T0314284

The Princeton Companion to Mathematics

The Princeton Companion to Mathematics

EDITOR

Timothy Gowers
University of Cambridge

ASSOCIATE EDITORS

June Barrow-Green
The Open University

Imre Leader
University of Cambridge

Princeton University Press
Princeton and Oxford

Published by Princeton University Press,
41 William Street, Princeton, New Jersey 08540

In the United Kingdom: Princeton University Press,
6 Oxford Street, Woodstock, Oxfordshire OX20 1TW

Library of Congress Cataloging-in-Publication Data

The Princeton companion to mathematics / Timothy Gowers, editor ;
  June Barrow-Green, Imre Leader, associate editors.
     p. cm.
  Includes bibliographical references and index.
  ISBN 978-0-691-11880-2 (hardcover : alk. paper)
   1. Mathematics—Study and teaching (Higher) 2. Princeton University.
  I. Gowers, Timothy. II. Barrow-Green, June, date– III. Leader, Imre.
QA11.2.P745 2008
510—dc22   2008020450

British Library Cataloging-in-Publication Data is available

Grateful acknowledgment is made for permission
to reprint the following illustrations in part VI:

Page 739. Portrait of René Descartes taken from *Pantheon
berühmter Menschen aller Zeiten* (Zwickau, 1830). Courtesy of
Niedersächsische Staats- und Universitätsbibliothek Göttingen.

Page 742. Portrait of Isaac Newton. By permission of
the Master and Fellows, Trinity College Cambridge.

Page 744. Copy after a portrait of Gottfried Leibniz by Andreas
Scheits (1703). Courtesy of Gottfried Wilhelm Leibniz
Bibliothek—Niedersächsische Landesbibliothek Hannover.

Page 748. Portrait of Leonhard Euler by J. F. A. Darbès (inv. no.
1829-8). Copyright: © Musée d'art et d'histoire, Ville de Genève.

Page 756. Portrait of Carl Friedrich Gauss. Courtesy of
Niedersächsische Staats- und Universitätsbibliothek Göttingen.

Page 775. Portrait of Bernhard Riemann. Courtesy of
Niedersächsische Staats- und Universitätsbibliothek Göttingen.

Page 786. Portrait of Henri Poincaré. Courtesy of
Henri Poincaré Archives (CNRS,UMR 7117, Nancy).

Page 788. Portrait of David Hilbert. Courtesy of
Niedersächsische Staats- und Universitätsbibliothek Göttingen.

This book has been composed in LucidaBright

Project management and composition
by T&T Productions Ltd, London

Printed on acid-free paper ∞
press.princeton.edu

Printed in the United States of America

20 19 18 17 16

Contents

Preface   ix
Contributors   xvii

Part I   Introduction

I.1   What Is Mathematics About?   1
I.2   The Language and Grammar of Mathematics   8
I.3   Some Fundamental Mathematical Definitions   16
I.4   The General Goals of Mathematical Research   47

Part II   The Origins of Modern Mathematics

II.1   From Numbers to Number Systems   77
II.2   Geometry   83
II.3   The Development of Abstract Algebra   95
II.4   Algorithms   106
II.5   The Development of Rigor in
       Mathematical Analysis   117
II.6   The Development of the Idea of Proof   129
II.7   The Crisis in the Foundations of Mathematics   142

Part III   Mathematical Concepts

III.1   The Axiom of Choice   157
III.2   The Axiom of Determinacy   159
III.3   Bayesian Analysis   159
III.4   Braid Groups   160
III.5   Buildings   161
III.6   Calabi–Yau Manifolds   163
III.7   Cardinals   165
III.8   Categories   165
III.9   Compactness and Compactification   167
III.10 Computational Complexity Classes   169
III.11 Countable and Uncountable Sets   170
III.12 C^*-Algebras   172
III.13 Curvature   172
III.14 Designs   172

III.15 Determinants   174
III.16 Differential Forms and Integration   175
III.17 Dimension   180
III.18 Distributions   184
III.19 Duality   187
III.20 Dynamical Systems and Chaos   190
III.21 Elliptic Curves   190
III.22 The Euclidean Algorithm and
        Continued Fractions   191
III.23 The Euler and Navier–Stokes Equations   193
III.24 Expanders   196
III.25 The Exponential and Logarithmic Functions   199
III.26 The Fast Fourier Transform   202
III.27 The Fourier Transform   204
III.28 Fuchsian Groups   208
III.29 Function Spaces   210
III.30 Galois Groups   213
III.31 The Gamma Function   213
III.32 Generating Functions   214
III.33 Genus   215
III.34 Graphs   215
III.35 Hamiltonians   215
III.36 The Heat Equation   216
III.37 Hilbert Spaces   219
III.38 Homology and Cohomology   221
III.39 Homotopy Groups   221
III.40 The Ideal Class Group   221
III.41 Irrational and Transcendental Numbers   222
III.42 The Ising Model   223
III.43 Jordan Normal Form   223
III.44 Knot Polynomials   225
III.45 K-Theory   227
III.46 The Leech Lattice   227
III.47 L-Functions   228
III.48 Lie Theory   229
III.49 Linear and Nonlinear Waves and Solitons   234
III.50 Linear Operators and Their Properties   239
III.51 Local and Global in Number Theory   241
III.52 The Mandelbrot Set   244
III.53 Manifolds   244
III.54 Matroids   244
III.55 Measures   246

III.56	Metric Spaces	247
III.57	Models of Set Theory	248
III.58	Modular Arithmetic	249
III.59	Modular Forms	250
III.60	Moduli Spaces	252
III.61	The Monster Group	252
III.62	Normed Spaces and Banach Spaces	252
III.63	Number Fields	254
III.64	Optimization and Lagrange Multipliers	255
III.65	Orbifolds	257
III.66	Ordinals	258
III.67	The Peano Axioms	258
III.68	Permutation Groups	259
III.69	Phase Transitions	261
III.70	π	261
III.71	Probability Distributions	263
III.72	Projective Space	267
III.73	Quadratic Forms	267
III.74	Quantum Computation	269
III.75	Quantum Groups	272
III.76	Quaternions, Octonions, and Normed Division Algebras	275
III.77	Representations	279
III.78	Ricci Flow	279
III.79	Riemann Surfaces	282
III.80	The Riemann Zeta Function	283
III.81	Rings, Ideals, and Modules	284
III.82	Schemes	285
III.83	The Schrödinger Equation	285
III.84	The Simplex Algorithm	288
III.85	Special Functions	290
III.86	The Spectrum	294
III.87	Spherical Harmonics	295
III.88	Symplectic Manifolds	297
III.89	Tensor Products	301
III.90	Topological Spaces	301
III.91	Transforms	303
III.92	Trigonometric Functions	307
III.93	Universal Covers	309
III.94	Variational Methods	310
III.95	Varieties	313
III.96	Vector Bundles	313
III.97	Von Neumann Algebras	313
III.98	Wavelets	313
III.99	The Zermelo–Fraenkel Axioms	314

Part IV Branches of Mathematics

IV.1	Algebraic Numbers	315
IV.2	Analytic Number Theory	332
IV.3	Computational Number Theory	348
IV.4	Algebraic Geometry	363
IV.5	Arithmetic Geometry	372
IV.6	Algebraic Topology	383
IV.7	Differential Topology	396
IV.8	Moduli Spaces	408
IV.9	Representation Theory	419
IV.10	Geometric and Combinatorial Group Theory	431
IV.11	Harmonic Analysis	448
IV.12	Partial Differential Equations	455
IV.13	General Relativity and the Einstein Equations	483
IV.14	Dynamics	493
IV.15	Operator Algebras	510
IV.16	Mirror Symmetry	523
IV.17	Vertex Operator Algebras	539
IV.18	Enumerative and Algebraic Combinatorics	550
IV.19	Extremal and Probabilistic Combinatorics	562
IV.20	Computational Complexity	575
IV.21	Numerical Analysis	604
IV.22	Set Theory	615
IV.23	Logic and Model Theory	635
IV.24	Stochastic Processes	647
IV.25	Probabilistic Models of Critical Phenomena	657
IV.26	High-Dimensional Geometry and Its Probabilistic Analogues	670

Part V Theorems and Problems

V.1	The ABC Conjecture	681
V.2	The Atiyah–Singer Index Theorem	681
V.3	The Banach–Tarski Paradox	684
V.4	The Birch–Swinnerton-Dyer Conjecture	685
V.5	Carleson's Theorem	686
V.6	The Central Limit Theorem	687
V.7	The Classification of Finite Simple Groups	687
V.8	Dirichlet's Theorem	689
V.9	Ergodic Theorems	689
V.10	Fermat's Last Theorem	691
V.11	Fixed Point Theorems	693
V.12	The Four-Color Theorem	696
V.13	The Fundamental Theorem of Algebra	698
V.14	The Fundamental Theorem of Arithmetic	699
V.15	Gödel's Theorem	700
V.16	Gromov's Polynomial-Growth Theorem	702
V.17	Hilbert's Nullstellensatz	703
V.18	The Independence of the Continuum Hypothesis	703
V.19	Inequalities	703
V.20	The Insolubility of the Halting Problem	706
V.21	The Insolubility of the Quintic	708
V.22	Liouville's Theorem and Roth's Theorem	710
V.23	Mostow's Strong Rigidity Theorem	711
V.24	The $\mathcal{P}$ versus $\mathcal{NP}$ Problem	713
V.25	The Poincaré Conjecture	714

V.26 The Prime Number Theorem and the
Riemann Hypothesis 714

V.27 Problems and Results in
Additive Number Theory 715

V.28 From Quadratic Reciprocity to
Class Field Theory 718

V.29 Rational Points on Curves and
the Mordell Conjecture 720

V.30 The Resolution of Singularities 722

V.31 The Riemann–Roch Theorem 723

V.32 The Robertson–Seymour Theorem 725

V.33 The Three-Body Problem 726

V.34 The Uniformization Theorem 728

V.35 The Weil Conjectures 729

Part VI Mathematicians

VI.1 Pythagoras (ca. 569 B.C.E.–ca. 494 B.C.E.) 733

VI.2 Euclid (ca. 325 B.C.E.–ca. 265 B.C.E.) 734

VI.3 Archimedes (ca. 287 B.C.E.–212 B.C.E.) 734

VI.4 Apollonius (ca. 262 B.C.E.–ca. 190 B.C.E.) 735

VI.5 Abu Ja'far Muhammad ibn Mūsā al-Khwārizmī
(800–847) 736

VI.6 Leonardo of Pisa (known as Fibonacci)
(ca. 1170–ca. 1250) 737

VI.7 Girolamo Cardano (1501–1576) 737

VI.8 Rafael Bombelli (1526–after 1572) 737

VI.9 François Viète (1540–1603) 737

VI.10 Simon Stevin (1548–1620) 738

VI.11 René Descartes (1596–1650) 739

VI.12 Pierre Fermat (160?–1665) 740

VI.13 Blaise Pascal (1623–1662) 741

VI.14 Isaac Newton (1642–1727) 742

VI.15 Gottfried Wilhelm Leibniz (1646–1716) 743

VI.16 Brook Taylor (1685–1731) 745

VI.17 Christian Goldbach (1690–1764) 745

VI.18 The Bernoullis (fl. 18th century) 745

VI.19 Leonhard Euler (1707–1783) 747

VI.20 Jean Le Rond d'Alembert (1717–1783) 749

VI.21 Edward Waring (ca. 1735–1798) 750

VI.22 Joseph Louis Lagrange (1736–1813) 751

VI.23 Pierre-Simon Laplace (1749–1827) 752

VI.24 Adrien-Marie Legendre (1752–1833) 754

VI.25 Jean-Baptiste Joseph Fourier (1768–1830) 755

VI.26 Carl Friedrich Gauss (1777–1855) 755

VI.27 Siméon-Denis Poisson (1781–1840) 757

VI.28 Bernard Bolzano (1781–1848) 757

VI.29 Augustin-Louis Cauchy (1789–1857) 758

VI.30 August Ferdinand Möbius (1790–1868) 759

VI.31 Nicolai Ivanovich Lobachevskii (1792–1856) 759

VI.32 George Green (1793–1841) 760

VI.33 Niels Henrik Abel (1802–1829) 760

VI.34 János Bolyai (1802–1860) 762

VI.35 Carl Gustav Jacob Jacobi (1804–1851) 762

VI.36 Peter Gustav Lejeune Dirichlet (1805–1859) 764

VI.37 William Rowan Hamilton (1805–1865) 765

VI.38 Augustus De Morgan (1806–1871) 765

VI.39 Joseph Liouville (1809–1882) 766

VI.40 Ernst Eduard Kummer (1810–1893) 767

VI.41 Évariste Galois (1811–1832) 767

VI.42 James Joseph Sylvester (1814–1897) 768

VI.43 George Boole (1815–1864) 769

VI.44 Karl Weierstrass (1815–1897) 770

VI.45 Pafnuty Chebyshev (1821–1894) 771

VI.46 Arthur Cayley (1821–1895) 772

VI.47 Charles Hermite (1822–1901) 773

VI.48 Leopold Kronecker (1823–1891) 773

VI.49 Georg Friedrich Bernhard Riemann
(1826–1866) 774

VI.50 Julius Wilhelm Richard Dedekind (1831–1916) 776

VI.51 Émile Léonard Mathieu (1835–1890) 776

VI.52 Camille Jordan (1838–1922) 777

VI.53 Sophus Lie (1842–1899) 777

VI.54 Georg Cantor (1845–1918) 778

VI.55 William Kingdon Clifford (1845–1879) 780

VI.56 Gottlob Frege (1848–1925) 780

VI.57 Christian Felix Klein (1849–1925) 782

VI.58 Ferdinand Georg Frobenius (1849–1917) 783

VI.59 Sofya (Sonya) Kovalevskaya (1850–1891) 784

VI.60 William Burnside (1852–1927) 785

VI.61 Jules Henri Poincaré (1854–1912) 785

VI.62 Giuseppe Peano (1858–1932) 787

VI.63 David Hilbert (1862–1943) 788

VI.64 Hermann Minkowski (1864–1909) 789

VI.65 Jacques Hadamard (1865–1963) 790

VI.66 Ivar Fredholm (1866–1927) 791

VI.67 Charles-Jean de la Vallée Poussin (1866–1962) 792

VI.68 Felix Hausdorff (1868–1942) 792

VI.69 Élie Joseph Cartan (1869–1951) 794

VI.70 Emile Borel (1871–1956) 795

VI.71 Bertrand Arthur William Russell (1872–1970) 795

VI.72 Henri Lebesgue (1875–1941) 796

VI.73 Godfrey Harold Hardy (1877–1947) 797

VI.74 Frigyes (Frédéric) Riesz (1880–1956) 798

VI.75 Luitzen Egbertus Jan Brouwer (1881–1966) 799

VI.76 Emmy Noether (1882–1935) 800

VI.77 Wacław Sierpiński (1882–1969) 801

VI.78 George Birkhoff (1884–1944) 802

VI.79 John Edensor Littlewood (1885–1977) 803

VI.80 Hermann Weyl (1885–1955) 805

VI.81 Thoralf Skolem (1887–1963) 806

VI.82 Srinivasa Ramanujan (1887–1920) 807

VI.83 Richard Courant (1888–1972) 808

VI.84 Stefan Banach (1892–1945) 809

VI.85 Norbert Wiener (1894–1964) 811

VI.86   Emil Artin (1898–1962)   812
VI.87   Alfred Tarski (1901–1983)   813
VI.88   Andrei Nikolaevich Kolmogorov (1903–1987)   814
VI.89   Alonzo Church (1903–1995)   816
VI.90   William Vallance Douglas Hodge (1903–1975)   816
VI.91   John von Neumann (1903–1957)   817
VI.92   Kurt Gödel (1906–1978)   819
VI.93   André Weil (1906–1998)   819
VI.94   Alan Turing (1912–1954)   821
VI.95   Abraham Robinson (1918–1974)   822
VI.96   Nicolas Bourbaki (1935–)   823

Part VII   The Influence of Mathematics

VII.1   Mathematics and Chemistry   827
VII.2   Mathematical Biology   837
VII.3   Wavelets and Applications   848
VII.4   The Mathematics of Traffic in Networks   862
VII.5   The Mathematics of Algorithm Design   871
VII.6   Reliable Transmission of Information   878
VII.7   Mathematics and Cryptography   887

VII.8   Mathematics and Economic Reasoning   895
VII.9   The Mathematics of Money   910
VII.10   Mathematical Statistics   916
VII.11   Mathematics and Medical Statistics   921
VII.12   Analysis, Mathematical and Philosophical   928
VII.13   Mathematics and Music   935
VII.14   Mathematics and Art   944

Part VIII   Final Perspectives

VIII.1   The Art of Problem Solving   955
VIII.2   "Why Mathematics?" You Might Ask   966
VIII.3   The Ubiquity of Mathematics   977
VIII.4   Numeracy   983
VIII.5   Mathematics: An Experimental Science   991
VIII.6   Advice to a Young Mathematician   1000
VIII.7   A Chronology of Mathematical Events   1010

Index   1015

Preface

1 What Is *The Companion*?

Bertrand Russell, in his book *The Principles of Mathematics*, proposes the following as a definition of pure mathematics.

> Pure Mathematics is the class of all propositions of the form "*p* implies *q*," where *p* and *q* are propositions containing one or more variables, the same in the two propositions, and neither *p* nor *q* contains any constants except logical constants. And logical constants are all notions definable in terms of the following: Implication, the relation of a term to a class of which it is a member, the notion of *such that*, the notion of relation, and such further notions as may be involved in the general notion of propositions of the above form. In addition to these, mathematics *uses* a notion which is not a constituent of the propositions which it considers, namely the notion of truth.

The Princeton Companion to Mathematics could be said to be about everything that Russell's definition leaves out.

Russell's book was published in 1903, and many mathematicians at that time were preoccupied with the logical foundations of the subject. Now, just over a century later, it is no longer a new idea that mathematics can be regarded as a formal system of the kind that Russell describes, and today's mathematician is more likely to have other concerns. In particular, in an era where so much mathematics is being published that no individual can understand more than a tiny fraction of it, it is useful to know not just which arrangements of symbols form grammatically correct mathematical statements, but also which of these statements deserve our attention.

Of course, one cannot hope to give a fully objective answer to such a question, and different mathematicians can legitimately disagree about what they find interesting. For that reason, this book is far less formal than Russell's and it has many authors with many different points of view. And rather than trying to give a precise answer to the question, "What makes a mathematical statement interesting?" it simply aims to present for the reader a large and representative sample of the ideas that mathematicians are grappling with at the beginning of the twenty-first century, and to do so in as attractive and accessible a way as possible.

2 The Scope of the Book

The central focus of this book is *modern, pure* mathematics, a decision about which something needs to be said. "Modern" simply means that, as mentioned above, the book aims to give an idea of what mathematicians are now doing: for example, an area that developed rapidly in the middle of the last century but that has now reached a settled form is likely to be discussed less than one that is still developing rapidly. However, mathematics carries its history with it: in order to understand a piece of present-day mathematics, one will usually need to know about many ideas and results that were discovered a long time ago. Moreover, if one wishes to have a proper perspective on today's mathematics, it is essential to have some idea of how it came to be as it is. So there is plenty of history in the book, even if the main reason for our including it is to illuminate the mathematics of today.

The word "pure" is more troublesome. As many have commented, there is no clear dividing line between pure and applied mathematics, and, just as a proper appreciation of modern mathematics requires some knowledge of its history, so a proper appreciation of pure mathematics requires some knowledge of applied mathematics and theoretical physics. Indeed, these areas have provided pure mathematicians with many fundamental ideas, which have given rise to some of the most interesting, important, and currently active branches of pure mathematics. This book is certainly not blind to the impact on pure mathematics of these other disciplines, nor does it ignore the practical and intellectual applications of pure mathematics. Nevertheless, the scope is narrower than it could be. At one stage it was suggested that a more accurate title would be "The Princeton Companion to Pure Mathematics": the only reason for rejecting this title was that it does not sound as good as the actual title.

Another thought behind the decision to concentrate on pure mathematics was that it would leave open the possibility of a similar book, a companion *Companion* so to speak, about applied mathematics and theoretical physics. Until such a book appears, *The Road to Reality*, by Roger Penrose (Knopf, New York, 2005), covers a very wide variety of topics in mathematical physics, written at a level fairly similar to that of this book, and Elsevier has recently brought out a five-volume *Encyclopedia of Mathematical Physics* (Elsevier, Amsterdam, 2006).

3   *The Companion* Is Not an Encyclopedia

The word "companion" is significant. Although this book is certainly intended as a useful work of reference, you should not expect too much of it. If there is a particular mathematical concept that you want to find out about, you will not necessarily be able to find out about it here, even if it is important; though the more important it is, the more likely it is to be included. In this respect, the book is like a human companion, complete with gaps in its knowledge and views on some topics that may not be universally shared. Having said that, we have at least aimed at some sort of balance: many topics are not covered, but those that are covered range very widely (much more so than one could reasonably expect of any single human companion). In order to achieve this kind of balance, we have been guided to some extent by "objective" indicators such as the American Mathematical Society's classification of mathematical topics, or the way that mathematics is divided into sections at the four-yearly International Congress of Mathematicians. The broad areas, such as number theory, algebra, analysis, geometry, combinatorics, logic, probability, theoretical computer science, and mathematical physics, are all represented, even if not all their subareas are. Inevitably, some of the choices about what to include, and at what length, were not the result of editorial policy, but were based on highly contingent factors such as who agreed to write, who actually submitted after having agreed, whether those who submitted stuck to their word limit, and so on. Consequently, there are some areas that are not as fully represented as

we would have liked, but the point came where it was better to publish an imperfect volume than to spend several more years striving for perfect balance. We hope that there will be future editions of *The Companion*: if so, there will be a chance to remedy any defects that there might be in this one.

Another respect in which this book differs from an encyclopedia is that it is arranged thematically rather than alphabetically. The advantage of this is that, although the articles can be enjoyed individually, they can also be regarded as part of a coherent whole. Indeed, the structure of the book is such that it would not be ridiculous to read it from cover to cover, though it would certainly be time-consuming.

4   The Structure of *The Companion*

What does it mean to say that *The Companion* is "arranged thematically"? The answer is that it is divided into eight parts, each with a different general theme and a different purpose. Part I consists of introductory material, which gives a broad overview of mathematics and explains, for the benefit of readers with less of a background in mathematics, some of the basic concepts of the subject. A rough rule of thumb is that a topic belongs in part I if it is part of the necessary background of all mathematicians rather than belonging to one specific area. GROUPS [I.3 §2.1] and VECTOR SPACES [I.3 §2.3] belong in this category, to take two obvious examples.

Part II is a collection of essays of a historical nature. Its aim is to explain how the distinctive style of modern mathematics came into being. What, broadly speaking, are the main differences between the way mathematicians think about their subject now and the way they thought about it 200 years ago (or more)? One is that there is a universally accepted standard for what counts as a proof. Closely related to this is the fact that mathematical analysis (calculus and its later extensions and developments) has been put on a rigorous footing. Other notable features are the extension of the concept of number, the abstract nature of algebra, and the fact that most modern geometers study *non-Euclidean* geometry rather than the more familiar triangles, circles, parallel lines, and the like.

Part III consists of fairly short articles, each one dealing with an important mathematical concept that has not appeared in part I. The intention is that this part of the book will be a very good place to look if there is a concept you do not know about but have often heard mentioned. If another mathematician, perhaps a colloquium speaker, assumes that you are familiar with a definition—for example, that of a SYMPLECTIC FORM [III.88], or the INCOMPRESSIBLE EULER EQUATION [III.23], or a SOBOLEV SPACE [III.29 §2.4], or the IDEAL CLASS GROUP [IV.1 §7]—and if you are too embarrassed to admit that in fact you are not, then you now have the alternative of looking these concepts up in *The Companion*.

The articles in part III would not be much use if all they gave was formal definitions: to understand a concept one wants to know what it means intuitively, why it is important, and why it was first introduced. Above all, if it is a fairly general concept, then one wants to know some good examples—ones that are not too simple and not too complicated. Indeed, it may well be that providing and discussing a well-chosen example is all that such an article needs to do, since a good example is much easier to understand than a general definition, and more experienced readers will be able to work out a general definition by abstracting the important properties from the example.

Another use of part III is to provide backup for part IV, which is the heart of the book. Part IV consists of twenty-six articles, considerably longer than those of part III, about different areas of mathematics. A typical part IV article explains some of the central ideas and important results of the area it treats, and does so as informally as possible, subject to the constraint that it should not be too vague to be informative. The original hope was for these articles to be "bedtime reading," that is, clear and elementary enough that one could read and understand them without continually stopping to think. For that reason, the authors were chosen with two priorities in mind, of equal importance: expertise and expository skill. But mathematics is not an easy subject, and in the end we had to regard the complete accessibility we originally hoped for as an ideal that we would strive toward, even if it was not achieved in every last subsection of every article. But even when the articles are tough going, they discuss what they discuss in a clearer and less formal way than a typical textbook, often with remarkable success. As with part III, several authors have achieved this by looking at illuminating examples, which they sometimes follow with more general theory and sometimes leave to speak for themselves.

Many part IV articles contain excellent descriptions of mathematical concepts that would otherwise have had articles devoted to them in part III. We originally planned to avoid duplication completely, and instead to include cross-references to these descriptions in part III. However, this risked irritating the reader, so we decided on the following compromise. Where a concept is adequately explained elsewhere, part III does not have a full article, but it does have a short description together with a cross-reference. This way, if you want to look a concept up quickly, you can use part III, and only if you need more detail will you be forced to follow the cross-reference to another part of the book.

Part V is a complement to part III. Again, it consists of short articles on important mathematical topics, but now these topics are the theorems and open problems of mathematics rather than the basic objects and tools of study. As with the book as a whole, the choice of entries in part V is necessarily far from comprehensive, and has been made with a number of criteria in mind. The most obvious one is mathematical importance, but some entries were chosen because it is possible to discuss them in an entertaining and accessible way, others because they have some unusual feature (an example is the FOUR-COLOR THEOREM [V.12], though this might well have been included anyway), some because the authors of closely related part IV articles felt that certain theorems should be discussed separately, and some because authors of several other articles wanted to assume them as background knowledge. As with part III, some of the entries in part V are not full articles but short accounts with cross-references to other articles.

Part VI is another historical section, about famous mathematicians. It consists of short articles, and the aim of each article is to give very basic biographical information (such as nationality and date of birth), together with an explanation of why the mathematician in question is famous. Initially, we planned to include living mathematicians, but in the end we came to the conclusion that it would be almost impossible to make a satisfactory selection of mathematicians working today, so we decided to restrict ourselves to mathematicians who had died, and moreover to mathematicians who were principally known for work carried out before 1950. Later mathematicians do of course feature in the book, since they are mentioned in other articles. They do not have their own entries, but one can get some idea of their achievements by looking them up in the index.

After six parts mainly about pure mathematics and its history, part VII finally demonstrates the great external impact that mathematics has had, both practically and intellectually. It consists of longer articles, some written by mathematicians with interdisciplinary interests and others by experts from other disciplines who make considerable use of mathematics.

The final part of the book contains general reflections about the nature of mathematics and mathematical life. The articles in this part are on the whole more accessible than the longer articles earlier in the book, so even though part VIII is the final part, some readers may wish to make it one of the first parts they look at.

The order of the articles within the parts is alphabetical in parts III and V and chronological in part VI. The decision to organize the articles about mathematicians in order of their dates of birth was carefully considered, and we made it for several reasons: it would encourage the reader to get a sense of the history of the subject by reading the part right through rather than just looking at individual articles; it would make it much clearer which mathematicians were contemporaries or near contemporaries; and after the slight inconvenience of looking up a mathematician by guessing his or her date of birth relative to those of other mathematicians, the reader would learn something small but valuable.

In the other parts, some attempt has been made to arrange the articles thematically. This applies in particular to part IV, where the ordering attempts to follow two basic principles: first, that articles about closely related branches of mathematics should be close to each other in the book; and second, that if it makes obvious sense to read article A before article B, then article A should come before article B in the book. This is easier said than done, since some branches are hard to classify: for instance, should arithmetic geometry count as algebra, geometry, or number theory? A case could be made for any of the three and it is artificial to decide on just one. So the ordering in part IV should not be taken as a classification scheme, but just as the best linear ordering we could think of.

As for the order of the parts themselves, the aim has been to make it the most natural one from a pedagogical point of view and to give the book some sense of direction. Parts I and II are obviously introductory, in different ways. Part III comes before part IV because in order to understand an area of mathematics one tends to start by grappling with new definitions. But part IV comes before part V because in order to appreciate a theorem it is a good idea to know how it fits into an area of mathematics. Part VI is placed after parts III–V because one can better appreciate the contribution of a famous mathematician after knowing some mathematics. Part VII is near the end for a similar reason: to understand the influence of mathematics, one should understand mathematics first. And the reflections of part VIII are a sort of epilogue, and therefore an appropriate way for the book to sign off.

5 Cross-References

From the start, it was planned that *The Companion* would have a large number of cross-references. One or two have even appeared in this preface, signalled by THIS FONT, together with an indication of where to find the relevant article. For example, the reference to a SYMPLECTIC FORM [III.88] indicated that symplectic forms are discussed in article number 88 of part III, and the reference to the IDEAL CLASS GROUP [IV.1 §7] pointed the reader to section 7 of article number 1 in part IV.

We have tried as hard as possible to produce a book that is a pleasure to read, and the aim is that cross-references should contribute to this pleasure. This may seem a rather strange thing to say, since it can be annoying to interrupt what one is reading in a book in order to spend a few seconds looking something up elsewhere. However, we have also tried to keep the articles as self-contained as is feasible. Thus, if you do not want to pursue the cross-references, then you will usually not have to. The main exception to this is that authors have been allowed to assume some knowledge of the concepts discussed in part I. If you do not know any university mathematics, then you would be well-advised to start by reading part I in full, as this will greatly reduce your need to look things up while reading later articles.

Sometimes a concept is introduced in an article and then explained in that article. The usual convention in mathematical writing is to italicize a term when it is being defined. We have stuck to something like that convention, but in an informal article it is not always clear what constitutes the moment of definition of a new or unfamiliar term. Our rough policy has been to italicize a term the first time it is used if that use is followed by a discussion that gives some kind of explanation of the term. We have also italicized terms that are not subsequently explained: this should be taken as a signal that the reader is not required to understand the term in order to understand the rest of the article in question. In more extreme cases of this kind, quotation marks may be used instead.

Many of the articles end with brief "Further Reading" sections. These are exactly that: suggestions for further reading. They should not be thought of as full-scale bibliographies such as one might find at the end of a survey article. Related to this is the fact that it is not a major concern of *The Companion* to give credit to all the mathematicians who made the discoveries that it discusses or to cite the papers where those discoveries appeared. The reader who is interested in original sources should be able to find them from the books and articles in the further reading sections, or from the Internet.

6 Who Is *The Companion* Aimed At?

The original plan for *The Companion* was that all of it should be accessible to anybody with a good background in high school mathematics (including calculus). However, it soon became apparent that this was an unrealistic aim: there are branches of mathematics that are so much easier to understand when one knows at least some university-level mathematics that it does not make good sense to attempt to explain them at a lower level. On the other hand, there are other parts of the subject that decidedly *can* be explained to readers without this extra experience. So in the end we abandoned the idea that the book should have a uniform level of difficulty.

Accessibility has, however, remained one of our highest priorities, and throughout the book we have tried to discuss mathematical ideas at the lowest level that is practical. In particular, the editors have tried very hard not to allow any material into the book that they do not themselves understand, which has turned out to be a very serious constraint. Some readers will find some articles too hard and other readers will find other articles too easy, but we hope that all readers from advanced high school level onwards will find that they enjoy a substantial proportion of the book.

What can readers of different levels hope to get out of *The Companion*? If you have embarked on a university-level mathematics course, you may find that you are presented with a great deal of difficult and unfamiliar material without having much idea why it is important and where it is all going. Then you can use *The Companion* to provide yourself with some perspective on the subject. (For example, many more people know what a ring is than can give a good reason for caring about rings. But there are very good reasons, which you can read about in RINGS, IDEALS, AND MODULES [III.81] and ALGEBRAIC NUMBERS [IV.1].)

If you are coming to the end of the course, you may be interested in doing research in mathematics. But undergraduate courses typically give you very little idea of what research is actually like. So how do you decide which areas of mathematics truly interest you at the research level? It is not easy, but the decision can make the difference between becoming disillusioned and ultimately not getting a Ph.D., and going on to a successful career in mathematics. This book, especially part IV, tells you what mathematicians of many different kinds are thinking about at the research level, and may help you to make a more informed decision.

If you are already an established research mathematician, then your main use for this book will probably be to understand better what your colleagues are up to. Most nonmathematicians are very surprised to learn how extraordinarily specialized mathematics has become. Nowadays it is not uncommon for a very good mathematician to be completely unable to understand the papers of another mathematician, even from an area that appears to be quite close. This is not a healthy state of affairs: anything that can be done to improve the level of communication among mathematicians is a good idea. The editors of this book have learned a huge amount from reading the articles carefully, and we hope that many others will avail themselves of the same opportunity.

7 What Does *The Companion* Offer That the Internet Does Not Offer?

In some ways the character of *The Companion* is similar to that of a large mathematical Web site such as the mathematical part of Wikipedia or Eric Weisstein's "Mathworld" (http://mathworld.wolfram.com/). In particular, the cross-references have something of the feel of hyperlinks. So is there any need for this book?

At the moment, the answer is yes. If you have ever tried to use the Internet to find out about a mathematical concept, then you will know that it is a hit-and-miss affair. Sometimes you find a good explanation that gives you the information you were looking for. But often you do not. The Web sites just mentioned are certainly useful, and recommended for material that is not covered here, but at the time of writing most of the online articles are written in a different style from the articles in this book: drier, and more concerned with giving the basic facts in an economical way than with reflecting on those facts. And one does not find long essays of the kind contained in parts I, II, IV, VII, and VIII of this book.

Some people will also find it advantageous to have a large collection of material in book form. As has already been mentioned, this book is organized not as a collection of isolated articles but as a carefully ordered sequence that exploits the linear structure that all books necessarily have and that Web sites do not have. And the physical nature of a book makes browsing through it a completely different experience from browsing a Web site: after reading the list of contents one can get a feel for the entire book, whereas with a large Web site one is somehow conscious only of the page one is looking at. Not everyone will agree with this or find it a significant advantage, but many undoubtedly will and it is for them that the book has been written. For now, therefore, *The Princeton Companion to Mathematics* does not have a serious online competitor: rather than competing with the existing Web sites, it complements them.

8 How *The Companion* Came into Being

The Princeton Companion to Mathematics was first conceived by David Ireland in 2002, who was at the time employed in the Oxford office of Princeton University Press. The most important features of the book—its title, its organization into sections, and the idea that one of these sections should consist of articles about major branches of mathematics—were all part of his original conception. He came to visit me in Cambridge to discuss his proposal, and when the moment came (it was clearly going to) for him to ask whether I would be prepared to edit it, I accepted more or less on the spot.

What induced me to make such a decision? It was partly because he told me that I would not be expected to do all the work on my own: not only would there be other editors involved, but also there would be considerable technical and administrative support. But a more fundamental reason was that the idea for the book was very similar to one that I had had myself in an idle moment as a research student. It would be wonderful, I thought then, if somewhere one could find a collection of well-written essays that presented for you the big themes of mathematical research in different areas of mathematics. Thus a little fantasy had been born, and suddenly I had the chance to turn it into a reality.

We knew from the outset that we wanted the book to contain a certain amount of historical reflection, and soon after this meeting David Ireland asked June Barrow-Green whether she was prepared to be another editor, with particular responsibility for the historical parts. To our delight, she accepted, and with her remarkable range of contacts she gave us access to more or less all the mathematical historians in the world.

There then began several meetings to plan the more detailed content of the volume, ending in a formal proposal to Princeton University Press. They sent it out to a team of expert advisers, and although some made the obvious point that it was a dauntingly huge project, all were enthusiastic about it. This enthusiasm was also evident at the next stage, when we began to find contributors. Many of them were very encouraging and said how glad they were that such a book was being produced, confirming what we already thought: that there was a gap in the market. During this stage, we benefited greatly from the advice and experience of Alison Latham, editor of *The Oxford Companion to Music*.

In the middle of 2003, David Ireland left Princeton University Press, and with it this project. This was a big blow, and we missed his vision and enthusiasm for the book: we hope that what we have finally produced is something like what he originally had in mind. However, there was a positive development at around the same time, when Princeton University Press decided to employ a small company called T&T Productions Ltd. The company was to be responsible for producing a book out of the files submitted by the contributors, as well as for doing a great deal of the day-to-day work such as sending out contracts, reminding contributors that their deadlines were approaching, receiving files, keeping a record of what had been done, and so on. Most of this work was done by Sam Clark, who is extraordinarily good at it and manages to be miraculously good-humored at the same time. In addition, he did a great deal of copy-editing as well, where that did not need too great a knowledge of mathematics (though as a former chemist he knows more than most people). With Sam's help we have not just a carefully edited book but one that is beautifully designed as well. Without him, I do not see how it would have ever been completed.

We continued to have regular meetings, to plan the book in more detail and to discuss progress on it. These meetings were now ably organized and chaired by Richard Baggaley, also from the Oxford office of Princeton University Press. He continued to do this until the summer of 2004, when Anne Savarese, Princeton's new reference editor, took over. Richard and

Anne have also been immensely useful, asking the editors the right awkward questions when we have been tempted to forget about the parts of the book that were not quite going to plan, and forcing on us a level of professionalism that, to me at least, does not come naturally.

In early 2004, at what we naively thought was a late stage in the preparation of the book, but which we now understand was actually near the beginning, we realized that, even with June's help, I had far too much to do. One person immediately sprang to mind as an ideal coeditor: Imre Leader, who I knew would understand what the book was trying to achieve and would have ideas about how to achieve it. He agreed, and quickly became an indispensable member of the editorial team, commissioning and editing several articles.

By the second half of 2007, we really were at a late stage, and by that time it had become clear that additional editorial help would make it much easier to complete the tricky tasks that we had been postponing and actually get the book finished. Jordan Ellenberg and Terence Tao agreed to help, and their contribution was invaluable. They edited some of the articles, wrote others, and enabled me to write several short articles on subjects that were outside my area of expertise, safe in the knowledge that they would stop me making serious errors. (I would have made several without their help, but take full responsibility for any that may have slipped through the net.) Articles by the editors have been left unattributed, but a note at the end of the contributor list explains which ones were written by which editor.

9 The Editorial Process

It is not always easy to find mathematicians with the patience and understanding to explain what they are doing to nonexperts or colleagues from other areas: too often they assume you know something that you do not, and it is embarrassing to admit that you are completely lost. However, the editors of this book have tried to help you by taking this burden of embarrassment on themselves. An important feature of the book has been that the editorial process has been a very active one: we have not just commissioned the articles and accepted whatever we have been sent. Some drafts have had to be completely discarded and new articles written in the light of editorial comments. Others have needed substantial changes, which have sometimes been made by the authors and sometimes by the editors. A few

articles were accepted with only trivial changes, but these were a very small minority.

The tolerance, even gratitude, with which almost all authors have allowed themselves to be subjected to this treatment has been a very welcome surprise and has helped the editors maintain their morale during the long years of preparation of this volume. We would like to express our gratitude in return, and we hope that they agree that the whole process has been worthwhile. To us it seems inconceivable that this amount of work could go into the articles *without* a substantial payoff. It is not my place to say how successful I think the outcome has been, but, given the number of changes that were made in the interests of accessibility, and given that interventionist editing of this type is rare in mathematics, I do not see how the book can fail to be unusual in a good way.

A sign of just how long everything has taken, and also of the quality of the contributors, is that a significant number of contributors have received major awards and distinctions since being invited to contribute. At least three babies have been born to authors in the middle of preparing articles. Two contributors, Benjamin Yandell and Graham Allan, have sadly not lived to see their articles in print, but we hope that in a small way this book will be a memorial to them.

10 Acknowledgments

An early part of the editorial process was of course planning the book and finding authors. This would have been impossible without the help and advice of several people. Donald Albers, Michael Atiyah, Jordan Ellenberg, Tony Gardiner, Sergiu Klainerman, Barry Mazur, Curt McMullen, Robert O'Malley, Terence Tao, and Avi Wigderson all gave advice that in one way or another had a beneficial effect on the shape of the book. June Barrow-Green has been greatly helped in her task by Jeremy Gray and Reinhard Siegmund-Schultze. In the final weeks, Vicky Neale very kindly agreed to proofread certain sections of the book and help with the index; she was amazing at this, picking up numerous errors that we would never have spotted ourselves and are very pleased to have corrected. And there is a long list of mathematicians and mathematical historians who have patiently answered questions from the editors: we would like to thank them all.

I am grateful to many people for their encouragement, including virtually all the contributors to this volume and many members of my immediate family, particularly my father, Patrick Gowers: this support has kept me going despite the mountainous appearance of the task ahead. I would also like to thank Julie Barrau for her less direct but equally essential help. During the final months of preparation of the book, she agreed to take on much more than her fair share of our domestic duties. Given that a son was born to us in November 2007, this made a huge difference to my life, as has she.

Timothy Gowers

Contributors

Graham Allan, *late Reader in Mathematics, University of Cambridge*
THE SPECTRUM [III.86]

Noga Alon, *Baumritter Professor of Mathematics and Computer Science, Tel Aviv University*
EXTREMAL AND PROBABILISTIC COMBINATORICS [IV.19]

George Andrews, *Evan Pugh Professor in the Department of Mathematics, The Pennsylvania State University*
SRINIVASA RAMANUJAN [VI.82]

Tom Archibald, *Professor, Department of Mathematics, Simon Fraser University*
THE DEVELOPMENT OF RIGOR IN MATHEMATICAL ANALYSIS [II.5], CHARLES HERMITE [VI.47]

Sir Michael Atiyah, *Honorary Professor, School of Mathematics, University of Edinburgh*
WILLIAM VALLANCE DOUGLAS HODGE [VI.90], ADVICE TO A YOUNG MATHEMATICIAN [VIII.6]

David Aubin, *Assistant Professor, Institut de Mathématiques de Jussieu*
NICOLAS BOURBAKI [VI.96]

Joan Bagaria, *ICREA Research Professor, University of Barcelona*
SET THEORY [IV.22]

Keith Ball, *Astor Professor of Mathematics, University College London*
THE EUCLIDEAN ALGORITHM AND CONTINUED FRACTIONS [III.22], OPTIMIZATION AND LAGRANGE MULTIPLIERS [III.64], HIGH-DIMENSIONAL GEOMETRY AND ITS PROBABILISTIC ANALOGUES [IV.26]

Alan F. Beardon, *Professor of Complex Analysis, University of Cambridge*
RIEMANN SURFACES [III.79]

David D. Ben-Zvi, *Associate Professor of Mathematics, University of Texas, Austin*
MODULI SPACES [IV.8]

Vitaly Bergelson, *Professor of Mathematics, The Ohio State University*
ERGODIC THEOREMS [V.9]

Nicholas Bingham, *Professor, Mathematics Department, Imperial College London*
ANDREI NIKOLAEVICH KOLMOGOROV [VI.88]

Béla Bollobás, *Professor of Mathematics, University of Cambridge and University of Memphis*
GODFREY HAROLD HARDY [VI.73], JOHN EDENSOR LITTLEWOOD [VI.79], ADVICE TO A YOUNG MATHEMATICIAN [VIII.6]

Henk Bos, *Honorary Professor, Department of Science Studies, Aarhus University; Professor Emeritus, Department of Mathematics, Utrecht University*
RENÉ DESCARTES [VI.11]

Bodil Branner, *Emeritus Professor, Department of Mathematics, Technical University of Denmark*
DYNAMICS [IV.14]

Martin R. Bridson, *Whitehead Professor of Pure Mathematics, University of Oxford*
GEOMETRIC AND COMBINATORIAL GROUP THEORY [IV.10]

John P. Burgess, *Professor of Philosophy, Princeton University*
ANALYSIS, MATHEMATICAL AND PHILOSOPHICAL [VII.12]

Kevin Buzzard, *Professor of Pure Mathematics, Imperial College London*
L-FUNCTIONS [III.47], MODULAR FORMS [III.59]

Peter J. Cameron, *Professor of Mathematics, Queen Mary, University of London*
DESIGNS [III.14], GÖDEL'S THEOREM [V.15]

Jean-Luc Chabert, *Professor, Laboratoire Amiénois de Mathématique Fondamentale et Appliquée, Université de Picardie*
ALGORITHMS [II.4]

Eugenia Cheng, *Lecturer, Department of Pure Mathematics, University of Sheffield*
CATEGORIES [III.8]

Clifford Cocks, *Chief Mathematician, Government Communications Headquarters, Cheltenham*
MATHEMATICS AND CRYPTOGRAPHY [VII.7]

Alain Connes, *Professor, Collège de France, IHES, and Vanderbilt University*
ADVICE TO A YOUNG MATHEMATICIAN [VIII.6]

Leo Corry, *Director, The Cohn Institute for History and Philosophy of Science and Ideas, Tel Aviv University*
THE DEVELOPMENT OF THE IDEA OF PROOF [II.6]

Wolfgang Coy, *Professor of Computer Science, Humboldt-Universität zu Berlin*
JOHN VON NEUMANN [VI.91]

Tony Crilly, *Emeritus Reader in Mathematical Sciences, Department of Economics and Statistics, Middlesex University*
ARTHUR CAYLEY [VI.46]

Serafina Cuomo, *Lecturer in Roman History, School of History, Classics and Archaeology, Birkbeck College*
PYTHAGORAS [VI.1], EUCLID [VI.2], ARCHIMEDES [VI.3], APOLLONIUS [VI.4]

Mihalis Dafermos, *Reader in Mathematical Physics, University of Cambridge*
GENERAL RELATIVITY AND THE EINSTEIN EQUATIONS [IV.13]

Partha Dasgupta, *Frank Ramsey Professor of Economics,*
University of Cambridge
MATHEMATICS AND ECONOMIC REASONING [VII.8]

Ingrid Daubechies, *Professor of Mathematics,*
Princeton University
WAVELETS AND APPLICATIONS [VII.3]

Joseph W. Dauben, *Distinguished Professor,*
Herbert H. Lehman College and City University of New York
GEORG CANTOR [VI.54], ABRAHAM ROBINSON [VI.95]

John W. Dawson Jr., *Professor of Mathematics, Emeritus,*
The Pennsylvania State University
KURT GÖDEL [VI.92]

Francois de Gandt, *Professeur d'Histoire des Sciences et*
de Philosophie, Université Charles de Gaulle, Lille
JEAN LE ROND D'ALEMBERT [VI.20]

Persi Diaconis, *Mary V. Sunseri Professor of Statistics and*
Mathematics, Stanford University
MATHEMATICAL STATISTICS [VII.10]

Jordan S. Ellenberg, *Associate Professor of Mathematics,*
University of Wisconsin
ELLIPTIC CURVES [III.21], SCHEMES [III.82],
ARITHMETIC GEOMETRY [IV.5]

Lawrence C. Evans, *Professor of Mathematics,*
University of California, Berkeley
VARIATIONAL METHODS [III.94]

Florence Fasanelli, *Program Director,*
American Association for the Advancement of Science
MATHEMATICS AND ART [VII.14]

Anita Burdman Feferman, *Independent Scholar and Writer,*
ALFRED TARSKI [VI.87]

Solomon Feferman, *Patrick Suppes Family Professor of*
Humanities and Sciences and Emeritus Professor of Mathematics
and Philosophy, Department of Mathematics, Stanford University
ALFRED TARSKI [VI.87]

Charles Fefferman, *Professor of Mathematics,*
Princeton University
THE EULER AND NAVIER–STOKES EQUATIONS [III.23],
CARLESON'S THEOREM [V.5]

Della Fenster, *Professor, Department of Mathematics and*
Computer Science, University of Richmond, Virginia
EMIL ARTIN [VI.86]

José Ferreirós, *Professor of Logic and Philosophy of Science,*
University of Seville
THE CRISIS IN THE FOUNDATIONS OF MATHEMATICS [II.7],
JULIUS WILHELM RICHARD DEDEKIND [VI.50],
GIUSEPPE PEANO [VI.62]

David Fisher, *Associate Professor of Mathematics,*
Indiana University, Bloomington
MOSTOW'S STRONG RIGIDITY THEOREM [V.23]

Terry Gannon, *Professor,*
Department of Mathematical Sciences, University of Alberta
VERTEX OPERATOR ALGEBRAS [IV.17]

A. Gardiner, *Reader in Mathematics and Mathematics Education,*
University of Birmingham
THE ART OF PROBLEM SOLVING [VIII.1]

Charles C. Gillispie, *Dayton-Stockton Professor of*
History of Science, Emeritus, Princeton University
PIERRE-SIMON LAPLACE [VI.23]

Oded Goldreich, *Professor of Computer Science,*
Weizmann Institute of Science, Israel
COMPUTATIONAL COMPLEXITY [IV.20]

Catherine Goldstein, *Directeur de Recherche,*
Institut de Mathématiques de Jussieu, CNRS, Paris
PIERRE FERMAT [VI.12]

Fernando Q. Gouvêa, *Carter Professor of Mathematics,*
Colby College, Waterville, Maine
FROM NUMBERS TO NUMBER SYSTEMS [II.1],
LOCAL AND GLOBAL IN NUMBER THEORY [III.51]

Andrew Granville, *Professor, Department of*
Mathematics and Statistics, Université de Montréal
ANALYTIC NUMBER THEORY [IV.2]

Ivor Grattan-Guinness, *Emeritus Professor of the*
History of Mathematics and Logic, Middlesex University
ADRIEN-MARIE LEGENDRE [VI.24], JEAN-BAPTISTE JOSEPH
FOURIER [VI.25], SIMÉON-DENIS POISSON [VI.27], AUGUSTIN-LOUIS
CAUCHY [VI.29], BERTRAND ARTHUR WILLIAM RUSSELL [VI.71],
FRIGYES (FRÉDÉRIC) RIESZ [VI.74]

Jeremy Gray, *Professor of History of Mathematics,*
The Open University
GEOMETRY [II.2], FUCHSIAN GROUPS [III.28],
CARL FRIEDRICH GAUSS [VI.26], AUGUST FERDINAND
MÖBIUS [VI.30], NICOLAI IVANOVICH LOBACHEVSKII [VI.31],
JÁNOS BOLYAI [VI.34], GEORG BERNHARD FRIEDRICH
RIEMANN [VI.49], WILLIAM KINGDON CLIFFORD [VI.55],
ÉLIE JOSEPH CARTAN [VI.69], THORALF SKOLEM [VI.81]

Ben Green, *Herchel Smith Professor of Pure Mathematics,*
University of Cambridge
THE GAMMA FUNCTION [III.31], IRRATIONAL AND
TRANSCENDENTAL NUMBERS [III.41], MODULAR
ARITHMETIC [III.58], NUMBER FIELDS [III.63],
QUADRATIC FORMS [III.73], TOPOLOGICAL SPACES [III.90],
TRIGONOMETRIC FUNCTIONS [III.92]

Ian Grojnowski, *Professor of Pure Mathematics,*
University of Cambridge
REPRESENTATION THEORY [IV.9]

Niccolò Guicciardini, *Associate Professor of History of Science,*
University of Bergamo
ISAAC NEWTON [VI.14]

Michael Harris, *Professor of Mathematics,*
Université Paris 7—Denis Diderot
"WHY MATHEMATICS?" YOU MIGHT ASK [VIII.2]

Ulf Hashagen, *Doctor, Munich Center for the History of Science*
and Technology, Deutsches Museum, Munich
PETER GUSTAV LEJEUNE DIRICHLET [VI.36]

Nigel Higson, *Professor of Mathematics,*
The Pennsylvania State University
OPERATOR ALGEBRAS [IV.15],
THE ATIYAH–SINGER INDEX THEOREM [V.2]

Andrew Hodges, *Tutorial Fellow in Mathematics,*
Wadham College, University of Oxford
ALAN TURING [VI.94]

F. E. A. Johnson, *Professor of Mathematics,*
University College London
BRAID GROUPS [III.4]

Mark Joshi, *Associate Professor,*
Centre for Actuarial Studies, University of Melbourne
THE MATHEMATICS OF MONEY [VII.9]

Kiran S. Kedlaya, *Associate Professor of Mathematics,*
Massachusetts Institute of Technology
FROM QUADRATIC RECIPROCITY TO CLASS FIELD THEORY [V.28]

Frank Kelly, *Professor of the Mathematics of Systems and*
Master of Christ's College, University of Cambridge
THE MATHEMATICS OF TRAFFIC IN NETWORKS [VII.4]

Sergiu Klainerman, *Professor of Mathematics,*
Princeton University
PARTIAL DIFFERENTIAL EQUATIONS [IV.12]

Jon Kleinberg, *Professor of Computer Science, Cornell University*
THE MATHEMATICS OF ALGORITHM DESIGN [VII.5]

Israel Kleiner, *Professor Emeritus,*
Department of Mathematics and Statistics, York University
KARL WEIERSTRASS [VI.44]

Jacek Klinowski, *Professor of Chemical Physics,*
University of Cambridge
MATHEMATICS AND CHEMISTRY [VII.1]

Eberhard Knobloch, *Professor, Institute for Philosophy, History*
of Science and Technology, Technical University of Berlin
GOTTFRIED WILHELM LEIBNIZ [VI.15]

János Kollár, *Professor of Mathematics, Princeton University*
ALGEBRAIC GEOMETRY [IV.4]

T. W. Körner, *Professor of Fourier Analysis,*
University of Cambridge
SPECIAL FUNCTIONS [III.85], TRANSFORMS [III.91],
THE BANACH–TARSKI PARADOX [V.3],
THE UBIQUITY OF MATHEMATICS [VIII.3]

Michael Krivelevich, *Professor of Mathematics,*
Tel Aviv University
EXTREMAL AND PROBABILISTIC COMBINATORICS [IV.19]

Peter D. Lax, *Professor, Courant Institute of*
Mathematical Sciences, New York University
RICHARD COURANT [VI.83]

Jean-François Le Gall, *Professor of Mathematics,*
Université Paris-Sud, Orsay
STOCHASTIC PROCESSES [IV.24]

W. B. R. Lickorish, *Emeritus Professor of Geometric Topology,*
University of Cambridge
KNOT POLYNOMIALS [III.44]

Martin W. Liebeck, *Professor of Pure Mathematics,*
Imperial College London
PERMUTATION GROUPS [III.68], THE CLASSIFICATION OF
FINITE SIMPLE GROUPS [V.7], THE INSOLUBILITY OF
THE QUINTIC [V.21]

Jesper Lützen, *Professor, Department of Mathematical Sciences,*
University of Copenhagen
JOSEPH LIOUVILLE [VI.39]

Des MacHale, *Associate Professor of Mathematics,*
University College Cork
GEORGE BOOLE [VI.43]

Alan L. Mackay, *Professor Emeritus,*
School of Crystallography, Birkbeck College
MATHEMATICS AND CHEMISTRY [VII.1]

Shahn Majid, *Professor of Mathematics,*
Queen Mary, University of London
QUANTUM GROUPS [III.75]

Lech Maligranda, *Professor of Mathematics,*
Luleå University of Technology, Sweden
STEFAN BANACH [VI.84]

David Marker, *Head of the Department of Mathematics,*
Statistics, and Computer Science, University of Illinois at Chicago
LOGIC AND MODEL THEORY [IV.23]

Jean Mawhin, *Professor of Mathematics,*
Université Catholique de Louvain
CHARLES-JEAN DE LA VALLÉE POUSSIN [VI.67]

Barry Mazur, *Gerhard Gade University Professor,*
Mathematics Department, Harvard University
ALGEBRAIC NUMBERS [IV.1]

Dusa McDuff, *Professor of Mathematics,*
Stony Brook University and Barnard College
ADVICE TO A YOUNG MATHEMATICIAN [VIII.6]

Colin McLarty, *Truman P. Handy Associate Professor of*
Philosophy and of Mathematics, Case Western Reserve University
EMMY NOETHER [VI.76]

Bojan Mohar, *Canada Research Chair in Graph Theory,*
Simon Fraser University; Professor of Mathematics,
University of Ljubljana
THE FOUR-COLOR THEOREM [V.12]

Peter M. Neumann, *Fellow and Tutor in Mathematics,*
The Queen's College, Oxford; University Lecturer in
Mathematics, University of Oxford
NIELS HENRIK ABEL [VI.33], ÉVARISTE GALOIS [VI.41],
FERDINAND GEORG FROBENIUS [VI.58], WILLIAM BURNSIDE [VI.60]

Catherine Nolan, *Associate Professor of Music,*
The University of Western Ontario
MATHEMATICS AND MUSIC [VII.13]

James Norris, *Professor of Stochastic Analysis,*
Statistical Laboratory, University of Cambridge
PROBABILITY DISTRIBUTIONS [III.71]

Brian Osserman, *Assistant Professor, Department of*
Mathematics, University of California, Davis
THE WEIL CONJECTURES [V.35]

Richard S. Palais, *Professor of Mathematics,*
University of California, Irvine
LINEAR AND NONLINEAR WAVES AND SOLITONS [III.49]

Marco Panza, *Directeur de Recherche, CNRS, Paris*
JOSEPH LOUIS LAGRANGE [VI.22]

Karen Hunger Parshall, *Professor of History and Mathematics,*
University of Virginia
THE DEVELOPMENT OF ABSTRACT ALGEBRA [II.3],
JAMES JOSEPH SYLVESTER [VI.42]

Gabriel P. Paternain, *Reader in Geometry and Dynamics,*
University of Cambridge
SYMPLECTIC MANIFOLDS [III.88]

Jeanne Peiffer, *Directeur de Recherche,*
CNRS, Centre Alexandre Koyré, Paris
THE BERNOULLIS [VI.18]

Birgit Petri, *Ph.D. Candidate,*
Fachbereich Mathematik, Technische Universität Darmstadt
LEOPOLD KRONECKER [VI.48], ANDRÉ WEIL [VI.93]

Carl Pomerance, *Professor of Mathematics, Dartmouth College*
COMPUTATIONAL NUMBER THEORY [IV.3]

Helmut Pulte, *Professor, Ruhr-Universität Bochum*
CARL GUSTAV JACOB JACOBI [VI.35]

Bruce Reed, *Canada Research Chair in Graph Theory,*
McGill University
THE ROBERTSON–SEYMOUR THEOREM [V.32]

Michael C. Reed, *Bishop-MacDermott Family Professor of*
Mathematics, Duke University
MATHEMATICAL BIOLOGY [VII.2]

Adrian Rice, *Associate Professor of Mathematics,*
Randolph-Macon College, Virginia
A CHRONOLOGY OF MATHEMATICAL EVENTS [VIII.7]

Eleanor Robson, *Senior Lecturer, Department of History and*
Philosophy of Science, University of Cambridge
NUMERACY [VIII.4]

Igor Rodnianski, *Professor of Mathematics, Princeton University*
THE HEAT EQUATION [III.36]

John Roe, *Professor of Mathematics,*
The Pennsylvania State University
OPERATOR ALGEBRAS [IV.15],
THE ATIYAH–SINGER INDEX THEOREM [V.2]

Mark Ronan, *Professor of Mathematics, University of*
Illinois at Chicago; Honorary Professor of Mathematics,
University College London
BUILDINGS [III.5], LIE THEORY [III.48]

Edward Sandifer, *Professor of Mathematics,*
Western Connecticut State University
LEONHARD EULER [VI.19]

Peter Sarnak, *Professor, Princeton University and*
Institute for Advanced Study, Princeton
ADVICE TO A YOUNG MATHEMATICIAN [VIII.6]

Tilman Sauer, *Doctor, Einstein Papers Project,*
California Institute of Technology
HERMANN MINKOWSKI [VI.64]

Norbert Schappacher, *Professor, Institut de Recherche*
Mathématique Avancée, Strasbourg
LEOPOLD KRONECKER [VI.48], ANDRÉ WEIL [VI.93]

Andrzej Schinzel, *Professor of Mathematics,*
Polish Academy of Sciences
WACŁAW SIERPIŃSKI [VI.77]

Erhard Scholz, *Professor of History of Mathematics, Department*
of Mathematics and Natural Sciences, Universität Wuppertal
FELIX HAUSDORFF [VI.68], HERMANN WEYL [VI.80]

Reinhard Siegmund-Schultze, *Professor,*
Faculty of Engineering and Science, University of Agder, Norway
HENRI LEBESGUE [VI.72], NORBERT WIENER [VI.85]

Gordon Slade, *Professor of Mathematics,*
University of British Columbia
PROBABILISTIC MODELS OF CRITICAL PHENOMENA [IV.25]

David J. Spiegelhalter, *Winton Professor of the Public*
Understanding of Risk, University of Cambridge
MATHEMATICS AND MEDICAL STATISTICS [VII.11]

Jacqueline Stedall, *Junior Research Fellow in Mathematics,*
The Queen's College, Oxford
FRANÇOIS VIÈTE [VI.9]

Arild Stubhaug, *Freelance Writer, Oslo*
SOPHUS LIE [VI.53]

Madhu Sudan, *Professor of Computer Science and Engineering,*
Massachusetts Institute of Technology
RELIABLE TRANSMISSION OF INFORMATION [VII.6]

Terence Tao, *Professor of Mathematics,*
University of California, Los Angeles
COMPACTNESS AND COMPACTIFICATION [III.9], DIFFERENTIAL
FORMS AND INTEGRATION [III.16], DISTRIBUTIONS [III.18],
THE FOURIER TRANSFORM [III.27], FUNCTION SPACES [III.29],
HAMILTONIANS [III.35], RICCI FLOW [III.78], THE SCHRÖDINGER
EQUATION [III.83], HARMONIC ANALYSIS [IV.11]

Jamie Tappenden, *Associate Professor of Philosophy,*
University of Michigan
GOTTLOB FREGE [VI.56]

C. H. Taubes, *William Petschek Professor of Mathematics,*
Harvard University
DIFFERENTIAL TOPOLOGY [IV.7]

Rüdiger Thiele, *Privatdozent, Universität Leipzig*
CHRISTIAN FELIX KLEIN [VI.57]

Burt Totaro, *Lowndean Professor of Astronomy and Geometry,*
University of Cambridge
ALGEBRAIC TOPOLOGY [IV.6]

Lloyd N. Trefethen, *Professor of Numerical Analysis,*
University of Oxford
NUMERICAL ANALYSIS [IV.21]

Dirk van Dalen, *Professor,*
Department of Philosophy, Utrecht University
LUITZEN EGBERTUS JAN BROUWER [VI.75]

Richard Weber, *Churchill Professor of Mathematics for*
Operational Research, University of Cambridge
THE SIMPLEX ALGORITHM [III.84]

Dominic Welsh, *Professor of Mathematics,*
Mathematical Institute, University of Oxford
MATROIDS [III.54]

Avi Wigderson, *Professor in the School of Mathematics,*
Institute for Advanced Study, Princeton
EXPANDERS [III.24], COMPUTATIONAL COMPLEXITY [IV.20]

Herbert S. Wilf, *Thomas A. Scott Professor of Mathematics,*
University of Pennsylvania
MATHEMATICS: AN EXPERIMENTAL SCIENCE [VIII.5]

David Wilkins, *Lecturer in Mathematics, Trinity College, Dublin*
WILLIAM ROWAN HAMILTON [VI.37]

Benjamin H. Yandell, *Pasadena, California (deceased)*
DAVID HILBERT [VI.63]

Eric Zaslow, *Professor of Mathematics, Northwestern University*
CALABI–YAU MANIFOLDS [III.6], MIRROR SYMMETRY [IV.16]

Doron Zeilberger, *Board of Governors Professor of Mathematics,*
Rutgers University
ENUMERATIVE AND ALGEBRAIC COMBINATORICS [IV.18]

Unattributed articles were written by the editors. In part III,
Imre Leader wrote the articles THE AXIOM OF CHOICE [III.1], THE
AXIOM OF DETERMINACY [III.2], CARDINALS [III.7], COUNTABLE
AND UNCOUNTABLE SETS [III.11], GRAPHS [III.34], JORDAN NORMAL
FORM [III.43], MEASURES [III.55], MODELS OF SET THEORY [III.57],
ORDINALS [III.66], THE PEANO AXIOMS [III.67], RINGS, IDEALS, AND
MODULES [III.81], and THE ZERMELO–FRAENKEL AXIOMS [III.99].
In part V, THE INDEPENDENCE OF THE CONTINUUM HYPOTHESIS
[V.18] is by Imre Leader and THE THREE-BODY PROBLEM [V.33]
is by June Barrow-Green. In part VI, June Barrow-Green wrote
all of the unattributed articles. All other unattributed articles
throughout the book were written by Timothy Gowers.

The Princeton Companion to Mathematics

Part I
Introduction

I.1 What Is Mathematics About?

It is notoriously hard to give a satisfactory answer to the question, "What is mathematics?" The approach of this book is not to try. Rather than giving a *definition* of mathematics, the intention is to give a good idea of what mathematics is by describing many of its most important concepts, theorems, and applications. Nevertheless, to make sense of all this information it is useful to be able to classify it somehow.

The most obvious way of classifying mathematics is by its subject matter, and that will be the approach of this brief introductory section and the longer section entitled SOME FUNDAMENTAL MATHEMATICAL DEFINITIONS [I.3]. However, it is not the only way, and not even obviously the best way. Another approach is to try to classify the kinds of questions that mathematicians like to think about. This gives a usefully different view of the subject: it often happens that two areas of mathematics that appear very different if you pay attention to their subject matter are much more similar if you look at the kinds of questions that are being asked. The last section of part I, entitled THE GENERAL GOALS OF MATHEMATICAL RESEARCH [I.4], looks at the subject from this point of view. At the end of that article there is a brief discussion of what one might regard as a third classification, not so much of mathematics itself but of the content of a typical article in a mathematics journal. As well as theorems and proofs, such an article will contain definitions, examples, lemmas, formulas, conjectures, and so on. The point of that discussion will be to say what these words mean and why the different kinds of mathematical output are important.

1 Algebra, Geometry, and Analysis

Although any classification of the subject matter of mathematics must immediately be hedged around with qualifications, there is a crude division that undoubtedly works well as a first approximation, namely the division of mathematics into algebra, geometry, and analysis. So let us begin with this, and then qualify it later.

1.1 Algebra versus Geometry

Most people who have done some high school mathematics will think of algebra as the sort of mathematics that results when you substitute letters for numbers. Algebra will often be contrasted with arithmetic, which is a more direct study of the numbers themselves. So, for example, the question, "What is 3×7?" will be thought of as belonging to arithmetic, while the question, "If $x + y = 10$ and $xy = 21$, then what is the value of the larger of x and y?" will be regarded as a piece of algebra. This contrast is less apparent in more advanced mathematics for the simple reason that it is very rare for numbers to appear without letters to keep them company.

There is, however, a different contrast, between algebra and *geometry*, which is much more important at an advanced level. The high school conception of geometry is that it is the study of shapes such as circles, triangles, cubes, and spheres together with concepts such as rotations, reflections, symmetries, and so on. Thus, the objects of geometry, and the processes that they undergo, have a much more visual character than the equations of algebra.

This contrast persists right up to the frontiers of modern mathematical research. Some parts of mathematics involve manipulating symbols according to certain rules: for example, a true equation remains true if you "do the same to both sides." These parts would typically be thought of as algebraic, whereas other parts are concerned with concepts that can be visualized, and these are typically thought of as geometrical.

However, a distinction like this is never simple. If you look at a typical research paper in geometry, will it be full of pictures? Almost certainly not. In fact, the methods used to solve geometrical problems very often involve a great deal of symbolic manipulation, although good powers of visualization may be needed to find and use these methods and pictures will typically underlie what is going on. As for algebra, is it "mere" symbolic manipulation? Not at all: very often one solves an algebraic problem by finding a way to visualize it.

As an example of visualizing an algebraic problem, consider how one might justify the rule that if a and b are positive integers then $ab = ba$. It is possible to approach the problem as a pure piece of algebra (perhaps proving it by induction), but the easiest way to convince yourself that it is true is to imagine a rectangular array that consists of a rows with b objects in each row. The total number of objects can be thought of as a lots of b, if you count it row by row, or as b lots of a, if you count it column by column. Therefore, $ab = ba$. Similar justifications can be given for other basic rules such as $a(b + c) = ab + ac$ and $a(bc) = (ab)c$.

In the other direction, it turns out that a good way of solving many geometrical problems is to "convert them into algebra." The most famous way of doing this is to use Cartesian coordinates. For example, suppose that you want to know what happens if you reflect a circle about a line L through its center, then rotate it through $40°$ counterclockwise, and then reflect it once more about the same line L. One approach is to visualize the situation as follows.

Imagine that the circle is made of a thin piece of wood. Then instead of reflecting it about the line you can rotate it through $180°$ about L (using the third dimension). The result will be upside down, but this does not matter if you simply ignore the thickness of the wood. Now if you look up at the circle from below while it is rotated counterclockwise through $40°$, what you will see is a circle being rotated *clockwise* through $40°$. Therefore, if you then turn it back the right way up, by rotating about L once again, the total effect will have been a clockwise rotation through $40°$.

Mathematicians vary widely in their ability and willingness to follow an argument like that one. If you cannot quite visualize it well enough to see that it is definitely correct, then you may prefer an algebraic approach, using the theory of linear algebra and matrices (which will be discussed in more detail in [I.3 §3.2]). To begin with, one thinks of the circle as the set of all pairs of numbers (x, y) such that $x^2 + y^2 \leqslant 1$. The two transformations, reflection in a line through the center of the circle and rotation through an angle θ, can both be represented by 2×2 matrices, which are arrays of numbers of the form $\left(\begin{smallmatrix} a & b \\ c & d \end{smallmatrix}\right)$. There is a slightly complicated, but purely algebraic, rule for multiplying matri-

ces together, and it is designed to have the property that if matrix A represents a transformation R (such as a reflection) and matrix B represents a transformation T, then the product AB represents the transformation that results when you first do T and then R. Therefore, one can solve the problem above by writing down the matrices that correspond to the transformations, multiplying them together, and seeing what transformation corresponds to the product. In this way, the geometrical problem has been converted into algebra and solved algebraically.

Thus, while one can draw a useful distinction between algebra and geometry, one should not imagine that the boundary between the two is sharply defined. In fact, one of the major branches of mathematics is even called ALGEBRAIC GEOMETRY [IV.4]. And as the above examples illustrate, it is often possible to translate a piece of mathematics from algebra into geometry or vice versa. Nevertheless, there is a definite difference between algebraic and geometric *methods of thinking*—one more symbolic and one more pictorial—and this can have a profound influence on which subjects a mathematician chooses to pursue.

1.2 Algebra versus Analysis

The word "analysis," used to denote a branch of mathematics, is not one that features at high school level. However, the word "calculus" is much more familiar, and differentiation and integration are good examples of mathematics that would be classified as analysis rather than algebra or geometry. The reason for this is that they involve *limiting processes*. For example, the derivative of a function f at a point x is the limit of the gradients of a sequence of chords of the graph of f, and the area of a shape with a curved boundary is defined to be the limit of the areas of rectilinear regions that fill up more and more of the shape. (These concepts are discussed in much more detail in [I.3 §5].)

Thus, as a first approximation, one might say that a branch of mathematics belongs to analysis if it involves limiting processes, whereas it belongs to algebra if you can get to the answer after just a finite sequence of steps. However, here again the first approximation is so crude as to be misleading, and for a similar reason: if one looks more closely one finds that it is not so much *branches* of mathematics that should be classified into analysis or algebra, but mathematical *techniques*.

Given that we cannot write out infinitely long proofs, how can we hope to prove anything about limiting processes? To answer this, let us look at the justification for the simple statement that the derivative of x^3 is $3x^2$. The usual reasoning is that the gradient of the chord of the line joining the two points (x, x^3) and $((x + h), (x + h)^3)$ is

$$\frac{(x + h)^3 - x^3}{x + h - x},$$

which works out as $3x^2 + 3xh + h^2$. As h "tends to zero," this gradient "tends to $3x^2$," so we say that the gradient at x *is* $3x^2$. But what if we wanted to be a bit more careful? For instance, if x is very large, are we really justified in ignoring the term $3xh$?

To reassure ourselves on this point, we do a small calculation to show that, whatever x is, the error $3xh + h^2$ can be made arbitrarily small, provided only that h is sufficiently small. Here is one way of going about it. Suppose we fix a small positive number ϵ, which represents the error we are prepared to tolerate. Then if $|h| \leqslant \epsilon/6x$, we know that $|3xh|$ is at most $\epsilon/2$. If in addition we know that $|h| \leqslant \sqrt{\epsilon/2}$, then we also know that $h^2 \leqslant \epsilon/2$. So, provided that $|h|$ is smaller than the minimum of the two numbers $\epsilon/6x$ and $\sqrt{\epsilon/2}$, the difference between $3x^2 + 3xh + h^2$ and $3x^2$ will be at most ϵ.

There are two features of the above argument that are typical of analysis. First, although the statement we wished to prove was about a limiting process, and was therefore "infinitary," the actual *work* that we needed to do to prove it was entirely finite. Second, the nature of that work was to find sufficient conditions for a certain fairly simple inequality (the inequality $|3xh + h^2| \leqslant \epsilon$) to be true.

Let us illustrate this second feature with another example: a proof that $x^4 - x^2 - 6x + 10$ is positive for every real number x. Here is an "analyst's argument." Note first that if $x \leqslant -1$ then $x^4 \geqslant x^2$ and $10 - 6x \geqslant 0$, so the result is certainly true in this case. If $-1 \leqslant x \leqslant 1$, then $|x^4 - x^2 - 6x|$ cannot be greater than $x^4 + x^2 + 6|x|$, which is at most 8, so $x^4 - x^2 - 6x \geqslant -8$, which implies that $x^4 - x^2 - 6x + 10 \geqslant 2$. If $1 \leqslant x \leqslant \frac{3}{2}$, then $x^4 \geqslant x^2$ and $6x \leqslant 9$, so $x^4 - x^2 - 6x + 10 \geqslant 1$. If $\frac{3}{2} \leqslant x \leqslant 2$, then $x^2 \geqslant \frac{9}{4}$, so $x^4 - x^2 = x^2(x^2 - 1) \geqslant \frac{9}{4} \cdot \frac{5}{4} > 2$. Also, $6x \leqslant 12$, so $10 - 6x \geqslant -2$. Therefore, $x^4 - x^2 - 6x + 10 > 0$. Finally, if $x \geqslant 2$, then $x^4 - x^2 = x^2(x^2 - 1) \geqslant 3x^2 \geqslant 6x$, from which it follows that $x^4 - x^2 - 6x + 10 \geqslant 10$.

The above argument is somewhat long, but each step consists in proving a rather simple inequality—this is the sense in which the proof is typical of analysis. Here, for contrast, is an "algebraist's proof." One simply points out that $x^4 - x^2 - 6x + 10$ is equal to $(x^2 - 1)^2 + (x - 3)^2$, and is therefore always positive.

This may make it seem as though, given the choice between analysis and algebra, one should go for algebra. After all, the algebraic proof was much shorter, and makes it obvious that the function is always positive. However, although there were several steps to the analyst's proof, they were all easy, and the brevity of the algebraic proof is misleading since no clue has been given about how the equivalent expression for $x^4 - x^2 - 6x + 10$ was found. And in fact, the general question of when a polynomial can be written as a sum of squares of other polynomials turns out to be an interesting and difficult one (particularly when the polynomials have more than one variable).

There is also a third, hybrid approach to the problem, which is to use calculus to find the points where $x^4 - x^2 - 6x + 10$ is minimized. The idea would be to calculate the derivative $4x^3 - 2x - 6$ (an algebraic process, justified by an analytic argument), find its roots (algebra), and check that the values of $x^4 - x^2 - 6x + 10$ at the roots of the derivative are positive. However, though the method is a good one for many problems, in this case it is tricky because the cubic $4x^3 - 2x - 6$ does not have integer roots. But one could use an analytic argument to find small intervals inside which the minimum must occur, and that would then reduce the number of cases that had to be considered in the first, purely analytic, argument.

As this example suggests, although analysis often involves limiting processes and algebra usually does not, a more significant distinction is that algebraists like to work with exact formulas and analysts use estimates. Or, to put it even more succinctly, algebraists like equalities and analysts like inequalities.

2   The Main Branches of Mathematics

Now that we have discussed the differences between algebraic, geometrical, and analytical thinking, we are ready for a crude classification of the subject matter of mathematics. We face a potential confusion, because the words "algebra," "geometry," and "analysis" refer *both* to specific branches of mathematics *and* to ways of thinking that cut across many different branches. Thus, it makes sense to say (and it is true) that some branches of analysis are more algebraic (or geometrical) than others; similarly, there is no paradox in the fact that algebraic topology is almost entirely algebraic and geometrical in character, even though the objects it studies, topological spaces, are part of analysis. In this section, we shall think primarily in terms of subject matter, but it is important to keep in mind the distinctions of the previous section and be aware that they are in some ways more fundamental. Our descriptions will be very brief: further reading about the main branches of mathematics can be found in parts II and IV, and more specific points are discussed in parts III and V.

2.1 Algebra

The word "algebra," when it denotes a branch of mathematics, means something more specific than manipulation of symbols and a preference for equalities over inequalities. Algebraists are concerned with number systems, polynomials, and more abstract structures such as groups, fields, vector spaces, and rings (discussed in some detail in SOME FUNDAMENTAL MATHEMATICAL DEFINITIONS [I.3]). Historically, the abstract structures emerged as generalizations from concrete instances. For instance, there are important analogies between the set of all integers and the set of all polynomials with rational (for example) coefficients, which are brought out by the fact that both sets are examples of algebraic structures known as *Euclidean domains*. If one has a good understanding of Euclidean domains, one can apply this understanding to integers and polynomials.

This highlights a contrast that appears in many branches of mathematics, namely the distinction between general, abstract statements and particular, concrete ones. One algebraist might be thinking about groups, say, in order to understand a particular rather complicated group of symmetries, while another might be interested in the general theory of groups on the grounds that they are a fundamental class of mathematical objects. The development of abstract algebra from its concrete beginnings is discussed in THE ORIGINS OF MODERN ALGEBRA [II.3].

A supreme example of a theorem of the first kind is THE INSOLUBILITY OF THE QUINTIC [V.21]—the result that there is no formula for the roots of a quintic polynomial in terms of its coefficients. One proves this theorem by analyzing symmetries associated with the roots of a polynomial, and understanding the group that these symmetries form. This concrete example of a group (or rather, class of groups, one for each polynomial) played a very important part in the development of the abstract theory of groups.

As for the second kind of theorem, a good example is THE CLASSIFICATION OF FINITE SIMPLE GROUPS [V.7], which describes the basic building blocks out of which any finite group can be built.

Algebraic structures appear throughout mathematics, and there are many applications of algebra to other areas, such as number theory, geometry, and even mathematical physics.

2.2 Number Theory

Number theory is largely concerned with properties of the set of positive integers, and as such has a considerable overlap with algebra. But a simple example that illustrates the difference between a typical question in algebra and a typical question in number theory is provided by the equation $13x - 7y = 1$. An algebraist would simply note that there is a one-parameter family of solutions: if $y = \lambda$ then $x = (1 + 7\lambda)/13$, so the general solution is $(x, y) = ((1 + 7\lambda)/13, \lambda)$. A number theorist would be interested in *integer* solutions, and would therefore work out for which integers λ the number $1 + 7\lambda$ is a multiple of 13. (The answer is that $1 + 7\lambda$ is a multiple of 13 if and only if λ has the form $13m + 11$ for some integer m.)

However, this description does not do full justice to modern number theory, which has developed into a highly sophisticated subject. Most number theorists are not directly trying to solve equations in integers; instead they are trying to understand structures that were originally developed to study such equations but which then took on a life of their own and became objects of study in their own right. In some cases, this process has happened several times, so the phrase "number theory" gives a very misleading picture of what some number theorists do. Nevertheless, even the most abstract parts of the subject can have down-to-earth applications: a notable example is Andrew Wiles's famous proof of FERMAT'S LAST THEOREM [V.10].

Interestingly, in view of the discussion earlier, number theory has two fairly distinct subbranches, known as ALGEBRAIC NUMBER THEORY [IV.1] and ANALYTIC NUMBER THEORY [IV.2]. As a rough rule of thumb, the study of equations in integers leads to algebraic number theory, while analytic number theory has its roots in the study of prime numbers, but the true picture is of course more complicated.

2.3 Geometry

A central object of study is the *manifold*, which is discussed in [I.3 §6.9]. Manifolds are higher-dimensional generalizations of shapes like the surface of a sphere: a small portion of a manifold looks flat, but the manifold as a whole may be curved in complicated ways. Most people who call themselves geometers are studying manifolds in one way or another. As with algebra, some will be interested in particular manifolds and others in the more general theory.

Within the study of manifolds, one can attempt a further classification, according to when two manifolds are regarded as "genuinely distinct." A topologist regards two objects as the same if one can be continuously deformed, or "morphed," into the other; thus, for example, an apple and a pear would count as the same for a topologist. This means that relative distances are not important to topologists, since one can change them by suitable continuous stretches. A *differential* topologist asks for the deformations to be "smooth" (which means "sufficiently differentiable"). This results in a finer classification of manifolds and a different set of problems. At the other, more "geometrical," end of the spectrum are mathematicians who are much more interested in the precise nature of the distances between points on a manifold (a concept that would not make sense to a topologist) and in auxiliary structures that one can associate with a manifold. See RIEMANNIAN METRICS [I.3 §6.10] and RICCI FLOW [III.78] for some indication of what the more geometrical side of geometry is like.

2.4   Algebraic Geometry

As its name suggests, algebraic geometry does not have an obvious place in the above classification, so it is easier to discuss it separately. Algebraic geometers also study manifolds, but with the important difference that their manifolds are defined using polynomials. (A simple example of this is the surface of a sphere, which can be defined as the set of all (x, y, z) such that $x^2 + y^2 + z^2 = 1$.) This means that algebraic geometry is algebraic in the sense that it is "all about polynomials" but geometric in the sense that the set of solutions of a polynomial in several variables is a geometric object.

An important part of algebraic geometry is the study of *singularities*. Often the set of solutions to a system of polynomial equations is similar to a manifold, but has a few exceptional, singular points. For example, the equation $x^2 = y^2 + z^2$ defines a (double) cone, which has its vertex at the origin $(0, 0, 0)$. If you look at a small enough neighborhood of a point x on the cone, then, provided x is not $(0, 0, 0)$, the neighborhood will resemble a flat plane. However, if x *is* $(0, 0, 0)$, then no matter how small the neighborhood is, you will still see the

vertex of the cone. Thus, $(0, 0, 0)$ is a singularity. (This means that the cone is not actually a manifold, but a "manifold with a singularity.")

The interplay between algebra and geometry is part of what gives algebraic geometry its fascination. A further impetus to the subject comes from its connections to other branches of mathematics. There is a particularly close connection with number theory, explained in ARITHMETIC GEOMETRY [IV.5]. More surprisingly, there are important connections between algebraic geometry and mathematical physics. See MIRROR SYMMETRY [IV.16] for an account of some of these.

2.5   Analysis

Analysis comes in many different flavors. A major topic is the study of PARTIAL DIFFERENTIAL EQUATIONS [IV.12]. This began because partial differential equations were found to govern many physical processes, such as motion in a gravitational field, for example. But partial differential equations arise in purely mathematical contexts as well—particularly in geometry—so they give rise to a big branch of mathematics with many subbranches and links to many other areas.

Like algebra, analysis has an abstract side as well. In particular, certain abstract structures, such as BANACH SPACES [III.62], HILBERT SPACES [III.37], C^*-ALGEBRAS [IV.15 §3], and VON NEUMANN ALGEBRAS [IV.15 §2], are central objects of study. These four structures are all infinite-dimensional VECTOR SPACES [I.3 §2.3], and the last two are "algebras," which means that one can multiply their elements together as well as adding them and multiplying them by scalars. Because these structures are infinite dimensional, studying them involves limiting arguments, which is why they belong to analysis. However, the extra algebraic structure of C^*-algebras and von Neumann algebras means that in those areas substantial use is made of algebraic tools as well. And as the word "space" suggests, geometry also has a very important role.

DYNAMICS [IV.14] is another significant branch of analysis. It is concerned with what happens when you take a simple process and do it over and over again. For example, if you take a complex number z_0, then let $z_1 = z_0^2 + 2$, and then let $z_2 = z_1^2 + 2$, and so on, then what is the limiting behavior of the sequence $z_0, z_1, z_2, \ldots$? Does it head off to infinity or stay in some bounded region? The answer turns out to depend in a complicated way on the original number z_0. Exactly *how* it depends on z_0 is a question in dynamics.

Sometimes the process to be repeated is an "infinitesimal" one. For example, if you are told the positions, velocities, and masses of all the planets in the solar system at a particular moment (as well as the mass of the Sun), then there is a simple rule that tells you how the positions and velocities will be different an instant later. Later, the positions and velocities have changed, so the calculation changes; but the basic rule is the same, so one can regard the whole process as applying the same simple infinitesimal process infinitely many times. The correct way to formulate this is by means of partial differential equations and therefore much of dynamics is concerned with the long-term behavior of solutions to these.

2.6 Logic

The word "logic" is sometimes used as a shorthand for all branches of mathematics that are concerned with fundamental questions about mathematics itself, notably SET THEORY [IV.22], CATEGORY THEORY [III.8], MODEL THEORY [IV.23], and logic in the narrower sense of "rules of deduction." Among the triumphs of set theory are GÖDEL'S INCOMPLETENESS THEOREMS [V.15] and Paul Cohen's proof of THE INDEPENDENCE OF THE CONTINUUM HYPOTHESIS [V.18]. Gödel's theorems in particular had a dramatic effect on philosophical perceptions of mathematics, though now that it is understood that not every mathematical statement has a proof or disproof most mathematicians carry on much as before, since most statements they encounter *do* tend to be decidable. However, set theorists are a different breed. Since Gödel and Cohen, many further statements have been shown to be undecidable, and many new axioms have been proposed that would make them decidable. Thus, decidability is now studied for *mathematical* rather than philosophical reasons.

Category theory is another subject that began as a study of the processes of mathematics and then became a mathematical subject in its own right. It differs from set theory in that its focus is less on mathematical objects themselves than on what is done to those objects—in particular, the maps that transform one to another.

A *model* for a collection of axioms is a mathematical structure for which those axioms, suitably interpreted, are true. For example, any concrete example of a group is a model for the axioms of group theory. Set theorists study models of set-theoretic axioms, and these are essential to the proofs of the famous theorems mentioned above, but the notion of a model is more widely applicable and has led to important discoveries in fields well outside set theory.

2.7 Combinatorics

There are various ways in which one can try to define combinatorics. None is satisfactory on its own, but together they give some idea of what the subject is like. A first definition is that combinatorics is about counting things. For example, how many ways are there of filling an $n \times n$ square grid with 0s and 1s if you are allowed at most two 1s in each row and at most two 1s in each column? Because this problem asks us to count something, it is, in a rather simple sense, combinatorial.

Combinatorics is sometimes called "discrete mathematics" because it is concerned with "discrete" structures as opposed to "continuous" ones. Roughly speaking, an object is discrete if it consists of points that are isolated from each other, and continuous if you can move from one point to another without making sudden jumps. (A good example of a discrete structure is the *integer lattice* $\mathbb{Z}^2$, which is the grid consisting of all points in the plane with integer coordinates, and a good example of a continuous one is the surface of a sphere.) There is a close affinity between combinatorics and theoretical computer science (which deals with the quintessentially discrete structure of sequences of 0s and 1s), and combinatorics is sometimes contrasted with analysis, though in fact there are several connections between the two.

A third view of combinatorics is that it is concerned with mathematical structures that have "few constraints." This idea helps to explain why number theory, despite the fact that it studies (among other things) the distinctly discrete set of all positive integers, is not considered a branch of combinatorics.

In order to illustrate this last contrast, here are two somewhat similar problems, both about positive integers.

(i) Is there a positive integer that can be written in a thousand different ways as a sum of two squares?
(ii) Let $a_1, a_2, a_3, \ldots$ be a sequence of positive integers, and suppose that each a_n lies between n^2 and $(n+1)^2$. Will there always be a positive integer that can be written in a thousand different ways as a sum of two numbers from the sequence?

The first question counts as number theory, since it concerns a very specific sequence—the sequence of squares—and one would expect to use properties of this special set of numbers in order to determine the answer, which turns out to be yes.[1]

The second question concerns a far less structured sequence. All we know about a_n is its rough size—it is fairly close to n^2—but we know nothing about its more detailed properties, such as whether it is a prime, or a perfect cube, or a power of 2, etc. For this reason, the second problem belongs to combinatorics. The answer is not known. If the answer turns out to be yes, then it will show that, in a sense, the number theory in the first problem was an illusion and that all that really mattered was the rough rate of growth of the sequence of squares.

2.8   Theoretical Computer Science

This branch of mathematics is described at considerable length in part IV, so we shall be brief here. Broadly speaking, theoretical computer science is concerned with efficiency of computation, meaning the amounts of various resources, such as time and computer memory, needed to perform given computational tasks. There are mathematical models of computation that allow one to study questions about computational efficiency in great generality without having to worry about precise details of how algorithms are implemented. Thus, theoretical computer science is a genuine branch of pure mathematics: in theory, one could be an excellent theoretical computer scientist and be unable to program a computer. However, it has had many notable applications as well, especially to cryptography (see MATHEMATICS AND CRYPTOGRAPHY [VII.7] for more on this).

2.9   Probability

There are many phenomena, from biology and economics to computer science and physics, that are so complicated that instead of trying to understand them in complete detail one tries to make probabilistic statements instead. For example, if you wish to analyze how a disease is likely to spread, you cannot hope to take account of all the relevant information (such as who will come into contact with whom) but you can build a mathematical model and analyze it. Such models can have

unexpectedly interesting behavior with direct practical relevance. For example, it may happen that there is a "critical probability" p with the following property: if the probability of infection after contact of a certain kind is above p then an epidemic may very well result, whereas if it is below p then the disease will almost certainly die out. A dramatic difference in behavior like this is called a *phase transition*. (See PROBABILISTIC MODELS OF CRITICAL PHENOMENA [IV.25] for further discussion.)

Setting up an appropriate mathematical model can be surprisingly difficult. For example, there are physical circumstances where particles travel in what appears to be a completely random manner. Can one make sense of the notion of a random continuous path? It turns out that one can—the result is the elegant theory of BROWNIAN MOTION [IV.24]—but the proof that one can is highly sophisticated, roughly speaking because the set of all possible paths is so complex.

2.10   Mathematical Physics

The relationship between mathematics and physics has changed profoundly over the centuries. Up to the eighteenth century there was no sharp distinction drawn between mathematics and physics, and many famous mathematicians could also be regarded as physicists, at least some of the time. During the nineteenth century and the beginning of the twentieth century this situation gradually changed, until by the middle of the twentieth century the two disciplines were very separate. And then, toward the end of the twentieth century, mathematicians started to find that ideas that had been discovered by physicists had huge mathematical significance.

There is still a big cultural difference between the two subjects: mathematicians are far more interested in finding rigorous proofs, whereas physicists, who use mathematics as a tool, are usually happy with a convincing argument for the truth of a mathematical statement, even if that argument is not actually a proof. The result is that physicists, operating under less stringent constraints, often discover fascinating mathematical phenomena long before mathematicians do.

Finding rigorous proofs to back up these discoveries is often extremely hard: it is far more than a pedantic exercise in certifying the truth of statements that no physicist seriously doubted. Indeed, it often leads to further mathematical discoveries. The articles VERTEX OPERATOR ALGEBRAS [IV.17], MIRROR SYMMETRY

---

1. Here is a quick hint at a proof. At the beginning of ANALYTIC NUMBER THEORY [IV.2] you will find a condition that tells you precisely which numbers can be written as sums of two squares. From this criterion it follows that "most" numbers cannot. A careful count shows that if N is a large integer, then there are many more expressions of the form $m^2 + n^2$ with both m^2 and n^2 less than N than there are numbers less than $2N$ that can be written as a sum of two squares. Therefore there is a lot of duplication.

[IV.16], GENERAL RELATIVITY AND THE EINSTEIN EQUA-
TIONS [IV.13], and OPERATOR ALGEBRAS [IV.15] describe
some fascinating examples of how mathematics and
physics have enriched each other.

I.2 The Language and Grammar of Mathematics

1 Introduction

It is a remarkable phenomenon that children can learn
to speak without ever being consciously aware of the
sophisticated grammar they are using. Indeed, adults
too can live a perfectly satisfactory life without ever
thinking about ideas such as parts of speech, subjects,
predicates, or subordinate clauses. Both children and
adults can easily recognize ungrammatical sentences,
at least if the mistake is not too subtle, and to do this
it is not necessary to be able to explain the rules that
have been violated. Nevertheless, there is no doubt that
one's understanding of language is hugely enhanced by
a knowledge of basic grammar, and this understanding
is essential for anybody who wants to do more with
language than use it unreflectingly as a means to a
nonlinguistic end.

The same is true of mathematical language. Up to
a point, one can do and speak mathematics without
knowing how to classify the different sorts of words
one is using, but many of the sentences of advanced
mathematics have a complicated structure that is much
easier to understand if one knows a few basic terms
of mathematical grammar. The object of this section
is to explain the most important mathematical "parts
of speech," some of which are similar to those of natural languages and others quite different. These are
normally taught right at the beginning of a university
course in mathematics. Much of *The Companion* can be
understood without a precise knowledge of mathematical grammar, but a careful reading of this article will
help the reader who wishes to follow some of the later,
more advanced parts of the book.

The main reason for using mathematical grammar is
that the statements of mathematics are supposed to
be completely precise, and it is not possible to achieve
complete precision unless the language one uses is free
of many of the vaguenesses and ambiguities of ordinary
speech. Mathematical sentences can also be highly complex: if the parts that made them up were not clear and
simple, then the unclarities would rapidly accumulate
and render the sentences unintelligible.

To illustrate the sort of clarity and simplicity that is
needed in mathematical discourse, let us consider the
famous mathematical sentence "Two plus two equals
four" as a sentence of English rather than of mathematics, and try to analyze it grammatically. On the face of it,
it contains three nouns ("two," "two," and "four"), a verb
("equals") and a conjunction ("plus"). However, looking
more carefully we may begin to notice some oddities.
For example, although the word "plus" resembles the
word "and," the most obvious example of a conjunction, it does not behave in quite the same way, as is
shown by the sentence "Mary and Peter love Paris." The
verb in this sentence, "love," is plural, whereas the verb
in the previous sentence, "equals," was singular. So the
word "plus" seems to take two objects (which happen
to be numbers) and produce out of them a new, single object, while "and" conjoins "Mary" and "Peter" in
a looser way, leaving them as distinct people.

Reflecting on the word "and" a bit more, one finds
that it has two very different uses. One, as above, is to
link two nouns, whereas the other is to join two whole
sentences together, as in "Mary likes Paris and Peter
likes New York." If we want the basics of our language
to be absolutely clear, then it will be important to be
aware of this distinction. (When mathematicians are at
their most formal, they simply outlaw the noun-linking
use of "and"—a sentence such as "3 and 5 are prime
numbers" is then paraphrased as "3 is a prime number
and 5 is a prime number.")

This is but one of many similar questions: anybody
who has tried to classify all words into the standard
eight parts of speech will know that the classification is
hopelessly inadequate. What, for example, is the role of
the word "six" in the sentence "This section has six sub-
sections"? Unlike "two" and "four" earlier, it is certainly
not a noun. Since it modifies the noun "subsection" it
would traditionally be classified as an adjective, but
it does not behave like most adjectives: the sentences
"My car is not very fast" and "Look at that tall building" are perfectly grammatical, whereas the sentences
"My car is not very six" and "Look at that six building"
are not just nonsense but ungrammatical nonsense. So
do we classify adjectives further into numerical adjectives and nonnumerical adjectives? Perhaps we do, but
then our troubles will be only just beginning. For example, what about possessive adjectives such as "my" and
"your"? In general, the more one tries to refine the classification of English words, the more one realizes how
many different grammatical roles there are.

2 Four Basic Concepts

Another word that famously has three quite distinct meanings is "is." The three meanings are illustrated in the following three sentences.

(1) 5 is the square root of 25.
(2) 5 is less than 10.
(3) 5 is a prime number.

In the first of these sentences, "is" could be replaced by "equals": it says that two objects, 5 and the square root of 25, are in fact one and the same object, just as it does in the English sentence "London is the capital of the United Kingdom." In the second sentence, "is" plays a completely different role. The words "less than 10" form an adjectival phrase, specifying a property that numbers may or may not have, and "is" in this sentence is like "is" in the English sentence "Grass is green." As for the third sentence, the word "is" there means "is an example of," as it does in the English sentence "Mercury is a planet."

These differences are reflected in the fact that the sentences cease to resemble each other when they are written in a more symbolic way. An obvious way to write (1) is $5 = \sqrt{25}$. As for (2), it would usually be written $5 < 10$, where the symbol "<" means "is less than." The third sentence would normally not be written symbolically because the concept of a prime number is not quite basic enough to have universally recognized symbols associated with it. However, it is sometimes useful to do so, and then one must invent a suitable symbol. One way to do it would be to adopt the convention that if n is a positive integer, then $P(n)$ stands for the sentence "n is prime." Another way, which does not hide the word "is," is to use the language of sets.

2.1 Sets

Broadly speaking, a *set* is a collection of objects, and in mathematical discourse these objects are mathematical ones such as numbers, points in space, or even other sets. If we wish to rewrite sentence (3) symbolically, another way to do it is to define P to be the collection, or set, of all prime numbers. Then we can rewrite it as "5 belongs to the set P." This notion of belonging to a set *is* sufficiently basic to deserve its own symbol, and the symbol used is "$\in$." So a fully symbolic way of writing the sentence is $5 \in P$.

The members of a set are usually called its *elements*, and the symbol "$\in$" is usually read "is an element of." So the "is" of sentence (3) is more like "$\in$" than "=."

Although one cannot directly substitute the phrase "is an element of" for "is," one can do so if one is prepared to modify the rest of the sentence a little.

There are three common ways to denote a specific set. One is to list its elements inside curly brackets: $\{2, 3, 5, 7, 11, 13, 17, 19\}$, for example, is the set whose elements are the eight numbers 2, 3, 5, 7, 11, 13, 17, and 19. The majority of sets considered by mathematicians are too large for this to be feasible—indeed, they are often infinite—so a second way to denote sets is to use dots to imply a list that is too long to write down: for example, the expressions $\{1, 2, 3, \ldots, 100\}$ and $\{2, 4, 6, 8, \ldots\}$ can be used to represent the set of all positive integers up to 100 and the set of all positive even numbers, respectively. A third way, and the way that is most important, is to define a set via a *property*: an example that shows how this is done is the expression $\{x : x \text{ is prime and } x < 20\}$. To read an expression such as this, one first reads the opening curly bracket as "The set of." Next, one reads the symbol that occurs before the colon. The colon itself one reads as "such that." Finally, one reads what comes after the colon, which is the property that determines the elements of the set. In this instance, we end up saying, "The set of x such that x is prime and x is less than 20," which is in fact equal to the set $\{2, 3, 5, 7, 11, 13, 17, 19\}$ considered earlier.

Many sentences of mathematics can be rewritten in set-theoretic terms. For example, sentence (2) earlier could be written as $5 \in \{n : n < 10\}$. Often there is no point in doing this (as here, where it is much easier to write $5 < 10$) but there are circumstances where it becomes extremely convenient. For example, one of the great advances in mathematics was the use of Cartesian coordinates to translate geometry into algebra and the way this was done was to define geometrical objects as sets of points, where points were themselves defined as pairs or triples of numbers. So, for example, the set $\{(x, y) : x^2 + y^2 = 1\}$ is (or represents) a circle of radius 1 with its center at the origin $(0, 0)$. That is because, by the Pythagorean theorem, the distance from $(0, 0)$ to (x, y) is $\sqrt{x^2 + y^2}$, so the sentence "$x^2 + y^2 = 1$" can be reexpressed geometrically as "the distance from $(0, 0)$ to (x, y) is 1." If all we ever cared about was which points were in the circle, then we could make do with sentences such as "$x^2 + y^2 = 1$," but in geometry one often wants to consider the entire circle as a single object (rather than as a multiplicity of points, or as a property that points might have), and then set-theoretic language is indispensable.

A second circumstance where it is usually hard to do without sets is when one is defining new mathematical objects. Very often such an object is a set together with a *mathematical structure* imposed on it, which takes the form of certain relationships among the elements of the set. For examples of this use of set-theoretic language, see sections 1 and 2, on number systems and algebraic structures, respectively, in SOME FUNDAMENTAL MATHEMATICAL DEFINITIONS [I.3].

Sets are also very useful if one is trying to do *metamathematics*, that is, to prove statements not about mathematical objects but about the process of mathematical reasoning itself. For this it helps a lot if one can devise a very simple language—with a small vocabulary and an uncomplicated grammar—into which it is in principle possible to translate all mathematical arguments. Sets allow one to reduce greatly the number of parts of speech that one needs, turning almost all of them into nouns. For example, with the help of the membership symbol "∈" one can do without adjectives, as the translation of "5 is a prime number" (where "prime" functions as an adjective) into "5 ∈ P" has already suggested.[1] This is of course an artificial process—imagine replacing "roses are red" by "roses belong to the set R"—but in this context it is not important for the formal language to be natural and easy to understand.

2.2 Functions

Let us now switch attention from the word "is" to some other parts of the sentences (1)–(3), focusing first on the phrase "the square root of" in sentence (1). If we wish to think about this phrase grammatically, then we should analyze what sort of role it plays in a sentence, and the analysis is simple: in virtually any mathematical sentence where the phrase appears, it is followed by the name of a number. If the number is n, then this produces the slightly longer phrase, "the square root of n," which is a noun phrase that denotes a number and plays the same grammatical role as a number (at least when the number is used as a noun rather than as an adjective). For instance, replacing "5" by "the square root of 25" in the sentence "5 is less than 7" yields a new sentence, "The square root of 25 is less than 7," that is still grammatically correct (and true).

One of the most basic activities of mathematics is to take a mathematical object and transform it into

another one, sometimes of the same kind and sometimes not. "The square root of" transforms numbers into numbers, as do "four plus," "two times," "the cosine of," and "the logarithm of." A nonnumerical example is "the center of gravity of," which transforms geometrical shapes (provided they are not too exotic or complicated to have a center of gravity) into points—meaning that if S stands for a shape, then "the center of gravity of S" stands for a point. A *function* is, roughly speaking, a mathematical transformation of this kind.

It is not easy to make this definition more precise. To ask, "What is a function?" is to suggest that the answer should be a *thing* of some sort, but functions seem to be more like processes. Moreover, when they appear in mathematical sentences they do not behave like nouns. (They are more like prepositions, though with a definite difference that will be discussed in the next subsection.) One might therefore think it inappropriate to ask what kind of object "the square root of" is. Should one not simply be satisfied with the grammatical analysis already given?

As it happens, no. Over and over again, throughout mathematics, it is useful to think of a mathematical phenomenon, which may be complex and very unthinglike, as a single object. We have already seen a simple example: a collection of infinitely many points in the plane or space is sometimes better thought of as a single geometrical shape. Why should one wish to do this for functions? Here are two reasons. First, it is convenient to be able to say something like, "The derivative of sin is cos," or to speak in general terms about some functions being differentiable and others not. More generally, functions can have *properties*, and in order to discuss those properties one needs to think of functions as things. Second, many algebraic structures are most naturally thought of as sets of functions. (See, for example, the discussion of groups and symmetry in [I.3 §2.1]. See also HILBERT SPACES [III.37], FUNCTION SPACES [III.29], and VECTOR SPACES [I.3 §2.3].)

If f is a function, then the notation $f(x) = y$ means that f turns the object x into the object y. Once one starts to speak formally about functions, it becomes important to specify exactly which objects are to be subjected to the transformation in question, and what sort of objects they can be transformed into. One of the main reasons for this is that it makes it possible to discuss another notion that is central to mathematics, that of *inverting* a function. (See [I.4 §1] for a discussion of why it is central.) Roughly speaking, the inverse of a function is another function that undoes it, and that it

---

undoes; for example, the function that takes a number n to $n - 4$ is the inverse of the function that takes n to $n + 4$, since if you add four and then subtract four, or vice versa, you get the number you started with.

Here is a function f that cannot be inverted. It takes each number and replaces it by the nearest multiple of 100, rounding up if the number ends in 50. Thus, $f(113) = 100$, $f(3879) = 3900$, and $f(1050) = 1100$. It is clear that there is no way of undoing this process with a function g. For example, in order to undo the effect of f on the number 113 we would need $g(100)$ to equal 113. But the same argument applies to every number that is at least as big as 50 and smaller than 150, and $g(100)$ cannot be more than one number at once.

Now let us consider the function that doubles a number. Can this be inverted? Yes it can, one might say: just divide the number by two again. And much of the time this would be a perfectly sensible response, but not, for example, if it was clear from the context that the numbers being talked about were positive integers. Then one might be focusing on the difference between even and odd numbers, and this difference could be encapsulated by saying that odd numbers are precisely those numbers n for which the equation $2x = n$ does *not* have a solution. (Notice that one can undo the doubling process by halving. The problem here is that the relationship is not symmetrical: there is no function that can be undone by doubling, since you could never get back to an odd number.)

To specify a function, therefore, one must be careful to specify two sets as well: the *domain*, which is the set of objects to be transformed, and the *range*, which is the set of objects they are allowed to be transformed into. A function f from a set A to a set B is a rule that specifies, for each element x of A, an element $y = f(x)$ of B. Not every element of the range needs to be used: consider once again the example of "two times" when the domain and range are both the set of all positive integers. The set $\{f(x) : x \in A\}$ of values actually taken by f is called the *image* of f. (Slightly confusingly, the word "image" is also used in a different sense, applied to the individual *elements* of A: if $x \in A$, then its image is $f(x)$.)

The following symbolic notation is used. The expression $f : A \to B$ means that f is a function with domain A and range B. If we then write $f(x) = y$, we know that x must be an element of A and y must be an element of B. Another way of writing $f(x) = y$ that is sometimes more convenient is $f : x \mapsto y$. (The bar on the arrow is to distinguish it from the arrow in $f : A \to B$, which has a very different meaning.)

If we want to undo the effect of a function $f : A \to B$, then we can, as long as we avoid the problem that occurred with the approximating function discussed earlier. That is, we can do it if $f(x)$ and $f(x')$ are different whenever x and x' are different elements of A. If this condition holds, then f is called an *injection*. On the other hand, if we want to find a function g that is undone by f, then we can do so as long as we avoid the problem of the integer-doubling function. That is, we can do it if every element y of B is equal to $f(x)$ for some element x of A (so that we have the option of setting $g(y) = x$). If this condition holds, then f is called a *surjection*. If f is both an injection and a surjection, then f is called a *bijection*. Bijections are precisely the functions that have inverses.

It is important to realize that not all functions have tidy definitions. Here, for example, is the specification of a function from the positive integers to the positive integers: $f(n) = n$ if n is a prime number, $f(n) = k$ if n is of the form 2^k for an integer k greater than 1, and $f(n) = 13$ for all other positive integers n. This function has an unpleasant, arbitrary definition but it is nevertheless a perfectly legitimate function. Indeed, "most" functions, though not most functions that one actually uses, are so arbitrary that they cannot be defined. (Such functions may not be useful as individual objects, but they are needed so that the set of all functions from one set to another has an interesting mathematical structure.)

2.3 Relations

Let us now think about the grammar of the phrase "less than" in sentence (2). As with "the square root of," it must always be followed by a mathematical object (in this case a number again). Once we have done this we obtain a phrase such as "less than n," which is importantly different from "the square root of n" because it behaves like an adjective rather than a noun, and refers to a property rather than an object. This is just how prepositions behave in English: look, for example, at the word "under" in the sentence "The cat is under the table."

At a slightly higher level of formality, mathematicians like to avoid too many parts of speech, as we have already seen for adjectives. So there is no symbol for "less than": instead, it is combined with the previous word "is" to make the phrase "is less than," which is

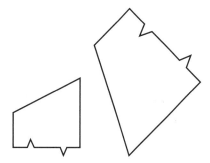

Figure 1 Similar shapes.

denoted by the symbol "<." The grammatical rules for this symbol are once again simple. To use "<" in a sentence, one should precede it by a noun and follow it by a noun. For the resulting grammatically correct sentence to make sense, the nouns should refer to numbers (or perhaps to more general objects that can be put in order). A mathematical "object" that behaves like this is called a *relation*, though it might be more accurate to call it a potential relationship. "Equals" and "is an element of" are two other examples of relations.

As with functions, it is important, when specifying a relation, to be careful about which objects are to be related. Usually a relation comes with a set A of objects that may or may not be related to each other. For example, the relation "<" might be defined on the set of all positive integers, or alternatively on the set of all real numbers; strictly speaking these are different relations. Sometimes relations are defined with reference to two sets A and B. For example, if the relation is "$\in$," then A might be the set of all positive integers and B the set of all sets of positive integers.

There are many situations in mathematics where one wishes to regard different objects as "essentially the same," and to help us make this idea precise there is a very important class of relations known as *equivalence relations*. Here are two examples. First, in elementary geometry one sometimes cares about shapes but not about sizes. Two shapes are said to be *similar* if one can be transformed into the other by a combination of reflections, rotations, translations, and enlargements (see figure 1); the relation "is similar to" is an equivalence relation. Second, when doing ARITHMETIC MODULO m [III.59], one does not wish to distinguish between two whole numbers that differ by a multiple of m: in this case one says that the numbers are *congruent* (mod m); the relation "is congruent (mod m) to" is another equivalence relation.

What exactly is it that these two relations have in common? The answer is that they both take a set (in the first case the set of all geometrical shapes, and in the second the set of all whole numbers) and split it into parts, called *equivalence classes*, where each part consists of objects that one wishes to regard as essentially the same. In the first example, a typical equivalence class is the set of all shapes that are similar to some given shape; in the second, it is the set of all integers that leave a given remainder when you divide by m (for example, if $m = 7$ then one of the equivalence classes is the set $\{\ldots, -16, -9, -2, 5, 12, 19, \ldots\}$).

An alternative definition of what it means for a relation $\sim$, defined on a set A, to be an equivalence relation is that it has the following three properties. First, it is *reflexive*, which means that $x \sim x$ for every x in A. Second, it is *symmetric*, which means that if x and y are elements of A and $x \sim y$, then it must also be the case that $y \sim x$. Third, it is *transitive*, meaning that if x, y, and z are elements of A such that $x \sim y$ and $y \sim z$, then it must be the case that $x \sim z$. (To get a feel for these properties, it may help if you satisfy yourself that the relations "is similar to" and "is congruent (mod m) to" both have all three properties, while the relation "<," defined on the positive integers, is transitive but neither reflexive nor symmetric.)

One of the main uses of equivalence relations is to make precise the notion of QUOTIENT [I.3 §3.3] constructions.

2.4 Binary Operations

Let us return to one of our earlier examples, the sentence "Two plus two equals four." We have analyzed the word "equals" as a relation, an expression that sits between the noun phrases "two plus two" and "four" and makes a sentence out of them. But what about "plus"? That also sits between two nouns. However, the result, "two plus two," is not a sentence but a noun phrase. That pattern is characteristic of *binary operations*. Some familiar examples of binary operations are "plus," "minus," "times," "divided by," and "raised to the power."

As with functions, it is customary, and convenient, to be careful about the set to which a binary operation is applied. From a more formal point of view, a binary operation on a set A is a function that takes pairs of elements of A and produces further elements of A from them. To be more formal still, it is a function with the set of all pairs (x, y) of elements of A as its domain and with A as its range. This way of looking at it is not reflected in the notation, however, since the symbol for the operation comes between x and y rather than before them: we write $x + y$ rather than $+(x, y)$.

There are four properties that a binary operation may have that are very useful if one wants to manipulate sentences in which it appears. Let us use the symbol $*$ to denote an arbitrary binary operation on some set A. The operation $*$ is said to be *commutative* if $x * y$ is always equal to $y * x$, and *associative* if $x * (y * z)$ is always equal to $(x * y) * z$. For example, the operations "plus" and "times" are commutative and associative, whereas "minus," "divided by," and "raised to the power" are neither (for instance, $9 - (5 - 3) = 7$ while $(9 - 5) - 3 = 1$). These last two operations raise another issue: unless the set A is chosen carefully, they may not always be defined. For example, if one restricts one's attention to the positive integers, then the expression $3 - 5$ has no meaning. There are two conventions one could imagine adopting in response to this. One might decide not to insist that a binary operation should be defined for every pair of elements of A, and to regard it as a desirable extra property of an operation if it *is* defined everywhere. But the convention actually in force is that binary operations *do* have to be defined everywhere, so that "minus," though a perfectly good binary operation on the set of all integers, is not a binary operation on the set of all positive integers.

An element e of A is called an *identity* for $*$ if $e * x = x * e = x$ for every element x of A. The two most obvious examples are 0 and 1, which are identities for "plus" and "times," respectively. Finally, if $*$ has an identity e and x belongs to A, then an *inverse* for x is an element y such that $x * y = y * x = e$. For example, if $*$ is "plus" then the inverse of x is $-x$, while if $*$ is "times" then the inverse is $1/x$.

These basic properties of binary operations are fundamental to the structures of abstract algebra. See FOUR IMPORTANT ALGEBRAIC STRUCTURES [I.3 §2] for further details.

3 Some Elementary Logic

3.1 Logical Connectives

A *logical connective* is the mathematical equivalent of a conjunction. That is, it is a word (or symbol) that joins two sentences to produce a new one. We have already discussed an example, namely "and" in its sentence-linking meaning, which is sometimes written by the symbol "$\wedge$," particularly in more formal or abstract mathematical discourse. If P and Q are statements (note here the mathematical habit of representing not just numbers but any objects whatsoever by single letters), then $P \wedge Q$ is the statement that is true if and only if both P and Q are true.

Another connective is the word "or," a word that has a more specific meaning for mathematicians than it has for normal speakers of the English language. The mathematical use is illustrated by the tiresome joke of responding, "Yes please," to a question such as, "Would you like your coffee with or without sugar?" The symbol for "or," if one wishes to use a symbol, is "$\vee$," and the statement $P \vee Q$ is true if and only if P is true or Q is true. This is taken to include the case when they are both true, so "or," for mathematicians, is always the so-called *inclusive* version of the word.

A third important connective is "implies," which is usually written "$\Rightarrow$." The statement $P \Rightarrow Q$ means, roughly speaking, that Q is a consequence of P, and is sometimes read as "if P then Q." However, as with "or," this does not mean quite what it would in English. To get a feel for the difference, consider the following even more extreme example of mathematical pedantry. At the supper table, my young daughter once said, "Put your hand up if you are a girl." One of my sons, to tease her, put his hand up on the grounds that, since she had not added, "and keep it down if you are a boy," his doing so was compatible with her command.

Something like this attitude is taken by mathematicians to the word "implies," or to sentences containing the word "if." The statement $P \Rightarrow Q$ is considered to be true under all circumstances except one: it is not true if P is true and Q is false. This is the *definition* of "implies." It can be confusing because in English the word "implies" suggests some sort of connection between P and Q, that P in some way causes Q or is at least relevant to it. If P causes Q then certainly P cannot be true without Q being true, but all a mathematician cares about is this logical consequence and not whether there is any reason for it. Thus, if you want to prove that $P \Rightarrow Q$, all you have to do is rule out the possibility that P could be true and Q false at the same time. To give an example: if n is a positive integer, then the statement "n is a perfect square with final digit 7" implies the statement "n is a prime number," not because there is any connection between the two but because no perfect square ends in a 7. Of course, implications of this kind are less interesting mathematically than more genuine-seeming ones, but the reward for accepting them is that, once again, one avoids being confused by some of the ambiguities and subtle nuances of ordinary language.

3.2 Quantifiers

Yet another ambiguity in the English language is exploited by the following old joke that suggests that our priorities need to be radically rethought.

(4) Nothing is better than lifelong happiness.

(5) But a cheese sandwich is better than nothing.

(6) Therefore, a cheese sandwich is better than lifelong happiness.

Let us try to be precise about how this play on words works (a good way to ruin any joke, but not a tragedy in this case). It hinges on the word "nothing," which is used in two different ways. The first sentence means "There is no single thing that is better than lifelong happiness," whereas the second means "It is better to have a cheese sandwich than to have nothing at all." In other words, in the second sentence, "nothing" stands for what one might call the null option, the option of having nothing, whereas in the first it does not (to have nothing is not better than to have lifelong happiness).

Words like "all," "some," "any," "every," and "nothing" are called *quantifiers*, and in the English language they are highly prone to this kind of ambiguity. Mathematicians therefore make do with just two quantifiers, and the rules for their use are much stricter. They tend to come at the beginning of sentences, and can be read as "for all" (or "for every") and "there exists" (or "for some"). A rewriting of sentence (4) that renders it unambiguous (but less like real English) is

(4′) For all x, lifelong happiness is at least as good as x.

The second sentence cannot be rewritten in these terms because the word "nothing" is not playing the role of a quantifier. (Its nearest mathematical equivalent is something like the *empty set*, that is, the set with no elements.)

Armed with "for all" and "there exists," we can be clear about the difference between the beginnings of the following sentences.

(7) Everybody likes at least one drink, namely water.

(8) Everybody likes at least one drink; I myself go for red wine.

The first sentence makes the point (not necessarily correctly) that there is one drink that everybody likes,

whereas the second claims merely that we all have something we like to drink, even if that something varies from person to person. The precise formulations that capture the difference are as follows.

(7′) There exists a drink D such that, for every person P, P likes D.

(8′) For every person P there exists a drink D such that P likes D.

This illustrates an important general principle: if you take a sentence that begins "for every x there exists y such that ..." and interchange the two parts so that it now begins "there exists y such that, for every x, ...," then you obtain a much stronger statement, since y is no longer allowed to depend on x. If the second statement is still true—that is, if you really can choose a y that works for all the x at once—then the first statement is said to hold *uniformly*.

The symbols $\forall$ and $\exists$ are often used to stand for "for all" and "there exists," respectively. This allows us to write quite complicated mathematical sentences in a highly symbolic form if we want to. For example, suppose we let P be the set of all primes, as we did earlier. Then the following symbols make the claim that there are infinitely many primes, or rather a slightly different claim that is equivalent to it.

(9) $\forall n \; \exists m \quad (m > n) \; \wedge \; (m \in P)$.

In words, this says that for every n we can find some m that is both bigger than n and a prime. If we wish to unpack sentence (9) further, we could replace the part $m \in P$ by

(10) $\forall a, b \quad ab = m \Rightarrow ((a = 1) \vee (b = 1))$.

There is one final important remark to make about the quantifiers "$\forall$" and "$\exists$." I have presented them as if they were freestanding, but actually a quantifier is always associated with a set (one says that it *quantifies over* that set). For example, sentence (10) would not be a translation of the sentence "m is prime" if a and b were allowed to be fractions: if $a = 3$ and $b = \frac{7}{3}$ then $ab = 7$ without either a or b equaling 1, but this does not show that 7 is not a prime. Implicit in the opening symbols $\forall a, b$ is the idea that a and b are intended to be *positive integers*. If this had not been clear from the context, then we could have used the symbol $\mathbb{N}$ (which stands for the set of all positive integers) and started sentence (10) with $\forall a, b \in \mathbb{N}$ instead.

3.3 Negation

The basic idea of negation in mathematics is very simple: there is a symbol, "¬," which means "not," and if P is any mathematical statement, then $\neg P$ stands for the statement that is true if and only if P is not true. However, this is another example of a word that has a slightly more restricted meaning to mathematicians than it has in ordinary speech.

To illustrate this phenomenon once again, let us take A to be a set of positive integers and ask ourselves what the negation is of the sentence "Every number in the set A is odd." Many people when asked this question will suggest, "Every number in the set A is even." However, this is wrong: if one thinks carefully about what exactly would have to happen for the first sentence to be false, one realizes that all that is needed is that *at least one* number in A should be even. So in fact the negation of the sentence is, "There exists a number in A that is even."

What explains the temptation to give the first, incorrect answer? One possibility emerges when one writes the sentence more formally, thus:

(11) $\forall n \in A \quad n$ is odd.

The first answer is obtained if one negates just the last part of this sentence, "n is odd"; but what is asked for is the negation of the *whole sentence*. That is, what is wanted is not

(12) $\forall n \in A \quad \neg(n$ is odd),

but rather

(13) $\neg(\forall n \in A \quad n$ is odd),

which is equivalent to

(14) $\exists n \in A \quad n$ is even.

A second possible explanation is that one is inclined (for psycholinguistic reasons) to think of the phrase "every element of A" as denoting something like a single, typical element of A. If that comes to have the feel of a particular number n, then we may feel that the negation of "n is odd" is "n is even." The remedy is not to think of the phrase "every element of A" on its own: it should always be part of the longer phrase, "*for* every element of A."

3.4 Free and Bound Variables

Suppose we say something like, "At time t the speed of the projectile is v." The letters t and v stand for real numbers, and they are called *variables*, because in the back of our mind is the idea that they are changing. More generally, a variable is any letter used to stand for a mathematical object, whether or not one thinks of that object as changing through time. Let us look once again at the formal sentence that said that a positive integer m is prime:

(10) $\forall a, b \quad ab = m \Rightarrow ((a = 1) \lor (b = 1))$.

In this sentence, there are three variables, a, b, and m, but there is a very important grammatical and semantic difference between the first two and the third. Here are two results of that difference. First, the sentence does not really make sense unless we already know what m is from the context, whereas it is important that a and b do *not* have any prior meaning. Second, while it makes perfect sense to ask, "For which values of m is sentence (10) true?" it makes no sense at all to ask, "For which values of a is sentence (10) true?" The letter m in sentence (10) stands for a fixed number, not specified in this sentence, while the letters a and b, because of the initial $\forall a, b$, do not *stand for* numbers—rather, in some way they search through all pairs of positive integers, trying to find a pair that multiply together to give m. Another sign of the difference is that you can ask, "What number is m?" but not, "What number is a?" A fourth sign is that the meaning of sentence (10) is completely unaffected if one uses different letters for a and b, as in the reformulation

(10′) $\forall c, d \quad cd = m \Rightarrow ((c = 1) \lor (d = 1))$.

One cannot, however, change m to n without establishing first that n denotes the same integer as m. A variable such as m, which denotes a specific object, is called a *free* variable. It sort of hovers there, free to take any value. A variable like a and b, of the kind that does not denote a specific object, is called a *bound* variable, or sometimes a *dummy* variable. (The word "bound" is used mainly when the variable appears just after a quantifier, as in sentence (10).)

Yet another indication that a variable is a dummy variable is when the sentence in which it occurs can be rewritten without it. For instance, the expression $\sum_{n=1}^{100} f(n)$ is shorthand for $f(1) + f(2) + \cdots + f(100)$, and the second way of writing it does not involve the letter n, so n was not really standing for anything in the first way. Sometimes, actual elimination is not possible, but one feels it could be done in principle. For instance, the sentence "For every real number x, x is either positive, negative, or zero" is a bit like putting together infinitely many sentences such as "t is either positive, negative, or zero," one for each real number t, none of which involves a variable.

4 Levels of Formality

It is a surprising fact that a small number of set-theoretic concepts and logical terms can be used to provide a precise language that is versatile enough to express all the statements of ordinary mathematics. There are some technicalities to sort out, but even these can often be avoided if one allows not just sets but also numbers as basic objects. However, if you look at a well-written mathematics paper, then much of it will be written not in symbolic language peppered with symbols such as $\forall$ and $\exists$, but in what appears to be ordinary English. (Some papers are written in other languages, particularly French, but English has established itself as the international language of mathematics.) How can mathematicians be confident that this ordinary English does not lead to confusion, ambiguity, and even incorrectness?

The answer is that the language typically used is a careful compromise between fully colloquial English, which would indeed run the risk of being unacceptably imprecise, and fully formal symbolism, which would be a nightmare to read. The ideal is to write in as friendly and approachable a way as possible, while making sure that the reader (who, one assumes, has plenty of experience and training in how to read mathematics) can see easily how what one writes could be made more formal if it became important to do so. And sometimes it does become important: when an argument is difficult to grasp it may be that the only way to convince oneself that it is correct is to rewrite it more formally.

Consider, for example, the following reformulation of the principle of mathematical induction, which underlies many proofs:

(15) Every nonempty set of positive integers has a least element.

If we wish to translate this into a more formal language we need to strip it of words and phrases such as "nonempty" and "has." But this is easily done. To say that a set A of positive integers is nonempty is simply

to say that there is a positive integer that belongs to A. This can be stated symbolically:

(16) $\exists n \in \mathbb{N} \quad n \in A.$

What does it mean to say that A has a least element? It means that there exists an element x of A such that every element y of A is either greater than x or equal to x itself. This formulation is again ready to be translated into symbols:

(17) $\exists x \in A \quad \forall y \in A \quad (y > x) \ \vee \ (y = x).$

Statement (15) says that (16) implies (17) for every set A of positive integers. Thus, it can be written symbolically as follows:

(18) $\forall A \subset \mathbb{N}$
$[(\exists n \in \mathbb{N} \quad n \in A)$
$\Rightarrow (\exists x \in A \ \forall y \in A \quad (y > x) \ \vee \ (y = x))].$

Here we have two very different modes of presentation of the same mathematical fact. Obviously (15) is much easier to understand than (18). But if, for example, one is concerned with the foundations of mathematics, or wishes to write a computer program that checks the correctness of proofs, then it is better to work with a greatly pared-down grammar and vocabulary, and then (18) has the advantage. In practice, there are many different levels of formality, and mathematicians are adept at switching between them. It is this that makes it possible to feel completely confident in the correctness of a mathematical argument even when it is *not* presented in the manner of (18)—though it is also this that allows mistakes to slip through the net from time to time.

I.3 Some Fundamental Mathematical Definitions

The concepts discussed in this article occur throughout so much of modern mathematics that it would be inappropriate to discuss them in part III—they are too basic. Many later articles will assume at least some acquaintance with these concepts, so if you have not met them, then reading this article will help you to understand significantly more of the book.

1 The Main Number Systems

Almost always, the first mathematical concept that a child is exposed to is the idea of numbers, and numbers retain a central place in mathematics at all levels.

However, it is not as easy as one might think to say what the word "number" means: the more mathematics one learns, the more uses of this word one comes to know, and the more sophisticated one's concept of number becomes. This individual development parallels a historical development that took many centuries (see FROM NUMBERS TO NUMBER SYSTEMS [II.1]).

The modern view of numbers is that they are best regarded not individually but as parts of larger wholes, called *number systems*; the distinguishing features of number systems are the arithmetical operations—such as addition, multiplication, subtraction, division, and extraction of roots—that can be performed on them. This view of numbers is very fruitful and provides a springboard into abstract algebra. The rest of this section gives a brief description of the five main number systems.

1.1 The Natural Numbers

The *natural numbers*, otherwise known as the *positive integers*, are the numbers familiar even to young children: 1, 2, 3, 4, and so on. It is the natural numbers that we use for the very basic mathematical purpose of counting. The set of all natural numbers is usually denoted $\mathbb{N}$. (Some mathematicians prefer to include 0 as a natural number as well: for instance, this is the usual convention in logic and set theory. Both conventions are to be found in this book, but it should always be clear which one is being used.)

Of course, the phrase "1, 2, 3, 4, and so on" does not constitute a formal definition, but it does suggest the following basic picture of the natural numbers, one that we tend to take for granted.

(i) Given any natural number n there is another, $n+1$, that comes next—known as the *successor* of n.
(ii) A list that starts with 1 and follows each number by its successor will include every natural number exactly once and nothing else.

This picture is encapsulated by THE PEANO AXIOMS [III.67].

Given two natural numbers m and n one can add them together or multiply them, obtaining in each case a new natural number. By contrast, subtraction and division are not always possible. If we want to give meaning to expressions such as $8 - 13$ or $\frac{5}{7}$, then we must work in a larger number system.

1.2 The Integers

The natural numbers are not the only whole numbers, since they do not include zero or negative numbers, both of which are indispensable to mathematics. One of the first reasons for introducing zero was that it is needed for the normal decimal notation of positive integers—how else could one conveniently write 1005? However, it is now thought of as much more than just a convenience, and the property that makes it significant is that it is an *additive identity*, which means that adding zero to any number leaves that number unchanged. And while it is not particularly interesting to do to a number something that has no effect, the property itself is interesting and distinguishes zero from all other numbers. An immediate illustration of this is that it allows us to think about negative numbers: if n is a positive integer, then the defining property of $-n$ is that when you add it to n you get zero.

Somebody with little mathematical experience may unthinkingly assume that numbers are for counting and find negative numbers objectionable because the answer to a question beginning "How many" is never negative. However, simple counting is not the only use for numbers, and there are many situations that are naturally modeled by a number system that includes both positive and negative numbers. For example, negative numbers are sometimes used for the amount of money in a bank account, for temperature (in degrees Celsius or Fahrenheit), and for altitude compared with sea level.

The set of all integers—positive, negative, and zero—is usually denoted $\mathbb{Z}$ (for the German word "Zahlen," meaning "numbers"). Within this system, subtraction is always possible: that is, if m and n are integers, then so is $m - n$.

1.3 The Rational Numbers

So far we have considered only whole numbers. If we form all possible fractions as well, then we obtain the *rational numbers*. The set of all rational numbers is denoted $\mathbb{Q}$ (for "quotients").

One of the main uses of numbers besides counting is *measurement*, and most quantities that we measure are ones that can vary continuously, such as length, weight, temperature, and velocity. For these, whole numbers are inadequate.

A more theoretical justification for the rational numbers is that they form a number system in which division is always possible—except by zero. This fact, together with some basic properties of the arithmetical operations, means that $\mathbb{Q}$ is a *field*. What fields are and why they are important will be explained in more detail later (section 2.2).

1.4 The Real Numbers

A famous discovery of the ancient Greeks, often attributed, despite very inadequate evidence, to the school of PYTHAGORAS [VI.1], was that the square root of 2 is not a rational number. That is, there is no fraction p/q such that $(p/q)^2 = 2$. The Pythagorean theorem about right-angled triangles (which was probably known at least a thousand years before Pythagoras) tells us that if a square has sides of length 1, then the length of its diagonal is $\sqrt{2}$. Consequently, there are lengths that cannot be measured by rational numbers.

This argument seems to give strong practical reasons for extending our number system still further. However, such a conclusion can be resisted: after all, we cannot make any measurements with infinite precision, so in practice we round off to a certain number of decimal places, and as soon as we have done so we have presented our measurement as a rational number. (This point is discussed more fully in NUMERICAL ANALYSIS [IV.21].)

Nevertheless, the *theoretical* arguments for going beyond the rational numbers are irresistible. If we want to solve polynomial equations, take LOGARITHMS [III.25 §4], do trigonometry, or work with the GAUSSIAN DISTRIBUTION [III.71 §5], to give just four examples from an almost endless list, then irrational numbers will appear everywhere we look. They are not used directly for the purposes of measurement, but they are needed if we want to reason theoretically about the physical world by describing it mathematically. This necessarily involves a certain amount of idealization: it is far more convenient to say that the length of the diagonal of a unit square is $\sqrt{2}$ than it is to talk about what would be observed, and with what degree of certainty, if one tried to measure this length as accurately as possible.

The real numbers can be thought of as the set of all numbers with a finite or infinite decimal expansion. In the latter case, they are defined not directly but by a process of successive approximation. For example, the squares of the numbers 1, 1.4, 1.41, 1.414, 1.4142, 1.41421, . . . , get as close as you like to 2, if you go far enough along the sequence, which is what we mean by saying that the square root of 2 is the infinite decimal 1.41421 . . . .

The set of all real numbers is denoted $\mathbb{R}$. A more abstract view of $\mathbb{R}$ is that it is an extension of the rational number system to a larger field, and in fact the only one possible in which processes of the above kind always give rise to numbers that themselves belong to $\mathbb{R}$.

Because real numbers are intimately connected with the idea of limits (of successive approximations), a true appreciation of the real number system depends on an understanding of mathematical analysis, which will be discussed in section 5.

1.5 The Complex Numbers

Many polynomial equations, such as the equation $x^2 = 2$, do not have rational solutions but can be solved in $\mathbb{R}$. However, there are many other equations that cannot be solved even in $\mathbb{R}$. The simplest example is the equation $x^2 = -1$, which has no real solution since the square of any real number is positive or zero. In order to get around this problem, mathematicians introduce a symbol, i, which they treat as a number, and they simply *stipulate* that i^2 is to be regarded as equal to -1. The *complex number system*, denoted $\mathbb{C}$, is the set of all numbers of the form $a + b i$, where a and b are real numbers. To add or multiply complex numbers, one treats i as a variable (like x, say), but any occurrences of i^2 are replaced by -1. Thus,

$$(a + b i) + (c + d i) = (a + c) + (b + d) i$$

and

$$(a + b i)(c + d i) = ac + bci + adi + bdi^2$$
$$= (ac - bd) + (bc + ad)i.$$

There are several remarkable points to note about this definition. First, despite its apparently artificial nature, it does not lead to any inconsistency. Secondly, although complex numbers do not directly count or measure anything, they are immensely useful. Thirdly, and perhaps most surprisingly, even though the number i was introduced to help us solve just one equation, it in fact allows us to solve *all* polynomial equations. This is the famous FUNDAMENTAL THEOREM OF ALGEBRA [V.13].

One explanation for the utility of complex numbers is that they provide a concise way to talk about many aspects of geometry, via *Argand diagrams*. These represent complex numbers as points in the plane, the number $a + b i$ corresponding to the point with coordinates (a, b). If $r = \sqrt{a^2 + b^2}$ and $\theta = \tan^{-1}(b/a)$, then

$a = r \cos\theta$ and $b = r \sin\theta$. It turns out that multiplying a complex number $z = x + y$i by $a + b$i corresponds to the following geometrical process. First, you associate z with the point (x, y) in the plane. Next, you multiply this point by r, obtaining the point (rx, ry). Finally, you rotate this new point counterclockwise about the origin through an angle of θ. In other words, the effect on the complex plane of multiplication by $a + b$i is to dilate it by r and then rotate it by θ. In particular, if $a^2 + b^2 = 1$, then multiplying by $a + b$i corresponds to rotating by θ.

For this reason, polar coordinates are at least as good as Cartesian coordinates for representing complex numbers: an alternative way to write $a + b$i is $r\mathrm{e}^{\mathrm{i}\theta}$, which tells us that the number has distance r from the origin and is positioned at an angle θ around from the positive part of the real axis (in a counterclockwise direction). If $z = r\mathrm{e}^{\mathrm{i}\theta}$ with $r > 0$, then r is called the *modulus* of z, denoted by $|z|$, and θ is the *argument* of z. (Since adding 2π to θ does not change $\mathrm{e}^{\mathrm{i}\theta}$, it is usually understood that $0 \leqslant \theta < 2\pi$, or sometimes that $-\pi \leqslant \theta < \pi$.) One final useful definition: if $z = x + \mathrm{i}y$ is a complex number, then its *complex conjugate*, written $\bar{z}$, is the number $x - y$i. It is easy to check that $z\bar{z} = x^2 + y^2 = |z|^2$.

2 Four Important Algebraic Structures

In the previous section it was emphasized that numbers are best thought of not as individual objects but as members of *number systems*. A number system consists of some objects (numbers) together with operations (such as addition and multiplication) that can be performed on those objects. As such, it is an example of an *algebraic structure*. However, there are many very important algebraic structures that are not number systems, and a few of them will be introduced here.

2.1 Groups

If S is a geometrical shape, then a *rigid motion* of S is a way of moving S in such a way that the distances between the points of S are not changed—squeezing and stretching are not allowed. A rigid motion is a *symmetry* of S if, after it is completed, S looks the same as it did before it moved. For example, if S is an equilateral triangle, then rotating S through $120°$ about its center is a symmetry; so is reflecting S about a line that passes through one of the vertices of S and the midpoint of the opposite side.

More formally, a symmetry of S is a function f from S to itself such that the distance between any two points x and y of S is the same as the distance between the transformed points $f(x)$ and $f(y)$.

This idea can be hugely generalized: if S is any mathematical structure, then a symmetry of S is a function from S to itself that preserves its structure. If S is a geometrical shape, then the mathematical structure that should be preserved is the distance between any two of its points. But there are many other mathematical structures that a function may be asked to preserve, most notably algebraic structures of the kind that will soon be discussed. It is fruitful to draw an analogy with the geometrical situation and regard any structure-preserving function as a sort of symmetry.

Because of its extreme generality, symmetry is an all-pervasive concept within mathematics; and wherever symmetries appear, structures known as *groups* follow close behind. To explain what these are and why they appear, let us return to the example of an equilateral triangle, which has, as it turns out, six possible symmetries.

Why is this? Well, let f be a symmetry of an equilateral triangle with vertices A, B, and C and suppose for convenience that this triangle has sides of length 1. Then $f(\mathrm{A})$, $f(\mathrm{B})$, and $f(\mathrm{C})$ must be three points of the triangle and the distances between these points must all be 1. It follows that $f(\mathrm{A})$, $f(\mathrm{B})$, and $f(\mathrm{C})$ are distinct vertices of the triangle, since the furthest apart *any* two points can be is 1 and this happens only when the two points are distinct vertices. So $f(\mathrm{A})$, $f(\mathrm{B})$, and $f(\mathrm{C})$ are the vertices A, B, and C in some order. But the number of possible orders of A, B, and C is 6. It is not hard to show that, once we have chosen $f(\mathrm{A})$, $f(\mathrm{B})$, and $f(\mathrm{C})$, the rest of what f does is completely determined. (For example, if X is the midpoint of A and C, then $f(\mathrm{X})$ must be the midpoint of $f(\mathrm{A})$ and $f(\mathrm{C})$ since there is no other point at distance $\frac{1}{2}$ from $f(\mathrm{A})$ and $f(\mathrm{C})$.)

Let us refer to these symmetries by writing down in order what happens to the vertices A, B, and C. So, for instance, the symmetry ACB is the one that leaves the vertex A fixed and exchanges B and C, which is achieved by reflecting the triangle in the line that joins A to the midpoint of B and C. There are three reflections like this: ACB, CBA, and BAC. There are also two rotations: BCA and CAB. Finally, there is the "trivial" symmetry, ABC, which leaves all points where they were originally. (The "trivial" symmetry is useful in much the same way as zero is useful for the algebra of integer addition.)

What makes these and other sets of symmetries into groups is that any two symmetries can be *composed*, meaning that one symmetry followed by another produces a third (since if two operations both preserve a structure then their combination clearly does too). For example, if we follow the reflection BAC by the reflection ACB, then we obtain the rotation CAB. To work this out, one can either draw a picture or use the following kind of reasoning: the first symmetry takes A to B and the second takes B to C, so the combination takes A to C, and similarly B goes to A, and C to B. Notice that the order in which we perform the symmetries matters: if we had started with the reflection ACB and then done the reflection BAC, then we would have obtained the rotation BCA. (If you try to see this by drawing a picture, it is important to think of A, B, and C as labels that stay where they are rather than moving with the triangle—they mark positions that the vertices can occupy.)

We can think of symmetries as "objects" in their own right, and of composition as an algebraic operation, a bit like addition or multiplication for numbers. The operation has the following useful properties: it is ASSOCIATIVE, the trivial symmetry is an IDENTITY ELEMENT, and every symmetry has an INVERSE [I.2 §2.4]. (For example, the inverse of a reflection is itself, since doing the same reflection twice leaves the triangle where it started.) More generally, any set with a binary operation that has these properties is called a group. It is *not* part of the definition of a group that the binary operation should be commutative, since, as we have just seen, if one is composing two symmetries then it often makes a difference which one goes first. However, if it is commutative then the group is called *Abelian*, after the Norwegian mathematician NIELS HENRIK ABEL [VI.33]. The number systems $\mathbb{Z}$, $\mathbb{Q}$, $\mathbb{R}$, and $\mathbb{C}$ all form Abelian groups with the operation of addition, or *under* addition, as one usually says. If you remove zero from $\mathbb{Q}$, $\mathbb{R}$, and $\mathbb{C}$, then they form Abelian groups under multiplication, but $\mathbb{Z}$ does not because of a lack of inverses: the reciprocal of an integer is not usually an integer. Further examples of groups will be given later in this section.

2.2 Fields

Although several number systems form groups, to regard them merely as groups is to ignore a great deal of their algebraic structure. In particular, whereas a group has just one binary operation, the standard number systems have two, namely addition and multiplication (from which further ones, such as subtraction and division, can be derived). The formal definition of a *field* is quite long: it is a set with two binary operations and there are several axioms that these operations must satisfy. Fortunately, there is an easy way to remember these axioms. You just write down all the basic properties you can think of that are satisfied by addition and multiplication in the number systems $\mathbb{Q}$, $\mathbb{R}$, and $\mathbb{C}$.

These properties are as follows. Both addition and multiplication are commutative and associative, and both have identity elements (0 for addition and 1 for multiplication). Every element x has an additive inverse $-x$ and a multiplicative inverse $1/x$ (except that 0 does not have a multiplicative inverse). It is the existence of these inverses that allows us to define subtraction and division: $x - y$ means $x + (-y)$ and x/y means $x \cdot (1/y)$.

That covers all the properties that addition and multiplication satisfy individually. However, a very general rule when defining mathematical structures is that if a definition splits into parts, then the definition as a whole will not be interesting *unless those parts interact.* Here our two parts are addition and multiplication, and the properties mentioned so far do not relate them in any way. But one final property, known as the *distributive law*, does this, and thereby gives fields their special character. This is the rule that tells us how to multiply out brackets: $x(y+z) = xy+xz$ for any three numbers x, y, and z.

Having listed these properties, one may then view the whole situation *abstractly* by regarding the properties as axioms and saying that a field is any set with two binary operations that satisfy all those axioms. However, when one works in a field, one usually thinks of the axioms not as a list of statements but rather as a general license to do all the algebraic manipulations that one can do when talking about rational, real, and complex numbers.

Clearly, the more axioms one has, the harder it is to find a mathematical structure that satisfies them, and it is indeed the case that fields are harder to come by than groups. For this reason, the best way to understand fields is probably to concentrate on examples. In addition to $\mathbb{Q}$, $\mathbb{R}$, and $\mathbb{C}$, one other field stands out as fundamental, namely $\mathbb{F}_p$, which is the set of integers modulo a prime p, with addition and multiplication also defined modulo p (see MODULAR ARITHMETIC [III.58]).

What makes fields interesting, however, is not so much the existence of these basic examples as the fact that there is an important process of *extension* that allows one to build new fields out of old ones. The idea is to start with a field $\mathbb{F}$, find a polynomial P that has no roots in $\mathbb{F}$, and "adjoin" a new element to $\mathbb{F}$ with the stipulation that it is a root of P. This produces an extended field $\mathbb{F}'$, which consists of everything that one can produce from this root and from elements of $\mathbb{F}$ using addition and multiplication.

We have already seen an important example of this process: in the field $\mathbb{R}$, the polynomial $P(x) = x^2 + 1$ has no root, so we adjoined the element i and let $\mathbb{C}$ be the field of all combinations of the form $a + b$i.

We can apply exactly the same process to the field $\mathbb{F}_3$, in which again the equation $x^2 + 1 = 0$ has no solution. If we do so, then we obtain a new field, which, like $\mathbb{C}$, consists of all combinations of the form $a + b$i, but now a and b belong to $\mathbb{F}_3$. Since $\mathbb{F}_3$ has three elements, this new field has nine elements. Another example is the field $\mathbb{Q}(\sqrt{2})$, which consists of all numbers of the form $a + b\sqrt{2}$, where now a and b are rational numbers. A slightly more complicated example is $\mathbb{Q}(\gamma)$, where γ is a root of the polynomial $x^3 - x - 1$. A typical element of this field has the form $a + b\gamma + c\gamma^2$, with a, b, and c rational. If one is doing arithmetic in $\mathbb{Q}(\gamma)$, then whenever γ^3 appears, it can be replaced by $\gamma + 1$ (because $\gamma^3 - \gamma - 1 = 0$), just as i^2 can be replaced by -1 in the complex numbers. For more on why field extensions are interesting, see the discussion of AUTOMORPHISMS in section 4.1.

A second very significant justification for introducing fields is that they can be used to form vector spaces, and it is to these that we now turn.

2.3 Vector Spaces

One of the most convenient ways to represent points in a plane that stretches out to infinity in all directions is to use Cartesian coordinates. One chooses an origin and two directions X and Y, usually at right angles to each other. Then the pair of numbers (a, b) stands for the point you reach in the plane if you go a distance a in direction X and a distance b in direction Y (where if a is a negative number such as -2, this is interpreted as going a distance $+2$ in the opposite direction to X, and similarly for b).

Another way of saying the same thing is this. Let $\boldsymbol{x}$ and $\boldsymbol{y}$ stand for the unit vectors in directions X and Y, respectively, so their Cartesian coordinates are $(1, 0)$

and $(0, 1)$. Then every point in the plane is a so-called *linear combination* $a\boldsymbol{x} + b\boldsymbol{y}$ of the *basis vectors* $\boldsymbol{x}$ and $\boldsymbol{y}$. To interpret the expression $a\boldsymbol{x} + b\boldsymbol{y}$, first rewrite it as $a(1, 0) + b(0, 1)$. Then a times the unit vector $(1, 0)$ is $(a, 0)$ and b times the unit vector $(0, 1)$ is $(0, b)$ and when you add $(a, 0)$ and $(0, b)$ coordinate by coordinate you get the vector (a, b).

Here is another situation where linear combinations appear. Suppose you are presented with the differential equation $(\mathrm{d}^2 y / \mathrm{d}x^2) + y = 0$, and happen to know (or notice) that $y = \sin x$ and $y = \cos x$ are two possible solutions. Then you can easily check that $y = a \sin x + b \cos x$ is a solution for any pair of numbers a and b. That is, any linear combination of the existing solutions $\sin x$ and $\cos x$ is another solution. It turns out that all solutions are of this form, so we can regard $\sin x$ and $\cos x$ as "basis vectors" for the "space" of solutions of the differential equation.

Linear combinations occur in many many contexts throughout mathematics. To give one more example, an arbitrary polynomial of degree 3 has the form $ax^3 + bx^2 + cx + d$, which is a linear combination of the four basic polynomials 1, x, x^2, and x^3.

A *vector space* is a mathematical structure in which the notion of linear combination makes sense. The objects that belong to the vector space are usually called *vectors*, unless we are talking about a specific example and are thinking of them as concrete objects such as polynomials or solutions of a differential equation. Slightly more formally, a vector space is a set V such that, given any two vectors $\boldsymbol{v}$ and $\boldsymbol{w}$ (that is, elements of V) and any two real numbers a and b, we can form the linear combination $a\boldsymbol{v} + b\boldsymbol{w}$.

Notice that this linear combination involves objects of two different kinds, the vectors $\boldsymbol{v}$ and $\boldsymbol{w}$ and the numbers a and b. The latter are known as *scalars*. The operation of forming linear combinations can be broken up into two constituent parts: addition and scalar multiplication. To form the combination $a\boldsymbol{v} + b\boldsymbol{w}$, first multiply the vectors $\boldsymbol{v}$ and $\boldsymbol{w}$ by the scalars a and b, obtaining the vectors $a\boldsymbol{v}$ and $b\boldsymbol{w}$, and then add these resulting vectors to obtain the full combination $a\boldsymbol{v} + b\boldsymbol{w}$.

The definition of linear combination must obey certain natural rules. Addition of vectors must be commutative and associative, with an identity (the *zero vector*) and an inverse for each $\boldsymbol{v}$ (written $-\boldsymbol{v}$). Scalar multiplication must obey a sort of associative law, namely that $a(b\boldsymbol{v})$ and $(ab)\boldsymbol{v}$ are always equal. We also need two distributive laws: $(a + b)\boldsymbol{v} = a\boldsymbol{v} + b\boldsymbol{v}$ and $a(\boldsymbol{v} + \boldsymbol{w}) =$

$a\boldsymbol{v} + a\boldsymbol{w}$ for any scalars a and b and any vectors $\boldsymbol{v}$ and $\boldsymbol{w}$.

Another context in which linear combinations arise, one that lies at the heart of the usefulness of vector spaces, is the solution of simultaneous equations. Suppose one is presented with the two equations $3x + 2y = 6$ and $x - y = 7$. The usual way to solve such a pair of equations is to try to eliminate either x or y by adding an appropriate multiple of one of the equations to the other: that is, by taking a certain linear combination of the equations. In this case, we can eliminate y by adding twice the second equation to the first, obtaining the equation $5x = 20$, which tells us that $x = 4$ and hence that $y = -3$. Why were we allowed to combine equations like this? Well, let us write L_1 and R_1 for the left- and right-hand sides of the first equation, and similarly L_2 and R_2 for the second. If, for some particular choice of x and y, it is true that $L_1 = R_1$ and $L_2 = R_2$, then clearly $L_1 + 2L_2 = R_1 + 2R_2$, as the two sides of this equation are merely giving different names to the same numbers.

Given a vector space V, a *basis* is a collection of vectors $\boldsymbol{v}_1, \boldsymbol{v}_2, \ldots, \boldsymbol{v}_n$ with the following property: every vector in V can be written in exactly one way as a linear combination $a_1 \boldsymbol{v}_1 + a_2 \boldsymbol{v}_2 + \cdots + a_n \boldsymbol{v}_n$. There are two ways in which this can fail: there may be a vector that cannot be written as a linear combination of $\boldsymbol{v}_1, \boldsymbol{v}_2, \ldots, \boldsymbol{v}_n$ or there may be a vector that can be so expressed, but in more than one way. If every vector is a linear combination then we say that the vectors $\boldsymbol{v}_1, \boldsymbol{v}_2, \ldots, \boldsymbol{v}_n$ *span* V, and if no vector is a linear combination in more than one way then we say that they are *independent*. An equivalent definition is that $\boldsymbol{v}_1, \boldsymbol{v}_2, \ldots, \boldsymbol{v}_n$ are independent if the only way of writing the zero vector as $a_1 \boldsymbol{v}_1 + a_2 \boldsymbol{v}_2 + \cdots + a_n \boldsymbol{v}_n$ is by taking $a_1 = a_2 = \cdots = a_n = 0$.

The number of elements in a basis is called the *dimension* of V. It is not immediately obvious that there could not be two bases of different sizes, but it turns out that there cannot, so the concept of dimension makes sense. For the plane, the vectors $\boldsymbol{x}$ and $\boldsymbol{y}$ defined earlier formed a basis, so the plane, as one would hope, has dimension 2. If we were to take more than two vectors, then they would no longer be independent: for example, if we take the vectors $(1, 2)$, $(1, 3)$, and $(3, 1)$, then we can write $(0, 0)$ as the linear combination $8(1, 2) - 5(1, 3) - (3, 1)$. (To work this out one must solve some simultaneous equations—this is typical of calculations in vector spaces.)

The most obvious n-dimensional vector space is the space of all sequences $(x_1, \ldots, x_n)$ of n real numbers. To add this to a sequence $(y_1, \ldots, y_n)$ one simply forms the sequence $(x_1 + y_1, \ldots, x_n + y_n)$ and to multiply it by a scalar c one forms the sequence $(cx_1, \ldots, cx_n)$. This vector space is denoted $\mathbb{R}^n$. Thus, the plane with its usual coordinate system is $\mathbb{R}^2$ and three-dimensional space is $\mathbb{R}^3$.

It is not in fact necessary for the number of vectors in a basis to be finite. A vector space that does not have a finite basis is called *infinite dimensional*. This is not an exotic property: many of the most important vector spaces, particularly spaces where the "vectors" are functions, are infinite dimensional.

There is one final remark to make about scalars. They were defined earlier as real numbers that one uses to make linear combinations of vectors. But it turns out that the calculations one does with scalars, in particular solving simultaneous equations, can all be done in a more general context. What matters is that they should belong to a field, so $\mathbb{Q}$, $\mathbb{R}$, and $\mathbb{C}$ can all be used as systems of scalars, as indeed can more general fields. If the scalars for a vector space V come from a field $\mathbb{F}$, then one says that V is a vector space *over* $\mathbb{F}$. This generalization is important and useful: see, for example, ALGEBRAIC NUMBERS [IV.1 §17].

2.4 Rings

Another algebraic structure that is very important is a *ring*. Rings are not quite as central to mathematics as groups, fields, or vector spaces, so a proper discussion of them will be deferred to RINGS, IDEALS, AND MODULES [III.81]. However, roughly speaking, a ring is an algebraic structure that has most, but not necessarily all, of the properties of a field. In particular, the requirements of the multiplicative operation are less strict. The most important relaxation is that nonzero elements of a ring are not required to have multiplicative inverses; but sometimes multiplication is not even required to be commutative. If it is, then the ring itself is said to be commutative—a typical example of a commutative ring is the set $\mathbb{Z}$ of all integers. Another is the set of all polynomials with coefficients in some field $\mathbb{F}$.

3 Creating New Structures Out of Old Ones

An important first step in understanding the definition of some mathematical structure is to have a supply of examples. Without examples, a definition is dry and abstract. With them, one begins to have a feeling for the structure that its definition alone cannot usually provide.

One reason for this is that it makes it much easier to answer basic questions. If you have a general statement about structures of a given type and want to know whether it is true, then it is very helpful if you can test it in a wide range of particular cases. If it passes all the tests, then you have some evidence in favor of the statement. If you are lucky, you may even be able to see why it is true; alternatively, you may find that the statement is true for each example you try, but always for reasons that depend on particular features of the example you are examining. Then you will know that you should try to avoid these features if you want to find a counterexample. If you *do* find a counterexample, then the general statement is false, but it may still happen that a modification to the statement is true and useful. In that case, the counterexample will help you to find an appropriate modification.

The moral, then, is that examples are important. So how does one find them? There are two completely different approaches. One is to build them from scratch. For example, one might define a group G to be the group of all symmetries of an icosahedron. Another, which is the main topic of this section, is to take some examples that have already been constructed and build new ones out of them. For instance, the group $\mathbb{Z}^2$, which consists of all pairs of integers (x, y), with addition defined by the obvious rule $(x, y) + (x', y') = (x + x', y + y')$, is a "product" of two copies of the group $\mathbb{Z}$. As we shall see, this notion of product is very general and can be applied in many other contexts. But first let us look at an even more basic method of finding new examples.

3.1   Substructures

As we saw earlier, the set $\mathbb{C}$ of all complex numbers, with the operations of addition and multiplication, forms one of the most basic examples of a field. It also contains many *subfields*: that is, subsets that themselves form fields. Take, for example, the set $\mathbb{Q}(i)$ of all complex numbers of the form $a + bi$ for which a and b are rational. This is a subset of $\mathbb{C}$ and is also a field. To show this, one must prove that $\mathbb{Q}(i)$ is *closed* under addition, multiplication, and the taking of inverses. That is, if z and w are elements of $\mathbb{Q}(i)$, then $z + w$ and zw must be as well, as must $-z$ and $1/z$ (this last requirement applying only when $z \neq 0$). Axioms

such as the commutativity and associativity of addition and multiplication are then true in $\mathbb{Q}(i)$ for the simple reason that they are true in the larger set $\mathbb{C}$.

Even though $\mathbb{Q}(i)$ is contained in $\mathbb{C}$, it is a more interesting field in some important ways. But how can this be? Surely, one might think, an object cannot become *more* interesting when most of it is taken away. But a moment's further thought shows that it certainly can: for example, the set of all prime numbers contains fascinating mysteries of a kind that one does not expect to encounter in the set of all positive integers. As for fields, THE FUNDAMENTAL THEOREM OF ALGEBRA [V.13] tells us that every polynomial equation has a solution in $\mathbb{C}$. This is very definitely not true in $\mathbb{Q}(i)$. So in $\mathbb{Q}(i)$, and in many other fields of a similar kind, we can ask which polynomial equations have solutions. This turns out to be a deep and important question that simply does not arise in the larger field $\mathbb{C}$.

In general, given an example X of an algebraic structure, a substructure of X is a subset Y that has relevant closure properties. For instance, groups have subgroups, vector spaces have subspaces, rings have subrings (and also IDEALS [III.81]), and so on. If the property defining the substructure Y is a sufficiently interesting one, then Y may well be significantly different from X and may therefore be a useful addition to one's stock of examples.

This discussion has focused on algebra, but interesting substructures abound in analysis and geometry as well. For example, the plane $\mathbb{R}^2$ is not a particularly interesting set, but it has subsets, such as the MANDELBROT SET [IV.14 §2.8], to give just one example, that are still far from fully understood.

3.2   Products

Let G and H be two groups. The *product group* $G \times H$ has as its elements all pairs of the form (g, h) such that g belongs to G and h belongs to H. This definition shows how to build the elements of $G \times H$ out of the elements of G and the elements of H. But to define a group we need to do more: we are given binary operations on G and H and we must use them to build a binary operation on $G \times H$. If g_1 and g_2 are elements of G, let us write $g_1 g_2$ for the result of applying G's binary operation to them, as is customary, and let us do the same for H. Then there is an obvious binary operation we can define on the pairs, namely

$$(g_1, h_1)(g_2, h_2) = (g_1 g_2, h_1 h_2).$$

That is, one applies the binary operation from G to the first coordinate and the binary operation from H to the second.

One can form products of vector spaces in a very similar way. If V and W are two vector spaces, then the elements of $V \times W$ are all pairs of the form (v, w) with v in V and w in W. Addition and scalar multiplication are defined by the formulas

$$(v_1, w_1) + (v_2, w_2) = (v_1 + v_2, w_1 + w_2)$$

and

$$\lambda(v, w) = (\lambda v, \lambda w).$$

The dimension of the resulting space is the sum of the dimensions of V and W. (It is actually more usual to denote this space by $V \oplus W$ and call it the *direct sum* of V and W. Nevertheless, it is a product construction.)

It is not always possible to define product structures in this simple way. For example, if $\mathbb{F}$ and $\mathbb{F}'$ are two fields, we might be tempted to define a "product field" $\mathbb{F} \times \mathbb{F}'$ using the formulas

$$(x_1, y_1) + (x_2, y_2) = (x_1 + x_2, y_1 + y_2)$$

and

$$(x_1, y_1)(x_2, y_2) = (x_1 x_2, y_1 y_2).$$

However, this definition does not give us a field. Most of the axioms hold, including the existence of additive and multiplicative identities—they are $(0, 0)$ and $(1, 1)$, respectively—but the nonzero element $(1, 0)$ does not have a multiplicative inverse, since the product of $(1, 0)$ and (x, y) is $(x, 0)$, which can never equal $(1, 1)$.

Occasionally we can define more complicated binary operations that do make the set $\mathbb{F} \times \mathbb{F}'$ into a field. For instance, if $\mathbb{F} = \mathbb{F}' = \mathbb{R}$, then we can define addition as above but define multiplication in a less obvious way as follows:

$$(x_1, y_1)(x_2, y_2) = (x_1 x_2 - y_1 y_2, x_1 y_2 + x_2 y_1).$$

Then we obtain $\mathbb{C}$, the field of complex numbers, since the pair (x, y) can be identified with the complex number $x + iy$. However, this is not a product field in the general sense we are discussing.

Returning to groups, what we defined earlier was the *direct product* of G and H. However, there are other, more complicated products of groups, which can be used to give a much richer supply of examples. To illustrate this, let us consider the *dihedral group* D_4, which is the group of all symmetries of a square, of which there are eight. If we let R stand for one of the reflections and T for a counterclockwise quarter turn, then

every symmetry can be written in the form $T^i R^j$, where i is 0, 1, 2, or 3 and j is 0 or 1. (Geometrically, this says that you can produce any symmetry by either rotating through a multiple of $90°$ or reflecting and then rotating.)

This suggests that we might be able to regard D_4 as a product of the group $\{I, T, T^2, T^3\}$, consisting of four rotations, with the group $\{I, R\}$, consisting of the identity I and the reflection R. We could even write (T^i, R^j) instead of $T^i R^j$. However, we have to be careful. For instance, $(TR)(TR)$ does not equal $T^2 R^2 = T^2$ but I. The correct rule for multiplication can be deduced from the fact that $RTR = T^{-1}$ (which in geometrical terms is saying that if you reflect the square, rotate it counterclockwise through $90°$, and reflect back, then the result is a *clockwise* rotation through $90°$). It turns out to be

$$(T^i, R^j)(T^{i'}, R^{j'}) = (T^{i + (-1)^j i'}, R^{j + j'}).$$

For example, the product of (T, R) with (T^3, R) is $T^{-2} R^2$, which equals T^2.

This is a simple example of a "semidirect product" of two groups. In general, given two groups G and H, there may be several interesting ways of defining a binary operation on the set of pairs (g, h), and therefore several potentially interesting new groups.

3.3 Quotients

Let us write $\mathbb{Q}[x]$ for the set of all polynomials in the variable x with rational coefficients: that is, expressions like $2x^4 - \frac{3}{2}x + 6$. Any two such polynomials can be added, subtracted, or multiplied together and the result will be another polynomial. This makes $\mathbb{Q}[x]$ into a commutative ring, but not a field, because if you divide one polynomial by another then the result is not (necessarily) a polynomial.

We will now convert $\mathbb{Q}[x]$ into a field in what may at first seem a rather strange way: by regarding the polynomial $x^3 - x - 1$ as "equivalent" to the zero polynomial. To put this another way, whenever a polynomial involves x^3 we will allow ourselves to replace x^3 by $x + 1$, and we will regard the new polynomial that results as equivalent to the old one. For example, writing "$\sim$" for "is equivalent to":

$$x^5 = x^3 x^2 \sim (x + 1)x^2 = x^3 + x^2$$
$$\sim x + 1 + x^2 = x^2 + x + 1.$$

Notice that in this way we can convert any polynomial into one of degree at most 2, since whenever the degree is higher, you can reduce it by taking out x^3 from the term of highest degree and replacing it by $x + 1$, just as we did above.

Notice also that whenever we do such a replacement, the difference between the old polynomial and the new one is a multiple of $x^3 - x - 1$. For example, when we replaced $x^3 x^2$ by $(x + 1)x^2$ the difference was $(x^3 - x - 1)x^2$. Therefore, what our process amounts to is this: two polynomials are equivalent if and only if their difference is a multiple of the polynomial $x^3 - x - 1$.

Now the reason $\mathbb{Q}[x]$ was not a field was that nonconstant polynomials do not have multiplicative inverses. For example, it is obvious that one cannot multiply x^2 by a polynomial and obtain the polynomial 1. However, we can obtain a polynomial that is *equivalent* to 1 if we multiply by $1 + x - x^2$. Indeed, the product of the two is

$$x^2 + x^3 - x^4 \sim x^2 + x + 1 - (x + 1)x = 1.$$

It turns out that *all* polynomials that are not equivalent to zero (that is, are not multiples of $x^3 - x - 1$) have multiplicative inverses in this generalized sense. (To find an inverse for a polynomial P one applies the generalized EUCLID ALGORITHM [III.22] to find polynomials Q and R such that $PQ + R(x^3 - x - 1) = 1$. The reason we obtain 1 on the right-hand side is that $x^3 - x - 1$ cannot be factorized in $\mathbb{Q}[x]$ and P is not a multiple of $x^3 - x - 1$, so their highest common factor is 1. The inverse of P is then Q.)

In what sense does this mean that we have a field? After all, the product of x^2 and $1 + x - x^2$ was not 1: it was merely equivalent to 1. This is where the notion of quotients comes in. We simply decide that when two polynomials are equivalent, we will regard them as equal, and we denote the resulting mathematical structure by $\mathbb{Q}[x]/(x^3 - x - 1)$. This structure turns out to be a field, and it turns out to be important as the smallest field that contains $\mathbb{Q}$ and also has a root of the polynomial $X^3 - X - 1$. What is this root? It is simply x. This is a slightly subtle point because we are now thinking of polynomials in two different ways: as elements of $\mathbb{Q}[x]/(x^3 - x - 1)$ (at least when equivalent ones are regarded as equal), and also as functions defined on $\mathbb{Q}[x]/(x^3 - x - 1)$. So the polynomial $X^3 - X - 1$ is not the zero polynomial, since for example it takes the value 5 when $X = 2$ and the value $x^6 - x^2 - 1 \sim (x + 1)^2 - x^2 - 1 \sim 2x$ when $X = x^2$.

You may have noticed a strong similarity between the discussion of the field $\mathbb{Q}[x]/(x^3 - x - 1)$ and the discussion of the field $\mathbb{Q}(y)$ at the end of section 2.2. And indeed, this is no coincidence: they are two different

ways of describing the same field. However, thinking of the field as $\mathbb{Q}[x]/(x^3 - x - 1)$ brings significant advantages, as it converts questions about a mysterious set of complex numbers into more approachable questions about polynomials.

What does it mean to "regard two mathematical objects as equal" when they are not equal? A formal answer to this question uses the notion of equivalence relations and equivalence classes (discussed in THE LANGUAGE AND GRAMMAR OF MATHEMATICS [I.2 §2.3]): one says that the elements of $\mathbb{Q}[x]/(x^3 - x - 1)$ are not in fact polynomials but *equivalence classes* of polynomials. However, to understand the notion of a quotient it is much easier to look at an example with which we are all familiar, namely the set $\mathbb{Q}$ of rational numbers. If we are trying to explain carefully what a rational number is, then we may start by saying that a typical rational number has the form a/b, where a and b are integers and b is not 0. And it is possible to define the set of rational numbers to be the set of all such expressions, with the rules

$$\frac{a}{b} + \frac{c}{d} = \frac{ad + bc}{bd}$$

and

$$\frac{a}{b}\frac{c}{d} = \frac{ac}{bd}.$$

However, there is one very important further remark we must make, which is that we do not regard all such expressions as different: for example, $\frac{1}{2}$ and $\frac{3}{6}$ are supposed to be the same rational number. So we define two expressions $\frac{a}{b}$ and $\frac{c}{d}$ to be equivalent if $ad = bc$ and we regard equivalent expressions as denoting the same number. Notice that the expressions can be genuinely different, but we think of them as denoting the same object.

If we do this, then we must be careful whenever we define functions and binary operations. For example, suppose we tried to define a binary operation "∘" on $\mathbb{Q}$ by the natural-looking formula

$$\frac{a}{b} \circ \frac{c}{d} = \frac{a + c}{b + d}.$$

This definition turns out to have a very serious flaw. To see why, let us apply it to the fractions $\frac{1}{2}$ and $\frac{1}{3}$. Then it gives us the answer $\frac{2}{5}$. Now let us replace $\frac{1}{2}$ by the equivalent fraction $\frac{3}{6}$ and apply the formula again. This time it gives us the answer $\frac{4}{9}$, which is different. Thus, although the formula defines a perfectly good binary operation on the set of *expressions* of the form $\frac{a}{b}$, it does not make any sense as a binary operation on the set of *rational numbers*.

In general, it is essential to check that if you put equivalent objects in then you get equivalent objects out. For example, when defining addition and multiplication for the field $\mathbb{Q}[x]/(x^3 - x - 1)$, one must check that if P and P' differ by a multiple of $x^3 - x - 1$, and Q and Q' also differ by a multiple of $x^3 - x - 1$, then so do $P + Q$ and $P' + Q'$, and so do PQ and $P'Q'$. This is an easy exercise.

An important example of a quotient construction is that of a *quotient group*. If G is a group and H is a subgroup of G, then it is natural to try to do what we did for polynomials and define g_1 and g_2 to be equivalent if $g_1^{-1}g_2$ (the obvious notion of the "difference" between g_1 and g_2) belongs to H. The equivalence class of an element g is easily seen to be the set of all elements gh such that $h \in H$, which is usually written gH. (It is called a *left coset* of H.)

There is a natural candidate for a binary operation $*$ on the set of all left cosets: $g_1H * g_2H = g_1g_2H$. In other words, given two left cosets, pick elements g_1 and g_2 from each, form the product g_1g_2, and take the left coset g_1g_2H. Once again, it is important to check that if you pick different elements from the original cosets, then you will still get the coset g_1g_2H. It turns out that this is not always the case: one needs the additional assumption that H is a *normal subgroup*, which means that if h is any element of H, then ghg^{-1} is an element of H for every element g of G. Elements of the form ghg^{-1} are called *conjugates* of h; thus, a normal subgroup is a subgroup that is "closed under conjugation."

If H is a normal subgroup, then the set of left cosets forms a group under the binary operation just defined. This group is written G/H and is called the quotient of G by H. One can regard G as a product of H and G/H (though it may be a somewhat complicated product), so if you understand both H and G/H, then for many purposes you understand G. Therefore, groups G that do not have normal subgroups (other than G itself and the subgroup that consists of just the identity element) have a special role, a bit like the role of prime numbers in number theory. They are called *simple groups*. (See THE CLASSIFICATION OF FINITE SIMPLE GROUPS [V.7].)

Why is the word "quotient" used? Well, a quotient is normally what you get when you divide one number by another, so to understand the analogy let us think about dividing 21 by 3. We can think of this as dividing up twenty-one objects into sets of three objects each and asking how many sets we get. This can be described in terms of equivalence as follows. Let us call two objects equivalent if they belong to the same one of the seven sets. Then there can be at most seven inequivalent objects. So when we regard equivalent objects as the same, we "divide out by the equivalence," obtaining a "quotient set" that has seven elements.

A rather different use of quotients leads to an elegant definition of the mathematical shape known as a *torus*: that is, the shape of the surface of a doughnut (of the kind that has a hole). We start with the plane, $\mathbb{R}^2$, and define two points (x, y) and (x', y') to be equivalent if $x - x'$ and $y - y'$ are both integers. Suppose that we regard any two equivalent points as the same and that we start at a point (x, y) and move right until we reach the point $(x + 1, y)$. This point is "the same" as (x, y), since the difference is $(1, 0)$. Therefore, it is as though the entire plane has been wrapped around a vertical cylinder of circumference 1 and we have gone around this cylinder once. If we now apply the same argument to the y-coordinate, noting that (x, y) is always "the same" point as $(x, y + 1)$, then we find that this cylinder is itself "folded around" so that if you go "upwards" by a distance of 1 then you get back to where you started. But that is what a torus is: a cylinder that is folded back into itself. (This is not the only way of defining a torus, however. For example, it can be defined as the product of two circles.)

Many other important objects in modern geometry are defined using quotients. It often happens that the object one starts with is extremely big, but that at the same time the equivalence relation is very generous, in the sense that it is easy for one object to be equivalent to another. In that case the number of "genuinely distinct" objects can be quite small. This is a rather loose way of talking, since it is not really the *number* of distinct objects that is interesting so much as the complexity of the set of these objects. It might be better to say that one often starts with a hopelessly large and complicated structure but "divides out most of the mess" and ends up with a quotient object that has a structure that is simple enough to be manageable while still conveying important information. Good examples of this are the FUNDAMENTAL GROUP [IV.6 §2] and the HOMOLOGY AND COHOMOLOGY GROUPS [IV.6 §4] of a topological space; an even better example is the notion of a MODULI SPACE [IV.8].

Many people find the idea of a quotient somewhat difficult to grasp, but it is of major importance throughout mathematics, which is why it has been discussed at some length here.

4 Functions between Algebraic Structures

One rule with almost no exceptions is that mathematical structures are not studied in isolation: as well as the structures themselves one looks at certain *functions* defined on those structures. In this section we shall see which functions are worth considering, and why. (For a discussion of functions in general, see THE LANGUAGE AND GRAMMAR OF MATHEMATICS [I.2 §2.2].)

4.1 Homomorphisms, Isomorphisms, and Automorphisms

If X and Y are two examples of a particular mathematical structure, such as a group, field, or vector space, then, as was suggested in the discussion of symmetry in section 2.1, there is a class of functions from X to Y of particular interest, namely the functions that "preserve the structure." Roughly speaking, a function $f : X \rightarrow Y$ is said to preserve the structure of X if, given any relationship between elements of X that is expressed in terms of that structure, there is a corresponding relationship between the images of those elements that is expressed in terms of the structure of Y. For example, if X and Y are groups and a, b, and c are elements of X such that $ab = c$, then, if f is to preserve the algebraic structure of X, $f(a)f(b)$ must equal $f(c)$ in Y. (Here, as is usual, we are using the same notation for the binary operations that make X and Y groups as is normally used for multiplication.) Similarly, if X and Y are fields, with binary operations that we shall write using the standard notation for addition and multiplication, then a function $f : X \rightarrow Y$ will be interesting only if $f(a) + f(b) = f(c)$ whenever $a + b = c$ and $f(a)f(b) = f(c)$ whenever $ab = c$. For vector spaces, the functions of interest are ones that preserve linear combinations: if V and W are vector spaces, then $f(a\boldsymbol{v} + b\boldsymbol{w})$ should always equal $af(\boldsymbol{v}) + bf(\boldsymbol{w})$.

A function that preserves structure is called a *homomorphism*, though homomorphisms of particular mathematical structures often have their own names: for example, a homomorphism of vector spaces is called a linear map.

There are some useful properties that a homomorphism may have if we are lucky. To see why further properties can be desirable, consider the following example. Let X and Y be groups and let $f : X \rightarrow Y$ be the function that takes every element of X to the identity element e of Y. Then, according to the definition above, f preserves the structure of X, since whenever $ab = c$, we have $f(a)f(b) = ee = e = f(c)$. However, it seems more accurate to say that f has *collapsed* the structure. One can make this idea more precise: although $f(a)f(b) = f(c)$ whenever $ab = c$, *the converse does not hold*: it is perfectly possible for $f(a)f(b)$ to equal $f(c)$ without ab equaling c, and indeed that happens in the example just given.

An *isomorphism* between two structures X and Y is a homomorphism $f : X \rightarrow Y$ that has an inverse $g : Y \rightarrow X$ that is also a homomorphism. For most algebraic structures, if f has an inverse g, then g is automatically a homomorphism; in such cases we can simply say that an isomorphism is a homomorphism that is also a BIJECTION [I.2 §2.2]. That is, f is a one-to-one correspondence between X and Y that preserves structure.[1]

If X and Y are fields, then these considerations are less interesting: it is a simple exercise to show that every homomorphism $f : X \rightarrow Y$ that is not identically zero is automatically an isomorphism between X and its image $f(X)$, that is, the set of all values taken by the function f. So structure cannot be collapsed without being lost. (The proof depends on the fact that the zero in Y has no multiplicative inverse.)

In general, if there is an isomorphism between two algebraic structures X and Y, then X and Y are said to be *isomorphic* (coming from the Greek words for "same" and "shape"). Loosely, the word "isomorphic" means "the same in all essential respects," where what counts as essential is precisely the algebraic structure. What is absolutely *not* essential is the nature of the objects that have the structure: for example, one group might consist of certain complex numbers, another of integers modulo a prime p, and a third of rotations of a geometrical figure, and they could all turn out to be isomorphic. The idea that two mathematical constructions can have very different constituent parts and yet in a deeper sense be "the same" is one of the most important in mathematics.

An *automorphism* of an algebraic structure X is an isomorphism from X to itself. Since it is hardly surprising that X is isomorphic to itself, one might ask what the point is of automorphisms. The answer is that automorphisms are precisely the algebraic symmetries

---

1. Let us see how this claim is proved for groups. If X and Y are groups, $f : X \rightarrow Y$ is a homomorphism with inverse $g : Y \rightarrow X$, and u, v, and w are elements of Y with $uv = w$, then we must show that $g(u)g(v) = g(w)$. To do this, let $a = g(u)$, $b = g(v)$, and $d = g(w)$. Since f and g are inverse functions, $f(a) = u$, $f(b) = v$, and $f(d) = w$. Now let $c = ab$. Then $w = uv = f(a)f(b) = f(c)$, since f is a homomorphism. But then $f(c) = f(d)$, which implies that $c = d$ (just apply the function g to $f(c)$ and $f(d)$). Therefore $ab = d$, which tells us that $g(u)g(v) = g(w)$, as we needed to show.

alluded to in our discussion of groups. An automorphism of X is a function from X to itself that preserves the structure (which now comes in the form of statements like $ab = c$). The composition of two automorphisms is clearly a third, and as a result the automorphisms of a structure X form a group. Although the individual automorphisms may not be of much interest, the group certainly is, as it often encapsulates what one really wants to know about a structure X that is too complicated to analyze directly.

A spectacular example of this is when X is a field. To illustrate, let us take the example of $\mathbb{Q}(\sqrt{2})$. If $f : \mathbb{Q}(\sqrt{2}) \to \mathbb{Q}(\sqrt{2})$ is an automorphism, then $f(1) = 1$. (This follows easily from the fact that 1 is the only multiplicative identity.) It follows that $f(2) = f(1 + 1) = f(1) + f(1) = 1 + 1 = 2$. Continuing like this, we can show that $f(n) = n$ for every positive integer n. Then $f(n) + f(-n) = f(n + (-n)) = f(0) = 0$, so $f(-n) = -f(n) = -n$. Finally, $f(p/q) = f(p)/f(q) = p/q$ when p and q are integers with $q \neq 0$. So f takes every rational number to itself. What can we say about $f(\sqrt{2})$? Well, $f(\sqrt{2})f(\sqrt{2}) = f(\sqrt{2} \cdot \sqrt{2}) = f(2) = 2$, but this implies only that $f(\sqrt{2})$ is $\sqrt{2}$ or $-\sqrt{2}$. It turns out that both choices are possible: one automorphism is the "trivial" one, $f(a+b\sqrt{2}) = a+b\sqrt{2}$, and the other is the more interesting one, $f(a + b\sqrt{2}) = a - b\sqrt{2}$. This observation demonstrates that there is no algebraic difference between the two square roots; in this sense, the field $\mathbb{Q}(\sqrt{2})$ does not know which square root of 2 is positive and which negative. These two automorphisms form a group, which is isomorphic to the group consisting of the elements ± 1 under multiplication, or the group of integers modulo 2, or the group of symmetries of an isosceles triangle that is not equilateral, or.... The list is endless.

The automorphism groups associated with certain field extensions are called *Galois groups*, and are a vital component of the proof of THE INSOLUBILITY OF THE QUINTIC [V.21], as well as large parts of ALGEBRAIC NUMBER THEORY [IV.1].

An important concept associated with a homomorphism ϕ between algebraic structures is that of a *kernel*. This is defined to be the set of all elements x of X such that $\phi(x)$ is the identity element of Y (where this means the additive identity if X and Y are structures that involve both additive and multiplicative binary operations). The kernel of a homomorphism tends to be a substructure of X with interesting properties. For instance, if G and K are groups, then the kernel of a homomorphism from G to K is a normal subgroup of

G; and conversely, if H is a normal subgroup of G, then the *quotient map*, which takes each element g to the left coset gH, is a homomorphism from G to the quotient group G/H with kernel H. Similarly, the kernel of any ring homomorphism is an IDEAL [III.81], and every ideal I in a ring R is the kernel of a "quotient map" from R to R/I. (This quotient construction is discussed in more detail in RINGS, IDEALS, AND MODULES [III.81].)

4.2 Linear Maps and Matrices

Homomorphisms between vector spaces have a distinctive geometrical property: they send straight lines to straight lines. For this reason they are called *linear maps*, as was mentioned in the previous subsection. From a more algebraic point of view, the structure that linear maps preserve is that of linear combinations: a function f from one vector space to another is a linear map if $f(a\boldsymbol{u} + b\boldsymbol{v}) = af(\boldsymbol{u}) + bf(\boldsymbol{v})$ for every pair of vectors $\boldsymbol{u}, \boldsymbol{v} \in V$ and every pair of scalars a and b. From this one can deduce the more general assertion that $f(a_1\boldsymbol{v}_1 + \cdots + a_n\boldsymbol{v}_n)$ is always equal to $a_1 f(\boldsymbol{v}_1) + \cdots + a_n f(\boldsymbol{v}_n)$.

Suppose that we wish to define a linear map from V to W. How much information do we need to provide? In order to see what sort of answer is required, let us begin with a similar but slightly easier question: how much information is needed to specify a point in space? The answer is that, once one has devised a sensible coordinate system, three numbers will suffice. If the point is not too far from Earth's surface then one might wish to use its latitude, its longitude, and its height above sea level, for instance. Can a linear map from V to W similarly be specified by just a few numbers?

The answer is that it can, at least if V and W are finite dimensional. Suppose that V has a basis $\boldsymbol{v}_1, \ldots, \boldsymbol{v}_n$, that W has a basis $\boldsymbol{w}_1, \ldots, \boldsymbol{w}_m$, and that $f : V \to W$ is the linear map we would like to specify. Since every vector in V can be written in the form $a_1\boldsymbol{v}_1 + \cdots + a_n\boldsymbol{v}_n$ and since $f(a_1\boldsymbol{v}_1 + \cdots + a_n\boldsymbol{v}_n)$ is always equal to $a_1 f(\boldsymbol{v}_1) + \cdots + a_n f(\boldsymbol{v}_n)$, once we decide what $f(\boldsymbol{v}_1), \ldots, f(\boldsymbol{v}_n)$ are we have specified f completely. But each vector $f(\boldsymbol{v}_j)$ is a linear combination of the basis vectors $\boldsymbol{w}_1, \ldots, \boldsymbol{w}_m$: that is, it can be written in the form

$$f(\boldsymbol{v}_j) = a_{1j}\boldsymbol{w}_1 + \cdots + a_{mj}\boldsymbol{w}_m.$$

Thus, to specify an individual $f(\boldsymbol{v}_j)$ needs m numbers, the scalars $a_{1j}, \ldots, a_{mj}$. Since there are n different vectors $\boldsymbol{v}_j$, the linear map is determined by the mn numbers a_{ij}, where i runs from 1 to m and j from 1 to n.

These numbers can be written in an array, as follows:

$$\begin{pmatrix} a_{11} & a_{12} & \cdots & a_{1n} \\ a_{21} & a_{22} & \cdots & a_{2n} \\ \vdots & \vdots & \ddots & \vdots \\ a_{m1} & a_{m2} & \cdots & a_{mn} \end{pmatrix}.$$

An array like this is called a *matrix*. It is important to note that a different choice of basis vectors for V and W would lead to a different matrix, so one often talks of the matrix of f *relative to* a given pair of bases (a basis for V and a basis for W).

Now suppose that f is a linear map from V to W and that g is a linear map from U to V. Then fg stands for the linear map from U to W obtained by doing first g, then f. If the matrices of f and g, relative to certain bases of U, V, and W, are A and B, then what is the matrix of fg? To work it out, one takes a basis vector $\boldsymbol{u}_k$ of U and applies to it the function g, obtaining a linear combination $b_{1k}\boldsymbol{v}_1 + \cdots + b_{nk}\boldsymbol{v}_n$ of the basis vectors of V. To this linear combination one applies the function f, obtaining a rather complicated linear combination of linear combinations of the basis vectors $\boldsymbol{w}_1, \ldots, \boldsymbol{w}_m$ of W.

Pursuing this idea, one can calculate that the entry in row i and column j of the matrix P of fg is $a_{i1}b_{1j} + a_{i2}b_{2j} + \cdots + a_{in}b_{nj}$. This matrix P is called the *product* of A and B and is written AB. If you have not seen this definition then you will find it hard to grasp, but the main point to remember is that there is a way of calculating the matrix for fg from the matrices A and B of f and g, and that this matrix is denoted AB. Matrix multiplication of this kind is associative but not commutative. That is, $A(BC)$ is always equal to $(AB)C$ but AB is not necessarily the same as BA. The associativity follows from the fact that composition of the underlying linear maps is associative: if A, B, and C are the matrices of f, g, and h, respectively, then $A(BC)$ is the matrix of the linear map "do h-then-g, then f" and $(AB)C$ is the matrix of the linear map "do h, then g-then-f," and these are the same linear map.

Let us now confine our attention to *automorphisms* from a vector space V to itself. These are linear maps $f : V \to V$ that can be inverted; that is, for which there exists a linear map $g : V \to V$ such that $fg(\boldsymbol{v}) = gf(\boldsymbol{v}) = \boldsymbol{v}$ for every vector $\boldsymbol{v}$ in V. These we can think of as "symmetries" of the vector space V, and as such they form a group under composition. If V is n dimensional and the scalars come from the field $\mathbb{F}$, then this group is called $\text{GL}_n(\mathbb{F})$. The letters "G" and "L" stand for "general" and "linear"; some of the most important and difficult problems in mathematics arise when one tries to understand the structure of the general linear groups (and related groups) for certain interesting fields $\mathbb{F}$ (see REPRESENTATION THEORY [IV.9 §§5,6]).

While matrices are very useful, many interesting linear maps are between infinite-dimensional vector spaces, and we close this section with two examples for the reader who is familiar with elementary calculus. (There will be a brief discussion of calculus later in this article.) For the first, let V be the set of all functions from $\mathbb{R}$ to $\mathbb{R}$ that can be differentiated and let W be the set of *all* functions from $\mathbb{R}$ to $\mathbb{R}$. These can be made into vector spaces in a simple way: if f and g are functions, then their sum is the function h defined by the formula $h(x) = f(x) + g(x)$, and if a is a real number then af is the function k defined by the formula $k(x) = af(x)$. (So, for example, we could regard the polynomial $x^2 + 3x + 2$ as a linear combination of the functions x^2, x, and the constant function 1.) Then differentiation is a linear map (from V to W), since the derivative $(af + bg)'$ is $af' + bg'$. This is clearer if we write $\text{D}f$ for the derivative of f: then we are saying that $\text{D}(af + bg) = a\,\text{D}f + b\,\text{D}g$.

A second example uses integration. Let V be another vector space of functions, and let u be a function of *two* variables. (The functions involved have to have certain properties for the definition to work, but let us ignore the technicalities.) Then we can define a linear map T on the space V by the formula

$$(Tf)(x) = \int u(x, y)f(y)\,\text{d}y.$$

Definitions like this one can be hard to take in, because they involve holding in one's mind three different levels of complexity. At the bottom we have real numbers, denoted by x and y. In the middle are functions like f, u, and Tf, which turn real numbers (or pairs of them) into real numbers. At the top is another function, T, but the "objects" that it transforms are themselves functions: it turns a function like f into a different function Tf. This is just one example where it is important to think of a function as a single, elementary "thing" rather than as a process of transformation. (See the discussion of functions in THE LANGUAGE AND GRAMMAR OF MATHEMATICS [I.2 §2.2].) Another remark that may help to clarify the definition is that there is a very close analogy between the role of the two-variable function $u(x, y)$ and the role of a matrix a_{ij} (which can itself be thought of as a function of the two integer variables i and j). Functions like u are sometimes called *kernels* (which should not be confused with kernels of homomorphisms). For more about linear maps between infinite-dimensional spaces, see OPERATOR ALGEBRAS [IV.15] and LINEAR OPERATORS [III.50].

4.3 Eigenvalues and Eigenvectors

Let V be a vector space and let $S : V \to V$ be a linear map from V to itself. An *eigenvector* of S is a nonzero vector $\boldsymbol{v}$ in V such that $S\boldsymbol{v}$ is proportional to $\boldsymbol{v}$; that is, $S\boldsymbol{v} = \lambda\boldsymbol{v}$ for some scalar λ. The scalar in question is called the *eigenvalue* corresponding to $\boldsymbol{v}$. This simple pair of definitions is extraordinarily important: it is hard to think of any branch of mathematics where eigenvectors and eigenvalues do not have a major part to play. But what is so interesting about $S\boldsymbol{v}$ being proportional to $\boldsymbol{v}$? A rather vague answer is that in many cases the eigenvectors and eigenvalues associated with a linear map contain all the information one needs about the map, and in a very convenient form. Another answer is that linear maps occur in many different contexts, and questions that arise in those contexts often turn out to be questions about eigenvectors and eigenvalues, as the following two examples illustrate.

First, imagine that you are given a linear map T from a vector space V to itself and want to understand what happens if you perform the map repeatedly. One approach would be to pick a basis of V, work out the corresponding matrix A of T, and calculate the powers of A by matrix multiplication. The trouble is that the calculation will be messy and uninformative, and it does not really give much insight into the linear map.

However, it often happens that one can pick a very special basis, consisting only of eigenvectors, and in that case understanding the powers of T becomes easy. Indeed, suppose that the basis vectors are $\boldsymbol{v}_1, \boldsymbol{v}_2, \ldots, \boldsymbol{v}_n$ and that each $\boldsymbol{v}_i$ is an eigenvector with corresponding eigenvalue λ_i. That is, suppose that $T(\boldsymbol{v}_i) = \lambda_i \boldsymbol{v}_i$ for every i. If $\boldsymbol{w}$ is any vector in V, then there is exactly one way of writing it in the form $a_1\boldsymbol{v}_1 + \cdots + a_n\boldsymbol{v}_n$, and then

$$T(\boldsymbol{w}) = \lambda_1 a_1 \boldsymbol{v}_1 + \cdots + \lambda_n a_n \boldsymbol{v}_n.$$

Roughly speaking, this says that T stretches the part of $\boldsymbol{w}$ in direction $\boldsymbol{v}_i$ by a factor of λ_i. But now it is easy to say what happens if we apply T not just once but m times to $\boldsymbol{w}$. The result will be

$$T^m(\boldsymbol{w}) = \lambda_1^m a_1 \boldsymbol{v}_1 + \cdots + \lambda_n^m a_n \boldsymbol{v}_n.$$

In other words, now the amount by which we stretch in the $\boldsymbol{v}_i$ direction is λ_i^m, and that is all there is to it.

Why should one be interested in doing linear maps over and over again? There are many reasons: one fairly convincing one is that this sort of calculation is exactly what Google does in order to put Web sites into a useful order. Details can be found in THE MATHEMATICS OF ALGORITHM DESIGN [VII.5].

The second example concerns the interesting property of the EXPONENTIAL FUNCTION [III.25] e^x: that its derivative is the same function. In other words, if $f(x) = e^x$, then $f'(x) = f(x)$. Now differentiation, as we saw earlier, can be thought of as a linear map, and if $f'(x) = f(x)$ then this map leaves the function f unchanged, which says that f is an eigenvector with eigenvalue 1. More generally, if $g(x) = e^{\lambda x}$, then $g'(x) = \lambda e^{\lambda x} = \lambda g(x)$, so g is an eigenvector of the differentiation map, with eigenvalue λ. Many linear differential equations can be thought of as asking for eigenvectors of linear maps defined using differentiation. (Differentiation and differential equations will be discussed in the next section.)

5 Basic Concepts of Mathematical Analysis

Mathematics took a huge leap forward in sophistication with the invention of calculus, and the notion that one can specify a mathematical object indirectly by means of better and better approximations. These ideas form the basis of a broad area of mathematics known as *analysis*, and the purpose of this section is to help the reader who is unfamiliar with them. However, it will not be possible to do full justice to the subject, and what is written here will be hard to understand without at least some prior knowledge of calculus.

5.1 Limits

In our discussion of real numbers (section 1.4) there was a brief discussion of the square root of 2. How do we know that 2 has a square root? One answer is the one given there: that we can calculate its decimal expansion. If we are asked to be more precise, we may well end up saying something like this. The real numbers 1, 1.4, 1.41, 1.414, 1.4142, 1.41421, ..., which have terminating decimal expansions (and are therefore rational), approach another real number $x = 1.4142135\ldots$. We cannot actually write down x properly because it has an infinite decimal expansion but we can at least explain how its digits are defined: for example, the third digit after the decimal point is a 4 because 1.414 is the largest multiple of 0.001 that squares to less than 2. It follows that the squares of the original numbers, 1, 1.96, 1.9881, 1.999396, 1.99996164, 1.9999899241,..., approach 2, and this is why we are entitled to say that $x^2 = 2$.

Suppose that we are asked to determine the length of a curve drawn on a piece of paper, and that we are given a ruler to help us. We face a problem: the ruler is straight and the curve is not. One way of tackling the problem is as follows. First, draw a few points $P_0, P_1, P_2, \ldots, P_n$ along the curve, with P_0 at one end and P_n at the other. Next, measure the distance from P_0 to P_1, the distance from P_1 to P_2, and so on up to P_n. Finally, add all these distances up. The result will not be an exactly correct answer, but if there are enough points, spaced reasonably evenly, and if the curve does not wiggle too much, then our procedure will give us a good notion of the "approximate length" of the curve. Moreover, it gives us a way to *define* what we mean by the "exact length": suppose that, as we take more and more points, we find that the approximate lengths, in the sense just defined, approach some number l. Then we say that l is the length of the curve.

In both these examples there is a number that we reach by means of better and better approximations. I used the word "approach" in both cases, but this is rather vague, and it is important to make it precise. Let $a_1, a_2, a_3, \ldots$ be a sequence of real numbers. What does it mean to say that these numbers approach a specified real number l?

The following two examples are worth bearing in mind. The first is the sequence $\frac{1}{2}, \frac{2}{3}, \frac{3}{4}, \frac{4}{5}, \ldots$. In a sense, the numbers in this sequence approach 2, since each one is closer to 2 than the one before, but it is clear that this is not what we mean. What matters is not so much that we get closer and closer, but that we get *arbitrarily close*, and the only number that is approached in this stronger sense is the obvious "limit," 1.

A second sequence illustrates this in a different way: $1, 0, \frac{1}{2}, 0, \frac{1}{3}, 0, \frac{1}{4}, 0, \ldots$. Here, we would like to say that the numbers approach 0, even though it is not true that each one is closer than the one before. Nevertheless, it is true that eventually the sequence gets as close as you like to 0 and remains at least that close.

This last phrase serves as a definition of the mathematical notion of a *limit*: the limit of the sequence of numbers $a_1, a_2, a_3, \ldots$ is l if eventually the sequence gets as close as you like to l and remains that close. However, in order to meet the standards of precision demanded by mathematics, we need to know how to translate English words like "eventually" into mathematics, and for this we need QUANTIFIERS [I.2 §3.2].

Suppose δ is a positive number (which one usually imagines as small). Let us say that a_n is δ-*close* to l if $|a_n - l|$, the difference between a_n and l, is less than δ. What would it mean to say that eventually the sequence gets δ-close to l and stays there? It means that from some point onwards, all the a_n are δ-close to l. And what is the meaning of "from some point onwards"? It is that there is some number N (the point in question) with the property that a_n is δ-close to l from N onwards—that is, for every n that is greater than or equal to N. In symbols:

$$\exists N \quad \forall n \geqslant N \quad a_n \text{ is } \delta\text{-close to } l.$$

It remains to capture the idea of "as close as you like." What this means is that the above sentence is true for any δ you might wish to specify. In symbols:

$$\forall \delta > 0 \quad \exists N \quad \forall n \geqslant N \quad a_n \text{ is } \delta\text{-close to } l.$$

Finally, let us stop using the nonstandard phrase "δ-close":

$$\forall \delta > 0 \quad \exists N \quad \forall n \geqslant N \quad |a_n - l| < \delta.$$

This sentence is not particularly easy to understand. Unfortunately (and interestingly in the light of the discussion in [I.2 §4]), using a less symbolic language does not necessarily make things much easier: "Whatever positive δ you choose, there is some number N such that for all bigger numbers n the difference between a_n and l is less than δ."

The notion of limit applies much more generally than just to real numbers. If you have any collection of mathematical objects and can say what you mean by the distance between any two of those objects, then you can talk of a sequence of those objects having a limit. Two objects are now called δ-close if the *distance* between them is less than δ, rather than the difference. (The idea of distance is discussed further in METRIC SPACES [III.56].) For example, a sequence of points in space can have a limit, as can a sequence of functions. (In the second case it is less obvious how to define distance—there are many natural ways to do it.) A further example comes in the theory of fractals (see DYNAMICS [IV.14]): the very complicated shapes that appear there are best defined as limits of simpler ones.

Two other ways of saying "the limit of the sequence $a_1, a_2, \ldots$ is l" are "a_n *converges to* l" and "a_n *tends to* l." One sometimes says that this happens *as n tends to infinity.* Any sequence that has a limit is called *convergent*. If a_n converges to l then one often writes $a_n \to l$.

5.2 Continuity

Suppose you want to know the approximate value of π^2. Perhaps the easiest thing to do is to press a π button on a calculator, which displays 3.1415927, and then an x^2 button, after which it displays 9.8696044. Of course, one knows that the calculator has not actually squared π: instead it has squared the number 3.1415927. (If it is a good one, then it may have secretly used a few more digits of π without displaying them, but not infinitely many.) Why does it not matter that the calculator has squared the wrong number?

A first answer is that it was only an *approximate* value of π^2 that was required. But that is not quite a complete explanation: how do we know that if x is a good approximation to π then x^2 is a good approximation to π^2? Here is how one might show this. If x is a good approximation to π, then we can write $x = \pi + \delta$ for some very small number δ (which could be negative). Then $x^2 = \pi^2 + 2\delta\pi + \delta^2$. Since δ is small, so is $2\delta\pi + \delta^2$, so x^2 is indeed a good approximation to π^2.

What makes the above reasoning work is that the function that takes a number x to its square is *continuous*. Roughly speaking, this means that if two numbers are close, then so are their squares.

To be more precise about this, let us return to the calculation of π^2, and imagine that we wish to work it out to a much greater accuracy—so that the first hundred digits after the decimal point are correct, for example. A calculator will not be much help, but what we might do is find a list of the digits of π (on the Internet you can find sites that tell you at least the first fifty million), use this to define a new x that is a much better approximation to π, and then calculate the new x^2 by getting a computer to do the necessary long multiplication.

How close to π do we need x to be for x^2 to be within 10^{-100} of π^2? To answer this, we can use our earlier argument. Let $x = \pi + \delta$ again. Then $x^2 - \pi^2 = 2\delta\pi + \delta^2$, and an easy calculation shows that this has modulus less than 10^{-100} if δ has modulus less than 10^{-101}. So we will be all right if we take the first 101 digits of π after the decimal point.

More generally, *however* accurate we wish our estimate of π^2 to be, we can achieve this accuracy if we are prepared to make x a sufficiently good approximation to π. In mathematical parlance, the function $f(x) = x^2$ is *continuous at π*.

Let us try to say this more symbolically. The statement "$x^2 = \pi^2$ to within an accuracy of ϵ" means that $|x^2 - \pi^2| < \epsilon$. To capture the phrase "however accurate," we need this to be true for every positive ϵ, so we should start by saying $\forall \epsilon > 0$. Now let us think about the words "if we are prepared to make x a sufficiently good approximation to π." The thought behind them is that there is some $\delta > 0$ for which the approximation is guaranteed to be accurate to within ϵ as long as x is within δ of π. That is, there exists a $\delta > 0$ such that if $|x - \pi| < \delta$ then it is guaranteed that $|x^2 - \pi^2| < \epsilon$. Putting everything together, we end up with the following symbolic sentence:

$$\forall \epsilon > 0 \quad \exists \delta > 0 \quad (|x - \pi| < \delta \Rightarrow |x^2 - \pi^2| < \epsilon).$$

To put that in words: "Given any positive number ϵ there is a positive number δ such that if $|x - \pi|$ is less than δ then $|x^2 - \pi^2|$ is less than ϵ." Earlier, we found a δ that worked when ϵ was chosen to be 10^{-100}: it was 10^{-101}.

What we have just shown is that the function $f(x) = x^2$ is continuous at the point $x = \pi$. Now let us generalize this idea: let f be any function and let a be any real number. We say that f is *continuous at a* if

$$\forall \epsilon > 0 \quad \exists \delta > 0 \quad (|x - a| < \delta \Rightarrow |f(x) - f(a)| < \epsilon).$$

This says that however accurate an estimate for $f(a)$ you wish $f(x)$ to be, you can achieve this accuracy if you are prepared to make x a sufficiently good approximation to a. The function f is said to be *continuous* if it is continuous at every a. Roughly speaking, what this means is that f has no "sudden jumps." (It also rules out certain kinds of very rapid oscillations that would also make accurate estimates difficult.)

As with limits, the idea of continuity applies in much more general contexts, and for the same reason. Let f be a function from a set X to a set Y, and suppose that we have two notions of distance, one for elements of X and the other for elements of Y. Using the expression $d(x, a)$ to denote the distance between x and a, and similarly for $d(f(x), f(a))$, one says that f is *continuous at a* if

$$\forall \epsilon > 0 \quad \exists \delta > 0 \quad (d(x, a) < \delta \Rightarrow d(f(x), f(a)) < \epsilon)$$

and that f is *continuous* if it is continuous at every a in X. In other words, we replace differences such as $|x - a|$ by distances such as $d(x, a)$.

Like homomorphisms (which are discussed in section 4.1 above), continuous functions can be regarded as preserving a certain sort of structure. It can be shown that a function f is continuous if and only if, whenever

$a_n \to x$, we also have $f(a_n) \to f(x)$. That is, continuous functions are functions that preserve the structure provided by convergent sequences and their limits.

5.3 Differentiation

The derivative of a function f at a value a is usually presented as a number that measures the rate of change of $f(x)$ as x passes through a. The purpose of this section is to promote a slightly different way of regarding it, one that is more general and that opens the door to much of modern mathematics. This is the idea of differentiation as *linear approximation*.

Intuitively speaking, to say that $f'(a) = m$ is to say that if one looks through a very powerful microscope at the graph of f in a tiny region that includes the point $(a, f(a))$, then what one sees is almost exactly a straight line of gradient m. In other words, in a sufficiently small neighborhood of the point a, the function f is approximately linear. We can even write down a formula for the linear function g that approximates f:

$$g(x) = f(a) + m(x - a).$$

This is the equation of the straight line of gradient m that passes through the point $(a, f(a))$. Another way of writing it, which is a little clearer, is

$$g(a + h) = f(a) + mh,$$

and to say that g approximates f in a small neighborhood of a is to say that $f(a + h)$ is *approximately* equal to $f(a) + mh$ when h is small.

One must be a little careful here: after all, if f does not jump suddenly, then, when h is small, $f(a + h)$ will be close to $f(a)$ and mh will be small, so $f(a + h)$ is approximately equal to $f(a) + mh$. This line of reasoning seems to work regardless of the value of m, and yet we wanted there to be something special about the choice $m = f'(a)$. What singles out that particular value is that $f(a + h)$ is not just close to $f(a) + mh$, but so close that the difference $\epsilon(h) = f(a + h) - f(a) - mh$ is small *compared with* h. That is, $\epsilon(h)/h \to 0$ as $h \to 0$. (This is a slightly more general notion of limit than the one discussed in section 5.1. It means that you can make $\epsilon(h)/h$ as small as you like if you make h small enough.)

The reason these ideas can be generalized is that the notion of a linear map is much more general than simply a function from $\mathbb{R}$ to $\mathbb{R}$ of the form $g(x) = mx + c$. Many functions that arise naturally in mathematics—and also in science, engineering, economics, and many other areas—are functions of *several variables*, and can

therefore be regarded as functions defined on a vector space of dimension greater than 1. As soon as we look at them this way, we can ask ourselves whether, in a small neighborhood of a point, they can be approximated by linear maps. It is very useful if they can: a general function can behave in very complicated ways, but if it can be approximated by a linear function, then at least in small regions of n-dimensional space its behavior is much easier to understand. In this situation one can use the machinery of linear algebra and matrices, which leads to calculations that are feasible, especially if one has the help of a computer.

Imagine, for instance, a meteorologist interested in how the direction and speed of the wind change as one looks at different parts of some three-dimensional region above Earth's surface. Wind behaves in complicated, chaotic ways, but to get some sort of handle on this behavior one can describe it as follows. To each point (x, y, z) in the region (think of x and y as horizontal coordinates and z as a vertical one) one can associate a vector (u, v, w) representing the velocity of the wind at that point: u, v, and w are the components of the velocity in the x-, y-, and z-directions.

Now let us change the point (x, y, z) very slightly by choosing three small numbers h, k, and l and looking at $(x + h, y + k, z + l)$. At this new point, we would expect the wind vector to be slightly different as well, so let us write it $(u + p, v + q, w + r)$. How does the small change (p, q, r) in the wind vector depend on the small change (h, k, l) in the position vector? Provided the wind is not too turbulent and h, k, and l are small enough, we expect the dependence to be roughly linear: that is how nature seems to work. In other words, we expect there to be some linear map T such that (p, q, r) is roughly $T(h, k, l)$ when h, k, and l are small. Notice that each of p, q, and r depends on each of h, k, and l, so nine numbers will be needed in order to specify this linear map. In fact, we can express it in matrix form:

$$\begin{pmatrix} p \\ q \\ r \end{pmatrix} = \begin{pmatrix} a_{11} & a_{12} & a_{13} \\ a_{21} & a_{22} & a_{23} \\ a_{31} & a_{32} & a_{33} \end{pmatrix} \begin{pmatrix} h \\ k \\ l \end{pmatrix}.$$

The matrix entries a_{ij} express individual dependencies. For example, if x and z are held fixed, then we are setting $h = l = 0$, from which it follows that the rate of change of u as just y varies is given by the entry a_{12}. That is, a_{12} is the *partial derivative* $\partial u / \partial y$ at the point (x, y, z).

This tells us how to calculate the matrix, but from the conceptual point of view it is easier to use vector notation. Write $\boldsymbol{x}$ for (x, y, z), $\boldsymbol{u}(\boldsymbol{x})$ for (u, v, w), $\boldsymbol{h}$ for (h, k, l), and $\boldsymbol{p}$ for (p, q, r). Then what we are saying is that

$$\boldsymbol{p} = T(\boldsymbol{h}) + \boldsymbol{\epsilon}(\boldsymbol{h})$$

for some vector $\boldsymbol{\epsilon}(\boldsymbol{h})$ that is small relative to $\boldsymbol{h}$. Alternatively, we can write

$$\boldsymbol{u}(\boldsymbol{x} + \boldsymbol{h}) = \boldsymbol{u}(\boldsymbol{x}) + T(\boldsymbol{h}) + \boldsymbol{\epsilon}(\boldsymbol{h}),$$

a formula that is closely analogous to our earlier formula $g(x + h) = g(x) + mh + \epsilon(h)$. This tells us that if we add a small vector $\boldsymbol{h}$ to $\boldsymbol{x}$, then $\boldsymbol{u}(\boldsymbol{x})$ will change by roughly $T(\boldsymbol{h})$.

More generally, let $\boldsymbol{u}$ be a function from $\mathbb{R}^n$ to $\mathbb{R}^m$. Then $\boldsymbol{u}$ is defined to be *differentiable* at a point $\boldsymbol{x} \in \mathbb{R}^n$ if there is a linear map $T : \mathbb{R}^n \to \mathbb{R}^m$ such that, once again, the formula

$$\boldsymbol{u}(\boldsymbol{x} + \boldsymbol{h}) = \boldsymbol{u}(\boldsymbol{x}) + T(\boldsymbol{h}) + \boldsymbol{\epsilon}(\boldsymbol{h})$$

holds, with $\boldsymbol{\epsilon}(\boldsymbol{h})$ small relative to $\boldsymbol{h}$. The linear map T is the *derivative of $\boldsymbol{u}$ at $\boldsymbol{x}$.*

An important special case of this is when $m = 1$. If $f : \mathbb{R}^n \to \mathbb{R}$ is differentiable at $\boldsymbol{x}$, then the derivative of f at $\boldsymbol{x}$ is a linear map from $\mathbb{R}^n$ to $\mathbb{R}$. The matrix of T is a row vector of length n, which is often denoted $\nabla f(x)$ and referred to as the *gradient* of f at x. This vector points in the direction in which f increases most rapidly and its magnitude is the rate of change in that direction.

5.4   Partial Differential Equations

Partial differential equations are of immense importance in physics, and have inspired a vast amount of mathematical research. Three basic examples will be discussed here, as an introduction to more advanced articles later in the volume (see, in particular, PARTIAL DIFFERENTIAL EQUATIONS [IV.12]).

The first is the *heat equation,* which, as its name suggests, describes the way the distribution of heat in a physical medium changes with time:

$$\frac{\partial T}{\partial t} = \kappa \left(\frac{\partial^2 T}{\partial x^2} + \frac{\partial^2 T}{\partial y^2} + \frac{\partial^2 T}{\partial z^2} \right).$$

Here, $T(x, y, z, t)$ is a function that specifies the temperature at the point (x, y, z) at time t.

It is one thing to read an equation like this and understand the symbols that make it up, but quite another to see what it really means. However, it is important to do so, since of the many expressions one could write down that involve partial derivatives, only a minority are of much significance, and these tend to be the ones

that have interesting interpretations. So let us try to interpret the expressions involved in the heat equation.

The left-hand side, $\partial T / \partial t$, is quite simple. It is the rate of change of the temperature $T(x, y, z, t)$ when the spatial coordinates x, y, and z are kept fixed and t varies. In other words, it tells us how fast the point (x, y, z) is heating up or cooling down at time t. What would we expect this to depend on? Well, heat takes time to travel through a medium, so although the temperature at some distant point (x', y', z') will eventually affect the temperature at (x, y, z), the way the temperature is changing *right now* (that is, at time t) will be affected only by the temperatures of points very close to (x, y, z): if points in the immediate neighborhood of (x, y, z) are hotter, on average, than (x, y, z) itself, then we expect the temperature at (x, y, z) to be increasing, and if they are colder then we expect it to be decreasing.

The expression in brackets on the right-hand side appears so often that it has its own shorthand. The symbol Δ, defined by

$$\Delta f = \frac{\partial^2 f}{\partial x^2} + \frac{\partial^2 f}{\partial y^2} + \frac{\partial^2 f}{\partial z^2},$$

is known as the *Laplacian.* What information does Δf give us about a function f? The answer is that it captures the idea in the last paragraph: it tells us how the value of f at (x, y, z) compares with the average value of f in a small neighborhood of (x, y, z), or, more precisely, with the limit of the average value in a neighborhood of (x, y, z) as the size of that neighborhood shrinks to zero.

This is not immediately obvious from the formula, but the following (not wholly rigorous) argument in one dimension gives a clue about why second derivatives should be involved. Let f be a function that takes real numbers to real numbers. Then to obtain a good approximation to the second derivative of f at a point x, one can look at the expression $(f'(x) - f'(x - h))/h$ for some small h. (If one substitutes $-h$ for h in the above expression, one obtains the more usual formula, but this one is more convenient here.) The derivatives $f'(x)$ and $f'(x - h)$ can themselves be approximated by $(f(x + h) - f(x))/h$ and $(f(x) - f(x - h))/h$, respectively, and if we substitute these approximations into the earlier expression, then we obtain

$$\frac{1}{h} \left(\frac{f(x + h) - f(x)}{h} - \frac{f(x) - f(x - h)}{h} \right),$$

which equals $(f(x + h) - 2f(x) + f(x - h))/h^2$. Dividing the top of this last fraction by 2, we obtain $\frac{1}{2}(f(x + h) +$

$f(x - h)) - f(x)$: that is, the difference between the value of f at x and the average value of f at the two surrounding points $x + h$ and $x - h$.

In other words, the second derivative conveys just the idea we want—a comparison between the value at x and the average value near x. It is worth noting that if f is linear, then the average of $f(x - h)$ and $f(x + h)$ will be *equal* to $f(x)$, which fits with the familiar fact that the second derivative of a linear function f is zero.

Just as, when defining the first derivative, we have to divide the difference $f(x + h) - f(x)$ by h so that it is not automatically tiny, so with the second derivative it is appropriate to divide by h^2. (This is appropriate, since, whereas the first derivative concerns linear approximations, the second derivative concerns *quadratic* ones: the best quadratic approximation for a function f near a value x is $f(x + h) \approx f(x) + hf'(x) + \frac{1}{2}h^2 f''(x)$, an approximation that one can check is exact if f was a quadratic function to start with.)

It is possible to pursue thoughts of this kind and show that if f is a function of three variables then the value of Δf at (x, y, z) does indeed tell us how the value of f at (x, y, z) compares with the average values of f at points nearby. (There is nothing special about the number 3 here—the ideas can easily be generalized to functions of any number of variables.) All that is left to discuss in the heat equation is the parameter κ. This measures the *conductivity* of the medium. If κ is small, then the medium does not conduct heat very well and ΔT has less of an effect on the rate of change of the temperature; if it is large then heat is conducted better and the effect is greater.

A second equation of great importance is the *Laplace equation*, $\Delta f = 0$. Intuitively speaking, this says of a function f that its value at a point (x, y, z) is always equal to the average value at the immediately surrounding points. If f is a function of just one variable x, this says that the second derivative of f is zero, which implies that f is of the form $ax + b$. However, for two or more variables, a function has more flexibility—it can lie above the tangent lines in some directions and below it in others. As a result, one can impose a variety of boundary conditions on f (that is, specifications of the values f takes on the boundaries of certain regions), and there is a much wider and more interesting class of solutions.

A third fundamental equation is the *wave equation*. In its one-dimensional formulation it describes the motion of a vibrating string that connects two points

A and B. Suppose that the height of the string at distance x from A and at time t is written $h(x, t)$. Then the wave equation says that

$$\frac{1}{v^2} \frac{\partial^2 h}{\partial t^2} = \frac{\partial^2 h}{\partial x^2}.$$

Ignoring the constant $1/v^2$ for a moment, the left-hand side of this equation represents the acceleration (in a vertical direction) of the piece of string at distance x from A. This should be proportional to the force acting on it. What will govern this force? Well, suppose for a moment that the portion of string containing x were absolutely straight. Then the pull of the string on the left of x would exactly cancel out the pull on the right and the net force would be zero. So, once again, what matters is how the height at x compares with the average height on either side: if the string lies above the tangent line at x, then there will be an upwards force, and if it lies below, then there will be a downwards one. This is why the second derivative appears on the right-hand side once again. How much force results from this second derivative depends on factors such as the density and tautness of the string, which is where the constant comes in. Since h and x are both distances, v^2 has dimensions of $(\text{distance}/\text{time})^2$, which means that v represents a speed, which is, in fact, the speed of propagation of the wave.

Similar considerations yield the three-dimensional wave equation, which is, as one might now expect,

$$\frac{1}{v^2} \frac{\partial^2 h}{\partial t^2} = \frac{\partial^2 h}{\partial x^2} + \frac{\partial^2 h}{\partial y^2} + \frac{\partial^2 h}{\partial z^2},$$

or, more concisely,

$$\frac{1}{v^2} \frac{\partial^2 h}{\partial t^2} = \Delta h.$$

One can be more concise still and write this equation as $\square^2 h = 0$, where $\square^2 h$ is shorthand for

$$\Delta h - \frac{1}{v^2} \frac{\partial^2 h}{\partial t^2}.$$

The operation $\square^2$ is called the *d'Alembertian*, after D'ALEMBERT [VI.20], who was the first to formulate the wave equation.

5.5   Integration

Suppose that a car drives down a long straight road for one minute, and that you are told where it starts and what its speed is during that minute. How can you work out how far it has gone? If it travels at the same speed for the whole minute then the problem is very simple indeed—for example, if that speed is thirty miles per hour then we can divide by sixty and see that it has gone half a mile—but the problem becomes more interesting if the speed varies. Then, instead of trying to give an exact answer, one can use the following technique to approximate it. First, write down the speed of the car at the beginning of each of the sixty seconds that it is traveling. Next, for each of those seconds, do a simple calculation to see how far the car would have gone during that second if the speed had remained exactly as it was at the beginning of the second. Finally, add up all these distances. Since one second is a short time, the speed will not change very much during any one second, so this procedure gives quite an accurate answer. Moreover, if you are not satisfied with this accuracy, then you can improve it by using intervals that are shorter than a second.

If you have done a first course in calculus, then you may well have solved such problems in a completely different way. In a typical question, one is given an explicit formula for the speed at time t—something like $at + u$, for example—and in order to work out how far the car has gone one "integrates" this function to obtain the formula $\frac{1}{2}at^2 + ut$ for the distance traveled at time t. Here, integration simply means the opposite of differentiation: to find the integral of a function f is to find a function g such that $g'(t) = f(t)$. This makes sense, because if $g(t)$ is the distance traveled and $f(t)$ is the speed, then $f(t)$ is indeed the rate of change of $g(t)$.

However, antidifferentiation is not the *definition* of integration. To see why not, consider the following question: what is the distance traveled if the speed at time t is e^{-t^2}? It is known that there is no nice function (which means, roughly speaking, a function built up out of standard ones such as polynomials, exponentials, logarithms, and trigonometric functions) with e^{-t^2} as its derivative, yet the question still makes good sense and has a definite answer. (It is possible that you have heard of a function $\Phi(t)$ that differentiates to $e^{-t^2/2}$, from which it follows that $\Phi(t\sqrt{2})/\sqrt{2}$ differentiates to e^{-t^2}. However, this does not remove the difficulty, since $\Phi(t)$ is defined as the integral of $e^{-t^2/2}$.)

In order to define integration in situations like this where antidifferentiation runs into difficulties, we must fall back on messy approximations of the kind discussed earlier. A formal definition along such lines was given by RIEMANN [VI.49] in the mid nineteenth century. To see what Riemann's basic idea is, and to see also that integration, like differentiation, is a procedure that can usefully be applied to functions of more than one variable, let us look at another physical problem.

Suppose that you have a lump of impure rock and wish to calculate its mass from its density. Suppose also that this density is not constant but varies rather irregularly through the rock. Perhaps there are even holes inside, so that the density is zero in places. What should you do?

Riemann's approach would be this. First, you enclose the rock in a cuboid. For each point (x, y, z) in this cuboid there is then an associated density $d(x, y, z)$ (which will be zero if (x, y, z) lies outside the rock or inside a hole). Second, you divide the cuboid into a large number of smaller cuboids. Third, in each of the small cuboids you look for the point of lowest density (if any point in the cuboid is not in the rock, then this density will be zero) and the point of highest density. Let C be one of the small cuboids and suppose that the lowest and highest densities in C are a and b, respectively, and that the volume of C is V. Then the mass of the part of the rock that lies in C must lie between aV and bV. Fourth, add up all the numbers aV that are obtained in this way, and then add up all the numbers bV. If the totals are M_1 and M_2, respectively, then the total mass of rock has to lie between M_1 and M_2. Finally, repeat this calculation for subdivisions into smaller and smaller cuboids. As you do this, the resulting numbers M_1 and M_2 will become closer and closer to each other, and you will have better and better approximations to the mass of the rock.

Similarly, his approach to the problem about the car would be to divide the minute up into small intervals and look at the minimum and maximum speeds during those intervals. For each interval, this would give him a pair of numbers a and b for which he could say that the car had traveled a distance of at least a and at most b. Adding up these sets of numbers, he could then say that over the full minute the car must have traveled a distance of at least D_1 (the sum of the as) and at most D_2 (the sum of the bs).

With both these problems we had a function (density/speed) defined on a set (the cuboid/a minute of time) and in a certain sense we wanted to work out the "total amount" of the function. We did so by dividing the set into small parts and doing simple calculations in those parts to obtain approximations to this amount from below and above. This process is what is known as (Riemann) *integration*. The following notation is common: if S is the set and f is the function, then the total amount of f in S, known as the *integral*, is written $\int_S f(x) \, dx$. Here, x denotes a typical element of S. If, as in the density example, the elements of S are points

(x, y, z), then vector notation such as $\int_S f(\boldsymbol{x}) \, \mathrm{d}\boldsymbol{x}$ can be used, though often it is not and the reader is left to deduce from the context that an ordinary "x" denotes a vector rather than a real number.

We have been at pains to distinguish integration from antidifferentiation, but a famous theorem, known as *the fundamental theorem of calculus*, asserts that the two procedures do, in fact, give the same answer, at least when the function in question has certain continuity properties that all "sensible" functions have. So it is usually legitimate to regard integration as the opposite of differentiation. More precisely, if f is continuous and $F(x)$ is defined to be $\int_a^x f(t) \, \mathrm{d}t$ for some a, then F can be differentiated and $F'(x) = f(x)$. That is, if you integrate a continuous function and differentiate it again, you get back to where you started. Going the other way around, if F has a continuous derivative f and $a < x$, then $\int_a^x f(t) \, \mathrm{d}t = F(x) - F(a)$. This almost says that if you differentiate F and then integrate it again, you get back to F. Actually, you have to choose an arbitrary number a and what you get is the function F with the constant $F(a)$ subtracted.

To get an idea of the sort of exceptions that arise if one does not assume continuity, consider the so-called *Heaviside step function* $H(x)$, which is 0 when $x < 0$ and 1 when $x \geqslant 0$. This function has a jump at 0 and is therefore not continuous. The integral $J(x)$ of this function is 0 when $x < 0$ and x when $x \geqslant 0$, and for almost all values of x we have $J'(x) = H(x)$. However, the gradient of J suddenly changes at 0, so J is not differentiable there and one cannot say that $J'(0) = H(0) = 1$.

5.6 Holomorphic Functions

One of the jewels in the crown of mathematics is *complex analysis*, which is the study of differentiable functions that take complex numbers to complex numbers. Functions of this kind are called *holomorphic*.

At first, there seems to be nothing special about such functions, since the definition of a derivative in this context is no different from the definition for functions of a real variable: if f is a function then the derivative $f'(z)$ at a complex number z is defined to be the limit as h tends to zero of $(f(z + h) - f(z))/h$. However, if we look at this definition in a slightly different way (one that we saw in section 5.3), we find that it is not altogether easy for a complex function to be differentiable. Recall from that section that differentiation means *linear approximation*. In the case of a complex function,

this means that we would like to approximate it by functions of the form $g(w) = \lambda w + \mu$, where λ and μ are complex numbers. (The approximation near z will be $g(w) = f(z) + f'(z)(w - z)$, which gives $\lambda = f'(z)$ and $\mu = f(z) - zf'(z)$.)

Let us regard this situation geometrically. If $\lambda \neq 0$ then the effect of multiplying by λ is to expand z by some factor r and to rotate it by some angle θ. This means that many transformations of the plane that we would ordinarily consider to be linear, such as reflections, shears, or stretches, are ruled out. We need two real numbers to specify λ (whether we write it in the form $a + bi$ or $re^{i\theta}$), but to specify a general linear transformation of the plane takes four (see the discussion of matrices in section 4.2). This reduction in the number of degrees of freedom is expressed by a pair of differential equations called the *Cauchy–Riemann equations*. Instead of writing $f(z)$ let us write $u(x + iy) + iv(x + iy)$, where x and y are the real and imaginary parts of z and $u(x + iy)$ and $v(x + iy)$ are the real and imaginary parts of $f(x + iy)$. Then the linear approximation to f near z has the matrix

$$\begin{pmatrix} \dfrac{\partial u}{\partial x} & \dfrac{\partial u}{\partial y} \\ \dfrac{\partial v}{\partial x} & \dfrac{\partial v}{\partial y} \end{pmatrix}.$$

The matrix of an expansion and rotation always has the form $\left(\begin{smallmatrix} a & b \\ -b & a \end{smallmatrix} \right)$, from which we deduce that

$$\frac{\partial u}{\partial x} = \frac{\partial v}{\partial y} \quad \text{and} \quad \frac{\partial u}{\partial y} = -\frac{\partial v}{\partial x}.$$

These are the Cauchy–Riemann equations. One consequence of these equations is that

$$\frac{\partial^2 u}{\partial x^2} + \frac{\partial^2 u}{\partial y^2} = \frac{\partial^2 v}{\partial x \partial y} - \frac{\partial^2 v}{\partial y \partial x} = 0.$$

(It is not obvious that the necessary conditions hold for the symmetry of the mixed partial derivatives, but when f is holomorphic they do.) Therefore, u satisfies the Laplace equation (which was discussed in section 5.4). A similar argument shows that v does as well.

These facts begin to suggest that complex differentiability is a much stronger condition than real differentiability and that we should expect holomorphic functions to have interesting properties. For the remainder of this subsection, let us look at a few of the remarkable properties that they do indeed have.

The first is related to the fundamental theorem of calculus (discussed in the previous subsection). Suppose that F is a holomorphic function and that we are given its derivative f and the value of $F(u)$ for some complex number u. How can we reconstruct F? An approximate method is as follows. Let w be another complex number and let us try to work out $F(w)$. We take a sequence of points $z_0, z_1, \ldots, z_n$ with $z_0 = u$ and $z_n = w$, and with the differences $|z_1 - z_0|, |z_2 - z_1|, \ldots, |z_n - z_{n-1}|$ all small. We can then approximate $F(z_{i+1}) - F(z_i)$ by $(z_{i+1} - z_i) f(z_i)$. It follows that $F(w) - F(u)$, which equals $F(z_n) - F(z_0)$, is approximated by the sum of all the $(z_{i+1} - z_i) f(z_i)$. (Since we have added together many small errors, it is not obvious that this approximation is a good one, but it turns out that it is.) We can imagine a number z that starts at u and follows a path P to w by jumping from one z_i to another in small steps of $\delta z = z_{i+1} - z_i$. In the limit as n goes to infinity and the steps δz go to zero we obtain a so-called *path integral*, which is denoted $\int_P f(z) \, dz$.

The above argument has the consequence that if the path P begins and ends at the same point u, then the path integral $\int_P f(z) \, dz$ is zero. Equivalently, if two paths P_1 and P_2 have the same starting point u and the same endpoint w, then the path integrals $\int_{P_1} f(z) \, dz$ and $\int_{P_2} f(z) \, dz$ are the same, since they both give the value $F(w) - F(u)$.

Of course, in order to establish this, we made the big assumption that f was the derivative of a function F. Cauchy's theorem says that the same conclusion is true if f is holomorphic. That is, rather than requiring f to be the derivative of another function, it asks for f itself to have a derivative. If that is the case, then any path integral of f depends only on where the path begins and ends. What is more, these path integrals can be used to define a function F that differentiates to f, so a function with a derivative automatically has an antiderivative.

It is not necessary for the function f to be defined on the whole of $\mathbb{C}$ for Cauchy's theorem to be valid: everything remains true if we restrict attention to a *simply connected domain*, which means an OPEN SET [III.90] with no holes in it. If there are holes, then two path integrals may differ if the paths go around the holes in different ways. Thus, path integrals have a close connection with the *topology* of subsets of the plane, an observation that has many ramifications throughout modern geometry. For more on topology, see section 6.4 of this article and ALGEBRAIC TOPOLOGY [IV.6].

A very surprising fact, which can be deduced from Cauchy's theorem, is that if f is holomorphic then it can be differentiated twice. (This is completely untrue of real-valued functions: consider, for example, the function f where $f(x) = 0$ when $x < 0$ and $f(x) = x^2$ when $x \geqslant 0$.) It follows that f' is holomorphic, so it too can be differentiated twice. Continuing, one finds that f can be differentiated any number of times. Thus, for complex functions differentiability implies infinite differentiability. (This property is what is used to establish the symmetry, and even the existence, of the mixed partial derivatives mentioned earlier.)

A closely related fact is that wherever a holomorphic function is defined it can be expanded in a power series. That is, if f is defined and differentiable everywhere on an open disk of radius R about w, then it will be given by a formula of the form

$$f(z) = \sum_{n=0}^{\infty} a_n (z - w)^n,$$

valid everywhere in that disk. This is called the *Taylor expansion* of f.

Another fundamental property of holomorphic functions, one that shows just how "rigid" they are, is that their entire behavior is determined just by what they do in a small region. That is, if f and g are holomorphic and they take the same values in some tiny disk, then they must take the same values everywhere. This remarkable fact allows a process of *analytic continuation*. If it is difficult to define a holomorphic function f everywhere you want it defined, then you can simply define it in some small region and say that elsewhere it takes the only possible values that are consistent with the ones that you have just specified. This is how the famous RIEMANN ZETA FUNCTION [IV.2 §3] is conventionally defined.

Finally, we mention a theorem of LIOUVILLE [VI.39], which states that if f is a holomorphic function defined on the whole complex plane, and if f is bounded (that is, if there is some constant C such that $|f(z)| \leqslant C$ for every complex number z), then f must be constant. Once again, this is obviously false for real functions. For example, the function $\sin(x)$ has no difficulty combining boundedness with very good behavior: it can be expanded in a power series that converges everywhere. (However, if you use the power series to define an extension of the function $\sin(x)$ to the complex plane, then the function you obtain is unbounded, as Liouville's theorem predicts.)

6   What Is Geometry?

It is not easy to do justice to geometry in this article because the fundamental concepts of the subject are either too simple to need explaining—for example, there is no need to say here what a circle, line, or plane is—or sufficiently advanced that they are better discussed in parts III and IV of the book. However, if you have not met the advanced concepts and have no idea what modern geometry is like, then you will get much more out of this book if you understand two basic ideas: the relationship between geometry and symmetry, and the notion of a manifold. These ideas' will occupy us for the rest of the article.

6.1 Geometry and Symmetry Groups

Broadly speaking, geometry is the part of mathematics that involves the sort of language that one would conventionally regard as geometrical, with words such as "point," "line," "plane," "space," "curve," "sphere," "cube," "distance," and "angle" playing a prominent role. However, there is a more sophisticated view, first advocated by KLEIN [VI.57], that regards *transformations* as the true subject matter of geometry. So, to the above list one should add words like "reflection," "rotation," "translation," "stretch," "shear," and "projection," together with slightly more nebulous concepts such as "angle-preserving map" or "continuous deformation."

As was discussed in section 2.1, transformations go hand in hand with groups, and for this reason there is an intimate connection between geometry and group theory. Indeed, given any group of transformations, there is a corresponding notion of geometry, in which one studies the phenomena that are unaffected by transformations in that group. In particular, two shapes are regarded as *equivalent* if one can be turned into the other by means of one of the transformations in the group. Different groups will of course lead to different notions of equivalence, and for this reason mathematicians frequently talk about geomet*ries*, rather than about a single monolithic subject called geometry. This subsection contains brief descriptions of some of the most important geometries and their associated groups of transformations.

6.2 Euclidean Geometry

Euclidean geometry is what most people would think of as "ordinary" geometry, and, not surprisingly given its name, it includes the basic theorems of Greek geometry that were the staple of geometers for over two millennia. For example, the theorem that the three

angles of a triangle add up to 180° belongs to Euclidean geometry.

To understand Euclidean geometry from a transformational viewpoint, we need to say how many dimensions we are working in, and we must of course specify a group of transformations. The appropriate group is the group of *rigid* transformations. These can be thought of in two different ways. One is that they are the transformations of the plane, or of space, or more generally of $\mathbb{R}^n$ for some n, that *preserve distance*. That is, T is a rigid transformation if, given any two points x and y, the distance between Tx and Ty is always the same as the distance between x and y. (In dimensions greater than 3, distance is defined in a way that naturally generalizes the Pythagorean formula. See METRIC SPACES [III.56] for more details.)

It turns out that every such transformation can be realized as a combination of rotations, reflections, and translations, and this gives us a more concrete way to think about the group. Euclidean geometry, in other words, is the study of concepts that do not change when you rotate, reflect, or translate, and these include points, lines, planes, circles, spheres, distance, angle, length, area, and volume. The rotations of $\mathbb{R}^n$ form an important group, the *special orthogonal group*, known as $\mathrm{SO}(n)$. The larger *orthogonal group* $\mathrm{O}(n)$ includes reflections as well. (It is not quite obvious how to define a "rotation" of n-dimensional space, but it is not too hard to do. An *orthogonal map* of $\mathbb{R}^n$ is a linear map T that preserves distances, in the sense that $d(Tx, Ty)$ is always the same as $d(x, y)$. It is a *rotation* if its DETERMINANT [III.15] is 1. The only other possibility for the determinant of a distance-preserving map is -1. Maps with determinant -1 are like reflections in that they turn space "inside out.")

6.3 Affine Geometry

There are many linear maps besides rotations and reflections. What happens if we enlarge our group from $\mathrm{SO}(n)$ or $\mathrm{O}(n)$ to include as many of them as possible? For a transformation to be part of a group it must be *invertible* and not all linear maps are, so the natural group to look at is the group $\mathrm{GL}_n(\mathbb{R})$ of all invertible linear transformations of $\mathbb{R}^n$, a group that we first met in section 4.2. These maps all leave the origin fixed, but if we want we can incorporate translations and consider a larger group that consists of all transformations of the form $x \mapsto Tx + b$, where b is a fixed vector and T is an invertible linear map. The resulting geometry is called *affine* geometry.

Since linear maps include stretches and shears, they preserve neither distance nor angle, so these are not concepts of affine geometry. However, points, lines, and planes remain as points, lines, and planes after an invertible linear map and a translation, so these concepts do belong to affine geometry. Another affine concept is that of two lines being parallel. (That is, although angles in general are not preserved by linear maps, angles of zero are.) This means that although there is no such thing as a square or a rectangle in affine geometry, one can still talk about a parallelogram. Similarly, one cannot talk of circles but one can talk of ellipses, since a linear map transformation of an ellipse is another ellipse (provided that one regards a circle as a special kind of ellipse).

6.4 Topology

The idea that the geometry associated with a group of transformations "studies the concepts that are preserved by all the transformations" can be made more precise using the notion of EQUIVALENCE RELATIONS [I.2 §2.3]. Indeed, let G be a group of transformations of $\mathbb{R}^n$. We might think of an n-dimensional "shape" as being a subset S of $\mathbb{R}^n$, but if we are doing G-geometry, then we do not want to distinguish between a set S and any other set we can obtain from it using a transformation in G. So in that case we say that the two shapes are *equivalent*. For example, two shapes are equivalent in Euclidean geometry if and only if they are congruent in the usual sense, whereas in two-dimensional affine geometry all parallelograms are equivalent, as are all ellipses. One can think of the basic objects of G-geometry as *equivalence classes* of shapes rather than the shapes themselves.

Topology can be thought of as the geometry that arises when we use a particularly generous notion of equivalence, saying that two shapes are equivalent, or *homeomorphic*, to use the technical term, if each can be "continuously deformed" into the other. For example, a sphere and a cube are equivalent in this sense, as figure 1 illustrates.

Because there are very many continuous deformations, it is quite hard to prove that two shapes are *not* equivalent in this sense. For example, it may seem obvious that a sphere (this means the surface of a ball rather than the solid ball) cannot be continuously deformed into a torus (the shape of the surface of a doughnut of the kind that has a hole in it), since they are fundamentally different shapes—one has a "hole" and the other

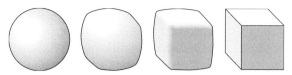

Figure 1 A sphere morphing into a cube.

does not. However, it is not easy to turn this intuition into a rigorous argument. For more on this kind of problem, see INVARIANTS [I.4 §2.2], ALGEBRAIC TOPOLOGY [IV.6], and DIFFERENTIAL TOPOLOGY [IV.7].

6.5 Spherical Geometry

We have been steadily relaxing our requirements for two shapes to be equivalent, by allowing more and more transformations. Now let us tighten up again and look at *spherical* geometry. Here the universe is no longer $\mathbb{R}^n$ but the *n-dimensional sphere* S^n, which is defined to be the surface of the $(n + 1)$-dimensional ball of radius 1, or, to put it more algebraically, the set of all points $(x_1, x_2, \ldots, x_{n+1})$ in $\mathbb{R}^{n+1}$ such that $x_1^2 + x_2^2 + \cdots + x_{n+1}^2 = 1$. Just as the surface of a three-dimensional ball is two dimensional, so this set is n dimensional. We shall discuss the case $n = 2$ here, but it is easy to generalize the discussion to larger n.

The appropriate group of transformations is SO(3): the group that consists of all rotations about axes that go through the origin. (One could allow reflections as well and take O(3).) These are symmetries of the sphere S^2, and that is how we regard them in spherical geometry, rather than as transformations of the whole of $\mathbb{R}^3$.

Among the concepts that make sense in spherical geometry are line, distance, and angle. It may seem odd to talk about a line if one is confined to the surface of a ball, but a "spherical line" is not a line in the usual sense. Rather, it is a subset of S^2 obtained by intersecting S^2 with a plane through the origin. This produces a *great circle*, that is, a circle of radius 1, which is as large as it can be given that it lives inside a sphere of radius 1.

The reason that a great circle deserves to be thought of as some sort of line is that the shortest path between any two points x and y in S^2 will always be along a great circle, *provided that the path is confined to S^2*. This is a very natural restriction to make, since we are regarding S^2 as our "universe." It is also a restriction of some practical relevance, since the shortest sensible route between two distant points on Earth's surface will not be the straight-line route that burrows hundreds of miles underground.

The *distance* between two points x and y is defined to be the length of the shortest path from x to y that lies entirely in S^2. (If x and y are opposite each other, then there are infinitely many shortest paths, all of length π, so the distance between x and y is π.) How about the *angle* between two spherical lines? Well, the lines are intersections of S^2 with two planes, so one can define it to be the angle between these two planes in the Euclidean sense. A more aesthetically pleasing way to view this, because it does not involve ideas external to the sphere, is to notice that if you look at a very small region about one of the two points where two spherical lines cross, then that portion of the sphere will be almost flat, and the lines almost straight. So you can define the angle to be the usual angle between the "limiting" straight lines inside the "limiting" plane.

Spherical geometry differs from Euclidean geometry in several interesting ways. For example, the angles of a spherical triangle always add up to *more* than $180°$. Indeed, if you take as the vertices the North Pole, a point on the equator, and a second point a quarter of the way around the equator from the first, then you obtain a triangle with three right angles. The smaller a triangle, the flatter it becomes, and so the closer the sum of its angles comes to $180°$. There is a beautiful theorem that gives a precise expression to this: if we switch to radians, and if we have a spherical triangle with angles α, β, and γ, then its area is $\alpha + \beta + \gamma - \pi$. (For example, this formula tells us that the triangle with three angles of $\frac{1}{2}\pi$ has area $\frac{1}{2}\pi$, which indeed it does as the surface area of a ball of radius 1 is 4π and this triangle occupies one-eighth of the surface.)

6.6 Hyperbolic Geometry

So far, the idea of defining geometries with reference to sets of transformations may look like nothing more than a useful way to view the subject, a unified approach to what would otherwise be rather different-looking aspects. However, when it comes to hyperbolic geometry, the transformational approach becomes indispensable, for reasons that will be explained in a moment.

The group of transformations that produces hyperbolic geometry is called $\mathrm{PSL}_2(\mathbb{R})$, the *projective special linear group* in two dimensions. One way to present this group is as follows. The *special linear group* $\mathrm{SL}_2(\mathbb{R})$ is the set of all matrices $\left(\begin{smallmatrix} a & b \\ c & d \end{smallmatrix}\right)$ with DETERMINANT [III.15]

$ad - bc$ equal to 1. (These form a group because the product of two matrices with determinant 1 again has determinant 1.) To make this "projective," one then regards each matrix A as *equivalent* to $-A$: for example, the matrices $\left(\begin{smallmatrix} 3 & -1 \\ -5 & 2 \end{smallmatrix}\right)$ and $\left(\begin{smallmatrix} -3 & 1 \\ 5 & -2 \end{smallmatrix}\right)$ are equivalent.

To get from this group to the geometry one must first interpret it as a group of transformations of some two-dimensional set of points. Once we have done this, we have what is called a *model* of two-dimensional hyperbolic geometry. The subtlety is that there is no single model of hyperbolic geometry that is clearly the most natural in the way that the sphere is the most natural model of spherical geometry. (One might think that the sphere was the *only* sensible model of spherical geometry, but this is not in fact the case. For example, there is a natural way of associating with each rotation of $\mathbb{R}^3$ a transformation of $\mathbb{R}^2$ with a "point at infinity" added, so the extended plane can be used as a model of spherical geometry.) The three most commonly used models of hyperbolic geometry are called the half-plane model, the disk model, and the hyperboloid model.

The *half-plane model* is the one most directly associated with the group $\mathrm{PSL}_2(\mathbb{R})$. The set in question is the upper half-plane of the complex numbers $\mathbb{C}$, that is, the set of all complex numbers $z = x + iy$ such that $y > 0$. Given a matrix $\left(\begin{smallmatrix} a & b \\ c & d \end{smallmatrix}\right)$, the corresponding transformation is the one that takes the point z to the point $(az + b)/(cz + d)$. (Notice that if we replace a, b, c, and d by their negatives, then we get the same transformation.) The condition $ad - bc = 1$ can be used to show that the transformed point will still lie in the upper half-plane, and also that the transformation can be inverted.

What this does not yet do is tell us anything about *distances*, and it is here that we need the group to "generate" the geometry. If we are to have a notion of distance d that is sensible from the perspective of our group of transformations, then it is important that the transformations should preserve it. That is, if T is one of the transformations and z and w are two points in the upper half-plane, then $d(T(z), T(w))$ should always be the same as $d(z, w)$. It turns out that there is essentially only *one* definition of distance that has this property, and that is the sense in which the group defines the geometry. (One could of course multiply all distances by some constant factor such as 3, but this would be like measuring distances in feet instead of yards, rather than a genuine difference in the geometry.)

This distance has some properties that at first seem odd. For example, a typical *hyperbolic line* takes the form of a semicircular arc with endpoints on the real axis. However, it is semicircular only from the point of view of the Euclidean geometry of $\mathbb{C}$: from a hyperbolic perspective it would be just as odd to regard a Euclidean straight line as straight. The reason for the discrepancy is that hyperbolic distances become larger and larger, relative to Euclidean ones, the closer you get to the real axis. To get from a point z to another point w, it is therefore shorter to take a "detour" away from the real axis, and the best detour turns out to be along an arc of the circle that goes through z and w and cuts the real axis at right angles. (If z and w are on the same vertical line, then one obtains a "degenerate circle," namely that vertical line.) These facts are no more paradoxical than the fact that a flat map of the world involves distortions of spherical geometry, making Greenland very large, for example. The half-plane model is like a "map" of a geometric structure, the hyperbolic plane, that in reality has a very different shape.

One of the most famous properties of two-dimensional hyperbolic geometry is that it provides a geometry in which Euclid's *parallel postulate* fails to hold. That is, it is possible to have a hyperbolic line L, a point x not on the line, and two different hyperbolic lines through x, neither of which meets L. All the other axioms of Euclidean geometry are, when suitably interpreted, true of hyperbolic geometry as well. It follows that the parallel postulate cannot be deduced from those axioms. This discovery, associated with GAUSS [VI.26], BOLYAI [VI.34], and LOBACHEVSKII [VI.31], solved a problem that had bothered mathematicians for over two thousand years.

Another property complements the result about the angle sums of spherical and Euclidean triangles. There is a natural notion of hyperbolic area, and the area of a hyperbolic triangle with angles α, β, and γ is $\pi - \alpha - \beta - \gamma$. Thus, in the hyperbolic plane $\alpha + \beta + \gamma$ is always *less* than π, and it almost equals π when the triangle is very small. These properties of angle sums reflect the fact that the sphere has positive CURVATURE [III.13], the Euclidean plane is "flat," and the hyperbolic plane has negative curvature.

The *disk model*, conceived in a famous moment of inspiration by POINCARÉ [VI.61] as he was getting into a bus, takes as its set of points the *open unit disk* in $\mathbb{C}$, that is, the set D of all complex numbers with modulus less than 1. This time, a typical transformation takes the following form. One takes a real number θ, and a complex number a from inside D, and sends each z in D to the point $e^{i\theta}(z - a)/(1 - \bar{a}z)$. It is not

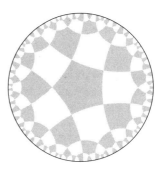

Figure 2 A tessellation of the hyperbolic disk.

completely obvious that these transformations form a group, and still less that the group is isomorphic to $\mathrm{PSL}_2(\mathbb{R})$. However, it turns out that the function that takes z to $-(iz + 1)/(z + i)$ maps the unit disk to the upper half-plane and vice versa. This shows that the two models give the same geometry and can be used to transfer results from one to the other.

As with the half-plane model, distances become larger, relative to Euclidean distances, as you approach the boundary of the disk: from a hyperbolic perspective, the diameter of the disk is infinite and it does not really have a boundary. Figure 2 shows a tessellation of the disk by shapes that are congruent in the sense that any one can be turned into any other by means of a transformation from the group. Thus, even though they do not look identical, within hyperbolic geometry they all have the same size and shape. Straight lines in the disk model are either arcs of (Euclidean) circles that meet the unit circle at right angles, or segments of (Euclidean) straight lines that pass through the center of the disk.

The *hyperboloid model* is the model that explains why the geometry is called hyperbolic. This time the set is the hyperboloid consisting of all points $(x, y, z) \in \mathbb{R}^3$ such that $z > 0$ and $x^2 + y^2 + 1 = z^2$. This is the hyperboloid of revolution about the z-axis of the hyperbola $x^2 + 1 = z^2$ in the plane $y = 0$. A general transformation in the group is a sort of "rotation" of the hyperboloid, and can be built up from genuine rotations about the z-axis, and "hyperbolic rotations" of the xz-plane, which have matrices of the form

$$\begin{pmatrix} \cosh\theta & \sinh\theta \\ \sinh\theta & \cosh\theta \end{pmatrix}.$$

Just as an ordinary rotation preserves the unit circle, one of these hyperbolic rotations preserves the hyperbola $x^2 + 1 = z^2$, moving points around inside it. Again, it is not quite obvious that this gives the same group of transformations, but it does, and the hyperboloid model is equivalent to the other two.

6.7   Projective Geometry

Projective geometry is regarded by many as an old-fashioned subject, and it is no longer taught in schools, but it still has an important role to play in modern mathematics. We shall concentrate here on the *real projective plane*, but projective geometry is possible in any number of dimensions and with scalars in any field. This makes it particularly useful to algebraic geometers.

Here are two ways of regarding the projective plane. The first is that the set of points is the ordinary plane, together with a "line at infinity." The group of transformations consists of functions known as *projections*. To understand what a projection is, imagine two planes P and P′ in space, and a point x that is not in either of them. We can "project" P onto P′ as follows. If a is a point in P, then its image $\phi(a)$ is the point where the line joining x to a meets P′. (If this line is parallel to P′, then $\phi(a)$ is a point on the line at infinity of P′.) Thus, if you are at x and a picture is drawn on the plane P, then its image under the projection ϕ will be the picture drawn on P′ that to you looks exactly the same. In fact, however, it will have been distorted, so the transformation ϕ has made a difference to the shape. To turn ϕ into a transformation of P itself, one can follow it by a rigid transformation that moves P′ back to where P is.

Such projections clearly do not preserve distances, but they do preserve other interesting concepts, such as *points*, *lines*, quantities known as *cross-ratios*, and, most famously, *conic sections*. A conic section is the intersection of a plane with a cone, and it can be a circle, an ellipse, a parabola, or a hyperbola. From the point of view of projective geometry, these are all the same kind of object (just as, in affine geometry, one can talk about ellipses but there is no special ellipse called a circle).

A second view of the projective plane is that it is the set of all lines in $\mathbb{R}^3$ that go through the origin. Since a line is determined by the two points where it intersects the unit sphere, one can regard this set as a sphere, but with the significant difference that *opposite points are regarded as the same*—because they correspond to the same line.

Under this view, a typical transformation of the projective plane is obtained as follows. Take any invertible linear map, and apply it to $\mathbb{R}^3$. This takes lines through the origin to lines through the origin, and can therefore be thought of as a function from the projective plane to itself. If one invertible linear map is a multiple of another, then they will have the same effect on all lines, so the resulting group of transformations is like $\mathrm{GL}_3(\mathbb{R})$, except that all nonzero multiples of any given matrix are regarded as equivalent. This group is called the *projective special linear group* $\mathrm{PSL}_3(\mathbb{R})$, and it is the three-dimensional equivalent of $\mathrm{PSL}_2(\mathbb{R})$, which we have already met. Since $\mathrm{PSL}_3(\mathbb{R})$ is bigger than $\mathrm{PSL}_2(\mathbb{R})$, the projective plane comes with a richer set of transformations than the hyperbolic plane, which is why fewer geometrical properties are preserved. (For example, we have seen that there is a useful notion of hyperbolic distance, but there is no obvious notion of projective distance.)

6.8   Lorentz Geometry

This is a geometry used in the theory of special relativity to model four-dimensional *spacetime*, otherwise known as *Minkowski space*. The main difference between it and four-dimensional Euclidean geometry is that, instead of the usual notion of distance between two points (t, x, y, z) and (t', x', y', z'), one considers the quantity

$$-(t - t')^2 + (x - x')^2 + (y - y')^2 + (z - z')^2,$$

which would be the square of the Euclidean distance were it not for the all-important minus sign before $(t - t')^2$. This reflects the fact that space and time are significantly different (though intertwined).

A *Lorentz transformation* is a linear map from $\mathbb{R}^4$ to $\mathbb{R}^4$ that preserves these "generalized distances." Letting g be the linear map that sends (t, x, y, z) to $(-t, x, y, z)$ and letting G be the corresponding matrix (which has $-1, 1, 1, 1$ down the diagonal and 0 everywhere else), we can define a Lorentz transformation abstractly as one whose matrix Λ satisfies $\Lambda^{\mathrm{T}} G \Lambda = G$, where I is the 4×4 identity matrix and Λ^{T} is the transpose of Λ. (The *transpose* of a matrix A is the matrix B defined by $B_{ij} = A_{ji}$.)

A point (t, x, y, z) is said to be *spacelike* if $-t^2 + x^2 + y^2 + z^2 > 0$, and *timelike* if $-t^2 + x^2 + y^2 + z^2 < 0$. If $-t^2 + x^2 + y^2 + z^2 = 0$, then the point lies in the *light cone*. All these are genuine concepts of Lorentzian geometry because they are preserved by Lorentz transformations.

Lorentzian geometry is also of fundamental importance to *general* relativity, which can be thought of as the study of *Lorentzian manifolds*. These are closely related to Riemannian manifolds, which are discussed in section 6.10. For a discussion of general relativity, see GENERAL RELATIVITY AND THE EINSTEIN EQUATIONS [IV.13].

6.9 Manifolds and Differential Geometry

To somebody who has not been taught otherwise, it is natural to think that Earth is flat, or rather that it consists of a flat surface on top of which there are buildings, mountains, and so on. However, we now know that it is in fact more like a sphere, appearing to be flat only because it is so large. There are various kinds of evidence for this. One is that if you stand on a cliff by the sea then you can see a definite horizon, not too far away, over which ships disappear. This would be hard to explain if Earth were genuinely flat. Another is that if you travel far enough in what feels like a straight line then you eventually get back to where you started. A third is that if you travel along a triangular route and the triangle is a large one, then you will be able to detect that its three angles add up to more than $180°$.

It is also very natural to believe that the geometry that best models that of the universe is three-dimensional Euclidean geometry, or what one might think of as "normal" geometry. However, this could be just as much of a mistake as believing that two-dimensional Euclidean geometry is the best model for Earth's surface.

Indeed, one can immediately improve on it by considering Lorentzian geometry as a model of spacetime, but even if there were no theory of special relativity, our astronomical observations would give us no particular reason to suppose that Euclidean geometry was the best model for the universe. Why should we be so sure that we would not obtain a better model by taking the three-dimensional surface of a very large four-dimensional ball? This might feel like "normal" space in just the way that the surface of Earth feels like a "normal" plane unless you travel large distances. Perhaps if you traveled far enough in a rocket without changing your course then you would end up where you started.

It is easy to describe "normal" space mathematically: one just associates with each point in space a triple of coordinates (x, y, z) in the usual way. How might we describe a huge "spherical" space? It is slightly harder, but not much: one can give each point *four* coordinates (x, y, z, w) but add the condition that these must satisfy the equation $x^2 + y^2 + z^2 + w^2 = R^2$ for some fixed R that we think of as the "radius" of the universe. This describes the three-dimensional surface of a four-dimensional ball of radius R in just the same way that the equation $x^2 + y^2 + z^2 = R^2$ describes the two-dimensional surface of a three-dimensional ball of radius R.

A possible objection to this approach is that it seems to rely on the rather implausible idea that the universe lives in some larger unobserved four-dimensional space. However, this objection can be answered. The object we have just defined, the *3-sphere* S^3, can also be described in what is known as an *intrinsic* way: that is, without reference to some surrounding space. The easiest way to see this is to discuss the 2-sphere first, in order to draw an analogy.

Let us therefore imagine a planet covered with calm water. If you drop a large rock into the water at the North Pole, a wave will propagate out in a circle of ever-increasing radius. (At any one moment, it will be a circle of constant latitude.) In due course, however, this circle will reach the equator, after which it will start to *shrink*, until eventually the whole wave reaches the South Pole at once, in a sudden burst of energy.

Now imagine setting off a three-dimensional wave in space—it could, for example, be a light wave caused by the switching on of a bright light. The front of this wave would now be not a circle but an ever-expanding spherical surface. It is logically possible that this surface could expand until it became very large and then contract again, not by shrinking back to where it started, but by turning itself inside out, so to speak, and shrinking to another point on the opposite side of the universe. (Notice that in the two-dimensional example, what you want to call the inside of the circle changes when the circle passes the equator.) With a bit of effort, one can visualize this possibility, and there is no need to appeal to the existence of a fourth dimension in order to do so. More to the point, this account can be turned into a mathematically coherent and genuinely three-dimensional description of the 3-sphere.

A different and more general approach is to use what is called an *atlas*. An atlas of the world (in the normal, everyday sense) consists of a number of flat pages, together with an indication of their *overlaps*: that is, of how parts of some pages correspond to parts of others. Now, although such an atlas is mapping out an external object that lives in a three-dimensional universe, the spherical geometry of Earth's surface can be read off from the atlas alone. It may be much less convenient to do this but it is possible: rotations, for example, might be described by saying that such-and-such a part of page 17 moved to a similar but slightly distorted part of page 24, and so on.

Not only is this possible, but one can *define* a surface by means of two-dimensional atlases. For example, there is a mathematically neat "atlas" of the 2-sphere that consists of just two pages, both of them circular. One is a map of the Northern Hemisphere plus a little bit of the Southern Hemisphere near the equator (to provide a small overlap) and the other is a map of the Southern Hemisphere with a bit of the Northern Hemisphere. Because these maps are flat, they necessarily involve some distortion, but one can specify what this distortion is.

The idea of an atlas can easily be generalized to three dimensions. A "page" now becomes a portion of three-dimensional space. The technical term is not "page" but "chart," and a three-dimensional atlas is a collection of charts, again with specifications of which parts of one chart correspond to which parts of another. A possible atlas of the 3-sphere, generalizing the simple atlas of the 2-sphere just discussed, consists of two solid three-dimensional balls. There is a correspondence between points toward the edge of one of these balls and points toward the edge of the other, and this can be used to describe the geometry: as you travel toward the edge of one ball you find yourself in the overlapping region, so you are also in the other ball. As you go further, you are off the map as far as the first ball is concerned, but the second ball has by that stage taken over.

The 2-sphere and the 3-sphere are basic examples of *manifolds*. Other examples that we have already met in this section are the torus and the projective plane. Informally, a *d*-dimensional manifold, or *d*-manifold, is any geometrical object M with the property that every point x in M is surrounded by what feels like a portion of *d*-dimensional Euclidean space. So, because small parts of a sphere, torus, or projective plane are very close to planar, they are all 2-manifolds, though when the dimension is two the word *surface* is more usual. (However, it is important to remember that a "surface" need not be the surface *of* anything.) Similarly, the 3-sphere is a 3-manifold.

The formal definition of a manifold uses the idea of atlases: indeed, one says that the atlas *is* a manifold. This is a typical mathematician's use of the word "is," and it should not be confused with the normal use. In practice, it is unusual to think of a manifold as a collection of charts with rules for how parts of them correspond, but the definition in terms of charts and atlases turns out to be the most convenient when

one wishes to reason about manifolds in general rather than discussing specific examples. For the purposes of this book, it may be better to think of a *d*-manifold in the "extrinsic" way that we first thought about the 3-sphere: as a *d*-dimensional "hypersurface" living in some higher-dimensional space. Indeed, there is a famous theorem of Nash that states that all manifolds arise in this way. Note, however, that it is not always easy to find a simple formula for defining such a hypersurface. For example, while the 2-sphere is described by the simple formula $x^2 + y^2 + z^2 = 1$ and the torus by the slightly more complicated and more artificial formula $(r - 2)^2 + z^2 = 1$, where r is shorthand for $\sqrt{x^2 + y^2}$, it is not easy to come up with a formula that describes a two-holed torus. Even the usual torus is far more easily described using quotients, as we did in section 3.3. Quotients can also be used to define a two-holed torus (see FUCHSIAN GROUPS [III.28]), and the reason one is confident that the result is a manifold is that every point has a small neighborhood that looks like a small part of the Euclidean plane. In general, a *d*-dimensional manifold can be thought of as any construction that gives rise to an object that is "locally like Euclidean space of *d* dimensions."

An extremely important feature of manifolds is that calculus is possible for functions defined on them. Roughly speaking, if M is a manifold and f is a function from M to $\mathbb{R}$, then to see whether f is differentiable at a point x in M you first find a chart that contains x (or a representation of it), and regard f as a function defined on the chart instead. Since the chart is a portion of the *d*-dimensional Euclidean space $\mathbb{R}^d$ and we can differentiate functions defined on such sets, the notion of differentiability now makes sense for f. Of course, for this definition to work for the manifold, it is important that if x belongs to two overlapping charts, then the answer will be the same for both. This is guaranteed if the function that gives the correspondence between the overlapping parts (known as a *transition function*) is itself differentiable. Manifolds with this property are called *differentiable manifolds*: manifolds for which the transition functions are continuous but not necessarily differentiable are called *topological manifolds*. The availability of calculus makes the theory of differentiable manifolds very different from that of topological manifolds.

The above ideas generalize easily from real-valued functions to functions from M to $\mathbb{R}^d$, or from M to M', where M' is another manifold. However, it is easier to judge whether a function defined on a manifold is differentiable than it is to say what the derivative is. The derivative at some point x of a function from $\mathbb{R}^n$ to $\mathbb{R}^m$ is a linear map, and so is the derivative of a function defined on a manifold. However, the domain of the linear map is not the manifold itself, which is not usually a vector space, but rather the so-called *tangent space* at the point x in question.

For more details on this and on manifolds in general, see DIFFERENTIAL TOPOLOGY [IV.7].

6.10 Riemannian Metrics

Suppose you are given two points P and Q on a sphere. How do you determine the distance between them? The answer depends on how the sphere is defined. If it is the set of all points (x, y, z) such that $x^2 + y^2 + z^2 = 1$ then P and Q are points in $\mathbb{R}^3$. One can therefore use the Pythagorean theorem to calculate the distance between them. For example, the distance between the points $(1, 0, 0)$ and $(0, 1, 0)$ is $\sqrt{2}$.

However, do we really want to measure the length of the line segment PQ? This segment does not lie in the sphere itself, so to use it as a means of defining length does not sit at all well with the idea of a manifold as an intrinsically defined object. Fortunately, as we saw earlier in the discussion of spherical geometry, there is another natural definition that avoids this problem: we can define the distance between P and Q as the length of the shortest path from P to Q that lies entirely within the sphere.

Now let us suppose that we wish to talk more generally about distances between points in manifolds. If the manifold is presented to us as a hypersurface in some bigger space, then we can use lengths of shortest paths as we did in the sphere. But suppose that the manifold is presented differently and all we have is a way of demonstrating that every point is contained in a chart—that is, has a neighborhood that can be associated with a portion of d-dimensional Euclidean space. (For the purposes of this discussion, nothing is lost if one takes d to be 2 throughout, in which case there is a correspondence between the neighborhood and a portion of the plane.) One idea is to define the distance between the two points to be the distance between the corresponding points in the chart, but this raises at least three problems.

The first is that the points P and Q that we are looking at might belong to different charts. This, however, is not too much of a problem, since all we actually need to do is calculate lengths of paths, and that can be done

provided we have a way of defining distances between points that are very close together, in which case we can find a single chart that contains them both.

The second problem, which is much more serious, is that for any one manifold there are many ways of choosing the charts, so this idea does not lead to a single notion of distance for the manifold. Worse still, even if one fixes one set of charts, these charts will overlap, and it may not be possible to make the notions of distance compatible where the overlap occurs.

The third problem is related to the second. The surface of a sphere is curved, whereas the charts of any atlas (in either the everyday or the mathematical sense) are flat. Therefore, the distances in the charts cannot correspond exactly to the lengths of shortest paths in the sphere itself.

The single most important moral to draw from the above problems is that if we wish to define a notion of distance for a given manifold, we have a great deal of choice about how to do so. Very roughly, a Riemannian metric is a way of making such a choice.

A little less roughly, a *metric* means a sensible notion of distance (the precise definition can be found in [III.56]). A Riemannian metric is a way of determining infinitesimal distances. These infinitesimal distances can be used to calculate lengths of paths, and then the distance between two points can be defined as the length of the shortest path between them. To see how this is done, let us first think about lengths of paths in the ordinary Euclidean plane. Suppose that (x, y) belongs to a path and $(x + \delta x, y + \delta y)$ is another point on the path, very close to (x, y). Then the distance between the two points is $\sqrt{\delta x^2 + \delta y^2}$. To calculate the length of a sufficiently smooth path, one can choose a large number of points along the path, each one very close to the next, and add up their distances. This gives a good approximation, and one can make it better and better by taking more and more points.

In practice, it is easier to work out the length using calculus. A path itself can be thought of as a moving point $(x(t), y(t))$ that starts when $t = 0$ and ends when $t = 1$. If δt is very small, then $x(t + \delta t)$ is approximately $x(t) + x'(t)\delta t$ and $y(t + \delta t)$ is approximately $y(t) + y'(t)\delta t$. Therefore, the distance between $(x(t), y(t))$ and $(x(t + \delta t), y(t + \delta t))$ is approximately $\delta t \sqrt{x'(t)^2 + y'(t)^2}$, by the Pythagorean theorem. Therefore, letting δt go to zero and integrating all the infinitesimal distances along the path, we obtain the formula

$$\int_0^1 \sqrt{x'(t)^2 + y'(t)^2} \, dt$$

for the length of the path. Notice that if we write $x'(t)$ and $y'(t)$ as dx/dt and dy/dt, then we can rewrite $\sqrt{x'(t)^2 + y'(t)^2}\,dt$ as $\sqrt{dx^2 + dy^2}$, which is the infinitesimal version of the expression $\sqrt{\delta x^2 + \delta y^2}$ that we had earlier. We have just defined a Riemannian metric, which is usually denoted by $dx^2 + dy^2$. This can be thought of as the square of the distance between the point (x, y) and the infinitesimally close point $(x + dx, y + dy)$.

If we want to, we can now prove that the shortest path between two points (x_0, y_0) and (x_1, y_1) is a straight line, which will tell us that the distance between them is $\sqrt{(x_1 - x_0)^2 + (y_1 - y_0)^2}$. (A proof can be found in VARIATIONAL METHODS [III.94].) However, since we could have just used this formula to begin with, this example does not really illustrate what is distinctive about Riemannian metrics. To do that, let us give a more precise definition of the disk model for hyperbolic geometry, which was discussed in section 6.6. There it was stated that distances become larger, relative to Euclidean distances, as one approaches the edge of the disk. A more precise definition is that the *open unit disk* is the set of all points (x, y) such that $x^2 + y^2 < 1$ and that the Riemannian metric on this disk is given by the expression $(dx^2 + dy^2)/(1 - x^2 - y^2)$. This is how we *define* the square of the distance between (x, y) and $(x + dx, y + dy)$. Equivalently, the length of a path $(x(t), y(t))$ with respect to this Riemannian metric is defined as

$$\int_0^1 \sqrt{\frac{x'(t)^2 + y'(t)^2}{1 - x(t)^2 - y(t)^2}}\,dt.$$

More generally, a *Riemannian metric* on a portion of the plane is an expression of the form

$$E(x, y)\,dx^2 + 2F(x, y)\,dx\,dy + G(x, y)\,dy^2$$

that is used to calculate infinitesimal distances and hence lengths of paths. (In the disk model we took $E(x, y)$ and $G(x, y)$ to be $1/(1 - x^2 - y^2)$ and $F(x, y)$ to be 0.) It is important for these distances to be positive, which will turn out to be the case provided that $E(x, y)G(x, y) - F(x, y)^2$ is always positive. One also needs the functions E, F, and G to satisfy certain smoothness conditions.

This definition generalizes straightforwardly to more dimensions. In n dimensions we must use an expression of the form

$$\sum_{i, j=1}^n F_{ij}(x_1, \ldots, x_n)\,dx_i\,dx_j.$$

to specify the squared distance between the points $(x_1, \ldots, x_n)$ and $(x_1 + dx_1, \ldots, x_n + dx_n)$. The numbers $F_{ij}(x_1, \ldots, x_n)$ form an $n \times n$ matrix that depends on the point $(x_1, \ldots, x_n)$. This matrix is required to be symmetric and positive definite: that is, $F_{ij}(x_1, \ldots, x_n)$ should always equal $F_{ji}(x_1, \ldots, x_n)$, and the expression that determines the squared distance should always be positive. It should also depend smoothly on the point $(x_1, \ldots, x_n)$.

Finally, now that we know how to define many different Riemannian metrics on portions of Euclidean space, we have many potential ways to define metrics on the charts that we use to define a manifold. A Riemannian metric on a *manifold* is a way of choosing compatible Riemannian metrics on the charts, where "compatible" means that wherever two charts overlap the distances should be the same. As mentioned earlier, once one has done this, one can define the distance between two points to be the length of a shortest path between them.

Given a Riemannian metric on a manifold, it is possible to define many other concepts, such as angles and volumes. It is also possible to define the important concept of *curvature*, which is discussed in RICCI FLOW [III.78]. Another important definition is that of a *geodesic*, which is the analogue for Riemannian geometry of a straight line in Euclidean geometry. A curve C is a geodesic if, given any two points P and Q on C that are sufficiently close, the shortest path from P to Q is part of C. For example, the geodesics on the sphere are the great circles.

As should be clear by now from the above discussion, on any given manifold there is a multitude of possible Riemannian metrics. A major theme in Riemannian geometry is to choose one that is "best" in some way. For example, on the sphere, if we take the obvious definition of the length of a path, then the resulting metric is particularly symmetric, and this is a highly desirable property. In particular, with this Riemannian metric the curvature of the sphere is the same everywhere. More generally, one searches for extra conditions to impose on Riemannian metrics. Ideally, these conditions should be strong enough that there is just one Riemannian metric that satisfies them, or at least that the family of such metrics should be very small.

I.4 The General Goals of Mathematical Research

The previous article introduced many concepts that appear throughout mathematics. This one discusses what mathematicians do with those concepts, and the sorts of questions they ask about them.

1 Solving Equations

As we have seen in earlier articles, mathematics is full of objects and structures (of a mathematical kind), but they do not simply sit there for our contemplation: we also like to *do* things to them. For example, given a number, there will be contexts in which we want to double it, or square it, or work out its reciprocal; given a suitable function, we may wish to differentiate it; given a geometrical shape, we may wish to transform it; and so on.

Transformations like these give rise to a never-ending source of interesting problems. If we have defined some mathematical process, then a rather obvious mathematical project is to invent techniques for carrying it out. This leads to what one might call *direct* questions about the process. However, there is also a deeper set of *inverse* questions, which take the following form. Suppose you are told what process has been carried out and what answer it has produced. Can you then work out what the mathematical object was that the process was applied to? For example, suppose I tell you that I have just taken a number and squared it, and that the result was 9. Can you tell me the original number?

In this case the answer is more or less yes: it must have been 3, except that if negative numbers are allowed, then another solution is -3.

If we want to talk more formally, then we say that we have been examining the equation $x^2 = 9$, and have discovered that there are two solutions. This example raises three issues that appear again and again.

- Does a given equation have any solutions?
- If so, does it have exactly one solution?
- What is the set in which solutions are required to live?

The first two concerns are known as the *existence* and the *uniqueness* of solutions. The third does not seem particularly interesting in the case of the equation $x^2 = 9$, but in more complicated cases, such as partial differential equations, it can be a subtle and important question.

To use more abstract language, suppose that f is a FUNCTION [I.2 §2.2] and that we are faced with a statement of the form $f(x) = y$. The direct question is to work out y given what x is. The inverse question is to work out x given what y is: this would be called solving the equation $f(x) = y$. Not surprisingly, questions about the solutions of an equation of this form are closely related to questions about the invertibility of the function f, which were discussed in [I.2]. Because x and y can be very much more general objects than numbers, the notion of solving equations is itself very general, and for that reason it is central to mathematics.

1.1 Linear Equations

The very first equations a schoolchild meets will typically be ones like $2x + 3 = 17$. To solve simple equations like this, one treats x as an unknown number that obeys the usual rules of arithmetic. By exploiting these rules one can transform the equation into something much simpler: subtracting 3 from both sides we learn that $2x = 14$, and dividing both sides of this new equation by 2 we then discover that $x = 7$. If we are very careful, we will notice that all we have shown is that *if* there is some number x such that $2x + 3 = 17$ *then* x must be 7. What we have not shown is that there is any such x. So strictly speaking there is a further step of checking that $2 \times 7 + 3 = 17$. This will obviously be true here, but the corresponding assertion is not always true for more complicated equations so this final step can be important.

The equation $2x + 3 = 17$ is called "linear" because the function f we have performed on x (to multiply it by 2 and add 3) is a linear one, in the sense that its graph is a straight line. As we have just seen, linear equations involving a single unknown x are easy to solve, but matters become considerably more sophisticated when one starts to deal with more than one unknown. Let us look at a typical example of an equation in two unknowns, the equation $3x + 2y = 14$. This equation has many solutions: for any choice of y you can set $x = (14 - 2y)/3$ and you have a pair (x, y) that satisfies the equation. To make it harder, one can take a second equation as well, $5x + 3y = 22$, say, and try to solve the two equations *simultaneously*. Then, it turns out, there is just one solution, namely $x = 2$ and $y = 4$. Typically, two linear equations in two unknowns have exactly one solution, just as these two do, which is easy to see if one thinks about the situation geometrically. An equation of the form $ax + by = c$ is the equation of a straight line in the xy-plane. Two lines normally meet in a single point, the exceptions being when they are identical, in which case they meet in infinitely many points, or parallel but not identical, in which case they do not meet at all.

If one has several equations in several unknowns, it can be conceptually simpler to think of them as one equation in one unknown. This sounds impossible, but it is perfectly possible if the new unknown is allowed to be a more complicated object. For example, the two equations $3x + 2y = 14$ and $5x + 3y = 22$ can be rewritten as the following single equation involving matrices and vectors:

$$\begin{pmatrix} 3 & 2 \\ 5 & 3 \end{pmatrix} \begin{pmatrix} x \\ y \end{pmatrix} = \begin{pmatrix} 14 \\ 22 \end{pmatrix}.$$

If we let A stand for the matrix, $\boldsymbol{x}$ for the unknown column vector, and $\boldsymbol{b}$ for the known one, then this equation becomes simply $A\boldsymbol{x} = \boldsymbol{b}$, which looks much less complicated, even if in fact all we have done is hidden the complication behind our notation.

There is more to this process, however, than sweeping dirt under the carpet. While the simpler notation conceals many of the specific details of the problem, it also *reveals* very clearly what would otherwise be obscured: that we have a linear map from $\mathbb{R}^2$ to $\mathbb{R}^2$ and we want to know which vectors $\boldsymbol{x}$, if any, map to the vector $\boldsymbol{b}$. When faced with a particular set of simultaneous equations, this reformulation does not make much difference—the calculations we have to do are the same—but when we wish to reason more generally, either directly about simultaneous equations or about other problems where they arise, it is much easier to think about a matrix equation with a single unknown vector than about a collection of simultaneous equations in several unknown numbers. This phenomenon occurs throughout mathematics and is a major reason for the study of high-dimensional spaces.

1.2   Polynomial Equations

We have just discussed the generalization of linear equations from one variable to several variables. Another direction in which one can generalize them is to think of linear functions as polynomials of degree 1 and consider functions of higher degree. At school, for example, one learns how to solve *quadratic* equations, such as $x^2 - 7x + 12 = 0$. More generally, a *polynomial equation* is one of the form

$$a_n x^n + a_{n-1} x^{n-1} + \cdots + a_2 x^2 + a_1 x + a_0 = 0.$$

To solve such an equation means to find a value of x for which the equation is true (or, better still, all such values). This may seem an obvious thing to say until one considers a very simple example such as the equation $x^2 - 2 = 0$, or equivalently $x^2 = 2$. The solution to

this is, of course, $x = \pm\sqrt{2}$. What, though, is $\sqrt{2}$? It is defined to be the positive number that squares to 2, but it does not seem to be much of a "solution" to the equation $x^2 = 2$ to say that x is plus or minus the positive number that squares to 2. Neither does it seem entirely satisfactory to say that $x = 1.4142135\ldots$, since this is just the beginning of a calculation that never finishes and does not result in any discernible pattern.

There are two lessons that can be drawn from this example. One is that what matters about an equation is often the *existence* and *properties* of solutions and not so much whether one can find a formula for them. Although we do not appear to learn anything when we are told that the solutions to the equation $x^2 = 2$ are $x = \pm\sqrt{2}$, this assertion does contain within it a fact that is not wholly obvious: that the number 2 has a square root. This is usually presented as a consequence of the *intermediate value theorem* (or another result of a similar nature), which states that if f is a continuous real-valued function and $f(a)$ and $f(b)$ lie on either side of 0, then somewhere between a and b there must be a c such that $f(c) = 0$. This result can be applied to the function $f(x) = x^2 - 2$, since $f(1) = -1$ and $f(2) = 2$. Therefore, there is some x between 1 and 2 such that $x^2 - 2 = 0$, that is, $x^2 = 2$. For many purposes, the mere existence of this x is enough, together with its defining properties of being positive and squaring to 2.

A similar argument tells us that all positive real numbers have positive square roots. But the picture changes when we try to solve more complicated quadratic equations. Then we have two choices. Consider, for example, the equation $x^2 - 6x + 7 = 0$. We could note that $x^2 - 6x + 7$ is -1 when $x = 4$ and 2 when $x = 5$ and deduce from the intermediate value theorem that the equation has some solution between 4 and 5. However, we do not learn as much from this as if we complete the square, rewriting $x^2 - 6x + 7$ as $(x-3)^2 - 2$. This allows us to rewrite the equation as $(x-3)^2 = 2$, which has the two solutions $x = 3 \pm \sqrt{2}$. We have already established that $\sqrt{2}$ exists and lies between 1 and 2, so not only do we have a solution of $x^2 - 6x + 7 = 0$ that lies between 4 and 5, but we can see that it is closely related to, indeed built out of, the solution to the equation $x^2 = 2$. This demonstrates a second important aspect of equation solving, which is that in many instances the explicit solubility of an equation is a *relative* notion. If we are given a solution to the equation $x^2 = 2$, we do not need any *new* input from the intermediate value theorem to solve the more complicated equation $x^2 - 6x + 7 = 0$: all we need is some algebra. The solution, $x = 3 \pm \sqrt{2}$, is given by an explicit expression, but inside that expression we have $\sqrt{2}$, which is *not* defined by means of an explicit formula but as a real number, with certain properties, that we can prove to exist.

Solving polynomial equations of higher degree is markedly more difficult than solving quadratics, and raises fascinating questions. In particular, there are complicated formulas for the solutions of cubic and quartic equations, but the problem of finding corresponding formulas for quintic and higher-degree equations became one of the most famous unsolved problems in mathematics, until ABEL [VI.33] and GALOIS [VI.41] showed that it could not be done. For more details about these matters see THE INSOLUBILITY OF THE QUINTIC [V.21]. For another article related to polynomial equations see THE FUNDAMENTAL THEOREM OF ALGEBRA [V.13].

1.3 Polynomial Equations in Several Variables

Suppose that we are faced with an equation such as

$$x^3 + y^3 + z^3 = 3x^2y + 3y^2z + 6xyz.$$

We can see straight away that there will be many solutions: if you fix x and y, then the equation is a cubic polynomial in z, and all cubics have at least one (real) solution. Therefore, for every choice of x and y there is some z such that the triple (x, y, z) is a solution of the above equation.

Because the formula for the solution of a general cubic equation is rather complicated, a precise specification of the set of all triples (x, y, z) that solve the equation may not be very enlightening. However, one can learn a lot by regarding this solution set as a geometric object—a two-dimensional surface in space, to be precise—and asking *qualitative* questions about it. One might, for instance, wish to understand roughly what shape it is. Questions of this kind can be made precise using the language and concepts of TOPOLOGY [I.3 §6.4].

One can of course generalize further and consider simultaneous solutions to several polynomial equations. Understanding the solution sets of such systems of equations is the province of ALGEBRAIC GEOMETRY [IV.4].

1.4 Diophantine Equations

As has been mentioned, the answer to the question of whether a particular equation has a solution varies according to where the solution is allowed to be. The equation $x^2 + 3 = 0$ has no solution if x is required to be real, but in the complex numbers it has the two solutions $x = \pm i\sqrt{3}$. The equation $x^2 + y^2 = 11$ has infinitely many solutions if we are looking for x and y in the real numbers, but none if they have to be integers.

This last example is a typical *Diophantine equation*, the name given to an equation if one is looking for integer solutions. The most famous Diophantine equation is the Fermat equation $x^n + y^n = z^n$, which is now known, thanks to Andrew Wiles, to have no positive integer solutions if n is greater than 2. (See FERMAT'S LAST THEOREM [V.10]. By contrast, the equation $x^2 + y^2 = z^2$ has infinitely many solutions.) A great deal of modern ALGEBRAIC NUMBER THEORY [IV.1] is concerned with Diophantine equations, either directly or indirectly. As with equations in the real and complex numbers, it is often fruitful to study the structure of sets of solutions to Diophantine equations: this investigation belongs to the area known as ARITHMETIC GEOMETRY [IV.5].

A notable feature of Diophantine equations is that they tend to be extremely difficult. It is therefore natural to wonder whether there could be a systematic approach to them. This question was the tenth in a famous list of problems asked by HILBERT [VI.63] in 1900. It was not until 1970 that Yuri Matiyasevich, building on work by Martin Davis, Julia Robinson, and Hilary Putnam, proved that the answer was no. (This is discussed further in THE INSOLUBILITY OF THE HALTING PROBLEM [V.20].)

An important step in the solution was taken in 1936, by CHURCH [VI.89] and TURING [VI.94]. This was to make precise the notion of a "systematic approach," by formalizing (in two different ways) the notion of an algorithm (see ALGORITHMS [II.4 §3] and COMPUTATIONAL COMPLEXITY [IV.20 §1]). It was not easy to do this in the pre-computer age, but now we can restate the solution of Hilbert's tenth problem as follows: there is no computer program that can take as its input any Diophantine equation, and without fail print "YES" if it has a solution and "NO" otherwise.

What does this tell us about Diophantine equations? We can no longer dream of a final theory that will encompass them all, so instead we are forced to restrict our attention to individual equations or special classes of equations, continually developing different methods for solving them. This would make them uninteresting after the first few, were it not for the fact that specific Diophantine equations have remarkable links with very general questions in other parts of mathematics. For example, equations of the form $y^2 = f(x)$, where $f(x)$ is a cubic polynomial in x, may look rather special, but in fact the ELLIPTIC CURVES [III.21] that they define are central to modern number theory, including the proof of Fermat's last theorem. Of course, Fermat's last theorem is itself a Diophantine equation, but its study has led to major developments in other parts of number theory. The correct moral to draw is perhaps this: solving a particular Diophantine equation is fascinating and worthwhile if, as is often the case, the result is more than a mere addition to the list of equations that have been solved.

1.5   Differential Equations

So far, we have looked at equations where the unknown is either a number or a point in n-dimensional space (that is, a sequence of n numbers). In order to generate these equations, we took various combinations of the basic arithmetic operations and applied them to our unknowns.

Here, for comparison, are two well-known differential equations, the first "ordinary" and the second "partial":

$$\frac{d^2 x}{dt^2} + k^2 x = 0,$$

$$\frac{\partial T}{\partial t} = \kappa \left(\frac{\partial^2 T}{\partial x^2} + \frac{\partial^2 T}{\partial y^2} + \frac{\partial^2 T}{\partial z^2} \right).$$

The first is the equation for simple harmonic motion, which has the general solution $x(t) = A \sin kt + B \cos kt$; the second is the heat equation, which was discussed in SOME FUNDAMENTAL MATHEMATICAL DEFINITIONS [I.3 §5.4].

For many reasons, differential equations represent a jump in sophistication. One is that the unknowns are *functions*, which are much more complicated objects than numbers or n-dimensional points. (For example, the first equation above asks what function x of t has the property that if you differentiate it twice then you get $-k^2$ times the original function.) A second is that the basic operations one performs on functions include differentiation and integration, which are considerably less "basic" than addition and multiplication. A third is that differential equations that can be solved in "closed form," that is, by means of a formula for the unknown function f, are the exception rather than the rule, even when the equations are natural and important.

Consider again the first equation above. Suppose that, given a function f, we write $\phi(f)$ for the function $(d^2 f / dt^2) + k^2 f$. Then ϕ is a linear map, in the sense that $\phi(f + g) = \phi(f) + \phi(g)$ and $\phi(af) = a\phi(f)$ for

any constant a. This means that the differential equation can be regarded as something like a matrix equation, but generalized to infinitely many dimensions. The heat equation has the same property: if we define $\psi(T)$ to be

$$\frac{\partial T}{\partial t} - \kappa \left(\frac{\partial^2 T}{\partial x^2} + \frac{\partial^2 T}{\partial y^2} + \frac{\partial^2 T}{\partial z^2} \right),$$

then ψ is another linear map. Such differential equations are called *linear*, and the link with linear algebra makes them markedly easier to solve. (A very useful tool for this is THE FOURIER TRANSFORM [III.27].)

What about the more typical equations, the ones that cannot be solved in closed form? Then the focus shifts once again toward establishing whether or not solutions *exist*, and if so what *properties* they have. As with polynomial equations, this can depend on what you count as an allowable solution. Sometimes we are in the position we were in with the equation $x^2 = 2$: it is not too hard to prove that solutions exist and all that is left to do is name them. A simple example is the equation $dy/dx = e^{-x^2}$. In a certain sense, this cannot be solved: it can be shown that there is no function built out of polynomials, EXPONENTIALS [III.25], and TRIGONOMETRIC FUNCTIONS [III.92] that differentiates to e^{-x^2}. However, in another sense the equation is easy to solve— all you have to do is integrate the function e^{-x^2}. The resulting function (when divided by $\sqrt{2\pi}$) is the NORMAL DISTRIBUTION [III.71 §5] function. The normal distribution is of fundamental importance in probability, so the function is given a name, Φ.

In most situations, there is no hope of writing down a formula for a solution, even if one allows oneself to integrate "known" functions. A famous example is the so-called THREE-BODY PROBLEM [V.33]: given three bodies moving in space and attracted to each other by gravitational forces, how will they continue to move? Using Newton's laws, one can write down some differential equations that describe this situation. NEWTON [VI.14] solved the corresponding equations for two bodies, and thereby explained why planets move in elliptical orbits around the Sun, but for three or more bodies they proved very hard indeed to solve. It is now known that there was a good reason for this: the equations can lead to chaotic behavior. (See DYNAMICS [IV.14] for more about chaos.) However, this opens up a new and very interesting avenue of research into questions of chaos and stability.

Sometimes there are ways of proving that solutions exist even if they cannot be easily specified. Then one may ask not for precise formulas, but for general descriptions. For example, if the equation has a time dependence (as, for instance, the heat equation and wave equations have), one can ask whether solutions tend to decay over time, or blow up, or remain roughly the same. These more qualitative questions concern what is known as *asymptotic behavior*, and there are techniques for answering some of them even when a solution is not given by a tidy formula.

As with Diophantine equations, there are some special and important classes of partial differential equations, including nonlinear ones, that *can* be solved exactly. This gives rise to a very different style of research: again one is interested in properties of solutions, but now these properties may be more algebraic in nature, in the sense that exact formulas will play a more important role. See LINEAR AND NONLINEAR WAVES AND SOLITONS [III.49].

2 Classifying

If one is trying to understand a new mathematical structure, such as a GROUP [I.3 §2.1] or a MANIFOLD [I.3 §6.9], one of the first tasks is to come up with a good supply of examples. Sometimes examples are very easy to find, in which case there may be a bewildering array of them that cannot be put into any sort of order. Often, however, the conditions that an example must satisfy are quite stringent, and then it may be possible to come up with something like an infinite list that includes every single one. For example, it can be shown that any VECTOR SPACE [I.3 §2.3] of dimension n over a field $\mathbb{F}$ is isomorphic to $\mathbb{F}^n$. This means that just one positive integer, n, is enough to determine the space completely. In this case our "list" will be $\{0\}, \mathbb{F}, \mathbb{F}^2, \mathbb{F}^3, \mathbb{F}^4, \ldots$. In such a situation we say that we have a *classification* of the mathematical structure in question.

Classifications are very useful because if we can classify a mathematical structure then we have a new way of proving results about that structure: instead of deducing a result from the axioms that the structure is required to satisfy, we can simply check that it holds for every example on the list, confident in the knowledge that we have thereby proved it in general. This is not always easier than the more abstract, axiomatic approach, but it certainly is sometimes. Indeed, there are several results proved using classifications that nobody knows how to prove in any other way. More generally, the more examples you know of a mathematical structure, the easier it is to think about that structure—testing hypotheses, finding counterexamples, and so

on. If you know *all* the examples of the structure, then for some purposes your understanding is complete.

2.1 Identifying Building Blocks and Families

There are two situations that typically lead to interesting classification theorems. The boundary between them is somewhat blurred, but the distinction is clear enough to be worth making, so we shall discuss them separately in this subsection and the next.

As an example of the first kind of situation, let us look at objects called *regular polytopes*. Polytopes are polygons, polyhedra, and their higher-dimensional generalizations. The regular polygons are those for which all sides have the same length and all angles are equal, and the regular polyhedra are those for which all faces are congruent regular polygons and every vertex has the same number of edges coming out of it. More generally, a higher-dimensional polytope is regular if it is as symmetrical as possible, though the precise definition of this is somewhat complicated. (Here, in three dimensions, is a definition that turns out to be equivalent to the one just given but easier to generalize. A *flag* is a triple (v, e, f) where v is a vertex of the polyhedron, e is an edge containing v, and f is a face containing e. A polyhedron is regular if for any two flags (v, e, f) and (v', e', f') there is a symmetry of the polyhedron that takes v to v', e to e', and f to f'.)

It is easy to see what the regular polygons are in two dimensions: for every k greater than 2 there is exactly one regular k-gon and that is all there is. In three dimensions, the regular polyhedra are the famous *Platonic solids*, that is, the tetrahedron, the cube, the octahedron, the dodecahedron, and the icosahedron. It is not too hard to see that there cannot be any more regular polyhedra, since there must be at least three faces meeting at each vertex, and the angles at that vertex must add up to less than $360°$. This constraint means that the only possibilities for the faces at a vertex are three, four, or five triangles, three squares, or three pentagons. These give the tetrahedron, the octahedron, the icosahedron, the cube, and the dodecahedron, respectively.

Some of the polygons and polyhedra just defined have natural higher-dimensional analogues. For example, if you take $n + 1$ points in $\mathbb{R}^n$ all at the same distance from one another, then they form the vertices of a *regular simplex*, which is an equilateral triangle or regular tetrahedron when $n = 2$ or 3. The set of all points $(x_1, x_2, \ldots, x_n)$ with $0 \leqslant x_i \leqslant 1$ for every i

forms the n-dimensional analogue of a unit square or cube. The octahedron can be defined as the set of all points (x, y, z) in $\mathbb{R}^3$ such that $|x| + |y| + |z| \leqslant 1$, and the analogue of this in n dimensions is the set of all points $(x_1, x_2, \ldots, x_n)$ such that $|x_1| + \cdots + |x_n| \leqslant 1$.

It is not obvious how the dodecahedron and icosahedron would lead to infinite families of regular polytopes, and it turns out that they do not. In fact, apart from three more examples in four dimensions, the above polytopes constitute a complete list. These three examples are quite remarkable. One of them has 120 "three-dimensional faces," each of which is a regular dodecahedron. It has a so-called dual, which has 600 regular tetrahedra as its "faces." The third example can be described in terms of coordinates: its vertices are the sixteen points of the form $(\pm 1, \pm 1, \pm 1, \pm 1)$, together with the eight points $(\pm 2, 0, 0, 0)$, $(0, \pm 2, 0, 0)$, $(0, 0, \pm 2, 0)$, and $(0, 0, 0, \pm 2)$.

The theorem that these are all the regular polytopes is significantly harder to prove than the result sketched above for three dimensions. The complete list was obtained by Schäfli in the mid nineteenth century; the first proof that there are no others was given by Donald Coxeter in 1969.

We therefore know that the regular polytopes in dimensions three and higher fall into three families—the n-dimensional versions of the tetrahedron, the cube, and the octahedron—together with five "exceptional" examples—the dodecahedron, the icosahedron, and the three four-dimensional polytopes just described. This situation is typical of many classification theorems. The exceptional examples, often called "sporadic," tend to have a very high degree of symmetry—it is almost as if we have no right to expect this degree of symmetry to be possible, but just occasionally by a happy chance it is. The families and sporadic examples that occur in different classification results are often closely related, and this can be a sign of deep connections between areas that do not at first appear to be connected at all.

Sometimes, instead of trying to classify all mathematical structures of a given kind, one identifies a certain class of "basic" structures out of which all the others can be built in a simple way. A good analogy for this is the set of primes, out of which all other integers can be built as products. Finite groups, for example, are all "products" of certain basic groups that are called *simple*. THE CLASSIFICATION OF FINITE SIMPLE GROUPS [V.7], one of the most famous theorems of twentieth-century mathematics, is discussed in part V.

For more on this style of classification theorem, see also LIE THEORY [III.48].

2.2 Equivalence, Nonequivalence, and Invariants

There are many situations in mathematics where two objects are, strictly speaking, different, but where we are not interested in the difference. In such situations we want to regard the objects as "essentially the same," or "equivalent." Equivalence of this kind is expressed formally by the notion of an EQUIVALENCE RELATION [I.2 §2.3].

For example, a topologist regards two shapes as essentially the same if one is a continuous deformation of the other, as we saw in [I.3 §6.4]. As pointed out there, a sphere is the same as a cube in this sense, and one can also see that the surface of a doughnut, that is, a torus, is essentially the same as the surface of a teacup. (To turn the teacup into a doughnut, let the handle expand while the cup part is gradually swallowed up into it.) It is equally obvious, intuitively speaking, that a sphere is *not* essentially the same as a torus, but this is much harder to prove.

Why should nonequivalence be harder to prove than equivalence? The answer is that in order to show that two objects are equivalent, all one has to do is find a single transformation that demonstrates this equivalence. However, to show that two objects are not equivalent, one must somehow consider *all possible* transformations and show that not one of them works. How can one rule out the existence of some wildly complicated continuous deformation that is impossible to visualize but happens, remarkably, to turn a sphere into a torus?

Here is a sketch of a proof. The sphere and the torus are examples of *compact orientable surfaces*, which means, roughly speaking, two-dimensional shapes that occupy a finite portion of space and have no boundary. Given any such surface, one can find an equivalent surface that is built out of triangles and is topologically the same. Here is a famous theorem of EULER [VI.19].

Let P be a polyhedron that is topologically the same as a sphere, and suppose that it has V vertices, E edges, and F faces. Then $V - E + F = 2$.

For example, if P is an icosahedron, then it has twelve vertices, thirty edges, and twenty faces, and $12 - 30 + 20$ is indeed equal to 2.

For this theorem, it is not in fact important that the triangles are flat: we can draw them on the original sphere, except that now they are spherical triangles. It is just as easy to count vertices, edges, and faces when we do this, and the theorem is still valid. A network of triangles drawn on a sphere is called a *triangulation* of the sphere.

Euler's theorem tells us that $V - E + F = 2$ regardless of what triangulation of the sphere we take. Moreover, the formula is still valid if the surface we triangulate is not a sphere but another shape that is topologically equivalent to the sphere, since triangulations can be continuously deformed without V, E, or F changing.

More generally, one can triangulate *any* surface, and evaluate $V - E + F$. The result is called the *Euler characteristic* of that surface. For this definition to make sense, we need the following fact, which is a generalization of Euler's theorem (and which is not much harder to prove than the original result).

(i) *Although a surface can be triangulated in many ways, the quantity $V - E + F$ will be the same for all triangulations.*

If we continuously deform the surface and continuously deform one of its triangulations at the same time, we can deduce that the Euler characteristic of the new surface is the same as that of the old one. In other words, fact (i) above has the following interesting consequence.

(ii) *If two surfaces are continuous deformations of each other, then they have the same Euler characteristic.*

This gives us a potential method for showing that surfaces are not equivalent: if they have different Euler characteristics then we know from the above that they are not continuous deformations of each other. The Euler characteristic of the torus turns out to be 0 (as one can show by calculating $V - E + F$ for any triangulation), and that completes the proof that the sphere and the torus are not equivalent.

The Euler characteristic is an example of an *invariant*. This means a function ϕ, the domain of which is the set of all objects of the kind one is studying, with the property that if X and Y are equivalent objects, then $\phi(X) = \phi(Y)$. To show that X is not equivalent to Y, it is enough to find an invariant ϕ for which $\phi(X)$ and $\phi(Y)$ are different. Sometimes the values ϕ takes are numbers (as with the Euler characteristic), but often they will be more complicated objects such as polynomials or groups.

It is perfectly possible for $\phi(X)$ to equal $\phi(Y)$ even when X and Y are not equivalent. An extreme example would be the invariant ϕ that simply took the value 0

for every object X. However, sometimes it is so hard to prove that objects are not equivalent that invariants can be considered useful and interesting even when they work only part of the time.

There are two main properties that one looks for in an invariant ϕ, and they tend to pull in opposite directions. One is that it should be as *fine* as possible: that is, as often as possible $\phi(X)$ and $\phi(Y)$ are different if X and Y are not equivalent. The other is that as often as possible one should actually be able to establish when $\phi(X)$ is different from $\phi(Y)$. There is not much use in having a fine invariant if it is impossible to calculate. (An extreme example would be the "trivial" invariant that simply mapped each X to its equivalence class. It is as fine as possible, but unless we have some independent means of specifying it, then it does not represent an advance on the original problem of showing that two objects are not equivalent.) The most powerful invariants therefore tend to be ones that can be calculated, but not very easily.

In the case of compact orientable surfaces, we are lucky: not only is the Euler characteristic an invariant that is easy to calculate, but it also classifies the compact orientable surfaces completely. To be precise, k is the Euler characteristic of a compact orientable surface if and only if it is of the form $2 - 2g$ for some nonnegative integer g (so the possible Euler characteristics are $2, 0, -2, -4, \ldots$), and two compact orientable surfaces with the same Euler characteristic are equivalent. Thus, if we regard equivalent surfaces as the same, then the number g gives us a complete specification of a surface. It is called the *genus* of the surface, and can be interpreted geometrically as the number of "holes" the surface has (so the genus of the sphere is 0 and that of the torus is 1).

For other examples of invariants, see ALGEBRAIC TOPOLOGY [IV.6] and KNOT POLYNOMIALS [III.44].

3 Generalizing

When an important mathematical definition is formulated, or theorem proved, that is rarely the end of the story. However clear a piece of mathematics may seem, it is nearly always possible to understand it better, and one of the most common ways of doing so is to present it as a special case of something more general. There are various different kinds of generalization, of which we discuss a few here.

3.1 Weakening Hypotheses and Strengthening Conclusions

The number 1729 is famous for being expressible as the sum of two cubes in two different ways: it is $1^3 + 12^3$ and also $9^3 + 10^3$. Let us now try to decide whether there is a number that can be written as the sum of four cubes in ten different ways.

At first this problem seems alarmingly difficult. It is clear that any such number, if it exists, must be very large and would be extremely tedious to find if we simply tested one number after another. So what can we do that is better than this?

The answer turns out to be that we should weaken our hypotheses. The problem we wish to solve is of the following general kind. We are given a sequence $a_1, a_2, a_3, \ldots$ of positive integers and we are told that it has a certain property. We must then prove that there is a positive integer that can be written as a sum of four terms of the sequence in ten different ways. This is perhaps an artificial way of thinking about the problem since the property we assume of the sequence is the property of "being the sequence of cubes," which is so specific that it is more natural to think of it as an *identification* of the sequence. However, this way of thinking encourages us to consider the possibility that the conclusion might be true for a much wider class of sequences. And indeed this turns out to be the case.

There are a thousand cubes less than or equal to 1 000 000 000. We shall now see that this property alone is sufficient to guarantee that there is a number that can be written as the sum of four cubes in ten different ways. That is, if $a_1, a_2, a_3, \ldots$ is *any* sequence of positive integers, and if none of the first thousand terms exceeds 1 000 000 000, then some number can be written as the sum of four terms of the sequence in ten different ways.

To prove this, all we have to do is notice that the number of different ways of choosing four distinct terms from the sequence $a_1, a_2, \ldots, a_{1000}$ is $1000 \times 999 \times 998 \times 997/24$, which is greater than $40 \times 1\,000\,000\,000$. The sum of any four terms of the sequence cannot exceed $4 \times 1\,000\,000\,000$. It follows that the average number of ways of writing one of the first 4 000 000 000 numbers as the sum of four terms of the sequence is at least ten. But if the average number of representations is at least ten, then there must certainly be numbers that have at least this number of representations.

Why did it help to generalize the problem in this way? One might think that it would be harder to prove a result if one assumed less. However, that is often not true. The less you assume, the fewer options you have when trying to use your assumptions, and that can speed up the search for a proof. Had we not generalized the problem above, we would have had too many options. For instance, we might have found ourselves trying to solve very difficult Diophantine equations involving cubes rather than noticing the easy counting argument. In a way, it was only once we had weakened our hypotheses that we understood the true nature of the problem.

We could also think of the above generalization as a strengthening of the conclusion: the problem asks for a statement about cubes, and we prove not just that but much more besides. There is no clear distinction between weakening hypotheses and strengthening conclusions, since if we are asked to prove a statement of the form $P \Rightarrow Q$, we can always reformulate it as $\neg Q \Rightarrow \neg P$. Then, if we weaken P we are weakening the hypotheses of $P \Rightarrow Q$ but strengthening the conclusion of $\neg Q \Rightarrow \neg P$.

3.2 Proving a More Abstract Result

A famous result in modular arithmetic, known as FERMAT'S LITTLE THEOREM [III.58], states that if p is a prime and a is not a multiple of p, then a^{p-1} leaves a remainder of 1 when you divide by p. That is, a^{p-1} is congruent to 1 mod p.

There are several proofs of this result, one of which is a good illustration of a certain kind of generalization. Here is the argument in outline. The first step is to show that the numbers $1, 2, \ldots, p - 1$ form a GROUP [I.3 §2.1] under multiplication mod p. (This means multiplication followed by taking the remainder on division by p. For example, if $p = 7$ then the "product" of 3 and 6 is 4, since 4 is the remainder when you divide 18 by 7.) The next step is to note that if $1 \leqslant a \leqslant p-1$ then the powers of a (mod p) form a subgroup of this group. Moreover, the size of the subgroup is the smallest positive integer m such that a^m is congruent to 1 mod p. One then applies *Lagrange's theorem*, which states that the size of a group is always divisible by the size of any of its subgroups. In this case, the size of the group is $p - 1$, from which it follows that $p - 1$ is divisible by m. But then, since $a^m = 1$, it follows that $a^{p-1} = 1$.

This argument shows that Fermat's little theorem is, when viewed appropriately, just one special case of Lagrange's theorem. (The word "just" is, however, a little misleading, because it is not wholly obvious that the integers mod p form a group in the way stated. This fact is proved using EUCLID'S ALGORITHM [III.22].)

Fermat could not have viewed his theorem in this way, since the concept of a group had not been invented when he proved it. Thus, the abstract concept of a group helps one to see Fermat's little theorem in a completely new way: it can be viewed as a special case of a more general result, but a result that cannot even be stated until one has developed some new, abstract concepts.

This process of abstraction has many benefits. Most obviously, it provides us with a more general theorem, one that has many other interesting particular cases. Once we see this, then we can prove the general result once and for all rather than having to prove each case separately. A related benefit is that it enables us to see connections between results that may originally have seemed quite different. And finding surprising connections between different areas of mathematics almost always leads to significant advances in the subject.

3.3   Identifying Characteristic Properties

There is a marked contrast between the way one defines $\sqrt{2}$ and the way one defines $\sqrt{-1}$, or i as it is usually written. In the former case one begins, if one is being careful, by proving that there is exactly one positive real number that squares to 2. Then $\sqrt{2}$ is defined to be this number.

This style of definition is impossible for i since there is no real number that squares to -1. So instead one asks the following question: if there were a number that squared to -1, what could one say about it? Such a number would not be a real number, but that does not rule out the possibility of *extending* the real number system to a larger system that contains a square root of -1.

At first it may seem as though we know precisely one thing about i: that $i^2 = -1$. But if we assume in addition that i obeys the normal rules of arithmetic, then we can do more interesting calculations, such as

$$(i + 1)^2 = i^2 + 2i + 1 = -1 + 2i + 1 = 2i,$$

which implies that $(i + 1)/\sqrt{2}$ is a square root of i.

From these two simple assumptions—that $i^2 = -1$ and that i obeys the usual rules of arithmetic—we can develop the entire theory of COMPLEX NUMBERS [I.3 §1.5] without ever having to worry about what i actually is. And in fact, once you stop to think about it,

the existence of $\sqrt{2}$, though reassuring, is not in practice anything like as important as *its* defining properties, which are very similar to those of i: it squares to 2 and obeys the usual rules of arithmetic.

Many important mathematical generalizations work in a similar way. Another example is the definition of x^a when x and a are real numbers with x positive. It is difficult to make sense of this expression in a direct way unless a is a positive integer, and yet mathematicians are completely comfortable with it, whatever the value of a. How can this be? The answer is that what really matters about x^a is not its numerical value but its *characteristic properties* when one thinks of it as a function of a. The most important of these is the property that $x^{a+b} = x^a x^b$. Together with a couple of other simple properties, this completely determines the function x^a. More importantly, it is these characteristic properties that one uses when reasoning about x^a. This example is discussed in more detail in THE EXPONENTIAL AND LOGARITHMIC FUNCTIONS [III.25].

There is an interesting relationship between abstraction and classification. The word "abstract" is often used to refer to a part of mathematics where it is more common to use characteristic properties of an object than it is to argue directly from a definition of the object itself (though, as the example of $\sqrt{2}$ shows, this distinction can be somewhat hazy). The ultimate in abstraction is to explore the consequences of a system of axioms, such as those for a group or a vector space. However, sometimes, in order to reason about such algebraic structures, it is very helpful to classify them, and the result of classification is to make them more concrete again. For instance, every finite-dimensional real vector space V is isomorphic to $\mathbb{R}^n$ for some nonnegative integer n, and it is sometimes helpful to think of V as the concrete object $\mathbb{R}^n$, rather than as an algebraic structure that satisfies certain axioms. Thus, in a certain sense, classification is the opposite of abstraction.

3.4   Generalization after Reformulation

Dimension is a mathematical idea that is also a familiar part of everyday language: for example, we say that a photograph of a chair is a two-dimensional representation of a three-dimensional object, because the chair has height, breadth, and depth, but the image just has height and breadth. Roughly speaking, the dimension of a shape is the number of independent directions one can move about in while staying inside the shape, and this rough conception can be made mathematically precise (using the notion of a VECTOR SPACE [I.3 §2.3]).

If we are given any shape, then its dimension, as one would normally understand it, must be a nonnegative integer: it does not make much sense to say that one can move about in 1.4 independent directions, for example. And yet there is a rigorous mathematical theory of *fractional* dimension, in which for every nonnegative real number d you can find many shapes of dimension d.

How do mathematicians achieve the seemingly impossible? The answer is that they *reformulate* the concept of dimension and only then do they generalize it. What this means is that they give a new definition of dimension with the following two properties.

(i) For all "simple" shapes the new definition agrees with the old one. For example, under the new definition a line will still be one dimensional, a square two dimensional, and a cube three dimensional.

(ii) With the new definition it is no longer obvious that the dimension of every shape must be a positive integer.

There are several ways of doing this, but most of them focus on the differences between length, area, and volume. Notice that a line segment of length 2 can be expressed as a union of two nonoverlapping line segments of length 1, a square of side-length 2 can be expressed as a union of four nonoverlapping squares of side-length 1, and a cube of side-length 2 can be expressed as a union of eight nonoverlapping cubes of side-length 1. It is because of this that if you enlarge a d-dimensional shape by a factor r, then its d-dimensional "volume" is multiplied by r^d. Now suppose that you would like to exhibit a shape of dimension 1.4. One way of doing it is to let $r - 2^{5/7}$, so that $r^{1.4} - 2$, and find a shape X such that if you expand X by a factor of r, then the expanded shape can be expressed as a union of two disjoint copies of X. Two copies of X ought to have twice the "volume" of X itself, so the dimension d of X ought to satisfy the equation $r^d = 2$. By our choice of r, this tells us that the dimension of X is 1.4. For more details, see DIMENSION [III.17].

Another concept that seems at first to make no sense is *noncommutative geometry*. The word "commutative" applies to BINARY OPERATIONS [I.2 §2.4] and therefore belongs to algebra rather than geometry, so what could "noncommutative geometry" possibly mean?

By now the answer should not be a surprise: one reformulates part of geometry in terms of a certain algebraic structure and then generalizes the algebra. The algebraic structure involves a commutative binary operation, so one can generalize the algebra by allowing the binary operation not to be commutative.

The part of geometry in question is the study of MANIFOLDS [I.3 §6.9]. Associated with a manifold X is the set $C(X)$ of all continuous complex-valued functions defined on X. Given two functions f, g in $C(X)$, and two complex numbers λ and μ, the linear combination $\lambda f + \mu g$ is another continuous complex-valued function, so it also belongs to $C(X)$. Therefore, $C(X)$ is a vector space. However, one can also *multiply* f and g to form the continuous function fg (defined by $(fg)(x) = f(x)g(x)$). This multiplication has various natural properties (for instance, $f(g + h) = fg + fh$ for all functions f, g, and h) that make $C(X)$ into an *algebra*, and even a C^*-ALGEBRA [IV.15 §3]. It turns out that a great deal of the geometry of a compact manifold X can be reformulated purely in terms of the corresponding C^*-algebra $C(X)$. The word "purely" here means that it is not necessary to refer to the manifold X in terms of which the algebra $C(X)$ was originally defined—all one uses is the fact that $C(X)$ is an algebra. This raises the possibility that there might be algebras that do *not* arise geometrically, but to which the reformulated geometrical concepts nevertheless apply.

An algebra has two binary operations: addition and multiplication. Addition is always assumed to be commutative, but multiplication is not: when multiplication is commutative as well, one says that the algebra is commutative. Since fg and gf are clearly the same function, the algebra $C(X)$ *is* a commutative C^*-algebra, so the algebras that arise geometrically are always commutative. However, many geometrical concepts, once they have been reformulated in algebraic terms, continue to make sense for noncommutative C^* algebras, and that is why the phrase "noncommutative" geometry is used. For more details, see OPERATOR ALGEBRAS [IV.15 §5].

This process of reformulating and then generalizing underlies many of the most important advances in mathematics. Let us briefly look at a third example. THE FUNDAMENTAL THEOREM OF ARITHMETIC [V.14] is, as its name suggests, one of the foundation stones of number theory: it states that every positive integer can be written in exactly one way as a product of prime numbers. However, number theorists like to look at enlarged number systems, and for most of these the obvious analogue of the fundamental theorem of arithmetic is no longer true. For example, in the RING [III.81 §1] of numbers of the form $a + b\sqrt{-5}$ (where a and b are required to be integers), the number 6 can be written either as 2×3 or as $(1 + \sqrt{-5}) \times (1 - \sqrt{-5})$. Since none of the numbers 2, 3, $1 + \sqrt{-5}$, or $1 - \sqrt{-5}$ can be decomposed further, the number 6 has two genuinely different prime factorizations in this ring.

There is, however, a natural way of generalizing the concept of "number" to include IDEAL NUMBERS [III.81 §2] that allow one to prove a version of the fundamental theorem of arithmetic in rings such as the one just defined. First, we must reformulate: we associate with each number γ the set of all its multiples $\delta\gamma$, where δ belongs to the ring. This set, which is denoted (γ), has the following closure property: if α and β belong to (γ) and δ and ϵ are any two elements of the ring, then $\delta\alpha + \epsilon\beta$ belongs to (γ).

A subset of a ring with that closure property is called an *ideal*. If the ideal is of the form (γ) for some number γ, then it is called a *principal ideal*. However, there are ideals that are not principal, so we can think of the set of ideals as generalizing the set of elements of the original ring (once we have reformulated each element γ as the principal ideal (γ)). It turns out that there are natural notions of addition and multiplication that can be applied to ideals. Moreover, it makes sense to define an ideal I to be "prime" if the only way of writing I as a product JK is if one of J and K is a "unit." In this enlarged set, unique factorization turns out to hold. These concepts give us a very useful way to measure "the extent to which unique factorization fails" in the original ring. For more details, see ALGEBRAIC NUMBERS [IV.1 §7].

3.5 Higher Dimensions and Several Variables

We have already seen that the study of polynomial equations becomes much more complicated when one looks not just at single equations in one variable, but at systems of equations in several variables. Similarly, we have seen that PARTIAL DIFFERENTIAL EQUATIONS [I.3 §5.4], which can be thought of as differential equations involving several variables, are typically much more difficult to analyze than ordinary differential equations, that is, differential equations in just one variable. These are two notable examples of a process that has generated many of the most important problems and results in mathematics, particularly over the last century or so: the process of generalization from one variable to several variables.

Suppose one has an equation that involves three real variables, x, y, and z. It is often useful to think of

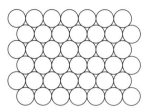

Figure 1 The densest possible packing of circles in the plane.

the triple (x, y, z) as an object in its own right, rather than as a collection of three numbers. Furthermore, this object has a natural interpretation: it represents a point in three-dimensional space. This geometrical interpretation is important, and goes a long way toward explaining why extensions of definitions and theorems from one variable to several variables are so interesting. If we generalize a piece of algebra from one variable to several variables, we can also think of what we are doing as generalizing from a one-dimensional setting to a higher-dimensional setting. This idea leads to many links between algebra and geometry, allowing techniques from one area to be used to great effect in the other.

4 Discovering Patterns

Suppose that you wish to fill the plane as densely as possible with nonoverlapping circles of radius 1. How should you do it? This question is an example of a so-called *packing problem*. The answer is known, and it is what one might expect: you should arrange the circles so that their centers form a triangular lattice, as shown in figure 1. In three dimensions a similar result is true, but much harder to prove: until recently it was a famous open problem known as the Kepler conjecture. Several mathematicians wrongly claimed to have solved it, but in 1998 a long and complicated solution, obtained with the help of a computer, was announced by Thomas Hales, and although his solution has proved very hard to check, the consensus is that it is probably correct.

Questions about packing of spheres can be asked in any number of dimensions, but they become harder and harder as the dimension increases. Indeed, it is likely that the best density for a ninety-seven-dimensional packing, say, will never be known. Experience with similar problems suggests that the best arrangement will almost certainly not have a simple structure such as one sees in two dimensions, so that the only method for finding it would be a "brute-force search" of some kind. However, to search for the best possible complicated structure is not feasible: even if one could somehow reduce the search to finitely many possibilities, there would be far more of them than one could feasibly check.

When a problem looks too difficult to solve, one should not give up completely. A much more productive reaction is to formulate related but more approachable questions. In this case, instead of trying to discover the very best packing, one can simply see how dense a packing one can find. Here is a sketch of an argument that gives a goodish packing in n dimensions, when n is large. One begins by taking a *maximal packing*: that is, one simply picks sphere after sphere until it is no longer possible to pick another one without it overlapping one of the spheres already chosen. Now let x be any point in $\mathbb{R}^n$. Then there must be a sphere in our collection such that the distance between x and its center is less than 2, since otherwise we could take a unit sphere about x and it would not overlap any of the other spheres. Therefore, if we take all the spheres in the collection and expand them by a factor of 2, then we cover all of $\mathbb{R}^n$. Since expanding an n-dimensional sphere by a factor of 2 increases its (n-dimensional) volume by a factor of 2^n, the proportion of $\mathbb{R}^n$ covered by the unexpanded spheres must be at least 2^{-n}.

Notice that in the above argument we learned nothing at all about the nature of the arrangements of spheres with density 2^{-n}. All we did was take a maximal packing, and that can be done in a very haphazard way. This is in marked contrast with the approach that worked in two dimensions, where we defined a specific pattern of circles.

This contrast pervades all of mathematics. For some problems, the best approach is to build a highly structured pattern that has the properties you need, while for others—usually problems for which there is no hope of obtaining an exact answer—it is better to look for less specific arrangements. "Highly structured" in this context often means "possessing a high degree of symmetry."

The triangular lattice is a rather simple pattern, but some highly structured patterns are much more complicated, and much more of a surprise when they are discovered. A notable example occurs in packing problems. By and large, the higher the dimension you are working in, the more difficult it is to find good patterns, but an exception to this general rule occurs at twenty-four dimensions. Here, there is a remarkable

construction, known as the *Leech lattice*, which gives rise to a miraculously dense packing. Formally, a *lattice* in $\mathbb{R}^n$ is a subset Λ with the following three properties.

(i) If x and y belong to Λ, then so do $x + y$ and $x - y$.

(ii) If x belongs to Λ, then x is *isolated*. That is, there is some $d > 0$ such that the distance between x and any other point of Λ is at least d.

(iii) Λ is not contained in any $(n - 1)$-dimensional subspace of $\mathbb{R}^n$.

A good example of a lattice is the set $\mathbb{Z}^n$ of all points in $\mathbb{R}^n$ with integer coordinates. If one is searching for a dense packing, then it is a good idea to look at lattices, since if you know that every nonzero point in a lattice has distance at least d from 0, then you know that *any* two points have distance at least d from each other. This is because the distance between x and y is the same as the distance between 0 and $y - x$, both of which lie in the lattice if x and y do. Thus, instead of having to look at the whole lattice, one can get away with looking at a small portion around 0.

In twenty-four dimensions it can be shown that there is a lattice Λ with the following additional properties, and that it is unique, in the sense that any other lattice with those properties is just a rotation of the first one.

(iv) There is a 24×24 matrix M with DETERMINANT [III.15] equal to 1 such that Λ consists of all integer combinations of the columns of M.

(v) If v is a point in Λ, then the square of the distance from 0 to v is an even integer.

(vi) The nonzero vector nearest to 0 is at distance 2. Thus, the balls of radius 1 about the points in Λ form a packing of $\mathbb{R}^{24}$.

The nonzero vector nearest to 0 is far from unique. In fact there are 196 560 of them, which is a remarkably large number considering that these points must all be at distance at least 2 from each other.

The Leech lattice also has an extraordinary degree of symmetry. To be precise, it has 8 315 553 613 086 720 000 rotational symmetries. (This number equals $2^{22} \cdot 3^9 \cdot 5^4 \cdot 7^2 \cdot 11 \cdot 13 \cdot 23$.) If you take the QUOTIENT [I.3 §3.3] of its symmetry group by the subgroup consisting of the identity and minus the identity, then you obtain the *Conway group* Co_1, which is one of the famous sporadic SIMPLE GROUPS [V.7]. The existence of so many symmetries makes it easier still to determine the smallest distance from 0 of any nonzero point of the lattice, since once you have checked one distance you have automatically checked lots of others (just as, in the triangular lattice, the six-fold rotational symmetry tells us that the distances from 0 to its six neighbors are all the same).

These facts about the Leech lattice illustrate a general principle of mathematical research: often, if a mathematical construction has one remarkable property, it will have others as well. In particular, a high degree of symmetry will often be related to other interesting features. So, although it is a surprise that the Leech lattice exists at all, it is not as surprising when one then discovers that it gives an extremely dense packing of $\mathbb{R}^{24}$. In fact, it was shown in 2004 by Henry Cohn and Abhinav Kumar that it gives the densest possible packing of spheres in twenty-four-dimensional space, at least among all packings derived from lattices. It is probably the densest packing of any kind, but this has not yet been proved.

5   Explaining Apparent Coincidences

The largest of all the sporadic finite simple groups is called the *Monster group*. Its name is partly explained by the fact that it has $2^{46} \cdot 3^{20} \cdot 5^9 \cdot 7^6 \cdot 11^2 \cdot 13^3 \cdot 17 \cdot 19 \cdot 23 \cdot 29 \cdot 31 \cdot 41 \cdot 47 \cdot 59 \cdot 71$ elements. How can one hope to understand a group of this size?

One of the best ways is to show that it is a group of symmetries of some other mathematical object (see the article on REPRESENTATION THEORY [IV.9] for much more on this theme), and the smaller that object is, the better. We have just seen that another large sporadic group, the Conway group Co_1, is closely related to the symmetry group of the Leech lattice. Might there be a lattice that played a similar role for the Monster group?

It is not hard to show that there will be at least *some* lattice that works, but more challenging is to find one of small dimension. It has been shown that the smallest possible dimension that can be used is 196 883.

Now let us turn to a different branch of mathematics. If you look at the article about ALGEBRAIC NUMBERS [IV.1 §8] you will see a definition of a function $j(z)$, called the *elliptic modular function*, of central importance in algebraic number theory. It is given as the sum of a series that starts

$$j(z) = \mathrm{e}^{-2\pi \mathrm{i} z} + 744 + 196\,884\mathrm{e}^{2\pi \mathrm{i} z}$$
$$+ 21\,493\,760\mathrm{e}^{4\pi \mathrm{i} z} + 864\,299\,970\mathrm{e}^{6\pi \mathrm{i} z} + \cdots.$$

Rather intriguingly, the coefficient of $\mathrm{e}^{2\pi \mathrm{i} z}$ in this series is 196 884, one more than the smallest possible dimension of a lattice that has the Monster group as its group of symmetries.

It is not obvious how seriously we should take this observation, and when it was first made by John McKay opinions differed about it. Some believed that it was probably just a coincidence, since the two areas seemed to be so different and unconnected. Others took the attitude that the function $j(z)$ and the Monster group are so important in their respective areas, and the number 196 883 so large, that the surprising numerical fact was probably pointing to a deep connection that had not yet been uncovered.

It turned out that the second view was correct. After studying the coefficients in the series for $j(z)$, McKay and John Thompson were led to a conjecture that related them all (and not just 196 884) to the Monster group. This conjecture was extended by John Conway and Simon Norton, who formulated the "Monstrous Moonshine" conjecture, which was eventually proved by Richard Borcherds in 1992. (The word "moonshine" reflects the initial disbelief that there would be a serious relationship between the Monster group and the j-function.)

In order to prove the conjecture, Borcherds introduced a new algebraic structure, which he called a VERTEX ALGEBRA [IV.17]. And to analyze vertex algebras, he used results from STRING THEORY [IV.17 §2]. In other words, he explained the connection between two very different-looking areas of pure mathematics with the help of concepts from theoretical physics.

This example demonstrates in an extreme way another general principle of mathematical research: if you can obtain the same series of numbers (or the same structure of a more general kind) from two different mathematical sources, then those sources are probably not as different as they seem. Moreover, if you can find one deep connection, you will probably be led to others. There are many other examples where two completely different calculations give the same answer, and many of them remain unexplained. This phenomenon results in some of the most difficult and fascinating unsolved problems in mathematics. (See the introduction to MIRROR SYMMETRY [IV.16] for another example.)

Interestingly, the j-function leads to a second famous mathematical "coincidence." There may not seem to be anything special about the number $\mathrm{e}^{\pi\sqrt{163}}$, but here is the beginning of its decimal expansion:

$$\mathrm{e}^{\pi\sqrt{163}}$$
$$= 262\,537\,412\,640\,768\,743.99999999999925\ldots,$$

which is astonishingly close to an integer. Again it is initially tempting to dismiss this as a coincidence, but one should think twice before yielding to the temptation. After all, there are not all that many numbers that can be defined as simply as $e^{\pi\sqrt{163}}$, and each one has a probability of less than one in a million million of being as close to an integer as $e^{\pi\sqrt{163}}$ is. In fact, it is not a coincidence at all: for an explanation see ALGEBRAIC NUMBERS [IV.1 §8].

6 Counting and Measuring

How many rotational symmetries are there of a regular icosahedron? Here is one way to work it out. Choose a vertex v of the icosahedron and let v' be one of its neighbors. An icosahedron has twelve vertices, so there are twelve places where v could end up after the rotation. Once we know where v goes, there are five possibilities for v' (since each vertex has five neighbors and v' must still be a neighbor of v after the rotation). Once we have determined where v and v' go, there is no further choice we can make, so the number of rotational symmetries is $5 \times 12 = 60$.

This is a simple example of a *counting argument*, that is, an answer to a question that begins "How many." However, the word "argument" is at least as important as the word "counting," since we do not put all the symmetries in a row and say "one, two, three, . . . , sixty," as we might if we were counting in real life. What we do instead is come up with a reason for the number of rotational symmetries being 5×12. At the end of the process, we understand more about those symmetries than merely how many there are. Indeed, it is possible to go further and show that the group of rotations of the icosahedron is A_5, the ALTERNATING GROUP [III.68] on five elements.

6.1 Exact Counting

Here is a more sophisticated counting problem. A *one-dimensional random walk* of n steps is a sequence of integers $a_0, a_1, a_2, \ldots, a_n$, such that for each i the difference $a_i - a_{i-1}$ is either 1 or -1. For example, $0, 1, 2, 1, 2, 1, 0, -1$ is a seven-step random walk. The number of n-step random walks that start at 0 is clearly 2^n, since there are two choices for each step (either you add 1 or you subtract 1).

Now let us try a slightly harder problem. How many walks of length $2n$ are there that start *and* end at 0? (We look at walks of length $2n$ since a walk that starts

and ends in the same place must have an even number of steps.)

In order to think about this problem, it helps to use the letters R and L (for "right" and "left") to denote adding 1 and subtracting 1, respectively. This gives us an alternative notation for random walks that start at 0: for example, the walk $0, 1, 2, 1, 2, 1, 0, -1$ would be rewritten as RRLRLLL. Now a walk will end at 0 if and only if the number of Rs is equal to the number of Ls. Moreover, if we are told the set of steps where an R occurs, then we know the entire walk. So what we are counting is the number of ways of choosing n of the $2n$ steps as the steps where an R will occur. And this is well-known to be $(2n)!/(n!)^2$.

Now let us look at a related quantity that is considerably less easy to determine: the number $W(n)$ of walks of length $2n$ that start and end at 0 and are never negative. Here, in the notation introduced for the previous problem, is a list of all such walks of length 6: RRRLLL, RRLRLL, RRLLRL, RLRRLL, and RLRLRL.

Now three of these five walks do not just start and end at 0 but visit it in the middle: RRLLRL visits it after four steps, RLRRLL after two, and RLRLRL after two and four. Suppose we have a walk of length $2n$ that is never negative and visits 0 for the first time after $2k$ steps. Then the remainder of the walk is a walk of length $2(n-k)$ that starts and ends at 0 and is never negative. There are $W(n-k)$ of these. As for the first $2k$ steps of such a walk, they must begin with R and end with L, and in between must never visit 0. This means that between the initial R and the final L they give a walk of length $2(k-1)$ that starts and ends at 1 and is never less than 1. The number of such walks is clearly the same as $W(k-1)$. Therefore, since the first visit to 0 must take place after $2k$ steps for some k between 1 and n, W satisfies the following slightly complicated recurrence relation:

$$W(n) = W(0)W(n-1) + \cdots + W(n-1)W(0).$$

Here, $W(0)$ is taken to be equal to 1.

This allows us to calculate the first few values of W. We have $W(1) = W(0)W(0) = 1$, which is easier to see directly: the only possibility is RL. Then $W(2) = W(1)W(0) + W(0)W(1) = 2$, and $W(3)$, which counts the number of such walks of length 6, equals $W(0)W(2) + W(1)W(1) + W(2)W(0) = 5$, confirming our earlier calculation.

Of course, it would not be a good idea to use the recurrence relation directly if one wished to work out $W(n)$ for large values of n such as 10^{10}. However, the recurrence is of a sufficiently nice form that it is amenable to treatment by GENERATING FUNCTIONS [IV.18 §§2.4, 3], as is explained in ENUMERATIVE AND ALGEBRAIC COMBINATORICS [IV.18 §3]. (To see the connection with that discussion, replace the letters R and L by the square brackets [and], respectively. A legal bracketing then corresponds to a walk that is never negative.)

The argument above gives an efficient way of calculating $W(n)$ exactly. There are many other exact counting arguments in mathematics. Here is a small further sample of quantities that mathematicians know how to count exactly without resorting to "brute force." (See the introduction to [IV.18] for a discussion of when one regards a counting problem as solved.)

(i) The number $r(n)$ of regions that a plane is cut into by n lines if no two of the lines are parallel and no three concurrent. The first four values of $r(n)$ are 2, 4, 7, and 11. It is not hard to prove that $r(n) = r(n-1) + n$, which leads to the formula $r(n) = \frac{1}{2}(n^2 + n + 2)$. This statement, and its proof, can be generalized to higher dimensions.

(ii) The number $s(n)$ of ways of writing n as a sum of four squares. Here we allow zero and negative numbers and we count different orderings as different (so, for example, $1^2 + 3^2 + 4^2 + 2^2$, $3^2 + 4^2 + 1^2 + 2^2$, $1^2 + (-3)^2 + 4^2 + 2^2$, and $0^2 + 1^2 + 2^2 + 5^2$ are considered to be four different ways of writing 30 as a sum of four squares). It can be shown that $s(n)$ is equal to 8 times the sum of all the divisors of n that are not multiples of 4. For example, the divisors of 12 are 1, 2, 3, 4, 6, and 12, of which 1, 2, 3, and 6 are not multiples of 4. Therefore $s(12) = 8(1 + 2 + 3 + 6) = 96$. The different ways are $1^2 + 1^2 + 1^2 + 3^2$, $0 + 2^2 + 2^2 + 2^2$, and the other expressions that can be obtained from these ones by reordering and replacing positive integers by negative ones.

(iii) The number of lines in space that meet a given four lines L_1, L_2, L_3, and L_4 when those four are in "general position." (This means that they do not have special properties such as two of them being parallel or intersecting each other.) It turns out that for any *three* such lines, there is a subset of $\mathbb{R}^3$ known as a *quadric surface* that contains them, and this quadric surface is unique. Let us take the surface for L_1, L_2, and L_3 and call it S.

The surface S has some interesting properties that allow us to solve the problem. The main one is that one can find a continuous family of lines (that is, a collection of lines $L(t)$, one for each real number t, that varies continuously with t) that, between them, make up the surface S and include each of the lines L_1, L_2, and L_3. But there is also *another* such continuous family of lines $M(s)$, each of which meets every line $L(t)$ in exactly one point. In particular, every line $M(s)$ meets all of L_1, L_2, and L_3, and in fact every line that meets all of L_1, L_2, and L_3 must be one of the lines $M(s)$.

It can be shown that L_4 intersects the surface S in exactly two points, P and Q. Now P lies in some line $M(s)$ from the second family, and Q lies in some other line $M(s')$ (which must be different, or else L_4 would equal $M(s)$ and intersect L_1, L_2, and L_3, contradicting the fact that the lines L_i are in general position). Therefore, the two lines $M(s)$ and $M(s')$ intersect all four of the lines L_i. But every line that meets all the L_i has to be one of the lines $M(s)$ and has to go through either P or Q (since the lines $M(s)$ lie in S and L_4 meets S at only those two points). Therefore, the answer is 2.

This question can be generalized very considerably, and answered by means of a technique known as *Schubert calculus*.

(iv) The number $p(n)$ of ways of expressing a positive integer n as a sum of positive integers. When $n = 6$ this number is 11, since $6 = 1 + 1 + 1 + 1 + 1 + 1 = 2 + 1 + 1 + 1 + 1 = 2 + 2 + 1 + 1 = 2 + 2 + 2 = 3 + 1 + 1 + 1 = 3 + 2 + 1 = 3 + 3 = 4 + 1 + 1 = 4 + 2 = 5 + 1 = 6$. The function $p(n)$ is called the *partition function*. A remarkable formula, due to HARDY [VI.73] and RAMANUJAN [VI.82], gives an approximation $\alpha(n)$ to $p(n)$ that is so accurate that $p(n)$ is always the nearest integer to $\alpha(n)$.

6.2 Estimates

Once we have seen example (ii) above, it is natural to ask whether it can be generalized. Is there a formula for the number $t(n)$ of ways of writing n as a sum of ten sixth powers, for example? It is generally believed that the answer to this question is no, and certainly no such formula has been discovered. However, as with packing problems, even if an exact answer does not seem to be forthcoming, it is still very interesting to obtain *estimates*. In this case, one can try to define an easily calculated function f such that $f(n)$ is always *approximately* equal to $t(n)$. If even that is too hard, one can try to find *two* easily calculated functions L and U such that $L(n) \leqslant t(n) \leqslant U(n)$ for every n. If we succeed, then we call L a *lower bound* for t and U an *upper bound*. Here are a few examples of quantities that nobody knows how to count exactly, but for which there are interesting approximations, or at least interesting upper and lower bounds.

(i) Probably the most famous approximate counting problem in all of mathematics is to estimate $\pi(n)$, the number of prime numbers less than or equal to n. For small values of n, we can of course compute $\pi(n)$ exactly: for example, $\pi(20) = 8$ since the primes less than or equal to 20 are 2, 3, 5, 7, 11, 13, 17, and 19. However, there does not seem to be a useful formula for $\pi(n)$, and although it is easy to think of a brute-force algorithm for computing $\pi(n)$—look at every number up to n, test whether it is prime, and keep count as you go along—such a procedure takes a prohibitively long time if n is at all large. Furthermore, it does not give us much insight into the nature of the function $\pi(n)$.

If, however, we modify the question slightly, and ask *roughly* how many primes there are up to n, then we find ourselves in the area known as ANALYTIC NUMBER THEORY [IV.2], a branch of mathematics with many fascinating results. In particular, the famous PRIME NUMBER THEOREM [V.26], proved by HADAMARD [VI.65] and DE LA VALLÉE POUSSIN [VI.67] at the end of the nineteenth century, states that $\pi(n)$ is approximately equal to $n/\log n$, in the sense that the ratio of $\pi(n)$ to $n/\log n$ converges to 1 as n tends to infinity.

This statement can be refined. It is believed that the "density" of primes close to n is about $1/\log n$, in the sense that a randomly chosen integer close to n has a probability of about $1/\log n$ of being prime. This would suggest that $\pi(n)$ should be about $\int_0^n \mathrm{d}t/\log t$, a function of n that is known as the *logarithmic integral* of n, or $\mathrm{li}(n)$.

How accurate is this estimate? Nobody knows, but the RIEMANN HYPOTHESIS [V.26], perhaps the most famous unsolved problem in mathematics, is equivalent to the statement that $\pi(n)$ and $\mathrm{li}(n)$ differ by at most $c\sqrt{n}\log n$ for some constant c. Since $\sqrt{n}\log n$ is much smaller than $\pi(n)$, this would tell us that $\mathrm{li}(n)$ was an extremely good approximation to $\pi(n)$.

(ii) A *self-avoiding walk* of length n in the plane is a sequence of points $(a_0, b_0), (a_1, b_1), (a_2, b_2), \ldots,$ (a_n, b_n) with the following properties.

- The numbers a_i and b_i are all integers.
- For each i, one obtains (a_i, b_i) from (a_{i-1}, b_{i-1}) by taking a horizontal or vertical step of length 1. That is, either $a_i = a_{i-1}$ and $b_i = b_{i-1} \pm 1$ or $a_i = a_{i-1} \pm 1$ and $b_i = b_{i-1}$.
- No two of the points (a_i, b_i) are equal.

The first two conditions tell us that the sequence forms a two-dimensional walk of length n, and the third says that this walk never visits any point more than once—hence the term "self-avoiding."

Let $S(n)$ be the number of self-avoiding walks of length n that start at $(0, 0)$. There is no known formula for $S(n)$, and it is very unlikely that such a formula exists. However, quite a lot is known about the way the function $S(n)$ grows as n grows. For instance, it is fairly easy to prove that $S(n)^{1/n}$ converges to a limit c. The value of c is not known, but it has been shown (with the help of a computer) to lie between 2.62 and 2.68.

(iii) Let $C(t)$ be the number of points in the plane with integer coordinates contained in a circle of radius t about the origin. That is, $C(t)$ is the number of pairs (a, b) of integers such that $a^2 + b^2 \leqslant t^2$. A circle of radius t has area πt^2, and the plane can be tiled by unit squares, each of which has a point with integer coordinates at its center. Therefore, when t is large it is fairly clear (and not hard to prove) that $C(t)$ is approximately πt^2. However, it is much less clear how good this approximation is.

To make this question more precise, let us set $\epsilon(t)$ to equal $|C(t) - \pi t^2|$. That is, $\epsilon(t)$ is the *error* in πt^2 as an estimate for $C(t)$. It was shown in 1915, by Hardy and Landau, that $\epsilon(t)$ must be at least $c\sqrt{t}$ for some constant $c > 0$, and this estimate, or something very similar, probably gives the right order of magnitude for $\epsilon(t)$. However, the best upper bound known, which was proved by Huxley in 2003 (the latest in a long line of successive improvements), is that $\epsilon(t)$ is at most $At^{131/208}(\log t)^{2.26}$ for some constant A.

6.3   Averages

So far, our discussion of estimates and approximations has been confined to problems where the aim is to count mathematical objects of a given kind. However, that is by no means the only context in which estimates can be interesting. Given a set of objects, one may wish to know not just how large the set is, but also what a typical object in the set looks like. Many questions of this kind take the form of asking what the average value is of some numerical parameter that is associated with each object. Here are two examples.

(i) What is the average distance between the starting point and the endpoint of a self-avoiding walk of length n? In this instance, the objects are self-avoiding walks of length n that start at $(0, 0)$, and the numerical parameter is the end-to-end distance.

Surprisingly, this is a notoriously difficult problem, and almost nothing is known. It is obvious that n is an upper bound for $S(n)$, but one would expect a typical self-avoiding walk to take many twists and turns and end up traveling much less far than n away from its starting point. However, there is no known upper bound for $S(n)$ that is substantially better than n.

In the other direction, one would expect the end-to-end distance of a typical self-avoiding walk to be greater than that of an ordinary walk, to give it room to avoid itself. This would suggest that $S(n)$ is significantly greater than $\sqrt{n}$, but it has not even been proved that it *is* greater.

This is not the whole story, however, and the problem will be discussed further in section 8.

(ii) Let n be a large randomly chosen positive integer and let $\omega(n)$ be the number of distinct prime factors of n. On average, how large will $\omega(n)$ be? As it stands, this question does not quite make sense because there are infinitely many positive integers, so one cannot choose one randomly. However, one can make the question precise by specifying a large integer m and choosing a random integer n between m and $2m$. It then turns out that the average size of $\omega(n)$ is around $\log \log n$.

In fact, much more is known than this. If all you know about a RANDOM VARIABLE [III.71 §4] is its average, then a great deal of its behavior is not determined, so for many problems calculating averages is just the beginning of the story. In this case, Hardy and Ramanujan gave an estimate for the STANDARD DEVIATION [III.71 §4] of $\omega(n)$, showing that it is about $\sqrt{\log \log n}$. Then Erdős and Kac went even further and gave a precise estimate for the probability that $\omega(n)$ differs from $\log \log n$ by more than $c\sqrt{\log \log n}$, proving the surprising fact that the distribution of ω is approximately GAUSSIAN [III.71 §5].

To put these results in perspective, let us think about the range of possible values of $\omega(n)$. At one extreme, n might be a prime itself, in which case it obviously has just one prime factor. At the other extreme, we can write the primes in ascending order as $p_1, p_2, p_3, \ldots$ and take numbers of the form $n = p_1 p_2 \cdots p_k$. With the help of the prime number theorem, one can show that the order of magnitude of k is $\log n / \log \log n$, which is much bigger than $\log \log n$. However, the results above tell us that such numbers are exceptional: a typical number has a few distinct prime factors, but nothing like as many as $\log n / \log \log n$.

6.4 Extremal Problems

There are many problems in mathematics where one wishes to maximize or minimize some quantity in the presence of various constraints. These are called *extremal problems*. As with counting questions, there are some extremal problems for which one can realistically hope to work out the answer exactly, and many more for which, even though an exact answer is out of the question, one can still aim to find interesting estimates. Here are some examples of both kinds.

(i) Let n be a positive integer and let X be a set with n elements. How many subsets of X can be chosen if none of these subsets is contained in any other?

A simple observation one can make is that if two different sets have the same size, then neither is contained in the other. Therefore, one way of satisfying the constraints of the problem is to choose all the sets of some particular size k. Now the number of subsets of X of size k is $n!/k!(n-k)!$, which is usually written $\binom{n}{k}$ (or $^{n}C_k$), and it is not hard to show that $\binom{n}{k}$ is largest when $k = n/2$ if n is even and when $k = (n \pm 1)/2$ if n is odd. For simplicity let us concentrate on the case when n is even. What we have just proved is that it is possible to pick $\binom{n}{n/2}$ subsets of an n-element set in such a way that none of them contains any other. That is, $\binom{n}{n/2}$ is a lower bound for the problem. A result known as *Sperner's theorem* states that it is an upper bound as well. That is, if you choose *more* than $\binom{n}{n/2}$ subsets of X, then, however you do it, one of these subsets will be contained in another. Therefore, the question is answered exactly, and the answer is $\binom{n}{n/2}$. (When n is odd, then the answer is $\binom{n}{(n+1)/2}$, as one might now expect.)

(ii) Suppose that the two ends of a heavy chain are attached to two hooks on the ceiling and that the chain is not supported anywhere else. What shape will the hanging chain take?

At first, this question does not look like a maximization or minimization problem, but it can be quickly turned into one. That is because a general principle from physics tells us that the chain will settle in the shape that minimizes its potential energy. We therefore find ourselves asking a new question: let A and B be two points at distance d apart, and let $\mathcal{C}$ be the set of all curves of length l that have A and B as their two endpoints. Which curve $C \in \mathcal{C}$ has the smallest potential energy? Here one takes the mass of any portion of the curve to be proportional to its length. The potential energy of the curve is equal to mgh, where m is the mass of the curve, g is the gravitational constant, and h is the height of the center of gravity of the curve. Since m and g do not change, another formulation of the question is: which curve $C \in \mathcal{C}$ has the smallest average height?

This problem can be solved by means of a technique known as *the calculus of variations*. Very roughly, the idea is this. We have a set, $\mathcal{C}$, and a function h defined on $\mathcal{C}$ that takes each curve $C \in \mathcal{C}$ to its average height. We are trying to minimize h, and a natural way to approach that task is to define some sort of derivative and look for a curve C at which this derivative is 0. Notice that the word "derivative" here does *not* refer to the rate of change of height as you move along the curve. Rather, it means the (linear) way that the average height of the entire curve changes in response to small perturbations of the curve. Using this kind of derivative to find a minimum is more complicated than looking for the stationary points of a function defined on $\mathbb{R}$, since $\mathcal{C}$ is an infinite-dimensional set and is therefore much more complicated than $\mathbb{R}$. However, the approach can be made to work, and the curve that minimizes the average height is known. (It is called a *catenary*, after the Latin word for chain.) Thus, this is another minimization problem that has been answered exactly.

For a typical problem in the calculus of variations, one is trying to find a curve, or surface, or more general kind of function, for which a certain quantity is minimized or maximized. If a minimum or maximum exists (which is by no means automatic when one is working with an infinite-dimensional set, so this can be an interesting and important question), the object that achieves it satisfies a system of PARTIAL DIFFERENTIAL EQUATIONS [I.3 §5.4] known as the *Euler–Lagrange equations*. For more about this style of minimization or maximization, see VARIATIONAL METHODS [III.94] (and also OPTIMIZATION AND LAGRANGE MULTIPLIERS [III.64]).

(iii) How many numbers can you choose between 1 and n if no three of them are allowed to lie in an arithmetic progression? If $n = 9$ then the answer is 5. To see this, note first that no three of the five numbers $1, 2, 4, 8, 9$ lie in an arithmetic progression. Now let us see if we can find six numbers that work.

If we make one of our numbers 5, then we must leave out either 4 or 6, or else we would have the progression $4, 5, 6$. Similarly, we must leave out one of 3 and 7, one of 2 and 8, and one of 1 and 9. But then we have left out four numbers. It follows that we cannot choose 5 as one of the numbers.

We must leave out one of 1, 2, and 3, and one of 7, 8, and 9, so if we leave out 5 then we must include 4 and 6. But then we cannot include 2 or 8. But we must also

leave out at least one of 1, 4, and 7, so we are forced to leave out at least four numbers.

An ugly case-by-case argument of this kind is feasible when $n = 9$, but as soon as n is at all large there are far too many cases for it to be possible to consider them all. For this problem, there does not seem to be a tidy answer that tells us exactly which is the largest set of integers between 1 and n that contains no arithmetic progression of length 3. So instead one looks for upper and lower bounds on its size. To prove a lower bound, one must find a good way of constructing a large set that does not contain any arithmetic progressions, and to prove an upper bound one must show that *any* set of a certain size must necessarily contain an arithmetic progression. The best bounds to date are very far apart. In 1947, Behrend found a set of size $n/e^{c\sqrt{\log n}}$ that contains no arithmetic progression, and in 1999 Jean Bourgain proved that every set of size $Cn\sqrt{\log \log n / \log n}$ contains an arithmetic progression. (If it is not obvious to you that these numbers are far apart, then consider what happens when $n = 10^{100}$, say. Then $e^{\sqrt{\log n}}$ is about $4\,000\,000$, while $\sqrt{\log n / \log \log n}$ is about 6.5.)

(iv) Theoretical computer science is a source of many minimization problems: if one is programming a computer to perform a certain task, then one wants it to do so in as short a time as possible. Here is an elementary-sounding example: how many steps are needed to multiply two n-digit numbers together?

Even if one is not too precise about what is meant by a "step," one can see that the traditional method, long multiplication, takes at least n^2 steps since, during the course of the calculation, each digit of the first number is multiplied by each digit of the second. One might imagine that this was necessary, but in fact there are clever ways of transforming the problem and dramatically reducing the time that a computer needs to perform a multiplication of this kind. The fastest known method uses THE FAST FOURIER TRANSFORM [III.26] to reduce the number of steps from n^2 to $Cn \log n \log \log n$. Since the logarithm of a number is much smaller than the number itself, one thinks of $Cn \log n \log \log n$ as being only just worse than a bound of the form Cn. Bounds of this form are called *linear*, and for a problem like this are clearly the best one can hope for, since it takes $2n$ steps even to read the digits of the two numbers.

Another question that is similar in spirit is whether there are fast algorithms for matrix multiplication. To multiply two $n \times n$ matrices using the obvious method one needs to do n^3 individual multiplications of the numbers in the matrices, but once again there are less obvious methods that do better. The main breakthrough on this problem was due to Strassen, who had the idea of splitting each matrix into four $n/2 \times n/2$ matrices and multiplying those together. At first it seems as though one has to calculate the products of eight pairs of $n/2 \times n/2$ matrices, but these products are related, and Strassen came up with *seven* such calculations from which the eight products could quickly be derived. One can then apply *recursion*: that is, use the same idea to speed up the calculation of the seven $n/2 \times n/2$ matrix products, and so on.

Strassen's algorithm reduces the number of numerical multiplications from about n^3 to about $n^{\log_2 7}$. Since $\log_2 7$ is less than 2.81, this is a significant improvement, but only when n is large. His basic divide-and-conquer strategy has been developed further, and the current record is better than $n^{2.4}$. In the other direction, the situation is less satisfactory: nobody has found a proof that one needs to use significantly more than n^2 multiplications.

For more problems of a similar kind, see COMPUTATIONAL COMPLEXITY [IV.20] and THE MATHEMATICS OF ALGORITHM DESIGN [VII.5].

(v) Some minimization and maximization problems are of a more subtle kind. For example, suppose that one is trying to understand the nature of the differences between successive primes. The smallest such difference is 1 (the difference between 2 and 3), and it is not hard to prove that there is no largest difference (given any integer n greater than 1, none of the numbers between $n! + 2$ and $n! + n$ is a prime). Therefore, there do not seem to be interesting maximization or minimization problems concerning these differences.

However, one can in fact formulate some fascinating problems if one first *normalizes* in an appropriate way. As was mentioned earlier in this section, the prime number theorem states that the density of primes near n is about $1/\log n$, so an average gap between two primes near n will be about $\log n$. If p and q are successive primes, we can therefore define a "normalized gap" to be $(q - p)/\log p$. The average value of this normalized gap will be 1, but is it sometimes much smaller and sometimes much bigger?

It was shown by Westzynthius in 1931 that even normalized gaps can be arbitrarily large, and it was widely believed that they could also be arbitrarily close to zero. (The famous twin prime conjecture—that there are infinitely many primes p for which $p + 2$ is also

a prime—implies this immediately.) However, it took until 2005 for this to be proved, by Goldston, Pintz, and Yıldırım. (See ANALYTIC NUMBER THEORY [IV.2 §§6–8] for a discussion of this problem.)

7   Determining Whether Different Mathematical Properties Are Compatible

In order to understand a mathematical concept, such as that of a group or a manifold, there are various stages one typically goes through. Obviously it is a good idea to begin by becoming familiar with a few representative examples of the structure, and also with techniques for building new examples out of old ones. It is also extremely important to understand the homomorphisms, or "structure-preserving functions," from one example of the structure to another, as was discussed in SOME FUNDAMENTAL MATHEMATICAL DEFINITIONS [I.3 §§4.1, 4.2].

Once one knows these basics, what is there left to understand? Well, for a general theory to be useful, it should tell us something about specific examples. For instance, as we saw in section 3.2, Lagrange's theorem can be used to prove Fermat's little theorem. Lagrange's theorem is a general fact about groups: that if G is a group of size n, then the size of any subgroup of G must be a factor of n. To obtain Fermat's little theorem, one applies Lagrange's theorem to the particular case when G is the multiplicative group of nonzero integers mod p. The conclusion one obtains—that a^p is always congruent to a—is far from obvious.

However, what if we want to know something about a group G that might not be true for all groups? That is, suppose that we wish to determine whether G has some property P that some groups have and others do not. Since we cannot prove that the property P follows from the group axioms, it might seem that we are forced to abandon the general theory of groups and look at the specific group G. However, in many situations there is an intermediate possibility: to identify some fairly general property Q that the group G has, and show that Q implies the more particular property P that interests us.

Here is an illustration of this sort of technique in a different context. Suppose we wish to determine whether the polynomial $p(x) = x^4 - 2x^3 - x^2 - 2x + 1$ has a real root. One method would be to study this particular polynomial and try to find a root. After quite a lot of effort we might discover that $p(x)$ can be factorized as $(x^2 + x + 1)(x^2 - 3x + 1)$. The first factor is always positive, but if we apply the quadratic formula to the second, we find that $p(x) = 0$ when $x = (3 \pm \sqrt{5})/2$. An alternative method, which uses a bit of general theory, is to notice that $p(1)$ is negative (in fact, it equals -3) and that $p(x)$ is large when x is large (because then the x^4 term is far bigger than anything else), and then to use the *intermediate value theorem*, the result that any continuous function that is negative somewhere and positive somewhere else must be zero somewhere in between.

Notice that, with the second approach, there was still some computation to do—finding a value of x for which $p(x)$ is negative—but that it was much easier than the computation in the first approach—finding a value of x for which $p(x)$ is zero. In the second approach, we established that p had the rather general property of *being negative somewhere*, and used the intermediate value theorem to finish off the argument.

There are many situations like this throughout mathematics, and as they arise certain general properties become established as particularly useful. For example, if you know that a positive integer n is prime, or that a group G is Abelian (that is, $gh = hg$ for any two elements g and h of G), or that a function taking complex numbers to complex numbers is HOLOMORPHIC [I.3 §5.6], then as a consequence of these general properties you know a lot more about the objects in question.

Once properties have established themselves as important, they give rise to a large class of mathematical questions of the following form: given a mathematical structure and a selection of interesting properties that it might have, which combinations of these properties imply which other ones? Not all such questions are interesting, of course—many of them turn out to be quite easy and others are too artificial—but some of them are very natural and surprisingly resistant to one's initial attempts to solve them. This is usually a sign that one has stumbled on what mathematicians would call a "deep" question. In the rest of this section let us look at a problem of this kind.

A group G is called *finitely generated* if there is some finite set $\{x_1, x_2, \ldots, x_k\}$ of elements of G such that all the rest can be written as products of elements in that set. For example, the group $SL_2(\mathbb{Z})$ consists of all 2×2 matrices $\left(\begin{smallmatrix} a & b \\ c & d \end{smallmatrix}\right)$ such that a, b, c, and d are integers and $ad - bc = 1$. This group is finitely generated: it is a nice exercise to show that every such matrix can be built from the four matrices $\left(\begin{smallmatrix} 1 & 1 \\ 0 & 1 \end{smallmatrix}\right)$, $\left(\begin{smallmatrix} 1 & -1 \\ 0 & 1 \end{smallmatrix}\right)$, $\left(\begin{smallmatrix} 1 & 0 \\ 1 & 1 \end{smallmatrix}\right)$, and $\left(\begin{smallmatrix} 1 & 0 \\ -1 & 1 \end{smallmatrix}\right)$ using matrix multiplication. (See [I.3 §3.2] for a

discussion of matrices. A first step toward proving this result is to show that $\left(\begin{smallmatrix} 1 & m \\ 0 & 1 \end{smallmatrix}\right)\left(\begin{smallmatrix} 1 & n \\ 0 & 1 \end{smallmatrix}\right) = \left(\begin{smallmatrix} 1 & m+n \\ 0 & 1 \end{smallmatrix}\right)$.)

Now let us consider a second property. If x is an element of a group G, then x is said to have *finite order* if there is some power of x that equals the identity. The smallest such power is called the *order of x*. For example, in the multiplicative group of nonzero integers mod 7, the identity is 1, and the order of the element 4 is 3, because $4^1 = 4$, $4^2 = 16 \equiv 2$ and $4^3 = 64 \equiv 1$ mod 7. As for 3, its first six powers are 3, 2, 6, 4, 5, 1, so it has order 6. Now some groups have the very special property that there is some integer n such that x^n equals the identity for every x—or, equivalently, the order of every x is a factor of n. What can we say about such groups?

Let us look first at the case where all elements have order 2. Writing e for the identity element, we are assuming that $a^2 = e$ for every element a. If we multiply both sides of this equation by the inverse a^{-1}, then we deduce that $a = a^{-1}$. The opposite implication is equally easy, so such groups are ones where every element is its own inverse.

Now let a and b be two elements of G. For any two elements a and b of any group we have the identity $(ab)^{-1} = b^{-1}a^{-1}$ (simply because $abb^{-1}a^{-1} = aa^{-1} = e$), and in our special group where all elements equal their inverses we can deduce from this that $ab = ba$. That is, G is automatically Abelian.

Already we have shown that one general property, that every element of G squares to the identity, implies another, that G is Abelian. Now let us add the condition that G is finitely generated, and let $x_1, x_2, \ldots, x_k$ be a *minimal* set of generators. That is, suppose that every element of G can be built up out of the x_i and that we need all of the x_i to be able to do this. Because G is Abelian and because every element is equal to its own inverse, we can rearrange products of the x_i into a *standard form*, where each x_i occurs at most once and the indices increase. For example, take the product $x_4x_3x_1x_4x_4x_1x_3x_1x_5$. Because G is Abelian, this equals $x_1x_1x_1x_3x_3x_4x_4x_4x_5$, and because each element is its own inverse this equals $x_1x_4x_5$, the standard form of the original expression.

This shows that G can have at most 2^k elements, since for each x_i we have the choice of whether or not to include it in the product (after it has been put in the form above). In particular, the properties "G is finitely generated" and "every nonidentity element of G has order 2" imply the third property "G is finite." It turns out to be fairly easy to prove that two elements whose standard forms are different are themselves different, so in fact G has exactly 2^k elements (where k is the size of a minimal set of generators).

Now let us ask what happens if n is some integer greater than 2 and $x^n = e$ for every element x. That is, if G is finitely generated and $x^n = e$ for every x, must G be finite? This turns out to be a much harder question, originally asked by BURNSIDE [VI.60]. Burnside himself showed that G must be finite if $n = 3$, but it was not until 1968 that his problem was solved, when Adian and Novikov proved the remarkable result that if $n \geqslant 4381$ then G does *not* have to be finite. There is of course a big gap between 3 and 4381, and progress in bridging it has been slow. It was only in 1992 that this was improved to $n \geqslant 13$, by Ivanov. And to give an idea of how hard the Burnside problem is, it is still not known whether a group with two generators such that the fifth power of every element is the identity must be finite.

8 Working with Arguments That Are Not Fully Rigorous

A mathematical statement is considered to be established when it has a proof that meets the high standards of rigor that are characteristic of the subject. However, nonrigorous arguments have an important place in mathematics as well. For example, if one wishes to apply a mathematical statement to another field, such as physics or engineering, then the truth of the statement is often more important than whether one has proved it.

However, this raises an obvious question: if one has not proved a statement, then what grounds could there be for believing it? There are in fact several different kinds of nonrigorous justification, so let us look at some of them.

8.1 Conditional Results

As was mentioned earlier in this article, the Riemann hypothesis is the most famous unsolved problem in mathematics. Why is it considered so important? Why, for example, is it considered more important than the twin prime conjecture, another problem to do with the behavior of the sequence of primes?

The main reason, though not the only one, is that it and its generalizations have a huge number of interesting consequences. In broad terms, the Riemann hypothesis tells us that the appearance of a certain degree of "randomness" in the sequence of primes is not misleading: in many respects, the primes really do behave like an appropriately chosen random set of integers.

If the primes behave in a random way, then one might imagine that they would be hard to analyze, but in fact randomness can be an advantage. For example, it is randomness that allows me to be confident that at least one girl was born in London on every day of the twentieth century. If the sex of babies were less random, I would be less sure: there could be some strange pattern such as girls being born on Mondays to Thursdays and boys on Fridays to Sundays. Similarly, if I know that the primes behave like a random sequence, then I know a great deal about their average behavior in the long term. The Riemann hypothesis and its generalizations formulate in a precise way the idea that the primes, and other important sequences that arise in number theory, "behave randomly." That is why they have so many consequences. There are large numbers of papers with theorems that are proved only under the assumption of some version of the Riemann hypothesis. Therefore, anybody who proves the Riemann hypothesis will change the status of all these theorems from conditional to fully proved.

How should one regard a proof if it relies on the Riemann hypothesis? One could simply say that the proof establishes that such and such a result is implied by the Riemann hypothesis and leave it at that. But most mathematicians take a different attitude. They believe the Riemann hypothesis, and believe that it will one day be proved. So they believe all its consequences as well, even if they feel more secure about results that can be proved unconditionally.

Another example of a statement that is generally believed and used as a foundation for a great deal of further research comes from theoretical computer science. As was mentioned in section 6.4 (iv), one of the main aims of computer science is to establish how quickly certain tasks can be performed by a computer. This aim splits into two parts: finding algorithms that work in as few steps as possible, and proving that every algorithm must take at least some particular number of steps. The second of these tasks is notoriously difficult: the best results known are far weaker than what is believed to be true.

There is, however, a class of computational problems, called *NP-complete* problems, that are known to be of *equivalent* difficulty. That is, if there were an efficient algorithm for one of these problems, then it could be converted into an efficient algorithm for any other.

However, largely for this very reason it is almost universally believed that there is in fact no efficient algorithm for any of the problems, or, as it is usually expressed, that "P does not equal NP." Therefore, if you want to demonstrate that no quick algorithm exists for some problem, all you have to do is prove that it is at least as hard as some problem that is already known to be NP-complete. This will not be a rigorous proof, but it will be a convincing demonstration, since most mathematicians are convinced that P does not equal NP. (See COMPUTATIONAL COMPLEXITY [IV.20] for much more on this topic.)

Some areas of research depend on several conjectures rather than just one. It is as though researchers in such areas have discovered a beautiful mathematical landscape and are impatient to map it out despite the fact that there is a great deal that they do not understand. And this is often a very good research strategy, even from the perspective of finding rigorous proofs. There is far more to a conjecture than simply a wild guess: for it to be accepted as important, it should have been subjected to tests of many kinds. For example, does it have consequences that are already known to be true? Are there special cases that one can prove? If it were true, would it help one solve other problems? Is it supported by numerical evidence? Does it make a bold, precise statement that would probably be easy to refute if it were false? It requires great insight and hard work to produce a conjecture that passes all these tests, but if one succeeds, one has not just an isolated statement, but a statement with numerous connections to other statements. This increases the chances that it will be proved, and greatly increases the chances that the proof of one statement will lead to proofs of others as well. Even a *counterexample* to a good conjecture can be extraordinarily revealing: if the conjecture is related to many other statements, then the effects of the counterexample will permeate the whole area.

One area that is full of conjectural statements is ALGEBRAIC NUMBER THEORY [IV.1]. In particular, the Langlands program is a collection of conjectures, due to Robert Langlands, that relate number theory to representation theory (it is discussed in REPRESENTATION THEORY [IV.9 §6]). Between them, these conjectures generalize, unify, and explain large numbers of other conjectures and results. For example, the Shimura–Taniyama–Weil conjecture, which was central to Andrew Wiles's proof of FERMAT'S LAST THEOREM [V.10], forms one small part of the Langlands program.

The Langlands program passes the tests for a good conjecture supremely well, and has for many years guided the research of a large number of mathematicians.

Another area of a similar nature is known as MIRROR SYMMETRY [IV.16]. This is a sort of DUALITY [III.19] that relates objects known as CALABI-YAU MANIFOLDS [III.6], which arise in ALGEBRAIC GEOMETRY [IV.4] and also in STRING THEORY [IV.17 §2], to other, dual manifolds. Just as certain differential equations can become much easier to solve if one looks at the FOURIER TRANSFORMS [III.27] of the functions in question, so there are calculations arising in string theory that look impossible until one transforms them into equivalent calculations in the dual, or "mirror," situation. There is at present no rigorous justification for the transformation, but this process has led to complicated formulas that nobody could possibly have guessed, and some of these formulas have been rigorously proved in other ways. Maxim Kontsevich has proposed a precise conjecture that would explain the apparent successes of mirror symmetry.

8.2   Numerical Evidence

The GOLDBACH CONJECTURE [V.27] states that every even number greater than or equal to 4 is the sum of two primes. It seems to be well beyond what anybody could hope to prove with today's mathematical machinery, even if one is prepared to accept statements such as the Riemann hypothesis. And yet it is regarded as almost certainly true.

There are two principal reasons for believing Goldbach's conjecture. The first is a reason we have already met: one would expect it to be true if the primes are "randomly distributed." This is because if n is a large even number, then there are many ways of writing $n = a + b$, and there are enough primes for one to expect that from time to time both a and b would be prime.

Such an argument leaves open the possibility that for some value of n that is not too large one might be unlucky, and it might just happen that $n - a$ was composite whenever a was prime. This is where numerical evidence comes in. It has now been checked that every even number up to 10^{14} can be written as a sum of two primes, and once n is greater than this, it becomes extremely unlikely that it could "just happen," by a fluke, to be a counterexample.

This is perhaps rather a crude argument, but there is a way to make it even more convincing. If one makes more precise the idea that the primes appear to be randomly distributed, one can formulate a stronger version of Goldbach's conjecture that says not only that every even number can be written as a sum or two primes, but also roughly how many ways there are of doing this. For instance, if a and $n - a$ are both prime, then neither is a multiple of 3 (unless one of them is equal to 3 itself). If n is a multiple of 3, then this merely says that a is not a multiple of 3, but if n is of the form $3m + 1$ then a cannot be of the form $3k + 1$ either (or $n - a$ would be a multiple of 3). So, in a certain sense, it is twice as easy for n to be a sum of two primes if it is a multiple of 3. Taking this kind of information into account, one can estimate in how many ways it "ought" to be possible to write n as a sum of two primes. It turns out that, for every even n, there should be many such representations. Moreover, one's predictions of *how* many are closely matched by the numerical evidence: that is, they are true for values of n that are small enough to be checked on a computer. This makes the numerical evidence much more convincing, since it is evidence not just for Goldbach's conjecture itself, but also for the more general principles that led us to believe it.

This illustrates a general phenomenon: the more precise the predictions that follow from a conjecture, the more impressive it is when they are confirmed by later numerical evidence. Of course, this is true not just of mathematics but of science more generally.

8.3 "Illegal" Calculations

In section 6.3 it was stated that "almost nothing is known" about the average end-to-end distance of an n-step self-avoiding walk. That is a statement with which theoretical physicists would strongly disagree. Instead, they would tell you that the end-to-end distance of a typical n-step self-avoiding walk is somewhere in the region of $n^{3/4}$. This apparent disagreement is explained by the fact that, although almost nothing has been rigorously proved, physicists have a collection of nonrigorous methods that, if used carefully, seem to give correct results. With their methods, they have in some areas managed to establish statements that go well beyond what mathematicians can prove. Such results are fascinating to mathematicians, partly because if one regards the results of physicists as mathematical conjectures then many of them are excellent conjectures, by the standards explained earlier: they are deep, completely unguessable in advance, widely believed to

be true, backed up by numerical evidence, and so on. Another reason for their fascination is that the effort to provide them with a rigorous underpinning often leads to significant advances in pure mathematics.

To give an idea of what the nonrigorous calculations of physicists can be like, here is a rough description of a famous argument of Pierre-Gilles de Gennes, which lies behind some of the results (or predictions, if you prefer to call them that) of physicists. In statistical physics there is a model known as the *n-vector model*, closely related to the Ising and Potts models described in PROBABILISTIC MODELS OF CRITICAL PHENOMENA [IV.25]. At each point of $\mathbb{Z}^d$ one places a unit vector in $\mathbb{R}^n$. This gives rise to a random configuration of unit vectors, with which one associates an "energy" that increases as the angles between neighboring vectors increase. De Gennes found a way of transforming the self-avoiding-walk problem so that it could be regarded as a question about the n-vector model in the case $n = 0$. The 0-vector problem itself does not make obvious sense, since there is no such thing as a unit vector in $\mathbb{R}^0$, but de Gennes was nevertheless able to take parameters associated with the n-vector model and show that if you let n converge to zero then you obtained parameters associated with self-avoiding walks. He proceeded to choose other parameters in the n-vector model to derive information about self-avoiding walks, such as the expected end-to-end distance.

To a pure mathematician, there is something very worrying about this approach. The formulas that arise in the n-vector model do not make sense when $n = 0$, so instead one has to regard them as limiting values when n tends to zero. But n is very clearly a positive integer in the n-vector model, so how can one say that it tends to zero? Is there some way of defining an n-vector model for more general n? Perhaps, but nobody has found one. And yet de Gennes's argument, like many other arguments of a similar kind, leads to remarkably precise predictions that agree with numerical evidence. There must be a good reason for this, even if we do not understand what it is.

The examples in this section are just a few illustrations of how mathematics is enriched by nonrigorous arguments. Such arguments allow one to penetrate much further into the mathematical unknown, opening up whole areas of research into phenomena that would otherwise have gone unnoticed. Given this, one might wonder whether rigor is important: if the results established by nonrigorous arguments are clearly true, then is that not good enough? As it happens, there are examples of statements that were "established" by nonrigorous methods and later shown to be false, but the most important reason for caring about rigor is that the understanding one gains from a rigorous proof is frequently deeper than the understanding provided by a nonrigorous one. The best way to describe the situation is perhaps to say that the two styles of argument have profoundly benefited each other and will undoubtedly continue to do so.

9 Finding Explicit Proofs and Algorithms

There is no doubt that the equation $x^5 - x - 13 = 0$ has a solution. After all, if we set $f(x) = x^5 - x - 13$, then $f(1) = -13$ and $f(2) = 17$, so somewhere between 1 and 2 there will be an x for which $f(x) = 0$.

That is an example of a *pure existence argument*—in other words, an argument that establishes that something exists (in this case, a solution to a certain equation), without telling us how to find it. If the equation had been $x^2 - x - 13 = 0$, then we could have used an argument of a very different sort: the formula for quadratic equations tells us that there are precisely two solutions, and it even tells us what they are (they are $(1 + \sqrt{53})/2$ and $(1 - \sqrt{53})/2$). However, there is no similar formula for quintic equations. (See THE INSOLUBILITY OF THE QUINTIC [V.21].)

These two arguments illustrate a fundamental dichotomy in mathematics. If you are proving that a mathematical object exists, then sometimes you can do so *explicitly*, by actually describing that object, and sometimes you can do so only *indirectly*, by showing that its nonexistence would lead to a contradiction.

There is also a spectrum of possibilities in between. As it was presented, the argument above showed merely that the equation $x^5 - x - 13 = 0$ has a solution between 1 and 2, but it also suggests a method for calculating that solution to any desired accuracy. If, for example, you want to know it to two decimal places, then run through the numbers $1, 1.01, 1.02, \ldots, 1.99, 2$ evaluating f at each one. You will find that $f(1.71)$ is approximately -0.0889 and that $f(1.72)$ is approximately 0.3337, so there must be a solution between the two (which the calculations suggest will be closer to 1.71 than to 1.72). And in fact there are much better ways, such as NEWTON'S METHOD [II.4 §2.3], of approximating solutions. For many purposes, a pretty formula for a solution is less important than a method of calculating or approximating it. (See NUMERICAL ANALYSIS

[IV.21 §1] for a further discussion of this point.) And if one has a method, its usefulness depends very much on whether it works quickly.

Thus, at one end of the spectrum one has simple formulas that define mathematical objects and can easily be used to find them, at the other one has proofs that establish existence but give no further information, and in between one has proofs that yield algorithms for finding the objects, algorithms that are significantly more useful if they run quickly.

Just as, all else being equal, a rigorous argument is preferable to a nonrigorous one, so an explicit or algorithmic argument is worth looking for even if an indirect one is already established, and for similar reasons: the effort to find an explicit argument very often leads to new mathematical insights. (Less obviously, as we shall soon see, finding *indirect* arguments can also lead to new insights.)

One of the most famous examples of a pure existence argument concerns TRANSCENDENTAL NUMBERS [III.41], which are real numbers that are not roots of any polynomial with integer coefficients. The first person to prove that such numbers existed was LIOUVILLE [VI.39], in 1844. He proved that a certain condition was sufficient to guarantee that a number was transcendental and demonstrated that it is easy to construct numbers satisfying his condition (see LIOUVILLE'S THEOREM AND ROTH'S THEOREM [V.22]). After that, various important numbers such as e and π were proved to be transcendental, but these proofs were difficult. Even now there are many numbers that are almost certainly transcendental but which have not been proved to be transcendental. (See IRRATIONAL AND TRANSCENDENTAL NUMBERS [III.41] for more information about this.)

All the proofs mentioned above were direct and explicit. Then in 1873 CANTOR [VI.54] provided a completely different proof of the existence of transcendental numbers, using his theory of COUNTABILITY [III.11]. He proved that the algebraic numbers were countable and the real numbers uncountable. Since countable sets are far smaller than uncountable sets, this showed that almost every real number (though not necessarily almost every real number you will actually meet) is transcendental.

In this instance, each of the two arguments tells us something that the other does not. Cantor's proof shows that there are transcendental numbers, but it does not provide us with a single example. (Strictly speaking, this is not true: one could specify a way of listing the algebraic numbers and then apply Cantor's famous diagonal argument to that particular list. However, the resulting number would be virtually devoid of meaning.) Liouville's proof is much better in that way, as it gives us a method of constructing several transcendental numbers with fairly straightforward definitions. However, if one knew only the explicit arguments such as Liouville's and the proofs that e and π are transcendental, then one might have the impression that transcendental numbers are numbers of a very special kind. The insight that is completely missing from these arguments, but present in Cantor's proof, is that a *typical* real number is transcendental.

For much of the twentieth century, highly abstract and indirect proofs were fashionable, but in more recent years, especially with the advent of the computer, attitudes have changed. (Of course, this is a very general statement about the entire mathematical community rather than about any single mathematician.) Nowadays, more attention is often paid to the question of whether a proof is explicit, and, if so, whether it leads to an efficient algorithm.

Needless to say, algorithms are interesting in themselves, and not just for the light they shed on mathematical proofs. Let us conclude this section with a brief description of a particularly interesting algorithm that has been developed by several authors over the last few years. It gives a way of computing the volume of a high-dimensional convex body.

A shape K is called *convex* if, given any two points x and y in K, the line segment joining x to y lies entirely inside K. For example, a square or a triangle is convex, but a five-pointed star is not. This concept can be generalized straightforwardly to n dimensions, for any n, as can the notions of area and volume.

Now let us suppose that an n-dimensional convex body K is specified for us in the following sense: we have a computer program that runs quickly and tells us, for each point $(x_1, \ldots, x_n)$, whether or not that point belongs to K. How can we estimate the volume of K? One of the most powerful methods for problems like this is *statistical*: you choose points at random and see whether they belong to K, basing your estimate of the volume of K on the frequency with which they do. For example, if you wanted to estimate π, you could take a circle of radius 1, enclose it in a square of side-length 2, and choose a large number of points randomly from the square. Each point has a probability $\pi/4$ (the ratio of the area π of the circle to the area 4 of the square)

of belonging to the circle, so we can estimate π by taking the proportion of points that fall in the circle and multiplying it by 4.

This approach works quite easily for very low dimensions but as soon as n is at all large it runs into a severe difficulty. Suppose for example that we were to try to use the same method for estimating the volume of an n-dimensional sphere. We would enclose that sphere in an n-dimensional cube, choose points at random in the cube, and see how often they belonged to the sphere as well. However, the ratio of the volume of an n-dimensional sphere to that of an n-dimensional cube that contains it is exponentially small, which means that the number of points you have to pick before even one of them lands in the sphere is exponentially large. Therefore, the method becomes hopelessly impractical.

All is not lost, though, because there is a trick for getting around this difficulty. You define a sequence of convex bodies, $K_0, K_1, \ldots, K_m$, each contained in the next, starting with the convex body whose volume you want to know, and ending with the cube, in such a way that the volume of K_i is always at least half that of K_{i+1}. Then for each i you estimate the ratio of the volumes of K_{i-1} and K_i. The product of all these ratios will be the ratio of the volume of K_0 to that of K_m. Since you know the volume of K_m, this tells you the volume of K_0.

How do you estimate the ratio of the volumes of K_{i-1} and K_i? You simply choose points at random from K_i and see how many of them belong to K_{i-1}. However, it is just here that the true subtlety of the problem arises: how do you choose points at random from a convex body K_i that you do not know much about? Choosing a random point in the n-dimensional cube is easy, since all you need to do is independently choose n random numbers $x_1, \ldots, x_n$, each between -1 and 1. But for a general convex body it is not easy at all.

There is a wonderfully clever idea that gets around this problem. It is to design carefully a random walk that starts somewhere inside the convex body and at each step moves to another point, chosen at random from just a few possibilities. The more random steps of this kind that are taken, the less can be said about where the point is, and if the walk is defined properly, it can be shown that after not too many steps, the point reached is almost purely random. However, the proof is not at all easy. (It is discussed further in HIGH-DIMENSIONAL GEOMETRY AND ITS PROBABILISTIC ANALOGUES [IV.26 §6].)

For further discussion of algorithms and their mathematical importance, see ALGORITHMS [II.4], COMPUTATIONAL NUMBER THEORY [IV.3], COMPUTATIONAL COMPLEXITY [IV.20], and THE MATHEMATICS OF ALGORITHM DESIGN [VII.5].

10 What Do You Find in a Mathematical Paper?

Mathematical papers have a very distinctive style, one that became established early in the twentieth century. This final section is a description of what mathematicians actually produce when they write.

A typical paper is usually a mixture of formal and informal writing. Ideally (but by no means always), the author writes a readable introduction that tells the reader what to expect from the rest of the paper. And if the paper is divided into sections, as most papers are unless they are quite short, then it is also very helpful to the reader if each section can begin with an informal outline of the arguments to follow. But the main substance of the paper has to be more formal and detailed, so that readers who are prepared to make a sufficient effort can convince themselves that it is correct.

The object of a typical paper is to establish *mathematical statements*. Sometimes this is an end in itself: for example, the justification for the paper may be that it proves a conjecture that has been open for twenty years. Sometimes the mathematical statements are established in the service of a wider aim, such as helping to explain a mathematical phenomenon that is poorly understood. But either way, mathematical statements are the main currency of mathematics.

The most important of these statements are usually called theorems, but one also finds statements called propositions, lemmas, and corollaries. One cannot always draw sharp distinctions between these kinds of statements, but in broad terms this is what the different words mean. A *theorem* is a statement that you regard as intrinsically interesting, a statement that you might think of isolating from the paper and telling other mathematicians about in a seminar, for instance. The statements that are the main goals of a paper are usually called theorems. A *proposition* is a bit like a theorem, but it tends to be slightly "boring." It may seem odd to want to prove boring results, but they can be important and useful. What makes them boring is that they do not surprise us in any way. They are statements that we need, that we expect to be true, and that we do not have much difficulty proving.

Here is a quick example of a statement that one might choose to call a proposition. The ASSOCIATIVE LAW FOR A BINARY OPERATION [I.2 §2.4] "$*$" states that $x * (y * z) = (x * y) * z$. One often describes this law informally by saying that "brackets do not matter." However, while it shows that we can write $x * y * z$ without fear of ambiguity, it does not show quite so obviously that we can write $a * b * c * d * e$, for example. How do we know that, just because the positions of brackets do not matter when you have three objects, they do not matter when you have more than three?

Many mathematics students go happily through university without noticing that this is a problem. It just seems obvious that the associative law shows that brackets do not matter. And they are basically right: although it is not completely obvious, it is certainly not a surprise and turns out to be easy to prove. Since we often need this simple result and could hardly call it a theorem, we might call it a proposition instead. To get a feel for how to prove it, you might wish to show that the associative law implies that

$$(a * ((b * c) * d)) * e = a * (b * ((c * d) * e)).$$

Then you can try to generalize what it is you are doing.

Often, if you are trying to prove a theorem, the proof becomes long and complicated, in which case if you want anybody to read it you need to make the structure of the argument as clear as possible. One of the best ways of doing this is to identify *subgoals*, which take the form of statements intermediate between your initial assumptions and the conclusion you wish to draw from them. These statements are usually called *lemmas*. Suppose, for example, that you are trying to give a very detailed presentation of the standard proof that $\sqrt{2}$ is irrational. One of the facts you will need is that every fraction p/q is equal to a fraction r/s with r and s not both even, and this fact requires a proof. For the sake of clarity, you might well decide to isolate this proof from the main proof and call the fact a lemma. Then you have split your task into two separate tasks: proving the lemma, and proving the main theorem using the lemma. One can draw a parallel with computer programming: if you are writing a complicated program, it is good practice to divide your main task into subtasks and write separate mini-programs for them, which you can then treat as "black boxes," to be called upon by other parts of the program whenever they are useful.

Some lemmas are difficult to prove and are useful in many different contexts, so the most important lemmas can be more important than the least important theorems. However, a general rule is that a result will be called a lemma if the main reason for proving it is in order to use it as a stepping stone toward the proofs of other results.

A *corollary* of a mathematical statement is another statement that follows easily from it. Sometimes the main theorem of a paper is followed by several corollaries, which advertise the strength of the theorem. Sometimes the main theorem itself is labeled a corollary, because all the work of the proof goes into proving a different, less punchy statement from which the theorem follows very easily. If this happens, the author may wish to make clear that the corollary is the main result of the paper, and other authors would refer to it as a theorem.

A mathematical statement is established by means of a *proof*. It is a remarkable feature of mathematics that proofs are possible: that, for example, an argument invented by EUCLID [VI.2] over two thousand years ago can still be accepted today and regarded as a completely convincing demonstration. It took until the late nineteenth and early twentieth centuries for this phenomenon to be properly understood, when the language of mathematics was *formalized* (see THE LANGUAGE AND GRAMMAR OF MATHEMATICS [I.2], and especially section 4, for an idea of what this means). Then it became possible to make precise the notion of a proof as well. From a logician's point of view a proof is a sequence of mathematical statements, each written in a formal language, with the following properties: the first few statements are the initial assumptions, or *premises*; each remaining statement in the sequence follows from earlier ones by means of logical rules that are so simple that the deductions are clearly valid (for instance rules such as "if $P \wedge Q$ is true then P is true," where "$\wedge$" is the logical symbol for "and"); and the final statement in the sequence is the statement that is to be proved.

The above idea of a proof is a considerable idealization of what actually appears in a normal mathematical paper under the heading "Proof." That is because a purely formal proof would be very long and almost impossible to read. And yet, the fact that arguments can in principle be formalized provides a very valuable underpinning for the edifice of mathematics, because it gives a way of resolving disputes. If a mathematician produces an argument that is strangely unconvincing, then the best way to see whether it is correct is to ask him or her to explain it more formally and in greater detail. This will usually either expose a mistake or make it clearer why the argument works.

Another very important component of mathematical papers is *definitions*. This book is full of them: see in particular part III. Some definitions are given simply because they enable one to speak more concisely. For example, if I am proving a result about triangles and I keep needing to consider the distances between the vertices and the opposite sides, then it is a nuisance to have to say "the distances from A, B, and C to the lines BC, AC, and AB, respectively," so instead I will probably choose a word like "altitude" and write, "Given a vertex of a triangle, define its *altitude* to be the distance from that vertex to the opposite side." If I am looking at triangles with obtuse angles, then I will have to be more careful: "Given a vertex A of a triangle ABC, define its *altitude* to be the distance from A to the unique line that passes through B and C." From then on, I can use the word "altitude" and the exposition of my proof will be much more crisp.

Definitions like this are mere definitions of convenience. When the need arises, it is pretty obvious what to do and one does it. But the really interesting definitions are ones that are far from obvious and that make you think in new ways once you know them. A very good example is the definition of the derivative of a function. If you do not know this definition, you will have no idea how to find out for which nonnegative x the function $f(x) = 2x^3 - 3x^2 - 6x + 1$ takes its smallest value. If you do know it, then the problem becomes a simple exercise. That is perhaps an exaggeration, since you also need to know that the minimum will occur either at 0 or at a point where the derivative vanishes, and you will need to know how to differentiate $f(x)$, but these are simple facts—propositions rather than theorems—and the real breakthrough is the concept itself.

There are many other examples of definitions like this, but interestingly they are more common in some branches of mathematics than in others. Some mathematicians will tell you that the main aim of their research is to find the right definition, after which their whole area will be illuminated. Yes, they will have to write proofs, but if the definition is the one they are looking for, then these proofs will be fairly straightforward. And yes, there will be problems they can solve with the help of the new definition, but, like the minimization problem above, these will not be central to the theory. Rather, they will demonstrate the power of the definition. For other mathematicians, the main purpose of definitions is to prove theorems, but even very theorem-oriented mathematicians will from time to time find that a good definition can have a major effect on their problem-solving prowess.

This brings us to mathematical problems. The main aim of an article in mathematics is usually to prove theorems, but one of the reasons for *reading* an article is to advance one's own research. It is therefore very welcome if a theorem is proved by a technique that can be used in other contexts. It is also very welcome if an article contains some good unsolved problems. By way of illustration, let us look at a problem that most mathematicians would not take all that seriously, and try to see what it lacks.

A number is called *palindromic* if its representation in base 10 is a palindrome: some simple examples are 22, 131, and 548 845. Of these, 131 is interesting because it is also a prime. Let us try to find some more prime palindromic numbers. Single-digit primes are of course palindromic, and two-digit palindromic numbers are multiples of 11, so only 11 itself is also a prime. So let us move quickly on to three-digit numbers. Here there turn out to be several examples: 101, 131, 151, 181, 191, 313, 353, 373, 383, 727, 757, 787, 797, 919, and 929. It is not hard to show that every palindromic number with an even number of digits is a multiple of 11, but the palindromic primes do not stop at 929—for example, 10 301 is the next smallest.

And now anybody with a modicum of mathematical curiosity will ask the question: are there infinitely many palindromic primes? This, it turns out, is an unsolved problem. It is believed (on the combined grounds that the primes should be sufficiently random and that palindromic numbers with an odd number of digits do not seem to have any particular reason to be factorizable) that there are, but nobody knows how to prove it.

This problem has the great virtue of being easy to understand, which makes it appealing in the way that FERMAT'S LAST THEOREM [V.10] and GOLDBACH'S CONJECTURE [V.27] are appealing. And yet, it is not a central problem in the way that those two are: most mathematicians would put it into a mental box marked "recreational" and forget about it.

What is the reason for this dismissive attitude? Are the primes not central objects of study in mathematics? Well, yes they are, but palindromic numbers are not. And the main reason they are not is that the definition of "palindromic" is extremely unnatural. If you know that a number is palindromic, what you know is less a feature of the number itself and more a feature of the particular way that, for accidental historical reasons,

we choose to represent it. In particular, the property depends on our choice of the number 10 as our base. For example, if we write 131 in base 3, then it becomes 11212, which is no longer the same when written backwards. By contrast, a prime number is prime however you write it.

Though persuasive, this is not quite a complete explanation, since there could conceivably be interesting properties that involved the number 10, or at least some artificial choice of number, in an essential way. For example, the problem of whether there are infinitely many primes of the form $2^n - 1$ *is* considered interesting, despite the use of the particular number 2. However, the choice of 2 can be justified here: $a^n - 1$ has a factor $a - 1$, so for any larger integer the answer would be no. Moreover, numbers of the form $2^n - 1$ have special properties that make them more likely to be prime. (See COMPUTATIONAL NUMBER THEORY [IV.3] for an explanation of this point.)

But even if we replace 10 by the "more natural" number 2 and look at numbers that are palindromic when written in binary, we still do not obtain a property that would be considered a serious topic for research. Suppose that, given an integer n, we define $r(n)$ to be the *reverse* of n—that is, the number obtained if you write n in binary and then reverse its digits. Then a palindromic number, in the binary sense, is a number n such that $n = r(n)$. But the function $r(n)$ is very strange and "unmathematical." For instance, the reverses of the numbers from 1 to 20 are 1, 1, 3, 1, 5, 3, 7, 1, 9, 5, 13, 3, 11, 7, 15, 1, 17, 9, 25, and 5, which gives us a sequence with no obvious pattern. Indeed, when one calculates this sequence, one realizes that it is even more artificial than it at first seemed. One might imagine that the reverse of the reverse of a number is the number itself, but that is not so. If you take the number 10, for example, it is 1010 in binary, so its reverse is 0101, which is the number 5. But this we would normally write as 101, so the reverse of 5 is not 10 but 5. But we cannot solve this problem by deciding to write 5 as 0101, since then we would have the problem that 5 was no longer palindromic, when it clearly ought to be.

Does this mean that nobody would be interested in a proof that there were infinitely many palindromic primes? Not at all. It can be shown quite easily that the number of palindromic numbers less than n is in the region of $\sqrt{n}$, which is a very small fraction indeed. It is notoriously hard to prove results about primes in sparse sets like this, so a solution to this conjecture would be a big breakthrough. However, the definition of "palindromic" is so artificial that there seems to be no way of using it in a detailed way in a mathematical proof. The only realistic hope of solving this problem would be to prove a much more general result, of which this would be just one of many consequences. Such a result would be wonderful, and undeniably interesting, but you will not discover it by thinking about palindromic numbers. Instead, you would be better off either trying to formulate a more general question, or else looking at a more natural problem of a similar kind. An example of the latter is this: are there infinitely many primes of the form $m^2 + 1$ for some positive integer m?

Perhaps the most important feature of a good problem is generality: the solution to a good problem should usually have ramifications beyond the problem itself. A more accurate word for this desirable quality is "generalizability," since some excellent problems may look rather specific. For example, the statement that $\sqrt{2}$ is irrational looks as though it is about just one number,

but once you know how to prove it, you will have no difficulty in proving that $\sqrt{3}$ is irrational as well, and in fact the proof can be generalized to a much wider class of numbers (see ALGEBRAIC NUMBERS [IV.1 §14]). It is quite common for a good problem to look uninteresting until you start to think about it. Then you realize that it has been asked for a reason: it might be the "first difficult case" of a more general problem, or it might be just one well-chosen example of a cluster of problems, all of which appear to run up against the same difficulty.

Sometimes a problem is just a question, but frequently the person who asks a mathematical question has a good idea of what the answer is. A *conjecture* is a mathematical statement that the author firmly believes but cannot prove. As with problems, some conjectures are better than others: as we have already discussed in section 8.1, the very best conjectures can have a major effect on the direction of mathematical research.

Part II
The Origins of Modern Mathematics

II.1 From Numbers to Number Systems
Fernando Q. Gouvêa

People have been writing numbers down for as long as they have been writing. In every civilization that has developed a way of recording information, we also find a way of recording numbers. Some scholars even argue that numbers came first.

It is fairly clear that numbers first arose as adjectives: they specified how many or how much of something there was. Thus, it was possible to talk about three apricots, say, long before it was possible to talk about the number 3. But once the concept of "threeness" is on the table, so that the same adjective specifies three fish and three horses, and once a written symbol such as "3" is developed that can be used in all of those instances, the conditions exist for 3 itself to emerge as an independent entity. Once it does, we are doing mathematics.

This process seems to have repeated itself many times when new kinds of numbers have been introduced: first a number is used, then it is represented symbolically, and finally it comes to be conceived as a thing in itself and as part of a system of similar entities.

1 Numbers in Early Mathematics

The earliest mathematical documents we know about go back to the civilizations of the ancient Middle East, in Egypt and in Mesopotamia. In both cultures, a scribal class developed. Scribes were responsible for keeping records, which often required them to do arithmetic and solve simple mathematical problems. Most of the mathematical documents we have from those cultures seem to have been created for the use of young scribes learning their craft. Many of them are collections of problems, provided with either answers or brief solutions: twenty-five problems about digging trenches in one tablet, twelve problems requiring the solution of a linear equation in another, problems about squares and their sides in a third.

Numbers were used both for counting and for measuring, so a need for fractional numbers must have come up fairly early. Fractions are complicated to write down, and computing with them can be difficult. Hence, the problem of "broken numbers" may well have been the first really challenging mathematical problem. How does one write down fractions? The Egyptians and the Mesopotamians came up with strikingly different answers, both of which are also quite different from the way we write them today.

In Egypt (and later in Greece and much of the Mediterranean world), the fundamental notion was "the nth part," as in "the third part of six is two." In this language, one would express the idea of dividing 7 by 3 as, "What is the third part of seven?" The answer is, "Two and the third." The process was complicated by an additional restriction: one never recorded a final result using more than one of the same kind of part. Thus, the number we would want to express as "two fifth parts" would have to be given as "the third and the fifteenth."

In Mesopotamia, we find a very different idea, which may have arisen to allow easy conversion between different kinds of units. First of all, the Babylonians had a way to generate symbols for all the numbers from 1 to 59. For larger numbers, they used a positional system much like the one we use today, but based on 60 rather than 10. So something like 1, 20 means one sixty and twenty units, that is, $1 \times 60 + 20 = 80$. The same system was then extended to fractions, so that one half was represented as thirty sixtieths. It is convenient to mark the beginning of the fractional part with a semicolon, though this and the comma are a modern convention that has no counterpart in the original texts. Then, for example, 1;24,36 means $1 + \frac{24}{60} + \frac{36}{60^2}$, which we would more usually write as $\frac{141}{100}$, or 1.41. The Mesopotamian way of writing numbers is called a *sexagesimal place-value system* by analogy with the system we use today, which is, of course, a *decimal* place-value system.

Neither of these systems is really equipped to deal well with complicated numbers. In Mesopotamia, for example, only *finite* sexagesimal expressions were employed, so the scribes were not able to write down an exact value for the reciprocal of 7 because there is no finite sexagesimal expression for $\frac{1}{7}$. In practice, this meant that to divide by 7 required finding an approximate answer. The Egyptian "parts" system, on the other hand, can represent any positive rational number, but doing so may require a sequence of denominators that to our eyes looks very complicated. One of the surviving papyri includes problems that look *designed* to produce just such complicated answers. One of these answers is "14, the 4th, the 56th, the 97th, the 194th, the 388th, the 679th, the 776th," which in modern notation is the fraction $14\frac{28}{97}$. It seems that the joy of computation for its own sake became well-established very early in the development of mathematics.

Mediterranean civilizations preserved both of these systems for a while. Most everyday numbers were specified using the system of "parts." On the other hand, astronomy and navigation required more precision, so the sexagesimal system was used in those fields. This included measuring time and angles. The fact that we still divide an hour into sixty minutes and a minute into sixty seconds goes back, via the Greek astronomers, to the Babylonian sexagesimal fractions; almost four thousand years later, we are still influenced by the Babylonian scribes.

2 Lengths Are Not Numbers

Things get more complicated with the mathematics of classical Greek and Hellenistic civilizations. The Greeks, of course, are famous for coming up with the first mathematical proofs. They were the first to attempt to do mathematics in a rigorously deductive way, using clear initial assumptions and careful statements. This, perhaps, is what led them to be very careful about numbers and their relations to other magnitudes.

Sometime before the fourth century B.C.E., the Greeks made the fundamental discovery of "incommensurable magnitudes." That is, they discovered that it is not always possible to express two given lengths as (integer) multiples of a third length. It is not just that lengths

and numbers are conceptually distinct things (though this was important too). The Greeks had found a *proof* that one cannot use numbers to represent lengths.

Suppose, they argued, you have two line segments. If their lengths are both given by numbers, then those numbers will at worst involve some fractions. By changing the unit of length, then, we can make sure that both of the lengths correspond to whole numbers. In other words, it must be possible to choose a unit length so that each of our segments consists of a whole number multiple of the unit. The two segments, then, could be "measured together," i.e., would be "commensurable."

Now here's the catch: the Greeks could *prove* that this was not always the case. Their standard example had to do with the side and the diagonal of a square. We do not know exactly how they first established that these two segments are not commensurable, but it might have been something like this: if you subtract the side from the diagonal, you will get a segment shorter than either of them; if both side and diagonal are measured by a common unit, then so is the difference. Now repeat the argument: take the remainder and subtract it from the side until we get a second remainder smaller than the first (it can be subtracted twice, in fact). The second remainder will also be measured by the common unit. It turns out to be quite easy to show that *this process will never terminate; instead, it will produce smaller and smaller remainder segments.* Eventually, the remainder segment will be smaller than the unit that supposedly measures it a whole number of times. That is impossible (no whole number is smaller than 1, after all), and hence we can conclude that the common unit does not, in fact, exist.

Of course, the diagonal does in fact have a length. Today, we would say that if the length of the side is one unit, then the length of the diagonal is $\sqrt{2}$ units, and we would interpret this argument as showing that the number $\sqrt{2}$ is not a fraction. The Greeks did not quite see in what sense $\sqrt{2}$ could be a number. Instead, it was a length, or, even better, the ratio between the length of the diagonal and the length of the side. Similar arguments could be applied to other lengths; for example, they knew that the side of a square of area 1 and a square of area 10 are incommensurable.

The conclusion, then, is that lengths are not numbers: instead, they are some other kind of magnitude. But now we are faced with a proliferation of magnitudes: numbers, lengths, areas, angles, volumes, etc. Each of these must be taken as a different kind of quantity, not comparable with the others.

This is a problem for geometry, particularly if we want to measure things. The Greeks solved this problem by relying heavily on the notion of a *ratio*. Two quantities of the same type have a ratio, and this ratio was allowed to be equal to the ratio of two quantities of another type: equality of two ratios was defined using Eudoxus's theory of proportion, the latter being one of the most important and deep ideas of Greek geometry. So, for example, rather than talking about a number called π, which to them would not be a number at all, they would say that "the ratio of the circle to the square on its radius is the same as the ratio of the circumference to the diameter." Notice that one of the two ratios is between two areas, the other between two lengths. The number π itself had no name in Greek mathematics, but the Greeks did compare it with ratios between numbers: ARCHIMEDES [VI.3] showed that it was just a little bit less than the ratio of 22 to 7 and just a little bit more than the ratio of 223 to 71.

Doing things this way seems ungainly to us, but it worked very well. Furthermore, it is philosophically satisfying to conceive of a great variety of magnitudes organized into various kinds (segments, angles, surfaces, etc.). Magnitudes of the same kind can be related to one another by ratios, and ratios can be compared with each other because they are relations perceived by our minds. In fact, the word for ratio, both in Greek and in Latin, is the same as the word for "reason" or "explanation" (*logos* in Greek, *ratio* in Latin). From the beginning, "irrational" (*alogos* in Greek) could mean both "without a ratio" and "unreasonable."

Inevitably, this austere theoretical system was somewhat disconnected from the everyday needs of people who needed to measure things such as lengths and angles. Astronomers kept right on using sexagesimal approximations, as did mapmakers and other scientists. There was some "leakage" of course: in the first century C.E., Heron of Alexandria wrote a book that reads like an attempt to apply the theoreticians' discoveries to practical measurement. It is to him, for example, that we owe the recommendation to use $\frac{22}{7}$ as an approximation for π. (Presumably, he chose Archimedes' upper bound because it was the simpler number.) In theoretical mathematics, however, the distinction between numbers and other kinds of magnitudes remained firm.

The history of numbers in the West over the fifteen hundred years that followed the classical Greek period can be seen as having two main themes: first, the Greek compartmentalization between different kinds of quantities was slowly demolished; second, in order to do this the notion of number had to be generalized over and over again.

3   Decimal Place Value

Our system for representing whole numbers goes back, ultimately, to the mathematicians of the Indian subcontinent. Sometime before (probably well before) the fifth century C.E., they created nine symbols to designate the numbers from one to nine and used the position of these symbols to indicate their actual value. So a 3 in the units position meant three, and a 3 in the tens position meant three tens, i.e., thirty. This, of course, is what we still do; the symbols themselves have changed, but not the principle. At about the same time, a place marker was developed to indicate an unoccupied space; this eventually evolved into our zero.

Indian astronomy made extensive use of sines, which are almost never whole numbers. To represent these, a Babylonian-style sexagesimal system was used, with each "sexagesimal unit" being represented using the decimal system. So "thirty-three and a quarter" might be represented as 33 15′, i.e., 33 units and 15 "minutes" (sixtieths).

Decimal place-value numeration was passed on from India to the Islamic world fairly early. In the ninth century C.E. in Baghdad, the recently established capital of the caliphate, one finds AL-KHWĀRIZMĪ [VI.5] writing a treatise on numeration in the Indian style, "using nine symbols." Several centuries later, al-Khwārizmī's treatise was translated into Latin. It was so popular and influential in late-medieval Europe that decimal numeration was often referred to as "algorism."

It is worth noting that in al-Khwārizmī's writing zero still had a special status: it was a place holder, not a number. But once we have a symbol, and we start doing arithmetic using these symbols, the distinction quickly disappears. We have to know how to add and multiply numbers by zero in order to multiply multi-digit numbers. In this way, "nothing" slowly became a number.

4   What People Want Is a Number

As Greek culture was displaced by other influences, the practical tradition became more important. One can see this in al-Khwārizmī's other famous book, whose title gave us the word "algebra." The book is actually a compendium of many different kinds of practical or semi-practical mathematics problems. Al-Khwārizmī opens the book with a declaration that tells us at once that we are no longer in the Greek mathematical world: "When I considered what people generally want in calculating, I found that it is always a number."

The first portion of al-Khwārizmī's book deals with quadratic equations and with the algebraic manipulations (done entirely in words, with no symbols whatsoever) needed to deal with them. His procedure is exactly the quadratic formula we still use, which of course requires extracting a square root. But in every example the number whose square root we need to find turns out to be a square, so that the square root is easily found—and al-Khwārizmī does get a number!

At other points in the book, however, we can see that al-Khwārizmī is beginning to think of irrational square roots as number-like entities. He teaches the reader how to manipulate symbols with square roots in them, and gives (in words, of course) examples such as $(20 - \sqrt{200}) + (\sqrt{200} - 10) = 10$. In the second part of the book, which deals with geometry and measurement, one even sees an approximation to a square root: "The product is one thousand eight hundred and seventy-five; take its root, it is the area; it is forty-three and a little."

The mathematicians of medieval Islam were influenced not only by the practical tradition represented by al-Khwārizmī, but also by the Greek tradition, especially EUCLID's [VI.2] *Elements*. One finds in their writing a mixture of Greek precision and a more practical approach to measurement. In Omar Khayyam's *Algebra*, for example, one sees both theorems in the Greek style and the desire for numerical solutions. In his discussion of cubic equations Khayyam manages to find solutions by means of geometric constructions but laments his inability to find numerical values.

Slowly, however, the realm of "number" began to grow. The Greeks might have insisted that $\sqrt{10}$ was not a number, but rather a name for a line segment, the side of a square whose area is 10, or a name for a ratio. Among the medieval mathematicians, both in Islam and in Europe, $\sqrt{10}$ started to behave more and more like a number, entering into operations and even appearing as the solution of certain problems.

5   Giving Equal Status to All Numbers

The idea of extending the decimal place-value system to include fractions was discovered by several mathe-maticians independently. The most influential of these was STEVIN [VI.10], a Flemish mathematician and engineer who popularized the system in a booklet called *De Thiende* ("The tenth"), which was first published in 1585. By extending place value to tenths, hundredths, and so on, Stevin created the system we still use today. More importantly, he explained how it simplified calculations that involved fractions, and gave many practical applications. The cover page, in fact, announces that the book is for "astrologers, surveyors, measurers of tapestries."

Stevin was certainly aware of some of the issues created by his move. He knew, for example, that the decimal expansion for $\frac{1}{3}$ was infinitely long; his discussion simply says that while it might be more correct to say that the full infinite expansion was the correct representation, in practice it made little difference if we truncated it.

Stevin was also aware that his system provided a way to attach a "number" (meaning a decimal expansion) to every single length. He saw little difference between 1.1764705882 (the beginning of the decimal expansion of $\frac{20}{17}$) and 1.4142135623 (the beginning of the decimal expansion of $\sqrt{2}$). In his *Arithmetic* he boldly declared that all (positive) numbers were squares, cubes, fourth powers, etc., and that roots were just numbers. He also says that "there are no absurd, irrational, irregular, inexplicable, or surd numbers." Those were all terms used for irrational numbers, i.e., numbers that are not fractions.

What Stevin was proposing, then, was to flatten the incredible diversity of "quantities" or "magnitudes" into one expansive notion of number, defined by decimal expansions. He was aware that these numbers could be represented as lengths along a line. This amounted to a fairly clear notion of what we now call the positive real numbers.

Stevin's proposal was made immensely more influential by the invention of logarithms. Like the sine and the cosine, these were practical computational tools. In order to be used, they needed to be tabulated, and the tables were given in decimal form. Very soon, everyone was using decimal representation.

It was only much later that it came to be understood what a bold leap this move represented. The positive real numbers are not just a larger number system; they are an *immensely* larger number system, whose internal complexity we still do not fully understand (see SET THEORY [IV.22]).

6 Real, False, Imaginary

Even as Stevin was writing, the next steps were being taken: under the pressure of the theory of equations, negative numbers and complex numbers began to be useful. Stevin himself was already aware of negative numbers, though he was clearly not quite comfortable with them. For example, he explained that the fact that -3 is a root of $x^2 + x - 6$ really means that 3 is a root of the associated polynomial $x^2 - x - 6$, obtained by replacing x by $-x$ everywhere.

This was an easy dodge, but cubic equations created more difficult problems. The work of several Italian mathematicians of the sixteenth century led to a method for solving cubic equations. As a crucial step, this method involved extracting a square root. The problem was that the number whose root was needed sometimes came out negative.

Up until then, it had always turned out that when an algebraic problem led to the extraction of the square root of a negative number, the problem simply had no solution. But the equation $x^3 = 15x + 4$ clearly *did* have a solution—indeed, $x = 4$ is one—it was just that applying the cubic formula required computing $\sqrt{-121}$.

It was BOMBELLI [VI.8], also a mathematician and engineer, who decided to bite the bullet and just see what happened. In his *Algebra*, published in 1572, he went ahead and computed with this "new kind of radical" and showed that he could find the solution of the cubic in this way. This showed that the cubic formula did indeed work in this case; more importantly, it showed that these strange new numbers could be useful.

It took a while for people to become comfortable with these new quantities. About fifty years later, we find both Albert Girard and DESCARTES [VI.11] saying that equations can have three sorts of roots: true (meaning positive), false (negative), and imaginary. It is not completely clear that they understood that these imaginary roots would be what we now call complex numbers; Descartes, at least, sometimes seems to be saying that an equation of degree n must have n roots, and that the ones that are neither "true" nor "false" must simply be imagined.

Slowly, however, complex numbers began to be used. They came up in the theory of equations, in debates about the logarithms of negative numbers, and in connection to trigonometry. Their connection with the sine and cosine functions (via the exponential) was turned into a powerful tool by EULER [VI.19] in the eighteenth century. By the middle of the eighteenth century, it was well-known that every polynomial had a complete set of roots in the complex numbers. This result became known as THE FUNDAMENTAL THEOREM OF ALGEBRA [V.13]; it was finally proved to everyone's satisfaction by GAUSS [VI.26]. Thus, the theory of equations did not seem to require any further extension of the notion of number.

7 Number Systems, Old and New

Since complex numbers are clearly different from real numbers, their presence stimulated people to begin classifying numbers into different kinds. Stevin's egalitarianism had its impact, but it could not quite erase the fact that whole numbers are nicer than decimals, and that fractions are generally easier to grasp than irrational numbers.

In the nineteenth century, all sorts of new ideas created the need for a more careful look at this classification. In number theory, Gauss and KUMMER [VI.40] started looking at subsets of the complex numbers that behaved in a way analogous to the integers, such as the set of all numbers $a + b\sqrt{-1}$ with a and b both integers. In the theory of equations, GALOIS [VI.41] pointed out that in order to do a careful analysis of the solvability of an equation one must start by agreeing on what numbers count as "rational." So, for example, he pointed out that in ABEL's [VI.33] theorem on the unsolvability of the quintic, "rational" meant "expressible as a quotient of polynomials in the symbols used as the coefficients of the equation," and he noted that the set of all such expressions obeyed the usual rules of arithmetic.

In the eighteenth century, Johann Lambert had established that e and π were irrational, and conjectured that in fact they were *transcendental*, that is, that they were not roots of any polynomial equation. Even the existence of transcendental numbers was not known at the time; LIOUVILLE [VI.39] proved that such numbers exist in 1844. Within a few decades, it was proved that both e and π were transcendental, and later in the century CANTOR [VI.54] showed that in fact the vast majority of real numbers were transcendental. Cantor's discovery highlighted, for the first time, that the system Stevin had popularized contained unexpected depths.

Perhaps the most important change in the concept of number, however, came after HAMILTON's [VI.37] discovery, in 1843, of a completely new number system. Hamilton had noticed that coordinatizing the plane using complex numbers (rather than simply using pairs of real numbers) vastly simplified plane geometry. He set out to find a similar way to parametrize three-dimensional space. This turned out to be impossible, but led Hamilton to a *four*-dimensional system, which he called the QUATERNIONS [III.76]. These behaved much like numbers, with one crucial difference: multiplication was not commutative, that is, if q and q' are quaternions, qq' and $q'q$ are usually *not* the same.

The quaternions were the first system of "hypercomplex numbers," and their appearance generated lots of new questions. Were there other such systems? What counts as a number system? If certain "numbers" can fail to satisfy the commutative law, can we make numbers that break other rules?

In the long run, this intellectual ferment led mathematicians to let go of the vague notion of "number" or "quantity" and to hold on, instead, to the more formal notion of an algebraic structure. Each of the number systems, in the end, is simply a set of entities on which we can do operations. What makes them interesting is that we can use them to parametrize, or coordinatize, systems that interest us. The whole numbers (or *integers*, to give them their latinized formal name), for example, formalize the notion of counting, while the real numbers parametrize the line and serve as the basis for geometry.

By the beginning of the twentieth century, there were many well-known number systems. The integers had pride of place, followed by a nested hierarchy consisting of the rational numbers (i.e., the fractions), the real numbers (Stevin's decimals, now carefully formalized), and the complex numbers. Still more general than the complex numbers were the quaternions. But these were by no means the only systems around. Number theorists worked with several different fields of algebraic numbers, subsets of the complex numbers that could be understood as autonomous systems. Galois had introduced finite systems that obeyed the usual rules of arithmetic, which we now call finite fields. Function theorists worked with fields of functions; they certainly did not think of these as numbers, but their analogy to number systems was known and exploited.

Early in the twentieth century, Kurt Hensel introduced the p-adic numbers [III.51], which were built from the rational numbers by giving a special role to a prime number p. (Since p can be chosen at will, Hensel in fact created infinitely many new number systems.) These too "obeyed the usual rules of arithmetic," in the sense that addition and multiplication behaved as

expected; in modern language, they were *fields*. The p-adics provided the first system of things that were recognizably numbers but that had no visible relation to the real or complex numbers—apart from the fact that both systems contained the rational numbers. As a result, they led Ernst Steinitz to create an abstract theory of fields.

The move to abstraction that appears in Steinitz's work had also occurred in other parts of mathematics, most notably the theory of groups and their representations and the theory of algebraic numbers. All of these theories were brought together into conceptual unity by NOETHER [VI.76], whose program came to be known as "abstract algebra." This left numbers behind completely, focusing instead on the abstract structure of sets with operations.

Today, it is no longer that easy to decide what counts as a "number." The objects from the original sequence of "integer, rational, real, and complex" are certainly numbers, but so are the p-adics. The quaternions are rarely referred to as "numbers," on the other hand, though they can be used to coordinatize certain mathematical notions. In fact, even stranger systems can show up as coordinates, such as Cayley's OCTONIONS [III.76]. In the end, whatever serves to parametrize or coordinatize the problem at hand is what we use. If the requisite system turns out not to exist yet, well, one just has to invent it.

Further Reading

Berlinghoff, W. P., and F. Q. Gouvêa. 2004. *Math through the Ages: A Gentle History for Teachers and Others*, expanded edn. Farmington, ME/Washington, DC: Oxton House/The Mathematical Association of America.

Ebbingaus, H.-D., et al. 1991. *Numbers*. New York: Springer.

Fauvel, J., and J. J. Gray, eds. 1987. *The History of Mathematics: A Reader*. Basingstoke: Macmillan.

Fowler, D. 1985. 400 years of decimal fractions. *Mathematics Teaching* 110:20–21.

———. 1999. *The Mathematics of Plato's Academy*, 2nd edn. Oxford: Oxford University Press.

Gouvêa, F. Q. 2003. *p-adic Numbers: An Introduction*, 2nd edn. New York: Springer.

Katz, V. J. 1998. *A History of Mathematics*, 2nd edn. Reading, MA: Addison-Wesley.

———, ed. 2007. *The Mathematics of Egypt, Mesopotamia, China, India, and Islam: A Sourcebook*. Princeton, NJ: Princeton University Press.

Mazur, B. 2002. *Imagining Numbers (Particularly the Square Root of Minus Fifteen)*. New York: Farrar, Straus, and Giroux.

Menninger, K. 1992. *Number Words and Number Symbols: A Cultural History of Numbers.* New York: Dover. (Translated by P. Broneer from the revised German edition of 1957/58: *Zahlwort und Ziffer. Eine Kulturgeschichte der Zahl.* Göttingen: Vandenhoeck und Ruprecht.)

Reid, C. 2006. *From Zero to Infinity: What Makes Numbers Interesting.* Natick, MA: A. K. Peters.

II.2 Geometry
Jeremy Gray

1 Introduction

The modern view of geometry was inspired by the novel geometrical theories of HILBERT [VI.63] and Einstein in the early years of the twentieth century, which built in their turn on other radical reformulations of geometry in the nineteenth century. For thousands of years, the geometrical knowledge of the Greeks, as set out most notably in EUCLID's [VI.2] *Elements*, was held up as a paradigm of perfect rigor, and indeed of human knowledge. The new theories amounted to the overthrow of an entire way of thinking. This essay will pursue the history of geometry, starting from the time of Euclid, continuing with the advent of non-Euclidean geometry, and ending with the work of RIEMANN [VI.49], KLEIN [VI.57], and POINCARÉ [VI.61]. Along the way, we shall examine how and why the notions of geometry changed so remarkably. Modern geometry itself will be discussed in later parts of this book.

2 Naive Geometry

Geometry generally, and Euclidean geometry in particular, is informally and rightly taken to be the mathematical description of what you see all around you: a space of three dimensions (left–right, up–down, forwards–backwards) that seems to extend indefinitely far. Objects in it have positions, they sometimes move around and occupy other positions, and all of these positions can be specified by measuring lengths along straight lines: this object is twenty meters from that one, it is two meters tall, and so on. We can also measure angles, and there is a subtle relationship between angles and lengths. Indeed, there is another aspect to geometry, which we do not see but which we reason about. Geometry is a mathematical subject that is full of *theorems*—the isosceles triangle theorem, the Pythagorean theorem, and so on—which collectively summarize what we can say about lengths, angles, shapes, and positions. What distinguishes this aspect

of geometry from most other kinds of science is its highly deductive nature. It really seems that by taking the simplest of concepts and thinking hard about them one can build up an impressive, deductive body of knowledge about space without having to gather experimental evidence.

But can we? Is it really as simple as that? Can we have genuine knowledge of space without ever leaving our armchairs? It turns out that we cannot: there are other geometries, also based on the concepts of length and angle, that have every claim to be useful, but that disagree with Euclidean geometry. This is an astonishing discovery of the early nineteenth century, but, before it could be made, a naive understanding of fundamental concepts, such as straightness, length, and angle, had to be replaced by more precise definitions—a process that took many hundreds of years. Once this had been done, first one and then infinitely many new geometries were discovered.

3 The Greek Formulation

Geometry can be thought of as a set of useful facts about the world, or else as an organized body of knowledge. Either way, the origins of the subject are much disputed. It is clear that the civilizations of Egypt and Babylonia had at least some knowledge of geometry—otherwise, they could not have built their large cities, elaborate temples, and pyramids. But not only is it difficult to give a rich and detailed account of what was known before the Greeks, it is difficult even to make sense of the few scattered sources that we have from before the time of Plato and Aristotle. One reason for this is the spectacular success of the later Greek writer, and author of what became the definitive text on geometry, Euclid of Alexandria (ca. 300 B.C.E.). One glance at his famous *Elements* shows that a proper account of the history of geometry will have to be about something much more than the acquisition of geometrical facts. The *Elements* is a highly organized, deductive body of knowledge. It is divided into a number of distinct themes, but each theme has a complex theoretical structure. Thus, whatever the origins of geometry might have been, by the time of Euclid it had become the paradigm of a logical subject, offering a kind of knowledge quite different from, and seemingly higher than, knowledge directly gleaned from ordinary experience.

Rather, therefore, than attempt to elucidate the early history of geometry, this essay will trace the high road of geometry's claim on our attention: the apparent certainty of mathematical knowledge. It is exactly this claim to a superior kind of knowledge that led eventually to the remarkable discovery of *non-Euclidean geometry*: there are geometries other than Euclid's that are every bit as rigorously logical. Even more remarkably, some of these turn out to provide better models of physical space than Euclidean geometry.

The *Elements* opens with four books on the study of plane figures: triangles, quadrilaterals, and circles. The famous theorem of Pythagoras is the forty-seventh proposition of the first book. Then come two books on the theory of ratio and proportion and the theory of similar figures (scale copies), treated with a high degree of sophistication. The next three books are about whole numbers, and are presumably a reworking of much older material that would now be classified as elementary number theory. Here, for example, one finds the famous result that there are infinitely many prime numbers. The next book, the tenth, is by far the longest, and deals with the seemingly specialist topic of lengths of the form $\sqrt{a} \pm \sqrt{b}$ (to write them as we would). The final three books, where the curious lengths studied in Book X play a role, are about three-dimensional geometry. They end with the construction of the five regular solids and a proof that there are no more. The discovery of the fifth and last had been one of the topics that excited Plato. Indeed, the five regular solids are crucial to the cosmology of Plato's late work the *Timaeus*.

Most books of the *Elements* open with a number of definitions, and each has an elaborate deductive structure. For example, to understand the Pythagorean theorem, one is driven back to previous results, and thence to even earlier results, until finally one comes to rest on basic definitions. The whole structure is quite compelling: reading it as an adult turned the philosopher Thomas Hobbes from incredulity to lasting belief in a single sitting. What makes the *Elements* so convincing is the nature of the arguments employed. With some exceptions, mostly in the number-theoretic books, these arguments use the axiomatic method. That is to say, they start with some very simple axioms that are intended to be self-evidently true, and proceed by purely logical means to deduce theorems from them.

For this approach to work, three features must be in place. The first is that *circularity* should be carefully avoided. That is, if you are trying to prove a statement P and you deduce it from an earlier statement, and deduce that from a yet earlier statement, and so on, then at no stage should you reach the statement

P again. That would not prove P from the axioms, but merely show that all the statements in your chain were equivalent. Euclid did a remarkable job in this respect.

The second necessary feature is that the rules of inference should be clear and acceptable. Some geometrical statements seem so obvious that one can fail to notice that they need to be proved: ideally, one should use no properties of figures other than those that have been clearly stated in their definitions, but this is a difficult requirement to meet. Euclid's success here was still impressive, but mixed. On the one hand, the *Elements* is a remarkable work, far outstripping any contemporary account of any of the topics it covers, and capable of speaking down the millennia. On the other, it has little gaps that from time to time later commentators would fill. For example, it is neither explicitly assumed nor proved in the *Elements* that two circles will meet if their centers lie outside each other and the sum of their radii is greater than the distance between their centers. However, Euclid is surprisingly clear that there are rules of inference that are of general, if not indeed universal, applicability, and others that apply to mathematics because they rely on the meanings of the terms involved.

The third feature, not entirely separable from the second, is adequate definitions. Euclid offered two, or perhaps three, sorts of definition. Book I opens with seven definitions of objects, such as "point" and "line," that one might think were primitive and beyond definition, and it has recently been suggested that these definitions are later additions. Then come, in Book I and again in many later books, definitions of familiar figures designed to make them amenable to mathematical reasoning: "triangle," "quadrilateral," "circle," and so on. The postulates of Book I form the third class of definition and are rather more problematic.

Book I states five "common notions," which are rules of inference of a very general sort. For example, "If equals be added to equals, the wholes are equals." The book also has five "postulates," which are more narrowly mathematical. For example, the first of these asserts that one may draw a straight line from any point to any point. One of these postulates, the fifth, became notorious: the so-called *parallel postulate*. It says that "If a straight line falling on two straight lines make the interior angles on the same side less than two right angles, the two straight lines, if produced indefinitely, meet on that side on which are the angles less than two right angles."

Parallel lines, therefore, are straight lines that do not meet. A helpful rephrasing of Euclid's parallel postulate was introduced by the Scottish editor, Robert Simson. It appears in his edition of Euclid's *Elements* from 1806. There he showed that the parallel postulate is equivalent, if one assumes those parts of the *Elements* that do not depend on it, to the following statement: given any line m in a plane, and any point P in that plane that does not lie on the line m, there is exactly one line n in the plane that passes through the point P and does not meet the line m. From this formulation it is clear that the parallel postulate makes two assertions: given a line and a point as described, a parallel line *exists* and it is *unique*.

It is worth noting that Euclid himself was probably well aware that the parallel postulate was awkward. It asserts a property of straight lines that seems to have made Greek mathematicians and philosophers uncomfortable, and this may be why its appearance in the *Elements* is delayed until proposition 29 of Book I. The commentator Proclus (fifth century C.E.), in his extensive discussion of Book I of the *Elements*, observed that the hyperbola and asymptote get closer and closer as they move outwards, but they never meet. If a line and a curve can do this, why not two lines? The matter needs further analysis. Unfortunately, not much of the *Elements* would be left if mathematicians dropped the parallel postulate and retreated to the consequences of the remaining definitions: a significant body of knowledge depends on it. Most notably, the parallel postulate is needed to prove that the angles in a triangle add up to two right angles—a crucial result in establishing many other theorems about angles in figures, including the Pythagorean theorem.

Whatever claims educators may have made about Euclid's *Elements* down the ages, a significant number of experts knew that it was an unsatisfactory compromise: a useful and remarkably rigorous theory could be had, but only at the price of accepting the parallel postulate. But the parallel postulate was difficult to accept on trust: it did not have the same intuitively obvious feel of the other axioms and there was no obvious way of verifying it. The higher one's standards, the more painful this compromise was. What, the experts asked, was to be done?

One Greek discussion must suffice here. In Proclus's view, if the truth of the parallel postulate was not obvious, and yet geometry was bare without it, then the only possibility was that it was true because it was a theorem. And so he gave it a proof. He argued as follows. Let two lines m and n cross a third line k at P and Q, respectively, and make angles with it that add up to two right angles. Now draw a line l that crosses m at P and enters the space between the lines m and n. The distance between l and m as one moves away from the point P continually increases, said Proclus, and therefore line l must eventually cross line n.

Proclus's argument is flawed. The flaw is subtle, and sets us up for what is to come. He was correct that the distance between the lines l and m increases indefinitely. But his argument assumes that the distance between lines m and n does not *also* increase indefinitely, and is instead bounded. Now Proclus knew very well that *if* the parallel postulate is granted, *then* it can be shown that the lines m and n are parallel and that the distance between them is a constant. But until the parallel postulate is proved, nothing prevents one saying that the lines m and n diverge. Proclus's proof does not therefore work unless one can show that lines that do not meet also do not diverge.

Proclus's attempt was not the only one, but it is typical of such arguments, which all have a standard form. They start by detaching the parallel postulate from Euclid's *Elements*, together with all the arguments and theorems that depend on it. Let us call what remains the "core" of the *Elements*. Using this core, an attempt is then made to derive the parallel postulate as a theorem. The correct conclusion to be derived from Proclus's attempt is not that the parallel postulate is a theorem, but rather that, given the core of the *Elements*, the parallel postulate is equivalent to the statement that lines that do not meet also do not diverge. Aganis, a writer of the sixth century C.E. about whom almost nothing is known, assumed, in a later attempt, that parallel lines are everywhere equidistant, and his argument showed only that, given the core, the Euclidean definition of parallel lines is equivalent to defining them to be equidistant.

Notice that one cannot even enter this debate unless one is clear which properties of straight lines belong to them by definition, and which are to be derived as theorems. If one is willing to add to the store of "common-sense" assumptions about geometry as one goes along, the whole careful deductive structure of the *Elements* collapses into a pile of facts.

This deductive character of the *Elements* is clearly something that Euclid regarded as important, but one can also ask what he thought geometry was *about*. Was it meant, for example, as a mathematical description of space? No surviving text tells us what he thought about this question, but it is worth noting that the most celebrated Greek theory of the universe, developed by Aristotle and many later commentators, assumed that space was finite, bounded by the sphere of the fixed stars. The mathematical space of the *Elements* is infinite, and so one has at least to consider the possibility that, for all these writers, mathematical space was not intended as a simple idealization of the physical world.

4 Arab and Islamic Commentators

What we think of today as Greek geometry was the work of a handful of mathematicians, mostly concentrated in a period of less than two centuries. They were eventually succeeded by a somewhat larger number of Arabic and Islamic writers, spread out over a much greater area and a longer time. These writers tend to be remembered as commentators on Greek mathematics and science, and for transmitting them to later Western authors, but they should also be remembered as creative, innovative mathematicians and scientists in their own right. A number of them took up the study of Euclid's *Elements*, and with it the problem of the parallel postulate. They too took the view that it was not a proper postulate, but one that could be proved as a theorem using the core alone.

Among the first to attempt a proof was Thābit ibn Qurra. He was a pagan from near Aleppo who lived and worked in Baghdad, where he died in 901. Here there is room to describe only his first approach. He argued that if two lines m and n are crossed by a third, k, and if they approach each other on one side of the line k, then they diverge indefinitely on the other side of k. He deduced that two lines that make equal alternate angles with a transversal (the marked angles in figure 1) cannot approach each other on one side of a transversal: the symmetry of the situation would imply that they approached on the other side as well, but he had shown that they would have to diverge on the other side. From this he deduced the Euclidean theory of parallels, but his argument was also flawed, since he had not considered the possibility that two lines could *diverge* in both directions.

The distinguished Islamic mathematician and scientist ibn al-Haytham was born in Basra in 965 and died in Egypt in 1041. He took a quadrilateral with two equal sides perpendicular to the base and dropped a perpendicular from one side to the other. He now attempted to prove that this perpendicular is equal to the base, and to do so he argued that as one of two original perpendiculars is moved toward the other, its tip sweeps

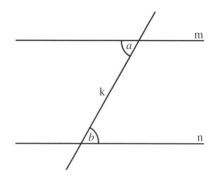

Figure 1 The lines m and n make equal alternate angles *a* and *b* with the transversal k.

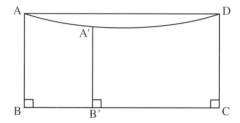

Figure 2 AB and CD are equal, the angle ADC is a right angle, A′B′ is an intermediate position of AB as it moves toward CD.

out a straight line, which will coincide with the perpendicular just dropped (see figure 2). This amounts to the assumption that the curve everywhere equidistant from a straight line is itself straight, from which the parallel postulate easily follows, and so his attempt fails. His proof was later heavily criticized by Omar Khayyam for its use of motion, which he found fundamentally unclear and alien to Euclid's *Elements*. It is indeed quite distinct from any use Euclid had for motion in geometry, because in this case the nature of the curve obtained is not clear: it is precisely what needs to be analyzed.

The last of the Islamic attempts on the parallel postulate is due to Naṣir al-Dīn al-Ṭūsī. He was born in Iran in 1201 and died in Baghdad in 1274. His extensive commentary is also one of our sources of knowledge of earlier Islamic mathematical work on this subject. Al-Ṭūsī focused on showing that if two lines begin to converge, then they must continue to do so until they eventually meet. To this end he set out to show that

(∗) if l and m are two lines that make an angle of less than a right angle, then every line perpendicular to l meets the line m.

He showed that if (∗) is true, then the parallel postulate follows. However, his argument for (∗) is flawed.

It is genuinely difficult to see what is wrong with some of these arguments if one uses only the techniques available to mathematicians of the time. Islamic mathematicians showed a degree of sophistication that was not to be surpassed by their Western successors until the eighteenth century. Unfortunately, however, their writings did not come to the attention of the West until much later, with the exception of a single work in the Vatican Library, published in 1594, which was for many years erroneously attributed to al-Ṭūsī (and which may have been the work of his son).

5   The Western Revival of Interest

The Western revival of interest in the parallel postulate came with the second wave of translations of Greek mathematics, led by Commandino and Maurolico in the sixteenth century and spread by the advent of printing. Important texts were discovered in a number of older libraries, and ultimately this led to the production of new texts of Euclid's *Elements*. Many of these had something to say about the problem of parallels, pithily referred to by Henry Savile as "a blot on Euclid." For example, the powerful Jesuit Christopher Clavius, who edited and reworked the *Elements* in 1574, tried to argue that parallel lines could be defined as equidistant lines.

The ready identification of physical space with the space of Euclidean geometry came about gradually during the sixteenth and seventeenth centuries, after the acceptance of Copernican astronomy and the abolition of the so-called sphere of fixed stars. It was canonized by NEWTON [VI.14] in his *Principia Mathematica*, which proposed a theory of gravitation that was firmly situated in Euclidean space. Although Newtonian physics had to fight for its acceptance, Newtonian cosmology had a smooth path and became the unchallenged orthodoxy of the eighteenth century. It can be argued that this identification raised the stakes, because any unexpected or counterintuitive conclusion drawn solely from the core of the *Elements* was now, possibly, a counterintuitive fact about space.

In 1663 the English mathematician John Wallis took a much more subtle view of the parallel postulate than any of his predecessors. He had been instructed by Halley, who could read Arabic, in the contents of the apocryphal edition of al-Ṭūsī's work in the Vatican Library, and he too gave an attempted proof. Unusually, Wallis also had the insight to see where his own argument was flawed, and commented that what it really showed was that, in the presence of the core, the parallel postulate was equivalent to the assertion that there exist similar figures that are not congruent.

Half a century later, Wallis was followed by the most persistent and thoroughgoing of all the defenders of the parallel postulate, Gerolamo Saccheri, an Italian Jesuit who published in 1733, the year of his death, a short book called *Euclid Freed of Every Flaw*. This little masterpiece of classical reasoning opens with a trichotomy. Unless the parallel postulate is known, the angle sum of a triangle may be either less than, equal to, or greater than two right angles. Saccheri showed that whatever happens in one triangle happens for them all, so there are apparently three geometries compatible with the core. In the first, every triangle has an angle sum less than two right angles (call this case L). In the second, every triangle has an angle sum equal to two right angles (call this case E). In the third, every triangle has an angle sum greater than two right angles (call this case G). Case E is, of course, Euclidean geometry, which Saccheri wished to show was the only case possible. He therefore set to work to show that each of the other cases independently self-destructed. He was successful with case G, and then turned to case L "which alone obstructs the truth of the [parallel] axiom," as he put it.

Case L proved to be difficult, and during the course of his investigations Saccheri established a number of interesting propositions. For example, if case L is true, then two lines that do not meet have just one common perpendicular, and they diverge on either side of it. In the end, Saccheri tried to deal with his difficulties by relying on foolish statements about the behavior of lines at infinity: it was here that his attempted proof failed.

Saccheri's work sank slowly, though not completely, into obscurity. It did, however, come to the attention of the Swiss mathematician Johann Lambert, who pursued the trichotomy but, unlike Saccheri, stopped short of claiming success in proving the parallel postulate. Instead the work was abandoned, and was published only in 1786, after his death. Lambert distinguished carefully between unpalatable results and impossibilities. He had a sketch of an argument to show that in case L the area of a triangle is proportional to the difference between two right angles and the angle sum of the triangle. He knew that in case L similar triangles had to be congruent, which would imply that the tables of trigonometric functions used in astronomy were not in fact valid and that different tables would have to be produced for every size of triangle. In particular, for every angle less than 60° there would be precisely one equilateral triangle with that given angle at each vertex. This would lead to what philosophers called an "absolute" measure of length (one could take, for instance, the length of the side of an equilateral triangle with angles equal to 30°), which LEIBNIZ's [VI.15] follower Wolff had said was impossible. And indeed it is counterintuitive: lengths are generally defined in relative terms, as, for instance, a certain proportion of the length of a meter rod in Paris, or of the circumference of Earth, or of something similar. But such arguments, said Lambert, "were drawn from love and hate, with which a mathematician can have nothing to do."

6   The Shift of Focus around 1800

The phase of Western interest in the parallel postulate that began with the publication of modern editions of Euclid's *Elements* started to decline with a further turn in that enterprise. After the French revolution, LEGENDRE [VI.24] set about writing textbooks, largely for the use of students hoping to enter the École Polytechnique, that would restore the study of elementary geometry to something like the rigorous form in which it appeared in the *Elements*. However, it was one thing to seek to replace books of a heavily intuitive kind, but quite another to deliver the requisite degree of rigor. Legendre, as he came to realize, ultimately failed in his attempt. Specifically, like everyone before him, he was unable to give an adequate defense of the parallel postulate. Legendre's *Éléments de Géométrie* ran to numerous editions, and from time to time a different attempt on the postulate was made. Some of these attempts would be hard to describe favorably, but the best can be extremely persuasive.

Legendre's work was classical in spirit, and he still took it for granted that the parallel postulate had to be true. But by around 1800 this attitude was no longer universally held. Not everybody thought that the postulate must, somehow, be defended, and some were prepared to contemplate with equanimity the idea that it might be false. No clearer illustration of this shift can be found than a brief note sent to GAUSS [VI.26] by F. K. Schweikart, a Professor of Law at the University of Marburg, in 1818. Schweikart described in a page the main results he had been led to in what he called "astral geometry," in which the angle sum of a triangle

was less than two right angles: squares had a particular form, and the altitude of a right-angled isosceles triangle was bounded by an amount Schweikart called "the constant." Schweikart went so far as to claim that the new geometry might even be the true geometry of space. Gauss replied positively. He accepted the results, and he claimed that he could do all of elementary geometry once a value for the constant was given. One could argue, somewhat ungenerously, that Schweikart had done little more than read Lambert's posthumous book—although the theorem about isosceles triangles is new. However, what is notable is the attitude of mind: the idea that this new geometry might be true, and not just a mathematical curiosity. Euclid's *Elements* shackled him no more.

Unfortunately, it is much less clear what Gauss himself thought. Some historians, bearing Gauss's remarkable mathematical originality in mind, have been inclined to interpret the evidence in such a way that Gauss emerges as the first person to discover non-Euclidean geometry. However, the evidence is slight, and it is difficult to draw firm conclusions from it. There are traces of some early investigations by Gauss of Euclidean geometry that include a study of a new definition of parallel lines; there are claims made by Gauss late in life that he had known this or that fact for many years; and there are letters he wrote to his friends. But there is no material in the surviving papers that allows us to reconstruct what Gauss knew or that supports the claim that Gauss discovered non-Euclidean geometry.

Rather, the picture would seem to be that Gauss came to realize during the 1810s that all previous attempts to derive the parallel postulate from the core of Euclidean geometry had failed and that all future attempts would probably fail as well. He became more and more convinced that there was another possible geometry of space. Geometry ceased, in his mind, to have the status of arithmetic, which was a matter of logic, and became associated with mechanics, an empirical science. The simplest accurate statement of Gauss's position through the 1820s is that he did not doubt that space might be described by a non-Euclidean geometry, and of course there was only one possibility: that of case L described above. It was an empirical matter, but one that could not be resolved by land-based measurements because any departure from Euclidean geometry was, evidently, very small. In this view he was supported by his friends, such as Bessel and Olbers, both professional astronomers. Gauss the scientist was convinced, but Gauss the mathematician may have retained a small degree of doubt, and certainly never developed the mathematical theory required to describe non-Euclidean geometry adequately.

One theory available to Gauss from the early 1820s was that of differential geometry. Gauss eventually published one of his masterworks on this subject, his *Disquisitiones Generales circa Superficies Curvas* (1827). In it he showed how to describe geometry on any surface in space, and how to regard certain features of the geometry of a surface as intrinsic to the surface and independent of how the surface was embedded into three-dimensional space. It would have been possible for Gauss to consider a surface of constant negative CURVATURE [III.78], and to show that triangles on such a surface are described by hyperbolic trigonometric formulas, but he did not do this until the 1840s. Had he done so, he would have had a surface on which the formulas of a geometry satisfying case L apply.

A surface, however, is not enough. We accept the validity of two-dimensional Euclidean geometry because it is a simplification of three-dimensional Euclidean geometry. Before a two-dimensional geometry satisfying the hypotheses of case L can be accepted, it is necessary to show that there is a plausible three-dimensional geometry analogous to case L. Such a geometry has to be described in detail and shown to be as plausible as Euclidean three-dimensional geometry. This Gauss simply never did.

7 Bolyai and Lobachevskii

The fame for discovering non-Euclidean geometry goes to two men, BOLYAI [VI.34] in Hungary and LOBACHEVSKII [VI.31] in Russia, who independently gave very similar accounts of it. In particular, both men described a system of geometry in two and three dimensions that differed from Euclid's but had an equally good claim to be the geometry of space. Lobachevskii published first, in 1829, but only in an obscure Russian journal, and then in French in 1837, in German in 1840, and again in French in 1855. Bolyai published his account in 1831, in an appendix to a two-volume work on geometry by his father.

It is easiest to describe their achievements together. Both men defined parallels in a novel way, as follows. Given a point P and a line m there will be some lines through P that meet m and others that do not. Separating these two sets will be two lines through P that do not quite meet m but which might come arbitrarily close, one to the right of P and one to the left. This situation is illustrated in figure 3: the two lines in question

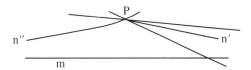

Figure 3 The lines n′ and n″ through P separate the lines through P that meet the line m from those that do not.

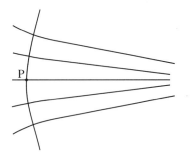

Figure 4 A curve perpendicular to a family of parallels.

are n′ and n″. Notice that lines on the diagram appear curved. This is because, in order to represent them on a flat, Euclidean page, it is necessary to distort them, unless the geometry is itself Euclidean, in which case one can put n′ and n″ together and make a single line that is infinite in both directions.

Given this new way of talking, it still makes sense to talk of dropping the perpendicular from P to the line m. The left and right parallels to m through P make equal angles with the perpendicular, called the *angle of parallelism*. If the angle is a right angle, then the geometry is Euclidean. However, if it is less than a right angle, then the possibility arises of a new geometry. It turns out that the size of the angle depends on the length of the perpendicular from P to m. Neither Bolyai nor Lobachevskii expended any effort in trying to show that there was not some contradiction in taking the angle of parallelism to be less than a right angle. Instead, they simply made the assumption and expended a great deal of effort on determining the angle from the length of the perpendicular.

They both showed that, given a family of lines all parallel (in the same direction) to a given line, and given a point on one of the lines, there is a curve through that point that is perpendicular to each of the lines (figure 4).

In Euclidean geometry the curve defined in this way is the straight line that is at right angles to the family of parallel lines and that passes through the given

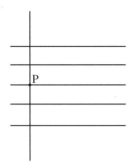

Figure 5 A curve perpendicular to a family of Euclidean parallels.

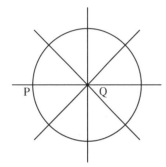

Figure 6 A curve perpendicular to a family of Euclidean lines through a point.

point (figure 5). If, again in Euclidean geometry, one takes the family of all lines through a common point Q and chooses another point P, then there will be a curve through P that is perpendicular to all the lines: the circle with center Q that passes through P (figure 6).

The curve defined by Bolyai and Lobachevskii has some of the properties of both these Euclidean constructions: it is perpendicular to all the parallels, but it is curved and not straight. Bolyai called such a curve an *L-curve*. Lobachevskii more helpfully called it a *horocycle*, and the name has stuck.

Their complicated arguments took both men into three-dimensional geometry. Here Lobachevskii's arguments were somewhat clearer than Bolyai's, and both men notably surpassed Gauss. If the figure defining a horocycle is rotated about one of the parallel lines, the lines become a family of parallel lines in three dimensions and the horocycle sweeps out a bowl-shaped surface, called the *F-surface* by Bolyai and the horosphere by Lobachevskii. Both men now showed that something remarkable happens. Planes through the horosphere cut it either in circles or in horocycles, and if a triangle

is drawn on a horosphere whose sides are horocycles, then the angle sum of such a triangle is two right angles. To put this another way, although the space that contains the horosphere is a three-dimensional version of case L, and is definitely not Euclidean, the geometry you obtain when you restrict attention to the horosphere is (two-dimensional) Euclidean geometry!

Bolyai and Lobachevskii also knew that one can draw spheres in their three-dimensional space, and they showed (though in this they were not original) that the formulas of spherical geometry hold independently of the parallel postulate. Lobachevskii now used an ingenious construction involving his parallel lines to show that a triangle on a sphere determines and is determined by a triangle in the plane, which also determines and is determined by a triangle on the horosphere. This implies that the formulas of spherical geometry must determine formulas that apply to the triangles on the horosphere. On checking through the details, Lobachevskii, and in more or less the same way Bolyai, showed that the triangles on the horosphere are described by the formulas of hyperbolic trigonometry.

The formulas for spherical geometry depend on the radius of the sphere in question. Similarly, the formulas of hyperbolic trigonometry depend on a certain real parameter. However, this parameter does not have a similarly clear geometrical interpretation. That defect apart, the formulas have a number of reassuring properties. In particular, they closely approximate the familiar formulas of plane geometry when the sides of the triangles are very small, which helps to explain how this geometry could have remained undetected for so long—it differs very little from Euclidean geometry in small regions of space. Formulas for length and area can be developed in the new setting: they show that the area of a triangle is proportional to the amount by which the angle sum of the triangle falls short of two right angles. Lobachevskii, in particular, seems to have felt that the very fact that there were neat and plausible formulas of this kind was enough reason to accept the new geometry. In his opinion, all geometry was about measurement, and theorems in geometry were unfailing connections between measurements expressed by formulas. His methods produced such formulas, and that, for him, was enough.

Bolyai and Lobachevskii, having produced a description of a novel three-dimensional geometry, raised the question of which geometry is true: is it Euclidean geometry or is it the new geometry for some value of the parameter that could presumably be determined experimentally? Bolyai left matters there, but Lobachevskii explicitly showed that measurements of stellar parallax might resolve the question. Here he was unsuccessful: such experiments are notoriously delicate.

By and large, the reaction to Bolyai and Lobachevskii's ideas during their lifetimes was one of neglect and hostility, and they died unaware of the success their discoveries would ultimately have. Bolyai and his father sent their work to Gauss, who replied in 1832 that he could not praise the work "for to do so would be to praise myself," adding, for extra measure, a simpler proof of one of Janos Bolyai's opening results. He was, he said, nonetheless delighted that it was the son of his old friend who had taken precedence over him. Janos Bolyai was enraged, and refused to publish again, thus depriving himself of the opportunity to establish his priority over Gauss by publishing his work as an article in a mathematics journal. Oddly, there is no evidence that Gauss knew the details of the young Hungarian's work in advance. More likely, he saw at once how the theory would go once he appreciated the opening of Bolyai's account.

A charitable interpretation of the surviving evidence would be that, by 1830, Gauss was convinced of the possibility that physical space might be described by non-Euclidean geometry, and he surely knew how to handle two-dimensional non-Euclidean geometry using hyperbolic trigonometry (although no detailed account of this survives from his hand). But the three-dimensional theory was known first to Bolyai and Lobachevskii, and may well not have been known to Gauss until he read their work.

Lobachevskii fared little better than Bolyai. His initial publication of 1829 was savaged in the press by Ostrogradskii, a much more established figure who was, moreover, in St Petersburg, whereas Lobachevskii was in provincial Kazan. His account in *Journal für die reine und angewandte Mathematik* (otherwise known as *Crelle's Journal*) suffered grievously from referring to results proved only in the Russian papers from which it had been adapted. His booklet of 1840 drew only one review, of more than usual stupidity. He did, however, send it to Gauss, who found it excellent and had Lobachevskii elected to the Göttingen Academy of Sciences. But Gauss's enthusiasm stopped there, and Lobachevskii received no further support from him.

Such a dreadful response to a major discovery invites analysis on several levels. It has to be said that the definition of parallels upon which both men depended was,

as it stood, inadequate, but their work was not criticized on that account. It was dismissed with scorn, as if it were self-evident that it was wrong: so wrong that it would be a waste of time finding the error it surely contained, so wrong that the right response was to heap ridicule upon its authors or simply to dismiss them without comment. This is a measure of the hold that Euclidean geometry still had on the minds of most people at the time. Even Copernicanism, for example, and the discoveries of Galileo drew a better reception from the experts.

8 Acceptance of Non-Euclidean Geometry

When Gauss died in 1855, an immense amount of unpublished mathematics was found among his papers. Among it was evidence of his support for Bolyai and Lobachevskii, and his correspondence endorsing the possible validity of non-Euclidean geometry. As this was gradually published, the effect was to send people off to look for what Bolyai and Lobachevskii had written and to read it in a more positive light.

Quite by chance, Gauss had also had a student at Göttingen who was capable of moving the matter decisively forward, even though the actual amount of contact between the two was probably quite slight. This was RIEMANN [VI.49]. In 1854 he was called to defend his Habilitation thesis, the postdoctoral qualification that was a German mathematician's license to teach in a university. As was the custom, he offered three titles and Gauss, who was his examiner, chose the one Riemann least expected: "On the hypotheses that lie at the foundation of geometry." The paper, which was to be published only posthumously, in 1867, was nothing less than a complete reformulation of geometry.

Riemann proposed that geometry was the study of what he called MANIFOLDS [I.3 §§6.9, 6.10]. These were "spaces" of points, together with a notion of distance that looked like Euclidean distance on small scales but which could be quite different at larger scales. This kind of geometry could be done in a variety of ways, he suggested, by means of the calculus. It could be carried out for manifolds of any dimension, and in fact Riemann was even prepared to contemplate manifolds for which the dimension was infinite.

A vital aspect of Riemann's geometry, in which he followed the lead of Gauss, was that it was concerned only with those properties of the manifold that were *intrinsic*, rather than properties that depended on some embedding into a larger space. In particular, the distance between two points x and y was defined to be the length of the shortest curve joining x and y that lay entirely within the surface. Such curves are called *geodesics*. (On a sphere, for example, the geodesics are arcs of great circles.)

Even two-dimensional manifolds could have different, intrinsic curvatures—indeed, a single two-dimensional manifold could have different curvatures in different places—so Riemann's definition led to infinitely many genuinely distinct geometries in each dimension. Furthermore, these geometries were best defined without reference to a Euclidean space that contained them, so the hegemony of Euclidean geometry was broken once and for all.

As the word "hypotheses" in the title of his thesis suggests, Riemann was not at all interested in the sorts of assumptions needed by Euclid. Nor was he much interested in the opposition between Euclidean and non-Euclidean geometry. He made a small reference at the start of his paper to the murkiness that lay at the heart of geometry, despite the efforts of Legendre, and toward the end he considered the three different geometries on two-dimensional manifolds for which the curvature is constant. He noted that one was spherical geometry, another was Euclidean geometry, and the third was different again, and that in each case the angle sums of all triangles could be calculated as soon as one knew the sum of the angles of any one triangle. But he made no reference to Bolyai or Lobachevskii, merely noting that if the geometry of space was indeed a three-dimensional geometry of constant curvature, then to determine which geometry it was would involve taking measurements in unfeasibly large regions of space. He did discuss generalizations of Gauss's curvature to spaces of arbitrary dimension, and he showed what METRICS [III.56] (that is, definitions of distance) there could be on spaces of constant curvature. The formula he wrote down is very general, but as with Bolyai and Lobachevskii it depended on a certain real parameter— the curvature. When the curvature is negative, his definition of distance gives a description of non-Euclidean geometry.

Riemann died in 1866, and by the time his thesis was published an Italian mathematician, Eugenio Beltrami, had independently come to some of the same ideas. He was interested in what the possibilities were if one wished to map one surface to another. For example, one might ask, for some particular surface S, whether it is possible to find a map from S to the plane such that the geodesics in S are mapped to straight lines in the plane. He found that the answer was yes if and only if

the space has constant curvature. There is, for example, a well-known map from the hemisphere to a plane with this property. Beltrami found a simple way of modifying the formula so that now it defined a map from a surface of constant *negative* curvature onto the interior of a disk, and he realized the significance of what he had done: his map defined a metric on the interior of the disk, and the resulting metric space obeyed the axioms for non-Euclidean geometry; therefore, those axioms would not lead to a contradiction.

Some years earlier, Minding, in Germany, had found a surface, sometimes called the pseudosphere, that had constant negative curvature. It was obtained by rotating a curve called the tractrix about its axis. This surface has the shape of a bugle, so it seemed rather less natural than the space of Euclidean plane geometry and unsuitable as a rival to it. The pseudosphere was independently rediscovered by LIOUVILLE [VI.39] some years later, and Codazzi learned of it from that source and showed that triangles on this surface are described by the formulas of hyperbolic trigonometry. But none of these men saw the connection to non-Euclidean geometry—that was left to Beltrami.

Beltrami realized that his disk depicted an infinite space of constant negative curvature, in which the geometry of Lobachevskii (he did not know at that time of Bolyai's work) held true. He saw that it related to the pseudosphere in a way similar to the way that a plane relates to an infinite cylinder. After a period of some doubt, he learned of Riemann's ideas and realized that his disk was in fact as good a depiction of the space of non-Euclidean geometry as any could be; there was no need to realize his geometry as that of a surface in Euclidean three-dimensional space. He thereupon published his essay, in 1868. This was the first time that sound foundations had been publicly given for the area of mathematics that could now be called non-Euclidean geometry.

In 1871 the young KLEIN [VI.57] took up the subject. He already knew that the English mathematician CAYLEY [VI.46] had contrived a way of introducing Euclidean metrical concepts into PROJECTIVE GEOMETRY [I.3 §6.7]. While studying at Berlin, Klein saw a way of generalizing Cayley's idea and exhibiting Beltrami's non-Euclidean geometry as a special case of projective geometry. His idea met with the disapproval of WEIERSTRASS [VI.44], the leading mathematician in Berlin, who objected that projective geometry was not a metrical geometry: therefore, he claimed, it could not generate metrical concepts. However, Klein persisted and in a series of three papers, in 1871, 1872, and 1873, showed that all the known geometries could be regarded as subgeometries of projective geometry. His idea was to recast geometry as the study of a group acting on a space. Properties of figures (subsets of the space) that remain invariant under the action of the group are the geometric properties. So, for example, in a projective space of some dimension, the appropriate group for projective geometry is the group of all transformations that map lines to lines, and the subgroup that maps the interior of a given conic to itself may be regarded as the group of transformations of non-Euclidean geometry: see the box on p. 94. (For a fuller discussion of Klein's approach to geometry, see [I.3 §6].)

In the 1870s Klein's message was spread by the first and third of these papers, which were published in the recently founded journal *Mathematische Annalen*. As Klein's prestige grew, matters changed, and by the 1890s, when he had the second of the papers republished and translated into several languages, it was this, the *Erlanger Programm*, that became well-known. It is named after the university where Klein became a professor, at the remarkably young age of twenty-three, but it was not his inaugural address. (That was about mathematics education.) For many years it was a singularly obscure publication, and it is unlikely that it had the effect on mathematics that some historians have come to suggest.

9 Convincing Others

Klein's work directed attention away from the *figures* in geometry and toward the *transformations* that do not alter the figures in crucial respects. For example, in Euclidean geometry the important transformations are the familiar rotations and translations (and reflections, if one chooses to allow them). These correspond to the motions of rigid bodies that contemporary psychologists saw as part of the way in which individuals learn the geometry of the space around them. But this theory was philosophically contentious, especially when it could be extended to another metrical geometry, non-Euclidean geometry. Klein prudently entitled his main papers "On the so-called non-Euclidean geometry," to keep hostile philosophers at bay (in particular Lotze, who was the well-established Kantian philosopher at Göttingen). But with these papers and the previous work of Beltrami the case for non-Euclidean geometry was made, and almost all mathematicians were persuaded. They believed, that is, that alongside Euclidean

geometry there now stood an equally valid mathematical system called non-Euclidean geometry. As for which one of these was true of space, it seemed so clear that Euclidean geometry was the sensible choice that there appears to have been little or no discussion. Lipschitz showed that it was possible to do all of mechanics in the new setting, and there the matter rested, a hypothetical case of some charm but no more. Helmholtz, the leading physicist of his day, became interested—he had known Riemann personally—and gave an account of what space would have to be if it was learned about through the free mobility of bodies. His first account was deeply flawed, because he was unaware of non-Euclidean geometry, but when Beltrami pointed this out to him he reworked it (in 1870). The reworked version also suffered from mathematical deficiencies, which were pointed out somewhat later by LIE [VI.53], but he had more immediate trouble from philosophers.

Their question was, "What sort of knowledge is this theory of non-Euclidean geometry?" Kantian philosophy was coming back into fashion, and in Kant's view knowledge of space was a fundamental pure a priori intuition, rather than a matter to be determined by experiment: without this intuition it would be impossible to have any knowledge of space at all. Faced with a rival theory, non-Euclidean geometry, neo-Kantian philosophers had a problem. They could agree that the mathematicians had produced a new and prolonged logical exercise, but could it be knowledge of the world? Surely the world could not have two kinds of geometry? Helmholtz hit back, arguing that knowledge of Euclidean geometry and non-Euclidean geometry would be acquired in the same way—through experience—but these empiricist overtones were unacceptable to the philosophers, and non-Euclidean geometry remained a problem for them until the early years of the twentieth century.

Mathematicians could not in fact have given a completely rigorous defense of what was becoming the accepted position, but as the news spread that there were two possible descriptions of space, and that one could therefore no longer be certain that Euclidean geometry was correct, the educated public took up the question: what was the geometry of space? Among the first to grasp the problem in this new formulation was POINCARÉ [VI.61]. He came to mathematical fame in the early 1880s with a remarkable series of essays in which he reformulated Beltrami's disk model so as to make it *conformal*: that is, so that angles in non-Euclidean geometry were represented by the same angles in the

Cross-ratios and distances in conics. A projective transformation of the plane sends four distinct points on a line, A, B, C, D, to four distinct collinear points, A′, B′, C′, D′, in such a way that the quantity

$$\frac{AB}{AD}\frac{CD}{CB}$$

is preserved: that is,

$$\frac{AB}{AD}\frac{CD}{CB} = \frac{A'B'}{A'D'}\frac{C'D'}{C'B'}.$$

This quantity is called the *cross-ratio* of the four points A, B, C, D, and is written CR(A, B, C, D).

In 1871, Klein described non-Euclidean geometry as the geometry of points inside a fixed conic, K, where the transformations allowed are the projective transformations that map K to itself and its interior to its interior (see figure 7). To define the distance between two points P and Q inside K, Klein noted that if the line PQ is extended to meet K at A and D, then the cross-ratio CR(A, P, D, Q) does not change if one applies a projective transformation: that is, it is a *projective invariant*. Moreover, if R is a third point on the line PQ and the points lie in the order P, Q, R, then CR(A, P, D, Q) CR(A, Q, D, R) = CR(A, P, D, R). Accordingly, he defined the distance between P and Q as $d(PQ) = -\frac{1}{2}\log CR(A, P, D, Q)$ (the factor of $-\frac{1}{2}$ is introduced to facilitate the later introduction of trigonometry). With this definition, distance is additive along a line: $d(PQ) + d(QR) = d(PR)$.

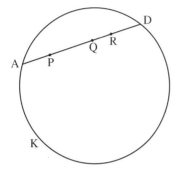

Figure 7 Three points, P, Q, and R, on a non-Euclidean straight line in Klein's projective model of non-Euclidean geometry.

model. He then used his new disk model to connect complex function theory, the theory of linear differential equations, RIEMANN SURFACE [III.79] theory, and non-Euclidean geometry to produce a rich new body of ideas. Then, in 1891, he pointed out that the disk model permitted one to show that any contradiction in non-Euclidean geometry would yield a contradiction in Euclidean geometry as well, and vice versa. Therefore, Euclidean geometry was consistent if and only if non-Euclidean geometry was consistent. A curious consequence of this was that if anybody *had* managed to derive the parallel postulate from the core of Euclidean geometry, then they would have inadvertently proved that Euclidean geometry was inconsistent!

One obvious way to try to decide which geometry described the actual universe was to appeal to physics. But Poincaré was not convinced by this. He argued in another paper (1902) that experience was open to many interpretations and there was no logical way of deciding what belonged to mathematics and what to physics. Imagine, for example, an elaborate set of measurements of angle sums of figures, perhaps on an astronomical scale. Something would have to be taken to be straight, perhaps the paths of rays of light. Suppose, finally, that the conclusion is that the angle sum of a triangle is indeed less than two right angles by an amount proportional to the area of the triangle. Poincaré said that there were two possible conclusions: light rays are straight and the geometry of space is non-Euclidean; or light rays are somehow curved, and space is Euclidean. Moreover, he continued, there was no logical way to choose between these possibilities. All one could do was to make a convention and abide by it, and the sensible convention was to choose the simpler geometry: Euclidean geometry.

This philosophical position was to have a long life in the twentieth century under the name of *conventionalism*, but it was far from accepted in Poincaré's lifetime. A prominent critic of conventionalism was the Italian Federigo Enriques, who, like Poincaré, was both a powerful mathematician and a writer of popular essays on issues in science and philosophy. He argued that one could decide whether a property was geometrical or physical by seeing whether we had any control over it. We cannot vary the law of gravity, but we can change the force of gravity at a point by moving matter around. Poincaré had compared his disk model to a metal disk that was hot in the center and got cooler as one moved outwards. He had shown that a simple law of cooling produced figures identical to those of non-Euclidean geometry. Enriques replied that heat was likewise something we can vary. A property such as Poincaré invoked, which was truly beyond our control, was not physical but geometric.

10 Looking Ahead

In the end, the question was resolved, but not in its own terms. Two developments moved mathematicians beyond the simple dichotomy posed by Poincaré. Starting in 1899, HILBERT [VI.63] began an extensive rewriting of geometry along axiomatic lines, which eclipsed earlier ideas of some Italian mathematicians and opened the way to axiomatic studies of many kinds. Hilbert's work captured very well the idea that if mathematics is sound, it is sound because of the nature of its reasoning, and led to profound investigations in mathematical logic. And in 1915 Einstein proposed his general theory of relativity, which is in large part a geometric theory of gravity. Confidence in mathematics was restored; our sense of geometry was much enlarged, and our insights into the relationships between geometry and space became considerably more sophisticated. Einstein made full use of contemporary ideas about geometry, and his achievement would have been unthinkable without Riemann's work. He described gravity as a kind of curvature in the four-dimensional manifold of spacetime (see GENERAL RELATIVITY AND THE EINSTEIN EQUATIONS [IV.13]). His work led to new ways of thinking about the large-scale structure of the universe and its ultimate fate, and to questions that remain unanswered to this day.

Further Reading

Bonola, R. 1955. *History of Non-Euclidean Geometry*, translated by H. S. Carslaw and with a preface by F. Enriques. New York: Dover.

Euclid. 1956. *The Thirteen Books of Euclid's Elements*, 2nd edn. New York: Dover.

Gray, J. J. 1989. *Ideas of Space: Euclidean, Non-Euclidean, and Relativistic*, 2nd edn. Oxford: Oxford University Press.

Gray, J. J. 2004. *Janos Bolyai, non-Euclidean Geometry and the Nature of Space*. Cambridge, MA: Burndy Library.

Hilbert, D. 1899. *Grundlagen der Geometrie* (many subsequent editions). Tenth edn., 1971, translated by L. Unger, *Foundations of Geometry*. Chicago, IL: Open Court.

Poincaré, H. 1891. Les géométries non-Euclidiennes. *Revue Générales des Sciences Pures et Appliquées* 2:769–74. (Reprinted, 1952, in *Science and Hypothesis*, pp. 35–50. New York: Dover.)

———. 1902. L'expérience et la géométrie. In *La Science et l'Hypothèse*, pp. 95–110. (Reprinted, 1952, in *Science and Hypothesis*, pp. 72–88. New York: Dover.)

II.3 The Development of Abstract Algebra
Karen Hunger Parshall

1 Introduction

What is algebra? To the high school student encountering it for the first time, algebra is an unfamiliar abstract language of x's and y's, a's and b's, together with rules for manipulating them. These letters, some of them variables and some constants, can be used for many purposes. For example, one can use them to express straight lines as equations of the form $y = ax + b$, which can be graphed and thereby visualized in the Cartesian plane. Furthermore, by manipulating and interpreting these equations, it is possible to determine such things as what a given line's root is (if it has one)— that is, where it crosses the x-axis—and what its slope is—that is, how steep or flat it appears in the plane relative to the axis system. There are also techniques for solving simultaneous equations, or equivalently for determining when and where two lines intersect (or demonstrating that they are parallel).

Just when there already seem to be a lot of techniques and abstract manipulations involved in dealing with lines, the ante is upped. More complicated curves like quadratics, $y = ax^2 + bx + c$, and even cubics, $y = ax^3 + bx^2 + cx + d$, and quartics, $y = ax^4 + bx^3 + cx^2 + dx + e$, enter the picture, but the same sort of notation and rules apply, and similar sorts of questions are asked. Where are the roots of a given curve? Given two curves, where do they intersect?

Suppose now that the same high school student, having mastered this sort of algebra, goes on to university and attends an algebra course there. Essentially gone are the by now familiar x's, y's, a's, and b's; essentially gone are the nice graphs that provide a way to picture what is going on. The university course reflects some brave new world in which the algebra has somehow become "modern." This *modern* algebra involves abstract structures—GROUPS [I.3 §2.1], RINGS [III.81 §1], FIELDS [I.3 §2.2], and other so-called objects—each one defined in terms of a relatively small number of axioms and built up of substructures like subgroups, ideals, and subfields. There is a lot of moving around between these objects, too, via maps like group homomorphisms and ring AUTOMORPHISMS [I.3 §4.1]. One objective of this new type of algebra is to understand the underlying structure of the objects and, in doing so, to build entire theories of groups or rings or fields. These abstract theories may then be applied in diverse settings where the basic axioms are satisfied but where it may not be at all apparent a priori that a group or a ring or a field may be lurking. This, in fact, is one of modern algebra's great strengths: once we have proved a general fact about an algebraic structure, there is no need to prove that fact separately each time we come across an instance of that structure. This abstract approach allows us to recognize that contexts that may look quite different are in fact importantly similar.

How is it that two endeavors—the high school analysis of polynomial equations and the modern algebra of the research mathematician—so seemingly different in their objectives, in their tools, and in their philosophical outlooks are both called "algebra"? Are they even related? In fact, they are, but the story of *how* they are is long and complicated.

2 Algebra before There Was Algebra: From Old Babylon to the Hellenistic Era

Solutions of what would today be recognized as first- and second-degree polynomial equations may be found in Old Babylonian cuneiform texts that date to the second millennium B.C.E. However, these problems were neither written in a notation that would be recognizable to our modern-day high school student nor solved using the kinds of general techniques so characteristic of the high school algebra classroom. Rather, particular problems were posed, and particular solutions obtained, from a series of recipe-like steps. No general theoretical justification was given, and the problems were largely cast geometrically, in terms of measurable line segments and surfaces of particular areas. Consider, for example, this problem, translated and transcribed from a clay tablet held in the British Museum (catalogued as BM 13901, problem 1) that dates from between 1800 and 1600 B.C.E.:

> The surface of my confrontation I have accumulated: $45'$ is it. 1, the projection, you posit. The moiety of 1 you break, $30'$ and $30'$ you make hold. $15'$ to $45'$ you append: by 1, 1 is equalside. $30'$ which you have made hold in the inside you tear out: $30'$ the confrontation.

This may be translated into modern notation as the equation $x^2 + 1x = \frac{3}{4}$, where it is important to notice that the Babylonian number system is base 60, so $45'$ denotes $\frac{45}{60} = \frac{3}{4}$. The text then lays out the following algorithm for solving the problem: take 1, the coefficient of the linear term, and halve it to get $\frac{1}{2}$. Square $\frac{1}{2}$

to get $\frac{1}{4}$. Add $\frac{1}{4}$ to $\frac{3}{4}$, the constant term, to get 1. This is the square of 1. Subtract from this the $\frac{1}{2}$ which you multiplied by to get $\frac{1}{2}$, the side of the square. The modern reader can easily see that this algorithm is equivalent to what is now called the quadratic formula, but the Babylonian tablet presents it in the context of a particular problem and repeats it in the contexts of other particular problems. There are no equations in the modern sense; the Babylonian writer is literally effecting a construction of plane figures. Similar problems and similar algorithmic solutions can also be found in ancient Egyptian texts such as the Rhind papyrus, believed to have been copied in 1650 B.C.E. from a text that was about a century and a half older.

There is a sharp contrast between the problem-oriented, untheoretical approach characteristic of texts from this early period and the axiomatic and deductive approach that EUCLID [VI.2] introduced into mathematics in around 300 B.C.E. in his magisterial, geometrical treatise, the *Elements*. (See GEOMETRY [II.2] for a further discussion of this work.) There, building on explicit definitions and a small number of axioms or self-evident truths, Euclid proceeded to deduce known—and almost certainly some hitherto unknown—results within a strictly geometrical context. Geometry done in this axiomatic context defined Euclid's standard of rigor. But what does this quintessentially geometrical text have to do with algebra? Consider the sixth proposition in Euclid's Book II, ostensibly a book on plane figures, and in particular quadrilaterals:

> If a straight line be bisected and a straight line be added to it in a straight line, the rectangle contained by the whole with the added straight line and the added straight line together with the square on the half is equal to the square on the straight line made up of the half and the added straight line.

While clearly a geometrical construction, it equally clearly describes two constructions—one a rectangle and one a square—that have equal areas. It therefore describes something that we should be able to write as an equation. Figure 1 gives the picture corresponding to Euclid's construction: he proves that the area of rectangle ADMK equals the sum of rectangles CDML and HMFG. To do this, he adds the square on CB—namely, square LHGE—to CDML and HMFG. This gives square CDFE. It is not hard to see that this is equivalent to the high school procedure of "completing the square" and to the algebraic equation $(2a + b)b + a^2 = (a + b)^2$, which we obtain by setting CB $= a$ and

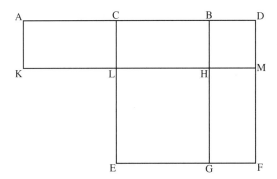

Figure 1 The sixth proposition from Euclid's Book II.

$BD = b$. Equivalent, yes, but for Euclid this is a specific *geometrical* construction and a particular *geometrical* equivalence. For this reason, he could not deal with anything but positive real quantities, since the *sides* of a geometrical figure could only be *measured* in those terms. Negative quantities did not and could not enter into Euclid's fundamentally geometrical mathematical world. Nevertheless, in the historical literature, Euclid's Book II has often been described as dealing with "geometrical algebra," and, because of our easy translation of the book's propositions into the language of algebra, it has been argued, albeit ahistorically, that Euclid *had* algebra but simply presented it geometrically.

Although Euclid's geometrical standard of rigor came to be regarded as a pinnacle of mathematical achievement, it was in many ways not typical of the mathematics of classical Greek antiquity, a mathematics that focused less on systematization and more on the clever and individualistic solution of particular problems. There is perhaps no better exemplar of this than ARCHIMEDES [VI.3], held by many to have been one of the three or four greatest mathematicians of all time. Still, Archimedes, like Euclid, posed and solved particular problems geometrically. As long as geometry defined the standard of rigor, not only negative numbers but also what we would recognize as polynomial equations of degree higher than three effectively fell outside the sphere of possible mathematical discussion. (As in the example from Euclid above, quadratic polynomials result from the geometrical process of completing the square; cubics could conceivably result from the geometrical process of completing the cube; but quartics and higher-degree polynomials could not be constructed in this way in familiar, three-dimensional space.) However, there was another mathematician of great importance to the present story, Diophantus of Alexandria (who was active in the middle of the third century C.E.). Like Archimedes, he posed particular problems, but he solved them in an algorithmic style much more reminiscent of the Old Babylonian texts than of Archimedes' geometrical constructions, and as a result he was able to begin to exceed the bounds of geometry.

In his text *Arithmetica*, Diophantus put forward general, indeterminate problems, which he then restricted by specifying that the solutions should have particular forms, before providing specific solutions. He expressed these problems in a very different way from the purely rhetorical style that held sway for centuries after him. His notation was more algebraic and was ultimately to prove suggestive to sixteenth-century mathematicians (see below). In particular, he used special abbreviations that allowed him to deal with the first *six* positive and negative powers of the unknown as well as with the unknown to the zeroth power. Thus, whatever his mathematics was, it was not the "geometrical algebra" of Euclid and Archimedes.

Consider, for example, this problem from Book II of the *Arithmetica*: "To find three numbers such that the square of any one of them minus the next following gives a square." In terms of modern notation, he began by restricting his attention to solutions of the form $(x + 1, 2x + 1, 4x + 1)$. It is easy to see that $(x+1)^2 - (2x+1) = x^2$ and $(2x+1)^2 - (4x+1) = 4x^2$, so two of the conditions of the problem are immediately satisfied, but he needed $(4x+1)^2 - (x+1) = 16x^2 + 7x$ to be a square as well. Arbitrarily setting $16x^2 + 7x = 25x^2$, Diophantus then determined that $x = \frac{7}{9}$ gave him what he needed, so a solution was $\frac{16}{9}, \frac{23}{9}, \frac{37}{9}$, and he was done. He provided no geometrical justification because in his view none was needed, a *single* numerical solution was all he required. He did not set up what we would recognize as a more general set of equations and try to find all possible solutions.

Diophantus, who lived more than four centuries after Archimedes' death, was doing neither geometry nor algebra in our modern sense, yet the kinds of problems and the sorts of solutions he obtained for them were very different from those found in the works of either Euclid or Archimedes. The extent to which Diophantus created a wholly new approach, rather than drawing on an Alexandrian tradition of what might be called "algorithmic algebraic," as opposed to "geometric algebraic," scholarship is unknown. It is clear that by the time Diophantus's ideas were introduced into the Latin West in the sixteenth century, they suggested new possibilities to mathematicians long conditioned to the authority of geometry.

3 Algebra before There Was Algebra: The Medieval Islamic World

The transmission of mathematical ideas was, however, a complex process. After the fall of the Roman Empire and the subsequent decline of learning in the West, both the Euclidean and the Diophantine traditions ultimately made their way into the medieval Islamic world. There they were not only preserved—thanks to the active translation initiatives of Islamic scholars—but also studied and extended.

AL-KHWĀRIZMĪ [VI.5] was a scholar at the royally funded House of Wisdom in Baghdad. He linked the kinds of geometrical arguments Euclid had presented in Book II of his *Elements* with the indigenous problem-solving algorithms that dated back to Old Babylonian times. In particular, he wrote a book on practical mathematics, entitled *al-Kitāb al-mukhtaṣar fī ḥisāb al-jabr wa'l-muqābala* ("The compendious book on calculation by completion and balancing"), beginning it with a theoretical discussion of what we would now recognize as polynomial equations of the first and second degrees. (The latinization of the word "al-jabr" or "completion" in his title gave us our modern term "algebra.") Because he employed neither negative numbers nor zero coefficients, al-Khwārizmī provided a systematization in terms of six separate kinds of examples where we would need just one, namely $ax^2 + bx + c = 0$. He considered, for example, the case when "a square and 10 roots are equal to 39 units," and his algorithmic solution in terms of multiplications, additions, and subtractions was in precisely the same form as the above solution from tablet BM 13901. This, however, was not enough for al-Khwārizmī. "It is necessary," he said, "that we should demonstrate geometrically the truth of the same problems which we have explained in numbers," and he proceeded to do this by "completing the square" in geometrical terms reminiscent of, but not as formal as, those Euclid used in Book II. (Abū Kāmil (ca. 850–930), an Egyptian Islamic mathematician of the generation after al-Khwārizmī, introduced a higher level of Euclidean formality into the geometric–algorithmic setting.) This juxtaposition made explicit how the relationships between geometrical areas and lines could be interpreted in terms of numerical multiplications, additions, and subtractions,

a key step that would ultimately suggest a move away from the *geometrical* solution of *particular problems* and toward an *algebraic* solution of *general types of equations*.

Another step along this path was taken by the mathematician and poet Omar Khayyam (ca. 1050–1130) in a book he entitled *Al-jabr* after al-Khwārizmī's work. Here he proceeded to systematize and solve what we would recognize, in the absence of both negative numbers and zero coefficients, as the cases of the cubic equation. Following al-Khwārizmī, Khayyam provided geometrical justifications, yet his work, even more than that of his predecessor, may be seen as closer to a general problem-solving technique for specific cases of equations, that is, closer to the notion of algebra.

The Persian mathematician al-Karajī (who flourished in the early eleventh century) also knew well and appreciated the geometrical tradition stemming from Euclid's *Elements*. However, like Abū-Kāmil, he was aware of the Diophantine tradition too, and synthesized in more general terms some of the procedures Diophantus had laid out in the context of specific examples in the *Arithmetica*. Although Diophantus's ideas and style were known to these and other medieval Islamic mathematicians, they would remain unknown in the Latin West until their rediscovery and translation in the sixteenth century. Equally unknown in the Latin West were the accomplishments of Indian mathematicians, who had succeeded in solving some quadratic equations algorithmically by the beginning of the eighth century and who, like Bragmagupta four hundred years later, had techniques for finding integer solutions to particular examples of what are today called Pell's equations, namely, equations of the form $ax^2 + b = y^2$, where a and b are integers and a is not a square.

4 Algebra before There Was Algebra: The Latin West

Concurrent with the rise of Islam in the East, the Latin West underwent a gradual cultural and political stabilization in the centuries following the fall of the Roman Empire. By the thirteenth century, this relative stability had resulted in the firm entrenchment of the Catholic Church as well as the establishment both of universities and of an active economy. Moreover, the Islamic conquest of most of the Iberian peninsula in the eighth century and the subsequent establishment there of an Islamic court, library, and research facility similar to the House of Wisdom in Baghdad brought the fruits of medieval Islamic scholarship to western Europe's doorstep. However, as Islam found its position on the Iberian peninsula increasingly compromised in the twelfth and thirteenth centuries, this Islamic learning, as well as some of the ancient Greek scholarship that the medieval Islamic scholars had preserved in Latin translation, began to filter into medieval Europe. In particular, FIBONACCI [VI.6], son of an influential administrator within the Pisan city state, encountered al-Khwārizmī's text and recognized not only the impact that the Arabic number system detailed there could have on accounting and commerce (Roman numerals and their cumbersome rules for manipulation were still widely in use) but also the importance of al-Khwārizmī's theoretical discussion, with its wedding of geometrical proof and the algorithmic solution of what we can interpret as first- and second-degree equations. In his 1202 book *Liber Abbaci*, Fibonacci presented al-Khwārizmī's work almost verbatim, and extolled all of these virtues, thus effectively introducing this knowledge and approach into the Latin West.

Fibonacci's presentation, especially of the practical aspects of al-Khwārizmī's text, soon became well-known in Europe. So-called abacus schools (named after Fibonacci's text and not after the Chinese calculating instrument) sprang up all over the Italian peninsula, particularly in the fourteenth and fifteenth centuries, for the training of accountants and bookkeepers in an increasingly mercantilistic Western world. The teachers in these schools, the "maestri d'abaco," built on and extended the algorithms they found in Fibonacci's text. Another tradition, the Cossist tradition—after the German word "Coss" connoting algebra, that is, "Kunstrechnung" or "artful calculation"—developed simultaneously in the Germanic regions of Europe and aimed to introduce algebra into the mainstream there.

In 1494 the Italian Luca Pacioli published (by now this is the operative word: Pacioli's text is one of the earliest *printed* mathematical texts) a compendium of all known mathematics. By this time, the geometrical justifications that al-Khwārizmī and Fibonacci had presented had long since fallen from the mathematical vernacular. By reintroducing them in his book, the *Summa*, Pacioli brought them back to the mathematical fore. Not knowing of Khayyam's work, he asserted that solutions had been discovered only in the six cases treated by both al-Khwārizmī and Fibonacci, even though there had been abortive attempts to solve the cubic and even

though he held out the hope that it could ultimately be solved.

Pacioli's book had highlighted a key unsolved problem: could algorithmic solutions be determined for the various cases of the cubic? And, if so, could these be justified geometrically with proofs similar in spirit to those found in the texts of al-Khwārizmī and Fibonacci?

Among several sixteenth-century Italian mathematicians who eventually managed to answer the first question in the affirmative was CARDANO [VI.7]. In his *Ars Magna*, or *The Great Art*, of 1545, he presented algorithms with geometric justifications for the various cases of the cubic, effectively completing the cube where al-Khwārizmī and Fibonacci had completed the square. He also presented algorithms that had been discovered by his student Ludovico Ferrari (1522–65) for solving the cases of the quartic. These intrigued him, because, unlike the algorithms for the cubic, they were not justified geometrically. As he put it in his book, "all those matters up to and including the cubic are fully demonstrated, but the others which we will add, either by necessity or out of curiosity, we do not go beyond barely setting out." An algebra was breaking out of the geometrical shell in which it had been encased.

5 Algebra Is Born

This process was accelerated by the rediscovery and translation into Latin of Diophantus's *Arithmetica* in the 1560s, with its abbreviated presentational style and ungeometrical approach. Algebra, as a general problem-solving technique, applicable to questions in geometry, number theory, and other mathematical settings, was established in RAPHAEL BOMBELLI's [VI.8] *Algebra* of 1572 and, more importantly, in VIÈTE's [VI.9] *In Artem Analyticem Isagoge*, or *Introduction to the Analytic Art*, of 1591. The aim of the latter was, in Viète's words, "to leave no problem unsolved," and to this end he developed a true notation—using vowels to denote variables and consonants to denote coefficients—as well as methods for solving equations in one unknown. He called his techniques "specious logistics."

Dimensionality—in the form of his so-called *law of homogeneity*—was, however, still an issue for Viète. As he put it, "[o]nly homogeneous magnitudes are to be compared to one another." The problem was that he distinguished two types of magnitudes: "ladder magnitudes"—that is, variables (A side) (or x in our modern notation), (A square) (or x^2), (A cube) (or x^3), etc.; and "compared magnitudes"—that is, coefficients (*B* length) of dimension one, (*B* plane) of dimension two, (*B* solid) of dimension three, etc. In the light of his law of homogeneity, then, Viète could legitimately perform the operation (*A* cube) + (*B* plane) (*A* side) (or $x^3 + bx$ in our notation), since the dimension of (*A* cube) is three, as is that of the product of the two-dimensional coefficient (*B* plane) and the one-dimensional variable (*A* side), but he could not legally add the three-dimensional variable (*A* cube) to the two-dimensional product of the one-dimensional coefficient (*B* length) and the one-dimensional variable (*A* side) (or, again, $x^3 + bx$ in our notation). Be this as it may, his "analytic art" still allowed him to add, subtract, multiply, and divide *letters* as opposed to specific numbers, and those letters, as long as they satisfied the law of homogeneity, could be raised to the second, third, fourth, or, indeed, any power. He had a rudimentary algebra, although he failed to apply it to curves.

The first mathematicians to do that were FERMAT [VI.12] and DESCARTES [VI.11] in their independent development of the analytic geometry so familiar to the high school algebra student of today. Fermat, and others like Thomas Harriot (ca. 1560–1621) in England, were influenced in their approaches by Viète, while Descartes not only introduced our present-day notational convention of representing variables by x's and y's and constants by a's, b's, and c's but also began the arithmetization of algebra. He introduced a unit that allowed him to interpret all geometrical magnitudes as line segments, whether they were x's, x^2's, x^3's, x^4's, or any higher power of x, thereby removing concerns about homogeneity. Fermat's main work in this direction was a 1636 manuscript written in Latin, entitled "Introduction to plane and solid loci" and circulated among the early seventeenth-century mathematical cognoscenti; Descartes's was *La Géométrie*, written in French as one of three appendices to his philosophical tract, *Discours de la Méthode*, published in 1637. Both were regarded as establishing the identification of geometrical curves with equations in two unknowns, or in other words as establishing analytic geometry and thereby introducing *algebraic* techniques into the solution of what had previously been considered *geometrical* problems. In Fermat's case, the curves were lines or conic sections—quadratic expressions in x and y; Descartes did this too, but he also considered equations more generally, tackling questions about the roots of polynomial equations that were connected with transforming and reducing the polynomials.

In particular, although he gave no proof or even general statement of it, Descartes had a rudimentary version of what we would now call THE FUNDAMENTAL THEOREM OF ALGEBRA [V.13], the result that a polynomial equation $x^n + a_{n-1}x^{n-1} + \cdots + a_1x + a_0$ of degree n has precisely n roots over the field $\mathbb{C}$ of complex numbers. For example, while he held that a given polynomial of degree n could be decomposed into n linear factors, he also recognized that the cubic $x^3 - 6x^2 + 13x - 10 = 0$ has three roots: the real root 2 and two complex roots. In his further exploration of these issues, moreover, he developed algebraic techniques, involving suitable transformations, for analyzing polynomial equations of the fifth and sixth degrees. Liberated from homogeneity concerns, Descartes was thus able to use his algebraic techniques freely to explore territory where the geometrically bound Cardano had clearly been reluctant to venture. NEWTON [VI.14] took the liberation of algebra from geometrical concerns a step further in his *Arithmetica Universalis* (or *Universal Arithmetic*) of 1707, arguing for the complete arithmetization of algebra, that is, for modeling algebra and algebraic operations on the real numbers and the usual operations of arithmetic.

Descartes's *La Géométrie* highlighted at least two problems for further algebraic exploration: the fundamental theorem of algebra and the solution of polynomial equations of degree greater than four. Although eighteenth-century mathematicians like D'ALEMBERT [VI.20] and EULER [VI.19] attempted proofs of the fundamental theorem of algebra, the first person to prove it rigorously was GAUSS [VI.26], who gave four distinct proofs over the course of his career. His first, an algebraic geometrical proof, appeared in his doctoral dissertation of 1799, while a second, fundamentally different proof was published in 1816, which in modern terminology essentially involved constructing the polynomial's splitting field. While the fundamental theorem of algebra established how many roots a given polynomial equation has, it did not provide insight into exactly what those roots were or how precisely to find them. That problem and its many mathematical repercussions exercised a number of mathematicians in the late eighteenth and nineteenth centuries and formed one of the strands of the mathematical thread that became modern algebra in the early twentieth century. Another emerged from attempts to understand the general behavior of systems of (one or more) polynomials in n unknowns, and yet another grew from efforts to approach number-theoretic questions algebraically.

6 The Search for the Roots of Algebraic Equations

The problem of finding roots of polynomials provides a direct link from the algebra of the high school classroom to that of the modern research mathematician. Today's high school student dutifully employs the quadratic formula to calculate the roots of second-degree polynomials. To derive this formula, one transforms the given polynomial into one that can be solved more easily. By more complicated manipulations of cubics and quartics, Cardano and Ferrari obtained formulas for the roots of those as well. It is natural to ask whether the same can be done for higher-degree polynomials. More precisely, are there formulas that involve just the usual operations of arithmetic—addition, subtraction, multiplication, and division—together with the extraction of roots? When there is such a formula, one says that the equation is *solvable by radicals.*

Although many eighteenth-century mathematicians (including Euler, Alexandre-Théophile Vandermonde (1735–96), WARING [VI.21], and Étienne Bézout (1730–83)) contributed to the effort to decide whether higher-order polynomial equations are solvable by radicals, it was not until the years from roughly 1770 to 1830 that there were significant breakthroughs, particularly in the work of LAGRANGE [VI.22], ABEL [VI.33], and Gauss.

In a lengthy set of "Réflexions sur la résolution algébrique des équations" (Reflections on the algebraic resolution of equations) published in 1771, Lagrange tried to determine principles underlying the resolution of algebraic equations in general by analyzing in detail the specific cases of the cubic and the quartic. Building on the work of Cardano, Lagrange showed that a cubic of the form $x^3 + ax^2 + bx + c = 0$ could always be transformed into a cubic with no quadratic term $x^3 + px + q = 0$ and that the roots of this could be written as $x = u + v$, where u^3 and v^3 are the roots of a certain quadratic polynomial equation. Lagrange was then able to show that if x_1, x_2, x_3 are the three roots of the cubic, the intermediate functions u and v could actually be written as $u = \frac{1}{3}(x_1 + \alpha x_2 + \alpha^2 x_3)$ and $v = \frac{1}{3}(x_1 + \alpha^2 x_2 + \alpha x_3)$, for α a primitive cube root of unity. That is, u and v could be written as rational expressions or resolvents in x_1, x_2, x_3. Conversely, starting with a linear expression $y = Ax_1 + Bx_2 + Cx_3$ in the roots x_1, x_2, x_3 and then permuting the roots in all possible ways yielded six expressions each of which was a root of a particular sixth-degree polynomial equation. An analysis of the latter equation (which involved the exploitation of properties of symmetric polynomials) yielded the same expressions for u and v in terms of x_1, x_2, x_3 and the cube root of unity α. As Lagrange showed, this kind of two-pronged analysis—involving intermediate expressions rational in the roots that are solutions of a solvable equation as well as the behavior of certain rational expressions under permutation of the roots—yielded the complete solution in the cases both of the cubic and the quartic. It was *one* approach that encompassed the solution of *both* types of equation. But could this technique be extended to the case of the quintic and higher-degree polynomials? Lagrange was unable to push it through in the case of the quintic, but by building on his ideas, first his student Paolo Ruffini (1765–1822) at the turn of the nineteenth century and then, definitively, the young Norwegian mathematician Abel in the 1820s showed that, in fact, the quintic is *not* solvable by radicals. (See THE INSOLUBILITY OF THE QUINTIC [V.21].) This negative result, however, still left open the questions of which algebraic equations *were* solvable by radicals and why.

As Lagrange's analysis seemed to underscore, the answer to this question in the cases of the cubic and the quartic involved in a critical way the cube and fourth roots of unity, respectively. By definition, these satisfy the particularly simple polynomial equations $x^3 - 1 = 0$ and $x^4 - 1 = 0$, respectively. It was thus natural to examine the general case of the so-called cyclotomic equation $x^n - 1 = 0$ and ask for what values n the nth roots of unity are actually constructible. To put this question in equivalent algebraic terms: for which n is it possible to find a formula for the nth roots of unity that expresses them in terms of integers using the usual arithmetical operations and extraction of square (but not higher) roots? This was one of the many questions explored by Gauss in his wide-ranging, magisterial, and groundbreaking 1801 treatise *Disquisitiones Arithmeticae.* One of his most famous results was that the regular 17-gon (or, equivalently, a 17th root of unity) was constructible. In the course of his analysis, he not only employed techniques similar to those developed by Lagrange but also developed key concepts such as MODULAR ARITHMETIC [III.58] and the properties of the modular "worlds" $\mathbb{Z}_p$, for p a prime, and, more generally, $\mathbb{Z}_n$, for $n \in \mathbb{Z}^+$, as well as the notion of a primitive element (a generator) of what would later be termed a cyclic group.

Although it is not clear how well he knew Gauss's work, in the years around 1830 GALOIS [VI.41] drew from the ideas both of Lagrange on the analysis of resolvents and of CAUCHY [VI.29] on permutations and substitutions to obtain a solution to the general problem of solvability of polynomial equations by radicals. Although his approach borrowed from earlier ideas, it was in one important respect fundamentally new. Whereas prior efforts had aimed at deriving an *explicit algorithm for calculating* the roots of a polynomial of a given degree, Galois formulated a theoretical process based on constructs more general than but derived from the given equation that allowed him to *assess whether or not that equation was solvable*.

To be more precise, Galois recast the problem into one in terms of two new concepts: fields (which he called "domains of rationality") and groups (or, more precisely, groups of substitutions). A polynomial equation $f(x) = 0$ of degree n was reducible over its domain of rationality—the ground field from which its coefficients were taken—if all n of its roots were in that ground field; otherwise, it was irreducible over that field. It could, however, be reducible over some larger field. Consider, for example, the polynomial $x^2 + 1$ as a polynomial over $\mathbb{R}$, the field of real numbers. While we know from high school algebra that this polynomial does not factor into a product of two real, linear factors (that is, there are no real numbers r_1 and r_2 such that $x^2 + 1 = (x - r_1)(x - r_2)$), it does factor over $\mathbb{C}$, the field of complex numbers, and, specifically, $x^2 + 1 = (x + \sqrt{-1})(x - \sqrt{-1})$. Thus, if we take all numbers of the form $a + b\sqrt{-1}$, where a and b belong to $\mathbb{R}$, then we enlarge $\mathbb{R}$ to a new field $\mathbb{C}$ in which the polynomial $x^2 + 1$ is reducible. If $\mathbb{F}$ is a field and x is an element of $\mathbb{F}$ that does not have an nth root in $\mathbb{F}$, then by a similar process we can adjoin an element y to $\mathbb{F}$ and stipulate that $y^n = x$. We call y a *radical*. The set of all polynomial expressions in y, with coefficients in $\mathbb{F}$, can be shown to form a larger field. Galois showed that if it was possible to enlarge $\mathbb{F}$ by successively adjoining radicals to obtain a field K in which $f(x)$ factored into n linear factors, then $f(x) = 0$ was solvable by radicals. He developed a process that hinged both on the notion of adjoining an element—in particular, a so-called primitive element—to a given ground field and on the idea of analyzing the internal structure of this new, enlarged field via an analysis of the (finite) group of substitutions (automorphisms of K) that leave invariant all rational relations of the n roots of $f(x) = 0$. The group-theoretic aspects of Galois's analysis were particularly potent; he introduced the notions, although not the modern terminology, of a normal subgroup of a group, a factor group, and a solvable group. Galois thus resolved the concrete problem of determining when a polynomial equation was solvable by radicals by examining it from the abstract perspective of groups and their internal structure.

Galois's ideas, although sketched in the early 1830s, did not begin to enter into the broader mathematical consciousness until their publication in 1846 in LIOUVILLE's [VI.39] *Journal des Mathématiques Pures et Appliquées*, and they were not fully appreciated until two decades later when first Joseph Serret (1819-85) and then JORDAN [VI.52] fleshed them out more fully. In particular, Jordan's *Traité des Substitutions et des Équations Algébriques* ("Treatise on substitutions and on algebraic equations") of 1870 not only highlighted Galois's work on the solution of algebraic equations but also developed the general structure theory of permutation groups as it had evolved at the hands of Lagrange, Gauss, Cauchy, Galois, and others. By the end of the nineteenth century, this line of development of group theory, stemming from efforts to solve algebraic equations by radicals, had intertwined with three others: the abstract notion of a group defined in terms of a group multiplication table, which was formulated by CAYLEY [VI.46], the structural work of mathematicians like Ludwig Sylow (1832-1918) and Otto Hölder (1859-1937), and the geometrical work of LIE [VI.53] and KLEIN [VI.57]. By 1893, when Heinrich Weber (1842-1914) codified much of this earlier work by giving the first actual abstract definitions of the notions both of group and field, thereby recasting them in a form much more familiar to the modern mathematician, groups and fields had been shown to be of central importance in a wide variety of areas, both mathematical and physical.

7 Exploring the Behavior of Polynomials in n Unknowns

The problem of solving algebraic equations involved finding the roots of polynomials in *one* unknown. At least as early as the late seventeenth century, however, mathematicians like LEIBNIZ [VI.15] had been interested in techniques for solving simultaneously systems of linear equations in more than two variables. Although his work remained unknown at the time, Leibniz considered three linear equations in three unknowns and determined their simultaneous solvability based on the value of a particular expression in the coefficients of the system. This expression, equivalent to what Cauchy would later call the DETERMINANT

[III.15] and which would ultimately be associated with an $n \times n$ square array or MATRIX [I.3 §4.2] of coefficients, was also developed and analyzed independently by Gabriel Cramer (1704–52) in the mid eighteenth century in the general context of the simultaneous solution of a system of n linear equations in n unknowns. From these beginnings, a theory of determinants, independent of the context of solving systems of linear equations, quickly became a topic of algebraic study in its own right, attracting the attention of Vandermonde, LAPLACE [VI.23], and Cauchy, among others. Determinants were thus an example of a new algebraic construct, the properties of which were then systematically explored.

Although determinants came to be viewed in terms of what SYLVESTER [VI.42] would dub matrices, a theory of matrices proper grew initially from the context not of solving simultaneous linear equations but rather of linearly transforming the variables of homogeneous polynomials in two, three, or more generally n variables. In the *Disquisitiones Arithmeticae*, for example, Gauss considered how binary and ternary quadratic forms with integer coefficients—expressions of the form $a_1 x^2 + 2a_2 xy + a_3 y^2$ and $a_1 x^2 + a_2 y^2 + a_3 z^2 + 2a_4 xy + 2a_5 xz + 2a_6 yz$, respectively—are affected by a linear transformation of their variables. In the ternary case, he applied the linear transformation $x = \alpha x' + \beta y' + \gamma z'$, $y = \alpha' x' + \beta' y' + \gamma' z'$, and $z = \alpha'' x' + \beta'' y' + \gamma'' z'$ to derive a new ternary form. He denoted the linear transformation of the variables by the square array

$$
\begin{array}{ccc}
\alpha, & \beta, & \gamma \\
\alpha', & \beta', & \gamma' \\
\alpha'', & \beta'', & \gamma''
\end{array}
$$

and, in the process of showing what the composition of two such transformations was, gave an explicit example of matrix multiplication. By the middle of the nineteenth century, Cayley had begun to explore matrices per se and had established many of the properties that the theory of matrices as a mathematical system in its own right enjoys. This line of algebraic thought was eventually reinterpreted in terms of the theory of algebras (see below) and developed into the independent area of linear algebra and the theory of VECTOR SPACES [I.3 §2.3].

Another theory that arose out of the analysis of linear transformations of homogeneous polynomials was the theory of invariants, and this too has its origins in some sense in Gauss's *Disquisitiones*. As in his study of ternary quadratic forms, Gauss began his study of binary forms by applying a linear transformation, specifically, $x = \alpha x' + \beta y'$, $y = \gamma x' + \delta y'$. The result was the new binary form $a_1'(x')^2 + 2a_2' x' y' + a_3'(y')^2$, where, explicitly, $a_1' = a_1 \alpha^2 + 2a_2 \alpha \gamma + a_3 \gamma^2$, $a_2' = a_1 \alpha \beta + a_2 (\alpha \delta + \beta \gamma) + a_3 \gamma \delta$, and $a_3' = a_1 \beta^2 + 2a_2 \beta \delta + a_3 \delta^2$. As Gauss noted, if you multiply the second of these equations by itself and subtract from this the product of the first and the third equations, you obtain the relation $a_2'^2 - a_1' a_3' = (a_2^2 - a_1 a_3)(\alpha \delta - \beta \gamma)^2$. To use language that Sylvester would develop in the early 1850s, Gauss realized that the expression $a_2^2 - a_1 a_3$ in the coefficients of the original binary quadratic form is an *invariant* in the sense that it remains unchanged up to a power of the determinant of the linear transformation. By the time Sylvester coined the term, the invariant phenomenon had also appeared in the work of the English mathematician BOOLE [VI.43], and had attracted Cayley's attention. It was not until after Cayley and Sylvester met in the late 1840s, however, that the two of them began to pursue a theory of invariants proper, which aimed to determine all invariants for homogeneous polynomials of degree m in n unknowns as well as simultaneous invariants for systems of such polynomials.

Although Cayley and (especially) Sylvester pursued this line of research from a purely algebraic point of view, invariant theory also had number-theoretic and geometric implications, the former explored by Gotthold Eisenstein (1823–52) and HERMITE [VI.47], the latter by Otto Hesse (1811–74), Paul Gordan (1837–1912), and Alfred Clebsch (1833–72), among others. It was of particular interest to understand how many "genuinely distinct" invariants were associated with a specific form, or system of forms. In 1868, Gordan achieved a fundamental breakthrough by showing that the invariants associated with any binary form in n variables can always be expressed in terms of a finite number of them. By the late 1880s and early 1890s, however, HILBERT [VI.63] brought new, abstract concepts associated with the theory of algebras (see below) to bear on invariant theory and, in so doing, not only reproved Gordan's result but also showed that the result was true for forms of degree m in n unknowns. With Hilbert's work, the emphasis shifted from the concrete calculations of his English and German predecessors to the kind of structurally oriented existence theorems that would soon be associated with abstract, modern algebra.

8   The Quest to Understand the Properties of "Numbers"

As early as the sixth century B.C.E., the Pythagoreans had studied the properties of numbers formally. For example, they defined the concept of a *perfect number*, which is a positive integer, such as $6 = 1 + 2 + 3$ and $28 = 1 + 2 + 4 + 7 + 14$, which is the sum of its divisors (excluding the integer itself). In the sixteenth century, Cardano and Bombelli had willingly worked with new expressions, complex numbers, of the form $a + \sqrt{-b}$, for real numbers a and b, and had explored their computational properties. In the seventeenth century, Fermat famously claimed that he could prove that the equation $x^n + y^n = z^n$, for n an integer greater than 2, had no solutions in the integers, except for the trivial cases when $z = x$ or $z = y$ and the remaining variable is zero. The latter result, known as FERMAT'S LAST THEOREM [V.10], generated many new ideas, especially in the eighteenth and nineteenth centuries, as mathematicians worked to find an actual proof of Fermat's claim. Central to their efforts were the creation and algebraic analysis of new types of number systems that extended the integers in much the same way that Galois had extended fields. This flexibility to create and analyze new number systems was to become one of the hallmarks of modern algebra as it would develop into the twentieth century.

One of the first to venture down this path was Euler. In the proof of Fermat's last theorem for the $n = 3$ case that he gave in his *Elements of Algebra* of 1770, Euler introduced the system of numbers of the form $a + b\sqrt{-3}$, where a and b are integers. He then blithely proceeded to factorize them into primes, without further justification, just as he would have factorized ordinary integers. By the 1820s and 1830s, Gauss had launched a more systematic study of numbers that are now called the *Gaussian integers*. These are all numbers of the form $a + b\sqrt{-1}$, for integers a and b. He showed that, like the integers, the Gaussian integers are closed under addition, subtraction, and multiplication; he defined the notions of unit, prime, and norm in order to prove an analogue of THE FUNDAMENTAL THEOREM OF ARITHMETIC [V.14] for them. He thereby demonstrated that there were whole new algebraic worlds to create and explore. (See ALGEBRAIC NUMBERS [IV.1] for more on these topics.)

Whereas Euler had been motivated in his work by Fermat's last theorem, Gauss was trying to generalize the LAW OF QUADRATIC RECIPROCITY [V.28] to a law of biquadratic reciprocity. In the quadratic case, the problem was the following. If a and m are integers with $m \geqslant 2$, then we say that a is a *quadratic residue* mod m if the equation $x^2 = a$ has a solution mod m; that is, if there is an integer x such that x^2 is congruent to a mod m. Now suppose that p and q are distinct odd primes. If you know whether p is a quadratic residue mod q, is there a simple way of telling whether q is a quadratic residue mod p? In 1785, Legendre had posed and answered this question—the status of q mod p will be the same as that of p mod q if at least one of p and q is congruent to 1 mod 4, and different if they are both congruent to 3 mod 4—but he had given a faulty proof. By 1796, Gauss had come up with the first rigorous proof of the theorem (he would ultimately give eight different proofs of it), and by the 1820s he was asking the analogous question for the case of two biquadratic equivalences $x^4 \equiv p \pmod{q}$ and $y^4 \equiv q \pmod{p}$. It was in his attempts to answer this new question that he introduced the Gaussian integers and signaled at the same time that the theory of residues of higher degrees would make it necessary to create and analyze still other new sorts of "integers." Although Eisenstein, DIRICHLET [VI.36], Hermite, KUMMER [VI.40], and KRONECKER [VI.48], among others, pushed these ideas forward in this Gaussian spirit, it was DEDEKIND [VI.50] in his tenth supplement to Dirichlet's *Vorlesungen über Zahlentheorie* (*Lectures on Number Theory*) of 1871 who fundamentally reconceptualized the problem by treating it not number theoretically but rather set theoretically and axiomatically. Dedekind introduced, for example, the general notions—if not what would become the precise axiomatic definitions—of fields, rings, IDEALS [III.81 §2], and MODULES [III.81 §3] and analyzed his number-theoretic setting in terms of these new, abstract constructs. His strategy was, from a philosophical point of view, not unlike that of Galois: translate the "concrete" problem at hand into new, more abstract terms in order to solve it more cleanly at a "higher" level. In the early twentieth century, NOETHER [VI.76] and her students, among them Bartel van der Waerden (1903–96), would develop Dedekind's ideas further to help create the structural approach to algebra so characteristic of the twentieth century.

Parallel to this nineteenth-century, number-theoretic evolution of the notion of "number" on the continent of Europe, a very different set of developments was taking place, initially in the British Isles. From the late eighteenth century, British mathematicians had debated not only the nature of number—questions such as,

"Do negative and imaginary numbers make sense?"—but also the meaning of algebra—questions like, "In an expression like $ax + by$, what values may a, b, x, and y legitimately take on and what precisely may '+' connote?" By the 1830s, the Irish mathematician HAMILTON [VI.37] had come up with a "unified" interpretation of the complex numbers that circumvented, in his view, the logical problem of adding a real number and an imaginary one, an apple and an orange. Given real numbers a and b, Hamilton conceived of the complex number $a + b\sqrt{-1}$ as the ordered pair (he called it a "couple") (a, b). He then defined addition, subtraction, multiplication, and division of such couples. As he realized, this also provided a way of representing numbers in the complex plane, and so he naturally asked whether he could construct algebraic, ordered triples so as to represent points in 3-space. After a decade of contemplating this question off and on, Hamilton finally answered it not for triples but for quadruples, the so-called QUATERNIONS [III.76], "numbers" of the form $(a, b, c, d) = a + b\mathrm{i} + c\mathrm{j} + d\mathrm{k}$, where a, b, c, and d are real and where i, j, k satisfy the relations ij $= -$ji $=$ k, jk $= -$kj $=$ i, ki $= -$ik $=$ j, i$^2 =$ j$^2 =$ k$^2 = -1$. As in the two-dimensional case, addition is defined component-wise, but multiplication, while definable in such a way that every nonzero element has a multiplicative inverse, is not commutative. Thus, this new number system did not obey all of the "usual" laws of arithmetic.

Although some of Hamilton's British contemporaries questioned the extent to which mathematicians were free to create such new mathematical worlds, others, like Cayley, immediately took the idea further and created a system of ordered 8-tuples, the octonions, the multiplication of which was neither commutative nor even, as was later discovered, associative. Several questions naturally arise about such systems, but one that Hamilton asked was what happens if the field of coefficients, the base field, is not the reals but rather the complexes? In that case, it is easy to see that the product of the two nonzero complex quaternions $(-\sqrt{-1}, 0, 1, 0) = -\sqrt{-1} + \mathrm{j}$ and $(\sqrt{-1}, 0, 1, 0) = \sqrt{-1} + \mathrm{j}$ is $1 + \mathrm{j}^2 = 1 + (-1) = 0$. In other words, the complex quaternions contain zero divisors—nonzero elements the product of which is zero—another phenomenon that distinguishes their behavior fundamentally from that of the integers. As it flourished in the hands of mathematicians like Benjamin Peirce (1809–80), FROBENIUS [VI.58], Georg Scheffers (1866–1945), Theodor Molien (1861–1941), CARTAN [VI.69], and Joseph H. M. Wedderburn (1882–1948), among others, this line of

thought resulted in a freestanding theory of algebras. This naturally intertwined with developments in the theory of matrices (the $n \times n$ matrices form an algebra of dimension n^2 over their base field) as it had evolved through the work of Gauss, Cayley, and Sylvester. It also merged with the not unrelated theory of n-dimensional vector spaces (n-dimensional algebras are n-dimensional vector spaces with a vector multiplication as well as a vector addition and scalar multiplication) that issued from ideas like those of Hermann Grassmann (1809–77).

9   Modern Algebra

By 1900, many new algebraic structures had been identified and their properties explored. Structures that were first isolated in one context were then found to appear, sometimes unexpectedly, in others: thus, these new structures were mathematically more general than the problems that had led to their discovery. In the opening decades of the twentieth century, algebraists (the term is not ahistorical by 1900) increasingly recognized these commonalities—these shared structures such as groups, fields, and rings—and asked questions at a more abstract level. For example, what are all of the finite simple groups? Can they be classified? (See THE CLASSIFICATION OF FINITE SIMPLE GROUPS [V.7].) Moreover, inspired by the set-theoretic and axiomatic work of CANTOR [VI.54], Hilbert, and others, they came to appreciate the common standard of analysis and comparison that axiomatization could provide. Coming from this axiomatic point of view, Ernst Steinitz (1871–1928), for example, laid the groundwork for an abstract theory of fields in 1910, while Abraham Fraenkel (1891–1965) did the same for an abstract theory of rings four years later. As van der Waerden came to realize in the late 1920s, these developments could be interpreted as dovetailing philosophically with results like Hilbert's in invariant theory and Dedekind's and Noether's in the algebraic theory of numbers. That interpretation, laid out in 1930 in van der Waerden's classic textbook *Moderne Algebra*, codified the structurally oriented "modern algebra" that subsumed the algebra of polynomials of the high school classroom and that continues to characterize algebraic thought today.

Further Reading

Bashmakova, I., and G. Smirnova. 2000. *The Beginnings and Evolution of Algebra*, translated by A. Shenitzer. Washington, DC: The Mathematical Association of America.

Corry, L. 1996. *Modern Algebra and the Rise of Mathematical Structures*. Science Networks, volume 17. Basel: Birkhäuser.

Edwards, H. M. 1984. *Galois Theory*. New York: Springer.

Heath, T. L. 1956. *The Thirteen Books of Euclid's Elements*, 2nd edn. (3 vols.). New York: Dover.

Høyrup, J. 2002. *Lengths, Widths, Surfaces: A Portrait of Old Babylonian Algebra and Its Kin*. New York: Springer.

Klein, J. 1968. *Greek Mathematical Thought and the Origin of Algebra*, translated by E. Brann. Cambridge, MA: The MIT Press.

Netz, R. 2004. *The Transformation of Mathematics in the Early Mediterranean World: From Problems to Equations*. Cambridge: Cambridge University Press.

Parshall, K. H. 1988. The art of algebra from al-Khwārizmī to Viète: A study in the natural selection of ideas. *History of Science* 26:129-64.

———. 1989. Toward a history of nineteenth-century invariant theory. In *The History of Modern Mathematics*, edited by D. E. Rowe and J. McCleary, volume 1, pp. 157-206. Amsterdam: Academic Press.

Sesiano, J. 1999. *Une Introduction à l'histoire de l'algèbre: Résolution des équations des Mésopotamiens à la Renaissance*. Lausanne: Presses Polytechniques et Universitaires Romandes.

Van der Waerden, B. 1985. *A History of Algebra from al-Khwārizmī to Emmy Noether*. New York: Springer.

Wussing, H. 1984. *The Genesis of the Abstract Group Concept: A Contribution to the History of the Origin of Abstract Group Theory*, translated by A. Shenitzer. Cambridge, MA: The MIT Press.

II.4　Algorithms
Jean-Luc Chabert

1　What Is an Algorithm?

It is not easy to give a precise definition of the word "algorithm." One can provide approximate synonyms: some other words that (sometimes) mean roughly the same thing are "rule," "technique," "procedure," and "method." One can also give good examples, such as long multiplication, the method one learns in high school for multiplying two positive integers together. However, although informal explanations and well-chosen examples do give a good idea of what an algorithm is, the concept has undergone a long evolution: it was not until the twentieth century that a satisfactory formal definition was achieved, and ideas about algorithms have evolved further even since then. In this article, we shall try to explain some of these developments and clarify the contemporary meaning of the term.

1.1　Abacists and Algorists

Returning to the example of multiplication, an obvious point is that how you try to multiply two numbers together is strongly influenced by how you represent those numbers. To see this, try multiplying the Roman numerals CXLVII and XXIX together without first converting them into their decimal counterparts, 147 and 29. It is difficult and time-consuming, and explains why arithmetic in the Roman empire was extremely rudimentary. A numeration system can be additive, as it was for the Romans, or *positional*, like ours today. If it is positional, then it can use one or several bases—for instance, the Sumerians used both base 10 and base 60.

For a long time, many processes of calculation used *abacuses*. Originally, these were lines traced on sand, onto which one placed stones (the Latin for small stone is *calculus*) to represent numbers. Later there were counting tables equipped with rows or columns onto which one placed tokens. These could be used to represent numbers to a given base. For example, if the base was 10, then a token would represent one unit, ten units, one hundred units, etc., according to which row or column it was in. The four arithmetic operations could then be carried out by moving the tokens according to precise rules. The Chinese counting frame can be regarded as a version of the abacus.

In the twelfth century, when the Arabic mathematical works were translated into Latin, the denary positional numeration system spread through Europe. This system was particularly suitable for carrying out the arithmetic operations, and led to new methods of calculation. The term *algoritmus* was introduced to refer to these, and to distinguish them from the traditional methods that used tokens on an abacus.

Although the signs for the numerals had been adapted from Indian practice, the numerals became known as Arabic. And the origin of the word "algorithm" is Arabic: it arose from a distortion of the name AL-KHWĀRIZMĪ [VI.5], who was the author of the oldest known work on algebra, in the first half of the ninth century. His treatise, entitled *al-Kitāb al-mukhtaṣar fī ḥisāb al-jabr wa'l-muqābala* ("The compendious book on calculation by completion and balancing"), gave rise to the word "algebra."

1.2　Finiteness

As we have just seen, in the Middle Ages the term "algorithm" referred to the processes of calculation based on the decimal notation for the integers. However, in the seventeenth century, according to D'ALEMBERT's [VI.20] *Encyclopédie*, the word was used in a more general sense, referring not just to arithmetic but also to methods in algebra and to other calculational procedures such as "the algorithm of the integral calculus" or "the algorithm of sines."

Gradually, the term came to mean any process of systematic calculation that could be carried out by means of very precise rules. Finally, with the growing role of computers, the important role of *finiteness* was fully understood: it is essential that the process stops and provides a result after a finite time. Thus one arrives at the following naive definition:

An algorithm is a set of finitely many rules for manipulating a finite amount of data in order to produce a result in a finite number of steps.

Note the insistence on finiteness: finiteness in the writing of the algorithm and finiteness in the implementation of the algorithm.

The formulation above is not of course a mathematical definition in the classical sense of the term. As we shall see later, it was important to formalize it further. But for now, let us be content with this "definition" and look at some classical examples of algorithms in mathematics.

2 Three Historical Examples

A feature of algorithms that we have not yet mentioned is *iteration*, or the repetition of simple procedures. To see why iteration is important, consider once again the example of long multiplication. This is a method that works for positive integers of any size. As the numbers get larger, the procedure takes longer, but—and this is of vital importance—the method is "the same": if you understand how to multiply two three-digit numbers together, then you do not need to learn any new principles in order to multiply two 137-digit numbers together (even if you might be rather reluctant to do the calculation). The reason for this is that the method for long multiplication involves a great deal of carefully structured repetition of much smaller tasks, such as multiplying two one-digit numbers together. We shall see that iteration plays a very important part in the algorithms to be discussed in this section.

2.1 Euclid's Algorithm: Iteration

One of the best, and most often used, examples to illustrate the nature of algorithms is EUCLID's ALGORITHM

[III.22], which goes back to the third century B.C.E. It is a procedure described by EUCLID [VI.2] to determine the *greatest common divisor* (gcd) of two positive integers a and b. (Sometimes the greatest common divisor is known as the *highest common factor* (hcf).)

When one first meets the concept of the greatest common divisor of a and b, it is usually defined to be the largest positive integer that is a divisor (or factor) of both a and b. However, for many purposes it is more convenient to think of it as the unique positive integer d with the following two properties. First, d is a divisor of a and b, and second, if c is any other divisor of a and b, then d is divisible by c. The method for determining d is provided by the first two propositions of Book VII of Euclid's *Elements*. Here is the first one: "Two unequal numbers being set out, and the less being continually subtracted in turn from the greater, if the number which is left never measures the one before it until a unit is left, the original numbers will be prime to one another." In other words, if by carrying out successive alternate subtractions one obtains the number 1, then the gcd of the two numbers is equal to 1. In this case one says that the numbers are *relatively prime* or *coprime*.

2.1.1 Alternate Subtractions

Let us describe Euclid's procedure in general. It is based on two simple observations:

(i) if $a = b$ then the gcd of a and b is b (or a);

(ii) d is a common divisor of a and b if and only if it is a common divisor of $a - b$ and b, which implies that the gcd of a and b is the same as the gcd of $a - b$ and b.

Now suppose that we wish to determine the gcd of a and b and suppose that $a \geqslant b$. If $a = b$ then observation (i) tells us that the gcd is b. Otherwise, observation (ii) tells us that the answer will be the same as it is for the two numbers $a - b$ and b. If we now let a_1 be the larger of these two numbers and b_1 the smaller (of course, if they are equal then we just set $a_1 = b_1 = b$), then we are faced with the same task that we started with—to determine the gcd of two numbers—but the larger of these two numbers, a_1, is smaller than a, the larger of the original two numbers. We can therefore repeat the process: if $a_1 = b_1$ then the gcd of a_1 and b_1, and hence that of a and b, is b_1, and otherwise we replace a_1 by $a_1 - b_1$ and reorganize the numbers $a_1 - b_1$ and b_1 so that if one of them is larger then it comes first.

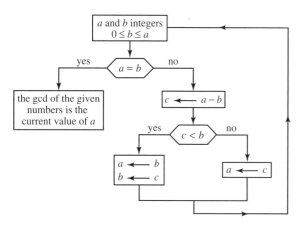

Figure 1 A flow chart for the
procedure in Euclid's algorithm.

One further observation is needed if we want to show that this procedure works. It is the following fundamental fact about the positive integers, sometimes known as the *well-ordering principle.*

(iii) A strictly decreasing sequence of positive integers $a_0 > a_1 > a_2 > \cdots$ must be finite.

Since the iterative procedure just described produces exactly such a strictly decreasing sequence, the iterations must eventually stop, which means that at some point a_k and b_k will be equal, and that value is thus the gcd of a and b (see figure 1).

2.1.2 Euclidean Divisions

Euclid's algorithm is usually described in a slightly different way. One makes use of a more complex procedure called *Euclidean division*—that is, division with remainder—which greatly reduces the number of steps that the algorithm takes. The basic fact underlying this procedure is that if a and b are two positive integers then there are (unique) integers q and r such that

$$a = bq + r \quad \text{and} \quad 0 \leqslant r < b.$$

The number q is called the *quotient* and r is the *remainder.* Remarks (i) and (ii) above are then replaced by the following ones:

(i′) if $r = 0$ then the gcd of a and b is equal to b;
(ii′) the gcd of a and b is the same as the gcd of b and r.

This time, at the first step, one replaces (a, b) by (b, r). If $r \neq 0$, then at the second step one replaces (b, r) by

(r, r_1), where r_1 is the remainder in the division of b by r, and so on. The sequence of remainders is strictly decreasing ($b > r > r_1 > r_2 \geqslant 0$), so the process stops and the gcd is the last nonzero remainder.

It is not hard to see that the two approaches are equivalent. Suppose, for example, that $a = 103\,438$ and $b = 37$. If you use the first approach, then you will repeatedly subtract 37 from 103 438 until you reach a number that is smaller than 37. This number will be the remainder when 103 438 is divided by 37, which is the first number you would calculate if you used the second approach. Thus, the reason for the second approach is that repeated subtraction can be a very inefficient way of calculating remainders. This efficiency gain is very important in practice: the second approach gives rise to a POLYNOMIAL-TIME ALGORITHM [IV.20 §2], while the time taken by the first is exponentially long.

2.1.3 Generalizations

Euclid's algorithm can be generalized to many other contexts where we have notions of addition, subtraction, and multiplication. For example, there is a variant of it that applies to the RING [III.81 §1] $\mathbb{Z}[i]$ of *Gaussian integers*, that is, numbers of the form $a + bi$, where a and b are ordinary integers. It can also be applied to the ring of all polynomials with real coefficients (or coefficients in any field, for that matter). The one requirement is that we should be able to find some analogue of the notion of division with remainder, after which the algorithm is virtually identical to the algorithm for positive integers. For example, we have the following statement for polynomials: given any two polynomials A and B with B not the zero polynomial, there are polynomials Q and R such that $A = BQ + R$ and either $R = 0$ or the degree of R is less than the degree of B.

As Euclid noticed (*Elements*, Book X, proposition 2), one may also carry out the procedure on pairs of numbers a and b that are not necessarily integers. It is easy to check that the process will stop if and only if the ratio a/b is a rational number. This observation leads to the concept of CONTINUED FRACTIONS [III.22], which are discussed in part III. They were not studied explicitly before the seventeenth century, but the roots of the idea can be traced back to ARCHIMEDES [VI.3].

2.2 The Method of Archimedes to Calculate π: Approximation and Finiteness

The ratio of the circumference of a circle to the diameter is a constant that has been denoted by π since the eighteenth century (see [III.70]). Let us see how Archimedes, in the third century B.C.E., obtained the classical approximation $\frac{22}{7}$ for this ratio. If one draws inscribed polygons (whose vertices lie on the circle) and circumscribed polygons (whose sides are tangent to the circle) and if one computes the length of these polygons, then one obtains lower and upper bounds for the value of π, since the circumference of the circle is greater than the length of any inscribed polygon and less than the length of any circumscribed polygon (figure 2). Archimedes started with regular hexagons, and then repeatedly doubled the number of sides, obtaining more and more precise bounds. He finished with ninety-six-sided polygons, obtaining the estimates

$$3 + \tfrac{10}{71} \leqslant \pi \leqslant 3 + \tfrac{1}{7}.$$

This process clearly involves iteration, but is it right to call it an algorithm? Strictly speaking it is not: however many sides you take for your polygon, all you will get is an approximation to π, so the process is not finite. However, what we do have is an algorithm that will calculate π to any desired accuracy: for example, if you demand an approximation that is correct to ten decimal places, then after a finite number of steps the algorithm will give you one. What matters now is that the process *converges*. That is, it is important that the values that come out of the iteration get arbitrarily close to π. The geometric origin of the method can be used to prove that this is indeed the case, and in 1609 in Germany Ludolph van Ceulen obtained an approximation accurate to thirty-five decimal places using polygons with 2^{62} sides.

Nevertheless, there is a clear difference between this algorithm for approximating π and Euclid's algorithm for calculating the gcd of two positive integers. Algorithms like Euclid's are often called *discrete algorithms*, and are contrasted with *numerical algorithms*, which are algorithms that are used to compute numbers that are not integers (see NUMERICAL ANALYSIS [IV.21]).

2.3 The Newton–Raphson Method: Recurrence Formulas

In around 1670, NEWTON [VI.14] devised a method for finding roots of equations, which he explained with reference to the example $x^3 - 2x - 5 = 0$. His explanation starts with the observation that the root x is approximately equal to 2. He therefore writes $x = 2 + p$ and obtains an equation for p by substituting $2 + p$ for x in the original equation. This new equation works out to be $p^3 + 6p^2 + 10p - 1 = 0$. Because x is close to 2, p is

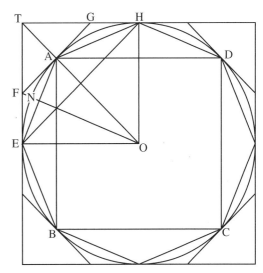

Figure 2 Approximation of π.

small, so he then estimates p by forgetting the terms p^3 and $6p^2$ (since these should be considerably smaller than $10p - 1$). This gives him the equation $10p - 1 = 0$, or $p = \frac{1}{10}$. Of course, this is not an exact solution, but it provides him with a new and better approximation, 2.1, for x. He then repeats the process, writing $x = 2.1 + q$, substituting to obtain an equation for q, solving this equation approximately, and refining his estimate still further. The estimate he obtains for q is -0.0054, so the next approximation for x is 2.0946.

How, though, can we be sure that this process really does converge to x? Let us examine the method more closely.

2.3.1 Tangents and Convergence

Newton's method can be interpreted geometrically in terms of the graph of a function f, though Newton himself did not do so. A root x of the equation $f(x) = 0$ corresponds to a point where the curve with equation $y = f(x)$ intersects the x-axis. If you start with an approximate value a for x and set $p = x - a$, as we did above, then when you substitute $a + p$ for x to obtain a new function $g(p)$, you are effectively moving the origin from $(0, 0)$ to the point $(a, 0)$. Then when you forget all powers of p other than the constant and linear terms, you are finding the best linear approximation to the function g—which, geometrically speaking, is the tangent line to g at the point $(0, g(0))$. Thus, the approximate value you obtain for p is the x-coordinate of the point where the tangent at $(0, g(0))$ crosses the

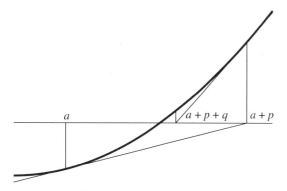

Figure 3 Newton's method.

x-axis. Adding a to this value returns the origin to $(0,0)$ and gives the new approximation to the root of f. This is why Newton's method is often called the *tangent method* (figure 3). And one can now see that the new approximation will definitely be better than the old one if the tangent to f at $(a, f(a))$ intersects the x-axis at a point that lies between a and the point where the curve $y = f(x)$ intersects the x-axis.

As it happens, this is not the case for Newton's choice of the value $a = 2$ above, but it is true for the approximate value 2.1 and for all subsequent ones. Geometrically, the favorable situation occurs if the point $(a, f(a))$ lies above the x-axis in a convex part of the curve that crosses the x-axis or below the x-axis in a concave part of the curve that crosses the x-axis. Under these circumstances, and provided the root is not a multiple one, the convergence is *quadratic*, meaning that the error at each stage is roughly the square of the error at the previous stage—or, equivalently, the approximation is valid to a number of decimal places that roughly doubles at each stage. This is enormously fast.

The choice of the initial approximation value is obviously important, and raises unexpectedly subtle questions. These are clearer if we look at *complex* polynomials and their complex roots. Newton's method can be easily adapted to this more general context. Suppose that z is a root of some complex polynomial and that z_0 is an initial approximation for z. Newton's method then gives us a sequence $z_0, z_1, z_2, \ldots$, which may or may not converge to z. We define the *domain of attraction*, denoted $A(z)$, to be the set of all complex numbers z_0 such that the resulting sequence does indeed converge to z. How do we determine $A(z)$?

The first person to ask this problem was CAYLEY [VI.46], in 1879. He noticed that the solution is easy

for quadratic polynomials but difficult as soon as the degree is 3 or more. For example, the domains of attraction of the roots ± 1 of the polynomial $z^2 - 1$ are the open half-planes bounded by the vertical axis, but the domains corresponding to the roots 1, ω, and ω^2 of $z^3 - 1$ are extremely complicated sets. They were described by Julia in 1918—such subsets are now called *fractal sets*. Newton's method and fractal sets are discussed further in DYNAMICS [IV.14].

2.3.2 Recurrence Formulas

At each stage of his method, Newton had to produce a new equation, but in 1690 Raphson noticed that this was not really necessary. For particular examples, he gave single formulas that could be used at each step, but his basic observation applies in general and leads to a general formula for every case, which one can easily obtain using the interpretation in terms of tangents. Indeed, the tangent to the curve $y = f(x)$ at the point of x-coordinate a has the equation $y - f(a) = f'(a)(x - a)$, and it cuts the x-axis at the point with x-coordinate $a - f(a)/f'(a)$. What we now call the *Newton–Raphson method* springs from this simple formula. One starts with an initial approximation $a_0 = a$ and then defines successive approximations using the recurrence formula

$$a_{n+1} = a_n - \frac{f(a_n)}{f'(a_n)}.$$

As an example, let us consider the function $f(x) = x^2 - c$. Here, Newton's method provides a sequence of approximations of the square root $\sqrt{c}$ of c, given by the recurrence formula $a_{n+1} = \frac{1}{2}(a_n + c/a_n)$ (which we obtain by substituting $x^2 + c$ for f in the general formula above). This method for approximating square roots was known by Heron of Alexandria in the first century. Note that if a_0 is close to $\sqrt{c}$, then c/a_0 is also close, $\sqrt{c}$ lies between them, and $a_1 = \frac{1}{2}(a_0 + c/a_0)$ is their arithmetic mean.

3 Does an Algorithm Always Exist?

3.1 Hilbert's Tenth Problem: The Need for Formalization

In 1900, at the Second International Congress of Mathematicians, HILBERT [VI.63] proposed a list of twenty-three problems. These problems, and Hilbert's works in general, had a huge influence on mathematics during the twentieth century (Gray 2000). We are interested here in *Hilbert's tenth problem*: given a Diophantine equation, that is, a polynomial equation with any number of indeterminates and with integer coefficients, "a process is sought by which it can be determined, in a finite number of operations, whether the equation is solvable in integral numbers." In other words, we have to find an algorithm which tells us, for any Diophantine equation, whether or not it has at least one integer solution. Of course, for many Diophantine equations it is easy to find solutions, or to prove that no solutions exist. However, this is by no means always the case: consider, for instance, the Fermat equation $x^n + y^n = z^n$ ($n \geqslant 3$). (Even before the solution of FERMAT'S LAST THEOREM [V.10] an algorithm was known for determining for any specific n whether this equation had a solution. However, one could not call it easy.)

If Hilbert's tenth problem has a positive answer, then one can demonstrate it by exhibiting a "process" of the sort that Hilbert asked for. To do this, it is not necessary to have a precise understanding of what a "process" is. However, if you want to give a *negative* answer, then you have to show that *no algorithm exists*, and for that you need to say precisely what counts as an algorithm. In section 1.2 we gave a definition that seems to be reasonably precise, but it is not precise enough to enable us to think about Hilbert's tenth problem. What kind of rules are we allowed to use in an algorithm? How can we be sure that no algorithm achieves a certain task, rather than just that we are unable to find one?

3.2 Recursive Functions: Church's Thesis

What we need is a *formal* definition of the notion of an algorithm. In the seventeenth century, LEIBNIZ [VI.15] envisaged a universal language that would allow one to reduce mathematical proofs to simple computations. Then, during the nineteenth century, logicians such as Charles Babbage, BOOLE [VI.43], FREGE [VI.56], and PEANO [VI.62] tried to formalize mathematical reasoning by an "algebraization" of logic. Finally, between 1931 and 1936, GÖDEL [VI.92], CHURCH [VI.89], and Stephen Kleene introduced the notion of *recursive functions* (see Davis (1965), which contains the original texts). Roughly speaking, a recursive function is one that can be calculated by means of an algorithm, but the *definition* of recursive functions is different, and is completely precise.

3.2.1 *Primitive Recursive Functions*

Another rough definition of a recursive function is as follows: a recursive function is one that has an induc-

tive definition. To give an idea of what this means, let us consider addition and multiplication as functions from $\mathbb{N} \times \mathbb{N}$ to $\mathbb{N}$. To emphasize this, we shall write $\mathrm{sum}(x, y)$ and $\mathrm{prod}(x, y)$ for $x + y$ and xy, respectively.

A familiar fact about multiplication is that it is "repeated addition." Let us examine this idea more precisely. We can define the function "prod" in terms of the function "sum" by means of the following two rules: $\mathrm{prod}(1, y)$ equals y and $\mathrm{prod}(x + 1, y)$ equals $\mathrm{sum}(\mathrm{prod}(x, y), y)$. Thus, if you know $\mathrm{prod}(x, y)$ and you know how to calculate sums, then you can work out $\mathrm{prod}(x + 1, y)$. Since you also know the "base case" $\mathrm{prod}(1, y)$, a simple inductive argument shows that these simple rules completely determine the function "prod."

We have just seen how one function can be "recursively defined" in terms of another. We now want to understand the class of *all* functions from $\mathbb{N}^n$ to $\mathbb{N}$ that can be built up in a few basic ways, of which recursion is the most important. We shall refer to functions from $\mathbb{N}^n$ to $\mathbb{N}$ as *n-ary functions*.

To begin with, we need an initial stock of functions out of which the rest will be built. It turns out that a very simple set of functions is enough. Most basic are the *constant functions*: that is, functions that take every n-tuple in $\mathbb{N}^n$ to some fixed positive integer c. Another very simple function, but the function that allows us to create much more interesting ones, is the *successor function*, which takes a positive integer n to the next one, $n + 1$. Finally, we have *projection functions*: the function U_k^n takes a sequence $(x_1, \ldots, x_n)$ in $\mathbb{N}^n$ and maps it to the kth coordinate x_k.

We then have two ways of constructing functions from other functions. The first is *substitution*. Given an m-ary function ϕ and m n-ary functions $\psi_1, \ldots, \psi_m$, one defines an n-ary function by

$$(x_1, \ldots, x_n) \mapsto \phi(\psi_1(x_1, \ldots, x_n), \ldots, \psi_m(x_1, \ldots, x_n)).$$

For example, $(x + y)^2 = \mathrm{prod}(\mathrm{sum}(x, y), \mathrm{sum}(x, y))$, so we can obtain the function $(x, y) \mapsto (x + y)^2$ from the functions "sum" and "prod" by means of substitution.

The second method of construction is called *primitive recursion*. This is a more general form of the inductive method we used above in order to construct the function "prod" from the function "sum." Given an $(n - 1)$-ary function ψ and an $(n + 1)$-ary function μ, one defines an n-ary function ϕ by saying that

$$\phi(1, x_2, \ldots, x_n) = \psi(x_2, \ldots, x_n)$$

and

$$\phi(k + 1, x_2, \ldots, x_n)$$
$$= \mu(k, \phi(k, x_2, \ldots, x_n), x_2, \ldots, x_n).$$

In other words, ψ tells you the "initial values" of ϕ (the values when the first coordinate is 1) and μ tells you how to work out $\phi(k + 1, x_2, \ldots, x_n)$ in terms of $\phi(k, x_2, \ldots, x_n), x_2, \ldots, x_n$ and k. (The sum-product example was simpler because we did not have a dependence on k.)

A *primitive recursive function* is any function that can be built from the initial stock of functions using the two operations, substitution and primitive recursion, that we have just described.

3.2.2 Recursive Functions

If you think for a while about primitive recursion and know a small amount about programming computers, you should be able to convince yourself that they are *effectively computable*: that is, that for any primitive recursive function there is an algorithm for computing it. (For example, the operation of primitive recursion can usually be realized in a rather direct way as a FOR loop.)

How about the converse? Are all computable functions primitive recursive? Consider, for example, the function that takes the positive integer n to p_n, the nth prime number. It is not hard to devise a simple algorithm for computing p_n, and it is then a good exercise (if you want to understand primitive recursion) to convert this algorithm into a proof that the function is primitive recursive.

However, it turns out that this function is not typical: there are computable functions that are not primitive recursive. In 1928, Wilhelm Ackermann defined a function, now known as the *Ackermann function*, that has a "doubly inductive" definition. The following function is not quite the same as Ackermann's, but it is very similar. It is the function $A(x, y)$ that is determined by the following recurrence rules:

(i) $A(1, y) = y + 2$ for every y;
(ii) $A(x, 1) = 2$ for every x;
(iii) $A(x+1, y+1) = A(x, A(x+1, y))$ whenever $x > 1$ and $y > 1$.

For example, $A(2, y + 1) = A(1, A(2, y)) = A(2, y) + 2$. From this and the fact that $A(2, 1) = 2$ it follows that $A(2, y) = 2y$ for every y. In a similar way one can show that $A(3, y) = 2^y$, and in general that for each x

the function that takes y to $A(x + 1, y)$ "iterates" the function that takes y to $A(x, y)$. This means that the values of $A(x, y)$ are extremely large even when x and y are fairly small. For example, $A(4, y + 1) = 2^{A(4,y)}$, so in general $A(4, y)$ is given by an "exponential tower" of height y. We have $A(4, 1) = 2$, $A(4, 2) = 2^2 = 4$, $A(4, 3) = 2^4 = 16$, $A(4, 4) = 2^{16} = 65\,536$, and $A(4, 5) = 2^{65\,536}$, which is too large a number for its decimal notation to be reproduced here.

It can be shown that for every primitive recursive function ϕ there is some x such that the function $A(x, y)$ grows faster than $\phi(y)$. This is proved by an inductive argument. To oversimplify slightly, if $\psi(y)$ and $\mu(y)$ have already been shown to grow more slowly than $A(x, y)$, then one can show that the function ϕ produced from them by primitive recursion also grows more slowly. This allows us to define a "diagonal" function $A(y) = A(y, y)$ that is not primitive recursive because it grows faster than any of the functions $A(x, y)$.

If we are trying to understand in a precise way which functions can be calculated algorithmically, then our definition will surely have to encompass functions like the Ackermann function, since they can in principle be computed. Therefore, we must consider a larger class of functions than just the primitive recursive ones. This is what Gödel, Church, and Kleene did, and they obtained in different ways the same class of *recursive functions*. For instance, Kleene added a third method of construction, which he called *minimization*. If f is an $(n + 1)$-ary function, one defines an n-ary function g by taking $g(x_1, \ldots, x_n)$ to be the smallest y such that $f(x_1, \ldots, x_n, y) = 0$. (If there is no such y, one regards g as undefined for $(x_1, \ldots, x_n)$. We shall ignore this complication in what follows.)

It turns out that, not only is the Ackermann function recursive, but so are all functions that one can write a computer program to calculate. So this gives us the formal definition of computability that we did not have before.

3.2.3 Effective Calculability

Once the notion of recursive functions had been formulated, Church claimed that the class of recursive functions was exactly the same as the class of "effectively calculable" functions. This claim is widely believed, but it is a conviction that cannot be proved since the notion of recursive function is a mathematically precise concept while that of an effectively calculable function is an intuitive notion, rather like that of "algorithm." Church's statement lies in the realm of metamathematics and is now called *Church's thesis*.

3.3 Turing Machines

One of the strongest pieces of evidence for Church's thesis is that in 1936 TURING [VI.94] found a very different-looking way of formalizing the notion of an algorithm, which he showed was equivalent. That is, every function that was computable in his new sense was recursive and vice versa. His approach was to define a notion that is now called a *Turing machine*, which can be thought of as an extremely primitive computer, and which played an important part in the development of actual computers. Indeed, functions that are computable by Turing machines are precisely those that can be programmed on a computer. The primitive architecture of Turing machines does not make them any less powerful: it merely means that in practice they would be too cumbersome to program or to implement in hardware. Since recursive functions are the same as Turing-computable functions, it follows that recursive functions too are those functions that can be programmed on a computer, so to disbelieve Church's thesis would be to maintain that there are some "effective procedures" that cannot be converted into computer programs—which seems rather implausible. A description of Turing machines can be found in COMPUTATIONAL COMPLEXITY [IV.20 §1].

Turing introduced his machines in response to a question that generalized Hilbert's tenth problem. The *Entscheidungsproblem*, or *decision problem*, was also asked by Hilbert, in 1922. He wanted to know whether there was a "mechanical process" by which one could determine whether any given mathematical statement could be proved. In order to think about this, Turing needed a precise notion of what constituted a "mechanical process." Once he had defined Turing machines, he was able to show by means of a fairly straightforward diagonal argument that the answer to Hilbert's question was no. His argument is outlined in THE INSOLUBILITY OF THE HALTING PROBLEM [V.20].

4 Properties of Algorithms

4.1 Iteration versus Recursion

As previously mentioned, we often encounter computation rules which define each element of a sequence in terms of the preceding elements. This gives rise to two different ways of carrying out the computation.

The first is *iteration*: one computes the first terms, then one obtains succeeding terms by means of a recurrence formula. The second is *recursion*, a procedure which seems circular at first because one defines a procedure in terms of itself. However, this is allowed because the procedure calls on itself with smaller values of the variables. The concept of recursion is subtle and powerful. Let us try to clarify the difference between recursion and iteration with some examples.

Suppose that we wish to compute $n! = 1 \cdot 2 \cdot 3 \cdots (n-1) \cdot n$. An obvious way of doing it is to note the recurrence relation $n! = n \cdot (n-1)!$ and the initial value $1! = 1$. Having done so, one could then compute successively the numbers $2!$, $3!$, $4!$, and so on until one reached $n!$, which would be the iterative approach. Alternatively, one could say that if $\text{fact}(n)$ is the result of a procedure that leads to $n!$, then $\text{fact}(n) = n \times \text{fact}(n-1)$, which would be a recursive procedure. The second approach says that to obtain $n!$ it suffices to know how to obtain $(n-1)!$, and to obtain $(n-1)!$ it suffices to know how to obtain $(n-2)!$, and so on. Since one knows that $1! = 1$, one can obtain $n!$. Thus, recursion is a bit like iteration but thought of "backwards."

In some ways this example is too simple to show clearly the difference between the two procedures. Moreover, if one wishes to compute $n!$, then iteration seems simpler and more natural than recursion. We now look at an example where recursion is far simpler than iteration.

4.1.1 The Tower of Hanoi

The Tower of Hanoi is a problem that goes back to Édouard Lucas in 1884. One is given n disks, all of different sizes and each with a hole in the middle, stacked on a peg A in order of size, with the largest one at the bottom. We also have two empty pegs B and C. The problem is to move the stack from peg A to peg B while obeying the following rules. One is allowed to move just one disk at a time, and each move consists in taking the top disk from one of the pegs and putting it onto another peg. In addition, no disk may ever be placed above a smaller disk.

The problem is easy if you have just three disks, but becomes rapidly harder as the number of disks increases. However, with the help of recursion one can see very quickly that an algorithm exists for moving the disks in the required way. Indeed, suppose that we know a procedure $H(n-1)$ that solves the problem for $n-1$ disks. Here is a procedure $H(n)$ for n disks: move the first $n - 1$ disks on top of A to C with the procedure $H(n - 1)$, then move the last disk on A to B, and finally apply once more the procedure $H(n - 1)$ to move all the disks from C to B. If we write $H_{AB}(n)$ for the procedure that moves n disks from peg A to peg B according to the rules, then we can represent this recursion symbolically as

$$H_{AB}(n) = H_{AC}(n - 1)H_{AB}(1)H_{BC}(n - 1).$$

Thus, $H_{AB}(n)$ is deduced from $H_{AC}(n-1)$ and $H_{BC}(n-1)$, which are clearly equivalent to $H_{AB}(n - 1)$. Since $H_{AB}(1)$ is certainly easy, we have the full recursion.

One can easily check by induction that this procedure takes $2^n - 1$ moves—moreover, it turns out that the task cannot be accomplished in fewer moves. Thus, the number of moves is an exponential function of n, so for large n the procedure will be very long.

Furthermore, the larger n is, the more memory one must use to keep track of where one is in the procedure. By contrast, if we wish to carry out an iteration during an iterative procedure, it is usually enough to know just the result of the previous iteration. Thus, the most we need to remember is the result of one iteration. There is in fact an iterative procedure for the Tower of Hanoi as well. It is easy to describe, but it is much less obvious that it actually solves the problem. It encodes the positions of the n disks as an n-bit sequence and at each step applies a very simple rule to obtain the next n-bit sequence. This rule makes no reference to how many steps have so far taken place, and therefore the amount of memory needed, beyond that required to store the positions of the disks, is very small.

4.1.2 The Extended Euclid Algorithm

Euclid's algorithm is another example that lends itself in a very natural way to a recursive procedure. Recall that if a and b are two positive integers, then we can write $a = qb + r$ with $0 \leqslant r < b$. The algorithm depended on the observation that $\gcd(a, b) = \gcd(b, r)$. Since the remainder r can be calculated easily from a and b, and since the pair (b, r) is smaller than the pair (a, b), this gives us a recursive procedure, which stops when we reach a pair of the form $(a, 0)$.

An important extension of Euclid's algorithm is *Bézout's lemma*, which states that for any pair of positive integers (a, b) there exist (not necessarily positive) integers u and v such that

$$ua + vb = d = \gcd(a, b).$$

How can we obtain such integers u and v? The answer is given by the *extended Euclid algorithm*, which again

can be defined using recursion. Suppose we can find a pair (u', v') that works for b and r: that is, $u'b + v'r = d$. Since $a = qb + r$, we can substitute $r = a - qb$ into this equation and deduce that $d = u'b + v'(a - qb) = v'a + (u' - v'q)b$. Thus, setting $u = v'$ and $v = u' - v'q$, we have $ua + vb = d$. Since a pair (u, v) that works for a and b can be easily calculated from a pair (u', v') that works for the smaller b and r, this gives us a recursive procedure. The "bottom" of the recursion is when $r = 0$, in which case we know that $1b + 0r = d$. Once we reach this, we can "run back up" through Euclid's algorithm, successively modifying our pair (u, v) according to the rule just given. Notice, incidentally, that the fact that this procedure exists is a proof of Bézout's lemma.

4.2 Complexity

So far we have considered algorithms in a theoretical way and ignored their obvious practical importance. However, the mere existence of an algorithm for carrying out a certain task does not guarantee that your computer can do it, because some algorithms take so many steps that no computer can implement them (unless you are prepared to wait billions of years for the answer). The *complexity* of an algorithm is, loosely speaking, the number of steps it takes to complete its task (as a function of the size of the input). More precisely, this is the *time complexity* of the algorithm. There is also its *space complexity*, which measures the maximum amount of memory a computer needs in order to implement it. *Complexity theory* is the study of the computational resources that are needed to carry out various tasks. It is discussed in detail in COMPUTATIONAL COMPLEXITY [IV.20]—here we shall give a hint of it by examining the complexity of one algorithm.

4.2.1 The Complexity of Euclid's Algorithm

The length of time that a computer will take to implement Euclid's algorithm is closely related to the number of times one needs to compute quotients and remainders: that is, to the number of times that the recursive procedure calls on itself. Of course, this number depends in turn on the size of the numbers a and b whose gcd is to be determined. An initial observation is that if $0 < b \leqslant a$, then the remainder in the division of a by b is less than $a/2$. To see this, notice that if $b \geqslant a/2$ then the remainder is $a - b$, which is at most $a/2$, whereas if $b \leqslant a/2$ then we know that the remainder is at most b and so is again at most $a/2$. It follows that after two steps of calculating the remainder, one arrives at a pair where the larger number is at most half what it was before. From this it is easy to show that the number of such calculations needed is at most $2 \log_2 a + 1$, which is roughly proportional to the number of digits of a. Since this number is far smaller than a itself, the algorithm can be used easily for very large numbers, which gives it great practical utility to go with its theoretical significance.

The number of divisions needed in the worst case does not appear to have been studied until the first half of the nineteenth century: the above bound of $2 \log_2 a + 1$ was given by Pierre-Joseph-Étienne Finck in 1841. It is in fact not hard to improve this result slightly and prove that the algorithm takes longest when a and b are consecutive Fibonacci numbers. This implies that the number of divisions needed is never more than $\log_\phi a + 1$, where ϕ is the golden ratio.

Euclid's algorithm also has low space complexity: once one has replaced a pair (a, b) by a new pair (b, r), one can forget the original pair, so at any stage one does not have to hold very much in one's memory (or store it in the memory of one's computer). By contrast, the extended Euclid algorithm appears to require one to remember the entire sequence of calculations that leads to the gcd d of a and b, so that one can make a series of substitutions and eventually find u and v such that $ua + vb = d$. However, a closer look at it reveals that one can perform it while keeping track of only a few numbers at any one time.

Let us see how this works with an example. We shall set $a = 38$, $b = 21$, and find integers u and v such that $38u + 21v = 1$. We begin by writing down the first step of Euclid's algorithm:

$$38 = 1 \times 21 + 17.$$

This tells us that $17 = 38 - 21$. Now we write down the second step:

$$21 = 1 \times 17 + 4.$$

We know how to write 17 in terms of 38 and 21, so let us do a substitution:

$$21 = 1 \times (38 - 21) + 4.$$

Rearranging this, we discover that $4 = 2 \times 21 - 38$. Now we write down the third step of Euclid's algorithm:

$$17 = 4 \times 4 + 1.$$

We know how to write 17 and 4 in terms of 38 and 21, so let us substitute again:

$$38 - 21 = 4 \times (2 \times 21 - 38) + 1.$$

Rearranging this, we find that $1 = 5 \times 38 - 9 \times 21$, and we have finished.

If you think about this procedure, you will see that at each stage one just has to keep track of how two numbers are expressed in terms of a and b. Thus, the space complexity of the extended Euclid algorithm is small if you implement it properly.

5 Modern Aspects of Algorithms

5.1 Algorithms and Chance

Earlier it was remarked that the notion of algorithm has continued to develop even since its formalization in the 1920s and 1930s. One of the main reasons for this has been the realization that *randomness* can be a very useful tool in algorithms. This may seem puzzling at first, since algorithms as we have described them are deterministic procedures; in a moment we shall give an example that illustrates how randomness can be used. A second reason is the recent development of the notion of a *quantum algorithm*: for more about this, see QUANTUM COMPUTATION [III.74].

The following simple example illustrates how chance can be useful. Given an integer n, we shall define a function $f(n)$ that is not too hard to calculate but is difficult to analyze. If n has d digits, then you approximate $\sqrt{n}$ to the point where the first d digits after the decimal point are correct (using Newton's method, say), and let $f(n)$ equal the dth digit. Now suppose that you wish to know roughly what proportion of numbers n between 10^{30} and 10^{31} have $f(n) = 0$. There does not seem to be a good way of determining this theoretically, but calculating it on a computer looks very hard, too, as there are so many numbers between 10^{30} and 10^{31}. However, if one chooses a random sample of 10000 numbers between 10^{30} and 10^{31} and does the calculation for just those numbers, then with high probability the proportion of those numbers with $f(n) = 0$ will be roughly the same as the proportion of all numbers in the range with $f(n) = 0$. Thus, if you do not demand absolute certainty but instead are satisfied with a very small error probability, then you can achieve your goal with much more modest computational resources.

5.1.1 *Pseudorandom Numbers*

How, though, does one use a deterministic computer to select ten thousand random numbers between 10^{30} and 10^{31}? The answer is that one does not in fact need to: it is almost always good enough to make a *pseudorandom* selection instead. The basic idea is well-illustrated by a method proposed by VON NEUMANN [VI.91] in the mid 1940s. One begins with a $2n$-digit integer a, called the "seed," calculates a^2, and extracts from a^2 a new $2n$-digit number b by taking all the digits of a^2 from the $(n + 1)$st to the $3n$th. One then repeats the procedure for b, and so on. Because of the way multiplication jumbles up digits, the resulting sequence of $2n$-digit numbers is very hard to distinguish from a truly random sequence, and can be used in randomized algorithms.

There are many other ways of producing pseudorandom sequences, and this raises an obvious question: what properties should a sequence have for us to regard it as pseudorandom? This turns out to be a complex question, and several different answers have been proposed. Randomized algorithms and pseudorandomness are discussed in depth in COMPUTATIONAL COMPLEXITY [IV.20 §§6, 7], and a formal definition of "pseudorandom generators" can be found there. (See also COMPUTATIONAL NUMBER THEORY [IV.3 §2] for an account of a famous randomized algorithm for testing whether a number is prime.) Here, let us discuss a similar question about infinite sequences of zeros and ones. When should we regard such a sequence as "random"?

Again, many different answers have been suggested. One idea is to consider simple statistical tests: we would expect that in the long run the frequency of zeros should be roughly the same as that of ones, and more generally that any small subsequence such as 00110 should appear with the "right" frequency (which for this sequence would be $\frac{1}{32}$ since it has length 5).

It is perfectly possible, however, for a sequence to pass these simple tests but to be generated by a deterministic procedure. If one is trying to decide whether a sequence of zeros and ones is *actually* random—that is, produced by some means such as tossing a coin—then we will be very suspicious of a sequence if we can identify an algorithm that produces the same sequence. For example, we would reject a sequence that was derived in a simple way from the digits of π, even if it passed the statistical tests. However, merely to ask that a sequence cannot be produced by a recursive procedure does not give a good test for randomness: for example, if one takes any such sequence and alternates the terms of that sequence with zeros, one then obtains a new sequence that is far from random, but which still cannot be produced recursively.

For this reason, von Mises suggested in 1919 that a sequence of zeros and ones should be called random if

it is not only the case that the limit of the frequency of ones is $\frac{1}{2}$, but also that the same is true for any subsequence that can be extracted "by means of a reasonable procedure." In 1940 Church made this more precise by translating "by means of a reasonable procedure" into "by means of a recursive function." However, even this condition is too weak: there are such sequences that do not satisfy the "law of the iterated logarithm" (something that a random sequence would satisfy). Currently, the so-called Martin–Löf thesis, formulated in 1966, is one of the most commonly used definitions of randomness: a random sequence is a sequence that satisfies all the "effective statistical sequential tests," a notion that we cannot formulate precisely here, but which uses in an essential manner the notion of recursive function. By contrast with Church's thesis, with which almost every mathematician agrees, the Martin–Löf thesis is still very much under discussion.

5.2 The Influence of Algorithms on Contemporary Mathematics

Throughout its history, mathematics has concerned itself with problems of existence. For example, are there TRANSCENDENTAL NUMBERS [III.41], that is, numbers that are not the root of any polynomial with integer coefficients? There are two kinds of answers to such questions: either one actually exhibits a number such as π and proves that it is transcendental (this was done by Carl Lindemann in 1873), or one gives an "indirect existence proof," such as CANTOR's [VI.54] demonstration that there are "far more" real numbers than there are roots of polynomials with integer coefficients (see COUNTABLE AND UNCOUNTABLE SETS [III.11]), which shows in particular that some real numbers must be transcendental.

5.2.1 Constructivist Schools

In around 1910, under the leadership of L. E. J. BROUWER [VI.75], the INTUITIONIST SCHOOL [II.7 §3.1] of mathematics arose, which rejected the principle of the excluded middle, the principle that every mathematical assertion is either true or false. In particular, Brouwer did not accept that the existence of a mathematical object such as a transcendental number is proved by the fact that its nonexistence would lead to a contradiction. This was the first of several "constructivist" schools, for which an object exists if and only if it can be constructed explicitly.

Not many working mathematicians have subscribed to these principles, but almost all would agree that there is an important difference between constructive proofs and indirect proofs of existence, a difference that has come to seem more important with the rise of computer science. This has added a further level of refinement: sometimes, even if you know that a mathematical object can be produced algorithmically, you still care whether the algorithm can be made to work in a reasonably short time.

5.2.2   Effective Results

In number theory there is an important distinction between "effective" and "ineffective" results. For example, MORDELL'S CONJECTURE [V.29], proposed in 1922 and finally proved by Faltings in 1983, states that a smooth rational plane curve of degree $n > 3$ has at most finitely many points with rational coefficients. Among its many consequences is that the Fermat equation $x^n + y^n = z^n$ has only finitely many integral solutions for each $n \geqslant 4$. (Of course, we now know that it has no nontrivial solutions, but the Mordell conjecture was proved before Fermat's last theorem, and it has many other consequences.) However, Faltings's proof is *ineffective*, which means that it does not give any information about how many solutions there are (except that there are not infinitely many), or how large they can be, so one cannot use a computer to find them all and know that one has finished the job. There are many other very important proofs in number theory that are ineffective, and replacing any one of them with an effective argument would be a major breakthrough.

A completely different set of issues was raised by another solution to a famous open problem, the FOUR-COLOR THEOREM [V.12], which was conjectured by Francis Guthrie, a student of DE MORGAN [VI.38], in 1852 and proved in 1976 by Appel and Haken, with a proof that made essential use of computers. They began with a theoretical argument that reduced the problem to checking finitely many cases, but the number of cases was so large that it could not be done by hand and was instead done by computers. But how should we judge such a proof? Can we be sure that the computer has been programmed correctly? And even if it has, how do we know with a computation of that size that the computer has operated correctly? And does a proof that relies on a computer really tell us *why* the theorem is true? These questions continue to be debated today.

Further Reading

Archimedes. 2002. *The Works of Archimedes*, translated by T. L. Heath. London: Dover. Originally published 1897, Cambridge University Press, Cambridge.

Chabert, J.-L., ed. 1999. *A History of Algorithms: From the Pebble to the Microchip*. Berlin: Springer

Davis, M., ed. 1965. *The Undecidable*. New York: The Raven Press.

Euclid. 1956. *The Thirteen Books of Euclid's Elements*, translated by T. L. Heath (3 vols.), 2nd edn. London: Dover. Originally published 1929, Cambridge University Press, Cambridge.

Gray, J. J. 2000. *The Hilbert Challenge*. Oxford: Oxford University Press.

Newton, I. 1969. *The Mathematical Papers of Isaac Newton*, edited by D. T. Whiteside, volume 3 (1670–73), pp. 43–47. Cambridge: Cambridge University Press.

II.5   The Development of Rigor in Mathematical Analysis
Tom Archibald

1   Background

This article is about how rigor came to be introduced into mathematical analysis. This is a complicated topic, since mathematical practice has changed considerably, especially in the period between the founding of the calculus (shortly before 1700) and the early twentieth century. In a sense, the basic criteria for what constitutes a correct and logical argument have not altered, but the circumstances under which one would require such an argument, and even to some degree the purpose of the argument, have altered with time. The voluminous and successful mathematical analysis of the 1700s, associated with names such as Johann and Daniel BERNOULLI [VI.18], EULER [VI.19], and LAGRANGE [VI.22], lacked foundational clarity in ways that were criticized and remedied in subsequent periods. By around 1910 a general consensus had emerged about how to make arguments in analysis rigorous.

Mathematics consists of more than techniques for calculation, methods for describing important features of geometric objects, and models of worldly phenomena. Nowadays, almost all working mathematicians are trained in, and concerned with, the production of rigorous arguments that justify their conclusions. These conclusions are usually framed as *theorems*, which are statements of fact, accompanied by an argument, or proof, that the theorem is indeed true. Here is a simple example: every positive whole number that is divisible by 6 is also divisible by 2. Running through the six times table (6, 12, 18, 24, ...) we see that each number is even, which makes the statement easy enough to believe. A possible justification of it would be to say that since 6 is divisible by 2, then every number divisible by 6 must also be divisible by 2.

Such a justification might or might not be thought of as a thorough proof, depending on the reader. For on hearing the justification we can raise questions: is it always true that if a, b, and c are three positive whole numbers such that c is divisible by b and b is divisible by a, then c is divisible by a? What is divisibility exactly? What is a whole number? The mathematician deals with such questions by precisely defining concepts (such as divisibility of one number by another), basing the definitions on a smallish number of undefined terms ("whole number" might be one, though it is possible to start even further back, with sets). For example, one could define a number n to be divisible by a number m if and only if there exists an integer q such that $qm = n$. Using this definition, we can give a more precise proof: if n is divisible by 6, then $n = 6q$ for some q, and therefore $n = 2(3q)$, which proves that n is divisible by 2. Thus we have used the definitions to show that the definition of divisibility by 2 holds whenever the definition of divisibility by 6 holds.

Historically, mathematical writers have been satisfied with varying levels of rigor. Results and methods have often been widely used without a full justification of the kind just outlined, particularly in bodies of mathematical thought that are new and rapidly developing. Some ancient cultures, the Egyptians for example, had methods for multiplication and division, but no justification of these methods has survived and it does not seem especially likely that formal justification existed. The methods were probably accepted simply because they worked, rather than because there was a thorough argument justifying them.

By the middle of the seventeenth century, European mathematical writers who were engaged in research were well-acquainted with the model of rigorous mathematical argument supplied by EUCLID's [VI.2] *Elements*. The kind of deductive, or synthetic, argument we illustrated earlier would have been described as a proof *more geometrico*—in the geometrical way. While Euclid's arguments, assumptions, and definitions are not wholly rigorous by today's standards, the basic idea was clear: one proceeds from clear definitions and generally agreed basic ideas (such as that the whole is greater than the part) to deduce theorems (also called

propositions) in a step-by-step manner, not bringing in anything extra (either on the sly or unintentionally). This classical model of geometric argument was widely used in reasoning about whole numbers (for example by FERMAT [VI.12]), in analytic geometry (DESCARTES [VI.11]), and in mechanics (Galileo).

This article is about rigor in *analysis*, a term which itself has had a shifting meaning. Coming from ancient origins, by around 1600 the term was used to refer to mathematics in which one worked with an unknown (something we would now write as x) to do a calculation or find a length. In other words, it was closely related to algebra, though the notion was imported into geometry by Descartes and others. However, over the course of the eighteenth century the word came to be associated with the calculus, which was the principal area of application of analytic techniques. When we talk about rigor in analysis it is the rigorous theory of the mathematics associated with differential and integral calculus that we are principally discussing. In the third quarter of the seventeenth century rival methods for the differential and integral calculus were devised by NEWTON [VI.14] and LEIBNIZ [VI.15], who thereby synthesized and extended a considerable amount of earlier work concerned with tangents and normals to curves and with the areas of regions bounded by curves. The techniques were highly successful, and were extended readily in a variety of directions, most notably in mechanics and in differential equations.

The key common feature of this research was the use of infinities: in some sense, it involved devising methods for combining infinitely many infinitely small quantities to get a finite answer. For example, suppose we divide the circumference of a circle into a (large) number of equal parts by marking off points at equal distances, then joining the points and creating triangles by joining the points to the center. Adding up the areas of the triangles approximates the circular area, and the more points we use the better the approximation. If we imagine infinitely many of these inscribed triangles, the area of each will be "infinitely small" or *infinitesimal*. But because the total involves adding up infinitely many of them, it may be that we get a finite positive total (rather than just 0, from adding up infinitely many zeros, or an infinite number, as we would get if we added the same finite number to itself infinitely many times). Many techniques for doing such calculations were devised, though the interpretation of what was taking place varied. Were the infinities involved "real" or merely "potential"? If something is "really" infinitesimal, is it just zero? Aristotelian writers had abhorred actual infinities, and complaints about them were common at the time.

Newton, Leibniz, and their immediate followers provided mathematical arguments to justify these methods. However, the introduction of techniques involving reasoning with infinitely small objects, limiting processes, infinite sums, and so forth meant that the founders of the calculus were exploring new ground in their arguments, and the comprehensibility of these arguments was frequently compromised by vague terminology or by the drawing of one conclusion when another might seem to follow equally well. The objects they were discussing included infinitesimals (quantities infinitely smaller than those we experience directly), ratios of vanishingly small quantities (i.e., fractions in, or approaching, the form 0/0), and finite sums of infinitely many positive terms. Taylor series representations, in particular, provoked a variety of questions. A function may be written as a series in such a way that the series, when viewed as a function, will have, at a given point $x = a$, the same value as the function, the same rate of change (or first derivative), and the same higher-order derivatives to arbitrary order:

$$f(x) = f(a) + f'(a)(x - a) + \tfrac{1}{2}f''(a)(x - a)^2 + \cdots .$$

For example, $\sin x = x - x^3/3! + x^5/5! + \cdots$, a fact already known to Newton though such series are now named after Newton's disciple BROOK TAYLOR [VI.16].

One problem with early arguments was that the terms being discussed were used in different ways by different writers. Other problems arose from this lack of clarity, since it concealed a variety of issues. Perhaps the most important of these was that an argument could fail to work in one context, even though a very similar argument worked perfectly well in another. In time, this led to serious problems in extending analysis. Eventually, analysis became fully rigorous and these difficulties were solved, but the process was a long one and it was complete only by the beginning of the twentieth century.

Let us consider some examples of the kinds of difficulties that arose from the very beginning, using a result of Leibniz. Suppose we have two variables, u and v, each of which changes when another variable, x, changes. An infinitesimal change in x is denoted dx, the differential of x. The differential is an infinitesimal quantity, thought of as a geometrical magnitude, such as a length, for example. This was imagined to be combined or compared with other magnitudes in the usual ways (two lengths can be added, have a ratio, and so on). When x changes to $x + dx$, u and v change to $u + du$ and $v + dv$, respectively. Leibniz concluded that the product uv would then change to $uv + u \, dv + v \, du$, so that $d(uv) = u \, dv + v \, du$. His argument is, roughly, that $d(uv) = (u + du)(v + dv) - uv$. Expanding the right-hand side using regular algebra and then simplifying gives $u \, dv + v \, du + du \, dv$. But the term $du \, dv$ is a second-order infinitesimal, vanishingly small compared with the first-order differentials, and is thus treated as equal to 0. Indeed, one aspect of the problems is that there appears to be an *inconsistency* in the way that infinitesimals are treated. For instance, if you want to work out the derivative of $y = x^2$, the calculation corresponding to the one just given (expanding $(x + dx)^2$, and so on) shows that $dy/dx = 2x + dx$. We then treat the dx on the right-hand side as zero, but the one on the left-hand side seems as though it ought to be an infinitesimal *nonzero* quantity, since otherwise we could not divide by it. So is it zero or not? And if not, how do we get around the apparent inconsistency?

At a slightly more technical level, the calculus required mathematicians to deal repeatedly with the "ultimate" values of ratios of the form dy/dx when the quantities in both numerator and denominator approach or actually reach 0. This phrasing uses, once again, the differential notation of Leibniz, though the same issues arose for Newton with a slightly different notational and conceptual approach. Newton generally spoke of variables as depending on time, and he sought (for example) the values approached when "evanescent increments"—vanishingly small time intervals—are considered. One long-standing set of confusions arose precisely from this idea that variable quantities were in the process of changing, whether with time or with changes in the value of another variable. This means that we talk about values of a variable approaching a given value, but without a clear idea of what this "approach" actually is.

2   Eighteenth-Century Approaches and Critiques

Of course, had the calculus not turned out to be an enormously fruitful field of endeavor, no one would have bothered to criticize it. But the methods of Newton and Leibniz were widely adopted for the solution of problems that had interested earlier generations (notably tangent and area problems) and for the posing and solution of problems that these techniques suddenly made far more accessible. Problems of areas, maxima and minima, the formulation and solution of differential equations to describe the shape of hanging chains or the positions of points on vibrating strings, applications to celestial mechanics, the investigation of problems having to do with the properties of functions (thought of for the most part as analytic expressions involving variable quantities)—all these fields and more were developed over the course of the eighteenth century by mathematicians such as Taylor, Johann and Daniel Bernoulli, Euler, D'ALEMBERT [VI.20], Lagrange, and many others. These people employed many virtuoso arguments of suspect validity. Operations with divergent series, the use of imaginary numbers, and manipulations involving actual infinities were used effectively in the hands of the most capable of these writers. However, the methods could not always be explained to the less capable, and thus certain results were not reliably reproducible—a very odd state for mathematics from today's standpoint. To do Euler's calculations, one needed to be Euler. This was a situation that persisted well into the following century.

Specific controversies often highlighted issues that we now see as a result of foundational confusion. In the case of infinite series, for example, there was confusion about the domain of validity of formal expressions. Consider the series

$$1 - 1 + 1 - 1 + 1 - 1 + 1 - \cdots .$$

In today's usual elementary definition (due to CAUCHY [VI.29] around 1820) we would now consider this series to be divergent because the sequence of partial sums $1, 0, 1, 0, \ldots$ does not tend to a limit. But in fact there was some controversy about the actual meaning of such expressions. Euler and Nicolaus Bernoulli, for example, discussed the potential distinction between the *sum* and the *value* of an infinite sum, Bernoulli arguing that something like $1 - 2 + 6 - 24 + 120 + \cdots$ has no sum but that this algebraic expression does constitute a value. Whatever may have been meant by this, Euler defended the notion that the sum of the series is the value of the finite expression that gives rise to the series. In his 1755 *Institutiones Calculi Differentialis*, he gives the example of $1 - x + x^2 - x^3 + \cdots$, which comes from $1/(1 + x)$, and later defended the view that this meant that $1 - 1 + 1 - 1 + \cdots = \frac{1}{2}$. His view was not universally accepted. Similar controversies arose in considering how to extend the values of functions outside their usual domain, for example with the logarithms of negative numbers.

Probably the most famous eighteenth-century critique of the language and methods of eighteenth-century analysis is due to the philosopher George Berkeley (1685–1753). Berkeley's motto, "To be is to be perceived," expresses his idealist stance, which was coupled with a strong view that the abstraction of individual qualities, for the purposes of philosophical discussion, is impossible. The objects of philosophy should thus be things that are perceived, and perceived in their entirety. The impossibility of perceiving infinitesimally small objects, combined with their manifestly abstracted nature, led him to attack their use in his 1734 treatise *The Analyst: Or, a Discourse Addressed to an Infidel Mathematician*. Referring sarcastically in 1734 to infinitesimals as the "ghosts of departed quantities," Berkeley argued that neglecting some quantity, no matter how small, was inappropriate in mathematical argument. He quoted Newton in this regard, to the effect that "in mathematical matters, errors are to be condemned, no matter how small." Berkeley continued, saying that "[n]othing but the obscurity of the subject" could have induced Newton to impose this kind of reasoning on his followers. Such remarks, while they apparently did not dissuade those enamored of the methods, contributed to a sentiment that aspects of the calculus required deeper explanation. Writers such as Euler, d'Alembert, Lazare Carnot, and others attempted to address foundational criticisms by clarifying what differentials were, and gave a variety of arguments to justify the operations of the calculus.

2.1 Euler

Euler contributed to the general development of analysis more than any other individual in the eighteenth century, and his approaches to justifying his arguments were enormously influential even after his death, owing to the success and wide use of his important textbooks. Euler's reasoning is sometimes regarded as rather careless since he operated rather freely with the notation of the calculus, and many of his arguments are certainly deficient by later standards. This is particularly true of arguments involving infinite series and products. A typical example is provided by an early version of his proof that

$$\sum_{n=1}^{\infty} \frac{1}{n^2} = \frac{\pi^2}{6} .$$

His method is as follows. Using the known series expansion for $\sin x$ he considered the zeros of

$$\frac{\sin \sqrt{x}}{\sqrt{x}} = 1 - \frac{x}{3!} + \frac{x^2}{5!} - \frac{x^3}{7!} + \cdots .$$

These lie at π^2, $(2\pi)^2$, $(3\pi)^2$, ... Applying (without argument) the factor theorem for *finite* algebraic equations he expressed this equation as

$$\frac{\sin \sqrt{x}}{\sqrt{x}} = \left(1 - \frac{x}{\pi^2}\right)\left(1 - \frac{x}{4\pi^2}\right)\left(1 - \frac{x}{9\pi^2}\right) \cdots .$$

Now, it can be seen that the coefficient of x in the infinite sum, $-\frac{1}{6}$, should equal the negative of the sum of the coefficients of x in the product. Euler apparently concluded this by imagining multiplying out the infinitely many terms and selecting the 1 from all but one of them. This gives

$$\frac{1}{\pi^2} + \frac{1}{4\pi^2} + \frac{\cdot\,1}{9\pi^2} + \cdots = \frac{1}{6},$$

and multiplying both sides by π^2 gives the required sum.

We now think of this approach as having several problems. The product of the infinitely many terms may or may not represent a finite value, and today we would specify conditions for when it does. Also, applying a result about (finite) polynomials to (infinite) power series is a step that requires justification. Euler himself was to provide alternative arguments for this result later in his life. But the fact that he may have known *counterexamples*—situations in which such usages would not work—was not, for him, a decisive obstacle. This view, in which one reasoned in a generic situation that might admit a few exceptions, was common at his time, and it was only in the late nineteenth century that a concerted effort was made to state the results of analysis in ways that set out precisely the conditions under which the theorems would hold.

Euler did not dwell on the interpretation of infinite sums or infinitesimals. Sometimes he was happy to regard differentials as actually equal to zero, and to derive the meaning of a ratio of differentials from the context of the problem:

> An infinitely small quantity is nothing but a vanishing quantity and therefore will be actually equal to 0.... Hence there are not so many mysteries hidden in this concept as there are usually believed to be. These supposed mysteries have rendered the calculus of the infinitely small quite suspect to many people.

This statement, from the *Institutiones Calculi Differentialis* of 1755, was followed by a discussion of proportions in which one of the ratios is $0/0$, and a justification of the fact that differentials may be neglected in calculations with ordinary numbers. This accurately describes a good deal of his practice—when he worked with differential equations, for example.

Controversial matters did arise, however, and debates about definitions were not unusual. The best-known example involves discussions connected with the so-called vibrating string problem, which involved Euler, d'Alembert, and Daniel Bernoulli. These were closely connected with the definition of FUNCTIONS [I.2 §2.2], and the question of which functions studied by analysis actually could be represented by series (in particular trigonometric series). The idea that a curve of arbitrary shape could serve as an initial position for a vibrating string extended the idea of function, and the work of FOURIER [VI.25] in the early nineteenth century made such functions analytically accessible. In this context, functions with broken graphs (a kind of *discontinuous* function) came under inspection. Later, how to deal with such functions would be a decisive issue for the foundations of analysis, as the more "natural" objects associated with algebraic operations and trigonometry gave way to the more general modern concept of function.

2.2   Responses from the Late Eighteenth Century

One significant response to Berkeley in Britain was that of Colin Maclaurin (1698–1746), whose 1742 textbook *A Treatise of Fluxions* attempted to clarify the foundations of the calculus and do away with the idea of infinitely small quantities. Maclaurin, a leading figure of the Scottish Enlightenment of the mid eighteenth century, was the most distinguished British mathematician of his time and an ardent proponent of Newton's methods. His work, unlike that of many of his British contemporaries, was read with interest on the Continent, especially his elaborations of Newtonian celestial mechanics. Maclaurin attempted to base his reasoning on the notion of the limits of what he termed "assignable" finite quantities. Maclaurin's work is famously obscure, though it did provide examples of calculating the limits of ratios. Perhaps his most important contribution to the clarification of the foundations of analysis was his influence on d'Alembert.

D'Alembert had read both Berkeley and Maclaurin and followed them in rejecting infinitesimals as real quantities. While exploring the idea of a differential as a limit, he also attempted to reconcile his idea with the idea that infinitesimals may be consistently regarded as being actually zero, perhaps in a nod to Euler's view. The main exposition of d'Alembert's views may be found in the *Encyclopédie*, in the articles on differentials (published in 1754) and on limits (1765). D'Alembert argued for the importance of geometric rather than algebraic limits. His meaning seems to have been that the quantities being investigated should not be treated merely formally, by substitution and simplification. Rather, a limit should be understood as the limit of a length (or collection of lengths), area, or other dimensioned quantity, in much the way that a circle may be seen as a limit of inscribed polygons. His aim seems primarily to have been to establish the reality of the objects described by existing algorithms, since the actual calculations he employs are carried out with differentials.

2.2.1 Lagrange

In the course of the eighteenth century, the differential and the integral calculus gradually distinguished themselves as a set of methods distinct from their applications in mechanics and physics. At the same time, the primary focus of the methods moved away from geometry, so that in work of the second half of the eighteenth century we increasingly see calculus treated as "algebraic analysis" of "analytic functions." The term "analytic" was used in a variety of senses. For many writers, such as Euler, it merely referred to a function (that is, a relationship between variable quantities) that is given by a single expression of the type used in analysis.

Lagrange provided a foundation for the calculus that was indebted to this algebraic viewpoint. Lagrange concentrated on power-series expansions as the basic entity of analysis, and through his work the term analytic function evolved toward its more recent meaning connected with the existence of a convergent Taylor series representation. His approach reached a full expression in his *Théorie des Fonctions Analytiques* of 1797. This was a version of his lectures at the École Polytechnique, a new institution for the elite training of military engineers in revolutionary France. Lagrange assumed that a function must necessarily be expressible as an infinite series of algebraic functions, basing this argument on the existence of expansions for known functions. He first sought to show that "in general" no negative or fractional powers would appear in the expansion, and from this he obtained a power-series representation. His arguments here are surprising, and somewhat ad hoc, and I use an example given by Fraser (1987). The slightly strange notation is based on that of Lagrange. Suppose that one seeks an expansion of $f(x) = \sqrt{x + i}$ in powers of i. In general, only

integer powers will be involved. Terms of the form $i^{m/n}$ do not make sense, says Lagrange, since the expression of the function $\sqrt{x + i}$ is only two-valued, while $i^{m/n}$ has n values. Hence the series

$$\sqrt{x + i} = \sqrt{x} + pi + qi^2 + \cdots + ti^k + \cdots$$

obtains its two values from the term $\sqrt{x}$, and all other powers must be integral. With fractional exponents set aside, Lagrange argued that $f(x + i) = f(x) + i^a P(x, i)$, with P finite for $i = 0$. Successive application of this result gave him the expansion

$$f(x + i) = f(x) + pi + qi^2 + ri^3 + \cdots,$$

where i was a small increment. The number p depends on x, so Lagrange defined a *derived function* $f'(x) = p(x)$. The French term *dérivée* is the origin of the term derivative, and in Lagrange's language f is the "primitive" of this derived function. Similar arguments can be made to relate the higher coefficients to the higher derivatives in the usual Taylor formula.

This approach, which seems oddly circular to modern eyes, relied on the eighteenth-century distinction between the "algebraic" infinite process of the series expansion on the one hand, and the use of differentials on the other. Lagrange did not see the original series expansion as based on the limit process. With the renewed emphasis on limits and modern definitions developed by Cauchy, this approach was soon to be regarded as untenable.

3 The First Half of the Nineteenth Century

3.1 Cauchy

Many writers contributed to discussions on rigor in analysis in the first decades of the nineteenth century. It was Cauchy who was to revive the limit approach to greatest effect. His aim was pedagogical, and his ideas were probably worked out in the context of preparing his introductory lectures for the École Polytechnique at the beginning of the 1820s. Although the students were the best in France in scholarly ability, many found the approach too difficult. As a result, while Cauchy himself continued to use his methods, other instructors held on to older approaches using infinitesimals, which they found more intuitively accessible for the students as well as better adapted to the solution of problems in elementary mechanics. Cauchy's self-imposed exile from Paris in the 1830s further limited the impact of his approach, which was initially taken up only by a few of his students.

Nonetheless, Cauchy's definitions of limit, of continuity, and of the derivative gradually came into general use in France, and were influential elsewhere as well, especially in Italy. Moreover, his methods of using these definitions in proofs, and particularly his use of mean value theorems in various forms, moved analysis from a collection of symbolic manipulations of quantities with special properties toward the science of argument about infinite processes using close estimation via the manipulation of inequalities.

In some respects, Cauchy's greatest contribution lay in his clear definitions. For earlier writers, the sum of an infinite series was a somewhat vague notion, sometimes interpreted by a kind of convergence argument (as with the sum of a geometric series such as $\sum_{n=0}^{\infty} 2^{-n}$) and sometimes as the value of the function from which the series was derived (as Euler, for example, often regarded it). Cauchy revised the definition to state that the sum of an infinite series was the limit of the sequence of partial sums. This provided a unified approach for series of numbers and series of functions, an important step in the move to base calculus and analysis on ideas about real numbers. This trend, eventually dominant, is often referred to as the "arithmetization of analysis." Similarly, a continuous function is one for which "an infinitely small increase of the variable produces an infinitely small increase of the function itself" (Cauchy 1821, pp. 34–35).

As we see from the example just given, Cauchy did not shy away from infinitely small quantities, nor did he analyze this notion further. The limit of a variable quantity is defined in a way that we would now regard as conversational, or heuristic:

> When the values that are successively assigned to a given variable approach a fixed value indefinitely, in such a way that it ends up differing from it as little as one wishes, this latter value is called the *limit* of all the others. Thus, for example, an irrational number is the limit of the various fractions that provide values that are closer and closer to it.
>
> Cauchy (1821, p. 4)

These ideas were not completely rigorous by modern standards, but he was able to use them to provide a unified foundation for the basic processes of analysis.

This use of infinitely small quantities appears, for example, in his definition of a continuous function. To paraphrase his definition, suppose that a function $f(x)$ is single-valued on some finite interval of the real line, and choose any value x_0 inside the interval. If the value

of x_0 is increased to $x_0 + a$, the function also changes by the amount $f(x_0 + a) - f(x_0)$. Cauchy says that the function f is continuous for this interval if, for each value of x_0 in that interval, the numerical value of the difference $f(x_0 + a) - f(x_0)$ decreases indefinitely to 0 with a. In other words, Cauchy defines continuity as a property *on an interval* rather than at a point, in essence by saying that on that interval infinitely small changes in the argument produce infinitely small changes in the function value. Cauchy appears to have considered continuity to be a property of a function on an interval.

This definition emphasizes the importance of jumps in the value of the function for the understanding of its properties, something that Cauchy had encountered early in his career when discussing THE FUNDAMENTAL THEOREM OF CALCULUS [I.3 §5.5]. In his 1814 memoir on definite integrals, Cauchy stated:

> If the function $\phi(z)$ increases or decreases in a continuous manner between $z = b'$ and $z = b''$, the value of the integral $[\int_{b'}^{b''} \phi'(z)\,dz]$ will ordinarily be represented by $\phi(b'') - \phi(b')$. But if … the function passes suddenly from one value to another sensibly different … the ordinary value of the integral must be diminished.
>
> *Oeuvres* (volume 1, pp. 402–3)

In his lectures, Cauchy assumed continuity when defining the definite integral. He considered first of all a division of the interval of integration into a finite number of subintervals on which the function is either increasing or decreasing. (This is not possible for all functions, but this appeared not to concern Cauchy.) He then defined the definite integral as the limit of the sum $S = (x_1 - x_0)f(x_0) + (x_2 - x_1)f(x_1) + \cdots + (X - x_{n-1})f(x_{n-1})$ as the number n becomes very large. Cauchy gives a detailed argument for the existence of this limit, using his theorem of the mean and the fact of continuity.

Versions of the main subjects of Cauchy's lectures were published in 1821 and 1823. Every student at the École Polytechnique would have been aware of them subsequently, and many would have used them explicitly. They were joined in 1841 by a version of the course elaborated by Cauchy's associate, the Abbé Moigno. They were referred to frequently in France and the definitions employed by Cauchy became standard there. We also know that the lectures were studied by others, notably by ABEL [VI.33] and DIRICHLET [VI.36], who spent time in Paris in the 1820s, and by RIEMANN [VI.49].

Cauchy's movement away from the formal approach of Lagrange rejected the "vagueness of algebra." Although he was clearly guided by intuition (both geometric and otherwise), he was well aware that intuition could be misleading, and produced examples to show the value of adhering to precise definitions. One famous example, the function that takes the value e^{-1/x^2} when $x \neq 0$ and zero when $x = 0$, is differentiable infinitely many times, yet it does not yield a Taylor series that converges to the function at the origin. Despite this example, which he mentioned in his lectures, Cauchy was not a specialist in counterexamples, and in fact the trend toward producing counterexamples for the purpose of clarifying definitions was a later development.

Abel famously drew attention to an error in Cauchy's work: his statement that a convergent series of continuous functions has a continuous sum. For this to be true, the series must be uniformly convergent, and in 1826 Abel gave as a counterexample the series

$$\sum_{k=1}^{\infty} (-1)^{k+1} \frac{\sin kx}{k},$$

which is discontinuous at odd multiples of π. Cauchy was led to make this distinction only much later, after the phenomenon had been identified by several writers. Historians have written extensively about this apparent error; one influential account, due to Bottazzini, proposes that for various reasons Cauchy would not have found Abel's example telling, even if he had known of it at the time (this account appears in Bottazzini (1990, p. LXXXV)).

Before leaving the time of Cauchy, we should note the related independent activity of BOLZANO [VI.28]. Bolzano, a Bohemian priest and professor whose ideas were not widely disseminated at the time, investigated the foundations of the calculus extensively. In 1817, for example, he gave what he termed a "purely analytic proof of the theorem that between any two values that possess opposite signs, at least one real root of the equation exists": the intermediate value theorem. Bolzano also studied infinite sets: what is now called the Bolzano–Weierstrass theorem states that for every bounded infinite set there is at least one point having the property that any disk about that point contains infinitely many points of the set. Such "limit points" were studied independently by WEIERSTRASS [VI.44]. By the 1870s, Bolzano's work became more broadly known.

3.2   Riemann, the Integral, and Counterexamples

Riemann is indelibly associated with the foundations of analysis because of the Riemann integral, which is part of every calculus course. Despite this, he was not always driven by issues involving rigor. Indeed he remains a standard example of the fruitfulness of nonrigorous intuitive invention. There are many points in Riemann's work at which issues about rigor arise naturally, and the wide interest in his innovations did much to direct the attention of researchers to making these insights precise.

Riemann's definition of the definite integral was presented in his 1854 *Habilitationschrift*—the "second thesis," which qualified him to lecture at a university for fees. He generalized Cauchy's notion to functions that are not necessarily continuous. He did this as part of an investigation of FOURIER SERIES [III.27] expansions. The extensive theory of such series was devised by Fourier in 1807 but not published until the 1820s. A Fourier series represents a function in the form

$$f(x) = a_0 + \sum_{n=1}^{\infty} (a_n \cos(nx) + b_n \sin(nx))$$

on a finite interval.

The immediate inspiration for Riemann's work was DIRICHLET [VI.36], who had corrected and developed earlier faulty work by Cauchy on the question of when and whether the Fourier series expansion of a function converges to the function from which it is derived. In 1829 Dirichlet had succeeded in proving such convergence for a function with period 2π that is integrable on an interval of that length, does not possess infinitely many maxima and minima there, and at jump discontinuities takes on the average value between the two limiting values on each side. As Riemann noted, following his professor Dirichlet, "this subject stands in the closest connection to the principles of infinitesimal calculus, and can therefore serve to bring these to greater clarity and definiteness" (Riemann 1854, p. 238). Riemann sought to extend Dirichlet's investigations to further cases, and was thus led to investigate in detail each of the conditions given by Dirichlet. Accordingly, he generalized the definition of a definite integral as follows:

We take between a and b an increasing sequence of values $x_1, x_2, \ldots, x_{n-1}$, and for brevity designate $x_1 - a$ by δ_1, $x_2 - x_1$ by δ_2, $\ldots$, $b - x_{n-1}$ by δ_n and by ϵ a positive proper fraction. Then the value of the sum

$$S = \delta_1 f(a + \epsilon_1 \delta_1) + \delta_2 f(x_1 + \epsilon_2 \delta_2)$$
$$+ \delta_3 f(x_2 + \epsilon_3 \delta_3) + \cdots + \delta_n f(x_{n-1} + \epsilon_n \delta_n)$$

depends on the choice of the intervals δ and the quantities ϵ. If it has the property that it approaches infinitely closely a fixed limit A no matter how the δ and ϵ are chosen, as δ becomes infinitely small, then we call this value $\int_a^b f(x)\,\mathrm{d}x$.

In connection with this definition of the integral, and in part to show its power, Riemann provided an example of a function that is discontinuous in any interval, yet can be integrated. The integral thus has points of nondifferentiability on each interval. Riemann's definition rendered problematic the inverse relationship between differentiation and integration, and his example brought this problem out clearly. The role of such "pathological" counterexamples in pushing the development of rigor, already apparent in Cauchy's work, intensified greatly around this time.

Riemann's definition was published only in 1867, following his death; an expository version due to Gaston Darboux appeared in French in 1873. The popularization and extension of Riemann's approach went hand in hand with the increasing appreciation of the importance of rigor associated with the Weierstrass school, discussed below. Riemann's approach focused attention on sets of points of discontinuities, and thus were seminal for CANTOR's [VI.54] investigations into point sets in the 1870s and afterwards.

The use of the *Dirichlet principle* serves as a further example of the way in which Riemann's work drew attention to problems in the foundations of analysis. In connection with his research into complex analysis, Riemann was led to investigate solutions to the so-called *Dirichlet problem*: given a function g, defined on the boundary of a closed region in the plane, does there exist a function f that satisfies the LAPLACE PARTIAL DIFFERENTIAL EQUATION [I.3 §5.4] in the interior and takes the same values as g on the boundary? Riemann asserted that the answer was yes. To demonstrate this, he reduced the question to proving the existence of a function that minimizes a certain integral over the region, and argued on physical grounds that such a minimizing function must always exist. Even before Riemann's death his assertion was questioned by WEIERSTRASS [VI.44], who published a counterexample in 1870. This led to attempts to reformulate Riemann's results and prove them by other

means, and ultimately to a rehabilitation of the Dirichlet principle through the provision of precise and broad hypotheses for its validity, which were expressed by HILBERT [VI.63] in 1900.

4 Weierstrass and His School

Weierstrass had a passion for mathematics as a student at Bonn and Münster, but his student career was very uneven. He spent the years from 1840 to 1856 as a high school teacher, undertaking research independently but at first publishing obscurely. Papers from 1854 onward in *Journal für die reine und angewandte Mathematik* (otherwise known as *Crelle's Journal*) attracted wide attention to his talent, and he obtained a professorship in Berlin in 1856. Weierstrass began to lecture regularly on mathematical analysis, and his approach to the subject developed into a series of four courses of lectures given cyclically between the early 1860s and 1890. The lectures evolved over time and were attended by a large number of important mathematical researchers. They also indirectly influenced many others through the circulation of unpublished notes. This circle included R. Lipschitz, P. du Bois-Reymond, H. A. Schwarz, O. Hölder, Cantor, L. Koenigsberger, G. Mittag-Leffler, KOVALEVSKAYA [VI.59], and L. Fuchs, to name only some of the most important. Through their use of Weierstrassian approaches in their own research, and their espousal of his ideas in their own lectures, these approaches became widely used well before the eventual publication of a version of his lectures late in his life. The account that follows is based largely on the 1878 version of the lectures. His approach was also influential outside Germany: parts of it were absorbed in France in the lectures of HERMITE [VI.47] and JORDAN [VI.52], for example.

Weierstrass's approach builds on that of Cauchy (though the detailed relationship between the two bodies of work has never been fully examined). The two overarching themes of Weierstrass's approach are, on the one hand, the banning of the idea of motion, or changing values of a variable, from limit processes, and, on the other, the representation of functions, notably of a complex variable. The two are intimately linked. Essential to the motion-free definition of a limit is Weierstrass's nascent investigation of what we would now call the topology of the real line or complex plane, with the idea of a limit point, and a clear distinction between local and global behavior. The central objects of study for Weierstrass are functions (of one or more real or complex variable quantities), but it should be borne in mind that set theory is not involved, so that functions are *not* to be thought of as sets of ordered pairs.

The lectures begin with a now-familiar subject: the development of rational, negative, and real numbers from the integers. For example, negative numbers are defined operationally by making the integers closed under the operation of subtraction. He attempted a unified approach to the definition of rational and irrational numbers which involved unit fractions and decimal expansions and now seems somewhat murky. Weierstrass's definition of the real numbers appears unsatisfactory to modern eyes, but the general path of *arithmetization* of analysis was established by this approach. In parallel to the development of number systems, he also developed different classes of functions, building them up from rational functions by using power-series representations. Thus, in Weierstrass's approach, a polynomial (called an integer rational function) is generalized to a "function of integer character," which means a function with a convergent power-series expansion everywhere. The Weierstrass factorization theorem asserts that any such function may be written as a (possibly infinite) product of certain "prime" functions and exponential functions with polynomial exponents of a certain type.

The limit definition given by Weierstrass has thoroughly modern features:

> That a variable quantity x becomes infinitely small simultaneously with another quantity y means: "After the assumption of an arbitrarily small quantity ϵ a bound δ for x may be found, such that for every value of x for which $|x| < \delta$, the corresponding value of $|y|$ will be less than ϵ."
>
> Weierstrass (1988, p. 57)

Weierstrass immediately used this definition to give a proof of continuity for rational functions of several variables, using an argument that could appear in a textbook today. The former notions of variables tending to given values were replaced by quantified statements about linked inequalities. The framing of hypotheses in terms of inequalities became a guiding motif in the work of Weierstrass's school: here we mention in passing the Lipschitz and Hölder conditions in the existence theory for differential equations. The clarity that this language gave to problems involving the interchange of limits, for example, meant that previously intractable problems could now be handled in

a routine way by those inculcated in the Weierstrass approach.

The fact that general functions were built from rational functions using series expansions gave the latter a key role in Weierstrass's work, and as early as 1841 he had identified the importance of uniform convergence. The distinction between uniform and point-wise convergence was made very clearly in his lectures. A series converges, as it does for Cauchy, if its sequence of partial sums converges, though now the convergence is phrased in the following terms: the series $\sum f_n(x)$ converges to s_0 at $x = x_0$ if, given an arbitrary positive ϵ, there is an integer N such that $|s_0 - (f_1(x_0) + f_2(x_0) + \cdots + f_n(x_0))| < \epsilon$ for every $n > N$. The convergence is uniform on a domain of the variable if the same N will work for that ϵ value for all x in the domain. Uniform convergence guarantees continuity of the sum, since these are series of rational, hence continuous, functions. From this point of view, then, uniform convergence is important well beyond the context of trigonometric series (important though those may be). Indeed, it is a central tool of the entire theory of functions.

Weierstrass's role as a critic of rigor in the work of others, notably Riemann, has already been noted. More than any other leading figure, he generated counterexamples to illustrate difficulties with received notions and to distinguish between different kinds of analytical behavior. One of his best-known examples was of an everywhere-continuous but nowhere-differentiable function, namely $f(x) = \sum b^n \cos(a^n x)$, which is uniformly convergent for $b < 1$ but fails to be differentiable at any x if $ab > 1 + \frac{3}{2}\pi$. Similarly he constructed functions for which the Dirichlet principle fails, examples of sets constituting "natural boundaries," that is, obstacles to continuing series expansions into larger domains, and so forth. The careful distinctions he encouraged, and the very procedure of seeking pathological rather than typical examples, threw the spotlight on the precision of hypotheses in analysis to an unprecedented degree. From the 1880s, with the maturity of this program, analysis no longer dealt with generic cases and looked instead for absolutely precise statements in a way that has for the most part endured to the present. This was also to become a pattern and an imperative in other areas of mathematics, though sometimes the passage from reasoning from generic examples to fully expressed hypotheses and definitions took decades. (Algebraic geometry provides a famous example, one in which reasoning with generic cases lasted until the 1920s.) In this sense the form of rigorous argument and exposition espoused by Weierstrass and his school was to become a pattern for mathematics generally.

4.1   The Aftermath of Weierstrass and Riemann

Analysis became the model subdiscipline for rigor for a variety of reasons. Of course, analysis was important for the sheer volume and range of application of its results. Not everyone agreed with the precise way in which Weierstrass approached foundational questions (through series, rational functions, and so on). Indeed, Riemann's more geometric approach had also attracted followers, if not exactly a school, and the insights his approach afforded were enthusiastically embraced. However, any subsequent discussion had to take place at a level of rigor comparable to that which Weierstrass had attained. While approaches to the foundations of analysis were to vary, the idea that limits should be rigorously handled in much the way that Weierstrass did was not to alter. Among the remaining central issues for rigor was the definition of the number systems.

For the real numbers, probably the most successful definition (in terms of its later use) was provided by DEDEKIND [VI.50]. Dedekind, like Weierstrass, took the integers as fundamental, and extended them to the rationals, noting that the algebraic properties satisfied by the latter are those satisfied by what we now call a FIELD [I.3 §2.2]. (This idea is also Dedekind's.) He then showed that the rational numbers satisfy a *trichotomy law*. That is, each rational number x divides the entire collection into three parts: x itself, rational numbers greater than x, and rational numbers less than x. He also showed that the rationals greater and less than a given number extend to infinity, and that any rational corresponds to a distinct point on the number line. However, he also observed that along that line there are infinitely many points that do not correspond to any rational. Using the idea that to every point on the line there should correspond a number, he constructed the remainder of the continuum (that is, the real line) by the use of *cuts*. These are ordered pairs (A_1, A_2) of nonempty sets of rational numbers such that every element of the first set is less than every element of the second, and such that taken together they contain all the rationals. Such cuts may obviously be produced by an element x, in which case x is either the greatest element of A_1 or the least element of A_2. But sometimes A_1 does not have a greatest element, or A_2 a least element, and in that case we can use the cut to define a

new number, which is necessarily irrational. The set of all such cuts may be shown to correspond to the points of the number line, so that nothing is left out. A critical reader might feel that this is begging the question, since the idea of the number line constituting a continuum in some way might seem to be a hidden premise.

Dedekind's construction stimulated a good deal of discussion, especially in Germany, about the best way to found the real numbers. Participants included Cantor, E. Heine, and the logician FREGE [VI.56]. Heine and Cantor, for example, considered real numbers as equivalence classes of Cauchy sequences of rationals, together with a machinery that permitted them to define the basic arithmetical operations. A very similar approach was proposed by the French mathematician Charles Méray. Frege, by contrast, in his 1884 *Die Grundlagen der Arithmetik*, sought to found the integers on logic. While his attempts to construct the reals along these lines did not bear fruit, he had an important role in his insistence that the various constructions should not merely be mathematically functional but should also be demonstrably free from internal contradiction.

Despite much activity on the foundations of the real numbers, infinite sets, and other basic notions for analysis, consensus remained elusive. For example, the influential Berlin mathematician LEOPOLD KRONECKER [VI.48] denied the existence of the reals, and held that all true mathematics was to be based on finite sets. Like Weierstrass, with whom he worked and whom he influenced, he emphasized the strong analogies between the integers and the polynomials, and sought to use this algebraic foundation to build all of mathematics. Hence for Kronecker the entire main path of research in analysis was anathema, and he opposed it ardently. These views were influential, both directly and indirectly, on a number of later writers, including BROUWER [VI.75], the intuitionist school around him, and the algebraist and number theorist Kurt Hensel.

All efforts to found analysis were based in one way or another on an underlying notion (not always made explicit) of quantity. The foundational framework of analysis, however, was to shift over the period from 1880 to 1910 toward the theory of sets. This had its origin in the work of Cantor, a student of Weierstrass who began studying discontinuities of Fourier series in the early 1870s. Cantor became concerned about how to distinguish between different types of infinite sets. His proofs that the rational numbers and the algebraic numbers are COUNTABLE [III.11] while the reals are not led him to a hierarchy of infinite sets of different car-dinality. The importance of this discovery for analysis was at first not widely recognized, though in the 1880s Mittag-Leffler and Hurwitz both made significant appli-cations of notions about derived sets (the set of limit points of a given set) and dense or nowhere-dense sets.

Cantor gradually came to the view that set theory could function as a foundational tool for all of math-ematics. As early as 1882 he wrote that the science of sets encompassed arithmetic, function theory, and geometry, combining them into a "higher unity" based on the idea of cardinality. However, this proposal was vaguely articulated and at first attracted no adher-ents. Nonetheless, sets began to find their way into the language of analysis, most notably through ideas of MEASURE [III.55] and measurability of a set. Indeed, one important route to the absorption of analysis by set theory was the path that sought to determine what kind of function could "measure" a set in an abstract sense. The work of LEBESGUE [VI.72] and BOREL [VI.70] around 1900 on integration and measurability tied set theory to the calculus in a very concrete and intimate way.

A further key step in the establishment of the foun-dations of analysis in the early twentieth century was a new emphasis on mathematical theories as axiomatic structures. This received enormous impetus from the work of Hilbert, who, beginning in the 1890s, had sought to provide a renewed axiomatization of geom-etry. PEANO [VI.62] in Italy headed a school with simi-lar aims. Hilbert redefined the reals on these axiomatic grounds, and his many students and associates turned to axiomatics with enthusiasm for the clarity the ap-proach could provide. Rather than proving the exis-tence of specific entities such as the reals, the math-ematician posits a system satisfying the fundamental properties they possess. A real number (or whatever object) is then defined by the set of axioms provided. As Epple has pointed out, such definitions were con-sidered to be ontologically neutral in that they did not provide methods for telling real numbers from other objects, or even state whether they existed at all (Epple 2003, p. 316). Hilbert's student Ernst Zermelo began work on axiomatizing set theory along these lines, pub-lishing his axioms in 1908 (see [IV.22 §3]). Problems with set theory had emerged in the form of paradoxes, the most famous due to RUSSELL [VI.71]: if S is the set of all sets that do not contain themselves, then it is not possible for S to be in S, nor can it not be in S. Zer-melo's axiomatics sought to avoid this difficulty, in part

by avoiding the definition of set. By 1910, WEYL [VI.80] was to refer to mathematics as the science of "∈," or set membership, rather than the science of quantity. Nonetheless, Zermelo's axioms as a foundational strat-egy were contested. For one thing, a consistency proof for the axioms was lacking. Such "meaning-free" axiom-atization was also contested on the grounds that it removed intuition from the picture.

Against the complex and rapidly developing back-ground of mathematics in the early twentieth century, these debates took on many dimensions that have implications well beyond the question of what consti-tutes rigorous argument in analysis. For the practicing analyst, however, as well as for the teacher of basic infinitesimal calculus, these discussions are marginal to everyday mathematical life and education, and are treated as such. Set theory is pervasive in the language used to describe the basic objects. Real-valued func-tions of one real variable are defined as sets of ordered pairs of real numbers, for example; a set-theoretic defi-nition of an ordered pair was given by WIENER [VI.85] in 1914, and the set-theoretic definition of functions may be dated from that time. However, research in analysis has been largely distinct from, and generally avoids, the foundational issues that may remain in connection with this vocabulary. This is not at all to say that contempo-rary mathematicians treat analysis in a purely formal way. The intuitive content associated with numbers and functions is very much a part of the way of thinking of most mathematicians. The axioms for the reals and for set theory form a framework to be referred to when necessary. But the essential objects of basic analysis, namely derivatives, integrals, series, and their existence or convergence behaviors, are dealt with along the lines of the early twentieth century, so that the ontologi-cal debates about the infinitesimal and infinite are no longer very lively.

A coda to this story is provided by the researches of ROBINSON [VI.95] (1918-74) into "nonstandard" analy-sis, published in 1961. Robinson was an expert in model theory: the study of the relationship between systems of logical axioms and the structures that may satisfy them. His differentials were obtained by adjoining to the regular real numbers a set of "differentials," which satisfied the axioms of an ordered field (in which there is ordinary arithmetic like that of the real numbers) but in addition had elements that were smaller than $1/n$ for every positive integer n. In the eyes of some, this creation eliminated many of the unpleasant fea-tures of the usual way of dealing with the reals, and realized the ultimate goal of Leibniz to have a theory of infinitesimals which was part of the same structure as that of the reals. Despite stimulating a flurry of activity, and considerable acclaim from some quarters, Robinson's approach has never been widely accepted as a working foundation for analysis.

Further Reading

Bottazzini, U. 1990. Geometrical rigour and "modern analysis": an introduction to Cauchy's *Cours d'Analyse*. In Cauchy (1821). Bologna: Editrice CLUB.

Cauchy, A.-L. 1821. *Cours d'Analyse de l'École Royale Polytechnique: Première Partie—Analyse Algébrique*. Paris: L'Imprimerie Royale. (Reprinted, 1990, by Editrice CLUB, Bologna.)

Epple, M. 2003. The end of the science of quantity: foundations of analysis, 1860–1910. In *A History of Analysis*, edited by H. N. Jahnke, pp. 291–323. Providence, RI: American Mathematical Society.

Fraser, C. 1987. Joseph Louis Lagrange's algebraic vision of the calculus. *Historia Mathematica* 14:38–53.

Jahnke, H. N., ed. 2003. *A History of Analysis*. Providence, RI: American Mathematical Society/London Mathematical Society.

Riemann, G. F. B. 1854. Ueber die Darstellbarkeit einer Function durch eine trigonometrische Reihe. *Königlichen Gesellschaft der Wissenschaften zu Göttingen* 13:87–131. Republished in Riemann's collected works (1990): *Gesammelte Mathematische Werke und Wissenschaftliche Nachlass und Nachträge*, edited by R. Narasimhan, 3rd edn., pp. 259–97. Berlin: Springer.

Weierstrass, K. 1988. *Einleitung in die Theorie der Analytischen Functionen: Vorlesung Berlin 1878*, edited by P. Ullrich. Braunschweig: Vieweg/DMV.

II.6   The Development of the Idea of Proof
Leo Corry

1   Introduction and Preliminary Considerations

In many respects the development of the idea of proof is coextensive with the development of mathematics as a whole. Looking back into the past, one might at first consider mathematics to be a body of scientific knowledge that deals with the properties of numbers, magnitudes, and figures, obtaining its justifications from proofs rather than, say, from experiments or inductive inferences. Such a characterization, however, is not without problems. For one thing, it immediately leaves out important chapters in the history of civilization that are more naturally associated with mathematics than with any other intellectual activity. For example, the Mesopotamian and Egyptian cultures developed elaborate bodies of knowledge that would most naturally be described as belonging to arithmetic or geometry, even though nothing is found in them that comes close to the idea of proof as it was later practiced in mathematics at large. To the extent that any justification is given, say, in the thousands of mathematical procedures found on clay tablets written in cuneiform script, it is inductive or based on experience. The tablets repetitively show—without additional explanation or attempts at general justifications—a given procedure to be followed whenever one is pursuing a certain type of result. Later on, in the context of Chinese, Japanese, Mayan, or Hindu cultures, one again finds important developments in fields naturally associated with mathematics. The extent to which these cultures pursued the idea of mathematical proof—a question that is debated among historians to this day—was undoubtedly not as great as it was in Greek tradition, and it certainly did not take the specific forms we typically associate with the latter. Should one nevertheless say that these are instances of mathematical knowledge, even though they are not justified on the basis of some kind of general, deductive proof? If so, then we cannot characterize mathematics as a body of knowledge that is backed up by proofs, as suggested above. However, this litmus test certainly provides a useful criterion—one that we do not want to give up too easily—for distinguishing mathematics from other intellectual endeavors.

Without totally ignoring these important questions, the present account focuses on a story that started, at some point in the past, usually taken to be before or around the fifth century B.C.E. in Greece, with the realization that there was a distinctive body of claims, mainly associated with numbers and with diagrams, whose truth could be and needed to be vindicated in a very special way—namely, by means of a general, deductive argument, or "proof." Exactly when and how this story began is unclear. Equally unclear are the direct historical sources of such a unique idea. Since the emphasis on the use of logic and reason in constructing an argument was well-entrenched in other spheres of public life in ancient Greece—such as politics, rhetoric, and law—much earlier than the fifth century B.C.E., it is possible that it is in those domains that the origins of mathematical proof are to be found.

The early stages of this story raise additional questions, both historical and methodological. For instance, Thales of Miletus, the first mathematician known by name (though he was also a philosopher and scientist), is reported to have *proved* several geometric theorems, such as, for instance, that the opposite angles between two intersecting straight lines are equal, or that if two vertices of a triangle are the endpoints of the diameter of a circle and the third is any other point on the circle then the triangle must be right angled. Even if we were to accept such reports at face value, several questions would immediately arise: in what sense can it be asserted that Thales "proved" these results? More specifically, what were Thales's initial assumptions and what inference methods did he take to be valid? We know very little about this. However, we do know that, as a result of a complex historical process, a certain corpus of knowledge eventually developed that comprised known results, techniques employed, and problems (both solved and yet requiring solution). This corpus gradually also incorporated the regulatory idea of proof: that is, the idea that some kind of general argument, rather than an example (or even many examples), was the necessary justification to be sought in all cases. As part of this development, the idea of proof came to be associated with *strictly deductive* arguments, as opposed to, say, dialogic (meaning "negotiated") or "probabilistically inferred" truth. It is an interesting and difficult historical question to establish why this was the case, and one that we will not address here.

EUCLID's [VI.2] *Elements* was compiled some time around the year 300 B.C.E. It stands out as the most successful and comprehensive attempt of its kind to organize the basic concepts, results, proofs, and techniques required by anyone wanting to master this increasingly complex body of knowledge. Still, it is important to stress that it was not the only such attempt within the Hellenic world. This endeavor was not just a matter of compilation, codification, and canonization, such as one can find in any other evolving field of learning at any point in time. Instead, the assertions it contained were of two different kinds, and the distinction was vitally important. On the one hand there were basic assumptions, or *axioms*, and on the other there were *theorems*, which were typically more elaborate statements, together with accounts of how they followed from the axioms—that is, proofs. The way that proof was conceived and realized in the *Elements* became the paradigm for centuries to come.

This article outlines the evolution of the idea of deductive proof as initially shaped in the framework of Euclidean-style mathematics and as subsequently practiced in the mainstream mathematical culture of ancient Greece, the Islamic world, Renaissance Europe, early modern European science, and then in the nineteenth century and at the turn of the twentieth. The main focus will be on geometry: other fields like arithmetic and algebra will be treated mainly in relation to it. This choice is amply justified by the subject matter itself. Indeed, much as mathematics stands out among the sciences for the unique way in which it relies on proof, so Euclidean-style geometry stood out—at least until well into the seventeenth century—among closely related disciplines such as arithmetic, algebra, and trigonometry.

Results in these other disciplines, or indeed the disciplines as a whole, were often regarded as fully legitimate only when they had been provided with a geometric (or geometric-like) foundation. However, important developments in nineteenth-century mathematics, mainly in connection with the rise of NON-EUCLIDEAN GEOMETRIES [II.2 §§6–10] and with problems in the FOUNDATIONS OF ANALYSIS [II.5], eventually led to a fundamental change of orientation, where arithmetic (and eventually SET THEORY [IV.22]) became the bastion of certainty and clarity from which other mathematical disciplines, geometry included, drew their legitimacy and their clarity. (See THE CRISIS IN THE FOUNDATIONS OF MATHEMATICS [II.7] for a detailed account of this development.) And yet, even before this fundamental change, Euclidean-style proof was not the only way in which mathematical proof was conceived, explored, and practiced. By focusing mainly on geometry, the present account will necessarily leave out important developments that eventually became the mainstream of legitimate mathematical knowledge. To mention just one important example in this regard, a fundamental question that will not be pursued here is how the principle of mathematical induction originated and developed, became accepted as a legitimate inference rule of universal validity, and was finally codified as one of the basic axioms of arithmetic in the late nineteenth century. Moreover, the evolution of the notion of proof involves many other dimensions that will not be treated here, such as the development of the internal organization of mathematics into subdisciplines, as well as the changing interrelations between mathematics and its neighboring disciplines. At a different level, it is related to how mathematics itself evolved as a socially institutionalized enterprise: we shall not discuss interesting questions about how proofs are produced, made public, disseminated, criticized, and often rewritten and improved.

2   Greek Mathematics

Euclid's *Elements* is the paradigmatic work of Greek mathematics, partly for what it has to say about the basic concepts, tools, results, and problems of synthetic geometry and arithmetic, but also for how it regards the role of a mathematical proof and the form that such a proof takes. All proofs appearing in the *Elements* have six parts and are accompanied by a diagram. I illustrate this with the example of proposition I.37. Euclid's text is quoted here in the classical translation of Sir Thomas Heath, and the meaning of some terms differs from current usage. Thus, two triangles are said to be "in the same parallels" if they have the same height and both their bases are contained in a single line, and any two figures are said to be "equal" if their areas are equal. For the sake of explanation, names of the parts of the proof have been added: these do not appear in the original. The proof is illustrated in figure 1.

Protasis (enunciation).   Triangles which are on the same base and in the same parallels are equal to one another.

Ekthesis (setting out).   Let ABC, DBC be triangles on the same base BC and in the same parallels AD, BC.

Diorismos (definition of goal).   I say that the triangle ABC is equal to the triangle DBC.

Kataskeue (construction).   Let AD be produced in both directions to E, F; through B let BE be drawn parallel to CA, and through C let CF be drawn parallel to BD.

Apodeixis (proof).   Then each of the figures EBCA, DBCF is a parallelogram; and they are equal, for they are on the same base BC and in the same parallels BC, EF. Moreover the triangle ABC is half of the parallelogram EBCA, for the diameter AB bisects it; and the triangle DBC is half of the parallelogram DBCF, for the diameter DC bisects it. Therefore the triangle ABC is equal to the triangle DBC.

Sumperasma (conclusion).   Therefore triangles which are on the same base and in the same parallels are equal to one another.

This is an example of a proposition that states a property of geometric figures. The *Elements* also includes propositions that express a task to be carried out. An

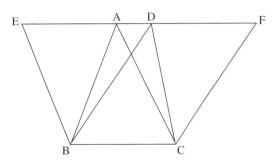

Figure 1 Proposition I.37 of Euclid's *Elements*.

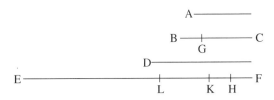

Figure 2 Proposition IX.35 of Euclid's *Elements*.

example is proposition I.1: "On a given finite straight line to construct an equilateral triangle." The same six parts of the proof and the diagram invariably appear in propositions of this kind as well. This formal structure is also followed in all propositions appearing in the three *arithmetic* books of the *Elements* and, most importantly, all of them are always accompanied by a diagram. Thus, for instance, consider proposition IX.35, which in its original version reads as follows:

> If as many numbers as we please be in continued proportion, and there be subtracted from the second and the last numbers equal to the first, then, as the excess of the second is to the first, so will the excess of the last be to all those before it.

This cumbersome formulation may prove incomprehensible on first reading. In more modern terms, an equivalent to this theorem would state that, given a geometric progression $a_1, a_2, \ldots, a_{n+1}$, we have

$$(a_{n+1} - a_1) : (a_1 + a_2 + \cdots + a_n) = (a_2 - a_1) : a_1.$$

This translation, however, fails to convey the spirit of the original, in which no formal symbolic manipulation is, or can be, made. More importantly, a modern algebraic proof fails to convey the ubiquity of diagrams in Greek mathematical proofs, even where they are not needed for a truly geometric construction. Indeed, the accompanying diagram for proposition IX.35 is shown as figure 2 and the first few lines of the proof are as follows:

> Let there be as many numbers as we please in contin-ued proportion A, BC, D, EF, beginning from A as least and let there be subtracted from BC and EF the num-bers BG, FH, each equal to A; I say that, as GC is to A, so is EH to A, BC, D. For let FK be made equal to BC and FL equal to D. . . .

This proposition and its proof provide good exam-ples of the capabilities, as well as the limitations, of ancient Greek practices of notation, and especially of how they managed without a truly symbolic language. In particular, they demonstrate that proofs were never conceived by the Greeks, even ideally, as purely logical constructs, but rather as specific kinds of arguments that one applied to a diagram. The diagram was not just a visual aid to the argumentation. Rather, through the *ekthesis* part of the proof, it embodied the idea referred to by the general character and formulation of the proposition.

Together with the centrality of diagrams, the six-part structure is also typical of most of Greek math-ematics. The constructions and diagrams that typi-cally appeared in Greek mathematical proofs were not of an arbitrary kind, but what we identify today as straightedge-and-compass constructions. The reason-ing in the *apodeixis* part could be either a direct deduc-tion or an argument by contradiction, but the result was always known in advance and the proof was a means to justify it. In addition, Greek geometric thinking, and in particular Euclid-style geometric proofs, strictly adhered to a principle of homogeneity. That is, magni-tudes were only compared with, added to, or subtracted from magnitudes of like kind—numbers, lengths, areas, or volumes. (See NUMBERS [II.1 §2] for more about this.)

Of particular interest are those Greek proofs con-cerned with lengths of curves, as well as with areas or volumes enclosed by curvilinear shapes. Greek mathe-maticians lacked a flexible notation capable of express-ing the gradual approximation of curves by polygons and an eventual passage to the infinite. Instead, they devised a special kind of proof that involved what can retrospectively be seen as an implicit passage to the limit, but which did so in the framework of a purely geo-metric proof and thus unmistakably followed the six-part proof-scheme described above. This implicit pas-sage to the infinite was based on the application of a continuity principle, later associated with ARCHIMEDES [VI.3]. In Euclid's formulation, for instance, the princi-

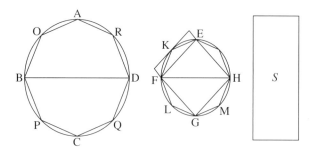

Figure 3 Proposition XII.2 of Euclid's *Elements*.

ple states that, given two unequal magnitudes of the same kind, *A*, *B* (be they two lengths, two areas, or two volumes), with *A* greater than *B*, and if we subtract from *A* a magnitude which is greater than *A*/2, and from the remainder we subtract a magnitude that is greater than its half, and if this process is iterated a sufficient number of times, then we will eventually remain with a magnitude that is smaller than *B*. Euclid used this prin-ciple to prove, for instance, that the ratio of the areas of two circles equals the ratio of the squares of their diameters (XII.2). The method used, later known as the *exhaustion method*, was based on a *double contradic-tion* that became standard for many centuries to come. This double contradiction is illustrated in figure 3, the accompanying diagram to the proposition.

If the ratio of the square on BD to the square on FH is not the same as the ratio of circle ABCD to circle EFGH, then it must be the same as the ratio of circle ABCD to an area *S* either larger or smaller than cir-cle EFGH. The curvilinear figures are approximated by polygons, since the continuity principle allows the dif-ference between the inscribed polygon and the circle to be as close as desired (e.g., closer than the differ-ence between *S* and EFGH). The "double contradiction" is reached if one assumes that *S* is either smaller or larger than EFGH.

Forms of proof and constructions other than those mentioned so far are occasionally found in Greek math-ematical texts. These include diagrams based on what is assumed to be the synchronized motion of two lines (e.g., the trisectrix, or Archimedes' spiral), mechanical devices of many sorts, or reasoning based on ideal-ized mechanical considerations. However, the Euclid-ean type of proof described above remained a model to be followed wherever possible. There is a famous Archimedes palimpsest that provides evidence of how less canonical methods, drawing on mechanical consid-erations (albeit of a highly idealized kind), were used to deduce results about areas and volumes. However, even this bears testimony to the primacy of the ideal model: there is a letter from Archimedes to Eratosthenes in which he displays the ingenuity of his mechanical methods but at the same time is at pains to stress their heuristic character.

3   Islamic and Renaissance Mathematics

Just as Euclid is now considered to represent an entire mainstream tradition of Greek mathematics, so AL-KHWĀRIZMĪ [VI.5] is regarded as a representative of Islamic mathematics. There are two main traits of his work that are relevant to the present account and that became increasingly central to the development of mathematics, starting with his works in the late eighth century and continuing until the works of CARDANO [VI.7] in sixteenth-century Italy. These traits are a pervasive "algebraization" of mathematical thinking, and a continued reliance on Euclidean-style geometric proof as the main way of legitimizing the validity of mathematical knowledge in general and of algebraic reasoning in mathematics in particular.

The prime example of this combination is found in al-Khwārizmī's seminal text *al-Kitāb al-mukhtaṣar fī ḥisāb al-jabr wa'l-muqābala* ("The compendious book on calculation by completion and balancing"), where he discusses the solutions of problems in which the unknown length appears in combination with numbers and squares (the side of which is an unknown). Since he only envisages the possibility of positive "coefficients" and positive rational solutions, al-Khwārizmī needs to consider six different situations each of which requires a different recipe for finding the unknown: the full-grown idea of a general quadratic equation and an algorithm to solve it in all cases does not appear in Islamic mathematical texts. For instance, the problem "squares and roots equal to numbers" (e.g., $x^2 + 10x = 39$, in modern notation) and the problem "roots and numbers equal to squares" (e.g., $3x + 4 = x^2$) are considered to be completely different ones, as are their solutions, and accordingly al-Khwārizmī treats them separately. In all cases, however, al-Khwārizmī *proves* the validity of the method described by translating it into geometric terms and then relying on Euclid-like geometric theorems built around a specific diagram. It is noteworthy, however, that the problems refer to specific numerical quantities associated with the magnitudes involved, and these measured magnitudes refer to the accompanying diagrams as well. In this way, al-Khwārizmī interestingly departs from the Euclidean style of proof. Still,

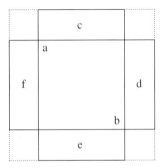

Figure 4 Al-Khwārizmī's geometric justification of the formula for a quadratic equation.

the Greek principle of homogeneity is essentially preserved, as the three quantities usually involved in the problem are all of the same kind, namely, areas.

Consider, for instance, the equation $x^2 + 10x = 39$, which corresponds to the following problem of al-Khwārizmī.

> What is the square which combined with ten of its roots will give a sum total of 39?

The recipe prescribes the following steps.

> Take one-half of the roots [5] and multiply them by itself [25]. Add this amount to 39 and obtain 64. Take the square root of this, which is eight, subtract from it half the roots, leaving three. The number three therefore represents one root of this square, which itself, of course, is nine.

The *justification* is provided by figure 4.

Here ab represents the said square, which for us is x^2, and the rectangles c, d, e, f represent an area of $\frac{10}{4}x$ each, so that all of them together equal $10x$, as in the problem. Thus, the small squares in the corners represent an area of 6.25 each, and we can "complete" the large square, being equal to 64, and whose side is therefore 8, thus yielding the solution 3 for the unknown.

Abu Kamil Shuja, just one generation after al-Khwārizmī, added force to this approach when he solved additional problems while specifically relying on theorems taken from the *Elements*, including the accompanying diagrams, in order to justify his method of solution. The primacy of the Euclidean-type proof, which was already accepted in geometry and arithmetic, thus also became associated with the algebraic methods that eventually turned into the main topic of interest in Renaissance mathematics. Cardano's 1545 *Ars*

Magna, the foremost example of this new trend, presented a complete treatment of the equations of third and fourth degree. Although the algebraic line of reasoning that he adopted and developed became increasingly abstract and formal, Cardano continued to justify his arguments and methods of solution by reference to Euclid-like geometric arguments based on diagrams.

4 Seventeenth-Century Mathematics

The next significant change in the conception of proof appears in the seventeenth century. The most influential development of mathematics in this period was the creation of the infinitesimal calculus simultaneously by NEWTON [VI.14] and LEIBNIZ [VI.15]. This momentous development was the culmination of a process that spanned most of the century, involving the introduction and gradual improvement of important techniques for determining areas and volumes, gradients of tangents, and maxima and minima. These developments included the elaboration of traditional points of view that went back to the Greek classics, as well as the introduction of completely new ideas such as the "indivisibles," whose status as a legitimate tool for mathematical proof was hotly debated. At the same time, the algebraic techniques and approaches that Renaissance mathematicians continued to expand upon, following on from their Islamic predecessors, now gained additional impetus and were gradually incorporated—starting with the work of FERMAT [VI.12] and DESCARTES [VI.11]—into the arsenal of tools available for proving geometric results. Underlying these various trends were different conceptions and practices of mathematical proof, which are briefly described and illustrated now.

Examples of how the classical Greek conception of geometric proof was essentially followed but at the same time fruitfully modified and expanded are found in the work of Fermat, as can be seen in his calculation of the area enclosed by a generalized hyperbola (in modern notation $(y/a)^m = (x/b)^n$ ($m, n \neq 1$)) and its asymptotes.

The quadratic hyperbola (i.e., a figure represented by $y = 1/x^2$), for instance, is defined here in terms of a purely geometric relationship on any two of its points, namely, that the ratio between the squares built on the abscissas equals the inverse ratio between the lengths of the ordinates. In its original version it is expressed as follows: $AG^2 : AH^2 :: IH : EG$ (see figure 5). It should be noticed that this is not an equation in the present sense

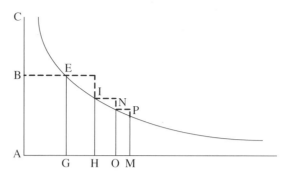

Figure 5 Diagram for Fermat's proof of the area under a hyperbola.

of the word, on which the standard symbolic manipulations can be directly performed. Rather, this is a four-term proportion to which the rules of Greek classical mathematics apply. Also, the proof was entirely geometric and indeed it essentially followed the Euclidean style. Thus, if the segments AG, AH, AO, etc., are chosen in continued proportion, then one can prove that the rectangles EH, IO, NM, etc., are also in continued proportion, and indeed that EH : IO :: IO : NM :: · · · :: AH : AG.

Fermat made use of proposition IX.35 of the *Elements* (mentioned above), which comprises an expression for the sum of any number of quantities in a geometric progression, namely (in more modern notation):

$$(a_{n+1} - a_1) : (a_1 + a_2 + \cdots + a_n) = (a_2 - a_1) : a_1.$$

But at this point his proof takes an interesting turn. He introduces the somewhat obscure concept of "*adequare*," which he found in the works of Diophantus, and which allows a kind of "approximate equality." Specifically, this idea allows him to bypass the cumbersome procedure of double contradiction typically used in Greek geometry as an implicit passage to the infinite. A figure bounded by GE, by the horizontal asymptote, and by the hyperbola will equal the infinite sum of rectangles obtained when the rectangle EH "will vanish and will be reduced to nothing." Further, proposition IX.35 implies that this sum equals the area of the rectangle BG. Significantly, Fermat still chose to rely on the authority of the ancients, hinting at the method of double contradiction when he declared that this result "would be easy to confirm by a more lengthy proof carried out in the manner of Archimedes."

Attempts to expand the accepted canon of geometric proof eventually led to the more progressive approaches associated with the idea of indivisibles, as practiced by Cavalieri, Roberval, and Torricelli. This is well illustrated by Torricelli's 1643 calculation of the volume of the infinite body created by rotating the hyperbola $xy = k^2$ around the y-axis, with values of x between 0 and a (as we would describe it in modern terms).

The essential idea of indivisibles is that areas are considered to be sums, or collections, of infinitely many line segments, and volumes are considered to be sums, or collections, of infinitely many areas. In this example, Torricelli calculated the volume of revolution by considering it to be a sum of the curved surfaces of an infinite collection of cylinders successively inscribed within each other and having radii ranging from 0 to a. In modern algebraic terms, the height of the inscribed cylinder with radius x is k^2/x, so the area of its curved surface is $2\pi x(k^2/x) = \pi(\sqrt{2}k)^2$, a constant value that is independent of x and equal to the area of a circle of radius $\sqrt{2}k$. Thus, in Torricelli's approach based on the geometry of indivisibles, the collection of all surfaces that, when taken together, comprise the infinite body can be equated to a collection of circles with area $2\pi k^2$, one for each x between 0 and a, or equivalently to a cylinder of volume $2\pi k^2 a$.

The rules of Euclid-like geometric proof were completely contravened in proofs of this kind and this made them unacceptable in the eyes of many. On the other hand, their fruitfulness was highly appealing, especially in cases like this one in which an infinite body was shown to have a finite volume, a result which Torricelli himself found extremely surprising. Both supporters and detractors alike, however, realized that techniques of this kind might lead to contradictions and inaccurate results. By the eighteenth century, with the accelerated development of the infinitesimal calculus and its associated techniques and concepts, techniques based on indivisibles had essentially disappeared.

The limits set by the classical paradigm of Euclidean geometric proof were then transgressed in a different direction by the all-embracing algebraization of geometry at the hands of Descartes. The fundamental step undertaken by Descartes was to introduce unit lengths as a key element in the diagrams used in geometric proofs. The radical innovation implied by this step, allowing the hitherto nonexistent possibility of defining operations with line segments, was explicitly stressed by Descartes in *La Géométrie* in 1637:

> Just as arithmetic consists of only four or five operations, namely addition, subtraction, multiplication,

division, and the extraction of roots, which may be considered a kind of division, so in geometry, to find required lines it is merely necessary to add or subtract other lines; or else, taking one line, which I shall call the unit in order to relate it as closely as possible to numbers, and which can in general be chosen arbitrarily, and having given two other lines, to find a fourth line which shall be to one of the given lines as the other is to the unit (which is the same as multiplication); or again, to find a fourth line which is to one of the given lines as the unit is to the other (which is equivalent to division); or, finally, to find one, two, or several mean proportionals between the unit and some other line (which is the same as extracting the square root, cube root, etc., of the given line).

Thus, for instance, given two segments BD, BE, the division of their lengths is represented by BC in figure 6, in which AB represents the unit length.

Although the proof was Euclid-like in appearance (because of the diagram and the use of the theory of similar triangles), the introduction of the unit length and its use for defining the operations with segments set it radically apart and opened completely new horizons for geometric proofs. Not only had measurements of length been absent from Euclidean-style proofs thus far, but also, as a consequence of the very existence of these operations, the essential dimensionality traditionally associated with geometric theorems lost its significance. Descartes used expressions such as $a - b$, a/b, a^2, b^3, and their roots, but he stressed that they should all be understood as "only simple lines, which, however, I name squares, cubes, etc., so that I make use of the terms employed in algebra." With the removal of dimensionality, the requirement of homogeneity also became unnecessary. Unlike his predecessors, who handled magnitudes only when they had a direct geometric significance, Descartes could not see any problem in forming an expression such as $a^2b^2 - b$ and then extracting its cube root. In order to do so, he said "we must consider the quantity a^2b^2 divided once by the unit, and the quantity b multiplied twice by the unit." Sentences of this kind would be simply incomprehensible to Greek geometers, as well as to their Islamic and Renaissance followers.

This algebraization of geometry, and particularly the newly created possibility of proving geometric facts via algebraic procedures, was strongly related to the recent consolidation of the idea of an algebraic equation, seen as an autonomous mathematical entity, for which formal rules of manipulation were well-known and could be systematically applied. This idea reached

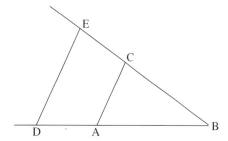

Figure 6 Descartes's geometric calculation
of the division of two given segments.

full maturity in the hands of VIÈTE [VI.9] only around
1591. But not all mathematicians in the seventeenth
century saw the important developments associated
with algebraic thinking either as a direction to be naturally adopted or as a clear sign of progress in the latter discipline. A prominent opponent of any attempt
to deviate from the classical Euclidean-style approach
in geometry was none other than NEWTON [VI.14], who,
in the *Arithmetica Universalis* (1707), was emphatic in
expressing his views:

> Equations are expressions of arithmetic computation
> and properly have no place in geometry, except in
> so far as truly geometrical quantities (lines, surfaces,
> solids and proportions) are thereby shown equal, some
> to others. Multiplications, divisions, and computations
> of that kind have recently been introduced into geom-
> etry, unadvisedly and against the first principle of this
> science.... Therefore these two sciences ought not to
> be confounded, and recent generations by confounding
> them have lost that simplicity in which all geometrical
> elegance consists.

Newton's *Principia* bears witness to the fact that
statements like this one were far from mere lip service, as Newton consistently preferred Euclidean-style
proofs, considering them to be the correct language for
presenting his new physics and for bestowing it with
the highest degree of certainty. He used his own calculus only where strictly necessary, and barred algebra
from his treatise entirely.

5 Geometry and Proof in Eighteenth-Century Mathematics

Mathematical analysis became the primary focus of
mathematicians in the eighteenth century. Questions
relating to the foundations of analysis arose immediately after the calculus began to be developed and were

not settled until the late nineteenth century. To a considerable extent these questions were about the nature
of legitimate mathematical proof, and debates about
them played an important role in undermining the long-
undisputed status of geometry as the basis for mathematical certainty and bestowing this status on arithmetic instead. The first important stage in this process
was EULER's [VI.19] reformulation of the calculus. Once
separated from its purely geometric roots, the calculus
came to be centered on the algebraically oriented concept of function. This trend for favoring algebra over
geometry was given further impetus by Euler's successors. D'ALEMBERT [VI.20], for instance, associated
mathematical certainty above all with algebra—because
of its higher degree of generality and abstraction—and
only subsequently with geometry and mechanics. This
was a clear departure from the typical views of Newton
and of his contemporaries. The trend reached a peak
and was transformed into a well-conceived program in
the hands of LAGRANGE [VI.22], who in the preface to
his 1788 *Méchanique Analitique* famously expressed
a radical view about how one could achieve certainty
in the mathematical sciences while distancing oneself
from geometry. He wrote as follows:

> One will not find figures in this work. The methods
> that I expound require neither constructions, nor geo-
> metrical or mechanical arguments, but only algebraic
> operations, subject to a regular and uniform course.

The details of these developments are beyond the scope
of this article. What is important to stress, however, is
that in spite of their very considerable impact, the basic
conceptions of proof in the more mainstream realm of
geometry did not change very much during the eighteenth century. An illuminating perspective on these
conceptions is offered by the views of contemporary
philosophers, especially Immanuel Kant.

Kant had a very profound knowledge of contemporary science, and particularly of mathematics. A
philosophical discussion of his views on mathematical
knowledge and proof need not concern us here. However, given his acquaintance with contemporary conceptions, they do provide an insightful *historical* perspective on proof as it was understood at the time. Of
particular interest is the contrast he draws between
a philosophical argument, on the one hand, and a
geometric proof, on the other. Whereas the former
deals with general concepts, the latter deals with concrete, yet nonempirical, concepts, by reference to "visualizable intuitions" (*Anschauung*). This difference is

epitomized in the following, famous passage from his *Critique of Pure Reason*.

> Suppose a philosopher be given the concept of a triangle and he is left to find out, in his own way, what relation the sum of its angles bears to a right angle. He has nothing but the concept of a figure enclosed by three straight lines, and possessing three angles. However long he meditates on this concept, he will never produce anything new. He can analyze and clarify the concept of a straight line or of an angle or of the number three, but he can never arrive at any properties not already contained in these concepts. Now let the geometrician take up these questions. He at once begins by constructing a triangle. Since he knows that the sum of two right angles is exactly equal to the sum of all the adjacent angles which can be constructed from a single point on a straight line, he prolongs one side of his triangle and obtains two adjacent angles, which together are equal to two right angles. He then divides the external angle by drawing a line parallel to the opposite side of the triangle, and observes that he has thus obtained an external adjacent angle which is equal to an internal angle—and so on. In this fashion, through a chain of inferences guided throughout by intuition, he arrives at a fully evident and universally valid solution of the problem.

In a nutshell, then, for Kant the nature of mathematical proof that sets it apart from other kinds of deductive argumentation (like philosophy) lies in the centrality of the diagrams and the role that they play. As in the *Elements*, this diagram is not just a heuristic guide for what is no more than abstract reasoning, but rather an "intuition," a singular embodiment of the mathematical idea that is clearly located not only in space, but rather in space and time. In fact,

> I cannot represent to myself a line, however small, without drawing it in thought, that is gradually generating all its parts from a point. Only in this way can the intuition be obtained.

This role played by diagrams as "visualizable intuitions" is what provides, for Kant, the explanation of why geometry is not just an empirical science, but also not just a huge tautology devoid of any synthetic content. According to him, geometric proof is constrained by logic but it is much more than just a purely logical analysis of the terms involved. This view was at the heart of a novel philosophical analysis whose starting point was the then-entrenched conception of what a mathematical proof is.

6   Nineteenth-Century Mathematics and the Formal Conception of Proof

The nineteenth century was full of important developments in geometry and other parts of mathematics, not just of the methods but also of the aims of the various subdisciplines. Logic, as a field of knowledge, also underwent significant changes and a gradual mathematization that entirely transformed its scope and methods. Consequently, by the end of the century the conception of proof and its role in mathematics had also been deeply transformed.

In Göttingen in 1854 RIEMANN [VI.49] gave his seminal talk "On the hypotheses which lie at the foundations of geometry." At around the same time, the works of BOLYAI [VI.34] and LOBACHEVSKII [VI.31] on non-Euclidean geometry, as well as the related ideas of GAUSS [VI.26], all dating from the 1830s, began to be more generally known. The existence of coherent, alternative geometries brought about a pressing need for the most basic, longstanding beliefs about the essence of geometric knowledge, including the role of proof and mathematical rigor, to be revised. Of even greater significance in this regard was the renewed interest in PROJECTIVE GEOMETRY [I.3 §6.7], which became a very active field of research with its own open research questions and foundational issues after the publication of Jean Poncelet's 1822 treatise. The addition of projective geometry to the many other possible geometric perspectives prompted a variety of attempts at unification and classification, the most significant of which were those based on group-theoretic ideas. Particularly notable were those of KLEIN [VI.57] and LIE [VI.53] in the 1870s. In 1882, Moritz Pasch published an influential treatise on projective geometry devoted to a systematic exploration of its axiomatic foundations and the interrelationships among its fundamental theorems. Pasch's book also attempted to close the many logical gaps that had been found in Euclidean geometry over the years. More systematically than any of his fellow nineteenth-century mathematicians, Pasch emphasized that all geometric results should be obtained from axioms by strict logical deduction, without relying on analytical means, and above all without appeal to diagrams or to properties of the figures involved. Thus, although in some ways he was consciously reverting to the canons of Euclid-like proof (which by then were somewhat loosened), his attitude toward diagrams was fundamentally different. Aware of the potential limitations of visualizing diagrams (and perhaps their misleading influence) he put a much greater emphasis on the pure logical structure of the proof than his predecessors had. Nevertheless, he was not led to an outright formalist view of geometry and geometric proof. Rather, he consistently adopted an empirical approach to the origins and meaning of geometry and fell short of claiming that diagrams were for heuristic use only:

> The basic propositions [of geometry] cannot be understood without corresponding drawings; they express what has been observed from certain, very simple facts. The theorems are not founded on observations, but rather, they are proved. Every inference performed during a deduction must find confirmation in a drawing, yet it is not justified by a drawing but from a certain preceding statement (or a definition).

Pasch's work definitely contributed to diagrams losing their central status in geometric proofs in favor of purely deductive relations, but it did not directly lead to a thorough revision of the status of the axioms of geometry, or to a change in the conception that geometry deals essentially with the study of our spatial, visualizable intuition (in the sense of *Anschauung*). The all-important nineteenth-century developments in geometry produced significant changes in the conception of proof only under the combined influence of additional factors.

Mathematical analysis continued to be a primary field of research, and the study of its foundations became increasingly identified with arithmetic, rather than geometric, rigor. This shift was provoked by the works of mathematicians like CAUCHY [VI.29], WEIERSTRASS [VI.44], CANTOR [VI.54], and DEDEKIND [VI.50], which aimed at eliminating intuitive arguments and concepts in favor of ever more elementary statements and definitions. (In fact, it was not until the work of Dedekind on the foundations of arithmetic, in the last third of the century, that the rigorous formulation pursued in these works was given any kind of axiomatic underpinning.) The idea of investigating the axiomatic basis of mathematical theories, whether geometry, algebra, or arithmetic, and of exploring alternative possible systems of postulates was indeed pursued during the nineteenth century by mathematicians such as George Peacock, Charles Babbage, John Herschel, and, in a different geographical and mathematical context, Hermann Grassmann. But such investigations were the exception rather than the rule, and they had only a fairly limited role in shaping a new conception of proof in analysis and geometry.

One major turning point, where the above trends combined to produce a new kind of approach to proof, is to be found in the works of GIUSEPPE PEANO [VI.62] and his Italian followers. Peano's mainstream activities were as a competent analyst, but he was also interested in artificial languages, and particularly in developing an artificial language that would allow a completely formal treatment of mathematical proofs. In 1889 his successful application of such a conceptual language to arithmetic yielded his famous POSTULATES FOR THE NATURAL NUMBERS [III.67]. Pasch's systems of axioms for projective geometry posed a challenge to Peano's artificial language, and he set out to investigate the relationship between the logical and the geometric terms involved in the deductive structure of geometry. In this context he introduced the idea of an independent set of axioms, and applied this concept to his own system of axioms for projective geometry, which were a slight modification of Pasch's. This view did not lead Peano to a formalistic conception of proof, and he still conceived geometry in terms very similar to his predecessors:

> Anyone is allowed to take a hypothesis and develop its logical consequences. However, if one wants to give this work the name of geometry it is necessary that such hypotheses or postulates express the result of simple and elementary observations of physical figures.

Under the influence of Peano, Mario Pieri developed a symbolism with which to handle abstract–formal theories. Unlike Peano and Pasch, Pieri consistently promoted the idea of geometry as a purely logical system, where theorems are deduced from hypothetical premises and where the basic terms are completely detached from any empirical or intuitive significance.

A new chapter in the history of geometry and of proof was opened at the end of the nineteenth century with the publication of HILBERT's [VI.63] *Grundlagen der Geometrie*, a work that synthesized and brought to completion the various trends of geometric research described above. Hilbert was able to achieve a comprehensive analysis of the logical interrelations among the fundamental results of projective geometry, such as the theorems of Desargues and Pappus, while paying particular attention to the role of continuity considerations within their proofs. His analysis was based on the introduction of a generalized analytic geometry, in which the coordinates may be taken from a variety of different NUMBER FIELDS [III.63], rather than from the real numbers alone. This approach created a purely

synthetic arithmetization of any given type of geometry, and thus helped to clarify the logical structure
of Euclidean geometry as a deductive system. It also
clarified the relationship between Euclidean geometry
and the various other kinds of known geometries—non-
Euclidean, projective, or non-Archimedean. This focus
on logic implied, among other things, that diagrams
should be relegated to a merely heuristic role. In fact,
although diagrams still appear in many proofs in the
Grundlagen, the entire purpose of the logical analysis
is to avoid being misled by diagrams. Proofs, and particularly geometric proofs, have thus become purely logical arguments, rather than arguments about diagrams.
And at the same time, the essence and the role of the
axioms from which the derivations in question start
also underwent a dramatic change.

Following Pasch's lead, Hilbert introduced a new system of axioms for geometry that attempted to close
the logical gaps inherent in earlier systems. These
axioms were of five kinds—axioms of incidence, of
order, of congruence, of parallels, and of continuity—
each of which expressed a particular way in which
spatial intuition manifests itself in our understanding.
They were formulated for three fundamental kinds of
object: points, lines, and planes. These remained undefined, and the system of axioms was meant to provide
an implicit definition of them. In other words, rather
than defining points or lines at the outset and then postulating axioms that are assumed to be valid for them,
a point and a line were not directly defined, except as
entities that satisfy the axioms postulated by the system. Further, Hilbert demanded that the axioms in a
system of this kind should be mutually independent,
and introduced a method for checking that this demand
is fulfilled; in order to do so, he constructed models
of geometries that fail to satisfy a given axiom of the
system but satisfy all the others. Hilbert also required
that the system be consistent, and that the consistency
of geometry could be made to depend, in his system,
on that of arithmetic. He initially assumed that proving the consistency of arithmetic would not present a
major obstacle and it was a long time before he realized
that this was not the case. Two additional requirements
that Hilbert initially introduced for axiomatic systems
were simplicity and completeness. Simplicity meant, in
essence, that an axiom should not contain more than
"a single idea." The demand that every axiom in a system be "simple," however, was never clearly defined or
systematically pursued in subsequent works of Hilbert
or any of his successors. The last requirement, com-

pleteness, meant for Hilbert in 1900 that any adequate
axiomatization of a mathematical domain should allow
for a derivation of *all* the known theorems of the discipline in question. Hilbert claimed that his axioms would
indeed yield all the known results of Euclidean geometry, but of course this was not a property that he could
formally prove. In fact, since this property of "completeness" cannot be formally checked for any given
axiomatic system, it did not become one of the standard requirements of an axiomatic system. It is important to note that the concept of completeness used by
Hilbert in 1900 is completely different from the currently accepted, model-theoretical one that appeared
much later. The latter amounts to the requirement that
in a given axiomatic system every true statement, be it
known or unknown, should be provable.

The use of undefined concepts and the concomitant
conception of axioms as implicit definitions gave enormous impetus to the view of geometry as a purely logical system, such as Pieri had devised it, and eventually
transformed the very idea of truth and proof in mathematics. Hilbert claimed on various occasions—echoing
an idea of Dedekind—that, in his system, "points, lines,
and planes" could be substituted by "chairs, tables, and
beer mugs," without thereby affecting in any sense the
logical structure of the theory. Moreover, in the light
of discussions about set-theoretical paradoxes, Hilbert
strongly emphasized the view that the logical consistency of a concept implicitly defined by axioms was the
essence of mathematical existence. Under the influence
of these views, of the new methodological tools introduced by Hilbert, and of the successful overview of the
foundations of geometry thus achieved, many mathematicians went on to promote new views of mathematics and new mathematical activities that in many
senses went beyond the views embodied in Hilbert's
approach. On the one hand, a trend that thrived in the
United States at the beginning of the twentieth century,
led by Eliakim H. Moore, turned the study of systems
of postulates into a mathematical field in its own right,
independent of direct interest in the field of research
defined by the systems in question. For instance, these
mathematicians defined the minimal set of independent postulates for groups, fields, projective geometry,
etc., without then proceeding to investigate of any of
these individual disciplines. On the other hand, prominent mathematicians started to adopt and develop
increasingly formalistic views of proof and of mathematical truth, and began applying them in a growing number of mathematical fields. The work of the

radically modernist mathematician FELIX HAUSDORFF [VI.68] provides important examples of this trend, as he was among the first to consistently associate Hilbert's achievement with a new, formalistic view of geometry. In 1904, for instance, he wrote:

> In all philosophical debates since Kant, mathematics, or at least geometry, has always been treated as heteronomous, as dependent on some external instance of what we could call, for want of a better term, intuition, be it pure or empirical, subjective or scientifically amended, innate or acquired. The most important and fundamental task of modern mathematics has been to set itself free from this dependency, to fight its way through from heteronomy to autonomy.

Hilbert himself would pursue such a point of view around 1918, when he engaged in the debates about the consistency of arithmetic and formulated his "finitist" program. This program did indeed adopt a strongly formalistic view, but it did so with the restricted aim of solving this particular problem. It is therefore important to stress that Hilbert's conceptions of geometry were, and remained, essentially empiricist and that he never regarded his axiomatic analysis of geometry as part of an overall formalistic conception of mathematics. He considered the axiomatic approach as a tool for the conceptual clarification of existing, well-elaborated theories, of which geometry provided only the most prominent example.

The implication of Hilbert's axiomatic approach for the concept of proof and of truth in mathematics provoked strong reactions from some mathematicians, and prominently so from FREGE [VI.56]. Frege's views are closely related to the changing status of logic at the turn of the twentieth century and its gradual process of mathematization and formalization. This process was an outcome of the successive efforts through the nineteenth century of BOOLE [VI.43], DE MORGAN [VI.38], Grassmann, Charles S. Peirce, and Ernst Schröder at formulating an algebra of logic. The most significant step toward a new, formal conception of logic, however, came with the increased understanding of the role of the logical QUANTIFIERS [I.2 §3.2] (universal, $\forall$, and existential, $\exists$) in the process of formulating a modern mathematical proof. This understanding emerged in an informal, but increasingly clear, fashion as part of the process of the rigorization of analysis and the distancing from visual intuition, especially at the hands of Cauchy, BOLZANO [VI.28], and Weierstrass. It was formally defined and systematically codified for the first

time by Frege in his 1879 *Begriffsschrift*. Frege's system, as well as similar ones proposed later by Peano and by RUSSELL [VI.71], brought to the fore a clear distinction between propositional connectives and quantifiers, as well as between logical symbols and algebraic or arithmetic ones.

Frege formulated the idea of a *formal system*, in which one defines in advance all the allowable symbols, all the rules that produce well-formed formulas, all axioms (i.e., certain preselected, well-formed formulas), and all the rules of inference. In such systems any deduction can be checked *syntactically*—in other words, by purely symbolic means. On the basis of such systems Frege aimed to produce theories with no logical gaps in their proofs. This would apply not only to analysis and to its arithmetic foundation—the mathematical fields that provided the original motivation for his work—but also to the new systems of geometry that were evolving at the time. On the other hand, in Frege's view the axioms of mathematical theories—even if they appear in the formal system merely as well-formed formulas—embody truths about the world. This is precisely the source of his criticism of Hilbert. It is the truth of the axioms, asserted Frege, that certifies their consistency, rather than the other way around, as Hilbert suggested.

We thus see how foundational research in two separate fields—geometry and analysis—was inspired by different methodologies and philosophical outlooks, but converged at the turn of the twentieth century to create an entirely new conception of mathematical proof. In this conception a mathematical proof is seen as a purely logical construct validated in purely syntactic terms, independently of any visualization through diagrams. This conception has dominated mathematics ever since.

Epilogue: Proof in the Twentieth Century

The new notion of proof that stabilized at the beginning of the twentieth century provided an idealized model—broadly accepted to this day—of what should constitute a valid mathematical argument. To be sure, actual proofs devised and published by mathematicians since that time are seldom presented as fully formalized texts. They typically present a clearly articulated argument in a language that is precise enough to convince the reader that it could—in principle, and perhaps with straightforward (if sustained) effort—be turned into one. Throughout the decades, however, some limitations of this dominant idea have gradually emerged and alternative conceptions of what should count as a valid mathematical argument have become increasingly accepted as part of current mathematical practice.

The attempt to pursue this idea systematically to its full extent led, early on and very unexpectedly, to a serious difficulty with the notion of a proof as a completely formalized and purely syntactic deductive argument. In the early 1920s, Hilbert and his collaborators developed a fully fledged mathematical theory whose subject matter was "proof," considered as an object of study in itself. This theory, which presupposed the formal conception of proof, arose as part of an ambitious program for providing a direct, *finitistic* consistency proof of arithmetic represented as a formalized system. Hilbert asserted that, just as the physicist examines the physical apparatus with which he carries out his experiments and the philosopher engages in a critique of reason, so the mathematician should be able to analyze mathematical proofs and do so strictly by mathematical means. About a decade after the program was launched, GÖDEL [VI.92] came up with his astonishing INCOMPLETENESS THEOREM [V.15], which famously showed that "mathematical truth" and "provability" were not one and the same thing. Indeed, in any consistent, sufficiently rich axiomatic system (including the systems typically used by mathematicians) there are true mathematical statements that cannot be proved. Gödel's work implied that Hilbert's finitistic program was too optimistic, but at the same time it also made clear the deep mathematical insights that could be obtained from Hilbert's proof theory.

A closely related development was the emergence of proofs that certain important mathematical statements were undecidable. Interestingly, these seemingly negative results have given rise to new ideas about the legitimate grounds for establishing the truth of such statements. For instance, in 1963 Paul Cohen established that the CONTINUUM HYPOTHESIS [IV.22 §5] can be neither proved nor disproved in the usual systems of axioms for set theory. Most mathematicians simply accept this idea and regard the problem as solved (even if not in the way that was originally expected), but some contemporary set theorists, notably Hugh Woodin, maintain that there are good reasons to believe that the hypothesis is *false*. The strategy they follow in order to justify this assertion is fundamentally different from the formal notion of proof: they devise new axioms, demonstrate that these axioms have very desirable properties, argue that they should therefore be accepted, and then show that they imply the negation of

the continuum hypothesis. (See SET THEORY [IV.22 §10] for further discussion.)

A second important challenge came from the ever-increasing length of significant proofs appearing in various mathematical domains. A prominent example was the CLASSIFICATION THEOREM FOR FINITE SIMPLE GROUPS [V.7], whose proof was worked out in many separate parts by a large numbers of mathematicians. The resulting arguments, if put together, would reach about ten thousand pages, and errors have been found since the announcement in the early 1980s that the proof was complete. It has always been relatively straightforward to fix the errors and the theorem is indeed accepted and used by group theorists. Nevertheless, the notion of a proof that is too long for a single human being to check is a challenge to our conception of when a proof should be accepted as such. The more recent, very conspicuous cases of FERMAT'S LAST THEOREM [V.10] and THE POINCARÉ CONJECTURE [V.25] were hard to survey for different reasons: not only were they long (though nowhere near as long as the classification of finite simple groups), but they were also very difficult. In both cases there was a significant interval between the first announcement of the proofs and their complete acceptance by the mathematical community because checking them required enormous efforts by the very few people qualified to do so. There is no controversy about either of these two breakthroughs, but they do raise an interesting sociological problem: if somebody claims to have proved a theorem and nobody else is prepared to check it carefully (perhaps because, unlike the two theorems just mentioned, this one is not important enough for another mathematician to be prepared to spend the time that it would take), then what is the status of the theorem?

Proofs based on probabilistic considerations have also appeared in various mathematical domains, including number theory, group theory, and combinatorics. It is sometimes possible to prove mathematical statements (see, for example, the discussion of random primality testing in COMPUTATIONAL NUMBER THEORY [IV.3 §2]), not with complete certainty, but in such a way that the probability of error is tiny—at most one in a trillion, say. In such cases, we may not have a formal proof, but the chances that we are mistaken in considering the given statement to be true are probably lower than, say, the chance that there is a significant mistake in one of the lengthy proofs mentioned above.

Another challenge has come from the introduction of computer-assisted methods of proof. For instance, in 1976 Kenneth Appel and Wolfgang Haken settled a famous old problem by proving the FOUR-COLOR THEOREM [V.12]. Their proof involved the checking of a huge number of different map configurations, which they did with the help of a computer. Initially, this raised debates about the legitimacy of their proof but it quickly became accepted and there are now several proofs of this kind. Some mathematicians even believe that computer-assisted and, more importantly, *computer-generated* proofs are the future of the entire discipline. Under this (currently minority) view, our present views about what counts as an acceptable mathematical proof will soon become obsolete.

A last point to stress is that many branches of mathematics now contain conjectures that seem to be both fundamentally important and out of reach for the foreseeable future. Mathematicians persuaded of the truth of such conjectures increasingly undertake the systematic study of their consequences, assuming that an acceptable proof will one day appear (or at least that the conjecture is true). Such conditional results are published in leading mathematical journals and doctoral degrees are routinely awarded for them.

These trends all raise interesting questions about existing conceptions of legitimate mathematical proofs, the status of truth in mathematics, and the relationship between "pure" and "applied" fields. The formal notion of a proof as a string of symbols that obeys certain syntactical rules continues to provide an ideal model for the principles that underlie what most mathematicians see as the essence of their discipline. It allows far-reaching mathematical analysis of the power of certain axiomatic systems, but at the same time it falls short of explaining the changing ways in which mathematicians decide what kinds of arguments they are willing to accept as legitimate in their actual professional practice.

Acknowledgments. I thank José Ferreirós and Reviel Netz for useful comments on previous versions of this text.

Further Reading

Bos, H. 2001. *Redefining Geometrical Exactness. Descartes' Transformation of the Early Modern Concept of Construction.* New York: Springer.

Ferreirós, J. 2000. *Labyrinth of Thought. A History of Set Theory and Its Role in Modern Mathematics.* Boston, MA: Birkhäuser.

Grattan-Guinness, I. 2000. *The Search for Mathematical Roots, 1870–1940: Logics, Set Theories and the Foundations of Mathematics from Cantor through Russell to Gödel.* Princeton, NJ: Princeton University Press.

Netz, R. 1999. *The Shaping of Deduction in Greek Mathematics: A Study in Cognitive History.* Cambridge: Cambridge University Press.

Rashed, R. 1994. *The Development of Arabic Mathematics: Between Arithmetic and Algebra*, translated by A. F. W. Armstrong. Dordrecht: Kluwer.

II.7 The Crisis in the Foundations of Mathematics
José Ferreirós

The foundational crisis is a celebrated affair among mathematicians and it has also reached a large non-mathematical audience. A well-trained mathematician is supposed to know something about the three viewpoints called "logicism," "formalism," and "intuitionism" (to be explained below), and about what GÖDEL'S INCOMPLETENESS RESULTS [V.15] tell us about the status of mathematical knowledge. Professional mathematicians tend to be rather opinionated about such topics, either dismissing the foundational discussion as irrelevant—and thus siding with the winning party—or defending, either as a matter of principle or as an intriguing option, some form of revisionist approach to mathematics. But the real outlines of the historical debate are not well-known and the subtler philosophical issues at stake are often ignored. Here we shall mainly discuss the former, in the hope that this will help bring the main conceptual issues into sharper focus.

The foundational crisis is usually understood as a relatively localized event in the 1920s, a heated debate between the partisans of "classical" (meaning late-nineteenth-century) mathematics, led by HILBERT [VI.63], and their critics, led by BROUWER [VI.75], who advocated strong revision of the received doctrines. There is, however, a second, and in my opinion very important, sense in which the "crisis" was a long and global process, indistinguishable from the rise of modern mathematics and the philosophical and methodological issues it created. This is the standpoint from which the present account has been written.

Within this longer process one can still pick out some noteworthy intervals. Around 1870 there were many discussions about the acceptability of non-Euclidean geometries, and also about the proper foundations of complex analysis and even the real numbers. Early in the twentieth century there were debates about set theory, about the concept of the continuum, and about the role of logic and the axiomatic method versus the role of intuition. By about 1925 there was a crisis in the proper sense, during which the main opinions in these debates were developed and turned into detailed mathematical research projects. And in the 1930s GÖDEL [VI.92] proved his incompleteness results, which could not be assimilated without some cherished beliefs being abandoned. Let us analyze some of these events and issues in greater detail.

1 Early Foundational Questions

There is evidence that in 1899 Hilbert endorsed the viewpoint that came to be known as *logicism*. Logicism was the thesis that the basic concepts of mathematics are definable by means of logical notions, and that the key principles of mathematics are deducible from logical principles alone.

Over time this thesis has become unclear, based as it seems to be on a fuzzy and immature conception of the scope of logical theory. But historically speaking logicism was a neat intellectual reaction to the rise of modern mathematics, and particularly to the set-theoretic approach and methods. Since the majority opinion was that set theory is just a part of (refined) logic,[1] this thesis was thought to be supported by the fact that the theories of natural and real numbers can be derived from set theory, and also by the increasingly important role of set-theoretic methods in algebra and in real and complex analysis.

Hilbert was following DEDEKIND [VI.50] in the way he understood mathematics. For us, the essence of Hilbert's and Dedekind's early logicism is their self-conscious endorsement of certain modern methods, however daring they seemed at the time. These methods had emerged gradually during the nineteenth century, and were particularly associated with Göttingen mathematics (GAUSS [VI.26] and DIRICHLET [VI.36]); they experienced a crucial turning point with RIE-MANN's [VI.49] novel ideas, and were developed further by Dedekind, CANTOR [VI.54], Hilbert, and other, lesser figures. Meanwhile, the influential Berlin school of mathematics had opposed this new trend, KRO-NECKER [VI.48] head-on and WEIERSTRASS [VI.44] more subtly. (The name of Weierstrass is synonymous with the introduction of rigor in real analysis, but in fact, as will be indicated below, he did not favor the more modern methods elaborated in his time.) Mathematicians in

Paris and elsewhere also harbored doubts about these new and radical ideas.

The most characteristic traits of the modern approach were:

(i) acceptance of the notion of an "arbitrary" function proposed by Dirichlet;

(ii) a wholehearted acceptance of infinite sets and the higher infinite;

(iii) a preference "to put thoughts in the place of calculations" (Dirichlet), and to concentrate on "structures" characterized axiomatically; and

(iv) a frequent reliance on "purely existential" methods of proof.

An early and influential example of these traits was Dedekind's approach (1871) to ALGEBRAIC NUMBER THEORY [IV.1]—his set-theoretic definition of NUMBER FIELDS [III.63] and IDEALS [III.81 §2], and the methods by which he proved results such as the fundamental theorem of unique decomposition. In a remarkable departure from the number-theoretic tradition, Dedekind studied the factorization properties of algebraic integers in terms of ideals, which are certain infinite sets of algebraic integers. Using this new abstract concept, together with a suitable definition of the product of two ideals, Dedekind was able to prove in full generality that, within any ring of algebraic integers, ideals possess a unique decomposition into prime ideals.

The influential algebraist Kronecker complained that Dedekind's proofs do not enable us to calculate, in a particular case, the relevant divisors or ideals: that is, the proof was *purely existential*. Kronecker's view was that this abstract way of working, made possible by the set-theoretic methods and by a concentration on the algebraic properties of the structures involved, was too remote from an algorithmic treatment—that is, from so-called *constructive* methods. But for Dedekind this complaint was misguided: it merely showed that he had succeeded in elaborating the principle "to put thoughts in the place of calculations," a principle that was also emphasized in Riemann's theory of complex functions. Obviously, concrete problems would require the development of more delicate computational techniques, and Dedekind contributed to this in several papers. But he also insisted on the importance of a general, conceptual theory.

The ideas and methods of Riemann and Dedekind became better known through publications of the period 1867–72. These were found particularly shocking

---

1. One should mention that key figures like Riemann and Cantor disagreed (see Ferreirós 1999). The "majority" included Dedekind, PEANO [VI.62], Hilbert, RUSSELL [VI.71], and others.

because of their very explicit defense of the view that mathematical theories *ought not* to be based upon formulas and calculations—they should always be based on clearly formulated *general concepts*, with analytical expressions or calculating devices relegated to the further development of the theory.

To explain the contrast, let us consider the particularly clear case of the opposition between the different approaches of Riemann and Weierstrass to function theory. Weierstrass explicitly represented analytic (or HOLOMORPHIC [I.3 §5.6]) functions as collections of power series of the form $\sum_{n=0}^{\infty} a_n(z-a)^n$, which were connected with each other by ANALYTIC CONTINUATION [I.3 §5.6]. Riemann chose a very different and more abstract approach, defining a function to be analytic if it satisfies the CAUCHY–RIEMANN DIFFERENTIABILITY CONDITIONS [I.3 §5.6].[2] This neat conceptual definition appeared objectionable to Weierstrass, as the class of differentiable functions had never been carefully characterized (in terms of series representations, for example). Exercising his famous critical abilities, Weierstrass offered examples of continuous functions that were nowhere differentiable.

It is worth mentioning that, in preferring infinite series as the key means for research in analysis and function theory, Weierstrass remained closer to the old eighteenth-century idea of a function as an analytical expression. On the other hand, Riemann and Dedekind were always in favor of Dirichlet's abstract idea of a function f as an "arbitrary" way of associating with each x some $y = f(x)$. (Previously it had been required that y should be expressed in terms of x by means of an explicit formula.) In his letters, Weierstrass criticized this conception of Dirichlet's as too general and vague to constitute the starting point for any interesting mathematical development. He seems to have missed the point that it was in fact just the right framework in which to define and analyze general concepts such as CONTINUITY [I.3 §5.2] and INTEGRATION [I.3 §5.5]. This framework came to be called the *conceptual approach* in nineteenth-century mathematics.

Similar methodological debates emerged in other areas too. In a letter of 1870, Kronecker went as far as saying that the Bolzano-Weierstrass theorem was

an "obvious sophism," promising that he would offer counterexamples. The Bolzano-Weierstrass theorem, which states that an infinite bounded set of real numbers has an accumulation point, was a cornerstone of classical analysis, and was emphasized as such by Weierstrass in his famous Berlin lectures. The problem for Kronecker was that this theorem rests entirely on the completeness axiom for the real numbers (which, in one version, states that every sequence of nonempty nested closed intervals in $\mathbb{R}$ has a nonempty intersection). The real numbers cannot be constructed in an elementary way from the rational numbers: one has to make heavy use of infinite sets (such as the set of all possible "Dedekind cuts," which are subsets $C \subset \mathbb{Q}$ such that $p \in C$ whenever p and q are rational numbers such that $p < q$ and $q \in C$). To put it another way: Kronecker was drawing attention to the problem that, very often, the accumulation point in the Bolzano-Weierstrass theorem cannot be constructed by elementary operations from the rational numbers. The classical idea of the set of real numbers, or "the continuum," already contained the seeds of the *nonconstructive* ingredient in modern mathematics.

Later on, in around 1890, Hilbert's work on invariant theory led to a debate about his purely existential proof of another basic result, the *basis theorem*, which states (in modern terminology) that every ideal in a polynomial ring is finitely generated. Paul Gordan, famous as the "king" of invariants for his heavily algorithmic work on the topic, remarked humorously that this was "theology," not mathematics! (He apparently meant that, because the proof was purely existential, rather than constructive, it was comparable with philosophical proofs of the existence of God.)

This early foundational debate led to a gradual clarification of the opposing viewpoints. Cantor's proofs in set theory also became quintessential examples of the modern methodology of existential proof. He offered an explicit defense of the higher infinite and modern methods in a paper of 1883, which was peppered with hidden attacks on Kronecker's views. Kronecker in turn criticized Dedekind's methods publicly in 1882, spoke privately against Cantor, and in 1887 published an attempt to spell out his foundational views. Dedekind replied with a detailed set-theoretic (and "thus," for him, logicistic) theory of the natural numbers in 1888.

The early round of criticism ended with an apparent victory for the modern camp, which enrolled new and powerful allies such as Hurwitz, MINKOWSKI [VI.64], Hilbert, Volterra, Peano, and HADAMARD [VI.65], and

---

2. Riemann determined particular functions by a series of *independent* traits such as the associated RIEMANN SURFACE [III.79] and the behavior at singular points. These traits determined the function via a certain variational principle (the "Dirichlet principle"), which was also criticized by Weierstrass, who gave a counterexample to it. Hilbert and Kneser would later reformulate and justify the principle.

which was defended by influential figures such as KLEIN [VI.57]. Although Riemannian function theory was still in need of further refinement, recent developments in real analysis, number theory, and other fields were showing the power and promise of the modern methods. During the 1890s, the modern viewpoint in general, and logicism in particular, enjoyed great expansion. Hilbert developed the new methodology into the axiomatic method, which he used to good effect in his treatment of geometry (1899 and subsequent editions) and of the real number system.

Then, dramatically, came the so-called logical paradoxes, discovered by Cantor, Russell, Zermelo, and others, which will be discussed below. These were of two kinds. On the one hand, there were arguments showing that assumptions that certain sets exist lead to contradictions. These were later called the *set-theoretic* paradoxes. On the other, there were arguments, later known as the *semantic* paradoxes, which showed up difficulties with the notions of truth and definability. These paradoxes completely destroyed the attractive view of recent developments in mathematics that had been proposed by logicism. Indeed, the heyday of logicism came *before* the paradoxes, that is, before 1900; it subsequently enjoyed a revival with Russell and his "theory of types," but by 1920 logicism was of interest more to philosophers than to mathematicians. However, the divide between advocates of the modern methods and constructivist critics of these methods was there to stay.

2 Around 1900

Hilbert opened his famous list of mathematical problems at the Paris International Congress of Mathematics of 1900 with Cantor's CONTINUUM PROBLEM [IV.22 §5], a key question in set theory, and with the problem of whether every set can be well-ordered. His second problem amounted to establishing the consistency of the notion of the set $\mathbb{R}$ of real numbers. It was not by chance that he began with these problems: rather, it was a way of making a clear statement about how mathematics should be in the twentieth century. Those two problems, and THE AXIOM OF CHOICE [III.1] employed by Hilbert's young colleague Zermelo to show that $\mathbb{R}$ (the *continuum*) can be well-ordered, are quintessential examples of the traits (i)-(iv) that were listed above. It is little wonder that less daring minds objected and revived Kronecker's doubts, as can be seen in many publications of 1905-6. This brings us to the next stage of the debate.

2.1 Paradoxes and Consistency

In a remarkable turn of events, the champions of modern mathematics stumbled upon arguments that cast new doubts on its cogency. In around 1896, Cantor discovered that the seemingly harmless concepts of the set of all ordinals and the set of all cardinals led to contradictions. In the former case the contradiction is usually called the *Burali-Forti paradox*; the latter is the *Cantor paradox*. The assumption that all transfinite ordinals form a set leads, by Cantor's previous results, to the result that there is an ordinal that is less than itself—and similarly for cardinals. Upon learning of these paradoxes, Dedekind began to doubt whether human thought is completely rational. Even worse, in 1901-2 Zermelo and Russell discovered a very elementary contradiction, now known as *Russell's paradox* or sometimes as the *Zermelo-Russell paradox*, which will be discussed in a moment. The untenability of the previous understanding of set theory as logic became clear, and there began a new period of instability. But it should be said that only logicists were seriously upset by these arguments: they were presented with contradictions in their theories.

Let us explain the importance of the Zermelo-Russell paradox. From Riemann to Hilbert, many authors accepted the principle that, given any well-defined logical or mathematical property, there exists a set of *all* objects satisfying that property. In symbols: given a well-defined property p, there exists another object, the set $\{x : p(x)\}$. For example, corresponding to the property of "being a real number" (which is expressed formally by Hilbert's axioms) there is the set of all real numbers; corresponding to the property of "being an ordinal" there is the set of all ordinals; and so on. This is called the *comprehension principle*, and it constitutes the basis for the logicistic understanding of set theory, often called naive set theory, although its naivete is only clear with hindsight. The principle was thought of as a basic logical law, so that all of set theory was merely a part of elementary logic.

The Zermelo-Russell paradox shows that the comprehension principle is contradictory, and it does so by formulating a property that seems to be as basic and purely logical as possible. Let $p(x)$ be the property $x \notin x$ (bearing in mind that negation and membership were assumed to be purely logical concepts). The comprehension principle yields the existence of the set $R = \{x : x \notin x\}$, but this leads quickly to a contradiction: if $R \in R$, then $R \notin R$ (by the definition of R), and similarly, if $R \notin R$, then $R \in R$. Hilbert (like his older colleague FREGE [VI.56]) was led to abandon logicism, and even wondered whether Kronecker might have been right all along. Eventually he concluded that set theory had shown the need to refine logical theory. It was also necessary to establish set theory axiomatically, as a basic *mathematical* theory based on mathematical (not logical) axioms, and Zermelo undertook this task.

Hilbert famously advocated that to claim that a set of mathematical objects exists is tantamount to proving that the corresponding axiom system is consistent— that is, free of contradictions. The documentary evidence suggests that Hilbert came to this celebrated principle in reaction to Cantor's paradoxes. His reasoning may have been that, instead of jumping directly from well-defined concepts to their corresponding sets, one had first to prove that the concepts are logically consistent. For example, before one could accept the set of all real numbers, one should prove the consistency of Hilbert's axiom system for them. Hilbert's principle was a way of removing any metaphysical content from the notion of mathematical existence. This view, that mathematical objects had a sort of "ideal existence" in the realm of thought rather than an independent metaphysical existence, had been anticipated by Dedekind and Cantor.

The "logical" paradoxes included not only the ones that go by the names of Burali-Forti, Cantor, and Russell, but also many semantic paradoxes formulated by Russell, Richard, König, Grelling, etc. (Richard's paradox will be discussed below.) Much confusion emerged from the abundance of different paradoxes, but one thing is clear: they played an important role in promoting the development of modern logic and convincing mathematicians of the need for strictly formal presentation of their theories. Only when a theory has been stated within a precise formal language can one disregard the semantic paradoxes, and even formulate the distinction between these and the set-theoretic ones.

2.2 Predicativity

When the books of Frege and Russell made the paradoxes of set theory widely known to the mathematical community in 1903, POINCARÉ [VI.61] used them to put forward criticisms of both logicism and formalism.

His analysis of the paradoxes led him to coin an important new notion, *predicativity*, and maintain that impredicative definitions should be avoided in mathematics. Informally, a definition is impredicative when

it introduces an element by reference to a totality that already contains that element. A typical example is the following: Dedekind defines the set $\mathbb{N}$ of natural numbers as the intersection of all sets that contain 1 and are closed under an injective function σ such that $1 \notin \sigma(\mathbb{N})$. (The function σ is called the *successor function*.) His idea was to characterize $\mathbb{N}$ as minimal, but in his procedure the set $\mathbb{N}$ is first introduced by appeal to a totality of sets that should already include $\mathbb{N}$ itself. This kind of procedure appeared unacceptable to Poincaré (and also to Russell), especially when the relevant object can be specified *only* by reference to the more embracing totality. Poincaré found examples of impredicative procedures in each of the paradoxes he studied.

Take, for instance, Richard's paradox, which is one of the linguistic or semantic paradoxes (where, as we said, the notions of truth and definability are prominent). One begins with the idea of *definable* real numbers. Because definitions must be expressed in a certain language by finite expressions, there are only countably many definable numbers. Indeed, we can explicitly count the definable real numbers by listing them in alphabetical order of their definitions. (This is known as the *lexicographic order*.) Richard's idea was to apply to this list a diagonal process, of the kind used by Cantor to prove that $\mathbb{R}$ is not COUNTABLE [III.11]. Let the definable numbers be $a_1, a_2, a_3, \ldots$. Define a new number r in a systematic way, making sure that the nth decimal digit of r is different from the nth decimal digit of a_n. (For example, let the nth digit of r be 2 unless the nth digit of a_n is 2, in which case let the nth digit of r be 4.) Then r cannot belong to the set of definable numbers. But in the course of this construction, the number r has just been defined in finitely many words! Poincaré would ban impredicative definitions and would therefore prevent the introduction of the number r, since it was defined with reference to the totality of all definable numbers.[3]

In this kind of approach to the foundations of mathematics, all mathematical objects (beyond the natural numbers) must be introduced by explicit definitions. If a definition refers to a presumed totality of which the object being defined is itself a member, we are involved in a circle: the object itself is then a constituent of its own definition. In this view, "definitions"

---

3. The modern solution is to establish mathematical definitions within a well-determined formal theory, whose language and expressions are fixed to begin with. Richard's paradox takes advantage of an ambiguity as to what the available means of definition are.

must be predicative: one refers only to totalities that have already been established before the object one is defining. Important authors such as Russell and WEYL [VI.80] accepted this point of view and developed it.

Zermelo was not convinced, arguing that impredicative definitions were often used unproblematically, not only in set theory (as in Dedekind's definition of $\mathbb{N}$, for example), but also in classical analysis. As a particular example, he cited CAUCHY's [VI.29] proof of THE FUNDAMENTAL THEOREM OF ALGEBRA [V.13],[4] but a simpler example of impredicative definition is the least upper bound in real analysis. The real numbers are not introduced separately, by explicit predicative definitions of each one of them; rather, they are introduced as a completed whole, and the particular way in which the least upper bound of an infinite bounded set of reals is singled out becomes impredicative. But Zermelo insisted that these definitions are innocuous, because the object being defined is not "created" by the definition; it is merely singled out (see his paper of 1908 in van Heijenoort (1967, pp. 183–98)).

Poincaré's idea of abolishing impredicative definitions became important for Russell, who incorporated it as the "vicious circle principle" in his influential *theory of types*. Type theory is a system of higher-order logic, with quantification over properties or sets, over relations, over sets of sets, and so on. Roughly speaking, it is based on the idea that the elements of any set should always be objects of a certain homogeneous type. For instance, we can have sets of "individuals," such as $\{a, b\}$, or sets of *sets* of individuals, such as $\{\{a\}, \{a, b\}\}$, but never a "mixed" set like $\{a, \{a, b\}\}$. Russell's version of type theory became rather complicated because of the so-called ramification he adopted in order to avoid impredicativity. This system, together with axioms of infinity, choice, and "reducibility" (a surprisingly ad hoc means to "collapse" the ramification), sufficed for the development of set theory and the number systems. Thus it became the logical basis for the renowned *Principia Mathematica* by Whitehead and Russell (1910–13), in which they carefully developed a foundation for mathematics.

Type theory remained the main logical system until about 1930, but under the form of *simple* type theory

(that is, without ramification), which, as Chwistek, Ramsey, and others realized, suffices for a foundation in the style of *Principia*. Ramsey proposed arguments that were aimed at eliminating worries about impredicativity, and he tried to justify the other existence axioms of *Principia*—the axiom of infinity and the axiom of choice—as logical principles. But his arguments were inconclusive. Russell's attempt to rescue logicism from the paradoxes remained unconvincing, except to some philosophers (especially members of the Vienna Circle).

Poincaré's suggestions also became a key principle for the interesting foundational approach proposed by Weyl in his book *Das Kontinuum* (1918). The main idea was to accept the theory of the natural numbers as they were conventionally developed using classical logic, but to work predicatively from there on. Thus, unlike Brouwer, Weyl accepted the principle of the excluded middle. (This, and Brouwer's views, will be discussed in the next section.) However, the full system of the real numbers was not available to him: in his system the set $\mathbb{R}$ was not complete and the Bolzano-Weierstrass theorem failed, which meant that he had to devise sophisticated replacements for the usual derivations of results in analysis.

The idea of *predicative foundations* for mathematics, in the style of Weyl, has been carefully developed in recent decades with noteworthy results (see Feferman 1998). Predicative systems lie between those that countenance all of the modern methodology and the more stringent constructivistic systems. This is one of several foundational approaches that do not fit into the conventional but by now outdated triad of logicism, formalism, and intuitionism.

2.3 Choices

As important as the paradoxes were, their impact on the foundational debate has often been overstated. One frequently finds accounts that take the paradoxes as the real starting point of the debate, in strong contrast with our discussion in section 1. But even if we restrict our attention to the first decade of the twentieth century, there was another controversy of equal, if not greater, importance: the arguments that surrounded the axiom of choice and Zermelo's proof of the well-ordering theorem.

Recall from section 2.1 that the association between sets and their defining properties was at the time deeply ingrained in the minds of mathematicians and logicians (via the contradictory principle of comprehension). The axiom of choice (AC) is the principle that,

---

4. Cauchy's reasoning was clearly nonconstructive, or "purely existential" as we have been saying. In order to show that the polynomial must have one root, Cauchy studied the absolute value of the polynomial, which has a global minimum σ. This global minimum is impredicatively defined. Cauchy assumed that it was positive, and from this he derived a contradiction.

given any infinite family of disjoint nonempty sets, there is a set, known as a *choice set*, that contains exactly one element from each set in the family. The problem with this, said the critics, is that it merely stipulates the existence of the choice set and does not give a defining property for it. Indeed, when it *is* possible to characterize the choice set explicitly, then the use of AC is avoidable! But in the case of Zermelo's well-ordering theorem it is essential to employ AC. The required well-ordering of $\mathbb{R}$ "exists" in the ideal sense of Cantor, Dedekind, and Hilbert, but it seemed clear that it was completely out of reach from any constructivist perspective.

Thus, the axiom of choice exacerbated obscurities in previous conceptions of set theory, forcing mathematicians to introduce much-needed clarifications. On the one hand, AC was nothing but an explicit statement of previous views about *arbitrary* subsets, and yet, on the other, it obviously clashed with strongly held views about the need to explicitly define infinite sets by properties. The stage was set for deep debate. The discussions about this particular topic contributed more than anything else to a clarification of the existential implications of modern mathematical methods. It is instructive to know that BOREL [VI.70], Baire, and LEBESGUE [VI.72], who became critics, had all relied on AC in less obvious ways in order to prove theorems of analysis. Not by chance, the axiom was suggested to Zermelo by an analyst, Erhard Schmidt, who was a student of Hilbert.[5]

After the publication of Zermelo's proof, an intense debate developed throughout Europe. Zermelo was spurred on to work out the foundations of set theory in an attempt to show that his proof could be developed within an unexceptionable axiom system. The outcome was his famous AXIOM SYSTEM [IV.22 §3], a masterpiece that emerged from careful analysis of set theory as it was historically given in the contributions of Cantor and Dedekind and in Zermelo's own theorem. With some additions due to Fraenkel and VON NEUMANN [VI.91] (the axioms of replacement and regularity) and the major innovation proposed by Weyl and SKOLEM [VI.81] (to formulate it within FIRST-ORDER LOGIC [IV.23 §1], i.e., quantifying over individuals, the sets, but not over their properties), the axiom system became in the 1920s the one that we now know.

The ZFC system (this stands for "Zermelo-Fraenkel with choice") codifies the key traits of modern mathematical methodology, offering a satisfactory framework for the development of mathematical theories and the conduct of proofs. In particular, it includes strong existence principles, allows impredicative definitions and arbitrary functions, warrants purely existential proofs, and makes it possible to define the main mathematical structures. It thus exhibits all the tendencies (i)–(iv) mentioned in section 1. Zermelo's own work was completely in line with Hilbert's informal axiomatizations of about 1900, and he did not forget to promise a proof of consistency. Axiomatic set theory, whether in the Zermelo-Fraenkel presentation or the von Neumann–Bernays–Gödel version, is the system that most mathematicians regard as the working foundation for their discipline.

As of 1910, the contrast between Russell's type theory and Zermelo's set theory was strong. The former system was developed within formal logic, and its point of departure (albeit later compromised for pragmatic reasons) was in line with predicativism; in order to derive mathematics, the system needed the existential assumptions of infinity and choice, but these were rhetorically treated as tentative hypotheses rather than outright axioms. The latter system was presented informally, adopted the impredicative standpoint wholeheartedly, and asserted as axioms strong existential assumptions that were sufficient to derive all of classical mathematics and Cantor's theory of the higher infinite. In the 1920s the separation diminished greatly, especially with respect to the first two traits just indicated. Zermelo's system was perfected and formulated within the language of modern formal logic. And the Russellians adopted simple type theory, thus accepting the impredicative and "existential" methodology of modern mathematics. This is often given the (potentially confusing) term "Platonism": the objects that the theory refers to are treated *as if* they were independent of what the mathematician can actually and explicitly define.

Meanwhile, back in the first decade of the twentieth century, a young mathematician in the Netherlands was beginning to find his way toward a philosophically colored version of constructivism. Brouwer presented his strikingly peculiar metaphysical and ethical views in 1905, and started to elaborate a corresponding foundation for mathematics in his thesis of 1907. His philosophy of "intuitionism" derived from the old metaphysical view that individual consciousness is the one and

---

5. One may still gain much insight by reading the letters exchanged by the French analysts in 1905 (see Moore 1982; Ewald 1996) and Zermelo's clever arguments in his second 1908 proof of well-ordering (van Heijenoort 1967).

only source of knowledge. This philosophy is perhaps of little interest in itself, so we shall concentrate here on Brouwer's constructivistic principles. In the years around 1910, Brouwer became a renowned mathematician, with crucial contributions to topology such as his FIXED POINT THEOREM [V.11]. By the end of World War I, he started to publish detailed elaborations of his foundational ideas, helping to create the famous "crisis," to which we now turn. He was also successful in establishing the customary (but misleading) distinction between formalism and intuitionism.

3 The *Crisis* in a Strict Sense

In 1921, the *Mathematische Zeitschrift* published a paper by Weyl in which the famous mathematician, who was a disciple of Hilbert, openly espoused intuitionism and diagnosed a "crisis in the foundations" of mathematics. The crisis pointed toward a "dissolution" of the old state of analysis, by means of Brouwer's "revolution." Weyl's paper was meant as a propaganda pamphlet to rouse the sleepers, and it certainly did. Hilbert answered in the same year, accusing Brouwer and Weyl of attempting a "putsch" aimed at establishing "dictatorship à la Kronecker" (see the relevant papers in Mancosu (1998) and van Heijenoort (1967)). The foundational debate shifted dramatically toward the battle between Hilbert's attempts to justify "classical" mathematics and Brouwer's developing reconstruction of a much-reformed intuitionistic mathematics.

Why was Brouwer "revolutionary"? Up to 1920 the key foundational issues had been the acceptability of the real numbers and, more fundamentally, of the impredicativity and strong existential assumptions of set theory, which supported the higher infinite and the unrestricted use of existential proofs. Set theory and, by implication, classical analysis had been criticized for their reliance on impredicative definitions and for their strong existential assumptions (in particular, the axiom of choice, of which extensive use was made by SIERPIŃSKI [VI.77] in 1918). Thus, the debate in the first two decades of the twentieth century was mainly about which principles to accept when it came to defining and establishing the existence of sets and subsets. A key question was, can one make rigorous the vague idea behind talk of "arbitrary subsets"? The most coherent reactions had been Zermelo's axiomatization of set theory and Weyl's predicative system in *Das Kontinuum*. (The *Principia Mathematica* of Whitehead and Russell was an unsuc-

cessful compromise between predicativism and classical mathematics.)

Brouwer, however, brought new and even more basic questions to the fore. No one had questioned the traditional ways of reasoning about the natural numbers: classical logic, in particular the use of quantifiers and the principle of the excluded middle, had been used in this context without hesitation. But Brouwer put forward principled critiques of these assumptions and started developing an alternative theory of analysis that was much more radical than Weyl's. In doing so, he came upon a new theory of the continuum, which finally enticed Weyl and made him announce the coming of a new age.

3.1 Intuitionism

Brouwer began systematically developing his views with two papers on "intuitionistic set theory," written in German and published in 1918 and 1919 by the *Verhandelingen* of the Dutch Academy of Sciences. These contributions were part of what he regarded as the "Second Act" of intuitionism. The "First Act" (from 1907) had been his emphasis on the intuitive foundations of mathematics. Already Klein and Poincaré had insisted that intuition has an inescapable role to play in mathematical knowledge: as important as logic is in proofs and in the development of mathematical theory, mathematics cannot be reduced to pure logic; theories and proofs are of course organized logically, but their basic principles (axioms) are grounded in intuition. But Brouwer went beyond them and insisted on the absolute independence of mathematics from language and logic.

From 1907, Brouwer rejected the principle of the excluded middle (PEM), which he regarded as equivalent to Hilbert's conviction that all mathematical problems are solvable. PEM is the logical principle that the statement $p \vee \neg p$ (that is, either p or not p) must always be true, whatever the proposition p may be. (For example, it follows from PEM that either the decimal expansion of π contains infinitely many sevens or it contains only finitely many sevens, even though we do not have a proof of which.) Brouwer held that our customary logical principles were abstracted from the way we dealt with subsets of a finite set, and that it was wrong to apply them to infinite sets as well. After World War I he started the systematic reconstruction of mathematics.

The intuitionist position is that one can only state "p or q" when one can give either a constructive proof of

p or a constructive proof of q. This standpoint has the consequence that proofs by contradiction (*reductio ad absurdum*) are not valid. Consider Hilbert's first proof of his basis theorem (section 1), achieved by *reductio*: he showed that one can derive a contradiction from the assumption that the basis is infinite, and from this he concluded that the basis is finite. The logic behind this procedure is that we start from a concrete instance of PEM, $p \vee \neg p$, show that $\neg p$ is untenable, and conclude that p must be true. But constructive mathematics asks for *explicit* procedures for constructing each object that is assumed to exist, and explicit constructions behind any mathematical statement. Similarly, we have mentioned before (section 2.1) Cauchy's proof of the fundamental theorem of algebra, as well as many proofs in real analysis that invoke the least upper bound. All of these proofs are invalid for a constructivist, and several mathematicians have tried to save the theorems by finding constructivist proofs for them. For instance, both Weyl and Kneser worked on constructivist proofs for the fundamental theorem of algebra.

It is easy to give instances of the use of PEM that a constructivist will not accept: one just has to apply it to any unsolved mathematical problem. For example, Catalan's constant is the number

$$K = \sum_{n=0}^{\infty} \frac{(-1)^n}{(2n+1)^2}.$$

It is not known whether K is transcendental, so if p is the statement "Catalan's constant is transcendental," then a constructivist will not accept that p is either true or false.

This may seem odd, or even obviously wrong, until one realizes that constructivists have a different view about what truth *is*. For a constructivist, to say that a proposition is true simply *means* that we can prove it in accordance with the stringent methods that we are discussing; to say that it is false *means* that we can actually exhibit a counterexample to it. Since there is no reason to suppose that every existence statement has either a strict constructivist proof or an explicit counterexample, there is no reason to believe PEM (with this notion of truth). Thus, in order to establish the existence of a natural number with a certain property, a proof by *reductio ad absurdum* is not enough. Existence must be shown by explicit determination or construction if you want to persuade a constructivist.

Notice also how this viewpoint implies that mathematics is not timeless or ahistorical. It was only in 1882 that Lindemann proved that π is a TRANSCENDEN-

TAL NUMBER [III.41]. Since that date, it has been possible to assign a truth value to statements that were neither true nor false before, according to intuitionists. This may seem paradoxical, but it was just right for Brouwer, since in his view mathematical objects are mental constructions and he rejected as "metaphysics" the assumption that they have an independent existence.

In 1918, Brouwer replaced the sets of Cantor and Zermelo by constructive counterparts, which he would later call "spreads" and "species." A *species* is basically a set that has been defined by a characteristic property, but with the proviso that *every* element has been previously and independently defined by an explicit construction. In particular, the definition of any given species will be strictly predicative.

The concept of a *spread* is particularly characteristic of intuitionism, and it forms the basis for Brouwer's definition of the continuum. It is an attempt to avoid idealization and do justice to the temporal nature of mathematical constructions. Suppose, for example, that we wish to define a sequence of rational numbers that gives better and better approximations to the square root of 2. In classical analysis, one conceives of such sequences as existing in their entirety, but Brouwer defined a notion that he called a *choice sequence*, which pays more attention to how they might be produced. One way to produce them is to give a rule, such as the recurrence relation $x_{n+1} = (x_n^2 + 2)/2x_n$ (and the initial condition $x_1 = 2$). But another is to make less rigidly determined choices that obey certain constraints: for instance, one might insist that x_n has denominator n and that x_n^2 differs from 2 by at most $100/n$, which does not determine x_n uniquely but does ensure that the sequence produces better and better approximations to $\sqrt{2}$.

A choice sequence is therefore not required to be completely specified from the outset, and it can involve choices that are freely made by the mathematician at different moments in time. Both these features make choice sequences very different from the sequences of classical analysis: it has been said that intuitionist mathematics is "mathematics in the making." By contrast, classical mathematics is marked by a kind of timeless objectivity, since its objects are assumed to be fully determined in themselves and independent of the thinking processes of mathematicians.

A spread has choice sequences as its elements—it is something like a law that regulates how the sequences are constructed.[6] For instance, one could take a spread that consisted of all choice sequences that began in some particular way, and such a spread would represent a segment—in general, spreads do not represent isolated elements, but continuous domains. By using spreads whose elements satisfy the Cauchy condition, Brouwer offered a new mathematical conception of the *continuum*: rather than being made up of points (or real numbers) with some previous Platonic existence, it was more genuinely "continuous." Interestingly, this view is reminiscent of Aristotle, who, twenty-three centuries earlier, had emphasized the priority of the continuum and rejected the idea that an extended continuum can be made up of unextended points.

The next stage in Brouwer's redevelopment of analysis was to analyze the idea of a function. Brouwer defined a function to be an assignment of values to the elements of a spread. However, because of the nature of spreads, this assignment had to be wholly dependent on an initial segment of the choice sequence in order to be constructively admissible. This threw up a big surprise: all functions that are everywhere defined are continuous (and even uniformly continuous). What, you might wonder, about the function f where $f(x) = 0$ when $x < 0$ and $f(x) = 1$ when $x \geq 0$? For Brouwer, this is not a well-defined function, and the underlying reason for this is that one can determine spreads for which we do not know (and may never know) whether they are positive, zero, or negative. For instance, one could let x_n be 1 if all the even numbers between 4 and $2n$ are sums of two primes, and -1 otherwise.

The rejection of PEM has the effect that intuitionistic negation differs in meaning from classical negation. Thus, intuitionistic arithmetic is also different from classical arithmetic. Nevertheless, in 1933 Gödel and Gentzen were able to show that the DEDEKIND–PEANO AXIOMS [III.67] of arithmetic are consistent *relative to* formalized intuitionistic arithmetic. (That is, they were able to establish a correspondence between the sentences of both formal systems, such that a contradiction in classical arithmetic yields a contradiction in its intuitionistic counterpart; thus, if the latter is consistent, the former must be as well.) This was a small triumph for the Hilbertians, though corresponding proofs

for systems of analysis or set theory have never been found.

Initially there had been hopes that the development of intuitionism would end in a simple and elegant presentation of pure mathematics. However, as Brouwer's reconstruction developed in the 1920s, it became more and more clear that intuitionistic analysis was extremely complicated and foreign. Brouwer was not worried, for, as he would say in 1933, "the spheres of truth are less transparent than those of illusion." But Weyl, although convinced that Brouwer had delineated the domain of mathematical intuition in a completely satisfactory way, remarked in 1925: "the mathematician watches with pain the largest part of his towering theories dissolve into mist before his eyes." Weyl seems to have abandoned intuitionism shortly thereafter. Fortunately, there was an alternative approach that suggested another way of rehabilitating classical mathematics.

3.2   Hilbert's Program

This alternative approach was, of course, Hilbert's program, which promised, in the memorable phrasing of 1928, "to eliminate from the world once and for all the skeptical doubts" as to the acceptability of the classical theories of mathematics. The new perspective, which he started to develop in 1904, relied heavily on formal logic and a combinatorial study of the formulas that are provable from given formulas (the axioms). With modern logic, proofs are turned into formal computations that can be checked mechanically, so that the process is purely constructivistic.

In the light of our previous discussion (section 1), it is interesting that the new project was to employ Kroneckerian means for a justification of modern, anti-Kroneckerian methodology. Hilbert's aim was to show that it is impossible to prove a contradictory formula from the axioms. Once this had been shown combinatorially or constructively (or, as Hilbert also said, finitarily), the argument can be regarded as a justification of the axiom system—even if we read the axioms as talking about non-Kroneckerian objects like the real numbers or transfinite sets.

Still, Hilbert's ideas at the time were marred by a deficient understanding of logical theory.[7] It was only in 1917–18 that Hilbert returned to this topic, now with

---

6. More precisely, a spread is defined by means of two laws; see Heyting (1956), or more recently van Atten (2003), for further details on this and other points. One can picture a spread as a subtree of the universal tree of natural numbers (consisting of all finite sequences of natural numbers), together with an assignment of previously available mathematical objects to the nodes. One law of the spread determines nodes in the tree, the other maps them to objects.

7. The logic he presented in 1905 lagged far behind Frege's system of 1879 or Peano's of the 1890s. We do not enter into the development of logical theory in this period (see, for example, Moore 1998).

a refined understanding of logical theory and a greater awareness of the considerable technical difficulties of his project. Other mathematicians played very significant parts in promoting this better understanding. By 1921, helped by his assistant Bernays, Hilbert had arrived at a very refined conception of the formalization of mathematics, and had perceived the need for a deeper and more careful probing into the logical structure of mathematical proofs and theories. His program was first clearly formulated in a talk at Leipzig late in 1922.

Here we will describe the mature form of Hilbert's program, as it was presented for instance in the 1925 paper "On the infinite" (see van Heijenoort 1967). The main goal was to establish, by means of syntactic consistency proofs, the logical acceptability of the principles and modes of inference of modern mathematics. Axiomatics, logic, and formalization made it possible to study mathematical theories from a purely mathematical standpoint (hence the name *metamathematics*), and Hilbert hoped to establish the consistency of the theories by employing very weak means. In particular, Hilbert hoped to answer all of the criticisms of Weyl and Brouwer, and thereby justify set theory, the classical theory of real numbers, classical analysis, and of course classical logic with its PEM (the basis for indirect proofs by *reductio ad absurdum*).

The whole point of Hilbert's approach was to make mathematical theories fully precise, so that it would become possible to obtain precise results about their properties. The following steps are indispensable for the completion of such a program.

(i) Finding suitable axioms and primitive concepts for a mathematical theory T, such as that of the real numbers.

(ii) Finding axioms and inference rules for classical logic, which makes the passage from given propositions to new propositions a purely syntactic, formal procedure.

(iii) Formalizing T by means of the formal logical calculus, so that propositions of T are just strings of symbols, and proofs are sequences of such strings that obey the formal rules of inference.

(iv) A *finitary* study of the formalized proofs of T that shows that it is impossible for a string of symbols that expresses a contradiction to be the last line of a proof.

In fact, steps (ii) and (iii) can be solved with rather simple systems formalized in first-order logic, like those studied in any introduction to mathematical logic, such as Dedekind–Peano arithmetic or Zermelo–Fraenkel set theory. It turns out that first-order logic is enough for codifying mathematical proofs, but, interestingly, this realization came rather late—after GÖDEL'S THEOREMS [V.15].

Hilbert's main insight was that, when theories are formalized, any proof becomes a *finite* combinatorial object: it is just an array of strings of symbols complying with the formal rules of the system. As Bernays said, this was like "projecting" the deductive structure of a theory T into the number-theoretic domain, and it became possible to express in this domain the consistency of T. These realizations raised hopes that a finitary study of formalized proofs would suffice to establish the consistency of the theory, that is, to prove the sentence expressing the consistency of T. But this hope, not warranted by the previous insights, turned out to be wrong.[8]

Also, a crucial presupposition of the program was that not only the logical calculus but also each of the axiomatic systems would be *complete*. Roughly speaking, this means that they would be sufficiently powerful to allow the derivation of all the relevant results.[9] This assumption turned out to be wrong for systems that contain (primitive recursive) arithmetic, as Gödel showed.

It remains to say a bit more about what Hilbert meant by *finitism* (for details, see Tait 1981). This is one of the points in which his program of the 1920s adopted to some extent the principles of intuitionists such as Poincaré and Brouwer and deviated strongly from the ideas Hilbert himself had considered in 1900. The key idea is that, contrary to the views of logicists like Frege and Dedekind, logic and pure thought require something that is given "intuitively" in our immediate experience: the signs and formulas.

In 1905, Poincaré had put forth the view that a formal consistency proof for arithmetic would be circular, as such a demonstration would have to proceed by induction on the length of formulas and proofs, and thus would rely on the same axiom of induction that it was supposed to establish. Hilbert replied in the 1920s that the form of induction required at the metamathematical level is much weaker than full arithmetical induction, and that this weak form is grounded on the

---

8. For further details, see, for example, Sieg (1999).

9. The notion of "relevant result" should of course be made precise: doing so leads to the notion either of syntactic completeness or of semantic completeness.

finitary consideration of signs that he took to be intuitively given. Finitary mathematics was not in need of any further justification or reduction.

Hilbert's program proceeded gradually by studying weak theories at first and proceeding to progressively stronger ones. The *metatheory* of a formal system studies properties such as consistency, completeness, and some others ("completeness" in the logical sense means that all true or *valid* formulas that can be represented in the calculus are formally deducible in it). Propositional logic was quickly proved to be consistent and complete. First-order logic, also known as *predicate logic*, was proved complete by Gödel in his dissertation of 1929. For all of the 1920s, the attention of Hilbert and coworkers was set on elementary arithmetic and its subsystems; once this had been settled, the project was to move on to the much more difficult, but crucial, cases of the theory of real numbers and set theory. Ackermann and von Neumann were able to establish consistency results for certain subsystems of arithmetic, but between 1928 and 1930 Hilbert was convinced that the consistency of arithmetic had already been established. Then came the severe blow of Gödel's incompleteness results (see section 4).

The name "formalism," as a description of this program, came from the fact that Hilbert's *method* consisted in formalizing each mathematical theory, and formally studying its proof structure. However, this name is rather one-sided and even confusing, especially because it is usually contrasted with intuitionism, a full-blown *philosophy* of mathematics. Like most mathematicians, Hilbert never viewed mathematics as a mere game played with formulas. Indeed, he often emphasized the meaningfulness of (informal) mathematical statements and the depth of conceptual content expressed in them.[10]

3.3 Personal Disputes

The crisis was unfolding not just at an intellectual level but also at a personal level. One should perhaps tell this story as a tragedy, in which the personalities of the main figures and the successive events made the final result quite inescapable.

Hilbert and Brouwer were very different personalities, though they were both extremely willful and clever men. Brouwer's worldview was idealistic and tended to solipsism. He had an artistic temperament and an eccentric private life. He despised the modern world, looking to the inner life of the self as the only way out (at least in principle, though not always in practice). He preferred to work in isolation, although he had good friends in the mathematical community, especially in the international group of topologists that gathered around him. Hilbert was typically modernist in his views and attitudes; full of optimism and rationalism, he was ready to lead his university, his country, and the international community into a new world. He was very much in favor of collaboration, and felt happy to join Klein's schemes for institutional development and power.

As a consequence of World War I, Germans in the early 1920s were not allowed to attend the International Congresses of Mathematicians. When the opportunity finally arose in 1928, Hilbert was eager to seize on it, but Brouwer was furious because of restrictions that were still imposed on the German delegation and sent a circular letter in order to convince other mathematicians. Their viewpoints were widely known and led to a clash between the two men. On another level, Hilbert had made important concessions to his opponents in the 1920s, hoping that he would succeed in his project of finding a consistency proof. Brouwer emphasized these concessions, accusing him of failing to recognize authorship, and demanded new concessions.[11] Hilbert must have felt insulted and perhaps even threatened by a man whom he regarded as perhaps the greatest mathematician of the younger generation.

The last straw came with an episode in 1928. Brouwer had since 1915 been a member of the editorial board of *Mathematische Annalen*, the most prestigious mathematics journal at the time, of which Hilbert had been the main editor since 1902. Ill with "pernicious anemia," and apparently thinking that he was close to the end, Hilbert feared for the future of his journal and decided it was imperative to remove Brouwer from the editorial board. When he wrote to other members of the board explaining his scheme, which he was already carrying out, Einstein replied saying that his proposal was unwise and that he wanted to have nothing to do with it. Other members, however, did not wish to upset the old and admired Hilbert. Finally, a dubious procedure was adopted, where the whole board was dissolved and created anew. Brouwer was greatly disturbed by this

---

10. This is very explicit, for example, in the lectures of 1919–20 edited by Rowe (1992), and also in the 1930 paper that bears exactly the same title (see *Gesammelte Abhandlungen*, volume 3).

11. See his "Intuitionistic reflections on formalism" of 1928 (in Mancosu 1998).

action, and as a result of it the journal lost Einstein and Carathéodory, who had previously been main editors (see van Dalen 2005).

After that, Brouwer ceased to publish for some years, leaving some book plans unfinished. With his disappearance from the scene, and with the gradual disappearance of previous political turbulences, the feelings of "crisis" began to fade away (see Hesseling 2003). Hilbert did not intervene much in the subsequent debates and foundational developments.

4 Gödel and the Aftermath

It was not only the *Annalen* war that Hilbert won: the mathematical community as a whole continued to work in the style of modern mathematics. And yet his program suffered a profound blow with the publication of Gödel's famous 1931 article in the *Monatshefte für Mathematik und Physik*. An extremely ingenious development of metamathematical methods—the arithmetization of metamathematics—allowed Gödel to prove that systems like axiomatic set theory or Dedekind–Peano arithmetic are incomplete (see GÖDEL'S THEOREM [V.15]). That is, there exist propositions P formulated strictly in the language of the system such that neither P nor $\neg P$ is formally provable in the system.

This theorem already presented a deep problem for Hilbert's endeavor, as it shows that formal proof cannot even capture arithmetical truth. But there was more. A close look at Gödel's arguments made it clear that this first metamathematical proof could itself be formalized, which led to "Gödel's second theorem"—that it is impossible to establish the consistency of the systems mentioned above by any proof that can be codified *within* them. Gödel's arithmetization of metamathematics makes it possible to build a sentence, in the language of formal arithmetic, that expresses the consistency of this same formal system. And this sentence turns out to be among those that are unprovable.[12] To express it contrapositively, a finitary formal proof (codifiable in the system of formal arithmetic) of the impossibility of proving $1 = 0$ could be transformed into a contradiction of the system! Thus, if the system is indeed consistent (as most mathematicians are convinced it is), then there is no such finitary proof.

According to what Gödel called at the time "the von Neumann conjecture" (namely, that if there is a finitary proof of consistency, then it can be formalized and codified within elementary arithmetic), the second theorem implies the failure of Hilbert's program (see Mancosu (1999, p. 38) and, for more on the reception, Dawson (1997, pp. 68 ff)). One should emphasize that Gödel's negative results are purely constructivistic and even finitistic, valid for all parties in the foundational debate. They were difficult to digest, but in the end they led to a reestablishment of the basic terms for foundational studies.

Mathematical logic and foundational studies continued to develop brilliantly with Gentzen-style proof theory, with the rise of MODEL THEORY [IV.23], etc.—all of which had their roots in the foundational studies of the first third of the twentieth century. Although the Zermelo–Fraenkel axioms suffice for giving a rigorous foundation to most of today's mathematics, and have a rather convincing intuitive justification in terms of the "iterative" conception of sets,[13] there is a general feeling that foundational studies, instead of achieving their ambitious goal, "found themselves attracted into the whirl of mathematical activity, and are now enjoying full voting rights in the mathematical senate."[14]

However, this impression is somewhat superficial. Proof theory has developed, leading to noteworthy reductions of classical theories to systems that can be regarded as constructive. A striking example is that analysis can be formalized in *conservative extensions* of arithmetic: that is, in systems that extend the language of arithmetic while including all theorems of arithmetic, but which are "conservative" in the sense that they have no new consequences in the language of arithmetic. Some parts of analysis can even be developed in conservative extensions of primitive recursive arithmetic (see Feferman 1998). This raises questions about the philosophical bases on which the admissibility of the relevant constructive theories can be founded. But for these systems the question is far less simple than it was for Hilbert's finitary mathematics; it seems

---

12. For further details, see, for example, Smullyan (2001), van Heijenoort (1967), and good introductions to mathematical logic. Both theorems were carefully proved in Hilbert and Bernays (1934/39). Bad expositions and faulty interpretations of Gödel's results abound.

13. The basic idea is to view the set-theoretic universe as a product of iterating the following basic operation: one starts with a basic domain V_0 (possibly finite or even equal to $\varnothing$) and forms all possible *sets of* elements in the domain; this gives a new domain V_1, and one iterates forming sets of $V_0 \cup V_1$, and so on (to infinity and beyond!). This produces an open-ended set-theoretic universe, masterfully described by Zermelo (1930). On the iterative conception, see, for example, the last papers in Bernacerraf and Putnam (1983).

14. To use the words of Gian-Carlo Rota in an essay of 1973.

fair to say that no general consensus has yet been reached.

Whatever its roots and justification may be, mathematics is a human activity. This truism is clear from the subsequent development of our story. The mathematical community refused to abandon "classical" ideas and methods; the constructivist "revolution" was aborted. In spite of its failure, formalism established itself in practice as the avowed ideology of twentieth-century mathematicians. Some have remarked that formalism was less a real faith than a Sunday refuge for those who spent their weekdays working on mathematical objects as something very real. The Platonism of working days was only abandoned, as a BOURBAKI [VI.96] member said, when a ready-made reply was needed to unwelcome philosophical questions concerning mathematical knowledge.

One should note that formalism suited very well the needs of a self-conscious, autonomous community of research mathematicians. It granted them full freedom to choose their topics and to employ modern methods to explore them. However, to reflective mathematical minds it has long been clear that it is not the answer. Epistemological questions about mathematical knowledge have not been "eliminated from the world"; philosophers, historians, cognitive scientists, and others keep looking for more adequate ways of understanding its content and development. Needless to say, this does not threaten the autonomy of mathematical researchers—if autonomy is to be a concern, perhaps we should worry instead about the pressures exerted on us by the market and other powers.

Both (semi-)constructivism and modern mathematics have continued to develop: the contrast between them has simply been consolidated, though in a very unbalanced way, since some 99% of practicing mathematicians are "modern." (But do statistics matter when it comes to the correct methods for mathematics?) In 1905, commenting on the French debate, HADAMARD [VI.65] wrote that "there are two conceptions of mathematics, two mentalities, in evidence." It has now come to be recognized that there is value in both approaches: they complement each other and can coexist peacefully. In particular, interest in effective methods, algorithms, and computational mathematics has grown markedly in recent decades—and all of these are closer to the constructivist tradition.

The foundational debate left a rich legacy of ideas and results, key insights and developments, including the formulation of axiomatic set theories and the rise

of intuitionism. One of the most important of these developments was the emergence of modern mathematical logic as a refinement of axiomatics, which led to the theories of recursion and computability in around 1936 (see ALGORITHMS [II.4 §3.2]). In the process, our understanding of the characteristics, possibilities, and limitations of formal systems was hugely clarified.

One of the hottest issues throughout the whole debate, and probably its main source, was the question of how to understand the continuum. The reader may recall the contrast between the set-theoretic understanding of the real numbers and Brouwer's approach, which rejected the idea that the continuum is "built of" points. That this is a labyrinthine question was further established by results on Cantor's continuum hypothesis (CH), which postulates that the cardinality of the set of real numbers is $\aleph_1$, the second transfinite cardinal, or equivalently that every infinite subset of $\mathbb{R}$ must biject with either $\mathbb{N}$ or with $\mathbb{R}$ itself. Gödel proved in 1939 that CH is consistent with axiomatic set theory, but Paul Cohen proved in 1963 that it is independent of its axioms (i.e., Cohen proved that the negation of CH is consistent with axiomatic set theory [IV.22 §5]). The problem is still alive, with a few mathematicians proposing alternative approaches to the continuum and others trying to find new and convincing set-theoretic principles that will settle Cantor's question (see Woodin 2001).

The foundational debate has also contributed in a definitive way to clarifying the peculiar style and methodology of modern mathematics, especially the so-called Platonism or existential character of modern mathematics (see the classic 1935 paper of Bernays in Benacerraf and Putnam (1983)), by which is meant (here at least) a methodological trait rather than any supposed implications of metaphysical existence. Modern mathematics investigates structures by considering their elements as given independently of human (or mechanical) capabilities of effective definition and construction. This may seem surprising, but perhaps this trait can be explained by broader characteristics of scientific thought and the role played by mathematical structures in the modeling of scientific phenomena.

In the end, the debate made it clear that mathematics and its modern methods are still surrounded by important philosophical problems. When a sizable amount of mathematical knowledge can be taken for granted, theorems can be established and problems can be solved with the certainty and clarity for which mathematics is celebrated. But when it comes to laying out the bare beginnings, philosophical issues cannot be avoided. The reader of these pages may have felt this at several places, especially in the discussion of intuitionism, but also in the basic ideas behind Hilbert's program, and of course in the problem of the relationship between formal mathematics and its informal counterpart, a problem that is brought into sharp focus by Gödel's theorems.

Acknowledgments. I thank Mark van Atten, Jeremy Gray, Paolo Mancosu, José F. Ruiz, Wilfried Sieg, and the editors for their helpful comments on a previous version of this paper.

Further Reading

It is impossible to list here all the relevant articles by Bernays, Brouwer, Cantor, Dedekind, Gödel, Hilbert, Kronecker, von Neumann, Poincaré, Russell, Weyl, Zermelo, etc. The reader can easily find them in the source books by van Heijenoort (1967), Benacerraf and Putnam (1983), Heinzmann (1986), Ewald (1996), and Mancosu (1998).

Benacerraf, P., and H. Putnam, eds. 1983. *Philosophy of Mathematics: Selected Readings*. Cambridge: Cambridge University Press.

Dawson Jr., J. W. 1997. *Logical Dilemmas: The Life and Work of Kurt Gödel*. Wellesley, MA: A. K. Peters.

Ewald, W., ed. 1996. *From Kant to Hilbert: A Source Book in the Foundations of Mathematics*, 2 vols. Oxford: Oxford University Press.

Feferman, S. 1998. *In the Light of Logic*. Oxford: Oxford University Press.

Ferreirós, J. 1999. *Labyrinth of Thought: A History of Set Theory and Its Role in Modern Mathematics*. Basel: Birkhäuser.

Heinzmann, G., ed. 1986. *Poincaré, Russell, Zermelo et Peano*. Paris: Vrin.

Hesseling, D. E. 2003. *Gnomes in the Fog: The Reception of Brouwer's Intuitionism in the 1920s*. Basel: Birkhäuser.

Heyting, A. 1956. *Intuitionism: An Introduction*. Amsterdam: North-Holland. Third revised edition, 1971.

Hilbert, D., and P. Bernays. 1934/39. *Grundlagen der Mathematik*, 2 vols. Berlin: Springer.

Mancosu, P., ed. 1998. *From Hilbert to Brouwer: The Debate on the Foundations of Mathematics in the 1920s*. Oxford: Oxford University Press.

———. 1999. Between Vienna and Berlin: the immediate reception of Gödel's incompleteness theorems. *History and Philosophy of Logic* 20:33–45.

Mehrtens, H. 1990. *Moderne—Sprache—Mathematik*. Frankfurt: Suhrkamp.

Moore, G. H. 1982. *Zermelo's Axiom of Choice*. New York: Springer.

———. 1998. Logic, early twentieth century. In *Routledge Encyclopedia of Philosophy*, edited by E. Craig. London: Routledge.

Rowe, D. 1992. *Natur und mathematisches Erkennen*. Basel: Birkhäuser.

Sieg, W. 1999. Hilbert's programs: 1917–1922. *The Bulletin of Symbolic Logic* 5:1–44.

Smullyan, R. 2001. *Gödel's Incompleteness Theorems*. Oxford: Oxford University Press.

Tait, W. W. 1981. Finitism. *Journal of Philosophy* 78:524–46.

van Atten, M. 2003. *On Brouwer*. Belmont, CA: Wadsworth.

van Dalen, D. 1999/2005. *Mystic, Geometer, and Intuitionist: The Life of L. E. J. Brouwer*. Volume I: *The Dawning Revolution*. Volume II: *Hope and Disillusion*. Oxford: Oxford University Press.

van Heijenoort, J., ed. 1967. *From Frege to Gödel: A Source Book in Mathematical Logic*. Cambridge, MA: Harvard University Press. (Reprinted, 2002.)

Weyl, H. 1918. *Das Kontinuum*. Leipzig: Veit.

Whitehead, N. R., and B. Russell. 1910/13. *Principia Mathematica*. Cambridge: Cambridge University Press. Second edition 1925/27. (Reprinted, 1978.)

Woodin, W. H. 2001. The continuum hypothesis, I, II. *Notices of the American Mathematical Society* 48:567–76, 681–90.

Part III
Mathematical Concepts

III.1 The Axiom of Choice

Consider the following problem: it is easy to find two irrational numbers a and b such that $a + b$ is rational, or such that ab is rational (in both cases one could take $a = \sqrt{2}$ and $b = -\sqrt{2}$), but is it possible for a^b to be rational? Here is an elegant proof that the answer is yes. Let $x = \sqrt{2}^{\sqrt{2}}$. If x is rational then we have our example. But $x^{\sqrt{2}} = \sqrt{2}^2 = 2$ is rational, so if x is irrational then again we have an example.

Now this argument certainly establishes that it is possible for a and b to be irrational and for a^b to be rational. However, the proof has a very interesting feature: it is *nonconstructive*, in the sense that it does not actually name two irrationals a and b that work. Instead, it tells us that either we can take $a = b = \sqrt{2}$ or we can take $a = \sqrt{2}^{\sqrt{2}}$ and $b = \sqrt{2}$. Not only does it not tell us which of these alternatives will work, it gives us absolutely no clue about how to find out.

Arguments of this kind have troubled some philosophers and philosophically inclined mathematicians, but as far as mainstream mathematics goes they are a fully accepted and important style of reasoning. Formally, we have appealed to the "law of the excluded middle." We have shown that the negation of the statement cannot be true, and deduced that the statement itself must be true. A typical reaction to the proof above is not that it is in any sense invalid, but merely that its nonconstructive nature is rather surprising.

Nevertheless, faced with a nonconstructive proof, it is very natural to ask whether there is a constructive proof. After all, an actual construction may give us more insight into the statement, which is an important point since we prove things not only to be sure they are true but also to gain an idea of *why* they are true. Of course, to ask whether there is a constructive proof is not to suggest that the nonconstructive proof is invalid, but just that it may be more informative to have a constructive proof.

The *axiom of choice* is one of several rules that we use for building sets out of other sets. Typical examples of such rules are the statement that for any set A we can form the set of all its subsets, and the statement that for any set A and any property p we can form the set of all elements of A that satisfy p (these are usually called the *power-set axiom* and the *axiom of comprehension*, respectively). Roughly speaking, the axiom of choice says that we are allowed to make an arbitrary number of unspecified choices when we wish to form a set.

Like the other axioms, the axiom of choice can seem so natural that one may not even notice that one is using it, and indeed it was applied by many mathematicians before it was first formalized. To get an idea of what it means, let us look at the well-known proof that the union of a countable family of COUNTABLE SETS [III.11] is countable. The fact that the family is countable allows us to write out the sets in a list $A_1, A_2, A_3, \ldots$, and then the fact that each individual set A_n is countable allows us to write its elements in a list $a_{n1}, a_{n2}, a_{n3}, \ldots$. We then finish the proof by finding some systematic way of counting through the elements a_{nm}.

Now in that proof we actually made an infinite number of unspecified choices. We were told that each A_n was countable and then for each A_n we "chose" a listing of the elements of A_n *without specifying the choice we had made.* Moreover, since we are told absolutely nothing about the sets A_n, it is clearly impossible to say how we choose to list them. This remark does not invalidate the proof, but it does show that it is nonconstructive. (Note, however, that if we are actually told what the sets A_n are, then we may well be able to specify listings of their elements and thereby give a constructive proof that the union of those particular sets is countable.)

Here is another example. A GRAPH [III.34] is *bipartite* if its vertices can be split into two classes X and Y in such a way that no two vertices in the same class are connected by an edge. For example, any even cycle (an even number of points arranged in a circle, with consecutive points joined) is bipartite, while no odd cycle is. Now, is an infinite disjoint union of even cycles bipartite? Of course it is: we just split each of the individual cycles C into two classes X_C and Y_C and then let X be the union of the sets X_C and Y be the union of the sets Y_C. But how do we choose for each cycle C which set to call X_C and which to call Y_C? Again, we cannot actually specify how we do this, so we are using the axiom of choice (even if we do not explicitly say so).

In general, the axiom of choice states that if we are given a family of nonempty sets X_i, then we may select an element x_i from each one. More precisely, it states that if the X_i are nonempty sets, where i ranges over some index set I, then there is a function f defined on I such that $f(i) \in X_i$ for all i. Such a function f is called a *choice function* for the family.

For one set we do not need any separate rule to do this: indeed, the statement that a set X_1 is nonempty is exactly the statement that there exists $x_1 \in X_1$. (More formally, we might say that the function f that takes 1 to x_1 is a choice function for the "family" that consists of the single set X_1.) For two sets, and indeed for any finite collection of sets, one can prove the existence of a choice function by induction on the number of sets. But for infinitely many sets it turns out that one cannot deduce the existence of a choice function from the other rules for building sets.

Why do people make a fuss about the axiom of choice? The main reason is that if it is used in a proof, then that part of the proof is automatically nonconstructive. This is reflected in the very statement of the axiom. For the other rules that we use, such as "one may take the union of two sets," the set whose existence is being asserted is uniquely defined by its properties (u is an element of $X \cup Y$ if and only if it is an element of X or of Y or of both). But this is not the case with the axiom of choice: the object whose existence is asserted (a choice function) is not uniquely specified by its properties, and there will typically be many choice functions.

For this reason, the general view in mainstream mathematics is that, although there is nothing wrong with using the axiom of choice, it is a good idea to signal that one has used it, to draw attention to the fact that one's proof is not constructive.

An example of a statement whose proof involves the axiom of choice is THE BANACH–TARSKI PARADOX [V.3]. This says that there is a way of dividing up a solid unit sphere into a finite number of subsets and then reassembling these subsets (using rotations, reflections, and translations) to form two solid unit spheres. The proof does not provide an explicit way of defining the subsets.

It is sometimes claimed that the axiom of choice has "undesirable" or "highly counterintuitive" consequences, but in almost all cases a little thought reveals that the consequence under consideration is actually not counterintuitive at all. For example, consider the Banach–Tarski paradox above. Why does it seem strange and paradoxical? It is because we feel that volume has not been preserved. And indeed, this feeling can be converted into a rigorous argument that the subsets used in the decomposition cannot all be sets to which one can meaningfully assign a volume. But that is not a paradox at all: we can say what we mean by the volume of a nice set such as a polyhedron, but there is no reason to suppose that we can give a sensible definition of volume for *all* subsets of the sphere. (The subject called measure theory can be used to give a volume to a very wide class of sets, called the MEASURABLE SETS [III.55], but there is no reason at all to believe that all sets should be measurable, and indeed it can be shown, again by a use of the axiom of choice, that there are sets that are not measurable.)

There are two forms of the axiom of choice that are more often used in daily mathematical life than the basic form we have been discussing. One is the *well-ordering principle*, which states that every set can be WELL-ORDERED [III.66]. The other is *Zorn's lemma*, which states that under certain circumstances "maximal" elements exist. For example, a basis for a vector space is precisely a maximal linearly independent set, and it turns out that Zorn's lemma applies to the collection of linearly independent sets in a vector space, which shows that every vector space has a basis.

These two statements are called forms of the axiom of choice because they are equivalent to it, in the sense that each one both implies the axiom of choice and may be deduced from it, in the presence of the other rules for building sets. A good way of seeing why these two other forms of the axiom have a nonconstructive feel to them is to spend a few minutes trying to find a well-ordering of the reals, or a basis for the vector space of all sequences of real numbers.

For more about the axiom of choice, and especially about its relationship to the other axioms of formal set theory, see SET THEORY [IV.22].

III.2 The Axiom of Determinacy

Consider the following "infinite game." Two players, A and B, take turns to name natural numbers, with A going first, say. By doing this, they generate an infinite sequence. A wins the game if this sequence is "eventually periodic," and B wins if it is not. (An eventually periodic sequence is one like $1, 56, 4, 5, 8, 3, 5, 8, 3, 5, 8, 3, 5, 8, 3, \ldots$: that is, one that settles down after a while to a recurring pattern.) It is not hard to see that B has a winning strategy for this game, since eventually periodic sequences are rather special. However, at any stage of the game it is always possible that A will win (if B plays sufficiently badly), since every finite sequence is the beginning of many eventually periodic sequences.

More generally, any collection S of infinite sequences of natural numbers gives rise to an infinite game: A's object is now to ensure that the sequence produced is one of the sequences in S, and B's object is to ensure the reverse. The resulting game is called *determined* if one of the two players has a winning strategy. As we have seen, the game is certainly determined when S is the set of eventually periodic sequences, and indeed for just about any set S that one writes down it is easy to see that the corresponding game is determined. Nevertheless, it turns out that there are games that are not determined. (It is an instructive exercise to see where the plausible-seeming argument, "If A does not have a winning strategy, then A cannot force a win, so B must have a winning strategy," breaks down.)

It is not too hard to construct nondetermined games, but the constructions use THE AXIOM OF CHOICE [III.1]: roughly speaking, one can well-order all possible strategies so that each one has fewer predecessors than there are infinite sequences, and select sequences to belong to S or its complement in a way that stops each strategy in turn from being a winning strategy for either player.

The *axiom of determinacy* states that all games are determined. It contradicts the axiom of choice, but it is a rather interesting axiom when it is added to THE ZERMELO–FRAENKEL AXIOMS [III.99] *without* choice. It turns out, for example, to imply that many sets of reals have surprisingly good properties, such as being Lebesgue measurable. Variants of the axiom of determinacy are closely connected with the theory of large cardinals. For more details, see SET THEORY [IV.22].

Banach Spaces

See NORMED SPACES AND BANACH SPACES [III.62]

III.3 Bayesian Analysis

Suppose you throw a pair of standard dice. The probability that the total is 10 is $\frac{1}{12}$ because there are thirty-six ways the dice can come up, of which three (4 and 6, 5 and 5, and 6 and 4) give 10. If, however, you look at the first die and see that it came up as a 6, then the *conditional probability* that the total is 10, given this information, is $\frac{1}{6}$ (since that is the probability that the other die comes up as a 4).

In general, the *probability of A given B* is defined to be the probability of A and B divided by the probability of B. In symbols, one writes

$$\mathbb{P}[A|B] = \frac{\mathbb{P}[A \wedge B]}{\mathbb{P}[B]}.$$

From this it follows that $\mathbb{P}[A \wedge B] = \mathbb{P}[A|B]\,\mathbb{P}[B]$. Now $\mathbb{P}[A \wedge B]$ is the same as $\mathbb{P}[B \wedge A]$. Therefore,

$$\mathbb{P}[A|B]\,\mathbb{P}[B] = \mathbb{P}[B|A]\,\mathbb{P}[A],$$

since the left-hand side is $\mathbb{P}[A \wedge B]$ and the right-hand side is $\mathbb{P}[B \wedge A]$. Dividing through by $\mathbb{P}[B]$ we obtain *Bayes's theorem*:

$$\mathbb{P}[A|B] = \frac{\mathbb{P}[B|A]\,\mathbb{P}[A]}{\mathbb{P}[B]},$$

which expresses the conditional probability of A given B in terms of the conditional probability of B given A.

A fundamental problem in statistics is to analyze random data given by an unknown PROBABILITY DISTRIBUTION [III.71]. Here, Bayes's theorem can make a significant contribution. For example, suppose you are told that some unbiased coins have been tossed and that three of them have come up heads. Suppose that you are told that the number of coins tossed is between 1 and 10, and that you wish to guess this number. Let H_3 stand for the event that three coins came up heads and let C be the number of coins. Then for each n between 1 and 10 it is not hard to calculate the conditional probability $\mathbb{P}[H_3|C = n]$, but we would like to know the reverse, namely $\mathbb{P}[C = n|H_3]$. Bayes's theorem tells us that it is

$$\mathbb{P}[H_3|C = n]\frac{\mathbb{P}[C = n]}{\mathbb{P}[H_3]}.$$

This would tell us the ratios between the various conditional probabilities $\mathbb{P}[C = n|H_3]$ if we knew what the probabilities $\mathbb{P}[C = n]$ were. Typically, one does *not* know this, but one makes some kind of guess, called a *prior distribution*. For example, one might guess, before knowing that three coins had come up heads, that for each n between 1 and 10 the probability that n coins had been chosen was $\frac{1}{10}$. *After* this information, one would use the calculation above to revise one's assessment and obtain a *posterior distribution*, in which the probability that $C = n$ would be proportional to $\frac{1}{10}\mathbb{P}[H_3 | C = n]$.

There is more to Bayesian analysis than simply applying Bayes's theorem to replace prior distributions by posterior distributions. In particular, as in the example just given, there is not always an obvious prior distribution to take, and it is a subtle and interesting mathematical problem to devise methods for choosing prior distributions that are "optimal" in different ways. For further discussion, see MATHEMATICS AND MEDICAL STATISTICS [VII.11] and MATHEMATICAL STATISTICS [VII.10].

III.4   Braid Groups
F. E. A. Johnson

Take two parallel planes, each punctured at n points. Label the holes 1 to n in each plane, and run a string from each hole in the first plane to one in the second, in such a way that no two strings go to the same hole. The result is an *n-braid*. Two different 3-braids, shown in two-dimensional projection in a similar manner to KNOT DIAGRAMS [III.44], are given in figure 1.

As the diagrams suggest, we insist that the strings go from left to right without "doubling back"; so, for example, a knotted string is not allowed.

A certain freedom is allowed when we describe a braid: provided that the string ends remain fixed and the strings do not break or pass through each other, one can stretch, contract, bend, and otherwise move the strings about in three dimensions and end up with the "same" braid. This notion of "sameness" is an EQUIVALENCE RELATION [I.2 §2.3] called *braid isotopy*.

Braids may be composed as follows: arrange a pair of braids end to end to abut in a common (middle) plane, join up the strings, and remove the middle plane. For the braids X and Y in figure 1, the composition XY is given in figure 2.

With this notion of composition, n-braids form a group B_n. In our example, $Y = X^{-1}$, since "pulling all the strings tight" shows that XY is isotopic to the *trivial* braid (figure 3), which acts as the identity.

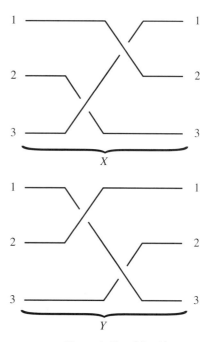

Figure 1 Two 3-braids.

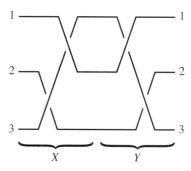

Figure 2 Braid composition.

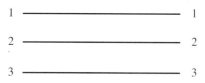

Figure 3 The trivial braid.

As a group, B_n is generated by elements $(\sigma_i)_{1 \leqslant i \leqslant n-1}$, where σ_i is formed from the trivial braid by crossing the ith string over the $(i + 1)$st as in figure 4. The reader may perceive a similarity between the σ_i and the adjacent transpositions that generate the group S_n of

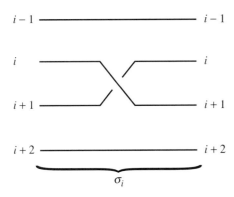

Figure 4 The generator σ_i.

PERMUTATIONS [III.68] of $\{1,\ldots,n\}$. Indeed, any braid determines a permutation by the rule

$i \mapsto$ right-hand label of ith string.

Ignoring everything except the behavior at the ends gives a surjective homomorphism $B_n \to S_n$, which maps σ_i to the transposition $(i, i+1)$. This is *not* an isomorphism, however, as B_n is infinite. In fact, σ_i has infinite order, whereas the transposition $(i, i+1)$ squares to the identity. In his celebrated 1925 paper "Theorie der Zöpfe," ARTIN [VI.86] showed that multiplication in B_n is completely described by the relations

$$\sigma_i \sigma_j = \sigma_j \sigma_i \quad (|i - j| \geqslant 2),$$
$$\sigma_i \sigma_{i+1} \sigma_i = \sigma_{i+1} \sigma_i \sigma_{i+1}.$$

These relations have subsequently acquired importance in statistical physics, where they are known as the Yang–Baxter equations.

In groups defined by generators and relations it is usually difficult (there being no method that works uniformly in all cases) to decide whether an arbitrary word in the generators represents the identity element (see GEOMETRIC AND COMBINATORIAL GROUP THEORY [IV.10]). For B_n, Artin solved this problem geometrically, by "combing the braid." An alternative algebraic method, due to Garside (1967), also decides when two elements in B_n are conjugate.

In relation to the decidability of such questions, and in many other respects, braid groups display close affinities with *linear groups*: that is, groups in which all elements behave as if they were invertible $N \times N$ matrices. Although such similarities suggested that it should be possible to prove that braid groups genuinely are linear, the problem of doing so remained unsolved for many years, until in 2001 a proof was eventually found by Bigelow and independently by Krammer.

The groups described here are, strictly speaking, braid groups of the plane, the plane being the object punctured. Other braid groups also occur, often in surprising contexts. The connection with statistical physics has already been mentioned. They also arise in algebraic geometry, when algebraic curves become punctured by discarding exceptional points. Thus, though originating in topology, braids may intervene significantly in areas such as "constructive Galois theory" that seem at first sight to be purely algebraic.

III.5 Buildings
Mark Ronan

The invertible linear transformations on a vector space form a group, called the *general linear group*. If n is the dimension of the vector space and K is the field of scalars, then it is denoted by $\mathrm{GL}_n(K)$, and if we pick a basis for the vector space, then each group element can be written as an $n \times n$ matrix whose DETERMINANT [III.15] is nonzero. This group and its subgroups are of great interest in mathematics, and can be studied "geometrically" in the following way. Instead of looking at the vector space V, where of course the origin plays a unique role and is fixed by the group, we use the PROJECTIVE SPACE [I.3 §6.7] associated with V: the points of projective space are the one-dimensional subspaces of V, the lines are the two-dimensional subspaces, the planes are the three-dimensional subspaces, and so on.

Several important subgroups of $\mathrm{GL}_n(K)$ can be obtained by imposing constraints on the linear maps (or matrices). For example, $\mathrm{SL}_n(K)$ consists of all linear transformations of determinant 1. The group $\mathrm{O}(n)$ consists of all linear transformations α of an n-dimensional real inner-product space such that $\langle \alpha v, \alpha w \rangle = \langle v, w \rangle$ for any two vectors v and w (or in matrix terms all real matrices A such that $AA^{\mathrm{T}} = I$); more generally, one can define many similar subgroups of $\mathrm{GL}_n(K)$ by taking all linear maps that preserve certain forms, such as bilinear or sesquilinear forms. These subgroups are called *classical groups*. The classical groups are either simple or close to simple (for example, we can often make them simple by quotienting out by the subgroup of scalar matrices). When K is the field of real or complex numbers, the classical groups are *Lie groups*.

Lie groups and their classification are discussed in LIE THEORY [III.48]: the simple Lie groups comprise the classical groups, which fall into one of four families, known as A_n, B_n, C_n, and D_n (where n is a natural number), along with other types known as E_6, E_7, E_8,

F_4, and G_2. The subscripts are related to the dimensions of the groups. For example, the groups of type A_n are the groups of invertible linear transformations in $n + 1$ dimensions.

These simple Lie groups have analogues over any field, where they are often referred to as *groups of Lie type*. For example, K can be a finite field, in which case the groups are finite. It turns out that almost all finite simple groups are of Lie type: see THE CLASSIFICATION OF FINITE SIMPLE GROUPS [V.7]. A geometric theory underlying the classical groups had been developed by the first half of the twentieth century. It used projective space and various subgeometries of projective space, which made it possible to provide analogues for the classical groups, but it did not provide analogues for the groups of types E_6, E_7, E_8, F_4, and G_2. For this reason, Jacques Tits looked for a geometric theory that would embrace all families, and ended up creating the theory of *buildings*.

The full abstract definition of a building is somewhat complicated, so instead we shall try to give some idea of the concept by looking at the building associated with the groups $GL_n(K)$ and $SL_n(K)$, which are of type A_{n-1}. This building is an *abstract simplicial complex*, which can be thought of as a higher-dimensional analogue of a GRAPH [III.34]. It consists of a collection of points called *vertices*; as in a graph, some pairs of vertices form *edges*; however, it is then possible for triples of vertices to form two-dimensional *faces*, and for sets of k vertices to form $(k-1)$-dimensional "simplexes." (The geometrical meaning of the word "simplex" is a convex hull of a finite set of points in general position: for instance, a three-dimensional simplex is a tetrahedron.) All faces of simplexes must also be included, so for example three vertices cannot form a two-dimensional face unless each pair is joined by an edge.

To form the building of type A_{n-1}, we start by taking all the 1-spaces, 2-spaces, 3-spaces, and so on (corresponding to points, lines, planes, and so on, in projective space), and treat them as "vertices." The simplexes are formed by all nested sequences of proper subspaces: for example, a 2-space inside a 4-space inside a 5-space will form a "triangle" whose vertices are these three subspaces. The simplexes of maximal dimension have $n - 1$ vertices: a 1-space inside a 2-space inside a 3-space, and so on. These simplexes are called *chambers*.

There are many subspaces, so a building is a huge object. However, buildings have important subgeometries called *apartments*, which in the A_{n-1} case are

obtained by taking a basis for the vector space, and then taking all subspaces generated by subsets of this basis. For example, in the A_3 case our vector space is four dimensional, so a basis has four elements; its subsets span four 1-spaces, six 2-spaces, and four 3-spaces. To visualize this apartment it helps to view the four 1-spaces as the vertices of a tetrahedron, the six 2-spaces as the midpoints of its edges, and the four 3-spaces as the midpoints of its faces. The apartment has twenty-four chambers, six for each face of the original tetrahedron, and they form a triangular tiling of the surface of the tetrahedron. This surface is topologically equivalent to a sphere, as are all apartments of this building: such buildings are called *spherical*. The buildings for the groups of Lie type are all spherical, and, just as A_3 is related to the tetrahedron, their apartments are related to the regular and semiregular polyhedra in n dimensions, where n is the subscript in the Lie notation given earlier.

Buildings have the following two noteworthy features. First, any two chambers lie in a common apartment: this is not obvious in the example above but it can be proved using linear algebra. Second, in any building all apartments are isomorphic and any two apartments intersect nicely: more precisely, if A and A' are apartments, then $A \cap A'$ is convex and there is an isomorphism from A to A' that fixes $A \cap A'$. These two features were originally used by Tits in defining buildings.

The theory of spherical buildings does not just give a pleasing geometric basis for the groups of Lie type: it can also be used to construct the ones of types E_6, E_7, E_8, and F_4, for an arbitrary field K, without the need for sophisticated machinery such as Lie algebras. Once the building has been constructed (and a construction can be given in a surprisingly simple manner), a theorem of Tits on the existence of automorphisms shows that the groups themselves must exist.

In a spherical building the apartments are tilings of a sphere, but other types of buildings also play significant roles. Of particular importance are *affine buildings*, in which the apartments are tilings of Euclidean space; such buildings arise in a natural way from groups, such as $GL_n(K)$, where K is a p-ADIC FIELD [III.51]. For such fields there are two buildings, one spherical and one affine, but the affine one carries more information and yields the spherical building as a structure "at infinity." Going beyond affine buildings, there are hyperbolic buildings, whose apartments are tilings of hyperbolic space; they arise naturally in the study of hyperbolic Kac–Moody groups.

III.6   Calabi–Yau Manifolds

Eric Zaslow

1   Basic Definition

Calabi–Yau manifolds, named after Eugenio Calabi and Shing-Tung Yau, arise in Riemannian geometry and algebraic geometry, and play a prominent role in string theory and mirror symmetry.

In order to explain what they are, we need first to recall the notion of orientability on a real MANI-FOLD [I.3 §6.9]. Such a manifold is *orientable* if you can choose coordinate systems at each point in such a way that any two systems $x = (x^1, \ldots, x^m)$ and $y = (y^1, \ldots, y^m)$ that are defined on overlapping sets give rise to a positive Jacobian: $\det(\partial y^i / \partial x^j) > 0$. The notion of a Calabi–Yau manifold is the natural complex analogue of this. Now the manifold is complex, and for each local coordinate system $z = (z^1, \ldots, z^n)$ one has a HOLOMORPHIC FUNCTION [I.3 §5.6] $f(z)$. It is vital that f should be *nonvanishing*: that is, it never takes the value 0. There is also a compatibility condition: if $\tilde{z}(z)$ is another coordinate system, then the corresponding function $\tilde{f}$ is related to f by the equation $f = \tilde{f} \det(\partial \tilde{z}^a / \partial z^b)$. Note that if we replace all complex terms by real terms in this definition, then we have the notion of a real orientation. So a Calabi–Yau manifold can be thought of informally as a complex manifold with complex orientation.

2   Complex Manifolds and Hermitian Structure

Before we go any further, a few words about complex and Kähler geometry are in order. A complex manifold is a structure that looks locally like $\mathbb{C}^n$, in the sense that one can find complex coordinates $z = (z^1, \ldots, z^n)$ near every point. Moreover, where two coordinate systems z and $\tilde{z}$ overlap, the coordinates $\tilde{z}^a$ are holomorphic when they are regarded as functions of the z^b. Thus, the notion of a holomorphic function on a complex manifold makes sense and does not depend on the coordinates used to express the function. In this way, the local geometry of a complex manifold does indeed look like an open set in $\mathbb{C}^n$, and the tangent space at a point looks like $\mathbb{C}^n$ itself.

On complex vector spaces it is natural to consider Hermitian INNER PRODUCTS [III.37] represented by HER-MITIAN MATRICES [III.50 §3] $g_{a\bar{b}}$ with respect to a basis e_a. On complex manifolds, a Hermitian inner product on the tangent spaces is called a "Hermitian metric,"

and is represented in a coordinate basis by a Hermitian matrix $g_{a\bar{b}}$, which depends on position.[1]

3   Holonomy, and Calabi–Yau Manifolds in Riemannian Geometry

On a RIEMANNIAN MANIFOLD [I.3 §6.10] one can move a vector along a path so as to keep it of constant length and "always pointing in the same direction." *Curvature* expresses the fact that the vector you wind up with at the end of the path depends on the path itself. When your path is a closed loop, the vector at the starting point comes back to a new vector at the same point. (A good example to think about is a path on a sphere that goes from the North Pole to the equator, then a quarter of the way around the equator, then back to the North Pole again. When the journey is completed, the "constant" vector that began by pointing south will have been rotated by 90°.) With each loop we associate a matrix operator called the *holonomy matrix*, which sends the starting vector to the ending vector; the group generated by all of these matrices is called the *holonomy group* of the manifold. Since the length of the vector does not change during the process of keeping it constant along the loop, the holonomy matrices all lie in the orthogonal group of length-preserving matrices, $O(m)$. If the manifold is oriented, then the holonomy group must lie in $SO(m)$, as one can see by transporting an oriented basis of vectors around the loop.

Every complex manifold of (complex) dimension n is also a real manifold of (real) dimension $m = 2n$, which one can think of as coordinatized by the real and imaginary parts of the complex coordinates z^j. Real manifolds that arise in this way have additional structure. For example, the fact that we can multiply complex coordinate directions by $i = \sqrt{-1}$ implies that there must be an operator on the real tangent space that squares to -1. This operator has eigenvalues $\pm i$, which can be thought of as "holomorphic" and "anti-holomorphic" directions. The Hermitian property states that these directions are orthogonal, and we say that the manifold is *Kähler* if they remain so after transport around loops. This means that the holonomy group is a subgroup of $U(n)$ (which itself is a subgroup of $SO(m)$: complex manifolds always have *real* orientations). There is a nice local characterization of the Kähler property: if $g_{a\bar{b}}$ are the components of

---

1. The notation $g_{a\bar{b}}$ indicates the conjugate-linear property of a Hermitian inner product.

the Hermitian metric in some coordinate patch, then there exists a function φ on that patch such that $g_{a\bar{b}} = \partial^2\varphi/\partial z^a\partial\bar{z}^b$.

Given a complex orientation—that is, the nonmetric definition of a Calabi–Yau manifold given above—a *compatible Kähler structure* leads to a holonomy that lies in $\mathrm{SU}(n) \subset \mathrm{U}(n)$, the natural analogue of the case of real orientation. This is the metric definition of a Calabi–Yau manifold.

4 The Calabi Conjecture

Calabi conjectured that, for any Kähler manifold of complex dimension n and any complex orientation, there exists a function u and a new Kähler metric $\tilde{g}$, given in coordinates by

$$\tilde{g}_{a\bar{b}} = g_{a\bar{b}} + \frac{\partial^2 u}{\partial z^a\partial\bar{z}^b},$$

that is compatible with the orientation. In equations, the compatibility condition states that

$$\det\left(g_{a\bar{b}} + \frac{\partial^2 u}{\partial z^a\partial\bar{z}^b}\right) = |f|^2,$$

where f is the holomorphic orientation function discussed above. Thus, the metric notion of a Calabi–Yau manifold amounts to a formidable nonlinear partial differential equation for u. Calabi proved the uniqueness and Yau proved the existence of a solution to this equation. So in fact the metric definition of a Calabi–Yau manifold is uniquely determined by its Kähler structure and its complex orientation.

Yau's theorem establishes that the space of metrics with holonomy group $\mathrm{SU}(n)$ on a manifold with complex orientation is in correspondence with the space of inequivalent Kähler structures. The latter space can easily be probed with the techniques of algebraic geometry.

5 Calabi–Yau Manifolds in Physics

Einstein's theory of gravity, general relativity, constructs equations that the metric of a Riemannian space-time manifold must obey (see GENERAL RELATIVITY AND THE EINSTEIN EQUATIONS [IV.13]). The equations involve three symmetric tensors: the metric, the RICCI CURVATURE [III.78] tensor, and the energy-momentum tensor of matter. A Riemannian manifold whose Ricci tensor vanishes is a solution to these equations when there is no matter, and is a special case of an *Einstein manifold*. A Calabi–Yau manifold with

its unique $\mathrm{SU}(n)$-holonomy metric has vanishing Ricci tensor, and is therefore of interest in general relativity.

A fundamental problem in theoretical physics is the incorporation of Einstein's theory into the quantum theory of particles. This enterprise is known as *quantum gravity*, and Calabi–Yau manifolds figure prominently in the leading theory of quantum gravity, STRING THEORY [IV.17 §2].

In string theory, the fundamental objects are one-dimensional "strings." The motion of the strings in space-time is described by two-dimensional trajectories, known as *worldsheets*, so every point on the worldsheet is labeled by the point in space-time where it sits. In this way, string theory is constructed from a quantum field theory of maps from two-dimensional RIEMANN SURFACES [III.79] to a space-time manifold M. The two-dimensional surface should be given a Riemannian metric, and there is an infinite-dimensional space of such metrics to consider. This means that we must solve quantum gravity in two dimensions— a problem that, like its four-dimensional cousin, is too hard. If, however, it happens that the two-dimensional worldsheet theory is conformal (invariant under local changes of scale), then just a finite-dimensional space of conformally inequivalent metrics remains, and the theory is well-defined.

The Calabi–Yau condition arises from these considerations. The requirement that the two-dimensional theory should be conformal, so that the string theory makes good sense, is in essence the requirement that the Ricci tensor of space-time should vanish. Thus, a two-dimensional condition leads to a space-time equation, which turns out to be exactly Einstein's equation without matter. We add to this condition the "phenomenological" criterion that the theory be endowed with "supersymmetry," which requires the space-time manifold M to be complex. The two conditions together mean that M is a complex manifold with holonomy group $\mathrm{SU}(n)$: that is, a Calabi–Yau manifold. By Yau's theorem, the choices of such M can easily be described by algebraic geometric methods.

We remark that there is a kind of distillation of string theory called "topological strings," which can be given a rigorous mathematical framework. Calabi–Yau manifolds are both symplectic and complex, and this leads to two versions of topological strings, called A and B, that one can associate with a Calabi–Yau manifold. Mirror symmetry is the remarkable phenomenon that the A version of one Calabi–Yau manifold is related to the B version of an entirely different "mirror partner." The mathematical consequences of such an equivalence are extremely rich. (See MIRROR SYMMETRY [IV.16] for more details. For other notions related to those discussed in this article, see SYMPLECTIC MANIFOLDS [III.88].)

The Calculus of Variations
See VARIATIONAL METHODS [III.94]

III.7   Cardinals

The cardinality of a set is a measure of how large that set is. More precisely, two sets are said to have the same cardinality if there is a bijection between them. So what do cardinalities look like?

There are finite cardinalities, meaning the cardinalities of finite sets: a set has "cardinality n" if it has precisely n elements. Then there are COUNTABLE [III.11] infinite sets: these all have the same cardinality (this follows from the definition of "countable"), usually written $\aleph_0$. For example, the natural numbers, the integers, and the rationals all have cardinality $\aleph_0$. However, the reals are uncountable, and so do not have cardinality $\aleph_0$. In fact, their cardinality is denoted by $2^{\aleph_0}$.

It turns out that cardinals can be added and multiplied and even raised to powers of other cardinals (so "$2^{\aleph_0}$" is not an isolated piece of notation). For details, and more explanation, see SET THEORY [IV.22 §2].

III.8   Categories
Eugenia Cheng

When we study GROUPS [I.3 §2.1] or VECTOR SPACES [I.3 §2.3], we pay particular attention to certain classes of maps between them: the important maps between groups are the group HOMOMORPHISMS [I.3 §4.1], and the important maps between vector spaces are the LINEAR MAPS [I.3 §4.2]. What makes these maps important is that they are the functions that "preserve structure": for example, if ϕ is a homomorphism from a group G to a group H, then it "preserves multiplication," in the sense that $\phi(g_1 g_2) = \phi(g_1)\phi(g_2)$ for any pair of elements g_1 and g_2 of G. Similarly, linear maps preserve addition and scalar multiplication.

The notion of a structure-preserving map applies far more generally than just to these two examples, and one of the purposes of category theory is to understand the general properties of such maps. For instance, if A, B, and C are mathematical structures of some given type, and f and g are structure-preserving maps from A to B and from B to C, respectively, then their composite $g \circ f$ is a structure-preserving map from A to C. That is, structure-preserving maps can be *composed* (at least if the range of one equals the domain of the other). We also use structure-preserving maps to decide when to regard two structures as "essentially the same": we call A and B *isomorphic* if there is a structure-preserving map from A to B with an inverse that also preserves structure.

A *category* is a mathematical structure that allows one to discuss properties such as these in the abstract. It consists of a collection of *objects*, together with *morphisms* between those objects. That is, if a and b are two objects in the category, then there is a collection of morphisms between a and b. There is also a notion of *composition* of morphisms: if f is a morphism from a to b and g is a morphism from b to c, then there is a *composite* of f and g, which is a morphism from a to c. This composition must be associative. In addition, for each object a there is an "identity morphism," which has the property that if you compose it with another morphism f then you get f.

As the earlier discussion suggests, an example of a category is the category of groups. The objects of this category are groups, the morphisms are group homomorphisms, and composition and the identity are defined in the way we are used to. However, it is by no means the case that all categories are like this, as the following examples show.

(i) We can form a category by taking the natural numbers as its objects, and letting the morphisms from n to m be all the $n \times m$ matrices with real entries. Composition of morphisms is the usual matrix multiplication. We would not normally think of an $n \times m$ matrix as a map from the number n to the number m, but the axioms for a category are nevertheless satisfied.

(ii) Any set can be turned into a category: the objects are the elements of the set, and a morphism from x to y is the assertion "$x = y$." We can also make an ordered set into a category by letting a morphism from x to y be the assertion "$x \leqslant y$." (The "composite" of "$x \leqslant y$" and "$y \leqslant z$" is "$x \leqslant z$.")

(iii) Any group G can be made into a category as follows: you have just one object, and the morphisms from that object to itself are the elements of the group, with the group multiplication defining the composition of two morphisms.

(iv) There is an obvious category where the objects are TOPOLOGICAL SPACES [III.90] and the morphisms are continuous functions. A less obvious category with the same objects takes as its morphisms not continuous functions but HOMOTOPY CLASSES [IV.6 §2] of continuous functions.

Morphisms are also called *maps*. However, as the above examples illustrate, the maps in a category do not have to be remotely map-like. They are also called *arrows*, partly to emphasize the more abstract nature of a general category, and partly because arrows are often used to represent morphisms pictorially.

The general framework and language of "objects and morphisms" enable us to seek and study structural features that depend only on the "shape" of the category, that is, on its morphisms and the equations they satisfy. The idea is both to make general arguments that are then applicable to all categories possessing particular structural features, and also to be able to make arguments in specific environments without having to go into the details of the structures in question. The use of the former to achieve the latter is sometimes referred to, endearingly or otherwise, as "abstract nonsense."

As we mentioned above, the morphisms in a category are generally depicted as arrows, so a morphism f from a to b is depicted as $a \xrightarrow{f} b$ and composition is depicted by concatenating the arrows $a \xrightarrow{f} b \xrightarrow{g} c$. This notation greatly eases complex calculations and gives rise to the so-called *commutative diagrams* that are often associated with category theory; an equality between composites of morphisms such as $g \circ f = t \circ s$ is expressed by asserting that the following diagram *commutes*, that is, that either of the two different paths from a to c yields the same composite:

Proving that one long string of compositions equals another then becomes a matter of "filling in" the space in between with smaller diagrams that are already known to commute. Furthermore, many important mathematical concepts can be described in terms of commutative diagrams: some examples are free groups, free rings, free algebras, quotients, products, disjoint unions, function spaces, direct and inverse limits, completion, compactification, and geometric realization.

Let us see how it is done in the case of disjoint unions. We say that a *disjoint union* of sets A and B is a set U equipped with morphisms $A \xrightarrow{p} U$ and $B \xrightarrow{q} U$ such that, given any set X and morphisms $A \xrightarrow{f} X$ and $B \xrightarrow{g} X$, there is a unique morphism $U \xrightarrow{h} X$ that makes the following diagram commute:

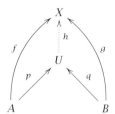

Here p and q tell us how A and B inject into the disjoint union. The "such that" part of the definition above is a *universal property*. It expresses the fact that giving a function from the disjoint union to another set is precisely the same as giving a function from each of the individual sets; this completely characterizes a disjoint union (which we regard as defined up to isomorphism). Another viewpoint is that the universal property expresses the fact that a disjoint union is the "most free" way of having two sets map into another set, neither adding any information nor collapsing any information. Universal properties are central to the way category theory describes structures that are somehow "canonical." (See also the discussion of free groups in GEOMETRIC AND COMBINATORIAL GROUP THEORY [IV.10].)

Another key concept in a category is that of an *isomorphism*. As one might expect, this is defined to be a morphism with a two-sided inverse. Isomorphic objects in a given category are thought of as "the same, as far as this particular category is concerned." Thus, categories provide a framework in which the most natural way of classifying objects is "up to isomorphism."

Categories are mathematical structures of a certain kind, and as such they themselves form a category (subject to size restrictions so as to avoid a Russell-type paradox). The morphisms, which are the structure-preserving maps for categories, are called *functors*. In other words, a functor F from a category X to a category Y takes the objects of X to the objects of Y and the morphisms of X to the morphisms of Y in such a way that the identity of a is taken to the identity of Fa and the composite of f and g is taken to the composite of Ff and Fg. An important example of a functor is the one that takes a topological space S with a "marked point" s to its fundamental group $\pi_1(S, s)$:

it is one of the basic theorems of algebraic topology that a continuous map between two topological spaces (that takes marked point to marked point) gives rise to a homomorphism between their fundamental groups.

Furthermore, there is a notion of morphism between functors called a *natural transformation*, which is analogous to the notion of homotopy between maps of topological spaces. Given continuous maps $F, G : X \rightarrow Y$, a homotopy from F to G gives us, for every point x in X, a path in Y from Fx to Gx; analogously, given functors $F, G : X \longrightarrow Y$, a natural transformation from F to G gives us, for every point x in X, a *morphism* in Y from Fx to Gx. There is also a commuting condition that is analogous to the fact that, in the case of homotopy, a path in X must have its image under F continuously transformed to its image under G without passing over any "holes" in the space Y. This avoidance of holes is expressed in the category case by the commutativity of certain squares in the target category Y, which is known as the "naturality condition."

One example of a natural transformation encodes the fact that every vector space is *canonically* isomorphic to its double dual; there is a functor from the category of vector spaces to itself that takes each vector space to its double dual, and there is an invertible natural transformation from this functor to the identity functor via the canonical isomorphisms. By contrast, every finite-dimensional vector space is isomorphic to its dual, but not canonically so because the isomorphism involves an arbitrary choice of basis; if we attempt to construct a natural transformation in this case, we find that the naturality condition fails. In the presence of natural transformations, categories actually form a 2-*category*, which is a two-dimensional generalization of a category, with objects, morphisms, and morphisms between morphisms. These last are thought of as two-dimensional morphisms; more generally an n-category has morphisms for each dimension up to n.

Categories and the language of categories are used in a wide variety of other branches of mathematics. Historically, the subject is closely associated with algebraic topology; the notions were first introduced in 1945 by Eilenberg and Mac Lane. Applications followed in algebraic geometry, theoretical computer science, theoretical physics, and logic. Category theory, with its abstract nature and lack of dependency on other fields of mathematics, can be thought of as "foundational." In fact, it has been proposed as an alternative candidate for the foundations of mathematics, with the notion of morphism as the basic one from which everything else is built up, instead of the relation of set membership that is used in SET-THEORETIC FOUNDATIONS [IV.22 §4].

Class Field Theory
See FROM QUADRATIC RECIPROCITY TO CLASS FIELD THEORY [V.28]

Cohomology
See HOMOLOGY AND COHOMOLOGY [III.38]

III.9 Compactness and Compactification
Terence Tao

In mathematics, it is well-known that the behavior of finite sets and the behavior of infinite sets can be rather different. For instance, each of the following statements is easily seen to be true whenever X is a finite set but false whenever X is an infinite set.

All functions are bounded. If $f : X \rightarrow \mathbb{R}$ is a real-valued function on X, then f must be bounded (i.e., there exists a finite number M such that $|f(x)| \leqslant M$ for all $x \in X$).

All functions attain a maximum. If $f : X \rightarrow \mathbb{R}$ is a real-valued function on X, then there must exist at least one point $x_0 \in X$ such that $f(x_0) \geqslant f(x)$ for all $x \in X$.

All sequences have constant subsequences. If $x_1, x_2, x_3, \cdots \in X$ is a sequence of points in X, then there must exist a subsequence $x_{n_1}, x_{n_2}, x_{n_3}, \ldots$ that is constant. In other words, $x_{n_1} = x_{n_2} = \cdots = c$ for some $c \in X$. (This fact is sometimes known as the *infinite pigeonhole principle*.)

The first statement—that all functions on a finite set are bounded—can be viewed as a very simple example of a *local-to-global principle*. The hypothesis is an assertion of "local" boundedness: it asserts that $|f(x)|$ is bounded for each point $x \in X$ separately, but with a bound that may depend on x. The conclusion is that of "global" boundedness: that $|f(x)|$ is bounded by a *single* bound M for all $x \in X$.

So far we have viewed the object X only as a set. However, in many areas of mathematics we like to endow our objects with additional structures, such as a TOPOLOGY [III.90], a METRIC [III.56], or a GROUP STRUCTURE [I.3 §2.1]. When we do this, it turns out that some objects exhibit properties similar to those of finite sets (in particular, they enjoy local-to-global principles), even though as sets they are infinite. In the categories of topological spaces and metric spaces, these "almost-finite" objects are known as *compact spaces*. (Other categories have "almost-finite" objects as well. For example, in the category of groups there is a notion of a *pro-finite group*; for LINEAR OPERATORS [III.50] between NORMED SPACES [III.62] the analogous notion is that of a *compact operator*, which is "almost of finite rank"; and so forth.)

A good example of a compact set is the closed unit interval $X = [0, 1]$. This is an infinite set, so the previous three assertions are all false as stated for X. But if we modify them by inserting topological concepts such as continuity and convergence, then we can restore these assertions for $[0, 1]$ as follows.

All continuous functions are bounded. If $f : X \to \mathbb{R}$ is a real-valued continuous function on X, then f must be bounded. (This is again a type of local-to-global principle: if a function does not vary too much locally, then it does not vary too much globally.)

All continuous functions attain a maximum. If $f : X \to \mathbb{R}$ is a real-valued continuous function on X, then there must exist at least one point $x_0 \in X$ such that $f(x_0) \geqslant f(x)$ for all $x \in X$.

All sequences have convergent subsequences. If $x_1, x_2, x_3, \cdots \in X$ is a sequence of points in X, then there must exist a subsequence $x_{n_1}, x_{n_2}, x_{n_3}, \ldots$ that converges to some limit $c \in X$. (This statement is known as the *Bolzano–Weierstrass theorem*.)

To these assertions we can add a fourth (which, like the others, has a rather trivial analogue for finite sets).

All open covers have finite subcovers. If $\mathcal{V}$ is a collection of open sets and the union of all these open sets contains X (in which case $\mathcal{V}$ is called an *open cover* of X), then there must exist a finite subcollection $V_{n_1}, V_{n_2}, \ldots, V_{n_k}$ of sets in $\mathcal{V}$ that still covers X.

All four of these topological statements are false for sets such as the open unit interval $(0, 1)$ or the real line $\mathbb{R}$, as one can easily check by constructing simple counterexamples. The *Heine–Borel theorem* asserts that when X is a subset of a Euclidean space $\mathbb{R}^n$, the above statements are all true when X is topologically closed and bounded, and all false otherwise.

The above four assertions are closely related to each other. For instance, if you know that all sequences in X contain convergent subsequences, then you can quickly deduce that all continuous functions have a maximum. This is done by first constructing a *maximizing sequence*—a sequence of points x_n in X such that $f(x_n)$ approaches the maximal value of f (or, more precisely, its supremum)—and then investigating a convergent subsequence of that sequence. In fact, given some fairly mild assumptions on the space X (e.g., that X is a metric space), one can deduce any of these four statements from any of the others.

To oversimplify a little, we say that a topological space X is *compact* if one (and hence all) of the above four assertions holds for X. Because the four assertions are not quite equivalent in general, the formal definition of compactness uses only the fourth version: that every open cover has a finite subcover. There are other notions of compactness, such as *sequential compactness*, for example, which is based on the third version, but the distinctions between these notions are technical and we shall gloss over them here.

Compactness is a powerful property of spaces, and it is used in many ways in many different areas of mathematics. One is via appeal to local-to-global principles: one establishes local control on a function, or on some other quantity, and then uses compactness to boost the local control to global control. Another is to locate maxima or minima of a function, which is particularly useful in the CALCULUS OF VARIATIONS [III.94]. A third is to partially recover the notion of a limit when dealing with nonconvergent sequences, by accepting the need to pass to a subsequence of the original sequence. (However, different subsequences may converge to different limits; compactness guarantees the existence of a limit point, but not its uniqueness.) Compactness of one object also tends to beget compactness of other objects; for instance, the image of a compact set under a continuous map is still compact, and the product of finitely many or even infinitely many compact sets continues to be compact. This last result is known as *Tychonoff's theorem*.

Of course, many spaces of interest are not compact. An obvious example is the real line $\mathbb{R}$, which is not compact, because it contains sequences such as $1, 2, 3, \ldots$ that are "trying to escape" the real line and that do not leave behind any convergent subsequences. However, one can often recover compactness by adding a few more points to the space: this process is known as *compactification*. For instance, one can compactify the real line by adding one point at each end: we call the added points $+\infty$ and $-\infty$. The resulting object, known as the *extended real line* $[-\infty, +\infty]$, can be given a topology in a natural way, which basically defines what it means to converge to $+\infty$ or to $-\infty$. The extended real line is compact: any sequence x_n of extended real numbers will have a subsequence that either converges to $+\infty$, or converges to $-\infty$, or converges to a finite number. Thus, by using this compactification of the real line, we can generalize the notion of a limit to one that no longer has to be a real number. While there are some drawbacks to dealing with extended reals instead of ordinary reals (for instance, one can always add two real numbers together, but the sum of $+\infty$ and $-\infty$ is undefined), the ability to take limits of what would otherwise be divergent sequences can be very useful, particularly in the theory of infinite series and improper integrals.

It turns out that a single noncompact space can have many different compactifications. For instance, by the device of *stereographic projection*, one can topologically identify the real line with a circle that has a single point removed. (For example, if one maps the real number x to the point $(x/(1 + x^2), x^2/(1 + x^2))$, then $\mathbb{R}$ maps to the circle of radius $\frac{1}{2}$ and center $(0, \frac{1}{2})$, with the north pole $(0, 1)$ removed.) If we then insert the missing point, we obtain the *one-point compactification* $\mathbb{R} \cup \{\infty\}$ of the real line. More generally, any reasonable topological space (e.g., a locally compact Hausdorff space) has a number of compactifications, ranging from the one-point compactification $X \cup \{\infty\}$, which is the "minimal" compactification as it adds only one point, to the *Stone–Čech compactification* βX, which is the "maximal" compactification and adds an enormous number of points. The Stone–Čech compactification $\beta\mathbb{N}$ of the natural numbers $\mathbb{N}$ is the space of *ultrafilters*, which are very useful tools in the more infinitary parts of mathematics.

One can use compactifications to distinguish between different types of divergence in a space. For instance, the extended real line $[-\infty, +\infty]$ distinguishes between divergence to $+\infty$ and divergence to $-\infty$. In a similar spirit, by using compactifications of the plane $\mathbb{R}^2$ such as the PROJECTIVE PLANE [I.3 §6.7], one can distinguish a sequence that diverges along (or near) the x-axis from a sequence that diverges along (or near) the y-axis. Such compactifications arise naturally in situations in which sequences that diverge in different ways exhibit markedly different behavior.

Another use of compactifications is to allow one to view one type of mathematical object rigorously as a limit of others. For instance, one can view a straight line in the plane as the limit of increasingly large circles by describing a suitable compactification of the space of circles that includes lines. This perspective allows us to deduce certain theorems about lines from analogous theorems about circles, and conversely to deduce certain theorems about very large circles from theorems about lines. In a rather different area of mathematics, the Dirac delta function is not, strictly speaking, a function, but it exists in certain (local) compactifications of spaces of functions, such as spaces of MEASURES [III.55] or DISTRIBUTIONS [III.18]. Thus, one can view the Dirac delta function as a limit of classical functions, and this can be very useful for manipulating it. One can also use compactifications to view the continuous as the limit of the discrete: for instance, it is possible to compactify the sequence $\mathbb{Z}/2\mathbb{Z}, \mathbb{Z}/3\mathbb{Z}, \mathbb{Z}/4\mathbb{Z}, \dots$ of cyclic groups in such a way that their limit is the circle group $\mathbb{T} = \mathbb{R}/\mathbb{Z}$. These simple examples can be generalized to much more sophisticated examples of compactifications, which have many applications in geometry, analysis, and algebra.

III.10   Computational Complexity Classes

One of the basic challenges of theoretical computer science is to determine what computational resources are necessary in order to perform a given computational task. The most basic resource is *time*, or equivalently (given the hardware) the number of steps needed to implement the most efficient algorithm that will carry out the task. Especially important is how this time scales up with the size of the input for the task: for instance, how much longer does it take to factorize an integer with $2n$ digits than an integer with n digits? Another resource connected with the feasibility of a computation is *memory*: one can ask how much storage space a computer will need in order to implement an algorithm, and how this can be minimized. A *complexity class* is a set of computational problems that can be performed with certain restrictions on the resources allowed. For instance, the complexity class $\mathcal{P}$ consists of all problems that can be performed in "polynomial time": that is, there is some positive integer k such that if the size of the problem is n (in the example above, the size was the number of digits of the integer to be factorized), then the computation can be carried out in at most n^k steps. A problem belongs to $\mathcal{P}$ if and only if the time taken to solve it scales up by at most a constant factor when the size of the input scales up by a constant factor. A good example of such a problem is multiplication of two n-digit numbers: if you use ordinary long multiplication, then replacing n by $2n$ increases the time taken by a factor of 4.

Suppose that you are presented with a positive integer x and told that it is a product of two primes p and q. How difficult is it to determine p and q? Nobody knows, but one thing is easy to see: if you are told p and q, then it is not hard (for a computer, at any rate) to check that pq really does equal x. Indeed, as we have just seen, long multiplication takes polynomial time, and comparing the answer with x is even easier. The complexity class $\mathcal{NP}$ consists of those computational tasks for which a correct answer can be *verified* in polynomial time, even if it cannot necessarily be *found* in polynomial time. Remarkably, although this is a fundamental distinction, nobody knows how to prove that $\mathcal{P} \neq \mathcal{NP}$: this problem is widely considered to be the most important in theoretical computer science.

We briefly mention two other important complexity classes. $\mathcal{PSPACE}$ consists of all problems that can be solved using an amount of memory that grows at most polynomially with the size of the input. It turns out to be the natural class associated with reasonable computational strategies for games such as chess. The complexity class $\mathcal{NC}$ is the set of all Boolean functions that can be computed by a "circuit of polynomial size and depth at most a polynomial in $\log n$." This last class is a model for the class of problems that can be solved very rapidly using parallel processing. In general, complexity classes are often surprisingly good at characterizing large families of problems with interesting and intuitively recognizable features in common. Another remarkable fact is that almost all complexity classes have "hardest problems" within them: that is, problems for which a solution can be converted into a solution for any other problem in the class. These problems are said to be *complete* for the class in question.

These issues, as well as several other complexity classes, are discussed in COMPUTATIONAL COMPLEXITY [IV.20]. A vast number of further classes can be found at

http://qwiki.stanford.edu/wiki/Complexity_Zoo

along with a brief definition of each.

Continued Fractions

See THE EUCLIDEAN ALGORITHM AND CONTINUED FRACTIONS [III.22]

III.11 Countable and Uncountable Sets

Infinite sets arise all the time in mathematics: the natural numbers, the squares, the primes, the integers, the rationals, the reals, and so on. It is often natural to try to compare the sizes of these sets: intuitively, one feels that the set of natural numbers is "smaller" than the set of integers (as it contains just the positive ones), and much larger than the set of squares (since a typical large integer is unlikely to be a square). But can we make comparisons of size in a precise way?

An obvious method of attack is to build on our intuition about finite sets. If A and B are finite sets, there are two ways we might go about comparing their sizes. One is to count their elements: we obtain two nonnegative integers m and n and just look at whether $m < n$, $m = n$, or $m > n$. But there is another important method, which does not require us to know the sizes of either A or B. This is to pair off elements from A with elements of B until one or other of the sets runs out of elements: the first one to run out is the smaller set, and if there is a dead heat, then the sets have the same size.

A suitable modification of this second method works for infinite sets as well: we can declare two sets to be of equal size if there is a one-to-one correspondence between them. This turns out to be an important and useful definition, though it does have some consequences that seem a little odd at first. For example, there is an obvious one-to-one correspondence between natural numbers and perfect squares: for each n we let n correspond to n^2. Thus, according to this definition there are "as many" squares as there are natural numbers. Similarly, we could show that there are as many primes as natural numbers by associating n with the nth prime number.[1]

What about $\mathbb{Z}$? It seems that it should be "twice as large" as $\mathbb{N}$, but again we can find a one-to-one correspondence between them. We just list the integers in the order $0, 1, -1, 2, -2, 3, -3, \ldots$ and then match the

---

1. For sufficiently nice sets of integers there is a definition of "density" that can be useful too. According to this definition, the even numbers have density $\frac{1}{2}$, while the squares and the primes have density 0, as one might expect. However, this is not the notion of size under discussion here.

natural numbers with them in the obvious way: 1 with 0, then 2 with 1, then 3 with −1, then 4 with 2, then 5 with −2, and so on.

An infinite set is called *countable* if it has the same size as the natural numbers. As the above example shows, this is exactly the same as saying that we can *list* the elements of the set. Indeed, if we have listed a set as $a_1, a_2, a_3, \ldots$, then our correspondence is just to send n to a_n. It is worth noting that there are of course many attempted listings that fail: for example, for $\mathbb{Z}$ we might have tried $-3, -2, -1, 0, 1, 2, 3, 4, \ldots$. So it is important to recognize that when we say that a set is countable we are not saying that *every* attempt to list it works, or even that the obvious attempt does: we are merely saying that there is *some* way of listing the elements. This is in complete contrast to finite sets, where if we attempt to match up two sets and find some elements of one set left over, then we know that the two sets cannot be in one-to-one correspondence. It is this difference that is mainly responsible for the "odd consequences" mentioned above.

Now that we have established that some sets that seem smaller or larger than $\mathbb{N}$, such as the squares or the integers, are actually countable, let us turn to a set that seems "much larger," namely $\mathbb{Q}$. How could we hope to list all the rationals? After all, between any two of them you can find infinitely many others, so it seems hard not to leave some of them out when you try to list them. However, remarkable as it may seem, it *is* possible to list the rationals. The key idea is that listing the rationals whose numerator and denominator are both smaller (in modulus) than some fixed number k is easy, as there are only finitely many of them. So we go through in order: first when both numerator and denominator are at most 1, then when they are at most 2, and so on (being careful not to relist any number, so that for example $\frac{1}{2}$ should not also appear as $\frac{2}{4}$ or $\frac{3}{6}$). This leads to an ordering such as $0, 1, -1, 2, -2, \frac{1}{2}, -\frac{1}{2},$ $3, -3, \frac{1}{3}, -\frac{1}{3}, \frac{2}{3}, -\frac{2}{3}, \frac{3}{2}, -\frac{3}{2}, 4, -4, \ldots$.

We could use the same idea to list sets that look even larger, such as, for example, the *algebraic* numbers (all real numbers, such as $\sqrt{2}$, that satisfy a polynomial equation with integer coefficients). Indeed, we note that each polynomial has only finitely many roots (which are therefore listable), so all we need to do is list the polynomials (as then we can go through them, in order, listing their roots). And we can do that by applying the same technique again: for each d we list those polynomials of degree at most d that we have not already listed, with coefficients that are at most d in modulus.

Based on the above examples, one might well guess that *every* infinite set is countable. But a beautiful argument of CANTOR [VI.54], called his "diagonal" argument, shows that the real numbers are not countable. We imagine that we have a list of all real numbers, say $r_1, r_2, r_3, \ldots$. Our aim is to show that this list cannot possibly contain all the reals, so we wish to construct a real that is not on this list. How do we accomplish this? We have each r_i written as an infinite decimal, say, and now we define a new number s as follows. For the first digit of s (after the decimal point), we choose a digit that is not the first digit of r_1. Note that this already guarantees that s cannot equal r_1. (To avoid coincidences with recurring 9s and the like, it is best to choose this first digit of s not to be 0 or 9 either.) Then, for the second digit of s, we choose a digit that is not the second digit of r_2; this guarantees that s cannot be equal to r_2. Continuing in this way, we end up with a real number s that is not on our list: whatever n is, the number s cannot be r_n, as s and r_n differ in the nth decimal place!

One can use similar arguments any time that we have "an infinite number of independent choices" to make in specifying an object (like the various digits of s). For example, let us use the same ideas to show that the set of all subsets of $\mathbb{N}$ is uncountable. Suppose we have listed all the subsets as $A_1, A_2, A_3, \ldots$. We will define a new set B that is not equal to any of the A_n. So we include the point 1 in B if and only if 1 does not belong to A_1 (this guarantees that B is not equal to A_1), and we include 2 in B if and only if 2 does not belong to A_2, and so on. It is amusing to note that one can write this set B down as $\{n \in \mathbb{N} : n \notin A_n\}$, which shows a striking resemblance to the set in Russell's paradox.

Countable sets are the "smallest" infinite sets. However, the set of real numbers is by no means the "largest" infinite set. Indeed, the above argument shows that no set X can be put into one-to-one correspondence with the set of all its subsets. So the set of all subsets of the real numbers is "strictly larger" than the set of real numbers, and so on.

The notion of countability is often a very fruitful one to bear in mind. For example, suppose we want to know whether or not all real numbers are algebraic. It is a genuinely hard exercise to write down a particular real that is TRANSCENDENTAL [III.41] (meaning not algebraic; see LIOUVILLE'S THEOREM AND ROTH'S THEOREM [V.22] for an idea of how it can be done), but the above notions make it utterly trivial that transcendental numbers exist. Indeed, the set of all real numbers is uncountable but the set of algebraic numbers is countable! Furthermore, this shows that "most" real numbers are transcendental: the algebraic numbers form only a tiny proportion of the reals.

III.12 C^*-Algebras

A BANACH SPACE [III.62] is both a VECTOR SPACE [I.3 §2.3] and a METRIC SPACE [III.56], and the study of Banach spaces is therefore a mixture of linear algebra and analysis. However, one can arrive at more sophisticated mixtures of algebra and analysis if one looks at Banach spaces that have more algebraic structure. In particular, while one can add two elements of a Banach space together, one cannot in general multiply them. However, sometimes one can: a vector space with a multiplicative structure is called an *algebra*, and if the vector space is also a Banach space, and if the multiplication has the property that $\|xy\| \leqslant \|x\| \|y\|$ for any two elements x and y, then it is called a *Banach algebra*. (This name does not really reflect historical reality, since the basic theory of Banach algebras was not worked out by Banach. A more appropriate name might have been Gelfand algebras.)

A C^*-algebra is a Banach algebra with an *involution*, which means a function that associates with each element x another element x^* in such a way that $x^{**} = x$, $\|x^*\| = \|x\|$, $(x + y)^* = x^* + y^*$, and $(xy)^* = y^* x^*$ for any elements x and y; this involution is required to satisfy the C^*-*identity* $\|xx^*\| = \|x\|^2$. A basic example of a C^*-algebra is the algebra $B(H)$ of all continuous linear maps T defined on a HILBERT SPACE [III.37] H. The norm of T is defined to be the smallest constant M such that $\|Tx\| \leqslant M\|x\|$ for every $x \in H$, and the involution takes T to its *adjoint*. This is a map T^* that has the property that $\langle x, Ty \rangle = \langle T^*x, y \rangle$ for every x and y in H. (It can be shown that there is exactly one map with this property.) If H is finite dimensional, then T can be thought of as an $n \times n$ matrix for some n, and T^* is then the complex conjugate of the transpose of T.

A fundamental theorem of Gelfand and Naimark states that every C^*-algebra can be represented as a subalgebra of $B(H)$ for some Hilbert space H. For more information, see OPERATOR ALGEBRAS [IV.15 §3].

III.13   Curvature

If you cut an orange in half, scoop out the inside, and try to flatten one of the resulting hemispheres of peel, then you will tear it. If you try to flatten a horse's saddle, or a soggy potato chip, then you will have the opposite problem: this time, there is "too much" of the surface to flatten and you will have to fold it over itself. If, however, you have a roll of wallpaper and wish to flatten it, then there is no difficulty: you just unroll it. Surfaces such as spheres are said to be *positively curved*, ones with a saddle-like shape are *negatively curved*, and ones like a piece of wallpaper are *flat*.

Notice that a surface can be flat in this sense even if it does not lie in a plane. This is because curvature is defined in terms of the *intrinsic geometry* of a surface, where distance is measured in terms of paths that lie inside the surface.

There are various ways of making the above notion of curvature precise, and also quantitative, so that with each point of a surface one can associate a number that tells you "how curved" it is at that point. In order to do this, the surface must have a RIEMANNIAN METRIC [I.3 §6.10] on it, which is used to determine the lengths of paths. The notion of curvature can also be generalized to higher dimensions, so that one can talk about the curvature of a point in a d-dimensional Riemannian manifold. However, when the dimension is higher than 2, the way that the manifold can curve at a point is more complicated, and is expressed not by a single number but by the so-called *Ricci tensor*. See RICCI FLOW [III.78] for more details.

Curvature is one of the fundamental concepts of modern geometry: not only the notion just described but also various alternative definitions that measure in other ways how far a geometric object deviates from being flat. It is also an integral part of the theory of general relativity (which is discussed in GENERAL RELATIVITY AND THE EINSTEIN EQUATIONS [IV.13]).

III.14   Designs
Peter J. Cameron

Block designs were first used in the design of experiments in statistics, as a method for coping with systematic differences in the experimental material. Suppose, for example, that we want to test seven different varieties of seed in an agricultural experiment, and that we have twenty-one plots of land available for the experiment. If the plots can be regarded as identical, then the best strategy is clearly to plant three plots with each variety. Suppose, however, that the available plots are on seven farms in different regions, with three plots on each farm. If we simply plant one variety on each farm, we lose information, because we cannot distinguish systematic differences between regions from

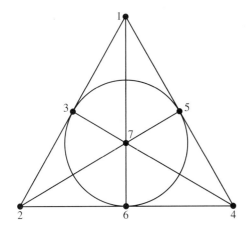

Figure 1 A block design.

differences in the seed varieties. It is better to follow a scheme like this: plant varieties 1, 2, 3 on the first farm; 1, 4, 5 on the second; and then 1, 6, 7; 2, 4, 6; 2, 5, 7; 3, 4, 7; and 3, 5, 6. This design is represented in figure 1.

This arrangement is called a *balanced incomplete-block design*, or BIBD for short. The blocks are the sets of seed varieties used on the seven farms. The blocks are "incomplete" because not every variety can be planted on every farm; the design is "balanced" because each pair of varieties occurs in the same block the same number of times (just once in this case). This is a $(7, 3, 1)$ design: there are seven varieties; each block contains three of them; and two varieties occur together in a block once. It is also an example of a finite *projective plane*. Because of the connection with geometry, varieties are usually called "points."

Mathematicians have developed an extensive theory of BIBDs and related classes of designs. Indeed, the study of such designs predates their use in statistics. In 1847, T. P. Kirkman showed that a $(v, 3, 1)$ design exists if and only if v is congruent to 1 or 3 mod 6. (Such designs are now called *Steiner triple systems*, although Steiner did not pose the problem of their existence until 1853.)

Kirkman also posed a more difficult problem. In his own words,

> Fifteen young ladies in a school walk out three abreast for seven days in succession: it is required to arrange them daily so that no two shall walk twice abreast.

The solution requires a $(15, 3, 1)$ Steiner triple system with the extra property that the thirty-five blocks can be partitioned into seven sets called "replicates," each replicate consisting of five blocks that partition the set of points. Kirkman himself gave a solution, but it was not until the late 1960s that Ray-Chaudhuri and Wilson showed that $(v, 3, 1)$ designs with this property exist whenever v is congruent to 3 mod 6.

For which v, k, λ do designs exist? Counting arguments show that, given k and λ, the values of v for which (v, k, λ) designs exist are restricted to certain congruence classes. (We noted above that $(v, 3, 1)$ designs exist only if v is congruent to 1 or 3 mod 6.) An asymptotic existence theory developed by Richard Wilson shows that this necessary condition is sufficient for the existence of a design, apart from finitely many exceptions, for each value of k and λ.

The concept of design has been further generalized: a t-(v, k, λ) design has the property that any t points are contained in exactly λ blocks. Luc Teirlinck showed that nontrivial t-designs exist for all t, but examples for $t > 3$ are comparatively rare.

The statisticians' concerns are a bit different. In our introductory example, if only six farms were available, we could not use a BIBD for the experiment, but would have to choose the most "efficient" possible design (allowing the most information to be obtained from the experimental results). A BIBD is most efficient if it exists; but not much is known in other cases.

There are other types of design; these can be important to statistics and also lead to new mathematics. Here, for example, is an *orthogonal array*: if you take any two rows of this matrix you obtain a 2×9 matrix in which each ordered pair of symbols from $\{0, 1, 2\}$ occurs exactly once as a column.

$$
\begin{matrix}
0 & 0 & 0 & 1 & 1 & 1 & 2 & 2 & 2 \\
0 & 1 & 2 & 0 & 1 & 2 & 0 & 1 & 2 \\
0 & 1 & 2 & 1 & 2 & 0 & 2 & 0 & 1 \\
0 & 2 & 1 & 1 & 0 & 2 & 2 & 1 & 0 \\
\end{matrix}
$$

It could be used if we had four different treatments, each of which could be applied at three different levels, and if we had nine plots available for testing.

Design theory is closely related to other combinatorial topics such as error-correcting codes; indeed, Fisher "discovered" the Hamming codes as designs five years before R. W. Hamming found them in the context of error correction. Other related subjects include packing and covering problems, and especially finite geometry, where many finite versions of classical geometries can be regarded as designs.

III.15 Determinants

The determinant of a 2×2 matrix

$$\begin{pmatrix} a & b \\ c & d \end{pmatrix}$$

is defined to be $ad - bc$. The determinant of a 3×3 matrix

$$\begin{pmatrix} a & b & c \\ d & e & f \\ g & h & i \end{pmatrix}$$

is defined to be $aei + bfg + cdh - afh - bdi - ceg$. What do these expressions have in common, how do they generalize, and why is the generalization significant?

To begin with the first question, let us make a few simple observations. Both expressions are sums and differences of products of entries from the matrix. Each one of these products contains exactly one element from each row of the matrix and also exactly one element from each column. In both cases, a minus sign seems to attach itself to the products for which the entries selected from the matrix "slope upward" rather than "downward."

Up to a point it is easy to see how to extend this definition to $n \times n$ matrices with $n \geqslant 4$. We simply take sums and differences of all possible products of n entries, where one entry from each row is used and one from each column. The difficulty comes in deciding which of these products to add and which to subtract. To do this we take one of the products and use it to define a permutation σ of the set $\{1, 2, \ldots, n\}$ as follows. For each $i \leqslant n$, the product contains exactly one entry in the ith row. If it belongs to the jth column then $\sigma(i) = j$. The product is added if this permutation is even and subtracted if it is odd (see PERMUTATION GROUPS [III.68]). So, for example, the permutation corresponding to the entry afh in the 3×3 determinant above sends 1 to 1, 2 to 3, and 3 to 2. This is an odd permutation, which is why afh receives a minus sign.

We still need to explain why the particular choice of products and minus signs that we have just defined is important. The reason is that it tells us something about the effect of a matrix when it is considered as a linear map. Let A be an $n \times n$ matrix. Then, as explained in [I.3 §3.2], A specifies a linear map α from $\mathbb{R}^n$ to $\mathbb{R}^n$. The determinant of A tells us what this linear map does to volumes. More precisely, if X is a subset of $\mathbb{R}^n$ with n-dimensional volume V, then αX, the result of transforming X using the linear map α, will have vol-

ume V times the determinant of A. We could write this symbolically as follows:

$$\text{vol}(\alpha X) = \det A \cdot \text{vol}(X).$$

For example, consider the 2×2 matrix

$$A = \begin{pmatrix} \cos\theta & -\sin\theta \\ \sin\theta & \cos\theta \end{pmatrix}.$$

The corresponding linear map is a rotation of $\mathbb{R}^2$ through an angle of θ. Since rotating a shape does not affect its volume, we should expect the determinant of A to be 1, and sure enough it is $\cos^2\theta + \sin^2\theta$, which is 1 by Pythagoras's theorem.

The above explanation is a slight oversimplification in one respect: determinants can be negative, but clearly volumes cannot. If the determinant of a matrix is -2, to give an example, it means that the linear map multiplies volumes by 2 but also "turns shapes inside out" by reflecting them.

Determinants have many useful properties, which become obvious once one knows the above interpretation in terms of volumes. (However, it is much less obvious that this interpretation is correct: in setting up the theory of determinants one must do some work somewhere.) Let us give three of these properties.

(i) Let V be a VECTOR SPACE [I.3 §2.3] and let $\alpha : V \to V$ be a linear map. Let $\boldsymbol{v}_1, \ldots, \boldsymbol{v}_n$ be a basis of V and let A be the matrix of α with respect to this basis. Now let $\boldsymbol{w}_1, \ldots, \boldsymbol{w}_n$ be another basis of V and let B be the matrix of α with respect to this different basis. Then A and B are different matrices, but since they both represent the linear map α, they must have the same effect on volumes. It follows that $\det(A) = \det(B)$. To put this another way: the determinant is better thought of as a property of linear maps rather than of matrices.

Two matrices that represent the same linear map in the above sense are called *similar*. It turns out that A and B are similar if and only if there is an invertible matrix P such that $P^{-1}AP = B$. (An $n \times n$ matrix P is *invertible* if there is a matrix Q such that PQ equals the $n \times n$ identity matrix, I_n, which turns out to imply that QP equals I_n as well. If this is true, then Q is called the *inverse* of P and is denoted P^{-1}.) What we have just shown is that similar matrices have the same determinant.

(ii) If A and B are any two $n \times n$ matrices, then they represent linear maps α and β of $\mathbb{R}^n$. The product AB represents the linear map $\alpha\beta$: that is, the linear map that results from doing β followed by α. Since β multiplies volumes by $\det B$ and α multiplies them by det A, $\alpha\beta$ multiplies them by det A det B. It follows that det(AB) = det A det B. (The determinant of a product equals the product of the determinants.)

(iii) If A is a matrix with determinant 0 and B is any other matrix, then AB will have determinant 0 as well, by the multiplicative property just discussed. It follows that AB cannot equal I_n, since I_n has determinant 1. Therefore a matrix with determinant 0 is not invertible. The converse of this turns out to be true as well: a matrix with nonzero determinant is invertible. Thus, the determinant gives us a way of finding out whether a matrix can be inverted.

III.16 Differential Forms and Integration
Terence Tao

It goes without saying that integration is one of the fundamental concepts of single-variable calculus. However, there are in fact *three* concepts of integration that appear in the subject: the *indefinite integral* $\int f$ (also known as the *antiderivative*), the *unsigned definite integral* $\int_{[a,b]} f(x)\,\mathrm{d}x$ (which one would use to find the area under a curve, or the mass of a one-dimensional object of varying density), and the *signed definite integral* $\int_a^b f(x)\,\mathrm{d}x$ (which one would use, for instance, to compute the work required to move a particle from a to b). For simplicity we shall restrict our attention here to functions $f : \mathbb{R} \to \mathbb{R}$ that are continuous on the entire real line (and similarly, when we come to differential forms, we shall discuss only forms that are continuous on the entire domain). We shall also informally use terminology such as "infinitesimal" in order to avoid having to discuss the (routine) "epsilon–delta" analytical issues that one must resolve in order to make these integration concepts fully rigorous.

These three concepts of integration are of course closely related to each other in single-variable calculus; indeed, THE FUNDAMENTAL THEOREM OF CALCULUS [I.3 §5.5] relates the signed definite integral $\int_a^b f(x)\,\mathrm{d}x$ to any one of the indefinite integrals $F = \int f$ by the formula

$$\int_a^b f(x)\,\mathrm{d}x = F(b) - F(a), \qquad (1)$$

while the signed and unsigned integrals are related by the simple identity

$$\int_a^b f(x)\,\mathrm{d}x = -\int_b^a f(x)\,\mathrm{d}x = \int_{[a,b]} f(x)\,\mathrm{d}x, \qquad (2)$$

which is valid whenever $a \leqslant b$.

When one moves from single-variable calculus to several-variable calculus, though, these three concepts begin to diverge significantly from each other. The indefinite integral generalizes to the notion of a *solution to a differential equation*, or to an *integral* of a connection, VECTOR FIELD [IV.6 §5], or BUNDLE [IV.6 §5]. The unsigned definite integral generalizes to the LEBESGUE INTEGRAL [III.55], or more generally to *integration on a measure space*. Finally, the signed definite integral generalizes to the *integration of forms*, which will be our focus here. While these three concepts are still related to each other, they are not as interchangeable as they are in the single-variable setting. The integration-of-forms concept is of fundamental importance in differential topology, geometry, and physics, and also yields one of the most important examples of COHOMOLOGY [IV.6 §4], namely *de Rham cohomology*, which (roughly speaking) measures the extent to which the fundamental theorem of calculus fails in higher dimensions and on general manifolds.

To provide some motivation for the concept, let us informally revisit one of the basic applications of the signed definite integral from physics, namely computing the amount of work required to move a one-dimensional particle from point a to point b in the presence of an external field. (For example, one might be moving a charged particle in an electric field.) At the infinitesimal level, the amount of work required to move a particle from a point $x_i \in \mathbb{R}$ to a nearby point $x_{i+1} \in \mathbb{R}$ is (up to a small error) proportional to the displacement $\Delta x_i = x_{i+1} - x_i$, with the constant of proportionality $f(x_i)$ depending on the initial location x_i of the particle. Thus, the total work required for this is approximately $f(x_i)\Delta x_i$. Note that we do not require x_{i+1} to be to the right of x_i, so the displacement Δx_i (or the infinitesimal work $f(x_i)\Delta x_i$) may well be negative. To return to the noninfinitesimal problem of computing the work required to move from a to b, we arbitrarily select a discrete path $x_0 = a, x_1, x_2, \ldots, x_n = b$ from a to b, and approximate the work as

$$\int_a^b f(x)\,\mathrm{d}x \approx \sum_{i=0}^{n-1} f(x_i)\Delta x_i. \qquad (3)$$

Again, we do *not* require x_{i+1} to be to the right of x_i; it is quite possible for the path to "backtrack" repeatedly: for instance, one might have $x_i < x_{i+1} > x_{i+2}$ for some i. However, it turns out that the effect of such backtracking eventually cancels itself out; regardless of what path we choose, the expression (3) above converges as the maximum step size tends to zero, and the limit is the signed definite integral

$$\int_a^b f(x)\,dx, \tag{4}$$

provided only that the total length $\sum_{i=0}^{n-1} |\Delta x_i|$ of the path (which controls the amount of backtracking involved) stays bounded. In particular, in the case when $a = b$, so that all paths are *closed* (i.e., $x_0 = x_n$), we see that the signed definite integral is zero:

$$\int_a^a f(x)\,dx = 0. \tag{5}$$

From this informal definition of the signed definite integral it is obvious that we have the concatenation formula

$$\int_a^c f(x)\,dx = \int_a^b f(x)\,dx + \int_b^c f(x)\,dx \tag{6}$$

regardless of the relative position of the real numbers a, b, and c. In particular (setting $a = c$ and using (5)) we conclude that

$$\int_a^b f(x)\,dx = -\int_b^a f(x)\,dx.$$

Thus, if we reverse a path from a to b to form a path from b to a, then the sign of the integral changes. This contrasts with the *unsigned definite integral* $\int_{[a,b]} f(x)\,dx$, since the set $[a,b]$ of numbers between a and b is exactly the same as the set of numbers between b and a. Thus we see that paths are not quite the same as sets: they carry an *orientation* which can be reversed, whereas sets do not.

Now let us move from one-dimensional integration to higher-dimensional integration: that is, from single-variable calculus to several-variable calculus. It turns out that there are *two* objects whose dimensions may increase: the "ambient space,"[1] which will now be $\mathbb{R}^n$ instead of $\mathbb{R}$, and the path, which will now become an oriented k-dimensional manifold S, over which the integration will take place. For example, if $n = 3$ and $k = 2$, then one is integrating over a surface that lives in $\mathbb{R}^3$.

Let us begin with the case $n \geq 1$ and $k = 1$. Here, we will be integrating over a continuously differentiable path (or *oriented rectifiable curve*) γ in $\mathbb{R}^n$ starting and ending at points a and b, respectively. (These points may or may not be distinct, depending on whether the path is open or closed.) From a physical point of view, we are still computing the work required to move from a to b, but now we are moving in several dimensions

instead of one. In the one-dimensional case, we did not need to specify exactly which path we used to get from a to b, because all backtracking canceled itself out. However, in higher dimensions, the exact choice of the path γ becomes important.

Formally, a path from a to b can be described (or *parametrized*) as a continuously differentiable function γ from the unit interval $[0, 1]$ to $\mathbb{R}^n$ such that $\gamma(0) = a$ and $\gamma(1) = b$. For instance, the line segment from a to b can be parametrized as $\gamma(t) = (1 - t)a + tb$. This segment also has many other parametrizations, such as $\tilde{\gamma}(t) = (1 - t^2)a + t^2 b$; however, as in the one-dimensional case, the exact choice of parametrization does not ultimately influence the integral. On the other hand, the reverse line segment $(-\gamma)(t) = ta + (1 - t)b$ from b to a is a genuinely different path; the integral along $-\gamma$ will turn out to be the negative of the integral along γ.

As in the one-dimensional case, we will need to approximate the continuous path γ by a discrete path

$$x_0 = \gamma(t_0), \ x_1 = \gamma(t_1), \ x_2 = \gamma(t_2), \ \ldots, \ x_n = \gamma(t_n),$$

where $\gamma(t_0) = a$ and $\gamma(t_n) = b$. Again, we allow some backtracking: t_{i+1} is not necessarily larger than t_i. The displacement $\Delta x_i = x_{i+1} - x_i \in \mathbb{R}^n$ from x_i to x_{i+1} is now a *vector* rather than a scalar. (Indeed, with an eye on the generalization to manifolds, one should think of Δx_i as an infinitesimal *tangent vector* to the ambient space $\mathbb{R}^n$ at the point x_i.) In the one-dimensional case, we converted the scalar displacement Δx_i into a new number $f(x_i)\Delta x_i$, which was linearly related to the original displacement by a proportionality constant $f(x_i)$ that depended on the position x_i. In higher dimensions, we again have a linear dependence, but this time, since the displacement is a vector, we must replace the simple constant of proportionality by a *linear transformation* ω_{x_i} from $\mathbb{R}^n$ to $\mathbb{R}$. Thus, $\omega_{x_i}(\Delta x_i)$ represents the infinitesimal "work" required to move from x_i to x_{i+1}. In technical terms, ω_{x_i} is a *linear functional* on the space of tangent vectors at x_i, and is thus a *cotangent vector* at x_i. By analogy with (3), the net work $\int_\gamma \omega$ required to move from a to b along the path γ is approximated by

$$\int_\gamma \omega \approx \sum_{i=0}^{n-1} \omega_{x_i}(\Delta x_i). \tag{7}$$

As in the one-dimensional case, one can show that the right-hand side of (7) converges if the maximum step size $\sup_{0 \leq i \leq n-1} |\Delta x_i|$ of the path converges to zero and the total length $\sum_{i=0}^{n-1} |\Delta x_i|$ of the path stays

---

1. We will start with integration on Euclidean spaces $\mathbb{R}^n$ for simplicity, although the true power of the integration-of-forms concept is much more apparent when we integrate on more general spaces, such as abstract n-dimensional manifolds.

bounded. The limit is written as $\int_\gamma \omega$. (Recall that we are restricting our attention to continuous functions. The existence of this limit uses the continuity of ω.)

The object ω, which continuously assigns[2] a cotangent vector to each point in $\mathbb{R}^n$, is called a 1-*form*, and (7) leads to a recipe for integrating any 1-form ω on a path γ. That is, to shift the emphasis slightly, it allows us to integrate the path γ "against" the 1-form ω. Indeed, it is useful to think of this integration as a *binary* operation (similar in some ways to the dot product) that takes the curve γ and the form ω as inputs, and returns a scalar $\int_\gamma \omega$ as output. There is in fact a "duality" between curves and forms; compare, for instance, the identity

$$\int_\gamma (\omega_1 + \omega_2) = \int_\gamma \omega_1 + \int_\gamma \omega_2,$$

which expresses (part of) the fundamental fact that integration of forms is a linear operation, with the identity

$$\int_{\gamma_1 + \gamma_2} \omega = \int_{\gamma_1} \omega + \int_{\gamma_2} \omega,$$

which generalizes (6) whenever the initial point of γ_2 is the final point of γ_1, where $\gamma_1 + \gamma_2$ is the *concatenation* of γ_1 and γ_2.[3]

Recall that if f is a differentiable function from $\mathbb{R}^n$ to $\mathbb{R}$, then its derivative at a point x is a linear map from $\mathbb{R}^n$ to $\mathbb{R}$ (see [I.3 §5.3]). If f is continuously differentiable, then this linear map depends continuously on x, and can therefore be thought of as a 1-form, which we denote by df, writing df_x for the derivative at x. This 1-form can be characterized as the unique 1-form such that one has the approximation

$$f(x + v) \approx f(v) + df_x(v)$$

for all infinitesimal v. (More rigorously, the condition is that $|f(x + v) - f(v) - df_x(v)|/|v| \to 0$ as $v \to 0$.)

The fundamental theorem of calculus (1) now generalizes to

$$\int_\gamma df = f(b) - f(a) \tag{8}$$

whenever γ is any oriented curve from a point a to a point b. In particular, if γ is closed, then $\int_\gamma df = 0$. Note that in order to interpret the left-hand side of the above equation, we are regarding it as a particular example of an integral of the form $\int_\gamma \omega$: in this case, ω happens to be the form df. Note also that, with this interpretation, df has an independent meaning (it is a 1-form) even if it does not appear under an integral sign.

A 1-form whose integral against every sufficiently small[4] closed curve vanishes is called *closed*, while a 1-form that can be written as df for some continuously differentiable function is called *exact*. Thus, the fundamental theorem implies that every exact form is closed. This turns out to be a general fact, valid for all manifolds. Is the converse true: that is, is every closed form exact? If the domain is a Euclidean space, or indeed any other *simply connected* manifold, then the answer is yes (this is a special case of the *Poincaré lemma*), but it is not true for general domains. In modern terminology, this demonstrates that the de Rham cohomology of such domains can be nontrivial.

As we have just seen, a 1-form can be thought of as an object ω that associates with each path γ a scalar, which we denote by $\int_\gamma \omega$. Of course, ω is not just any old function from paths to scalars: it must satisfy the concatenation and reversing rules discussed earlier, and this, together with our continuity assumptions, more or less forces it to be associated with some kind of continuously varying linear function that can be used, in combination with γ, to define an integral. Now let us see if we can generalize this basic idea from paths to k-dimensional sets with $k > 1$. For simplicity we shall stick to the two-dimensional case, that is, to integration of forms on (oriented) surfaces in $\mathbb{R}^n$, since this already illustrates many features of the general case.

Physically, such integrals arise when one is computing a *flux* of some field (e.g., a magnetic field) across a surface. We parametrized one-dimensional oriented curves as continuously differentiable functions γ from the interval $[0, 1]$ to $\mathbb{R}^n$. It is thus natural to parametrize two-dimensional oriented surfaces as continuously differentiable functions ϕ defined on the unit square $[0, 1]^2$. This does not in fact cover all possible surfaces one wishes to integrate over, but it turns out that one can cut up more general surfaces into pieces that can be parametrized using "nice" domains such as $[0, 1]^2$.

In the one-dimensional case, we cut up the oriented interval $[0, 1]$ into infinitesimal oriented intervals from t_i to $t_{i+1} = t_i + \Delta t$, which led to infinitesimal curves from $x_i = \gamma(t_i)$ to $x_{i+1} = \gamma(t_{i+1}) = x_i + \Delta x_i$. Note that

---

2. More precisely, one can think of ω as a *section* of the *cotangent bundle*.

3. This duality is best understood using the abstract, and much more general, formalism of homology and cohomology. In particular, one can remove the requirement that γ_2 begins where γ_1 leaves off by generalizing the notion of an integral to cover not just integration on paths, but also integration on *formal sums or differences* of paths. This makes the duality between curves and forms more symmetric.

4. The precise condition needed is that the curve should be *contractible*, which means that it can be continuously shrunk down to a point.

Δx_i and Δt are related by the approximation $\Delta x_i \approx y'(t_i)\Delta t_i$. In the two-dimensional case, we will cut up the unit square $[0, 1]^2$ into infinitesimal squares in an obvious way.[5] A typical one of these will have corners of the form (t_1, t_2), $(t_1 + \Delta t, t_2)$, $(t_1, t_2 + \Delta t)$, $(t_1 + \Delta t, t_2 + \Delta t)$. The surface described by ϕ can then be partitioned into regions, with corners $\phi(t_1, t_2)$, $\phi(t_1 + \Delta t, t_2)$, $\phi(t_1, t_2 + \Delta t)$, $\phi(t_1 + \Delta t, t_2 + \Delta t)$, each of which carries an orientation. Since ϕ is differentiable, it is approximately linear at small distance scales, so this region is approximately an oriented parallelogram in $\mathbb{R}^n$ with corners x, $x + \Delta_1 x$, $x + \Delta_2 x$, $x + \Delta_1 x + \Delta_2 x$, where $x = \phi(t_1, t_2)$ and $\Delta_1 x$ and $\Delta_2 x$ are the infinitesimal vectors

$$\Delta_1 x = \frac{\partial \phi}{\partial t_1}(t_1, t_2)\Delta t, \qquad \Delta_2 x = \frac{\partial \phi}{\partial t_2}(t_1, t_2)\Delta t.$$

Let us refer to this object as the infinitesimal parallelogram with *dimensions* $\Delta_1 x \wedge \Delta_2 x$ and *base point* x. For now, we will think of the symbol "$\wedge$" as a mere notational convenience and not try to interpret it. In order to integrate in a manner analogous with integration on curves, we now need some sort of functional ω_x at this base point that depends continuously on x. This functional should take the above infinitesimal parallelogram and return an infinitesimal number $\omega_x(\Delta_1 x \wedge \Delta_2 x)$, which one can think of as the amount of "flux" passing through this parallelogram.

As in the one-dimensional case, we expect ω_x to have certain properties. For instance, if you double $\Delta_1 x$, you double one of the sides of the infinitesimal parallelogram, so (by the continuity of ω) the "flux" passing through the parallelogram should double. More generally, $\omega_x(\Delta_1 x \wedge \Delta_2 x)$ should depend linearly on each of $\Delta_1 x$ and $\Delta_2 x$: in other words, it is *bilinear*. (This generalizes the linear dependence in the one-dimensional case.)

Another important property is that

$$\omega_x(\Delta_2 x \wedge \Delta_1 x) = -\omega_x(\Delta_1 x \wedge \Delta_2 x). \qquad (9)$$

That is, the bilinear form ω_x is *antisymmetric*. Again, this has an intuitive explanation: the parallelogram represented by $\Delta_2 x \wedge \Delta_1 x$ is the same as that represented by $\Delta_1 x \wedge \Delta_2 x$ except that it has had its orientation reversed, so the "flux" now counts negatively where it used to count positively, and vice versa. Another way of seeing this is to note that if $\Delta_1 x = \Delta_2 x$, then the parallelogram is degenerate and there should be no flux.

---

5. One could also use infinitesimal oriented rectangles, parallelograms, triangles, etc.; this leads to an equivalent concept of the integral.

Antisymmetry follows from this and the bilinearity. A 2-*form* ω is a continuous assignment of a functional ω_x with these properties to each point x.

If ω is a 2-form and $\phi : [0, 1]^2 \to \mathbb{R}^n$ is a continuously differentiable function, we can now define the integral $\int_\phi \omega$ of ω "against" ϕ (or, more precisely, the integral against the image under ϕ of the oriented square $[0, 1]^2$) by the approximation

$$\int_\phi \omega \approx \sum_i \omega_{x_i}(\Delta x_{1,i} \wedge \Delta x_{2,i}), \qquad (10)$$

where the image of ϕ is (approximately) partitioned into parallelograms of dimensions $\Delta x_{1,i} \wedge \Delta x_{2,i}$ based at points x_i. We do not need to decide what order these parallelograms should be arranged in, because addition is both commutative and associative. One can show that the right-hand side of (10) converges to a unique limit as one makes the partition of parallelograms "increasingly fine," though we will not make this precise here.

We have thus shown how to integrate 2-forms against oriented two-dimensional surfaces. More generally, one can define the concept of a k-form on an n-dimensional manifold (such as $\mathbb{R}^n$) for any $0 \leqslant k \leqslant n$ and integrate this against an oriented k-dimensional surface in that manifold. For instance, a 0-form on a manifold X is the same thing as a scalar function $f : X \to \mathbb{R}$, whose integral on a positively oriented point x (which is zero dimensional) is $f(x)$, and on a negatively oriented point x is $-f(x)$. A k-form tells us how to assign a value to an infinitesimal k-dimensional parallelepiped with dimensions $\Delta x_1 \wedge \cdots \wedge \Delta x_k$, and hence to a portion of k-dimensional "surface," in much the same way as we have seen when $k = 2$. By convention, if $k \neq k'$, the integral of a k-dimensional form on a k'-dimensional surface is understood to be zero. We refer to 0-forms, 1-forms, 2-forms, etc. (and formal sums and differences thereof), collectively as *differential forms*.

There are three fundamental operations that one can perform on scalar functions: addition $(f, g) \mapsto f + g$, pointwise product $(f, g) \mapsto fg$, and differentiation $f \mapsto \mathrm{d}f$, although the last of these is not especially useful unless f is continuously differentiable. These operations have various relationships with each other. For instance, the product is *distributive* over addition,

$$f(g + h) = fg + fh,$$

and differentiation is a *derivation* with respect to the product:

$$\mathrm{d}(fg) = (\mathrm{d}f)g + f(\mathrm{d}g).$$

It turns out that one can generalize all three of these operations to differential forms. Adding a pair of forms is easy: if ω and η are two k-forms and ϕ : $[0,1]^k \to \mathbb{R}^n$ is a continuously differentiable function, then $\int_\phi (\omega + \eta)$ is defined to be $\int_\phi \omega + \int_\phi \eta$. One multiplies forms using the so-called *wedge product*. If ω is a k-form and η is an l-form, then $\omega \wedge \eta$ is a $(k + l)$-form. Roughly speaking, given a $(k + l)$-dimensional infinitesimal parallelepiped with base point x and dimensions $\Delta x_1 \wedge \cdots \wedge \Delta x_{k+l}$, one evaluates ω and η at the parallelepipeds with base point x and dimensions $\Delta x_1 \wedge \cdots \wedge \Delta x_k$ and $\Delta x_{k+1} \wedge \cdots \wedge \Delta x_{k+l}$, respectively, and multiplies the results together.

As for differentiation, if ω is a continuously differentiable k-form, then its derivative $d\omega$ is a $(k + 1)$-form that measures something like the "rate of change" of ω. To see what this might mean, and in particular to see why $d\omega$ is a $(k + 1)$-form, let us think how we might answer a question of the following kind. We are given a spherical surface in $\mathbb{R}^3$ and a flow, and we would like to know the net flux out of the surface: that is, the difference between the amount of flux coming out and the amount going in. One way to do this would be to approximate the surface of the sphere by a union of tiny parallelograms, to measure the flux through each one, and to take the sum of all these fluxes. Another would be to approximate the solid sphere by a union of tiny parallelepipeds, to measure the *net* flux out of each of these, and to add up the results. If a parallelepiped is small enough, then we can closely approximate the net flux out of it by looking at the difference, for each pair of opposite faces, between the amount coming out of the parallelepiped through one and the amount going into it through the other, and this will depend on the rate of change of the 2-form.

The process of summing up the net fluxes out of the parallelepipeds is more rigorously described as integrating a 3-form over the solid sphere. In this way, one can see that it is natural to expect that information about how a 2-form varies should be encapsulated in a 3-form.

The exact construction of these operations requires a little bit of algebra and is omitted here. However, we remark that they obey similar laws to their scalar counterparts, except that there are some sign changes that are ultimately due to the antisymmetry (9). For instance, if ω is a k-form and η is an l-form, the commutative law for multiplication becomes

$$\omega \wedge \eta = (-1)^{kl} \eta \wedge \omega,$$

basically because kl swaps are needed to interchange k dimensions with l dimensions; and the derivation rule for differentiation becomes

$$d(\omega \wedge \eta) = (d\omega) \wedge \eta + (-1)^k \omega \wedge (d\eta).$$

Another rule is that the differentiation operator d is nilpotent:

$$d(d\omega) = 0. \tag{11}$$

This may seem rather unintuitive, but it is fundamentally important. To see why it might be expected, let us think about differentiating a 1-form twice. The original 1-form associates a scalar with each small line segment. Its derivative is a 2-form that associates a scalar with each small parallelogram. This scalar essentially measures the sum of the scalars given by the 1-form as you go around the four edges of the parallelogram, though to get a sensible answer when you pass to the limit you have to divide by the area of the parallelogram. If we now repeat the process, we are looking at a sum of the six scalars associated with the six faces of a parallelepiped. But each of these scalars in turn comes from a sum of the scalars associated with the four directed edges around the corresponding face, and each edge is therefore counted twice (as it belongs to two faces), once in each direction. Therefore, the contributions from each edge cancel and the sum of all contributions is zero.

The description given earlier of the relationship between integrating a 2-form over the surface of a sphere and integrating its derivative over the solid sphere can be thought of as a generalization of the fundamental theorem of calculus, and can itself be generalized considerably: *Stokes's theorem* is the assertion that

$$\int_S d\omega = \int_{\partial S} \omega \tag{12}$$

for any oriented manifold S and form ω, where ∂S is the oriented boundary of S (which we will not define here). Indeed one can view this theorem as a definition of the derivative operation $\omega \mapsto d\omega$; thus, differentiation is the *adjoint* of the boundary operation. (For instance, the identity (11) is dual to the geometric observation that the boundary ∂S of an oriented manifold itself has no boundary: $\partial(\partial S) = \varnothing$.) As a particular case of Stokes's theorem, we see that $\int_S d\omega = 0$ whenever S is a *closed* manifold, i.e., one with no boundary. This observation lets one extend the notions of closed and exact forms to general differential forms, which (together with (11)) allows one to fully set up *de Rham cohomology*.

We have already seen that 0-forms can be identified with scalar functions. Also, in Euclidean spaces one can use the inner product to identify linear functionals with vectors, and therefore 1-forms can be identified with vector fields. In the special (but very physical) case of three-dimensional Euclidean space $\mathbb{R}^3$, 2-forms can *also* be identified with vector fields via the famous *right-hand rule*,[6] and 3-forms can be identified with scalar functions by a variant of this rule. (This is an example of a concept known as *Hodge duality*.) In this case, the differentiation operation $\omega \mapsto d\omega$ can be identified with the *gradient* operation $f \mapsto \nabla f$ when ω is a 0-form, with the *curl* operation $X \mapsto \nabla \times X$ when ω is a 1-form, and with the *divergence* operation $X \mapsto \nabla \cdot X$ when ω is a 2-form. Thus, for instance, the rule (11) implies that $\nabla \times \nabla f = 0$ and $\nabla \cdot (\nabla \times X)$ for any suitably smooth scalar function f and vector field X, while various cases of Stokes's theorem (12), with this interpretation, become the various theorems about integrals of curves and surfaces in three dimensions that you may have seen referred to as "the divergence theorem," "Green's theorem," and "Stokes's theorem" in a course on several-variable calculus.

Just as the signed definite integral is connected to the unsigned definite integral in one dimension via (2), there is a connection between integration of differential forms and the Lebesgue (or Riemann) integral. On the Euclidean space $\mathbb{R}^n$ one has the n standard coordinate functions $x_1, x_2, \ldots, x_n : \mathbb{R}^n \to \mathbb{R}$. Their derivatives $dx_1, \ldots, dx_n$ are then 1-forms on $\mathbb{R}^n$. Taking their wedge product, one obtains an n-form $dx_1 \wedge \cdots \wedge dx_n$. We can multiply this by any (continuous) scalar function $f : \mathbb{R}^n \to \mathbb{R}$ to obtain another n-form $f(x) \, dx_1 \wedge \cdots \wedge dx_n$. If Ω is any open bounded domain in $\mathbb{R}^n$, we then have the identity

$$\int_\Omega f(x) \, dx_1 \wedge \cdots \wedge dx_n = \int_\Omega f(x) \, dx,$$

where on the left-hand side we have an integral of a differential form (with Ω viewed as a positively oriented n-dimensional manifold) and on the right-hand side we have the Riemann or Lebesgue integral of f on Ω. If we give Ω the negative orientation, we have to reverse the sign of the left-hand side. This correspondence generalizes (2).

There is one last operation on forms that is worth pointing out. Suppose we have a continuously differentiable map $\Phi : X \to Y$ from one manifold to another (we allow X and Y to have different dimensions). Then

of course every point x in X *pushes forward* to a point $\Phi(x)$ in Y. Similarly, if we let $v \in T_x X$ be an infinitesimal tangent vector to X based at x, then this tangent vector also *pushes forward* to a tangent vector $\Phi_* v \in T_{\Phi(x)} Y$ based at $\Phi(x)$; informally speaking, $\Phi_* v$ can be defined by requiring the infinitesimal approximation $\Phi(x + v) = \Phi(x) + \Phi_* v$. One can write $\Phi_* v = D\Phi(x)(v)$, where $D\Phi : T_x X \to T_{\Phi(x)} Y$ is the *derivative* of the several-variable map Φ at x. Finally, any k-dimensional oriented manifold S in X also pushes forward to a k-dimensional oriented manifold $\Phi(S)$ in X, although in some cases (e.g., if the image of Φ has dimension less than k) this pushed-forward manifold may be degenerate.

We have seen that integration is a duality pairing between manifolds and forms. Since manifolds push forward under Φ from X to Y, we expect forms to *pull back* from Y to X. Indeed, given any k-form ω on Y, we can define the *pullback* $\Phi^* \omega$ as the unique k-form on X such that we have the *change-of-variables formula*

$$\int_{\Phi(S)} \omega = \int_S \Phi^*(\omega).$$

In the case of 0-forms (i.e., scalar functions), the pullback $\Phi^* f : X \to \mathbb{R}$ of a scalar function $f : Y \to \mathbb{R}$ is given explicitly by $\Phi^* f(x) = f(\Phi(x))$, while the pullback of a 1-form ω is given explicitly by the formula

$$(\Phi^* \omega)_x(v) = \omega_{\Phi(x)}(\Phi_* v).$$

Similar definitions can be given for other differential forms. The pullback operation enjoys several nice properties: for instance, it respects the wedge product,

$$\Phi^*(\omega \wedge \eta) = (\Phi^* \omega) \wedge (\Phi^* \eta),$$

and the derivative,

$$d(\Phi^* \omega) = \Phi^*(d\omega).$$

By using these properties, one can recover rather painlessly the change-of-variables formulas in several-variable calculus. Moreover, the whole theory carries over effortlessly from Euclidean spaces to other manifolds. It is because of this that the theory of differential forms and integration is an indispensable tool in the modern study of manifolds, and especially in DIFFERENTIAL TOPOLOGY [IV.7].

III.17 Dimension

What is the difference between a two-dimensional set and a three-dimensional set? A rough answer that one might give is that a two-dimensional set lives inside a plane, while a three-dimensional set fills up a portion of

---

6. This is an entirely arbitrary convention; one could just as easily have used the left-hand rule to provide this identification, and apart from some harmless sign changes here and there, one gets essentially the same theory as a consequence.

space. Is this a good answer? For many sets it does seem to be: triangles, squares, and circles can be drawn in a plane, while tetrahedra, cubes, and spheres cannot. But how about the surface of a sphere? This we would normally think of as two dimensional, contrasting it with the solid sphere, which is three dimensional. But the surface of a sphere does not live inside a plane.

Does this mean that our rough definition was incorrect? Not exactly. From the perspective of linear algebra, the set $\{(x, y, z) : x^2 + y^2 + z^2 = 1\}$, which is the surface of a sphere of radius 1 in $\mathbb{R}^3$ centered at the origin, *is* three dimensional, precisely because it is not contained in a plane. (One can express this in algebraic language by saying that the affine subspace generated by the sphere is the whole of $\mathbb{R}^3$.) However, this sense of "three dimensional" does not do justice to the rough idea that the surface of a sphere has no thickness. Surely there ought to be another sense of dimension in which the surface of a sphere is two dimensional?

As this example illustrates, dimension, though very important throughout mathematics, is not a single concept. There turn out to be many natural ways of generalizing our ideas about the dimensions of simple sets such as squares and cubes, and they are often incompatible with one another, in the sense that the dimension of a set may vary according to which definition you use. The remainder of this article will set out a few different definitions.

One very basic idea we have about the dimension of a set is that it is "the number of coordinates you need to specify a point." We can use this to justify our instinct that the surface of a sphere is two dimensional: you can specify any point by giving its longitude and latitude. It is a little tricky to turn this idea into a rigorous mathematical definition because you can in fact specify a point of the sphere by means of just *one* number if you do not mind doing it in a highly artificial way. This is because you can take any two numbers and interleave the digits to form a single number from which the original two numbers can be recovered. For instance, from the two numbers $\pi = 3.141592653\ldots$ and e $= 2.718281828\ldots$ you can form the number $32.174118529821685238\ldots$, and by taking alternate digits you get back π and e again. It is even possible to find a *continuous* function f from the closed interval $[0, 1]$ (that is, the set of all real numbers between 0 and 1, inclusive) to the surface of a sphere that takes every value.

We therefore have to decide what we mean by a "natural" coordinate system. One way of making this deci-

sion leads to the definition of a *manifold*, a very important concept that is discussed in [I.3 §6.9] and also in DIFFERENTIAL TOPOLOGY [IV.7]. This is based on the idea that every point in the sphere is contained in a neighborhood N that "looks like" a piece of the plane, in the sense that there is a "nice" one-to-one correspondence ϕ between N and a subset of the Euclidean plane $\mathbb{R}^2$. Here, "nice" can have different meanings: typical ones are that ϕ and its inverse should both be continuous, or differentiable, or infinitely differentiable.

Thus, the intuitive notion that a d-dimensional set is one where you need d numbers to specify a point can be developed into a rigorous definition that tells us, as we had hoped, that the surface of a sphere is two dimensional. Now let us take another intuitive notion and see what we can get from it.

Suppose I want to cut a piece of paper into two pieces. The boundary that separates the pieces will be a curve, which we would normally like to think of as one dimensional. Why is it one dimensional? Well, we could use the same reasoning: if you cut a curve into two pieces then the part where the two pieces meet each other is a single point (or pair of points if the curve is a loop), which is zero dimensional. That is, there appears to be a sense in which a $(d - 1)$-dimensional set is needed if you want to cut a d-dimensional set into two.

Let us try to be slightly more precise about this idea. Suppose that X is a set and x and y are points in X. Let us call a set Y a *barrier* between x and y if there is no continuous path from x to y that avoids Y. For example, if X is a solid sphere of radius 2, x is the center of X, and y is a point on the boundary of X, then one possible barrier between x and y is the surface of a sphere of radius 1. With this terminology in place, we can make the following inductive definition. A finite set is zero dimensional, and in general we say that X is *at most d dimensional* if between any two points in X there is a barrier that is at most $(d - 1)$ dimensional. We also say that X is *d dimensional* if it is at most d dimensional but not at most $(d - 1)$ dimensional.

The above definition makes sense, but it runs into difficulties: one can construct a pathological set X that acts as a barrier between any two points in the plane, but contains no segment of any curve. This makes X zero dimensional and therefore makes the plane one dimensional, which is not satisfactory. A small modification to the above definition eliminates such pathologies and gives a definition that was put forward by BROUWER [VI.75]. A complete METRIC SPACE [III.56] X is said to have dimension at most d if, given any pair

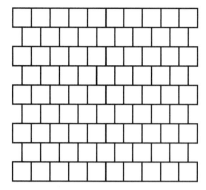

Figure 1 How to cover with squares
so that no four overlap.

of disjoint closed sets A and B, you can find disjoint open sets U and V with $A \subset U$ and $B \subset V$ such that the complement Y of $U \cup V$ (that is, everything in X that does not belong to either U or V) has dimension at most $d - 1$. The set Y is the barrier—the main difference is that we have now asked for it to be closed. The induction starts with the empty set, which has dimension -1. Brouwer's definition is known as the *inductive dimension* of a set.

Here is another basic idea that leads to a useful definition of dimension, proposed by LEBESGUE [VI.72]. Suppose you want to cover an open interval of real numbers (that is, an interval that does not contain its endpoints) with shorter open intervals. Then you will be forced to make the shorter ones overlap, but you can do it in such a way that no point is contained in more than two of your intervals: just start each new interval close to the end of the previous one.

Now suppose that you want to cover an open square (that is, one that does not contain its boundary) with smaller open squares. Again you will be forced to make the smaller squares overlap, but this time the situation is slightly worse: some points will have to be contained in three squares. However, if you take squares arranged like bricks, as in figure 1, and expand them slightly, then you can do the covering in such a way that no four squares overlap. In general, it seems that to cover a typical d-dimensional set with small open sets, you need to have overlaps of $d + 1$ sets but you do not need to have overlaps greater than this.

The precise definition that this leads to is surprisingly general: it makes sense not just for subsets of $\mathbb{R}^n$ but even for an arbitrary TOPOLOGICAL SPACE [III.90]. We say that a set X is *at most d dimensional* if, however you cover X with a finite collection of open sets

$U_1, \ldots, U_n$, you can find a finite collection of open sets $V_1, \ldots, V_m$ with the following properties:

  (i) the sets V_i also cover the whole of X;
 (ii) every V_i is a subset of at least one U_i;
(iii) no point is contained in more than $d + 1$ of the V_i.

If X is a metric space, then we can choose our U_i to have small diameter, thereby forcing the V_i to be small. So this definition is basically saying that it is possible to cover X with open sets with no $d + 2$ of them overlapping, and that these open sets can be as small as you like.

We then define the *topological dimension* of X to be the smallest d such that X is at most d dimensional. And again it can be shown that this definition assigns the "correct" dimension to the familiar shapes of elementary geometry.

A fourth intuitive idea leads to concepts known as *homological* and *cohomological* dimension. Associated with any suitable topological space X, such as a manifold, are sequences of groups known as HOMOLOGY AND COHOMOLOGY GROUPS [IV.6 §4]. Here we will discuss homology groups, but a very similar discussion is possible for cohomology. Roughly speaking, the nth homology group tells you how many interestingly different continuous maps there are from closed n-dimensional manifolds M to X. If X is a manifold of dimension less than n, then it can be shown that the nth homology group is trivial: in a sense, there is not enough room in X to define any map that is interestingly different from a constant map. On the other hand, the nth homology group of the n-sphere itself is $\mathbb{Z}$, which says that one can classify the maps from the n-sphere to itself by means of an integer parameter.

It is therefore tempting to say that a space is at least n dimensional if there is room inside it for interesting maps from n-dimensional manifolds. This thought leads to a whole class of definitions. The homological dimension of a structure X is defined to be the largest n for which some substructure of X has a nontrivial nth homology group. (It is necessary to consider substructures, because homology groups can also be trivial when there is too *much* room: it then becomes easy to deform a continuous map and show that it is equivalent to a constant map.) However, homology is a very general concept and there are many different homology theories, so there are many different notions of homological dimension. Some of these are geometric, but there are also homology theories for algebraic structures: for example, using suitable theories, one can define the homological dimension of algebraic structures such as RINGS [III.81 §1] or GROUPS [I.3 §2.1]. This is a very good example of geometrical ideas having an algebraic payoff.

Now let us turn to a fifth and final (for this article at least) intuitive idea about dimension, namely the way it affects how we measure *size*. If you want to convey how big a shape X is, then a good way of doing so is to give the length of X if X is one dimensional, the area if it is two dimensional, and the volume if it is three dimensional. Of course, this presupposes that you already know what the dimension is, but, as we shall see, there is a way of deciding which measure is the most appropriate *without* determining the dimension in advance. Then the tables are turned: we can actually *define* the dimension to be the number that corresponds to the best measure.

To do this, we use the fact that length, area, and volume scale in different ways when you expand a shape. If you take a curve and expand it by a factor of 2 (in all directions), then its length doubles. More generally, if you expand by a factor of C, then the length multiplies by C. However, if you take a two-dimensional shape and expand it by C, then its area multiplies by C^2. (Roughly speaking, this is because each little portion of the shape expands by C "in two directions" so you have to multiply the area by C twice.) And the volume of a three-dimensional shape multiplies by C^3: for instance, the volume of a sphere of radius 3 is twenty-seven times the volume of a sphere of radius 1.

It may look as though we still have to decide in advance whether we will talk about length, area, or volume before we can even begin to think about how the measurement scales when we expand the shape. But this is not the case. For instance, if we expand a square by a factor of 2, then we obtain a new square that can be divided up into four congruent copies of the original square. So, without having decided in advance that we are talking about area, we can say that the size of the new square is four times that of the old square.

This observation has a remarkable consequence: there are sets to which it is natural to assign a dimension that is not an integer! Perhaps the simplest example is a famous set first defined by CANTOR [VI.54] and now known as the *Cantor set*. This set is produced as follows. You start with the closed interval $[0, 1]$, and call it X_0. Then you form a set X_1 by removing the middle third of X_0: that is, you remove all points between $\frac{1}{3}$ and $\frac{2}{3}$, but leave $\frac{1}{3}$ and $\frac{2}{3}$ themselves. So X_1 is the union of the closed intervals $[0, \frac{1}{3}]$ and $[\frac{2}{3}, 1]$. Next, you

remove the middle thirds of these two closed intervals to produce a set X_2, so X_2 is the union of the intervals $[0, \frac{1}{9}]$, $[\frac{2}{9}, \frac{1}{3}]$, $[\frac{2}{3}, \frac{7}{9}]$, and $[\frac{8}{9}, 1]$.

In general, X_n is a union of closed intervals, and X_{n+1} is what you get by removing the middle thirds of each of these intervals—so X_{n+1} consists of twice as many intervals as X_n, but they are a third of the size. Once you have produced the sequence $X_0, X_1, X_2, \ldots$, you define the Cantor set to be the intersection of all the X_i: that is, all the real numbers that remain, no matter how far you go with the process of removing middle thirds of intervals. It is not hard to show that these are precisely the numbers whose ternary expansions consist just of 0s and 2s. (There are some numbers that have two different ternary expansions. For instance, $\frac{1}{3}$ can be written either as 0.1 or as 0.02222.... In such cases we take the recurring expansion rather than the terminating one. So $\frac{1}{3}$ belongs to the Cantor set.) Indeed, when you remove middle thirds for the nth time, you are removing all numbers that have a 1 in the nth place after the "decimal" (in fact, ternary) point.

The Cantor set has many interesting properties. For example, it is UNCOUNTABLE [III.11], but it also has MEASURE [III.55] zero. Briefly, the first of these assertions follows from the fact that there is a different element of the Cantor set for every subset A of the natural numbers (just take the ternary number $0.a_1a_2a_3\ldots$, where $a_i = 2$ whenever $i \in A$ and $a_i = 0$ otherwise), and there are uncountably many subsets of the natural numbers. To justify the second, note that the total length of the intervals making up X_n is $(\frac{2}{3})^n$ (since one removes a third of X_{n-1} to produce X_n). Since the Cantor set is contained in every X_n, its measure must be smaller than $(\frac{2}{3})^n$, whatever n is, which means that it must be zero. Thus, the Cantor set is very large in one respect and very small in another.

A further property of the Cantor set is that it is *self-similar*. The set X_1 consists of two intervals, and if you look at just one of these intervals as the middle thirds are repeatedly removed, then what you see is just like the construction of the whole Cantor set, but scaled down by a factor of 3. That is, the Cantor set consists of two copies of itself, each scaled down by a factor of 3. From this we deduce the following statement: if you expand the Cantor set by a factor of 3, then you can divide the expanded set up into two congruent copies of the original, so it is "twice as big."

What consequence should this have for the dimension of the Cantor set? Well, if the dimension is d, then the expanded set ought to be 3^d times as big. Therefore, 3^d should equal 2. This means that d should be $\log 2/\log 3$, which is roughly 0.63.

Once one knows this, the mystery of the Cantor set is lessened. As we shall see in a moment, a theory of fractional dimension can be developed with the useful property that a countable union of sets of dimension at most d has dimension at most d. Therefore, the fact that the Cantor set has dimension greater than 0 implies that it cannot be countable (since single points have dimension 0). On the other hand, because the dimension of the Cantor set is less than 1, it is *much* smaller than a one-dimensional set, so it is no surprise that its measure is zero. (This is a bit like saying that a surface has no volume, but now the two dimensions are 0.63 and 1 instead of 2 and 3.)

The most useful theory of fractional dimension is one developed by HAUSDORFF [VI.68]. One begins with a concept known as Hausdorff measure, which is a natural way of assessing the "d-dimensional volume" of a set, even if d is not an integer. Suppose you have a curve in $\mathbb{R}^3$ and you want to work out its length by considering how easy it is to cover it with spheres. A first idea might be to say that the length was the smallest you could make the sum of the diameters of the spheres. But this does not work: you might be lucky and find that a long curve was tightly wrapped up, in which case you could cover it with a single sphere of small diameter.

However, this would no longer be possible if your spheres were required to be small. Suppose, therefore, that we require all the diameters of the spheres to be at most δ. Let $L(\delta)$ be the smallest we can then get the sum of the diameters to be. The smaller δ is, the less flexibility we have, so the larger $L(\delta)$ will be. Therefore, $L(\delta)$ tends to a (possibly infinite) limit L as δ tends to 0, and we call L the length of the curve.

Now suppose that we have a smooth surface in $\mathbb{R}^3$ and want to deduce its area from information about covering it with spheres. This time, the area that you can cover with a very small sphere (so small that it meets only one portion of the surface and that portion is almost flat) will be roughly proportional to the *square* of the diameter of the sphere. But that is the only detail we need to change: let $A(\delta)$ be the smallest we can make the sum of the squares of the diameters of a set of spheres that cover the surface, if all those spheres have diameter at most δ. Then declare the area of the surface to be the limit of $A(\delta)$ as δ tends to 0. (Strictly speaking, we ought to multiply this limit by $\pi/4$, but then we get a definition that does not generalize easily.)

We have just given a way of defining length and area, for shapes in $\mathbb{R}^3$. The only difference between the two was that for length we considered the sum of the diameters of small spheres, while for area we considered the sum of the *squares* of the diameters of small spheres. In general, we define the *d-dimensional Hausdorff measure* in a similar way, but considering the sum of the dth powers of the diameters.

We can use the concept of Hausdorff measure to give a rigorous definition of fractional dimension. It is not hard to show that for any shape X there will be exactly one appropriate d, in the following sense: if c is less than d, then the c-dimensional Hausdorff measure of X is infinite, while if c is greater than d, then it is 0. (For instance, the c-dimensional Hausdorff measure of a smooth surface is 0 if $c < 2$ and infinite if $c > 2$.) This d is called the *Hausdorff dimension* of the set X. Hausdorff dimension is very useful for analyzing fractal sets, which are discussed further in DYNAMICS [IV.14].

It is important to realize that the Hausdorff dimension of a set need not equal its topological dimension. For example, the Cantor set has topological dimension zero and Hausdorff dimension $\log 2/\log 3$. A larger example is a very wiggly curve known as the *Koch snowflake*. Because it is a curve (and a single point is enough to cut it into two) it has topological dimension 1. However, because it is very wiggly, it has infinite length, and its Hausdorff dimension is in fact $\log 4/\log 3$.

III.18 Distributions
Terence Tao

A function is normally defined to be an object $f : X \to Y$ which assigns to each point x in a set X, known as the *domain*, a point $f(x)$ in another set Y, known as the *range* (see THE LANGUAGE AND GRAMMAR OF MATHEMATICS [I.2 §2.2]). Thus, the definition of functions is set-theoretic and the fundamental operation that one can perform on a function is *evaluation*: given an element x of X, one evaluates f at x to obtain the element $f(x)$ of Y.

However, there are some fields of mathematics where this may not be the best way of describing functions. In geometry, for instance, the fundamental property of a function is not necessarily how it acts on points, but rather how it *pushes forward* or *pulls back* objects that are more complicated than points (e.g., other functions, BUNDLES [IV.6 §5] and sections, SCHEMES [IV.5 §3] and sheaves, etc.). Similarly, in analysis, a function need not necessarily be defined by what it does to points, but may instead be defined by what it does to objects of different kinds, such as sets or other functions; the former leads to the notion of a *measure*; the latter to that of a *distribution*.

Of course, all these notions of function and function-like objects are related. In analysis, it is helpful to think of the various notions of a function as forming a spectrum, with very "smooth" classes of functions at one end and very "rough" ones at the other. The smooth classes of functions are very restrictive in their membership: this means that they have good properties, and there are many operations that one can perform on them (such as, for example, differentiation), but it also means that one cannot necessarily ensure that the functions one is working with belong to this category. Conversely, the rough classes of functions are very general and inclusive: it is easy to ensure that one is working with them, but the price one pays is that the number of operations one can perform on these functions is often sharply reduced (see FUNCTION SPACES [III.29]).

Nevertheless, the various classes of functions can often be treated in a unified manner, because it is often possible to approximate rough functions arbitrarily well (in an appropriate TOPOLOGY [III.90]) by smooth ones. Then, given an operation that is naturally defined for smooth functions, there is a good chance that there will be exactly one natural way to extend it to an operation on rough functions: one takes a sequence of better and better smooth approximations to the rough functions, performs the operation on them, and passes to the limit.

Distributions, or *generalized functions*, belong at the rough end of the spectrum, but before we say what they are, it will be helpful to begin by considering some smoother classes of functions, partly for comparison and partly because one obtains rough classes of functions from smooth ones by a process known as *duality*: a *linear functional* defined on a space E of functions is simply a linear map ϕ from E to the scalars $\mathbb{R}$ or $\mathbb{C}$. Typically, E is a normed space, or at least comes with a topology, and the *dual space* is the space of *continuous* linear functionals.

The class $C^{\omega}[-1, 1]$ of analytic functions.   These are in many ways the "nicest" functions of all, and include many familiar functions such as $\exp(x)$, $\sin(x)$, polynomials, and so on. However, we shall not discuss them further, because for many purposes they form too rigid a class to be useful. (For example, if an analytic func-

tion is zero everywhere on an interval, then it is forced to be zero everywhere.)

The class $C_{\mathrm{c}}^{\infty}[-1, 1]$ of test functions.   These are the smooth (that is, infinitely differentiable) functions f, defined on the interval $[-1, 1]$, that vanish on neighborhoods of 1 and -1. (That is, one can find $\delta > 0$ such that $f(x) = 0$ whenever $x > 1 - \delta$ or $x < -1 + \delta$.) They are more numerous than analytic functions and therefore more tractable for analysis. For instance, it is often useful to construct smooth "cutoff functions," which are functions that vanish outside some small set but do not vanish inside it. Also, all the operations from calculus (differentiation, integration, composition, convolution, evaluation, etc.) are available for these functions.

The class $C^0[-1, 1]$ of continuous functions.   These functions are regular enough for the notion of evaluation, $x \mapsto f(x)$, to make sense for every $x \in [-1, 1]$, and one can integrate such functions and perform algebraic operations such as multiplication and composition, but they are not regular enough that operations such as differentiation can be performed on them. Still, they are usually considered among the smoother examples of functions in analysis.

The class $L^2[-1, 1]$ of square-integrable functions. These are measurable functions $f : [-1, 1] \to \mathbb{R}$ for which the Lebesgue integral $\int_{-1}^{1} |f(x)|^2 \, \mathrm{d}x$ is finite. Usually one regards two such functions f and g as equal if the set of x such that $f(x) \neq g(x)$ has measure zero. (Thus, from the set-theoretic point of view, the object in question is really an EQUIVALENCE CLASS [I.2 §2.3] of functions.) Since a singleton $\{x\}$ has measure zero, we can change the value of $f(x)$ without changing the function. Thus, the notion of evaluation does not make sense for a square-integrable function $f(x)$ at any specific point x. However, two functions that differ on a set of measure zero have the same LEBESGUE INTEGRAL [III.55], so integration does make sense.

A key point about this class is that it is *self-dual* in the following sense. Any two functions in this class can be paired together by the *inner product* $\langle f, g \rangle = \int_{-1}^{1} f(x)g(x) \, \mathrm{d}x$. Therefore, given a function $g \in L^2[-1, 1]$, the map $f \mapsto \langle f, g \rangle$ defines a linear functional on $L^2[-1, 1]$, which turns out to be continuous. Moreover, given any continuous linear functional ϕ on $L^2[-1, 1]$, there is a unique function $g \in L^2[-1, 1]$ such that $\phi(f) = \langle f, g \rangle$ for every f. This is a special case of one of the *Riesz representation theorems*.

The class $C^0[-1, 1]^*$ of finite Borel measures. Any finite Borel MEASURE [III.55] μ gives rise to a continuous linear functional on $C^0[-1, 1]$ defined by $f \mapsto \langle \mu, f \rangle = \int_{-1}^{1} f(x) \, d\mu$. Another of the Riesz representation theorems says that every continuous linear functional on $C^0[-1, 1]$ arises in this way, so one could in principle define a finite Borel measure to be a continuous linear functional on $C^0[-1, 1]$.

The class $C_c^\infty[-1, 1]^*$ of distributions. Just as measures can be viewed as continuous linear functionals on $C^0[-1, 1]$, a *distribution* μ is a continuous linear functional on $C_c^\infty[-1, 1]$ (with an appropriate topology). Thus, a distribution can be viewed as a "virtual function": it cannot itself be directly evaluated, or even integrated over an open set, but it can still be paired with any test function $g \in C_c^\infty[-1, 1]$, producing a number $\langle \mu, g \rangle$. A famous example is the *Dirac distribution* δ_0, defined as the functional which, when paired with any test function g, returns the evaluation $g(0)$ of g at zero: $\langle \delta_0, g \rangle = g(0)$. Similarly, we have the derivative of the Dirac distribution, $-\delta_0'$, which, when paired with any test function g, returns the derivative $g'(0)$ of g at zero: $\langle -\delta_0', g \rangle = g'(0)$. (The reason for the minus sign will be given later.) Since test functions have so many operations available to them, there are many ways to define continuous linear functionals, so the class of distributions is quite large. Despite this, and despite the indirect, virtual nature of distributions, one can still define many operations on them; we shall discuss this later.

The class $C^\omega[-1, 1]^*$ of hyperfunctions. There are classes of functions more general still than distributions. For instance, there are hyperfunctions, which roughly speaking one can think of as linear functionals that can be tested only against analytic functions $g \in C^\omega[-1, 1]$ rather than against test functions $g \in C^\infty[-1, 1]$. However, as the class of analytic functions is so sparse, hyperfunctions tend not to be as useful in analysis as distributions.

At first glance, the concept of a distribution has limited utility, since all a distribution μ is empowered to do is to be tested against test functions g to produce inner products $\langle \mu, g \rangle$. However, using this inner product, one can often take operations that are initially defined only on test functions, and *extend* them to distributions by duality. A typical example is differentiation. Suppose one wants to know how to define the derivative μ' of a distribution, or in other words how to define $\langle \mu', g \rangle$

for any test function g and distribution μ. If μ is itself a test function $\mu = f$, then we can evaluate this using integration by parts (recalling that test functions vanish at -1 and 1). We have

$$
\begin{aligned}
\langle f', g \rangle &= \int_{-1}^{1} f'(x) g(x) \, dx \\
&= -\int_{-1}^{1} f(x) g'(x) \, dx = -\langle f, g' \rangle.
\end{aligned}
$$

Note that if g is a test function, then so is g'. We can therefore generalize this formula to arbitrary distributions by defining $\langle \mu', g \rangle = -\langle \mu, g' \rangle$. This is the justification for the differentiation of the Dirac distribution: $\langle \delta_0', g \rangle = -\langle \delta_0, g' \rangle = -g'(0)$.

More formally, what we have done here is to compute the adjoint of the differentiation operation (as defined on the dense space of test functions). Then we have taken adjoints again to define the differentiation operation for general distributions. This procedure is well-defined and also works for many other concepts; for instance, one can add two distributions, multiply a distribution by a smooth function, convolve two distributions, and compose distributions on both the left and the right with suitably smooth functions. One can even take Fourier transforms of distributions. For instance, the Fourier transform of the Dirac delta δ_0 is the constant function 1, and vice versa (this is essentially the Fourier inversion formula), while the distribution $\sum_{n \in \mathbb{Z}} \delta_0(x - n)$ is its own Fourier transform (this is essentially the Poisson summation formula). Thus, the space of distributions is quite a good space to work in, in that it contains a large class of functions (e.g., all measures and integrable functions), and is also closed under a large number of common operations in analysis. Because the test functions are dense in the space of distributions, the operations as defined on distributions are usually compatible with those on test functions. For instance, if f and g are test functions and $f' = g$ in the sense of distributions, then $f' = g$ will also be true in the classical sense. This often allows one to manipulate distributions as if they were test functions without fear of confusion or inaccuracy. The main operations one has to be careful about are evaluation and pointwise multiplication of distributions, both of which are usually not well-defined (e.g., the square of the Dirac delta distribution is not well-defined as a distribution).

Another way to view distributions is as the *weak limit* of test functions. A sequence of functions f_n is said to *converge weakly* to a distribution μ if $\langle f_n, g \rangle \to \langle \mu, g \rangle$

for all test functions g. For instance, if φ is a test function with total integral $\int_{-1}^{1} \varphi = 1$, then the test functions $f_n(x) = n\varphi(nx)$ can be shown to converge weakly to the Dirac delta distribution δ_0, while the functions $f'_n(x) = n^2\varphi'(nx)$ converge weakly to the derivative δ'_0 of the Dirac delta. On the other hand, the functions $g_n(x) = \cos(nx)\varphi(x)$ converge weakly to zero (this is a variant of the *Riemann–Lebesgue lemma*). Thus weak convergence has some unusual features not present in stronger notions of convergence, in that severe oscillations can sometimes "disappear" in the limit. One advantage of working with distributions instead of smoother functions is that one often has some compactness in the space of distributions under weak limits (e.g., by the Banach–Alaoglu theorem). Thus, distributions can be thought of as asymptotic extremes of behavior of smoother functions, just as real numbers can be thought of as limits of rational numbers.

Because distributions can be easily differentiated, while still being closely connected to smoother functions, they have been extremely useful in the study of partial differential equations (PDEs), particularly when the equations are linear. For instance, the general solution to a linear PDE can often be described in terms of its *fundamental solution*, which solves the PDE in the sense of distributions. More generally, distribution theory (together with related concepts, such as that of a *weak derivative*) gives an important (though certainly not the only) means to define *generalized solutions* of both linear and nonlinear PDEs. As the name suggests, these generalize the concept of smooth (or *classical*) solutions by allowing the formation of singularities, shocks, and other nonsmooth behavior. In some cases the easiest way to construct a smooth solution to a PDE is first to construct a generalized solution and then to use additional arguments to show that the generalized solution is in fact smooth.

III.19 Duality

Duality is an important general theme that has manifestations in almost every area of mathematics. Over and over again, it turns out that one can associate with a given mathematical object a related, "dual" object that helps one to understand the properties of the object one started with. Despite the importance of duality in mathematics, there is no single definition that covers all instances of the phenomenon. So let us look at a few examples and at some of the characteristic features that they exhibit.

1 Platonic Solids

Suppose you take a cube, draw points at the centers of each of its six faces, and let those points be the vertices of a new polyhedron. The polyhedron you get will be a regular octahedron. What happens if you repeat the process? If you draw a point at the center of each of the eight faces of the octahedron, you will find that these points are the eight vertices of a cube. We say that the cube and the octahedron are *dual* to one another. The same can be done for the other Platonic solids: the dodecahedron and the icosahedron are dual to one another, while the dual of a tetrahedron is again a tetrahedron.

The duality just described does more than just split up the five Platonic solids into three groups: it allows us to associate statements about a solid with statements about its dual. For instance, two faces of a dodecahedron are *adjacent* if they share an edge, and this is so if and only if the corresponding vertices of the dual icosahedron are linked by an edge. And for this reason there is also a correspondence between edges of the dodecahedron and edges of the icosahedron.

2 Points and Lines in the Projective Plane

There are several equivalent definitions of the PROJECTIVE PLANE [I.3 §6.7]. One, which we shall use here, is that it is the set of all lines in $\mathbb{R}^3$ that go through the origin. These lines we call the "points" of the projective plane. In order to visualize this set as a geometrical object and to make its "points" more point-like, it is helpful to associate each line through the origin with the pair of points in $\mathbb{R}^3$ at which it intersects the unit sphere: indeed, one can define the projective plane as the unit sphere with opposite points identified.

A typical "line" in the projective plane is the set of all "points" (that is, lines through the origin) that lie in some plane through the origin. This is associated with the great circle in which that plane intersects the unit sphere, once again with opposite points identified.

There is a natural association between lines and points in the projective plane: each point P is associated with the line L that consists of all points orthogonal to P, and each line L is associated with the single point P that is orthogonal to all points in L. For example, if P is the z-axis, then the associated projective line L is the set of all lines through the origin that lie in the xy-plane, and vice versa. This association has the following basic property: if a point P belongs to a line L, then the line associated with P contains the point associated with L.

This allows us to translate statements about points and lines into logically equivalent statements about lines and points. For example, three points are collinear (that is, they all lie in a line) if and only if the corresponding lines are concurrent (that is, there is some point that is contained in all of them). In general, once you have proved a theorem in projective geometry, you get another, dual, theorem for free (unless the dual theorem turns out to be the same as the original one).

3   Sets and Their Complements

Let X be a set. If A is any subset of X, then the *complement* of A, written A^c, is the set of all elements of X that do not belong to A. The complement of the complement of A is clearly A, so there is a kind of duality between sets and their complements. *De Morgan's laws* are the statements that $(A \cap B)^c = A^c \cup B^c$ and $(A \cup B)^c = A^c \cap B^c$: they tell us that complementation "turns intersections into unions," and vice versa. Notice that if we apply the first law to A^c and B^c, then we find that $(A^c \cap B^c)^c = A \cup B$. Taking complements of both sides of this equality gives us the second law.

Because of de Morgan's laws, any identity involving unions and intersections remains true when you interchange them. For example, one useful identity is $A \cup (B \cap C) = (A \cup B) \cap (A \cup C)$. Applying this to the complements of the sets and using de Morgan's laws, it is straightforward to deduce the equally useful identity $A \cap (B \cup C) = (A \cap B) \cup (A \cap C)$.

4   Dual Vector Spaces

Let V be a VECTOR SPACE [I.3 §2.3], over $\mathbb{R}$, say. The *dual space* V^* is defined to be the set of all *linear functionals* on V: that is, linear maps from V to $\mathbb{R}$. It is not hard to define appropriate notions of addition and scalar multiplication and show that these make V^* into a vector space as well.

Suppose that T is a LINEAR MAP [I.3 §4.2] from a vector space V to a vector space W. If we are given an element w^* of the dual space W^*, then we can use T and w^* to create an element of V^* as follows: it is the map that takes v to the real number $w^*(Tv)$. This map, which is denoted by T^*w^*, is easily checked to be linear. The function T^* is itself a linear map, called the *adjoint* of T, and it takes elements of W^* to elements of V^*.

This is a typical feature of duality: a function f from object A to object B very often gives rise to a function g from the dual of B to the dual of A.

Suppose that T^* is a surjection. Then if $v \neq v'$, we can find v^* such that $v^*(v) \neq v^*(v')$, and then $w^* \in W^*$ such that $T^*w^* = v^*$, so that $T^*w^*(v) \neq T^*w^*(v')$, and therefore $w^*(Tv) \neq w^*(Tv')$. This implies that $Tv \neq Tv'$, which proves that T is an injection. We can also prove that if T^* is an injection, then T is a surjection. Indeed, if T is not a surjection, then TV is a proper subspace of W, which allows us to find a nonzero linear functional w^* such that $w^*(Tv) = 0$ for every $v \in V$, and hence such that $T^*w^* = 0$, which contradicts the injectivity of T^*. If V and W are finite dimensional, then $(T^*)^* = T$, so in this case we find that T is an injection if and only if T^* is a surjection, and vice versa. Therefore, we can use duality to convert an existence problem into a uniqueness problem. This conversion of one kind of problem into a different kind is another characteristic and very useful feature of duality.

If a vector space has additional structure, the definition of the dual space may well change. For instance, if X is a real BANACH SPACE [III.62], then X^* is defined to be the space of all *continuous* linear functionals from X to $\mathbb{R}$, rather than the space of *all* linear functionals. This space is also a Banach space: the norm of a continuous linear functional f is defined to be $\sup\{|f(x)| : x \in X, \|x\| \leqslant 1\}$. If X is an explicit example of a Banach space (such as one of the spaces discussed in FUNCTION SPACES [III.29]), it can be extremely useful to have an explicit description of the dual space. That is, one would like to find an explicitly described Banach space Y and a way of associating with each nonzero element y of Y a nonzero continuous linear functional ϕ_y defined on X, in such a way that every continuous linear functional is equal to ϕ_y for some $y \in Y$.

From this perspective, it is more natural to regard X and Y as having the same status. We can reflect this in our notation by writing $\langle x, y \rangle$ instead of $\phi_y(x)$. If we do this, then we are drawing attention to the fact that the map $\langle \cdot, \cdot \rangle$, which takes the pair (x, y) to the real number $\langle x, y \rangle$, is a continuous bilinear map from $X \times Y$ to $\mathbb{R}$.

More generally, whenever we have two mathematical objects A and B, a set S of "scalars" of some kind, and a function $\beta : A \times B \to S$ that is a structure-preserving map in each variable separately, we can think of the elements of A as elements of the dual of B, and vice versa. Functions like β are called *pairings*.

5   Polar Bodies

Let X be a subset of $\mathbb{R}^n$ and let $\langle \cdot, \cdot \rangle$ be the standard INNER PRODUCT [III.37] on $\mathbb{R}^n$. Then the *polar* of X, denoted X°, is the set of all points $y \in \mathbb{R}^n$ such that $\langle x, y \rangle \leqslant 1$ for every $x \in X$. It is not hard to check that X° is closed and convex, and that if X is closed and convex, then $(X^\circ)^\circ = X$. Furthermore, if $n = 3$ and X is a Platonic solid centered at the origin, then X° is (a multiple of) the dual Platonic solid, and if X is the "unit ball" of a normed space (that is, the set of all points of norm at most 1), then X° is (easily identified with) the unit ball of the dual space.

6   Duals of Abelian Groups

If G is an Abelian group, then a *character* on G is a homomorphism from G to the group $\mathbb{T}$ of all complex numbers of modulus 1. Two characters can be multiplied together in an obvious way, and this multiplication makes the set of all characters on G into another Abelian group, called the *dual group*, $\hat{G}$, of the group G. Again, if G has a topological structure, then one usually imposes an additional continuity condition.

An important example is when the group is itself $\mathbb{T}$. It is not hard to show that the continuous homomorphisms from $\mathbb{T}$ to $\mathbb{T}$ all have the form $e^{i\theta} \mapsto e^{in\theta}$ for some integer n (which may be negative or zero). Thus, the dual of $\mathbb{T}$ is (isomorphic to) $\mathbb{Z}$.

This form of duality between groups is called *Pontryagin duality*. Note that there is an easily defined pairing between G and $\hat{G}$: given an element $g \in G$ and a character $\psi \in \hat{G}$, we define $\langle g, \psi \rangle$ to be $\psi(g)$.

Under suitable conditions, this pairing extends to *functions* defined on G and $\hat{G}$. For instance, if G and $\hat{G}$ are finite, and $f : G \to \mathbb{C}$ and $F : \hat{G} \to \mathbb{C}$, then we can define $\langle f, F \rangle$ to be the complex number $|G|^{-1} \sum_{g \in G} \sum_{\psi \in \hat{G}} f(g) F(\psi)$. In general, one obtains a pairing between a complex HILBERT SPACE [III.37] of functions on G and a Hilbert space of functions on $\hat{G}$.

This extended pairing leads to another important duality. Given a function in the Hilbert space $L^2(\mathbb{T})$, its *Fourier transform* is the function $\hat{f} \in \ell_2(\mathbb{Z})$ that is defined by the formula

$$\hat{f}(n) = \frac{1}{2\pi} \int_0^{2\pi} f(e^{i\theta}) e^{-in\theta} \, d\theta.$$

The Fourier transform, which can be defined similarly for functions on other Abelian groups, is immensely useful in many areas of mathematics. (See, for example, FOURIER TRANSFORMS [III.27] and REPRESENTATION THEORY [IV.9].) By contrast with some of the previous examples, it is *not* always easy to translate a statement about a function f into an equivalent statement about its Fourier transform $\hat{f}$, but this is what gives the Fourier transform its power: if you wish to understand a function f defined on $\mathbb{T}$, then you can explore its properties by looking at both f and $\hat{f}$. Some properties will follow from facts that are naturally expressed in terms of f and others from facts that are naturally expressed in terms of $\hat{f}$. Thus, the Fourier transform "doubles one's mathematical power."

7   Homology and Cohomology

Let X be a compact n-dimensional MANIFOLD [I.3 §6.9]. If M and M' are an i-dimensional submanifold and an $(n - i)$-dimensional submanifold of X, respectively, and if they are well-behaved and in sufficiently general position, then they will intersect in a finite set of points. If one assigns either 1 or -1 to each of these points in a natural way that takes account of how M and M' intersect, then the sum of the numbers at the points is an invariant called the *intersection number* of M and M'. This number turns out to depend only on the HOMOLOGY CLASSES [IV.6 §4] of M and M'. Thus, it defines a map from $H_i(X) \times H_{n-i}(X)$ to $\mathbb{Z}$, where we write $H_r(X)$ for the rth homology group of X. This map is a group homomorphism in each variable separately, and the resulting pairing leads to a notion of duality called *Poincaré duality*, and ultimately to the modern theory of *cohomology*, which is dual to homology. As with some of our other examples, many concepts associated with homology have dual concepts: for example, in homology one has a *boundary map*, whereas in cohomology there is a *coboundary map* (in the opposite direction). Another example is that a continuous map from X to Y gives rise to a homomorphism from the homology group $H_i(X)$ to the homology group $H_i(Y)$, and also to a homomorphism from the cohomology group $H^i(Y)$ to the cohomology group $H^i(X)$.

8   Further Examples Discussed in This Book

The examples above are not even close to a complete list: even in this book there are several more. For instance, the article on DIFFERENTIAL FORMS [III.16] discusses a pairing, and hence a duality, between k-forms and k-dimensional surfaces. (The pairing is given by integrating the form over the surface.) The article on

DISTRIBUTIONS [III.18] shows how to use duality to give rigorous definitions of function-like objects such as the Dirac delta function. The article on MIRROR SYMMETRY [IV.16] discusses an astonishing (and still largely conjectural) duality between CALABI–YAU MANIFOLDS [III.6] and so-called "mirror manifolds." Often the mirror manifold is much easier to understand than the original manifold, so this duality, like the Fourier transform, makes certain calculations possible that would otherwise be unthinkable. And the article on REPRESENTATION THEORY [IV.9] discusses the "Langlands dual" of certain (non-Abelian) groups: a proper understanding of this duality would solve many major open problems.

III.20   Dynamical Systems and Chaos

From a scientific point of view, a dynamical system is a physical system, such as a collection of planets or the water in a canal, that changes over time. Typically, the positions and velocities of the parts of such a system at a time t depend only on the positions and velocities of those parts just before that time, which means that the behavior of the system is governed by a system of PARTIAL DIFFERENTIAL EQUATIONS [I.3 §5.3]. Often, a very simple collection of partial differential equations can lead to very complicated behavior of the physical system.

From a mathematical point of view, a dynamical system is any mathematical object that evolves in time according to a precise rule that determines the behavior of the system at time t from its behavior just beforehand. Sometimes, as above, "just beforehand" refers to a time infinitesimally earlier, which is why calculus is involved. But there is also a vigorous theory of *discrete* dynamical systems, where the "time" t takes integer values, and the "time just before t" is $t − 1$. If f is the function that tells us how the system at time t depends on the system at time $t − 1$, then the system as a whole can be thought of as the process of *iterating* f: that is, applying f over and over again.

As with continuous dynamical systems, a very simple function f can lead to very complicated behavior if you iterate it enough times. In particular, some of the most interesting dynamical systems, both discrete ones and continuous ones, exhibit an extreme sensitivity to initial conditions, which is known as *chaos*. This is true, for example, of the equations that govern weather. One cannot hope to specify exactly the wind speed at every point on the Earth's surface (not to mention high above

it), which means that one has to make do with approximations. Because the relevant equations are chaotic, the resulting inaccuracies, which may be small to start with, rapidly propagate and overwhelm the system: you could start with a different, equally good approximation and find that after a fairly short time the system had evolved in a completely different way. This is why accurate forecasting is impossible more than a few days in advance.

For more about dynamical systems and chaos, see DYNAMICS [IV.14].

III.21   Elliptic Curves
Jordan S. Ellenberg

An elliptic curve over a field K can be defined as an algebraic curve of genus 1 over K, endowed with a point defined over K. If this definition is too abstract for your tastes, then an equivalent definition is the following: an elliptic curve is a curve in the plane determined by an equation of the form

$$y^2 + a_1xy + a_3y = x^3 + a_2x^2 + a_4x + a_6. \quad (1)$$

When the characteristic of K is not 2, we can transform this equation into the simpler form $y^2 = f(x)$, for some cubic polynomial f. In this sense, an elliptic curve is a rather concrete object. However, this definition has given rise to a subject of seemingly inexhaustible mathematical interest, which has provided a tremendous fund of ideas, examples, and problems in number theory and algebraic geometry. This is in part because there are many values of "X" for which it is the case that "the simplest interesting example of X is an elliptic curve."

For instance, the points of an elliptic curve E with coordinates in K naturally form an Abelian group, which we call $E(K)$. The connected projective VARIETIES [III.95] that admit a group law of this kind are called *Abelian varieties*; and elliptic curves are just the Abelian varieties that are one dimensional. The Mordell–Weil theorem tells us that, when K is a number field and A is an Abelian variety, $A(K)$ is actually a *finitely generated* Abelian group, called a *Mordell–Weil group*; these Abelian groups are much studied but have retained much of their mystery (see RATIONAL POINTS ON CURVES AND THE MORDELL CONJECTURE [V.29]). Even when A is an elliptic curve, in which case we would call it E instead, there is a great deal that we do not know, though THE BIRCH–SWINNERTON-DYER CONJECTURE [V.4] offers a conjectural formula for the rank of the group $E(K)$. For much more on the topic of rational points on elliptic curves, see ARITHMETIC GEOMETRY [IV.5].

Since $E(K)$ forms an Abelian group, given any prime p one can look at the subgroup of elements P such that $pP = 0$. This subgroup is called $E(K)[p]$. In particular, we can take the algebraic closure $\bar{K}$ of K and look at $E(\bar{K})[p]$. It turns out that, when K is a NUMBER FIELD [III.63] (or, for that matter, any field of characteristic not equal to p), this group is isomorphic to $(\mathbb{Z}/p\mathbb{Z})^2$, no matter what choice of E we started with. If the group is the same for all elliptic curves, why is it interesting? Because it turns out that the GALOIS GROUP [V.21] $\mathrm{Gal}(\bar{K}/K)$ permutes the set $E(\bar{K})[p]$. In fact, the action of $\mathrm{Gal}(\bar{K}/K)$ on the group $(\mathbb{Z}/p\mathbb{Z})^2$ gives rise to a REPRESENTATION [III.77] of the Galois group. This is a foundational example in the theory of *Galois representations*, which has become central to contemporary number theory. Indeed, the proof of FERMAT'S LAST THEOREM [V.10] by Andrew Wiles is in the end a theorem about the Galois representations that arise from elliptic curves. And what Wiles proved about these special Galois representations is itself a small special case of the family of conjectures known as the *Langlands program*, which proposes a thoroughgoing correspondence between Galois representations and *automorphic forms*, which are generalized versions of the classical analytic functions called MODULAR FORMS [III.59].

In another direction, if E is an elliptic curve over $\mathbb{C}$, then the set of points of E with complex coordinates, which we denote $E(\mathbb{C})$, is a COMPLEX MANIFOLD [III.88 §3]. It turns out that this manifold can always be expressed as the quotient of the complex plane by a certain group Λ of transformations. What is more, these transformations are just translations: each map sends z to $z + c$ for some complex number c. (This expression of $E(\mathbb{C})$ as a quotient is carried out with the help of ELLIPTIC FUNCTIONS [V.31].) Each elliptic curve gives rise in this way to a subset—indeed, a subgroup—of the complex numbers; the elements of this subgroup are called *periods* of the elliptic curve. This construction can be regarded as the very beginning of *Hodge theory*, a powerful branch of algebraic geometry with a reputation for extreme difficulty. (The *Hodge conjecture*, a central question in the theory, is one of the Clay Institute's million-dollar-prize problems.)

Yet another point of view is presented by the MODULI SPACE [IV.8] of elliptic curves, denoted $M_{1,1}$. This is itself a curve, but not an elliptic one. (In fact, if I am completely honest, I should say that $M_{1,1}$ is not quite a curve at all—it is an object called, depending on whom you ask, an ORBIFOLD [IV.4 §7] or an *algebraic stack*—you can think of it as a curve from which someone has removed a few points, folded the points in half or into thirds, and then glued the folded-up points back in. You might find it reassuring to know that even professionals in the subject find this process rather difficult to visualize.) The curve $M_{1,1}$ is a "simplest example" in two ways: it is the simplest *modular curve*, and simultaneously the simplest moduli space of curves.

III.22 The Euclidean Algorithm and Continued Fractions
Keith Ball

1 The Euclidean Algorithm

THE FUNDAMENTAL THEOREM OF ARITHMETIC [V.14], which states that every integer can be factored into primes in a unique way, has been known since antiquity. The usual proof depends upon what is known as the Euclidean algorithm, which constructs the highest common factor (h, say) of two numbers m and n. In doing so, it shows that h can be written in the form $am + bn$ for some pair of integers a, b (not necessarily positive). For example, the highest common factor of 17 and 7 is 1, and sure enough we can express 1 as the combination $1 = 5 \times 17 - 12 \times 7$.

The algorithm works as follows. Assume that m is larger than n and start by dividing m by n to yield a quotient q_1 and a nonnegative remainder r_1 that is less than n. Then we have

$$m = q_1 n + r_1. \tag{1}$$

Now since $r_1 < n$ we may divide n by r_1 to obtain a second quotient and remainder:

$$n = q_2 r_1 + r_2. \tag{2}$$

Continue in this way, dividing r_1 by r_2, r_2 by r_3, and so on. The remainders get smaller each time but cannot go below zero. So the process must stop at some point with a remainder of 0: that is, with a division that comes out exactly. For instance, if $m = 165$ and $n = 70$, the algorithm generates the sequence of divisions

$$165 = 2 \times 70 + 25, \tag{3}$$
$$70 = 2 \times 25 + 20, \tag{4}$$
$$25 = 1 \times 20 + 5, \tag{5}$$
$$20 = 4 \times 5 + 0. \tag{6}$$

The process guarantees that the last nonzero remainder, 5 in this case, is the highest common factor of m and n. On the one hand, the last line shows that 5 is a factor of the previous remainder 20. Now the last-but-one line shows that 5 is also a factor of the remainder 25 that occurred one step earlier, because 25 is expressed as a combination of 20 and 5. Working back up the algorithm we conclude that 5 is a factor of both $m = 165$ and $n = 70$. So 5 is certainly *a* common factor of m and n.

On the other hand, the last-but-one line shows that 5 can be written as a combination of 25 and 20 with integer coefficients. Since the previous line shows that 20 can be written as a combination of 70 and 25 we can write 5 in terms of 70 and 25:

$$5 = 25 - 20 = 25 - (70 - 2 \times 25) = 3 \times 25 - 70.$$

Continuing back up the algorithm we can express 25 in terms of 165 and 70 and conclude that

$$5 = 3 \times (165 - 2 \times 70) - 70 = 3 \times 165 - 7 \times 70.$$

This shows that 5 is the *highest* common factor of 165 and 70 because any factor of 165 and 70 would automatically be a factor of $3 \times 165 - 7 \times 70$: that is, a factor of 5. Along the way we have shown that the highest common factor can be expressed as a combination of the two original numbers m and n.

2 Continued Fractions for Numbers

During the 1500 years following Euclid, it was realized by mathematicians of the Indian and Arabic schools that the application of the Euclidean algorithm to a pair of integers m and n could be encoded in a formula for the ratio m/n. The equation (1) can be written

$$\frac{m}{n} = q_1 + \frac{r_1}{n} = q_1 + \frac{1}{F},$$

where $F = n/r_1$. Now equation (2) expresses F as

$$F = q_2 + \frac{r_2}{r_1}.$$

The next step of the algorithm will produce an expression for r_1/r_2 and so on. If the algorithm stops after k steps, then we can put these expressions together to get what is called the *continued fraction* for m/n:

$$\frac{m}{n} = q_1 + \cfrac{1}{q_2 + \cfrac{1}{q_3 + \cfrac{\cdots}{\cdots + \cfrac{1}{q_k}}}}.$$

For example,

$$\frac{165}{70} = 2 + \cfrac{1}{2 + \cfrac{1}{1 + \frac{1}{4}}}.$$

The continued fraction can be constructed directly from the ratio $165/70 = 2.35714\ldots$ without reference to the integers 165 and 70. We start by subtracting from $2.35714\ldots$ the largest whole number we can: namely 2. Now we take the reciprocal of what is left: $1/0.35714\ldots = 2.8$. Again we subtract off the largest integer we can, 2, which tells us that $q_2 = 2$. The reciprocal of 0.8 is 1.25, so $q_3 = 1$ and then, finally, $1/0.25 = 4$, so $q_4 = 4$ and the continued fraction stops.

The mathematician John Wallis, who worked in the seventeenth century, seems to have been the first to give a systematic account of continued fractions and to recognize that continued-fraction expansions exist for all numbers (not only rational numbers), provided that we allow the continued fraction to have infinitely many levels. If we start with any positive number, we can build its continued fraction in the same way as for the ratio $2.35714\ldots$. For example, if the number is $\pi = 3.14159265\ldots$, we start by subtracting 3, then take the reciprocal of what is left: $1/0.14159\ldots = 7.06251\ldots$. So for π we get that the second quotient is 7. Continuing the process we build the continued fraction

$$\pi = 3 + \cfrac{1}{7 + \cfrac{1}{15 + \cfrac{1}{1 + \cfrac{1}{292 + \frac{1}{1 + \cdots}}}}}. \qquad (7)$$

The numbers 3, 7, 15, and so on, that appear in the fraction are called the *partial quotients* of π.

The continued fraction for a real number can be used to approximate it by rational numbers. If we truncate the continued fraction after several steps, we are left with a finite continued fraction which is a rational number: for example, by truncating the fraction (7), one level down we get the familiar approximation $\pi \approx 3 + 1/7 = 22/7$; at the second level we get the approximation $3 + 1/(7 + 1/15) = 333/106$. The truncations at different levels thus generate a sequence of rational approximations: the sequence for π begins

$$3, \; 22/7, \; 333/106, \; 355/113, \; \ldots.$$

Whatever positive real number x we start with, the sequence of continued-fraction approximations will approach x as we move further down the fraction. Indeed, the formal interpretation of the equation (7) is precisely that the successive truncations of the fraction approach π.

Naturally, in order to get better approximations to a number x we need to take more "complicated" fractions—fractions with larger numerator and denominator. The continued-fraction approximations to x are *best* approximations to x in the following sense: if p/q is one of these fractions, then it is impossible to find any fraction r/s that is closer than p/q to x and that has denominator s smaller than q.

Moreover, if p/q is one of the approximations coming from the continued fraction for x, then the error $x - p/q$ cannot be too large relative to the size of the denominator q; specifically, it is always true that

$$\left| x - \frac{p}{q} \right| \leqslant \frac{1}{q^2}. \tag{8}$$

This error estimate shows just how special the continued-fraction approximations are: if you pick a denominator q without thinking, and then select the numerator p that makes p/q closest to x, the only thing you can guarantee is that x lies between $(p - 1/2)/q$ and $(p + 1/2)/q$. So the error could be as large as $1/(2q)$, which is much bigger than $1/(q^2)$ if q is a large integer.

Sometimes a continued-fraction approximation to x can have even smaller error than is guaranteed by (8). For example, the approximation $\pi \approx 355/113$ that we get by truncating (7) at the third level is exceptionally accurate, the reason being that the next partial quotient, 292, is rather large. So we are not changing the fraction much by ignoring the tail $1/(292 + 1/(1 + \cdots))$. In this sense, the most difficult number to approximate by fractions is the one with the smallest possible partial quotients, i.e., the one with all its partial quotients equal to 1. This number,

$$1 + \cfrac{1}{1 + \cfrac{1}{1 + \cdots}}, \tag{9}$$

can be easily calculated because the sequence of partial quotients is periodic: it repeats itself. If we call the number ϕ, then $\phi - 1$ is $1/(1 + 1/(1 + \cdots))$. The reciprocal of this number is exactly the continued fraction (9) for ϕ. Hence

$$\frac{1}{\phi - 1} = \phi,$$

which in turn implies that $\phi^2 - \phi = 1$. The roots of this quadratic equation are $(1 + \sqrt{5})/2 = 1.618\ldots$ and $(1 - \sqrt{5})/2 = -0.618\ldots$. Since the number we are trying to find is positive, it is the first of these roots: the so-called *golden ratio*.

It is quite easy to show that, just as (9) represents the positive solution of the equation $x^2 - x - 1 = 0$, any other periodic continued fraction represents a root of a quadratic equation. This fact seems to have been understood already in the sixteenth century. It is quite a lot trickier to prove the converse: that the continued fraction of any quadratic surd is periodic. This was established by LAGRANGE [VI.22] during the eighteenth century and is closely related to the existence of units in quadratic NUMBER FIELDS [III.63].

3   Continued Fractions for Functions

Several of the most important functions in mathematics are most easily described using infinite sums. For example, the EXPONENTIAL FUNCTION [III.25] has the infinite series

$$e^x = 1 + x + \frac{x^2}{2} + \cdots + \frac{x^n}{n!} + \cdots.$$

There are also a number of functions that have simple continued-fraction expansions: continued fractions involving a variable like x. These are probably the most important continued fractions historically.

For example, the function $x \mapsto \tan x$ has the continued fraction

$$\tan x = \cfrac{x}{1 - \cfrac{x^2}{3 - \cfrac{x^2}{5 - \cdots}}}, \tag{10}$$

valid for any value of x other than the odd multiples of $\pi/2$, where the tangent function has a vertical asymptote.

Whereas the infinite series of a function can be truncated to provide *polynomial* approximations to the function, truncation of the continued fraction provides approximations by *rational functions*: functions that are ratios of polynomials. For instance, if we truncate the fraction for the tangent after one level, then we get the approximation

$$\tan x \approx \frac{x}{1 - x^2/3} = \frac{3x}{3 - x^2}.$$

This continued fraction, and the rapidity with which its truncations approach $\tan x$, played the central role in the proof that π is irrational: that π is not the ratio of two whole numbers. The proof was found by Johann Lambert in the 1760s. He used the continued fraction to show that if x is a rational number (other than 0), then $\tan x$ is not. But $\tan \pi/4 = 1$ (which certainly is rational), so $\pi/4$ cannot be.

III.23   The Euler and Navier–Stokes Equations
Charles Fefferman

The Euler and Navier–Stokes equations describe the motion of an idealized fluid. They are important in science and engineering, yet they are very poorly understood. They present a major challenge to mathematics.

To state the equations we work in Euclidean space $\mathbb{R}^d$, with d equal to 2 or 3. Suppose that, at position $x = (x_1, \ldots, x_d) \in \mathbb{R}^d$ and at time $t \in \mathbb{R}$, the fluid is moving with a velocity vector $u(x, t) = (u_1(x, t), \ldots, u_d(x, t)) \in \mathbb{R}^d$, and the pressure in the fluid is $p(x, t) \in \mathbb{R}$. The Euler equation is

$$\left(\frac{\partial}{\partial t} + \sum_{j=1}^{d} u_j \frac{\partial}{\partial x_j} \right) u_i(x, t) = \frac{-\partial p}{\partial x_i}(x, t) \quad (i = 1, \ldots, d)$$

(1)

for all (x, t); and the Navier-Stokes equation is

$$\left(\frac{\partial}{\partial t} + \sum_{j=1}^{d} u_j \frac{\partial}{\partial x_j} \right) u_i(x, t)$$
$$= \nu \left(\sum_{j=1}^{d} \frac{\partial^2}{\partial x_j^2} \right) u_i(x, t) - \frac{\partial p}{\partial x_i}(x, t) \quad (i = 1, \ldots, d)$$

(2)

for all (x, t). Here, $\nu > 0$ is a coefficient of friction called the "viscosity" of the fluid.

In this article we restrict our attention to incompressible fluids, which means that, in addition to requiring that they satisfy (1) or (2), we also demand that

$$\operatorname{div} u \equiv \sum_{j=1}^{d} \frac{\partial u_j}{\partial x_j} = 0 \qquad (3)$$

for all (x, t). The Euler and Navier-Stokes equations are nothing but Newton's law $F = ma$ applied to an infinitesimal portion of the fluid. In fact, the vector

$$\left(\frac{\partial}{\partial t} + \sum_{j=1}^{d} u_j \frac{\partial}{\partial x_j} \right) u$$

is easily seen to be the acceleration experienced by a molecule of fluid that finds itself at position x at time t.

The forces F leading to the Euler equation arise entirely from pressure gradients (e.g., if the pressure increases with height, then there is a net force pushing the fluid down). The additional term

$$\nu \left(\sum_{j=1}^{d} \frac{\partial^2}{\partial x_j^2} \right) u$$

in (2) arises from frictional forces.

The Navier-Stokes equations agree very well with experiments on real fluids under many and varied circumstances. Since fluids are important, so are the Navier-Stokes equations.

The Euler equation is simply the limiting case $\nu = 0$ of Navier-Stokes. However, as we shall see, solutions of the Euler equation behave very differently from solutions of the Navier-Stokes equation, even when ν is small.

We want to understand the solutions of the Euler equations (1) and (3), or the Navier-Stokes equations (2) and (3), together with an initial condition

$$u(x) = u^0(x) \quad \text{for all } x \in \mathbb{R}^d, \qquad (4)$$

where $u^0(x)$ is a given initial velocity, i.e., a vector-valued function on $\mathbb{R}^d$. For consistency with (3), we assume that

$$\operatorname{div} u^0(x) = 0 \quad \text{for all } x \in \mathbb{R}^d.$$

Also, to avoid physically unreasonable conditions, such as infinite energy, we demand that $u^0(x)$, as well as $u(x, t)$ for each fixed t, should tend to zero "fast enough" as $|x| \to \infty$. We will not specify here exactly what is meant by "fast enough," but we assume from now on that we are dealing only with such rapidly decreasing velocities.

A physicist or engineer would want to know how to calculate efficiently and accurately the solution to the Navier-Stokes equations (2)-(4), and to understand how that solution behaves. A mathematician asks first whether a solution exists, and, if so, whether there is only one solution. Although the Euler equation is 250 years old and the Navier-Stokes equation well over 100 years old, there is no consensus among experts as to whether Navier-Stokes or Euler solutions exist for all time, or whether instead they "break down" at a finite time. Definitive answers supported by rigorous proofs seem a long way off.

Let us state more precisely the problem of "breakdown" for the Euler and Navier-Stokes equations. Equations (1)-(3) refer to the first and second derivatives of $u(x, t)$. It is natural to suppose that the initial velocity $u^0(x)$ in (4) has derivatives

$$\partial^\alpha u^0(x) = \left(\frac{\partial}{\partial x_1} \right)^{\alpha_1} \cdots \left(\frac{\partial}{\partial x_d} \right)^{\alpha_d} u^0(x)$$

of all orders, and that these derivatives tend to zero "fast enough" as $|x| \to \infty$. We then ask whether the Navier-Stokes equations (2)-(4), or the Euler equations (1), (3), and (4), have solutions $u(x, t)$, $p(x, t)$, defined for all $x \in \mathbb{R}^d$ and $t > 0$, such that the derivatives

$$\partial_{x,t}^\alpha u(x, t) = \left(\frac{\partial}{\partial t} \right)^{\alpha_0} \left(\frac{\partial}{\partial x_1} \right)^{\alpha_1} \cdots \left(\frac{\partial}{\partial x_d} \right)^{\alpha_d} u(x, t)$$

and $\partial_{x,t}^\alpha p(x, t)$ of all orders exist for all $x \in \mathbb{R}^d$, $t \in [0, \infty)$ (and tend to zero "fast enough" as $|x| \to \infty$). A pair u and p with these properties is called a "smooth" solution for the Euler or Navier-Stokes equations. No one knows whether such solutions exist (in the three-dimensional case). It is known that, for some positive time $T = T(u^0) > 0$ depending on the initial velocity

u^0 in (4), there exist smooth solutions $u(x,t)$, $p(x,t)$ to the Euler or Navier-Stokes equations, defined for $x \in \mathbb{R}^d$ and $t \in [0,T)$.

In two space dimensions (one speaks of "2D Euler" or "2D Navier-Stokes"), we can take $T = +\infty$; in other words, there is no "breakdown" for 2D Euler or 2D Navier-Stokes. In three space dimensions, no one can rule out the possibility that, for some finite $T = T(u^0)$ as above, there is an Euler or Navier-Stokes solution $u(x,t)$, $p(x,t)$, which is defined and smooth on

$$\Omega = \{(x,t) : x \in \mathbb{R}^3,\ t \in [0,T)\},$$

such that some derivative $|\partial_{x,t}^\alpha u(x,t)|$ or $|\partial_{x,t}^\alpha p(x,t)|$ is unbounded on Ω. This would imply that there is no smooth solution past time T. (We say that the 3D Navier-Stokes or Euler solution "breaks down" at time T.) Perhaps this can actually happen for 3D Euler and/or Navier-Stokes. No one knows what to believe.

Many computer simulations of the 3D Navier-Stokes and Euler equations have been carried out. Navier-Stokes simulations exhibit no evidence of breakdown, but this may mean only that initial velocities u^0 that lead to breakdown are exceedingly rare. Solutions of 3D Euler behave very wildly, so that it is hard to decide whether a given numerical study indicates a breakdown. Indeed, it is notoriously hard to perform a reliable numerical simulation of the 3D Euler equations.

It is useful to study how a Navier-Stokes or Euler solution behaves if one assumes that there is a breakdown. For instance, if there is a breakdown at time $T < \infty$ for the 3D Euler equation, then a theorem of Beale, Kato, and Majda asserts that the "vorticity"

$$\omega(x,t) = \mathrm{curl}(u(x,t))$$
$$= \left(\frac{\partial u_2}{\partial x_3} - \frac{\partial u_3}{\partial x_2}, \frac{\partial u_3}{\partial x_1} - \frac{\partial u_1}{\partial x_3}, \frac{\partial u_1}{\partial x_2} - \frac{\partial u_2}{\partial x_1} \right) \quad (5)$$

grows so large as $t \to T$ that the integral

$$\int_0^T \left(\max_{x \in \mathbb{R}^3} |\omega(x,t)| \right) dt$$

diverges. This has been used to invalidate some plausible computer simulations that allegedly indicated a breakdown for 3D Euler. It is also known that the direction of the vorticity vector $\omega(x,t)$ must vary wildly with x, as t approaches a finite breakdown time T.

The vector ω in (5) has a natural physical meaning: it indicates how the fluid is rotating about the point x at time t. A small pinwheel placed in the fluid in position x at time t with its axis of rotation oriented parallel to $\omega(x,t)$ would be turned by the fluid at an angular velocity $|\omega(x,t)|$.

For the 3D Navier-Stokes equation, a recent result of V. Sverak shows that if there is a breakdown, then the pressure $p(x,t)$ is unbounded, both above and below.

A promising idea, pioneered by J. Leray in the 1930s, is to study "weak solutions" of the Navier-Stokes equations. The idea is as follows. At first glance, the Navier-Stokes equations (2) and (3) make sense only when $u(x,t)$, $p(x,t)$ are sufficiently smooth: for example, one would like the second derivatives of u with respect to the x_j to exist. However, a formal calculation shows that (2) and (3) are apparently equivalent to conditions that we shall call (2′) and (3′), which make sense even when $u(x,t)$ and $p(x,t)$ are very rough. Let us first see how to derive (2′) and (3′), and then we will discuss their use.

The starting point is the observation that a function F on $\mathbb{R}^n$ is equal to zero if and only if $\int_{\mathbb{R}^n} F\theta\, dx = 0$ for every smooth function θ. Applying this remark to the 3D Navier-Stokes equations (2) and (3) and performing a simple formal computation (an integration by parts), we find that (2) and (3) are equivalent to the following equations:

$$\iint_{\mathbb{R}^3 \times (0,\infty)} \left\{ -\sum_{i=1}^3 u_i \frac{\partial \theta_i}{\partial t} - \sum_{i,j=1}^3 u_i u_j \left(\frac{\partial \theta_i}{\partial x_j} \right) \right\} dx\, dt$$
$$= \iint_{\mathbb{R}^3 \times (0,\infty)} \left\{ \nu \sum_{i,j=1}^3 \left(\frac{\partial^2}{\partial x_j^2} \theta_i \right) u_i + \left(\sum_{i=1}^3 \frac{\partial \theta_i}{\partial x_i} \right) p \right\} dx\, dt$$
$$(2')$$

and

$$\iint_{\mathbb{R}^3 \times (0,\infty)} \left\{ \sum_{i=1}^3 u_i \frac{\partial \varphi}{\partial x_i} \right\} dx\, dt = 0. \quad (3')$$

More precisely, given any smooth functions $u(x,t)$ and $p(x,t)$, equations (2) and (3) hold if and only if (2′) and (3′) are satisfied for arbitrary smooth functions $\theta_1(x,t)$, $\theta_2(x,t)$, $\theta_3(x,t)$, and $\varphi(x,t)$ that vanish outside a compact subset of $\mathbb{R}^3 \times (0,\infty)$.

We call θ_1, θ_2, θ_3, and ϕ *test functions*, and we say that u and p form a *weak solution* of 3D Navier-Stokes. Since all the derivatives in (2′) and (3′) are applied to smooth test functions, equations (2′) and (3′) make sense even for very rough functions u and p. To summarize, we have the following conclusion.

A smooth pair (u, p) solves 3D Navier-Stokes if and only if it is a weak solution. However, the idea of a weak solution makes sense even for rough (u, p).

We hope to use weak solutions, by carrying out the following plan.

Step (i): prove that suitable weak solutions exist for 3D Navier–Stokes on all of $\mathbb{R}^3 \times (0, \infty)$.

Step (ii): prove that any suitable weak solution of 3D Navier–Stokes must be smooth.

Step (iii): conclude that the suitable weak solution constructed in step (i) is in fact a smooth solution of the 3D Navier–Stokes equations on all of $\mathbb{R}^3 \times (0, \infty)$.

Here, "suitable" means "not too big"; we omit the precise definition.

Analogues of the above plan have succeeded for interesting partial differential equations. But for 3D Navier–Stokes, the plan has been only partly carried out. It has been known for a long time how to construct suitable weak solutions of 3D Navier–Stokes, but the uniqueness of these solutions has not been proved. Thanks to the work of Sheffer, of Lin, and of Caffarelli, Kohn, and Nirenberg, it is known that any suitable weak solution to 3D Navier–Stokes must be smooth (i.e., it must possess derivatives of all orders), outside a set $E \subset \mathbb{R}^3 \times (0, \infty)$ of small FRACTAL DIMENSION [III.17]. In particular, E cannot contain a curve. To rule out a breakdown, one would have to show that E is the empty set.

For the Euler equation, weak solutions again make sense, but examples due to Sheffer and Shnirelman show that they can behave very strangely. A two-dimensional fluid that is initially at rest and subject to no outside forces can suddenly start moving in a bounded region of space and then return to rest. Such behavior can occur for a weak solution of 2D Euler.

The Navier–Stokes and Euler equations give rise to a number of fundamental problems in addition to the breakdown problem discussed above. We finish this article with one such problem. Suppose that we fix an initial velocity $u^0(x)$ for the 3D Navier–Stokes or Euler equation. The energy E_0 at time $t = 0$ is given by

$$E_0 = \frac{1}{2} \int_{\mathbb{R}^3} |u(x, 0)|^2 \, \mathrm{d}x.$$

For $v \geqslant 0$, let $u^{(v)}(x, t) = (u_1^{(v)}, u_2^{(v)}, u_3^{(v)})$ denote the Navier–Stokes solution with initial velocity u^0 and with viscosity v. (If $v = 0$, then $u^{(0)}$ is an Euler solution.) We assume that $u^{(v)}$ exists for all time, at least when $v > 0$. The energy for $u^{(v)}(x, t)$ at time $t \geqslant 0$ is given by

$$E^{(v)}(t) = \frac{1}{2} \int_{\mathbb{R}^3} |u^{(v)}(x, t)|^2 \, \mathrm{d}x.$$

An elementary calculation based on (1)–(3) (we multiply (1) or (2) by $u_i(x)$, sum over i, integrate over all $x \in \mathbb{R}^3$,

and integrate by parts) shows that

$$\frac{\mathrm{d}}{\mathrm{d}t} E^{(v)}(t) = -\frac{1}{2} v \int_{\mathbb{R}^3} \sum_{i,j=1}^{3} \left(\frac{\partial u_i^{(v)}}{\partial x_j} \right)^2 \mathrm{d}x. \tag{6}$$

In particular, for the Euler equation we have $v = 0$, and (6) shows that the energy is equal to E_0, independently of time, as long as the solution exists.

Now suppose that v is small but nonzero. From (6) it is natural to guess that $|(\mathrm{d}/\mathrm{d}t) E^{(v)}(t)|$ is small when v is small, so that the energy remains almost constant for a long time. However, numerical and physical experiments suggest strongly that this is not the case. Instead, it seems that there exists $T_0 > 0$, depending on u^0 but independent of v, such that the fluid loses at least half of its initial energy by time T_0, regardless of how small v is (provided that $v > 0$).

It would be very important if one could prove (or disprove) this assertion. We need to understand why a tiny viscosity dissipates a lot of energy.

III.24 Expanders
Avi Wigderson

---

1 The Basic Definition

An expander is a special sort of GRAPH [III.34] that has remarkable properties and many applications. Roughly speaking, it is a graph that is very hard to disconnect because every set of vertices in the graph is joined by many edges to its complement. More precisely, we say that a graph with n vertices is a *c-expander* if for every $m \leqslant \frac{1}{2} n$ and every set S of m vertices there are at least cm edges between S and the complement of S.

This definition is particularly interesting when G is sparse: in other words, when G has few edges. We shall concentrate on the important special case where G is *regular of degree d* for some fixed constant d that is independent of the number n of vertices: this means that every vertex is joined to exactly d others. When G is regular of degree d, the number of edges from S to its complement is obviously at most dm, so if c is some fixed constant (that is, not tending to zero with n), then the number of edges between any set of vertices and its complement is within a constant of the largest number possible. As this comment suggests, we are usually interested not in single graphs but in infinite families of graphs: we say that an infinite family of d-regular graphs is a *family of expanders* if there is a constant $c > 0$ such that each graph in the family is a c-expander.

2 The Existence of Expanders

The first person to prove that expanders exist was Pinkser, who proved that if n is large and $d \geqslant 3$, then almost every d-regular graph with n vertices is an expander. That is, he proved that there is a constant $c > 0$ such that for every fixed $d \geqslant 3$, the proportion of d-regular graphs with n vertices that are *not* expanders tends to zero as n tends to infinity. This proof was an early example of the PROBABILISTIC METHOD [IV.19 §3] in combinatorics. It is not hard to see that if a d-regular graph is chosen uniformly at random, then the *expected* number of edges leaving a set S is $d|S|(n - |S|)/n$, which is at least $(\frac{1}{2}d)|S|$. Standard "tail estimates" are then used to prove that, for any fixed S, the probability that the number of edges leaving S is significantly different from its expected value is extremely small: so small that if we add up the probabilities for all sets, then even the sum is small. So with high probability all sets S have at least $c|S|$ edges to their complement. (In one respect this description is misleading: it is not a straightforward matter to discuss probabilities of events concerning random d-regular graphs because the edges are not independently chosen. However, Bollobás has defined an equivalent model for random regular graphs that allows them to be handled.)

Note that this proof does not give us an explicit description of any expander: it merely proves that they exist in abundance. This is a drawback to the proof, because, as we shall see later, there are applications for expanders that depend on some kind of explicit description, or at least on an efficient method of producing expanders. But what exactly is an "explicit description" or an "efficient method"? There are many possible answers to this question, of which we shall discuss two. The first is to demand that there is an algorithm that can list, for any integer n, all the vertices and edges of a d-regular c-expander with around n vertices (we could be flexible about this and ask for the number of vertices to be between n and n^2, say) in a time that is polynomial in n. (See COMPUTATIONAL COMPLEXITY [IV.20 §2] for a discussion of polynomial-time algorithms.) Descriptions of this kind are sometimes called "mildly explicit."

To get an idea of what is "mild" about this, consider the following graph. Its vertices are all 01 sequences of length k, and two such sequences are joined by an edge if they differ in exactly one place. This graph is sometimes called the *discrete cube* in k dimensions. It has 2^k vertices, so the time taken to list all the vertices and edges will be huge compared with k. However, for many purposes we do not actually need such a list: what matters is that there is a concise way of representing each vertex, and an efficient algorithm for listing the (representations of the) neighbors of any given vertex. Here the 01 sequence itself is a very concise representation, and given such a sequence σ it is very easy to list, in a time that is polynomial in k rather than 2^k, the k sequences that can be obtained by altering σ in one place. Graphs that can be efficiently described in this way (so that listing the neighbors of a vertex takes a time that is polynomial in the *logarithm* of the number of vertices) are called *strongly explicit*.

The quest for explicitly constructed expanders has been the source of some beautiful mathematics, which has often used ideas from fields such as number theory and algebra. The first explicit expander was discovered by Margulis. We give his construction and another one; we stress that although these constructions are very simple to describe, it is rather less easy to prove that they really are expanders.

Margulis's construction gives an 8-regular graph G_m for every integer m. The vertex set is $\mathbb{Z}_m \times \mathbb{Z}_m$, where $\mathbb{Z}_m$ is the set of all integers mod m. The neighbors of the vertex (x, y) are $(x + y, y)$, $(x - y, y)$, $(x, y + x)$, $(x, y - x)$, $(x + y + 1, y)$, $(x - y + 1, y)$, $(x, y + x + 1)$, $(x, y - x + 1)$ (all operations are mod m). Margulis's proof that G_m is an expander was based on REPRESENTATION THEORY [IV.9] and did not provide any specific bound on the expansion constant c. Gabber and Galil later derived such a bound using HARMONIC ANALYSIS [IV.11]. Note that this family of graphs is strongly explicit.

Another construction provides, for each prime p, a 3-regular graph with p vertices. This time the vertex set is $\mathbb{Z}_p$, and a vertex x is connected to $x + 1$, $x - 1$, and x^{-1} (where this is the inverse of x mod p, and we define the inverse of 0 to be 0). The proof that these graphs are expanders depends on a deep result in number theory, called the Selberg $3/16$ theorem. This family is only mildly explicit, since we are at present unable to generate large primes deterministically.

Until recently, the only known methods for explicitly constructing expanders were algebraic. However, in 2002 Reingold, Vadhan, and Wigderson introduced the so-called zigzag product of graphs, and used it to give a combinatorial, iterative construction of expanders.

3   Expanders and Eigenvalues

The condition that a graph should be a c-expander involves all subsets of the vertices. Since there are exponentially many subsets, it would seem on the face of it that checking whether a graph is a c-expander is an exponentially long task. And, indeed, this problem turns out to be CO-NP COMPLETE [IV.20 §§3, 4]. However, we shall now describe a closely related property that can be checked in polynomial time, and which is in some ways more natural.

Given a graph G with n vertices, its *adjacency matrix* A is the $n \times n$ matrix where A_{uv} is defined to be 1 if u is joined to v and 0 otherwise. This matrix is real and symmetric, and therefore has n real EIGENVALUES [I.3 §4.3] $\lambda_1, \lambda_2, \ldots, \lambda_n$, which we name in such a way that $\lambda_1 \geqslant \lambda_2 \geqslant \cdots \geqslant \lambda_n$. Moreover, EIGENVECTORS [I.3 §4.3] with distinct eigenvalues are orthogonal.

It turns out that these eigenvalues encode a great deal of useful information about G. But before we come to this, let us briefly consider how A acts as a linear map. If we are given a function f, defined on the vertices of G, then Af is the function whose value at u is the sum of $f(v)$ over all neighbors v of u. From this we see immediately that if G is d-regular and f is the function that is 1 at every vertex, then Af is the function that is d at every vertex. In other words, a constant function is an eigenvector of A with eigenvalue d. It is also not hard to see that this is the largest possible eigenvalue λ_1, and that if the graph is connected, then the second largest eigenvalue λ_2 will be strictly less than d.

In fact, the relationship between λ_2 and connectivity properties of the graph is considerably deeper than this: roughly speaking, the further away λ_2 is from d, the bigger the expansion parameter c of the graph. More precisely, it can be shown that c lies between $\frac{1}{2}(d - \lambda_2)$ and $\sqrt{2d(d - \lambda_2)}$. From this it follows that an infinite family of d-regular graphs is a family of expanders if and only if there is some constant $a > 0$ such that the *spectral gaps* $d - \lambda_2$ are at least a for every graph in the family. One of the many reasons these bounds on c are important is that although, as we have remarked, it is hard to test whether a graph is a c-expander, its second largest eigenvalue can be computed in polynomial time. So we can at least obtain estimates for how good the expansion properties of a graph are.

Another important parameter of a d-regular graph G is the largest absolute value of any eigenvalue apart from λ_1; this parameter is denoted by $\lambda(G)$. If $\lambda(G)$ is small, then G behaves in many respects like a random d-regular graph. For example, let A and B be two disjoint sets of vertices. If G were random, a small calculation shows that we would expect the number $E(A, B)$ of edges from A to B to be about $d|A|\,|B|/n$. It can be shown that, for any two disjoint sets in any d-regular graph G, $E(A, B)$ will differ from this expected amount by at most $\lambda(G)\sqrt{|A|\,|B|}$. Therefore, if $\lambda(G)$ is a small fraction of d, then between any two reasonably large sets A and B we get roughly the number of edges that we expect. This shows that graphs for which $\lambda(G)$ is small "behave like random graphs."

It is natural to ask how small $\lambda(G)$ can be in d-regular graphs. Alon and Boppana proved that it was always at least $2\sqrt{d - 1} - g(n)$ for a certain function g that tends to zero as n increases. Friedman proved that almost all d-regular graphs G with n vertices have $\lambda(G) \leqslant 2\sqrt{d - 1} + h(n)$, where $h(n)$ tends to zero, so a typical d-regular graph comes very close to matching the best possible bound for $\lambda(G)$. The proof was a tour de force. Even more remarkably, it is possible to match the lower bound with *explicit* constructions: the famous Ramanujan graphs of Lubotzky, Philips, and Sarnak, and, independently, Margulis. They constructed, for each d such that $d - 1$ is a prime power, a family of d-regular graphs G with $\lambda(G) = 2\sqrt{d - 1}$.

4   Applications of Expanders

Perhaps the most obvious use for expanders is in communication networks. The fact that expanders are highly connected means that such a network is highly "fault tolerant," in the sense that one cannot cut off part of the network without destroying a large number of individual communication lines. Further desirable properties of such a network, such as a small diameter, follow from an analysis of random walks on expanders.

A *random walk* of length m on a d-regular graph G is a path $v_0, v_1, \ldots, v_m$, where each v_i is a randomly chosen neighbor of v_{i-1}. Random walks on graphs can be used to model many phenomena, and one of the questions one frequently asks about a random walk is how rapidly it "mixes." That is, how large does m have to be before the probability that $v_m = v$ is approximately the same for all vertices v?

If we let $p_k(v)$ be the probability that $v_k = v$, then it is not hard to show that $p_{k+1} = d^{-1} A p_k$. In other words, the *transition matrix* T of the random walk, which tells you how the distribution after $k + 1$ steps depends on the distribution after k steps, is d^{-1} times the adjacency matrix A. Therefore, its largest eigenvalue is 1, and if $\lambda(G)$ is small then all other eigenvalues are small.

Suppose that this is the case, and let p be any PROBABILITY DISTRIBUTION [III.71] on the vertices of G. Then we can write p as a linear combination $\sum_i u_i$, where u_i is an eigenvector of T with eigenvalue $d^{-1}\lambda_i$. If T is applied k times, then the new distribution will be $\sum_i (d^{-1}\lambda_i)^k u_i$. If $\lambda(G)$ is small, then $(d^{-1}\lambda_i)^k$ tends rapidly to zero, except that it equals 1 when $i = 1$. In other words, after a short time, the "nonconstant part" of p goes to zero and we are left with the uniform distribution.

Thus, random walks on expanders mix rapidly. This property is at the heart of some of the applications of expanders. For example, suppose that V is a large set, f is a function from V to the interval $[0,1]$, and we wish to estimate quickly and accurately the average of f. A natural idea is to choose a random sample $v_1, v_2, \dots, v_k$ of points in V and calculate the average $k^{-1}\sum_{i=1}^{k} f(v_i)$. If k is large and the v_i are chosen independently, then it is not too hard to prove that this sample average will almost certainly be close to the true average: the probability that they differ by more than ϵ is at most $e^{-\epsilon^2 k}$.

This idea is very simple, but actually implementing it requires a source of randomness. In theoretical computer science, randomness is regarded as a resource, and it is desirable to use less of it if one can. The above procedure needed about $\log(|V|)$ bits of randomness for each v_i, so $k\log(|V|)$ bits in all. Can we do better? Ajtai, Komlós, and Szemerédi showed that the answer is yes: big time! What one does is associate V with the vertices of an explicit expander. Then, instead of choosing $v_1, v_2, \dots, v_k$ independently, one chooses them to be the vertices of a random walk in this expanding graph, starting at a random point v_1 of V. The randomness needed for this is far smaller: $\log(|V|)$ bits for v_1 and $\log(d)$ bits for each further v_i, making $\log(|V|) + k\log(d)$ bits in all. Since V is very large and d is a fixed constant, this is a big saving: we essentially pay only for the first sample point.

But is this sample any good? Clearly there is a heavy dependence between the v_i. However, it can be shown that *nothing* is lost in accuracy: again, the probability that the estimate differs from the true mean by more than ϵ is at most $e^{-\epsilon^2 k}$. Thus, there are no costs attached to the big saving in randomness.

This is just one of a huge number of applications of expanders, which include both practical applications and applications in pure mathematics. For instance, they were used by Gromov to give counterexamples to certain variants of the famous BAUM–CONNES CONJECTURE [IV.15 §4.4]. And certain bipartite graphs called "lossless expanders" have been used to produce linear codes with efficient decodings. (See RELIABLE TRANSMISSION OF INFORMATION [VII.6] for a description of what this means.)

III.25 The Exponential and Logarithmic Functions

1 Exponentiation

The following is a very well-known mathematical sequence: 2, 4, 8, 16, 32, 64, 128, 256, 512, 1024, ... Each term in this sequence is twice the term before, so, for instance, 128, the seventh term in the sequence, is equal to $2 \times 2 \times 2 \times 2 \times 2 \times 2 \times 2$. Since repeated multiplications of this kind occur throughout mathematics, it is useful to have a less cumbersome notation for them, so $2 \times 2 \times 2 \times 2 \times 2 \times 2 \times 2$ is normally written as 2^7, which we read as "2 to the power 7" or just "2 to the 7." More generally, if a is any real number and m is any positive integer, then a^m stands for $a \times a \times \cdots \times a$, where there are m as in the product. This product is called "a to the m," and numbers of the form a^m are called the *powers* of a.

The process of raising a number to a power is known as *exponentiation*. (The number m is called the *exponent*.) A fundamental fact about exponentiation is the following identity:

$$a^{m+n} = a^m \cdot a^n$$

This says that exponentiation "turns addition into multiplication." It is easy to see why this identity must be true if one looks at a small example and temporarily reverts to the old, cumbersome notation. For instance,

$$2^7 = 2 \times 2 \times 2 \times 2 \times 2 \times 2 \times 2$$
$$= (2 \times 2 \times 2) \times (2 \times 2 \times 2 \times 2)$$
$$= 2^3 \times 2^4.$$

Suppose now that we are asked to evaluate $2^{3/2}$. At first sight, the question seems misconceived: an essential part of the definition of 2^m that has just been given was that m was a positive integer. The idea of multiplying one-and-a-half 2s together does not make sense. However, mathematicians like to generalize, and even if we cannot immediately make sense of 2^m except when

m is a positive integer, there is nothing to stop us inventing a meaning for it for a wider class of numbers.

The more natural we make our generalization, the more interesting and useful it is likely to be. And the way we make it natural is to ensure that at all costs we keep the property of "turning addition into multiplication." This, it turns out, leaves us with only one sensible choice for what $2^{3/2}$ should be. If the fundamental property is to be preserved, then we must have

$$2^{3/2} \cdot 2^{3/2} = 2^{3/2+3/2} = 2^3 = 8.$$

Therefore, $2^{3/2}$ has to be $\pm\sqrt{8}$. It turns out to be convenient to take $2^{3/2}$ to be positive, so we define $2^{3/2}$ to be $\sqrt{8}$.

A similar argument shows that 2^0 should be defined to be 1: if we wish to keep the fundamental property, then

$$2 = 2^1 = 2^{1+0} = 2^1 \cdot 2^0 = 2 \cdot 2^0.$$

Dividing both sides by 2 gives the answer $2^0 = 1$.

What we are doing with these kinds of arguments is solving a *functional equation*, that is, an equation where the unknown is a function. So that we can see this more clearly, let us write $f(t)$ for 2^t. The information we are given is the fundamental property $f(t + u) = f(t)f(u)$ together with one value, $f(1) = 2$, to get us started. From this we wish to deduce as much as we can about f.

It is a nice exercise to show that the two conditions we have placed on f determine the value of f at every rational number, at least if f is assumed to be positive. For instance, to show that $f(0)$ should be 1, we note that $f(0)f(1) = f(1)$, and we have already shown that $f(3/2)$ must be $\sqrt{8}$. The rest of the proof is in a similar spirit to these arguments, and the conclusion is that $f(p/q)$ must be the qth root of 2^p. More generally, the only sensible definition of $a^{p/q}$ is the qth root of a^p.

We have now extracted everything we can from the functional equation, but we have made sense of a^t only if t is a rational number. Can we give a sensible definition when t is irrational? For example, what would be the most natural definition of $2^{\sqrt{2}}$? Since the functional equation alone does not determine what $2^{\sqrt{2}}$ should be, the way to answer a question like this is to look for some natural additional property that f might have that would, together with the functional equation, specify f uniquely. It turns out that there are two obvious choices, both of which work. The first is that f should be an *increasing* function: that is, if s is less than t, then

$f(s)$ is less than $f(t)$. Alternatively, one can assume that f is CONTINUOUS [I.3 §5.2].

Let us see how the first property can in principle be used to work out $2^{\sqrt{2}}$. The idea is not to calculate it directly but to obtain better and better *estimates*. For instance, since $1.4 < \sqrt{2} < 1.5$ the order property tells us that $2^{\sqrt{2}}$ should lie between $2^{7/5}$ and $2^{3/2}$, and in general that if $p/q < \sqrt{2} < r/s$ then $2^{\sqrt{2}}$ should lie between $2^{p/q}$ and $2^{r/s}$. It can be shown that if two rational numbers p/q and r/s are very close to each other, then $2^{p/q}$ and $2^{r/s}$ are also close. It follows that as we choose fractions p/q and r/s that are closer and closer together, so the resulting numbers $2^{p/q}$ and $2^{r/s}$ converge to some limit, and this limit we call $2^{\sqrt{2}}$.

2   The Exponential Function

One of the hallmarks of a truly important concept in mathematics is that it can be defined in many different but equivalent ways. The exponential function $\exp(x)$ very definitely has this property. Perhaps the most basic way to think of it, though for most purposes not the best, is that $\exp(x) = e^x$, where e is a number whose decimal expansion begins 2.7182818. Why do we focus on this number? One property that singles it out is that if we differentiate the function $\exp(x) = e^x$, then we obtain e^x again—and e is the only number for which that is true. Indeed, this leads to a second way of defining the exponential function: it is the only solution of the differential equation $f'(x) = f(x)$ that satisfies the initial condition $f(0) = 1$.

A third way to define $\exp(x)$, and one that is often chosen in textbooks, is as the limit of a power series:

$$\exp(x) = 1 + x + \frac{x^2}{2!} + \frac{x^3}{3!} + \cdots,$$

known as the *Taylor series* of $\exp(x)$. It is not immediately obvious that the right-hand side of this definition gives us some number raised to the power x, which is why we are using the notation $\exp(x)$ rather than e^x. However, with a bit of work one can verify that it yields the basic properties $\exp(x+y) = \exp(x)\exp(y)$, $\exp(0) = 1$, and $(d/dx)\exp(x) = \exp(x)$.

There is yet another way to define the exponential function, and this one comes much closer to telling us what it really means. Suppose you wish to invest some money for ten years and are given the following choice: either you can add 100% to your investment (that is, double it) at the end of the ten years, or each year you can take whatever you have and increase it by 10%. Which would you prefer?

The second is the better investment because in the second case the interest is *compounded*: for instance, if you start with $100, then after a year you will have $110 and after two years you will have $121. The increase of $11 in the second year breaks down as 10% interest on the original $100 plus a further dollar, which is 10% interest on the interest earned in the first year. Under the second scheme, the amount of money you end up with is $100 times $(1.1)^{10}$, since each year it multiplies by 1.1. The approximate value of $(1.1)^{10}$ is 2.5937, so you will get almost $260 instead of $200.

What if you compounded your interest monthly? Instead of multiplying your investment by $1\frac{1}{10}$ ten times, you would multiply it by $1\frac{1}{120}$ 120 times. By the end of ten years your $100 would have been multiplied by $(1 + \frac{1}{120})^{120}$, which is approximately 2.707. If you compounded it daily, you could increase this to approximately 2.718, which is suspiciously close to e. In fact, e can be defined as the limit, as n tends to infinity, of the number $(1 + \frac{1}{n})^n$.

It is not instantly obvious that this expression really does tend to a limit. For any fixed power m, the limit of $(1 + \frac{1}{n})^m$ as n tends to infinity is 1, while for any fixed n, the limit as m tends to infinity is ∞. When it comes to $(1 + \frac{1}{n})^n$, the increase in the power just compensates for the decrease in the number $1 + \frac{1}{n}$ and we get a limit between 2 and 3. If x is any real number, then $(1 + \frac{x}{n})^n$ also converges to a limit, and this we define to be $\exp(x)$.

Here is a sketch of an argument that shows that if we define $\exp(x)$ in this way, then we obtain the main property that we need if our definition is to be a good one, namely $\exp(x)\exp(y) = \exp(x + y)$. Let us take

$$\left(1 + \frac{x}{n}\right)^n \left(1 + \frac{y}{n}\right)^n,$$

which equals

$$\left(1 + \frac{x}{n} + \frac{y}{n} + \frac{xy}{n^2}\right)^n.$$

Now the ratio of $1 + x/n + y/n + xy/n^2$ to $1 + x/n + y/n$ is smaller than $1 + xy/n^2$, and $(1 + xy/n^2)^n$ can be shown to converge to 1 (as here the increase in n is not enough to compensate for the rapid decrease in xy/n^2). Therefore, for large n the number we have is very close to

$$\left(1 + \frac{x + y}{n}\right)^n.$$

Letting n tend to infinity, we deduce the result.

3 Extending the Definition to Complex Numbers

If we think of $\exp(x)$ as e^x, then the idea of generalizing the definition to complex numbers seems hopeless: our intuition tells us nothing, the functional equation does not help, and we cannot use continuity or order relations to determine it for us. However, both the power series and the compound-interest definitions can be generalized easily. If z is a complex number, then the most usual definition of $\exp(z)$ is

$$1 + z + \frac{z^2}{2!} + \frac{z^3}{3!} + \cdots.$$

Setting $z = i\theta$, for a real number θ, and splitting the resulting expression into its real and imaginary parts, we obtain

$$1 - \frac{\theta^2}{2!} + \frac{\theta^4}{4!} + \cdots + i\left(\theta - \frac{\theta^3}{3!} + \frac{\theta^5}{5!} - \cdots\right),$$

which, using the power-series expansions for $\cos(\theta)$ and $\sin(\theta)$, tells us that $\exp(i\theta) = \cos(\theta) + i\sin(\theta)$, the formula for the point with argument θ on the unit circle in the complex plane. In particular, if we take $\theta = \pi$, we obtain the famous formula $e^{i\pi} = -1$ (since $\cos(\pi) = -1$ and $\sin(\pi) = 0$).

This formula is so striking that one feels that it ought to hold for a good reason, rather than being a mere fact that one notices after carrying out some formal algebraic manipulations. And indeed there is a good reason. To see it, let us return to the compound-interest idea and define $\exp(z)$ to be the limit of $(1 + z/n)^n$ as n tends to infinity. Let us concentrate just on the case where $z = i\pi$: why should $(1 + i\pi/n)^n$ be close to -1 when n is very large?

To answer this, let us think geometrically. What is the effect on a complex number of multiplying it by $1 + i\pi/n$? On the Argand diagram this number is very close to 1 and vertically above it. Because the vertical line through 1 is tangent to the circle, this means that the number is very close indeed to a number that lies on the circle and has argument π/n (since the argument of a number on the circle is the length of the circular arc from 1 to that number, and in this case the circular arc is almost straight). Therefore, multiplication by $1 + i\pi/n$ is very well approximated by rotation through an angle of π/n. Doing this n times results in a rotation by π, which is the same as multiplication by -1. The same argument can be used to justify the formula $\exp(i\theta) = \cos(\theta) + i\sin(\theta)$.

Continuing in this vein, let us see why the derivative of the exponential function is the exponential function.

We know already that $\exp(z + w) = \exp(z)\exp(w)$, so the derivative of exp at z is the limit as w tends to zero of $\exp(z)(\exp(w) - 1)/w$. It is therefore enough to show that $\exp(w) - 1$ is very close to w when w is small. To get a good idea of $\exp(w)$ we should take a large n and consider $(1 + w/n)^n$. It is not hard to prove that this is indeed close to $1 + w$, but here is an informal argument instead. Suppose that you have a bank account that offers a tiny rate of interest over a year, say 0.5%. How much better would you do if you could compound this interest monthly? The answer is not very much: if the total amount of interest is very small, then the interest on the interest is negligible. This, in essence, is why $(1 + w/n)^n$ is approximately $1 + w$ when w is small.

One can extend the definition of the exponential function yet further. The main ingredients one needs are addition, multiplication, and the possibility of limiting arguments. So, for example, if x is an element of a BANACH ALGEBRA [III.12] A, then $\exp(x)$ makes sense. (Here, the power series definition is the easiest, though not necessarily the most enlightening.)

4   The Logarithm Function

Natural logarithms, like exponentials, can be defined in many ways. Here are three.

(i) The function log is the inverse of the function exp. That is, if t is a positive real number, then the statement $u = \log(t)$ is equivalent to the statement $t = \exp(u)$.

(ii) Let t be a positive real number. Then
$$\log(t) = \int_1^t \frac{\mathrm{d}x}{x}.$$

(iii) If $|x| < 1$ then $\log(1 + x) = x - \frac{1}{2}x^2 + \frac{1}{3}x^3 - \cdots$. This defines $\log(t)$ for $0 < t < 2$. If $t \geqslant 2$ then $\log(t)$ can be defined as $-\log(1/t)$.

The most important feature of the logarithmic function is a functional equation that is the reverse of the functional equation for exp, namely $\log(st) = \log(s) + \log(t)$. That is, whereas exp turns addition into multiplication, log turns multiplication into addition. A more formal way of putting this is that $\mathbb{R}$ forms a group under addition, and $\mathbb{R}_+$, the set of positive real numbers, forms a group under multiplication. The function exp is an isomorphism from $\mathbb{R}$ to $\mathbb{R}_+$, and log, its inverse, is an isomorphism from $\mathbb{R}_+$ to $\mathbb{R}$. Thus, in a sense the two groups have the same structure, and the exponential and logarithmic functions demonstrate this.

Let us use the first definition of log to see why $\log(st)$ must equal $\log(s) + \log(t)$. Write $s = \exp(a)$ and $t = \exp(b)$. Then $\log(s) = a$, $\log(t) = b$, and
$$\log(st) = \log(\exp(a)\exp(b))$$
$$= \log(\exp(a + b))$$
$$= a + b.$$

The result follows.

In general, the properties of log closely follow those of exp. However, there is one very important difference, which is a complication that arises when one tries to extend log to the complex numbers. At first it seems quite easy: every complex number z can be written as $re^{i\theta}$ for some nonnegative real number r and some θ (the modulus and argument of z, respectively). If $z = re^{i\theta}$ then $\log(z)$, one might think, should be $\log(r) + i\theta$ (using the functional equation for log and the fact that log inverts exp). The problem with this is that θ is not uniquely determined. For instance, what is $\log(1)$? Normally we would like to say 0, but we could, perversely, say that $1 = e^{2\pi i}$ and claim that $\log(1) = 2\pi i$.

Because of this difficulty, there is no single best way to define the logarithmic function on the entire complex plane, even if 0, a number that does not have a logarithm however you look at it, is removed. One convention is to write $z = re^{i\theta}$ with $r > 0$ and $0 \leqslant \theta < 2\pi$, which can be done in exactly one way, and *then* define $\log(z)$ to be $\log(r) + i\theta$. However, this function is not continuous: as you cross the positive real axis, the argument jumps by 2π and the logarithm jumps by $2\pi i$.

Remarkably, this difficulty, far from being a blow to mathematics, is an entirely positive phenomenon that lies behind several remarkable theorems in complex analysis, such as Cauchy's residue theorem, which allows one to evaluate very general path integrals.

III.26   The Fast Fourier Transform

If $f : \mathbb{R} \to \mathbb{R}$ is a periodic function with period 1, then one can obtain a great deal of useful information about f by calculating its Fourier coefficients (see THE FOURIER TRANSFORM [III.27] for a discussion of why). This is true for both theoretical and practical reasons, and because of the latter it is highly desirable to have a good way of computing Fourier coefficients quickly. A method for doing this was discovered by Cooley and Tukey in 1965 (though it turned out that Gauss had anticipated them over 150 years earlier).

The rth Fourier coefficient of f is given by the formula

$$\hat{f}(r) = \int_0^1 f(x)e^{-2\pi i r x}\, dx.$$

If we do not have an explicit formula for the integral (as would be the case, for instance, if f were derived from some physical signal rather than a mathematical formula), then we will want to approximate this integral numerically, and a natural way to do that is to *discretize* it: that is, turn it into a sum of the form $N^{-1}\sum_{n=0}^{N-1} f(n/N)e^{-2\pi i r n/N}$. If f is not too wildly oscillating and r is not too big, then this should be a good approximation.

The sum above will be unchanged if we add a multiple of N to r, so we now care only about the values of f at points of the form n/N. Moreover, the periodicity of f tells us that adding a multiple of N to n also makes no difference. So we can regard both n and r as belonging to the group $\mathbb{Z}_N$ of integers mod N (see MODULAR ARITHMETIC [III.58]). Let us change our notation to one that reflects this. Given a function g defined on $\mathbb{Z}_N$ we define the *discrete Fourier transform* of g to be the function $\hat{g}$, also defined on $\mathbb{Z}_N$, which is given by the formula

$$\hat{g}(r) = N^{-1} \sum_{n \in \mathbb{Z}_N} g(n)\omega^{-rn}, \qquad (1)$$

where we are writing ω for $e^{2\pi i/N}$, so that $\omega^{-rn} = e^{-2\pi i r n/N}$. Note that the sum over n could be regarded as a sum from 0 to $N-1$ just as above; the other notational change is that we have written $g(n)$ instead of $f(n/N)$.

The discrete Fourier transform can be thought of as multiplying a column vector (corresponding to the function g) by an $N \times N$ matrix (with entries $N^{-1}\omega^{-rn}$ for each r and n). Therefore it can be calculated using about N^2 arithmetical operations. The fast Fourier transform arises from the observation that the sum in (1) has symmetry properties that allow it to be calculated much more efficiently. This is most easily seen when N is a power of 2, and to make it even easier we shall look at the case $N = 8$. The sums to be evaluated are then

$$g(0) + \omega^{-r}g(1) + \omega^{-2r}g(2) + \cdots + \omega^{-7r}g(7)$$

for each r between 0 and 7. Now a sum like this can be rewritten as

$$g(0) + \omega^{-2r}g(2) + \omega^{-4r}g(4) + \omega^{-6r}g(6)$$
$$+ \omega^{-r}(g(1) + \omega^{-2r}g(3) + \omega^{-4r}g(5) + \omega^{-6r}g(7)),$$

which is interesting because

$$g(0) + \omega^{-2r}g(2) + \omega^{-4r}g(4) + \omega^{-6r}g(6)$$

and

$$g(1) + \omega^{-2r}g(3) + \omega^{-4r}g(5) + \omega^{-6r}g(7)$$

are themselves values of discrete Fourier transforms. For instance, if we set $h(n) = g(2n)$ for $0 \leqslant n \leqslant 3$, and write ψ for $\omega^2 = e^{2\pi i/4}$, then the first expression equals $h(0) + \psi^{-r}h(1) + \psi^{-2r}h(2) + \psi^{-3r}h(3)$. If we think of h as being defined on $\mathbb{Z}_4$, then this is precisely the formula for $\hat{h}(r)$.

A similar remark applies to the second expression, so if we can calculate the discrete Fourier transforms of the "even part" of g and the "odd part" of g, then it will be very straightforward to obtain each value of the Fourier transform of g itself: it will be a linear combination of values of the transforms of the two parts of g. Thus, if N is even and we write $F(N)$ for the number of operations needed to calculate the discrete Fourier transform of a function defined on $\mathbb{Z}_N$, we obtain a recurrence of the form

$$F(N) = 2F(N/2) + CN.$$

The interpretation of this is that in order to work out the N values of the transform of a function on $\mathbb{Z}_N$, it is enough to work out two such transforms for functions on $\mathbb{Z}_{N/2}$ and work out N linear combinations, each of which takes a constant number of steps.

If N is a power of 2, then we can iterate this: $F(N/2)$ will be at most $2F(N/4) + CN/2$, and so on. It is not hard to show as a result that $F(N)$ is at most $CN \log N$ for some constant C, a considerable improvement on CN^2. If N is not a power of 2, then the above argument does not work, but there are modifications of the method that do, and that lead to similar efficiency gains. (Indeed, this is true for the Fourier transform on an arbitrary finite Abelian group.)

Once we can calculate Fourier transforms efficiently, there are other calculations that immediately become easy as well. A simple example is the *inverse* Fourier transform, which has a formula very similar to that of the Fourier transform and can therefore be calculated in a similar way. Another calculation that becomes easy is the *convolution* of two sequences, which is defined as follows. If $a = (a_0, a_1, a_2, \ldots, a_m)$ and $b = (b_0, b_1, b_2, \ldots, b_n)$ are two sequences, then their convolution is the sequence $c = (c_0, c_1, c_2, \ldots, c_{m+n})$, where each c_r is defined to be $a_0 b_r + a_1 b_{r-1} + \cdots + a_r b_0$. This sequence is denoted by $a * b$. One of the most important properties of Fourier transforms is that they "convert convolutions into multiplication." That is, if we find a suitable way of regarding a and b as functions on $\mathbb{Z}_N$, then the Fourier transform of $a * b$ is the function $r \mapsto \hat{a}(r)\hat{b}(r)$. Therefore, to work out $a * b$ we can work out $\hat{a}$ and $\hat{b}$, multiply them together for each r, and take the inverse Fourier transform of the result. All stages of this calculation are quick, so calculating convolutions is quick.

This immediately leads to a quick way of multiplying the two polynomials $a_0 + a_1x + \cdots + a_mx^m$ and $b_0 + b_1x + \cdots + b_nx^n$ together, since the coefficients of the product are given by the sequence $c = a * b$. If all the a_i are between 0 and 9, it is a quick process to evaluate the product polynomial at $x = 10$ (since none of the coefficients c_r will have many digits), so we also have a method of multiplying two n-digit integers together that is far faster than long multiplication. These are two of the huge number of applications of the fast Fourier transform. A more direct source of applications occurs in engineering, where one frequently wishes to analyze a signal by looking at its Fourier transform. A very surprising application is to QUANTUM COMPUTATION [III.74]: a famous result of Peter Shor is that one can use a quantum computer to factorize large integers very quickly; this algorithm depends in an essential way on the fast Fourier transform, but uses the power of quantum computing in an almost miraculous way to divide the $N \log N$ steps into N lots of $\log N$ steps that can be carried out "in parallel."

III.27 The Fourier Transform
Terence Tao

Let f be a function from $\mathbb{R}$ to $\mathbb{R}$. Typically, there is not much that one can say about f, but certain functions have useful symmetry properties. For instance, f is called *even* if $f(-x) = f(x)$ for every x, and it is called *odd* if $f(-x) = -f(x)$ for every x. Furthermore, every function f can be written as a *superposition* of an even part, f_e, and an odd part, f_o. For instance, the function $f(x) = x^3 + 3x^2 + 3x + 1$ is neither even nor odd, but it can be written as $f_e(x) + f_o(x)$, where $f_e(x) = 3x^2 + 1$ and $f_o(x) = x^3 + 3x$. For a general function f, the decomposition is unique and is given by the formulas

$$f_e(x) = \tfrac{1}{2}(f(x) + f(-x))$$

and

$$f_o(x) = \tfrac{1}{2}(f(x) - f(-x)).$$

What are the symmetry properties enjoyed by even and odd functions? A useful way to regard them is as follows. We have a group of two transformations of the real line: one is the identity map $\iota : x \mapsto x$ and the other is the reflection $\rho : x \mapsto -x$. Now any transformation ϕ of the real line gives rise to a transformation of the functions defined on the real line: given a function f, the transformed function is the function $g(x) = f(\phi(x))$. In the case at hand, if $\phi = \iota$ then the transformed function is just $f(x)$, while if $\phi = \rho$ then it is $f(-x)$. If f is either even or odd, then both the transformed functions are *scalar multiples* of the original function f. In particular, when $\phi = \rho$, the transformed function is $f(x)$ when f is even (so the scalar multiple is 1) and $-f(x)$ when f is odd (so the scalar multiple is -1).

The procedure just described can be thought of as a very simple prototype of the general notion of a Fourier transform. Very broadly speaking, a Fourier transform is a systematic way to decompose "generic" functions into a superposition of "symmetric" functions. These symmetric functions are usually quite explicitly defined: for instance, one of the most important examples is a decomposition into the TRIGONOMETRIC FUNCTIONS [III.92] $\sin(nx)$ and $\cos(nx)$. They are also often related to physical concepts such as frequency or energy. The symmetry will usually be associated with a GROUP [I.3 §2.1] G, which is usually Abelian. (In the case considered above, it is the two-element group.) Indeed, the Fourier transform is a fundamental tool in the study of groups, and more precisely in the REPRESENTATION THEORY [IV.9] of groups, which concerns different ways in which a group can be regarded as a group of symmetries. It is also related to topics in linear algebra, such as the representation of a vector as linear combinations of an ORTHONORMAL BASIS [III.37], or as linear combinations of EIGENVECTORS [I.3 §4.3] of a matrix or LINEAR OPERATOR [III.50].

For a more complicated example, let us fix a positive integer n and let us define a systematic way of decomposing functions from $\mathbb{C}$ to $\mathbb{C}$, that is, complex-valued functions defined on the complex plane. If f is such a function and j is an integer between 0 and $n - 1$, then we say that f is a *harmonic of order j* if it has the following property. Let $\omega = e^{2\pi i/n}$, so that ω is a primitive nth root of 1 (meaning that $\omega^n = 1$ but no smaller positive power of ω gives 1). Then $f(\omega z) = \omega^j f(z)$ for every $z \in \mathbb{C}$. Notice that if $n = 2$, then $\omega = -1$, so when $j = 0$ we recover the definition of an even function and when $j = 1$ we recover the definition of an odd function. In fact, inspired by this, we can give a general formula for a decomposition of f into harmonics, which again turns out to be unique. If we define

$$f_j(z) = \frac{1}{n} \sum_{k=0}^{n-1} f(\omega^k z) \omega^{-jk},$$

then it is a simple exercise to prove that

$$f(z) = \sum_{j=0}^{n-1} f_j(z)$$

for every z (use the fact that $\sum_j \omega^{-jk} = n$ if $k = 0$ and 0 otherwise), and that $f_j(\omega z) = \omega^j f_j(z)$ for every z. Thus, f can be decomposed as a sum of harmonics. The group associated with this Fourier transform is the multiplicative group of the nth roots of unity $1, \omega, \ldots, \omega^{n-1}$, or the cyclic group of order n. The root ω^j is associated with the rotation of the complex plane through an angle of $2\pi j/n$.

Now let us consider infinite groups. Let f be a complex-valued function defined on the unit circle $\mathbb{T} = \{z \in \mathbb{C} : |z| = 1\}$. To avoid technical issues we shall assume that f is *smooth*—that is, it is infinitely differentiable. Now if f is a function of the simple form $f(z) = cz^n$ for some integer n and some constant c, then f will have rotational symmetry of order n. That is, if $\omega = e^{2\pi i/n}$ again, then $f(\omega z) = f(z)$ for all complex numbers z. After our earlier examples, it should come as no surprise that an arbitrary smooth function f can be expressed as a superposition of such rotationally symmetric functions. Indeed, one can write

$$f(z) = \sum_{n=-\infty}^{\infty} \hat{f}(n) z^n,$$

where the numbers $\hat{f}(n)$, called the *Fourier coefficients* of f at the *frequencies* n, are given by the formula

$$\hat{f}(n) = \frac{1}{2\pi} \int_0^{2\pi} f(e^{i\theta}) e^{-in\theta} \, d\theta.$$

This formula can be thought of as the limiting case $n \to \infty$ of the previous decomposition, restricted to the unit circle. It can also be regarded as a generalization of the Taylor series expansion of a HOLOMORPHIC FUNCTION [I.3 §5.6]. If f is holomorphic on the closed unit disk $\{z \in \mathbb{C} : |z| \leqslant 1\}$, then one can write

$$f(z) = \sum_{n=0}^{\infty} a_n z^n,$$

where the *Taylor coefficient* a_n is given by the formula

$$a_n = \frac{1}{2\pi i} \int_{|z|=1} \frac{f(z)}{z^{n+1}} \, dz.$$

In general, there are very strong links between Fourier analysis and complex analysis.

If f is smooth, then its Fourier coefficients decay to zero very quickly and it is easy to show that the Fourier series $\sum_{n=-\infty}^{\infty} \hat{f}(n) z^n$ converges. The issue becomes more subtle if f is not smooth (for instance, if it is merely continuous). Then one must be careful to specify the precise sense in which the series converges. In fact, a significant portion of HARMONIC ANALYSIS [IV.11] is devoted to questions of this kind, and to developing tools for answering them.

The group of symmetries associated with this version of Fourier analysis is the circle group $\mathbb{T}$. (Notice that we can think of the number $e^{i\theta}$ both as a point in the circle and as a rotation through an angle of θ. Thus, the circle can be identified with its own group of rotational symmetries.) But there is a second group that is important here as well, namely the additive group $\mathbb{Z}$ of all integers. If we take two of our basic symmetric functions, z^m and z^n, and multiply them together, then we obtain the function z^{m+n}, so the map $n \to z^n$ is an isomorphism from $\mathbb{Z}$ to the set of all these functions under multiplication. The group $\mathbb{Z}$ is known as the *Pontryagin dual* of $\mathbb{T}$.

In the theory of partial differential equations and in related areas of harmonic analysis, the most important Fourier transform is defined on the Euclidean space $\mathbb{R}^d$. Among all functions $f : \mathbb{R}^d \to \mathbb{C}$, the ones considered to be "basic" are the *plane waves* $f(x) = c_\xi e^{2\pi i x \cdot \xi}$, where $\xi \in \mathbb{R}^d$ is a vector (known as the *frequency* of the plane wave), $x \cdot \xi$ is the dot product between the position x and the frequency ξ, and c_ξ is a complex number (whose magnitude is the *amplitude* of the plane wave). Notice that sets of the form $H_\lambda = \{x : x \cdot \xi = \lambda\}$ are (hyper)planes orthogonal to ξ, and on each such set the value of $f(x)$ is constant. Moreover, the value taken by f on H_λ is always equal to the value taken on $H_{\lambda+2\pi}$. This explains the name "plane waves." It turns out that if a function f is sufficiently "nice" (e.g., smooth and rapidly decreasing as x gets large), then it can be represented uniquely as the superposition of plane waves, where a "superposition" is now interpreted as an integral rather than a summation. More precisely, we have the formulas[1]

$$f(x) = \int_{\mathbb{R}^d} \hat{f}(\xi) e^{2\pi i x \cdot \xi} \, d\xi,$$

---

1. In some texts, the Fourier transform is defined slightly differently, with factors such as 2π and -1 being moved to other places. These notational differences have some minor benefits and drawbacks, but they are all equivalent to each other.

where

$$\hat{f}(\xi) = \int_{\mathbb{R}^d} f(x) e^{-2\pi i x \cdot \xi} \, dx.$$

The function $\hat{f}(\xi)$ is known as the *Fourier transform* of f, and the second formula is known as the *Fourier inversion formula*. These two formulas show how to determine the Fourier-transformed function from the original function and vice versa. One can view the quantity $\hat{f}(\xi)$ as the extent to which the function f contains a component that oscillates at frequency ξ. As it turns out, there is no difficulty in justifying the convergence of these integrals when f is sufficiently nice, though the issue again becomes more subtle for functions that are somewhat rough or slowly decaying. In this case, the underlying group is the Euclidean group $\mathbb{R}^d$ (which can also be thought of as the group of d-dimensional translations); note that both the position variable x and the frequency variable ξ are contained in $\mathbb{R}^d$, so $\mathbb{R}^d$ is also the Pontryagin dual group in this setting.[2]

One major application of the Fourier transform lies in understanding various linear operations on functions, such as, for instance, the Laplacian on $\mathbb{R}^d$. Given a function $f : \mathbb{R}^d \to \mathbb{C}$, its Laplacian Δf is defined by the formula

$$\Delta f(x) = \sum_{j=1}^{d} \frac{\partial^2 f}{\partial x_j^2},$$

where we think of the vector x in coordinate form, $x = (x_1, \ldots, x_d)$, and of f as a function $f(x_1, \ldots, x_d)$ of d real variables. To avoid technicalities let us consider only those functions that are smooth enough for the above formula to make sense without any difficulty.

In general, there is no obvious relationship between a function f and its Laplacian Δf. But when f is a plane wave such as $f(x) = e^{2\pi i x \cdot \xi}$, there is a very simple relationship:

$$\Delta e^{2\pi i x \cdot \xi} = -4\pi^2 |\xi|^2 e^{2\pi i x \cdot \xi}.$$

That is, the effect of the Laplacian on the plane wave $e^{2\pi i x \cdot \xi}$ is to multiply it by the scalar $-4\pi^2 |\xi|^2$. In other words, the plane wave is an eigenfunction[3] for the Laplacian Δ, with eigenvalue $-4\pi^2 |\xi|^2$. (More generally, plane waves will be eigenfunctions for any linear operation that commutes with translations.) Therefore, the Laplacian, when viewed through the lens of the

Fourier transform, is very simple: the Fourier transform lets one write an arbitrary function as a superposition of plane waves, and the Laplacian has a very simple effect on each plane wave. To be explicit about it,

$$\begin{aligned} \Delta f(x) &= \Delta \int_{\mathbb{R}^d} \hat{f}(\xi) e^{2\pi i x \cdot \xi} \, d\xi \\ &= \int_{\mathbb{R}^d} \hat{f}(\xi) \Delta e^{2\pi i x \cdot \xi} \, d\xi \\ &= \int_{\mathbb{R}^d} (-4\pi^2 |\xi|^2) \hat{f}(\xi) e^{2\pi i x \cdot \xi} \, d\xi, \end{aligned}$$

which gives us a formula for the Laplacian of a general function. Here we have interchanged the Laplacian Δ with an integral; this can be rigorously justified for suitably nice f, but we omit the details.

This formula represents Δf as a superposition of plane waves. But any such representation is unique, and the Fourier inversion formula tells us that

$$\Delta f(x) = \int_{\mathbb{R}^d} \widehat{\Delta f}(\xi) e^{2\pi i x \cdot \xi} \, d\xi.$$

Therefore,

$$\widehat{\Delta f}(\xi) = (-4\pi^2 |\xi|^2) \hat{f}(\xi),$$

a fact that can also be derived directly from the definition of the Fourier transform using integration by parts. This identity shows that the Fourier transform *diagonalizes* the Laplacian: the operation of taking the Laplacian, when viewed using the Fourier transform, is nothing more than multiplication of a function $F(\xi)$ by the *multiplier* $-4\pi^2 |\xi|^2$. The quantity $-4\pi^2 |\xi|^2$ can be interpreted as the *energy level* associated[4] with the frequency ξ. In other words, the Laplacian can be viewed as a *Fourier multiplier*, meaning that to calculate the Laplacian you take the Fourier transform, multiply by the multiplier, and then take the inverse Fourier transform again. This viewpoint allows one to manipulate the Laplacian very easily. For instance, we can iterate the above formula to compute higher powers of the Laplacian:

$$\widehat{\Delta^n f}(\xi) = (-4\pi^2 |\xi|^2)^n \hat{f}(\xi) \quad \text{for } n = 0, 1, 2, \ldots.$$

Indeed, we are now in a position to develop more general functions of the Laplacian. For instance, we can take a square root as follows:

$$\widehat{\sqrt{-\Delta} f}(\xi) = 2\pi |\xi| \hat{f}(\xi).$$

This leads to the theory of fractional differential operators (which are in turn a special case of *pseudodifferential operators*), as well as the more general theory

---

2. This is because of our reliance on the dot product; if one did not want to use this dot product, the Pontryagin dual would instead be $(\mathbb{R}^d)^*$, the dual vector space to $\mathbb{R}^d$. But this subtlety is not too important in most applications.

3. Strictly speaking, this is a *generalized* eigenfunction, as plane waves are not square-integrable on $\mathbb{R}^d$.

4. When taking this view, it is customary to replace Δ by $-\Delta$ in order to make the energies positive.

of FUNCTIONAL CALCULUS [IV.15 §3.1], in which one starts with a given operator (such as the Laplacian) and then studies various functions of that operator, such as square roots, exponentials, inverses, and so forth.

As the above discussion shows, the Fourier transform can be used to develop a number of interesting operations, which have particular importance in the theory of differential equations. To analyze these operations effectively, one needs various *estimates* on the Fourier transform. For instance, it is often important to know how the size of a function f, as measured by some norm, relates to the size of its Fourier transform, as measured by a possibly different norm. For a further discussion of this point, see FUNCTION SPACES [III.29]. One particularly important and striking estimate of this type is the *Plancherel identity*,

$$\int_{\mathbb{R}^d} |f(x)|^2 \, dx = \int_{\mathbb{R}^d} |\hat{f}(\xi)|^2 \, d\xi,$$

which shows that the L_2-norm of a Fourier transform is actually *equal* to the L_2-norm of the original function. The Fourier transform is therefore a unitary operation, so one can view the frequency-space representation of a function as being in some sense a "rotation" of the physical-space representation.

Developing further estimates related to the Fourier transform and associated operators is a major component of harmonic analysis. A variant of the Plancherel identity is the *convolution formula*:

$$\int_{\mathbb{R}^d} f(y)g(x-y) \, dy = \int_{\mathbb{R}^d} \hat{f}(\xi)\hat{g}(\xi)e^{2\pi ix\cdot\xi} \, d\xi.$$

This formula allows one to analyze the *convolution*

$$f * g(x) = \int_{\mathbb{R}^d} f(y)g(x-y) \, dy$$

of two functions f and g in terms of their Fourier transforms; in particular, if the Fourier coefficients of f or g are small, then we expect the convolution $f * g$ to be small as well. This relationship means that the Fourier transform controls certain *correlations* of a function with itself and with other functions, which makes the Fourier transform an important tool in understanding the randomness and uniform distribution properties of various objects in probability theory, harmonic analysis, and number theory. For instance, one can pursue the above ideas to establish the central limit theorem, which asserts that the sum of many independent random variables will eventually resemble a GAUSSIAN DISTRIBUTION [III.71 §5]; one can even use such methods to establish VINOGRADOV'S THEOREM [V.27], that every sufficiently large odd number is the sum of three primes.

There are many directions in which to generalize the above set of ideas. For instance, one can replace the Laplacian by a more general operator and the plane waves by (generalized) eigenfunctions of that operator. This leads to the subject of SPECTRAL THEORY [III.86] and functional calculus; one can also study the algebra of Fourier multipliers (and of convolution) more abstractly, which leads to the theory of C^*-ALGEBRAS [IV.15 §3]. One can also go beyond the theory of linear operators and study bilinear, multilinear, or even fully nonlinear operators. This leads in particular to the theory of *paraproducts*, which are generalizations of the pointwise product operation $(f(x), g(x)) \mapsto fg(x)$ that are of importance in differential equations. In another direction, one can replace Euclidean space $\mathbb{R}^d$ by a more general group, in which case the notion of a plane wave is replaced by the notion of a *character* (if the group is Abelian) or a *representation* (if the group is non-Abelian). There are other variants of the Fourier transform, such as the Laplace transform or the Mellin transform (for more about other transforms, see the article TRANSFORMS [III.91]), which are very similar algebraically to the Fourier transform and play similar roles (for instance, the Laplace transform is also useful in analyzing differential equations). We have already seen that Fourier transforms are connected to Taylor series; there is also a connection to some other important series expansions, notably Dirichlet series, as well as expansions of functions in terms of SPECIAL POLYNOMIALS [III.85] such as orthogonal polynomials or SPHERICAL HARMONICS [III.87].

The Fourier transform decomposes a function exactly into many components, each of which has a precise frequency. In some applications it is more useful to adopt a "fuzzier" approach, in which a function is decomposed into fewer components but each component has a range of frequencies rather than consisting purely of a single frequency. Such decompositions can have the advantage of being less constrained by the *uncertainty principle*, which asserts that it is impossible for both a function and its Fourier transform to be concentrated in very small regions of $\mathbb{R}^d$. This leads to some variants of the Fourier transform, such as WAVELET TRANSFORMS [VII.3], which are better suited to a number of problems in applied and computational mathematics, and also to certain questions in harmonic analysis and differential equations. The uncertainty principle, being fundamental to quantum mechanics, also connects the Fourier transform to mathematical physics, and in particular to the connections between classical and quantum physics, which can be studied rigorously using the methods of geometric quantization and microlocal analysis.

III.28 Fuchsian Groups
Jeremy Gray

One of the most basic objects in geometry is the *torus*: a surface that has the shape of the surface of a bagel. If you want to construct one, you can do so by taking a square and gluing opposite edges together. When you glue the top and bottom edges together you have a cylinder, and when you glue the other two edges together, which have now become circles, you obtain your torus.

A more mathematical way of making a torus is as follows. We start with the usual (x, y) coordinate plane and the square in it with vertices at $(0,0)$, $(1,0)$, $(1,1)$, and $(0,1)$, which consists of the points whose coordinates satisfy $0 \leqslant x \leqslant 1$, $0 \leqslant y \leqslant 1$. This square can be moved around horizontally and vertically. If we shift it m units horizontally and n units vertically, where m and n are integers, we get the square that consists of the points whose coordinates satisfy $m \leqslant x \leqslant m + 1$, $n \leqslant y \leqslant n+1$. As m and n run through all the integers, we see that the copies of the square cover the whole plane, with four squares coming together at each point with integer coordinates. The plane is said to be *tiled* or *tessellated* (from the Latin word for a marble chip in a mosaic), and it is easy to see that you can color the squares alternately black and white and get an infinite checkerboard pattern.

To make the torus we "identify" points. We say that the points (x, y) and (x', y') correspond to the same point in a certain new figure if $x - x'$ and $y - y'$ are both integers. To see what the new figure looks like, we observe that any point in the plane corresponds to a point inside, or on the edge of, our original square. Moreover, the point (x, y) corresponds to exactly one point inside the square provided that neither x nor y is an integer. So our new space looks a lot like our original square. But what about the points $(\frac{1}{4}, 0)$ and $(\frac{1}{4}, 1)$? They correspond to the same point in our new space, as do any corresponding pairs of points on the upper and lower edges of our square. So those edges are identified in our new space. By a similar argument, so too are the left and right edges. The result is that, after points are identified according to our rule, we obtain the torus.

If we make the torus in this way, we can draw small figures on it just by drawing them in the original square;

lengths in the square will then correspond exactly to lengths on the torus. This is how old-fashioned printing on a drum works: an inked figure on a cylinder is rolled over the paper to make exact copies of the figure. Thus, as far as small figures are concerned, the geometry of the torus is exactly like Euclidean geometry. In mathematical language we say that the geometry on the torus is induced from the geometry on the plane, and therefore that it is *locally Euclidean*. Globally, of course, it is different, because one can draw curves on the torus that cannot be shrunk to a point, whereas one cannot do so on the plane.

Notice, too, that we have brought in a group to do the bulk of the work for us. In this case the group is the set of all pairs (m, n) where m and n are integers, with $(m, n) + (m', n')$ defined to be $(m + m', n + n')$.

The torus and the sphere are but two of an infinite class of surfaces that are closed (they have no boundary) and compact (they do not in any sense go off to infinity). Other examples include the two-holed torus, and more generally the n-holed torus (the surfaces of genus $2, 3, 4, \dots$). To create these in a similar way, we need *Fuchsian groups*.

It is natural to expect that we can get other surfaces by using polygons with more than four sides. It turns out that if you use a polygon with eight sides, for example a regular octagon, and glue sides 1 and 3 together, 2 and 4 together, 5 and 7 together, and 6 and 8 together, you get the two-holed torus. How can we use a group to achieve the same result, as we did with the torus? For that we need a way of fitting lots of copies of the octagon together so that they overlap only along edges. The problem is that one cannot tile the plane with octagons: the angles of an octagon are $135°$, and that is far too big because we need eight octagons to fit together at each vertex.

The way forward here is to use HYPERBOLIC GEOMETRY [I.3 §6.6] instead of Euclidean geometry. But we can also work with our bare hands. Take the unit disk in the complex plane, $\mathbb{D} = \{z : |z| \leqslant 1\}$. Take the group of what are called *Möbius transformations*, which are maps of the form $z \mapsto (az + b)/(cz + d)$. It is a routine calculation to show that these maps send circles and straight lines to circles and straight lines (they mix the two types up, sometimes sending a circle to a straight line and vice versa) and that they map angles to equal angles, just like the more familiar Euclidean rotations. If we now select just those Möbius transformations that map $\mathbb{D}$ to itself, then we have a group that we shall call G. Indeed, we very nearly have a Fuchsian group.

We need to find a shape that will play the role that the square played in the Euclidean plane. Our group G has the property that it maps diameters of $\mathbb{D}$ and arcs of circles perpendicular to the boundary of $\mathbb{D}$ to diameters of $\mathbb{D}$ and arcs of circles perpendicular to the boundary of $\mathbb{D}$, so we let these play the role of straight lines and use eight of them as the edges of a (non-Euclidean) octagon. We find that we can do this in many ways, so we pick one with the highest degree of symmetry to make things easy for ourselves. That is, we draw a "regular octagon" centered on the center of the disk $\mathbb{D}$. This still leaves us with some choice: the bigger the octagon, the smaller its angles. So we draw the octagon with angles of $\pi/4$, which allows eight of them to cluster at each vertex, and then we can fit them together as we want. If we identify points that lie in corresponding places in different copies of the polygon, then the resulting space is a RIEMANN SURFACE [III.79] of genus 2.

A Fuchsian group is a subgroup of the group G (of Möbius transformations that map $\mathbb{D}$ to itself) that moves some polygon around "en bloc" and thereby tiles the disk. Just as with the torus, we have a notion of equivalent points (ones that are in the corresponding place in different tiles) and when we identify equivalent points we get the space that we would also have obtained by identifying the edges of the polygon in pairs, which is the space we wanted.

All this can be described in the language of hyperbolic geometry. The *disk model* is defined by means of a RIEMANNIAN METRIC [I.3 §6.10] on $\mathbb{D}$, the differential of which is given by

$$\mathrm{d}s = \frac{|\mathrm{d}z|}{\sqrt{1 - |z|^2}}.$$

The elements of G move figures around in $\mathbb{D}$ in a way that preserves hyperbolic distances. It follows that the geometry on the surface that we obtain by identifying points in the manner just described is *locally hyperbolic*, just as that of the torus was locally Euclidean.

It turns out that if we carry out the above construction starting with a regular $4n$-sided figure (with $n > 2$), then we obtain a Riemann surface of genus n. But mathematicians can do much more. If you go back to the plane and start not with a square but with a rectangle, or still more generally a parallelogram, it is reasonably easy to see that the same construction can be carried out. Indeed, if you just watch the original construction from an appropriate angle, instead of from

vertically above the plane, then the square will turn into any parallelogram you choose (possibly enlarged or contracted). When you use a parallelogram, you again obtain a torus, but it differs from the original one in the same way that the square and the parallelogram differ: angles are distorted. It is a not entirely trivial exercise to show that the only angle-preserving maps from one parallelogram to another are similarities (uniform scaling by the same amount in two, and therefore all, directions). So the resulting tori have a different sense of what angles are: that is, they have different *conformal structures*.

The same happens in the hyperbolic disk. If one picks a $4n$-sided polygon (its sides are parts of geodesics) whose edges come in pairs of equal length, and one finds a group that moves this polygon around en bloc and matches the edges exactly, then a Riemann surface is once again obtained, but if the polygons are not conformally equivalent, then neither are the corresponding surfaces; they have the same genus, n, but different conformal structures. We can even go further and allow some of the vertices of the polygon to lie on the boundary of the disk, in which case the corresponding sides of the polygon are infinitely long with respect to the hyperbolic metric. The space we then construct is a "punctured" Riemann surface, and again mathematicians can vary its conformal structure.

The fundamental importance of Fuchsian groups derives from the *uniformization theorem*, which says that all but the simplest Riemann surfaces arise from some Fuchsian group in the fashion described above. This includes every Riemann surface of genus greater than 1, and those of genus 1 with at least one puncture, with any possible conformal structure.

The name Fuchsian group was given to these groups by POINCARÉ [VI.61] in 1881, who discovered them in the course of work on the hypergeometric equation and related differential equations, which had been inspired by the work of the German mathematician Lazarus Fuchs. KLEIN [VI.57] protested to him that a better procedure might have been to name them after Schwarz, and Poincaré was willing to agree once he read the relevant paper by Schwarz, but by then Fuchs had given his approval to the name. When Klein protested too much (in Poincaré's view), Poincaré publicly gave the name *Kleinian groups* to the analogous class of groups that arise in the study of conformal transformations of the three-dimensional unit ball. The names have stuck ever since, but the study of Kleinian groups is much more difficult and neither Poincaré nor Klein could do much with the concept. However, the idea that every Riemann surface might arise from either the sphere, the Euclidean plane, or the hyperbolic plane was something they both came to conjecture. Rigorous proofs of this statement, the uniformization theorem, were to be given only in 1907, by Poincaré and Koebe independently.

The formal definition of a Fuchsian group is as follows. A subgroup H of the group of all Möbius transformations is said to act *discontinuously* if, for every compact set K in the disk $\mathbb{D}$ the sets $h(K)$ and K are disjoint except for finitely many $h \in H$. A *Fuchsian group* is a subgroup H of the group of all Möbius transformations that acts discontinuously on the disk $\mathbb{D}$.

III.29 Function Spaces
Terence Tao

1 What Is a Function Space?

When one works with real or complex numbers, there is a natural notion of the *magnitude* of a number x, namely its modulus $|x|$. One can also use this notion of magnitude to define a distance $|x - y|$ between two numbers x and y and thereby say in a quantitative way which pairs of numbers are close and which ones are far apart.

The situation becomes more complicated, however, when one deals with objects with more degrees of freedom. Consider for instance the problem of determining the "magnitude" of a three-dimensional rectangular box. There are several candidates for such a magnitude: length, width, height, volume, surface area, diameter (the length of a long diagonal), eccentricity, and so forth. Unfortunately, these magnitudes do not give equivalent comparisons: for example, box A may be longer and have a greater volume than box B, but box B may be wider and have a greater surface area. Because of this, one abandons the idea that there should be only one notion of "magnitude" for boxes, and instead accepts that there is a multiplicity of such notions and that they can all be useful: for some applications one may wish to distinguish the large-volume boxes from the small-volume boxes, while in others one may wish to distinguish the eccentric boxes from the round boxes. Of course, there are several relationships between the different notions of magnitude (e.g., the ISOPERIMETRIC INEQUALITY [IV.26] allows one to place an upper limit on the possible volume if one knows the surface area), so the situation is not as disorganized as it may at first appear.

Now let us turn to functions with a fixed domain and range. (A good case to have in mind is functions $f : [-1, 1] \to \mathbb{R}$ from the interval $[-1, 1]$ to the real line $\mathbb{R}$.) These objects have infinitely many degrees of freedom, so it should not be surprising that there are now infinitely many distinct notions of "magnitude," which all provide different answers to the question "how large is a given function f?" (or to the closely related question "how close together are two functions f and g?"). In some cases, certain functions may have infinite magnitude by one measure and finite magnitude by another (similarly, a pair of functions may be very close by one measure and very far apart by another). Again, this situation may seem chaotic, but it simply reflects the fact that functions have many distinct characteristics—some are tall, some are broad, some are smooth, some are oscillatory, and so forth—and that, depending on the application at hand, one may need to give more weight to one of these characteristics than to others. In analysis, these characteristics are embodied in a variety of standard *function spaces* and their associated *norms*, which are available to describe functions both qualitatively and quantitatively.

Formally, a function space is a NORMED SPACE [III.62] X, the elements of which are functions (with some fixed domain and range). A majority (but certainly not all) of the standard function spaces considered in analysis are not just normed spaces but also BANACH SPACES [III.62]. The norm $\|f\|_X$ of a function f in X is the function space's way of measuring how large f is. It is common, though not universal, for the norm to be defined by a simple formula and for the space X to consist precisely of those functions f for which the resulting definition $\|f\|_X$ makes sense and is finite. Thus, the mere fact that a function f belongs to a function space X can already convey some qualitative information about that function. For example, it may imply some regularity,[1] decay, boundedness, or integrability on the function f. The actual value of the norm $\|f\|_X$ makes this information quantitative. It may tell us *how* regular f is, *how much* decay it has, *by which constant* it is bounded, or *how large* its integral is.

2 Examples of Function Spaces

We now present a sample of commonly used function spaces. For simplicity we shall consider only spaces of functions from $[-1, 1]$ to $\mathbb{R}$.

---

1. The more smoothly a function varies, the more "regular" it is considered to be.

2.1 $C^0[-1, 1]$

This space consists of all CONTINUOUS FUNCTIONS [I.3 §5.2] from $[-1, 1]$ to $\mathbb{R}$, and is sometimes denoted $C[-1, 1]$. Continuous functions are regular enough to allow one to avoid many of the technical subtleties associated with very rough functions. Continuous functions on a COMPACT [III.9] interval such as $[-1, 1]$ are bounded, so the most natural norm to place on this space is the *supremum norm*, denoted $\|f\|_\infty$, which is the largest possible value of $|f(x)|$. (Formally, it is defined to be $\sup\{|f(x)| : x \in [-1, 1]\}$, but for continuous functions on $[-1, 1]$ the two definitions are equivalent.)

The supremum norm is the norm associated with uniform convergence: a sequence $f_1, f_2, \ldots$ converges uniformly to f if and only if $\|f_n - f\|_\infty$ tends to 0 as n tends to ∞. The space $C^0[-1, 1]$ has the useful property that one can multiply functions together as well as adding them. This makes it a basic example of a *Banach algebra*.

2.2 $C^1[-1, 1]$

This is a space that has a more restricted membership than $C^0[-1, 1]$: not only must a function f in $C^1[-1, 1]$ be continuous but it must also have a derivative that is continuous. The supremum norm here is no longer a natural one, because a sequence of continuously differentiable functions can converge in this norm to a nondifferentiable function. Instead, the right norm here is the C^1-*norm* $\|f\|_{C^1[-1,1]}$, which is defined to be $\|f\|_\infty + \|f'\|_\infty$.

Notice that the C^1-norm measures both the size of a function *and* the size of its derivative. (Merely controlling the latter would be unsatisfactory, since it would give constant functions a norm of zero.) Thus it is a norm that forces a greater degree of regularity than the supremum norm. One can similarly define the space $C^2[-1, 1]$ of twice continuously differentiable functions, and so forth, all the way up to the space $C^\infty[-1, 1]$ of infinitely differentiable functions. (There are also "fractional" versions of these spaces, such as $C^{0,\alpha}[-1, 1]$, the space of α-Hölder continuous functions. We will not discuss these variants here.)

2.3   The Lebesgue Spaces $L^p[-1, 1]$

The supremum norm $\|f\|_\infty$ mentioned earlier gives simultaneous control on the size of $|f(x)|$ for all $x \in [-1, 1]$. However, this means that if there is a tiny set

of x for which $|f(x)|$ is very large, then $\|f\|_\infty$ is very large, even if a typical value of $|f(x)|$ is much smaller. It is sometimes more advantageous to work with norms that are less influenced by the values of a function on small sets. The L^p-*norm* of a function f is

$$\|f\|_p = \left(\int_{-1}^1 |f(x)|^p \, dx \right)^{1/p}.$$

This is defined for $1 \leqslant p < \infty$ and for any measurable f. The function space $L^p[-1, 1]$ is the class of measurable functions for which the above norm is finite. The norm $\|f\|_\infty$ of a measurable function f is its *essential supremum*: roughly speaking this means the largest value of $|f(x)|$ if you ignore sets of measure zero. It turns out to be the limit of the norms $\|f\|_p$ as p tends to infinity. The space $L^\infty[-1, 1]$ consists of those measurable functions f for which $\|f\|_\infty$ is finite. While the L^∞ norm is concerned solely with the "height" of a function, the L^p norms are instead concerned with a combination of the "height" and "width" of a function.

Particularly important among these norms is the L^2-norm, since $L^2[-1, 1]$ is a HILBERT SPACE [III.37]. This space is exceptionally rich in symmetries: there is a wide variety of *unitary transformations*, that is, invertible linear maps T defined on $L^2[-1, 1]$ such that $\|Tf\|_2 = \|f\|_2$ for every function $f \in L^2[-1, 1]$.

2.4   The Sobolev Spaces $W^{k,p}[-1, 1]$

The Lebesgue norms control, to some extent, the height and width of a function, but say nothing about regularity; there is no reason why a function in L^p should be differentiable or even continuous. To incorporate such information one often turns to the *Sobolev norms* $\|f\|_{W^{k,p}[-1,1]}$, defined for $1 \leqslant p \leqslant \infty$ and $k \geqslant 0$ by

$$\|f\|_{W^{k,p}[-1,1]} = \sum_{j=0}^k \left\| \frac{d^j f}{dx^j} \right\|_p.$$

The *Sobolev space* $W^{k,p}[-1, 1]$ is the space of functions for which this norm is finite. Thus, a function lies in $W^{k,p}[-1, 1]$ if it and its first k derivatives all belong to $L^p[-1, 1]$. There is one subtlety: we do not require f to be k times differentiable in the usual sense, but in the weaker sense of DISTRIBUTIONS [III.18]. For instance, the function $f(x) = |x|$ is not differentiable at zero, but it does have a natural weak derivative: the function $f'(x)$ which is -1 when $x < 0$ and $+1$ when $x > 0$. This function lies in $L^\infty[-1, 1]$ (since the set $\{0\}$ has measure zero, we do not need to specify $f'(0)$), and therefore f lies in $W^{1,\infty}[-1, 1]$ (which turns out to be the space of *Lipschitz-continuous* functions). We need to consider these generalized differentiable functions because without them the space $W^{k,p}[-1,1]$ would not be complete.

Sobolev norms are particularly natural and useful in the analytical study of partial differential equations and mathematical physics. For instance, the $W^{1,2}$ norm can be interpreted as (the square root of) an "energy" associated with a function.

3   Properties of Function Spaces

There are many ways in which knowledge of the structure of function spaces can assist in the study of functions. For instance, if one has a good basis for the function space, so that every function in the space is a (possibly infinite) linear combination of basis elements, and one has some quantitative estimates on how this linear combination converges to the original function, then this allows one to represent that function efficiently in terms of a number of coefficients, and also allows one to approximate that function by smoother functions. For instance, one basic result about $L^2[-1,1]$ is the *Plancherel theorem*, which asserts, among other things, that there are numbers $(a_n)_{n=-\infty}^\infty$ such that

$$\left\| f - \sum_{n=-N}^N a_n e^{\pi i n x} \right\|_2 \to 0 \quad \text{as } N \to \infty.$$

This shows that any function in $L^2[-1,1]$ can be approximated to any desired accuracy in L^2 by a *trigonometric polynomial*: that is, an expression of the form $\sum_{n=-N}^N a_n e^{\pi i n x}$. The number a_n is the nth *Fourier coefficient* $\hat{f}(n)$ of f. It is given by the formula

$$\hat{f}(n) = \frac{1}{2} \int_{-1}^1 f(x) e^{-\pi i n x} \, \mathrm{d}x.$$

One can regard this result as saying that the functions $e^{\pi i n x}$ form a very good basis for $L^2[-1,1]$. (They are in fact an *orthonormal basis*: they have norm 1 and the inner product of two different ones is always zero.)

Another very basic fact about function spaces is that certain function spaces embed into others, so that a function from one space automatically also belongs to other spaces. Furthermore, there is often some inequality that gives an upper bound for one norm in terms of another. For instance, a function in a high-regularity space such as $C^1[-1,1]$ automatically belongs to a low-regularity space such as $C^0[-1,1]$, and a function in a high-integrability space such as $L^\infty[-1,1]$ automatically belongs to a low-integrability space such as $L^1[-1,1]$. (This statement is no longer

true if one replaces the interval $[-1,1]$ by a set of infinite measure, such as the real line $\mathbb{R}$.) These inclusions cannot be reversed; however, one does have the *Sobolev embedding theorem*, which allows one to "trade" regularity for integrability. This result tells us that spaces with lots of regularity but low integrability can be embedded into spaces with low regularity but high integrability. A sample estimate of this type is

$$\|f\|_\infty \leqslant \|f\|_{W^{1,1}[-1,1]},$$

which tells us that if the integrals of $|f(x)|$ and $|f'(x)|$ are both finite, then f must be bounded (which is a far stronger integrability condition than the finiteness of $\|f\|_1$).

Another very useful concept is that of DUALITY [III.19]. Given a function space X, one can define the dual space X^*, which is formally defined as the class of all *continuous linear functionals* on X, or more precisely all maps $\omega : X \to \mathbb{R}$ (or $\omega : X \to \mathbb{C}$, if the function space is complex valued) that are linear and continuous with respect to the norm of X. For example, it turns out that every linear functional ω on the space $L^p[-1,1]$ is of the form

$$\omega(f) = \int_{-1}^1 f(x) g(x) \, \mathrm{d}x$$

for some function g in $L^q[-1,1]$, where q is the *dual* or *conjugate exponent* of p, defined by the equation $1/p + 1/q = 1$.

One can sometimes analyze functions in a function space by looking instead at how the linear functionals in the dual space act on those functions. Similarly, one can often analyze a continuous linear operator $T : X \to Y$ from one function space to another by first considering the *adjoint operator* $T^* : Y^* \to X^*$, defined for all linear functionals $\omega : Y \to \mathbb{R}$ by letting $T^*\omega$ be the functional on X defined by the formula $T^*\omega(x) = \omega(Tx)$.

We mention one more important fact about function spaces, which is that certain function spaces X "interpolate" between two other function spaces X_0 and X_1. For example, there is a natural sense in which the spaces $L^p[-1,1]$ with $1 < p < \infty$ "lie between" the spaces $L^1[-1,1]$ and $L^\infty[-1,1]$. The precise definition of interpolation is too technical for this article, but its usefulness lies in the fact that the "extreme" spaces X_0 and X_1 are often easier to deal with than the "intermediate" spaces X. For this reason, it is sometimes possible to prove difficult results about X by proving much easier results about X_0 and X_1 and "interpolating" between them. For instance, it can be used to give a short proof of *Young's inequality*, which is the following statement. Let $1 \leqslant p, q, r \leqslant \infty$ satisfy the equation $1/p + 1/q = 1/r + 1$, let f and g belong to $L^p(\mathbb{R})$ and $L^q(\mathbb{R})$, respectively, and let $f * g$ be the *convolution* of f and g: that is, $f * g(x) = \int_{-\infty}^{\infty} f(y)g(x-y)\,\mathrm{d}y$. Then

$$\left(\int_{-\infty}^{\infty} |f * g(x)|^r \,\mathrm{d}x \right)^{1/r}$$
$$\leqslant \left(\int_{-\infty}^{\infty} |f(x)|^p \,\mathrm{d}x \right)^{1/p} \left(\int_{-\infty}^{\infty} |g(x)|^q \,\mathrm{d}x \right)^{1/q}.$$

Interpolation is useful here because the inequality is easy to prove in the extreme cases when $p = 1$, when $q = 1$, or when $r = \infty$. It is much harder to prove this result without the help of interpolation theory.

III.30   Galois Groups

Given a polynomial function f with rational coefficients, the *splitting field* of f is defined to be the smallest FIELD [I.3 §2.2] that contains all rational numbers and all the roots of f. The *Galois group* of f is the group of all AUTOMORPHISMS [I.3 §4.1] of the splitting field. Each such automorphism permutes the roots of f, so the Galois group can be thought of as a subset of the group of all PERMUTATIONS [III.68] of these roots. The structure and properties of the Galois group are closely connected with the solubility of the polynomial: in particular, the Galois group can be used to show that not all polynomials are *solvable by radicals* (that is, solvable by means of a formula that involves the usual arithmetic operations together with the extraction of roots). This theorem, spectacular as it is, is by no means the only application of Galois groups: they play a central role in modern algebraic number theory.

For more details, see THE INSOLUBILITY OF THE QUINTIC [V.21] and ALGEBRAIC NUMBERS [IV.1 §20].

III.31   The Gamma Function
Ben Green

If n is a positive integer, then its *factorial*, written $n!$, is the number $1 \times 2 \times \cdots \times n$: that is, the product of all positive integers up to n. For example, the first eight factorials are $1, 2, 6, 24, 120, 720, 5040,$ and $40\,320$. (The exclamation mark was introduced by Christian Kramp 200 years ago as a convenience to the printer: it is perhaps also intended to convey some alarm at the rapidity with which $n!$ grows. An obsolete notation, which can still be found in some twentieth-century texts, is $\lfloor\underline{n}$.) From this definition, it might appear to be

impossible to make sense of the idea of the factorial of a number that is not a positive integer, but, as it turns out, it is not just possible to do so, but also extremely useful.

The *gamma function*, written Γ, is a function that agrees with the factorial function at positive integer values, but that makes sense for any real number, and even for any complex number. Actually, for various reasons it is natural to define Γ so that $\Gamma(n) = (n-1)!$ for $n = 2, 3, \ldots$. Let us start by writing

$$\Gamma(s) = \int_0^{\infty} x^{s-1} \mathrm{e}^{-x} \,\mathrm{d}x, \qquad (1)$$

without paying too much attention to whether the integral converges. If we integrate by parts, then we find that

$$\Gamma(s) = [-x^{s-1}\mathrm{e}^{-x}]_0^{\infty} + \int_0^{\infty} (s-1)x^{s-2}\mathrm{e}^{-x} \,\mathrm{d}x. \qquad (2)$$

As x tends to infinity, $x^{s-1}\mathrm{e}^{-x}$ tends to zero, and if s is, for example, a real number greater than 1, then $x^{s-1} = 0$ when $x = 0$. Therefore, for such s, we can ignore the first term in the above expression. But the second one is simply the formula for $\Gamma(s-1)$, so we have shown that $\Gamma(s) = (s-1)\Gamma(s-1)$, which is just what we need if we want to think of $\Gamma(s)$ as something like $(s-1)!$.

It is not hard to show that the integral is in fact convergent whenever s is a *complex* number and $\operatorname{Re}(s)$ (the real part of s) is positive. Moreover, it defines a HOLO-MORPHIC FUNCTION [I.3 §5.6] in that region. When the real part of s is negative, the integral does not converge at all, and so the formula (1) cannot be used to define the gamma function in its entirety. However, we can instead use the property $\Gamma(s) = (s-1)\Gamma(s-1)$ to *extend* the definition. For example, when $-1 < \operatorname{Re}(s) \leqslant 0$, we know that the definition does not work directly, but it does work for $s + 1$, since $\operatorname{Re}(s+1) > 0$. We would like $\Gamma(s+1)$ to equal $s\Gamma(s)$, so it makes sense to define $\Gamma(s)$ to be $\Gamma(s+1)/s$. Once we have done this, we can turn our attention to values of s with $-2 < \operatorname{Re}(s) \leqslant -1$, and so on.

The reader may object that in defining $\Gamma(0)$ (for example), we have divided by zero. This is perfectly permissible, however, if all we require of Γ is that it should be MEROMORPHIC [V.31], because meromorphic functions are allowed to take the "value" ∞. Indeed, it is not hard to see that Γ, as we have defined it, has simple poles at $0, -1, -2, \ldots$.

There are in fact many functions that share the useful properties of Γ. (For instance, because $\cos(2\pi s) = \cos(2\pi(s+1))$ for any s, and $\cos(2\pi n) = 1$ for every integer n, the function $F(s) = \Gamma(s)\cos(2\pi s)$ also has the property $F(s) = (s-1)F(s-1)$ and $F(n) = (n-1)!$.) Nevertheless, for a variety of reasons, the function Γ, as we have defined it, is the most natural meromorphic extension of the factorial function. The most persuasive reason is the fact that it arises so often in natural contexts, but it is also, in a certain sense, the smoothest interpolation of the factorial function to all positive real values. In fact, if $f : (0,\infty) \to (0,\infty)$ is such that $f(x+1) = xf(x)$, $f(1) = 1$, and $\log f$ is convex, then $f = \Gamma$.

There are many interesting formulas involving Γ, such as $\Gamma(s)\Gamma(1-s) = \pi/\sin(\pi s)$. There is also the famous result $\Gamma(\frac{1}{2}) = \sqrt{\pi}$, which is essentially equivalent to the fact that the area under the "normal distribution curve" $h(x) = (1/\sqrt{2\pi})e^{-x^2/2}$ is 1 (this can be seen by making the substitution $x = u^2/2$ in (1)). A very important result concerning Γ is the Weierstrass product expansion, which states that

$$\frac{1}{\Gamma(z)} = ze^{\gamma z} \prod_{n=1}^{\infty} \left(1 + \frac{z}{n}\right)e^{-z/n}$$

for all complex z, where γ is Euler's constant:

$$\gamma = \lim_{n\to\infty}\left(1 + \frac{1}{2} + \cdots + \frac{1}{n} - \log n\right).$$

This formula makes it clear that Γ never vanishes, and that it has simple poles at 0 and the negative integers.

Why is the gamma function important? A simple reason is that it occurs frequently in many parts of mathematics, but one can still ask why this should be so. One reason is that Γ, as defined in (1), is the *Mellin transform* of the unarguably natural function $f(x) = e^{-x}$. The Mellin transform is a type of FOURIER TRANSFORM [III.27], but it is defined for functions on the group $(\mathbb{R}^+, \times)$ rather than $(\mathbb{R}, +)$ (which is the habitat of the most familiar type of Fourier transform). For this reason, Γ is often seen in number theory, particularly ANALYTIC NUMBER THEORY [IV.2], where multiplicatively defined functions are often studied by taking Fourier transforms.

One appearance of Γ in a number-theoretical context is in the functional equation for the RIEMANN ZETA FUNCTION [IV.2 §3], namely,

$$\Xi(s) = \Xi(1-s),$$

where

$$\Xi(s) = \Gamma(s/2)\pi^{-s/2}\zeta(s). \tag{3}$$

The ζ function has a well-known product representation

$$\zeta(s) = \prod_{p}(1 - p^{-s})^{-1},$$

where the product is over primes and the representation is valid for $\text{Re}(s) > 1$. The extra factor $\Gamma(s/2)\pi^{-s/2}$ in (3) may be regarded as coming from the "prime at infinity" (a term which may be rigorously defined).

Stirling's formula is a very useful tool in dealing with the gamma function: it provides a rather accurate estimate for $\Gamma(z)$ in terms of simpler functions. A very rough (but often useful) approximation for $n!$ is $(n/e)^n$, which tells us that $\log(n!)$ is about $n(\log n - 1)$. Stirling's formula is a sharper version of this crude estimate. Let $\delta > 0$ and suppose that z is a complex number that has modulus at least 1 and argument between $-\pi + \delta$ and $\pi - \delta$. (This second condition keeps z away from the negative real axis, where the poles are.) Then Stirling's formula states that

$$\log\Gamma(z) = (z - \tfrac{1}{2})\log z - z + \tfrac{1}{2}\log 2\pi + E,$$

where the error E is at most $C(\delta)/|z|$. Here, $C(\delta)$ stands for a certain positive real number that depends on δ. (The smaller you make δ, the larger you have to make $C(\delta)$.) Using this, one may confirm that Γ decays exponentially as $\text{Im}\,z \to \infty$ in any fixed vertical strip in the complex plane. In fact, if $\alpha < \sigma < \beta$, then

$$|\Gamma(\sigma + it)| \leqslant C(\alpha, \beta)|t|^{\beta-1}e^{-\pi|t|/2}$$

for all $|t| > 1$, uniformly in σ.

III.32 Generating Functions

Suppose that you have defined a combinatorial structure, and for each nonnegative integer n you wish to understand how many examples of this structure there are of size n. If a_n denotes this number, then the object that you are trying to analyze is the sequence $a_0, a_1, a_2, a_3, \ldots$. If the structure is quite complicated, then this may be a very hard problem, but one can sometimes make it easier by considering a different object, the *generating function* of the sequence, which contains the same information.

To define this function, one simply regards the sequence a_n as the sequence of coefficients in a power series. That is, the generating function f of the sequence is given by the formula

$$f(x) = a_0 + a_1 x + a_2 x^2 + a_3 x^3 + \cdots.$$

The reason this can be useful is that one can sometimes derive a succinct expression for f and analyze it without reference to the individual numbers a_n. For example, one important generating function has the formula $f(x) = (1 - \sqrt{1 - 4x})/2x$. In such cases, one can deduce properties of the sequence $a_0, a_1, a_2, \ldots$ from properties of f, rather than the other way round.

For more on generating functions, see ENUMERATIVE AND ALGEBRAIC COMBINATORICS [IV.18] and TRANSFORMS [III.91].

III.33   Genus

The *genus* is a topological invariant of surfaces: that is, a quantity associated with a surface that does not change when the surface is continuously deformed. Roughly speaking, it corresponds to the number of holes of that surface, so a sphere has genus 0, a torus has genus 1, a pretzel shape (that is, the surface of a blown-up figure eight) has genus 2, and so on. If one triangulates an orientable surface and counts the vertices, edges, and faces in the triangulation, denoting their numbers by V, E, and F, respectively, then the *Euler characteristic* is defined to be $V - E + F$. It can be shown that if g is the genus and χ is the Euler characteristic, then $\chi = 2 - 2g$. See [I.4 §2.2] for a fuller discussion.

A famous result of POINCARÉ [VI.61] states that for every nonnegative integer g there is precisely one orientable surface of genus g. (Moreover, genus can also be defined for nonorientable surfaces, where a similar result holds.) See DIFFERENTIAL TOPOLOGY [IV.7 §2.3] for more about this theorem.

One can associate an orientable surface, and therefore a genus, with a smooth algebraic curve. An ELLIPTIC CURVE [III.21] can be defined as a smooth curve of genus 1. See ALGEBRAIC GEOMETRY [IV.4 §10] for more details.

III.34   Graphs

A graph is one of the simplest of all mathematical structures: it consists of some elements called *vertices* (of which there are usually just finitely many), some pairs of which are deemed to be "joined" or "adjacent." It is customary to represent the vertices by points in a plane and to join adjacent points by a line. The line is referred to as an *edge* (though how the line is drawn or visualized is irrelevant: all that is important is whether or not two points are joined).

For example, the rail network of a country can be represented by a graph: we can use vertices to represent the stations, and we can join two vertices if they represent consecutive stations along some railroad. Another example is provided by the Internet: the vertices are all the world's computers, and two are adjacent if there is a direct link between them.

Many questions in graph theory take the form of asking what some structural property of a graph can tell you about its other properties. For example, suppose that we are trying to find a graph with n vertices that does not contain a triangle (defined to be a set of three vertices that are mutually joined). How many edges can the graph have? Clearly $\frac{1}{4}n^2$ is possible, at least if n is even, since one can then divide up the n vertices into two equal classes and join all vertices in one class to all vertices in the other. But can there be more edges than that?

Here is another example of a typical question about graphs. Let k be a positive integer. Must there exist an n such that every graph with n vertices always contains either k vertices that are all joined to each other or k vertices no two of which are joined to each other? This question is quite easy for $k = 3$ (where $n = 6$ suffices), but already for $k = 4$ it is not obvious that such an n exists.

For more on these problems (the first is the founding problem of "extremal graph theory," while the second is the founding problem of "Ramsey theory") and on the study of graphs in general, see EXTREMAL AND PROBABILISTIC COMBINATORICS [IV.19].

III.35   Hamiltonians
Terence Tao

At first glance, the many theories and equations of modern physics exhibit a bewildering diversity: for instance, compare classical mechanics with quantum mechanics, or nonrelativistic physics with relativistic physics, or particle physics with statistical mechanics. However, there are strong unifying themes connecting all of these theories. One of these is the remarkable fact that in all of them the evolution of a physical system over time (as well as the steady states of that system) is largely controlled by a single object, the *Hamiltonian* of that system, which can often be interpreted as describing the total energy of any given state in that system. Roughly speaking, each physical phenomenon (e.g., electromagnetism, atomic bonding, particles in a potential well, etc.) may correspond to a single Hamiltonian H, while each type of mechanics (classical, quantum, statistical, etc.) corresponds to a different way of using that Hamiltonian to describe a physical system. For instance, in classical physics, the Hamiltonian is a function $(q, p) \mapsto H(q, p)$ of the positions q and momenta p of the system, which then evolve according to Hamilton's equations:

$$\frac{\mathrm{d}q}{\mathrm{d}t} = \frac{\partial H}{\partial p}, \qquad \frac{\mathrm{d}p}{\mathrm{d}t} = -\frac{\partial H}{\partial q}.$$

In (nonrelativistic) quantum mechanics, the Hamiltonian H becomes a LINEAR OPERATOR [III.50] (which is often a formal combination of the position operators q and momentum operators p), and the wave function ψ of the system then evolves according to THE SCHRÖDINGER EQUATION [III.83]:

$$i\hbar \frac{\mathrm{d}}{\mathrm{d}t} \psi = H\psi.$$

In statistical mechanics, the Hamiltonian H is a function of the microscopic state (or *microstate*) of a system, and the probability that a system at a given temperature T will lie in a given microstate is proportional to $e^{-H/kT}$. And so on and so forth.

Many fields of mathematics are closely intertwined with their counterparts in physics, and so it is not surprising that the concept of a Hamiltonian also appears in pure mathematics. For instance, motivated by classical physics, Hamiltonians (as well as generalizations of Hamiltonians, such as *moment maps*) play a major role in dynamical systems, differential equations, Lie group theory, and symplectic geometry. Motivated by quantum mechanics, Hamiltonians (as well as generalizations, such as *observables* or *pseudo-differential operators*) are similarly prominent in operator algebras, spectral theory, representation theory, differential equations, and microlocal analysis.

Because of their presence in so many areas of physics and mathematics, Hamiltonians are useful for building bridges between seemingly unrelated fields: for instance, between classical mechanics and quantum mechanics, or between symplectic mechanics and operator algebras. The properties of a given Hamiltonian often reveal much about the physical or mathematical objects associated with that Hamiltonian. For example, the symmetries of a Hamiltonian often induce corresponding symmetries in objects described using that Hamiltonian. While not every interesting feature of a mathematical or physical object can be read off directly from its Hamiltonian, this concept is still fundamental to understanding the properties and behavior of such objects.

See also VERTEX OPERATOR ALGEBRAS [IV.17 §2.1], MIRROR SYMMETRY [IV.16 §§2.1.3, 2.2.1], and SYMPLECTIC MANIFOLDS [III.88 §2.1].

III.36   The Heat Equation
Igor Rodnianski

The heat equation was first proposed by FOURIER [VI.25] as a mathematical description of the transfer of heat in solid bodies. Its influence has subsequently been felt in many corners of mathematics: it provides explanations for such disparate phenomena as the formation of ice (the *Stefan problem*), the theory of incompressible viscous fluids (the NAVIER–STOKES EQUATION [III.23]), geometric flows (e.g., curve shortening, and the harmonic-map heat flow problem), BROWNIAN MOTION [IV.24], liquid filtration in porous media (the *Hele-Shaw problem*), index theorems (e.g., the *Gauss–Bonnet–Chern formula*), the price of stock options (the BLACK–SCHOLES FORMULA [VII.9 §2]), and the topology of three-dimensional manifolds (THE POINCARÉ CONJECTURE [V.25]). But the bright future of the heat equation could have been predicted at its birth: after all, another small event that accompanied it was the creation of FOURIER ANALYSIS [III.27].

The propagation of heat is based on a simple continuity principle. The change in the quantity of heat u in a small volume ΔV over a small interval of time Δt is approximately

$$CD \frac{\partial u}{\partial t} \Delta t \Delta V,$$

where C is the heat capacity of the substance and D is its density; but it is also given by the amount of heat entering and exiting through ΔV, which is approximately

$$K \Delta t \int_{\partial \Delta V} \frac{\partial u}{\partial \boldsymbol{n}},$$

where K is the heat conductivity constant and $\boldsymbol{n}$ is the unit normal to the boundary of ΔV.

Thus, setting the values of all physical constants to 1, dividing through by Δt and ΔV, and letting them tend to zero, we find that the evolution of the amount of heat (that is, the temperature) in a three-dimensional solid Ω is governed by the following classical heat equation, where $u(t, \boldsymbol{x})$ is the temperature at time t at the point $\boldsymbol{x} = (x, y, z)$:

$$\frac{\partial}{\partial t} u(t, \boldsymbol{x}) - \Delta u(t, \boldsymbol{x}) = 0. \qquad (1)$$

Here

$$\Delta = \frac{\partial^2}{\partial x^2} + \frac{\partial^2}{\partial y^2} + \frac{\partial^2}{\partial z^2}$$

is the three-dimensional Laplacian; Δu is the limit as the diameter of ΔV tends to zero of the quantity

$$\frac{1}{\Delta V} \int_{\partial \Delta V} \frac{\partial u}{\partial \mathbf{n}}.$$

To determine $u(t, \mathbf{x})$, equation (1) needs to be complemented by the *initial distribution* $u_0(\mathbf{x}) = u(0, \mathbf{x})$ and *boundary conditions* on the solid interface $\partial \Omega$. For example, for a solid unit cube C with surface maintained at zero temperature, the heat equation is considered as a problem with Dirichlet boundary conditions and, as was proposed by Fourier, $u(t, \mathbf{x})$ can be found by the method of separation of variables by expanding $u_0(\mathbf{x})$ into its Fourier series

$$u_0(x, y, z) = \sum_{k,m,l=0}^{\infty} C_{kml} \sin(\pi k x) \\ \times \sin(\pi m y) \sin(\pi l z),$$

which leads to the solution

$$u(t, x, y, z) = \sum_{k,m,l=0}^{\infty} e^{-\pi^2(k^2 + m^2 + l^2)t} C_{kml} \sin(\pi k x) \\ \times \sin(\pi m y) \sin(\pi l z).$$

This simple example already illuminates a fundamental property of the heat equation: the tendency of its solutions to converge to an equilibrium state. In this case it reflects a physically intuitive fact that the temperature $u(t, \mathbf{x})$ converges to the constant distribution $u^*(\mathbf{x}) = C_{000}$.

Propagation of heat in an insulated body corresponds to the choice of the *Neumann* boundary conditions, in which the normal derivative of u (normal, that is, to the boundary $\partial \Omega$) is set to vanish. Its solutions can be constructed in a similar fashion.

The reason that Fourier analysis is intimately connected with the heat equation is that the trigonometric functions are EIGENFUNCTIONS [I.3 §4.3] of the Laplacian. A variety of more general heat equations can be obtained if one replaces the Laplacian by a more general linear, SELF-ADJOINT [III.50 §3.2], nonnegative HAMILTONIAN [III.35] H with a discrete set of eigenvalues λ_n and corresponding eigenfunctions ψ_n. That is, one considers the heat flow

$$\frac{\partial}{\partial t} u + H u = 0.$$

The solution $u(t)$ is given by the formula $u(t) = e^{-tH} u_0$, where e^{-tH} is the *heat semigroup* generated by H, which also takes the more explicit form

$$u(t, \mathbf{x}) = \sum_{n=0}^{\infty} e^{-\lambda_n t} C_n \psi_n(\mathbf{x}).$$

Here the coefficients C_n are the Fourier coefficients of u_0 relative to H: that is, they are the coefficients that arise when we write u_0 as a sum $\sum_{n=0}^{\infty} C_n \psi_n$. (The existence of such a decomposition follows from the SPECTRAL THEOREM [III.50 §3.4] for self-adjoint operators. In a similar way, heat flows can also be generated by self-adjoint operators with a continuous spectrum.) In particular, the asymptotic behavior of $u(t, \mathbf{x})$ as $t \to +\infty$ is completely determined by the spectrum of H.

Although explicit, representations like this do not provide very good quantitative descriptions of the behavior of the heat equation. To obtain such descriptions one has to abandon the idea of constructing solutions explicitly and look instead for principles and methods that apply to general classes of solutions while also being sufficiently robust to be useful in the analysis of more complicated heat equations.

The first methods of this type are called *energy identities*. To derive an energy identity, one multiplies the heat equation by a certain quantity, which may depend on the given solution, and integrates by parts. The simplest two identities of this type are the *conservation of total heat* of an insulated body,

$$\frac{\mathrm{d}}{\mathrm{d}t} \int_\Omega u(t, \mathbf{x}) \, \mathrm{d}\mathbf{x} = 0,$$

and the energy identity,

$$\int_\Omega u^2(t, \mathbf{x}) \, \mathrm{d}\mathbf{x} + 2 \int_0^t \int_\Omega |\nabla u(s, \mathbf{x})|^2 \, \mathrm{d}\mathbf{x} \, \mathrm{d}s \\ = \int_\Omega u^2(0, \mathbf{x}) \, \mathrm{d}x.$$

The second identity already captures a fundamental smoothing property of the heat equation: since all three integrands are nonnegative and the first and third integrals are finite, the average of the mean-square gradient of u is finite, even if the initial mean-square gradient is infinite, and it even decreases to zero with t. In fact, away from the boundary of Ω an arbitrary amount of smoothing takes place, and not just on average but at *every* time $t > 0$.

The second fundamental principle of the heat equation is the *global maximum principle*

$$\max_{\mathbf{x} \in \Omega, 0 \leqslant t \leqslant T} u(t, \mathbf{x}) \\ \leqslant \max\left(u(0, \mathbf{x}), \max_{\mathbf{x} \in \partial\Omega, 0 \leqslant t \leqslant T} u(t, \mathbf{x})\right),$$

which tells us the familiar fact that the hottest spot in the body, over all time, is either on its boundary or in the initial distribution.

Finally, the diffusive properties of the heat equation in $\mathbb{R}^n$ are captured by the *Harnack inequality* for nonnegative solutions u. It tells us that

$$\frac{u(t_2, \boldsymbol{x}_2)}{u(t_1, \boldsymbol{x}_1)} \geq \left(\frac{t_1}{t_2}\right)^{n/2} e^{-|\boldsymbol{x}_2 - \boldsymbol{x}_1|^2/4(t_2 - t_1)}$$

when $t_2 > t_1$. This tells us that if the temperature at $\boldsymbol{x}_1$ at time t_1 takes a certain value, then the temperature at $\boldsymbol{x}_2$ at time t_2 cannot be too much smaller.

This form of the Harnack inequality features a very important object in the study of the heat equation, called the *heat kernel*:

$$p(t, \boldsymbol{x}, \boldsymbol{y}) = \frac{1}{(4\pi t)^{n/2}} e^{-|\boldsymbol{x} - \boldsymbol{y}|^2/4t}.$$

One of its many uses is that it allows one to construct solutions of the heat equation in the whole of space (that is, in $\mathbb{R}^n$) from initial data u_0, by the formula

$$u(t, \boldsymbol{x}) = \int_{\mathbb{R}^n} p(t, \boldsymbol{x}, \boldsymbol{y}) u_0(\boldsymbol{y}) \, d\boldsymbol{y}.$$

It also shows that after a time t initial point disturbances become distributed in a ball of radius $\sqrt{t}$ around the point of the original disturbance. This sort of relation between spatial scales and timescales is the characteristic *parabolic scaling* of the heat equation.

As was shown by Einstein, the heat equation is intimately connected with the diffusion process of Brownian motion. In fact, the mathematical description of Brownian motion is in terms of a random process B_t with transitional probability densities given by the heat kernel $p(t, \boldsymbol{x}, \boldsymbol{y})$. For the n-dimensional Brownian motion $B_t^{\boldsymbol{x}}$ starting at $\boldsymbol{x}$, the function

$$u(t, \boldsymbol{x}) = \mathbb{E}[u_0(\sqrt{2} B_t^{\boldsymbol{x}})]$$

computed with the help of expectation value $\mathbb{E}$ is precisely the solution of the heat equation in $\mathbb{R}^n$ with initial data $u_0(\boldsymbol{x})$. This connection is the start of a mutually beneficial relationship between the theory of the heat equation and probability. Among the most profitable applications of this relationship is the *Feynman–Kac formula*

$$u(t, \boldsymbol{x}) = \mathbb{E}\left[\exp\left(-\int_0^t V(\sqrt{2} B_s^{\boldsymbol{x}}) \, ds\right) u_0(\sqrt{2} B_t^{\boldsymbol{x}}) \right],$$

which connects Brownian motion with solutions of the heat equation

$$\frac{\partial}{\partial t} u(t, \boldsymbol{x}) - \Delta u(t, \boldsymbol{x}) + V(\boldsymbol{x}) u(t, \boldsymbol{x}) = 0$$

with initial data $u_0(\boldsymbol{x})$.

The three fundamental principles of the heat equation described above are remarkably robust, in the sense that they, or weaker versions of them, hold even for very general variants of the classical equation. For instance, they can be applied to the question of the continuity of solutions of the heat equation

$$\frac{\partial}{\partial t} u - \sum_{i,j=1}^n \frac{\partial}{\partial x_i}\left(a_{ij}(\boldsymbol{x}) \frac{\partial}{\partial x_j} u\right) = 0,$$

where all that is assumed of the coefficients a_{ij} is that they are bounded and that they satisfy the *ellipticity condition* $\lambda |\xi|^2 \leq \sum_{i,j} a_{ij} \xi^i \xi^j \leq \Lambda |\xi|^2$. One can even look at the equations in "nondivergence form":

$$\frac{\partial}{\partial t} u - \sum_{i,j=1}^n a_{ij}(\boldsymbol{x}) \frac{\partial}{\partial x_i} \frac{\partial}{\partial x_j} u = 0.$$

Here, the connection between the heat equation and the corresponding stochastic diffusion process turns out to be particularly helpful. This analysis has led to beautiful applications in the CALCULUS OF VARIATIONS [III.94] and in fully nonlinear problems.

The same principles also hold for the heat equations on RIEMANNIAN MANIFOLDS [I.3 §6.10]. The appropriate analogue of the Laplacian for a manifold M is the *Laplace–Beltrami operator* Δ_M, and the heat equation for M is

$$\frac{\partial}{\partial t} u - \Delta_M u = 0.$$

If the Riemannian metric is g, then in local coordinates Δ_M takes the form

$$\Delta_M = \frac{1}{\sqrt{\det g(\boldsymbol{x})}} \sum_{i,j=1}^n \frac{\partial}{\partial x_i}\left(g^{ij}(\boldsymbol{x}) \sqrt{\det g(\boldsymbol{x})} \frac{\partial}{\partial x_j}\right).$$

In this case, a version of the Harnack inequality holds for the heat equation on a manifold that has RICCI CURVATURE [III.78] bounded from below. Interest in the heat equations on manifolds is in part motivated by nonlinear geometric flows and attempts to understand their long-term behavior. One of the earliest geometric flows was the *harmonic map flow*

$$\frac{\partial}{\partial t} \Phi - \Delta_M^N \Phi = 0,$$

which describes a deformation of the map $\Phi(t, \cdot)$ between two compact Riemannian manifolds M and N. The operator Δ_M^N is a nonlinear Laplacian that is constructed by projecting Δ_M onto the tangent space of N. This is a *gradient flow* associated with the energy

$$E[U] = \frac{1}{2} \int_M |dU|_N^2;$$

it measures the stretching of the map U between M and N. Under the assumption that the *sectional curvature* of N is nonpositive, it can be shown that the harmonic map heat flow is regular and converges, as $t \to +\infty$, to a harmonic map between M and N, which is a critical point of the energy functional $E[U]$. This heat equation is used to establish the existence of harmonic maps and to construct a continuous deformation of a given map $\Phi(0, \cdot)$ to a harmonic map $\Phi(+\infty, \cdot)$. The curvature assumption on the target manifold N is responsible for the crucial *monotonicity* properties of the harmonic map heat flow, which come to light through the use of the energy estimates.

An even more spectacular application of a deformation principle of this kind appears in the three-dimensional RICCI FLOW [III.78]

$$\frac{\partial}{\partial t} g_{ij} = -2\mathrm{Ric}_{ij}(g).$$

This is a *quasilinear* heat evolution of a family of metrics $g_{ij}(t)$ on a given manifold M. In this case the flow is not necessarily regular; nonetheless, it can be extended as a flow with "surgeries" in such a way that the structure of the surgeries and the long-term behavior of the flow can be precisely analyzed. This analysis shows in particular that any three-dimensional simply connected manifold is diffeomorphic to a three-dimensional sphere, which gives the proof of the Poincaré conjecture.

The long-term behavior of the heat equation is also important in the analysis of *reaction–diffusion systems* and associated biological phenomena. This was suggested already in the work of TURING [VI.94] in his attempt to understand *morphogenesis* (the formation of inhomogeneous patterns such as animal-coat patterns from a nearly homogeneous initial state) by means of exponential instabilities in the reaction-diffusion equations

$$\frac{\partial}{\partial t} u = \mu \Delta u + f(u, v), \qquad \frac{\partial}{\partial t} v = \nu \Delta v + g(u, v).$$

These examples emphasize the long-term behavior of the heat equation, and in particular the tendency of its solutions to converge to an equilibrium, or alternatively to develop exponential instabilities. However, it turns out that the short-term behavior of the heat equation on a manifold M is of the utmost importance in connection with the geometry and topology of M. This connection is twofold: first, one seeks to establish a relationship between the spectrum of Δ_M and the geometry of M; second, one can use an analysis of the short-term behavior to prove *index theorems*. The former aspect, in the context of planar domains, is captured by Marc Kac's well-known question, "Can one hear the shape of a drum?" For manifolds it begins with the *Weyl formula*

$$\sum_{i=0}^{\infty} e^{-t\lambda_i} = \frac{1}{(4\pi t)^{n/2}} (\mathrm{Vol}(M) + O(t))$$

as t tends to 0. The left-hand side of the identity is the trace of the heat kernel of Δ_M. That is,

$$\sum_{i=0}^{\infty} e^{-t\lambda_i} = \mathrm{tr}\, e^{-t\Delta_M} = \int_M p(t, x, x)\, dx,$$

where $p(t, x, y)$ is such that any solution of the heat equation $\partial u / \partial t - \Delta_M u = 0$ with $u(0, x) = u_0(x)$ is given by the expression

$$u(t, x) = \int_M p(t, x, y) u_0(y)\, dy.$$

The right-hand side of the Weyl identity reflects the short-term asymptotics of the heat kernel $p(t, x, y)$.

The heat-flow approach to the proof of the index theorems can be viewed as a refinement of both sides of the Weyl identity. The trace on the left-hand side is replaced by a more complicated "super-trace," while the right-hand side involves full asymptotics of the heat kernel, which requires one to understand subtle cancelations. The simplest example of this kind is the *Gauss–Bonnet formula*

$$\chi(M) = 2\pi \int_M R,$$

which connects the Euler characteristic of a two-dimensional manifold M and the integral of its scalar curvature. The Euler characteristic $\chi(M)$ arises from a linear combination of traces of the heat flows associated with the *Hodge Laplacian* $(d + d^*)^2$ restricted to the space of exterior differential 0-forms, 1-forms, and 2-forms. A proof of a general ATIYAH–SINGER INDEX THEOREM [V.2] involves heat flows associated with an operator given by the square of a *Dirac operator*.

III.37 Hilbert Spaces

The theory of VECTOR SPACES [I.3 §2.3] and LINEAR MAPS [I.3 §4.2] underpins a large part of mathematics. However, angles cannot be defined using vector space concepts alone, since linear maps do not in general preserve angles. An *inner product space* can be thought of as a vector space with just enough extra structure for the notion of angle to make sense.

The simplest example of an inner product on a vector space is the standard scalar product defined on $\mathbb{R}^n$, the space of all real sequences of length n, as follows. If $v = (v_1, \ldots, v_n)$ and $w = (w_1, \ldots, w_n)$ are two such sequences, then their scalar product, denoted $\langle v, w \rangle$, is the sum $v_1 w_1 + v_2 w_2 + \cdots + v_n w_n$. (For example, the scalar product of $(3, 2, -1)$ and $(1, 4, 4)$ is $3 \times 1 + 2 \times 4 + (-1) \times 4 = 7$.)

Among the properties that the scalar product has are the following two.

(i) It is linear in each variable separately. That is, $\langle \lambda u + \mu v, w \rangle = \lambda \langle u, w \rangle + \mu \langle v, w \rangle$ for any three vectors u, v, and w and any two scalars λ and μ, and similarly $\langle u, \lambda v + \mu w \rangle = \lambda \langle u, v \rangle + \mu \langle u, w \rangle$.

(ii) The scalar product $\langle v, v \rangle$ of any vector v with itself is always a nonnegative real number, and is zero only if v is zero.

In a general vector space, any function $\langle v, w \rangle$ of pairs of vectors v and w that has these two properties is called an inner product, and a vector space with an inner product is called an inner product space. If the vector space has complex scalars, then instead of (i) one must use the following modification.

(i′) For any three vectors u, v, and w and any two scalars λ and μ, $\langle \lambda u + \mu v, w \rangle = \lambda \langle u, w \rangle + \mu \langle v, w \rangle$, and $\langle u, \lambda v + \mu w \rangle = \bar{\lambda} \langle u, v \rangle + \bar{\mu} \langle u, w \rangle$. That is, the inner product is *conjugate-linear* in the second variable.

The reason this has anything to do with angles is that in $\mathbb{R}^2$ and $\mathbb{R}^3$ the scalar product of two vectors v and w works out as the length of v times the length of w times the cosine of the angle between them. In particular, since a vector v makes an angle of zero with itself, $\langle v, v \rangle$ is the square of the length of v.

This gives us a natural way to *define* length and angle in an inner product space. The length, or *norm*, of a vector v, denoted $\|v\|$, is $\sqrt{\langle v, v \rangle}$. Given two vectors v and w, the angle between them is defined by the fact that it lies between 0 and π (or 180°) and its cosine is $\langle v, w \rangle / \|v\| \|w\|$. Once length has been defined, we can also talk about distance: the distance $d(v, w)$ between v and w is the length of their difference, or $\|v - w\|$. This definition of distance satisfies the axioms for a METRIC SPACE [III.56]. From the notion of angle, we can say what it is for v and w to be orthogonal to each other: this simply means that $\langle v, w \rangle = 0$.

The usefulness of inner product spaces goes far beyond their ability to represent the geometry of two- and three-dimensional space. Where they really come into their own is if they are infinite dimensional. Then it becomes convenient if they satisfy the additional property of *completeness*, which is briefly discussed at the end of [III.62]. A complete inner product space is called a *Hilbert space*.

Two important examples of Hilbert spaces are the following.

(i) ℓ_2 is the natural infinite-dimensional generalization of $\mathbb{R}^n$ with the standard scalar product. It is the set of all sequences $(a_1, a_2, a_3, \dots)$ such that the infinite sum $|a_1|^2 + |a_2|^2 + |a_3|^2 + \cdots$ converges. The inner product of $(a_1, a_2, a_3, \dots)$ and $(b_1, b_2, b_3, \dots)$ is $a_1 b_1 + a_2 b_2 + a_3 b_3 + \cdots$ (which can be shown to converge by the CAUCHY–SCHWARZ INEQUALITY [V.19]).

(ii) $L_2[0, 2\pi]$ is the set of all functions f defined on the interval $[0, 2\pi]$ of all real numbers between 0 and 2π, such that the integral $\int_0^{2\pi} |f(x)|^2 \, dx$ makes sense and is finite. The inner product of two functions f and g in $L_2[0, 2\pi]$ is defined to be $\int_0^{2\pi} f(x) g(x) \, dx$. (For technical reasons, this definition is not quite accurate, as a nonzero function can have norm zero, but this problem can easily be dealt with.)

The second of these examples is central to Fourier analysis. A *trigonometric function* is a function of the form $\cos(mx)$ or $\sin(nx)$. The inner product of any two different trigonometric functions is zero, so they are all orthogonal. Even more importantly, the trigonometric functions serve as a coordinate system for the space $L_2[0, 2\pi]$, in that every function f in the space can be represented as an (infinite) linear combination of trigonometric functions. This allows Hilbert spaces to model sound waves: if the function f represents a sound wave, then the trigonometric functions are the pure tones that are its constituent parts.

These properties of trigonometric functions illustrate a very important general phenomenon in the theory of Hilbert spaces: that every Hilbert space has an *orthonormal basis*. This means a set of vectors e_i with the following three properties:

• $\|e_i\| = 1$ for every i;
• $\langle e_i, e_j \rangle = 0$ whenever $i \neq j$; and
• every vector v in the space can be expressed as a convergent sum of the form $\sum_i \lambda_i e_i$.

The trigonometric functions do not quite form an orthonormal basis of $L_2[0, 2\pi]$ but suitable multiples of them do. There are many contexts besides Fourier analysis where one can obtain useful information about a vector by decomposing it in terms of a given orthonormal basis, and many general facts that can be deduced from the existence of such bases.

Hilbert spaces (with complex scalars) are also central to quantum mechanics. The vectors of a Hilbert space can be used to represent possible states of a quantum mechanical system, and observable features of that system correspond to certain linear maps.

For this and other reasons, the study of LINEAR OPERATORS [III.50] on Hilbert spaces is a major branch of mathematics: see OPERATOR ALGEBRAS [IV.15].

III.38 Homology and Cohomology

If X is a TOPOLOGICAL SPACE [III.90], then one can associate with it a sequence of groups $H_n(X, R)$, where R is a commutative RING [III.81 §1] such as $\mathbb{Z}$ or $\mathbb{C}$. These groups, the *homology groups* of X (with coefficients in R), are a powerful invariant: powerful because they contain a great deal of information about X but are nevertheless easy to compute, at least compared with some other invariants. The closely related *cohomology groups* $H^n(X, R)$ are more useful still because they can be made into a ring: to oversimplify slightly, an element of the cohomology group $H^n(X)$ is an EQUIVALENCE CLASS [I.2 §2.3] [Y] of a subspace Y of codimension n. (Of course, for this to make true sense X should be a fairly nice space such as a MANIFOLD [I.3 §6.9].) Then, if [Y] and [Z] belong to $H^n(X, R)$ and $H^m(X, R)$, respectively, their product is [$Y \cap Z$]. Since $Y \cap Z$ "typically" has codimension $n + m$, the equivalence class [$Y \cap Z$] belongs to $H^{n+m}(X, R)$. Homology and cohomology groups are described in more detail in ALGEBRAIC TOPOLOGY [IV.6].

The concepts of homology and cohomology have become far more general than the above discussion suggests, and are no longer tied to topological spaces: for instance, the notion of group cohomology is of great importance in algebra. Even within topology, there are many different homology and cohomology theories. In 1945, Eilenberg and Steenrod devised a small number of axioms that greatly clarified the area: a homology theory is any association of groups with topological spaces that satisfies these axioms, and the fundamental properties of homology theories follow from the axioms.

III.39 Homotopy Groups

If X is a TOPOLOGICAL SPACE [III.90], then a *loop* in X is a path that begins and ends at the same point; or, more formally, a continuous function $f : [0, 1] \to X$ such that $f(0) = f(1)$. The point where the path begins and ends is called the *base point*. If two loops have the same base point, they are called *homotopic* if one can be continuously deformed to the other, with all the intermediate

paths living in X and beginning and ending at the given base point. For example, if X is the plane $\mathbb{R}^2$, then any two paths that begin and end at $(0, 0)$ are homotopic, whereas if X is the plane with the origin removed, then whether or not two paths (that begin and end at some other point) are homotopic depends on whether or not they go around the origin the same number of times.

Homotopy is an EQUIVALENCE RELATION [I.2 §2.3], and the equivalence classes of paths with base point x form the *fundamental group* of X, relative to x, which is denoted by $\pi_1(X, x)$. If X is connected, then this does not depend on x and we can write $\pi_1(X)$ instead. The group operation is "concatenation": given two paths that begin and end at x, their "product" is the combined path that goes along one and then the other, and the product of equivalence classes is then defined to be the equivalence class of the product. This group is a very important invariant (see for instance GEOMETRIC AND COMBINATORIAL GROUP THEORY [IV.10 §7]); it is the first in a sequence of higher-dimensional homotopy groups, which are described in ALGEBRAIC TOPOLOGY [IV.6 §§2, 3].

III.40 The Ideal Class Group

THE FUNDAMENTAL THEOREM OF ARITHMETIC [V.14] asserts that every positive integer can be written in exactly one way (apart from reordering) as a product of primes. Analogous theorems are true in other contexts as well: for example, there is a unique factorization theorem for polynomials, and another one for *Gaussian integers*, that is, numbers of the form $a + ib$ where a and b are integers.

However, for most NUMBER FIELDS [III.63], the associated "ring of integers" does not have the unique-factorization property. For example, in the RING [III.81 §1] of numbers of the form $a + b\sqrt{-5}$ with a and b integers, one can factorize 6 either as 2×3 or as $(1 + \sqrt{-5})(1 - \sqrt{-5})$.

The ideal class group is a way of measuring how badly unique factorization fails. Given any ring of integers of a number field, one can define a multiplicative structure on its set of IDEALS [III.81 §2], for which unique factorization holds. The elements of the ring itself correspond to so-called "principal ideals," so if every ideal is principal, then unique factorization holds for the ring. If there are nonprincipal ideals, then one can define a natural EQUIVALENCE RELATION [I.2 §2.3] on them in such a way that the equivalence classes, which are called *ideal classes*, form a GROUP [I.3 §2.1]. This group is the ideal class group. All principal ideals belong to the class that forms the identity of this group, so the larger and more complex the ideal group is, the further the ring is from having the unique-factorization property. For more details, see ALGEBRAIC NUMBERS [IV.1], and in particular section 7.

III.41   Irrational and Transcendental Numbers
Ben Green

An irrational number is one that cannot be written as a/b with both a and b integers. A great many naturally occurring numbers, such as $\sqrt{2}$, e, and π, are irrational. The following proof that $\sqrt{2}$ is irrational is one of the best-known arguments in all of mathematics. Suppose that $\sqrt{2} = a/b$; since common factors can be canceled, we may assume that a and b have no common factor; we have $a^2 = 2b^2$, which means that a must be even; write $a = 2c$; but then $4c^2 = 2b^2$, which implies that $2c^2 = b^2$, and hence b must be even too; this, however, is contrary to our assumption that a and b were coprime.

There are several famous conjectures in mathematics that ask whether certain specific numbers are rational or not. For example, $\pi + e$ and π^e are not known to be irrational, and neither is Euler's constant:

$$\gamma = \lim_{n \to \infty} \left(1 + \frac{1}{2} + \cdots + \frac{1}{n} - \log n \right) \approx 0.577215\ldots.$$

It is known that $\zeta(3) = 1 + 2^{-3} + 3^{-3} + \cdots$ is irrational. Almost certainly, $\zeta(5), \zeta(7), \zeta(9), \ldots$ are all irrational as well. However, although it has been shown that infinitely many of these numbers are irrational, no specific one is known to be.

A classic proof is that of the irrationality of e. If

$$e = \sum_{j=0}^{\infty} \frac{1}{j!}$$

were equal to p/q, then we would have

$$p(q-1)! = \sum_{j=0}^{\infty} \frac{q!}{j!}.$$

The left-hand side and the terms of the sum with $j \leqslant q$ are all integers. Therefore the quantity

$$\sum_{j \geqslant q+1} \frac{q!}{j!} = \frac{1}{q+1} + \frac{1}{(q+1)(q+2)} + \cdots$$

is also an integer. But it is not hard to show that this quantity lies strictly between 0 and 1, a contradiction.

The principle used here, that a nonzero integer must have absolute value at least one, is surprisingly powerful in the theory of irrational and transcendental numbers.

Some numbers are more irrational than others. In a sense, the most irrational number is $\tau = \frac{1}{2}(1 + \sqrt{5})$, the golden ratio, because the best rational approximations to it, which are ratios of consecutive Fibonacci numbers, approach it rather slowly. There is also a very elegant proof that τ is irrational. This is based on the observation that the $\tau \times 1$ rectangle R may be divided into a square of side 1 and a $1/\tau \times 1$ rectangle. If τ were rational, then we would be able to create a rectangle with integer sides that was similar to R. From this we could remove a square, and we would be left with a smaller rectangle with integer sides that would still be similar to R. We could continue this process ad infinitum, which is clearly impossible.

A *transcendental* number is one which is not *algebraic*, that is to say, is not the root of a polynomial equation with integer coefficients. Thus, $\sqrt{2}$ is not transcendental, since it solves $x^2 - 2 = 0$, and neither is $\sqrt{7 + \sqrt{17}}$.

Are there, in fact, any transcendental numbers? This question was answered by LIOUVILLE [VI.39] in 1844, who showed that various numbers were transcendental, of which

$$\kappa = \sum_{n \geqslant 1} 10^{-n!}$$

$$= 0.11000100000000000000000010\ldots$$

is a well-known example. This is not algebraic, because it can be approximated by rationals more accurately than any algebraic number can. For example, the rational approximation $110001/1000000$ is very close indeed to κ, but its denominator is not particularly large.

Liouville showed that if α is a root of a polynomial of degree n, then

$$\left| \alpha - \frac{a}{q} \right| > \frac{C}{q^n}$$

for all integers a and q and for some constant C depending on α. In words, α cannot be too well approximated by rationals. Roth later proved that the exponent n here can actually be replaced by $2 + \varepsilon$ for any $\varepsilon > 0$. (For more on these topics, see LIOUVILLE'S THEOREM AND ROTH'S THEOREM [V.22].)

A completely different approach to the existence of transcendental numbers was discovered by CANTOR [VI.54] thirty years later. He proved that the set of algebraic numbers is COUNTABLE [III.11], which means, roughly speaking, that they may be listed in order. More precisely, there is a surjective map from $\mathbb{N}$, the set of natural numbers, to the set of algebraic numbers. By contrast, the real numbers $\mathbb{R}$ are not countable. Cantor's famous proof of this uses a diagonalization argument to show that any listing of all the real numbers must be incomplete. There must, therefore, be real numbers that are not algebraic.

It is generally rather difficult to prove that a specific number is transcendental. For instance, it is by no means the case that all transcendental numbers are very well approximated by rationals; this merely provides a useful sufficient condition. There are other ways to establish that numbers are transcendental. Both e and π are known to be transcendental, and it is known that $|e - a/b| > C(\varepsilon)/b^{2+\varepsilon}$ for all $\varepsilon > 0$, so e is not all that well approximated by rationals. Since $\zeta(2m)$ is always a rational multiple of π^{2m}, it follows that the numbers $\zeta(2), \zeta(4), \ldots$ are all transcendental.

The modern theory of transcendental numbers contains a wealth of beautiful results. An early one is the Gel'fond–Schneider theorem, which says that α^β is transcendental if $\alpha \neq 0, 1$ is algebraic, and if β is algebraic but not rational. In particular, $\sqrt{2}^{\sqrt{2}}$ is transcendental. There is also the *six-exponentials theorem*, which states that if x_1, x_2 are two linearly independent complex numbers, and if y_1, y_2, y_3 are three linearly independent complex numbers, then at least one of the six numbers

$$e^{x_1 y_1}, e^{x_1 y_2}, e^{x_1 y_3}, e^{x_2 y_1}, e^{x_2 y_2}, e^{x_2 y_3}$$

is transcendental. Related to this is the (as yet unsolved) *four-exponentials conjecture*: if x_1 and x_2 are two linearly independent complex numbers, and if y_1 and y_2 are linearly independent, then at least one of the four exponentials

$$e^{x_1 y_1}, e^{x_1 y_2}, e^{x_2 y_1}, e^{x_2 y_2}$$

is transcendental.

III.42 The Ising Model

The Ising model is one of the fundamental models of statistical physics. It was originally designed as a model for the behavior of a ferromagnetic material when it is heated up, but it has since been used to model many other phenomena.

The following is a special case of the model. Let G_n be the set of all pairs of integers with absolute value

at most n. A *configuration* is a way of assigning to each point x in G_n a number σ_x, which equals 1 or -1. The points represent atoms and $\sigma(x)$ represents whether x has "spin up" or "spin down." With each configuration σ we associate an "energy" $E(\sigma)$, which equals $-\sum \sigma_x \sigma_y$, where the sum is taken over all pairs of neighboring points x and y. Thus, the energy is high if many points have different signs from some of their neighbors, and low if G_n is divided into large clusters of points with the same sign.

Each configuration is assigned a probability, which is proportional to $e^{-E(\sigma)/T}$. Here, T is a positive real number that represents temperature. The probability of a given configuration is therefore higher when it has small energy, so there is a tendency for a typical configuration to have clusters of points with the same sign. However, as the temperature T increases, this clustering effect becomes smaller since the probabilities become more equal.

The *two-dimensional Ising model with zero potential* is the limit of this model as n tends to infinity. For a more detailed discussion of the general model and of the *phase transition* associated with it, see PROBABILISTIC MODELS OF CRITICAL PHENOMENA [IV.25 §5].

III.43 Jordan Normal Form

Suppose that you are presented with an $n \times n$ real or complex MATRIX [I.3 §4.2] A and would like to understand it. You might ask how it behaves as a LINEAR MAP [I.3 §4.2] on $\mathbb{R}^n$ or $\mathbb{C}^n$, or you might wish to know what the powers of A are. In general, answering these questions is not particularly easy, but for some matrices it is very easy. For example, if A is a *diagonal* matrix (that is, one whose nonzero entries all lie on the diagonal), then both questions can be answered immediately: if x is a vector in $\mathbb{R}^n$ or $\mathbb{C}^n$, then Ax will be the vector obtained by multiplying each entry of x by the corresponding diagonal element of A, and to compute A^m you just raise each diagonal entry to the power m.

So, given a linear map T (from $\mathbb{R}^n$ to $\mathbb{R}^n$ or from $\mathbb{C}^n$ to $\mathbb{C}^n$), it is very nice if we can find a basis with respect to which T has a diagonal matrix; if this can be done, then we feel that we "understand" the linear map. Saying that such a basis exists is the same as saying that there is a basis consisting of EIGENVECTORS [I.3 §4.3]: a linear map is called *diagonalizable* if it has such a basis. Of course, we may apply the same terminology to a matrix (since a matrix A determines a linear map on $\mathbb{R}^n$ or $\mathbb{C}^n$, by mapping x to Ax). So a matrix is also called diagonalizable if it has a basis of eigenvectors, or equivalently if there is an invertible matrix P such that $P^{-1}AP$ is diagonal.

Is every matrix diagonalizable? Over the reals, the answer is no for uninteresting reasons, since there need not even be any eigenvectors: for example, a rotation in the plane clearly has no eigenvectors. So let us restrict our attention to matrices and linear maps over the complex numbers.

If we have a matrix A, then its *characteristic polynomial*, namely $\det(A - tI)$, certainly has a root, by THE FUNDAMENTAL THEOREM OF ALGEBRA [V.13]. If λ is such a root, then standard facts from linear algebra tell us that $A - \lambda I$ is singular, and therefore that there is a vector x such that $(A - \lambda I)x = 0$, or equivalently that $Ax = \lambda x$. So we do have at least one eigenvector. Unfortunately, however, there need not be enough eigenvectors to form a basis. For example, consider the linear map T that sends $(1,0)$ to $(0,1)$ and $(0,1)$ to $(0,0)$. The matrix of this map (with respect to the obvious basis) is $\left(\begin{smallmatrix} 0 & 0 \\ 1 & 0 \end{smallmatrix}\right)$. This matrix is not diagonalizable. One way of seeing why not is the following. The characteristic polynomial turns out to be t^2, of which the only root is 0. An easy computation reveals that if $Ax = 0$ then x has to be a multiple of $(0,1)$, so we cannot find two linearly independent eigenvectors. A rather more elegant method of proof is to observe that T^2 is the zero matrix (since it maps each of $(1,0)$ and $(0,1)$ to $(0,0)$), so that if T were diagonalizable, then its diagonal matrix would have to be zero (since any nonzero diagonal matrix has a nonzero square), and therefore T would have to be the zero matrix, which it is not.

The same argument shows that *any* matrix A such that $A^k = 0$ for some k (such matrices are called *nilpotent*) must fail to be diagonalizable, unless A is itself the zero matrix. This applies, for example, to any matrix that has all of its nonzero entries below the main diagonal.

What, then, *can* we say about our nondiagonalizable matrix T above? In a sense, one feels that $(1,0)$ is "nearly" an eigenvector, since we do have $T^2(1,0) = (0,0)$. So what happens if we extend our point of view by allowing such vectors? One would say that a vector x is a *generalized eigenvector* of T, with eigenvalue λ, if some power of $T - \lambda$ maps x to zero. For instance, in our example above the vector $(1,0)$ is a generalized eigenvector with eigenvalue 0. And, just as we have an "eigenspace" associated with each eigenvalue λ (defined to be the space of all eigenvectors with eigenvalue λ), we also have a "generalized eigenspace,"

which consists of all generalized eigenvectors with eigenvalue λ.

Diagonalizing a matrix corresponds exactly to decomposing the vector space ($\mathbb{C}^n$) into eigenspaces. So it is natural to hope that one could decompose the vector space into generalized eigenspaces for *any* matrix. And this turns out to be true. The way of breaking up the space is called *Jordan normal form*, which we shall now describe in more detail.

Let us pause for a moment and ask: what is the very simplest situation in which we get a generalized eigenvector? It would surely be the obvious generalization of the above example to n dimensions. In other words, we have a linear map T that sends e_1 to e_2, e_2 to e_3, and so on, until e_{n-1} is sent to e_n, with e_n itself mapped to zero. This corresponds to the matrix

$$\begin{pmatrix} 0 & 0 & 0 & \cdots & 0 & 0 \\ 1 & 0 & 0 & \cdots & 0 & 0 \\ 0 & 1 & 0 & \cdots & 0 & 0 \\ \vdots & \vdots & \vdots & \ddots & \vdots & \vdots \\ 0 & 0 & 0 & \cdots & 1 & 0 \end{pmatrix}.$$

Although this matrix is not diagonalizable, its behavior is at least very easy to understand.

The Jordan normal form of a matrix will be a diagonal sum of matrices that are easily understood in the way that this one is. Of course, we have to consider eigenvalues other than zero: accordingly, we define a *block* to be any matrix of the form

$$\begin{pmatrix} \lambda & 0 & 0 & \cdots & 0 & 0 \\ 1 & \lambda & 0 & \cdots & 0 & 0 \\ 0 & 1 & \lambda & \cdots & 0 & 0 \\ \vdots & \vdots & \vdots & \ddots & \vdots & \vdots \\ 0 & 0 & 0 & \cdots & 1 & \lambda \end{pmatrix}.$$

Note that this matrix A, with λI subtracted, is precisely the matrix above, so that $(A - \lambda I)^n$ is indeed zero. Thus, a block represents a linear map that is indeed easy to understand, and all its vectors are generalized eigenvectors with the same eigenvalue. The Jordan normal form theorem tells us that every matrix can be decomposed into such blocks: that is, a matrix is in Jordan normal form if it is of the form

$$\begin{pmatrix} B_1 & 0 & \cdots & 0 \\ 0 & B_2 & \cdots & 0 \\ \vdots & \vdots & \ddots & \vdots \\ 0 & 0 & \cdots & B_k \end{pmatrix}.$$

Here, the B_i are blocks, which can have different sizes, and the 0s represent submatrices of the matrix with sizes depending on the block sizes. Note that a block of size 1 simply consists of an eigenvector.

Once a matrix A is put into Jordan normal form, we have broken up the space into subspaces on which it is easy to understand the action of A. For example, suppose that A is the matrix

$$\begin{pmatrix} 4 & 0 & 0 & 0 & 0 & 0 & 0 \\ 1 & 4 & 0 & 0 & 0 & 0 & 0 \\ 0 & 1 & 4 & 0 & 0 & 0 & 0 \\ 0 & 0 & 0 & 4 & 0 & 0 & 0 \\ 0 & 0 & 0 & 1 & 4 & 0 & 0 \\ 0 & 0 & 0 & 0 & 0 & 2 & 0 \\ 0 & 0 & 0 & 0 & 0 & 1 & 2 \end{pmatrix},$$

which is made out of three blocks, of sizes 3, 2, and 2. Then we can instantly read off a great deal of information about A. For instance, consider the eigenvalue 4. Its algebraic multiplicity (its multiplicity as a root of the characteristic polynomial) is 5, since it is the sum of the sizes of all the blocks with eigenvalue 4, while its geometric multiplicity (the dimension of its eigenspace) is 2, since it is the *number* of such blocks (because in each block we only have one actual eigenvector). And even the minimal polynomial of the matrix (the smallest-degree polynomial $P(t)$ such that $P(A) = 0$) is easy to write down. The minimal polynomial of each block can be written down instantly: if the block has size k and generalized eigenvalue λ, then it is $(t - \lambda)^k$. The minimal polynomial of the whole matrix is then the "lowest common multiple" of the polynomials for the individual blocks. For the matrix above, we get $(t - 4)^3$, $(t-4)^2$, and $(t-2)^2$ for the three blocks, so the minimal polynomial of the whole matrix is $(t - 4)^3 (t - 2)^2$.

There are some generalizations of Jordan normal form, away from the context of linear maps acting on vector spaces. For example, there is an analogue of the theorem that applies to Abelian groups, which turns out to be the statement that every finite Abelian group can be decomposed as a direct product of cyclic groups.

III.44   Knot Polynomials
W. B. R. Lickorish

1   Knots and Links

A *knot* is a curve in three-dimensional space that is closed (in other words, it stops where it began) and never meets itself along its way. A *link* is several such curves, all disjoint from one another, which are called the *components* of the link. Some simple examples of knots and links are the following:

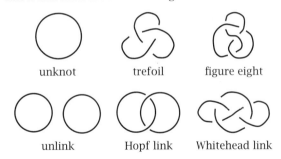

unknot   trefoil   figure eight

unlink   Hopf link   Whitehead link

Two knots are equivalent or "the same" if one can be moved continuously, never breaking the "string," to become the other. *Isotopy* is the technical term for such movement. For example, the following knots are the same:

The first problem in knot theory is how to decide whether two knots are the same. Two knots may appear to be very different but how does one *prove* that they are different? In classical geometry two triangles are the same (or *congruent*) if one can be moved rigidly on to the other. Numbers that measure side-lengths and angles are assigned to each triangle to help determine whether this is the case. Similarly, mathematical entities called *invariants* can be associated with knots and links in such a way that if two links have different invariants, then they cannot be the same link. Many invariants relate to the geometry or topology of the complement of a link in three-dimensional space. The FUNDAMENTAL GROUP [IV.6 §2] of this complement is an excellent invariant, but algebraic techniques are then needed to distinguish the groups. The polynomial of J. W. Alexander (published in 1926) is a link invariant derived from distinguishing such groups. Although rooted in ALGEBRAIC TOPOLOGY [IV.6], the Alexander polynomial has long been known to satisfy a skein relation (see below). The HOMFLY polynomial of 1984 generalizes the Alexander polynomial and can be based on the simple combinatorics of skein theory alone.

1.1   The HOMFLY Polynomial

Suppose that links are oriented so that directions, indicated by arrows, are given to all components. To each oriented link L is assigned its HOMFLY polynomial

$P(L)$, a polynomial with integer coefficients in two variables v and z (allowing both positive and negative powers of v and z). The polynomials are such that

$$P(\text{unknot}) = 1 \qquad (1)$$

and there is a linear *skein relation*

$$v^{-1}P(L_+) - vP(L_-) = zP(L_0). \qquad (2)$$

This means that equation (2) holds whenever three links have identical diagrams except near one crossing where they are as follows

then this equation holds.

This turns out to be good notation, although one could in principle use x and y in place of v^{-1} and $-v$. Although Alexander's polynomial satisfied a particular instance of (2), it took almost sixty years and the discovery of the Jones polynomial for it to be realized that this general linear relation can be used. Note that there are two possible types of crossing in a diagram of an oriented link. A crossing is *positive* if, when approaching the crossing along the under-passing arc in the direction of the arrow, the other directed arc is seen to cross over from left to right. If the over-passing arc crosses from right to left, the crossing is *negative*. When interpreting the skein relation at a crossing of a link L, it is vital that L be regarded as L_+ if the crossing is positive and as L_- if it is negative.

The theorem that underpins this theory, which is not at all obvious, is that it is possible to assign such polynomials to oriented links in a coherent fashion, uniquely, independent of any choice of a link's diagram. A proof of this is given in Lickorish (1997).

1.2 HOMFLY Calculations

In a diagram of a knot it is always possible to change some of the crossings, from over to under, to achieve a diagram of the unknot. Links can be undone similarly. Using this, the polynomial of any link can be calculated from the above equations, though the length of the calculation is exponential in the number of crossings. The following is a calculation of $P(\text{trefoil})$. Firstly, consider the following instance of the skein relation:

$$v^{-1}P(\text{⬭⬭}) - vP(\text{⬭⬭}) = zP(\text{◯◯}).$$

Substituting the polynomial 1 for the polynomials of the two unknots, this shows that the HOMFLY polynomial of the two-component unlink is $z^{-1}(v^{-1} - v)$. A

second usage of the skein relation is

$$v^{-1}P\!\left(\text{⬭}\right) - vP\!\left(\text{⬭}\right) = zP\!\left(\text{⬭}\right).$$

Substituting the previous answer for the unlink shows that the HOMFLY polynomial of the Hopf link is equal to $z^{-1}(v^{-3} - v^{-1}) - zv^{-1}$. Finally, consider the following instance of the skein relation:

$$v^{-1}P\!\left(\text{⬭}\right) - vP\!\left(\text{⬭}\right) = zP\!\left(\text{⬭}\right).$$

Substitution of the polynomial already calculated for the Hopf link and of course the value 1 for the unknot shows that

$$P(\text{trefoil}) = -v^{-4} + 2v^{-2} + z^2 v^{-2}.$$

A similar calculation shows that

$$P(\text{figure eight}) = v^2 - 1 + v^{-2} - z^2.$$

The trefoil and the figure eight thus have different polynomials; this *proves* they are different knots. Experimentally, if a trefoil is actually made from a necklace (using the clasp to join the ends together) it is indeed found to be impossible to move it to the configuration of a figure eight knot. Note that the polynomial of a knot is not dependent on the choice of its orientation (but this is not so for links).

Reflecting a knot in a mirror is equivalent to changing every crossing in a diagram of the knot from an over-crossing to an under-crossing and vice versa (consider the plane of the diagram to be the mirror). The polynomial of the reflection is always the same as that of the original knot *except* that every occurrence of v must be replaced by one of $-v^{-1}$. Thus the trefoil and its reflection,

have polynomials

$$-v^{-4} + 2v^{-2} + z^2 v^{-2} \quad \text{and} \quad -v^4 + 2v^2 + z^2 v^2.$$

As these polynomials are not the same, the trefoil and its reflection are different knots.

2 Other Polynomial Invariants

The HOMFLY polynomial was inspired by the discovery in 1984 of the polynomial of V. F. R. Jones. For an oriented link L, the Jones polynomial $V(L)$ has just one variable t (together with t^{-1}). It is obtained from $P(L)$ by substituting $v = t$ and $z = t^{1/2} - t^{-1/2}$, where

$t^{1/2}$ is just a formal square root of t. The Alexander polynomial is obtained by the substitution $v = 1$, $z = t^{-1/2} - t^{1/2}$. This latter polynomial is well understood in terms of topology, by way of the fundamental group, covering spaces, and homology theory, and can be calculated by various methods involving determinants. It was J. H. Conway who, in discussing in 1969 his normalized version of the Alexander polynomial (the polynomial in one variable z obtained by substituting $v = 1$ into the HOMFLY polynomial), first developed the theory of skein relations.

There is one more polynomial (due to L. H. Kauffman) based on a linear skein relation. The relation involves four links with unoriented diagrams differing as follows:

There are examples of pairs of knots that the Kauffman polynomial but not the HOMFLY polynomial can distinguish and vice versa; some pairs are not distinguished by any of these polynomials.

2.1 Application to Alternating Knots

For the Jones polynomial there is a particularly simple formulation, by means of "Kauffman's bracket polynomial," that leads to an easy proof that the Jones (but *not* the HOMFLY) polynomial is coherently defined. This approach has been used to give the first rigorous confirmation of P. G. Tait's (1898) highly believable proposal that a reduced alternating diagram of a knot has the minimal number of crossings for any diagram of that knot. Here "alternating" means that in going along the knot the crossings go: ... over, under, over, under, over, ... Not every knot has such a diagram. "Reduced" means that there are, adjacent to each crossing, four *distinct* regions of the diagram's planar complement. Thus, for example, any nontrivial reduced alternating diagram is not a diagram of the unknot. Also, the figure eight knot certainly has no diagram with only three crossings.

2.2 Physics

Unlike that of Alexander, the HOMFLY polynomial has no known interpretation in terms of classical algebraic topology. It can, however, be reformulated as a collection of state sums, summing over certain labelings of a knot diagram. This recalls ideas from statistical mechanics; an elementary account is given in Kauffman

(1991). An amplification of the whole HOMFLY polynomial theory leads into a version of conformal field theory called topological quantum field theory.

Further Reading

Kauffman, L. H. 1991. *Knots and Physics*. Singapore: World Scientific.
Lickorish, W. B. R. 1997. *An Introduction to Knot Theory*. Graduate Texts in Mathematics, volume 175. New York: Springer.
Tait, P. G. 1898. On knots. In *Scientific Papers*, volume I, pp. 273–347. Cambridge: Cambridge University Press.

III.45 *K*-Theory

K-theory concerns one of the most important invariants of a TOPOLOGICAL SPACE [III.90] X, a pair of groups called the *K-groups* of X. To form the group $K^0(X)$ one takes all (equivalence classes of) vector bundles on X, and uses the direct sum as the group operation. This leads not to a group but to a semigroup. However, from the semigroup one can easily construct a group in the same way that one constructs $\mathbb{Z}$ out of $\mathbb{N}$: by taking equivalence classes of expressions of the form $a - b$. If i is a positive integer, then there is a natural way of defining a group $K^{-i}(X)$: it is closely related to the group $K^0(S^i \times X)$. The very important *Bott periodicity theorem* says that $K^i(X)$ depends only on the parity of i, so there are in fact just two distinct *K*-groups, $K^0(X)$ and $K^1(X)$. See ALGEBRAIC TOPOLOGY [IV.6 §6] for more details.

If X is a topological space such as a compact manifold, then one can associate with it the C^*-algebra $C(X)$ of all continuous functions from X to $\mathbb{C}$. It turns out to be possible to define the *K*-groups in terms of this algebra in such a way that it applies to algebras that are not of the form $C(X)$. In particular, it applies to algebras where multiplication is not commutative. For instance, *K*-theory provides important invariants of C^*-algebras. See OPERATOR ALGEBRAS [IV.15 §4.4].

Lagrange Multipliers
See OPTIMIZATION AND LAGRANGE MULTIPLIERS [III.64]

III.46 The Leech Lattice

To define a *lattice* in $\mathbb{R}^d$ one chooses d linearly independent vectors $v_1, \dots, v_d$ and takes all combinations of the form $a_1 v_1 + \cdots + a_d v_d$, where $a_1, \ldots, a_d$ are integers. For example, to define the *hexagonal lattice* in $\mathbb{R}^2$ one can take v_1 and v_2 to be $(1, 0)$ and $(\frac{1}{2}, \frac{\sqrt{3}}{2})$, respectively. Notice that v_2 is v_1 rotated by $\pi/3$, and also that $v_2 - v_1$ is v_2 rotated by $\pi/3$. Continuing this process, one can generate all the points in a regular hexagon about the origin.

The hexagonal lattice is unusual, among lattices in $\mathbb{R}^2$, in that it has a rotational symmetry of order 6. This makes it the "best" lattice in many ways. (For example, bees arrange their hives in hexagonal lattices, soap bubbles of similar sizes naturally organize themselves into hexagonal lattices, and so on.) The Leech lattice plays a similar role in twenty-four dimensions: it is the "most symmetrical" of all twenty-four-dimensional lattices, with a degree of symmetry that is quite extraordinary. It is discussed in more detail in THE GENERAL GOALS OF MATHEMATICAL RESEARCH [I.4 §4].

III.47 L-Functions
Kevin Buzzard

1 How Can We "Package" a Sequence of Numbers?

Suppose we are given a sequence of numbers such as

$$\pi, \ \sqrt{2}, \ 6.023 \times 10^{23}, \ \ldots.$$

How can we package up this sequence into *one* object that remembers everything about the sequence, and that might even give us new insights into the sequence? One standard technique is to use a GENERATING FUNCTION [III.32], but here is another way, which has proved very fruitful in number theory and elsewhere. Given a sequence $a_1, a_2, a_3, \ldots$, we define the *Dirichlet series*

$$L(s) = \frac{a_1}{1^s} + \frac{a_2}{2^s} + \frac{a_3}{3^s} + \cdots$$
$$= \sum_{n \geqslant 1} a_n / n^s.$$

Here, s could be a positive integer, or a real number, for example. As long as our sequence $a_1, a_2, \ldots$ does not grow too quickly (which we shall henceforth assume), the series $L(s)$ will converge for all sufficiently large values of s. Moreover, it may be a very "rich" object, even if the initial sequence is simple. For example, if $a_n = 1$ for all n, then the resulting function $L(s)$ is the famous RIEMANN ZETA FUNCTION [IV.2 §3] $\zeta(s) = 1^{-s} + 2^{-s} + 3^{-s} + \cdots$, which converges when $s > 1$ and was shown by Euler to satisfy the following identities,

among others (there is one for each even number):

$$\zeta(2) = \pi^2/6, \qquad \zeta(4) = \pi^4/90,$$
$$\zeta(12) = \frac{691 \pi^{12}}{638\,512\,875}.$$

Thus, even a sequence as simple as $1, 1, 1, \ldots$ leads us to some natural questions that cry out to be answered.

The zeta function is the prototypical example of an *L-function*. However, not every Dirichlet series deserves to be called an *L*-function. We will mention below some "good" properties that the zeta function has: roughly speaking, a Dirichlet series is considered to be an *L*-function if it has these good properties. This is not a formal definition of course, but in fact there is no formal definition of "an *L*-function." (People have tried to give one, but there is no real consensus about what the right definition should be.) What happens in practice is that a mathematician finds a way of associating a sequence $a_1, a_2, \ldots$ of numbers with a mathematical object X, and if evidence then emerges to suggest that the associated Dirichlet series $L(s)$ shares the good properties of the zeta function, then $L(s)$ will be called the *L*-function of X.

2 What Good Properties Might $L(s)$ Have?

One can check that the zeta function can also be expressed as an infinite product over primes $\zeta(s) = \prod_p (1 - p^{-s})^{-1}$. The product is usually referred to as an *Euler product*, and if a Dirichlet series is to deserve the title of *L*-function, then it should have some kind of analogous product expansion. The existence of such an expansion is closely related to, but a little stronger than, the property that the sequence $a_1, a_2, \ldots$ should be *multiplicative*, which means that $a_{mn} = a_m a_n$ whenever m and n are coprime.

To go further we must expand our horizons. It is not hard to show that our definition of $L(s)$ makes sense even when s is a *complex* number, as long as it has a sufficiently large real part. Moreover, it defines a HOLOMORPHIC FUNCTION [I.3 §5.6] in the region of the complex plane where the sum converges. For example, the Dirichlet series defining the zeta function converges for every s such that $\mathrm{Re}(s) > 1$. A standard fact about the zeta function is that it has a unique extension to a holomorphic function of s for *any* complex number $s \neq 1$. This phenomenon is known as *meromorphic continuation* of the zeta function. It is similar to the fact that the infinite sum $1 + x + x^2 + x^3 + \cdots$ converges only when $|x| < 1$ but, when rewritten as $1/(1 - x)$, has a natural interpretation for any complex number x other than 1.

A meromorphic continuation is another of the properties that one would expect of a general L-function. It is important to stress, however, that extending a Dirichlet series to a function on the whole complex plane is *not* a "purely formal" technique: for a random sequence $a_1, a_2, \ldots$ there is no reason at all for the associated Dirichlet series $L(s)$ to have a natural extension beyond the region where the series converges. The existence of a meromorphic continuation is somehow a rigorous way of asserting the existence of subtle symmetries in the series.

While on the subject of meromorphic continuation, we should briefly mention the RIEMANN HYPOTHESIS [V.26], a conjecture which states that, once one has extended $\zeta(s)$ to a function on the whole complex plane, the complex numbers s such that $0 < \mathrm{Re}(s) < 1$ and $\zeta(s) = 0$ all have real part equal to $\frac{1}{2}$. There are analogous Riemann hypotheses for many L-functions, almost all of which are open problems.

The final property we shall emphasize is that there is a relatively simple formula relating $\zeta(s)$ and $\zeta(1 - s)$. This relation is called the *functional equation* of the zeta function, and any Dirichlet series worthy of the name L-function should also have an analogous property. (In general one looks for a relation between $L(s)$ and $\bar{L}(k - s)$, where k is some real number and $\bar{L}(s)$ is the Dirichlet series associated with the series of complex conjugates $\overline{a_1}, \overline{a_2}, \ldots$.)

There are many examples of Dirichlet series arising in number theory that do have, or are at least conjectured to have, these three key properties: an Euler product, meromorphic continuation, and a functional equation. These are the Dirichlet series that have come to be known as L-functions. For example, if A and B are integers such that the three roots of the cubic polynomial $x^3 + Ax + B$ are distinct, then the equation

$$y^2 = x^3 + Ax + B \qquad (1)$$

defines an ELLIPTIC CURVE [III.21], and associated with it there is a natural sequence $a_1, a_2, \ldots$ (where a_n is related to the number of solutions of (1) modulo n, at least when n is prime—see ARITHMETIC GEOMETRY [IV.5 §5.1] for more details). However, it was an open problem for years to establish the existence of a meromorphic continuation of the associated Dirichlet series $L(s)$ to the complex plane: it is now known to exist (and indeed to have no poles) as a consequence of the work of Wiles, Taylor, and others that grew out of the proof of FERMAT'S LAST THEOREM [V.10].

3 What Is the Point of L-Functions?

One of the first uses of L-functions was by DIRICHLET [VI.36] himself, who used them to prove that there are infinitely many primes in a general arithmetic progression (see ANALYTIC NUMBER THEORY [IV.2 §4]). In fact, although the Riemann hypothesis is still an open problem, even partial results about the locations of the zeros of the Riemann zeta function have deep consequences in the theory of a distribution of prime numbers.

However, over the last hundred years mathematicians have realized a second use for them: if X is a mathematical object and $L(s)$ is its associated L-function, then there are deep conjectures relating the arithmetic of X to the values that $L(s)$ assumes, typically at points where the Dirichlet series defining $L(s)$ does not converge! Hence, one can investigate X by investigating its L-function. One basic example of this phenomenon is THE BIRCH–SWINNERTON-DYER CONJECTURE [V.4], a weak form of which states that the L-function associated with equation (1) should vanish at $s = 1$ if and only if (1) has infinitely many solutions such that both x and y are rational numbers. Much is known about this conjecture, and it has been vastly generalized by work of Deligne, Belinson, Bloch, and Kato. However, at the time of writing it remains open.

III.48 Lie Theory
Mark Ronan

1 Lie Groups

Why are groups important in mathematics? One major reason is that it is often possible to understand a mathematical structure by understanding its symmetries, and the symmetries of a given mathematical structure form a group. Some mathematical structures are so symmetrical that they have not just a finite number of symmetries, but a continuous family of them. When this is the case, we find ourselves in the realms of Lie groups and Lie theory.

One of the simplest "continuous" groups is the group $\mathrm{SO}(2)$, which consists of all rotations of the plane $\mathbb{R}^2$ about the origin. With each element of $\mathrm{SO}(2)$ one can associate an angle θ: the angle of the rotation in question. If we write R_θ for the counterclockwise rotation by θ, then the group operation is given by $R_\theta R_\varphi = R_{\theta + \varphi}$, where $R_{2\pi}$ is understood to equal R_0, the identity element of the group.

The group $SO(2)$ is not just a continuous group, but also a *Lie group*. Roughly speaking, this means that it is a group in which one can meaningfully define the concept of a smooth curve (that is, a curve that is not just continuous but differentiable as well). Given any two elements R_θ and R_φ of $SO(2)$, one can easily define a smooth path from R_θ to R_φ by smoothly modifying θ until it becomes φ. (The most obvious such path would be given in parametric form by $R_{(1-t)\theta+t\varphi}$, as t goes from 0 to 1.) It is not always the case that every pair of points in a Lie group can be connected by a path: when they can, the Lie group is said to be *connected*. An example of a Lie group that is not connected is $O(2)$, which consists of $SO(2)$ together with all reflections of the plane about lines through the origin. Any two rotations can be linked by a path, as can any two reflections, but there is no continuous way of changing a rotation into a reflection.

Lie groups were introduced by SOPHUS LIE [VI.53] in order to create an analogue of GALOIS THEORY [V.21] for differential equations. Lie groups that consist of invertible linear transformations of $\mathbb{R}^n$ or $\mathbb{C}^n$, like the examples above, are called *linear* Lie groups, and they are an important subclass. For linear Lie groups it is fairly easy to work out what terms such as "continuous," "differentiable," or "smooth" should mean. However, one can also consider more abstract Lie groups (both real and complex), with elements that are not given as linear transformations. In order to give a proper definition of Lie groups in their full generality, one needs the concept of a smooth MANIFOLD [I.3 §6.9]. However, for simplicity we shall mostly restrict attention to linear Lie groups.

A very common way to create a Lie group is to collect all transformations of a given space that preserve one or more specified geometric structures. For instance, the *general linear group* $GL_n(\mathbb{R})$ is defined to be the group of all invertible linear transformations from $\mathbb{R}^n$ to $\mathbb{R}^n$. Inside this group is the *special linear group* $SL_n(\mathbb{R})$, in which we retain only those linear transformations that preserve volume and orientation (or equivalently those with DETERMINANT [III.15] equal to 1). If instead we retain the linear transformations that preserve distance, then we obtain the *orthogonal group* $O(n)$; if we retain linear transformations that preserve both distance and orientation we obtain the *special orthogonal group* $SO(n)$, which is easily seen to equal $SL_n(\mathbb{R}) \cap O(n)$. The *Euclidean group* $E(n)$ of rigid motions of $\mathbb{R}^n$ (that is, all transformations that preserve distances and angles, such as rotations, reflec-

tions, and translations) is generated by the orthogonal group $O(n)$, together with the group of translations (which is isomorphic to $\mathbb{R}^n$). There are analogues of all of the above groups in which the real numbers $\mathbb{R}$ are replaced by the complex numbers $\mathbb{C}$. For instance, $GL_n(\mathbb{C})$ is the group of all invertible complex-linear transformations of $\mathbb{C}^n$, and the complex analogue of the orthogonal group $O(n)$ is the *unitary group* $U(n)$. There are also the *symplectic groups* $Sp(2n)$, which are analogues of $O(n)$ and $U(n)$ over the QUATERNIONS [III.76]. These are all manifestly linear Lie groups except for $E(n)$, and in fact it is not difficult to describe a linear Lie group that is isomorphic to $E(n)$ as well.

Many important examples of Lie groups are finite dimensional, which roughly means that they can be described using a finite number of continuous parameters. (Infinite-dimensional Lie groups, while important, are more difficult to handle and will not be discussed in detail here.) For example, the group $SO(3)$, of rotations of $\mathbb{R}^3$ that fix the origin, is three dimensional. Each rotation can be specified using three parameters, which could, for instance, be taken as rotations around the x-axis, y-axis, and z-axis. These particular parameters are known to airline pilots as roll, pitch, and yaw, where the x-axis is in the direction of the airplane. Another way of specifying each rotation is by its axis and angle of rotation. Two parameters are needed to specify the axis (using spherical coordinates for example), and one parameter is needed to specify the angle of rotation. Let us take this angle to be between 0 and π (a rotation by an angle greater than π has the same effect as a rotation by an angle less than π from the opposite direction).

We can represent $SO(3)$ geometrically as follows. Let B be a ball of radius π centered at the origin. Given any noncentral point P of B, associate with it the rotation of $\mathbb{R}^3$ about the axis OP (where O is the origin) through an angle that is given in radians by the distance from O to P. With O itself we associate the identity map, so the only ambiguity is that a rotation through π radians is associated with two opposite points P and P' on the surface of B. We can remove this ambiguity by gluing all such pairs of points together. This tells us what $SO(3)$ looks like as a TOPOLOGICAL SPACE [III.90]: it is equivalent to the three-dimensional PROJECTIVE SPACE [I.3 §6.7] $\mathbb{RP}^3$. The group $SO(2)$, by comparison, is much simpler, and is topologically equivalent to a circle.

Lie groups arise naturally in any subject that involves continuous motion. For instance, they appear in applied topics such as the design of guidance systems and also in very pure topics such as geometry or differential equations. Lie groups, and the closely related Lie algebras discussed below, also frequently arise in many types of algebra, particularly in the algebraic structures that appear in quantum mechanics and other related branches of physics.

2 Lie Algebras

As the examples above show, Lie groups are often "curved" and have some nontrivial topology. However, one can profitably analyze a Lie group by associating with it a flat space known as a *Lie algebra*. This idea is similar to the idea of studying a symmetric object such as a sphere by first studying its relationship to one of its tangent planes. The Lie algebra uses the tangent space to the Lie group at the identity element, and one can view it as a "logarithm" of the Lie group.

To see how Lie algebras arise, let us consider a linear Lie group. The elements of the group can be viewed as linear transformations on a vector space, or equivalently (when we have selected a coordinate basis) as square matrices. In general, two matrices A and B do not commute (that is, AB does not have to equal BA), but the situation becomes simpler if one looks at matrices that are very close to the identity matrix I. If $A = I + \epsilon X$ and $B = I + \epsilon Y$ for some very small positive ϵ and two fixed matrices X and Y, then

$$AB = I + \epsilon(X + Y) + \epsilon^2 XY$$

and

$$BA = I + \epsilon(X + Y) + \epsilon^2 YX.$$

Thus, if we ignore the terms containing ϵ^2, we see that A and B "almost commute," and that multiplication of A and B "almost corresponds to" addition of X and Y: indeed, one can view X and Y as analogous to "logarithms" of A and B respectively.

Let us now informally define the *Lie algebra* $\mathfrak{g}$ of a linear Lie group G to be the space of all matrices X such that, for sufficiently small ϵ, the matrix $I + \epsilon X$ lies in G, up to errors of size ϵ^2. For example, the Lie algebra $\mathfrak{gl}_n(\mathbb{C})$ of the general linear group $GL_n(\mathbb{C})$ is the space of all $n \times n$ complex matrices. One can view the Lie algebra as describing all possible instantaneous directions and speeds within the group G, and a more precise definition is the collection of all derivatives R'_0 of smooth curves $\epsilon \mapsto R_\epsilon$ in G that pass through the identity element R_0. This definition can also be extended to more abstract Lie groups without much difficulty. (To return to the example of the airplane pilot, an element of the

Lie group SO(3) could be used to describe the current orientation of the aircraft with respect to a fixed coordinate system, whereas an element of the Lie algebra $\mathfrak{so}(3)$ could be used to describe the current rate of roll, pitch, and yaw that the pilot is applying to the aircraft to smoothly change its orientation.)

As we have just seen, the Lie algebra $\mathfrak{gl}_n(\mathbb{C})$ of the general linear group $GL_n(\mathbb{C})$ is the space of all $n \times n$ complex matrices. The Lie algebra $\mathfrak{sl}_n(\mathbb{C})$ of the special linear group $SL_n(\mathbb{C})$ is the subspace of all matrices with trace zero. This is because $\det(I + \epsilon X) = 1 + \epsilon \operatorname{tr} X$, up to errors of size ϵ^2, so if $\epsilon \mapsto I + \epsilon X$ is a path in the group, then $\operatorname{tr} X = 0$. The Lie algebra $\mathfrak{so}(n)$ of SO(n) is equal to the Lie algebra $\mathfrak{o}(n)$ of O(n), and both are equal to the space of all antisymmetric matrices. Similarly, both the Lie algebra $\mathfrak{su}(n)$ of SU(n) and the Lie algebra $\mathfrak{u}(n)$ of U(n) are equal to the space of skew-Hermitian matrices. (A matrix is skew-Hermitian if it equals minus the complex conjugate of its transpose.)

The fact that a Lie group is closed under multiplication can be used to show that its Lie algebra is closed under addition. Thus, a Lie algebra is a (real) vector space. However, it has some additional structure that makes it far more than *just* a vector space. For instance, let A and B be two elements of the Lie group G that are very close to the identity. Then we can write $A \approx I + \epsilon X$ and $B \approx I + \epsilon Y$ for some very small ϵ and some elements X and Y of the Lie algebra $\mathfrak{g}$. A little matrix algebra shows that the *commutator* $ABA^{-1}B^{-1}$ of A and B, which is the element of G that measures the extent to which A and B fail to commute, can be approximated by $I + \epsilon^2[X, Y]$, where $[X, Y] = XY - YX$. This quantity $[X, Y]$ is called the *Lie bracket* of X and Y. Informally, it represents the net direction of motion if one first moves an infinitesimal amount in the X direction, then in the Y direction, then back in the X direction and back in the Y direction, in that order. The resulting new direction may be quite different from the original directions X and Y.

The Lie bracket obeys a number of nice identities, such as the antisymmetric identity $[X, Y] = -[Y, X]$ and the *Jacobi identity*

$$[[X, Y], Z] + [[Y, Z], X] + [[Z, X], Y] = 0.$$

One can in fact use such identities to define Lie algebras in a completely abstract fashion, without any reference to matrices or Lie groups, in much the same way that other algebraic objects such as groups, rings, and fields can be defined using a handful of algebraic identities as axioms, but we shall not focus on the abstract approach to Lie algebras here. A familiar example of a Lie algebra is $\mathbb{R}^3$ with the Lie bracket $[x, y]$ defined to be the cross-product $x \times y$. Notice that the Lie bracket does not satisfy the associative law (unless it is trivial).

We have seen that a linear Lie group G naturally generates the bracket operation $[\cdot, \cdot]$ on its Lie algebra $\mathfrak{g}$. Conversely, if the Lie group is connected, one can almost reconstruct it from the Lie algebra, with its addition, scalar multiplication, and Lie bracket operation. More precisely, every element A of the Lie group can be written as an EXPONENTIAL [III.25] $\exp(X)$ of an element X of the Lie algebra. For example, if the Lie group is SO(2), then we can identify it with the unit circle in $\mathbb{C}$. The tangent to this circle at 1 is a vertical line, so we can identify the Lie algebra with the set $i\mathbb{R}$ of purely imaginary numbers. (Normally, however, we would just say that the Lie algebra is $\mathbb{R}$.) The rotation through an angle θ can then be written as $\exp(i\theta)$. Note that this representation is not unique, since $\exp(i\theta) = \exp(i(\theta + 2\pi))$. It is not hard to see that the Lie group $\mathbb{R}$ also has $\mathbb{R}$ as its Lie algebra (to make sense of this it helps to replace $\mathbb{R}$ by the multiplicative group of positive real numbers, which is isomorphic to $\mathbb{R}$), and that in this case the representation of a group element as an exponential *is* unique. In general, if two connected Lie groups have the same Lie algebra, then those Lie groups share the same universal cover, and are therefore closely related to one another.

In the case of linear Lie groups, the exponential can be described by the familiar formula

$$\exp(X) = \lim_{n \to \infty} \left(I + \frac{1}{n}X \right)^n.$$

For more abstract Lie groups, the exponential is best described in the language of ordinary differential equations,[1] using a suitable generalization of the identity

$$\frac{\mathrm{d}}{\mathrm{d}t}\mathrm{e}^{tX} = X\mathrm{e}^{tX}$$

from single-variable calculus. However, owing to the noncommutativity of the Lie group, it is not quite true that $\exp(X + Y)$ equals $\exp(X)\exp(Y)$; instead, the correct identity is the *Baker–Campbell–Hausdorff formula*

$$\exp(X)\exp(Y) = \exp(X + Y + \tfrac{1}{2}[X, Y] + \cdots),$$

where the missing terms consist of a moderately complicated infinite series involving the Lie bracket. The exponential map that connects Lie algebras and Lie groups is closely related to the Lie bracket, and because of this it is possible to study and classify Lie groups by first studying and classifying Lie algebras with their Lie bracket operation.

3   Classification

It is always of interest when a mathematical structure can be classified, but especially so if the structure is important and the classification is not straightforward. By these criteria, the results that have been obtained concerning the classification of Lie algebras are undeniably interesting, and they are regarded as one of the great mathematical achievements from around the turn of the twentieth century.

It turns out to be easier to classify *complex* Lie algebras: that is, Lie algebras such as $\mathfrak{sl}_n(\mathbb{C})$ that have the structure of a complex vector space. Each real Lie algebra embeds in a complex Lie algebra of twice the (real) dimension, known as the *complexification* of the original algebra. However, a complex Lie algebra may arise as the complexification of several different real Lie algebras (known as *real forms* of the complex Lie algebra).

In classifying Lie groups and Lie algebras, the first step is to restrict attention to *simple* Lie groups and Lie algebras; these are analogous to prime numbers in the sense that they cannot be "factored" into smaller components. For instance, the Euclidean group E(n) contains the translation group $\mathbb{R}^n$ as a connected normal subgroup. If we factor out this group, then we obtain the orthogonal group O(n), so E(n) is not simple. More formally, a Lie group is *simple* if it contains no proper connected normal subgroups, and a Lie algebra is *simple* if it contains no proper IDEALS [III.81 §2]. In this sense, the Lie group SL$_n(\mathbb{C})$ and its Lie algebra $\mathfrak{sl}_n(\mathbb{C})$ are simple for every n. Finite-dimensional, complex, simple Lie algebras were classified by Wilhelm Killing and Élie CARTAN [VI.69] in 1888–94.

This classification is often placed in the context of so-called *semisimple* Lie algebras, which can be factored in a unique way (up to rearrangement) as a direct sum of simple Lie algebras, just as a natural number can be factored uniquely as a product of prime numbers. Furthermore, a theorem of Levi shows that a general finite-dimensional Lie algebra $\mathfrak{g}$ can be expressed as a combination (or, more precisely, a "semidirect product") of a

---

1. Indeed, Lie groups and Lie algebras are an excellent tool for describing the algebraic aspects of ordinary and partial differential equations; the evolution of such equations through time can be modeled using a Lie group, and the differential operators used to describe an equation can be modeled on the associated Lie algebra. However, we will not discuss this important connection between Lie theory and differential equations here.

semisimple algebra (called a *Levi subalgebra* of $\mathfrak{g}$) and a solvable subalgebra (known as the *radical* of $\mathfrak{g}$). Solvable Lie algebras, which are related to the concept of a SOLVABLE GROUP [V.21] in group theory, are difficult to classify, but in many applications one can restrict attention to semisimple Lie algebras, and hence to simple Lie algebras.

A simple Lie algebra $\mathfrak{g}$ splits into smaller subalgebras, which are not ideals but which are related to one another in particularly nice ways. The case of $\mathfrak{sl}_{n+1}$ is typical and we shall use it to explain the general theory. It comprises all $(n+1) \times (n+1)$ matrices of trace zero, and splits as a direct sum in the following way:

$$\mathfrak{sl}_{n+1} = \mathfrak{n}_+ \oplus \mathfrak{h} \oplus \mathfrak{n}_-,$$

where $\mathfrak{h}$ is the set of diagonal matrices of trace zero, and $\mathfrak{n}_+$ and $\mathfrak{n}_-$ are, respectively, the sets of upper and lower triangular matrices with 0s on the diagonal. Two diagonal matrices X and Y commute with one another, so their Lie bracket $[X, Y] = XY - YX$ is 0. In other words, if X and Y belong to $\mathfrak{h}$, then $[X, Y] = 0$. A Lie algebra in which $[X, Y] = 0$ for any two elements X and Y is called *Abelian*.

Each simple Lie algebra $\mathfrak{g}$ has a similar decomposition where the subspace $\mathfrak{h}$ is a maximal Abelian subalgebra called a *Cartan subalgebra*. (For Lie algebras that are not simple, the definition of Cartan subalgebras is more complicated.) Cartan subalgebras are important because their action on the rest of the Lie algebra can be simultaneously diagonalized. What this means is that a complement to $\mathfrak{h}$ can be split up into one-dimensional components $\mathfrak{g}_\alpha$, known as *root spaces*, that are invariant under the action of $\mathfrak{h}$. To put this another way, if X belongs to $\mathfrak{h}$, and Y belongs to a root space, then $[X, Y]$ is a scalar multiple of Y. (The diagonalization requires THE FUNDAMENTAL THEOREM OF ALGEBRA [V.13], which is why we need to work with complex Lie algebras.)

For $\mathfrak{sl}_{n+1}$ this works as follows. Each root space $\mathfrak{g}_{ij}$ is the one-dimensional space of matrices whose entries are 0 except for a single entry in the ith row and jth column. If $X \in \mathfrak{h}$ (that is, if X is a diagonal matrix of trace zero) and $Y \in \mathfrak{g}_{ij}$, then it is not hard to check that $[X, Y]$ also lies in $\mathfrak{g}_{ij}$. In fact,

$$[X, Y] = (X_{ii} - X_{jj})Y.$$

If we identify the diagonal matrix X with the vector whose n coordinates appear down its diagonal, and if we write e_i for the vector that is 1 in the ith position and 0 elsewhere, then $X_{ii} - X_{jj}$ can be rewritten

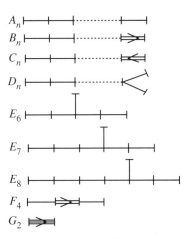

Figure 1 Dynkin diagrams.

as $\langle e_i - e_j, X \rangle$. We refer to the vectors $e_i - e_j$ as *root vectors*.

In general, a complex semisimple Lie algebra $\mathfrak{g}$ can be completely described by its root vectors α and corresponding root spaces $\mathfrak{g}_\alpha$. The *rank* of $\mathfrak{g}$ equals the dimension of the Cartan subalgebra $\mathfrak{h}$, and also equals the dimension of the vector space spanned by the root vectors. For example, $\mathfrak{sl}_{n+1}$ has rank n, and its root vectors are the vectors $e_i - e_j$, as we have just seen. Sets of root vectors are far from arbitrary: they must obey some simple but quite restrictive geometric properties. For instance, if a root vector α is reflected in the hyperplane perpendicular to another root vector β, the result must be a third root vector $s_\beta(\alpha)$, where s_β is the reflection concerned. (To make the notion of "perpendicular" precise, one needs to define a special inner product on the Cartan subalgebra, known as the *Killing form*, but we shall not discuss this here.) The group generated by these reflections is called the *Weyl group* of the Lie algebra.

The root vectors form what is called a *root system*, and the geometric properties mentioned above allow one to classify all root systems, and hence all complex semisimple Lie algebras. This classification is given by some very simple diagrams called *Dynkin diagrams*, which are shown in figure 1.

The nodes of the diagram correspond to so-called *simple roots*. Every root is a linear combination of simple roots with coefficients that are either all nonnegative or all nonpositive. The nature of the bond (or lack thereof) between two nodes determines the inner product of the corresponding simple roots. If there is no bond, then the inner product is 0; if there is a single bond, then the root vectors have the same length and the angle between them is 120°. In diagrams that have only single bonds, the root vectors span a set of lines in $\mathbb{R}^n$ in which the angle between any two lines is either 90° or 60°. In the diagrams B_n, C_n, F_4, and G_2 there are arrows between certain pairs of nodes. The direction of an arrow is from a long root to a short root: the ratio of the root lengths is $\sqrt{2}$ in the first three cases and $\sqrt{3}$ in the case of G_2. In these cases there are exactly two root lengths, but in the single-bond cases all roots have the same length.

The A_n diagram is the one for $\mathfrak{sl}_{n+1}$. The simple roots are $e_i - e_{i+1}$ for $1 \leqslant i \leqslant n$, going from left to right on the diagram. Notice that the inner product of two simple roots is 0 unless they are adjacent on the diagram, in which case it is -1. Each root $e_i - e_j$ is a sum of simple roots with coefficients all 1 or all -1 on a connected segment of the diagram.

The four infinite families A_n B_n, C_n, and D_n correspond to the *classical Lie algebras*, of which $\mathfrak{sl}_{n+1}(\mathbb{R})$, $\mathfrak{so}(2n + 1)$, $\mathfrak{sp}(2n)$, and $\mathfrak{so}(2n)$ are real forms. These are the algebras associated with the *classical Lie groups* $\mathrm{SL}_{n+1}(\mathbb{R})$, $\mathrm{SO}(2n+1)$, $\mathrm{Sp}(2n)$, and $\mathrm{SO}(2n)$, respectively.

As mentioned earlier, a simple Lie algebra $\mathfrak{g}$ of rank n decomposes as the direct sum of a Cartan subalgebra of dimension n plus a set of one-dimensional root spaces, one for each root. It follows that

$$\dim \mathfrak{g} = \text{the rank of } \mathfrak{g} + \text{the number of roots.}$$

Here are the dimensions of the simple Lie algebras:

$$\dim A_n = n + n(n + 1) = n(n + 2),$$
$$\dim B_n = n + 2n^2 = n(2n + 1),$$
$$\dim C_n = n + 2n^2 = n(2n + 1),$$
$$\dim D_n = n + 2n(n - 1) = n(2n - 1),$$
$$\dim G_2 = 2 + 12 = 14,$$
$$\dim F_4 = 4 + 48 = 52,$$
$$\dim E_6 = 6 + 72 = 78,$$
$$\dim E_7 = 7 + 126 = 133,$$
$$\dim E_8 = 8 + 240 = 248.$$

Each node of the diagram corresponds to a simple root, and hence to a reflection across the hyperplane perpendicular to that root. This set of reflections generates the Weyl group W in a particularly elegant way. If s_i denotes the reflection corresponding to node i, then W is generated by elements s_i of order 2, subject only to the relations

$$(s_i s_j)^{m_{ij}} = 1,$$

where m_{ij} is the order of $s_i s_j$ (see [IV.10 §2] for a discussion of generators and relations). These orders are determined by the diagram according to the following rules:

(i) $s_i s_j$ has order 2 if there is no bond;
(ii) $s_i s_j$ has order 3 if there is a single bond;
(iii) $s_i s_j$ has order 4 if there is a double bond; and
(iv) $s_i s_j$ has order 6 if there is a triple bond.

For example, the Weyl group of type A_n is isomorphic to the SYMMETRIC GROUP [III.68] S_{n+1}, and one can take $s_1, \ldots, s_n$ to be the transpositions $(1\ 2), (2\ 3), \ldots, (n\ n + 1)$. Notice that the Dynkin diagrams for the B_n and C_n root systems yield the same Weyl group.

In principle, this classification of root systems leads to a classification of all semisimple finite-dimensional Lie algebras and Lie groups. However, there are many fundamental questions about simple Lie algebras and Lie groups that remain only partly understood. For instance, one particularly important aim of Lie theory is to understand the linear representations of a given Lie group or Lie algebra; roughly speaking, a linear representation is a way of interpreting an abstract Lie group or Lie algebra as a linear Lie group or Lie algebra by assigning a matrix to each of its elements. While the representations of all the simple Lie algebras and Lie groups have been classified and described explicitly, these descriptions are not always easy to work with, and answering basic questions (such as how a given representation decomposes into simpler representations) often requires some sophisticated tools from algebraic combinatorics.

The theory of root systems outlined above can also be extended to an important class of infinite-dimensional Lie algebras, namely the Kac–Moody algebras. Such algebras arise in several areas of physics (such as are described in VERTEX OPERATOR ALGEBRAS [IV.17]) and algebraic combinatorics.

III.49 Linear and Nonlinear Waves and Solitons
Richard S. Palais

1 John Scott Russell and the Great Wave of Translation

To the world at large, John Scott Russell is known as the naval architect who designed *The Great Eastern*, a steamship larger than any built before. But long after *The Great Eastern* has been forgotten, Russell will be remembered by mathematicians as the man who, despite limited mathematical training and background, was the first person to recognize the highly important mathematical concept known as a *soliton*, which he referred to as "the great wave of translation." Here is his oft-quoted passage in which he describes how he first became acquainted with it:

> I was observing the motion of a boat which was rapidly drawn along a narrow channel by a pair of horses, when the boat suddenly stopped—not so the mass of water in the channel which it had put in motion; it accumulated round the prow of the vessel in a state of violent agitation, then suddenly leaving it behind, rolled forward with great velocity, assuming the form of a large solitary elevation, a rounded, smooth and well-defined heap of water, which continued its course along the channel apparently without change of form or diminution of speed. I followed it on horseback, and overtook it still rolling on at a rate of some eight or nine miles an hour, preserving its original figure some thirty feet long and a foot to a foot and a half in height. Its height gradually diminished, and after a chase of one or two miles I lost it in the windings of the channel. Such, in the month of August 1834, was my first chance interview with that singular and beautiful phenomenon which I have called the Wave of Translation.
>
> Russell (1844)

You may feel that there is nothing unusual about what Russell describes here, and indeed many before and since have watched this same scenario play out without noticing anything out of the ordinary. But Russell was very familiar with wave phenomena and had a scientist's keenly observant eye. What struck him was the remarkable *stability* of the bow wave as it traveled over a long distance. He knew that if one tried to create a traveling water wave on, say, a calm lake, it would soon disperse into a train of smaller wavelets—it would *not* just go marching along as a single "heap" over a long distance. There was clearly something very special about water waves traveling in a narrow and shallow channel.

Russell became fascinated—even a little obsessed—with his discovery. He built a wave tank behind his home and proceeded to do extensive experiments, recording the results as data and sketches in his notebooks. He found, for example, that the speed of a soliton depended on its height, and he was even able to discover the correct formula for the speed as a function of height. More surprising still, in Russell's notebooks one finds remarkable sketches of a two-soliton interaction—something that would evoke amazement when

it was rediscovered as a rigorous solution to the KdV equation (see section 3 below) more than a hundred years later.

However, as we shall see, solitons are very much a nonlinear phenomenon, and when some of the best mathematicians of Russell's day, notably Stokes and Airy, tried to understand Russell's observations using the linearized theory of water waves that was then available, they failed to find any trace of soliton-like behavior and expressed doubts that what Russell had seen was real.

It was not until after Russell's death, with the more sophisticated nonlinear mathematical treatment by Boussinesq in 1871 and by Korteweg and de Vries in 1895, that Russell's careful observations and experiments were at last seen to be in complete agreement with mathematical theory. And it took another seventy years before the full importance of the great wave of translation was recognized, after which it became an object of intensive study for the rest of the twentieth century.

2   The Korteweg–de Vries Equation

Korteweg and de Vries were the first people to derive the appropriate differential equation to describe the motion of a wave in a shallow channel. We can write their equation, usually called the *KdV equation*, in a succinct form as follows:

$$u_t + uu_x + \delta^2 u_{xxx} = 0.$$

Here, u is a function of two variables, x and t, which represent space and time, respectively. "Space" is one dimensional, so x is a real number, and $u(x, t)$ represents the height of the wave at x at time t. The notation u_t is shorthand for $\partial u / \partial t$; similarly, u_x stands for $\partial u / \partial x$ and u_{xxx} stands for $\partial^3 u / \partial x^3$.

This is an example of an *evolution equation*: if, for each t, we write $u(t)$ for the function from $\mathbb{R}$ to $\mathbb{R}$ that takes x to $u(x, t)$, then it describes how the function $u(t)$ "evolves" over time. The *Cauchy problem* for an evolution equation is the problem of determining this evolution from knowledge of its initial value $u(0)$.

2.1   Some Model Equations

To put the KdV equation into perspective, it is useful to think briefly about three other evolution equations. The first is the classic WAVE EQUATION [I.3 §5.4]

$$u_{tt} - c^2 u_{xx} = 0.$$

To solve the Cauchy problem for this equation, we factor the wave operator $(\partial^2/\partial t^2) - c^2(\partial^2/\partial x^2)$ as a product $((\partial/\partial t) - c(\partial/\partial x))((\partial/\partial t) + c(\partial/\partial x))$. Then we transform to so-called characteristic coordinates $\xi = x - ct$, $\eta = x + ct$. The equation becomes $\partial^2 u/\partial\xi\partial\eta = 0$, which clearly has the general solution $u(\xi,\eta) = F(\xi) + G(\eta)$. Transforming back to "laboratory coordinates" x, t, the general solution is $u(x,t) = F(x - ct) + G(x + ct)$. If the initial shape of the wave is $u(x,0) = u_0(x)$ and its initial velocity is $u_t(x,0) = v(x,0) = v_0(x)$, then an easy algebraic computation gives the following very explicit formula:

$$u(x,t) = \tfrac{1}{2}[u_0(x - ct) + u_0(x + ct)]$$
$$+ \frac{1}{2c}\int_{x-ct}^{x+ct} v_0(\xi)\,\mathrm{d}\xi,$$

known as "d'Alembert's solution" of the Cauchy problem for the wave equation.

Note the geometric interpretation in the important "plucked string" case, $v_0 = 0$; the initial profile u_0 breaks up into the sum of two "traveling waves," both with the same profile $\tfrac{1}{2}u_0$, one traveling to the right, and the other to the left, both with speed c. It is an easy exercise to derive d'Alembert's solution using the following hint: since $u_0(x) = F(x) + G(x)$, $u_0'(x) = F'(x) + G'(x)$, while $v_0(x) = u_t(x,0) = -cF'(x) + cG'(x)$.

The next equation to think about is

$$u_t = -u_{xxx}, \tag{1}$$

which we can obtain from the KdV equation if we drop the nonlinear term uu_x. This equation is not just linear but also translation invariant (meaning that if $u(x,t)$ is a solution, then so is $u(x - x_0, t - t_0)$ for any constants x_0 and t_0). Such equations can be solved using THE FOURIER TRANSFORM [III.27]. Let us try to find a "plane-wave" solution of the form $u(x,t) = \mathrm{e}^{\mathrm{i}(kx-\omega t)}$. If we substitute this into (1), then we obtain the equation

$$-\mathrm{i}\omega\mathrm{e}^{\mathrm{i}(kx-\omega t)} = \mathrm{i}k^3\mathrm{e}^{\mathrm{i}(kx-\omega t)},$$

and therefore the simple algebraic equation $\omega + k^3 = 0$. This is called the *dispersion relation* of (1): with the help of the Fourier transform it is not hard to show that every solution is a superposition of solutions of the form $\mathrm{e}^{\mathrm{i}(kx-\omega t)}$, and the dispersion relation tells us how the "wave number" k is related to the "angular frequency" ω in each of these elementary solutions.

The function $\mathrm{e}^{\mathrm{i}(kx-\omega t)}$ represents a wave that travels at a speed of ω/k, which we have just shown to be equal to $-k^2$. Therefore, the different plane-wave components of the solution travel at different speeds: the

higher the angular frequency, the greater the speed. For this reason, the equation (1) is called *dispersive*.

What happens if instead we omit the u_{xxx} term from the KdV equation? Then we obtain the *inviscid Burgers equation*

$$u_t + uu_x = 0. \tag{2}$$

The term uu_x can be rewritten as $(\partial/\partial x)(\tfrac{1}{2}u^2)$. Let us consider the integral $\int_{-\infty}^{\infty} u(x,t)\,\mathrm{d}x$, which is a function of t. The derivative of this function is $\int_{-\infty}^{\infty} u_t\,\mathrm{d}x$, which equation (2) tells us is equal to

$$-\int_{-\infty}^{\infty} \frac{\partial}{\partial x}(\tfrac{1}{2}u^2)\,\mathrm{d}x,$$

which equals $[-\tfrac{1}{2}u(x,t)^2]_{-\infty}^{\infty}$. Therefore, if $\tfrac{1}{2}u(x,t)^2$ vanishes at infinity, then $\int_{-\infty}^{\infty} u(x,t)\,\mathrm{d}x$ is a "constant of the motion." We say that the inviscid Burgers equation is a *conservation law*. (The argument we have just used can be used for any equation of the form $u_t = (F(u))_x$, where F is a smooth function of u and its partial derivatives with respect to x. This is known as the *general conservation law*. For example, taking $F(u) = -(\tfrac{1}{2}u^2 + \delta^2 u_{xx})$ gives rise to the KdV equation.)

The inviscid Burgers equation (and other conservation laws where F is a function just of u) can be solved using the *method of characteristics*. The idea of this method is to look for smooth curves $(x(s),t(s))$ in the xt-plane along which the solution to the Cauchy problem is constant. Suppose that s_0 is such that $t(s_0) = 0$, and write x_0 for $x(s_0)$. Then the constant value that the solution $u(x,t)$ will have to take along this curve is $u(x_0,0)$, which we also write as $u_0(x_0)$. The derivative of u along this so-called *characteristic curve* is $(\mathrm{d}/\mathrm{d}s)u(x(s),t(s)) = u_x x' + u_t t'$, so if we want the solution to be constant along the curve, then we need this to be 0. Therefore, using the fact that $u_t = -uu_x$, we find that

$$\frac{\mathrm{d}x}{\mathrm{d}t} = \frac{x'(s)}{t'(s)} = -\frac{u_t}{u_x} = u(x(s),t(s)) = u_0(x_0),$$

so the characteristic curve is a straight line of slope $u_0(x_0)$. In other words, u has the constant value $u_0(x_0)$ along the line $x = x_0 + u_0(x_0)t$.

Note the following geometric interpretation of this last result: to find the wave profile at time t (i.e., the graph of the map $x \mapsto u(x,t)$), we translate each point $(x,u_0(x))$ of the initial profile to the right by the amount $u_0(x)t$. Suppose we look at a portion of the initial profile where u_0 is decreasing. Then the earlier, and higher, parts of the initial wave are translated at a greater speed (since $u_0(x)$ is larger), so that the negative slope of the wave becomes more negative. Indeed, after a finite time the earlier part of the wave "catches up" with the later part, which means that we no longer have a graph of a function. The first time at which this sort of problem happens is called the "breaking time," since one can visualize it as the breaking of a wave. This process is usually referred to as *shock formation*, or *steepening and breaking of the wave profile*: once again, the phenomenon occurs for many other conservation laws.

2.2 Split-Stepping

Now let us return to the KdV equation itself, in the form $u_t = -uu_x - u_{xxx}$. Why is it that this equation gives rise to the remarkable stability of the solutions that was observed experimentally by Russell? Intuitively, the reason is that there is a balance between the dispersing effect of the u_{xxx} term and the shock-forming effect of the uu_x term.

There turns out to be a very general technique for analyzing balances of this kind. In the pure-mathematics community it is usually called the *Trotter product formula*, while in the applied-mathematics and numerical-analysis communities it is called *split-stepping*. The rough idea is simple: as t increases to $t + \Delta t$, you first change u to $u - u_{xxx}\Delta t$, as would be required by the equation $u_t = -u_{xxx}$, and then you take a further step to $u - u_{xxx}\Delta t - uu_x\Delta t$, the small change required by the equation $u_t = -uu_x$. To work out the function $u(t, x)$, you start at the initial function u_0 and take a succession of alternating small steps of this form. You then take the limit as the step size tends to zero.

Split-stepping suggests a way to understand the mechanism by which dispersion from u_{xxx} balances shock formation from uu_x in KdV. If we imagine the evolution of the wave profile as made up of a succession of pairs of small steps in this way, then when u, u_x, and u_{xxx} are not too large, the steepening mechanism will dominate. But as the time t approaches the breaking time T_B, u remains bounded (since it is made out of horizontally translated parts of u_0). It is not hard to prove that the maximum slope (that is, the maximum value of u_x) blows up like the function $(T_B - t)^{-1}$, while at the same place u_{xxx} blows up like the function $(T_B - t)^{-5}$. Thus, near the breaking time, and breaking point, the u_{xxx} term will dwarf the nonlinearity and will disperse the incipient shock. Thus, the stability is caused by a kind of negative feedback. Computer simulations show just such a scenario playing out.

3 Solitons and Their Interactions

We have just seen that the KdV equation expresses a balance between dispersion from its third-derivative term and the shock-forming tendency of its nonlinear term, and in fact many models of one-dimensional physical systems that exhibit mild dispersion and weak nonlinearity lead to KdV as the controlling equation at some level of approximation.

In their 1894 paper, Korteweg and de Vries introduced the KdV equation and gave a convincing mathematical argument that this was the equation that governed wave motion in a shallow canal. They also showed by explicit computation that it admitted traveling-wave solutions that had exactly the properties that had been described by Russell, including the relation of height to speed that Russell had determined experimentally with the help of his wave tank.

But it was only much later that further remarkable properties of the KdV equation became evident. In 1954, Fermi, Pasta, and Ulam (FPU) used one of the very first digital computers to perform numerical experiments on an elastic string with a nonlinear restoring force, and their results contradicted the then current expectations of how energy should distribute itself among the normal modes of such a system. A decade later, Zabusky and Kruskal reexamined the FPU results in a famous paper in which they showed that the FPU string was well approximated by the KdV equation. They then did their own computer experiments, solving the Cauchy problem for KdV with initial conditions corresponding to those used in the FPU experiments. In the results of these simulations they observed the first example of a "soliton," a term that they coined to describe a remarkable particle-like behavior (elastic scattering) exhibited by certain KdV solutions. Zabusky and Kruskal showed how the coherence of solitons explained the anomalous results observed by Fermi, Pasta, and Ulam. But in solving that mystery they had uncovered a larger one: the behavior of KdV solitons was unlike anything seen before in applied mathematics, and the search for an explanation of their remarkable behavior led to a series of discoveries that changed the course of applied mathematics for the next thirty years. We shall now fill in some of the mathematical details behind the above sketch, beginning with a discussion of explicit solutions to the KdV equation.

It is straightforward to find the traveling-wave solutions of KdV. First, we substitute a traveling wave $u(x, t) = f(x - ct)$ into KdV, obtaining the ordinary differential equation $-cf' + 6ff' + f''' = 0$. If we add as a boundary condition that f should vanish at infinity, then a routine computation leads to the following two-parameter family of traveling-wave solutions:

$$u(x,t) = 2a^2 \operatorname{sech}^2(a(x - 4a^2t + d)).$$

These are the solitary waves seen by Russell, and they are now usually referred to as the 1-*soliton solutions* of KdV. Note that their amplitude, $2a^2$, is just half their speed, $4a^2$, while their "width" is proportional to a^{-1}. Thus, taller solitary waves are thinner and move faster.

Next, following Toda, we will "derive"[1] the 2-soliton solutions of KdV. Rewrite the 1-soliton solution as $u(x,t) = 2(\partial^2/\partial x^2) \log \cosh(a(x - 4a^2t + \delta))$, or $u(x,t) = 2(\partial^2/\partial x^2) \log K(x,t)$, where $K(x,t) = (1 + e^{2a(x-4a^2t+\delta)})$. We now try to generalize, looking for solutions of the form $u(x,t) = 2(\partial^2/\partial x^2) \log K(x,t)$, with $K(x,t) = 1 + A_1 e^{2\eta_1} + A_2 e^{2\eta_2} + A_3 e^{2(\eta_1+\eta_2)}$, where $\eta_i = a_i(x - 4a_i^2t + d_i)$, and we shall choose the A_i and d_i by substituting into KdV and seeing what works. One can check that KdV is satisfied for $u(x,t)$ of this form and arbitrary A_1, A_2, a_1, a_2, d_1, d_2, provided that we define $A_3 = ((a_2 - a_1)/(a_1 + a_2))^2 A_1 A_2$, and solutions of KdV arising in this way are called the KdV 2-*soliton solutions*.

It can now be shown that for these choices of a_1 and a_2,

$$u(x,t) = 12 \frac{3 + 4\cosh(2x - 8t) + \cosh(4x - 64t)}{[\cosh(3x - 36t) + 3\cosh(x - 28t)]^2}.$$

In particular, $u(x,0) = 6 \operatorname{sech}^2(x)$, $u(x,t)$ is asymptotically equal to $2 \operatorname{sech}^2(x - 4t - \phi) + 8 \operatorname{sech}^2(x - 16t + \frac{1}{2}\phi)$ when t is large and negative, and $u(x,t)$ is asymptotically equal to $2 \operatorname{sech}^2(x - 4t + \phi) + 8 \operatorname{sech}^2(x - 16t - \frac{1}{2}\phi)$ when t is large and positive, where $\phi = \frac{1}{3}\log(3)$.

Note what this says. If we follow the evolution from $-T$ to T (where T is large and positive), we first see the superposition of two 1-solitons: a larger and thinner one to the left of, and catching up with, a shorter, fatter, and slower-moving one to the right. Around $t = 0$ they merge into a single lump (with the shape $6 \operatorname{sech}^2(x)$), and then they separate again, with their original shapes restored—but now the taller and thinner one is to the right. It is almost as if they had passed right through each other. The only effect of their interaction is the pair of phase shifts: the slower one is retarded slightly from where it would have been, and the faster one is slightly ahead of where it would have been. Except for these phase shifts, the final result is what we might

expect from a linear interaction. It is only if we look closely at the interaction as the two solitons meet that we can detect its highly nonlinear nature. (Note, for example, that at time $t = 0$, the maximum amplitude, 6, of the combined wave is actually less than the maximum amplitude, 8, of the taller wave when they are separated.) But of course the really striking fact is the resilience of the two individual solitons: their ability to put themselves back together after the collision. Not only is no energy radiated away, but their actual shapes are preserved. (Remarkably, Russell (1844, p. 384) gives a sketch of a 2-soliton interaction experiment that he had carried out in his wave tank!)

Now back to the computer experiment of Zabusky and Kruskal. For numerical reasons, they chose to deal with the case of periodic boundary conditions: in effect, studying the KdV equation $u_t + uu_x + \delta^2 u_{xxx} = 0$ (which they label (1)) on the circle instead of on the line. For their published report, they chose $\delta = 0.022$ and used the initial condition $u(x,0) = \cos(\pi x)$. With the above background in mind, it is interesting to read the following extract from their 1965 report, which contains the first use of the term "soliton":

> (I) Initially the first two terms of Eq. (1) dominate and the classical overtaking phenomenon occurs; that is u steepens in regions where it has negative slope. (II) Second, after u has steepened sufficiently, the third term becomes important and serves to prevent the formation of a discontinuity. Instead, oscillations of small wavelength (of order δ) develop on the left of the front. The amplitudes of the oscillations grow, and finally *each* oscillation achieves an almost steady amplitude (that increases linearly from left to right) and has the shape of an individual solitary-wave of (1). (III) Finally, each "solitary wave pulse" or *soliton* begins to move uniformly at a rate (relative to the background value of u from which the pulse rises) which is linearly proportional to its amplitude. Thus, the solitons spread apart. Because of the periodicity, two or more solitons eventually overlap spatially and interact nonlinearly. Shortly after the interaction they reappear virtually unaffected in size or shape. In other words, solitons "pass through" one another without losing their identity. *Here we have a nonlinear physical process in which interacting localized pulses do not scatter irreversibly.*
>
> Zabusky and Kruskal (1965)

Further Reading

Lax, P. D. 1996. *Outline of a Theory of the KdV Equation in Recent Mathematical Methods in Nonlinear Wave Propagation.* Lecture Notes in Mathematics, volume 1640, pp. 70–102. New York: Springer.

---

1. This is a complete swindle! Only knowledge of the form of the solutions allows us to make the clever choice of K.

Palais, R. S. 1997. The symmetries of solitons. *Bulletin of the American Mathematical Society* 34:339–403.

Russell, J. S. 1844. Report on waves. In *Report of the 14th Meeting of the British Association for the Advancement of Science*, pp. 311–90. London: John Murray.

Toda, M. 1989. *Nonlinear Waves and Solitons.* Dordrecht: Kluwer.

Zabusky, N. J., and M. D. Kruskal. 1965. Interaction of solitons in a collisionless plasma and the recurrence of initial states. *Physics Review Letters* 15:240–43.

III.50   Linear Operators and Their Properties

1   Some Examples of Linear Operators

A LINEAR MAP [I.3 §4.2] between two VECTOR SPACES [I.3 §2.3] V and W is a function $T : V \to W$ that satisfies the condition $T(\lambda_1 v_1 + \lambda_2 v_2) = \lambda_1 T v_1 + \lambda_2 T v_2$. Two phrases that are used almost interchangeably with "linear map" are "linear transformation" and "linear operator." The former is often used when one wishes to draw attention to the effect of a linear map on some other object; for example, one might well choose to use the word "transformation" to describe geometrical operations such as reflections or rotations. As for "operator," it tends to be the word of choice when the linear map is between infinite-dimensional spaces, especially when it is just one of an ensemble of linear maps that form an algebra. It is these maps that we shall discuss here.

Let us begin with some examples of linear operators.

(i) If X is a BANACH SPACE [III.62] whose elements are infinite sequences, then we can define a "shift" S from X to X, which takes the sequence $(a_1, a_2, a_3, \dots)$ to the sequence $(0, a_1, a_2, a_3, \dots)$. (In other words, it puts a 0 at the beginning and shifts the other values of the sequence one place to the right.) The map S is linear, and if the norm on X is not too pathological, then S will be a continuous function from X to X.

(ii) If X is a SPACE OF FUNCTIONS [III.29] defined on the closed interval $[0, 1]$ and w is some fixed function, then the map M that takes the function f to the product fw (which is shorthand for the function $x \mapsto f(x)w(x)$) is linear, and, provided w is small enough in some appropriate sense, M is a continuous linear map from X to X. Such maps are called *multipliers*. (Note that the property of "being a multiplier" depends not just on the space X and the map M but also on the way we choose to represent X as a space of functions, so it is not an intrinsic property of the map itself.)

(iii) Another important way of defining linear operators on function spaces is to use a *kernel*. This is a function K of two variables, which can be used to define a linear map in a way that is similar to the way a matrix can be used to define a map between finite-dimensional vector spaces. The following formula uses K to define a linear map T:

$$Tf(x) = \int K(x, y) f(y) \, \mathrm{d}y. \tag{1}$$

Note the formal similarity between this and the formula

$$(Av)_i = \sum_j A_{ij} v_j,$$

which defines the product of a matrix with a column vector. Once again, K will have to satisfy appropriate conditions in order for (1) to define a continuous linear map.

A good example of a linear operator defined by a kernel is THE FOURIER TRANSFORM [III.27] $\mathcal{F}$, which takes a function in $L^2(\mathbb{R})$ to another such function. It is defined by the formula

$$(\mathcal{F}f)(\alpha) = \int_{-\infty}^{\infty} f(x) \mathrm{e}^{-\mathrm{i}\alpha x} \, \mathrm{d}x.$$

The kernel in this case is the function $K(\alpha, x) = \mathrm{e}^{-\mathrm{i}\alpha x}$.

(iv) If f is a differentiable function defined on $\mathbb{R}$, say, and we write $\mathrm{D}f$ for its derivative, then we can think of D as a linear map, since $\mathrm{D}(\lambda f + \mu g) = \lambda \mathrm{D}f + \mu \mathrm{D}g$. In order to regard D as an operator, we need to require f to belong to a suitable function space. The best way of doing this varies from context to context: choosing a good function space can be very important and can raise subtle questions. One way is not to insist that D is defined for every function in the space and not to require D to be continuous: sometimes it is enough if D is discontinuous but defined on a dense set of functions.

Similarly, many partial differential operators, such as the GRADIENT [I.3 §5.3] and the LAPLACIAN [I.3 §5.4], are linear operators when viewed appropriately.

2   Algebras of Operators

Although individual operators can be important, linear operators would not be as interesting as they are if it were not for the fact that they can be formed into *families*. If X is a Banach space, then the set $B(X)$ of all continuous linear operators from X to itself forms a structure known as a *Banach algebra*. Roughly speaking, this means that it is a Banach space (the norm of an operator T is defined to be the supremum of $\|Tx\|$ over all x such that $\|x\| \leqslant 1$) in which the elements can be multiplied as well as added. The product of T_1 and T_2 is defined to be the composition $T_1 T_2$, and it is easily seen to satisfy the inequality $\|T_1 T_2\| \leqslant \|T_1\| \|T_2\|$. This algebra is particularly important when X is a HILBERT SPACE [III.37] H: subalgebras of $B(H)$ have a very rich structure, which is discussed in OPERATOR ALGEBRAS [IV.15].

3 Properties of Operators Defined on a Hilbert Space

Unlike a general Banach space, a Hilbert space H has an inner product. It is therefore natural to ask that a continuous linear operator from H to H should relate to the inner product somehow. This basic idea leads to several different definitions, each of which picks out an important class of operators.

3.1 Unitary and Orthogonal Maps

Perhaps the most obvious condition one might require of an operator T is that it should *preserve* the inner product, in the sense that $\langle Tx, Ty \rangle$ should equal $\langle x, y \rangle$ for any two vectors x and y. In particular, this implies that $\|Tx\| = \|x\|$ for every x, and therefore that T is an *isometry* (that is, a map that preserves distances). If in addition, T is invertible, which it will be if its image is the whole of H, then T is a *unitary* map. The unitary maps form a group. If H is n dimensional, then this group is an important LIE GROUP [III.48 §1] called $U(n)$. If H is a real Hilbert space (as opposed to a complex one), then the word "orthogonal" is used instead of "unitary" and the corresponding Lie group is called $O(n)$. When $n = 3$, orthogonal maps are rotations and reflections, so $O(n)$ is the generalization of the group of rotations and reflections to n dimensions.

3.2 Hermitian and Self-Adjoint Maps

Given any operator T from H to H, there is an operator T^* from H to H with the property that $\langle Tx, y \rangle = \langle x, T^*y \rangle$ for every x and y. This operator is unique, and it is called the *adjoint* of T. A second property that T can have is that of equaling its own adjoint, which is the case if and only if $\langle Tx, y \rangle = \langle x, Ty \rangle$ for every x and y. Such operators are called *Hermitian* or, when the scalars are real, *self-adjoint*. A simple source of examples of Hermitian maps is multipliers on the space $L^2[0, 1]$, where the function one multiplies by is bounded and real-valued. As we shall see in a moment, there is a sense in which these are the only examples.

3.3 Properties of Matrices

If H is a finite-dimensional space with an orthonormal basis, then we can form the matrix A of T with respect to that basis. The various properties of T discussed above then turn out to be equivalent to properties of the matrix A. The *transpose* of A is the matrix A^{T} defined by $(A^{\mathrm{T}})_{ij} = A_{ji}$, and the *conjugate transpose* is the matrix A^* defined by $(A^*)_{ij} = \overline{A_{ji}}$. An $n \times n$ matrix A is *unitary* if AA^* is the identity, *orthogonal* if A is real and AA^{T} is the identity, *Hermitian* if $A = A^*$, and *self-adjoint* if $A = A^{\mathrm{T}}$ (in which case we say that A is *symmetric*). The operator T has one of these four properties if and only if its matrix A has the corresponding property.

3.4 The Spectral Theorem

Notice that the adjoint of a unitary operator is the *inverse* of that operator. In particular, both unitary and Hermitian operators commute with their adjoints. An operator with this property is called *normal*. Normal operators are important because of the famous spectral theorem. If T is a normal operator on a finite-dimensional space H, then the spectral theorem asserts that H has an ORTHONORMAL BASIS [III.37] of eigenvectors of T. In other words, there is a basis of H consisting of orthogonal unit vectors, with the property that the matrix of T with respect to this basis is diagonal. This is an extremely useful theorem in linear algebra. In general, if T is a normal operator on a Hilbert space H, then the spectral theorem tells us that there is something like a "basis" for H, with respect to which T is a multiplier. To put this slightly differently, there is an isometric isomorphism ϕ from H to a Hilbert space H' of functions that are square-integrable with respect to some MEASURE [III.55], and the map $\phi T \phi^{-1}$ is a multiplier on H'.

3.5 Projections

Another important class of maps on a Hilbert space is the set of *orthogonal projections*. In general, an element T of an algebra is an *idempotent* if it has the property that $T^2 = T$. If the algebra is an algebra of operators on a space X, then T is called a *projection*. To see why this name is appropriate, note that every x is mapped to the subspace TX of X, and all points in that subspace are left fixed by T (since $T(Tx) = T^2x = Tx$). A projection is *orthogonal* if Tx is always orthogonal to $x - Tx$. This tells us that T is a projection on to some subspace Y of

H, and that it takes each vector to the nearest point in *Y*, so that the vector $x - Tx$ is orthogonal to the whole of the subspace *Y*.

III.51 Local and Global in Number Theory
Fernando Q. Gouvêa

Analogy is a powerful tool. When one can see parallels between two different theories, this often allows one to transport insights from one to the other. The idea of studying something "locally" comes from the theory of functions. Imported into number theory by way of an analogy between functions and numbers, it leads us to a whole new kind of number, the *p*-adic numbers, and to the *local–global principle*, which has become one of the guiding ideas of modern number theory.

1   Studying Functions Locally

Suppose that we have a polynomial such as

$$f(x) = -18 + 21x - 26x^2 + 22x^3 - 8x^4 + x^5.$$

From the very way the polynomial is written down, we can see certain things about it. For example, we can see at once that if we plug in $x = 0$, we get $f(0) = -18$. Other things are less apparent. For example, to determine the values of $f(2)$ or $f(3)$, we would have to do some arithmetic. But if we were to rewrite the polynomial as

$$f(x) = 5(x - 2) - 6(x - 2)^2 - 2(x - 2)^3 \\ + 2(x - 2)^4 + (x - 2)^5,$$

we could see at once that $f(2) = 0$. (Of course, one needs to check that those two expressions really are equal!) Similarly, we can check that

$$f(x) = 10(x - 3)^2 + 16(x - 3)^3 + 7(x - 3)^4 + (x - 3)^5$$

and see at once that $f(3)$ is also zero, and in fact that the polynomial has a double root at $x = 3$.

One way to think about this is to describe the first expression as "local at $x = 0$," because it privileges the value 0 over all others. Then the other two expressions are local at 2 and local at 3, respectively. On the other hand, a formula like

$$f(x) = (x - 2)(x - 3)^2(x^2 + 1)$$

(which is also correct) is clearly more "global." It tells us where all the roots are: at 2, 3, and $\pm\sqrt{-1}$, with the 3 being a double root.

The same ideas extend to functions that are not polynomials, as long as we allow the expressions to be infinite. So, for example, let us take

$$g(x) = \frac{x^2 - 5x + 2}{x^3 - 2x^2 + 2x - 4}.$$

Locally at 0, we can write this as

$$g(x) = -\tfrac{1}{2} + x + \tfrac{1}{2}x^2 - \tfrac{3}{8}x^3 - \tfrac{3}{16}x^4 + \tfrac{7}{32}x^5 + \cdots.$$

Or we can write it locally at 2:

$$g(x) = -\tfrac{2}{3}(x - 2)^{-1} + \tfrac{5}{18} + \tfrac{5}{54}(x - 2) \\ - \tfrac{35}{324}(x - 2)^2 + \tfrac{55}{972}(x - 2)^3 \\ - \tfrac{115}{5832}(x - 2)^4 + \tfrac{65}{17496}(x - 2)^5 + \cdots.$$

Notice that this time we had to use a *negative* power of $(x - 2)$, because plugging in $x = 2$ makes the denominator zero. Nevertheless, the expansion tells us that the "badness" at 2 is not too bad. Specifically, we can see that while $g(2)$ is undefined, $(x - 2)g(2)$ makes sense and is equal to $-\tfrac{2}{3}$.

It is easy to keep going. To handle general functions locally at *a*, we may sometimes need to use fractional powers of $(x - a)$, but it does not get much worse than that. Such expansions are a very powerful tool in the theory of functions. One of the motivations for the discovery of the *p*-adic numbers was to find a similarly powerful tool for the study of numbers.

2   Numbers Are Like Functions

It was DEDEKIND [VI.50] and Heinrich Weber who first realized that an analogy could be drawn between numbers and functions. In their scheme, positive whole numbers were compared to polynomials, while fractions were analogous to quotients of polynomials such as the function $g(x)$ above. More complicated functions were like more complicated kinds of number. ELLIPTIC FUNCTIONS [V.31], for example, were similar to certain kinds of algebraic number. On the other hand, functions like $\sin(x)$ were more like TRANSCENDENTAL NUMBERS [III.41] such as e or π.

Dedekind and Weber pushed the idea that "functions are like numbers" in order to understand functions better. In particular, they showed that the techniques developed to study algebraic numbers could be used to study a whole class of functions, which came to be known as algebraic functions. It was Kurt Hensel, however, who saw that if functions are like numbers, then numbers must be like functions. In particular, he set out to find an analogue, for numbers, of the local expansions that were so useful in the theory of functions.

To get to Hensel's idea, let us start by noticing that the way we usually represent numbers already points in the right direction. After all, an expression like 34 291 really means

$$34\,291 = 1 + 9 \cdot 10 + 2 \cdot 10^2 + 4 \cdot 10^3 + 3 \cdot 10^4.$$

If we allow ourselves to think of 10 as being something like the variable x, this looks exactly like a polynomial. What is more, just as we can expand a polynomial in terms of different expressions $(x - a)$, we can write numbers in other bases. For example,

$$34\,291 = 4 + 4 \cdot 11 + 8 \cdot 11^2 + 3 \cdot 11^3 + 2 \cdot 11^4.$$

It is easy to see how to find this expansion. First, divide 34 291 by 11, and look at the remainder. It is 4. That is our first term. Next, subtract 4 from the original number to get something divisible by 11:

$$34\,291 - 4 = 34\,287 = 3117 \cdot 11.$$

Now divide 3117 by 11 to find the next remainder, which will give the second term. Keep repeating this process, and you will find the base-11 expansion.

That sounds very promising, but there is one little insight missing. The fact is that 10 is not really like $(x - 2)$, because 10 *can be factored*, while $(x - 2)$ cannot. So expanding a number in base 10 is a little like trying to express a polynomial in powers of $(x^2 - 3x + 2)$, which factors as $(x-1)(x-2)$. Such an expansion is not really local, since it is looking at two possible values of x at once. Similarly, the base-10 expansion mixes information about 2 and information about 5. The upshot is that we should always use a *prime number* as our base.

Just to fix ideas, let us choose $p = 11$. We already know that we can write positive numbers in base 11, i.e., as "polynomials in powers of 11." What happens if we try it with a fraction? Let us take $\frac{1}{2}$. The first step is to find the remainder, that is, to find a number r (between 0 and 10) such that $\frac{1}{2} - r$ is divisible by 11. Well, $\frac{1}{2} - 6 = -\frac{11}{2} = -\frac{1}{2} \cdot 11$. So the first term is 6. (To see what is meant by divisibility here, consider what would have happened if we had taken $r = 4$. Then $\frac{1}{2} - r$ would have been $-\frac{7}{2}$, and if we divide that by 11 we get $-\frac{7}{22}$, which has a factor of 11 in the denominator. It is this that is not allowed and that does not happen when $r = 6$.)

Now we repeat with the quotient, which was $-\frac{1}{2}$. We see that $-\frac{1}{2} - 5 = -\frac{11}{2} = -\frac{1}{2} \cdot 11$. So the second term will be $5 \cdot 11$. But now we find ourselves having to do $-\frac{1}{2}$ again! So we will do this again and again, and *all*

of the remaining terms will have coefficient 5. In other words,

$$\tfrac{1}{2} = 6 + 5 \cdot 11 + 5 \cdot 11^2 + 5 \cdot 11^3 + 5 \cdot 11^4 + 5 \cdot 11^5 + \cdots.$$

It is not clear quite what the equals sign means here, but in any case we have obtained an infinite expansion in powers of 11. It is called the 11-*adic expansion* of $\frac{1}{2}$. Furthermore, the expansion "works" when we do arithmetic with it. For example, if we multiply it by 2 and do all the rearranging ($2 \times 6 = 12 = 1 + 11$, so carry a 1, etc.) we do end up with 1.

Hensel showed that one can do this with all algebraic numbers as long as one allows infinite expansions, a finite number of negative powers of 11 (so that one can handle $\frac{5}{33}$ and similar things), and, in certain cases, fractional powers of 11. He argued that we should view such expansions as giving information "locally at 11." The same happens with all of the prime numbers. So if we have a prime number p we can consider our numbers "locally at p" by taking their expansions in powers of p. These we call their p-*adic expansions*. Just as in the case of functions, such expansions immediately tell us how divisible by p a number is, while hiding all the information about other primes; in that sense, they are truly "local."

3 p-adic Numbers

The best answers always raise new questions. Having discovered that any rational number has a p-adic expansion, and that one can "do arithmetic" directly with the expansions, it is inevitable to ask whether we have therefore enlarged the world of numbers under consideration. Once we have chosen the prime p, any rational number gives us a p-adic expansion. But does every such expansion come from a rational number?

Not a chance. It is easy to see that the set of all expansions is much bigger than the set of all rational numbers. Hensel's next move, then, was to point out that the set $\mathbb{Q}_p$ of all possible p-adic expansions is a new realm of numbers, which he called the p-*adic numbers*. It includes not only all the rational numbers, but also a lot more.

The best way to think of $\mathbb{Q}_p$ is by analogy with the set $\mathbb{R}$ of all real numbers. Real numbers are usually given by their decimal expansions. When we write $e = 2.718\ldots$, what we mean is that

$$e = 2 + 7 \cdot 10^{-1} + 1 \cdot 10^{-2} + 8 \cdot 10^{-3} + \cdots.$$

The set of all such expansions is the set of all real numbers. It contains all the rational numbers, but is much bigger.

Of course, except for the fact that both contain the rationals, these two realms are almost completely different. For example, in both $\mathbb{Q}_p$ and $\mathbb{R}$ there is a natural notion of "distance between two numbers." But these distances are completely different, even when the numbers in question are rational. So, in the reals, 2 is very close to 2001/1000. In the 5-adics, however, the distance between these two numbers is quite large!

It turns out that we can do calculus with p-adic numbers, just as we do it with reals. Many other mathematical ideas also extend. So Hensel's ideas led to a system of "parallel (numerical) universes"—one for each prime, plus the real numbers—in which we can do mathematics.

4 The Local–Global Principle

At first, most mathematicians seem to have found Hensel's new numbers interesting in a formal way, but also to have wondered what the point of them was. One does not adopt a new number system just for fun; it needs to be useful for something. Hensel was fascinated by his numbers and kept writing about them, but to begin with he had trouble demonstrating their usefulness. He showed, for example, that they could be used to develop the basics of algebraic number theory in a new way—but most folks seemed happy with the old way.

One can demonstrate the power of a new idea by giving a beautiful and easy proof of a difficult result. Hensel wrote a paper purporting to do just that: he gave an easy and elegant p-adic proof that the number e is transcendental. This did get people's attention. Unfortunately, when they looked hard at the proof they realized that it contained a subtle error. As a result, mathematicians' attitude of suspicion about Hensel's strange new numbers was reinforced.

The tide was turned by Helmut Hasse. He had been studying in Göttingen. At one point, he walked into a used bookstore and found a copy of Hensel (1913), a book written a few years earlier. Hasse was fascinated, and moved to Marburg to study with Hensel. A couple of years later, in 1920, he found the idea that was to make the p-adic numbers a crucial tool for number theorists.

What Hasse showed was that it was possible to answer some questions in number theory by answering them "locally." Here is a (not very important, but fairly easy to follow) example. Suppose x is a rational number that is a square of some other rational number

y, so $x = y^2$. Since all rational numbers are also p-adic, it is true that *for every prime number p the number x, thought of as a p-adic number, is a square.* And similarly, the real number x is a square. In other words, the rational number y is a kind of "global" square root, in that it serves as a square root in each local setting.

So far, so boring. But now reverse the thing. Suppose that we know that *for every prime number p the number x, thought of as a p-adic number, is the square of some p-adic number (which may depend on p)*, and also that x, thought of as a real number, is the square of some real number. A priori, these local square roots of x could all be different! But it turns out that under these assumptions x must be the square of some *rational* number, so that in fact all the local roots must come from a "global" root.

This leads us to think of the rational numbers as "global" and of the various $\mathbb{Q}_p$ and of $\mathbb{R}$ as "local." Then the previous paragraph claims that the property of "being a square" is true globally if and only if it is true "everywhere locally." This turns out to be a powerful and illuminating idea, and it has become known as the *Hasse principle* or the *local–global principle*.

Our example, of course, demonstrates the principle in its strongest case: solve a problem locally in all cases, and you have solved it globally. That is often too much to hope for. Nevertheless, attacking a problem locally and then putting the local pieces together has become a fundamental technique in modern number theory. It has been used to simplify older proofs, as in CLASS FIELD THEORY [V.28], and also to obtain new results, as in Wiles's proof of FERMAT'S LAST THEOREM [V.10]. So Hensel was right after all: his new numbers have earned their place along with the real numbers in every number theorist's heart.

Further Reading

Gouvêa, F. Q. 2003. *p-adic Numbers: An Introduction*, revised 3rd printing of the 2nd edn. New York: Springer.

Hasse, H. 1962. Kurt Hensels entscheidener Anstoss zur Entdeckung des Lokal–Global-Prinzips. *Journal für die reine und angewandte Mathematik* 209:3–4.

Hensel, K. 1913. *Zahlentheorie*. Leipzig: G. J. Göschenische.

Roquette, P. 2002. History of valuation theory. I. In *Valuation Theory and Its Applications*, volume I, pp. 291–355. Providence, RI: American Mathematical Society.

Ullrich, P. 1995. On the origins of p-adic analysis. *Proceedings of the 2nd Gauss Symposium. Conference A: Mathematics and Theoretical Physics, Munich, 1993*, pp. 459–73. Symposia Gaussiana. Berlin: Walter de Gruyter.

Ullrich, P. 1998. The genesis of Hensel's p-adic numbers. In *Charlemagne and His Heritage. 1200 Years of Civilization and Science in Europe*, volume 2, pp. 163–78. Turnhout: Brepols.

The Logarithmic Function

See THE EXPONENTIAL AND LOGARITHMIC FUNCTIONS [III.25]

III.52 The Mandelbrot Set

Suppose we have a complex polynomial f defined by a formula $f(z) = z^2 + C$ for some complex number C. Then for any choice of complex number z_0 we can form a sequence $z_0, z_1, z_2, \ldots$ by *iterating*, that is, repeatedly applying, the function f. So we let $z_1 = f(z_0)$, $z_2 = f(z_1)$, and so on. Sometimes the resulting sequence will tend to infinity, but sometimes it remains bounded—that is, it stays within a fixed distance from 0. For example, if we take $C = 2$ and start with $z_0 = 1$, then the sequence goes $1, 3, 11, 123, 15131, \ldots$ and clearly tends to infinity, whereas if we start with z_0 equal to $\frac{1}{2}(1 - i\sqrt{7})$, then we find that $z_1 = z_0^2 + 2 = z_0$ so the sequence is bounded since all its terms are equal to z_0. The *Julia set* associated with the constant C is the set of all z_0 for which the sequence remains bounded. Julia sets often have a fractal shape (see [IV.14 §2.5]).

To define a Julia set, one fixes C and considers different possibilities for z_0. What happens if one fixes z_0 and considers different possibilities for C? The result is the *Mandelbrot set*. The precise definition is that it is the set of all C such that the sequence is bounded if you take $z_0 = 0$. (One could consider other values of z_0, but the resulting sets are not interestingly different because they are related to each other by a simple change of variables.)

The Mandelbrot set also has an intricate fractal shape—one that has captured the popular imagination. The detailed geometry of the Mandelbrot set is not yet fully understood; some of the resulting open problems are of major importance because they encode very general information about dynamical systems. See DYNAMICS [IV.14 §2.8] for more details.

III.53 Manifolds

The surface of a sphere has the property that if you look at a very small portion of it then that portion will look like part of a plane. More generally, a *d-dimensional manifold*, or *d-manifold*, is a geometrical object that looks "locally" like d-dimensional EUCLIDEAN SPACE [I.3 §6.2]. Thus, 2-manifolds are smooth surfaces such as those of a sphere or a torus. Higher-dimensional manifolds are harder to visualize, but are a major topic of research. The basics of manifolds are set out in SOME FUNDAMENTAL MATHEMATICAL DEFINITIONS [I.3 §§6.9, 6.10]. More advanced ideas are discussed in DIFFERENTIAL TOPOLOGY [IV.7] and ALGEBRAIC TOPOLOGY [IV.6]. See also ALGEBRAIC GEOMETRY [IV.4], MODULI SPACES [IV.8], and RICCI FLOW [III.78]. (Even this is far from a complete list of articles in which manifolds feature.)

III.54 Matroids
Dominic Welsh

The original aim of Hassler Whitney when he introduced the concept of a matroid in 1935 was to produce an abstract notion that would capture the main ingredients of the structure of a set of vectors in a VECTOR SPACE [I.3 §2.3], while avoiding any explicit mention of linear independence.

To do this he singled out two fundamental properties and postulated that any family of subsets that possessed these properties was the collection of "independent sets" of a "matroid." The first of these properties was an obvious one: any subset of a linearly independent set is also linearly independent. The second property was more subtle: if A and B are two linearly independent sets and B contains more elements than A, then there exists some element of B that is not in A but which, when added to A, gives a set that is still linearly independent. Finally, in order to avoid trivialities he insisted that in every matroid the empty set must be independent.

Thus, formally, a *matroid* is defined to be a finite set E together with a family of subsets of E which are called the *independent sets* and which satisfy the following axioms.

  (i) The empty set is independent.
 (ii) Every subset of an independent set is independent.
(iii) If A and B are independent sets, with the number of elements of A being one less than the number of elements of B, then there is some x in B that is not in A such that $A \cup \{x\}$ is also independent.

Property (iii) is called the *exchange axiom*. The most fundamental example of a matroid is a set of vectors

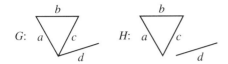

Figure 1 Two graphs giving rise to the same matroid.

in a vector space with the "independent sets" being the usual linearly independent ones: in this case the exchange axiom is known as Steinitz's exchange lemma. However, there are many examples of matroids that are not subsets of vector spaces.

Here, for example, is an important class of matroids that arise from graph theory. A *cycle* in a graph is a collection of edges of the form (v_1, v_2), (v_2, v_3), ..., (v_{k-1}, v_k), (v_k, v_1), where the v_i are distinct vertices. Take any graph and call a subset of edges "independent" if it contains no cycle.

So here we are thinking of a cycle among the edges as being in some way similar to a linear dependence among some vectors. It is obvious that any subset of an independent set will also not contain a cycle, so condition (ii) is satisfied. Slightly less obvious is that if A and B are sets of t and $t + 1$ edges, respectively, neither containing a cycle, then there will be at least one edge in B but not in A which can be added to A without creating a cycle. So we see that this is another example of a matroid, even though it arises in a very different context from the vector space one.

As it turns out, there is a way of identifying the edges of a graph with a set of vectors in a vector space over the field $\mathbb{F}_2$ of integers mod 2 (see MODULAR ARITHMETIC [III.58]). If G has n vertices and one associates with each vertex a basis element of $\mathbb{F}_2^n$, then one can associate with each edge the vector that is given by the sum of the basis elements corresponding to its two endpoints. A set of edges will then be independent if and only if the corresponding vectors in $\mathbb{F}_2^n$ are linearly independent. However, as we shall see, there are important examples of matroids that are not even *isomorphic* to sets of vectors.

Note that the collection of the independent sets (in a graph) conveys part of the information present in the graph, but by no means all of it. For example, consider the graphs G and H in figure 1. As graphs, G and H are distinct, but both give the same matroid on the set $\{a, b, c, d\}$ (the independent sets are all subsets of size less than or equal to 3, except for $\{a, b, c\}$). Note that this matroid is also the same as the matroid formed by

the columns of the matrix

$$A = \begin{pmatrix} 1 & 0 & 1 & 1 \\ 0 & 1 & 1 & 1 \\ 0 & 0 & 0 & 1 \end{pmatrix}.$$

with column labels a, b, c, d.

However, it turns out that most matroids do not come from either graphs or matrices.

Although a matroid is defined by very simple axioms, many basic results from both linear algebra and graph theory can be extended to the wider setting of matroids. For example, suppose that G is a connected graph. It is not hard to prove that if B is a maximal independent set of the matroid on G, then B is a tree which is incident with every vertex of G. Such a tree is called a *spanning tree* of G. All spanning trees of a connected graph have the same number of edges, namely, one less than the number of vertices. Similarly, in a vector space, or indeed in any subset of vectors, all maximal linearly independent sets have the same size. Both of these are special cases of the general result that in any matroid all maximal independent sets have the same size. This common size is called the *rank* of the matroid and, by analogy with vector spaces, a maximal independent set in a matroid is called a *basis*.

Matroids arise naturally in many parts of mathematics, and they often turn up unexpectedly. For example, consider the *minimum connector problem*: a company needs to connect a number of cities by links, such as railways or phone cables, and wishes to minimize the total cost. This is clearly equivalent to the following problem. Given a connected graph G, with each edge e having a nonnegative weight $w(e)$, find a set of edges that has the minimum total weight but that connects all the vertices of G. It is not hard to see that this problem reduces to finding a spanning tree of minimum weight.

For this there is a classical algorithm. It is the simplest possible algorithm one could imagine for the problem, and it works as follows. Start by choosing an edge of minimum weight, and at each subsequent step add an edge of minimum weight to your chosen set provided that at no stage a cycle is formed.

For example, consider the graph in figure 2. The algorithm would successively select the edges (a, b), (b, c), (d, f), (e, f), (c, d), giving a spanning tree of total weight $1 + 2 + 3 + 5 + 7 = 18$. Because of the way it works, the algorithm is known as a *greedy algorithm*.

At first sight, it seems rather unlikely that this algorithm could work, as it denies the possibility that choosing a suboptimal edge now might have a payoff

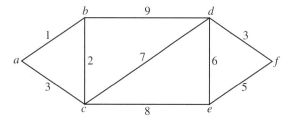

Figure 2 A graph with edge-weights.

later. However, it is not hard to show that the algorithm is actually correct. In fact, it extends in almost exactly the same way to matroids in general: what it gives is a (rather fast) algorithm for selecting a basis of minimum weight in a matroid in which each element has a nonnegative weight.

Somewhat more surprisingly, matroids are the only structures for which the greedy algorithm works. More precisely, suppose that $\mathcal{I}$ is a family of subsets of a set E with the property that if $A \in \mathcal{I}$ and $B \subseteq A$, then $B \in \mathcal{I}$. Now let w be any weight function and suppose that the problem is to select a member B of $\mathcal{I}$ which has maximum weight, where the weight of a set is just the sum of the weights of its elements. As above, the greedy algorithm starts by choosing an element e of maximum weight and then successively picks elements of maximum weight from the remaining elements subject to the proviso that at each stage, the set of elements chosen is a member of $\mathcal{I}$. It turns out that the following is true: *the greedy algorithm works on $\mathcal{I}$ for all weight functions w if and only if $\mathcal{I}$ is the collection of independent sets of a matroid.* Thus, matroids form a "natural home" for many optimization problems. Moreover, the concept is genuinely useful, since many of the matroids that arise in such problems are not derived from either vector spaces or graphs.

III.55 Measures

To understand measure theory, and to see why it is useful and important, it is instructive to start with a problem about lengths. Suppose that we have a sequence of intervals in $[0, 1]$ (the closed interval from 0 to 1), of total length less than 1. Can they cover $[0, 1]$? In other words, given intervals $[a_1, b_1], [a_2, b_2], \ldots$, with $\sum (b_n - a_n) < 1$, is it possible that their union equals $[0, 1]$?

One is tempted to answer "no, as the total length is too small." But this is just to restate the question. After all, why should "total length less than 1" actually imply

that the intervals cannot cover $[0, 1]$? Another tempting answer is to say "just rearrange the intervals so that they go from the left to the right, and then we never get to the right-hand end of $[0, 1]$." In other words, if the nth interval has length $b_n - a_n = d_n$, then just translate the intervals to be the intervals $[0, d_1], [d_1, d_1 + d_2], \ldots$. In this rearrangement it is indeed true that we never cover any point beyond $\sum d_n$, and so do not cover all of $[0, 1]$, but why does this imply that the original intervals do not cover $[0, 1]$?

It is quite easy to see that this rearrangement argument works for a *finite* number of intervals, but it does not work in general. Indeed, suppose we ask our original question again, but this time for the rationals: that is, let us replace the interval $[0, 1]$ by the rational interval $[0, 1] \cap \mathbb{Q}$. If our intervals have lengths $\frac{1}{4}, \frac{1}{8}, \frac{1}{16}, \ldots$, for example, so that the total length is only $\frac{1}{2}$, then certainly the left-to-right intervals will cover only the interval $[0, \frac{1}{2}] \cap \mathbb{Q}$, but it *is* possible for the original intervals to cover all of $[0, 1] \cap \mathbb{Q}$, since we can just enumerate the rationals as $q_1, q_2, \ldots$ (see COUNTABLE AND UNCOUNTABLE SETS [III.11]), and then put an interval of length $\frac{1}{4}$ around q_1, one of length $\frac{1}{8}$ around q_2, and so on.

This observation shows that the answer to our problem must involve properties of the reals that are not shared by the rationals, which wrecks any kind of "it is obvious" argument. In fact, the result *is* true for the reals, but its proof is a good exercise.

Why is this an important fact? It stems from a wish to define "length" for general sets of reals (for simplicity, we will concentrate on $[0, 1]$, just to avoid some technicalities about "infinite length"). What should the "length" of a set be? For intervals the answer is clear, and it is also clear for finite unions of intervals. But what about sets like $\{\frac{1}{2}, \frac{1}{3}, \frac{1}{4}, \ldots\}$, or $\mathbb{Q}$ itself?

A natural first attempt would be to use finite unions of intervals: one could take the length of a set A to be the least value of the length of a finite union of intervals that covers A. More precisely, one could define the length of A to be the infimum of $(b_1 - a_1) + \cdots + (b_n - a_n)$, taken over all finite unions of intervals $[a_1, b_1] \cup \cdots \cup [a_n, b_n]$ that cover A.

Unfortunately, this definition has some very undesirable properties. For example, the length of the set of all rational numbers in the interval $[0, 1]$ would then be 1, as would the length of all irrational numbers in $[0, 1]$. We would thus have two disjoint sets (and very natural ones at that) such that the length of their union is not the sum of their lengths. So this form of "length" is not really well-behaved for such sets.

What we want is a notion of length that applies to all the sets we know and are used to, and is *additive*, meaning that the length of $A \cup B$ is the sum of the lengths of A and B whenever A and B are disjoint. Remarkably, this *can* be achieved, and the key idea is to allow *countable* covers. That is, we modify the above definition as follows: the length (or *measure*, to give it its usual name) of a set A is the infimum of $(b_1 - a_1) + (b_2 - a_2) + \cdots$, taken over all unions of intervals $[a_1, b_1] \cup [a_2, b_2] \cup \cdots$ that cover A. Note that, thanks to the puzzle discussed earlier, the measure of the interval $[a, b]$ is $b - a$, just as we would hope.

It is also not hard to see that the measure of the set of rationals in $[0, 1]$ is zero, and it turns out that the measure of the irrationals in $[0, 1]$ is 1. Indeed, any countable set has measure zero. In many contexts, sets of measure zero are regarded as "negligible" or "of no importance." It is worth mentioning that there are also sets of measure zero that are uncountable (an example is the CANTOR SET [III.17]).

It turns out that, even with this definition, there are pairs of disjoint sets A and B such that the measure of $A \cup B$ is not the sum of the measures of A and B. However, it can be shown that for all "reasonable" sets the measure is additive. More precisely, one says that a subset of $[0, 1]$ is *measurable* if the measures of it and its complement add up to 1, as they should. If A and B are disjoint measurable sets, then the measure of their union *is* the sum of their measures.

This is a very useful fact, since it can be shown that every set that arises naturally in mathematics, or that has an explicit definition, is measurable: intervals, finite unions of intervals, countable unions of intervals, Cantor sets, things involving rationals or irrationals, and so on. In fact, the union of any countable family of measurable sets is again measurable: one says that the measurable sets form a *sigma-algebra*. Even better, for measurable sets the measure is *countably* additive, meaning that the measure of a disjoint union of countably many measurable sets is the sum of the measures of the individual sets.

More generally, in many other settings, one wants to end up with a sigma-algebra, containing all the sets one is interested in, on which we can define a countably additive measure, or "length function." The above example is called *Lebesgue measure on* $[0, 1]$. In general, whenever one wishes to define a countably additive measure, one always needs a result like the puzzle above in order to get started.

An important sigma-algebra is the algebra of all *Borel sets*. This is the smallest sigma-algebra that contains all open and closed intervals. Roughly speaking, it is the collection of all sets that you can build out of open and closed intervals using countable unions and intersections. (However, this masks the fact that the building up process can be very complicated: there is in fact a transfinite hierarchy of Borel sets.) The sigma-algebra of all Borel sets is smaller than the sigma-algebra of all Lebesgue-measurable sets because an arbitrary set of measure zero does not have to be a Borel set. Borel sets are one of the basic notions of DESCRIPTIVE SET THEORY [IV.22 §9]: in a certain technical sense they are "easily describable."

Here is another example of a sigma-algebra with a countably additive measure: we could work in $[0, 1]^2$ (the unit square in the plane), and base our ideas upon rectangles instead of intervals. So we would define the measure of a set as the least total area of a sequence of rectangles that covers the set. This gives an elegant and powerful approach to integration: the integral of a function f (defined on $[0, 1]$, say, and taking values in $[0, 1]$) is just defined to be the "area under its graph": that is, the measure of the set $\{(x, y) : y \leqslant f(x)\}$. Many complicated-looking functions can now be integrated: for example, the function f that is 1 on the rationals and 0 on the irrationals is easily checked to have an integral, namely 0, whereas in earlier theories such as Riemann integration that function would be too rapidly varying to be integrable.

This approach to integration gives rise to the so-called *Lebesgue integral* (further discussed in the article on LEBESGUE [VI.72]), which is one of the fundamental concepts in mathematics. It allows one to integrate a wide range of functions that are not Riemann integrable, but the main reason for its importance is not so much this as the fact that the Lebesgue integral has very good limiting properties that the Riemann integral lacks. For example, if $f_1, f_2, \ldots$ is a sequence of Lebesgue-integrable functions from $[0, 1]$ to $[0, 1]$ and $f_n(x)$ converges to $f(x)$ for every x, then f is Lebesgue-integrable and the Lebesgue integrals of the functions f_n converge to the Lebesgue integral of f.

III.56 Metric Spaces

There are many contexts in mathematics, especially in analysis, where one would like to say that two mathematical objects are close, and understand precisely what that means. If the two objects are the points (x_1, x_2) and (y_1, y_2) in a plane, then the task is straightforward: the distance between them is

$$\sqrt{(y_1 - x_1)^2 + (y_2 - x_2)^2},$$

by the Pythagorean theorem, and it makes sense to say that the points are close if this distance is small.

Now suppose that we have two points in n-dimensional space, $(x_1, \ldots, x_n)$ and $(y_1, \ldots, y_n)$. It is a simple matter to generalize the formula just given when $n = 2$ and define the distance between them to be

$$\sqrt{(y_1 - x_1)^2 + (y_2 - x_2)^2 + \cdots + (y_n - x_n)^2}.$$

Of course, the fact that the formula can be easily generalized is not in itself a guarantee that the resulting notion is a sensible definition of distance. And this raises the question of what properties we would like a definition to have for it to count as sensible. A metric space is an abstract notion that results from thinking about this question.

Let X be a set of "points." Suppose that, given any two of these points, x and y say, we have a way of assigning a real number $d(x, y)$ that we wish to regard as the distance between them. The following three properties are ones that it would be highly desirable for this idea of distance to have.

(P1) $d(x, y) \geqslant 0$ with equality if and only if $x = y$.
(P2) $d(x, y) = d(y, x)$ for any two points x and y.
(P3) $d(x, y) + d(y, z) \geqslant d(x, z)$ for any three points x, y, and z.

The first of these properties says that the distance between two points is always positive, except when the two points are the same, when it is zero. The second says that distance is a *symmetric* notion: the distance from x to y is the same as the distance from y to x. The third is called the *triangle inequality*: if you imagine x, y, and z as the vertices of a triangle, it says that the length of any side never exceeds the sum of the lengths of the other two sides.

A function d defined on pairs of points (x, y) from a set X is called a *metric* if it has properties (P1)–(P3) above. In that case, X and d together form a *metric space*. This abstraction of the usual notion of distance is very useful, and there are many important examples of metrics that are not necessarily derived from the Pythagorean theorem. Here are a few examples.

(i) Let X be n-dimensional space, that is, the set $\mathbb{R}^n$ of all sequences $(x_1, \ldots, x_n)$ of n real numbers. It

can be shown that the formula derived above from the Pythagorean theorem gives a notion of distance that does indeed satisfy properties (P1)–(P3). This metric is called the *Euclidean distance* and the resulting metric space is called *Euclidean space*. Euclidean spaces are perhaps the single most basic and important class of metric spaces in mathematics.

(ii) Nowadays, information is often transmitted digitally in the form of a string of 0s and 1s, such as 000111010010. The *Hamming distance* between two such strings is defined to be the number of places where the strings are different. For example, the Hamming distance between the strings 00110100 and 00100101 is 2, since the strings differ in the fourth and eighth places only. This idea of distance also satisfies properties (P1)–(P3).

(iii) If you are driving from one town to another, then the distance you care about is not the distance as the crow flies but the length of the shortest route along the network of available roads. Similarly, if you wish to travel from London to Sydney, then what matters is the length of the shortest path (known as a *geodesic*) along the Earth's surface, rather than the "actual" distance through the Earth itself. Many useful metrics come from this general idea of a shortest route, which guarantees that property (P3) will hold.

(iv) An important feature of Euclidean distance is its rotational symmetry: in other words, rotating the plane, or space, does not alter the Euclidean distances between points. There are other metrics that also have a great deal of symmetry, and these have great geometrical significance. In particular, the discovery of the HYPERBOLIC METRIC [I.3 §§6.6, 6.10] in the early nineteenth century demonstrated that the parallel postulate could not be proved using Euclid's other axioms. This resolved a question that had been open for thousands of years. See RIEMANNIAN METRICS [I.3 §6.10].

III.57 Models of Set Theory

A model of set theory is, roughly speaking, a structure in which the usual AXIOMS OF SET THEORY [IV.22 §3.1] (that is, the axioms of ZF or ZFC) hold. To explain what this means, let us think first about groups. The axioms of group theory mention certain operations (such as multiplication and inversion), and a model of group theory is a set, equipped with such operations, such that the axioms hold. In other words, a model of group theory is nothing other than a group. So what does a "model of ZF" mean? The axioms of ZF mention one relation, namely "is an element of," or "∈." A model of ZF is a set M, on which there is a relation E, such that all the axioms of ZF hold in S if we replace "∈" by "E."

However, there is one very important difference between these two sorts of model. When one first meets groups, one starts with some very simple examples, such as cyclic groups, or groups of symmetries of regular polygons, and one then builds up to more sophisticated examples such as the SYMMETRIC AND ALTERNATING GROUPS [III.68], and beyond. But this gentle process is not available for models of ZF. Indeed, since all of mathematics can be formulated in the language of ZF, it follows that *every* model of ZF has to contain a "copy" of the whole world of mathematics. This makes studying models of ZF rather difficult.

One aspect that is often found puzzling is the fact that a model of ZF is a *set*. This might seem to mean that there is a "universal" set (a set that has every set as a member), but from RUSSELL'S PARADOX [II.7 §2.1] it is easy to see that there can be no such set. The answer to this apparent problem is that the model M is indeed a set in the real mathematical universe, but that inside the model there is no universal set—in other words, there is no element x of M such that yEx for every element y of M. Thus, from the perspective of the model, the statement "there is no universal set" is true.

See MODEL THEORY [IV.23] for more about models in general, and SET THEORY [IV.22] for more about models of set theory.

III.58 Modular Arithmetic
Ben Green

Is there a square number whose decimal expansion ends … 7? Is 438 345 divisible by 9? For which positive integers n is $n^2 - 5$ a power of two? Is $n^7 - 77$ ever a Fibonacci number?

These questions, and more, can be answered using modular arithmetic. Let us look at the first question. Listing the first few squares, $1, 4, 9, 16, \ldots$, one does not find any whose final digit is 7. In fact, writing down just the final digits, one gets the sequence

$$1, 4, 9, 6, 5, 6, 9, 4, 1, 0, 1, 4, 9, 6, 5, 6, \ldots,$$

which seems to repeat (and thus never contain the number 7).

An explanation of this phenomenon is as follows. Let n be a number to be squared. We can always write n as a multiple of 10 plus a remainder; that is, $n = 10q + r$, where $r \in \{0, 1, \ldots, 9\}$. Now, if we square n we get

$$n^2 = (10q + r)^2$$
$$= 100q^2 + 20qr + r^2$$
$$= 10(10q^2 + 2qr) + r^2.$$

The only part of this expression that affects the final digit is the r^2, which immediately explains why the sequence of last digits of squares repeats with period 10, and hence contains no 7s.

Modular arithmetic is essentially just a notation for writing down arguments of this sort. If two numbers (like n and r) leave the same remainder on division by 10, then we say that they are *congruent modulo* 10 and write $n \equiv r \mod 10$. What we proved above is the statement that, if $n \equiv r \mod 10$, then $n^2 \equiv r^2 \mod 10$.

Everything we have just said applies equally well if we replace 10 by an arbitrary *modulus* m: if n and r leave the same remainder on division by m, then we say that n and r are *congruent modulo* m and we write $n \equiv r \mod m$. Equivalently, n and r are congruent modulo m if m divides $n - r$. (An integer a is said to *divide* another integer b if b is an integer multiple of a.) The above argument is just one instance of the following general fact, which is not hard to prove: if $a \equiv a' \mod m$ and $b \equiv b' \mod m$, then $ab \equiv a'b' \mod m$ and $a + b \equiv a' + b' \mod m$.

Now observe that $10 \equiv 1 \mod 9$. It follows that $10 \times 10 \equiv 1 \times 1 \equiv 1 \mod 9$, and in fact that $10^d \equiv 1 \mod 9$ for any $d \in \mathbb{N}$. Suppose that we have a number N whose decimal expansion is $a_d a_{d-1} \cdots a_2 a_1 a_0$. This means that

$$N = a_d 10^d + a_{d-1} 10^{d-1} + \cdots + a_1 10 + a_0.$$

Applying the rules of modular arithmetic, we get

$$N \equiv a_d + a_{d-1} + \cdots + a_1 + a_0 \mod 9.$$

This gives the well-known test for divisibility by 9: simply add up the digits of the number in base 10, and see if the result is divisible by 9. For the example $N = 438\,345$ the sum of the digits is 27, which is divisible by 9. So N is a multiple of 9 (in fact $N = 9 \times 48\,705$).

If m is a modulus and n is an integer, then there is precisely one value of r between 0 and $m - 1$ such that $n \equiv r \mod m$. This number r is often called the least residue or simply the *residue* of n to the modulus m.

Now let us consider the third question posed at the beginning of this article, namely the matter of when $n^2 - 5$ is a power of two. When $n = 3$, $3^2 - 5 = 4$ *is* a power of two, but a little experimentation does not reveal any further examples. What aspect of the problem changes as n becomes larger than 3? The key observation is that $n^2 - 5$ is now greater than 4, and so if it were a power of 2, then it would have to be divisible by 8. That would mean that $n^2 \equiv 5 \mod 8$, but this is never the case. Indeed, the residues of the first eight squares are 1, 4, 1, 0, 1, 4, 1, 0, and we know that the sequence will repeat with period 8 (actually, it repeats with period 4). Thus, it never contains a 5.

Modular arithmetic should be used with care. Although the rules for addition and subtraction are simple, division is somewhat more subtle. For example, if we are given that $ac \equiv bc \mod m$, it is not, in general, permissible to divide by c and conclude that $a \equiv b \mod m$: consider, for instance, the case $a = 2$, $b = 4$, $c = 3$, $m = 6$.

Let us examine what has just gone wrong. To say that $ac \equiv bc \mod m$ means that m divides $ac - bc = (a-b) \times c$. But this clearly does not mean that m divides $a - b$, since m could divide c (or at least have a common factor with it). However, if m has no factor in common with c, then it must divide $a - b$, so in this case we do indeed have $a \equiv b \mod m$. In particular, for any prime number p we have the very useful *cancelation law*: if $ac \equiv bc \mod p$ and $c \not\equiv 0 \mod p$, then $a \equiv b \mod p$.

The examples so far may have suggested that the principal uses of modular arithmetic are to do with specific moduli such as 10 and 8. However, this is far from true, and the subject really comes into its own when one looks at more general m. For example, one of the basic results in number theory is *Fermat's little theorem*, which states that if p is a prime and $a \not\equiv 0 \mod p$, then $a^{p-1} \equiv 1 \mod p$. Let us quickly prove this. Consider the numbers $a, 2a, 3a, \ldots, (p-1)a \mod p$. If $ra \equiv sa \mod p$, then from the cancelation law we can deduce that $r \equiv s \mod p$, from which it follows that $a, 2a, \ldots, (p-1)a$ are all different modulo p. Furthermore, none of these numbers is $0 \mod p$. We are thus forced to conclude that the numbers $a, 2a, 3a, \ldots, (p-1)a \mod p$ are simply a rearrangement of the numbers $1, 2, 3, \ldots, p-1 \mod p$. In particular, the products

of the numbers in these two sets are the same, which implies that

$$a^{p-1}(p-1)! \equiv (p-1)! \mod p.$$

Since $(p-1)!$ is not a multiple of p, we can apply the cancelation law again and divide both sides by $(p-1)!$. This implies the result.

Euler's theorem is a generalization of Fermat's little theorem to composite moduli. It states that if m is a positive integer and a is another positive integer that is *coprime to* m (this means that a and m have no common factor), then $a^{\phi(m)} \equiv 1 \mod m$. Here ϕ is *Euler's totient function*: $\phi(m)$ is the number of integers less than m that are coprime to m. For instance, if $m = 9$, then the integers less than m and coprime to m are 1, 2, 4, 5, 7, and 8, so $\phi(9) = 6$ and we can deduce from Euler's theorem that $5^6 \equiv 1 \mod 9$. Let us check this directly: $5^6 = 15\,625$, so the sum of its digits is 19, which is indeed congruent to $1 \mod 9$. For further discussion of the Fermat-Euler theorem, see MATHEMATICS AND CRYPTOGRAPHY [VII.7], COMPUTATIONAL NUMBER THEORY [IV.3], and THE WEIL CONJECTURES [V.35].

The final question from above—whether $n^7 - 77$ is ever a Fibonacci number—is left as an exercise to the reader.

III.59 Modular Forms
Kevin Buzzard

1 A Lattice in the Complex Numbers

When one first learns about the complex numbers, one is taught to think of them as a two-dimensional space, with one real and one imaginary dimension: a complex number $z = x + iy$ has real part x and imaginary part y, where i is a square root of -1.

Now let us consider what the complex numbers that have *integers* for their real and imaginary parts look like. These complex numbers, such as $3 + 4i$ or $-23i$, form a "lattice" in the complex plane (see figure 1).

By definition, every element of this lattice is of the form $m + ni$ for some pair of integers m and n. We say that the lattice is *generated* by 1 and i, and use the notation $\mathbb{Z} + \mathbb{Z}i$ for it. Note that this lattice can be generated in plenty of other ways. For example, it is also generated by the pair $(1, -i)$, the pair $(1, 100+i)$ or even the pair $(101 + i, 100 + i)$. In fact, one can easily check that this lattice is generated by the pair $(a + bi, c + di)$ (meaning that every element of the lattice is an integer combination of $a + bi$ and $c + di$) if and only if $a, b, c,$ and d are integers and $ad - bc = \pm 1$.

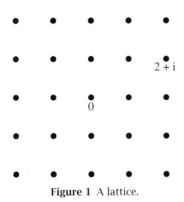

Figure 1 A lattice.

2 More General Lattices

Now let v and w be *any* two complex numbers and consider the set of complex numbers of the form $av + bw$, again with a and b integers (see figure 2).

A lattice is exactly such a thing: a grid $\mathbb{Z}v + \mathbb{Z}w$ in the complex plane generated by two complex numbers v and w, with the provisos that neither v nor w is zero and that v/w is not real (this is just to ensure that v and w do not both lie on one line).

If $\tau = x + iy$ is a complex number with $y \neq 0$, then there is a standard lattice associated with τ, namely $\mathbb{Z}\tau + \mathbb{Z}$. We call this lattice Λ_τ and note that $\Lambda_\tau = \Lambda_{-\tau}$. In general, however, distinct complex numbers τ give rise to distinct lattices—and furthermore there are plenty of lattices that are not equal to Λ_τ for any τ, for the simple reason that 1 belongs to Λ_τ for every τ.

3 Relations between Lattices

If Λ is a lattice generated by v and w, and α is a nonzero complex number, then one can multiply the entire situation by α and deduce that $\alpha\Lambda$ is the lattice generated by αv and αw. Geometrically, this says that one can rotate and rescale lattices.

If Λ is a lattice generated by v and w, and we scale it by dividing everything by w, then we get a new lattice $(1/w)\Lambda$, which is generated by v/w and $w/w = 1$. In particular, this new lattice is equal to Λ_τ for the complex number $\tau = v/w$.

It may seem like an odd thing to do, but one can apply this scaling trick to Λ_τ itself. The lattice Λ_τ is generated by $(\tau, 1)$ but also by any pair $(v, w) = (a\tau + b, c\tau + d)$, if a, b, c, and d are integers such that $ad - bc = \pm 1$. If we divide by $c\tau + d$ and set $\sigma = (a\tau + b)/(c\tau + d)$, then we see that

$$\frac{1}{c\tau + d}\Lambda_\tau = \Lambda_\sigma. \tag{1}$$

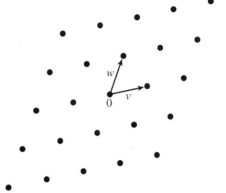

Figure 2 A general lattice.

4 Modular Forms as Functions on Lattices

The formal definition of a modular form is rather unenlightening: it is a function that obeys certain boundedness conditions and transformation properties. One way of seeing where the transformation properties come from is to think about lattices. If k is an integer, then a *modular form of weight k* is a function f that associates a complex number $f(\Lambda)$ with any lattice Λ, and has the property that

$$f(\alpha\Lambda) = \alpha^{-k}f(\Lambda). \tag{2}$$

The function also has to satisfy some other properties (a differentiability condition and a boundedness condition), but the crucial property is the one above. If k is even and at least 4, then an example of a modular form of weight k is the *Eisenstein series* G_k defined by the formula

$$G_k(\Lambda) = \sum_{0 \neq \lambda \in \Lambda} \lambda^{-k}.$$

The assumption that k is at least 4 guarantees that the sum converges, and the evenness of k ensures that the function is nonzero.

We have seen that any lattice can be scaled so that it takes the form Λ_τ for some τ, so (2) implies that a modular form will be determined by its values on such lattices. If $\mathcal{H}$ denotes the complex numbers with positive imaginary part, then, because $\Lambda_\tau = \Lambda_{-\tau}$, a modular form is in fact determined by its values on Λ_τ for $\tau \in \mathcal{H}$.

However, an arbitrary function on $\mathcal{H}$ does not give us a modular form: equation (1) tells us that if f is a modular form and F is the function on $\mathcal{H}$ defined by $F(\tau) = f(\Lambda_\tau)$, then F must satisfy the equation

$$F\left(\frac{a\tau + b}{c\tau + d}\right) = (c\tau + d)^k F(\tau) \tag{3}$$

for every $a, b, c, d \in \mathbb{Z}$ such that $ad - bc = 1$. (The reason we exclude the case $ad - bc = -1$ is that $(a\tau + b)/(c\tau + d)$ would not be in the upper half-plane in this case.) This is the equation at the heart of the definition of a modular form.

Over the years, mathematicians have isolated other desirable properties that F should have in order to give a useful theory. Nowadays, modular forms are required to obey the additional properties that F is HOLOMORPHIC [I.3 §5.6] and that $F(x + iy)$ does not grow too quickly as y goes to $+\infty$; these assumptions imply that the vector space of weight k modular forms is finite dimensional. The Eisenstein series above do have these additional properties, and are the first basic examples of modular forms.

5 Why Modular Forms?

Modular forms have connections with arithmetic, geometry, representation theory, and even physics. Modular forms also played a key role in the Taylor–Wiles proof of FERMAT'S LAST THEOREM [V.10]. Why is this? One general reason is that there are links between modular forms and other mathematical objects: here we briefly explain one of the links.

Lattices in the complex plane are related to ELLIPTIC CURVES [III.21]: the quotient of the complex numbers by a lattice is an elliptic curve, and every elliptic curve arises in this way. Hence to study elliptic curves, or families of elliptic curves, one can instead study families of lattices. One way of studying an object is by studying the functions on that object, and a modular form is *precisely* that: a function on the collection of all lattices. And indeed, automorphic forms, which are generalizations of modular forms, have been used to great effect in studying a wide variety of families of algebraic objects in this way.

III.60 Moduli Spaces

An important general problem in mathematics is *classification* (see THE GENERAL GOALS OF MATHEMATICAL RESEARCH [I.4 §2]). Often, one has a set of mathematical structures and a notion of equivalence, and one would like to describe the EQUIVALENCE CLASSES [I.2 §2.3]. For example, two (compact, orientable) surfaces are often regarded as equivalent if each can be continuously deformed into the other. Each equivalence class is then fully described by the GENUS [III.33], or "number of holes," in the surface.

Topological equivalence is rather "crude," in the sense that it is relatively easy for two surfaces to be equivalent. As a result, the equivalence classes are parametrized by a fairly simple set: the set of all positive integers. But there are many geometrical contexts in which finer notions of equivalence are important. For example, in several contexts one wishes to regard two two-dimensional LATTICES [III.59] as equivalent if one is a rotation and enlargement of the other. Equivalence relations such as this one often lead to parameter sets that themselves have an interesting geometrical structure. Such sets are called *moduli spaces*. For details, see [IV.8] and also [V.23].

III.61 The Monster Group

THE CLASSIFICATION OF FINITE SIMPLE GROUPS [V.7] is one of the landmarks of twentieth-century mathematics. As its name suggests, it gives a complete description of all finite simple groups, which can be thought of as the building blocks for all finite groups. It states that each finite simple group belongs to one of eighteen infinite families, or else is one of twenty-six "sporadic" examples. The Monster group is the largest of these sporadic groups, with 808 017 424 794 512 875 886 459 904 961 710 757 005 754 368 000 000 000 elements.

As well as having a starring role in the classification theorem, the Monster group has remarkable and deep connections with other areas of mathematics. Most notably, the smallest dimension of a faithful REPRESENTATION [IV.9] of the Monster group is 196 883, while the coefficient of $e^{2\pi i z}$ in the important and famous "elliptic modular function" (see ALGEBRAIC NUMBERS [IV.1 §8]) is 196 884. Far from being an amusing coincidence, the fact that these two numbers differ by just 1 is a manifestation of a very deep connection between the two. See VERTEX OPERATOR ALGEBRAS [IV.17 §4.2] for further details.

The Navier–Stokes Equation
> *See* THE EULER AND NAVIER–STOKES EQUATIONS [III.23]

III.62 Normed Spaces and Banach Spaces

It is often useful to approximate a function f by a polynomial P. For example, if you are designing a pocket calculator and want it to calculate LOGARITHMS

[III.25 §4], you cannot expect it to do so exactly, since a calculator cannot handle infinitely many digits, so instead you will get it to calculate a different function $P(x)$ that approximates $\log(x)$ well. Polynomials are a good choice, because they can be built up from the basic operations of addition and multiplication. This idea raises two questions: which functions can you hope to approximate, and what counts as a good approximation?

Clearly, the answer to the second question determines the answer to the first, but there is no single right answer to the second: it is up to you what you would like to declare to be a good approximation. However, not all decisions are equally natural. Suppose that P and Q are polynomials, f and g are more general functions, and x is a real number. If $P(x)$ is close to $f(x)$ and $Q(x)$ is close to $g(x)$, then $P(x) + Q(x)$ will be close to $f(x) + g(x)$. Also, if λ is a real number and $P(x)$ is close enough to $f(x)$, then $\lambda P(x)$ will be close to $\lambda f(x)$. This informal argument suggests that the functions that we can approximate well will form a VECTOR SPACE [I.3 §2.3].

We have arrived, by one of many possible routes, at the following general situation: we are given a vector space V (consisting, in our case, of certain functions) and we would like to be able to say in a precise way what it is for two elements of the vector space to be *close*.

The idea of closeness is formally captured by the notion of a METRIC SPACE [III.56], so the obvious approach is to define a metric d on the vector space V. Now a general principle, when putting two structures together (in this case, the linear structure of the vector space and the distance structure of the metric), is that the two structures should *relate* to one another in a natural way. In our case, there are two natural properties that one should ask for. The first is *translation invariance*. If u and v are two vectors and we translate them by adding w to both, then their distance should not change: that is, $d(u + w, v + w) = d(u, v)$. The second is that the metric should *scale correctly*. For example, if one doubles two vectors u and v, then the distance between them should double. More generally, if one multiplies u and v by a scalar λ, then the distance between them should multiply by $|\lambda|$: that is, $d(\lambda u, \lambda v) = |\lambda| d(u, v)$.

If a metric has the first of these properties, then, setting $w = -u$, we find that $d(u, v) = d(0, v - u)$. It follows that if we know distances from 0, then we know all distances. Let us write $\|v\|$ instead of $d(0, v)$. Then what we have just shown is that $d(u, v) = \|v - u\|$. The

expression $\|\cdot\|$ is called a *norm*, and $\|v\|$ is the *norm of* v. The following two properties of norms are easy to deduce from the fact that d is a metric that scales properly.

(i) For any vector v, $\|v\| \geqslant 0$. Moreover, $\|v\| = 0$ only if $v = 0$.

(ii) For any vector v and any scalar λ, $\|\lambda v\| = |\lambda| \|v\|$.

We also have the so-called *triangle inequality*.

(iii) $\|u + v\| \leqslant \|u\| + \|v\|$ for any two vectors u and v.

This follows from translation invariance and the triangle inequality for metric spaces, since

$$\|u + v\| = d(0, u + v) \leqslant d(0, u) + d(u, u + v)$$
$$= d(0, u) + d(0, v) = \|u\| + \|v\|.$$

In general, any function $\|\cdot\|$ on a vector space V that has properties (i)–(iii) is called a norm on V. A vector space with a norm on it is called a *normed space*. Given a normed space V, we can say that two vectors u and v are close if their distance $\|v - u\|$ is small.

There are many important examples of normed spaces, several of which are discussed elsewhere in this volume. One class of examples that stands out is that of HILBERT SPACES [III.37], which can be thought of as norms given by distances that stay the same not just when you translate but also when you rotate. Other examples are discussed in FUNCTION SPACES [III.29].

Let us return to the problem of how to discuss approximation by polynomials. The most commonly given answers to the two questions that arose earlier are as follows. The functions that one can approximate well are all continuous functions defined on some closed interval $[a, b]$ of real numbers. These functions form a vector space which is denoted $C[a, b]$. To make the notion of good approximation precise, we introduce a norm on this space: $\|f\|$ is defined to be the largest value of $|f(x)|$ for any x in the interval (that is, for any x between a and b). With this definition, the distance $\|f - g\|$ between two functions f and g will be small if and only if $|f(x) - g(x)|$ is small for every x in the interval. In this situation one says that f *uniformly approximates* g. It is not obvious that every continuous function on $[a, b]$ can be uniformly approximated by a polynomial: the statement that it can is called the *Weierstrass approximation theorem*.

Here is a different way in which normed spaces arise. For most PARTIAL DIFFERENTIAL EQUATIONS [I.3 §5.4] it is not possible to write down a tidy formula that solves them. However, there are many techniques for proving that solutions *exist*, and they usually involve limiting arguments. For example, sometimes one can generate a sequence of functions $f_1, f_2, \ldots$ and show that these functions "converge" to some "limiting function" f, which, owing to the way we constructed the sequence $f_1, f_2, \ldots$, must be a solution to the equation. Again, if we want to make sense of this, we must know what it is for two functions to be close, which means that the functions f_n should belong to a normed space.

How can we show that these functions converge to a limit f if we cannot already describe f? The answer is that most interesting normed spaces, including Hilbert spaces and most important function spaces, have an additional property, called *completeness*, which guarantees, under certain conditions, that limits do indeed exist. Informally, it says that if the vectors in a sequence $v_1, v_2, \ldots$ all get very close to each other when you go far enough along the sequence, then they must converge to a limit, v, that belongs to the normed space as well. A complete normed space is known as a *Banach space*, after the Polish mathematician STEFAN BANACH [VI.84], who developed much of the general theory of such spaces. Banach spaces have many useful properties that normed spaces do not have in general: the completeness property can be thought of as ruling out pathological examples.

The theory of Banach spaces is sometimes known as *linear analysis*, since by mixing vector spaces and metric spaces it mixes linear algebra and analysis. Banach spaces arise throughout modern analysis: see, for example, the articles in this volume on PARTIAL DIFFERENTIAL EQUATIONS [IV.12], HARMONIC ANALYSIS [IV.11], and OPERATOR ALGEBRAS [IV.15].

III.63 Number Fields
Ben Green

A *number field* K is a "finite-degree field extension" of $\mathbb{Q}$, the field of rational numbers. This means that K is a FIELD [I.3 §2.2] that is finite dimensional when one regards it as a VECTOR SPACE [I.3 §2.3] over $\mathbb{Q}$. The following alternative description is somewhat more concrete. Take finitely many algebraic numbers $\alpha_1, \ldots, \alpha_k$ (that is, roots of polynomials with integer coefficients) and consider the field K of all rational functions in the α_i. (In other words, K consists of numbers like $\alpha_1^2 \alpha_3 / (\alpha_2^2 + 7)$.) Then it turns out that K is a num-

ber field (the one thing that is not completely obvious is that it has finite degree over $\mathbb{Q}$), which we denote by $\mathbb{Q}(\alpha_1, \ldots, \alpha_k)$. Conversely, every number field is of this form.

The simplest number fields are perhaps the *quadratic fields*. These are fields of the form $\mathbb{Q}(\sqrt{d}) = \{a + b\sqrt{d} : a, b \in \mathbb{Q}\}$, where d is an integer (which, it is important to stress, may be negative) that is square-free. This last condition tells us that d has no nontrivial square factors. It is there for convenience so that all the $\mathbb{Q}(\sqrt{d})$ will be distinct. (For example, $\mathbb{Q}(\sqrt{12})$, if we were to allow it, would equal $\mathbb{Q}(\sqrt{3})$, since $\sqrt{12} = 2\sqrt{3}$.) Among the other important number fields are the *cyclotomic fields*. Here we take a primitive mth root of unity ζ_m (which, for concreteness, one could take to be $e^{2\pi i/m}$) and "adjoin" it to $\mathbb{Q}$, obtaining the field $\mathbb{Q}(\zeta_m)$.

Why consider number fields? Historically, an important reason is that they allow us to factorize certain Diophantine equations. For example, the Ramanujan–Nagell equation $x^2 = 2^n - 7$ may be factorized as

$$(x + \sqrt{-7})(x - \sqrt{-7}) = 2^n$$

if we allow coefficients in the field $\mathbb{Q}(\sqrt{-7})$, while the Fermat equation $x^n + y^n = z^n$ is equivalent to

$$x^n = (z - y)(z - \zeta_n y) \cdots (z - \zeta_n^{n-1} y) \qquad (1)$$

if we allow coefficients in the field $\mathbb{Q}(\zeta_n)$.

Before one can start thinking about whether such factorizations are useful, it is necessary to understand the notion of an *integer* in a number field K. A number $\alpha \in K$ is an (algebraic) integer if it is a root of a *monic* polynomial with coefficients in $\mathbb{Z}$: that is, a polynomial with leading coefficient 1. For simple fields like $\mathbb{Q}(\sqrt{d})$ with d squarefree, the integers can be described quite explicitly. They are all the numbers of the form $a + b\sqrt{d}$ for integers a and b, unless $d \equiv 1 \pmod 4$, in which case we must include more numbers: we get all numbers of the form $a + b(\frac{1}{2}(1 + \sqrt{d}))$, again for integers a and b. The set of integers in K is often denoted by $\mathcal{O}_K$, and it forms a RING [III.81 §1].

Unfortunately, factorizations such as (1) are not as helpful as they seem at first sight: $\mathcal{O}_K$ turns out not to be OK, at least if one expects familiar properties of the ring $\mathbb{Z}$ to carry over unchanged. In particular, unique factorization into primes fails to hold: for example, $2 \cdot 3 = (1 + \sqrt{-5})(1 - \sqrt{-5})$ in the field $\mathbb{Q}(\sqrt{-5})$. The numbers on both sides are integers in this field, and it is not possible to decompose any of them any further.

Amazingly, unique factorization may be restored by embedding $\mathcal{O}_K$ into a larger set, which consists of objects called IDEALS [III.81 §2]. There is a natural EQUIVALENCE RELATION [I.2 §2.3] that one can place on these ideals, and the number of equivalence classes, called the *class number* and written $h(K)$, is one of the most important invariants in number theory: in a certain sense, it measures "the extent to which unique factorization fails" in the number field K. (See ALGEBRAIC NUMBERS [IV.1 §7] for more details.) The fact that it is finite is one of the two basic *finiteness theorems* in algebraic number theory.

When $h(K) = 1$, the integers $\mathcal{O}_K$ themselves enjoy unique factorization, without the need for extra ideals. This does not happen particularly often; among the fields $\mathbb{Q}(\sqrt{-d})$ with d positive and squarefree, only nine have this property, namely $d = 1, 2, 3, 7, 11, 19, 43, 67$, and 163. The problem of determining these numbers was posed by GAUSS [VI.26] and finally solved by Heegner in 1952.

The fact that $h(\mathbb{Q}(\sqrt{-163})) = 1$ is closely related to some remarkable facts. For example, the polynomial $x^2 + x + 41$ takes prime values when $x = 0, 1, \ldots, 39$ (observe that $4 \times 41 = 163 + 1$), and the number $e^{\pi\sqrt{163}}$ is within 10^{-12} of an integer.

It is a well-known open problem to decide whether or not there are infinitely many fields $\mathbb{Q}(\sqrt{d})$, $d > 0$, with class number 1. Gauss and many subsequent authors have conjectured that there are.

The second basic finiteness result in algebraic number theory is *Dirichlet's unit theorem*. A *unit* is simply some $x \in \mathcal{O}_K$ such that there exists $y \in \mathcal{O}_K$ with $xy = 1$. The numbers 1 and -1 are always units, but there can certainly be others: for example, $17 - 12\sqrt{2}$ is a unit in $\mathbb{Q}(\sqrt{2})$ (since its reciprocal is $17 + 12\sqrt{2}$). The units form an Abelian group $\mathcal{U}_K$ under multiplication. Dirichlet's theorem states that this group has finite rank, which means that it is generated by finitely many of its elements.

If $d > 0$ is squarefree and if $K = \mathbb{Q}(\sqrt{d})$, then $\mathcal{U}_K$ has rank 1. When $d \not\equiv 1 \pmod 4$, the fact that it has rank *at least* 1 is equivalent to the statement that the Pell equation $x^2 - dy^2 = 1$ always has a nontrivial solution. This is because the Pell equation factors as $(x - y\sqrt{d})(x + y\sqrt{d}) = 1$. The unit $17 - 12\sqrt{2}$ in $\mathbb{Q}(\sqrt{2})$ corresponds to the solution $x = 17$, $y = 12$ of the equation $x^2 - 2y^2 = 1$.

For more about some of the topics discussed in this article, see FERMAT'S LAST THEOREM [V.10].

III.64 Optimization and Lagrange Multipliers
Keith Ball

1 Optimization

Soon after being introduced to calculus, most students learn of its application to *optimization*: that is, to the problem of finding the largest or smallest value of a given differentiable function, which is usually referred to as the *objective function*. A very helpful observation is that if f is an objective function that is maximized or minimized at x, then the tangent to the graph at the point $(x, f(x))$ will be horizontal, since otherwise we can find some value x' close to x for which $f(x')$ is higher. This means that we can narrow down the search for the maximum and minimum values of f by looking just at the values of $f(x)$ for which $f'(x) = 0$.

Now suppose that we have an objective function of more than one variable, such as, for example, the function

$$F(x, y) = 2x + 10y - x^2 + 2xy - 3y^2.$$

The "graph" of F is obtained by plotting the values $F(x, y)$ of F as heights above the corresponding points (x, y) of the plane, so now it is a surface instead of a curve. A smooth surface possesses not a tangent *line* at each point, but a tangent *plane*. If F has a maximum value, it will occur at a point where the tangent plane is horizontal.

The tangent plane at each point (x, y) is the graph of the linear function that best approximates F near (x, y). For small values of h and k, $F(x + h, y + k)$ will be approximately equal to $F(x, y)$ plus a function of the form

$$(h, k) \mapsto ah + bk,$$

that is, $F(x, y)$ plus a linear function of h and k. As explained in SOME FUNDAMENTAL MATHEMATICAL DEFINITIONS [I.3 §5.3], the derivative of F at (x, y) is this linear map. The map can be represented by the pair of numbers (a, b), which can in turn be thought of as a vector in $\mathbb{R}^2$. This derivative vector is usually called the *gradient* of the function F at the point (x, y) and is written $\nabla F(x, y)$. In vector notation (writing $\boldsymbol{x}$ for (x, y) and $\boldsymbol{h}$ for (h, k)), the approximation to F near (x, y) is

$$F(\boldsymbol{x} + \boldsymbol{h}) \approx F(\boldsymbol{x}) + \boldsymbol{h} \cdot \nabla F. \tag{1}$$

Thus, ∇F points in the direction in which F increases most rapidly if you start at $\boldsymbol{x}$, and the magnitude of ∇F is the slope of the "graph" of F in this direction.

The components a and b of the gradient can be calculated using partial differentiation. The number a tells us how quickly $F(x, y)$ changes as we vary x while keeping y fixed: so to find a, we differentiate $F(x, y) = 2x + 10y - x^2 + 2xy - 3y^2$ with respect to x, treating y as a constant. In this case we get the partial derivative

$$a = \frac{\partial F(x, y)}{\partial x} = 2 - 2x + 2y.$$

Similarly,

$$b = \frac{\partial F(x, y)}{\partial y} = 10 + 2x - 6y.$$

Now, if we want to locate points where the tangent plane is horizontal, then we want to find the points at which the gradient is zero: that is, the points at which the vector (a, b) is the zero vector. So we solve the pair of simultaneous equations

$$2 - 2x + 2y = 0,$$
$$10 + 2x - 6y = 0$$

to get $x = 4$, $y = 3$. Thus, the only candidate for the maximum is the point $(4, 3)$, where F takes the value 19. It can be checked that 19 is indeed the maximum value of F.

2 The Gradient and Contours

One of the most common ways of representing surfaces (landscapes on maps, for example) is by means of *contour lines*, or curves of constant height. In the xy-plane, we plot several curves of the form $F(x, y) = V$, for various "representative" values of V. For the function we considered earlier,

$$F(x, y) = 2x + 10y - x^2 + 2xy - 3y^2,$$

the values 0, 8, 14, 18, 19 yield the contour plot shown in figure 1. The 14 contour, for example, contains all the points at which the surface has height 14. The figure indicates that this particular surface is an elliptical hump whose peak occurs at $(4, 3)$ and has height 19.

There is a simple geometrical relationship between the contour lines and the gradient vector. The vector equation (1) shows that the direction $\boldsymbol{h}$ in which F is instantaneously constant is the direction which makes the scalar product $\boldsymbol{h} \cdot \nabla F$ equal to 0: the direction perpendicular to ∇F. At each point, the gradient vector is perpendicular to the contour through that point. This fact underlies the method of Lagrange multipliers that we shall discuss in the next section.

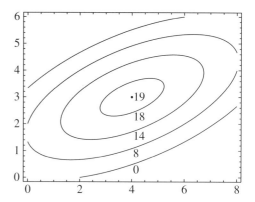

Figure 1 A contour plot.

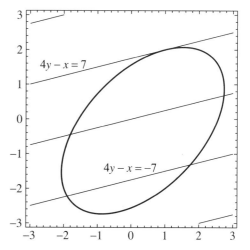

Figure 2 Constrained optimization.

3 Constrained Optimization and Lagrange Multipliers

It often happens that we are interested in the maximum or minimum value of an objective function that depends upon several variables whose values are constrained to satisfy certain equations or inequalities. Consider, for example, the following problem.

Find the maximum value of

$$F(x, y) = 4y - x$$

over all pairs (x, y) satisfying the constraint

$$G(x, y) = x^2 - xy + y^2 - x + y - 4 = 0. \quad (2)$$

Figure 2 shows the curve in the xy-plane defined by $G(x, y) = 0$ (an ellipse), and also a number of contour lines of the function $4y - x$. Our aim is to find the largest that $4y - x$ can be if (x, y) is a point on the curve. So we want to find the largest value of V for which the corresponding contour $4y - x = V$ contains a point on the curve. The value of V increases as the lines move up the diagram, and the uppermost line that touches the curve is the one labeled $4y - x = 7$. So the maximum value we are looking for is 7, and it occurs at the point where the line $4y - x = 7$ touches the curve. It is easy to check that this point is $(1, 2)$.

How could we locate this point algebraically, rather than by drawing? The important thing to notice is that the optimizing line is tangent to the curve: the line and the curve are parallel at their common point. The *line* was chosen to be a contour of the function F. The *curve* is also a contour: the 0 contour of G. From the discussion in the previous section we know that these contours are perpendicular to the gradients of F and G, respectively (at the point in question). So the two gradient vectors are parallel to one another and are therefore multiples of one another: $\nabla F = \lambda \nabla G$, say.

We thus have a way to hunt for solutions to the constrained optimization problem

maximize $F(x, y)$ subject to $G(x, y) = 0$.

We look for a point (x, y) and a number λ such that

$$\nabla F(x, y) = \lambda \nabla G(x, y) \quad \text{and} \quad G(x, y) = 0. \quad (3)$$

For our example (2), the gradient equation gives two partial derivative equations,

$$-1 = \lambda(2x - y - 1), \qquad 4 = \lambda(-x + 2y + 1),$$

from which we conclude that

$$x = \frac{2 + \lambda}{3\lambda}, \qquad y = \frac{7 - \lambda}{3\lambda}. \quad (4)$$

Substituting these values into the equation $G(x, y) = 0$, we obtain

$$\frac{13(1 - \lambda^2)}{3\lambda^2} = 0,$$

which has two solutions: $\lambda = 1$ and $\lambda = -1$. If we substitute $\lambda = 1$ into (4), we get the point $(1, 2)$ where F is at its maximum. (Taking $\lambda = -1$ yields the minimum.)

The number λ that we introduced to solve the problem is called a *Lagrange multiplier*. It is possible to reformulate the problem by defining the *Lagrangian*

$$\mathcal{L}(x, y, \lambda) = F(x, y) - \lambda G(x, y)$$

and then condensing the equations (3) into a single equation

$$\nabla \mathcal{L} = 0.$$

The reason this works is that if we differentiate $\mathcal{L}$ with respect to λ, then we obtain $G(x, y)$, so requiring this partial derivative to be zero is equivalent to requiring $G(x, y)$ to be zero. And asking for the other two partial derivatives to be zero is equivalent to requiring that $\nabla F = \lambda \nabla G$. The remarkable fact about this reformulation is that it has turned a *constrained* optimization problem involving x and y into an *unconstrained* problem involving x, y, and λ.

4 The General Method of Lagrange Multipliers

In real problems we may wish to optimize a function F of many variables $x_1, \ldots, x_n$ under many constraints $G_1(x_1, \ldots, x_n) = 0$, $G_2(x_1, \ldots, x_n) = 0$, $\ldots$, $G_m(x_1, \ldots, x_n) = 0$. In this case we introduce a Lagrange multiplier for each constraint and define the Lagrangian $\mathcal{L}$ by the formula

$$\mathcal{L}(x_1, \ldots, x_n, \lambda_1, \ldots, \lambda_m)$$
$$= F(x_1, \ldots, x_n) - \sum_1^m \lambda_i G_i(x_1, \ldots, x_n).$$

The partial derivative of $\mathcal{L}$ with respect to λ_i is zero if and only if $G_i(x_1, \ldots, x_n) = 0$. And the partial derivatives with respect to the x_i will all be zero if and only if $\nabla F = \sum_1^m \lambda_i \nabla G_i$. This tells us that any direction that is perpendicular to all the gradients ∇G_i (and therefore lies in all their "contour hypersurfaces") will be perpendicular to the gradient ∇F as well, so we cannot find a direction in which F increases while all the constraints are satisfied.

Problems of this kind occur frequently in economics, where the objective function F is a cost (which we are probably trying to minimize), and the constraints force us to allocate spending among different items so as to satisfy certain overall demands. For instance, we might want to minimize the total cost of supplies of various different foodstuffs that between them had to satisfy various nutritional demands. In this case, the Lagrange multipliers have an interpretation as "notional prices." As we have just seen, at the optimum point we have an equation of the form $\nabla F = \sum_1^m \lambda_i \nabla G_i$. This tells us how much F will vary as we vary the G_i by small amounts: that is, it tells us the costs associated with increasing the various demands.

For a further use of Lagrange multipliers, see THE MATHEMATICS OF TRAFFIC IN NETWORKS [VII.4].

III.65 Orbifolds

If you take a QUOTIENT [I.3 §3.3] of the plane $\mathbb{R}^2$ by a group of symmetries, then you may obtain a MANIFOLD

[I.3 §6.9]. For instance, if the group consists of all translations by an integer vector, then two points (x, y) and (z, w) are equivalent if and only if $z - x$ and $w - y$ are both integers, and the quotient space is a torus. However, if you take instead the group of all rotations about the origin through a multiple of $\pi/3$, then every point apart from the origin is equivalent to exactly five others, while the origin is equivalent only to itself. The result in this case is not a manifold, because the exceptional behavior at the origin results in a singularity. However, it is a well-understood kind of singularity. An *orbifold* is, roughly speaking, just like a manifold, except that whereas manifolds are locally like $\mathbb{R}^n$, orbifolds are locally like quotients of $\mathbb{R}^n$ by groups of symmetries, and can therefore have a few singularities. See ALGEBRAIC GEOMETRY [IV.4 §7] and also MIRROR SYMMETRY [IV.16 §7].

III.66   Ordinals

Loosely speaking, the ordinals are what we get if, starting with 0, we use the following two procedures. We can add 1 to whatever we have, and we can "collect together" (or "take the limit of") whatever we have so far. So from 0 we would get 1, then 2, then 3, and so on. After all of those, we could take their "limit" (i.e., the limit of $0, 1, 2, 3, \ldots$), which is called ω. Then we can add 1, obtaining $\omega + 1$, then $\omega + 2$, and so on. And then we can take the limit of all of *those*, to obtain an ordinal we could write as $\omega + \omega$. And so on. Note that this final "and so on" carries quite a lot inside it. For example, the ordinals do not just consist of finite sums of ωs and natural numbers, since we can take the limit of $\omega, \omega + \omega, \omega + \omega + \omega, \ldots$, which we might call ω^2.

Ordinals arise in two ways (which turn out to be closely related). First, they give a measure of the "size" of a *well-ordering*. A well-ordering on a set is an ordering in which every (nonempty) subset has a least element. For example, the set $\{\frac{1}{2}, \frac{2}{3}, \frac{3}{4}, \ldots\} \cup \{\frac{3}{2}, \frac{5}{3}, \frac{7}{4}, \ldots\}$ in the reals is well-ordered, while the set $\{\ldots, \frac{1}{4}, \frac{1}{3}, \frac{1}{2}\}$ is not. The first set is *order isomorphic* to the ordinals less than $\omega + \omega$, meaning that there is a bijection that preserves the order. So one says that that set has *order type* $\omega + \omega$.

Ordinals also commonly arise when one wishes to index transfinite processes. Here "transfinite" means "going beyond finite." As an example, suppose that we wish to "count, in increasing order" the elements of the well-ordered set above. How would we do it? We would start with $\frac{1}{2}$, then $\frac{2}{3}$, then $\frac{3}{4}$, and so on. But, at the end of all time, we would still not have reached elements like $\frac{3}{2}$ or $\frac{5}{3}$. So we would start again: "at time ω" we would count $\frac{3}{2}$, then at time $\omega + 1$ we would count $\frac{5}{3}$, and so on. Thus our counting is complete by time $\omega + \omega$.

For a more detailed explanation of ordinals, including more examples and more on how they arise in mathematics, see SET THEORY [IV.22 §2].

III.67   The Peano Axioms

Everyone knows what the natural numbers are: 0, 1, 2, 3, and so on. But how would we make that "and so on" precise? Can we look at the way that we reason about natural numbers and isolate a few basic principles, or *axioms*, whose consequences do complete justice to our intuitive picture of what the natural numbers should be? To put it another way, when we are proving something about the natural numbers, what assumptions do we need in order to get started?

To answer this question, let us strip things down to the bare minimum: we have an object called 0, and an operation s, called the *successor function*, which we think of intuitively as "adding 1." In this pared-down language, we would like to say two things: that all the numbers $0, s(0), s(s(0)), \ldots$ are distinct natural numbers, and that there are no others.

One simple way is to use the following two axioms. The first says that 0 is not a successor:

(i) For all x, $s(x) \neq 0$.

The second says that distinct elements stay distinct when you take their successors:

(ii) For all x and y, if $x \neq y$, then $s(x) \neq s(y)$.

Note that this implies, for example, that $s(s(s(0))) \neq s(0)$, for if they *were* equal, then, from rule (ii), we could deduce that $s(s(0)) = 0$, contradicting rule (i).

Now, how can we say that there are no other natural numbers? One would like to say that, for every x, either $x = 0$ or $x = s(0)$ or $x = s(s(0))$ or $\cdots$, but that is an infinitely long statement, and those are definitely not allowed. After the failure of that very natural attempt, one might guess that there is no way to achieve the goal, but in fact there is a brilliant solution: induction. Here is an axiom that expresses the principle of induction.

(iii) Let A be any subset of the natural numbers with the following properties: $0 \in A$, and $s(x) \in A$ whenever $x \in A$. Then A must be the set of all natural numbers.

Note that this does express our intuitive idea that there are no "extra" natural numbers, since we can take A to be the set of all the numbers $0, s(0), s(s(0)), \ldots$ that were on our list.

Rules (i), (ii), and (iii) are called the *Peano axioms* for the natural numbers. As explained above, they "characterize" the natural numbers, in the sense that all reasoning about the natural numbers may be reduced or rewritten in such a way that the only assumptions one needs are the Peano axioms.

There is a related system used in logic, called the *first-order Peano axioms*. The idea here is that we want to express the Peano axioms in the language of FIRST-ORDER LOGIC [IV.23 §1]. This means that we are allowed variables (that are interpreted as ranging over the natural numbers), as well as the symbols 0 and s, logical connectives, and the like, but nothing more: so there is no "member of" symbol, and no sets are allowed. (However, for technical reasons one does allow symbols for "plus" and "times.")

To give an idea of what is allowed and what is not, consider the statements "there are infinitely many perfect squares" and "every infinite set of positive integers contains either infinitely many odd numbers or infinitely many even numbers." With a little effort, we can express the first of these statements in first-order logic, as follows:

$$(\forall m)(\exists n)(\exists x) \quad xx = m + n.$$

In words, this says that for every m you can find a perfect square of the form $m + n$ (which is how we express the fact that it is larger than m). However, in order to express the second statement, we find ourselves wanting to write $(\forall A)$, where A ranges over all possible *subsets* of the natural numbers, rather than all possible *elements*: this is the main thing that is not allowed in first-order logic.

By this criterion, rules (i) and (ii) are fine, but rule (iii) is not. Instead, we have to use an "axiom scheme," which is an infinite set of axioms, one for each first-order statement $p(x)$. So our version of rule (iii) is this: for each statement $p(x)$, we have an axiom saying that if $p(0)$ is true, and $p(x)$ implies $p(s(x))$, then $p(x)$ is true for all x.

Note that these axioms do not have the full strength of the usual Peano axioms. For instance, there are only countably many possible formulas $p(x)$, whereas there are uncountably many sets A. It turns out that in fact there are "nonstandard" models of these axioms, meaning structures other than the natural numbers that satisfy the axioms of first-order Peano arithmetic.

Actually, one also allows *parameters* in the statements $p(x)$; for example, $p(x)$ could be the statement "there exists z with $x = y + z$," which would correspond to the set of all natural numbers greater than or equal to y, and would therefore depend on y. And one also adds some axioms saying how plus and times behave (for example, commutativity of addition). This whole collection of axioms is known as Peano arithmetic, or PA for short.

See MODEL THEORY [IV.23] for more on some of the topics discussed in this article.

III.68 Permutation Groups
Martin W. Liebeck

Let S be a set. A *permutation* of S is a function from S to S that is both injective and surjective—in other words, a function that "rearranges" the elements of S. For example, if $S = \{1, 2, 3\}$, then the function $a : S \to S$ that sends 1 to 3, 2 to 1, and 3 to 2 is a permutation of S; so is the function b that sends 1 to 3, 2 to 2, and 3 to 1; whereas the function c that sends 1 to 3, 2 to 1, and 3 to 1 is not a permutation. An example of a permutation of the set of real numbers $\mathbb{R}$ is the function $x \mapsto 8 - 2x$.

From the point of view of finite group theory, the most important permutations to study are those of the set $I_n = \{1, 2, \ldots, n\}$, where n is a positive integer. Let S_n denote the set of all permutations of I_n. So, for example, the permutations a and b defined in the previous paragraph lie in S_3. To count how many permutations there are altogether in S_n, observe that, for a permutation $f : I_n \to I_n$, there are n choices for $f(1)$, then $n - 1$ choices for $f(2)$ (we can choose anything different from $f(1)$), then $n - 2$ for $f(3)$, and so on, until we have just 1 choice for $f(n)$. Therefore the total number of permutations in S_n is $n(n-1)(n-2) \cdots 1 = n!$.

If f and g are permutations of a set S, their composition $f \circ g$ is defined by $f \circ g(s) = f(g(s))$ for all $s \in S$, and it is quite easy to see that $f \circ g$ is also a permutation of S. It is usual to drop the "$\circ$" symbol and write just fg instead of $f \circ g$. For example, if $a, b \in S_3$ are as in the first paragraph, then $ab \in S_3$ sends 1 to 2, 2 to 1, and 3 to 3, while ba sends 1 to 1, 2 to 3, and 3 to 2. Notice that $ab \neq ba$.

For any set S, the *identity* function $\iota : S \to S$, defined by $\iota(s) = s$ for all $s \in S$, is a permutation of S; and if f

is a permutation of S, then there is an *inverse* permutation f^{-1} that sends everything back to where it came from and therefore satisfies $ff^{-1} = f^{-1}f = \iota$. For example, the inverse of the above permutation $a \in S_3$ is the permutation that sends 1 to 2, 2 to 3, and 3 to 1. Also, for any permutations f, g, h of S, we have $f(gh) = (fg)h$, since both sides send any $s \in S$ to $f(g(h(s)))$.

Thus, the set of all permutations of S, together with the BINARY OPERATION [I.2 §2.4] of composition, satisfies the axioms for a GROUP [I.3 §2.1]. In particular, S_n is a finite group of size $n!$, known as the *symmetric group of degree n*.

There is a neat way of representing permutations succinctly, known as the *cycle notation*. It is best explained with an example. Let $d \in S_6$ be the permutation $1 \mapsto 3$, $2 \mapsto 5$, $3 \mapsto 6$, $4 \mapsto 4$, $5 \mapsto 2$, $6 \mapsto 1$. We can represent this more economically by writing $1 \mapsto 3 \mapsto 6 \mapsto 1$, $2 \mapsto 5 \mapsto 2$, and $4 \mapsto 4$. We say the symbols 1, 3, 6 form a *cycle* of d (of length 3); similarly, 2, 5 form a cycle of length 2, and 4 a cycle of length 1. We then compress our notation even further and write $d = (1\,3\,6)\,(2\,5)\,(4)$, indicating that each number 1, 3, 6 in the first cycle is sent to the next one, except for the last which is sent back to the first, and likewise for the second and third cycles. This is the cycle notation for d; notice that the cycles have no symbols in common—they are called *disjoint* cycles. It is not too hard to see that every permutation in S_n can be expressed as a product of disjoint cycles; this is what we mean by the cycle notation for a permutation. For example, in cycle notation, the six permutations of S_3 are ι, $(1\,2)\,(3)$, $(1\,3)\,(2)$, $(2\,3)\,(1)$, $(1\,2\,3)$, and $(1\,3\,2)$. (The permutations a and b in the first paragraph are $(1\,3\,2)$ and $(1\,3)\,(2)$, respectively.) You might like to while away a few minutes by working out the multiplication table of S_3.

The *cycle-shape* of a permutation g is the sequence of numbers we get by writing down the lengths of the disjoint cycles in the cycle notation for g, in decreasing order. For example, the cycle-shape of the permutation $(1\,6\,3)\,(2\,4)\,(5\,8)\,(7)\,(9)$ in S_9 is $(3, 2, 2, 1, 1)$, or more succinctly $(3, 2^2, 1^2)$.

One can define the *powers* of a permutation $f \in S_n$ in a natural way—namely, $f^1 = f$, $f^2 = ff$, $f^3 = f^2f$, and so on. For example, if $e = (1\,2\,3\,4) \in S_4$, then $e^2 = (1\,3)\,(2\,4)$, $e^3 = (1\,4\,3\,2)$, and $e^4 = \iota$. The *order* of a permutation $f \in S_n$ is defined to be the smallest positive integer r such that $f^r = \iota$: that is, the smallest number of times we have to do f to send everything back to where it came from. So the order of the 4-cycle

e above is 4. In general, the order of an r-cycle (i.e., a cycle of length r) is equal to r, and the order of a permutation in cycle notation is equal to the least common multiple of the lengths of the (disjoint) cycles.

It is often useful to be able to work out the order of a permutation. Here is one such instance. Suppose we shuffle a pack of eight cards in the following way: the pack is divided into two equal parts and then "interlaced," so that if the original order was $1, 2, 3, 4, \ldots$, the new order is $1, 5, 2, 6, \ldots$. How many times must this shuffle be repeated before the cards are again in the original order? Well, the shuffle gives the permutation of the eight card positions sending 1 to 1, 2 to 5, 3 to 2, 4 to 6, and so on, which in cycle notation is $(1)\,(2\,5\,3)\,(4\,6\,7)\,(8)$. This has order 3, so the cards return to their original order after three shuffles. Things get quite interesting if we consider the same problem for different numbers of cards—you might like to try it yourself with fifty-two cards, for instance.

There is one slightly more subtle aspect of permutations which is important for group theory: namely, the theory of *even* and *odd* permutations. Again, this is best illustrated by example. Take $n = 3$, and let x_1, x_2, x_3 be three variables. Let us think of the permutations in S_3 as moving these variables around rather than the numbers 1, 2, and 3. So, for instance, we shall take the permutation $(1\,3\,2)$ to send x_1 to x_3, x_2 to x_1, and x_3 to x_2. Now let Δ be the expression $\Delta = (x_1 - x_2)(x_1 - x_3)(x_2 - x_3)$. We can apply permutations in S_3 to Δ in an obvious way: for example, $(1\,2\,3)$ sends Δ to $(x_2 - x_3)(x_2 - x_1)(x_3 - x_1)$. Notice that this is just the expression for Δ with two of the brackets, $(x_1 - x_2)$ and $(x_1 - x_3)$, reversed. So $(1\,2\,3)$ sends Δ to Δ. However, if we apply $(1\,2)\,(3)$ to Δ, we get $(x_2 - x_1)(x_2 - x_3)(x_1 - x_3) = -\Delta$. You can see that each permutation in S_3 sends Δ to either $+\Delta$ or $-\Delta$. Call those permutations that send Δ to $+\Delta$ *even permutations* and those that send Δ to $-\Delta$ *odd permutations*. Check that ι, $(1\,2\,3)$, and $(1\,3\,2)$ are even, while $(1\,2)\,(3)$, $(1\,3)\,(2)$, and $(2\,3)\,(1)$ are odd.

The definition of even and odd permutations for general n is very similar to this example. Let $x_1, \ldots, x_n$ be variables, and regard the permutations in S_n as moving these variables around rather than the symbols $1, 2, \ldots, n$. Define Δ to be the product of all $x_i - x_j$ for $i < j$. Just as in the example, we can apply any permutation $g \in S_n$ to Δ, and the result will be either $+\Delta$ or $-\Delta$. Define the *signature* of g to be the number $\text{sgn}(g) \in \{+1, -1\}$ such that $g(\Delta) = \text{sgn}(g)\Delta$. This defines the signature function $\text{sgn} : S_n \to \{+1, -1\}$.

Then a permutation $g \in S_n$ is *even* if $\mathrm{sgn}(g) = +1$, and is *odd* if $\mathrm{sgn}(g) = -1$.

It follows easily from the definition that

$$\mathrm{sgn}(gh) = \mathrm{sgn}(g)\,\mathrm{sgn}(h)$$

for any $g, h \in S_n$, and also that the signature of any 2-cycle is -1. Since an r-cycle $(a_1\, a_2\, \cdots\, a_r)$ can be expressed as a product $(a_1\, a_r)(a_1\, a_{r-1}) \cdots (a_1\, a_2)$ of 2-cycles, the signature of the r-cycle is $(-1)^{r-1}$. Hence, if $g \in S_n$ has cycle-shape $(r_1, r_2, \ldots, r_k)$, then

$$\mathrm{sgn}(g) = (-1)^{r_1 - 1}(-1)^{r_2 - 1} \cdots (-1)^{r_k - 1}.$$

This makes it easy to work out the signature of any permutation. For example, the even permutations in S_5 are those that have cycle-shape (1^5), $(2^2, 1)$, $(3, 1^2)$, or (5). If you count these, you will find that there are sixty even permutations in S_5 altogether, which is exactly half of the total of $5! = 120$ permutations in S_5. In general, the number of even permutations in S_n is $\frac{1}{2}n!$.

So what is the point of this complicated definition? The answer is that the set of all even permutations in S_n forms a subgroup of size $\frac{1}{2}n!$, known as the *alternating group of degree n*, and written as A_n. The alternating groups are very important examples of finite groups, because of the fact that, for $n \geqslant 5$, A_n is a *simple group*—that is, its only NORMAL SUBGROUPS [I.3 §3.3] are the identity subgroup and A_n itself (see THE CLASSIFICATION OF FINITE SIMPLE GROUPS [V.7]). For example, A_5 is a simple group of size 60, and in fact is the smallest non-Abelian finite simple group.

III.69 Phase Transitions

If you heat up a block of ice, then it turns into water. This very familiar phenomenon is actually rather mysterious, because it shows that the properties of the chemical H_2O do not depend continuously on temperature: the block of ice goes straight from a solid to a liquid, rather than doing so by a process of gradual softening.

This is an example of a *phase transition*. Phase transitions tend to occur in systems that involve a large number of particles with "local" interactions—that is, where the behavior of one particle is directly influenced only by the particles in its immediate vicinity.

Such systems can be modeled mathematically, and the study of these models belongs to the area known as *statistical physics*. For further discussion of such models, see PROBABILISTIC MODELS OF CRITICAL PHENOMENA [IV.25].

III.70 π

What makes one number more fundamental and important, mathematically speaking, than another? Why, for instance, would almost everybody agree that 2 is more important than $\frac{43}{32}$? One possible answer is that what really matters about a number is its properties, and in particular any interesting properties it might have that distinguish it from all other numbers. Of course, we now have to decide what counts as an interesting property: for example, why do we not regard it as interesting that $\frac{43}{32}$ is the only number that gives you $\frac{43}{16}$ when you double it? An obvious reason is that there is an analogous property for *every* number x you might care to choose: x is the only number that gives you $2x$ when you double it. By contrast, the property "is the smallest prime number" does not mention any specific number and is easily stated in terms of a concept, that of "prime number," whose importance is itself easy to explain. This property must apply to exactly one number, so it is likely that that number will have an important part to play in mathematics, and indeed it does. (As it happens, $\frac{43}{32}$ is conjectured to be an important critical exponent in statistical physics, which means that it *can* be singled out as an interesting number, though still nothing like as fundamental as 2.)

Everybody agrees that π is one of the most important numbers in mathematics, and it is easy to justify this assessment by the criterion of the previous paragraph, because π has an abundance of properties—so many that when π appears unexpectedly in a calculation, one is not unduly surprised. For example, the following is a famous theorem of EULER [VI.19]:

$$\sum_{n=1}^{\infty} \frac{1}{n^2} = 1 + \frac{1}{4} + \frac{1}{9} + \frac{1}{16} + \frac{1}{25} + \cdots = \frac{\pi^2}{6}.$$

What on earth, one might wonder, has π to do with adding up reciprocals of squares? This is a perfectly legitimate question, but the idea that there could in principle be a connection is not, to an experienced mathematician, a surprise. A very common way to prove mathematical identities is to show that the two sides of the identity are different ways of evaluating the same quantity. In this case, one can use a basic fact from FOURIER ANALYSIS [III.27], known as *Plancherel's identity*, which states the following. If $f : \mathbb{R} \to \mathbb{C}$ is a periodic function with period 2π, and for every integer n (positive or negative) we define its nth Fourier coefficient a_n by the formula

$$a_n = \frac{1}{2\pi} \int_{-\pi}^{\pi} f(x)e^{inx}\, dx,$$

then

$$\frac{1}{2\pi} \int_{-\pi}^{\pi} |f(x)|^2\, dx = \sum_{n=-\infty}^{\infty} |a_n|^2.$$

If you now take as f the function that is 1 whenever x is between $(2n - \frac{1}{2})\pi$ and $(2n + \frac{1}{2})\pi$ for some integer n, and 0 otherwise, then you find that the left-hand side works out as $\frac{1}{2}$. You also find, after a small calculation, that $|a_n|^2 = 1/(\pi n)^2$ when n is odd, that $|a_0|^2 = \frac{1}{4}$, and that $|a_n|^2 = 0$ whenever n is even and nonzero. Therefore,

$$\frac{1}{2} = \frac{1}{4} + \frac{1}{\pi^2} \sum_{n\,\text{odd}} \frac{1}{n^2}.$$

Bearing in mind that $n^2 = (-n)^2$, we can deduce easily that

$$\frac{\pi^2}{8} = 1 + \frac{1}{3^2} + \frac{1}{5^2} + \frac{1}{7^2} + \cdots.$$

This closely resembles the identity we were trying to prove, which we can get by noticing that the right-hand side is equal to $\sum_n 1/n^2 - \sum_n 1/(2n)^2$, which is three quarters of $\sum_n 1/n^2$. Therefore, $\sum_n 1/n^2 = \pi^2/6$.

Now we have a reason for the appearance of π: it comes up in the formula for the Fourier coefficients. What is more, its appearance there can be explained as well. A periodic function on $\mathbb{R}$ is more naturally thought of as a function defined on the unit circle. The Fourier coefficient a_n is a certain average defined on the unit circle, so we have to divide by the length of the circle, which is 2π.

What, then, is π? Well, we have just seen what is perhaps the most elementary definition: it is the ratio of the circumference of a circle to its diameter. But what makes π so interesting is that it has many different defining properties. Here are a few more of them.

(i) Define a function $\sin x$ to be equal to the sum of the power series

$$x - \frac{x^3}{3!} + \frac{x^5}{5!} - \cdots.$$

Then π is the smallest positive number x such that $\sin x = 0$. (For more on $\sin x$, see TRIGONOMETRIC FUNCTIONS [III.92].)

(ii) $\pi = \int_{-1}^{1} \frac{dx}{\sqrt{1-x^2}}$.

(iii) $\frac{\pi}{2} = \int_{-1}^{1} \sqrt{1-x^2}\, dx$.

(iv) $\frac{\pi}{4} = \left(1 - \frac{1}{3} + \frac{1}{5} - \frac{1}{7} + \frac{1}{9} - \cdots\right)$.

(v) $\sqrt{2\pi} = \int_{-\infty}^{\infty} e^{-x^2/2}\, dx$.

(vi) $\pi = \sum_{k=0}^{\infty} \frac{1}{16^k} \left(\frac{4}{8k+1} - \frac{2}{8k+4} - \frac{1}{8k+5} - \frac{1}{8k+6}\right)$.

The integrals on the right-hand sides of the second and third properties above are expressions for half the circumference of the unit circle and half its area, respectively. So those definitions are analytical expressions of the geometrical facts that a unit circle has circumference 2π and area π, respectively.

The fifth property tells us what constant to put in front of $e^{-x^2/2}$ to make it into the famous NORMAL DISTRIBUTION [III.71 §5]. (Why should π come into it? One can give several reasons. One is that the function $e^{-x^2/2}$ has a special role in Fourier analysis, and so does π. Another fundamental property of $e^{-x^2/2}$ is that the function $f(x,y) = e^{-(x^2+y^2)/2}$ is *rotationally invariant*, and rotations involve circles, which involve π.)

The last formula above is a remarkable recent discovery of David Bailey, Peter Borwein, and Simon Plouffe. The presence of the factor $1/16^k$ leads to a way of calculating hexadecimal digits of π (that is, digits to base 16), without needing to work out all the earlier digits first. It has been used to work out digits that are astonishingly far along the hexadecimal expansion: for example, it is known that the trillionth hexadecimal digit is 8. (See MATHEMATICS: AN EXPERIMENTAL SCIENCE [VIII.5 §7] for a further discussion of this formula.)

A fact that seems paradoxical to many nonmathematicians is that a number as natural as π turns out to be IRRATIONAL, and also TRANSCENDENTAL [III.41]. However, this is not surprising at all: the defining properties of π are simple, but they do not lead to solutions of polynomial equations, so it would be extraordinary if π were *not* transcendental. Similarly, it would be a major surprise if one could find any pattern in the decimal digits of π. Indeed, π is conjectured to be *normal to base* 10, meaning that every sequence of digits occurs with about the frequency you would expect: for example, if you look at pairs of consecutive digits, then you expect 35 to occur about a hundredth of the time. However, this conjecture seems to be very hard, and it has not even been proved that the decimal expansion of π contains all the digits from 0 to 9 infinitely often.

III.71   Probability Distributions
James Norris

1   Discrete Distributions

When we toss a coin, we have no idea whether it will land heads or tails. However, there is a different sense in which the behavior of the coin is highly predictable: if it is tossed many times, then the proportion of heads is very likely to be close to $\frac{1}{2}$.

In order to study this phenomenon mathematically, we need to model it, and this is done by defining a *sample space*, which represents the set of possible outcomes, and a *probability distribution* on that space, which tells you their probabilities. In the case of a coin, the natural sample space is the set $\{H, T\}$, and the obvious distribution assigns the number $\frac{1}{2}$ to each element. Alternatively, since we are interested in the number of heads, we could use the set $\{0, 1\}$ instead: after one toss, there is a probability of $\frac{1}{2}$ that the number of heads is 0 and a probability of $\frac{1}{2}$ that it is 1. More generally, a (discrete) sample space is simply a set Ω, and a probability distribution on Ω is a way of assigning a nonnegative real number to each element of Ω in such a way that the sum of all these numbers is 1. The number assigned to a particular element of Ω is then interpreted as the probability that some corresponding outcome will occur, the total probability being 1.

If Ω is a set of size n, then the *uniform distribution* on Ω is the probability distribution that assigns a probability of $1/n$ to each element of Ω. However, it is often more appropriate to assign different probabilities to different outcomes. For example, given any real number p between 0 and 1, the *Bernoulli distribution with parameter p* on the set $\{0, 1\}$ is the distribution that assigns the number p to 1 and $1 - p$ to 0. This can be used to model the toss of a biased coin.

Suppose now that we toss an unbiased coin n times. If we are interested in the outcome of every toss, then we would choose the sample space consisting of all possible sequences of 0s and 1s of length n. For instance, if $n = 5$, a typical element of the sample space is 01101. (This particular element represents the outcome tails, heads, heads, tails, heads, in that order.) Since there are 2^n such sequences and they are all equally likely, the appropriate distribution on this space will be the uniform one, which assigns a probability of $1/2^n$ to each sequence.

But what if we are interested not in the particular sequence of heads and tails but just in the *total number of heads*? In that case, we could take as our sample space the set $\{0, 1, 2, \ldots, n\}$. The probability that the total number of heads is k is 2^{-n} times the number of sequences of 0s and 1s that contain exactly k 1s. This number is

$$\binom{n}{k} = \frac{n!}{k!(n-k)!},$$

so the probability we assign to k is $p_k = \binom{n}{k}2^{-n}$.

More generally, for a sequence of n independent experiments, each with the same probability p of success, the probability of a given sequence of k successes and $n - k$ failures is $p^k(1-p)^{n-k}$. So, the probability of having exactly k successes is $p_k = \binom{n}{k}p^k(1-p)^{n-k}$. This is called the *binomial distribution* with parameters n and p. It models the number of heads if you toss a biased coin n times, for example.

Suppose we perform such experiments for as long as we need to in order to obtain one success. When k experiments are performed, the probability of getting $k - 1$ failures followed by a success is $p_k = (1-p)^{k-1}p$. Therefore, this formula gives us the distribution of the number of experiments up to the first success. It is called the *geometric distribution* of parameter p. In particular, the number of tosses of a fair coin needed to get the first head has a geometric distribution of parameter $\frac{1}{2}$. Notice that our sample space is now the set of all nonnegative integers—in particular, it is infinite. So in this case the condition that the probabilities add up to 1 is requiring that a certain infinite series (the series $\sum_{k=1}^{\infty} p_k$) converges to 1.

Now let us imagine a somewhat more complicated experiment. Suppose we have a radioactive source that occasionally emits an alpha particle. It is often reasonable to suppose that these emissions are independent and equally likely to occur at any time. If the average number of emissions per minute is λ, say, then what is the probability that during any given minute there will be k particles emitted?

One way to think about this question is to divide up the minute into n equal intervals, for some large n. If n is large enough, then the probability of two emissions occurring in the same interval is so small that it can be ignored, and therefore, since the average number of emissions per minute is λ, the probability of an emission during any given interval must be approximately λ/n. Let us call this number p. Since the emissions are independent, we can now regard the number of emissions as the number of successes when we do n trials, each with probability p of success. That is, we have the binomial distribution with parameters n and p, where $p = \lambda/n$.

Notice that as n gets larger, p gets smaller. Also, the approximations just made become better and better. It is therefore natural to let n tend to infinity and study the resulting "limiting distribution." It can be checked that, in the limit as $n \to \infty$, the binomial probabilities converge to $p_k = e^{-\lambda}\lambda^k/k!$. These numbers define a distribution on the set of all nonnegative integers, known as the *Poisson distribution* of parameter λ.

2 Probability Spaces

Suppose that I throw a dart at a dartboard. Not being very good at darts, I am not able to say very much about where the dart will land, but I can at least try to model it probabilistically. The obvious sample space to take consists of a circular disk, the points of which represent where the dart lands. However, now there is a problem: if I look at any particular point in the disk, the probability that the dart will land at precisely that point is zero. So how do I define a probability distribution?

A clue to the answer lies in the fact that it seems to be perfectly easy to make sense of a question such as "What is the probability that I will hit the bull's-eye?" In order to hit the bull's-eye, the dart has to land in a certain region of the board, and the probability of this happening does not have to be zero. It might, for instance, be equal to the area of the bull's-eye region divided by the total area of the board.

What we have just observed is that even if we cannot assign probabilities to individual *points* in the sample space, we can still hope to give probabilities to *subsets*. That is, if Ω is a sample space and A is a subset of Ω, we can try to assign a number $\mathbb{P}(A)$ between 0 and 1 to the set A. This represents the probability that the random outcome belongs to the set A, and can be thought of as something like a notion of "mass" for the set A.

For this to work, we need $\mathbb{P}(\Omega)$ to be 1 (since the probability of getting *something* in the sample space must be 1). Also, if A and B are disjoint subsets of Ω, then $\mathbb{P}(A \cup B)$ should be $\mathbb{P}(A) + \mathbb{P}(B)$. From this it follows that if $A_1, \ldots, A_n$ are all disjoint, then $\mathbb{P}(A_1 \cup \cdots \cup A_n)$ is equal to $\mathbb{P}(A_1) + \cdots + \mathbb{P}(A_n)$. Actually, it turns out to be important that this should be true not just for finite unions but even for COUNTABLY INFINITE [III.11] ones as well. (Related to this point is the fact that one does not attempt to define $\mathbb{P}(A)$ for *every* subset A of Ω but just for MEASURABLE SUBSETS [III.55]. For our purposes,

it is sufficient to regard $\mathbb{P}(A)$ as given whenever A is a set we can actually define.)

A *probability space* is a sample space Ω together with a function $\mathbb{P}$, defined on all "sensible" subsets A of Ω, that satisfies the conditions mentioned in the previous two paragraphs. The function $\mathbb{P}$ itself is known as a *probability measure* or *probability distribution*. The term *probability distribution* is often preferred when we specify $\mathbb{P}$ concretely.

3 Continuous Probability Distributions

There are three particularly important distributions defined on subsets of $\mathbb{R}$, of which two will be discussed in this section. The first is the *uniform distribution* on the interval $[0, 1]$. We would like to capture the idea that "all points in $[0, 1]$ are equally likely." In view of the problems mentioned above, how should we do this?

A good way is to take seriously the "mass" metaphor. Although we cannot calculate the mass of an object by adding up the masses of all the infinitely small points that make up the object, we can assign to those points a *density* and integrate it. That is exactly what we shall do here. We assign a *probability density* of 1 to each point in the interval $[0, 1]$. Then we determine the probability of a subinterval, $[\frac{1}{3}, \frac{1}{2}]$ say, by calculating the integral $\mathbb{P}([\frac{1}{3}, \frac{1}{2}]) = \int_{1/3}^{1/2} 1 \, dx = \frac{1}{6}$. More generally, the probability associated with an interval $[a, b]$ will just be its length $b - a$. The probability of a union of disjoint intervals will then be the sum of the lengths of those intervals, and so on.

This "continuous" uniform distribution sometimes arises naturally from requirements of symmetry, just like its discrete counterpart. It can also arise as a limiting distribution. For instance, suppose that a hermit lives deep in a cave, away from any clocks or sources of natural light, and that each "day" he spends lasts for a random length of time between twenty-three and twenty-five hours. To start with, he will have some idea of what the time is, and be able to make statements such as, "I'm having lunch now, so it's probably light outside," but after a few weeks of this regime, he will no longer have any idea: any outside time will be just as likely as any other.

Now let us look at a rather more interesting density function, which depends on the choice of a positive constant λ. Consider the density function $f(x) = \lambda e^{-\lambda x}$, defined on the set of all nonnegative real numbers. To work out the probability associated with an interval

$[a, b]$, we now calculate

$$\int_a^b f(x) \, dx = \int_a^b \lambda e^{-\lambda x} \, dx = e^{-\lambda a} - e^{-\lambda b}.$$

The resulting probability distribution is called the *exponential distribution with parameter* λ. The exponential distribution is appropriate if we are modeling the time T of a spontaneous event, such as the time it takes for a radioactive nucleus to decay, or for the next spam email to arrive. The reason for this is based on the assumption of *memorylessness*: for example, if we know that the nucleus remains intact at time s, the probability that it will remain intact until a later time $s + t$ is the same as the original probability that it would remain intact to time t. Let $G(t)$ represent the probability that the nucleus remains intact up to time t. Then the probability that it remains intact up to time $s + t$ given that it has remained intact up to time s is $G(s + t)/G(s)$, so this has to equal $G(t)$. Equivalently, $G(s + t) = G(s)G(t)$. The only decreasing functions that have this property are EXPONENTIAL FUNCTIONS [III.25], that is, functions of the form $G(t) = e^{-\lambda t}$ for some positive λ. Since $1 - G(t)$ represents the probability that the nucleus decays before time t, this should equal $\int_0^t f(x) \, dx$, from which it is easy to deduce that $f(x) = \lambda e^{-\lambda x}$.

We shall come to the third, and most important, distribution below.

4   Random Variables, Mean, and Variance

Given a probability space, an *event* is defined to be a (sufficiently nice) subset of that space. For example, if the probability space is the interval $[0, 1]$ with the uniform distribution, then the interval $[\frac{1}{2}, 1]$ is an event: it represents a randomly chosen number between 0 and 1 turning out to be at least $\frac{1}{2}$. It is often useful to think not just about random events, but also about random *numbers* associated with a probability space. For example, let us look once again at a sequence of tosses of a biased coin that has probability p of coming up heads. The natural sample space associated with this experiment is the set Ω of all sequences ω of 0s and 1s. Earlier, we showed that the probability of obtaining k heads is $p_k = \binom{n}{k} p^k (1 - p)^{n-k}$, and we described that as a distribution on the sample space $\{0, 1, 2, \dots, n\}$. However, it is in many ways more natural, and often far more convenient, to regard the original set Ω as the sample space and to define a function X from Ω to $\mathbb{R}$ to represent the number of heads: that is, $X(\omega)$ is the number of 1s in the sequence ω. We then write

$$\mathbb{P}(X = k) = p_k = \binom{n}{k} p^k (1 - p)^{n-k}.$$

A function like this is called a *random variable*. If X is a random variable and it takes values in a set Y, then the *distribution* of X is the function P defined on subsets of Y by the formula

$$P(A) = \mathbb{P}(X \in A) = \mathbb{P}(\{\omega \in \Omega : X(\omega) \in A\}).$$

It is not hard to see that P is indeed a probability distribution on Y.

For many purposes, it is enough to know the distribution of a random variable. However, the notion of a random variable defined on a sample space captures our intuition of a random quantity, and it allows us to ask further questions. For example, if we were to ask for the probability that there were k heads given that the first and last tosses had the same outcome, then the distribution of X would not provide the answer, whereas our richer model of regarding X as a function defined on sequences would do so. Furthermore, we can talk of *independent* random variables, $X_1, \dots, X_n$ say, meaning that the subset of Ω where $X_i(\omega) \in A_i$ for all i has probability given by the product $\mathbb{P}(X_1 \in A_1) \times \cdots \times \mathbb{P}(X_n \in A_n)$ for all possible sets of values A_i.

Associated with a random variable X are two important numbers that begin to characterize it, called the *mean* or *expectation* $\mathbb{E}(X)$ and the *variance* $\text{var}(X)$. Both these numbers are determined by the distribution of X. If X takes integer values, with distribution $\mathbb{P}(X = k) = p_k$, then

$$\mathbb{E}(X) = \sum_k k p_k, \qquad \text{var}(X) = \sum_k (k - \mu)^2 p_k,$$

where $\mu = \mathbb{E}(X)$. The mean tells us how big X is on average. The variance, or more precisely its square root, the *standard deviation* $\sigma = \sqrt{\text{var}(X)}$, tells us how far away X lies, typically, from its mean. It is not hard to derive the following useful alternative formula for the variance:

$$\text{var}(X) = \mathbb{E}(X^2) - \mathbb{E}(X)^2.$$

To understand the meaning of the variance, consider the following situation. Suppose that one hundred people take an exam and you are told that their average mark is 75%. This gives you some useful information, but by no means a complete picture of how the marks are distributed. For example, perhaps the exam consisted of four questions of which three were very easy and one almost impossible, so that all the marks were clustered around 75%. Or perhaps about fifty people got full marks and fifty got around half marks. To model this situation let the sample space Ω consist of the hundred people and let the probability distribution be the uniform distribution. Given a random person ω, let $X(\omega)$ be that person's mark. Then in the first situation, the variance will be small, since almost everybody's mark is close to the mean of 75%; whereas in the second it is close to $25^2 = 625$, since almost everybody's mark was about 25 away from the mean. Thus, the variance helps us to understand the difference between the two situations.

As we discussed at the start of this article, it is known from experience that the "expected" number of heads in a sequence of n tosses of a fair coin is around $\frac{1}{2}n$, in the sense that the proportion is usually close to $\frac{1}{2}$. It is not hard to work out that, if X models the number of heads in n tosses, that is, if X is binomially distributed with parameters n and $\frac{1}{2}$, then $\mathbb{E}(X) = \frac{1}{2}n$. The variance of X is $\frac{1}{4}n$, so the natural distance scale with which to measure the spread of the distribution is $\sigma = \frac{1}{2}\sqrt{n}$. This allows us to see that X/n is close to $\frac{1}{2}$ with probability close to 1 for large n, in accordance with experience.

More generally, if $X_1, X_2, \ldots, X_n$ are independent random variables, then $\mathrm{var}(X_1 + \cdots + X_n) = \mathrm{var}(X_1) + \cdots + \mathrm{var}(X_n)$. It follows that if all the X_i have the same distribution with mean μ and variance σ^2, then the variance of the *sample average* $\bar{X} = n^{-1}(X_1 + \cdots + X_n)$ is $n^{-2}(n\sigma^2) = \sigma^2/n$, which tends to zero as n tends to infinity. This observation can be used to prove that, for any $\epsilon > 0$, the probability that $|\bar{X} - \mu|$ is greater than ϵ tends to zero as n tends to infinity. Thus, the sample average "converges in probability" to the mean μ.

This result is called the *weak law of large numbers*. The argument sketched above implicitly assumes that the random variables have finite variance, but this assumption turns out not to be necessary. There is also a *strong law of large numbers*, which states that, with probability 1, the sample average of the first n variables converges to μ as n tends to infinity. As its name suggests, the strong law is stronger than the weak law, in the sense that the weak law can be deduced from the strong law. Notice that these laws make long-term predictions of a statistical kind about the real events that we have chosen to model using probability theory. Moreover, these predictions can be checked experimentally, and the experimental evidence confirms them. This provides a convincing scientific justification for our models.

5   The Normal Distribution and the Central Limit Theorem

As we have seen, for the binomial distribution with parameters n and p, the probability p_k is given by the formula $\binom{n}{k}p^k(1-p)^{n-k}$. If n is large and you plot the points (k, p_k) on a graph, then you will notice that they lie in a bell-shaped curve that has a sharp peak around the mean np. The width of the tall part of the curve has order of magnitude $\sqrt{np(1-p)}$, the standard deviation of the distribution. Let us assume for simplicity that np is an integer, and define a new probability distribution q_k by $q_k = p_{k+np}$. The points (k, q_k) peak at $k = 0$. If you now rescale the graph, compressing horizontally by a factor of $\sqrt{np(1-p)}$ and expanding vertically by the same factor, then the points will all lie close to the graph of

$$f(x) = \frac{1}{\sqrt{2\pi}} e^{-x^2/2}.$$

This is the density function of a famous distribution known as the *standard normal distribution* on $\mathbb{R}$. It is also often called the *Gaussian distribution*.

To put this differently, if you toss a biased coin a large number of times, then the number of heads, minus its mean and divided by its standard deviation, is close to a standard normal random variable.

The function $(1/\sqrt{2\pi})e^{-x^2/2}$ occurs in a huge variety of mathematical contexts, from probability theory to FOURIER ANALYSIS [III.27] to quantum mechanics. Why should this be? The answer, as it is for many such questions, is that there are properties that this function has that are shared by no other function.

One such property is *rotational invariance*. Suppose once again that we are throwing a dart at a dartboard and aiming for the bull's-eye. We could model this as the result of adding two independent normal distributions at right angles to each other: one for the x-coordinate and one for the y-coordinate (each having mean 0 and variance 1, say). If we do this, then the two-dimensional "density function" is given by the formula $(1/2\pi)e^{-x^2/2}e^{-y^2/2}$, which can conveniently be written as $(1/2\pi)e^{-r^2/2}$, where r denotes the length of (x, y). In other words, the density function depends only on the distance from the origin. (This is why it is called "rotationally invariant.") This very appealing property holds in more dimensions as well. And it turns out to be quite easy to check that $(1/2\pi)e^{-r^2/2}$ is the *only* such function: more precisely, it is the only rotation-invariant density function that makes the coordinates

x and y into independent random variables of variance 1. Thus, the normal distribution has a very special symmetry property.

Properties like this go some way toward explaining the ubiquity of the normal distribution in mathematics. However, the normal distribution has an even more remarkable property, which leads to its appearance wherever mathematics is used to model disorder in the real world. The *central limit theorem* states that, for any sequence of independent and identically distributed random variables $X_1, X_2, \ldots$ (with finite mean μ and nonzero finite variance σ^2), we have

$$\lim_{n \to \infty} \mathbb{P}(X_1 + \cdots + X_n \leqslant n\mu + \sqrt{n}\sigma x)$$
$$= \int_{-\infty}^{x} \frac{1}{\sqrt{2\pi}} e^{-y^2/2} \, dy$$

for every real number x. The expected value of $X_1 + \cdots + X_n$ is $n\mu$ and its standard deviation is $\sqrt{n}\sigma$, so another way of thinking about this is to let $Y_n = (X_1 + \cdots + X_n - n\mu)/\sqrt{n}\sigma$. This rescales $X_1 + \cdots + X_n$ to have mean 0 and variance 1, and the probability becomes the probability that $Y_n \leqslant x$. Thus, *whatever* distribution we start with, the limiting distribution of the sum of many independent copies is normal (after appropriate rescaling). Many natural processes can realistically be modeled as accumulations of small independent random effects, and this is why many distributions that one observes, such as the distribution of heights of adults in a given town, have a familiar bell-shaped curve.

A useful application of the central limit theorem is to simplify what look like impossibly complicated calculations. For example, when the parameter n is large, the calculation of binomial probabilities becomes prohibitively complicated. But if X is a binomial random variable, with parameters n and $\frac{1}{2}$, for instance, then we can write X as a sum $Y_1 + \cdots + Y_n$, where $Y_1, \ldots, Y_n$ are independent Bernoulli random variables with parameter $\frac{1}{2}$. Then, by the central limit theorem,

$$\lim_{n \to \infty} \mathbb{P}(X \leqslant \tfrac{1}{2}n + \tfrac{1}{2}\sqrt{n}x) = \int_{-\infty}^{x} \frac{1}{\sqrt{2\pi}} e^{-y^2/2} \, dy.$$

III.72 Projective Space

The *real projective plane* can be defined in various ways. One way is to use three *homogeneous coordinates*: a typical point is represented as (x, y, z), where not all of x, y, and z are equal to 0, with the convention that if λ is a nonzero constant, then (x, y, z) and $(\lambda x, \lambda y, \lambda z)$ are

regarded as equal. Notice that for each (x, y, z) the set of all points of the form $(\lambda x, \lambda y, \lambda z)$ is the line through the origin and (x, y, z), and indeed a more geometrical definition of the real projective plane is that it is the set of all lines in $\mathbb{R}^3$ that pass through the origin. Each such line meets the unit sphere in exactly two points, which are opposite each other, and a third way of defining the real projective plane is to define opposite points in the unit sphere to be equivalent and to take the QUOTIENT [I.3 §3.3] of the unit sphere by this EQUIVALENCE RELATION [I.2 §2.3]. A fourth way to define the projective plane is to start with the usual Euclidean plane and to add one "point at infinity" for each possible slope that a line can have. With an appropriate topology, this defines the projective plane as a COMPACTIFICATION [III.9] of the Euclidean plane.

Taking the third definition, a *line* in the projective plane is defined to be a great circle with its opposite points identified. It is then not hard to see that any two lines meet in exactly one point (since any two great circles meet in exactly two opposite points) and that any two points are contained in exactly one line. This property can be used to define much more abstract generalizations of the notion of a projective plane.

There are similar definitions for other fields besides $\mathbb{R}$, and also in higher dimensions. For instance, *complex projective n-space* is the set of all points of the form $(z_1, z_2, \ldots, z_{n+1})$, where not every z_i is 0, with $(z_1, z_2, \ldots, z_{n+1})$ equivalent to $(\lambda z_1, \lambda z_2, \ldots, \lambda z_{n+1})$ if λ is a nonzero complex scalar. This is the set of all "complex lines" in $\mathbb{C}^{n+1}$ that pass through the origin. See SOME FUNDAMENTAL MATHEMATICAL DEFINITIONS [I.3 §6.7] for more details about projective geometry.

III.73 Quadratic Forms
Ben Green

A quadratic form is a homogeneous polynomial of degree 2 in some finite set of unknowns $x_1, x_2, \ldots, x_n$: an example is $q(x_1, x_2, x_3) = x_1^2 - 3x_1x_2 + 4x_3^2$. Here, the coefficients 1, -3, and 4 are integers, but the idea generalizes straightforwardly from $\mathbb{Z}$ to any ring R. Since linear functions are undeniably important and 2 is the next positive integer after 1, one might expect quadratic forms to be important as well, and indeed they are, in many different branches of mathematics, including linear algebra itself.

Here are two theorems about quadratic forms.

Theorem 1. *If x, y, and z are three points in $\mathbb{R}^d$, then the distances between them satisfy the triangle inequality*

$$|x - z| \leqslant |x - y| + |y - z|.$$

Theorem 2. *An odd prime p can be written as the sum of two squares if and only if it leaves remainder 1 on division by 4.*

It is not at first sight clear why theorem 1 has anything to do with quadratic forms. The reason is that the square of the *Euclidean distance*

$$|x| = \sqrt{x_1^2 + \cdots + x_d^2}$$

is a quadratic form over the real numbers $\mathbb{R}$ (here, the x_i are the coordinates of x). This form is derived from the *inner product*

$$\langle x, y \rangle = x_1 y_1 + \cdots + x_d y_d$$

by taking $|x|^2$ to be $\langle x, x \rangle$. The inner product satisfies the relations

(i) $\langle x, x \rangle \geqslant 0$ for all $x \in \mathbb{R}^d$, with equality if and only if $x = 0$.

(ii) $\langle x, y + z \rangle = \langle x, y \rangle + \langle x, z \rangle$ for all $x, y, z \in \mathbb{R}^d$.

(iii) $\langle \lambda x, y \rangle = \langle x, \lambda y \rangle = \lambda \langle x, y \rangle$ for all $\lambda \in \mathbb{R}$ and $x, y \in \mathbb{R}^d$.

(iv) $\langle x, y \rangle = \langle y, x \rangle$ for all $x, y \in \mathbb{R}^d$.

More generally, any function $\phi(x, y)$ that satisfies these relations is called an inner product. The triangle inequality is a consequence of arguably the most important inequality in mathematics, the CAUCHY–SCHWARZ INEQUALITY [V.19]

$$|\langle x, y \rangle| \leqslant |x|\,|y|.$$

Not all quadratic forms on $\mathbb{R}^d$ come from inner products, but they do all come from symmetric bilinear forms $g : \mathbb{R}^d \times \mathbb{R}^d \to \mathbb{R}$. These are functions of two variables that satisfy all the axioms of an inner product except possibly (i), the positivity criterion. Given a quadratic form $q(x) = g(x, x)$, one may recover g using the *polarization identity*

$$g(x, y) = \tfrac{1}{2}(q(x + y) - q(x) - q(y)).$$

This correspondence between quadratic forms and symmetric bilinear forms works just as well when $\mathbb{R}$ is replaced by any field k, except that there are some serious technical issues when k has characteristic two (due to the presence of the fraction $\frac{1}{2}$ in the above formula). In linear algebra one often *defines* quadratic forms by first discussing symmetric bilinear forms. The

advantage of this more abstract approach over the concrete definition we gave at the beginning is that it is not necessary to specify a basis for $\mathbb{R}^d$.

If one makes a good choice of basis, then the quadratic form can be made to look particularly pleasant: we may always choose a basis in such a way that

$$q(x) = x_1^2 + \cdots + x_s^2 - x_{s+1}^2 - \cdots - x_t^2$$

for some s and t satisfying $0 \leqslant s \leqslant t \leqslant d$. Here $x_1, \ldots, x_t$ are the coefficients of x with respect to the basis we have carefully chosen. The quantity $s - t$ is called the *signature* of the form. When $s = d$ (as is the case for the form defining the Euclidean distance) the form is said to be *positive definite*. Forms that are not positive definite occur very commonly. For example, the form $x^2 + y^2 + z^2 - t^2$ is used to define MINKOWSKI SPACE [I.3 §6.8], which plays a key role in special relativity.

We turn now to examples of quadratic forms in number theory, beginning with two very famous theorems about quadratic forms over the integers $\mathbb{Z}$. The first is theorem 2, mentioned at the start of the article. It is due to FERMAT [VI.12]. There are many related results for other binary quadratic forms such as $x^2 + 2y^2$ and $x^2 + 3y^2$. In general, however, the question of which primes are represented by $x^2 + ny^2$ is extremely subtle and interesting, and leads one to CLASS FIELD THEORY [V.28].

In 1770 LAGRANGE [VI.22] showed that every number n can be written as a sum of four squares. In fact, the number of such representations of n, $r_4(n)$, is given by the formula

$$r_4(n) = \sum_{\substack{d \mid n \\ 4 \nmid d}} d.$$

This formula can be explained using the theory of MODULAR FORMS [III.59], one of the most important topics in number theory. Indeed, the generating series

$$f(z) = \sum_{n=0}^{\infty} r_4(n) e^{2\pi i n z}$$

is a *theta series*, as a result of which it satisfies certain transformations that identify it as a modular form.

A remarkable theorem of Conway and Schneeberger states that if a quadratic form $a_1 x_1^2 + a_2 x_2^2 + a_3 x_3^2 + a_4 x_4^2$ with $a_1, \ldots, a_4 \in \mathbb{N}$ represents all the positive integers less than or equal to 15, then it represents *all* positive integers. RAMANUJAN [VI.82] listed fifty-five such forms; actually, one of his forms did not represent 15, but the remaining fifty-four forms constitute the complete list. For example, every positive integer can be written as $x_1^2 + 2x_2^2 + 4x_3^2 + 13x_4^2$.

Quadratic forms in three variables are more difficult to treat. GAUSS [VI.26] proved that $n = x_1^2 + x_2^2 + x_3^2$ if and only if n does not have the form $4^t(8k + 7)$ for integers t and k. It is still not known exactly which integers can be written as $x_1^2 + x_2^2 + 10x_3^2$ (this is known as *Ramanujan's ternary form*).

From the point of view of prime number theory, quadratic forms in *one* variable are the hardest to understand. For example, are there infinitely many primes of the form $x^2 + 1$?

Let us mention one final topic, where quadratic forms over $\mathbb{R}$ are studied but where the unknowns $x_1, \ldots, x_n$ are replaced by integers. In particular, let us mention a beautiful result of Margulis, which confirmed a conjecture of Oppenheim. One instance of the result is the following: for any $\epsilon > 0$, one may find integers x_1, x_2, and x_3 such that

$$0 < |x_1^2 + x_2^2\sqrt{2} - x_3^2\sqrt{3}| < \epsilon.$$

The proof uses techniques from ERGODIC THEORY [V.9], which in related contexts are proving very influential at the forefront of research today. No explicit bounds are known on how large x_1, x_2, and x_3 need to be.

III.74   Quantum Computation

A quantum computer is a theoretical device that makes use of the phenomenon of "superposition" in quantum mechanics to carry out certain computations in a way that is fundamentally different from any known classical methods, and in a few important cases remarkably efficient. In classical physics, if there is some property that a particle could have, then either it has it or it does not. But according to quantum mechanics, it can exist in a sort of indeterminate state that is a linear combination of several states, in some of which it might have the property in question and in others not. The coefficients in this linear combination are called *probability amplitudes*: the modulus squared of the coefficient associated with a state tells you the probability of finding that the particle is in that state if you do a measurement.

Exactly what happens when you take a measurement is puzzling, and the subject of much debate among physicists and philosophers. Fortunately, however, one can understand quantum computation without solving the measurement problem, as it is called: indeed, one

can get away with not understanding quantum mechanics at all. (Similarly, and for similar reasons, one could in principle do significant work in theoretical computer science without having the slightest idea what a transistor is or how it works.)

To understand quantum computation it is helpful to look at two other models of computation. The notion of a *classical* computation is a mathematical distillation of what actually goes on inside your computer. The "state" of a computer at any one time is modeled by an *n-bit string*: that is, a sequence of 0s and 1s of length n. Let us write σ for a typical string and $\sigma_1, \sigma_2, \ldots, \sigma_n$ for the bits that make it up. A "computation" is a sequence of very simple operations performed on the initial string. For example, one operation might be to choose three numbers i, j, and k, all less than n, and change the kth bit σ_k of the current state σ to 1 if $\sigma_i = \sigma_j = 1$ and to 0 otherwise. What makes an operation such as this "simple" is that it is *local* in character: what it does to σ depends on, and affects, just a bounded number of bits of σ (in this case it depends on two bits and affects one). The "state space" of a classical computer, in this model, is the set $\{0, 1\}^n$ of all possible n-bit strings, which we shall denote by Q_n.

After a certain number of stages, we declare the computation to have finished. At this point we perform a simple sequence of "measurements" on the final state, which consist in looking at the bits of the string we have ended up with. If our problem is a "decision problem," then we will typically organize the computation so that all we need to look at is a single bit: if it is 0 then the answer is no and if it is 1 then the answer is yes.

If the ideas of the last two paragraphs are unfamiliar to you, then you are strongly advised to read the first few sections of COMPUTATIONAL COMPLEXITY [IV.20] before continuing with this article.

The next model we shall consider is *probabilistic computation*. This is just like classical computation except that at each stage we are allowed to toss a (possibly biased) coin and let the simple operation we perform depend on the outcome of the toss. For instance, we might again choose three numbers i, j, and k, but this time proceed as follows: with probability $\frac{2}{3}$ we perform the operation described earlier, and with probability $\frac{1}{3}$ we change σ_k to $1 - \sigma_k$. Remarkably, introducing randomness into algorithms can be extremely helpful. (Equally remarkably, there are strong theoretical reasons for believing that all algorithms that use randomness can in fact be "derandomized." See [IV.20 §7.1] for details.)

Suppose that we allow our randomized probabilistic computation to run for k steps and that we do not examine the result. How should we model the current state of the computer? We could use exactly the same definition as in the classical case—a state is an n-bit string—and simply say that the computation is in a state that we cannot know until we do a measurement. But the state of the computer is not a complete mystery: for each n-bit string σ there will be some probability p_σ that the state is σ. In other words, it is better to think of the state of the computer as a PROBABILITY DISTRIBUTION [III.71] on Q_n. This probability distribution will depend on the initial string, and therefore it can in principle give us useful information about that string.

Here is how to use a randomized computation to solve a decision problem. Let us write $P(\sigma)$ for the probability that a certain bit (without loss of generality the first) is 1 at the end of the computation, when the initial string is σ. Suppose we can arrange for $P(\sigma)$ to be at least a for all strings σ for which the answer is yes, and at most some smaller number b for all strings σ for which the answer is no. Let c be the average of a and b. Now run the computation m times for some large m. With very high probability, if the answer is yes then when we have finished the first bit will have been 1 more than cm times, and if the answer is no then it will have been 1 fewer than cm times. So we can solve the decision problem, not with certainty, but at least with a negligibly small chance of error.

The "state space" of a probabilistic computer consists of all possible probability distributions on Q_n, or equivalently all possible functions $p : Q_n \to [0, 1]$ such that $\sum_{\sigma \in Q_n} p_\sigma = 1$. The state space of a *quantum* computer also consists of functions defined on Q_n, but there are two differences. First, they can take complex as well as real values. Second, if $\lambda : Q_n \to \mathbb{C}$ is a state, then the requirement on the size of λ is that $\sum_{\sigma \in Q_n} |\lambda|^2 = 1$. In other words, λ is a unit vector in the HILBERT SPACE [III.37] $\ell_2(Q_n, \mathbb{C})$ rather than a nonnegative unit vector in the BANACH SPACE [III.62] $\ell_1(Q_n, \mathbb{R})$. The scalars λ_σ are the probability amplitudes mentioned earlier. We shall explain what this means later.

Among the possible states of a quantum computer are the "basis states," which are the functions that take the value 1 at one string and 0 everywhere else. It is customary to use Dirac's "bra" and "ket" notation for these, writing $|\sigma\rangle$ if the string in question is σ. Other "pure states" are then linear combinations of these, and Dirac's notation is again used. For instance, if $n = 5$,

then one fairly simple state that the computer could be in is $|\psi\rangle = (1/\sqrt{2})|01101\rangle + (i/\sqrt{2})|11001\rangle$.

To get from one state to another, we again apply "local" operations, but adapted to the new, Hilbert space context. Suppose first that we have a basis state $|\sigma\rangle$. Again we look at a very small number of bits. If, for instance, we look at three bits, at i, j, and k, then there are eight possibilities for the triple $\tau = (\sigma_1, \sigma_2, \sigma_3)$, which we could think of as the basis states in a much smaller state space: the space of all functions $\mu : Q_3 \to \mathbb{C}$ such that $\sum_{\tau \in Q_3} |\mu_\tau|^2 = 1$. The obvious operations that take unit vectors to unit vectors in a complex Hilbert space are the UNITARY MAPS [III.50 §3.1], and these are indeed what are used.

Let us illustrate this with an example. Suppose that $n = 5$, and that i, j, and k are 1, 2, and 4. One possible operation on these three bits would send $|000\rangle$ to $(|000\rangle + i|111\rangle)/\sqrt{2}$ and $|111\rangle$ to $(i|000\rangle + |111\rangle)/\sqrt{2}$, leaving all other three-bit sequences as they are. If our initial basis state is $|01000\rangle$, then the first, second, and fourth bits are in the state $|000\rangle$, so the resulting state at the end of the operation would be $(|01000\rangle + i|11110\rangle)/\sqrt{2}$.

Now that we have explained what a basic operation does to a basis state, we have in fact explained what it does in general, since the basis states form a basis of the state space. In other words, if you start with a linear combination (or superposition) of basis states, you apply the operation described above to each basis state and take the corresponding linear combination of the results.

Thus, an elementary operation of quantum computation consists in acting on the state space by means of a very special sort of unitary map. If the operation is on k bits (where k is typically very small indeed), then the matrix of this map will be a diagonal sum of 2^{n-k} copies of the $2^k \times 2^k$ unitary matrix used to manipulate those k bits (if the basis elements are appropriately ordered). A quantum computation is a sequence of these elementary operations.

Measuring the result of a quantum computation is more mysterious. The basic idea is simple: we do a certain number of elementary operations and then look at one of the bits of the resulting state. But what does this mean, when the state is not a basis state but rather a superposition of such states? The answer is that when we "measure" the rth bit of the output, we are doing a probabilistic process that is somewhat different from the measurement of a probabilistic computation: if the output state is $\sum_{\sigma \in Q_n} \lambda_\sigma |\sigma\rangle$, then the probability that we observe 1 is the sum of all $|\lambda_\sigma|^2$ such that the kth bit of σ is 1, and the probability that we observe 0 is the same sum but over those σ for which the kth bit is 0. This is why the numbers λ_σ are called probability amplitudes. In order to get a useful answer from a quantum computation, one runs it several times, just as with a probabilistic computation.

Note the following two important differences between a quantum computation and a probabilistic computation. We described the state of a probabilistic computation as a probability distribution on Q_n, which one could also call a convex combination of basis states. But this probability distribution is not telling us what is in the computer: that is a basis state. Rather, it is describing our *knowledge* about what is in the computer. By contrast, the state of a quantum computer *really is* a unit vector in a 2^n-dimensional Hilbert space. So in a certain sense a huge amount of computation can go on in parallel: this is what gives quantum computation its power. Although we cannot know much about the computation, since a single measurement causes it to "collapse," we can hope to organize it so that different parts of it "interfere" with each other. This "interference" is related to the second main difference, which is the fact that we deal with probability amplitudes rather than probabilities. Roughly speaking, a quantum computation can "split up" and "reassemble itself," whereas once a probabilistic computation splits up it stays split up. Crucial to the reassembly process in a quantum computation is *cancellation* of probability amplitudes: to give an extreme example, if you multiply a typical unitary matrix by its inverse, then there is a huge amount of cancellation to get all the off-diagonal entries of the resulting matrix to be zero.

All this raises two obvious questions: what are quantum computers good for, and can they actually be built? It turns out that a quantum computer can carry out classical and probabilistic computations, so the first question is asking whether they can do anything further.[1] One might think so, since the state space is so much bigger than it is for a classical computation (it is 2^n dimensional rather than merely n dimensional), and the reassembly process means that we can potentially afford to visit remote parts of the state space, where all coefficients might be of very similar (and small) magnitudes, and come back again to a state where a useful

measurement can be made. However, the very vastness of this space means that most states are completely inaccessible unless one is prepared to use a vast number of basic operations. Additionally, it is important that at the end of the computation the output should not be a "typical" state, since only very special states give rise to useful measurements.

These arguments show that if a quantum computation is to be useful, then it will have to be very carefully (and cleverly) organized. However, there is a spectacular example of just such a computation: Peter Shor's use of a quantum computer to calculate FAST FOURIER TRANSFORMS [III.26] extremely rapidly. The fast Fourier transform has a symmetry that allows the calculation to be split up and carried out "in parallel" (it might be better to say "in superposition") in a way that is ideally suited to a quantum computer. A super-fast Fourier transform can then be used to solve (by classical methods) some famous computational problems, such as the discrete logarithm problem and the factorization of large integers. The latter can then be used to break a public-key cryptosystem, the encryption method that lies at the heart of modern computer security. (See MATHEMATICS AND CRYPTOGRAPHY [VII.7 §5] and COMPUTATIONAL NUMBER THEORY [IV.3 §3] for further discussion of these problems.)

Can a machine be built that would actually be able to do this? There are formidable problems to overcome, arising from a phenomenon in quantum mechanics known as "decoherence," which makes it very hard to stop a complicated state from "collapsing" to a simpler one that is no longer of use. Some progress has been made, but it is too early to say whether, or when, a quantum computer will be built that can factorize large numbers quickly.

Nevertheless, the theoretical challenges raised by the notion of a quantum computer are fascinating. Perhaps the most interesting one is very simple: find an application of quantum computers that is significantly different from the few that have already been found. The fact that quantum computers can factorize large numbers is strong evidence that they are more powerful, but it would be good to have a better understanding of why. (It is known that quantum computers are better for some other uses, such as COMMUNICATION COMPLEXITY [IV.20 §5.1.4].) Is there a much simpler task that is easy for quantum computers and difficult for classical computers, at least if some well-known plausible hypothesis is true about what classical computers cannot do? Can quantum computers solve NP-COMPLETE

---

1. It is also possible to simulate a quantum computation classically, but it would take an absurdly long time to do so: quantum computers cannot calculate noncomputable functions, but they may be far more efficient at calculating some computable ones.

[IV.20 §4] problems? The majority opinion is that they cannot, and indeed the statement that they cannot is becoming another of the many "plausible hypotheses" of complexity theory, but it would be good to have stronger reasons for believing in this statement, such as a proof subject to already-known plausible hypotheses in classical computation.

III.75 Quantum Groups
Shahn Majid

There are at least three different paths that lead to the objects known today as quantum groups. They could be summarized briefly as quantum geometry, quantum symmetry, and self-duality. Any one of them would be a great reason to invent quantum groups and each of them had a role in the development of the modern theory.

1 Quantum Geometry

One of the great discoveries in physics in the last century was that classical mechanics should be replaced by quantum mechanics, in which the space of possible positions and momenta of a particle is replaced by the formulation of position and momentum as mutually noncommuting operators. This noncommutativity underlies Heisenberg's "uncertainty principle," but it also suggests the need for a more general notion of geometry in which coordinates need not commute. One approach to noncommutative geometry is discussed in OPERATOR ALGEBRAS [IV.15 §5]. However, another approach is to note that geometry really grew out of examples such as spheres, tori, and so forth, which are LIE GROUPS [III.48 §1] or objects closely related to Lie groups. If one wants to "quantize" geometry, one should first think about how to generalize basic examples like this: in other words, one should try to define "quantum Lie groups" and associated "quantum" homogeneous spaces.

The first step is to consider geometrical structures not so much in terms of their points but in terms of corresponding *algebras*. For example, the group $SL_2(\mathbb{C})$ is defined as the set of 2×2 matrices $\left(\begin{smallmatrix} \alpha & \beta \\ \gamma & \delta \end{smallmatrix}\right)$ of complex numbers such that $\alpha\delta - \beta\gamma = 1$. We can think of this as a subset of $\mathbb{C}^4$, and indeed not just a subset but a VARIETY [III.95]. The natural class of functions associated with this variety is the set of polynomials in four variables (which are defined on $\mathbb{C}^4$) restricted to the variety. However, if two polynomials take equal values on the

variety, then we identify them. In other words, we take the algebra of polynomials in four variables a, b, c, and d and QUOTIENT [I.3 §3.3] by the IDEAL [III.81 §2] generated by the polynomial $ad - bc - 1$. (This construction is discussed in detail in ARITHMETIC GEOMETRY [IV.5 §3.2].) Let us call the resulting algebra $\mathbb{C}[SL_2]$.

We can do the same for any subset $X \subset \mathbb{C}^n$ that is defined by polynomial relations. This gives us a precise one-to-one correspondence between subsets of this type and certain commutative algebras equipped with n generators. Let us write $\mathbb{C}[X]$ for the algebra that corresponds to X. As with many similar constructions (see, for example, the discussion of adjoint maps in DUALITY [III.19]), a suitable map from X to Y gives rise to a map from $\mathbb{C}[Y]$ to $\mathbb{C}[X]$. More precisely, the map ϕ from X to Y has to be polynomial (in a suitable sense) and the resulting map from $\mathbb{C}[Y]$ to $\mathbb{C}[X]$ is an algebra homomorphism ϕ^* that satisfies the formula $\phi^*(p)(x) = p(\phi x)$ for every $x \in X$ and $p \in \mathbb{C}[Y]$.

Going back to our example, the set $SL_2(\mathbb{C})$ has a group structure $SL_2(\mathbb{C}) \times SL_2(\mathbb{C}) \to SL_2(\mathbb{C})$ defined by the matrix product. The set $SL_2(\mathbb{C}) \times SL_2(\mathbb{C})$ is a variety in $\mathbb{C}^8$ and the matrix product depends in a polynomial way on the entries in the matrices, so we obtain an algebra homomorphism $\Delta : \mathbb{C}[SL_2] \to \mathbb{C}[SL_2] \otimes \mathbb{C}[SL_2]$, which is known as the *coproduct*. (The algebra $\mathbb{C}[SL_2] \otimes \mathbb{C}[SL_2]$ is isomorphic to $\mathbb{C}[SL_2 \times SL_2]$.) It turns out that Δ can be expressed by the formula

$$\Delta \begin{pmatrix} a & b \\ c & d \end{pmatrix} = \begin{pmatrix} a & b \\ c & d \end{pmatrix} \otimes \begin{pmatrix} a & b \\ c & d \end{pmatrix}.$$

This formula needs a word or two of explanation: the variables a, b, c, and d are the four generators of the algebra of polynomials in four variables (and hence of its quotient by $ad - bc - 1$), and the right-hand side is a shorthand way of saying that $\Delta a = a \otimes a + b \otimes c$, and so on. Thus, Δ is defined on the generators by a sort of mixture of TENSOR PRODUCTS [III.89] and matrix multiplication.

One can then show that the associativity of matrix multiplication in SL_2 is equivalent to the assertion that $(\Delta \otimes id)\Delta = (id \otimes \Delta)\Delta$. To understand what these expressions mean, bear in mind that Δ takes elements of $\mathbb{C}[SL_2]$ to elements of $\mathbb{C}[SL_2] \otimes \mathbb{C}[SL_2]$. Thus, when we apply the map $(\Delta \otimes id)\Delta$, for example, we begin by applying Δ, and thereby creating an element of $\mathbb{C}[SL_2] \otimes \mathbb{C}[SL_2]$. This element will be a linear combination of elements of the form $p \otimes q$, each of which will then be replaced by $\Delta p \otimes q$.

Similarly, one can express the rest of the group structure of $SL_2(\mathbb{C})$ equivalently in terms of the algebra $\mathbb{C}[SL_2]$. There is a *counit* map $\epsilon : \mathbb{C}[SL_2] \to k$, which corresponds to the group identity, and an *antipode* map $S : \mathbb{C}[SL_2] \to \mathbb{C}[SL_2]$, which corresponds to the group inversion. The group axioms appear as equivalent properties of these maps, making $\mathbb{C}[SL_2]$ into a "Hopf algebra" or "quantum group." The formal definition is as follows.

Definition. A *Hopf algebra* over a field k is a quadruple (H, Δ, ϵ, S), where

(i) H is a unital algebra over k;

(ii) $\Delta : H \to H \otimes H$, $\epsilon : H \to k$ are algebra homomorphisms such that $(\Delta \otimes \mathrm{id})\Delta = (\mathrm{id} \otimes \Delta)\Delta$ and $(\epsilon \otimes \mathrm{id})\Delta = (\mathrm{id} \otimes \epsilon)\Delta = \mathrm{id}$;

(iii) $S : H \to H$ is a linear map such that $m(\mathrm{id} \otimes S)\Delta = m(S \otimes \mathrm{id})\Delta = 1\epsilon$, where m is the product operation on H.

There are two great things about this formulation. The first is that the notion of a Hopf algebra makes sense over any field. The second is that nowhere did we demand that H was commutative. Of course, if H is derived from a group, then it certainly is commutative (since multiplying two polynomials is commutative), so if we can find a noncommutative Hopf algebra, then we have obtained a strict generalization of the notion of a group. The great discovery of the past two decades is that there are indeed many natural noncommutative examples.

For example, the quantum group $\mathbb{C}_q[SL_2]$ is defined as the free associative *noncommutative* algebra on symbols a, b, c, and d modulo the relations

$$ba - qab, \quad bc - cb, \quad ca - qac, \quad dc - qcd,$$
$$db = qbd, \quad da = ad + (q - q^{-1})bc, \quad ad - q^{-1}bc = 1.$$

This forms a Hopf algebra with Δ given by the same formula as it is for $\mathbb{C}[SL_2]$ and with suitable maps ϵ and S. Here q is a nonzero element of $\mathbb{C}$, and as $q \to 1$ one obtains $\mathbb{C}[SL_2]$. This example generalizes to canonical examples $\mathbb{C}_q[G]$ for all complex simple Lie groups G.

Much of group theory and Lie group theory can be generalized to quantum groups. For example, Haar integration is a linear map $\int : H \to k$ that is translation invariant in a certain sense that involves Δ. If it exists, it is unique up to a scalar multiple, and it does indeed exist in most cases of interest, including all finite-dimensional Hopf algebras. Likewise, the notion of a complex of DIFFERENTIAL FORMS [III.16] (Ω, d) makes sense over any algebra H as a proxy for a differential structure. Here, $\Omega = \bigoplus_n \Omega^n$ is required to be an associative algebra generated by $\Omega^0 = H$ and Ω^1, but one does not assume that it is graded-commutative as in the classical case. When H is a Hopf algebra one can ask that Ω is translation invariant, again in a certain sense that involves the coproduct Δ. In this case both Ω and its COHOMOLOGY [IV.6 §4] as a complex are super (or graded) quantum groups. The axioms of a (graded) Hopf algebra were originally introduced by Heinz Hopf in 1947 precisely to express the structure of the cohomology ring of a group, so this result brings us back full circle to the origins of the subject. For most quantum groups, including all the $\mathbb{C}_q[G]$, one has a natural minimal complex (Ω, d). Thus, a "quantum group" is not merely a Hopf algebra but has additional structure analogous to that of a Lie group.

There are many other quantum groups that are not related to q-deformations. There are also applications of the theory to finite groups. If G is a finite group, one has a corresponding algebra $k(G)$ of all functions on G with pointwise product and a coproduct $(\Delta f)(g, h) = f(gh)$ for $f \in k(G)$ and $g, h \in G$. Here we identify $k(G) \otimes k(G)$ and $k(G \times G)$, which makes Δf into a function of two variables, and one may check even more simply that this is a Hopf algebra. There can never be an interesting classical differential structure on a finite set, but if we use the methods developed for quantum groups, then we have one or more translation-invariant complexes (Ω^1, d) on any finite group. Applying further parts of the theory of quantum group differential geometry, one finds, for example, that the alternating group A_4 is naturally Ricci-flat, while the symmetric group S_3 naturally has constant CURVATURE [III.13], much like a 3-sphere.

2   Quantum Symmetry

Symmetry in mathematics is usually expressed as the action of a group or Lie algebra of finite or infinitesimal transformations of some structure. If you have a collection of transformations that is closed under inversion and composition, then you necessarily have an ordinary group. So how might one generalize this? The answer is that one begins by observing that a group G can act on several objects at the same time. If a group acts on two objects X and Y, then it also acts on their direct product $X \times Y$, with $g(x, y) = (gx, gy)$. Here we are making implicit use of a diagonal or "duplication" map $\Delta : G \to G \times G$, which duplicates a group element so that one copy can act on the first object and the other on the second object. In order to generalize this it once again pays to replace the notion of a group G by that of an algebra. This time we use the *group algebra* kG, which is the set of all formal linear combinations $\sum_i \lambda_i g_i$, where the g_i are elements of G and the λ_i are scalars from the field k. The elements of G (considered as particularly simple linear combinations of this kind) form a basis of kG and we multiply them as we would in G itself. One then extends this definition to products of more general linear combinations in the obvious way. We also extend Δ linearly from $\Delta g = g \otimes g$ on the basis elements to a map from kG to $kG \otimes kG$. Together with some associated maps ϵ and S, this makes kG into a Hopf algebra. Note that this is a completely different use of the coproduct from the one in the previous section, since the group product has already gone into the algebra. One has a similar story for the "enveloping algebra" $U(\mathfrak{g})$ associated with any Lie algebra $\mathfrak{g}$; this is generated by a basis of $\mathfrak{g}$ with certain relations and becomes a Hopf algebra with the coproduct $\Delta \xi = \xi \otimes 1 + 1 \otimes \xi$ "sharing out" an element $\xi \in \mathfrak{g}$ for the purposes of acting on a tensor product of objects on which $\mathfrak{g}$ acts.

Extrapolating from these two examples, a general "quantum symmetry" means an algebra H equipped with further structure Δ that allows one to form a tensor product $V \otimes W$ of any two representations V, W of the algebra in an associative manner. An element $h \in H$ acts as $h(v \otimes w) = (\Delta h)(v \otimes w)$, where one part of Δh acts on $v \in V$ and another part on $w \in W$. This is a second route to the Hopf algebra axioms we gave in the previous section.

Note that, in the examples just given, Δ has had a symmetric output. As a consequence, if V and W are representations of a group or Lie algebra, then $V \otimes W$ and $W \otimes V$ are isomorphic via the obvious map that takes $v \otimes w$ to $w \otimes v$. In general, however, $V \otimes W$ and $W \otimes V$ may be unrelated, so it is now the tensor product that is being made noncommutative. In nice examples it may be the case that $V \otimes W \cong W \otimes V$, but not necessarily by the obvious map. Instead, there may be a nontrivial isomorphism for every pair V, W, which may nevertheless obey some reasonable conditions. This happens for a large class of examples, denoted by $U_q(\mathfrak{g})$ and associated with all complex simple Lie algebras. For these examples, the isomorphism obeys the braid or Yang–Baxter relations among any three representations (see BRAID GROUPS [III.4]). As a result, these quantum groups lead to KNOT AND 3-MANIFOLD INVARIANTS

[III.44] (the Jones knot invariant comes from the example $U_q(\mathfrak{sl}_2)$, where $\mathfrak{sl}_2$ is the Lie algebra of the group $SL_2(\mathbb{C})$). The parameter q can usefully be regarded here as a formal variable, and these examples can be thought of as some kind of deformation of the classical enveloping algebras $U(\mathfrak{g})$. They arose originally in work of Drinfeld and of Jimbo in the theory of quantum integrable systems.

3 Self-duality

A third point of view is that Hopf algebras are the next simplest CATEGORY [III.8] after Abelian groups of structures that admit a FOURIER TRANSFORM [III.27]. It is not immediately obvious, but the axioms (i)–(iii) in the definition we gave earlier have a certain symmetry. One can write out the requirement (i) of a unital algebra H in terms of linear maps $m : H \otimes H \to k$ and $\eta : k \to H$ (here η specifies the identity element of H as the image of $1 \in k$) that have to obey some straightforward commutative diagrams. If you reverse all the arrows in these diagrams, then you have the axioms displayed in (ii), obtaining what could be called a "coalgebra." The requirement that the coalgebra structures Δ and ϵ are algebra maps is given by a collection of diagrams that is invariant under arrow reversal. Finally, the axioms in (iii), as commutative diagrams, are invariant under arrow reversal in the above sense.

Thus, the axioms of a Hopf algebra have the special property of being symmetric under arrow reversal. A practical consequence is that if H is a finite-dimensional Hopf algebra, then so is H^*, with all structure maps defined as the adjoints of those of H (which necessarily reverses arrows). In the infinite-dimensional case one needs a suitable topological dual, or one can just speak of two Hopf algebras as dually paired to each other. For instance, $\mathbb{C}_q[SL_2]$ and $U_q(\mathfrak{sl}_2)$ above are dually paired, while if G is finite then $(kG)^* = k(G)$, the Hopf algebra of functions on G.

As an application, let H be finite dimensional with basis $\{e_a\}$, let H^* have a dual basis $\{f^a\}$, and let $\int$ denote a right-translation-invariant integral on H. The Fourier transform $\mathcal{F} : H \to H^*$ is defined as

$$\mathcal{F}(h) = \sum_a \left(\int e_a h \right) f^a$$

and has many remarkable properties. A special case is a Fourier transform $\mathcal{F} : k(G) \to kG$ for any finite group G, *which does not have to be Abelian*. If G happens to be Abelian, then $kG \cong k(\hat{G})$, where $\hat{G}$ is the group of characters, and we recover the usual Fourier transform

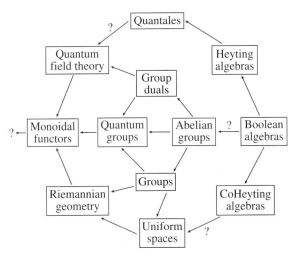

Figure 1 Putting quantum groups in context. Self-dual categories are shown on the horizontal axis.

for finite Abelian groups. The point is that in the non-Abelian case, kG is not commutative and hence not the algebra of functions on any usual "Fourier dual" space.

This point of view is responsible for the second main class of genuine quantum groups to have been discovered, namely the "bicrossproduct" ones of self-dual form. They are simultaneously "coordinate" and "symmetry" algebras, and are truly connected with quantum mechanics. An example, which is written

$$\mathbb{C}[\mathbb{R}^3 \rtimes \mathbb{R}]_\lambda \blacktriangleright\!\!\triangleleft U(\mathfrak{so}(1,3)),$$

is the so-called *Poincaré quantum group* of a certain noncommutative spacetime with coordinates x, y, z, t, where t does not commute with the other variables. This quantum group can also be interpreted as the quantization of a particle moving in a curved geometry with black-hole-like features. In essence, the self-duality of quantum groups provides a paradigm for "toy models" of the unification of gravity (as spacetime geometry) and quantum theory.

This is part of a wider picture indicated in figure 1. A category of objects with a coherent notion of "tensor product" is called a *monoidal* (or *tensor*) *category*, and we have seen that this is the case for representations of quantum groups. There, one also has a "forgetful functor" to the category of vector spaces, which forgets the quantum group action. This embeds quantum groups into the next most general self-dual category (in a representation-theoretic sense), namely that of functors between monoidal categories. Over on the right, I have included Boolean algebras as primitive struc-

tures with (de Morgan) duality. However, the connection between duality here and the other dualities is speculative.

Further Reading

Majid, S. 2002. *A Quantum Groups Primer*. London Mathematical Society Lecture Notes, volume 292. Cambridge: Cambridge University Press.

III.76 Quaternions, Octonions, and Normed Division Algebras

Mathematics took a leap forward in sophistication with the introduction of the COMPLEX NUMBERS [I.3 §1.5]. To define these, one suspends one's disbelief, introduces a new number i, and declares that $i^2 = -1$. A typical complex number is of the form $a + ib$, and the arithmetic of complex numbers is easy to deduce from the normal rules of arithmetic for real numbers. For example, to calculate the product of $1 + 2i$ and $2 + i$ one simply expands some brackets:

$$(1 + 2i)(2 + i) = 2 + 5i + 2i^2 = 5i,$$

the last equality following from the fact that $i^2 = -1$. One of the great advantages of the complex numbers is that, if complex roots are allowed, every polynomial can be factorized into linear factors: this is the famous FUNDAMENTAL THEOREM OF ALGEBRA [V.13].

Another way to define a complex number is to say that it is a pair of real numbers. That is, instead of writing $a + ib$ one writes simply (a, b). To add two complex numbers is simple, and exactly what one does when adding two vectors: $(a, b) + (c, d) = (a + c, b + d)$. However, it is less obvious how to multiply: the product of (a, b) and (c, d) is $(ac - bd, ad + bc)$, which seems an odd definition unless one goes back to thinking of (a, b) and (c, d) as $a + ib$ and $c + id$.

Nevertheless, the second definition draws our attention to the fact that the complex numbers are formed out of the two-dimensional VECTOR SPACE [I.3 §2.3] $\mathbb{R}^2$ with a carefully chosen definition of multiplication. This immediately raises a question: could we do the same for higher-dimensional spaces?

As it stands, this question is not wholly precise, since we have not been clear about what "the same" means. To make it precise, we must ask what properties this multiplication should have. So let us return to $\mathbb{R}^2$ and think about why it would be a bad idea to define the product of (a, b) and (c, d) in a simple-minded way as

(ac, bd). Of course, part of the reason is that the product of $a + ib$ and $c + id$ is not $ac + ibd$, but why should we not also be interested in other ways of multiplying vectors in $\mathbb{R}^2$?

The trouble with this alternative definition is that it allows *zero divisors*, that is, pairs of nonzero numbers that multiply together to give zero. For example, it gives us $(1, 0)(0, 1) = (0, 0)$. If we have zero divisors, then we cannot have multiplicative inverses, since if every nonzero number in a number system has a multiplicative inverse, and if $xy = 0$, then either $x = 0$ or $y = x^{-1}xy = x^{-1}0 = 0$. And if we do not have multiplicative inverses, then we cannot define a useful notion of division.

Let us return then to the usual definition of the complex numbers and try to think how we can go beyond it. One way we might try to "do the same" as we did before is to do to the complex numbers what we did to the real numbers. That is, why not define a "super-complex" number to be an ordered pair (z, w) of complex numbers? Since we still want to have a vector space, we will continue to define the sum of (z, w) and (u, v) to be $(z + u, w + v)$, but we need to think about the best way of defining their product. An obvious guess is to use precisely the expression that worked before, namely $(zu - wv, zv + wu)$. But if we do that, then the product of $(1, i)$ and $(1, -i)$ works out to be $(1 + i^2, i - i) = (0, 0)$, so we have zero divisors.

This example came from the following thought. The *modulus* of a complex number $z = a + ib$, which measures the length of the vector (a, b), is the real number $|z| = \sqrt{a^2 + b^2}$. This can also be written as $\sqrt{\bar{z}z}$, where $\bar{z}$ is the *complex conjugate* $a - ib$ of z. Now if a and b are allowed to take *complex* values, then there is no reason for $a^2 + b^2$ to be nonnegative, so we may not be able to take its square root. Moreover, if $a^2 + b^2 = 0$ it does not follow that $a = b = 0$. The example above came from taking $a = 1$ and $b = i$ and multiplying the number $(1, i)$ by its "conjugate" $(1, -i)$.

There is, nevertheless, a natural way to define the modulus of a pair (z, w) that works even when z and w are complex numbers. The number $|z|^2 + |w|^2$ *is* guaranteed to be nonnegative, so we can take its square root. Moreover, if $z = a + ib$ and $w = c + id$, then we will obtain the number $(a^2 + b^2 + c^2 + d^2)^{1/2}$, which is the length of the vector (a, b, c, d).

This observation leads to another: the complex conjugate of a real number is the number itself, so, if we want to "use the same formula" for the complex numbers as we used for the reals, we are free to introduce

complex conjugates into that formula. Before we try to do that, let us think about what we might mean by the "conjugate" of a pair (z, w). We expect $(z, 0)$ to behave like the complex number z, so its conjugate should be $(\bar{z}, 0)$. Similarly, if z and w are real, then the conjugate of (z, w) should be $(z, -w)$. This leaves us with two reasonable possibilities for a general pair (z, w): either $(\bar{z}, -\bar{w})$ or $(\bar{z}, -w)$. Let us consider the second of these.

We would like the product of (z, w) and its conjugate, which we are defining as $(\bar{z}, -w)$, to be $(|z|^2 + |w|^2, 0)$. We want to achieve this by introducing complex conjugates into the formula

$$(z, w)(u, v) = (zu - wv, zv + wu).$$

An obvious way of getting the result we want is to take

$$(z, w)(u, v) = (zu - \bar{w}v, \bar{z}v + wu),$$

and this modified formula, it turns out, defines an ASSOCIATIVE BINARY OPERATION [I.2 §2.4] on the set of pairs (z, w). If you try the other definition of conjugate, you will find that you end up with zero divisors. (A first indication of trouble is that, under the other definition, the pair $(0, i)$ is its own conjugate.)

We have just defined the *quaternions*, a set $\mathbb{H}$ of "numbers" that form a four-dimensional real vector space, or alternatively a two-dimensional complex vector space. (The letter "H" is in honor of William Rowan Hamilton, their discoverer. See HAMILTON [VI.37] for the story of how the discovery was made.) But why should we have wished to do that? This question becomes particularly pressing when we notice that the notion of multiplication that we have defined is not commutative. For example, $(0, 1)(i, 0) = (0, i)$, while $(i, 0)(0, 1) = (0, -i)$.

To answer it, let us take a step back and think about the complex numbers again. The most obvious justification for introducing those is that one can use them to solve all polynomial equations, but that is by no means the only justification. In particular, complex numbers have an important *geometrical* interpretation, as rotations and enlargements. This connection becomes particularly clear if we choose yet another way of writing the complex number $a + ib$, as the matrix $\left(\begin{smallmatrix} a & -b \\ b & a \end{smallmatrix}\right)$. Multiplication by the complex number $a + ib$ can be thought of as a LINEAR MAP [I.3 §4.2] on the plane $\mathbb{R}^2$, and this is the matrix of that linear map. For example, the complex number i corresponds to the matrix $\left(\begin{smallmatrix} 0 & -1 \\ 1 & 0 \end{smallmatrix}\right)$, which is the matrix of a counterclockwise rotation through $\frac{1}{2}\pi$ about the origin, and this rotation is exactly what multiplying by i does to the complex plane.

If complex numbers can be thought of as linear maps from $\mathbb{R}^2$ to $\mathbb{R}^2$, then quaternions should have an interpretation as linear maps from $\mathbb{C}^2$ to $\mathbb{C}^2$. And indeed they do. Let us associate with the pair (z, w) the matrix $\left(\begin{smallmatrix} z & \bar{w} \\ -w & \bar{z} \end{smallmatrix}\right)$. Now let us consider the product of two such matrices:

$$\begin{pmatrix} z & \bar{w} \\ -w & \bar{z} \end{pmatrix} \begin{pmatrix} u & \bar{v} \\ -v & \bar{u} \end{pmatrix} = \begin{pmatrix} zu - \bar{w}v & z\bar{v} + \bar{w}\bar{u} \\ -\bar{z}v - wu & \bar{z}\bar{u} - w\bar{v} \end{pmatrix}.$$

This product is precisely the matrix associated with the pair $(zu - w\bar{v}, zv + w\bar{u})$, which is the quaternionic product of (z, w) and (u, v)! As an immediate corollary, we have a proof of a fact mentioned earlier: that quaternionic multiplication is associative. Why? Because matrix multiplication is associative. (And *that* is true because the composition of functions is associative: see [I.3 §3.2].)

Notice that the DETERMINANT [III.15] of the matrix $\left(\begin{smallmatrix} z & \bar{w} \\ -w & \bar{z} \end{smallmatrix}\right)$ is $|z|^2 + |w|^2$, so the modulus of the pair (z, w) (which is defined to be $\sqrt{|z|^2 + |w|^2}$) is just the determinant of the associated matrix. This proves that the modulus of the product of two quaternions is the product of their moduli (since the determinant of a product is the product of determinants). Notice also that the adjoint of the matrix (that is, the complex conjugate of the transpose matrix) is $\left(\begin{smallmatrix} \bar{z} & -\bar{w} \\ w & z \end{smallmatrix}\right)$, which is the matrix associated with the conjugate pair $(\bar{z}, -w)$. Finally, notice that if $|z|^2 + |w|^2 = 1$, then

$$\begin{pmatrix} z & \bar{w} \\ -w & \bar{z} \end{pmatrix} \begin{pmatrix} \bar{z} & -\bar{w} \\ w & z \end{pmatrix} = \begin{pmatrix} 1 & 0 \\ 0 & 1 \end{pmatrix},$$

which tells us that the matrix is UNITARY [III.50 §3.1]. Conversely, any unitary 2×2 matrix with determinant 1 can easily be shown to have the form $\left(\begin{smallmatrix} z & \bar{w} \\ -w & \bar{z} \end{smallmatrix}\right)$. Therefore, the unit quaternions (that is, the quaternions of modulus 1) have a geometrical interpretation: they correspond to the "rotations" of $\mathbb{C}^2$ (that is, the unitary maps of determinant 1), just as the unit complex numbers correspond to the rotations of $\mathbb{R}^2$.

The group of unitary transformations of $\mathbb{C}^2$ of determinant 1 is an important LIE GROUP [III.48 §1] called the *special unitary group* SU(2). Another important Lie group is the group SO(3), of rotations of $\mathbb{R}^3$. Surprisingly, the unit quaternions can be used to describe this group as well. To see this, it is convenient to present the quaternions in another, more conventional, way.

Quaternions, as they are usually introduced, are a system of numbers where we introduce not just one square root of -1 but three, called i, j, and k (together with their negatives). Once one knows that $i^2 = j^2 = k^2 = -1$, and also that ij = k, jk = i, and ki = j, one has

all the information one needs to multiply two quaternions. For example, ji = jjk = $-$k. A typical quaternion takes the form $a + ib + jc + kd$, which corresponds to the pair of complex numbers $(a + ic, b + id)$ in our previous way of thinking about quaternions. Now if we want, we can think of this quaternion as a pair $(a, \boldsymbol{v})$, where a is a real number and $\boldsymbol{v}$ is the vector (b, c, d) in $\mathbb{R}^3$. The product of $(a, \boldsymbol{v})$ and $(b, \boldsymbol{w})$ then works out to be $(ab - \boldsymbol{v} \cdot \boldsymbol{w}, a\boldsymbol{w} + b\boldsymbol{v} + \boldsymbol{v} \wedge \boldsymbol{w})$, where $\boldsymbol{v} \cdot \boldsymbol{w}$ and $\boldsymbol{v} \wedge \boldsymbol{w}$ are the scalar and vector products of $\boldsymbol{v}$ and $\boldsymbol{w}$.

If $q = (a, \boldsymbol{u})$ is a quaternion of modulus 1, then $a^2 + \|\boldsymbol{u}\|^2 = 1$, so we can write q in the form $(\cos\theta, \boldsymbol{v}\sin\theta)$ with $\boldsymbol{v}$ a unit vector. This quaternion corresponds to a counterclockwise rotation R about an axis in direction $\boldsymbol{v}$ through an angle of 2θ. This angle is not what one might at first expect, and neither is the way the correspondence works. If $\boldsymbol{w}$ is another vector, we can represent it as the quaternion $(0, \boldsymbol{w})$. We would now like a neat expression for the quaternion $(0, R\boldsymbol{w})$; it turns out that $(0, R\boldsymbol{w}) = q(0, \boldsymbol{w})q^*$, where q^* is the conjugate $(\cos\theta, -\boldsymbol{v}\sin\theta)$ of q, which is also its multiplicative inverse, as q has modulus 1. So to do the rotation R, you do not multiply by q but rather you *conjugate* by q. (This is a different meaning of the word "conjugate," referring to multiplying on one side by q and on the other side by q^{-1}.) Now if q_1 and q_2 are quaternions corresponding to rotations R_1 and R_2, respectively, then

$$q_2 q_1(0, \boldsymbol{w})q_1^* q_2^* = q_2 q_1(0, \boldsymbol{w})(q_2 q_1)^*,$$

from which it follows that $q_2 q_1$ corresponds to the rotation $R_2 R_1$. This tells us that quaternionic multiplication corresponds to composition of rotations.

The unit quaternions form a group, as we have already seen—it is SU(2). It might appear that we have shown that SU(2) is the same as the group SO(3) of rotations of $\mathbb{R}^3$. However, we have not quite done this, because for each rotation of $\mathbb{R}^3$ there are *two* unit quaternions that give rise to it. The reason is simple: a counterclockwise rotation through θ about a vector $\boldsymbol{v}$ is the same as a counterclockwise rotation through $-\theta$ about $-\boldsymbol{v}$. In other words, if q is a unit quaternion, then q and $-q$ give rise to the same rotation of $\mathbb{R}^3$. So SU(2) is not isomorphic to SO(3); rather, it is a *double cover* of SO(3). This fact has important ramifications in mathematics and physics. In particular, it lies behind the notion of the "spin" of an elementary particle.

Let us return to the question we were considering earlier: for which n is there a good way of multiplying vectors in $\mathbb{R}^n$? We now know that we can do it for $n = 1$,

2, or 4. When $n = 4$ we had to sacrifice commutativity, but we were amply rewarded for this, since quaternion multiplication gives a very concise way of representing the important groups SU(2) and SO(3). These groups are not commutative, so it was essential to our success that quaternion multiplication should also not be commutative.

One obvious thing we can do is continue the process that led to the quaternions. That is, we can consider pairs (q, r) of quaternions, and multiply these pairs together using the formula

$$(q, r)(s, t) = (qs - r^*t, q^*t + rs).$$

Since the conjugate q^* of a quaternion q is the analogue of the complex conjugate $\bar{z}$ of a complex number z, this is basically the same formula that we used for multiplication of pairs of complex numbers—that is, for quaternions.

However, we need to be careful: multiplication of quaternions is not commutative, so there are in fact many formulas we could write down that would be "basically the same" as the earlier one. Why choose the above one, rather than, say, replacing q^*t by tq^*?

It turns out that the formula suggested above leads to zero divisors. For example, $(j, i)(l, k)$ works out to be $(0, 0)$. However, the modified formula

$$(q, r)(s, t) = (qs - tr^*, q^*t + sr),$$

which one can discover fairly quickly if one bears in mind that one would like $(q, r)(q^*, -r)$ to work out as $(|q|^2 + |r|^2, 0)$, does produce a useful number system. It is denoted $\mathbb{O}$ and its elements are called the *octonions* (or sometimes the *Cayley numbers*). Unfortunately, multiplication of octonions is not even associative, but it does have two very good properties: every nonzero octonion has a multiplicative inverse, and two nonzero octonions never multiply together to give zero. (Because octonion multiplication is not associative, these two properties are no longer obviously equivalent. However, any subalgebra of the octonions generated by two elements *is* associative, and this is enough to prove the equivalence.)

So now we have number systems when $n = 1, 2, 4$, or 8. It turns out that these are the *only* dimensions with good notions of multiplication. Of course, "good" has a technical meaning here: matrix multiplication, which is associative but gives zero divisors, is for many purposes "better" than octonion multiplication, which has no zero divisors but is not associative. So let us finish by seeing more precisely what it is that is special about dimensions 1, 2, 4, and 8.

All the number systems constructed above have a notion of size given by a NORM [III.62]. For real and complex numbers z, the norm of z is just its modulus. For a quaternion or octonion x, it is defined to be $\sqrt{x^*x}$, where x^* is the conjugate of x (a definition that works for real and complex numbers as well). If we write $\|x\|$ for the norm of x, then the norms constructed have the property that $\|xy\| = \|x\| \|y\|$ for every x and y. This property is extremely useful: for example, it tells us that the elements of norm 1 are closed under multiplication, a fact that we used many times when discussing the geometric importance of complex numbers and quaternions.

The feature that distinguishes dimensions $1, 2, 4,$ and 8 from all other dimensions is that these are the only dimensions for which one can define a norm $\| \cdot \|$ and a notion of multiplication with the following properties.

(i) There is a multiplicative identity: that is, a number 1 such that $1x = x1 = x$ for every x.

(ii) Multiplication is *bilinear*, meaning that $x(y + z) = xy + xz$ for every x, y, and z, and $x(ay) = a(xy)$ whenever a is a real number, and similarly for multiplication on the right.

(iii) For any x and y, $\|xy\| = \|x\| \|y\|$ (and therefore there are no zero divisors).

A *normed division algebra* is a vector space $\mathbb{R}^n$ together with a norm and a method of multiplying vectors that satisfy the above properties. So normed division algebras exist only in dimensions 1, 2, 4, and 8. Furthermore, even in these dimensions, $\mathbb{R}$, $\mathbb{C}$, $\mathbb{H}$, and $\mathbb{O}$ are the only examples.

There are various ways to prove this fact, which is known as *Hurwitz's theorem*. Here is a very brief sketch of one of them. The idea is to prove that if a normed division algebra A contains one of the above examples, then either it *is* that example, or it contains the next one in the sequence. So either A is one of $\mathbb{R}$, $\mathbb{C}$, $\mathbb{H}$, and $\mathbb{O}$ or A contains the algebra produced by doing to $\mathbb{O}$ the process we used to construct $\mathbb{H}$ from $\mathbb{C}$ and $\mathbb{O}$ from $\mathbb{H}$, a process known as the *Cayley–Dickson construction*. However, if one applies the Cayley–Dickson construction to $\mathbb{O}$, one obtains an algebra with zero divisors.

To see how such an argument might work, let us imagine, for the sake of example, that A contains $\mathbb{O}$ as a proper subalgebra. It turns out that the norm on A must be a EUCLIDEAN NORM [III.37]—that is, a norm derived from an inner product. (Roughly speaking, this is because multiplication by an element of norm 1 does not change the norm, which gives A so many symmetries that the norm on A has to be the most symmetric of all, namely Euclidean.) Let us call an element of A *imaginary* if it is orthogonal to the element 1. Then we can define a conjugation operation on A by taking 1^* to be 1 and x^* to be $-x$ when x is imaginary, and extending linearly. This operation can be shown to have all the properties one would like. In particular, $aa^* = a^*a = \|a\|^2$ for every element a of A. Let us choose a norm-1 element of A that is orthogonal to all of $\mathbb{O}$ and call it i. Then $i^* = -i$, so $1 = i^*i = -i^2$, so $i^2 = -1$. Now take the algebra generated by i and the copy of $\mathbb{O}$ that lies in A. With some algebraic manipulation, one can demonstrate that this consists of elements of the form $x + iy$, with x and y belonging to $\mathbb{O}$. Moreover, the product of $x + iy$ and $z + iw$ turns out to be $xz - wy^* + i(x^*w + zy)$, which is exactly what the Cayley–Dickson construction gives.

For further details about quaternions and octonions, two excellent sources are a discussion by John Baez at http://math.ucr.edu/home/baez/octonions and a book, *On Quaternions and Octonions: Their Geometry, Arithmetic, and Symmetry*, by J. H. Conway and D. A. Smith (2003; Wellesley, MA: AK Peters).

III.77   Representations

A *linear representation* of a finite GROUP [I.3 §2.1] G is a way of associating a linear map T_g, from some VECTOR SPACE [I.3 §2.3] V to itself, with each element g of G. Of course, this association must reflect the group structure of G, so T_gT_h should equal T_{gh}, and if e is the identity of G, then T_e should be the identity map on V.

One useful aspect of linear representations is that the dimension of the vector space V may be considerably smaller than the size of G. If this is the case, then the representation packages the information about G in a particularly efficient way. For example, the ALTERNATING GROUP [III.68] A_5, which has sixty elements, is isomorphic to the group of rotational symmetries of an icosahedron, and can therefore be thought of as a group of transformations of $\mathbb{R}^3$ (or, equivalently, of 3×3 matrices).

A more fundamental reason for representations being useful is that every representation can be decomposed into building blocks known as *irreducible* representations. It turns out that a great deal of information about G can be deduced from a few basic facts about its irreducible representations.

These ideas can be generalized to infinite groups as well, and are particularly important in the case of

LIE GROUPS [III.48 §1]. Since Lie groups have a differentiable structure, the representations of interest are those where the homomorphism $g \mapsto T_g$ reflects this structure (for example, by being differentiable).

Representations are discussed in much greater detail in REPRESENTATION THEORY [IV.9]. See also OPERATOR ALGEBRAS [IV.15 §2].

III.78   Ricci Flow
Terence Tao

Ricci flow is a technique that allows one to take an arbitrary RIEMANNIAN MANIFOLD [I.3 §6.10] and smooth out the geometry of that manifold to make it look more symmetric. It has proven to be a very useful tool in understanding the topology of such manifolds.

Ricci flow can be defined for Riemannian manifolds of any dimension, but for the sake of exposition we restrict ourselves here to two-dimensional manifolds (i.e., surfaces) as they are easy to visualize. From our everyday experience with three-dimensional space $\mathbb{R}^3$, we are familiar with many surfaces, such as spheres, cylinders, planes, tori (the shape of the surface of a doughnut), and so forth. This is an *extrinsic* way to think about surfaces: as subsets of a larger *ambient space*, which in this case is three-dimensional Euclidean space $\mathbb{R}^3$. On the other hand, one can think about surfaces in a more abstract *intrinsic* manner: by considering how the points in the surface stand in relation to each other, but not in relation to any external space. (For instance, the Klein bottle makes perfect sense as a surface from an intrinsic viewpoint, but cannot be viewed extrinsically in three-dimensional Euclidean space $\mathbb{R}^3$, although it can be viewed extrinsically in four-dimensional Euclidean space $\mathbb{R}^4$.) It turns out that the two viewpoints are mostly equivalent to each other, but it will be more convenient here to adopt the intrinsic perspective.

A good example of a surface is the surface of Earth. Extrinsically, this is a subset of a three-dimensional space $\mathbb{R}^3$. But we can also view this surface two dimensionally by using an *atlas*: a collection of *maps* or *charts* that describe various regions of this surface by identifying them with a subset of a two-dimensional plane. As long as we have enough charts to cover the original surface, this atlas is sufficient to describe the surface. This way of thinking of a surface is not completely intrinsic, because there is more than one atlas that one could associate with this surface, and they may differ in various minor ways. For instance, in one atlas the city of Los Angeles might be on the boundary of one of the charts, whereas in another atlas it might be in the interior of every chart that it appears in. However, there are many facts one can deduce from an atlas that do not depend on the choice of atlas; for instance, using any accurate atlas of Earth one can see that it is impossible to travel from Los Angeles to Sydney without crossing at least one ocean. If a fact regarding a surface does not depend on which atlas one uses, we say that it is *intrinsic* or *coordinate-independent*. It will turn out that Ricci flow is an intrinsic flow on surfaces; it can be defined without any knowledge of charts or of some external space.

We have informally described the mathematical concept of a surface, or two-dimensional manifold. But to describe Ricci flow we need the more sophisticated concept of a *Riemannian surface* (or two-dimensional Riemannian manifold). This is a surface M with an additional (intrinsic) object, a *Riemannian metric* g, which specifies the distance $d(x, y)$ between any two points x, y on the surface. This metric allows one to define the angle $\angle y_1, y_2$ that any two curves y_1, y_2 on the surface make where they intersect; for instance, the Earth's equator intersects any longitude at right angles. And it can also be used to define the area $|A|$ of any given set A on the surface (e.g., the area of Australia). There are a number of properties that these concepts of distance, angle, and area have to satisfy, but the most important property can be stated informally as follows: *the geometry of a Riemannian surface has to be very close to the geometry of the Euclidean plane at small length scales.*

To give an example of what the above statement means, take any point x in the surface M, and pick any positive radius r. Because the Riemannian metric g specifies a notion of distance, we can define the *disk $B(x, r)$ of radius r centered at x* to be the set of all points y whose distance $d(x, y)$ to x is less than r. Because the Riemannian metric g defines a notion of area, we can then discuss the area of this disk $B(x, r)$. In the Euclidean plane, this area would of course be πr^2. In a Riemannian surface, this need not be the case: for instance, the total area of the surface of Earth (and hence of all disks within this surface) is finite, even though πr^2 can be arbitrarily large as r goes to infinity. However, we do require that, when r is very small, the area of the disk $B(x, r)$ becomes increasingly close to πr^2; more precisely, we require that the ratio between the area and πr^2 converges to 1 in the limit as r tends to 0.

This brings us to the notion of *scalar curvature $R(x)$*. In some cases, such as on the sphere, the area $|B(x, r)|$ of a small disk $B(x, r)$ is actually a little bit smaller than πr^2; when this is the case, we say that the surface has *positive scalar curvature* at x. In some other cases, such as on a saddle, the area $|B(x, r)|$ of a small disk $B(x, r)$ is a bit larger than πr^2; then we say that the surface has *negative scalar curvature* at x. In other cases again, such as on a cylinder, the area $|B(x, r)|$ of a small disk $B(x, r)$ is equal (or very nearly equal) to πr^2; in this case we say the surface has *vanishing scalar curvature* at x. (This is despite the cylinder being "curved" when viewed extrinsically as a subset of three-dimensional space.) Note that on a complicated surface it is perfectly possible to have positive scalar curvature at some points of the surface and negative or vanishing scalar curvature at other points. The scalar curvature $R(x)$ at any given point x can be defined more precisely by the formula

$$R(x) = \lim_{r \to 0} \frac{\pi r^2 - |B(x, r)|}{\pi r^4 / 24}.$$

(For surfaces in an external space, this intrinsic concept of scalar curvature is almost identical to the extrinsic concept of *Gauss curvature*, which we will not discuss here.)

One can refine this notion to that of *Ricci curvature* $\mathrm{Ric}(x)(v, v)$. Consider now an angular sector $A(x, r, \theta, v)$ inside a small disk $B(x, r)$ of small angular aperture θ (measured in radians) about some direction v (a unit vector) emanating from x. This sector is well-defined, basically because the Riemannian metric gives us the appropriate notions of distance and angle. In Euclidean space, the area $|A(x, r, \theta, v)|$ of this sector is $\frac{1}{2}\theta r^2$. But on a surface, the area $|A(x, r, \theta, v)|$ may be slightly less (respectively, slightly more) than $\frac{1}{2}\theta r^2$. In these cases we say that the surface has positive (respectively, negative) Ricci curvature at x in the direction v. More precisely, we have

$$\mathrm{Ric}(x)(v, v) = \lim_{r \to 0} \lim_{\theta \to 0} \frac{\frac{1}{2}\theta r^2 - |A(x, r, \theta, v)|}{\theta r^4 / 24}.$$

Now it turns out that for surfaces, this more complicated notion of curvature is in fact equal to half the scalar curvature: $\mathrm{Ric}(x)(v, v) = \frac{1}{2}R(x)$. In particular, the direction v plays no role in Ricci curvature in two dimensions. However, it is possible to extend all of the above concepts to other dimensions. (For instance, to define scalar and Ricci curvature for three-dimensional manifolds, one would use balls and solid sectors instead of disks and angular sectors, as well as making other necessary adjustments, such as replacing the expression πr^2 with $\frac{4}{3}\pi r^3$.) In higher dimensions it turns out that the Ricci curvature is more complicated than the scalar curvature. For instance, in three dimensions it is possible for a point x to have positive Ricci curvature in one direction but negative Ricci curvature in another; intuitively, this means that narrow sectors in the former direction "curve inward," whereas narrow sectors in the latter direction "curve outward."

Now we can describe *Ricci flow* informally as the process of *stretching* the metric g in directions of negative Ricci curvature, and *contracting* the metric in directions of positive Ricci curvature. The stronger the curvature, the faster the stretching or contracting of the metric. The concepts of stretching and contracting will not be defined formally here, but they increase or decrease the distance between points along these directions. By changing the notion of distance, one also affects the notions of angle and volume (though it turns out that Ricci flow in two dimensions is *conformal*, which means that the notion of angle remains unaffected by the flow; this fact is closely related to the previously mentioned fact that in two dimensions the Ricci curvature is the same in all directions). Ricci flow can be described succinctly and precisely by the equation

$$\frac{d}{dt} g = -2\,\mathrm{Ric},$$

although we will not define here exactly what it means to differentiate the metric g with respect to the time variable t, or what it means for that derivative to equal the Ricci curvature multiplied by -2.

In principle, one could perform Ricci flow on a manifold for as long a period of time as one wished. In practice, however, it is possible (especially in the presence of positive curvature) for the Ricci flow to cause a manifold to develop *singularities*: points where it ceases to look like a manifold, and where the geometry may stop resembling Euclidean geometry even at very small scales. For example, if one starts with a perfectly round sphere and performs Ricci flow, what happens is that the sphere contracts at a steady rate until it becomes a point, which is no longer a two-dimensional manifold. In three dimensions, more complicated singularities are possible: for instance, one can have a *neck pinch*, in which a cylinder-like "neck" of the manifold shrinks under Ricci flow, until at one or more places along the neck, the cylinder has tapered down to a point. The types of possible singularity formations for three-dimensional Ricci flow were only classified completely in a recent and very important paper of Grigori Perelman.

Some years ago, Richard Hamilton made the fundamental observation that Ricci flow is an excellent tool for simplifying the structure of a manifold: generally speaking, it compresses all the positive-curvature parts of the manifold into nothingness, while expanding the negative-curvature parts of the manifold until they become very homogeneous, in the sense that the manifold begins to look much the same no matter which vantage point one selects inside it. Indeed, the flow seems to separate the manifold into extremely symmetric components. For instance, in two dimensions the Ricci flow always ends up endowing the manifold with a metric of constant curvature, which could be positive (as in the sphere), zero (as in the cylinder), or negative (as in *hyperbolic space*); the fact that such a constant-curvature metric can always be found is known as the UNIFORMIZATION THEOREM [V.34] and is of fundamental importance in the theory of surfaces. In higher dimensions, the Ricci flow can develop singularities before perfect symmetry is attained, but it turns out that it is possible to perform "surgeries" (see DIFFERENTIAL TOPOLOGY [IV.7 §§2.3, 2.4]) on the singularities that develop this way, so that the manifold becomes smooth again and one can restart the Ricci flow process. (The surgery may, however, change the topology of the manifold: for instance, it can convert a connected manifold into two disconnected pieces.) In three dimensions it has recently been shown by Perelman that Ricci flow, when augmented by surgery to remove the singularities, does indeed convert an arbitrary manifold (obeying some mild assumptions) into a finite union of some very symmetric and explicitly describable pieces; the precise statement of this conclusion was known as the *geometrization conjecture* of Thurston. One consequence of this conjecture, which is now a rigorous theorem proved by Perelman, is the POINCARÉ CONJECTURE [V.25]: any compact three-dimensional manifold that is *simply connected* (meaning that any closed loop on the manifold can be contracted smoothly to a point without ever leaving the manifold) can in fact be smoothly deformed into a 3-sphere (which is to four-dimensional Euclidean space as the usual two-dimensional sphere is to three-dimensional Euclidean space). The proof of Poincaré's conjecture is one of the most impressive recent achievements of modern mathematics.

III.79 Riemann Surfaces
Alan F. Beardon

Let D be a *region* (that is, a connected open set) in the complex plane. If f is a complex-valued function defined on D, then we can define its derivative just as we would for real-valued functions defined on subsets of $\mathbb{R}$: the derivative of f at w is the limit as z tends to w of the "difference quotient" $(f(z) - f(w))/(z - w)$. Of course, this limit does not necessarily exist, but if it exists for every w in D, then f is said to be *analytic*, or *holomorphic*, on D. Analytic functions have amazing properties; for example, if a function is analytic in a region, then it automatically has a Taylor-series expansion at each point of the region, from which one can deduce that it is infinitely differentiable. This is in stark contrast to the theory of real functions of a real variable, where, for example, a function may be once differentiable but not twice differentiable at some point x, yet three-times differentiable at some other point y. *Complex analysis* is the study of analytic functions. Perhaps more than any other mathematical topic, it is both immensely useful in a practical sense and profound and beautiful in a theoretical sense. (Some of the basic results of complex analysis are described in [I.3 §5.6].)

Just as group theorists do not generally distinguish between isomorphic groups, and topologists do not distinguish between homeomorphic topological spaces, complex analysts do not distinguish between two regions D and D' if there is an analytic bijection between D and D'. When this is the case, we say that D and D' are *conformally equivalent*. Conformal equivalence is, as its name suggests, an EQUIVALENCE RELATION [I.2 §2.3]: the proof depends on the surprising fact that if f is an analytic bijection from D to D', then its inverse $f^{-1} : D' \to D$ is also analytic. Again, this contrasts with real analysis. If D and D' are conformally equivalent, then "interesting" properties of analytic functions on D are transferred automatically to corresponding properties of analytic functions defined on D'. Indeed, this statement can almost be taken as a definition of "interesting" properties (although admittedly this conflicts with the numerical side of complex analysis, because purely numerical statements do not usually transfer under such maps). Naturally, we would like to know which properties of analytic functions are "interesting" in this sense. One such property is that (except at certain isolated points) the angle between two intersecting curves in D is preserved under an analytic map: this is the origin of the term "conformal." It is less well-known that if a bijection (which is not assumed to be differentiable) preserves the angles between curves (that is, both their magnitude and whether they are measured clockwise or counterclockwise), then it is analytic. Thus, loosely speaking, the preservation of angles implies the existence of a Taylor series!

The impact of complex analysis on other topics is so great that it is natural to try to find the most general type of surface on which we can study analytic functions. This leads to the definition of a *Riemann surface* (after BERNHARD RIEMANN [VI.49], who introduced the idea in his doctoral dissertation). In order to put a coordinate system on a surface S we try to map S bijectively onto a plane region D; if we succeed, then we can transfer the coordinates from D to S. For many surfaces (for example, a sphere) it is not possible to find such a map, and we have to be satisfied with *local coordinates*. This means that at each point w of S, we map a neighborhood N of w onto a plane region, and so obtain coordinates that are restricted to N. As there are usually infinitely many ways to do this, we are forced to consider the class of *transition maps*; that is, the maps from one coordinate system at w to another. The surface is a *Riemann surface* precisely when each such transition map is an analytic bijection. This definition resembles that of a two-dimensional MANIFOLD [I.3 §6.9], but the requirement that the transition maps should be analytic is much stronger, so by no means every 2-manifold is a Riemann surface.

It is not difficult to construct Riemann surfaces. Consider, for example, a sphere S resting on a horizontal table. If we imagine a light source at the highest point P of the sphere, then each point of S except P casts a "shadow" on the table: since the table has a simple coordinate system, we can use these "shadows" to define a coordinate system on all of S except the point P. Similarly, a light source at the point Q of tangency with the table casts a shadow onto the (horizontal) tangent plane at P, and this gives a coordinate system valid throughout S except at Q. It can be shown that if the second coordinate system is composed with a reflection, then the sphere does have the structure of a Riemann surface. This is an extremely important example, because it allows one to handle questions involving infinity in a satisfactory way; it is known as the *Riemann sphere*.

For another example, consider a cube C, and (for simplicity only) remove the eight vertices. Given a face F of C (without its bounding edges), we can find a Euclidean rigid motion that maps F into $\mathbb{C}$, so we can easily define a coordinate system on F. If w is an interior point of an edge E of C, we can "open" the two faces that meet at E to make a planar region that contains E, and then map this region into $\mathbb{C}$ by a Euclidean rigid motion. In this way we see that C (less its vertices) is a Riemann surface. The problem with the vertices can be solved by technical means, and this method can then be generalized to show that any polyhedron (even one with holes, such as a "square" torus) is a Riemann surface. These are known as *compact surfaces*. It is a deep but fascinating classical result that each such surface corresponds bijectively to an irreducible polynomial $P(z, w)$ in two complex variables. To give an idea of how the correspondence works, let us consider an equation such as $w^3 + wz + z^2 = 0$. For each z this can be solved to give three values of w, say w_1, w_2, and w_3; as we allow z to vary in $\mathbb{C}$, the values w_j vary, and as they do so they create a Riemann surface W, which can be shown to be connected. This surface can be thought of as lying "above" $\mathbb{C}$, and for all but a finite set of z in $\mathbb{C}$ there are exactly three points on W that are "above" z.

As we have mentioned, Riemann surfaces are important because they are the most general surfaces on which one can study analytic functions, with all of their remarkable properties. It is easy to define what we mean by an analytic function f on a Riemann surface R. Given a coordinate system on part of R, we can think of f as a function of the coordinates, and we then regard f as analytic if and only if it depends analytically on the coordinates. Because the transition maps are analytic, f will be analytic with respect to one coordinate system if and only if it is analytic with respect to all the other coordinate systems defined at the point in question.

This simple property—that if something holds in one coordinate system, then it holds in all of them—is one of the crucial features of the theory. For example, suppose that we have two curves crossing on an (abstract) Riemann surface. If we transfer the two curves to plane regions using different local coordinate systems at the crossing point, and then measure the angle of intersection in each case, we must get the same result (since the transition from one coordinate system to another preserves angles). It follows that the angle between intersecting curves on an abstract Riemann surface is a well-defined concept.

It turns out that analysis on Riemann surfaces goes beyond analytic functions. *Harmonic functions* (solutions of LAPLACE'S EQUATION [I.3 §5.4]) are intimately connected to analytic functions, since the real part of an analytic function is harmonic and any harmonic function is (locally) the real part of an analytic function. Thus, on a Riemann surface, complex analysis merges almost imperceptibly with potential theory (which is the study of harmonic functions).

Perhaps the most profound theorem of all about Riemann surfaces is the UNIFORMIZATION THEOREM [V.34]. Roughly speaking, this says that every Riemann surface is obtained from either Euclidean, spherical, or hyperbolic geometry (see SOME FUNDAMENTAL MATHEMATICAL DEFINITIONS [I.3 §§6.2, 6.5, 6.6]) by taking a polygon in that geometry and gluing its sides together, in the same way that one obtains a torus by gluing opposite sides of a rectangle together. (See also FUCHSIAN GROUPS [III.28].) Remarkably, only very few Riemann surfaces come from the Euclidean or spherical geometries; essentially, every Riemann surface can be constructed in this way from (and only from) the hyperbolic plane. This means that virtually every region in the complex plane comes equipped with a natural and intrinsic geometry whose character is hyperbolic and *not*, as one might expect, Euclidean. The Euclidean character of a generic plane region comes from its embedding in $\mathbb{C}$, and not from its own intrinsic hyperbolic geometry.

III.80 The Riemann Zeta Function

The *Riemann zeta function* ζ is a function defined on the complex numbers that encapsulates in a remarkable way many of the most important properties about the distribution of prime numbers. If s is a complex number with real part greater than 1, then $\zeta(s)$ is defined to be $\sum_{n=1}^{\infty} n^{-s}$. The condition that $\mathrm{Re}(s) > 1$ is needed to ensure that this series converges. However, because the resulting function is HOLOMORPHIC [I.3 §5.6], it is possible to extend the definition by means of analytic continuation. The result is a function that is defined everywhere on the complex plane (though it takes the value ∞ at 1).

A first clue to why this function is related to the distribution of primes is *Euler's product formula*:

$$\zeta(s) = \prod_p (1 - p^{-s})^{-1}.$$

Here, the product on the right-hand side is over all primes. The formula can be proved by writing $(1 - p^{-s})^{-1}$ as $1 + p^{-s} + p^{-2s} + \cdots$, expanding out the product, and using THE FUNDAMENTAL THEOREM OF ARITHMETIC [V.14]. Deeper connections were discovered by

RIEMANN [VI.49], who formulated the famous RIEMANN HYPOTHESIS [IV.2 §3].

The Riemann zeta function is just one of a family of functions that encode important number-theoretic information. For example, the *Dirichlet L-functions* are closely related to the distribution of primes in arithmetic progressions. For more details about these and about the Riemann zeta function itself, see ANALYTIC NUMBER THEORY [IV.2]. Some more sophisticated zeta functions are described in THE WEIL CONJECTURES [V.35]. See also L-FUNCTIONS [III.47].

III.81   Rings, Ideals, and Modules

1   Rings

A ring, like a GROUP [I.3 §2.1] or a FIELD [I.3 §2.2], is an algebraic structure that satisfies certain axioms. To remember the axioms for both rings and fields at the same time, it is helpful to think of two simple examples: with the two operations of addition and multiplication, the set $\mathbb{Z}$ of all integers forms a ring and the set $\mathbb{Q}$ of all rational numbers forms a field. In general, a ring is a set R with two BINARY OPERATIONS [I.2 §2.4], denoted by "+" and "×", which satisfies all the field axioms apart from the one that says that nonzero elements have multiplicative inverses.

Although the integers are the prototypical example of a ring, the notion arose historically as an abstraction from several sources, one of which was polynomials. Like integers, polynomials (with real coefficients, say) can be added and multiplied, and these operations have all the properties one might expect, such as the fact that multiplication is distributive over addition, so the space of such polynomials forms a ring. Other examples include the integers modulo n (for any positive integer n), the rationals (or indeed any other field), and the set $\mathbb{Z}[i]$ of all complex numbers $a + bi$ such that a and b are integers.

Sometimes the assumptions that multiplication is commutative and has an identity element are dropped. This leads to a more complicated theory, but it does encompass important examples such as the set of all $n \times n$ matrices (with elements in a given field, or even just a ring).

As with other algebraic structures, there are several ways of forming new rings from old ones: for instance, we can take subrings and direct products of two rings. Slightly less obviously, we can start with a ring R and form the ring of all polynomials with coefficients in R.

We can also take QUOTIENTS [I.3 §3.3], but in order to discuss these we must introduce the notion of an ideal.

2   Ideals

A typical quotient construction for an algebraic structure A will identify some substructure B and regard two elements of A as "equivalent" if they "differ by an element of B." If A is a group or a VECTOR SPACE [I.3 §2.3], then B will be a subgroup or a subspace. However, the situation for rings is slightly different.

We can see why if we think about quotients in another way: as images of HOMOMORPHISMS [I.3 §4.1]. The substructures that we like to quotient by are the kernels of these homomorphisms, so we should ask ourselves what the kernel of a ring homomorphism (that is, the set of elements that map to 0) will be like.

If $\phi : R \to R'$ is a homomorphism between two rings, and $\phi(a) = \phi(b) = 0$, then $\phi(a + b) = 0$. Also, if r is any element of R, then $\phi(ra) = \phi(r)\phi(a) = 0$. Thus, the kernel of a homomorphism is closed under addition, and also under multiplication by any element of the ring. These two properties define the notion of an *ideal*. For example, the set of all even integers is an ideal in $\mathbb{Z}$. In interesting cases, ideals are not subrings, since if an ideal contains 1 then it must contain r for every r in the ring. (An example that makes the difference very clear is the subset of the ring of all polynomials that consists of all constant polynomials. The constants form a subring, but they certainly do not form an ideal.)

It is not hard to show that for any ideal I in a ring R there is a homomorphism that has I as its kernel, namely the quotient map from R to the quotient R/I. Here R/I is a construction that as usual we think of as "R, but with two elements considered the same if they differ by an element of I."

Quotients of rings are extremely useful in ALGEBRAIC NUMBER THEORY [IV.1] because they allow us to rephrase questions about algebraic numbers as questions about polynomials. To get an idea of how this is done, consider the ring $\mathbb{Z}[X]$ of all polynomials with integer coefficients, and the ideal that consists of all multiples (by integer polynomials) of the polynomial $X^2 + 1$. In the quotient of $\mathbb{Z}[X]$ by this ideal, we regard two polynomials as the same if they differ by a multiple of $X^2 + 1$. In particular, X^2 is the same as -1. In other words, in this quotient ring we have a square root of -1, and in fact the quotient ring is isomorphic to the ring $\mathbb{Z}[i]$ that we met earlier.

One of the things we like to do to integers is factorize them, and we can try to do the same in rings as well. However, it turns out that, while it is usually possible to factorize an element of a ring into "irreducible" ones that cannot be factorized further (like the primes in $\mathbb{Z}$), in many cases the factorization is not unique. At first, this might be rather unexpected, and indeed it was a stumbling block for many early workers (in the eighteenth and nineteenth centuries). Here is an example: in the ring $\mathbb{Z}[\sqrt{-3}]$, which consists of all complex numbers $a + b\sqrt{-3}$, where a and b are integers, the number 4 may be factorized as 2×2 and also as $(1 + \sqrt{-3}) \times (1 - \sqrt{-3})$.

3 Modules

Modules are to rings as vector spaces are to fields. In other words, they are algebraic structures where the basic operations are addition and scalar multiplication, but now the scalars are allowed to come from a ring rather than a field. For an example of a module over a ring that is not a field, take any Abelian group G. This can be turned into a module over $\mathbb{Z}$, with addition given by the group operation and scalar multiplication defined in the obvious way: for instance, $3g$ means $g + g + g$, and $-2g$ means the inverse of $g + g$.

The simplicity of this definition masks the fact that the structure of modules is in general far more subtle than that of vector spaces. For example, we can define a *basis* of a module to be a linearly independent set of elements that spans the module. However, many useful facts about bases in vector spaces do not hold for modules. For instance, in $\mathbb{Z}$, which we may consider as a module over itself, the set $\{2, 3\}$ spans the module but does not contain a basis, and similarly the set $\{2\}$ is linearly independent but cannot be extended to a basis. In fact, modules may be very far from having a basis: for example, if we consider the integers modulo n as a module over $\mathbb{Z}$, then even a single element x fails to be linearly independent, since $nx = 0$.

The following example of a module is an important one. Let V be a complex vector space and let α be a linear map from V to V. This can be made into a module over the ring $\mathbb{C}[X]$: if $v \in V$ and P is a complex polynomial, then Pv is defined to be $P(\alpha)v$. (For instance, if P is the polynomial $x^2 + 1$, then $Pv = \alpha^2 v + v$.) Applying general structural results about modules to this example, one obtains a proof of the JORDAN NORMAL FORM THEOREM [III.43].

III.82 Schemes
Jordan S. Ellenberg

One frequently finds in the history of mathematics that a definition thought to be completely general was in fact too restrictive to treat certain problems of interest. The notion of "number," for instance, has been expanded again and again—most notably to incorporate irrationalities and complex numbers, the former arising from problems in geometry and the latter needed in order to describe solutions to arbitrary algebraic equations. In a similar way, algebraic geometry, which was once understood as the study of *algebraic varieties*, or solution sets of algebraic equations in some finite-dimensional space, has grown to encompass more general objects known as "schemes." As a very meager example one might consider the two equations $x + y = 0$ and $(x + y)^2 = 0$. The two equations have the same set of solutions in the plane, so they describe the same variety; but the schemes attached to the two objects are completely different. The reformulation of algebraic geometry in the language of schemes was a tremendous project spearheaded by Alexander Grothendieck in the 1960s. As the above example suggests, the scheme-theoretic viewpoint tends to emphasize the algebraic aspects of the subject (equations) rather than the traditionally geometric ones (solution sets of equations). This viewpoint has made a reality of the long-hoped-for unification of ALGEBRAIC NUMBER THEORY [IV.1] and algebraic geometry, and, indeed, much recent progress in number theory would have been impossible without the geometric insight supplied by the theory of schemes.

Even schemes are not enough to handle all the problems of current interest, and still more general notions (stacks, "noncommutative varieties," derived categories of sheaves, etc.) are brought to bear when necessary. These can appear exotic, but to our successors they will no doubt be second nature, just as schemes are to us. For more on algebraic geometry in general, see ALGEBRAIC GEOMETRY [IV.4]. Schemes are discussed at greater length in ARITHMETIC GEOMETRY [IV.5].

III.83 The Schrödinger Equation
Terence Tao

In mathematical physics, the Schrödinger equation (and the closely related Heisenberg equation) are the most fundamental equations in nonrelativistic quantum mechanics, playing the same role as Hamilton's laws of motion (and the closely related Poisson equation) in nonrelativistic classical mechanics. (In relativistic quantum mechanics, the equations of quantum field theory take over the role of Heisenberg's equation, while Schrödinger's equation does not have a natural direct analogue.) In pure mathematics, the Schrödinger equation, together with its variants, is one of the basic equations studied in the field of PARTIAL DIFFERENTIAL EQUATIONS [IV.12], and has applications to geometry, to spectral and scattering theory, and to integrable systems.

The Schrödinger equation can be used to describe the quantum dynamics of many-particle systems under the influence of a variety of forces, but for simplicity let us consider just a single particle, of mass $m > 0$, moving about in n-dimensional space $\mathbb{R}^n$ subject to the influence of a potential, which we shall take to be a function $V : \mathbb{R}^n \to \mathbb{R}$. To avoid technicalities we shall assume that all the functions we discuss are smooth.

In classical mechanics, this particle would have a specific position $q(t) \in \mathbb{R}^n$ and a specific momentum $p(t) \in \mathbb{R}^n$ for each time t. (Eventually we shall observe the familiar law $p(t) = mv(t)$, where $v(t) = q'(t)$ is the velocity of the particle.) Thus, the state of this system at any given time t is described by the element $(q(t), p(t))$ of the space $\mathbb{R}^n \times \mathbb{R}^n$, which is known as *phase space*. The *energy* of this state is described by the HAMILTONIAN FUNCTION [III.35] $H : \mathbb{R}^n \times \mathbb{R}^n \to \mathbb{R}$ on phase space, defined in this case by

$$H(q, p) = \frac{|p|^2}{2m} + V(q).$$

(Physically, the quantity $|p|^2/2m = \frac{1}{2}m|v|^2$ represents kinetic energy, while $V(q)$ represents potential energy.) The system then evolves according to *Hamilton's equations of motion*:

$$q'(t) = \frac{\partial H}{\partial p}, \qquad p'(t) = -\frac{\partial H}{\partial q}, \qquad (1)$$

where we keep in mind that p and q are vectors, so that these derivatives are GRADIENTS [I.3 §5.3]. Hamilton's equations of motion are valid for any classical system, but in our specific case of a particle in a "potential well," they become

$$q'(t) = \frac{1}{m}p(t), \qquad p'(t) = -\nabla V(q). \qquad (2)$$

The first equation is asserting that $p = mv$, while the second equation is basically Newton's second law of motion.

From (1) we can easily derive *Poisson's equation of motion*

$$\frac{\mathrm{d}}{\mathrm{d}t}A(q(t), p(t)) = \{H, A\}(q(t), p(t)) \qquad (3)$$

for any *classical observable* $A : \mathbb{R}^n \times \mathbb{R}^n \to \mathbb{R}$, where

$$\{H, A\} = \frac{\partial H}{\partial p}\frac{\partial A}{\partial q} - \frac{\partial A}{\partial p}\frac{\partial H}{\partial q}$$

is the *Poisson bracket* of H and A. Setting $A = H$, we have in particular the *conservation-of-energy law*:

$$H(q(t), p(t)) = E \qquad (4)$$

for all $t \in \mathbb{R}$ and some quantity E independent of t.

Now we analyze the quantum mechanical analogue of the above classical system. We need a small[1] parameter $\hbar > 0$, known as *Planck's constant*. The state of the particle at a time t is no longer described by a single point $(q(t), p(t))$ in phase space, but is instead described by a *wave function*, which is a complex-valued function of position that evolves over time: that is, for each t we have a function $\psi(t)$ from $\mathbb{R}^n$ to $\mathbb{C}$. It is required to obey the normalization condition $\langle \psi(t), \psi(t) \rangle = 1$, where $\langle \cdot, \cdot \rangle$ denotes the inner product

$$\langle \phi, \psi \rangle = \int_{\mathbb{R}^n} \phi(q)\overline{\psi(q)}\,\mathrm{d}q.$$

Unlike a classical particle, a wave function $\psi(t)$ does not necessarily have a specific position $q(t)$. However, it does have an *average position* $\langle q(t) \rangle$, defined as

$$\langle q(t) \rangle = \langle Q\psi(t), \psi(t) \rangle = \int_{\mathbb{R}^n} q|\psi(t, q)|^2\,\mathrm{d}q.$$

Here, we have written $\psi(t, q)$ for the value of $\psi(t)$ at the point q, and Q is the *position operator*, defined by $(Q\psi)(t, q) = q\psi(t, q)$: that is, Q is the operator that multiplies pointwise by q. Similarly, while ψ does not have a specific momentum $p(t)$, it does have an *average momentum* $\langle p(t) \rangle$, defined as

$$\langle p(t) \rangle = \langle P\psi(t), \psi(t) \rangle = \frac{\hbar}{\mathrm{i}} \int_{\mathbb{R}^n} (\nabla_q \psi(t, q))\overline{\psi(t, q)}\,\mathrm{d}q,$$

where the *momentum operator* P is defined by *Planck's law*

$$P\psi(t, q) = \frac{\hbar}{\mathrm{i}}\nabla_q \psi(t, q).$$

Note that the vector $\langle p(t) \rangle$ is real-valued because all the components of P are SELF-ADJOINT [III.50 §3.2]. More generally, given any *quantum observable*, by which we mean a self-adjoint OPERATOR [III.50] A acting on the space $L^2(\mathbb{R}^n)$ of complex-valued square integrable functions, we can define the *average value* $\langle A(t) \rangle$ of A

---

1. In many applications it is convenient to normalize $\hbar$ (and m) to equal 1.

at time t by the formula

$$\langle A(t) \rangle = \langle A\psi(t), \psi(t) \rangle.$$

The analogue of Hamilton's equations of motion (1) is now the *time-dependent Schrödinger equation*:

$$i\hbar \frac{\partial \psi}{\partial t} = H\psi, \qquad (5)$$

where H is now a quantum observable rather than a classical one. More precisely,

$$H = \frac{|P|^2}{2m} + V(Q).$$

In other words, we have

$$i\hbar \frac{\partial \psi}{\partial t}(t, q) = H\psi(t, q)$$
$$= -\frac{\hbar^2}{2m}\Delta_q \psi(t, q) + V(q)\psi(t, q),$$

where

$$\Delta_q \psi = \sum_{j=1}^{n} \frac{\partial^2 \psi}{\partial q_j^2}$$

is the *Laplacian* of ψ. The analogue of Poisson's equation of motion (3) is the *Heisenberg equation*

$$\frac{\mathrm{d}}{\mathrm{d}t}\langle A(t) \rangle = \left\langle \frac{\mathrm{i}}{\hbar}[H(t), A(t)] \right\rangle \qquad (6)$$

for any observable A, where $[A, B] = AB - BA$ is the *commutator* or *Lie bracket* of A and B. (The quantity $(\mathrm{i}/\hbar)[A, B]$ is occasionally referred to as the *quantum Poisson bracket* of A and B.)

If the quantum state ψ oscillates in time according to the formula $\psi(t, q) = \mathrm{e}^{(E/\mathrm{i}\hbar)t}\psi(0, q)$ for some real number E (known as the *energy level* or *eigenvalue*), then one has the *time-independent Schrödinger equation*:

$$H\psi(t) = E\psi(t) \quad \text{for all times } t \qquad (7)$$

(compare this with (4)). More generally, the important subject of *spectral theory* provides many links between the time-dependent equation (5) and the time-independent equation (7).

There are several strong analogies between the equations of classical mechanics and those of quantum mechanics. For instance, from (6) one has the equations

$$\frac{\mathrm{d}}{\mathrm{d}t}\langle q(t) \rangle = \frac{1}{m}\langle p(t) \rangle, \qquad \frac{\mathrm{d}}{\mathrm{d}t}\langle p(t) \rangle = -\langle \nabla_q V(q)(t) \rangle,$$

which should be compared with (2). Also, given any classical solution $t \mapsto (q(t), p(t))$ to Hamilton's equation of motion, one can construct a corresponding family of *approximate* solutions $\psi(t)$ to Schrödinger's

equation, for instance by the formula[2]

$$\psi(t, q) = \mathrm{e}^{(\mathrm{i}/\hbar)L(t)}\mathrm{e}^{(\mathrm{i}/\hbar)p(t)\cdot(q - q(t))}\varphi(q - q(t)),$$

where

$$L(t) = \int_0^t \frac{p(s)^2}{2m} - V(q(s))\,\mathrm{d}s$$

is the *classical action* and φ is any slowly varying function that is normalized in the sense that

$$\int_{\mathbb{R}^n} |\varphi(q)|^2\,\mathrm{d}q = 1.$$

One can verify that ψ solves (5) except for some errors that are small when $\hbar$ is small. In physics, this fact is an example of the *correspondence principle*, which asserts that classical mechanics can be used to approximate quantum mechanics accurately if Planck's constant is small and one is working at macroscopic scales (which is what allows us to use slowly varying functions φ). In mathematics (and more precisely in the fields of *microlocal analysis* and *semi-classical analysis*), there are a number of formalizations of this principle that allow us to use knowledge about the behavior of Hamilton's equations of motion in order to analyze the Schrödinger equation. For example, if the classical equations of motion have periodic solutions, then the Schrödinger equation often has nearly periodic solutions, whereas if the classical equations have very chaotic solutions, then the Schrödinger equation typically does as well (this phenomenon is known as *quantum chaos* or *quantum ergodicity*).

There are many aspects of the Schrödinger equation that are of interest. We mention just one of them here for illustration, namely that of *scattering theory*. If the potential function V decays sufficiently quickly at infinity, and $k \in \mathbb{R}^n$ is a nonzero frequency vector, then, setting the energy level as $E = \hbar^2|k|^2/2m$, the time-independent Schrödinger equation $H\psi = E\psi$ admits solutions $\psi(q)$ that behave asymptotically (as $|q| \to \infty$) as

$$\psi(q) \approx \mathrm{e}^{\mathrm{i}k\cdot q} + f\left(\frac{q}{|q|}, k\right)\frac{\mathrm{e}^{\mathrm{i}|k||q|}}{r^{(n-1)/2}}$$

for some canonical function $f : S^{n-1} \times \mathbb{R}^n \to \mathbb{C}$, which is known as the *scattering amplitude function*. This scattering amplitude depends (in a nonlinear fashion) on

---

2. Intuitively, this function $\psi(t, q)$ is localized in position near $q(t)$ and localized in momentum near $p(t)$, and is thus localized near $(q(t), p(t))$ in phase space. Such a localized function, exhibiting such "particle-like" behavior as having a reasonably well-defined position and velocity, is sometimes known as a "wave packet." A typical solution of the Schrödinger equation does not behave like a wave packet, but can be decomposed as a *superposition* or *linear combination* of wave packets; such decompositions are a useful tool in analyzing general solutions of such equations.

the potential V, and the map from V to f is known as the *scattering transform*. The scattering transform can be viewed as a nonlinear variant of THE FOURIER TRANSFORM [III.27]; it is connected to many areas of partial differential equations, such as the theory of *integrable systems*.

There are many generalizations and variants of the Schrödinger equation; one can generalize to many-particle systems, or add other forces such as magnetic fields or even nonlinear terms. One can also couple this equation to other physical equations such as MAXWELL'S EQUATIONS [IV.13 §1.1] of electromagnetism, or replace the domain $\mathbb{R}^n$ by another space such as a torus, a discrete lattice, or a manifold. Alternatively, one could place some impenetrable obstacles in the domain (thus effectively removing those regions of space from the domain). The study of all of these variants leads to a vast and diverse field in both pure mathematics and in mathematical physics.

III.84 The Simplex Algorithm
Richard Weber

1 Linear Programming

The simplex algorithm is the preeminent tool for solving some of the most important mathematical problems arising in business, science, and technology. In these problems, which are called linear programs, we are to maximize (or minimize) a linear function subject to linear constraints. An example is the diet problem posed by the U.S. Air Force in 1947: find quantities of seventy-seven differently priced foodstuffs (cheese, spinach, etc.) to satisfy a man's minimum daily requirements for nine nutrients (protein, iron, etc.) at least cost. Further applications occur in choosing the elements of an investment portfolio, rostering an airline's crew, and finding optimal strategies in two-person games. The study of linear programming has inspired many of the central ideas of optimization theory, such as DUALITY [III.19], the importance of convexity, and COMPUTATIONAL COMPLEXITY [IV.20].

The input data of a linear program (LP) consists of two vectors $b \in \mathbb{R}^m$ and $c \in \mathbb{R}^n$, and an $m \times n$ matrix $A = (a_{ij})$. The problem is to find values for n nonnegative decision variables, $x_1, \dots, x_n$, to maximize the *objective function* $c_1 x_1 + \cdots + c_n x_n$, subject to m constraints, $a_{i1} x_1 + \cdots + a_{in} x_n \leqslant b_i$, $i = 1, \dots, m$. In the diet problem, $n = 77$ and $m = 9$. In the following simple example (not a diet problem), $n = 2$ and $m = 3$. In

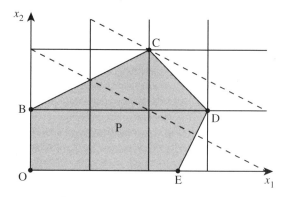

Figure 1 Feasible region "P" of an LP.

serious real-life problems, n and m can be greater than 100 000.

$$\text{Maximize} \quad x_1 + 2x_2$$
$$\text{subject to} \quad -x_1 + 2x_2 \leqslant 2,$$
$$x_1 + x_2 \leqslant 4,$$
$$2x_1 - x_2 \leqslant 5,$$
$$x_1, x_2 \geqslant 0.$$

The constraints define a feasible region for (x_1, x_2), a convex polygon that is depicted by the shaded region "P" in figure 1. The two dotted lines mark those x where the value of the objective function value is 4 and where it is 6. Clearly, it is maximized at point C.

The general story is similar to that of the example. If the feasible region $P = \{x : Ax \leqslant b, x \geqslant 0\}$ is nonempty, then it is a convex polytope in $\mathbb{R}^n$, and an optimal solution can be found at one of its vertices. It is helpful to introduce "slack variables" $x_3, x_4, x_5 \geqslant 0$ to take up the slack on the left of the inequality constraints. We can write

$$-x_1 + 2x_2 + x_3 \qquad = 2,$$
$$x_1 + x_2 \qquad + x_4 \qquad = 4,$$
$$2x_1 - x_2 \qquad + x_5 = 5.$$

We now have three equations in five variables, so we can set any two of the variables $x_1, \dots, x_5$ equal to 0, and solve the equations for the other three variables (or solve a perturbation of them if they happen not to be independent). There are ten ways to choose two variables from five. Not all of the ten corresponding solutions satisfy $x_1, x_2, x_3, x_4, x_5 \geqslant 0$, but five of them do. These are called *basic feasible solutions* (BFSs), and correspond to the vertices of P marked O, B, C, D, E.

2 How the Algorithm Works

George Dantzig invented the simplex algorithm in 1947 as a means of solving the Air Force's diet problem mentioned at the start. The word "program" was not yet used to mean computer code, but was a military term for a logistic plan or schedule. The fundamental fact on which the algorithm relies is that if an LP has a bounded optimal solution, then the optimum value is attained at a BFS, i.e., at a vertex (or so-called "extreme point") of the polytope of feasible points, P. Another name for the feasible polytope is "simplex," which is where the algorithm gets its name. It works as follows.

Step 0. Pick a BFS.

Step 1. Test whether this BFS is optimal.

If so, stop. If not, go to step 2.

Step 2. Find a better BFS.

Repeat from step 1.

Since there are only finitely many BFSs (i.e., vertices of P), the algorithm must stop.

Now that we have an overview, let us look at the details. Suppose that at step 0 we pick the BFS of $x = (x_1, x_2, x_3, x_4, x_5) = (0, 0, 2, 4, 5)$, corresponding to vertex O. At step 1 we wish to know whether the objective function can be increased if x_1 or x_2 is increased from 0. So we write x_3, x_4, x_5, and the objective function $c^T x$ in terms of x_1 and x_2, and display this as dictionary 1.

Dictionary 1
$x_3 = 2 + x_1 - 2x_2,$
$x_4 = 4 - x_1 - x_2,$
$x_5 = 5 - 2x_1 + x_2,$
$c^T x = x_1 + 2x_2.$

The last equation in the dictionary shows that we can increase the value of $c^T x$ by increasing either x_1 or x_2 from 0. Suppose that we increase x_2. The first and second equations show that x_3 and x_4 must decrease, and we cannot increase x_2 beyond 1, at which point $x_3 = 0$ and $x_4 = 3$, $x_5 = 6$. Increasing x_2 as much as possible, we complete step 2 and arrive at a new BFS of $x = (0, 1, 0, 3, 6)$, which is vertex B. Now we are ready for step 1 again, and so we write x_2, x_4, x_5, and $c^T x$ in terms of the variables that are now zero, namely x_1, x_3, to give dictionary 2.

Dictionary 2
$x_2 = 1 + \frac{1}{2}x_1 - \frac{1}{2}x_3,$
$x_4 = 3 - \frac{3}{2}x_1 + \frac{1}{2}x_3,$
$x_5 = 6 - \frac{3}{2}x_1 - \frac{1}{2}x_3,$
$c^T x = 2 + 2x_1 - x_3.$

Dictionary 3
$x_1 = 2 + \frac{1}{3}x_3 - \frac{2}{3}x_4,$
$x_2 = 2 - \frac{1}{3}x_3 - \frac{1}{3}x_4,$
$x_5 = 3 - x_3 + x_4,$
$c^T x = 6 - \frac{1}{3}x_3 - \frac{4}{3}x_4.$

This shows that $c^T x$ can be increased by increasing x_1 from 0, but that x_1 can increase no further than 2 because at that point $x_4 = 0$. This brings us to a new solution $(2, 2, 0, 0, 3)$, which is vertex C. Once more, we are ready for step 1, and so compute dictionary 3, now writing things in terms of x_3 and x_4, which are 0. The algorithm now stops because, as we require $x_3, x_4 \geqslant 0$, the bottom line of dictionary 3 proves that $c^T x \leqslant 6$ for all feasible x.

There is other important information in the final dictionary. If b is changed to $b + \epsilon$, for small $\epsilon^T = (\epsilon_1, \epsilon_2, \epsilon_3)$, then the maximum value of $c^T x$ will change to $6 + \frac{1}{3}\epsilon_1 + \frac{4}{3}\epsilon_2$. The coefficient $\frac{1}{3}$ is called a "shadow price," because it is what we should be willing to pay per unit increase in b_1.

3 How the Algorithm Performs

In running the simplex algorithm the serious work comes in computing the dictionaries. To find dictionary 2, we could use the first equation of dictionary 1 to rewrite x_2 in terms of x_1 and x_3, and then substitute for x_2 in the other equations. Versions of the simplex algorithm have been invented that reduce the computing effort by taking advantage of special structure in the matrix A, such as the fact that most of its entries are zero. The dictionary data is often held in a so-called *tableau* of coefficients.

There are many other practical and theoretical issues. One concerns the selection of the *pivot*, that is, the variable that is to be increased from 0. Starting at O, and depending on which of x_1 and x_2 we choose as the first variable to increase from zero, the path to C can be O, E, D, C or O, B, C. There is no known way to guarantee that the algorithm takes the shortest path.

The question of how many steps the simplex algorithm really needs is related to the famous *Hirsch conjecture*: that for any bounded n-dimensional polytope with m faces, the diameter (defined as the maximum number of edges on the shortest edge-traversing path between any two vertices) is at most $m - n$. If this were true, it would suggest that some version of the simplex algorithm might run in a number of steps that grows only linearly in the numbers of variables and constraints. However, Klee and Minty (1972) have given an example based on a perturbed n-dimensional cube ($m = 2n$ faces and diameter n), in which if the algorithm selects among possible pivots by choosing the one for which the objective function increases at the greatest rate per unit increase in that variable, then it visits all 2^n vertices before reaching the optimum. Indeed, for most deterministic pivot selection rules, examples are known in which the number of steps grows exponentially in n.

Fortunately, things are usually much better in practical problems than in worst-case examples. Typically, only $O(m)$ steps are needed to solve a problem with m constraints. Moreover, Khachian (1979) proved (by analysis of the so-called *ellipsoid algorithm*) that linear programs can in principle be solved by an algorithm whose running time grows only polynomially in n. Thus linear programming is much easier than "integer linear programming," in which $x_1, \ldots, x_n$ are required to be integers and for which no algorithm with polynomial running time is known.

Karmarkar (1984) pioneered development of "interior" methods for linear programming problems. These move through the interior of the polytope P, rather than among its vertices, and can sometimes solve large LPs more quickly than the simplex algorithm. Modern computer software uses both methods and can easily solve LPs with millions of variables and constraints.

Further Reading

Dantzig, G. 1963. *Linear Programming and Extensions*. Princeton, NJ: Princeton University Press.

Karmarkar, N. 1984. A new polynomial-time algorithm for linear programming. *Combinatorica* 4:373–95.

Khachian, L. G. 1979. A polynomial algorithm in linear programming. *Soviet Mathematics Doklady* 20:191–94.

Klee, V., and G. Minty. 1972. How good is the simplex algorithm? In *Inequalities III*, edited by O. Shisha, volume 16, pp. 159–75. New York: Academic Press.

Solitons

> *See* LINEAR AND NONLINEAR WAVES AND SOLITONS [III.49]

III.85 Special Functions
T. W. Körner

Suppose that the only functions we have come across are quotients of polynomials and that we are asked to solve the differential equation

$$f'(x) = 1/x \qquad (1)$$

for all $x > 0$, subject to the condition $f(1) = 0$.

If we try $f(x) = P(x)/Q(x)$, where P and Q are polynomials with no common factors, then we find that

$$x(Q(x)P'(x) - P(x)Q'(x)) = Q(x)^2.$$

By comparing coefficients we can show that $Q(0) = P(0) = 0$, which shows that, contrary to our assumptions, both $P(x)$ and $Q(x)$ are divisible by x. Thus, we cannot solve equation (1) in terms of known functions. However, THE FUNDAMENTAL THEOREM OF CALCULUS [I.3 §5.5] tells us that equation (1) does indeed have a solution, namely

$$F(x) = \int_1^x \frac{1}{t}\, dt.$$

Further study shows that the function F has many useful properties. For example, using the substitution $u = t/a$, we find that

$$\begin{aligned}
F(ab) &= \int_1^a \frac{1}{t}\, dt + \int_a^{ab} \frac{1}{t}\, dt \\
&= \int_1^a \frac{1}{t}\, dt + \int_1^b \frac{1}{u}\, du \\
&= F(a) + F(b),
\end{aligned}$$

and, using the formula for differentiating an inverse function, we find that F^{-1} is the solution of the differential equation

$$g'(x) = g(x).$$

We therefore give the function a name (the *logarithm*) and add it to our list of standard functions.

At a more advanced level, integration by parts shows that the GAMMA FUNCTION [III.31] (introduced by EULER [VI.19])

$$\Gamma(x) = \int_0^\infty t^{x-1} e^{-t}\, dt,$$

defined for all $x > 0$, has the property that

$$\Gamma(x) = (x - 1)\Gamma(x - 1)$$

for all $x > 1$, and therefore $\Gamma(n) = (n - 1)!$ for all integers $n \geqslant 1$ (since $\Gamma(1) = 1$). As one might expect from its association with factorials, the gamma function turns out to be very useful in number theory and statistics.

In practice, a "special function" is any function that, like the logarithm and the gamma function, has been extensively studied and has turned out to be useful. Some authors use the phrase "special functions" in a more restricted sense, meaning something like "functions that turn up in the solution of physical problems" or "functions other than those generally provided by a pocket calculator," but these restrictions do not seem to be very useful.

In spite of this apparent generality, the theory of special functions is linked in the minds of many mathematicians to a collection of particular ideas and methods. Indeed, it is often linked to particular books like Whittaker and Watson's *A Course of Modern Analysis* (which was first published in 1902 and is still selling well) and Abramowitz and Stegun's *Handbook of Mathematical Functions*. These connections may simply be accidents of history, but the phrase "special functions" is often associated with other phrases like "equations of mathematical physics," "beautiful formulas," and "sheer ingenuity." We illustrate this and other themes in the particular case of *Legendre polynomials*. (The next paragraph involves more advanced mathematics and glosses over several long calculations, but the reader may simply glance over its contents and resume careful reading thereafter.)

Suppose that we wish to examine the gravitational potential ψ of Earth by looking at solutions of LA-PLACE'S EQUATION [I.3 §5.4] $\Delta\psi = 0$. Since Earth is an oblate spheroid that is nearly spherical, we use spherical polar coordinates (r, θ, ϕ) and, noting that Earth is symmetric about its axis of rotation, we may suppose that ψ depends only on r and θ. Under these assumptions, Laplace's equation takes the form

$$\sin\theta \frac{\partial}{\partial r}\left(r^2 \frac{\partial \psi}{\partial r}\right) + \frac{\partial}{\partial \theta}\left(\sin\theta \frac{\partial \psi}{\partial \theta}\right) = 0. \quad (2)$$

Following the standard technique of separation of variables, we look for solutions of the form $\psi(r, \theta) = R(r)\Theta(\theta)$. After a little calculation, equation (2) yields

$$\frac{1}{R(r)}\frac{d}{dr}(r^2 R'(r)) = -\frac{1}{\sin\theta\,\Theta(\theta)}\frac{d}{d\theta}(\sin\theta\,\Theta'(\theta)) \quad (3)$$

Since one side of equation (3) depends on r alone and the other on θ alone, both sides must equal some constant k. The equation

$$\frac{1}{R(r)}\frac{d}{dr}(r^2 R'(r)) = k$$

has the solution $R(r) = r^l$ whenever $l(l+1) = k$. The corresponding equation for Θ is then

$$\frac{1}{\sin\theta\,\Theta(\theta)}\frac{d}{d\theta}(\sin\theta\,\Theta'(\theta)) = -l(l+1). \quad (4)$$

We now make the substitution $x = \cos\theta$, $y(x) = \Theta(\theta)$ to convert (4) to *Legendre's equation*

$$(1 - x^2)y''(x) - 2xy'(x) + l(l+1)y(x) = 0. \quad (5)$$

Routine equating of coefficients reveals that, if we seek nontrivial solutions of the form $f(x) = \sum_{j=0}^{\infty} a_j x^j$, then, unless l is an integer, $f(x)$ is unbounded as x approaches 1 (that is, as θ approaches 0), so these solutions are not useful physically. However, if l is a positive integer, then there is a polynomial solution of degree l. (If l is a negative integer, the same polynomials reappear.) In fact, we have the following stronger statement: if l is a positive integer, then there exists a unique polynomial P_l of degree l satisfying Legendre's equation (5) such that $P_l(1) = 1$. We call P_l the lth Legendre polynomial. Returning to our original problem, we see that it has solutions of the form

$$\psi(r, \theta) = \sum_{n=0}^{\infty} A_n \frac{P_n(\cos\theta)}{r^{n+1}}.$$

It is obvious to the physicist, and can be proved by the mathematician, that this is the most general solution if we also demand that $\phi(r, \theta) \to 0$ as $r \to \infty$. Notice that if r is large, then only the first few terms will contribute much to the final answer.

There are many different ways of obtaining the Legendre polynomials. The reader is invited to verify that, if we define Q_n inductively by setting $Q_0(x) = 1$ and $Q_1(x) = x$, and using the "three-term recurrence relation"

$$(n+1)Q_{n+1}(x) - (2n+1)xQ_n(x) + nQ_{n-1}(x) = 0,$$

then $Q_n(1) = 1$ and Q_n is a polynomial that satisfies Legendre's equation (5) (with $l = n$), from which it follows that Q_n is the Legendre polynomial of degree n.

If we set $v_n(x) = (x^2 - 1)^n$, then

$$(x^2 - 1)v_n'(x) = 2nxv(x).$$

Differentiating both sides of this equation $n + 1$ times using Leibniz's rule, we see that $v_n^{(n)}$ satisfies Legendre's equation (5) with $l = n$. Differentiating $v_n(x) = (x-1)^n(x+1)^n$ n times using Leibniz's rule and noting that all but one of the resulting terms vanish when $x = 1$, we see that v_n^n is a polynomial with $v_n^{(n)}(1) = 2^n n!$. Putting all this information together, we obtain *Rodriguez's formula*

$$P_n(x) = \frac{1}{2^n n!}v_n^{(n)}(x) = \frac{1}{2^n n!}\frac{d}{dx}(x^2 - 1)^n.$$

Equation (5) is an example of a *Sturm–Liouville equation*. Setting $l = n$ and $y = P_n$ and rewriting slightly, we obtain the equation

$$\frac{d}{dx}((1 - x^2)P_n'(x)) + n(n+1)P_n(x) = 0. \quad (6)$$

If m and n are positive integers, then, using (6) and integrating by parts, we obtain

$$-n(n+1) \int_{-1}^{1} P_n(x) P_m(x) \, dx$$

$$= \int_{-1}^{1} \left(\frac{d}{dx} ((1-x^2) P_n'(x)) \right) P_m(x) \, dx$$

$$= [(1-x^2) P_n'(x) P_m(x)]_{-1}^{1}$$

$$+ \int_{-1}^{1} (1-x^2) P_n'(x) P_m'(x) \, dx$$

$$= \int_{-1}^{1} (1-x^2) P_n'(x) P_m'(x) \, dx.$$

Thus, by symmetry,

$$n(n+1) \int_{-1}^{1} P_n(x) P_m(x) \, dx$$

$$= m(m+1) \int_{-1}^{1} P_n(x) P_m(x) \, dx,$$

and, if $m \neq n$,

$$\int_{-1}^{1} P_n(x) P_m(x) \, dx = 0. \tag{7}$$

The "orthogonality relation" given by (7) has important consequences. Since P_r is a polynomial of degree exactly r, we know that any polynomial Q of degree $n-1$ or less can be written

$$Q(x) = \sum_{r=0}^{n-1} a_r P_r(x)$$

and so

$$\int_{-1}^{1} P_n(x) Q(x) \, dx = \sum_{r=0}^{n-1} a_r \int_{-1}^{1} P_n(x) P_r(x) \, dx = 0. \tag{8}$$

Thus, P_n is orthogonal to all polynomials of lower degree.

Suppose that the points where $P_n(x)$ changes sign in the interval $[-1, 1]$ are $\alpha_1, \ldots, \alpha_m$. Then, if we write

$$Q(x) = (x - \alpha_1)(x - \alpha_2) \cdots (x - \alpha_m),$$

we know that $P(x) Q(x)$ does not change sign on $[-1, 1]$ and so

$$\int_{-1}^{1} P_n(x) Q(x) \, dx \neq 0.$$

By equation (8) this means that the degree m of Q is at least n and so (since a polynomial of degree n can have at most n zeros) P_n must have exactly n distinct zeros and they must all lie in the interval $[-1, 1]$.

GAUSS [VI.26] made use of these facts to obtain a powerful method of numerical integration. Suppose that $x_1, x_2, \ldots, x_{n+1}$ are distinct points on $[-1, 1]$. If we set

$$e_j(x) = \prod_{i \neq j} \frac{x - x_i}{x_j - x_i},$$

then $e_j(x)$ is a polynomial of degree n that takes the value 1 when $x = x_j$ and 0 when $x = x_k$ with $k \neq j$. Thus, if R is any polynomial of degree at most n, the polynomial Q given by

$$Q(x) = R(x_1) e_1(x) + R(x_2) e_2(x) + \cdots$$
$$+ R(x_{n+1}) e_{n+1}(x) - R(x)$$

has degree at most n, and $R - Q$ vanishes at the $n+1$ points x_j. It follows that $R = Q$, so

$$R(x) = R(x_1) e_1(x) + R(x_2) e_2(x) + \cdots$$
$$+ R(x_{n+1}) e_{n+1}(x).$$

If we write $a_j = \int_{-1}^{1} e_j(x) \, dx$, then

$$\int_{-1}^{1} R(x) \, dx = a_1 R(x_1) + a_2 R(x_2) + \cdots + a_{n+1} R(x_{n+1}).$$

It is natural to hope that the approximation

$$\int_{-1}^{1} f(x) \, dx \approx a_1 f(x_1) + a_2 f(x_2) + \cdots + a_{n+1} f(x_{n+1}), \tag{9}$$

which is an exact equality when f is a polynomial of degree n or less, will work well for other well-behaved functions.

Gauss observed that we can make a major improvement by taking the x_j to be the $n+1$ roots of the $(n+1)$st Legendre polynomial. Suppose that P is a polynomial of degree at most $2n+1$. Then we can write

$$P(x) = Q(x) P_{n+1}(x) + R(x),$$

where Q and R are polynomials of degree at most n and P_{n+1} is the $(n+1)$st Legendre polynomial. Now P_{n+1} is orthogonal to polynomials over lower degree (and, in particular, to Q), $P_{n+1}(x_j) = 0$ by the definition of x_j, and the approximation (9) is an equality for R. Thus,

$$\int_{-1}^{1} P(x) \, dx = \int_{-1}^{1} P_{n+1}(x) Q(x) \, dx + \int_{-1}^{1} R(x) \, dx$$

$$= 0 + \sum_{j=1}^{n+1} a_j R(x_j)$$

$$= \sum_{j=1}^{n+1} a_j (P_{n+1}(x_j) Q(x_j) + R(x_j))$$

$$= \sum_{j=1}^{n+1} a_j P(x_j).$$

We have shown that the "quadrature formula" (9) is actually exact for all polynomials of degree at most $2n+1$, provided we choose the x_j to be the numbers suggested by Gauss. Unsurprisingly, this choice gives an extremely good way of estimating integrals numerically. "Gaussian quadrature" is one of the two main methods used to evaluate integrals on computers today.

We conclude with a brief look at a few other special functions.

Consider de Moivre's formula

$$\cos n\theta + i\sin n\theta = (\cos\theta + i\sin\theta)^n.$$

Using the binomial expansion, we see that

$$\cos n\theta + i\sin n\theta = \sum_{r=0}^{n}\binom{n}{r}i^r\cos^{n-r}\theta\sin^r\theta,$$

and, taking real parts,

$$\cos n\theta = \sum_{r=0}^{\lfloor n/2\rfloor}\binom{n}{2r}(-1)^r\cos^{n-2r}\theta\sin^{2r}\theta.$$

Since $\sin^2\theta = 1 - \cos^2\theta$, we have

$$\cos n\theta = \sum_{r=0}^{\lfloor n/2\rfloor}\binom{n}{2r}(-1)^r\cos^{n-2r}\theta(1-\cos^2\theta)^r$$

$$= T_n(\cos\theta),$$

where T_n is a polynomial of degree n called the nth *Chebyshev polynomial*. The Chebyshev polynomials play an important role in numerical analysis.

The next collection of functions requires us to calculate with infinite sums. Readers may treat our calculations as plausible or justify them rigorously according to taste. Observe first that

$$h(x) = \sum_{n=-\infty}^{\infty}\frac{1}{(x-n\pi)^2}$$

is well-defined for each real x that is not a multiple of π. Note also that $h(x+\pi) = h(x)$ and $h(\frac{1}{2}\pi - x) = h(\frac{1}{2}\pi + x)$. Set $f(x) = h(x) - \operatorname{cosec}^2(\pi x)$. By showing that there are constants K_1 and K_2 such that

$$0 < \sum_{n=1}^{\infty}\frac{1}{(x-n\pi)^2} < K_1$$

and

$$0 < \operatorname{cosec}^2 x - \frac{1}{x^2} < K_2$$

for all $0 < x \leqslant \frac{1}{2}\pi$, we deduce that there is a constant K such that $|f(x)| < K$ for all $0 < x < \pi$. Simple calculations show that

$$f(x) = \tfrac{1}{4}(f(\tfrac{1}{2}x) + f(\tfrac{1}{2}(x+\pi))). \tag{10}$$

A single application of (10) shows that $|f(x)| < \frac{1}{2}K$ for all $0 < x < \pi$, and repeated applications show that $f(x) = 0$. Thus

$$\operatorname{cosec}^2 x = \sum_{n=-\infty}^{\infty}\frac{1}{(x-n\pi)^2}$$

for all real noninteger x.

If we seek analogues in the complex plane, we are led to functions of the type

$$F(z) = \sum_{n=-\infty}^{\infty}\sum_{m=-\infty}^{\infty}\frac{1}{(z-n-mi)^3}.$$

Observe that, while the real function $\operatorname{cosec}^2 x$ satisfies $\operatorname{cosec}^2(x+\pi) = \operatorname{cosec}^2(x)$ and is periodic with period π, the complex function F just defined satisfies

$$F(z+1) - F(z), \qquad F(z+i) - F(z)$$

and is *doubly periodic* with periods 1 and i. Functions like F are called *elliptic functions* and have a theory that parallels that of the TRIGONOMETRIC FUNCTIONS [III.92].

The function $E(x) = (2\pi)^{-1/2}e^{-x^2/2}$ is called the *Gaussian* or *normal* function and appears in probability and the study of diffusion processes (see [III.71 §5] and [IV.24]). The partial differential equation

$$\frac{\partial^2\phi}{\partial x^2}(x,t) = K\frac{\partial\phi}{\partial t}(x,t)$$

with x corresponding to distance and t to time provides a reasonable model for diffusion. It is easy to check that $\phi(x,t) = \psi(x,t) = (Kt)^{-1/2}E(x(Kt)^{-1/2})$ is a solution. By sketching a graph of $\psi(x,t)$ as a function of x for various values of t, readers will see that ψ can be considered as the response to a disturbance at $x = 0$ when $t = 0$. By considering the behavior of $\psi(x,t)$ as a function of t for a given value of x, they will see that "the effect at x of a disturbance at the origin becomes noticeable only after a time of the order $x^{1/2}$." Living cells depend on diffusion processes and the result just given suggests (correctly) that such processes are very slow over long distances. It is plausible that this sets a limit on the size of a single cell: a large organism must be multi-celled.

Statisticians constantly use the related *error function*

$$\operatorname{erf}(x) = \frac{2}{\pi^{1/2}}\int_0^x \exp(-t^2)\,dt.$$

There is a famous theorem of LIOUVILLE [VI.39] that shows that $\operatorname{erf}(x)$ cannot be expressed as a composition of elementary functions (such as quotients of polynomials, trigonometric functions, and EXPONENTIAL FUNCTIONS [III.25]).

We have been able to look at only a few properties of a few special functions in this article, but even this small sample shows how much interesting mathematics arises when we study one function or a class of particular functions rather than functions in general.

III.86 The Spectrum
G. R. Allan

In the theory of LINEAR MAPS [I.3 §4.2], or *operators*, on a VECTOR SPACE [I.3 §2.3], the notions of EIGEN-VALUE AND EIGENVECTOR [I.3 §4.3] play an important role. Recall that if V is a vector space (over $\mathbb{R}$ or $\mathbb{C}$) and if $T : V \to V$ is a linear mapping, then an *eigenvector* of T is a nonzero vector e in V such that $T(e) = \lambda e$ for some scalar λ; then λ is the *eigenvalue* corresponding to the eigenvector e. If V is finite dimensional, then the eigenvalues are also the roots of the *characteristic polynomial* $\chi(t) = \det(tI - T)$ of T. Because every nonconstant complex polynomial has a root (the so-called FUNDAMENTAL THEOREM OF ALGEBRA [V.13]), it follows that every linear operator on a finite-dimensional, complex vector space has at least one eigenvalue. If the scalar field is $\mathbb{R}$, then not all operators have eigenvectors (e.g., consider a rotation about the origin in $\mathbb{R}^2$).

The linear operators that arise in analysis usually act on infinite-dimensional spaces (see [III.50]). We consider *continuous linear operators* acting on a complex BANACH SPACE [III.62]; these will be referred to simply as *operators* (even though not all linear operators on an infinite-dimensional Banach space are continuous). We shall now see that, for X infinite dimensional, not every such operator has an eigenvalue.

Example 1. Let X be the Banach space $C[0, 1]$, consisting of all continuous, complex-valued functions on the closed interval $[0, 1]$ of the real line. The vector-space structure is the "natural" one (e.g., for $f, g \in X$ the sum $f + g$ is defined by setting $(f + g)(t) = f(t) + g(t)$ for each t and the norm is the *supremum norm*, that is, the largest value of any $|f(t)|$).

Now let u be a continuous complex-valued function on $[0, 1]$. We can associate with it a *multiplication operator* M_u on $C[0, 1]$ as follows. Given a function f, let $M_u(f)$ be the function that takes t to $u(t)f(t)$. It is clear that M_u is linear and continuous. We shall see that whether M_u has an eigenvalue depends on the choice of u. We consider two simple cases.

(i) Let u be the constant function $u(t) \equiv k$. Then evidently M_u has the single eigenvalue k and every (nonzero) function f in X is an eigenvector.

(ii) Let $u(t) = t$ for all t. Suppose that the complex number λ is an eigenvalue of M_u. Then there is some $f \in C[0, 1]$, not identically zero, such that $u(t)f(t) = \lambda f(t)$ and so $(t - \lambda)f(t) = 0$ for all

t. But then $f(t) = 0$ for all $t \neq \lambda$, so that, since f is continuous, $f(t) \equiv 0$, contrary to hypothesis. So, for this choice of u, the operator M_u has no eigenvalue.

Let X be a complex Banach space and let T be an operator on X. Then T is said to be *invertible* if and only if there is some operator S on X for which $ST = TS = I$ (here, ST is the composition of S and T, and I is the identity operator on X). It can be shown that T is invertible if and only if T is both *injective* (i.e., $T(x) = 0$ only for $x = 0$) and *surjective* (i.e., $T(X) = X$). The part here that is not just simple algebra is to show that if T is both injective and surjective, then the linear inverse T^{-1} is a *continuous* operator. A complex number λ is an eigenvalue of T precisely if $T - \lambda I$ is *not* injective.

If V is a *finite-dimensional* space, then an injective operator $T : V \to V$ is necessarily also surjective, and hence invertible. For X infinite dimensional this implication is no longer valid.

Example 2. Let H be the HILBERT SPACE [III.37] ℓ^2 that consists of all sequences $(\xi_n)_{n \geqslant 1}$ of complex numbers such that $\sum_{n \geqslant 1} |\xi_n|^2 < \infty$. Let S be the "right-shift" operator defined by $S(\xi_1, \xi_2, \xi_3, \dots) = (0, \xi_1, \xi_2, \dots)$. Then S is injective but not surjective. The "reverse shift" S^*, defined by $S^*(\xi_1, \xi_2, \xi_3, \dots) = (\xi_2, \xi_3, \dots)$, is surjective but not injective.

With this example in mind, we make the following definition.

Definition 3. Let X be a complex Banach space and let T be an operator on X. The *spectrum* of T, denoted by $\mathrm{Sp}\, T$ (or $\sigma(T)$), is the set of all complex numbers λ such that $T - \lambda I$ is not invertible.

The following remarks should be clear.

(i) If X is finite dimensional, then $\mathrm{Sp}\, T$ is just the set of eigenvalues of T.

(ii) For general X, $\mathrm{Sp}\, T$ includes the set of eigenvalues of T, but may be larger (e.g., in example 2, 0 is not an eigenvalue of S, but 0 does belong to the spectrum of S).

It is easy to show that the spectrum is always a *bounded* and *closed* (i.e., COMPACT [III.9]) subset of $\mathbb{C}$. A rather deeper fact is that it is never empty: that is, there will always be some λ for which $T - \lambda I$ is not invertible. That is proved by applying LIOUVILLE'S THEOREM [I.3 §5.6] to the analytic operator-valued function $\lambda \mapsto (\lambda I - T)^{-1}$, defined for λ not in the spectrum of T.

Example 1 continued. We have already seen that not all multiplication operators have eigenvalues. However, they do have an easily described spectrum. Let M_u be such an operator and let S be the set of all values $u(t)$ taken by the function u. Let $\mu = u(t_0)$ be one of these values and consider the operator $M_u - \mu I$. Given any function f in $C[0, 1]$, the value of $(M_u - \mu I)f$ at t_0 is $u(t_0)f(t_0) - \mu f(t_0) = 0$. It follows that $M_u - \mu I$ is not surjective (for instance, the range of $M_u - \mu I$ does not contain any nonzero constant function) and therefore μ belongs to the spectrum of M_u. Thus, S is contained in the spectrum of M_u; it is not hard to show that the two are in fact equal.

We may easily generalize this example to show that if K is *any* nonempty compact subset of $\mathbb{C}$, then there is a linear operator T with K as its spectrum. Let X be the space of continuous complex-valued functions defined on K, for each $z \in K$, let $u(z) = z$, and let T be the multiplication operator M_u, defined as it was when K was the set $[0, 1]$.

The spectrum is central to most aspects of operator theory. We shall briefly mention a result about Hilbert space operators, known as the spectral theorem (there are a number of variations).

Let H be a Hilbert space with inner product $\langle x, y \rangle$. A continuous linear operator T on H is called *Hermitian* if $\langle Tx, y \rangle = \langle x, Ty \rangle$ for every x and y in H.

Examples 4.

(i) If H is finite dimensional, then a linear operator S on H is Hermitian if and only if, with respect to some (and hence *every*) ORTHONORMAL BASIS [III.37], S is represented by a Hermitian matrix (i.e., a matrix A with $A = \bar{A}^{\mathrm{T}}$).

(ii) On the Hilbert space $L_2[0, 1]$, let M_u be the operator of multiplication by a continuous function u (just as in example 1, but now we apply M_u to functions in $L_2[0, 1]$ rather than just $C[0, 1]$). Then M_u is Hermitian if and only if u is real-valued.

If H is finite dimensional and T is a Hermitian operator on H, then H has an orthonormal basis consisting of eigenvectors of T (a "diagonal basis"). Equivalently, $T = \sum_{j=1}^{k} \lambda_j P_j$, where $\{\lambda_1, \ldots, \lambda_k\}$ are the *distinct* eigenvalues of T and P_j is the orthogonal projection of H onto the eigenspace $E_j \equiv \{x \in H : Tx = \lambda_j x\}$.

If H is infinite dimensional and T is a Hermitian operator on H, then it is *not* generally true that H has a basis of eigenvectors. But, very importantly, the representation $T = \sum \lambda_j P_j$ *does* generalize to a representation $T = \int \lambda \, dP$, a kind of integral with respect to a "projection-valued measure" on the spectrum of T.

There is an intermediate case, for so-called *compact Hermitian operators*, "compactness" being a kind of strong continuity, of great importance in applications. The technicalities are much simpler than in the general case, involving an infinite sum, rather than an integral. A very readable introduction may be found in Young (1988).

Further Reading

Young, N. 1988. *An Introduction to Hilbert Space.* Cambridge: Cambridge University Press.

III.87 Spherical Harmonics

The starting point for FOURIER ANALYSIS [III.27] is the observation that a wide class of periodic functions $f(\theta)$ with period 2π can be decomposed as infinite linear combinations of the TRIGONOMETRIC FUNCTIONS [III.92] $\sin n\theta$ and $\cos n\theta$, or, equivalently, as sums of the form $\sum_{n=-\infty}^{\infty} a_n e^{in\theta}$.

A useful way to think of a periodic function f defined on the real line is as an equivalent function F defined on $\mathbb{T}$, the unit circle in the complex plane. A typical point on the circle has the form $e^{i\theta}$, and we define $F(e^{i\theta})$ to be $f(\theta)$. (Note that if we add 2π to θ then $F(e^{i\theta})$ does not change because $e^{i\theta} = e^{i(\theta+2\pi)}$ and $f(\theta)$ does not change because f is periodic with period 2π.)

If $f(\theta) = \sum_{n=-\infty}^{\infty} a_n e^{in\theta}$ and we write z for $e^{i\theta}$, then $F(z) = \sum_{n=-\infty}^{\infty} a_n z^n$. Therefore, if we consider functions defined on $\mathbb{T}$ rather than periodic functions defined on $\mathbb{R}$, then Fourier analysis decomposes our functions into infinite linear combinations of the functions z^n, where n can be any integer.

What is special about the functions z^n? The answer is that they are the *characters* of $\mathbb{T}$, which means that they are the only nonzero continuous complex-valued functions defined on $\mathbb{T}$ that satisfy the relation $\phi(zw) = \phi(z)\phi(w)$ for every z and w in $\mathbb{T}$.

Now imagine that F is a function defined not on $\mathbb{T}$ but on the two-dimensional set S^2, which is the unit sphere in $\mathbb{R}^3$ (defined as the set of points (x, y, z) such that $x^2 + y^2 + z^2 = 1$). More generally, how about functions F defined on S^{d-1} (defined as the set of points $(x_1, \ldots, x_d)$ such that $x_1^2 + \cdots + x_d^2 = 1$)? Is there a natural way of decomposing such an F, at least if it is sufficiently nice? That is, is there a good way of generalizing Fourier analysis to higher-dimensional spheres?

There is an important and initially discouraging difference between the sphere S^2 and the circle $S^1 = \mathbb{T}$. We defined $\mathbb{T}$ as a set of complex numbers rather than as a set of points in the plane $\mathbb{R}^2$ because that way it forms a multiplicative group. The sphere, by contrast, does not have a useful group structure (for a clue about why, see QUATERNIONS, OCTONIONS, AND NORMED DIVISION ALGEBRAS [III.76]), so we cannot talk about characters. This makes it less obvious what the "nice" functions should be, into which we might hope to decompose more general functions.

However, there is another way of explaining why the trigonometric functions arise naturally, one that does not involve complex numbers. We can write a typical point in S^1 as (x, y) with $x^2 + y^2 = 1$, or equivalently as $(\cos \theta, \sin \theta)$ for some real number θ. Then our basic functions, if we wish to avoid complex numbers, are $\cos n\theta$ and $\sin n\theta$, but these can also be written in terms of x and y. For instance, $\cos \theta$ and $\sin \theta$ are x and y, respectively, $\cos 2\theta = \cos^2 \theta - \sin^2 \theta = x^2 - y^2$, and so on. (Note that $x^2 - y^2 = 2x^2 - 1 = 1 - 2y^2$, since $x^2 + y^2 = 1$.) In general, $\cos n\theta$ and $\sin n\theta$ can always be written as polynomials in $\cos \theta$ and $\sin \theta$, so the basic trigonometric functions can be thought of as restrictions to the unit circle of certain polynomials.

What are these polynomials? It turns out that they are *harmonic* and *homogeneous*. A harmonic polynomial $p(x, y)$ is one that satisfies the LAPLACE EQUATION [I.3 §5.4] $\Delta p = 0$, where Δp stands for

$$\frac{\partial^2 p}{\partial x^2} + \frac{\partial^2 p}{\partial y^2}.$$

For instance, if $p(x, y) = x^2 - y^2$, then $\partial^2 p / \partial x^2 = 2$ and $\partial^2 p / \partial y^2 = -2$, so $x^2 - y^2$ is, as we would hope, a harmonic polynomial. Since the Laplacian Δ is a linear operator, the harmonic polynomials form a vector space. A homogeneous polynomial of degree n is one in which the total degree of each term is n, or equivalently a polynomial $p(x, y)$ such that $p(\lambda x, \lambda y)$ is always equal to $\lambda^n p(x, y)$. For example, $x^3 - 3xy^2$ is homogeneous of degree 3 (and also harmonic). The homogeneous harmonic polynomials of degree n form a subspace of the space of all harmonic polynomials. It has dimension 1 when $n = 0$ and 2 when $n > 0$. (When $n > 0$ it corresponds to the space of functions of the form $A \cos n\theta + B \sin n\theta$. The polynomial $x^3 - 3xy^2$, for instance, corresponds to the function $\cos 3\theta$.)

The notion of a harmonic polynomial generalizes very easily to higher dimensions. For example, in three dimensions a harmonic polynomial is a polynomial $p(x, y, z)$ such that

$$\frac{\partial^2 p}{\partial x^2} + \frac{\partial^2 p}{\partial y^2} + \frac{\partial^2 p}{\partial z^2} = 0.$$

A *spherical harmonic* of *order n* and *dimension d* is the restriction to the sphere S^{d-1} of a harmonic polynomial in d variables that is homogeneous of degree n.

Here are some of the properties of spherical harmonics that make them particularly useful and closely analogous to the trigonometric polynomials on the circle. We shall fix a dimension d and use the notation $\mathrm{d}\mu$ to denote *Haar measure* on the unit sphere $S = S^{d-1}$. Basically, this means that if f is an integrable function from S to $\mathbb{R}$, then $\int_S f(x) \, \mathrm{d}\mu$ is its average.

(i) **Orthogonality.** If p and q are spherical harmonics of dimension d, and if they have different degrees, then $\int_S p(x)q(x) \, \mathrm{d}\mu = 0$.

(ii) **Completeness.** Every function $f : S \to \mathbb{R}$ that belongs to $L^2(S, \mu)$ (meaning that $\int_S |f(x)|^2 \, \mathrm{d}\mu$ exists and is finite) can be written as a sum $\sum_{n=0}^{\infty} H_n$ (with convergence in $L^2(S, \mu)$), where H_n is a spherical harmonic of order n.

(iii) **Finite-dimensionality of decomposition.** For each d and n, the vector space of spherical harmonics of dimension d and order n is finite dimensional.

From these three properties it is easy to deduce that $L^2(S, \mu)$ has an ORTHONORMAL BASIS [III.37] consisting of spherical harmonics.

Why are spherical harmonics natural, and why are they useful? Both questions can be given several answers: here is one for each.

The Laplace operator Δ, which operates on functions defined on $\mathbb{R}^n$, can be generalized to functions defined on any RIEMANNIAN MANIFOLD [I.3 §6.10] M. The generalization, denoted Δ_M, is called the *Laplace–Beltrami operator* for M, and its behavior gives one a great deal of information about the geometry of M. In particular, the Laplace–Beltrami operator can be defined for the sphere S^{d-1}, where it is called simply the *Beltrami operator*. It turns out that the spherical harmonics are the EIGENVECTORS [I.3 §4.3] of the Beltrami operator. More precisely, a spherical harmonic of dimension d and order n is an eigenvector with eigenvalue $-n(n + d - 2)$. (Notice that the second derivative of $\cos n\theta$ is $-n^2 \cos n\theta$, which corresponds to the case $d = 2$.) This gives an alternative, more natural (but less elementary) definition of spherical harmonics. This definition, combined with the fact that the Laplace operator is self-adjoint, explains many of the important properties of spherical harmonics. (See LINEAR OPERATORS AND THEIR PROPERTIES [III.50 §3] for an amplification of this remark.)

One reason for the importance of Fourier analysis is that many important linear operators become diagonal, and hence particularly easy to understand, when they are applied to the Fourier transform of a function. For example, if f is a smooth periodic function and we write it as $\sum_{n \in \mathbb{Z}} a_n e^{in\theta}$, then the derivative of f is $\sum_{n \in \mathbb{Z}} na_n e^{in\theta}$. Writing $\hat{f}(n)$ and $\widehat{f'}(n)$ for the nth Fourier coefficients of f and f', respectively, we deduce that $\widehat{f'}(n) = n\hat{f}(n)$, which tells us that to differentiate a function f all we have to do is multiply its Fourier transform pointwise by the function $g(n) = n$. This provides a very useful technique for solving differential equations.

As has already been mentioned, spherical harmonics are eigenvectors of the Laplacian, but they also diagonalize several other linear operators. A good example is the *spherical Radon transform*, which is defined as follows. If f is a function from S^{d-1} to $\mathbb{R}$, then its spherical Radon transform Rf is another function from S^{d-1} to $\mathbb{R}$, and the value of Rf at a point x is the average value of f over all points y that are orthogonal to x. This is closely related to the more usual Radon transform, which replaces a function defined on the plane by its averages over lines; inverting the Radon transform is important for creating images from the outputs of medical scanners. The spherical harmonics turn out to be eigenfunctions for the spherical Radon transform. More generally, any transform T of the form $Tf(x) = \int_S w(x \cdot y) f(y) \, d\mu(y)$, where w is a suitable function (or generalized function), is diagonalized by spherical harmonics. The eigenvalue associated with a given spherical harmonic can be calculated by the so-called *Funk–Hecke formula*.

Spherical harmonics give a way of linking CHEBYSHEV AND LEGENDRE POLYNOMIALS [III.85], and showing that both of them are natural concepts. The Chebyshev polynomials are those polynomials in x that are also spherical harmonics of dimension 2: that is, that are equal on S^1 to homogeneous harmonic polynomials in two variables. For instance, because $x^2 + y^2 = 1$ for every (x, y) in the circle S^1, the function $x^3 - 3xy^2$ that we considered earlier is equal on S^1 to the function $4x^3 - 3x$, so $4x^3 - 3x$ is a Chebyshev polynomial. The Legendre polynomials are those polynomials

in x that are equal to spherical harmonics of dimension 3. For example, if $p(x, y, z) = 2x^2 - y^2 - z^2$ then $\Delta p = 0$, and $p(x, y, z) = 3x^2 - 1$ everywhere on S^2, since $x^2 + y^2 + z^2 = 1$. Therefore, $3x^2 - 1$ is a Legendre polynomial.

Here is a sketch of a proof that these polynomials are equal to the Chebyshev and Legendre polynomials as they are usually defined. The usual definition is that they are sequences of polynomials, one for each degree, that are uniquely determined by certain orthogonality relations. Because spherical harmonics of different orders are orthogonal, the polynomials just described also satisfy certain orthogonality relations. When one works out what these are, one discovers that they are precisely the relations that define the Chebyshev and Legendre polynomials.

III.88 Symplectic Manifolds
Gabriel P. Paternain

Symplectic geometry is the geometry that governs classical physics, and more generally plays an important role in helping us to understand the actions of groups on manifolds. It shares some features with Riemannian geometry and complex geometry, and there is an important special class of manifolds, the *Kähler manifolds*, in which all three geometric structures are unified.

1 Symplectic Linear Algebra

Just as RIEMANNIAN GEOMETRY [I.3 §6.10] is based on EUCLIDEAN GEOMETRY [I.3 §6.2], symplectic geometry is based on the geometry of the so-called *linear symplectic space* $(\mathbb{R}^{2n}, \omega_0)$.

Given two vectors $v = (q, p)$ and $v' = (q', p')$ in the plane $\mathbb{R}^2$, the *signed area* $\omega_0(v, v')$ of the parallelogram spanned by v and v' is given by the formula

$$\omega_0(v, v') = \det \begin{pmatrix} q' & q \\ p' & p \end{pmatrix} = pq' - qp'.$$

It can also be written using matrices and inner products as $\omega_0(v, v') = v' \cdot Jv$, where J is the 2×2 matrix

$$J = \begin{pmatrix} 0 & 1 \\ -1 & 0 \end{pmatrix}.$$

If a linear transformation $A : \mathbb{R}^2 \to \mathbb{R}^2$ is area preserving and orientation preserving, then $\omega_0(Av, Av') = \omega_0(v, v')$ for every v and v'.

Symplectic geometry studies two-dimensional signed area measurements like this, as well as transformations that preserve these measurements, but it applies to general spaces of dimension $2n$ rather than just to the plane.

If we split $\mathbb{R}^{2n}$ up as $\mathbb{R}^n \times \mathbb{R}^n$, then we can write a vector v in $\mathbb{R}^{2n}$ as $v = (q, p)$, where q and p each belong to $\mathbb{R}^n$. The *standard symplectic form* $\omega_0 : \mathbb{R}^{2n} \times \mathbb{R}^{2n} \to \mathbb{R}$ is defined by the formula

$$\omega_0(v, v') = p \cdot q' - q \cdot p',$$

where "$\cdot$" denotes the usual inner product in $\mathbb{R}^n$. Geometrically, $\omega_0(v, v')$ can be interpreted as the sum of the signed areas of the parallelograms spanned by the projections of v and v' to the $q_i p_i$-planes. In terms of matrices, we can write

$$\omega_0(v, v') = v' \cdot J v, \tag{1}$$

where J is the $2n \times 2n$ matrix

$$J = \begin{pmatrix} 0 & I \\ -I & 0 \end{pmatrix} \tag{2}$$

and I is the $n \times n$ identity matrix.

A linear map $A : \mathbb{R}^{2n} \to \mathbb{R}^{2n}$ that preserves the product ω_0 of any two vectors (that is, $\omega_0(Av, Av') = \omega_0(v, v')$ for all $v, v' \in \mathbb{R}^{2n}$) is called a *symplectic linear transformation*; equivalently, a $2n \times 2n$ matrix A is symplectic if and only if $A^{\mathrm{T}} J A = J$, where A^{T} is the transpose of A. Symplectic linear transformations are to symplectic geometry as rigid motions are to Euclidean geometry. The set of all symplectic linear transformations of $(\mathbb{R}^{2n}, \omega_0)$ is one of the classical LIE GROUPS [III.48 §1] and is denoted by $\mathrm{Sp}(2n)$. One can show that symplectic matrices $A \in \mathrm{Sp}(2n)$ always have DETERMINANT [III.15] 1, and are thus volume preserving. However, the converse does not hold when $n \geqslant 2$. For instance, if $n = 2$, the linear map

$$(q_1, q_2, p_1, p_2) \mapsto (aq_1, q_2/a, ap_1, p_2/a)$$

has determinant 1 for any $a \neq 0$, but it is symplectic only if $a^2 = 1$.

The standard symplectic form ω_0 has three properties worth noting. First, it is *bilinear*: the expression $\omega_0(v, v')$ varies linearly in v when v' is held fixed, and vice versa. Second, it is *antisymmetric*: we have $\omega_0(v, v') = -\omega_0(v', v)$ for all v and v', and in particular $\omega_0(v, v) = 0$. Finally, it is *nondegenerate*, which means that for every nonzero v there is a nonzero v' such that $\omega_0(v, v') \neq 0$. The standard symplectic form ω_0 is not the only form that obeys these three properties; however, it turns out that any form with

these three properties can be converted into the standard form ω_0 after an invertible linear change of variables. (This is a special case of *Darboux's theorem*.) Thus $(\mathbb{R}^{2n}, \omega_0)$ is essentially the "only" linear symplectic geometry in $2n$ dimensions. There are no symplectic forms in odd-dimensional spaces.

2   Symplectic Diffeomorphisms of $(\mathbb{R}^{2n}, \omega_0)$

In Euclidean geometry, all rigid motions are automatically linear (or affine) transformations. However, in symplectic geometry there are many more symplectic maps than just the symplectic linear transformations. These nonlinear symplectic maps in $(\mathbb{R}^{2n}, \omega_0)$ are one of the principal objects of study in symplectic geometry.

Let $U \subset \mathbb{R}^{2n}$ be an open set. Recall that a map $\phi : U \to \mathbb{R}^{2n}$ is called *smooth* if it has continuous partial derivatives of all orders. A *diffeomorphism* is a smooth map with smooth inverse.

A smooth nonlinear map $\phi : U \to \mathbb{R}^{2n}$ is said to be *symplectic* if, for every $x \in U$, the *Jacobian matrix* $\phi'(x)$ of first derivatives of ϕ is a symplectic linear transformation. Informally, a symplectic map is one that behaves like a symplectic linear transformation at infinitesimally small scales. Since symplectic linear transformations have determinant 1, we can conclude using several-variable calculus that a symplectic map is always locally volume preserving and locally invertible; roughly speaking, this means that the map $\phi : A \to \phi(A)$ is invertible whenever A is a sufficiently small subset of U, and $\phi(A)$ has the same volume as A. However, the converse is not true when $n \geqslant 2$; the class of symplectic maps is much more restricted than that of volume-preserving maps. In fact, Gromov's nonsqueezing theorem (see below) shows how striking this difference can be.

Symplectic maps have been around for quite a long time in Hamiltonian mechanics under the name of *canonical transformations*. We briefly explain this in the next subsection.

2.1   Hamilton's Equations

How can we produce nonlinear symplectic maps? Let us begin by exploring a familiar example. Consider the motion of a simple pendulum with length l and mass m and let $q(t)$ be the angle it makes with the vertical at time t. The equation of motion is

$$\frac{\mathrm{d}^2 q}{\mathrm{d}t^2} + \frac{g}{l} \sin q = 0,$$

where g is the acceleration due to gravity. If we define the *momentum* p as $p = ml^2\dot{q}$, then we may transform this second-order differential equation into a first-order system in the *phase plane* $\mathbb{R}^2$, namely

$$\frac{d}{dt}(q, p) = X(q, p), \qquad (3)$$

where the *vector field* $X : \mathbb{R}^2 \to \mathbb{R}^2$ is given by the formula $X(q, p) = (p/ml^2, -mgl\sin q)$. For each $(q(0), p(0)) \in \mathbb{R}^2$ there is a unique solution $(q(t), p(t))$ to (3) with initial condition $(q(0), p(0))$. Then for any fixed time t we obtain an *evolution map* (or flow) $\phi_t : \mathbb{R}^2 \to \mathbb{R}^2$ given by $\phi_t(q(0), p(0)) = (q(t), p(t))$, which has the remarkable property of being *area preserving*. This can be deduced from the observation that X is *divergence free*, or in other words that

$$\frac{d}{dq}\frac{p}{ml^2} + \frac{d}{dp}(-mgl\sin q) = 0.$$

In fact, for every time t, ϕ_t is a symplectic map on $(\mathbb{R}^2, \omega_0)$.

More generally, any system in classical mechanics with finitely many degrees of freedom can be reformulated in a similar fashion, so that the evolution maps ϕ_t are always symplectic maps; in this context, they are also known as canonical transformations. The Irish mathematician WILLIAM ROWAN HAMILTON [VI.37] showed us how to do this in general more than 170 years ago. Given any smooth function $H : \mathbb{R}^{2n} \to \mathbb{R}$ (called the *Hamiltonian*), the system of first-order differential equations given by

$$\frac{dq_i}{dt} = \frac{\partial H}{\partial p_i}, \qquad i = 1, \ldots, n, \qquad (4)$$

$$\frac{dp_i}{dt} = -\frac{\partial H}{\partial q_i}, \qquad i = 1, \ldots, n, \qquad (5)$$

will (under some mild growth assumptions on H, which we ignore here) give rise to evolution operators $\phi_t : \mathbb{R}^{2n} \to \mathbb{R}^{2n}$, which are symplectic maps on $(\mathbb{R}^{2n}, \omega_0)$ for every time t. To see the connection with the symplectic form ω_0, observe that we may rewrite (4) and (5) in the following equivalent form:

$$\frac{dx}{dt} = J\nabla H(x), \qquad (6)$$

where ∇H is the usual GRADIENT [I.3 §5.3] of H and J was defined in (2). From (6), (1), and the antisymmetry property of ω_0, it is then not difficult to verify that ϕ_t is symplectic for every t (the main trick is to compute the derivative of $\omega_0(\phi_t'(x)v, \phi_t'(x)v')$ in t and check that it equals zero).

We have already pointed out that symplectic maps are volume preserving. The preservation of volume by

Hamiltonian systems (a result known as *Liouville's theorem*) attracted considerable attention in the nineteenth century and it was a driving force in the development of ERGODIC THEORY [V.9], which studies recurrence properties of measure-preserving transformations.

Symplectic maps or canonical transformations play an important role in classical physics, as they allow one to replace a complicated system by an equivalent system that is simpler to analyze.

2.2   Gromov's Nonsqueezing Theorem

What is the difference between a symplectic map and a volume-preserving map? In order to answer this question, suppose that we have two connected open sets U and V in $\mathbb{R}^{2n}$ and that we wish to embed one into the other using a symplectic map. This means that we are looking for a symplectic map $\phi : U \to V$ such that ϕ is a homeomorphism onto its image. We know such a ϕ must be volume preserving, so we clearly have the restriction that the volume of U should be at most the volume of V, but is this restriction all that matters? Consider the open ball $B(R) = \{x \in \mathbb{R}^{2n} : |x| < R\}$, which has radius R and center at the origin, and clearly has finite volume. It is not hard to embed it symplectically into the infinite-volume cylinder given by

$$C(r) = \{(q, p) \in \mathbb{R}^{2n} : q_1^2 + q_2^2 < r^2\},$$

whatever the values of R and r. Indeed, the linear symplectic map

$$(q, p) \mapsto (aq_1, aq_2, q_3, \ldots, q_n, p_1/a, p_2/a, p_3, \ldots, p_n)$$

will do the trick when a is sufficiently small and positive. However, the situation is radically different if instead we consider the infinite-volume cylinder

$$Z(r) = \{(q, p) \in \mathbb{R}^{2n} : q_1^2 + p_1^2 < r^2\}.$$

We could try with a similar linear map like

$$(q, p) \mapsto (aq_1, q_2/a, q_3, \ldots, q_n, ap_1, p_2/a, p_3, \ldots, p_n).$$

This map is volume preserving (it has determinant 1) and for a small it embeds $B(R)$ into $Z(r)$. However, it is symplectic only if $a = 1$, so it will give a symplectic embedding only if $R \leqslant r$. One is tempted to think that if $R > r$, then there should still be a nonlinear symplectic embedding squeezing $B(R)$ into $Z(r)$, but a remarkable theorem of Gromov from 1985 asserts that it is not possible to find such a map.

In spite of this deep result of Gromov, and other results that followed it, we still do not know much about how sets in $\mathbb{R}^{2n}$ embed into one another.

3 Symplectic Manifolds

Recall from DIFFERENTIAL TOPOLOGY [IV.7] that a *manifold* of dimension d is a TOPOLOGICAL SPACE [III.90] such that each point has a neighborhood that is homeomorphic to an open set in Euclidean space $\mathbb{R}^d$. One can think of $\mathbb{R}^d$ as a *local model* for this manifold, in the sense that it describes what the manifold looks like at very small distance scales. Recall also that a *smooth* manifold is one for which the "transition functions" are smooth. This means that if $\psi : U \to \mathbb{R}^d$ and $\varphi : V \to \mathbb{R}^d$ are coordinate charts, then the transition function $\psi \circ \varphi^{-1}$ between the open sets $\varphi(U \cap V)$ and $\psi(U \cap V)$ is smooth.

A symplectic manifold is defined similarly, but now the local model is the linear symplectic space $(\mathbb{R}^{2n}, \omega_0)$. More precisely, a symplectic manifold M is a manifold of dimension $2n$ that can be covered with domains of coordinate charts whose transition functions are symplectic diffeomorphisms of $(\mathbb{R}^{2n}, \omega_0)$.

Of course, any open set in $(\mathbb{R}^{2n}, \omega_0)$ is a symplectic manifold. An example of a compact symplectic manifold is the torus $\mathbb{T}^{2n}$, which is obtained as the quotient of $\mathbb{R}^{2n}$ by the action of $\mathbb{Z}^{2n}$. In other words, two points $x, y \in \mathbb{R}^{2n}$ are equivalent if $x - y$ has integer coordinates. Other important examples of symplectic manifolds include RIEMANN SURFACES [III.79], complex PROJECTIVE SPACE [III.72], and cotangent BUNDLES [IV.6 §5]. However, it is a wide open problem to determine, given a compact manifold, whether it can be assigned a system of coordinate charts that makes it symplectic.

We have seen that in $(\mathbb{R}^{2n}, \omega_0)$, one can assign an "area" $\omega_0(v, v')$ to any parallelogram in the space $\mathbb{R}^{2n}$. In a symplectic manifold M, one can similarly assign an area $\omega_p(v, v')$, but only to *infinitesimal* parallelograms based at a point $p \in M$. The axes of such a parallelogram are two infinitesimal vectors (or more precisely *tangent vectors*) v and v'. There is a unique way to do this so that all the coordinate charts for M are symplectic diffeomorphisms. In the language of DIFFERENTIAL FORMS [III.16], the map $p \mapsto \omega_p$ is an antisymmetric nondegenerate 2-form, which can then be used to compute the "area" $\int_S \omega$ of noninfinitesimal two-dimensional surfaces S in M. One can show that for any sufficiently small closed surface S, the integral $\int_S \omega$ vanishes, so ω is a *closed* form. Indeed, one can define a symplectic manifold more abstractly (without reference to charts) as a smooth manifold equipped with a closed, antisymmetric nondegenerate 2-form ω; a clas-

sical theorem of Darboux asserts that this abstract definition is equivalent to the more concrete definition using coordinate charts.

Finally, a special class of symplectic manifolds is given by *Kähler manifolds*. These are symplectic manifolds that are also *complex manifolds*, in such a way that the two structures are naturally compatible, a condition that generalizes the relationship (1). Observe that if one identifies points (q, p) in $\mathbb{R}^{2n}$ with points $p + iq$ in $\mathbb{C}^n$, then the linear transformation $J : \mathbb{R}^{2n} \to \mathbb{R}^{2n}$ becomes the operation of multiplication by i:

$$J : (z_1, \ldots, z_n) \mapsto (iz_1, \ldots, iz_n).$$

Thus the identity (1) relates the symplectic structure (as given by ω_0), the complex structure (as given by J), and the Riemannian structure (as given by the dot product "·"). A *complex manifold* is a manifold that at small distance scales looks like regions of $\mathbb{C}^n$, with the transition functions required to be HOLOMORPHIC [I.3 §5.6]. (A smooth map $f : U \subset \mathbb{C}^n \to \mathbb{C}^n$ is said to be holomorphic if each coordinate component of $f(z_1, \ldots, z_n)$ is holomorphic in each variable z_k.) On a complex manifold we can multiply tangent vectors by i. This gives us at each point $p \in M$ a linear map J_p such that $J_p^2 v = -v$ for all tangent vectors v at p. A Kähler manifold is a complex manifold M with a symplectic structure ω (which computes signed areas of infinitesimal parallelograms) and a Riemannian metric g (which computes an inner product $g_p(v, v')$ of any two tangent vectors v, v' at p); these two structures are linked together by the analogue of (1), namely

$$\omega_p(v, v') = g_p(v', J_p v).$$

Examples of Kähler manifolds include complex vector spaces $\mathbb{C}^n$, Riemann surfaces, and complex projective spaces $\mathbb{CP}^n$.

An example of a compact symplectic manifold that is not Kähler can be obtained by taking the quotient of $\mathbb{R}^4$ by a symplectic action of a group that looks like $\mathbb{Z}^4$ but with a group operation that differs from the usual one. The change in the group structure manifests itself as a topological property (an odd first Betti number) that prevents the quotient being Kähler.

Further Reading

Arnold, V. I. 1989. *Mathematical Methods of Classical Mechanics*, 2nd edn. Graduate Texts in Mathematics, volume 60. New York: Springer.

McDuff, D., and D. Salamon. 1998. *Introduction to Symplectic Topology*, 2nd edn. Oxford Mathematical Monographs. Oxford: Clarendon Press/Oxford University Press.

III.89   Tensor Products

If U, V, and W are VECTOR SPACES [I.3 §2.3] over some field, then a *bilinear map* from $U \times V$ to W is a map ϕ obeying the rules

$$\phi(\lambda u + \mu u', v) = \lambda \phi(u, v) + \mu \phi(u', v)$$

and

$$\phi(u, \lambda v + \mu v') = \lambda \phi(u, v) + \mu \phi(u, v').$$

That is, it is linear in each variable separately.

Many important maps, such as INNER PRODUCTS [III.37], are bilinear. The *tensor product $U \otimes V$* of two vector spaces U and V is a way of capturing the idea of the "most general" bilinear map that we can define on $U \times V$. To get an idea of what this might mean, let us try to imagine a "completely arbitrary" bilinear map from $U \times V$ to a "completely arbitrary" vector space W, and let us use the notation $u \otimes v$ instead of $\phi(u, v)$. Now because our linear map is perfectly general, all we know about it is what we can deduce from the fact that it is bilinear. For example, we know that $u \otimes v_1 + u \otimes v_2 = u \otimes (v_1 + v_2)$. This example might suggest that all elements of $U \otimes V$ are of the form $u \otimes v$, but that is certainly not the case: for instance, in general there is no way of simplifying an expression such as $u_1 \otimes v_1 + u_2 \otimes v_2$. (This reflects the fact that the set of values taken by a bilinear map from $U \times V$ to W is not in general a subspace of W.)

Thus, a typical element of $U \otimes V$ is a *linear combination* of elements of the form $u \otimes v$, with the rule that different linear combinations give the same element of $U \otimes V$ whenever they are forced to by the bilinearity property: for instance, $(u_1 + 2u_2) \otimes (v_1 - v_2)$ will always be equal to

$$u_1 \otimes v_1 + 2u_2 \otimes v_1 - u_1 \otimes v_2 - 2u_2 \otimes v_2.$$

A more formal way of expressing the above ideas is to say that $U \otimes V$ has a *universal property*. (See GEOMETRIC AND COMBINATORIAL GROUP THEORY [IV.10] for some other examples of universal properties. See also CATEGORIES [III.8].) The property in question is the following: given any bilinear map ϕ from $U \times V$ to a space W, we can find a *linear* map α from $U \otimes V$ to W such that $\phi(u, v) = \alpha(u \otimes v)$ for every u and v. That is, every bilinear map ϕ defined on $U \times V$ is naturally associated with a linear map defined on $U \otimes V$. (This linear map takes $u \otimes v$ to $\phi(u, v)$: the identifications made in the definition of the tensor product ensure that we can

extend this to linear combinations of such elements in a consistent way.)

It is not hard to show that if U and V are finite dimensional, with bases $u_1, \dots, u_m$ and $v_1, \dots, v_n$, then the vectors $u_i \otimes v_j$ form a basis for $U \otimes V$. Other important properties of the tensor product are that it is commutative and associative, in the sense that $U \otimes V$ is naturally isomorphic to $V \otimes U$ and $U \otimes (V \otimes W)$ is naturally isomorphic to $(U \otimes V) \otimes W$.

We have been discussing tensor products of vector spaces, but the definition can easily be generalized to any algebraic structure for which some notion of bilinearity makes sense, such as a MODULE [III.81 §3] or a C^*-ALGEBRA [IV.15 §3]. Sometimes the tensor product of two structures is not what you would immediately expect. For instance, let $\mathbb{Z}_n$ be the set of integers mod n, and consider both $\mathbb{Z}_n$ and $\mathbb{Q}$ as modules over $\mathbb{Z}$. Then their tensor product is zero. This reflects the fact that every bilinear map from $\mathbb{Z}_n \times \mathbb{Q}$ must be the zero map.

Tensor products occur in many mathematical contexts. For a good example, see QUANTUM GROUPS [III.75].

Transcendental Numbers

See IRRATIONAL AND TRANSCENDENTAL
NUMBERS [III.41]

III.90   Topological Spaces
Ben Green

A topological space is the most basic context in which one can understand the notion of a CONTINUOUS FUNCTION [I.3 §5.2].

Let us recall a standard definition of what it means for a function $f : \mathbb{R} \to \mathbb{R}$ to be continuous. Suppose that $f(x) = y$. Then f is continuous at x provided that $f(x')$ is close to y whenever x' is close to x. Of course, to make this a mathematically rigorous notion we have to be precise about the meaning of "close." We could say that $f(x')$ is close to y if $|f(x') - f(x)| < \varepsilon$, where $\varepsilon > 0$ is some small positive constant. And we could deem x to be close to x' whenever $|x - x'| < \delta$, where δ is another positive constant.

We say that f is *continuous at x* if an appropriate δ can be found, regardless of how small ε was chosen to be (δ is allowed, of course, to depend on ε). And f is said to be *continuous* if it is continuous at every point x on the real line.

How might we generalize this notion, replacing $\mathbb{R}$ by an arbitrary set X? Our existing definition makes sense only if we can decide when two points $x, x' \in X$ are close. For a general set, which might not be nicely embedded in Euclidean space, this is impossible without the addition of further structure. (When such structure is added one has the notion of a METRIC SPACE [III.56]: metric spaces are less general than topological spaces.)

If the notion of closeness is unavailable, how should one define continuity? The answer may be found in the notion of an *open set*. A set $U \subset \mathbb{R}$ is said to be *open* if for any point x in U there is an interval (a, b) that contains x (that is, $a < x < b$) and is contained in U.

It is an amusing exercise to check that if $f : \mathbb{R} \to \mathbb{R}$ is continuous, and if U is open, then $f^{-1}(U)$ is open. Conversely, if $f^{-1}(U)$ is open for every open set U, then f is continuous. Thus, at least for functions from $\mathbb{R}$ to $\mathbb{R}$, one may characterize continuity purely in terms of open sets. The notion of closeness is used only when it comes to defining what an open set is.

We now turn to the formal definition. A *topological space* is a set X together with a collection $\mathcal{U}$ of subsets of X (called the "open sets") satisfying the following axioms.

- The empty set $\varnothing$ and the set X are both open.
- $\mathcal{U}$ is closed under taking arbitrary unions (so if $(U_i)_{i \in I}$ is a collection of open sets, then so is $\bigcup_{i \in I} U_i$).
- $\mathcal{U}$ is closed under taking finite intersections (so if $U_1, \ldots, U_k$ are open sets, then so is $U_1 \cap \cdots \cap U_k$).

The collection $\mathcal{U}$ is called a *topology* on X. It is easy to verify that the usual open subsets of $\mathbb{R}$ satisfy the above axioms: thus, $\mathbb{R}$ forms a topological space with these sets.

A subset of a topological space is called *closed* if and only if its complement is open. Note that "closed" does not mean "not open": for example, in the space $\mathbb{R}$, the half-open interval $[0, 1)$ is neither open nor closed, and the empty set is both open and closed.

Note that we do not demand many properties from our open sets: this makes the notion of topological space a rather general one. Indeed, under many circumstances the concept is a little *too* general: then it can be convenient to assume that a topological space has further properties. For instance, a topological space X is called *Hausdorff* if, for any two distinct points x_1 and x_2 in X, there are disjoint open sets U_1 and U_2 that contain x_1 and x_2, respectively. Hausdorff topological spaces (of which $\mathbb{R}$ is an obvious example) have many useful properties that general topological spaces do not necessarily have.

We saw earlier that for functions from $\mathbb{R}$ to $\mathbb{R}$ the notion of continuity could be formulated entirely in terms of open sets. This means that we can define continuity for functions between topological spaces: if X and Y are two topological spaces and if $f : X \to Y$ is a function between them, then we simply define f to be continuous if $f^{-1}(U)$ is open for every open set $U \subset Y$. Remarkably, we have found a useful definition of continuity that does not rely on a notion of distance.

A continuous map that has a continuous inverse is known as a *homeomorphism*. If there is a homeomorphism between two spaces X and Y, then they are regarded as equivalent from the point of view of topology. In topology texts one will often see it said that a topologist is unable to distinguish between a doughnut and a teacup because each can be continuously deformed into the other (imagine that they are both made of modeling clay).

If X is a topological space, then a very useful way of describing the topology on X is by giving a *basis* for it. This is a subcollection $\mathcal{B} \subseteq \mathcal{U}$ with the property that every open set (that is, every element of $\mathcal{U}$) is a union of open sets in $\mathcal{B}$. A basis for $\mathbb{R}$ with the usual topology is the collection of open intervals $\{(a, b) : a < b\}$, and a basis for $\mathbb{R}^2$ is the collection of *open balls*: that is, sets of the form $\{B_\delta(x) = \{y : |x - y| < \delta\}\}$.

Let us give some examples.

The discrete topology. Let X be any set whatsoever, and take $\mathcal{U}$ to be the collection of all subsets of X. It is a simple matter to check that the axioms for a topological space are satisfied.

Euclidean spaces. Let $X = \mathbb{R}^d$, and let $\mathcal{U}$ contain all sets that are open in the Euclidean metric. That is, $U \subseteq X$ is open if, for every $u \in U$, there is $\delta > 0$ such that $B_\delta(u)$ is contained in U. It is only slightly more taxing to check that the axioms are satisfied in this case. More generally, for any metric space the open sets can be defined in a similar way and they form a topological space.

Subspace topology. If X is a topological space and if $S \subseteq X$, then we may make S a topological space. We declare the open sets in S to be all sets of the form $S \cap U$, where $U \in \mathcal{U}$ is an open set in X.

The Zariski topology. This is used in ALGEBRAIC GEOMETRY [IV.4]. It is specified by giving its closed sets (and hence, by complementation, its open sets)—these are the zero loci of systems of polynomial equations. On $\mathbb{C}^2$, for example, these closed sets are precisely the sets of the form

$$\{(z_1, z_2) : f_1(z_1, z_2) = f_2(z_1, z_2)$$
$$= \cdots = f_k(z_1, z_2) = 0\},$$

where $f_1, \ldots, f_k$ are polynomials. To show that this defines a topology is somewhat nontrivial, the difficulty being to show that an arbitrary intersection of closed sets is closed (which is equivalent to the assertion that an arbitrary union of open sets is open). This is a consequence of Hilbert's basis theorem.

The notion of topological space is a very good example of the power of abstraction in mathematics. The definition is simple and covers a wide variety of natural situations, yet it has enough content that one can make interesting definitions and prove theorems purely within the world of topological spaces. It is often fun to take a familiar concept, that applies to $\mathbb{R}$ or $\mathbb{R}^2$, say, and try to find an analogue of it in the world of general topological spaces. We give two examples.

Connectedness. The rough idea of connectedness is that a connected set is one that does not break up into pieces in an obvious way. Most people would imagine that they could discern, from a list of pictures of reasonably sensible subsets of $\mathbb{R}^2$, which were connected and which were not. But can one give a precise mathematical definition that applies to all sets, including potentially very wild ones, and says whether they are connected or not? For example, is the space

$$S = ((\mathbb{Q} \times \mathbb{R}) \cup (\mathbb{R} \times \mathbb{Q})) \setminus (\mathbb{Q} \times \mathbb{Q}),$$

which consists of all points with exactly one rational coordinate, connected or not (with the subspace topology)? It turns out that a definition can indeed be given, and moreover that it applies not just to $\mathbb{R}^2$ but to general topological spaces. We say that a space X is *connected* if there is no decomposition $X = U_1 \cup U_2$ of X into two disjoint, nonempty, open sets. We leave it to the reader to decide whether S is connected or not.

Compactness. This is one of the most important concepts in all of mathematics, but it can appear strange at first sight. It comes from attempting to abstract the notion of a closed and bounded set (in $\mathbb{R}^2$, say) to a general topological space. We say that X is *compact* if, given any collection C of open sets U that cover X (i.e.,

whose union is X), we may find a finite subcollection $\{U_1, \ldots, U_k\} \subseteq C$ that still covers X. Specializing this definition to $\mathbb{R}^2$ with the usual topology, it can indeed be proved that a set $S \subseteq \mathbb{R}^2$ is compact (in the subspace topology) if and only if it is closed and bounded. See COMPACTNESS AND COMPACTIFICATION [III.9] for more information.

III.91   Transforms
T. W. Körner

If we have a finite sequence $a_0, a_1, \ldots, a_n$ of real numbers (written briefly as $\boldsymbol{a}$), then we can look at the polynomial

$$P_{\boldsymbol{a}}(t) = a_0 + a_1 t + \cdots + a_n t^n.$$

Conversely, given a polynomial Q of degree $m \leqslant n$, we can recover a unique sequence $b_0, b_1, \ldots, b_n$ such that

$$Q(t) = b_0 + b_1 t + \cdots + b_n t^n$$

by, for example, taking $b_k = Q^{(k)}(0)/k!$.

We observe that if $a_0, a_1, \ldots, a_n$ and $b_0, b_1, \ldots, b_n$ are finite sequences, then

$$P_{\boldsymbol{a}}(t) P_{\boldsymbol{b}}(t) = P_{\boldsymbol{a} * \boldsymbol{b}}(t),$$

where $\boldsymbol{a} * \boldsymbol{b} = \boldsymbol{c}$ is a sequence $c_0, c_1, \ldots, c_{2n}$ given by

$$c_k = a_0 b_k + a_1 b_{k-1} + \cdots + a_k b_0,$$

where we interpret a_i and b_i as 0 if $i > n$. This sequence is called the *convolution* of the sequences $\boldsymbol{a}$ and $\boldsymbol{b}$.

To see the kind of use that one can make of this observation, consider what happens when we throw two dice, the first of which has probability a_u of showing u and the second of which has probability b_v of showing v. The probability that their sum is k is given by the number c_k defined above. If we take both a_u and b_u to be the probability of throwing u with an ordinary fair die (so they are equal to $\frac{1}{6}$ if $1 \leqslant u \leqslant 6$, and 0 otherwise), then

$$P_{\boldsymbol{c}}(t) = P_{\boldsymbol{a}}(t) P_{\boldsymbol{b}}(t)$$
$$= (\tfrac{1}{6}(t + t^2 + \cdots + t^6))^2.$$

This polynomial can be rewritten as

$$\tfrac{1}{36}(t(t+1)(t^4 + t^2 + 1))(t(t^2 + t + 1)(t^3 + 1))$$
$$= \tfrac{1}{36}(t(t+1)(t^2 + t + 1))(t(t^4 + t^2 + 1)(t^3 + 1))$$
$$= P_A(t) P_B(t),$$

where A and B are two different sequences, given by $A_1 = A_4 = \frac{1}{6}$, $A_2 = A_3 = \frac{2}{6}$, and $A_u = 0$ otherwise, and $B_1 = B_3 = B_4 = B_5 = B_6 = B_8 = \frac{1}{6}$, and $B_v = 0$

otherwise. Thus, if we take two fair dice A and B and number A so that it has 2 on two faces, 3 on two faces, 1 on one face, and 4 on the remaining face, and we number B so that it has 1, 3, 4, 5, 6, and 8 on its faces, then the probability of throwing a sum k is the same as with an ordinary pair of dice. It is not hard to show, by considering the roots of the polynomial $t + t^2 + \cdots + t^6$, that this is the only nonstandard labeling of dice with strictly positive integers that has this property.

These general ideas are easily extended to infinite sequences. If $\boldsymbol{a}$ is the sequence $a_0, a_1, \ldots$, we can define an "infinite polynomial" $(\mathcal{G}\boldsymbol{a})(t)$ to be $\sum_{r=0}^{\infty} a_r t^r$. For the moment, we shall proceed formally, without worrying in what sense the sum exists. Observe that, much as before,

$$(\mathcal{G}\boldsymbol{a})(t)(\mathcal{G}\boldsymbol{b})(t) = (\mathcal{G}(\boldsymbol{a} * \boldsymbol{b}))(t),$$

where the infinite sequence $\boldsymbol{c} = \boldsymbol{a} * \boldsymbol{b}$ is given by

$$c_k = a_0 b_k + a_1 b_{k-1} + \cdots + a_k b_0.$$

(Again, we call this the convolution of $\boldsymbol{a}$ and $\boldsymbol{b}$.)

There is a well-known problem in which we are asked how many ways there are of making change for r units of currency using notes of given denominations. (For example, we can ask how many ways there are of making \$43 out of \$1 and \$5 bills.) If we can make r units in a_r ways using one set of denominations and b_r ways using a completely different set, then it is not hard to see that, if we are allowed to use both sets of denominations, we can make up k units in c_k ways, where c_k is again the number defined earlier.

Let us see how this applies in the simple case where a_r is the number of ways of making up r dollars using \$1 bills and b_r is the number of ways of making up r dollars using \$2 bills. We observe that

$$(\mathcal{G}\boldsymbol{a})(t) = \sum_{r=0}^{\infty} t^r = \frac{1}{1-t},$$

$$(\mathcal{G}\boldsymbol{b})(t) = \sum_{r=0}^{\infty} t^{2r} = \frac{1}{1-t^2},$$

and so, using partial fractions,

$$(\mathcal{G}\boldsymbol{c})(t) = (\mathcal{G}(\boldsymbol{a} * \boldsymbol{b}))(t) = (\mathcal{G}\boldsymbol{a})(t)(\mathcal{G}\boldsymbol{b})(t)$$

$$= \frac{1}{(1-t)(1-t^2)} = \frac{1}{(1-t)^2(1+t)}$$

$$= \frac{1}{2(1-t)^2} + \frac{1}{4(1+t)} + \frac{1}{4(1-t)}$$

$$= \frac{1}{2}\sum_{r=0}^{\infty}(r+1)t^r + \frac{1}{4}\sum_{r=0}^{\infty}(-1)^r t^r + \frac{1}{4}\sum_{r=0}^{\infty} t^r$$

$$= \sum_{r=0}^{\infty} \frac{2r + 3 + (-1)^r}{4} t^r.$$

Thus we can make change for r dollars in $\frac{1}{2}(r+1)$ ways when r is odd and $\frac{1}{2}(r+2)$ ways when r is even. In this simple case it is easy to obtain the result directly but the method indicated works automatically in all cases. (The calculations can be made easier if we allow ourselves to work with complex roots.)

We have produced a "generating function transform" or "$\mathcal{G}$-transform," which takes a sequence $a_0, a_1, \ldots$ into a Taylor series $\sum_{r=0}^{\infty} a_r x^r$. (These names are not standard: most mathematicians would simply talk about GENERATING FUNCTIONS [IV.18 §§2.4, 3].) The next two examples show how we can use $\mathcal{G}$-transforms to restate problems about sequences as problems about Taylor series. Consider first the problem of finding a sequence u_n such that $u_0 = 0$, $u_1 = 1$, and

$$u_{n+2} - 5u_{n+1} + 6u_n = 0$$

for all $n \geqslant 0$. Observe that we must have

$$u_{n+2}t^{n+2} - 5u_{n+1}t^{n+2} + 6u_n t^{n+2} = 0$$

for all $n \geqslant 0$, so that summing over all $n \geqslant 0$ yields

$$((\mathcal{G}\boldsymbol{u})(t) - u_1 t - u_0) - 5(t(\mathcal{G}\boldsymbol{u})(t) - u_0) + 6t^2(\mathcal{G}\boldsymbol{u})(t) = 0.$$

Recalling that $u_0 = 0$, $u_1 = 1$, and rearranging, we obtain

$$(6t^2 - 5t + 1)(\mathcal{G}\boldsymbol{u})(t) = t.$$

Thus, using partial fractions, we obtain

$$(\mathcal{G}\boldsymbol{u})(t) = \frac{t}{6t^2 - 5t + 1} = \frac{t}{(1 - 2t)(1 - 3t)}$$

$$= \frac{-1}{1 - 2t} + \frac{1}{1 - 3t}$$

$$= -\sum_{r=0}^{\infty}(2t)^r + \sum_{r=0}^{\infty}(3t)^r$$

$$= \sum_{r=0}^{\infty}(3^r - 2^r)t^r.$$

It follows that $u_r = 3^r - 2^r$.

Next consider the rather trivial problem of finding a sequence u_n such that $u_0 = 1$ and

$$(n + 1)u_{n+1} + u_n = 0$$

for all $n \geqslant 0$. For every t we have

$$(n + 1)u_{n+1}t^n + u_n t^n = 0$$

and so, summing over all n and assuming that the usual laws of differentiation apply to infinite sums, we obtain

$$(\mathcal{G}\boldsymbol{u})'(t) + (\mathcal{G}\boldsymbol{u})(t) = 0.$$

This differential equation gives $(\mathcal{G}\boldsymbol{u})(t) = Ae^{-t}$ for some constant A. Setting $t = 0$, we obtain

$$1 = u_0 = (\mathcal{G}\boldsymbol{u})(0) = Ae^0 = A.$$

Thus

$$(\mathcal{G}\boldsymbol{u})(t) = \mathrm{e}^{-t} = \sum_{r=0}^{\infty} \frac{(-1)^r}{r!} t^r,$$

so $u_r = (-1)^r / r!$.

We can write down some of the correspondences between sequences and their $\mathcal{G}$-transforms:

$$(a_0, a_1, a_2, \dots) \longleftrightarrow (\mathcal{G}\boldsymbol{a})(t),$$
$$(a_0 + b_0, a_1 + b_1, a_2 + b_2, \dots) \longleftrightarrow (\mathcal{G}\boldsymbol{a})(t) + (\mathcal{G}\boldsymbol{b})(t),$$
$$\boldsymbol{a} * \boldsymbol{b} \longleftrightarrow (\mathcal{G}\boldsymbol{a})(t)(\mathcal{G}\boldsymbol{b})(t),$$
$$(0, a_0, a_1, a_2, \dots) \longleftrightarrow t(\mathcal{G}\boldsymbol{a})(t),$$
$$(a_1, 2a_2, 3a_2, \dots) \longleftrightarrow (\mathcal{G}\boldsymbol{a})'(t).$$

It is also important that we can recover the sequence $\boldsymbol{a}$ from its $\mathcal{G}$-transform. One way of seeing this is to note that

$$a_r = \frac{(\mathcal{G}\boldsymbol{a})^{(r)}(0)}{r!}.$$

We can use these rules, as in the examples above, to convert problems about sequences into problems about functions and vice versa. In textbooks and examinations, the effect of such a transformation is to make things simpler. In real life, it will usually convert your problem into a more complicated problem. However, occasionally you strike lucky and it is these occasions that make transforms such a valuable weapon in the mathematician's armory.

Up to now we have handled $\mathcal{G}$-transforms formally. However, if we wish to use the methods of analysis, we need to know that $\sum_{r=0}^{\infty} a_r t^r$ converges, at least when $|t|$ is small. Provided that the a_r do not increase too rapidly, this will always be the case. However, we run into difficulties when we try to extend our ideas to "two-sided sequences" (a_r), where r runs through all integers rather than just the nonnegative ones, and to the resulting sums $\sum_{r=-\infty}^{\infty} a_r t^r$. If $|t|$ is small, then $|t^r|$ is large when r is large and negative, while if $|t|$ is large, then $|t^r|$ is large when r is large and positive. In many cases, the best we can hope for is that $\sum_{r=-\infty}^{\infty} a_r t^r$ might converge when $t = -1$ and $t = 1$. It is not very useful to talk about functions defined at only two points, but we save the situation by moving from $\mathbb{R}$ to $\mathbb{C}$.

If we have a well-behaved sequence (a_r) of complex numbers where r runs through all integers, then we consider the sum $\sum_{r=-\infty}^{\infty} a_r z^r$, where the complex number z belongs to the unit circle (or, in other words, $|z| = 1$). Since any such z can be written

$$z = \mathrm{e}^{\mathrm{i}\theta} = \cos\theta + \mathrm{i}\sin\theta$$

with $\theta \in \mathbb{R}$, it is more usual to talk about the 2π-periodic function $\sum_{r=-\infty}^{\infty} a_r \mathrm{e}^{\mathrm{i}r\theta}$. We thus have the "Fourier series transform" (once again, the name is nonstandard) $\mathcal{H}$ given by

$$(\mathcal{H}\boldsymbol{a})(\theta) = \sum_{r=-\infty}^{\infty} a_r \mathrm{e}^{\mathrm{i}r\theta}.$$

The $\mathcal{H}$-transform takes a two-sided sequence $\boldsymbol{a}$ to a 2π-periodic complex-valued function $f = \mathcal{H}\boldsymbol{a}$ on the real line, but historically mathematicians have been more interested in reversing the process and obtaining $\boldsymbol{a}$ from f. If

$$f(\theta) = \sum_{r=-\infty}^{\infty} a_r \mathrm{e}^{\mathrm{i}r\theta},$$

then, arguing formally,

$$\frac{1}{2\pi} \int_{-\pi}^{\pi} f(\theta) \mathrm{e}^{-\mathrm{i}k\theta} \, \mathrm{d}\theta = \frac{1}{2\pi} \int_{-\pi}^{\pi} \sum_{r=-\infty}^{\infty} a_r \mathrm{e}^{\mathrm{i}(r-k)\theta} \, \mathrm{d}\theta$$
$$= \sum_{r=-\infty}^{\infty} \frac{a_r}{2\pi} \int_{-\pi}^{\pi} \mathrm{e}^{\mathrm{i}(r-k)\theta} \, \mathrm{d}\theta$$
$$= \sum_{r=-\infty}^{\infty} \frac{a_r}{2\pi} \int_{-\pi}^{\pi} \cos(r-k)\theta + \mathrm{i}\sin(r-k)\theta \, \mathrm{d}\theta = a_k.$$

If we write

$$\hat{f}(k) = \frac{1}{2\pi} \int_{-\pi}^{\pi} f(\theta) \mathrm{e}^{-\mathrm{i}k\theta} \, \mathrm{d}\theta,$$

then we obtain the celebrated Fourier sum formula

$$f(\theta) = \sum_{r=-\infty}^{\infty} \hat{f}(r) \mathrm{e}^{\mathrm{i}r\theta}. \qquad (1)$$

DIRICHLET [VI.36] proved that this formula holds in its natural interpretation for reasonably well-behaved functions, but the question of the appropriate interpretation and proof for wider classes of functions took much longer to settle (see CARLESON'S THEOREM [V.5]). Aspects of the question are still open today.

It is worth noting that we can obtain qualitative information about a sequence from its $\mathcal{H}$-transform and vice versa. For example, if $a_r r^{m+3}$ forms a bounded sequence, then the rules for term-by-term differentiation show that $\mathcal{H}\boldsymbol{a}$ is continuously m times differentiable, and if f is m times continuously differentiable, then repeated integration by parts shows that the numbers $r^m \hat{f}(r)$ form a bounded sequence.

Suppose that f represents a signal fed into a "black box," such as a telephone system, which gives rise to a resultant signal Tf. Many important black boxes in physics and engineering have the "infinite linearity" property that

$$T\left(\sum_{r=-\infty}^{\infty} c_r g_r \right)(\theta) = \sum_{r=-\infty}^{\infty} c_r T g_r(\theta)$$

for all well-behaved function g_r and constants c_r. Many such systems also have the key property that

$$T e_k(\theta) = \gamma_k e_k(\theta)$$

for some constant γ_k, where we have written $e_k(\theta)$ for the quantity $\mathrm{e}^{-\mathrm{i}k\theta}$. In other words, the functions e_k are EIGENFUNCTIONS [I.3 §4.3] for T. We can use the Fourier sum formula to obtain the formula

$$Tf(\theta) = \left(\sum_{r=-\infty}^{\infty} \hat{f}(r) T e_r \right)(\theta)$$
$$= \sum_{r=-\infty}^{\infty} \gamma_r \hat{f}(r) e_r(\theta).$$

In this context, it makes sense to think of f as the weighted sum of simple signals e_k of frequency k.

Mathematicians are always interested to see what happens if sums are replaced by integrals. In this case we obtain the classical Fourier transform. If F is a reasonably well-behaved function $F : \mathbb{R} \to \mathbb{C}$, then we define its *Fourier transform* $\mathcal{F}F$ by the formula

$$\mathcal{F}F(\lambda) = \int_{-\infty}^{\infty} F(t) \mathrm{e}^{-\mathrm{i}\lambda s} \, \mathrm{d}s.$$

Much of the analysis that is typically taught in the first year or two of a university mathematics course was developed in the context of this transform and related topics. Using that analysis, it is not hard to obtain the correspondences

$$F(t) \longleftrightarrow (\mathcal{F}F)(\lambda),$$
$$F(t) + G(t) \longleftrightarrow (\mathcal{F}F)(\lambda) + (\mathcal{F}G)(\lambda),$$
$$F * G(t) \longleftrightarrow (\mathcal{F}F)(\lambda)(\mathcal{F}G)(\lambda),$$
$$F(t+u) \longleftrightarrow \mathrm{e}^{-\mathrm{i}u\lambda}(\mathcal{F}F)(\lambda),$$
$$F'(t) \longleftrightarrow \mathrm{i}\lambda \mathcal{F}F(\lambda).$$

In this context we define the convolution of F and G by

$$F * G(t) = \int_{-\infty}^{\infty} F(t-s) G(s) \, \mathrm{d}s.$$

There is an element of truth in the saying that the importance of the Fourier transform is that it converts convolution to multiplication and the importance of convolution is that it is the operation that is transformed to multiplication by the Fourier transform. Just as we can use the $\mathcal{G}$-transform to solve difference equations, we can use the $\mathcal{F}$-transform to solve important classes of PARTIAL DIFFERENTIAL EQUATIONS [I.3 §5.4] that occur in physics and some parts of probability theory. For more on the Fourier transform, see [III.27].

By rescaling the Fourier sum formula (1), we obtain the formula

$$F(t) = \sum_{r=-\infty}^{\infty} \frac{1}{2\pi N} \int_{-\pi N}^{\pi N} F(s) \mathrm{e}^{-\mathrm{i}rs/N} \, \mathrm{d}s \mathrm{e}^{\mathrm{i}rt/N}$$

when $|t| < \pi N$. If we let $N \to \infty$, we obtain, more or less formally,

$$F(t) = \frac{1}{2\pi} \int_{-\infty}^{\infty} (\mathcal{F}F)(s) \mathrm{e}^{\mathrm{i}st} \, \mathrm{d}s,$$

which translates to the marvelous formula

$$(\mathcal{F}\mathcal{F}F)(t) = 2\pi F(-t).$$

Like the Fourier sum formula, this *Fourier inversion formula* can be proved under a wide range of circumstances, though often at the price of reinterpreting the formula in novel ways.

Beautiful though the Fourier inversion formula is, it should be noted that, both in practice and in theory, we often need only the observation that $\mathcal{F}F = \mathcal{F}G$ implies $F = G$. The *uniqueness of the Fourier transform* is often easier to prove and more convenient to use, and it holds over a wider range of conditions than the inversion formula. A similar observation holds for other transforms.

When we talked about the Fourier sums associated with 2π-periodic functions, we said that $\hat{f}(r)$ measured the proportion of the signal f with frequency $2\pi r$. In the same way, $(\mathcal{F}F)(\lambda)$ gives a measure of the proportion of F composed of frequencies close to λ. There is a family of inequalities, known generically as *Heisenberg uncertainty principles*, which say, in effect, that if most of $\mathcal{F}F$ is concentrated in a narrow band, then the signal F must be very spread out. This fact places strong restrictions on our ability to manipulate signals and occupies a central place in quantum theory.

At the beginning of this article we talked about transformations of sequences and saw that it was easier to handle one-sided sequences than two-sided sequences. In the same way, we can apply Fourier transforms to a wider range of functions $F : \mathbb{R} \to \mathbb{C}$ if we know that $F(t) = 0$ for $t < 0$. More specifically, if F is such a one-sided function, and if it does not grow too fast, then we can compute the *Laplace transform*

$$(\mathcal{L}F)(x + \mathrm{i}y) = \int_{-\infty}^{\infty} F(s) \mathrm{e}^{-(x+\mathrm{i}y)s} \, \mathrm{d}s$$
$$= \int_{0}^{\infty} F(s) \mathrm{e}^{-(x+\mathrm{i}y)s} \, \mathrm{d}s$$

whenever x and y are real and x is sufficiently large. If we use the more natural notation

$$(\mathcal{L}F)(z) = \int_{-\infty}^{\infty} F(s) \mathrm{e}^{-zs} \, \mathrm{d}s,$$

we see that $\mathcal{L}F$ can be considered as a weighted average of HOLOMORPHIC [I.3 §5.6] (that is, complex differentiable) functions and this can be used to show that $\mathcal{L}F$ is holomorphic. The Laplace transform shares many of the properties of the Fourier transform and we can use these, as well as the extensive collection of results on holomorphic functions, whenever we manipulate Laplace transforms. Many of the deepest results in number theory, such as the PRIME NUMBER THEOREM [V.26], are most easily obtained by clever uses of the Laplace transform.

The transforms we have discussed all belong to the same family, as is indicated by the fact that they all take convolution to multiplication. The general idea of a transform has been developed in several different directions, generally by concentrating on some aspects of the "classical transforms" and being prepared to lose others.

One of the most important of these new transforms is the *Gelfand transform*, which gives a concrete representation of the abstract *commutative Banach algebras*. It is discussed in OPERATOR ALGEBRAS [IV.15 §3.1]. Other *integral transforms* extend the integral definition of the Fourier transform by setting up a correspondence

$$F(t) \longmapsto \int_{-\infty}^{\infty} F(s)K(\lambda - s)\,\mathrm{d}s$$

or, more generally,

$$F(t) \longmapsto \int_{-\infty}^{\infty} F(s)\kappa(s, \lambda)\,\mathrm{d}s.$$

Another interesting transform is the *Radon* or *x-ray transform*. We shall consider the three-dimensional case and talk very informally. Suppose we shine a beam of radiation through a body in direction $\boldsymbol{u}$. Suppose also that f is a function defined on $\mathbb{R}^3$ that represents how much radiation is absorbed by different parts of the body. What we can measure is the amount of radiation absorbed along any given straight line. We might present some of this information in the form of a two-dimensional image, which could represent the amount absorbed by all lines in the direction $\boldsymbol{u}$. In general, we can use f to define a new function

$$(\mathcal{R}f)(\boldsymbol{u}, \boldsymbol{v}) = \int_{-\infty}^{\infty} f(t\boldsymbol{u} + \boldsymbol{v})\,\mathrm{d}t,$$

which tells us how much radiation is absorbed along the line in direction $\boldsymbol{u}$ that goes through a vector $\boldsymbol{v}$ perpendicular to $\boldsymbol{u}$. The *tomography problem* deals with the recovery of f from $\mathcal{R}f$.

Because the idea of a transform has been developed in so many different directions, any attempt to give a general definition results in something too general to be useful. The most that we can say about the various transforms is that they present a more or less distant analogy to the classical Fourier transforms and that this analogy has been found useful by those who developed them. (See also THE FOURIER TRANSFORM [III.27], SPHERICAL HARMONICS [III.87], REPRESENTATION THEORY [IV.9 §3], and WAVELETS AND APPLICATIONS [VII.3].)

III.92 Trigonometric Functions
Ben Green

The basic trigonometric functions "sin" and "cos," as well as the four related functions "tan," "cot," "sec," and "cosec," will probably be familiar to most readers in some form. One way to define the sine function sin : $\mathbb{R} \to [-1, 1]$ is as follows.

In almost all branches of mathematics one measures angles using *radians*, which are defined in terms of arclength: to say that the angle $\angle$AOB in figure 1 is θ radians is to say that the arc AB of the circle has length θ. This definition makes sense when $0 \leqslant \theta < 2\pi$. One then defines $\sin\theta$ to be the length PB, where P is the foot of the perpendicular from B to OA. It is very important that this length be taken with the correct *sign*. If $0 < \theta < \pi$ then we take the positive sign, whereas if $\pi < \theta < 2\pi$ we take the negative sign. In other words, $\sin\theta$ is the y-coordinate of the point B.

The sine function is, at the moment, defined on the interval $[0, 2\pi)$. To define it on all of $\mathbb{R}$ one simply insists that it be periodic with period 2π (that is, that it satisfies the relation $\sin\theta = \sin(2\pi n + \theta)$ for any integer n).

There is one problem with our definition of sine. What do we mean by the *length* of the arc AB? The only really satisfactory way of understanding this is to use calculus. The equation of the unit circle is $y = \sqrt{1 - x^2}$, at least if (x, y) lies in the upper-right quadrant. (Otherwise one must be careful about sign.) The formula for the arc-length of a curve $y = f(x)$ between $y = a$ and $y = b$ is

$$S = \int_a^b \sqrt{1 + (\mathrm{d}x/\mathrm{d}y)^2}\,\mathrm{d}y.$$

(This may be thought of as a definition, though the motivation for the definition comes from pictures.) For the circle, $\sqrt{1 + (\mathrm{d}x/\mathrm{d}y)^2} = 1/\sqrt{1 - y^2}$. Since the arc-length of the circle between the points P $= (x, \sin\theta)$

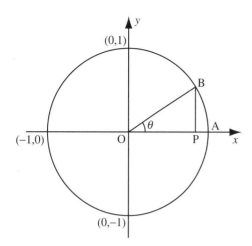

Figure 1 Interpreting trigonometric functions geometrically.

and A = $(1,0)$ is θ, this gives the formula

$$\int_0^{\sin\theta} \frac{\mathrm{d}y}{\sqrt{1-y^2}} = \theta \tag{1}$$

for $0 \leqslant \theta \leqslant \pi/2$ (we do not care about what x is). This can be regarded as giving a precise, even if implicit, definition of $\sin\theta$ for $0 \leqslant \theta \leqslant \pi/2$.

As with many of the most natural concepts in mathematics, sin may be defined in a multitude of equivalent ways. Another definition (whose equivalence to the first one is not obvious) is

$$\sin z = z - \frac{z^3}{3!} + \frac{z^5}{5!} - \frac{z^7}{7!} + \cdots. \tag{2}$$

This infinite series converges for all real z. The resulting definition has a distinct advantage over (1), in that it also makes sense when z is an arbitrary *complex* number (that is why we replaced the letter θ by z). It therefore allows us to extend sin to a HOLOMORPHIC FUNCTION [I.3 §5.6] on $\mathbb{C}$.

If the sine function is analytic, then what is its derivative? The answer is the cosine function $\cos z$, which may be defined in much the same way as sin: either geometrically or using a power series. The power series is

$$\cos(z) = 1 - \frac{z^2}{2!} + \frac{z^4}{4!} - \frac{z^6}{6!} + \cdots, \tag{3}$$

which may be obtained by differentiating the series for sin term by term (naturally, this is an operation that must be properly justified, but it can be).

If one differentiates again, one gets the formula $(\mathrm{d}^2/\mathrm{d}z^2)\sin z = -\sin z$. In fact, it is possible to define

$\sin : \mathbb{R} \to [-1,1]$ as the unique solution y to the differential equation $y'' = -y$ that also satisfies the initial value conditions $y(0) = 0$, $y'(0) = 1$. This is a very sensible way of proving that the two definitions (1) and (2) are equivalent (it is a good calculus exercise to prove that $\sin'' = -\sin$ using (1)).

Ultimately, the power series expansions (2) and (3) display the most important side of sin and cos, which is their relation with the EXPONENTIAL FUNCTION [III.25]:

$$\mathrm{e}^z = 1 + z + \frac{z^2}{2!} + \frac{z^3}{3!} + \cdots.$$

Comparing this with (2) and (3), one gets the famous formula

$$\mathrm{e}^{\mathrm{i}\theta} = \cos\theta + \mathrm{i}\sin\theta.$$

The exponential functions $\theta \mapsto \mathrm{e}^{\mathrm{i}n\theta}$ are *characters*, that is, HOMOMORPHISMS [I.3 §4.1] from $\mathbb{R}/2\pi\mathbb{Z}$ to the unit circle S^1 (which form groups under addition mod 2π and multiplication, respectively). This makes them the natural objects with which to do a FOURIER ANALYSIS [III.27] of 2π-periodic functions on $\mathbb{R}$. Because sin and cos are real-valued, it is convenient to try to decompose such a function $f(x)$ not into exponentials, but as a series

$$a_0 + a_1\cos x + b_1\sin x + a_2\cos 2x + b_2\sin 2x + \cdots.$$

Under favorable circumstances (if the function f is sufficiently smooth, say) one can recover the coefficients a_i, b_i by using *orthogonality relations* such as

$$\frac{1}{\pi}\int_0^{2\pi}\cos nx\cos mx\,\mathrm{d}x$$
$$= \begin{cases} 0 & \text{for all } n, m \geqslant 0, \ n \neq m, \\ 1 & n = m, \end{cases}$$

and

$$\frac{1}{\pi}\int_0^{2\pi}\cos nx\sin mx\,\mathrm{d}x = 0 \quad \text{for all } n, m \geqslant 0.$$

Thus, for example, we have

$$a_n = \frac{1}{\pi}\int_0^{2\pi}f(x)\cos nx\,\mathrm{d}x.$$

Such decompositions into trigonometric functions ultimately underlie devices like compact disk players and mobile phones.

Let us conclude by remarking that there is a whole zoo of formulas concerning sin, cos, and the other four trigonometric functions (which we have not discussed here), as well as integrals involving these functions. It is these formulas that make the trigonometric functions an indispensable tool in classical Euclidean geometry. There are many further formulas in that setting. To mention just one beautiful example, the area of a triangle inscribed in a unit circle with angles A, B, and C is exactly $2 \sin A \sin B \sin C$.

Uncountable Sets

See COUNTABLE AND UNCOUNTABLE SETS
[III.11]

III.93   Universal Covers

Let X be a TOPOLOGICAL SPACE [III.90]. A *loop* in X can be defined as a continuous function f from the closed interval $[0, 1]$ to X such that $f(0) = f(1)$. A *continuous family* of loops is a continuous function F from $[0, 1]^2$ to X such that $F(t, 0) = F(t, 1)$ for every t; the idea is that for each t we can define a loop f_t by taking $f_t(s)$ to be $F(t, s)$, and if we do this then the loops f_t "vary continuously" with t. A loop f is *contractible* if it can be continuously shrunk to a point: more formally, there should be a continuous family of loops $F(t, s)$ with $F(0, s) = f(s)$ for every s and with all values of $F(1, s)$ equal. If all loops are contractible, then X is said to be *simply connected*. For instance, a sphere is simply connected, but a torus is not because there are loops that "go around" the torus and therefore cannot be contracted (since any continuous deformation of a loop that goes around the torus goes around the same number of times).

Given any sufficiently nice path-connected space (that is, a space X such as a MANIFOLD [I.3 §6.9], with the property that any two points in X are linked by a continuous path), we can define a closely related *simply connected* space $\tilde{X}$ as follows. First, we pick an arbitrary "base point" x_0 in X. We then take the set of all continuous paths f from $[0, 1]$ to X such that $f(0) = x_0$ (but we do not necessarily ask for $f(1)$ to be x_0). Next, we regard two of these paths f and g as *equivalent*, or *homotopic*, if $f(1) = g(1)$ and there is a continuous family of paths that begins with f and ends with g and always has the same beginning point and endpoint. That is, f and g are homotopic if there is a continuous function F from $[0, 1]^2$ to X such that $F(t, 0) = x_0$ and $F(t, 1) = f(1) = g(1)$ for every t, and $F(0, s) = f(s)$ and $F(1, s) = g(s)$ for every s. Finally, we define the *universal cover* $\tilde{X}$ of X to be the space of all homotopy classes of paths: that is, it is the QUOTIENT [I.3 §3.3] of the space of all continuous paths that start at x_0 by the EQUIVALENCE RELATION [I.2 §2.3] of homotopy.

Let us see how this works in practice. As mentioned earlier, the torus is not simply connected, so what is its universal cover? To answer this question, it helps to think of the torus in a slightly artificial way: fix a point x_0 and define the torus to be the set of all continuous paths that begin at x_0, with two of these paths regarded as equivalent if they have the same endpoint. If we do this, then for each path "all we care about" is where it ends, and the set of endpoints is clearly the torus itself. But this was not the definition of the universal cover. There we cared not just about the endpoint of a path but also about *how we reached* the endpoint. For instance, if the path happens to be a loop, in which case the endpoint is x_0 itself, then we care about how many times that loop goes around the torus, and in what manner it goes around.

The torus can be defined as the quotient of $\mathbb{R}^2$ by the equivalence relation where we define two points as equivalent if their difference belongs to $\mathbb{Z}^2$. Then any point in $\mathbb{R}^2$ maps to a point in the torus (by the quotient map). Any continuous path on the torus then "lifts" uniquely to the plane in the following sense. Fix a point u_0 in $\mathbb{R}^2$ that maps to x_0 in the torus. Then if you trace out any continuous path in the torus that starts at x_0, there will be exactly one way of tracing out a path in $\mathbb{R}^2$ such that each point in that path maps to the appropriate point in the path in the torus.

Now suppose that we have two paths in the torus that start at x_0 and end at the same point x_1. Then the "lifts" of those paths both start at u_0, but all we know about their endpoints is that they are *equivalent*: we do not know that they are the same. Indeed, if the first path is a contractible loop and the second is a loop that goes once around the torus, then their lifts will end at *different* points. It turns out (and if you try to visualize this then you will see that the result is very natural and plausible) that the "lifts" of two paths will end at the same point if and only if the original paths are homotopic. In other words, there is a one-to-one correspondence between homotopy classes of paths in the torus and points in $\mathbb{R}^2$. This shows that $\mathbb{R}^2$ is the universal cover of the torus. In a sense, the operation of passing from a space to its universal cover "unfolds" the quotienting operation that we use to get from the universal cover to the space.

A fruitful way to think of this example is to think about the natural GROUP ACTION [IV.9 §2] of $\mathbb{Z}^2$ on $\mathbb{R}^2$. This associates with each element (m, n) of $\mathbb{Z}^2$ the translation $(x, y) \mapsto (x + m, y + n)$. We can then regard the torus as the quotient of $\mathbb{Z}^2$ by this action. That is, the elements of the torus are the *orbits* of the action (which are sets of the form $\{(x + m, y + n) : (m, n) \in \mathbb{Z}^2\}$) with the quotient topology (which basically means that two translates of $\mathbb{Z}^2$ are close when you think they are close). The action of $\mathbb{Z}^2$ on $\mathbb{R}^2$ is *free* and *discrete*, which means that each nonzero element of $\mathbb{Z}^2$ moves a small neighborhood of each point entirely off itself. It turns out that every sufficiently nice space X arises as the quotient of its universal cover by a similar group action: this group is the FUNDAMENTAL GROUP [IV.6 §2] of X.

As its name suggests, the universal cover has a universal property. Roughly speaking, a *cover* of a space X is a space Y and a continuous surjection from Y to X such that the inverse image of a small neighborhood in X is a disjoint union of small neighborhoods in Y. If U is the universal cover of X and Y is any other cover of X, then U can be made into a cover of Y in a natural way. For instance, one can define a cover of the torus by an infinite cylinder by wrapping the cylinder around, and the cylinder can in turn be covered by the plane. Thus, all connected covers of X are quotients of the universal cover. What is more, each is the space of orbits for the action on the universal cover of a subgroup of the fundamental group of X. This observation sets up a correspondence between conjugacy classes of subgroups of the fundamental group of X and equivalence classes of covers. This "Galois correspondence" has many analogues elsewhere in mathematics, most classically in the theory of field extensions (see THE INSOLUBILITY OF THE QUINTIC [V.21]).

An example of the use of universal covers can be found in GEOMETRIC AND COMBINATORIAL GROUP THEORY [IV.10 §§7, 8].

III.94 Variational Methods
Lawrence C. Evans

The *calculus of variations* is both a theory in itself and a toolbox of techniques for studying certain kinds of (often extremely nonlinear) ordinary and partial differential equations. These equations, which arise when we seek critical points of appropriate "energy" functionals, are usually far more tractable than other nonlinear problems.

1 Critical Points

Let us begin with a simple observation from first-year calculus, where we learn that if $f = f(t)$ is a smooth function defined on the real line $\mathbb{R}$ and if f

has a local minimum (or maximum) at a point t_0, then $(\mathrm{d}f/\mathrm{d}t)(t_0) = 0$.

The calculus of variations vastly extends this insight. The basic object to be considered is a *functional F*, which is applied not to real numbers but to functions, or rather to certain admissible classes of functions. That is, F takes functions u to real numbers $F(u)$. If u_0 is a minimizer of F (that is, $F(u_0) \leqslant F(u)$ for all admissible functions u), then we can expect that "the derivative of F at u_0 is zero." Of course, this idea has to be made precise, which one might expect to be tricky since the space of admissible functions is infinite dimensional. But in practice these so-called variational methods end up using just standard calculus, and they provide deep insights into the nature of minimizing functions u_0.

2 One-Dimensional Problems

The simplest situation to which variational techniques apply involves functions of a single variable. Let us see why minimizers of appropriate functionals in this setting must automatically satisfy certain ordinary differential equations.

2.1 Shortest Distance

As a warmup problem, we shall show that the shortest path between two points in the plane is a line segment. Of course, this is obvious, but the methods we develop can be applied to much more interesting situations.

Suppose, then, that we are given two points P and Q in the plane. We take as our class of admissible functions all smooth, real-valued functions u, defined on some interval $I = [a, b]$, such that $u(a) = P$ and $u(b) = Q$. The length of this curve is

$$F[u] = \int_I (1 + (u')^2)^{1/2} \, \mathrm{d}x, \tag{1}$$

where $u = u(x)$ and a prime denotes differentiation with respect to x. Now suppose that some particular curve u_0 minimizes the length. We want to deduce that the graph of u_0 is a line segment, which we will do by "setting the derivative of F to zero" at the minimizer u_0.

To make sense of this idea, select any other smooth function w that is defined on our interval I and that vanishes at its endpoints. For each t define $f(t)$ to be $F[u_0 + tw]$. Since the graph of the function $u_0 + tw$ connects the given endpoints, and since u_0 gives the minimum length, it follows that the function f, which is just an ordinary function from $\mathbb{R}$ to $\mathbb{R}$, has a minimum at $t = 0$. Therefore, $(\mathrm{d}f/\mathrm{d}t)(0) = 0$. But we can explicitly compute $(\mathrm{d}f/\mathrm{d}t)(0)$ by differentiating under the integral sign and then integrating by parts. This gives

$$\int_I \frac{u_0' w'}{(1 + (u_0')^2)^{1/2}}\, \mathrm{d}x = -\int_I \left(\frac{u_0'}{(1 + (u_0')^2)^{1/2}}\right)' w\, \mathrm{d}x.$$

This identity holds for all functions w with the properties specified above, and consequently

$$\left(\frac{u_0'}{(1 + (u_0')^2)^{1/2}}\right)' = \frac{u_0''}{(1 + (u_0')^2)^{3/2}} = 0 \qquad (2)$$

everywhere on the interval I.

To summarize the discussion so far: if the graph of u_0 minimizes the distance between the given endpoints, then u_0'' identically equals zero, and therefore the shortest path is a line segment. This conclusion may not seem too exciting, but even this simple case has an interesting feature. The calculus of variations automatically focuses our attention on the expression

$$\kappa = \frac{u''}{(1 + (u')^2)^{3/2}},$$

which turns out to be the *curvature* of the graph of u. The graph of the minimizer u_0 has zero curvature everywhere.

2.2 Generalization: The Euler–Lagrange Equations

It turns out that the technique we used for the previous example is extremely powerful and can be vastly generalized.

One useful extension is to replace the length functional (1) by a more general functional of the form

$$F[u] = \int_I L(u', u, x)\, \mathrm{d}x, \qquad (3)$$

where $L = L(v, z, x)$ is a given function, sometimes called the *Lagrangian*. Then $F[u]$ can be interpreted as the "energy" (or "action") of a given function u defined on the interval I.

Suppose next that a particular curve u_0 is a minimizer of F, subject to certain fixed boundary conditions. We want to extract information about the behavior of u_0, and to do so we proceed as in our first example. We select a smooth function w as above, define $f(t) = F[u_0 + tw]$, observe that f has a minimum at $t = 0$, and consequently deduce that $(\mathrm{d}f/\mathrm{d}t)(0) = 0$. As in the previous calculation, we then explicitly compute this derivative:

$$\frac{\mathrm{d}f}{\mathrm{d}t}(0) = \int_I L_v w' + L_z w\, \mathrm{d}x = \int_I (-(L_v)' + L_z) w\, \mathrm{d}x.$$

Here, L_v and L_z stand for the partial derivatives $\partial L/\partial v$ and $\partial L/\partial z$, evaluated at (u_0', u_0, x). This expression

equals zero for all functions w satisfying the given conditions. Therefore,

$$-(L_v(u_0', u_0, x))' + L_z(u_0', u_0, x) = 0 \qquad (4)$$

everywhere on the interval I. This nonlinear ordinary differential equation for the function u_0 is called the *Euler–Lagrange equation*. The key point is that any minimizer of our functional F must be a solution of this differential equation, which often contains important geometrical or physical information.

For example, take $L(v, z, x) = \frac{1}{2}mv^2 - W(z)$, which we interpret as the difference between the kinetic energy and the potential energy W of a particle of mass m moving along the real line. The Euler–Lagrange equation (4) is then

$$mu_0'' = -W'(u_0),$$

which is *Newton's second law of motion*. The calculus of variations provides us with an elegant derivation of this fundamental law of physics.

2.3 Systems

We can generalize further, by taking

$$F[\boldsymbol{u}] = \int_I L(\boldsymbol{u}', \boldsymbol{u}, x)\, \mathrm{d}x, \qquad (5)$$

where now we are taking vector-valued functions $\boldsymbol{u}$ that map the interval I into $\mathbb{R}^m$. If $\boldsymbol{u}_0$ is a minimizer in some appropriate class of functions, then one can compute the Euler–Lagrange equation using ideas similar to those discussed above. We obtain the equations

$$-(L_{v^k}(\boldsymbol{u}_0', \boldsymbol{u}_0, x))' + L_{z^k}(\boldsymbol{u}_0', \boldsymbol{u}_0, x) = 0, \qquad (6)$$

one for each k. Here L_{v^k} and L_{z^k} represent the partial derivatives of L with respect to the kth variables of $\boldsymbol{u}'$ and $\boldsymbol{u}$, evaluated at $(\boldsymbol{u}_0', \boldsymbol{u}_0, x)$. These equations form a *system* of coupled ordinary differential equations for the components of $\boldsymbol{u}_0 = (u_0^1, \ldots, u_0^m)$.

For a geometric example, put

$$L(v, z, x) = \left(\sum_{i,j=1}^m g_{ij}(z) v^i v^j\right)^{1/2},$$

so that $F[\boldsymbol{u}]$ is the length of the curve $\boldsymbol{u}$ in the RIEMANNIAN METRIC [I.3 §6.10] determined by the g_{ij}. When $\boldsymbol{u}_0$ is a curve of constant unit speed, the Euler–Lagrange system of equations (6) can be rewritten, after some work, to read

$$(u_0^k)'' + \sum_{i,j=1}^m \Gamma_{ij}^k (u_0^i)'(u_0^j)' = 0 \quad (k = 1, \ldots, m)$$

for certain expressions Γ_{ij}^k, called *Christoffel symbols*, that can be computed in terms of the g_{ij}. Solutions of this system of ordinary differential equations are called *geodesics*. Thus, we have deduced that *length-minimizing curves are geodesics*.

A physical example is $L(v, z, x) = \frac{1}{2} m |v|^2 - W(z)$, for which the Euler-Lagrange equation is

$$m \boldsymbol{u}_0'' = -\nabla W(\boldsymbol{u}_0).$$

This is Newton's second law of motion for a particle in $\mathbb{R}^m$ moving under the influence of the potential energy W.

3 Higher-Dimensional Problems

Variational methods also apply to expressions involving functions of several variables, in which case the resulting Euler-Lagrange equations are *partial* differential equations (PDEs).

3.1 Least Area

A first example extends our earlier examination of shortest curves. For this problem we are given a region U in the plane, with boundary ∂U, and a real-valued function g defined on the boundary. We then look at a class of admissible real-valued functions u, defined on U, with the condition that u should equal g on the boundary. We can think of the graph of u as a two-dimensional curved surface with a boundary equal to the graph of g. The area of this surface is

$$F[u] = \int_U (1 + |\nabla u|^2)^{1/2} \, dx. \tag{7}$$

Let us assume that a particular function u_0 minimizes the area among all other surfaces with the given boundary. What can we deduce about the geometric behavior of this so-called *minimal surface*?

Yet again we proceed by writing $f(t) = F[u_0 + tw]$, differentiating with respect to t, and so on. After some calculation we eventually discover that

$$\operatorname{div}\left(\frac{\nabla u_0}{(1 + |\nabla u_0|^2)^{1/2}} \right) = 0 \tag{8}$$

within the region U, where "div" denotes the divergence operator. This nonlinear PDE is the *minimal surface equation*. The left-hand side turns out to be a formula for (twice) the *mean curvature* of the graph of u_0. Consequently, we have shown that *a minimal surface has zero mean curvature everywhere*.

Minimal surfaces are sometimes regarded physically as the surfaces formed by soap films when they are stretched between a fixed wire frame that traces out the boundary specified by the function g.

3.2 Generalization: The Euler-Lagrange Equations

It is now straightforward, and sometimes very profitable, to replace the area functional (7) by the general expression

$$F[u] = \int_U L(\nabla u, u, x) \, dx, \tag{9}$$

in which we now take U to be a region in $\mathbb{R}^n$. Assuming that u_0 is a minimizer, subject to given boundary conditions, we deduce the *Euler-Lagrange equation*

$$-\operatorname{div}(\nabla_v L(\nabla u_0, u_0, x)) + L_z(\nabla u_0, u_0, x) = 0. \tag{10}$$

This is a nonlinear PDE that a minimizer must satisfy. A given PDE is called *variational* if it has this form.

If, for example, we take $L(v, z, x) = \frac{1}{2} |v|^2 + G(z)$, the corresponding Euler-Lagrange equation is the *nonlinear Poisson equation*

$$\Delta u = g(u),$$

where $g = G'$ and $\Delta u = \sum_{k=1}^n u_{x_k x_k}$ is the LAPLA-CIAN [I.3 §5.4] of u. We have shown that this important PDE is variational. This is a valuable insight, since we can then find solutions by constructing minimizers (or other critical points) of the functional $F[u] = \int_U \frac{1}{2} |\nabla u|^2 + G(u) \, dx$.

4 Further Issues in the Calculus of Variations

Our examples have shown pretty convincingly how useful our simple method, called computing the *first variation*, can be when applied to the right geometrical and physical problems. And indeed, variational principles and methods appear in several branches of both mathematics and physics. Many of the objects that mathematicians consider most important have an underlying variational principle of some kind. The list is impressive and, besides the examples we have discussed, includes Hamilton's equations, the Yang-Mills and Selberg-Witten equations, various nonlinear wave equations, Gibbs states in statistical physics, and dynamic programming equations from optimal control theory.

Many issues remain. For example, if $f = f(t)$ has a local minimum at a point t_0, then we know not only that $(df/dt)(t_0) = 0$, but also that $(d^2f/dt^2)(t_0) \geqslant 0$. The attentive reader will correctly guess that a generalization of this observation, called computing the *second variation*, is important for the calculus of variations. It provides an insight into appropriate convexity conditions that are needed to ensure that critical points are in fact stable minimizers. Even more fundamental is the question of the existence of minimizers or other critical points. Here mathematicians have devoted great ingenuity to designing appropriate function spaces within which "generalized" solutions can be found. But these weak solutions need not be smooth, and so the further question of their regularity and/or possible singularities must then be addressed.

However, these are all highly technical mathematical issues, far beyond the scope of this article. We end our discussion here, in the hope that our excessive demands upon the reader's attention have been minimized.

III.95 Varieties

Two simple examples of varieties are the circle and the parabola, which can be defined by the polynomial equations $x^2 + y^2 = 1$ and $y = x^2$, respectively. With one qualification, a variety is the solution set of a system of polynomial equations. The qualification is that there are certain examples that we do not want to include. For instance, the set of solutions to the equation $x^2 - y^2 = 0$ is the union of the two lines $x = y$ and $x = -y$, which naturally splits into two pieces. So the solution set to a system of polynomial equations is called an *algebraic set*, and it is called a *variety* if it cannot be written as a union of smaller algebraic sets.

The examples just given were subsets of the plane $\mathbb{R}^2$. However, the concept is much more general: varieties can live in $\mathbb{R}^n$ for any n, and also in $\mathbb{C}^n$ for any n. Indeed, the definitions make sense, and are interesting and important, in $\mathbb{F}^n$, where $\mathbb{F}$ can be any field.

The varieties defined so far have been *affine* varieties. For many purposes it is more convenient to deal with *projective* varieties. The definition is similar, but now they live inside a PROJECTIVE SPACE [III.72], and the polynomials used to define them must be homogeneous—that is, any multiple of a solution must still be a solution.

See ALGEBRAIC GEOMETRY [IV.4] and ARITHMETIC GEOMETRY [IV.5] for more information.

III.96 Vector Bundles

Let X be a TOPOLOGICAL SPACE [III.90]. A vector bundle over X is, roughly speaking, a way of associating a vector space with each point x of X in such a way that these spaces "vary continuously" as you vary x. As an example, consider a smooth surface X in $\mathbb{R}^3$. Associated

with each point x is the *tangent plane* at x, which varies continuously with x and can be identified in a natural way with a two-dimensional vector space. A more precise definition is as follows: a *vector bundle of rank n* over X is a topological space E, together with a continuous map $p : E \to X$, such that the inverse image $p^{-1}(x)$ of each point x (that is, the set of points in E that map to x) is an n-dimensional vector space. Moreover, for every sufficiently small region U of X, the inverse image of U is homeomorphic to $\mathbb{R}^n \times U$ (this property is called *local triviality*). The most obvious vector bundle of rank n over X is the space $\mathbb{R}^n \times X$ with the map $p(v, x) = x$; this is called the *trivial bundle*. However, the interesting bundles are the nontrivial ones, such as the tangent bundle of the 2-sphere. One can learn a great deal about a topological space by understanding its vector bundles. For this reason, vector bundles are central to algebraic topology. See ALGEBRAIC TOPOLOGY [IV.6 §5] for more details.

III.97 Von Neumann Algebras

A *unitary representation* of a GROUP [I.3 §2.1] G is a HOMOMORPHISM [I.3 §4.1] that associates with each element g of G a UNITARY MAP [III.50 §3.1] U_g defined on some HILBERT SPACE [III.37] H. A von Neumann algebra is a special kind of C^*-ALGEBRA [III.12], intimately connected with the theory of unitary representations. There are several equivalent ways of defining von Neumann algebras. One is as follows. It can be checked that, given any unitary representation, its *commutant*, defined to be the set of all OPERATORS [III.50] in $B(H)$ that commute with every single unitary map U_g in the representation, forms a C^*-algebra. Von Neumann algebras are algebras that arise in this way. They can also be defined abstractly as follows: a C^*-algebra A is a von Neumann algebra if there is a BANACH SPACE [III.62] X such that the DUAL [III.19 §4] of X is A (when A is itself considered as a Banach space).

The basic building blocks of von Neumann algebras are special kinds of von Neumann algebras called *factors*. The classification of factors is a major topic of research, which includes some of the most celebrated theorems of the second half of the twentieth century. See OPERATOR ALGEBRAS [IV.15 §2] for more details.

III.98 Wavelets

If you wish to send a black and white picture from one computer to another, then an obvious way of doing it is to encode it pixel by pixel: 0 for black and 1 for white. However, for certain pictures this would obviously be extremely inefficient. For instance, if the picture is a square, the left half of which is entirely white and the right half of which is entirely black, then it is clearly much better to send a set of instructions for reconstructing the picture than a list of every single pixel. Furthermore, the precise details of the pixels usually do not matter: if you want a patch of gray, then it is enough to put in black and white pixels in the right proportion and make sure that they are evenly distributed.

However, finding a good way of encoding pictures is difficult, and an important area of research in engineering. A picture can be thought of as a function from a rectangle to $\mathbb{R}$. The set of all such functions forms a VECTOR SPACE [I.3 §2.3], and a natural way to try to come up with a good encoding is to find a good basis for this space. Here "good" means that the functions one is interested in (that is, ones that correspond to the kinds of pictorial representations one might wish to send) are determined by just a few of their coefficients, apart from minor variations that are not detectable by the human eye.

Wavelets are a particularly good basis for many purposes. In some ways they are like FOURIER TRANSFORMS [III.27], but they are much better suited to encoding details such as sharp boundaries, and patterns that are "localized," rather than spread throughout the picture. For more details, see WAVELETS AND APPLICATIONS [VII.3].

III.99   The Zermelo–Fraenkel Axioms

The Zermelo–Fraenkel, or ZF, axioms are a collection of axioms that provide a foundation for set theory. They may be viewed in two ways. The first is as a list of the "allowed operations" on sets. For example, there is an axiom that states that, given sets x and y, there exists a "pair set," whose members are precisely x and y.

One of the reasons the ZF axioms are important is that it is possible to reduce all of mathematics to set theory, so the ZF axioms can be regarded as a foundation for mathematics as a whole. Of course, for this to be the case it is vital that the operations allowed by the ZF axioms do indeed allow one to perform all of the usual mathematical constructions. Some of the axioms are rather subtle as a result.

The other way to view the ZF axioms is as giving us just what we need to "build up" the world of all sets, starting with just the empty set. One can look at the various ZF axioms and see that each one plays an essential role as we create the set-theoretic universe. Equivalently, they are "closure rules" that any universe of sets, or more precisely any model of set theory, should obey. So, for example, there is an axiom that says that every set has a power set (the set of all its subsets), and this axiom allows us to build up a huge collection of sets starting with just the empty set: one obtains the power set of the empty set, the power set of the power set of the empty set, and so on. Indeed, the universe of all sets could (in a sense) be described as the closure of the empty set under all the allowable operations of ZF.

The ZF axioms are written in the language of FIRST-ORDER LOGIC [IV.23 §1]. So each axiom can mention variables (which are interpreted as ranging over all sets), as well as the usual logical operations, and also one "primitive relation," namely membership. For example, the pair-set axiom above would be formally written as

$$(\forall x)(\forall y)(\exists z)(\forall t)(t \in z \iff t = x \text{ or } t = y).$$

By convention, the ZF axioms do not include the AXIOM OF CHOICE [III.1]; when one does include the axiom of choice, the axioms are usually called the "ZFC axioms."

For a more detailed discussion of the ZF axioms see SET THEORY [IV.22 §3.1].

Part IV
Branches of Mathematics

IV.1 Algebraic Numbers
Barry Mazur

The roots of our subject go back to ancient Greece while its branches touch almost all aspects of contemporary mathematics. In 1801 the *Disquisitiones Arithmeticae* of CARL FRIEDRICH GAUSS [VI.26] was first published, a "founding treatise," if ever there was one, for the modern attitude toward number theory. Many of the still unachieved aims of current research can be seen, at least in embryonic form, as arising from Gauss's work.

This article is meant to serve as a companion to the reader who might be interested in learning, and thinking about, some of the classical theory of algebraic numbers. Much can be understood, and much of the beauty of algebraic numbers can be appreciated, with a minimum of theoretical background. I recommend that readers who wish to begin this journey carry in their backpacks Gauss's *Disquisitiones Arithmeticae* as well as Davenport's *The Higher Arithmetic* (1992), which is one of the gems of exposition of the subject, and which explains the founding ideas clearly and in depth using hardly anything more than high school mathematics.

1 The Square Root of 2

The study of algebraic numbers and algebraic integers begins with, and constantly reverts back to, the study of ordinary rational numbers and ordinary integers. The first algebraic irrationalities occurred not so much as *numbers* but rather as *obstructions* to simple answers to questions in geometry.

That the ratio of the diagonal of a square to the length of its side cannot be expressed as a ratio of whole numbers is purported to be one of the vexing discoveries of the early Pythagoreans. But this very ratio, when squared, is 2:1. So we might—and later mathematicians certainly did—deal with it algebraically. We can think of this ratio as a cipher, about which we know nothing

beyond the fact that its square is 2 (a viewpoint taken toward algebraic numbers by KRONECKER [VI.48], as we shall see below). We can write $\sqrt{2}$ in various forms, e.g.,

$$\sqrt{2} = |1 - i|, \tag{1}$$

and we can think of $1 - i = 1 - e^{2\pi i/4}$ as the world's simplest trigonometric sum; we shall see generalizations of this for all quadratic surds below. We can also view $\sqrt{2}$ as a limit of various infinite sequences, one of which is given by the elegant CONTINUED FRACTION [III.22]

$$\sqrt{2} = 1 + \cfrac{1}{2 + \cfrac{1}{2 + \cdots}}. \tag{2}$$

Directly connected to this continued fraction (2) is the Diophantine equation

$$2X^2 - Y^2 = \pm 1 \tag{3}$$

known as the *Pell equation*. There are infinitely many pairs of integers (x, y) satisfying this equation, and the corresponding fractions y/x are precisely what you get by truncating the expression in (2). For example, the first few solutions are $(1, 1)$, $(2, 3)$, $(5, 7)$, and $(12, 17)$, and

$$\left.\begin{aligned} \tfrac{3}{2} &= 1 + \tfrac{1}{2} = 1.5, \\[1ex] \tfrac{7}{5} &= 1 + \cfrac{1}{2 + \tfrac{1}{2}} = 1.4, \\[1ex] \tfrac{17}{12} &= 1 + \cfrac{1}{2 + \cfrac{1}{2 + \tfrac{1}{2}}} = 1.416\ldots. \end{aligned}\right\} \tag{4}$$

Replace the ± 1 on the right-hand side of (3) by *zero* and you get $2X^2 - Y^2 = 0$, an equation all of whose positive real-number solutions (X, Y) have the ratio $Y/X = \sqrt{2}$, so it is easy to see that the sequence of fractions (4) (these being alternately larger and smaller than $\sqrt{2} = 1.414\ldots$) converges to $\sqrt{2}$ in the limit. Even more striking is that (4) is a list of fractions that best approximate $\sqrt{2}$. (A rational number a/d is said to be a *best approximant* to a real number α if a/d is closer to α than any rational number of denominator smaller than or equal to d.) To deepen the picture, consider another important infinite expression,

the conditionally convergent series

$$\frac{\log(\sqrt{2}+1)}{\sqrt{2}} = 1 - \frac{1}{3} - \frac{1}{5} + \frac{1}{7} + \frac{1}{9} + \cdots \pm \frac{1}{n} + \cdots. \quad (5)$$

Here the n range over positive odd numbers, and the sign of the term $\pm 1/n$ is *plus* if n has a remainder of 1 or 7 when divided by 8, and it is *minus* if n has a remainder of 3 or 5. This elegant formula (5), which you are invited to "check out" at least to one digit accuracy with a calculator, is an instance of the powerful and general theory of *analytic formulas for special values of L*-FUNCTIONS [III.47], which plays the role of a bridge between the more algebraic and the more analytic sides of the story. When we allude to this, below, we will call it "the analytic formula," for short.

2   The Golden Mean

If you are looking for quadratic irrationalities that have been the subject of geometric fascination through the ages, then $\sqrt{2}$ has a strong competitor in the number $\frac{1}{2}(1+\sqrt{5})$, known as the *golden mean*. The ratio $\frac{1}{2}(1+\sqrt{5})$:1 gives the proportions of a rectangle with the property that when you remove a square from it, as in figure 1, you are left with a smaller rectangle whose sides are in the same proportion. Its corresponding trigonometric sum description is

$$\frac{1}{2}(1+\sqrt{5}) = \frac{1}{2} + \cos\frac{2}{5}\pi - \cos\frac{4}{5}\pi. \quad (6)$$

Its continued-fraction expansion is

$$\frac{1}{2}(1+\sqrt{5}) = 1 + \cfrac{1}{1+\cfrac{1}{1+\cdots}}, \quad (7)$$

where the sequence of fractions obtained by successive truncations of this continued fraction,

$$\frac{y}{x} = \frac{1}{1}, \frac{2}{1}, \frac{3}{2}, \frac{5}{3}, \frac{8}{5}, \frac{13}{8}, \frac{21}{13}, \frac{34}{21}, \dots, \quad (8)$$

is a sequence of best rational-number approximants to

$$\tfrac{1}{2}(1+\sqrt{5}) = 1.618033988749894848\dots,$$

where "best" has the sense already mentioned. For example, the fraction

$$\frac{34}{21} = 1 + \cfrac{1}{1+\cfrac{1}{1+\cfrac{1}{1+\cfrac{1}{1+\cfrac{1}{1+\cfrac{1}{1+\frac{1}{1}}}}}}}$$

equals $1.619047619047619047\dots$ and is closer to the golden mean than any fraction with denominator less than 21.

Figure 1 The outer rectangle has its height-to-width ratio equal to the golden mean. If you remove a square from it as indicated in the figure, you are left with a rectangle that has the golden mean as its width-to-height ratio. This procedure is of course repeatable.

Nevertheless, the exclusive appearance of 1s in this continued fraction[1] can be used to show that, among all irrational real numbers, the golden mean is the number that is, in a specific technical sense, *least well approximated by rational numbers*.

Readers familiar with the sequence of Fibonacci numbers will recognize them in the successive denominators of (8), and in the numerators as well. The analogue to equation (3) is

$$X^2 + XY - Y^2 = \pm 1. \quad (9)$$

This time, if you replace the ± 1 on the right-hand side of the equation by 0, you get the equation $X^2 + XY - Y^2 = 0$, whose positive real-number solutions (X, Y) have the ratio $Y/X = \frac{1}{2}(1+\sqrt{5})$, that is, the golden mean. And now the numerators and denominators y, x that appear in (8) run through the positive integral solutions of (9). The analogue of formula (5) (the "analytic formula") for the golden mean is the conditionally convergent infinite sum

$$\frac{2\log(\frac{1}{2}(1+\sqrt{5}))}{\sqrt{5}} = 1 - \frac{1}{2} - \frac{1}{3} + \frac{1}{4} + \frac{1}{6} + \cdots \pm \frac{1}{n} + \cdots, \quad (10)$$

---

1. The continued-fraction expansion of any real quadratic algebraic number has an eventually recurring pattern in its entries, as is vividly exhibited by the two examples (2) and (7) given above.

where the n range over positive integers not divisible by 5, and the sign of $\pm 1/n$ is *plus* if n has a remainder of ± 1 when divided by 5, and *minus* otherwise.

What governs the choice of the plus terms and minus terms is whether or not n is a *quadratic residue modulo* 5. Here is a brief explanation of this terminology. If m is an integer, two integers a, b are said to be *congruent modulo m* (in symbols we write $a \equiv b \mod m$) if the difference $a - b$ is an integral multiple of m; if a, b, and m are positive numbers, it is equivalent to ask that a and b have the same "remainder" (sometimes also called "residue") when each is divided by m (see MODULAR ARITHMETIC [III.58]). An integer a relatively prime to m is called a *quadratic residue* modulo m if a is congruent to the square of some integer, modulo m; otherwise it is called a *quadratic nonresidue* modulo m. So, $1, 4, 6, 9, \dots$ are quadratic residues modulo 5, while $2, 3, 7, 8, \dots$ are quadratic nonresidues modulo 5.

A generalization of equations (5) and (10) (the "analytic formula for the L-function attached to quadratic Dirichlet characters") gives a very surprising formula for the conditionally convergent sum of terms $\pm 1/n$, where n runs through positive integers relatively prime to a fixed integer and the sign of $\pm 1/n$ corresponds to whether n is a quadratic residue, or nonresidue modulo that integer.

3 Quadratic Irrationalities

The quadratic formula

$$X = \frac{-b \pm \sqrt{b^2 - 4ac}}{2a}$$

gives the solutions (usually two) to the general quadratic polynomial equation $aX^2 + bX + c = 0$ as a rational expression of the number $\sqrt{D}$, where $D = b^2 - 4ac$ is known as the *discriminant* of the polynomial $aX^2 + bX + c$, or, equivalently, of the corresponding homogeneous QUADRATIC FORM [III.73] $aX^2 + bXY + cY^2$. This formula introduces many irrational numbers: Plato's dialogue "Theaetetus" has the young Theaetetus credited with the discovery that $\sqrt{D}$ is irrational whenever D is a natural number that is not a perfect square. The curious switch, from initially perceiving an *obstruction* to a problem to eventually embodying this obstruction as a *number* or an *algebraic object of some sort* that we can effectively study, is repeated over and over again, in different contexts, throughout mathematics. Much later, *complex* quadratic irrationalities also made their appearance. Again these were not at first regarded as "numbers as such," but rather as *obstructions* to the

solution of problems. Nicholas Chuquet, for example, in his 1484 manuscript, *Le Triparty*, raised the question of whether or not there is a number whose triple is four plus its square and he comes to the conclusion that there is no such number because the quadratic formula applied to this problem yields "impossible" numbers, i.e., complex quadratic irrationalities in our terminology.[2]

For any real quadratic ("integral") irrationality there is a discussion along similar lines to the ones we have just given (expressions (1)–(5) for $\sqrt{2}$ and expressions (6)–(10) for $\frac{1}{2}(1 + \sqrt{5})$). For complex irrationalities, there is also such a theory, but with interesting twists. For one thing, we do not have anything directly comparable to continued-fraction expansions for a complex quadratic irrationality. In fact, the simple, but true, answer to the problem of how to find an infinite number of rational numbers that converge to such an irrationality is that you cannot! Correspondingly, the analogue of the Pell equation has only finitely many solutions. As a consolation, however, the appropriate "analytic formula" has a simpler sum, as we will see below.

Let d be any square-free integer, positive or negative. Associated with d is a particularly important number τ_d, defined as follows. If d is congruent to 1 mod 4 (that is, if $d - 1$ is a multiple of 4), then $\tau_d = \frac{1}{2}(1 + \sqrt{d})$; otherwise, $\tau_d = \sqrt{d}$. We will refer to these quadratic irrationalities τ_d as *fundamental algebraic integers of degree* 2. The general notion of an "algebraic integer" is defined in section 11. An algebraic integer of degree two is simply a root of a quadratic polynomial of the form $X^2 + aX + b$ with a, b ordinary integers. In the first case (when $d \equiv 1$ modulo 4), τ_d is a root of the polynomial $X^2 - X + \frac{1}{4}(1 - d)$ and in the second it is a root of $X^2 - d$. The reason special names are given to these quadratic irrationalities is that *any* quadratic algebraic integer is a linear combination (with ordinary integers as coefficients) of 1 and one of these fundamental quadratic algebraic integers.

4 Rings and Fields

I think that one of the big early advances in mathematics is the now-current, universal recognition of the importance of studying the properties of *collections* of

---

2. BOMBELLI [VI.8], in the sixteenth century, would refer to irrational square roots, of positive or of negative numbers, as "deaf" (reminiscent of the word *surd* that is still in use) and as "numbers impossible to name."

mathematical objects, and not just the objects in isolation. A *ring R* of complex numbers is a collection of them that contains 1 and is closed under the operations of addition, subtraction, and multiplication. That is, if a, b are any two numbers in R, $a \pm b$ and ab must also be in R. If such a ring R has the further property that it is closed under division by nonzero elements (i.e., if a/b is again in R whenever a and b are, and $b \neq 0$), then we say that R is a *field*. (These concepts are discussed further in FIELDS [I.3 §2.2] and RINGS, IDEALS, AND MODULES [III.81].) The ring $\mathbb{Z}$ of ordinary integers, $\{0, \pm 1, \pm 2, \dots\}$ is our "founding example" of a ring; visibly, it is the smallest ring of complex numbers.

The collection of all real or complex numbers that are integral linear combinations of 1 and τ_d is closed under addition, subtraction, and multiplication, and is therefore a ring, which we denote by R_d. That is, R_d is the set of all numbers of the form $a + b\tau_d$ where a and b are ordinary integers. These rings R_d are our first, basic, examples of *rings of algebraic integers* beyond that prototype, $\mathbb{Z}$, and they are the most important rings that are receptacles for quadratic irrationalities. Every quadratic irrational algebraic integer is contained in exactly one R_d.

For example, when $d = -1$ the corresponding ring R_{-1}, usually referred to as the ring of *Gaussian integers*, consists of the set of complex numbers whose real and imaginary parts are ordinary integers. These complex numbers may be visualized as the vertices of the infinite tiling of the complex plane by squares whose sides have length 1 (see figure 2).

When $d = -3$ the complex numbers in the corresponding ring R_{-3} may be visualized as the vertices of a tiling of the complex plane by equilateral triangles (see figure 3).

With the rings R_d in hand, we may ask ring-theoretic questions about them, and here is some of the standard vocabulary useful for this. A *unit u* in a given ring R of complex numbers is a number in R whose reciprocal $1/u$ is also in R; an *irreducible* element in R is a nonunit that cannot be written as the product of two nonunits in R. A ring of complex numbers R has the *unique factorization property* if every nonzero, nonunit, algebraic number in R can be expressed as a product of irreducible elements in exactly one way (where two factorizations are counted as the same if one can be obtained from the other by rearranging the order in which the irreducible elements appear and multiplying them by units).

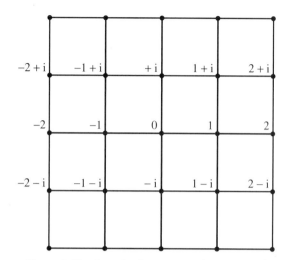

Figure 2 The Gaussian integers are the vertices of this lattice of squares tiling the complex plane.

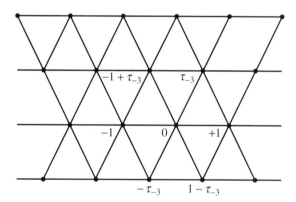

Figure 3 The elements of the ring R_{-3} are the vertices of this lattice of equilateral triangles tiling the complex plane.

In the prototype ring $\mathbb{Z}$ of ordinary integers, the only units are ± 1 and the irreducible elements are all numbers of the form $\pm p$ with p prime. The fundamental fact that any ordinary integer greater than 1 can be uniquely expressed as a product of (positive) prime numbers (that is, that $\mathbb{Z}$ enjoys the unique factorization property) is crucial for much of the number theory done with ordinary integers. That this unique factorization property for integers actually required proof was itself a hard-won realization of Gauss, who also provided its proof (see THE FUNDAMENTAL THEOREM OF ARITHMETIC [V.14]).

It is easy to see that there are only four units in the ring R_{-1} of Gaussian integers, namely ± 1 and $\pm i$;

multiplication by any of these units effects a *symmetry* of the infinite square tiling (figure 2 above). There are only six units in the ring R_{-3}, namely ± 1, $\pm\frac{1}{2}(1 + \sqrt{-3})$ and $\pm\frac{1}{2}(1 - \sqrt{-3})$; multiplication by any of these units results in a symmetry of the infinite triangular tiling illustrated in figure 3.

Fundamental to understanding the arithmetic of R_d is the following question: which ordinary prime numbers p are irreducible elements of R_d and which ones factorize as products of irreducible elements in R_d? We will see shortly that if a prime number does factorize in R_d, it must be expressible as the product of precisely two irreducible factors. For example, in the ring of Gaussian integers, R_{-1}, we have the factorizations

$$2 = (1 + i)(1 - i),$$
$$5 = (1 + 2i)(1 - 2i),$$
$$13 = (2 + 3i)(2 - 3i),$$
$$17 = (1 + 4i)(1 - 4i),$$
$$29 = (2 + 5i)(2 - 5i),$$
$$\vdots$$

where all the Gaussian integer factors in brackets above are irreducible elements of the ring of Gaussian integers.

Let us say that an odd prime p *splits* in R_{-1} if it factorizes into a product of at least two primes and *remains prime* if it does not do so. As we shall soon see, the officially agreed-upon definitions of splitting and remaining prime for more general rings of algebraic integers (even ones of the form R_d) are worded slightly, but very significantly, differently from the way we have just defined these concepts in the ring R_{-1} of Gaussian integers. (Note that we have excluded the prime $p = 2$ from the above dichotomy. This is because 2 *ramifies* in R_{-1}; for a discussion of this concept see section 7 below.) In any event, there is an elementary computable *rule* that tells us, for any R_d, which primes p split and which remain prime in this agreed sense. The rule depends upon the residue of p modulo $4d$: the reader is invited to guess it for the ring of Gaussian integers given the data just displayed above. In general, an elementary computable rule that says which primes split and which do not in a ring of algebraic integers such as R_d is referred to as a *splitting law* for the ring of algebraic integers in question.

5   The Rings R_d of Quadratic Integers

There is a very important "symmetry," or AUTOMOR-PHISM [I.3 §4.1], defined on the ring R_d. It sends $\sqrt{d}$ to $-\sqrt{d}$, keeps all ordinary integers fixed, and more generally, for rational numbers u and v, it sends $\alpha = u + v\sqrt{d}$ to what we may call its *algebraic conjugate* $\alpha' = u - v\sqrt{d}$. (The word "algebraic" is to remind you that this is not necessarily the same as the complex-conjugate symmetry of the complex numbers!)

You can immediately work out the formulas for this algebraic conjugation operation on the fundamental quadratic irrationalities τ_d: if d is not congruent to 1 modulo 4, then $\tau_d = \sqrt{d}$, so obviously $\tau_d' = -\tau_d$, while if d is congruent to 1 modulo 4, then $\tau_d = \frac{1}{2}(1 + \sqrt{d})$ and $\tau_d' = \frac{1}{2}(1 - \sqrt{d}) = 1 - \tau_d$. This symmetry $\alpha \mapsto \alpha'$ respects all algebraic formulas. For example, to work out the algebraic conjugate of a polynomial expression like $\alpha\beta + 2\gamma^2$, where α, β, and γ are numbers in R_d, you just replace each individual number by its algebraic conjugate, obtaining the expression $\alpha'\beta' + 2\gamma'^2$.

The most telling integer quantity attached to a number $\alpha = x + y\tau_d$ in R_d is its *norm* $N(\alpha)$, which is defined to be the product $\alpha\alpha'$. This equals $x^2 - dy^2$ when $\tau_d = \sqrt{d}$ and $x^2 + xy - \frac{1}{4}(d-1)y^2$ when $\tau_d = \frac{1}{2}(1 + \sqrt{d})$. The norm turns out to be *multiplicative*, meaning that $N(\alpha\beta) = N(\alpha)N(\beta)$, as you can directly check by multiplying out the formula for the norm of each factor and comparing with the norm of the product. This gives us a useful tactic for trying to factorize algebraic numbers in R_d, and offers criteria for determining whether a number α in R_d is a unit, and whether it is prime in R_d. In fact, an element $\alpha \in R_d$ is a unit if and only if $N(\alpha) = \alpha\alpha' = \pm 1$; in other words, the units are given by the integral solutions to the equations

$$X^2 - dY^2 = \pm 1 \tag{11}$$

or

$$X^2 + XY - \tfrac{1}{4}(d - 1)Y^2 = \pm 1 \tag{12}$$

following the two cases. Here is the proof of this. If $\alpha = x + y\tau_d$ is a unit in R_d, then its reciprocal, $\beta = 1/\alpha$, must also be in R_d, and, of course, we have $\alpha\beta = 1$. Applying the norm to both sides of this equation and using the multiplicative property discussed above, we see that $N(\alpha)$ and $N(\beta)$ are reciprocal ordinary integers. Therefore, they are either both equal to $+1$ or both equal to -1. This shows that (x, y) is a solution to whichever of equation (11) or (12) is appropriate. In the other direction, if $N(\alpha) = \alpha\alpha' = \pm 1$, then the reciprocal of α is simply $\pm\alpha'$. This is in R_d so α is indeed a unit in R_d.

These homogeneous quadratic forms, the left-hand sides of equations (11) and (12) (which generalize formulas (3) and (9)), play an important role; let us refer to whichever of them is relevant to R_d as the *fundamental quadratic form* for R_d, and to its discriminant D as the *fundamental discriminant*. (D is equal to d if d is congruent to 1 modulo 4 and to $4d$ otherwise.) When d is negative there are only finitely many units (if $d < -3$ the only ones are ± 1) but when d is positive, so that R_d consists entirely of real numbers, there are infinitely many. The ones that are greater than 1 are powers of a smallest such unit, ε_d, and this is called the *fundamental unit.*

For example, when $d = 2$ the fundamental unit, ε_2, is $1 + \sqrt{2}$, and when $d = 5$ it is the golden mean, $\varepsilon_5 = \frac{1}{2}(1 + \sqrt{5})$. Since any power of a unit is again a unit, we immediately have a machine for producing infinitely many units from any single one. For example, taking powers of the golden mean, we get

$$\varepsilon_5 = \tfrac{1}{2}(1 + \sqrt{5}), \qquad \varepsilon_5^2 = \tfrac{1}{2}(3 + \sqrt{5}),$$
$$\varepsilon_5^3 = 2 + \sqrt{5}, \qquad \varepsilon_5^4 = \tfrac{1}{2}(7 + 3\sqrt{5}),$$
$$\varepsilon_5^5 = \tfrac{1}{2}(11 + 5\sqrt{5}),$$

all of which are units in R_5. The study of these fundamental units was already under way in the twelfth century in India, but in general their detailed behavior as d varies still holds mysteries for us today. For example, there is a deep theorem of Hua (1942) that tells us that $\varepsilon_d < (4e^2 d)^{\sqrt{d}}$ (for a proof of it along with a historical discussion of such estimates, see chapters 3 and 8 in Narkiewicz (1973)). There are examples of d that come close to attaining that bound, but we still do not know whether or not there is a positive number η and an infinity of square-free d for which $\varepsilon_d > d^{d^\eta}$. (The answer to this question would be yes if, for example, there were an infinity of R_d satisfying the unique factorization property! This follows from a famous theorem of Brauer (1947) and Siegel (1935); for a proof of the Brauer–Siegel theorem, see theorem 8.2 of chapter 8 in Narkiewicz (1973) or Lang (1970).)

6 Binary Quadratic Forms and the Unique Factorization Property

The principle of unique factorization is an all-important fact for the ring of ordinary integers $\mathbb{Z}$. The question of whether this principle does or does not hold for a given ring R_d is central to the algebraic number theory. There are helpful, analyzable, *obstructions* to the validity of unique factorization in R_d. These obstructions, in turn, connect with profound arithmetic issues, and have become the focus of important study in their own right. One such mode of expressing the

obstruction to unique factorization is already prominent in Gauss's *Disquisitiones Arithmeticae* (1801), in which much of the basic theory of R_d was already laid down.

This "obstruction" has to do with how many "essentially different" binary quadratic forms $aX^2 + bXY + cY^2$ there are with discriminant equal to the fundamental discriminant D of R_d. (Recall that the discriminant of $aX^2 + bXY + cY^2$ is $b^2 - 4ac$, and that D equals $4d$ unless $d \equiv 1 \bmod 4$, in which case it equals d.)

In order to define a binary quadratic form $aX^2 + bXY + cY^2$ of discriminant D, what you need to provide is simply a triplet of coefficients (a, b, c) such that $b^2 - 4ac = D$. Given such a form, one can use it to define other ones. For example, if we make a small linear change of the variables, replacing X by $X - Y$ and keeping Y fixed, then we get $a(X - Y)^2 + b(X - Y)Y + cY^2$, which simplifies to $aX^2 + (b - 2a)XY + (c - b + a)Y^2$. That is, we get a new binary quadratic form whose triplet of coefficients is $(a, b - 2a, c - b + a)$, and which (as can easily be checked) has the same discriminant D. We can "reverse" this change by replacing X by $X + Y$ and keeping Y fixed. If we do this reversal and perform the corresponding simplification then we get back our original binary quadratic form. Because of this reversibility, these two quadratic forms take exactly the same set of integer values as X and Y vary: it is therefore reasonable to think of them as *equivalent*.

More generally, then, one says that two binary quadratic forms are equivalent if one can be turned into the other (or minus the other) by any "reversible" linear change of variables with integer coefficients. That is, one chooses integers r, s, u, v such that $rv - su = \pm 1$, replaces X and Y by the linear combinations $X' = rX + sY$, $Y' = uX + vY$, and simplifies the resulting expression to get a new triplet of coefficients. The condition $rv - su = \pm 1$ guarantees that by a similar operation we can get back to our original binary quadratic form, and also that the new binary quadratic form has the same discriminant D as the old one. So when we talk of "essentially different" binary quadratic forms of discriminant D we mean that we cannot turn one into the other by this kind of change of variables.

Here is the surprising obstruction to unique factorization that Gauss discovered.

The unique factorization principle is valid in R_d if and only if every homogeneous quadratic form $aX^2 + bXY + cY^2$ with discriminant equal to the fundamental discriminant of R_d is equivalent to the fundamental quadratic form of R_d.

Furthermore, the collection of inequivalent quadratic forms whose discriminant is the fundamental discriminant of R_d expresses in concrete terms the degree to which R_d "enjoys unique factorization."

If you have never seen this theory of binary quadratic forms before, try your hand at working with quadratic forms in the case where $D = -23$. The idea is to start with some particular quadratic form $aX^2 + bXY + cY^2$ of your choice with discriminant $D = b^2 - 4ac = -23$. Then, using a sequence of carefully chosen linear changes of variables you reduce the size of the coefficients $a, b,$ and c until you can go no further. Eventually you should end up with one of the two (inequivalent) quadratic forms that there are with discriminant -23: the fundamental form $X^2 + XY + 6Y^2$, or the form $2X^2 + XY + 3Y^2$. For example, can you see that the binary quadratic form $X^2 + 3XY + 8Y^2$ is equivalent to $X^2 + XY + 6Y^2$?

This type of exercise offers a small hint of the role that the *geometry of numbers* will play in the eventual theory. As you might expect from the venerability of these ideas, elegant streamlined methods have been discovered for making such calculations. Nevertheless, it is an open secret that any working mathematician, contemporary or ancient, engaged in this subject or nearby subjects, has done a myriad of straightforward simple hand computations along the lines of the above exercise.

If you try a few examples of this exercise, as I hope you do, here is one way of organizing your calculations. First, find a simple reversible linear change of variables to turn your form into an equivalent one with $a, b, c \geqslant 0$. (You may also have to multiply the whole form by -1.)

The cleanest way of writing down all binary quadratic forms given by triplets (a, b, c) of discriminant -23 is to list the triplets in increasing order of b, which will now be an odd positive integer. For each value of b you can then choose a and c in such a way that their product is $\frac{1}{4}(b^2 + 23)$. At this point the aim is to build up a repertoire of moves that tend to decrease b (which will keep a and c within bounds as well). A big clue, and aid, here is that for any pair of relatively prime integers x, y if you evaluate your quadratic form $aX^2 + bXY + cY^2$ at $(X, Y) = (x, y)$ to get the integer $a' = ax^2 + bxy + cy^2$, you can find, for appropriate b' and c', a quadratic form $a'X^2 + b'XY + c'Y^2$ equivalent to yours, with first coefficient a'. So, one tactic is to look for small integers represented by your quadratic form. Also the "example" linear change of variables $X \mapsto X - Y, Y \mapsto Y$ will

lead you to be able to reduce the coefficient b to an integer smaller than $2a$. Can you check that $X^2 + XY + 6Y^2$ and $2X^2 + XY + 3Y^2$ are inequivalent?

Now, as we have just discussed, it follows from the general theory that R_{-23} does not have the unique factorization property. We can also see this directly. For example,

$$\tau_{-23} \cdot \tau'_{-23} = 2 \cdot 3,$$

and all four of the factors in this equation are irreducible in R_{-23}. To be a faithful companion, I should at this point give at least a hint at what connection there might be between this specific "failure of unique factorization" and the previous discussion. It may become a bit clearer in the next paragraph, but the underlying tension in the equation $\tau_{-23} \cdot \tau'_{-23} = 2 \cdot 3$ is that all the factors in our ring are prime: we are *missing* any elements in our ring R_{-23} that could factorize it further. We lack, for example, elements that play the role of the *greatest common divisor* of factors of this equation. The general theory regarding these matters (which we are not entering into here, but see EUCLID'S ALGORITHM [III.22]) tells us that what is missing is some element γ in R_{-23} that is both a linear combination of the numbers τ_{-23} and 2 (with coefficients in the ring R_{-23}) and also a common divisor of τ_{-23} and 2 in the ring R_{-23}, i.e., such that τ_{-23}/γ and $2/\gamma$ are both in R_{-23}. There is no such element, for its norm must divide $N(\tau_{-23}) = 6$ and $N(2) = 4$, and therefore be equal to 2, which can easily be shown to be impossible. But we are interested, rather, in the phenomenon that *inequivalence* of certain binary quadratic forms will indeed show this, so let us go on.

First, check that any linear combination

$$\alpha \cdot \tau_{-23} + \beta \cdot 2$$

with α, β elements of R_{-23} can also be written as $u \cdot \tau_{-23} + v \cdot 2$, where u and v are ordinary integers. Now compute the binary quadratic form given by systematically taking the norms of these linear combinations, and viewing these norms as functions of the integer coefficients u, v:

$$N(u \cdot \tau_{-23} + v \cdot 2) = (\tau_{-23}u + 2v)(\tau'_{-23}u + 2v)$$
$$= 6u^2 + 2uv + 4v^2.$$

Viewing the u and the v as *variables*, and dubbing them U and V to emphasize their status as variables, we can say that the *norm quadratic form* obtained from the collection of linear combinations of τ_{-23} and 2 is

$$6U^2 + 2UV + 4V^2 = 2 \cdot (3U^2 + UV + 2V^2).$$

Now suppose that, contrary to fact, there *were* a common divisor, y, as above; in particular, the multiples of y in the ring R_{-23} would then be precisely the linear combinations of the numbers τ_{-23} and 2. We would then have another way of describing those linear combinations; namely, for any pair of ordinary integers (u, v) there would be a pair of ordinary integers (r, s) such that

$$u \cdot \tau_{-23} + v \cdot 2 = y \cdot (r\tau_{-23} + s) = ry\tau_{-23} + sy.$$

Taking norms, as above, we would get

$$N(y \cdot (r\tau_{-23} + s)) = N(ry\tau_{-23} + sy)$$
$$= N(y)(6r^2 + rs + s^2).$$

Again, thinking of r and s as variables and renaming them R and S we would have the corresponding norm quadratic form:

$$N(y) \cdot (6R^2 + RS + S^2) = 2 \cdot (6R^2 + RS + S^2).$$

Given the above facts—dependent, of course, on the contrary-to-fact hypothesis that there is a y as above—the key idea is that there would be linear changes of variables from (U, V) to (R, S) and back that would establish an equivalence between the two quadratic forms $2 \cdot (3U^2 + UV + 2V^2)$ and $2 \cdot (6R^2 + RS + S^2)$. But these quadratic forms are not equivalent! Their inequivalence therefore shows that the putative y does not exist and factorization in the ring R_{-23} is not unique.

7 Class Numbers and the Unique Factorization Property

In the previous section we saw that the collection of inequivalent quadratic forms of discriminant equal to the fundamental discriminant provides us with an obstruction to unique factorization. Somewhat later, a more articulated version of this obstruction arose, known as the *ideal class group* H_d of R_d. As its name implies, to describe this we must use the vocabulary of IDEALS [III.81 §2] and GROUPS [I.3 §2.1]. A subset I of R_d is an *ideal* if it has the following closure properties: if α belongs to I, so do $-\alpha$ and $\tau_d\alpha$, and if α and β belong to I, so does $\alpha + \beta$. (The first and third properties imply together that any integer combination of α and β belongs to I.) The basic example of such an ideal is the set of all multiples of some fixed, nonzero element y of R_d, where by a *multiple* of y we mean the product of y and an element of R_d. We denote this set tersely as (y), or, slightly more expressively, as $y \cdot R_d$. An ideal of this sort, i.e., one that can be expressed as the set of all multiples of a single nonzero element y, is

called a *principal ideal*. For example, the ring R_d itself is an ideal (it consists, after all, of all linear combinations of 1 and τ_d) and is even a principal ideal: in our laconic terminology, it can be denoted $(1) = 1 \cdot R_d = R_d$. Strictly speaking, the singleton $\{0\}$ is also an ideal, but the ones that will interest us are the *nonzero ideals*.

As a direct counterpart to the obstruction principle involving binary quadratic forms that was described in the previous section, we have the following obstruction principle involving ideals.

The unique factorization principle is valid in R_d if and only if every ideal in R_d is principal.

Reflecting on this, you can get a sense of why the word "ideal" might have been chosen. Every principal ideal in R_d is of the form $y \cdot R_d$ for some number y in R_d (which is uniquely determined apart from multiplication by units), but sometimes there are more general ideals. These arise if you ever have two elements of R_d (think of τ_{-23} and 2, as in the previous section) such that the set of all their integer combinations *cannot* be expressed as the set of multiples of some fixed number y in R_d. This phenomenon is a sign that we may be missing numbers in R_d that provide fine enough factorizations to make the arithmetic in R_d as smooth going as one might hope for. Just as a principal ideal $y \cdot R_d$ corresponds to the number y, ideals of this more general kind (think of the set of all integer combinations of τ_{-23} and 2) can be thought of as corresponding to "ideal numbers" that should, "by rights," be present in our ring, but happen not to be.

Once we think of ideals as standing for ideal numbers it makes some sense to try to multiply them: if I, J are two ideals in R_d, we let $I \cdot J$ denote the set of all finite sums of products $\alpha \cdot \beta$ in which α is in I and β is in J. The product of two principal ideals $(y_1) \cdot (y_2)$ is the principal ideal $(y_1 \cdot y_2)$ so, just as one would hope, multiplication of principal ideals corresponds to multiplication of the corresponding numbers. Multiplication of any ideal I by the ideal (1) leaves I unchanged: $(1) \cdot I = I$; we therefore refer to the ideal (1) as *the unit ideal*. With this new notion of *multiplication of ideals* we can now give the general definition of what it means for a prime number p to split or to remain prime in a ring R_d, the definition we promised in section 4.

The idea behind the definition is to use multiplication of ideals rather than of numbers. So if we are thinking about a prime p, the first thing we do is turn our attention to the principal ideal (p) in R_d. If this can be factorized as a product of two different ideals (*not* necessarily principal ideals, this is the whole point) in R_d, and if neither of these is the unit ideal $(1) = R_d$, then we say that p *splits* in R_d. If, on the other hand, no factorization of the ideal (p) can be made without one of the factors being the ideal $(1) = R_d$, then we say that p *remains prime* in R_d. There is also a third important definition: if the principal ideal (p) can be expressed as the square of another ideal I, then we say that p *ramifies* in R_d. Continuing with the momentum of this definition, we may say that an ideal P is a *prime ideal* if P cannot be "factorized" as the product of two ideals neither of which is the unit ideal. This definition makes sense whether or not P is principal, so we are subtly shifting our attention from the multiplicative arithmetic of the numbers in R_d to the ideals.

By definition, two ideals are in the same *ideal class* if when you multiply each by an appropriate principal ideal you get the same ideal as a result. This is a natural EQUIVALENCE RELATION [I.2 §2.3] on ideals. It is also one that *respects products*, meaning that if I and J are two ideals, then the ideal class of their product $I \cdot J$ depends only on the ideal classes of I and J. (In other words, if I' is in the same ideal class as I and J' is in the same ideal class as J', then $I' \cdot J'$ is in the same ideal class as $I \cdot J$.) We can therefore say what we mean by *multiplication of ideal classes*: to multiply two classes, pick an ideal from each, multiply those, and take the ideal class of the resulting product. The set H_d of ideal classes of R_d, given this operation of multiplication, forms an Abelian group, in the sense that the multiplication law we have just defined is associative and commutative, and there are inverses. The identity element is the principal ideal R_d itself. This group H_d, the *ideal class group*, directly measures the extent to which the ideals of the ring R_d are principal: roughly speaking it is what you get if you take the multiplicative structure of all ideals and "divide out" by the principal ones.

As was mentioned in section 6, there is a close connection between ideal classes and binary quadratic forms. To begin to see this, take an ideal I of R_d and write it as the set of all integer combinations of two elements α, β of R_d. Then consider the norm function on the elements of I, that is,

$$N(x\alpha + y\beta) = (x\alpha + y\beta)(x\alpha' + y\beta')$$
$$= \alpha\alpha' x^2 + (\alpha\beta' + \alpha'\beta)xy + \beta\beta' y^2.$$

This is a binary quadratic form in the variable coefficients x and y. If you start with a different choice of α, β that generate I you get a different form, but the two forms are scalar multiples of two forms with discriminant D that are equivalent to one another. Even better, the equivalence class of these forms depends only on the ideal class of I.

It can be shown that there are only a finite number of distinct ideal classes of R_d; that is, the ideal class group H_d is finite. The number of its elements is denoted h_d and called the *class number* of R_d. So, the obstruction to unique factorization of R_d is given by the nontriviality of the group H_d; equivalently, unique factorization holds for R_d if and only if its class number is 1. But whether or not H_d is trivial, its detailed group-theoretic structure is profoundly related to the arithmetic of R_d.

The class number enters into the generalizations of formulas (5) and (10) of section 1; that is, the *analytic formulas* we alluded to in that section. These formulas represent just the beginning of one of the ongoing chapters of our subject, and form a bridge between the world of discrete arithmetical issues and that of calculus, infinite series, and volumes of spaces, all of which can be attacked by the methods of COMPLEX ANALYSIS [I.3 §5.6]. Here is a sample of them.

(i) If $d > 0$ is a square-free integer and D is either d or $4d$ according to whether d is congruent to 1 modulo 4 or not, then

$$h_d \cdot \frac{\log \varepsilon_d}{\sqrt{D}} = \sum_{n>0} \pm \frac{1}{n},$$

where the integers n run through those that are relatively prime to D and the signs $\pm$ are chosen in a way that depends only on the residue class of n modulo D.

(ii) If $d < 0$ we have a somewhat simpler formula: there is no fundamental unit ε_d in R_d to contend with, but when $d = -1$ or -3, there are more roots of unity than merely ± 1. If w_d denotes the number of roots of unity in R_d, then $w_{-1} = 4$, $w_{-3} = 6$ and otherwise $w_d = 2$, and then one has a formula of the following type:

$$\frac{h_d}{w_d\sqrt{|D|}} = \sum_{n>0} \pm \frac{1}{n}.$$

As d tends to $-\infty$ the class number h_d tends to infinity.

We have effective lower bounds for the growth of h_d but these lower bounds are probably still far from the actual growth (cf. Goldfeld 1985). The effective lower bounds that are known are exceedingly weak. They follow, however, from beautiful work of Goldfeld, and of Gross and Zagier: for every real number

$r < 1$ there is a computable constant $C(r)$ such that $h_d > C(r) \log |D|^r$. Here is a sample:

$$h_d > \frac{1}{55} \prod_{p|D} \left(1 - \frac{2\sqrt{p}}{p+1}\right) \cdot \log |D|$$

if $(D, 5077) = 1$.

It is a striking lacuna in our theory that, even today, nobody knows how to prove that there are infinitely many values of $d > 0$ for which R_d enjoys the unique factorization property—particularly since we expect that more than three quarters of them do! Our expectations are even more precise than that, thanks to Henri Cohen and Hendrik Lenstra, who make use of certain probabilistic expectations (now known as the *Cohen–Lenstra heuristics*) to conjecture that the density of positive fundamental discriminants of class number 1 among all positive fundamental discriminants is $0.75446\ldots$.

8 The Elliptic Modular Function and the Unique Factorization Property

A different obstruction to unique factorization in R_d is available when d is negative. Now R_d may be thought of as a lattice in the complex plane (see figure 3), which makes a wonderful tool available for us: the classical *elliptic modular function* of KLEIN [VI.57],

$$j(z) = e^{-2\pi i z} + 744 + 196\,884\,e^{2\pi i z}$$
$$+ 21\,493\,760\,e^{4\pi i z} + 864\,299\,970\,e^{6\pi i z} + \cdots .$$
$$(13)$$

This function, also colloquially referred to as the "j-function," converges for complex numbers $z = x + iy$ with $y > 0$. If $z = x + iy$ and $z' = x' + iy'$ are two such complex numbers, then $j(z) = j(z')$ if and only if the lattice generated by z and 1 in the complex plane is the same as the lattice generated by z' and 1 (or, equivalently, $z' = (az + b)/(cz + d)$, where a, b, c, and d are ordinary integers such that $ad - bc = 1$). We can paraphrase this by saying that the value $j(z)$ depends only on, and characterizes, the lattice generated by z and 1.

It turns out (by a theorem of Schneider) that if an algebraic number $\alpha = x + iy$ with $y > 0$ has the property that $j(\alpha)$ is also algebraic, then α is a (complex) quadratic irrationality; and the converse is also true. In particular, since $\alpha = \tau_d$ is such a complex quadratic irrationality when d is negative, the value, $j(\tau_d)$, of the j-function on τ_d is an algebraic number—in fact, an algebraic integer. This will be of some importance for

our story. First, since the ring R_d as situated in the complex plane is simply the lattice generated by τ_d and 1, it follows from the previous paragraph that this value $j(\tau_d)$ will be the same if we replace τ_d by *any* element α of R_d, as long as the lattice generated by α and 1 is the entire ring R_d. More importantly, $j(\tau_d)$ is an algebraic integer of degree roughly comparable with the class number of R_d. In particular, it is an ordinary integer if and only if the ring R_d has the unique factorization property. (This result is one of the great applications of a classical theory known as *complex multiplication*.) In brief, here is yet another answer to the question of when the unique factorization principle holds for R_d when d is negative: if $j(\tau_d)$ is an ordinary integer, the answer is *yes*; otherwise it is *no*.

The search for the full list of negative values of d for which R_d has the unique factorization property makes a marvelous tale: there are precisely nine values of d for which it occurs (see below), but for over two decades number theorists, while knowing these nine, could prove only that there were no more than ten. The history of how the *nonexistence* of a possible tenth value of d was established, and reestablished, is one of the thrilling chapters in our subject. K. Heegner, in an article published in 1952, provided what he claimed was a proof of the nonexistence of the possible *tenth value of d*. However, Heegner's proof was framed in somewhat unfamiliar language and was not understood by the mathematicians of the time. His paper and his purported proof were largely forgotten until the late 1960s, when the nonexistence of the tenth field was established (to the mathematical community's satisfaction) by Stark (1967) and independently, via a different method, by Baker (1971). It was only then that mathematicians took a second and closer look at Heegner's original article and discovered that he had indeed proven exactly what he claimed. Moreover, his proof offered an elegant direct conceptual road to an understanding of the underlying issue.

Here are the nine values of d:

$$d = -1, -2, -3, -7, -11, -19, -43, -67, -163.$$

And here are the corresponding nine values of $j(\tau_d)$:

$$j(\tau_d) = 2^6 3^3,\ 2^6 5^3,\ 0,\ -3^3 5^3,\ -2^{15},\ -2^{15} 3^3,$$
$$-2^{18} 3^3 5^3,\ -2^{15} 3^3 5^3 11^3,\ -2^{18} 3^3 5^3 23^3 29^3.$$

As Stark once pointed out, if, for some of these values of d, you simply "plug" τ_d into the power-series expansion for j, you get rather surprising formulas. For example, when $d = -163$, then

$$e^{-2\pi i \tau_d} = -e^{\pi\sqrt{163}}$$

is the first term of the power series for $j(\tau_{-163})$ (see formula (13)). Since $j(\tau_{-163}) = -2^{18}3^3 5^3 23^3 29^3$ and since all the terms $e^{2\pi n \tau_d}$ ($n > 0$) that appear in the power series for the j-function are relatively small, we find that $e^{\pi\sqrt{163}}$ is incredibly close to an integer. Indeed, it is $2^{18}3^3 5^3 23^3 29^3 + 744 + \cdots$, which works out as $262\,537\,412\,640\,768\,744 - \epsilon$, where the error term ϵ is less than 7.5×10^{-13}.

9 Representations of Prime Numbers by Binary Quadratic Forms

More often than you might expect, it turns out to be possible to translate difficult and/or somewhat artificial problems about ordinary integers into natural and tractable problems about larger rings of algebraic integers. My favorite elementary example of this type is the theorem due to FERMAT [VI.12] that if a prime number p may be expressed as a sum of two squares, $p = a^2 + b^2$ with $0 < a \leqslant b$, then it has only one such expression. (For example, $1^2 + 10^2$ is the only way of expressing the prime number 101 as the sum of two squares.) Moreover, a prime number p can be expressed as a sum of two squares if and only if $p = 2$ or p is of the form $4k + 1$. (The "only if" part of this is easy to see: since any square is congruent either to 0 or to 1 mod 4, an odd integer that is a sum of two squares is necessarily congruent to 1 mod 4.) These statements about ordinary integers can be translated into basic statements about the ring of Gaussian integers. For if we write $a^2 + b^2 = (a + ib)(a - ib)$, with $i = \sqrt{-1}$, then we can view $a^2 + b^2$ as the norm of the (conjugate) elements $a + ib$ in the ring of Gaussian integers. So, if p is a prime number that admits an expression as a sum of squares, $p = a^2 + b^2$, it follows that each of the elements $a \pm ib$ has norm a prime integer. It is easy to deduce that $a \pm ib$ is itself a prime in the ring of Gaussian integers. Indeed, any factorization of $a \pm ib$ into a product of two Gaussian integers would have the property that the norms of the factors are ordinary integers which multiply out to be the prime p, and this severely limits their possibilities: one of them has to be a unit.

In other words, whenever $p = a^2 + b^2$, then

$$p = (a + ib)(a - ib)$$

is a factorization of the ordinary integer prime p into a product of two Gaussian integer primes. The uniqueness part of Fermat's theorem then follows from (in fact, it is readily seen to be equivalent to) the unique factorization property of the ring R_{-1} of Gaussian integers. That any prime number p of the form $4k + 1$ admits such an expression as a sum of two squares follows from the *splitting law* for primes p in the ring of Gaussian integers: an odd prime number p is a norm, and hence splits into the product of two distinct primes, in the ring of Gaussian integers if and only if p is congruent to 1 mod 4. This result is just the beginning of an immense chapter of arithmetic.

10 Splitting Laws and the Race between Residues and Nonresidues

The simple *splitting law* for ordinary prime integers p in the ring of Gaussian integers, which states that p splits if $p \equiv 1 \bmod 4$ and not if $p \equiv -1 \bmod 4$, invites us to ask how often each of these cases occurs (see figure 4). DIRICHLET [VI.36] proved a famous theorem that says that there are infinitely many primes in the arithmetic progression $c, m + c, 2m + c, \ldots$ if the integers m and c are relatively prime. A more precise version of his result gives a clear asymptotic answer to the question we have just asked: as x goes to infinity, the ratio of the number of primes less than x that split to the number that do not tends to 1. (See ANALYTIC NUMBER THEORY [IV.2 §4] for a further discussion of Dirichlet's theorem.)

For fun, one might ask a fussier question: which type of prime less than x is actually in greater abundance, the nonsplit primes or the split ones (see figure 4)? To put some perspective on this, let us widen our query: for q equal either to 4 or to an odd prime, let $A(x)$ be the number of primes $\ell < x$ that are quadratic residues modulo q and let $B(x)$ be the number of primes $\ell < x$ that are quadratic nonresidues modulo q. Let $D(x) = A(x) - B(x)$ be the difference; what does $D(x)$ look like?

For an absorbing account of the history and status of this problem, see Granville and Martin (2006).

11 Algebraic Numbers and Algebraic Integers

Now that we have seen the algebraic integers $j(\tau_d)$ for negative values of d, and have touched on trigonometric sums, we have a few hints that, as with ordinary integers, the deep structure of these rings of quadratic integers may be better understood within a larger context of algebraic numbers. So now let us deal with algebraic numbers in full generality.

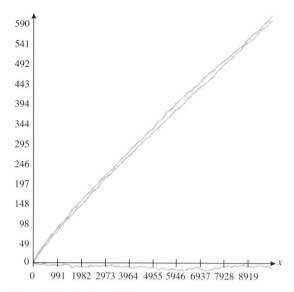

Figure 4 The higher of the two graphs in the figure represents the number of primes less than X that *remain prime* in the ring of Gaussian integers, and the lower represents the number of primes less than X that *split* in the ring of Gaussian integers. The third graph hovering around the x-axis represents the difference between the two numbers. We thank William Stein for this data.

By a *monic* polynomial, we mean a polynomial of the form

$$P(X) = X^n + a_1 X^{n-1} + \cdots + a_{n-1}X + a_n,$$

i.e., a polynomial of degree n such that the coefficient of X^n is 1. In general, the other coefficients are just assumed to be complex numbers. If $P(X) = X^n + a_1 X^{n-1} + \cdots + a_{n-1}X + a_n$ is such a polynomial, and if Θ is a complex number such that $P(\Theta) = 0$, or, equivalently, if Θ satisfies the polynomial equation

$$\Theta^n + a_1 \Theta^{n-1} + \cdots + a_{n-1}\Theta + a_n = 0,$$

we say that Θ is a *root* of the polynomial $P(X)$. THE FUNDAMENTAL THEOREM OF ALGEBRA [V.13], initially proved by Gauss, guarantees that any such polynomial of degree n factors into a product of n linear polynomials. That is,

$$P(X) = (X - \Theta_1)(X - \Theta_2) \cdots (X - \Theta_n)$$

for some complex numbers $\Theta_1, \Theta_2, \ldots, \Theta_n$ that are in fact precisely the roots of the polynomial $P(X)$.

If Θ is a root of such a polynomial $P(X) = X^n + a_1 X^{n-1} + \cdots + a_{n-1}X + a_n$ and if in addition the coefficients a_i are rational numbers, then Θ is called an *algebraic number*. If the coefficients are not just rational but are in fact integers, then Θ is called an *algebraic integer*. So, for example, the square root of any rational number is an algebraic number and the square root of any "ordinary" integer is an algebraic integer. The same holds true for nth roots of ordinary integers, or of algebraic integers, for any natural number n. For an example of a different sort, we have already mentioned the theorem that the values of the j-function on complex quadratic irrational integers are algebraic integers. For a (random) particular case of that theorem, the complex number $j(\tau_{-23})$ is a root of the monic polynomial

$$X^3 + 3\,491\,750X^2 - 5\,151\,296\,875X$$
$$+ 12\,771\,880\,859\,375.$$

An exercise: show that any algebraic number can be expressed as an algebraic integer divided by an ordinary integer.

12 Presentation of Algebraic Numbers

In dealing with any mathematical concept, we confront, in one way or another, the dual problem of the various forms in which it comes to us when it arises in our work, and the various ways we can *present* it so as to deal with it effectively. We have already seen a bit of this at the outset of this article, in our discussion of quadratic surds, and we will continue to see it in our treatment of them below, where the various modes in which quadratic surds can be presented—as *radicals*, as *eventually recurrent continued fractions*, or as *trigonometric sums*—come together, all contributing to their unified theory.

This issue of presentation is all the more of a problem with algebraic numbers in general, which may come to us in a multitude of ways. For example, they can arise as the coordinates of points on specific algebraic varieties whose defining equations may not be easily available, or as special values of functions like the j-function. It is natural, then, to look for some uniform way of presenting algebraic numbers, and the history of the subject shows how much effort has been devoted to such a search. For example, consider the focus on iterated radical expressions, as in the famous formula for the solution to the general cubic equation $X^3 = bX + c$ given by

$$X = \left(\frac{c}{2} + \sqrt{\frac{c^2}{2} - \frac{b^3}{27}}\right)^{1/3} + \left(\frac{c}{2} - \sqrt{\frac{c^2}{2} - \frac{b^3}{27}}\right)^{1/3}, \quad (14)$$

or the corresponding general solution to the fourth-degree equation. These were major achievements of sixteenth-century Italian algebra, and they culminated in the proof that the general fifth-degree algebraic number could *not* be so expressed, which was a major achievement of the early nineteenth century (see THE INSOLUBILITY OF THE QUINTIC [V.21]). The challenge to give *some* analytic expression for such fifth-degree algebraic numbers was the source of a classic book by Klein, *The Icosahedron*, written in the late nineteenth century. Kronecker wrote that it was the "dream of his youth" (his *Jugendtraum*) to establish a uniform mode of presentation for a class of algebraic numbers that interested him, by expressing them as values of certain analytic functions.

13   Roots of Unity

A central role in the theory of algebraic numbers is played by the *roots of unity*, that is, the n complex solutions of the equation $X^n = 1$, or equivalently the n roots of the polynomial $X^n - 1$. If we let $\zeta_n = e^{2\pi i/n}$, then these roots are precisely ζ_n and its powers, so in particular they are algebraic integers. They give us the factorization

$$X^n - 1 = (X - 1)(X - \zeta_n)(X - \zeta_n^2) \cdots (X - \zeta_n^{n-1}).$$

Now the powers of ζ_n form the vertices of a regular n-gon in the complex plane, centered at the origin. This has the following consequence, noticed by Gauss in his youth. It can be shown that compass and straightedge constructions allow us, in effect, to extract square roots, so whenever ζ_n can be given as an expression built out of just square roots and the usual arithmetical operations, we have, implicitly, a ruler-and-compass construction of the regular n-gon, and conversely.

To get some idea of why square roots are so closely connected with these constructions, consider this. If we have given ourselves a *unit measure*, which we can view as the distance between the numbers 0 and 1 in the (complex) plane, and if we have already constructed, by whatever device, a specific point, x say, between 0 and 1 on the horizontal axis of the plane, we can first "construct" $x/4$ by straightedge and compass, and then go on to form a right-angled triangle with hypotenuse of length $1 + x/4$ and one of its other sides of length $1 - x/4$ (again using a straightedge and compass). The Pythagorean theorem gives us that the third side of that triangle is of length $\sqrt{x}$. If one follows this line of thought (but adapts it to deal with complex quantities as well as the real number x as in the example we have

just discussed), then one can see that the equations

$$\zeta_3 = \tfrac{1}{2}(1 + i\sqrt{3}),$$
$$\zeta_4 = \sqrt{i},$$
$$\zeta_5 = \tfrac{1}{4}(\sqrt{5} - 1) + i\tfrac{1}{8}\left(\sqrt{5 + \sqrt{5}}\right),$$
$$\zeta_6 = -\tfrac{1}{2}(1 + i\sqrt{3})$$

provide (implicit) constructions of the equilateral triangle, the square, the regular pentagon, and the regular hexagon, respectively. By contrast, ζ_7 cannot be expressed solely in terms of the arithmetical operations and square roots (it is the root of a quadratic equation with coefficients that are rational expressions in the roots of the irreducible *cubic* polynomial $X^3 - \tfrac{7}{3}X + \tfrac{7}{27}$), which already suggests that the regular heptagon might fail to be constructible by the standard classical means—and indeed it does fail without some act of "angle trisection." (In principle, though, the reader can work out an expression for ζ_7 in terms of square roots and cube roots by means of the information provided in the parenthetical phrase above, together with equation (14).)

Gauss showed that if $n > 2$ is a prime number then the regular n-gon is classically constructible if and only if n is a *Fermat prime*, that is, a prime number of the form $2^{2^a} + 1$. So, for example, the 11-gon and 13-gon are not constructible by classical means, but since ζ_{17} is expressible as nested rational expressions of square roots, the 17-gon is, famously, constructible.

So, not all roots of unity can be expressed as iterated rational expressions of square roots. However, this inhospitability is not mutual, since all square roots of integers can be expressed as integer combinations of roots of unity. More mysteriously, the elusive fundamental units ε_d (for d positive), for which there is no known formula, are intimately related to a unit c_d in R_d which is an explicit rational expression of roots of unity. (See below: it is called a *circular unit*.) This satisfies the elegant formula

$$c_d = \varepsilon_d^{h_d}, \tag{15}$$

which establishes yet another explicit test of unique factorization: the equality $c_d = \varepsilon_d$ is a "litmus" requirement for the unique factorization principle to hold in R_d.

To give the flavor of the formulas involved, let p be an odd prime number and let a be an integer not divisible by p. Then define $\sigma_p(a)$ to be $+1$ if a is a *quadratic residue modulo p*, that is, if a is congruent to the square of an integer modulo p, and -1 if not. The simple trigonometric sums of (1) and (6) generalize to *quadratic Gauss sums*:

$$\pm i^{(p-1)/2}\sqrt{p} = \zeta_p + \sigma_p(2)\zeta_p^2 + \sigma_p(3)\zeta_p^3 + \cdots$$
$$+ \sigma_p(p-2)\zeta_p^{p-2} + \sigma_p(p-1)\zeta_p^{p-1}. \tag{16}$$

This formula is not too hard to prove, apart from determining which sign is correct in the initial $\pm$, but after considerable efforts Gauss managed to work this out too. To see the connection between, say, formula (6) and (16) note that when $p = 5$, the left-hand side of (16) is $\sqrt{5}$ and the right-hand side is

$$\zeta_5 - \zeta_5^2 - \zeta_5^{-2} + \zeta_5^{-1} = 2\cos\tfrac{2}{5}\pi - 2\cos\tfrac{4}{5}\pi.$$

As for the circular unit c_p, it is defined to be

$$\prod_{a=1}^{(p-1)/2}(\zeta_p^a - \zeta_p^{-a})^{\sigma_p(a)} = \prod_{a=1}^{(p-1)/2}\sin(\pi a/p)^{\sigma_p(a)},$$

and this leads to further formulas. For example, when $p = 5$, we have $\varepsilon_p = \tau_5 = \tfrac{1}{2}(1 + \sqrt{5})$, and since $h_5 = 1$, formula (6) for $p = 5$ tells us that

$$\frac{1+\sqrt{5}}{2} = \frac{\zeta_5 - \zeta_5^{-1}}{\zeta_5^2 - \zeta_5^{-2}} = \frac{\sin\tfrac{1}{5}\pi}{\sin\tfrac{2}{5}\pi}.$$

14 The Degree of an Algebraic Number

If Θ is an algebraic integer that is also a rational number, then Θ is an "ordinary" integer. Here is the proof of this fact. If Θ is a rational number, then we may write $\Theta = C/D$ as a fraction in lowest terms. If Θ is also an algebraic integer, then it is the root of a monic polynomial with rational integer coefficients, $\Theta^n + a_1\Theta^{n-1} + \cdots + a_n$, so we have an equation

$$(C/D)^n + a_1(C/D)^{n-1} + \cdots + a_{n-1}(C/D) + a_n = 0.$$

Multiplying through by D^n we get

$$C^n + a_1C^{n-1}D + \cdots + a_{n-1}CD^{n-1} + a_nD^n = 0,$$

where all terms are (ordinary) integers, and all but the first one is divisible by D. If $D > 1$ then it has some prime factor p, so all terms apart from the first are also divisible by p. Since the terms add up to zero, it follows that p divides C^n, which implies that p divides C, which contradicts the assertion that the fraction C/D is in its lowest terms. This in turn contradicts the hypothesis that Θ can be expressed as a ratio of whole numbers in the first place. As the reader may like to verify, this fact implies the result attributed to Theaetetus above, that $\sqrt{A}$ is irrational if and only if A is not a perfect square.

The *degree* of an algebraic number Θ is defined to be the smallest degree, n, of any polynomial relation $\Theta^n + a_1\Theta^{n-1} + \cdots + a_{n-1}\Theta + a_n = 0$ that Θ satisfies, where the coefficients a_i are rational numbers. The corresponding polynomial, $P(X) = X^n + a_1X^{n-1} + \cdots + a_{n-1}X + a_n$ is unique, since if there were two of them then their difference would be of smaller degree and would also have Θ as a root. (One could make it monic by dividing it through by the leading coefficient.) Let us call $P(X)$ the *minimal polynomial* of Θ. The minimal polynomial is *irreducible* over the field of rational numbers: that is, it cannot be factored as a product of two polynomials, each of smaller degree and having rational numbers as coefficients. (If it could, then it would not be of minimal degree, since one of its factors would have Θ as a root.) The minimal polynomial $P(X)$ of Θ is a factor of any monic polynomial $G(X)$ with rational coefficients that has Θ as root. (The greatest common divisor of P and G is another monic polynomial with rational coefficients that has Θ as a root, so it cannot be of degree smaller than that of P and it must therefore *be P*.) The minimal polynomial $P(X)$ of Θ has distinct roots. (If $P(X)$ had multiple roots, then a little elementary calculus shows that it would share a nontrivial factor with its derivative, $P'(X)$. Since the derivative is of lower degree than $P(X)$ and again has rational coefficients, the greatest common divisor of P and P' would provide a nontrivial factorization of $P(X)$, contradicting its irreducibility.)

A fundamental result due to Gauss is that the nth root of unity $\zeta_n = e^{2\pi i/n}$ is an algebraic integer of degree precisely $\phi(n)$, where ϕ is Euler's ϕ-function. For example, if p is prime, the minimal polynomial of ζ_p is

$$\frac{X^p - 1}{X - 1} = X^{p-1} + X^{p-2} + \cdots + X + 1,$$

which is of degree $\phi(p) = p - 1$.

15 Algebraic Numbers as Ciphers Determined by Their Minimal Polynomials

We have expressly insisted that our algebraic numbers are complex numbers (of a certain sort). But another possible attitude toward an algebraic number, Θ, an attitude at times promoted by Kronecker, among others, is to deal with Θ as an unknown satisfying only the algebraic relations implied by the fact that it is a root of its (unique monic) minimal polynomial with rational coefficients. For example, if the minimal polynomial of Θ is $P(X) = X^3 - X - 1$, then, according to this view,

Θ is just an algebraic symbol that comes with the rule that any occurrence of Θ^3 may be replaced by $\Theta + 1$ (rather as the complex number i can be regarded as a symbol with the property that i^2 may be replaced by -1). Any root of the minimal polynomial of Θ satisfies all the same polynomial relations with rational coefficients that Θ satisfies; these roots are called *conjugates* of Θ. If Θ is an algebraic number of degree n, then Θ has n *distinct* conjugates, all of them again, of course, algebraic numbers.

16   A Few Remarks about the Theory of Polynomials

Central to the theory of polynomials in one variable—and, therefore, particularly to the theory of algebraic numbers—is the general relationship that *roots* have to *coefficients*:

$$\prod_{i=1}^{n}(X - T_i) = X^n + \sum_{j=1}^{n}(-1)^j A_j(T_1, T_2, \ldots, T_n)X^{n-j}.$$

The polynomial $A_j(T_1, T_2, \ldots, T_n)$ is homogeneous of degree j (this means that every monomial in it has total degree j), has integer coefficients, and is symmetric in (i.e., unchanged by any permutation of) the variables $T_1, T_2, \ldots, T_n$.

The constant term is given by the product of the roots:

$$A_n(T_1, T_2, \ldots, T_n) = T_1 \cdot T_2 \cdots T_n,$$

which is known as the *norm* form. The coefficient of X^{n-1} is given by the sum of the roots:

$$A_1(T_1, T_2, \ldots, T_n) = T_1 + T_2 + \cdots + T_n,$$

and this is the *trace* form.

When $n = 2$ the norm and trace are all the symmetric polynomials in the list. For $n = 3$, beyond the norm and trace we also have the symmetric polynomial of degree two:

$$A_2(T_1, T_2, T_3) = T_1 T_2 + T_2 T_3 + T_3 T_1$$
$$= \tfrac{1}{2}\{(T_1 + T_2 + T_3)^2 - (T_1^2 + T_2^2 + T_3^2)\}.$$

It is of major importance to this theory, and more specifically to GALOIS THEORY [V.21], that the symmetry properties of the conjugate roots are nicely reflected in these symmetric polynomials. In particular, we have the fundamental result that *any* symmetric polynomial in $T_1, T_2, \ldots, T_n$ with rational coefficients can be expressed as a polynomial with rational coefficients in

the symmetric polynomials $A_j(T_1, T_2, \ldots, T_n)$, and similarly with integral coefficients. For example, the equation above shows that $T_1^2 + T_2^2 + T_3^2$ can be expressed as

$$A_1(T_1, T_2, T_3)^2 - 2A_2(T_1, T_2, T_3).$$

17   Fields of Algebraic Numbers and Rings of Algebraic Integers

The inverse of a nonzero algebraic number is again an algebraic number; the sum, difference, and product of two algebraic numbers are algebraic numbers; the sum, difference, and product of two algebraic integers are algebraic integers. The neat proofs of these (latter) facts are a good demonstration of the power of linear algebra, and in particular of *Cramer's rule*. This states that any matrix with integer coefficients (and therefore also any linear transformation of a finite-dimensional vector space that preserves an integer lattice) satisfies a monic polynomial identity with integer coefficients.

To see just how useful this remark is for finding polynomial relations, and more specifically for showing that the collections of algebraic numbers and algebraic integers are closed under sums and products, try your hand at showing that $\sqrt{2} + \sqrt{3}$ is an algebraic integer. One way to do it is to search for the monic fourth-degree polynomial equation that it satisfies. But this is hardly a beautiful calculation! If, however, you are familiar with linear algebra, then a less painful route is to form the four-dimensional vector space over the rational numbers, generated by $1, \sqrt{2}, \sqrt{3},$ and $\sqrt{6}$ (which are linearly independent when the scalars are rational). Multiplication by $\sqrt{2} + \sqrt{3}$ defines a linear transformation T of this vector space, and one can compute its characteristic polynomial P. The *Cayley–Hamilton theorem* says that $P(T) = 0$, and this translates into the statement that $\sqrt{2} + \sqrt{3}$ is a root of P.

These "closure properties" we have just discussed lead us to study, in complete generality, fields of algebraic numbers and rings of algebraic integers. A *number field* is a field that is generated (as a field) by finitely many algebraic numbers. A standard result tells us that any number field K can in fact be generated by a single carefully chosen algebraic number. The degree of this algebraic number equals the *degree* of K, which is defined to be the dimension of K when K is viewed as a vector space over the field $\mathbb{Q}$ of rational numbers. One of the main introductory observations of Galois theory is that if K is a number field of degree n, then there are exactly n distinct ring homomorphisms ("embeddings") $\iota : K \to \mathbb{C}$ from K into the field of complex numbers. (This means that ι sends 1 to 1 and respects the addition and multiplication laws within K. That is, $\iota(x + y) = \iota(x) + \iota(y)$ and $\iota(x \cdot y) = \iota(x) \cdot \iota(y)$.) From these embeddings, we can construct some very useful rational-valued functions on K. For any element x in K, we form the n complex numbers $x_1, x_2, \ldots, x_n$ that are the images of x under the n different embeddings of K into $\mathbb{C}$. We then let

$$a_j(x) = A_j(x_1, x_2, \ldots, x_n),$$

where $A_j(X_1, X_2, \ldots, X_n)$ is the jth symmetric polynomial of section 14 above. (Because the polynomials A_j are symmetric, we do not have to worry about the order of the images $x_1, x_2, \ldots, x_n$ in the above expression.) It is not immediately obvious that the values of a_j are rational numbers, but there is a theorem that tells us this.

If an algebraic number Θ in K generates K (as a field), then the rational numbers $a_j(\Theta)$ are the coefficients of its minimal polynomial; in general they are the coefficients of a power of its minimal polynomial. The most prominent of these functions are the multiplicative function $a_n(x) = x_1 \cdot x_2 \cdot \ldots \cdot x_n$, called the *norm* function, usually denoted $x \mapsto N_{K/\mathbb{Q}}(x)$, and the additive function $a_1(x) = x_1 + x_2 + \cdots + x_n$, called the *trace* function, usually denoted $x \mapsto \mathrm{trace}_{K/\mathbb{Q}}(x)$.

The trace function can be used to define a fundamental symmetric bilinear form on the $\mathbb{Q}$-vector space K,

$$\langle x, y \rangle = \mathrm{trace}_{K/\mathbb{Q}}(x \cdot y),$$

which turns out to be nondegenerate. This nondegeneracy, together with the fact that if x, y are both algebraic integers, then $\langle x, y \rangle$ is an ordinary integer, can be used to show that the ring $\mathcal{O}(K)$ of *all* algebraic integers in K is finitely generated as an additive group. More specifically, there is a *basis* of algebraic integers in K, that is, a finite set $\{\Theta_1, \Theta_2, \ldots, \Theta_n\}$, such that any other algebraic integer in K can be expressed as an "ordinary" integer combination of the numbers Θ_i.

Let us summarize this structure. The number field K is a finite-dimensional vector space over $\mathbb{Q}$ and comes equipped with a nondegenerate bilinear symmetric form $(x, y) \mapsto \langle x, y \rangle$, and also with a lattice $\mathcal{O}(K) \subset K$. Moreover, the restriction of the bilinear form to $\mathcal{O}(K)$ takes on integral values.

The *discriminant* of K, denoted $D(K)$, is defined to be the DETERMINANT [III.15] of the matrix whose ij-entry is $\langle \Theta_i, \Theta_j \rangle$, for $\{\Theta_1, \Theta_2, \ldots, \Theta_n\}$ a basis of the lattice

$\mathcal{O}(K)$; this determinant does not depend on the basis chosen.

The discriminant represents important information about the number field K. For one thing, there is a natural generalization to any number field of the notions of *splitting* and *ramification* that we discussed for quadratic fields, and the prime divisors p of $D(K)$ are precisely those prime numbers that ramify in the field extension K. By a theorem of MINKOWSKI [VI.64], the absolute value of the discriminant $D(K)$ of a number field K of degree n is always greater than

$$\left(\frac{\pi}{4} \right)^n \cdot \left(\frac{n^n}{n!} \right)^2 .$$

This is greater than 1 unless K is the field of rational numbers. It follows that any nontrivial extension of the field of rational numbers has some prime that ramifies in it, a result that would be very hard to prove without the help of the algebraic structures we have just defined. This integer $D(K)$ really is quite a discriminating "tag" for our number field K, for, by a theorem of HERMITE [VI.47], given any integer D there are only finitely many different number fields with discriminant equal to D. (Not all integers can be discriminants: as is true for quadratic number fields, the integers D that are discriminants are either divisible by 4 or else congruent to 1 modulo 4.)

18   On the Size(s) of the Absolute Values of All Conjugates of an Algebraic Integer

As we have just seen, the coefficients of the minimal polynomial for an algebraic integer Θ are given by the ordinary integers $a_j(\Theta_1, \Theta_2, \ldots, \Theta_n)$, where the numbers Θ_i are all the conjugates of Θ. The sizes of all these coefficients must therefore all be less than some universal number M that depends only on the degree of Θ and the largest absolute value of any of its conjugates. As a consequence, given any n and any positive number B, there are only finitely many algebraic integers Θ of degree less than n such that the absolute values of Θ and its conjugates are all less than B. (This is because for any n and M there are only finitely many polynomials of degree less than or equal to n with the absolute values of all their integer coefficients at most M.) This finiteness result is the key to the following observation, due to Kronecker: if Θ is an algebraic number and if the absolute values of Θ and of all its conjugates are equal to 1, then Θ is a root of unity. Indeed, all the powers of Θ have degree at most that of Θ, and they enjoy the same property: their absolute value, and that of all their conjugates, is equal to 1. Consequently, there are only finitely many such algebraic numbers, from which it follows that there must be at least one coincidence of the form $\Theta^a = \Theta^b$ for different a and b. But this can happen only if Θ is a root of unity.

19   Weil Numbers

To follow this thread for just a bit, let us generalize the hypothesis of Kronecker's observation, and define a *Weil number*[3] of absolute value r to be a nonzero algebraic integer such that it and all of its conjugates have the same absolute value r. By the discussion in the previous section there are only finitely many distinct Weil numbers of given degree and absolute value. By Kronecker's theorem, which we have just described, the Weil numbers of absolute value 1 are precisely the roots of unity. Here are further basic facts that you might try to prove. First, the quadratic Weil numbers ω are precisely those quadratic algebraic integers such that $|\operatorname{trace}(\omega)| \leqslant 2\sqrt{|N(\omega)|} = 2\sqrt{|\omega\omega'|}$, where ω' is the (algebraic) conjugate of ω. Second, if p is prime then a quadratic Weil number ω of absolute value $\sqrt{p}$ is a prime element of the (unique) ring of quadratic integers R_d that contains ω, and therefore gives a prime factorization $\omega\omega' = \pm p$ of the integer p in that ring.

Weil numbers of absolute value $p^{\nu/2}$, where p is again a prime number and ν is a natural number, are extremely important in arithmetic: they hold the key to counting numbers of rational solutions of systems of polynomial equations over finite fields. For just one concrete example, the Gaussian integer $\omega = -1 + i$ and its algebraic conjugate (which, in this instance, is also its complex conjugate) $\bar{\omega} = -1 - i$ are Weil numbers (of absolute value 2) that control the number of solutions of the equation $y^2 - y = x^3 - x$ over all finite fields of size a power of 2. Specifically, the number of solutions of that equation over a field of order 2^ν is given by the formula

$$2^\nu - (-1 - i)^\nu - (-1 + i)^\nu$$

(which is an ordinary integer). This leads to another immense chapter of mathematics.

20   Epilogue

The single symmetry $\alpha \mapsto \alpha'$, the algebraic conjugation in the rings R_d that we have discussed, gave birth, thanks to ABEL [VI.33] and GALOIS [VI.41] in the beginning of the nineteenth century, to the rich study of (Galois) groups of symmetries of general number fields (see THE INSOLUBILITY OF THE QUINTIC [V.21]). This study continues with great intensity, since these Galois groups and their linear representations hold the key to a very detailed understanding of number fields. In its modern dress, algebraic number theory is closely connected with what is often called ARITHMETIC GEOMETRY [IV.5]. Kronecker's dream of getting explicit control of a wealth of algebraic number theoretic material by expressing algebraic numbers in terms of natural analytic functions has not yet been fully realized. Nevertheless, the scope of this dream (and, one might also add, the supply of natural analytic and algebraic functions) has expanded substantially: the full range of algebraic geometry and group representation theory is now being brought to bear on it. This is done, for example, by the *Langlands program*, which among other things works with objects known as *Shimura varieties*. On the one hand, these varieties have close connections with the theory of group representations and classical algebraic geometry, which greatly helps us to understand them. On the other hand, they are a rich source of concrete linear representations of Galois groups of number fields. This program, one of the glories of current mathematics, will, I expect, make a terrific chapter for a *Companion to Mathematics* to be written at the beginning of the next century.

Further Reading

Basic Texts

First, I list three classics that require a minimum of background.

Davenport, H. 1992. *The Higher Arithmetic: An Introduction to the Theory of Numbers*. Cambridge: Cambridge University Press.
Gauss, C. F. 1986. *Disquisitiones Arithmeticae*, English edn. New York: Springer.
Hardy, G. H., and E. M. Wright. 1980. *An Introduction to the Theory of Numbers*, 5th edn. Oxford: Oxford University Press.

At a more advanced level, the following are extraordinary expository books.

Borevich, Z. I., and I. R. Shafarevich. 1966. *Number Theory*. New York: Academic Press.
Cassels, J., and A. Fröhlich. 1967. *Algebraic Number Theory*. New York: Academic Press.

---

3. This is a weaker condition than is usually required for Weil numbers but our deviation from standard usage should not be the cause of too much confusion.

Cohen, H. 1993. *A Course in Computational Algebraic Number Theory*. New York: Springer.

Ireland, K., and M. Rosen. 1982. *A Classical Introduction to Modern Number Theory*, 2nd edn. New York: Springer.

Serre, J.-P. 1973. *A Course in Arithmetic*. New York: Springer.

Technical Articles and Books

Baker, A. 1971. Imaginary quadratic fields with class number 2. *Annals of Mathematics (2)* 94:139–52.

Brauer, R. 1950. On the Zeta-function of algebraic number fields. I. *American Journal of Mathematics* 69:243–50.

Brauer, R. 1950. On the Zeta-function of algebraic number fields. II. *American Journal of Mathematics* 72:739–46.

Goldfeld, D. 1985. Gauss's class number problem for imaginary quadratic fields. *Bulletin of the American Mathematical Society* 13:23–37.

Granville, A., and G. Martin. 2006. Prime number races. *American Mathematical Monthly* 113:1–33.

Gross, B., and D. Zagier. 1986. Heegner points and derivatives of *L*-series. *Inventiones Mathematicae* 84:225–320.

Heegner, K. 1952. Diophantische Analysis und Modulfunktionen. *Mathematische Zeitschrift* 56:227–53.

Hua, L.-K. 1942. On the least solution of Pell's equation. *Bulletin of the American Mathematical Society* 48:731–35.

Lang, S. 1970. *Algebraic Number Theory*. Reading, MA: Addison-Wesley.

Narkiewicz, W. 1973. *Algebraic Numbers*. Warsaw: Polish Scientific Publishers.

Siegel, C. L. 1935. Über die Classenzahl quadratischer Zahlörper. *Acta Arithmetica* 1:83–86.

Stark, H. 1967. A complete determination of the complex quadratic fields of class-number one. *Michigan Mathematical Journal* 14:1–27.

IV.2 Analytic Number Theory
Andrew Granville

1 Introduction

What is number theory? One might have thought that it was simply the study of numbers, but that is too broad a definition, since numbers are almost ubiquitous in mathematics. To see what distinguishes number theory from the rest of mathematics, let us look at the equation $x^2 + y^2 = 15\,925$, and consider whether it has any solutions. One answer is that it certainly does: indeed, the solution set forms a circle of radius $\sqrt{15\,925}$ in the plane. However, a number theorist is interested in *integer* solutions, and now it is much less obvious whether any such solutions exist.

A useful first step in considering the above question is to notice that $15\,925$ is a multiple of 25: in fact, it is

25×637. Furthermore, the number 637 can be decomposed further: it is 49×13. That is, $15\,925 = 5^2 \times 7^2 \times 13$. This information helps us a lot, because if we can find integers a and b such that $a^2 + b^2 = 13$, then we can multiply them by $5 \times 7 = 35$ and we will have a solution to the original equation. Now we notice that $a = 2$ and $b = 3$ works, since $2^2 + 3^2 = 13$. Multiplying these numbers by 35, we obtain the solution $70^2 + 105^2 = 15\,925$ to the original equation.

As this simple example shows, it is often useful to decompose positive integers multiplicatively into components that cannot be broken down any further. These components are called *prime numbers*, and THE FUNDAMENTAL THEOREM OF ARITHMETIC [V.14] states that every positive integer can be written as a product of primes in exactly one way. That is, there is a one-to-one correspondence between positive integers and finite products of primes. In many situations we know what we need to know about a positive integer once we have decomposed it into its prime factors and understood those, just as we can understand a lot about molecules by studying the atoms of which they are composed. For example, it is known that the equation $x^2 + y^2 = n$ has an integer solution if and only if every prime of the form $4m + 3$ occurs an even number of times in the prime factorization of n. (This tells us, for instance, that there are no integer solutions to the equation $x^2 + y^2 = 13\,475$, since $13\,475 = 5^2 \times 7^2 \times 11$, and 11 appears an odd number of times in this product.)

Once one begins the process of determining which integers are primes and which are not, it is soon apparent that there are many primes. However, as one goes further and further, the primes seem to consist of a smaller and smaller proportion of the positive integers. They also seem to come in a somewhat irregular pattern, which raises the question of whether there is any formula that describes all of them. Failing that, can one perhaps describe a large class of them? We can also ask whether there are infinitely many primes. If there are, can we quickly determine how many there are up to a given point? Or at least give a good estimate for this number? Finally, when one has spent long enough looking for primes, one cannot help but ask whether there is a quick way of recognizing them. This last question is discussed in COMPUTATIONAL NUMBER THEORY [IV.3]; the rest motivate the present article.

Now that we have discussed what marks number theory out from the rest of mathematics, we are ready to make a further distinction: between *algebraic* and *analytic* number theory. The main difference is that in algebraic number theory (which is the main topic of ALGEBRAIC NUMBERS [IV.1]) one typically considers questions with answers that are given by exact formulas, whereas in analytic number theory, the topic of this article, one looks for *good approximations*. For the sort of quantity that one estimates in analytic number theory, one does not expect an exact formula to exist, except perhaps one of a rather artificial and unilluminating kind. One of the best examples of such a quantity is one we shall discuss in detail: the number of primes less than or equal to x.

Since we are discussing approximations, we will need terminology that allows us to give some idea of the quality of an approximation. Suppose, for example, that we have a rather erratic function $f(x)$ but are able to show that, once x is large enough, $f(x)$ is never bigger than $25x^2$. This is useful because we understand the function $g(x) = x^2$ quite well. In general, if we can find a constant c such that $|f(x)| \leqslant cg(x)$ for every x, then we write $f(x) = O(g(x))$. A typical usage occurs in the sentence "the average number of prime factors of an integer up to x is $\log \log x + O(1)$"; in other words, there exists some constant $c > 0$ such that |the average $- \log \log x| \leqslant c$ once x is sufficiently large.

We write $f(x) \sim g(x)$ if $\lim_{x \to \infty} f(x)/g(x) = 1$; and also $f(x) \approx g(x)$ when we are being a little less precise, that is, when we want to say that $f(x)$ and $g(x)$ come close when x is sufficiently large, but we cannot be, or do not want to be, more specific about what we mean by "come close."

It is convenient for us to use the notation $\sum$ for sums and $\prod$ for product. Typically we will indicate beneath the symbol what terms the sum, or product, is to be taken over. For example, $\sum_{m \geqslant 2}$ will be a sum over all integers m that are greater than or equal to 2, whereas $\prod_{p \text{ prime}}$ will be a product over all primes p.

2   Bounds for the Number of Primes

Ancient Greek mathematicians knew that there were infinitely many primes. Their beautiful proof by contradiction goes as follows. Suppose that there are only finitely many primes, say k of them, which we will denote by $p_1, p_2, \ldots, p_k$. What are the prime factors of $p_1 p_2 \cdots p_k + 1$? Since this number is greater than 1 it must have at least one prime factor, and this must be p_j for some j (since *all* primes are contained among $p_1, p_2, \ldots, p_k$). But then p_j divides both $p_1 p_2 \cdots p_k$ and $p_1 p_2 \cdots p_k + 1$, and hence their difference, 1, which is impossible.

Many people dislike this proof, since it does not actually exhibit infinitely many primes: it merely shows that there cannot be finitely many. It is more or less possible to correct this deficiency by defining the sequence $x_1 = 2$, $x_2 = 3$, and $x_{k+1} = x_1 x_2 \cdots x_k + 1$ for each $k \geqslant 2$. Then each x_k must contain at least one prime factor, q_k say, and these prime factors must be distinct, since if $k < \ell$, then q_k divides x_k which divides $x_\ell - 1$, while q_ℓ divides x_ℓ. This gives us an infinite sequence of primes.

In the eighteenth century EULER [VI.19] gave a different proof that there are infinitely many primes, one that turned out to be highly influential in what was to come later. Suppose again that the list of primes is $p_1, p_2, \ldots, p_k$. As we have mentioned, the fundamental theorem of arithmetic implies that there is a one-to-one correspondence between the set of all integers and the set of products of the primes, which, if those are the only primes, is the set $\{p_1^{a_1} p_2^{a_2} \cdots p_k^{a_k} : a_1, a_2, \ldots, a_k \geqslant 0\}$. But, as Euler observed, this implies that a sum involving the elements of the first set should equal the analogous sum involving the elements of the second set:

$$\sum_{\substack{n \geqslant 1 \\ n \text{ a positive integer}}} \frac{1}{n^s}$$

$$= \sum_{a_1, a_2, \ldots, a_k \geqslant 0} \frac{1}{(p_1^{a_1} p_2^{a_2} \cdots p_k^{a_k})^s}$$

$$= \left(\sum_{a_1 \geqslant 0} \frac{1}{(p_1^{a_1})^s} \right) \left(\sum_{a_2 \geqslant 0} \frac{1}{(p_2^{a_2})^s} \right) \cdots \left(\sum_{a_k \geqslant 0} \frac{1}{(p_k^{a_k})^s} \right)$$

$$= \prod_{j=1}^{k} \left(1 - \frac{1}{p_j^s} \right)^{-1}.$$

The last equality holds because each sum in the second-last line is the sum of a geometric progression. Euler then noted that if we take $s = 1$, the right-hand side equals some rational number (since each $p_j > 1$) whereas the left-hand side equals ∞. This is a contradiction, so there cannot be finitely many primes. (To see why the left-hand side is infinite when $s = 1$, note that $(1/n) \geqslant \int_n^{n+1} (1/t) \, dt$ since the function $1/t$ is decreasing, and therefore $\sum_{n=1}^{N-1} (1/n) \geqslant \int_1^N (1/t) \, dt = \log N$ which tends to ∞ as $N \to \infty$.)

During the proof above, we gave a formula for $\sum n^{-s}$ under the false assumption that there are only finitely many primes. To correct it, all we have to do is rewrite it in the obvious way without that assumption:

$$\sum_{\substack{n \geqslant 1 \\ n \text{ a positive integer}}} \frac{1}{n^s} = \prod_{p \text{ prime}} \left(1 - \frac{1}{p^s} \right)^{-1}. \qquad (1)$$

Now, however, we need to be a little careful about whether the two sides of the formula converge. It is safe to write down such a formula when both sides are absolutely convergent, and this is true when $s > 1$. (An infinite sum or product is *absolutely convergent* if the value does not change when we take the terms in any order we want.)

Like Euler, we want to be able to interpret what happens to (1) when $s = 1$. Since both sides converge and are equal when $s > 1$, the natural thing to do is consider their common limit as s tends to 1 from above. To do this we note, as above, that the left-hand side of (1) is well approximated by

$$\int_1^\infty \frac{\mathrm{d}t}{t^s} = \frac{1}{s-1},$$

so it diverges as $s \to 1^+$. We deduce that

$$\prod_{p \text{ prime}} \left(1 - \frac{1}{p}\right) = 0. \tag{2}$$

Taking logarithms and discarding negligible terms, we then find that

$$\sum_{p \text{ prime}} \frac{1}{p} = \infty. \tag{3}$$

So how numerous are the primes? One way to get an idea is to determine the behavior of the sum analogous to (3) for other sequences of integers. For instance, $\sum_{n \geqslant 1} 1/n^2$ converges, so the primes are, in this sense, more numerous than the squares. This argument works if we replace the power 2 by any $s > 1$, since then, as we have just observed, the sum $\sum_{n \geqslant 1} 1/n^s$ is about $1/(s-1)$ and in particular converges. In fact, since $\sum_{n \geqslant 1} 1/n(\log n)^2$ converges, we see that the primes are in the same sense more numerous than the numbers $\{n(\log n)^2 : n \geqslant 1\}$, and hence there are infinitely many integers x for which the number of primes less than or equal to x is at least $x/(\log x)^2$.

Thus, there seem to be primes in abundance, but we would also like to verify our observations, made from calculations, that the primes constitute a smaller and smaller proportion of the integers as the integers become larger and larger. The easiest way to see this is to try to count the primes using the "sieve of Eratosthenes." In the sieve of Eratosthenes one starts with all the positive integers up to some number x. From these, one deletes the numbers 4, 6, 8 and so on—that is, all multiples of 2 apart from 2 itself. One then takes the first undeleted integer greater than 2, which is 3, and deletes all *its* multiples—again, not including the number 3 itself. Then one removes all multiples of 5 apart

from 5, and so on. By the end of this process, one is left with the primes up to x.

This suggests a way to guess at how many there are. After deleting every second integer up to x other than 2 (which we call "sieving by 2") one is left with roughly half the integers up to x; after sieving by 3, one is left with roughly two thirds of those that had remained; continuing like this we expect to have about

$$x \prod_{p \leqslant y} \left(1 - \frac{1}{p}\right) \tag{4}$$

integers left by the time we have sieved with all the primes up to y. Once $y = \sqrt{x}$ the undeleted integers are 1 and the primes up to x, since every composite has a prime factor no bigger than its square root. So, is (4) a good approximation for the number of primes up to x when $y = \sqrt{x}$?

To answer this question, we need to be more precise about what the formula in (4) is estimating. It is supposed to approximate the number of integers up to x that have no prime factors less than or equal to y, plus the number of primes up to y. The so-called *inclusion-exclusion principle* can be used to show that the approximation given in (4) is accurate to within 2^k, where k is the number of primes less than or equal to y. Unless k is very small, this error term of 2^k is far larger than the quantity we are trying to estimate, and the approximation is useless. It is quite good if k is less than a small constant times $\log x$, but, as we have seen, this is far less than the number of primes we expect up to y if $y \approx \sqrt{x}$. Thus it is not clear whether (4) can be used to obtain a good estimate for the number of primes up to x. What we *can* do, however, is use this argument to give an upper bound for the number of primes up to x, since the number of primes up to x is never more than the number of integers up to x that are free of prime factors less than or equal to y, plus the number of primes up to y, which is no more than 2^k plus the expression in (4).

Now, by (2), we know that as y gets larger and larger the product $\prod_{p \leqslant y}(1 - 1/p)$ converges to zero. Therefore, for any small positive number ε we can find a y such that $\prod_{p \leqslant y}(1 - 1/p) < \varepsilon/2$. Since every term in this product is at least $1/2$, the product is at least $1/2^k$. Hence, for any $x \geqslant 2^{2k}$ our error term, 2^k, is no bigger than the quantity in (4), and therefore the number of primes up to x is no larger than twice (4), which, by our choice of y, is less than εx. Since we were free to make ε as small as we liked, the primes are indeed a vanishing proportion of all the integers, as we predicted.

Even though the error term in the inclusion–exclusion principle is too large for us to use that method to estimate (4) when $y = \sqrt{x}$, we can still hope that (4) is a good approximation for the number of primes up to x: perhaps a different argument would give us a much smaller error term. And this turns out to be the case: in fact, the error never gets much bigger than (4). However, when $y = \sqrt{x}$ the number of primes up to x is actually about 8/9 times (4). So why does (4) not give a good approximation? After sieving with prime p we supposed that roughly 1 in every p of the remaining integers were deleted: a careful analysis yields that this can be justified when p is small, but that this becomes an increasingly poor approximation of what really happens for larger p; in fact (4) *does not* give a correct approximation once y is bigger than a fixed power of x. So what goes wrong? In the hope that the proportion is roughly $1/p$ lies the unspoken assumption that the consequences of sieving by p are independent of what happened with the primes smaller than p. But if the primes under consideration are no longer small, then this assumption is false. This is one of the main reasons that it is hard to estimate the number of primes up to x, and indeed similar difficulties lie at the heart of many related problems.

One can refine the bounds given above but they do not seem to yield an asymptotic estimate for the primes (that is, an estimate which is correct to within a factor that tends to 1 as x gets large). The first good guesses for such an estimate emerged at the beginning of the nineteenth century, none better than what emerges from an observation of GAUSS [VI.26], made when studying tables of primes up to three million at sixteen years of age, that "the density of primes at around x is about $1/\log x$." Interpreting this, we guess that the number of primes up to x is about

$$\sum_{n=2}^{x} \frac{1}{\log n} \approx \int_{2}^{x} \frac{dt}{\log t}.$$

Let us compare this prediction (rounded to the nearest integer) with the latest data on numbers of primes, discovered by a mixture of ingenuity and computational power. Table 1 shows the actual numbers of primes up to various powers of 10 together with the difference between these numbers and what Gauss's formula gives. The differences are far smaller than the numbers themselves, so his prediction is amazingly accurate. It does seem always to be an overcount, but since the width of the last column is about half that of the

Table 1 Primes up to various x, and the overcount in Gauss's prediction.

x	$\pi(x) = \#\{\text{primes} \leqslant x\}$	Overcount: $\int_{2}^{x} \frac{dt}{\log t} - \pi(x)$
10^8	5 761 455	753
10^9	50 847 534	1 700
10^{10}	455 052 511	3 103
10^{11}	4 118 054 813	11 587
10^{12}	37 607 912 018	38 262
10^{13}	346 065 536 839	108 970
10^{14}	3 204 941 750 802	314 889
10^{15}	29 844 570 422 669	1 052 618
10^{16}	279 238 341 033 925	3 214 631
10^{17}	2 623 557 157 654 233	7 956 588
10^{18}	24 739 954 287 740 860	21 949 554
10^{19}	234 057 667 276 344 607	99 877 774
10^{20}	2 220 819 602 560 918 840	222 744 643
10^{21}	21 127 269 486 018 731 928	597 394 253
10^{22}	201 467 286 689 315 906 290	1 932 355 207

central one it appears that the difference is something like $\sqrt{x}$.

In the 1930s, Harald Cramér, the great probability theorist, gave a probabilistic way of interpreting Gauss's prediction. We can represent the primes as a sequence of 0s and 1s. If we start with 3 and put a 1 each time we encounter a prime and 0 otherwise, then we obtain the sequence $1, 0, 1, 0, 1, 0, 0, 0, 1, 0, 1, \ldots$. Cramér's idea is to suppose that this sequence, which represents the primes, has the same properties as a "typical" sequence of 0s and 1s, and to use this principle to make precise conjectures about the primes. More precisely, let $X_3, X_4, \ldots$ be an infinite sequence of RANDOM VARIABLES [III.71 §4] taking the values 0 or 1, and let the variable X_n equal 1 with probability $1/\log n$ (so that it equals 0 with probability $1 - 1/\log n$). Assume also that the variables are independent, so for each m knowledge about the variables other than X_m tells us nothing about X_m itself. Cramér's suggestion was that any statement about the distribution of 1s in the sequence that represents the primes will be true if and only if it is true with probability 1 for his random sequences. Some care is needed in interpreting this statement: for example, with probability 1 a random sequence will contain infinitely many even numbers. However, it is possible to formulate a general principle that takes account of such examples.

Here is an example of a use of the Gauss–Cramér model. With the help of the CENTRAL LIMIT THEOREM

[III.71 §5] one can prove that, with probability 1, there are

$$\int_2^x \frac{dt}{\log t} + O(\sqrt{x} \log x)$$

1s among the first x terms in our sequence. The model tells us that the same should be true of the sequence representing primes, and so we predict that

$$\#\{\text{primes up to } x\} = \int_2^x \frac{dt}{\log t} + O(\sqrt{x} \log x), \quad (5)$$

just as the table suggests.

The Gauss–Cramér model provides a beautiful way to think about distribution questions concerning the prime numbers, but it does not give proofs, and it does not seem likely that it can be made into such a tool; so for proofs we must look elsewhere. In analytic number theory one attempts to count objects that appear naturally in arithmetic, yet which resist being counted easily. So far, our discussion of the primes has concentrated on upper and lower bounds that follow from their basic definition and a few elementary properties—notably the fundamental theorem of arithmetic. Some of these bounds are good and some not so good. To improve on these bounds we shall do something that seems unnatural at first, and reformulate our question as a question about complex functions. This will allow us to draw on deep tools from analysis.

3 The "Analysis" in Analytic Number Theory

These analytic techniques were born in an 1859 memoir of RIEMANN [VI.49], in which he looked at the function that appears in the formula (1) of Euler, but with one crucial difference: now he considered *complex* values of s. To be precise, he defined what we now call the *Riemann zeta function* as follows:

$$\zeta(s) = \sum_{n \geqslant 1} \frac{1}{n^s}.$$

It can be shown quite easily that this sum converges whenever the real part of s is greater than 1, as we have already seen in the case of real s. However, one of the great advantages of allowing complex values of s is that the resulting function is HOLOMORPHIC [I.3 §5.6], and we can use a process of *analytic continuation* to make sense of $\zeta(s)$ for every s apart from 1. (A similar but more elementary example of this phenomenon is the infinite series $\sum_{n \geqslant 0} z^n$, which converges if and only if $|z| < 1$. However, when it does converge, it equals $1/(1 - z)$, and this formula defines a holomorphic function that is defined everywhere except $z = 1$.) Riemann proved the remarkable fact that confirming

Gauss's conjecture for the number of primes up to x is equivalent to gaining a good understanding of the zeros of the function $\zeta(s)$, that is, of the values of s for which $\zeta(s) = 0$. Riemann's deep work gave birth to our subject, so it seems worthwhile to at least sketch the key steps in the argument linking these seemingly unconnected topics.

Riemann's starting point was Euler's formula (1). It is not hard to prove that this formula is valid when s is complex, as long as its real part is greater than 1, so we have

$$\zeta(s) = \prod_{p \text{ prime}} \left(1 - \frac{1}{p^s}\right)^{-1}.$$

If we take the logarithm of both sides and then differentiate, we obtain the equation

$$-\frac{\zeta'(s)}{\zeta(s)} = \sum_{p \text{ prime}} \frac{\log p}{p^s - 1} = \sum_{p \text{ prime}} \sum_{m \geqslant 1} \frac{\log p}{p^{ms}}.$$

We need some way to distinguish between primes $p \leqslant x$ and primes $p > x$; that is, we want to count those primes p for which $x/p \geqslant 1$, but not those with $x/p < 1$. This can be done using the *step function* that takes the value 0 for $y < 1$ and the value 1 for $y > 1$ (so that its graph looks like a step). At $y = 1$, the point of discontinuity, it is convenient to give the function the average value, $\frac{1}{2}$. Perron's formula, one of the big tools of analytic number theory. describes this step function by an integral, as follows. For any $c > 0$,

$$\frac{1}{2\pi i} \int_{s:\text{Re}(s)=c} \frac{y^s}{s} \, ds = \begin{cases} 0 & \text{if } 0 < y < 1, \\ \frac{1}{2} & \text{if } y = 1, \\ 1 & \text{if } y > 1. \end{cases}$$

The integral is a *path integral* along a vertical line in the complex plane: the line consisting of all points $c + it$ with $t \in \mathbb{R}$. We apply Perron's formula with $y = x/p^m$, so that we count the term corresponding to p^m when $p^m < x$, but not when $p^m > x$. To avoid the "$\frac{1}{2}$," assume that x is not a prime power. In that case we obtain

$$\sum_{\substack{p \text{ prime, } m \geqslant 1 \\ p^m \leqslant x}} \log p$$

$$= \frac{1}{2\pi i} \sum_{p \text{ prime, } m \geqslant 1} \log p \int_{s:\text{Re}(s)=c} \left(\frac{x}{p^m}\right)^s \frac{ds}{s}$$

$$= -\frac{1}{2\pi i} \int_{s:\text{Re}(s)=c} \frac{\zeta'(s)}{\zeta(s)} \frac{x^s}{s} \, ds. \quad (6)$$

We can justify swapping the order of the sum and the integral if c is taken large enough, since everything then converges absolutely. Now the left-hand side of the above equation is not counting the number of primes up to x but rather a "weighted" version: for each prime p we add a weight of $\log p$ to the count. It turns out, though, that Gauss's prediction for the number of primes up to x follows so long as we can show that x is a good estimate for this weighted count when x is large. Notice that the sum in (6) is exactly the logarithm of the lowest common multiple of the integers less than or equal to x, which perhaps explains why this weighted counting function for the primes is a natural function to consider. Another explanation is that if the density of primes near p is indeed about $1/\log p$, then multiplying by a weight of $\log p$ makes the density everywhere about 1.

If you know some complex analysis, then you will know that *Cauchy's residue theorem* allows one to evaluate the integral in (6) in terms of the "residues" of the integrand $(\zeta'(s)/\zeta(s))(x^s/s)$, that is, the poles of this function. Moreover, for any function f that is analytic except perhaps at finitely many points, the poles of $f'(s)/f(s)$ are the zeros and poles of f. Each pole of $f'(s)/f(s)$ has order 1, and the residue is simply the order of the corresponding zero, or minus the order of the corresponding pole, of f. Using these facts we can obtain the *explicit formula*

$$\sum_{\substack{p \text{ prime, } m \geqslant 1 \\ p^m \leqslant x}} \log p = x - \sum_{\rho:\zeta(\rho)=0} \frac{x^\rho}{\rho} - \frac{\zeta'(0)}{\zeta(0)}. \quad (7)$$

Here the zeros of $\zeta(s)$ are counted with multiplicity: that is, if ρ is a zero of $\zeta(s)$ of order k, then there are k terms for ρ in the sum. It is astonishing that there can be such a formula, an exact expression for the number of primes up to x in terms of the zeros of a complicated function: you can see why Riemann's work stretched people's imagination and had such an impact.

Riemann made another surprising observation which allows us to easily determine the values of $\zeta(s)$ on the left-hand side of the complex plane (where the function is not naturally defined). The idea is to multiply $\zeta(s)$ by some simple function so that the resulting product $\xi(s)$ satisfies the *functional equation*

$$\xi(s) = \xi(1-s) \quad \text{for all } s. \quad (8)$$

He determined that this can be done by taking $\xi(s) = \frac{1}{2}s(s-1)\pi^{-s/2}\Gamma(\frac{1}{2}s)\zeta(s)$. Here $\Gamma(s)$ is the famous GAMMA FUNCTION [III.31], which equals the factorial function at positive integers (that is, $\Gamma(n) = (n-1)!$), and is well-defined and continuous for all other s.

A careful analysis of (1) reveals that there are no zeros of $\zeta(s)$ with $\operatorname{Re}(s) > 1$. Then, with the help of

(8), we can deduce that the only zeros of $\zeta(s)$ with $\operatorname{Re}(s) < 0$ lie at the negative even integers $-2, -4, \ldots$ (the "trivial zeros"). So, to be able to use (7), we need to determine the zeros inside the *critical strip*, the set of all s such that $0 \leqslant \operatorname{Re}(s) \leqslant 1$. Here Riemann made yet another extraordinary observation which, if true, would allow us tremendous insight into virtually every aspect of the distribution of primes.

The Riemann hypothesis. If $0 \leqslant \operatorname{Re}(s) \leqslant 1$ and $\zeta(s) = 0$, then $\operatorname{Re}(s) = \frac{1}{2}$.

It is known that there are infinitely many zeros on the line $\operatorname{Re}(s) = \frac{1}{2}$, crowding closer and closer together as we go up the line. The Riemann hypothesis has been verified computationally for the ten billion zeros of lowest height (that is, with $|\operatorname{Im}(s)|$ smallest), it can be shown to hold for at least 40% of all zeros, and it fits nicely with many different heuristic assertions about the distribution of primes and other sequences. Yet, for all that, it remains an unproved hypothesis, perhaps the most famous and tantalizing in all of mathematics.

How did Riemann think of his "hypothesis"? Riemann's memoir gives no hint as to how he came up with such an extraordinary conjecture, and for a long time afterwards it was held up as an example of the great heights to which humankind could ascend by pure thought alone. However, in the 1920s Siegel and WEIL [VI.93] got hold of Riemann's unpublished notes and from these it is evident that Riemann had been able to determine the lowest few zeros to several decimal places through extensive hand calculations—so much for "pure thought alone"! Nevertheless, the Riemann hypothesis is a mammoth leap of imagination and to have come up with an algorithm to calculate zeros of $\zeta(s)$ is a remarkable achievement. (See COMPUTATIONAL NUMBER THEORY [IV.3] for a discussion of how zeros of $\zeta(s)$ can be calculated.)

If the Riemann hypothesis is true, then it is not hard to prove the bound

$$\left| \frac{x^\rho}{\rho} \right| \leqslant \frac{x^{1/2}}{|\operatorname{Im}(\rho)|}.$$

Inserting this into (7) one can deduce that

$$\sum_{\substack{p \text{ prime} \\ p \leqslant x}} \log p = x + O(\sqrt{x}\log^2 x). \quad (9)$$

This, in turn, can be "translated" into (5). In fact these estimates hold if and only if the Riemann hypothesis is true.

The Riemann hypothesis is not an easy thing to understand, nor to fully appreciate. The equivalent, (5), is perhaps easier. Another version, which I prefer, is that, for every $N \geqslant 100$,

$$|\log(\mathrm{lcm}[1, 2, \ldots, N]) - N| \leqslant \sqrt{N}(\log N)^2.$$

To focus on the overcount in Gauss's guesstimate for the number of primes up to x, we use the following approximation, which can be deduced from (7) if, and only if, the Riemann hypothesis is true:

$$\frac{\int_2^x (1/\log t)\,\mathrm{d}t - \#\{\text{primes} \leqslant x\}}{\sqrt{x}/\log x}$$
$$\approx 1 + 2 \sum_{\substack{\text{all real numbers } \gamma > 0 \\ \text{such that } \frac{1}{2}+i\gamma \\ \text{is a zero of } \zeta(s)}} \frac{\sin(\gamma \log x)}{\gamma}. \quad (10)$$

The right-hand side here is the overcount in Gauss's prediction for the number of primes up to x, divided by something that grows like $\sqrt{x}$. When we looked at the table of primes it seemed that this quantity should be roughly constant. However, that is not quite true as we see upon examining the right-hand side. The first term on the right-hand side, the "1," corresponds to the contribution of the squares of the primes in (7). The subsequent terms correspond to the terms involving the zeros of $\zeta(s)$ in (7); these terms have denominator γ so the most significant terms in this sum are those with the smallest values of γ. Moreover, each of these terms is a sine wave, which oscillates, half the time positive and half the time negative. Having the "$\log x$" in there means that these oscillations happen slowly (which is why we hardly notice them in the table above), but they do happen, and indeed the quantity in (10) does eventually get negative. No one has yet determined a value of x for which this is negative (that is, a value of x for which there are more than $\int_2^x (1/\log t)\,\mathrm{d}t$ primes up to x), though our best guess is that the first time this happens is for

$$x \approx 1.398 \times 10^{316}.$$

How does one arrive at such a guess given that the table of primes extends only up to 10^{22}? One begins by using the first thousand terms of the right-hand side of (10) to approximate the left-hand side; wherever it looks as though it could be negative, one approximates with more terms, maybe a million, until one becomes pretty certain that the value is indeed negative.

It is not uncommon to try to understand a given function better by representing it as a sum of sines and cosines like this; indeed this is how one studies the harmonics in music, and (10) becomes quite compelling from this perspective. Some experts suggest that (10)

tells us that "the primes have music in them" and thus makes the Riemann hypothesis believable, even desirable.

To prove unconditionally that

$$\#\{\text{primes} \leqslant x\} \sim \int_2^x \frac{\mathrm{d}t}{\log t},$$

the so-called *prime number theorem*, we can take the same approach as above but, since we are not asking for such a strong approximation to the number of primes up to x, we need to show only that the zeros near to the line $\mathrm{Re}(s) = 1$ do not contribute much to the formula (7). By the end of the nineteenth century this task had been reduced to showing that there are no zeros actually *on* the line $\mathrm{Re}(s) = 1$: this was eventually established by DE LA VALLÉE POUSSIN [VI.67] and HADAMARD [VI.65] in 1896.

Subsequent research has provided wider and wider subregions of the critical strip without zeros of $\zeta(s)$ (and thus improved approximations to the number of primes up to x), without coming anywhere near to proving the Riemann hypothesis. This remains as an outstanding open problem of mathematics.

A simple question like "How many primes are there up to x?" deserves a simple answer, one that uses elementary methods rather than all of these methods of complex analysis, which seem far from the question at hand. However, (7) tells us that the prime number theorem is true *if and only if* there are no zeros of $\zeta(s)$ on the line $\mathrm{Re}(s) = 1$, and so one might argue that it is inevitable that complex analysis must be involved in such a proof. In 1949 Selberg and Erdős surprised the mathematical world by giving an elementary proof of the prime number theorem. Here, the word "elementary" does not mean "easy" but merely that the proof does not use advanced tools such as complex analysis—in fact, their argument is a complicated one. Of course their proof must somehow show that there is no zero on the line $\mathrm{Re}(s) = 1$, and indeed their combinatorics cunningly masks a subtle complex analysis proof beneath the surface (read Ingham's discussion (1949) for a careful examination of the argument).

4   Primes in Arithmetic Progressions

After giving good estimates for the number of primes up to x, which from now on we shall denote by $\pi(x)$, we might ask for the number of such primes that are congruent to $a \bmod q$. (If you do not know what this means, see MODULAR ARITHMETIC [III.58].) Let us write $\pi(x; q, a)$ for this quantity. To start with, note that there is only one prime congruent to 2 mod 4, and indeed there can be no more than one prime in any arithmetic progression $a, a+q, a+2q, \ldots$ if a and q have a common factor greater than 1. Let $\phi(q)$ denote the number of integers a, $1 \leqslant a \leqslant q$, such that $(a,q) = 1$. (The notation (a,q) stands for the highest common factor of a and q.) Then all but a small finite number of the infinitely many primes belong to the $\phi(q)$ arithmetic progressions $a, a+q, a+2q, \ldots$ with $1 \leqslant a < q$ and $(a,q) - 1$. Calculation reveals that the primes seem to be pretty evenly split between these $\phi(q)$ arithmetic progressions, so we might guess that in the limit the proportion of primes in each of them is $1/\phi(q)$. That is, whenever $(a,q) = 1$, we might conjecture that, as $x \to \infty$,

$$\pi(x;q,a) \sim \frac{\pi(x)}{\phi(q)}. \tag{11}$$

It is far from obvious even that the number of primes congruent to a mod q is infinite. This is a famous theorem of DIRICHLET [VI.36]. To begin to consider such questions we need a systematic way to identify integers n that are congruent to a mod q, and this Dirichlet provided by introducing a class of functions now known as *(Dirichlet) characters*. Formally, a *character* mod q is a function χ from $\mathbb{Z}$ to $\mathbb{C}$ with the following three properties (in ascending order of interest):

(i) $\chi(n) = 0$ whenever n and q have a common factor greater than 1;

(ii) χ is *periodic* mod q (that is, $\chi(n+q) = \chi(n)$ for every integer n);

(iii) χ is *multiplicative* (that is, $\chi(mn) = \chi(m)\chi(n)$ for any two integers m and n).

An easy but important example of a character mod q is the *principal character* χ_q, which takes the value 1 if $(n,q) = 1$ and 0 otherwise. If q is prime, then another important example is the *Legendre symbol* $(\frac{\cdot}{q})$: one sets $(\frac{n}{q})$ to be 0 if n is a multiple of q, 1 if n is a quadratic residue mod q, and -1 if n is a quadratic nonresidue mod q. (An integer n is called a *quadratic residue* mod q if n is congruent mod q to a perfect square.) If q is composite, then a function known as the *Legendre–Jacobi symbol* $(\frac{\cdot}{q})$, which generalizes the Legendre symbol, is also a character. This too is an important example that helps us, in a slightly less direct way, to recognize squares mod q.

These characters are all real-valued, which is the exception rather than the rule. Here is an example of a genuinely complex-valued character in the case $q = 5$. Set $\chi(n)$ to be 0 if $n \equiv 0 \pmod 5$, i if $n \equiv$ 2, -1 if $n \equiv 4$, $-$i if $n \equiv 3$, and 1 if $n \equiv 1$. To see that this is a character, note that the powers of 2 mod 5 are $2,4,3,1,2,4,3,1,\ldots$, while the powers of i are $i, -1, -i, 1, i, -1, -i, 1, \ldots$.

It can be shown that there are precisely $\phi(q)$ distinct characters mod q. Their usefulness to us comes from the properties above, together with the following formula, in which the sum is over all characters mod q and $\bar{\chi}(a)$ denotes the complex conjugate of $\chi(a)$:

$$\frac{1}{\phi(q)} \sum_{\chi} \bar{\chi}(a)\chi(n) = \begin{cases} 1 & \text{if } n \equiv a \pmod q, \\ 0 & \text{otherwise.} \end{cases}$$

What is this formula doing for us? Well, understanding the set of integers congruent to a mod q is equivalent to understanding the function that takes the value 1 if $n \equiv a \pmod q$ and 0 otherwise. This function appears on the right-hand side of the formula. However, it is not a particularly nice function to deal with, so we write it as a linear combination of characters, which are much nicer functions because they are multiplicative. The coefficient associated with the character χ in this linear combination is the number $\bar{\chi}(a)/\phi(q)$.

From the formula, it follows that

$$\sum_{\substack{p \text{ prime, } m \geqslant 1 \\ p^m \leqslant x \\ p^m \equiv a \, (\text{mod } q)}} \log p$$
$$= \frac{1}{\phi(q)} \sum_{\chi \, (\text{mod } q)} \bar{\chi}(a) \sum_{\substack{p \text{ prime, } m \geqslant 1 \\ p^m \leqslant x}} \chi(p^m) \log p.$$

The sum on the left-hand side is a natural adaptation of the sum we considered earlier when we were counting all primes. And we can estimate it if we can get good estimates for each of the sums

$$\sum_{\substack{p \text{ prime, } m \geqslant 1 \\ p^m \leqslant x}} \chi(p^m) \log p.$$

We approach these sums much as we did before, obtaining an explicit formula, analogous to (7), (10), now in terms of the zeros of the *Dirichlet L-function*:

$$L(s,\chi) = \sum_{n \geqslant 1} \frac{\chi(n)}{n^s}.$$

This function turns out to have properties closely analogous to the main properties of $\zeta(s)$. In particular, it is here that the multiplicativity of χ is all-important, since it gives us a formula similar to (1):

$$\sum_{n \geqslant 1} \frac{\chi(n)}{n^s} = \prod_{p \text{ prime}} \left(1 - \frac{\chi(p)}{p^s}\right)^{-1}. \tag{12}$$

That is, $L(s, \chi)$ has an *Euler product*. We also believe the "generalized Riemann hypothesis" that all zeros ρ of $L(\rho, \chi) = 0$ in the critical strip satisfy $\mathrm{Re}(\rho) = \frac{1}{2}$. This would imply that the number of primes up to x that are congruent to $a \bmod q$ can be estimated as

$$\pi(x; q, a) = \frac{\pi(x)}{\phi(q)} + O(\sqrt{x} \log^2(qx)). \qquad (13)$$

Therefore, the generalized Riemann hypothesis implies the estimate we were hoping for (formula (11)), provided that x is a little bigger than q^2.

In what range can we prove (11) unconditionally— that is, without the help of the generalized Riemann hypothesis? Although we can more or less translate the proof of the prime number theorem over into this new setting, we find that it gives (11) only when x is very large. In fact, x has to be bigger than an exponential in a power of q, which is a lot bigger than the "x is a little larger than q^2" that we obtained from the generalized Riemann hypothesis. We see a new type of problem emerging here, in which we are asking for a good starting point for the range of x for which we obtain good estimates, as a function of the modulus q; this does not have an analogy in our exploration of the prime number theorem. By the way, even though this bound "x is a little larger than q^2" is far out of reach of current methods, it still does not seem to be the best answer; calculations reveal that (11) seems to hold when x is just a little bigger than q. So even the Riemann hypothesis and its generalizations are not powerful enough to tell us the precise behavior of the distribution of primes.

Throughout the twentieth century much thought was put in to bounding the number of zeros of Dirichlet L-functions near to the 1-line. It turns out that one can make enormous improvements in the range of x for which (11) holds (to "halfway between polynomial in q and exponential in q") provided there are no *Siegel zeros*. These putative zeros β of $L(s, (\frac{\cdot}{q}))$ would be real numbers with $\beta > 1 - c/\sqrt{q}$; they can be shown to be extremely rare if they exist at all.

That Siegel zeros are rare is a consequence of the *Deuring–Heilbronn phenomenon*: that zeros of L-FUNCTIONS [III.47] repel each other, rather like similarly charged particles. (This phenomenon is akin to the fact that different algebraic numbers repel one another, part of the basis of the subject of Diophantine approximation.)

How big is the smallest prime congruent to $a \bmod q$ when $(a, q) = 1$? Despite the possibility of the existence of Siegel zeros, one can prove that there is always such

a prime less than $q^{5.5}$ if q is sufficiently large. Obtaining a result of this type is not difficult when there are no Siegel zeros. If there are Siegel zeros, then we go back to the explicit formula, which is similar to (7) but now concerns zeros of $L(s, \chi)$. If β is a Siegel zero, then it turns out that in the explicit formula there are now two obviously large terms: $x/\phi(q)$ and $-(\frac{a}{q})x^\beta/\beta\phi(q)$. When $(\frac{a}{q}) = 1$ it appears that they might almost cancel (since β is close to 1), but with more care we obtain

$$x - \frac{a}{q}\frac{x^\beta}{\beta} = (x - x^\beta) + x^\beta\left(1 - \frac{1}{\beta}\right) \sim x(1 - \beta)\log x.$$

This is a smaller main term than before, but it is not too hard to show that it is bigger than the contributions of all of the other zeros combined, because the Deuring–Heilbronn phenomenon implies that the Siegel zero repels those zeros, forcing them to be far to the left. When $(\frac{a}{q}) = -1$, the same two terms tell us that if $(1 - \beta)\log x$ is small, then there are twice as many primes as we would expect up to x that are congruent to $a \bmod q$.

There is a close connection between Siegel zeros and *class numbers*, which are defined and discussed in ALGEBRAIC NUMBERS [IV.1 §7]. Dirichlet's *class number formula* states that $L(1, (\frac{\cdot}{q})) = \pi h_{-q}/\sqrt{q}$ for $q > 6$, where h_{-q} is the class number of the field $\mathbb{Q}(\sqrt{-q})$. A class number is always a positive integer, so this result immediately implies that $L(1, (\frac{\cdot}{q})) \geqslant \pi/\sqrt{q}$. Another consequence is that h_{-q} is small if and only if $L(1, (\frac{\cdot}{q}))$ is small. The reason this gives us information about Siegel zeros is that one can show that the derivative $L'(\sigma, (\frac{\cdot}{q}))$ is positive (and not too small) for real numbers σ close to 1. This implies that $L(1, (\frac{\cdot}{q}))$ is small if and only if $L(s, (\frac{\cdot}{q}))$ has a real zero close to 1, that is, a Siegel zero β. When $h_{-q} = 1$, the link is more direct: it can be shown that the Siegel zero β is approximately $1 - 6/(\pi\sqrt{q})$. (There are also more complicated formulas for larger values of h_{-q}.)

These connections show that getting good lower bounds on h_{-q} is equivalent to getting good bounds on the possible range for Siegel zeros. Siegel showed that for any $\varepsilon > 0$ there exists a constant $c_\varepsilon > 0$ such that $L(1, (\frac{\cdot}{q})) \geqslant c_\varepsilon q^{-\varepsilon}$. His proof was unsatisfactory because by its very nature one cannot give an explicit value for c_ε. Why not? Well, the proof comes in two parts. The first assumes the generalized Riemann hypothesis, in which case an explicit bound follows easily. The second obtains a lower bound *in terms of the first counterexample* to the generalized Riemann hypothesis. So if the generalized Riemann hypothesis is true but remains unproved, then Siegel's proof cannot be exploited to give explicit bounds. This dichotomy, between what can be proved with an explicit constant and what cannot be, is seen far and wide in analytic number theory—and when it appears it usually stems from an application of Siegel's result, and especially its consequences for the range in which the estimate (11) is valid.

A polynomial with integer coefficients cannot always take on prime values when we substitute in an integer. To see this, note that if p divides $f(m)$ then p also divides $f(m + p), f(m + 2p), \ldots$ However, there are some prime-rich polynomials, a famous example being the polynomial $x^2 + x + 41$, which is prime for $x = 0, 1, 2, \ldots, 39$. There are almost certainly quadratic polynomials that take on more consecutive prime values, though their coefficients would have to be very large. If we ask the more restricted question of when the polynomial $x^2 + x + p$ is prime for $x = 0, 1, 2, \ldots, p-2$, then the answer, given by Rabinowitch, is rather surprising: it happens if and only if $h_{-q} = 1$, where $q = 4p - 1$. Gauss did extensive calculations of class numbers and predicted that there are just nine values of q with $h_{-q} = 1$, the largest of which is $163 = 4 \times 41 - 1$. Using the Deuring–Heilbronn phenomenon researchers showed, in the 1930s, that there is at most one q with $h_{-q} = 1$ that is not already on Gauss's list; but as usual with such methods, one could not give a bound on the size of the putative extra counterexample. It was not until the 1960s that Baker and Stark proved that there was no tenth q, both proofs involving techniques far removed from those here (in fact Heegner gave what we now understand to have been a correct proof in the 1950s but he was so far ahead of his time that it was difficult for mathematicians to appreciate his arguments and to believe that all of the details were correct). In the 1980s Goldfeld, Gross, and Zagier gave the best result to date, showing that $h_{-q} \geqslant \frac{1}{7700} \log q$ this time using the Deuring–Heilbronn phenomenon with the zeros of yet another type of L-function to repel the zeros of $L(s, (\frac{\cdot}{q}))$.

This idea that primes are well-distributed in arithmetic progressions except for a few rare moduli was exploited by Bombieri and Vinogradov to prove that (11) holds "almost always" when x is a little bigger than q^2 (that is, in the same range that we get "always" from the generalized Riemann hypothesis). More precisely, for given large x we have that (11) holds for "almost all" q less than $\sqrt{x}/(\log x)^2$ and for all a such that $(a, q) = 1$. "Almost all" means that, out of all q less

than $\sqrt{x}/(\log x)^2$, the proportion for which (11) does not hold for every a with $(a, q) = 1$ tends to 0 as $x \to \infty$. Thus, the possibility is not ruled out that there are infinitely many counterexamples. However, since this would contradict the generalized Riemann hypothesis, we do not believe that it is so.

The *Barban–Davenport–Halberstam theorem* gives a weaker result, but it is valid for the whole feasible range: for any given large x, the estimate (11) holds for "almost all" pairs q and a such that $q \leqslant x/(\log x)^2$ and $(a, q) = 1$.

5   Primes in Short Intervals

Gauss's prediction referred to the primes "around" x, so it perhaps makes more sense to interpret his statement by considering the number of primes in short intervals at around x. If we believe Gauss, then we might expect the number of primes between x and $x + y$ to be about $y/\log x$. That is, in terms of the prime-counting function π, we might expect that

$$\pi(x + y) - \pi(x) \sim \frac{y}{\log x} \tag{14}$$

for $|y| \leqslant x/2$. However, we have to be a little careful about the range for y. For example, if $y = \frac{1}{2} \log x$, then we certainly cannot expect to have half a prime in each interval. Obviously we need y to be large enough that the prediction can be interpreted in a way that makes sense; indeed, the Gauss–Cramér model suggests that (14) should hold when $|y|$ is a little bigger than $(\log x)^2$.

If we attempt to prove (14) using the same methods we used in the proof of the prime number theorem, we find ourselves bounding differences between ρth powers as follows:

$$\left| \frac{(x + y)^\rho - x^\rho}{\rho} \right| = \left| \int_x^{x+y} t^{\rho-1} \, \mathrm{d}t \right|$$
$$\leqslant \int_x^{x+y} t^{\mathrm{Re}(\rho)-1} \, \mathrm{d}t$$
$$\leqslant y(x + y)^{\mathrm{Re}(\rho)-1}.$$

With bounds on the density of zeros of $\zeta(s)$ well to the right of $\frac{1}{2}$, it has been shown that (14) holds for y a little bigger than $x^{7/12}$; but there is little hope, even assuming the Riemann hypothesis, that such methods will lead to a proof of (14) for intervals of length $\sqrt{x}$ or less.

In 1949 Selberg showed that (14) is true for "almost all" x when $|y|$ is a little bigger than $(\log x)^2$, assuming the Riemann hypothesis. Once again, "almost all" means with density tending to 1, rather than "all," and it is feasible that there are infinitely many counterexamples, though at that time it seemed highly unlikely.

It therefore came as a surprise when Maier showed, in 1984, that, for any fixed $A > 0$, the estimate (14) fails for infinitely many integers x, with $y = (\log x)^A$. His ingenious proof rests on showing that the small primes do not always have as many multiples in an interval as one might expect.

Let $p_1 = 2 < p_2 = 3 < \cdots$ be the sequence of primes. We are now interested in the size of the gaps $p_{n+1} - p_n$ between consecutive primes. Since there are about $x/\log x$ primes up to x, the average difference is $\log x$ and we might ask how often the difference between consecutive primes is about average, whether the differences can get really small, and whether the differences can get really large. The Gauss–Cramér model suggests that the proportion of n for which the gap between consecutive primes is more than λ times the average, that is $p_{n+1} - p_n > \lambda \log p_n$, is approximately $e^{-\lambda}$; and, similarly, the proportion of intervals $[x, x + \lambda \log x]$ containing exactly k primes is approximately $e^{-\lambda}\lambda^k/k!$, a suggestion which, as we shall see, is supported by other considerations. By looking at the tail of this distribution, Cramér conjectured that $\limsup_{n\to\infty}(p_{n+1} - p_n)/(\log p_n)^2 = 1$, and the evidence we have *seems* to support this (see table 2).

The Gauss–Cramér model does have a big drawback: it does not "know any arithmetic." In particular, as we noted earlier, it does not predict divisibility by small primes. One manifestation of this failing is that it predicts that there should be just about as many gaps of length 1 between primes as there are of length 2. However, there is only one gap of length 1, since if two primes differ by 1, then one of them must be even, whereas there are many examples of pairs of primes differing by 2, and there are believed to be infinitely many. For the model to make correct conjectures about prime pairs, we must consider divisibility by small primes in the formulation of the model, which makes it rather more complicated. Since there are these glaring errors in the simpler model, Cramér's conjecture for the largest gaps between consecutive primes must be treated with a degree of suspicion. And in fact, if one corrects the model to account for divisibility by small primes, one is led to conjecture that $\limsup_{n\to\infty}(p_{n+1} - p_n)/(\log p_n)^2$ is greater than $\frac{9}{8}$.

Finding large gaps between primes is equivalent to finding long sequences of composite numbers. How about trying to do this explicitly? For example, we know that $n! + j$ is composite for $2 \leqslant j \leqslant n$, as it is divisible by j. Therefore we have a gap of length at least n between consecutive primes, the first of which is

Table 2 The largest known gaps between primes.

p_n	$p_{n+1} - p_n$	$\dfrac{p_{n+1} - p_n}{\log^2 p_n}$
113	14	0.6264
1 327	34	0.6576
31 397	72	0.6715
370 261	112	0.6812
2 010 733	148	0.7026
20 831 323	210	0.7395
25 056 082 087	456	0.7953
2 614 941 710 599	652	0.7975
19 581 334 192 423	766	0.8178
218 209 405 436 543	906	0.8311
1 693 182 318 746 371	1132	0.9206

the largest prime less than or equal to $n! + 1$. However, this observation is not especially helpful, since the average gap between primes around $n!$ is $\log(n!)$, which is approximately equal to $n \log n$, whereas we are looking for gaps that are *larger* than the average. However, it is possible to generalize this argument and show that there are indeed long sequences of consecutive integers, each with a small prime factor. In the 1930s, Erdős reformulated the question as follows. Fix a positive integer z, and for each prime $p \leqslant z$ choose an integer a_p in such a way that, for as large an integer y as possible, every positive integer $n \leqslant y$ satisfies at least one of the congruences $n \equiv a_p \pmod{p}$. Now let X be the product of all the primes up to z (which means, by the prime number theorem, that $\log X$ is about z), and let x be the integer between X and $2X$ such that $x \equiv -a_p \pmod{p}$ for every $p \leqslant z$. (This integer exists, by the *Chinese remainder theorem*.) If m is an integer between $x + 1$ and $x + y$, then $m - x$ is a positive integer less than y, so $m - x \equiv a_p \pmod{p}$ for some prime $p \leqslant z$. Since $x \equiv -a_p \pmod{p}$, it follows that m is divisible by p. Thus, all the integers from $x + 1$ to $x + y$ are composite. Using this basic idea, it can be shown that there are infinitely many primes p_n for which $p_{n+1} - p_n$ is about $(\log p_n)(\log\log p_n)$, which is significantly larger than the average but nowhere close to Cramér's conjecture.

6   Gaps between Primes That Are Smaller Than the Average

We have just seen how to show that there are infinitely many pairs of consecutive primes whose difference is much bigger than the average: that is, $\limsup_{n\to\infty}(p_{n+1} - p_n)/(\log p_n) = \infty$. We would now like to show that there are infinitely many pairs of consecutive primes whose difference is much smaller than the average: that is, $\liminf_{n\to\infty}(p_{n+1} - p_n)/(\log p_n) = 0$. Of course, it is believed that there are infinitely many pairs of primes that differ by 2, but this question seems intractable for now.

Until recently researchers had very little success with the question of small gaps; the best result before 2000 was that there are infinitely many gaps of size less than one-quarter of the average. However, a recent method of Goldston, Pintz, and Yıldırım, which counts primes in short intervals with simple weighting functions, proves that $\liminf_{n\to\infty}(p_{n+1} - p_n)/(\log p_n) = 0$, and even that there are infinitely many pairs of consecutive primes with difference no larger than about $\sqrt{\log p_n}$. Their proof, rather surprisingly, rests on estimates for primes in arithmetic progressions; in particular, that (11) holds for almost all q up to $\sqrt{x}$ (as discussed earlier). Moreover, they obtain a conditional result of the following kind: if in fact (11) holds for almost all q up to a little larger than $\sqrt{x}$, then it follows that there exists an integer B such that $p_{n+1} - p_n \leqslant B$ for infinitely many primes p_n.

7 Very Small Gaps between Primes

There appear to be many pairs of primes that differ by two, like 3 and 5, 5 and 7, . . . , the so-called *twin primes*, though no one has yet proved that there are infinitely many. In fact, for every even integer $2k$ there seem to be many pairs of primes that differ by $2k$, but again no one has yet proved that there are infinitely many. This is one of the outstanding problems in the subject.

In a similar vein is Goldbach's conjecture from the 1760s: is it true that every even integer greater than 2 is the sum of two primes? This is still an open question, and indeed a publisher recently offered a million dollars for its solution. We know it is true for almost all integers, and it has been computer tested for every even integer up to 4×10^{14}. The most famous result on this question is due to Chen (1966), who showed that every even integer can be written as the sum of a prime and a second integer that has *at most two* prime factors (that is, it could be a prime or an "almost-prime").

In fact, GOLDBACH [VI.17] never asked this question. He asked Euler, in a letter in the 1760s, whether every integer greater than 1 can be written as the sum of at most three primes, which would imply what we now call the "Goldbach conjecture." In the 1920s Vinogradov showed that every sufficiently large odd integer can be written as the sum of three primes (and thus every sufficiently large even integer can be written as the sum of four primes). We actually believe that every odd integer greater than 5 is the sum of three primes but the known proofs only work once the numbers involved are large enough. In this case we can be explicit about "sufficiently large"—at the moment the proof needs them to be at least e^{5700}, but it is rumored that this may soon be substantially reduced, perhaps even to 7.

To guess at the precise number of prime pairs q, $q + 2$ with $q \leqslant x$ we proceed as follows. If we do not consider divisibility by the small primes, then the Gauss–Cramér model suggests that a random integer up to x is prime with probability roughly $1/\log x$, so we might expect $x/(\log x)^2$ prime pairs q, $q + 2$ up to x. However, we do have to account for the small primes, as the q, $q + 1$ example shows, so let us consider 2-divisibility. The proportion of random pairs of integers that are both odd is $\frac{1}{4}$, whereas the proportion of random q such that q and $q + 2$ are both odd is $\frac{1}{2}$. Thus we should adjust our guess $x/(\log x)^2$ by a factor $(\frac{1}{2})/(\frac{1}{4}) = 2$. Similarly, the proportion of random pairs of integers that are both not divisible by 3 (or indeed by any given odd prime p) is $(\frac{2}{3})^2$ (and $(1 - 1/p)^2$, respectively), whereas the proportion of random q such that q and $q + 2$ are both not divisible by 3 (or by prime p) is $\frac{1}{3}$ (and $(1 - 2/p)$, respectively). Adjusting our formula for each prime p we end up with the prediction

$$\#\{q \leqslant x : q \text{ and } q + 2 \text{ both prime}\}$$
$$\sim 2 \prod_{p \text{ an odd prime}} \frac{(1 - 2/p)}{(1 - 1/p)^2} \frac{x}{(\log x)^2}.$$

This is known as the *asymptotic twin prime conjecture*. Despite its plausibility there do not seem to be any practical ideas around for turning the heuristic argument above into something rigorous. The one good unconditional result known is that the number of twin primes less than or equal to x is never more than four times the quantity we have just predicted. One can make a more precise prediction replacing $x/(\log x)^2$ by $\int_2^x (1/(\log t)^2)\,dt$, and then we expect that the difference between the two sides is no more than $c\sqrt{x}$ for some constant $c > 0$, a guesstimate that is well supported by computational evidence.

A similar method allows us to make predictions for the number of primes in any polynomial-type patterns. Let $f_1(t), f_2(t), \ldots, f_k(t) \in \mathbb{Z}[t]$ be distinct irreducible polynomials of degree greater than or equal to 1 with positive leading coefficient, and define $\omega(p)$ to be the number of integers $n \pmod{p}$ for which p divides

$f_1(n)f_2(n)\cdots f_k(n)$. (In the case of twin primes above we have $f_1(t) = t$, $f_2(t) = t + 2$ with $\omega(2) = 1$ and $\omega(p) = 2$ for all odd primes p.) If $\omega(p) = p$ then p always divides at least one of the polynomial values, so they can be simultaneously prime just finitely often (an example of this is when $f_1(t) = t$, $f_2(t) = t + 1$, in which case $\omega(2) = 2$). Otherwise we have an *admissible set* of polynomials for which we predict that the number of integers n less than x for which all of $f_1(n), f_2(n), \ldots, f_k(n)$ are prime is about

$$\prod_{p \text{ prime}} \frac{(1 - \omega(p)/p)}{(1 - 1/p)^k}$$
$$\times \frac{x}{\log|f_1(x)| \log|f_2(x)| \cdots \log|f_k(x)|} \quad (15)$$

once x is sufficiently large. One can use a similar heuristic to make predictions in Goldbach's conjecture, that is, for the number of pairs of primes p, q for which $p + q = 2N$. Again, these predictions are very well matched by the computational evidence.

There are just a few cases of conjecture (15) that have been proved. Modifications of the proof of the prime number theorem give such a result for admissible polynomials $qt + a$ (in other words, for primes in arithmetic progressions) and for admissible $at^2 + btu + cu^2 \in \mathbb{Z}[t, u]$ (as well as some other polynomials in two variables of degree two). It is also known for a certain type of polynomial in n variables of degree n (the admissible "norm-forms").

There was little improvement on this situation during the twentieth century until quite recently, when, by very different methods, Friedlander and Iwaniec broke through this stalemate showing such a result for the polynomial $t^2 + u^4$, and then Heath-Brown did so for any admissible homogeneous polynomial in two variables of degree three.

Another truly extraordinary breakthrough occurred recently with a result of Green and Tao, proved in 2004, which states that for every k there are infinitely many k-term arithmetic progressions of primes: that is, pairs of integers a, d such that $a, a+d, a+2d, \ldots, a+(k-1)d$ are all prime. Green and Tao are currently hard at work attempting to show that the number of k-term arithmetic progressions of primes is indeed well approximated by (15). They are also extending their results to other families of polynomials.

8   Gaps between Primes Revisited

In the 1970s Gallagher deduced from the conjectured prediction (15) (with $f_j(t) = t + a_j$) that the propor-

tion of intervals $[x, x + \lambda \log x]$ which contain exactly k primes is close to $e^{-\lambda}\lambda^k/k!$ (as was also deduced, in section 5 above, from the Gauss–Cramér heuristics). This has recently been extended to support the prediction that, as we vary x from X to $2X$, the number of primes in the interval $[x, x + y]$ is normally distributed with mean $\int_x^{x+y}(1/\log t)\,dt$ and variance $(1 - \delta)y/\log x$, where δ is some constant strictly between 0 and 1 and we take y to be x^δ.

When $y > \sqrt{x}$ the Riemann zeta function supplies information on the distribution of primes in intervals $[x, x + y)$ via the explicit formula (7). Indeed, when we compute the "variance"

$$\frac{1}{X}\int_X^{2X}\left(\sum_{\substack{p \text{ prime,} \\ x < p \leqslant x+y}} \log p - y\right)^2 dx$$

using the explicit formula we obtain a sum of terms of the form $\int_X^{2X} x^{i(\gamma_j - \gamma_k)}\,dx$. Here we are assuming the Riemann hypothesis and writing the zeros of $\zeta(s)$ as $\frac{1}{2} \pm i\gamma_n$ with $0 < \gamma_1 < \gamma_2 < \cdots$. This sum is dominated by the terms corresponding to those pairs γ_j, γ_k for which $|\gamma_j - \gamma_k|$ is small (in which case there is little cancellation in the integral). Therefore, in order to understand the variance for the distribution of primes in short intervals we need to understand the distribution of the zeros of $\zeta(s)$ in short intervals. In 1973 Montgomery investigated this and suggested that the proportion of pairs of zeros of $\zeta(s)$ whose difference is less than α times the average gap between consecutive zeros is given by the integral

$$\int_0^\alpha \left(1 - \left(\frac{\sin \pi\theta}{\pi\theta}\right)^2\right) d\theta, \quad (16)$$

and he proved an equivalent form of this in a limited range. If the zeros were placed "randomly," then (16) would be replaced by α. In fact (16) is about $\frac{1}{9}\alpha^3$ for small α, which is far smaller than α. This means that there are far fewer pairs of zeros of $\zeta(s)$ that are close together than one might expect, which we express informally by saying that the zeros of $\zeta(s)$ *repel* one another.

In a now-famous conversation that took place at the Institute for Advanced Study in Princeton, Montgomery mentioned his ideas to the physicist Freeman Dyson. Dyson immediately recognized (16) as a function that comes up in modeling energy levels in quantum chaos. Believing that this was unlikely to be a coincidence, he suggested that the zeros of the Riemann zeta function are distributed, *in all aspects*, like energy levels, which are in turn modeled on the distribution of EIGENVALUES

[I.3 §4.3] of random HERMITIAN MATRICES [III.50 §3]. There is now substantial computational and theoretical evidence that Dyson's suggestion is correct and can be extended to Dirichlet L-functions, as well as other types of L-functions, and even to other statistics about L-functions.

One note of caution. Few of the conjectured consequences of this new "random matrix theory" have been unconditionally proved, or seem likely to be in the foreseeable future. It simply provides a tool to make predictions where that was too difficult to do before. However, there is at least one key question about which we still cannot make a well-substantiated prediction: how big does $\zeta(s)$ get on the $\frac{1}{2}$-line? One can show that $\log|\zeta(\frac{1}{2}+it)|$ gets larger than $\sqrt{\log T}$ for values of t close to T, and that it gets no larger than $\log T$. However, it is unclear, even if we do not insist on a rigorous proof, whether the true maximal order is nearer the upper or lower bound.

9   Sieve Methods

Almost all of our discussion so far has been about developments of Riemann's approach to counting primes. This approach is very delicate and not as adaptable as one might wish to many natural questions (such as counting k-tuples of primes $n + a_1, n + a_2, \ldots, n + a_k$). However, one can go back to *sieve methods*, which are modifications of the sieve of Eratosthenes, and at least get upper bounds. For example, suppose we want to find an upper bound for the number of prime pairs n, $n + 2$ with $N < n \leqslant 2N$. One possibility would be to fix a number y and determine for how many pairs n, $n + 2$ with $N < n \leqslant 2N$ it is the case that neither n nor $n + 2$ has a prime factor less than y. If we took y to be $(2N)^{1/2}$, then this method would exactly count the twin primes, but it seems to be far too difficult to implement. But it turns out that if instead we take y to be a small power of N, then the calculations become much easier and there are methods of obtaining good bounds. (However, the bounds given by these methods become less accurate as the power gets closer to $\frac{1}{2}$.)

In the 1920s Brun showed how to make the principle of inclusion–exclusion into a useful tool in this type of question. This principle is best exhibited when counting the number of integers n in a set S that are coprime to given integer m. We begin with the number of integers in S, which is obviously more than the quantity we seek. Next, we subtract, for each prime p dividing m, the number of integers in S that are divisible by p. If

$n \in S$ is divisible by exactly r prime factors of m, then we have counted $1 + r \times (-1)$ for the contribution of n so far, which is less than or equal to 0, and less than 0 for $r \geqslant 2$; whereas we wanted to count 0 when $r \geqslant 2$ (since n is not coprime to m). Thus we obtain a number that is less than the quantity we seek. To compensate for that, we add back in the number of integers in S divisible by pq for each pair of primes $p < q$ which divide m. We have now counted $1 + r \times (-1) + \binom{r}{2} \times 1$ for the contribution of n, which is greater than or equal to 0, and greater than 0 for $r \geqslant 3$. Similarly, we subtract the number of integers divisible by pqr, etc.

For each $n \in S$ we end up counting $(1 - 1)^r$ for n, where r is the number of distinct prime factors of (m, n). Expanding this sum with the binomial theorem we may reexpress this identity as follows. Let $\chi_m(n) = 1$ if $(n, m) = 1$ and 0 otherwise. Then

$$\chi_m(n) = \sum_{d|(m,n)} \mu(d),$$

where $\mu(m)$, the Möbius function, equals 0 if m is divisible by the square of a prime and equals $(-1)^{\omega(m)}$ otherwise, where $\omega(m)$ is the number of distinct prime factors of m.

The inclusion–exclusion inequalities just discussed may be obtained from

$$\sum_{\substack{d|(m,n) \\ \omega(d) \leqslant 2k+1}} \mu(d) \leqslant \chi_m(n) \leqslant \sum_{\substack{d|(m,n) \\ \omega(d) \leqslant 2k}} \mu(d),$$

which holds for any $k \geqslant 0$, by summing over all $n \in S$.

The reason for using these abbreviated sums rather than the complete sum is that there are far fewer terms and thus, when one sums over values of n, there will be far fewer rounding errors (remember that it was rounding errors that sank our attempt to estimate the number of primes up to x using the sieve of Eratosthenes). On the other hand, they have the disadvantage that they cannot possibly give the exact answer, since they are missing many appropriate terms. However, with a judicious choice of k the missing terms do not contribute much to the complete sum and we get a good answer.

Minor variants work well for many questions. In the "combinatorial sieve" one selects which d are part of the upper and lower bound sums, not by counting the total number of prime factors they contain but instead using other criteria, such as the numbers of prime factors of d in each of several intervals. Using such a method, Brun showed that there cannot be too many twin primes p, $p + 2$; indeed, the sum of $1/p$, over all primes p for which $p + 2$ is also prime, converges, in contrast with (3).

In the "Selberg upper bound sieve" one comes up with some numbers λ_d that are nonzero only when $d \leqslant D$ (where D is chosen to be not too large), with the property that

$$\chi_m(n) \leqslant \left(\sum_{d \mid n} \lambda_d \right)^2 \quad \text{for all } n.$$

Summing over the appropriate n one then finds the optimal solution by minimizing the resulting quadratic form. Lower bounds can also be obtained out of Selberg's methods. It was by using such methods that Chen was able to prove there are infinitely many primes p for which $p + 2$ has at most two prime factors, and that Goldston, Pintz, and Yıldırım were able to establish that there are sometimes short gaps between primes. These methods are also an essential ingredient in the work of Green and Tao. One can also get good upper bounds on the number of primes in arithmetic progressions and short intervals:

- the number of primes in any interval of length y is never greater than $2y / \log y$;
- the number of primes less than x in an arithmetic progression mod q is never greater than $2x / \phi(q) \log(x/q)$.

Notice that in each case the log in the denominator is of the number of integers being considered (y and x/q, respectively), not $\log x$ as expected, though this will only make a significant difference if the number of integers being considered is small. Otherwise these inequalities are bigger than the expected quantity by a factor of 2. Can this "2" be improved? It will be difficult because we showed earlier that if there are Siegel zeros then we get twice as many primes as expected in certain arithmetic progressions. Therefore, if we can improve the "2" in these two formulas, then we can deduce that there are no Siegel zeros!

10   Smooth Numbers

An integer is *y-smooth* if all of its prime factors are less than or equal to y. A proportion $1 - \log 2$ of the integers up to x are $\sqrt{x}$-smooth, and indeed, for any fixed $u > 1$ there exists some number $\rho(u) > 0$ such that if $x = y^u$, then a proportion $\rho(u)$ of the integers up to x are y-smooth. This proportion does not seem to have any easy definition in general. For $1 \leqslant u \leqslant 2$ we have $\rho(u) = 1 - \log u$, but for larger u it is best defined as

$$\rho(u) = \frac{1}{u} \int_0^1 \rho(u - t) \, dt,$$

an *integral delay equation*. Such an equation is typical when we give precise estimates for questions that arise in sieve theory.

Questions about the distribution of smooth numbers arise frequently in the analysis of algorithms, and have consequently been the focus of a lot of recent research. (See COMPUTATIONAL NUMBER THEORY [IV.3 §3] for an example of the use of smooth numbers.)

11   The Circle Method

Another method of analysis that plays a prominent role in this subject is the so-called *circle method*, which goes back to HARDY [VI.73] and LITTLEWOOD [VI.79]. This method uses the fact that, for any integer n,

$$\int_0^1 e^{2i\pi nt} \, dt = \begin{cases} 1 & \text{if } n = 0, \\ 0 & \text{otherwise.} \end{cases}$$

For example, if we wish to count the number, $r(n)$, of solutions to the equation $p + q = n$ with p and q prime, we can express it as an integral as follows:

$$r(n) = \sum_{\substack{p,q \leqslant n \\ \text{both prime}}} \int_0^1 e^{2i\pi(p+q-n)t} \, dt$$

$$= \int_0^1 e^{-2i\pi nt} \left(\sum_{p \text{ prime, } p \leqslant n} e^{2i\pi pt} \right)^2 dt.$$

The first equality holds because the integrand is 0 when $p + q \neq n$ and 1 otherwise, and the second is easy to check.

At first sight it looks more difficult to estimate the integral than it is to estimate $r(n)$ directly, but this is not the case. For instance, the prime number theorem for arithmetic progressions allows us to estimate $P(t) = \sum_{p \leqslant n} e^{2i\pi pt}$ when t is a rational ℓ/m with m small. For in this case,

$$P\left(\frac{\ell}{m}\right) = \sum_{(a,m)=1} e^{2i\pi a\ell/m} \sum_{\substack{p \leqslant n, \\ p \equiv a \pmod{m}}} 1$$

$$\approx \sum_{(a,m)=1} e^{2i\pi a\ell/m} \frac{\pi(n)}{\phi(m)} = \mu(m) \frac{\pi(n)}{\phi(m)}.$$

If t is sufficiently close to ℓ/m, then $P(t) \approx P(\ell/m)$; such values of t are called the *major arcs* and we believe that the integral over the major arcs gives, in total, a very good approximation to $r(n)$; indeed, we get something very close to the quantity one predicts from something like (15). Thus to prove the Goldbach conjecture we need to show that the contribution to the integral from the other values of t (that is, from the *minor arcs*) is small. In many problems one can successfully do this, but no one has yet succeeded in doing so for the Goldbach problem. Also useful is the "discrete analogue" of the above: using the identity

$$\frac{1}{m} \sum_{j=0}^{m-1} e^{2i\pi jn/m} \, dt = \begin{cases} 1 & \text{if } n \equiv 0 \pmod{m}, \\ 0 & \text{otherwise} \end{cases}$$

(which holds for any given integer $m \geqslant 1$), we have that

$$r(n) = \sum_{\substack{p,q \leqslant n \\ \text{both prime}}} \frac{1}{m} \sum_{j=0}^{m-1} e^{2i\pi j(p+q-n)/m}$$

$$= \sum_{j=0}^{m-1} e^{-2i\pi jn/m} P(j/m)^2$$

provided $m > n$. A similar analysis can be used here but working mod m sometimes has advantages, as it allows us to use properties of the multiplicative group mod m.

Sums like $P(j/m)$ in the paragraph above or more simple sums like $\sum_{n \leqslant N} e^{2i\pi n^k/m}$ are called *exponential sums*. They play a central role in many of the calculations one does in analytic number theory. There are several techniques for investigating them.

(1) It is easy to calculate the sum $\sum_{n \leqslant N} e^{2i\pi n/m}$, since it is a geometric progression. With higher-degree polynomials one can often reduce to this case; for example, by writing $n_1 - n_2 = h$ we have

$$\left| \sum_{n \leqslant N} e^{2i\pi n^2/m} \right|^2$$

$$= \sum_{n_1, n_2 \leqslant N} e^{2i\pi(n_1^2 - n_2^2)/m}$$

$$= \sum_{|h| \leqslant N} e^{2i\pi h^2/m} \sum_{\substack{\max\{0, -h\} < n_2 \\ \leqslant \min\{N, N-h\}}} e^{4i\pi hn_2/m},$$

and the inner sum is now a geometric progression.

(2) The work of Weil and Deligne, which gives very accurate results on the number of solutions to equations mod p, is ideally suited to many applications in analytic number theory. For example, the "Kloosterman sum" $\sum_{a_1 a_2 \cdots a_k \equiv b \pmod{p}} e^{2i\pi(a_1 + a_2 + \cdots + a_k)/p}$, where the a_i run over the integers mod p and $(b, p) = 1$, appears naturally in many questions; Deligne showed that it has absolute value less than or equal to $kp^{(k-1)/2}$, an extraordinary amount of cancellation in this sum which has about p^{k-1} summands, each of absolute value 1. (See THE WEIL CONJECTURES [V.35].)

(3) We discussed earlier the fact that the values of $\zeta(s)$ satisfy a symmetry about the line $\mathrm{Re}(s) = \frac{1}{2}$, given by the "functional equation." There are other functions (called "modular functions") that also have symme-

tries in the complex plane; typically the value of the function at s is related to the value of the function at $(\alpha s + \beta)/(\gamma s + \delta)$, for some integers $\alpha, \beta, \gamma, \delta$ satisfying $\alpha\delta - \beta\gamma = 1$. Sometimes an exponential sum can be related to the value of a modular function, and subsequently to the value of that modular function at another point, using the symmetry of the function.

12 More *L*-Functions

There are many types of *L*-functions beyond Dirichlet *L*-functions, some of which are well understood, some not (see *L*-FUNCTIONS [III.47]). The type that has received the most attention recently is a class of *L*-functions that can be associated with elliptic curves (see ARITHMETIC GEOMETRY [IV.5 §5.1]). An *elliptic curve E* is given by an equation of the form $y^2 = x^3 + ax + b$, where the *discriminant* $4a^3 + 27b^2$ is nonzero. The associated *L*-function $L(E, s)$ is most easily described in terms of its Euler product:

$$L(E, s) = \prod_p \left(1 - \frac{a_p}{p^s} + \frac{p}{p^{2s}} \right)^{-1}. \tag{17}$$

Here a_p is an integer which, for primes p not dividing $4a^3 + 27b^2$, is defined to be p minus the number of solutions $(x, y) \pmod{p}$ to the equation $y^2 \equiv x^3 + ax + b \pmod{p}$. It can be shown that each $|a_p|$ is less than $2\sqrt{p}$, so the Euler product above converges absolutely when $\mathrm{Re}(s) > \frac{3}{2}$. Therefore, (17) is a good definition for these values of s. Can we now extend it to the whole of the complex plane, as we did for $\zeta(s)$? This is a very deep problem—the answer is yes; in fact, it is the celebrated theorem of Andrew Wiles that implied FERMAT'S LAST THEOREM [V.10].

Another interesting question is to understand the distribution of values of $a_p/2\sqrt{p}$ as we range over primes p. These all lie in the interval $[-1, 1]$. One might expect them to be uniformly distributed in the interval, but in fact this is never the case. As discussed in ALGEBRAIC NUMBERS [IV.1] one can write $a_p = \alpha_p + \bar{\alpha}_p$, where $|\alpha_p| = \sqrt{p}$, and α_p is called the Weil number. If we write $\alpha = \sqrt{p} e^{\pm i\theta_p}$, then $a_p = 2\sqrt{p}\cos(\theta_p)$ for some angle $\theta_p \in [0, \pi]$. We can then think of θ_p as belonging to the upper half of a circle. The surprise is that for almost all elliptic curves the θ_p are not uniformly distributed, which would mean the proportion in a certain arc would be proportional to the length of that arc. Rather, they are distributed in such a way that the proportion of them in any given arc is proportional to the area under that arc. This is a recent result of Richard Taylor.

The correct analogue of the Riemann hypothesis for $L(E, s)$ turns out to be that all the nontrivial zeros lie on the line $\text{Re}(s) = 1$. This is believed to be true. Moreover, it is believed that they, like the zeros of $\zeta(s)$, are distributed according to the rules that govern the eigenvalues of randomly chosen matrices.

These L-functions often have zeros at $s = 1$ (which is linked to THE BIRCH–SWINNERTON-DYER CONJECTURE [V.4]) and these zeros repel zeros of Dirichlet L-functions (which is what was used by Goldfeld, Gross, and Zagier, as mentioned in section 4, to get their lower bound on h_{-q}).

L-functions arise in many areas of arithmetic geometry, and their coefficients typically describe the number of points satisfying certain equations mod p. The *Langlands program* seeks to understand these connections at a deep level.

It seems that every "natural" L-function has many of the same analytic properties as those discussed in this article. Selberg has proposed that this phenomenon should be even more general. Consider sums $A(s) = \sum_{n \geqslant 1} a_n / n^s$ that

- are well-defined when $\text{Re}(s) > 1$,
- have an Euler product $\prod_p (1 + b_p / p^s + b_{p^2} / p^{2s} + \cdots)$ in this (or an even smaller) region,
- have coefficients a_n that are smaller than any given power of n, once n is sufficiently large,
- satisfy $|b_n| < \kappa n^\theta$ for some constants $\theta < \frac{1}{2}$ and $\kappa > 0$.

Selberg conjectures that we should be able to give a good definition to $A(s)$ on the whole complex plane, and that $A(s)$ should have a symmetry connecting the value of $A(s)$ with $A(1-s)$. Furthermore, he conjectures that the Riemann hypothesis should hold for $A(s)$!

The current wishful thinking is that Selberg's family of L-functions is precisely the same as those considered by Langlands.

13 Conclusion

In this article we have described current thinking on several key questions about the distribution of primes. It is frustrating that after centuries of research so little has been proved, the primes guarding their mysteries so jealously. Each new breakthrough seems to require brilliant ideas and extraordinary technical prowess. As EULER [VI.19] wrote in 1770:

> Mathematicians have tried in vain to discover some order in the sequence of prime numbers but we have every reason to believe that there are some mysteries which the human mind will never penetrate.

Further Reading

Hardy and Wright's classic book (1980) stands alone among introductory number theory texts for the quality of its discussion of analytic topics. The best introduction to the heart of analytic number theory is the masterful book by Davenport (2000). Everything you have ever wanted to know about the Riemann zeta function is in Titchmarsh (1986). Finally, there are two recently released books by modern masters of the subject (Iwaniec and Kowalski 2004; Montgomery and Vaughan 2006) that introduce the reader to the key issues of the subject.

The reference list below includes several papers, significant for this article, whose content is not discussed in any of the listed books.

Davenport, H. 2000. *Multiplicative Number Theory*, 3rd edn. New York: Springer.

Deligne, P. 1977. Applications de la formule des traces aux sommes trigonométriques. In *Cohomologie Étale* (SGA 4 1/2). Lecture Notes in Mathematics, volume 569. New York: Springer.

Green, B., and T. Tao. 2008. The primes contain arbitrarily long arithmetic progressions. *Annals of Mathematics* 167: 481–547.

Hardy, G. H., and E. M. Wright. 1980. *An Introduction to the Theory of Numbers*, 5th edn. Oxford: Oxford University Press.

Ingham, A. E. 1949. Review 10,595c (MR0029411). *Mathematical Reviews*. Providence, RI: American Mathematical Society.

Iwaniec, H., and E. Kowalski. 2004. *Analytic Number Theory*. AMS Colloquium Publications, volume 53. Providence, RI: American Mathematical Society.

Montgomery, H. L., and R. C. Vaughan. 2006. *Multiplicative Number Theory I: Classical Theory*. Cambridge: Cambridge University Press.

Soundararajan, K. 2007. Small gaps between prime numbers: the work of Goldston-Pintz-Yıldırım. *Bulletin of the American Mathematical Society* 44:1–18.

Titchmarsh, E. C. 1986. *The Theory of the Riemann Zeta-Function*, 2nd edn. Oxford: Oxford University Press.

IV.3 Computational Number Theory
Carl Pomerance

1 Introduction

Historically, computation has been a driving force in the development of mathematics. To help measure the sizes of their fields, the Egyptians invented geometry. To help predict the positions of the planets, the Greeks invented trigonometry. Algebra was invented to deal with equations that arose when mathematics was used to model the world. The list goes on, and it is not just historical. If anything, computation is more important than ever. Much of modern technology rests on algorithms that compute quickly: examples range from the WAVELETS [VII.3] that allow CAT scans, to the numerical extrapolation of extremely complex systems in order to predict weather and global warming, and to the combinatorial algorithms that lie behind Internet search engines (see THE MATHEMATICS OF ALGORITHM DESIGN [VII.5 §6]).

In pure mathematics we also compute, and many of our great theorems and conjectures are, at root, motivated by computational experience. It is said that GAUSS [VI.26], who was an excellent computationalist, needed only to work out a concrete example or two to discover, and then prove, the underlying theorem. While some branches of pure mathematics have perhaps lost contact with their computational origins, the advent of cheap computational power and convenient mathematical software has helped to reverse this trend.

One mathematical area where the new emphasis on computation can be clearly felt is number theory, and that is the main topic of this article. A prescient call-to-arms was issued by Gauss as long ago as 1801:

> The problem of distinguishing prime numbers from composite numbers, and of resolving the latter into their prime factors, is known to be one of the most important and useful in arithmetic. It has engaged the industry and wisdom of ancient and modern geometers to such an extent that it would be superfluous to discuss the problem at length. Nevertheless we must confess that all methods that have been proposed thus far are either restricted to very special cases or are so laborious and difficult that even for numbers that do not exceed the limits of tables constructed by estimable men, they try the patience of even the practiced calculator. And these methods do not apply at all to larger numbers.... Further, the dignity of the science itself seems to require that every possible means be explored for the solution of a problem so elegant and so celebrated.

Factorization into primes is a very basic issue in number theory, but essentially all branches of number theory have a computational component. And in some areas there is such a robust computational literature that we discuss the algorithms involved as mathematically interesting objects in their own right. In this article we will briefly present a few examples of the computational spirit: in analytic number theory (the distribution of primes and the Riemann hypothesis); in Diophantine equations (Fermat's last theorem and the ABC conjecture); and in elementary number theory (primality and factorization). A secondary theme that we shall explore is the strong and constructive interplay between computation, heuristic reasoning, and conjecture.

2 Distinguishing Prime Numbers from Composite Numbers

The problem is simple to state. Given an integer $n > 1$, decide if n is prime or composite. And we all know an algorithm. Divide n by each positive integer in turn. Either we find a proper factor, in which case we know that n is composite, or we do not, in which case we know that n is prime. For example, take $n = 269$. It is odd, so it has no even divisors. It is not a multiple of 3, so it has no divisor which is a multiple of 3. Continuing, we rule out 5, 7, 11, and 13. The next possibility, 17, has a square that is greater than 269, which means that if 269 were a multiple of 17, then it would also have to be a multiple of some number less than 17. Since we have ruled that out, we can stop our trial division at 13 and conclude that 269 is prime. (If we were actually carrying out the algorithm, we might try dividing 269 by 17, in which case we would discover that $269 = 15 \times 17 + 14$. At that point we would notice that the quotient, 15, is less than 17, which is what would tell us that 17^2 was greater than 269. Then we could stop.) In general, since a composite number n has a proper factor d with $d \leqslant \sqrt{n}$, one can give up on the trial dividing once one passes $\sqrt{n}$, at which point we know that n is prime.

This straightforward method is excellent for mental computation with small numbers, and for machine computation for somewhat larger numbers. But it scales poorly, in that if you double the number of digits of n, then the time for the worst case is squared; it is therefore an "exponential-time" algorithm. One might tolerate such an algorithm for twenty-digit inputs, but think how long it would take to establish the primality of a forty-digit number! And you can forget about numbers with hundreds or thousands of digits. The issue of how the running time of an algorithm scales when one goes to larger inputs is absolutely paramount in measuring one algorithm against another. In contrast to the exponential time it takes to use trial division to recognize primes, consider the problem of multiplying two numbers. The school method of multiplication is to take each digit of one number in turn and multiply it by the other number, forming a parallelogram array.

One then performs an addition to obtain the answer. If you now double the number of digits in each number, then the parallelogram becomes twice as large in each dimension, so the running time grows by a factor of about 4. Multiplication of two numbers is an example of a "polynomial time" algorithm; its running time scales by a constant factor when the input length is doubled.

One might then rephrase Gauss's call to arms as follows. Is there a polynomial-time algorithm that distinguishes prime numbers from composite numbers? Is there a polynomial-time algorithm that can produce a nontrivial factor of a composite number? It might not be apparent at this point that these are two different questions, since trial division does both. We will see, though, that it is convenient to separate them, as did Gauss.

Let us focus on recognizing primes. What we would like is a simply computed criterion that primes satisfy and composites do not, or vice versa. An old theorem of Wilson might just fit the bill. Note that $6! = 720$, which is just one less than a multiple of 7. Wilson's theorem asserts that if n is prime, then $(n-1)! \equiv -1$ (mod n). (The meaning of this and similar statements is explained in MODULAR ARITHMETIC [III.58].) This cannot hold when n is composite, for if p is a prime factor of n and is smaller than n, then it is a factor of $(n-1)!$, so it cannot possibly be a factor of $(n-1)! + 1$. Thus, we have an ironclad criterion for primality. However, the Wilson criterion does not meet the standard of being simply computed, since we know no especially rapid way of computing factorials modulo another number. For example, Wilson predicts that $268! \equiv -1$ (mod 269), as we have already seen that 269 is prime. But if we did not know this already, how in the world could we quickly find the remainder when $268!$ is divided by 269? We can work out the product $268!$ one factor at a time, but this would take many more steps than trying divisors up to 17. It is hard to prove that something *cannot* be done, and in fact there is no theorem that says we cannot compute $a! \bmod b$ in polynomial time. We do know some ways of speeding up the computation over the totally naive method, but all methods known so far take exponential time. So, Wilson's theorem initially seems promising, but in fact it is no help at all unless we can find a fast way to compute $a! \bmod b$.

How about FERMAT'S LITTLE THEOREM [III.58]? Note that $2^7 = 128$, which is 2 more than a multiple of 7. Or take $3^5 = 243$, which is 3 mod 5. Fermat's little theorem tells us that if n is prime and a is any integer, then $a^n \equiv a$ (mod n). If computing a large factorial modulo n is hard, perhaps it is also hard to compute a large power modulo n.

It cannot hurt to try it out for some moderate example to see if any ideas pop up. Take $a = 2$ and $n = 91$, so that we are trying to compute $2^{91} \bmod 91$. A powerful idea in mathematics is that of reduction. Can we reduce this computational problem to a smaller one? Notice that if we had already computed $2^{45} \bmod 91$, obtaining a remainder r_1, say, then $2^{91} \equiv 2r_1^2$ (mod 91). That is, it is just a short additional calculation to get to our goal, yet the power 45 is only half as big. How to continue is clear: we further reduce to the exponent 22, which is less than half of 45. If $2^{22} \bmod 91 = r_2$, then $2^{45} \equiv 2r_2^2$ (mod 91). And of course 2^{22} is the square of 2^{11}, and so on. It is not so hard to "automate" this procedure: the exponent sequence

$$1, \ 2, \ 5, \ 11, \ 22, \ 45, \ 91$$

can be read directly from the binary (base 2) representation of 91 as 1011011, since the above sequence in binary is

$$1, \ 10, \ 101, \ 1011, \ 10110, \ 101101, \ 1011011.$$

These are the initial strings from the left of 1011011. And it is plain that the transition from one term to the next is either the double or the double plus 1.

This procedure scales nicely. When the number of digits of n is doubled, so is the sequence of exponents, and the time it takes to get from one exponent to the next, being a modular multiplication, is multiplied by 4. (As with naive multiplication, naive divide-with-remainder also takes four times as long when the size of the problem is doubled.) Thus, the overall time is multiplied by 8, yielding a polynomial time method. We call this the "powermod" algorithm.

So, let us try to illustrate Fermat's little theorem, taking $a = 2$ and $n = 91$. Our sequence of powers is

$$2^1 \equiv 2, \qquad 2^2 \equiv 4, \qquad 2^5 \equiv 32, \qquad 2^{11} \equiv 46,$$
$$2^{22} \equiv 23, \qquad 2^{45} \equiv 57, \qquad 2^{91} \equiv 37,$$

where each congruence is modulo 91, and each term in the sequence is found by squaring the prior one mod 91 or squaring and multiplying by 2 mod 91.

Wait a second: does Fermat's little theorem not say that we are supposed to get 2 for the final residue? Well, yes, but this is guaranteed only if n is prime. And as you have probably already noticed, 91 is composite. In fact, the computation proves this.

Quite remarkably, here is an example of a computation that proves that n is composite, yet it does not reveal any nontrivial factorization!

You are invited to try out the powermod algorithm as above, but to change the base of the power from 2 to 3. The answer you should come to is that $3^{91} \equiv 3$ (mod 91): that is, the congruence for Fermat's little theorem holds. Since you already know that 91 is composite, I am sure you would not jump to the false conclusion that it is prime! So, as it stands, Fermat's little theorem can sometimes be used to recognize composites, but it cannot be used to recognize primes.

There are two interesting further points to be made regarding Fermat's little theorem. First, on the negative side, there are some composites, such as $n = 561$, where the Fermat congruence holds for *every* integer a. These numbers n are called *Carmichael numbers*, and unfortunately (from the point of view of testing primality) there are infinitely many of them, a result due to Alford, Granville, and me. But, on the positive side, if one were to choose randomly among all pairs a, n for which $a^n \equiv a$ (mod n), with $a < n$ and n bounded by a large number x, almost certainly (as x grows) you would choose a pair with n prime, a result of Erdős and myself.

It is possible to combine Fermat's little theorem with another elementary property of (odd) prime numbers. If n is an odd prime, there are exactly two solutions to the congruence $x^2 \equiv 1$ (mod n), namely ± 1. Actually, some composites have this property as well, but composites divisible by two different odd primes do not.

Now let us suppose that n is an odd number and that we wish to determine whether it is prime. Suppose that we pick some number a with $1 \leqslant a \leqslant n-1$ and discover that $a^{n-1} \equiv 1$ (mod n). If we set $x = a^{(n-1)/2}$, then $x^2 = a^{n-1} \equiv 1$ (mod n); so, by the simple property of primes just mentioned, if n is prime, then x must be ± 1. Therefore, if we calculate $a^{(n-1)/2}$ and discover that it is not congruent to ± 1 (mod n), then n must be composite.

Let us try this idea with $a = 2$, $n = 561$. We know already that $2^{560} \equiv 1$ (mod 561), so what is 2^{280} mod 561? This too turns out to be 1, so we have not shown that 561 is composite. However, we can go further, since now we know that 2^{140} is also a square root of 1 and computing this we find that $2^{140} \equiv 67$ (mod 561). So now we have found a square root of 1 that is not ± 1, which proves that 561 is composite. (Of course, for this particular number, it is obviously divisible by 3, so there was not really any mystery about whether it was prime or composite. But the method can be used in much less obvious cases.) In practice, there is no need to backtrack from a higher exponent to a smaller one. Indeed, in order to calculate 2^{560} mod 561 by the efficient method outlined earlier, one calculates the numbers 2^{140} and 2^{280} along the way, so that this generalization of the earlier test is both quicker and stronger.

Here is the general principle that we have illustrated. Suppose that n is an odd prime and let a be an integer not divisible by n. Write $n - 1 = 2^s t$, where t is odd. Then

$$\text{either } a^t \equiv 1 \pmod{n} \quad \text{or} \quad a^{2^i t} \equiv -1 \pmod{n}$$

for some $i = 0, 1, \ldots, s - 1$. Call this the *strong Fermat congruence*. The wonderful thing here is that, as proved independently by Monier and Rabin, there is no analogue of a Carmichael number. They showed that if n is an odd composite, then the strong Fermat congruence fails for at least three quarters of the choices for a with $1 \leqslant a \leqslant n - 1$.

If you want only to be able to distinguish between primes and composites in practice, and you do not insist on proof, then you have read enough. Namely, given a large odd number n, choose twenty values of a at random from $[1, n - 1]$, and begin trying to verify the strong Fermat congruence with these bases a. If it should ever fail, you may stop: the number n must be composite. And if the strong Fermat congruence holds, we might surmise that n is actually prime. Indeed, if n were composite, the Monier–Rabin theorem says that the chance that the strong Fermat congruence would hold for twenty random bases is at most 4^{-20}, which is less than one chance in a trillion. Thus we have a remarkable *probabilistic* test for primality. If it tells us that n is composite, then we know for sure that n is composite; if it tells us that n is prime, then the chances that n is not prime are so small as to be more or less negligible.

If three quarters of the numbers a in $[1, n - 1]$ provide the key to an easily checkable proof that the odd composite number n is indeed composite, surely it should not be so hard to find just one! How about checking small numbers a, in order, until one is found? Excellent, but when do we stop? Let us think about this for a moment. We have given up the power of randomness and are forcing ourselves to choose sequentially among small numbers for the trial bases a. Can we argue heuristically that they continue to behave as if they were random choices? Well, there *are* some connections among them. For example, if taking $a = 2$ does not result in a proof that n is composite, then neither will taking any power of 2. It is theoretically possible for 2 and 3 not to give proofs that n is composite but for 6 to work just fine, but this turns out not to be very common. So let us amend the heuristic and assume that we have independence for *prime* values of a. Up to $\log n \log \log n$ there are about $\log n$ primes (via the PRIME NUMBER THEOREM [V.26] discussed later in this article); so, heuristically, the probability that n is composite, but that none of these primes help us to prove it, is about $4^{-\log n} < n^{-4/3}$. Since the infinite sum $\sum n^{-4/3}$ converges, perhaps a stopping point of $\log n \log \log n$ is sufficient, at least for large n.

Miller was able to prove the slightly weaker result that a stopping point of $c(\log n)^2$ is adequate, but his proof assumes a generalization of the RIEMANN HYPOTHESIS [V.26]. (We discuss the Riemann hypothesis below; the generalization that Miller assumes is beyond the scope of this article.) In further work, Bach was able to show that we may take $c = 2$ in this last result. Summarizing, if this generalized Riemann hypothesis holds, and if the strong Fermat congruence holds for every positive integer $a \leqslant 2(\log n)^2$, then n is prime. So, provided that a famous unproved hypothesis in another field of mathematics is correct, one can decide in polynomial time, via a deterministic algorithm, whether n is prime or composite. (It has been tempting to *use* this conditional test, for if it should ever lie to you and tell you that a particular composite number is prime, then this failure—if you were able to detect it—would be a disproof of one of the most famous conjectures in mathematics. Perhaps this is not too disastrous a failure!)

After Miller's test in the 1970s, the question continually challenging us was whether it is possible to test for primality in polynomial time without assuming unproved hypotheses. Recently, Agrawal et al. (2004) answered this question with a resounding yes. Their idea begins with a combination of the binomial theorem and Fermat's little theorem. Given an integer a, consider the polynomial $(x + a)^n$ and expand it in the usual way through the binomial theorem. Each intermediate term between the leading x^n and the trailing a^n has the coefficient $n!/(j!(n - j)!)$ for some j between 1 and $n - 1$. If n is prime, then this coefficient, which is an integer, is divisible by n because n appears as a factor in the numerator that is not canceled by any factors in the denominator. That is, the coefficient is

0 (mod n). For example, $(x + 1)^7$ is equal to

$$x^7 + 7x^6 + 21x^5 + 35x^4 + 35x^3 + 21x^2 + 7x + 1,$$

and we see each internal coefficient is a multiple of 7. Thus, we have $(x + 1)^7 \equiv x^7 + 1 \pmod 7$. (Two polynomials are congruent mod n if corresponding coefficients are congruent mod n.) In general, if n is prime and a is any integer, then via this binomial-theorem idea and Fermat's little theorem we have

$$(x + a)^n \equiv x^n + a^n \equiv x^n + a \pmod n.$$

It is an easy exercise to show that this congruence in the simple case $a = 1$ is actually equivalent to primality. But as with the Wilson criterion we know no way of quickly verifying that all these coefficients are indeed divisible by n.

However, one can do more with polynomials than raise them to powers. We can also divide one polynomial by another to find a quotient and a remainder, just as we do with integers. It makes sense, for example, to say that $g(x) \equiv h(x) \pmod{f(x)}$, meaning that $g(x)$ and $h(x)$ leave the same remainder when divided by $f(x)$. We will write $g(x) \equiv h(x) \pmod{n, f(x)}$ if the remainders upon division by $f(x)$ are congruent mod n. As with the powermod algorithm for integer congruences, we can quickly compute $g(x)^n \pmod{n, f(x)}$, provided the degree of $f(x)$ is not too big. This is exactly what Agrawal et al. propose. They have an auxiliary polynomial $f(x)$ of not-too-high degree such that, if

$$(x + a)^n \equiv x^n + a \pmod{n, f(x)}$$

for each $a = 1, 2, \ldots, B$, for a not-too-high bound B, then n must be in a set that contains the primes and certain composites that are easily recognized as composites. (Not all composites are hard to recognize as such, e.g., any number with a small prime factor is easy to recognize.) These ideas put together form the primality test of Agrawal et al. To give the argument in full detail one has to specify the auxiliary polynomial $f(x)$ that is used and what the bound B is, and one has to prove rigorously that it is exactly the primes which pass the test.

Agrawal et al. (2004) show that the auxiliary polynomial $f(x)$ can be taken to be the beautifully simple $x^r - 1$, with an elementary upper bound for r of about $(\log n)^5$. Doing this leads to a time bound of about $(\log n)^{10.5}$ for the algorithm. Using a numerically ineffective tool, they bring the time bound down to $(\log n)^{7.5}$. Recently, Lenstra and I presented a not-so-simple but numerically effective method of bringing the exponent on $\log n$ down to 6. We did this by expanding the set of polynomials used beyond those of the form $x^r - 1$: in particular we used polynomials that are related to Gauss's famous algorithm for construction of certain regular n-gons with straightedge and compass (see ALGEBRAIC NUMBERS [IV.1 §13]). It was indeed satisfying to us to bring in a famous tool of Gauss to say something about his problem of distinguishing prime numbers from composite numbers.

Are the new polynomial-time primality tests good in practice? So far, the answer is no, the competition is just too tough. For example, using the arithmetic of ELLIPTIC CURVES [III.21] we can come up with bona fide proofs of primality for huge numbers. This algorithm is conjectured to run in polynomial time but we have not even proved that it always terminates. If, at the end of the day, or in this case the end of the run, we have a legitimate proof, then perhaps we can tolerate the situation of not being sure that it would work out when we started! The method, pioneered by Atkin and Morain, has recently proved the primality of a number that has over 20 000 decimal digits, and is not of some special form such as $2^n - 1$ that makes testing for primality easier. The record for the new breed of polynomial-time tests is a measly 300 digits.

For numbers of certain special forms there are much faster primality tests. Mersenne primes comprise the most famous of these forms; these are primes that are 1 less than a power of 2. It is suspected that there are infinitely many examples, but we seem to be very far from a proof of this. Just forty-three Mersenne primes are known, the record example being $2^{30\,402\,457} - 1$, a prime with more than 9.15 million decimal digits.

For much more on primality testing, and for references to various other sources, see Crandall and Pomerance (2005).

3 Factoring Composite Numbers

Compared with what we know about testing primality, our ability to factor large numbers is still in the dark ages. In fact this imbalance between the two problems forms the bulwark for the security of electronic commerce on the Internet. (See MATHEMATICS AND CRYPTOGRAPHY [VII.7] for an account of why.) This is a very important application of mathematics, but also an odd one, and not something to brag about, since it depends on the inability of mathematicians to efficiently solve a basic problem!

Nevertheless, we do have our tricks. Part of the landscape is EUCLID'S ALGORITHM [III.22] for computing the greatest common divisor (GCD) of two numbers. One might naively think that, to find the GCD of two positive integers m and n, one should find all of their divisors and pick the largest one common to the two. But Euclid's algorithm is much more efficient: the number of arithmetic steps is bounded by the logarithm of the smaller number, so not only does it run in polynomial time, it is in fact quite speedy.

So, if we can build up a special number m that may be likely to have a nontrivial factor in common with n, we can use Euclid's algorithm to discover this factor. For example, Pollard and Strassen (independently) used this idea, together with fast subroutines for multiplication and polynomial evaluation, to enhance the trial division method discussed in the last section. Somewhat miraculously, one can take the integers up to $n^{1/2}$, break them into $n^{1/4}$ subintervals of length $n^{1/4}$, and for each subinterval calculate the GCD of n with the product of all the integers in the subinterval, spending only about $n^{1/4}$ elementary steps in total. If n is composite, then at least one GCD will be larger than 1, and then a search over the first such subinterval will locate a nontrivial factor of n. To date, this algorithm is the fastest rigorous and deterministic method of factoring that we know.

Most practical factoring algorithms are based on unproved but reasonable-seeming hypotheses about the natural numbers. Although we may not know how to prove rigorously that these methods will always produce a factorization, or do so quickly, in practice they do. This situation resembles the experimental sciences, where hypotheses are tested against experiments. Our experience with certain factoring algorithms is now so overwhelming that a scientist might claim that a physical law is involved. As mathematicians, we still search for proof, but fortunately the numbers we factor do not feel the need to wait for us.

I often mention a contest problem from my high school years: factor 8051. The trick is to notice that $8051 = 90^2 - 7^2 = (90 - 7)(90 + 7)$, from which the factorization $83 \cdot 97$ can be read off. In fact every odd composite can be factored as the difference of two squares, an idea that goes back to FERMAT [VI.12]. Indeed, if n has the nontrivial factorization ab, then let $u = \frac{1}{2}(a + b)$ and $v = \frac{1}{2}(a - b)$, so that $n = u^2 - v^2$, and $a = u + v$, $b = u - v$. This method works very well if n has a divisor very close to $n^{1/2}$, as $n = 8051$ does, but in the worst case, the Fermat method is slower than trial division.

My quadratic sieve method (which follows work of Kraitchik, Brillhart–Morrison, and Schroeppel) tries to efficiently extend Fermat's idea to all odd composites. For example, take $n = 1649$. We start just above $n^{1/2}$ with $j = 41$, and consider the numbers $j^2 - 1649$. As j runs, we will eventually hit a value where $j^2 - 1649$ is a square, and so be able to use Fermat's method. Let's try it:

$$41^2 - 1649 = 32,$$
$$42^2 - 1649 = 115,$$
$$43^2 - 1649 = 200,$$
$$\vdots$$

Well, no squares yet, which is not surprising, since the Fermat method is often very poor. But wait, do the first and third lines not multiply together to give a square? Yes they do, $32 \cdot 200 = 80^2$. So, multiplying the first and third lines, and treating them as congruences mod 1649, we have

$$(41 \cdot 43)^2 \equiv 80^2 \pmod{1649}.$$

That is, we have a pair u, v with $u^2 \equiv v^2 \pmod{1649}$. This is not quite the same as having $u^2 - v^2 = 1649$, but we do have 1649 a divisor of $u^2 - v^2 = (u - v)(u + v)$. Now maybe 1649 divides one of these factors, but if it does not, then it is split between them, and so a computation of the GCD of $u - v$ (or $u + v$) with 1649 will reveal a proper factor. Now $v = 80$ and $u = 41 \cdot 43 \equiv 114 \pmod{1649}$, and so we see instantly that $u \not\equiv \pm v \pmod{1649}$, so we are in business. The GCD of $114 - 80 = 34$ with 1649 is 17. Dividing, we see that $1649 = 17 \cdot 97$, and we are done.

Can we generalize this? In trying to factor $n = 1649$ we considered consecutive values of the quadratic polynomial $f(j) = j^2 - n$ for j starting just above $\sqrt{n}$, and viewed these as congruences $j^2 \equiv f(j) \pmod{n}$. Then we found a set $\mathcal{M}$ of numbers j with $\prod_{j \in \mathcal{M}} f(j)$ equal to a square, say v^2. We then let $u = \prod_{j \in \mathcal{M}} j$, so that $u^2 \equiv v^2 \pmod{n}$. Since $u \not\equiv \pm v \pmod{n}$, we could split n via the GCD of $u - v$ and n.

There is another lesson that we can learn from our small example with $n = 1649$. We used 32 and 200 to form our square, but we ignored 115. If we had thought about it, we might have noticed from the start that 32 and 200 were more likely to be useful than 115. The reason is that 32 and 200 are *smooth numbers* (meaning that they have only small prime factors), while 115 is not smooth, having the relatively large prime factor 23. Say you have $k + 1$ positive integers that involve

in their prime factorizations only the first k primes. It is an easy theorem that some nonempty subset of these numbers has product a square. The proof has us associate with each of these numbers, which can be written in the form $p_1^{a_1} p_2^{a_2} \cdots p_k^{a_k}$, an *exponent vector* $(a_1, a_2, \ldots, a_k)$. Since squares are detected by all even exponents, we really only care whether the exponents a_i are odd or even. Thus, we think of these vectors as having coordinates 0 and 1, and when we add them (which corresponds to multiplying the underlying numbers), we do so mod 2. Since we have $k + 1$ vectors, each with only k coordinates, an easy matrix calculation leads quickly to a nonempty subset that adds up to the 0-vector. The product of the corresponding integers is then a square.

In our toy example with $n = 1649$, the first and third numbers, which are $32 = 2^5 3^0 5^0$ and $200 = 2^3 3^0 5^2$, have exponent vectors $(5, 0, 0)$ and $(3, 0, 2)$, which reduce to $(1, 0, 0)$ and $(1, 0, 0)$, so we see that the sum of them is $(0, 0, 0)$, which indicates that we have a square. We were lucky that we could make do with just two vectors, instead of the four that the above argument shows would be sufficient.

In general with the quadratic sieve, one finds smooth numbers in the sequence $j^2 - n$, forms the exponent vectors mod 2, and then uses a matrix to find a nonempty subset which adds up to the 0-vector, which then corresponds to a set $\mathcal{M}$ for which $\prod_{j \in \mathcal{M}} f(j)$ is a square.

In addition, the "sieve" in the quadratic sieve comes in with the search for smooth values of $f(j) = j^2 - n$. These numbers are the consecutive values of a (quadratic) polynomial, so those divisible by a given prime can be found in regular places in the sequence. For example, in our illustration, $j^2 - 1649$ is divisible by 5 precisely when $j \equiv 2$ or $3 \pmod 5$. A sieve very much like the sieve of Eratosthenes can then be used to efficiently find the special numbers j where $j^2 - n$ is smooth. A key issue, though, is how smooth a value $f(j)$ has to be for us to decide to accept it. If we choose a smaller bound for the primes involved, we do not have to find all that many of them to use the matrix method. But such very smooth values might be very rare. If we use a larger bound for the primes involved, then smooth values of $f(j)$ may be more common, but we will need many of them. Somewhere between smaller and larger is just right! In order to make the choice, it would help to know how frequently values of an irreducible quadratic polynomial are smooth. Unfortunately, we do not have a theorem that tells us, but we can still make a good choice by assuming that this frequency is about that for a random number of the same size, an assumption that is probably correct even if it is hard to prove.

Finally, note that if the final GCD yields only a trivial factor with n, one can continue just a bit longer and find more linear dependencies, each with a fresh chance at splitting n.

These thoughts lead us to a time bound of about

$$\exp\left(\sqrt{\log n \, \log \log n}\,\right)$$

for the quadratic sieve to factor n. Instead of being exponential in the number of digits of n, as with trial division, this is exponential in about the square root of the number of digits of n. This is certainly a huge improvement, but it is still a far cry from polynomial time.

Lenstra and I actually have a rigorous random factoring method with the same time complexity as that above for the quadratic sieve. (It is random in the sense that a coin is flipped at various junctures, and decisions on what to do next depend on the outcomes of these flips. Through this process, we expect to get a bona fide factorization within the advertised time bound.) However, the method is not so computer practical, and if you had to choose in practice between the two, then you should go with the nonrigorous quadratic sieve. A triumph for the quadratic sieve was the 1994 factorization of the 129-digit RSA cryptographic challenge first published in Martin Gardner's column in *Scientific American* in 1977.

The *number field sieve*, which is another sieve-based factoring algorithm, was discovered in the late 1980s by Pollard for integers close to powers, and later developed by Buhler, Lenstra, and me for general integers. The method is similar in spirit to the quadratic sieve, but assembles its squares from the product of certain sets of algebraic integers. The number field sieve has a conjectured time complexity of the type

$$\exp(c(\log n)^{1/3}(\log \log n)^{2/3}),$$

for a value of c slightly below 2. For composite numbers beyond 100 digits or so that have no small prime factor, it is the method of choice, with the current record being 200 decimal digits.

The sieve-based factorization methods share the property that if you use them, then all composite numbers of about the same size are equally hard to factor. For instance, factoring n will be about as difficult if n is a product of five primes each roughly near the fifth root of n as it will be if n is a product of two primes roughly near the square root of n. This is quite unlike trial division, which is happiest when there is a small prime factor. We will now describe a famous factorization method due to Lenstra that detects small prime factors before large ones, and beyond baby cases is much superior to trial dividing. This is his *elliptic curve method*.

Just as the quadratic sieve searches for a number m with a nontrivial GCD with n, so does the elliptic curve method. But where the quadratic sieve painstakingly builds up to a successful m from many small successes, the elliptic curve method hopes to hit upon m with essentially one lucky choice.

Choosing random numbers m and testing their GCD with n can also have instant success, but you can well imagine that if n has no small prime factors, then the expected time for success would be enormous. Instead, the elliptic curve method involves considerably more cleverness.

Consider first the "$p - 1$ method" of Pollard. Suppose you have a number n you wish to factor and a certain large number k. Unbeknownst to you, n has a prime factor p with $p - 1$ a divisor of k, and another prime factor q with $q - 1$ not a divisor of k. You can use this imbalance to split n. First of all, by Fermat's little theorem there are many numbers u with $u^k \equiv 1 \pmod{p}$ and $u^k \not\equiv 1 \pmod{q}$. Say you have one of these, and let m be $u^k - 1$ reduced mod n. Then the GCD of m and n is a nontrivial factor of n; it is divisible by p but not by q. Pollard suggests taking k as the least common multiple of the integers to some moderate bound so that it has many divisors and perhaps a decent chance that it is divisible by $p - 1$. The best case of Pollard's method is when n has a prime factor p with $p - 1$ smooth (has all small prime factors—see the quadratic sieve discussion above). But if n has no prime factors p with $p - 1$ smooth, Pollard's method fares poorly.

What is going on here is that corresponding to the prime p we have the multiplicative GROUP [I.3 §2.1] of the $p - 1$ nonzero residues mod p. Furthermore, when doing arithmetic mod n with numbers relatively prime to n, we are, whether we realize it or not, doing arithmetic in this group. We are exploiting the fact that u^k is the group identity mod p, but not mod q.

Lenstra had the brilliant idea of using the Pollard method in the context of elliptic curve groups. There are many elliptic curve groups associated with the prime p, and therefore many chances to hit upon one where the number of elements is smooth. Of great importance here are theorems of Hasse and Deuring. An ELLIPTIC CURVE [III.21] mod p (for $p > 3$) can be taken as the set of solutions to the congruence $y^2 \equiv x^3 + ax + b \pmod{p}$, for given integers a, b with the property that $x^3 + ax + b$ does not have repeated roots mod p. There is one additional "point at infinity" thrown in (see below). A fairly simple addition law (but not as simple as adding coordinatewise!) makes the elliptic curve into a group, with the point at infinity as the identity (see RATIONAL POINTS ON CURVES AND THE MORDELL CONJECTURE [V.29]). Hasse, in a result later generalized by WEIL [VI.93] with his famous proof of the "Riemann hypothesis for curves," showed us that the number of elements in the elliptic curve group always lies between $p + 1 - 2\sqrt{p}$ and $p + 1 + 2\sqrt{p}$ (see THE WEIL CONJECTURES [V.35])). And Deuring proved that every number in this range is indeed associated with some elliptic curve mod p.

Say we randomly choose integers x_1, y_1, a, and then choose b so that y_1^2 is congruent to $x_1^3 + ax_1 + b$ (mod n). This gives us the curve with coefficients a, b and a point $P = (x_1, y_1)$ on the curve. One can then mimic the Pollard strategy, with a number k as before with many divisors, and with the point P playing the role of u. Let kP denote the k-fold sum of P added to itself using elliptic curve addition. If kP is the point at infinity on the curve considered mod p (which it will be if the number of points on the curve is a divisor of k), but not on the curve considered mod q, then this gives us a number m whose GCD with n is divisible by p and not by q. We will have factored n.

To see where m comes from it is convenient to consider the curve projectively: we take solutions (x, y, z) of the congruence $y^2 z \equiv x^3 + axz^2 + bz^3 \pmod{p}$. The triple (cx, cy, cz) when $c \neq 0$ is considered to be the same as (x, y, z). The mysterious point at infinity is now demystified; it is just $(0, 1, 0)$. And our point P is $(x_1, y_1, 1)$. (This is the mod p version of classical PROJECTIVE GEOMETRY [I.3 §6.7].) Say we work mod n and compute the point $kP = (x_k, y_k, z_k)$. Then the candidate for the number m is just z_k. Indeed, if kP is the point at infinity mod p, then $z_k \equiv 0 \pmod{p}$, and if it is not the point at infinity mod q, then $z_k \not\equiv 0 \pmod{q}$.

When Pollard's $p - 1$ method fails, our only recourse is to raise k or give up. With the elliptic curve method, if things do not work for our randomly chosen curve, we can pick another. Corresponding to the hidden prime p in n, we are actually picking new elliptic curve groups mod p, and so gaining a fresh chance for the number of elements in the group to be smooth. The elliptic curve

method has been quite successful in factoring numbers which have a prime factor up to about fifty decimal digits, and occasionally even somewhat larger primes have been discovered.

We conjecture that the expected time for the elliptic curve method to find the least prime factor p of n is about

$$\exp\left(\sqrt{2 \log p \, \log \log p}\,\right)$$

arithmetic operations mod n. What is holding us back from proving this conjecture is not lack of knowledge about elliptic curves, but rather lack of knowledge of the distribution of smooth numbers.

For more on these and other factorization methods, the reader is referred to Crandall and Pomerance (2005).

4   The Riemann Hypothesis and the Distribution of the Primes

As a teenager looking at a modest table of primes, Gauss conjectured that their frequency decays logarithmically and that $\mathrm{li}(x) = \int_2^x (1/\log t) \, \mathrm{d}t$ should be a good approximation for $\pi(x)$, the number of primes between 1 and x. Sixty years later, RIEMANN [VI.49] showed how Gauss's conjecture can be proved if one assumes that the Riemann zeta function $\zeta(s) = \sum_n n^{-s}$ has no zeros in the complex half-plane where the real part of s is greater than $\frac{1}{2}$. The series for $\zeta(s)$ converges only for $\mathrm{Re}\, s > 1$, but it may be analytically continued to $\mathrm{Re}\, s > 0$, with a simple pole at $s = 1$. (For a brief description of the process of analytic continuation, see SOME FUNDAMENTAL MATHEMATICAL DEFINITIONS [I.3 §5.6].) This continuation may be seen quite concretely via the identity $\zeta(s) = s/(s-1) - s \int_1^\infty \{x\} x^{-s-1} \, \mathrm{d}x$, with $\{x\}$ the fractional part of x (so that $\{x\} = x - [x]$): note that this integral converges quite nicely in the half-plane $\mathrm{Re}\, s > 0$. In fact, via Riemann's functional equation mentioned below, $\zeta(s)$ can be continued to a meromorphic function in the whole complex plane, with the single pole at $s = 1$.

The assertion that $\zeta(s) \neq 0$ for $\mathrm{Re}\, s > \frac{1}{2}$ is known as the RIEMANN HYPOTHESIS [IV.2 §3]; arguably it is the most famous unsolved problem in mathematics. Though HADAMARD [VI.65] and DE LA VALLÉE POUSSIN [VI.67] were able in 1896 to prove (independently) a weak form of Gauss's conjecture known as the PRIME NUMBER THEOREM [V.26], the apparent breathtaking strength of the approximation $\mathrm{li}(x)$ to $\pi(x)$ is uncanny. For example, take $x = 10^{22}$. We have

$$\pi(10^{22}) = 201\,467\,286\,689\,315\,906\,290$$

exactly, and, to the nearest integer, we have

$$\mathrm{li}(10^{22}) \approx 201\,467\,286\,691\,248\,261\,497.$$

As you can plainly see, Gauss's guess is right on the money!

The numerical computation of $\mathrm{li}(x)$ is simple via numerical methods for integration, and it is directly obtainable in various mathematics computing packages. However, the computation of $\pi(10^{22})$ (due to Gourdon) is far from trivial. It would be far too laborious to count these approximately 2×10^{20} primes one by one, so how are they counted? In fact, we have various combinatorial tricks to count without listing everything. For example, one does not need to count one by one to see that there are exactly $2[10^{22}/6] + 1$ integers in the interval from 1 to 10^{22} that are relatively prime to 6. Rather, one thinks of these numbers grouped in blocks of six, with two in each block coprime to 6. (The "+1" comes from the partial block at the end.) Building on early ideas of Meissel and Lehmer, Lagarias, Miller, and Odlyzko presented an elegant combinatorial method for computing $\pi(x)$ that takes about $x^{2/3}$ elementary steps. The method was refined by Deléglise and Rivat, and then Gourdon found a way to distribute the computation to many computers.

From work of von Koch, and later Schoenfeld, we know that the Riemann hypothesis is *equivalent* to the assertion that

$$|\pi(x) - \mathrm{li}(x)| < \sqrt{x}\log x \qquad (1)$$

for all $x \geqslant 3$ (see Crandall and Pomerance 2005, exercise 1.37). Thus, the mammoth calculation of $\pi(10^{22})$ might be viewed as computational evidence for the Riemann hypothesis—in fact, if the count had turned out to violate (1), we would have had a disproof.

It may not be obvious what (1) has to do with the location of the zeros of $\zeta(s)$. To understand the connection, let us first dismiss the so-called "trivial" zeros, which occur at each negative even integer. The nontrivial zeros ρ are known to be infinite in number, and, as mentioned above, are conjectured to satisfy $\mathrm{Re}\,\rho \leqslant \frac{1}{2}$. There are certain symmetries among these zeros: indeed, if ρ is a zero, then so are $\bar{\rho}$, $1 - \rho$, and $1 - \bar{\rho}$. Therefore, the Riemann hypothesis is the assertion that every nontrivial zero has real part equal to $\frac{1}{2}$. (The symmetry with ρ and $1 - \rho$, which follows from Riemann's functional equation $\zeta(1 - s) = 2(2\pi)^{-s}\cos(\frac{1}{2}\pi s)\Gamma(s)\zeta(s)$, perhaps provides some heuristic support for the Riemann hypothesis.)

The connection to prime numbers begins with THE FUNDAMENTAL THEOREM OF ARITHMETIC [V.14], which yields the identity

$$\zeta(s) = \sum_{n=1}^{\infty} n^{-s} = \prod_{p \text{ prime}} \sum_{j=0}^{\infty} p^{-js}$$
$$= \prod_{p \text{ prime}} (1 - p^{-s})^{-1},$$

a product that converges when $\mathrm{Re}\,s > 1$. Thus, taking the logarithmic derivative (that is, taking the logarithm of both sides and then differentiating), we have

$$\frac{\zeta'(s)}{\zeta(s)} = -\sum_{p \text{ prime}} \frac{\log p}{p^s - 1} = -\sum_{p \text{ prime}} \sum_{j=1}^{\infty} \frac{\log p}{p^{js}}.$$

That is, if we define $\Lambda(n)$ to be $\log p$ if $n = p^j$ for a prime p and an integer $j \geqslant 1$, and $\Lambda(n) = 0$ if n is not of this form, then we have the identity

$$\sum_{n=1}^{\infty} \frac{\Lambda(n)}{n^s} = -\frac{\zeta'(s)}{\zeta(s)}.$$

Through various relatively routine calculations, one can then relate the function

$$\psi(x) = \sum_{n \leqslant x} \Lambda(n)$$

to the residues at the poles of ζ'/ζ, which correspond to the zeros (and single pole) of ζ. In fact, as Riemann showed, we have the following beautiful formula:

$$\psi(x) = x - \sum_{\rho} \frac{x^{\rho}}{\rho} - \log(2\pi) - \tfrac{1}{2}\log(1 - x^{-2})$$

if x itself is not a prime or prime power, and where the sum over the nontrivial zeros ρ of ζ is to be understood in the symmetric sense where we sum over those ρ with $|\mathrm{Im}\,\rho| < T$ and let $T \to \infty$. Through elementary manipulations, an understanding of the function $\psi(x)$ readily gives an equivalent understanding of $\pi(x)$, and it should be clear now that $\psi(x)$ is intimately connected to the nontrivial zeros ρ of ζ.

The function $\psi(x)$ defined above has a simple interpretation. It is the logarithm of the least common multiple of the integers in the interval $[1, x]$. As with (1) we have an elementary translation of the Riemann hypothesis: it is equivalent to the assertion that

$$|\psi(x) - x| < \sqrt{x}\log^2 x$$

for all $x \geqslant 3$. This inequality involves only the elementary concepts of least common multiple, natural logarithm, absolute value, and square root, yet it is equivalent to the Riemann hypothesis.

A number of nontrivial zeros ρ of $\zeta(s)$ have actually been calculated and it has been verified that they lie on the line $\operatorname{Re} s = \frac{1}{2}$. One might wonder how someone can computationally verify that a complex number ρ has $\operatorname{Re} \rho = \frac{1}{2}$. For example, suppose that we are carrying calculations to (an unrealistically large) 10^{10} significant digits, and suppose we come across a zero with real part $\frac{1}{2} + 10^{-10^{100}}$. It would be far beyond the precision of the calculation to be able to distinguish this number from $\frac{1}{2}$ itself. Nevertheless, we do have a method for seeing if particular zeros ρ satisfy $\operatorname{Re} \rho = \frac{1}{2}$. There are two ideas involved, one of which comes from elementary calculus. If we have a continuous real-valued function $f(x)$ defined on the real numbers, we can sometimes use the intermediate value theorem to count zeros. For example, say $f(1) > 0$, $f(1.7) < 0$, $f(2.3) > 0$. Then we know for sure that f has at least one zero between 1 and 1.7, and at least one zero between 1.7 and 2.3. If we know for other reasons that f has exactly two zeros, then we have accounted for both of them. To locate zeros of the complex function $\zeta(s)$, a real-valued function $g(t)$ is constructed with the property that $\zeta(\frac{1}{2} + it) = 0$ if and only if $g(t) = 0$. By looking at sign changes for $g(t)$ for $0 < t < T$, we can get a *lower bound* for the number of zeros ρ of ζ with $\operatorname{Re} \rho = \frac{1}{2}$ and $0 < \operatorname{Im} \rho < T$. In addition, we can use the so-called *argument principle* from complex analysis to count the *exact number* of zeros with $0 < \operatorname{Im} \rho < T$. If we are lucky and this exact count is equal to our lower bound, then we have accounted for all of ζ's zeros here, showing that they all have real part $\frac{1}{2}$ (and, in addition, that they are all simple zeros). If the counts did not match, it would not be a disproof of the Riemann hypothesis, but certainly it would indicate a region where we should be checking the data more closely. So far, whenever we have tried this approach, the counts have matched, though sometimes we have been forced to evaluate $g(t)$ at very closely spaced points.

The first few nontrivial zeros were computed by Riemann himself. The famous cryptographer and early computer scientist ALAN TURING [VI.94] also computed some zeta zeros. The current record for this kind of calculation is held by Gourdon, who has shown that the first 10^{13} zeta zeros with positive imaginary part all have real part equal to $\frac{1}{2}$, as predicted by Riemann. Gourdon's method is a modification of that pioneered by Odlyzko and Schönhage (1988), who ushered in the modern age of zeta-zero calculations.

Explicit zeta-function calculations can lead to highly useful explicit prime number estimates. If p_n is the nth prime, then the prime number theorem implies that $p_n \sim n \log n$ as $n \to \infty$. Actually, there is a sec-

ondary term of order $n \log \log n$, and so for all sufficiently large n, we have $p_n > n \log n$. By using explicit zeta estimates, Rosser was able to put a numerical bound on the "sufficiently large" in this statement, and then, by checking small cases, was able to prove that in fact $p_n > n \log n$ for every n. The paper of Rosser and Schoenfeld (1962) is filled with highly useful and numerically explicit inequalities of this kind.

Let us imagine for a moment that the Riemann hypothesis had been proved. Mathematics is never "used up," as there is always that next problem around the bend. Even if we know that all of zeta's nontrivial zeros lie on the line $\operatorname{Im} s = \frac{1}{2}$, we can still ask how they are distributed on this line. We have a fairly concise understanding of how many zeros there should be up to a given height T. In fact, as already found by Riemann, this count is about $(1/2\pi)T \log T$. Thus, on average, the zeros would tend to get closer and closer with about $(1/2\pi) \log T$ of them in a unit interval near height T.

This tells us the average distance, or spacing, between one zeta zero and the next, but there is much more that one can ask about how these spacings are distributed. In order to discuss this question, it is very convenient to "normalize" the spacings, so that the average (normalized) gap between consecutive zeros is 1. By Riemann's result, together with our assumption of the Riemann hypothesis, this can be done if we multiply a gap near T by $(1/2\pi) \log T$, or, equivalently, if for each zero ρ we replace its imaginary part $t = \operatorname{Im} \rho$ by $(1/2\pi)t \log t$. In this way we arrive at a sequence $\delta_1, \delta_2, \ldots$ of normalized gaps between consecutive zeros, which on average are about 1.

Checking numerically, we see that some δ_n are large, with others close to 0; it is just the average that is 1. Mathematics is well equipped to study random phenomena, and we have names for various PROBABILITY DISTRIBUTIONS [III.71], such as Poisson, Gaussian, etc. Is this what is happening here? These zeta zeros are not random at all, but perhaps thinking in terms of randomness has promise.

In the early twentieth century, HILBERT [VI.63] and Pólya suggested that the zeros of the zeta function might correspond to the EIGENVALUES [I.3 §4.3] of some OPERATOR [III.50]. Now this is provocative! But what operator? Some fifty years later in a now famous conversation between Dyson and Montgomery at the Institute for Advanced Study, it was conjectured that the nontrivial zeros behave like the eigenvalues of

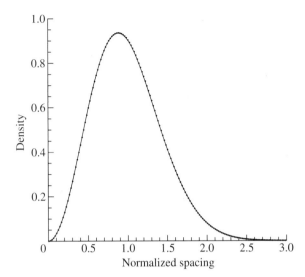

Figure 1 Nearest-neighbor spacing
and the Gaudin distribution.

a random matrix from the so-called *Gaussian unitary ensemble*. This conjecture, now known as the GUE conjecture, can be numerically tested in various ways. Odlyzko has done this, and found persuasive evidence for the conjecture: the higher the batches of zeros one looks at, the more closely their distribution corresponds to what the GUE conjecture predicts.

For example, take the $1\,041\,417\,089$ numbers δ_n with n starting at $10^{23} + 17\,368\,588\,794$. (The imaginary parts of these zeros are around 1.3×10^{22}.) For each interval $(j/100, (j+1)/100]$ we can compute the proportion of these normalized gaps that lie in this interval, and plot it. If we were dealing with eigenvalues from a random matrix from the GUE, we would expect these statistics to converge to a certain distribution known as the *Gaudin distribution* (for which there is no closed formula, but which is easily computable). Odlyzko has kindly supplied me with the graph in figure 1, which plots the Gaudin distribution against the data just described (but leaves out every second data point to avoid clutter). Like pearls on a necklace! The fit is absolutely remarkable.

The vital interplay of thought experiments and numerical computation has taken us to what we feel is a deeper understanding of the zeta function. But where do we go next? The GUE conjecture suggests a connection to random matrix theory, and pursuing further connections seems promising to many. It may be that random matrix theory will allow us only to formulate

great conjectures about the zeta function, and will not lead to great theorems. But then again, who can deny the power of a glimpse at the truth? We await the next chapter in this development.

5   Diophantine Equations and the ABC Conjecture

Let us move now from the Riemann hypothesis to FERMAT'S LAST THEOREM [V.10]. Until the last decade it too was one of the most famous unsolved problems in mathematics, once even having a mention on an episode of *Star Trek*. The assertion is that the equation $x^n + y^n = z^n$ has no solutions in positive integers x, y, z, n, where $n \geqslant 3$. This conjecture had remained unproved for three and a half centuries until Andrew Wiles published a proof in 1995. In addition, perhaps more important than the solution of this particular Diophantine equation (that is, an equation where the unknowns are restricted to the integers), the centuries-long quest for a proof helped establish the field of ALGEBRAIC NUMBER THEORY [IV.1]. And the proof itself established a long-sought and wonderful connection between MODULAR FORMS [III.59] and elliptic curves.

But do you know why Fermat's last theorem is true? That is, just in case you are not an expert on all of the intricacies of the proof, are you surprised that there are in fact no solutions? In fact, there is a fairly simple heuristic argument that supports the assertion. First note that the case $n = 3$, namely $x^3 + y^3 = z^3$, can be handled by elementary methods, and this in fact had already been done by EULER [VI.19]. So, let us focus on the cases when $n \geqslant 4$.[1] Let S_n be the set of positive nth powers of integers. How likely is it that the sum of two members of S_n is itself a member of S_n? Well, not at all likely, since Wiles has proved that this never occurs! But recall that we are trying to think naively.

Let us try to mimic our situation by replacing the set S_n with a random set. In fact, we will throw all of the powers together into one set. Following an idea of Erdős and Ulam (1971) we create a set $\mathcal{R}$ by a random process: each integer m is considered independently, and the chance it gets thrown into $\mathcal{R}$ is proportional to $m^{-3/4}$. This process would typically give us about $x^{1/4}$ numbers in $\mathcal{R}$ in the interval $[1, x]$, or at least this would be the order of magnitude. Now the total number of fourth and higher powers between 1 and x is also about $x^{1/4}$,

---

1. Actually, Fermat himself had a simple proof in the case $n = 4$, but we ignore this.

so we can take our random set $\mathcal{R}$ as modeling the situation for these powers, namely the union of all sets S_n for $n \geqslant 4$. We ask how likely it is to have $a + b = c$ where a, b, and c all come from $\mathcal{R}$.

The probability that a number m may be represented as $a + b$ with $0 < a < b < m$ and $a, b \in \mathcal{R}$ is proportional to $\sum_{0 < a < m/2} a^{-3/4}(m - a)^{-3/4}$, since for each a less than m the probability that a and $m - a$ both lie in $\mathcal{R}$ is $a^{-3/4}(m-a)^{-3/4}$. Actually, there is a minor caveat when m is even, since then $a = m - a$ when $a = \frac{1}{2}m$: to cover this, we add the single term $(\frac{1}{2}m)^{-3/4}$ to the above sum. Replacing each $m - a$ in the sum with $\frac{1}{2}m$, we get a larger sum that is easy to estimate and turns out to be proportional to $m^{-1/2}$. That is, the chance that a random number m is a sum of two members of $\mathcal{R}$ is at most a certain quantity that is proportional to $m^{-1/2}$. Now the events that would have to occur for m to be given as such a sum involve numbers smaller than m, so the event that m itself is in $\mathcal{R}$ is independent of these. Therefore, the probability that m is not only the sum of two members of $\mathcal{R}$, but also itself a member of $\mathcal{R}$, is at most a quantity proportional to $m^{-1/2}m^{-3/4} = m^{-5/4}$. So now we can count how many times we should expect a sum of two members of $\mathcal{R}$ to itself be a member of $\mathcal{R}$. This is at most a constant times $\sum_m m^{-5/4}$. But this sum is convergent, so we expect only finitely many examples. Further, since the tail of a convergent series is tiny, we do not expect any large examples.

Thus, this argument suggests that there are at most finitely many positive integer solutions to

$$x^u + y^v = z^w, \qquad (2)$$

where the exponents u, v, w are at least 4. Since Fermat's last theorem is the special case when $u = v = w$, we would have at most finitely many counterexamples to that as well.

This seems tidy enough, but now we get a surprise! There are actually *infinitely many solutions* to (2) in positive integers with u, v, w all at least 4. For example, note that $17^4 + 34^4 = 17^5$. This is the case $a = 1$, $b = 2$, $u = 4$ of a more general identity: if a, b are positive integers, and $c = a^u + b^u$, we have $(ac)^u + (bc)^u = c^{u+1}$. Another way to get infinitely many examples is to build on the possible existence of just one example. If x, y, z, u, v, w are positive integers satisfying (2), then with the same exponents, we may replace x, y, z with $a^{vw}x$, $a^{uw}y$, $a^{uv}z$ for any integer a, and so get infinitely many solutions.

The point is that events of the kind that we are considering—that a given integer is a power—are not quite independent. For instance, if A and B are both uth powers, then so is AB, and this idea is exploited in the infinite families just mentioned.

So how do we neatly bar these trivialities and come to the rescue of our heuristic argument? One simple way to do this is to insist that the numbers x, y, z in (2) be relatively prime. This gives no restriction whatsoever in the Fermat case of equal exponents, since a solution to $x^n + y^n = z^n$ with d the greatest common divisor of x, y, z leads to the coprime solution $(x/d)^n + (y/d)^n = (z/d)^n$.

Concerning Fermat's last theorem, one might ask how far it had actually been verified before the final proof by Wiles. The paper by Buhler et al. (1993) reports a verification for all exponents n up to $4\,000\,000$. This type of calculation, which is far from trivial, has its roots in nineteenth-century work of KUMMER [VI.40] and early-twentieth-century work of Vandiver. In fact, Buhler et al. (1993) also verify in the same range a related conjecture of Vandiver dealing with cyclotomic fields, but this conjecture may in fact be false in general.

The probabilistic thinking above, combined with computation of small cases, can carry us deeply into some very provocative conjectures. The above probabilistic argument can easily be extended to suggest that (2) has at most finitely many relatively prime solutions x, y, z over all possible exponent triples u, v, w with $1/u + 1/v + 1/w < 1$. This conjecture has come to be known as the Fermat–Catalan conjecture, since it contains within it essentially Fermat's last theorem and also the Catalan conjecture (recently proved by Mihăilescu) that 8 and 9 are the only consecutive powers.

It is good that we do allow for the possibility that there are *some* solutions, and this is where our main topic of computing comes in. For example, since $1 + 8 = 9$, we have a solution to $x^7 + y^3 = z^2$, where $x = 1$, $y = 2$, and $z = 3$. (The exponent 7 is chosen to insure that the reciprocal sum of the exponents is less than 1. Of course, we could replace 7 by any larger integer, but since in each case the power involved is the number 1, they should all together be considered as just one example.) Here are the known solutions to (2):

$$1^n + 2^3 = 3^2,$$
$$2^5 + 7^2 = 3^4,$$
$$13^2 + 7^3 = 2^9,$$

$$2^7 + 17^3 = 71^2,$$
$$3^5 + 11^4 = 122^2,$$
$$33^8 + 1\,549\,034^2 = 15\,613^3,$$
$$1414^3 + 2\,213\,459^2 = 65^7,$$
$$9262^3 + 15\,312\,283^2 = 113^7,$$
$$17^7 + 76\,271^3 = 21\,063\,928^2,$$
$$43^8 + 96\,222^3 = 30\,042\,907^2.$$

The larger members were found in an exhaustive computer search by Beukers and Zagier. Perhaps this is the complete list of all solutions, or perhaps not—we have no proof.

However, for particular choices of u, v, w, more can be said. Using results from a famous paper of Faltings, Darmon and Granville (1995) have shown that for any fixed choice of u, v, w with reciprocal sum at most 1, there are at most finitely many coprime triples x, y, z solving (2). For a particular choice of exponents, one might try to actually find all of the solutions. If it can be handled at all, this task can involve a delicate interplay between ARITHMETIC GEOMETRY [IV.5], effective methods in transcendental number theory, and good hard computing. In particular, the exponent triple sets $\{2,3,7\}$, $\{2,3,8\}$, $\{2,3,9\}$, and $\{2,4,5\}$ are known to have all their solutions in the above table. See Poonen et al. (2007) for the treatment of the case $\{2,3,7\}$ and links to other work.

THE ABC CONJECTURE [V.1] of Oesterlé and Masser is deceptively simple. It involves positive integer solutions to the equation $a + b = c$, hence the name. To put some meaning into $a + b = c$, we define the *radical* of a nonzero integer n as the product of the primes that divide n, denoting this as $\mathrm{rad}(n)$. So, for example, $\mathrm{rad}(10) = 10$, $\mathrm{rad}(72) = 6$, and $\mathrm{rad}(65\,536) = 2$. In particular, high powers have small radicals in comparison to the number itself, and so do many other numbers. Basically, the ABC conjecture asserts that if $a + b = c$, then the radical of abc cannot be too small. More specifically we have the following.

The ABC conjecture. *For each $\varepsilon > 0$ there are at most finitely many relatively prime positive integer triples a, b, c with $a + b = c$ and $\mathrm{rad}(abc) < c^{1-\varepsilon}$.*

Note that the ABC conjecture immediately solves the Fermat–Catalan problem. Indeed, if u, v, w are positive integers with $1/u + 1/v + 1/w < 1$, then it is easily found that we must have $1/u + 1/v + 1/w \leqslant 41/42$. Suppose we have a coprime solution to (2). Then $x \leqslant$

$z^{w/u}$ and $y \leqslant z^{w/v}$, so that
$$\mathrm{rad}(x^u y^v z^w) \leqslant xyz \leqslant (z^w)^{41/42}.$$

Thus, the ABC conjecture with $\varepsilon = 1/42$ implies that there are at most finitely many solutions.

The ABC conjecture has many other marvelous consequences; for a delightful survey, see Granville and Tucker (2002). In fact, the ABC conjecture and its generalizations can be used to prove so many things that I have joked that it is beginning to resemble a false statement, since a false statement implies everything. But probably the ABC conjecture is true. Indeed, though a bit harder to see, the Erdős–Ulam probabilistic argument can be modified to provide heuristic evidence for it too.

Basic to this argument is a perfectly rigorous result on the distribution of integers n for which $\mathrm{rad}(n)$ is below some bound. These ideas, which lead to a more explicit version of the ABC conjecture, are worked through in the thesis of van Frankenhuijsen and by Stewart and Tenenbaum. Here is a slightly weaker statement: if $a + b = c$ are relatively prime positive integers and c is sufficiently large, then we have
$$\mathrm{rad}(abc) > c^{1 - 1/\sqrt{\log c}}. \tag{3}$$

One might like to know how the numerical evidence stacks up against (3). This inequality asserts that if $\mathrm{rad}(abc) = r$, then $\log(c/r)/\sqrt{\log c} < 1$. So, let $T(a,b,c)$ denote the test statistic $\log(c/r)/\sqrt{\log c}$. There is a Web site maintained by Nitaj that contains a wealth of information about the ABC conjecture (www.math.unicaen.fr/~nitaj/abc.html). Checking the data, there are quite a few examples with $T(a,b,c) \geqslant 1$, the champion so far being

$$a = 7^2 \cdot 41^2 \cdot 311^3 = 2\,477\,678\,547\,239$$
$$b = 11^{16} \cdot 13^2 \cdot 79 = 613\,474\,843\,408\,551\,921\,511$$
$$c = 2 \cdot 3^3 \cdot 5^{23} \cdot 953 = 613\,474\,845\,886\,230\,468\,750$$
$$r = 2 \cdot 3 \cdot 5 \cdot 7 \cdot 11 \cdot 13 \cdot 41 \cdot 79 \cdot 311 \cdot 953$$
$$= 28\,828\,335\,646\,110,$$

so that
$$T(a,b,c) = \frac{\log(c/r)}{\sqrt{\log c}} = 2.43886\ldots.$$

Is it always true that $T(a,b,c) < 2.5$?

One can get carried away with heuristics, forgetting that one is not actually proving a theorem, but making a guess. Heuristics are often based on the idea of randomness, and all bets are off if there is some underlying structure. But how do we know that there is no underlying structure? Consider the case of an "*abcd* conjecture." Here we consider integers a, b, c, and d with $a + b + c + d = 0$. The condition that the terms be relatively prime now takes on two possible meanings: pairwise relatively prime or no nontrivial common divisor of all four numbers. The first condition seems more in the spirit of the three-term conjecture, but may be a tad too strong in that it disallows using any even numbers. So say we take the four terms with no pair having a common factor greater than 2. Under this condition, our heuristics seem to suggest that for each $\varepsilon > 0$, we have

$$\mathrm{rad}(abcd)^{1+\varepsilon} < \max\{|a|, |b|, |c|, |d|\} \qquad (4)$$

for at most finitely many cases. But consider the polynomial identity

$$(x + 1)^5 = (x - 1)^5 + 10(x^2 + 1)^2 - 8$$

(suggested to me by Granville). If we take x as a multiple of 10, the four terms involved in the identity are pairwise relatively prime except for the last two, which have a common factor of 2. Let $x = 11^k - 1$, which is a multiple of 10. The largest of the four terms is 11^{5k}, and the radical of the product of the four terms is at most

$$110(11^k - 2)((11^k - 1)^2 + 1) < 110 \cdot 11^{3k}.$$

The heuristics are saying that this cannot be, yet here it is right before our eyes!

What is happening is that the polynomial identity is supplying an underlying structure. For the four-term *abcd* conjecture, Granville conjectures that for each $\varepsilon > 0$, all counterexamples to (4) come from at most finitely many polynomial families. And the number of polynomial families grows to infinity as ε shrinks to 0.

We have looked here at only a small portion of the field of Diophantine equations, and then we have looked mainly at the dynamic relationship between heuristics and computational searches for small solutions. For much more on the subject of computational Diophantine methods, see Smart (1998).

Heuristic arguments often assume that the objects of study behave as if they were random, and we have visited several cases where it is useful to think this way. Other examples include the twin prime conjecture (there are infinitely many primes p such that $p + 2$ is prime), Goldbach's conjecture (every even number larger than 2 is the sum of two primes), and countless other conjectures in number theory. Often the computational evidence for the probabilistic view is striking,

even overwhelming, and we become convinced of the truth of our model. But on the other hand, if it is this pseudo-proof that is all we have to go on, we may still be very far from the truth. Nevertheless, the interplay of computations and heuristic thinking forms an indispensable part of our arsenal, and mathematics is the richer for it.

Remarks and Acknowledgments. I would like to recommend to the reader the book by Cohen (1993) for a discussion of computational algebraic number theory, a subject that is neglected in this article. I am grateful to the following people, who generously shared their expertise: X. Gourdon, A. Granville, A. Odlyzko, E. Schaefer, K. Soundararajan, C. Stewart, R. Tijdeman, and M. van Frankenhuijsen. I am also thankful to A. Granville and D. Pomerance for helpful suggestions with the exposition. I was supported in part by NSF grant DMS-0401422.

Further Reading

Agrawal, M., N. Kayal, and N. Saxena. 2004. PRIMES is in P. *Annals of Mathematics* 160:781–93.

Buhler, J., R. Crandall, R. Ernvall, and T. Metsänkylä. 1993. Irregular primes and cyclotomic invariants to four million. *Mathematics of Computation* 61:151–53.

Cohen, H. 1993. *A Course in Computational Algebraic Number Theory*. Graduate Texts in Mathematics, volume 138. New York: Springer.

Crandall, R., and C. Pomerance. 2005. *Prime Numbers: A Computational Perspective*, 2nd edn. New York: Springer.

Darmon, H., and A. Granville. 1995. On the equations $z^m = F(x, y)$ and $Ax^p + By^q = Cz^r$. *Bulletin of the London Mathematical Society* 27:513–43.

Erdős, P., and S. Ulam. 1971. Some probabilistic remarks on Fermat's last theorem. *Rocky Mountain Journal of Mathematics* 1:613–16.

Granville, A., and T. J. Tucker. 2002. It's as easy as *abc*. *Notices of the American Mathematical Society* 49:1224–31.

Odlyzko, A. M., and A. Schönhage. 1988. Fast algorithms for multiple evaluations of the Riemann zeta function. *Transactions of the American Mathematical Society* 309:797–809.

Poonen, B., E. Schaefer, and M. Stoll. 2007. Twists of $X(7)$ and primitive solutions to $x^2 + y^3 = z^7$. *Duke Mathematics Journal* 137:103–58.

Rosser, J. B., and L. Schoenfeld. 1962. Approximate formulas for some functions of prime numbers. *Illinois Journal of Mathematics* 6:64–94.

Smart, N. 1998. *The Algorithmic Resolution of Diophantine Equations*. London Mathematical Society Student Texts, volume 41. Cambridge: Cambridge University Press.

IV.4 Algebraic Geometry
János Kollár

1 Introduction

Succinctly put, algebraic geometry is the study of geometry using polynomials and the investigation of polynomials using geometry.

Many of us were taught the beginnings of algebraic geometry in high school, under the name "analytic geometry." When we say that $y = mx + b$ is the equation of a line L, or that $x^2 + y^2 = r^2$ describes a circle C of radius r, we establish a basic connection between geometry and algebra.

If we want to find the points where the line L and the circle C intersect, we just substitute $mx + b$ for y in the circle equation to get $x^2 + (mx + b)^2 = r^2$ and solve the resulting quadratic equation to obtain the x coordinates of the two intersection points.

This simple example encapsulates the method of algebraic geometry: a geometric problem is translated into algebra, where it is readily solvable; conversely, we get insight into algebra problems by using geometry. It is hard to guess the solutions of systems of polynomial equations, but once a corresponding geometric picture is drawn, we start to have a qualitative understanding of them. The precise quantitative answer is then provided by algebra.

2 Polynomials and Their Geometry

Polynomials are the expressions one can put together from variables and numbers by addition and multiplication. The most familiar are one-variable polynomials such as $x^3 - x + 4$, but we can use two or three variables to get, for instance, $2x^5 - 3xy^2 + y^3$ (which has degree 5 in two variables) or $x^5 - y^7 + x^2z^8 - xyz + 1$ (which has degree 10 in three variables). In general, one can use n variables, in which case they are frequently denoted by $x_1, x_2, \ldots, x_n$, and we write $f(x_1, \ldots, x_n)$, $f(\boldsymbol{x})$ or simply f to denote an unspecified polynomial.

Polynomials are the only functions that computers can work with. (Although your pocket calculator is likely to have a button for logarithms, it is secretly computing a polynomial whose value at a number b agrees with $\log b$ up to many decimal places.)

We can slightly rewrite the equations we gave earlier for the line L and the circle C: as $y - mx - b = 0$ and $x^2 + y^2 - r^2 = 0$. We can then describe L and C as *zero*

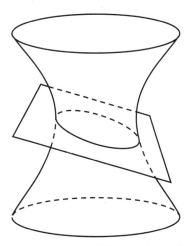

Figure 1 A hyperboloid intersecting a plane.

sets: L is the zero set of $y - mx - b$ (that is, the set of all points (x, y) such that $y - mx - b = 0$) and C is the zero set of $x^2 + y^2 - r^2$.

Similarly, the zero set of $2x^2 + 3y^2 - z^2 - 7$ in 3-space is a hyperboloid, the zero set of $z - x - y$ in 3-space is a plane, and the common zero set of these two equations in 3-space is the intersection of the hyperboloid and the plane, which is an ellipse (see figure 1).

The set of common zeros of a system of polynomial equations in any number of variables is called an *algebraic set*. These are the basic objects of algebraic geometry.

Most people feel that geometry ends in 3-space. Very few have a feeling for 4-space, also called *space-time*, and 5-space is by and large inconceivable to almost everyone. So what is the meaning of geometry in many variables?

Algebra comes to our rescue here. While I have great difficulty visualizing what a four-dimensional sphere of radius r in 5-space should be, I can easily write down its equation,

$$x_1^2 + x_2^2 + x_3^2 + x_4^2 + x_5^2 - r^2 = 0,$$

and work with it. This equation is also something a computer can handle, which is immensely useful in applications.

I will, nonetheless, stick to two or three variables for the rest of this article. This is where all geometry starts and there are plenty of interesting questions and results.

The importance of algebraic geometry derives from the fact that significant interactions between algebra and geometry happen very frequently. Let us look at two examples, just for illustration.

3 Most Shapes Are Algebraic

Shapes that occur frequently enough to have their own name, for instance, lines, planes, circles, ellipses, hyperbolas, parabolas, hyperboloids, paraboloids, ellipsoids, are almost all algebraic. Even the more esoteric conchoid (or shell curve) of Dürer, the trident of NEWTON [VI.14], and the folium of Kepler are algebraic.

Some shapes cannot be described by polynomial equations, but they can be described by polynomial inequalities. For instance, the inequalities $0 \leqslant x \leqslant a$ and $0 \leqslant y \leqslant b$ together describe a rectangle with side lengths a, b. Shapes described by polynomial inequalities are called *semialgebraic*, and every polyhedron is semialgebraic.

Not everything is an algebraic set, though. Look, for example, at the graph of the sine function $y = \sin x$. This crosses the x-axis infinitely many times (at multiples of π). If $f(x)$ is any polynomial, then it has at most as many roots as its degree, so $y = f(x)$ will never look like $y = \sin x$.

We can, however, get very close to $\sin x$ with a polynomial if we concentrate on values of x that are not too large. For instance, the degree-7 Taylor polynomial

$$x - \tfrac{1}{6}x^3 + \tfrac{1}{120}x^5 - \tfrac{1}{5040}x^7$$

differs from $\sin x$ by an error of at most 0.1 for $-\pi < x < \pi$. This is a very special case of a basic theorem of Nash that says that every "reasonable" geometric shape is algebraic if we ignore what happens very far from the origin. So, what is reasonable? Certainly not everything. Fractals seem profoundly nonalgebraic. The nicest shapes are MANIFOLDS [I.3 §6.9], and all of these can be described by polynomials.

Nash's theorem. *Let M be any manifold in $\mathbb{R}^n$. Fix any large number R. Then there is a polynomial f whose zero set is as close to M as we want, at least inside a ball of radius R around the origin.*

4 Codes and Finite Geometries

Consider the equation $x^2 + y^2 = z^2$, which describes a double cone in 3-space (see figure 4). If we confine ourselves to natural numbers, then the solutions of $x^2 + y^2 = z^2$ are the *Pythagorean triples*, corresponding to right-angled triangles where all sides have integer lengths, of which the two best-known examples are $(3, 4, 5)$ and $(5, 12, 13)$.

Let us now look at the same equation, but declare that we care only about the *parities* of the two sides (that is, whether they are even or odd). For instance, $3^2 + 15^2$ and 4^2 are both even, so we say that $3^2 + 15^2 \equiv 4^2 \pmod 2$. The parities of $x^2 + y^2$ and of z^2 depend only on those of x, y, and z, so we can pretend that x, y, and z are all either 0 (the even case) or 1 (the odd case). Our equation modulo 2 therefore has four solutions:

$$000, \ 011, \ 101, \ 110.$$

These look like code words in a computer message. It was quite a surprise when it was discovered that using polynomials and their solutions modulo 2 is a great—probably the best—way of constructing ERROR-CORRECTING CODES [VII.6 §§3–5].

There is something very substantial and new happening here. Let us think for a moment about what 3-space is for us. For many it is an amorphous everything, but for algebraic geometers (with DESCARTES [VI.11] as our ancestor) it is simply a collection of points described by three numbers, the x, y, and z coordinates. Let us make a jump here, and declare that "3-space modulo 2" is the collection of all "points" given by three coordinates modulo 2. Four of these are listed above, and there are four more. The beauty of algebra is that suddenly we can talk about lines, planes, spheres, cones in this "3-space having only eight points."

We do not need to stop here, and one can work modulo any integer. For example, working modulo 7, we have 0, 1, 2, 3, 4, 5, 6 as possible coordinates, and so "3-space modulo 7" has $7^3 = 343$ points.

Talking about geometry in these spaces is very intriguing, but also technically difficult. Its great reward is that one can view this process as a "discretization" of ordinary space. Working modulo n for large n (especially when n is a prime number) gets very close to the usual geometry.

This approach is especially fruitful in number-theoretic questions. It was, for instance, instrumental in Wiles's proof of Fermat's last theorem.

For more on these topics, see ARITHMETIC GEOMETRY [IV.5].

5 Snapshots of Polynomials

Consider the equation $x^2 + y^2 = R$. If $R > 0$, then the real solutions form a circle of radius $\sqrt{R}$; if $R = 0$, we get only the origin; and if $R < 0$, we get the empty set. Thus, if $R > 0$, then the geometry of the solution set determines what R is, but otherwise it does not. We can of course look at complex solutions, and the complex solutions always determine R. (For instance, the intersection points with the x-axis are $(\pm\sqrt{R}, 0)$.)

If R is a rational number, we can ask about rational solutions of $x^2 + y^2 = R$, and if R is an integer, we can also look for solutions in the "plane modulo m" for any m.

One can even look for solutions where $x = x(t)$, $y = y(t)$ are themselves polynomials in a variable t. (Most generally, we can ask for solutions where x, y are elements of any ring containing the number R.)

To my mind, the polynomial is the central object, and each time we look at solution sets we are taking a "snapshot" of the polynomial. Some snapshots are good (like the above real snapshot for $R > 0$) and some are bad (like the above real snapshot for $R < 0$).

How good can snapshots be? Can we determine a polynomial from its snapshots?

One frequently talks about "the" equation of a hyperbola, but "an" equation would be more correct. Indeed, the hyperbola $x^2 - y^2 - R = 0$ can also be given by an equation $cx^2 - cy^2 - cR = 0$, for any $c \neq 0$. We can also use the equation $(x^2 - y^2 - R)^2 = 0$, which we may well not recognize in its expanded form. Higher powers can also be used. What about the equation $f(x, y) = (x^2 - y^2 - R)(x^2 + y^2 + R^2) = 0$? If we look only at real solutions, this is still just the hyperbola since $x^2 + y^2 + R^2$ is always positive for x, y real. However, as with one-variable polynomials, one should look at all complex roots to understand everything. Then we see that $f(\sqrt{-1}R, 0) = 0$, but the complex point $(\sqrt{-1}R, 0)$ is not on the hyperbola $x^2 - y^2 - R = 0$. In general, as long as $R \neq 0$, we get that if f is a polynomial that has exactly the same complex roots as $x^2 - y^2 - R$, then $f(x, y) = c(x^2 - y^2 - R)^m$ for some m and $c \neq 0$.

Why is the $R = 0$ case different? The reason is that for $R \neq 0$ the polynomial $x^2 - y^2 - R$ is *irreducible* (that is, it cannot be written as the product of other polynomials), while $x^2 - y^2 = (x + y)(x - y)$ is reducible with *irreducible factors* $x + y$ and $x - y$. In the latter case one gets that if $g(x, y)$ is a polynomial that has exactly the same complex roots as $x^2 - y^2$, then $f = c \cdot (x + y)^m (x - y)^n$ for some m, n and $c \neq 0$.

The analogous question for systems of equations is answered by the fundamental theorem of algebraic geometry. It is sometimes called Hilbert's theorem on the zeros, but its German name is used most of the time. For simplicity, we state only the case of one equation.

Hilbert's Nullstellensatz. *Two complex polynomials f and g have the same complex solutions if and only if they have the same irreducible factors.*

We can do even better for polynomials with integer coefficients. For instance, $x^2 - y^2 - 1 = 0$ and $2(x^2 - y^2 - 1) = 0$ have the same solutions over the real or complex numbers, and the same solutions modulo p for any odd prime p, but they have different solutions modulo 2. The general result in this case is easy and simple.

Arithmetic Nullstellensatz. *Two polynomials with integer coefficients f and g have the same solutions modulo m for every m if and only if $f = \pm g$.*

6 Bézout's Theorem and Intersection Theory

If $h(x)$ is a polynomial of degree n, then it has n complex roots, at least when they are counted with multiplicity. What happens with a system $f(x, y) = g(x, y) = 0$? Geometrically we see two curves in the plane, so we expect that there will typically be finitely many intersection points.

If f, g are both linear, we have two lines in the plane. These usually intersect in a single point, but they can be parallel and they can coincide. The first case leads to the classical declaration that "parallel lines meet at infinity" and the definition of projective planes and PROJECTIVE SPACES [III.72]. (The introduction of projective spaces and the corresponding projective varieties is a key step in algebraic geometry. It is somewhat technical so we shall skip it here, but it is indispensable even at the most basic level.)

Next, consider two polynomials of degree 2, that is, two plane conics. Two smooth conics usually intersect in at most four points (just try this by drawing two ellipses). There are also some rather degenerate cases. Two conics may coincide, or, if they are both reducible, they can have a common line. In any case, we are ready to formulate a basic result, dating back to 1779.

Bézout's theorem. *Let $f_1(\mathbf{x}), \ldots, f_n(\mathbf{x})$ be n polynomials in n variables, and for each i let d_i be the degree of f_i. Then either*

(i) *the equation(s) $f_1(\mathbf{x}) = \cdots = f_n(\mathbf{x}) = 0$ have at most $d_1 d_2 \cdots d_n$ solutions; or*

(ii) *the f_i vanish identically on an algebraic curve C, and so there is a continuous family of solutions.*

As an example, the second alternative happens for the system of equations $xz - y^2 = y^3 - z^2 = x^3 - z = 0$, which has (t, t^2, t^3) as a solution for any t. This case is actually quite rare. If we pick the coefficients of the polynomials f_i randomly, then the first alternative happens with probability 1.

Ideally, we would like to make the stronger claim that if the first alternative happens, then there are *exactly* $d_1 d_2 \cdots d_n$ solutions, but counted "with multiplicity." This actually works, and gives us our first example of an extremely useful feature of algebraic geometry. Even in very degenerate situations it is possible to define and count the multiplicities easily. This is frequently of great help since the typical (or "generic") cases are usually very hard to compute. To get around this problem, we can sometimes find a special, degenerate case where we know that the answer will be the same, but the computations are much easier.

There are two ways to think about multiplicity: one algebraic and one geometric. The algebraic definition is computationally very efficient, but somewhat technical. The geometric interpretation is easier to explain, so that is the one we shall give here, but it would be hard to compute with in practice.

If $\boldsymbol{x} = \boldsymbol{p}$ is an isolated solution of the equations $f_1(\boldsymbol{x}) = \cdots = f_n(\boldsymbol{x}) = 0$ *with multiplicity* m, then the perturbed system

$$f_1(\boldsymbol{x}) + \epsilon_1 = \cdots = f_n(\boldsymbol{x}) + \epsilon_n = 0$$

has exactly m solutions near $\boldsymbol{x} = \boldsymbol{p}$ for almost all small values of the ϵ_i.

Intersection theory is the branch of algebraic geometry that deals with generalizations of Bézout's theorem. Above, we looked at intersections of *hypersurfaces*—that is, of zero sets of single polynomials—but we may wish to look at intersections of more general algebraic sets. Also, even when the second alternative holds, we may want to count the number of isolated intersection points; this can be very tricky but also very useful.

7   Varieties, Schemes, Orbifolds, and Stacks

Consider the system $xz = yz = 0$ in 3-space. It consists of two pieces, the $z = 0$ plane and the $x = y = 0$ line. It is easy to see that neither the plane nor the line can be written as the union of algebraic sets (except by nitpickers who point out that the line is the union of the line itself and of any point on the line). In general, any algebraic set can be written in exactly one way as the union of smaller algebraic sets that in turn cannot be decomposed further. These basic building blocks are called *irreducible* algebraic sets or *algebraic varieties*.

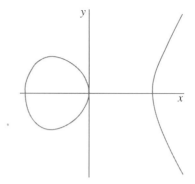

Figure 2 A smooth cubic: $y^2 = x^3 - x$.

Sometimes this is not exactly what one would naively expect. For instance, the curve in figure 2 has two connected components. The two parts are, however, not algebraic sets.

An explanation is provided by looking at the *complex* solutions of this equation. We shall see later that these form a connected set, namely a torus (with a missing point at infinity). We see two components when we look at the real solutions because we are taking a cross-section of this torus.

In general, the zero set $f = 0$ is irreducible as an algebraic set if and only if f is irreducible as a polynomial (or if it is the power of an irreducible polynomial). The implication in one direction is easy to see: if $f = gh$, then the zero set of f is the union of the zero set of g and of the zero set of h.

For many questions, keeping track only of the zero set is not enough. For instance, look at the polynomial $f = x^2(x - 1)(x - 2)^3$. It has degree 6 and three roots at $x = 0, 1, 2$. These roots behave differently, however, and one usually says that f has a double root at $x = 0$ and a triple root at $x = 2$. If we perturb f by adding a small number ϵ to it, then the perturbed equation $f(x) + \epsilon = 0$ has two (complex) solutions near 0, one solution near 1 and three (complex) solutions near 2. Thus, these multiplicities carry important geometric meaning about the perturbation of the equation.

Similarly, it is natural to say that while $x^2 y = 0$ and $xy^3 = 0$ define the same algebraic set (consisting of the two axes), the first "assigns multiplicity 2" to the y-axis and the other "assigns multiplicity 3" to the x-axis.

More complicated things can happen for systems of equations. Consider the systems $x = y^2 = 0$ and $x^3 = y = 0$ in 3-space. Both define the z-axis and it is reasonable to say that the first does so with multiplicity 2, the second with multiplicity 3. There is, however, a further difference. In the first case the multiplicity seems to "go in the y-direction" and in the second case it seems to go in the x-direction. We can also look at other systems, like $x - cy = y^3 = 0$, if we want to see more complicated behavior.

Roughly speaking, a *scheme* is an algebraic set where we also keep track of the multiplicities and of the directions they occur in.

Consider the xy-plane and consider the map that reflects across the origin. Thus a point (x, y) is mapped to $(-x, -y)$. Let us try to glue each point (x, y) to its image $(-x, -y)$. What do we get? The right half-plane $x \geqslant 0$ is mapped to the left half-plane $x \leqslant 0$, so it is enough to work out what happens with the right half-plane. The positive y-axis is glued to the negative y-axis, and the resulting surface is a dunce cap (but less pointy).

Algebraically, it is one half of the cone $z^2 = x^2 + y^2$. This cone looks nice and smooth except at the vertex. There it is more complicated, but the above construction shows that it can be obtained from a plane by a reflection across a point. More generally, suppose we take the n-dimensional space $\mathbb{R}^n$ and finitely many symmetries of it. If we glue together points that move into each other, we again get an algebraic variety, most of whose points are smooth, but some of which are more complicated. A variety made up of pieces like these is called an *orbifold*. (When this is defined more precisely, we also keep track of which symmetries have been used.) In practice, such varieties occur frequently; that is why they deserve a separate name.

Finally, if we marry a scheme to an orbifold, the outcome is a *stack*. The study of stacks is strongly recommended to people who would have been flagellants in earlier times.

8 Curves, Surfaces, Threefolds

As with any geometric object, one of the simplest questions one can ask about a variety is: what is its dimension? As expected, a curve in the plane has dimension 1, and a surface in 3-space has dimension 2. This seems quite simple until one writes down examples like $S = (x^4 + y^4 + z^4 = 0)$, which is only the origin in $\mathbb{R}^3$. This example is, nonetheless, still two dimensional: the explanation is that we were looking at the wrong snapshot. Using complex numbers we can solve the equation as $z = \sqrt[4]{-x^4 - y^4}$, so the complex solutions of $x^4 + y^4 + z^4 = 0$ can be described by two independent variables x, y and a dependent variable z. Thus, it is quite reasonable to say that S is two dimensional.

This idea works more generally. If X is any variety in some complex space $\mathbb{C}^n$, then choose a random set of n independent directions to serve as a basis, or coordinate system, for $\mathbb{C}^n$, and hence for X. With probability 1 (i.e., except in degenerate cases) one finds that there is some d such that the first d coordinates of a point x in X can vary independently, while the rest depend on them. This number d depends on X only and is called the *dimension* (or, to be precise, the *algebraic dimension*) of X.

If X is a variety and f is a polynomial, then the intersection $X \cap (f = 0)$ has dimension one less than $\dim X$ (unless f vanishes identically on X or never takes the value zero on X).

If X is a subset of $\mathbb{R}^n$ defined by real equations, and if it is smooth (see the next section for a discussion of smoothness), then its TOPOLOGICAL DIMENSION [III.17] is the same as its algebraic dimension.

For complex varieties, the topological dimension is twice the algebraic dimension. Thus, for an algebraic geometer, $\mathbb{C}^n$ has dimension n. In particular, for us $\mathbb{C}$ is the "complex line," whereas everybody else calls this the "complex plane." Our "complex plane" is, of course, $\mathbb{C}^2$.

A variety of dimension 1 is called a *curve*. A *surface* is a variety of dimension 2, and a *threefold* is a variety of dimension 3.

The theory of algebraic curves is a very well developed and beautiful subject. We shall see later how one can start to get an overview of all algebraic curves. Surfaces have been intensively studied for the last century, and now we have reached a reasonably complete understanding of them. This is a much more complicated theory than for curves. Still very little is known for varieties of dimension 3 and up. At least conjecturally, all these dimensions behave in roughly the same way. Despite some progress, especially in dimension 3, many questions are wide open.

9 Singularities and Their Resolutions

If we look at the simplest examples of algebraic curves in figure 3, we see that most points of a curve are smooth, but that there may be a finite set of more complicated singular points. Let us compare these with the curve in figure 2.

All three curves pass through the origin, since their equation has no constant term. The equation of figure 2 has a linear term and the curve looks nice and smooth at the origin, whereas the equations of figure 3 contain

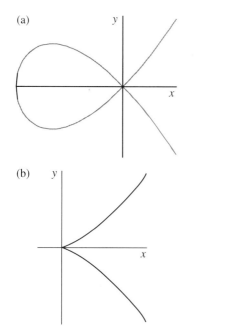

Figure 3 Singular cubics: (a) $y^2 = x^3 + x^2$ and (b) $y^2 = x^3$.

no linear term and the curves are more complicated at the origin. This is not an accident. For small values of x, the higher powers $x^2, x^3, \ldots$ are much smaller than x in absolute value, so near the origin the linear terms dominate. If we have only linear terms $ax + by = 0$, we get a line through the origin, and an algebraic curve $ax + by + cx^2 + gxy + ey^2 + \cdots = 0$ is close to the line $ax + by = 0$, at least for very small values of x and y.

The study of a curve near another point with coordinates (p, q) can be reduced to the case $(p, q) = (0, 0)$ via the coordinate change $(x, y) \mapsto (x - p, y - q)$.

In general, if $f(\mathbf{0}) = 0$ and f has a (nonzero) linear term $L(f)$, the hypersurface $f = 0$ is very close to the hyperplane $L(f) = 0$. This is the so-called *implicit function theorem*. Such points are called *smooth*. Points that are not smooth are called *singular*. One can easily show that the singular points of X form an algebraic set, defined by the vanishing of all partial derivatives $\partial f / \partial x_i$. A random hypersurface will, with probability 1, be smooth, but there are many singular hypersurfaces as well.

The smooth and singular points of an arbitrary variety of dimension d can be defined analogously by comparing X with d-dimensional linear subspaces.

Singularities also occur in other geometric fields, such as topology and differential geometry, but by and large these fields shy away from their study (with the

notable exception of catastrophe theory). By contrast, algebraic geometry provides very powerful tools for their investigation.

Let us start with singularities of hypersurfaces, or equivalently with *critical points* of functions. When thinking about these it is natural to work not just with polynomials but with more general power series, that is, functions $f(x_1, \ldots, x_n)$ that can be written as "polynomials of infinite degree." For simplicity of notation we shall assume that $f(\mathbf{0}) = 0$. Two functions f, g are considered to be *equivalent* if there is a coordinate change $x_i \mapsto \phi_i(\mathbf{x})$, where each ϕ_i is given by a power series, such that $f(\phi_1(\mathbf{x}), \ldots, \phi_n(\mathbf{x})) = g(\mathbf{x})$.

In the one-variable case, any f can be written as

$$f = x^m (a_m + a_{m+1}x + \cdots),$$

where $a_m \neq 0$. The (inverse of the) substitution

$$x \mapsto x \sqrt[m]{a_m + a_{m+1}x + \cdots}$$

then shows that f is equivalent to x^m. The functions x^m are inequivalent for different values of m, so in this particular case the lowest-degree monomial occurring in f determines f up to equivalence. (Note that even if f is a polynomial, the above change of variable involves an infinite power series: it is because we cannot invert polynomials, even locally, that it is more convenient to consider general power series.)

In general, the lowest-degree terms of a power series do not determine the singularity, but taking more terms is usually enough to do so, because of the following result.

Algebraization of analytic singularities. *Given a power series f, let $f_{\leqslant N}$ denote the polynomial obtained from f by deleting all monomials of degree greater than N. If $\mathbf{0}$ is an isolated singular point of the hypersurface $(f = 0)$, then f is equivalent to $f_{\leqslant N}$ for sufficiently large N.*

To see an example of a nonisolated singularity at $\mathbf{0}$, take

$$g(x, y, z) = \left(y + \frac{x}{1 - x}\right)^2 - z^3$$
$$= (y + x + x^2 + x^3 + \cdots)^2 - z^3.$$

It has singular points not just at $\mathbf{0}$, but everywhere along the curve $y + (x/(1 - x)) = z = 0$. On the other hand, one can easily check that all truncations $g_{\leqslant N}$ do have an isolated singular point at $\mathbf{0}$.

If we have two power series, f and g, we can view functions of the form $f + \epsilon g$ as perturbations of f. A very fruitful question of singularity theory asks:

what can we say about the perturbations of a given polynomial or power series f?

For instance, in the one-variable case, the polynomial x^m can be perturbed as $x^m + \epsilon x^r$, which is equivalent to x^r if $r < m$. Every perturbation contains x^m, so if $r > m$, then no perturbation of x^m will be equivalent to x^r (because near the origin x^m will be much larger than x^r). Hence, up to equivalence, the set of all possible perturbations of x^m is $\{x^r : r \leqslant m\}$.

On the other hand, it is not hard to see that for any given ϵ, there are only twenty-four different values of η for which the polynomials $xy(x^2 - y^2) + \epsilon y^2(x^2 - y^2)$ and $xy(x^2 - y^2) + \eta y^2(x^2 - y^2)$ are equivalent. (Indeed, both polynomials describe four lines through the origin. The first one gives the lines $y = 0$, $x = y$, $x = -y$, and $x = -\epsilon y$, and the second gives the same lines except that η replaces ϵ. The linear part of any supposed equivalence gives a linear transformation mapping the first set of four lines to the second. There are twenty-four ways to assign which line goes to which line.) Thus $xy(x^2 - y^2)$ has a continuous family of inequivalent perturbations.

Simple singularities. *Suppose that the polynomial or power series $f(x_1, \ldots, x_n)$ has only finitely many inequivalent perturbations. Then f is equivalent to one of the following normal forms:*

$$
\begin{array}{lll}
A_m & x_1^{m+1} + x_2^2 + \cdots + x_n^2 & (m \geqslant 1), \\
D_m & x_1^2 x_2 + x_2^{m-1} + x_3^2 + \cdots + x_n^2 & (m \geqslant 4), \\
E_6 & x_1^3 + x_2^4 + x_3^2 + \cdots + x_n^2, & \\
E_7 & x_1^3 + x_1 x_2^3 + x_3^2 + \cdots + x_n^2, & \\
E_8 & x_1^3 + x_2^5 + x_3^2 + \cdots + x_n^2. &
\end{array}
$$

The names should bring to mind the CLASSIFICATION OF LIE GROUPS [III.48]. The connections are numerous but not easy to explain. When $n = 3$, these are also called *Du Val singularities* or *rational double points*.

Consider again the cone $z^2 = x^2 + y^2$. Earlier, we described a two-to-one parametrization of it. Here is another, and for many purposes better, parametrization over the real numbers.

In the (u, v, w)-space consider the smooth cylinder $u^2 + v^2 = 1$. The map $(u, v, w) \mapsto (uw, vw, w)$ maps the cylinder onto the cone (see figure 4). The map is one-to-one away from the vertex, the preimage of which is the circle $u^2 + v^2 = 1$ in the $(w = 0)$-plane.

(Sharp-eyed readers will have noticed that this map is not so nice if we use complex numbers. In general, we want parametrizations that work both for real and

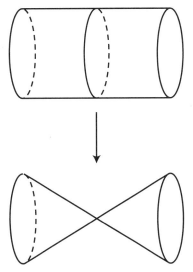

Figure 4 A resolution of the cone.

complex numbers, but that would be quite a bit more complicated to describe.)

The advantage of the cylinder over the cone is that it does not have a singularity. Parametrizations of varieties in terms of smooth varieties are very useful, and there is a major result that tells us that they always exist, at least when the varieties are real or complex. (The corresponding result is still unknown for the finite geometries considered earlier.)

Resolution of singularities (Hironaka). *For any variety X there is another smooth variety Y and a polynomially defined surjective map $\pi : Y \to X$ such that π is invertible at all smooth points of X.*

(In the cone example above, one can take the whole cylinder, but the cylinder minus finitely many points in the collapsed circle would also work. In order to avoid such silly cases, we require π to be surjective in a very strong sense: if a sequence of smooth points $x_i \in X$ converges to a limit in X, then a subsequence of their preimages $\pi^{-1}(x_i)$ converges to a limit in Y.)

10 Classification of Curves

In order to get an idea of how the classification of algebraic varieties should proceed, let us look at hypersurfaces of degree d in n-space. These are given by a degree-d polynomial $f(x_1, \ldots, x_n) = 0$. The set of all polynomials of degree at most d forms a vector space $V_{n,d}$. Thus hypersurfaces have two obvious discrete invariants, the dimension and the degree, and one can move between hypersurfaces of the same dimension and degree by varying the coefficients of f continuously. Moreover, the entire set $V_{n,d}$ is itself an algebraic variety. Our aim is to develop a similar understanding for all varieties, which can be done in two steps.

The first step is to define some integers, naturally attached to varieties, which stay the same if we change a variety continuously. Such integers are called *discrete invariants*. The simplest example is the dimension.

The second is to show that the set of all varieties with the same discrete invariant is parametrized by another algebraic variety, called the MODULI SPACE [IV.8]. Moreover, we would like the variety used for this parametrization to be chosen as economically as possible. We will look at this in more detail in the next section.

Let us see how it is accomplished for curves. Here there is only one more discrete invariant besides the dimension, known as the *genus* of the curve. This has many different definitions: one of the simplest is through topology. Let E be a smooth curve and let us look at its complex points. Locally, this set looks like $\mathbb{C}$, so it is a topological surface. After patching up some holes at infinity, we get a compact surface. Multiplication by $\sqrt{-1}$ gives an orientation, so basic topology tells us that we get a sphere with a certain number of handles attached (see DIFFERENTIAL TOPOLOGY [IV.7]). The genus of the curve is defined to be the number of these handles (that is, the genus of the corresponding surface). To see what this means in practice, let us look at some examples.

A line in 2-space is like the complex numbers, which can be viewed as a sphere minus a point. This sphere, $\mathbb{C}$ plus the point at infinity, is also called the *Riemann sphere*. So the genus is zero.

Next, we look at conics. Here it is better to use some projective geometry. Take any tangent of the conic and move this so that it becomes the line at infinity. Then we get a parabola, which, in suitable coordinates, is given by an equation $y = x^2$. The polynomial map $t \mapsto (t, t^2)$, with its inverse $(x, y) \mapsto x$, shows that this parabola is isomorphic to a line, so again has genus 0.

Cubics are quite a bit more complicated. A first warning is that $y = x^3$ is the wrong cubic to look at. It is smooth (and has genus 0) but it is singular at infinity. (The earlier expediency of keeping silent about projective geometry starts to bite us!) In any case, the correct thing to do is to choose the tangent line of the cubic at an inflection point and move that to infinity. After some computation we obtain a much-simplified

equation $y^2 = f(x)$, where f has degree 3. What is the genus?

Consider the special case $y^2 = x(x - 1)(x - 2)$. We try to understand the two-to-one projection to the (complex) x-axis, but it is better to do this when the x-axis has already had the point at infinity added, so that it is the Riemann sphere. If we remove the interval $0 \leqslant x \leqslant 1$ and the half line $2 \leqslant x \leqslant +\infty$ from the Riemann sphere, then the function $y = \sqrt{x(x-1)(x-2)}$ has two branches. (This means that y takes two different values for each x, the positive and negative square roots of $x(x-1)(x-2)$, but if one moves x about, one can let y vary in a continuous way.) The sphere minus two slits is topologically like a cylinder, hence the complex cubic is glued together from two cylinders. So we get the torus and the genus is 1.

It turns out that a smooth plane curve of degree d has genus $\frac{1}{2}(d-1)(d-2)$, but I find this hard to see directly topologically.

It is a (probably hopeless) dream of algebraic geometers to give a similarly simple description of the discrete invariants for higher-dimensional varieties. Unfortunately, the topological invariants of the complex points are not good enough, and they probably mislead more than help.

As a further illustration of the approach to the classification of curves, here is a list of all curves of low genus.

Genus 0. There is only one curve of genus 0. As we saw, it can be realized as a line or as a conic in the plane.

Genus 1. Every curve of genus 1 is a plane cubic, and it can be given by an equation of the form $y^2 = f(x)$, where f has degree 3. Genus-1 curves are usually called ELLIPTIC CURVES [III.21], since they first appeared (in the guise of elliptic integrals) in connection with the arc length of ellipses. We look at these in more detail later.

Genus 2. Every curve of genus 2 can be given by an equation of the form $y^2 = f(x)$, where f has degree 5. (These curves are singular at infinity.) More generally, if f has degree $2g + 1$ or $2g + 2$, then the curve $y^2 = f(x)$ has genus g. For $g \geqslant 3$, such curves, called *hyperelliptic*, are rather special.

Genus 3. Every curve of genus 3 can be realized as a plane curve of degree 4 (or it is hyperelliptic).

Genus 4. Every curve of genus 4 can be presented as a space curve given by two equations of degrees 2 and 3 (or it is hyperelliptic).

It should be emphasized that hyperelliptic curves do not form a separate family. One can move continuously from any hyperelliptic curve to a general curve of the kind described above. This can be seen through more-complicated representations.

One can continue in this manner a bit longer, up to about genus 10, but no such explicit construction is possible when the genus is large.

11 Moduli Spaces

Let us go back to plane cubics, which we parametrized by the vector space $V_{2,3}$ of degree-3 polynomials in two variables. This is not very economical. For instance, $x^3 + 2y^3 + 1$ and $3x^3 + 6y^3 + 3$ are different polynomials, but define the same curve. Furthermore, there is not much reason to distinguish $x^3 + 2y^3 + 1$ from $2x^3 + y^3 + 1$, since they are obtained from each other by switching the two coordinate axes. More generally, as we have seen in the previous section, any cubic curve can be transformed into one given by an equation $y^2 = f(x)$, where $f = ax^3 + bx^2 + cx + d$.

This is better but not yet optimal, and there are two more steps to take. First, one can set the leading coefficient of f to be 1. Indeed, substitute $y = \sqrt{a}\, y_1$ and then divide the whole equation by a to get $y_1^2 = x^3 + \cdots$. Second, we can make a substitution $x = ux_1 + v$ to get another elliptic curve with equation $y^2 = f(ux_1 + v) = f_1(x_1)$, where f_1 is easy to write down explicitly. One can see that these are the only coordinate changes that we can make without messing up the form $y^2 = $ (cubic polynomial).

It is still not very clear what happens. To get a better answer, look at the three roots of f, so $f(x) = (x - r_1)(x - r_2)(x - r_3)$. (Again, complex numbers inevitably appear.) If we make the substitution $x \mapsto (r_2 - r_1)x + r_1$, we get a new polynomial $f_1(x)$, two of whose roots are 0 and 1. Thus our elliptic curve is transformed into $y^2 = x(x - 1)(x - \lambda)$. So instead of the four unknown coefficients of f, we are down to only one unknown, λ.

This form is still not completely unique. In our transformation we sent r_1, r_2 to 0, 1, but we could have used any two roots. For instance, we can substitute $x \mapsto 1 - x$, sending $\lambda \mapsto 1 - \lambda$, or $x \mapsto \lambda x$, sending $\lambda \mapsto \lambda^{-1}$. All together, the six values

$$\lambda, \ \frac{1}{\lambda}, \ 1 - \lambda, \ \frac{1}{1 - \lambda}, \ \frac{-\lambda}{1 - \lambda}, \ \frac{1 - \lambda}{-\lambda}$$

give "the same" elliptic curve. Most of the time these six values are different, but there may be coincidences. For instance, we get only three different values if

$\lambda = -1$. This corresponds to the fact that the elliptic curve $y^2 = x(x - 1)(x + 1)$ has four symmetries: $(x, y) \mapsto (-x, \pm\sqrt{-1}\,y)$ and $(x, y) \mapsto (x, \pm y)$. (An unusual feature of elliptic curves is that they all have the second pair of symmetries. At $\lambda = 1$ we pick up 4/2 new symmetries, which corresponds to halving the number of different values above.)

The best way to think about it is to view this as an action of the symmetric group S_3 (the group of permutations of a three-element set) on the set $\mathbb{C} \setminus \{0, 1\}$.

It is not at all obvious that we have run out of tricks, but we have in fact reached the final result.

Moduli of elliptic curves. *The set of all elliptic curves is in a natural one-to-one correspondence with the points of the quotient orbifold $(\mathbb{C} \setminus \{0, 1\})/S_3$. The orbifold points correspond to the elliptic curves with extra automorphisms.*

This is the simplest illustration of a general phenomenon.

Moduli principle. *In most cases of interest, the set of all algebraic varieties with fixed discrete invariants is in a natural one-to-one correspondence with the points of an orbifold. The orbifold points correspond to the varieties with extra automorphisms.*

The moduli orbifold (also called the moduli space) of smooth curves of genus g is denoted by $\mathcal{M}_g$. These are among the most intensely studied orbifolds in algebraic geometry, especially since the recent discovery of their fundamental position in STRING THEORY [IV.17 §2] and MIRROR SYMMETRY [IV.16].

12 Effective Nullstellensatz

In order to show that there are still interesting elementary questions in algebraic geometry, let us try to decide when m given polynomials $f_1, \ldots, f_m$ have no common complex zero. The classical answer is given by the following result, which tells us that an obviously necessary condition is in fact sufficient.

Weak Nullstellensatz. *The polynomials $f_1, \ldots, f_m$ have no common complex zero if and only if there are polynomials $g_1, \ldots, g_m$ such that*

$$g_1 f_1 + \cdots + g_m f_m = 1.$$

Let us now make a guess that we can find g_j with degree at most 100. We can then write

$$g_j = \sum_{i_1 + \cdots + i_n \leqslant 100} a_{j, i_1, \ldots, i_n} x_1^{i_1} \cdots x_n^{i_n},$$

where the $a_{j,i_1,\dots,i_n}$ are indeterminates. If we write $g_1 f_1 + \cdots + g_m f_m$ as a polynomial in the variables $x_1, \dots, x_n$, then all the coefficients must vanish, save the constant term which must equal 1. Thus we get a system of *linear* equations in the indeterminates $a_{j,i_1,\dots,i_n}$. The solvability of systems of linear equations is well-known (with good computer implementations). Thus we can decide if there is a solution with $\deg g_j \leqslant 100$. Of course it is possible that 100 was too small a guess, and we may have to repeat the process with larger and larger degree bounds. Will this ever end? The answer is given by the following result, which was proved only recently.

Effective Nullstellensatz. *Let $f_1, \dots, f_m$ be polynomials of degree less than or equal to d in n variables, where $d \geqslant 3$, $n \geqslant 2$. If they have no common zero, then $g_1 f_1 + \cdots + g_m f_m = 1$ has a solution such that $\deg g_j \leqslant d^n - d$.*

For most systems, one can find solutions such that $\deg g_j \leqslant (n-1)(d-1)$, but in general the upper bound $d^n - d$ cannot be improved.

As explained above, this provides a computational method for deciding whether or not a system of polynomial equations has a common solution. Unfortunately, this is rather useless in practice as we end up with exceedingly large linear systems. We still do not have a computationally effective and foolproof method.

13 So, What Is Algebraic Geometry?

To me algebraic geometry is a belief in the unity of geometry and algebra. The most exciting and profound developments arise from the discovery of new connections. We have seen hints of some of these; many more were left unmentioned. Born with Cartesian coordinates, algebraic geometry is now intertwined with coding theory, number theory, computer-aided geometric design, and theoretical physics. Several of these connections have emerged in the last decade, and I hope to see many more in the future.

Further Reading

Most of the algebraic geometry literature is very technical. A notable exception is *Plane Algebraic Curves* (Birkhäuser, Boston, MA, 1986), by E. Brieskorn and H. Knörrer, which starts with a long overview of algebraic curves through arts and sciences since antiquity,

with many nice pictures and reproductions. *A Scrapbook of Complex Curve Theory* (American Mathematical Society, Providence, RI, 2003), by C. H. Clemens, and *Complex Algebraic Curves* (Cambridge University Press, Cambridge, 1992), by F. Kirwan, also start at an easily accessible level, but then delve more quickly into advanced subjects.

The best introduction to the techniques of algebraic geometry is *Undergraduate Algebraic Geometry* (Cambridge University Press, Cambridge, 1988), by M. Reid. For those wishing for a general overview, *An Invitation to Algebraic Geometry* (Springer, New York, 2000), by K. E. Smith, L. Kahanpää, P. Kekäläinen, and W. Traves, is a good choice, while *Algebraic Geometry* (Springer, New York, 1995), by J. Harris, and *Basic Algebraic Geometry*, volumes I and II (Springer, New York, 1994), by I. R. Shafarevich, are suitable for more systematic readings.

IV.5 Arithmetic Geometry
Jordan S. Ellenberg

1 Diophantine Problems, Alone and in Teams

Our goal is to sketch some of the essential ideas of arithmetic geometry; we begin with a problem which, on the face of it, involves no geometry and only a bit of arithmetic.

Problem. Show that the equation

$$x^2 + y^2 = 7z^2 \tag{1}$$

has no solution in nonzero rational numbers x, y, z.

(Note that it is only in the coefficient 7 that (1) differs from the Pythagorean equation $x^2 + y^2 = z^2$, which we know has *infinitely* many solutions. It is a feature of arithmetic geometry that modest changes of this kind can have drastic effects!)

Solution. Suppose x, y, z are rational numbers satisfying (1); we will derive from this a contradiction.

If n is the least common denominator of x, y, z, we can write

$$x = a/n, \quad y = b/n, \quad z = c/n$$

such that a, b, c, and n are integers. Our original equation (1) now becomes

$$\left(\frac{a}{n}\right)^2 + \left(\frac{b}{n}\right)^2 = 7\left(\frac{c}{n}\right)^2,$$

and multiplying through by n^2 one has

$$a^2 + b^2 = 7c^2. \tag{2}$$

If a, b, and c have a common factor m, then we can replace them by a/m, b/m, and c/m, and (2) still holds for these new numbers. We may therefore suppose that a, b, and c are integers with no common factor.

We now reduce the above equation modulo 7 (see MODULAR ARITHMETIC [III.58]). Denote by $\bar{a}$ and $\bar{b}$ the reductions of a and b modulo 7. The right-hand side of (2) is a multiple of 7, so it reduces to 0. We are left with

$$\bar{a}^2 + \bar{b}^2 = 0. \tag{3}$$

Now there are only seven possibilities for $\bar{a}$, and seven possibilities for $\bar{b}$. So the analysis of the solutions of (3) amounts to checking the forty-nine choices of $\bar{a}$, $\bar{b}$ and seeing which ones satisfy the equation. A few minutes of calculation are enough to convince us that (3) is satisfied only if $\bar{a} = \bar{b} = 0$.

But saying that $\bar{a} = \bar{b} = 0$ is the same as saying that a and b are both multiples of 7. This being the case, a^2 and b^2 are both multiples of 49. It follows that their sum, $7c^2$, is a multiple of 49 as well. Therefore, c^2 is a multiple of 7, and this implies that c itself is a multiple of 7. In particular, a, b, and c share a common factor of 7. We have now arrived at the desired contradiction, since we chose a, b, and c to have no common factor. Thus, the hypothesized solution leads us to a contradiction, so we are forced to conclude that there is not, in fact, any solution to (1) consisting of nonzero rational numbers.[1]

In general, the determination of rational solutions to a polynomial equation like (2) is called a *Diophantine problem*. We were able to dispose of (2) in a paragraph, but that turns out to be the exception: in general, Diophantine problems can be extraordinarily difficult. For instance, we might modify the exponents in (2) and consider the equation

$$x^5 + y^5 = 7z^5. \tag{4}$$

I do not know whether (4) has any solutions in nonzero rational numbers or not; one can be sure, though, that determining the answer would be a substantial piece of work, and it is quite possible that the most powerful techniques available to us are insufficient to answer this simple question.

More generally, one can take an arbitrary commutative RING [III.81] R, and ask whether a certain polynomial equation has solutions in R. For instance, does (2) have a solution with x, y, z in the polynomial ring $\mathbb{C}[t]$? (The answer is yes. We leave it as an exercise

to find some solutions.) We call the problem of solving a polynomial equation over R a *Diophantine problem over R*. The subject of arithmetic geometry has no precise boundary, but to a first approximation one may say that it concerns the solution of Diophantine problems over subrings of NUMBER FIELDS [III.63]. (To be honest, a problem is usually called Diophantine *only* when R is a subring of a number field. However, the more general definition suits our current purposes.)

With any particular equation like (2), one can associate *infinitely many* Diophantine problems, one for each commutative ring R. A central insight—in some sense the basic insight—of modern algebraic geometry is that this whole gigantic ensemble of problems can be treated as a single entity. This widening of scope reveals structure that is invisible if we consider each problem on its own. The aggregate we make of all these Diophantine problems is called a *scheme*. We will return to schemes later, and will try, without giving precise definitions, to convey some sense of what is meant by this not very suggestive term.

A word of apology: I will give only the barest sketch of the immense progress that has taken place in arithmetic geometry in recent decades—there is simply too much to cover in an article of the present scope. I have chosen instead to discuss at some length the idea of a scheme, assuming, I hope, minimal technical knowledge on the part of the reader. In the final section, I shall discuss some outstanding problems in arithmetic geometry with the help of the ideas developed in the body of the article. It must be conceded that the theory of schemes, developed by Grothendieck and his collaborators in the 1960s, belongs to algebraic geometry as a whole, and not to arithmetic geometry alone. I think, though, that in the arithmetic setting, the use of schemes, and the concomitant extension of geometric ideas to contexts that seem "nongeometric" at first glance, is particularly central.

2   Geometry without Geometry

Before we dive into the abstract theory of schemes, let us splash around a little longer among the polynomial equations of degree 2. Though it is not obvious from our discussion so far, the solution of Diophantine problems is properly classified as part of geometry. Our goal here will be to explain why this is so.

Suppose we consider the equation

$$x^2 + y^2 = 1. \tag{5}$$

---

1. Exercise: why does our argument not obtain a contradiction from the solution $x = y = z = 0$?

One can ask: which values of $x, y \in \mathbb{Q}$ satisfy (5)? This problem has a flavor very different from that of the previous section. There we looked at an equation with *no* rational solutions. We shall see in a moment that (5), by contrast, has *infinitely* many rational solutions. The solutions $x = 0$, $y = 1$ and $x = \frac{3}{5}$, $y = -\frac{4}{5}$ are representative examples. (The four solutions $(\pm 1, 0)$ and $(0, \pm 1)$ are the ones that would be said, in the usual mathematical parlance, to be "staring you in the face.")

Equation (5) is, of course, immediately recognizable as "the equation of a circle." What, precisely, do we mean by that assertion? We mean that the set of pairs of real numbers (x, y) satisfying (5) forms a circle when plotted in the Cartesian plane.

So geometry, as usually construed, makes its entrance in the figure of the circle. Now suppose that we want to find more solutions to (5). One way to proceed is as follows. Let P be the point $(1, 0)$, and let L be a line through P of slope m. Then we have the following geometric fact.

(G) The intersection of a line with a circle consists of either zero, one, or two points; the case of a single point occurs only when the line is tangent to the circle.

From (G) we conclude that, unless L is the tangent line to the circle at P, there is exactly one point other than P where the line intersects the circle. In order to find solutions (x, y) to (5), we must determine coordinates for this point. So suppose L is the line through $(1, 0)$ with slope m, which is to say it is the line L_m whose equation is $y = m(x - 1)$. Then in order to find the x-coordinates of the points of intersection between L_m and the circle, we need to solve the simultaneous equations $y = m(x - 1)$ and $x^2 + y^2 = 1$; that is, we need to solve $x^2 + m^2(x - 1)^2 = 1$ or, equivalently,

$$(1 + m^2)x^2 - 2m^2 x + (m^2 - 1) = 0. \qquad (6)$$

Of course, (6) has the solution $x = 1$. How many other solutions are there? The geometric argument above leads us to believe that there is at most one solution to (6). Alternatively, we can use the following algebraic fact, which is analogous[2] to the geometric fact (G).

(A) The equation $(1 + m^2)x^2 - 2m^2 x + (m^2 - 1) = 0$ has either zero, one, or two solutions in x.

---

2. Note that (A), unlike (G), contains no mention of tangency; that is because the notion of tangency is more subtle in the algebraic setting, as we will see in section 4 below.

Of course, the conclusion of statement (A) holds for *any* nontrivial quadratic equation in x, not just (6); it is a consequence of the factor theorem.

In this case, it is not really necessary to appeal to any theorem; one can find by direct computation that the solutions of (6) are $x = 1$ and $x = (m^2 - 1)/(m^2 + 1)$. We conclude that the intersection between the unit circle and L_m consists of $(1, 0)$ and the point P_m with coordinates

$$\left(\frac{m^2 - 1}{m^2 + 1}, \frac{-2m}{m^2 + 1} \right). \qquad (7)$$

Equation (7) establishes a correspondence $m \mapsto P_m$, which associates with each slope m a solution P_m to (5). What is more, since every point on the circle, other than $(1, 0)$ itself, is joined to $(1, 0)$ by a unique line, we find that we have established a one-to-one correspondence between slopes m and solutions, other than $(1, 0)$, to equation (5).

A very nice feature of this construction is that it allows us to construct solutions to (5) not only over $\mathbb{R}$ but over smaller fields, like $\mathbb{Q}$: it is evident that, when m is rational, so are the coordinates of the solution yielded by (7). For example, taking $m = 2$ yields the solution $(\frac{3}{5}, -\frac{4}{5})$. In fact, not only does (7) show us that (5) admits infinitely many solutions over $\mathbb{Q}$, it also gives us an explicit way to *parametrize* the solutions in terms of a variable m. We leave it as an exercise to prove that the solutions of (5) over $\mathbb{Q}$, apart from $(1, 0)$, are in one-to-one correspondence with rational values of m. Alas, rare is the Diophantine problem whose solutions can be parametrized in this way! Still, polynomial equations like (5) with solutions that can be parametrized by one or more variables play a special role in arithmetic geometry; they are called *rational varieties* and constitute by any measure the best-understood class of examples in the subject.

I want to draw your attention to one essential feature of this discussion. We relied on geometric intuition (e.g., our knowledge of facts like (G)) to give us ideas about how to construct solutions to (5). On the other hand, now that we have erected an algebraic justification for our construction, we can kick away our geometric intuition as needless scaffolding. It was a geometric fact about lines and circles that *suggested* to us that (6) should have only one solution other than $x = 1$. However, once one has had that thought, one can *prove* that there is at most one such solution by means of the purely algebraic statement (A), which involves no geometry whatsoever.

The fact that our argument can stand without any reference to geometry means that it can be applied in situations that might not, at first glance, seem geometric. For instance, suppose we wished to study solutions to (5) over the finite field $\mathbb{F}_7$. Now this solution set would not seem rightfully to be called "a circle" at all—it is just a finite set of points! Nonetheless, our geometrically inspired argument still works perfectly. The possible values of m in $\mathbb{F}_7$ are 0, 1, 2, 3, 4, 5, 6, and the corresponding solutions P_m are $(-1, 0)$, $(0, -1)$, $(2, 2)$, $(5, 5)$, $(5, 2)$, $(2, 5)$, $(0, 1)$. These seven points, together with $(1, 0)$, form the whole solution set of (5) over $\mathbb{F}_7$.

We have now started to reap the benefits of considering a whole bundle of Diophantine problems at once; in order to find the solutions to (5) over $\mathbb{F}_7$, we used a method that was inspired by the problem of finding solutions to (5) over $\mathbb{R}$. Similarly, in general, methods suggested by geometry can help us solve Diophantine problems. And these methods, once translated into purely algebraic form, still apply in situations that do not appear to be geometric.

We must now open our minds to the possibility that the purely algebraic appearance of certain equations is deceptive. Perhaps there could be a sense of "geometry" that was general enough to include entities like the solution set of (5) over $\mathbb{F}_7$, and in which this particular example had every right to be called a "circle." And why not? It has properties a circle has: most importantly for us, it has either zero, one, or two intersection points with any line. Of course, there are features of "circleness" which this set of points lacks: infinitude, continuity, roundness, etc. But these latter qualities turn out to be inessential when we are doing arithmetic geometry. From our viewpoint the set of solutions of (5) over $\mathbb{F}_7$ has every right to be called the unit circle.

To sum up, you might think of the modern point of view as an upending of the traditional story of Cartesian space. There, we have geometric objects (curves, lines, points, surfaces) and we ask questions such as, "What is the equation of this curve?" or "What are the coordinates of that point?" The underlying object is the geometric one, and the algebra is there to tell us about its properties. For us, the situation is exactly reversed: the underlying object is the *equation*, and the various geometric properties of solution sets of the equation are merely tools that tell us about the equation's algebraic properties. For an arithmetic geometer, "the unit circle" *is* the equation $x^2 + y^2 = 1$. And the round thing on the page? That is just a *picture* of the solutions to the equation over $\mathbb{R}$. It is a distinction that makes a remarkable difference.

3   From Varieties to Rings to Schemes

In this section, we will attempt to give a clearer answer to the question, "What is a scheme?" Instead of trying to lay out a precise definition—which requires more algebraic apparatus than would fit comfortably here—we will approach the question by means of an analogy.

3.1   Adjectives and Qualities

So let us think about adjectives. Any adjective, such as "yellow" for instance, picks out a set of nouns to which the adjective applies. For each adjective A, we might call this set of nouns $\Gamma(A)$. For instance, $\Gamma("yellow")$ is an infinite set that might look like {lemon, school bus, banana, sun, ... }.[3] And anyone would agree that $\Gamma(A)$ is an important thing to know about A.

Now suppose that, moved by a desire for lexical parsimony, a theoretician among us suggested that adjectives could in fact be dispensed with entirely. If, instead of A, we spoke only of $\Gamma(A)$, we could get by with a grammatical theory involving only nouns.

Is this a good idea? Well, there are certainly some obvious ways that things could go wrong. For instance, what if lots of different adjectives were sent to the same set of nouns? Then our new viewpoint would be less precise than the old one. But it certainly seems that if two adjectives apply to *exactly* the same set of nouns, then it is fair to say that the adjectives are the same, or at least synonymous.

What about relationships between adjectives? For instance, we can ask of two adjectives whether one is *stronger* than another, in the way that "gigantic" is stronger than "large." Is this relationship between adjectives still visible on the level of sets of nouns? The answer is yes: it seems fair to say that A is "stronger than" B precisely when $\Gamma(A)$ is a subset of $\Gamma(B)$. In other words, what it means to say that "gigantic" is stronger than "large" is that all gigantic things are large, though some large things may not be gigantic.

So far, so good. We have paid a price in technical difficulty: it is much more cumbersome to speak of infinite sets of nouns than it was to use simple, familiar adjectives. But we have gained something, too:

---

3. Of course, in real life, there are nouns whose relationship with "yellow" is not so clear-cut, but since our goal is to make this look like mathematics, let us pretend that every object in the world is either definitively yellow or definitively not yellow.

the opportunity for generalization. Our theoretician—whom we may now call a "set-theoretic grammarian"—observes that there is, perhaps, nothing special about the sets of nouns that happen to be of the form $\Gamma(A)$ for some already known adjective A. Why not take a conceptual leap and *redefine* the word "adjective" to mean "a set of nouns"? To avoid confusion with the usual meaning of "adjective," the theoretician might even use a new term, like "quality," to refer to his new objects of study.

Now we have a whole new world of qualities to play with. For example, there is a quality {"school bus", "sun"} which is stronger than "yellow," and a quality {"sun"} (not the same thing as the *noun* "sun"!) which is stronger than the qualities "yellow," "gigantic," "large," and {"school bus", "sun"}.

I may not have convinced you that, on balance, this reconception of the notion of "adjective" is a good idea. In fact, it probably is not, which is why set-theoretic grammar is not a going concern. The corresponding story in algebraic geometry, however, is quite a different matter.

3.2   Coordinate Rings

A warning: the next couple of sections will be difficult going for those not familiar with rings and ideals—such readers can either skip to section 4, or try to follow the discussion after reading RINGS, IDEALS, AND MODULES [III.81] (see also ALGEBRAIC NUMBERS [IV.1]).

Let us recall that a *complex affine variety* (from now on, just "variety") is the set of solutions over $\mathbb{C}$ to some finite set of polynomial equations. For instance, one variety V we could define is the set of points (x, y) in $\mathbb{C}^2$ satisfying our favorite equation

$$x^2 + y^2 = 1. \tag{8}$$

Then V is what we called in the previous section "the unit circle," though in fact the shape of the set of complex solutions of (8) is a sphere with two points removed. (This is not supposed to be obvious.) It is a question of general interest, given some variety X, to understand the ring of polynomial functions that take points on X to complex numbers. This ring is called the *coordinate ring* of X, and is denoted $\Gamma(X)$.

Certainly, given any polynomial in x and y, we can regard it as a function defined on our particular variety V. So is the coordinate ring of V just the polynomial ring $\mathbb{C}[x, y]$? Not quite. Consider, for instance, the function $f = 2x^2 + 2y^2 + 5$. If we evaluate this function

at various points on V,

$$f(0, 1) = 7, \ f(1, 0) = 7,$$
$$f(1/\sqrt{2}, 1/\sqrt{2}) = 7, \ f(\mathrm{i}, \sqrt{2}) = 7, \ \ldots,$$

we notice that f keeps taking the same value; indeed, since $x^2 + y^2 = 1$ for all $(x, y) \in V$, we see that $f = 2(x^2 + y^2) + 5$ takes the value 7 at *every* point on V. So $2x^2 + 2y^2 + 5$ and 7 are just different names for the same function on V.

So $\Gamma(V)$ is smaller than $\mathbb{C}[x, y]$; it is the ring obtained from $\mathbb{C}[x, y]$ by declaring two polynomials f and g to be the same function whenever they take the same value at every point of V. (More formally, we are defining an EQUIVALENCE RELATION [I.2 §2.3] on the set of complex polynomials in two variables.) It turns out that f and g have this property precisely when their difference is a multiple of $x^2 + y^2 - 1$. Thus, the ring of polynomial functions on V is the quotient of $\mathbb{C}[x, y]$ by the ideal generated by $x^2 + y^2 - 1$. This ring is denoted by $\mathbb{C}[x, y]/(x^2 + y^2 - 1)$.

We have shown how to attach a ring of functions to any variety. It is not hard to show that, if X and Y are two varieties, and if their coordinate rings $\Gamma(X)$ and $\Gamma(Y)$ are ISOMORPHIC [I.3 §4.1], then X and Y are in a sense the "same" variety. It is a short step from this observation to the idea of abandoning the study of varieties entirely in favor of the study of rings. Of course, we are here in the position of the set-theoretic grammarian in the parable above, with "variety" playing the part of "adjective" and "coordinate ring" the part of "set of nouns."

Happily, we can recover the geometric properties of a variety from the algebraic properties of its coordinate ring; if this were not the case, the coordinate ring would not be such a useful object! The relationship between geometry and algebra is a long story—and much of it belongs to algebraic geometry in general, not arithmetic geometry in particular—but to give the flavor, let us discuss some examples.

A straightforward geometric property of a variety is *irreducibility*. We say a variety X is *reducible* if X can be expressed as the union of two varieties X_1 and X_2, neither of which is the whole of X. For example, the variety

$$x^2 = y^2 \tag{9}$$

in $\mathbb{C}^2$ is the union of the lines $x = y$ and $x = -y$. A variety is called *irreducible* if it is not reducible. All varieties are thus built up from irreducible varieties: the relationship between irreducible varieties and general varieties is rather like the relationship between prime numbers and general positive integers.

Moving from geometry to algebra, we recall that a ring R is called an *integral domain* if, whenever f, g are nonzero elements of R, their product fg is also nonzero; the ring $\mathbb{C}[x, y]$ is a good example.

Fact. A variety X is irreducible if and only if $\Gamma(X)$ is an integral domain.

Experts will note that we are glossing over issues of "reducedness" here.

We will not prove this fact, but the following example is illustrative: consider the two functions $f = x - y$ and $g = x + y$ on the variety X defined by (9). Neither of these functions is the zero function; note, for instance, that $f(1, -1)$ is nonzero, as is $g(1, 1)$. Their product, however, is $x^2 - y^2$, which is equal to zero on X; so $\Gamma(X)$ is not an integral domain. Notice that the functions f and g that we chose are closely related to the decomposition of X as the union of two smaller varieties.

Another crucial geometric notion is that of functions from one variety to another. (It is common practice to call such functions "maps" or "morphisms"; we will use the three words interchangeably.) For instance, suppose that W is the variety in $\mathbb{C}^3$ determined by the equation $xyz = 1$. Then the map $F : \mathbb{C}^3 \to \mathbb{C}^2$ defined by

$$F(x, y, z) = \left(\frac{1}{2}(x + yz), \frac{1}{2\mathrm{i}}(x - yz) \right)$$

maps points of W to points of V.

It turns out that knowing the coordinate rings of varieties makes it very easy to see the maps between the varieties. We merely observe that if $G : V_1 \to V_2$ is a map between varieties V_1 and V_2, and if f is a polynomial function on V_2, then we have a polynomial function on V_1 that sends every point v to $f(G(v))$. This function on V_1 is denoted by $G^*(f)$. For example, if f is the function $x + y$ on V, and F is the map above, $F^*(f) = \frac{1}{2}(x + yz) + \frac{1}{2\mathrm{i}}(x - yz)$. It is easy to check that G^* is a $\mathbb{C}$-algebra homomorphism (that is, a homomorphism of rings that sends each element of $\mathbb{C}$ to itself) from $\Gamma(V_2)$ to $\Gamma(V_1)$. What is more, one has the following theorem.

Fact. For any pair of varieties V, W, the correspondence sending G to G^* is a bijection between the polynomial functions sending W to V and the $\mathbb{C}$-algebra homomorphisms from $\Gamma(V)$ to $\Gamma(W)$.

You would not be far off in thinking of the statement "there is an injective map from V to W" as analogous to "quality A is stronger than quality B."

The move to transform geometry into algebra is not something one undertakes out of sheer love of abstraction, or hatred of geometry. Instead, it is part of the universal mathematical instinct to unify seemingly disparate theories. I cannot put it any better than Dieudonné (1985) does in his *History of Algebraic Geometry*:

> ... from [the 1882 memoirs of] Kronecker and Dedekind–Weber dates the awareness of the profound analogies between algebraic geometry and the theory of algebraic numbers, which originated at the same time. Moreover, this conception of algebraic geometry is the most simple and most clear for us, trained as we are in the wielding of "abstract" algebraic notions: rings, ideals, modules, etc. But it is precisely this "abstract" character that repulsed most contemporaries, disconcerted as they were by not being able to recover the corresponding geometric notions easily. Thus the influence of the algebraic school remained very weak up until 1920. ... It certainly seems that Kronecker was the first to dream of one vast algebraico-geometric construction comprising these two theories at once; this dream has begun to be realized only recently, in our era, with the theory of schemes.

Let us therefore move on to schemes.

3.3   Schemes

We have seen that each variety X gives rise to a ring $\Gamma(X)$, and furthermore that the algebraic study of these rings can stand in for the geometric study of varieties. But just as not every set of nouns corresponds to an adjective, not every ring arises as the coordinate ring of a variety. For example, the ring $\mathbb{Z}$ of integers is not the coordinate ring of a variety, as we can see by the following argument: for every complex number a and every variety V, the constant function a is a function on V, and therefore $\mathbb{C} \subset \Gamma(V)$ for every variety V. Since $\mathbb{Z}$ does not contain $\mathbb{C}$ as a subring, it is not the coordinate ring of any variety.

Now we are ready to imitate the set-theoretic grammarian's coup de grâce. We know that some, but not all, rings arise from geometric objects (varieties); and we know that the geometry of these varieties is described by algebraic properties of these special rings. Why not, then, just consider *every* ring R to be a "geometric object" whose geometry is determined by algebraic properties of R? The grammarian needed to invent a new word, "quality," to describe his generalized adjectives; we are in the same position with our rings-that-are-not-coordinate-rings; we will call them *schemes*.

So, after all this work, the definition of scheme is rather prosaic—schemes are rings! (In fact, we are hiding some technicalities; it is correct to say that *affine schemes* are rings. Restricting our attention to affine schemes will not interfere with the phenomena that we are aiming to explain.) More interesting is to ask how we can carry out the task whose difficulty "disconcerted" the early algebraic geometers—how can we identify "geometric" features of arbitrary rings?

For instance, if R is supposed to be an arbitrary geometric object, it ought to have "points." But what are the "points" of a ring? Clearly we cannot mean by this the *elements* of the ring; for in the case $R = \Gamma(X)$, the elements of R are *functions* on X, not points on X. What we need, given a point p on X, is some entity attached to the ring R that corresponds to p.

The key observation is that we can think of p as a map from $\Gamma(X)$ to $\mathbb{C}$: given a function f from $\Gamma(X)$ we map it to the complex number $f(p)$. This map is a homomorphism, called the *evaluation homomorphism at* p. Since points on X give us homomorphisms on $\Gamma(X)$, a natural way to define the word "point" for the ring $R = \Gamma(X)$, without using geometry, is to say that a "point" is a homomorphism from R to $\mathbb{C}$. It turns out that the kernel of such a homomorphism is a maximal ideal, i.e., a proper ideal in R which is contained in no larger ideal except R itself. Moreover, every maximal ideal of R arises from a point p of X. So a very concise way to describe the points of X might be to say that they are the maximal ideals of R. A modern algebraic geometer would say that all *prime* ideals correspond to points, not only the maximal ones. The "points" corresponding to the nonmaximal ideals are not points in the usual sense of the term; for instance, the point corresponding to the zero ideal (when it is prime) is the "generic point," which is in one sense everywhere on X at once, and in another sense nowhere in particular at all. This description sounds rather woolly, but on the algebraic side the zero ideal is something quite concrete—and in fact, having a precise notion of "generic point" turns out quite often to be useful in making a certain species of vague geometric argument into a rigorous proof.

The definition we have arrived at makes sense for *all* rings R, and not just those of the form $R = \Gamma(X)$. So we might define the "points" of a ring R to be its prime ideals. The set of prime ideals of R is given the name

Spec R, and it is Spec R that we call the *scheme associated with* R. (More precisely, Spec R is defined to be a "locally ringed topological space" whose points are the prime ideals of R, but we will not need the full power of this definition for our discussion here.)

We are now in a position to elucidate our claim, made in the first section, that a scheme incorporates into one package Diophantine problems over many different rings. Suppose, for instance, that R is the ring $\mathbb{Z}[x, y]/(x^2 + y^2 - 1)$. We are going to catalog the homomorphisms $f : R \to \mathbb{Z}$. To specify f, I merely have to tell you the values of $f(x)$ and $f(y)$ in $\mathbb{Z}$. But I cannot choose these values arbitrarily: since $x^2 + y^2 - 1 = 0$ in R, it must be the case that $f(x)^2 + f(y)^2 - 1 = 0$ in $\mathbb{Z}$. In other words, the pair $(f(x), f(y))$ constitutes a solution over $\mathbb{Z}$ to the Diophantine equation $x^2 + y^2 = 1$. What is more, the same argument shows that, for *any* ring S, a homomorphism $f : R \to S$ yields a solution over S to $x^2 + y^2 = 1$, and vice versa. In summary,

> for each S, there is a one-to-one correspondence between the set of ring homomorphisms from R to S, and solutions over S to $x^2 + y^2 = 1$.

This behavior is what we have in mind when we say that the ring R "packages" information about Diophantine equations over different rings.

It turns out, just as one might hope, that every interesting geometric property of varieties can be computed by means of the coordinate ring, which means it can be defined not only for varieties but also for general schemes. We have already seen, for instance, that a variety X is irreducible if and only if $\Gamma(X)$ is an integral domain. Thus, we say in general that a scheme Spec R is irreducible if and only if R is an integral domain (or, more precisely, if the quotient of R by its nilradical is an integral domain). One can speak of the connectedness of a scheme, its dimension, whether it is smooth, and so forth. All these geometric properties turn out, like irreducibility, to have purely algebraic descriptions. In fact, to the arithmetic geometer's way of thinking, all these *are*, at bottom, algebraic properties.

3.4 Example: Spec $\mathbb{Z}$, the Number Line

The first ring we encounter in our mathematical education—and the ring that is the ultimate subject of number theory—is $\mathbb{Z}$, the ring of integers. How does it fit into our picture? The scheme Spec $\mathbb{Z}$ has as its points the set of prime ideals of $\mathbb{Z}$, which come in two flavors: there are the principal ideals (p), with p a prime number; and there is the zero ideal.

We are supposed to think of $\mathbb{Z}$ as the ring of "functions" on Spec $\mathbb{Z}$. How can an integer be a function? Well, I merely need to tell you how to evaluate an integer n at a point of Spec $\mathbb{Z}$. If the point is a nonzero prime ideal (p), then the evaluation homomorphism at (p) is precisely the homomorphism whose kernel is (p); so the value of n at (p) is just the reduction of n modulo p. At the point (0), the evaluation homomorphism is the identity map $\mathbb{Z} \to \mathbb{Z}$; so the value of n at (0) is just n.

4   How Many Points Does a Circle Have?

We now return to the method of section 2, paying particular attention to the case where the equation $x^2 + y^2 = 1$ is considered over a finite field $\mathbb{F}_p$.

Let us write V for the scheme of solutions of $x^2 + y^2 = 1$. For any ring R, we will denote by $V(R)$ the set of solutions of $x^2 + y^2 = 1$.

If R is a finite field $\mathbb{F}_p$, the set $V(\mathbb{F}_p)$ is a subset of $\mathbb{F}_p^2$. In particular, it is a *finite* set. So it is natural to wonder how large this set is: in other words, how many points does a circle have?

In section 2, guided by our geometric intuition, we observed that, for every $m \in \mathbb{Q}$, the point

$$P_m = \left(\frac{m^2 - 1}{m^2 + 1}, \frac{-2m}{m^2 + 1} \right)$$

lies on V.

The algebraic computation showing that P_m satisfies the equation $x^2 + y^2 = 1$ is no different over a finite field. So we might be inclined to think that $V(\mathbb{F}_p)$ consists of $p + 1$ points: namely, the points P_m for each $m \in \mathbb{F}_p$, together with $(1, 0)$.

But this is not right: for instance, when $p = 5$ it is easy to check that the four points $(0, 1)$, $(0, -1)$, $(1, 0)$, $(-1, 0)$ make up all of $V(\mathbb{F}_5)$. Computing P_m for various m, we quickly discover the problem; when m is 2 or 3, the formula for P_m does not make sense, because the denominator $m^2 + 1$ is zero! This is a wrinkle we did not see over $\mathbb{Q}$, where $m^2 + 1$ was always positive.

What is the geometric story here? Consider the intersection of the line L_2, that is, the line $y = 2(x - 1)$, with V. If (x, y) belongs to this intersection, then $x^2 + (2(x-1))^2 = 1$, so $5x^2 - 8x + 3 = 0$. Since $5 = 0$ and $8 = 3$ in $\mathbb{F}_5$, this equation can be written as $3 - 3x = 0$; in other words, $x = 1$, which in turn implies that $y = 0$. In other words, the line L_2 intersects the circle V at only one point!

We are left with two possibilities, both disturbing to our geometric intuition. We might declare that L_2 is tangent to V; but this means that V would have multiple

tangents at $(1, 0)$, since the vertical line $x = 1$ should surely still be considered a tangent. The alternative is to declare that L_2 is *not* tangent to V; but then we are in the equally unsavory situation of having a line which, while not tangent to the circle V, intersects it at only one point. You are now beginning to see why I did not include an algebraic definition of "tangent" in statement (A) above!

This quandary illustrates the nature of arithmetic geometry nicely. When we move into novel contexts, like geometry over $\mathbb{F}_p$, some features stay fixed (such as "a line intersects a circle in at most two points"), while others have to be discarded (such as "there exists exactly one line, which we may call the tangent line to the circle at $(1, 0)$, that intersects the circle at $(1, 0)$ and no other point"[4]).

Notwithstanding these subtleties, we are now ready to compute the number of points in $V(\mathbb{F}_p)$. First of all, when $p = 2$ one can check directly that $(0, 1)$ and $(1, 0)$ are the only two points in $V(\mathbb{F}_2)$. Having treated this case, we assume for the rest of this section that p is odd. It follows from basic number theory that the equation $m^2 + 1 = 0$ has a solution in $\mathbb{F}_p$ if and only if $p \equiv 1 \pmod 4$, in which case there are exactly two such m. So, if $p \equiv 3 \pmod 4$, then every line L_m intersects the circle at a point other than $(1, 0)$, and we have $p + 1$ points in all. If $p \equiv 1 \pmod 4$, there are two choices of m for which L_m intersects V only at $(1, 0)$; eliminating these two choices of m yields a total of $p - 1$ points in $V(\mathbb{F}_p)$.

We conclude that $|V(\mathbb{F}_p)|$ is equal to 2 when $p = 2$, to $p - 1$ when $p \equiv 1 \pmod 4$, and to $p + 1$ when $p \equiv 3 \pmod 4$. The interested reader will find the following exercises useful: how many solutions are there to $x^2 + 3y^2 = 1$ over $\mathbb{F}_p$? What about $x^2 + y^2 = 0$?

More generally, let X be the scheme of solutions of *any* system of equations

$$F_1(x_1, \ldots, x_n) = 0, \quad F_2(x_1, \ldots, x_n) = 0, \ldots, \quad (10)$$

where the F_i are polynomials with integral coefficients. Then one can associate with F a list of integers $N_2(X), N_3(X), N_5(X), \ldots$, where $N_p(X)$ is the number of solutions to (10) with $x_1, \ldots, x_n \in \mathbb{F}_p$. This list of integers turns out to contain a surprising amount of geometric information about the scheme X; even for the simplest schemes, the analysis of these lists is a deep problem of intense current interest, as we will see in the next section.

---

4. In this case, the right attitude to adopt is that L_2 is not tangent to V, but that there are certain nontangent lines that intersect the circle at a single point.

5 Some Problems in Classical and Contemporary Arithmetic Geometry

In this section I will try to give an impression of a few of arithmetic geometry's great successes, and to gesture at some problems of current interest for researchers in the area.

A word of warning is in order. In what follows, I will be trying to give brief and nontechnical descriptions of some mathematics of extreme depth and complexity. Consequently, I will feel very free to oversimplify. I will try to avoid making assertions that are actually false, but I will often use definitions (like that of the L-function attached to an elliptic curve) that do not exactly agree with those in the literature.

5.1   From Fermat to Birch–Swinnerton-Dyer

The world is not lacking in expositions of the proof of FERMAT'S LAST THEOREM [V.10] and I will not attempt to give another one here, although it is without question the most notable contemporary achievement in arithmetic geometry. (Here I am using the mathematician's sense of "contemporary," which, as the old joke goes, means "theorems proved since I entered graduate school." The shorthand for "theorems proved before I entered graduate school" is "classical.") I will content myself with making some comments about the structure of the proof, emphasizing connections with the parts of arithmetic geometry we have discussed above.

Fermat's last theorem (rightly called "Fermat's conjecture," since it is almost impossible to imagine that FERMAT [VI.12] proved it) asserts that the equation

$$A^\ell + B^\ell = C^\ell, \qquad (11)$$

where ℓ is an odd prime, has no solutions in positive integers A, B, C.

The proof uses the crucial idea, introduced independently by Frey and Hellegouarch, of associating with any solution (A, B, C) of (11) a certain variety $X_{A,B}$, namely the curve described by the equation

$$y^2 = x(x - A^\ell)(x + B^\ell).$$

What can we say about $N_p(X_{A,B})$? We begin with a simple heuristic. There are p choices for x in $\mathbb{F}_p$. For each choice of x, there are either zero, one, or two choices for y, depending on whether $x(x - A^\ell)(x + B^\ell)$ is a quadratic nonresidue, zero, or a quadratic residue in $\mathbb{F}_p$. Since there are equally many quadratic residues and nonresidues in $\mathbb{F}_p$, we might guess that those two cases arise equally often. If so, there would on average be one choice of y for each of the p choices of x, which

inclines us to make the estimate $N_p(X_{A,B}) \sim p$. Define a_p to be the error in this estimate: $a_p = p - N_p(X_{A,B})$. It is worth remembering that when X was the scheme attached to $x^2 + y^2 = 1$, the behavior of $p - N_p(X)$ was very regular; in particular, this quantity took the value 1 at primes congruent to 1 mod 4 and -1 at primes congruent to 3 mod 4. (We note, in particular, that the heuristic estimate $N_p(X) \sim p$ is quite good in this case.) Might one hope that a_p displays the same kind of regularity?

In fact, the behavior of the a_p is very *irregular*, as a famous theorem of Mazur shows; not only do the a_p fail to vary periodically, even their reductions modulo various primes are irregular!

Fact (Mazur). Suppose that ℓ is a prime greater than 3, and let b be a positive integer. It is not the case that a_p takes the same value (mod ℓ) for all primes p congruent to 1 (mod b).[5]

On the other hand—if I may compress a 200-page paper into a slogan—Wiles proved that, when A, B, C is a solution to (11), the reductions mod ℓ of the a_p *necessarily* behaved periodically, contradicting Mazur's theorem when $\ell > 3$. The case $\ell = 3$ is an old theorem of EULER [VI.19]. This completes the proof of Fermat's conjecture and, I hope, bolsters our assertion that the careful study of the values $N_p(X)$ is an interesting way to study a variety X!

But the story does not end with Fermat. In general, if $f(x)$ is a cubic polynomial with coefficients in $\mathbb{Z}$ and no repeated roots, the curve E defined by the equation

$$y^2 = f(x) \qquad (12)$$

is called an ELLIPTIC CURVE [III.21] (note well that an elliptic curve is not an ellipse). The study of rational points on elliptic curves (that is, pairs of rational numbers satisfying (12)) has been occupying arithmetic geometers since before our subject existed as such; a decent treatment of the story would fill a book, as indeed it does fill the book of Silverman and Tate (1992). We can define $a_p(E)$ to be $p - N_p(E)$ as above. First of all, if our heuristic $N_p(E) \sim p$ is a good estimate, we might expect that $a_p(E)$ is small compared with p; and, in fact, a theorem of Hasse from the 1930s shows that $a_p(E) \leqslant 2\sqrt{p}$ for all but finitely many p.

---

5. The theorem proved by Mazur is stated by him in a very different and much more general way: he proves that certain *modular curves* do not possess any rational points. This implies that a version of the fact above is true, not only for $X_{A,B}$, but for *any* equation of the form $y^2 = f(x)$, where f is a cubic polynomial without repeated roots. We will leave it to the other able treatments of Fermat to develop that point of view.

It turns out that some elliptic curves have infinitely many rational points, and some only finitely many. One might expect that an elliptic curve with many points over $\mathbb{Q}$ would tend to have more points over finite fields as well, since the coordinates of a rational point can be reduced mod p to yield a point over the finite field $\mathbb{F}_p$. Conversely, one might imagine that, by knowing the list of numbers a_p, one could draw conclusions about the points of E over $\mathbb{Q}$.

In order to draw such conclusions, one needs a nice way to package the information of the infinite list of integers a_p. Such a package is given by the *L*-FUNCTION [III.47] of the elliptic curve, defined to be the following function of a variable s:

$$L(E, s) = \prod_p{}' (1 - a_p p^{-s} + p^{1-2s})^{-1}. \qquad (13)$$

The notation $\prod'$ means that this product is evaluated over all primes apart from a finite set, which is easy to determine from the polynomial f. (As is often the case, we are oversimplifying; what I have written here differs in some irrelevant-to-us respects from what is usually called $L(E, s)$ in the literature.) It is not hard to check that (13) is a convergent product when s is a real number greater than $\frac{3}{2}$. Not much deeper is the fact that the right-hand side of (13) is well-defined when s is a complex number whose real part exceeds $\frac{3}{2}$. What *is* much deeper—following from the theorem of Wiles, together with later theorems of Breuil, Conrad, Diamond, and Taylor—is that we can extend $L(E, s)$ to a HOLOMORPHIC FUNCTION [I.3 §5.6] defined for *every* complex number s.

A heuristic argument might suggest the following relationship between the values of $N_p(E)$ and the value of $L(E, 1)$. If the a_p are typically negative (corresponding to the $N_p(E)$ typically being greater than p) the terms in the infinite product tend to be smaller than 1; when the a_p are positive, the terms in the product tend to be larger than 1. In particular, one might expect the value of $L(E, 1)$ to be closer to 0 when E has many rational points. Of course, this heuristic should be taken with a healthy pinch of salt, given that $L(E, 1)$ is not in fact defined by the infinite product on the right-hand side of (13)! Nonetheless, THE BIRCH–SWINNERTON-DYER CONJECTURE [V.4], which makes precise the heuristic prediction above, is widely believed, and supported by many partial results and numerical experiments. We do not have the space here to state the conjecture in full generality. However, the following conjecture would follow from Birch–Swinnerton-Dyer.

Conjecture. The elliptic curve E has infinitely many points over $\mathbb{Q}$ if and only if $L(E, 1) = 0$.

Kolyvagin proved one direction of this conjecture in 1988: that E has finitely many rational points if $L(E, 1) \neq 0$. (To be precise, he proved a theorem that yields the assertion here once combined with the later theorems of Wiles and others.) It follows from a theorem of Gross and Zagier that E has infinitely many rational points if $L(E, s)$ has a *simple* zero at $s = 1$. That more or less sums up our present knowledge about the relationship between L-functions and rational points on elliptic curves. This lack of knowledge has not, however, prevented us from constructing a complex of ever more rarefied conjectures in the same vein, of which the Birch–Swinnerton-Dyer conjecture is only a tiny and relatively down-to-earth sliver.

Before we leave the subject of counting points behind, we will pause and point out one more beautiful result: the theorem of ANDRÉ WEIL [VI.93] bounding the number of points on a curve over a finite field. (Because we have not introduced projective geometry, we will satisfy ourselves with a somewhat less beautiful formulation than the usual one.) Let $F(x, y)$ be an irreducible polynomial in two variables, and let X be the scheme of solutions of $F(x, y) = 0$. Then the complex points of X define a certain subset of $\mathbb{C}^2$, which we call an *algebraic curve*. Since X is obtained by imposing one polynomial condition on the points of $\mathbb{C}^2$, we expect that X has complex dimension 1, which is to say it has real dimension 2. Topologically speaking, $X(\mathbb{C})$ is, therefore, a surface. It turns out that, for almost all choices of F, the surface $X(\mathbb{C})$ will have the topology of a "g-holed doughnut" with d points removed, for some nonnegative integers g and d. In this case we say that X is a *curve of genus g*.

In section 2 we saw that the behavior of schemes over finite fields seemed to "remember" facts arising from our geometric intuition over $\mathbb{R}$ and $\mathbb{C}$: our example there was the fact that circles and lines intersect in at most two points.

The theorem of Weil reveals a similar, though much deeper, phenomenon.

Fact. Suppose the scheme X of solutions of $F(x, y)$ is a curve of genus g. Then, for all but finitely many primes p, the number of points of X over $\mathbb{F}_p$ is at most $p + 1 + 2g\sqrt{p}$ and at least $p + 1 - 2g\sqrt{p} - d$.

Weil's theorem illustrates the startlingly close bonds between geometry and arithmetic. The more complicated the topology of $X(\mathbb{C})$, the further the number of

$\mathbb{F}_p$-points can vary from the "expected" answer of p. What is more, it turns out that knowing the size of the set $X(\mathbb{F}_q)$ for every finite field $\mathbb{F}_q$ allows us to determine the genus of X. In other words, the *finite sets of points* $X(\mathbb{F}_q)$ somehow "remember" the topology of the space of complex points $X(\mathbb{C})$! In modern language, we say that there is a theory applying to general schemes, called *étale cohomology*, which mimics the theory of cohomology applying to the topology of varieties over $\mathbb{C}$.

Let us return for a moment to our favorite curve, by taking the polynomial $F(x, y) = x^2 + y^2 - 1$. In this case, it turns out that $X(\mathbb{C})$ has $g = 0$ and $d = 2$: our previous result that $X(\mathbb{F}_p)$ contains either $p + 1$ or $p - 1$ points therefore conforms exactly with the Weil bounds. We also remark that elliptic curves always have genus 1; so the theorem of Hasse alluded to above is a special case of Weil's theorem as well.

Recall from section 2 that the solutions to $x^2 + y^2 = 1$, over $\mathbb{R}$, over $\mathbb{Q}$, or over various finite fields, could be parametrized by the variable m. It was this parametrization that enabled us to determine a simple formula for the size of $X(\mathbb{F}_p)$ in this case. We remarked earlier that most schemes could not be so parametrized; now we can make that statement a bit more precise, at least for algebraic curves.

Fact. If X is a genus-0 curve, then the points of X can be parametrized by a single variable.

The converse of this fact is more or less true as well (though stating it properly requires us to say more than we can here about "singular curves"). In other words, a thoroughly algebraic question—whether the solutions of a Diophantine equation can be parametrized—is hereby given a geometric answer.

5.2 Rational Points on Curves

As we said above, some elliptic curves (which are curves of genus 1) have finitely many rational points, and others have infinitely many. What is the situation for algebraic curves of other flavors?

We have already encountered a curve of genus 0 with infinitely many points: namely, the curve $x^2 + y^2 = 1$. On the other hand, the curve $x^2 + y^2 = 7$ also has genus 0, and a simple modification of the argument of the first section shows that this curve has *no* rational points. It turns out these are the only two possibilities.

Fact. If X is a curve of genus 0, then $X(\mathbb{Q})$ is either empty or infinite.

Genus-1 curves are known to fall into a similar dichotomy, thanks to the theorem of Mazur we alluded to earlier.

Fact. If X is a genus-1 curve, then either X has at most sixteen rational points or it has infinitely many rational points.

What about curves of higher genus? In the early 1920s, Mordell made the following conjecture.

Conjecture. If X is a curve of genus greater than 2, then X has finitely many rational points.

This conjecture was proved by Faltings in 1983; in fact, he proved a more general theorem of which this conjecture is a special case. It is worth remarking that the work of Faltings involves a great deal of importation of geometric intuition to the study of the scheme $\operatorname{Spec}\mathbb{Z}$.

When you prove that a set is finite, it is natural to wonder whether you can bound its size. For example, if $f(x)$ is a degree 6 polynomial with no repeated roots, the curve $y^2 = f(x)$ turns out to have genus 2; so by Faltings's theorem there are only finitely many pairs of rational numbers (x, y) satisfying $y^2 = f(x)$.

Question. Is there a constant B such that, for all degree 6 polynomials with coefficients in $\mathbb{Q}$ and no repeated roots, the equation $y^2 = f(x)$ has at most B solutions?

This question remains open, and I do not think there is a strong consensus about whether the answer will be yes or no. The current world record is held by the curve

$$y^2 = 378\,371\,081x^2(x^2 - 9)^2 - 229\,833\,600(x^2 - 1)^2,$$

which was constructed by Keller and Kulesz and has 588 rational points.

Interest in the above question comes from its relation to a conjecture of Lang, which involves points on higher-dimensional varieties. Caporaso, Harris, and Mazur showed that Lang's conjecture implies a positive answer to the question above. This suggests a natural attack on the conjecture: if one can find a way to construct an infinite sequence of degree 6 polynomials $f(x)$ so that the equations $y = f(x)$ have ever more numerous rational solutions, then one has a disproof of Lang's conjecture! No one has yet been successful at this task. If one could *prove* that the answer to the question above was affirmative, it would probably bolster our faith in the correctness of Lang's conjecture, though of course it would bring us no nearer to turning the conjecture into a theorem.

In this article we have seen only a glimpse of the modern theory of arithmetic geometry, and perhaps I have overemphasized mathematicians' successes at the expense of the much larger territory of questions, like Lang's conjecture above, about which we remain wholly ignorant. At this stage in the history of mathematics, we can confidently say that the schemes attached to Diophantine problems *have geometry*. What remains is to say as much as we can about *what this geometry is like*, and in this respect, despite the progress described here, our understanding is still quite unsatisfactory when compared with our knowledge of more classical geometric situations.

Further Reading

Dieudonné, J. 1985. *History of Algebraic Geometry.* Monterey, CA: Wadsworth.

Silverman, J., and J. Tate. 1992. *Rational Points on Elliptic Curves.* New York: Springer.

IV.6   Algebraic Topology
Burt Totaro

Introduction

Topology is concerned with the properties of a geometric shape that are unchanged when we continuously deform it. In more technical terms, topology tries to classify TOPOLOGICAL SPACES [III.90], where two spaces are considered the same if they are homeomorphic. Algebraic topology assigns numbers to a topological space, which can be thought of as the "number of holes" in that space. These holes can be used to show that two spaces are not homeomorphic: if they have different numbers of holes of some kind, then one cannot be a continuous deformation of the other. In the happiest cases, we can hope to show the converse statement: that two spaces with the same number of holes (in some precise sense) *are* homeomorphic.

Topology is a relatively new branch of mathematics, with its origins in the nineteenth century. Before that, mathematics usually sought to solve problems exactly: to solve an equation, to find the path of a falling body, to compute the probability that a game of dice will lead to bankruptcy. As the complexity of mathematical problems grew, it became clear that most problems would never be solved by an exact formula: a classic example is the problem, known as THE THREE-BODY PROBLEM [V.33], of computing the future movements of Earth, the Sun, and the Moon under the influence of gravity. Topology allows the possibility of making qualitative predictions when quantitative ones are impossible. For example, a simple topological fact is that a trip from New York to Montevideo must cross the equator at some point, although we cannot say exactly where.

1   Connectedness and Intersection Numbers

Perhaps the simplest topological property is one called *connectedness*. This can be defined in various ways, as we shall see in a moment, but once we have a notion of what it means for a space to be connected we can then divide a topological space up into connected pieces, called *components*. The number of these pieces is a simple but useful INVARIANT [I.4 §2.2]: if two spaces have different numbers of connected components, then they are not homeomorphic.

For nice topological spaces, the different definitions of connectedness are equivalent. However, they can be generalized to give ways of measuring the number of holes in a space; these generalizations are interestingly different and all of them are important.

The first interpretation of connectedness uses the notion of a *path*, which is defined to be a continuous mapping f from the unit interval $[0, 1]$ to a given space X. (We think of f as a path from $f(0)$ to $f(1)$.) Let us declare two points of X to be equivalent if there is a path from one to the other. The set of EQUIVALENCE CLASSES [I.2 §2.3] is called the set of *path components* of X and is written $\pi_0(X)$. This is a very natural way of defining the "number of connected pieces" into which X breaks up. One can generalize this notion by considering mappings into X from other standard spaces such as spheres: this leads to the notion of homotopy groups, which will be the topic of section 2.

A different way of thinking about connectedness is based on functions from X to the real line rather than functions from a line segment into X. Let us assume that we are in a situation where it makes sense to differentiate functions on X. For example, X could be an open subset of some Euclidean space, or more generally a SMOOTH MANIFOLD [I.3 §6.9]. Consider all the real-valued functions on X whose derivative is everywhere equal to zero: these functions form a real VECTOR SPACE [I.3 §2.3], which we call $H^0(X, \mathbb{R})$ (the "zeroth cohomology group of X with real coefficients"). Calculus tells us that if a function defined on an interval has derivative zero, then it must be constant, but that is not true when the domain has several connected pieces: all we can say then is that the function is constant on each connected piece of X. The number of degrees of freedom of such a function is therefore equal to the number of connected pieces, so the dimension of the vector space $H^0(X, \mathbb{R})$ is another way to describe the number of connected components of X. This is the simplest example of a cohomology group. Cohomology will be discussed in section 4.

We can use the idea of connectedness to prove a serious theorem of algebra: every real polynomial of odd degree has a real root. For example, there must be some real number x such that $x^3 + 3x - 4 = 0$. The basic observation is that when x is a large positive number or a highly negative number, the term x^3 is much bigger (in absolute value) than the other terms of the polynomial. Since this top term is an odd power of x, we have $f(x) > 0$ for some positive number x and $f(x) < 0$ for some negative number x. If f were never equal to zero, then it would be a continuous mapping from the real line into the real line minus the origin. But the real line is connected, while the real line minus the origin has two connected components, the positive and negative numbers. It is easy to show that a continuous map from a connected space X to another space Y must map X into just one connected component of Y: in our case, this contradicts the fact that f takes both positive and negative values. Therefore f must be equal to zero at some point, and the proof is complete.

This argument can be phrased in terms of the "intermediate value theorem" of calculus, which is indeed one of the most basic topological theorems. An equivalent reformulation of this theorem states that a continuous curve that goes from the lower half-plane to the upper half-plane must cross the horizontal axis at some point. This idea leads to *intersection numbers*, one of the most useful concepts in topology. Let M be a smooth oriented manifold. (Roughly speaking, a manifold is oriented if you cannot continuously slide a shape about inside it and end up with a reflection of that shape. The simplest nonoriented manifold is a Möbius strip: to reflect a shape, slide it around the strip an odd number of times.) Let A and B be two closed oriented submanifolds of M with dimensions adding up to the dimension of M. Finally, suppose that A and B intersect transversely, so that their intersection has the "correct" dimension, namely 0, and is therefore a collection of separated points.

Now let p be one of these points. There is a way of assigning a weight of $+1$ or -1 to p, which depends

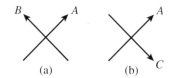

Figure 1 Intersection numbers: (a) $A \cdot B = 1$; (b) $A \cdot C = -1$.

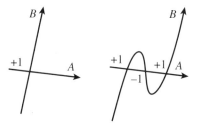

Figure 2 Moving a submanifold.

in a natural way on the relationship between the orientations of A, B, and M (see figure 1). For example, if M is a sphere, A is the equator of M, B is a closed curve, and appropriate directions are given to A and B, then the weight of p will tell you whether B crosses A upwards or downwards at p. If A and B intersect in only finitely many points, then we can define the intersection number of A and B, written $A \cdot B$, to be the sum of the weights ($+1$ or -1) at all the intersection points. In particular, this will happen if M is COMPACT [III.9] (that is, we can think of it as a closed bounded subset of $\mathbb{R}^N$ for some N).

The important point about the intersection number is that it is an *invariant*, in the following sense: if you move A and B about in a continuous way, ending up with another pair of transverse submanifolds A' and B', then the intersection number $A' \cdot B'$ is the same as $A \cdot B$, even though the number of intersection points can change. To see why this might be true, consider again the case where A and B are curves and M is two dimensional: if A and B meet at a point with weight 1, we can wiggle one of them to turn that point into three points with weights 1, -1, and 1, but the total contribution to the intersection number is unchanged. This is illustrated in figure 2. As a result, the intersection number $A \cdot B$ is defined for *any* two submanifolds of complementary dimension: if they do not intersect transversely, one can move them until they do and use the definition we have just given.

In particular, if two submanifolds have nonzero intersection number, then they can never be moved to

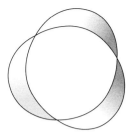

Figure 3 A surface bounded by a knot.

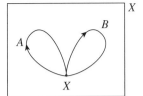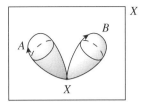

Figure 4 Multiplication in the fundamental group and in higher homotopy groups.

be disjoint from each other. This is another way to describe the earlier arguments about connectedness. It is easy to write down one curve from New York to Montevideo whose intersection number with the equator is equal to 1. Therefore, no matter how we move that curve (provided that we keep the endpoints fixed: more generally, if either A or B has a boundary, then that boundary should be kept fixed), its intersection number with the equator will always be 1, and in particular it must meet the equator in at least one point.

One of many applications of intersection numbers in topology is the idea of *linking numbers*, which comes from KNOT THEORY [III.44]. A *knot* is a path in space that begins and ends at the same point, or, more formally, a closed connected one-dimensional submanifold of $\mathbb{R}^3$. Given any knot K, it is always possible to find a surface S in $\mathbb{R}^3$ with K as its boundary (see figure 3). Now let L be a knot that is disjoint from K. The linking number of K with L is defined to be the intersection number of L with the surface S. The properties of intersection numbers imply that if the linking number of K with L is nonzero, then the knots K and L are "linked," in the sense that it is impossible to pull them apart.

2   Homotopy Groups

If we remove the origin from the plane $\mathbb{R}^2$, then we obtain a new space that is different from the plane in a fundamental way: it has a hole in it. However, we cannot detect this difference by counting components, since both the plane and the plane without the origin are connected. We begin this section by defining an invariant called the *fundamental group*, which does detect this kind of hole.

As a first approximation, one could say that the elements of the fundamental group of a space X are *loops*, which can be formally defined as continuous functions f from $[0,1]$ to X such that $f(0) = f(1)$. However, this is not quite accurate, for two reasons. The first reason, which is extremely important, is that two loops

are regarded as equivalent if one can be continuously deformed to the other while all the time staying inside X. If this is the case, we say that they are *homotopic*. To be more formal about this, let us suppose that f_0 and f_1 are two loops. Then a *homotopy* between f_0 and f_1 is a collection of loops f_s in X, one for each s between 0 and 1, such that the function $F(s,t) = f_s(t)$ is a continuous function from $[0,1]^2$ to X. Thus, as s increases from 0 to 1, the loop f_s moves continuously from f_0 to f_1. If two loops are homotopic, then we count them as the same. So the elements of the homotopy group are not actually loops but equivalence classes, or *homotopy classes*, of loops.

Even this is not quite correct, because for technical reasons we need to impose an extra condition on our loops: that they all start from (and therefore end at) a given point, called the *base point*. If X is connected, it turns out not to matter what this base point is, but we need it to be the same for all loops. The reason for this is that it gives us a way to multiply two loops: if x is the base point and A and B are two loops that start and end at x, then we can define a new loop by going around A and then going around B. This is illustrated in figure 4. We regard this new loop as the product of the loops A and B. It is not hard to check that the homotopy class of this product depends only on the homotopy classes of A and B, and that the resulting binary operation turns the set of homotopy classes of loops into a GROUP [I.3 §2.1]. It is this group that we call the fundamental group of X. It is denoted $\pi_1(X)$.

The fundamental group can be computed for most of the spaces we are likely to encounter. This makes it an important way to distinguish one space from another. First of all, for any n the fundamental group of $\mathbb{R}^n$ is the trivial group with just one element, because any loop in $\mathbb{R}^n$ can be continuously shrunk to its base point. On the other hand, the fundamental group of $\mathbb{R}^2 \setminus \{0\}$, the plane with the origin removed, is isomorphic to the group $\mathbb{Z}$ of the integers. This tells us that we can associate with any loop in $\mathbb{R}^2 \setminus \{0\}$ an integer that does not change if we modify the loop in a continuous way. This integer is known as the *winding number*. Intuitively, the winding number measures the total number of times that the mapping goes around the origin, with counterclockwise circuits counting positively and clockwise ones negatively. Since the fundamental group of $\mathbb{R}^2 \setminus \{0\}$ is not the trivial group, $\mathbb{R}^2 \setminus \{0\}$ cannot be homeomorphic to the plane. (It is an interesting exercise to try to find an elementary proof of this result—that is, a proof that does not use, or implicitly reconstruct, any of the machinery of algebraic topology. Such proofs do exist, but it is tricky to find them.)

A classic application of the fundamental group is to prove THE FUNDAMENTAL THEOREM OF ALGEBRA [V.13], which states that every nonconstant polynomial with complex coefficients has a complex root. (The proof is sketched in the article just cited, though the fundamental group is not explicitly mentioned there.)

The fundamental group tells us about the number of "one-dimensional holes" that a space has. A basic example is given by the circle, which has fundamental group $\mathbb{Z}$, just as $\mathbb{R}^2 \setminus \{0\}$ does, and for essentially the same reason: given a path in the circle that begins and ends at the same point, we can see how many times it goes around the circle. In the next section we shall see some more examples.

Before we think about higher-dimensional holes, we first need to discuss one of the most important topological spaces: the n-dimensional sphere. For any natural number n, this is defined to be the set of points in $\mathbb{R}^{n+1}$ at distance 1 from the origin. It is denoted S^n. Thus, the 0-sphere S^0 consists of two points, the 1-sphere S^1 is the circle, and the 2-sphere S^2 is the usual sphere, like the surface of Earth. Higher-dimensional spheres take a little bit of getting used to, but we can work with them in the same way that we can with lower-dimensional spheres. For example, we can construct the 2-sphere from a closed two-dimensional disk by identifying all the points on the boundary circle with each other. In the same way, the 3-sphere can be obtained from a solid three-dimensional ball by identifying all the points on the boundary 2-sphere. A related picture is to think of the 3-sphere as being obtained from our familiar three-dimensional space $\mathbb{R}^3$ by adding one point "at infinity."

Now let us think about the familiar sphere S^2. This has trivial fundamental group, since any loop drawn on the sphere can be shrunk to a point. However, this does not mean that the topology of S^2 is trivial. It just means that in order to detect its interesting properties

we need a different invariant. And it is possible to base such an invariant on the observation that even if loops can always be shrunk, there are other maps that cannot. Indeed, the sphere itself cannot be shrunk to a point. To say this more formally, the identity map from the sphere to itself is not homotopic to a map from the sphere to just one point.

This idea leads to the notion of higher-dimensional homotopy groups of a topological space X. The rough idea is to measure the number of "n-dimensional holes" in X, for any natural number n, by considering all the continuous mappings from the n-sphere to X. We want to see whether any of these spheres wrap around a hole in X. Once again, we consider two mappings from S^n to X to be equivalent if they are homotopic. And the elements of the nth homotopy group $\pi_n(X)$ are again defined to be the homotopy classes of these mappings.

Let f be a continuous map from $[0,1]$ to X with $f(0) = f(1) = x$. If we like we can turn the interval $[0,1]$ into the circle S^1 by "identifying" the points 0 and 1: then f becomes a map from S^1 to X, with one specified point in S^1 mapping to x. In order to be able to define a group operation for mappings from a higher-dimensional S^n, we similarly fix a point s in S^n and a base point x in X and look just at maps that send s to x.

Let A and B be two continuous mappings from S^n to X with this property. The "product" mapping $A \cdot B$ from S^n to X is defined as follows. First "pinch" the equator of S^n down to a point. When $n = 1$, the equator consists of just two points and the result is a figure eight. Similarly, for general n, we end up with two copies of S^n that touch each other, one made out of the northern hemisphere and one out of the southern hemisphere of the original unpinched copy of S^n. We now use the map A to map the bottom half into X and the map B to map the top half into X, with the equator mapping to the base point x. (For both halves, the pinched equator is playing the part of the point s.)

As in the one-dimensional case, this operation makes the set $\pi_n(X)$ into a group, and this group is the nth homotopy group of the space X. One can think of it as measuring how many "n-dimensional holes" a space has.

These groups are the beginning of "algebraic" topology: starting from any topological space, we construct an algebraic object, in this case a group. If two spaces are homeomorphic, then their fundamental groups (and higher homotopy groups) must be isomorphic. This is richer than the original idea of just measuring the *number* of holes, since a group contains more information than just a number.

Any continuous function from S^n into $\mathbb{R}^m$ can be continuously shrunk to a point in a straightforward way. This shows that all the higher homotopy groups of $\mathbb{R}^m$ are also trivial, which is a precise formulation of the vague idea that $\mathbb{R}^m$ has no holes.

Under certain circumstances one can show that two different topological spaces X and Y must have the same number of holes of all types. This is clearly true if X and Y are homeomorphic, but it is also true if X and Y are equivalent in a weaker sense, known as *homotopy equivalence*. Let X and Y be topological spaces and let f_0 and f_1 be continuous maps from X to Y. A homotopy from f_0 to f_1 is defined more or less as it was for spheres: it is a continuous family of continuous maps from X to Y that starts with f_0 and ends with f_1. As then, if such a homotopy exists, we say that f_0 and f_1 are homotopic. Next, a homotopy equivalence from a space X to a space Y is a continuous map $f : X \to Y$ such that there is another continuous map $g : Y \to X$ with the property that the composition $g \circ f : X \to X$ is homotopic to the identity map on X, and $f \circ g : Y \to Y$ is homotopic to the identity map on Y. (Notice that if we replaced the word "homotopic" with "equal," we would obtain the definition of a homeomorphism.) When there is a homotopy equivalence from X to Y, we say that X and Y are *homotopy equivalent*, and also that X and Y have the same *homotopy type*.

A good example is when X is the unit circle and Y is the plane with the origin removed. We have already observed that these have the same fundamental group, and commented that it was "for essentially the same reason." Now we can be more precise. Let $f : X \to Y$ be the map that takes (x, y) to (x, y) (where the first (x, y) belongs to the circle and the second to the plane). Let $g : Y \to X$ be the map that takes (u, v) to

$$\left(\frac{u}{\sqrt{u^2 + v^2}}, \frac{v}{\sqrt{u^2 + v^2}} \right).$$

(Note that $u^2 + v^2$ is never zero because the origin is not contained in Y.) Then $g \circ f$ is easily seen to equal the identity on the unit circle, so it is certainly homotopic to the identity. As for $f \circ g$, it is given by the same formula as g itself. More geometrically, it takes the points on each radial line to the point where that line intersects the unit circle. It is not hard to show that this map is homotopic to the identity on Y. (The basic idea is to "shrink the radial lines down" to the points where they intersect the circle.)

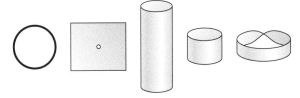

Figure 5 Some spaces that are homotopy equivalent to the circle.

Very roughly speaking, two spaces are homotopy equivalent if they have the same number of holes of all types. This is a more flexible notion of "having the same shape" than the notion of homeomorphism. For example, Euclidean spaces of different dimensions are not homeomorphic to each other, but they are all homotopy equivalent. Indeed, they are all homotopy equivalent to a point: such spaces are called *contractible*, and one thinks of them as the spaces that have no hole of any sort. The circle is not contractible, but it is homotopy equivalent to many other natural spaces: the plane $\mathbb{R}^2$ minus the origin (as we have seen), the cylinder $S^1 \times \mathbb{R}$, the compact cylinder $S^1 \times [0, 1]$, and even the Möbius strip (see figure 5). Most invariants in algebraic topology (such as homotopy groups and cohomology groups) are the same for any two spaces that are homotopy equivalent. Thus, knowing that the fundamental group of the circle is isomorphic to the integers tells us that the same is true for the various homotopy equivalent spaces just mentioned. Roughly speaking, this says that all these spaces have "one basic one-dimensional hole."

3   Calculations of the Fundamental Group and Higher Homotopy Groups

To give some more feeling for the fundamental group, let us review what we already know and look at a few more examples. The fundamental group of the 2-sphere, or indeed of any higher-dimensional sphere, is trivial. The two-dimensional torus $S^1 \times S^1$ has fundamental group $\mathbb{Z}^2 = \mathbb{Z} \times \mathbb{Z}$. Thus, a loop in the torus determines two integers, which measure how many times it winds around in the meridian direction and how many in the longitudinal direction.

The fundamental group can also be non-Abelian; that is, we can have $ab \neq ba$ for some elements a and b of the fundamental group. The simplest example is a space X built out of two circles that meet at a single point (see figure 6). The fundamental group of X is the FREE GROUP [IV.10 §2] on two generators a and

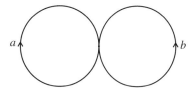

Figure 6 One-point union of two circles.

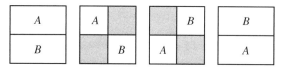

Figure 7 Proof that π_2 of any space is Abelian.

b. Roughly speaking, an element of this group is any product you can write down using the generators and their inverses, such as $abaab^{-1}a$, except that if a and a^{-1} or b and b^{-1} appear next to each other, you cancel them first. (So instead of $abb^{-1}bab^{-1}$ one would simply write $abab^{-1}$, for example.) The generators correspond to loops around each of the two circles. The free group is in a sense the most highly non-Abelian group. In particular, ab is not equal to ba, which in topological terms tells us that going around loop a and then loop b in the space X is not homotopic to the loop that goes around loop b and then loop a.

This space may seem somewhat artificial, but it is homotopy equivalent to the plane with two points removed, which appears in many contexts. More generally, the fundamental group of the plane with d points removed is the free group on d generators: this is a precise sense in which the fundamental group measures the number of holes.

In contrast with the fundamental group, the higher homotopy groups $\pi_n(X)$ are Abelian when n is at least 2. Figure 7 gives a "proof without words" in the case $n = 2$, the proof being the same for any larger n. In the figure, we view the 2-sphere as the square with its boundary identified to a point. So any elements A and B of $\pi_2(X)$ are represented by continuous maps of the square to X that map the boundary of the square to the base point x. The figure exhibits (several steps of) a homotopy from AB to BA, with the shaded regions and the boundary of the square all mapping to the base point x. The picture is reminiscent of the simplest nontrivial braid, in which one string is twisted around another; this is the beginning of a deep connection between algebraic topology and BRAID GROUPS [III.4].

The fundamental group is especially powerful in low dimensions. For example, every compact connected surface (or two-dimensional manifold) is homeomorphic to one of those on a standard list (see DIFFERENTIAL TOPOLOGY [IV.7 §2.3]), and we compute that all the manifolds on this list have different (nonisomorphic) fundamental groups. So, when you capture a closed surface in the wild, computing its fundamental group tells you exactly where it fits in the classification. Moreover, the geometric properties of the surface are closely tied to its fundamental group. The surfaces with a RIEMANNIAN METRIC [I.3 §6.10] of positive CURVATURE [III.13] (the 2-sphere and REAL PROJECTIVE PLANE [I.3 §6.7]) are exactly the surfaces with finite fundamental group; the surfaces with a metric of curvature zero (the torus and Klein bottle) are exactly the surfaces with a fundamental group that is infinite but "almost Abelian" (there is an Abelian subgroup of finite index); and the remaining surfaces, those that have a metric of negative curvature, have "highly non-Abelian" fundamental group, like the free group (see figure 8).

After more than a century of studying three-dimensional manifolds, we now know, thanks to the advances of Thurston and Perelman, that the picture is almost the same for these as it is for 2-manifolds: the fundamental group controls the geometric properties of the 3-manifold almost completely (see DIFFERENTIAL TOPOLOGY [IV.7 §2.4]). But this is completely untrue for 4-manifolds and in higher dimensions: there are many different *simply connected* manifolds, meaning manifolds with trivial fundamental group, and we need more invariants to be able to distinguish between them. (To begin with, the 4-sphere S^4 and the product $S^2 \times S^2$ are both simply connected. More generally, we can take the connected sum of any number of copies of $S^2 \times S^2$, obtained by removing 4-balls from these manifolds and identifying the boundary 3-spheres. These 4-manifolds are all simply connected, and yet no two of them are homeomorphic or even homotopy equivalent.)

An obvious way in which we might try to distinguish different spaces is to use *higher* homotopy groups, and indeed this works in simple cases. For example, π_2 of the connected sum of r copies of $S^2 \times S^2$ is isomorphic to $\mathbb{Z}^{2r}$. Also, we can show that the sphere S^n of any dimension is not contractible (although it is simply connected for $n \geqslant 2$) by computing that $\pi_n(S^n)$ is isomorphic to the integers (rather than the trivial group). Thus, each continuous map from the n-sphere to itself determines an integer, called the *degree* of the map,

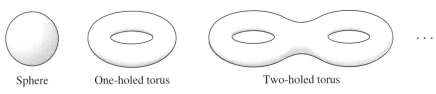

| Sphere | One-holed torus | Two-holed torus |

Figure 8 A sphere, a torus, and a surface of genus 2.

which generalizes the notion of winding number for maps from the circle to itself.

In general, however, the homotopy groups are not a practical way of distinguishing one space from another, because they are amazingly hard to compute. A first hint of this was Hopf's 1931 discovery that $\pi_3(S^2)$ is isomorphic to the integers: it is clear that the 2-sphere has a two-dimensional hole, as measured by $\pi_2(S^2) \cong \mathbb{Z}$, but in what sense does it have a three-dimensional hole? This does not correspond to our naive view of what such a hole should be. The problem of computing the homotopy groups of spheres turns out to be one of the hardest in all of mathematics: some of what we know is shown in table 1, but despite massive efforts the homotopy groups $\pi_i(S^2)$, for example, are known only for $i \leqslant 64$. There are tantalizing patterns in these calculations, with a number-theoretic flavor, but it seems impossible to formulate a precise guess for the homotopy groups of spheres in general. And computing the homotopy groups for spaces more complex than spheres is even more complicated.

To get an idea of the difficulties involved, let us define the so-called *Hopf map* from S^3 to S^2, which turns out to represent a nonzero element of $\pi_3(S^2)$. There are in fact several equivalent definitions. One of them is to regard a point (x_1, x_2, x_3, x_4) in S^3 as a pair of complex numbers (z_1, z_2) such that $|z_1|^2 + |z_2|^2 = 1$. This we do by setting $z_1 = x_1 + \mathrm{i}x_2$ and $z_2 = x_3 + \mathrm{i}x_4$. We then map the pair (z_1, z_2) to the complex number z_1/z_2. This may not look like a map to S^2, but it is because z_2 may be zero, so in fact the image of the map is not $\mathbb{C}$ but the *Riemann sphere* $\mathbb{C} \cup \infty$, which can be identified with S^2 in a natural way.

Another way of defining the Hopf map is to regard points (x_1, x_2, x_3, x_4) in S^3 as unit quaternions. In the article on quaternions in this volume [III.76], it is shown that each unit quaternion can be associated with a rotation of the sphere. If we fix some point s in the sphere and map each unit quaternion to the image of s under the associated rotation, then we get a map from S^3 to S^2 that is homotopic to the map defined in the previous paragraph.

The Hopf map is an important construction, and will reappear more than once later in this article.

4 Homology Groups and the Cohomology Ring

Homotopy groups, then, can be rather mysterious and very hard to calculate. Fortunately, there is a different way to measure the number of holes in a topological space: homology and cohomology groups. The definitions are more subtle than the definition of homotopy groups, but the groups turn out to be easier to compute and are for this reason much more commonly used.

Recall that elements of the nth homotopy group $\pi_n(X)$ of a topological space X are represented by continuous maps from the n-sphere to X. Let X be a manifold, for simplicity. There are two key differences between homotopy groups and homology groups. The first is that the basic objects of homology are more general than n-dimensional spheres: *every* closed oriented n-dimensional submanifold A of X determines an element of the nth homology group of X, $H_n(X)$. This might make homology groups seem much bigger than homotopy groups, but that is not the case, because of the second major difference between homotopy and homology. As with homotopy, the elements of the homology groups are not the submanifolds themselves but equivalence classes of submanifolds, but the definition of the equivalence relation for homology makes it much easier for two of these submanifolds to be equivalent than it is for two spheres to be homotopic.

We shall not give a formal definition of homology, but here are some examples that convey some of its flavor. Let X be the plane with the origin removed and let A be a circle that goes around the origin. If we continuously deform this circle, we will obtain a new curve that is homotopic to the original circle, but with homology we can do more. For instance, we can start with a continuous deformation that causes two of its points to touch and turns it into a figure eight. One half of this figure eight will have to contain the origin, but we can leave

Table 1 The first few homotopy groups of spheres.

	S^1	S^2	S^3	S^4	S^5	S^6	S^7	S^8	S^9
π_1	$\mathbb{Z}$	0	0	0	0	0	0	0	0
π_2	0	$\mathbb{Z}$	0	0	0	0	0	0	0
π_3	0	$\mathbb{Z}$	$\mathbb{Z}$	0	0	0	0	0	0
π_4	0	$\mathbb{Z}/2$	$\mathbb{Z}/2$	$\mathbb{Z}$	0	0	0	0	0
π_5	0	$\mathbb{Z}/2$	$\mathbb{Z}/2$	$\mathbb{Z}/2$	$\mathbb{Z}$	0	0	0	0
π_6	0	$\mathbb{Z}/4\times\mathbb{Z}/3$	$\mathbb{Z}/4\times\mathbb{Z}/3$	$\mathbb{Z}/2$	$\mathbb{Z}/2$	$\mathbb{Z}$	0	0	0
π_7	0	$\mathbb{Z}/2$	$\mathbb{Z}/2$	$\mathbb{Z}\times\mathbb{Z}/4\times\mathbb{Z}/3$	$\mathbb{Z}/2$	$\mathbb{Z}/2$	$\mathbb{Z}$	0	0
π_8	0	$\mathbb{Z}/2$	$\mathbb{Z}/2$	$\mathbb{Z}/2\times\mathbb{Z}/2$	$\mathbb{Z}/8\times\mathbb{Z}/3$	$\mathbb{Z}/2$	$\mathbb{Z}/2$	$\mathbb{Z}$	0
π_9	0	$\mathbb{Z}/3$	$\mathbb{Z}/3$	$\mathbb{Z}/2\times\mathbb{Z}/2$	$\mathbb{Z}/2$	$\mathbb{Z}/8\times\mathbb{Z}/3$	$\mathbb{Z}/2$	$\mathbb{Z}/2$	$\mathbb{Z}$
π_{10}	0	$\mathbb{Z}/3\times\mathbb{Z}/5$	$\mathbb{Z}/3\times\mathbb{Z}/5$	$\mathbb{Z}/8\times\mathbb{Z}/3\times\mathbb{Z}/3$	$\mathbb{Z}/2$	0	$\mathbb{Z}/8\times\mathbb{Z}/3$	$\mathbb{Z}/2$	$\mathbb{Z}/2$

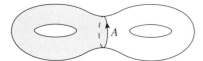

Figure 9 The circle A represents zero
in the homology of the surface.

that still and slide the other part away. The result is then two closed curves, with the origin inside one and outside the other. This pair of curves, which together form a 1-manifold with two components, is equivalent to the original circle. It can be seen as a continuous deformation of a more general kind.

A second example shows how natural it is to include other manifolds in the definition of homology. This time let X be $\mathbb{R}^3$ with a circle removed, and let A be a sphere that contains the circle in its interior. Suppose that the circle is in the XY-plane and that both it and the sphere A are centered at the origin. Then we can pinch the top and bottom of A toward the origin until they just touch. If we do so, then we obtain a shape that looks like a torus, except that the hole in the middle has been shrunk to zero. But we can open up this hole with the help of a further continuous deformation and obtain a genuine torus, which is a "tube" around the original circle. From the point of view of homology, this torus is equivalent to the sphere A.

A more general rule is that if X is a manifold and B is a compact oriented $(n + 1)$-dimensional submanifold of X with a boundary, then this boundary ∂B will be equivalent to zero (which is the same as saying that $[\partial B] = 0$ in $H_n(X)$): see figure 9.

The group operation is easy to define: if A and B are two disjoint submanifolds of X, giving rise to homology classes $[A]$ and $[B]$, then $[A] + [B]$ is the homol-ogy class of $[A \cup B]$. (More generally, the definition of homology allows us to add up any collection of sub-manifolds, whether or not they overlap.) Here are some simple examples of homology groups, which, unlike the fundamental group, are always Abelian. The homology groups of a sphere, $H_i(S^n)$, are isomorphic to the integers $\mathbb{Z}$ for $i = 0$ and for $i = n$, and 0 otherwise. This contrasts with the complicated homotopy groups of the sphere, and better reflects the naive idea that the n-sphere has one n-dimensional hole and no other holes. Note that the fundamental group of the circle, the group of integers, is the same as its first homology group. More generally, for any path-connected space, the first homology group is always the "Abelianization" of the fundamental group (which is formally defined to be its largest Abelian quotient). For example, the fundamental group of the plane with two points removed is the free group on two generators, while the first homology group is the free *Abelian* group on two generators, or $\mathbb{Z}^2$.

The homology groups of the two-dimensional torus $H_i(S^1\times S^1)$ are isomorphic to $\mathbb{Z}$ for $i = 0$, to $\mathbb{Z}^2$ for $i = 1$, and to $\mathbb{Z}$ for $i = 2$. All of this has geometric meaning. The zeroth homology group of any space is isomorphic to $\mathbb{Z}^r$ for a space X with r connected components. So the fact that the zeroth homology group of the torus is isomorphic to $\mathbb{Z}$ means that the torus is connected. Any closed loop in the torus determines an element of the first homology group $\mathbb{Z}^2$, which measures how many times the loop winds around the meridian and longitudinal directions of the torus. And finally, the homology of the torus in dimension 2 is isomorphic to $\mathbb{Z}$ because the torus is a closed orientable manifold. That tells us that the whole torus defines an element of the second homology group of the torus, which is in fact a generator of that group. By contrast, the homotopy group

$\pi_2(S^1 \times S^1)$ is the trivial group: there are no interesting maps from the 2-sphere to the 2-torus, but homology shows that there are interesting maps from other closed 2-manifolds to the 2-torus.

As we have mentioned, calculating homology groups is much easier than calculating homotopy groups. The main reason for this is the existence of results that tell you the homology groups of a space that is built up from smaller pieces in terms of the homology groups of those pieces and their intersections. Another important property of homology groups is that they are "functorial" in the sense that a continuous map f from a space X to a space Y leads in a natural way to a homomorphism f_* from $H_i(X)$ to $H_i(Y)$ for each i: $f_*([A])$ is defined to be $[f(A)]$. In other words, $f_*([A])$ is the equivalence class of the image of A under f.

We can define the closely related idea of "cohomology" simply by a different numbering. Let X be a closed oriented n-dimensional manifold. Then we define the ith *cohomology group* $H^i(X)$ to be the homology group $H_{n-i}(X)$. Thus, one way to write down a cohomology class (an element of $H^i(X)$) is by choosing a closed oriented submanifold S of codimension i in X. (This means that the dimension of S is $n - i$.) We write $[S]$ for the corresponding cohomology class.

For more general spaces than manifolds, cohomology is not just a simple renumbering of homology. Informally, if X is a topological space, then we think of an element of $H^i(X)$ as being represented by a codimension-i subspace of X that can move around freely in X. For example, suppose that f is a continuous map from X to an i-dimensional manifold. If X is a manifold and f is sufficiently "well-behaved," then the inverse image of a "typical" point in the manifold will be an i-codimensional submanifold of X, and as we move the point about, this submanifold will vary continuously, and will do so in a way that is similar to the way that a circle became two circles and a sphere became a torus earlier. If X is a more general topological space, the map f still determines a cohomology class in $H^i(X)$, which we think of as being represented by the inverse image in X of any point in the manifold.

However, even when X is an oriented n-dimensional manifold, cohomology has distinct advantages over homology. This may seem odd, since the cohomology groups are the homology groups with different names. However, this renumbering allows us to give very useful extra algebraic structure to the cohomology groups of X: not only can we add cohomology classes, we can multiply them as well. Furthermore, we can do so in such a

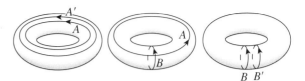

Figure 10 $A^2 = A \cdot A' = 0$, $A \cdot B = [\text{point}]$, and $B^2 = B \cdot B' = 0$.

way that, taken together, the cohomology groups of X form a RING [III.81 §1]. (Of course, we could do this for the homology groups, but the cohomology groups form a so-called *graded* ring. In particular, if $[A] \in H^i(X)$ and $[B] \in H^j(X)$, then $[A] \cdot [B] \in H^{i+j}(X)$.)

The multiplication of cohomology classes has a rich geometric meaning, especially on manifolds: it is given by the *intersection* of two submanifolds. This generalizes our discussion of intersection numbers in section 1: there we considered zero-dimensional intersections of submanifolds, whereas we are now considering (cohomology classes of) higher-dimensional intersections. To be precise, let S and T be closed oriented submanifolds of X, of codimension i and j, respectively. By moving S slightly (which does not change its class in $H^i(X)$) we can assume that S and T intersect transversely, which implies that the intersection of S and T is a smooth submanifold of codimension $i + j$ in X. Then the product of the cohomology classes $[S]$ and $[T]$ is simply the cohomology class of the intersection $S \cap T$ in $H^{i+j}(X)$. (In addition, the submanifold $S \cap T$ inherits an orientation from S, T, and X: this is needed to define the associated cohomology class.)

As a result, to compute the cohomology ring of a manifold, it is enough to specify a basis for the cohomology groups (which, as we have already discussed, are relatively easy to determine) using some submanifolds and to see how these submanifolds intersect. For example, we can compute the cohomology ring of the 2-torus as shown in figure 10. For another example, it is not hard to show that the cohomology of the COMPLEX PROJECTIVE PLANE [III.72] $\mathbb{CP}^2$ has a basis given by three basic submanifolds: a point, which belongs to $H^4(\mathbb{CP}^2)$ because it is a submanifold of codimension 4; a complex projective line $\mathbb{CP}^1 = S^2$, which belongs to $H^2(\mathbb{CP}^2)$; and the whole manifold $\mathbb{CP}^2$, which is in $H^0(\mathbb{CP}^2)$ and represents the identity element 1 of the cohomology ring. The product in the cohomology ring is described by saying that $[\mathbb{CP}^1][\mathbb{CP}^1] = [\text{point}]$, because any two distinct lines $\mathbb{CP}^1$ in the plane meet transversely in a single point.

This calculation of the cohomology ring of the complex projective plane, although very simple, has several strong consequences. First of all, it implies Bézout's theorem on intersections of complex algebraic curves (see ALGEBRAIC GEOMETRY [IV.4 §6]). An algebraic curve of degree d in $\mathbb{CP}^2$ represents d times the class of a line $\mathbb{CP}^1$ in $H^2(\mathbb{CP}^2)$. Therefore, if two algebraic curves D and E of degrees d and e meet transversely, then the cohomology class $[D \cap E]$ equals

$$[D] \cdot [E] = (d[\mathbb{CP}^1])(e[\mathbb{CP}^1]) = de[\text{point}].$$

For complex submanifolds of a complex manifold, intersection numbers are always $+1$, not -1, and so this means that D and E meet in exactly de points.

We can also use the computation of the cohomology ring of $\mathbb{CP}^2$ to prove something about the homotopy groups of spheres. It turns out that $\mathbb{CP}^2$ can be constructed as the union of the 2-sphere and the closed four-dimensional ball, with each point of the boundary S^3 of the ball identified with a point in S^2 by the Hopf map, which was defined in the previous section.

A constant map from one space to another, or a map homotopic to a constant map, gives rise to the zero homomorphism between the homology groups H_i, at least when $i > 0$. The Hopf map $f : S^3 \rightarrow S^2$ also induces the zero homomorphism because the nonzero homology groups of S^3 and S^2 are in different dimensions. Nonetheless, we will show that f is not homotopic to the constant map. If it were, then the space $\mathbb{CP}^2$ obtained by attaching a 4-ball to the 2-sphere using the map f would be homotopy equivalent to the space obtained by attaching a 4-ball to the 2-sphere using a constant map. The latter space Y is the union of S^2 and S^4 identified at one point. But in fact Y is not homotopy equivalent to the complex projective plane, because their cohomology rings are not isomorphic. In particular, the product of any element of $H^2(Y)$ with itself is zero, unlike what happens in $\mathbb{CP}^2$ where $[\mathbb{CP}^1][\mathbb{CP}^1] = [\text{point}]$. Therefore f is nonzero in $\pi_3(S^2)$. A more careful version of this argument shows that $\pi_3(S^2)$ is isomorphic to the integers, and the Hopf map $f : S^3 \rightarrow S^2$ is a generator of this group.

This argument shows some of the rich relations between all the basic concepts of algebraic topology: homotopy groups, cohomology rings, manifolds, and so on. To conclude, here is a way to visualize the nontriviality of the Hopf map $f : S^3 \rightarrow S^2$. Look at the subset of S^3 that maps to any given point of the 2-sphere. These inverse images are all circles in the 3-sphere. To draw them, we can use the fact that S^3 minus a point

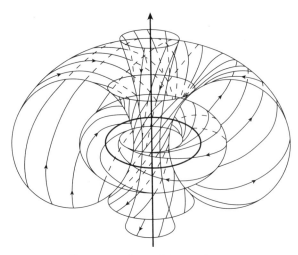

Figure 11 Fibers of the Hopf map.

is homeomorphic to $\mathbb{R}^3$; so these inverse images form a family of disjoint circles that fills up three-dimensional space, with one circle being drawn as a line (the circle through the point we removed from S^3). The striking feature of this picture is that any two of this huge family of circles have linking number 1 with each other: there is no way to pull any two of them apart (see figure 11).

5   Vector Bundles and Characteristic Classes

We now introduce another major topological idea: fiber bundles. If E and B are topological spaces, x is a point in B, and $p : E \rightarrow B$ is a continuous map, then the *fiber* of p *over* x is the subspace of E that maps to x. We say that p is a *fiber bundle*, with fiber F, if every fiber of p is homeomorphic to the same space F. We call B the *base space* and E the *total space*. For example, any product space $B \times F$ is a fiber bundle over B, called the trivial F-bundle over B. (The continuous map in this case is the map that takes (x, y) to x.) But there are many nontrivial fiber bundles. For example, the Möbius strip is a fiber bundle over the circle with fiber a closed interval. This example helps to explain the old name "twisted product" for fiber bundles. Another example: the Hopf map makes the 3-sphere the total space of a circle bundle over the 2-sphere.

Fiber bundles are a fundamental way to build up complicated spaces from simple pieces. We will focus on the most important special case: vector bundles. A *vector bundle* over a space B is a fiber bundle $p : E \rightarrow B$ whose fibers are all real vector spaces of some dimension n.

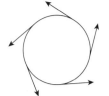 

Figure 12 Trivializations of the tangent
bundle for the circle and the torus.

Figure 13 The hairy ball theorem.

This dimension is called the *rank* of the vector bundle. A *line bundle* means a vector bundle of rank 1; for example, we can view the Möbius strip (not including its boundary) as a line bundle over the circle S^1. It is a *nontrivial* line bundle; that is, it is not isomorphic to the trivial line bundle $S^1 \times \mathbb{R}$. (There are many ways of constructing it: one is to take the strip $\{(x, y) : 0 \leqslant x \leqslant 1\}$ and identify each point $(0, y)$ with the point $(1, -y)$. The base space of this line bundle is the set of all points $(x, 0)$, which is a circle since $(0, 0)$ and $(1, 0)$ have been identified.)

If M is a smooth manifold of dimension n, its *tangent bundle* $TM \to M$ is a vector bundle of rank n. We can easily define this bundle by considering M as a submanifold of some Euclidean space $\mathbb{R}^N$. (Every smooth manifold can be embedded into Euclidean space.) Then TM is the subspace of $M \times \mathbb{R}^N$ of pairs (x, v) such that the vector v is tangent to M at the point x; the map $TM \to M$ sends a pair (x, v) to the point x. The fiber over x then has the form of the set of all pairs (x, v) with v belonging to an affine subspace of $\mathbb{R}^N$ of dimension equal to that of M. For any fiber bundle, a *section* means a continuous map from the base space B to the total space E that maps each point x in B to some point in the fiber over x. A section of the tangent bundle of a manifold is called a *vector field*. We can draw a vector field on a given manifold by putting an arrow (possibly of zero length) at every point of the manifold.

In order to classify smooth manifolds, it is important to study their tangent bundles, and in particular to see whether they are trivial or not. Some manifolds, like the circle S^1 and the torus $S^1 \times S^1$, do have trivial tangent bundle. The tangent bundle of an n-manifold M is trivial if and only if we can find n vector fields that are linearly independent at every point of M. So we can prove that the tangent bundle is trivial just by writing down such vector fields; see figure 12 for the circle or the torus. But how can we show that the tangent bundle of a given manifold is nontrivial?

One way is to use intersection numbers. Let M be a closed oriented n-manifold. We can identify M with the image of the "zero-section" inside the tangent bundle TM, the section that assigns to every point of M the zero vector at that point. Since the dimension of TM is precisely double that of M, the discussion of intersection numbers in section 1 gives a well-defined integer $M^2 = M \cdot M$, the self-intersection number of M inside TM; this is called the *Euler characteristic* $\chi(M)$. By the definition of intersection numbers, for any vector field v on M that meets the zero-section transversely, the Euler characteristic of M is equal to the number of zeros of v, counted with signs.

As a result, if the Euler characteristic of M is not zero, then every vector field on M must meet the zero-section; in other words, every vector field on M must equal zero somewhere. The simplest example occurs when M is the 2-sphere S^2. We can easily write down a vector field (for example, the one pointing toward the east along circles of latitude, which vanishes at the north and south poles) whose intersection number with the zero-section is 2. Therefore the 2-sphere has Euler characteristic 2, and so every vector field on the 2-sphere must vanish somewhere. This is a famous theorem of topology known as the "hairy ball theorem": it is impossible to comb the hair on a coconut (see figure 13).

This is the beginning of the theory of *characteristic classes*, which measure how nontrivial a given vector bundle is. There is no need to restrict ourselves to the tangent bundle of a manifold. For any oriented vector bundle E of rank n on a topological space X, we can define a cohomology class $\chi(E)$ in $H^n(X)$, the *Euler class*, which vanishes if the bundle is trivial. Intuitively, the Euler class of E is the cohomology class represented by the zero set of a general section of E, which (for example, if X is a manifold) should be a codimension-n submanifold of X, since X has codimension n in E. If X is a closed oriented n-manifold, then the Euler class of the tangent bundle in $H^n(X) = \mathbb{Z}$ is the Euler characteristic of X.

One of the inspirations for the theory of characteristic classes was the Gauss–Bonnet theorem, generalized to all dimensions in the 1940s. The theorem expresses the Euler characteristic of a closed manifold with a Riemannian metric as the integral over the manifold of a certain curvature function. More broadly, a central goal of differential geometry is to understand how the geometric properties of a Riemannian manifold such as its curvature are related to the topology of the manifold.

The characteristic classes for *complex* vector bundles (that is, bundles where the fibers are complex vector spaces) turn out to be particularly convenient: indeed, real vector bundles are often studied by constructing the associated complex vector bundle. If E is a complex vector bundle of rank n over a topological space X, the *Chern classes* of E are a sequence $c_1(E), \ldots, c_n(E)$ of cohomology classes on X, with $c_i(E)$ belonging to $H^{2i}(X)$, which all vanish if the bundle is trivial. The top Chern class, $c_n(E)$, is simply the Euler class of E: thus, it is the first obstruction to finding a section of E that is everywhere nonzero. The more general Chern classes have a similar interpretation. For any $1 \leqslant j \leqslant n$, choose j general sections of E. The subset of X over which these sections become linearly dependent will have codimension $2(n + 1 - j)$ (assuming, for example, that X is a manifold). The Chern class $c_{n+1-j}(E)$ is precisely the cohomology class of this subset. Thus the Chern classes measure in a natural way the failure of a given complex vector bundle to be trivial. The *Pontryagin classes* of a real vector bundle are defined to be the Chern classes of the associated complex vector bundle.

A triumph of differential topology is Sullivan's 1977 theorem that there are only finitely many smooth closed simply connected manifolds of dimension at least 5 with any given homotopy type and given Pontryagin classes of the tangent bundle. This statement fails badly in dimension 4, as Donaldson discovered in the 1980s (see DIFFERENTIAL TOPOLOGY [IV.7 §2.5]).

6 K-Theory and Generalized Cohomology Theories

The effectiveness of vector bundles in geometry led to a new way of measuring the "holes" in a topological space X: looking at how many different vector bundles over X there are. This idea gives a simple way to define a cohomology-like ring associated to any space, known as K-theory (after the German word "Klasse," since the theory involves equivalence classes of vector bundles). It turns out that K-theory gives a very useful new angle by which to look at topological spaces. Some problems that could be solved only with enormous effort using ordinary cohomology became easy with K-theory. The idea was created in algebraic geometry by Grothendieck in the 1950s and then brought into topology by Atiyah and Hirzebruch in the 1960s.

The definition of K-theory can be given in a few lines. For a topological space X, we define an Abelian group $K^0(X)$, the K-theory of X, whose elements can be written as formal differences $[E] - [F]$, where E and F are any two complex vector bundles over X. The only relations we impose in this group are that $[E \oplus F] = [E] + [F]$ for any two vector bundles E and F over X. Here $E \oplus F$ denotes the *direct sum* of the two bundles; if E_x and F_x denote the fibers at a given point x in X, the fiber of $E \oplus F$ at x is simply $E_x \times F_x$.

This simple definition leads to a rich theory. First of all, the Abelian group $K^0(X)$ is in fact a ring: we multiply two vector bundles on X by forming the TENSOR PRODUCT [III.89]. In this respect, K-theory behaves like ordinary cohomology. The analogy suggests that the group $K^0(X)$ should form part of a whole sequence of Abelian groups $K^i(X)$, for integers i, and indeed these groups can be defined. In particular, $K^{-i}(X)$ can be defined as the subgroup of those elements of $K^0(S^i \times X)$ whose restriction to $K^0(\text{point} \times X)$ is zero.

Then a miracle occurs: the groups $K^i(X)$ turn out to be *periodic* of order 2: $K^i(X) = K^{i+2}(X)$ for all integers i. This is a famous phenomenon known as *Bott periodicity*. So there are really only two different K-groups attached to any topological space: $K^0(X)$ and $K^1(X)$.

This may suggest that K-theory contains less information than ordinary cohomology, but that is not so. Neither K-theory nor ordinary cohomology determines the other, although there are strong relations between them. Each brings different aspects of the shape of a space to the fore. Ordinary cohomology, with its numbering, shows fairly directly the way a space is built up from pieces of different dimensions. K-theory, having only two different groups, looks cruder at first (and is often easier to compute as a result). But geometric problems involving vector bundles often involve information that is subtle and hard to extract from ordinary cohomology, whereas this information is brought to the surface by K-theory.

The basic relation between K-theory and ordinary cohomology is that the group $K^0(X)$ constructed from the vector bundles on X "knows" something about all the even-dimensional cohomology groups of X. To be precise, the rank of the Abelian group $K^0(X)$ is the sum of the ranks of all the even-dimensional cohomology groups $H^{2i}(X)$. This connection comes from associating with a given vector bundle on X its Chern classes. The odd K-group $K^1(X)$ is related in the same way to the odd-dimensional ordinary cohomology.

As we have already hinted, the precise group $K^0(X)$, as opposed to just its rank, is better adapted to some geometric problems than ordinary cohomology. This phenomenon shows the power of looking at geometric problems in terms of vector bundles, and thus ultimately in terms of linear algebra. Among the classic applications of K-theory is the proof, by Bott, Kervaire, and Milnor, that the 0-sphere, the 1-sphere, the 3-sphere, and the 7-sphere are the only spheres whose tangent bundles are trivial. This has a deep algebraic consequence, in the spirit of the fundamental theorem of algebra: the only dimensions in which there can be a real division algebra (not assumed to be commutative or even associative) are 1, 2, 4, and 8. There are indeed division algebras of all four types: the real numbers, complex numbers, quaternions, and octonions (see QUATERNIONS, OCTONIONS, AND NORMED DIVISION ALGEBRAS [III.76]).

Let us see why the existence of a real division algebra of dimension n implies that the $(n-1)$-sphere has trivial tangent bundle. In fact, let us merely assume that we have a finite-dimensional real vector space V with a bilinear map $V \times V \to V$, which we call the "product," such that if x and y are vectors in V with $xy = 0$, then either $x = 0$ or $y = 0$. For convenience, let us also assume that there is an identity element 1 in V, so $1 \cdot x = x \cdot 1 = x$ for all $x \in V$; one can, however, do without this assumption. If V has dimension n, then we can identify V with $\mathbb{R}^n$. Then, for each point x in the sphere S^{n-1}, left multiplication by x gives a linear isomorphism from $\mathbb{R}^n$ to itself. By scaling the output to have length 1, left multiplication by x gives a diffeomorphism from S^{n-1} to itself which maps the point 1 (scaled to have length 1) to x. Taking the derivative of this diffeomorphism at the point 1 gives a linear isomorphism from the tangent space of the sphere at the point 1 to the tangent space at x. Since the point x on the sphere is arbitrary, a choice of basis for the tangent space of the sphere at the point 1 determines a trivialization of the whole tangent bundle of the $(n-1)$-sphere.

Among other applications, K-theory provides the best "explanation" for the low-dimensional homotopy groups of spheres, and in particular for the number-theoretic patterns that are seen there. Notably, denom-inators of Bernoulli numbers appear among those groups (such as $\pi_{n+3}(S^n) \cong \mathbb{Z}/24$ for n at least 5), and this pattern was explained using K-theory by Milnor, Kervaire, and Adams.

THE ATIYAH–SINGER INDEX THEOREM [V.2] provides a deep analysis of linear differential equations on closed manifolds using K-theory. The theorem has made K-theory important for gauge theories and string theories in physics. K-theory can also be defined for noncommutative rings, and is in fact the central concept in "noncommutative geometry" (see OPERATOR ALGEBRAS [IV.15 §5]).

The success of K-theory led to a search for other "generalized cohomology theories." There is one other theory that stands out for its power: *complex cobordism*. The definition is very geometric: the complex cobordism groups of a manifold M are generated by mappings of manifolds (with a complex structure on the tangent bundle) into M. The relations say that any manifold counts as zero if it is the boundary of some other manifold. For example, the union of two circles would count as zero if you could find a cylinder whose ends were those circles.

It turns out that complex cobordism is much richer than either K-theory or ordinary cohomology. It sees far into the structure of a topological space, but at the cost of being difficult to compute. Over the past thirty years, a whole series of cohomology theories, such as elliptic cohomology and Morava K-theories, have been constructed as "simplifications" of complex cobordism: there is a constant tension in topology between invariants that carry a lot of information and invariants that are easy to compute. In one direction, complex cobordism and its variants provide the most powerful tool for the computation and understanding of the homotopy groups of spheres. Beyond the range where Bernoulli numbers appear, we see deeper number theory such as MODULAR FORMS [III.59]. In another direction, the geometric definition of complex cobordism makes it useful in algebraic geometry.

7 Conclusion

The line of thought introduced by pioneering topologists like RIEMANN [VI.49] is simple but powerful. Try to translate any problem, even a purely algebraic one, into geometric terms. Then ignore the details of the geometry and study the underlying shape or topology of the problem. Finally, go back to the original problem and see how much has been gained. The fundamental topological ideas such as cohomology are used throughout mathematics, from number theory to string theory.

Further Reading

From the definition of topological spaces to the fundamental group and a little beyond, I like M. A. Armstrong's *Basic Topology* (Springer, New York, 1983). The current standard graduate textbook is A. Hatcher's *Algebraic Topology* (Cambridge University Press, Cambridge, 2002). Two of the great topologists, Bott and Milnor, are also brilliant writers. Every young topologist should read R. Bott and L. Tu's *Differential Forms in Algebraic Topology* (Springer, New York, 1982), J. Milnor's *Morse Theory* (Princeton University Press, Princeton, NJ, 1963), and J. Milnor and J. Stasheff's *Characteristic Classes* (Princeton University Press, Princeton, NJ, 1974).

IV.7 Differential Topology
C. H. Taubes

1 Smooth Manifolds

This article is about classifying certain objects called smooth manifolds, so I need to start by telling you what they are. A good example to keep in mind is the surface of a smooth ball. If you look at a small portion of it from very close up, then it looks like a portion of a flat plane, but of course it differs in a radical way from a flat plane on larger distance scales. This is a general phenomenon: a smooth manifold can be very convoluted, but must be quite regular in close-up. This "local regularity" is the condition that each point in a manifold belongs to a neighborhood that looks like a portion of standard Euclidean space in some dimension. If the dimension in question is d for every point of the manifold, then the manifold itself is said to have dimension d. A schematic of this is shown in figure 1.

What does it mean to say that a neighborhood "looks like a portion of standard Euclidean space"? It means that there is a "nice" one-to-one map ϕ from the neighborhood into $\mathbb{R}^d$ (with its usual notion of distance). One can think of ϕ as "identifying" points in the neighborhood with points in $\mathbb{R}^d$: that is, x is identified with $\phi(x)$. If we do this, then the function ϕ is called a *coordinate chart* of the neighborhood, and any chosen basis for the linear functions on the Euclidean space is called a *coordinate system*. The reason for this is that ϕ allows us to use the coordinates in $\mathbb{R}^d$ to label points in the neighborhood: if x belongs to the neighborhood,

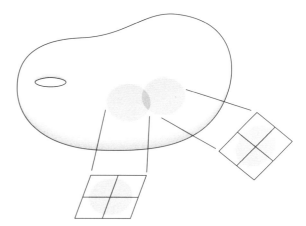

Figure 1 Small portions of a manifold resemble regions in a Euclidean space.

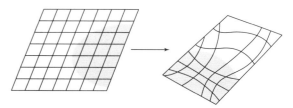

Figure 2 A transition function from a rectangular grid to a distorted rectangular grid.

then one can label it with the coordinates of $\phi(x)$. For example, Europe is part of the surface of a sphere. A typical map of Europe identifies each point in Europe with a point in flat, two-dimensional Euclidean space, that is, a square grid labeled with latitude and longitude. These two numbers give us a coordinate system for the map, which can also be transferred to a coordinate system for Europe itself.

Now, here is a straightforward but central observation. Suppose that M and N are two neighborhoods that intersect, and suppose that functions $\phi : M \to \mathbb{R}^d$ and $\psi : N \to \mathbb{R}^d$ are used to give them each a coordinate chart. Then the intersection $M \cap N$ is given *two* coordinate charts, and this gives us an identification between the open regions $\phi(M \cap N)$ and $\psi(M \cap N)$ of $\mathbb{R}^d$: given a point x in the first region, the corresponding point in the second is $\psi(\phi^{-1}(x))$. This composition of maps is called a *transition function*, and it tells you how the coordinates from one of the charts on the intersecting region relate to those of the other. The transition function is a HOMEOMORPHISM [III.90] between the regions $\phi(M \cap N)$ and $\psi(M \cap N)$.

Suppose that you take a rectangular grid in the first Euclidean region and use the transition function $\psi\phi^{-1}$ to map it to the second one. It is possible that the image will again be a rectangular grid, but in general it will be somewhat distorted. An illustration is given in figure 2.

The proper term for a space whose points are surrounded by regions that can be identified with parts of Euclidean space is a *topological manifold*. The word "topological" is used in order to indicate that there are no constraints on the coordinate-chart transition functions apart from the basic one that they should be continuous. However, some continuous functions are quite unpleasant, so one typically introduces extra constraints in order to limit the distorting effect that the transition functions can have on a rectangular coordinate grid.

Of prime interest here is the case where the transition functions are required to be differentiable to all orders. If a manifold has a collection of charts for which all the transition functions are infinitely differentiable, then it is said to have a *smooth structure*, and it is called a *smooth manifold*. Smooth manifolds are especially interesting because they are the natural arena for calculus. Roughly speaking, they are the most general context in which the notion of differentiation to any order makes intrinsic sense.

A function f, defined on a manifold, is said to be *differentiable* if, given any of its coordinate charts $\phi : N \to \mathbb{R}^d$, the function $g(y) = f(\phi^{-1}(y))$ (which is defined on a region of $\mathbb{R}^d$) is DIFFERENTIABLE [I.3 §5.3]. Calculus is impossible on a manifold if it does not admit charts with differentiable transition functions, because a function that might appear differentiable in one chart will not, in general, be differentiable when viewed from a neighboring chart.

Here is a one-dimensional example to illustrate this point. Consider the following two coordinate charts of a neighborhood of the origin in the real line. The first is the obvious chart that simply represents a real number x by itself. (Formally speaking, one is taking the function ϕ to be defined by the simple formula $\phi(x) = x$.) The second represents x by the point $x^{1/3}$. (Here the cube root of a negative number x is defined to be minus the cube root of $-x$.) What is the transition function between these two charts? Well, if t is a point in the region of $\mathbb{R}$ used for the first chart, then $\phi^{-1}(t) = t$, so $\psi(\phi^{-1}(t)) = \psi(t) = t^{1/3}$. This is a continuous function of t but it is not differentiable at the origin.

Now consider the simplest possible function defined on the region used for the second chart, the function

$h(s) = s$, and let us work out the corresponding function f on the manifold itself. The value of f at x should be the value of h at the point s corresponding to x. This point is $\psi(x) = x^{1/3}$, so $f(x) = h(x^{1/3}) = x^{1/3}$. Finally, since the point x in the manifold corresponds to the point $t = \phi(x) = x$ in the first region, the corresponding function on the first region is $g(t) = t^{1/3}$. (This is the same function as f only because ϕ happens to be the very special map that takes each number to itself.) Thus, the eminently differentiable function h on one coordinate chart translates into the continuous but not differentiable function g on the other.

Suppose one is given a topological manifold M with two sets of charts, both of which have infinitely differentiable transition functions. Then each set of charts gives us a smooth structure on the manifold. Of great importance is the fact that these two smooth structures can be fundamentally different.

To see what this means, let us call the sets of charts K and L. Given a function f, let us call it K-*differentiable* if it is differentiable from the viewpoint of K, and L-*differentiable* if it is differentiable from the viewpoint of L. It may easily happen that a function is K-differentiable without being L-differentiable or vice versa. However, we can say that K and L *give the same smooth structure* on M when there is a map, F, from M to itself with the following three properties. First, F is invertible and both F and F^{-1} are continuous. Second, the composition of F with any function that is K-differentiable is L-differentiable. Third, the composition of the inverse of F^{-1} with any function that is L-differentiable is K-differentiable. Loosely speaking, F turns the K-differentiable functions into L-differentiable ones and F^{-1} turns them back again. If no such function F exists, then the smooth structures given by K and L are considered to be genuinely different.

To see how this plays out, let us look at the one-dimensional example again. As noted previously, the functions that you deem to be differentiable if you use the ϕ-chart are not the same as those you deem to be differentiable if you use the ψ-chart. For example, the function $x \mapsto x^{1/3}$ is not ϕ-differentiable but it is ψ-differentiable. Even so, the ϕ-differentiable and ψ-differentiable sets of functions define the *same* smooth structure for the line, since any ψ-differentiable function becomes ϕ-differentiable once you compose it with the self-map $F : t \mapsto t^3$.

It is very far from obvious that any manifold can have more than one smooth structure, but this turns out to be the case. There are also manifolds that are entirely lacking in smooth structures. These two facts lead directly to the central concern of this essay, the long-sought quest for the two holy grails of differential topology.

- A list of all smooth structures on any given topological manifold.
- An algorithm to identify any given smooth structure on any given topological manifold with the corresponding structure from the list.

2   What Is Known about Manifolds?

Much has been accomplished as of the writing of this article with respect to the two points listed above. This said, the task for this part of the article is to summarize the state of affairs at the beginning of the twenty-first century. Various examples of manifolds are described along the way.

The story here requires a brief, preliminary digression to set the stage. If you have two manifolds and you set them side by side without their touching, then technically speaking they can be regarded as a single manifold that happens to have two components. In such a case, one can study the components individually. Therefore, in this article I shall talk exclusively about *connected* manifolds: that is, manifolds with just one component. In a connected manifold, one can get from any point to any other point without ever leaving the manifold.

A second technical point is that it is useful to distinguish between manifolds such as the sphere, which are bounded in extent, and manifolds such as the plane, which go off to infinity. More precisely, I am talking about the distinction between COMPACT [III.9] and noncompact manifolds: a compact manifold can be thought of as one that can be expressed as a closed bounded subset of $\mathbb{R}^n$ for some n. The discussion that follows will be almost entirely about compact manifolds. As some of the examples below will demonstrate, the story for compact manifolds is less convoluted than the analogous story for noncompact ones. For simplicity I shall often use the word "manifold" to mean "compact manifold"; it will be clear from the context if noncompact manifolds are also being discussed.

2.1   Dimension 0

There is only one dimension-0 manifold. It is a single point. The period at the end of this sentence looks,

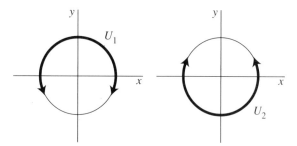

Figure 3 Two charts that cover the circle.

from afar, like a connected, dimension-0 manifold. Note that the distinction between topological and smooth is irrelevant here.

2.2   Dimension 1

There is only one compact, connected, one-dimensional topological manifold, namely the circle. Moreover, the circle has just one smooth structure. Here is one way to represent this structure. Take as a representative circle the unit circle in the xy-plane, that is, the set of all points (x, y) with $x^2 + y^2 = 1$. This can be covered by two overlapping intervals, each of which covers slightly more than half of the circle. The intervals U_1 and U_2 are drawn in figure 3. Each interval constitutes a coordinate chart. The one on the left, U_1, can be parametrized in a continuous fashion by taking the angle of a given point as measured counterclockwise from the positive x-axis. For example, the point $(1, 0)$ has angle 0, and the point $(-1, 0)$ has angle π. In order to parametrize U_2 by angle, you will have to start with angle π at the negative x-axis. If you move around U_2, varying this angle continuously, then when you reach the point $(1, 0)$ you will have parametrized it as a point in U_2 using the angle 2π.

As you can see, the arcs U_1 and U_2 intersect in two separated, smaller arcs; these are labeled V_1 and V_2 in figure 4. The transition function on V_1 is the identity map, since the U_1 angle of any given point in V_1 is the same as its U_2 angle. By contrast, the U_2 angle of a point in V_2 is obtained from the U_1 angle by adding 2π. Thus, the transition function on V_2 is not the identity map but the map that adds 2π to the coordinate function.

This one-dimensional example brings up a number of important issues, all related to a particularly troubling question. To state it, consider first that there are lots of closed loops in the plane that can be taken as model circles. Indeed, the word "lots" considerably understates the situation. Moreover, why should we restrict our

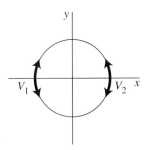

Figure 4 The intersection of the arcs U_1 and U_2.

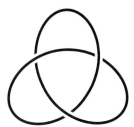

Figure 5 A knotted loop in 3-space.

attention to circles in the plane? There are closed loops galore in 3-space too: see figure 5, for example. For that matter, any manifold of dimension greater than 1 has smooth loops. Earlier, it was asserted that there is just one smooth, compact, connected, one-dimensional manifold, so all of these loops must be considered the "same." Why is this?

Here is the answer. We often think of a manifold as it might appear were it sitting in some larger space. For example, we might imagine a circle sitting in the plane, or sitting knotted in three-dimensional Euclidean space. However, the notion of "smooth manifold" introduced above is an *intrinsic* one, in the sense that it does not depend on how the manifold is placed inside a higher-dimensional space. Indeed, it is not even necessary for there to be a higher-dimensional space at all. In the case of the circle, this can be said in the following way. The circle can be placed as a loop in the plane, or as a knot in 3-space, or whatever. Each view of the circle in a higher-dimensional Euclidean space defines a collection of functions that are considered differentiable: one just takes the differentiable functions of the coordinates of the big Euclidean space and restricts them to the circle. As it turns out, any one such collection defines the same smooth structure on the circle as any other. Thus, the smooth structures that are provided by these different views of a circle are all the same, even though

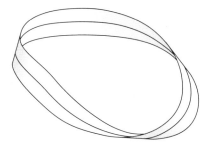

Figure 6 A Möbius strip has just one side.

there are many interesting ways of placing a circle in a given higher-dimensional space. (In fact, the classification of knots in 3-space is a fascinating, vibrant topic in its own right: see KNOT POLYNOMIALS [III.44].)

How is it proved that there is only one smooth structure for the circle? For that matter, how is it proved that there is but a single compact topological manifold in dimension 1? Since this article is not meant to provide proofs, these questions are left as serious exercises with the following advice. Think hard about the definitions and, for the smooth-manifold question, use some calculus.

2.3   Dimension 2

The story for two-dimensional, connected, compact manifolds is much richer than that for dimension 1. In the first place, there is a basic dichotomy between two kinds of manifold: orientable and nonorientable. Roughly speaking, this is the distinction between manifolds that have two sides and those that have just one. To give a more formal definition, a two-dimensional manifold is called *orientable* if every loop in the manifold that does not cross itself or have any kinks has two distinct sides. This is to say that there is no path from one side of the loop to the other that avoids the loop yet remains very close to it. The Möbius strip (see figure 6) is not orientable because there are paths from one side of the central loop to the other that do not cross the central loop yet remain very close to it. The orientable, compact, connected, topological, two-dimensional manifolds are in one-to-one correspondence with a collection of fundamental foods: the apple, the doughnut, the two-holed pretzel, the three-holed pretzel, the four-holed pretzel, and so on (see figure 7). Technically, they are classified by an integer, called the *genus*. This is 0 for the sphere, 1 for the torus, 2 for the two-holed torus, etc. The genus counts the number of holes that appear in a given example from

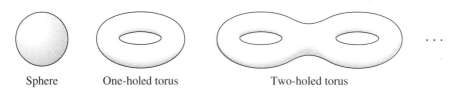

Sphere   One-holed torus   Two-holed torus

Figure 7 The orientable manifolds of dimension 2.

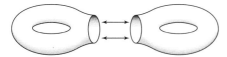

Figure 8 Cutting and gluing.

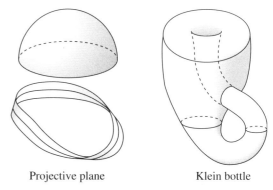

Projective plane   Klein bottle

Figure 9 Two nonorientable surfaces. To form the projective plane, one identifies the boundary of the Möbius strip with the boundary of the hemisphere.

figure 7. To say that this classifies them is to say that two such manifolds are the same if and only if they have the same genus. This is a theorem due to POINCARÉ [VI.61].

As it turns out, every topological two-dimensional manifold has exactly one smooth structure, so the list in figure 7 is the same as the list of the *smooth* orientable two-dimensional manifolds. Here one should keep in mind that the notion of a smooth manifold is intrinsic, and therefore independent of how the manifold is represented as a surface in 3-space, or in any other space. For example, the surfaces of an orange, a banana, and a watermelon each represent embedded images of the two-dimensional sphere, the leftmost example in figure 7.

The shapes illustrated in figure 7 suggest an idea that plays a key role when it comes to classifying manifolds of higher dimensions. Notice that the two-holed torus can be viewed as the result of taking two one-holed tori, cutting disks out of both, gluing the results together across their boundary circles, and then smoothing the corners. This operation is depicted in figure 8. This sort of cutting and gluing operation is an example of what is called a *surgery*. The analogous surgery can also be done with a one-holed torus and a two-holed torus to obtain a three-holed torus. And so on. Thus, all of the oriented two-dimensional manifolds can be built using standard surgeries on copies of just two fundamental building blocks: the one-holed torus and the sphere. Here is a nice exercise to test your understanding of this process. Suppose that you perform a surgery, as in figure 8, on a sphere and another manifold M. Prove that the resulting manifold is the same, with regard to its topological and smooth structure, as M.

As it turns out, all of the nonorientable two-dimensional manifolds can be built using a version of surgery

that first cuts a disk out of an orientable two-dimensional manifold and then glues on a Möbius strip. To be more precise, note that the Möbius strip has a circle as its boundary. Cut a disk out of any given orientable, two-dimensional manifold and the result also has a circular boundary. Glue the latter circular boundary to the Möbius strip boundary, smooth the corners, and the result is a smooth manifold that is nonorientable. Every nonorientable, topological (and thus every nonorientable, smooth), two-dimensional manifold is obtained in this way. Moreover, the manifold you get depends only on the number of holes (the genus) of the orientable manifold that is used.

The manifold obtained from the surgery of a Möbius strip with a sphere is called the *projective plane*. The one that uses the Möbius strip and the torus is called the *Klein bottle*. These shapes are illustrated in figure 9. No nonorientable example can be put into three-dimensional Euclidean space in a clean way; any such placement is forced to have portions that pass through other portions, as can be seen in the illustration of the Klein bottle.

How does one prove that the list given above exhausts all two-dimensional manifolds? One method uses versions of the geometric techniques that are discussed below in the three-dimensional context.

2.4 Dimension 3

There is now a complete classification of all smooth, three-dimensional manifolds; however, this is a very recent achievement. There has been for some time a conjectured list of all three-dimensional manifolds, and a conjectured procedure for telling one from the other. The proof of these conjectures was recently completed by Grigori Perelman; this is a much-celebrated event in the mathematics community. The proof uses geometry about which more is said in the final part of this article. Here I shall concentrate on the classification scheme.

Before getting to the classification scheme, it is necessary to introduce the notion of a *geometric structure* on a manifold. Roughly speaking, this means a rule for defining the lengths of paths on the manifold. This rule must satisfy the following conditions. The constant path that simply stays at one point has length 0, but any path that moves at all has positive length. Second, if one path starts where another ends, the length of their concatenation (that is, the result of putting the two paths together) is the sum of their lengths.

Note that a rule of this sort for path lengths leads naturally to a notion of distance $d(x, y)$ between any two points x and y on the manifold: one takes the length of the shortest path between them. It turns out to be particularly interesting when $d(x, y)^2$ varies as a smooth function of x and y.

As it happens, there is nothing special about having a geometric structure. Manifolds have them in spades. The following are three very useful geometric structures for the interior of the ball of radius 2 about the origin in n-dimensional Euclidean space. In these formulas, the given path is viewed as if drawn in real time by some hyper-dimensional artist, with $x(t)$ denoting the position of the pencil tip on the path at time t. Here, t ranges over some interval of the real line:

$$\left. \begin{aligned} \text{length} &= \int |\dot{x}(t)|\,\mathrm{d}t; \\[4pt] \text{length} &= \int |\dot{x}(t)|\,\frac{1}{1 + \frac{1}{4}|x(t)|^2}\,\mathrm{d}t; \\[4pt] \text{length} &= \int |\dot{x}(t)|\,\frac{1}{1 - \frac{1}{4}|x(t)|^2}\,\mathrm{d}t. \end{aligned} \right\} \quad (1)$$

In these formulas, $\dot{x}$ denotes the time-derivative of the path $t \to x(t)$.

The first of these geometric structures leads to the standard Euclidean distance between pairs of points. For this reason it is called the *Euclidean geometry* for the ball. The second defines what is called *spherical geometry* because the distance between any two points

is the angle between certain corresponding points in the sphere of radius 1 in $(n + 1)$-dimensional Euclidean space. The correspondence comes from an $(n + 1)$-dimensional version of the stereographic projection that is used for maps of the Earth's polar regions. The third distance function defines what is called the *hyperbolic geometry* on the ball. This arises when the ball of radius 2 in n-dimensional Euclidean space is identified in a certain way with a particular hyperbola in $(n + 1)$-dimensional Euclidean space.

The geometric structures that are depicted in (1) turn out to be symmetrical with respect to rotations and certain other transformations of the unit ball. (You can read more about Euclidean, spherical, and hyperbolic geometry in SOME FUNDAMENTAL MATHEMATICAL DEFINITIONS [I.3 §§6.2, 6.5, 6.6].)

As was remarked above, there are very many geometric structures on any given manifold and so one might hope to find one that has some particularly desirable properties. With this goal in mind, suppose that I have specified some "standard" geometric structure S for the ball in $\mathbb{R}^n$ to serve as a model of an exceptionally desirable structure. This could be one of the ones I have just defined or some other favorite. This leads to a corresponding notion of the structure S for a compact manifold. Roughly speaking, one says that a geometric structure on a manifold is of the type S if every point in the manifold feels as though it belongs to the unit ball with the structure S, that is, if one can use the structure S on the ball to provide coordinate charts that respect the geometric structure on the manifold. To be more precise, suppose that I am defining a coordinate system in a small neighborhood N of x by means of a function $\phi : N \to \mathbb{R}^d$. If I can always do this in such a way that the image $\phi(N)$ lies inside the ball, and such that the distance between any two points x and y in N equals the distance between their images $\phi(x)$ and $\phi(y)$, defined in terms of the structure S on the ball, then I will say that the manifold has structure of type S. In particular, a geometric structure is said to be *Euclidean*, *spherical*, or *hyperbolic* when the structure on the ball is Euclidean, spherical, or hyperbolic, respectively.

For example, the sphere in any dimension has a spherical geometric structure (as it should!). As it turns out, every two-dimensional manifold has a geometric structure that is either spherical, Euclidean, or hyperbolic. Moreover, if it has a structure of one of these types, then it cannot have one of a different type. In particular, the sphere has a spherical structure, but not a Euclidean or hyperbolic structure. Meanwhile, the torus in dimension 2 has a Euclidean geometric structure but only a Euclidean one, and all of the other manifolds listed in figure 7 have hyperbolic geometric structures and only hyperbolic ones.

William Thurston had the great insight to realize that three-dimensional manifolds might be classifiable using geometric structures. In particular, he made what was known as the *geometrization conjecture*, which says, roughly speaking, that every three-dimensional manifold is made up of "nice" pieces:

Every smooth three-dimensional manifold can be cut in a canonical fashion along a predetermined set of two-dimensional spheres and one-holed tori so that each of the resulting parts has precisely one of a list of eight possible geometric structures.

The eight possible structures include the spherical, Euclidean, and hyperbolic ones. These plus the other five are, in a sense that can be made precise, those that are maximally symmetric. The other five are associated with various LIE GROUPS [III.48 §1], as are the listed three.

Since its proof by Perelman, the geometrization conjecture has come to be known as the geometrization theorem. As I shall explain in a moment, this provides a satisfactory resolution of the three-dimensional part of the quest set out at the end of section 1. This is because a manifold with one of the eight geometric structures can be described in a canonical fashion using group theory. As a result, the geometrization theorem turns the classification issue for manifolds into a question that group theory can answer. What follows is an indication of how this comes about.

Each of the eight geometric structures has an associated *model space* which has the given geometric structure. For example, in the case of the spherical structure, the model space is the three-dimensional sphere. For the Euclidean structure, the model space is the three-dimensional Euclidean space. For the hyperbolic structure, it is the hyperbola in the four-dimensional Euclidean space, where the coordinates (x, y, z, t) obey $t^2 = 1 + x^2 + y^2 + z^2$. In all of the eight cases, the model space has a canonical group of self-maps that preserve the distance between any two pairs of points. In the Euclidean case, this group is the group of translations and rotations of the three-dimensional Euclidean space. In the spherical case, it is the group of rotations of the four-dimensional Euclidean space, and in the hyperbolic case, it is the group of Lorentz transformations of four-dimensional Minkowski space. The associated group of self-maps is called the *isometry group* for the given geometric structure.

The connection between manifolds and group theory arises because a certain set of discrete subgroups of the isometry group of any one of the eight model spaces determines a compact manifold with the corresponding geometric structure. (A subgroup is called *discrete* if every point in the subgroup is isolated, meaning that it belongs to a neighborhood that contains no other points from the subgroup.) This compact manifold is obtained as follows. Two points x and y in the model space are declared to be *equivalent* if there is an isometry T, belonging to the subgroup, such that $Tx = y$. In other words, x is equivalent to all its images under isometries from the subgroup. It is easy to check that this notion of equivalence is a genuine EQUIVALENCE RELATION [I.2 §2.3]. The equivalence classes are then in one-to-one correspondence with the points of the associated compact manifold.

Here is a one-dimensional example of how this works. Think of the real line as a model space whose isometry group is the group of translations. The set of translations by integer multiples of 2π forms a discrete subgroup of this group. Given a point t in the real line, the possible images under translations from the subgroup are all the numbers of the form $t + 2n\pi$, where n is an integer, so one regards two real numbers as equivalent if they differ by a multiple of 2π, and the equivalence class of t is $\{t + 2n\pi : n \in \mathbb{Z}\}$. One can associate with this equivalence class the point $(x, y) = (\cos t, \sin t)$ in the circle, since adding a multiple of 2π to t does not affect either its sine or its cosine. (Intuitively speaking, if you regard each t as equivalent to $t + 2\pi$, then you are wrapping the real line around and around a circle.)

This association between certain subgroups of the isometry group and compact manifolds with the given geometric structure goes in the other direction as well. That is, the subgroup can be recovered from the manifold in a relatively straightforward fashion using the fact that each point in the manifold lies in a coordinate chart where its distance function is the same as that of the associated model space.

Even before Perelman's work there was a tremendous amount of evidence for the validity of the geometrization conjecture, much of it supplied by Thurston. In order to discuss this evidence, a small digression is required to give some of the background. First, I need to bring in the notion of a *link* in the three-dimensional sphere. A link is the name given to a finite disjoint

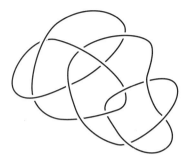

Figure 10 A link formed out of two knots.

union of knots. Figure 10 depicts an example of one that is made out of two knots.

I also need the notion of *surgery on a link*. To this end, thicken the link so as to view it as a union of knotted, solid tubes. (Think of the knot as the copper in an insulated wire and view the solid tube as the copper plus the surrounding insulation.) Notice that the boundary of any given component tube is really a copy of our one-holed torus from figure 7. Therefore, removing any one of the tubes leaves a tubular-shaped missing region from the three-dimensional sphere whose boundary is a torus.

Now, to define a surgery, imagine removing a knotted tube and then gluing it back in a different way. That is, imagine gluing the boundary of the tube to the boundary of the resulting missing region using an identification that is *not* the same as the original. For example, take the "unknot," a standard round circle in a given plane, here viewed as living inside a coordinate chart of the three-dimensional sphere. Take out the solid tube around it, and then replace the tube by gluing the boundary in the "wrong" way, as follows. Consider the leftmost torus in figure 11 as the boundary of the complement of the tube in $\mathbb{R}^3$. Consider the middle torus as the inside of the tube. The "wrong" gluing identifies the circles marked "R" and "L" on the leftmost torus with their counterparts on the middle torus. The resulting space is a three-dimensional manifold which turns out to be the product of the circle with the two-dimensional sphere. That is to say, it is the set of ordered pairs (x, y), where x is a point in the circle and y is a point in the two-dimensional sphere. There are many other possible ways to glue the boundary torus, and almost all of the corresponding surgeries give rise to distinct three-dimensional manifolds. One of these is illustrated in the rightmost part of figure 11.

In general, given any link one can construct a countably infinite set of distinct, smooth three-dimensional

manifolds by using surgeries on it. Furthermore, Raymond Lickorish proved that *every* three-dimensional manifold can be obtained by using surgery on *some* link in the three-dimensional sphere. Unfortunately, this characterization of three-dimensional manifolds via surgeries on links does not provide a satisfactory resolution to the central quest of classifying smooth structures because the process is far from unique: for any given manifold there is a bewildering assortment of links and surgeries that can be used to produce it. Moreover, as of this writing, there is no known way to classify knots and links in the three-dimensional sphere.

In any event, here is a taste of Thurston's evidence for his geometrization conjecture. Given any link, all but finitely many of the three-dimensional manifolds you can produce from it by surgery satisfy the conclusions of the geometrization conjecture. Thurston also proved that, given any knot apart from the unknot, all but finitely many surgeries on it produce a manifold with a hyperbolic geometric structure.

By the way, Perelman's proof of the geometrization theorem gives as a special case a proof of the *Poincaré conjecture*, proposed by Poincaré in 1904. To state this we need the notion of a *simply connected* manifold. This is a manifold with the property that any closed loop in it can be shrunk down to a point. To be more precise, designate a point in the manifold as the "base point." Then any path in the manifold that starts and ends at the chosen base point can be continuously deformed in such a way that at each stage of the deformation the path still starts and ends at the base point, and so that the end result is the trivial path that starts at the base point and just stays there. For example, the two-dimensional sphere is simply connected, but the torus is not, since a loop that goes "once around" the torus (for example, any of the loops R or L in the various tori of figure 11) cannot be shrunk to a point. In fact, a sphere is the only two-dimensional manifold that is simply connected, and spheres are simply connected in all dimensions greater than 1.

The Poincaré conjecture. Every compact, simply connected, three-dimensional manifold is the three-dimensional sphere.

2.5   Dimension 4

This is the weird dimension. Nobody has managed to formulate a useful and viable conjecture for the classification of smooth, compact, four-dimensional manifolds. On the other hand, the classification story for

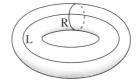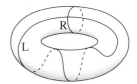

Figure 11 Different ways of gluing a tube into a tube-shaped hole.

many categories of topological four-dimensional manifolds is well-understood. For the most part, this work is by Michael Freedman.

Some of the topological manifolds in dimension 4 do not admit smooth structures. The so-called "$\frac{11}{8}$ conjecture" proposes necessary and sufficient conditions for a four-dimensional, topological manifold to have at least one smooth structure. The fraction $\frac{11}{8}$ here refers to the absolute value of the ratio of the rank to the signature of a certain symmetric, bilinear form that appears in the four-dimensional story. The case $\frac{0}{0}$ excepted, the conjecture asserts that a smooth structure exists if and only if this ratio is at least $\frac{11}{8}$. The bilinear form in question is obtained by counting with signed weights the intersection points between various two-dimensional surfaces inside the given four-dimensional manifold. In this regard, note that a typical pair of two-dimensional surfaces in four dimensions will intersect at finitely many points. This is a higher-dimensional analogue of a fact that is rather easier to visualize: that a typical pair of loops in the two-dimensional plane will intersect at finitely many points. Not surprisingly, the bilinear form here is called the *intersection form*; it plays a prominent role in Freedman's classification theorems.

Meanwhile, the problem of listing all smooth structures is wide open in four dimensions: there are no cases of a topological manifold with at least one smooth structure where the list of distinct structures is known to be complete. Some topological four-dimensional manifolds are known to have (countably) infinitely many distinct smooth structures. For others there is only one known structure. For example, the four-dimensional sphere has one obvious smooth structure and this is the only one known. However, the underlying topological manifold may, for all anyone knows, have many distinct smooth structures. By the way, the story for noncompact manifolds in dimension 4 is truly bizarre. For example, it is known that there are uncountably many smooth manifolds that are homeomorphic to the standard, four-dimensional Euclidean space. But even here, our understanding is

less than optimal since there is no known explicit construction of a single one of these "exotic" smooth structures.

Simon Donaldson provided a set of geometric invariants that have the power to distinguish smooth structures on a given topological 4-manifold. Donaldson's invariants were recently superseded by a suite of more computable invariants; these were proposed by Edward Witten and are called the *Seiberg–Witten invariants*. More recently still, Peter Oszvath and Zoltan Szabo designed a possibly equivalent set of invariants that are even easier to use. Do the Seiberg–Witten invariants (broadly defined) distinguish all smooth structures? No one knows. A bit more is said about these invariants in the final part of this article.

Note that Freedman's results include the topological version of the four-dimensional Poincaré conjecture that follows.

The four-dimensional sphere is the only compact, topological 4-manifold with the following property: every based map from either a one-dimensional circle or a two-dimensional sphere can be continuously deformed so that the result maps onto the base point.

The smooth version of this conjecture has not been resolved.

Is there a four-dimensional version of the geometrization conjecture/theorem?

2.6 Dimensions 5 and Greater

Surprisingly enough, the issues raised at the end of the first section have more or less been resolved in all dimensions that are greater than 4. This was done some time ago by Stephen Smale with input from John Stallings. In these higher dimensions it is also possible to say what conditions need to hold in order for a topological manifold to admit a smooth structure. For example, John Milnor and others determined that the respective number of smooth structures on the spheres of dimensions 5–18 are as follows: 1, 1, 28, 2, 8, 6, 992, 1, 3, 2, 16 256, 2, 16, 16.

At first sight, it is surprising that the dimensions greater than 4 are easier to deal with than dimensions 3 and 4. However, there is a good reason for this. It turns out that there is more room to maneuver in these higher-dimensional spaces and this extra room makes all the difference. To get a sense for this, let n be a positive integer, and let S^n denote the n-dimensional sphere. To make this more concrete, view S^n as the set of points $(x_1, \ldots, x_{n+1})$ in the Euclidean space $\mathbb{R}^n$ such that $x_1^2 + \cdots + x_{n+1}^2 = 1$. Now consider the product manifold, $S^n \times S^n$. This is the set of pairs of points (x, y), where x is in one copy of S^n and y is in another. This product manifold has dimension $2n$. A standard picture of $S^n \times S^n$ has two distinguished copies of S^n inside it, one consisting of all points of the form (x, y) with $y = (1, 0, \ldots)$ and the other consisting of all points (x, y) with $x = (1, 0, \ldots)$. Let us call the first copy S_R and the second one S_L. Of particular interest here is the fact that S_R and S_L intersect in precisely one point, the point $((1, 0, \ldots), (1, 0, \ldots))$.

By the way, in the $n = 1$ case, the space $S^1 \times S^1$ is the doughnut in figure 7. The one-dimensional spheres S_R and S_L inside it are the circles that are drawn in the leftmost diagram in figure 11.

If you are with me so far, suppose now that an advanced alien en route from Arcturus to the galactic center kidnaps you and drops you into some unknown, $2n$-dimensional manifold. You suspect that it is $S^n \times S^n$, but are not sure. One reason that you suspect this to be the case is that you have found a pair of n-dimensional spheres in it, one you call M_R and the other you call M_L. Unfortunately, they intersect in $2N + 1$ points, where $N > 0$. You would be less nervous about things if you could find a pair of different spheres that intersect precisely once. So you wonder whether perhaps you can push M_L around a bit so as to remove the $2N$ unwanted intersection points.

The surprise here is that the issue of removing intersection points in any dimension concerns only certain zero-, one-, and two-dimensional manifolds that live inside your $2n$-dimensional one. This is an old observation due to Hassler Whitney. In particular, Whitney discovered that in the $2n$-dimensional manifold you must be able to find a disk of dimension two whose boundary loop lies half in M_L and half in M_R. This boundary loop must hit two of the intersection points (one when it passes from M_L to M_R and one when it passes back again). The disk must also stick out orthogonally to M_L and M_R where it touches them. If its interior is disjoint from both M_L and M_R, and if there are no points where

the disk comes back to intersect itself, then you can push the part of M_L that is very near the disk along the disk while stretching the remaining part to keep things from tearing. If you extend the disk a bit past M_R, then you will have removed two of the intersection points when you have pushed past the end of the disk. Figure 12 is a schematic of this. This pushing operation (the *Whitney trick*) can be performed in any manifold of any dimension if you can find the required disk. The problem is to find the disk. Figure 13 is a drawing of a cross-sectional slice showing a "good" disk on the left and some badly chosen disks in the middle and on the right. If you have a badly chosen disk that nevertheless satisfies the required boundary conditions, then you might hope to find a tiny wiggle of its interior that makes it better. You would like the new disk to have no self-intersection points and you would like its interior to be disjoint from both M_L and M_R. No wiggle along a direction that is parallel to the disk itself will help, for any such wiggle only changes the position of the intersection point in the disk. Likewise, a wiggle in a direction parallel to the offending M_L or M_R is useless since it only changes the position of the intersection point in the latter space. Thus, $2 + n$ of the $2n$ dimensions are useless when it comes to wiggling a disk. However, there are $2n - (n + 2) = n - 2$ remaining dimensions to work with, which is a positive number when $2n > 4$. In fact, when this is true a generic wiggle in any of these extra dimensions does the trick.

Now, when $2n = 4$ (so $n = 2$) there are no extra dimensions, and, consequently, no small wiggle can make a new disk without intersection points. So if a given candidate disk intersects M_R, then the Whitney trick just trades the old pair of intersection points for a new collection. If the disk intersects either itself or M_L, then the new version of M_L has self-intersection points: that is, points where one part has come around to intersect another.

This failure of the Whitney trick is the bane of four-dimensional topology. Thus, a major lemma for Michael Freedman's classification theorem about topological four-dimensional manifolds describes ubiquitous circumstances where a topologically (but not smoothly!) embedded disk can be found for use in the Whitney trick.

3 How Geometry Enters the Fray

Much of our current understanding about smooth manifolds in dimensions 4 or less has come via what

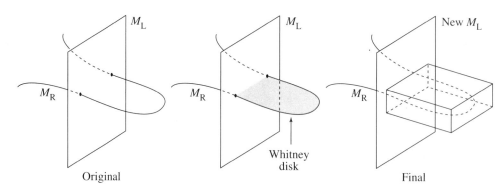

Figure 12 The Whitney trick.

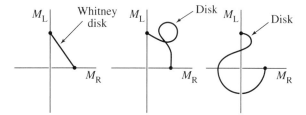

Figure 13 Some possible Whitney disks.

might be called geometric techniques. The search for a canonical geometric structure on a given three-dimensional manifold is an example. Perelman's proof of the geometrization theorem proceeds in this manner. The idea is to choose any convenient geometric structure on a given three-dimensional manifold and then continuously deform it by some well-defined rule. If one views the deformation as a time-dependent process, then the goal is to design the deformation rule to make the geometric structure ever more symmetric as time goes on.

A rule introduced and much studied by Richard Hamilton and then used by Perelman specifies the time-derivative of the geometric structure at any given time in terms of certain of its properties at that time. It is a nonlinear version of the classical HEAT EQUATION [I.3 §5.4]. For those unfamiliar with the latter, the simplest version modifies functions on the real line and will now be described. Let τ denote the time parameter, and let $f(x)$ denote a given function on the line, representing the initial distribution of heat. The resulting time-dependent family of functions associates with any given positive value for τ a function, $F_\tau(x)$, which represents the distribution of heat at time τ. The partial derivative of $F_\tau(x)$ with respect to τ is equal to its second partial derivative with respect to x, and the initial

condition is that $F_0(x) = f(x)$. If the initial function f is zero outside some interval, then one can write down a formula for F_τ:

$$F_\tau(x) = \frac{1}{(2\pi\tau)^{1/2}} \int_{-\infty}^{\infty} e^{-(x-y)^2/2\tau} f(y)\,dy. \quad (2)$$

One can see from (2) that $F_\tau(x)$ tends uniformly to zero in x as τ tends to infinity. In particular, this limit is completely ignorant of the starting function f; and, being identically zero, it is also the most symmetric function possible. The representation for F_τ in (2) indicates how this comes about. The value of F_τ at any given point is a weighted average of the values of the original function. Moreover, as τ increases, this average looks more like the standard average over ever-larger regions of the line. Physically this is very plausible as well: the heat spreads itself out more and more thinly as time goes on.

The time-dependent family of geometric structures that Hamilton introduced and Perelman used is defined by an equation that relates the time-derivative of the geometric structure at any given time to its *Ricci curvature*, a certain natural substitute in the context of geometric structures for the second derivatives that enter the heat equation for the functions F_τ above. The idea much studied by Hamilton and then by Perelman is to let the evolving geometric structure decompose the manifold into the canonical pieces that are predicted to exist by the geometrization conjecture. Perelman proved that the pieces required by the geometrization conjecture emerge as regions whose points stay relatively close together (as measured by a certain rescaling of the distance function) while the points in distinct regions move farther and farther apart.

The equation used by Perelman and Hamilton for the time-evolution of a geometric structure is rather complicated. Its standard incarnation involves the notion of a RIEMANNIAN METRIC [I.3 §6.10]. This appears in any given coordinate chart on an n-dimensional manifold as a symmetric, positive-definite $n \times n$ matrix whose entries are functions of the coordinates. The various components of this matrix are traditionally written as $\{g_{ij}\}_{1 \leqslant i, j \leqslant n}$. The matrix determines the geometric structure and can in turn be derived from it.

Hamilton and Perelman study a time-dependent family of Riemannian metrics, $\tau \rightarrow g_\tau$, where the rule for the time dependence is obtained using an equation for the τ-derivative of g_τ that has the schematic form $\partial_\tau (g_\tau)_{ij} = -2R_{ij}[g_\tau]$, where $\{R_{ij}\}_{1 \leqslant i, j \leqslant n}$ are the components of the aforementioned Ricci curvature, a certain symmetric matrix that is determined at any given τ by the metric g_τ. Every Riemannian metric has a Ricci curvature; its components are standard (nonlinear) functions of the components of the matrix and their first- and second-order partial derivatives in the coordinate directions. The Ricci curvatures for the metrics that define the respective Euclidean, spherical, and hyperbolic geometries have the particularly simple form $R_{ij} = cg_{ij}$, where c is 0, 1, or -1, respectively. For more about these ideas, see RICCI FLOW [III.78].

As was mentioned at the beginning of this part of the article, geometry has also played a central role in the developments in the classification program for smooth, four-dimensional manifolds. In this case, geometrically defined data are used to distinguish smooth structures on topologically equivalent manifolds. What follows is a very brief sketch of how this is done.

To begin with, the idea is to introduce a geometric structure on the manifold and then to use the latter to define a canonical system of partial differential equations. In any given coordinate chart, these equations are for a particular set of functions. The equations state that certain linear combinations of the collection of first derivatives of the functions from the set are equal to terms that are linear and quadratic in the values of the functions themselves. In the case of the Donaldson invariants, and also of the newer Seiberg–Witten invariants, the relevant equations are nonlinear generalizations of the MAXWELL EQUATIONS [IV.13 §1.1] for electricity and magnetism.

In any event, one then counts the solutions with algebraic weights. The purpose of the algebraic weighting of the count is to obtain an INVARIANT [I.4 §2.2], that is, a count that does not change if the given geometric structure is changed. The point here is that the naive count will typically depend on the structure, but a suitably weighted count will not. Imagine, for example, that one has a continuously varying family of geometric structures, and that new solutions appear and old ones disappear only in pairs, where one solution has been assigned weight $+1$ and the other -1.

The following toy model illustrates this appearance and disappearance phenomenon. The equation in question is for a single function on the circle. That is, it will concern a function, f, of one variable, x, that is periodic with period 2π. For example, take the equation $\partial f / \partial x + \tau f - f^3 = 0$, where τ is a constant that is specified in advance. Varying τ can now be viewed as a model for the variation of the geometric structure. When $\tau > 0$ there are exactly three solutions: $f \equiv 0$, $f \equiv \tau$, and $f \equiv -\tau$. However, when $\tau \leqslant 0$, the only solution is $f \equiv 0$. Thus, the number of solutions changes as τ crosses zero. Even so, a suitable weighted count is independent of τ.

Let us return now to the four-dimensional story. If the weighted sum is independent of the chosen *geometric* structure, then it depends only on the underlying *smooth* structure. Therefore, if two geometric structures on a given topological manifold provide distinct sums, then the underlying smooth structures must be distinct.

As I remarked earlier, Oszvath and Szabo have defined invariants for four-dimensional manifolds that are easier to use than the Seiberg–Witten invariants, but probably equivalent to them. These are also defined as the number of solutions to a particular system of differential equations, counted in a creative way. In this case, the equations are analogues of the CAUCHY–RIEMANN EQUATIONS [I.3 §5.6], and the arena is a space that can be defined after cutting the 4-manifold into simpler pieces. There are myriad ways to slice a 4-manifold in the prescribed manner, but a suitably creative, algebraic count of solutions provides the same number for each.

With hindsight, one can see that the use of differential equations to distinguish smooth structures on a given topological manifold makes good sense, since a smooth structure is needed to take a derivative in the first place. Even so, this author is constantly amazed by the fact that the Donaldson/Seiberg-Witten/Oszvath-Szabo strategy of algebraically counting differential equation solutions yields counts that are both tractable and useful. (Getting the same count in all cases is no help at all.)

Further Reading

Those who wish to learn more about manifolds in general can consult J. Milnor's book *Topology from the Differentiable Viewpoint* (Princeton University Press, Princeton, NJ, 1997) or the book *Differential Topology* (Prentice Hall, Englewood Cliffs, NJ, 1974), by V. Guillemin and A. Pollack. A good introduction to the classification problem in dimensions 2 and 3 is the book *Three-Dimensional Geometry and Topology* (Princeton University Press, Princeton, NJ, 1997), by W. Thurston. This book also has a nice discussion of geometric structures. A full account of Perelman's proof of the Poincaré conjecture can be found in *Ricci Flow and the Poincaré Conjecture*, by J. Morgan and G. Tian (American Mathematical Society, Providence, RI, 2007). The story for topological 4-manifolds is told in the book by M. Freedman and F. Quinn titled *Topology of 4-Manifolds* (Princeton University Press, Princeton, NJ, 1990). There are no books available that serve as general introductions to the smooth 4-manifold story. A book that does introduce the Seiberg–Witten invariants is *The Seiberg–Witten Equations and Applications to the Topology of Smooth Four-Manifolds* (Princeton University Press, Princeton, NJ, 1995), by J. Morgan. Meanwhile, the Donaldson invariants are discussed in detail in the book by Donaldson and P. Kronheimer titled *Geometry of Four-Manifolds* (Oxford University Press, Oxford, 1990). Finally, parts of the story for dimensions greater than 4 are told in *Lectures on the h-Cobordism Theorem* (Princeton University Press, Princeton, NJ, 1965), by J. Milnor, and *Foundational Essays on Topological Manifolds, Smoothings and Triangulations* (Princeton University Press, Princeton, NJ, 1977), by R. Kirby and L. Siebenman.

IV.8 Moduli Spaces
David D. Ben-Zvi

Many of the most important problems in mathematics concern CLASSIFICATION [I.4 §2]. One has a class of mathematical objects and a notion of when two objects should count as equivalent. It may well be that two equivalent objects look superficially very different, so one wishes to describe them in such a way that equivalent objects have the same description and inequivalent objects have different descriptions.

Moduli spaces can be thought of as *geometric* solutions to *geometric* classification problems. In this article we shall illustrate some of the key features of mod-

uli spaces, with an emphasis on the moduli spaces of RIEMANN SURFACES [III.79]. In broad terms, a *moduli problem* consists of three ingredients.

Objects: which geometric objects would we like to describe, or *parametrize*?
Equivalences: when do we identify two of our objects as being isomorphic, or "the same"?
Families: how do we allow our objects to vary, or modulate?

In this article we will discuss what these ingredients signify, as well as what it means to *solve* a moduli problem, and we will give some indications as to why this might be a good thing to do.

Moduli spaces arise throughout ALGEBRAIC GEOMETRY [IV.4], differential geometry, and ALGEBRAIC TOPOLOGY [IV.6]. (Moduli spaces in topology are often referred to as *classifying spaces*.) The basic idea is to give a geometric structure to the *totality* of the objects we are trying to classify. If we can understand this geometric structure, then we obtain powerful insights into the geometry of the objects themselves. Furthermore, moduli spaces are rich geometric objects in their own right. They are "meaningful" spaces, in that any statement about their geometry has a "modular" interpretation, in terms of the original classification problem. As a result, when one investigates them one can often reach much further than one can with other spaces. Moduli spaces such as the moduli of ELLIPTIC CURVES [III.21] (which we discuss below) play a central role in a variety of areas that have no immediate link to the geometry being classified, in particular in ALGEBRAIC NUMBER THEORY [IV.1] and algebraic topology. Moreover, the study of moduli spaces has benefited tremendously in recent years from interactions with physics (in particular with STRING THEORY [IV.17 §2]). These interactions have led to a variety of new questions and new techniques.

1 Warmup: The Moduli Space of Lines in the Plane

Let us begin with a problem that looks rather simple, but that nevertheless illustrates many of the important ideas of moduli spaces.

Problem. Describe the collection of all lines in the real plane $\mathbb{R}^2$ that pass through the origin.

To save writing, we are using the word "line" to mean "line that passes through the origin." This classification problem is easily solved by assigning to each line L an essential parameter, or *modulus*, a quantity that we can calculate for each line and that will help us tell different lines apart. All we have to do is take standard Cartesian coordinates x, y on the plane and measure the angle $\theta(L)$ between the line L and the x-axis, taken in counterclockwise fashion. We find that the possible values of θ are those for which $0 \leqslant \theta < \pi$, and that for every such θ there is exactly one line L that makes an angle of θ with the x-axis. So as a *set*, we have a complete solution to our classification problem: the set of lines L, known as *the real projective line* $\mathbb{RP}^1$, is in one-to-one correspondence with the half-open interval $[0, \pi)$.

However, we are seeking a *geometric* solution to the classification problem. What does this entail? We have a natural notion of when two lines are near each other, which our solution should capture—in other words, the collection of lines has a natural TOPOLOGY [III.90]. So far, our solution does not reflect the fact that lines L for which the angle $\theta(L)$ is close to π are almost horizontal: they are therefore close to the x-axis (for which $\theta = 0$) and to the lines L with $\theta(L)$ close to zero. We need to find some way of "wrapping around" the interval $[0, \pi)$ so that π becomes close to 0.

One way to do this is to take not the half-open interval $[0, \pi)$ but the closed interval $[0, \pi]$, and then to "identify" the points 0 and π. (This idea can easily be made formal by defining an appropriate EQUIVALENCE RELATION [I.2 §2.3].) If π and 0 are regarded as the same, then numbers close to π are close to numbers close to 0. This is a way of saying that if you attach the two ends of a line segment together, then, topologically speaking, you obtain a circle.

A more natural way of achieving the same end is suggested by the following geometric construction of $\mathbb{RP}^1$. Consider the unit circle $S^1 \subset \mathbb{R}^2$. To each point $s \in S^1$, there is an obvious way of assigning a line $L(s)$: take the line that passes through s and the origin. Thus, we have a *family of lines parametrized by S^1*, that is, a map (or function) $s \mapsto L(s)$ that takes points in S^1 to lines in our set $\mathbb{RP}^1$. What is important about this is that we already know what it means for two points in S^1 to be close to each other, and the map $s \mapsto L(s)$ is continuous. However, this map is a two-to-one function rather than a bijection, since s and $-s$ always give the same line. To remedy this, we can identify each s in the circle S^1 with its antipodal point $-s$. We then have a one-to-one correspondence between $\mathbb{RP}^1$ and the resulting QUOTIENT SPACE [I.3 §3.3] (which again is topologically

a circle), and this correspondence is continuous in both directions.

The key feature of the space $\mathbb{RP}^1$, considered as the *moduli space* of lines in the plane, is that it captures the ways in which lines can *modulate*, or vary continuously in families. But when do families of lines arise? A good example is provided by the following construction. Whenever we have a continuous curve $C \subset \mathbb{R}^2 \setminus 0$ in the plane, we can assign to each point c in C the line $L(c)$ that passes through 0 and c. This gives us a family of lines parametrized by C. Moreover, the function that takes c to $L(c)$ is a continuous function from C to $\mathbb{RP}^1$, so the parametrization is a continuous one.

Suppose, for example, that C is a copy of $\mathbb{R}$ realized as the set of points $(x, 1)$ at height 1. Then the map from C to $\mathbb{RP}^1$ gives an isomorphism between $\mathbb{R}$ and the set $\{L : \theta(L) \neq 0\}$, which is the subset of $\mathbb{RP}^1$ consisting of all lines apart from the x-axis. Put more abstractly, we have an intuitive notion of what it means for a collection of lines through the origin to depend continuously on some parameters, and this notion is captured precisely by the geometry of $\mathbb{RP}^1$: for instance, if you tell me you have a continuous 37-parameter family of lines in $\mathbb{R}^2$, this is the same as saying that you have a continuous map from $\mathbb{R}^{37}$ to $\mathbb{RP}^1$, which sends a point $v \in \mathbb{R}^{37}$ to a line $L(v) \in \mathbb{RP}^1$. (More concretely, we could say that the real function $v \mapsto \theta(L(v))$ on $\mathbb{R}^{37}$ is continuous away from the locus where θ is close to π. Near this locus we could use instead the function ϕ that measures the angle from the y-axis.)

1.1 Other Families

The idea of families of lines leads to various other geometric structures on the space $\mathbb{RP}^1$, and not just its topological structure. For example, we have the notion of a *differentiable* family of lines in the plane, which is a family of lines for which the angles vary differentiably. (The same ideas apply if we replace "differentiable" by "measurable," "C^∞," "real analytic," etc.) To parametrize such a family appropriately, we would like $\mathbb{RP}^1$ to be a DIFFERENTIABLE MANIFOLD [I.3 §6.9], so that we can calculate derivatives of functions on it. Such a structure on $\mathbb{RP}^1$ can be specified by using the angle functions θ and ϕ defined in the previous section. The function θ gives us a coordinate for lines that are not too close to the x-axis, and ϕ gives us a coordinate for lines that are not too close to the y-axis. We can calculate derivatives of functions on $\mathbb{RP}^1$ by writing them in terms of these coordinates. One can justify this differentiable structure on $\mathbb{RP}^1$ by checking that for any differentiable curve $C \subset \mathbb{R}^2 \setminus 0$ the map $c \mapsto L(c)$ comes out as differentiable. This means that if $L(c)$ is not close to the x-axis, then the function $x \mapsto \theta(L(x))$ is differentiable at $x = c$, and similarly for ϕ and the y-axis. The functions $x \mapsto \theta(L(x))$ and $\mapsto \phi(L(x))$ are called *pullbacks*, because they are the result of converting, or "pulling back," θ and ϕ from functions defined on $\mathbb{RP}^1$ to functions defined on C.

We now can state the fundamental property of $\mathbb{RP}^1$ as a differentiable space.

A differentiable family of lines in $\mathbb{R}^2$ parametrized by a differentiable manifold X is the same thing as a function from X to $\mathbb{RP}^1$, taking a point x to a line $L(x)$, such that the pullbacks $x \mapsto \theta(L(x))$ and $x \mapsto \phi(L(x))$ of the functions θ, ϕ are differentiable functions.

We say that $\mathbb{RP}^1$ (with its differentiable structure) is the *moduli space* of (differentiably varying families of) lines in $\mathbb{R}^2$. This means that $\mathbb{RP}^1$ carries the *universal differentiable family of lines*. From the very definition, we have assigned to each point of $\mathbb{RP}^1$ a line in $\mathbb{R}^2$, and these lines vary differentiably as we vary the point. The above assertion says that *any* differentiable family of lines, parametrized by a space X, is described by giving a map $f : X \rightarrow \mathbb{RP}^1$ and assigning to $x \in X$ the line $L(f(x))$.

1.2 Reformulation: Line Bundles

It is interesting to reformulate the notion of a (continuous or differentiable) family of lines as follows. Let X be a space and let $x \mapsto L(x)$ be an assignment of lines to points in X. For each point $x \in X$, we place a copy of $\mathbb{R}^2$ at x; in other words, we consider the Cartesian product $X \times \mathbb{R}^2$. We may now visualize the line $L(x)$ as living in the copy of $\mathbb{R}^2$ that lies over x. This gives us a continuously varying collection of lines $L(x)$ parametrized by $x \in X$, otherwise known as a *line bundle* over X. Moreover, this line bundle is embedded in the "trivial" VECTOR BUNDLE [IV.6 §5] $X \times \mathbb{R}^2$, which is the constant assignment that takes each x to the plane $\mathbb{R}^2$. In the case when X is $\mathbb{RP}^1$ itself, we have a "tautological" line bundle: to each point $s \in \mathbb{RP}^1$, which we can think of as a line L_s in $\mathbb{R}^2$, it assigns that very same line L_s.

Proposition. *For any topological space X there is a natural bijection between the following two sets:*

(i) *the set of continuous functions $f : X \rightarrow \mathbb{RP}^1$; and*
(ii) *the set of line bundles on X that are contained in the trivial vector bundle $X \times \mathbb{R}^2$.*

This bijection sends a function f to the corresponding pullback of the tautological line bundle on $\mathbb{RP}^1$. That is, the function f is mapped to the line bundle $x \mapsto L_{f(x)}$. (This is a pullback because it converts L from a function defined on $\mathbb{RP}^1$ to a function defined on X.)

Thus, the space $\mathbb{RP}^1$ carries the *universal* line bundle that sits in the trivial $\mathbb{R}^2$ bundle—any time we have a line bundle sitting in the trivial $\mathbb{R}^2$ bundle, we can obtain it by pulling back the universal (tautological) example on $\mathbb{RP}^1$.

1.3 Invariants of Families

Associated with any continuous function f from the circle S^1 to itself is an integer known as its *degree*. Roughly speaking, the degree of f is the number of times $f(x)$ goes around the circle when x goes around once. (If it goes backwards n times, then we say that the degree is $-n$.) Another way to think of the degree is as the number of times a typical point in S^1 is passed by $f(x)$ as x goes around the circle, where we count this as $+1$ if it is passed in the counterclockwise direction and -1 if it is passed in the clockwise direction.

Earlier, we showed that the circle S^1, which we obtained by identifying the endpoints of the closed interval $[0, \pi]$, could be used to parametrize the moduli space $\mathbb{RP}^1$ of lines. Combining this with the notion of degree, we can draw some interesting conclusions. In particular, we can define the notion of *winding numbers*. Suppose that we are given a continuous function y from the circle S^1 into the plane $\mathbb{R}^2$ and suppose that it avoids 0. The image of this map will be a closed loop C (which may cross itself). This defines for us a map from S^1 to itself: first do y to obtain a point c in C, then work out $L(c)$, which belongs to $\mathbb{RP}^1$, and finally use the parametrization of $\mathbb{RP}^1$ to associate with $L(c)$ a point in S^1 again. The degree of the resulting composite map will be *twice* the number of times that y, and hence C, winds around 0, so half this number is defined to be the winding number of y.

More generally, given a family of lines in $\mathbb{R}^2$ parametrized by some space X, we would like to measure the "manner in which X winds around the circle." To be precise, given a function ϕ from X to $\mathbb{RP}^1$, which defines the parametrized family of lines, we would like to be able to say, for any map $f : S^1 \rightarrow X$, what the winding number is of the composition ϕf, which takes a point x in S^1 to its image $f(x)$ in X and from there to the corresponding line $\phi(f(x))$ in the family. Thus, the map ϕ gives us a way of assigning to each function $f : S^1 \to X$ an integer, the winding number of ϕf. The way this assignment works does not change if ϕ is continuously deformed: that is, it is a topological invariant of ϕ. What it does depend on is the class that ϕ belongs to in the first COHOMOLOGY GROUP [IV.6 §4] of X, $H^1(X, \mathbb{Z})$. Equivalently, to any line bundle on a space X which is contained in the trivial $\mathbb{R}^2$-bundle, we have associated a cohomology class, known as the *Euler class* of the bundle. This is the first example of a CHARACTERISTIC CLASS [IV.6 §5] for vector bundles. It demonstrates that if we understand the topology of moduli spaces of classes of geometric objects, then we can define topological invariants for families of those objects.

2   The Moduli of Curves and Teichmüller Spaces

We now turn our attention to perhaps the most famous examples of moduli spaces, the moduli spaces of curves, and their first cousins, the *Teichmüller spaces*. These moduli spaces are the geometric solution to the problem of classification of compact Riemann surfaces, and can be thought of as the "higher theory" of Riemann surfaces. The moduli spaces are "meaningful spaces," in that each of their points stands for a Riemann surface. As a result, any statement about their geometry tells us something about the geometry of Riemann surfaces.

We turn first to the objects. Recall that a *Riemann surface* is a topological surface X (connected and oriented) to which a *complex structure* has been given. Complex structures can be described in many ways, and they enable us to do complex analysis, geometry, and algebra on the surface X. In particular, they enable us to define HOLOMORPHIC [I.3 §5.6] (complex-analytic) and MEROMORPHIC FUNCTIONS [V.31] on open subsets of X. To be precise, X is a two-dimensional manifold, but the charts are thought of as open subsets of $\mathbb{C}$ rather than of $\mathbb{R}$, and the maps that glue them together are required to be holomorphic. An equivalent notion is that of a *conformal structure* on X, which is the structure needed to make it possible to define angles between curves in X. Yet another important equivalent notion is that of *algebraic structure* on X, making X into a *complex-algebraic curve* (leading to the persistent confusion in terminology: a Riemann surface is two dimensional, and therefore a surface, from the point of view of topology or the real numbers, but one dimensional, and therefore a curve, from the point of view of

complex analysis and algebra). An algebraic structure is what allows us to speak of polynomial, rational, or algebraic functions on X, and is usually specified by realizing X as the set of solutions to polynomial equations in complex PROJECTIVE SPACE [III.72] $\mathbb{CP}^2$ (or $\mathbb{CP}^n$).

In order to speak of a classification problem, let alone a moduli space, for Riemann surfaces we must next specify when we regard two Riemann surfaces as equivalent. (We postpone the discussion of the final ingredient, the notion of families of Riemann surfaces, to section 2.2.) To do this, we must give a notion of *isomorphism* between Riemann surfaces: when should two Riemann surfaces X and Y be "identified," or thought of as giving two equivalent realizations of the same underlying object of our classification? This issue was hidden in our toy example of classifying lines in the plane: there we simply identified two lines if and only if they were *equal* as lines in the plane. This naive option is not available to us with the more abstractly defined Riemann surfaces. If we considered Riemann surfaces realized concretely as subsets of some larger space—for example, as solution sets to algebraic equations in complex projective space—we could similarly choose to identify surfaces only if they were equal as subsets. However, this is too fine a classification for most applications: what we care about is the *intrinsic geometry* of Riemann surfaces, and not incidental features that result from the particular way we choose to realize them.

At the other extreme, we might choose to ignore the extra geometric structure that makes a surface into a Riemann surface. That is, we could identify two Riemann surfaces X and Y if they are topologically equivalent, or homeomorphic (the "coffee mug is a doughnut" perspective). The classification of compact Riemann surfaces up to topological equivalence is captured by a single positive integer, the genus g ("number of holes") of the surface. Any surface of genus zero is homeomorphic to the Riemann sphere $\mathbb{CP}^1 \simeq S^2$, any surface of genus 1 is homeomorphic to a torus $S^1 \times S^1$, and so on. Thus, in this case there is no issue of "modulation"— the classification is solved by giving a list of possible values of a single discrete invariant.

However, if we are interested in Riemann surfaces *as Riemann surfaces* rather than simply as topological manifolds, then this classification is too crude: it completely ignores the complex structure. We would now like to refine our classification to remedy this defect. To this end, we say that two Riemann surfaces X and Y are (conformally, or holomorphically) *equivalent* if there is a topological equivalence between them that preserves the geometry, i.e., a homeomorphism that preserves the angles between curves, or takes holomorphic functions to holomorphic functions, or takes rational functions to rational functions. (These conditions are all equivalent.) Note that we still have at our disposal our discrete invariant—the genus of a surface. However, as we shall see, this invariant is not fine enough to distinguish between all inequivalent Riemann surfaces. In fact, it is possible to have families of inequivalent Riemann surfaces that are parametrized by *continuous* parameters (but we cannot make proper sense of this idea until we have said precisely what is meant by a family of Riemann surfaces). Thus, the next step is to fix our discrete invariant and to try to classify all the different isomorphism classes of Riemann surfaces with the same genus by assembling them in a natural geometric fashion.

An important step toward this classification is the UNIFORMIZATION THEOREM [V.34]. This states that any simply connected Riemann surface is holomorphically isomorphic to one of the following three: the Riemann sphere $\mathbb{CP}^1$, the complex plane $\mathbb{C}$, or the upper half-plane $\mathbb{H}$ (equivalently, the unit disk D). Since the UNIVERSAL COVERING SPACE [III.93] of any Riemann surface is a simply connected Riemann surface, the uniformization theorem provides an approach to classifying arbitrary Riemann surfaces. For instance, any COMPACT [III.9] Riemann surface of genus zero is simply connected, and in fact homeomorphic to the Riemann sphere, so the uniformization theorem already solves our classification problem in genus zero: up to equivalence, $\mathbb{CP}^1$ is the *only* Riemann surface of genus zero, and so in this case the topological and conformal classifications agree.

2.1 Moduli of Elliptic Curves

Next, we consider Riemann surfaces whose universal cover is $\mathbb{C}$, which is the same as saying that they are quotients of $\mathbb{C}$. For example, we can look at a quotient of $\mathbb{C}$ by $\mathbb{Z}$, which means that we regard two complex numbers z and w as equivalent if $z - w$ is an integer. This has the effect of "wrapping $\mathbb{C}$ around" into a cylinder. Cylinders are not compact, but to get a compact surface we could take a quotient by $\mathbb{Z}^2$ instead: that is, we could regard z and w as equivalent if their difference is of the form $a + b$i, where a and b are both integers. Now $\mathbb{C}$ is wrapped around in two directions and the result is a torus with a complex (or, equivalently, conformal or algebraic) structure. This is a compact Riemann surface of genus 1. More generally, we

can replace $\mathbb{Z}^2$ by any lattice L, regarding z and w as equivalent if $z - w$ belongs to L. (A *lattice L* in $\mathbb{C}$ is an additive subgroup of $\mathbb{C}$ with two properties. First, it is not contained in any line. Second, it is *discrete*, which means that there is a constant $d > 0$ such that the distance between any two points in L is at least d. Lattices are also discussed in THE GENERAL GOALS OF MATHEMATICAL RESEARCH [I.4 §4]. A *basis* for a lattice L is a pair of complex numbers u and v belonging to L such that every z in L can be written in the form $au + bv$ with a and b integers. Such a basis will not be unique: for example, if $L = \mathbb{Z} \oplus \mathbb{Z}$, then the obvious basis is $u = 1$ and $v = $ i, but $u = 1$ and $v = 1 + $ i would do just as well.) If we take a quotient of $\mathbb{C}$ by a lattice, then we again obtain a torus with complex structure. It turns out that any compact Riemann surface of genus 1 can be produced in this way.

From a topological point of view, any two tori are the same, but once we consider the complex structure we start to find that different choices of lattice may lead to different Riemann surfaces. Certain changes to L do *not* have an effect: for example, if we multiply a lattice L by some nonzero complex number λ, then the quotient surface $\mathbb{C}/L$ will not be affected. That is, $\mathbb{C}/L$ is naturally isomorphic to $\mathbb{C}/\lambda L$. Therefore, we need only worry about the difference between lattices when one is not a multiple of the other. Geometrically, this says that one cannot be obtained from the other by a combination of rotation and dilation.

Notice that by taking the quotient $\mathbb{C}/L$ we obtain not just a "naked" Riemann surface, but one equipped with an "origin," that is, a distinguished point $e \in E$, which is the image of the origin $0 \in \mathbb{C}$. In other words, we obtain an *elliptic curve*:

Definition. An elliptic curve (over $\mathbb{C}$) is a Riemann surface E of genus 1, equipped with a marked point $e \in E$. Elliptic curves, up to isomorphism, are in bijection with lattices $L \subset \mathbb{C}$ up to rotation.

Remark. In fact, since $L \subset \mathbb{C}$ is a *subgroup* of the Abelian group $\mathbb{C}$, the elliptic curve $E = \mathbb{C}/L$ is naturally an Abelian group, with e as its identity element. This is an important motivation for keeping e as part of the data that defines an elliptic curve. A more subtle reason for remembering the location of e when we speak of E is that it helps us to define E more uniquely. This is useful, because any surface E of genus 1 has lots of symmetries, or AUTOMORPHISMS [I.3 §4.1]: there is always a holomorphic automorphism of E taking any point x to any other given point y. (If we think of E

as a group, these are achieved by translations.) Thus, if someone hands us another genus-1 surface E', there may be no way to identify E with E', or there may be infinitely many ways: we can always compose a given isomorphism between them with a self-symmetry of E. As we will discuss later, automorphisms haunt almost every moduli problem, and are crucial when we consider the behavior of families. It is usually convenient to "rigidify" the situation somewhat, so that the possible isomorphisms between different objects are less "floppy" and more uniquely determined. In the case of elliptic curves, distinguishing the point e achieves this by reducing the symmetry of E. Once we do that, there is usually at most one way to identify two elliptic curves (one way, that is, that takes origin to origin).

We see that Riemann surfaces of genus 1 (with the choice of a marked point) can be described by concrete "linear algebra data": a lattice $L \subset \mathbb{C}$, or rather the equivalence class consisting of all nonzero scalar multiples λL of L. This is the ideal setting to study a classification, or moduli, problem. The next step is to find an explicit parametrization of the collection of all lattices, up to multiplication, and to decide in what sense we have obtained a geometric solution to the classification problem.

In order to parametrize the collection of lattices, we follow a procedure used for all moduli problems: first parametrize lattices together with the choice of some additional structure, and then see what happens when we forget this choice. For every lattice L we choose a basis $\omega_1, \omega_2 \in L$: that is, we represent L as the set of all integer combinations $a\omega_1 + b\omega_2$. We do this in an *oriented* fashion: we require that the *fundamental parallelogram* spanned by ω_1 and ω_2 is positively oriented. (That is, the numbers $0, \omega_1, \omega_1 + \omega_2$, and ω_2 list the vertices of the parallelogram in a counterclockwise order. From the geometric point of view of the elliptic curve E, L is the FUNDAMENTAL GROUP [IV.6 §2] of E, and the orientation condition says that we generate L by two loops, or "meridians," $A = \omega_1$, $B = \omega_2$, which are oriented, in that their oriented intersection number $A \cap B$ is equal to $+1$ rather than -1.) Since we are interested in lattices only up to multiplication, we can multiply L by a complex number so as to turn ω_1 into 1 and hence ω_2 into $\omega = \omega_2/\omega_1$. The orientation condition now says that ω is in the upper half-plane $\mathbb{H}$: i.e., its imaginary part is positive, Im $\omega > 0$. Conversely, any complex number $\omega \in \mathbb{H}$ in the upper half-plane determines a unique oriented lattice $L = \mathbb{Z}1 \oplus \mathbb{Z}\omega$ (that is,

the set of all integer combinations $a + b\omega$ of 1 and ω) and no two of these lattices are related by a rotation.

What does this tell us about elliptic curves? We saw earlier that an elliptic curve is defined by a lattice L and an identity e. Now we have seen that if we give L some extra structure, namely an oriented basis, then we can parametrize it by a complex number $\omega \in \mathbb{H}$. This makes precise for us the "additional structure" that we want to place on elliptic curves. We say that a *marked* elliptic curve is an elliptic curve E, e together with the choice of an oriented basis ω_1, ω_2 for the associated lattice (fundamental group) L of E. The point is that any lattice has infinitely many different bases, which lead to many automorphisms of E. By "marking" one of these bases, we stop them being automorphisms.

2.2 Families and Teichmüller Spaces

With our new definition, we can summarize the earlier discussion by saying that marked elliptic curves are in bijection with points $\omega \in \mathbb{H}$ of the upper half-plane. The upper half-plane is, however, much more than just a *set* of points: it carries a host of geometric structures, in particular a topology and a complex structure. In what sense do these structures reflect geometric properties of marked elliptic curves? In other words, in what sense is the complex manifold $\mathbb{H}$, known in this context as the *Teichmüller space* $\mathcal{T}_{1,1}$ of genus-1 Riemann surfaces with one marked point, a geometric solution to the problem of classifying marked elliptic curves?

In order to answer this question, we need the notion of a continuous family of Riemann surfaces, and also the notion of a complex-analytic family. A *continuous family of Riemann surfaces* parametrized by a topological space S, such as the circle S^1, for example, is a "continuously varying" assignment of a Riemann surface X_s to every point s of S. In our example of the moduli of lines in the plane, a continuous family of lines was characterized by the property that the angles between the lines and the x-axis or y-axis defined continuous functions of the parameters. Geometrically defined collections of lines, such as those produced by a curve C in the plane, then gave rise to continuous families. More abstractly, a continuous family of lines defined a line *bundle* over the parameter space. A good criterion for a family of Riemann surfaces is likewise that any "reasonably defined" geometric quantity that we can calculate for every Riemann surface should vary continuously in the family. For example, a classical construction of Riemann surfaces of genus g comes from taking $4g$-gons and gluing opposite sides together. The resulting Riemann surface is fully determined by the edge-lengths and angles of the polygon. Therefore, a continuous family of Riemann surfaces described in this fashion should be precisely a family such that the edge-lengths and angles give *continuous* functions of the parameter set.

In more abstract topological terms, if we have a collection $\{X_s, \ s \in S\}$ of Riemann surfaces depending on points in a space S and we wish to make it into a continuous family, then we should give the union $\bigcup_{s \in S} X_s$ itself the structure of a topological space X, which should simultaneously extend the topology on each individual X_s. The result is called a *Riemann surface bundle*. Associated with X is the map that takes each point x to the particular s for which x belongs to X_s. We should demand that this map is continuous, and perhaps more (it could be a fibration, or fiber bundle). This definition has the advantage of great flexibility. For example, if S is a complex manifold, then in just the same way we can speak of a *complex-analytic family of Riemann surfaces* $\{X_s, \ s \in S\}$ parametrized by S: now we ask for the union of the X_s to carry not just a topology but a complex structure (i.e., it should form a complex manifold), extending the complex structure on the fibers and mapping holomorphically to the parameter set. The same holds with "complex-analytic" replaced by "algebraic." These abstract definitions have the property that if our Riemann surfaces are described in a concrete way—cut out by equations, glued from coordinate patches, etc.— then the coefficients of our equations or gluing data will vary as complex-analytic functions in our family precisely when the family is complex analytic (and likewise for continuous or algebraic families).

As a reality check, note that a (continuous, analytic, or other) family of Riemann surfaces parametrized by a single *point* $s = S$ is indeed just a single Riemann surface X_s. Just as in this simple case we wish to consider Riemann surfaces only up to equivalence, so there is a notion of equivalence or isomorphism of two analytic families $\{X_s\}$ and $\{X_s'\}$ parametrized by the same space S. We simply regard the families as equivalent if the surfaces X_s and X_s' are isomorphic for every s, and if the isomorphism depends analytically on s.

Armed with the notion of family, we can now formulate the characteristic property that the upper half-plane possesses when we think of it as the moduli space of marked elliptic curves. We define a continuous or analytic family of marked elliptic curves to be a family where the underlying genus-1 surfaces vary continuously or analytically, while the choice of basepoint $e_s \in E_s$ and the basis of the lattice L_s vary continuously.

The upper half-plane $\mathbb{H}$ plays a role for marked elliptic curves that is similar to the role played by $\mathbb{RP}^1$ for lines in the plane. The following theorem makes this statement precise.

Theorem. *For any topological space S, there is a one-to-one correspondence between continuous maps from S to $\mathbb{H}$ and isomorphism classes of continuous families of marked elliptic curves parametrized by S. Similarly, there is a one-to-one correspondence between analytic maps from any complex manifold S to $\mathbb{H}$ and isomorphism classes of analytic families of marked elliptic curves parametrized by S.*

If we apply the theorem in the case where S is a single point, it simply tells us that the points of $\mathbb{H}$ are in bijection with the isomorphism classes of marked elliptic curves, as we already knew. However, it contains more information: it says that $\mathbb{H}$, with its topology and complex structure, *embodies the structure* of marked elliptic curves and the ways in which they can modulate. At the other extreme, we could take $S = \mathbb{H}$ itself, mapping S to $\mathbb{H}$ by the identity map. This expresses the fact that $\mathbb{H}$ itself carries a family of marked elliptic curves, i.e., the collection of Riemann surfaces defined by $\omega \in \mathbb{H}$ fit together into a complex manifold fibering over $\mathbb{H}$ with elliptic curve fibers. This family is called the *universal family*, since by the theorem any family is "deduced" (or pulled back) from this one universal example.

2.3 From Teichmüller Spaces to Moduli Spaces

We have arrived at a complete and satisfying picture for the classification of elliptic curves when we choose in addition a marking (that is, an oriented basis of the associated lattice $L = \pi_1(E)$). What can we say about elliptic curves themselves, without the choice of marking? We somehow need to "forget" the marking, by regarding two points of $\mathbb{H}$ as equivalent if they correspond to two different markings of the same elliptic curve.

Now, given any two bases of the group (or lattice) $\mathbb{Z} \oplus \mathbb{Z}$, there is an invertible 2×2 matrix with integer entries that takes one basis to the other. If the two bases are *oriented*, then this matrix will have determinant 1, which means that it is an element

$$A = \begin{pmatrix} a & b \\ c & d \end{pmatrix} \in \mathrm{SL}_2(\mathbb{Z})$$

of the group of invertible unimodular matrices over $\mathbb{Z}$. Similarly, given any two oriented bases (ω_1, ω_2) and (ω_1', ω_2') of a lattice L, which can be thought of as oriented identifications of L with $\mathbb{Z} \oplus \mathbb{Z}$, there is a matrix $A \in SL_2(\mathbb{Z})$ such that $\omega_1' = a\omega_1 + b\omega_2$ and $\omega_2' = c\omega_1 + d\omega_2$. If we now consider the normalized bases $(1, \omega)$ and $(1, \omega')$, where $\omega = \omega_1/\omega_2$ and $\omega' = \omega_1'/\omega_2'$, then we obtain a transformation of the upper half-plane. It is given by the formula

$$\omega' = \frac{a\omega + b}{c\omega + d}.$$

That is, the group $SL_2(\mathbb{Z})$ is acting on the upper half-plane by linear fractional (or Möbius) transformations with integer coefficients, and two points in the upper half-plane correspond to the same elliptic curve if one can be turned into the other by means of such a transformation. If this is the case, then we should regard the two points as equivalent: that is how we formalize the idea of "forgetting" the marking. Note also that the scalar matrix $-\mathrm{Id}$ in $SL_2(\mathbb{Z})$, which negates both ω_1 and ω_2, acts trivially on the upper half-plane, so that we in fact get an action of $PSL_2(\mathbb{Z}) = SL_2(\mathbb{Z})/\{\pm \mathrm{Id}\}$ on $\mathbb{H}$.

So we come to the conclusion that *elliptic curves (up to isomorphism) are in bijection with orbits of $PSL_2(\mathbb{Z})$ on the upper half-plane, or equivalently with points of the quotient space $\mathbb{H}/PSL_2(\mathbb{Z})$*. This quotient space has a natural quotient topology, and in fact can be given a complex-analytic structure, which, it turns out, identifies it with the complex plane $\mathbb{C}$ itself. To see this one uses the classical MODULAR FUNCTION [IV.1 §8] $j(z)$, a complex-analytic function on $\mathbb{H}$ which is invariant under the modular group $PSL_2(\mathbb{Z})$ and which therefore defines a natural coordinate $\mathbb{H}/PSL_2(\mathbb{Z}) \to \mathbb{C}$.

It appears that we have solved the moduli problem for elliptic curves: we have a topological, and even complex-analytic, space $\mathfrak{M}_{1,1} = \mathbb{H}/PSL_2(\mathbb{Z})$ whose points are in one-to-one correspondence with isomorphism classes of elliptic curves. This already qualifies $\mathfrak{M}_{1,1}$ as the *coarse moduli space* for elliptic curves, which means it is as good a moduli space as we can hope for. However, $\mathfrak{M}_{1,1}$ fails an important test for a moduli space that $\mathcal{T}_{1,1}$ passed (as we saw in section 2.2): it is *not* true, even for the circle $S = S^1$, that every continuous family of elliptic curves over S corresponds to a map from S to $\mathfrak{M}_{1,1}$.

The reason for this failure is the problem of automorphisms. These are equivalences from E to itself: that is, complex-analytic maps from E to E that preserve the basepoint e. Equivalently, they are given by complex-analytic self-maps of $\mathbb{C}$ that preserve 0 and the lattice

L. Such a map must be a rotation: that is, multiplication by some complex number λ of modulus 1. It is easy to check that for most lattices L in the plane, the only rotation that sends L to itself is multiplication by $\lambda = -1$. Note that this is the same -1 that we quotiented out by to pass from $SL_2(\mathbb{Z})$ to $PSL_2(\mathbb{Z})$. However, there are two special lattices that have greater symmetry. These are the *square lattice* $L = \mathbb{Z} \cdot 1 \oplus \mathbb{Z} \cdot i$, corresponding to the fourth root of unity i, and the *hexagonal lattice* $L = \mathbb{Z} \cdot 1 \oplus \mathbb{Z} \cdot e^{2\pi i/6}$, corresponding to a sixth root of unity. (Note that the hexagonal lattice is also represented by the point $\omega = e^{2\pi i/3}$.) The square lattice, which corresponds to the elliptic curve formed by gluing the opposite sides of a square, has as its symmetries the group $\mathbb{Z}/4\mathbb{Z}$ of rotational symmetries of the square. The hexagonal lattice, which corresponds to the elliptic curve formed by gluing the opposite sides of a regular hexagon, has as its symmetries the group $\mathbb{Z}/6\mathbb{Z}$ of rotational symmetries of a hexagon.

We see that the number of automorphisms of an elliptic curve jumps discontinuously at the special points $\omega = i$ and $\omega = e^{2\pi i/6}$. This already suggests that something might be wrong with $\mathfrak{M}_{1,1}$ as a moduli space. Note that we avoided this problem with the moduli $\mathcal{T}_{1,1}$ of *marked* elliptic curves, since there are no automorphisms of an elliptic curve that also preserve the marking. Another place we might have observed this problem with $\mathfrak{M}_{1,1}$ is when we passed to the quotient $\mathbb{H}/PSL_2(\mathbb{Z})$. We avoided the automorphism $\lambda = -1$ by quotienting by $PSL_2(\mathbb{Z})$ rather than $SL_2(\mathbb{Z})$. However, the two special points i and $e^{2\pi i/6}$ are preserved by integer Möbius transformations of $\mathbb{H}$ other than the identity, and they are the only points with that property. This means that the quotient $\mathbb{H}/PSL_2(\mathbb{Z})$ naturally comes with conical singularities at the points corresponding to these two orbits: one looks like a cone with angle π, and the other like a cone with angle $\frac{2}{3}\pi$. (To see why this is plausible, imagine the following simpler instance of the same phenomenon. If for every complex number z you identify z with $-z$, then the result is to wrap the complex plane around into a cone with a singularity at 0. The reason 0 is singled out is that it is preserved by the transformation $z \mapsto -z$. Here the angle would be π because the identification of points is two-to-one away from the singularity and π is half of 2π.) It is possible to massage these singularities away using the j-function, but they are indicating a basic difficulty.

So why do automorphisms form an obstacle to the existence of "good" moduli spaces? We can demonstrate the difficulty by considering an interesting continuous family of marked elliptic curves parametrized by the circle $S = S^1$. Let $E(\mathrm{i})$ be the "square" elliptic curve that we considered earlier, based on the lattice of integer combinations of 1 and i. Next, for every t between 0 and 1, let E_t be a copy of $E(\mathrm{i})$. Thus, we have taken the constant, or "trivial," family of elliptic curves over the closed unit interval $[0, 1]$, where every curve in the family is $E(\mathrm{i})$. Now we identify the elliptic curves at the two ends of this family, not in the obvious way, but by using the automorphism given by a 90° rotation, or multiplication by i. This means that we are looking at the family of elliptic curves over the circle where each member of the family is a copy of the elliptic curve $E(\mathrm{i})$, but these copies twist by 90° as we go around the circle.

It is easy to see that there is no way to capture this family of elliptic curves by means of a map from S^1 to the space $\mathfrak{M}_{1,1}$. Since all of the members of the family are isomorphic, each point of the circle should map to the same point in $\mathfrak{M}_{1,1}$ (the equivalence class of i in $\mathbb{H}$). But the constant map $S^1 \to \{\mathrm{i}\} \in \mathfrak{M}_{1,1}$ classifies the *trivial* family $S^1 \times E_i$ of elliptic curves over S^1, that is, the family where every curve is equal to $E(\mathrm{i})$ but the curves *do not* twist as we go around! Thus, there are more families of elliptic curves than there are maps to $\mathfrak{M}_{1,1}$; the quotient space $\mathbb{H}/\mathrm{PSL}_2(\mathbb{Z})$ cannot handle the complications caused by automorphisms. A variant of this construction applies to complex-analytic families with S^1 replaced by $\mathbb{C}^\times$. This is a very general phenomenon in moduli problems: whenever objects have nontrivial automorphisms, we can imitate the construction above to get nontrivial families over an interesting parameter set, all of whose members are the same. As a result, they cannot be classified by a map to the set of all isomorphism classes.

What do we do about this problem? One approach is to resign ourselves to having coarse moduli spaces, which have the right points and right geometry but do not quite classify arbitrary families. Another approach is the one that leads to $\mathcal{T}_{1,1}$: we can fix markings of one kind or another, which "kill" all automorphisms. In other words, we choose enough extra structure on our objects so that there do not remain any (nontrivial) automorphisms that preserve all this decoration. In fact, one can be far more economical than picking a basis of the lattice L and obtaining the infinite covering $\mathcal{T}_{1,1}$ of $\mathfrak{M}_{1,1}$: one can fix a basis of L only up to some congruence (for example, of $L/2L$). Finally, we can simply learn to come to terms with the automorphisms, keeping them as part of the data, resulting in "spaces" where points have internal symmetries. This is

the notion of an ORBIFOLD [IV.4 §7], or STACK [IV.4 §7], which is flexible enough to deal with essentially all moduli problems.

3   Higher-Genus Moduli Spaces and Teichmüller Spaces

We would now like to generalize as much as possible of the picture of elliptic curves and their moduli to higher-genus Riemann surfaces. For each g we would like to define a space $\mathfrak{M}_g$, called the *moduli space of curves of genus g*, that classifies compact Riemann surfaces of genus g and tells us how they modulate. Thus, the points of $\mathfrak{M}_g$ should correspond to our objects, compact Riemann surfaces of genus g, or, to be more accurate, equivalence classes of such surfaces, where two surfaces are considered to be equivalent if there is a complex-analytic isomorphism between them. In addition, we would like $\mathfrak{M}_g$ to do the best it can to embody the structure of continuous families of genus-g surfaces. Likewise, there are spaces $\mathfrak{M}_{g,n}$ parametrizing "n-punctured" Riemann surfaces of genus g. This means we consider not "bare" Riemann surfaces, but Riemann surfaces together with a "decoration" or "marking" by n distinct labeled points (punctures). Two of these are considered to be equivalent if there is a complex-analytic isomorphism between them that takes punctures to punctures and preserves labels. Since there are Riemann surfaces with automorphisms, we do not expect $\mathfrak{M}_g$ to be able to classify all families of Riemann surfaces: that is, we will expect examples similar to the twisted square-lattice construction discussed earlier. However, if we consider Riemann surfaces with enough extra markings, then we will be able to obtain a moduli space in the strongest sense. One way to choose such markings is to consider $\mathfrak{M}_{g,n}$ with n large enough (for fixed g). Another approach will be to mark generators of the fundamental group, leading to the Teichmüller spaces $\mathcal{T}_g$ and $\mathcal{T}_{g,n}$. We now outline this process.

To construct the space $\mathfrak{M}_g$, we return to the uniformization theorem. Any compact surface X of genus $g > 1$ has as its universal cover the upper half-plane $\mathbb{H}$, so it is represented as a quotient $X = \mathbb{H}/\Gamma$, where Γ is a representation of the fundamental group of X as a subgroup of conformal self-maps of $\mathbb{H}$. The group of all conformal automorphisms of $\mathbb{H}$ is $\mathrm{PSL}_2(\mathbb{R})$, the group of linear fractional transformations with real coefficients. The fundamental groups of all compact genus-g Riemann surfaces are isomorphic to a fixed abstract group

Γ_g, with $2g$ generators A_i, B_i ($i = 1, \ldots, g$) and one relation: that the product of all commutators $A_i B_i A_i^{-1} B_i^{-1}$ is the identity. A subgroup $\Gamma \subset \mathrm{PSL}_2(\mathbb{R})$ that acts on $\mathbb{H}$ in such a way that the quotient $\mathbb{H}/\Gamma$ is a Riemann surface (technically, the action should have no fixed points and should be properly discontinuous) is known as a FUCHSIAN GROUP [III.28]. Thus, the analogue of the representation of elliptic curves by lattices $L \simeq \mathbb{Z} \oplus \mathbb{Z}$ in the plane is the representation of higher-genus Riemann surfaces as $\mathbb{H}/\Gamma$, where Γ is a Fuchsian group.

The Teichmüller space $\mathcal{T}_g$ of genus-g Riemann surfaces is the space that solves the moduli problem for genus-g surfaces when they come with a marking of their fundamental group. This means that our objects are genus-g surfaces X plus a set of generators A_i, B_i of $\pi_1(X)$, which give an isomorphism between $\pi_1(X)$ and Γ_g, up to conjugation.[1] Our equivalences are complex-analytic maps that preserve the markings. Finally, our continuous (respectively, complex-analytic) families are continuous (complex-analytic) families of Riemann surfaces with continuously varying markings of the fundamental group. In other words, we are asserting the existence of a topological space/complex manifold $\mathcal{T}_g$, with a complex-analytic family of marked Riemann surfaces over it, and the following strong property.

The characteristic property of $\mathcal{T}_g$. *For any topological space (respectively, complex manifold) S, there is a bijection between continuous maps (respectively, holomorphic maps) $S \to \mathcal{T}_g$ and isomorphism classes of continuous (respectively, complex-analytic) families of marked genus-g surfaces parametrized by S.*

3.1   Digression: "Abstract Nonsense"

It is interesting to note that, while we have yet to see why such a space exists, it follows from general, nongeometric principles—CATEGORY THEORY [III.8] or "abstract nonsense"—that it is completely and uniquely determined, both as a topological space and as a complex manifold, by this characteristic property. In a very abstract way, every topological space M can be uniquely reconstructed from its set of points, the set of paths between these points, the set of surfaces spanning these paths, and so on. To put it differently, we can

think of M as a "machine" that assigns to any topological space S the set of continuous maps from S to M. This machine is known as the "functor of points of M." Similarly, a complex manifold M provides a machine that assigns to any other complex manifold S the set of complex-analytic maps from S to M. A curious discovery of category theory (the *Yoneda lemma*) is that for very general reasons (having nothing to do with geometry), these machines (or functors) uniquely determine M as a space, or a complex manifold.

Any moduli problem in the sense we have described (giving objects, equivalences, and families) also gives such a machine, where to S we assign the set of all families over S, up to isomorphism. So *just by setting up the moduli problem* we have already uniquely determined the topology and complex structure on Teichmüller space. The interesting part then is to know whether or not there *actually exists* a space giving rise to the same machine we have constructed, whether we can construct it explicitly, and whether we can use its geometry to learn interesting facts about Riemann surfaces.

3.2   Moduli Spaces and Representations

Coming back to earth, we discover that we have a fairly concrete model of Teichmüller space at our disposal. Once we have fixed the marking $\pi_1(X) \simeq \Gamma_g$, we are simply looking at all ways to represent Γ_g as a Fuchsian subgroup of $\mathrm{PSL}_2(\mathbb{R})$. Ignoring the Fuchsian condition for a moment, this means finding $2g$ real matrices (up to $\pm \mathrm{Id}$) $A_i, B_i \in \mathrm{PSL}_2(\mathbb{R})$ satisfying the commutator relation of Γ_g. This gives an explicit set of (algebraic!) equations for the entries of the $2g$ matrices, which determine the space of all representations $\Gamma_g \to \mathrm{PSL}_2(\mathbb{R})$. We must now quotient out by the action of $\mathrm{PSL}_2(\mathbb{R})$ that simultaneously conjugates all $2g$ matrices to obtain the *representation variety* $\mathrm{Rep}(\Gamma_g, \mathrm{PSL}_2(\mathbb{R}))$. This is analogous to considering lattices in $\mathbb{C}$ up to rotation, and is motivated by the fact that the quotients of $\mathbb{H}$ by two conjugate subgroups of $\mathrm{PSL}_2(\mathbb{R})$ will be isomorphic.

Once we have described the space of all representations of Γ_g into $\mathrm{PSL}_2(\mathbb{R})$, we can then single out Teichmüller space as the subset of the representation variety that consists of Fuchsian representations of Γ_g into $\mathrm{PSL}_2(\mathbb{R})$. Luckily this subset is *open* in the representation variety, which gives a nice realization of $\mathcal{T}_g$ as a topological space—in fact, $\mathcal{T}_g$ is homeomorphic to $\mathbb{R}^{6g-6}$. (This can be seen very explicitly in terms of the *Fenchel–Nielsen* coordinates, which parametrize a

---

1. Note that while the fundamental group of X depends on the choice of a basepoint, $\pi_1(X, x)$ and $\pi_1(X, y)$ may be identified by choosing a path from x to y, and the different choices are related by conjugation by a loop. Thus, if we are willing to identify sets of generators A_i, B_i when they differ only by a conjugation, then we can ignore the choice of a basepoint.

surface in $\mathcal{T}_g$ via a cut-and-paste procedure involving $3g - 3$ lengths and $3g - 3$ angles.) We may now try to "forget" the marking $\pi_1(X) \cong \Gamma_g$, to obtain the moduli space $\mathfrak{M}_g$ of unmarked Riemann surfaces. In other words, we would like to take $\mathcal{T}_g$ and identify any two points that represent the same underlying Riemann surface with different markings. This identification is achieved by the action of a group, the genus-g *mapping class group* MCG_g or *Teichmüller modular group*, on $\mathcal{T}_g$, which generalizes the modular group $\mathrm{PSL}_2(\mathbb{Z})$ that acts on $\mathbb{H} = \mathcal{T}_{1,1}$. (The mapping class group is defined as the group of all self-diffeomorphisms of a genus-g surface—remember that all such surfaces are topologically the same—modulo those diffeomorphisms that act trivially on the fundamental group.) As in the case of elliptic curves, Riemann surfaces with automorphisms correspond to points in $\mathcal{T}_g$ fixed by some subgroup of MCG_g, and give rise to singular points in the quotient $\mathfrak{M}_g = \mathcal{T}_g / \mathrm{MCG}_g$.

Representation varieties, or moduli spaces of representations, are an important and concrete class of moduli spaces that arise throughout geometry, topology, and number theory. Given any (discrete) group Γ, we ask (for example) for a space that parametrizes homomorphisms of Γ into the group of $n \times n$ matrices. The notion of equivalence is given by conjugation by GL_n, and that of families by continuous (or analytic, or algebraic, etc.) families of matrices. This problem is interesting even when the group Γ is $\mathbb{Z}$. Then we are simply considering invertible $n \times n$ matrices (the image of $1 \in \mathbb{Z}$) up to conjugacy. It turns out that there is no moduli space for this problem, even in the coarse sense, unless we consider only "nice enough" matrices: for example, matrices that consist of only a single Jordan block. This is a good example of a ubiquitous phenomenon in moduli problems: one is often forced to throw out some "bad" (unstable) objects in order to have any chance of obtaining a moduli space. (See the paper by Mumford and Suominen (1972) for a detailed discussion.)

3.3 Moduli Spaces and Jacobians

The upper half-plane $\mathbb{H} = \mathcal{T}_{1,1}$, together with the action of $\mathrm{PSL}_2(\mathbb{Z})$, gives an appealingly complete picture of the moduli problem for elliptic curves and its geometry. The same cannot be said, unfortunately, for the picture of $\mathcal{T}_g$ as an open subset of the representation variety. In particular, the representation variety does not even carry a natural complex structure, so we cannot

see from this description the geometry of $\mathcal{T}_g$ as a complex manifold. This failure reflects some of the ways in which the study of moduli spaces is more complicated for genus greater than 1. In particular, the moduli spaces of higher-genus surfaces are not described purely by linear algebra plus data about orientation, as is the case in genus 1.

Part of the blame for this complexity lies with the fact that the fundamental group $\Gamma_g \simeq \pi_1(X)$ ($g > 1$) is no longer Abelian, and in particular it is no longer equal to the first homology group $\mathrm{H}_1(X, \mathbb{Z})$. A related problem is that X is no longer a group. A beautiful solution to this problem is given by the construction of the Jacobian $\mathrm{Jac}(X)$, which shares with elliptic curves the properties of being a torus (homeomorphic to $(S^1)^{2g}$), an Abelian group, and a complex (in fact complex-algebraic) manifold. (The Jacobian of an elliptic curve is the elliptic curve itself.) The Jacobian captures the "Abelian" or "linear" aspects of the geometry of X. There is a moduli space $\mathcal{A}_g$ for such complex-algebraic tori (known as *Abelian varieties*), which does share all of the nice properties and linear algebraic description of the moduli of elliptic curves $\mathfrak{M}_{1,1} = \mathcal{A}_1$. The good news—the Torelli theorem—is that by assigning to each Riemann surface X its Jacobian we embed $\mathfrak{M}_g$ as a closed, complex-analytic subset of $\mathcal{A}_g$. The *interesting* news—the Schottky problem—is that the image is quite complicated to characterize intrinsically. In fact, solutions to this problem have come from as far afield as the study of nonlinear partial differential equations!

3.4 Further Directions

In this section we give hints at some interesting questions about, and applications of, moduli spaces.

Deformations and degenerations. Two of the main topics in moduli spaces ask which objects are very near to a given one, and what lies very far away. Deformation theory is the calculus of moduli spaces: it describes their infinitesimal structure. In other words, given an object, deformation theory is concerned with describing all its small perturbations (see Mazur (2004) for a beautiful discussion of this). At the other extreme, we can ask what happens when our objects degenerate? Most moduli spaces, for example the moduli of curves, are not compact, so there are families "going off to infinity." It is important to find "meaningful" compactifications of moduli spaces, which classify the possible degenerations of our objects. Another advantage of compactifying moduli spaces is that we can then calculate integrals over the completed space. This is crucial for the next item.

Invariants from moduli spaces. An important application of moduli spaces in geometry and topology is inspired by quantum field theory, where a particle, rather than following the "best" classical path between two points, follows all paths with varying probabilities (see MIRROR SYMMETRY [IV.16 §2.2.4]). Classically, one calculates many topological invariants by picking a geometric structure (such as a metric) on a space, calculating some quantity using this structure, and finally proving that the result of the calculation did not depend on the structure we chose. The new alternative is to look at *all* such geometric structures, and integrate some quantity over the space of all choices. The result, if we can show convergence, will manifestly not depend on any choices. String theory has given rise to many important applications of this idea, in particular by giving a rich structure to the collection of integrals obtained in this way. Donaldson and Seiberg-Witten theories use this philosophy to give topological invariants of four-manifolds. Gromov-Witten theory applies it to the topology of SYMPLECTIC MANIFOLDS [III.88], and to counting problems in algebraic geometry, such as, How many rational plane curves of degree 5 pass through fourteen points in general position? (Answer: 87 304.)

Modular forms. One of the most profound ideas in mathematics, the Langlands program, relates number theory to function theory (harmonic analysis) on very special moduli spaces, generalizing the moduli space of elliptic curves. These moduli spaces (Shimura varieties) are expressible as quotients of symmetric spaces (such as $\mathbb{H}$) by arithmetic groups (such as $\mathrm{PSL}_2(\mathbb{Z})$). MODULAR FORMS [III.59] and automorphic forms are special functions on these moduli spaces, described by their interaction with the large symmetry groups of the spaces. This is an extremely exciting and active area of mathematics, which counts among its recent triumphs the proof of FERMAT'S LAST THEOREM [V.10] and the Shimura-Taniyama-Weil conjecture (Wiles, Taylor-Wiles, Breuil-Conrad-Diamond-Taylor).

Further Reading

For historical accounts and bibliographies on moduli spaces, the following articles are highly recommended.

A beautiful and accessible overview of moduli spaces, with an emphasis on the notion of deformations, is given by Mazur (2004). The articles by Hain (2000) and Looijenga (2000) give excellent introductions to the study of the moduli spaces of curves, perhaps the oldest and most important of all moduli problems. The article by Mumford and Suominen (1972) introduces the key ideas underlying the study of moduli spaces in algebraic geometry.

Hain, R. 2000. Moduli of Riemann surfaces, transcendental aspects. In *School on Algebraic Geometry, Trieste, 1999*, pp. 293-353. ICTP Lecture Notes Series, no. 1. Trieste: The Abdus Salam International Centre for Theoretical Physics.

Looijenga, E. 2000. A minicourse on moduli of curves. In *School on Algebraic Geometry, Trieste, 1999*, pp. 267-91. ICTP Lecture Notes Series, no. 1. Trieste: The Abdus Salam International Centre for Theoretical Physics.

Mazur, B. 2004. Perturbations, deformations and variations (and "near-misses") in geometry. Physics and number theory. *Bulletin of the American Mathematical Society* 41(3):307-36.

Mumford, D., and K. Suominen. 1972. Introduction to the theory of moduli. In *Algebraic Geometry, Oslo, 1970: Proceedings of the Fifth Nordic Summer School in Mathematics*, edited by F. Oort, pp. 171-222. Groningen: Wolters-Noordhoff.

IV.9 Representation Theory
Ian Grojnowski

1 Introduction

It is a fundamental theme in mathematics that many objects, both mathematical and physical, have symmetries. The goal of GROUP [I.3 §2.1] theory in general, and representation theory in particular, is to study these symmetries. The difference between representation theory and general group theory is that in representation theory one restricts one's attention to symmetries of VECTOR SPACES [I.3 §2.3]. I will attempt here to explain why this is sensible and how it influences our study of groups, causing us to focus on groups with certain nice structures involving *conjugacy classes*.

2 Why Vector Spaces?

The aim of representation theory is to understand how the *internal* structure of a group controls the way it acts *externally* as a collection of symmetries. In the other direction, it also studies what one can learn about a group's internal structure by regarding it as a group of symmetries.

We begin our discussion by making more precise what we mean by "acts as a collection of symmetries." The idea we are trying to capture is that if we are given a group G and an object X, then we can associate with each element g of G some symmetry of X, which we call $\phi(g)$. For this to be sensible, we need the composition of symmetries to work properly: that is, $\phi(g)\phi(h)$ (the result of applying $\phi(h)$ and then $\phi(g)$) should be the same symmetry as $\phi(gh)$. If X is a set, then a symmetry of X is a particular kind of PERMUTATION [III.68] of its elements. Let us denote by $\mathrm{Aut}(X)$ the group of *all* permutations of X. Then an *action* of G on X is defined to be a homomorphism from G to $\mathrm{Aut}(X)$. If we are given such a homomorphism, then we say that G *acts* on X.

The image to have in mind is that G "does things" to X. This idea can often be expressed more conveniently and vividly by forgetting about ϕ in the notation: thus, instead of writing $\phi(g)(x)$ for the effect on x of the symmetry associated with g, we simply think of g itself as a permutation and write gx. However, sometimes we do need to talk about ϕ as well: for instance, we might wish to compare two different actions of G on X.

Here is an example. Take as our object X a square in the plane, centered at the origin, and let its vertices be A, B, C, and D (see figure 1). A square has eight symmetries: four rotations by multiples of $90°$ and four reflections. Let G be the group consisting of these eight symmetries; this group is often called D_8, or the *dihedral group* of order 8. By definition, G acts on the square. But it also acts on the set of *vertices* of the square: for instance, the action of the reflection through the y-axis is to switch A with B and C with D. It might seem as though we have done very little here. After all, we defined G as a group of symmetries so it does not take much effort to associate a symmetry with each element of G. However, we did not define G as a group of permutations of the set $\{A, B, C, D\}$, so we have at least done something.

To make this point clearer, let us look at some other sets on which G acts, which will include any set that we can build sufficiently naturally from the square. For instance, G acts not only on the set of vertices $\{A, B, C, D\}$, but on the set of edges $\{AB, BC, CD, DA\}$ and on the set of cross-diagonals $\{AC, BD\}$ as well. Notice in the latter case that some of the elements of G act in the same way: for example, a clockwise rotation through $90°$ interchanges the two diagonals, as does a counterclockwise rotation through $90°$. If all the elements of G act differently, then the action is called *faithful*.

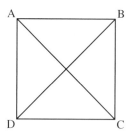

Figure 1 A square and its diagonals.

Notice that the operations on the square ("reflect through the y-axis," "rotate through $90°$," and so on) can be applied to the whole Cartesian plane $\mathbb{R}^2$. Therefore, $\mathbb{R}^2$ is another (and much larger) set on which G acts. To call $\mathbb{R}^2$ a set, though, is to forget the very interesting fact that the elements in $\mathbb{R}^2$ can be added together and multiplied by real numbers: in other words, $\mathbb{R}^2$ is a *vector space*. Furthermore, the action of G is well-behaved with respect to this extra structure. For instance, if g is one of our symmetries and v_1 and v_2 are two elements of $\mathbb{R}^2$, then g applied to the sum $v_1 + v_2$ yields the sum $g(v_1) + g(v_2)$. Because of this, we say that G acts *linearly* on the vector space $\mathbb{R}^2$. When V is a vector space, we denote by $\mathrm{GL}(V)$ the set of invertible linear maps from V to V. If V is the vector space $\mathbb{R}^n$, this group is the familiar group $\mathrm{GL}_n(\mathbb{R})$ of invertible $n \times n$ matrices with real entries; similarly, when $V = \mathbb{C}^n$ it is the group of invertible matrices with complex entries.

Definition. A *representation* of a group G on a vector space V is a homomorphism from G to $\mathrm{GL}(V)$.

In other words, a group action is a way of regarding a group as a collection of permutations, while a representation is the special case where these permutations are invertible linear maps. One sometimes sees representations referred to, for emphasis, as *linear* representations. In the representation of D_8 on $\mathbb{R}^2$ that we described above, the homomorphism from G to $\mathrm{GL}_2(\mathbb{R})$ took the symmetry "clockwise rotation through $90°$" to the matrix $\left(\begin{smallmatrix} 0 & 1 \\ -1 & 0 \end{smallmatrix}\right)$ and the symmetry "reflection through the y-axis" to the matrix $\left(\begin{smallmatrix} -1 & 0 \\ 0 & 1 \end{smallmatrix}\right)$.

Given one representation of G, we can produce others using natural constructions from linear algebra. For example, if ρ is the representation of G on $\mathbb{R}^2$ described above, then its DETERMINANT [III.15] $\det \rho$ is a homomorphism from G to $\mathbb{R}^*$ (the group of nonzero real numbers under multiplication), since

$$\det(\rho(gh)) = \det(\rho(g)\rho(h)) = \det(\rho(g))\det(\rho(h)),$$

by the multiplicative property of determinants. This makes det ρ a one-dimensional representation, since each nonzero real number t can be thought of as the element "multiply by t" of $GL_1(\mathbb{R})$. If ρ is the representation of D_8 just discussed, then under det ρ we find that rotations act as the identity and reflections act as multiplication by -1.

The definition of "representation" is formally very similar to the definition of "action," and indeed, since every linear automorphism of V is a permutation on the set of vectors in V, the representations of G on V form a subset of the actions of G on V. But the set of representations is in general a much more interesting object. We see here an instance of a general principle: if a set comes equipped with some extra structure (as a vector space comes with the ability to add elements together), then it is a mistake not to make use of that structure; and the more structure the better.

In order to emphasize this point, and to place representations in a very favorable light, let us start by considering the general story of actions of groups on sets. Suppose, then, that G is a group that acts on a set X. For each x, the set of all elements of the form gx, as g ranges over G, is called the *orbit* of x. It is not hard to show that the orbits form a partition of X.

Example. Let G be the dihedral group D_8 acting on the set X of *ordered pairs* of vertices of the square, of which there are sixteen. Then there are three orbits of G on X, namely $\{AA, BB, CC, DD\}$, $\{AB, BA, BC, CB, CD, DC, DA, AD\}$, and $\{AC, CA, BD, DB\}$.

An action of G on X is called *transitive* if there is just one orbit. In other words, it is transitive if for every x and y in X you can find an element g such that $gx = y$. When an action is *not* transitive, we can consider the action of G on each orbit separately, which effectively breaks up the action into a collection of transitive actions on disjoint sets. So in order to study *all* actions of G on sets it suffices to study *transitive* actions; you can think of actions as "molecules" and transitive actions as the "atoms" into which they can be decomposed. We shall see that this idea of *decomposing into objects that cannot be further decomposed* is fundamental to representation theory.

What are the possible transitive actions? A rich source of such actions comes from subgroups H of G. Given a subgroup H of G, a *left coset* of H is a set of the form $\{gh : h \in H\}$, which is commonly denoted by gH. An elementary result in group theory is that the left cosets form a partition of G (as do the right cosets,

if you prefer them). There is an obvious action of G on the set of left cosets of H, which we denote by G/H: if g' is an element of G, then it sends the coset gH to the coset $(g'g)H$.

It turns out that every transitive action is of this form! Given a transitive action of G on a set X, choose some $x \in X$ and let H_x be the subgroup of G consisting of all elements h such that $hx = x$. (This set is called the *stabilizer* of x.) Then one can check that the action of G on X is the same[1] as that of G on the left cosets of H_x. For example, the action of D_8 on the first orbit above is isomorphic to the action on the left cosets of the two-element subgroup H generated by a reflection of the square through its diagonal. If we had made a different choice of x, for example the point $x' = gx$, then the subgroup of G fixing x' would just be gH_xg^{-1}. This is a so-called *conjugate subgroup*, and it gives a different description of the same orbit, this time as left cosets of gH_xg^{-1}.

It follows that there is a one-to-one correspondence between transitive actions of G and conjugacy classes of subgroups (that is, collections of subgroups conjugate to some given subgroup). If G acts on our original set X in a nontransitive way, then we can break X up into a union of orbits, each of which, as a result of this correspondence, is associated with a conjugacy class of subgroups. This gives us a convenient "bookkeeping" mechanism for describing the action of G on X: just keep track of how many times each conjugacy class of subgroups arises.

Exercise. Check that in the example earlier the three orbits correspond (respectively) to a two-element subgroup R generated by reflection through a diagonal, the trivial subgroup, and another copy of the group R.

This completely solves the problem of how groups act on sets. The internal structure that controls the action is the *subgroup* structure of G.

In a moment we will see the corresponding solution to the problem of how groups act on vector spaces. First, let us just stare at sets for a while and see why, though we have answered our question, we should not feel too happy about it.[2]

The problem is that the subgroup structure of a group is *just horrible*.

---

1. By "the same" we mean "isomorphic as sets with *G*-action." The casual reader may read this as "the same," while the more careful reader should stop here and work out, or look up, precisely what is meant.

2. Exercise: go back to the example of D_8 and list all the possible transitive actions.

For example, any finite group of order n is a subgroup of the SYMMETRIC GROUP [III.68] S_n (this is "Cayley's theorem," which follows by considering the action of G on itself), so in order to list the conjugacy classes of subgroups of the symmetric group S_n one must understand all finite groups of size less than n.[3] Or consider the cyclic group $\mathbb{Z}/n\mathbb{Z}$. The subgroups correspond to the divisors of n, a subtle property of n that makes the cyclic groups behave quite differently as n varies. If n is prime, then there are very few subgroups, while if n is a power of 2 there are quite a few. So number theory is involved even if all we want to do is understand the subgroup structure of a group as simple as a cyclic group.

With some relief we now turn our attention back to linear representations. We will see that, just as with actions on sets, one can decompose representations into "atomic" ones. But, by contrast with the case of sets, these atomic representations (called "irreducible" representations, or sometimes simply "irreducibles") turn out to exhibit quite beautiful regularities.

The nice properties of representation theory come largely from the following fact. While elements of the symmetric group S_n can be multiplied together, elements of $GL(V)$, being matrices, can be *added* as well as multiplied. (But beware: the sum of two elements of $GL(V)$ is not necessarily an element of $GL(V)$, because it may not be invertible. It is, however, an element of the endomorphism algebra $\mathrm{End}(V)$. When $V = \mathbb{C}^n$, $\mathrm{End}(V)$ is just the familiar algebra of all $n \times n$ matrices with complex entries, both invertible and not.)

To see the difference it makes to be able to add, consider the cyclic group $G = \mathbb{Z}/n\mathbb{Z}$. For each $\omega \in \mathbb{C}$ with $\omega^n = 1$, we get a representation χ_ω of G on $\mathbb{C}$ by associating the element $r \in \mathbb{Z}/n\mathbb{Z}$ with multiplication by ω^r, which we think of as a linear map from the one-dimensional space $\mathbb{C}$ to itself. This gives us n different one-dimensional representations, one for each nth root of unity, and it turns out that there are no others. Moreover, if $\rho : G \to GL(V)$ is any representation of $\mathbb{Z}/n\mathbb{Z}$, then we can write it as a direct sum of these representations by imitating the formula for finding the Fourier mode of a function. Using the representation ρ, we associate with each r in $\mathbb{Z}/n\mathbb{Z}$ a linear map $\rho(r)$. Now let us define a linear map $p_\omega : V \to V$ by the formula

$$p_\omega = \frac{1}{n} \sum_{0 \leqslant r < n} \omega^{-r} \rho(r).$$

Then p_ω is an element of $\mathrm{End}(V)$, and one can check that it is actually a PROJECTION [III.50 §3.5] onto a subspace V_ω of V. In fact, this subspace is an EIGENSPACE [I.3 §4.3]: it consists of all vectors v such that $\rho(1)v = \omega v$, which implies, since ρ is a representation, that $\rho(r)v = \omega^r v$. The projection p_ω should be thought of as the analogue of the nth FOURIER COEFFICIENT [III.27] $a_n(f)$ of a function $f(\theta)$ on the circle; note the formal similarity of the above formula to the Fourier expansion formula $a_n(f) = \int e^{-2\pi i n\theta} f(\theta)\, d\theta$.

Now the interesting thing about the Fourier series of f is that, under favorable circumstances, it adds up to f itself: that is, it decomposes f into TRIGONOMETRIC FUNCTIONS [III.92]. Similarly, what is interesting about the subspaces V_ω is that we can use them to decompose the representation ρ. The composition of any two distinct projections p_ω is 0, from which it can be shown that

$$V = \bigoplus_\omega V_\omega.$$

We can write each subspace V_ω as a sum of one-dimensional spaces, which are copies of $\mathbb{C}$, and the restriction of ρ to any one of these is just the simple representation χ_ω defined earlier. Thus, ρ has been decomposed as a combination of very simple "atoms" χ_ω.[4]

This ability to add matrices has a very useful consequence. Let a finite group G act on a complex vector space V. A subspace W of V is called G-*invariant* if $gW = W$ for every $g \in G$. Let W be a G-invariant subspace, and let U be a complementary subspace (that is, one such that every element v of V can be written in exactly one way as $w + u$ with $w \in W$ and $u \in U$). Let ϕ be an arbitrary projection onto U. Then it is a simple exercise to show that the linear map $1/|G| \sum_{g \in G} g\phi$ is also a projection onto a complementary subspace, but with the added advantage that it is G-invariant. This latter fact follows because applying an element g' to the sum just rearranges its terms.

The reason this is so useful is that it allows us to decompose an arbitrary representation into a direct sum of *irreducible representations*, which are representations without a G-invariant subspace. Indeed, if ρ is

---

3. THE CLASSIFICATION OF FINITE SIMPLE GROUPS [V.7] does at least allow us to estimate the *number* γ_n of subgroups of S_n up to conjugacy: it is a result of Pyber that $2^{((1/16)+o(1))n^2} \leqslant \gamma_n \leqslant 24^{((1/6)+o(1))n^2}$. Equality is expected for the lower bound.

4. To summarize the rest of this article: the similarity to the Fourier transform is not just analogy—decomposing a representation into its irreducible summands is a notion that includes both this example and the Fourier transform.

not irreducible, then there is a G-invariant subspace W. By the above remark, we can write $G = W \oplus W'$ with W' also G-invariant. If either W or W' has a further G-invariant subspace, then we can decompose it further, and so on. We have just seen this done for the cyclic group: in that case the irreducible representations were the one-dimensional representations χ_ω.

The irreducible representations are the basic building blocks of arbitrary complex representations, just as the basic building blocks for actions on sets are the transitive actions. It raises the question of what the irreducible representations are, a question that has been answered for many important examples, but which is not yet solvable by any general procedure.

To return to the difference between actions and representations, another important observation is that any action of a group G on a finite set X can be *linearized* in the following sense. If X has n elements, then we can look at the HILBERT SPACE [III.37] $L^2(X)$ of all complex-valued functions defined on X. This has a natural basis given by the "delta functions" δ_x, which send x to 1 and all other elements of X to 0. Now we can turn the action of G on X into an action of G on the basis in an obvious way: we just define $g\delta_x$ to be δ_{gx}. We can extend this definition by linearity, since an arbitrary function f is a linear combination of the basis functions δ_x. This gives us an action of G on $L^2(X)$, which can be defined by a simple formula: if f is a function in $L^2(X)$, then gf is the function defined by $(gf)(x) = f(g^{-1}x)$. Equivalently, gf does to gx what f does to x. Thus, an action on sets can be thought of as an assignment of a very special matrix to every group element, namely a matrix with only 0s and 1s and precisely one 1 in each row and each column. (Such matrices are called *permutation matrices*.) By contrast, a general representation assigns an *arbitrary* invertible matrix.

Now, even when X itself is a single orbit under the action of G, the above representation on $L^2(X)$ can break up into pieces. For an extreme example of this phenomenon, consider the action of $\mathbb{Z}/n\mathbb{Z}$ on itself by multiplication. We have just seen that, by means of the "Fourier expansion" above, this breaks up into a sum of n one-dimensional representations.

Let us now consider the action of an arbitrary group G on itself by multiplication, or, to be more precise, left multiplication. That is, we shall associate with each element g the permutation of G that takes each h in G to gh. This action is obviously transitive. As an action on a *set* it cannot be decomposed any further. But when we *linearize* this action to a representation of G on the

vector space $L^2(G)$, we have much greater flexibility to decompose the action. It turns out that, not only does it break up into a direct sum of many irreducible representations, but *every* irreducible representation ρ of G occurs as one of the summands in this direct sum, and the number of times that ρ appears is equal to the dimension of the subspace on which it acts.

The representation we have just discussed is called the *left regular representation* of G. The fact that every irreducible representation occurs in it so regularly makes it extremely useful. Notice that it is easier to decompose representations on complex vector spaces than on real vector spaces, since every automorphism of a complex vector space has an eigenvector. So it is simplest to begin by studying complex representations.

The time has now come to state the fundamental theorem about complex representations of finite groups. This theorem tells us how many irreducible representations there are for a finite group, and, more colorfully, that representation theory is a "non-Abelian analogue of Fourier decomposition."

Let $\rho : G \to \mathrm{End}(V)$ be a representation of G. The *character* χ_ρ of ρ is defined to be its trace: that is, χ_ρ is a function from G to $\mathbb{C}$ and $\chi_\rho(g) = \mathrm{tr}(\rho(g))$ for each g in G. Since $\mathrm{tr}(AB) = \mathrm{tr}(BA)$ for any two matrices A and B, we have $\chi_\rho(hgh^{-1}) = \chi_\rho(g)$. Therefore, χ_V is very far from an arbitrary function on G: it is a function that is constant on each *conjugacy class*. Let K_G denote the vector space of all complex-valued functions on G with this property; it is called the *representation ring* of G.

The characters of the irreducible representations of a group form a very important set of data about the group, which it is natural to organize into a matrix. The columns are indexed by the conjugacy classes, the rows by the irreducible representations, and each entry is the value of the character of the given representation at the given conjugacy class. This array is called the *character table* of the group, and it contains all the important information about representations of the group: it is our periodic table. The basic theorem of the subject is that this array is a *square*.

Theorem (the character table is square). *Let G be a finite group. Then the characters of the irreducible representations form an orthonormal basis of K_G.*

When we say that the basis of characters is *orthonormal* we mean that the Hermitian inner product defined by

$$\langle \chi, \psi \rangle = |G|^{-1} \sum_{g \in G} \chi(g)\overline{\psi(g)}$$

is 1 when $\chi = \psi$ and 0 otherwise. The fact that it is a basis implies in particular that there are exactly as many irreducible representations as there are conjugacy classes in G, and the map from isomorphism classes of representations to K_G that sends each ρ to its character is an injection. That is, an arbitrary representation is determined up to isomorphism by its character.

The internal structure of a group G that controls how it can act on vector spaces is the structure of conjugacy classes of elements of G. This is a much gentler structure than the set of all conjugacy classes of *subgroups* of G. For example, in the symmetric group S_n two permutations belong to the same conjugacy class if and only if they have the same cycle type. Therefore, in that group there is a bijection between conjugacy classes and partitions of n.[5]

Furthermore, whereas it is completely unclear how to count subgroups, conjugacy classes are much easier to handle. For instance, since they partition the group, we have the formula $|G| = \sum_{C \text{ a conjugacy class}} |C|$. On the representation side, there is a similar formula, which arises from the decomposition of the regular representation $L^2(G)$ into irreducibles: $|G| = \sum_{V \text{ irreducible}} (\dim V)^2$. It is inconceivable that there might be a similarly simple formula for sums over all subgroups of a group.

We have reduced the problem of understanding the general structure of the representations of a finite group G to the problem of determining the character table of G. When $G = \mathbb{Z}/n\mathbb{Z}$, our description of the n irreducible representations above implies that all the entries of this matrix are roots of unity. Here are the character tables for D_8 (on the left), the group of symmetries of the square, and, just for contrast, for the group $\mathbb{Z}/3\mathbb{Z}$ (on the right):

1	1	1	1	1		1	1	1
1	1	1	-1	-1		1	z	z^2
1	1	-1	1	-1		1	z^2	z
1	1	-1	-1	1				
2	-2	0	0	0				

where $z = \exp(2\pi i/3)$.

The obvious question—Where did the first table come from?—indicates the main problem with the theorem: though it tells us the shape of the character table, it leaves us no closer to understanding what the actual

character values are. We know *how many* representations there are, but not *what* they are, or even what their dimensions are. We do not have a general method for constructing them, a kind of "non-Abelian Fourier transform." This is the central problem of representation theory.

Let us see how this problem can be solved for the group D_8. Over the course of this article, we have already encountered three irreducible representations of this group. The first is the "trivial" one-dimensional representation: the homomorphism $\rho : D_8 \to \mathrm{GL}_1$ that takes every element of D_8 to the identity. The second is the two-dimensional representation we wrote down in the first section, where each element of D_8 acts on $\mathbb{R}^2$ in the obvious way. The determinant of this representation is a one-dimensional representation that is *not* trivial: it sends the rotations to 1 and the reflections to -1. So we have constructed three rows of the character table above. There are five conjugacy classes in D_8 (trivial, reflection through axis, reflection through diagonal, $90°$ rotation, $180°$ rotation), so we know that there are just two more rows.

The equality $|G| = 8 = 2^2 + 1 + 1 + (\dim V_4)^2 + (\dim V_5)^2$ implies that these missing representations are one dimensional. One way of getting the missing character values is to use orthogonality of characters.

A slightly (but only slightly) less ad hoc way is to decompose $L^2(X)$ for small X. For example when X is the pair of diagonals $\{AC, BD\}$, we have $L^2(X) = V_4 \oplus \mathbb{C}$, where $\mathbb{C}$ is the trivial representation.

We are now going to start pointing the way toward some more modern topics in representation theory. Of necessity, we will use language from fairly advanced mathematics: the reader who is familiar with only some of this language should consider browsing the remaining sections, since different discussions have different prerequisites.

In general, a good, but not systematic, way of finding representations is to find objects on which G acts, and "linearize" the action. We have seen one example of this: when G acts on a set X we can consider the linearized action on $L^2(X)$. Recall that the irreducible G-sets are all of the form G/H, for H some subgroup of G. As well as looking at $L^2(G/H)$, we can consider, for every representation W of H, the vector space $L^2(G/H, W) = \{f : G \to W \mid f(gh) = h^{-1}f(g), \, g \in G, \, h \in H\}$; in geometric language, for those who prefer it, this is the space of sections of the associated W-bundle on G/H. This representation of G is called the *induced representation* of W from H to G.

---

5. Not only is the set of all partitions a sensible combinatorial object, it is far smaller than the set of all subgroups of S_n: HARDY [VI.73] and RAMANUJAN [VI.82] showed that the number of partitions of n is about $(1/4n\sqrt{3})e^{\pi\sqrt{(2n/3)}}$.

Other linearizations are also important. For example, if G acts continuously on a topological space X, we can consider how it acts on homology classes and hence on the HOMOLOGY GROUPS [IV.6 §4] of X.[6] The simplest case of this is the map $z \to \bar{z}$ of the circle S^1. Since this map squares to the identity map, it gives us an action of $\mathbb{Z}/2\mathbb{Z}$ on S^1, which becomes a representation of $\mathbb{Z}/2\mathbb{Z}$ on $H_1(S^1) = \mathbb{R}$ (which represents the identity as multiplication by 1 and the other element of $\mathbb{Z}/2\mathbb{Z}$ as multiplication by -1).

Methods like these have been used to determine the character tables of all finite SIMPLE GROUPS [I.3 §3.3], but they still fall short of a uniform description valid for all groups.

There are many arithmetic properties of the character table that hint at properties of the desired non-Abelian Fourier transform. For example, the size of a conjugacy class divides the order of the group, and in fact the dimension of a representation also divides the order of the group. Pursuing this thought leads to an examination of the values of the characters mod p, relating them to the so-called *p-local subgroups*. These are groups of the form $N(Q)/Q$, where Q is a subgroup of G, the number of elements of Q is a power of p, and $N(Q)$ is the *normalizer* of Q (defined to be the largest subgroup of G that contains Q as a normal subgroup). When the so-called "p-Sylow subgroup" of G is Abelian, beautiful conjectures of Broué give us an essentially complete picture of the representations of G. But in general these questions are at the center of a great deal of contemporary research.

3 Fourier Analysis

We have justified the study of group actions on vector spaces by explaining that the theory of representations has a nice structure that is not present in the theory of group actions on sets. A more historically based account would start by saying that spaces of functions very often come with natural actions of some group G, and many problems of traditional interest can be related to the decomposition of these representations of G.

In this section we will concentrate on the case where G is a compact LIE GROUP [III.48 §1]. We will see that in this case many of the nice features of the representation theory of finite groups persist.

The prototypical example is the space $L^2(S^1)$ of square-integrable functions on the circle S^1. We can think of the circle as the unit circle in $\mathbb{C}$, and thereby identify it with the group of rotations of the circle (since multiplication by $e^{i\theta}$ rotates the circle by θ). This action linearizes to an action on $L^2(S^1)$: if f is a square-integrable function defined on S^1 and w belongs to the circle, then $(w \cdot f)(z)$ is defined to be $f(w^{-1}z)$. That is, $w \cdot f$ does to wz what f does to z.

Classical Fourier analysis expands functions in the space $L^2(S^1)$ in terms of a basis of trigonometric functions: the functions z^n for $n \in \mathbb{Z}$. (These look more "trigonometric" if one writes $e^{i\theta}$ for z and $e^{in\theta}$ for z^n.) If we fix w and write $\phi_n(z) = z^n$, then $(w \cdot \phi_n)(z) = \phi_n(w^{-1}z) = w^{-n}\phi_n(z)$. In particular, $w \cdot \phi_n$ is a multiple of ϕ_n for each w, so the one-dimensional subspace generated by ϕ_n is invariant under the action of S^1. In fact, *every* irreducible representation of S^1 is of this form, as long as we restrict attention to continuous representations.

Now let us consider an innocuous-looking generalization of the above situation: we shall replace 1 by n and try to understand $L^2(S^n)$, the space of complex-valued square-integrable functions on the n-sphere S^n. The n-sphere is acted on by the group of rotations $SO(n+1)$. As usual, this can be converted into a representation of $SO(n+1)$ on the space $L^2(S^n)$, which we would like to decompose into irreducible representations; equivalently, we would like to decompose $L^2(S^n)$ into a direct sum of minimal $SO(n+1)$-invariant subspaces.

This turns out to be possible, and the proof is very similar to the proof for finite groups. In particular, a compact group such as $SO(n+1)$ has a natural PROBABILITY MEASURE [III.71 §2] on it (called *Haar measure*) in terms of which we can define averages. Roughly speaking, the only difference between the proof for $SO(n+1)$ and the proof in the finite case is that we have to replace a few sums by integrals.

The general result that one can prove by this method is the following. If G is a compact group that acts continuously on a compact space X (in the sense that each permutation $\phi(g)$ of X is continuous, and also that $\phi(g)$ varies continuously with g), then $L^2(X)$ splits up into an orthogonal direct sum of finite-dimensional minimal G-invariant subspaces; equivalently, the linearized action of G on $L^2(X)$ splits up into an orthogonal direct sum of irreducible representations, all of which are finite dimensional. The problem of finding a

---

6. The homology groups discussed in the article just referred to consist of formal sums of homology classes with integer coefficients. Here, where a vector space is required, we are taking real coefficients.

Hilbert space basis of $L^2(X)$ then splits into two sub-problems: we must first determine the irreducible representations of G, a problem which is independent of X, and then determine how many times each of these irreducible representations occurs in $L^2(X)$.

When $G = S^1$ (which we identified with SO(2)) and $X = S^1$ as well, we saw that these irreducible representations were one dimensional. Now let us look at the action of the compact group SO(3) on S^2. It can be shown that the action of G on $L^2(S^2)$ commutes with the *Laplacian*, the differential operator Δ on $L^2(S^2)$ defined by

$$\Delta = \frac{\partial^2}{\partial x^2} + \frac{\partial^2}{\partial y^2} + \frac{\partial^2}{\partial z^2}.$$

That is, $g(\Delta f) = \Delta(gf)$ for any $g \in G$ and any (sufficiently smooth) function f. In particular, if f is an eigenfunction for the Laplacian (which means that $\Delta f = \lambda f$ for some $\lambda \in \mathbb{C}$), then for each $g \in \text{SO}(3)$ we have

$$\Delta gf = g\Delta f = g\lambda f = \lambda gf,$$

so gf is also an eigenfunction for Δ. Therefore, the space V_λ of all eigenvectors for the Laplacian with eigenvalue λ is G-invariant. In fact, it turns out that if V_λ is nonzero then the action of G on V_λ is an irreducible representation. Furthermore, each irreducible representation of SO(3) arises exactly once in this way. More precisely, we have a Hilbert space direct sum,

$$L^2(S^2) = \bigoplus_{n \geqslant 0} V_{2n(2n+2)},$$

and each eigenspace $V_{2n(2n+2)}$ has dimension $2n + 1$. Note that this is a case where the set of eigenvalues is *discrete*. (These eigenspaces are discussed further in SPHERICAL HARMONICS [III.87].)

The nice feature that each irreducible representation appears at most once is rather special to the example $L^2(S^n)$. (For an example where this does not happen, recall that with the regular representation $L^2(G)$ of a finite group G each irreducible representation ρ occurs $\dim \rho$ times in $L^2(G)$.) However, other features are more generic: for example, when a compact Lie group acts differentiably on a space X, then the sum of all the G-invariant subspaces of $L^2(X)$ corresponding to a particular representation is always equal to the set of common eigenvectors of some family of commuting differential operators. (In the example above, there was just one operator, the Laplacian.)

Interesting SPECIAL FUNCTIONS [III.85], such as solutions of certain differential equations, often admit representation-theoretic meaning, for example as matrix coefficients. Their properties can then easily be deduced from general results in functional analysis and representation theory rather than from any calculation. Hypergeometric equations, Bessel equations, and many integrable systems arise in this way.

There is more to say about the similarities between the representation theory of compact groups and that of finite groups. Given a compact group G and an irreducible representation ρ of G, we can again take its trace (since it is finite dimensional) and thereby define its character χ_ρ. Just as before, χ_ρ is constant on each conjugacy class. Finally, "the character table is square," in the sense that the characters of the irreducible representations form an orthonormal basis of the Hilbert space of all square-integrable functions that are conjugation invariant in this sense. (Now, though, the "square matrix" is infinite.) When $G = S^1$ this is the Fourier theorem; when G is finite this is the theorem of section 2.

4 Noncompact Groups, Groups in Characteristic p, and Lie Algebras

The "character table is square" theorem focuses our attention on groups with nice conjugacy-class structure. What happens when we take such a group but relax the requirement that it be compact?

A paradigmatic noncompact group is the real numbers $\mathbb{R}$. Like S^1, $\mathbb{R}$ acts on itself in an obvious way (the real number t is associated with the translation $s \mapsto s + t$), so let us linearize that action in the usual way and look for a decomposition of $L^2(\mathbb{R})$ into $\mathbb{R}$-invariant subspaces.

In this situation we have a *continuous family* of irreducible one-dimensional representations: for each real number λ we can define the function χ_λ by $\chi_\lambda(x) = e^{2\pi i \lambda x}$. These functions are not square integrable, but despite this difficulty classical Fourier analysis tells us that we can write an L^2-function in terms of them. However, since the Fourier modes now vary in a continuous family, we can no longer decompose a function as a sum: rather we must use an integral. First, we define the Fourier transform $\hat{f}$ of f by the formula $\hat{f}(\lambda) = \int f(x)e^{2\pi i \lambda x}\,dx$. The desired decomposition of f is then $f(x) = \int \hat{f}(\lambda)e^{-2\pi i \lambda x}\,d\lambda$. This, the *Fourier inversion formula*, tells us that f is a weighted integral of the functions χ_λ. We can also think of it as something like a decomposition of $L^2(\mathbb{R})$ as a "direct integral" (rather than direct sum) of the one-dimensional subspaces generated by the functions χ_λ. However, we must treat this picture with due caution since the functions χ_λ do not belong to $L^2(\mathbb{R})$.

This example indicates what we should expect in general. If X is a space with a measure and G acts continuously on it in a way that preserves the measures of subsets of X (as translations did with subsets of $\mathbb{R}$), then the action of G on X gives rise to a measure μ_X defined on the set of all irreducible representations, and $L^2(X)$ can be decomposed as the integral over all irreducible representations with respect to this measure. A theorem that explicitly describes such a decomposition is called a *Plancherel* theorem for X.

For a more complicated but more typical example, let us look at the action of $SL_2(\mathbb{R})$ (the group of real 2×2 matrices with determinant 1) on $\mathbb{R}^2$ and see how to decompose $L^2(\mathbb{R}^2)$. As we did when we looked at functions defined on S^2, we shall make use of a differential operator. This involves the small technicality that we should look at smooth functions, and we do not ask for them to be defined at the origin. The appropriate differential operator this time turns out to be the Euler vector field $x(\partial/\partial x) + y(\partial/\partial y)$. It is not hard to check that if f satisfies the condition $f(tx, ty) = t^s f(x, y)$ for every x, y, and $t > 0$, then f is an eigenfunction of this operator with eigenvalue s, and indeed all functions in the eigenspace with this eigenvalue, which we shall denote by W_s, are of this form. We can also split W_s up as $W_s^+ \oplus W_s^-$, where W_s^+ and W_s^- consist of the even and odd functions in W_s, respectively.

The easiest way of analyzing the structure of W_s is to compute the action of the LIE ALGEBRA [III.48 §2] $\mathfrak{sl}_2$. For those readers unfamiliar with Lie algebras, we will say only that the Lie algebra of a Lie group G keeps track of the action of elements of G that are "infinitesimally close to the identity," and that in this case the Lie algebra $\mathfrak{sl}_2$ can be identified with the space of 2×2 matrices of trace 0, with $\left(\begin{smallmatrix} a & b \\ c & -a \end{smallmatrix}\right)$ acting as the differential operator $(-ax - by)(\partial/\partial x) + (-cx + ay)(\partial/\partial y)$.

Every element of W_s is a function on $\mathbb{R}^2$. If we restrict these functions to the unit circle, then we obtain a map from W_s to the space of smooth functions defined on S^1, which turns out to be an isomorphism. We already know that this space has a basis of Fourier modes z^m, which we can now think of as $(x + iy)^m$, defined when $x^2 + y^2 = 1$. There is a unique extension of this from a function defined on S^1 to a function in W_s, namely the function $w_m(x, y) = (x + iy)^m (x^2 + y^2)^{(s-m)/2}$. One can then check the following actions of simple matrices on these functions (to do so, recall the association of the matrices with differential operators given in the

previous paragraph):

$$\begin{pmatrix} 0 & -i \\ i & 0 \end{pmatrix} \cdot w_m = m w_m,$$

$$\begin{pmatrix} 1 & i \\ i & -1 \end{pmatrix} \cdot w_m = (m - s) w_{m+2},$$

$$\begin{pmatrix} 1 & -i \\ -i & -1 \end{pmatrix} \cdot w_m = (-m - s) w_{m-2}.$$

It follows that if s is not an integer, then from any function w_m in W_s^+ we can produce all the others using the action of $SL_2(\mathbb{R})$. Therefore, $SL_2(\mathbb{R})$ acts irreducibly on W_s^+. Similarly, it acts irreducibly on W_s^-. We have therefore encountered a significant difference between this and the finite/compact case: when G is not compact, irreducible representations of G can be infinite dimensional.

Looking more closely at the formulas for W_s when $s \in \mathbb{Z}$, we see more disturbing differences. In order to understand these, let us distinguish carefully between representations that are *reducible* and representations that are *decomposable*. The former are representations that have nontrivial G-invariant subspaces, whereas the latter are representations where one can decompose the space on which G acts into a direct sum of G-invariant subspaces. Decomposable representations are obviously reducible. In the finite/compact case, we used an averaging process to show that reducible representations are decomposable. Now we do not have a natural probability measure to use for the averaging, and it turns out that there can be reducible representations that are not decomposable.

Indeed, if s is a nonnegative integer, then the subspaces W_s^+ and W_s^- give us an example of this phenomenon. They are indecomposable (in fact, this is true even when s is a negative integer not equal to -1) but they contain an invariant subspace of dimension $s + 1$. Thus, we cannot write the representation as a direct sum of irreducible representations. (One can do something a little bit weaker, however: if we quotient out by the $(s + 1)$-dimensional subspace, then the quotient representation can be decomposed.)

It is important to understand that in order to produce these indecomposable but reducible representations we worked not in the space $L^2(\mathbb{R}^2)$ but in the space of smooth functions on $\mathbb{R}^2$ with the origin removed. For instance, the functions w_m above are not square integrable. If we look just at representations of G that act on subspaces of $L^2(X)$, then we *can* split them up into a direct sum of irreducibles: given a G-invariant subspace, its orthogonal complement is also G-invariant.

It might therefore seem best to ignore the other, rather subtle representations and just look at these ones. But it turns out to be easier to study *all* representations and only later ask which ones occur inside $L^2(X)$. For $SL_2(\mathbb{R})$, the representations we have just constructed (which were subquotients of $W_s^{\pm}$) exhaust all the irreducible representations,[7] and there is a Plancherel formula for $L^2(\mathbb{R}^2)$ that tells us which ones appear in $L^2(\mathbb{R}^2)$ and with what multiplicity:

$$L^2(\mathbb{R}^2) = \int_{-\infty}^{\infty} W_{-1+it} e^{it} \, dt.$$

To summarize: if G is not compact, then we can no longer take averages over G. This has various consequences:

Representations occur in continuous families. The decomposition of $L^2(X)$ takes the form of a direct integral, not a direct sum.

Representations do not split up into a direct sum of irreducibles. Even when a representation admits a finite composition series, as with the action of $SL^2(\mathbb{R})$ on $W_s^{\pm}$, it need not split up into a direct sum. So to describe all representations we need to do more than just describe the irreducibles—we also need to describe the glue that holds them together.

So far, the theory of representations of a noncompact group G seems to have *none* of the pleasant features of the compact case. But one thing does survive: there is still an analogue of the theorem that the character table is square. Indeed, we can still define characters in terms of the traces of group elements. But now we must be careful, since the irreducible representation may be on an infinite-dimensional vector space, so that its trace cannot be defined so easily. In fact, characters are not functions on G, but only DISTRIBUTIONS [III.18]. The character of a representation determines the *semisimplification* of a representation ρ: that is, it tells us which irreducible representations are part of ρ, but not how they are glued together.[8]

These phenomena were discovered by Harish-Chandra in the 1950s in an extraordinary series of works that completely described the representation theory of Lie groups such as the ones we have discussed (the precise condition is that they should be real and reductive—a concept that will be explained later in this article) and the generalizations of classical theorems of Fourier analysis to this setting.[9]

Independently and slightly earlier, Brauer had investigated the representation theory of *finite* groups on finite-dimensional vector spaces over fields of characteristic p. Here, too, reducible representations need not decompose as direct sums, though in this case the problem is not lack of compactness (obviously, since everything is finite) but an inability to *average* over the group: we would like to divide by $|G|$, but often this is zero. A simple example that illustrates this is the action of $\mathbb{Z}/p\mathbb{Z}$ on the space $\mathbb{F}_p^2$ that takes x to the 2×2 matrix $\left(\begin{smallmatrix} 1 & x \\ 1 & 0 \end{smallmatrix}\right)$. This is reducible, since the column vector $\left(\begin{smallmatrix} 1 \\ 0 \end{smallmatrix}\right)$ is fixed by the action, and therefore generates an invariant subspace. However, if one could decompose the action, then the matrices $\left(\begin{smallmatrix} 1 & x \\ 1 & 0 \end{smallmatrix}\right)$ would all be diagonalizable, which they are not.

It is possible for there to be infinitely many indecomposable representations, which again may vary in families. However, as before, there are only finitely many *irreducible* representations, so there is some chance of a "character table is square" theorem in which the rows of the square are parametrized by characters of irreducible representations. Brauer proved just such a theorem, pairing the characters with p-*semisimple* conjugacy classes in G: that is, conjugacy classes of elements whose order is not divisible by p.

We will draw two crude morals from the work of Harish-Chandra and of Brauer. The first is that the category of representations of a group is always a reasonable object, but when the representations are infinite dimensional it requires serious technical work to set it up. Objects in this category do not necessarily decompose as a direct sum of irreducibles (one says that the category is not *semisimple*), and can occur in infinite families, but irreducible objects pair off in some precise way with certain "diagonalizable" conjugacy classes in the group—there is always some kind of analogue of "the character table is square" theorem.

It turns out that when we consider representations in more general contexts—Lie algebras acting on vector spaces, quantum groups, p-adic groups on infinite-dimensional complex or p-adic vector spaces, etc.—these qualitative features stay the same.

---

7. To make this precise requires some care about what we mean by "isomorphic." Because many different topological vector spaces can have the same underlying $\mathfrak{sl}_2$-module, the correct notion is of *infinitesimal* equivalence. Pursuing this notion leads to the category of *Harish-Chandra modules*, a category with good finiteness properties.

8. It is a major theorem of Harish-Chandra that the distribution that defines a character is given by *analytic* functions on a dense subset of the semisimple elements of the group.

9. The problem of determining the irreducible *unitary* representations for real reductive groups has still not been solved; the most complete results are due to Vogan.

The second moral is that we should always hope for some "non-Abelian Fourier transform": that is, a set that parametrizes irreducible representations and a description of the character values in terms of this set.

In the case of real reductive groups Harish-Chandra's work provides such an answer, generalizing the Weyl character formula for compact groups; for arbitrary groups no such answer is known. For special classes of groups, there are partially successful general principles (the orbit method, Broué's conjecture), of which the deepest are the extraordinary circle of conjectures known as the Langlands program, which we shall discuss later.

5   Interlude: The Philosophical Lessons of "The Character Table Is Square"

Our basic theorem ("the character table is square") tells us to expect that the category of all irreducible representations of G is interesting when the conjugacy-class structure of G is in some way under control. We will finish this essay by explaining a remarkable family of examples of such groups—the rational points of *reductive* algebraic groups—and their conjectured representation theory, which is described by the *Langlands program*.

An *affine algebraic group* is a subgroup of some group GL_n that is defined by polynomial equations in the matrix coefficients. For example, the determinant of a matrix is a polynomial in the matrix coefficients, so the group SL_n, which consists of all matrices in GL_n with determinant 1, is such a group. Another is SO_n, which is the set of matrices with determinant 1 that satisfy the equation $AA^T = I$.

The above notation did not specify what sort of coefficients we were allowing for the matrices. That vagueness was deliberate. Given an algebraic group G and a field k, let us write $G(k)$ for the group where the coefficients are taken to have values in k. For example, $SL_n(\mathbb{F}_q)$ is the set of $n \times n$ matrices with coefficients in the finite field $\mathbb{F}_q$ and determinant 1. This group is finite, as is $SO_n(\mathbb{F}_q)$, while $SL_n(\mathbb{R})$ and $SO_n(\mathbb{R})$ are Lie groups. Moreover, $SO_n(\mathbb{R})$ is compact, while $SL_n(\mathbb{R})$ is not. So among affine algebraic groups over fields one already finds all three types of groups we have discussed: finite groups, compact Lie groups, and noncompact Lie groups.

We can think of $SL_n(\mathbb{R})$ as the set of matrices in $SL_n(\mathbb{C})$ that are equal to their complex conjugates. There is another involution on $SL_n(\mathbb{C})$ that is a sort of "twisted" form of complex conjugation, where we send a matrix A to the complex conjugate of $(A^{-1})^T$. The fixed points of this new involution (that is, the determinant-1 matrices A such that A equals the complex conjugate of $(A^{-1})^T$) form a group called $SU_n(\mathbb{R})$. This is also called a *real form* of $SL_n(\mathbb{C})$,[10] and it is compact.

The groups $SL_n(\mathbb{F}_q)$ and $SO_n(\mathbb{F}_q)$ are almost simple groups;[11] the classification of finite simple groups tells us, mysteriously, that all but twenty-six of the finite simple groups are of this form. A much, much easier theorem tells us that the *connected compact* groups are also of this form.

Now, given an algebraic group G, we can also consider the instances $G(\mathbb{Q}_p)$, where $\mathbb{Q}_p$ is the field of p-adic numbers, and also $G(\mathbb{Q})$. For that matter, we may consider $G(k)$ for any other field k, such as the FUNCTION FIELD OF AN ALGEBRAIC VARIETY [V.30]. The lesson of section 4 is that we may hope for all of these many groups to have a good representation theory, but that to obtain it there will be serious "analytic" or "arithmetic" difficulties to overcome, which will depend strongly on the properties of the field k.

Lest the reader adopt too optimistic a viewpoint, we point out that not every affine algebraic group has a nice conjugacy-class structure. For example, let V_n be the set of upper triangular matrices in GL_n with 1s along the diagonal, and let k be $\mathbb{F}_q$. For large n, the conjugacy classes in $V_n(\mathbb{F}_q)$ form large and complex families: to parametrize them sensibly one needs more than n parameters (in other words, they belong to families of dimension greater than n, in an appropriate sense), and it is not in fact known how to parametrize them even for a smallish value of n, such as 11. (It is not obvious that this is a "good" question though.)

More generally, solvable groups tend to have horrible conjugacy-class structure, even when the groups themselves are "sensible." So we might expect their representation theory to be similarly horrible. The best we can hope for is a result that describes the entries of the character table *in terms of* this horrible structure—some kind of non-Abelian Fourier integral. For certain p-groups Kirillov found such a result in the 1960s, as

---

10. When we say that $SL_n(\mathbb{R})$ and $SU_n(\mathbb{R})$ are both "real forms" of $SL_n(\mathbb{C})$, what is meant more precisely is that in both cases the group can be described as a subgroup of some group of real matrices that consists of all solutions to a set of polynomial equations, and that when the same set of equations is applied instead to the group of *complex* matrices the result is isomorphic to $SL_n(\mathbb{C})$.

11. Which is to say that the quotient of these groups by their center is simple.

an example of the "orbit method," but the general result is not yet known.

On the other hand, groups that are similar to connected compact groups do have a nice conjugacy-class structure: in particular, finite simple groups do. An algebraic group is called *reductive* if $G(\mathbb{C})$ has a compact real form. So, for instance, SL_n is reductive by the existence of the real form $SU_n(\mathbb{R})$. The groups GL_n and SO_n are also reductive, but V_n is not.[12]

Let us examine the conjugacy classes in the group SU_n. Every matrix in $SU_n(\mathbb{R})$ can be diagonalized, and two conjugate matrices have the same eigenvalues, up to reordering. Conversely, any two matrices in $SU_n(\mathbb{R})$ with the same eigenvalues are conjugate. Therefore, the conjugacy classes are parametrized by the quotient of the subgroup of all diagonal matrices by the action of S_n that permutes the entries.

This example can be generalized. Any compact connected group has a *maximal torus* T, that is, a maximal subgroup isomorphic to a product of circles. (In the previous example it was the subgroup of diagonal matrices.) Any two maximal tori are conjugate in G, and any conjugacy class in G intersects T in a unique W-orbit on T, where W is the *Weyl group*, the finite group $N(T)/T$ (where $N(T)$ is the normalizer of T).

The description of conjugacy classes in $G(\bar{k})$, for an algebraically closed field $\bar{k}$, is only a little more complicated. Any element $g \in G(\bar{k})$ admits a JORDAN DECOMPOSITION [III.43]: it can be written as $g = su = us$, where s is conjugate to an element of $T(\bar{k})$ and u is unipotent when considered as an element of $GL_n(\bar{k})$. (A matrix A is *unipotent* if some power of $A - I$ is zero.) Unipotent elements never intersect compact subgroups. When $G = GL_n$ this is the usual Jordan decomposition; conjugacy classes of unipotent elements are parametrized by partitions of n, which, as we mentioned in section 2, are precisely the conjugacy classes of $W = S_n$. For general reductive groups, unipotent conjugacy classes are again almost the same thing as conjugacy classes in W.[13] In particular, there are finitely many, independent of $\bar{k}$.

Finally, when k is not algebraically closed, one describes conjugacy classes by a kind of Galois descent;

for example, in $GL_n(k)$, semisimple classes are still determined by their characteristic polynomial, but the fact that this polynomial has coefficients in k constrains the possible conjugacy classes.

The point of describing the conjugacy-class structure in such detail is to describe the representation theory in analogous terms. A crude feature of the conjugacy-class structure is the way it decouples the field k from finite combinatorial data that is attached to G but independent of k—things like W, the lattice defining T, roots, and weights.

The "philosophy" suggested by the theorem that the character table is square suggests that the representation theory should also admit such a decoupling: it should be built out of the representation theory of k^*, which is the analogue of the circle, and out of the combinatorial structure of $G(\bar{k})$ (such as the finite groups W). Moreover, representations should have a "Jordan decomposition":[14] the "unipotent" representations should have some kind of combinatorial complexity but little dependence on k, and compact groups should have no unipotent representations.

The Langlands program provides a description along the lines laid out above, but it goes beyond any of the results we have suggested in that it also describes the entries of the character table. Thus, for this class of examples, it gives us (conjecturally) the hoped-for "non-Abelian Fourier transform."

6   Coda: The Langlands Program

And so we conclude by just hinting at statements. If $G(k)$ is a reductive group, we want to describe an appropriate category of representations for $G(k)$, or at least the character table, which we may think of as a "semisimplification" of that category.

Even when k is finite, it is too much to hope that conjugacy classes in $G(k)$ parametrize irreducible representations. But something not so far off is conjectured, as follows.

To a reductive group G over an algebraically closed field, Langlands attaches another reductive group ${}^L G$, the *Langlands dual*, and conjectures that representations of $G(k)$ will be parametrized by conjugacy classes

---

12. The miracle, not relevant for this discussion, is that compact connected groups can be easily classified. Each one is essentially a product of circles and non-Abelian simple compact groups. The latter are parametrized by DYNKIN DIAGRAMS [III.48 §3]. They are SU_n, Sp_{2n}, SO_n, and five others, denoted E_6, E_7, E_8, F_4, and G_2. That is it!

13. They are different, but related. Precisely, they are given by combinatorial data, Lusztig's *two-sided cells* for the corresponding affine Weyl group.

14. The first such theorems were proved for $GL_n(\mathbb{F}_q)$ by Green and Steinberg. However, the notion of Jordan decomposition for characters originates with Brauer, in his work on modular representation theory. It is part of his modular analogue of the "character table is square" theorem, which we mentioned in section 3.

in $^L G(\mathbb{C})$.[15] However, these are not conjugacy classes of *elements* of $^L G(\mathbb{C})$, as before, but of *homomorphisms* from the Galois group of k to $^L G$. The Langlands dual was originally defined in a combinatorial manner, but there is now a conceptual definition. A few examples of pairs $(G, ^L G)$ are $(\mathrm{GL}_n, \mathrm{GL}_n)$, $(\mathrm{SO}_{2n+1}, \mathrm{Sp}_{2n})$, and $(\mathrm{SL}_n, \mathrm{PGL}_n)$.

In this way the Langlands program describes the representation theory as built out of the structure of G and the arithmetic of k.

Although this description indicates the flavor of the conjectures, it is not quite correct as stated. For instance, one has to modify the Galois group[16] in such a way that the correspondence is true for the group $\mathrm{GL}_1(k) = k^*$. When $k = \mathbb{R}$, we get the representation theory of $\mathbb{R}^*$ (or its compact form S^1), which is Fourier analysis; on the other hand, when k is a p-adic local field, the representation theory of k^* is described by local class field theory. We already see an extraordinary aspect of the Langlands program: it precisely unifies and generalizes harmonic analysis and number theory.

The most compelling versions of the Langlands program are "equivalences of derived categories" between the category of representations and certain geometric objects on the spaces of Langlands parameters. These conjectural statements are the hoped-for Fourier transforms.

Though much progress has been made, a large part of the Langlands program remains to be proved. For finite reductive groups, slightly weaker statements have been proved, mostly by Lusztig. As all but twenty-six of the finite simple groups arise from reductive groups, and as the sporadic groups have had their character tables computed individually, this work already determines the character tables of all the finite simple groups.

For groups over $\mathbb{R}$, the work of Harish-Chandra and later authors again confirms the conjectures. But for other fields, only fragmentary theorems have been proved. There is much still to be done.

Further Reading

A nice introductory text on representation theory is Alperin's *Local Representation Theory* (Cambridge University Press, Cambridge, 1993). As for the Langlands program, the 1979 American Mathematical Society volume titled *Automorphic Forms, Representations, and L-functions* (but universally known as "The Corvallis Proceedings") is more advanced, and as good a place to start as any.

IV.10 Geometric and Combinatorial Group Theory
Martin R. Bridson

---

1   What Are Combinatorial and Geometric Group Theory?

Groups and geometry are ubiquitous in mathematics, groups because the symmetries (or AUTOMORPHISMS [I.3 §4.1]) of any mathematical object in any context form a group and geometry because it allows one to think intuitively about abstract problems and to organize families of objects into spaces from which one may gain some global insight.

The purpose of this article is to introduce the reader to the study of infinite, discrete groups. I shall discuss both the combinatorial approach to the subject that held sway for much of the twentieth century and the more geometric perspective that has led to an enormous flowering of the subject in the last twenty years. I hope to convince the reader that the study of groups is a concern for all of mathematics rather than something that belongs particularly to the domain of algebra.

The principal focus of *geometric group theory* is the interaction of geometry/topology and group theory, through group actions and through suitable translations of geometric concepts into group theory. One wants to develop and exploit this interaction for the benefit of both geometry/topology and group theory. And, in keeping with our assertion that groups are important throughout mathematics, one hopes to illuminate and solve problems from elsewhere in mathematics by encoding them as problems in group theory.

Geometric group theory acquired a distinct identity in the late 1980s but many of its principal ideas have their roots in the end of the nineteenth century. At that time, low-dimensional topology and *combinatorial group theory* emerged entwined. Roughly speaking, combinatorial group theory is the study of groups defined in terms of *presentations*, that is, by means of generators and relations. In order to follow the rest of this introduction the reader must first understand what these terms mean. Since their definitions would require

---

15. The $\mathbb{C}$ here is because we are looking at representations on *complex* vector spaces; if we were looking at representations on vector spaces over some field $\mathbb{F}$, we would take $^L G(\mathbb{F})$.

16. The appropriately modified Galois group is called the Weil-Deligne group.

an unacceptably long break in the flow of our discussion, I will postpone them to the next section, but I strongly advise the reader who is unfamiliar with the meaning of the expression $\Gamma = \langle a_1, \ldots, a_n \mid r_1, \ldots, r_m \rangle$ to pause and read that section before continuing with this one.

The rough definition of combinatorial group theory just given misses the point that, like many parts of mathematics, it is a subject defined more by its core problems and its origins than by its fundamental definitions. The initial impetus for the subject came from the description of discrete groups of hyperbolic isometries and, most particularly, the discovery of the FUNDAMENTAL GROUP [IV.6 §2] of a MANIFOLD [I.3 §6.9] by POINCARÉ [VI.61] in 1895. The group-theoretic issues that emerged were brought into sharp focus by the work of Tietze and Dehn in the first decade of the twentieth century and drove much of combinatorial group theory for the remainder of the century.

Not all of the epoch-defining problems came from topology: other areas of mathematics threw up fundamental questions as well. Here are some of the forms they took: Does there exist a group of the following type? Which groups have the following property? What are the subgroups of …? Is the following group infinite? When can one determine the structure of a group from its finite quotients? In the sections that follow I shall attempt to illustrate the mathematical culture associated with questions of this kind, but let me immediately mention some easily stated but difficult classical problems. (i) Let G be a group that is finitely generated and suppose that there is some positive integer n such that $x^n = 1$ for every x in G. Must G be finite? (ii) Is there a finitely presented group Γ and a surjective homomorphism $\phi : \Gamma \to \Gamma$ such that $\phi(y) = 1$ for some $y \neq 1$? (iii) Does there exist a finitely presented, infinite, SIMPLE GROUP [I.3 §3.3]? (iv) Is every countable group isomorphic to a subgroup of a finitely generated group, or even a finitely presented group?

The first of these questions was asked by Burnside in 1902 and the second by Hopf in connection with his study of degree-1 maps between manifolds. I shall present the answers to all four questions (in section 5) to illustrate an important aspect of both combinatorial and geometric group theory: one develops techniques that allow the construction of *explicit groups* with prescribed properties. Such constructions are of particular interest when they illustrate the diversity of possible phenomena in other branches of mathematics.

Another kind of question that raises basic issues in combinatorial group theory takes the form: Does there exist an algorithm to determine whether or not a group (or given elements of a group) has such-and-such a property? For example, does there exist an algorithm that can take any finite presentation and decide in a finite number of steps whether or not the group presented is trivial? Questions of this type led to a profound and mutually beneficial interaction between group theory and logic, given full voice by the Higman embedding theorem, which we shall discuss in section 6. Moreover, via the conduit of combinatorial group theory, logic has influenced topology as well: one uses group-theoretic constructions to show, for example, that there is no algorithm to determine which pairs of compact triangulated manifolds are homeomorphic in dimensions 4 and above. This shows that certain kinds of classification results that have been obtained in two and three dimensions do not have higher-dimensional analogues.

One might reasonably regard combinatorial group theory as the attempt to develop algebraic techniques to solve the types of questions described above, and in the course of doing so to identify classes of groups that are worthy of particular study. This last point, the question of which groups deserve our attention, is tackled head-on in the final section of this article.

Some of the triumphs of combinatorial group theory are intrinsically combinatorial in nature, but many more have had their true nature revealed by the introduction of geometric techniques in the past twenty years. A fine example of this is the way in which Gromov's insights have connected algorithmic problems in group theory to so-called filling problems in Riemannian geometry. Moreover, the power of geometric group theory is by no means confined to improving the techniques of combinatorial group theory: it naturally leads one to think about many other issues of fundamental importance. For example, it provides a context in which one can illuminate and vastly extend classical RIGIDITY THEOREMS [V.23], such as that of Mostow. The key to applications such as this is the idea that finitely generated groups can usefully be regarded as geometric objects in their own right. This idea has its origins in the work of CAYLEY [VI.46] (1878) and Dehn (1905) but its full force was recognized and promoted by Gromov, starting in the 1980s. It is the key idea that underpins the later sections of this article.

2   Presenting Groups

How should one describe a group? An example will illustrate the standard way of doing so and give some idea of why it is often appropriate.

Consider the familiar tiling of the Euclidean plane by equilateral triangles. How might you describe the full group Γ_Δ of symmetries of this tiling, i.e., the rigid motions of the plane that send tiles to tiles? Let us focus on a single tile T and a particular edge e of T, and use this to pick out three symmetries. The first, which we shall call α, is the reflection of the plane in the line that contains e and the other two, β and γ, are the reflections in the lines that join the endpoints of e to the midpoints of the opposite edges in T. With some effort one can convince oneself that every symmetry of the tiling can be obtained by performing these three operations repeatedly in a suitable order. One expresses this by saying that the set $\{\alpha, \beta, \gamma\}$ *generates* the group Γ_Δ.

A further useful observation is that if one performs the operation α twice, the tiling is returned to its original position: that is, $\alpha^2 = 1$. Likewise, $\beta^2 = \gamma^2 = 1$. One can also verify that $(\alpha\beta)^6 = (\alpha\gamma)^6 = (\beta\gamma)^3 = 1$.

It turns out that the group Γ_Δ is completely determined by these facts alone, a statement that we summarize by the notation

$$\Gamma_\Delta = \langle \alpha, \beta, \gamma \mid \alpha^2, \beta^2, \gamma^2, (\alpha\beta)^6, (\alpha\gamma)^6, (\beta\gamma)^3 \rangle.$$

The aim of the rest of this section is to say in more detail what this means.

To begin with, notice that from the facts we are given we can deduce others: for example, bearing in mind that $\beta^2 = \gamma^2 = (\beta\gamma)^3 = 1$, we can show that

$$(\gamma\beta)^3 = (\gamma\beta)^3(\beta\gamma)^3 = 1$$

as well (where the last equality follows after repeatedly canceling pairs of the form $\beta\beta$ or $\gamma\gamma$). We wish to convey the idea that in Γ_Δ there are no relationships between the generators except those that follow from the facts above by this kind of argument.

Now let us try to say this more formally. We define a *set of generators* for a group Γ to be a subset $S \subset \Gamma$ such that every element of Γ is equal to some product of elements of S and their inverses. That is, every element can be written in the form $s_1^{\varepsilon_1} s_2^{\varepsilon_2} \cdots s_n^{\varepsilon_n}$, where each s_i is an element of S and each ε_i is 1 or -1. We then call a product of this kind a *relation* if it is equal to the identity in Γ.

There is an awkward ambiguity here. When we talk about "the product" of some elements of Γ, it sounds as though we are referring to another element of Γ, but

we certainly did not mean this at the end of the last paragraph: a relation is not the identity element of Γ but rather a *string of symbols* such as $ab^{-1}a^{-1}bc$ that yields the identity in Γ when you interpret a, b, and c as generators in the set S. In order to be clear about this, it is useful to define another group, known as the *free group* $F(S)$.

For concreteness we shall describe the free group with three generators, taking our set S to be $\{a, b, c\}$. A typical element is a "word" in the elements of S and their inverses, such as the expression $ab^{-1}a^{-1}bc$ considered in the previous paragraph. However, we sometimes regard two words as the same: for instance, $abcc^{-1}ac$ and $abab^{-1}bc$ are the same because they become identical when we cancel out the inverse pairs cc^{-1} and $b^{-1}b$. More formally, we define two such words to be *equivalent* and say that the elements of the free group are the EQUIVALENCE CLASSES [I.2 §2.3]. To multiply words together, we just concatenate them: for instance, the product of ab^{-1} and $bcca$ is $ab^{-1}bcca$, which we can shorten to $acca$. The identity is the "empty word." This is the free group on three generators a, b, and c. It should be clear how to generalize it to an arbitrary set S, though we shall continue to discuss the set $S = \{a, b, c\}$.

A more abstract way of characterizing the free group on a, b, and c is to say that it has the following *universal property*: if G is any group and ϕ is any function from $S = \{a, b, c\}$ to G, then there is a unique homomorphism Φ from $F(S)$ to G that takes a to $\phi(a)$, b to $\phi(b)$, and c to $\phi(c)$. Indeed, if we want Φ to have these properties, then our definition is forced upon us: for example, $\Phi(ab^{-1}ca)$ will have to be $\phi(a)\phi(b)^{-1}\phi(c)\phi(a)$, by the definition of a homomorphism. So the uniqueness is obvious. The rough reason that this definition really does give rise to a well-defined homomorphism is that the only equations that are true in $F(S)$ are ones that are true in all groups: in order for Φ not to be a homomorphism, one would need a relation to hold in $F(S)$ that did not hold in G, but this is impossible.

Now let us return to our example Γ_Δ. We would like to prove that it is (isomorphic to) the "freest" group with generators α, β, and γ that satisfies the relations $\alpha^2 = \beta^2 = \gamma^2 = (\alpha\beta)^6 = (\alpha\gamma)^6 = (\beta\gamma)^3 = 1$. But what exactly is this "freest" group that we are claiming is isomorphic to Γ_Δ?

To avoid confusion about the meaning of α, β, and γ (are they elements of Γ_Δ or of the group that we are trying to construct that will turn out to be isomorphic to Γ_Δ?) we shall use the letters a, b, and c

when we answer this question. Thus, we are trying to build the "freest" group with generators a, b, and c that satisfies the relations $a^2 = b^2 = c^2 = (ab)^6 = (ac)^6 = (bc)^3 = 1$, which we denote by $G = \langle a, b, c \mid a^2, b^2, c^2, (ab)^6, (ac)^6, (bc)^3 \rangle$.

There are two ways of going about this task. One is to imitate the above discussion of the free group itself, except that now we say that two words are equivalent if you can get from one to the other by inserting or deleting not just inverse pairs but also one of the words a^2, b^2, c^2, $(ab)^6$, $(ac)^6$, or $(bc)^3$. For example, ab^2c is equivalent to ac in this group. G is then defined to be the set of equivalence classes of words with the product coming from concatenation.

A neater way to obtain G is more conceptual and exploits the universal property of the free group. As G is to be generated by a, b, and c, the universal property of the free group $F(S)$ tells us that there will have to be a unique homomorphism Φ from $F(S)$ to G such that $\Phi(a) = a$, $\Phi(b) = b$, and $\Phi(c) = c$. Moreover, we require that all of a^2, b^2, c^2, $(ab)^6$, $(ac)^6$, and $(bc)^3$ must map to the identity element in G. It follows that the KERNEL [I.3 §4.1] of Φ is a NORMAL SUBGROUP [I.3 §3.3] of $F(S)$ that contains the set $R = \{a^2, b^2, c^2, (ab)^6, (ac)^6, (bc)^3\}$. Let us write $\langle\!\langle R \rangle\!\rangle$ for the smallest normal subgroup of $F(S)$ that contains R (or equivalently the intersection of all normal subgroups of $F(S)$ that contain R). Then there is a surjective homomorphism from the QUOTIENT [I.3 §3.3] $F(S)/\langle\!\langle R \rangle\!\rangle$ to *any* group that is generated by a, b, and c and satisfies the relations $a^2 = b^2 = c^2 = (ab)^6 = (ac)^6 = (bc)^3 = 1$. This quotient itself is the group we are looking for: it is the largest group generated by a, b, and c that satisfies the relations in R.

Our assertion about Γ_Δ is that it is isomorphic to the group $G = \langle a, b, c \mid a^2, b^2, c^2, (ab)^6, (ac)^6, (bc)^3 \rangle$ that we have just described (in two ways). More precisely, the map from $F(S)/\langle\!\langle R \rangle\!\rangle$ to Γ_Δ that takes a to α, b to β, and c to γ is an isomorphism.

The above construction is very general. If we are given a group Γ, then a *presentation* of Γ is a set S that generates Γ, together with a set $R \subset F(S)$ of relations, such that Γ is isomorphic to the quotient $F(S)/\langle\!\langle R \rangle\!\rangle$. If both S and R are finite sets, one says that the presentation is finite. A group is *finitely presented* if it has a finite presentation.

We can also define presentations in the abstract, without mentioning a group Γ in advance: given any set S and any subset $R \subset F(S)$, we just define $\langle S \mid R \rangle$ to be the group $F(S)/\langle\!\langle R \rangle\!\rangle$. This is the "freest" group

generated by S that satisfies the relations in R: the only relations that hold in $\langle S \mid R \rangle$ are the ones that can be deduced from the relations R.

A psychological advantage of switching to this more abstract setting is that, whereas previously we began with a group Γ and asked how we might present it, we can now write down group presentations at will, starting with any set S and prescribing a set of words R in the symbols $S^{\pm 1}$. This gives us a very flexible way of constructing a wide variety of groups. We might, for example, use a group presentation to encode a question from elsewhere in mathematics. We could then ask about the properties of the group thus defined, and see what they had to tell us about our original problem.

3 Why Study Finitely Presented Groups?

Groups arise throughout the whole of mathematics as *groups of automorphisms*. These are maps from an object to itself that preserve all of the defining structure: two examples are the invertible LINEAR MAPS [I.3 §4.2] from a VECTOR SPACE [I.3 §2.3] to itself, and the homeomorphisms from a TOPOLOGICAL SPACE [III.90] to itself. Groups encapsulate the essence of symmetry and for this reason demand our attention. We are driven to understand their general nature, identify groups that deserve particular attention, and develop techniques for constructing new groups (from old ones, or from new ideas). And, reversing the process of abstraction, when *given* a group, we want to find concrete instances of it. For example, we might like to realize it as the group of automorphisms of some interesting object, with the aim of illuminating the nature of both the object and the group. (See the article on REPRESENTATION THEORY [IV.9] for more on this theme.)

3.1 Why Present Groups in Terms of Generators and Relations?

The short answer is that this is the form in which groups often "appear in nature." This is particularly true in topology. Before looking at a general result that illustrates this point, let us examine a simple example. Consider the group D of all isometries of $\mathbb{R}$ that are generated by the reflections at the points 0, 1, and 2: that is, the group generated by the three functions α_0, α_1, and α_2, which take x to $-x$, $2-x$, and $4-x$, respectively. You may recognize this group to be the infinite dihedral group, and you may notice that the generator α_2 is superfluous, since it can be generated from α_0

and α_1. But let us close our eyes to these observations as we let a presentation emerge from the action.

To this end, we choose an open interval U with the property that the images of U under the maps in D cover the whole of the real line, say $U = (-\frac{1}{2}, \frac{3}{2})$. Now let us record two pieces of data: the only elements of D (apart from the identity) that fail to move U completely off itself are α_0 and α_1, and, among all products of length at most 3 in those two letters, the only nontrivial ones that act as the identity on $\mathbb{R}$ are α_0^2 and α_1^2. You may like to prove that $\langle \alpha_0, \alpha_1 \mid \alpha_0^2, \alpha_1^2 \rangle$ is a presentation of D.

This is in fact a special case of a general result, which we now state. (The proof of it is somewhat involved.) Let X be a topological space that is both PATH CONNECTED [IV.6 §1] and SIMPLY CONNECTED [III.93], and let Γ be a group of homeomorphisms from X to itself. Then any choice of path-connected open subset $U \subset X$ such that the images of U cover all of X gives rise to a presentation $\Gamma = \langle S \mid R \rangle$, where $S = \{ \gamma \in \Gamma \mid \gamma(U) \cap U \neq \varnothing \}$ and R consists of all words $w \in F(S)$ of length at most 3 such that $w = 1$ in Γ. Thus, the identification of a suitable subset U provides one with a presentation of Γ, and the task of a group theorist is to determine the nature of the group from this information.

To see how difficult this task is, you might like to consider the groups

$$G_n = \langle a_1, \ldots, a_n \mid a_i^{-1} a_{i+1} a_i a_{i+1}^{-2}, \ i = 1, \ldots, n \rangle,$$

where we interpret $i + 1$ as 1 when $i = n$. One of G_3 and G_4 is trivial and the other is infinite. Can you decide which is which?

To illustrate a more subtle point, let us consider a finitely presented group that we perhaps feel we understand: the group Γ_Δ that we were discussing earlier. If we want to describe this group to a blind friend unfamiliar with the triangular tiling of the plane, what can we say to make her understand the group, or at least convince her that we understand the group?

Our friend might reasonably ask us to list the elements of our group, so we begin to describe them as products (words) in the given generators. But as we begin to do so we hit a problem: we do not want to list any element more than once and in order to avoid redundancy we have to know which pairs of words w_1, w_2 represent the same element of Γ_Δ; equivalently, we must be able to recognize which words $w_1^{-1} w_2$ are relations in the group. Determining which words are relations is called the *word problem* for the group. Even

in Γ_Δ this takes some work, and in the groups G_n we quickly find ourselves at a loss.

Note that as well as allowing one to list the elements of the group effectively, a solution to the word problem also allows one to determine the multiplication table, since deciding whether $w_1 w_2 = w_3$ is the same as deciding whether $w_1 w_2 w_3^{-1} = 1$.

3.2 Why *Finitely* Presented Groups?

The packaging of infinite objects into finite amounts of data arises throughout mathematics in the various guises of COMPACTNESS [III.9]. Finite presentation is basically a compactness condition: a group can be finitely presented if and only if it is the fundamental group of a reasonable compact space, as we shall see later.

Another good reason for studying finitely presented groups is that the Higman embedding theorem (to be discussed later) allows us to encode questions about arbitrary TURING MACHINES [IV.20 §1.1] as questions about such groups and their subgroups.

4 The Fundamental Decision Problems

In exploring the geometry and topology of low-dimensional manifolds at the beginning of the twentieth century, Max Dehn saw that many of the problems that he was wrestling with could be "reduced" to questions about finitely presented groups. For example, he gave a simple formula for associating with a KNOT DIAGRAM [III.44] a finite presentation of a group. There was one relation for each crossing in the diagram and he argued that the resulting group would be isomorphic to $\mathbb{Z}$ if and only if the knot was the unknot: that is, if and only if it could be continuously deformed into a circle. It is extremely hard to tell by staring at a knot diagram whether it is actually the unknot, so this seems like a useful reduction until one realizes that it can be just as hard to tell whether a finitely presented group is isomorphic to $\mathbb{Z}$. For example, here is the presentation of $\mathbb{Z}$ that Dehn's recipe associates with one of smallest possible pictures of the unknot, namely a diagram with just four crossings:

$$\langle a_1, a_2, a_3, a_4, a_5 \mid$$
$$a_1^{-1} a_3 a_4^{-1}, a_2 a_3^{-1} a_1, a_3 a_4^{-1} a_2^{-1}, a_4 a_5^{-1} a_4 a_3^{-1} \rangle.$$

Thus Dehn's investigations led him to understand how difficult it is to extract information from a group presentation. In particular, he was the first to identify the fundamental role of the word problem, which we alluded to earlier, and he was one of the first to begin to understand that there are fundamental problems associated with the challenge of developing *algorithms* that extract knowledge from well-defined objects such as group presentations. In his famous article of 1912 Dehn writes:

> The general discontinuous group is given by n generators and m relations between them. ... Here *there are above all three fundamental problems* whose solution is very difficult and which will not be possible without a penetrating study of the subject.
>
> 1. **The identity [word] problem:** An element of the group is given as a product of generators. One is required to give a method whereby it may be decided in a finite number of steps whether this element is the identity or not.
> 2. **The transformation [conjugacy] problem:** Any two elements S and T of the group are given. A method is sought for deciding the question whether S and T can be transformed into each other, i.e., whether there is an element U of the group satisfying the relation
> $$S = UTU^{-1}.$$
> 3. **The isomorphism problem:** Given two groups, one is to decide whether they are isomorphic or not (and further, whether a given correspondence between the generators of one group and elements of the other is an isomorphism or not).

We shall take these problems as the starting point for three lines of enquiry. First, we shall work toward an outline of the proof that all of these problems are, in a strict sense, unsolvable for general finitely presented groups.

The second use that we shall make of Dehn's problems is to hold them up as fundamental measures of complexity for each of the classes of groups that we subsequently encounter. If we can prove, for example, that the isomorphism problem is solvable in one class of groups but not in another, then we will have given genuine substance to previously vague assertions to the effect that the second class is "harder."

Finally, I want to make the point that geometry lies at the heart of the fundamental issues in combinatorial group theory: it may not be immediately obvious, but its implicit presence is nonetheless a fundamental trait of group theory and not something imposed for reasons of taste. To illustrate this point I shall explain how the study of the large-scale geometry of least-area disks in RIEMANNIAN MANIFOLDS [I.3 §6.10] is intimately connected with the study of the complexity of word problems in arbitrary finitely presented groups.

5 New Groups from Old

Suppose that you have two groups, G_1 and G_2, and want to combine them to form a new group. The first method that is taught in a typical course on group theory is to take the Cartesian product $G_1 \times G_2$: a typical element has the form (g, h) with $g \in G_1$ and $h \in G_2$, and the product of (g, h) with (g', h') is defined to be (gg', hh'). The set of elements of the form (g, e) (where e is the identity of G_2) is a copy of G_1 inside $G_1 \times G_2$, and similarly the set of elements of the form (e, h) is a copy of G_2.

These copies have nontrivial relations between their elements: for example, $(e, h)(g, e) = (g, e)(e, h)$. We would now like to take two groups Γ_1 and Γ_2 and combine them in a different way to form a group called the *free product* $\Gamma_1 * \Gamma_2$, which contains copies of Γ_1 and Γ_2 and as few additional relations as possible. That is, we would like there to be embeddings $i_j : \Gamma_j \hookrightarrow \Gamma_1 * \Gamma_2$ so that $i_1(\Gamma_1)$ and $i_2(\Gamma_2)$ generate $\Gamma_1 * \Gamma_2$ but they are not intertwined in any way. This requirement is neatly encapsulated by the following universal property: given any group G and any two homomorphisms $\phi_1 : \Gamma_1 \to G$ and $\phi_2 : \Gamma_2 \to G$, there should be a unique homomorphism $\Phi : \Gamma_1 * \Gamma_2 \to G$ such that $\Phi \circ i_j = \phi_j$ for $j = 1, 2$. (Less formally, Φ behaves like ϕ_1 on the copy of Γ_1 and behaves like ϕ_2 on the copy of Γ_2.)

It is easy to check that this property characterizes $\Gamma_1 * \Gamma_2$ up to isomorphism, but it leaves open the question of whether $\Gamma_1 * \Gamma_2$ actually exists. (These are the standard pros and cons of defining an object by means of a universal property.) In the present setting, existence is easily established using presentations: let $\langle A_1 \mid R_1 \rangle$ be a presentation of Γ_1 and let $\langle A_2 \mid R_2 \rangle$ be a presentation of Γ_2, with A_1 and A_2 disjoint, and then define $\Gamma_1 * \Gamma_2$ to be $\langle A_1 \sqcup A_2 \mid R_1 \sqcup R_2 \rangle$ (where $\sqcup$ denotes a union of disjoint sets).

More intuitively, one can define $\Gamma_1 * \Gamma_2$ to be the set of alternating sequences $a_1 b_1 \cdots a_n b_n$ with each a_i belonging to Γ_1 and each b_j belonging to Γ_2, with the extra condition that none of the a_i and b_j equals the identity, except possibly a_1 or b_n. The group operations in Γ_1 and Γ_2 extend to this set in an obvious way: for example, $(a_1 b_1 a_2)(a_1' b_1') = a_1 b_1 a_2' b_1'$, where $a_2' = a_2 a_1'$, except that if $a_2 a_1' = 1$ then the product cancels down to $a_1 b_2'$, where $b_2' = b_1 b_1'$.

Free products occur naturally in topology: if one has topological spaces X_1, X_2 with marked points $p_1 \in X_1$, $p_2 \in X_2$, then the FUNDAMENTAL GROUP [IV.6 §2] of the space $X_1 \vee X_2$ obtained from $X_1 \sqcup X_2$ by making the identification $p_1 = p_2$ is the free product of $\pi_1(X_1, p_1)$ and $\pi_1(X_2, p_2)$. The Seifert–van Kampen theorem tells one how to present the fundamental group of a space obtained by gluing X_1 and X_2 along larger subspaces. If the inclusion of the subspaces gives rise to an injection of fundamental groups, then one can express the fundamental group of the resulting space as an *amalgamated free product*, which we now define.

Let Γ_1 and Γ_2 be two groups. If some other group contains copies of Γ_1 and Γ_2, then the intersection of those copies must contain the identity element. The free product $\Gamma_1 * \Gamma_2$ was the freest group we could build that was subject to this minimal constraint. Now we shall insist that the copies of Γ_1 and Γ_2 intersect nontrivially, specify which of their subgroups must lie in the intersection, and build the freest group that satisfies this constraint.

Suppose, then, that A_1 is a subgroup of Γ_1 and that ϕ is an isomorphism from A_1 to a subgroup A_2 of Γ_2. As in the example of the free product, one can define the "freest product that identifies A_1 and A_2" by means of a universal property. Again, one can establish the existence of such a group using presentations: if $\Gamma_1 = \langle S_1 \mid R_1 \rangle$ and $\Gamma_2 = \langle S_2 \mid R_2 \rangle$, the group we seek takes the form

$$\langle S_1 \sqcup S_2 \mid R_1 \sqcup R_2 \sqcup T \rangle.$$

Here, $T = \{u_a v_a^{-1} \mid a \in A_1\}$, where u_a is some word that represents a in (the presentation of) Γ_1 and v_a is a word that represents $\phi(a)$ in Γ_2.

This group is called the *amalgamated free product of Γ_1 and Γ_2 along A_1 and A_2*. It is often described by the casual and ambiguous notation $\Gamma_1 *_{A_1 = A_2} \Gamma_2$, or even $\Gamma_1 *_A \Gamma_2$, where $A \cong A_j$ is an abstract group.

Unlike with free products, it is no longer obvious that the maps $\Gamma_i \to \Gamma_1 *_A \Gamma_2$ implicit in this construction are injective, but they do turn out to be, as was shown by Schreier in 1927.

A related construction of Higman, Neumann, and Neumann in 1949 answers the following question: given a group Γ and an isomorphism $\psi : B_1 \to B_2$ between subgroups of Γ, can one always embed Γ in a bigger group so that ψ becomes the restriction to B_1 of a conjugation?

By now, having seen the idea in the context of both free products and amalgamated free products, the reader may guess how one goes about answering this question: one writes down the presentation of a universal candidate for the desired enveloping group, denoted $\Gamma *_\psi$, and then one sets about proving that the natural map from Γ to $\Gamma *_\psi$ (which takes each word

to itself) is injective. Thus, given $\Gamma = \langle A \mid R \rangle$, we introduce a symbol $t \notin A$ (usually called the *stable letter*), we choose for each $b \in B_1$ words $\hat{b}, \tilde{b} \in F(A)$ with $\hat{b} = b$ and $\tilde{b} = \psi(b)$ in Γ, and we define

$$\Gamma *_\psi = \langle A, t \mid R, t\hat{b}t^{-1}\tilde{b}^{-1} \ (b \in B_1) \rangle.$$

This is the freest group we can build from Γ by adjoining a new element t and requiring it to satisfy all the equations we want it to, namely $t\hat{b}t^{-1} = \tilde{b}$ for every $b \in B_1$ (which we can think of as saying that $tbt^{-1} = \psi(b)$). This group is called an *HNN extension* of Γ (after Higman, Neumann, and Neumann).

Now we must show that the natural map from Γ to $\Gamma *_\psi$ is injective. That is, if you take an element y of Γ and regard it as an element of $\Gamma *_\psi$, you should not be able to use t and the relations in $\Gamma *_\psi$ to cancel y down to the identity. This is proved with the help of the following more general result known as *Britton's lemma*. Suppose that w is a word in the free group $F(A, t)$. Then the only circumstances under which it can give rise to the identity in the group $\Gamma *_\psi$ are if either it does not involve t and represents the identity in Γ or it involves t but can be simplified in an obvious way by containing a "pinch." A pinch is a subword of the form $t b t^{-1}$, where b is a word in $F(A)$ that represents an element of B_1 (in which case we can replace it by $\psi(b)$), or one of the form $t^{-1} b' t$, where b' represents an element of B_2 (in which case we can replace it by $\psi^{-1}(b')$). Thus, if you are given a word that involves t and contains no pinches, then you know that it cannot be canceled down to the identity.

A similar noncancellation result holds for the amalgamated free product $\Gamma_1 *_{A_1 = A_2} \Gamma_2$. If $g_1, \dots, g_n$ belong to Γ_1 but not to A_1 and $h_1, \dots, h_n$ belong to Γ_2 but not to A_2, then the word $g_1 h_1 g_2 h_2 \cdots g_n h_n$ cannot equal the identity in $\Gamma_1 *_{A_1 = A_2} \Gamma_2$.

These noncancellation results do far more than show that the natural homomorphisms we have been considering are injective: they also demonstrate further aspects of freeness in amalgamated free products and HNN extensions. For example, suppose that in the amalgamated free product $\Gamma_1 *_{A_1 = A_2} \Gamma_2$ we can find an element g of Γ_1 that generates an infinite group that intersects A_1 in the identity and an element h of Γ_2 that does the same for A_2. Then the subgroup of $\Gamma_1 *_{A_1 = A_2} \Gamma_2$ generated by g and h is the free group on those two generators. With a little more effort, one can deduce that any finite subgroup of $\Gamma_1 *_{A_1 = A_2} \Gamma_2$ has to be conjugate to a subgroup of the obvious copy of either Γ_1 or Γ_2. Similarly, the finite subgroups of $\Gamma *_\psi$ are conjugates of subgroups of Γ. We shall exploit these facts in the constructions that follow.

There are many ways of combining groups that I have not mentioned here. I have chosen to focus on amalgamated free products and HNN extensions partly because they lead to transparent solutions of the basic problems discussed below but more because of their primitive appeal and the way in which they arise naturally in the calculation of fundamental groups. They also mark the beginning of *arboreal group theory*, which we will discuss later. If space allowed, I would go on to describe semidirect and wreath products, which are also indispensable tools of the group theorist.

Before turning to some applications of HNN extensions and amalgamated free products, I want to return to the Burnside problem, which asks if there exist finitely generated infinite groups all of whose elements have a given finite order. This question generated important developments throughout the twentieth century, particularly in Russia. It is appropriate to mention it here because it provides another illustration of the fact that it can be useful to study a universal object in order to solve a general question.

5.1 The Burnside Problem

Given an exponent m, one clarifies the problem at hand by considering the *free Burnside group* $B_{n,m}$ given by the presentation $\langle a_1, \ldots, a_n \mid R_m \rangle$, where R_m consists of all mth powers in the free group $F(a_1, \ldots, a_n)$. It is clear that $B_{n,m}$ maps onto any group with at most n generators in which every element has order dividing m. Therefore, there exists a finitely generated infinite group with all elements of the same finite order if and only if, for suitable values of n and m, the group $B_{n,m}$ is infinite. Thus, a question that takes the form, Does there exist a group such that ...?, becomes a question about just one group.

Novikov and Adian showed in 1968 that $B_{n,m}$ is infinite when $n \geqslant 2$ and $m \geqslant 667$ is odd. Determining the exact range of values for which $B_{n,m}$ is infinite is an active area of research. Of far greater interest is the open question of whether there exist finitely presented infinite groups that are quotients of $B_{n,m}$. Zelmanov was awarded the Fields Medal for proving that each $B_{n,m}$ has only finitely many finite quotients.

5.2 Every Countable Group Can Be Embedded in a Finitely Generated Group

Given a countable group G we can list its elements, $g_0, g_1, g_2, \ldots$, taking g_0 to be the identity. We can then take a free product of G with an infinite cyclic group $\langle s \rangle \cong \mathbb{Z}$. Let Σ_1 be the set of all elements of $G * \mathbb{Z}$ of the form $s_n = g_n s^n$ with $n \geqslant 1$. Then the subgroup $\langle \Sigma_1 \rangle$ generated by Σ_1 is isomorphic to the free group $F(\Sigma_1)$. Similarly, if we let $\Sigma_2 = \{s_2, s_3, \ldots\}$ (so it is Σ_1 with the element $s_1 = g_1 s$ removed), then $\langle \Sigma_2 \rangle$ is isomorphic to $F(\Sigma_2)$. It follows that the map $\psi(s_n) = s_{n+1}$ gives rise to an isomorphism from $\langle \Sigma_1 \rangle$ to $\langle \Sigma_2 \rangle$. Now take the HNN extension $(G * \mathbb{Z}) *_\psi$, whose stable letter we denote by t. This group contains a copy of G, as we noted before. Moreover, since we have ensured that $t s_n t^{-1} = s_{n+1}$ for every $n \geqslant 1$, it can be generated by just the three elements $s_1, s,$ and t. Thus, we have embedded an arbitrary countable group into a group with three generators. (We leave the reader to think about how one can vary this construction to produce a group with two generators.)

5.3 There Are Uncountably Many Nonisomorphic Finitely Generated Groups

This was proved by B. H. Neumann in 1932. Since there are infinitely many primes, there are uncountably many nonisomorphic groups of the form $\bigoplus_{p \in P} \mathbb{Z}_p$, where P is an infinite set of primes. We have seen that each of these groups can be embedded in a finitely generated group, and our earlier comments on finite subgroups of HNN extensions show that no two of the resulting finitely generated groups are isomorphic.

5.4 An Answer to Hopf's Question

A group G is called *Hopfian* if every surjective homomorphism from G to G is an isomorphism. Most familiar groups have this property: for example, finite groups obviously do, as do $\mathbb{Z}^n$ (as you can prove using linear algebra) and free groups. So too do groups of matrices such as $SL_n(\mathbb{Z})$, as we shall discuss in a moment. A simple example of a non-Hopfian group is the group consisting of all infinite sequences of integers (under pointwise addition), since the function that takes $(a_1, a_2, a_3, \ldots)$ to $(a_2, a_3, a_4, \ldots)$ is a surjective homomorphism that contains $(1, 0, 0, \ldots)$ in its kernel. But is there a finitely presented example? The answer is yes, and Higman was the first to construct one. The following examples are due to Baumslag and Solitar.

Let $p \geqslant 2$ be an integer and identify $\mathbb{Z}$ with the free group $\langle a \rangle$ generated by a single generator a. Then the subgroups $p\mathbb{Z}$ and $(p+1)\mathbb{Z}$ of $\mathbb{Z}$ are identified with the powers of a^p and a^{p+1}, respectively. Let ψ be the isomorphism between these subgroups that takes a^p to a^{p+1} and consider the corresponding HNN extension

B. This has presentation $B = \langle a, t \mid ta^{-p}t^{-1}a^{p+1} \rangle$. The homomorphism $\psi : B \to B$ defined by $t \mapsto t, a \mapsto a^p$ is clearly a surjection but its kernel contains, for example, the element $c = ata^{-1}t^{-1}a^{-2}tat^{-1}a$, which does not contain a pinch and is therefore not equal to the identity, by Britton's lemma. (If you want to convince yourself how useful this lemma is, set $p = 3$ and try to prove directly that c is not equal to the identity in the group B just defined.)

5.5   A Group That Has No Faithful Linear Representation

One can show that a finitely generated group G of matrices over any field is *residually finite*, which means that for each nontrivial element $g \in G$ there exists a finite group Q and a homomorphism $\pi : G \to Q$ with $\pi(g) \neq 1$. For example, if you are given an element $g \in \mathrm{SL}_n(\mathbb{Z})$, then you can pick an integer m bigger than the absolute values of all the entries in g (which is an $n \times n$ matrix) and consider the homomorphism from $\mathrm{SL}_n(\mathbb{Z})$ to $\mathrm{SL}_n(\mathbb{Z}/m\mathbb{Z})$ that reduces the matrix entries mod m. The image of g in the finite group $\mathrm{SL}_n(\mathbb{Z}/m\mathbb{Z})$ is clearly nontrivial.

Non-Hopfian groups are not residually finite, and hence are not isomorphic to a group of matrices over any field. One can see that the non-Hopfian group B defined above is not residually finite by considering what happens to the nontrivial element c. We saw that there was a surjective homomorphism $\psi : B \to B$ with $\psi(c) = 1$. Let c_n be an element such that $\psi^n(c_n) = c$ (which exists since ψ is a surjection). If there were a homomorphism π from B to a finite group Q with $\pi(c) \neq 1$, then we would have infinitely many distinct homomorphisms from B to Q, namely the compositions $\pi \circ \psi^n$; these are distinct because $\pi \circ \psi^m(c_n) = 1$ if $m > n$ and $\pi \circ \psi^n(c_n) = \pi(c) \neq 1$. This is a contradiction, since a homomorphism π from a finitely generated group to a finite group is determined by what it does to the generators, so there can only be finitely many such homomorphisms.

5.6   Infinite Simple Groups

Britton's lemma actually tells us more than that $c \neq 1$: the subgroup Λ of B generated by t and c is in fact a free group on those generators. Thus we may form the amalgamated free product Γ of two copies of B, denoted B_1 and B_2, by gluing together the two copies of Λ with the isomorphism $c_1 \mapsto t_2, t_1 \mapsto c_2$. We have

seen that in any finite quotient of $\Gamma = B_1 *_\Lambda B_2$, the elements $c_1 (= t_2)$ and $c_2 (= t_1)$ must have trivial image, and it is easy to deduce from this that in fact the quotient must be trivial. Thus Γ is an infinite group with no finite quotients. It follows that the quotient of Γ by any maximal proper normal subgroup is also infinite (and it is simple by maximality).

The simple group that we have constructed is infinite and finitely generated but it is not finitely presentable. Finitely presented infinite simple groups do exist, but they are much harder to construct.

6   Higman's Theorem and Undecidability

We have seen that there are uncountably many (non-isomorphic) finitely generated groups. But as there are only countably many finitely *presented* groups, only countably many finitely generated groups can be subgroups of finitely presented groups. Which ones are they?

A complete answer to this question is provided by a beautiful and deep theorem proved by Graham Higman in 1961, which says, roughly, that the groups that arise are all those that are algorithmically describable. (If you have no idea what this means, even roughly, then you might like to read THE INSOLUBILITY OF THE HALTING PROBLEM [V.20] before continuing with this section.)

A set S of words over a finite alphabet A is called *recursively enumerable* if there is some algorithm (or more formally, Turing machine) that can produce a complete list of the elements of S. A case of particular interest is when A is just a singleton, in which case a word is determined by its length and we can think of S as a set of nonnegative integers. The elements of S need not be listed in a sensible order, so having an algorithm that produces an exhaustive list of S does not mean that one can use the algorithm to determine that some given word w does *not* belong to S: if you imagine standing by your computer as it enumerates S, there will not in general come a time when you can say to yourself, "If it was going to appear, then it would have done so by now," and therefore be certain that it is not in S. If you want an algorithm with this further property, then you need the stronger notion of a *recursive set*, which is a set S such that S and its complement are *both* recursively enumerable. Then you can list all the elements that belong to S and you can also list all the elements that do not belong to S.

A finitely generated group is said to be *recursively presentable* if it has a presentation with a finite number of generators and a recursively enumerable set of defining relations. In other words, such a group is not necessarily finitely presented, but at least the presentation of the group is "nice" in the sense that it can be generated by some algorithm.

Higman's embedding theorem states that *a finitely generated group G is recursively presentable if and only if it is isomorphic to a subgroup of a finitely presented group.*

To get a feeling for how nonobvious this is, you might consider the following presentation of the group of all rationals under addition, in which the generator a_n corresponds to the fraction $1/n!$:

$$Q = \langle a_1, a_2, \cdots \mid a_n^n = a_{n-1} \ \forall n \geqslant 2 \rangle.$$

Higman's theorem tells us that Q can be embedded in a finitely presented group, but no truly explicit embedding is known.

The power of Higman's theorem is illustrated by the ease with which it implies the celebrated undecidability results that were rightly regarded as watersheds of twentieth-century mathematics. In order to make this case convincingly, I shall give a complete proof (except that I shall assume some of the facts mentioned earlier) that there exist finitely presented groups with unsolvable word problems, and also that there are sequences of finitely presented groups among which one cannot decide isomorphism. We shall also see how these group-theoretic results can be used to translate undecidability phenomena into topology.

The basic seed of undecidability comes from the fact that there are recursively enumerable subsets $S \subset \mathbb{N}$ that are not recursive. Using this fact one can readily construct finitely generated groups with an unsolvable word problem: given such a set of integers S we consider

$$J = \langle a, b, t \mid t(b^n a b^{-n}) t^{-1} = b^n a b^{-n} \ \forall n \in S \rangle.$$

This is the HNN extension of the free group $F(a, b)$ associated with the identity map $L \to L$, where L is the subgroup generated by $\{b^n a b^{-n} : n \in S\}$. Britton's lemma tells us that the word

$$w_m = t(b^m a b^{-m}) t^{-1} (b^m a^{-1} b^{-m})$$

equals $1 \in J$ if and only if $m \in S$, and by definition there is no algorithm to decide if $m \in S$, so we cannot decide which of the w_m are relations. Thus J has an unsolvable word problem.

That there exist finitely presented groups for which the word problem is unsolvable is a much deeper fact, but with Higman's embedding theorem at hand the proof becomes almost trivial: Higman tells us that J can be embedded in a finitely presented group Γ, and it is a relatively straightforward exercise to show that if one cannot decide which words in the generators of J represent the identity, then one cannot decide for arbitrary words in the generators of Γ either.

Once one has a finitely presented group with an unsolvable word problem, it is easy to translate undecidability into all manner of other problems. For example, suppose that $\Gamma = \langle A \mid R \rangle$ is a finitely presented group with an unsolvable word problem, where $A = \{a_1, \ldots, a_n\}$ and no a_i equals the identity in Γ. For each word w made out of the letters in A and their inverses, define a group Γ_w to have presentation

$$\langle A, s, t \mid R, \ t^{-1}(s^i a_i s^{-i}) t (s^i w s^{-i}), \ i = 1, \ldots, n \rangle.$$

It is not hard to show that if $w = 1$ in Γ then Γ_w is the free group generated by s and t. If $w \neq 1$, then Γ_w is an HNN extension. In particular, it contains a copy of Γ, and hence has an unsolvable word problem, which means that it cannot be a free group. Thus, since there is no algorithm to decide whether $w = 1$ in Γ, one cannot decide which of the groups Γ_w are isomorphic to which others.

A variant of this argument shows that there is no algorithm to determine whether or not a given finitely presented group is trivial.

We shall see in a moment that every finitely presented group G is the fundamental group of some compact four-dimensional manifold. By following a standard proof of this theorem with considerable care, Markov proved in 1958 that in dimensions 4 and above there is no algorithm to decide which compact manifolds (presented as simplicial complexes, for example) are homeomorphic. His basic idea was to show that if there were an algorithm to determine which triangulated 4-manifolds are homeomorphic, then one could use it to determine which finitely presented groups are trivial, which we know is impossible. In order to implement this idea one has to be careful to arrange that the 4-manifolds associated with different presentations of the trivial group are homeomorphic: this is the delicate part of the argument.

Strikingly, there does exist an algorithm to decide which compact three-dimensional manifolds are isomorphic. This is an extremely deep theorem that relies in particular on Perelman's solution to THURSTON'S GEOMETRIZATION CONJECTURE [IV.7 §2.4].

7 Topological Group Theory

Let us change perspective now and look at the symbols $P \equiv \langle a_1, \ldots, a_2 \mid r_1, \ldots, r_m \rangle$ through the eyes of a topologist. Instead of interpreting P as a recipe for constructing a group, we regard it as a recipe for constructing a TOPOLOGICAL SPACE [III.90], or more specifically a *two-dimensional complex*. Such spaces consist of points, called *vertices*, some of which are linked by directed paths, called *edges*, or 1-*cells*. If a collection of such 1-cells forms a cycle, then it can be filled in with a *face*, or 2-cell: topologically speaking, each face is a disk with a directed cycle as its boundary.

To see what this complex is, let us first consider the standard presentation $P \equiv \langle a, b \mid aba^{-1}b^{-1} \rangle$ of $\mathbb{Z}^2$. (This is generated by a and b and the relation tells us that $ab = ba$.) We begin with a graph K^1 that has a single vertex and two edges (which are loops) that are directed and labeled a and b. Next, we take a square $[0,1] \times [0,1]$, the sides of which are directed and labeled a, b, a^{-1}, b^{-1} as we proceed around the boundary. Imagine gluing the boundary of the square to the graph so as to respect the labeling of edges: with a bit of thought, you should be able to see that the result is a torus, that is, a surface in the shape of a bagel. An observation that turns out to be important is that the fundamental group of the torus is $\mathbb{Z}^2$, the group we started with.

The idea of "gluing" is made precise by the use of *attaching maps*: we take a continuous map ϕ from the boundary of the square S to the graph K^1 that sends the corners of the square to the vertex of K^1 and sends each side (minus its vertices) homeomorphically onto an open edge. The torus is then the quotient of $K^1 \sqcup S$ by the equivalence relation that identifies each x in the boundary of the square with its image $\phi(x)$.

With this more abstract language in hand, it is easy to see how the above construction generalizes to arbitrary presentations: given a presentation $P \equiv \langle a_1, \ldots, a_n \mid r_1, \ldots, r_m \rangle$, one takes a graph with a single vertex and n oriented loops, which are labeled $a_1, \ldots, a_n$. Then for each r_j one attaches a polygonal disk by gluing its boundary circuit to the sequence of oriented edges that traces out the word r_j.

In general, the result will not be a surface as it was for $\langle a, b \mid aba^{-1}b^{-1} \rangle$. Rather, it will be a two-dimensional complex with singularities along the edges and at the vertex. You may find it instructive to do some more examples. From $\langle a \mid a^2 \rangle$ one gets the projective plane; from $\langle a, b, c, d \mid aba^{-1}b^{-1}, cdc^{-1}d \rangle$ one gets a torus and a Klein bottle stuck together at a point. Picturing the 2-complex for $\langle a, b \mid a^2, b^3, (ab)^3 \rangle$ is already rather difficult.

The construction of $K(P)$ is the beginning of *topological group theory*. The Seifert–van Kampen theorem (mentioned earlier) implies that the fundamental group of $K(P)$ is the group presented by P. But the group no longer sits inertly in the form of an inscrutable presentation—now it acts on the UNIVERSAL COVERING [III.93] of $K(P)$ by homeomorphisms known as "deck transformations." Thus, through the simple construction of $K(P)$ (and the elegant theory of covering spaces in topology) we achieve our aim of realizing an abstract finitely presented group as the group of symmetries of an object with a potentially rich structure, on which we can bring global geometric and topological techniques to bear.

To obtain an improved topological model for our group, we can embed $K(P)$ in $\mathbb{R}^5$ (just as one can embed a finite GRAPH [III.34] in $\mathbb{R}^3$) and consider the compact four-dimensional manifold M obtained by taking all points that are a small fixed distance from the image. (I am assuming that the embedding is suitably "tame," which one can arrange.) The mental picture to strive for here is a higher-dimensional analogue of the surface (sleeve) one gets by taking the points in $\mathbb{R}^3$ that are a small fixed distance from an embedded graph. The fundamental group of M is again the group presented by P, so now we have our arbitrary finitely presented group acting on a manifold (the universal cover of M). This allows us to use the tools of analysis and differential geometry.

The constructions of $K(P)$ and M establish the more difficult implication of the theorem, promised earlier, that a group can be finitely presented if and only if it is the fundamental group of a compact cell complex and of a compact 4-manifold. This result raises several natural questions. First, are there better, more informative, topological models for an arbitrary finitely presented group Γ? And if not, then what can one say about the classes of groups defined by the natural constraints that arise when one tries to improve the model? For example, we would like to construct a lower-dimensional manifold with fundamental group Γ, enabling us to exploit our physical insight into three-dimensional geometry. But it turns out that the fundamental groups of compact three-dimensional manifolds are very special; this observation lies near the heart of a great deal of mathematics at the end of the twentieth century. Other interesting fields open up when one asks which groups arise as the fundamental groups of compact spaces satisfying CURVATURE [III.13] conditions, or constraints coming from complex geometry.

A particularly rich set of constraints comes from the following question. Can one arrange for an arbitrary finitely presented group to be the fundamental group of a compact space (a complex or manifold, perhaps) whose universal cover is CONTRACTIBLE [IV.6 §2]? This is a natural question from the point of view of topology because a space with a contractible universal cover is, up to HOMOTOPY [IV.6 §2], completely determined by its fundamental group. If the fundamental group is Γ, then such a space is called a *classifying space* for Γ and its homotopy-invariant properties provide a rich array of invariants for the group Γ (getting away from the gross dependence that $K(P)$ has on P rather than Γ).

If our earlier discussion of how hard it is to recognize Γ from P has left you very skeptical about whether this dependence can actually be removed, then your skepticism is well-founded: there are many obstructions to the construction of compact classifying spaces for an arbitrary finitely presented group; the study of them (under the generic name *finiteness conditions*) is a rich area at the interface of modern group theory, topology, and homological algebra.

One aspect of this area is the search for natural conditions that ensure the *existence* of compact classifying spaces (not necessarily manifolds). This is one of several places where manifestations of nonpositive curvature play a fundamental role in modern group theory. More combinatorial conditions also arise. For example, Lyndon proved that for any presentation $P \equiv \langle A \mid r \rangle$ where the single defining relation $r \in F(A)$ is not a nontrivial power, the universal cover of $K(P)$ is contractible.

A neighboring and highly active area of research concerns questions of uniqueness and rigidity for classifying spaces. (Here, as is common, the word *rigidity* is used to describe a situation in which requiring two objects to be equivalent in an apparently weak sense forces them to be equivalent in an apparently stronger sense.) For example, the (open) *Borel conjecture* asserts that if two compact manifolds have isomorphic fundamental groups and contractible universal covers, then those manifolds must be homeomorphic.

I have been talking mostly about realizing groups as fundamental groups, which led to certain free actions. That is, we could interpret the elements of the group as symmetries of a topological space and none of these symmetries had any fixed points. Before moving on to geometric group theory I should point out that there are many situations in which the most illuminating actions of a group are not free: one instead allows well-understood stabilizers. (The *stabilizer* of a point is the set of all symmetries in the group that leave that point fixed.) For example, the natural way in which to study Γ_Δ is by its action on the triangulated plane, each vertex of which is left unmoved by twelve symmetries.

A deeper illustration of the merits of seeking insight into algebraic structure through nonfree actions on suitable topological spaces comes from the Bass–Serre theory of groups acting on trees, which subsumes the theory of amalgamated free products and HNN extensions, whose potency we saw earlier. (This theory and its extensions often go under the heading of *arboreal group theory*.)

A *tree* is a connected graph that has no circuits in it. It is helpful to regard it as a METRIC SPACE [III.56] in which each edge has length 1. The group actions that one allows on trees are those that take edges to edges isometrically, never flipping an edge.

If a group Γ acts on a set X (in other words, if it can be regarded as a group of symmetries of X), then the *orbit* of a point $x \in X$ is the set of all its images gx with $g \in \Gamma$. A group Γ can be expressed as an amalgamated free product $A *_C B$ if and only if it acts on a tree in such a way that there are two orbits of vertices, one orbit of edges, and stabilizers A, B, C (where A and B are the stabilizers of adjacent vertices and intersect in C, which is the edge stabilizer). HNN extensions correspond to actions with one orbit of vertices and one orbit of edges. Thus, amalgamated free products and HNN extensions appear as *graphs of groups*, which are the basic objects of Bass–Serre theory. These objects allow one to recover groups acting on trees from the quotient data of the action, i.e., the quotient space (which is a graph) and the pattern of edge and vertex stabilizers.

An early benefit of Bass–Serre theory is a transparent and instructive proof that any finite subgroup of $A *_C B$ is conjugate to a subgroup of either A or B: given any set V of vertices in a tree, there is a unique vertex or midpoint x minimizing $\max\{d(x, v) \mid v \in V\}$; one applies this observation with V an orbit of the finite subgroup; x provides a fixed point for the action of the subgroup; and any point stabilizer is conjugate to a subgroup of either A or B.

Arboreal group theory goes much deeper than this first application suggests. It is the basis for a decomposition theory of finitely presented groups from which it emerges, for example, that there is an essentially canonical maximal splitting of an arbitrary finitely presented group as a graph of groups with cyclic edge stabilizers. This provides a striking parallel with the decomposition theory of 3-manifolds, a parallel that extends far beyond a mere analogy and accounts for much of the deepest work in geometric group theory in the past ten years. If you want to learn more about this, search the literature for *JSJ decompositions*. You may also want to search for *complexes of groups*, which provide the appropriate higher-dimensional analogue for graphs of groups.

8   Geometric Group Theory

Let us refresh the image of $K(P)$ in our mind's eye by thinking again about the presentation $P \equiv \langle a, b \mid aba^{-1}b^{-1} \rangle$ of $\mathbb{Z}$. The complex $K(P)$, as we saw earlier, is a torus. Now the torus can be defined as the quotient of the Euclidean plane $\mathbb{R}^2$ by the action of the group $\mathbb{Z}^2$ (where the point $(m, n) \in \mathbb{Z}^2$ acts as the translation $(x, y) \mapsto (x + m, y + n)$): in fact, $\mathbb{R}^2$, with an appropriate square tiling, is the universal cover of the torus. If we look at the orbit of the point 0 under this action, it forms a copy of $\mathbb{Z}^2$, and one can thereby see the large-scale geometry of $\mathbb{Z}^2$ laid out for us. We can make the idea of the "geometry of $\mathbb{Z}^2$" precise by decreeing that edges of the tiling have length 1 and defining the *graph distance* between vertices to be the length of the shortest path of edges connecting them.

As this example shows, the construction of $K(P)$ involves the two main (intertwined) strands of geometric group theory. In the first and more classical strand, one studies actions of groups on metric and topological spaces in order to elucidate the structures of both the space and the group (as with the action of $\mathbb{Z}^2$ on the plane in our example, or the action of the fundamental group of $K(P)$ on its universal cover in general). The quality of the insights that one obtains varies according to whether the action has or does not have certain desirable properties. The action of $\mathbb{Z}^2$ on $\mathbb{R}^2$ consists of isometries on a space with a fine geometric structure, and the quotient (the torus) is compact. Such actions are in many ways ideal, but sometimes one accepts weaker admission criteria in order to obtain a more diverse class of groups, and sometimes one demands even more structure in order to narrow the focus and study groups and spaces of an exceptional, but for that reason interesting, character.

This first strand of geometric group theory mingles with the second. In the second strand, one regards finitely generated groups as geometric objects in their own right equipped with *word metrics*, which are defined as follows. Given a finite generating set S for a group Γ, one defines the *Cayley graph* of Γ by joining each element $\gamma \in \Gamma$ by an edge to each element of the form γs or γs^{-1} with $s \in S$ (which is the same as the graph formed by the edges of the universal covering of $K(P)$). The distance $d_S(\gamma_1, \gamma_2)$ between γ_1 and γ_2 is then the length of the shortest path from γ_1 to γ_2 if all edges have length 1. Equivalently, it is the length of the shortest word in the free group on S that is equal to $\gamma_1^{-1}\gamma_2$ in Γ.

The word metric and the Cayley graph depend on the choice of generating set but their large-scale geometry does not. In order to make this idea precise, we introduce the notion of a *quasi-isometry*. This is an equivalence relation that identifies spaces that are similar on a large scale. If X and Y are two metric spaces, then a quasi-isometry from X to Y is a function $\phi : X \to Y$ with the following two properties. First, there are positive constants c, C, and ϵ such that $cd(x, x') - \epsilon \leqslant d(\phi(x), \phi(x')) \leqslant Cd(x, x') + \epsilon$: this says that ϕ distorts sufficiently large distances by at most a constant factor. Second, there is a constant C' such that for every $y \in Y$ there is some $x \in X$ for which $d(\phi(x), y) \leqslant C'$: this says that ϕ is a "quasi-surjection" in the sense that every element of Y is close to the image of an element of X.

Consider for example the two spaces $\mathbb{R}^2$ and $\mathbb{Z}^2$, where the metric on $\mathbb{Z}^2$ is given by the graph distance defined earlier. In this case the map $\phi : \mathbb{R}^2 \to \mathbb{Z}^2$ that takes (x, y) to $(\lfloor x \rfloor, \lfloor y \rfloor)$ (where $\lfloor x \rfloor$ denotes the largest integer less than or equal to x) is easily seen to be a quasi-isometry: if the Euclidean distance d between two points (x, y) and (x', y') is at least 10, say, then the graph distance between $(\lfloor x \rfloor, \lfloor y \rfloor)$ and $(\lfloor x' \rfloor, \lfloor y' \rfloor)$ will certainly lie between $\frac{1}{2}d$ and $2d$. Notice how little we care about the local structure of the two spaces: the map ϕ is a quasi-isometry despite not even being continuous.

It is not hard to check that if ϕ is a quasi-isometry from X to Y, then there is a quasi-isometry ψ from Y to X that "quasi-inverts" ϕ, in the sense that every x in X is at most a bounded distance from $\psi\phi(x)$ and every y in Y is at most a bounded distance from $\phi\psi(y)$. Once one has established this, it is easy to see that quasi-isometry is an equivalence relation.

Returning to Cayley graphs and word metrics, it turns out that if you take two different sets of generators for the same group, then the resulting Cayley graphs will be quasi-isometric. Thus, any property of a Cayley graph that is invariant under quasi-isometry will be a property not just of the graph but of the group itself. When dealing with such invariants we are free to think of Γ itself as a space (since we do not care which Cayley graph we form), and we can replace it by any metric space that is quasi-isometric to it, such as the universal cover of a closed Riemannian manifold with fundamental group Γ (whose existence we discussed earlier). Then the tools of analysis can be brought to bear on it.

A fundamental fact, discovered independently by many people and often called the *Milnor–Švarc lemma*, provides a crucial link between the two main strands of geometric group theory. Let us call a metric space X a *length space* if the distance between each pair of points is the infimum of the lengths of paths joining them. The Milnor–Švarc lemma states that if a group Γ acts "properly discontinuously" as a set of isometries of a length space X, and if the quotient is compact, then Γ is finitely generated and quasi-isometric to X (for any choice of word metric).

We have seen an example of this already: $\mathbb{Z}^2$ is quasi-isometric to the Euclidean plane. Less obviously, the same is true of Γ_Δ. (Consider the map that takes each element α of Γ_Δ to the point of $\mathbb{Z}^2$ nearest $\alpha(0)$.)

The fundamental group of a compact Riemannian manifold is quasi-isometric to the universal cover of that manifold. Therefore, from the point of view of quasi-isometry invariants, the study of such manifolds is equivalent to the study of arbitrary finitely presented groups. In a moment we will discuss some nontrivial consequences of this equivalence. But first let us reflect on the fact that, when finitely generated groups are considered as metric objects in the framework of large-scale geometry, they present us with a new challenge: we should *classify finitely generated groups up to quasi-isometry*.

This is an impossible task, of course, but nevertheless serves as a beacon in modern geometric group theory, one that has guided us toward many beautiful theorems, particularly under the general heading of rigidity. For example, suppose that you come across a finitely generated group Γ that is reminiscent of $\mathbb{Z}^n$ on a large scale: in other words, quasi-isometric to it. We are not necessarily given any algebraically defined map between this mystery group and $\mathbb{Z}^n$, and yet it transpires that such a group must contain a copy of $\mathbb{Z}^n$ as a subgroup of finite index.

At the heart of this result is *Gromov's polynomial-growth theorem*, a landmark theorem published in 1981. This theorem concerns the number of points within a distance r of the identity in a finitely generated group Γ. This will be a function $f(r)$, and Gromov was interested in how the function $f(r)$ grows as r tends to infinity, and what that tells us about the group Γ.

If Γ is an Abelian group with d generators, then it is not hard to see that $f(r)$ is at most $(2r + 1)^d$ (since each generator is raised to a power between $-r$ and r). Thus, in this case $f(r)$ is bounded above by a polynomial in r. At the other extreme, if Γ is a free group with two generators a and b, say, then $f(r)$ is exponentially large, since all sequences of length r that consist of as and bs (and not their inverses) give different elements of Γ.

Given this sharp contrast in behavior, one might wonder whether requiring $f(r)$ to be bounded above by a polynomial forces Γ to exhibit a great deal of commutativity. Fortunately, there is a much-studied definition that makes this idea precise. Given any group G and any subgroup H of G, the *commutator* $[G, H]$ is the subgroup generated by all elements of the form $ghg^{-1}h^{-1}$, where g belongs to G and h belongs to H. If G is Abelian, then $[G, H]$ contains just the identity. If G is not Abelian, then $[G, G]$ forms a group G_1 that contains other elements besides the identity, but it may be that $[G, G_1]$ is trivial. In that case, one says that G is a two-step nilpotent group. In general, a *k-step nilpotent* group G is one where, if you form a sequence by setting $G_0 = G$ and $G_{i+1} = [G, G_i]$ for each i, then you eventually reach the trivial group, and the first time you do so is at G_k. A *nilpotent* group is a group that is k-step nilpotent for some k.

Gromov's theorem states that a group has polynomial growth if and only if it has a nilpotent subgroup of finite index. This is a quite extraordinary fact: the polynomial-growth condition is easily seen to be independent of the choice of word metric and to be an invariant of quasi-isometry. Thus the seemingly rigid and purely algebraic condition of having a nilpotent subgroup of finite index is in fact a quasi-isometry invariant, and therefore a flabby, robust characteristic of the group.

In the past fifteen years quasi-isometric rigidity theorems have been established for many other classes of groups, including lattices in semisimple Lie groups and the fundamental groups of compact 3-manifolds (where the classification up to quasi-isometry involves more than algebraic equivalences), as well as various classes defined in terms of their graph of group decompositions. In order to prove theorems of this type, one must identify nontrivial invariants of quasi-isometry that allow one to distinguish and relate various classes of spaces. In many cases such invariants come from the development of suitable analogues of the tools of algebraic topology, modified so that they behave well with respect to quasi-isometries rather than continuous maps.

9 The Geometry of the Word Problem

It is time to explain the comments I made earlier about the geometry inherent in the basic decision problems of combinatorial group theory. I shall concentrate exclusively on the geometry of the word problem.

Gromov's *filling theorem* describes a startlingly intimate connection between the highly geometric study of disks with minimal area in RIEMANNIAN GEOMETRY [I.3 §6.10] and the study of word problems, which seems to belong more to algebra and logic.

On the geometric side, the basic object of study is the *isoperimetric function* $\mathrm{Fill}_M(l)$ of a complete Riemannian manifold M. Given any contractible closed path of length l, there is a disk of minimal area that is bounded by that path. The largest such area, over all closed paths of length l, is defined to be $\mathrm{Fill}_M(l)$. Thus, the isoperimetric function is the smallest function of which it is true to say that every closed path of length l can be filled by a disk of area at most $\mathrm{Fill}_M(l)$.

The image to have in mind here is that of a soap film: if one twists a circular wire of length l in Euclidean space and dips it in soap, the film that forms has area at most $l^2/4\pi$, whereas if one performs the same experiment in HYPERBOLIC SPACE [I.3 §6.6], the area of the film is bounded by a linear function of l. Correspondingly, the isoperimetric functions of $\mathbb{E}^n$ and $\mathbb{H}^n$ (and quotients of them by groups of isometries) are quadratic and linear, respectively. In a moment we shall discuss what types of isoperimetric functions arise when one considers other geometries (more precisely, compact Riemannian manifolds).

To state the filling theorem we need to think about the algebraic side as well. Here, we identify a function that measures the complexity of a direct attack on the word problem for an arbitrary finitely presented group $\Gamma = \langle A \mid R \rangle$. If we wish to know whether a word w equals the identity in Γ and do not have any further insight into the nature of Γ, then there is not much we can do other than repeatedly insert or remove the given relations $r \in R$.

Consider the simple example $\Gamma = \langle a, b \mid b^2 a, baba \rangle$. In this group aba^2b represents the identity. How do we prove this? Well,

$$aba^2b = a(b^2a)ba^2b = ab(baba)ab$$
$$= abab = a(baba)a^{-1} = aa^{-1} = 1.$$

Now let us think about the proof geometrically, via the Cayley graph. Since $aba^2b = 1$ in the group Γ, we obtain a cycle in this graph if we start at the identity and go along edges labeled a, b, a, a, b, in that order (in which case we visit the vertices $1, a, ab, aba, aba^2$, $aba^2b = 1$). The equalities in the proof can be thought of as a way of "contracting" this cycle down to the identity by means of inserting or deleting small loops: for instance, we could insert b, a, b, a into the list of edge directions, since $baba$ is a relation, or we could delete a trivial loop of the form a, a^{-1}. This contraction can be given a more topological character if we turn our Cayley graph into a two-dimensional complex by filling in each small loop with a *face*. Then the contraction of the original cycle consists in gradually moving it across these small faces.

Thus, the difficulty of demonstrating that a word w equals the identity is intimately connected with the *area* of w, denoted $\mathrm{Area}(w)$, which can be thought of algebraically as the smallest sequence of relations you need to insert and delete to turn w into the identity, or geometrically as the smallest number of faces you need to make a disk that fills the cycle represented by w.

The *Dehn function* $\delta_\Gamma : \mathbb{N} \to \mathbb{N}$ bounds $\mathrm{Area}(w)$ in terms of the length $|w|$ of the word w: $\delta_\Gamma(n)$ is the largest area of any word of length at most n that equals 1 in Γ. If the Dehn function grows rapidly, then the word problem is hard, since there are short words that are equal to the identity, but their area is very large, so that any demonstration that they are equal to the identity has to be very long. Results bounding the Dehn function are called *isoperimetric inequalities*.

The subscript on δ_Γ is somewhat misleading since different finite presentations of the same group will in general yield different Dehn functions. This ambiguity is tolerated because it is tightly controlled: if the groups defined by two finite presentations are isomorphic, or just quasi-isometric, then the corresponding Dehn functions have similar growth rates. More precisely, they are *equivalent*, with respect to what is sometimes called the *standard equivalence relation* "$\simeq$" of geometric group theory: given two monotone functions $f, g : [0, \infty) \to [0, \infty)$, one writes $f \preccurlyeq g$ if there exists a constant $C > 0$ such that $f(l) \leqslant Cg(Cl+C) + Cl + C$ for all $l \geqslant 0$, and $f \simeq g$ if $f \preccurlyeq g$ and $g \preccurlyeq f$; and one extends this relation to include functions from $\mathbb{N}$ to $[0, \infty)$.

You will have noticed a resemblance between the definitions of $\text{Fill}_M(l)$ and $\delta_\Gamma(n)$. The filling theorem relates them precisely: it states that *if M is a smooth compact manifold, then $\text{Fill}_M(l) \simeq \delta_\Gamma(l)$, where Γ is the fundamental group $\pi_1 M$ of M*.

For example, since $\mathbb{Z}^2$ is the fundamental group of the torus $T = \mathbb{R}^2/\mathbb{Z}^2$, which has Euclidean geometry, $\delta_{\mathbb{Z}^2}(l)$ is quadratic.

9.1   What Are the Dehn Functions?

We have seen that the complexity of word problems is related to the study of isoperimetric problems in Riemannian and combinatorial geometry. Such insights have, in the last fifteen years, led to great advances in the understanding of the nature of Dehn functions. For example, one can ask for which numbers ρ the function n^ρ is a Dehn function. The set of all such numbers, which can be shown to be countable, is known as the *isoperimetric spectrum*, denoted IP, and it is now largely understood.

Following work by many authors, Brady and Bridson proved that the closure of IP is $\{1\} \cup [2, \infty)$. The finer structure of IP was described by Birget, Rips, and Sapir in terms of the time functions of Turing machines. A further result by the same authors and Ol'shanskii explains how fundamental Dehn functions are to understanding the complexity of arbitrary approaches to the word problem for finitely generated groups Γ: the word problem for Γ lies in NP if and only if Γ is a subgroup of a finitely presented group with polynomial Dehn function. (Here, NP is the class of problems in the famous "P versus NP" question: see COMPUTATIONAL COMPLEXITY [IV.20 §3] for a description of this class.)

The structure of IP raises an obvious question: What can one say about the two classes of groups singled out as special—those with linear Dehn functions and those with quadratic ones? The true nature of the class of groups with a quadratic Dehn function remains obscure for the moment but there is a beautifully definitive description of those with a linear Dehn function: they are the *word hyperbolic groups*, which we shall discuss in the next section.

Not all Dehn functions are of the form n^α: there are Dehn functions such as $n^\alpha \log n$, for example, and others that grow more quickly than any iterated

exponential, for example that of

$$\langle a, b \mid aba^{-1}bab^{-1}a^{-1}b^{-2} \rangle.$$

If the word problem for Γ is unsolvable, then $\delta_\Gamma(n)$ will grow faster than any recursive function (indeed this serves as a definition of such groups).

9.2   The Word Problem and Geodesics

A *closed geodesic* on a Riemannian manifold is a loop that locally minimizes distance, such as a loop formed by an elastic band when released on a perfectly smooth surface. Examples such as the great circles on a sphere or the waist of an hourglass show that manifolds may contain closed geodesics that are *null-homotopic*: that is, they can be moved continuously until they are reduced to a point. But can one construct a compact topological manifold with the property that no matter what metric one puts on it there will always be infinitely many such geodesics? (Technically, if you go around a geodesic loop n times, then you get a geodesic; we avoid this by counting only "primitive" geodesics.)

From a purely geometric point of view this is a daunting problem: all specific metric information has been stripped away and one has to deal with an arbitrary metric on the floppy topological object left behind. But group theory provides a solution: *if the Dehn function of the fundamental group $\pi_1 M$ grows at least as fast as 2^{2^n}, then in any Riemannian metric on M there will be infinitely many closed geodesics that are null-homotopic*. The proof of this is too technical to sketch here.

10   Which Groups Should One Study?

Several special classes of groups have emerged from our previous discussion, such as nilpotent groups, 3-manifold groups, groups with linear Dehn functions, and groups with a single defining relation. Now we shall change viewpoint and ask which groups present themselves for study as we set out to explore the universe of all finitely presented groups, starting with the easiest ones.

The trivial group comes first, of course, followed by the finite groups. Finite groups are discussed in various other places in this volume, so I shall ignore them in what follows and adopt the approach of large-scale geometry, blurring the distinction between groups that have a common subgroup of finite index.

The first infinite group is surely $\mathbb{Z}$, but what comes next is open to debate. If one wants to retain the safety of commutativity, then finitely generated Abelian groups come next. Then, as one slowly relinquishes commutativity and control over growth and constructibility, one passes through the progressively larger classes of nilpotent, polycyclic, solvable, and elementary amenable groups. We have already met nilpotent groups in our discussion of Gromov's polynomial-growth theorem. They crop up in many contexts as the most natural generalization of Abelian groups and much is known about them, not least because one can prove a great deal by induction on the k for which they are k-step nilpotent. One can also exploit the fact that G is built from the finitely generated Abelian groups G_i/G_{i+1} in a very controlled way. The larger class of polycyclic groups is built in a similar way, while finitely generated solvable groups are built in a finite number of steps from Abelian groups that need not be finitely generated. This last class is not only larger but wilder; the isomorphism problem is solvable among polycyclic groups, for example, but unsolvable among solvable groups. By definition a group G is solvable if its *derived series*, defined inductively by $G^{(n)} = [G^{(n-1)}, G^{(n-1)}]$ with $G^{(0)} = G$, terminates in a finite number of steps.

The concept known as *amenability* forms an important link between geometry, analysis, and group theory. Solvable groups are amenable but not vice versa. It is not quite the case that a finitely presented group is amenable if and only if it does not contain a free subgroup of rank 2, but for a novice this serves as a good rule of thumb.

Now, let us return to $\mathbb{Z}$ in a more adventurous frame of mind, throw away the security of commutativity, and start taking free products instead. In this more liberated approach, finitely generated free groups appear after $\mathbb{Z}$ as the first groups in the universe. What comes next? Thinking geometrically, we might note that free groups are precisely those groups that have a tree as a Cayley graph and then ask which groups have Cayley graphs that are *tree-like*.

A key property of a tree is that all of its triangles are degenerate: if you take any three points in the tree and join them by shortest paths, then every point in one of these paths is contained in at least one other path as well. This is a manifestation of the fact that trees are spaces of infinite negative curvature. To get a feeling for why, consider what happens when one rescales the metric on a space of bounded negative curvature such as the hyperbolic plane $\mathbb{H}^2$. If we replace the standard distance function $d(x, y)$ by $(1/n)d(x, y)$ and let n tend to ∞, then the curvature of this space (in

the classical sense of differential geometry) tends to $-\infty$. This is captured by the fact that triangles look increasingly degenerate: there is a constant $\delta(n)$, with $\delta(n) \to 0$ as $n \to \infty$, such that any side of a triangle in the scaled hyperbolic space $(\mathbb{H}^2, (1/n)d)$ is contained in the $\delta(n)$-neighborhood of the union of the other two sides. More colloquially, triangles in $\mathbb{H}^2$ are *uniformly thin* and get increasingly thin as one rescales the metric.

With this picture in mind, one might move a little away from trees by asking which groups have Cayley graphs in which all triangles are uniformly thin. (It makes little sense to specify the thinness constant δ since it will change when one changes generating set.) The answer is Gromov's *hyperbolic groups*. This is a fascinating class of groups that has many equivalent definitions and arises in many contexts. For example, we have already met it as the class of groups that have linear Dehn functions. (It is not at all obvious that these two definitions are equivalent.)

Gromov's great insight is that because the thin-triangles condition encapsulates so much of the essence of the large-scale geometry of negatively curved manifolds, hyperbolic groups share many of the rich properties enjoyed by the groups that act nicely by isometries on such spaces. Thus, for example, hyperbolic groups have only finitely many conjugacy classes of finite subgroups, contain no copy of $\mathbb{Z}^2$, and (after accounting for torsion) have compact classifying spaces. Their conjugacy problems can be solved in less than quadratic time, and Sela showed that one can even solve the isomorphism problem among torsion-free hyperbolic groups. In addition to their many fascinating properties and natural definition, a further source of interest in hyperbolic groups is the fact that in a precise statistical sense, a *random finitely presented group* will be hyperbolic.

Spaces of negative and nonpositive curvature have played a central role in many branches of mathematics in the last twenty years. There is no room even to begin to justify this assertion here but it does guide us in where to look for natural enlargements of the class of hyperbolic groups: we want *nonpositively curved groups*, defined by requiring that their Cayley graphs enjoy a key geometric feature that cocompact groups of isometries inherit from simply connected spaces of nonpositive curvature ("CAT(0) spaces"). But in contrast to the hyperbolic case, the class of groups that one obtains varies considerably when one perturbs the definition, and delineating the resulting classes and their (rich) properties has been the subject of much research.

The added complications that one encounters when one moves from negative to nonpositive curvature are exemplified by the fact that the isomorphism problem is unsolvable in one of the most prominent classes that arises: the so-called *combable groups*.

Let us now return to free groups and ask which hyperbolic groups are the *immediate* neighbors of free groups. Remarkably, this vague question has a convincing answer.

One of the great triumphs of arboreal group theory is the proof that there is a finite description of the set $\text{Hom}(G, F)$ of homomorphisms from an arbitrary finitely generated group G to a free group F. The basic building blocks in this description are what Sela calls *limit groups*. One of the many ways of defining a limit group L is that for each finite subset $X \subset L$ there should exist a homomorphism to a finitely generated free group that is injective on X.

Limit groups can also be defined as those whose FIRST-ORDER LOGIC [IV.23 §1] resembles that of a free group in a precise sense. To see how first-order logic can be used to say something nontrivial about a group, consider the sentence

$$\forall x, y, z$$
$$(xy \neq yx) \vee (yz \neq zy) \vee (xz = zx) \vee (y = 1).$$

A group with this property is *commutative transitive*: if x commutes with $y \neq 1$, and y commutes with z, then x commutes with z. Free groups and Abelian groups have this property but a direct product of non-Abelian free groups, for example, does not.

It is a simple exercise to show that free Abelian groups are limit groups. But if one restricts attention to groups that have precisely the same first-order logic as free groups, one gets a smaller class consisting only of hyperbolic groups. The groups in this class are the subject of intense scrutiny at the moment. They all have negatively curved two-dimensional classifying spaces, built from graphs and hyperbolic surfaces in a hierarchical manner. The fundamental groups Σ_g of closed surfaces of genus $g \geqslant 2$ lie in this class, lending substance to the traditional opinion in combinatorial group theory that, among nonfree groups, it is the groups Σ_g that resemble free groups F_n most closely.

Incorporating this opinion into our earlier discussion, we arrive at the view that the groups $\mathbb{Z}^n$, the free groups F_n, and the groups Σ_g are the most basic of infinite groups. This is the start of a rich vein of ideas involving the automorphisms of these groups. In particular, there are many striking parallels between their outer automorphism groups $\text{GL}_n(\mathbb{Z})$, $\text{Out}(F_n)$, and $\text{Mod}_g \cong \text{Out}(\Sigma_g)$ (the mapping class group). These three classes of groups play a fundamental role across a broad spectrum of mathematics. I have mentioned them here in order to make the point that, beyond the search for knowledge about natural classes of groups, there are certain "gems" in group theory that merit a deep and penetrating study in their own right. Other groups that people might suggest for this category include Coxeter groups (generalized reflection groups, for which Γ_Δ is a prototype) and Artin groups (particularly BRAID GROUPS [III.4], which again crop up in many branches of mathematics).

I have thrown classes of groups at you thick and fast in this last section. Even so, there are many fascinating classes of groups and important issues that I have ignored completely. But so it must be, for as Higman's theorem assures us, the challenges, joys, and frustrations of finitely presented groups can never be exhausted.

Further Reading

Bridson, M. R., and A. Haefliger. 1999. *Metric Spaces of Non-Positive Curvature*. Grundlehren der Mathematischen Wissenschaften, volume 319. Berlin: Springer.

Gromov, M. 1984. Infinite groups as geometric objects. In *Proceedings of the International Congress of Mathematicians, Warszawa, Poland, 1983*, volume 1, pp. 385-92. Warsaw: PWN.

——. 1993. Asymptotic invariants of infinite groups. In *Geometric Group Theory*, volume 2. London Mathematical Society Lecture Note Series, volume 182. Cambridge: Cambridge University Press.

Lyndon, R. C., and P. E. Schupp. 2001. *Combinatorial Group Theory*. Classics in Mathematics. Berlin: Springer.

IV.11 Harmonic Analysis
Terence Tao

1 Introduction

Much of analysis tends to revolve around the study of general classes of FUNCTIONS [I.2 §2.2] and OPERATORS [III.50]. The functions are often real-valued or complex-valued, but may take values in other sets, such as a VECTOR SPACE [I.3 §2.3] or a MANIFOLD [I.3 §6.9]. An operator is itself a function, but at a "second level," because its domain and range are themselves spaces of functions: that is, an operator takes a function (or perhaps more than one function) as its input and returns a transformed function as its output. Harmonic analysis focuses in particular on the *quantitative* properties of such functions, and how these quantitative properties change when various operators are applied to them.[1]

What is a "quantitative property" of a function? Here are two important examples. First, a function is said to be *uniformly bounded* if there is some real number M such that $|f(x)| \leqslant M$ for every x. It can often be useful to know that two functions f and g are "uniformly close," which means that their difference $f - g$ is uniformly bounded with a small bound M. Second, a function is called *square integrable* if the integral $\int |f(x)|^2 \, dx$ is finite. The square integrable functions are important because they can be analyzed using the theory of HILBERT SPACES [III.37].

A typical question in harmonic analysis might then be the following: if a function $f : \mathbb{R}^n \to \mathbb{R}$ is square integrable, its gradient ∇f exists, and all the n components of ∇f are also square integrable, does this imply that f is uniformly bounded? (The answer is yes when $n = 1$, and no, but only just, when $n = 2$; this is a special case of the *Sobolev embedding theorem*, which is of fundamental importance in the analysis of PARTIAL DIFFERENTIAL EQUATIONS [IV.12].) If so, what are the precise bounds one can obtain? That is, given the integrals of $|f|^2$ and $|(\nabla f)_i|^2$, what can you say about the uniform bound M that you obtain for f?

Real and complex functions are of course very familiar in mathematics, and one meets them in high school. In many cases one deals primarily with SPECIAL FUNCTIONS [III.85]: polynomials, exponentials, trigonometric functions, and other very concrete and explicitly defined functions. Such functions typically have a very rich algebraic and geometric structure, and many questions about them can be answered exactly using techniques from algebra and geometry.

However, in many mathematical contexts one has to deal with functions that are not given by an explicit formula. For example, the solutions to ordinary and partial differential equations often cannot be given in an explicit algebraic form (as a composition of familiar functions such as polynomials, EXPONENTIAL FUNCTIONS [III.25], and TRIGONOMETRIC FUNCTIONS [III.92]). In such cases, how does one think about a function? The

answer is to focus on its *properties* and see what can be deduced from them: even if the solution of a differential equation cannot be described by a useful formula, one may well be able to establish certain basic facts about it and be able to derive interesting consequences from those facts. Some examples of properties that one might look at are measurability, boundedness, continuity, differentiability, smoothness, analyticity, integrability, or quick decay at infinity. One is thus led to consider interesting *general classes* of functions: to form such a class one chooses a property and takes the set of all functions with that property. Generally speaking, analysis is much more concerned with these general classes of functions than with individual functions. (See also FUNCTION SPACES [III.29].)

This approach can in fact be useful even when one is analyzing a single function that is very structured and has an explicit formula. It is not always easy, or even possible, to exploit this structure and formula in a purely algebraic manner, and then one must rely (at least in part) on more analytical tools instead. A typical example is the *Airy function*

$$\text{Ai}(x) = \int_{-\infty}^{\infty} e^{i(x\xi + \xi^3)} \, d\xi.$$

Although this is defined explicitly as a certain integral, if one wants to answer such basic questions as whether $\text{Ai}(x)$ is always a convergent integral, and whether this integral goes to zero as $x \to \pm\infty$, it is easiest to proceed using the tools of harmonic analysis. In this case, one can use a technique known as the *principle of stationary phase* to answer both these questions affirmatively, although there is the rather surprising fact that the Airy function decays almost exponentially fast as $x \to +\infty$, but only polynomially fast as $x \to -\infty$.

Harmonic analysis, as a subfield of analysis, is particularly concerned not just with qualitative properties like the ones mentioned earlier, but also with *quantitative bounds* that relate to those properties. For instance, instead of merely knowing that a function f is bounded, one may wish to know *how* bounded it is. That is, what is the *smallest* $M \geqslant 0$ such that $|f(x)| \leqslant M$ for all (or almost all) $x \in \mathbb{R}$; this number is known as the *sup norm* or L^∞-*norm* of f, and is denoted $\|f\|_{L^\infty}$. Or instead of assuming that f is square integrable one can quantify this by introducing the L^2-*norm* $\|f\|_{L^2} = (\int |f(x)|^2 \, dx)^{1/2}$; more generally one can quantify pth-power integrability for $0 < p < \infty$ via the L^p-*norm* $\|f\|_{L^p} = (\int |f(x)|^p \, dx)^{1/p}$. Similarly, most of the other qualitative properties mentioned above can be quantified by a variety of NORMS [III.62],

---

1. Strictly speaking, this sentence describes the field of *real-variable harmonic analysis*. There is another field called *abstract harmonic analysis*, which is primarily concerned with how real- or complex-valued functions (often on very general domains) can be studied using symmetries such as translations or rotations (for instance, via the Fourier transform and its relatives); this field is of course related to real-variable harmonic analysis, but is perhaps closer in spirit to representation theory and functional analysis, and will not be discussed here.

which assign a nonnegative number (or $+\infty$) to any given function and which provide some quantitative measure of one characteristic of that function. Besides being of importance in pure harmonic analysis, quantitative estimates involving these norms are also useful in applied mathematics, for instance in performing an error analysis of some numerical algorithm.

Functions tend to have infinitely many degrees of freedom, and it is thus unsurprising that the number of norms one can place on a function is infinite as well: there are many ways of quantifying how "large" a function is. These norms can often differ quite dramatically from each other. For instance, if a function f is very large for just a few values, so that its graph has tall, thin "spikes," then it will have a very large L^∞-norm, but $\int |f(x)| \, dx$, its L^1-norm, may well be quite small. Conversely, if f has a very broad and spread-out graph, then it is possible for $\int |f(x)| \, dx$ to be very large even if $|f(x)|$ is small for every x: such a function has a large L^1-norm but a small L^∞-norm. Similar examples can be constructed to show that the L^2-norm sometimes behaves very differently from either the L^1-norm or the L^∞-norm. However, it turns out that the L^2-norm lies "between" these two norms, in the sense that if one controls *both* the L^1-norm *and* the L^∞-norm, then one also automatically controls the L^2-norm. Intuitively, the reason is that if the L^∞-norm is not too large then one eliminates all the spiky functions, and if the L^1-norm is small then one eliminates most of the broad functions; the remaining functions end up being well-behaved in the intermediate L^2-norm. More quantitatively, we have the inequality

$$\|f\|_{L^2} \leqslant \|f\|_{L^1}^{1/2} \|f\|_{L^\infty}^{1/2},$$

which follows easily from the trivial algebraic fact that if $|f(x)| \leqslant M$, then $|f(x)|^2 \leqslant M|f(x)|$. This inequality is a special case of HÖLDER'S INEQUALITY [V.19], which is one of the fundamental inequalities in harmonic analysis. The idea that control of two "extreme" norms automatically implies further control on "intermediate" norms can be generalized tremendously and leads to very powerful and convenient methods known as *interpolation*, which is another basic tool in this area.

The study of a single function and all its norms eventually gets somewhat tiresome, though. Nearly all fields of mathematics become a lot more interesting when one considers not just objects, but also *maps* between objects. In our case, the objects in question are functions, and, as was mentioned in the introduction, a map that takes functions to functions is usually

referred to as an *operator*. (In some contexts it is also called a TRANSFORM [III.91].) Operators may seem like fairly complicated mathematical objects—their inputs and outputs are functions, which in turn have inputs and outputs that are usually numbers—but they are in fact a very natural concept since there are many situations where one wants to transform functions. For example, *differentiation* can be thought of as an operator, which takes a function f to its derivative df/dx. This operator has a well-known (partial) inverse, *integration*, which takes f to the function F that is defined by the formula

$$F(x) = \int_{-\infty}^{x} f(y) \, dy.$$

A less intuitive, but particularly important, example is THE FOURIER TRANSFORM [III.27]. This takes f to a function $\hat{f}$, given by the formula

$$\hat{f}(x) = \int_{-\infty}^{\infty} e^{-2\pi i x y} f(y) \, dy.$$

It is also of interest to consider operators that take two or more inputs. Two particularly common examples are the *pointwise product* and *convolution*. If f and g are two functions, then their pointwise product fg is defined in the obvious way:

$$(fg)(x) = f(x)g(x).$$

The convolution, denoted $f * g$, is defined as follows:

$$f * g(x) = \int_{-\infty}^{\infty} f(y)g(x - y) \, dy.$$

This is just a very small sample of interesting operators that one might look at. The original purpose of harmonic analysis was to understand the operators that were connected to Fourier analysis, real analysis, and complex analysis. Nowadays, however, the subject has grown considerably, and the methods of harmonic analysis have been brought to bear on a much broader set of operators. For example, they have been particularly fruitful in understanding the solutions of various linear and nonlinear partial differential equations, since the solution of any such equation can be viewed as an operator applied to the initial conditions. They are also very useful in analytic and combinatorial number theory, when one is faced with understanding the oscillation present in various expressions such as exponential sums. Harmonic analysis has also been applied to analyze operators that arise in geometric measure theory, probability theory, ergodic theory, numerical analysis, and differential geometry.

A primary concern of harmonic analysis is to obtain both qualitative and quantitative information about the effects of these operators on generic functions. A typical example of a quantitative estimate is the inequality

$$\|f * g\|_{L^\infty} \leqslant \|f\|_{L^2} \|g\|_{L^2},$$

which is true for all $f, g \in L^2$. This result, which is a special case of *Young's inequality*, is easy to prove: one just writes out the definition of $f * g(x)$ and applies the CAUCHY–SCHWARZ INEQUALITY [V.19]. As a consequence, one can draw the qualitative conclusion that the convolution of two functions in L^2 is always continuous. Let us briefly sketch the argument, since it is an instructive one.

A fundamental fact about functions in L^2 is that any such function f can be approximated arbitrarily well (in the L^2-norm) by a function $\tilde{f}$ that is continuous and *compactly supported*. (The second condition means that $\tilde{f}$ takes the value zero everywhere outside some interval $[-M, M]$.) Given any two functions f and g in L^2, let $\tilde{f}$ and $\tilde{g}$ be approximations of this kind. It is an exercise in real analysis to prove that $\tilde{f} * \tilde{g}$ is continuous, and it follows easily from the inequality above that $\tilde{f} * \tilde{g}$ is close to $f * g$ in the L^∞-norm, since

$$f * g - \tilde{f} * \tilde{g} = f * (g - \tilde{g}) + (f - \tilde{f}) * \tilde{g}.$$

Therefore, $f * g$ can be approximated arbitrarily well in the L^∞-norm by continuous functions. A standard result in basic real analysis (that a uniform limit of continuous functions is continuous) now tells us that $f * g$ is continuous.

Notice the general structure of this argument, which occurs frequently in harmonic analysis. First, one identifies a "simple" class of functions for which one can easily prove the result one wants. Next, one proves that every function in a much wider class can be approximated in a suitable sense by simple functions. Finally, one uses this information to deduce that the result holds for functions in the wider class as well. In our case, the simple functions were the continuous functions of finite support, the wider class consisted of square-integrable functions, and the suitable sense of approximation was closeness in the L^2-norm.

We shall give some further examples of qualitative and quantitative analysis of operators in the next section.

2   Example: Fourier Summation

To illustrate the interplay between quantitative and qualitative results, we shall now sketch some of the basic theory of summation of Fourier series, which historically was one of the main motivations for studying harmonic analysis.

In this section, we shall consider functions f that are *periodic* with period 2π: that is, functions such that $f(x + 2\pi) = f(x)$ for all x. An example of such a function is $f(x) = 3 + \sin(x) - 2\cos(3x)$. A function like this, which can be written as a finite linear combination of functions of the form $\sin(nx)$ and $\cos(nx)$, is called a *trigonometric polynomial*. The word "polynomial" is used here because any such function can be expressed as a polynomial in $\sin(x)$ and $\cos(x)$, or alternatively, and somewhat more conveniently, as a polynomial in e^{ix} and e^{-ix}. That is, it can be written as $\sum_{n=-N}^{N} c_n e^{inx}$ for some N and some choice of coefficients $(c_n : -N \leqslant n \leqslant N)$. If we know that f can be expressed in this form, then we can work out the coefficient c_n quite easily: it is given by the formula

$$c_n = \frac{1}{2\pi} \int_0^{2\pi} f(x) e^{-inx} \, dx.$$

It is a remarkable and very important fact that we can say something similar about a much wider class of functions—if, that is, we now allow *infinite* linear combinations. Suppose that f is a periodic function that is also continuous (or, more generally, that f is *absolutely integrable*, meaning that the integral of $|f(x)|$ between 0 and 2π is finite). We can then define the *Fourier coefficients* $\hat{f}(n)$ of f, using exactly the formula we had above for c_n:

$$\hat{f}(n) = \frac{1}{2\pi} \int_0^{2\pi} f(x) e^{-inx} \, dx.$$

The example of trigonometric polynomials now suggests that one should have the identity

$$f(x) - \sum_{n=-\infty}^{\infty} \hat{f}(n) e^{inx},$$

expressing f as a sort of "infinite trigonometric polynomial," but this is not always true, and even when it is true it takes some effort to justify it rigorously, or even to say precisely what the infinite sum means.

To make the question more precise, let us introduce for each natural number N the *Dirichlet summation operator* S_N. This takes a function f to the function $S_N f$ that is defined by the formula

$$S_N f(x) = \sum_{n=-N}^{N} \hat{f}(n) e^{inx}.$$

The question we would like to answer is whether $S_N f$ converges to f as $N \to \infty$. The answer turns out to be surprisingly complicated: not only does it depend on the assumptions that one places on the function f, but it also depends critically on how one defines "convergence." For example, if we assume that f is continuous and ask for the convergence to be uniform, then the answer is very definitely no: there are examples of continuous functions f for which $S_N f$ does not even converge pointwise to f. However, if we ask for a weaker form of convergence, the answer is yes: $S_N f$ will necessarily converge to f in the L^p topology for any $0 < p < \infty$, and even though it does not have to converge pointwise, it will converge *almost everywhere*, meaning that the set of x for which $S_N f(x)$ does not converge to x has MEASURE [III.55] zero. If instead one assumes only that f is absolutely integrable, then it is possible for the partial sums $S_N f$ to diverge at every single point x, as well as being divergent in the L^p topology for every p such that $0 < p \leqslant \infty$. The proofs of all of these results ultimately rely on very quantitative results in harmonic analysis, and in particular on various L^p-type estimates on the Dirichlet sum $S_N f(x)$, as well as estimates connected with the closely related *maximal operator*, which takes f to the function $\sup_{N>0} |S_N f(x)|$.

As these results are a little tricky to prove, let us first discuss a simpler result, in which the Dirichlet summation operators S_N are replaced by the *Fejér summation operators* F_N. For each N, the operator F_N is the average of the first N Dirichlet operators: that is, it is given by the formula

$$F_N = \frac{1}{N}(S_0 + \cdots + S_{N-1}).$$

It is not hard to show that if $S_N f$ converges to f, then so does $F_N f$. However, by averaging the $S_N f$ we allow cancellations to take place that sometimes make it possible for $F_N f$ to converge to f even when $S_N f$ does not. Indeed, here is a sketch of a proof that $F_N f$ converges to f whenever f is continuous and periodic—which, as we have seen, is far from true of $S_N f$.

In its basic structure, the argument is similar to the one we used when showing that the convolution of two functions in L^2 is continuous. Note first that the result is easy to prove when f is a trigonometric polynomial, since then $S_N f = f$ for every N from some point onward. Now the *Weierstrass approximation theorem* says that every continuous periodic function f can be uniformly approximated by trigonometric polynomials: that is, for every $\varepsilon > 0$ there is a trigonometric polynomial such that $\|f - g\|_{L^\infty} \leqslant \varepsilon$. We know that $F_N g$ is close to g for large N (since g is a trigonometric polynomial), and would like to deduce the same for f.

The first step is to use some routine trigonometric manipulation to prove the identity

$$F_N f(x) = \int_{-\pi}^{\pi} \frac{\sin^2(\frac{1}{2}Ny)}{N\sin^2(\frac{1}{2}y)} f(x-y)\, dy.$$

The precise form of this expression is less important than two properties of the function

$$u(y) = \frac{\sin^2(\frac{1}{2}Ny)}{N\sin^2(\frac{1}{2}y)}$$

that we shall use. One is that $u(y)$ is always nonnegative and the other is that $\int_{-\pi}^{\pi} u(y)\, dy = 1$. These two facts allow us to say that

$$F_N h(x) = \int_{-\pi}^{\pi} u(y)h(x-y)\, dy$$
$$\leqslant \|h\|_{L^\infty} \int_{-\pi}^{\pi} u(y)\, dy = \|h\|_{L^\infty}.$$

That is, $\|F_N h\|_{L^\infty} \leqslant \|h\|_{L^\infty}$ for any bounded function h.

To apply this result, we choose a trigonometric polynomial g such that $\|f - g\|_{L^\infty} \leqslant \varepsilon$ and let $h = f - g$. Then we find that $\|F_N h\|_{L^\infty} = \|F_N f - F_N g\|_{L^\infty} \leqslant \varepsilon$ as well. As mentioned above, if we choose N large enough, then $\|F_N g - g\|_{L^\infty} \leqslant \varepsilon$, and then we use the TRIANGLE INEQUALITY [V.19] to say that

$$\|F_N f - f\|_{L^\infty}$$
$$\leqslant \|F_N f - F_N g\|_{L^\infty} + \|F_N g - g\|_{L^\infty} + \|g - f\|_{L^\infty}.$$

Since each term on the right-hand side is at most ε, this shows that $\|F_N f - f\|_{L^\infty}$ is at most 3ε. And since ε can be made arbitrarily small, this shows that $F_N f$ converges to f.

A similar argument (using MINKOWSKI'S INTEGRAL INEQUALITY [V.19] instead of the triangle inequality) shows that $\|F_N f\|_{L^p} \leqslant \|f\|_{L^p}$ for all $1 \leqslant p \leqslant \infty$, $f \in L^p$, and $N \geqslant 1$. As a consequence, one can modify the above argument to show that $F_N f$ converges to f in the L^p topology for every $f \in L^p$. A slightly more difficult result (relying on a basic result in harmonic analysis known as the *Hardy–Littlewood maximal inequality*) asserts that, for every $1 < p \leqslant \infty$, there exists a constant C_p such that one has the inequality $\|\sup_N |F_N f|\|_{L^p} \leqslant C_p \|f\|_{L^p}$ for all $f \in L^p$; as a consequence, one can show that $F_N f$ converges to f almost everywhere for every $f \in L^p$ and $1 < p \leqslant \infty$. A slight modification of this argument also allows one to treat the endpoint case when f is merely assumed to be absolutely integrable; see the discussion on the Hardy–Littlewood maximal inequality at the end of this article.

Now let us return briefly to Dirichlet summation. Using some fairly sophisticated techniques in harmonic analysis (such as Calderón–Zygmund theory) one can show that when $1 < p < \infty$ the Dirichlet operators S_N are bounded in L^p uniformly in N. In other words, for every p in this range there exists a positive real number C_p such that $\|S_N f\|_{L^p} \leqslant C_p \|f\|_{L^p}$ for every function f in L^p and every nonnegative integer N. As a consequence, one can show that $S_N f$ converges to f in the L^p topology for all f in L^p and every p such that $1 < p < \infty$. However, the quantitative estimate on S_N fails at the endpoints $p = 1$ and $p = \infty$, and from this one can also show that the convergence result also fails at these endpoints (either by explicitly constructing a counterexample or by using general results such as the so-called *uniform boundedness principle*).

What happens if we ask for $S_N f$ to converge to f almost everywhere? Almost-everywhere convergence does not follow from convergence in L^p when $p < \infty$, so we cannot use the above results to prove it. It turns out to be a much harder question, and was a famous open problem, eventually answered by CARLESON'S THEOREM [V.5] and an extension of it by Hunt. Carleson proved that one has an estimate of the form $\|\sup_N |S_N f|\|_{L^p} \leqslant C_p \|f\|_{L^p}$ in the case $p = 2$, and Hunt generalized the proof to cover all p with $1 < p < \infty$. This result implies that the Dirichlet sums of an L^p function do indeed converge almost everywhere when $1 < p \leqslant \infty$. On the other hand, this estimate fails at the endpoint $p = 1$, and there is in fact an example due to KOLMOGOROV [VI.88] of an absolutely integrable function whose Dirichlet sums are everywhere divergent. These results require a lot of harmonic analysis theory. In particular they use many decompositions of both the spatial variable and the frequency variable, keeping the Heisenberg uncertainty principle in mind. They then carefully reassemble the pieces, exploiting various manifestations of orthogonality.

To summarize, quantitative estimates such as L^p estimates on various operators provide an important route to establishing qualitative results, such as convergence of certain series or sequences. In fact there are a number of principles (notably the uniform boundedness principle and a result known as *Stein's maximal principle*) which assert that in certain circumstances this is the *only* route, in the sense that a quantitative estimate must exist in order for the qualitative result to be true.

3   Some General Themes in Harmonic Analysis: Decomposition, Oscillation, and Geometry

One feature of harmonic analysis methods is that they tend to be *local* rather than *global*. For instance, if one is analyzing a function f it is quite common to decompose it as a sum $f = f_1 + \cdots + f_k$, with each function f_i "localized" in the sense that its support (the set of values x for which $f_i(x) \neq 0$) has a small diameter. This would be called localization in the *spatial variable*. One can also localize in the *frequency variable* by applying the process to the Fourier transform $\hat{f}$ of f. Having split f up like this, one can carry out estimates for the pieces separately and then recombine them later. One reason for this "divide and conquer" strategy is that a typical function f tends to have many different features—for example, it may be very "spiky," "discontinuous," or "high frequency" in some places, and "smooth" or "low frequency" in others—and it is difficult to treat all of these features at once. A well-chosen decomposition of the function f can isolate these features from each other, so that each component has only one salient feature that could cause difficulty: the spiky part can go into one f_i, the high-frequency part into another, and so on. In reassembling the estimates from the individual components, one can use crude tools such as the triangle inequality or more refined tools, for instance those relying on some sort of orthogonality, or perhaps a clever algorithm that groups the components into manageable clusters. The main drawback of the decomposition method (other than an aesthetic one) is that it tends to give bounds that are not quite optimal; however, in many cases one is content with an estimate that differs from the best possible one by a multiplicative constant.

To give a simple example of the method of decomposition, let us consider the Fourier transform $\hat{f}(\xi)$ of a function $f : \mathbb{R} \to \mathbb{C}$, defined (for suitably nice functions f) by the formula

$$\hat{f}(\xi) = \int_{\mathbb{R}} f(x) e^{-2\pi i x \xi} \, dx.$$

What we can say about the size of $\hat{f}$, as measured by suitable norms, if we are given information about the size of f, as measured by other norms?

Here are two simple observations in response to this question. First, since the modulus of $e^{-2\pi i x \xi}$ is always equal to 1, it follows that $|\hat{f}(\xi)|$ is at most $\int_{\mathbb{R}} |f(x)| \, dx$. This tells us that $\|\hat{f}\|_{L^\infty} \leqslant \|f\|_{L^1}$, at least if $f \in L^1$. In particular, $\hat{f} \in L^\infty$. Secondly, the Plancherel theorem, a very basic fact of Fourier analysis, tells us that $\|\hat{f}\|_{L^2}$ is equal to $\|f\|_{L^2}$ if $f \in L^2$. Therefore, if f belongs to L^2 then so does $\hat{f}$.

We would now like to know what happens if f lies in an intermediate L^p space. In other words, what happens if $1 < p < 2$? Since L^p is not contained in either L^1 or L^2, one cannot use either of the above two results directly. However, let us take a function $f \in L^p$ and consider what the difficulty is. The reason f may not lie in L^1 is that it may decay too slowly: for instance, the function $f(x) = (1 + |x|)^{-3/4}$ tends to zero more slowly than $1/x$ as $x \to \infty$, so its integral is infinite. However, if we raise f to the power $3/2$ we obtain the function $(1 + |x|)^{-9/8}$ which decays quickly enough to have a finite integral, so f does belong to $L^{3/2}$. Similar examples show that the reason f may fail to belong to L^2 is that it can have places where it tends to infinity slowly enough for the integral of $|f|^p$ to be finite but not slowly enough for the integral of $|f|^2$ to be finite.

Notice that these two reasons are completely different. Therefore, we can try to decompose f into two pieces, one consisting of the part where f is large and the other consisting of the part where f is small. That is, we can choose some threshold λ and define $f_1(x)$ to be $f(x)$ when $|f(x)| < \lambda$ and 0 otherwise, and define $f_2(x)$ to be $f(x)$ when $|f(x)| \geqslant \lambda$ and 0 otherwise. Then $f_1 + f_2 = f$, and f_1 and f_2 are the "small part" and "large part" of f, respectively.

Because $|f_1(x)| < \lambda$ for every x, we find that

$$|f_1(x)|^2 = |f_1(x)|^{2-p}|f_1(x)|^p < \lambda^{2-p}|f_1(x)|^p.$$

Therefore, f_1 belongs to L^2 and $\|f_1\|_{L^2} \leqslant \lambda^{2-p}\|f_1\|_{L^p}$. Similarly, because $|f_2(x)| \geqslant \lambda$ whenever $f_2(x) \neq 0$, we have the inequality $|f_2(x)| \leqslant |f_2(x)|^p/\lambda^{p-1}$ for every x, which tells us that f_2 belongs to L^1 and that $\|f_2\|_{L^1} \leqslant \|f_2\|_{L^p}/\lambda^{p-1}$.

From our knowledge about the L^2-norm of f_1 and the L^1-norm of f_2 we can obtain upper bounds for the L^2-norm of $\hat{f}_1$ and the L^∞-norm of $\hat{f}_2$, by our remarks above. By using this strategy for every λ and combining the results in a clever way, one can obtain the *Hausdorff–Young inequality*, which is the following assertion. Let p lie between 1 and 2 and let p' be the *dual exponent* of p, which is the number $p/(p-1)$. Then there is a constant C_p such that, for every function $f \in L^p$, one has the inequality $\|\hat{f}\|_{L^{p'}} \leqslant C_p\|f\|_{L^p}$. The particular decomposition method we have used to obtain this result is formally known as the method of *real interpolation*. It does not give the best possible value of C_p, which turns out to be $p^{1/2p}/(p')^{1/2p'}$, but that requires more delicate methods.

Another basic theme in harmonic analysis is the attempt to quantify the elusive phenomenon of *oscillation*. Intuitively, if an expression oscillates wildly, then we expect its average value to be relatively small in magnitude, since the positive and negative parts, or in the complex case the parts with a wide range of different arguments, will cancel out. For instance, if a 2π-periodic function f is smooth, then for large n the Fourier coefficient

$$\hat{f}(n) = \frac{1}{2\pi}\int_{-\pi}^{\pi} f(x)\mathrm{e}^{-\mathrm{i}nx}$$

will be very small since $\int_{-\pi}^{\pi} \mathrm{e}^{-\mathrm{i}nx} = 0$ and the comparatively slow variation in $f(x)$ is not enough to stop the cancellation occurring. This assertion can easily be proved rigorously by repeated integration by parts. Generalizations of this phenomenon include the so-called *principle of stationary phase*, which among other things allows one to obtain precise control on the Airy function $\mathrm{Ai}(x)$ discussed earlier. It also yields the Heisenberg uncertainty principle, which relates the decay and smoothness of a function to the decay and smoothness of its Fourier transform.

A somewhat different manifestation of oscillation lies in the principle that if one has a sequence of functions that oscillate in different ways, then their sum should be significantly smaller than the bound that follows from the triangle inequality. Again, this is the result of cancellation that is simply not noticed by the triangle inequality. For instance, the Plancherel theorem in Fourier analysis implies, among other things, that a trigonometric polynomial $\sum_{n=-N}^{N} c_n\mathrm{e}^{\mathrm{i}nx}$ has an L^2-norm of

$$\left(\frac{1}{2\pi}\int_0^{2\pi}\left|\sum_{n=-N}^{N} c_n\mathrm{e}^{\mathrm{i}nx}\right|^2\right)^{1/2} = \left(\sum_{n=-N}^{N}|c_n|^2\right)^{1/2}.$$

This bound (which can also be proved by direct calculation) is smaller than the upper bound of $\sum_{n=-N}^{N}|c_n|$ that would be obtained if we simply applied the triangle inequality to the functions $c_n\mathrm{e}^{\mathrm{i}nx}$. This identity can be viewed as a special case of the Pythagorean theorem, together with the observation that the harmonics $\mathrm{e}^{\mathrm{i}nx}$ are all *orthogonal* to each other with respect to the INNER PRODUCT [III.37]

$$\langle f,g\rangle = \frac{1}{2\pi}\int_0^{2\pi} f(x)\overline{g(x)}\,\mathrm{d}x.$$

This concept of orthogonality has been generalized in a number of ways. For instance, there is a more general and robust concept of "almost orthogonality," which roughly speaking means that the inner products of a collection of functions are small but not necessarily 0.

Many arguments in harmonic analysis will, at some point, involve a combinatorial statement about certain types of geometric objects such as cubes, balls, or boxes. For instance, one useful such statement is the *Vitali covering lemma*, which asserts that, given any collection $B_1, \dots, B_k$ of balls in Euclidean space $\mathbb{R}^n$, there will be a subcollection $B_{i_1}, \dots, B_{i_m}$ of balls that are disjoint, but that nevertheless contain a significant fraction of the volume covered by the original balls. To be precise, one can choose the disjoint balls so that

$$\text{vol}\left(\bigcup_{j=1}^{m} B_{i_j} \right) \geqslant 5^{-n} \, \text{vol}\left(\bigcup_{j=1}^{k} B_j \right).$$

(The constant 5^{-n} can be improved, but this will not concern us here.) This result is obtained by a "greedy algorithm": one picks balls one by one, at each stage choosing the largest ball among the B_j that is disjoint from all the balls already selected.

One consequence of the Vitali covering lemma is the *Hardy–Littlewood maximal inequality*, which we will briefly describe. Given any function $f \in L^1(\mathbb{R}^n)$, any $x \in \mathbb{R}^n$, and any $r > 0$, we can calculate the average of $|f|$ in the n-dimensional sphere $B(x, r)$ of center x and radius r. Next, we can define the *maximal function* F of f by letting $F(x)$ be the largest of all these averages as r ranges over all positive real numbers. (More precisely, one takes the supremum.) Then, for each positive real number λ one can define a set X_λ to be the set of all x such that $F(x) > \lambda$. The Hardy–Littlewood maximal inequality asserts that the volume of X_λ is at most $5^n \|f\|_{L^1} / \lambda.$[2]

To prove it, one observes that X_λ can be covered by balls $B(x, r)$ on each of which the integral of $|f|$ is at least $\lambda \, \text{vol}(B(x, r))$. To this collection of balls one can then apply the Vitali covering lemma, and the result follows. The Hardy–Littlewood maximal inequality is a quantitative result, but it has as a qualitative consequence the *Lebesgue differentiation theorem*, which asserts the following. If f is any absolutely integrable function defined on $\mathbb{R}^n$, then for almost every $x \in \mathbb{R}^n$ the averages

$$\frac{1}{\text{vol}(B(x, r))} \int_{B(x, r)} f(y) \, dy$$

of f over the Euclidean balls about x tend to $f(x)$ as $r \to 0$. This example demonstrates the impor-

tance of the underlying geometry (in this case, the combinatorics of metric balls) in harmonic analysis.

Further Reading

Stein, E. M. 1970. *Singular Integrals and Differentiability Properties of Functions*. Princeton, NJ: Princeton University Press.

——. 1993. *Harmonic Analysis*. Princeton, NJ: Princeton University Press.

Wolff, T. H. 2003. *Lectures on Harmonic Analysis*, edited by I. Łaba and C. Shubin. University Lecture Series, volume 29. Providence, RI: American Mathematical Society.

IV.12 Partial Differential Equations
Sergiu Klainerman

Introduction

Partial differential equations (or PDEs) are an important class of *functional equations*: they are equations, or systems of equations, in which the unknowns are functions of more than one variable. As a very crude analogy, PDEs are to functions as polynomial equations (such as $x^2 + y^2 = 1$, for example) are to numbers. The distinguishing feature of PDEs, as opposed to more general functional equations, is that they involve not only unknown functions, but also various *partial derivatives* of those functions, in algebraic combination with each other and with other, fixed, functions. Other important kinds of functional equations are *integral equations*, which involve various integrals of the unknown functions, and *ordinary differential equations* (ODEs), in which the unknown functions depend on only one independent variable (such as a time variable t) and the equation involves only ordinary derivatives $d/dt, d^2/dt^2, d^3/dt^3, \dots$ of these functions.

Given the immense scope of the subject the best I can hope to do is to give a very crude perspective on some of the main issues and an even cruder idea of the multitude of current research directions. The difficulty one faces in trying to describe the subject of PDEs starts with its very definition. Is it a unified area of mathematics, devoted to the study of a clearly defined set of objects (in the way that algebraic geometry studies solutions of polynomial equations or topology studies manifolds, for example), or is it rather a collection of separate fields, such as general relativity, several complex variables, or hydrodynamics, each one vast in its own right and centered on a particular, very difficult, equation or class of equations? I will attempt to argue

---

2. This version of the Hardy–Littlewood inequality looks somewhat different from the one mentioned briefly in the previous section, but one can deduce that inequality from this one by the real interpolation method discussed earlier.

below that, even though there are fundamental difficulties in formulating a general theory of PDEs, one can nevertheless find a remarkable unity between various branches of mathematics and physics that are centered on individual PDEs or classes of PDEs. In particular, certain ideas and methods in PDEs have turned out to be extraordinarily effective across the boundaries of these separate fields. It is thus no surprise that the most successful book ever written about PDEs did not mention PDEs in its title: it was *Methods of Mathematical Physics* by COURANT [VI.83] and HILBERT [VI.63].

As it is impossible to do full justice to such a huge subject in such limited space I have been forced to leave out many topics and relevant details; in particular, I have said very little about the fundamental issue of breakdown of solutions, and there is no discussion of the main open problems in PDEs. A longer and more detailed version of the article, which includes these topics, can be found at

http://press.princeton.edu/titles/8350.html

1 Basic Definitions and Examples

The simplest example of a PDE is the LAPLACE EQUATION [I.3 §5.4]

$$\Delta u = 0. \tag{1}$$

Here, Δ is the *Laplacian*, that is, the *differential operator* that transforms functions $u = u(x_1, x_2, x_3)$ defined from $\mathbb{R}^3$ to $\mathbb{R}$ according to the rule

$$\Delta u(x_1, x_2, x_3)$$
$$= \partial_1^2 u(x_1, x_2, x_3) + \partial_2^2 u(x_1, x_2, x_3) + \partial_3^2 u(x_1, x_2, x_3),$$

where ∂_1, ∂_2, ∂_3 are standard shorthand for the partial derivatives $\partial/\partial x_1$, $\partial/\partial x_2$, $\partial/\partial x_3$. (We will use this shorthand throughout the article.) Two other fundamental examples (also described in [I.3 §5.4]) are the *heat equation* and the *wave equation*:

$$-\partial_t u + k\Delta u = 0, \tag{2}$$
$$-\partial_t^2 u + c^2 \Delta u = 0. \tag{3}$$

In each case one is asked to find a function u that satisfies the corresponding equations. For the Laplace equation u will depend on x_1, x_2, and x_3, and for the other two it will depend on t as well. Observe that equations (2) and (3) again involve the symbol Δ, but also partial derivatives with respect to the time variable t. The constants k (which is positive) and c are fixed and represent the rate of diffusion and the speed of light, respectively. However, from a mathematical point of

view they are not important, since if $u(t, x_1, x_2, x_3)$ is a solution of (3), for example, then $v(t, x_1, x_2, x_3) = u(t, x_1/c, x_2/c, x_3/c)$ satisfies the same equation with $c = 1$. Thus, when one is studying the equations one can set these constants to be 1. Both equations are called *evolution equations* because they are supposed to describe the change of a particular physical object as the time parameter t varies. Observe that (1) can be interpreted as a particular case of both (2) and (3): if $u = u(t, x_1, x_2, x_3)$ is a solution of either (2) or (3) that is independent of t, then $\partial_t u = 0$, so u must satisfy (1).

In all three examples mentioned above, we tacitly assume that the solutions we are looking for are sufficiently differentiable for the equations to make sense. As we shall see later, one of the important developments in the theory of PDEs was the study of more refined notions of solutions, such as DISTRIBUTIONS [III.18], which require only *weak* versions of differentiability.

Here are some further examples of important PDEs. The first is THE SCHRÖDINGER EQUATION [III.83],

$$i\partial_t u + k\Delta u = 0, \tag{4}$$

where u is a function from $\mathbb{R} \times \mathbb{R}^3$ to $\mathbb{C}$. This equation describes the quantum evolution of a massive particle, $k = \hbar/2m$, where $\hbar > 0$ is Planck's constant and m is the mass of the particle. As with the heat equation, one can set k to equal 1 after a simple change of variables. Though the equation is formally very similar to the heat equation, it has very different qualitative behavior. This illustrates an important general point about PDEs: that small changes in the form of an equation can lead to very different properties of solutions.

A further example is the Klein–Gordon equation

$$-\partial_t^2 u + c^2 \Delta u - \left(\frac{mc^2}{\hbar}\right)^2 u = 0. \tag{5}$$

This is the relativistic counterpart to the Schrödinger equation: the parameter m has the physical interpretation of mass and mc^2 has the physical interpretation of rest energy (reflecting Einstein's famous equation $E = mc^2$). One can normalize the constants c and $mc^2/\hbar$ so that they both equal 1 by applying a suitable change of variables to time and space.

Though all five equations mentioned above first appeared in connection with specific physical phenomena, such as heat transfer for (2) and propagation of electromagnetic waves for (3), they have, miraculously, a range of relevance far beyond their original applications. In particular there is no reason to restrict their study to three space dimensions: it is very easy to generalize them to similar equations in n variables $x_1, x_2, \ldots, x_n$.

All the PDEs listed so far obey a simple but fundamental property called the *principle of superposition*: if u_1 and u_2 are two solutions to one of these equations, then any linear combination $a_1 u_1 + a_2 u_2$ of these solutions is also a solution. In other words, the space of all solutions is a VECTOR SPACE [I.3 §2.3]. Equations that obey this property are known as *homogeneous linear equations*. If the space of solutions is an affine space (that is, a translate of a vector space) rather than a vector space, we say that the PDE is an *inhomogeneous linear equation*; a good example is *Poisson's equation*:

$$\Delta u = f, \tag{6}$$

where $f : \mathbb{R}^3 \to \mathbb{R}$ is a function that is given to us and $u : \mathbb{R}^3 \to \mathbb{R}$ is the unknown function. Equations that are neither homogeneous linear nor inhomogeneous linear are known as *nonlinear*. The following equation, the MINIMAL SURFACE EQUATION [III.94 §3.1], is manifestly nonlinear:

$$\partial_1 \left(\frac{\partial_1 u}{(1 + |\partial_1 u|^2 + |\partial_2 u|^2)^{1/2}} \right)$$
$$+ \partial_2 \left(\frac{\partial_2 u}{(1 + |\partial_1 u|^2 + |\partial_y u|^2)^{1/2}} \right) = 0. \tag{7}$$

The graphs of solutions $u : \mathbb{R}^2 \to \mathbb{R}$ of this equation are area-minimizing surfaces (like soap films).

Equations (1), (2), (3), (4), (5) are not just linear: they are all examples of *constant-coefficient linear equations*. This means that they can be expressed in the form

$$\mathcal{P}[u] = 0, \tag{8}$$

where $\mathcal{P}$ is a differential operator that involves linear combinations, with constant real or complex coefficients, of mixed partial derivatives of u. (Such operators are called *constant-coefficient linear differential operators*.) For instance, in the case of the Laplace equation (1), $\mathcal{P}$ is simply the Laplacian Δ, while for the wave equation (3), $\mathcal{P}$ is the *d'Alembertian*

$$\mathcal{P} = \Box = -\partial_t^2 + \partial_1^2 + \partial_2^2 + \partial_3^2.$$

The characteristic feature of linear constant-coefficient operators is *translation invariance*. Roughly speaking, this means that if you translate a function u, then you translate $\mathcal{P}u$ in the same way. More precisely, if $v(x)$ is defined to be $u(x - a)$ (so the value of u at x becomes the value of v at $x + a$; note that x and a belong to $\mathbb{R}^3$ here), then $\mathcal{P}v(x)$ is equal to $\mathcal{P}u(x - a)$. As a consequence of this basic fact we infer that solutions to the homogeneous, linear, constant-coefficient equation (8) are still solutions when translated.

Since symmetries play such a fundamental role in PDEs we should stop for a moment to make a general definition. A symmetry of a PDE is any invertible operation $T : u \mapsto T(u)$ from functions to functions that preserves the space of solutions, in the sense that u solves the PDE if and only if $T(u)$ solves the same PDE. A PDE with this property is then said to be *invariant* under the symmetry T. The symmetry T is often a linear operation, though this does not have to be the case. The composition of two symmetries is again a symmetry, as is the inverse of a symmetry, and so it is natural to view a collection of symmetries as forming a GROUP [I.3 §2.1] (which is typically a finite- or infinite-dimensional LIE GROUP [III.48 §1]).

Because the translation group is intimately connected with THE FOURIER TRANSFORM [III.27] (indeed, the latter can be viewed as the representation theory of the former), this symmetry strongly suggests that Fourier analysis should be a useful tool to solve constant-coefficient PDEs, and this is indeed the case.

Our basic constant-coefficient linear operators, the Laplacian Δ and the d'Alembertian $\Box$, are formally similar in many respects. The Laplacian is fundamentally associated with the geometry of EUCLIDEAN SPACE [I.3 §6.2] $\mathbb{R}^3$ and the d'Alembertian is similarly associated with the geometry of MINKOWSKI SPACE [I.3 §6.8] $\mathbb{R}^{1+3}$. This means that the Laplacian commutes with all the rigid motions of the Euclidean space $\mathbb{R}^3$, while the d'Alembertian commutes with the corresponding class of *Poincaré transformations* of Minkowski spacetime. In the former case this simply means that invariance applies to all transformations of $\mathbb{R}^3$ that preserve the Euclidean distances between points. In the case of the wave equation, the Euclidean distance has to be replaced by the *spacetime distance* between points (which would be called *events* in the language of relativity): if $P = (t, x_1, x_2, x_3)$ and $Q(s, y_1, y_2, y_3)$, then the distance between them is given by the formula

$$d_M(P, Q)^2$$
$$= -(t - s)^2 + (x_1 - y_1)^2 + (x_2 - y_2)^2 + (x_3 - y_3)^2.$$

As a consequence of this basic fact we infer that all solutions to the wave equation (3) are invariant under translations and LORENTZ TRANSFORMATIONS [I.3 §6.8].

Our other evolution equations (2) and (4) are clearly invariant under rotations of the space variables $x = (x^1, x^2, x^3) \in \mathbb{R}^3$, when t is fixed. They are also *Galilean invariant*, which means, in the particular case of the Schrödinger equation (4), that whenever $u =$

$u(t, x)$ is a solution so is the function $u_v(t, x) = e^{i(x \cdot v)} e^{it|v|^2}(t, x - vt)$ for any vector $v \in \mathbb{R}^3$.

Poisson's equation (6), on the other hand, is an example of a *constant-coefficient inhomogeneous linear equation*, which means that it takes the form

$$\mathcal{P}[u] = f \qquad (9)$$

for some constant-coefficient linear differential operator $\mathcal{P}$ and known function f. To solve such an equation requires one to understand the invertibility or otherwise of the linear operator $\mathcal{P}$: if it is invertible then u will equal $\mathcal{P}^{-1}f$, and if it is not invertible then either there will be no solution or there will be infinitely many solutions. Inhomogeneous equations are closely related to their homogeneous counterpart; for instance, if u_1, u_2 both solve the inhomogeneous equation (9) with the same inhomogeneous term f, then their difference $u_1 - u_2$ solves the corresponding homogeneous equation (8).

Linear homogeneous PDEs satisfy the principle of superposition but they do not have to be translation invariant. For example, suppose that we modify the heat equation (2) so that the coefficient k is no longer constant but rather an arbitrary, positive, smooth function of (x_1, x_2, x_3). Such an equation models the flow of heat in a medium in which the rate of diffusion varies from point to point. The corresponding space of solutions is not translation invariant (which is not surprising as the medium in which the heat flows is not translation invariant). Equations like this are called *linear equations with variable coefficients*. It is more difficult to solve them and describe their qualitative features than it is for constant-coefficient equations. (See, for example, STOCHASTIC PROCESSES [IV.24 §5.2] for an approach to equations of type (2) with variable k.) Finally, nonlinear equations such as (7) can often still be written in the form (8), but the operator $\mathcal{P}$ is now a *nonlinear* differential operator. For instance, the relevant operator for (7) is given by the formula

$$\mathcal{P}[u] = \sum_{i=1}^{2} \partial_i \left(\frac{1}{(1 + |\partial u|^2)^{1/2}} \partial_i u \right),$$

where $|\partial u|^2 = (\partial_1 u)^2 + (\partial_2 u)^2$. Operators such as these are clearly not linear. However, because they are ultimately constructed from algebraic operations and partial derivatives, both of which are "local" operations, we observe the important fact that $\mathcal{P}$ is at least still a "local" operator. More precisely, if u_1 and u_2 are two functions that agree on some open set D, then the expressions $\mathcal{P}[u_1]$ and $\mathcal{P}[u_2]$ also agree on this set. In

particular, if $\mathcal{P}[0] = 0$ (as is the case in our example), then whenever u vanishes on a domain, $\mathcal{P}[u]$ will also vanish on that domain.

So far we have tacitly assumed that our equations take place in the whole of a space such as $\mathbb{R}^3$, $\mathbb{R}^+ \times \mathbb{R}^3$, or $\mathbb{R} \times \mathbb{R}^3$. In reality one is often restricted to a fixed domain of that space. Thus, for example, equation (1) is usually studied on a bounded open domain of $\mathbb{R}^3$ subject to a specified *boundary condition*. Here are some basic examples of boundary conditions.

Example. The *Dirichlet problem* for Laplace's equation on an open domain of $D \subset \mathbb{R}^3$ is the problem of finding a function u that behaves in a prescribed way on the boundary of D and obeys the Laplace equation inside.

More precisely, one specifies a continuous function $u_0 : \partial D \to \mathbb{R}$ and looks for a continuous function u, defined on the closure $\bar{D}$ of D, that is twice continuously differentiable inside D and solves the equations

$$\left. \begin{array}{ll} \Delta u(x) = 0 & \text{for all } x \in D, \\ u(x) = u_0(x) & \text{for all } x \in \partial D. \end{array} \right\} \qquad (10)$$

A basic result in PDEs asserts that if the domain D has a sufficiently smooth boundary, then there is exactly one solution to the problem (10) for any prescribed function u_0 on the boundary ∂D.

Example. The *Plateau problem* is the problem of finding the surface of minimal total area that bounds a given curve.

When the surface is the graph of a function u on some suitably smooth domain D, in other words a set of the form $\{(x, y, u(x, y)) : (x, y) \in D\}$, and the bounding curve is the graph of a function u_0 over the boundary ∂D of D, then this problem turns out to be equivalent to the Dirichlet problem (10), but with the linear equation (1) replaced by the nonlinear equation (7). For the above equations, it is also often natural to replace the Dirichlet boundary condition $u(x) = u_0(x)$ on the boundary ∂D with another boundary condition, such as the *Neumann boundary condition* $n(x) \cdot \nabla_x u(x) = u_1(x)$ on ∂D, where $n(x)$ is the outward normal (of unit length) to D at x. Generally speaking, Dirichlet boundary conditions correspond to "absorbing" or "fixed" barriers in physics, whereas Neumann boundary conditions correspond to "reflecting" or "free" barriers.

Natural boundary conditions can also be imposed for our evolution equations (2)–(4). The simplest one is to prescribe the values of u when $t = 0$. We can think of this more geometrically. We are prescribing the values of u at each spacetime point of form $(0, x, y, z)$, and the set of all such points is a hyperplane in $\mathbb{R}^{1+3}$: it is an example of an *initial time surface*.

Example. The *Cauchy problem* (or *initial value problem*, sometimes abbreviated to IVP) for the heat equation (2) asks for a solution $u : \mathbb{R}^+ \times \mathbb{R}^3 \to \mathbb{R}$ on the spacetime domain $\mathbb{R}^+ \times \mathbb{R}^3 = \{(t, x) : t > 0,\ x \in \mathbb{R}^3\}$, which equals a prescribed function $u_0 : \mathbb{R}^3 \to \mathbb{R}$ on the initial time surface $\{0\} \times \mathbb{R}^3 = \partial(\mathbb{R}^+ \times \mathbb{R}^3)$.

In other words, the Cauchy problem asks for a sufficiently smooth function u, defined on the closure of $\mathbb{R}^+ \times \mathbb{R}^3$ and taking values in $\mathbb{R}$, that satisfies the conditions

$$\left. \begin{aligned} -\partial_t u(t, x) + k\Delta u(t, x) = 0 \\ \text{for every } (t, x) \in \mathbb{R}^+ \times \mathbb{R}^3, \\ u(0, x) = u_0(x) \quad \text{for every } x \in \mathbb{R}^3. \end{aligned} \right\} \quad (11)$$

The function u_0 is often referred to as the *initial conditions*, or *initial data*, or just *data*, for the problem. Under suitable smoothness and decay conditions, one can show that this equation has exactly one solution u for each choice of data u_0. Interestingly, this assertion fails if one replaces the *future* domain $\mathbb{R}^+ \times \mathbb{R}^3 = \{(t, x) : t > 0,\ x \in \mathbb{R}^3\}$ by the *past* domain $\mathbb{R}^- \times \mathbb{R}^3 = \{(t, x) : t < 0,\ x \in \mathbb{R}^3\}$.

A similar formulation of the IVP holds for the Schrödinger equation (4), though in this case we can solve both to the past and to the future. However, in the case of the wave equation (3) we need to specify not just the initial *position* $u(0, x) = u_0(x)$ on the initial time surface $t = 0$, but also an initial *velocity* $\partial_t u(0, x) = u_1(x)$, since equation (3) (unlike (2) or (4)) cannot formally determine $\partial_t u$ in terms of u. One can construct unique smooth solutions (both to the future and to the past of the initial hyperplane $t = 0$) to the IVP for (3) for very general smooth initial conditions u_0, u_1.

Many other boundary-value problems are possible. For instance, when analyzing the evolution of a wave in a bounded domain D (such as a sound wave), it is natural to work with the spacetime domain $\mathbb{R} \times D$ and prescribe *both* Cauchy data (on the initial boundary $0 \times D$) *and* Dirichlet or Neumann data (on the spatial boundary $\mathbb{R} \times \partial D$). On the other hand, when the physical problem under consideration is the evolution of a wave outside a bounded obstacle (for example, an electromagnetic wave), one considers instead the evolution in $\mathbb{R} \times (\mathbb{R}^3 \setminus D)$ with a boundary condition on D.

The choice of boundary condition and initial conditions for a given PDE is very important. For equations of physical interest these arise naturally from the context in which they are derived. For example, in the case of a vibrating string, which is described by solutions of the one-dimensional wave equation $\partial_t^2 u - \partial_x^2 u = 0$ in the domain $(a, b) \times \mathbb{R}$, the initial conditions $u = u_0$ and $\partial_t u = u_1$ at $t = t_0$ amount to specifying the original position and velocity of the string. The boundary condition $u(a) = u(b) = 0$ is what tells us that the two ends of the string are fixed.

So far we have considered just *scalar* equations. These are equations where there is only one unknown function u, which takes values either in the real numbers $\mathbb{R}$ or in the complex numbers $\mathbb{C}$. However, many important PDEs involve either multiple unknown scalar functions or (equivalently) functions that take values in a multidimensional vector space such as $\mathbb{R}^m$. In such cases, we say that we have a *system* of PDEs. An important example of a system is that of the CAUCHY–RIEMANN EQUATIONS [I.3 §5.6]:

$$\partial_1 u_2 - \partial_2 u_1 = 0, \qquad \partial_1 u_1 + \partial_2 u_2 = 0, \qquad (12)$$

where $u_1, u_2 : \mathbb{R}^2 \to \mathbb{R}$ are real-valued functions on the plane. It was observed by CAUCHY [VI.29] that a complex function $w(x + iy) = u_1(x, y) + iu_2(x, y)$ is HOLOMORPHIC [I.3 §5.6] if and only if its real and imaginary parts u_1, u_2 satisfy the system (12). This system can still be represented in the form of a constant-coefficient linear PDE (8), but u is now a vector $\left(\begin{smallmatrix} u_1 \\ u_2 \end{smallmatrix} \right)$, and $\mathcal{P}$ is not a scalar differential operator, but rather a *matrix* of operators $\left(\begin{smallmatrix} -\partial_2 & \partial_1 \\ \partial_1 & \partial_2 \end{smallmatrix} \right)$.

The system (12) contains two equations and two unknowns. This is the standard situation for a *determined system*. Roughly speaking, a system is called *overdetermined* if it contains more equations than unknowns and *underdetermined* if it contains fewer equations than unknowns. Underdetermined equations typically have infinitely many solutions for any given set of prescribed data; conversely, overdetermined equations tend to have no solutions at all, unless some additional *compatibility conditions* are imposed on the prescribed data.

Observe also that the Cauchy–Riemann operator $\mathcal{P}$ has the following remarkable property:

$$\mathcal{P}^2[u] = \mathcal{P}[\mathcal{P}[u]] = \begin{pmatrix} \Delta u_1 \\ \Delta u_2 \end{pmatrix}.$$

Thus $\mathcal{P}$ can be viewed as a square root of the two-dimensional Laplacian Δ. One can define a similar type of square root for the Laplacian in higher dimensions and, more surprisingly, even for the d'Alembertian operator $\Box$ in $\mathbb{R}^{1+3}$. To achieve this we need to have four 4×4 complex matrices $\gamma^1, \gamma^2, \gamma^3, \gamma^4$ that satisfy the property

$$\gamma^\alpha \gamma^\beta + \gamma^\beta \gamma^\alpha = -2m^{\alpha\beta} I.$$

Here, I is the unit 4×4 matrix and $m^{\alpha\beta} = \frac{1}{2}$ when $\alpha = \beta = 1$, $-\frac{1}{2}$ when $\alpha = \beta \neq 1$, and 0 otherwise. Using the γ matrices we can introduce the *Dirac operator* as follows. If $u = (u_1, u_2, u_3, u_4)$ is a function in $\mathbb{R}^{1+3}$ with values in $\mathbb{C}^4$, then we set $\mathcal{D}u = i\gamma^\alpha \partial_\alpha u$. It is easy to check that, indeed, $\mathcal{D}^2 u = \Box u$. The equation

$$\mathcal{D}u = ku \tag{13}$$

is called the *Dirac equation* and it is associated with a free, massive, relativistic particle such as an electron.

One can extend the concept of a PDE further to cover unknowns that are not, strictly speaking, functions taking values in a vector space, but are instead sections of a VECTOR BUNDLE [IV.6 §5], or perhaps a map from one MANIFOLD [I.3 §6.9] to another; such generalized PDEs play an important role in geometry and modern physics. A fundamental example is given by the EINSTEIN FIELD EQUATIONS [IV.13]. In the simplest, "vacuum," case, they take the form

$$\text{Ric}(g) = 0, \tag{14}$$

where $\text{Ric}(g)$ is the RICCI CURVATURE [III.78] tensor of the spacetime manifold $M = (M, g)$. In this case the spacetime metric itself is the unknown to be solved for. One can often reduce such equations *locally* to more traditional PDE systems by selecting a suitable choice of coordinates, but the task of selecting a "good" choice of coordinates, and working out how different choices are compatible with each other, is a nontrivial and important one. Indeed, the task of selecting a good set of coordinates in order to solve a PDE can end up being a significant PDE problem in its own right.

PDEs are ubiquitous throughout mathematics and science. They provide the basic mathematical framework for some of the most important physical theories: elasticity, hydrodynamics, electromagnetism, general relativity, and nonrelativistic quantum mechanics, for example. The more modern relativistic quantum field theories lead, in principle, to equations in an infinite number of unknowns, which lie beyond the scope of PDEs. Yet, even in that case, the basic equations preserve the locality property of PDEs. Moreover, the starting point of a QUANTUM FIELD THEORY [IV.17 §2.1.4] is always a classical field theory, which is described by

systems of PDEs. This is the case, for example, in the standard model of weak and strong interactions, which is based on the so-called Yang–Mills–Higgs field theory. If we also include the ordinary differential equations of classical mechanics, which can be viewed as one-dimensional PDEs, we see that essentially all of physics is described by differential equations. As examples of PDEs underlying some of our most basic physical theories we refer to the articles that discuss THE EULER AND NAVIER–STOKES EQUATIONS [III.23], THE HEAT EQUATION [III.36], THE SCHRÖDINGER EQUATION [III.83], and THE EINSTEIN EQUATIONS [IV.13].

An important feature of the main PDEs is their apparent universality. Thus, for example, the wave equation, first introduced by D'ALEMBERT [VI.20] to describe the motion of a vibrating string, was later found to be connected with the propagation of sound and electromagnetic waves. The heat equation, first introduced by FOURIER [VI.25] to describe heat propagation, appears in many other situations in which dissipative effects play an important role. The same can be said about the Laplace equation, the Schrödinger equation, and many other basic equations.

It is even more surprising that equations that were originally introduced to describe specific physical phenomena have played a fundamental role in several areas of mathematics that are considered to be "pure," such as complex analysis, differential geometry, topology, and algebraic geometry. Complex analysis, for example, which studies the properties of holomorphic functions, can be regarded as the study of solutions to the Cauchy–Riemann equations (12) in a domain of $\mathbb{R}^2$. Hodge theory is based on studying the space of solutions to a class of linear systems of PDEs on manifolds that generalize the Cauchy–Riemann equations: it plays a fundamental role in topology and algebraic geometry. THE ATIYAH–SINGER INDEX THEOREM [V.2] is formulated in terms of a special class of linear PDEs on manifolds, related to the Euclidean version of the Dirac operator. Important geometric problems can be reduced to finding solutions to specific PDEs, typically nonlinear. We have already seen one example: the Plateau problem of finding surfaces of minimal total area that pass through a given curve. Another striking example is the UNIFORMIZATION THEOREM [V.34] in the theory of surfaces, which takes a compact Riemannian surface S (a two-dimensional surface with a RIEMANNIAN METRIC [I.3 §6.10]) and, by solving the PDE

$$\Delta_S u + e^{2u} = K \tag{15}$$

(which is a nonlinear variant of the Laplace equation (1)), *uniformizes* the metric so that it is "equally curved" at all points on the surface (or, more precisely, has constant SCALAR CURVATURE [III.78]) without changing the *conformal class* of the metric (i.e., without distorting any of the angles subtended by curves on the surface). This theorem is of fundamental importance to the theory of such surfaces: in particular, it allows one to give a topological classification of compact surfaces in terms of a single number $\chi(S)$, which is called the EULER CHARACTERISTIC [I.4 §2.2] of the surface S. The three-dimensional analogue of the uniformization theorem, the GEOMETRIZATION CONJECTURE [IV.7 §2.4] of Thurston, has recently been established by Perelman, who did so by solving yet another PDE; in this case, the equation is the RICCI FLOW [III.78] equation

$$\partial_t g = 2\operatorname{Ric}(g), \qquad (16)$$

which can be transformed into a nonlinear version of the heat equation (2) after a carefully chosen change of coordinates. The proof of the geometrization conjecture is a decisive step toward the total classification of all three-dimensional compact manifolds, in particular establishing the well-known POINCARÉ CONJECTURE [IV.7 §2.4]. To overcome the many technical details in establishing this conjecture, one needs to make a detailed qualitative analysis of the behavior of solutions to the Ricci flow equation, a task which requires just about all the advances made in geometric PDEs in the last hundred years.

Finally, we note that PDEs arise not only in physics and geometry but also in many fields of applied science. In engineering, for example, one often wants to *control* some feature of the solution u to a PDE by carefully selecting whatever components of the given data one can directly influence; consider, for instance, how a violinist controls the solution to the vibrating string equation (closely related to (3)) by modulating the force and motion of a bow on that string in order to produce a beautiful sound. The mathematical theory dealing with these types of issues is called *control theory*.

When dealing with complex physical systems, one cannot possibly have complete information about the state of the system at any given time. Instead, one often makes certain randomness assumptions about various factors that influence it. This leads to the very important class of equations called *stochastic differential equations* (SDEs), where one or more components of the equation involve a RANDOM VARIABLE [III.71 §4] of some sort. An example of this is in the BLACK–SCHOLES MODEL [VII.9 §2] in mathematical finance. A general discussion of SDEs can be found in STOCHASTIC PROCESSES [IV.24 §6].

The plan for the rest of this article is as follows. In section 2 I shall describe some of the basic notions and achievements of the general theory of PDEs. The main point I want to make here is that, in contrast with ordinary differential equations, for which a general theory is both possible and useful, partial differential equations do not lend themselves to a useful general theoretical treatment because of some important obstructions that I shall try to describe. One is thus forced to discuss special classes of equations such as *elliptic, parabolic, hyperbolic,* and *dispersive equations.* In section 3 I will try to argue that, despite the impossibility of developing a useful general theory that encompasses all, or most, of the important examples, there is nevertheless an impressive unifying body of concepts and methods for dealing with various basic equations, and this gives PDEs the feel of a well-defined area of mathematics. In section 4 I develop this further by trying to identify some common features in the derivation of the main equations that are dealt with in the subject. An additional source of unity for PDEs is the central role played by the issues of *regularity* and *breakdown* of solutions, which is discussed only briefly here. In the final section we shall discuss some of the main goals that can be identified as driving the subject.

2   General Equations

One might expect, after looking at other areas of mathematics such as algebraic geometry or topology, that there was a very general theory of PDEs that could be specialized to various specific cases. As I shall argue below, this point of view is seriously flawed and very much out of fashion. It does, however, have important merits, which I hope to illustrate in this section. I shall avoid giving formal definitions and focus instead on representative examples. The reader who wants more precise definitions can consult the online version of this article.

For simplicity we shall look mostly at *determined* systems of PDEs. The simplest distinction, which we have already made, is between scalar equations, such as (1)–(5), which consist of only one equation and one unknown, and systems of equations, such as (12) and (13). Another simple but important concept is that of the *order* of a PDE, which is defined to be the highest derivative that appears in the equation; this concept is analogous to that of the *degree* of a polynomial. For instance, the five basic equations (1)–(5) listed earlier are second order in space, although some (such as (2) or (4)) are only first order in time. Equations (12) and (13), as well as the Maxwell equations, are first order.[1]

We have seen that PDEs can be divided into linear and nonlinear equations, with the linear equations being divided further into constant-coefficient and variable-coefficient equations. One can also divide nonlinear PDEs into several further classes depending on the "strength" of the nonlinearity. At one end of the scale, a *semilinear* equation is one in which all the nonlinear components of the equation have strictly lower order than the linear components. For instance, equation (15) is semilinear, because the nonlinear component e^u is of zero order, i.e., it contains no derivatives, whereas the linear component $\Delta_S u$ is of second order. These equations are close enough to being linear that they can often be effectively viewed as perturbations of a linear equation. A more strongly nonlinear class of equations is that of *quasilinear equations*, in which the highest-order derivatives of u appear in the equation only in a linear manner but the coefficients attached to those derivatives may depend in some nonlinear manner on lower-order derivatives. For instance, the second-order equation (7) is quasilinear, because if one uses the product rule to expand the equation, then it takes the quasilinear form

$$F_{11}(\partial_1 u, \partial_2 u)\partial_1^2 u + F_{12}(\partial_1 u, \partial_2 u)\partial_1\partial_2 u$$
$$+ F_{22}(\partial_1 u, \partial_2 u)\partial_2^2 u = 0$$

for some explicit algebraic functions F_{11}, F_{12}, F_{22} of the lower-order derivatives of u. While quasilinear equations can still sometimes be analyzed by perturbative techniques, this is generally more difficult to accomplish than it is for an analogous semilinear equation. Finally, we have *fully nonlinear equations*, which exhibit no linearity properties whatsoever. A typical example is the *Monge–Ampère equation*

$$\det(\mathrm{D}^2 u) = F(x, u, \mathrm{D}u),$$

where $u : \mathbb{R}^n \to \mathbb{R}$ is the unknown function, $\mathrm{D}u$ is the GRADIENT [I.3 §5.3] of u, $\mathrm{D}^2 u = (\partial_i \partial_j u)_{1 \leqslant i,j \leqslant n}$ is the *Hessian matrix* of u, and $F : \mathbb{R}^n \times \mathbb{R} \times \mathbb{R}^n \to \mathbb{R}$ is a given function. This equation arises in many geometric contexts, ranging from manifold-embedding problems

to the complex geometry of CALABI–YAU MANIFOLDS [III.6]. Fully nonlinear equations are among the most difficult and least well-understood of all PDEs.

Remark. Most of the basic equations of physics, such as the Einstein equations, are quasilinear. However, fully nonlinear equations arise in the theory of characteristics of linear PDEs, which we discuss below, and also in geometry.

2.1 First-Order Scalar Equations

It turns out that first-order scalar PDEs in any number of dimensions can be reduced to systems of first-order ODEs. As a simple illustration of this important fact consider the following equation in two space dimensions:

$$a^1(x^1, x^2)\partial_1 u(x^1, x^2) + a^2(x^1, x^2)\partial_2 u(x^1, x^2)$$
$$= f(x^1, x^2), \quad (17)$$

where a^1, a^2, f are given real functions in the variables $x = (x^1, x^2) \in \mathbb{R}^2$. We associate with (17) the first-order 2×2 system

$$\left.\begin{array}{l} \dfrac{\mathrm{d}x^1}{\mathrm{d}s}(s) = a^1(x^1(s), x^2(s)), \\[2mm] \dfrac{\mathrm{d}x^2}{\mathrm{d}s} = a^2(x^1(s), x^2(s)). \end{array}\right\} \quad (18)$$

To simplify matters, let us assume that $f = 0$.

Suppose now that $x(s) = (x^1(s), x^2(s))$ is a solution of (18), and let us consider how $u(x^1(s), x^2(s))$ varies as s varies. By the chain rule we know that

$$\frac{\mathrm{d}}{\mathrm{d}s} u = \partial_1 u \frac{\mathrm{d}}{\mathrm{d}s}\frac{\mathrm{d}x^1}{\mathrm{d}s} + \partial_2 u \frac{\mathrm{d}x^2}{\mathrm{d}s},$$

and equations (17) and (18) imply that this equals zero (by our assumption that $f = 0$). In other words, any solution $u = u(x^1, x^2)$ of (17) with $f = 0$ is constant along any parametrized curve of the form $x(s) = (x^1(s), x^2(s))$ that satisfies (18).

Thus, in principle, if we know the solutions to (18), which are called *characteristic curves* for the equation (17), then we can find all solutions to (17). I say "in principle" because, in general, the nonlinear system (18) is not so easy to solve. Nevertheless, ODEs are simpler to deal with, and the fundamental theorem of ODEs, which we will discuss later in this section, allows us to solve (18) at least locally and for a small interval in s.

The fact that u is constant along characteristic curves allows us to obtain important qualitative information even when we cannot find explicit solutions. For example, suppose that the coefficients a^1, a^2 are smooth (or

---

1. There is a simple trick, well-known in ordinary differential equations, for converting higher-order equations into a lower-order (or even first-order) system of equations by increasing the number of unknowns. See the discussion in DYNAMICS [IV.14 §1.2].

real analytic) and that the initial data is smooth (or real analytic) everywhere on the set $\mathcal{H}$ where it is defined, except at some point x_0 where it is discontinuous. Then the solution u remains smooth (or real analytic) at all points except along the characteristic curve Γ that starts at x_0, or, in other words, along the solution to (18) that satisfies the initial condition $x(0) = x_0$. That is, the discontinuity at x_0 propagates precisely along Γ. We see here the simplest manifestation of an important principle, which we shall explain in more detail later: *singularities of solutions to PDEs propagate along characteristics* (or, more generally, hypersurfaces).

One can generalize equation (17) to allow the coefficients a_1, a_2, and f to depend not only on $x = (x^1, x^2)$ but also on u:

$$a^1(x, u(x))\partial_1 u(x) + a^2(x, u(x))\partial_2 u(x) = f(x, u(x)). \tag{19}$$

The associated *characteristic system* becomes

$$\left.\begin{array}{l} \dfrac{dx^1}{ds}(s) = a^1(x(s), u(s, x(s))), \\[2mm] \dfrac{dx^2}{ds}(s) = a^2(x(s), u(s, x(s))). \end{array}\right\} \tag{20}$$

As a special example of (19) consider the scalar equation in two space dimensions,

$$\partial_t u + u\partial_x u = 0, \quad u(0, x) = u_0(x), \tag{21}$$

which is called the *Burgers equation*. Here we have set $a^1(x, u(x)) = 1$ and $a^2(x, u(x)) = u(x)$. With this choice of a^1, a^2, we can take $x^1(s)$ to be s in (20). Then, renaming $x^2(s)$ as $x(s)$, we derive the *characteristic equation* in the form

$$\frac{dx}{ds}(s) = u(s, x(s)). \tag{22}$$

For any given solution u of (21) and any characteristic curve $(s, x(s))$ we have $(d/ds)u(s, x(s)) = 0$. Thus, in principle, knowing the solutions to (22) should allow us to determine the solutions to (21). However, this argument seems worryingly circular, since u itself appears in (22).

To see how this difficulty can be circumvented, consider the IVP for (21): that is, look for solutions that satisfy $u(0, x) = u_0(x)$. Consider an associated characteristic curve $x(s)$ such that, initially, $x(0) = x_0$. Then, since u is constant along the curve, we must have $u(s, x(s)) = u_0(x_0)$. Hence, going back to (22), we infer that $dx/ds = u_0(x_0)$ and thus $x(s) = x_0 + su_0(x_0)$. We thus deduce that

$$u(s, x_0 + su_0(x_0)) = u_0(x_0), \tag{23}$$

which implicitly gives us the form of the solution u. We see once more, from (23), that if the initial data is smooth (or real analytic) everywhere except at a point x_0 of the line $t = 0$, then the corresponding solution is also smooth (or real analytic) everywhere in a small neighborhood V of x_0, except along the characteristic curve that begins at x_0. The smallness of V is necessary here because new singularities can form at large scales. Indeed, u has to be constant along the lines $x + su_0(x)$, whose slopes depend on $u_0(x)$. At a point where these lines cross we would obtain different values of u, which is impossible unless u *becomes singular by this point*. This blow-up phenomenon occurs for any smooth, nonconstant initial data u_0.

Remark. There is an important difference between the linear equation (17) and the quasilinear equation (19). The characteristics of the first depend only on the coefficients $a^1(x)$, $a^2(x)$, while the characteristics of the second depend explicitly on a particular solution u of the equation. In both cases, singularities can only propagate along the characteristic curves of the equation. For nonlinear equations, however, new singularities can form at large distance scales, whatever the smoothness of the initial data.

The above procedure extends to fully nonlinear scalar equations in $\mathbb{R}^d$ such as the *Hamilton–Jacobi equation*

$$\partial_t u + H(x, Du) = 0, \quad u(0, x) = u_0(x), \tag{24}$$

where $u : \mathbb{R} \times \mathbb{R}^n \to \mathbb{R}$ is the unknown function, Du is the gradient of u, and the HAMILTONIAN [III.35] $H : \mathbb{R}^d \times \mathbb{R}^d \to \mathbb{R}$ and the initial data $u_0 : \mathbb{R}^d \to \mathbb{R}$ are given. For instance, the *eikonal equation* $\partial_t u = |Du|$ is a special instance of a Hamilton–Jacobi equation. We associate with (24) the ODE system

$$\left.\begin{array}{l} \dfrac{dx^i}{dt} = \dfrac{\partial}{\partial p_i}H(x(t), p(t)), \\[2mm] \dfrac{dp_i}{dt} = -\dfrac{\partial}{\partial x^i}H(x(t), p(t)), \end{array}\right\} \tag{25}$$

where i runs from 1 to d. The equations (25) are known as a *Hamiltonian system* of ODEs. The relationship between this system and the corresponding Hamilton–Jacobi equation is a little more involved than in the cases discussed above. Briefly, we can construct a solution u to (24) based only on the knowledge of the solutions $(x(t), p(t))$ to (25), which are called the *bicharacteristic curves* of the nonlinear PDE. Once again, singularities can only propagate along bicharacteristic curves (or hypersurfaces). As in the case of the Burgers equation, singularities will occur for more or less any smooth data. Thus, a classical, continuously differentiable solution can only be constructed locally in time. Both Hamilton–Jacobi equations and Hamiltonian systems play a fundamental role in classical mechanics as well as in the theory of the propagation of singularities in linear PDEs. The deep connection between Hamiltonian systems and first-order Hamilton–Jacobi equations played an important role in the introduction of the Schrödinger equation into quantum mechanics.

2.2   The Initial Value Problem for ODEs

Before we can continue with our general presentation of PDEs we need first to discuss, for the sake of comparison, the IVP for ODEs. Let us start with a first-order ODE

$$\partial_x u(x) = f(x, u(x)) \tag{26}$$

subject to the initial condition

$$u(x_0) = u_0. \tag{27}$$

Let us also assume for simplicity that (26) is a scalar equation and that f is a well-behaved function of x and u, such as $f(x, u) = u^3 - u + 1 + \sin x$. From the initial data u_0 we can determine $\partial_x u(x_0)$ by substituting x_0 into (26). If we now differentiate the equation (26) with respect to x and apply the chain rule, we derive the equation

$$\partial_x^2 u(x) = \partial_x f(x, u(x)) + \partial_u f(x, u(x)) \partial_x u(x),$$

which for the example just defined works out to be $\cos x + 3u^2(x)\partial_x u(x) - \partial_x u(x)$. Hence,

$$\partial_x^2 u(x_0) = \partial_x f(x_0, u_0) + \partial_u f(x_0, u_0)\partial_x u_0,$$

and since $\partial_x u(x_0)$ has already been determined we find that $\partial_x^2 u(x_0)$ can also be explicitly calculated from the initial data u_0. This calculation also involves the function f and its first partial derivatives. Taking higher derivatives of the equation (26) we can recursively determine $\partial_x^3 u(x_0)$, as well as all other higher derivatives of u at x_0. Therefore, one can in principle determine $u(x)$ with the help of the Taylor series

$$u(x) = \sum_{k \geqslant 0} \frac{1}{k!} \partial_x^k u(x_0)(x - x_0)^k$$
$$= u(x_0) + \partial_x u(x_0)(x - x_0)$$
$$\qquad + \frac{1}{2!}\partial_x^2(x_0)(x - x_0)^2 + \cdots.$$

We say "in principle" because there is no guarantee that the series converges. There is, however, a very important theorem, called the *Cauchy–Kovalevskaya theorem*, which asserts that if the function f is real analytic, as is certainly the case for our function $f(x, u) = u^3 - u + 1 + \sin x$, then there will be some neighborhood J of x_0 where the Taylor series converges to a real-analytic solution u of the equation. It is then easy to show that the solution thus obtained is the unique solution to (26) that satisfies the initial condition (27). To summarize: if f is a well-behaved function, then the initial value problem for ODEs has a solution, at least in some time interval, and that solution is unique.

The same result does not always hold if we consider a more general equation of the form

$$a(x, u(x))\partial_x u = f(x, u(x)), \quad u(x_0) = u_0. \tag{28}$$

Indeed, the recursive argument outlined above breaks down in the case of the scalar equation $(x - x_0)\partial_x u = f(x, u)$ for the simple reason that we cannot even determine $\partial_x u(x_0)$ from the initial condition $u(x_0) = u_0$. A similar problem occurs for the equation $(u - u_0)\partial_x u = f(x, u)$. An obvious condition that allows us to extend our previous recursive argument to (28) is to insist that $a(x_0, u_0) \neq 0$. Otherwise, we say that the IVP (28) is *characteristic*. If both a and f are also real analytic, the Cauchy–Kovalevskaya theorem applies again and we obtain a unique, real-analytic solution of (28) in a small neighborhood of x_0. In the case of an $N \times N$ system,

$$A(x, u(x))\partial_x u = F(x, u(x)), \quad u(x_0) = u_0,$$

$A = A(x, u)$ is an $N \times N$ matrix, and the *noncharacteristic condition* becomes

$$\det A(x_0, u_0) \neq 0. \tag{29}$$

It turns out, and this is extremely important in the development of the theory of ODEs, that, while the nondegeneracy condition (29) is essential to obtain a unique solution of the equation, the analyticity condition is not at all important: it can be replaced by a simple *local Lipschitz condition* for A and F. It suffices to assume, for example, that their first partial derivatives exist and that they are locally bounded. This is always the case if the first derivatives of A and F are continuous.

Theorem (the fundamental theorem of ODEs). *If the matrix $A(x_0, u_0)$ is invertible and if A and F are continuous and have locally bounded first derivatives, then there is some time interval $J \subset \mathbb{R}$ that contains x_0, and a* unique solution[2] u defined on J that satisfies the initial conditions $u(x_0) = u_0$.

The proof of the theorem is based on the *Picard iteration method*. The idea is to construct a sequence of approximate solutions $u_{(n)}(x)$ that converge to the desired solution. Without loss of generality we can assume A to be the identity matrix.[3] One starts by setting $u_{(0)}(x) = u_0$ and then defines, recursively,

$$\partial_x u_{(n)}(x) - \Gamma(x, u_{(n-1)}(x)), \quad u_{(n-1)}(x_0) = u_0.$$

Observe that at every stage all we need to solve is a very simple linear problem, which makes Picard iteration easy to implement numerically. As we shall see below, variations of this method are also used for solving nonlinear PDEs.

Remark. In general, the local existence theorem is sharp, in the sense that its conditions cannot be relaxed. We have seen that the invertibility condition for $A(x_0, u_0)$ is necessary. Also, it is not always possible to extend the interval J in which the solution exists to the whole of the real line. As an example, consider the nonlinear equation $\partial_x u = u^2$ with initial data $u = u_0$ at $x = 0$, for which the solution $u = u_0/(1 - xu_0)$ becomes infinite in finite time: in the terminology of PDEs, it *blows up*.

In view of the fundamental theorem and the example mentioned above, one can define the main goals of the mathematical theory of ODEs as follows.

(i) Find criteria for global existence. In the case of blow-up describe the limiting behavior.

(ii) In the case of global existence describe the asymptotic behavior of solutions and families of solutions.

Though it is impossible to develop a general theory that achieves both goals (in practice one is forced to restrict oneself to special classes of equations motivated by applications), the general local existence and uniqueness theorem mentioned above provides a powerful unifying theme. It would be very helpful if a similar situation were to hold for general PDEs.

2.3   The Initial Value Problem for PDEs

In the one-dimensional situation one specifies initial conditions at a point. The natural higher-dimensional analogue is to specify them on hypersurfaces $\mathcal{H} \subset \mathbb{R}^d$, that is, $(d-1)$-dimensional subsets (or, to be more precise, submanifolds). For a general equation of order k, that is, one that involves k derivatives, we need to specify the values of u and of its first $k-1$ derivatives in the direction normal to $\mathcal{H}$. For example, in the case of the second-order wave equation (3) and the initial hyperplane $t = 0$ we need to specify initial data for u and $\partial_t u$.

If we wish to use initial data of this kind to start obtaining a solution, it is important that the data should not be degenerate. (We have already seen this in the case of ODEs.) For this reason, we make the following general definition.

Definition. Suppose that we have a kth-order quasilinear system of equations, and the initial data comes in the form of the first $k-1$ normal derivatives that a solution u must satisfy on a hypersurface $\mathcal{H}$. We say that the system is *noncharacteristic* at a point x_0 of $\mathcal{H}$ if we can use the initial data to determine formally all the other higher partial derivatives of u at x_0, in terms of the data.

As a very rough picture to have in mind, it may be helpful to imagine an infinitesimally small neighborhood of x_0. If the hypersurface $\mathcal{H}$ is smooth, then its intersection with this neighborhood will be a piece of a $(d-1)$-dimensional affine subspace. The values of u and the first $k-1$ normal derivatives on this intersection are given by the initial data, and the problem of determining the other partial derivatives is a problem in linear algebra (because everything is infinitesimally small). To say that the system is noncharacteristic at x_0 is to say that this linear algebra problem can be uniquely solved, which is the case provided that a certain matrix is invertible. This is the nondegeneracy condition referred to earlier.

To illustrate the idea, let us look at first-order equations in two space dimensions. In this case $\mathcal{H}$ is a curve Γ, and since $k - 1 = 0$ we must specify the restriction of u to $\Gamma \subset \mathbb{R}^2$ but we do not have to worry about any derivatives. Thus, we are trying to solve the system

$$a^1(x, u(x))\partial_1 u(x) + a^2(x, u(x))\partial_2 u(x)$$
$$= f(x, u(x)), \quad u|_\Gamma = u_0, \quad (30)$$

where a^1, a^2, and f are real-valued functions of x (which belongs to $\mathbb{R}^2$) and u. Assume that in a small neighborhood of a point p the curve Γ is described parametrically as the set of points $x = (x^1(s), x^2(s))$. We denote by $n(s) = (n_1(s), n_2(s))$ a unit normal to Γ.

---

2. Since we are not assuming that A and F are analytic, the solution may not be analytic, but it does have continuous first derivatives.

3. Since A is invertible we can multiply both sides of the equation by the inverse matrix A^{-1}.

As in the case of ODEs, which we looked at earlier, we would like to find conditions on Γ such that for a given point in Γ we can determine all derivatives of u from the data u_0, the derivatives of u along Γ, and the equation (30). Out of all possible curves Γ we distinguish in particular the *characteristic* ones we have already encountered above (see (20)):

$$\left.\begin{array}{l} \dfrac{\mathrm{d}x^1}{\mathrm{d}s} = a^1(x(s), u(x(s))), \\[2mm] \dfrac{\mathrm{d}x^2}{\mathrm{d}s} = a^2(x(s), u(x(s))), \end{array}\right\} \quad x(0) = p.$$

One can prove the following fact:

Along a characteristic curve, the equation (30) is degenerate. That is, we cannot determine the first-order derivatives of u uniquely in terms of the data u_0.

In terms of the rough picture above, at each point there is a direction such that if the hypersurface, which in this case is a line, is along that direction, then the resulting matrix is singular. If you follow this direction, then you travel along a characteristic curve.

Conversely, if the nondegeneracy condition

$$a^1(p, u(p))n_1(p) + a_2(p, u(p))n_2(p) \neq 0 \quad (31)$$

is satisfied at some point $p = x(0) \in \Gamma$, then we can determine all higher derivatives of u at x_0 uniquely in terms of the data u_0 and its derivatives along Γ. If the curve Γ is given by the equation $\psi(x^1, x^2) = 0$, with nonvanishing gradient $D\psi(p) \neq 0$, then the condition (31) takes the form

$$a^1(p, u(p))\partial_1\psi(p) + a^2(p, u(p))\partial_2\psi(p) \neq 0.$$

With a little more work one can extend the above discussion to higher-order equations in higher dimensions, and even to systems of equations. Particularly important is the case of a second-order scalar equation in $\mathbb{R}^d$,

$$\sum_{i,j=1}^{d} a^{ij}(x)\partial_i\partial_j u = f(x, u(x)), \quad (32)$$

together with a hypersurface $\mathcal{H}$ in $\mathbb{R}^d$ defined by the equation $\psi(x) = 0$, where ψ is a function with non-vanishing gradient $D\psi$. Define the unit normal at a point $x_0 \in \mathcal{H}$ to be $n = D\psi/|D\psi|$, or, in component form, $n_i = \partial_i\psi/|\partial\psi|$. As initial conditions for (32) we prescribe the values of u and its normal derivative $n[u](x) = n_1(x)\partial_1 u(x) + n_2(x)\partial_2 u(x) + \cdots + n_d(x)\partial_d u(x)$ on $\mathcal{H}$:

$$u(x) = u_0(x), \quad n[u](x) = u_1(x), \quad x \in \mathcal{H}.$$

It can be shown that $\mathcal{H}$ is noncharacteristic (with respect to equation (32)) at a point p (that is, we can determine all derivatives of u at p in terms of the initial data u_0, u_1) if and only if

$$\sum_{i,j=1}^{d} a^{ij}(p)\partial_i\psi(p)\partial_j\psi(p) \neq 0. \quad (33)$$

On the other hand, $\mathcal{H}$ is a characteristic hypersurface for (32) if

$$\sum_{i,j=1}^{d} a^{ij}(x)\partial_i\psi(x)\partial_j\psi(x) = 0 \quad (34)$$

for every x in $\mathcal{H}$.

Example. If the coefficients a of (32) satisfy the condition

$$\sum_{i,j=1}^{d} a^{ij}(x)\xi_i\xi_j > 0, \quad \forall \xi \in \mathbb{R}^d, \ \forall x \in \mathbb{R}^d, \quad (35)$$

then clearly, by (34), no surface in $\mathbb{R}^d$ can be characteristic. This is the case, in particular, for the Laplace equation $\Delta u = f$. Consider also the minimal surface equation (7) written in the form

$$\sum_{i,j=1,2} h^{ij}(\partial u)\partial_i\partial_j u = 0, \quad (36)$$

with $h^{11}(\partial u) = 1 + (\partial_2 u)^2$, $h^{22}(\partial u) = 1 + (\partial_1 u)^2$, $h^{12}(\partial u) = h^{21}(\partial u) = -\partial_1 u\partial_2 u$. It is easy to check that the quadratic form associated with the symmetric matrix $h^{ij}(\partial u)$ is positive definite for every ∂u. Indeed,

$$h^{ij}(\partial u)\xi_i\xi_j$$
$$= (1 + |\partial u|^2)^{-1/2}(|\xi|^2 - (1 + |\partial u|^2)^{-1}(\xi \cdot \partial u)^2) > 0.$$

Thus, even though (36) is not linear, we see that all surfaces in $\mathbb{R}^2$ are noncharacteristic.

Example. Consider the wave equation $\Box u = f$ in $\mathbb{R}^{1+d}$. All hypersurfaces of the form $\psi(t, x) = 0$ for which

$$(\partial_t\psi)^2 = \sum_{i=1}^{d} (\partial_i\psi)^2 \quad (37)$$

are characteristic. This is the famous eikonal equation, which plays a fundamental role in the study of wave propagation. Observe that it splits into two Hamilton–Jacobi equations (see (24)):

$$\partial_t\psi = \pm\left(\sum_{i=1}^{d} (\partial_i\psi)^2\right)^{1/2}. \quad (38)$$

The bicharacteristic curves of the associated Hamiltonians are called bicharacteristic curves of the wave equation. As particular solutions of (37) we find $\psi_+(t, x) =$

$(t - t_0) + |x - x_0|$ and $\psi_-(t, x) = (t - t_0) - |x - x_0|$, whose level surfaces $\psi_\pm = 0$ correspond to forward and backward light cones with their vertex at $p = (t_0, x_0)$. These represent, physically, the union of *all light rays emanating from a point source at p*. The light rays are given by the equation $(t - t_0)\omega = (x - x_0)$, for $\omega \in \mathbb{R}^3$ with $|\omega| = 1$, and are precisely the (t, x) components of the bicharacteristic curves of the Hamilton–Jacobi equations (38). More generally, the characteristics of the linear wave equation

$$a^{00}(t, x)\partial_t^2 u - \sum_{i,j} a^{ij}(t, x)\partial_i \partial_j u = 0, \qquad (39)$$

with $a^{00} > 0$ and a^{ij} satisfying (35), are given by the Hamilton–Jacobi equations:

$$-a^{00}(t, x)(\partial_t \psi)^2 + a^{ij}(x)\partial_i \psi \partial_j \psi = 0$$

or, equivalently,

$$\partial_t \psi = \pm \left((a^{00})^{-1} \sum_{i,j} a^{ij}(x)\partial_i \psi \partial_j \psi \right)^{1/2}. \qquad (40)$$

The bicharacteristics of the corresponding Hamiltonian systems are called bicharacteristic curves of (39).

Remark. In the case of the first-order scalar equations (17) we have seen how knowledge of characteristics can be used to find, implicitly, general solutions. We have also seen that singularities propagate only along characteristics. In the case of second-order equations the characteristics are not sufficient to solve the equations, but they continue to provide important information, such as how the singularities propagate. For example, in the case of the wave equation $\Box u = 0$ with smooth initial data u_0, u_1 everywhere except at a point $p = (t_0, x_0)$, the solution u has singularities present at all points of the light cone $-(t - t_0)^2 + |x - x_0|^2 = 0$ with vertex at p. A more refined version of this fact shows that the singularities propagate along bicharacteristics. The general principle here is that *singularities propagate along characteristic hypersurfaces of a PDE*. Since this is a very important principle, it pays to give it a more precise formulation that extends to general boundary conditions, such as the Dirichlet condition for (1).

Propagation of singularities. *If the boundary conditions or the coefficients of a PDE are singular at some point p, and otherwise smooth (or real analytic) everywhere in some small neighborhood V of p, then a solution of the equation cannot be singular in V except along a characteristic hypersurface passing through p.*

In particular, if there are no such characteristic hypersurfaces, then any solution of the equation must be smooth (or real analytic) at every point of V other than p.

Remarks. (i) The heuristic principle mentioned above is invalid, in general, at large scales. Indeed, as we have shown in the case of the Burgers equation, solutions to nonlinear evolution equations can develop new singularities whatever the smoothness of the initial conditions. Global versions of the principle can be formulated for linear equations based on the bicharacteristics of the equation. See (iii) below.

(ii) According to the principle, it follows that any solution of the equation $\Delta u = f$, satisfying the boundary condition $u|_{\partial D} = u_0$ with a boundary value u_0 that merely has to be continuous, is automatically smooth everywhere in the interior of D provided that f itself is smooth there. Moreover, the solution is real analytic if f is real analytic.

(iii) More precise versions of this principle, which plays a fundamental role in the general theory, can be given for linear equations. In the case of the general wave equation (39), for example, one can show that singularities propagate along bicharacteristics. These are the bicharacteristic curves associated with the Hamilton–Jacobi equation (40).

2.4   The Cauchy–Kovalevskaya Theorem

In the case of ODEs we have seen that a noncharacteristic IVP always admits solutions locally (that is, in some time interval about a given point). Is there a higher-dimensional analogue of this fact? The answer is yes, provided that we restrict ourselves to the real-analytic situation, which is covered by an appropriate extension of the Cauchy–Kovalevskaya theorem. More precisely, one can consider general quasilinear equations, or systems, with real-analytic coefficients, real-analytic hypersurfaces $\mathcal{H}$, and appropriate real-analytic initial data on $\mathcal{H}$.

Theorem (Cauchy–Kovalevskaya (CK)). *If all the real-analyticity conditions made above are satisfied and if the initial hypersurface $\mathcal{H}$ is noncharacteristic at x_0,[4] then in some neighborhood of x_0 there is a unique real-analytic solution $u(x)$ that satisfies the system of equations and the corresponding initial conditions.*

---

4. For second-order equations of the kind of (32), this is precisely condition (33).

In the special case of linear equations, an important companion theorem, due to Holmgren, asserts that the analytic solution given by the CK theorem is unique in the class of all smooth solutions and smooth noncharacteristic hypersurfaces $\mathcal{H}$. The CK theorem shows that, given the noncharacteristic condition and the analyticity assumptions, the following straightforward way of finding solutions works: look for a formal expansion of the kind $u(x) = \sum_\alpha C_\alpha (x - x_0)^\alpha$ by determining the constants C_α recursively from simple algebraic formulas arising from the equation and initial conditions on $\mathcal{H}$. More precisely, the theorem ensures that the naive expansion obtained in this way converges in a small neighborhood of $x_0 \in \mathcal{H}$.

It turns out, however, that the analyticity conditions required by the CK theorem are much too restrictive, and therefore the apparent generality of the result is misleading. A first limitation becomes immediately apparent when we consider the wave equation $\Box u = 0$. A fundamental feature of this equation is *finite speed of propagation*, which means, roughly speaking, that if at some time t a solution u is zero outside some bounded set, then the same must be true at all later times. However, analytic functions cannot have this property unless they are identically zero (see SOME FUNDAMENTAL MATHEMATICAL DEFINITIONS [I.3 §5.6]). Therefore, it is impossible to discuss the wave equation properly within the class of real-analytic solutions. A related problem, first pointed out by HADAMARD [VI.65], concerns the impossibility of solving the Cauchy problem, in many important cases, for arbitrary smooth nonanalytic data. Consider, for example, the Laplace equation $\Delta u = 0$ in $\mathbb{R}^d$. As we have established above, any hypersurface $\mathcal{H}$ is noncharacteristic, yet the Cauchy problem $u|_{\mathcal{H}} = u_0$, $n[u]|_{\mathcal{H}} = u_1$, for arbitrary smooth initial conditions u_0, u_1, may admit no local solutions in a neighborhood of any point of $\mathcal{H}$. Indeed, take $\mathcal{H}$ to be the hyperplane $x_1 = 0$ and assume that the Cauchy problem can be solved for given nonanalytic smooth data in a domain that includes a closed ball B centered at the origin. The corresponding solution can also be interpreted as the solution to the Dirichlet problem in B, with the values of u prescribed on the boundary ∂B. But this, according to our heuristic principle (which can easily be made rigorous in this case), must be real analytic everywhere in the interior of B, contradicting our assumptions about the initial data.

On the other hand, the Cauchy problem for the wave equation $\Box u = 0$ in $\mathbb{R}^{d+1}$ has a unique solution for any smooth initial data u_0, u_1 that is prescribed on

a *spacelike hypersurface*. This means a hypersurface $\psi(t, x) = 0$ such that at every point $p = (t_0, x_0)$ that belongs to it the normal vector at p lies inside the light cone (either in the future direction or in the past direction). To say this analytically,

$$|\partial_t \psi(p)| > \left(\sum_{i=1}^d |\partial_i \psi(p)|^2 \right)^{1/2}. \tag{41}$$

This condition is clearly satisfied by a hyperplane of the form $t = t_0$, but any other hypersurface close to this is also spacelike. By contrast, the IVP is *ill-posed* for a timelike hypersurface, i.e., a hypersurface for which

$$|\partial_t \psi(p)| < \left(\sum_{i=1}^d |\partial_i \psi(p)|^2 \right)^{1/2}.$$

That is, we cannot, for general non-real-analytic initial conditions, find a solution of the IVP. An example of a timelike hypersurface is given by the hyperplane $x^1 = 0$. Let us explain the term "ill-posed" more precisely.

Definition. A given problem for a PDE is said to be well-posed if both existence and uniqueness of solutions can be established for arbitrary data that belongs to a specified large space of functions, which includes the class of smooth functions.[5] Moreover, the solutions must depend continuously on the data. A problem that is not well-posed is called ill-posed.

The continuous dependence on the data is very important. Indeed, the IVP would be of little use if very small changes in the initial conditions resulted in very large changes in the corresponding solutions.

2.5 Standard Classification

The different behavior of the Laplace and wave equations mentioned above illustrates the fundamental difference between ODEs and PDEs and the illusory generality of the CK theorem. Given that these two equations are so important in geometric and physical applications, it is of great interest to find the broadest classes of equations with which they share their main properties. The equations modeled by the Laplace equation are called *elliptic*, while those modeled by the wave equation are called *hyperbolic*. The other two important models are the heat equation (see (2)) and the Schrödinger equation (see (4)). The general classes of equations that they resemble are called *parabolic* and *dispersive*, respectively.

---

5. Here we are necessarily vague. A precise space can be specified in each given case.

Elliptic equations are the most robust and the easiest to characterize: they are the ones that admit no characteristic hypersurfaces.

Definition. A linear, or quasilinear, $N \times N$ system with no characteristic hypersurfaces is called elliptic.

Equations of type (32) whose coefficients a^{ij} satisfy condition (35) are clearly elliptic. The minimal surface equation (7) is also elliptic. It is also easy to verify that the Cauchy–Riemann system (12) is elliptic. As was pointed out by Hadamard, the IVP is not well-posed for elliptic equations. The natural way of parametrizing the set of solutions to an elliptic PDE is to prescribe conditions for u, and some of its derivatives (the number of derivatives will be roughly half the order of the equation) at the boundary of a domain $D \subset \mathbb{R}^n$. These are called *boundary-value problems* (BVPs). A typical example is the Dirichlet boundary condition $u|_{\partial D} = u_0$ for the Laplace equation $\Delta u = 0$ in a domain $D \subset \mathbb{R}^n$. One can show that, if the domain D satisfies certain mild regularity assumptions and the boundary value u_0 is continuous, then this problem admits a unique solution that depends continuously on u_0. We say that the Dirichlet problem for the Laplace equation is well-posed. Another well-posed problem for the Laplace equation is given by the Neumann boundary condition $n[u]|_{\partial D} = f$, where n is the exterior unit normal to the boundary. This problem is well-posed for all continuous functions f defined on ∂D with zero mean average. A typical problem of general theory is to classify all well-posed BVPs for a given elliptic system.

As a consequence of our propagation-of-singularities principle, we deduce, heuristically at least, the following general fact:

Classical solutions of elliptic equations with smooth (or real-analytic) coefficients in a regular domain D are smooth (or real analytic) in the interior of D, whatever the degree of smoothness of the boundary conditions.[6]

Hyperbolic equations are, essentially, those for which the IVP is well-posed. In that sense, they provide the natural class of equations for which one can prove a result similar to the local existence theorem for ODEs. More precisely, for each sufficiently regular set of initial conditions there is a unique solution. We can

thus think of the Cauchy problem as a natural way of parametrizing the set of all solutions to the equations.

The definition of hyperbolicity depends, however, on the particular hypersurface we are considering as the initial hypersurface. Thus, in the case of the wave equation $\Box u = 0$, the standard IVP

$$u(0, x) = u_0(x), \quad \partial_t u(0, x) = u_1$$

is well-posed. This means that for any smooth initial data u_0, u_1 we can find a unique solution of the equation, which depends continuously on u_0, u_1. As we have already mentioned, the IVP for $\Box u = 0$ remains well-posed if we replace the initial hypersurface $t = 0$ by any spacelike hypersurface $\psi(t, x) = 0$ (see (41)). However, it fails to be well-posed for timelike hypersurfaces, for which there may be no solution with prescribed, nonanalytic, Cauchy data.

It is more difficult to give algebraic conditions for hyperbolicity. Roughly speaking, hyperbolic equations are at the opposite end of the spectrum from elliptic equations: whereas elliptic equations have no characteristic hypersurfaces, hyperbolic equations have as many as possible passing through any given point. One of the most useful classes of hyperbolic equations, which includes most of the important known examples, consists of equations of the form

$$A^0(t, x, u)\partial_t u + \sum_{i=1}^d A_i(t, x, u)\partial_i u = F(t, x, u),$$
$$u|_{\mathcal{H}} = u_0, \quad (42)$$

where all the coefficients $A^0, A^1, \dots, A^d$ are symmetric $N \times N$ matrices and $\mathcal{H}$ is given by $\psi(t, x) = 0$. Such a system is well-posed provided that the matrix

$$A^0(t, x, u)\partial_t \psi(t, x) + \sum_{i=1}^d A_i(t, x, u)\partial_i \psi(t, x) \quad (43)$$

is positive definite. A system (42) that satisfies these conditions is called *symmetric hyperbolic*. In the particular case when $\psi(t, x) = t$, the condition (43) becomes

$$(A^0 \xi, \xi) \geqslant c|\xi|^2 \quad \forall \xi \in \mathbb{R}^N.$$

The following is a fundamental result in the theory of general hyperbolic equations. It is called the local existence and uniqueness of solutions for symmetric hyperbolic systems.

Theorem (fundamental theorem for hyperbolic equations). *The IVP (42) is locally well-posed for symmetric hyperbolic systems with sufficiently smooth A, F, and $\mathcal{H}$ and sufficiently smooth initial conditions u_0. In*

---

6. Provided that the boundary condition under consideration is well-posed. Moreover, this heuristic principle holds, in general, only for classical solutions of a nonlinear equation. There are in fact examples of well-posed BVPs, for certain nonlinear elliptic systems, with no classical solutions.

other words, if the appropriate smoothness conditions are satisfied, then for any point $p \in \mathcal{H}$ there is a small neighborhood $\mathcal{D}$ of p[7] inside which there is a unique, continuously differentiable solution u.

Remarks. (i) The local character of the theorem is essential, just as it was for the general propagation-of-singularities principle discussed earlier, since the result cannot be globalized in the particular case of the Burgers equation (21), which fits trivially into the framework of general nonlinear symmetric hyperbolic systems. A precise version of the theorem above gives a lower bound on how large $\mathcal{D}$ can be.

(ii) The proof of the theorem is based on a variation of the Picard iteration method that we encountered earlier for ODEs. One starts by taking $u_{(0)} = u_0$ in a neighborhood of $\mathcal{H}$. Then one defines functions $u_{(n)}$ recursively as follows:

$$A^0(t, x, u_{(n-1)})\partial_t u_{(n)} + \sum_{i=1}^{d} A_i(t, x, u_{(n-1)})\partial_i u_{(n)}$$
$$= F(t, x, u_{(n-1)}), \quad u_{(n)}|\mathcal{H} = u_0.$$

Notice that at each stage of the iteration we have to solve a linear equation. Linearization is an extremely important tool in studying nonlinear PDEs. We can almost never understand their behavior without linearizing them around important special solutions. Thus, almost invariably, hard problems in nonlinear PDEs reduce to understanding specific problems in linear PDEs.

(iii) To implement the Picard iteration method we need to get precise estimates concerning $u_{(n)}$ in terms of $u_{(n-1)}$. This step requires *energy type a priori estimates*, which we will discuss in section 3.3.

Another important property of hyperbolic equations (which is not shared by elliptic, parabolic, or dispersive equations) is *finite speed of propagation*, which was mentioned earlier in the case of the wave equation (3). Consider this simple case again. The IVP can be solved explicitly by the so-called *Kirchhoff formula*. The formula allows us to conclude that if the initial data at $t = 0$ is zero outside a ball $B_a(x_0)$ of radius $a > 0$ centered at $x_0 \in \mathbb{R}^3$, then at time $t > 0$ the solution u is zero outside the ball $B_{a+ct}(x_0)$. In general, finite speed of propagation can best be formulated in terms of domains of dependence and influence of

hyperbolic equations (see the online version for general definitions).

Hyperbolic PDEs play a fundamental role in physics, as they are intimately tied to the relativistic nature of the modern theory of fields. Equations (3), (5), (13) are the simplest examples of *linear field theories*, and they are manifestly hyperbolic. Other basic examples appear in *gauge field theories* such as MAXWELL'S EQUATIONS [IV.13 §1.1] $\partial^\alpha F_{\alpha\beta} = 0$ or the *Yang–Mills equations* $D^\alpha F_{\alpha\beta} = 0$. Finally, the Einstein equations (14) are also hyperbolic.[8] Other important examples of hyperbolic equations arise in the physics of elasticity and inviscid fluids. As examples of the latter, the Burgers equation (21) and the compressible Euler equation are hyperbolic.

Elliptic equations, on the other hand, appear naturally in describing time-independent, or more generally *steady-state*, solutions of hyperbolic equations. Elliptic equations can also be derived, directly, by well-defined VARIATIONAL PRINCIPLES [III.94].

Finally, a few words about parabolic equations and Schrödinger-type equations, which are intermediate between the elliptic and hyperbolic ones. Large classes of useful equations of these types are given by

$$\partial_t u - Lu = f \tag{44}$$

and

$$i\partial_t u + Lu = f, \tag{45}$$

respectively, where L is an elliptic second-order operator. One looks for solutions $u = u(t, x)$, defined for $t \geqslant t_0$, with the prescribed initial condition

$$u(t_0, x) = u_0(x) \tag{46}$$

on the hypersurface $t = t_0$. Strictly speaking, this hypersurface is characteristic, since the order of the equation is 2 and we cannot determine $\partial_t^2 u$ at $t = t_0$ directly from the equation. Yet this is not a serious problem; we can still determine $\partial_t^2 u$ formally by differentiating the equation with respect to ∂_t. Thus, the IVP (44) (or (45)) with initial condition (46) is well-posed, but not quite in the same sense as for hyperbolic equations. For example, the heat equation $-\partial_t u + \Delta u$ is well-posed for positive t but ill-posed for negative t. The heat equation may also not have unique solutions for the IVP unless we make assumptions about how fast the initial data is allowed to grow at infinity. One can also show

---

7. By "point" we mean that p is a spacetime point $(t, x) \in \mathbb{R}^{1+d}$. Similarly, $\mathcal{D}$ is a set of spacetime points.

8. For gauge theories and Einstein equations the notion of hyperbolicity depends on the choice of gauge or coordinates. In the case of the Yang–Mills equations, for example, one obtains a well-defined system of nonlinear wave equations only in the Lorentz gauge.

that the characteristic hypersurfaces of the equation (44) are all of the form, and therefore parabolic equations are quite similar to elliptic equations. For example, one can show that if the coefficients a^{ij} and f are smooth (or real analytic), then the solution u must be smooth (or real analytic in x) for $t > t_0$ even if the initial data u_0 is not smooth, which is consistent with our propagation-of-singularities principle. The heat equation smooths out initial conditions. It is for this reason that the heat equation is useful in many applications. In physics, parabolic PDEs arise whenever diffusion or dissipation phenomena are important, while in geometry and calculus of variations, parabolic PDEs often arise as gradient flows of positive-definite functionals. Ricci flow (16) can also be viewed as a parabolic PDE, after a suitable change of coordinates.

Dispersive PDEs, of which the Schrödinger equation (4) is a fundamental example, are evolution equations that behave analogously to hyperbolic PDEs in many respects. For instance, the IVP tends to be locally well-posed both forward and backward in time. However, solutions to dispersive PDEs do not propagate along characteristic surfaces. Instead, they move at speeds that are determined by their spatial frequency; in general, high-frequency waves tend to propagate at much greater speeds than low-frequency waves, which eventually leads to a *dispersion* of the solution into increasingly large areas of space. In fact, the speed of propagation of solutions is typically infinite. This behavior also differs from that of parabolic equations, which tend to *dissipate* the high-frequency components of a solution (sending them to zero) rather than dispersing them. In physics, dispersive equations arise in quantum mechanics: they are the *nonrelativistic limit* $c \rightarrow \infty$ of relativistic equations and they are also approximations to model certain types of fluid behavior. For instance, the KORTEWEG–DE VRIES EQUATION [III.49],

$$\partial_t u + \partial_x^3 u = 6u \partial_x u,$$

is a dispersive PDE that models the behavior of small-amplitude waves in a shallow canal.

2.6 Special Topics for Linear Equations

The greatest successes of the general theory have been in connection with linear equations, especially those with constant coefficients, for which Fourier analysis provides an extremely powerful tool. While the related issues of classification, well-posedness, and propagation of singularities have dominated the study of lin-

ear equations, there are other issues of interest as well, including the following.

2.6.1 Local Solvability

This is the problem of determining the conditions on a linear operator P and given data f under which the equation (9) is locally solvable. The Cauchy–Kovalevskaya theorem gives a criterion for local solvability when f and the coefficients of P are real analytic, but it is a remarkable phenomenon that when one relaxes this assumption slightly, asking for f to be smooth rather than real analytic, serious obstructions to local solvability appear. For instance, the *Lewy operator*

$$P[u](t,z) = \frac{\partial u}{\partial \bar{z}}(t,z) - \mathrm{i}z\frac{\partial u}{\partial t}(t,z),$$

defined on complex-valued functions $u : \mathbb{R} \times \mathbb{C} \rightarrow \mathbb{C}$, has the property that equation (9) is locally solvable for real-analytic f but not for "most" smooth f. The Lewy operator is intimately connected to the tangential Cauchy–Riemann equations on the Heisenberg group in $\mathbb{C}^2$. It was discovered in the study of the restriction of the two-dimensional analogue of the Cauchy–Riemann operator P to a quadric in $\mathbb{C}^2$. This example was the starting point for the theory of *local solvability*, whose goal is to characterize linear equations that are locally solvable. The theory of Cauchy–Riemann manifolds—which has its origin in the study of restrictions of the Cauchy–Riemann equations (in higher dimensions) to real hypersurfaces, each of which comes with an associated "tangential Cauchy–Riemann complex"—is another extremely rich source of examples of interesting linear PDEs, which do not fit into the standard classification.

2.6.2 Unique Continuation

This concerns various ill-posed problems where solutions may not always exist, but one still has uniqueness. A fundamental example is that of *analytic continuation*: two holomorphic functions on a connected domain D that agree on a nondiscrete set (such as a disk or an interval) must necessarily agree everywhere on D. This fact can be viewed as a unique continuation result for the Cauchy–Riemann equations (12). Another example in a similar spirit is *Holmgren's theorem*, which asserts that solutions to a linear PDE (9) that has real-analytic coefficients and data are unique, even in the class of smooth functions. More generally, the study of ill-posed problems (such as the wave equation with pre-scribed data on a timelike surface rather than a space-like one) arises naturally in connection with control theory.

2.6.3 Spectral Theory

There is no way I can even begin to give an account of this theory, which is of fundamental importance not only to quantum mechanics and other physical theories, but also to geometry and ANALYTIC NUMBER THEORY [IV.2]. Just as a matrix A can often be analyzed through its EIGENVALUES AND EIGENVECTORS [I.3 §4.3] by the tools of linear algebra, one can learn much about a linear differential operator $\mathcal{P}$ and its associated PDE by understanding that operator's SPECTRUM [III.86] and eigenfunctions with the help of tools from FUNCTIONAL ANALYSIS [IV.15]. A typical problem in spectral theory is the *eigenvalue problem* in $\mathbb{R}^d$:

$$-\Delta u(x) + V(x)u(x) = \lambda u(x).$$

A function u that is localized in space (for example, by being bounded in the $L^2(\mathbb{R}^d)$-norm) and that satisfies this equation is mapped by the linear operator $-\Delta + V$ to the function λu: we say that u is an *eigenfunction* with *eigenvalue* λ.

Suppose that we have an eigenfunction u and let $\phi(t, x) = e^{-i\lambda t}u(x)$. It is easy to check that ϕ is a solution of the Schrödinger equation

$$i\partial_t\phi + \Delta\phi - V\phi = 0. \tag{47}$$

Moreover, it has a very special form. Such solutions are called *bound states* of the physical system described by (47). The eigenvalues λ, which form a discrete set, correspond to the quantum energy levels of the sys-tem. They are very sensitive to the choice of potential V. The *inverse spectral problem* is also important: can one determine the potential V from knowledge of the corresponding eigenvalues? The eigenvalue problem can be studied in considerable generality by replacing the operator $-\Delta + V$ with other elliptic operators. For instance, in geometry it is important to study the eigen-value problem for the *Laplace–Beltrami operator*, which is the natural generalization of the Laplace operator from $\mathbb{R}^n$ to general RIEMANNIAN MANIFOLDS [I.3 §6.10]. When the manifold has some arithmetic structure (for instance, if it is the quotient of the upper half-plane by a discrete arithmetic group), this problem is of major importance in number theory, leading, for instance, to the theory of *Hecke–Maas forms*. A famous problem in differential geometry ("can you hear the shape of a drum?") is to characterize the metric on a compact surface from the spectral properties of the associated Laplace–Beltrami operator.

2.6.4 Scattering Theory

This theory formalizes the intuition from quantum mechanics that a potential which is small or localized is largely unable to "trap" a quantum particle, which is therefore likely to escape to infinity in a manner resem-bling that of a free particle. In the case of equation (47), solutions that scatter are those that behave freely as $t \to \infty$. That is, they behave like solutions to the free Schrödinger equation $i\partial_t\psi + \Delta\psi = 0$. A typical prob-lem in scattering theory is to show that, if $V(x)$ tends to zero sufficiently fast as $|x| \to \infty$, all solutions, except the bound states, scatter as $t \to \infty$.

2.7   Conclusions

In the analytic case, the CK theorem allows us to solve the IVP locally for very general classes of PDEs. We have a general theory of characteristic hypersurfaces of PDEs and a good general understanding of how they relate to propagation of singularities. We can also distinguish in considerable generality the fundamental classes of elliptic and hyperbolic equations and can define gen-eral parabolic and dispersive equations. The IVP for a large class of nonlinear hyperbolic systems can be solved locally in time, for sufficiently smooth initial conditions. Similar local-in-time results hold for gen-eral classes of nonlinear parabolic and dispersive equa-tions. For linear equations a lot more can be done. We have satisfactory results concerning the regularity of solutions for elliptic and parabolic equations and a good understanding of the propagation of singular-ities for a large class of hyperbolic equations. Some aspects of spectral theory and scattering theory and problems of unique continuation can also be studied in considerable generality.

The main defect of the general theory concerns the passage from local to global. Important global features of special equations are too subtle to fit into a general scheme. Rather, each important PDE requires special treatment. This is particularly true for nonlinear equa-tions: the long-term behavior of solutions is very sen-sitive to the special features of the equation at hand. Moreover, general points of view may obscure, through unnecessary technical complications, the main proper-ties of the important special cases. A useful general framework is one that provides a simple and elegant treatment of a particular phenomenon, as is the case for symmetric hyperbolic systems and the phenomenon of local well-posedness and finite speed of propagation. However, it turns out that symmetric hyperbolic systems are simply too general for the study of more refined questions about the important examples of hyperbolic equations.

3   General Ideas

As one turns away from the general theory, one may be inclined to accept the pragmatic point of view described earlier, according to which PDEs is not a real subject but is rather a collection of subjects such as hydrodynamics, general relativity, several complex variables, elasticity, etc., each organized around a special equation. However, this rather widespread viewpoint has its own serious drawbacks. Even though specific equations have specific properties, the tools that are used to derive them are intimately related. In fact, there is an impressive body of knowledge relevant to all important equations, or at least large classes of them. Lack of space does not allow me to do anything more than enumerate them below.[9]

3.1   Well-Posedness

As is clear from the previous section, well-posed problems are at the heart of the modern theory of PDEs. Recall that these are problems that admit unique solutions for given smooth initial or boundary conditions, and that the corresponding solutions have to depend continuously on the data. It is this condition that leads to the classification of PDEs into elliptic, hyperbolic, parabolic, and dispersive equations. The first step in the study of a nonlinear evolution equation is a proof of a local-in-time existence and uniqueness theorem, similar to the one for ODEs. *Ill-posedness*, the counterpart of well-posedness, is also important in many applications. The Cauchy problem for the wave equation (3), with data on the timelike hypersurface $z = 0$, is a typical example. Ill-posed problems appear naturally in control theory, as we have mentioned, and also in inverse scattering.

3.2   Explicit Representations and Fundamental Solutions

Our basic equations (2)–(5) can be solved explicitly. For example, the solution to the IVP for the heat equation in $\mathbb{R}^{1+d}_+$, that is, the problem of finding a function u that satisfies

$$-\partial_t u + \Delta u = 0, \quad u(0, x) = u_0(x),$$

for $t \geqslant 0$, is given by

$$u(t, x) = \int_{\mathbb{R}^d} E_d(t, x - y) u_0(y) \, dy$$

for a certain function E_d, which is called the *fundamental solution* of the heat operator $-\partial_t + \Delta$. This function can be defined explicitly: when $t \leqslant 0$ it is 0, and when $t > 0$ it is given by the formula $E_d(t, x) = (4\pi t)^{-d/2} e^{-|x|^2/4t}$. Observe that E_d satisfies the equation $(-\partial_t + \Delta)E = 0$ in both regions $t < 0$ and $t > 0$, but it has a singularity at $t = 0$, which prevents it from satisfying the equation in the whole of $\mathbb{R}^{1+d}$. In fact, we can check that for any function[10] $\phi \in C_0^\infty(\mathbb{R}^{d+1})$, we have

$$\int_{\mathbb{R}^{d+1}} E_d(t, x)(\partial_t \phi(t, x) + \Delta \phi(t, x)) \, dt \, dx = \phi(0, 0).$$
$$(48)$$

In the language of DISTRIBUTION THEORY [III.18], formula (48) means that E_d, as a distribution, satisfies the equation $(-\partial_t + \Delta)E_d = \delta_0$, where δ_0 is the *Dirac distribution* in $\mathbb{R}^{1+d}$ supported at the origin. That is, $\delta_0(\phi) = \phi(0, 0)$, $\forall \phi \in C_0^\infty(\mathbb{R}^{d+1})$. A similar notion of fundamental solution can be defined for the Poisson, wave, Klein–Gordon, and Schrödinger equations.

A powerful method of solving linear PDEs with constant coefficients is based on THE FOURIER TRANSFORM [III.27]. For example, consider the heat equation $\partial_t - \Delta u = 0$ in one space dimension, with initial condition $u(0, x) = u_0$. Define $\hat{u}(t, \xi)$ to be the Fourier transform of u relative to the space variable:

$$\hat{u}(t, \xi) = \int_{-\infty}^{+\infty} e^{-ix\xi} u(t, x) \, dx.$$

It is easy to see that $\hat{u}(t, \xi)$ satisfies the differential equation

$$\partial_t \hat{u}(t, \xi) = -\xi^2 \hat{u}(t, \xi), \quad \hat{u}(0, \xi) = \hat{u}_0(\xi).$$

This can be solved by a simple integration, which results in the formula $\hat{u}(t, \xi) = \hat{u}_0(\xi) e^{-t|\xi|^2}$. Thus, with

---

9. I fail to mention in the few examples given above some of the important functional analytic tools connected to Hilbert space methods, compactness, the implicit function theorems, etc. I also fail to mention the importance of probabilistic methods and the development of topological methods for dealing with global properties of elliptic PDEs.

---

10. That is, any function that is smooth and has compact support in $\mathbb{R}^{1+d}$.

the help of the inverse Fourier transform, we derive a formula for $u(t, x)$:

$$u(t, x) = (2\pi)^{-1} \int_{-\infty}^{+\infty} e^{ix\xi} e^{-t|\xi|^2} \hat{u}_0(\xi) \, d\xi.$$

Similar formulas can be derived for our other basic evolution equations. For example, in the case of the wave equation $-\partial_t^2 u + \Delta u = 0$ in three dimensions, subject to the initial data $u(0, x) = u_0$, $\partial_t u(0, x) = 0$, we find that

$$u(t, x) = (2\pi)^{-3} \int_{\mathbb{R}^3} e^{ix\xi} \cos(t|\xi|) \hat{u}_0(\xi) \, d\xi. \quad (49)$$

After some work, one can reexpress formula (49) in the form

$$u(t, x) = \partial_t \left((4\pi t)^{-1} \int_{|x-y|=t} u_0(y) \, da(y) \right), \quad (50)$$

where da is the area element of the sphere $|x - y| = t$ of radius t centered at x. This is the well-known *Kirchhoff formula*. By contrast with (49), the integration here is with respect to the physical variables t and x only. It is instructive to compare these two formulas. Using the Plancherel identity it is very easy to deduce from (49) the L^2 bound

$$\int_{\mathbb{R}^3} |u(t, x)|^2 \, dx \leqslant C \|u_0\|_{L^2(\mathbb{R}^3)}^2,$$

while the possibility of obtaining such a bound from (50) seems unlikely since the formula involves a derivative. On the other hand, (50) is perfect for giving us information about the domain of influence. Indeed, we can see immediately from the formula that if u_0 is zero outside the ball $B_a = \{|x - x_0| \leqslant a\}$, then $u(t, x)$ is zero outside the ball $B_{a+|t|}$ for any time t. This fact does not seem at all transparent in the Fourier-based formula (49). The fact that different representations of solutions have different, even opposite, strengths and weaknesses has important consequences for constructing approximate solutions, or *parametrices*, for more complicated equations, such as linear equations with variable coefficients or nonlinear wave equations. There are two possible types of constructions: those in physical space, which mimic the physical-space formula (50), and those in Fourier space, which mimic the formula (49).

3.3 A Priori Estimates

Most equations cannot be solved explicitly. However, if we are interested in *qualitative* information about a solution, then it is not necessary to derive it from an exact formula. But how else, one might wonder, can we

extract such information? A priori estimates are a very important technique for doing this.

The best-known examples are *energy estimates, the maximum principle*, and *monotonicity arguments*. The simplest example of the first type is the following identity (which is a very simple example of a so-called *Bochner-type identity*):

$$\int_{\mathbb{R}^d} |\partial^2 u(x)|^2 \, dx = \int_{\mathbb{R}^d} |\Delta u(x)|^2 \, dx.$$

The left-hand side is shorthand for

$$\int_{\mathbb{R}^d} \sum_{1 \leqslant i,j \leqslant d} |\partial_i \partial_j u(x)|^2 \, dx$$

and the identity holds for all functions u that are twice continuously differentiable and tend to zero as $|x| \to \infty$. This formula can be justified fairly simply by integrating by parts. As a consequence of the Bochner identity, we obtain the a priori estimate that if u is a smooth solution to the Poisson equation (6) with square-integrable data f, and if it tends to zero at infinity, then the square integral of its second derivatives is bounded:

$$\int_{\mathbb{R}^d} |\partial^2 u(x)|^2 \, dx \leqslant \int_{\mathbb{R}^d} |f(x)|^2 \, dx < \infty. \quad (51)$$

Thus we obtain the qualitative fact that, on average (in a mean-square sense), u has "two more degrees of regularity" than f.[11] This is called an energy-type estimate because, in physical situations, the square of the L^2-norm can often be interpreted as some type of kinetic energy.

The Bochner identity can be extended to more general Riemannian manifolds than $\mathbb{R}^d$, although one then picks up some additional lower-order terms involving the curvature of those manifolds. Such identities play a major role in the theory of geometric PDEs on these manifolds.

Energy-type identities and estimates also exist for parabolic, dispersive, and hyperbolic PDEs. For instance, they play a fundamental role in demonstrating the local existence, uniqueness, and finite speed of propagation for hyperbolic PDEs with smooth initial data. Energy estimates become particularly powerful when combined with inequalities such as the *Sobolev embedding inequality*, which allows one to convert

---

11. A crucial fact, about which one can read more in the online version, is that the L^2-norms in (51) can be replaced by L^p-norms, $1 < p < \infty$, or Hölder-type norms. The first case corresponds to *Calderon–Zygmund estimates*, while the second corresponds to *Schauder estimates*. Both are extremely important in the study of regularity properties for solutions to second-order elliptic PDEs.

the "L^2" information provided by these estimates into pointwise (or "L^∞") type information (see FUNCTION SPACES [III.29 §§2.4, 3]).

While energy identities and L^2 estimates (which, as in the above example, come from integration by parts) apply to all, or at least major classes of, PDEs, the *maximum principle* can be applied only to elliptic and parabolic PDEs. The following theorem is the simplest manifestation of it. Note that the theorem provides us with important quantitative information about solutions to the Laplace equation even in the absence of any explicit representation for them.

Theorem (maximum principle). *Assume that u is a solution to the Laplace equation (1) on a bounded connected domain $D \in \mathbb{R}^d$ with a smooth boundary ∂D. Assume also that u is continuous on the closure of D and has continuous first and second partial derivatives in the interior of D. Then u must achieve its maximum and minimum values on the boundary. Moreover, if the maximum or minimum is also achieved at an interior point of D, then u must be constant in D.*

The method is very robust and can easily be extended to a large class of second-order elliptic equations. It can also be extended to parabolic equations and systems, and plays a crucial role in, for example, the study of Ricci flow.

Let us briefly mention some other important classes of a priori estimates. The Sobolev inequalities, which are of prime importance in elliptic equations, have several counterparts in linear and nonlinear hyperbolic and dispersive equations, including the *Strichartz estimates* and *bilinear estimates*. In connection with ill-posed problems and unique continuation, *Carleman estimates* play a fundamental role. Finally, several a priori estimates arising from monotonicity formulas[12]—such as *virial identities*, *Pohozaev identities*, or *Morawetz inequalities*—can be used to establish the breakdown of regularity or the *blow-up* of solutions to some nonlinear equations, and to guarantee global existence and decay of solutions to others.

To summarize, it is not much of an exaggeration to say that a priori estimates play a fundamental role in more or less every aspect of the modern theory of PDEs.

---

12. Perhaps the most familiar example of a monotonicity phenomenon is the *second law of thermodynamics* from physics, which asserts that, for many physical systems, the total *entropy* of the system is an increasing function of time.

3.4 Bootstrap and Continuity Arguments

The *bootstrap* argument is a method, or rather a powerful general philosophy, to derive a priori estimates for nonlinear equations. According to this philosophy we start by making educated assumptions about the solutions we are trying to describe. These assumptions allow us to think of the original nonlinear problem as a linear one whose coefficients satisfy properties consistent with the assumptions. We may then use linear methods, based on other a priori estimates that we already know, to try to show that the solutions to this linear problem behave as well as we have postulated—in fact, even better. One can characterize this powerful method, which allows us to use linear theory without actually having to linearize the equation, as a *conceptual linearization*. It can also be regarded as a continuity argument relative to some parameter, which might be the natural time parameter of an evolution problem, but it could also be an artificial parameter which we have the freedom to introduce ourselves. This latter situation is typical of applications to nonlinear elliptic equations. In the online version of this article we provide a few examples to illustrate the method in both cases.

3.5 The Method of Generalized Solutions

Since a PDE involves differentiation, it might seem obvious that in any discussion of PDEs we should restrict our attention to differentiable functions. However, it is possible to generalize the notion of differentiation so that it makes sense for a wider class of functions, and even for function-like objects, such as distributions, that are not functions at all. This allows us to make sense of a PDE in a broader context, and admits the possibility of *generalized solutions*.

The best way to introduce generalized solutions in PDEs and explain why they are important is through the *Dirichlet principle*. This originates in the observation that, out of all functions that are defined on a bounded domain $D \subset \mathbb{R}^d$, that satisfy prescribed Dirichlet boundary condition $u|_{\partial D} = f$, and that live in an appropriate functional space X, the functions u that minimize the Dirichlet integral (or Dirichlet functional)

$$\|u\|_{Dr}^2 = \frac{1}{2} \int_D |\nabla u|^2 = \frac{1}{2} \sum_{i=1}^d \int_D |\partial_i u|^2 \quad (52)$$

are the harmonic functions (that is, solutions of the equation $\Delta u = 0$). It was RIEMANN [VI.49] who first had the idea of trying to use this fact to solve the Dirichlet problem: in order to find a solution u to the problem

$$\Delta u = 0, \quad u|_{\partial D} = u_0, \tag{53}$$

one should find (by some means other than solving the Dirichlet problem) a function u that minimizes the Dirichlet integral while equaling u_0 on ∂D. To do this, one must specify the set by functions, or rather the function space, over which the minimization is taking place. The history of how this choice was made is a fascinating one. A natural choice is $X = C^1(\bar{D})$, the space of continuously differentiable functions on $\bar{D}$, where the norm of a function v is

$$\|v\|_{C^1(\bar{D})} = \sup_{x \in D}(|v(x)| + |\partial v(x)|).$$

In particular, the Dirichlet norm $\|v\|_{Dr}$ is finite when v belongs to this space. In fact, Riemann chose $X = C^2(\bar{D})$ (a similar space but designed for twice continuously differentiable functions). This bold but flawed attempt was followed by a penetrating criticism by WEIERSTRASS [VI.44], who showed that the functional does not have to achieve its minimum in either $C^2(\bar{D})$ or $C^1(\bar{D})$. However, Riemann's basic idea was revived, and it eventually triumphed after a long and inspiring process that involved defining appropriate function spaces, introducing the notion of generalized solutions, and developing a *regularity theory* for them. (The precise formulation of the Dirichlet principle also requires the definition of SOBOLEV SPACES [III.29 §2.4].)

Let us briefly summarize the method, which has since been vastly extended so that it can be applied to a large class of linear[13] and nonlinear elliptic and parabolic equations. It is based on two steps. In the first step one applies a minimization procedure. Although, as Weierstrass discovered, the natural function spaces may not contain functions that achieve the minimum, one can use such a procedure to find a *generalized* solution instead. This may not seem very interesting, since we were looking for a *function* that solves the Dirichlet problem (or one of the other problems to which the method can be applied). But this is where the second step comes in: it is sometimes possible to show that the generalized solution must in fact be a classical solution (that is, an appropriately smooth function) after all. This is the "regularity theory" mentioned earlier. In some situations, however, the generalized solution may turn out to have singularities and therefore not be regular. Then the challenge is to understand the

nature of these singularities and to prove realistic *partial* regularity results. For instance, it is sometimes possible to prove that the generalized solution is smooth everywhere apart from in a small "exceptional set."

Though generalized solutions are at their most effective for elliptic problems, their range of applicability encompasses all PDEs. For example, we have already seen that the fundamental solutions to the basic linear equations have to be interpreted as distributions, which are examples of generalized solutions.

The notion of generalized solutions has also proved successful for nonlinear evolution problems, such as systems of conservation laws in one space dimension. An excellent example is provided by the Burgers equation (21). As we have seen, solutions to $\partial_t u + u \partial_x u = 0$ develop singularities in finite time no matter how smooth the initial conditions are. It is natural to ask whether solutions continue to make sense, as generalized solutions, even beyond the time when these singularities form. A natural notion of generalized solution is a function u such that

$$\int_{\mathbb{R}^{1+1}} (\partial_t u + u \partial_x u)\phi = 0$$

for every smooth function ϕ that is zero outside a bounded set, since one can make sense of the integral even when u is not a differentiable function. Integrating this by parts (the first term with respect to t and the second with respect to x) one obtains the following formulation:

$$\int_{\mathbb{R}^{1+1}} u \partial_t \phi + \frac{1}{2} \int_{\mathbb{R}^{1+1}} u^2 \partial_x \phi = 0 \quad \forall \phi \in C_0^\infty(\mathbb{R}^{1+1}).$$

It can be shown that, under additional conditions called *entropy conditions*, the IVP for the Burgers equation admits a unique generalized solution that is *global*: that is, valid for every $t \in \mathbb{R}$. Today we have a satisfactory theory of global solutions to a large class of hyperbolic systems of one-dimensional "conservation laws." These systems, for which the above-mentioned theory applies, are called *strictly hyperbolic*.

For more complicated nonlinear evolution equations, the question of what constitutes a good concept of a generalized solution, though fundamental, is far murkier. For higher-dimensional evolution equations the first concept of a *weak solution* was introduced by Leray. Let us call a generalized solution *weak* if one cannot prove any type of uniqueness for it. This unsatisfactory situation may be temporary, i.e., the result of our technical inabilities, or unavoidable, in the sense that the concept itself is flawed. Leray was able to produce, by a compactness method, a weak solution of the IVP

---

13. A notable example for applications in geometry is Hodge theory.

for the NAVIER–STOKES EQUATIONS [III.23]. The great advantage of the compactness method (and its modern extensions, which can, in some cases, cleverly circumvent lack of compactness) is that it produces global solutions for all data. This is particularly important for supercritical or critical nonlinear evolution equations, which we will discuss later. For these we expect classical solutions to develop singularities in a finite time. The problem, however, is that one has very little control over such solutions. In particular, we do not know how to prove their uniqueness.[14] Similar types of solutions were later introduced for other important nonlinear evolution equations. In most of the interesting cases of supercritical evolution equations, such as the Navier–Stokes equations, the usefulness of the types of weak solutions discovered so far remains undecided.

3.6 Microlocal Analysis, Parametrices, and Paradifferential Calculus

One of the fundamental difficulties of hyperbolic and dispersive equations is the interplay between geometric properties, which concern the physical space, and other properties, intimately tied to oscillations, that are best seen in Fourier space. *Microlocal analysis* is a general still-developing philosophy according to which one isolates the main difficulties by careful localizations in physical space or Fourier space or both. An important application of this point of view is the construction of parametrices for linear hyperbolic equations and their use in proving results about the propagation of singularities. Parametrices, as we have already mentioned, are approximate solutions of linear equations with variable coefficients, with error terms that are smoother. The *paradifferential calculus* is an extension of microlocal analysis to nonlinear equations. It allows one to manipulate the form of a nonlinear equation by taking account of how large and small frequencies interact, and it has achieved a remarkable technical versatility.

3.7 Scaling Properties of Nonlinear Equations

A PDE is said to have a *scaling property* if, whenever one rescales a solution in an appropriate way, one obtains another solution. Essentially, all basic nonlinear equations have well-defined scaling properties. Take, for example, the Burgers equation (21), $\partial_t u + u \partial_x u = 0$. If

u is a solution of this equation, then so is the function u_λ defined by $u_\lambda(t, x) = u(\lambda t, \lambda x)$. Similarly, if u is a solution of the cubic nonlinear Schrödinger equation in $\mathbb{R}^d$,

$$i\partial_t u + \Delta u + c|u|^2 u = 0, \tag{54}$$

then so is $u_\lambda(t, x) = \lambda u(\lambda^2 t, \lambda x)$. The relationship between the nonlinear scaling of the equation and the a priori estimates available for solutions to the equations leads to an extremely useful classification of equations into subcritical, critical, and supercritical equations. This will be discussed in more detail in the next section. For the moment it suffices to say that subcritical equations are those for which the nonlinearity can be controlled by the existing a priori estimates of the equation, while supercritical equations are those for which the nonlinearity appears to be stronger. Critical equations are borderline. The definition of criticality and its relationship with the issue of regularity play a very important heuristic role in nonlinear PDEs. One expects supercritical equations to develop singularities and subcritical equations not to.

4 The Main Equations

In the previous section we argued that, while there is no hope of finding a general theory of all PDEs, there is nevertheless a wealth of general ideas and techniques that are relevant to the study of almost all important equations. In this section we indicate how it may be possible to identify the features that characterize the equations we call important.

Most of our basic PDEs can be derived from simple geometric principles, which happen to coincide with some of the underlying geometric principles of modern physics. These simple principles provide a unifying framework[15] for the subject and help endow it with a sense of purpose and cohesion. They also explain why a very small number of linear differential operators, such as the Laplacian and the d'Alembertian, are all-pervasive.

Let us begin with the operators. The Laplacian is the simplest differential operator that is invariant under rigid motions of Euclidean space—a fact that we noted at the beginning of this article. This is important mathematically and physically: mathematically because it

---

14. Leray was very concerned about this point. Though, like all other researchers after him, he was unable to prove uniqueness of his weak solution, he managed to show that it must coincide with a classical one as long as the latter does not develop singularities.

15. The scheme sketched below is only an attempt to show that, in spite of the enormous number of PDEs studied by mathematicians, physicists, and engineers, there are nevertheless simple basic principles that unite them. I do not want, by any means, to imply that the equations discussed below are the only ones worthy of our attention.

results in many symmetry properties and physically because many physical laws are themselves invariant under rigid motions. The d'Alembertian is, similarly, the simplest differential operator that is invariant under the natural symmetries, or Poincaré transformations, of Minkowski space.

Now let us turn to the equations. From the point of view of physics, the heat equation is basic because it is the simplest paradigm for diffusive phenomena, while the Schrödinger equation can be viewed as the Newtonian limit of the Klein–Gordon equation. The geometric framework of the former is Galilean space, which itself is simply the Newtonian limit of Minkowski space.[16]

From a mathematical point of view, the heat, Schrödinger, and wave equations are basic because the corresponding differential operators $\partial_t - \Delta$, $(1/i)\partial_t - \Delta$, and $\partial_t^2 - \Delta$ are the simplest evolution operators that can be built out of Δ. The wave operator, as just discussed, is basic in a deeper way because of the association between $\Box = -\partial_t^2 + \Delta$ and the geometry of Minkowski space $\mathbb{R}^{1+n}$. As for Laplace's equation, one can view solutions to $\Delta\phi = 0$ as special time-independent solutions to $\Box\phi = 0$. Appropriate invariant and local definitions of square roots of Δ and $\Box$, or $\Box - k^2$, corresponding to "spinorial representations" of the Lorentz group, lead to the associated Dirac operators (see (13)). In the same vein we can associate with every Riemannian or Lorentzian manifold the operator Δ_g or $\Box_g$, respectively, or the corresponding Dirac operators. These equations inherit in a straightforward way the symmetries of the spaces on which they are defined.

4.1 Variational Equations

There is a general and extremely effective method for generating equations with prescribed symmetries that plays a fundamental role in both physics and geometry. One starts with a scalar quantity, called a *Lagrangian*, such as

$$\mathcal{L}[\phi] = \sum_{\mu,\nu=0}^{3} m^{\mu\nu}\partial_\mu\phi\partial_\nu\phi - V(\phi), \qquad (55)$$

with ϕ a real-valued function defined on $\mathbb{R}^{1+3}$ and V some real function of ϕ such as, for example, $V(\phi) = \phi^3$. Here ∂_μ denotes the partial derivatives with respect to the coordinates x^μ, $\mu = 0, 1, 2, 3$, and $m^{\mu\nu} = m_{\mu\nu}$, as earlier, denotes the 4×4 diagonal matrix with diagonal entries $(-1, 1, 1, 1)$, associated with the Minkowski

---

16. This is done by starting with the Minkowski metric $m = \mathrm{diag}(-1/c^2, 1, 1, 1)$, where c corresponds to the velocity of light, and letting $c \to \infty$.

metric. We associate with $\mathcal{L}[\phi]$ the so-called *action integral*:

$$S[\phi] = \int_{\mathbb{R}^{3+1}} \mathcal{L}[\phi].$$

Notice that both $\mathcal{L}[\phi]$ and $S[\phi]$ are invariant under translations and Lorentz transformations. In other words, if $T : \mathbb{R}^{1+3} \to \mathbb{R}^{1+3}$ is a function that does not change the metric and we define a new function by $\psi(t, x) = \phi(T(t, x))$, then $\mathcal{L}[\phi] = \mathcal{L}[\psi]$ and $S[\phi] = S[\psi]$.

We shall consider a function ϕ that minimizes the action integral. From this we wish to deduce that the derivative of S at ϕ, in some appropriate sense, is zero, and hence to deduce other properties about ϕ. But ϕ is a function that lives in an infinite-dimensional space, so we cannot talk about derivatives in a completely straightforward way. To deal with this problem, we define a *compact variation* of ϕ to be a smooth one-parameter family of functions $\phi^{(s)} : \mathbb{R}^{1+3} \to \mathbb{R}$, defined for each s in some interval $(-\epsilon, \epsilon)$, such that $\phi^{(0)}(x) = \phi(x)$ for every $x \in \mathbb{R}^3$ and $\phi^{(s)}(x) = \phi(x)$ for every (s, x) outside some bounded subset of $\mathbb{R}^{1+3}$. This allows us to differentiate with respect to s.

Given such a variation, we denote the derivative $\mathrm{d}\phi^{(s)}/\mathrm{d}s|_{s=0}$ by $\dot\phi$.

Definition. A field ϕ is said to be *stationary* with respect to S if, for *any* compact variation $\phi^{(s)}$ of ϕ, we have

$$\left.\frac{\mathrm{d}}{\mathrm{d}s}S[\phi^{(s)}]\right|_{s=0} = 0.$$

The variational principle. *The variational principle, or principle of least action, states that an acceptable solution of a given physical system must be stationary with respect to the action integral associated with the Lagrangian of the system.*

The variational principle enables us to associate with the given Lagrangian a system of PDEs, obtained from the fact that ϕ is stationary, called the *Euler–Lagrange equations*. We illustrate this by showing that the nonlinear wave equation in $\mathbb{R}^{1+3}$, namely

$$\Box\phi - V'(\phi) = 0, \qquad (56)$$

is the Euler–Lagrange equation associated with the Lagrangian (55). Given a compact variation $\phi^{(s)}$ of ϕ, we set $S(s) = S[\phi^{(s)}]$. Integration by parts gives

$$\left.\frac{\mathrm{d}}{\mathrm{d}s}S(s)\right|_{s=0} = \int_{\mathbb{R}^{3+1}} [-m^{\mu\nu}\partial_\mu\dot\phi\partial_\nu\phi - V'(\phi)\dot\phi]$$

$$= \int_{\mathbb{R}^{3+1}} \dot\phi[\Box\phi - V'(\phi)].$$

In view of the action principle and the arbitrariness of $\dot{\phi}$ we infer that ϕ must satisfy equation (56). Thus (56) is indeed the Euler–Lagrange equation associated with the Lagrangian $\mathcal{L}[\phi] = m^{\mu\nu}\partial_\mu\phi\partial_\nu\phi - V(\phi)$.

One can similarly show that the Maxwell equations of electromagnetism—along with their beautiful extensions to the Yang–Mills equations, wave maps, and the Einstein equations of general relativity—are also variational. That is, they too can be derived from a Lagrangian.

Remark. The variational principle asserts only that the acceptable solutions of a given system are stationary: in general, we have no reason to expect that the desired solutions minimize or maximize the action integral. Indeed, this fails to be the case for systems that have a time dependence, such as the Maxwell equations, Yang–Mills equations, wave maps, and Einstein equations.

However, there is a large class of variational problems, corresponding to time-independent physical systems or geometric problems, for which the desired solutions *do* turn out to be extremal. The simplest example is that of geodesics in a Riemannian manifold M, which are minimizers[17] with respect to length. More precisely, the *length functional* takes a curve y that passes through two fixed points of M and associates with it its length $L(y)$, which plays the role of an action integral. In this case a geodesic is not just a stationary point for the functional but a minimum. We also saw earlier that, according to the Dirichlet principle, solutions to the Dirichlet problem (53) minimize the Dirichlet integral (52). Another example is provided by the minimal surface equation (7), the solutions of which are minimizers of the area integral.

The study of minimizers of various functionals, i.e., action integrals, is a venerable subject in mathematics that goes under the name of *calculus of variations* (see VARIATIONAL METHODS [III.94] for further discussion).

Associated with the variational principle is another fundamental principle. A *conservation law* for an evolution PDE is a law that says that some quantity, typically an integral quantity depending on the solution, must remain constant over time, for every solution of the equation.

Noether's principle. *To any continuous one-parameter group of symmetries of the Lagrangian there corresponds a conservation law for the associated Euler–Lagrange PDE.*

Examples of such conservation laws are the familiar laws of conservation of energy, conservation of momentum, and conservation of angular momentum, all of which have important physical meaning. (The one-parameter group of symmetries for energy, for example, is just translations in time.) In the case of equation (56), the law of conservation of energy takes the form

$$E(t) = E(0), \tag{57}$$

where the quantity $E(t)$, which equals

$$\int_{\Sigma_t} \left(\tfrac{1}{2}(\partial_t\phi)^2 + \frac{1}{2}\sum_{i=1}^{3}(\partial_i\phi)^2 + V(\phi) \right) \mathrm{d}x, \tag{58}$$

is called the *total energy* at time t. (We write Σ_t for the set of all points (t, x, y, z) as (x, y, z) ranges over $\mathbb{R}^3$.) Observe that (57) provides an extremely important a priori estimate for solutions to (56) in the case when $V \geqslant 0$. Indeed, if the energy of the initial data at $t = 0$ is finite (that is, if $E(0) < \infty$), then

$$\int_{\Sigma_t} \left(\tfrac{1}{2}(\partial_t\phi)^2 + \frac{1}{2}\sum_{i=1}^{3}(\partial_i\phi)^2 \right) \mathrm{d}x \leqslant E(0).$$

We say that the energy identity (57) is *coercive*, which means that it leads to an absolute bound on all solutions with finite initial energy.

4.2 The Issue of Criticality

For the most basic evolution equations of mathematical physics, there are typically no better a priori estimates known than those provided by the energy. Taking into account the scaling properties of the corresponding equations as well, one is led to the very important classification of our basic equations, mentioned earlier, into *subcritical*, *critical*, and *supercritical* equations. To see how this is done, consider again the nonlinear scalar equation $\Box\phi - V'(\phi) = 0$, and take $V(\phi)$ to be $(1/(p+1))|\phi|^{p+1}$. Recall that the energy integral is given by (58). If we assign to the spacetime variables the dimension of length, L, then the spacetime derivatives have dimension L^{-1} and therefore $\Box$ has the dimension of L^{-2}. To be able to balance the left- and right-hand sides of the equation $\Box\phi = |\phi|^{p-1}\phi$, we need to assign a length scale to ϕ; we find this to be $L^{2/(1-p)}$. Thus the energy integral,

$$E(t) = \int_{\mathbb{R}^d} (2^{-1}|\partial\phi|^2 + |\phi|^{p+1}) \, \mathrm{d}x,$$

---

17. This is true, in general, only for sufficiently short geodesics, i.e., ones that pass through two points close to each other.

has the dimension L^c, $c = d - 2 + (4/(1 - p))$, with d corresponding to the volume element $dx = dx^1 dx^2 \cdots dx^d$, which scales like L^d. We say that the equation is *subcritical* if $c < 0$, *critical* if $c = 0$, and *supercritical* if $c > 0$. Thus, for example, $\Box\phi - \phi^5 = 0$ is critical in dimension $d = 3$. The same sort of dimensional analysis can be done for all our other basic equations. An evolutionary PDE is said to be *regular* if all smooth finite-energy initial conditions lead to global smooth solutions. It is conjectured that all subcritical equations are regular, but one expects supercritical equations to develop singularities. Critical equations are important borderline cases. The heuristic reason for this is that the nonlinearity tends to produce singularities while the coercive estimates prevent it. In subcritical equations the coercive estimates are stronger, while for supercritical equations it is the nonlinearity that is stronger. However, there may be other, more subtle a priori estimates that are not accounted for by our crude heuristic argument. Thus, some supercritical equations, such as the Navier–Stokes equations, may still be regular.

4.3 Other Equations

Many other familiar equations can be derived from the variational ones described above by the following procedures.

4.3.1 Symmetry Reductions

Sometimes a PDE is very hard to solve but becomes much easier if one places additional symmetry constraints on solutions. For example, if the PDE is rotation invariant and we look just for rotation-invariant solutions $u(t, x)$, then we can regard these solutions as functions of t and $r = |x|$, effectively reducing the dimension of the problem. By this procedure of *symmetry reduction* one can then derive a new PDE that is much simpler than the original one. Another, somewhat more general, way of obtaining simpler equations is to look for solutions that satisfy some further property. For instance, one can assume that they are stationary (that is, that they do not depend on the time variable), spherically symmetric, *self-similar* (which means that $u(t, x)$ depends only on x/t^a), or *traveling waves* (which means that $u(t, x)$ depends only on $x - vt$ for some fixed velocity vector v). Typically, the equations obtained by such reductions have a variational structure themselves. In fact, the symmetry reduction can be applied directly to the original Lagrangian.

4.3.2 The Newtonian Approximation and Other Limits

We can derive a large class of new equations as *limits* of the basic ones described above by taking one or more characteristic speeds to infinity. The most important example is the *Newtonian limit*, which is obtained by letting the velocity of light go to infinity. As we have already mentioned, the Schrödinger equation can be derived in this way from the linear Klein–Gordon equation. Similarly, we can derive the Lagrangians for the equations of nonrelativistic elasticity, fluid dynamics, or magnetohydrodynamics. It is an interesting fact that the nonrelativistic equations tend to look more messy than the relativistic ones. The simple geometric structure of the original equations gets lost in the limit. The remarkable simplicity of the relativistic equations is a powerful example of the importance of relativity as a unifying principle.

Once we are in the familiar world of Newtonian physics we can perform other well-known limiting procedures. The famous INCOMPRESSIBLE EULER EQUATIONS [III.23] are obtained by taking the limit of the general nonrelativistic fluid equations as the speed of sound tends to infinity. Various other limits are obtained relative to other characteristic speeds of the system or in connection with specific boundary conditions, such as the boundary-layer approximation in fluids. For example, in the limit as all characteristic speeds tend to infinity, the equations of elasticity turn into the familiar equations of a rigid body in classical mechanics.

4.3.3 Phenomenological Assumptions

Even after taking various limits and making symmetry reductions, the equations may still remain intractable. However, in various applications it makes sense to assume that certain quantities are sufficiently small to be neglected. This leads to simplified equations that could be called *phenomenological*[18] in the sense that they are not derived from first principles.

Phenomenological equations are "toy equations" that are used to illustrate and isolate important physical phenomena in complicated systems. A typical way of generating interesting phenomenological equations is to try to write down the simplest model equation that

---

18. I use this term here quite freely; it is typically used in a somewhat different context. Also, some of the equations that I call phenomenological below, e.g., dispersive equations, can be given formal asymptotic derivations.

still exhibits a particular feature of the original system. For instance, the self-focusing plane-wave effects of compressible fluids or elasticity can be illustrated by the simple-minded Burgers equation $u_t + uu_x = 0$. Nonlinear dispersive phenomena, typical of fluids, can be illustrated by the famous Korteweg–de Vries equation $u_t + uu_x + u_{xxx} = 0$. The nonlinear Schrödinger equation (54) provides a good model problem for nonlinear dispersive effects in optics.

If it is well chosen, a model equation can lead to basic insights into the original equation itself. For this reason, simplified model problems are also essential in the day-to-day work of the rigorous researcher into PDEs, who tests ideas on carefully selected model problems. It is crucial to emphasize that good results concerning the basic physical equations are rare; a very large percentage of important rigorous work in PDEs deals with simplified equations selected, for technical reasons, to isolate and focus our attention on some specific difficulties present in the basic equations.

In the above discussion we have not mentioned diffusive equations[19] such as the Navier–Stokes equations. These are in fact not variational, and therefore do not quite fit into the above description. Though they could be viewed as phenomenological equations, they can also be derived from basic microscopic laws such as those governing the Newtonian-mechanical interactions of a very large number of particles N. In principle,[20] the equations of continuum mechanics, such as the Navier–Stokes equations, could be derived by letting the number of particles $N \to \infty$.

Diffusive equations also turn out to be very useful in connection with geometric problems. Geometric flows such as mean curvature, inverse mean curvature, harmonic maps, Gauss curvature, and Ricci flow are some of the best-known examples. Diffusive equations can often be interpreted as the gradient flow for an associated elliptic variational problem. They can be used to construct nontrivial stationary solutions to the corresponding stationary systems, in the limit as $t \to \infty$, or to produce foliations with remarkable properties, such as one that was used recently in the proof of a famous conjecture of Penrose. As we have already mentioned, this idea has recently found an extraordinary application in the work of Perelman, who has used Ricci flow to settle the three-dimensional Poincaré conjecture. One

of his main new ideas was to interpret Ricci flow as a gradient flow.

4.4   Regularity or Breakdown

An additional source of unity for the subject of PDEs is the central role played by the problem of *regularity or breakdown* of solutions to the basic equations. It is intimately tied to the fundamental mathematical question of understanding what we actually mean by solutions and, from a physical point of view, to the issue of understanding the limits of validity of the corresponding physical theories. Thus, in the case of the Burgers equation, for example, the problem of singularities can be tackled by extending our concept of solutions to accommodate *shock waves*, which are solutions that are discontinuous across certain curves in the (t, x)-space. In this case one can define a function space of generalized solutions in which the IVP has unique, global solutions. Though the situation for more realistic physical systems is far less clear and far from being satisfactorily solved, the generally held opinion is that shock-wave-type singularities can be accommodated without breaking the boundaries of the physical theory at hand. The situation for singularities in general relativity is radically different. The singularities one expects there are such that no continuation of solutions is possible without altering the physical theory itself. The prevailing opinion here is that only a gravitational quantum field theory could achieve this.

5   General Conclusions

What, then, is the modern theory of PDEs? As a first approximation, one could say that it is the pursuit of the following main goals.

(i) *Understand the problem of evolution for the basic equations of mathematical physics*. The most pressing issue in this regard is to understand *when and how the local[21] (with respect to time) smooth solutions of the basic equations develop singularities*. A simple-minded criterion for distinguishing between regular theories and those that may admit singular solutions is given by the distinction between subcritical and supercritical equations. As mentioned earlier, it is widely believed

---

19. That is, equations where some of the basic physical quantities, such as energy, are not conserved and may in fact decrease in time. These are typically of parabolic type.

20. To establish this rigorously remains a major challenge.

21. One of the important achievements of the past century of mathematics was the establishment of a general procedure that guarantees the existence and uniqueness of a local-in-time solution to broad classes of initial conditions and large classes of nonlinear equations, including all those we have already mentioned above.

that *subcritical equations are regular and that supercritical equations are not.* Indeed, many subcritical equations have been proved to be regular even though we lack a general procedure for establishing regularity results of this kind. The situation with supercritical equations is far more subtle. To start with, an equation that we now call supercritical[22] may in fact turn out to be critical, or even subcritical, upon the discovery of additional a priori estimates. Thus an important question concerning the issue of criticality, and consequently that of singular behavior, is: are there other, stronger, local a priori bounds that cannot be derived from Noether's principle? The discovery of such a bound would be a major event in both mathematics and physics.

Once we understand that the presence of singularities in our basic evolution equations is unavoidable, we have to face the question of whether they can somehow be accommodated by a more general concept of what a solution is or whether their structure is such that the equation itself, indeed the physical theory that it underlies, becomes meaningless. An acceptable concept of a generalized solution should, of course, preserve the deterministic nature of the equations: in other words, it should be uniquely determined from its Cauchy data.

Finally, once an acceptable concept of generalized solutions is found, we would like to use it to determine some important qualitative features, such as long-term asymptotic behavior. One can formulate a limitless number of such questions, the answers to which will vary from equation to equation.

(ii) *Understand in a rigorous mathematical fashion the range of validity of various approximations.* The equations obtained by various limiting procedures or phenomenological assumptions can of course be studied in their own right, as the examples that we have referred to above are. However, they present us with additional problems to do with the mechanics of how they are derived from equations that we regard as more fundamental. It is entirely possible, for example, that the dynamics of a derived system of equations leads to behavior *that is incompatible with the assumptions made in its derivation.* Alternatively, a particular simplifying assumption, such as spherical symmetry in general relativity or zero vorticity for compressible fluids, may turn out to be unstable at large

scales and therefore not a reliable predictor of the general case. These and other similar situations lead to important dilemmas: should we persist in studying the approximate equations even when, in many cases, we face formidable mathematical difficulties (some which may turn out to be quite pathological and are perhaps related to the nature of the approximation), or should we abandon them in favor of the original system or a more suitable approximation? Whatever one may feel about this in any specific situation, it is clear that the problem of understanding, rigorously, the range of validity of various approximations is one of the fundamental goals in PDEs.

(iii) *Devise and analyze the right equation for studying the specific geometric or physical problem at hand.* This last goal is equally important even though it is necessarily vague. The enormously important role played by PDEs in various branches of mathematics is more evident than ever. One looks in awe at how equations such as the Laplace, heat, wave, Dirac, KdV, Maxwell, Yang–Mills, and Einstein equations, which were originally introduced in specific physical contexts, turned out to have very deep applications to seemingly unrelated problems in areas such as geometry, topology, algebra, and combinatorics. Other PDEs appear naturally in geometry when we look for embedded objects with optimal geometric shapes, such as solutions to isoperimetric problems, minimal surfaces, surfaces of least distortion or minimal curvature, or, more abstractly, connections, maps, or metrics with distinguished properties. They are variational in character, just like the main equations of mathematical physics. Other equations have been introduced with the goal of allowing one to deform a general object, such as a map, connection, or metric, to an optimal one. They usually arise in the form of geometric, parabolic flows. The most famous example of this is Ricci flow, first introduced by Richard Hamilton, who hoped to use it to deform Riemannian metrics into Einstein metrics. Similar ideas were used earlier to construct, for example, stationary harmonic maps with the help of a harmonic heat flow, and self-dual Yang–Mills connections with the help of a Yang–Mills flow. In addition to the successful use of Ricci flow to settle the Poincaré conjecture in three dimensions, another remarkable recent example of the usefulness of geometric flows is that of the inverse mean flow, first introduced by Geroch, to settle the so-called Riemannian version of the Penrose inequality.

---

22. What we call supercritical depends on the strongest a priori coercive estimate available.

Further Reading

Brezis, H., and F. Browder. 1998. Partial differential equations in the 20th century. *Advances in Mathematics* 135: 76–144.

Constantin, P. 2007. On the Euler equations of incompressible fluids. *Bulletin of the American Mathematical Society* 44:603–21.

Evans, L. C. 1998. *Partial Differential Equations*. Graduate Studies in Mathematics, volume 19. Providence, RI: American Mathematical Society.

John, F. 1991. *Partial Differential Equations*. New York: Springer.

Klainerman, S. 2000. PDE as a unified subject. In *GAFA 2000*, *Visions in Mathematics—Towards 2000* (special issue of *Geometric and Functional Analysis*), part 1, pp. 279–315.

Wald, R. M. 1984. *General Relativity*. Chicago, IL: Chicago University Press.

IV.13 General Relativity and the Einstein Equations
Mihalis Dafermos

Einstein's formulation of general relativity represents one of the great triumphs of modern physics and provides the currently accepted classical theory that unifies gravitation, inertia, and geometry. The *Einstein equations* are the mathematical embodiment of this theory.

The definitive form of the equations,

$$R_{\mu\nu} - \tfrac{1}{2} R g_{\mu\nu} = 8\pi T_{\mu\nu}, \qquad (1)$$

was attained in November 1915; this was the final act of Einstein's eight-year struggle to generalize his *principle of relativity* so as to encompass gravitation, which had been described in the earlier "Newtonian" theory by the *Poisson equation*

$$\frac{\partial^2 \phi}{\partial x^2} + \frac{\partial^2 \phi}{\partial y^2} + \frac{\partial^2 \phi}{\partial z^2} = 4\pi\mu \qquad (2)$$

for the potential ϕ and mass density μ.

An obvious contrast between the Einstein equations (1) and the Poisson equation (2) is that the mysterious notation of the former makes it far less obvious what they even mean. This has given the subject of general relativity a reputation for difficulty and impenetrability. However, this reputation is to some extent unwarranted. Both (1) and (2) represent the culmination of revolutionary theories whose formulations presuppose a complicated conceptual framework. For better or for worse, however, the structure necessary to formulate Poisson's equation has been incorporated into our traditional mathematical notation and school education. As a result, $\mathbb{R}^3$, with its Cartesian coordinate system, and notions such as functions, partial derivatives, masses, forces, and so on, are familiar to people with a general mathematical background, while the conceptual structure of general relativity is much less so, both with respect to its basic physical notions and with respect to the mathematical objects that are needed to model them. However, once one comes to terms with these, the equations turn out to be more natural and, one might even dare say, simpler.

Thus, the first task of this article is to explain in more detail the conceptual structure of general relativity. Our aim will be to make it clear what the equations (1) actually denote, and, moreover, why they are in a certain sense the simplest equations one can write down, given the general framework of the theory. This in turn will require us to review *special relativity* and its implications for the structure of matter, which will bring us to the unified concept of *stress–energy–momentum*, described by a *tensorial object* T. Finally, we will join Einstein in his inspired leap to the notion of a general four-dimensional *Lorentzian manifold* $(\mathcal{M}, g)$ that represents our space-time continuum. We shall see that equation (1) expresses a relationship between the tensor T and the *geometry* of g as expressed in its so-called *curvature*.

There is more to truly understanding a theory than merely knowing how to write down its governing equations. General relativity is associated with some of the most spectacular predictions of twentieth-century physics: *gravitational collapse*, *black holes*, *space-time singularities*, the *expansion of the universe*. These phenomena (which were completely unknown in 1915 and thus played no role in the formulation of the equations (1)) revealed themselves only when the conceptual issues surrounding the problem of global *dynamics* of solutions were understood. This took a surprisingly long time, though the story is not as well-known as the heroic struggle to attain (1). The article will conclude with a very brief glimpse into the fascinating dynamics of the Einstein equations.

1 Special Relativity

1.1 Einstein, 1905

Einstein's 1905 formulation of special relativity stipulated that all fundamental laws of physics should be invariant under *Lorentz transformations* of the *frame of reference* defined by x, y, z, and t. A Lorentz transformation is any composition of translations, rotations, and the *Lorentz boost*, which is given by the formulas

$$\left.\begin{array}{ll} \tilde{x} = \dfrac{x - vt}{\sqrt{1 - v^2/c^2}}, & \tilde{y} = y, \\[3ex] \tilde{t} = \dfrac{t - vx/c^2}{\sqrt{1 - v^2/c^2}}, & \tilde{z} = z, \end{array}\right\} \qquad (3)$$

where c is a certain constant and $|v| < c$. Thus, Einstein's stipulation was that if one changes coordinates by means of a Lorentz transformation, then the form of all fundamental equations will remain the same. This set of transformations had already been identified in the context of the study of the vacuum *Maxwell equations* for the electric field E and magnetic field B:

$$\left.\begin{array}{ll} \nabla \cdot E = 0, & \nabla \cdot B = 0, \\[1ex] c^{-1}\partial_t B + \nabla \times E = 0, & c^{-1}\partial_t E - \nabla \times B = 0. \end{array}\right\} \quad (4)$$

Indeed, the Lorentz transformations are precisely the transformations that keep the form of the above equations invariant if we also transform E and B appropriately. Their significance was emphasized by POINCARÉ [VI.61]. However, it was Einstein's profound insight to elevate this invariance to the status of fundamental physical principle, despite its incompatibility with what we now usually call *Galilean relativity*, which corresponds to taking $c \to \infty$ in (3). A surprising consequence of Lorentz invariance is that the notion of simultaneity is not absolute but depends on the observer: given two distinct events that occur at (t, x, y, z) and (t, x', y', z'), it is easy to find a Lorentz transformation such that the transformed events no longer have the same t-coordinate.

It follows from a celebrated result in partial differential equations known as the *strong Huygens principle*, applied to (4), that electromagnetic disturbances in vacuum propagate with speed c, which we thus identify as the speed of light. In view of Lorentz invariance, this statement is independent of the frame! A further postulate of the principle of relativity is that physical theories should not allow massive particles to move at speeds (as measured in any frame) greater than or equal to c.

1.2   Minkowski, 1908

Einstein's understanding of special relativity was "algebraic." It was MINKOWSKI [VI.64] who first understood

its underlying geometric structure, namely, that the content of the principle was contained in the *metric element*

$$-c^2\,\mathrm{d}t^2 + \mathrm{d}x^2 + \mathrm{d}y^2 + \mathrm{d}z^2 \qquad (5)$$

defined on $\mathbb{R}^4$ with coordinates (t, x, y, z). We call $\mathbb{R}^4$ endowed with the metric (5) *Minkowski space-time* and denote it $\mathbb{R}^{3+1}$. Points of $\mathbb{R}^{3+1}$ are referred to as *events*. The expression (5) is classical notation for the *inner product* defined on tangent vectors $v = (c^{-1}v^0, v^1, v^2, v^3)$, $w = (c^{-1}w^0, w^1, w^2, w^3)$ on $\mathbb{R}^4$ by

$$\langle v, w \rangle = -v^0 w^0 + v^1 w^1 + v^2 w^2 + v^3 w^3. \quad (6)$$

The Lorentz transformations constitute precisely the *symmetry group* of the geometry defined by (5). Einstein's principle of relativity could now be understood as the principle that the fundamental equations of physics must refer to space-time only through geometric quantities: that is, quantities that can be defined purely in terms of the metric. For example, from this point of view the reason that the notion of absolute simultaneity is not allowed is that it depends on a privileged hyperplane through any given point of $\mathbb{R}^{3+1}$. But there are Lorentz transformations that preserve the metric and send this hyperplane to another one through the given point, so nothing in the metric can pick out one particular hyperplane. Note that if a physical theory makes use of geometric quantities only, then it is automatically invariant under Lorentz transformations: this observation renders many complicated calculations unnecessary.

Let us explore this geometric point of view further. Note that nonzero vectors v are naturally classified by the inner product $\langle \cdot, \cdot \rangle$ into three types, called *timelike*, *null*, and *spacelike*, according to whether $\langle v, v \rangle < 0$, $\langle v, v \rangle = 0$, or $\langle v, v \rangle > 0$, respectively. Idealized point particles traverse curves γ through space-time; these are called the *world lines* of the corresponding particles. The postulate (referred to earlier) that speed in any frame of reference is bounded by the speed of light c can now be formulated as the following statement: *if γ is the world line of a particle, then the vector $\mathrm{d}\gamma/\mathrm{d}s$ must be timelike*. (Null lines correspond to light rays in the geometric optics limit of (4).) This statement is independent of the parameter s of γ, but for world lines we shall always assume that $\mathrm{d}t/\mathrm{d}s > 0$. To phrase this more geometrically, $\langle \mathrm{d}\gamma/\mathrm{d}s, (c^{-1}, 0, 0, 0) \rangle < 0$, which we interpret as the statement that γ is *future-directed*.

We can now define the "length" of the world line of a particle by

$$L(\boldsymbol{\gamma}) = \int_{s_1}^{s_2} \sqrt{-\langle \dot{\boldsymbol{\gamma}}, \dot{\boldsymbol{\gamma}} \rangle}\, ds$$

$$= \int_{s_1}^{s_2} \sqrt{c^2 \left(\frac{dt}{ds}\right)^2 - \left(\frac{dx}{ds}\right)^2 - \left(\frac{dy}{ds}\right)^2 - \left(\frac{dz}{ds}\right)^2}\, ds. \tag{7}$$

Classically, the above expression would have been written simply as

$$L(\boldsymbol{\gamma}) = \int_{\gamma} \sqrt{-(-c^2\, dt^2 + dx^2 + dy^2 + dz^2)},$$

which explains the notation (5). We refer to the quantity $c^{-1}L(\boldsymbol{\gamma})$ as *proper time*. This is the time that is relevant in local physical processes; in particular, if *you* are the particle traversing the world line $\boldsymbol{\gamma}$, then $c^{-1}L(\boldsymbol{\gamma})$ is the time that you will *feel*.

The metric (5) contains three-dimensional Euclidean geometry

$$dx^2 + dy^2 + dz^2,$$

restricted to $t = 0$, say. More interestingly, it also contains *non-Euclidean geometry*

$$\left(1 - \frac{x}{r}\right) dx^2 + \left(1 - \frac{y}{r}\right) dy^2 + \left(1 - \frac{z}{r}\right) dz^2$$

when it is restricted to the hypersurface $t = c^{-1}r = c^{-1}\sqrt{x^2 + y^2 + z^2}$. It is hard to overestimate how revolutionary the notion was that the time of physical processes (including our very sensations) and the length of measuring rods are two interdependent aspects of a geometric structure that naturally lives on a four-dimensional space-time continuum. Indeed, even Einstein initially rejected Minkowski space-time, preferring to retain the independent reality of a definite "space," albeit a space with a relative notion of simultaneity. Only as a result of his search for general relativity did he realize that this view is fundamentally untenable. We shall return to this in section 3.

2   Relativistic Dynamics and the Unification of Energy, Momentum, and Stress

Besides the space-time concept and its geometrization, the principle of relativity led to a profound rearrangement and unification of the fundamental concepts of dynamics: mass, energy, and momentum. Einstein's celebrated relation between mass and energy in the rest frame,

$$E_0 = mc^2, \tag{8}$$

is the best-known expression of one aspect of this unification. This relation arises naturally when one attempts to generalize Newton's second law $m(d\boldsymbol{v}/dt) = \boldsymbol{f}$ to a relation between 4-vectors in Minkowski space.

General relativity has to be formulated in terms of *fields* rather than particles. As a first step toward understanding it, let us look at continuous media. Now, instead of particles we consider *matter fields*; the unification of dynamical concepts encompasses what is known as *stress*, and its complete expression is embodied by the so-called *stress–energy–momentum tensor* $\boldsymbol{T}$. This tensor is fundamental to general relativity, so we have no choice but to familiarize ourselves with it. It will be the key to the form of the Einstein equations (1) as well as to the object on their right-hand side.

For each point $\boldsymbol{q} \in \mathbb{R}^{3+1}$, the stress–energy–momentum tensor field $\boldsymbol{T}$ gives us a map

$$\boldsymbol{T} : \mathbb{R}_{\boldsymbol{q}}^4 \times \mathbb{R}_{\boldsymbol{q}}^4 \to \mathbb{R} \tag{9}$$

defined by the formula

$$\boldsymbol{T}(\boldsymbol{w}, \tilde{\boldsymbol{w}}) = \sum_{\alpha,\beta=0}^{3} T_{\alpha\beta} w^\alpha \tilde{w}^\beta.$$

Here, $T_{\alpha\beta} = T_{\beta\alpha}$ for each α and β. By $\mathbb{R}_{\boldsymbol{q}}^4$ we mean the space of vectors *at* $\boldsymbol{q}$. (In Minkowski coordinates, we often identify $\mathbb{R}^4$ with $\mathbb{R}_{\boldsymbol{q}}^4$, but it will be important to distinguish between the two when considering arbitrary coordinates in section 3.2.) Bilinear maps of the form (9) are known as *covariant 2-tensors*.

If the only matter present is described by what is known as a *perfect fluid*, then the components of $\boldsymbol{T}$ are given by

$$T_{00} = (\rho + p)u^0 u^0 - p, \qquad T_{0i} = (\rho + p)u^i u^0,$$

$$T_{ij} = (\rho + p)u^i u^j + p\delta^{ij},$$

where $\boldsymbol{u}$ is the 4-velocity, a timelike vector normalized such that $\langle \boldsymbol{u}, \boldsymbol{u} \rangle = -c^2$, ρ is the *mass–energy*, p is the *pressure*, and where $\delta_{ij} = 1$ if $i = j$, 0 if $i \neq j$, and i and j range over 1, 2, 3. Greek indices will range over 0, 1, 2, 3. We identify T_{00} with *energy*, T_{0i} with *momentum*, and T_{ij} with *stress*. These notions are clearly frame-dependent. Finally, observe that $\boldsymbol{T}(\boldsymbol{u}, \boldsymbol{u}) = \rho c^2$. This is the field-theoretic version of the famous equation (8).

In general, $\boldsymbol{T}$ is derived from the totality of all the matter fields by constitutive functions that depend on the nature of the matter fields and their interactions. We need not worry here about such things. But, regardless of the nature of the matter fields involved, we always postulate that the following equations are satisfied:

$$-\partial_0 T_{0\alpha} + \sum_{i=1}^{3} \partial_i T_{i\alpha} = 0.$$

Defining $\nabla^0 = -\partial_0$, $\nabla^i = \partial_i$, and introducing the *Einstein summation convention*, under which summation is implicit when an index appears both upstairs and downstairs, we may rewrite this as

$$\nabla^\mu T_{\mu\nu} = 0. \tag{10}$$

These equations are Lorentz invariant.

The above relations embody the *conservation of stress-energy-momentum* at a differential level. Integrating (10) between homologous hypersurfaces and applying the Minkowski-space version of the divergence theorem, one obtains global balance laws. If one assumes that $T_{\alpha\beta}$ is compactly supported, then, integrating between $t = t_1$ and $t = t_2$, one obtains

$$\int_{t=t_2} T_{0\alpha} \, dx^1 \, dx^2 \, dx^3 = \int_{t=t_1} T_{0\alpha} \, dx^1 \, dx^2 \, dx^3. \tag{11}$$

With respect to the chosen Lorentz frame, the zeroth component of the above equation represents the *conservation of total energy*, while the remaining components represent *conservation of total momentum*.

In the case of a perfect fluid, if we close the system (10) by adjoining a conservation law for particle number

$$\nabla^\alpha (n\boldsymbol{u}_\alpha) = 0$$

and postulate constitutive relations between ρ, p, particle number density n, and entropy per particle s, compatible with the laws of thermodynamics, then we arrive at the so-called *relativistic Euler equations*.

3 From Special to General Relativity

With the elements of special relativity at hand, together with their deep implications for the nature of energy, momentum, and stress, we can now pass to the formulation of general relativity.

3.1 The Equivalence Principle

Einstein understood as early as 1907 that the most profound aspect of the gravitational force could not be described within the relativity principle as he had formulated it in 1905. This aspect is what he called *the equivalence principle.*

The easiest setting in which to understand this principle is that of the "test particle" with velocity $\boldsymbol{v}(t)$ in a fixed gravitational field ϕ. In this case, we have that the classical *gravitational force* is given by $\boldsymbol{f} = -m\nabla\phi$, and we may rewrite Newton's second law $m(d\boldsymbol{v}/dt) = \boldsymbol{f}$ as

$$\frac{d\boldsymbol{v}}{dt} = -\nabla\phi. \tag{12}$$

Notice that the mass m has dropped out! Thus, the gravitational field accelerates all objects at a given position in the same way. This explains the fact, recorded already in late antiquity by Ioannes Philoponus and popularized in Western Europe by Galileo, that the time it takes objects to fall from a given height is independent of their weight.

It was Einstein who first interpreted this property as a sort of covariance with respect to transformations to *noninertial*, that is to say *accelerated*, frames. For instance, in the case of a constant gravitational field, which corresponds to the case $\phi(z) = fz$, we can pass to the accelerated frame

$$\tilde{z} = z + \tfrac{1}{2} f t^2$$

and write (12) as

$$\frac{d\boldsymbol{v}}{dt} = 0. \tag{13}$$

Similarly, one can reverse the argument to "simulate" a gravitational field when none is present by expressing (13) in an accelerated frame.

3.2 Vectors, Tensors, and Equations in General Coordinates

Exactly what the equivalence principle means in general is somewhat obscure and has been the subject of debate ever since Einstein introduced it. Nevertheless, the above considerations suggest that, even in the absence of gravity, it would be useful to know how various objects and equations appear when expressed in arbitrary coordinate systems. That is to say, let us change from our Minkowski coordinates x^0, x^1, x^2, x^3 to the most general coordinate system, which we shall write as $\bar{x}^{\bar{\mu}} = \bar{x}^{\bar{\mu}}(x^0, x^1, x^2, x^3)$, where $\bar{\mu}$ ranges over 0, 1, 2, 3.

Expressing scalar functions in arbitrary coordinates poses no problem. But what about vector fields? If $\boldsymbol{v}$ is a vector field expressed in Minkowski coordinates as (v^0, v^1, v^2, v^3), how do we express $\boldsymbol{v}$ in our new coordinates $\bar{x}^{\bar{\mu}}$?

One has to think a bit about what a vector field actually *is*. The correct point of view is to consider a vector field $\boldsymbol{v}$ as a first-order differential operator defined (using Einstein's summation convention) by $\boldsymbol{v}(f) = v^\mu \partial_\mu f$. So we seek $v^{\bar{\mu}}$ such that $\boldsymbol{v}(f) = v^{\bar{\mu}} \partial_{\bar{\mu}} f$ for all functions f. The chain rule then gives us our answer:

$$v^{\bar{\mu}} = \frac{\partial \bar{x}^{\bar{\mu}}}{\partial x^\nu} v^\nu. \tag{14}$$

What about tensors, such as the stress–energy–momentum tensor T? In view of the definition (9), we seek

$T_{\bar{\mu}\bar{\nu}}$ such that

$$T(\boldsymbol{u},\boldsymbol{v}) = T_{\bar{\mu}\bar{\nu}}u^{\bar{\mu}}v^{\bar{\nu}}, \tag{15}$$

where the numbers $u^{\bar{\mu}}$ are the components of $\boldsymbol{u}$ with respect to the coordinates $\bar{x}^{\bar{\mu}}$ as we have just calculated them above. (Note that these components depend on the point $\boldsymbol{q}$. This is why it is now essential to distinguish $\mathbb{R}^4_{\boldsymbol{q}}$ from $\mathbb{R}^4$.) Again, the chain rule gives us the answer:

$$T_{\bar{\mu}\bar{\nu}} = T_{\mu\nu}\frac{\partial x^{\nu}}{\partial \bar{x}^{\bar{\nu}}}\frac{\partial x^{\mu}}{\partial \bar{x}^{\bar{\mu}}}.$$

Classically, we write

$$T = T_{\bar{\mu}\bar{\nu}}\,\mathrm{d}\bar{x}^{\bar{\mu}}\,\mathrm{d}\bar{x}^{\bar{\nu}} = T_{\mu\nu}\,\mathrm{d}x^{\mu}\,\mathrm{d}x^{\nu}.$$

One can interpret the above as a shorthand notation for (15), but it also tells us how to compute $T_{\bar{\mu}\bar{\nu}}$ from $T_{\mu\nu}$ by formally applying the chain rule to $\mathrm{d}\bar{x}^{\bar{\mu}}$.

There is another covariant symmetric 2-tensor besides T that is relevant here. This is the Minkowski metric itself. Indeed, the classical form of the Minkowski metric (5) corresponds to the representation

$$\eta_{\mu\nu}\,\mathrm{d}x^{\mu}\,\mathrm{d}x^{\nu},$$

where the $\eta_{\mu\nu}$ for Minkowski coordinates x^{μ} are given by $\eta_{00} = -1$, $\eta_{0i} = 0$, $\eta_{ij} = 1$ if $i = j$, and $\eta_{ij} = 0$ if $i \neq j$. To avoid the cumbersome notation $\langle \cdot, \cdot \rangle$, let us refer to the Minkowski metric as $\boldsymbol{\eta}$. Following the above, we may express $\boldsymbol{\eta}$ in general coordinates $\bar{x}^{\bar{\mu}}$ by

$$\eta_{\bar{\mu}\bar{\nu}}\,\mathrm{d}\bar{x}^{\bar{\mu}}\,\mathrm{d}\bar{x}^{\bar{\nu}},$$

where $\eta_{\bar{\mu}\bar{\nu}}$ is computed by formal application of the chain rule.

It is clear that if one tries to transform an equation such as (10) into general coordinates, then the components of $\boldsymbol{\eta}$ and their derivatives will appear in the equations. Einstein (always thinking "algebraically") was seeking laws of motion for both matter and the gravitational field that would have the same *form* in all coordinate systems. As he understood it, this meant that all objects that appear should transform as tensors and should be considered a priori "unknown." He referred to this principle as "general covariance." This suggests that $\boldsymbol{\eta}$ should be replaced by an *unknown* symmetric 2-tensor. Let us call this 2-tensor $\boldsymbol{g}$. One can of course try to write down an equation for the "unknown" $\boldsymbol{g}$ that forces it to be the "known" Minkowski metric $\boldsymbol{\eta}$. Thus, "general covariance" per se does not force one to abandon $\boldsymbol{\eta}$. But in view of the fact that $\boldsymbol{g}$ and T have the same number of components, it was a natural step to consider $\boldsymbol{g}$ as the embodiment of the gravitational field and to try to look for an equation that related $\boldsymbol{g}$ and T directly. In this way, the framework of general relativity was born.

3.3   Lorentzian Geometry

The profound insight of replacing the fixed Minkowski $\boldsymbol{\eta}$ with a dynamic $\boldsymbol{g}$ brought Einstein to what we now call *Lorentzian geometry*. Lorentzian geometry generalizes Minkowski geometry following the blueprint of RIEMANN [VI.49]. That is, we replace the Minkowski metric $\boldsymbol{\eta}$ by a general map

$$\boldsymbol{g} : \mathbb{R}^4_{\boldsymbol{q}} \times \mathbb{R}^4_{\boldsymbol{q}} \to \mathbb{R}.$$

In other words, we replace $\boldsymbol{\eta}$ by a symmetric covariant 2-tensor, which is expressed in arbitrary coordinates x^{μ} by

$$g_{\mu\nu}\,\mathrm{d}x^{\mu}\,\mathrm{d}x^{\nu}.$$

Moreover, we require that at each point $\boldsymbol{q}$ the bilinear form $\boldsymbol{g}(\cdot, \cdot)$ can be diagonalized to the Minkowski form (6). Loosely speaking, a Lorentzian metric is one that "looks locally like the Minkowski metric," just as a RIEMANNIAN METRIC [I.3 §6.10] looks locally like the Euclidean metric.

Just as with the Minkowski metric, the bilinear form $\boldsymbol{g}$ permits us to classify nonzero vectors $\boldsymbol{v}_{\boldsymbol{q}}$ at a point $\boldsymbol{q}$ as *timelike*, *null*, or *spacelike* and to define proper times of world lines $y(s) = (x^0(s), x^1(s), x^2(s), x^3(s))$ by the formula (7), but with $\langle \dot{y}, \dot{y} \rangle$ replaced by $g_{\mu\nu}\dot{x}^{\mu}\dot{x}^{\nu}$. It is in this sense that we can speak of the *geometry* of $\boldsymbol{g}$.

In view of Minkowski's formulation of the special relativity principle as the statement that the equations of physics refer to space-time only through geometric quantities associated with the Minkowski metric, it is natural to look for a generalization of this principle, and indeed a suitable version immediately suggests itself. It is the principle that *the equations of physics refer to the space-time coordinates only via geometric quantities naturally associated with $\boldsymbol{g}$.*

We saw earlier that the kinematic constraint on "test particles," as formulated geometrically for the Minkowski metric, was that $\mathrm{d}\boldsymbol{y}/\mathrm{d}s$ should be timelike; this makes sense for an arbitrary Lorentzian metric. But how does one formulate differential equations? For instance, how does one formulate an analogue of (10) that refers only to $\boldsymbol{g}$?

It turned out that in the Riemannian case, a set of natural geometric concepts suitable for the task had already been developed in the nineteenth and early twentieth centuries by Riemann, Bianchi, Christoffel, Ricci, and Levi-Civita. These carry over directly to the Lorentzian case.

One begins by defining the so-called *Christoffel symbols* $\Gamma^\lambda_{\mu\nu}$ by

$$\Gamma^\lambda_{\mu\nu} = \tfrac{1}{2} g^{\lambda\rho}(\partial_\mu g_{\rho\nu} + \partial_\nu g_{\mu\rho} - \partial_\rho g_{\mu\nu}).$$

Here, the numbers $g^{\mu\nu}$ are the components of the "inverse metric" of g: that is, they are the unique solution to the equation $g^{\mu\nu}g_{\nu\lambda} = \delta^\mu_\lambda$, where, as usual, $\delta^\mu_\lambda = 1$ if $\lambda = \mu$ and 0 otherwise. (It turns out that $g^{\mu\nu}$ is very useful for the calculational gymnastics that are typical of tensor analysis when it exploits the Einstein summation convention.)

One can then define a differential operator ∇_μ called a *connection*, which acts on vector fields by

$$\nabla_\mu v^\nu = \partial_\mu v^\nu + \Gamma^\nu_{\mu\lambda} v^\lambda \qquad (16)$$

and on covariant 2-tensors by

$$\nabla_\lambda T_{\mu\nu} = \partial_\lambda T_{\mu\nu} - \Gamma^\sigma_{\lambda\mu} T_{\sigma\nu} - \Gamma^\sigma_{\lambda\nu} T_{\mu\sigma}. \qquad (17)$$

The left-hand sides of (16) and (17) define tensors that can be expressed in any coordinate system by a formal application of the chain rule.

With the help of this differential operator, one could now write the analogue of equations (10) for an arbitrary metric g as

$$\nabla^\mu T_{\mu\nu} = 0, \qquad (18)$$

where $\nabla^\mu = g^{\mu\nu}\nabla_\nu$ refers to the connection associated with g.

If we consider a limit as the matter field becomes concentrated at a point, or rather as the stress–energy-momentum tensor $T_{\mu\nu}$ is nonzero only on a world line, then this curve will be a *geodesic* of g: that is, a curve that locally maximizes the proper time defined by g. These are the analogues of straight timelike lines in Minkowski space. In this limit, the motion of the matter does not depend on the nature of the stress–energy-momentum tensor, but only on the geometry of the metric that defines geodesics. Thus, all objects fall in the same way. These considerations give a concrete realization to the equivalence principle in general relativity.

Finally, it is important to remark that for a general metric g, the identity (18) *does not* imply global conservation laws (11) for "total energy" and "total momentum." Such laws hold only if g has symmetries. The fact that the fundamental conservation laws survive in general only at the infinitesimal level is an important insight into the nature of these principles in physics.

3.4   Curvature and the Einstein Equations

It remains, then, to give a set of equations for the metric g that relate it to T. In anticipation of a Newtonian limit, we expect these equations to be second order, and we expect them to implement "general covariance" in the simplest way possible: they should refer to no other structure but g itself and T.

Again, Riemannian geometry provides ready-made tensorial objects that are invariantly associated with g. One can define the *Riemann curvature tensor*

$$R_{\mu\nu\lambda\rho}\, dx^\mu\, dx^\nu\, dx^\lambda\, dx^\rho$$

with components given by

$$R_{\mu\nu\lambda\rho} = g_{\mu\sigma}(\partial_\rho \Gamma^\sigma_{\nu\lambda} - \partial_\lambda \Gamma^\sigma_{\nu\rho} + \Gamma^\tau_{\nu\lambda}\Gamma^\sigma_{\tau\rho} - \Gamma^\tau_{\nu\rho}\Gamma^\sigma_{\tau\lambda}).$$

One can also define the *Ricci curvature*

$$R_{\mu\nu}\, dx^\mu\, dx^\nu,$$

a covariant symmetric 2-tensor with components given by

$$R_{\mu\nu} = g^{\lambda\rho}R_{\mu\nu\lambda\rho},$$

and the *scalar curvature*

$$R = g^{\mu\nu}R_{\mu\nu}.$$

If g were the induced (Riemannian) metric on a 2-surface in $\mathbb{R}^3$, then R would just be twice the *Gauss curvature K*. The above expressions should be thought of as complicated tensorial generalizations of Gauss curvature to several dimensions.

The final piece of the puzzle for the formulation of the Einstein equations (1) is provided by the following constraint that Einstein demanded: whatever the equation relating the metric and the stress–energy-momentum tensor of matter, (18) (the infinitesimal conservation of stress–energy-momentum) should hold *as a consequence*. Now, it turns out that for *any* metric g, the so-called *Bianchi identities* imply that

$$\nabla^\mu(R_{\mu\nu} - \tfrac{1}{2}g_{\mu\nu}R) = 0. \qquad (19)$$

It is thus natural to postulate a linear relation between $T_{\mu\nu}$ and the tensor $R_{\mu\nu} - \tfrac{1}{2}g_{\mu\nu}R$. The form

$$R_{\mu\nu} - \tfrac{1}{2}g_{\mu\nu}R = 8\pi G c^{-4} T_{\mu\nu} \qquad (20)$$

is then uniquely determined by the requirement that it should give the correct Newtonian limit when one makes the identifications

$$g_{00} \sim 1 + 2\phi/c^2, \quad g_{0j} \sim 0, \quad g_{ij} \sim (1 - 2\phi/c^2)\delta_{ij}.$$

The form (1) corresponds to the usual units $G = c = 1$. Note that (1), when written out explicitly, is nonlinear in the metric components $g_{\mu\nu}$.

Einstein did not stop at the Newtonian limit. By considering geodesic motion in solutions of the linearized equations (20), Einstein was able to determine the correct value for the *anomalous precession of the perihelion of Mercury*, an effect that Newtonian theory was unable to explain. Since (20) had no adjustable parameters after determining the Newtonian limit, this was a genuine *test* of the theory. A few years later the gravitational "bending" of light was observed. This had been calculated theoretically in the context of the geometric optics approximation where light rays follow null geodesics in a fixed space-time background. Post-Newtonian predictions of (1) have now been verified by various solar system tests, confirming general relativity in this regime to a high degree of accuracy.

One special case of (20) is when we postulate that $T_{\mu\nu} = 0$. The equations then simplify to

$$R_{\mu\nu} = 0. \qquad (21)$$

These are known as the *vacuum equations*. The Minkowski metric (5) is a particular solution (but not the only one!).

The vacuum equations can be derived formally as the EULER–LAGRANGE EQUATIONS [III.94] corresponding to the so-called *Hilbert Lagrangian*:

$$\mathcal{L}(\boldsymbol{g}) = \int R \sqrt{-g}\, \mathrm{d}x^0 \, \mathrm{d}x^1 \, \mathrm{d}x^2 \, \mathrm{d}x^3.$$

(The expression $\sqrt{-g}\, \mathrm{d}x^0 \, \mathrm{d}x^1 \, \mathrm{d}x^2 \, \mathrm{d}x^3$ denotes the natural *volume form* associated with $\boldsymbol{g}$.) HILBERT [VI.63], who was following closely Einstein's struggle to formulate a theory of gravity with a dynamic metric $\boldsymbol{g}$, arrived at his Lagrangian (actually a more general version of the above yielding the coupled Einstein–Maxwell system) very shortly before Einstein obtained the general equations (20).

Many of the most interesting phenomena that come from the equations (20) are already present in the vacuum case (21). This is somewhat ironic, because it was the forms of T and (10) that dictated (20). Note, in contrast, that in the Newtonian theory (2), the "vacuum" equations $\mu = 0$ and standard boundary conditions at infinity imply $\phi = 0$. Thus, the Newtonian theory of the vacuum is trivial.

The part of the curvature tensor $R_{\mu\nu\lambda\rho}$ that is not forced to vanish from (21) is known as the *Weyl curvature*. This curvature measures the "tidal" distortion of families of geodesics. Thus, the "local strength" of gravitational fields in vacuum regions is related in the Newtonian limit to the tidal forces on macroscopic test matter, not the norm of the gravitational force.

3.5 The Manifold Concept

We have been able to get this far without really addressing the question of *where the metric $\boldsymbol{g}$ is defined*. In passing from the Minkowski metric to a general $\boldsymbol{g}$, Einstein did not originally have in mind replacing the domain $\mathbb{R}^4$. But it is clear in the Riemannian case from the theory of surfaces that the natural object for a metric to live on is not necessarily $\mathbb{R}^2$ but a general surface. For instance, the metric $\mathrm{d}\theta^2 + \sin\theta\, \mathrm{d}\phi^2$ naturally lives on the sphere $\mathbb{S}^2$. In saying this, we are to understand that one requires several coordinate systems of the type (θ, ϕ) to cover all of $\mathbb{S}^2$. The n-dimensional generalization of the object where Riemannian or Lorentzian metrics naturally live is a MANIFOLD [I.3 §6.9]. Manifolds are the structures obtained by consistently smoothly pasting together local coordinate systems.

Thus, general relativity allows the space-time continuum not to be $\mathbb{R}^4$ but instead to be a general manifold $\mathcal{M}$, which may very well be topologically inequivalent to $\mathbb{R}^4$, just as $\mathbb{S}^2$ is inequivalent to $\mathbb{R}^2$. We call the pair $(\mathcal{M}, \boldsymbol{g})$ a *Lorentzian manifold*. Properly put, the unknown in the Einstein equations is not just $\boldsymbol{g}$ but the pair $(\mathcal{M}, \boldsymbol{g})$.

It is interesting that this fundamental fact, namely that the topology of space-time is not a priori determined by the equations, arises almost as an afterthought. Moreover, it was a thought that took many years to be clarified.

3.6 Waves, Gauges, and Hyperbolicity

When written out explicitly in arbitrary coordinates (try it!), the Einstein equations do not appear to be of any usual type, such as elliptic (like THE POISSON EQUATION [IV.12 §1]), parabolic (like THE HEAT EQUATION [I.3 §5.4]), or hyperbolic (like THE WAVE EQUATION [I.3 §5.4]; see [IV.12 §2.5] for more about these different classes of PDEs). This is related to the fact that, given a solution, one can form a "new" solution by composing the old solution with a coordinate transformation. We can do this for new coordinate systems whose coordinate transformations differ from the identity only in a ball. This fact, known as the *hole argument*, confused Einstein and his mathematical collaborator Marcel Grossmann, who were thinking algebraically in terms of the form of the equations in coordinates, and temporarily led them to reject "general covariance." The resulting backtracking delayed the final correct formulation of (1) by about two years. The geometric interpretation of the theory immediately suggests the resolution to the dilemma: such solutions are to be considered "the same" because they are the same from the point of view of all geometric measurements. In modern language, a solution to the Einstein vacuum equations (say) is an EQUIVALENCE CLASS [I.2 §2.3] of space-times $(\mathcal{M}, g)$, where two space-times are equivalent if there exists a diffeomorphism ϕ between them such that in any open set the metric has the same coordinate form when one identifies local coordinates by ϕ.

It turns out that once these conceptual issues are overcome, the Einstein equations can be viewed as hyperbolic. The easiest way to do this is to impose a *gauge*: that is to say, a certain restriction on the coordinate system. Specifically, one requires the coordinate functions x^α to satisfy the wave equation $\Box_g x^\alpha = 0$, where the *d'Alembertian* operator is defined by the formula

$$\Box_g = \frac{1}{\sqrt{-g}} \partial_\mu (\sqrt{-g} g^{\mu\nu} \partial_\nu).$$

Such coordinates always exist locally and they are traditionally called *harmonic coordinates*, although the term *wave coordinates* would perhaps be more appropriate. The Einstein equation can then be written as a system

$$\Box_g g_{\mu\nu} = N_{\mu\nu}(\{g_{\alpha\beta}\}, \{\partial_\gamma g_{\alpha\beta}\}),$$

where $N_{\mu\nu}$ is a nonlinear expression that is quadratic in the $\partial_\gamma g_{\alpha\beta}$. In view of the Lorentzian signature of the metric, the above system constitutes what is known as *a second-order nonlinear (but quasilinear) hyperbolic system*.

At this point, it is instructive to make a comparison with the Maxwell equations. Suppose we are given an electric field E and a magnetic field B defined on Minkowski space. A 4-*potential* is a vector field A such that $E_i = -\partial_i A_0 - c^{-1} \partial_t A_i$, and $B_i = \sum_{j,k=1}^3 \epsilon_{ijk} \partial_j A_k$. (Here $\epsilon_{123} = 1$, and ϵ_{ijk} is totally antisymmetric, i.e., it transforms to its negative under permutation of any two indices.) If one wishes to view A as the fundamental physical object, then one notices that if A is replaced by the field $\tilde{A}$, defined by the formula

$$\tilde{A} = A + (-c^{-1} \partial_t \psi, \partial_1 \psi, \partial_2 \psi, \partial_3 \psi),$$

where ψ is an *arbitrary* function, then $\tilde{A}$ is also a 4-potential for E and B. One can expect a determined equation for A only if one imposes further conditions on it: that is, if one "fixes the gauge." (The terminology "gauge" is originally due to WEYL [VI.80].) In the so-called *Lorentz gauge*

$$\nabla^\mu A_\mu = 0,$$

the Maxwell equations can be written

$$\Box A_\mu = -c^{-2} \partial_t^2 A_\mu + \sum_i \partial_i^2 A_\mu = 0,$$

from which the wave properties are completely manifest. The gauge-symmetric point of view lived on to later twentieth century glory: the *Yang–Mills equations*, which are a nonlinear generalization of the Maxwell equations with a similar gauge symmetry, are the central part of the so-called *standard model* for particle physics.

The hyperbolicity property of the Einstein equations has two important repercussions. The first is that there should exist *gravitational waves*. This was noted by Einstein at least as early as 1918, essentially as a result of a linearized version of the considerations in the above discussion. The second is that there is a WELL-POSED INITIAL VALUE PROBLEM [IV.12 §2.4] for the Einstein equations (1) with the domain-of-dependence property, when these are coupled with appropriate matter equations. In particular, this is true in the vacuum case (21). The proper conceptual framework to formulate the latter problem took a long time to get right, and was only completely understood through work of Choquet-Bruhat and Geroch in the 1950s and 1960s, based on the fundamental concept of *global hyperbolicity* due to Leray. Well-posedness means that one could associate a unique solution (in the vacuum case, a Lorentzian 4-manifold $(\mathcal{M}, g)$ satisfying (21)) with a suitable notion of initial data. Of course, "initial data" does not mean "data at time $t = 0$," since the concept of $t = 0$ is not geometric. Instead, the data take the form of some Riemannian 3-manifold $(\Sigma, \bar{g})$ with a symmetric covariant 2-tensor K. The triple $(\Sigma, \bar{g}, K)$ has to satisfy the so-called *Einstein constraint equations*. But with this notion, the fundamental problem of general relativity, despite its revolutionary conceptual structure, is thoroughly classical: to determine the relation of the solution to initial data, that is to say, to determine the future from knowledge of the "present." This is the problem of *dynamics*.

4 The Dynamics of General Relativity

In this final section we give a taste of our current mathematical understanding of the dynamics of the Einstein equations.

4.1 Stability of Minkowski Space and the Nonlinearity of Gravitational Radiation

In any physical theory in which one can formulate the problem of dynamics, the most basic question is the stability of the trivial solution. In other words, if we make a small change to the "initial conditions," will the resulting change to the solution be small as well? In the case of general relativity, this is the question of stability of the Minkowski space-time $\mathbb{R}^{3+1}$. This fundamental result was proven for the vacuum equations (21) in 1993 by Christodoulou and Klainerman.

The proof of the stability of Minkowski space made it possible to formulate the *laws of gravitational radiation* rigorously. Gravitational radiation is yet to be observed directly, but it has been inferred, originally by Hulse and Taylor, from the energy loss of a binary system. This work gave them the only Nobel prize (1993) directly associated with the Einstein equations! The blueprint for the mathematical formulation of the radiation problem is based on work of Bondi and later Penrose. One associates with the space-time $(\mathcal{M}, g)$ an ideal boundary "at infinity," known as *null infinity* and denoted $\mathcal{I}^+$. Physically, the points of $\mathcal{I}^+$ correspond to observers who are far away from the isolated self-gravitating system but who are receiving its signals. Gravitational radiation can be identified with certain tensors defined on $\mathcal{I}^+$ from rescaled boundary limits of various geometric quantities. As Christodoulou was to discover, the laws of gravitational radiation are themselves nonlinear, and the nonlinearity is potentially relevant for observation.

4.2   Black Holes

Perhaps no prediction of general relativity is better known today than that of black holes.

The story of black holes begins with the so-called *Schwarzschild* metric:

$$-\left(1 - \frac{2m}{r}\right)dt^2 + \left(1 - \frac{2m}{r}\right)^{-1}dr^2$$
$$+ r^2(d\theta^2 + \sin^2\theta\, d\phi^2). \quad (22)$$

The parameter m here is a positive constant. This is a solution of the vacuum Einstein equations (21) that was found in 1916. The original interpretation of (22) was that it modeled the gravitational field in a vacuum region outside a star. That is to say, (22) was considered only in some coordinate range $r > R_0$, for an $R_0 > 2m$, and the metric was matched at $r = R_0$ to a "static" interior metric satisfying the coupled Einstein–Euler system in the coordinate range $r \leqslant R_0$. (This latter metric is again of the form (22), but with $m = m(r)$ such that $m \to 0$ as $r \to 0$.)

From the theoretical point of view, a natural problem poses itself. Suppose we do away with the star alto-

gether and try to consider (22) for *all* values of r. What happens then to the metric (22) at $r = 2m$? In the (r, t) coordinates, the metric element appears to be singular. But this turns out to be an illusion! By a simple change of coordinates, one can easily extend the metric regularly as a solution of (21) beyond $r = 2m$. That is, there exists a manifold $\mathcal{M}$ that contains both a region $r > 2m$ and a region $0 < r < 2m$, separated by a regular (null) hypersurface $\mathcal{H}^+$. The metric element (22) is valid everywhere except on $\mathcal{H}^+$, where it must be rewritten in regular coordinates.

It turns out that the hypersurface $\mathcal{H}^+$ can be characterized by an exceptional global property: it defines the boundary of the region of space-time that can send signals to null infinity $\mathcal{I}^+$, or, in the physical interpretation, to distant observers. In general, the set of points that *cannot* send signals to null infinity $\mathcal{I}^+$ is known as the *black hole* region of space-time. Thus, the region $0 < r < 2m$ is the black hole region of $\mathcal{M}$, and $\mathcal{H}^+$ is known as the *event horizon*.

These issues took a long time to be sorted out, partly because the language of global Lorentzian geometry was developed long after the original formulation of the Einstein equations. The global geometry of the extended space-time $\mathcal{M}$ was clarified by Synge in around 1950 and finally by Kruskal in 1960. The name "black hole" is due to the imaginative physicist John Wheeler. From their beginnings as a theoretical curiosity, black holes have become part of the accepted astrophysical explanation for a wide variety of phenomena, and in particular are thought to represent the end-state for the gravitational collapse of many stars.

4.3   Space-Time Singularities

A second natural problem poses itself in relation to the Schwarzschild metric (22), now considered in the region $r < 2m$ of the extended space-time $\mathcal{M}$: *what happens at $r = 0$?*

A computation reveals that as $r \to 0$, the Kretchmann scalar $R_{\mu\nu\lambda\rho}R^{\mu\nu\lambda\rho}$ blows up. Since this expression is a geometric invariant, it follows that, unlike the situation at $r = 2m$, the space-time is *not* regularly extendable beyond 0. Moreover, timelike geodesics (freely falling observers in the test particle approximation) entering the black hole region reach $r = 0$ in finite proper time, so they are "incomplete" in the sense that they cannot be continued indefinitely. They thus "observe" the breakdown of the geometry of the space-time metric. Moreover, macroscopic observers approaching $r = 0$ are torn apart by the gravitational "tidal forces."

In the early years of the subject, it was thought that this seemingly pathological behavior was connected to the high degree of symmetry of the Schwarzschild metric and that "generic" solutions would not exhibit such phenomena. That this is not the case was shown by Penrose's celebrated *incompleteness theorem* of 1965. This states that solutions to the initial value problem for the Einstein equations coupled to appropriate matter will *always* contain such incomplete timelike or null geodesics if the initial data hypersurface is noncompact and contains what is known as a closed trapped surface. The Schwarzschild case may appear to suggest that such incomplete geodesics are associated with the curvature blowing up. However, the situation can in fact be very different, as is apparent in the celebrated *Kerr* solutions, a remarkable two-parameter family of solutions to the vacuum equations (21), discovered only in 1963, which are rotating versions of (22). In the Kerr solutions, incomplete timelike geodesics meet a so-called *Cauchy horizon*, a smooth boundary of the region of space-time that is uniquely determined by initial data.

The theorem of Penrose gives rise to two important conjectures. The first, known as *weak cosmic censorship*, says roughly that for generic physically plausible initial data for suitable Einstein-matter systems, geodesic incompleteness, if it occurs, is always confined to black hole regions. The second, *strong cosmic censorship*, says roughly that for generic admissible initial data, incompleteness of the solution is always associated with a local obstruction to extendability, such as the blow-up of curvature. The latter conjecture would ensure that the unique solution of the initial value problem is the only classical space-time that can arise from the data. That is to say, it would imply that classical determinism holds for the Einstein equations.

Both conjectures are false if we drop the assumption that the initial data are generic, and this is one reason for their difficulty. Indeed, Christodoulou has constructed spherically symmetric solutions of the coupled Einstein-scalar field system (arising from regular initial data) that are geodesically incomplete but do not contain black hole regions. Such space-times are said to contain *naked singularities*.

Naked singularities are easy to construct if one does not require that they arise from the collapse of regular initial data. An example is the Schwarzschild metric (22) for $m < 0$. This metric, however, does not admit a complete asymptotically flat Cauchy hypersurface. This

fact is related to the celebrated *positive energy theorem* of Schoen and Yau.

4.4 Cosmology

The space-times $(\mathcal{M}, \boldsymbol{g})$ discussed previously are all idealized representations of isolated systems. The "rest of the universe" is excised and replaced by an "asymptotically flat end"; far-away observers are placed at an ideal boundary "at infinity." But what if we are more ambitious and consider our space-time $(\mathcal{M}, \boldsymbol{g})$ as representing the whole universe? The study of this latter problem is known as *cosmology*.

Observations suggest that on very large scales the universe is approximately homogeneous and isotropic. This is sometimes known as the *Copernican principle*. Interestingly, one cannot solve the Poisson equation (2) with a constant $\nabla \phi$ and constant nonzero μ on $\mathbb{R}^4$. Thus, in Newtonian physics, cosmology never became a rational science.[1] General relativity, on the other hand, does admit homogeneous and isotropic solutions as well as their perturbations. Indeed, cosmological solutions of the Einstein equations were studied by Einstein himself, de Sitter, Friedmann, and Lemaitre in the early years of the subject.

When general relativity was formulated, the prevailing view was that the universe should be static. This led Einstein to add a term $\Lambda g_{\mu\nu}$ to the left-hand side of his equations, fine-tuned so as to allow for such a solution. The constant Λ is known as the *cosmological constant*. The expansion of the universe is now considered to be an observational fact, beginning with the fundamental discoveries of Hubble. Expanding universes can be modeled to a first approximation by so-called Friedmann–Lemaitre solutions to the Einstein–Euler system, with various values of Λ. In the past direction, these solutions are singular: this singular behavior is often given the suggestive name "the big bang."

4.5 Future Developments

The plethora of exact solutions of the Einstein equations gives us a taste of what the qualitative behavior of more general solutions may be. But a true qualitative understanding of the nature of general solutions has been achieved only in a neighborhood of the very

---

1. One can study "Newtonian cosmology" by modifying the foundations of the Newtonian theory so as to describe the theory with a nonmetric connection on, say, $\mathbb{T}^3 \times \mathbb{R}$. But this step is of course inspired by general relativity (see section 3.5).

simplest solutions. The question of the stability of the black hole solutions described above remains unanswered, as do the cosmic censorship conjectures and the nature of the singularities that occur generically in general relativity. Yet these questions are fundamental to the physical interpretation of the theory, and indeed to assessing its very validity.

How likely is it that these questions can ever be answered by rigorous mathematics? Problems concerning the singular behavior of nonlinear hyperbolic partial differential equations are notoriously difficult. The rich geometric structure of the Einstein equations appears at first as a formidable additional complication, but it may also turn out to be a blessing. One can only hope that the Einstein equations will continue to reveal beautiful mathematical structure that answers fundamental questions about our physical world.

Further Reading

Christodoulou, D. 1999. On the global initial value problem and the issue of singularities. *Classical Quantum Gravity* 16:A23–A35.

Hawking, S. W., and G. F. R. Ellis. 1973. *The Large Scale Structure of Space-Time*. Cambridge Monographs on Mathematical Physics, number 1. Cambridge: Cambridge University Press.

Penrose, R. 1965. Gravitational collapse and space-time singularities. *Physical Review Letters* 14:57–59.

Rendall, A. 2008. *Partial Differential Equations in General Relativity*. Oxford: Oxford University Press.

Weyl, H. 1919. *Raum, Zeit, Materie*. Berlin: Springer. (Also published in English, in 1952, as *Space, Time, Matter*. New York: Dover.)

IV.14 Dynamics
Bodil Branner

---

1 Introduction

Dynamical systems are used to describe the way systems evolve in time, and have their origin in the laws of nature that NEWTON [VI.14] formulated in *Principia Mathematica* (1687). The associated mathematical discipline, the theory of dynamics, is closely related to many parts of mathematics, in particular analysis, topology, measure theory, and combinatorics. It is also highly influenced and stimulated by problems from the natural sciences, such as celestial mechanics, hydrodynamics, statistical mechanics, meteorology, and other parts of mathematical physics,

as well as reaction chemistry, population dynamics, and economics.

Computer simulations and visualizations play an important role in the development of the theory; they have changed our views about what should be considered typical, rather than special and atypical.

There are two main branches of dynamical systems: continuous and discrete. The main focus of this paper will be *holomorphic dynamics*, which concerns discrete dynamical systems of a special kind. These systems are obtained by taking a HOLOMORPHIC FUNCTION [I.3 §5.6] f defined on the complex numbers and applying it repeatedly. An important example is when f is a quadratic polynomial.

1.1 Two Basic Examples

It is interesting to note that both types of dynamical system, continuous and discrete, can be well illustrated by examples that date back to Newton.

(i) *The N-body problem* models the motion in the solar system of the sun and $N - 1$ planets, and does so in terms of differential equations. Each body is represented by a single point, namely its center of mass, and the motion is determined by Newton's *universal law of gravitation*—also called the *inverse square law*. This says that the gravitational force between two bodies is proportional to each of their masses and inversely proportional to the square of the distance between them. Let r_i denote the position vector of the ith body, m_i its mass, and g the universal gravitational constant. Then the force on the ith body due to the jth has magnitude $g m_i m_j / \|r_j - r_i\|^2$, and its direction is along the line from r_i to r_j. We can work out the total force on the ith body by adding up all these forces for $j \neq i$. Since a unit vector in the direction from r_i to r_j is $(r_j - r_i) / \|r_j - r_i\|$, we obtain a force of

$$g \sum_{j \neq i} m_i m_j \frac{r_j - r_i}{\|r_j - r_i\|^3}.$$

(There is a cube on the bottom rather than a square in order to compensate for the magnitude of $r_j - r_i$.) A solution to the N-body problem is a set of differentiable vector functions $(r_1(t), \ldots, r_N(t))$, depending on time t, that satisfy the N differential equations

$$m_i r_i''(t) = g \sum_{j \neq i} m_i m_j \frac{r_j(t) - r_i(t)}{\|r_j(t) - r_i(t)\|^3},$$

which result from Newton's second law, which states that force = mass × acceleration.

Newton was able to solve the two-body problem explicitly. By neglecting the influence of other planets, he derived the laws formulated by Johannes Kepler, which describe how each planet moves in an elliptic orbit around the sun. However, the jump to $N > 2$ makes an enormous difference to the complication of the problem: except in very special cases, the system of equations can no longer be solved explicitly (see THE THREE-BODY PROBLEM [V.33]). Nevertheless, Newton's equations are of great practical importance when it comes to guiding satellites and other space missions.

(ii) NEWTON'S METHOD [II.4 §2.3] for solving equations is quite different and does not involve differential equations. We consider a differentiable function f of one real variable and wish to determine a zero of f, that is, a solution to the equation $f(x) = 0$. Newton's idea was to define a new function:

$$N_f(x) = x - \frac{f(x)}{f'(x)}.$$

To put this more geometrically, $N_f(x)$ is the x-coordinate of the point where the tangent line to the graph $y = f(x)$ at the point $(x, f(x))$ crosses the x-axis. (If $f'(x) = 0$, then this tangent line is horizontal and $N_f(x)$ is not defined.)

Under many circumstances, if x is close to a zero of f, then $N_f(x)$ is significantly closer. Therefore, if we start with some value x_0 and form the sequence obtained by repeated application of N_f, that is, the sequence $x_0, x_1, x_2, \ldots$, where $x_1 = N_f(x_0)$, $x_2 = N_f(x_1)$, and so on, we can expect that this sequence will converge to a zero of f. And this is true: if the initial value x_0 is sufficiently close to a zero, then the sequence does indeed converge toward that zero, and does so extremely quickly, basically doubling the number of correct digits in each step. This rapid convergence makes Newton's method very useful for numerical computations.

1.2 Continuous Dynamical Systems

We can think of a *continuous dynamical system* as a system of first-order differential equations, which determine how the system evolves in time. A solution is called an *orbit* or *trajectory*, and is parametrized by a number t, which one usually thinks of as time, that takes real values and varies continuously: hence the name "continuous" dynamical system. A *periodic orbit* of *period T* is a solution that repeats itself after time T, but not earlier.

The differential equation $x''(t) = -x(t)$ is of second order, but it is nevertheless a continuous dynamical system because it is equivalent to the system of two first-order differential equations $x_1'(t) = x_2(t)$ and $x_2'(t) = -x_1(t)$. In a similar way, the system of differential equations of the N-body problem can be brought into standard form by introducing new variables. The equations are equivalent to a system of $6N$ first-order differential equations in the variables of the position vectors $r_i = (x_{i1}, x_{i2}, x_{i3})$ and the velocity vectors $r_i' = (y_{i1}, y_{i2}, y_{i3})$. Thus, the N-body problem is a good example of a continuous dynamical system.

In general, if we have a dynamical system consisting of n equations, then we can write the ith equation in the form

$$x_i'(t) = f_i(x_1(t), \ldots, x_n(t)),$$

or alternatively we can write all the equations at once in the form $\boldsymbol{x}'(t) = \boldsymbol{f}(\boldsymbol{x}(t))$, where $\boldsymbol{x}(t)$ is the vector $(x_1(t), \ldots, x_n(t))$ and $\boldsymbol{f} = (f_1, \ldots, f_n)$ is a function from $\mathbb{R}^n$ to $\mathbb{R}^n$. Note that $\boldsymbol{f}$ is assumed not to depend on t. If it does, then the system can be brought into standard form by adding the variable $x_{n+1} = t$ and the differential equation $x_{n+1}'(t) = 1$, which increases the dimension of the system from n to $n + 1$.

The simplest systems are *linear* ones, where $\boldsymbol{f}$ is a linear map: that is, $\boldsymbol{f}(\boldsymbol{x})$ is given by $A\boldsymbol{x}$ for some constant $n \times n$ matrix A. The system above, $x_1'(t) = x_2(t)$ and $x_2'(t) = -x_1(t)$, is an example of a linear system. Most systems, however, including the one for the N-body problem, are *nonlinear*. If the function $\boldsymbol{f}$ is "nice" (for instance, differentiable), then *uniqueness* and *existence* of solutions are guaranteed for any initial point $\boldsymbol{x}_0$. That is, there is exactly one solution that passes through the point $\boldsymbol{x}_0$ at time $t = 0$. For example, in the N-body problem there is exactly one solution for any given set of initial position vectors and initial velocity vectors. It also follows from uniqueness that any pair of orbits must either coincide or be totally disjoint. (Bear in mind that the word "orbit" in this context does not mean the set of positions of a single point mass, but rather the evolution of the vector that represents all the positions and velocities of all the masses.)

Although it is seldom possible to express solutions to nonlinear systems explicitly, we know that they exist, and we call the dynamical system *deterministic* since solutions are completely determined by their initial conditions. For a given system and given initial conditions it is therefore theoretically possible to predict its entire future evolution.

1.3   Discrete Dynamical Systems

A *discrete dynamical system* is a system that evolves in jumps: "time," in such a system, is best represented by an integer rather than a real number. A good example is Newton's method for solving equations. In this instance, the sequence of points we saw earlier, $x_0, x_1, \ldots, x_k, \ldots$, where $x_k = N_f(x_{k-1})$, is called the *orbit* of x_0. We say that it is obtained by *iteration* of the function N_f, i.e., by repeated application of the function.

This idea can easily be generalized to other mappings $F : X \to X$, where X could be the real axis, an interval in the real axis, the plane, a subset of the plane, or some more complicated space. The important thing is that the output $F(x)$ of any input x can be used as the next input. This guarantees that the orbit of any x_0 in X is defined for all future times. That is, we can define a sequence, $x_0, x_1, \ldots, x_k, \ldots$, where $x_k = F(x_{k-1})$ for every k. If the function F has an inverse F^{-1}, then we can iterate both forwards and backwards and obtain the *full orbit* of x_0 as the bi-infinite sequence $\ldots, x_{-2}, x_{-1}, x_0, x_1, x_2, \ldots$, where $x_k = F(x_{k-1})$ and, equivalently, $x_{k-1} = F^{-1}(x_k)$, for all integer values.

The orbit of x_0 is *periodic* of period k if it repeats itself after time k, but not earlier, i.e., if $x_k = x_0$, but $x_j \neq x_0$ for $j = 1, \ldots, k-1$. The orbit is called *pre-periodic* if it is eventually periodic, in other words if there exist $\ell \geqslant 1$ and $k \geqslant 1$ such that x_ℓ is periodic of period k, but none of the x_j for $0 \leqslant j < \ell$ are periodic. The notion of pre-periodicity has no counterpart in continuous dynamics.

A discrete dynamical system is deterministic, since the orbit of any given initial point x_0 is completely determined once you know x_0.

1.4   Stability

The modern theory of dynamics was greatly influenced by the work of POINCARÉ [VI.61], and in particular by his prize-winning memoir on the three-body problem, succeeded by three more elaborate volumes on celestial mechanics, all from the late nineteenth century. The memoir was written in response to a competition where one of the proposed problems concerned stability of the solar system. Poincaré introduced the so-called *restricted three-body problem*, where the third body is assumed to have an infinitely small mass: it does not influence the motion of the other two bodies but it is influenced by them. Poincaré's work became the prelude to *topological dynamics*, which focuses on topological properties of solutions to dynamical systems and takes a qualitative approach to them.

Of special interest is the long-term behavior of a system. A periodic orbit is called *stable* if all orbits through points sufficiently close to it stay close to it at all future times. It is called *asymptotically stable* if all sufficiently close orbits approach it as time tends to infinity. Let us illustrate this by two linear examples in discrete dynamics. For the real function $F(x) = -x$, all points have a periodic orbit: 0 has period 1 and all other x have period 2. Every orbit is stable, but none is asymptotically stable. The real function $G(x) = \frac{1}{2}x$ has only one periodic orbit, namely 0. Since $G(0) = 0$, this orbit has period 1, and we call it a *fixed point*. If you take any number and repeatedly divide it 2, then the resulting sequence will approach 0, so the fixed point 0 is asymptotically stable.

One of the methods introduced by Poincaré during his study of the three-body problem was a reduction from a continuous dynamical system, in dimension n, say, to an associated discrete dynamical system, a mapping in dimension $n - 1$. The idea is as follows. Suppose we have a periodic orbit of period $T > 0$ in some continuous system. Choose a point $\boldsymbol{x}_0$ on the orbit and a hypersurface Σ through $\boldsymbol{x}_0$, for instance part of a hyperplane, such that the orbit cuts through Σ at $\boldsymbol{x}_0$. For any point in Σ that is sufficiently close to $\boldsymbol{x}_0$, one can follow its orbit around and see where it next intersects Σ. This defines a transformation, known as the *Poincaré map*, which takes the original point to the next point of intersection of its orbit with Σ. It follows from the fact that dynamical systems have unique solutions that every Poincaré map is injective in the neighborhood of $\boldsymbol{x}_0$ (within Σ) for which the Poincaré map is defined. One can perform both forwards and backwards iterations. Note that the periodic orbit of $\boldsymbol{x}_0$ in the continuous system is stable (respectively, asymptotically stable) exactly when the fixed point $\boldsymbol{x}_0$ of the Poincaré map in the discrete system is stable (respectively, asymptotically stable).

1.5   Chaotic Behavior

The notion of *chaotic dynamics* arose in the 1970s. It has been used in different settings, and there is no single definition that covers all uses of the term. However, the property that best characterizes chaos is the phenomenon of *sensitive dependence on initial conditions*.

Poincaré was the first to observe sensitivity to initial conditions in his treatment of the three-body problem.

Instead of describing his observations let us look at a much simpler example from discrete dynamics. Take as a dynamical space X the half-open unit interval $[0, 1)$, and let F be the function that doubles a number and reduces it modulo 1. That is, $F(x) = 2x$ when $0 \leqslant x < \frac{1}{2}$ and $F(x) = 2x - 1$ when $\frac{1}{2} \leqslant x < 1$. Let x_0 be a number in X and let its iterates be $x_1 = F(x_0)$, $x_2 = F(x_1)$, and so on. Then x_k is the fractional part of $2^k x_0$. (The fractional part of a real number t is what you get when you subtract the largest integer less than t.)

A good way to understand the behavior of the sequence $x_0, x_1, x_2, \ldots$ of iterates is to consider the binary expansion of x_0. Suppose, for example, that this begins $0.110100010100111\ldots$. To double a number when it is written in binary, all you have to do is shift every digit to the left (just as one does in the decimal system when multiplying by 10). So $2x_0$ will have a binary expansion that begins $1.10100010100111\ldots$. To obtain $F(x_0)$, we have to take the fractional part of this, which we do by subtracting the initial 1. This gives us $x_1 = 0.10100010100111\ldots$. Repeating the process we find that $x_2 = 0.0100010100111\ldots$, $x_3 = 0.100010100111\ldots$, and so on. (Notice that when we calculated x_3 from x_2 there was no need to subtract 1, since the first digit after the "decimal point" was a 0.) Now consider a different choice of initial number, $x_0' = 0.110100010110110\ldots$. The first nine digits after the decimal point are the same as the first nine digits of x_0, so x_0' is very close to x_0. However, if we apply F ten times to x_0 and x_0', then their respective eleventh digits have shifted leftwards and become the first digits of $x_{10} = 0.00111\ldots$ and $x_{10}' = 0.10110\ldots$. These two numbers differ by almost $\frac{1}{2}$, so they are not at all close.

In general, if we know x_0 to an accuracy of k binary digits and no more, then after k iterations of the map F we have lost all information: x_k could lie anywhere in the interval $[0, 1)$. Therefore, even though the system is deterministic, it is impossible to predict its long-term behavior without knowing x_0 with perfect accuracy.

This is true in general: it is impossible to make long-term predictions in any part of a dynamical system that shows sensitivity to initial conditions unless the initial conditions are known exactly. In practical applications this is never the case. For instance, when applying a mathematical model to perform weather forecasts, one does not know the initial conditions exactly, and this is why reliable long-term forecasting is impossible.

Sensitivity is also important in the notion of so-called *strange attractors*. A set A is called an *attractor* if all orbits that start in A stay in A and if all orbits through nearby points get closer and closer to A. In continuous systems, some simple sets that can be attractors are equilibrium points, periodic orbits (limit cycles), and surfaces such as a torus. In contrast to these examples, strange attractors have both complicated geometry and complicated dynamics: the geometry is *fractal* and the dynamics sensitive. We shall see examples of fractals later on.

The best-known strange attractor is the *Lorenz attractor*. In the early 1960s, the meteorologist Edward N. Lorenz studied a three-dimensional continuous dynamical system that gave a simplified model of heat flow. While doing so, he noticed that if he restarted his computer with its initial conditions chosen as the output of an earlier calculation, then the trajectory started to diverge from the one he had previously observed. The explanation he found was that the computer used more precision in its internal calculations than it showed in its output. For this reason, it was not immediately apparent that the initial conditions were in fact very slightly different from before. Because the system was sensitive, this tiny difference eventually made a much bigger difference. He coined the poetic phrase "the butterfly effect" to describe this phenomenon, suggesting that a small disturbance such as a butterfly flickering its wings could in time have a dramatic effect on the long-term evolution of the weather and trigger a tornado thousands of miles away. Computer simulations of the Lorenz system indicate that solutions are attracted to a complicated set that "looks like" a strange attractor. The question of whether it actually was one remained open for a long time. It is not obvious how trustworthy computer simulations are when one is studying sensitive systems, since the computer rounds off the numbers in each step. In 1998 Warwick Tucker gave a computer-assisted proof that the Lorenz attractor is in fact a strange attractor. He used *interval arithmetic*, where numbers are represented by intervals and estimates can be made precise.

For topological reasons, sensitivity to initial conditions is possible for continuous dynamical systems only when the dimension is at least 3. For discrete systems where the map F is injective, the dimension must be at least 2. However, for noninjective mappings, sensitivity can occur for one-dimensional systems, as we saw with the example given earlier. This is one of the reasons that discrete one-dimensional dynamical systems have been intensively studied.

1.6 Structural Stability

Two dynamical systems are said to be *topologically equivalent* if there is a homeomorphism (a continuous map with continuous inverse) that maps the orbits of one system onto the orbits of the other, and vice versa. Roughly speaking, this means that there is a continuous change of variables that turns one system into the other.

As an example, consider the discrete dynamical system given by the real quadratic polynomial $F(x) = 4x(1-x)$. Suppose we were to make the substitution $y = -4x + 2$. How could we describe the system in terms of y? Well, if we apply F, then we change x to $4x(1-x)$, which means that $y = -4x+2$ changes $F(x)$ to $-4F(x) + 2 = -16x(1-x) + 2$. But

$$-16x(1-x) + 2 = 16x^2 - 16x + 2$$
$$= (-4x + 2)^2 - 2$$
$$= y^2 - 2.$$

Therefore, the effect of applying the polynomial function F to x is to apply a different polynomial function to y, namely $Q(y) = y^2 - 2$. Since the change of variables from x to $-4x + 2$ is continuous and invertible, one says that the functions F and Q are *conjugate*.

Because F and Q are conjugate, the orbit of any x_0 under F becomes, after the change of variables, the orbit of the corresponding point $y_0 = -4x_0 + 2$ under Q. That is, for every k we have $y_k = -4x_k+2$. The two systems are topologically equivalent: if you want to understand the dynamics of one of them, you can if you study the other, since its dynamics will be qualitatively the same.

For continuous dynamical systems the notion of equivalence is slightly looser in that we allow a homeomorphism between two topologically equivalent systems to map one orbit onto another without respecting the exact time evolution, but for discrete dynamical systems we must demand that the time evolution is respected as in the example above: in other words, we insist on conjugacy.

The term *dynamical system* was coined by Stephen Smale in the 1960s and has taken off since then. Smale evolved the theory of *robust* systems, also named *structurally stable* systems, a notion that was introduced in the 1930s by Alexander A. Andronov and Lev S. Pontryagin. A dynamical system is called structurally stable if all systems sufficiently close to it, belonging to

some specified family of systems, are in fact topologically equivalent to it. We say that they all have the same qualitative behavior. An example of the kind of family one might consider is the set of all real quadratic polynomials of the form $x^2 + a$. This family is parametrized by a, and the systems close to a given polynomial $x^2 + a_0$ are all the polynomials $x^2 + a$ for which a is close to a_0. We shall return to the question of structural stability when we discuss holomorphic dynamics later.

If a family of dynamical systems parametrized by a variable a is not structurally stable, it may still be that the system with parameter a_0 is topologically equivalent to all systems with parameter a in some region that contains a_0. A major goal of research into dynamics is to understand not just the qualitative structure of each system in the family, but also the structure of the *parameter space*, that is, how it is divided up into such regions of stability. The boundaries that separate these regions form what is called the *bifurcation set*: if a_0 belongs to this set, then there will be parameters a arbitrarily close to a_0 for which the corresponding system has a different qualitative behavior.

A description and classification of structurally stable systems and a classification of possible bifurcations is not within reach for general dynamical systems. However, one of the success stories in the subject, holomorphic dynamics, studies a special class of dynamical systems for which many of these goals have been attained. It is time to turn our attention to this class.

2 Holomorphic Dynamics

Holomorphic dynamics is the study of discrete dynamical systems where the map to be iterated is a HOLOMORPHIC FUNCTION [I.3 §5.6] of the COMPLEX NUMBERS [I.3 §1.5]. Complex numbers are typically denoted by z. In this article, we shall consider iterations of complex polynomials and rational functions (that is, functions like $(z^2+1)/(z^3+1)$ that are ratios of polynomials), but much of what we shall say about them is true for more general holomorphic functions, such as EXPONENTIAL [III.25] and TRIGONOMETRIC [III.92] functions.

Whenever one restricts attention to a special kind of dynamical system, there will be tools that are specially adapted to that situation. In holomorphic dynamics these tools come from complex analysis. When we concentrate on rational functions, there are more special tools, and if we restrict further to polynomials, then there are yet others, as we shall see.

Why might one be interested in iterating rational functions? One answer arose in 1879, when CAYLEY [VI.46] had the idea of trying to find roots of complex polynomials by extending Newton's method, which we discussed in the introduction, from real numbers to complex numbers. Given any polynomial P, the corresponding Newton function N_P is a rational function, given by the formula

$$N_P(z) = z - \frac{P(z)}{P'(z)} = \frac{zP'(z) - P(z)}{P'(z)}.$$

To apply Newton's method, one iterates this rational function.

The study of the iteration of rational functions flourished at the beginning of the twentieth century, thanks in particular to work of Pierre Fatou and Gaston Julia (who independently obtained many of the same results). Part of their work concerned the study of the local behavior of functions in the neighborhoods of a fixed point. But they were also concerned about global dynamical properties and were inspired by the theory of so-called *normal families*, then recently established by Paul Montel. However, research on holomorphic dynamics almost came to a stop around 1930, because the fractal sets that lay behind the results were so complicated as to be almost beyond imagination. The research came back to life in around 1980 with the vastly extended calculating powers of computers, and in particular the possibility of making sophisticated graphic visualizations of these fractal sets. Since then, holomorphic dynamics has attracted a lot of attention. New techniques continue to be developed and introduced.

To set the scene, let us start by looking at one of the simplest of polynomials, namely z^2.

2.1 The Quadratic Polynomial z^2

The dynamics of the simplest quadratic polynomial, $Q_0(z) = z^2$, plays a fundamental role in the understanding of the dynamics of any quadratic polynomial. Moreover, the dynamical behavior of Q_0 can be analyzed and understood completely.

If $z = re^{i\theta}$, then $z^2 = r^2e^{2i\theta}$, so squaring a complex number squares its modulus and doubles its argument. Therefore, the unit circle (the set of complex numbers of modulus 1) is mapped by Q_0 to itself, while a circle of radius $r < 1$ is mapped onto a circle closer to the origin, and a circle of radius $r > 1$ is mapped onto a circle farther away.

Let us look more closely at what happens to the unit circle. A typical point in the circle, $e^{i\theta}$, can be parametrized by its argument θ, which we can take to lie in the interval $[0, 2\pi)$. When we square this number, we obtain $e^{2i\theta}$, which is parametrized by the number 2θ if $2\theta < 2\pi$, but if $2\theta \geqslant 2\pi$, then we subtract 2π so that the argument, $2\theta - 2\pi$, still lies in $[0, 2\pi)$. This is strongly reminiscent of the dynamical system we considered in section 1.5. In fact, if we replace the argument θ by *its modified argument* $\theta/2\pi$, which amounts to writing $e^{2\pi i\theta}$ instead of $e^{i\theta}$, then it becomes exactly the same system. Therefore, the behavior of z^2 on the unit circle is chaotic.

As for the rest of the complex plane, the origin is an asymptotically stable fixed point, $Q_0(0) = 0$. For any point z_0 inside the unit circle the iterates z_k converge to 0 as k tends to infinity. For any point z_0 outside the unit circle the distance $|z_k|$ between the iterates z_k and the origin tends to infinity as k tends to infinity. The set of initial points z_0 with bounded orbit is equal to the closed unit disk, i.e., all points for which $|z_0| \leqslant 1$. Its boundary, the unit circle, divides the complex plane into two domains with qualitatively different dynamical behavior.

Some orbits of Q_0 are periodic. In order to determine which ones, we first notice that the only possibility outside the unit circle is the fixed point at the origin, since all other points, when you repeatedly square them, either get steadily closer and closer to the origin, or get steadily farther and farther away. So now let us look at the unit circle, and consider the point $e^{2\pi i\theta_0}$, with modified argument θ_0. If this point is periodic with period k, we must have $2^k\theta_0 = \theta_0 \pmod{1}$: that is, $(2^k - 1)\theta_0$ must be an integer. Because of this, it is convenient to parametrize a point on the unit circle by its modified argument. From now on, when we say "the point θ," we shall mean the point $e^{2\pi i\theta}$, and when we say "argument" we shall mean modified argument.

We have just established that the point θ is periodic with period k only if $(2^k - 1)\theta$ is an integer. It follows that there is one point of period 1, namely $\theta_0 = 0$. There are two points of period 2, forming one orbit, namely $\frac{1}{3} \mapsto \frac{2}{3} \mapsto \frac{1}{3}$. There are six points for period 3, forming two orbits, namely $\frac{1}{7} \mapsto \frac{2}{7} \mapsto \frac{4}{7} \mapsto \frac{1}{7}$ and $\frac{3}{7} \mapsto \frac{6}{7} \mapsto \frac{5}{7} \mapsto \frac{3}{7}$. (At each stage, we double the number we have, and subtract 1 if that is needed to get us back into the interval $[0, 1)$.) The points of period 4 are fractions with denominator 15, but the converse is not true: the fractions $\frac{3}{15} = \frac{1}{5}$ and $\frac{6}{15} = \frac{2}{5}$ have the lower period 2. The periodic points on the unit circle are *dense* in the unit circle, meaning that arbitrarily close to any point is a periodic point. This follows from the observation that all repeating binary expansions, such as 0.110001100011000110001100011000... are periodic, and any finite sequence of 0s and 1s is the start of a repeating sequence. One can, in fact, show that the periodic points on the unit circle are exactly the points whose argument is a fraction p/q in $[0, 1)$ with q odd. Any fraction with even denominator can be written in the form $p/(2^\ell q)$ for some odd number q. After ℓ iterations, such a fraction will land on a periodic point, so the initial point is pre-periodic. Points with rational argument in $[0, 1)$ have a finite orbit, while points with irrational argument have an infinite orbit. The reason for taking modified arguments is now justified: the behavior of the dynamics depends on whether θ_0 is rational or irrational.

When θ_0 is irrational its orbit may or may not be dense in $[0, 1)$. This is another fact that is easy to see if one considers binary expansions. For instance, a very special example of a θ_0 with a dense orbit is given by the binary expansion

$$\theta_0 = 0.0100011011000001010011100101110111\ldots,$$

where one obtains this expansion by simply listing all finite binary sequences in turn: first the blocks of length one, 0 and 1, then the blocks of length two, 00, 01, 10, and 11, and so on. When we iterate, this binary expansion shifts to the left and all possible finite sequences appear at some time or another at the beginning of some iterate θ_k.

2.2   Characterization of Periodic Points

Let z_0 be a fixed point of a holomorphic map F. How do the iterates of points near z_0 behave? The answer depends crucially on a number ρ, called the *multiplier* of the fixed point, which is defined to be $F'(z_0)$. To see why this is relevant, notice that if z is very close to z_0, then $F(z)$ is, to a first-order approximation, equal to $F(z_0) + F'(z_0)(z - z_0) = z_0 + \rho(z - z_0)$. Thus, when you apply F to a point near z_0, its difference from z_0 approximately multiplies by ρ. If $|\rho| < 1$, then nearby points will get closer to z_0, in which case z_0 is called an *attracting* fixed point. If $\rho = 0$, then this happens very quickly and z_0 is called *super-attracting*. If $|\rho| > 1$, then nearby points get farther away and z_0 is called *repelling*. Finally, if $|\rho| = 1$, then one says that z_0 is *indifferent*.

If z_0 is indifferent, then its multiplier will take the form $\rho = e^{2\pi i \theta}$, and near z_0 the map F will be approx-

imately a rotation about z_0 by an angle of $2\pi\theta$. The behavior of the system depends very much on the precise value of θ. We call the fixed point *rationally* or *irrationally indifferent* if θ is rational or irrational, respectively. The dynamics is not yet completely understood in all irrational cases.

A periodic point z_0 of period k will be a fixed point of the kth iterate $F^k = F \circ \cdots \circ F$ of F. For this reason we define its multiplier by $\rho = (F^k)'(z_0)$. It follows from the chain rule that

$$(F^k)'(z_0) = \prod_{j=0}^{k-1} F'(z_j)$$

and therefore that the derivative of F^k is the same at all points of the periodic orbit. This formula also implies that a super-attracting periodic orbit must contain a critical point (that is, a point where the derivative of F is zero): if $(F^k)'(z_0) = 0$, then at least one $F'(z_j)$ must be 0.

Note that 0 is a super-attracting fixed point of Q_0, and that any periodic orbit of Q_0 of period k on the unit circle has multiplier 2^k. All periodic orbits on the unit circle are therefore repelling.

2.3   A One-Parameter Family of Quadratic Polynomials

The quadratic polynomial Q_0 sits at the center of the one-parameter family of quadratic polynomials of the form $Q_c(z) = z^2 + c$. (We considered this family earlier, but then z and c were real rather than complex.) For each fixed complex number c we are interested in the dynamics of the polynomial Q_c under iteration. The reason we do not need to study more general quadratic polynomials is that they can be brought into this form by a simple substitution $w = az + b$, similar to the substitution in the real example in section 1.6. For any given quadratic polynomial P we can find exactly one substitution $w = az + b$ and one c such that

$$a(P(z)) + b = (az + b)^2 + c \quad \text{for all } z.$$

Therefore, if we understand the dynamics of the polynomials Q_c, then we understand the dynamics of all quadratic polynomials.

There are other representative families of quadratic polynomials that can be useful. One example is the family $F_\lambda(z) = \lambda z + z^2$. The substitution $w = z + \frac{1}{2}\lambda$ changes F_λ into Q_c, where $c = \frac{1}{2}\lambda - \frac{1}{4}\lambda^2$. We shall return to the expression of c in terms of λ later on. In the family of polynomials Q_c, the parameter $c = Q_c(0)$ coincides with the only *critical value* of Q_c in the plane:

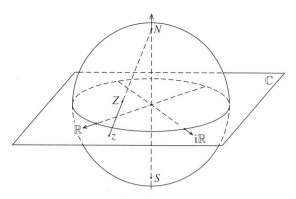

Figure 1 The Riemann sphere.

as we shall see later, critical orbits play an essential role in the analysis of the global dynamics. In the family of polynomials F_λ the parameter λ is equal to the multiplier of the fixed point at the origin of F_λ, which sometimes makes this family more convenient.

2.4 The Riemann Sphere

To understand further the dynamics of polynomials it is best to regard them as a special case of rational functions. Since a rational function can sometimes be infinite, the natural space to consider is not the complex plane $\mathbb{C}$ but the *extended complex plane*, which is the complex plane together with the point "∞." This space is denoted $\hat{\mathbb{C}} = \mathbb{C} \cup \{\infty\}$. A geometrical picture (see figure 1) is obtained by identifying the extended complex plane with the *Riemann sphere*. This is simply the unit sphere $\{(x_1, x_2, x_3) : x_1^2 + x_2^2 + x_3^2 = 1\}$ in three-dimensional space. Given a number z in the complex plane, the straight line joining z to the north pole $N = (0, 0, 1)$ intersects this sphere in exactly one place (apart from N itself). This place is the point in the sphere that is associated with z. Notice that the bigger $|z|$ is, the closer the associated point is to N. We therefore regard N as corresponding to the point ∞.

Let us now think of $Q_0(z) = z^2$ as a function from $\hat{\mathbb{C}}$ to $\hat{\mathbb{C}}$. We have seen that 0 is a super-attracting fixed point of Q_0. What about ∞, which is a fixed point as well? The classification we gave in terms of multipliers does not work at ∞, but a standard trick in this situation is to "move" ∞ to 0. If one wishes to understand the behavior of a function f with a fixed point at ∞, one can look instead at the function $g(z) = 1/f(1/z)$, which has a fixed point at 0 (since $1/f(1/0) = 1/f(\infty) = 1/\infty = 0$). When $f(z) = z^2$, $g(z)$ is also z^2, so ∞ is also a super-attracting fixed point of Q_0.

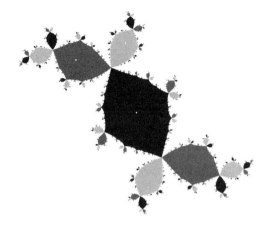

Figure 2 The *Douady rabbit*. The filled Julia set of Q_{c_0} where c_0 is the one root of the polynomial $(c^2 + c)^2 + c$ that has positive imaginary part. This corresponds to one of the three possible c values for which the critical orbit $0 \mapsto c \mapsto c^2 + c \mapsto (c^2 + c)^2 + c = 0$ is periodic of period 3. The critical orbit is marked with three white dots inside the filled Julia set: 0 in the black, c_0 in the light gray, and $c_0^2 + c_0$ in the gray. The corresponding three attracting basins of $Q_{c_0}^3$ are marked in black, light gray, and gray, respectively. The Julia set is the common boundary of the black, light gray, and gray basins of attraction as well as of $A_{c_0}(\infty)$.

In general, if P is any nonconstant polynomial, then it is natural to define $P(\infty)$ to be ∞. Applying the above trick, we obtain a rational function. For example, if $P(z) = z^2 + 1$, then $1/P(1/z) = z^2/(z^2 + 1)$. If P has degree at least 2, then ∞ is a super-attracting fixed point.

The connection between $\hat{\mathbb{C}}$ and rational functions is expressed by the following fact: a function $F : \hat{\mathbb{C}} \to \hat{\mathbb{C}}$ is holomorphic everywhere (with suitable definitions at ∞) if and only if it is a rational function. This is not obvious, but is typically proved in a first course in complex analysis. Among the rational functions, the polynomials are the ones for which $F(\infty) = \infty = F^{-1}(\infty)$.

A polynomial P of degree d has $d - 1$ critical points in the plane (not including ∞). These are the roots of the derivative P', counted with multiplicity. The critical point at ∞ has multiplicity $d - 1$, as can again be seen by looking at the map $1/P(1/z)$. In particular, quadratic polynomials have exactly one critical point in the plane. The degree of a rational function P/Q (where P and Q have no common roots) is defined to be the maximal degree of the polynomials P and Q. A rational function of degree d has $2d - 2$ critical points in $\hat{\mathbb{C}}$, as we have just seen for polynomials.

2.5   Julia Sets of Polynomials

It can be shown that the only invertible holomorphic maps from $\mathbb{C}$ to $\mathbb{C}$ are polynomials of degree 1, that is, functions of the form $az + b$ with $a \neq 0$. The dynamical behavior of these maps is easy to analyze, simple, and hence not interesting.

From now on, therefore, we shall consider only polynomials P of degree at least 2. For all such polynomials, ∞ is a super-attracting fixed point, from which it follows that the plane is split into two disjoint sets with qualitatively different dynamics, one consisting of points that are attracted to ∞ and the other consisting of points that are not. The *attracting basin* of ∞, denoted by $A_P(\infty)$, consists of all initial points z such that $P^k(z) \to \infty$ as $k \to \infty$. (Here, $P^k(z)$ stands for the result of applying P to z k times.) The complement of $A_P(\infty)$ is called the *filled Julia set*, and is denoted by K_P. It can be defined as the set of all points z such that the sequence $z, P(z), P^2(z), P^3(z), \ldots$ is bounded. (It is not hard to show that sequences of this kind either tend to ∞ or are bounded.)

The attracting basin of ∞ is an open set and the filled Julia set is a closed, bounded set (i.e., a COMPACT SET [III.9]). The attracting basin of ∞ is always connected. For this reason the boundary of K_P is equal to the boundary of $A_P(\infty)$. The common boundary is called the *Julia set* of P and is denoted by J_P. The three sets K_P, $A_P(\infty)$, and J_P are completely invariant, i.e., $P(K_P) = K_P = P^{-1}(K_P)$, and so on. If we replace P by any iterate P^k, then the filled Julia set, the attracting basin of ∞, and the Julia set of P^k are the same sets as those of P.

For the polynomial Q_0, we showed earlier that the filled Julia set is the closed unit disk, $\{z : |z| \leqslant 1\}$; the attracting basin of ∞ is its complement, $\{z : |z| > 1\}$; and the Julia set is the unit circle, $\{z : |z| = 1\}$.

The name "filled Julia set" refers to the fact that K_P is equal to J_P with all its holes (or, more formally, the bounded components of its complement) filled in. The complement of the Julia set is called the *Fatou set* and any connected component of it is called a *Fatou component*.

Figures 2–6 show different examples of Julia sets of quadratic polynomials Q_c. For simplicity we set $K_{Q_c} = K_c$, $A_{Q_c}(\infty) = A_c(\infty)$, and $J_{Q_c} = J_c$. Note that all Julia sets J_c are symmetric around 0, owing to the symmetry in the formula: $Q_c(-z) = Q_c(z)$, which implies that if a point z belongs to J_c, then so does $-z$.

Figure 3   The Julia set of $Q_{1/4}$. Every point inside the Julia set (including the critical point 0) is attracted (under repeated applications of $Q_{1/4}$) to the rationally indifferent fixed point $\frac{1}{2}$ with multiplier $\rho = 1$, which belongs to $J_{1/4}$.

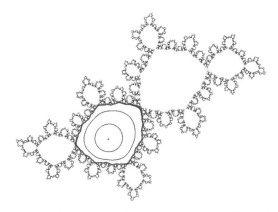

Figure 4   The Julia set of Q_c with a so-called *Siegel disk* around an irrationally indifferent fixed point of multiplier $\rho = e^{2\pi i(\sqrt{5}-1)/2}$. The corresponding c-value is equal to $\frac{1}{2}\rho - \frac{1}{4}\rho^2$. In the Siegel disk, the Fatou component containing the fixed point, the action of Q_c can, after a suitable change of variables, be expressed as $w \mapsto \rho w$. The fixed point is marked and so are some orbits of points in its vicinity. The critical orbit is dense in the boundary of the Siegel disk.

2.6   Properties of Julia Sets

In this section we shall list several common properties of Julia sets. The proofs of these, which are beyond the scope of this article, mostly depend on the theory of *normal families*.

- The Julia set is the set of points for which the system displays sensitivity to initial conditions, i.e., the chaotic subset of the dynamical system.

- The repelling orbits belong to the Julia set and form a dense subset of the set. That is, any point in the Julia set can be approximated arbitrarily well by a repelling point. This is the definition originally used by Julia. (Of course, the name "Julia set" was used only later.)

- For any point z in the Julia set, the set of iterated preimages $\bigcup_{k=1}^{\infty} F^{-k}(z)$ forms a dense subset of the Julia set. This property is used when one is making computer pictures of Julia sets.

- In fact, for any point z in $\hat{\mathbb{C}}$ (with at most one or two exceptions), the closure of the set of iterated preimages contains the Julia set.

- For any point z in the Julia set and any neighborhood U_z of z, the iterated images $F^k(U_z)$ cover all of $\hat{\mathbb{C}}$ except at most one or two exceptional points. This property demonstrates an extreme sensitivity to initial conditions.

- If Ω is a union of Fatou components that is completely invariant (that is, $F(\Omega) = \Omega = F^{-1}(\Omega)$), then the boundary of Ω coincides with the Julia set. This justifies the definition of the Julia set of a polynomial as the boundary of the attracting basin of ∞. Compare also with figure 2, where the attracting basins of $Q_{c_0}^3$ and $A_{c_0}(\infty)$ are examples of such completely invariant sets.

- The Julia set is either connected or consists of uncountably many connected components. An example of the latter is shown in figure 6.

- The Julia set is typically a fractal: when one zooms in on it, one finds that the complication of the set is repeated at all scales. It is also *self-similar*, in the following sense: for any noncritical point z in the Julia set, any sufficiently small neighborhood U_z of z is mapped bijectively onto $F(U_z)$, a neighborhood of $F(z)$. The Julia set in U_z and the Julia set in $F(U_z)$ look alike.

All but the last two properties can easily be verified in the example Q_0. In this case the exceptional points are 0 and ∞.

2.7 Böttcher Maps and Potentials

2.7.1 Böttcher Maps

Consider the quadratic polynomial $Q_{-2}(z) = z^2 - 2$. If z belongs to the interval $[-2, 2]$, then z^2 belongs to the interval $[0, 4]$, so $Q_{-2}(z)$ also belongs to the interval $[-2, 2]$. It follows that this interval is contained in the filled Julia set K_{-2}.

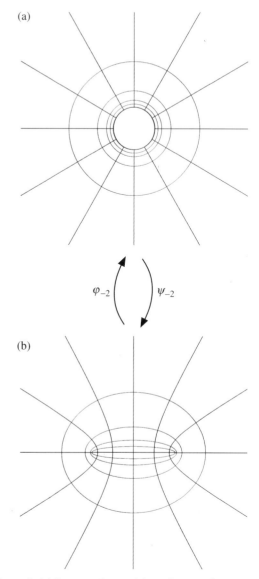

Figure 5 (a) Some equipotentials and external rays $\mathcal{R}_0(\theta)$ of Q_0 in $A_0(\infty)$, the set of complex numbers of modulus greater than 1. (b) The corresponding equipotentials and external rays $\mathcal{R}_{-2}(\theta)$ of Q_{-2} in $A_{-2}(\infty)$, the set of complex numbers not in $K_{-2} = J_{-2} = [-2, 2]$. The external rays that are drawn have arguments $\theta = \frac{1}{12}p$, where $p = 0, 1, \ldots, 11$.

The polynomial $Q_{-2}(z)$ is not topologically equivalent to $Q_0(w) = w^2$, but when z is big enough, it behaves in a similar way, since 2 is small compared with z^2. We can express this similarity with an appropriate holomorphic change of variables. Indeed, suppose that $z = w + 1/w$. Then when w changes to w^2,

z changes to $w^2 + 1/w^2$. But this equals

$$(w + 1/w)^2 - 2 = z^2 - 2 = Q_{-2}(z).$$

The reason this does not show that Q_0 and Q_{-2} are equivalent is that the change of variables cannot be inverted. However, in a suitable region it can. If $z = w + 1/w$, then $w^2 - wz + 1 = 0$. Solving this quadratic equation we find that $w = \frac{1}{2}(z \pm \sqrt{z^2 - 4})$, which leaves us with the problem of which square root to take. It can be shown that for one choice $|w| < 1$ and for the other choice $|w| > 1$, as long as z does not lie in the interval $[-2, 2]$. If we always choose the square root for which $|w| > 1$, then it turns out that the resulting function of z is a continuous function (in fact, holomorphic) from the set $\mathbb{C} \setminus [-2, 2]$ of complex numbers not in $[-2, 2]$ to the set $\{w : |w| > 1\}$ of complex numbers of modulus greater than 1.

Once this is established, it follows that the behavior of Q_{-2} on the set $\mathbb{C} \setminus [-2, 2]$ is topologically the same as the behavior of Q_0 on the set $\{w : |w| > 1\}$. In particular, points outside $\mathbb{C} \setminus [-2, 2]$ have orbits that tend to infinity under iteration by Q_{-2}. Therefore, the attracting basin $A_{-2}(\infty)$ of Q_{-2} is $\mathbb{C} \setminus [-2, 2]$, and the filled Julia set K_{-2} and the Julia set J_{-2} are both equal to $[-2, 2]$.

Let us write $\psi_{-2}(w)$ for $w + 1/w$. The function ψ_{-2}, which we used to change variables, maps circles of radius greater than 1 onto ellipses, and takes radial lines $\mathcal{R}_0(\theta)$ that consists of all complex numbers of some given argument θ and modulus greater than 1 to half-branches of hyperbolas. Since the ratio of $\psi_{-2}(w)$ to w tends to 1 as $w \to \infty$, each radial line will be the asymptote of the corresponding hyperbola half-branch (see figure 5).

It turns out that what we have just done for the polynomial Q_{-2} can be done for any quadratic polynomial Q_c. That is, for sufficiently large complex numbers there is a holomorphic function, denoted φ_c, called the *Böttcher map*, that changes variables in such a way that Q_c turns into Q_0, in the sense that $\varphi_c(Q_c(z)) = \varphi_c(z)^2$. (The map ψ_{-2} described above is the *inverse* of the Böttcher map in the case $c = -2$, rather than the map itself.) After the change of variables, the new coordinates are called *Böttcher coordinates*.

More generally, for all monic polynomials P (i.e., polynomials with leading coefficient 1) there is a unique holomorphic change of variables φ_P that converts P into the function $z \mapsto z^d$ for large enough z, in the sense that $\varphi_P(P(z)) = \varphi_P(z)^d$, and has the property

that $(\varphi_P(z)/z) \to 1$ as $z \to \infty$. The inverse of φ_P is written ψ_P.

2.7.2 Potentials

As we have noted already, if one repeatedly squares a complex number z of modulus greater than 1, then it will escape to infinity. The larger the modulus of z, the faster the iterates will tend to infinity. If instead of squaring, one applies a monic polynomial P of degree d, then for large enough z it is again true that the iterates $z, P(z), P^2(z), \ldots$ tend to infinity. It follows from the formula $\varphi_P(P(z)) = \varphi_P(z)^d$ that $\varphi_P(P^k(z)) = \varphi_P(z)^{d^k}$. Therefore, the speed at which the iterates tend to infinity depends not on $|z|$ but on $|\varphi_P(z)|$: the larger the value of $|\varphi_P(z)|$, the faster the convergence. For this reason, the level sets of $|\varphi_P|$, that is, sets of the form $\{z \in \mathbb{C} : |\varphi_P(z)| = r\}$, are important.

For many purposes it is useful to look not at the function φ_P itself but at the function $g_P(z) = \log|\varphi_P(z)|$. This function is called the *potential*, or *Green's function*. It has the same level sets as $|\varphi_P(z)|$, but the advantage that it is a HARMONIC FUNCTION [IV.24 §5.1].

Clearly, g_P is defined whenever φ_P is defined. But we can in fact extend the definition of g_P to the whole of the attracting basin $A_P(\infty)$. Given any z for which the iterates $P^k(z)$ tend to infinity, one chooses some k such that $\varphi_P(P^k(z))$ is defined and one sets $g_P(z)$ to be $d^{-k} \log|\varphi_P(P^k(z))|$. Notice that $\varphi_P(P^{k+1}(z)) = \varphi_P(P^k(z))^d$, so $\log|\varphi_P(P^{k+1}(z))| = d\log|\varphi_P(P^k(z))|$, from which it is easy to deduce that the value of $d^{-k} \log|\varphi_P(P^k(z))|$ does not depend on the choice of k.

The level sets of g_P are called *equipotentials*. Notice that the equipotential of potential $g_P(z)$ is mapped by P onto the equipotential of potential $g_P(P(z)) = dg_P(z)$. As we shall see, useful information about the dynamics of the polynomial P can be deduced from information about its equipotentials.

If ψ_P is defined everywhere on the circle C_r of radius r, for some $r > 1$, then it maps it to $\{z : |\varphi_P(z)| = r\}$, which is the equipotential of potential $\log r$. For large enough r, this equipotential is a simple closed curve encircling K_P, and it shrinks as r decreases. It is possible for two parts of this curve to come together so that it forms a figure-of-eight shape and then splits into two, like an amoeba dividing, but this can happen only if the curve crosses a critical point of P. Therefore, if all the critical points of P belong to

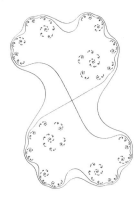

Figure 6 The Julia set of a quadratic polynomial Q_c for which the critical point 0 escapes to infinity under iteration. The Julia set is totally disconnected. The figure-of-eight-shaped curve with 0 at its intersection point is the equipotential through 0. The simple closed curve surrounding it is the equipotential through the critical value c.

the filled Julia set K_P (as in the example Q_{-2}, where $0 \in K_{-2} = [-2, 2]$), then it cannot happen. In this case, the Böttcher map φ_P can be defined on the whole of the attracting basin $A_P(\infty)$, and it is a bijection from $A_P(\infty)$ to the attracting basin $A_0(\infty) = \{w \in \mathbb{C} : |w| > 1\}$ of the polynomial z^d. There are equipotentials of potential t for every $t > 0$ and they are all simple closed curves. (Compare with figure 5.) As t approaches 0, the equipotential of potential t, together with its interior, forms a shape that gets closer and closer to the filled Julia set K_P. It follows that K_P is a connected set, as is the Julia set J_P.

On the other hand, if at least one of the critical points in the plane belongs to $A_P(\infty)$, then at a certain point the image of C_r splits into two or more pieces. In particular, the equipotential containing the fastest escaping critical point (i.e., the critical point with the highest value of the potential g_P) has at least two loops, as is illustrated in figure 6. The inside of each loop is mapped by P onto the inside of the equipotential of the corresponding critical value, which is a simple closed curve (since the potential of the critical value is greater than the potential of any critical point). Inside each loop there must be points from the filled Julia set K_P, so this set must be disconnected. The Böttcher map can always be defined on the outside of the equipotential of the fastest escaping critical point and can therefore always be applied to the fastest escaping critical value.

If Q_c is a quadratic polynomial for which 0 escapes to infinity under iteration, then the filled Julia set turns

out to be *totally disconnected*, which means that the connected components of K_c are points. None of these points is isolated: they can all be obtained as limits of sequences of other points of K_c. A set which is compact, totally disconnected, and with no isolated points is called a CANTOR SET [III.17], since such a set is homeomorphic to Cantor's middle-thirds set. Note that in this case $K_c = J_c$. For Q_c we have the following dichotomy: the Julia set J_c is connected if 0 has a bounded orbit, and it is totally disconnected if 0 escapes to infinity under iteration. We shall return to this dichotomy when we come to define the Mandelbrot set later in this article.

2.7.3 External Rays of Polynomials with Connected Julia Set

We have just obtained information by looking at the images under ψ_P of circles of radius greater than 1. We can obtain complementary information from the images of *radial lines*, which cut all these circles at right angles. If the Julia set is connected, then, as we saw in the discussion of potentials, the Böttcher map φ_P is a bijection from the attracting basin $A_P(\infty)$ to the attracting basin of z^d, which is the complement $\{w : |w| > 1\}$ of the closed unit disk. As before, let $\mathcal{R}_0(\theta)$ denote the half-line that consists of all complex numbers of argument θ and modulus greater than 1. Because $(\varphi_P(z)/z) \to 1$ as $z \to \infty$, the image of $\mathcal{R}_0(\theta)$ under ψ_P is a half-infinite curve consisting of points with arguments getting closer and closer to θ. This curve is denoted by $\mathcal{R}_P(\theta)$, and is known as the *external ray of argument θ* of P. Note that $\mathcal{R}_0(\theta)$ is the external ray of argument θ of z^d.

One can think of equipotentials as contour lines of the potential function, and of external rays as the lines of steepest ascent. Between the two of them, they provide a parametrization of the attracting basin, just as modulus and argument provide a parametrization of $\{z : |z| > 1\}$: if you know the potential at a certain complex number z, and you also know which external ray it lies on, then you know what z is. Moreover, a ray of argument θ is mapped by P onto the ray of argument $d\theta$, just as, when a number z lies on the half-line $\mathcal{R}_0(\theta)$, then z^d lies on the half-line $\mathcal{R}_0(d\theta)$.

We say that an external ray *lands* if $\psi_P(re^{2\pi i\theta})$ converges to a limit as $r \searrow 1$. If this happens, then the limit is called the *landing point*. However, it may happen that the end of the ray oscillates so much that there is a continuum of different limit points. In this case the ray is *nonlanding.* It can be shown that all rational rays land. Since a rational ray is either periodic or pre-periodic under iteration by P, the landing point of a rational ray must be either a periodic or a pre-periodic point in the Julia set. Much of the structure of the Julia set can be picked up from knowledge about common landing points. In the example illustrated in figure 2, the closures of the three Fatou components containing the critical orbit have one point in common. This point is a repelling fixed point and the common landing point of the rays of argument $\frac{1}{7}, \frac{2}{7}, \frac{4}{7}$. The rays of argument $\frac{1}{7}$ and $\frac{2}{7}$ are adjacent to the Fatou component containing the critical value c_0. These two arguments will show up again in the parameter plane and tell us where c_0 is situated.

2.7.4 Local Connectedness

In the example illustrated in figure 5 the inverse of the Böttcher map (the function ψ_{-2}) is defined on the set $\{w : |w| > 1\}$ of all complex numbers w of modulus greater than 1. However, it can be continuously extended to a function defined on the larger set $\{w : |w| \geqslant 1\}$. If we use the formula $\psi_{-2}(w) = w + 1/w$, then we have $\psi_{-2}(e^{2\pi i\theta}) = 2\cos(2\pi\theta)$, which is the landing point of the external ray $\mathcal{R}_{-2}(\theta)$. For an arbitrary connected filled Julia set K_P, we have the following result of Carathéodory: the inverse ψ_P of the Böttcher map has a continuous extension from $\{w : |w| > 1\}$ to $\{w : |w| \geqslant 1\}$ if and only if K_P is *locally connected*. To understand what this means, imagine a set that is shaped like a comb. From any point in this set to any other point there is a continuous path that lies in the set, but it is possible for the two points to be very close and for the shortest path to be very long. This happens, for example, if the two points are the ends of neighboring teeth of the comb. A connected set X is called locally connected if every point has arbitrarily small connected neighborhoods. It is possible to build comb-like sets (with infinitely many teeth) that contain points for which all connected neighborhoods have to be large. The filled Julia sets in the examples in figures 2–5 are locally connected, but there are examples of filled Julia sets that are not locally connected. When K_P is locally connected, then all external rays land, and the landing point is a continuous function of the argument. Under these circumstances, we have a natural and useful parametrization of the Julia set J_P.

2.8 The Mandelbrot Set M

We shall now restrict our attention to quadratic polynomials of the form Q_c. These are parametrized by the complex number c, and in this context we shall refer to the complex plane as the *parameter plane*, or *c-plane*. We would like to understand the family of dynamical systems that arise when we iterate the polynomials Q_c. Our goal will be to do this by dividing the c-plane into regions that correspond to polynomials with qualitatively the same dynamics. These regions will be separated by their boundaries, which together form the so-called *bifurcation set*. This consists of "unstable" c-values: that is, values of c for which there are other values arbitrarily nearby that give rise to qualitatively different dynamical behavior. In other words, a parameter c belongs to the bifurcation set if a small perturbation of c can make an important difference to the dynamics.

Recall the dichotomy that we stated earlier: the Julia set J_c is connected if the critical point 0 belongs to the filled Julia set K_c and is totally disconnected if 0 belongs to the attracting basin $A_c(\infty)$. This dichotomy motivates the following definition: the *Mandelbrot set M* consists of the c-values for which J_c is connected. That is,

$$M = \{c \in \mathbb{C} \mid Q_c^k(0) \not\to \infty \text{ as } k \to \infty\}.$$

Since the Julia set represents the chaotic part of the dynamical system given by Q_c, the dynamical behavior is certainly qualitatively affected by whether c belongs to M or not. We have therefore made a start toward our goal, but the division of the plane into M and $\mathbb{C} \setminus M$ is very coarse, and it does not obviously give us the complete understanding we are looking for.

The important set is in fact not M, but its boundary ∂M, which is illustrated in figure 7. Notice that this set has a number of "holes" (in fact, infinitely many). The Mandelbrot set itself is obtained by filling in all these holes. More precisely, the complement of ∂M consists of an infinite collection of connected components, of which one, the outside of the set, stretches off to infinity, while all the others are bounded. The "holes" are the bounded components.

This definition is similar to the definition of the Julia set of a polynomial. It is easy to define the filled Julia set, and the Julia set is then defined as its boundary. The Julia set provides a lot of structure in the dynamical plane, the z-plane. The Mandelbrot set is similarly easy to define, and its boundary provides a lot of structure in the c-plane. Remarkably, even though each Julia

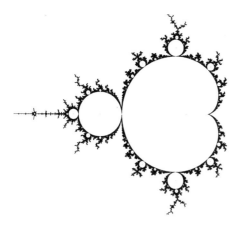

Figure 7 The boundary ∂M of the Mandelbrot set.

set concerns just one dynamical system, while the Mandelbrot set concerns an entire family of systems, there are close analogies between them, as will become clear.

Pioneering work on holomorphic dynamics in general and quadratic polynomials in particular was carried out in the early 1980s by Adrien Douady and John H. Hubbard. They introduced the name "Mandelbrot set" and proved several results about it. In particular, they defined a sort of Böttcher map, denoted by Φ_M, for the Mandelbrot set, which is a map from the complement of the Mandelbrot set to the complement of the closed unit disk.

The definition of Φ_M is actually quite simple: for each c let $\Phi_M(c)$ equal $\varphi_c(c)$, where φ_c is the Böttcher map for the parameter c. However, Douady and Hubbard did more than merely define Φ_M: they proved that it is a holomorphic bijection with holomorphic inverse.

Once we have Φ_M we can make further definitions, just as we did with the Böttcher map. For instance, we can define a *potential* G on the complement of the Mandelbrot set by setting $G(c) = g_c(c) = \log|\Phi_M(c)|$. An *equipotential* is then a level set of Φ_M (that is, a set of the form $\{c \in \mathbb{C} : |\Phi_M(c)| = r\}$ for some $r > 1$) and the *external ray of argument* θ is the set $\{c \in \mathbb{C} : \arg(\Phi_M(c)) = 2\pi\theta\}$ (that is, the inverse image of a radial line $\mathcal{R}_0(\theta)$). The latter is denoted by $\mathcal{R}_M(\theta)$ and it is asymptotic to the radial line of argument θ. The rational external rays are known to land (see figure 8).

It follows from the above that as t approaches zero, the equipotential of potential t, together with its interior, gets closer and closer to M: that is, M is the intersection of all such sets. Hence, M is a connected, closed, bounded subset of the plane.

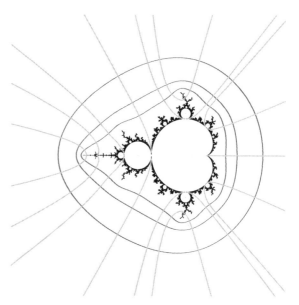

Figure 8 Some equipotentials of M and the external rays of arguments θ of periods 1, 2, 3, and 4. In counterclockwise direction the arguments between 0 and $\frac{1}{2}$ are 0, $\frac{1}{15}$, $\frac{2}{15}$, $\frac{1}{7}$, $\frac{3}{15}$, $\frac{4}{15}$, $\frac{2}{7}$, $\frac{1}{3}$, $\frac{6}{15}$, $\frac{3}{7}$, and $\frac{7}{15}$; and symmetrically in clockwise direction they are $1 - \theta$ with θ as above. The external rays of argument $\frac{1}{7}$ and $\frac{2}{7}$ are landing at the root point of the hyperbolic component that has c_0, the parameter value of the Douady rabbit in figure 2, as its center. The rays of argument $\frac{3}{15}$ and $\frac{4}{15}$ are landing at the root point of the copy of M shown in figure 9.

2.8.1 J-Stability

As we have mentioned and as figure 7 suggests, the complement of ∂M has infinitely many connected components. These components are of great dynamical significance: if c and c' are two parameters taken from the same component, then the dynamical systems arising from Q_c and $Q_{c'}$ can be shown to be essentially the same. To be precise, they are *J-equivalent*, which means that there is a continuous change of variables that converts the dynamics on one Julia set to the dynamics on the other. If c belongs to the boundary ∂M, then there are parameter values c' arbitrarily close to c for which Q_c and $Q_{c'}$ are not J-equivalent, so ∂M is the "bifurcation set with respect to J-stability." We shall comment on the global structural stability later.

2.8.2 Hyperbolic Components

From now on, we shall use the word "component" to refer to the holes of the Mandelbrot set—that is, to the bounded components of the complement of ∂M.

We start by considering the component containing $c = 0$, the central component $\mathcal{H}_0$. Recall from section 2.3 that, after a suitable change of variables, one can change the polynomial $F_\lambda(z) = \lambda z + z^2$ into the polynomial Q_c, where the parameters λ and c are related by the equation $c = \frac{1}{2}\lambda - \frac{1}{4}\lambda^2$. The parameter λ has a dynamical meaning: the origin is a fixed point of F_λ and λ is its multiplier. This knowledge tells us that the corresponding Q_c has a fixed point of multiplier λ; we denote the fixed point by α_c. For $|\lambda| < 1$ the fixed point is attracting.

The unit disk $\{\lambda : |\lambda| < 1\}$ corresponds to the central component $\mathcal{H}_0$, and the function that takes a parameter c in $\mathcal{H}_0$ to the corresponding parameter λ in the unit disk is called the *multiplier map*, and is denoted by $\rho_{\mathcal{H}_0}$. Thus, $\rho_{\mathcal{H}_0}(c)$ is the multiplier of the fixed point α_c of the polynomial Q_c. The multiplier map $\rho_{\mathcal{H}_0}$ is a holomorphic isomorphism from $\mathcal{H}_0$ to the unit disk. As we have just seen, the inverse map is given by $\rho_{\mathcal{H}_0}^{-1}(\lambda) = \frac{1}{2}\lambda - \frac{1}{4}\lambda^2$. This map extends continuously to the unit circle, and thereby gives us a parametrization of the boundary of the central component $\mathcal{H}_0$ by points λ of modulus 1. The image of the unit circle under the map $\lambda \mapsto \frac{1}{2}\lambda - \frac{1}{4}\lambda^2$ is a *cardioid*. This explains the heart-like shape of the largest part of the Mandelbrot set, which can be seen in figure 7.

Any quadratic polynomial has two fixed points if we count with multiplicity (in fact, two distinct ones unless $c = \frac{1}{4}$). The central component $\mathcal{H}_0$ is characterized as the component of c-values for which Q_c has an attracting fixed point. For any c outside the cardioid, Q_c has two repelling fixed points, but it may have an attracting periodic orbit of a period greater than 1. It is an important fact that the attracting basin of an attracting periodic orbit always contains a critical orbit. Therefore, for any quadratic polynomial there can be at most one attracting periodic orbit.

We call a component $\mathcal{H}$ of the Mandelbrot set a *hyperbolic component* if, for every parameter c in $\mathcal{H}$, the polynomial Q_c has an attracting periodic orbit. For any given hyperbolic component, the periods of the attracting periodic orbits will be the same. There is a corresponding multiplier map $\rho_{\mathcal{H}}$, from $\mathcal{H}$ to the unit disk, which assigns to each parameter c in $\mathcal{H}$ the multiplier of the attracting periodic orbit. This multiplier map is always a holomorphic isomorphism that extends continuously to the boundary $\partial\mathcal{H}$ of $\mathcal{H}$.

The points $\rho_{\mathcal{H}}^{-1}(0)$ and $\rho_{\mathcal{H}}^{-1}(1)$ are called the *center* and the *root* of $\mathcal{H}$. The center of $\mathcal{H}$ is the unique c in $\mathcal{H}$ for which the periodic orbit of Q_c is super-attracting.

As for the root, if the period of the component is k, then it will be the landing point for a pair of external rays of periodic arguments of period k. (For the central component $\mathcal{H}_0$ there is only one ray assigned.) Conversely, every external ray with such an argument lands at the root point of a hyperbolic component of period k. Thus, the arguments of these rays give addresses to the hyperbolic components. This can be seen in figure 8, from which one can read off the mutual positions of all the components of periods 1–4.

As a consequence of the above, the number of hyperbolic components corresponding to a certain period k can be determined both as the number of roots in the polynomial $Q_c^k(0)$ that are not roots in $Q_c^\ell(0)$ for some $\ell < k$ and also as the number of pairs of rational arguments with denominator $2^k - 1$ that cannot be expressed with denominator $2^\ell - 1$ for some $\ell < k$.

For any component $\mathcal{H}$ with center c_0 let $\mathcal{R}_M(\theta_-)$ and $\mathcal{R}_M(\theta_+)$ be the pair of rays landing at the root point. Then, in the dynamical plane of Q_{c_0}, the pair of rays $\mathcal{R}_{c_0}(\theta_-)$ and $\mathcal{R}_{c_0}(\theta_+)$ are adjacent to the Fatou component of Q_{c_0} containing c_0, and they land at the root point of that Fatou component.

2.8.3   Structural Stability

Suppose that Q_c has a super-attracting periodic orbit of period k, and let z_0 be a point in this orbit. Then $Q_c^k(z_0) = z_0$, and the derivative of Q_c^k at z_0 is 0. It follows from the chain rule that there is at least one z_i in the orbit at which the derivative of Q_c is 0: that is, 0 belongs to the orbit. Therefore, the center of a hyperbolic component cannot be structurally stable, since the critical orbit of the center-polynomial is finite, but it is infinite for all nearby polynomials. However, if we remove from the complex plane not just ∂M but also all the centers of hyperbolic components, then we obtain the splitting we have been looking for: any connected component of the remaining set forms a structurally stable region. For any pair of parameter values c and c' in such a component, Q_c and $Q_{c'}$ are conjugate, meaning that there is a continuous change of variables in the plane that converts the dynamics of one polynomial into those of the other.

2.8.4   Conjectures

The above discussion raises an obvious question: we have a good understanding of the hyperbolic components of the complement of ∂M, but are there components that are *not* hyperbolic? The following conjecture expresses a widely held belief, but it is as yet unproved.

The hyperbolicity conjecture. *All the bounded components of the complement of ∂M are hyperbolic.*

The hyperbolicity conjecture can be stated in greater generality for rational functions, where it says that every rational function can be approximated arbitrarily closely by a *hyperbolic rational function.* Here, "hyperbolic" means that the dynamics is expanding on the Julia set. We shall not go further into this, but only mention that the dynamics on the Julia set is expanding for every Q_c with c in a hyperbolic component of M, and also in the unbounded component, the complement of M. The Julia set J_c can in these cases be thought of as a "strange repeller": the dynamics is chaotic and the geometry is fractal (except for $c = 0$).

The main conjecture about the Mandelbrot set is, however, the following.

The local connectivity conjecture. *The Mandelbrot set is locally connected.*

This conjecture, often referred to as MLC, is important for many reasons. To begin with, it is known that it implies the hyperbolicity conjecture. Second, if M is locally connected, then Ψ_M, the inverse of Φ_M, which is a holomorphic bijection from the set outside the closed unit disk to the complement of the Mandelbrot set, has a continuous extension to the unit circle, and all external rays land in a continuous manner. This would give us a useful parametrization of ∂M. One can then give a beautifully simple abstract combinatorial description of M, despite the fact that ∂M is a complicated fractal. (Mitsuhiro Shishikura has proved that the HAUSDORFF DIMENSION [III.17] of ∂M is the maximum possible in the plane, namely 2.)

2.9 Universality of *M*

The Mandelbrot set is remarkably ubiquitous. For example, homeomorphic copies of M appear inside M itself, as is apparent from figure 9. Inside other families of holomorphic mappings that depend holomorphically on some parameter, we again find homeomorphic copies of M. For this reason, M is said to be *universal.* Douady and Hubbard have captured the reason behind the phenomenon of universality by defining a notion of a *quadratic-like mapping.* The kth iterate of a quadratic polynomial is globally a polynomial

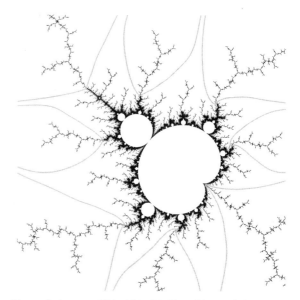

Figure 9 A copy of M within M. The address of the copy is given by the arguments of the two external rays that+ land at the cusp, the root point of the copy. Here the arguments are $\frac{3}{15}$ and $\frac{4}{15}$. Compare with figure 8. The rays are drawn to indicate where the "decorations" should be cut off in order to have the bare copy of M.

of degree 2^k, but locally it may behave like a quadratic polynomial. The same is true for a rational function or an iterate of it. By a quadratic-like mapping we mean a triple (f, V, W) where V and W are open simply connected domains (that is, connected open sets without holes), $\bar{V} \subset W$, and f is a holomorphic map that maps V onto W with degree 2. (This means that every point in W has two preimages, up to multiplicity, in V.) Such a map f has a single critical point ω in V, and behaves in many ways like a quadratic polynomial. The filled Julia set K_f is defined as the set of points z in V for which the iterates $f^k(z)$ stay in V for all $k \geq 0$. A dichotomy similar to the one for quadratic polynomials holds for quadratic-like mappings as well: K_f is connected if and only if the critical point ω is contained in K_f. For any quadratic-like mapping with a connected filled Julia set, Douady and Hubbard have defined a strategy, called *straightening,* which associates with the mapping a unique c-value in M. For a family of quadratic-like mappings $\{f_\lambda\}_{\lambda \in \Lambda}$ the Mandelbrot set M_Λ is defined as the set of λ for which K_{f_λ} is connected. We obtain through straightening a mapping $\Xi : M_\Lambda \to M$, which takes λ to the uniquely associated c-value.

In the copy of M shown in figure 9, the "center" associated with $c = 0$ in M corresponds to a polynomial Q_{c_0} for which the critical point 0 is periodic of period 4, and for which a suitable restriction of the fourth iterate $f_{c_0} = Q_{c_0}^4$ is quadratic-like from V_0 to its image W_0. Moreover, there is a neighborhood $\mathcal{V}_0$ of c_0 in the c-plane such that for any c in $\mathcal{V}_0$ the restriction of $f_c = Q_c^4$ to V_0 is a quadratic-like map from V_0 to its image W_c, and such that the map Ξ is a homeomorphism from $M_{\mathcal{V}_0}$ to M.

The infinitely many copies of M that appear inside M may suggest that M has a self-similarity property. However, there is another phenomenon that pulls in the opposite direction. The c-values for which the critical point 0 is pre-periodic form a dense subset of ∂M. If $\tilde{c}$ is one of these special c-values, then there are two contexts in which one may look at magnifications of smaller and smaller neighborhoods of $\tilde{c}$: the first is the Julia set $J_{\tilde{c}}$ of the polynomial $Q_{\tilde{c}}$ in neighborhoods of $z = \tilde{c}$, and the second is the Mandelbrot set in neighborhoods of $c = \tilde{c}$. It turns out that the pictures are *asymptotically similar*, which means that the greater the magnification, and the smaller the neighborhood, the more similar the two pictures become.

This is an extraordinary fact. Indeed, it may even seem to be impossible, since in any neighborhood of $\tilde{c}$ the Mandelbrot set contains infinitely many copies of itself, while the Julia set is known to contain no such copies. The explanation for the apparent paradox is that the copies of the Mandelbrot set get smaller very quickly as their distance to $\tilde{c}$ decreases. Hence, if one magnifies a small enough neighborhood, the copies that are there are practically invisible.

2.10   Newton's Method Revisited

Let us return briefly to Newton's method for polynomials. Consider any polynomial P of degree $d \geqslant 2$ that has only simple roots. Then the Newton function N_P is a rational function of degree d, and each simple root of P is a super-attracting fixed point of N_P. For quadratic polynomials the number of roots of P coincides with the number of critical points of N_P (since $2d - 2 = 2$ when $d = 2$). For polynomials of degree $d > 2$ there are more critical points than the roots can account for.

Cayley considered Newton's method for quadratic polynomials with two distinct roots $P(z) = (z - r_1)(z - r_2)$. He showed that the function $\mu(z) = (z - r_1)/(z - r_2)$, which maps the root r_1 onto 0 and the root r_2 onto ∞, provides a change of variables that turns

N_P into the quadratic polynomial Q_0 on the Riemann sphere $\hat{\mathbb{C}}$. When one translates the dynamics of Q_0 to the dynamics of Newton's method one finds that the unit circle corresponds to the bisector of r_1 and r_2 and that all points in the half-plane containing r_i, $i = 1, 2$, are therefore attracted to r_i under iteration by N_P.

Cayley announced that he would write about Newton's iteration for cubic polynomials. However, it took about a hundred years before any such paper appeared. For a cubic polynomial P with three simple roots the Newton function N_P has three super-attracting fixed points, each of which gives rise to an attracting basin. The Julia set of N_P is the common boundary of these three basins, and is therefore a complicated fractal set. Moreover, N_P has an extra critical point since $2d - 2 = 4$ for $d = 3$. The extra critical point may be attracted to one of the roots under iteration, or it can have its own independent behavior. In order to catch the behavior of all cubic polynomials under Newton's iteration (except the one with one root of multiplicity three) it is sufficient to consider the one-parameter family of polynomials $P_\lambda(z) = (z - 1)(z - \frac{1}{2} - \lambda)(z - \frac{1}{2} + \lambda)$. The extra critical point for the corresponding Newton function N_λ then turns out to be at the origin. Suppose that we associate three colors, for instance red, blue, and green, with the three roots 1, $\frac{1}{2} + \lambda$, $\frac{1}{2} - \lambda$. We can then color the λ-plane, which is the parameter plane in this context, as follows. A parameter value λ is colored red, blue, or green if the critical point 0 is attracted under iteration by N_λ to the root of that color. If it is not attracted to any of the three roots, then we color with a fourth color, yellow, say. The universality of the Mandelbrot set is thereby demonstrated: in the λ-plane one can observe yellow copies of it, which one can explain by showing that families of suitably restricted iterates of N_λ are quadratic-like.

3   Concluding Remarks

We have illustrated several results in holomorphic dynamics through examples, including the transferring of definitions and results from the dynamical planes to the parameter plane. The structures of the filled Julia sets and the Mandelbrot set are partly understood through analysis of their complements, linked together via the Böttcher maps φ_c and Φ_M. The functions that are used for changing variables in J-stability and structural stability are examples of so-called *quasiconformal mappings*. This is a concept that was introduced into holomorphic dynamics in the early 1980s by

Dennis Sullivan. They are indispensable for discussing change of *complex structure, straightening, holomorphic motion, surgery*, and many other phenomena. The interested reader is referred to the books listed below. The first two contain expository papers, the third is a graduate textbook, and the fourth is a collection of papers. They all contain many further references.

Acknowledgments. The computer drawings in this article were obtained from a program written by Christian Henriksen.

Further Reading

Devaney, R. L., and L. Keen, eds. 1989. *Chaos and Fractals. The Mathematics Behind the Computer Graphics.* Proceedings of Symposia in Applied Mathematics, volume 39. Providence, RI: American Mathematical Society.

———. 1994. *Complex Dynamical Systems. The Mathematics Behind the Mandelbrot and Julia Sets.* Proceedings of Symposia in Applied Mathematics, volume 49. Providence, RI: American Mathematical Society.

Lei, T., ed. 2000. *The Mandelbrot Set, Theme and Variations.* London Mathematical Society Lecture Note Series, volume 274. Cambridge: Cambridge University Press.

Milnor, J. 1999. *Dynamics in One Complex Variable.* Weisbaden: Vieweg.

IV.15 Operator Algebras
Nigel Higson and John Roe

1 The Beginnings of Operator Theory

We can ask two basic questions about any equation, or system of equations: is there a solution, and, if there is, is it unique? Experience with finite systems of linear equations indicates that the two questions are interconnected. Consider for instance the equations

$$2x + 3y - 5z = a,$$
$$x - 2y + z = b,$$
$$3x + y - 4z = c.$$

Notice that the left-hand side of the third equation is the sum of the left-hand sides of the first two. As a result, no solution to the system exists unless $a + b = c$. But if $a + b = c$, then any solution of the first two equations is also a solution of the third; and in any linear system involving more unknowns than equations, solutions, when they exist, are never unique. In the present case, if (x, y, z) is a solution, then so is $(x + t, y + t, z + t)$, for any t. Thus the same phenomenon (a linear relation among the equations) that

prevents the system from admitting solutions in some cases also prevents solutions from being unique in other cases.

To make the relation between existence and uniqueness of solutions more precise, consider a general system of linear equations of the form

$$k_{11}u_1 + k_{12}u_2 + \cdots + k_{1n}u_n = f_1,$$
$$k_{21}u_1 + k_{22}u_2 + \cdots + k_{2n}u_n = f_2,$$
$$\vdots$$
$$k_{n1}u_1 + k_{n2}u_2 + \cdots + k_{nn}u_n = f_n$$

consisting of n equations in n unknowns. The scalars k_{ji} form a matrix of coefficients and the problem is to solve for the u_i in terms of the f_j. The general theorem illustrated by our particular numerical example above is that the number of linear conditions that the f_j must satisfy if a solution is to exist is equal to the number of arbitrary constants appearing in the general solution when a solution does exist. To use a more technical vocabulary, the dimension of the KERNEL [I.3 §4.1] of the matrix $K = \{k_{ji}\}$ is equal to the dimension of its cokernel. In the example, these numbers are both 1.

A little more than a hundred years ago, FREDHOLM [VI.66] made a study of *integral equations* of the type

$$u(y) - \int k(y, x)u(x)\,\mathrm{d}x = f(y).$$

These arose from questions in theoretical physics, and the problem was to solve for the function u in terms of the function f. Since an integral can be thought of as a limit of finite sums, Fredholm's equation is an infinite-dimensional counterpart of the finite-dimensional linear systems considered above, in which vectors with n components are replaced by functions with values at infinitely many different points x. (Strictly speaking, Fredholm's equation is analogous to a matrix equation of the type $u - Ku = f$ rather than $Ku = f$. The altered form of the left-hand side has no effect on the overall behavior of the matrix equation, but it does considerably alter the behavior of the integral equation. As we shall see, Fredholm was fortunate to work with a class of equations whose behavior mirrors that of matrix equations very closely.)

A very simple example is

$$u(y) - \int_0^1 u(x)\,\mathrm{d}x = f(y).$$

To solve this equation, it helps to observe that the quantity $\int_0^1 u(x)\,\mathrm{d}x$, when thought of as a function of y, is a *constant*. Thus in the homogeneous case ($f \equiv 0$), the only possible solutions for $u(y)$ are the constant functions. On the other hand, for a general function f, solutions exist if and only if the single linear condition $\int_0^1 f(y)\,\mathrm{d}y = 0$ is satisfied. So in this example the dimension of the kernel and the dimension of the cokernel are both 1. Fredholm set out on a systematic exploration of the analogy between matrix theory and integral equations that this example suggests. He was able to prove that, for equations of his type, the dimensions of the kernel and of the cokernel are always finite and equal.

Fredholm's work sparked the imagination of HILBERT [VI.63], who made a detailed study of the *integral operators* that transform $u(y)$ into $\int k(y,x)u(x)\,\mathrm{d}x$, in the special case where the real-valued function k is *symmetric*, meaning that $k(x,y) = k(y,x)$. The finite-dimensional counterpart of Hilbert's theory is the theory of real symmetric matrices. Now if K is such a matrix, then a standard result from linear algebra asserts that there is an orthonormal basis consisting of EIGENVECTORS [I.3 §4.3] for K, or equivalently that there is a *unitary* matrix U such that $U^{-1}TU$ is diagonal. (*Unitary* means that U is invertible and preserves the lengths of vectors: $\|Uv\| = \|v\|$ for all vectors v.) Hilbert obtained an analogous theory for all symmetric integral operators. He showed that there exist functions $u_1(y), u_2(y), \ldots$ and real numbers $\lambda_1, \lambda_2, \ldots$ such that

$$\int k(y,x)u_n(x)\,\mathrm{d}x = \lambda_n u_n(y).$$

Thus $u_n(y)$ is an *eigenfunction* for the integral operator, with eigenvalue λ_n.

In most cases it is hard to calculate u_n and λ_n explicitly, but calculation *is* possible when $k(x,y) = \phi(x - y)$ for some periodic function ϕ. If the range of integration is $[0,1]$ and the period of ϕ is 1, then the eigenfunctions are $\cos(2k\pi y)$, $k = 0, 1, 2, \ldots$, and $\sin(2k\pi y)$, $k = 1, 2, \ldots$. In this case, the theory of FOURIER SERIES [III.27] tells us that a general function $f(y)$ on $[0,1]$ can be expanded as the sum of a series $\sum(a_k \cos 2k\pi y + b_k \sin 2k\pi y)$ of cosines and sines. Hilbert showed that, in general, there is an analogous expansion

$$f(y) = \sum a_n u_n(y)$$

in terms of the eigenfunctions for *any* symmetric integral operator. In other words, the eigenfunctions form a *basis*, just as in the finite-dimensional case. Hilbert's result is now called the *spectral theorem* for symmetric integral operators.

1.1   From Integral Equations to Functional Analysis

Hilbert's spectral theorem led to an explosion of activity, since integral operators arise in many different areas of mathematics (including, for example, the DIRICHLET PROBLEM [IV.12 §1] in partial differential equations and the REPRESENTATION THEORY OF COMPACT GROUPS [IV.9 §3]). It was soon recognized that these operators are best viewed as linear transformations on the HILBERT SPACE [III.37] of all functions $u(y)$ such that $\int |u(y)|^2\,\mathrm{d}y < \infty$. Such functions are called *square-integrable*, and the collection of all of them is denoted $L^2[0,1]$.

With the important concept of Hilbert space available, it became convenient to examine a much broader range of operators than the integral operators initially considered by Fredholm and Hilbert. Since Hilbert spaces are VECTOR SPACES [I.3 §2.3] and METRIC SPACES [III.56], it made sense to look first at operators from a Hilbert space to itself that are both linear and continuous: these are usually called *bounded* linear operators. The analogue of the symmetry condition $k(x,y) = k(y,x)$ on integral operators is the condition that a bounded linear operator T be *self-adjoint*, which is to say that $\langle Tu, v \rangle = \langle u, Tv \rangle$ for all vectors u and v in the Hilbert space (the angle brackets denote the inner product). A simple example of a self-adjoint operator is the *multiplication operator* by a real-valued function $m(y)$; this is the operator M defined by the formula $(Mu)(y) = m(y)u(y)$. (The finite-dimensional counterpart to a multiplication operator is a diagonal matrix K, which multiplies the jth component of the vector by the matrix entry k_{jj}.)

Hilbert's spectral theorem for symmetric integral operators tells us that every such operator can be given a particularly nice form: with respect to a suitable "basis" of $L^2[0,1]$, namely a basis of eigenfunctions, it will have an infinite diagonal matrix. Moreover, the basis vectors can be chosen to be orthogonal to each other. For a general self-adjoint operator, this is not true. Consider, for instance, the multiplication operator from $L^2[0,1]$ to itself that takes each square-integrable function $u(y)$ to the function $yu(y)$. This operator has no EIGENVECTORS [I.3 §4.3], since if λ is an EIGENVALUE [I.3 §4.3], then we need $yu(y) = \lambda u(y)$ for every y, which implies that $u(y) = 0$ for every y not equal to λ, and hence that $\int |u(y)|^2\,\mathrm{d}y = 0$. However, this example is not particularly worrying, since a multiplication operator of this kind is a sort of continuous analogue of the operator defined by a diagonal matrix.

It turns out that if we enlarge our concept of "diagonal" to include multiplication operators, then all self-adjoint operators are "diagonalizable," in the sense that, after a suitable "change of basis," they become multiplication operators.

To make this statement precise, we need the notion of the SPECTRUM [III.86] of an operator T. This is the set of complex numbers λ for which the operator $T - \lambda I$ does not have a bounded inverse (here I is the identity operator on Hilbert space). In finite dimensions the spectrum is precisely the set of eigenvalues, but in infinite dimensions this is not always so. Indeed, whereas every symmetric matrix has at least one eigenvalue, a self-adjoint operator, as we have just seen, need not. As a result of this, the spectral theorem for bounded self-adjoint operators is phrased not in terms of eigenvalues but in terms of the spectrum. One way of formulating it is to state that any self-adjoint operator T is *unitarily equivalent* to a multiplication operator $(Mu)(y) = m(y)u(y)$, where the closure of the range of the function $m(y)$ is the spectrum of T. Just as in the finite-dimensional case, a *unitary* operator is an invertible operator U that preserves the lengths of vectors. To say that T and M are unitarily equivalent is to say that there is some unitary map U, which we can think of as an analogue of a change-of-basis matrix, such that $T = U^{-1}MU$. This generalizes the statement that any real symmetric matrix is unitarily equivalent to a diagonal matrix with the eigenvalues along the diagonal.

1.2   The Mean Ergodic Theorem

A beautiful application of the spectral theorem was found by VON NEUMANN [VI.91]. Imagine a checkerboard on which are distributed a certain number of checkers. Imagine that for each square there is designated a "successor" square (in such a way that no two squares have the same successor), and that every minute the checkers are rearranged by moving each one to its successor square. Now focus attention on a single square and each minute record with a 1 or 0 whether or not there is a piece on the square. This produces a succession of readings $R_1, R_2, R_3, \ldots$ like this:

$$00100110010110100100 \cdots.$$

We might expect that over time, the average number of positive readings $R_j = 1$ will converge to the number of pieces on the board divided by the number of squares. If the rearrangement rule is not complicated enough, then this will not happen. For example, in the most extreme case, if the rule designates each square

as its own successor, then the readout will be either $00000 \cdots$ or $111111 \cdots$, depending on whether or not we chose a square with a piece on it to begin with. But if the rule is sufficiently complicated, then the "time average" $(1/n) \sum_{j=1}^{n} R_j$ will indeed converge to the number of pieces on the board divided by the number of squares, as expected.

The checkerboard example is elementary, since in fact the only "sufficiently complicated" rules in this finite case are cyclic permutations of the squares of the board, and thus *all* the squares move past our observation post in succession. However, there are related examples where one observes only a small fraction of the data. For instance, replace the set of squares on a checkerboard with the set of points on a circle, and in place of the checkers, imagine that a subset S of a circle is marked as occupied. Let the rearrangement rule be the rotation of points on the circle through some irrational number of degrees. Stationed at a point x of the circle, we record whether x belongs to S, the first rotated copy of S, the second rotated copy of S, and so on to obtain a sequence of 0 or 1 readings as before. One can show that (for nearly every x) the time average of our observations will converge to the proportion of the circle occupied by S.

Similar questions about the relationship between time and space averages had arisen in thermodynamics and elsewhere, and the expectation that time and space averages should agree when the rearrangement rule is sufficiently complex became known as the *ergodic hypothesis*.

Von Neumann brought operator theory to bear on this question in the following way. Let H be the Hilbert space of functions on the squares of the checkerboard, or the Hilbert space of square-integrable functions on the circle. The rearrangement rule gives rise to a unitary operator U on H by means of the formula

$$(Uf)(y) = f(\phi^{-1}(y)),$$

where ϕ is the function describing the rearrangement. Von Neumann's ergodic theorem asserts that if no nonconstant function in H is fixed by U (this is one way of saying that the rearrangement rule is "sufficiently complicated"), then, for every function $f \in H$, the limit

$$\lim_{n \to \infty} \frac{1}{n} \sum_{j=1}^{n} U^j f$$

exists and is equal to the constant function whose value everywhere is the average value of f. (To apply this to our examples, take $f(x)$ to be the function that is 1 if the point x is occupied and 0 otherwise.)

Von Neumann's theorem can be deduced from a spectral theorem for unitary operators that is analogous to the spectral theorem for self-adjoint operators. Every unitary operator can be reduced to a multiplication operator, not by real-valued functions but by functions whose values are complex numbers of absolute value 1. The key to the proof then becomes a statement about complex numbers of absolute value 1: if z is such a complex number, different from 1, then the expression $(1/n) \sum_{j=1}^{n} z^j$ approaches zero as $n \to \infty$. This in turn is easily proved using the formula for the sum of a geometric series, $\sum_{j=1}^{n} z^j = z(1 - z^n)/(1 - z)$. (More detail can be found in ERGODIC THEOREMS [V.9].)

1.3   Operators and Quantum Theory

Von Neumann realized that Hilbert spaces and their operators provide the correct mathematical tools to formalize the laws of quantum mechanics, introduced in the 1920s by Heisenberg and Schrödinger.

The *state* of a physical system at any given instant is the list of all the information needed to determine its future behavior. If, for instance, the system consists of a finite number of particles, then classically its state consists of the list of the position and momentum vectors of all the constituent particles. By contrast, in von Neumann's formulation of quantum mechanics one associates with each physical system a Hilbert space H, and a state of the system is represented by a unit vector u in H. (If u and v are unit vectors and v is a scalar multiple of u, then u and v determine the same state.)

Associated with each observable quantity (perhaps the total energy of the system, or the momentum of one particle within the system) is a self-adjoint operator Q on H whose spectrum is the set of all observed values of that quantity (hence the origin of the term "spectrum"). States and observables are related as follows: when a system is in the state described by a unit vector $u \in H$, the *expected value* of the observable quantity corresponding to a given self-adjoint operator Q is the inner product $\langle Qu, u \rangle$. This may not be a value that is ever actually measured: rather, it is the average of values that are obtained from many repeated experiments with the system when it is in the given state u. The relation between states and observables reflects the paradoxical behavior of quantum mechanics: it is possible, and in fact typical, for a system to exist in a "superposed" state, under which repeated identical experiments produce distinct outcomes. A measure-

ment of an observable quantity will produce a determinate outcome if and only if the state of the system is an eigenvector for the operator associated with that quantity.

A distinctive feature of quantum theory is that the operators associated with different observables typically do not commute with one another. If two operators do not commute, then they will typically have no eigenvectors in common, and, as a result, simultaneous measurements of two different observables will typically not result in determinate values for both of them. A famous example is provided by the operators P and Q associated with the position and momentum of a particle moving along a line. They satisfy the *Heisenberg commutation relation*

$$QP - PQ = i\hbar I,$$

where $\hbar$ is a certain physical constant. (This is an instance of a general principle which relates the noncommutativity of observables in quantum mechanics to the *Poisson bracket* of the corresponding observables in classical mechanics: see MIRROR SYMMETRY [IV.16 §§2.1.3, 2.2.1].) As a result, it is impossible for the particle simultaneously to have a determinate momentum and position. This is the *uncertainty principle*.

It turns out that there is an essentially unique way of representing the Heisenberg commutation relation using self-adjoint operators on Hilbert space: the Hilbert space H must be $L^2(\mathbb{R})$; the operator P must be $-i\hbar \, d/dx$; and the operator Q must be multiplication by x. This theorem allows one to determine explicitly the observable operators for simple physical systems. For example, in a system consisting of a particle on a line subject to a force directed toward the origin which is proportional to the distance from the origin (as if the particle were attached to a spring, anchored at the origin), the operator for total energy is

$$E = -\frac{\hbar^2}{2m}\frac{d^2}{dx^2} + \frac{k}{2}x^2,$$

where k is a constant which determines the overall strength of the force. The spectrum of this operator is the set

$$\{(n + \tfrac{1}{2})\hbar(k/m)^{1/2} : n = 0, 1, 2, \dots\}.$$

These are therefore the possible values for the total energy of the system. Notice that the energy can assume only a discrete set of values. This is another characteristic and fundamental feature of quantum theory.

Another important example is the operator of total energy for the hydrogen atom. Like the operator above, this may be realized as a certain explicit partial differential operator. It can be shown that the eigenvalues of this operator form a sequence proportional to $\{-1, -\frac{1}{4}, -\frac{1}{9}, \dots\}$. A hydrogen atom, when disturbed, may release a photon, resulting in a drop in its total energy. The released photon will have energy equal to the difference between the energies of the initial and final states of the atom, and therefore it is proportional to a number of the form $1/n^2 - 1/m^2$. When light from hydrogen is passed through a prism or diffraction grating, bright lines are indeed observed at wavelengths corresponding to these possible energies. Spectral observations of this sort provide experimental confirmation for quantum mechanical predictions.

So far we have discussed states of a quantum system only at a single instant. However, quantum systems evolve in time, just as classical systems do: to describe this evolution we need a law of motion. The time evolution of a quantum system is represented by a family of unitary operators $U_t : H \to H$, parametrized by the real numbers. If the system is in an initial state u, it will be in the state $U_t u$ after t units of time. Because the passage of s units of time followed by t further units is the same as the passage of $s + t$ units, the unitary operators U_t satisfy the *group law* $U_s U_t = U_{s+t}$. An important theorem of Marshall Stone asserts that there is a one-to-one correspondence between unitary groups $\{U_t\}$ and self-adjoint operators E given by the formula

$$\mathrm{i}E = \left(\frac{\mathrm{d}U_t}{\mathrm{d}t}\right)_{t=0} = \lim_{t \to 0} \frac{1}{t}(U_t - I).$$

The quantum law of motion is that the *generator E* corresponding in this way to time evolution is the operator associated with the observable "total energy." When E is realized as a differential operator on a Hilbert space of functions (as in the examples above), this statement becomes a differential equation, the *Schrödinger equation*.

1.4   The GNS Construction

The time-evolution operators U_t of quantum mechanics satisfy the law $U_s U_t = U_{s+t}$. More generally, we define a *unitary representation* of a GROUP [I.3 §2.1] G to be a family of unitary operators U_g, one for each $g \in G$, satisfying the law $U_{g_1 g_2} = U_{g_1} U_{g_2}$ for all $g_1, g_2 \in G$. Originally introduced by FROBENIUS [VI.58] as a tool for the study of finite groups, REPRESENTATION THEORY [IV.9] has become indispensable in mathematics and physics wherever the symmetries of a system must be taken into account.

If U is a unitary representation of G and v is a vector, then $\sigma : g \mapsto \langle U_g v, v \rangle$ is a function defined on G. The law $U_{g_1 g_2} = U_{g_1} U_{g_2}$ implies that σ has an important positivity property, namely

$$\sum_{g_1, g_2 \in G} \overline{a_{g_1}} \, a_{g_2} \sigma(g_1^{-1} g_2) = \left\| \sum a_g U_g v \right\|^2 \geqslant 0,$$

for any scalars $a_g \in \mathbb{C}$. A function defined on G and having this positivity property is said to be *positive definite*. Conversely, from a positive-definite function one can build a unitary representation. This *GNS construction* (in honor of Israel Gelfand, Mark Naimark, and Irving Segal) begins by considering the group elements themselves as basis vectors in an abstract vector space. We can attempt to define an inner product on this vector space by means of the formula

$$\langle g_1, g_2 \rangle = \sigma(g_1^{-1} g_2).$$

The resulting object may differ from a genuine Hilbert space in two respects. First, there may be nonzero vectors whose length, as measured by the inner product, is zero (although the hypothesis that σ is positive definite does rule out the possibility that there might be vectors of *negative* length). Second, the COMPLETENESS AXIOM [III.62] of Hilbert space theory may not be satisfied. However, there is a "completion" procedure which fixes both these deficiencies. Applied in the present case, it produces a Hilbert space H_σ that carries a unitary representation of G.

Versions of the GNS construction arise in several areas of mathematics. They have the advantage that the functions on which the constructions are based are easy to manipulate. For instance, convex combinations of positive-definite functions are again positive definite, and this allows geometrical methods to be applied to the study of representations.

1.5   Determinants and Traces

The original works of Fredholm and Hilbert borrowed heavily from traditional concepts of linear algebra, and in particular the theory of DETERMINANTS [III.15]. In view of the complicated definition of the determinant even for finite matrices, it is perhaps not surprising that the infinite-dimensional situation presented extraordinary challenges. Very soon, much simpler alternative approaches were found that avoided determinants altogether. But it is interesting to note that the determinant, or to be more exact the related notion of the trace, has played an important role in recent developments on which we will report later in this article.

The *trace* of an $n \times n$ matrix is the sum of its diagonal entries. As with the determinant, the trace of a matrix A is equal to the trace of BAB^{-1} for any invertible matrix B. In fact, the trace is related to the determinant by the formula $\det(\exp(A)) = \exp(\operatorname{tr}(A))$ (because of the invariance properties of trace and determinant, it is enough to check this for diagonal matrices, where it is easy). In infinite dimensions the trace need not make sense since the sum of the diagonal entries of an $\infty \times \infty$ matrix may not converge. (The trace of the identity operator is a case in point: the diagonal entries are all 1, and if there are infinitely many of them, then their sum is not well-defined.) One way to address this problem is to limit oneself to operators for which the sum *is* well-defined. An operator T is said to be of *trace class* if, for every two sequences $\{u_j\}$ and $\{v_j\}$ of pairwise orthogonal vectors of length 1, the sum $\sum_{j=1}^{\infty} \langle T u_j, v_j \rangle$ is absolutely convergent. A trace-class operator T has a well-defined and finite trace, namely the sum $\sum_{j=1}^{\infty} \langle T u_j, u_j \rangle$ (which is independent of the choice of orthonormal basis $\{u_j\}$).

Integral operators such as those appearing in Fredholm's equation provide natural examples of trace-class operators. If $k(y, x)$ is a smooth function, then the operator $Tu(y) = \int k(y, x) u(x) \, dx$ is of trace class, and its trace is equal to $\int k(x, x) \, dx$, which can be regarded as the "sum" of the diagonal elements of the "continuous matrix" k.

2   Von Neumann Algebras

The *commutant* of a set S of bounded linear operators on a Hilbert space H is the collection S' of all operators on H that commute with every operator in the set S. The commutant of any set is an *algebra* of operators on H. That is, if T_1 and T_2 are in the commutant, then so are $T_1 T_2$ and any linear combination $a_1 T_1 + a_2 T_2$.

As mentioned in the previous section, a *unitary representation* of a group G on a Hilbert space H is a collection of unitary operators U_g, labeled by elements of G, with the property that for any two group elements g_1 and g_2 the composition $U_{g_1} U_{g_2}$ is equal to $U_{g_1 g_2}$. A *von Neumann algebra* is any algebra of operators on a complex Hilbert space H which is the commutant of some unitary representation of a group on H. Every von Neumann algebra is closed under adjoints and under limits of nearly every sort. For example, it is closed under pointwise limits: if $\{T_n\}$ is a sequence of operators in a von Neumann algebra M, and if $T_n v \to Tv$, for every vector $v \in H$, then $T \in M$.

It is easy to check that every von Neumann algebra M is equal to its own double commutant M'' (the commutant of the commutant of M). Von Neumann proved that if a self-adjoint algebra M of operators is closed under pointwise limits, then M is equal to the commutant of the group of unitary operators in its commutant, and is therefore a von Neumann algebra.

2.1   Decomposing Representations

Let $g \to U_g$ be a unitary representation of a group G on a Hilbert space H. If a closed subspace H_0 of H is mapped into itself by all the operators U_g, then it is said to be an *invariant subspace* for the representation. If H_0 is invariant, then since the operators U_g map H_0 to itself, their restrictions to H_0 constitute another representation of G, called a *subrepresentation* of the original.

A subspace H_0 is invariant for a representation, and so determines a subrepresentation, if and only if the orthogonal projection operator $P : H \to H_0$ belongs to the commutant of that representation. This points to a close connection between subrepresentations and von Neumann algebras. In fact, von Neumann algebra theory can be thought of as the study of the ways in which unitary representations can be decomposed into subrepresentations.

A representation is *irreducible* if it has no nontrivial invariant subspace. A representation that does have a nontrivial invariant subspace H_0 can be divided into two subrepresentations: those associated with H_0 and those associated with its orthogonal complement $H_0^{\perp}$. Unless both the representations H_0 and $H_0^{\perp}$ are irreducible, we will be able to divide one or both of them into still smaller pieces by repeating the process that was just carried out for H. If the initial Hilbert space H is finite dimensional, then continuing in this way we will eventually decompose it into irreducible subrepresentations. In the language of matrices, we will obtain a basis for H with respect to which all the operators in the group are simultaneously block diagonal, in such a way that each block represents an irreducible group of unitary operators on a smaller Hilbert space.

Reducing a unitary representation on a finite-dimensional Hilbert space into irreducible subrepresentations is a bit like decomposing an integer into a product of prime factors. As with prime factorization, the decomposition process for a finite-dimensional unitary representation has only one possible end: there is, up to ordering, a unique list of irreducible representations into which a given unitary representation decomposes.

But in infinite dimensions the decomposition process faces a number of difficulties, the most surprising of which is that there may be two decompositions of the same representation into entirely different sets of irreducible subrepresentations.

In the face of this, a different form of decomposition suggests itself, which is roughly analogous to the factorization of an integer into prime powers instead of individual primes. Let us refer to the prime powers into which an integer is decomposed as its *components*. They have two characteristic properties: no two components share a common factor, and any two (proper) factors of the same component *do* share a common factor. Similarly, one can decompose a unitary representation into *isotypical components*, which have analogous properties: no two distinct isotypical components share a common (meaning isomorphic) subrepresentation, and any two subrepresentations of the same isotypical component have themselves a common sub-subrepresentation. Any unitary representation (finite dimensional or not) can be decomposed into isotypical components, and this decomposition is unique.

In finite dimensions, every isotypical representation decomposes into a (finite) number of *identical* irreducible subrepresentations (like the prime factors of a prime power). In infinite dimensions this is not so. In effect, much of von Neumann algebra theory is concerned with analyzing the many possibilities that arise.

2.2   Factors

The commutant of an isotypical unitary representation is called a *factor*. Concretely, a factor is a von Neumann algebra M whose *center*, the set of all operators in M that commute with every member of M, consists of nothing more than scalar multiples of the identity operator. This is because projections in the center of M correspond to projections onto combinations of isotypical subrepresentations. Every von Neumann algebra can be uniquely decomposed into factors.

A factor is said to be of *type I* if it arises as the commutant of an isotypical representation that is a multiple of a single irreducible representation. Every type I factor is isomorphic to the algebra of all bounded operators on a Hilbert space. In finite dimensions, every factor is of type I, since as we already noted every isotypical representation decomposes into a multiple of one irreducible representation.

The existence of unitary representations with more than one decomposition into irreducible components is

related to the existence of factors that are *not* of type I. Von Neumann, together with Francis Murray, investigated this possibility in a series of papers that mark the foundation of operator algebra theory. They introduced an order structure on the collection of subrepresentations of a given isotypical representation or, to put it in terms of the commutant, on the collection of projections in a given factor. If H_0 and H_1 are subrepresentations of the isotypical representation H, then we write $H_0 \preceq H_1$ if H_0 is isomorphic to a subrepresentation of H_1. Murray and von Neumann proved that this is a total ordering: either $H_0 \preceq H_1$; or $H_1 \preceq H_0$; or both, in which case H_0 and H_1 are isomorphic. For example, in a finite-dimensional type I situation, where H is a multiple of n copies of a single irreducible representation, each subrepresentation is the sum of $m \leqslant n$ copies of the irreducible representation, and the order structure of the (isomorphism classes of) subrepresentations is the same as the order structure of the integers $\{0, 1, \ldots, n\}$.

Murray and von Neumann showed that the only order structures that can arise from factors are the following very simple ones:

Type I, $\{0, 1, 2, \ldots, n\}$ or $\{0, 1, 2, \ldots, \infty\}$;

Type II, $[0, 1]$ or $[0, \infty]$;

Type III, $\{0, \infty\}$.

The *type* of a factor is determined from the order structure of its projections according to this table.

In the case of factors of type II, the order structure is that of an interval of *real numbers*, not integers. Any subrepresentation of an isotypical representation of type II can be divided into yet smaller subrepresentations: we shall never reach an irreducible "atom." Nevertheless, subrepresentations can still be compared in size by means of the "real-valued dimension" provided by Murray and von Neumann's theorem.

A notable example of a factor of type II may be obtained as follows. Let G be a group and let $H = \ell^2(G)$ be a Hilbert space having basis vectors $[g]$ corresponding to the elements $g \in G$. Then there is a natural representation of G on H derived from the group multiplication law, called the *regular representation*: given an element g of G, the corresponding unitary map U_g is the linear operator that takes each basis vector $[g']$ in $\ell^2(G)$ to the basis vector $[gg']$. The commutant of this representation is a von Neumann algebra M. If G is a commutative group, then all the operators U_g are in the center of M; but if G is far enough from commutativity (for instance, if it is a free group), then M

will have trivial center and will therefore be a factor. It can be shown that this factor is of type II. There is a simple explicit formula for the real-valued dimension of a subrepresentation corresponding to an orthogonal projection $P \in M$. Represent P by an infinite matrix relative to the basis $\{[g]\}$ of H. Because P commutes with the representation, it is easy to see that the diagonal elements of P are all the same, equal to some real number between 0 and 1. This real number is the dimension of the subrepresentation corresponding to P.

More recently, the Murray–von Neumann dimension theory has found unexpected applications in TOPOLOGY [I.3 §6.4]. Many important topological concepts, such as Betti numbers, are defined as the (integer-valued) dimensions of certain vector spaces. Using von Neumann algebras, one can define real-valued counterparts of these quantities that have useful additional properties. In this way, one can use von Neumann algebra theory to obtain topological conclusions. The von Neumann algebras used here are typically obtained by the construction of the previous paragraph from the FUNDAMENTAL GROUP [IV.6 §2] of some compact space.

2.3   Modular Theory

Type III factors remained rather mysterious for a long time; indeed, Murray and von Neumann were at first unable to determine whether any such factors existed. They eventually managed to do so, but the fundamental breakthrough in the area came well after their pioneering work, when it was realized that each von Neumann algebra has a special family of symmetries, its so-called *modular automorphism group*.

To explain the origins of modular theory, let us consider once again the von Neumann algebra obtained from the regular representation of a group G. We defined the operators U_g on $\ell^2(G)$ by multiplying *on the left* by elements of G; but we could equally well have considered a representation defined by multiplying *on the right*. This would have yielded a different von Neumann algebra.

So long as we deal only with discrete groups G this difference is unimportant, because the map $S : [g] \mapsto [g^{-1}]$ is a unitary operator on H that interchanges the left and right regular representations. But for certain *continuous* groups the problem arises that the function $f(g)$ may be square-integrable while $f(g^{-1})$ is not. In this situation there is no simple unitary isomorphism analogous to the one for discrete groups. To remedy this, one must introduce a correction factor called the *modular function* of G.

The project of modular theory is to show that something analogous to the modular function can be constructed for any von Neumann algebra. This object then serves as an invariant for all factors of type III, whether or not they are explicitly derived from groups.

Modular theory exploits a version of the GNS construction (section 1.4). Let M be a self-adjoint algebra of operators. A linear functional $\phi : M \to \mathbb{C}$ is called a *state* if it is positive in the sense that $\phi(T^*T) \geqslant 0$, for every $T \in M$ (this terminology is derived from the connection described earlier between Hilbert space theory and quantum mechanics). For the purposes of modular theory we restrict attention to *faithful* states, those for which $\phi(T^*T) = 0$ implies $T = 0$. If ϕ is a state, then the formula

$$\langle T_1, T_2 \rangle = \phi(T_1^* T_2)$$

defines an inner product on the vector space M. Applying the GNS procedure, we obtain a Hilbert space H_M. The first important fact about H_M is that every operator T in M determines an operator on H_M. Indeed, a vector $V \in H_M$ is a limit $V = \lim_{n \to \infty} V_n$ of elements in M, and we can apply an operator $T \in M$ to the vector V using the formula

$$TV = \lim_{n \to \infty} TV_n,$$

where on the right-hand side we use multiplication in the algebra M. Because of this observation, we can think of M as an algebra of operators on H_M, rather than as an algebra of operators on whatever Hilbert space we began with.

Next, the adjoint operation equips the Hilbert space H_M with a natural "antilinear" operator $S : H_M \to H_M$ by the formula[1] $S(V) = V^*$. Since $U_g^* = U_{g^{-1}}$ for the regular representation, this is indeed analogous to the operator S we encountered in our discussion of continuous groups. The important theorem of Minoru Tomita and Masamichi Takesaki asserts that, as long as the original state ϕ satisfies a continuity condition, the *complex powers* $U_t = (S^*S)^{it}$ have the property that $U_t M U_{-t} = M$, for all t.

The transformations of M given by the formula $T \mapsto U_t T U_{-t}$ are called the *modular automorphisms* of M. Alain Connes proved that they depend only in a rather inessential way on the original faithful state ϕ. To be precise, changing ϕ changes the modular automorphisms only by *inner automorphisms*, that is, transformations of the form $T \mapsto UTU^{-1}$, where U is a unitary

---

1. The interpretation of this formula on the completion H_M of M is a delicate matter.

operator in M itself. The remarkable conclusion is that every von Neumann algebra M has a canonical one-parameter group of "outer automorphisms," which is determined by M alone and not by the state ϕ that is used to define it.

The modular group of a type I or type II factor consists only of the identity transformation; however, the modular group of a type III factor is much more complex. For example, the set

$$\{t \in \mathbb{R} : T \mapsto U_t T U_{-t} \text{ is an inner automorphism}\}$$

is a subgroup of $\mathbb{R}$ and an invariant of M that can be used to distinguish between uncountably many different type III factors.

2.4 Classification

A crowning achievement of von Neumann algebra theory is the classification of factors that are *approximately finite dimensional*. These are the factors that are in a certain sense limits of finite-dimensional algebras. Besides the range of the dimension function, which separates factors into types, the sole invariant is the *module*. This is a flow on a certain space that is assembled from the modular automorphism group.

A lot of attention is currently being given to the long-standing problem of distinguishing among the type II factors associated with the regular representations of groups. Of special interest is the case of FREE GROUPS [IV.10 §2], around which has flourished the subject of free probability theory. Despite intensive effort, some fundamental questions remain open: at the time of writing it is unknown whether the factors associated with the free groups on two and on three generators are isomorphic.

Another important development has been *subfactor theory*, which attempts to classify the ways in which factors can be realized within other factors. A remarkable and surprising theorem of Vaughan Jones shows that, in the type II situation, where continuous values of dimensions are the norm, the dimensions of subfactors can in certain situations assume only a discrete range of values. The combinatorics associated with this result have also appeared in other apparently quite unrelated parts of mathematics, notably KNOT THEORY [III.44].

3 C*-Algebras

Von Neumann algebra theory helps describe the structure of a single representation of a group on a Hilbert space. But in many situations it is of interest to gain an understanding of all possible unitary representations. To shed some light on this problem we turn to a related but different part of operator algebra theory.

Consider the collection $\mathcal{B}(H)$ of all bounded operators on a Hilbert space H. It has two very different structures: *algebraic* operations, such as addition, multiplication, and formation of adjoints; and *analytic* structures, such as the operator norm

$$\|T\| = \sup\{\|Tu\| : \|u\| \leqslant 1\}.$$

These structures are not independent of one another. Suppose, for instance, that $\|T\| < 1$ (an analytic hypothesis). Then the geometric series

$$S = I + T + T^2 + T^3 + \cdots$$

converges in $\mathcal{B}(H)$, and its limit S satisfies

$$S(I - T) = (I - T)S = I.$$

It follows that $I - T$ is invertible in $\mathcal{B}(H)$ (an algebraic conclusion). One can easily deduce from this that the *spectral radius* $r(T)$ of any operator T (defined to be the greatest absolute value of any complex number in the spectrum of T) is less than or equal to its norm.

The remarkable *spectral radius formula* goes much further in the same direction. It asserts that $r(T) = \lim_{n \to \infty} \|T^n\|^{1/n}$. If T is *normal* ($TT^* = T^*T$), and in particular if T is self-adjoint, then it may be shown that $\|T^n\| = \|T\|^n$. As a result, the spectral radius of T is precisely equal to the norm of T. There is therefore a very close connection between the algebraic structure of $\mathcal{B}(H)$, particularly algebraic structure related to the adjoint operation, and the analytic structure.

Not all the properties of $\mathcal{B}(H)$ are relevant to this connection between algebra and analysis. A *C*-algebra* A is an abstract structure that has enough properties for the argument of the previous two paragraphs to remain valid. A detailed definition would be out of place here, but it is worth mentioning that a crucial condition relating norm, multiplication, and $*$-operation is

$$\|a^*a\| = \|a\|^2, \quad a \in A,$$

called the *C*-identity* for A. We also note that special classes of operators on Hilbert space (unitaries, orthogonal projections, and so on) all have their counterparts in a general C^*-algebra. For example, a *unitary* $u \in A$ satisfies $uu^* = u^*u = 1$, and a *projection* p satisfies $p = p^2 = p^*$.

A simple example of a C^*-algebra is obtained by starting with a single operator $T \in \mathcal{B}(H)$. The collection of all operators $S \in \mathcal{B}(H)$ that can be obtained as limits of polynomials in T and T^* is a C^*-algebra said to be *generated* by T. The C^*-algebra generated by T is commutative if and only if T is normal; this is one reason for the importance of normal operators.

3.1   Commutative C^*-Algebras

If X is a COMPACT [III.9] TOPOLOGICAL SPACE [III.90], then the collection $C(X)$ of continuous functions $f : X \to \mathbb{C}$ comes with natural algebraic operations (inherited from the usual ones on $\mathbb{C}$) and a norm $\|f\| = \sup\{|f(x)| : x \in X\}$. In fact, these operations make $C(X)$ into a C^*-algebra. The multiplication in $C(X)$ is *commutative*, because the multiplication of complex numbers is commutative.

A basic result of Gelfand and Naimark asserts that *every* commutative C^*-algebra is isomorphic to some $C(X)$. Given a commutative C^*-algebra A, one constructs X as the collection of all algebra homomorphisms $\xi : A \to \mathbb{C}$, and the *Gelfand transform* then associates with $a \in A$ the function $\xi \mapsto \xi(a)$ from X to $\mathbb{C}$.

The Gelfand–Naimark theorem is a foundational result of operator theory. For example, a modern proof of the spectral theorem might proceed as follows. Let T be a self-adjoint or normal operator on a Hilbert space H, and let A be the commutative C^*-algebra generated by T. By the Gelfand–Naimark theorem, A is isomorphic to $C(X)$ for some space X, which may in fact be identified with the spectrum of T. If v is a unit vector in H, then the formula $S \mapsto \langle Sv, v \rangle$ defines a state ϕ on A. The GNS space associated with this state is a Hilbert space of functions on X, and elements of $A = C(X)$ act as multiplication operators. In particular, T acts as a multiplication operator. A small additional argument shows that T is unitarily equivalent to this multiplication operator, or at least to a direct sum of such operators (which is itself a multiplication operator on a larger space).

Continuous functions can be composed: if f and g are continuous functions (with the range of g contained in the domain of f), then $f \circ g$ is also a continuous function. Since the Gelfand–Naimark theorem tells us that any self-adjoint element of a C^*-algebra A sits inside an algebra isomorphic to the continuous functions on the spectrum of a, we conclude that if $a \in A$ is self-adjoint, and if f is a continuous function defined on the spectrum of a, then an operator $f(a)$ exists in A. This *functional calculus* is a key technical tool in C^*-algebra theory. For example, suppose that $u \in A$ is unitary and $\|u - 1\| < 2$. Then the spectrum of u is a subset of the

unit circle in $\mathbb{C}$ that does not contain -1. One can define a continuous branch of the complex logarithm function on such a subset, and it follows that there is an element $a = \log u$ of the algebra such that $a = -a^*$ and $u = e^a$. The path $t \mapsto e^{ta}, 0 \leqslant t \leqslant 1$, is then a continuous path of unitaries in A connecting u to the identity. Thus every unitary sufficiently close to the identity is connected to the identity by a unitary path.

3.2   Further Examples of C^*-Algebras

3.2.1   The Compact Operators

An operator on a Hilbert space has *finite rank* if its range is a finite-dimensional subspace. The operators of finite rank form an algebra, and its closure is a C^*-algebra called the algebra of *compact* operators and denoted $\mathcal{K}$. One can also view $\mathcal{K}$ as a "limit" of matrix algebras

$$M_1(\mathbb{C}) \to M_2(\mathbb{C}) \to M_3(\mathbb{C}) \to \cdots ,$$

where each matrix algebra is included in the next by

$$A \mapsto \begin{pmatrix} A & 0 \\ 0 & 0 \end{pmatrix}.$$

Many natural operators are compact, including the integral operators that arose in Fredholm's theory. The identity operator on a Hilbert space is compact if and only if that Hilbert space is finite dimensional.

3.2.2   The CAR Algebra

The presentation of $\mathcal{K}$ as a limit of matrix algebras leads one to consider other "limits" of a similar sort. (We shall not attempt a formal definition of these limits here, but it is important to note that the limit of a sequence $A_1 \to A_2 \to A_3 \to \cdots$ depends on the homomorphisms $A_i \to A_{i+1}$ as well as on the algebras A_i.) One particularly important example is obtained as the limit

$$M_1(\mathbb{C}) \to M_2(\mathbb{C}) \to M_4(\mathbb{C}) \to \cdots ,$$

where each matrix algebra is included in the next by

$$A \mapsto \begin{pmatrix} A & 0 \\ 0 & A \end{pmatrix}.$$

This is called the *CAR algebra*, because it contains elements that represent the *canonical anticommutation relations* that arise in quantum theory. C^*-algebras find several applications to quantum field theory and quantum statistical mechanics which extend von Neumann's formulation of quantum theory in terms of Hilbert space.

3.2.3   Group C^*-Algebras

If G is a group and $g \mapsto U_g$ is a unitary representation of G on a Hilbert space H, we can consider the smallest C^*-algebra of operators on H containing all the U_g; this is called the C^*-algebra *generated* by the representation. An important example is the *regular representation* on the Hilbert space $\ell^2(G)$ generated by G, which we defined in section 2.2. The C^*-algebra that it generates is denoted $C_r^*(G)$. The subscript "r" refers to the regular representation. Considering other representations leads to other, potentially different, group C^*-algebras.

Consider, for example, the case $G = \mathbb{Z}$. Since this is a commutative group, its C^*-algebra is also commutative, and thus it is isomorphic to $C(X)$ for a suitable X, by the Gelfand–Naimark theorem. In fact, X is the unit circle S^1, and the isomorphism

$$C(S^1) \cong C_r^*(\mathbb{Z})$$

takes a function on the circle to its Fourier series.

States defined on group C^*-algebras correspond to positive-definite functions defined on groups, and hence to unitary group representations. In this way new representations may be constructed and studied. For example, using states of group C^*-algebras it is possible to give to the set of irreducible representations of G the structure of a topological space.

3.2.4   The Irrational Rotation Algebra

The algebra $C^*(\mathbb{Z})$ is generated by a single unitary element U (corresponding to $1 \in \mathbb{Z}$). Moreover, it is the *universal example* of such a C^*-algebra, which is to say that given any C^*-algebra A and unitary $u \in A$, there is one and only one homomorphism $C^*(\mathbb{Z}) \to A$ sending U to u. In fact, this is nothing other than the functional calculus homomorphism for the unitary u.

If instead we consider the universal example of a C^*-algebra generated by *two* unitaries U, V subject to the relation

$$UV = e^{2\pi i \alpha} VU,$$

where α is irrational, we obtain a noncommutative C^*-algebra called the *irrational rotation algebra* A_α. The irrational rotation algebras have been studied intensively from a number of points of view. Using K-theory (see below) it has been shown that A_{α_1} is isomorphic to A_{α_2} if and only if $\alpha_1 \pm \alpha_2$ is an integer.

It can be shown that the irrational rotation algebra is *simple*, which implies that *any* pair of unitaries U, V satisfying the commutation relation above will generate

a copy of A_α. (Note the contrast with the case of a single unitary: 1 is a unitary operator, but it does not generate a copy of $C^*(\mathbb{Z})$.) This allows us to give a concrete representation of A_α on the Hilbert space $L^2(S^1)$, where U is the rotation through $2\pi\alpha$ and V is multiplication by $z : S^1 \to \mathbb{C}$.

4   Fredholm Operators

A *Fredholm operator* between Hilbert spaces is defined to be a bounded operator T for which the kernel and cokernel are finite dimensional. This means that the homogeneous equation $Tu = 0$ admits only finitely many linearly independent solutions, while the inhomogeneous equation $Tu = v$ admits a solution if v satisfies a finite number of linear conditions. The terminology arises from Fredholm's original work on integral equations; he showed that if K is an integral operator, then $I + K$ is a Fredholm operator.

For the operators that Fredholm considered, the dimensions of the kernel and cokernel must be equal, but in general this need not be so. The *unilateral shift operator* S, which maps the infinite "row vector" $(a_1, a_2, a_3, \dots)$ to $(0, a_1, a_2, \dots)$, is an example. The equation $Su = 0$ has only the zero solution, but the equation $Su = v$ has a solution only if the first coordinate of the vector v is zero.

The *index* of a Fredholm operator is defined to be the integer difference

$$\text{index}(T) = \dim(\ker(T)) - \dim(\text{coker}(T)).$$

For example, every invertible operator is a Fredholm operator of index 0, whereas the unilateral shift is a Fredholm operator of index -1.

4.1   Atkinson's Theorem

Consider the two systems of linear equations

$$\left.\begin{cases} 2.1x + y = 0 \\ 4x + 2y = 0 \end{cases}\right\} \quad \text{and} \quad \left.\begin{cases} 2x + y = 0 \\ 4x + 2y = 0 \end{cases}\right\}.$$

Although the coefficients of these equations are very close, the dimensions of their kernels are quite different: the left-hand system has only the zero solution, whereas the right-hand system has the nontrivial solutions $(t, -2t)$. Thus the dimension of the kernel is an *unstable* invariant of the system of equations. A similar remark applies to the dimension of the cokernel. By contrast, the index is stable, despite its definition as the difference of two unstable quantities.

An important theorem of Frederick Atkinson gives precise expression to these stability properties. Atkinson's theorem asserts that an operator T is Fredholm if and only if it is invertible modulo compact operators. This implies that any operator that is sufficiently close to a Fredholm operator is itself a Fredholm operator with the same index, and that if T is a Fredholm operator and K is a compact operator, then $T + K$ is a Fredholm operator with the same index as T. Notice that, since integral operators are compact operators, this contains Fredholm's original theorem as a special case.

4.2 The Toeplitz Index Theorem

TOPOLOGY [I.3 §6.4] studies those properties of mathematical systems that remain the same when the system is (continuously) perturbed. Atkinson's theorem tells us that the Fredholm index is a topological quantity. In many contexts it is possible to obtain a formula for the index of a Fredholm operator in terms of other, apparently quite different, topological quantities. Formulas of this sort often indicate deep connections between analysis and topology and often have powerful applications.

The simplest example involves the *Toeplitz operators*. A Toeplitz operator has a matrix with the special form

$$T = \begin{pmatrix} b_0 & b_1 & b_2 & b_3 & \cdots \\ b_{-1} & b_0 & b_1 & b_2 & \cdots \\ b_{-2} & b_{-1} & b_0 & b_1 & \cdots \\ b_{-3} & b_{-2} & b_{-1} & b_0 & \cdots \\ \vdots & \vdots & \vdots & \vdots & \ddots \end{pmatrix}.$$

In other words, as you go down each diagonal of the matrix, the entries remain constant. The sequence of coefficients $\{b_n\}_{n=-\infty}^{\infty}$ defines a function $f(z) = \sum_{n=-\infty}^{\infty} b_n z^{-n}$ on the unit circle in the complex plane, called the *symbol* of the Toeplitz operator. It can be shown that a Toeplitz operator whose symbol is a continuous function which is never zero is Fredholm. What is its index?

The answer is given by thinking about the symbol as a mapping from the unit circle to the nonzero complex numbers: in other words, as a closed path in the nonzero complex plane. The fundamental topological invariant of such a path is its *winding number*: the number of times it "goes around" the origin in the counterclockwise direction. It can be proved that the index of a Toeplitz operator with nonzero symbol f is minus the winding number of f. For example, if f is the function $f(z) = z$ (with winding number $+1$),

then the associated Toeplitz operator is the unilateral shift S that we encountered earlier (with index -1). The Toeplitz index theorem is a very special case of THE ATIYAH–SINGER INDEX THEOREM [V.2], which gives a topological formula for the indices of various Fredholm operators that arise in geometry.

4.3 Essentially Normal Operators

Atkinson's theorem suggests that compact perturbations of an operator are in some sense "small." This leads to the study of properties of an operator that are preserved by compact perturbation. For instance, the *essential spectrum* of an operator T is the set of complex numbers λ for which $T - \lambda I$ fails to be Fredholm (that is, invertible modulo compact operators). Two operators T_1 and T_2 are *essentially equivalent* if there is a unitary operator U such that $U T_1 U^*$ and T_2 differ by a compact operator. A beautiful theorem originally due to WEYL [VI.80] asserts that two self-adjoint or normal operators are essentially equivalent if and only if they have the same essential spectrum.

One might argue that the restriction to normal operators in this theorem is inappropriate. Since we are concerned with properties that are preserved by compact perturbation, would it not be more appropriate to consider *essentially normal* operators—that is, operators T for which $T^*T - TT^*$ is compact? This apparently modest variation leads to an unexpected result. The unilateral shift S is an example of an essentially normal operator. Its essential spectrum is the unit circle, as is the essential spectrum of its adjoint; however, S and S^* cannot be essentially equivalent, because S has index -1 and S^* has index $+1$. Thus some new ingredient, beyond the essential spectrum, is needed to classify essentially normal operators. In fact, it follows easily from Atkinson's theorem that if essentially normal operators T_1 and T_2 are to be essentially equivalent, then not only must they have the same essential spectrum but also, for every λ not in the essential spectrum, the Fredholm index of $T_1 - \lambda I$ must be equal to the Fredholm index of $T_2 - \lambda I$. The converse of this statement was proved by Larry Brown, Ron Douglas, and Peter Fillmore in the 1970s, using entirely novel techniques that led to a new era of interaction between C^*-algebra theory and topology.

4.4 *K*-Theory

A remarkable feature of the Brown–Douglas–Fillmore work was the appearance within it of tools from

ALGEBRAIC TOPOLOGY [IV.6], notably *K-theory*. Remember that, according to the Gelfand–Naimark theorem, the study of (suitable) topological spaces and the study of commutative C^*-algebras are one and the same; all the techniques of topology can be transferred, via the Gelfand–Naimark isomorphism, to commutative C^*-algebras. Having made this observation, it is natural to ask which of these techniques can be extended further, to provide information about *all* C^*-algebras, commutative or not. The first and best example is *K*-theory.

In its most basic form, *K*-theory associates with each C^*-algebra A an Abelian group $K(A)$, and with each homomorphism of C^*-algebras a corresponding homomorphism of Abelian groups. The building blocks for $K(A)$ can be thought of as generalized Fredholm operators associated with A; the generalization is that these operators act on "Hilbert spaces" in which the complex scalars are replaced by elements of the C^*-algebra A. The group $K(A)$ itself is defined to be the collection of connected components of the space of all such generalized Fredholm operators. Thus if $A = \mathbb{C}$, for instance (so that we are dealing with classical Fredholm operators), then $K(A) = \mathbb{Z}$. This follows from the fact that two Fredholm operators are connected by a path of Fredholm operators if and only if they have the same index.

One of the great strengths of *K*-theory is that one can construct *K*-theory classes from a variety of different ingredients. For example, every projection $p \in A$ defines a class in $K(A)$ which can be thought of as a "dimension" for the range of p. This connects *K*-theory to the classification of factors (section 2.2), and has become an important tool in the effort to classify various families of C^*-algebras, such as the irrational rotation algebras. (It was at one time thought that the irrational rotation algebras might not contain any nontrivial projections at all: the construction of such projections by Marc Rieffel was an important step in the development of C^*-algebra *K*-theory.) Another beautiful example is George Elliott's classification theorem for locally finite-dimensional C^*-algebras like the CAR algebra; they are completely determined by *K*-theoretic invariants.

The problem of computing the *K*-theory groups of noncommutative C^*-algebras, particularly group C^*-algebras, has turned out to have important connections with topology. In fact, some key advances in topology have come from C^*-algebra theory in this way, thereby allowing operator algebraists to repay some of the debt they owe to the topologists for *K*-theory. The principal organizing problem in this area is the *Baum–Connes conjecture*, which proposes a description of the *K*-theory of group C^*-algebras in terms of invariants familiar in algebraic topology. Most of the progress on the conjecture to date is the result of work of Gennadi Kasparov, who dramatically broadened the original discoveries of Brown, Douglas, and Fillmore to cover not just single essentially normal operators but also noncommuting systems of operators, that is, C^*-algebras. Kasparov's work is now a central component of operator algebra theory.

5 Noncommutative Geometry

DESCARTES's [VI.11] invention of coordinates showed that one can do geometry by thinking about coordinate functions rather than directly thinking about points in space and their interrelationships: these coordinate functions are the familiar x, y, and z. The Gelfand–Naimark theorem can be viewed as one expression of this idea of passing from the "point picture" of a space X to the "field picture" of the algebra $C(X)$ of functions on it. The success of *K*-theory in operator algebras invites us to ponder whether the field picture might be *more powerful* than the point picture, since *K*-theory can be applied to noncommutative C^*-algebras which may not have any "points" (homomorphisms to $\mathbb{C}$) at all.

One of the most exciting research frontiers in operator algebra theory is reached along a path which develops these thoughts. The *noncommutative geometry* program of Connes takes seriously the idea that a general C^*-algebra should be thought of as an algebra of functions on a "noncommutative space," and goes on to develop "noncommutative" versions of many ideas from geometry and topology, as well as completely new constructions that have no commutative counterpart. Noncommutative geometry begins with the creative reformulation of ideas from ordinary geometry in ways that involve only operators and functions, but not points.

Consider, for instance, the circle S^1. The algebra $C(S^1)$ reflects all the topological properties of S^1, but to incorporate its *metric* (distance-related) properties as well we look not just at $C(S^1)$ but at the pair consisting of the algebra $C(S^1)$ and the operator $D = \mathrm{i}\,\mathrm{d}/\mathrm{d}\theta$ on the Hilbert space $H = L^2(S^1)$. Notice that if f is a function on the circle (considered as a multiplication operator on H), then the commutator $Df - fD$ is also a multiplication operator, this time by $\mathrm{i}\,\mathrm{d}f/\mathrm{d}\theta$. It follows that ordinary measurements of angular distance between points on the circle can be recovered from $C(S^1)$ and D by the formula

$$d(p, q) = \max\{|f(p) - f(q)| : \|Df - fD\| \leqslant 1\}.$$

Connes argues that operator $|D|^{-1}$ plays the role of the "unit of arc-length ds" in this and many other, more complicated situations.[2]

Another feature of the examples Connes considers, also of central importance in noncommutative geometry, is the fact that the operator $|D|^{-k}$ is a trace-class operator (see section 1.5) when k is large enough. In the case of the circle, k needs to be bigger than 1. Computations with traces connect noncommutative geometry to COHOMOLOGY THEORY [IV.6 §4]. We now have two kinds of "noncommutative algebraic topology," namely K-theory and a new variant of homology called *cyclic cohomology*; the connection between the two is provided by a very general index theorem.

There are several procedures that produce noncommutative C^*-algebras (to which Connes's methods can be applied) from classical geometric data. The irrational rotation algebras A_θ are examples; the classical picture to which they apply is the QUOTIENT SPACE [I.3 §3.3] of the circle by the group of rotations through multiples of θ. Classical methods of geometry and topology are unable to handle this quotient space, but the noncommutative approach via A_θ is much more successful.

An exciting but speculative possibility is that the basic laws of physics should be addressed from the perspective of noncommutative geometry. The transition to noncommutative C^*-algebras can be viewed as analogous to the transition from classical to quantum mechanics. However, Connes has argued that noncommutative C^*-algebras play a role in describing the physical world even before the transition is made to quantum physics.

Further Reading

Connes, A. 1995. *Noncommutative Geometry*. Boston, MA: Academic Press.
Davidson, K. 1996. C^*-*Algebras by Example*. Providence, RI: American Mathematical Society.
Fillmore, P. 1996. *A User's Guide to Operator Algebras*. Canadian Mathematical Society Series of Monographs and Advanced Texts. New York: John Wiley.
Halmos, P. R. 1963. What does the spectral theorem say? *American Mathematical Monthly* 70:241–47.

2. The operator D is not quite invertible since it vanishes on constant functions. A small modification must therefore be made before considering inverse operators. The operator $|D|$ is by definition the positive square root of D^2.

IV.16   Mirror Symmetry
Eric Zaslow

1   What Is Mirror Symmetry?

Mirror symmetry is a phenomenon found in theoretical physics that has had profound mathematical applications. It burst onto the mathematical scene after Candelas, de la Ossa, Green, and Parkes exploited the physical phenomenon to make precise predictions about certain sequences of numbers describing geometric spaces. The sequence predicted by those authors began 2875, 609 250, 317 206 375, . . . , and was far beyond the scope of calculation at the time. The phenomenon of mirror symmetry is that some physical theories have equivalent, "mirror" theories that lead to the same predictions. If some prediction requires a hard calculation but is easy to perform in the mirror theory, then you can get the answer for free! These physical theories do not have to be realistic models of physics. For instance, beginning students of physics often study point particles on frictionless planes. Although they are unrealistic, such toy models can bring the physical concepts into focus and their analysis can give rise to very interesting mathematics.

1.1   Exploiting Equivalences

Children at school in the 1950s used log tables to exploit the equivalence of multiplication of positive numbers with addition of real numbers. Given the problem of multiplying two large numbers a and b, they would use a table to look up the logarithms $\log(a)$ and $\log(b)$ (to a certain number of significant figures), then add them by hand. They would then use the same table to find which number had a logarithm equal to $\log(a) + \log(b)$. The answer is ab.

College students sometimes exploit the equivalence defined by FOURIER TRANSFORMS [III.27] to solve differential equations. Basically, the Fourier transform is a rule that maps one function $f(x)$ to a new function $\hat{f}(p)$. What is nice is that the transform of the derivative $f'(x)$ relates in a very simple way to $\hat{f}(p)$: it is $ip\hat{f}(p)$, where i is the imaginary number $\sqrt{-1}$. If you want to solve a differential equation such as $f'(x) + 2f(x) = h(x)$, where $h(x)$ is a given function and you are trying to find f, you can map the equation to its Fourier transform equation $ip\hat{f}(p) + 2\hat{f}(p) = \hat{h}(p)$. This is much easier: it is an algebraic equation rather than a differential equation, and has the solution $\hat{f}(p) = \hat{h}(p)/(2 + \mathrm{i}p)$. The solution $f(x)$ is then the function which has $\hat{h}(p)/(2 + \mathrm{i}p)$ as its Fourier transform.

Mirror symmetry is like a fancy Fourier transform, mapping much more information than is contained in a single function. Every aspect of a physical theory is involved.

This article will (eventually) focus on the mathematics of mirror symmetry, but it is crucial to understand its physical origins. We therefore begin with a brief guide to physics. (For a further discussion of mathematical physics, see VERTEX OPERATOR ALGEBRAS [IV.17 §2].) This is in no way an adequate treatment—a separate *Companion to Physics* would be needed—but we hope to give enough of the flavor of the subject to help the reader with the later sections. (A reader familiar with physical theories may wish to skip the next section and refer back as needed.)

2 Theories of Physics

2.1 Formulations of Mechanics and Action Principles

2.1.1 Newtonian Physics

Newton's second law states that a particle moving through space accelerates[1] in proportion to the force it experiences: $F = m\ddot{x}$. The force is itself the (negative) gradient of a gravitational potential $V(x)$, so this equation can be written $m\ddot{x} + \nabla V(x) = 0$. Stationary particles sit at minima of the potential: examples are a ball in equilibrium at the end of a spring, or a pea at the bottom of a bowl. In stable situations, there is a restoring force proportional to some displacement distance. This means that in some appropriate coordinate, $F \sim -x$, so $V(x) = kx^2/2$, for some k. The solutions are oscillatory, with angular frequency $\omega = \sqrt{k/m}$. This model is called the *simple harmonic oscillator*.

2.1.2 The Least Action Principle

Every major theory can also be formulated by means of an idea known as the *least action principle*. Let us see how it works for the equations of Newtonian mechanics. Consider an arbitrary path of a particle $x(t)$ and

form the quantity

$$S(x) = \int [\tfrac{1}{2}m\dot{x}^2 - V(x)]\,\mathrm{d}t.$$

Here and below, the notation x may represent more than one coordinate. If x is used as a point in space-time, it will include the time coordinate, if that is not otherwise noted. Likewise, we omit component notation on most vectors. The notation should be clear from the context. The quantity $S(x)$, which is known as the *action*, equals the kinetic energy minus the potential energy. One then considers which paths minimize this action. That is, we ask which paths $x(t)$ have the property that, when they are perturbed by a small amount $\delta x(t)$, the action is unchanged, to leading order. (So in fact we require only that the action is unchanged to first order, and not that it is actually minimized. Solutions of saddle-point type are allowed.) The answer turns out to be precisely those paths that satisfy $m\ddot{x} + \nabla V(x) = 0$.[2]

For example, consider the simple harmonic oscillator in two dimensions. We can model x as a complex number and set $V(x) = k|x|^2$. The action is then $\int \tfrac{1}{2}[m|\dot{x}|^2 - k|x|^2]$. Note that a phase rotation $x \to \mathrm{e}^{\mathrm{i}\theta}x$ leaves the action invariant, and is therefore a symmetry of the equations of motion.

Lesson. Physical solutions extremize the action.

The principle of least action applies to many other physical situations, as we shall see below. First, though, we describe another formulation of mechanics.

2.1.3 The Hamiltonian Formulation of Mechanics

HAMILTON's [VI.37] formulation of the equations of motion also deserves mention. It leads to first-order equations. Let S be the action and define L by $S = \int L\,\mathrm{d}t$, and consider the (typical) case where L is a function of coordinates x and their time derivatives $\dot{x}$. Then set $p = \mathrm{d}L/\mathrm{d}\dot{x}$, a function that can depend both on x and on $\dot{x}$. (In the example $L = \tfrac{1}{2}m\dot{x}^2 - V(x)$ that we have already considered, we find that $p = m\dot{x}$, or $\dot{x} = p/m$.) Now let us consider the function $H = p\dot{x} - L$, which is called the HAMILTONIAN [III.35], and change variables from $(x, \dot{x})$ to (x, p) so as to remove all mention of $\dot{x}$. In the example, H works out to be

$$\frac{p^2}{m} - \left(\frac{p^2}{2m} - V(x)\right) = \frac{p^2}{2m} + V(x),$$

---

1. Acceleration is the second derivative of position with respect to time. We denote position by x, which is shorthand for a three-component position vector, and we denote time derivatives by dots, so acceleration is denoted by $\ddot{x}$.

2. To see this, replace x by $x + \delta x$ in the action and keep only the linear terms in δx and its time derivative. For V the linear terms are $(\nabla V)\delta x$. One then has to integrate by parts to remove the time derivative of δx and isolate it as a factor in the integrand. The integral will be zero for arbitrary variations δx only when the term multiplying it vanishes. This gives the equation. Try it!

which is the total energy. For the simple harmonic oscillator, $H = p^2/2m + kx^2/2$.

The equations $\dot{x} = \partial H/\partial p$ and $\dot{p} = -\partial H/\partial x$ are the equations of motion in the Hamiltonian formulation; they can be shown to be equivalent to those obtained from the action principle. In the example, $\dot{x} = p/m$ and $\dot{p} = -\nabla V$. Using the first equation to replace p by $m\dot{x}$ in the second, we recover the equation $m\ddot{x} + \nabla V(x) = 0$. More generally, one can consider the time derivative of some quantity $f(x, p)$ constructed from p and x and prove—using the chain rule and the equations of motion—that

$$\dot{f} = \frac{\partial f}{\partial x}\frac{\partial H}{\partial p} - \frac{\partial f}{\partial p}\frac{\partial H}{\partial x} = \{H, f\}.$$

The term in the middle is called the *Poisson bracket* of H and f, denoted $\{H, f\}$.

Lesson. The Hamiltonian controls time dependence through the Poisson bracket.

Notice that when we plug the coordinates x and p themselves into the bracket, we derive the identity

$$\{x, p\} = -1. \tag{1}$$

It is also possible to begin with the Hamiltonian viewpoint. One considers a space endowed with a bracket operation on functions, such that there are coordinate functions (not uniquely determined) obeying $\{x, p\} = -1$. The mechanical model is defined by a function $H(x, p)$, which determines the dynamics.

2.1.4 Symmetry

A brief remark on symmetry is in order. NOETHER [VI.76] proved that in the action formulation of mechanics, a symmetry of the action results in a conserved quantity. The prototypical example is translational or rotational symmetry, where the potential of a particle is invariant under some direction of translation or rotation: the corresponding conserved quantity is then momentum or angular momentum. In the example above, $V(x) = k|x|^2/2$ is independent of θ, the phase of x. The equation of motion determined by varying θ is $\mathrm{d}(m|x|^2\dot{\theta})/\mathrm{d}t = 0$, so in this case it is the angular momentum $m|x|^2\dot{\theta}$ that is conserved. In the Hamiltonian formulation, since a conserved quantity $f(x, p)$ does not change with time, it must have zero Poisson bracket with the Hamiltonian: $\{H, f\} = 0$. In particular, the Hamiltonian itself is conserved.

2.1.5 Action Functions for Other Theories

Returning now to action principles, we shall see how different physical theories are described through different actions. In electricity and magnetism, MAXWELL'S EQUATIONS [IV.13 §1.1] can be formulated in the form $\delta S = 0$, where now the action S takes the form of an integral over space and time of the electric (E) and magnetic (B) fields. In the case where there are no sources, the action is written

$$S = \frac{1}{8\pi e^2}\int [E^2 - B^2]\,\mathrm{d}x\,\mathrm{d}t, \tag{2}$$

where e is the electric charge of an electron. There is one important difference from the previous example, which is that the variations of the action must be taken with respect to the fundamental fields, and E and B are not fundamental as they are derived from the electromagnetic potential $A = (\phi, A)$ by the equations $E = \nabla\phi - \dot{A}$, $B = \nabla \times A$. If you rewrite S in terms of A, vary A by δA, and set $\delta S = 0$, then you recover Maxwell's equations from the least action principle.

It is clear that the electromagnetic action merely changes sign under the replacement $E \to B$, $B \to -E$, and therefore any solution $\delta S = 0$ remains a solution under the transformation. This is an example of an equivalence of a classical theory of physics. In fact, this symmetry extends to the case where there are sources (such as electrons) if we also interchange electric and magnetic sources. (No magnetic sources have been observed in the universe, but a theory with such objects still makes sense.)

Lesson. Physical equivalences act on fields and their sources.

Electricity and magnetism is a "field theory," which means that the degrees of freedom involve functions that depend on position in space. Contrast this with Newtonian mechanics, where the spatial degrees of freedom are just the coordinates of the particle(s). However, there is not much conceptual distance between the two, as can be seen in the following toy model.

We will consider the simplest example: a scalar field, ϕ. That is, ϕ is just a function that takes numerical values. Now imagine that space has just one dimension, not three, and further that that dimension is a circle, which we can describe with an angular coordinate, θ. At any fixed point in time we can use FOURIER SERIES [III.27] to write the scalar field as $\phi(\theta) = \sum_n c_n \exp(\mathrm{i}n\theta)$, where the c_n are the Fourier coefficients, and if we want the values of ϕ to be real numbers then we must insist that $c_{-n} = c_n^*$. We can then think of $\phi(\theta)$ not as a function but as an infinite-dimensional vector $(c_0, c_1, \ldots)$. The spatial dependence of ϕ is completely determined by the coefficients c_n. If we now wish to consider time dependence, then all we have to do is use time-dependent components $(c_0(t), c_1(t), \ldots)$, which looks a lot like an infinite set of quantum-mechanical particles c_n. Thus, the function ϕ has the Fourier expansion $\phi(\theta, t) = \sum_n c_n(t) \exp(in\theta)$.

The simplest action for a scalar field ϕ that allows wave-like solutions of the equations of motion serves as a natural analogue of equation (2):

$$S = \int \frac{1}{2\pi} [(\dot\phi)^2 - (\phi')^2] \, d\theta \, dt, \qquad (3)$$

where $\phi' = \partial\phi/\partial\theta$. When we plug the Fourier expansion into the action and perform the θ integration, we get

$$S = \int \sum_n [|\dot c_n|^2 - n^2 |c_n|^2] \, dt. \qquad (4)$$

Note that the term in brackets is just the action for a particle c_n in a quadratic potential, as in section 2.1.2. We simply have an infinite number of harmonic oscillators (with the exception of the c_0 degree of freedom, which corresponds to a free particle in no potential).

Lesson. Field theory is like point particle theory with an infinite number of particles. The particles correspond to the degrees of freedom of the field. When the action is just quadratic in the derivatives, the particles have an interpretation as simple harmonic oscillators.

Even GENERAL RELATIVITY [IV.13] fits into this framework as a field theory. For a space-time M, the field is the RIEMANNIAN METRIC [I.3 §6.10] on space-time. The metric is what determines the lengths of paths between points—so a stretching of space-time, for example, is represented by a rescaled metric. The action is then constructed as the integral of the Riemannian curvature scalar $\mathcal{R}$ over space-time: $S = \int_M \mathcal{R}.$[3]

2.2 Quantum Theory

Mirror symmetry is an equivalence of quantum theories, so we must develop an understanding of what a quantum theory is and what an equivalence looks like. There are two formulations of quantum mechanics: the operator formulation and Feynman's path-integral formulation.

Both formulations are probabilistic, meaning that you cannot predict exactly what will be observed in

a single measurement, but you can make precise predictions about what will be observed after multiple, repeated measurements in the same environment. For instance, your experimental apparatus may involve a beam of electrons hitting a screen and making a mark. The beam will contain millions of electrons, so the pattern of marks on the screen can be predicted with great accuracy. However, we cannot say what will happen to a single, given electron—all we can do is assign probabilities to the outcomes of various measurements. These probabilities are encoded in the so-called "wave function" Ψ of the particle.

2.2.1 Hamiltonian Formulation

In the operator formulation of quantum mechanics, the positions and momenta of classical mechanics (and any quantity formed from them) are converted into OPERATORS [III.50] acting on a HILBERT SPACE [III.37] according to the following rule: *replace the Poisson bracket* $\{\cdot, \cdot\}$ *by* $i/\hbar[\cdot, \cdot]$, where $[A, B] = AB - BA$ is the *commutator bracket* and $\hbar$ is Planck's constant. Thus, for example, we get from equation (1) the relation $[x, p] = i\hbar$. The state of a particle (or system) is now defined not as a set of values of x and p but as a vector Ψ in the Hilbert space. Once again, time evolution is determined by the Hamiltonian, H, but now H is an operator. The basic dynamical equation is

$$H\Psi = i\hbar \frac{d}{dt} \Psi. \qquad (5)$$

This is called the *Schrödinger equation.*

Lesson. To quantize a classical theory, replace ordinary degrees of freedom by operators on a vector space; replace Poisson brackets by commutator brackets.

In the case where we have a particle on the real line $\mathbb{R}$, the Hilbert space is the space of square-integrable functions $L^2(\mathbb{R})$, so we write Ψ as $\Psi(x)$. The commutation relation is obeyed if we think of x as the operator that sends the function $\Psi(x)$ to the function $x\Psi(x)$. Now the relation $[x, p] = i\hbar$ means that we should represent p as the operator $-i\hbar(d/dx)$. The values of the classical quantity associated with an operator correspond to the EIGENVALUES [I.3 §4.3] of that operator, so for example a state with momentum p has the form $\Psi \sim \exp(ipx/\hbar)$. Unfortunately, this is not square-integrable on the real line, but it would become so if we identified x and $x + 2\pi R$, for some number (radius) $R > 0$. Topologically, this COMPACTIFIES [III.9] $\mathbb{R}$ to a circle, but note that Ψ will be single-valued only if

---

3. In 3-space, the paraboloid $z = \frac{1}{2}ax^2 + \frac{1}{2}by^2$ has curvature ab at the origin.

$p = n\hbar/R$, where n is an integer. Thus, momentum is "quantized" in units of $\hbar/R$.[4] The integer label of the c_n of equation (4) can therefore also be thought of as a momentum.

In the above example, $\mathbb{R}$ is the degree of freedom of the classical coordinate x. In other examples, there is a copy of $L^2(\mathbb{R})$ for each real degree of freedom, whether or not it represents a geometric location.

Another novelty is that position and momentum do not commute as operators in quantum mechanics, meaning they cannot be simultaneously diagonalized: you cannot specify the position and momentum simultaneously. This is a form of Heisenberg's uncertainty principle (see OPERATOR ALGEBRAS [IV.15 §1.3]).

2.2.2   Symmetry

As the rules of quantization would suggest, a symmetry of a quantum theory is an operator A such that $[H, A] = 0$. That is, A commutes with the Hamiltonian, and therefore respects the dynamics.

2.2.3   Example: The Simple Harmonic Oscillator

We now discuss an example that will be useful later on for understanding quantum field theory and mirror symmetry: the simple harmonic oscillator in quantum mechanics. Suppose that the constants are chosen so that the Hamiltonian is given by $H = x^2 + p^2$. If one defines $a = (x + ip)/\sqrt{2}$ and $a^\dagger = (x - ip)/\sqrt{2}$, then one can show that $a^\dagger$ raises the energy of a state by one unit[5] and a lowers the energy by one unit. Invoking the physical argument that there is a ground state Ψ_0 of lowest energy, this state must obey $a\Psi_0 = 0$. One then finds that all states can be written in terms of the basis vectors $\Psi_n = (a^\dagger)^n \Psi_0$ with energy $n + \frac{1}{2}$. Note that Ψ_0 has energy $\frac{1}{2}$.[6] The basis $\{\Psi_n\}$ is called the *occupation number* basis, since the interpretation is that Ψ_n has n energy "quanta" above the ground state.

---

4. We shall occasionally choose our units to make $\hbar$ equal to 1. For example, we could work in the fictitious time unit of "sqeconds," one second equals $\hbar$ sqeconds.

5. Here is the calculation: $[a, a^\dagger] = 1$ and $H = a^\dagger a + \frac{1}{2}$. Further, $[H, a^\dagger] = a^\dagger$ and $[H, a] = -a$. These equations have the following interpretation. Suppose Ψ is an eigenvector of H with eigenvalue (energy) E. Then $H\Psi = E\Psi$. Consider $a^\dagger\Psi$. One quickly finds that

$$H(a^\dagger\Psi) = (Ha^\dagger - a^\dagger H + a^\dagger H)\Psi = ([H, a^\dagger] + a^\dagger H)\Psi$$

$$= (a^\dagger + a^\dagger E)\Psi = (E + 1)(a^\dagger\Psi).$$

We learn that $a^\dagger\Psi$ has eigenvalue $E + 1$, so $a^\dagger$ has "raised" the energy by one unit.

6. It is instructive to write these equations in terms of the operators defined by x and p.

2.2.4   Path-Integral Formulation

Feynman's path integral formulation of quantum mechanics builds on the idea of the least action principle. In this formulation, the probability of an experiment is calculated through an average over *all* paths of particles, and not just the ones which extremize the action. Each path $x(t)$ is weighted by the factor $\exp(iS(x)/\hbar)$, where $S(x)$ is the action of the path $x(t)$ and $\hbar$ is Planck's constant, which is very small compared with macroscopic action scales. This average can be an imaginary number, but the probability of the process is the square of its absolute value.

Note that $\exp(iS/\hbar) = \cos(S/\hbar) + i\sin(S/\hbar)$, so if S changes appreciably when we vary $x(t)$, then the real and imaginary parts will oscillate rapidly, since $\hbar$ is small. Then, when we integrate over paths $x(t)$, the positive and negative oscillations will roughly cancel. As a result, the main contributions to the weighted sum over paths will come from those paths for which S does not vary when the path does: the classical paths! However, if the variations are sufficiently small compared with $\hbar$, then nonclassical paths can contribute appreciably. One typically separates the degrees of freedom into the classical trajectory piece and the quantum fluctuations near it. Then one can organize the path integral in a perturbation theory around the parameter $\hbar$.

We have not yet discussed the integrand of the path integral, and will not go into the details of this. The main point is that the theory makes a prediction about the likelihood of measuring a physical process. Each process determines a possible integrand. For example, from our discussion above we learn that the integrand for measuring the likelihood of a quantum-mechanical particle going from the point x_0 at time t_0 to the point x_1 at time t_1 gives nonzero weight—determined by the exponentiated action—to all paths that go from x_0 to x_1 as t goes from t_0 to t_1, and zero weight to all other paths.

It is illustrative to consider a toy model of a path integral on a "space-time" that consists of just a single point. Then the possible "paths" of a scalar field, say, are simply the values that the field can take at the point, so they are real numbers. The action is then an ordinary function $S(x)$ on $\mathbb{R}$. For the purposes of this example, let us consider the case where $iS/\hbar = -\frac{1}{2}x^2 + \lambda x^3$. The possible integrands are (sums of) powers of x, so the basic path integrals to perform are $\int x^k \exp(-\frac{1}{2}x^2 + \lambda x^3)\,dx$, which we denote by $\langle x^k \rangle$. The value at $\lambda = 0$

is easily calculated.[7] For small λ we expand $e^{\lambda x^3}$ as $1 + \lambda x^3 + \lambda^2 x^6/2 + \cdots$, and evaluate each term by the same methods as for $\lambda = 0$. This is how we construct a well-defined perturbation theory, even when the integral is not calculable.

As we see from this example, path integrals are easiest when the action is only quadratic in the variables, just as we found in the operator formulation of quantum mechanics. The mathematical reason for this is that Gaussian integrals (exponentials of squares) can be done explicitly, while integrals involving exponentials of cubics or higher are difficult or impossible. For quadratic actions, the path integral can be evaluated exactly, but when cubic or higher terms appear, the perturbation series is necessary.

2.2.5 Quantum Field Theory

The generalization to field theories follows our earlier pattern. We think of quantum field theories, then, as being like quantum mechanics with infinite numbers of particles. In fact, the quantum field theories in which the fields Φ and their derivatives do not have more than quadratic terms in the action are easily understood in this way—we had a preview of this in equation (4). The Fourier components correspond to particles indexed by their momenta. Each one looks like a simple harmonic oscillator at some frequency, which will depend on the Fourier coefficient. The quantum Hilbert space is then a (tensor) product of lots of different "occupation number Hilbert spaces," one for each Fourier component of each field. Since the occupation number basis is also an energy eigenbasis, these states have a simple time evolution under the Hamiltonian H. That is, if $H = E$ on some state $\Psi(t = 0)$, then that state evolves like

$$\Psi(t) = \exp(iEt/\hbar)\Psi(0).$$

However, if the action includes terms that are *cubic or higher*, then things get interesting: particles can decay! This can be seen, for example, from the scalar field of equation (3) if we include a term ϕ^3 in the action, and therefore also the Hamiltonian. If we write this using Fourier components, we get terms involving three oscillators, such as $a_3^\dagger a_4^\dagger a_7$. To see this, recall that after we quantize the real field ϕ, the Fourier components

c_n act as harmonic oscillators, and we have written a_n for the associated creation and annihilation operators. Since the Hamiltonian governs time evolution according to equation (5), this means that over time one particle (the 7 mode) can decay into two others (the 3 and the 4). Such decay processes occur in real life, and it is a great triumph of quantum field theory that it can predict such events with astounding accuracy.

In fact, because the space of paths of fields is infinite dimensional, the path integral in quantum field theory is not usually defined in a mathematically rigorous way. However, the perturbation series for producing predictions can be defined just as for quantum mechanics, and this is how physicists make their predictions in practice. This perturbation series is organized in terms of *Feynman diagrams* (which are discussed in VERTEX OPERATOR ALGEBRAS [IV.17]). These diagrams, and the rules for computing them, completely solve the perturbation problem.

As in the example of quantum mechanics, different integrands of the path integral correspond to different predictions. If Φ is some function of the fields of some quantum field theory, we write $\langle\Phi\rangle$ for the path integral with Φ as an integrand (as we did for $\langle x^k\rangle$ in the previous section). We call such a term a "correlation function." If $\Phi = \phi_1(x_1) \cdots \phi_n(x_n)$, the answer will depend on the action of the theory, the fields ϕ_i, and the space-time points x_i.

One might wonder if a symmetry of a classical theory always remains a symmetry of the same theory after quantization. The answer is sometimes no. Such a case is known as an "anomaly." Roughly speaking, this is because the measure of integration of the path integral is not preserved under the symmetry, but this is a somewhat heuristic explanation because the path integral has no rigorous definition in general.

Returning to our cubic example, if the interaction term ϕ^3 has a coefficient λ, so that it is $\lambda\phi^3$, then we organize the perturbation series as a power series in λ. In terms of paths, probabilities of decay processes can be evaluated by considering paths that split into two—like the letter Y—with each leg carrying the label of the appropriate particle.

2.2.6 String Theory

Feynman's perturbation theory has an important generalization in *string theory*. String theory considers particles not as points but as loops. Instead of paths of particles through space-time, we get paths of loops,

---

7. Consider
$$\int \exp(-\tfrac{1}{2}x^2 + Jx)\,dx = \int \exp(-\tfrac{1}{2}(x+J)^2)\exp(\tfrac{1}{2}J^2)\,dx$$
$$= \sqrt{2\pi}\exp(\tfrac{1}{2}J^2).$$

Now if we differentiate this answer with respect to J, and set $J = 0$, we get $\langle x\rangle$. Taking k derivatives gives $\langle x^k\rangle$, and the theory is solved.

which look like two-dimensional surfaces. String theory amplitudes are computed by summing over all surfaces. These sums are organized in a perturbation series in powers of the so-called *string coupling constant*, λ_g. The power of λ_g in the perturbation series depends on the number of holes in the surface.

The surfaces are called worldsheets. At each point of the worldsheet, its location in space-time is determined by coordinates X^i. These coordinates themselves depend on the location on the worldsheet. In effect, we get an *auxiliary* theory: a field theory of coordinates on the two-dimensional surface! In string theory, even this two-dimensional field theory must be considered as a quantum field theory. The fields of the two-dimensional theory are maps from the surface to actual space-time. However, from the point of view of the worldsheet, the worldsheet itself is a two-dimensional space-time and the maps are fields on *this* space-time with values in some other (target) space.

Mirror symmetry was discovered as a result of the study of these quantum field theories on two-dimensional surfaces. Subsequently, the same phenomenon was discovered in the case where the strings were not closed loops but filaments with endpoints. Both cases play an important role below.

3 Equivalence in Physics

Mirror symmetry is a particular type of equivalence of quantum field theories. As we have seen, quantum field theories are rules for producing probabilities of physical processes. In the path-integral formulation, probabilities are computed from correlation functions of fields. According to Feynman, these correlation functions can be thought of as being averages over all paths of fields. Each path is weighted by $\exp(iS/\hbar)$, where S is the action of the path and $\hbar$ is Planck's constant. Let us denote the correlation function of some integrand Φ in theory A as $\langle\Phi\rangle_A$. Recall that Φ can depend on various fields ϕ_i and points of space-time x_i, and the correlation function will depend on all these and the action of theory A.

Equivalence, then, is a map from all possible fields ϕ_i in a theory A to corresponding fields $\tilde{\phi}_i$ in a theory B such that
$$\langle\Phi\rangle_A = \langle\tilde{\Phi}\rangle_B.$$

(For the moment, we deliberately neglect to notate the dependence on the points x_i.) One special correlation function is $\langle 1 \rangle$, which we call the *partition function* and denote by Z. As the field 1 always gets mapped to 1, we

derive the corollary that the partition functions must be equal: $Z_A = Z_B$.

Of course, this all has a description in the operator formulation of the quantum theory. Each state Ψ and each operator a in one theory must get mapped to a corresponding state $\tilde{\Psi}$ and operator $\tilde{a}$ in the mirror theory, in such a way that corresponding operators map corresponding states to states which themselves correspond. Here one sees the sharp analogy with the slide rule and the operations of multiplication and addition of numbers.

Each theory is typically described through some mathematical model, so an equivalence implies a host of mathematical identities between quantities constructed from corresponding models.

The particular case of mirror symmetry refers to an equivalence of quantum field theories on a two-dimensional surface. The most typical example of mirror symmetry is the physical theory whose fields are maps φ from a two-dimensional RIEMANN SURFACE [III.79] Σ to some target space, M. Such a theory is called a *sigma model*. As we saw above, in string theory M plays the role of actual space-time, but for our purposes we can even consider the case where M is the real line $\mathbb{R}$, so that φ is an ordinary function. This case has already been studied in section 2.1.5. The action is given in equation (4). We can then write the partition function as
$$Z = \langle 1 \rangle = \int [\mathcal{D}\varphi]e^{iS(\varphi)/\hbar},$$

where $[\mathcal{D}\varphi]$ represents the measure of integration over all paths.[8]

One approach to evaluating the partition function Z is through a process known as *Wick rotation*. One first Euclideanizes the time coordinate by writing $\tau = it$ (this is the Wick rotation), which leads to an imaginary Euclidean action iS_E. One then tries to evaluate the path integral in this framework, hoping that the answer will be HOLOMORPHIC [I.3 §5.6]. If it is, then one can use analytic continuation to work out the answer for ordinary time. The advantage is that the Euclidean exponential weighting becomes $\exp(-S_E/\hbar)$, so the minima of S_E receive the greatest weighting and the integral might converge. The nonconstant minima of the Euclidean action are called *instantons*. After Euclideanizing equation (4), the action becomes the "energy" S_E of the

---

8. Warning: these expressions represent only the "bosonic" part of a theory with "supersymmetry," meaning, in particular, that there are "fermionic" terms that complete the theory. We omit the fermionic completions for ease of notation and exposition.

map φ:

$$S_{\mathrm{E}} = \int_{\Sigma} |\nabla\varphi|^2.$$

The energy of a map has a *conformal symmetry*, meaning that it is independent of local scale transformations on the Riemann surface, that is, transformations that can be locally approximated by a combination of rotations and dilations. Invariance under rescaling by a positive number λ can easily be seen: each of the two derivatives in $|\nabla\varphi|^2$ decreases by a factor of λ, while the area element increases by λ^2. Rotational invariance is clear from the form of $|\nabla\varphi|^2$. The combination of the two, along with the fact that this argument did not depend on the derivatives of the scaling parameter λ, leads to the statement of local scale invariance.

The conformal symmetry of the action is an example of a classical symmetry of the action that is not necessarily maintained in the quantum theory. However, the quantum theory has no anomaly—meaning that the symmetry *is* preserved—if M is chosen to be a complex, CALABI–YAU MANIFOLD [III.6].

The Calabi-Yau condition can be thought of as a complex notion of orientation. Recall that for an oriented manifold one can continuously choose, on each patch, a basis for the tangent space such that, when we move from patch to patch, the determinant of the change-of-basis matrix is equal to one. The same is true on a Calabi-Yau manifold, but now we consider complex bases for the complex tangent spaces.

When the target manifold is a Calabi-Yau manifold, the instantons are complex analytic maps from the two-dimensional surface. Instantons are not "close" to the constant paths; their effects are therefore not accessible by perturbative methods such as Feynman diagrams. They are therefore "nonperturbative" phenomena. An example from quantum mechanics would be a particle in a double-well potential such as $(x^2 - 1)^2$. The zero-energy minima are the two constant (stationary) paths at $x = \pm 1$. An instanton path could go from $x = -1$ to $x = +1$, or vice versa. Such trajectories occur and are known as "quantum tunneling."

Lesson. Inaccessible by perturbation theory, instantonic effects are notoriously challenging to calculate.

3.1 Mirror Pairs

In the setting above, we considered maps from a two-dimensional surface Σ to a target (Calabi-Yau) space. Let us denote this quantum field theory by $Q(M)$, which

is shorthand for the collection of all fields and all possible correlation functions created from them. In this setup, we say that the Calabi-Yau manifolds M and W are "mirror pairs" if $Q(M)$ is equivalent to $Q(W)$. Through the magic of mirror symmetry, hard problems in $Q(M)$ involving instantons can be answered in $Q(W)$ by considering only the much simpler constant paths.

4 Mathematical Distillation

A physical theory contains a tremendous amount of information. For example, correlation functions can involve any number of fields, each evaluated at different points on the two-dimensional surface. This is typically too unwieldy a situation to approach mathematically. Instead, equipped with a symmetry of the theory called "supersymmetry," a mathematical distillation can be performed. The distillation procedure is called *topological twisting*, and the resulting "topological field theory" has correlation functions that are independent of the positions of points. Because of this independence, the correlation functions are certain characteristic numbers associated with the underlying geometric setup. In fact, there are two types of twisting, typically called A and B, which capture different aspects of the manifold in question.

4.1 Complex and Symplectic Geometry

4.1.1 Complex Geometry

To get a feel for the geometric aspect captured by topological twisting, recall that we can construct the circle S^1 from the real line $\mathbb{R}$ by identifying the points θ and $\theta + 2\pi$, and therefore also $\theta + 2\pi n$, where n is any integer. What we have done is identified points related by a *lattice of integer translations*. We could choose the lattice to consist of multiples of some other real number r, but since any two such lattices differ only by an overall scaling of $\mathbb{R}$, we would effectively get the same space. In the complex plane $\mathbb{C}$, we can do the same thing with a two-dimensional lattice of translations generated by two complex numbers λ_1 and λ_2, as long as the quotient λ_2/λ_1 is not real. This space is called a *torus* and has the same topology as any two-dimensional surface with one hole. It has more structure, however, because it can be covered by regions described by a complex coordinate—with different regions related by complex analytic maps. The pairs (λ_1, λ_2) and $(\lambda_1, \lambda_2 + \lambda_1)$ generate the same lattice of translations, as do the pairs (λ_1, λ_2) and $(\lambda_2, -\lambda_1)$. In fact, lattices related by a complex rescaling of $\mathbb{C}$ are equivalent, so a better parametrization of the lattice is the ratio $\tau = \lambda_2/\lambda_1$.

By redefining the direction of one of the λs, we can assume that the imaginary part of τ is positive, so τ takes values in the upper half of the complex plane. By the reasoning above, we note that τ and $\tau + 1$, as well as $-1/\tau$, all come from the same lattice. The number τ can also be thought of in the following way. The torus has two distinct loops, one generated by a straight path from z to $z + \lambda_1$, and one generated by a straight path from z to $z + \lambda_2$. Then λ_1 and λ_2 are both the result of the line integral of the complex differential dz over the loop. In fact, the loop did not even need to be straight to lead to this conclusion. The values of such integrals over subspaces without boundaries (the loops, here) are more generally called *periods*.

Although any two tori are topologically equivalent, one can show that there is no *complex analytic* map between two complex tori described by genuinely different values of τ. The parameter τ therefore determines the complex geometry of the space. Roughly speaking, we think of this parameter as describing the *shape* of the torus. (See MODULI SPACES [IV.8 §2.1] for a further discussion of this.)

The topological B-model depends only on the complex geometry of the target space M. That is, the theory depends, continuously, only on the parameter τ.

4.1.2   Symplectic Geometry

Another aspect of geometry is the *size* of the torus, which is described simply by an area element. Let us recall that, topologically, all tori look like $\mathbb{R}^2$ with points identified by the lattice of integer horizontal and vertical translations (but not necessarily in a way that would respect any complex geometry). The points of the torus can be thought of as the unit square with opposite sides glued together. An area element in $\mathbb{R}^2$ looks like $\rho\,dx\,dy$, which then determines the area ρ of the unit square. These notions of two-dimensional area generalize to two-dimensional subspaces in higher-dimensional spaces. The study of such structures is called SYMPLECTIC GEOMETRY [III.88], and so we call ρ the *symplectic parameter*.

The topological A-model depends only on the symplectic geometry of the target space M. That is, the theory depends, continuously, only on the parameter ρ.

4.2   Cohomological Theories

As you might imagine, the passage from an ordinary theory to a topological theory involves identifying many aspects of the physical theory that were previously distinct, such as different point values of a single field. Mathematically, a well-established method of producing topological aspects of a structure—and one that involves making identifications—is through a COHOMOLOGY THEORY [IV.6 §4]. Cohomology theories follow the pattern of having an operator δ obeying the equation $\delta \circ \delta = 0$. We think of this equation as the statement image$(\delta) \subset$ ker(δ). The cohomology group $H(\delta)$ is formed as the quotient $H(\delta) = \text{ker}(\delta)/\text{image}(\delta)$, which means that we identify any two vectors u and v satisfying $\delta u = \delta v = 0$, so long as the difference $u - v$ can be written as δw for some w. Then $H(\delta)$ is just the space of all such vectors, up to identifications.

The topological twisting of physical theories is similar. The operator δ is a physical operator acting on a Hilbert space of states. The presence of supersymmetry in our theories ensures that δ exists and squares to zero. The vector states of the topological theory are just the elements of $H(\delta)$, i.e., states in the original theory Ψ obeying $\delta\Psi = 0$, up to identification. In many cases, these states can be identified with ground states.

It is crucial that supersymmetry is a symmetry that contains the complex translations of points on the two-dimensional surface. This means that the value of a field operator $\phi(z)$ at one point is identified with its value $\phi(z')$ at another. In other words, the physics of the topological theory is independent of the positions of the operators! In the path-integral formulation, this means that the correlation functions are independent of the positions of the fields inserted into the integrand. What can they depend on, then? They depend on the particular field or combination of fields inserted, and they depend on the geometric parameter (such as ρ or τ) of the space M.

4.2.1   The A-Model and the B-Model

Given a Calabi–Yau space, one can actually construct *two* operators, δ_A and δ_B, each of which squares to zero. There are therefore two distinct corresponding topological twistings and two distinct topological theories that can be constructed from a Calabi–Yau space.

If M and W are mirror Calabi–Yau pairs, you might wonder if the topological models constructed from them will still be equivalent theories. The answer is a most interesting form of yes: the resulting A-model of one Calabi–Yau manifold M is equivalent to the B-model of the mirror W, and vice versa! The complex and symplectic aspects of the theories get interchanged under mirror symmetry! In particular, a hard symplectic question of M might get mapped to an easy computation involving the complex geometry of W.

We emphasize here that the two manifolds may be completely topologically distinct. For example, the Euler characteristic of one is the negative of the other.

5 Basic Example: T-Duality

Although the circle is not complex, it provides a very illustrative entry into mirror symmetry that can be studied quite easily. We will find an equivalence between two theories constructed from circles. The equivalence will be very nontrivial, however, as states of very different kinds will be shown to correspond.

Consider the case where the two-dimensional surface is a cylinder, with spatial dimension a unit circle, and one dimension of time, and let us look at the sigma model (these were introduced in section 3). Suppose also that the target space is a circle of radius R, which we denote by S_R^1. We think of S_R^1 as the real line, with two points identified if they differ by a multiple of $2\pi R$. Maps from one circle to another can be classified by their *winding number*, an integer that tells you how many (net) times the image of a point goes around the second circle when the point goes once around the first. The map $\theta \mapsto mR\theta$ from the circle to S_R^1 has winding number m. This allows us to write the field $\varphi(\theta)$ as a winding piece, $mR\theta$, plus an honest Fourier series (no winding): $\varphi(\theta) = mR\theta + x + \sum_{n \neq 0} c_n \exp(in\theta)$. Here we have singled out the constant mode $x = c_0$ of the Fourier series. We have expanded just the θ dependence in a series, so every continuous parameter (x and the c_n) should be thought of as a function of time, as well.

The energy, or Hamiltonian, of such a map is computed as in section 2.1.3:

$$H = (mR)^2 + \dot{x}^2 + \sum_n |\dot{c}_n|^2 + n^2 |c_n|^2.$$

Comparing this with the harmonic oscillator Hamiltonian of section 2.1.3, we can see that each degree of freedom $c_n(t)$ plays the role of a (complex) quantum-mechanical particle in a simple harmonic oscillator potential. There is an occupation-mode basis for describing the quantum mechanics of each mode.[9] The full Hilbert space of the quantum theory is the (tensor) product of each of these, plus parts involving the

constant mode and winding number, which we now discuss. (Remember, each degree of freedom of the classical theory becomes a *particle* in the quantum field theory.)

The constant mode x has energy $\dot{x}^2$, and therefore has no associated potential (it can be anywhere on the circle). This mode represents a free quantum-mechanical particle on the circle. Recall that the momentum of the x particle is represented by the operator $-i(d/dx)$. This operator has eigenfunctions e^{ipx}. The requirement that these eigenfunctions are invariant under the translation $x \to x + 2\pi R$ means that the eigenvalues of momentum are "quantized," and have the form $p = n/R$.

In contrast to momentum, the integer winding number (m) is really a classical label for the possible maps from a circle to a circle. Although integral, it is clearly on a different footing from the integer n of momentum. Still, it is also an important label on the Hilbert space. For each m, we have a space of m-winding configurations which gets quantized to become the mth sector of the Hilbert space. Roughly, this sector $\mathcal{H}_m$ comprises the functions of all the degrees of freedom of all the m-winding maps. We can consider the winding number as an operator by simply declaring that the states with winding number m have eigenvalue mR.

Ignoring the oscillator modes for the moment, the state of momentum n/R with winding m has energy $(n/R)^2 + (mR)^2$. In particular, the energy is unchanged if we make the simultaneous switches $(m, n) \leftrightarrow (n, m)$ and $R \leftrightarrow 1/R$. Since the oscillator modes a_n have energies that are independent of R, and since the modes are noninteracting particles, this symmetry can be extended to a full equivalence of the theories with targets S_R^1 and $S_{1/R}^1$, with momentum in one theory corresponding to winding number in the other.

In this example, the target space S^1 is neither complex nor symplectic. As a result, we cannot construct the topological A- and B-models. Nevertheless, we have demonstrated the stronger statement that the two sigma models with target space S_R^1 and $S_{1/R}^1$ are equivalent. The theories are mirror pairs. In the special case of circles, mirror symmetry is referred to as T-duality. In fact, the entire phenomenon of mirror symmetry—even for noncircles—can be deduced from T-duality.

5.1 Tori

If we take the product of two circles $S_{R_1}^1 \times S_{R_2}^1$, we get a torus. We can think of the torus as a circle family of circles, since for each point in $S_{R_2}^1$ we have a circle $S_{R_1}^1$. As

---

9. Each $a_n^\dagger = [\text{Re}(\dot{c}_n) - in \, \text{Re}(c_n)]/\sqrt{2n}$ is a raising operator, and similarly for the imaginary parts of the c_n.

we have seen in section 4.1.1, this space is complex—specifically, it is the complex plane $\mathbb{C}$ quotiented by a lattice of translations. A particularly simple lattice is the one generated by the translations $z \to z + R_1$ and $z \to z + iR_2$. As discussed in section 4.1.1 above, the lattice is determined by the complex number $\tau = iR_2/R_1$, equal to the ratio of integrals ("periods") of the complex form $\mathrm{d}z$ over the two nontrivial loops of the torus.

The symplectic data is captured by the area element. Recall that we can choose coordinates x and y such that the identifications look like unit translations in each direction. Then the (normalized) area element of the torus with radii R_1 and R_2 is $R_1 R_2 \, \mathrm{d}x \, \mathrm{d}y$, which integrates to $R_1 R_2$ on the unit square. Let us define the symplectic parameter $\rho = iR_1 R_2$. We now perform T-duality for the first circle $R_1 \to 1/R_1$. We see that under this substitution, the complex and symplectic parameters get interchanged:[10]

$$\tau \longleftrightarrow \rho.$$

Lessons. Mirror symmetry interchanges complex and symplectic parameters. Mirror symmetry is T-duality.

5.2   The General Case

The torus is the only compact one-dimensional Calabi–Yau space and is therefore the simplest one, but the discussion above is part of a more general picture. The Calabi–Yau condition ensures a unique complex volume element, or orientation ($\mathrm{d}z$, above), whose "periods" determine, and in turn vary with, the complex parameters. Though the A- and B-models both turn out to be rather simple in the case of the torus, what is important in general is that the B-model is completely determined by how the periods of the complex volume element (which were λ_1 and λ_2 in section 4.1.1) change with the parameters of the theory (of which there was just one in section 4.1.1, namely τ). Again, the relation $\tau = \lambda_2/\lambda_1$ is quite simple for the torus, but more complicated in general. In any case, this data gives all the information of the B-model. The reason for all of this is that the instantons of the B-model turn out to be just the constant maps. Each point of the target space determines a constant map, and as a result the B-model is reduced to (classical) complex geometry of the target space. This is determined by the periods.

This state of affairs is to be compared with the A-model. The A-model depends on the symplectic parameters ρ, i.e., the areas of two-dimensional surfaces

inside the target space. In contrast to the B-model, however, the dependence on ρ is very complicated, in general. The reason for this is that the instantons of the A-model are area-minimizing surfaces inside the target space, and their enumeration is a notoriously challenging problem. (The problem is not terribly challenging for the torus, however.) Mathematically, the A-model instantons are described by the theory of Gromov–Witten invariants, the subject to which we now turn.

6   Mirror Symmetry and Gromov–Witten Theory

As we mentioned above, the B-model on W is explained entirely by the classical complex geometry of W. The only relevant maps for B-model computations are the constant ones, so the space of such maps is equal to W itself, and correlators reduce to (classical) integrals over W. In fact, one of the integrands to be integrated is the complex volume element. Let us call the parameter for all possible complex volume elements τ. B-model correlation functions are then determined by τ-dependent integrals over W. In particular, the partition function $Z_{\mathrm{B}}^{(W)}$ of the B-model on W depends on τ, so we write it as $Z_{\mathrm{B}}^{(W)}(\tau)$.

The main point about topological twisting is that local variations of the fields are all identified, as they are related by the operator δ. In particular, varying the point on the worldsheet is a trivial operation in the topological theory. It turns out that, for the B-model on W, *only* the constant maps contributed, but for the A-model the situation is a bit more subtle. To give a feel for the geometry, consider again the winding of a map from a circle to a circle. Maps with different windings can never be deformed continuously into one another. The winding number is a measure of how the first circle "wraps" (or winds) around the target, according to the map. Because it is a discrete parameter it cannot change under continuous variations. Likewise, when M is a higher-dimensional space, the two-dimensional surface Σ can "wrap" around two-dimensional subspaces of M by different amounts. The parameters for wrapping are again discrete. A map φ can wrap Σ around the basic surfaces C_i in M by different integer amounts, k_i. We say that $k = k_i$ labels the "class" of the map φ. (More precisely, $\varphi(\Sigma)$ is a closed 2-cycle when Σ is compact, and k labels its homology class.) Different classes k contribute through different (Euclidean) actions $S_k(\rho)$, which depend on the areas ρ and the

---

10. The parameters τ and ρ can also have real parts, but we neglect the details for simplicity.

class k but not on the continuous details of the map φ_k. The partition function can have contributions from all classes. Different classes may contribute differently not only through the exponential weighting, but also in accordance with how many *minimal surfaces* they contain. (A good example of a minimal surface in three-dimensional space is a soap film. If you fix the boundary with a wire, the soap film will seek to find the minimum-area surface with that boundary.) In our examples, the space M is actually complex; the minimal surfaces we speak of in Gromov–Witten theory are complex analytic maps from Σ. That is, if you have a complex coordinate for Σ, then the complex coordinates for the surfaces M can be written as complex analytic functions of Σ.

The difference between the A-model and the B-model comes from the fact that the topological model is constructed from an operator δ, which was guaranteed to exist by the presence of supersymmetry in our theories. For the different models, the relevant supersymmetry operators δ_A and δ_B are simply different. As we saw above, the maps relevant to the A-model are the instantons, or complex analytic maps from Σ to M. Roughly, then, A-model correlation functions on M, and in particular the partition function $Z_A^{(M)}$, are sums over classes k of surfaces in M and sums over instantons in each class, each one weighted by its instanton action $\exp(-S_k(\rho))$. We have explicitly written the dependence on the parameter for the symplectic structure ρ. For Calabi–Yau manifolds, such maps should be discrete, and it is a conjecture, true in all known cases, that they are finite in number if we fix the class, k. All this data is packaged in a function of ρ, and based on what we have argued, the partition function must take the general form

$$Z_A^{(M)}(\rho) = \sum_k n_k \exp(-S_k(\rho)).$$

The coefficients n_k are called *Gromov–Witten invariants*.[11]

Putting things together, if (M, A) is mirror to (W, B), and if we can identify for each complex parameter τ for W a corresponding symplectic parameter $\rho(\tau)$ for M, then we have

$$Z_A^{(M)}(\rho) = Z_A^{(M)}(\rho(\tau)) = Z_B^{(W)}(\tau). \qquad (6)$$

The first equality means we should rewrite ρ in terms of τ, and the second says that the answer should be given by the corresponding B-model on W. Therefore, all of the information about complex analytic surfaces in M, which is encapsulated in the coefficients n_k, is completely determined by the classical geometry of W!

This remarkable predictive power—the computation of an infinite number of difficult Gromov–Witten invariants through equations such as (6)—is what led to such intense interest in mirror symmetry at its inception.

7 Orbifolds and Nongeometric Phases

7.1 Nongeometric Theories

Mirror symmetry is about an equivalence of quantum field theories, and not every such field theory has the geometric content of a target space as in the sigma model. The structure involved in mirror symmetry—or at least its topological version—begins with a quantum theory with a supersymmetry algebra that allows for the passage to a topological theory. That is, there is a Hilbert space of states, a Hamiltonian operator, and a particular algebra of symmetries, i.e., operators that commute with the Hamiltonian. There are no dictates as to how one constructs such a setup, and the sigma model of maps to a target space is only one such way. Other methods abound. The geometric case is merely the one most suited for mathematicization (and exposition), which is why we have focused on the theory with a target space.

As an intermediate case—possibly geometric, possibly not—we will discuss the so-called orbifold theories.

7.2 Orbifolds

When space-time is a cylinder $S^1 \times \mathbb{R}$, with a circle S^1 as its spatial dimension, there is a fascinating construction in quantum field theory known as an *orbifold theory*. This is defined as follows. Suppose there is a finite group G of symmetries (such as a reflection symmetry). That is, each group element acts as an operator on the Hilbert space, so if $g \in G$ then it sends a state Ψ to a state $g\Psi$. Then one defines a new theory by identifying states related by the symmetry. To construct the theory, let us first consider the ground state Ψ_0 of the original theory. This is assumed to be invariant under the group: that is, $g\Psi_0 = \Psi_0$ for all group elements, g.[12] One then constructs the space $\mathcal{H}_0$ of all

---

11. Though our discussion makes it seem as though the n_k are integers, in fact they are only rational numbers. They can be expressed in terms of more basic integers, however. These integers are the ones referred to at the beginning of this article.

12. In the case where there are flat directions of a potential, as in a free particle on a circle (no potential at all), the ground state may be a superposition of classical values of the field. For the circle, the constant wave function $\Psi = 1$ is not associated with a single, classical location. It is still invariant under any group of rotations, however.

invariant states. This is known as the *untwisted sector*, and Ψ_0 is the ground state of the untwisted sector. In the case where G is commutative, a *twisted sector* is then constructed for every group element $g \in G$.[13] To construct the twisted sector, first think of the spatial dimension S^1 as being an interval $[0, 1]$ with endpoints 0 and 1 identified. Recall that the Hilbert space of states is constructed from (functions of) all the degrees of freedom of the possible configurations of fields. The twisted sector $\mathcal{H}_g$ corresponds to additional field configurations Φ that are related at the two ends by the action of g: so $\Phi(1) = g\Phi(0)$. Such field configurations represent configurations on the circle S^1 since left and right ends are related by the group, and therefore get identified. These additional configurations are thus part of the orbifold theory. One constructs a sector $\mathcal{H}_g$ of the Hilbert space by taking all such states Ψ_g that also obey the invariance condition $h\Psi_g = \Psi_g$ for all group elements h.

Orbifolds may be geometric, as they are in the case of the sigma model to a manifold X on which a discrete group G acts. For example, rotations act on the plane, and we can consider the four-element group generated by a right-angle rotation. The quotient of the plane by these rotations looks like a cone. As another example, the finite groups of symmetries of the platonic solids (tetrahedron, cube, etc.) act on the two-dimensional sphere by rotations. When we take $X = S^2$ and G a platonic group, we get an interesting orbifold. In fact, if we simply take the space of orbits of the group G, it is topologically just a sphere again, but not a smooth one—it has cone points. These cone points would be troublesome in a quantum field theory, but the "stringy" orbifold is perfectly "smooth."

The orbifold theory itself carries a symmetry. For example, if G is the commutative group with two elements, then there is an untwisted sector and a unique twisted sector. There is a symmetry corresponding to multiplication by 1 in the untwisted sector and by -1 in the twisted sector. This symmetry is not geometric. Orbifold theories with symmetries can often themselves be orbifolded in such a way as to recover the original theory. In fact, the theory and its orbifold are also often mirror pairs! Greene and Plesser used such a construction to create the first examples of mirror pairs. Furthermore, they used ways of ascribing geometric interpretations to some nongeometrically constructed

theories so as to identify mirror Calabi–Yau spaces. To be precise, they took the space of all nonzero complex 5-vectors $X = (X_1, X_2, X_3, X_4, X_5)$ satisfying the equation

$$X_1^5 + X_2^5 + X_3^5 + X_4^5 + X_5^5 + \tau X_1 X_2 X_3 X_4 X_5 = 0,$$

identifying X with λX for any nonzero complex number λ. (If X is a solution, then so is λX.) The equation actually defines a family of complex spaces, since $\tau \in \mathbb{C}$ is a parameter. The orbifold theory is defined from the finite group of phase transformations

$$(X_1, X_2, X_3, X_4, X_5)$$
$$\mapsto (\omega^{n_1} X_1, \omega^{n_2} X_2, \omega^{n_3} X_3, \omega^{n_4} X_4, \omega^{n_5} X_5),$$

where $\omega = \mathrm{e}^{2\pi i/5}$ and $\sum_{i=1}^{5} n_i$ is a multiple of 5. This space and its orbifold are actually the mirror pair about which Candelas et al. made their famous predictions.

8   Boundaries and Categories

The entire story of mirror symmetry becomes much richer when we allow the strings to have endpoints. Strings with ends are called "open strings," while "closed strings" refers to loops. Mathematically, allowing ends corresponds to adding boundaries to the worldsheet surfaces. With this addition, we would like to perform the same topological twisting. To do so, we must first ensure that some supersymmetry condition persists when we put the boundary conditions on the fields. If we begin with a Calabi–Yau target manifold, we can ask to preserve the conditions that allow either the A-twisting or the B-twisting (but not both: the boundary condition will destroy some symmetry, much as pinning a rope will constrain its degrees of freedom). After the twist, the boundary topological theory will depend on symplectic or complex information, respectively.

For the A-model, the endpoints or boundaries must lie on a Lagrangian subspace. The Lagrangian condition constrains half the coordinates; for linear spaces it is like a restriction to the real part of a complex vector space. For the B-model the boundaries must lie on a complex space. Locally, a complex space looks like $\mathbb{C}^n$ and a complex subspace is described by complex analytic equations in the coordinates. A boundary condition that preserves supersymmetry and allows a chosen topological twisting is called a *brane*. (The terminology mimics the word "membrane," but applies to any dimension.) In short, A-branes are Lagrangian; B-branes are complex.

---

13. The twisted sectors are properly labeled by conjugacy classes, which are the same as group elements when G is commutative.

To package all the information of the topological boundary theory, one appeals to the mathematical notion of a CATEGORY [III.8]. A category is a way of talking about structure: it consists of *objects*, and for any pair of objects there is a space of *morphisms* from one object to the other. Often the objects are mathematical structures of some kind and the morphisms from one object to another are the functions that preserve the relevant structure. For example, if the objects are (i) SETS [I.3 §2.1], (ii) TOPOLOGICAL SPACES [III.90], (iii) GROUPS [I.3 §2.1], (iv) VECTOR SPACES [I.3 §2.3], or (v) chain complexes, then the morphisms are, respectively, (i) MAPS [I.2 §2.2], (ii) CONTINUOUS MAPS [III.90], (iii) HOMOMORPHISMS [I.3 §4.1], (iv) LINEAR MAPS [I.3 §4.2], or (v) chain maps. The morphism spaces between objects should be thought of as some kind of relational data. Morphisms themselves interact with one another, as they can be composed when the end object of one morphism is the start object of another. The composition is associative, so whether you compute abc as $(ab)c$ or $a(bc)$ does not matter. A useful image is a directed graph, which is a category with vertices as objects and paths between two vertices as morphisms. Composition is defined in this category by concatenating paths.

In the case of a two-dimensional field theory with boundary conditions, we construct a category whose objects are branes (i.e., boundary conditions). The morphisms between two branes α and β are the ground states $\mathcal{H}_{\alpha\beta}$ of the boundary field theory defined on the infinite strip $[0,1] \times \mathbb{R}$, where we put the boundary condition α on the left boundary $\{0\} \times \mathbb{R}$ and the condition β on the right boundary $\{1\} \times \mathbb{R}$. Morphisms are composed by gluing boundaries together, and associativity is guaranteed by topological invariance.[14]

Mirror symmetry with boundary conditions then becomes the following statement: two manifolds M and W are mirror pairs if the brane category of the A-twisting of M is equivalent to the brane category of the B-twisting of W (and vice versa). The mathematical translation of this statement is called the *homological mirror symmetry conjecture*, due to Kontsevich. On the A-model side, the brane category is the so-called *Fukaya category*, and is governed by complex analytic maps from surfaces with boundaries, where the boundaries must be mapped to Lagrangian branes. On the B-model side, the branes form a category determined by complex subspaces, together with complex analytic VECTOR BUNDLES [IV.6 §5] on them. A complex vector bundle associates a complex vector space to every point. For example, the complex circle $\{x^2 + y^2 = 1\}$ in $\mathbb{C}^2$ has a complex tangent space at every point. "Complex analytic" means that this subspace of $\mathbb{C}^2$ changes in a complex analytic way. For the complex circle, the space of tangent vectors at (x, y) consists of all multiples of the vector $(-y, x)$, an assignment which is clearly complex analytic. Physically, the bundles arise from allowing charges on the endpoints of strings.

Kontsevich's conjecture asserts that these two categories of branes are equivalent. That statement is natural from the physics point of view, but by identifying the precise categories that correspond to the physical picture, this conjecture is a major contribution to the translation of mirror symmetry from physics into rigorous mathematics. The equivalence of categories means that not only is there a corresponding Lagrangian A-brane of M for every complex B-brane of W, but that the *relationships*, or morphisms, between branes are also in correspondence.

8.1 Example: Torus

Kontsevich's conjecture can be proven and easily illustrated in the example of a 2-torus. Think of the now-familiar symplectic two-torus as being the two-dimensional plane, with integer lattice translations identified. We take the torus to have area element $A \, dx \, dy$, so that the symplectic parameter is the imaginary number $\rho = iA$, as in section 4.1.2. Now consider straight lines on the plane. These will correspond to closed circles on the torus as long as they have rational slope: $m = d/r$, with d and r relatively prime integers. They are Lagrangian branes of the A-model boundary theory. The minimal-energy open strings connecting one line of slope $m = d/r$ to another of slope $m' = d'/r'$ are those that have zero length. They are therefore the points of intersection. It is an easy exercise to show that there are $|dr' - rd'|$ such points.

On the mirror side, we again have a torus, but with a complex parameter τ, and for the two tori to be mirror pairs, we should set $\tau = \rho$. The objects of the B-model brane category are complex vector bundles. It is a theorem that the basic bundles are classified by their

---

14. We speak of associativity of the topological states, which are themselves cohomology classes. At the "chain" level, before the topological twisting, there is no associativity. The notion of a category with morphisms that have a cohomology and compose only "up to cohomology" is called an A_∞ category. One can also imagine a categorical definition that captures the structure of surfaces with handles and holes. Indeed, the proper mathematical framework for a complete understanding of mirror symmetry is still under construction.

rank r and degree d, two integers.[15] It is customary to organize these two numbers into what is known as a "slope," $m = d/r$ (the nomenclature preceded this application), and basic bundles must have d and r relatively prime.

We can now easily guess that under the mirror correspondence we have

$$\text{slope} \longleftrightarrow \text{slope}.$$

This means that a Lagrangian brane of slope m on the torus with symplectic parameter ρ should correspond to a *complex* vector bundle with slope m in the mirror torus with complex parameter ρ. Now suppose we have the B-model version of our example above, so we take two vector bundles of slope m and m'. In fact, the minimum-energy open strings between two complex analytic bundles of slope m and m' correspond to complex maps between the bundles, and the RIEMANN-ROCH FORMULA [V.31] counts this number as $|dr' - rd'|$. This is the same result as for our A-model calculation above! Therefore, corresponding objects relate in a corresponding way. Beyond the morphism spaces, one checks finally that the compositions of corresponding morphisms correspond, just as for logarithms and slide rules. Doing so proves Kontsevich's conjecture.

8.2   Definition and Conjecture

In fact, Kontsevich's definition of mirror symmetry is really a conjecture stating that the boundary notion of mirror symmetry as an equivalence of categories is compatible with, and even implies, the traditional notion of mirror symmetry that relates Gromov–Witten theory and complex structures.

One way to show this is to try to reconstruct the Gromov–Witten invariants from the boundary theory. A heuristic, geometric approach to doing so involves looking at the diagonal boundary condition in two copies of a space. A disk mapping into two copies of a space is described by two maps of a disk into the space. Further, if the boundary condition is diagonal, this means that the maps have to agree on the boundary. What we have, then, is two disks inside a space which agree on the boundary. That is exactly what a sphere is: two disks (or cups) glued together! The disks are the two hemispheres, and they are glued

along the equator. Now the minimal disks are instantons for the open string (with boundary), and by gluing them together along a common boundary, we have constructed a minimal sphere, or closed-string instanton. Thus the open string on this double theory should recover the closed string on the original theory.

A more algebraic approach sees the closed-string deformations as deformations of the category of branes. That is, a change in bulk (nonboundary) theory induces a change in boundary theory. But once equipped with a category, one can classify its deformations intrinsically. That is, if one views a category as a fancy algebra,[16] then, as the deformations of an algebra are easily classified through a notion called Hochschild cohomology, the deformations of a category can be treated similarly. One arrives at the maxim that the closed string is the Hochschild cohomology of the open string. By computing the Hochschild cohomology of a brane category, one can, in principle, check this maxim, establish Kontsevich's conjecture, and then prove the connection to traditional mirror symmetry and Gromov–Witten theory.

9   Unifying Themes

How does one find mirror pairs (M, W)? What is the construction? Although mirror symmetry has spawned many results and proofs, these basic questions continue to vex.

On the one hand, Hori and Vafa have given a physics proof of mirror symmetry, which constructs mirror pairs but not through an evident mathematical channel. Of course, one can attempt to mathematicize the physical argument, but that does not seem to lead to insights into the construction—perhaps because path integrals and other methods of quantum field theory such as renormalization are not very well understood mathematically.

Batyrev has devised a procedure for constructing mirror pairs within the context of toric geometry. This method is a generalization, to a wide class of examples, of the original construction of Greene and Plesser. The recipe has been extremely successful in producing examples of every stripe. However, the underlying meaning behind the construction is unclear.

As for a geometric construction of mirror pairs, there is a physical argument that makes contact with mathematics, but it has not yet been made rigorous. The argument uses T-duality. Start with the B-model on M and consider a point P of M as a zero-dimensional

---

15. A vector bundle assigns a vector space to each point of the torus. The rank is the dimension of that space. The degree is roughly a measure of the complexity of the bundle. For example, if we have a two-dimensional surface and consider the bundle that assigns to each point the tangent space at that point, the degree is equal to $2 - 2g$, where g is the number of holes on the surface.

16. An algebra is a category with one object.

complex subspace. Then the choice of point P on M is parametrized by M itself. By mirror symmetry, there should be a corresponding Lagrangian brane T on the mirror manifold W. Furthermore, the choices of T must equal the choices of P, i.e., the manifold M. Therefore, if we can find the brane T on W, we can parametrize the choices of T, and recover M. So we can find the mirror M of W from W itself.

This construction is geometric and has something to say about the structure of the Calabi–Yau spaces involved in mirror symmetry. Specifically, the choices of a Lagrangian brane always look like a family of tori. Therefore, M itself should look like a family of tori. Further, one can argue that by performing T-duality in families of tori (in a similar way to how one does it for a single torus), one arrives back at the mirror manifold, W. This is what we did for the torus, thought of as a circle family $(S^1_{R_2})$ of circles $S^1_{R_1}$. When we T-dualized each member of the family, we found the mirror torus. So mirror symmetry is T-duality, and Calabi–Yau spaces of mirror symmetry should look like families of tori. This approach also relates to the homological mirror symmetry construction. Though promising, it remains mathematically elusive.

Various points of view on mirror symmetry are helpful for different applications. To date, no unified understanding of the phenomenon has been achieved. To some extent, we are still "feeling the elephant."

10   Applications to Physics and Mathematics

As a computational tool in string theory, mirror symmetry is unparalleled in its power. When combined with other physical equivalences, its power is multiplied. For example, there are certain equivalences in physics that relate one type of string theory to another.

Without going into the details of string theory, we can get a flavor of its complexity by returning to mirror symmetry. Recall that the B-model was able to compute the difficult instantons on the A-model, yielding a great simplification of the two-dimensional quantum field theory on the worldsheet. But this whole quantum field theory was just an *auxiliary* tool for computing some Feynman diagram for the perturbation theory of the full string theory! Unfortunately, a satisfactory description of the full string theory path integral is, at the time of writing, way out of reach. String theory instanton effects are mostly unknown to us, unless a string equivalence or other argument can relate them to a perturbative effect in a *different* string theory. The perturbative string calculation

in that other theory may then be performed by exploiting mirror symmetry. Tracing through chains of equivalences in such a manner, many different phenomena in string theory can ultimately be calculated via mirror symmetry.

In principle, one should be able to calculate *all* nonperturbative and perturbative aspects of a single theory by outsourcing the calculations to equivalent theories and exploiting mirror symmetry. The barriers to doing this at the time of writing are largely technological, not conceptual.

Beyond physics, the rich texture of mirror symmetry means that there is interesting mathematics to be discovered in the proper formulation of the problem. For example, defining the precise categories of branes in full generality remains a challenge.

Yet there are also direct applications to mathematical questions. We have already discussed how enumerative geometry has been revolutionized by mirror symmetry and the counting of instantons. Results in symplectic geometry have also been obtained. Occasionally, two objects may be proven to be equivalent as B-model branes. If the A-model mirrors can then be found, one has the result that the corresponding Lagrangian subspaces of the mirror symplectic space are also equivalent. Of course, to make such an argument, one must first prove Kontsevich's version of mirror symmetry for the mirror pair considered. As a final recent example, Kapustin and Witten have found a relation of mirror symmetry to the geometric Langlands program in representation theory. This program, loosely stated, is a correspondence between objects associated with two-dimensional surfaces and Lie groups. From a surface Σ and a gauge group G, one constructs the space $\mathcal{M}_H$ of solutions to Hitchin's equations. Central to that program are complex analytic objects on $\mathcal{M}_H$ that behave nicely under the action of an algebra of operations. The Langlands correspondence relates two sets of such objects: one easy to calculate and the other more difficult. In fact $\mathcal{M}_H$ is itself a family of tori, and the easy objects correspond to points. Mirror symmetry states that the points should turn into the tori under T-duality, so the hard objects should correspond to the tori themselves! It is an appealing proposition, and making it precise mathematics will be difficult—but the gauntlet has been thrown down.

The discovery that mirror symmetry relates to the geometric Langlands program has elicited great excitement among researchers and reveals yet another facet of this fascinating phenomenon.

Further Reading

The article "Physmatics" (which can be found online at www.claymath.org/library/senior_scholars/zaslow_physmatics.pdf) is a general discussion of the relationship between mathematics and physics, and may serve as a complement to this article. Readers with a university-level mathematics background who want to learn about mirror symmetry in more detail could try consulting the book *Mirror Symmetry* (Clay Mathematics Monographs, volume 1, edited by K. Hori and others (American Mathematical Society, Providence, RI, 2003)).

IV.17 Vertex Operator Algebras
Terry Gannon

1 Introduction

Algebra is the mathematics that places more emphasis on abstract structure than on intrinsic meaning. The conceptual simplifications that can result when context is stripped away from structure give algebra a special power and clarity compared with other areas: compare, for example, the difficulty of visualizing four-dimensional space with the triviality of manipulating quadruples (x_1, x_2, x_3, x_4) of real numbers. However, this abstractness can also blind us. For instance, basic identities like $ab = ba$ and $a(bc) = (ab)c$ that are obeyed by numbers can be modified in countless directions, and each modification defines a new algebraic structure, but it is hard to guess from a purely abstract perspective which of these modifications will give rise to a rich, accessible, and interesting theory. For guidance, algebra has traditionally turned to geometry. For example, over a century ago LIE [VI.53] suggested that the identities $ab = -ba$ and $a(bc) = (ab)c + b(ac)$ were worth studying for geometrical reasons: the resulting structures are now called LIE ALGEBRAS [III.48 §2]. More recently, as we shall see, physics has joined geometry in this guiding role and has had spectacular success.

The renowned physicist and mathematician Edward Witten believes that a major theme of twenty-first-century mathematics will be its reconciliation with the branch of physics known as quantum field theory. Conformal field theory (the quantum field theory that underlies string theory) is an especially symmetric and well-behaved class of quantum field theories. When this notion is translated into algebra, the result is a structure known as a *vertex operator algebra* (VOA). This article sketches where VOAs come from, what they are, and what they are good for.

To aim to explain a VOA in a few pages is almost as absurd as to aim to explain quantum field theory in a few pages, but, undaunted, I shall try to do both. Obviously it will be necessary to gloss over many important technicalities and to commit major simplifications; without question this exposition will raise the ire of experts and the eyebrows of knowledgeable amateurs, but I hope that it will at least convey the essence of this important and beautiful area. Vertex operator algebras are the algebra of string theory: they should be thought of as the same sort of gift to the twenty-first century that Lie algebras were to the twentieth.

2 Where VOAs Come From

The two most revolutionary developments in physics in the early twentieth century are usually held to be relativity and quantum mechanics. They are revolutionary not just because they have consequences that are extremely counterintuitive, but also because they provide very general frameworks that can potentially affect all physical theories: one can take a theory from classical physics, such as the theory of the harmonic oscillator or the theory of electrostatic force, for example, and one can try to make it "relativistic," so that it becomes compatible with relativity, or to "quantize" it, so that it becomes compatible with quantum mechanics.

Unfortunately, nobody knows how to make relativity fully compatible with quantum mechanics. To put this another way, the ultimate concern of relativity is gravitation, and a direct application to gravity of the usual quantizing techniques fails. This ought to mean that a fundamentally new physics arises at small distance scales that we are ignoring. Indeed, naive calculations suggest that the space-time "continuum" at distance scales of around 10^{-35} m should deteriorate into some sort of "quantum foam," whatever that might mean. (10^{-35} m is extremely small: for instance, the order of magnitude of the size of an atom is 10^{-10} m.)

Perhaps the most popular and controversial approach to quantum gravity is string theory. The electron is a *particle*, i.e., in principle it can be localized to a point. In string theory, the fundamental object is a *string*, a finite curve of length approximately 10^{-35} m. In place of the dozens of kinds of fundamental particles in the generally accepted quantum field theory, there is only one string, whose precise physical properties (mass, charge, etc.) depend on its current "vibrational mode."

As the string moves, it traces out a surface called a worldsheet. For reasons that we will sketch below, much of string theory reduces to studying conformal field theory, which is the induced quantum theory on these surfaces. Probably no other structures have affected so many areas of "pure" mathematics in so short a time as string theory and, what is essentially the same thing, conformal field theory. Indeed, five of the twelve Fields Medals awarded in the 1990s (namely, those to Drinfel'd, Jones, Witten, Borcherds, and Kontsevich) were for such work. We shall focus in this article on their algebraic impact; see MIRROR SYMMETRY [IV.16] for some geometrical implications.

2.1 Physics 101

A quick overview of physics will be useful for the discussion. Further details can be found in MIRROR SYMMETRY [IV.16 §2].

2.1.1 States, Observables, and Symmetries

A physical theory is a set of laws that govern the behavior of some kind of physical system. A *state* of that system is a complete mathematical description of the system at a particular time: for instance, if the system consists of a single particle, then we could take its state to be its position x and momentum $p = m(\mathrm{d}/\mathrm{d}t)x$ (where m is its mass). An *observable* is a physically measurable quantity such as position, momentum, or energy. It is through observables that a theory is compared with experiment. Of course, for this to be true we also need to know what an observable is from a theoretical point of view.

In classical physics, an observable is just a numerical function of the state. For example, our single particle has energy E, which depends on the position and momentum via a formula of the form $E = (1/2m)p^2 + V(x)$. (This gives us the kinetic energy plus the potential energy.) Classical states at different times are related by the equations of motion, which are usually expressed as differential equations. However, string theory and conformal field theory (CFT) are quantum theories, which are significantly different from classical theories: one can think of them as "applied linear algebra." Whereas a classical state was given by a collection of a few numbers (two, in the case of the particle above), a quantum state is an element of a HILBERT SPACE [III.37], which for the purposes of discussion we can think of as a column vector with infinitely many complex entries. As for a quantum observable, it is a HERMITIAN OPERATOR [III.50 §3.2] on the Hilbert space,

which we can think of as an $\infty \times \infty$ matrix $\hat{A}$ that acts on the states by matrix multiplication. As in classical physics, one of the most important observables is energy, which is given by the *Hamiltonian operator* $\hat{H}$.

It is far from obvious how a linear operator that takes states to states has anything to do with the notion of a physical observation, and indeed the relationship between observables and observation is a major difference between classical and quantum theories. If $\hat{A}$ is an observable, then the SPECTRAL THEOREM [III.50 §3.4] tells us that the Hilbert space has an ORTHONORMAL BASIS [III.37] of EIGENVECTORS [I.3 §4.3]. When we do the experiment that is modeled by the observable $\hat{A}$, the answer we obtain will be one of the eigenvalues of $\hat{A}$. However, this answer is usually not fully determined by the state v. Instead, it is given by a probability distribution: the probability of obtaining a particular eigenvalue is proportional to the square of the norm of the projection of v into the corresponding eigenspace. Thus, the only circumstances under which the answer is determined in advance are if the state v is an eigenvector of $\hat{A}$.

There are two independent ways in which a quantum state can evolve in time: a deterministic evolution between measurements, governed by the famous SCHRÖDINGER EQUATION [III.83], and a probabilistic and discontinuous one that occurs at the instant when a measurement is made. For our purposes, only the deterministic evolution will be relevant.

The symmetries of CFT are extremely rich, as we shall see. Symmetries in physical theories are highly desirable because of two consequences that they have. First, they lead by NOETHER'S THEOREM [IV.12 §4.1] to *conserved quantities*, i.e., quantities independent of time. For example, the equations of motion of our particles are usually invariant under translation: for instance, the gravitational force between two particles depends only on the difference between their positions. The corresponding conservation law in this case is the conservation of momentum. A second consequence of symmetries in quantum theories is that infinitesimal generators of the symmetries act on the state space $\mathcal{H}$ (the Hilbert space to which the states belong), forming a representation of the Lie algebra. Both consequences are important to CFT.

2.1.2 The Lagrangian Formulation and Feynman Diagrams

We will need two of the languages in which physics is written. One is the *Lagrangian* formalism, which is responsible for the relationship between string theory and CFT, as well as for the appearance of modular functions in string theory. The other is the *Hamiltonian* or *Poisson bracket* formalism, which is where algebra arises. Vertex operator algebras try to explain the "miracle" that these two formalisms cohere.

The Lagrangian formalism can be expressed classically through Hamilton's *action principle*. When there are no forces present, particles travel in straight lines, which are the curves of shortest length. Hamilton's principle explains how this idea generalizes to arbitrary forces: instead of minimizing length, the particle minimizes a related quantity S called the *action*.

The quantum version of Hamilton's principle is due to Feynman. He expresses the probability of measuring the system in some final (eigen)state $|\text{out}\rangle$, given that it was originally in some initial state $|\text{in}\rangle$, using a "path integral" of $e^{iS/\hbar}$ over all possible histories that connect $|\text{in}\rangle$ and $|\text{out}\rangle$. The details are not important for us (and in any case are mathematically dubious in general). The intuition behind the path integral formulation is that the particle simultaneously follows every one of those histories, and each of them contributes to the probability. $\hbar$ is called *Planck's constant*; in the "classical limit" as $\hbar \to 0$, the contribution from the path that satisfies Hamilton's principle dominates everything else.

The main use of Feynman's path integral is in perturbation theory. Finding exact solutions in physics is typically impossible and rarely useful. In practice, it suffices to find the first few terms in some Taylor expansion of the solution. This so-called "perturbative" approach to quantum theories is particularly transparent in Feynman's formalism, where each term of the expansion can be represented pictorially as a graph. See figure 1(a) for typical examples. The graphs contributing to the nth-order term in this Taylor expansion will involve n vertices. Feynman's rules describe how to convert these graphs into integral expressions for computing the individual terms in the Taylor expansion.

In this article we are interested in perturbative string theory. The string Feynman diagrams (see figure 1(b) for three equivalent ones) are surfaces called *worldsheets*; the need for quantum foam is avoided because these surfaces are much less singular than the particle graphs (which have singularities at each vertex), and this is also largely why the mathematics of strings is so nice. To cut a long story short, each term in the perturbative expression for probabilities in string theory can be calculated from a quantity called a "correlation function" in a CFT that lives on the corresponding

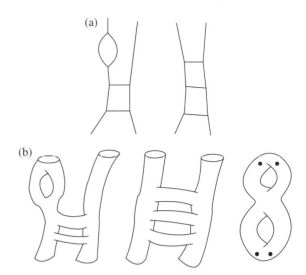

Figure 1 Some Feynman diagrams of (a) particles and (b) strings.

worldsheet. Feynman's path integral here amounts to the integral of a quantity that CFT can compute, over some MODULI SPACE [IV.8] of surfaces.

The vertices in a Feynman diagram represent places where one particle absorbs or emits another. The corresponding rules of string theory tell us that we should dissect the worldsheet into "tubular Y-shapes," or spheres with three legs, as in figure 2. Since these spheres with legs play the role of vertices in the Feynman diagram, the factor they contribute to the integrand of the path integral is called a *vertex operator*, and now it describes the absorption or emission of one *string* by another. A vertex operator algebra is the "algebra" of these vertex operators.

2.1.3 The Hamiltonian Formulation and Algebra

The Poisson bracket $\{A, B\}_{\mathrm{P}}$ of two classical observables A and B is defined to be

$$\frac{\partial A}{\partial x}\frac{\partial B}{\partial p} - \frac{\partial B}{\partial x}\frac{\partial A}{\partial p}.$$

Note that $\{A, B\}_{\mathrm{P}} = -\{B, A\}_{\mathrm{P}}$: in other words, the Poisson bracket is *anti-commutative*. It also satisfies the *Jacobi identity*

$$\{A, \{B, C\}_{\mathrm{P}}\}_{\mathrm{P}} + \{B, \{C, A\}_{\mathrm{P}}\}_{\mathrm{P}} + \{C, \{A, B\}_{\mathrm{P}}\}_{\mathrm{P}} = 0,$$

and therefore defines a Lie algebra. The Hamiltonian formulation of classical physics expresses the evolution of an observable A by means of the differential equation $\dot{A} = \{A, H\}_{\mathrm{P}}$, where H is the HAMILTONIAN

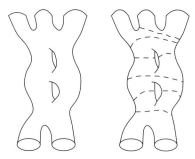

Figure 2 Dissecting a surface.

[III.35]: that is, the energy observable. The quantum version of this picture is due to Heisenberg and Dirac: the observables are now linear operators rather than smooth functions, and the Poisson bracket is replaced by the *commutator* $[\hat{A}, \hat{B}] = \hat{A} \circ \hat{B} - \hat{B} \circ \hat{A}$ of operators. This again has the anti-commuting property $[\hat{A}, \hat{B}] = -[\hat{B}, \hat{A}]$ and again satisfies the Jacobi identity, so the process of "quantization" gives rise to a homomorphism of Lie algebras. The derivative with respect to time of a quantum observable $\hat{A}$ is then the natural analogue of the classical case: it is proportional to $[\hat{A}, \hat{H}]$, where $\hat{H}$ is the Hamiltonian operator. Thus the Hamiltonian has a dual role: as the energy observable and as the controller of time evolution. All of physics is stored in the action of the observables on state space $\mathcal{H}$, as well as the commutators of these observables with $\hat{H}$.

Let us illustrate this picture with the *quantum spring*, also known as the *harmonic oscillator*. The position and momentum observables $\hat{x}$, $\hat{p}$ are operators acting on the infinite-dimensional space $\mathcal{H}$ of possible spring-states. It is more convenient to work with certain combinations of them called $\hat{a}$ and $\hat{a}^{\dagger}$ (the dagger denotes the "Hermitian adjoint," or complex-conjugate transpose), which obey the commutator relation $[\hat{a}, \hat{a}^{\dagger}] = I$, where I is the identity operator. It turns out that all other observables can be built from $\hat{a}$ and $\hat{a}^{\dagger}$. For example, the Hamiltonian $\hat{H}$ is $l(\hat{a}^{\dagger} \hat{a} + \frac{1}{2})$ for some positive constant l. The *vacuum*, which is denoted $|0\rangle$, is the state of minimum energy. In other words, the state $|0\rangle$ is an eigenvector of $\hat{H}$ with smallest possible eigenvalue: $\hat{H}|0\rangle = E_0|0\rangle$ for some $E_0 \in \mathbb{R}$ and all other eigenvalue E of $\hat{H}$ are greater than E_0. It follows from this that $\hat{a}|0\rangle = 0$. To see why, consider the effect of $\hat{H}$ on $\hat{a}|0\rangle$:

$$\hat{H}\hat{a}|0\rangle = l(\hat{a}^{\dagger}\hat{a} + \tfrac{1}{2})\hat{a}|0\rangle = l(\hat{a}\hat{a}^{\dagger} - \tfrac{1}{2})\hat{a}|0\rangle$$
$$= \hat{a}l(\hat{a}^{\dagger}\hat{a} - \tfrac{1}{2})|0\rangle = \hat{a}(\hat{H} - l)|0\rangle = (E_0 - l)\hat{a}|0\rangle.$$

Here, we have used the fact that $\hat{a}^{\dagger}\hat{a} = \hat{a}\hat{a}^{\dagger} - I$. (The observables $\hat{a}$ and $\hat{a}^{\dagger}$ are called *creation and annihila-tion operators* because, as we shall see later, they can be interpreted as adding or removing a particle from a certain n-particle state. Showing this uses the fact that they produce $\pm I$ when you interchange their order.) This calculation shows that if $\hat{a}|0\rangle$ is not zero, then it is an eigenvector of $\hat{H}$ with an eigenvalue smaller than E_0, which is a contradiction.

Since $\hat{a}|0\rangle = 0$, it follows that $\hat{H}|0\rangle = \frac{1}{2}l|0\rangle$, so $E_0 = \frac{1}{2}l$. We now define, for each positive integer n, a state $|n\rangle$ to be $(\hat{a}^{\dagger})^n|0\rangle \in \mathcal{H}$. Similar calculations to the one just given show that $|n\rangle$ has energy $E_n = (2n + 1)E_0$. For example,

$$\hat{H}|1\rangle = l(\hat{a}^{\dagger}\hat{a} + \tfrac{1}{2})\hat{a}^{\dagger}|0\rangle = l(\hat{a}^{\dagger}(\hat{a}^{\dagger}\hat{a} + I) + \tfrac{1}{2}\hat{a}^{\dagger})|0\rangle$$
$$= \tfrac{3}{2}l\hat{a}^{\dagger}|0\rangle = E_1|1\rangle.$$

(Note that we used the fact that $a|0\rangle = 0$ in the penultimate equality above.) We think of the vacuum as the ground state, and $|n\rangle$ as being the state with n *quantum particles*. These states $|n\rangle$ span all of the state space $\mathcal{H}$. To see how some observable acts on some state, one writes the observable in terms of the basic observables $\hat{a}$, $\hat{a}^{\dagger}$ and the state in terms of the basic states $|n\rangle$. In this algebraic way we can recover all of the physics.

This idea of building up the whole space $\mathcal{H}$ from the vacuum and the operators is a fruitful one in mathematics as well: something similar happens for the most important modules of most of the important Lie algebras.

2.1.4 Fields

A classical *field* is a function of space and time. Its values can be numbers or vectors, which represent quantities such as air temperature or the current in a river. The values taken by a *quantum* field are operators; furthermore, a quantum field is not a *function* of space and time, but a more general object called a DISTRIBUTION [III.18]. The prototypical example of a distribution is the *Dirac delta function* $\delta(x - a)$. Despite its name, this is not a function: rather, it is defined by the property that

$$\int f(x)\,\delta(x - a)\,dx = f(a) \qquad (1)$$

for any sufficiently well-behaved function $f(x)$. Even though $\delta(x - a)$ is not a function, one can informally interpret it as the derivative of a step function, and one can visualize it as equaling 0 everywhere except at $x = a$, where it is infinite, in such a way that the infinitely tall and infinitely thin rectangle under the graph has area 1. However, it really only makes sense inside an integral, as in (1). Similar remarks apply to distributions in general, so a quantum field can really only be evaluated inside an integral of space and time, applied to some "test function" like f above. The value of such an integral will be an operator on the state space $\mathcal{H}$.

Dirac deltas appear in classical mechanics when one takes Poisson brackets of classical fields. Similarly, commutators of quantum fields involve delta functions too. For example, in the simplest cases the quantum fields φ satisfy

$$\left.\begin{aligned} [\varphi(x,t),\varphi(x',t)] &= 0, \\ \left[\varphi(x,t),\frac{\partial}{\partial t}\varphi(x',t)\right] &= \mathrm{i}\hbar\delta(x-x'). \end{aligned}\right\} \quad (2)$$

This is a mathematical way of expressing, in the context of quantum field theory, the cherished physical principle called *locality*.[1] the only way we can *directly* affect something is by nudging it. In order to influence something not touching us, we must propagate a disturbance from us to it, such as a ripple in water. The main purpose of both classical and quantum fields is that they provide a natural vehicle for realizing locality. Locality is also at the heart of vertex operator algebras.

An important aspect of modern physics is that many of the central concepts of classical physics become less central, and are instead *derived* quantities. For example, the basic object of GENERAL RELATIVITY [IV.13] is a Lorentzian manifold, and familiar physical quantities such as mass and gravitational force are, from the point of view of this manifold, just names (that are not wholly precise) given to certain of its geometrical features.

Particles are obviously essential to classical physics, but we have not mentioned them in our brief sketch of quantum field theory. They arise through the so-called *modes* of quantum fields φ, which play the role of the operators $\hat{a}$, $\hat{a}^{\dagger}$ that we met in section 2.1.3. A mode is the operator that results from hitting the quantum field with an appropriate test function and integrating—just as one does when working out a Fourier coefficient, in which case the test functions are TRIGONOMETRIC FUNCTIONS [III.92]. In fact, when viewed appropriately, modes actually *are* Fourier coefficients of a certain kind. The commutators of these modes can be obtained from the commutators of the fields. Now, recall that the vertex operators of string theory are related to the

emission and absorption of strings. As we shall see shortly, these vertex operators are the quantum fields in a quantum field theory of point particles (namely, the associated conformal field theory); the modes of these vertex operators generate the "particles" (or in more conventional language, the *states*) in that conformal field theory. Equivalently, they generate the various vibrational states of a single string in that string theory.

2.2   Conformal Field Theory

A *conformal field theory* (CFT) is a quantum field theory with a two-dimensional space-time whose symmetries include all *conformal transformations*. We shall explain what this means in the next paragraphs, but for now it is enough to know that a CFT is a particularly symmetrical kind of quantum field theory. A CFT lives on the worldsheet Σ traced by a set of strings as they evolve, sometimes colliding and separating, through time. In this subsection we shall informally sketch their basic theory; in section 3.1 we shall be more precise.

CFT, like any quantum field theory in two dimensions, has two almost independent halves. This is easiest to see in the context of string theory: the ripples on the string are responsible for the physical properties (charge, mass, etc.) of the corresponding state, but they can move (at the speed of light) either clockwise or counterclockwise around the string. When they do so, they just pass through each other without interacting. These two alternatives, clockwise and counterclockwise, yield the two *chiral halves* of CFT. To study a CFT, one first analyzes its chiral halves and then splices them together to form the "bichiral" physical quantities. Almost all attention in CFT by mathematicians has focused on the chiral (as opposed to physical) data, and indeed that is where vertex operator algebras live. For ease of presentation, we will usually suppress one of the chiral halves.

A conformal transformation is a transformation that preserves angles. The simplest reason one can give for why two dimensions are so special for CFT is that there are far more conformal transformations in two dimensions than there are in higher dimensions. When $n > 2$ the only examples are the obvious ones: combinations of translations, rotations, and enlargements. This means that the space of all local conformal transformations in $\mathbb{R}^n$ is $\binom{n+2}{2}$ dimensional. However, when $n = 2$ the space of local conformal transformations is far richer: it is *infinite* dimensional. Indeed, if you identify $\mathbb{R}^2$ with the complex plane $\mathbb{C}$, then any HOLOMORPHIC FUNCTION [I.3 §5.6] $f(z)$ that does not have

---

1. More precisely, for quantum fields, locality takes the form that if not even light can connect two given space-time points, then the quantum fields at those points must be causally independent. In particular, measurements at such points can be performed simultaneously with arbitrary precision. In quantum theories, this requires those operators to commute. Equation (2) is a generous way to satisfy locality.

zero derivative at a point z_0 is conformal near z_0. Since a CFT is invariant under conformal transformations and there are many conformal transformations, a CFT is especially symmetrical: this is what makes CFTs so interesting mathematically.

Lie algebras arise naturally whenever one has local symmetries, and indeed one can form an infinite-dimensional Lie algebra out of the infinitesimal conformal transformations. This algebra has a basis l_n, $n \in \mathbb{Z}$, that obeys the Lie-bracket relations

$$[l_m, l_n] = (m - n)l_{m+n}. \tag{3}$$

The algebraic interpretation of the conformal symmetry of CFT turns out to be that these basis elements l_n act naturally on all the quantities in the theory, as we shall explain below.

The basic example that underlies all the others is when space-time Σ is a semi-infinite cylinder corresponding to an incoming string. It is parametrized by time $t < 0$ and the angle $0 \leqslant \theta < 2\pi$ around the string. We can conformally map the cylinder to the punctured disk in $\mathbb{C}$ by $z = e^{t-i\theta}$, so $t = -\infty$ corresponds to $z = 0$. This allows us to say what we mean by conformal symmetries of the cylinder.

The quantum fields $\varphi(z)$ of CFT are the vertex operators of string theory. As always, these quantum fields φ are "operator-valued distributions" on space-time Σ, acting on the space $\mathcal{H}$ of states. Now it is possible for a field φ to be "holomorphic," in the following sense. First, you calculate its modes φ_n, one for each $n \in \mathbb{Z}$, which are linear maps from the state space $\mathcal{H}$ to itself, given by the formula

$$\varphi_n = \int \varphi(z) z^{n-1} \, dz,$$

where the integral is around a small circle about the origin. Then you take these modes as the coefficients of a formal power series $\sum_{n \in \mathbb{Z}} \varphi_n z^n$. We call φ holomorphic if this formal power series can be identified with φ, in a sense that we shall discuss more in section 3.1. A typical field $\varphi(z)$ is not holomorphic: rather, it is a combination of holomorphic and anti-holomorphic fields, which make up the two chiral halves of CFT. We will focus on the space of holomorphic fields $\varphi(z)$, which we call $\mathcal{V}$. This turns out to form a vertex operator algebra (as do the anti-holomorphic fields).

For example, the most important vertex operator comes directly from the conformal symmetry: the *stress-energy tensor* $T(z) \in \mathcal{V}$ is the "conserved current" that Noether's theorem associates with the conformal symmetry. Labeling its modes (Noether's

"conserved charges" here) by $L_n = \int T(z) z^{-n-3} \, dz$, so that $T(z) = \sum_n L_n z^{-n-2}$, we find that they *almost* realize the conformal algebra: instead of (3), however, they obey the slightly more complicated relations

$$[L_m, L_n] = (m - n)L_{m+n} + \delta_{n,-m} \frac{m(m^2 - 1)}{12} cI, \tag{4}$$

where I is the identity. In other words, the operators L_n and I form an extension of the conformal algebra by I. The resulting infinite-dimensional Lie algebra is called the *Virasoro algebra* $\mathfrak{Vir}$. The number c appearing in (4) is called the *central charge* of the CFT and is a rough measure of its size.

The operators L_n do not precisely represent the conformal algebra (3). Instead, they form a so-called *projective representation*. Projective representations of symmetries, such as (4), are common in quantum theories. The fact that they are not true representations is not a problem, since one can turn them into true representations by extending the algebra. In our case, the state space $\mathcal{H}$ carries inside it a true representation of the Virasoro algebra $\mathfrak{Vir}$, which is useful as it means $\mathfrak{Vir}$ can be used to organize $\mathcal{H}$.

Any quantum field theory has what is called a *state-field correspondence*: with each field φ one associates its incoming state, which is the limit as the time t tends to $-\infty$ of $\varphi|0\rangle$ (as always, $|0\rangle$ is the vacuum state in $\mathcal{H}$ and φ acts on states). CFT is unusual in that the state-field correspondence is a bijection. This means we can identify $\mathcal{H}$ and $\mathcal{V}$ and use states to label all fields.

We want to make $\mathcal{V}$ into some sort of algebra, but the obvious direct approach of taking products $\varphi_1(z)\varphi_2(z)$ fails, since distributions, unlike true functions, cannot in general be multiplied. For example, the Dirac delta $\delta(x - a)$ cannot be squared without causing problems in (1). However, even if the product $\varphi_1(z)\varphi_2(z)$ does not make sense, one can make sense of $\varphi_1(z_1)\varphi_2(z_2)$ as an operator-valued distribution on Σ^2. It is then possible to recover most of the physics of CFT by studying the singular terms as $z_2 \to z_1$. By the *operator product expansion*, we mean expanding products $\varphi_1(z_1)\varphi_2(z_2)$ as sums of the form $\sum_h (z_1 - z_2)^h O_h(z_1)$. The set $\mathcal{V}$ is closed under this product in the sense that each coefficient $O_h(z)$ lies in $\mathcal{V}$. A typical example is

$$T(z_1)T(z_2) = \tfrac{1}{2}c(z_1 - z_2)^{-4}I + 2(z_1 - z_2)^{-2}T(z_1)$$
$$+ (z_1 - z_2)\frac{d}{dz}T(z_1) + \cdots.$$

Physicists call $\mathcal{V}$ a *chiral algebra*; for us it is the prototypical example of a vertex operator algebra. It is not an algebra in the conventional sense though, since, given vertex operators $\varphi_1(z)$ and $\varphi_2(z)$, we have not just a single product $\varphi_1(z) * \varphi_2(z)$ in $\mathcal{V}$ but infinitely many products $\varphi_1(z) *_h \varphi_2(z) = O_h(z)$, all belonging to $\mathcal{V}$.

The Hamiltonian plays a crucial role in any quantum field theory; here it turns out to be proportional to the mode L_0 discussed earlier. Being an observable, L_0 is diagonalizable on $\mathcal{H}$, which means that any state $v \in \mathcal{H}$ can be written as a sum $\sum_h v_h$, where $v_h \in \mathcal{H}$ has energy h: that is, $L_0 v_h = h v_h$.

There is a special class of CFT that is particularly well-behaved. Let $\bar{\mathcal{V}}$ denote the space of all anti-holomorphic fields in the CFT—it is the other chiral half. Recall that the full CFT consists of $\mathcal{V}$ and $\bar{\mathcal{V}}$ spliced together. We call the CFT *rational* if $\mathcal{V} \oplus \bar{\mathcal{V}}$ is so large that it has finite index, in an appropriate sense, in the full space of quantum fields in the CFT. The name "rational" arises because the central charge c and other parameters in a rational CFT have to be rational numbers.

The mathematics of rational CFT is especially rich. Let us briefly look at one example. (We will use several words that will be unfamiliar to most readers, but at least it will give some idea of which areas are touched by CFT.) As with everything else, the quantum probabilities arising in CFT are found by first computing chiral quantities and splicing them together. These chiral quantities are called *conformal* or *chiral blocks*, and are found using simple Feynman-like rules applied to dissections like figure 2. In rational CFT we get a finite-dimensional space $\mathcal{F}_{g,n}$ of chiral blocks for any worldsheet Σ, i.e., for any choice of genus g and number n of punctures. These spaces carry projective representations of the mapping class group $\Gamma_{g,n}$ (defined to be the fundamental group π_1 of the moduli space $\mathcal{M}_{g,n}$). This $\Gamma_{g,n}$-representation is the source, for instance, of Jones's relation of the BRAID GROUP [III.4] (and hence KNOTS [III.44]) to subfactors, Borcherds's explanation of "Monstrous Moonshine," the Drinfel'd–Kohno monodromy theorem, and the modularity of affine Kac–Moody characters. Some of this we will touch on in section 4.

The most important example here is the torus, where the chiral blocks are *modular functions*, a class of functions of fundamental mathematical importance. A modular function is a meromorphic function (that is, a function that is holomorphic except at a few "poles" where it can tend to infinity) $f(\tau)$ that is defined on the upper half-plane $\mathbb{H} = \{\tau \in \mathbb{C} \mid \mathrm{Im}\,\tau > 0\}$ and that is "symmetric" with respect to the group $\mathrm{SL}_2(\mathbb{Z})$ of 2×2 matrices with integer entries and determinant 1, in the sense that for any such matrix $\left(\begin{smallmatrix} a & b \\ c & d \end{smallmatrix}\right)$ the function $f(\tau)$ is closely related (though not necessarily exactly equal) to the function $f((a\tau + b)/(c\tau + d))$. We shall discuss this further in section 3.2.

The appearance of modularity can be understood by recalling from section 2.1.2 that Feynman's path integral in string theory is an integral over moduli spaces. The moduli space $\mathcal{M}_{1,0}$ for the torus can be written as the quotient of the half-plane $\mathbb{H}$ by the action of $\mathrm{SL}_2(\mathbb{Z})$. Therefore, if one lifts the integrand of Feynman's integral from $\mathcal{M}_{1,0}$ to $\mathbb{H}$, one obtains a function $\mathcal{Z}(\tau)$ that is invariant under $\mathrm{SL}_2(\mathbb{Z})$ and hence modular. This integrand $\mathcal{Z}(\tau)$ is a quadratic combination of the chiral blocks for the torus.

3 What VOAs Are

It is possible to give a fully axiomatic definition of vertex operator algebras. However, when one first encounters this definition (and not just the first time either) it can seem very complicated and arbitrary, and one is given no feel for the importance of VOAs. Our treatment below will be much more informal: this will clarify their importance even if it hides much of their complexity. Thanks to the previous section, it is possible to give a quick justification for VOAs: if you concede that CFT (or equivalently, perturbative string theory) is important, and if you have seen how closely related CFT is to VOAs, then you must concede that VOAs are important. However, this is not the whole story, as we shall see.

3.1 Their Definition

Let us begin by defining them in terms of other concepts that must themselves be defined: a vertex operator algebra is an algebra of vertex operators, or in other words the chiral algebra $\mathcal{V}$ of a conformal field theory.

The most important thing to understand in this definition is that a vertex operator is a quantum field, which, as we have seen, is an "operator-valued distribution of space-time." So we can think of it informally as a matrix-valued function of space-time, where the matrix is $\infty \times \infty$ and its entries can be generalized functions like the Dirac delta (1). However, we shall give a much better description of these vertex operators shortly.

By "space-time" we mean the unit disk in $\mathbb{C}$ punctured at $z = 0$. Recall from section 2.2 that string-theoretically this set corresponds to a semi-infinite cylinder parametrized by the angle $-\pi < \theta \leqslant \pi$ running around the string as well as the time $-\infty < t < 0$

running along the axis: the map from this to the punctured disk was $(\theta, t) \mapsto z = e^{t-i\theta}$. We want to restrict our attention to quantum fields that depend holomorphically on z. However, it is not obvious what "holomorphic" means for distributions. We touched on this question in section 2.2: now we shall look at it in more detail.

To do this, we need a more concrete description of a vertex operator. The key idea is a very convenient algebraic interpretation of holomorphic distributions. Consider the sum

$$d(z) = \sum_{n=-\infty}^{\infty} z^n. \tag{5}$$

Multiply it by $f(z) = 3z^{-2} - 5z^3$, say. This gives us

$$f(z)d(z) = 3 \sum_{n=-\infty}^{\infty} z^{n-2} - 5 \sum_{n=-\infty}^{\infty} z^{n+3}$$

$$= 3 \sum_{n=-\infty}^{\infty} z^n - 5 \sum_{n=-\infty}^{\infty} z^n = -2d(z).$$

A few more examples like this will convince you that $f(z)d(z) = f(1)d(z)$ for *any* polynomial function f of z and z^{-1}. Therefore, $d(z)$ behaves exactly like the Dirac delta $\delta(z - 1)$, at least for polynomial test functions f. Note that $d(z)$ cannot converge for any z: the positive powers have a convergent sum only for $|z| < 1$, and the negative powers only for $|z| > 1$. The "function" $d(z)$ is an example of a *formal power series*: any series $\sum_{n=-\infty}^{\infty} a_n z^n$, where the coefficients a_n can be anything and we ignore all convergence issues.

By inspection, these formal power series are "holomorphic" throughout the punctured plane: after all, holomorphic just means that the complex derivative d/dz exists, and the derivative $\sum_n na_n z^{n-1}$ of a formal power series clearly remains a formal power series. (By contrast, nonholomorphic series would involve the complex conjugate $\bar{z}$.)

So that is what a vertex operator looks like: a formal power series $\sum_{n=-\infty}^{\infty} a_n z^n$, where each coefficient a_n is now an operator (endomorphism) on the space $\mathcal{V}$ of states, which is an infinite-dimensional vector space. Since the vertex operators are in one-to-one correspondence with the states (we called this the "state–field correspondence" above), we can label these vertex operators with states: the standard convention is to denote the vertex operator corresponding to state $v \in \mathcal{V}$ by

$$Y(v, z) = \sum_{n=-\infty}^{\infty} v_n z^{-n-1}. \tag{6}$$

The symbol "Y" should remind you of the sphere with three legs, which as we know is the vertex of string

theory. These coefficients v_n are the modes: as in any quantum field theory, all observables and states in the theory are built up from them.

The most important state in the theory is the vacuum $|0\rangle$. It corresponds to the identity vertex operator: $Y(|0\rangle, z) = I$. From the physical point of view, the vertex operator $Y(v, z)$ is the field that created the state v at time $t = -\infty$, i.e., $Y(v, 0)|0\rangle$ exists and equals v. (Recall that in our model $z = 0$ corresponds to $t = -\infty$.) Among other things, this means that $v_{-1}(|0\rangle) = v$, so indeed the modes applied to $|0\rangle$ generate $\mathcal{V}$, as is required in any quantum field theory.

The most important observable in the theory is the Hamiltonian, or energy operator, which we denote by L_0. It is diagonalizable (so $\mathcal{V}$ can be written as a sum of L_0-eigenspaces) and all of its eigenvalues must be integers. For example, the vacuum $|0\rangle$ has 0 energy: $L_0|0\rangle = 0$. Since $|0\rangle$ should have the minimum energy, the L_0-decomposition of $\mathcal{V}$ is then $\mathcal{V} = \bigoplus_{n=0}^{\infty} \mathcal{V}_n$, where $\mathcal{V}_0 = \mathbb{C}|0\rangle$. Each space $\mathcal{V}_n$ turns out to be finite dimensional, and we can think of L_0 as defining a $\mathbb{Z}_+$-grading on state space $\mathcal{V}$.

The most important vertex operator in the theory is the stress-energy tensor $T(z)$. The corresponding state is called the *conformal vector* ω: $Y(\omega, z) = T(z)$. This means that ω has modes $\omega_n = L_{n-1}$ that form a representation (4) of the Virasoro algebra $\mathfrak{Vir}$. (This is the algebraic expression for the requirement of conformal symmetry.) The conformal vector has energy 2: $\omega \in \mathcal{V}_2$.

So far our theory is seriously underdetermined. The most important axiom to help us to pin it down further is locality. With a little work, one can show that this reduces to the condition that the commutator $[Y(u, z), Y(v, w)]$ of two vertex operators should be a finite linear combination of the Dirac delta $\delta(z - w) = z^{-1} \sum_{n=-\infty}^{\infty} (w/z)^n$ and its derivatives $(\partial^k/\partial w^k)\delta(z - w)$. Now, $(z - w)^{k+1}(\partial^k/\partial w^k)\delta(z - w) = 0$. To see this, look at the case $k = 1$:

$$(z - w)^2 \frac{\partial}{\partial w} \delta(z - w)$$

$$= \sum_{n=-\infty}^{\infty} (nw^{n-1}z^{-n+1} - 2nw^n z^{-n} + nw^{n+1} z^{-n-1})$$

$$= \sum_{n=-\infty}^{\infty} ((n + 1) - 2n + (n - 1))w^n z^{-n} = 0.$$

The proof for general k is similar. Therefore, locality can be recast in an equivalent form as follows: given any $u, v \in \mathcal{V}$, there is a positive number N such that

$$(z - w)^N [Y(u, z), Y(v, w)] = 0. \tag{7}$$

This equation may look strange. Why can we not simply divide out the $(z - n)^N$ and get that all vertex operators commute? The reason is that when formal power series are involved, there can be zero divisors. For example, it is easy to check that $(z - 1) \sum_{n \in \mathbb{Z}} z^n = 0$. Locality in the form (7) is at the heart of VOAs; for instance, one can express it as a triply infinite sequence of identities that the modes must obey, and this emphasizes just how restrictive a condition it is, and how correspondingly interesting it is to find examples of VOAs.

This completes the definition of a VOA. A consequence of these properties is that the modes u_n respect the L_0-grading that we mentioned earlier. This means that if u has energy k and v has energy l, then $u_n(v)$ has energy $k + l - n - 1$. The definition followed here is sometimes called a VOA *of CFT-type*, for obvious reasons. Sometimes in the literature some of these conditions are weakened or dropped. For example, much of the theory is independent of the existence of the conformal vector ω, although to us it will be crucial, for reasons that will be explained in the next subsection.

A VOA is simultaneously a physical and a mathematical object. We have emphasized their physical origins in order to help explain the motivation for studying them. We know they should be valuable to mathematics, simply because CFT is, and indeed this is the case, as we shall see in section 4. But from a purely mathematical point of view, they might appear somewhat ad hoc, as though we had a list of mathematical ingredients and said to ourselves, "Let's consider this, and then have some of these, oh, and perhaps one of those too, but with the following extra assumption: ...." Fortunately, there are more abstract formulations of VOAs that make them appear much less arbitrary as mathematical structures. For example, Huang has shown that they can be regarded as "two-dimensionalized" Lie algebras, in the following sense. If you want to keep track of the Lie brackets in an expression such as $[a, [[b, c], d]]$ (which is important since the Lie bracket is not an associative operation), you can do so with the help of a binary tree, and in fact it is easy to formulate Lie algebras in the language of such trees. If one then replaces binary trees by diagrams made out of spheres with legs, as we did with Feynman diagrams earlier, one obtains a structure that is equivalent to a VOA. (Of course, this is very far from a full explanation of what Huang did: his proof is extremely long.)

3.2 Basic Properties

We see from the definition sketched in the last subsection that a VOA is an infinite-dimensional $\mathbb{Z}_+$-graded vector space with infinitely many products (namely $u *_n v = u_n(v)$), which obey infinitely many identities. Needless to say, it is not an easy definition, and there are no easy examples.

However, if we ignore the conformal symmetry (i.e., the conformal vector ω), then there are some simple, though uninteresting, examples. The easiest is the one-dimensional algebra $\mathcal{V} = \mathbb{C}|0\rangle$. More generally, a VOA $\mathcal{V}$ that obeys (7) with $N = 0$ is a commutative associative algebra with a unit $1 = |0\rangle$. It also has a *derivation* $T = L_{-1}$, with respect to the product $u * v = u_{-1}(v)$: this means a linear map that obeys the product rule satisfied by derivatives, namely $T(u * v) = (Tu) * v + u * (Tv)$. The converse of this statement is true too: any such algebra is a VOA that obeys (7) with $N = 0$. In these simple examples, the role of the derivation T is to recover the z-dependence of the vertex operator.

Therefore, we need N not to be zero in (7) if we want interesting examples. Likewise, the vertex operators $Y(u, z)$ must be distributions (that is, they must involve doubly infinite sums) or again the VOA reduces to a commutative associative algebra.

It is also easy to show that in any VOA (again the existence of the conformal vector is not needed), the space $\mathcal{V}_1$ is a Lie algebra, with Lie bracket given by $[uv] = u_0(v)$. This is important because each $\mathcal{V}_n$ will carry a representation of this Lie algebra, and $\mathcal{V}_1$ generates continuous symmetries of the VOA (at least when $\mathcal{V}_1 \neq \{0\}$). For a typical VOA $\mathcal{V}$ these Lie algebras are very familiar. For instance, for the VOAs associated with rational CFT, they are *reductive*, which means that they are a direct sum of copies of the trivial Lie algebra $\mathbb{C}$ with simple Lie algebras.

The existence of the conformal vector becomes important when one starts to consider the representation theory of VOAs. A $\mathcal{V}$-*module* is defined in a natural way. We shall not give full details here, but, roughly speaking, it is a space on which $\mathcal{V}$ acts in such a way that as much as possible of the VOA structure is respected. For example, $\mathcal{V}$ will automatically be a module for itself, just as a group acts on itself in a simple way. (See REPRESENTATION THEORY [IV.9 §2] for an explanation of the latter.) A *rational* VOA is defined to be one that has the simplest representation theory: it has only finitely many irreducible $\mathcal{V}$-modules, and any $\mathcal{V}$-module is a direct sum of irreducible ones. They are called rational VOAs because they are the VOAs that come from rational CFT. For these VOAs, $\mathcal{V}$ acts irreducibly on itself.

Assume now that $\mathcal{V}$ is rational. Any irreducible $\mathcal{V}$-module M will inherit from $\mathcal{V}$ an L_0-grading by rational numbers, $M = \bigoplus_h M_h$, into finite-dimensional spaces M_h. The *character* $\chi_M(\tau)$ is defined by

$$\chi_M(\tau) = \sum_h \dim M_h e^{2\pi i \tau (h - c/24)}, \qquad (8)$$

where c is the central charge. This definition arises naturally in CFT as well as in Lie theory (or *affine Kac-Moody algebras*), although the curious "$c/24$," needed for (9) below, is mysterious in Lie theory. (In CFT it has a natural explanation as a certain topological effect.) These characters converge for any τ in the upper half-plane $\mathbb{H}$. They carry a representation of the modular group $\mathrm{SL}_2(\mathbb{Z})$:

$$\chi_M\left(\frac{a\tau + b}{c\tau + d}\right) = \sum_{N \in \mathrm{Irr}(V)} \rho\begin{pmatrix} a & b \\ c & d \end{pmatrix}_{MN} \chi_N(\tau), \quad (9)$$

where $\mathrm{Irr}(V)$ denotes the (finite) set of irreducible V-modules, and $\rho\left(\begin{smallmatrix} a & b \\ c & d \end{smallmatrix}\right)$ is a matrix with complex entries, whose rows and columns are labeled by $M, N \in \mathrm{Irr}(V)$. Equation (9) holds for any $\left(\begin{smallmatrix} a & b \\ c & d \end{smallmatrix}\right)$ in $\mathrm{SL}_2(Z)$, i.e., for any integers a, b, c, d satisfying $ad - bc = 1$. The lengthy proof of (9), by Zhu, is perhaps the high point of VOA theory, and owes much to the intuitions of rational CFT. In the next section, we shall get some idea of why it is so important.

4 What Are VOAs Good For?

This section describes what are probably the two most significant applications of VOAs. But let us begin by listing (without any explanations) a few others. Inspired by the geometry of string theory, vertex operator (super)algebras have been assigned to manifolds, resulting in a powerful, though complicated, algebraic invariant of those manifolds that generalizes and enriches more classical data such as de Rham cohomology. VOAs associated with affine Kac-Moody algebras at "degenerate" levels k are deeply related to the geometric Langlands program. The modularity of affine algebra characters, as well as that of, for example, lattice theta functions, are all special cases of Zhu's theorem, which places these modularities in a much broader context.

4.1 The Mathematical Formulation of CFT

Since the 1970s quantum field theory has had considerable success, especially in geometry, by studying classical structures using infinite-dimensional methods; this is a theme in particular of Atiyah's school. Conformal field theories are a class of exceptionally symmetric quantum field theories, and they are also among the simplest nontrivial quantum field theories known. In the past two decades mathematics has feasted on this combination of symmetry and (relative) simplicity, often by "looping" or "complexifying" more classical structures, and the impact of CFT (or, equivalently, of string theory) has been especially significant and broad. In hindsight the importance of CFT to mathematics is not surprising: it is a coherent and intricate structure that straddles several disparate areas of mathematics, sprawling across geometry, number theory, analysis, combinatorics, and indeed algebra.

From this point of view, a crucial application of VOA theory has been to CFT itself. Quantum field theories are notoriously difficult to put on a rigorous mathematical footing. But the successful applications suggest that these difficulties are a symptom of mathematical profundity and subtlety rather than of irreparable mathematical incoherence. In this sense the situation is highly reminiscent of the deep conceptual challenges to eighteenth-century mathematicians that were raised by calculus. The definition of a VOA by Richard Borcherds makes the chiral algebra of a CFT completely rigorous, as well as concepts like the operator product expansion. Subsequent work (especially by Huang and Zhu) reconstructs from the VOA more and more of the CFT, in arbitrary genus. The resulting clarity makes the whole subject more accessible to, and hence exploitable by, mathematicians. Quantum field theories are here to stay in mathematics, and thanks to VOAs mathematicians are absorbing a large class of them completely and explicitly.

4.2 Monstrous Moonshine

In 1978 McKay noticed that $196\,884 \approx 196\,883$. Why was this an interesting observation? Well, the number on the left is the first meaningful coefficient of the j-FUNCTION [IV.1 §8]

$$j(\tau) = q^{-1} + (744+)\,196\,884q + 21\,493\,760q^2 \\ + 864\,299\,970q^3 + \cdots, \quad (10)$$

the generator of all modular functions for $\mathrm{SL}_2(\mathbb{Z})$. Recall that a modular function is a function $f(\tau)$ that is meromorphic in the upper half-plane $\mathbb{H}$ and invariant under the usual action of $\mathrm{SL}_2(\mathbb{Z})$. It should also be meromorphic at the boundary points $\mathbb{Q} \cup \{i\infty\}$, which are called *cusps*; we did not mention this condition earlier. The j-function generates these functions in the sense that any such modular function $f(\tau)$ can be written as a rational function $\text{poly}(j(\tau))/\text{poly}(j(\tau))$. In other words, $j(\tau)$ is a uniformizing function that identifies $(\mathbb{H} \cup \mathbb{Q} \cup \{i\infty\})/\text{SL}_2(\mathbb{Z})$ with the Riemann sphere $\mathbb{C} \cup \infty$. We bracketed the constant term 744 in (10) because although 744 was the traditional choice it can be freely replaced with any other number, including 0.

The number on the right in McKay's observation is the dimension of the smallest nontrivial representation of the Monster, the most exceptional of the FINITE SIMPLE GROUPS [V.7]. This relation between modular functions and the Monster was completely unexpected, as they seem to occupy completely independent spots in the mathematical universe. Conway, Norton, and others fleshed out and expanded McKay's original observation by making a number of conjectures, collectively called *Monstrous Moonshine*. For instance, with every pair (g, h) of commuting elements in the Monster (a group of size about 8×10^{53}), we expect there to be associated a function $j_{(g,h)}(\tau)$ that generates all modular functions for some discrete subgroup $\Gamma_{(g,h)}$ of $\text{SL}_2(\mathbb{Z})$. The j-function would be assigned in the case $g = h = \text{identity}$.

The first major step toward proving these Moonshine conjectures was made by Frenkel, Lepowsky, and Meurman in the mid 1980s. They constructed an infinite-dimensional vector space $V^\natural$ out of formal power series. They were motivated on the one hand by the vertex operators of string theory, and on the other by the formally similar distributions used in constructing affine algebra representations. This seemed a promising direction since for both string theory and affine algebra representations modular functions arise naturally. Together with a rich algebraic structure that came from these "vertex operators," $V^\natural$ was also acted on in a natural way by the Monster group. Moreover, although $V^\natural$ is infinite dimensional, it comes packaged into finite-dimensional pieces $V^\natural = \bigoplus_{n=-1}^{\infty} V_n^\natural$, and the "graded dimension" $\sum_n \dim(V_n^\natural)q^n$ equals $j - 744$. The action of the Monster sends each $V_n^\natural$ to itself; that is, each space $V_n^\natural$ itself carries a representation of the Monster. Frenkel, Lepowsky, and Meurman proposed that $V^\natural$ lies at the heart of the Monstrous Moonshine conjectures.

Borcherds was struck by the formal similarity between $V^\natural$ and the chiral algebras of CFTs, and by

abstracting out their important algebraic properties he defined a new structure called a vertex (operator) algebra. His axioms clarified their relationship with (generalizations of) Kac–Moody algebras, and by 1992 he had proved the main Conway–Norton conjecture (which corresponds to the case where g is arbitrary but h is the identity in the conjecture given earlier). Although his definition of VOAs required a deep understanding of the physics of CFT, his elaborate proof of this Moonshine conjecture is purely algebraic.

We would now call $V^\natural$ a rational VOA with only one irreducible module (namely itself); its symmetry group is the Monster and its character (8) is $j(\tau) - 744$. The removal of the constant term 744 from (10) is significant as it says that the Lie algebra $V_1^\natural$ is trivial—this is necessary if the symmetry group is to be finite. It is conjectured that $V^\natural$ is the unique VOA with central charge $c = 24$, trivial $\mathcal{V}_1$, and only one irreducible module. This is meant to be reminiscent of the LEECH LATTICE [I.4 §4], which is known to be the unique twenty-four-dimensional even self-dual lattice with no vectors of length $\sqrt{2}$. Indeed, the Leech lattice plays a crucial role in the construction of $V^\natural$.

Most of the Moonshine conjectures are still open and this deep connection between modular functions and the Monster is still somewhat mysterious. At the time of writing, however, VOAs still provide the only serious approach to the Moonshine conjectures.

Borcherds defined VOAs to clarify the chiral algebra of CFT and to tackle Monstrous Moonshine. For this work, he was awarded a Fields Medal in 1998.

Further Reading

Borcherds, R. E. 1986. Vertex algebras, Kac–Moody algebras, and the Monster. *Proceedings of the National Academy of Sciences of the USA* 83:3068–71.

——. 1992. Monstrous Moonshine and monstrous Lie superalgebras. *Inventiones Mathematicae* 109:405–44.

Di Francesco, P., P. Mathieu, and D. Sénéchal. 1996. *Conformal Field Theory*. New York: Springer.

Gannon, T. 2006. *Moonshine Beyond the Monster: The Bridge Connecting Algebra, Modular Forms and Physics*. Cambridge: Cambridge University Press.

Kac, V. G. 1998. *Vertex Algebras for Beginners*, 2nd edn. Providence, RI: American Mathematical Society.

Lepowsky, J., and H. Li. 2004. *Introduction to Vertex Operator Algebras and their Representations*. Boston, MA: Birkhäuser.

IV.18 Enumerative and Algebraic Combinatorics

Doron Zeilberger

1 Introduction

Enumeration, otherwise known as *counting*, is the oldest mathematical subject, while algebraic combinatorics is one of the youngest. Some cynics claim that algebraic combinatorics is not really a new *subject* but just a new *name* given to enumerative combinatorics in order to enhance its (former) poor image, but algebraic combinatorics is in fact the synthesis of two opposing trends: *abstraction of the concrete* and *concretization of the abstract*. The former trend dominated the first half of the twentieth century, starting with Hilbert's "theological" proof of the fundamental theorem of invariants, in which he showed by abstract means that certain invariants existed, but not how to find them. The latter trend is dominating contemporary mathematics, thanks to the omnipresence of The Mighty Computer.

The abstraction trend consists of the *categorization*, *conceptualization*, *structuralization*, and *fancification* (in short, "BOURBAKIZATION" [VI.96]) of mathematics. Enumeration did not escape this trend, and in the hands of such giants as Gian-Carlo Rota and Richard Stanley in America and Marco Schützenberger and Dominique Foata in France, classical, enumerative combinatorics became more conceptual, structural, and algebraic. However, as algebraic combinatorics has established itself as a fully fledged and separate mathematical speciality, the more recent trend toward the *explicit*, *concrete*, and *constructive* has left its mark as well. It has revealed that many algebraic structures have hidden combinatorial underpinnings; the attempts to unearth these have led to many fascinating discoveries and unsolved problems.

1.1 Enumeration

The fundamental theorem of enumeration, independently discovered by several anonymous cave dwellers, states that

$$|A| = \sum_{a \in A} 1.$$

In words: the number of elements in A is the sum over all elements of A of the constant function 1.

While this formula is still useful after all these years, enumerating specific finite sets is no longer considered mathematics. A genuine mathematical fact has to incorporate *infinitely* many facts, and the generic enumeration problem is to enumerate not just one set but all the sets in an infinite family.

To be precise, given an infinite sequence of sets $\{A_n\}_{n=0}^{\infty}$, where each set A_n consists of objects satisfying some combinatorial specifications that depend on the parameter n, answer the question: How many elements does A_n have?

In a moment we shall look at some examples. But before we can learn how to *answer* this kind of question, let us consider a meta-question: What is an answer?

This was posed, and beautifully answered, by Herbert Wilf. To give some background to Wilf's meta-answer, let us examine answers to some famous instances of enumeration questions.

In the list below, when we are given a set A_n (which will change from example to example), we shall write a_n instead of $|A_n|$. That is, a_n will stand for the number of elements of A_n.

(i) I Ching. If A_n is the set of all subsets of $\{1, \dots, n\}$, then $a_n = 2^n$.

(ii) Rabbi Levi Ben Gerson. If A_n is the set of PERMUTATIONS [III.68] on $\{1, \dots, n\}$, then $a_n = n!$.

(iii) Catalan. If A_n is the set of legal bracketings with n opening brackets and n closing brackets, then $a_n = (2n)!/(n+1)!n!$. (A *legal bracketing* is a sequence of n opening brackets and n closing brackets such that at no point in the sequence has the number of closing brackets exceeded the number of opening brackets. For instance, when $n = 2$ the legal bracketings are [ ][ ] and [[ ]].)

(iv) LEONARDO OF PISA [VI.6]. Let A_n be the set of finite sequences that consist only of 1s and 2s and that sum to n. (For example, when $n = 4$ the possible sequences are 1111, 112, 121, 211, and 22.) In this case, we have *three* equivalent answers as follows.

(i)
$$a_n = \frac{1}{\sqrt{5}} \left(\left(\frac{1+\sqrt{5}}{2} \right)^{n+1} - \left(\frac{1-\sqrt{5}}{2} \right)^{n+1} \right).$$

(ii)
$$a_n = \sum_{k=0}^{\lfloor n/2 \rfloor} \binom{n-k}{k}.$$

(iii) $a_n = F_{n+1}$, where F_n is the sequence *defined* by the recurrence $F_n = F_{n-1} + F_{n-2}$, subject to the initial conditions $F_0 = 0$, $F_1 = 1$.

(v) CAYLEY [VI.46]. If A_n is the set of labeled trees on n vertices, then $a_n = n^{n-2}$. (A *tree* is a connected GRAPH [III.34] without cycles, and it is *labeled* if the vertices have distinct names.)

(vi) If A_n is the set of labeled simple graphs with n vertices, then $a_n = 2^{n(n-1)/2}$. (A graph is *simple* if it has neither loops nor multiple edges.)

(vii) If A_n is the set of labeled *connected* simple graphs on n vertices (that is, graphs for which every vertex can be reached from every other by a path), then a_n is $n!$ times the coefficient of x^n in the power-series expansion of

$$\log\left(\sum_{k=0}^{\infty} \frac{2^{k(k-1)/2}}{k!} x^k\right).$$

(viii) If A_n is the set of Latin squares of size n ($n \times n$ matrices each of whose rows and columns is a permutation of $\{1, \ldots, n\}$), then not even a good approximation for a_n is known.

In 1982, Wilf defined an answer as follows.

Definition. An *answer* is a *polynomial-time algorithm* (in n) for computing a_n.

Wilf arrived at this definition after he refereed a paper proposing a "formula" for the answer to question (viii), and realized that its "computational complexity" exceeds that of the caveman's formula of direct counting.

What is a "formula"? It is really an algorithm that inputs n and outputs a_n. For example, $a_n = 2^n$ is shorthand for the recursive algorithm

if $n = 0$ then $a_n = 1$,
else $a_n = 2 \cdot a_{n-1}$,

which takes $O(n)$ steps. However, using the algorithm

if $n = 0$ then $a_n = 1$,
else if n is odd, then $a_n = 2a_{n-1}$,
else $a_n = a_{n/2}^2$

takes $O(\log n)$ steps, much faster than Wilf demands. In other cases, like enumerating self-avoiding walks, the best algorithm known is exponential, $O(c^n)$, and any lowering of the constant c is a major advance. (A *self-avoiding walk* is a sequence of points $x_0, x_1, \ldots, x_n$ in the two-dimensional integer lattice, where each x_i is one of the four neighbors of x_{i-1} and no two of the x_i are equal.) Notwithstanding these exceptions, Wilf's meta-answer is a very useful general guideline for evaluating answers.

Traditionally, the main customers of enumeration were probability and statistics. In fact, discrete probability is almost synonymous with enumerative combinatorics, since the probability of an event E occurring is the ratio of the number of successful cases divided by the total number. Also, statistical physics is, by and large, weighted enumeration of lattice models (see PHASE TRANSITIONS AND UNIVERSALITY [IV.25]). About fifty years ago, another important customer came along: computer science. Here one is interested in the COMPUTATIONAL COMPLEXITY [IV.20] of algorithms: that is, in the number of steps it takes to execute them.

2   Methods

The following tools are indispensable to the enumerative combinatorialist.

2.1   Decomposition

$$|A \cup B| = |A| + |B| \quad (\text{if } A \cap B = \varnothing).$$

In words: the size of the union of two disjoint sets equals the sum of their sizes.

$$|A \times B| = |A| \cdot |B|.$$

In words: the size of the Cartesian product of two sets (that is, the set of all pairs (a, b), where $a \in A$ and $b \in B$) equals the product of their sizes.

$$|A^B| = |A|^{|B|}.$$

In words: the size of the set of functions from B to A equals the size of A raised to the power the size of B. For example, the number of 0–1 sequences of length n, which can be viewed as functions from $\{1, 2, \ldots, n\}$ to $\{0, 1\}$, equals 2^n.

2.2   Refinement

If

$$A_n = \bigcup_k B_{nk} \quad (\text{disjoint union}),$$

and if b_{nk}, the number of elements of B_{nk}, is "nice" (and even if it is not), then

$$a_n = \sum_k b_{nk}.$$

The idea here is that it may be possible to take a set A_n that is difficult to count, and split it up into disjoint sets B_{nk} that are easier to count. For example, consider the set A_n of example (iv). This can be split into a disjoint union of subsets B_{nk}, where each B_{nk} consists of the sequences in A_n that have exactly k 2s. If there are k 2s, then there must be $n - 2k$ 1s, so $b_{nk} = \binom{n-k}{k}$. This yields answer (ii).

2.3 Recursion

Suppose that A_n can be decomposed in such a way that it is a combination of fundamental operations applied to the sets $A_{n-1}, A_{n-2}, \ldots, A_0$. Then a_n satisfies a recurrence relation of the form

$$a_n = P(a_{n-1}, a_{n-2}, \ldots, a_0).$$

For example, let A_n be the set of example (iv). If a sequence in A_n starts with a 1, then the rest of the sequence must add up to $n - 1$, and if it starts with a 2, then the rest must add up to $n - 2$. Since when $n \geqslant 2$ exactly one of these possibilities occurs and both are possible, we can decompose A_n into $1A_{n-1}$ and $2A_{n-2}$, where $1A_{n-1}$ is shorthand for the set of all sequences that begin with a 1 and continue with a sequence in A_{n-1}, and $2A_{n-2}$ is defined similarly. Since the sizes of $1A_{n-1}$ and $2A_{n-2}$ are clearly a_{n-1} and a_{n-2}, it follows that $a_n = a_{n-1} + a_{n-2}$, which yields answer (iii).

If A_n is the set of legal bracketings with n pairs (example (iii)), then a typical legal bracketing can be written recursively as $[L_1]L_2$, where L_1 and L_2 are smaller (possibly empty) legal bracketings. For example, if the bracketing is $[[][]][[]][[[]]]$ then $L_1 = []\,[]$ and $L_2 = [[]][[][]]$. If L_1 has k pairs, then L_2 has $n - 1 - k$ pairs. It follows that A_n can be identified with the union $\bigcup_{k=0}^{n-1} A_k \times A_{n-1-k}$, and, taking cardinalities, $a_n = \sum_{k=0}^{n-1} a_k a_{n-1-k}$. This is a *nonlinear* (in fact, quadratic) and *nonlocal* recurrence, but it is nevertheless one that satisfies Wilf's dictum.

2.4 Generatingfunctionology

According to Wilf, who coined this neologism by making it the title of his classic book (a free download from his Web site, even though it is still in print!):

> A generating function is a clothesline on which we hang up a sequence of numbers for display.

The method of generating functions is one of the most useful tools of the trade of enumeration. The generating function of a sequence, sometimes called its *z-transform*, is a discrete analogue of the LAPLACE TRANSFORM [III.91], and indeed goes back to LAPLACE [VI.23] himself. If the sequence is $(a_n)_{n=0}^\infty$, then its generating function $f(x)$ is defined to be $\sum_{n=0}^\infty a_n x^n$. In other words, the terms of the sequence are regarded as the coefficients of a power series in x.

Generating functions are so useful because information about the sequence (a_n) translates to information about $f(x)$ that is often easier to process, and after some manipulations one often gets additional information about $f(x)$ that can be translated back into information about the sequence. For example, if $a_0 = a_1 = 1$ and $a_n = a_{n-1} + a_{n-2}$ when $n \geqslant 2$, then we can do the following manipulations on $f(x)$:

$$
\begin{aligned}
f(x) &= \sum_{n=0}^\infty a_n x^n = a_0 + a_1 x + \sum_{n=2}^\infty a_n x^n \\
&= 1 + x + \sum_{n=2}^\infty (a_{n-1} + a_{n-2}) x^n \\
&= 1 + x + \sum_{n=2}^\infty a_{n-1} x^n + \sum_{n=2}^\infty a_{n-2} x^n \\
&= 1 + x + x \sum_{n=2}^\infty a_{n-1} x^{n-1} + x^2 \sum_{n=2}^\infty a_{n-2} x^{n-2} \\
&= 1 + x + x(f(x) - 1) + x^2 f(x) \\
&= 1 + (x + x^2) f(x).
\end{aligned}
$$

It follows that

$$f(x) = \frac{1}{1 - x - x^2}.$$

If one performs a partial-fraction decomposition, and expands the two resulting terms in a Taylor series, then one can obtain answer (i) to example (iv).

3 Weight Enumeration

According to the modern approach, pioneered by Pólya, Tutte, and Schützenberger, generating functions are neither "generating," nor are they functions. Rather, they are *formal power series* that are *weight enumerators* of combinatorial sets. (Usually, but not always, these sets are infinite: for a finite set the corresponding "power series" has only finitely many nonzero terms and is therefore a polynomial.)

A power series $\sum_{n=0}^\infty a_n x^n$ is called *formal* when one sheds its analytical connotation as a Taylor series of a function, and thereby obviates the need to worry about convergence. For example, the sum $\sum_{n=0} n!^{n!} x^n$ is perfectly legal as a formal power series even though it converges only when $x = 0$.

As for weight enumerators, consider the following situation. Suppose that we want to study the age distribution of a finite population. One way of doing this is to ask 121 questions. For each i between 0 and 120, we ask those whose age is i to raise their hand. Then we count each of these age-groups one by one, compiling a table of a_i ($0 \leqslant i \leqslant 120$), and finally computing the generating function

$$f(x) = \sum_{i=0}^{120} a_i x^i.$$

But if the size of the population is much less than 120, it is much more efficient, because fewer questions would be needed, to ask every person their age and then to declare the *weight* of a person of age i to be x^i. Then the generating function is the sum of these weights. That is,

$$f(x) = \sum_{\text{persons}} x^{\text{age(person)}},$$

which is a natural extension of the caveman's formula of naive counting. Once we know $f(x)$ we can easily compute statistically interesting quantities, like the *average* and the *variance*, which work out to be $\mu = f'(1)/f(1)$ and $\sigma^2 = f''(1)/f(1) + \mu - \mu^2$, respectively.

The general scenario is that we have an *interesting* (finite or infinite) combinatorial set, let us call it A, and a certain numerical *attribute*, $\alpha : A \to \mathbb{N}$, which assigns to each element of A a natural number. (Here we allow 0 as a natural number.) Then the *weight enumerator* of A with respect to α is defined by the formula

$$f(x) = \sum_{a \in A} x^{\alpha(a)}.$$

We shall also use the notation $|A|_x$ for $f(x)$. Obviously, this equals

$$\sum_{n=0}^{\infty} a_n x^n,$$

where a_n is the number of members of A whose α equals n. Hence if we have some kind of explicit expression for $f(x)$, we immediately have an "explicit" expression for the actual sequence a_n assuming, that is, that one considers the operations needed to calculate the nth coefficient a_n of $f(x)$ as constituting an explicit expression for a_n. Even if one does not, then it is still often possible to get a "nice" formula for a_n, or, failing this, to extract the asymptotics.

The fundamental operations for naive counting also hold for *weighted counting*: just replace $|\cdot|$ by $|\cdot|_x$. For example,

$$|A \cup B|_x = |A|_x + |B|_x$$

(if $A \cap B = \varnothing$) and

$$|A \times B|_x = |A|_x \cdot |B|_x.$$

Let us quickly see why the second of these is true. If the members of A and B are endowed with numerical attributes α and β, respectively, and one defines an attribute γ on $A \times B$ by letting $\gamma(a, b)$ equal $\alpha(a) + \beta(b)$, then

$$
\begin{aligned}
|A \times B|_x &= \sum_{(a,b) \in A \times B} x^{\gamma(a,b)} \\
&= \sum_{(a,b) \in A \times B} x^{\alpha(a) + \beta(b)} \\
&= \sum_{(a,b) \in A \times B} x^{\alpha(a)} \cdot x^{\beta(b)} \\
&= \sum_{a \in A} \sum_{b \in B} x^{\alpha(a)} \cdot x^{\beta(b)} \\
&= \left(\sum_{a \in A} x^{\alpha(a)} \right) \cdot \left(\sum_{b \in B} \cdot x^{\beta(b)} \right) \\
&= |A|_x \cdot |B|_x.
\end{aligned}
$$

Let us see how these facts can be useful. First, consider the *infinite* set A, of all (finite) sequences of 1s and 2s, and let the attribute be "sum of entries." Then the weight of 1221 is x^6, and, in general, the weight of a sequence $(a_1 \cdots a_r)$ is $x^{a_1 + \cdots + a_k}$. The set A can be naturally decomposed as

$$A = \{\phi\} \cup 1A \cup 2A,$$

where ϕ is the empty word, and $1A$ is short for the set of all sequences obtained by prefixing a 1 to members of A, and analogously for $2A$. Applying $|\cdot|_x$, we get

$$|A|_x = 1 + x|A|_x + x^2|A|_x,$$

which, in this simple case, can be solved *explicitly*, to yield, once again

$$|A|_x = \frac{1}{1 - x - x^2}.$$

A legal bracketing L is either empty (in which case the weight is $x^0 = 1$), or else, as we have already noted, it can be written as $L = [L_1]L_2$, where L_1 and L_2 are (shorter) legal bracketings. Conversely, whenever L_1 and L_2 are legal bracketings, so is $[L_1]L_2$. Let $\mathcal{L}$ be the (infinite) set of *all* legal bracketings, and define the weight of a legal bracketing to be x^n, where n is the number of bracket pairs $[\ ]$. For example, the weight of $[\ ]$ is x and the weight of $[[\ ][[\ ][\ ]]]$ is x^5. The set $\mathcal{L}$ decomposes naturally as follows:

$$\mathcal{L} = \{\phi\} \cup ([\mathcal{L}] \times \mathcal{L}),$$

where ϕ denotes the empty word and $[\mathcal{L}] \times \mathcal{L}$ denotes the set of all words of the form $[L_1]L_2$ with L_1 and L_2 in $\mathcal{L}$. This leads to the *nonlinear* (in fact, quadratic) equation

$$|\mathcal{L}|_x = 1 + x|\mathcal{L}|_x^2,$$

which yields, thanks to the Babylonians, the explicit expression

$$|\mathcal{L}|_x = \frac{1 - \sqrt{1 - 4x}}{2x}.$$

This in turn gives us the answer to example (iii) above, via Newton's binomial theorem.

Legal bracketings are equivalent to so-called *binary trees*, that is, unlabeled ordered trees where every vertex has either no children or exactly two children. For instance, when we write the legal bracketing [[ ] [ ]] [[ ]] [[ ] [[ ]]] in the form [L_1]L_2 we can think of [[ ] [ ]] [[ ]] [[ ] [[ ]]] as the parent, with children $L_1 = $ [ ] [ ] and $L_2 = $ [[ ]] [[ ] [[ ]]]. Then L_1's children are ϕ and [ ], while L_2's are [ ] and [[ ] [[ ]]]. This process continues until we have reached ϕ down every branch of the family.

If we try to count *penta-trees* instead, where each vertex may only have exactly zero or five children, then the generating function, alias weight-enumerator, satisfies the quintic equation

$$f = x + f^5,$$

which, according to ABEL [VI.33] and GALOIS [VI.41], is not *solvable by radicals* (see THE INSOLUBILITY OF THE QUINTIC [V.21]). However, solvability by radicals is not everything. More than 200 years ago, LAGRANGE [VI.22] devised a beautiful and extremely useful formula for extracting the coefficients of the generating function from the equation it satisfies, now called the *Lagrange inversion formula*. Using it one can easily show that the number of complete k-ary trees with $(k-1)m+1$ leaves is

$$\frac{(km)!}{((k-1)m+1)!m!}.$$

A multivariate generalization of the Lagrange inversion formula, discovered by the great Bayesian probabilist I. J. Good, enables one to enumerate *colored* trees and many other extensions.

3.1 Enumeration Ansatzes

If one wants to turn enumerative combinatorics into a *theory* rather than a collection of solved problems, one needs to introduce *classification*, and *enumeration paradigms* for counting sequences. But since "paradigm" is such a pretentious word, let us use the much humbler German word "ansatz," which roughly means "form of solution."

Let $(a_n)_{n=0}^{\infty}$ be a sequence, and let

$$f(x) = \sum_{n=0}^{\infty} a_n x^n$$

be its generating function. If we know the "form" of a_n, we can often deduce the form of $f(x)$ (and vice versa).

(i) If a_n is a polynomial in n, then $f(x)$ has the form

$$f(x) = \frac{P(x)}{(1-x)^{d+1}},$$

where P is a polynomial function and d is the degree of the polynomial that describes a_n.

(ii) If a_n is a *quasi-polynomial* in n (i.e., there exists an integer N such that for each $r = 0, \ldots, N-1$, the function $m \mapsto a_{mN+r}$ is a polynomial in m), then, for some (finite) sequence of integers $d_1, d_2, \ldots$ and some polynomial function P,

$$f(x) = \frac{P(x)}{(1-x)^{d_1}(1-x^2)^{d_2}(1-x^3)^{d_3} \cdots}.$$

(iii) If a_n is *C-recursive*, that is, if it satisfies a linear recurrence equation with constant coefficients

$$a_n = c_1 a_{n-1} + c_2 a_{n-2} + \cdots + c_d a_{n-d}$$

(a good example is the Fibonacci sequence), then $f(x)$ is a *rational* function of x: that is, $f(x) = P(x)/Q(x)$, where P and Q are polynomials.

(iv) If a_n satisfies a linear recurrence equation of the form

$$c_0(n)a_n = c_1(n)a_{n-1} + c_2(n)a_{n-2}$$
$$+ \cdots + c_d(n)a_{n-d},$$

where the coefficients $c_i(n)$ are polynomial in n, then it is said to be *P-recursive*. (For example, $a_n = n!$ is P-recursive since we have the recurrence $a_n = na_{n-1}$.) If this is the case, then $f(x)$ is *D-finite*, which means that it satisfies a linear differential equation with polynomial coefficients (in x).

In the case of $a_n = n!$ the recurrence $a_n = na_{n-1}$ is *first order*. A natural example of a P-recursive sequence satisfying a higher-order linear recurrence with polynomial coefficients is the sequence that counts the number of involutions on $\{1, \ldots, n\}$. (An involution is a permutation that equals its inverse.) Let us call this number w_n. The sequence (w_n) satisfies the recurrence relation

$$w_n = w_{n-1} + (n-1)w_{n-2}.$$

This recurrence follows from the fact that in the permutation n belongs either to a 1-cycle or to a 2-cycle. The former case accounts for w_{n-1} of the involutions, and the latter for $(n-1)w_{n-2}$ of them. (There are $n-1$ ways of choosing the cycle-mate, i, say, of n, and deleting the resulting cycle leaves an involution of the $n-2$ elements $\{1, \ldots, i-1, i+1, \ldots, n-1\}$.)

4   Bijective Methods

This last argument was a simple example of a *bijective proof*, in this case, of a recurrence for the number of involutions on n objects. Contrast it with the following proof.

The number of involutions of $\{1, \ldots, n\}$ with exactly k 2-cycles is

$$\binom{n}{2k} \frac{(2k)!}{k!2^k},$$

because we must first choose the $2k$ elements that will participate in the k 2-cycles, and then match them up into (unordered) pairs, which can be done in

$$(2k-1)(2k-3)\cdots 1 = \frac{(2k)!}{k!2^k}$$

ways. Hence

$$w_n = \sum_k \binom{n}{2k} \frac{(2k)!}{k!2^k}.$$

Nowadays such sums can be handled completely *automatically*, and if one inputs this sum to the Maple package EKHAD (downloadable from my Web site), one would get the recurrence $w_n = w_{n-1} + (n-1)w_{n-2}$ as the output, together with a (completely rigorous!) proof. While the so-called Wilf–Zeilberger (WZ) method is able to handle many such problems, there are many other cases where one still needs a human proof. In either case such proofs involve (algebraic, and sometimes analytic) *manipulations*. The great combinatorialist Adriano Garsia derogatorily calls such proofs "manipulatorics," and *real enumerators do not manipulate*, or at least try to avoid it whenever possible. The preferred method of proof is by BIJECTION [I.2 §2.2].

Suppose one has to prove that $|A_n| = |B_n|$ for every n, where A_n and B_n are combinatorial families. The "ugly way" is to get, by some means or other, algebraic or analytic expressions for $a_n = |A_n|$ and $b_n = |B_n|$. Then one *manipulates* a_n, getting another expression a_n', which in turn leads to yet another expression a_n'', and if one is patient enough, and clever enough, and in luck, or if the problem is not too deep, one eventually arrives at b_n, and the result follows.

On the other hand, the *nice* way of proving that $|A_n| = |B_n|$ is by constructing a (preferably nice) *bijection* $T_n : A_n \to B_n$, which immediately implies, as a corollary, that $|A_n| = |B_n|$.

In addition to being more *aesthetically* pleasing, a bijective proof is also *philosophically* more satisfactory. In fact, the notion of (cardinal) *number* is a highly sophisticated *derived* notion based on the much more basic notion of *being in bijection*. Indeed, according to FREGE [VI.56], the cardinal numbers are *equivalence classes*, where the EQUIVALENCE RELATION [I.2 §2.3] is "is in bijective correspondence with." Saharon Shelah said that people have been exchanging objects, in a one-to-one way, since long before they started to count. Also, a bijective proof *explains why* the two sets are equinumerous, as opposed to just certifying the formal correctness of this fact.

For example, suppose that Noah had wanted to prove that there were as many male as female creatures in his Ark. One way of proving this would have been to count the males and count the females, and check that the two resulting numbers were indeed the same. But a much better, conceptual, proof would have been to note that there is an obvious one-to-one correspondence between the set M of males and the set F of females: the function $w : M \to F$ defined by $w(x) = \text{WifeOf}(x)$ is a bijection, with inverse $h : F \to M$ defined by $h(y) = \text{HusbandOf}(y)$.

A classic example of a bijective proof is Glaisher's proof of EULER's [VI.19] "odd equals distinct" partition theorem. A *partition* of an integer n is a way of writing it as a sum of positive integers, where order does not matter. For example, 6 has eleven partitions: 6, 51, 42, 411, 33, 321, 3111, 222, 2211, 21111, 111111. (Here 3111 is shorthand for the sum $3 + 1 + 1 + 1$, and so on. Since order does not matter, we count 3111 as the same partition of 6 as 1311, 1131, and 1113. It is convenient to write the partitions with their numbers in decreasing order, as we have done.)

A partition is called *odd* if all its parts are odd, and it is called *distinct* if all its parts are distinct. Let $\text{Odd}(n)$ and $\text{Dis}(n)$ be the sets of odd and distinct partitions of n, respectively. For example, $\text{Odd}(6) = \{51, 33, 3111, 111111\}$ and $\text{Dis}(6) = \{6, 51, 42, 321\}$. Euler proved that $|\text{Odd}(n)| = |\text{Dis}(n)|$ for all n. His "manipulatorics" proof goes as follows. Let $o(n)$ and $d(n)$ be the number of odd and distinct partitions of n, respectively, and let us define the *generating functions*

$$f(q) = \sum_{n=0}^{\infty} o(n)q^n \quad \text{and} \quad g(q) = \sum_{n=0}^{\infty} d(n)q^n.$$

With the help of the "multiplication principle" for weighted counting, Euler showed that

$$f(q) = \prod_{i=0}^{\infty} \frac{1}{1 - q^{2i+1}} \quad \text{and} \quad g(q) = \prod_{i=0}^{\infty} (1 + q^i).$$

Using the algebraic identity $1 + y = (1 - y^2)/(1 - y)$, we have

$$\prod_{i=0}^{\infty}(1 + q^i) = \prod_{i=0}^{\infty}\frac{1 - q^{2i}}{1 - q^i}$$

$$= \frac{\prod_{i=0}^{\infty}(1 - q^{2i})}{\prod_{i=0}^{\infty}(1 - q^{2i})\prod_{i=0}^{\infty}(1 - q^{2i+1})}$$

$$= \prod_{i=0}^{\infty}\frac{1}{1 - q^{2i+1}}.$$

Hence $g(q) = f(q)$, and the identity $o(n) = d(n)$ follows by extracting the coefficient of q^n.

For a very long time, these kinds of manipulation were considered to belong to the realm of *analysis*, and in order to justify the manipulations of the infinite series and products, one talked about the "region of convergence," usually $|q| < 1$, and every step had to be justified by the appropriate analytical theorem. Only relatively recently did people come to realize that no analysis need be involved: everything makes sense in the *completely elementary* and much more rigorous (from the philosophical viewpoint) algebra of *formal power series*. One still needs to worry about convergence, so as to exclude, for example, an infinite product like $\prod_{i=0}^{\infty}(1 + x)$, but the notion of convergence in the ring of formal power series is much more user-friendly than its analytical namesake.

Even though invoking analysis was a red herring, Euler's proof, while purely algebraic and elementary, is nevertheless still manipulatorics. It would be much nicer to find a direct bijection between the sets $\mathrm{Dis}(n)$ and $\mathrm{Odd}(n)$. Such a bijection was given by Glaisher. Given a distinct partition, write each of its parts as $2^r \cdot s$, where s is odd, and replace it by 2^r copies of s. (For example, $12 = 4 \cdot 3$, so we would replace 12 by $3 + 3 + 3 + 3$.) The output is obviously a partition of the same integer n, but now into odd parts. For example, the partition $(10, 5, 4)$ is transformed to the new partition $(5, 5, 5, 1, 1, 1, 1)$. To define the inverse transformation, take an odd part a and count how many times it shows up. If it shows up m times, then write m in binary notation, $m = 2^{s_1} + \cdots + 2^{s_k}$, and replace the m copies of a by the k parts: $2^{s_1}a, \ldots, 2^{s_k}a$. It is not hard to check that if you do the first transformation to a partition in $\mathrm{Dis}(n)$ and then do the second transformation, you get back to the partition you started with.

When we perform algebraic (and logical, and even analytical) manipulations, we are really rearranging and combining symbols, and hence we are doing combinatorics in disguise. In fact, *everything is combinatorics.*

All we need to do is to take the combinatorics out of the closet, and make it explicit. The plus sign turns into (disjoint) union, the multiplication sign becomes Cartesian product, and induction turns into recursion. But what about the combinatorial counterpart of the minus sign? In 1982, Garsia and Steven Milne filled this gap by producing an ingenious "involution principle" that enables one to translate the implication

$$a = b \quad \text{and} \quad c = d \quad \Rightarrow \quad a - c = b - d$$

into a bijective argument, in the sense that if $C \subset A$ and $D \subset B$, and there are natural bijections $f : A \to B$ and $g : C \to D$ establishing that $|A| = |B|$, and $|C| = |D|$, then it is possible to construct an explicit bijection between $A \backslash C$ and $B \backslash D$. Let us define it in terms of people. Suppose that in a certain village all the adults are married, with the result that there is a natural bijection from the set of married men to the set of married women, $m \mapsto \mathrm{WifeOf}(m)$, with its inverse $w \mapsto \mathrm{HusbandOf}(w)$. In addition, some of the people have extramarital affairs, but only one per person, and all within the village. There is a natural bijection from the set of cheating men to the set of cheating women, called $m \mapsto \mathrm{MistressOf}(m)$, with its inverse $w \mapsto \mathrm{LoverOf}(w)$. It follows that there are as many faithful men as there are faithful women. But how do we match them up? (One might imagine, for example, that each faithful man wants a faithful woman to go to church with him.)

Here is how it is done. A faithful man first asks his wife to come with him. If she is faithful, she agrees. If she is not, she has a lover, and that lover has a wife. So she tells her husband: "Sorry, hubby, I am going to the pub with my lover, but my lover's wife may be free." If this happens, then the man asks the wife of the lover of his wife to go with him, and if she is faithful, she agrees. If she is not he keeps asking the wife of the lover of the woman who has just rejected his proposal. Since the village is finite, he will eventually get to a faithful woman.

The reaction of the combinatorial enumeration community to the involution principle was mixed. On the one hand it had the universal appeal of a general principle, one that should be useful in many attempts to find bijective proofs of combinatorial identities. On the other hand, its universality is also a major drawback, since involution-principle proofs usually do not give any insight into the *specific* structures involved, and one feels a bit cheated. Such a proof answers the *letter* of the question, but it misses its *spirit*. Given a proof of this kind, one still hopes for a *really* natural, "involution-principle-free proof." This is the case, for instance, with the celebrated Rogers–Ramanujan identity, which states that the number of partitions of an integer into parts that leave remainder 1 or 4 when divided by 5 equals the number of partitions of that integer with the property that the difference between any two parts is at least 2. For example, if $n = 7$ the cardinalities of $\{61, 4111, 1111111\}$ and $\{7, 61, 52\}$ are the same. Garsia and Milne invented their notorious principle in order to give a Rogers–Ramanujan bijection, thereby winning a \$50 prize from George Andrews. However, finding a *really nice* bijective proof is still an open problem.

A quintessential example of a bijective proof is Prüfer's proof of CAYLEY's [VI.46] celebrated result that there are n^{n-2} labeled trees on n vertices (example (v) earlier). Recall that a labeled tree is a labeled connected simple graph without cycles. Every tree has at least two vertices with only one neighbor (these are called *leaves*). A certain mapping called the *Prüfer bijection* associates with every labeled tree T a vector of integers $(a_1, \ldots, a_{n-2})$, with $1 \leqslant a_i \leqslant n$ for each i. This vector is called its *Prüfer code*. Since there are n^{n-2} such vectors, Cayley's formula follows once we have defined the mapping $f :$ Trees $\rightarrow$ Codes and proved that it is indeed a bijection. This really needs four steps: defining f, defining its alleged inverse map g, and proving that $g \circ f$ and $f \circ g$ are the identity maps on their respective domains.

The mapping f is defined recursively as follows. If the tree has 2 vertices, then its code is the empty sequence. Otherwise, let a_1 be the (sole) neighbor of the smallest leaf and let $(a_2, \ldots, a_{n-2})$ be the code of the smaller tree obtained by deleting that leaf.

5 Exponential Generating Functions

So far, when we have discussed generating functions, we have been talking about *ordinary generating functions* (or OGFs). These are ideally suited for counting ordered structures like integer partitions, ordered trees, and words. But many combinatorial families are really *sets*, where the order is immaterial. For these the natural concept is that of an *exponential generating function* (or EGF).

The EGF of a sequence $\{a(n)\}_{n=0}^{\infty}$ is defined to be

$$\sum_{n=0}^{\infty} \frac{a(n)}{n!} x^n.$$

Labeled objects can be often viewed as sets of smaller *irreducible* objects. For example, a permutation is the

disjoint union of *cycles*, a set partition is the disjoint union of *nonempty sets*, a (labeled) forest is the disjoint union of *labeled trees*, and so on.

Suppose that we have two combinatorial families A and B, and suppose that there are $a(n)$ labeled objects of size n in the A family, and $b(n)$ in the B family. We can construct a new set of labeled objects $C = A \times B$, where the labels are disjoint and distinct, and define the size of a pair to be the sum of the sizes of the components. We have

$$c(n) = \sum_{k=0}^{n} \binom{n}{k} a(k) b(n-k),$$

since we must

(i) decide the size of the first component, k (an integer between 0 and n), which forces the size of the second component to be $n - k$,

(ii) decide which of the n labels go to the first component ($\binom{n}{k}$ ways), and

(iii) pick the objects for each component from the A and B families, respectively, using the available labels ($a(k) b(n-k)$ ways).

Multiplying both sides by $x^n / n!$ and summing from $n = 0$ to $n = \infty$ yields

$$\sum_{n=0}^{\infty} \frac{c(n)}{n!} x^n = \sum_{n=0}^{\infty} \sum_{k=0}^{n} \frac{a(k)}{k!} x^k \frac{b(n-k)}{(n-k)!} x^{n-k}$$

$$= \left(\sum_{k=0}^{\infty} \frac{a(k)}{k!} x^k \right) \left(\sum_{n-k=0}^{\infty} \frac{b(n-k)}{(n-k)!} x^{n-k} \right).$$

Hence $\mathrm{EGF}(C) = \mathrm{EGF}(A)\, \mathrm{EGF}(B)$. Iterating, we get

$$\mathrm{EGF}(A_1 \times A_2 \times \cdots \times A_k) = \mathrm{EGF}(A_1) \cdots \mathrm{EGF}(A_k).$$

In particular, if all the A_i are the same, we have that the EGF of ordered k-tuples, A^k, equals $[\mathrm{EGF}(A)]^k$. But if "order does not matter," then the EGF of k-sets of A-objects is $[\mathrm{EGF}(A)]^k / k!$, since there are exactly $k!$ ways of arranging a k-set into an ordered array (since all labels are distinct, all these objects are different). Summing from $k = 0$ to $k = \infty$ we get the "fundamental theorem of exponential generating functions."

If B is a labeled combinatorial family that can be viewed as sets of "connected components" that belong to a combinatorial family A, then

$$\mathrm{EGF}(B) = \exp[\mathrm{EGF}(A)].$$

This useful theorem was part of the physics folklore for many years, and was also implicit in many older combinatorial proofs. However, it was explicated only in the early 1970s. It was fully "categorized" by means of Joyal's theory of species, which grew to be a beautiful theory of enumeration in the hands of the *école Québecoise* (the Labelle and Bergeron *frères*, Leroux, and others).

Here are some venerable examples. Let us try to find the EGF of set partitions. That is, let us try to figure out an expression for

$$\sum_{n=0}^{\infty} \frac{b(n)}{n!} x^n,$$

where $b(n)$ (so-called Bell numbers) denotes the number of set partitions of an n-element set.

Recall that a *set partition* of a set A is a set of pairwise-disjoint *nonempty* subsets of A, $\{A_1, \ldots, A_r\}$, such that the union of all the A_i equals A. For example, the set partitions of the 2-element set $\{1, 2\}$ are $\{\{1\}, \{2\}\}$ and $\{\{1, 2\}\}$.

The atomic objects in this example are *nonempty sets*. (We think of a set A as being the "trivial" partition of itself into just one set.) Let $a(n)$ be the number of ways of partitioning a set of size n into one nonempty set. Clearly, when $n = 0$ this cannot be done, so $a(0) = 0$. When $n \geqslant 1$ there is exactly one way of doing it, so the EGF of the sequence $a(n)$ is

$$A(x) = 0 + \sum_{n=1}^{\infty} \frac{1}{n!} x^n = e^x - 1.$$

It follows immediately from the fundamental theorem that

$$\sum_{n=0}^{\infty} \frac{b(n)}{n!} x^n = e^{e^x - 1}, \tag{1}$$

an identity of Bell. Nowadays, with computer algebra systems, this can be used immediately to crank out the first 100 terms of the sequence $b(n)$. For example, in Maple one simply types

```
taylor(exp(exp(x)-1),x=0,101);
```

so this is definitely an answer in the Wilfian sense. We can also easily derive *recurrences* (albeit ones that need at least $O(n)$ memory), by differentiating both sides of (1) and comparing coefficients.

That was really easy, so let us go on and prove something much deeper. How about an EGF-style proof of Levi Ben Gerson's celebrated formula for the number of permutations on n objects, $n!$ (example (ii) earlier)? Every permutation can be decomposed into a disjoint union of cycles, so the atomic objects are now *cycles*. How many n-cycles are there? The answer is of course $(n - 1)!$, since $(a_1, a_2, \ldots, a_n)$ is the

same as $(a_2, a_3, \ldots, a_n, a_1)$, which is the same as $(a_3, \ldots, a_n, a_1, a_2)$, etc., which means that we can pick the first entry arbitrarily, after which we have $(n - 1)!$ choices for placing the remaining entries. The EGF for cycles is therefore

$$\sum_{n=1}^{\infty} \frac{(n-1)!}{n!} x^n = \sum_{n=1}^{\infty} \frac{1}{n} x^n$$
$$= -\log(1 - x) = \log(1 - x)^{-1}.$$

Using the fundamental theorem, we get that the EGF of permutations is

$$\exp(\log(1 - x)^{-1}) = (1 - x)^{-1} = \sum_{n=0}^{\infty} x^n = \sum_{n=0}^{\infty} \frac{n!}{n!} x^n,$$

and voilà we have a beautiful new proof that the number of permutations on n objects is $n!$.

This argument may not look very impressive. But a slight modification leads immediately to the (ordinary) generating function for the number of permutations on $\{1, \ldots, n\}$ with exactly k cycles, which we shall denote by $c(n, k)$. Here we are fixing n and letting k vary, so the generating function is $C_n(\alpha) = \sum_{k=0}^{n} c(n, k) \alpha^k$. All we have to do to calculate this is go from *naive* counting to *weighted* counting, and assign to each permutation the weight $\alpha^{\#\text{cycles}}$. The fundamental theorem of exponential generating functions carries over word-for-word to weighted counting. The weighted EGF for cycles is $\alpha \log(1 - x)^{-1}$, so the weighted EGF for permutations is

$$\exp(\alpha \cdot \log(1 - x)^{-1}) = (1 - x)^{-\alpha} = \sum_{n=0}^{\infty} \frac{(\alpha)_n}{n!} x^n,$$

where

$$(\alpha)_n = \alpha(\alpha + 1) \cdots (\alpha + n - 1)$$

is the so-called *rising factorial*. We have therefore derived the far less trivial result that the number of permutations of $\{1, \ldots, n\}$ with exactly k cycles equals the coefficient of α^k in $(\alpha)_n$.

About ten years ago (Ehrenpreis and Zeilberger 1994) I used this technique to give a combinatorial proof of the Pythagorean theorem in the form

$$\sin^2 z + \cos^2 z = 1.$$

The functions $\sin z$ and $\cos z$ are the weighted EGFs for *increasing sequences* of odd and even lengths, respectively, with weight $(-1)^{\lfloor \text{length}/2 \rfloor}$. Hence the left-hand side is the weighted EGF for ordered pairs of increasing sequences

$$a_1 < \cdots < a_k, \qquad b_1 < \cdots < b_r,$$

such that k and r have the same parity, the sets $\{a_1, \ldots, a_k\}$ and $\{b_1, \ldots, b_r\}$ are disjoint, and the union of the two sets is $\{1, 2, \ldots, k + r\}$. There is a killer involution on these sets of pairs defined as follows.

If $a_k < b_r$ then map the pair to

$$a_1 < \cdots < a_k < b_r, \qquad b_1 < \cdots < b_{r-1}.$$

and otherwise map it to

$$a_1 < \cdots < a_{k-1}, \qquad b_1 < \cdots < b_r < a_k.$$

For example, the pair

$$1, 3, 5, 6 \qquad 2, 4, 7, 8, 9, 10, 11, 12,$$

whose sign is $(-1)^2 \cdot (-1)^4 = 1$, goes to the pair

$$1, 3, 5, 6, 12 \qquad 2, 4, 7, 8, 9, 10, 11,$$

whose sign is $(-1)^2 \cdot (-1)^3 = -1$ (and vice versa).

Since this mapping changes the sign, and is an involution, all such pairs can be paired up into mutually canceling pairs. But this mapping is undefined for one special pair, namely the pair (empty, empty), whose weight is 1. Therefore, the EGF for the sum of the weights of all pairs is 1, which explains the right-hand side.

Yet another application of this method is a proof of André's generating function for the number of *up-down* permutations. A permutation of $a_1 \cdots a_n$ is called up–down (or sometimes *zigzag*) if $a_1 < a_2 > a_3 < a_4 > a_5 < \cdots$. Let a_n be the number of up–down permutations. Then

$$\sum_{n=0}^{\infty} \frac{a(n)}{n!} x^n = \sec x + \tan x.$$

This is equivalent to saying that

$$\cos x \cdot \left(\sum_{n=0}^{\infty} \frac{a(n)}{n!} x^n \right) = 1 + \sin x.$$

Can you find the appropriate set and the killer involution?

6 Pólya–Redfield Enumeration

Often in enumeration it is easy enough to count *labeled* objects, but what about unlabeled ones? For example, the number of labeled (simple) graphs on n vertices (example (vi)) is trivially $2^{n(n-1)/2}$, but how many unlabeled graphs are there on n vertices? This is much harder, and in general there are no "nice" answers, but the best known way is via a powerful technique initiated by Pólya, which was largely anticipated by Redfield. Pólya enumeration lends itself very efficiently to counting chemical isomers, since, for example, all the

carbon atoms "look the same." Indeed, counting isomers was Pólya's initial motivation (see MATHEMATICS AND CHEMISTRY [VII.1 §2.3]).

The main idea is to view *unlabeled* objects as equivalence classes of easy-to-count *labeled* objects, and to count these equivalence classes. But what is the equivalence? The answer is that there is always a SYMMETRY GROUP [I.3 §2.1] involved, and it leads to a natural equivalence relation. Let the symmetry group be G, and let the set of labeled objects be A. Then two objects a and b of A are regarded as *equivalent* if $b = g(a)$ for some member g of the group G. This means that there is some symmetry g in the group G that transforms a to b. This is easily seen to be an equivalence relation and the equivalence classes are the sets

$$\text{Orbit}(a) = \{g(a) \mid g \in G\}, \quad a \in A,$$

which are known as *orbits*. Calling each orbit a "family," we have the task of counting the number of families. Note that G is a subgroup of the group of permutations of the finite set A.

Suppose that there is a picnic consisting of many families and we want to count the number of families. One way would be to define some "canonical head" of each family, say "mother," and count the number of mothers. But some daughters look like mothers, so this is not so easy. On the other hand, you cannot just count everybody, since then you would count each family several times. The problem is that "naive" counting of people (or objects) is giving a credit of 1 to each person, and this is inappropriate if we are trying to count families. If instead we were to ask each person "How big is your family?" and add to our count the reciprocal of that number, then the calculation would come out just right, since a family of size k would get a credit of $1/k$ for each of its members, and would therefore have been counted exactly once by the end. Going back to counting orbits, we see by the same reasoning that their number is

$$\sum_{a \in A} \frac{1}{|\text{Orbit}(a)|}.$$

The conceptual opposite of "orbit of a" is the subgroup of members of G that fix a:

$$\text{Fix}(a) = \{g \in G \mid g(a) = a\}.$$

(This is sometimes known as the *stabilizer* of a.) To each element $b = ga$ in the orbit of a, we can associate the left coset $g\,\text{Fix}(a)$ of $\text{Fix}(a)$. This association turns out to be a well-defined one-to-one correspondence between the orbit of a and the cosets of $\text{Fix}(a)$

in G, from which it follows that the size of Orbit(a) is $|G/\text{Fix}(a)|$. We can therefore substitute $|\text{Fix}(a)|/|G|$ for $1/|\text{Orbit}(a)|$ in the previous formula, which implies that the number of orbits is

$$\frac{1}{|G|} \sum_{a \in A} |\text{Fix}(a)|.$$

Let us use the notation χ(statement) to stand for 1 if the statement is true and 0 if it is false. Then

$$\frac{1}{|G|} \sum_{a \in A} |\text{Fix}(a)| = \frac{1}{|G|} \sum_{a \in A} \sum_{g \in G} \chi(g(a) = a)$$

$$= \frac{1}{|G|} \sum_{g \in G} \sum_{a \in A} \chi(g(a) = a)$$

$$= \frac{1}{|G|} \sum_{g \in G} \text{fix}(g),$$

where $\text{fix}(g)$ is the number of fixed points of g (when g is viewed as a permutation of A). We have just proved what used to be called *Burnside's lemma*, but it goes back to CAUCHY [VI.29] and FROBENIUS [VI.58]. It states that the total number of orbits equals the average number of fixed points of g, over all transformations g in G. If the group G is the full symmetric group of all the permutations of A, then the average number of fixed points equals 1 (since in this trivial case there is only one orbit!).

Enter Pólya. The objects that he was interested in counting (e.g., chemical isomers, or colorings of the faces of the cube) were all naturally *functions* from an *underlying set* to a set of *colors* (or atoms). Let us call the underlying set U and the set of colors C. A symmetry of U gives rise in a natural way to a transformation of the set of functions $f : U \to C$. Given a function f one defines a new function gf by $g(f)(u) = f(g(u))$. (If we think of f as a coloring, then gf is the new coloring that assigns to u the color that f assigned to $g(u)$.) Now let us think about the number of fixed points of g in the set of C-colorings of U. Such a fixed point is a coloring f that equals gf: that is, $f(u) = f(gu)$ for every u. But then $f(u) = f(gu) = f(g^2u) = \cdots$, which means that, given any cycle of g, f must assign the same color to all members of that cycle. It follows that the number of fixed colorings of g is $c^{\#\text{cycles}(g)}$, where $c = |C|$ is the number of colors.

Applying Burnside's lemma, we may deduce that the number of different colorings of U (up to G-equivalence) is

$$\frac{1}{|G|} \sum_{g \in G} c^{\#\text{cycles}(g)},$$

since an equivalence class of colorings is simply an orbit of one of the colorings in that class.

Here is a simple application. How many necklaces (without a clasp) are there that consist of p beads (where p is a prime) and that use a different colors? The underlying set is $\{0, \ldots, p-1\}$, and the symmetry group is $\mathbb{Z}_p$, the cyclic group of order p. As usual, regard the elements of the symmetry group as permutations of the set of beads. Since p is a prime, there are $p - 1$ elements of $\mathbb{Z}_p$ with one cycle (of length p), and one element (the identity permutation) with p cycles (all of length 1). It follows that the number of necklaces is

$$\frac{1}{p}((p-1) \cdot a + 1 \cdot a^p) = a + \frac{a^p - a}{p}.$$

In particular, since this number is necessarily an integer, we get as a bonus a combinatorial proof of FERMAT'S LITTLE THEOREM [III.58]: that $a^p - a$ is always a multiple of p. Perhaps one day there will be an equally nice combinatorial proof of Fermat's *last* theorem. All one has to do is to prove that there is no bijection from the union of the set of straight necklaces of size n using x colors, and the set of such necklaces using y colors, to the set of necklaces using z colors (with $n > 2$, of course).

If one wants to keep track of how many beads there are of each color, one simply replaces straight counting by weighted counting, and $c^{\#\text{cycles}(g)}$ is replaced by

$$(x_1 + \cdots + x_c)^{\alpha_1} \cdot (x_1^2 + \cdots + x_c^2)^{\alpha_2} \cdots$$

(assuming that g has α_1 1-cycles, α_2 2-cycles, etc.). The resulting expression is the celebrated *cycle-index polynomial*.

6.1 The Principle of Inclusion–Exclusion and Möbius Inversion

Another pillar of enumeration is the principle of inclusion–exclusion (nicknamed PIE). Suppose that there are n sins, $s_1, \ldots, s_n$, that a person may succumb to, and suppose that for each set of sins S, A_S is the set of people who have all the sins in S (and possibly others). Then the number of good people (without sins) is

$$\sum_S (-1)^{|S|} |A_S|.$$

For example, if the set A is the set of all permutations π of $\{1, \ldots, n\}$ and the ith sin is having $\pi[i] = i$, then $|A_S| = (n - |S|)!$, and we get that the number of *derangements* (permutations without fixed points) is

$$\sum_{k=0}^{n} (-1)^k \binom{n}{k} (n-k)! = n! \sum_{k=0}^{n} (-1)^k \frac{1}{k!},$$

which yields the *answer*: "closest integer to $n!/e$." This is sometimes called the "umbrella problem": if on a rainy day n absent-minded people go to a party and leave an umbrella by the door, and if on their departure they each take a random umbrella, then the probability that nobody ends up with the right umbrella is about $1/e$.

The PIE is a special case of *Möbius inversion* on general partially ordered sets (posets) where the poset happens to be the Boolean lattice. This realization was published in a seminal paper by Rota (1964) and reprinted in his collected works. It is considered by many to be the big bang that started modern algebraic combinatorics. Möbius's original inversion formula is recovered when the partially ordered set is $\mathbb{N}$ and the partial order is divisibility.

A contemporary account of enumeration from the "algebraic" point of view can be found in a marvelous two-volume set by Stanley (2000), which I strongly recommend.

7   Algebraic Combinatorics

So far I have described one of the routes to algebraic combinatorics: abstraction and conceptualization of classical enumeration. The other route, "concretization of the abstract," is almost everywhere dense in mathematics, and cannot be described in a few pages. Let me quote from the preface of the excellent *New Perspectives in Algebraic Combinatorics* by Billera et al. (1999).

> Algebraic combinatorics involves the use of techniques from algebra, topology, and geometry in the solution of combinatorial problems, or the use of combinatorial methods to attack problems in these areas. Problems amenable to the methods of algebraic combinatorics arise in these or other areas of mathematics or from diverse parts of applied mathematics. Because of this interplay with many fields of mathematics, algebraic combinatorics is an area in which a wide variety of ideas and methods come together.

7.1   Tableaux

An interesting class of objects that initially came up in group representation theory, but that turned out to be useful in many other areas—such as, for example, the theory of algorithms—are *Young tableaux*. They were first used by Reverend Alfred Young to construct *explicit* bases for the IRREDUCIBLE REPRESENTATIONS [IV.9 §2] of the SYMMETRIC GROUP [III.68]. For any partition $\lambda = \lambda_1 \cdots \lambda_k$ of n, a Young tableau of shape λ is an array of k left-justified rows with λ_1 entries in the first row, λ_2 entries in the second row, and so on, such that every row and every column is increasing, and the set of entries is $\{1, 2, \ldots, n\}$. For example, there are two standard Young tableaux whose shape is 22,

$$\begin{array}{cc} 1 & 2 \\ 3 & 4 \end{array} \qquad \begin{array}{cc} 1 & 3 \\ 2 & 4 \end{array},$$

and three of shape 31,

$$\begin{array}{ccc} 1 & 2 & 3 \\ 4 & & \end{array} \qquad \begin{array}{ccc} 1 & 2 & 4 \\ 3 & & \end{array} \qquad \begin{array}{ccc} 1 & 3 & 4 \\ 2 & & \end{array}.$$

Let f_λ be the number of standard Young tableaux of shape λ. For example, for $n = 4$: $f_4 = 1$, $f_{31} = 3$, $f_{22} = 2$, $f_{211} = 3$, and $f_{1111} = 1$. The sum of the squares of these numbers is $1^2 + 3^2 + 2^2 + 3^2 + 1^2 = 24 = 4!$.

The number f_λ is the dimension of the irreducible representation parametrized by λ. It follows by a result in REPRESENTATION THEORY [IV.9] known as *Frobenius reciprocity* that the same is true for all n. In other words,

$$\sum_{\lambda \vdash n} f_\lambda^2 = n!,$$

a result known as the *Young–Frobenius identity*. A gorgeous *bijective* proof of this identity, which has many beautiful properties, was given by Gilbert Robinson and Craige Schensted and later extended by Donald Knuth, and is now known as the Robinson–Schensted–Knuth correspondence. It inputs a permutation $\pi = \pi_1 \pi_2 \cdots \pi_n$, and outputs a pair of Young tableaux of the same shape, thereby proving the identity.

Algebraic combinatorics is currently a very active field, and as mathematics is becoming more and more concrete, constructive, and algorithmic, there are going to be many more combinatorial structures discovered in all areas of mathematics (and science!) and this will guarantee that algebraic combinatorialists will stay very busy for a long time to come.

Further Reading

Billera, L. J., A. Bjorner, C. Greene, R. E. Simion, and R P. Stanley, eds. 1999. *New Perspectives in Algebraic Combinatorics*. Cambridge: Cambridge University Press.

Ehrenpreis, L., and D. Zeilberger. 1994. Two EZ proofs of $\sin^2 z + \cos^2 z = 1$. *American Mathematical Monthly* 101: 691.

Rota, G.-C. 1964. On the foundations of combinatorial theory. I. Theory of Möbius functions. *Zeitschrift für Wahrscheinlichkeitstheorie und Verwandte Gebiete* 2:340–68.

Stanley, R. P. 2000. *Enumerative Combinatorics*, volumes 1 and 2. Cambridge: Cambridge University Press.

IV.19 Extremal and Probabilistic Combinatorics

Noga Alon and Michael Krivelevich

1 Combinatorics: An Introduction

1.1 Examples

It is hard to give a rigorous definition of combinatorics. Instead, let us start with a few examples to illustrate what the area is about.

(i) In the course of an examination of friendship between children some fifty years ago, the Hungarian sociologist Sandor Szalai observed that among any group of about twenty children he checked he could always find four children any two of whom were friends, or else four children no two of whom were friends. Despite the temptation to try to draw sociological conclusions, Szalai realized that this might well be a mathematical phenomenon rather than a sociological one. Indeed, a brief discussion with the mathematicians Erdős, Turán, and Sós convinced him this was the case. If X is any set of size 18 or more, and R is some symmetric RELATION [I.2 §2.3] on X, then there is always a subset S of X of size 4 with the following property: either xRy for any two distinct elements x, y of S, or $xR y$ for no two distinct elements x, y of S. In this case, X is a set of children and R is the relation "is friends with." This mathematical fact is a special case of *Ramsey's theorem*, which was proved by the economist and mathematician Frank Plumpton Ramsey in 1930. Ramsey's theorem led to the development of Ramsey theory, a branch of *extremal combinatorics*, which will be discussed in the next section.

(ii) In 1916, Schur was studying FERMAT'S LAST THEOREM [V.10]. It is sometimes possible to prove that a Diophantine equation has no solutions by showing that it has no solutions mod p for some prime p. However, Schur proved that for every integer k and every sufficiently large prime p, there are three integers a, b, and c, none of them congruent to 0 mod p, such that $a^k + b^k$ is congruent to c^k. Although this is a result in number theory, it has a relatively simple and purely combinatorial proof, which is another example of the many applications of Ramsey theory.

(iii) When studying the number of real zeros of random polynomials, LITTLEWOOD [VI.79] and Offord investigated in 1943 the following problem. Let $z_1, z_2, \ldots, z_n$ be n not-necessarily-distinct complex numbers, each of modulus at least 1. One can form 2^n sums by taking some subset of these numbers and adding them together (with the convention that if one takes the empty set, then the sum is 0). Littlewood and Offord wanted to know how many of these sums there could conceivably be such that the difference between any two of them had modulus less than 1. When $n = 2$ the answer is easily seen to be at most 2. There are four sums: 0, z_1, z_2, and $z_1 + z_2$. You cannot choose both of the first two or both of the last two or you will have a difference of z_1, which has modulus at least 1. Kleitman and Katona proved that in general the maximum is $\binom{n}{\lfloor n/2 \rfloor}$. Notice that a simple construction proves that this maximum can be achieved. Indeed, let $z_1 = z_2 = \cdots = z_n$ and choose all sums of precisely $\lfloor n/2 \rfloor$ of them. There are $\binom{n}{\lfloor n/2 \rfloor}$ such sums and they are all equal. The proof that one cannot do better than this uses tools from another area of extremal combinatorics, where the basic objects studied are systems of finite sets.

(iv) Consider a school in which there are m teachers $T_1, T_2, \ldots, T_m$ and n classes $C_1, C_2, \ldots, C_n$. The teacher T_i has to teach the class C_j for a specified number p_{ij} of lessons. What is the minimum possible number of periods in a complete timetable? Let d_i denote the total number of lessons the teacher T_i has to teach, and let c_j denote the total number of lessons the class C_j has to be taught. Clearly, the number of periods required for a complete schedule is at least as big as any d_i or c_j, and thus at least as big as the maximum of all these numbers, which we denote by d. It turns out that this obvious lower bound of d is also an upper bound: it is always possible to fit all the lessons that need to be taught into d periods. This is a consequence of *König's theorem*, which is a basic result in graph theory. Suppose now that the situation is not so simple: for every teacher T_i and every class C_j there is some specified set of d periods in which the teaching has to take place. Can we always find a feasible timetable with these more complicated constraints? Recent breakthroughs from a subject known as *list coloring* of graphs imply that it is always possible.

(v) Given a map with several countries represented, how many colors do you need if you want to color the countries without giving any two adjacent countries the same color? Here we assume that each country forms a connected region in the plane. Of course, at least four colors may be necessary: think of Belgium, France, Germany, and Luxembourg, out of which any two have a common border. The FOUR-COLOR THEOREM

[V.12], proved by Appel and Haken in 1976, asserts that you never need *more* than four colors. The study of this problem led to numerous interesting questions and results about graph coloring.

(vi) Let S be an arbitrary subset of the two-dimensional lattice $\mathbb{Z}^2$. For any two finite subsets $A, B \subset \mathbb{Z}$ we can think of the Cartesian product $A \times B$ as a sort of "combinatorial rectangle." This set has size $|A|\,|B|$ (where $|X|$ denotes the size of a set X), and we can define an obvious notion of the *density* $d_S(A, B)$ of S in $A \times B$ by the formula $d_S(A, B) = |S \cap (A \times B)|/|A|\,|B|$, which measures what proportion of the elements of $A \times B$ belong to S. For each k, let $d(S, k)$ be the largest possible value of $d_S(A, B)$ if $|A| = |B| = k$. What can we say about $d(S, k)$ as k tends to infinity? One might guess that almost any behavior is possible, but, remarkably, basic results in extremal graph theory (about the so-called Turán numbers of complete bipartite graphs) imply that $d(S, k)$ must always tend to 0 or 1.

(vii) Suppose that n basketball teams compete in a tournament and any two teams play each other exactly once. The organizers wish to award k prizes at the end of the tournament. It would be embarrassing if there ended up being a team that had not won a prize despite beating all the teams that had won a prize. However, unlikely though it might sound, it is quite possible that this will be the case *whatever* k teams they choose, at least if n is large enough. To demonstrate this is easy if one uses the *probabilistic method*, which is one of the most powerful techniques in combinatorics. For any fixed k, and all sufficiently large n, if the results of all the games are chosen randomly (and uniformly and independently), then there is a very high probability that for any k teams there is another team that beats all of them. Probabilistic combinatorics, which is one of the most active areas in modern combinatorics, started with the realization that probabilistic reasoning often provides simple solutions to problems of this type, problems that are often very hard to solve in any other way.

(viii) If G is a finite group of n elements, and H is a subgroup of size k in G, then there are n/k left cosets and n/k right cosets of H. Is there always a set of n/k elements of G that contains a single representative of each right coset and a single representative of each left coset? Hall's theorem, a basic result in graph theory, implies that there is. In fact, if H' is another subgroup of size k in G, then there is always a set of n/k elements of G that contains a single representative of each right coset of H and a single representative of each left coset

of H'. This may sound like a result in group theory, but it is really a (simple) result in combinatorics.

1.2 Topics

The examples described above illustrate some of the main themes of combinatorics. The subject, sometimes also called *discrete mathematics*, is a branch of mathematics that focuses on the study of discrete objects (as opposed to continuous ones) and their properties. Although combinatorics is probably as old as the human ability to count, the field has experienced tremendous growth during the last fifty years and has matured into a thriving area with its own set of problems, approaches, and methodology.

The examples above suggest that combinatorics is a basic mathematical discipline that plays a crucial role in the development of many other mathematical areas. In this essay we discuss some of the main aspects of this modern field, focusing on extremal and probabilistic combinatorics. (An account of combinatorial problems with a rather different flavor can be found in ALGEBRAIC AND ENUMERATIVE COMBINATORICS [IV.18].) It is, of course, impossible to cover the area fully in such a short article. A detailed account of the subject can be found in Graham, Grötschel, and Lovász (1995). Our main intention is to give a glimpse of the topics, methods, and applications illustrated by representative examples. The topics we discuss include extremal graph theory, Ramsey theory, the extremal theory of set systems, combinatorial number theory, combinatorial geometry, random graphs, and probabilistic combinatorics. The methods applied in the area include combinatorial techniques, probabilistic methods, tools from linear algebra, spectral techniques, and topological methods. We also discuss the algorithmic aspects and some of the many fascinating open problems in the area.

2 Extremal Combinatorics

Extremal combinatorics deals with the problem of determining or estimating the maximum or minimum possible size of a collection of finite objects that satisfies certain requirements. Such problems are often related to other areas, including computer science, information theory, number theory, and geometry. This branch of combinatorics has developed spectacularly over the last few decades (see, for example, Bollobás (1978), Jukna (2001), and their many references).

2.1 Extremal Graph Theory

A GRAPH [III.34] is one of the very basic combinatorial structures. It consists of a set of points, called *vertices*, some of which are linked by *edges*. One can represent a graph visually by drawing the vertices as points in the plane and the edges as lines (or curves). However, formally a graph is more abstract: it is just a set together with a collection of pairs taken from the set. More precisely, it consists of a set V, called the *vertex set*, and a set E, called the *edge set*; the elements of E (the edges) are sets of the form $\{u, v\}$, where u and v are distinct elements of V. If $\{u, v\}$ is an edge, we say that u and v are *adjacent*. The *degree* $d(v)$ of a vertex v is the number of vertices adjacent to it.

Here are a number of simple definitions associated with graphs that have emerged as important. A *path* of length k from u to v in G is a sequence of distinct vertices $u = v_0, v_1, \ldots, v_k = v$, where v_i and v_{i+1} are adjacent for all $i < k$. If $v_0 = v_k$ (but all vertices v_i for $i < k$ are distinct), this is called a *cycle* of length k, and is usually denoted by C_k. A graph G is *connected* if for any two vertices u, v of G there is a path from u to v. A *complete graph* K_r is a graph with r vertices such that any two of them are adjacent. A *subgraph* of a graph G is a graph that contains some of the vertices of G and some of its edges. A *clique* in G is a set of vertices in G such that any two of them are adjacent. The maximum size of a clique in G is called the *clique number* of G. Similarly, an *independent set* in G is a set of vertices in G with *no* two of them adjacent, and the *independence number* of G is the maximum size of an independent set in it.

Extremal graph theory deals with quantitative connections between various parameters of a graph, such as its numbers of vertices and edges, its clique number, or its independence number. In many cases a certain optimization problem involving these parameters has to be solved (for example, determining how big one parameter can be if another one is at most some given size), and its optimal solutions are the *extremal graphs* for this problem. Many important optimization problems that do not explicitly mention graphs can be reformulated, using the definitions above, as problems about extremal graphs.

2.1.1 Graph Coloring

Let us return to the map-coloring example discussed in the introduction. To translate the problem into mathematics, we can describe the map-coloring problem in terms of a graph G, as follows. The vertices of G correspond to the countries on the map, and two vertices are connected by an edge in G if and only if the corresponding countries share a common border. It is not hard to show that one can draw such a graph in such a way that no two edges cross each other: such graphs are called *planar*. Conversely, any planar graph arises in this way. Therefore, our problem is equivalent to the following: if you want to color the vertices of a planar graph so that no two adjacent vertices receive the same color, then how many colors do you need? (One can make the problem yet more mathematical by removing the nonmathematical notion of color. For example, one can assign to each vertex a positive integer instead.) Such a coloring is called *proper*. In this language, the four-color theorem states that every planar graph can be properly colored with four colors.

Here is another example of a graph-coloring problem. Suppose we must schedule meetings of several parliament committees. We do not wish to have two committees meeting at the same time if some parliament member belongs to both, so how many sessions do we need?

Again we can model this situation by using a graph G. The vertices of G represent the committees, with two vertices adjacent if and only if the corresponding committees share a member. A *schedule* is a function f that assigns to each committee one of k time slots. More mathematically, we can think of it as just a function from V to the set $\{1, 2, \ldots, k\}$. Let us call a schedule *valid* if no two adjacent vertices are assigned the same number. This corresponds to no two committees being assigned the same time slot if they share a member. The question then becomes, "What is the minimal value of k for which a valid schedule exists?"

The answer is called the *chromatic number* of the graph G, denoted $\chi(G)$: it is the smallest number of colors in any proper coloring of G. Notice that a coloring of a graph G is proper if and only if for each color the set of vertices of that color is independent. Therefore, $\chi(G)$ can also be defined as the smallest number of independent sets into which it is possible to partition the vertices of G. A graph is called *k-colorable* if it admits a k-coloring, or, equivalently, if it can be partitioned into k independent sets. Thus, $\chi(G)$ is the minimum k for which G is k-colorable.

Two simple examples are in order. If G is a complete graph K_n on n vertices, then obviously in any coloring of G all vertices get distinct colors, and thus n colors are necessary. Of course, n colors are also sufficient, so $\chi(K_n) = n$. If G is a cycle C_{2n+1} on $2n + 1$ vertices, then easy parity arguments show that at least three colors are needed, and three colors are enough: color the vertices along the cycle alternately by colors 1 and 2, and then color the last vertex by color 3. Thus, $\chi(C_{2n+1}) = 3$.

It is not hard to prove that G is 2-colorable if and only if it does not contain a cycle of odd length. Graphs that are 2-colorable are usually called *bipartite*, since they split into two parts, with all the edges going from one part to the other. The easy characterization ends here, and no simple criterion equivalent to k-colorability is available for $k \geqslant 3$. This is related to the fact that for each fixed $k \geqslant 3$ the computational problem of deciding whether a given graph is k-colorable is NP-hard, a notion discussed in COMPUTATIONAL COMPLEXITY [IV.20].

Coloring is one of the most fundamental notions of graph theory, as a huge array of problems in this field and in related areas like computer science and operations research can be formulated in terms of graph coloring. Finding an optimal coloring of a graph is known to be a very hard task, both theoretically and practically.

There are two simple yet fundamental lower bounds on the chromatic number. First, as every color class in a proper coloring of a graph G forms an independent set, it cannot be bigger than the independence number of G, which is denoted by $\alpha(G)$. Therefore, at least $|V(G)|/\alpha(G)$ colors are necessary. Secondly, if G contains a clique of size k, then k colors are needed to color that clique alone, and thus $\chi(G) \geqslant k$. This implies that $\chi(G) \geqslant \omega(G)$, where $\omega(G)$ is the clique number of G.

What about upper bounds on the chromatic number? One of the simplest approaches to coloring a graph is to do it *greedily*: put the vertices in some order and color them one by one, assigning to each one the smallest positive integer that has not already been assigned to one of its neighbors. While the greedy algorithm can sometimes be very inefficient (for example, it can color bipartite graphs in an unbounded number of colors, even though two colors are sufficient), it often works quite well. Observe that when applying the greedy algorithm, a color given to a vertex v is at most one more than the number of the neighbors of v preceding it in the chosen order, and is thus at most $d(v) + 1$, where $d(v)$ is the degree of v in G. It follows that if $\Delta(G)$ is the maximum degree of G, then the greedy algorithm uses at most $\Delta(G) + 1$ colors. Therefore $\chi(G) \leqslant \Delta(G) + 1$. This bound is tight for complete graphs and odd cycles,

and, as shown by Brooks in 1941, those are the only cases: if G is a graph of maximum degree Δ, then $\chi(G) \leqslant \Delta$ unless G contains a clique $K_{\Delta+1}$, or $\Delta = 2$ and G contains an odd cycle.

It is also possible to color the *edges* of a graph, rather than the vertices. In this case a proper coloring is defined to be one where no two edges that meet at a vertex are given the same color. The *chromatic index* of G, denoted by $\chi'(G)$, is the minimum k for which G admits a proper edge-coloring with k colors. For example, if G is the complete graph K_{2n}, then $\chi'(G) = 2n-1$. This turns out to be equivalent to the fact that it is possible to organize a round-robin tournament with $2n$ teams and fit it into $2n - 1$ rounds: just ask the manager of a soccer league. It is also not hard to show that $\chi'(K_{2n-1}) = 2n - 1$. Since in any proper edge-coloring of G all edges of G that are incident to a vertex v get distinct colors, the chromatic index is obviously at least as big as the maximum degree. Equality holds for bipartite graphs, as proved by König in 1931, which implies the existence of a complete timetable using d periods in the problem of teachers and classes discussed in the introduction.

Remarkably, this trivial lower bound of $\chi'(G) \geqslant \Delta(G)$ is very close to the true behavior of $\chi'(G)$. A fundamental theorem of Vizing from 1964 states that $\chi'(G)$ is always equal either to the maximum degree $\Delta(G)$ or to $\Delta(G)+1$. Thus, the chromatic index of G is much easier to approximate than its chromatic number.

2.1.2   Excluded Subgraphs

If a graph G has n vertices and contains no triangle (that is, three vertices all joined to each other) then how many edges can it contain? If n is even, then you can split the vertex set into two equal parts A and B of size $n/2$ and join every vertex in A to every vertex in B. The resulting graph G contains no triangles and has $n^2/4$ edges. Moreover, adding another edge will automatically create a triangle (in fact, several triangles). But is this the densest possible triangle-free graph? A hundred years ago the answer was shown to be yes by Mantel. (A similar theorem holds when n is odd, but now A and B must have nearly equal sizes $(n + 1)/2$ and $(n - 1)/2$.)

Let us look at a more general problem, where the role of the triangle is played by an arbitrary graph. More precisely, let H be any graph, with m vertices, say, and when $n \geqslant m$ let us define $ex(n, H)$ to be the maximum possible number of edges in a graph with n vertices that does not contain H as a subgraph. (The notation "ex" stands for "exclude.") The function $\text{ex}(n, H)$ is usually called the Turán number of H, for reasons that will become clear, and finding good approximations for it has been a central problem in extremal graph theory.

What kind of examples of graphs that do not contain H can we think of? One observation that gets us started is that if H has chromatic number r, then it cannot be a subgraph of a graph G with chromatic number less than r. (Why not? Because a proper $(r - 1)$-coloring of G provides us with a proper $(r - 1)$-coloring of any subgraph of G.) So a promising approach is to look for a graph G with n vertices, chromatic number $r - 1$, and as many edges as possible. This is easy to find. Our constraint is that the vertices can be partitioned into $r - 1$ independent sets. Once we have done that, we may as well include all edges between those sets. The result is a *complete $(r - 1)$-partite* graph. A routine calculation shows that in order to maximize the number of edges, one should partition into sets that have sizes as nearly equal as possible. (For example, if $n = 10$ and $r = 4$, then we would partition into three sets of sizes 3, 3, and 4.)

The graph that satisfies this condition is called the *Turán graph* $T_{r-1}(n)$ and the number of edges it contains is denoted by $t_{r-1}(n)$. We have just argued that $\text{ex}(n, H) \geqslant t_{r-1}(n)$, which can be shown to be at least as big as $(1 - 1/(r - 1))\binom{n}{2}$.

Turán's contribution to this area was to give an exact solution, in 1941, for the most important case, when H is the complete graph K_r on r vertices. He proved that $\text{ex}(n, K_r)$ is not just at least $t_{r-1}(n)$, but is actually equal to $t_{r-1}(n)$. Moreover, the only K_r-free graph with n vertices and $\text{ex}(n, K_r)$ edges is the Turán graph $T_{r-1}(n)$. Turán's paper is generally considered the starting point of extremal graph theory.

Later, Erdős, Stone, and Simonovits extended Turán's theorem by proving that the above simple lower bound for $\text{ex}(n, H)$ is *asymptotically* tight for any fixed H with chromatic number at least 3. That is, if r is the chromatic number of H, then the ratio of $\text{ex}(n, H)$ to $t_{r-1}(n)$ tends to 1 as n tends to infinity.

Thus, the function $\text{ex}(n, H)$ is well-understood for all nonbipartite graphs. Bipartite graphs are rather different, because their Turán numbers are much smaller: if H is bipartite, then $\text{ex}(n, H)/n^2$ tends to zero. Determining the asymptotics of $\text{ex}(n, H)$ in this case remains a challenging open problem with many unsettled questions. Indeed, the full story is unknown even for the very simple case when H is a cycle. Partial results

obtained so far use a variety of techniques from different fields, including probability theory, number theory, and algebraic geometry.

2.1.3   Matchings and Cycles

Let G be a graph. A *matching* in G is a collection of edges in G of which no two share a vertex. A matching M in G is called *perfect* if every vertex belongs to one of the edges in M. (The idea is that the edges determine a "match" for each vertex: the match for x is the vertex y for which xy is an edge of M.) Of course, for G to have a perfect matching it must have an even number of vertices.

One of the best-known theorems in graph theory is Hall's theorem, which provides a necessary and sufficient condition for the existence of a perfect matching in a bipartite graph. What kind of condition can this be? It is very easy to write down a trivial *necessary* condition, as follows. Let G be a bipartite graph with vertex sets A and B of equal size. (If they do not have equal size, then clearly there is no perfect matching.) Given any subset S of A, let $N(S)$ denote the set of all vertices in B that are joined to at least one vertex in S. If there is to be a matching, then it must be possible to assign to each vertex in S a distinct "match," so obviously $N(S)$ must have at least as many elements as S. Hall's theorem, proved in 1935, asserts that, remarkably, this obvious necessary condition is also sufficient. That is, if $N(S)$ is always at least as big as S, then there will be a perfect matching. More generally, if A is smaller than B, then the same condition guarantees that one can find a matching that includes every vertex in A (but leaves some vertices in B unmatched).

There is a useful reformulation of Hall's theorem in terms of set systems. Let $S_1, S_2, \ldots, S_n$ be a collection of sets, and suppose that we would like to find a system of *distinct representatives*: that is, a sequence $x_1, x_2, \ldots, x_n$ such that x_i is an element of S_i and no two of the x_i are the same. Obviously this cannot be done if the union of some k of the sets S_i has size less than k. Again, this obvious necessary condition is sufficient. It is not hard to show that this assertion is equivalent to Hall's theorem: let S be the union of the S_i and define a bipartite graph with vertex sets $\{1, 2, \ldots, n\}$ and S, joining i to x if and only if $x \in S_i$. Then a matching that includes all of the set $\{1, 2, \ldots, n\}$ picks out a system of distinct representatives: x_i is the element of S that is matched with i.

Hall's theorem can be applied to solve the problem of finding a system of representatives for the right and left cosets of a subgroup H, mentioned in section 1.1. Define a bipartite graph F, whose two sides (of size n/k each) are the left and right cosets of H. A left coset $g_1 H$ is connected by an edge of F to a right coset $H g_2$ if they share a common element. It is not difficult to show that F satisfies the Hall condition, and hence it has a perfect matching M. Choosing for each edge $(g_i H, H g_j)$ of M a common element of $g_i H$ and $H g_j$, we obtain the required family of representatives.

There is also a necessary and sufficient condition for the existence of a perfect matching in a general (not-necessarily-bipartite) graph G. This is a theorem of Tutte, which we shall not state here.

Recall that C_k denotes a cycle of length k. A cycle is a very basic graph structure, and, as one might expect, there are many extremal results concerning cycles.

Suppose that G is a connected graph with no cycles. If you pick a vertex and look at its neighbors and then the neighbors of its neighbors, and so on, you will see that it has a tree-like structure. Indeed, such graphs are called *trees*. An easy exercise shows that any tree with n vertices has exactly $n - 1$ edges. It follows that every graph G on n vertices with at least n edges has a cycle. If you want to guarantee that this cycle has certain extra properties, then you may need more edges. For example, the theorem of Mantel mentioned earlier implies that a graph G with n vertices and more than $n^2/4$ edges contains a triangle $C_3 = K_3$. One can also prove that a graph $G = (V, E)$ with $|E| > \frac{1}{2}k(|V| - 1)$ has a cycle of length longer than k (and this is in fact a sharp result).

A *Hamilton cycle* in a graph G is a cycle that visits every vertex of G. This term originated in a game, invented by HAMILTON [VI.37] in 1857, the objective of which was to complete a Hamilton cycle in the graph of the dodecahedron. A graph containing a Hamilton cycle is called *Hamiltonian*. This concept is strongly related to the well-known TRAVELING SALESMAN PROBLEM [VII.5 §2]: you are given a graph with positive weights assigned to the edges, and you must find a Hamilton cycle for which the sum of the weights of its edges is minimized. There are many sufficient criteria for a graph to be Hamiltonian, quite a few of which are based on the sequence of degrees. For example, Dirac proved in 1952 that a graph on $n \geqslant 3$ vertices all of whose degrees are at least $n/2$ is Hamiltonian.

2.2 Ramsey Theory

Ramsey theory is a systematic study of the following general phenomenon. Surprisingly often, a large struc-ture of a certain kind has to contain a fairly large highly organized substructure, even if the structure itself is completely arbitrary and apparently chaotic. As succinctly put by the mathematician T. S. Motzkin, "Complete disorder is impossible." One might expect that the simple and very general form of this paradigm ensures that it has many diverse manifestations in different mathematical areas, and this is indeed the case. (One should, however, bear in mind that some natural statements of this kind are false for nonobvious reasons.)

A very simple statement, which can be regarded as a basic prototype for what follows, is the *pigeonhole principle*. This states that if a set X of n objects is colored with s colors, then there must be a subset of X of size at least n/s that uses just one color. Such a subset is called *monochromatic*.

The situation becomes more interesting if the set X has some additional structure. It then becomes natural to ask for a monochromatic subset that keeps some of the structure of X. However, it also becomes much less obvious whether such a subset exists. Ramsey theory consists of problems and theorems of this general kind. Although several Ramsey-type theorems had appeared before, Ramsey theory is traditionally regarded as having started with *Ramsey's theorem*, proved in 1930. Ramsey took as his set X the set of all the edges in a complete graph, and the monochromatic subset he obtained consisted of all the edges of some complete subgraph. A precise statement of his theorem is as follows. Let k and l be integers greater than 1. Then there exists an integer n such that, however you color the edges of the complete graph with n vertices, using the two colors red and blue, there will either be k vertices such that all edges between them are red or l vertices such that all edges between them are blue. That is, a sufficiently large complete graph colored with two colors contains a largish complete subgraph that is monochromatic. Let $R(k, l)$ denote the minimum number n with this property. In this language, the observation of Szalai, mentioned in the introduction, is that $R(4, 4) \leqslant 20$ (in fact, $R(4, 4) = 18$). Actually, Ramsey's theorem was more general, in that he allowed any number of colors, and the objects colored could be r-tuples of elements rather than just pairs, as one has when coloring graphs. The exact computation of small Ramsey numbers turns out to be a notoriously difficult task: even the value of $R(5, 5)$ is unknown at present.

The second cornerstone of Ramsey theory was laid by Erdős and Szekeres, who in 1935 wrote a paper containing several important Ramsey-type results. In particular, they proved the recursion $R(k,l) \leqslant R(k-1,l) + R(k,l-1)$. Combined with the easy boundary conditions $R(2,l) = l$, $R(k,2) = k$, the recursion leads to the estimate $R(k,l) \leqslant \binom{k+l-2}{k-1}$. In particular, for the so-called diagonal case $k = l$ we obtain $R(k,k) < 4^k$. Remarkably, no improvement in the exponent of the latter estimate has been found so far. That is, nobody has found an upper bound of the form C^k for some $C < 4$. The best lower bound known, which we shall discuss in section 3.2, is roughly $R(k,k) \geqslant 2^{k/2}$, so there is a rather substantial gap.

Another Ramsey-type statement, proved by Erdős and Szekeres, is of a geometric nature. They showed that for every $n \geqslant 3$ there exists a positive integer N such that, given any configuration of N points in the plane in general position (i.e., no three of them are on a line), there are n that form a convex n-gon. (It is instructive to prove that if $n = 4$ then N can be taken to be 5.) There are several proofs of this theorem, some using the general Ramsey theorem. It is conjectured that the smallest value of N that will do in order to ensure a convex n-gon is $2^{n-2} + 1$.

The classic Erdős–Szekeres paper also contains the following Ramsey-type result: any sequence of $n^2 + 1$ distinct numbers contains a monotone (increasing or decreasing) subsequence of length $n+1$.

This provides a quick lower bound of $\sqrt{n}$ for a well-known problem of Ulam, asking for the typical length of a longest increasing subsequence of a random sequence of length n. A detailed description of the distribution of this length has recently been given by Baik, Deift, and Johansson.

In 1927 van der Waerden proved what became known as van der Waerden's theorem: for all positive integers k and r there exists an integer W such that for every coloring of the set of integers $\{1, 2, \ldots, W\}$ using r colors, one of the colors contains an arithmetic progression of length k. The minimum W for which this is true is denoted by $W(k,r)$. Van der Waerden's bounds for $W(k,r)$ are enormous: they grow like an Ackermann-type function. A new proof of his theorem was found by Shelah in 1987, and yet another proof was given by Gowers in 2000, while he was studying the (much deeper) "density version" of the theorem, which will be described in section 2.4. These recent proofs provided improved upper bounds for $W(k,r)$, but the best-known lower bound for this number, which is only exponential in k for each fixed r, is much smaller.

Even before van der Waerden, Schur proved in 1916 that for any positive integer r there exists an integer $S(r)$ such that for every r-coloring of $\{1, \ldots, S(r)\}$ one of the colors contains a solution of the equation $x + y = z$. The proof can be derived rather easily from the general Ramsey theorem. Schur applied this statement to prove the following result, mentioned in section 1.1: for every k and all sufficiently large primes p, the equation $a^k + b^k = c^k$ has a nontrivial solution in the integers modulo p. To prove this result, assume that $p \geqslant S(k)$ and consider the FIELD [I.3 §2.2] $\mathbb{Z}_p$ of integers mod p. The nonzero elements of $\mathbb{Z}_p$ form a GROUP [I.3 §2.1] under multiplication. Let H be the subgroup of this group consisting of all kth powers: that is, $H = \{x^k : x \in \mathbb{Z}_p^*\}$. It is not hard to show that the index r of H is the highest common factor of k and $p - 1$, and in particular is at most k. The partition of $\mathbb{Z}_p^*$ into the cosets of H can be thought of as an r-coloring of $\mathbb{Z}_p^*$. By Schur's theorem there exist $x, y, z \in \{1, \ldots, p-1\}$ that all have the same color—that is, they all belong to the same coset of H. In other words, there exists a residue $d \in \mathbb{Z}_p^*$ such that $x = da^k$, $y = db^k$, $z = dc^k$, and $da^k + db^k = dc^k$ modulo p. The desired result follows if we multiply both sides by d^{-1}.

Many additional Ramsey-type results can be found in Graham, Rothschild, and Spencer (1990) or in Graham, Grötschel, and Lovász (1995, chapter 25).

2.3 Extremal Theory of Set Systems

Graphs are one of the fundamental structures studied by combinatorialists, but there are others too. An important branch of the subject is the study of *set systems*. Most often, these are simply collections of subsets of some n-element set. For example, the collection of all subsets of the set $\{1, 2, \ldots, n\}$ of size at most $n/3$ is a good example of a set system. An extremal problem in this area is any problem where the aim is to determine, or estimate, the maximum number of sets there can be in a set system that satisfies certain conditions. For example, one of the first results in the area was proved by Sperner in 1928. He looked at the following question: how large a collection of subsets can one choose from an n-element set in such a way that no set from the collection is a subset of any other? A simple example of a set system satisfying this condition is the collection of all sets of size r, for some r. From this it immediately follows that we can obtain a collection as large as the largest binomial coefficient, which is $\binom{n}{n/2}$ if n is even and $\binom{n}{(n+1)/2}$ if n is odd.

Sperner showed that this is indeed the maximum possible size of such a collection. This result supplies a quick solution to the real analogue of the problem of Littlewood and Offord described in section 1.1. Suppose that $x_1, x_2, \ldots, x_n$ are n not-necessarily-distinct real numbers, each of modulus at least 1. A first observation is that we may assume that all the x_i are positive, since if we replace a negative x_i by $-x_i$ (which is positive), then we end up with exactly the same set of sums, but shifted by $-x_i$. (To see this, compare a sum that used to involve x_i with the corresponding sum that does not involve $-x_i$, and vice versa.) But now, if A is a proper subset of B, then some x_i belongs to B and not to A, so

$$\sum_{i \in B} x_i - \sum_{i \in A} x_i \geqslant x_i \geqslant 1.$$

Therefore, the total number of subset sums you can find with any two differing by less than 1 is at most $\binom{n}{\lfloor n/2 \rfloor}$, by Sperner's theorem.

A set system is called an *intersecting family* if any two sets in the system intersect. Since a set and its complement cannot both belong to an intersecting family of subsets of $\{1, 2, \ldots, n\}$, we see immediately that such a family can have size at most 2^{n-1}. Moreover, this bound is achieved by, for example, the collection of all sets that contain the element 1. But what happens if we fix a k and assume in addition that all our sets have size k? We may assume that $n \geqslant 2k$, as otherwise the solution is trivial. Erdős, Ko, and Rado proved that the maximum is $\binom{n-1}{k-1}$. Here is a beautiful proof discovered later by Katona. Suppose you arrange the elements randomly around a circle. Then there are n ways of choosing k elements that are consecutive in this arrangement, and it is quite easy to convince yourself that at most k of these can intersect (if $n \geqslant 2k$). So out of these n sets of size k, only k of them can belong to any given intersecting family. Now it is also easy to show that every set has an equal chance of being one of these n sets, and this proves (by a simple double-counting argument) that the largest possible proportion of sets in the family is k/n. Therefore, the family itself has size at most $(k/n)\binom{n}{k}$, which equals $\binom{n-1}{k-1}$. The original proof of Erdős, Ko, and Rado is more complicated than this, but it is important because it introduced a technique known as *compression*, which was used to solve many other extremal problems.

Let $n > 2k$ be two positive integers. Suppose that you wish to color all subsets of the set $\{1, 2, \ldots, n\}$ of size k in such a way that any two sets with the same color intersect each other. What is the smallest number of

colors you can use? It is not difficult to see that $n - 2k + 2$ colors suffice. Indeed, one color class can be the family of all subsets of $\{1, 2, \ldots, 2k - 1\}$, which is clearly an intersecting family. And then, for each i such that $2k \leqslant i \leqslant n$, you can take the family of all subsets whose largest element is i. There are $n - 2k + 1$ such families, and any set of size k belongs either to one of them or to the first family. Therefore, $n - 2k + 2$ colors are enough.

Kneser conjectured in 1955 that this bound was tight: in other words, that if you have fewer than $n - 2k + 2$ colors then you will have to give the same color to some pair of disjoint sets. This conjecture was proved by Lovász in 1978. His proof is topological, and uses the Borsuk–Ulam theorem. Several simpler proofs have been found since, but they are all based on the topological idea in the first proof. Since Lovász's breakthrough, topological arguments have become an important part of the armory of researchers in combinatorics.

2.4   Combinatorial Number Theory

Number theory is one of the oldest branches of mathematics. At its core are problems about integers, but a sophisticated array of techniques has been developed to deal with those problems, and these techniques have often themselves been the basis for further study (see, for example, ALGEBRAIC NUMBERS [IV.1], ANALYTIC NUMBER THEORY [IV.2], and ARITHMETIC GEOMETRY [IV.5]). However, some problems in number theory have yielded to the methods of combinatorics. Some of these problems are extremal problems with a combinatorial flavor, while others are classical problems in number theory where the existence of a combinatorial solution has been quite surprising. We describe below a few examples. Many more can be found in chapter 20 of Graham, Grötschel, and Lovász (1995), in Nathanson (1996), and in Tao and Vu (2006).

A simple but important notion in the area is that of a *sumset*. If A and B are two sets of integers, or more generally are two subsets of an ABELIAN GROUP [I.3 §2.1], then the sumset $A + B$ is defined to be $\{a + b : a \in A, b \in B\}$. For instance, if $A = \{1, 3\}$ and $B = \{5, 6, 12\}$, then $A + B = \{6, 7, 8, 9, 13, 15\}$. There are many results relating the size and structure of $A + B$ to those of A and B. For example, the *Cauchy–Davenport theorem*, which has numerous applications in additive number theory, is the statement that if p is a prime, and A, B are two nonempty subsets of $\mathbb{Z}_p$, then the size of $A + B$ is at least the minimum of p and $|A| + |B| - 1$. (Equality occurs if A and B are arithmetic progressions with the same common difference.) CAUCHY [VI.29] proved this theorem in 1813, and applied it to give a new proof of a lemma that LAGRANGE [VI.22] had proved as part of his well-known 1770 paper that shows that every positive integer is a sum of four squares. Davenport formulated the theorem as a discrete analogue of a related conjecture of Khinchin about densities of sums of sequences of integers. The proofs given by Cauchy and by Davenport are combinatorial, but there is also a more recent algebraic proof, based on some properties of roots of polynomials. The advantage of the latter is that it provides many variants that do not seem to follow from the combinatorial approach. For example, let us define $A \oplus B$ to be the set of all $a + b$ such that $a \in A$, $b \in B$, and $a \neq b$. Then the smallest possible size of $A \oplus B$, given the sizes of A and B, is the minimum of p and $|A| + |B| - 2$. Further extensions can be found in Nathanson (1996) and in Tao and Vu (2006).

The theorem of van der Waerden mentioned in section 2.2 implies that, however you color the positive integers with some finite number r of colors, there must be some color that contains arithmetic progressions of every length. Erdős and Turán conjectured in 1936 that this always holds for the "most popular" color class. More precisely, they conjectured that for any positive integer k and for any real number $\epsilon > 0$, there is a positive integer n_0 such that if $n > n_0$, any set of at least ϵn positive integers between 1 and n contains a k-term arithmetic progression. (Setting $\epsilon = r^{-1}$ one can easily deduce van der Waerden's theorem from this.) After several partial results, this conjecture was proved by Szemerédi in 1975. His deep proof is combinatorial, and applies techniques from Ramsey theory and extremal graph theory. Furstenberg gave another proof in 1977, based on techniques of ERGODIC THEORY [V.9]. In 2000 Gowers gave a new proof, combining combinatorial arguments with tools from analytic number theory. This proof supplied a much better quantitative estimate. A related very recent spectacular result of Green and Tao asserts that there are arbitrarily long arithmetic progressions of prime numbers. Their proof combines number-theoretic techniques with the ergodic theory approach. Erdős conjectured that any infinite sequence n_i for which the sum $\sum_i (1/n_i)$ diverges contains arbitrarily long arithmetic progressions. This conjecture would imply the theorem of Green and Tao.

2.5   Discrete Geometry

Let P be a set of points and let L be a set of lines in the plane. Let us define an *incidence* to be a pair (p, ℓ),

where p is a point in P, ℓ is a line in L, and the point p lies on the line ℓ. Suppose that P contains m distinct points and L contains n distinct lines. How many incidences can there be? This is a geometrical problem, but again it has a strong flavor of extremal combinatorics. As such, it is typical of the area known as *discrete* (or *combinatorial*) geometry.

Let us write $I(m, n)$ for the maximum number of incidences there can be between m points and n lines. Szemerédi and Trotter determined the asymptotic behavior of this quantity, up to a constant factor, for all possible values of m and n. There are two absolute positive constants c_1, c_2 such that, for all m, n,

$$c_1(m^{2/3}n^{2/3} + m + n) \leqslant I(m, n)$$
$$\leqslant c_2(m^{2/3}n^{2/3} + m + n).$$

If $m > n^2$ or $n > m^2$ then one can establish the lower bound by taking all m points on a single line, or all n lines through a single point, respectively. In the harder cases when m and n are closer to each other, one can prove it by letting P contain all the points of a $\lfloor \sqrt{m} \rfloor$ by $\lfloor \sqrt{m} \rfloor$ grid, and by taking the n most "popular" lines: that is, the n lines that contain the most points of P. Establishing the upper bound is more difficult. The most elegant proof of it is due to Székely, and is based on the fact that, however you draw a graph with m vertices and more than $4m$ edges, you must have many pairs of edges that cross each other. (This is a rather simple consequence of the famous Euler formula connecting the numbers of vertices, edges, and regions in any drawing of a planar graph.) To bound the number of incidences between a set of points P and a set of lines L in the plane, one considers the graph whose vertices are the points P, and whose edges are all segments between consecutive points along a line in L. The desired bound is obtained by observing that the number of crossings in this graph does not exceed the number of pairs of lines in L, and yet should be large if there are many incidences.

Similar ideas can be used to give a partial answer to the following question: if you take n points in the plane, how many pairs (x, y) of these points can there be with the distance from x to y equal to 1? It is not surprising that the two problems are related: the number of such pairs is the number of incidences between the given n points and the n unit circles that are centered at these points. Here, however, there is a large gap between the best known upper bound, which is $cn^{4/3}$ for some absolute constant c, and the best known lower bound, which is only $n^{1+c'/\log\log n}$ for some constant $c' > 0$.

A fundamental theorem of Helly asserts that if you have a finite family $\mathcal{F}$ of at least $d + 1$ convex sets in $\mathbb{R}^d$, and if any $d + 1$ of them have a point in common, then all sets in the family have a common point. Now let us start with a weaker assumption: given any p of the sets, some $d + 1$ of those p sets have a point in common. (Here p is some integer greater than $d + 1$.) Can one then find a set X of at most C points such that each set in $\mathcal{F}$ contains a point in X, with C a constant that depends on p but not on the number of convex sets in the family? This question was raised by Hadwiger and Debrunner in 1957 and solved by Kleitman and Alon in 1992. The proof combines a "fractional version" of Helly's theorem with the duality of LINEAR PROGRAMMING [III.84] and various additional geometric results. Unfortunately, it gives a very poor estimate for C: even in two dimensions and with $p = 4$ it is not known what the best possible value of C is.

This is just a small sample of problems and results in discrete geometry. Such results have been applied extensively in computational geometry and in combinatorial optimization in recent decades. Two good books on the subject are Pach and Agarwal (1995) and Matoušek (2002).

2.6   Tools

Many of the basic results in extremal combinatorics were obtained mainly by ingenuity and detailed reasoning. However, the subject has grown out of this early stage: several deep tools have been developed that have been essential to much of the recent progress in the area. In this subsection, we include a very brief description of some of these tools.

Szemerédi's regularity lemma is a result in graph theory that has numerous applications in various areas, including combinatorial number theory, computational complexity, and, mainly, extremal graph theory. The precise statement of the lemma, which can be found, for example, in Bollobás (1978), is somewhat technical. The rough statement is that the vertex set of any large graph can be partitioned into a constant number of pieces of nearly equal size, so that the bipartite graphs between most pairs of pieces behave like random bipartite graphs. The strength of this lemma is that it applies to any graph, providing a rough approximation of its structure that enables one to extract a lot of information about it. A typical application is that a graph with "few" triangles can be "well-approximated" by a graph with no triangles. More precisely, for any $\epsilon > 0$ there

exists $\delta > 0$ such that if G is a graph with n vertices and at most δn^3 triangles, then one can remove at most ϵn^2 edges from G and make it triangle free. This innocent-looking statement turns out to imply the case $k = 3$ of Szemerédi's theorem that was mentioned earlier.

Tools from linear and multilinear algebra play an essential role in extremal combinatorics. The most fruitful technique of this kind, which is possibly also the simplest, is the so-called *dimension argument*. In its simplest form, the method can be described as follows. In order to bound the cardinality of a discrete structure A, one maps its elements to distinct vectors in a VECTOR SPACE [I.3 §2.3], and proves that those vectors are linearly independent. It then follows that the size of A is at most the dimension of the vector space in question. An early application of this argument was found by Larman, Rogers, and Seidel in 1977. They wanted to know how many points it was possible to find in $\mathbb{R}^n$ that determine at most two distinct differences. An example of such a system is the set of all points whose coordinates consist of $n - 2$ 0s and two 1s. Notice, however, that these points all lie in the hyperplane of points whose coordinates add up to 2. So this actually provides us with an example in $\mathbb{R}^{n-1}$. Therefore, we have a simple lower bound of $n(n + 1)/2$. Larman, Rogers, and Seidel matched this with an upper bound of $(n + 1)(n + 4)/2$. They did this by associating with each point of such a set a polynomial in n variables, and by showing that these polynomials are linearly independent and all lie in a space of dimension $(n + 1)(n + 4)/2$. This has been improved by Blokhuis to $(n + 1)(n + 2)/2$. He did this by finding $n + 1$ further polynomials that lie in the same space in such a way that the augmented set of polynomials is still linearly independent. More applications of the dimension argument can be found in Graham, Grötschel, and Lovász (1995, chapter 31).

Spectral techniques, that is, an analysis of EIGENVECTORS AND EIGENVALUES [I.3 §4.3], have been used extensively in graph theory. The link comes through the notion of an *adjacency matrix* of a graph G. This is defined to be the matrix A with entries $a_{u,v}$ for each pair of (not-necessarily-distinct) vertices u and v, where $a_{u,v} = 1$ if u and v are joined by an edge, and $a_{u,v} = 0$ otherwise. This matrix is symmetric, and therefore, by standard results in linear algebra, it has real eigenvalues and an ORTHONORMAL BASIS [III.37] of eigenvectors. It turns out that there is a tight relationship between the eigenvalues of the adjacency matrix A and several structural properties of the graph G, and these properties can often be useful in the study of various extremal problems. Of particular interest is the second largest eigenvalue of a regular graph. Suppose that every vertex of a graph G has degree d. Then the vector for which every entry is 1 is easily seen to be an eigenvector with eigenvalue d, and this is the largest eigenvalue. If all other eigenvalues have modulus much smaller than d, then it turns out that G behaves in many ways like a random d-regular graph. In particular, the number of edges inside any set of k of the vertices is roughly the same (provided k is not too small) as one would expect with a random graph. It follows easily that any set of vertices that is not too big has many neighbors among the vertices outside that set. Graphs with the latter property are called EXPANDERS [III.24] and have numerous applications in theoretical computer science. Constructing such graphs explicitly is not an easy matter and was at one time a major open problem. Now, however, several constructions are known, based on algebraic tools. See chapter 9 of Alon and Spencer (2000), and its references, for more details.

The application of topological methods in the study of combinatorial objects such as partially ordered sets, graphs, and set systems has already become part of the mathematical machinery commonly used in combinatorics. An early example is Lovász's proof of Kneser's conjecture, mentioned in section 2.3. Another example is a result of which the following is a representative special case. Suppose you have a piece of string with 10 red beads, 15 blue beads, and 20 yellow beads on it. Then, no matter what order the beads come in, you can cut the string in at most 12 places and place the resulting segments of beaded string into five piles, each of which contains two red beads, three blue beads, and four yellow beads. The number 12 is obtained by multiplying 4, the number of piles minus 1, by 3, the number of colors. The general case of this result was proved by Alon using a generalization of Borsuk's theorem. Many additional examples of topological proofs appear in Graham, Grötschel, and Lovász (1995, chapter 34).

3 Probabilistic Combinatorics

A wonderful development took place in twentieth-century mathematics when it was realized that it is sometimes possible to use probabilistic reasoning to prove mathematical statements that do not have an obvious probabilistic nature. For example, in the first half of the century, Paley, Zygmund, Erdős, Turán, Shannon, and others used probabilistic reasoning to obtain striking results in analysis, number theory, combinatorics,

and information theory. It soon became clear that the so-called *probabilistic method* is a very powerful tool for proving results in discrete mathematics. The early results combined combinatorial arguments with fairly elementary probabilistic techniques, but in recent years the method has been greatly developed, and now it often requires one to apply much more sophisticated techniques. A recent text dealing with the subject is Alon and Spencer (2000).

The applications of probabilistic techniques in discrete mathematics were initiated by Paul Erdős, who contributed to the development of the method more than anyone else. One can classify them into three groups.

The first deals with the study of certain classes of random combinatorial objects, like random graphs or random matrices. The results here are essentially results in probability theory, although most of them are motivated by problems in combinatorics. A typical problem is the following: if we pick a graph "at random," what is the probability that it contains a Hamilton cycle?

The second group consists of applications of the following idea. Suppose you want to prove that a combinatorial structure exists with certain properties. Then one possible method is to choose a structure randomly (from a probability distribution that you are free to specify) and estimate the probability that it has the properties you want. If you can show that this probability is greater than 0, then such a structure exists. Surprisingly often it is much easier to prove this than it is to give an example of a structure that works. For instance, is there a graph with large girth (meaning it has no short cycles) and large chromatic number? Even if "large" means "at least 7," it is very hard to come up with an example of such a graph. But their existence is a fairly easy consequence of the probabilistic method.

The third group of applications is perhaps the most striking of all. There are many examples of statements that appear to be completely deterministic (even when one is used to the idea of using probability to give existence proofs) but that nevertheless yield to probabilistic reasoning. In the remainder of this section we shall briefly describe some typical examples of each of these three kinds of application.

3.1 Random Structures

The systematic study of random graphs was initiated by Erdős and Rényi in 1960. The most common way of defining a random graph is to fix a probability p and then to join each pair of vertices with an edge with probability p, with all the choices made independently. The resulting graph is denoted $G(n, p)$. (Formally speaking, $G(n, p)$ is not a graph but a probability distribution, but one often talks about it as though it is a graph that has been produced in a random way.) Given any property, such as "contains no triangles," we can study the probability that $G(n, p)$ has that property.

A striking discovery of Erdős and Rényi was that many properties of graphs "emerge very suddenly." Some examples are "contains a Hamilton cycle," "is not planar," and "is connected." These properties are all *monotone*, which means that if a graph G has the property and you add an edge to G, then the resulting graph still has the property. Let us take one of these properties and define $f(p)$ to be the probability that the random graph $G(n, p)$ has it. Because the property is monotone, $f(p)$ increases as p increases. What Erdős and Rényi discovered was that almost all of this increase happens in a very short time. That is, $f(p)$ is almost 0 for small p and then suddenly changes very rapidly and becomes almost 1.

Perhaps the most famous and illustrative example of this swift change is the sudden appearance of the so-called *giant component*. Let us look at $G(n, p)$ when p has the form c/n. If $c < 1$, then with high probability all the connected components of $G(n, p)$ have size at most logarithmic in n. However, if $c > 1$, then $G(n, p)$ almost certainly has one component of size linear in n (the giant component), while all the rest have logarithmic size. This is related to the phenomenon of *phase transitions* in mathematical physics, which are discussed in PROBABILISTIC MODELS OF CRITICAL PHENOMENA [IV.25]. A result of Friedgut shows that the phase transition for a graph property that is "global," in a sense that can be made precise, is sharper than the one for a "local" property.

Another interesting early discovery in the study of random graphs was that many of the basic parameters of graphs are highly "concentrated." A striking example that illustrates what this means is the fact that, for any fixed value of p and for most values of n, almost all graphs $G(n, p)$ have the same clique number. That is, there exists some r (depending on p and n) such that with high probability, when n is large, the clique number of $G(n, p)$ is equal to r. Such a result cannot hold for all n, for continuity reasons, but in the exceptional cases there is still some r such that the clique number is almost certainly equal either to r or to $r + 1$. In both cases, r is roughly $2 \log n / \log(1/p)$. The proof

of this result is based on the so-called *second moment method*: one estimates the expectation and the variance of the number of cliques of a given size contained in $G(n, p)$, and applies well-known inequalities of Markov and CHEBYSHEV [VI.45].

The chromatic number of the random graph $G(n, p)$ is also highly concentrated. Its typical behavior for values of p that are bounded away from 0 was determined by Bollobás. A more general result, in which p is allowed to tend to 0 as $n \to \infty$, was proved by Shamir, Spencer, Łuczak, Alon, and Krivelevich. In particular, it can be shown that for every $\alpha < \frac{1}{2}$ and every integer-valued function $r(n) < n^\alpha$, there exists a function $p(n)$ such that the chromatic number of $G(n, p(n))$ is precisely $r(n)$ almost surely. However, determining the precise degree of concentration of the chromatic number of $G(n, p)$, even in the most basic and important case $p = \frac{1}{2}$ (in which all labeled graphs on n vertices occur with equal probability), remains an intriguing open problem.

Many additional results on random graphs can be found in Janson, Łuczak, and Ruciński (2000).

3.2 Probabilistic Constructions

One of the first applications of the probabilistic method in combinatorics was a lower bound given by Erdős for the Ramsey number $R(k, k)$, which was defined in section 2.2. He proved that if

$$\binom{n}{k} 2^{1 - \binom{k}{2}} < 1,$$

then $R(k, k) > n$. That is, there is a red/blue coloring of the edges of the complete graph on n vertices such that no clique of size k is completely red or completely blue. Notice that the number $n = \lfloor 2^{k/2} \rfloor$ satisfies the above inequality for all $k \geqslant 3$, so Erdős's result gives an exponential lower bound for $R(k, k)$. The proof is simple: if you color the edges randomly and independently, then the probability that any fixed set of k vertices has all its edges of the same color is twice $2^{-\binom{k}{2}}$. Thus, the expected number of cliques with this property is

$$\binom{n}{k} 2^{1 - \binom{k}{2}}.$$

If this is less than 1, then there must be at least *one* coloring for which there are no cliques with this property, and the result is proved.

Note that this proof is completely nonconstructive, in the sense that it merely proves the existence of such a coloring, but gives no efficient way of actually constructing one.

A similar computation yields a solution for the tournament problem mentioned in section 1.1. If the results of the tournament are random, then the probability, for any particular k teams, that no other team beats them all is $(1 - (1/2^k))^{n-k}$. From this it follows that if

$$\binom{n}{k}\left(1 - \frac{1}{2^k}\right)^{n-k} < 1,$$

then there is a nonzero probability that for every choice of k teams, there is another team that beats them all. In particular, it is possible for this to happen. If n is larger than about $k^2 2^k \log 2$, then the above inequality holds.

Probabilistic constructions have been very powerful in supplying lower bounds for Ramsey numbers. Besides the bound for $R(k,k)$ mentioned above, there is a subtle probabilistic proof, due to Kim, that $R(3,k) \geqslant ck^2 / \log k$, for some $c > 0$. This is known to be tight up to a constant factor, as proved by Ajtai, Komlós, and Szemerédi, who also used probabilistic methods.

3.3 Proving Deterministic Theorems

Suppose that you color the integers with k colors. Let us call a set S *multicolored* if all k colors appear in S. Straus conjectured that for every k there is an m with the following property: given any set S with m elements, there is a coloring of the integers with k colors such that all translates of S are multicolored. This conjecture was proved by Erdős and Lovász. The proof is probabilistic, and applies a tool called the *Lovász local lemma*, which, unlike many probabilistic techniques, allows one to show that certain events hold with nonzero probability even when this probability is extremely small. The assertion of this lemma, which has numerous additional applications, is, roughly, that for any finite collection of "nearly independent" low-probability events, there is a positive probability that none of the events holds. Note that the statement of Straus's conjecture has nothing to do with probability, and yet its proof relies on probabilistic arguments.

A graph G is k-colorable, as we have said, if you can properly color its vertices with k colors. Suppose now that instead of trying to use k colors in total, you have a separate list of k colors for each vertex, and this time you want to find a proper coloring of G where each vertex gets a color from its own list. If you can always do so, no matter what the lists are, then G is called k-*choosable*, and the smallest k for which G is k-choosable is called the *choice number* ch(G). If all the lists are the same, then one obtains a k-coloring,

so ch(G) must be at least as big as $\chi(G)$. One might expect ch(G) to be equal to $\chi(G)$, since it seems as though using different lists of k colors for different vertices would make it easier to find a proper coloring than using the same k colors for all vertices. However, this turns out to be far from true. It can be proved that for any constant c there is a constant C such that any graph with average degree at least C has choice number at least c. Such a graph might easily be bipartite (and therefore have chromatic number 2), so it follows that ch(G) can be much bigger than $\chi(G)$. Somewhat surprisingly, the proof of this result is probabilistic.

An interesting application of this fact concerns a graph that arises in Ramsey theory. Its vertices are all the points in the plane, with two vertices joined by an edge if and only if the distance between them is 1. The choice number of this graph is infinite, by the above result, but the chromatic number is known to be between 4 and 7.

A typical problem in Ramsey theory asks for a substructure of some kind that is entirely colored with one color. Its cousin, *discrepancy theory*, merely asks that the numbers of times the colors are used are not too close to each other. Probabilistic arguments have proved extremely useful in numerous problems of this general kind. For example, Erdős and Spencer proved that in any red/blue coloring of the edges of the complete graph K_n there is a subset V_0 of vertices such that the difference between the number of red edges inside V_0 and the number of blue edges inside V_0 is at least $cn^{3/2}$, for some absolute constant $c > 0$. This problem is a convincing manifestation of the power of probabilistic methods, since they can be used in the other direction as well, to prove that the result is tight up to a constant factor. Additional examples of such results can be found in Alon and Spencer (2000).

4 Algorithmic Aspects and Future Challenges

As we have seen, it is one matter to prove that a certain combinatorial structure exists, and quite another to construct an example. A related question is whether an example can be generated by means of an EFFICIENT ALGORITHM [IV.20 §2.3], in which case we call it *explicit*. This question has become increasingly important because of the rapid development of theoretical computer science, which has close connections with discrete mathematics. It is particularly interesting when the structures in question have been proved to exist by means of probabilistic arguments. Efficient algorithms for producing them are not just interesting on their own, but also have important applications in other areas. For example, explicit constructions of error-correcting codes that are as good as random ones are of major interest in CODING AND INFORMATION THEORY [VII.6], and explicit constructions of certain Ramsey-type colorings may have applications in DERANDOMIZATION [IV.20 §7.1.1] (the process of converting randomized algorithms into deterministic ones).

It turns out, however, that the problem of finding a good explicit construction is often very difficult. Even the simple proof of Erdős, described in section 3.2, that there are red/blue colorings of graphs with $\lfloor 2^{k/2} \rfloor$ vertices containing no monochromatic clique of size k leads to an open problem that seems very difficult. Can we construct, explicitly, such a graph with $n \geqslant (1 + \epsilon)^k$ vertices in time that is polynomial in n? Here we allow ϵ to be any constant, as long as it is positive. This problem is still wide open, despite considerable efforts from many mathematicians.

The application of other advanced tools, such as algebraic and analytic techniques, spectral methods, and topological proofs, also tends to lead in many cases to nonconstructive proofs. The conversion of these to algorithmic arguments may well be one of the main future challenges of the area.

Another interesting recent development is the increased appearance of computer-aided proofs in combinatorics, starting with the proof of the FOUR-COLOR THEOREM [V.12]. To incorporate such proofs into the area, without threatening its special beauty and appeal, is a further challenge.

These challenges, the fundamental nature of the area, its tight connection to other disciplines, and its many fascinating open problems ensure that combinatorics will continue to play an essential role in the general development of mathematics and science in the future.

Further Reading

Alon, N., and J. H. Spencer. 2000. *The Probabilistic Method*, 2nd edn. New York: John Wiley.

Bollobás, B. 1978. *Extremal Graph Theory*. New York: Academic Press.

Graham, R. L., M. Grötschel, and L. Lovász, eds. 1995. *Handbook of Combinatorics*. Amsterdam: North-Holland.

Graham, R. L., B. L. Rothschild, and J. H. Spencer. 1990. *Ramsey Theory*, 2nd edn. New York: John Wiley.

Janson, S., T. Łuczak, and A. Ruciński. 2000. *Random Graphs*. New York: John Wiley.

Jukna, S. 2001. *Extremal Combinatorics*. New York: Springer.

Matoušek, J. 2002. *Lectures on Discrete Geometry*. New York: Springer.

Nathanson, M. 1996. *Additive Number Theory: Inverse Theorems and the Geometry of Sumsets*. New York: Springer.

Pach, J., and P. Agarwal. 1995. *Combinatorial Geometry*. New York: John Wiley.

Tao, T., and V. H. Vu. 2006. *Additive Combinatorics*. Cambridge: Cambridge University Press.

IV.20   Computational Complexity
Oded Goldreich and Avi Wigderson

1   Algorithms and Computation

This article is concerned with what can be computed efficiently, and what cannot. We will introduce several important concepts and research areas, such as formal models of computation, measures of efficiency, the $\mathcal{P}$ versus $\mathcal{NP}$ question, NP-completeness, circuit complexity, proof complexity, randomized computation, pseudorandomness, probabilistic proof systems, cryptography, and more. Underlying them all are the related notions of *algorithms* and *computation*, and we begin by discussing these.

1.1   What Is an Algorithm?

Suppose that you are presented with a large positive integer N and asked to determine whether it is prime. What should you do? One possibility would be to apply the method of *trial division*. That is, first see whether N is even, then whether it is a multiple of 3, then whether it is a multiple of 4, and so on through all the numbers up to $\sqrt{N}$. If N is composite, then it has a factor between 2 and $\sqrt{N}$, so it is prime if and only if the answer to all these questions is no.

The trouble with this method is that it is highly *inefficient*. Suppose, for instance, that N has 101 digits. Then $\sqrt{N}$ is at least 10^{50}, so in order to carry the method out one would have to answer 10^{50} questions of the form, "Is K a factor of N?" This would take far longer than a human lifetime, even if all the world's computers devoted themselves to the task. What, then, is an "efficient procedure"? This question divides into two parts: what is a procedure, and what counts as efficient? We shall look at these two questions in turn.

Two very obvious conditions that a method should satisfy if it is to count as a procedure for solving this problem are *finiteness*—that the procedure should have a finite description (so, for example, one cannot simply look up the answer in an infinite list of integers and their factorizations)—and *correctness*—that, for every N, it correctly tells you whether N is prime.

There is also a third, more subtle, condition, which goes to the heart of what is meant by the word "algorithm." It is that it should consist of *simple steps*. This is needed in order to rule out ridiculous "procedures" such as, "See whether N has any nontrivial factors; declare N to be prime if and only if it does not." The problem with this is that we cannot see, just like that, whether N has nontrivial factors. By contrast, all that the method of trial division asks of us is that we should do basic arithmetic, such as increasing integers by 1, comparing them, and doing long division. Moreover, the procedures of basic arithmetic can be broken down into yet simpler steps: for instance, it is possible to do long division by a succession of elementary operations applied to single digits at a time.

In order to understand this simplicity condition better, and to prepare ourselves for a formal definition of the notion of algorithms, let us look at long division in slightly more detail. Suppose that you have a piece of paper in front of you and you want to divide 5 959 578 by 857. You will write the two numbers down, and then, as the calculation proceeds, you will write other numbers as well. For instance, you may wish to start by writing out all the multiples of 857 up to 9×857. At some point early on you will probably find yourself comparing $5999 = 7 \times 857$ with 5959: this you do by scanning the numbers from left to right and comparing individual digits. In this case, a difference is first detected in the third digit. You then write 5142 (which is 6×857) underneath the 5959, subtract (again by scanning numbers from left to right and performing single-digit operations), write down the difference 817, "bring down" the next digit, 5, of 5 959 578, and repeat the process with the number 8175.

At each stage in this calculation you are modifying the piece of paper in front of you. As you do so you need to keep track of which stage of the procedure you are at (whether you are writing out the initial table of multiples of 857, or seeing which one is the largest that does not exceed another number, or subtracting one number from another, or bringing down a digit, etc.), and which symbols on the page you are currently dealing with. What is remarkable is that this information has a *fixed size*, in the sense that it does not increase as the size of the input (that is, the two numbers to be divided) increases.

Therefore, the procedure can be regarded as making *local changes* to some "environment," using repeated applications of a fixed rule that does not depend on the input. (This rule will typically have some internal structure, such as a list of simpler rules together with specifications of the circumstances under which they should be applied.) In general, this is what we mean by a *computation*: it modifies an environment by means of repeated applications of a fixed rule. The rule is usually referred to as an *algorithm*. Notice that this description applies to many scientific theories of dynamic evolution in nature (of weather, chemical reactions, or biological processes, for example). Thus, these can be regarded as computational processes, of sorts. Some of these dynamical systems also demonstrate well the fact that simple, local rules can result in a very complex modification of the environment if they are iterated many times. (See DYNAMICS [IV.14] for further discussion of this phenomenon.)

Thoughts such as these lie behind the idea of a *Turing machine*, TURING's [VI.94] famous formalization of the notion of an algorithm. It is interesting that he came up with his formalization *before* computers existed. Indeed, this abstraction and central features of it, most notably the existence of a "universal" machine, greatly influenced the actual construction of computers.

It is very important to know that the idea of an algorithm can be formalized, so that one can talk precisely about whether there are algorithms that will perform particular tasks, how many steps they need for a given size of input, and so on. However, there are many ways of doing this, which all turn out to be equivalent, and for the purposes of understanding this article it is not necessary to go into the details of any particular method. (You can, if you like, think of an algorithm as any procedure that can be programmed on a real computer—slightly idealized so that it has unlimited storage space—and a step of an algorithm as any change of one of the bits of that computer from a 0 to a 1 or vice versa.) Nevertheless, just to show roughly how it is done, here is a brief description of the basic features of the Turing machine model.

To begin with, one makes the observation that all computational problems can be encoded as operations on sequences of 0s and 1s. (This observation is not just theoretically useful but also very important for the actual building of computers.) For example, all numbers that occur in the course of a computation can be converted into their binary representations; one can also use 1 to stand for "true" and 0 to stand for "false" and thereby perform the basic logical operations; and so on. For this reason we can define a very simple "environment" for a Turing machine: it is a "tape," infinitely long in both directions, that consists of a row of "cells," each of which contains either a 0 or a 1. Before the computation starts, a certain prespecified portion of this tape is filled with the *input*, which is a sequence of 0s and 1s. The algorithm is a little control mechanism. At any one time, this mechanism can be in one of a finite set of states, and it is located at one of the cells of the tape. According to the state it is in and the value, 0 or 1, that it sees at the cell it has reached, it makes three decisions: whether to change the value in the cell, whether to move left or right by one cell, and which state it should next be in.

One of the states of this control mechanism is "halt." If this state is reached, then the mechanism stops doing anything and is said to have halted. At that point, a certain prespecified portion of the tape will be regarded as the output of the machine. An algorithm can be thought of as any Turing machine that halts for every possible input. And the number of steps of the algorithm is the number of steps taken by that Turing machine. Remarkably, this very simple computational model is enough to capture the full power of computation: in theory one could build a Turing machine, out of clockwork, say, that would be able to do whatever a modern supercomputer can do. (However, it would take too long over each step to be practical for anything but the very simplest of computations.)

1.2   What Does an Algorithm Compute?

A Turing machine converts a sequence of 0s and 1s into another sequence of 0s and 1s. If we wish to use mathematical language to discuss this, then we need to give a name to the set of $\{0, 1\}$-sequences. To be precise, we consider the set of all *finite* sequences of 0s and 1s, and we call this set I. It is also useful to write I_n for the set of all $\{0, 1\}$-sequences of length n. If x is a sequence in I, then we write $|x|$ for its length: for instance, if x is the string 0100101, then $|x| = 7$. To say that a Turing machine converts a sequence of 0s and 1s into another such sequence (if it halts) is to say that it naturally defines a function from I to I. If M is the Turing machine and f_M is the corresponding function, then we say that M *computes* f_M.

Thus, every function $f : I \to I$ gives rise to a computational task, namely that of computing f. We say that f is *computable* if this is possible: that is, if there exists a Turing machine M such that the corresponding

function f_M is equal to f. A central early result (due to Turing and independently to CHURCH [VI.89]) is that some natural functions are *not* computable. (For more details, see THE INSOLUBILITY OF THE HALTING PROBLEM [V.20].) However, complexity theory deals only with computable functions, and studies which of these can be computed *efficiently*.

Using the notation we have just introduced, we can formally describe various different kinds of computational tasks, of which two major examples are *search problems* and *decision problems*. The aim of a search problem is, informally speaking, to find a mathematical object with certain properties: for instance, one might wish to find a solution to a system of equations, and this solution might not be unique. We can model this by means of a binary RELATION [I.2 §2.3] R on the set I: for a pair (x, y) of strings in I, we say that y *is a valid solution of problem instance x* if xRy. (This notation means that x is related to y in the way specified by R; another common notation for the same thing is $(x, y) \in R$.) For example, we might let x and y be binary expansions of positive integers N and K, respectively, and say that xRy if and only if N is a composite number and K is a nontrivial factor of N. Informally, this search problem would be, "Find a nontrivial factor of N." If M is an algorithm that computes a certain function $f_M : I \to I$, then we say that M *solves the search problem R* if $f_M(x)$ is a valid solution of x for every problem instance x that has a solution. For example, it solves the search problem just defined if, for every composite number N with binary expansion x, $f_M(x)$ is the binary expansion of a nontrivial factor K of N.

Notice that in the above example we were interested in positive integers, but formally speaking an algorithm is a function of binary strings. This was not a problem, because there is a convenient and natural way to *encode* integers as binary strings—via their usual binary expansions. For the rest of this article, we shall feel free to blur the distinction between the mathematical objects we wish to investigate and the strings we use to represent them in a computation. For instance, it is simpler to think of the algorithm M in the previous paragraph as computing a function $f_M : \mathbb{N} \to \mathbb{N}$, and solving the search problem if, for every composite number N, $f_M(N)$ is a nontrivial factor of N. We stress that the representation of objects by strings is a rather succinct one: it takes only $\lceil \log_2 N \rceil$ bits to represent the number N, so the number N is exponentially larger than the length of its representation.

Now let us turn to decision problems. These are simply problems where one is looking for a yes/no answer. The problem with which we opened this article—Is N a prime number?—is a classic example of a decision problem. Notice that here and in the paragraph before last we are using the word "problem" in a slightly unusual way, to mean a general class of questions rather than just one. In this example, the question, "Is 443 a prime number?" would be called an *instance* of the problem, "Is N a prime number?"

Modeling decision problems is very simple: they are subsets of I. The idea is that a subset S of I consists of all the strings where the answer is yes. So if the problem is to determine primality, then S would consist of all binary expansions of prime numbers, at least if we chose the obvious encoding of the problem. When do we say that a machine M solves the decision problem S? We would like it to compute a function f that says yes when the input x belongs to S and says no otherwise. That is, we say that M *solves the problem S* if the associated function f_M is a function from I to the set $\{0, 1\}$ such that $f_M(x) = 1$ whenever $x \in S$ and $f_M(x) = 0$ otherwise.

Most of this article will be focused on decision problems, but the reader should bear in mind that computational tasks that seem more complicated, including search problems, can in fact usually be reduced to sequences of decision problems. For example, if you can solve all decision problems and you want to factorize a large composite number N, then you can proceed as follows. First, determine whether the smallest prime factor of the number ends in a 1 (in its binary expansion). If the answer is yes, you can look at the next digit by asking if this factor ends in 11; if it is no, then you can ask if it ends in 10. You can continue this process, extending your knowledge of the smallest prime factor by one bit at a time. The number of queries you will need to make will be at most the number of digits of N.

2 Efficiency and Complexity

Near the beginning of this article we asked what was meant by the phrase "efficient procedure." We have now discussed the word "procedure" in some depth, but we have yet to say what we mean by "efficient," beyond pointing out that trial division takes too long to be practical if we have a very large integer and want to determine whether it is prime.

2.1 Complexity of Algorithms

How can we describe mathematically what it means for a procedure to "take too long to be practical"? The Turing-machine formalization is particularly useful for answering questions like this, because we can say precisely what a step of a Turing-machine computation is and this allows us to give a precise definition: an algorithm is a Turing machine, and its *complexity* is defined to be the number of steps the machine takes before halting.

If we look at this definition carefully, we see that what it defines is not just one number but a function. The time taken by a Turing machine depends on the input, so, given a Turing machine M and a string x, we can define $t_M(x)$ to be the number of steps M takes before halting when x is the input. The function $t_M : \text{I} \to \mathbb{N}$ is the *complexity function* of the machine M.

Most of the time, we are interested not so much in the full detail of this complexity function, but in the *worst-case complexity* of the machine M. This is a function $T_M : \mathbb{N} \to \mathbb{N}$ defined as follows. Given a positive integer n, $T_M(n)$ is the maximum value of $t_M(x)$ over all input strings x of length n. In other words, we want to know the longest possible time that our machine might take when faced with an input of length n. And usually we do not look for an exact formula for $T_M(n)$: for most purposes it is enough to have a good upper bound.

The function $t_M(x)$ is more accurately called the *time complexity* of the algorithm M, since it measures how long M takes given x as its input. But time is not the only resource that matters in computer science. Another is how much memory an algorithm uses, beyond that needed to store the input, and this too can be captured in our formal model. Given a Turing machine M and an input x, we can define $s_M(x)$ to be the number of cells, other than input cells, that are visited before the machine halts, under the extra condition that the input cells must be left unchanged.

2.2 Intrinsic Complexity of Problems

Much of this article will be concerned with a very general analysis of the power of computation. In particular, we shall discuss a central subfield of theoretical computer science known as *computational complexity* (or *complexity theory*). The aim of this area is to understand the *intrinsic complexity* of computational tasks.

Notice that we said "computational tasks" rather than "algorithms." This is an important distinction and it involves a change of focus. Returning to our example of primality testing, it is not too hard to estimate how long various algorithms take, and indeed we had no trouble in seeing that trial division would take a very long time indeed. But does that mean that the task of primality testing is *intrinsically* hard? Not necessarily, since there may be other algorithms that do the job much more quickly.

This idea fits neatly into our formal scheme. What would be a good definition of the complexity of a computational task? Roughly speaking, the complexity of such a task should be the smallest complexity of any algorithm M that solves it. A convenient way of saying this is as follows. If $T : \mathbb{N} \to \mathbb{N}$ is some integer function, we say that the task has *complexity at most T* if there is an algorithm M that solves the task such that $T_M \leqslant T$ (i.e., $T_M(n) \leqslant T(n)$ for every n).

If you want to show that a computational task is not intrinsically hard, then all you have to do is devise an algorithm with low complexity that solves this task. But what if you want to show that this task *is* intrinsically hard? Then you have to prove, for *every possible* low-complexity algorithm M, that M does not solve this task. This is much harder: even after half a century of intensive work, the best results that are known are very weak. Notice a big difference between the two kinds of research: one can find algorithms without knowing how the concept of "algorithm" is formalized, but to analyze *all* algorithms with a certain property, it is essential to have a precise definition of what an algorithm is. Fortunately, with Turing's formalization, we have one.

2.3   Efficient Computation and $\mathcal{P}$

Now we have ways of measuring the complexity of algorithms and computational tasks. But we have not yet addressed the question of when we should regard an algorithm as *efficient*, or a computational task as efficiently solvable. We shall propose a definition of efficiency that seems somewhat arbitrary and then explain why it is in fact a surprisingly good one.

If M is an algorithm, then we regard it as efficient if and only if it *terminates in polynomial time*. This means that there are constants c and k such that the worst-case complexity T_M always satisfies the inequality $T_M(n) \leqslant cn^k$. In other words, the time taken by the algorithm is bounded above by a polynomial function of the length of the input string. It is not hard to convince yourself that the familiar methods for adding or multiplying two n-digit numbers termi-

nate in polynomial time, whereas trial division for primality testing does not. Other familiar examples of tasks with efficient algorithms are putting a set of numbers in increasing order, computing the DETERMINANT [III.15] of a matrix (provided one uses row operations rather than substituting the entries directly into the formula), solving linear equations by Gaussian elimination, finding the shortest path in a given network, and more.

Since we are interested in the intrinsic complexity of computational tasks, we now define such a task to be *efficiently computable* if there is an efficient algorithm M that solves it. In our discussion of efficient computability, we shall focus on decision problems and consider the class of *all* decision problems that have efficient algorithms. Understanding it is *the* major goal of computational complexity theory. Here is a formal definition. We shall use the following convenient piece of notation: if M is a Turing machine and x is an input, then $M(x)$ is the output of x. (Earlier we wrote $f_M(x)$ for this function.) Since we are considering decision problems, $M(x)$ will be 0 or 1.

Definition. A decision problem $S \subseteq \mathbf{I}$ is *solvable in polynomial time* if there is a Turing machine M, terminating in polynomial time, such that $M(x) = 1$ if and only if $x \in S$.

The class of decision problems that are solvable in polynomial time is our first example of a *complexity class*. It is denoted $\mathcal{P}$.

The *asymptotic analysis* of running time, i.e., estimating the running time as a function of the input length, turns out to be crucial for revealing structure in the theory of efficient computation. The choice of polynomial time as the standard for efficiency may seem arbitrary, and theories could be developed with other choices, but it has amply justified itself. The main reason for this is that the class of polynomials (or functions bounded above by a polynomial) is closed under various operations that arise naturally in computation. In particular, the sum, product, or composition of two polynomials is again a polynomial. This allows us, for example, to think of long division as a basic, one-step operation when we are investigating the efficiency of algorithms for primality testing. In fact, long division takes more than one step, but it is in $\mathcal{P}$ so the time it takes does not affect whether an algorithm that uses it is itself in $\mathcal{P}$. In general, if we use the basic programming technique of *subroutines*, and if our subroutines are in $\mathcal{P}$, then we will preserve the efficiency of the algorithm as a whole.

Almost all computer programs that are used in practice turn out to be efficient in this theoretical sense. Of course, the converse is not true: an algorithm that runs in time n^{100} is completely useless despite the fact that n^{100} is a polynomial. However, this seems not to matter. It is unusual to discover even an n^{10}-time algorithm for a natural problem, and on the rare occasions when this happens, improvements to n^3- or n^2-time, which border on the practical, almost always follow.

It is important to contrast $\mathcal{P}$ with the class $\mathcal{EXP}$. A problem belongs to $\mathcal{EXP}$ if there is an algorithm that solves it in at most $\exp(p(n))$ steps for any input of length n, where p is some polynomial. (Roughly speaking, $\mathcal{EXP}$ consists of problems that can be solved in exponential time: the polynomial p makes the definition more robust and less dependent on the precise nature of encodings, etc.)

If you use trial division to test the primality of a number N with n digits in its binary expansion, then you have to do $\sqrt{N}$ long-division calculations. Since $\sqrt{N}$ is about $2^{n/2}$, this is an exponential-time procedure. Exponential running time is considered blatantly *inefficient*, and if the problem has no faster algorithm, then it is deemed intractable. It is known (via a basic technique called *diagonalization*) that $\mathcal{P} \neq \mathcal{EXP}$; furthermore, some problems in $\mathcal{EXP}$ really do require exponential time. Almost all problems and classes considered in this paper can easily be shown to belong to $\mathcal{EXP}$ via trivial, "brute-force" algorithms such as the trial division just discussed: the main question will be whether much faster algorithms can be devised for them.

3 The $\mathcal{P}$ versus $\mathcal{NP}$ Question

In this section we discuss the famous $\mathcal{P}$ versus $\mathcal{NP}$ question, which is usually formulated in terms of decision problems, but which also has an interpretation in terms of search problems. We shall start with the latter.

3.1 Finding versus Checking

Can you rearrange the letters CHAIRMITTE to form an English word? To solve a puzzle like this, one has to search among many possibilities (all permutations of those letters), perhaps building up fragments of words and hoping that inspiration will strike. Now consider the following question: can the letters of CHAIRMITTE be rearranged to form the word "arithmetic"? It is very easy (if slightly boring) to check that the answer is yes.

This informal example illustrates an important feature of many search problems: that once you find a solution, it is easy to recognize that it *is* a solution. The hard part is to find the solution in the first place. Or at least, so it seems. But actually proving that search problems of this kind are hard is a famous unsolved problem, the $\mathcal{P}$ *versus* $\mathcal{NP}$ *question.*

Another search problem with this quality, which is in fact quite general and has a natural appeal to mathematicians, is the task of finding proofs for valid mathematical statements. Again it seems to be far easier to check that an argument *is* a valid proof than it is to find the argument in the first place. Since finding a proof is a process that requires considerable creativity (as, in a much smaller way, is finding an anagram), the $\mathcal{P}$ versus $\mathcal{NP}$ question is, in a sense, asking whether this kind of creativity can be automated.

In section 3.2 we shall define the class $\mathcal{NP}$ formally. Informally, it corresponds to the set of all search problems for which it is easy to check whether you have found what you are searching for. Another example of such a problem is that of finding a factor of a large composite integer N. If you are told that K is a factor, then it is an easy task for you (or your computer) to verify that this is true: all you have to do is a single instance of long division.

A vast number of problems in science (such as creating theories to explain various natural phenomena) and engineering (such as creating designs under various physical and economic constraints) have the same property that success is much easier to recognize than to achieve in the first place. This gives some indication of the importance of this class of problems.

3.2 Deciding versus Verifying

For the purposes of theoretical analysis, it is actually more convenient to define $\mathcal{NP}$ as a class of *decision* problems. For instance, consider the decision problem, "Is N composite?" What makes this a problem in $\mathcal{NP}$ is that, whenever N is composite, there is a *short proof* of this fact. Such a proof consists of a factor of N, and is easy to check that this proof is correct. That is, it is easy to devise a polynomial-time algorithm M that takes as input a pair (N, K) of positive integers and outputs 1 if K is a nontrivial factor of N and 0 otherwise. If N is prime, then $M(N, K) = 0$ for every K, while if N is composite there will always exist an integer K such that $M(N, K) = 1$. Moreover, in this case the string that encodes K will be at most as long as the string that encodes N, though all we really care about is that it should not be too much longer. These properties we now encapsulate in a formal definition.

Definition (the complexity class $\mathcal{NP}$[1]). A decision problem $S \subset \mathtt{I}$ belongs to $\mathcal{NP}$ if there is a subset $R \subset \mathtt{I} \times \mathtt{I}$ with the following three properties.

(i) There is a polynomial function p such that $|y| \leqslant p(|x|)$ whenever $(x, y) \in R$.

(ii) x belongs to S if and only if there is some y such that (x, y) belongs to R.

(iii) The problem of determining whether a pair (x, y) belongs to R is in $\mathcal{P}$.

When such a y exists, it is called a *proof* (or *witness*) of the fact that x belongs to S. The polynomial-time algorithm for determining whether a pair (x, y) belongs to R is called a *verification procedure* for determining whether x belongs to S.

Notice that every problem S in the class $\mathcal{P}$ is also in $\mathcal{NP}$, since we can simply forget about the candidate proof y and use the efficient test for whether x belongs to S. On the other hand, every problem in $\mathcal{NP}$ is trivially in $\mathcal{EXP}$, because we can enumerate all possible ys (in exponential time) and check for each one whether it works. (This is more or less what we do with trial division.) Can this trivial algorithm be improved? Sometimes it can, even in very nonobvious cases. In fact, recently it was proved that the problem of determining whether a number N is composite belongs to $\mathcal{P}$. (Further details can be found in COMPUTATIONAL NUMBER THEORY [IV.3 §2].) However, we would like to know whether for *every* problem in $\mathcal{NP}$ one can do much better than the trivial algorithm.

3.3   The Big Conjecture

The $\mathcal{P}$ versus $\mathcal{NP}$ problem asks whether or not $\mathcal{P}$ equals $\mathcal{NP}$. In terms of decision problems, this question is asking *whether the existence of an efficient verification procedure for some set implies the existence of an efficient decision procedure for it.* In other words, if there is a polynomial-time algorithm for checking whether proofs that $x \in S$ are correct (as in the definition of $\mathcal{NP}$ just given), does it follow that there is a polynomial-time algorithm for deciding whether $x \in S$?

As our earlier examples suggest, the problem can also be formulated as a question about search problems. Suppose we have a set $R \subset \mathtt{I} \times \mathtt{I}$ satisfying properties (i) and (iii) of the definition of $\mathcal{NP}$. For instance, R might correspond to all pairs of integers (N, K) such that K is a nontrivial factor of N. Then the corresponding search problem, "Given a composite number N find a nontrivial factor K," is closely related to the integer factorization problem. In general, any such relation R gives rise to a search problem, "Given a string x, find a string y such that (x, y) belongs to R (if such a y exists)." Now the $\mathcal{P}$ versus $\mathcal{NP}$ problem asks the following: "Are all such search problems solvable in polynomial time?"

If the answer is yes, then the *mere fact* that it can be checked in polynomial time whether K is a nontrivial factor of N would imply that such a factor could actually be found in polynomial time.[2] Similarly, the mere fact that a short proof of a mathematical statement existed would be enough to guarantee that it could be found in a short time by a purely mechanical process. The apparent difference between the difficulty of discovering solutions and the ease of checking them once discovered would be entirely illusory.

This would be very strange, and almost all experts believe that it is not the case. However, nobody has managed to prove it. So the big conjecture is that $\mathcal{P}$ does not equal $\mathcal{NP}$. That is, finding is harder than checking, and efficient verification procedures do not necessarily lead to efficient algorithms for decision problems. This conjecture is strongly supported by our intuition, which has been developed over many centuries of dealing with search and decision problems in a wide variety of human activities. Further empirical evidence in favor of the conjecture is given by the fact that there are literally thousands of $\mathcal{NP}$ problems, from many mathematical and scientific disciplines, that are not known to be solvable in polynomial time, despite the fact that researchers have tried very hard to discover efficient procedures for solving them.

The $\mathcal{P} \neq \mathcal{NP}$ conjecture is certainly the most important open problem in computer science, and one of the most significant in all of mathematics. Our later section on circuit complexity (section 5.1) is devoted to attempts to prove it. There we shall discuss some partial results and limits of the techniques used so far.

---

1. The acronym NP stands for nondeterministic polynomial-time, where a *nondeterministic machine* is a *fictitious* computing device used in an alternative definition of the class $\mathcal{NP}$. The nondeterministic moves of such a machine correspond to guessing a "proof" in this definition.

---

2. Despite the fact that there is a polynomial-time algorithm for determining whether a number is composite, no such algorithm is known for actually finding its factors, and it is widely believed that no efficient algorithm exists for this.

3.4 $\mathcal{NP}$ versus co $\mathcal{NP}$

Another important class, known as co $\mathcal{NP}$, is the class of *complements* of sets (or decision problems) in $\mathcal{NP}$. For example, the problem "Is N prime?" belongs to co $\mathcal{NP}$ because there is an efficient verification procedure for showing that a given positive integer N is *not* prime, namely, exhibiting some factors. Equivalently, the set of primes belongs to co $\mathcal{NP}$ because its complement belongs to $\mathcal{NP}$.

Does $\mathcal{NP}$ equal co $\mathcal{NP}$? That is, if you have an efficient verification procedure for determining membership of a set S, do you also have one for determining *nonmembership*? Again, intuition would suggest not, or at least not necessarily. For instance, if a jumble of letters can be rearranged to form a word, then that word serves as a short demonstration. But suppose a jumble of letters *cannot* be rearranged to form a word. One could demonstrate this by looking at all possible rearrangements and noting that none of them is a word, but this is a very long demonstration and there does not seem to be a systematic way of finding a truly short one.

Here again intuition from mathematics is extremely relevant: to verify that a set of logical constraints is mutually *inconsistent*, that a family of polynomial equations has *no* common root, or that a set of regions in space has *empty* intersection seems far harder than to verify the opposite (exhibiting a consistent valuation, a common root, or a point that belongs to all the regions). Indeed, only when rare extra mathematical structure is available, such as DUALITY [III.19] theorems or complete systems of invariants, are we able to show that a set and its complement are computationally equivalent. So another big conjecture is that $\mathcal{NP}$ is not equal to co $\mathcal{NP}$. The section on proof complexity (section 5.3) looks further at this conjecture and at attempts to resolve it.

Surprisingly, it is not hard to show that the problem, "Is N composite?" which obviously belongs to $\mathcal{NP}$, actually belongs to co $\mathcal{NP}$ as well. To prove this, one uses the following fact from elementary number theory: p is prime if and only if there is an integer $a < p$ such that $a^{p-1} \equiv 1 \pmod{p}$ and $a^r \not\equiv 1$ whenever r is a factor of $p - 1$. Thus, to verify that p is prime it is enough to exhibit such an integer a. However, to check that a works, one needs to know the prime factorization of $p - 1$, and one must give a short proof that it really is a factorization into primes. This takes us back to the problem we started with, but the numbers are smaller so one can give a recursive argument. (We men-

tion again that the set of primes is actually in $\mathcal{P}$, but this is harder to prove.)

4 Reducibility and NP-Completeness

One sign that a mathematical problem is fundamental is that it has many equivalent formulations. This is true to a quite extraordinary extent for the $\mathcal{P}$ versus $\mathcal{NP}$ problem, as we shall see in this section. Fundamental to our discussion will be the notion of *polynomial-time reducibility*. Roughly speaking, one computational problem is polynomially reducible to another if any polynomial-time algorithm for the second can be converted into a polynomial-time algorithm for the first. Let us see an example of this, and then we will define the notion formally.

First, here is a famous problem in $\mathcal{NP}$, called SAT. Consider the logical formula

$$(p \vee q \vee \bar{r}) \wedge (\bar{p} \vee q) \wedge (p \vee \bar{q} \vee r) \wedge (\bar{p} \vee \bar{r}).$$

Here, p, q, and r are *propositions*, each of which can be true or false. The symbols "$\vee$" and "$\wedge$" stand for OR and AND, respectively, and $\bar{p}$ (read as "NOT-p") is the proposition that is true if and only if p is false.

Suppose now that p is true, q is true, and r is false. Then the first subformula $p \vee q \vee \bar{r}$ is true because at least one of p, q, and $\bar{r}$ is true. Similarly, one can check that all the other subformulas are true, which means that the entire formula is true. We call our choice of truth values for p, q, and r a *satisfying assignment* for the formula, and we say that the formula is *satisfiable*. A natural computation problem that arises is the following.

SAT: *given a propositional formula, is it satisfiable?*

In the example above, the formula was a conjunction of subformulas, called *clauses*. In their turn, these subformulas were disjunctions of propositions or their negations, which are called *literals*. (The *conjunction* of some formulas $\phi_1, \ldots, \phi_k$ is the formula $\phi_1 \wedge \cdots \wedge \phi_k$ and their *disjunction* is $\phi_1 \vee \cdots \vee \phi_k$.)

3SAT: *given a propositional formula that consists of a conjunction of clauses that contain at most three literals each, is the formula satisfiable?*

Notice that SAT and 3SAT are in $\mathcal{NP}$, since it is an easy matter to check whether a given truth assignment to the variables is a satisfying assignment for the formula.

Let us now turn to a second problem in $\mathcal{NP}$.

3-colorability: *given a planar map (such as one might find in an atlas), can its regions be colored with three colors,* Red, Blue, *and* Green, *such that no two adjacent countries have the same color?*[3]

We shall now "reduce" 3-colorability to 3SAT: that is, show how an algorithm that solves 3SAT can be used to solve 3-colorability as well. Suppose, then, that we have a map with n regions. We shall need $3n$ propositions, which we shall call $R_1, \ldots, R_n$, $B_1, \ldots, B_n$, and $G_1, \ldots, G_n$, and we would like to define a logical formula in such a way that a satisfying assignment of the formula will correspond to a 3-coloring of the graph. In the back of our minds, we shall think of R_i as the statement, "Region i of the map is colored Red," and similarly for B_i and G_i. We then take as our clauses some statements that tell us that every region receives a single color and no two adjacent regions receive the same color.

This is easy to do: to guarantee that region i receives a color, we take the clause $R_i \vee B_i \vee G_i$, and if regions i and j are adjacent, then to guarantee that they do not receive the same color we take the three clauses $\overline{R_i} \vee \overline{R_j}$, $\overline{B_i} \vee \overline{B_j}$, and $\overline{G_i} \vee \overline{G_j}$. (To ensure that no region is assigned more than one color, we can also add clauses of the form $\overline{R_i} \vee \overline{B_i}$, $\overline{B_i} \vee \overline{G_i}$, and $\overline{G_i} \vee \overline{R_i}$. Alternatively, we can allow multiple colors and finish by picking one of the assigned colors for each region.)

It is not hard to see that the conjunction of all these clauses is satisfiable if and only if there is a 3-coloring of the map. Furthermore, the conversion process is a simple one that can be carried out in a time that is polynomial in the number of regions in the map. Thus, we have our hoped-for polynomial-time reduction.

Now let us give a formal description of what we have just done.

Definition (polynomial-time reducibility). Let S and T be subsets of I. We say that S is *polynomial-time reducible* to T if there exists a polynomial-time computable function $h : \text{I} \to \text{I}$ such that $x \in S$ if and only if $h(x) \in T$.

If S is polynomial-time reducible to T, then the following algorithm can be used to decide membership of S: given x, compute $h(x)$ (in polynomial time), then decide whether $h(x) \in T$. Therefore, if membership of T can be decided in polynomial time, so can membership of S. An equivalent, and important, way of saying

this is that if membership of S cannot be decided in polynomial time, then neither can membership of T. In short, if S is hard, then T is hard.

Now let us give a very important definition based on the notion of polynomial-time reducibility.

Definition (NP-completeness). A decision problem S is *NP-complete* if S is in $\mathcal{NP}$ and every decision problem in $\mathcal{NP}$ is polynomial-time reducible to S.

That is, if S has a polynomial-time algorithm, then so do *all other* problems in $\mathcal{NP}$. Thus, an NP-complete (decision) problem is in a certain sense "universal" among all problems in $\mathcal{NP}$.

At first this may seem a peculiar definition, because it is far from obvious that there are any NP-complete problems! However, in 1971, it was proved that SAT is NP-complete, and since then thousands of problems have been proved to be NP-complete as well. (Hundreds of them are listed in Garey and Johnson (1979).) Other examples are 3SAT and 3-colorability. The significance of 3SAT is that it is one of the most basic of all NP-complete problems. (It is not too hard to show that, by contrast, 2SAT and 2-colorability have polynomial-time algorithms.) In order to prove that a decision problem S is NP-complete, one starts with a known NP-complete problem S' and finds a polynomial-time reduction from S' to S. It now follows that if S has a polynomial-time algorithm, then so does S' and hence so do all other problems in $\mathcal{NP}$. Sometimes these reductions are quite simple, like our reduction of 3-colorability to 3SAT. But sometimes they need a great deal of ingenuity.

Here are two further NP-complete problems.

Subset sum: *given a sequence of integers* $a_1, \ldots, a_n$ *and another integer b, does there exist a set J such that* $\sum_{i \in J} a_i = b$?

Traveling salesman problem: *given a finite* GRAPH [III.34] G, *does there exist a* Hamilton cycle? *That is, can one find a cycle of edges that visits each vertex of the graph exactly once?*

Interestingly, almost all natural problems in $\mathcal{NP}$ that are not obviously in $\mathcal{P}$ turn out to be NP-complete. However, there are two important examples that have not been shown to be NP-complete and are strongly believed not to be. The first is a problem we have already discussed: integer factorization. More precisely, consider the following decision problem.

---

3. Recall that the celebrated FOUR-COLOR THEOREM [V.12] asserts that this can always be done with four colors.

`Factor in interval`: *given* x, a, b, *does* x *have a prime factor* y *such that* $a \leqslant y \leqslant b$?

A polynomial-time algorithm for this can be combined with a simple binary search to find a prime factor if it exists. The reason this problem is unlikely to be NP-complete is that it also belongs to co $\mathcal{NP}$. (Roughly speaking, this is true because one can exhibit the prime factorization of x and demonstrate in polynomial time that it really is a prime factorization.) If it were NP-complete, then it would follow that $\mathcal{NP} \subset \text{co} \, \mathcal{NP}$, and hence, by symmetry, that $\mathcal{NP} = \text{co} \, \mathcal{NP}$.

The second example is the following.

`Graph isomorphism`: *given two graphs* G *and* H *with* n *vertices, is there a function* ϕ *from the vertex set of* G *to the vertex set of* H *such that* $\phi(x)\phi(y)$ *is an edge of* H *when, and only when,* xy *is an edge of* G?

Notice with these two examples how surprising it is that they can be reduced in polynomial time to problems such as `3SAT` or `3-colorability`. This is particularly true of the first, which has nothing to do with graphs or satisfiability of logical formulas.

If $\mathcal{P} \neq \mathcal{NP}$, then no NP-complete problem has a polynomial-time decision procedure. Consequently, the corresponding search problems cannot be solved in polynomial time. Thus, a proof that a problem is NP-complete is often taken as *evidence* that this problem is hard: if we could solve it, then we could also efficiently solve a multitude of other problems. But thousands of researchers (and tens of thousands of engineers) have, over several decades, tried and failed to find such procedures.

NP-completeness has more positive aspects as well. Sometimes it is possible to prove a fact about all sets in $\mathcal{NP}$ by establishing it only for some NP-complete set (and noting that polynomial-time reductions preserve the claimed property). Famous examples include the existence of "zero-knowledge proofs," established first for 3-coloring (see section 6.3.2), and the so-called *PCP theorem*, established first for 3SAT (see section 6.3.3).

5 Lower Bounds

As we mentioned earlier, it is very much harder to prove that certain problems *cannot* be solved efficiently than it is to find efficient algorithms (when they exist). In this section, we shall survey some of the basic methods that have been developed for finding lower bounds for the complexity of natural computational problems. That is,

we shall discuss results that say that no algorithm can run in fewer than a given number of steps.

In particular, we shall introduce the theories of *circuit complexity* and *proof complexity*. The first is defined with the long-term goal of proving that $\mathcal{P} \neq \mathcal{NP}$, and the second is a program that is aimed at proving that $\mathcal{NP} \neq \text{co} \, \mathcal{NP}$. Both of these theories use the notion of a *directed acyclic graph*, which models the flow of information in a computation or a proof, and the sequence of derivations of each new piece of information from previous ones.

A directed graph is a graph for which each edge is given a direction. One can visualize it as a graph with arrows along the edges. A *directed cycle* is a sequence of vertices $v_1, \ldots, v_t$ such that for every i between 1 and $t - 1$ there is an edge pointing from v_i toward v_{i+1} and there is also an edge pointing from v_t back to v_1. If a directed graph G has no directed cycle, then it is called *acyclic*. We shall abbreviate the phrase "directed acyclic graph" by writing DAG.

It is not hard to see that in every DAG there will be some vertices with no incoming edges and some with no outgoing edges. These are called *inputs* and *outputs*, respectively. If u and v are vertices of a DAG and there is an edge from u to v, then we say that u is a *predecessor* of v. The basic idea of the DAG model is that you place information at each input, and at each vertex v you have a very simple rule that derives some information at v from the information at all the predecessors of v. Starting at the inputs, you gradually move through the graph, working out the information at a vertex once you have worked out the information for all its predecessors, until you have reached all the outputs.

5.1 Boolean Circuit Complexity

A Boolean circuit is a DAG in which all the values at the inputs, outputs, and intermediate vertices are *bits*. That is, each vertex may take the value 0 or 1. We have to specify simple rules for determining the value at a vertex from the values of its predecessors, and the usual choice is to allow three logical operations: AND, OR, and NOT. We call a vertex v an *AND gate* if the following rule applies: the value at v is 1 if all its predecessors have value 1 and is otherwise 0. At an *OR gate* we have a similar rule: the value at v is 1 if and only if at least one of its predecessors has value 1. Finally, v is a *NOT gate* if it has exactly one predecessor u, and v takes the value 1 if and only if u takes the value 0.

Given any Boolean circuit with n inputs $u_1, \ldots, u_n$ and m outputs $v_1, \ldots, v_m$ one can associate with it a function f from I_n to I_m as follows. Given a $\{0, 1\}$-string $x = (x_1, \ldots, x_n)$ of length n, let each u_i take the value x_i. Then use the gates of the circuit to find the values at the outputs $v_1, \ldots, v_m$. If these are $y_1, \ldots, y_m$, then $f(x_1, \ldots, x_n) = (y_1, \ldots, y_m)$.

It is not hard to prove that *any* function from I_n to I_m can be computed in this way. Thus, we say that AND, OR, and NOT gates, or more briefly "$\wedge$", "$\vee$", and "$\neg$", form a *complete basis*. Moreover, this is true even if we restrict attention to DAGs where every vertex has at most two predecessors. In fact, we shall now assume that our DAGs have this property unless we say otherwise. There are other choices of gates that are complete bases, but we shall stick with "$\wedge$", "$\vee$", and "$\neg$" since this does not affect our discussion in an essential way.

It may be easy to show that every Boolean function f can be computed by means of a circuit, but as soon as one asks how large the circuit needs to be, one comes up against fascinating and very difficult questions. Thus, the following definition is central to the subject of circuit complexity.

Definition. Let f be a function from I_n to I_m. Then $S(f)$ is the size of the smallest Boolean circuit that computes f, where this is measured by the number of vertices in the corresponding DAG.

To see what this has to do with the $\mathcal{P}$ versus $\mathcal{NP}$ question, consider an NP-complete decision problem such as 3SAT. This can be coded as a function f from I to $\{0, 1\}$, with $f(x)$ taking the value 1 if and only if the formula corresponding to x is satisfiable. Now we cannot find a circuit to compute f for the simple reason that I is an infinite set. However, if we restrict attention to formulas that can be encoded as strings of length n, then we obtain a function $f_n : I_n \to \{0, 1\}$, and we can try to estimate $S(f_n)$.

If we do this for every n, then we obtain an estimate for the growth rate of $S(f_n)$ as n tends to infinity. Writing f for the infinite sequence of functions $(f_1, f_2, \ldots)$, let us define $S(f)$ to be the function that takes n to $S(f_n)$.

This is an important definition because of the following fact: if there is a polynomial-time algorithm for computing f, then the function $S(f)$ is bounded above by a polynomial. More generally, given any function $f : I \to I$, let f_n stand for the restriction of f to I_n. If f has Turing complexity T (as defined in section 2.1), then $S(f_n)$ is bounded above by a polynomial function of $T(n)$. That is, there is a sequence of circuits that computes the function f, and takes a time not significantly different from the time taken by the Turing machine.

This provides us with a potential method of proving lower bounds on computational complexity, since if we can prove that $S(f_n)$ grows very rapidly with n, then we have proved that the Turing complexity of f is very large. If f is a problem in $\mathcal{NP}$, then this proves that $\mathcal{P} \neq \mathcal{NP}$.

The circuit model of computation is finite rather than infinite, which raises an issue called *uniformity*. When we build a family of circuits from a Turing machine, the circuits are all in a certain sense "the same." More precisely, there is an algorithm that can generate these circuits, and the time it takes to generate each one is polynomial in its size. A uniform family of circuits is one that can be generated in this way.

However, by no means all families of circuits are uniform. Indeed, there are functions f that cannot be computed by Turing machines at all (let alone in a reasonable amount of time), despite having circuits of *linear* size. This extra power comes from the fact that these families of circuits do not have a succinct ("effective") description; that is, there is no single algorithm that can generate them. Such families are called *nonuniform*.

If there are many families of circuits that do not arise from Turing machines, then it would seem that proving good lower bounds for circuit complexity should be much harder than proving lower bounds for Turing complexity, since now one must rule out many more potential ways of computing a function. However, there is a strong sentiment that the extra power provided by nonuniformity is irrelevant to the $\mathcal{P}$ versus $\mathcal{NP}$ question: it is believed that for a natural problem such as 3SAT, nonuniformity does not help. Therefore, we have another big conjecture of theoretical computer science: that NP-complete sets do not have polynomial-size circuits. Why do we believe this conjecture? It would be nice to be able to say that its falsehood implied that $\mathcal{P} = \mathcal{NP}$.

We do not quite know that, but we do know that if it is false then "the polynomial-time hierarchy collapses." Roughly speaking, this means that a whole system of complexity classes, which appear to be distinct, would in fact all be the same, which would be very unexpected. In any case, it is hard to imagine that there might be a sequence of polynomial-sized circuits computing an NP-complete problem without its being possible to generate such a sequence by an efficient algorithm.

Even if we grant that nonuniformity does not help solve NP-complete problems, what is the point of replacing the Turing machine model by the more powerful model of circuit families? The main reason is that circuits are simpler mathematical objects than Turing machines, and have the great advantage of being *finite*. The hope is that, while abstracting away the uniformity condition, which ought to be irrelevant, circuits provide us with a model that can be analyzed using combinatorial techniques.

It is also worth mentioning that Boolean circuits are a natural computational model of "hardware complexity," so their study is of independent interest. Moreover, some of the techniques for analyzing Boolean functions have found applications elsewhere: for example, in computational learning theory, combinatorics, and game theory.

5.1.1   Basic Results and Questions

We have already mentioned several basic facts about Boolean circuits, in particular the fact that they can efficiently simulate Turing machines. Another basic fact is that *most Boolean functions require exponential-size circuits*. This can be proved by a simple counting argument: the number of small circuits is far smaller than the number of functions. More precisely, let the number of inputs be n. The number of possible functions defined on the set of all n-bit sequences is precisely 2^{2^n}. On the other hand, it is not hard to show that the number of circuits of size m is bounded above by around m^{m^2}. It follows easily that we cannot compute all functions unless $m > 2^{n/2}/n$. Furthermore, the proportion of functions that can be computed by a circuit of size at most m is tiny.

Thus, hard functions (for circuits and consequently for Turing machines) abound. However, this hardness is proved via a counting argument, which does not give us a way of actually exhibiting a hard function. That is, we cannot prove such hardness for any *explicit* function f, where "explicit" means that we place some algorithmic restriction on f, such as belonging to $\mathcal{NP}$ or $\mathcal{EXP}$. In fact, the situation is even worse: no *nontrivial* lower bound is known for any explicit function. For any function f on n bits (assuming that it depends on all its inputs), we trivially must have $S(f) \geqslant n$, just to read the inputs. A major open problem of circuit complexity is beating this trivial bound by more than a constant factor.

Open problem. *Find an explicit Boolean function f (or even a length-preserving function f) for which $S(f)$ is superlinear: that is, not bounded above by cn for any constant c.*

A particularly basic special case of this problem is the question of whether addition is easier than multiplication. Let ADD and MULT denote, respectively, the addition and multiplication functions defined on pairs of integers (presented in binary). For addition, the usual procedure one learns at school gives rise to a linear-time algorithm, which implies a linear upper bound for $S(\text{ADD})$ as well. For multiplication, the standard school algorithm runs in *quadratic* time: that is, the number of steps is proportional to n^2. This can be greatly improved (via FAST FOURIER TRANSFORMS [III.26]) to an algorithm that yields $S(\text{MULT}) < n(\log n)^2$. Since $\log n$ grows very slowly with n, this is only slightly superlinear. And now the question is whether this can be improved further. In particular, do there exist linear-size circuits for multiplication?

How can circuit complexity be a thriving subject if no nontrivial bounds are known for any explicit functions? The answer is that there have been some remarkable successes in proving lower bounds under natural extra assumptions on the circuits. We shall now describe the most important of these extra assumptions.

5.1.2   Monotone Circuits

As we have seen, general Boolean circuits can compute every Boolean function, and can do it at least as efficiently as general algorithms. Now some functions have additional properties that might lead one to expect that they could be computed with Boolean circuits of a particular kind. For example, consider the function CLIQUE, defined on the set of all graphs as follows. If G is a graph with n vertices, then a *clique* in G is defined to be a set of vertices such that any two are joined by an edge. Let us define CLIQUE(G) to be 1 if G contains a clique of size at least $\sqrt{n}$ and 0 otherwise.

Notice that if we add an edge to G, then either CLIQUE(G) changes from 0 to 1 or it stays the same. What it will *not* do is change from 1 to 0: adding an edge obviously cannot *destroy* a clique.

We can encode G as a string x of $\binom{n}{2}$ bits, one for each pair of vertices, assigning 1 to a bit if the corresponding pair of vertices is joined by an edge and 0 otherwise. If we then set CLIQUE(x) to equal CLIQUE(G), we find that changing any bit of x from a 0 to a 1 cannot change CLIQUE(x) from 1 to 0. Boolean functions with this property are called *monotone*.

When considering the complexity of monotone functions, it is extremely natural to restrict the circuits by allowing only AND and OR gates, and disallowing NOT gates. Notice that "∧" and "∨" are monotone operations, in the sense that changing an input bit from 0 to 1 will not change the output of the gate from 1 to 0, whereas "¬" is certainly not monotone in this sense. A circuit that uses just "∧" and "∨" is called a *monotone circuit*. It is not hard to show that every monotone function $f : \mathrm{I}_n \to \mathrm{I}_m$ can be computed by a monotone circuit, and that almost all monotone functions need exponential-sized circuits.

Does the extra restriction on the circuits make it easier to prove lower bounds? For over forty years the answer seemed to be not much: nobody could prove a super-polynomial lower bound for the monotone complexity of any explicit monotone function. But then, in 1985, a new technique called the *approximation method* was invented to prove the remarkable theorem that CLIQUE has super-polynomial monotone complexity. This technique eventually led to the following even stronger result.

Theorem. CLIQUE *requires monotone circuits of exponential size.*

Very roughly speaking, the approximation method works as follows. Assume that CLIQUE can be computed with a small monotone circuit. Then replace the occurrences of "∧" and "∨" in this circuit with other gates that are cleverly chosen (and complex to describe), denoting these by "∧̃" and "∨̃," respectively. The new gates are chosen to satisfy two key properties.

(i) Replacing one particular gate has only a "small" effect on the output of the circuit (where "small" is defined in terms of a certain natural but nontrivial measure of distance). Consequently, if a circuit has few gates, then replacing all of them yields a new circuit that approximates the original circuit for "most" choices of inputs.

(ii) On the other hand, *every* circuit (regardless of its size) containing only the approximating gates "∧̃" and "∨̃" computes a function that can be shown to be "far" from CLIQUE, in the sense that it disagrees with CLIQUE on many inputs.

CLIQUE is a well-known NP-complete problem, so the above theorem provides us with an explicit monotone function, conjectured not to be in $\mathcal{P}$, that cannot be computed by small monotone circuits. It is natural at this point to wonder whether every monotone function that *is* in $\mathcal{P}$ can be computed by a small monotone circuit. If so, we would be able to deduce that $\mathcal{P} \neq \mathcal{NP}$. However, the same method yields a *super-polynomial* lower bound for the size of monotone circuits that compute the PERFECT MATCHING function, which is monotone and is in $\mathcal{P}$. Given a graph G, this function outputs 1 if one can pair up the vertices in such a way that every pair is connected by an edge and 0 otherwise. Furthermore, exponential-size lower bounds are known for other monotone functions in $\mathcal{P}$, so general circuits are known to be substantially more powerful than monotone circuits, even for computing monotone functions.

5.1.3 Bounded-Depth Circuits

To understand the motivation for our next model, consider the following basic question: "Can one speed up computation by using several computers in parallel?" For instance, suppose that a certain task can be performed by one computer in t steps. Can it be performed by t (or even t^2) cooperating computers in constant time (or just in $\sqrt{t}$ time)? The common wisdom is that the answer depends on the task in question: if a single person can dig at a rate of one cubic meter per hour, then in one hour a hundred people can dig a ditch that is 100 m long, but not a hole 100 m deep. Determining which computational tasks can be "parallelized" when many processors are available and which are "inherently sequential" is a basic question for both practical and theoretical reasons.

A very good feature of the circuit model is that it can easily be used to study questions of this kind. Let us define the *depth* of a DAG to be the length of the longest directed path in it: that is, the longest sequence of vertices where there is an edge from each one to the next. This notion of depth models the *parallel time* needed to compute the function: if you put a separate processor at each gate of a circuit of depth d, and at each phase you evaluate all gates for which the inputs have already been evaluated, then the number of phases you need is d. Parallel time is another important computational resource. Here again our knowledge is scarce—we do not know how to disprove the statement that every explicit function can be computed by a circuit of polynomial size *and* logarithmic depth.

Thus, we will restrict d to be a constant. It then becomes necessary to allow our gates to have *unbounded fan-in*, meaning that the AND and OR gates are allowed to have any number of incoming edges. (If we do not allow this, then each output bit can depend only on a constant number of input bits.) With this very stringent restriction on circuit depth, it is possible to prove lower bounds for the complexity of explicit functions. For example, let PAR(x) (for "parity") equal 1 if and only if the binary string has an odd number of 1s, and let MAJ(x) (for "majority") equal 1 if and only if there are more 1s than 0s in x.

Theorem. *For any constant d, the functions* PAR *and* MAJ *cannot be computed by a polynomial-sized family of circuits of depth d.*

This result is due to another fundamental proof technique: the *random restriction method*. The idea is to fix at random (with judiciously chosen parameters) most of the input variables, by assigning them random values. Note that this simultaneously restricts the function as well as the circuit. This "restriction" should satisfy the following two properties.

(i) The restricted circuit becomes very simple: for instance, it may depend on only a small subset of the remaining, unfixed input variables.

(ii) The restricted function remains complex: for instance, it may depend on all remaining input variables.

For PAR the second property is easily seen to hold, and of course the heart of the matter is analyzing the effect of random restrictions on shallow circuits.

Interestingly, MAJ remains hard for constant-depth polynomial-size circuits even if the circuits are also allowed (unbounded fan-in) PAR-gates. However the "converse" does not hold; that is, PAR has constant-depth polynomial-size circuits with (unbounded fan-in) MAJ-gates. Indeed, the latter class seems to be quite powerful: nobody has managed to prove that there are functions in $\mathcal{NP}$ that cannot be computed by such circuits, even if the depth is restricted to 3.

5.1.4 Formula Size

Formulas are perhaps the most standard way in which mathematicians express functions. For example, given a quadratic polynomial $at^2 + bt + c$ with $b^2 > 4ac$, the larger of its two roots is represented in terms of its (input) coefficients a, b, and c by the formula $(-b + \sqrt{b^2 - 4ac})/2a$. This is an arithmetic formula. In Boolean formulas the logical operations "$\neg$", "$\wedge$", "$\vee$"

replace the arithmetic operations above. For example, if $x = (x_1, x_2)$ is a Boolean string of length 2, then PAR(x) is given by the formula $(\neg x_1 \wedge x_2) \vee (x_1 \wedge \neg x_2)$.

Any formula can be represented by a circuit, but this circuit has the additional property that its underlying DAG is a *tree*. Intuitively, this means that the computation is not allowed to reuse a previously computed partial result (unless it recomputes it). A natural size measure for formulas is the number of occurrences of variables in them, which is the same as the number of gates, to within a factor of 2.

Formulas are natural not only because of their prevalence in mathematics, but also because their size can be related to the depth of circuits and to the *memory* requirements of Turing machines (i.e., their space complexity).

By recursively using the above formula for PAR, that is, by using the fact that PAR($x_1, \ldots, x_{2n}$) is equal to PAR(PAR($x_1, \ldots, x_n$), PAR($x_{n+1}, \ldots, x_{2n}$)), we obtain a formula for the parity of n variables that has size n^2. Given the fact that PAR has a simple circuit of linear size, one might wonder if there are smaller formulas as well. One of the oldest results in circuit complexity gives a negative answer.

Theorem. *Boolean formulas for* PAR *and* MAJ *must have at least quadratic size.*

The proof follows a simple combinatorial (or information-theoretic) argument. By contrast, there are linear-size circuits for both functions. This is very easy to show for PAR, but not for MAJ.

Can we give super-polynomial lower bounds on formula size? One of the cleanest methods suggested so far is the *communication complexity method*, which provides an information-theoretic setting for studying this computational problem. The power of this approach has been demonstrated mainly in the context of monotone formulas, where it yields an exponential lower bound for the PERFECT MATCHING problem (defined in section 5.1.2).

Suppose that two players play the following game. One player is given a graph G with n vertices that contains no perfect matching, and the other is given a graph H, with the same vertices, that does contain a perfect matching. Then there must be some pair of vertices that are joined by an edge in H but not joined in G. The aim of the two players is to find such a pair by sending each other bit strings, which each thinks of as encoding messages according to some prearranged scheme. Of course, the player with graph G could simply send enough messages to specify the entire graph, but the question is whether there is some protocol that would enable them to find a pair of the desired kind with far fewer bits being exchanged. The smallest number of bits needed (in the worst case) is called the *monotone communication complexity* of the problem.

It has been shown that the monotone communication complexity must be at least linear in n, and this leads to the exponential lower bound just mentioned. More generally, if $f : I_n \rightarrow \{0, 1\}$ is a monotone function, then the monotone communication complexity of f is the smallest number of bits that must be exchanged, in the worst case, to find a place i where $x_i = 0$ and $y_i = 1$, if $f(x) = 0$ and $f(y) = 1$. If f is not monotone, then one simply asks to find i such that x_i and y_i differ, and the smallest number of exchanges needed is the *communication complexity* of f. It can be shown that the monotone formula size of f is at least $\exp(cm)$ for a positive constant c if and only if the monotone communication complexity of f is at least $c'm$ for a positive constant c'. The corresponding statement also holds for general formula size and general communication complexity.

5.1.5   *Why Is It so Difficult to Prove Lower Bounds?*

As we have seen, complexity theory has developed quite a few powerful techniques, which have been useful for proving strong lower bounds, at least in restricted models of computation. But they all fall well short of providing nontrivial lower bounds for *general* circuits. Is there a fundamental reason for this failure? The same may be asked about any long-standing mathematical problem, such as the RIEMANN HYPOTHESIS [V.26], for example, and the typical answer would be rather vague: that it seems that the current tools and ideas do not suffice.

Remarkably, for circuit complexity this vague feeling has been made into a precise theorem. Thus, there is a "formal excuse" for our failure so far. Roughly speaking, a very general class of arguments, called *natural proofs*, has been defined and shown to include all known proofs of lower bounds for restricted circuits. In fact, so broad is the class of arguments that it is very hard to envisage what an "unnatural" proof might be like. On the other hand, it has also been shown that if there is a natural proof that $\mathcal{P} \neq \mathcal{NP}$, then there are fairly efficient (not quite polynomial-time, but significantly faster than known) algorithms for various problems, including integer factorization. So if, like most

complexity theorists, you believe that these problems do *not* have efficient algorithms, then you also believe that *there is no natural proof that $\mathcal{P} \neq \mathcal{NP}$.*

The connection between natural proofs that $\mathcal{P} \neq \mathcal{NP}$ and some notoriously hard problems is through the notion of *pseudorandomness*, which is discussed in section 7.1.

One interpretation of this result is that it shows that general circuit lower bounds are "independent" of a certain natural fragment of PEANO ARITHMETIC [III.67]. This gives a hint that the $\mathcal{P}$ versus $\mathcal{NP}$ question may be independent of all of Peano arithmetic, or even of THE AXIOMS OF ZFC [IV.22 §3.1], although few believe the latter to be the case.

5.2   Arithmetic Circuits

As mentioned earlier, directed acyclic graphs can be used in various different contexts. We shall now leave Boolean functions and operations and look instead at arithmetical operations and functions that take numerical values, by which we mean values in $\mathbb{Q}$ or $\mathbb{R}$ or indeed in any FIELD [I.3 §2.2]. If F is a field, then we can consider a DAG in which the inputs are now elements of F and the gates are the field operations "+" and "×" (including multiplication by fixed field elements such as -1). Then, just as with Boolean circuits, once we know the inputs we can assign values to all vertices of the DAG: at each vertex one just applies the corresponding arithmetical operation to the values assigned to its predecessors, once these have been calculated. An arithmetic circuit computes a polynomial function $p : F^n \rightarrow F^m$, and every homogeneous polynomial function is computed by some circuit. To allow the computation of inhomogeneous polynomials, we augment the model by allowing a special input vertex whose value is the constant "1" of the field.

Let us consider a couple of examples. The polynomial $x^2 - y^2$, which as written requires two multiplications and one addition, can be computed by the circuit $(x + y)(x - y)$ which requires instead one multiplication and two additions. The polynomial x^d, which is defined using $d - 1$ multiplications, may in fact be computed with only $2 \log d$ multiplications: first compute $x, x^2, x^4, \ldots$ (each term in the sequence squaring the previous one), and then multiply together the appropriate subset of these powers to get the exponent d.

We denote by $S_F(p)$ the smallest possible size of a circuit that computes p. When we give no subscript, we shall assume that $F = \mathbb{Q}$, the field of rational numbers.

We do not count multiplication by a fixed field element as contributing to the size of a circuit: for example, when we said that $(x + y)(x - y)$ involves one multiplication, we were not counting the multiplication of y by -1. The reader may wonder about division. However, we will be mainly interested in computing polynomials, and for computing polynomials (over infinite fields) division can be efficiently emulated by the other operations. As usual, we will be interested in sequences of polynomials, one for every input size, and will study size asymptotically.

It is easy to see that, for any *fixed* finite field F, arithmetic circuits over F can simulate Boolean circuits (on Boolean inputs) with only a constant factor increase in size. Thus, lower bounds for such arithmetic circuits yield corresponding lower bounds for Boolean circuits. Therefore, if we want to avoid the extreme difficulty with which we are already familiar, it makes sense to focus more on infinite fields, where lower bounds may perhaps be easier to obtain.

As in the Boolean case, the mere existence of hard polynomials is easy to establish.[4] But, as before, we will be interested in *explicit* (families of) polynomials. The notion of explicitness is more delicate here, but it can be formally defined (and, for example, polynomials with algebraically independent coefficients are not considered explicit).

An important parameter, which is absent in the Boolean model, is the *degree* of the polynomial(s) being computed. For example, a polynomial of degree d, even in one variable, requires size at least $\log d$. Let us briefly consider the one-variable, or *univariate*, case first, in which the degree is the main parameter of interest, since this case already contains striking and important problems. Then we shall move to the general *multivariate* case, in which n, the number of inputs, will be the main parameter.

5.2.1   Univariate Polynomials

How tight is the $\log d$ lower bound for the size of an arithmetic circuit computing a polynomial of degree d? A simple dimension argument shows that for most degree-d polynomials p, $S(p)$ is proportional to d. However, we know of no explicit polynomial with this

property. (Of course, this is shorthand for "explicit family of polynomials, one for each degree d.") In fact, considerably less is known even than this.

Open problem.   *Find an explicit polynomial p of degree d, such that $S(p)$ is not bounded above by $c \log d$ for some constant c.*

Two concrete examples are illuminating. Let $p_d(x) = x^d$, and $q_d(x) = (x + 1)(x + 2) \cdots (x + d)$. We have already seen that $S(p_d) \leqslant 2 \log d$, so the trivial lower bound is relatively tight. On the other hand, it is a major open problem to determine $S(q_d)$, and the conjecture is that $S(q_d)$ grows more quickly than any power of $\log d$. This question is particularly important because of the following result. If $S(q_d)$ *is bounded above by a power of* $\log d$, *then integer factorization has polynomial-size circuits.*

5.2.2   Multivariate Polynomials

Now let us return to polynomials with n variables. It is convenient to make n our only input size parameter, so we shall restrict ourselves to polynomials of total degree at most n, even when we do not mention this restriction.

For almost every polynomial p in n variables, $S(p)$ is at least $\exp(n/2)$. Again, this follows from an easy dimension argument, but again we would like to find explicit (families of) polynomials that are hard to compute. Unlike in the Boolean world, here there are lower bounds that slightly exceed the trivial ones. The following theorem is proved using elementary tools from algebraic geometry.

Theorem.   *There is a positive constant c such that $S(x_1^n + x_2^n + \cdots + x_n^n) \geqslant cn \log n$.*

The same techniques extend to prove lower bounds of similar strength for other natural polynomials such as the symmetric polynomials and the DETERMINANT [III.15] (which can be regarded as a polynomial in the entries of the matrix). Establishing a stronger lower bound for some explicit polynomial is a major open problem. Another is obtaining a superlinear lower bound for any polynomial map of *constant total degree*. Outstanding candidates for the latter are the *linear* maps that compute the discrete Fourier transform over the complex numbers or the Walsh transform over the rationals. For both these transformations algorithms of time complexity $O(n \log n)$ are known.

---

4. A counting argument over infinite fields is inadequate (e.g., for every $a, b \in F$ the circuit $ax+b$ has size two, and so there are infinitely many circuits of size 2). Instead, a "dimension" argument is used, showing that the set of polynomials that are computable by small circuits forms a vector space of lower dimension than the set of all polynomials of adequate degree.

Now let us focus on specific polynomials of central importance. The most natural and well-studied candidate for the last open problem is MATRIX MULTIPLICATION [I.3 §4.2]: given two $m \times m$ matrices A, B, how many operations are needed to compute their product? The obvious algorithm, which follows from the definition of matrix product, requires about m^3 operations. Can this be beaten? It turns out that what really matters here is the number of multiplications. The first hint that one can improve on the obvious algorithm comes from the first nontrivial case (i.e., $m = 2$). While the usual algorithm uses eight multiplications, one can in fact reorganize the calculation and get away with only seven. This leads to a recursive argument: given a $2m \times 2m$ matrix, think of it as a 2×2 matrix, each entry of which is itself an $m \times m$ matrix. It follows that doubling the size of the matrix increases the number of multiplications needed by a factor of at most 7. This argument leads to an algorithm with only $m^{\log_2 7}$ multiplications (and roughly as many additions).

These ideas have been developed and extended to yield the following strong, but not quite linear, upper bound, where we denote by $n = m^2$ the natural input size, and by MM the matrix multiplication function.

Theorem. *For every field F there is a constant c such that $S_F(\text{MM}) \leqslant cn^{1.19}$.*

So what is the complexity of MM (even if one counts only multiplication gates)? Is it linear, or almost linear (something like $n \log n$, say), or is $S(\text{MM})$ at least n^α for some $\alpha > 1$? This is a famous open problem.

We next consider two polynomials in the $n = m^2$ variables representing an $m \times m$ matrix. We have already mentioned the determinant, but we shall also look at the *permanent*, which is defined by the determinant formula, except that now all the signs are positive. (In other words, one simply adds up $m!$ products instead of adding some and subtracting others.) We shall denote these by DET and PER, respectively.

While DET plays a major role in classical mathematics, PER is somewhat esoteric (though it appears in statistical mechanics and quantum mechanics). In the context of complexity theory both polynomials are of great importance, because they are representative of natural complexity classes. DET has relatively low complexity (and is related to the class of polynomials having polynomial-sized arithmetic formulas), while PER seems to have high complexity (indeed, it is complete for a complexity class of counting problems denoted $\#\mathcal{P}$, which extends $\mathcal{NP}$). Thus, it is natural

to conjecture that PER is *not polynomial-time reducible to* DET.

One restricted type of reduction that makes sense in this algebraic context is called *projection*. Suppose we wish to find an algorithm for computing the permanent of an $m \times m$ matrix A. One approach might be to construct an $M \times M$ matrix B such that each of its entries is either a (variable) entry of A or a fixed element of the field, and to do so in such a way that the determinant of B equals the permanent of A. Then, as long as M is not too much larger than m, we can use the efficient algorithm for DET to give us an efficient algorithm for PER. A projection of this kind is known to exist with $M = 3^m$, but this is nothing like good enough. Therefore we ask the following question.

Open problem. *Can the permanent of an $m \times m$ matrix be expressed as the determinant of an $M \times M$ matrix, with M bounded above by a polynomial in m?*

If so, then $\mathcal{P} = \mathcal{NP}$: therefore, the answer is likely to be no. Conversely, if the answer could be shown to be no, then this would provide a significant step toward proving that $\mathcal{P} \neq \mathcal{NP}$, though it would probably not imply it.

5.3   Proof Complexity

The concept of *proof* distinguishes mathematics from all other fields of human inquiry. Mathematicians have gathered millennia of experience to attribute such adjectives to proofs as "insightful," "original," "deep," and, most notably, "difficult." Can one quantify mathematically the difficulty of proving various theorems? This is exactly the task undertaken in proof complexity. It seeks to classify theorems according to the difficulty of proving them, much as circuit complexity seeks to classify functions according to the difficulty of computing them. In proofs, just as in computation, there will be a number of models, called *proof systems*, that capture the power of reasoning that is allowed to the prover.

The types of statements, theorems, and proofs we shall deal with are best illustrated by the following example. We warn the reader in advance that the theorem we are about to discuss may seem too trivial to give us any insight into the nature of proofs: however, it turns out to be highly relevant.

The theorem in question is the well-known *pigeonhole principle*, which states that if you have more pigeons than holes then at least two pigeons will have to share a hole. More formally, there is no INJECTION

[I.2 §2.2] f from a finite set X to a smaller finite set Y. Let us reformulate this theorem and then discuss the complexity of proving it. First, we turn it into a sequence of finite statements. For each $m > n$ let PHP_n^m stand for the statement, "You cannot fit m pigeons into n holes if each pigeon needs a hole to itself." A convenient way of formulating this mathematically is to use an $m \times n$ matrix of Boolean variables x_{ij}. This can be used to describe a hypothetical mapping if we interpret $x_{ij} = 1$ to mean that the ith pigeon is placed in the jth hole. The pigeonhole principle states that either some pigeon is not mapped anywhere or two pigeons are mapped to the same hole. In terms of the matrix, this says that either there is some i such that $x_{ij} = 0$ for every j, or we can find $i \neq i'$ and j such that $x_{ij} = x_{i'j} = 1$.[5] These conditions are easily expressible as a *propositional formula* in the variables x_{ij} (that is, an expression built out of the x_{ij} using "$\wedge$", "$\vee$", and "$\neg$"), and the pigeonhole principle is the statement that this formula is a *tautology*: that is, it is satisfied by *every* assignment of true or false values (or equivalently 1 or 0) to the variables.

How can we prove this tautology to someone who can read our proof and perform simple, efficient computations? Here are a few possibilities which differ from each other in a number of ways.

- The standard proof uses symmetry and induction. It reduces PHP_n^m to PHP_{n-1}^{m-1} by saying that once the first pigeon has been assigned a hole, the task that is left is to place the remaining $n - 1$ pigeons into $m - 1$ holes. Notice that these holes may not be the *first* $n - 1$ holes, so for such an argument to become a formal proof one must argue by symmetry. Our proof system must be strong enough to capture this symmetry (which amounts to a renaming of the variables), and it must also allow us to use induction.

- At the other extreme, one can obtain a trivial proof, which requires only "mechanical reasoning," by simply presenting an evaluation of the formula for every possible input. As there are mn variables, the proof length is 2^{mn}, which is exponential in the size of the formula describing the assertion PHP_n^m.

- A more sophisticated ("mechanical") proof uses counting. Assume for a contradiction that there

exists an assignment of truth values to the variables that falsifies the formula. Since each pigeon is mapped to some hole, the assignment must have at least m 1s. But since each hole contains at most one pigeon, the assignment must contain at most n 1s. Therefore, $m \leqslant n$, which contradicts the assumption that $m > n$. For this proof to be admissible, our system has to allow inferences powerful enough to do counting of this kind.

The lesson from the above example is that proofs and their length depend on the underlying proof system. But what exactly is a proof system, and how do we measure the complexity of a proof? It is to this question that we now turn. Here are the salient features that we expect from any such system.

Completeness: every true statement has a proof.
Soundness: no false statement has a proof.
Verification efficiency: given a mathematical statement T and a purported proof for it π, it can be easily checked whether π does indeed prove T in the system.[6]

Actually, even the first two requirements are too much to expect from strong proof systems, as GÖDEL [VI.92] famously proved in his INCOMPLETENESS THEOREM [V.15]. However, we are considering just propositional formulas with finite proofs, and for these there are proof systems. In this context, the above conditions are concisely captured by the following definition.

Definition. A (*propositional*) *proof system* is a polynomial-time Turing machine M with the property that T is a tautology if and only if there exists a ("*proof*") π such that $M(\pi, T) = 1$.[7]

As a simple example, consider the following "truth-table" proof system M_{TT}, which corresponds to the trivial proof in the foregoing example. Basically, this machine will declare a formula T to be a theorem if evaluating T on each possible input makes T true. A bit more formally, for any formula T in n variables, $M_{\text{TT}}(\pi, T) = 1$ if and only if π is a list of all binary strings of length n, and for each such string σ we have $T(\sigma) = 1$.

---

5. Note that we have not ruled out the possibility that some pigeon is mapped to more than one hole—we could do so, but the principle remains valid even if we do not.

6. Here, efficiency of the verification procedure refers to its running time measured in terms of the *total length of the alleged theorem and proof*. In contrast, in sections 3.2 and 6.3, we consider the running time as a function of the *length of the alleged theorem* (or, alternatively, allow only proofs of a priori bounded length).

7. In agreement with standard formalisms (see below), the proof is seen as coming before the theorem.

Notice that M_{TT} runs in polynomial time in its input length. The point, of course, is that for typical interesting formulas such as the pigeonhole principle, whose size depends polynomially on the number of variables, the input length is extremely long, since the proof π has length exponential in the length of the formula. This leads us to the definition of the efficiency (or complexity) of a general propositional proof system M: it is the length of the shortest proof of each tautology. That is, if T is a tautology, we define its complexity $\mathcal{L}_M(T)$ to be the length of the shortest string π such that $M(\pi, T) = 1$. We then measure the efficiency of the proof system itself (i.e., M) by defining $\mathcal{L}_M(n)$ to be the maximum of $\mathcal{L}_M(T)$ over all tautologies T of length n.

Is there a propositional proof system which has polynomial-size proofs for all tautologies? The following theorem provides a basic connection between this question and computational complexity, and in particular with the major question of section 3.4. It follows quite easily from the NP-completeness of SAT, the problem of satisfying propositional formulas (and the fact that a formula is satisfiable if and only if its negation is not a tautology).

Theorem. *There exists a proof system M such that $\mathcal{L}_M$ is polynomial if and only if $\mathcal{NP} = \text{co}\,\mathcal{NP}$.*

To start attacking this formidable problem it makes good sense to begin by considering simpler (and thus weaker) proof systems before moving on to more and more complex ones. Moreover, there are tautologies and proof systems that naturally suggest themselves as good ones to study, systems in which certain basic forms of reasoning are allowed while others are not. In the rest of this section we shall focus on some of these restricted proof systems.

If a typical proof in a branch of mathematics such as algebra, geometry, or logic is written out in full, then it starts with some axioms and proceeds to a conclusion using a set of very simple and transparent *deduction rules*. Each line of the proof consists of a mathematical statement, or formula, which follows from earlier statements by means of one of these rules.[8] This deductive approach goes right back to EUCLID [VI.2] and perfectly fits our DAG model: the inputs can be labeled by the axioms, every other vertex is assigned a deduction rule, and the statement associated with each vertex is the statement that follows from its predecessors by means of the specified rule.

There is an equivalent and somewhat more convenient view of (simple) proof systems, namely as (simple) *refutation* systems. These encapsulate the idea of a proof by contradiction. We assume the negation of the tautology T we wish to prove, and use the rules of the system to derive a contradiction—that is, a statement that is identically FALSE. It is often easy to write the negation of a tautology T as a conjunction of mutually contradicting formulas (e.g., a set of clauses with no common truth assignment, a system of polynomials with no common root, a collection of half-spaces with empty intersection, etc). Assuming, for a contradiction, that all these *are* simultaneously satisfiable by some σ (which could be an assignment, root, or point, respectively), we derive more and more formulas that must also be satisfied by σ because of the soundness of the derivation rules, until eventually we reach a blatant contradiction (such as $\neg x \wedge x$, $1 = 0$, or $1 < 0$, respectively). We will use the refutation viewpoint throughout, and often exchange "tautology" and its negation, "contradiction."

So we turn to studying the proof length $\mathcal{L}_\Pi(T)$ of tautologies T in proof systems Π. The first observation, which reveals a major difference between proof complexity and circuit complexity, is that the trivial counting argument *fails*. The reason is that, while the number of functions on n bits is 2^{2^n}, there are at most 2^n tautologies of length n. Thus, in proof complexity, even the *existence* of a hard tautology, let alone an explicit one, would be of interest. As we shall see, however, most known lower bounds (in restricted proof systems) apply to very natural tautologies.

5.3.1   Logical Proof Systems

The proof systems in this section will all have lines that are Boolean formulas. The differences between the systems will be in the structural limits that are imposed on these formulas.

The most basic proof system, called the *Frege system*, puts no restriction on the formulas manipulated by the proof. It has just one derivation rule, called the *cut rule*: from the two formulas $(A \vee C)$, $(B \vee \neg C)$ we can derive $A \vee B$. Different basic books in logic have slightly different ways of describing this system. However, from a computational perspective they are all equivalent, in the sense that (up to polynomial factors) the length of the shortest proofs is independent of which variant you pick.

---

8. General proof systems as we defined them can also be adapted to this formalism, by considering a deduction rule that corresponds to a single step of the machine M. However, the deduction rules considered below are even simpler, and more importantly they are natural.

The counting-based proof of the pigeonhole principle can be carried out efficiently in the Frege system (but this is not a trivial fact), which tells us that $\mathcal{L}_{\text{Frege}}(\text{PHP}_n^{n+1})$ is polynomial in n. The major open problem in proof complexity is to find any tautology (as usual we mean a family of tautologies) that has no polynomial-size proof in the Frege system.

Open problem. *Establish a super-polynomial lower bound for the Frege system.*

As it seems to be very hard to find lower bounds for Frege systems, we turn to natural and interesting subsystems. The most widely studied system is called *resolution*. Its importance stems from its use by most propositional (as well as first-order) *automated theorem provers.*[9] The formulas allowed in resolution refutations are simply clauses (disjunctions), so the cut rule defined earlier simplifies to the *resolution rule*: from two clauses $(A \vee x)$, $(B \vee \neg x)$ we can derive $A \vee B$, where A, B are clauses and x is a variable.

A major result of proof complexity is that proving the pigeonhole principle is hard in the resolution system.

Theorem. $\mathcal{L}_{\text{resolution}}(\text{PHP}_n^{n+1}) = 2^{\Omega(n)}$

The proof of this result is related in an interesting way to the circuit lower bounds for the parity and majority functions discussed in section 5.1.3.

5.3.2 Algebraic Proof Systems

Just as a natural contradiction in the Boolean setting is an unsatisfiable collection of clauses, a natural contradiction in the algebraic setting is a system of polynomials without a common root.[10]

How would you prove that the system $\{f_1 = xy + 1, f_2 = 2yz - 1, f_3 = xz + 1, f_4 = x + y + z - 1\}$ has no common root (over any field)? A quick way is to observe that $zf_1 - xf_2 + yf_3 - f_4 \equiv 1$. Clearly, a common root of the system would be a root of this linear combination, which is a contradiction because the constant 1 function has no root. Can we always use such proofs?

A famous theorem known as HILBERT'S NULLSTELLENSATZ [V.17] tells us that the answer is yes. It states that if $f_1, f_2, \ldots, f_n$ are polynomials (with any number of variables) that have no common root, then there exist polynomials $g_1, \ldots, g_n$ such that $\sum_i g_i f_i \equiv 1$. How efficient are such proofs? Can we always have proofs (i.e., g_is) of length polynomial in the description of the f_is? Unfortunately not: the shortest explicit description of the g_is may be of exponential length, though proving this fact is highly nontrivial.

Another natural proof system, which is related both to Hilbert's Nullstellensatz and to computations of Gröbner bases in symbolic algebra programs, is *polynomial calculus* (PC). The lines in this system are polynomials, represented explicitly by all their coefficients, and it has two deduction rules: for any two polynomials g, h, we can derive their sum, $g + h$, and for any polynomial g and any variable x_i, we can derive the product $x_i g$. PC is known to be exponentially stronger than the proof system underlying Hilbert's Nullstellensatz. However, strong size lower bounds (obtained from degree lower bounds) are known for this system as well. For example, encoding the pigeonhole principle as a contradicting set of constant degree polynomials, we have the following theorem.

Theorem. *For every n and every $m > n$, $\mathcal{L}_{\text{PC}}(\text{PHP}_n^m) \geqslant 2^{n/2}$, over every field.*

5.3.3 Geometric Proof Systems

Yet another natural way to represent contradictions is by sets of regions in space that have empty intersection. For instance, many important problems in *combinatorial optimization* concern systems of linear inequalities in $\mathbb{R}^n$ and their relationship to the Boolean cube $\{0, 1\}^n$. Each inequality defines a half-space, and the problem is to decide whether the intersection of all these half-spaces contains a point with coordinates all equal to 0 or 1.

The most basic proof system is called Cutting Planes (CP). A line of a proof is a linear inequality with integer coefficients. The deduction rules are that you can add two inequalities, and, less obviously, that you can divide the coefficients by a constant and do some rounding, taking advantage of the fact that the points of the solution space have integer coordinates.

While PHP_n^m is easy in this system, exponential lower bounds are known for other tautologies. They are obtained from the monotone circuit lower bounds of section 5.1.2.

---

9. These are algorithms that attempt to generate proofs for given tautologies. These tautologies may be boring mathematically but of great practical importance, such as the statement that a computer chip or communication protocol functions correctly. Interestingly, popular applications also include a variety of theorems that *are* mathematically interesting, such as results in basic number theory.

10. Moreover, polynomials can easily encode propositional formulas. First, one puts such a formula into *conjunctive normal form*, or CNF: that is, one expresses it as the conjunction of a collection of clauses. CNF formulas can easily be converted to a system of polynomials, one per clause, over any field. One often adds the polynomials $x_i^2 - x_i$, which ensure Boolean values.

6   Randomized Computation

Up to now, the computations we have considered have all been *deterministic*: that is, the output is completely determined by the inputs and the rules governing the computations. In this section we shall continue to focus on polynomial-time computations, but now we shall allow our computing devices to make *probabilistic*, or *randomized*, choices.

6.1   Randomized Algorithms

A famous example of such an algorithm is one that tests for primality. If N is the positive integer to be tested, then the algorithm randomly chooses k numbers less than N, and repeatedly performs a simple test using each of the chosen numbers in turn. If N is composite, then the probability that the test detects this is at least $\frac{3}{4}$. Therefore, the probability that the algorithm fails to detect it for any of the k numbers is at most $(\frac{1}{4})^k$, which is very small indeed for even modestly large values of k. Details of how the test works can be found in COMPUTATIONAL NUMBER THEORY [IV.3 §2].

It is not hard to give a rigorous definition of a randomized Turing machine, but we shall not need the precise details here. The main point is that if M is a randomized Turing machine and x is an input string, then $M(x)$ is not a fixed output string, but rather a RANDOM VARIABLE [III.71 §4]. If, for example, the output is a single bit, then we shall make statements such as, "The probability that $M(x) = 1$ is p." The actual value of $M(x)$ will depend on the particular random choices made by the machine M when it runs.

If we are using a randomized algorithm to solve a decision problem S, then we would like $M(x)$ to give the correct answer with high probability *whatever the input* x. (The correct answer is 1 if $x \in S$ and 0 otherwise.) This leads to the definition of the complexity class $\mathcal{BPP}$ (for *bounded error, probabilistic polynomial time*).

Definition ($\mathcal{BPP}$). A Boolean function f is in $\mathcal{BPP}$ if there exists a probabilistic polynomial-time machine M such that $\Pr[M(x) \neq f(x)] \leqslant \frac{1}{3}$ for every $x \in \mathrm{I}$.

The error bound $\frac{1}{3}$ is arbitrary, and can be made much smaller if one runs the algorithm several times and takes a majority vote of the answers. (We stress that the random moves in the various runs are independent.) Standard probabilistic estimates show that, for any k, the error probability can be reduced to 2^{-k} if one runs the algorithm $O(k)$ times.

Because randomness is believed to be "available" and an exponentially small chance of failure is of no practical importance, the class $\mathcal{BPP}$ is in many ways a better model for efficient computation than $\mathcal{P}$, which it trivially contains. Let us mention some relations of this class $\mathcal{BPP}$ to other complexity classes we have seen already. It is easy to see that $\mathcal{BPP} \subseteq \mathcal{EXP}$; if the machine tosses m coins, we could enumerate all 2^m possible outcomes of these coin tosses and take a majority vote. The relation of $\mathcal{BPP}$ to $\mathcal{NP}$ is not known, but it is known that if $\mathcal{P} = \mathcal{NP}$ then $\mathcal{P} = \mathcal{BPP}$ as well. Finally, nonuniformity can replace randomness: every function in $\mathcal{BPP}$ has polynomial-size circuits. But the fundamental question is whether or not randomized algorithms are genuinely more powerful than deterministic ones (for decision problems).

Open problem.   *Does* $\mathcal{P} = \mathcal{BPP}$?

As we mentioned earlier, a deterministic polynomial-time algorithm was recently discovered for primality testing, though in practice the randomized algorithm is much more efficient. However, there are quite a few problems[11] that are known to be in $\mathcal{BPP}$ but not known to be in $\mathcal{P}$. Indeed, for most of these problems randomness gives an exponential improvement over the best deterministic algorithms that are known. Is this evidence that randomness increases our power to solve decision problems? Surprisingly, a completely different kind of evidence (discussed in section 7.1) suggests the opposite, namely that $\mathcal{P} = \mathcal{BPP}$.

6.2   Counting at Random

One important general question regarding $\mathcal{NP}$ search problems is that of determining *how many* solutions a particular instance has. This includes a host of interesting problems from various disciplines: for example, counting the number of solutions to a system of multivariate polynomials, counting the number of perfect matchings of a graph (or, equivalently, computing the permanent of a $\{0, 1\}$ matrix), computing the volume of a polytope (defined by linear inequalities) in high dimension (see [I.4 §9] for more about this problem), computing various parameters of physical systems, etc.

For most of these problems, even approximate counting is good enough. Clearly, an approximate count of the number of solutions will in particular allow one to

---

11. A central example is Identity Testing: given an arithmetic circuit over Q, decide if it computes the identically zero polynomial.

determine whether a solution exists at all. For example, if one knows the approximate number of satisfying assignments for a given propositional formula, then one certainly knows whether this number is at least 1. This tells us whether the formula is satisfiable and solves an instance of SAT. Interestingly, the converse is also true: if one can solve SAT, then one can use this ability to produce a randomized algorithm for approximating the *number* of solutions, to within any constant factor greater than 1. More precisely, there is an efficient probabilistic algorithm that can produce such an approximate count if it is allowed to make free use of a subroutine that solves SAT instances. It turns out that analogous statements holds for all NP-complete problems.

For some problems, approximate counting can be done *without* the SAT subroutine. There are polynomial-time probabilistic algorithms for approximating the permanent of positive matrices, approximating the volume of polytopes, and more. These algorithms use a connection between approximate counting and another natural algorithmic problem: that of randomly generating a solution in such a way that all correct solutions are equally likely to occur. The basic technique is to construct a Markov chain on the space of solutions with uniform stationary distribution and to analyze the rate of convergence of the chain to this distribution (see Hochbaum 1996, chapter 12).

What about *exact* counting? It is believed that this *cannot* be done by an efficient probabilistic algorithm, even if it can make free use of a SAT subroutine. A remarkable "complete" problem for this class of counting problems is counting the number of perfect matchings in a graph. What is surprising about it is that there is an efficient algorithm for finding a perfect matching in a graph, if one exists, and yet counting such matchings is complete in the sense that an efficient algorithm for doing this can be turned into an efficient algorithm for the counting version of *any other* problem in $\mathcal{NP}$.

6.3 Probabilistic Proof Systems

As we saw earlier, proof systems are defined in terms of their verification procedure. In section 5.3, we considered verification procedures that run in time that is polynomial in the combined length of the assertion and its alleged proof. Here (as in section 3.2), we restrict our attention to verification procedures that run in time that is polynomial in the length of the assertion. Such proof systems are related to the class $\mathcal{NP}$, since sets S

in $\mathcal{NP}$ are those with the following property: there is a polynomial-time algorithm M such that x belongs to S if and only if there exists a string y of length polynomial in x with $M(x, y) = 1$. In other words, we can regard y as a concise proof (verifiable by M) that x belongs to S.

What if we now allow M to be a *randomized* algorithm? Then we obtain a *probabilistic proof system*. Such systems are not put forward as a substitute for the notion of mathematical proof, but rather as an interesting extension of the notion of efficient verifiability in situations where a tiny amount of error can be tolerated. As we shall see, various types of probabilistic proof systems yield enormous advantages in computer science. We shall exhibit three remarkable manifestations of this. The first shows that we can use it to prove many more theorems, the second that we can do so without revealing *anything* in our proof, and the third that alleged proofs can be written in such a way that verifiers need only look at a tiny handful of bits in order to decide whether they are correct.

6.3.1 Interactive Proof Systems

Recall the graph isomorphism problem from section 4. Given two graphs G and H, it asks whether H is obtained from G by simply permuting the vertices. This problem is clearly in $\mathcal{NP}$, since one can just exhibit a permutation that transforms G into H.

We can look at this as a protocol involving a verifier, who can do polynomial-time computations, and a prover, who has unlimited computational resources. The verifier wishes to be convinced that G and H are isomorphic, so the prover sends a permutation and the verifier checks (in polynomial time) that it is valid.

Suppose that we now look at the graph *non*isomorphism problem. Is there any way for a prover to convince a verifier that two graphs G and H are not isomorphic? Obviously there will be for some pairs of graphs (G, H), but there does not seem to be a systematic method of demonstration that works for *all* nonisomorphic pairs. Yet, remarkably, if we allow *randomness* and *interaction*, then there is a simple way for the verifier to be convinced.[12]

Here is how it works. The verifier chooses at random one of the two graphs G and H, randomly permutes its vertices, and sends it to the prover. The prover then

---

12. We note that allowing interaction without randomness does not yield any gain; that is, such interactive (but deterministic) proof systems are exactly as powerful as $\mathcal{NP}$.

sends back a message saying whether this permuted graph is G or H.

If G and H are not isomorphic, then the permuted graph is isomorphic to exactly one of G and H, so the prover can determine which and thereby get the right answer. But if G and H *are* isomorphic, then the prover has no way of knowing which graph has been permuted, and therefore has a 50% chance of getting the right answer.

So now, to become convinced, the verifier repeats the procedure k times. If the graphs are not isomorphic, the prover will always get the right answer. If they are isomorphic, then with probability $1 - 2^{-k}$ the prover will make at least one mistake. If k is large, this becomes a near-certainty, so if the prover never makes a mistake, then the verifier will be convinced that the graphs are not isomorphic.

That was an example of an *interactive proof system*. Given a decision problem S, an interactive proof system for S is a protocol involving an interacting verifier and prover, with the property that if $x \in S$ then the verifier will eventually output 1, while if $x \notin S$ then there is a probability of at least $\frac{1}{2}$ that the verifier will output 0. As in the example, the verifier can then repeat the protocol several times, thereby replacing $\frac{1}{2}$ by a probability very close to 1. Also as in the example, the verifier is allowed polynomial-time randomized computations and the prover has unlimited computational power. Finally, the number of rounds of the interaction must be at most polynomial in the size of the input x, so that the entire verification procedure is efficient. The class of decision problems for which an interactive proof system exists is denoted $\mathcal{IP}$.

One can view the protocol as an "interrogation" by a persistent student, who asks the teacher "tough" questions in order to be convinced of correctness. Interestingly, it turns out that asking "tough" questions is no better than asking random questions! That is, every set that has an interactive proof system also has one in which the verifier only asks random questions that are uniformly and independently distributed in some predetermined set.

It turns out that for *every* decision problem S that belongs to $\mathcal{NP}$ there is an interactive proof system that can be used to demonstrate that $x \notin S$. It works by demonstrating the *nonexistence of an NP-proof that x is in S*. The proof of this result, which tells us that co$\mathcal{NP} \subset \mathcal{IP}$, involves an arithmetization of Boolean formulas. Furthermore, a complete characterization of the power of interactive proofs is known.

Let $\mathcal{PSPACE}$ be the class of all problems solvable in polynomial *space* (or *memory*). Although solving problems in $\mathcal{PSPACE}$ may require *exponential* time, they all have interactive proofs.

Theorem. $\mathcal{IP} = \mathcal{PSPACE}$.

While it is not known if $\mathcal{NP} \neq \mathcal{PSPACE}$, it is widely believed to be the case, and so it seems that interactive proofs are much more powerful than standard noninteractive and deterministic proofs (that is, NP-proofs).

6.3.2   Zero-Knowledge Proof Systems

A typical mathematical proof not only guarantees the truth of a statement, but also *teaches* you something about it. In this section we shall discuss a kind of proof that teaches you absolutely nothing, beyond the fact that the statement is true. Since this seems impossible, let us give an example.

Suppose a prover wants to convince you that a certain map (in the geography sense) can be colored with three colors in such a way that no two adjacent regions have the same color. The most obvious approach is actually to show you a coloring, but this teaches you something—a particular coloring—which you would not otherwise be able to find easily, even knowing that it existed (since this search problem is NP-complete). Is there any way the prover can convince you without giving you this extra knowledge?

Here is a way of doing it. Given any coloring of the map, with red, blue, and green, say, one can produce other colorings by permuting the colors: for instance, one might change all the red regions into blue and all the blue ones into red. Let the prover take six copies of the map and color them in six different ways, one for each permutation of the three colors. Now we have a sequence of rounds. In each round the prover randomly chooses one of the six colored maps, you randomly choose a pair of adjacent regions, and the prover allows you to check that they have different colors, *but does not allow you to look at the rest of the map*. If the graph cannot be properly colored with three colors and the prover tries to cheat, then after enough rounds (a polynomial number suffices) you will discover the deception by hitting upon two adjacent regions that have been given the same color (or perhaps one of them has not been colored at all). However, at each stage, all you learn about the two regions you look at is that they have different colors—you have no idea what those colors are in the coloring the prover started with. So you end up with no knowledge beyond the fact that the map can (almost certainly) be properly colored.

Similarly, a "zero-knowledge proof" that a certain formula is satisfiable should not reveal a satisfying assignment, or even any partial information (such as the truth value of one of the variables), or irrelevant information that is hard to compute (such as how to factorize an integer that happens to be encoded by the formula). In general, a zero-knowledge proof is an interactive proof that does not help you (the verifier) to make *any* computations that you were not able to make efficiently already.

Which theorems have zero-knowledge proofs? Obviously, if the verifier can determine the answer with no help, then the theorem has a trivial zero-knowledge proof, in which the prover does nothing at all. Thus, any set in $\mathcal{BPP}$ has a zero-knowledge proof. The zero-knowledge proof outlined for 3-colorability depended on noncomputational procedures, such as the prover watching carefully to make sure that you just look at two regions. Implementing the protocol in full on a computer takes some care, but a method of doing it has been devised, which depends on the hardness of integer factorization. The result is a *zero-knowledge proof system*. Combining this with the NP-completeness of 3-colorability, one can prove that zero-knowledge proof systems exist for *every* set in $\mathcal{NP}$. More generally, we have the following theorem.

Theorem. *If one-way functions exist (these are defined in section 7), then every set in $\mathcal{NP}$ has a zero-knowledge proof system. Moreover, this proof system can be efficiently derived from the standard $\mathcal{NP}$ proof.*

This theorem has a dramatic effect on the design of cryptographic protocols (see section 7.2). Furthermore, under the same assumption, an even stronger result holds: any set that has an interactive proof system also has a zero-knowledge interactive proof system.

6.3.3 Probabilistically Checkable Proofs

In this section we turn to one of the deepest and most surprising discoveries about the power of probabilistic proofs. Here, as in the case of standard (noninteractive) proofs, the verifier receives a complete written proof. The catch is that the verifier may read only a very small, randomly selected, part of this proof.

A good analogy is to imagine that you are refereeing a paper and trying to decide the correctness of a long proof by reading just a few random lines. If the proof has a single (but crucial) mistake, then you will probably not read the relevant line so you will not notice the mistake. But this is true only for the "natural" way of writing down proofs. It turns out that there are ways of writing proofs "robustly" (with a certain amount of redundancy) so that any mistake will manifest itself in many different places. (This may remind you of ERROR-CORRECTING CODES [VII.6]. There is indeed an important analogy here, and cross-fertilization between the two areas has been very significant.) Such a robust proof system is called a PCP, which stands for "probabilistically checkable proof."

Loosely speaking, a *PCP system* for a set S consists of a probabilistic polynomial-time verifier who has access to individual bits in a string that represents the (alleged) proof. The verifier tosses coins and, depending on the outcome, accesses only a *constant* number of the bits in the alleged proof. It should output 1 whenever x belongs to S (and an adequate proof is provided), while if x does not belong to S, then (no matter which false proof is provided) it should output 0 with probability at least $\frac{1}{2}$.

Theorem (the PCP theorem). *Every set in $\mathcal{NP}$ has a PCP system. Furthermore, there exists a polynomial-time procedure for converting any NP-proof to the corresponding PCP.*

In particular, it follows that the (robust) PCP has length that is polynomial in the length of the input. In fact, this PCP is itself an NP-proof.[13]

On top of its direct conceptual appeal, the PCP theorem (and its variants) has a major application to complexity theory: it allows us to prove that several natural approximation problems are hard (assuming that $\mathcal{P} \neq \mathcal{NP}$).

For example, suppose we are given n linear equations over the two-element field $\mathbb{F}_2$. If we choose random values for the variables, then any given equation will be satisfied with probability $\frac{1}{2}$, so it is clearly possible to satisfy at least half the equations. Also, by linear algebra one can quickly determine whether it is possible to satisfy all the equations simultaneously. However, it turns out that if $\mathcal{P} \neq \mathcal{NP}$ then there is no polynomial-time algorithm that will output 1 if 99% of the equations can be satisfied simultaneously and 0 if it is impossible

---

13. Here we take advantage of the fact that PCP systems are defined to be error free when $x \in S$ and the fact that the verifier in the PCP theorem uses only a logarithmic number of coin tosses, so one can efficiently check all possible outcomes.

to satisfy more than 51% of them. That is, even *approximately* determining the number of equations that can be satisfied simultaneously is hard.

To see the connection between such approximation problems and PCP, note that a PCP system for any set S gives rise to an optimization problem as follows. Suppose we are given an input x. Then for any alleged proof that $x \in S$, which is presented as a string y, there is a certain probability that the verifier accepts y. What is the maximum of this probability over all alleged proofs y? If we could answer this question to within a factor of 2, then we would be able to tell whether x belongs to S. Hence, if S is an NP-complete decision problem, the PCP theorem implies that this optimization problem is NP-hard (that is, at least as hard as any problem in $\mathcal{NP}$). One can now use reductions, capitalizing on the fact that the verifier reads only a constant number of bits in the alleged proof, to obtain similar results for many natural optimization problems.

This is of great theoretical interest, but some practical disappointment: in many cases, approximate solutions would have been just as useful as exact ones, but they turn out to be just as hard to obtain.

6.4   Weak Random Sources

We now turn to the question of how to obtain the randomness for all the probabilistic computations discussed in this section. Although randomness seems to be present in the world (e.g., the perceived randomness in the weather, Geiger counters, Zener diodes, real coin flips, etc.), it does not seem to be in the perfect form of the unbiased and independent coin tosses we have postulated. If we actually want to use randomized procedures, then we need to convert weak sources of randomness into almost perfect ones, because this is what probabilistic computations were defined to work with.

Algorithms that convert imperfect randomness into a stream of almost completely independent and unbiased bits are called *randomness extractors*, and near optimal ones have been constructed. This large body of work is surveyed in Shaltiel (2002), for example. The questions that arise turn out to be related to certain types of pseudorandom generators (see section 7.1) as well as to combinatorics and coding theory.

To illustrate the nature of the problem of randomness extraction, we consider three relatively simple models of weak random sources. Imagine first that you are in possession of a biased coin that has probability

p of coming up Heads, where $\frac{1}{3} < p < \frac{2}{3}$, but you do not know the bias. Can you produce a uniformly distributed binary value using such a coin? A simple solution consists of tossing the coin twice, outputting 1 if the result is Heads followed by Tails and 0 if the result is Tails followed by Heads, and otherwise continuing to the next attempt. This way we can generate a perfect coin toss by tossing the biased coin an expected number $((1 - p)p)^{-1}$ of times.

A more challenging setting arises if you are given n different biased coins, with unknown biases $p_1, \ldots, p_n$, each in the interval $(\frac{1}{3}, \frac{2}{3})$, and you are asked to generate an almost uniformly distributed binary value by tossing each of these coins *exactly once*. Here a good solution consists of tossing all coins and outputting the parity of the number of Heads. It can be shown that the outcome will be 1 with a probability that is exponentially (in n) close to $\frac{1}{2}$.

Finally, consider a situation in which the devil designs the coins in the latter example, but does so after seeing the outcome of previous coin tosses. That is, you are tossing n different coins, but the bias of the ith coin (i.e., p_i) may depend on the outcome of the previous $i - 1$ coin tosses (but still lies between $\frac{1}{3}$ and $\frac{2}{3}$). It can be shown that in this case you cannot do better than simply outputting the outcome of the first coin. However, if you are allowed to use just a few genuinely random bits, then you can do much better: given just $O(\log(n/\epsilon))$ perfectly random coin tosses, together with the n biased coin tosses, you can output a string of length proportional to n that is "ϵ-close" to being uniformly distributed.

7   The Bright Side of Hardness

If $\mathcal{P} \neq \mathcal{NP}$, as almost everybody believes, then there are computational problems of great interest that are inherently intractable. This is bad news, but there is a bright side to the matter: computational hardness has many fascinating conceptual consequences as well as important practical applications.

The hardness assumption we shall make is the existence of *one-way functions*; namely, functions that are easy to compute but hard to invert. For example, the product of two integers is of course easy to compute, but its "inverse"—factoring the resulting product—is the integer factorization problem, widely believed to be intractable. For our purposes, we shall need the inverse to be hard not just in the worst case, but hard *on average*. For example, for factoring it is believed that the product of two random primes of length n cannot be factored in polynomial time, even with some small constant probability of success. In general, we shall say that a function $f : I_n \to I_n$ is a *one-way function* if it is easy to evaluate (i.e., there exists a polynomial-time algorithm that returns $f(x)$ when you input x) but hard to invert in the following average-case sense: any polynomial-time algorithm M will fail to invert f correctly for at least half the input strings $x \in I_n$. That is, for at least half the strings x, if you input $y = f(x)$ into M, then the output will not be a string x' such that $f(x') = y$.

Do one-way functions exist? It is easy to see that if $\mathcal{P} = \mathcal{NP}$ then the answer is no. The converse is an important open problem: *If $\mathcal{P} \neq \mathcal{NP}$, does it follow that one-way functions exist?*

Below, we discuss the connections between computational difficulty (in the form of one-way functions), and two important computational complexity theories: the theory of *pseudorandomness* and the theory of *cryptography*.

7.1 Pseudorandomness

What is randomness? When should we say that a mathematical or physical object behaves randomly? These are fundamental questions that have been thought about for centuries. When the objects are probability distributions, on n-bit sequences, say, there is consensus about one point at least: the *uniform* distribution (in which each n-bit string appears with probability 2^{-n}) is "the most random" one. More generally, it seems reasonable to say that any distribution that is statistically close to the uniform distribution should also be regarded as having "good randomness" properties.[14]

One of the great insights of computational complexity theory is that there are distributions that are extremely far from the uniform distribution, but which are nevertheless "effectively random." The reason is that they are *computationally indistinguishable* from the uniform distribution.

Let us try to formalize this idea. Suppose we can randomly sample n-bit strings chosen according to a probability distribution P_n, and suppose that we want to know whether P_n is in fact the uniform distribution. One way to try to tell is to fix an efficiently computable function $f : I_n \to \{0, 1\}$ and consider two experiments:

one of the probability that $f(x) = 1$ when x is chosen with probability $P_n(x)$, and the other of the probability that $f(x) = 1$ when x is chosen with the uniform probability 2^{-n}. If there is a noticeable discrepancy between these two probabilities, then certainly P_n is not uniform. However, the converse is not true: it may be that P_n is far from uniform, but *no* efficiently computable function f can help us detect this. In that case, we say that P_n is *pseudorandom*.

This definition is both general and pragmatic. It refers to any efficient procedure that may be employed in an attempt to tell two distributions apart. And it is pragmatic because for any practical purpose a pseudorandom distribution is as good as a random one, for reasons we shall now explain.

Notice first that the behavior of any efficient probabilistic algorithm will be virtually unaffected if we replace its random source with a pseudorandom one. Why? Because if its behavior changed, then the algorithm itself would have efficiently distinguished between the random and pseudorandom sources, contradicting the definition of pseudorandomness!

Replacing uniform distributions by pseudorandom distributions is beneficial provided we can generate the latter using fewer resources. In this context, the resource we are trying hardest to save on is randomness. Suppose we have an efficiently computable function $\phi : I_m \to I_n$ and suppose that $n > m$. Then we can define a probability distribution on n-bit strings by choosing a random m-bit string x and computing $\phi(x)$. If this distribution is pseudorandom, then ϕ is called a *pseudorandom generator*. The random string x is called the *seed*, and if the generator stretches m-bit long seeds into strings of length $n = \ell(m)$, then we call the function ℓ the *stretch measure* of the generator. The larger the stretch measure, the better the generator is considered to be.

Of course, all this raises an important question: Do pseudorandom generators exist? It is to this question that we now turn.

7.1.1 Hardness versus Randomness

There is an obvious connection between pseudorandom generators and computational difficulty, since the main property of a pseudorandom generator is that its output should be computationally hard to distinguish from a purely random string, even though the two distributions are significantly different. However, there is a much less obvious connection as well.

---

14. Two probability distributions p_1 and p_2 are *statistically close* if they assign roughly the same probabilities: that is, if $p_1(E) \approx p_2(E)$ for every event E.

Theorem. *Pseudorandom generators exist if and only if one-way functions exist. Furthermore, if pseudorandom generators exist then they exist for any stretch measure that is a polynomial.*[15]

This theorem converts computational difficulty, or *hardness*, into pseudorandomness, and vice versa. Furthermore, its proof links computational indistinguishability to computational unpredictability, hinting that the computational difficulty is linked to randomness, or at least to the appearance of randomness.

The existence of pseudorandom generators has the remarkable consequence that probabilistic algorithms can be partially or even wholly *derandomized*. The basic idea is this. Suppose you have a probabilistic algorithm that computes a function f and requires n^c random bits (where n denotes the length of the input). Suppose that this algorithm outputs $f(x)$ with probability at least $\frac{2}{3}$. If you replace the random bits with n^c *pseudorandom* bits, generated from a seed of size m, then the behavior of the algorithm will hardly be affected. Therefore, if m is small, then you can do the same computation with only a small amount of randomness. If m is as small as $O(\log n)$, then it becomes feasible to check through *all* possible seeds. For close to two thirds of these, the algorithm outputs $f(x)$. But this means we can compute $f(x)$ deterministically and efficiently by taking a majority vote!

Can this actually be done? Can we use hardness to achieve the ultimate derandomization result, that $\mathcal{BPP} = \mathcal{P}$? The theory has developed to give essentially optimal answers to this question. Notice that if we wish to achieve an exponential stretch measure, we do not mind if the algorithm that performs the stretch takes exponential time (in the length of the seed). Such pseudorandom generators exist under very plausible hardness assumptions, such as the assumption that $\mathcal{NP}$-complete problems require exponential-size Boolean circuits. More generally, we have the following theorem.

Theorem. *If, for some constant $\epsilon > 0$, $S(\mathsf{SAT}) > 2^{\epsilon n}$, then $\mathcal{BPP} = \mathcal{P}$. Moreover, SAT can be replaced by any problem computable in $2^{O(n)}$-time.*

7.1.2   Pseudorandom Functions

Pseudorandom generators allow you to generate long pseudorandom sequences efficiently from short ran-

dom seeds. Pseudorandom *functions* are even more powerful: if you are given a random seed of n bits, they provide you with an efficient way of computing a function $f : \mathsf{I}_n \to \{0, 1\}$ that is computationally indistinguishable from a random function. Thus, with just n bits of randomness, one has efficient access to 2^n bits that appear random. (Note that it is inefficient to scan through all these bits—what we are given is the ability to look at any one of them in polynomial time.)

It turns out that *pseudorandom functions can be constructed given any pseudorandom generator*, and that they have many applications (most notably in cryptography).

7.2   Cryptography

Cryptography has existed for millennia, but whereas in the past it was focused on one basic problem—that of providing secret communications—the modern computational theory of cryptography is interested in *all* tasks that involve several agents who each wish to obtain some information while preserving the secrecy of other information. An important priority besides *privacy* (that is, keeping secrets) is *resilience*: one would like guaranteed privacy even if one is not certain that the other participants are behaving honestly.

A good example to illustrate these difficulties is playing a game of poker over the telephone or e-mail. You are encouraged to ponder seriously how this might be done, and realize to what extent standard poker relies on human vision, physical implements like cards with opaque backs, etc., to protect privacy and prevent cheating.

The general goal of cryptography is to construct schemes, called *protocols*, that maintain *any* desired functionality (rules, privacy requirements, etc.), even in the face of malicious attempts to make them deviate from this functionality. As with pseudorandomness, there are two key assumptions underlying the new theory. First, it is assumed that all parties, including the malicious adversaries, are computationally limited. Second, it is assumed that there are hard functions. Sometimes these are one-way functions, and sometimes they are yet stronger functions called "trapdoor permutations," which also exist if integer factorization is hard.

This goal is an ambitious one, but it has been achieved. There is a result that says, roughly speaking, that *every functionality can be securely implemented.*

---

15. In other words, if you can achieve a stretch measure $\ell(m) = m + 1$, then you can also achieve a stretch measure of $\ell(m) = m^c$ for any $c > 1$.

This includes highly complex tasks such as playing poker over the phone, but also very basic ones such as secure communication, digital signatures (a digital analogue of handwritten signatures), collective coin flipping, auctions, elections, and the famous *millionaires' problem*: how can two people interact to determine who is richer, without either of them learning anything further about the other's wealth?

Let us very briefly hint at connections between cryptography and matters that we have already discussed. First of all, consider the very definition of the central notion of cryptography: that of a *secret*. If you have an n-bit string, then when should we say that it is completely secret? A natural definition would be that it is secret if nobody else has any information about it: that is, from anybody else's point of view it is equally likely to be any of the 2^n-bit strings. However, in the new computational complexity theory, this is not the definition taken, since a *pseudorandom* n-bit string will, for all practical purposes, be just as secret.

The difference between the two definitions of a secret is huge. The point of cryptography is not just to have secrets (that is easy, just select a string at random) but actually to *use* them without giving away information. At first this seems impossible, since any nontrivial use of a secret n-bit string will cut down the set of possible strings that it might be, and therefore give away genuine information. However, if the new probability distribution over the possible strings (after the information has been given away) is pseudorandom, then this information *cannot feasibly be used*, since no efficient algorithm can tell the difference between a string that gives rise to the information you have revealed and a truly random string.

A famous example of this idea is given by the so-called *public-key encryption schemes*, such as RSA, which are described in detail in MATHEMATICS AND CRYPTOGRAPHY [VII.7] and in Goldreich (2004, chapter 5). In the RSA scheme, if a user, say Alice, wants to receive messages, she publishes a number N, called a *public key*, which is a product of two primes P and Q. If you know N then you can encrypt any message, but to decrypt it you need to know P and Q. Thus, if integer factorization is hard, then only Alice can feasibly decrypt messages, even though P and Q are completely determined by N.

The generic problem about using secrets is one in which there are k parties, and each party has a string of bits. They are interested in the value of some efficiently computable function f that depends on all the strings

of bits, but they would like to ascertain this without giving away any information about their own strings beyond what follows from the value of f. For example, in the case of the millionaires' problem, there are two parties, each with a string that encodes their wealth. They would like a protocol that provides them with a single bit that tells them who is richer, but gives them no information beyond this. The precise formulation of this condition is an extension of the formulation of zero-knowledge proofs (presented in section 6.3.2). As hinted at earlier in this section, assuming the existence of trapdoor permutations, *every such multiparty computation can be performed without yielding anything beyond the designated outputs.*

Finally, we come to the issue of cheating. In the foregoing discussion, we did not worry about malicious behavior and focused on what participants may learn from the transcript of their interaction. But how can a player, Bob, say, be forced to act "as specified," when his actions may depend partly on his secrets, which he does not want to reveal? The answer is closely related to zero-knowledge proofs. Essentially, each player whose turn it is to perform some computation is asked to prove to the others that he has acted as specified. This is a (mathematically boring) theorem and the standard proof is obvious (i.e., revealing all his secrets). But as we saw in our discussion of zero-knowledge proof systems in section 6.3.2, if a proof exists, then a zero-knowledge proof can be efficiently derived from it. Thus, *Bob can convince the others of his proper behavior without revealing anything about his secrets.*

8 The Tip of an Iceberg

Even within the topics reviewed above, many important notions and results have not been discussed, for space reasons. Furthermore, other important topics and even wide areas have not been mentioned at all.

The $\mathcal{P}$ versus $\mathcal{NP}$ question, as well as most of the discussion so far, focuses on a simplified view of the goals of (efficient) computations. Specifically, we have insisted on efficient procedures that *always* give the *exact* answer. However, in practice one may be content with less. For example, one may be happy with an efficient procedure that gives the correct answer for a large fraction of the instances. This will be useful if all instances are equally interesting, but that is typically not the case. On the other hand, demanding success under all input distributions gives back worst-case complexity. Between these two extremes is a useful and appealing theory of *average-case complexity* (see Goldreich 1997): one demands that algorithms succeed with high probability on every possible input distribution that can be *efficiently sampled*.

Another possible relaxation is settling for approximate answers. This can mean many things, and the best notion of approximation varies from context to context. For search problems, we may be satisfied with a solution that is close in some METRIC [III.56] to being valid (see Hochbaum (1996) and THE MATHEMATICS OF ALGORITHM DESIGN [VII.5]). For decision problems, we might ask how close the input is (again in some natural metric) to an instance in the set (see Ron 2001). And there is also approximate counting, which was discussed in section 6.2.

In this article we have focused on the *running time* of procedures. This is arguably the most important complexity measure, but it is not the only one. Another is the amount of *work space* consumed during the computation (see Sipser 1997). Another important issue is the extent to which a computation can be performed in parallel; that is, speeding up the computation by splitting the work among several computing devices, which are viewed as components of the same (parallel) machine and are provided with direct access to the same memory module. In addition to the parallel *time*, a fundamentally important complexity measure in such a case is the number of parallel computing devices used (see Karp and Ramachandran 1990).

Finally, there are several computational models that we have not discussed here. Models of *distributed computing* refer to distant computing devices, each given a local input, which may be viewed as a part of a global input. In typical studies one wishes to minimize the amount of communication between these devices (and certainly avoid the communication of the entire input). In addition to measures of communication complexity, a central issue is asynchrony (see Attiya and Welch 1998). The *communication complexity* of two-argument (and many-argument) functions is a measure of their "complexity" (see Kushilevitz and Nisan 1996), but in these studies communication proportional to the length of the input is not ruled out (but rather appears frequently). While being "information theoretic" in nature, this model has many connections to complexity theory. Altogether different types of computational problems are investigated in the context of *computational learning theory* (see Kearns and Vazirani 1994) and of *online* algorithms (see Borodin and El-Yaniv 1998). Finally, QUANTUM COMPUTATION [III.74] investigates the possibility of using quantum mechanics to speed up computation (see Kitaev et al. 2002).

9   Concluding Remarks

We hope that this ultra-brief survey conveys the fascinating flavor of the concepts, results, and open problems that dominate the field of computational complexity. One important feature of the field we did not do justice to is the remarkable web of (often surprising) connections between different subareas, and its impact on progress.

For further details on sections 1–4 the reader is referred to standard textbooks such as Garey and Johnson (1979) and Sipser (1997). For further details on sections 5.1–5.3 the reader is referred to Boppana and Sipser (1990), Strassen (1990), and Beame and Pitassi (1998), respectively. For further details on sections 6 and 7 the reader is referred to Goldreich (1999) (and also to Goldreich (2001, 2004)).

Further Reading

Attiya, H., and J. Welch. 1998. *Distributed Computing: Fundamentals, Simulations and Advanced Topics.* Columbus, OH: McGraw-Hill.

Beame, P., and T. Pitassi. 1998. Propositional proof complexity: past, present, and future. *Bulletin of the European Association for Theoretical Computer Science* 65:66–89.

Boppana, R., and M. Sipser. 1990. The complexity of finite functions. In *Handbook of Theoretical Computer Science*, volume A, *Algorithms and Complexity*, edited by J. van Leeuwen. Cambridge, MA: MIT Press/Elsevier.

Borodin, A., and R. El-Yaniv. 1998. *On-line Computation and Competitive Analysis.* Cambridge: Cambridge University Press.

Garey, M. R., and D. S. Johnson. 1979. *Computers and Intractability: A Guide to the Theory of NP-Completeness.* New York: W. H. Freeman.

Goldreich, O. 1997. Notes on Levin's theory of average-case complexity. *Electronic Colloquium on Computational Complexity*, TR97-058.

———. 1999. *Modern Cryptography, Probabilistic Proofs and Pseudorandomness.* Algorithms and Combinatorics Series, volume 17. New York: Springer.

———. 2001. *Foundation of Cryptography*, volume 1: *Basic Tools.* Cambridge: Cambridge University Press.

———. 2004. *Foundation of Cryptography*, volume 2: *Basic Applications.* Cambridge: Cambridge University Press.

———. 2008. *Computational Complexity: A Conceptual Perspective.* Cambridge: Cambridge University Press.

Hochbaum, D., ed. 1996. *Approximation Algorithms for NP-Hard Problems.* Boston, MA: PWS.

Karp, R. M., and V. Ramachandran. 1990. Parallel algorithms for shared-memory machines. In *Handbook of Theoretical Computer Science*, volume A, *Algorithms and Complexity*, edited by J. van Leeuwen. Cambridge, MA: MIT Press/Elsevier.

Kearns, M. J., and U. V. Vazirani. 1994. *An Introduction to Computational Learning Theory*. Cambridge, MA: MIT Press.

Kitaev, A., A. Shen, and M. Vyalyi. 2002. *Classical and Quantum Computation*. Providence, RI: American Mathematical Society.

Kushilevitz, E., and N. Nisan. 1996. *Communication Complexity*. Cambridge: Cambridge University Press.

Ron, D. 2001. Property testing (a tutorial). In *Handbook on Randomized Computing*, volume II. Dordrecht: Kluwer.

Shaltiel, R. 2002. Recent developments in explicit constructions of extractors. *Bulletin of the European Association for Theoretical Computer Science* 77:67–95.

Sipser, M. 1997. *Introduction to the Theory of Computation*. Boston, MA: PWS.

Strassen, V. 1990: Algebraic complexity theory. In *Handbook of Theoretical Computer Science*, volume A, *Algorithms and Complexity*, edited by J. van Leeuwen. Cambridge, MA: MIT Press/Elsevier.

IV.21 Numerical Analysis

Lloyd N. Trefethen

1 The Need for Numerical Computation

Everyone knows that when scientists and engineers need numerical answers to mathematical problems, they turn to computers. Nevertheless, there is a widespread misconception about this process.

The power of numbers has been extraordinary. It is often noted that the scientific revolution was set in motion when Galileo and others made it a principle that everything must be measured. Numerical measurements led to physical laws expressed mathematically, and, in the remarkable cycle whose fruits are all around us, finer measurements led to refined laws, which in turn led to better technology and still finer measurements. The day has long since passed when an advance in the physical sciences could be achieved, or a significant engineering product developed, without numerical mathematics.

Computers certainly play a part in this story, yet there is a misunderstanding about what their role is. Many people imagine that scientists and mathematicians generate formulas, and then, by inserting numbers into these formulas, computers grind out the necessary results. The reality is nothing like this. What

really goes on is a far more interesting process of execution of *algorithms*. In most cases the job could not be done even in principle by formulas, for most mathematical problems cannot be solved by a finite sequence of elementary operations. What happens instead is that fast algorithms quickly converge to "approximate" answers that are accurate to three or ten digits of precision, or a hundred. For a scientific or engineering application, such an answer may be as good as exact.

We can illustrate the complexities of exact versus approximate solutions by an elementary example. Suppose we have one polynomial of degree 4,

$$p(z) = c_0 + c_1 z + c_2 z^2 + c_3 z^3 + c_4 z^4,$$

and another of degree 5,

$$q(z) = d_0 + d_1 z + d_2 z^2 + d_3 z^3 + d_4 z^4 + d_5 z^5.$$

It is well-known that there is an explicit formula that expresses the roots of p in terms of radicals (discovered by Ferrari around 1540), but no such formula for the roots of q (as shown by Ruffini and ABEL [VI.33] more than 250 years later; see THE INSOLUBILITY OF THE QUINTIC [V.21] for more details). Thus, in a certain philosophical sense the root-finding problems for p and q are utterly different. Yet in practice they hardly differ at all. If a scientist or a mathematician wants to know the roots of one of these polynomials, he or she will turn to a computer and get an answer to sixteen digits of precision in less than a millisecond. Did the computer use an explicit formula? In the case of q, the answer is certainly no, but what about p? Maybe, maybe not. Most of the time, the user neither knows nor cares, and probably not one mathematician in a hundred could write down formulas for the roots of p from memory.

Here are three more examples of problems that can be solved in principle by a finite sequence of elementary operations, like finding the roots of p.

(i) Linear equations: solve a system of n linear equations in n unknowns.

(ii) Linear programming: minimize a linear function of n variables subject to m linear constraints.

(iii) Traveling salesman problem: find the shortest tour between n cities.

And here are five that, like finding the roots of q, cannot generally be solved in this manner.

(iv) Find an EIGENVALUE [I.3 §4.3] of an $n \times n$ matrix.

(v) Minimize a function of several variables.

(vi) Evaluate an integral.

(vii) Solve an ordinary differential equation (ODE).

(viii) Solve a partial differential equation (PDE).

Can we conclude that (i)–(iii) will be easier than (iv)–(viii) in practice? Absolutely not. Problem (iii) is usually very hard indeed if n is, say, in the hundreds or thousands. Problems (vi) and (vii) are usually rather easy, at least if the integral is in one dimension. Problems (i) and (iv) are of almost exactly the same difficulty: easy when n is small, like 100, and often very hard when n is large, like 1 000 000. In fact, in these matters philosophy is such a poor guide to practice that, for each of the three problems (i)–(iii), when n and m are large one often ignores the exact solution and uses approximate (but fast!) methods instead.

Numerical analysis is the study of algorithms for solving the problems of continuous mathematics, by which we mean problems involving real or complex variables. (This definition includes problems like linear programming and the traveling salesman problem posed over the real numbers, but not their discrete analogues.) In the remainder of this article we shall review some of its main branches, past accomplishments, and possible future trends.

2   A Brief History

Throughout history, leading mathematicians have been involved with scientific applications, and in many cases this has led to the discovery of numerical algorithms still in use today. GAUSS [VI.26], as usual, is an outstanding example. Among many other contributions, he made crucial advances in least-squares data fitting (1795), systems of linear equations (1809), and numerical quadrature (1814), as well as inventing THE FAST FOURIER TRANSFORM [III.26] (1805), though the last did not become widely known until its rediscovery by Cooley and Tukey in 1965.

Around 1900, the numerical side of mathematics started to become less conspicuous in the activities of research mathematicians. This was a consequence of the growth of mathematics generally and of great advances in fields in which, for technical reasons, mathematical rigor had to be the heart of the matter. For example, many advances of the early twentieth century sprang from mathematicians' new ability to reason rigorously about infinity, a subject relatively far from numerical calculation.

A generation passed, and in the 1940s the computer was invented. From this moment numerical mathemat-ics began to explode, but now mainly in the hands of specialists. New journals were founded such as *Mathematics of Computation* (1943) and *Numerische Mathematik* (1959). The revolution was sparked by hardware, but it included mathematical and algorithmic developments that had nothing to do with hardware. In the half-century from the 1950s, machines sped up by a factor of around 10^9, but so did the best algorithms known for some problems, generating a combined increase in speed of almost incomprehensible scale.

Half a century on, numerical analysis has grown into one of the largest branches of mathematics, the specialty of thousands of researchers who publish in dozens of mathematical journals as well as applications journals across the sciences and engineering. Thanks to the efforts of these people going back many decades, and thanks to ever more powerful computers, we have reached a point where most of the classical mathematical problems of the physical sciences can be solved numerically to high accuracy. Most of the algorithms that make this possible were invented since 1950.

Numerical analysis is built on a strong foundation: the mathematical subject of *approximation theory*. This field encompasses classical questions of interpolation, series expansions, and HARMONIC ANALYSIS [IV.11] associated with NEWTON [VI.14], FOURIER [VI.25], Gauss, and others; semiclassical problems of polynomial and rational minimax approximation associated with names such as CHEBYSHEV [VI.45] and Bernstein; and major newer topics, including splines, radial basis functions, and WAVELETS [VII.3]. We shall not have space to address these subjects, but in almost every area of numerical analysis it is a fact that, sooner or later, the discussion comes down to approximation theory.

3   Machine Arithmetic and Rounding Errors

It is well-known that computers cannot represent real or complex numbers exactly. A quotient like $1/7$ evaluated on a computer, for example, will normally yield an inexact result. (It would be different if we designed machines to work in base 7!) Computers approximate real numbers by a system of *floating-point arithmetic*, in which each number is represented in a digital equivalent of scientific notation, so that the scale does not matter unless the number is so huge or tiny as to cause overflow or underflow. Floating-point arithmetic was invented by Konrad Zuse in Berlin in the 1930s, and by the end of the 1950s it was standard across the computer industry.

Until the 1980s, different computers had widely different arithmetic properties. Then, in 1985, after years of discussion, the IEEE (Institute of Electrical and Electronics Engineers) standard for binary floating-point arithmetic was adopted, or *IEEE arithmetic* for short. This standard has subsequently become nearly universal on processors of many kinds. An IEEE (double precision) real number consists of a 64-bit word divided into 53 bits for a signed fraction in base 2 and 11 bits for a signed exponent. Since $2^{-53} \approx 1.1 \times 10^{-16}$, IEEE numbers represent the numbers of the real line to a relative accuracy of about 16 digits. Since $2^{\pm 2^{10}} \approx 10^{\pm 308}$, this system works for numbers up to about 10^{308} and down to about 10^{-308}.

Computers do not merely represent numbers, of course; they perform operations on them such as addition, subtraction, multiplication, and division, and more complicated results are obtained from sequences of these elementary operations. In floating-point arithmetic, the computed result of each elementary operation is almost exactly correct in the following sense: if "$*$" is one of these four operations in its ideal form and "$\circledast$" is the same operation as realized on the computer, then for any floating-point numbers x and y, assuming that there is no underflow or overflow,

$$x \circledast y = (x * y)(1 + \varepsilon).$$

Here ε is a very small quantity, no greater in absolute value than a number known as *machine epsilon*, denoted by $\varepsilon_{\text{mach}}$, that measures the accuracy of the computer. In the IEEE system, $\varepsilon_{\text{mach}} = 2^{-53} \approx 1.1 \times 10^{-16}$.

Thus, on a computer, the interval $[1, 2]$, for example, is approximated by about 10^{16} numbers. It is interesting to compare the fineness of this discretization with that of the discretizations of physics. In a handful of solid or liquid or a balloonful of gas, the number of atoms or molecules in a line from one point to another is on the order of 10^8 (the cube root of Avogadro's number). Such a system behaves enough like a continuum to justify our definitions of physical quantities such as density, pressure, stress, strain, and temperature. Computer arithmetic, however, is more than a million times finer than this. Another comparison with physics concerns the precision to which fundamental constants are known, such as (roughly) 4 digits for the gravitational constant G, 7 digits for Planck's constant h and the elementary charge e, and 12 digits for the ratio $\mu_{\text{e}}/\mu_{\text{B}}$ of

the magnetic moment of the electron to the Bohr magneton. At present, almost nothing in physics is known to more than 12 or 13 digits of accuracy. Thus IEEE numbers are orders of magnitude more precise than any number in science. (Of course, purely mathematical quantities like π are another matter.)

In two senses, then, floating-point arithmetic is far closer to its ideal than is physics. It is a curious phenomenon that, nevertheless, it is floating-point arithmetic rather than the laws of physics that is widely regarded as an ugly and dangerous compromise. Numerical analysts themselves are partly to blame for this perception. In the 1950s and 1960s, the founding fathers of the field discovered that inexact arithmetic can be a source of danger, causing errors in results that "ought" to be right. The source of such problems is *numerical instability*: that is, the amplification of rounding errors from microscopic to macroscopic scale by certain modes of computation. These men, including VON NEUMANN [VI.91], Wilkinson, Forsythe, and Henrici, took great pains to publicize the risks of careless reliance on machine arithmetic. These risks are very real, but the message was communicated all too successfully, leading to the current widespread impression that the main business of numerical analysis is coping with rounding errors. In fact, the main business of numerical analysis is designing algorithms that converge quickly; rounding-error analysis, while often a part of the discussion, is rarely the central issue. If rounding errors vanished, 90% of numerical analysis would remain.

4 Numerical Linear Algebra

Linear algebra became a standard topic in undergraduate mathematics curriculums in the 1950s and 1960s, and has remained there ever since. There are several reasons for this, but I think one is at the bottom of it: the importance of linear algebra has exploded since the arrival of computers.

The starting point of this subject is *Gaussian elimination*, a procedure that can solve n linear equations in n unknowns using on the order of n^3 arithmetic operations. Equivalently, it solves equations of the form $Ax = b$, where A is an $n \times n$ matrix and x and b are column vectors of size n. Gaussian elimination is invoked on computers around the world almost every time a system of linear equations is solved. Even if n is as large as 1000, the time required is well under a second on a typical 2008 desktop machine. The idea of elimination was first discovered by Chinese scholars about 2000 years ago, and more recent contributors include LAGRANGE [VI.22], Gauss, and JACOBI [VI.35]. The modern way of describing such algorithms, however, was apparently introduced as late as the 1930s. Suppose that, say, α times the first row of A is subtracted from the second row. This operation can be interpreted as the multiplication of A on the left by the lower-triangular matrix M_1 consisting of the identity with the additional nonzero entry $m_{21} = -\alpha$. Further analogous row operations correspond to further multiplications on the left by lower-triangular matrices M_j. If k steps convert A to an upper-triangular matrix U, then we have $MA = U$ with $M = M_k \cdots M_2 M_1$, or, upon setting $L = M^{-1}$,

$$A = LU.$$

Here L is unit lower-triangular, that is, lower-triangular with all its diagonal entries equal to 1. Since U represents the target structure and L encodes the operations carried out to get there, we can say that Gaussian elimination is a process of *lower-triangular upper-triangularization*.

Many other algorithms of numerical linear algebra are also based on writing a matrix as a product of matrices that have special properties. To borrow a phrase from biology, we may say that this field has a central dogma:

algorithms $\longleftrightarrow$ matrix factorizations.

In this framework we can quickly describe the next algorithm that needs to be considered. Not every matrix has an LU factorization; a 2×2 counterexample is the matrix

$$A = \begin{pmatrix} 0 & 1 \\ 1 & 0 \end{pmatrix}.$$

Soon after computers came into use it was observed that even for matrices that do have LU factorizations, the pure form of Gaussian elimination is unstable, amplifying rounding errors by potentially large amounts. Stability can be achieved by interchanging rows during the elimination in order to bring maximal entries to the diagonal, a process known as *pivoting*. Since pivoting acts on rows, it again corresponds to a multiplication of A by other matrices on the left. The matrix factorization corresponding to Gaussian elimination with pivoting is

$$PA = LU,$$

where U is upper-triangular, L is unit lower-triangular, and P is a permutation matrix, i.e., the identity matrix

with permuted rows. If the permutations are chosen to bring the largest entry below the diagonal in column k to the (k, k) position before the kth elimination step, then L has the additional property $|\ell_{ij}| \leqslant 1$ for all i and j.

The discovery of pivoting came quickly, but its theoretical analysis has proved astonishingly hard. In practice, pivoting makes Gaussian elimination almost perfectly stable, and it is routinely done by almost all computer programs that need to solve linear systems of equations. Yet it was realized in around 1960 by Wilkinson and others that for certain exceptional matrices, Gaussian elimination is still unstable, even with pivoting. The lack of an explanation of this discrepancy represents an embarrassing gap at the heart of numerical analysis. Experiments suggest that the fraction of matrices (for example, among random matrices with independent normally distributed entries) for which Gaussian elimination amplifies rounding errors by a factor greater than $\rho n^{1/2}$ is in a certain sense exponentially small as a function of ρ as $\rho \to \infty$, where n is the dimension, but a theorem to this effect has never been proved.

Meanwhile, beginning in the late 1950s, the field of numerical linear algebra expanded in another direction: the use of algorithms based on ORTHOGONAL [III.50 §3] or UNITARY [III.50 §3] matrices, that is, real matrices with $Q^{-1} = Q^{\mathrm{T}}$ or complex ones with $Q^{-1} = Q^*$, where Q^* denotes the conjugate transpose. The starting point of such developments is the idea of *QR factorization*. If A is an $m \times n$ matrix with $m \geqslant n$, a QR factorization of A is a product

$$A = QR,$$

where Q has orthonormal columns and R is upper-triangular. One can interpret this formula as a matrix expression of the familiar idea of *Gram–Schmidt orthogonalization*, in which the columns $q_1, q_2, \ldots$ of Q are determined one after another. These column operations correspond to multiplication of A on the right by elementary upper-triangular matrices. One could say that the Gram–Schmidt algorithm aims for Q and gets R as a by-product, and is thus a process of *triangular orthogonalization*. A big event was when Householder showed in 1958 that a dual strategy of *orthogonal triangularization* is more effective for many purposes. In this approach, by applying a succession of elementary matrix operations each of which reflects $\mathbb{R}^m$ across a hyperplane, one reduces A to upper-triangular form via orthogonal operations: one aims at R and gets Q

as a by-product. The Householder method turns out to be more stable numerically, because orthogonal operations preserve norms and thus do not amplify the rounding errors introduced at each step.

From the QR factorization sprang a rich collection of linear algebra algorithms in the 1960s. The QR factorization can be used by itself to solve least-squares problems and construct orthonormal bases. More remarkable is its use as a step in other algorithms. In particular, one of the central problems of numerical linear algebra is the determination of the eigenvalues and eigenvectors of a square matrix A. If A has a complete set of eigenvectors, then by forming a matrix X whose columns are these eigenvectors and a diagonal matrix D whose diagonal entries are the corresponding eigenvalues, we obtain

$$AX = XD,$$

and hence, since X is nonsingular,

$$A = XDX^{-1},$$

the *eigenvalue decomposition*. In the special case in which A is HERMITIAN [III.50 §3], a complete set of orthonormal eigenvectors always exists, giving

$$A = QDQ^*,$$

where Q is unitary. The standard algorithm for computing these factorizations was developed in the early 1960s by Francis, Kublanovskaya, and Wilkinson: the *QR algorithm*. Because polynomials of degree 5 or more cannot be solved by a formula, we know that eigenvalues cannot generally be computed in closed form. The QR algorithm is therefore necessarily an iterative one, involving a sequence of QR factorizations that is in principle infinite. Nevertheless, its convergence is extraordinarily rapid. In the symmetric case, for a typical matrix A, the QR algorithm converges *cubically*, in the sense that at each step the number of correct digits in one of the eigenvalue–eigenvector pairs approximately triples.

The QR algorithm is one of the great triumphs of numerical analysis, and its impact through widely used software products has been enormous. Algorithms and analysis based on it led in the 1960s to computer codes in Algol and Fortran and later to the software library EISPACK ("Eigensystem Package") and its descendant LAPACK. The same methods have also been incorporated in general-purpose numerical libraries such as the NAG, IMSL, and *Numerical Recipes* collections, and in problem-solving environments such as MATLAB, Maple, and Mathematica. These developments

have been so successful that the computation of matrix eigenvalues long ago became a "black box" operation for virtually every scientist, with nobody but a few specialists knowing the details of how it is done. A curious related story is that EISPACK's relative LINPACK for solving linear systems of equations took on an unexpected function: it became the original basis for the benchmarks that all computer manufacturers run to test the speed of their computers. If a supercomputer is lucky enough to make the TOP500 list, updated twice a year since 1993, it is because of its prowess in solving certain matrix problems $Ax = b$ of dimensions ranging from 100 into the millions.

The eigenvalue decomposition is familiar to all mathematicians, but the development of numerical linear algebra has also brought its younger cousin onto the scene: the *singular value decomposition* (SVD). The SVD was discovered by Beltrami, JORDAN [VI.52], and SYLVESTER [VI.42] in the late nineteenth century, and made famous by Golub and other numerical analysts beginning in around 1965. If A is an $m \times n$ matrix with $m \geqslant n$, an SVD of A is a factorization

$$A = U\Sigma V^*,$$

where U is $m \times n$ with orthonormal columns, V is $n \times n$ and unitary, and Σ is diagonal with diagonal entries $\sigma_1 \geqslant \sigma_2 \geqslant \cdots \geqslant \sigma_n \geqslant 0$. One could compute the SVD by relating it to the eigenvalue problems for AA^* and A^*A, but this proves numerically unstable; a better approach is to use a variant of the QR algorithm that does not square A. Computing the SVD is the standard route to determining the NORM [III.62] $\|A\| = \sigma_1$ (here $\| \cdot \|$ is the HILBERT SPACE [III.37] or "2" norm), the norm of the inverse $\|A^{-1}\| = 1/\sigma_n$ in the case where A is square and nonsingular, or their product, known as the *condition number*,

$$\kappa(A) = \|A\| \, \|A^{-1}\| = \sigma_1/\sigma_n.$$

It is also a step in an extraordinary variety of further computational problems including rank-deficient least-squares, computation of ranges and nullspaces, determination of ranks, "total least-squares," low-rank approximation, and determination of angles between subspaces.

All the discussion above concerns "classical" numerical linear algebra, born in the period 1950–75. The ensuing quarter-century brought in a whole new set of tools: methods for large-scale problems based on *Krylov subspace iterations*. The idea of these iterations is as follows. Suppose a linear algebra problem is given that involves a matrix of large dimension, say $n \gg 1000$. The solution may be characterized as the vector $x \in \mathbb{R}^n$ that satisfies a certain variational property such as minimizing $\frac{1}{2}x^{\mathrm{T}}Ax - x^{\mathrm{T}}b$ (for solving $Ax = b$ if A is symmetric positive definite) or being a stationary point of $(x^{\mathrm{T}}Ax)/(x^{\mathrm{T}}x)$ (for solving $Ax = \lambda x$ if A is symmetric). Now if K_k is a k-dimensional subspace of $\mathbb{R}^n$ with $k \ll n$, then it may be possible to solve the same variational problem much more quickly in that subspace. The magical choice of K_k is a *Krylov subspace*

$$K_k(A, q) = \mathrm{span}(q, Aq, \ldots, A^{k-1}q)$$

for an initial vector q. For reasons that have fascinating connections with approximation theory, solutions in these subspaces often converge very rapidly to the exact solution in $\mathbb{R}^n$ as k increases, if the eigenvalues of A are favorably distributed. For example, it is often possible to solve a matrix problem involving 10^5 unknowns to ten-digit precision in just a few hundred iterations. The speedup compared with the classical algorithms may be a factor of thousands.

Krylov subspace iterations originated with the conjugate gradient and Lanczos iterations published in 1952, but in those years computers were not powerful enough to solve problems of a large enough scale for the methods to be competitive. They took off in the 1970s with the work of Reid and Paige and especially van der Vorst and Meijerink, who made famous the idea of *preconditioning*. In preconditioning a system $Ax = b$, one replaces it by a mathematically equivalent system such as

$$MAx = Mb$$

for some nonsingular matrix M. If M is well chosen, the new problem involving MA may have favorably distributed eigenvalues and a Krylov subspace iteration may solve it quickly.

Since the 1970s, preconditioned matrix iterations have emerged as an indispensable tool of computational science. As one indication of their prominence we may note that in 2001, Thomson ISI announced that the most heavily cited article in all of mathematics in the 1990s was the 1989 paper by van der Vorst introducing Bi-CGStab, a generalization of conjugate gradients for nonsymmetric matrices.

Finally, we must mention the biggest unsolved problem in numerical analysis. Can an arbitrary $n \times n$ matrix A be inverted in $O(n^\alpha)$ operations for every $\alpha > 2$? (The problems of solving a system $Ax = b$ or computing a matrix product AB are equivalent.) Gaussian elimination has $\alpha = 3$, and the exponent shrinks as

far as 2.376 for certain recursive (though impractical) algorithms published by Coppersmith and Winograd in 1990. Is there a "fast matrix inverse" in store for us?

5   Numerical Solution of Differential Equations

Long before much attention was paid to linear algebra, mathematicians developed numerical methods to solve problems of analysis. The problem of numerical integration or *quadrature* goes back to Gauss and NEWTON [VI.14], and even to ARCHIMEDES [VI.3]. The classic quadrature formulas are derived from the idea of interpolating data at $n + 1$ points by a polynomial of degree n, then integrating the polynomial exactly. Equally spaced interpolation points give the *Newton–Cotes formulas*, which are useful for small degrees but diverge at a rate as high as 2^n as $n \to \infty$: the *Runge phenomenon*. If the points are chosen optimally, then the result is *Gauss quadrature*, which converges rapidly and is numerically stable. It turns out that these optimal points are roots of Legendre polynomials, which are clustered near the endpoints. (A proof is sketched in SPECIAL FUNCTIONS [III.85].) Equally good for most purposes is *Clenshaw–Curtis quadrature*, where the interpolation points become $\cos(j\pi/n)$, $0 \leqslant j \leqslant n$. This quadrature method is also stable and rapidly convergent, and unlike Gauss quadrature can be executed in $O(n \log n)$ operations by the fast Fourier transform. The explanation of why clustered points are necessary for effective quadrature rules is related to the subject of potential theory.

Around 1850 another problem of analysis began to get attention: the solution of ODEs. The *Adams formulas* are based on polynomial interpolation in equally spaced points, which in practice typically number fewer than ten. These were the first of what are now called *multistep methods* for the numerical solution of ODEs. The idea here is that for an initial value problem $u' = f(t, u)$ with independent variable $t > 0$, we pick a small time step $\Delta t > 0$ and consider a finite set of time values

$$t_n = n\Delta t, \quad n \geqslant 0.$$

We then replace the ODE by an algebraic approximation that enables us to calculate a succession of approximate values

$$v^n \approx u(t_n), \quad n \geqslant 0.$$

(The superscript here is just a superscript, not a power.) The simplest such approximate formula, going back to EULER [VI.19], is

$$v^{n+1} = v^n + \Delta t f(t_n, v^n),$$

or, using the abbreviation $f^n = f(t_n, v^n)$,

$$v^{n+1} = v^n + \Delta t f^n.$$

Both the ODE itself and its numerical approximation may involve one equation or many, in which case $u(t,x)$ and v^n become vectors of an appropriate dimension. The Adams formulas are higher-order generalizations of Euler's formula that are much more efficient at generating accurate solutions. For example, the fourth-order Adams–Bashforth formula is

$$v^{n+1} = v^n + \tfrac{1}{24}\Delta t(55f^n - 59f^{n-1} + 37f^{n-2} - 9f^{n-3}).$$

The term "fourth-order" reflects a new element in the numerical treatment of problems of analysis: the appearance of questions of convergence as $\Delta t \to 0$. The formula above is of fourth order in the sense that it will normally converge at the rate $O((\Delta t)^4)$. The orders employed in practice are most often in the range 3–6, enabling excellent accuracy for all kinds of computations, typically in the range of 3–10 digits, and higher-order formulas are occasionally used when still more accuracy is needed.

Most unfortunately, the habit in the numerical analysis literature is to speak not of the *convergence* of these magnificently efficient methods, but of their *error*, or more precisely their *discretization* or *truncation error* as distinct from rounding error. This ubiquitous language of error analysis is dismal in tone, but seems ineradicable.

At the turn of the twentieth century, the second great class of ODE algorithms, known as *Runge–Kutta* or *one-step methods*, was developed by Runge, Heun, and Kutta. For example, here are the formulas of the famous fourth-order Runge–Kutta method, which advance a numerical solution (again scalar or system) from time step t_n to t_{n+1} with the aid of four evaluations of the function f:

$$
\begin{aligned}
a &= \Delta t f(t_n, v^n),\\
b &= \Delta t f(t_n + \tfrac{1}{2}\Delta t, v^n + \tfrac{1}{2}a),\\
c &= \Delta t f(t_n + \tfrac{1}{2}\Delta t, v^n + \tfrac{1}{2}b),\\
d &= \Delta t f(t_n + \Delta t, v^n + c),\\
v^{n+1} &= v^n + \tfrac{1}{6}(a + 2b + 2c + d).
\end{aligned}
$$

Runge–Kutta methods tend to be easier to implement but sometimes harder to analyze than multistep formulas. For example, for any s, it is a trivial matter to derive the coefficients of the s-step Adams–Bashforth formula, which has order of accuracy $p = s$. For Runge–Kutta methods, by contrast, there is no simple relationship between the number of "stages" (i.e., function evaluations per step) and the attainable order of accuracy. The classical methods with $s = 1, 2, 3, 4$ were known to Kutta in 1901 and have order $p = s$, but it was not until 1963 that it was proved that $s = 6$ stages are required to achieve order $p = 5$. The analysis of such problems involves beautiful mathematics from graph theory and other areas, and a key figure in this area since the 1960s has been John Butcher. For orders $p = 6, 7, 8$ the minimal numbers of stages are $s = 7, 9, 11$, while for $p > 8$ exact minima are not known. Fortunately, these higher orders are rarely needed for practical purposes.

When computers began to be used to solve differential equations after World War II, a phenomenon of the greatest practical importance appeared: once again, *numerical instability*. As before, this phrase refers to the unbounded amplification of local errors by a computational process, but now the dominant local errors are usually those of discretization rather than rounding. Instability typically manifests itself as an oscillatory error in the computed solution that blows up exponentially as more numerical steps are taken. One mathematician concerned with this effect was Germund Dahlquist. Dahlquist saw that the phenomenon could be analyzed with great power and generality, and some people regard the appearance of his 1956 paper as one of the events marking the birth of modern numerical analysis. This landmark paper introduced what might be called the *fundamental theorem of numerical analysis*:

$$\text{consistency} + \text{stability} = \text{convergence}.$$

The theory is based on precise definitions of these three notions along the following lines. *Consistency* is the property that the discrete formula has locally positive order of accuracy and thus models the right ODE. *Stability* is the property that errors introduced at one time step cannot grow unboundedly at later times. *Convergence* is the property that as $\Delta t \to 0$, in the absence of rounding errors, the numerical solution converges to the correct result. Before Dahlquist's paper, the idea of an equivalence of stability and convergence was perhaps in the air in the sense that practitioners realized that if a numerical scheme was not unstable, then it would probably give a good approximation to the right answer. His theory gave rigorous form to that idea for a wide class of numerical methods.

As computer methods for ODEs were being developed, the same was happening for the much bigger subject of PDEs. Discrete numerical methods for solving PDEs had been invented around 1910 by Richardson for applications in stress analysis and meteorology, and further developed by Southwell; in 1928 there was also a theoretical paper on finite-difference methods by COURANT [VI.83], Friedrichs, and Lewy. But although the Courant–Friedrichs–Lewy work later became famous, the impact of these ideas before computers came along was limited. After that point the subject developed quickly. Particularly influential in the early years was the group of researchers around von Neumann at the Los Alamos laboratory, including the young Peter Lax.

Just as for ODEs, von Neumann and his colleagues discovered that some numerical methods for PDEs were subject to catastrophic instabilities. For example, to solve the linear wave equation $u_t = u_x$ numerically we may pick space and time steps Δx and Δt for a regular grid,

$$x_j = j\Delta x, \quad t_n = n\Delta t, \quad j, n \geqslant 0,$$

and replace the PDE by algebraic formulas that compute a succession of approximate values:

$$v_j^n \approx u(t_n, x_j), \quad j, n \geqslant 0.$$

A well-known discretization for this purpose is the *Lax-Wendroff formula*:

$$v_j^{n+1} = v_j^n + \tfrac{1}{2}\lambda(v_{j+1}^n - v_{j-1}^n) + \tfrac{1}{2}\lambda^2(v_{j+1}^n - 2v_j^n + v_{j-1}^n),$$

where $\lambda = \Delta t / \Delta x$, which can be generalized to nonlinear systems of hyperbolic conservation laws in one dimension. For $u_t = u_x$, if λ is held fixed at a value less than or equal to 1, the method will converge to the correct solution as $\Delta x, \Delta t \to 0$ (ignoring rounding errors). If λ is greater than 1, on the other hand, it will explode. Von Neumann and others realized that the presence or absence of such instabilities could be tested, at least for linear constant-coefficient problems, by discrete FOURIER ANALYSIS [III.27] in x: "von Neumann analysis." Experience indicated that, as a practical matter, a method would succeed if it was not unstable. A theory soon appeared that gave rigor to this observation: the *Lax equivalence theorem*, published by Lax and Richtmyer in 1956, the same year as Dahlquist's paper. Many details were different—this theory was restricted to linear equations whereas Dahlquist's theory for ODEs also applied to nonlinear ones—but broadly speaking the new result followed the same pattern of equating convergence to consistency plus stability. Mathematically, the key point was the uniform boundedness principle.

In the half-century since von Neumann died, the Lax-Wendroff formula and its relatives have grown into a breathtakingly powerful subject known as *computational fluid dynamics*. Early treatments of linear and nonlinear equations in one space dimension soon moved to two dimensions and eventually to three. It is now a routine matter to solve problems involving millions of variables on computational grids with hundreds of points in each of three directions. The equations are linear or nonlinear; the grids are uniform or nonuniform, often adaptively refined to give special attention to boundary layers and other fast-changing features; the applications are everywhere. Numerical methods were used first to model airfoils, then whole wings, then whole aircraft. Engineers still use wind tunnels, but they rely more on computations.

Many of these successes have been facilitated by another numerical technology for solving PDEs that emerged in the 1960s from diverse roots in engineering and mathematics: finite elements. Instead of approximating a differential operator by a difference quotient, finite-element methods approximate the solution itself by functions f that can be broken up into simple pieces. For instance, one might partition the domain of f into elementary sets such as triangles or tetrahedra and insist that the restriction of f to each piece is a polynomial of small degree. The solution is obtained by solving a variational form of the PDE within the corresponding finite-dimensional subspace, and there is often a guarantee that the computed solution is optimal within that subspace. Finite-element methods have taken advantage of tools of functional analysis to develop to a very mature state. These methods are known for their flexibility in handling complicated geometries, and in particular they are entirely dominant in applications in structural mechanics and civil engineering. The number of books and articles that have been published about finite-element methods is in excess of 10 000.

In the vast and mature field of numerical solution of PDEs, what aspect of the current state of the art would most surprise Richardson or Courant, Friedrichs, and Lewy? I think it is the universal dependence on exotic algorithms of linear algebra. The solution of a large-scale PDE problem in three dimensions may require a system of a million equations to be solved at each time step. This may be achieved by a GMRES matrix iteration that utilizes a finite-difference preconditioner implemented by a Bi-CGStab iteration relying on another multigrid preconditioner. Such stacking of tools was surely not imagined by the early computer pioneers. The need for it ultimately traces to numerical instability, for as Crank and Nicolson first noted in 1947, the crucial tool for combating instability is the use of *implicit formulas*, which couple together unknowns at the new time step t_{n+1}, and it is in implementing this coupling that solutions of systems of equations are required.

Here are some examples that illustrate the successful reliance of today's science and engineering on the numerical solution of PDEs: chemistry (the SCHRÖDINGER EQUATION [III.83]); structural mechanics (the equations of elasticity); weather prediction (the geostrophic equations); turbine design (the NAVIER–STOKES EQUATIONS [III.23]); acoustics (the Helmholtz equation); telecommunications (MAXWELL'S EQUATIONS [IV.13 §1.1]); cosmology (the Einstein equations); oil discovery (the migration equations); groundwater remediation (Darcy's law); integrated circuit design (the drift diffusion equations); tsunami modeling (the shallow-water equations); optical fibers (the NONLINEAR WAVE EQUATIONS [III.49]); image enhancement (the Perona–Malik equation); metallurgy (the Cahn-Hilliard equation); pricing financial options (the BLACK–SCHOLES EQUATION [VII.9 §2]).

6 Numerical Optimization

The third great branch of numerical analysis is optimization, that is, the minimization of functions of several variables and the closely related problem of solution of nonlinear systems of equations. The development of optimization has been somewhat independent of that of the rest of numerical analysis, carried forward in part by a community of scholars with close links to operations research and economics.

Calculus students learn that a smooth function may achieve an extremum at a point of zero derivative, or at a boundary. The same two possibilities characterize the two big strands of the field of optimization. At one end there are problems of finding interior zeros and minima of unconstrained nonlinear functions by methods related to multivariate calculus. At the other are problems of linear programming, where the function to be minimized is linear and therefore easy to understand, and all the challenge is in the boundary constraints.

Unconstrained nonlinear optimization is an old subject. Newton introduced the idea of approximating functions by the first few terms of what we now call their Taylor series; indeed, Arnol'd has argued that Taylor series were Newton's "main mathematical discovery." To find a zero x_* of a function F of a real variable x, everyone knows the idea of *Newton's method*: at the kth step, given an estimate $x^{(k)} \approx x_*$, use the derivative $F'(x^{(k)})$ to define a linear approximation from which to derive a better estimate $x^{(k+1)}$:

$$x^{(k+1)} = x^{(k)} - F(x^{(k)})/F'(x^{(k)}).$$

Newton (1669) and Raphson (1690) applied this idea to polynomials, and Simpson (1740) generalized it to other functions F and to systems of two equations. In today's language, for a system of n equations in n unknowns, we regard F as an n-vector whose derivative at a point $x^{(k)} \in \mathbb{R}^n$ is the $n \times n$ Jacobian matrix with entries

$$J_{ij}(x^{(k)}) = \frac{\partial F_i}{\partial x_j}(x^{(k)}), \quad 1 \leqslant i, j \leqslant n.$$

This matrix defines a linear approximation to $F(x)$ that is accurate for $x \approx x^{(k)}$. Newton's method then takes the matrix form

$$x^{(k+1)} = x^{(k)} - (J(x^{(k)}))^{-1}F(x^{(k)}),$$

which in practice means that to get $x^{(k+1)}$ from $x^{(k)}$, we solve a linear system of equations:

$$J(x^{(k)})(x^{(k+1)} - x^{(k)}) = -F(x^{(k)}).$$

As long as J is Lipschitz continuous and nonsingular at x_* and the initial guess is good enough, the convergence of this iteration is quadratic:

$$\|x^{(k+1)} - x_*\| = O(\|x^{(k)} - x_*\|^2). \tag{1}$$

Students often think it might be a good idea to develop formulas to enhance the exponent in this estimate to 3 or 4. However, this is an illusion. Taking two steps at a time of a quadratically convergent algorithm yields a quartically convergent one, so the difference in efficiency between quadratic and quartic is at best a constant factor. The same goes if the exponent 2, 3, or 4 is replaced by any other number greater than 1. The true distinction is between all of these algorithms that converge *superlinearly*, of which Newton's method is the prototype, and those that converge *linearly* or *geometrically*, where the exponent is just 1.

From the point of view of multivariate calculus, it is a small step from solving a system of equations to minimizing a scalar function f of a variable $x \in \mathbb{R}^n$: to find a (local) minimum, we seek a zero of the gradient $g(x) = \nabla f(x)$, an n-vector. The derivative of g is the Jacobian matrix known as the *Hessian* of f, with entries

$$H_{ij}(x^{(k)}) = \frac{\partial^2 f}{\partial x_i \partial x_j}(x^{(k)}), \quad 1 \leqslant i, j \leqslant n,$$

and one may utilize it just as before in a Newton iteration to find a zero of $g(x)$, the new feature being that a Hessian is always symmetric.

Though the Newton formulas for minimization and finding zeros were already established, the arrival of computers created a new field of numerical optimization. One of the obstacles quickly encountered was that Newton's method often fails if the initial guess is not good. This problem has been comprehensively addressed both practically and theoretically by the algorithmic technologies known as *line searches* and *trust regions*.

For problems with more than a few variables, it also quickly became clear that the cost of evaluating Jacobians or Hessians at every step could be exorbitant. Faster methods were needed that might make use of inexact Jacobians or Hessians and/or inexact solutions of the associated linear equations, while still achieving superlinear convergence. An early breakthrough of this kind was the discovery of *quasi-Newton methods* in the 1960s by Broyden, Davidon, Fletcher, and Powell, in which partial information is used to generate steadily improving estimates of the true Jacobian or Hessian or its matrix factors. An illustration of the urgency of this subject at the time is the fact that in 1970 the optimal rank-two symmetric positive-definite quasi-Newton updating formula was published independently by no fewer than four different authors, namely Broyden, Fletcher, Goldfarb, and Shanno; their discovery has been known ever since as the *BFGS formula*. In subsequent years, as the scale of tractable problems has increased exponentially, new ideas have also become important, including *automatic differentiation*, a technology that enables derivatives of computed functions to be determined automatically: the computer program itself is "differentiated," so that as well as producing numerical outputs it also produces their derivatives. The idea of automatic differentiation is an old one, but for various reasons, partly related to advances in sparse linear algebra and to the development of "reverse mode" formulations, it did not become fully practical until the work of Bischof, Carle, and Griewank in the 1990s.

Unconstrained optimization problems are relatively easy, but they are not typical; the true depth of this field is revealed by the methods that have been developed for dealing with constraints. Suppose a function $f : \mathbb{R}^n \to \mathbb{R}$ is to be minimized subject to certain equality constraints $c_j(x) = 0$ and inequality constraints $d_j(x) \geqslant 0$, where $\{c_j\}$ and $\{d_j\}$ are also functions from $\mathbb{R}^n$ to $\mathbb{R}$. Even the problem of stating local optimality conditions for solutions to such problems is nontrivial, a matter involving LAGRANGE MULTIPLIERS [III.64] and a distinction between active and inactive constraints. This problem was solved by what are now known as the *KKT conditions*, introduced by Kuhn and Tucker in 1951 and also twelve years earlier, it was subsequently realized, by Karush. Development of algorithms for constrained nonlinear optimization continues to be an active research topic today.

The problem of constraints brings us to the other strand of numerical optimization, linear programming. This subject was born in the 1930s and 1940s with Kantorovich in the Soviet Union and Dantzig in the United States. As an outgrowth of his work for the U.S. Air Force in the war, Dantzig invented in 1947 the famous SIMPLEX ALGORITHM [III.84] for solving linear programs. A linear program is nothing more than a problem of minimizing a linear function of n variables subject to m linear equality and/or inequality constraints. How can this be a challenge? One answer is that m and n may be large. Large-scale problems may arise through discretization of continuous problems and also in their own right. A famous early example was Leontiev's theory of input–output models in economics, which won him the Nobel Prize in 1973. Even in the 1970s the Soviet Union used an input–output computer model involving thousands of variables as a tool for planning the economy.

The simplex algorithm made medium- and large-scale linear programming problems tractable. Such a problem is defined by its *objective function*, the function $f(x)$ to be minimized, and its *feasible region*, the set of vectors $x \in \mathbb{R}^n$ that satisfy all the constraints. For a linear program the feasible region is a polyhedron, a closed domain bounded by hyperplanes, and the optimal value of f is guaranteed to be attained at one of the vertex points. (A point is called a *vertex* if it is the unique solution of some subset of the equations that define the constraints.) The simplex algorithm proceeds by moving systematically downhill from one vertex to another until an optimal point is reached. All of the iterates lie on the boundary of the feasible region.

In 1984, an upheaval occurred in this field, triggered by Narendra Karmarkar at AT&T Bell Laboratories. Karmarkar showed that one could sometimes do much better than the simplex algorithm by working in the interior of the feasible region instead. Once a connection was shown between Karmarkar's method and the logarithmic barrier methods popularized by Fiacco and

McCormick in the 1960s, new interior methods for linear programming were devised by applying techniques previously viewed as suitable only for nonlinear problems. The crucial idea of working in tandem with a pair of primal and dual problems led to today's powerful primal–dual methods, which can solve continuous optimization problems with millions of variables and constraints. Starting with Karmarkar's work, not only has the field of linear programming changed completely, but the linear and nonlinear sides of optimization are seen today as closely related rather than essentially different.

7 The Future

Numerical analysis sprang from mathematics; then it spawned the field of computer science. When universities began to found computer science departments in the 1960s, numerical analysts were often in the lead. Now, two generations later, most of them are to be found in mathematics departments. What happened? A part of the answer is that numerical analysts deal with continuous mathematical problems, whereas computer scientists prefer discrete ones, and it is remarkable how wide a gap this can be.

Nevertheless, the computer science side of numerical analysis is of crucial importance, and I would like to end with a prediction that emphasizes this aspect of the subject. Traditionally one might think of a numerical algorithm as a cut-and-dried procedure, a loop of some kind to be executed until a well-defined termination criterion is satisfied. For some computations this picture is accurate. On the other hand, beginning with the work of de Boor, Lyness, Rice and others in the 1960s, a less deterministic kind of numerical computing began to appear: *adaptive algorithms*. In an adaptive quadrature program of the simplest kind, two estimates of the integral are calculated on each portion of a certain mesh and then compared to produce an estimate of the local error. Based on this estimate, the mesh may then be refined locally to improve the accuracy. This process is carried out iteratively until a final answer is obtained that aims to be accurate to a tolerance specified in advance by the user. Most such computations come with no guarantee of accuracy, but an exciting ongoing development is the advance of more sophisticated techniques of a posteriori error control that sometimes do provide guarantees. When these are combined with techniques of *interval arithmetic*, there is even the prospect of accuracy guaranteed with respect to rounding as well as discretization error.

First, computer programs for quadrature became adaptive; then programs for ODEs did as well. For PDEs, the move to adaptive programs is happening on a longer timescale. More recently there have been related developments in the computation of Fourier transforms, optimization, and large-scale numerical linear algebra, and some of the new algorithms adapt to the computer architecture as well as the mathematical problem. In a world where several algorithms are known for solving every problem, we increasingly find that the most robust computer program will be one that has diverse capabilities at its disposal and deploys them adaptively on the fly. In other words, numerical computation is increasingly embedded in intelligent control loops. I believe this process will continue, just as has happened in so many other areas of technology, removing scientists further from the details of their computations but offering steadily growing power in exchange. I expect that most of the numerical computer programs of 2050 will be 99% intelligent "wrapper" and just 1% actual "algorithm," if such a distinction makes sense. Hardly anyone will know how they work, but they will be extraordinarily powerful and reliable, and will often deliver results of guaranteed accuracy.

This story will have a mathematical corollary. One of the fundamental distinctions in mathematics is between linear problems, which can be solved in one step, and nonlinear ones, which usually require iteration. A related distinction is between forward problems (one step) and inverse problems (iteration). As numerical algorithms are increasingly embedded in intelligent control loops, almost every problem will be handled by iteration, regardless of its philosophical status. Problems of algebra will be solved by methods of analysis; and between linear and nonlinear, or forward and inverse, the distinctions will fade.

8 Appendix: Some Major Numerical Algorithms

The list in table 1 attempts to identify some of the most significant algorithmic (as opposed to theoretical) developments in the history of numerical analysis. In each case some of the key early figures are cited, more or less chronologically, and a key early date is given. Of course, any brief sketch of history like this must be an oversimplification. Distressing omissions of names occur throughout the list, including many early contributors in fields such as finite elements, preconditioning, and automatic differentiation, as well as more

Table 1 Some algorithmic developments in the history of numerical analysis.

Year	Development	Key early figures
263	Gaussian elimination	Liu, Lagrange, Gauss, Jacobi
1671	Newton's method	Newton, Raphson, Simpson
1795	Least-squares fitting	Gauss, Legendre
1814	Gauss quadrature	Gauss, Jacobi, Christoffel, Stieltjes
1855	Adams ODE formulas	Euler, Adams, Bashforth
1895	Runge–Kutta ODE formulas	Runge, Heun, Kutta
1910	Finite differences for PDE	Richardson, Southwell, Courant, von Neumann, Lax
1936	Floating-point arithmetic	Torres y Quevedo, Zuse, Turing
1943	Finite elements for PDE	Courant, Feng, Argyris, Clough
1946	Splines	Schoenberg, de Casteljau, Bezier, de Boor
1947	Monte Carlo simulation	Ulam, von Neumann, Metropolis
1947	Simplex algorithm	Kantorovich, Dantzig
1952	Lanczos and conjugate gradient iterations	Lanczos, Hestenes, Stiefel
1952	Stiff ODE solvers	Curtiss, Hirschfelder, Dahlquist, Gear
1954	Fortran	Backus
1958	Orthogonal linear algebra	Aitken, Givens, Householder, Wilkinson, Golub
1959	Quasi-Newton iterations	Davidon, Fletcher, Powell, Broyden
1961	QR algorithm for eigenvalues	Rutishauser, Kublanovskaya, Francis, Wilkinson
1965	Fast Fourier transform	Gauss, Cooley, Tukey, Sande
1971	Spectral methods for PDE	Chebyshev, Lanczos, Clenshaw, Orszag, Gottlieb
1971	Radial basis functions	Hardy, Askey, Duchon, Micchelli
1973	Multigrid iterations	Fedorenko, Bakhvalov, Brandt, Hackbusch
1976	EISPACK, LINPACK, LAPACK	Moler, Stewart, Smith, Dongarra, Demmel, Bai
1976	Nonsymmetric Krylov iterations	Vinsome, Saad, van der Vorst, Sorensen
1977	Preconditioned matrix iterations	van der Vorst, Meijerink
1977	MATLAB	Moler
1977	IEEE arithmetic	Kahan
1982	Wavelets	Morlet, Grossmann, Meyer, Daubechies
1984	Interior methods in optimization	Fiacco, McCormick, Karmarkar, Megiddo
1987	Fast multipole method	Rokhlin, Greengard
1991	Automatic differentiation	Iri, Bischof, Carle, Griewank

than half of the authors of the EISPACK, LINPACK, and LAPACK libraries. Even the dates can be questioned; the fast Fourier transform is listed as 1965, for example, since that is the year of the paper that brought it to the world's attention, though Gauss made the same discovery 160 years earlier. Nor should one imagine that the years from 1991 to the present have been a blank! No doubt in the future we shall identify developments from this period that deserve a place in the table.

Further Reading

Ciarlet, P. G. 1978. *The Finite Element Method for Elliptic Problems*. Amsterdam: North-Holland.

Golub, G. H., and C. F. Van Loan. 1996. *Matrix Computations*, 3rd edn. Baltimore, MD: Johns Hopkins University Press.

Hairer, E., S. P. Nørsett (for volume I), and G. Wanner. 1993, 1996. *Solving Ordinary Differential Equations*, volumes I and II. New York: Springer.

Iserles, A., ed. 1992–. *Acta Numerica* (annual volumes). Cambridge: Cambridge University Press.

Nocedal, I., and S. J. Wright. 1999. *Numerical Optimization*. New York: Springer.

Powell, M. J. D. 1981. *Approximation Theory and Methods*. Cambridge: Cambridge University Press.

Richtmyer, R. D., and K. W. Morton. 1967. *Difference Methods for Initial-Value Problems*. New York: Wiley Interscience.

IV.22   Set Theory
Joan Bagaria

1   Introduction

Among all mathematical disciplines, set theory occupies a special place because it plays two very different roles at the same time: on the one hand, it is an area of mathematics devoted to the study of abstract sets and their properties; on the other, it provides mathematics with its foundation. This second aspect of set theory gives it philosophical as well as mathematical significance. We shall discuss both aspects of the subject in this article.

2 The Theory of Transfinite Numbers

Set theory began with the work of CANTOR [VI.54]. In 1874 he proved that there are more real numbers than there are algebraic ones, thus showing that infinite sets can be of different sizes. This also provided a new proof of the existence of TRANSCENDENTAL NUMBERS [III.41]. Recall that a real number is called *algebraic* if it is the solution of some polynomial equation

$$a_n X^n + a_{n-1} X^{n-1} + \cdots + a_1 X + a_0 = 0,$$

where the coefficients a_i are integers (and $a_n \neq 0$). Thus, numbers like $\sqrt{2}$, $\frac{3}{4}$, and the golden ratio, $\frac{1}{2}(1 + \sqrt{5})$, are algebraic. A transcendental number is one that is not algebraic.

What does it mean to say that there are "more" real numbers than algebraic ones, when there are infinitely many of both? Cantor defined two sets A and B to have the same size, or *cardinality*, if there is a bijection between them: that is, if there is a one-to-one correspondence between the elements of A and the elements of B. If there is no bijection between A and B, but there is a bijection between A and a *subset* of B, then A is of *smaller cardinality* than B. So what Cantor in fact showed was that the set of algebraic numbers had smaller cardinality than that of all real numbers.

In particular, Cantor distinguished between two different kinds of infinite set: COUNTABLE AND UNCOUNTABLE [III.11]. A countable set is one that can be put into one-to-one correspondence with the natural numbers. In other words, it is a set that we can "enumerate," assigning a different natural number to each of its elements. Let us see how this can be done for the algebraic numbers. Given a polynomial equation as above, let the number

$$|a_n| + |a_{n-1}| + \cdots + |a_0| + n$$

be called its *index*. It is easy to see that for every $k > 0$ there are only a finite number of equations of index k. For instance, there are only four equations of index 3 with strictly positive a_n, namely, $X^2 = 0$, $2X = 0$, $X + 1 = 0$, and $X - 1 = 0$, which have as solutions 0, -1, and 1. Thus, we can enumerate the algebraic numbers by first enumerating all solutions of equations of index

1, then all solutions of equations of index 2 that we have not already enumerated, and so on. Therefore, the algebraic numbers are countable. Note that from this proof we also see that the sets $\mathbb{Z}$ and $\mathbb{Q}$ are countable.

Cantor discovered that, surprisingly, the set $\mathbb{R}$ of real numbers is not countable. Here is Cantor's original proof. Suppose, aiming for a contradiction, that $r_0, r_1, r_2, \ldots$ is an enumeration of $\mathbb{R}$. Let $a_0 = r_0$. Choose the least k such that $a_0 < r_k$ and put $b_0 = r_k$. Given a_n and b_n, choose the least l such that $a_n < r_l < b_n$, and put $a_{n+1} = r_l$. And choose the least m such that $a_{n+1} < r_m < b_n$, and put $b_{n+1} = r_m$. Thus, we have $a_0 < a_1 < a_2 < \cdots < b_2 < b_1 < b_0$. Now let a be the limit of the a_n. Then a is a real number different from r_n, for all n, contradicting our assumption that the sequence $r_0, r_1, r_2, \ldots$ enumerates all real numbers.

Thus it was established for the first time that there are at least two genuinely different kinds of infinite sets. Cantor also showed that there are bijections between any two of the sets $\mathbb{R}^n$, $n \geqslant 1$, and even $\mathbb{R}^{\mathbb{N}}$, the set of all infinite sequences $r_0, r_1, r_2, \ldots$ of real numbers, so all these sets have the same (uncountable) cardinality.

From 1879 to 1884 Cantor published a series of works that constitute the origin of set theory. An important concept that he introduced was that of infinite, or "transfinite," *ordinals*. When we use the natural numbers to count a collection of objects, we assign a number to each object, starting with 1, continuing with 2, 3, etc., and stopping when we have counted each object exactly once. When this process is over we have done two things. The more obvious one is that we have obtained a number n, the last number in the sequence, that tells us how many objects there are in the collection. But that is not all we have done: as we count we also define an *ordering* on the objects that we were counting, namely the order in which we count them. This reflects two different ways in which we can think about the set $\{1, 2, \ldots, n\}$. Sometimes all we care about is its size. Then, if we have a set X in one-to-one correspondence with $\{1, 2, \ldots, n\}$, we conclude that X has cardinality n. But sometimes we also take note of the natural ordering on the set $\{1, 2, \ldots, n\}$, in which case we observe that our one-to-one correspondence provides us with an ordering on X too. If we adopt the first point of view, then we are regarding n as a *cardinal*, and if we adopt the second, then we are regarding it as an ordinal.

If we have a countably infinite set, then we can think of that from the ordinal point of view too. For instance, if we define a one-to-one correspondence between $\mathbb{N}$ and $\mathbb{Z}$ by taking $0, 1, 2, 3, 4, 5, 6, 7, \ldots$ to $0, 1, -1, 2, -2, 3, -3, \ldots$, then we have not only shown that $\mathbb{N}$ and $\mathbb{Z}$ have the same cardinality, but also used the obvious ordering on $\mathbb{N}$ to define an ordering on $\mathbb{Z}$.

Suppose now that we want to count the points in the unit interval $[0, 1]$. Cantor's argument given above shows that no matter how we assign numbers in this interval to the numbers 0, 1, 2, 3, etc., we will run out of natural numbers before we have counted all points. However, when this happens, nothing prevents us from simply setting aside the numbers we have already counted and starting again. This is where transfinite ordinals come in: they are a continuation of the sequence $0, 1, 2, 3, \ldots$ "beyond infinity," and they can be used to count bigger infinite sets.

To start with, we need an ordinal number that represents the first position in the sequence that comes straight after all the natural numbers. This is the first infinite ordinal number, which Cantor denoted by ω. In other words, after $0, 1, 2, 3, \ldots$ comes ω. The ordinal ω has a different character from the previous ordinals, because although it has predecessors, it has no immediate predecessor (unlike 7, say, which has immediate predecessor 6). We say that ω is a *limit ordinal*. But once we have ω, we can continue the ordinal sequence in a very simple way, just by adding 1 repeatedly. Thus, the sequence of ordinal numbers begins as follows:

$$0, 1, 2, 3, 4, 5, 6, 7, \ldots, \omega, \omega + 1, \ \omega + 2, \ \omega + 3, \ldots.$$

After this comes the next limit ordinal, which it seems natural to call $\omega + \omega$, and which we can write as $\omega \cdot 2$. The sequence continues as

$$\omega \cdot 2, \omega \cdot 2 + 1, \omega \cdot 2 + 2, \ldots, \omega \cdot n, \ldots, \omega \cdot n + m, \ldots.$$

As this discussion indicates, there are two basic rules for generating new ordinals: adding 1 and passing to the limit. What we mean by "passing to the limit" is "assigning a new ordinal number to the position in the ordinal sequence that comes straight after all the ordinals obtained so far." For example, after all the ordinals $\omega \cdot n + m$ comes the next limit ordinal, which we write as $\omega \cdot \omega$, or ω^2, and we obtain

$$\omega^2, \omega^2 + 1, \ldots, \omega^2 + \omega, \ldots, \omega^2 + \omega \cdot n, \ldots, \omega^2 \cdot n, \ldots.$$

Eventually, we reach ω^3 and the sequence continues as

$$\omega^3, \omega^3 + 1, \ldots, \omega^3 + \omega, \ldots, \omega^3 + \omega^2, \ldots, \omega^3 \cdot n, \ldots.$$

The next limit ordinal is ω^4, and so on. The first limit ordinal after all the ω^n is ω^ω. And after $\omega^\omega, \omega^{\omega^\omega}$,

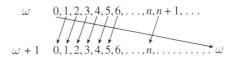

Figure 1 ω and $\omega + 1$ have the same cardinality.

$\omega^{\omega^{\omega^\omega}}, \ldots$ comes the limit ordinal denoted by ε_0. And on and on it goes.

In set theory, one likes to regard all mathematical objects as sets. For ordinals this can be done in a particularly simple way: we represent 0 by the empty set, and the ordinal number α is then identified with the set of all its predecessors. For instance, the natural number n is identified with the set $\{0, 1, \ldots, n - 1\}$ (which has cardinality n) and the ordinal $\omega + 3$ is identified with the set $\{0, 1, 2, 3, \ldots, \omega, \omega + 1, \omega + 2\}$. If we think of ordinals in this way, then the ordering on the set of ordinals becomes set membership: if α comes before β in the ordinal sequence, then α is one of the predecessors of β and therefore an element of β. A critically important property of this ordering is that each ordinal is a *well-ordered set*, which means that every nonempty subset of it has a least element.

As we said earlier, cardinal numbers are used for measuring the sizes of sets, while ordinal numbers indicate the position in an ordered sequence. This distinction is much more apparent for infinite numbers than for finite ones, because then it is possible for two different ordinals to have the same size. For example, the ordinals ω and $\omega + 1$ are different but the corresponding sets $\{0, 1, 2, \ldots\}$ and $\{0, 1, 2, \ldots, \omega\}$ have the same cardinality, as figure 1 shows. In fact, all sets that can be counted using the infinite ordinals we have described so far are countable. So in what sense are different ordinals different? The point is that although two sets such as $\{0, 1, 2, \ldots\}$ and $\{0, 1, 2, \ldots, \omega\}$ have the same cardinality, they are not *order isomorphic*: that is, you cannot find a bijection ϕ from one set to the other such that $\phi(x) < \phi(y)$ whenever $x < y$. Thus, they are the same "as sets" but not "as ordered sets."

Informally, the cardinal numbers are the possible sizes of sets. A convenient formal definition of a cardinal number is that it is an ordinal number that is bigger than all its predecessors. Two important examples of such ordinals are ω, the first infinite ordinal, and the set of all countable ordinals, which Cantor denoted by ω_1. The second of these is the first uncountable ordinal: uncountable since it cannot include itself as an element, and the first one because all its elements are countable. (If this seems paradoxical, consider the ordinal ω: it is infinite, but all its elements are finite.) Therefore, it is also a cardinal number, and when we consider this aspect of it rather than its order structure we call it $\aleph_1$, again following Cantor. Similarly, when we think of ω as a cardinal, we call it $\aleph_0$.

The process used to define $\aleph_1$ can be repeated. The set of all ordinals of cardinality $\aleph_1$ (or equivalently the set of all ordinals that can be put in one-to-one correspondence with the first uncountable ordinal ω_1) is the smallest ordinal that has cardinality greater than $\aleph_1$. As an ordinal it is called ω_2 and as a cardinal it is called $\aleph_2$. We can continue, generating a whole sequence of ordinals $\omega_1, \omega_2, \omega_3, \ldots$ of larger and larger cardinality. Moreover, using limits as well, we can continue this sequence transfinitely: for example, the ordinal ω_ω is the limit of all the ordinals ω_n. As we do this, we also produce the sequence of infinite, or transfinite, cardinals:

$$\aleph_0, \aleph_1, \ldots, \aleph_\omega, \aleph_{\omega+1}, \ldots, \aleph_{\omega^\omega}, \ldots,$$
$$\aleph_{\omega_1}, \ldots, \aleph_{\omega_2}, \ldots, \aleph_{\omega_\omega}, \ldots.$$

Given two natural numbers, we can calculate their sum and product. A convenient set-theoretic way to define these binary operations is as follows. Given two natural numbers m and n, take any two disjoint sets A and B of size m and n, respectively; $m + n$ is then the size of the union $A \cup B$. As for the product, it is the size of the set $A \times B$, the set of all ordered pairs (a, b) with $a \in A$ and $b \in B$. (For this set, which is called the *Cartesian product*, we do not need A and B to be disjoint.)

The point of these definitions is that they apply just as well to infinite cardinal numbers: just replace m and n in the above definitions by two infinite cardinals κ and λ. The resulting arithmetic of transfinite cardinals is very simple, however. It turns out that for all transfinite cardinals $\aleph_\alpha$ and $\aleph_\beta$,

$$\aleph_\alpha + \aleph_\beta = \aleph_\alpha \aleph_\beta = \max(\aleph_\alpha, \aleph_\beta) = \aleph_{\max(\alpha,\beta)}.$$

However, it is also possible to define cardinal exponentiation, and for this the picture changes completely. If κ and λ are two cardinals, then κ^λ is defined as the cardinality of the Cartesian product of λ copies of any set of cardinality κ. Equivalently, it is the cardinality of the set of all functions from a set of cardinality λ into a set of cardinality κ. Again, if κ and λ are *finite* numbers, this gives us the usual definition: for instance, the number of functions from a set of size 3 to a set of size 4 is 4^3. What happens if we take the simplest nontrivial transfinite example, $2^{\aleph_0}$? Not only is this question

extremely hard, there is a sense in which it cannot be resolved, as we shall see later.

The most obvious set of cardinality $2^{\aleph_0}$ is the set of functions from $\mathbb{N}$ to the set $\{0, 1\}$. If f is such a function, then we can regard it as giving the binary expansion of the number

$$x = \sum_{n \in \mathbb{N}} f(n) 2^{-(n+1)},$$

which belongs to the closed interval $[0, 1]$. (The power is $2^{-(n+1)}$ rather than 2^{-n} because we are using the convention, standard in set theory, that 0 is the first natural number rather than 1.) Since every point in $[0, 1]$ has at most two different binary representations, it follows easily that $2^{\aleph_0}$ is also the cardinality of $[0, 1]$, and therefore also the cardinality of $\mathbb{R}$. Thus, $2^{\aleph_0}$ is uncountable, which means that it is greater than or equal to $\aleph_1$. Cantor conjectured that it is exactly $\aleph_1$. This is the famous *continuum hypothesis*, which will be discussed at length in section 5 below.

It is not immediately obvious, but there are many mathematical contexts in which transfinite ordinals occur naturally. Cantor himself devised his theory of transfinite ordinals and cardinals as a result of his attempts, which were eventually successful, to prove the continuum hypothesis for closed sets. He first defined the *derivative* of a set X of real numbers to be the set you obtain when you throw out all the "isolated" points of X. These are points x for which you can find a small neighborhood around x that contains no other points in X. For example, if X is the set $\{0\} \cup \{1, \frac{1}{2}, \frac{1}{3}, \ldots\}$, then all points in X are isolated except for 0, so the derivative of X is the set $\{0\}$.

In general, given a set X, we can take its derivative repeatedly. If we set $X^0 = X$, then we obtain a sequence $X^0 \supseteq X^1 \supseteq X^2 \supseteq \cdots$, where X^{n+1} is the derivative of X^n. But the sequence does not stop here: we can take the intersection of all the X^n and call it X^ω, and if we do that, then we can define $X^{\omega+1}$ to be the derivative of X^ω, and so on. Thus, the reason that ordinals appear naturally is that we have *two* operations, taking the derivative and taking the intersection of everything so far, which correspond to successors and limits in the ordinal sequence. Cantor initially regarded superscripts such as $\omega + 1$ as "tags" that marked the transfinite stages of the derivation. These tags later became the countable ordinal numbers.

Cantor proved that for every closed set X there must be a countable ordinal α (which could be finite) such that $X^\alpha = X^{\alpha+1}$. It is easy to show that each X^β in the sequence of derivatives is closed, and that it contains all but countably many points of the original set X. Therefore, X^α is a closed set that contains no isolated points. Such sets are called *perfect sets* and it is not too hard to show that they are either empty or have cardinality $2^{\aleph_0}$. From this it follows that X is either countable or of cardinality $2^{\aleph_0}$.

The intimate connection, discovered by Cantor, between transfinite ordinals and cardinals and the structure of the continuum was destined to leave its mark on the entire subsequent development of set theory.

3 The Universe of All Sets

In the discussion so far we have taken for granted that every set has a cardinality, or in other words that for every set X there is a unique cardinal number that can be put into one-to-one correspondence with X. If κ is such a cardinal and $f : X \rightarrow \kappa$ is a bijection (recall that we identify κ with the set of all its predecessors), then we can define an ordering on X by taking $x < y$ if and only if $f(x) < f(y)$. Since κ is a well-ordered set, this makes X into a well-ordered set. But it is far from obvious that every set can be given a well-ordering: indeed, it is not obvious even for the set $\mathbb{R}$. (If you need convincing of this, then try to find one.)

Thus, to make full use of the theory of transfinite ordinals and cardinals and to solve some of the fundamental problems—such as computing where in the aleph hierarchy of infinite cardinals the cardinal of $\mathbb{R}$ is—one must appeal to the *well-ordering principle*: the assertion that every set can be well-ordered. Without this assertion, one cannot even make sense of the questions. The well-ordering principle was introduced by Cantor, but he was unable to prove it. HILBERT [VI.63] listed proving that $\mathbb{R}$ could be well-ordered as part of the first problem in his celebrated list of twenty-three unsolved mathematical problems presented in 1900 at the Second International Congress of Mathematicians in Paris. Four years later, Ernst Zermelo gave a proof of the well-ordering principle that drew a lot of criticism for its use of THE AXIOM OF CHOICE [III.1] (AC), a principle that had been tacitly used for many years but which was now brought into focus by Zermelo's result. AC states that *for every set X of pairwise-disjoint nonempty sets there is a set that contains exactly one element from each set in X*. In a second, much more detailed, proof published in 1908, Zermelo spells out some of the principles or axioms involved in his proof of the well-ordering principle, including AC.

In that same year, Zermelo published the first axiomatization of set theory, the main motivation being the need to continue with the development of set theory while avoiding the logical traps, or paradoxes, that originated in the careless use of the intuitive notion of a set (see THE CRISIS IN THE FOUNDATIONS OF MATHEMATICS [II.7]). For instance, it seems intuitively clear that every property determines a set, namely, the set of those objects that have that property. But then consider the property of *being an ordinal number*. If this property determined a set, this would be the set of all ordinal numbers. But a moment of reflection shows that there cannot be such a set, since it would be well-ordered and would therefore correspond to an ordinal greater than all ordinals, which is absurd. Similarly, the property of *being a set that is not an element of itself* cannot determine a set, for otherwise we fall into Russell's paradox, that if A is such a set, then A is an element of A if and only if A is not an element of A, which is absurd. Thus, not every collection of objects, not even those that are defined by some property, can be taken to be a set. So what is a set? Zermelo's 1908 axiomatization provides the first attempt to capture our intuitive notion of set in a short list of basic principles. It was later improved through contributions from SKOLEM [VI.81], Abraham Fraenkel, and VON NEUMANN [VI.91], becoming what is now known as *Zermelo-Fraenkel set theory with the axiom of choice*, or ZFC.

The basic idea behind the axioms of ZFC is that there is a "universe of all sets" that we would like to understand, and the axioms give us the tools we need to build sets out of other sets. In usual mathematical practice we take sets of integers, sets of real numbers, sets of functions, etc., but also sets of sets (such as sets of open sets in a TOPOLOGICAL SPACE [III.90]), sets of sets of sets (such as sets of open covers), and so on. Thus, the universe of all sets should consist not only of sets of objects, but also of sets of sets of objects, etc. Now it turns out that it is much more convenient to dispense with "objects" altogether and consider only sets whose elements are sets, whose elements are also sets, etc. Let us call those sets "pure sets." The restriction to pure sets is technically advantageous and yields a more elegant theory. Moreover, it is possible to model traditional mathematical concepts such as real numbers using pure sets, so one does not lose any mathematical power. Pure sets are built from nothing, i.e., the empty set, by successively applying the "set of" operation. A simple example is $\{\varnothing, \{\varnothing, \{\varnothing\}\}\}$: to build this we start by forming $\{\varnothing\}$, then $\{\varnothing, \{\varnothing\}\}$, and putting these two sets together gives us $\{\varnothing, \{\varnothing, \{\varnothing\}\}\}$. Thus, at every stage we form all the sets whose elements are sets

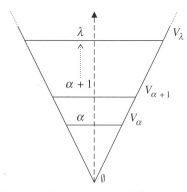

Figure 2 The universe V of all pure sets.

already obtained in the previous stages. Once again, this can be continued transfinitely: at limit stages we collect into a set all the sets obtained so far, and keep going. The universe of all (pure) sets, represented by the letter V and usually drawn as a V-shape with a vertical axis representing the ordinals (see figure 2), therefore forms a cumulative well-ordered hierarchy, indexed by the ordinal numbers, beginning with the empty set $\varnothing$. That is, we let

$$V_0 = \varnothing,$$

$$V_{\alpha+1} = \mathcal{P}(V_\alpha), \quad \text{the set of all subsets of } V_\alpha,$$

$$V_\lambda = \bigcup_{\beta < \lambda} V_\beta, \quad \text{the union of all the } V_\beta, \beta < \lambda,$$

if λ is a limit ordinal.

The universe of all sets is then the union of all the sets V_α such that α is an ordinal. More concisely,

$$V = \bigcup_\alpha V_\alpha.$$

3.1 The Axioms of ZFC

The ZFC axioms, stated informally, are the following.

(i) Extensionality. If two sets have the same elements, they are equal.

(ii) Power set. For every set x there is a set $\mathcal{P}(x)$ whose elements are all the subsets of x.

(iii) Infinity. There is an infinite set.

(iv) Replacement. If x is a set and ϕ is a *function-class*[1] restricted to x, then there is a set $y = \{\phi(u) : u \in x\}$.

(v) Union. For every set x, there is a set $\bigcup x$ whose elements are all the elements of the elements of x.

---

1. A function-class can be thought of as a function that is given as a definition rather than an object that has to exist as a set. The concept will be made precise in section 3.2.

(vi) Regularity. Every set x belongs to V_α, for some ordinal α.

(vii) Axiom of choice (AC). For every set X of pairwise-disjoint nonempty sets there is a set that contains exactly one element from each set in X.

Usually a further axiom appears on this list, called the *pairing axiom*. It asserts that for any two sets A and B the set $\{A, B\}$ exists. In particular, $\{A\}$ exists. Applying the union axiom to the set $\{A, B\}$ one then gets the union $A \cup B$ of A and B. But pairing can be derived from the other axioms. Another important axiom that appeared in Zermelo's original list, one that is both natural and very useful, is the *axiom of separation*. It states that for every set A and every *definable property* P, the set of elements of A that have the property P is also a set. But this axiom is a consequence of the axiom of replacement, so there is no need to include it in the list. Using the axiom of separation one can easily prove the existence of the empty set $\varnothing$, as well as the intersection $A \cap B$ and difference $A - B$ of any two sets A and B. The axiom of regularity is also known as the *axiom of foundation* and it is usually stated as follows: every nonempty set X has an $\in$-minimal element, i.e., an element that no element of X belongs to. In the presence of the other axioms the two formulations are equivalent. We chose the formulation in terms of the V_αs to stress the fact that this is a natural axiom based on the construction of the universe of all sets. But it is important to notice that the notions of "ordinal" and the "cumulative hierarchy of V_αs" need not appear in the formulation of the axioms of ZFC.

The axioms of ZFC lead a kind of double life. On the one hand, they tell us the things we can do with sets. In this sense, ZFC is just like any other collection of axioms for algebraic structures, e.g., the axioms for GROUPS [I.3 §2.1], or FIELDS [I.3 §2.2]: in both cases they give rules for creating new objects from old ones, though there are more rules for sets than there are for group or field elements and they are more complicated. Thus, just as one studies abstract groups, i.e., algebraic structures that satisfy the axioms for groups, so one can study the mathematical structures that satisfy the axioms of ZFC. These are called *models of ZFC*. Since, for reasons to be explained below, models of ZFC are not easy to come by, one is also interested in models of fragments of ZFC: that is, of axiom systems A that consist of just some of the axioms of ZFC. A model of a fragment A of ZFC is defined to be a pair $\langle M, E \rangle$, where M is a nonempty set and E is a binary relation on M, such that all axioms of A are true when the elements of M are interpreted as the sets and E is interpreted as the membership relation. For example, if A includes the union axiom, then for every element x of M there must be an element y of M such that zEy if and only if there exists w such that zEw and wEx. (If we replaced E by $\in$ and "element of M" by "set" in the last sentence, then we would recover the usual union axiom.)

The set $\langle V_\omega, \in \rangle$ is a model of all the axioms of ZFC except infinity, and $\langle V_{\omega+\omega}, \in \rangle$ is a model of ZFC except replacement. (To see why replacement fails, let x be the set ω and define a function ϕ on x by letting $\phi(n)$ equal $\omega + n$. The range of ϕ belongs to $V_{\omega+\omega+1}$ but not to $V_{\omega+\omega}$ because the ordinal $\omega + \omega$ does not belong to any set $V_{\omega+n}$ and $V_{\omega+\omega}$ is the union of the sets $V_{\omega+n}$.) For both these models, we took E to be $\in$, but one can also look at a completely different relation E on a set M, and see whether it happens to satisfy some of the axioms of ZFC. For example, take the pair $\langle \mathbb{N}, E \rangle$, where mEn if and only if the mth digit (counting from right to left) in the binary expansion of n is 1. This is a model of ZFC without the axiom of infinity, as the reader may care to check.

The other way of thinking of the ZFC axioms is that they tell us how to build up the hierarchy of the V_αs. Axiom (i), the axiom of extensionality, states that a set is something entirely determined by its elements. Axioms (ii)–(v) are tailored to construct V. The power-set axiom is what we use to get from V_α to $V_{\alpha+1}$. The axiom of infinity allows the construction to go into the transfinite. Indeed, in the context of the other ZFC axioms, this axiom is equivalent to the assertion that ω exists. The axiom of replacement is used to continue the construction of V at limit stages λ. To see this, consider the function defined by $F(x) = y$ if and only if x is an ordinal and $y = V_x$. The range of F restricted to λ then consists of all V_β with $\beta < \lambda$. By the axiom of replacement these sets form a set. Now, by an application of the union axiom to this set one obtains V_λ. Finally, the axiom of regularity states that all sets are obtained in this way: that is, the universe of all sets is precisely V. This rules out pathologies, such as sets that belong to themselves. The point is that for every set X there is a first α such that $X \in V_{\alpha+1}$. This α is called the *rank* of X and it marks the stage of the cumulative hierarchy where X was formed. So X could not possibly be an element of itself, since all elements of X must have a rank strictly smaller than the rank of X. The axiom of choice is equivalent, in the context of the other ZFC axioms, to the well-ordering principle.

3.2   Formulas and Models

The ZFC axioms can be formalized using the language of *first-order logic for sets*. The symbols of first-order logic are *variables* such as $x, y, z, \ldots$; the *quantifiers* "$\forall$" (for all) and "$\exists$" (there exists); the *logical connectives* "$\neg$" (not), "$\wedge$" (and), "$\vee$" (or), "$\rightarrow$" (if ..., then ...), and "$\leftrightarrow$" (if and only if); the equality symbol "$=$"; and parentheses. To make this first-order logic *for sets* we add one other symbol, "$\in$," standing for "is an element of," and the quantifiers are understood to range over sets. Here is how the axiom of extensionality is expressed in this language:

$$\forall x \forall y (\forall z (z \in x \leftrightarrow z \in y) \rightarrow x = y).$$

This reads as: for every set x and every set y, if every set z belongs to x if and only if it belongs to y (i.e., if x and y have the same elements), then x and y are equal. It is an example of a *formula* in our language. Formulas can be defined inductively as follows. The *atomic formulas* are $x = y$ and $x \in y$. Using quantifiers and logical connectives one can build up more complicated formulas using the following rules: if φ and ψ are formulas, then so are $\neg\varphi$, $(\varphi \wedge \psi)$, $(\varphi \vee \psi)$, $(\varphi \rightarrow \psi)$, $(\varphi \leftrightarrow \psi)$, $\forall x \varphi$, and $\exists x \varphi$. Thus, formulas are the formal counterpart of sentences in English (or in any other natural language) that talk only about sets and the membership relation. (For another discussion of formal languages, see LOGIC AND MODEL THEORY [IV.23 §1].)

Conversely, any formula of the formal language can be interpreted as a sentence (in English) about sets, and it makes sense to ask whether the interpreted sentence is true or not. Usually, by "true" we mean "true in the universe V of all sets," but it also makes sense to ask about the truth or falsity of a formula in any structure of the form $\langle M, E \rangle$, where E is a binary relation on M. For example, the formula $\forall x \exists y \; x \in y$ is true in all models $\langle M, E \rangle$ of ZFC, while the formula $\exists x \forall y \; y \in x$ is false (because of the axiom of regularity). Any formula that can be deduced from the axioms of ZFC is true in all models of ZFC.

Once we have defined what a formula is, we are in a position to make many statements precise that would otherwise not be. For example, the axiom of replacement involves the notion of a *function-class*. To make proper sense of it one formulates it in terms of first-order formulas. For example, the operation that takes each set a to the singleton $\{a\}$ is definable, and this depends on the fact that the statement $y = \{x\}$ can be expressed by the formula $\forall z (z \in y \leftrightarrow z = x)$. It is not a function, since it is defined on all sets, and the universe of all sets is not a set. This is why we use the different phrase "function-class." In addition, we sometimes allow *parameters* in our definitions of function-classes. For example, the function-class that, for a fixed set b, takes each set a to the set $a \cap b$ is defined by the formula $\forall z(z \in y \leftrightarrow z \in x \wedge z \in b)$, which depends on the set b: we call b a parameter and we say that the function-class is *definable with parameters*. More generally, a function-class is a function on sets given by a formula. But the function itself may not exist as a set, since its domain may contain all sets, or all ordinals, etc. Since the axiom of replacement is a statement about all function-classes, it is not in fact a single axiom but rather an "axiom scheme," consisting of one axiom for each function-class.

An important consequence of the fact that ZFC can be formalized in first-order logic is that it is subject to a remarkable theorem of Löwenheim and Skolem. The Löwenheim–Skolem theorem is a general result about first-order formal languages; in the particular case of ZFC, it says that if ZFC has a model, then it has a countable model. More precisely, given any model $M = \langle M, E \rangle$ of ZFC, there is a model N of ZFC contained in M that is countable and that satisfies exactly the same sentences as M. At first, this may seem paradoxical, for how can ZFC have a countable model if one can prove in ZFC that there are uncountable sets? Does the theorem not lead to a contradiction and therefore imply that there are no models of ZFC? Not quite. Suppose that we have a countable model N of ZFC and a set a in N. If we want to show that the statement "a is countable" is true in N, then we must show that *in* N there is a surjective map from ω to a. But it is possible for such a map to exist in V, or in some model M that is larger than N, without existing in N, because V and M contain more sets, and therefore more functions, than N does. In such a case, a is uncountable from the point of view of N but countable from the point of view of M or V.

Far from presenting a problem, the relativity of certain set-theoretic notions, like being countable or having a certain cardinality, with respect to different models of ZFC is an important phenomenon which, even if a bit disconcerting at first, may be used to great advantage in consistency proofs (see section 5 below).

It is not difficult to see that all the axioms of ZFC are true in V, which is hardly surprising since they were designed for that to happen. But the ZFC axioms may conceivably hold in some smaller universes. That is, there may be some class M properly contained in V,

or even some set M, and therefore by the Löwenheim–Skolem theorem also some countable set M, which is a model of ZFC. As we shall see, while the existence of models of ZFC cannot be proved in ZFC, the fact that one can consistently assume that they exist—provided ZFC is consistent, of course—is of the greatest importance for set theory.

4   Set Theory and the Foundation of Mathematics

As we have seen, we can use ZFC to develop the theory of transfinite numbers. But it turns out that all standard mathematical objects may be viewed as sets, and all classical mathematical theorems can be proved from ZFC using the usual logical rules of proof. For example, real numbers can be defined as certain sets of rational numbers, which can be defined as EQUIVALENCE CLASSES [I.2 §2.3] of ordered pairs of integers. The ordered pair (m, n) can be defined as the set $\{m, \{m, n\}\}$, integers can be defined as equivalence classes of ordered pairs of positive integers, and positive integers can be thought of as finite ordinals, which as we have seen can be defined as sets. Tracing back, one finds that a real number can be regarded as a set of sets of sets of sets of sets of finite ordinals. Similarly, all the usual mathematical objects—such as algebraic structures, vector spaces, topological spaces, smooth manifolds, dynamical systems, and so on—can be shown to exist in ZFC. Theorems concerning these objects can be expressed in the formal language of ZFC, as can their proofs. Of course, writing out a complete proof using the formal language would be extremely laborious, and the result would not only be very long but also virtually impossible to understand. It is important, however, to convince oneself that in principle it can be done. It is the fact that all standard mathematics can be formulated and developed within the axiomatic system of ZFC that makes *metamathematics* possible, that is, the rigorous mathematical study of mathematics itself. For example, it allows us to think about whether a mathematical statement has a proof: once we have rigorous definitions of "mathematical statement" and "proof," the question of whether a proof exists becomes a mathematical one with a determinate answer.

4.1   Undecidable Statements

In mathematics the truth of a mathematical statement φ is established by means of a proof from basic principles or axioms. Similarly, the falsity of φ is established by a proof of $\neg\varphi$. It is tempting to believe that there must always be a proof of either φ or $\neg\varphi$, but in 1931 GÖDEL [VI.92] proved in his famous INCOMPLETENESS THEOREMS [V.15] that this is not the case. The first incompleteness theorem says that in every axiomatic formal system that is consistent and rich enough to develop basic arithmetic there are *undecidable* statements: that is, statements such that neither they nor their negations are provable in the system. In particular, there are statements of the formal language of set theory that are neither provable nor disprovable from the ZFC axioms, supposing, that is, that ZFC is consistent.

But is ZFC consistent? The statement that asserts the consistency of ZFC, usually written as CON(ZFC), is the translation into the language of set theory of:

$$0 = 1 \text{ is not provable in ZFC.}$$

This statement asserts that the sequence of symbols $0 = 1$ is not the last step of any formal proof from ZFC. One can encode a formal proof as a finite sequence of natural numbers that satisfies certain arithmetical properties, and thereby regard the above statement as an arithmetical one. Gödel's second incompleteness theorem says that in any consistent axiomatic formal system that is rich enough to develop basic arithmetic, the arithmetical statement that asserts the consistency of the system cannot be proved. Thus, if ZFC is consistent, then its consistency can neither be proved nor disproved in ZFC.

ZFC is currently accepted as the standard formal system in which to develop mathematics. Thus, the truth of a mathematical statement is firmly established if its translation into the language of set theory is provable in ZFC. But what about undecidable statements? Since ZFC embodies all standard mathematical methods, the fact that a given mathematical statement φ is undecidable in ZFC means that the truth or falsity of φ cannot be established by means of usual mathematical practice. If all undecidable statements were like CON(ZFC), this would probably not be a cause of worry, since they seem not to directly affect the kind of mathematical problems that people are usually interested in. But for better or worse this is not so. As we will see, there are many statements of mathematical interest that are undecidable in ZFC.

There is an obvious way of showing that a mathematical statement has a proof: you just find one. But how can it be possible to prove, mathematically, that a given mathematical statement φ is undecidable in ZFC? This question has a short but far-reaching answer. If we can find a model M of ZFC in which φ is false, then there cannot be a proof of φ (because that proof would show that φ was true in M). Therefore, if we can find models M and N of ZFC with φ true in M and false in N, we can conclude that φ is undecidable.

Unfortunately, a consequence of Gödel's second incompleteness theorem is that it is not possible to prove in ZFC the existence of a model of ZFC. This is because another theorem of Gödel, called the *completeness theorem* for first-order logic, asserts that ZFC is consistent if and only if it has a model. However, we can get around this difficulty by splitting the proof of the undecidability of φ into two *relative consistency* proofs: the first is a proof that if ZFC is consistent, then so is ZFC plus φ; and the second is a proof that if ZFC is consistent, then so is ZFC plus the negation of φ. That is, one assumes that there is a model M of ZFC and proves the existence of two models of ZFC: one where φ holds, and one where it fails. One can then conclude that either φ and its negation are both unprovable in ZFC, or ZFC is inconsistent, in which case everything is provable.

One of the most surprising results of twentieth-century mathematics is that the continuum hypothesis is undecidable in ZFC.

5   The Continuum Hypothesis

Cantor's continuum hypothesis (CH), first formulated in 1878, states that every infinite set of real numbers is either countable or has the same cardinality as $\mathbb{R}$. In ZFC, since AC implies that every set, and in particular every infinite set of real numbers, can be put into one-to-one and onto correspondence with a cardinal number, one can easily see that CH is equivalent to the assertion that the cardinality of $\mathbb{R}$ is $\aleph_1$, or equivalently, that $2^{\aleph_0} = \aleph_1$, the version of the statement that we mentioned earlier.

Solving CH was the first problem in Hilbert's famous list of twenty-three unsolved problems, and has been one of the main driving forces for the development of set theory. In spite of many attempts at proving CH by Cantor himself and by many leading mathematicians of the first third of the twentieth century, no major progress was made until, sixty years after its formulation, Gödel was able to prove its consistency with ZFC.

5.1   The Constructible Universe

In 1938, Gödel found a way to construct, starting with a model M of ZFC, another model of ZFC, contained in M, where CH holds. He thereby proved the relative consistency of CH with ZFC. Gödel's model is known as the *constructible universe* and is represented by the letter L. Since M is a model of ZFC, we may view M as the universe V of all sets. Then L is built inside M in a way that is similar to how we built V, but with the following important difference. When we passed from V_α to $V_{\alpha+1}$ we took *all* subsets of V_α, but to go from L_α to $L_{\alpha+1}$ one takes only those subsets of L_α that are *definable* in L_α. That is, $L_{\alpha+1}$ consists of all sets of the form $\{a : a \in L_\alpha$ and $\varphi(a)$ holds in $L_\alpha\}$, where $\varphi(x)$ is a formula of the language of set theory that may mention elements of L_α. If λ is a limit ordinal, then L_λ is just the union of all the L_α, $\alpha < \lambda$, and L is the union of all the L_α, α an ordinal. Of course, we can also build L inside V. This is the *real L*, the universe of all constructible sets.

One important observation is that to build L it is not necessary to use AC, and so we do not require AC to hold in M. But once L is constructed it can be verified that AC holds in L, as do the other axioms of ZFC. The verification of AC is based on the fact that every element of L is defined at some stage α, and so it is uniquely determined by a formula and some ordinals. Therefore, any sensible well-ordering of all the formulas will naturally yield a well-ordering of L, and thus of every set in L. This shows that if ZF (i.e., ZFC minus AC) is consistent, then so is ZFC. In other words, if we add AC to the ZF axioms, then no contradiction is introduced into the system. This is very reassuring, for although AC has many desirable consequences it also has some that at first sight can appear counterintuitive, such as THE BANACH–TARSKI PARADOX [V.3].

That CH holds in L is due to the fact that in L every real number appears at some countable stage of the construction, i.e., in some L_α, where α is countable in L. To prove this, one shows first that every real r belongs to some L_β that satisfies a finite number of axioms of ZFC that are sufficient to build L, where β is an ordinal that is not necessarily countable. Then, with the help of the Löwenheim–Skolem theorem, one can show that there is a countable subset X of L_β that contains r and satisfies the same axioms as L_β. And then one shows that X must be isomorphic to L_α for some countable ordinal α, via an isomorphism that is the identity on r; this finishes the proof that r appears at a countable stage. But since there are only $\aleph_1$ countable ordinals, and L_α is countable for each countable ordinal α, there can be only $\aleph_1$ real numbers.

Since, for each ordinal α, L_α contains only the sets that are strictly necessary, namely those that were explicitly definable in one of the previous stages, L is the smallest possible model of ZFC containing all the ordinals, and in it the cardinality of $\mathbb{R}$ is also the smallest possible, namely $\aleph_1$. In fact, in L the *generalized continuum hypothesis* (GCH) holds: that is, for every ordinal α, $2^{\aleph_\alpha}$ has the smallest possible value, namely, $\aleph_{\alpha+1}$.

The theory of constructible sets went through an extraordinary development in the hands of Ronald Jensen. He showed that in L a well-known conjecture called Suslin's hypothesis was false (see section 10 below) and isolated two important combinatorial principles, known as $\diamond$ (diamond) and $\square$ (square), that hold in L. These two principles, which will not be defined here, enable us to carry out constructions of uncountable mathematical structures by induction on the ordinals in such a way that the construction does not break down at limit stages. This is extremely useful, because it allows one to prove consistency results without going to the trouble of analyzing constructible sets: if you can deduce a statement φ from $\diamond$ or $\square$, then it holds in L, since, by Jensen's results, $\diamond$ and $\square$ hold in L; it follows that φ is consistent with ZFC.

There is also an important generalization of the notion of constructibility, called *inner model theory*. Given any set A it is possible to build the *constructible closure* of A, which is the smallest model of ZF that contains all ordinals and A. This model, called $L(A)$, is built in the same way as L, but instead of beginning with the empty set one begins with the *transitive closure* of A, which consists of A, the elements of A, the elements of the elements of A, and so on. Models of this sort are examples of *inner models*: that is, models of ZF that contain all the ordinals and all the elements of their elements. Especially prominent are the inner models $L(r)$, where r is a real number, and $L(\mathbb{R})$, the constructible closure of the set of real numbers. Also very important are the inner models of large-cardinal axioms, which will be discussed in section 6 below.

After the result of Gödel, and given the repeated failed attempts to prove CH in ZFC, the idea started to take shape that maybe it was undecidable. To prove this, it was necessary to find a way to build a model of ZFC in which CH is false. This was finally accomplished twenty-five years later, in 1963, by Paul Cohen, using a revolutionary new technique called *forcing*.

5.2 Forcing

The forcing technique is an extremely flexible and powerful tool for building models of ZFC. It allows one to construct models with the most diverse properties and with great control over the statements that will hold in the model being constructed. It has made it possible to prove the consistency of many statements with ZFC that were not previously known to be consistent, and this has led to many undecidability results.

In a manner reminiscent of the way one passes from a field K to an algebraic extension $K[a]$, one goes from a model M of ZFC to a *forcing extension* $M[G]$ that is also a model of ZFC. However, the forcing method is far more complex, both conceptually and technically, involving set-theoretic, combinatorial, topological, logical, and metamathematical aspects.

To give an idea of how it works, let us consider Cohen's original problem of starting from a model M of ZFC and obtaining from it a model where CH fails. The only thing we know about M is that it is a model of ZFC, and as far as we know CH may hold in it. In fact, for all we know, M might be the constructible universe L: perhaps when we build L inside M we obtain the whole of M. Therefore, when we extend M we shall have to add to it some new real numbers to ensure that in the extension $M[G]$ there will be at least $\aleph_2$ of them. More precisely, we need the model $M[G]$ to satisfy the sentence that says that there are at least $\aleph_2$-many real numbers. However, the "real numbers" in $M[G]$ may not be real numbers in the actual universe V: all that matters is that in $M[G]$ they satisfy sentences that say "I am a real number." Similarly, the element of $M[G]$ that plays the role of the cardinal $\aleph_2$ need not be the actual cardinal $\aleph_2$ in V.

In order to explain the method, let us consider the simpler problem of adding to M just a single new real number r. To make things even simpler, let us think of r as just the binary representation of a real in $[0, 1]$. In other words, r is an infinite binary sequence in the real world V.

A first difficulty is that M may already contain all infinite binary sequences, in which case we will not be able to find one to add. However, by the Löwenheim–Skolem theorem, every model M of ZFC has a countable submodel N that satisfies exactly the same sentences of the language of set theory as M. Let us emphasize that N is countable in the real world, that is, in V; so there is, outside N, a function that enumerates all its elements. Nevertheless, N will contain sets x for which the sentence that says "x is uncountable" is true in N. Since M was a model of ZFC, so is N. So, since we really do not care about the size of M, but only that it is a model of ZFC, we may as well assume that $M = N$, so that M

itself is countable. And now, since there are uncountably many infinite binary sequences, there are plenty of them that do not belong to M.

So, can we just pick any one of them and add it to M? Well, no. The problem is that there are some binary sequences that have a great influence on any model that contains them. For example, we can encode any countable ordinal α as a real number as follows. First let f be a bijection from $\mathbb{N}$ to α and define a subset $A \subset \mathbb{N}^2$ to be $\{(m, n) \in \mathbb{N}^2 : f(m) < f(n)\}$. Now choose a bijection g from $\mathbb{N}$ to $\mathbb{N}^2$ and let $c(n) = 1$ if and only if $g(n) \in A$. If g is sufficiently explicit (as it can easily be chosen to be), then any model M that contains the infinite binary sequence c must contain the ordinal α, since α can be built out of c using the axioms of ZFC.

To see why this matters, suppose that M is of the form L_α, as constructed in V, where α is a countable ordinal in V. The existence of models of ZFC of this form follows, for instance, from the existence of large cardinals (see section 6 below), so we certainly cannot rule out this possibility. Since we want to build a model $M[c]$ of ZFC that contains a new infinite binary sequence c and all the elements of M, it will have to contain $L_\alpha(c)$, i.e., all sets that can be constructed in fewer than α steps starting with c. But if c is a sequence that encodes α, as above, then $M[c]$ cannot *equal* $L_\alpha(c)$ and still be a model of ZFC, since this would imply that $L_\alpha(c)$ contained itself. If we try to circumvent the problem by adding more sets to $M[c]$ so that it becomes a model of ZFC, then we may end up with $M[c] = L_\gamma$ for some ordinal γ greater than α. And this is not good for our purposes since CH holds in all models of ZFC of the form L_γ. The conclusion is that we cannot just pick an *arbitrary* c that is not in M: we will have to choose it very carefully.

The key idea is that c should be "generic," meaning that it should have no special property that singles it out. The reason for this is that if, as before, $M = L_\alpha$, and we want to ensure that $M[c] = L_\alpha(c)$ is still a model of ZFC, then we do not want c to have any special property that would interfere in the construction of $M[c]$ and cause some ZFC axiom not to hold any more. To accomplish this we build c little by little so that it avoids all the special properties that could possibly have any undesirable effect on $M[c]$. For example, if we do not want c to encode the ordinal α in the manner sketched above, we simply set some $c(n)$ equal to 0 for some n such that $g(n) \in A$.

Of course, if we have built up the first N binary digits of c and φ is a property that holds for all real numbers that begin with those N digits, then we cannot avoid φ without undoing our previous work. Let us call a property *avoidable* if every finite binary sequence p can be extended to a finite binary sequence q such that no infinite sequence that extends q has the property. For instance, the property "all terms in the sequence are zero" is avoidable, while the property "there are ten consecutive ones in the sequence" is not avoidable.

A real number c is called *generic*, or *Cohen*, over M if it avoids all avoidable properties that can be defined in M, that is, properties that can be defined by means of formulas that may mention sets in M. It is easy to see that c cannot belong to M, since if it did then the property "is equal to c" would be definable in M, and it is certainly avoidable.

Why should a generic real number exist? Once again, we use the fact that M is countable. From this it follows that there are only countably many avoidable properties. If we enumerate them as $\varphi_1, \varphi_2, \ldots$, then we can pick a finite sequence q_1 such that no infinite extension of q_1 satisfies φ_1. Then we can extend q_1 to q_2 such that no infinite extension of q_2 satisfies φ_2. Continuing in this way we create an infinite binary sequence c that does not have any of the properties φ_i. In other words, it is generic.

Now let $M[c]$ be the set of all sets that can be constructed, using c and the elements of M as parameters, in as many steps as the ordinals of M. For instance, if M were of the form L_α, then $M[c]$ would just be $L_\alpha(c)$. The model $M[c]$ is called a *Cohen-generic extension* of M.

It turns out that, miraculously, $M[c]$ is a model of ZFC. Moreover, it has the same ordinals as M and, therefore, it is not of the form L_γ, for any ordinal γ. In particular, when we build L inside $M[c]$, c does not belong to it. These statements are by no means easy to prove, but very roughly what Cohen showed was that a formula φ is true in $M[c]$ if and only if there is an initial segment p of c that "forces" φ to be true. Moreover, the relation "p forces φ to be true," which relates finite binary sequences to formulas and is written $p \Vdash \varphi$, can be defined in M. Therefore, to know whether a statement φ is true in $M[c]$ one just needs to check whether there is an initial segment p of c such that $p \Vdash \varphi$. In particular, using this result one can prove that $M[c]$ satisfies the ZFC axioms.

In order to build a model where CH fails, one adds not just one generic real number but $\aleph_2^M$ of them, where $\aleph_2^M$ is the ordinal that plays the role of $\aleph_2$ in M. That is, it is the second uncountable cardinal in M. This

need not be the real $\aleph_2$, and indeed it will not be if, for instance, M is of the form L_α for some countable ordinal α in V. Adding $\aleph_2^M$ generic real numbers can be done by finitely approximating any finite number of them while avoiding all avoidable properties they could have. Thus, instead of finite binary sequences we now work with finite sets of finite binary sequences indexed by ordinals less than $\aleph_2^M$. A generic object will be a sequence $\langle c_\alpha : \alpha < \aleph_2^M \rangle$ of Cohen reals over M, all different, and so CH is false in the generic extension $M[\langle c_\alpha : \alpha < \aleph_2^M \rangle]$.

However, there is an important point that needs to be addressed. When we add the new real numbers to M, it is important that the $\aleph_2$ of the new expanded model is the same as $\aleph_2^M$. Otherwise, CH might hold in the expanded model and our work would have been wasted. Fortunately, this is true, but again we must use the facts about forcing to prove it.

The same kind of forcing argument allows one to construct models where the cardinality of $\mathbb{R}$ is $\aleph_3$, or $\aleph_{27}$, or any other cardinal of uncountable *cofinality*, i.e., any uncountable cardinal that is not the least upper bound of countably many smaller cardinals. The cardinality of the continuum is, therefore, undetermined by ZFC. Furthermore, since CH holds in Gödel's constructible universe L and fails in the model constructed by Cohen using forcing, it is undecidable in ZFC.

Cohen also used forcing to prove that AC is independent of ZF. Since AC holds in L, this amounted to constructing a model of ZF in which AC was false. He did this by adding a countable collection $\langle c_n : n \in \mathbb{N} \rangle$ of generic real numbers to a countable model M of ZF. To see why this works, let N be the smallest submodel of $M[\langle c_n : n \in \mathbb{N} \rangle]$ that contains all the ordinals and the unordered set $A = \{c_n : n \in \mathbb{N}\}$. Thus, N is just $L(A)$, as built inside $M[\langle c_n : n \in \mathbb{N} \rangle]$. One can then show that N is a model of ZF, but that in N there is no well-ordering of A. The reason is that any well-ordering of A would be definable in $L(A)$ with a finite number of ordinals and finitely many elements of A as parameters, and then each one of the c_n would in its turn be definable by indicating its ordinal position in the well-ordering. But since the whole sequence of c_ns is generic over L, no formula can distinguish one of the c_ns from another unless they appear as parameters in the formula. Since we can choose two different c_ns that do not appear as parameters in the definition of the well-ordering of A, and that well-ordering distinguishes all the c_ns from each other, we have a contradiction. Therefore, the set A cannot be well-ordered, so AC does not hold.

Immediately after Cohen's proof of the independence of AC from ZF and of CH from ZFC, a result for which he got the Fields Medal in 1966, many set theorists started developing the forcing technique in its full generality (notably Azriel Lévy, Dana Scott, Joseph Shoenfield, and Robert Solovay) and began to apply it to other well-known mathematical problems. For instance, Solovay constructed a model of ZF in which every set of real numbers is LEBESGUE MEASURABLE [III.55], thereby showing that AC is necessary for the existence of nonmeasurable sets. He also constructed a model of ZFC where every *definable* set of real numbers is Lebesgue measurable; therefore, nonmeasurable sets, although they can be proved to exist (see the example in section 6.1 below), cannot be explicitly given; Solovay and Stanley Tennenbaum developed the theory of iterated forcing and used it to prove the consistency of Suslin's hypothesis (see section 10 below); Adrian Mathias proved the consistency of the infinitary form of RAMSEY'S THEOREM [IV.19 §2.2]; Saharon Shelah proved the undecidability of the Whitehead problem in group theory; and Richard Laver proved the consistency of the Borel conjecture; to cite just a few remarkable examples from the 1970s.

The forcing technique now pervades all of set theory. It continues to be a research area of great interest, very sophisticated from the technical point of view and of great beauty. It keeps producing important results, with applications in many areas of mathematics, such as topology, combinatorics, and analysis. Especially influential has been the development over the last twenty-five years of the theory of *proper forcing*, introduced by Shelah. Proper forcing has proved very useful in the context of forcing iterations, and in the formulation and study of new *forcing axioms*, which will be dealt with in section 10, as well as in the analysis of *cardinal invariants* of the continuum. These are uncountable cardinals associated with various topological or combinatorial properties of the real line that can consistently take different values in different models obtained by forcing. An example of a cardinal invariant is the least number of null sets needed to cover the real line. Another important development has been the use of *class forcing* by Anthony Dodd and Ronald Jensen for coding the universe into a single real number, which shows that, amazingly, one can always use forcing to turn any model M into a model of the form $L(r)$ for some real number r. A more recent contribution is the invention by W. Hugh Woodin of new powerful forcing notions associated with the theory of large cardinals (see the next section), which have provided new insights into the continuum hypothesis (see the end of section 10).

The large number of independence results obtained by forcing have made very clear that the axioms of ZFC are insufficient to answer many fundamental mathematical questions. Thus, it is desirable to find new axioms that, once added to ZFC, will provide a solution to some of those questions. We shall discuss some candidates in the next few sections.

6   Large Cardinals

As we have already seen, the collection of all ordinal numbers cannot form a set. But if it did, then to that set there would correspond an ordinal number κ. This ordinal would coincide with the κth cardinal $\aleph_\kappa$, since otherwise $\aleph_\kappa$ would be a larger ordinal. Moreover, V_κ would be a model of ZFC. We cannot prove in ZFC that there is an ordinal κ with these properties, for then we would have proved in ZFC that ZFC has a model, which is impossible by Gödel's second incompleteness theorem. So, why do we not add to ZFC the axiom that says that there is a cardinal κ such that V_κ is a model of ZFC?

This axiom, with the further requirement that κ be *regular*, that is, not the limit of fewer than κ smaller cardinals, was proposed in 1930 by SIERPIŃSKI [VI.77] and TARSKI [VI.87], and it is the first of the *large-cardinal axioms*. A cardinal κ with those properties is called *inaccessible*.

Other notions of large cardinals, which implied inaccessibility, kept appearing during the twentieth century. Some of them originated in generalizations to uncountable sets of the infinite version of Ramsey's theorem, which states that if each (unordered) pair of elements of ω (i.e., of natural numbers) is painted either red or blue, then there is an infinite subset X of ω such that all pairs of elements of X have the same color. The natural generalization of the theorem to ω_1 turns out to be false. However, on the positive side, Paul Erdős and Richard Rado proved that for every cardinal $\kappa > 2^{\aleph_0}$, if each pair of elements of κ is painted either red or blue, then there is a subset X of κ of size ω_1 such that all pairs of elements of X have the same color. This is one of the landmark results of the *partition calculus*, an important area of combinatorial set theory developed mainly by the Hungarian school, led by Erdős and András Hajnal. The problem of whether Ramsey's theorem can be generalized to some uncountable cardinal leads naturally to cardinals that are called *weakly compact*. A cardinal κ is weakly compact if it is uncountable and satisfies the strongest possible Ramsey-type theorem: whenever all pairs of elements of κ are painted either red or blue, there is a subset X of κ of size κ such that all pairs of elements of X have the same color. Weakly compact cardinals are inaccessible, so their existence cannot be proved in ZFC. Moreover, it turns out that below the first weakly compact cardinal, assuming it exists, there are many inaccessible cardinals, so the existence of a weakly compact cardinal cannot be proved even if one assumes the existence of inaccessible cardinals.

The most important large cardinals, the *measurable cardinals*, are much larger than the weakly compact ones, and were discovered in 1930 by Stanisław Ulam.

6.1   Measurable Cardinals

A set A of real numbers is a BOREL SET [III.55] if it can be obtained in countably many steps starting from the open intervals and applying the two operations of taking complements and countable unions. It is *null*, or has *measure zero*, if for every $\varepsilon > 0$ there is a sequence of open intervals $I_0, I_1, I_2, \ldots$ such that $A \subseteq \bigcup_n I_n$ and $\sum_n |I_n| < \varepsilon$. It is *Lebesgue measurable* if it is almost a Borel set, that is, if it differs from a Borel set by a null set. To each measurable set A corresponds a number $\mu(A) \in [0, \infty]$, its *measure*, that is invariant under translation of A and is *countably additive*, that is, the measure of a countable union of measurable pairwise-disjoint sets is the sum of their measures. Moreover, the measure of an interval is its length. (See MEASURES [III.55] for a fuller discussion of these concepts.)

One can prove in ZFC that there exist non-Lebesgue-measurable sets of real numbers. For example, the following set was discovered in 1905 by Giuseppe Vitali. Define two elements of the closed interval $[0, 1]$ to be equivalent if they differ by a rational, and let A be a subset of $[0, 1]$ that contains precisely one element from each equivalence class. This requires one to make a large number of choices, which can be done by AC. To see that A is not measurable, consider for each rational p the set $A_p = \{x + p : x \in A\}$. Any two of these sets are disjoint, because of the way we built A. Let B be the union of all A_p over all rational numbers p in the interval $[-1, 1]$. A cannot have measure zero, for then B itself would have measure zero, and this is impossible because $[0, 1] \subseteq B$. On the other hand, A cannot have positive measure either, since then B would

have infinite measure, and this is impossible because $B \subseteq [-1, 2]$.

Since measurable sets are closed under taking complements and countable unions, all Borel sets are measurable. In 1905 LEBESGUE [VI.72] showed that there are measurable sets that are not Borel. While reading Lebesgue's work, Mikhail Suslin noticed that Lebesgue had made a mistake in claiming that continuous images of Borel sets are Borel. Indeed, Suslin soon found a counterexample, which led eventually to the discovery of a new natural hierarchy of sets of reals beyond the Borel sets, the so-called *projective sets*. These are the sets that can be obtained from the Borel sets by taking continuous images and complements (see section 9 below). In 1917 Nikolai Luzin showed that all continuous images of Borel sets, the *analytic sets*, are also measurable. If a set is measurable, then so is its complement, so all complements of analytic sets, the *coanalytic* sets, are also Lebesgue measurable. It is therefore natural to ask whether we can continue like this. In particular, are continuous images of coanalytic sets, or Σ_2^1 sets, as they are known, also measurable? The answer to this question turns out to be undecidable in ZFC: in L there are Σ_2^1 sets that are not Lebesgue measurable, and with forcing one can construct models where all Σ_2^1 sets are measurable.

The proof given above of the existence of a non-Lebesgue-measurable set of reals hinges on the fact that Lebesgue measure is translation invariant. In fact, the proof shows that there cannot be any countably additive translation-invariant measure that extends Lebesgue measure and measures all sets of reals. Thus, a natural question, known as *the measure problem*, is whether, if one drops the requirement of translation invariance, there can exist some countably additive measure that extends Lebesgue measure and measures all sets of reals. If such a measure exists, then the cardinality of the continuum cannot be $\aleph_1$, nor $\aleph_2$, nor any $\aleph_n$ with $n < \omega$, etc. In fact, Ulam proved in 1930 that a positive solution to the measure problem implies that the cardinality of $\mathbb{R}$ is extremely large: it is greater than or equal to the least uncountable regular cardinal that is a limit of smaller cardinals. He also proved that the existence of a nontrivial countably additive measure on *any* set implies either a positive solution to the measure problem, or that there exists an uncountable cardinal κ with a (nontrivial) $\{0, 1\}$-valued κ-additive measure that measures all its subsets. Such a cardinal is called *measurable*. If κ is measurable, then it is weakly compact, and therefore inaccessible. In fact, the set of weakly compact cardinals smaller than κ has measure 1, and so κ is itself the κth weakly compact cardinal. It follows that the existence of a measurable cardinal cannot be proved in ZFC, even if one adds the axiom that inaccessible, or weakly compact, cardinals exist (unless, of course, ZFC plus the existence of such cardinals is inconsistent). A complete clarification of the measure problem was finally provided by Solovay, who showed that if the solution is positive, then there is an inner model with a measurable cardinal. Conversely, if there is a measurable cardinal, then one can build a forcing extension where the measure problem has a positive solution.

An unexpected consequence of the existence of a measurable cardinal is that the universe V cannot be L: that is, there are nonconstructible sets, and even nonconstructible real numbers. In fact, if there is a measurable cardinal, then V is much larger than L. For instance, the first uncountable cardinal, $\aleph_1$, is an inaccessible cardinal in L.

After the invention of forcing and the subsequent avalanche of independence results, the hope arose that axioms asserting the existence of large cardinals, like measurable cardinals, would settle some of the questions that, thanks to the forcing technique, had been proved undecidable in ZFC. It was soon shown, however, by Lévy and Solovay, that large-cardinal axioms could not settle CH, as one could easily use forcing to change the cardinality of the continuum and make CH hold or fail without destroying the large cardinals. But Solovay proved in 1969 that, surprisingly, if there exists a measurable cardinal, then all Σ_2^1 sets of real numbers are Lebesgue measurable. So, while the axiom that asserts the existence of a measurable cardinal cannot settle the size of the continuum, it has a profound effect on its structure. It is indeed astonishing that measurable cardinals, so far away from the sets of real numbers in the universe V, have such a strong influence on their basic properties. While the relationship between large cardinals and the structure of the continuum is not yet fully understood, great progress has been made in the last thirty years through the work done in *descriptive set theory* and *determinacy*, which will be described in sections 8 and 9 below.

Some of the deepest and most technically difficult work in set theory is currently devoted to the construction and analysis of canonical inner models for large cardinals. These are analogues of L for large cardinals, that is, they are models built in some canonical way that contain all the ordinals and are transitive (i.e., they contain all elements of their elements), and in which certain large cardinals exist. The larger the cardinal, the more difficult it is to build the model. This work is known as the *inner model program*.

One of the striking consequences of the inner model program is that it provides a way of measuring the *consistency strength* of virtually any set-theoretic statement φ, using large cardinals. That is, there are large-cardinal axioms A_1 and A_2 such that the consistency of ZFC plus φ implies that of ZFC plus A_1 and is implied by the consistency of ZFC plus A_2. We refer to A_1 as a *lower bound* for the consistency of φ and to A_2 as an *upper bound*. In the fortunate cases when the lower and upper bounds coincide, we obtain an exact measure of the consistency strength of φ. An upper bound A_2 is usually obtained by forcing over a model of ZFC plus A_2, whereas a lower bound A_1 is obtained by inner model theory. Earlier in this section we saw that the consistency strength of a positive solution to the measure problem is exactly that of the existence of a measurable cardinal. We shall see another important example in the next section.

Knowing upper and lower bounds for the consistency strength of set-theoretic statements—or, even better, knowing their exact consistency strength—is extremely useful for comparing them. Indeed, if the lower bound for a sentence φ is greater than the upper bound for another sentence ψ, then we can conclude, via Gödel's incompleteness theorem, that ψ does not imply φ.

7   Cardinal Arithmetic

Beyond the continuum hypothesis, understanding the behavior of the exponential function 2^κ for arbitrary infinite cardinals κ has been a motivating force in set theory. Cantor proved that $2^\kappa > \kappa$ for all κ, and Dénes König proved that the cofinality of 2^κ is always greater than κ: that is, 2^κ is not the limit of fewer than κ smaller cardinals. The GCH, which, as we saw, holds in the constructible universe L, states precisely that 2^κ has the least possible value, namely, the least cardinal greater than κ, usually denoted by κ^+. One might think that, as in the case of $2^{\aleph_0}$, by forcing it should be possible to build models of ZFC where 2^κ takes any prescribed value, subject only to the necessary requirement that its cofinality should be greater than κ. This is true for cardinals κ that are *regular*, that is, not the limit of fewer than κ smaller cardinals. Indeed, William Easton showed that for any function F on the regular cardinals such that $\kappa \leqslant \lambda$ implies $F(\kappa) \leqslant F(\lambda)$

and $F(\kappa)$ has cofinality greater than κ, there is a forcing extension of L in which $2^\kappa = F(\kappa)$, for all regular κ. So, for instance, one can build a model of ZFC where $2^{\aleph_0} = \aleph_7$, $2^{\aleph_1} = \aleph_{20}$, $2^{\aleph_2} = \aleph_{20}$, $2^{\aleph_3} = \aleph_{101}$, etc. This shows that the behavior of the exponential function for infinite regular cardinals is totally undetermined in ZFC, and anything possible can be attained by forcing.

But how about nonregular cardinals? Nonregular cardinals are called *singular*. Thus, an infinite cardinal κ is *singular* if it is the supremum of fewer than κ smaller cardinals. For instance, $\aleph_\omega$, being the supremum of the $\aleph_n$, $n \in \mathbb{N}$, is the first singular cardinal. Determining the possible values of the exponential function at singular cardinals is a very hard problem that has generated much important research and involves, quite surprisingly, the necessary use of large cardinals.

Using a *supercompact cardinal*, which is a measurable cardinal with certain further properties that make it much larger than ordinary measurable cardinals, Matthew Foreman and Woodin built a model of ZFC in which GCH fails everywhere, i.e., $2^\kappa > \kappa^+$ for all cardinals κ. But curiously, the value of the exponential function at a singular cardinal of *uncountable* cofinality is somehow determined by its values at smaller regular cardinals. Indeed, in 1975, Jack Silver proved that if κ is a singular cardinal of uncountable cofinality and $2^\alpha = \alpha^+$ for all $\alpha < \kappa$, then $2^\kappa = \kappa^+$. That is, if the GCH holds below κ, then it also holds at κ. That this is also the case for singular cardinals of *countable* cofinality is a consequence of the *singular cardinal hypothesis* (SCH), a general principle weaker than the GCH that completely determines singular cardinal exponentiation, relative to exponentiation for regular cardinals. A special case of SCH is the following. *If* $2^{\aleph_n} < \aleph_\omega$ *for all finite n, then* $2^{\aleph_\omega} = \aleph_{\omega+1}$. So, in particular, if the GCH holds below $\aleph_\omega$, then it must hold at $\aleph_\omega$. Shelah used his powerful "PCF theory" to obtain the unexpected result that if $2^{\aleph_n} < \aleph_\omega$ for all n, then $2^{\aleph_\omega} < \aleph_{\omega_4}$. So, if GCH holds below $\aleph_\omega$, then there is a bound (in ZFC!) on the possible values of $2^{\aleph_\omega}$. But can this value actually be greater than the least possible one, namely $\aleph_{\omega+1}$? In particular, can the GCH first fail at $\aleph_\omega$? The answer is yes, but large cardinals are needed. Indeed, on the one hand Menachem Magidor proved the consistency of the first failure of GCH at $\aleph_\omega$, assuming the consistency of the existence of a supercompact cardinal. Thus, the existence of a supercompact cardinal is an *upper bound* for the failure of SCH. On the other hand, using inner model theory, Dodd and Jensen showed that large car-

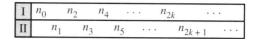

I	n_0		n_2		n_4	$\cdots$		n_{2k}		$\cdots$
II		n_1		n_3		n_5	$\cdots$		n_{2k+1}	$\cdots$

Figure 3 A run of the infinite game associated with a set $A \subseteq [0, 1]$.

dinals are required for this to happen. An exact measure of the consistency strength of the failure of SCH was later established by Moti Gitik.

8   Determinacy

It turns out that the existence of very large cardinals, such as supercompact cardinals, has a dramatic effect on the properties of sets of real numbers, especially when they can be defined in some simple way. The link between the two appears through the analysis of certain infinite two-player games that are associated with sets of real numbers. Given a subset A of $[0, 1]$, consider the following infinite game associated with A: there are two players, I and II, who alternately choose a number n_i that equals either 0 or 1. To begin with, player I plays n_0, then player II plays n_1, to which I answers by playing n_2, and so on. A run of the game is displayed in figure 3. At the end of the run, the players have produced an infinite binary sequence: $n_0, n_1, n_2, \ldots$. This sequence can be regarded as the binary expansion of a real number r in $[0, 1]$. Player I wins the game if r belongs to A and player II wins otherwise.

For example, if A is the interval $[0, \frac{1}{2}]$, then a winning strategy for player I is simply to start by playing 0, whereas if $A = [0, \frac{1}{4})$, then player II wins the game by playing 1 in her first move. But for most games, the question of who wins is not decided after any finite number of moves. For instance, if A is the set of rational points of $[0, 1]$, then one can easily see that player II has a strategy for winning the game (for example, whatever player I does, player II will win if she plays $01001000100001\ldots$), but she will not win at any finite stage of the run.

The game is *determined* if one of the two players has a winning strategy. Formally, a *strategy* for player II is a function f that assigns 0 or 1 to each finite binary sequence of odd length. It is a *winning strategy* if player II always wins the game if she plays $f(n_0, n_1, \ldots, n_{2k})$ in her kth turn, whatever moves are made by player I. Similarly, one can define a winning strategy for I. We say that the set A is *determined* if the game associated with A is determined. One might guess that every game is determined, but actually it is quite easy, using AC, to prove the existence of a game that is not determined.

It turns out that the determinacy of the games associated with certain classes of sets of reals implies that all sets in the class have properties similar to those of the Borel sets. For example, the *axiom of determinacy* (AD), which asserts that all sets of reals are determined, implies that every set of reals is Lebesgue measurable, has the property of Baire (i.e., differs from an open set by a set of first category), and has the perfect set property (i.e., contains a perfect set if it is uncountable). To give the flavor of a typical argument, let us indicate why every set A of reals is Lebesgue measurable.

First, one observes that it is enough to show that if all measurable subsets of A are null, then A itself must be null. And for this one plays, for every $\varepsilon > 0$, the *covering game* for A and ε. In this game, player I plays so that the sequence $a = \langle n_0, n_2, n_4, \dots \rangle$ represents an element of A, and player II plays (binary encodings of) finite unions of rational intervals, with measures adding up to at most ε, while attempting to cover a. It can be shown that if every measurable subset of A is null, then player I cannot have a winning strategy. So by AD there must be a winning strategy for II. Using this strategy one can show that the outer measure of A is at most ε. And since this works for all $\varepsilon > 0$, A must be null.

While AD rules out the existence of badly behaved sets of reals, it implies the negation of AC, so AD is inconsistent with ZFC. However, weaker versions of AD are compatible with, and even follow from, ZFC. Indeed, Donald Martin proved in 1975 that ZFC implies that every Borel set is determined. Moreover, if there exists a measurable cardinal, then every analytic set, and therefore also every coanalytic set, is determined. A natural question, therefore, is whether the existence of larger cardinals implies the determinacy of more complex sets such as the Σ_2^1 sets.

The intimate connection between large cardinals and the determinacy of simple sets of reals was first made explicit by Leo Harrington, who showed that the determinacy of all analytic sets is in fact equivalent to a large-cardinal principle slightly weaker than the existence of a measurable cardinal. As we shall shortly see, large cardinals imply the determinacy of certain simply definable sets of reals, the so-called projective sets, while the determinacy of those sets implies in turn the existence of the same kind of large cardinals in some inner models.

9   Projective Sets and Descriptive Set Theory

As we have seen, very basic questions about sets of real numbers can be extremely hard to answer. However, it often turns out to be possible to answer them for sets that occur "in nature," or that can be explicitly described. This raises the hope that one might be able to prove facts about definable sets of reals that cannot be proved for arbitrary sets.

The study of the structure of definable sets of reals is the subject of *descriptive set theory*. Examples of such sets are the Borel sets, and also the *projective* sets, which are sets that can be obtained from Borel sets by taking continuous images and complements. An equivalent definition of the projective sets is that they are subsets of $\mathbb{R}$ that can be obtained from closed subsets of $\mathbb{R}^n$ by a mixture of projecting to a lower dimension and taking complements. To see how this relates to definability, consider projecting a subset $A \subset \mathbb{R}^2$ down to the x-axis. The result will be the set of all x such that there exists y with $(x, y) \in A$. Thus, projection corresponds to existential quantification. Taking complements corresponds to negation, so one can combine the two and obtain universal quantification as well. One can therefore think of a projective set as a set that is definable from a closed set.

Since analytic sets are continuous images of Borel sets, they are projective. And so are the complements of the analytic sets, the coanalytic sets, and the continuous images of coanalytic sets, the Σ_2^1 sets. More complex projective sets are obtained by taking complements of Σ_2^1 sets, the so-called Π_2^1 sets, their continuous images, called Σ_3^1, etc. The projective sets form a hierarchy of increasing complexity, in accordance with the number of steps (always finite) that are necessary to obtain them from the Borel sets. Many sets of reals that appear naturally in usual mathematical practice are projective. Moreover, the results and techniques of descriptive set theory, although originally developed for the study of sets of reals, also apply to definable sets in any *Polish space* (a separable and complete-metrizable space). These include basic examples such as $\mathbb{R}^n$, $\mathbb{C}$, separable BANACH SPACES [III.62], etc., where projective sets arise in a very natural way. For example, in the space $C[0, 1]$ of continuous real-valued functions on $[0, 1]$ with the sup norm, the set of everywhere differentiable functions is coanalytic, and the set of functions that satisfy the mean value theorem is Π_2^1. Thus, since descriptive set theory deals with rather natural sets in Polish spaces of general mathematical interest, it is not surprising that it has found many applications in other areas of mathematics such as harmonic analysis, group actions, ergodic theory, and dynamical systems.

Classical results of descriptive set theory are that all analytic sets, and hence also all coanalytic sets, are Lebesgue measurable and have the Baire property, and that all uncountable analytic sets contain a perfect set. However, as we have already pointed out, one cannot prove in ZFC that all Σ_2^1 sets have those properties, since in L there are counterexamples. By contrast, if there exists a measurable cardinal, then they do have them. But what about more complex projective sets?

The theory of projective sets is closely tied to large cardinals. On the one hand, Solovay showed that if the existence of an inaccessible cardinal is consistent, then so is the statement that every projective set of reals is Lebesgue measurable, has the Baire property, etc. On the other hand, Shelah showed, quite unexpectedly, that the inaccessible cardinal is necessary, in the sense that if all Σ_3^1 sets are Lebesgue measurable, then $\aleph_1$ is an inaccessible cardinal in L.

Nearly all the classical properties of Borel and analytic sets are shared by the projective sets, assuming that they are determined. So since the determinacy of all projective sets cannot be proved in ZFC and since it allows for the extension of the theory of Borel and analytic sets to all projective sets in a very elegant and satisfactory way, it constitutes an excellent candidate for a new set-theoretic axiom. This axiom is known as *projective determinacy* (PD). It implies, for instance, that every projective set is Lebesgue measurable, has the Baire property, and has the perfect set property. In particular, since every uncountable perfect set has the same cardinality as $\mathbb{R}$, it implies that there is no projective counterexample to CH.

One of the most remarkable advances in set theory over the last twenty years is the proof that PD follows from the existence of large cardinals. Martin and John Steel proved in 1988 that if there exist infinitely many so-called *Woodin cardinals*, then PD holds. Woodin cardinals lie between measurable and supercompact in the hierarchy of large cardinals. Subsequently, Woodin showed that, surprisingly, the hypothesis that for each n it is consistent that there exist n Woodin cardinals is necessary in order to obtain the consistency of PD. Thus the existence of infinitely many Woodin cardinals is a sufficient, and essentially necessary, assumption for extending the classical theory of Borel and analytic

sets to all projective sets of reals, and more generally to all projective sets in Polish spaces.

In spite of the enormous success of the known large-cardinal axioms, not only in descriptive set theory but also in many other areas of mathematics, their status as true axioms of set theory is still a matter of debate. This is more so in the case of very large cardinals such as the supercompact ones, the reason being that there is as yet no inner model theory available for them, which means that there is not even strong evidence for their consistency. However, it should be noted that, as Harvey Friedman has shown, large cardinals are necessary even for proving quite simple-looking and rather natural statements about finite functions on the integers, which provides evidence for their essential role in even the most basic parts of mathematics. Another shortcoming of the known large-cardinal axioms is that they cannot decide some fundamental questions. The most conspicuous is CH, but there are others.

10 Forcing Axioms

Another old and basic question about the continuum that the known large-cardinal axioms cannot solve is *Suslin's hypothesis* (SH). Cantor had proved that every linearly ordered set that is dense (i.e., any two distinct elements have another element in between), complete (i.e., every nonempty subset with an upper bound has a supremum), separable (i.e., contains a dense countable subset), and without endpoints is order-isomorphic to the real line. In 1920 Suslin conjectured that if instead of separability one assumes the weaker *countable chain condition*, or CCC, which demands that every pairwise-disjoint collection of open intervals should be at most countable, then it must still be isomorphic to $\mathbb{R}$. The importance of SH for the development of set theory is that it led to the discovery of a new class of axioms, the so-called *forcing axioms*.

In 1967, Solovay and Tennenbaum used forcing to construct a model in which SH holds. The idea is to use the forcing to destroy any counterexamples that there might be to SH. But when one does this one may create new ones, and the result is that one needs to force again and again, transfinitely many times. The iteration of forcing is technically cumbersome and difficult to control, for many unwanted things can happen at the limit stages. For instance, ω_1 may be "collapsed," i.e., it may become countable.

Fortunately, these difficulties can be dealt with. In general, a forcing argument involves a partially ordered set. (In the case we looked at earlier, it was the set of all finite binary sequences, with $p < q$ if p was a proper initial segment of q.) If one starts with a model where GCH holds, uses only partial orderings that are CCC—that is, in which every set of incompatible elements is countable—and takes so-called *direct limits* at the limit stages, then in ω_2 steps one can destroy all counterexamples so that SH holds in the final model. On the other hand, Jensen proved in 1968 that a counterexample to SH exists in L, thereby proving the undecidability of SH in ZFC.

From the construction of Solovay and Tennenbaum, Martin isolated a new principle now known as *Martin's axiom* (MA), which generalizes the well-known *Baire category theorem*. The latter states that in every compact Hausdorff topological space, the intersection of a countable collection of dense open sets is nonempty. MA says the following:

In every compact Hausdorff CCC topological space, the intersection of $\aleph_1$ dense open sets is nonempty.

The condition that the space be CCC (i.e., every collection of pairwise-disjoint open sets is countable) is necessary, for without it the statement is false. It is easy to see that MA implies the negation of CH, for if there are only $\aleph_1$ real numbers, then the intersection of the $\aleph_1$ dense open sets $\mathbb{R} \setminus \{r\}$, as r ranges over all the real numbers, is empty. However, MA does not decide the cardinality of $\mathbb{R}$.

MA has been used with great success to solve many questions that are undecidable in ZFC. For example, it implies SH and that every Σ_2^1 set is Lebesgue measurable. But is MA really an axiom? In what sense, if any, is it a natural, or at least plausible, assumption about sets? Is the fact that it decides many ZFC undecidable questions sufficient for it to be accepted as being on a par with the ZFC axioms or the axioms of large cardinals? We shall come back to this.

MA has many different equivalent formulations. The original formulation of Martin was more closely connected with forcing—hence the term *forcing axiom*. Roughly speaking it said that if you have a CCC partial order, then you can avoid $\aleph_1$ avoidable properties, and not just countably many. This allows one to prove the existence of generic subsets of the partial order, over models M of size $\aleph_1$.

Stronger forcing axioms can be obtained by expanding the class of partial orderings to which MA applies while keeping the axiom consistent. An important such strengthening is the *proper forcing axiom* (PFA), which

is formulated for partial orderings that are *proper*. Properness is a property weaker than the CCC that was discovered by Shelah and is particularly useful when working with complicated forcing iterations. The strongest possible forcing axiom of this type was discovered by Foreman, Magidor, and Shelah in 1988. It is called *Martin's maximum* (MM) and is consistent with ZFC, assuming the consistency of a supercompact cardinal.

Both MM and PFA have striking consequences. For example, PFA, and therefore also MM, implies the axiom of projective determinacy (PD), the singular cardinal hypothesis (SCH), and that the cardinality of $\mathbb{R}$ is $\aleph_2$.

An advantage of forcing axioms is that one can apply them without having to go into the details of forcing, just as $\diamondsuit$ and $\square$ save one from having to go into the details of constructible sets. A very good example of this is PFA and some combinatorial principles derived from it, like the so-called *open coloring axiom*, which have been used with great success by Stevo Todorcevic to solve many outstanding problems in general topology and infinite combinatorics.

As we have already pointed out, forcing axioms are not as intuitively evident as the ZFC axioms, or even the axioms of large cardinals, so one can ask to what extent they should be considered as true axioms of set theory rather than just useful principles for showing that certain statements are consistent with ZFC. In the case of MA and some weaker forms of PFA and MM, some justification for their being taken as true axioms is based on the fact that they are equivalent to principles of *generic absoluteness*. That is, they assert, under certain restrictions that are necessary to avoid inconsistency, that *everything that might exist, does exist*. More precisely, if some set having certain properties could be forced to exist over V, then a set having the same properties already exists (in V). So, like the axioms of large cardinals, they are maximality principles, i.e., they attempt to make V as large as possible.

For example, MA is equivalent to the assertion that if a set X having some properties that depend exclusively on subsets of ω_1 could be forced to exist over V using a CCC partial ordering $\mathbb{P}$, then such an X already exists in V. This characterization of MA in terms of generic absoluteness provides some justification for regarding MA as a true axiom of set theory. The analogous principle of generic absoluteness, but for proper partial orderings instead of CCC, is known as the *bounded proper forcing axiom* (BPFA). Although weaker than PFA, BPFA is strong enough to decide many questions that the large-cardinal axioms are unable to settle. Most notably, Justin Moore has recently proved, following a series of results by Woodin, David Asperó, and Todorcevic, that BPFA implies that the cardinality of $\mathbb{R}$ is $\aleph_2$.

To finish, we briefly mention some deep results that establish strong underlying connections between large cardinals, inner models, determinacy, forcing axioms, generic absoluteness, and the continuum. These results hold under the assumption that for every ordinal α there exists a Woodin cardinal greater than α.

The first one, due to Shelah and Woodin, is that the theory of $L(\mathbb{R})$ is generically absolute. That is, all sentences with real numbers as parameters that would hold in the $L(\mathbb{R})$ of *any* generic extension of V are already true in the real $L(\mathbb{R})$. This kind of generic absoluteness implies that all sets of reals in $L(\mathbb{R})$, and in particular the projective sets, are Lebesgue measurable, have the Baire property, etc. Furthermore, by refining the Martin–Steel result that large cardinals imply PD, Woodin showed that in $L(\mathbb{R})$ every set of reals is determined.

Another result of Woodin is that there is an axiom, which he calls $(*)$, that is intended to play the role for subsets of ω_1 that PD plays for sets of natural numbers, in the sense that it decides "practically all" questions about those sets. Of course, no consistent axiom can really decide *all* questions that refer only to subsets of ω_1, since by Gödel's incompleteness theorem there will always be undecidable arithmetical statements. So, to formulate precisely the notion of *deciding practically all questions*, Woodin introduces a new logic, called Ω-*logic*, that strengthens ordinary first-order logic. One of the main features of Ω-logic is that the valid statements in Ω-logic are generically absolute. Under suitable large-cardinal hypotheses, $(*)$ is consistent in Ω-logic and decides in Ω-logic all questions that refer only to subsets of ω_1. The main open problem is the Ω-*conjecture*, whose formulation is quite technical and beyond the scope of this article. If the Ω-conjecture is true, then *any* axiom compatible with the existence of large cardinals that decides all questions that depend exclusively on subsets of ω_1 in Ω-logic must imply the negation of CH. Thus, the theories ZFC plus CH and ZFC plus not-CH are not equally reasonable from the point of view of Ω-logic, since in the presence of large cardinals CH puts an unnecessary limitation on the possibility of settling all natural questions about subsets of ω_1.

11   Final Remarks

In this short account of set theory, we have reviewed some of the key developments since its beginnings in the late nineteenth century. What started in the hands of Cantor as a mathematical theory of transfinite numbers has developed to become a general theory of infinite sets and a foundation for mathematics. The fact that it has been possible to unify all of classical mathematics into one single theoretical framework, the ZFC axiom system, is certainly remarkable. But beyond this, and most importantly, the techniques developed by set theory, such as constructibility, forcing, infinite combinatorics, the theory of large cardinals, determinacy, the descriptive theory of definable sets in Polish spaces, etc., have turned it into a discipline of great depth and beauty, with fascinating results that stimulate and challenge our imagination, and with numerous applications in areas such as algebra, topology, real and complex analysis, functional analysis, and measure theory. In the twenty-first century, the ideas and techniques generated within set theory will surely continue to contribute to the solution of outstanding mathematical problems, old as well as new, and will help mathematicians gain an ever deeper insight into the complexities and vastness of the mathematical universe.

Further Reading

Foreman, M., and A. Kanamori, eds. 2008. *Handbook of Set Theory*. New York: Springer.

Friedman, S. D. 2000. *Fine Structure and Class Forcing*. De Gruyter Series in Logic and Its Applications, volume 3. Berlin: Walter de Gruyter.

Hrbacek, K., and T. Jech. 1999. *Introduction to Set Theory*, 3rd edn., revised and expanded. New York: Marcel Dekker.

Jech, T. 2003. *Set Theory*, 3rd edn. New York: Springer.

Kanamori, A. 2003. *The Higher Infinite*, 2nd edn. Springer Monographs in Mathematics. New York: Springer.

Kechris, A. S. 1995. *Classical Descriptive Set Theory*. Graduate Texts in Mathematics. New York: Springer.

Kunen, K. 1980. *Set Theory: An Introduction to Independence Proofs*. Amsterdam: North-Holland.

Shelah, S. 1998. *Proper and Improper Forcing*, 2nd edn. New York: Springer.

Woodin, W. H. 1999. *The Axiom of Determinacy, Forcing Axioms, and the Nonstationary Ideal*. De Gruyter Series in Logic and Its Applications, volume 1. Berlin: Walter de Gruyter.

Zeman, M. 2001. *Inner Models and Large Cardinals*. De Gruyter Series in Logic and Its Applications, volume 5. Berlin: Walter de Gruyter.

IV.23 Logic and Model Theory
David Marker

1 Languages and Theories

Mathematical logic is the study of formal languages that are used to describe mathematical structures and what these can tell us about the structures themselves. We can learn a lot about a formal language by investigating which of its sentences are true for the structure it describes, and we can learn a lot about the structure by investigating the subsets of it that can be defined using the language. In this article, we shall see several examples of languages and the structures that they are used to describe. We shall also see instances of the remarkable phenomenon that theorems in logic can sometimes be used to prove "purely mathematical" results that seem to have nothing to do with logic. This introductory section briefly introduces some of the basic ideas that will be needed to understand the later sections.

All the formal languages that we consider will be extensions of a basic logical language that we shall denote by $\mathcal{L}_0$. The statements, or *formulas*, of this language are made up of the following components: *variables*, which are denoted by letters of the alphabet such as x or y, or letters with subscripts such as $v_1, v_2, \dots$; the *parentheses* "(" and ")"; the *equality symbol* "="; the *logical connectives* $\wedge$, $\vee$, $\neg$, $\rightarrow$, $\leftrightarrow$, which we read as "and," "or," "not," "implies," and "if and only if"; and the *quantifiers* $\exists$ and $\forall$, which we read as "there exists" and "for all." (If these symbols are unfamiliar to you, then you should read THE LANGUAGE AND GRAMMAR OF MATHEMATICS [I.2] before attempting to read this article.) Here are a couple of formulas of $\mathcal{L}_0$:

(i) $\forall x\ \forall y\ \exists z\ (z \neq x \wedge z \neq y)$;
(ii) $\forall x\ (x = y \vee x = z)$.

The first of these says that if any object exists at all then there are at least three objects, and the second says that y and z are the only objects. There is an important difference between the two formulas: the variables x, y, and z that occur in the first formula are all *bound* variables, which means that they are all attached to quantifiers, whereas in the second formula, only the variable x is bound, while the variables y and z are *free*. This means that the first formula expresses a statement about some mathematical structure, while the second

is a statement about not just a structure but also the particular elements y and z.

There are various rules that allow one to build larger formulas out of smaller ones. We will not give them all, but for example if ϕ and ψ are formulas, then $\neg\phi$, $\phi \vee \psi$, $\phi \wedge \psi$, $\phi \rightarrow \psi$, and $\phi \leftrightarrow \psi$ are all formulas. In general, if ϕ is built out of smaller formulas $\phi_1, \dots, \phi_n$ using logical connectives (and parentheses), then we call ϕ a *Boolean combination* of $\phi_1, \dots, \phi_n$. Another important way to modify a formula is quantification: if $\phi(x)$ is a formula involving a free variable x, then $\forall x \phi(x)$ and $\exists x \phi(x)$ are both formulas.

The formulas just discussed are "purely logical," which makes them not very useful for describing interesting mathematical structures. Suppose, for example, that we wanted to study real solutions to algebraic and exponential equations over the FIELD [I.3 §2.2] of real numbers. We can think of this as studying the "mathematical structure"

$$\mathbb{R}_{\exp} = (\mathbb{R}, +, \cdot, \exp, <, 0, 1),$$

where the right-hand side is a septuple that consists of the set $\mathbb{R}$ of real numbers, the binary operations of addition and multiplication, the EXPONENTIAL FUNCTION [III.25], the "less than" relation, and the real numbers 0 and 1.

The various components of this structure are of course related to each other in many ways, but we cannot express these relationships unless we are prepared to extend the basic language $\mathcal{L}_0$. For example, if we wanted to write, in a formal way, the statement that the exponential function turns addition into multiplication, then the obvious thing to write down would be

(i) $\forall x \forall y\ \exp(x) \cdot \exp(y) = \exp(x + y)$.

Here we have two quantifiers, two bound variables x and y, and the equals sign, but the rest of the formula involves extraneous elements such as "+", "$\cdot$", and "exp". Thus, to discuss the structure $\mathbb{R}_{\exp}$, we extend the language $\mathcal{L}_0$ to a language $\mathcal{L}_{\exp}$, by adding in the symbols "+", "$\cdot$", "exp", "<", "0", and "1". Of course, these come with various syntactic rules that reflect the fact that "+" is a binary operation, "exp" is a function, and so on. For instance, these rules would allow us to write $\exp(x + y) = z$ but would forbid us to write $\exp(x = y) + z$.

Here are three more $\mathcal{L}_{\exp}$-formulas:

(ii) $\forall x\ (x > 0 \rightarrow \exists y\ \exp(y) = x)$;
(iii) $\exists x\ x^2 = -1$;
(iv) $\exists y\ y^2 = x$.

We interpret these formulas as the assertions "for all positive x, there is a y such that $e^y = x$," "-1 is a square," and "x is a square." The first three formulas above are declarative statements about the structure $\mathbb{R}_{\exp}$. Formulas (i) and (ii) are true in $\mathbb{R}_{\exp}$, while (iii) is false. Formula (iv) is different because x is a free variable: thus, it expresses a property of x. (For instance, it is true if $x = 8$, but false if $x = -7$.) A *sentence* is defined to be a formula with no free variables. If ϕ is an $\mathcal{L}_{\exp}$-sentence, then ϕ is either true or false in $\mathbb{R}_{\exp}$.

If ϕ is a formula with free variables $x_1, \ldots, x_n$, and $a_1, \ldots, a_n$ are real numbers, then we write $\mathbb{R}_{\exp} \vDash \phi(a_1, \ldots, a_n)$ if the formula ϕ is true for the particular sequence $(a_1, \ldots, a_n)$. We think of the formula as defining the set

$$\{(a_1, \ldots, a_n) \in \mathbb{R}^n : \mathbb{R}_{\exp} \vDash \phi(a_1, \ldots, a_n)\},$$

that is, the set of all sequences $(a_1, \ldots, a_n)$ for which the formula is true when you set x_i to equal a_i for every i. For example, the formula

$$\exists z\, (x = z^2 + 1 \ \wedge \ y = z \cdot \exp(\exp(z)))$$

defines the parametrized curve

$$\{(t^2 + 1, te^{e^t}) : t \in \mathbb{R}\}.$$

For another example, one that illustrates an important point, let us consider the structure $(\mathbb{Z}, +, \cdot, 0, 1)$: that is, the integers, with addition, multiplication, 0, and 1. The language used to describe this structure is the *language of rings*, $\mathcal{L}_{\mathrm{rng}} = \mathcal{L}(+, \cdot, 0, 1)$. (The notation here lists the symbols that we add to the basic language $\mathcal{L}_0$.) The language $\mathcal{L}_{\mathrm{rng}}$ has no symbol for the usual ordering on $\mathbb{Z}$, but, surprisingly, this ordering can nevertheless be defined in terms of $\mathcal{L}_{\mathrm{rng}}$. (To appreciate the nonobviousness of this fact, the reader is encouraged to try to work out why it is true before reading on.)

The trick is to use a well-known theorem due to LAGRANGE [VI.22], which asserts that every nonnegative integer is a sum of four squares. It follows that the statement $x \geqslant 0$ can be defined by the formula

$$\exists y_1 \exists y_2 \exists y_3 \exists y_4 \quad x = y_1^2 + y_2^2 + y_3^2 + y_4^2.$$

(Of course, we are also using the fact that a negative integer cannot be written as a sum of four squares. Note too that a similar trick would work even if all one knew was that every nonnegative integer was a sum of a hundred squares.) Once one has a way of expressing the statement that x is nonnegative, it is easy to define the symbol "$<$". The interesting aspect of this is that

the reformulation was not obvious—it depended on a genuine mathematical theorem.

It is important to understand that formulas are restricted in several ways, of which two stand out in particular.

- Formulas are finite. We do not allow formulas like

$$\forall x > 0\ (x < 1 \vee x < 1 + 1 \vee x < 1 + 1 + 1 \vee \cdots),$$

  which would express the fact that $\mathbb{R}$ has the so-called Archimedean property. (If we did, then it would be much easier to define "$<$" above.)

- Quantifiers range over *elements* of the structure, and not subsets. This rules out a "second-order" formula such as

$$\forall S \subseteq \mathbb{R} \quad \text{(if S is bounded above,}$$
$$\text{then S has a least upper bound),}$$

  which would express the completeness of $\mathbb{R}$ by quantifying over all subsets S of $\mathbb{R}$. Since we look just at "first-order" formulas, what we are studying is often called *first-order logic*.

Now that we have seen some examples of languages, let us discuss them more generally. A *language* is basically something like $\mathcal{L}_{\exp}$ or $\mathcal{L}_{\mathrm{rng}}$ above: that is, a set of symbols (combined with the basic logical symbols) together with some rules concerning their use. If $\mathcal{L}$ is a language, then an $\mathcal{L}$-*structure* is a mathematical structure in which all the sentences of $\mathcal{L}$ can be interpreted. (This concept will become clearer in a moment, when we give a couple of examples.) An $\mathcal{L}$-*theory* T is just a set of $\mathcal{L}$-sentences, which one can think of as axioms that an $\mathcal{L}$-structure might or might not satisfy. A *model* of T is then an $\mathcal{L}$-structure $\mathcal{M}$ in which all the sentences of T, suitably interpreted, are true. For instance, the structure was a model for the formulas (i) and (ii) of the language $\mathcal{L}_{\exp}$ that we discussed earlier. (Another model for the same two formulas would be one in which we replaced the exponential function by the function 2^x and interpreted "exp" as referring to that function instead.)

The justification for the word "theory" is clearer in another example, the language of GROUPS [I.3 §2.1], $\mathcal{L}_{\mathrm{grp}} = \mathcal{L}(\circ, e)$. Here, $\circ$ is a binary operation symbol and e is a constant. We might look at the theory T_{grp} consisting of the sentences

(i) $\forall x \forall y \forall z\ x \circ (y \circ z) = (x \circ y) \circ z$;
(ii) $\forall x\ x \circ e = e \circ x = x$;
(iii) $\forall x \exists y\ x \circ y = y \circ x = e$;

which are the usual axioms for groups.

In order to interpret this language in some mathematical structure $\mathcal{M}$ we need $\mathcal{M}$ to consist of a set M, a binary operation $f : M^2 \to M$, and an element $a \in M$. We then interpret "$\circ$" as referring to f, "e" as referring to the element a, and quantification as being over the set M. Thus, for example, the interpretation of (iii) is that for every x in M there exists a y in M such that $f(x, y) = a$. Under this interpretation of the symbols of $\mathcal{L}_{\text{grp}}$, the structure $\mathcal{M}$ becomes an $\mathcal{L}_{\text{grp}}$-structure. This $\mathcal{L}_{\text{grp}}$-structure is a model of T_{grp} if in addition the sentences (i), (ii), and (iii) are all true. Since sentences (i)–(iii) are the axioms for groups, a model of T_{grp} is nothing other than a group.

We say that an $\mathcal{L}$-sentence ϕ is a *logical consequence* of a theory T, and write $T \models \phi$, if ϕ is true in every model of T. That is, $T \models \phi$ if ϕ is true in every structure in which all the sentences of T are true. Thus, the symbol "$\models$" has two different meanings, according to whether there is a structure or a theory on the left-hand side. However, these two meanings are closely related in that they are both concerned with truth in models: $\mathcal{M} \models \phi$ means that ϕ is true in the model $\mathcal{M}$, and $T \models \phi$, as we have just said, means that ϕ is true in every possible model of T. Either way, the symbol "$\models$" stands for a "semantic" notion of entailment.

Returning to the example of groups, if ϕ is a sentence in $\mathcal{L}_{\text{grp}}$, then $T_{\text{grp}} \models \phi$ if and only if ϕ is true for every group. So, for instance,

$$T_{\text{grp}} \models \forall x \forall y \forall z \ (xy \neq xz \lor y = z),$$

because if x, y, and z are elements of any group and $xy = xz$, then we can multiply both sides on the left by the inverse of x to deduce that $y = z$.

We can now describe some of the basic problems in logic.

(i) Given an $\mathcal{L}$-theory T, can we decide if a sentence ϕ is a logical consequence of T, and if so how?

(ii) Given an interesting mathematical structure, like $\mathbb{R}_{\text{exp}}$, or $(\mathbb{N}, +, \cdot, 0, 1)$, or the complex field, and a language $\mathcal{L}$ that describes the structure, can we determine which $\mathcal{L}$-sentences are true of the structure?

(iii) Given a structure described by a language, do the subsets of the structure that can be defined in the language have special properties? Are they in some sense "simple"? For example, earlier we saw how to use $\mathcal{L}_{\text{exp}}$ to define a certain curve in the plane. Now consider a very complicated set such

as a CANTOR SET [III.17] or the MANDELBROT SET [IV.14 §2.8]. Is it possible to prove that these sets *cannot* be defined in $\mathcal{L}_{\text{exp}}$ because they are "too complex" in some sense?

2   Completeness and Incompleteness

Let T be an $\mathcal{L}$-theory and let ϕ be an $\mathcal{L}$-sentence. To show that $T \models \phi$, we must show that ϕ holds in every model of T. Checking all models of T sounds like a daunting task, but fortunately it is not necessary, since instead we can use a *proof*. One of the first tasks in mathematical logic is to say precisely what this means.

Suppose, then, that $\mathcal{L}$ is some language and that T is a set of sentences in $\mathcal{L}$, i.e., an $\mathcal{L}$-theory. Suppose also that ϕ is a formula of $\mathcal{L}$. Informally speaking, a proof of ϕ assumes the statements of T and ends up establishing ϕ. We express this idea formally as follows. A *proof of ϕ from T* is a finite sequence of $\mathcal{L}$-formulas $\psi_1, \ldots, \psi_m$ (which one can think of as the lines of the proof) with the following properties:

(i) each ψ_i is either a logical axiom, or a sentence of T, or a formula that follows from the previous formulas $\psi_1, \ldots, \psi_{i-1}$ by means of simple logical rules;

(ii) $\psi_m = \phi$.

We shall not say precisely what a "simple logical rule" is, but three examples are

- from ϕ and ψ it follows that $\phi \land \psi$;
- from $\phi \land \psi$ it follows that ϕ;
- from $\phi(x)$ it follows that $\exists v \ \phi(v)$.

The other possible rules are similarly elementary.

There are three points about proofs that need to be stressed. The first is that they are finite, which may seem too obvious to mention but is important because it has a number of consequences that are not obvious. The second is that proof systems have to be *sound*: if there is a proof of ϕ from T, then ϕ is true in every model of T. To put this more succinctly, let us introduce the notation $T \vdash \phi$ for the statement that there is a proof of ϕ from T. Then soundness is the assertion that if $T \vdash \phi$ then $T \models \phi$. This is why we can prove that ϕ is true in every model of T by finding a proof rather than by looking at all the models. The third point is that it is easy to check whether a sequence of sentences is a proof. More precisely, there is an algorithm that can look at a sequence $\psi_1, \ldots, \psi_m$ and decide whether it really is a proof of ϕ from T.

It is not too surprising that if ϕ can be proved from T, then ϕ is true in all models of T. Much more remarkable is that the converse is also true: if ϕ cannot be proved from T, then there must be a model of T in which ϕ is false. This tells us that two very different notions—the finitistic, syntactic notion of "proof" and the semantic notion of "logical consequence," which concerns truth in models—always agree. This result is known as Gödel's completeness theorem. Here is its formal statement.

Theorem. *Let T be an $\mathcal{L}$-theory and let ϕ be an $\mathcal{L}$-sentence. Then $T \vDash \phi$ if and only if $T \vdash \phi$.*

Suppose that T is a simple theory like T_{grp}, where there is an algorithm to decide whether a sentence is in T. (In the case of T_{grp} this algorithm is particularly simple, but some theories might have infinitely many sentences.) We could write a computer program which, given a formula ϕ as its input, would systematically generate all possible proofs σ from T and check to see whether σ was a proof of ϕ. If such a program finds a proof of ϕ, then it halts and tells us that $T \vDash \phi$. We say that $\{\phi : T \vDash \phi\}$ is *recursively enumerable*.

However, one might hope for more. If $T \nvDash \phi$, our program above will go on searching forever, so it will never tell us that there is no proof of ϕ. We say that an $\mathcal{L}$-theory T is *decidable* if there is a computer program which, when given an $\mathcal{L}$-sentence ϕ as input, will always halt and tell us, one way or another, whether $T \vDash \phi$. Such a program would have to be cleverer than the one that just checks all possible proofs σ, and unfortunately such a program does not have to exist: as GÖDEL [VI.92] proved in his famous INCOMPLETENESS THEOREM [V.15], many important theories are undecidable. Here is a first version of his theorem, concerning the *theory of the natural numbers* (or theory of $\mathbb{N}$ for short), which means the set of all sentences in the language $\mathcal{L}_{\mathrm{rng}}$ that are true of the structure $(\mathbb{N}, +, \cdot, 0, 1)$.

Theorem. *The theory of the natural numbers is undecidable.*

At first, this might seem rather strange: after all, if T is the theory of $\mathbb{N}$, then T contains all true sentences about $\mathbb{N}$. So a sentence ϕ is provable from T if and only if it has a one-line proof (the line being ϕ itself). However, this does not make ϕ decidable, because the theory T is very complicated and there is no algorithm for deciding whether ϕ belongs to T.

One approach to proving the incompleteness theorem is to associate a natural number with each computer program in such a way that statements about programs can be recast as statements about natural numbers. The theory of $\mathbb{N}$ then determines whether a program P halts on input x, thus solving what is known as the *halting problem*. Since the halting problem was shown by TURING [VI.94] to be undecidable (a sketch of the proof can be found in THE INSOLUBILITY OF THE HALTING PROBLEM [V.20]), it follows that the theory of $\mathbb{N}$ is undecidable.

How can we understand the theory of $\mathbb{N}$? One might hope to find a much smaller theory that yielded the same true sentences. That is, we could try to find a simple set of axioms about $\mathbb{N}$ that we know are true and hope that every true sentence follows from these axioms. A good candidate is *first-order Peano arithmetic*, or PA. This is a theory in the language $\mathcal{L}(+, \cdot, 0, 1)$ that involves a few simple axioms about addition and multiplication, such as

$$\forall x \forall y \; x \cdot (y + 1) = x \cdot y + x,$$

together with axioms for induction.

Why do we need more than one axiom of induction? The reason is that the obvious statement that expresses the principle of mathematical induction, namely

$$\forall A \; (0 \in A \wedge \forall x \; x \in A \to x + 1 \in A) \to \forall x \; x \in A,$$

is not a first-order sentence, because the quantifier is applied to all subsets A of $\mathbb{N}$. (It is also not a sentence in $\mathcal{L}_{\mathrm{rng}}$ since it uses the symbol "$\in$", but this is a less fundamental problem.) To get around this difficulty, one has a separate axiom of induction for each formula ϕ. It is the assertion that

$$[\phi(0) \wedge \forall x \; (\phi(x) \to \phi(x + 1))] \to \forall x \; \phi(x).$$

In words, this says that if $\phi(0)$ is true and $\phi(x + 1)$ is true whenever $\phi(x)$ is true, then $\phi(x)$ is true for every x in $\mathbb{N}$.

Most of number theory can be formalized in PA and one might hope that PA $\vdash \phi$ for every ϕ that is true in $\mathbb{N}$. Sadly, this is not true. Here is a second version of Gödel's incompleteness theorem. Recall that the notation $\mathbb{N} \vDash \psi$ means simply that ψ is true in $\mathbb{N}$.

Theorem. *There is a sentence ψ such that $\mathbb{N} \vDash \psi$ but PA $\nvdash \psi$.*

Another way to state this result is to say that there is a sentence ψ such that PA $\nvdash \psi$ and PA $\nvdash \neg\psi$. To see that this is an equivalent statement, let ψ be any sentence. Then precisely one of ψ and $\neg\psi$ is true. Therefore, if the theorem is false, then PA must prove either ψ or $\neg\psi$. But this means that we can decide which by simply going through all possible proofs in PA until we find a proof of ψ or a proof of $\neg\psi$.

Gödel's original example of a true but unprovable sentence was a self-referential sentence that effectively asserted

"I am not provable from PA."

More precisely, he found a sentence ψ for which he was able to show that ψ is true in $\mathbb{N}$ if and only if ψ is not provable from PA. With more work he showed that there is a sentence that asserts

"PA is consistent"

that is unprovable from PA. The somewhat artificial and metamathematical nature of these sentences might lead one to hope that all "mathematically interesting" sentences about $\mathbb{N}$ are settled by PA. However, more recent work has shown that even this is a forlorn hope, since there are undecidable statements related to RAMSEY'S THEOREM [IV.19 §2.2] in finite combinatorics.

Undecidability also appears in number theory in a very basic way. *Hilbert's tenth problem* asked if there is an algorithm to decide whether a polynomial $p(X_1, \ldots, X_n)$ with integer coefficients has an integer zero. Davis, Matijasevic, Putnam, and Robinson showed that the answer is no.

Theorem. *For any recursively enumerable $S \subseteq \mathbb{N}$ there is $n > 0$ and $p(X, Y_1, \ldots, Y_n) \in \mathbb{Z}[X, Y_1, \ldots, Y_n]$ such that $m \in S$ if and only if $p(m, Y_1, \ldots, Y_n)$ has an integer zero.*

Since the halting problem provides an undecidable recursively enumerable set, the answer to Hilbert's tenth problem is no. An important open question is whether there is an algorithm to decide if a polynomial with *rational* coefficients has a *rational* zero. Hilbert's tenth problem is also discussed in THE INSOLUBILITY OF THE HALTING PROBLEM [V.20], and other interesting examples of undecidability can be found in GEOMETRIC AND COMBINATORIAL GROUP THEORY [IV.10].

3   Compactness

A theory T is called *satisfiable* if there are structures that satisfy all of the sentences in T (that is, if T has a model), and we call T *consistent* if we cannot derive a contradiction from T. Since our proof system is sound,

any satisfiable theory is consistent. On the other hand if T is not satisfiable, then every sentence ϕ is a logical consequence of T, for the trivial reason that there are no models of T in which ϕ is required to be true. But the completeness theorem then tells us that $T \vdash \phi$ for every ϕ. Choosing ϕ to be some contradictory statement, of the form $\psi \wedge \neg\psi$, for instance, we see that T is inconsistent. This way of reformulating the completeness theorem has the following simple consequence, called the *compactness theorem*, which turns out to be surprisingly important, as we shall see.

Theorem. *If every finite subset of T is satisfiable, then T is satisfiable.*

The reason this is true is that if T is not satisfiable then it is inconsistent (as we have just seen), which means that a contradiction can be proved from T. Since this proof, like all proofs, must be finite, it involves only finitely many sentences from T. Therefore, T has a finite subset that implies a contradiction, which contradicts our assumption that all finite subsets of T are satisfiable.

Although the compactness theorem is an easy consequence of the completeness theorem, it has many immediate intriguing consequences and lies at the heart of many constructions in model theory. Here are two simple applications that show that theories have many models that you might not expect. If $\mathcal{M}$ is some $\mathcal{L}$-structure, let us write $\text{Th}(\mathcal{M})$ for *the theory of* $\mathcal{M}$: that is, for the set of all $\mathcal{L}$-sentences that are true in $\mathcal{M}$. We also extend our earlier notation $\mathcal{M} \vDash \phi$ from single formulas to collections of formulas, so if $\mathcal{M}$ is an $\mathcal{L}$-structure and T is an $\mathcal{L}$-theory, then $\mathcal{M} \vDash T$ means that every sentence of T is true in M, or in other words that $\mathcal{M}$ is a model of T.

Corollary. *There exists an $\mathcal{L}_{\exp}$-structure $\mathcal{M}$ containing an infinite element a (which means that $a > 1$, $a > 1 + 1$, $a > 1 + 1 + 1$, etc.), such that $\mathcal{M} \vDash \text{Th}(\mathbb{R}_{\exp})$.*

That is, there is a structure $\mathcal{M}$ in which all the true first-order statements about the structure $\mathbb{R}_{\exp}$ are still true, but $\mathcal{M}$ is different from $\mathbb{R}_{\exp}$ because it contains an infinite element. To prove this, we add one more constant symbol c to our language and consider the theory T that consists of all the statements of $\text{Th}(\mathbb{R}_{\exp})$ (that is, all true statements about $\mathbb{R}_{\exp}$), together with the infinite sequence of statements $c > 1$, $c > 1 + 1$, $c > 1 + 1 + 1$, and so on. If Δ is any finite subset of T, then we can make $\mathbb{R}$ a model of Δ simply by interpreting

c as a sufficiently large real number—large enough to satisfy all the statements of the form $c > 1 + 1 + \cdots + 1$ that belong to Δ. Since we can model every finite subset Δ of T, the compactness theorem tells us that we can model T itself. If $\mathcal{M} \vDash T$, then the element named by c must be infinite.

The element $1/a$ will be an *infinitesimal* element of $\mathcal{M}$ (which means that it satisfies statements that effectively say that it is smaller than $1/n$ for every positive integer n). This observation is the first step toward a rigorous development of calculus with infinitesimals.

For another example, let $\mathcal{L}_{\text{rng}} = \mathcal{L}(+, \cdot, 0, 1)$ be the language of rings. Let T be the set of $\mathcal{L}$-sentences that are true in every finite field. We call T the *theory of finite fields*. Recall that a field is said to have *characteristic p* if p is the smallest positive integer (which has to be prime) such that $1 + 1 + \cdots + 1 = 0$ in the field, where the number of 1s in the sum is p. If there is no such p, then the field is said to have *characteristic zero*. Thus, the fields $\mathbb{Q}$, $\mathbb{R}$, and $\mathbb{C}$ all have characteristic zero.

Corollary. *There is a field F with characteristic zero such that $F \vDash T$.*

This result tells us that there is no possible set of axioms that characterizes the finite fields: given any set of statements that are true in all finite fields, there is an infinite field in which they are also all true. To prove it, we look at the theory T' that consists of T together with the statements $1 + 1 \neq 0$, $1 + 1 + 1 \neq 0$, and so on. Any finite set of statements in T' will be true of a finite field of sufficiently large characteristic, and thus satisfiable. By the compactness theorem T' is satisfiable, but a model of T clearly has to have characteristic zero.

The compactness theorem can sometimes be used to show the existence of interesting algebraic bounds. The next result allows us to deduce from HILBERT'S NULL-STELLENSATZ [V.17] a stronger "quantitative version." It is our first example of a statement that does not appear to be logical in nature but which can be proved using logic. Recall that a field is *algebraically closed* if every polynomial with coefficients in the field has a root in the field. (THE FUNDAMENTAL THEOREM OF ALGEBRA [V.13] is the assertion that $\mathbb{C}$ is an algebraically closed field.)

Proposition. *For any three positive integers n, m, d there is a positive integer l such that if K is an algebraically closed field and $f_1, \ldots, f_m$ are polynomials in n variables with coefficients in K, degree at most d and no common zero, then there are polynomials $g_1, \ldots, g_m$ of degree at most l such that $\sum g_i f_i = 1$.*

Hilbert's Nullstellensatz itself is the same statement but without the extra information about the degrees of the polynomials g_i.

To see how the proposition is proved, we will restrict our attention to the case $n = d = 2$. This is just for notational simplicity: the proof is almost identical in larger cases. For each i between 1 and m let

$$F_i = a_i X^2 + b_i Y^2 + c_i XY + d_i X + e_i Y + f_i.$$

For each k write down a formula ϕ_k that asserts that there are no polynomials $G_1, \ldots, G_m$ with degree at most k such that $1 = \sum F_i G_i$. Let T be the theory of algebraically closed fields with the formulas $\phi_1, \phi_2, \ldots$ and the assertion that the polynomials $F_1, \ldots, F_m$ have no common zero. If there is no positive integer l satisfying the conclusion of the proposition, then every finite subset of T is satisfiable. Hence, by the compactness theorem, T is satisfiable. If $K \vDash T$, then $F_1, \ldots, F_m$ are polynomials over an algebraically closed field with no common zero, but it is impossible to find polynomials $G_1, \ldots, G_m$ such that $\sum G_i F_i = 1$. This contradicts Hilbert's Nullstellensatz.

Notice that in the above argument we did not say anything about the dependence of l on n, m, and d. This is because the proof does not actually find a bound: it merely shows that some sort of bound must exist. However, good explicit bounds were recently discovered—see ALGEBRAIC GEOMETRY [IV.4] for more details.

4 The Complex Field

A surprising counterpoint to Gödel's incompleteness theorem is a result of TARSKI [VI.87], which states that the theories of the fields of real and complex numbers *are* decidable. The key to these results is a method known as *quantifier elimination*. If we have a formula without quantifiers that concerns the natural numbers, then it is easy to decide whether it is true or false. The negative solution to Hilbert's tenth problem shows that as soon as we start adding existential quantifiers (as we do if, for example, we assert that a polynomial has a zero), then we leave the realm of decidability.

Thus, if we want to show that a formula is decidable, it will be very useful if we can find an equivalent formula that does not have quantifiers. And in some settings, this turns out to be possible. For example, let $\phi(a, b, c)$ be the formula

$$\exists x \ ax^2 + bx + c = 0.$$

The usual rule for solving quadratics tells us that, as long as $a \neq 0$, this is true in $\mathbb{R}$ if and only if $b^2 \geqslant 4ac$. Therefore, $\mathbb{R} \vDash \phi(a, b, c)$ if and only if

$$[(a \neq 0 \wedge b^2 - 4ac \geqslant 0) \vee (a = 0 \wedge (b \neq 0 \vee c = 0))].$$

As for the complex numbers, it is easy to see that $\mathbb{C} \vDash \phi(a, b, c)$ if and only if

$$a \neq 0 \vee b \neq 0 \vee c = 0.$$

In either case, ϕ is equivalent to a formula with no quantifiers.

For a second example, let $\phi(a, b, c, d)$ be the formula

$$\exists x \exists y \exists u \exists v \, (xa + yc = 1 \; \wedge \; xb + yd = 0$$
$$\wedge \; ua + vc = 0 \; \wedge \; ub + vd = 1).$$

The formula $\phi(a, b, c, d)$ is the obvious way of asserting that the matrix $\left(\begin{smallmatrix} a & b \\ c & d \end{smallmatrix}\right)$ is invertible. However, by the DETERMINANT [III.15] test, we know that, for any field F, $F \vDash \phi(a, b, c, d)$ if and only if $ad - bc \neq 0$. Thus the existence of an inverse can be expressed by the quantifier-free formula $ad - bc \neq 0$.

Tarski proved that we can *always* eliminate quantifiers in algebraically closed fields.

Theorem. *For any $\mathcal{L}_{\mathrm{rng}}$-formula ϕ there is a quantifier-free formula ψ such that ϕ is equivalent to ψ in every algebraically closed field.*

Furthermore, Tarski gave an explicit algorithm for eliminating the quantifiers.

The equivalent quantifier-free formulas above were both finite Boolean combinations of formulas of the form $p(v_1, \ldots, v_n) = q(v_1, \ldots, v_n)$, where p and q are polynomials in n variables with integer coefficients. It is not hard to see that this is true of any quantifier-free $\mathcal{L}_{\mathrm{rng}}$-formula. It follows that a quantifier-free $\mathcal{L}_{\mathrm{rng}}$-*sentence* is particularly simple: if no free variables are allowed and no quantifiers are allowed, then there cannot be any variables! Therefore, the polynomials p and q have to be constant, which means that a quantifier-free $\mathcal{L}_{\mathrm{rng}}$-*sentence* is a finite Boolean combination of formulas of the form $k = l$ (where this should be regarded as an abbreviation for $1 + 1 + \cdots + 1 = 1 + 1 + \cdots + 1$, with k 1s on the left-hand side and l 1s on the right-hand side).

This leads to the decidability result. If we want to know whether $\mathbb{C} \vDash \phi$, then we use Tarski's algorithm to convert ϕ into an equivalent quantifier-free sentence. But the very simple form of such sentences makes their truth or falsity easy to decide.

In the remainder of this section, we shall discuss a number of other consequences of Tarski's theorem.

The first is that sentences in the language $\mathcal{L}_{\mathrm{rng}}$ cannot distinguish between different algebraically closed fields of the same characteristic. That is, if ϕ is any $\mathcal{L}_{\mathrm{rng}}$-sentence that is true for some algebraically closed field of characteristic p (where p is allowed to be zero), then it is true in every algebraically closed field of characteristic p.

To see why this is true, let K and F be two algebraically closed fields of characteristic p, and suppose that $K \vDash \phi$ (or in other words that ϕ is true of K). Let k be the field $\mathbb{Q}$ if the characteristic is zero and the field with p elements otherwise. Tarski's theorem tells us that there is a quantifier-free sentence ψ that is equivalent to ϕ in all algebraically closed fields of characteristic p. However, the extremely simple nature of the quantifier-free sentences of $\mathcal{L}_{\mathrm{rng}}$ means that their truth or falsity in any given field depends only on the elements 0, 1, $1 + 1$, and so on. Therefore,

$$K \vDash \psi \; \Leftrightarrow \; k \vDash \psi \; \Leftrightarrow \; F \vDash \psi.$$

Since $K \vDash \phi$ and ϕ and ψ are equivalent in all algebraically closed fields of characteristic p, it follows that $F \vDash \phi$ as well.

A consequence of this theorem is that an $\mathcal{L}_{\mathrm{rng}}$-sentence ϕ is true of the complex numbers if and only if it is true of the algebraic numbers $\mathbb{Q}^{\mathrm{alg}}$. (Recall that these are all roots of polynomials with integer coefficients. As one would expect, the algebraic numbers form an algebraically closed field, though this is not a wholly obvious fact.) Thus, rather surprisingly, if we wish to prove something about $\mathbb{Q}^{\mathrm{alg}}$, we have the option of working in $\mathbb{C}$ and using the methods of complex analysis; similarly, if we want to prove something about $\mathbb{C}$ we can, if it makes things easier, work in $\mathbb{Q}^{\mathrm{alg}}$ and use number theoretic methods.

Combining these ideas with the completeness theorem gives another useful tool. If ϕ is any $\mathcal{L}_{\mathrm{rng}}$-sentence, then the following are equivalent:

(i) ϕ is true in every algebraically closed field of characteristic zero;

(ii) for some $m > 0$, ϕ is true in every algebraically closed field of characteristic $p > m$;

(iii) there are arbitrarily large p such that ϕ is true in some algebraically closed field of characteristic p.

Let us see why this is so. Suppose first that ϕ is true in every algebraically closed field of characteristic 0. The completeness theorem then implies that there is a *proof* of ϕ from the axioms for algebraically closed fields combined with the sentences $1 \neq 0$, $1 + 1 \neq 0$, $1+1+1 \neq 0$, and so on. Since proofs are finite sequences of formulas, there must be some m such that the proof used only the first m of these sentences (not necessarily all of them). If p is some prime bigger than m, then this proof shows that ϕ holds in algebraically closed fields of characteristic p, since all the sentences we used are true in such fields.

We have just shown that (i) implies (ii). It is obvious that (ii) implies (iii). To see that (iii) implies (i), let us suppose that (i) fails, so that there is an algebraically closed field of characteristic zero in which $\neg\phi$ is true. Then, by the principle we proved earlier, $\neg\phi$ is true in *every* algebraically closed field of characteristic zero. Thus, since (i) implies (ii), there is an m such that $\neg\phi$ is true in every algebraically closed field of characteristic $p > m$. Therefore (iii) fails.

An interesting application of this theorem was found by Ax. It is another example of a statement that has nothing to do with logic, but which can be proved using logical tools. It is perhaps more striking than the previous example because in this case one does not even feel with hindsight that the statement did after all have some logical content.

Theorem. *If a polynomial map from $\mathbb{C}^n$ to $\mathbb{C}^n$ is an injection, then it must also be a surjection.*

The basic thought behind the proof of this result is very simple indeed: what is remarkable is that it is of any help. It is the observation that if k is a finite field, then every injective polynomial map from k^n to k^n is a surjection. This is true because every injection from a finite set to itself is automatically a surjection.

How do we exploit this observation? Well, the previous results tell us that, in several situations, statements are true for one field if and only if they are true for another. We shall use these results to transfer our problem from $\mathbb{C}$, where it is hard, to a finite field k, where it is trivial. The first step is a routine exercise: one shows that for each positive integer d there is a sentence ϕ_d in $\mathcal{L}_{\mathrm{rng}}$ that expresses the fact that every injective polynomial map from F^n to F^n, with the n polynomials all of degree at most d, is surjective. We would like to prove that all the sentences ϕ_d are true when $F = \mathbb{C}$.

The equivalences in the previous theorem imply that it is enough to prove that the sentences ϕ_d are true when F is the field $\mathbb{F}_p^{\mathrm{alg}}$, the algebraic closure of the

p-element field. (It can be shown that any field F is contained in an algebraically closed field. Roughly speaking, the *algebraic closure* of F is the smallest algebraically closed field that contains F.) Suppose, then, that some ϕ_d fails for $\mathbb{F}_p^{\mathrm{alg}}$. Then there must be an injective polynomial map f from $(\mathbb{F}_p^{\mathrm{alg}})^n$ to $(\mathbb{F}_p^{\mathrm{alg}})^n$ that is not surjective. Since every finite subset of $\mathbb{F}_p^{\mathrm{alg}}$ is contained in a finite subfield, there is a finite subfield k such that all the n polynomials used to define f have coefficients in k, from which it follows that f maps k^n to k^n. Moreover, by enlarging k if necessary, we can ensure that there is an element of k^n that is not in the image of f. But now we have succeeded in transferring ourselves to a finite field: this function $f : k^n \to k^n$ is an injection between finite sets that is not a surjection, which is a contradiction.

Quantifier elimination has other useful applications. Let F be a field, let K be a subfield of F, let $\psi(v_1, \ldots, v_n)$ be a quantifier-free formula, and let $a_1, \ldots, a_n$ be elements of K. Since, as we have already mentioned, quantifier-free formulas are just Boolean combinations of equalities between polynomials, the statement $\psi(a_1, \ldots, a_n)$ involves just the elements of K, and is therefore true in K if and only if it is true in F. By quantifier elimination, if K and F are algebraically closed, then the same is true for *all* formulas ψ, and not just those that are quantifier free. From this observation we can prove the "weak version" of Hilbert's Nullstellensatz. (For the proof, we shall need to assume a certain degree of familiarity with the basics of RING THEORY [III.81]. We shall also write $K[X]$ for the polynomial ring $K[X_1, \ldots, X_n]$ and $\bar{v}$ for the n-tuple $(v_1, \ldots, v_n)$.)

Proposition. *Suppose that K is an algebraically closed field, P is a prime ideal in $K[X]$, and g is a polynomial in $K[X]$ that does not belong to P. Then there is some $a = (a_1, \ldots, a_n)$ in K^n such that $f(a) = 0$ for every f that belongs to P, and such that $g(a) \neq 0$.*

Proof. Let F be the algebraic closure of the fraction field of the integral domain $K[X]/P$. We can view F as an extension field of K with a natural homomorphism $\eta : K[X] \to F$. Let $b_i = \eta(X_i)$ and let $b \in F^n$ be the element $(b_1, \ldots, b_n)$. Then $f(b) = 0$ for all $f \in P$ and $g(b) \neq 0$. We would like to find such an element in K. Since ideals in polynomial rings are finitely generated, we can find polynomials $f_1, \ldots, f_m$ that generate P. The sentence

$$\exists v_1 \cdots \exists v_n (f_1(\bar{v}) = \cdots = f_m(\bar{v}) = 0 \,\wedge\, g(\bar{v}) \neq 0)$$

is true in F. Thus it is also true in K and we can find $a \in K^n$ such that each $f \in P$ vanishes at a but $g(a) \neq 0$. $\square$

Notice that the above proof has the same basic structure as the result about polynomial maps on $\mathbb{C}^n$. The idea was to come up with a different field, in this case F, where the result was easy to prove, and use logical ideas to deduce the result for the field we were originally interested in, in this case K.

5 The Reals

Quantifier elimination in the language of rings does not work in the field of real numbers. For instance, the formula

$$\exists y \; x = y \cdot y,$$

which asserts "x is a square," is not equivalent to a quantifier-free formula in the language of rings. Of course, x is a square if and only if $x \geqslant 0$. So we *could* eliminate this quantifier if we were prepared to add a symbol for the ordering to our language. An amazing result of Tarski shows that this is the only obstruction to quantifier elimination.

Let $\mathcal{L}_{\mathrm{or}}$ be the language of ordered rings, which is the language of rings with the addition of the symbol "$<$" for an ordering. Which $\mathcal{L}_{\mathrm{or}}$-sentences are true in the real field? Some of the properties of $\mathbb{R}$ that we can formalize in $\mathcal{L}_{\mathrm{or}}$ include:

(i) the axioms for ordered fields, such as the sentence

$$\forall x \forall y \; (x > 0 \wedge y > 0) \to x \cdot y > 0;$$

(ii) the intermediate-value property for polynomials, which states that if $p(x)$ is a polynomial and there exist a and b such that $a < b$ and $p(a) < 0 < p(b)$, then there exists a real number c such that $a < c < b$ and $p(c) = 0$.

The intermediate-value property is expressed not by just one sentence, but by the infinite sequence of sentences

$$\forall d_0 \cdots \forall d_n \forall a \forall b$$
$$\left(\sum d_i a^i < 0 < \sum d_i b^i \to \exists c \sum d_i c^i = 0 \right),$$

one for each positive integer n.

An ordered field that satisfies the intermediate-value property is called a *real closed* field. It turns out that an equivalent way of axiomatizing real closed fields is as ordered fields for which every positive element is a square and every polynomial of odd degree has a zero. Tarski's theorem is the following statement.

Theorem. *For any $\mathcal{L}_{\mathrm{or}}$-formula ϕ there is a quantifier-free $\mathcal{L}_{\mathrm{or}}$-formula ψ such that ϕ and ψ are equivalent in every real closed field.*

What are the quantifier-free formulas of $\mathcal{L}_{\mathrm{or}}$? It turns out (and is not hard to show) that they are finite Boolean combinations of formulas of the form $p(v_1, \ldots, v_n) = q(v_1, \ldots, v_n)$ and formulas of the form $p(v_1, \ldots, v_n) < q(v_1, \ldots, v_n)$, where, as in the case of $\mathcal{L}_{\mathrm{rng}}$, p and q are polynomials in n and m variables, respectively, with integer coefficients. As for quantifier-free *sentences*, they are Boolean combinations of sentences of the form $k = l$ and sentences of the form $k < l$.

One consequence of quantifier elimination is the following result, which tells us that every $\mathcal{L}_{\mathrm{or}}$ statement that is true in $\mathbb{R}$ can be proved from the real-closed-field axioms. One says that these axioms *completely axiomatize* the theory of the real field.

Corollary. *Let K be a real closed field and let ϕ be an $\mathcal{L}_{\mathrm{or}}$-sentence. Then $K \vDash \phi$ if and only if $\mathbb{R} \vDash \phi$.*

To prove this, first use Tarski's theorem to find a quantifier-free sentence ψ such that ϕ and ψ are equivalent in any real closed field. Every ordered field has characteristic zero and contains the rational numbers as an ordered subfield. Therefore $\mathbb{Q}$ is a subfield of both K and $\mathbb{R}$. But the very simple nature of quantifier-free sentences in $\mathcal{L}_{\mathrm{or}}$ means that

$$K \vDash \psi \; \Leftrightarrow \; \mathbb{Q} \vDash \psi \; \Leftrightarrow \; \mathbb{R} \vDash \psi.$$

Since ϕ and ψ are equivalent in all real closed fields, it follows that $K \vDash \phi$ if and only if $\mathbb{R} \vDash \phi$.

By the completeness theorem, ϕ is true in every real closed field if and only if we can prove ϕ from the axioms for real closed fields, and ϕ is false in every real closed field if and only if we can prove $\neg\phi$ from the axioms for real closed fields. It follows that the $\mathcal{L}_{\mathrm{or}}$-theory of the real field is decidable. Indeed, if ϕ is true in $\mathbb{R}$, then by the corollary above, it is true in every real closed field, so it has a proof. If ϕ is false in $\mathbb{R}$, then $\neg\phi$ is true in $\mathbb{R}$, so for the same reason $\neg\phi$ has a proof. Therefore, to decide whether ϕ is true, one can search through all possible proofs from the axioms of real closed fields until one proves either ϕ or $\neg\phi$.

Let $\mathcal{M}$ be a mathematical structure consisting of a set M and various other parts such as functions and binary operations. A subset X of M is called *definable*, with respect to some language $\mathcal{L}$ that describes $\mathcal{M}$, if there is an $\mathcal{L}$-formula ϕ with a free variable x such that $X = \{x \in M : \phi(x)\}$. Quantifier elimination gives us a good geometric understanding of the definable sets. If K is an ordered field, we say that $X \subseteq K^n$ is *semialgebraic* if it is a finite Boolean combination of sets of the form

$$\{x \in K^n : p(x) = 0\} \quad \text{and} \quad \{x \in K^n : q(x) > 0\},$$

where $p, q \in K[X_1, \dots, X_n]$. By quantifier elimination, the definable sets in a real closed field are easily shown to be exactly the semialgebraic sets.

A simple application of this fact is that if A is a semialgebraic subset of $\mathbb{R}^n$, then the closure of A is also semialgebraic. Indeed, the closure of A is, by definition, the set

$$\left\{ x \in \mathbb{R}^n : \forall \epsilon > 0 \ \exists y \in A \ \sum_{i=1}^{n} (x_i - y_i)^2 < \epsilon \right\}.$$

This is a definable set, and hence a semialgebraic set.

Semialgebraic subsets of the real line are particularly simple. For any real polynomial f in one variable, the set $\{x \in \mathbb{R} : f(x) > 0\}$ is a finite union of open intervals. Therefore, any semialgebraic subset of $\mathbb{R}$ is a finite union of points and intervals. This simple fact is the starting point of the modern model-theoretic approach to $\mathbb{R}$. Let $\mathcal{L}^*$ be a language extending $\mathcal{L}_{\mathrm{or}}$ and let $\mathbb{R}^*$ denote the reals considered as an $\mathcal{L}^*$-structure. For example, below we will be interested in the case where $\mathcal{L}^* = \mathcal{L}_{\exp}$ and $\mathbb{R}^* = \mathbb{R}_{\exp}$. We say that $\mathbb{R}^*$ is *o-minimal* if every subset of $\mathbb{R}$ definable using $\mathcal{L}^*$-formulas is a finite union of points and intervals. The "o" in "o-minimal" stands for "ordered." $\mathbb{R}^*$ is o-minimal if every definable subset of $\mathbb{R}$ can be defined using only the ordering.

Pillay and Steinhorn introduced o-minimality, generalizing an earlier idea of van den Dries. It turned out to be a key definition, because although o-minimality is defined in terms of the one-dimensional set $\mathbb{R}$, it has remarkably strong consequences for definable subsets of $\mathbb{R}^n$ when $n > 1$.

To explain this, we inductively define a collection of basic sets called *cells* as follows.

- A subset X of $\mathbb{R}$ is a cell if and only if it is either a point or an interval.
- If X is a cell in $\mathbb{R}^n$ and f is a continuous definable function from X to $\mathbb{R}$, then the graph of f (which is a subset of $\mathbb{R}^{n+1}$) is a cell.
- If X is a cell in $\mathbb{R}^n$ and f and g are continuous definable functions from X to $\mathbb{R}$ such that $f(x) > g(x)$ for every $x \in X$, then $\{(x, y) : x \in X \text{ and } f(x) > y > g(x)\}$ is a cell, as are $\{(x, y) : x \in X \text{ and } f(x) > y\}$ and $\{(x, y) : x \in X \text{ and } y > f(x)\}$.

Cells are topologically simple definable sets that play the role of open intervals in $\mathbb{R}$. It is not hard to see that any cell is homeomorphic to $(0, 1)^n$ for some n. Remarkably, all definable sets can be decomposed into cells. The following theorem is a precise version of this statement.

Theorem.

(i) *If $\mathbb{R}^*$ is an o-minimal structure, then every definable set X can be partitioned into finitely many disjoint cells.*

(ii) *If $f : X \to \mathbb{R}$ is a definable function, then there is a partition of X into finitely many cells such that f is continuous on each cell.*

This is just the beginning. In any o-minimal structure, definable sets have many of the good topological and geometric properties of the semialgebraic sets. For example:

- Any definable set has finitely many connected components.
- Definable bounded sets can be definably triangulated.
- Suppose that X is a definable subset of $\mathbb{R}^{n+m}$. For each $a \in \mathbb{R}^m$, let X_a be the "cross-section" $\{x \in \mathbb{R}^n : (x, a) \in X\}$. Then there are only finitely many different homeomorphism types for the sets X_a.

As these results were known for semialgebraic sets, the real interest is in finding new o-minimal structures. The most interesting example is $\mathbb{R}_{\exp}$. It is known that $\mathbb{R}_{\exp}$ does not have quantifier elimination in the language $\mathcal{L}_{\exp}$. Wilkie showed that the next best thing is true. We say that $\mathbb{R}^n$ is an *exponential variety* if it is the zero set of a finite system of exponential terms. For example, the set $\{(x, y, z) : x = \exp(y)^2 - z^3 \wedge \exp(\exp(z)) = y - x\}$ is an exponential variety.

Theorem. *Every $\mathcal{L}_{\exp}$-definable subset of $\mathbb{R}^n$ is of the form*

$$\{x \in \mathbb{R}^n : \exists y \in \mathbb{R}^m \ (x, y) \in V\}$$

for some exponential variety $V \subseteq \mathbb{R}^{n+m}$.

In other words, the definable sets, though not exponential varieties themselves, are projections of exponential varieties, which makes them tractable. Indeed, a theorem from real analytic geometry, due to Khovanskii, states that every exponential variety has a finite number of connected components. Since this property is preserved by projections, it follows that every definable set has a finite number of connected components, and also that every definable subset of the real line is a finite union of points and intervals. Thus $\mathbb{R}_{\exp}$ is o-minimal and all of the results above about definable sets in o-minimal structures apply.

Tarski asked if the theory of $\mathbb{R}_{\exp}$ is decidable. This question remains open, but the answer is known to follow from the following conjecture of Schanuel in transcendental number theory.

Conjecture. *Suppose that $\lambda_1, \ldots, \lambda_n$ are complex numbers that are linearly independent over $\mathbb{Q}$. Then the field $\mathbb{Q}(\lambda_1, \ldots, \lambda_n, e^{\lambda_1}, \ldots, e^{\lambda_n})$ has transcendence degree at least n.*

Macintyre and Wilkie have shown that if Schanuel's conjecture is true, then the theory of $\mathbb{R}_{\exp}$ is decidable.

6 The Random Graph

Model-theoretic methods give interesting information about random GRAPHS [III.34]. Suppose we construct a graph as follows. The vertex set is the set $\mathbb{N}$ of all natural numbers $\mathbb{N}$. To decide whether we will have an edge between x and y (with $x \neq y$) we flip a coin, putting an edge there if and only if we get heads. Although these constructions are random, we will show below that, with probability 1, any two such graphs are isomorphic.

The proof depends on the following extension property. Let A and B be disjoint finite subsets of $\mathbb{N}$, and suppose that they have sizes n and m, respectively. We would like to find a vertex $x \in \mathbb{N}$ that is joined to every element of A and to no element of B. Now for any particular x, the probability that it does *not* have the desired property is $p = 1 - 2^{-(n+m)}$. Therefore, if we look at N different vertices, the probability that none of them has the desired property is p^N. Since this converges to zero with N, the probability that at least one $x \in \mathbb{N}$ has the property is 1. Moreover, since there are only countably many disjoint pairs (A, B) of finite sets, with probability 1 it is the case that for *every* such pair (A, B) one can find a vertex x that is joined to every vertex in A and to no vertex in B.

We can formalize this observation in a model-theoretic way. Let $\mathcal{L}_g = \mathcal{L}(\sim)$, where "$\sim$" is a binary relation symbol (which we read as "is joined to") and we let T be the $\mathcal{L}_g$-theory:

(i) $\forall x \forall y \; x \sim y \to y \sim x$;
(ii) $\forall x \; \neg(x \sim x)$;
(iii) $\Phi_{n,m}$ for $n, m \geqslant 0$.

Here $\Phi_{n,m}$ is the sentence

$$\forall x_1 \cdots \forall x_n \forall y_1 \cdots \forall y_m$$
$$\bigwedge_{i=1}^{n} \bigwedge_{j=1}^{m} x_i \neq y_j \to \exists z \left(\left(\bigwedge_{i=1}^{n} x_i \sim z \right) \wedge \left(\bigwedge_{i=1}^{m} \neg(y_i \sim z) \right) \right).$$

The first two sentences tell us that the relation "$\sim$" defines a graph, and for each pair (n, m) the sentence $\Phi_{n,m}$ tells us that the extension property holds for all pairs of disjoint sets A and B with A of size n and B of size m. Thus, a model of T is a graph for which the extension property holds for any pair of disjoint finite sets of vertices.

The argument above shows that with probability 1 the random graphs we constructed are models of T. Now let us see why they are isomorphic (again with probability 1). This will be an immediate consequence of the following theorem.

Theorem. *If G_1 and G_2 are any two countable models of T, then G_1 is isomorphic to G_2.*

Recall that an *isomorphism* between G_1 and G_2 means a bijection f from the vertex set of G_1 to the vertex set of G_2 such that x is joined to y in G_1 if and only if $f(x)$ is joined to $f(y)$ in G_2. The proof, which we shall now sketch, is a "back-and-forth" argument that gradually builds up an isomorphism between G_1 and G_2. First, let $a_0, a_1, \ldots$ be an enumeration of the vertices of G_1 and let $b_0, b_1, \ldots$ be an enumeration of the vertices of G_2. Let us set $f(a_0)$ to be b_0. Next, we choose an image for a_1: if a_1 is joined to a_0 then we need to find some vertex that is joined to b_0 and if a_1 is not joined to a_0 then we need to find a vertex that is not joined to b_0. Either way, we can do it because G is a model of T, so it satisfies the extension property. (The particular cases we use here are $\Phi_{1,0}$ and $\Phi_{0,1}$.)

It is tempting to continue finding images for a_2, a_3, and so on, in each case using the extension property to make sure that the images are joined to each other if and only if the original vertices are. The trouble with this is that we may not end up with a bijection, since for any particular b_j there is no guarantee that we will ever choose it as the image of some a_j. However, we can remedy this by alternately choosing an image for the first a_i that does not yet have an image, and a preimage for the first b_j that does not yet have a preimage. In this way we build the desired isomorphism.

It was not essential to use model theory to prove the above result. However, it has the following very nice model-theoretic consequence.

Corollary. *For any $\mathcal{L}_g$-sentence ϕ either ϕ is true in every model of T or $\neg \phi$ is true in every model of T. Moreover, there is an algorithm that will tell us which of ϕ or $\neg \phi$ is true in every model of T.*

To prove this, one first applies a slight strengthening of the compactness theorem, which allows one to conclude that if the result is false then there are *countable* models G_1 and G_2 of T such that ϕ is true in G_1 and $\neg\phi$ is true in G_2. But this shows that G_1 and G_2 are not isomorphic, and therefore directly contradicts the previous theorem.

To decide which of ϕ or $\neg\phi$ is true in every model of T, one searches through all possible proofs from the sentences of T. By the completeness theorem, one or other of the statements has a proof, so we will eventually find either a proof of ϕ or a proof of $\neg\phi$. At that point we will know which of ϕ and $\neg\phi$ is true in every model of T.

The theory T also gives us information about random finite graphs. Let $\mathcal{G}_N$ be the set of all graphs with vertices $\{1, 2, \ldots, N\}$. We consider the probability measure on $\mathcal{G}_N$ in which we make all graphs equally likely. This is the same as constructing a random graph on N vertices, where for each i and j we toss an unbiased coin in order to decide whether i is joined to j. For any $\mathcal{L}_g$-sentence ϕ, let us write $p_N(\phi)$ for the probability that a random graph on N vertices satisfies ϕ.

An easy variant of the argument for infinite graphs shows that for each extension axiom $\Phi_{n,m}$, the probability $p_N(\Phi_{n,m})$ tends to 1. Therefore, for any fixed M, if N is sufficiently large, then with very high probability a random graph on N vertices satisfies all the axioms $\Phi_{n,m}$ with $n, m \leqslant M$.

This observation allows us to use the theory T to get a good understanding of the asymptotic properties of random graphs. The following result is called a *zero-one law*.

Theorem. *Given any $\mathcal{L}_g$-sentence ϕ, the probability $p_N(\phi)$ either tends to 0 or tends to 1 as $N \to \infty$. Moreover, T axiomatizes the set of statements ϕ such that the limit is 1, called the* almost sure theory of graphs, *which is a decidable theory.*

This follows from our previous results. We saw earlier that either ϕ is true in every model of T or $\neg\phi$ is true in every model of T. In the first case, by the completeness theorem there must be a proof of ϕ from T. Since proofs are finite, this proof can use only finitely many of the statements $\Phi_{n,m}$. Therefore, there exists some M such that if $G \vDash \Phi_{M,M}$, then $G \vDash \phi$. But if G is a random graph on N vertices, then the probability that $G \vDash \Phi_{M,M}$ tends to 1, and therefore the probability $p_N(\phi)$ that $G \vDash \phi$ tends to 1 as well. The same argument holds if $\neg\phi$ is true in every model of T and shows that $p_N(\neg\phi)$ tends to 1, which implies that $p_N(\phi)$ tends to 0.

Note the following interesting consequence of this result. It is not hard to prove that the probability that a random graph contains at least $\frac{1}{2}\binom{N}{2}$ edges converges to $\frac{1}{2}$ as N tends to infinity. Combining this simple observation with the theorem we can deduce that the property "contains at least as many edges as nonedges" cannot be expressed by a first-order formula in $\mathcal{L}_g$. This is a purely syntactic result, but to prove it we made essential use of model theory.

Further Reading

Shoenfield (2001) is an excellent introduction to logic including the completeness and incompleteness theorems, basic computability theory, and elementary model theory.

The examples described here give only a small part of the flavor for modern model theory. Hodges (1993), Marker (2002), and Poizat (2000) are comprehensive introductions. Marker et al. (1995) contains several introductory articles on the model theory of fields.

In addition to providing tools for analyzing definability in particular structures, a major goal in model theory is proving structure theorems for wide classes of mathematical structures. A key feature is the development by Shelah of notions of dependence generalizing linear dependence in vector spaces and algebraic dependence in fields. Led by Hrushovski and Zilber, model theorists have studied the geometry of dependence and found that frequently it can be used to detect hidden algebraic structure.

In recent years, abstract model theory has found interesting applications in classical mathematics. Hrushovski used these ideas to give a model-theoretic proof of the Mordell-Lang conjecture for function fields in Diophantine geometry. Bouscaren (1998) is an excellent collection of survey articles leading up to Hrushovski's proof.

Bouscaren, E., ed. 1998. *Model Theory and Algebraic Geometry. An Introduction to E. Hrushovski's Proof of the Geometric Mordell-Lang Conjecture.* New York: Springer.

Hodges, W. 1993. *Model Theory.* Encyclopedia of Mathematics and Its Applications, volume 42. Cambridge: Cambridge University Press.

Marker, D. 2002. *Model Theory: An Introduction.* New York: Springer.

Marker, D., M. Messmer, and A. Pillay. 1995. *Model Theory of Fields.* New York: Springer.

Poizat, B. 2000. *A Course in Model Theory. An Introduction to Contemporary Mathematical Logic.* New York: Springer.

Shoenfield, J. 2001. *Mathematical Logic.* Natick, MA: A. K. Peters.

IV.24 Stochastic Processes
Jean-François Le Gall

1 Historical Introduction

Stochastic processes are one of the major themes of modern probability theory. Roughly speaking, they are mathematical models that describe the evolution of random phenomena as time goes by. In this article, we shall introduce and illustrate the fundamental ideas of the theory of stochastic processes by concentrating on the single most important example: Brownian motion. We start with a brief historical introduction, in order to provide some motivation for the mathematical theory that follows.

In 1828, the British botanist Robert Brown observed the very irregular and wiggly motion of small particles of pollen suspended in water. Brown pointed out the unpredictable character of the motion, which appeared to obey no known physical rule. During the nineteenth century, several physicists tried to understand the origin of this "Brownian motion," which turned out to be present in many other physical phenomena. Several theories were proposed, some of them rather fanciful: perhaps Brownian particles were living microscopic animals, or perhaps the motion was due to electrostatic forces. By the end of the century, however, physicists had concluded that the constant changes of direction in Brownian motion could be explained by the impacts on a particle from the molecules of the surrounding medium. If the particle was sufficiently light, then these numerous collisions could have a macroscopic influence on its displacement. This explanation was also consistent with the experimental observation that the motion became faster if the temperature of the water, and thus the thermal agitation of its molecules, increased.

Albert Einstein, in one of his three famous 1905 papers, was responsible for a major step forward in the understanding of Brownian motion. He worked out that if a Brownian particle starts at the origin, then after a fixed time t its position should be randomly distributed according to the (three-dimensional) GAUSSIAN DISTRIBUTION [III.71 §5] with mean 0 and variance $\sigma^2 t$, where σ^2 is a constant, called the *diffusion constant*, that measures how quickly the distribution spreads out with time. (One can think of this loosely as the speed of the Brownian motion, but we shall see later

that the word "speed" is not really appropriate.) Einstein's method was based on considerations of statistical physics, which led him to THE HEAT EQUATION [I.3 §5.4] and then to the Gaussian density that solves this equation (see section 5.2).

A few years before Einstein, the French mathematician Louis Bachelier, in his work about the mathematical modeling of stock markets, had already noticed the Gaussian distribution of Brownian motion. However, Bachelier was dealing not with the physical phenomenon known as Brownian motion, but rather with random walks where the step size was very small. As we shall see in sections 2 and 3, the two concepts are essentially equivalent from a mathematical viewpoint. Bachelier pointed out what we call today the *Markov property* of Brownian motion: if we wish to predict the displacement after time t of a Brownian particle, then knowledge of the path followed by the particle before time t does not help us any more than just knowing the position at time t. Bachelier's arguments were not completely satisfactory, and his ideas were not fully appreciated in his time.

How does one go about modeling a particle that moves in a random way? A first remark is that the position of the particle at time t will be a RANDOM VARIABLE [III.71 §4] B_t. But these random variables will depend on each other: if you know where the particle is at time t, it will affect your knowledge of how likely it is to be in a certain region at some later time. These two considerations can be accommodated if we take as our basic model a set of random variables B_t, one for each nonnegative real number, all defined on the same underlying probability space. This, formally speaking, is what a stochastic process is.

This may seem a rather simple definition, but in order for a stochastic process to be interesting it needs to have additional properties, and difficult mathematical questions arise as soon as one tries to obtain them. Let us write Ω for the underlying probability space. Then each of the random variables B_t is a function from Ω to $\mathbb{R}^3$, and therefore we associate a point in $\mathbb{R}^3$ with each pair (t, ω) (where t is a positive real number and ω belongs to Ω). So far we have thought about the probability distribution of B_t, so we have been focusing on what happens when we fix t and let ω vary. However, we must also consider what happens when we look at a "single instance" of a stochastic process, by fixing ω and letting t vary. For fixed ω, the function that takes t to $B_t(\omega)$ is called a *sample path*. If we want a rigorous mathematical theory of Brownian motion, then a very important property it should satisfy is that all the sample paths are continuous: that is, for fixed ω the point $B_t(\omega)$ depends continuously on t.

Physical observations, as well as the contributions of Einstein and Bachelier described above, suggested a few other properties that Brownian motion should satisfy. It then became a substantial mathematical problem to prove that there existed a stochastic process with those properties. Wiener was the first person to establish this, which he did in 1923, and for this reason the mathematical concept of Brownian motion is sometimes called the *Wiener process*.

The most famous names of probability theory in the twentieth century, including KOLMOGOROV [VI.88], Lévy, Itô, and Doob, all made important contributions to the study of Brownian motion. Detailed properties of the sample paths have received particular attention, ever since the physicist Jean Perrin observed that these functions are nowhere differentiable (despite Wiener's later result that they were continuous). The nondifferentiability of Brownian trajectories led Itô to introduce a differential calculus for functions of Brownian motion and more general stochastic processes. This Itô stochastic calculus, which will be briefly presented in section 4, has found many applications in many different areas of modern probability theory.

2 Coin Tossing and Random Walks

One of the easiest ways to understand Brownian motion is via another important concept of probability: that of *random walks*. Suppose you were to play a game where you repeatedly tossed a coin, winning €1 if it came up heads, and losing €1 if it came up tails. One could then define a sequence of random variables $S_0, S_1, S_2, \ldots$, where S_n represented your total gain (which could well be negative) after n tosses of the coin. Two simple properties of this sequence are that S_0 must be 0 and that S_n and S_{n-1} always differ by 1. One can see this in figure 1, which plots a graph of the sequence in the case where the coin tosses are HTTTHTHHHHTHHHTH ....

A third property becomes clear if one defines another sequence of random variables $\varepsilon_1, \varepsilon_2, \ldots$, representing the outcome of each toss of the coin. These are independent, and each ε_n takes the value 1 with probability $\frac{1}{2}$ and -1 with probability $\frac{1}{2}$. Moreover, for each n we can write $S_n = \varepsilon_1 + \cdots + \varepsilon_n$. The distribution of sums of this kind depends in a very simple way on the well-known BINOMIAL DISTRIBUTION [III.71 §1]. (To be pre-

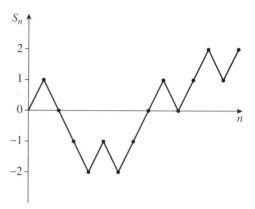

Figure 1 The accumulated gain in coin tossing.

cise, the binomial distribution tells you that the probability that the number of heads after n tosses is k is $2^{-n}\binom{n}{k}$. If it is k, then $S_n = k - (n - k) = 2k - n$.) What is more, if $m > 0$ then $S_{m+n} - S_m = \varepsilon_{m+1} + \cdots + \varepsilon_{m+n}$, which is also a sum of n of the ε_i, so the distribution of $S_{m+n} - S_m$ is the same as that of S_n. Note too that it is independent of the values of $S_0, S_1, \ldots, S_m$.

The name "random walk" comes from the fact that we can think of the sequence $S_0, S_1, S_2, \ldots$ as taking a succession of random steps, each of either 1 or -1. Brownian motion can be thought of as the limit of this process as the number of steps gets larger and larger and the sizes of the steps get correspondingly smaller.

To see what "correspondingly" means here, we appeal to the CENTRAL LIMIT THEOREM [III.71 §5], which tells us about the limiting behavior of the distribution of S_n when n gets large. Or rather, it tells us about the distribution of $(1/\sqrt{n})S_n$: the reason it is appropriate to divide by $\sqrt{n}$ is that $\sqrt{n}$ is the STANDARD DEVIATION [III.71 §4] of S_n. This one can think of as its "typical size": thus, when we divide by it, the "renormalized" distribution will have "typical size" 1 (and therefore we will get the same typical size for each n).

The precise information that the central limit theorem gives us is that for any real numbers a and b with $a < b$, the probability that $a < (1/\sqrt{n})S_n < b$ tends to

$$\frac{1}{\sqrt{2\pi}} \int_a^b e^{-x^2/2} \, \mathrm{d}x$$

as n tends to ∞. That is, the limiting behavior of the distribution of $(1/\sqrt{n})S_n$ is Gaussian with mean 0 and standard deviation 1. Since the distribution of $S_{m+n} - S_m$ is the same as that of S_n (as we saw earlier), this also tells us the limiting behavior of the distribution of $(1/\sqrt{n})(S_{m+n} - S_m)$ for any m.

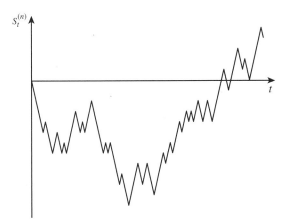

Figure 2 The rescaled random walk $S^{(n)}$ for $n = 100$.

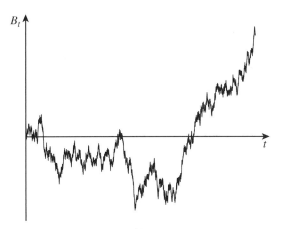

Figure 3 Simulation of linear Brownian motion.

3 From Random Walks to Brownian Motion

In the previous section, we looked at a sequence of random variables $S_0, S_1, S_2, \ldots$. This is another stochastic process, except that "time" is now represented by a positive integer. (One says that it is a *discrete-time* process.) Now let us try to do justice to the idea that Brownian motion is something like a random walk with infinitely many infinitesimally small steps. (We are now looking at one-dimensional Brownian motion, rather than the three-dimensional Brownian motion discussed right at the beginning of this article.)

It will be slightly simpler to think about a Brownian motion B_t that runs just for times t between 0 and 1. We hope that the distributions of B_t, and in particular of B_1, will be Gaussian, and the results from the last section suggest that this is exactly what we should expect if they are appropriately scaled limits of the distributions of the S_n. To be precise, suppose we have a graph like that of figure 1 but with some large number of steps n. Then the x-axis will go from 1 to n and the standard deviation of the height of the end of the graph will be $\sqrt{n}$. Therefore, if we shrink the graph horizontally by a factor of n and vertically by a factor of $\sqrt{n}$ we will obtain the graph of a random function $S^{(n)}$ from $[0, 1]$ to $\mathbb{R}$, and the standard deviation of $S^{(n)}(1)$ will be 1. Effectively, we are shrinking the time between the steps of the random walk from 1 to $1/n$ and shrinking the step size from 1 to $1/\sqrt{n}$. Also, so that the functions $S^{(n)}$ are defined everywhere, we "join the dots" of the graph with straight lines, just as we did in figure 1. A rescaled random walk of this kind is shown in figure 2.

At this point, we shall simply assume that the distributions of these rescaled random walks converge, in

an appropriate sense, to a stochastic process with continuous sample paths. This stochastic process is the Brownian motion B_t. The graph of a typical sample path is illustrated in figure 3. Notice how similar its general behavior is to that of the graph in figure 2.

If we want to approximate a Brownian motion that goes on forever rather than stopping at 1, all we have to do is let the rescaled random walk go on forever, rather than stopping after n steps.

Now let us give a more precise definition. A *linear Brownian motion starting at x* is a collection $(B_t)_{t \geqslant 0}$ of real-valued random variables with the following properties.

- $B_0 = x$. (In other words, $B_0(\omega) = x$ for every ω in the underlying probability space.)
- The sample paths are continuous.
- Given any $s < t$ the distribution of $B_t - B_s$ is Gaussian with mean 0 and variance $t - s$.
- Moreover, $B_t - B_s$ is independent of the process up to time s. (This implies the Markov property mentioned in section 1.)

Each of these properties has its counterpart for random walks, as we saw in the previous section. Therefore, even though it is not easy to prove that Brownian motion exists, the result is nevertheless highly plausible. (It turns out to be easy to construct a stochastic process that satisfies all the properties above apart from the second; the difficulty is in obtaining the continuity of the sample paths.) Another important remark is that the above properties characterize Brownian motion: any two stochastic processes with those properties are essentially the same.

We have not yet said what it means for the rescaled random walks $S^{(n)}$ to "converge" to Brownian motion. Rather than defining this notion precisely, we shall merely remark that any "reasonable" function that we can define on the processes $S^{(n)}$ will converge to the "corresponding" function of the limiting Brownian motion B_t. For example, as we have already seen, the probability that $S^{(n)}(1)$ lies between a and b converges to

$$\frac{1}{\sqrt{2\pi}} \int_a^b e^{-x^2/2}\, dx.$$

But B_1 is governed by the Gaussian distribution, so this is also the probability that B_1 lies between a and b.

A more interesting example is the proportion X_n of times t between 0 and 1 for which $S^{(n)}(t)$ is positive, or rather the way that this proportion (which is a random variable that depends on the walk $S^{(n)}$) is distributed. This "converges in distribution" to the distribution of the corresponding proportion X for Brownian motion. That is, for any $a < b$, the probability that the proportion X_n lies between a and b converges to the probability that the proportion X lies between a and b. The probability distribution for X is known explicitly, and is called *Paul Lévy's arcsine law*:

$$P[a \leqslant X \leqslant b] = \int_a^b \frac{dx}{\pi\sqrt{x(1-x)}}.$$

Perhaps surprisingly, X is more likely to be close to 0 or 1 than to $\frac{1}{2}$. The basic reason for this is that if s and t are two different times, then the events $B_s > 0$ and $B_t > 0$ are positively correlated.

The convergence of random walks to Brownian motion is just one special case of a much more general phenomenon (see, for example, Billingsley 1968). For instance, we can allow other probability distributions for the individual steps of the random walk. A typical result is that if each individual step has mean 0 (as is the case when we have +1 or −1 with probability $\frac{1}{2}$) and finite variance, then the limiting process will always be a simple rescaling of Brownian motion. In this sense Brownian motion appears as a universal object: it is the continuous limit of a wide range of discrete models. (See the introduction to PROBABILISTIC MODELS OF CRITICAL PHENOMENA [IV.25] for a discussion of universality.)

Now that we have discussed one-dimensional Brownian motion, let us think about how to model random continuous paths in three dimensions. An obvious way of doing it would be to take three independent Brownian motions, B_t^1, B_t^2, and B_t^3, and let these be the three coordinates of a point in a random path in $\mathbb{R}^3$. And

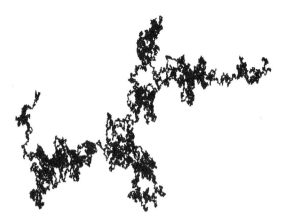

Figure 4 Simulation of planar Brownian motion.

indeed, this is how three-dimensional Brownian motion is defined. However, it is not quite so obvious that this is a good definition. In particular, it seems to depend on our choice of coordinate system, which is worrying if we want a good model for physical Brownian motion.

However, a central property of higher-dimensional Brownian motion (the definition just given clearly generalizes to any dimension d) is *rotational invariance*. That is, if we choose a different ORTHONORMAL BASIS [III.37] as our coordinate system, then we obtain the same stochastic process. The proof of this is a simple deduction from the basic fact that the DENSITY FUNCTION [III.71 §3] of a vector made up of d independent one-dimensional Gaussian random variables is

$$\frac{1}{(2\pi)^{d/2}} e^{-(x_1^2 + \cdots + x_d^2)/2}.$$

Since the quantity $x_1^2 + \cdots + x_d^2$ is just the square of the distance from 0 to $(x_1, \ldots, x_d)$, the density does not change when you rotate.

In the planar case $d = 2$, there is a much deeper invariance property, which we shall explain in section 5.3.

It is not hard to incorporate the notion of a diffusion constant into our model. (This is the constant σ^2 mentioned in section 1 that measures how quickly the Brownian motion tends to spread out.) All one has to do is rescale from B_t to $B_{\sigma^2 t}$.

As one might expect, higher-dimensional Brownian motions are limits of higher-dimensional random walks. This helps to explain why mathematical Brownian motion is a good model for the physical phenomenon observed by Brown: the erratic displacements caused by collisions with molecules resemble the steps of a random walk with very small step size. See figure 4 for a simulation of the curve of a planar Brownian motion over the time interval $[0, 1]$.

4   Itô's Formula and Martingales

Let f be a real-valued differentiable function. Suppose that we are told the values of $f'(x)$ at $0, 1/n, 2/n, \ldots,$ $(n - 1)/n$ for some large positive integer n and are asked to estimate $f(1) - f(0)$. If the derivative f' did not vary too rapidly, then we would expect the difference $f((j + 1)/n) - f(j/n)$ to be approximately $(1/n)f'(j/n)$, so a good approximation ought to be

$$\frac{1}{n}\left(f'(0) + f'\left(\frac{1}{n}\right) + f'\left(\frac{2}{n}\right) + \cdots + f'\left(\frac{n-1}{n}\right)\right).$$

THE FUNDAMENTAL THEOREM OF CALCULUS [I.3 §5.5] implies that this argument is indeed correct if the derivative f' is continuous.

Now let us look at a setup that is superficially similar. This time, let us suppose that the numbers $x_0, x_1, x_2, \ldots, x_n$ are the positions of a random walk with step size $1/\sqrt{n}$. Suppose that f is a function with a well-behaved derivative, and that we know the values of $f'(x)$ at $x_0, x_1, \ldots, x_{n-1}$. This time, let us think about estimating $f(x_n) - f(x_0)$.

If we follow the lines of our previous argument, then we will comment that $f(x_{j+1}) - f(x_j)$ is approximately $(x_{j+1} - x_j)f'(x_j)$, which would lead to an estimate of

$$(x_1 - x_0)f'(x_0) + (x_2 - x_1)f'(x_1)$$
$$+ \cdots + (x_n - x_{n-1})f'(x_{n-1}).$$

Now it is not obvious that this will still be a good estimate. The reason is that, typically, the random walk will have gone backwards and forwards, covering the same ground several times before reaching its eventual destination x_n, and this gives the errors in the approximations a chance to accumulate. To see that this is a serious problem, consider the very well-behaved function $f(x) = x^2$ and let $x_0 = 0$. In this case,

$$f(x_{j+1}) - f(x_j) = x_{j+1}^2 - x_j^2$$

and a simple calculation shows that this is equal to

$$(x_{j+1} - x_j)2x_j + (x_{j+1} - x_j)^2.$$

The first term here equals $(x_{j+1} - x_j)f'(x_j)$ and is therefore the approximation that we are considering, so the error we have to worry about is $(x_{j+1} - x_j)^2$, which is the square of the step size of the random walk. In other words, it is $1/n$. But there are n steps to the walk, so the total error (all of which is positive) is 1. Since the order of magnitude of x_n, and hence x_n^2, is typically

about 1, this is a significant fraction of $f(x_n) - f(x_0)$, and therefore our estimate is not a good one.

Remarkably, this turns out to be the "only" problem that can occur, and we can get around it rather easily. All we have to do is use one more term in the Taylor expansion. That is, we use the slightly more refined approximation

$$f(x_{j+1}) - f(x_j) = (x_{j+1} - x_j)f'(x_j)$$
$$+ \tfrac{1}{2}(x_{j+1} - x_j)^2 f''(x_j).$$

(Of course, now we are assuming that the *second* derivative f'' exists and is continuous.) Notice that in the example $f(x) = x^2$ just considered, $f''(x) = 2$ for every x, and so if we add up all the above approximations we get exactly the right answer. In general, as this observation would suggest, one can show that $f(x_n) - f(x_0)$ is well-approximated by

$$\sum_{j=0}^{n-1}(x_{j+1} - x_j)f'(x_j) + \frac{1}{2}\sum_{j=0}^{n-1}(x_{j+1} - x_j)^2 f''(x_j).$$

Now let us think about what happens to these two sums if we allow our random walks to converge to a Brownian motion B_t. A relatively straightforward argument, based on the fact that $(x_{j+1} - x_j)^2$ is just the reciprocal of the number of steps, shows that the limiting distribution of the second sum exists and is given by the integral $\frac{1}{2}\int_0^t f''(B_s)\,\mathrm{d}s$. This suggests that the first sum should also converge to a limit, which indeed it does: the limit is called the *stochastic integral* and is written $\int_0^t f'(B_s)\,\mathrm{d}B_s$. More precisely, one ends up with the formula

$$f(B_t) = f(B_0) + \int_0^t f'(B_s)\,\mathrm{d}B_s + \frac{1}{2}\int_0^t f''(B_s)\,\mathrm{d}s, \quad (1)$$

which is known as *Itô's formula*. Note the similarity to the fundamental theorem of calculus. The main difference is the extra term involving the second derivative, the so-called *Itô term*.

Why, one might wonder, is this interesting? If we wish to estimate the difference between two values of a function by integrating its derivative, why not choose a smooth path rather than a very wiggly one? The point, however, is that we are not interested in just one path. For any fixed sample path, the two sides of the above formula are just numbers, but if we think of B_t as a random variable, then they too become random variables. And since both sides are defined for all $t \geqslant 0$, they are actually stochastic processes. So what we are discussing is a way of integrating one stochastic process to produce another.

The reason Itô's formula is so useful is that stochastic integrals have properties that allow one to prove many facts about them. In particular, if we view the stochastic integral $\int_0^t f'(B_s)\,\mathrm{d}B_s$ as a collection of random variables indexed by the parameter t, then we have a stochastic process of an especially nice sort called a *martingale*. A martingale is a stochastic process $(M_t)_{t \geqslant 0}$ with the property that, whenever $s \leqslant t$, the expected value of M_t, conditional on the values of M_r for all $r \leqslant s$, is just M_s.

Brownian motion is a particularly simple kind of martingale, but martingales are much more general because $M_t - M_s$ is not *independent* of the values of M_r with $r \leqslant s$: all one knows is that the expectation of $M_t - M_s$, given those values, is zero. Here is an example that illustrates the difference: start running Brownian motion at 0; when it first reaches 1 (if it ever does), continue with Brownian motion but at double the speed (or rather, double the diffusion constant). In this case, the behavior of $M_t - M_s$ certainly depends on what has happened up to s, but its expectation is nevertheless zero.

In a certain sense, the stochastic integral term in Itô's formula behaves like a Brownian motion "run at a varying speed," rather like the example just given. The precise result is that there exists another Brownian motion $\beta = (\beta_t)_{t \geqslant 0}$ such that, for every $t \geqslant 0$,

$$\int_0^t f'(B_s)\,\mathrm{d}B_s = \beta_{\int_0^t f'(B_s)^2\,\mathrm{d}s}.$$

This is in fact true for any continuous martingale—not just one given by a stochastic integral—and the relevant time change is a quantity called the *quadratic variation* of the martingale. Therefore, the graph of a continuous martingale is obtained from that of a Brownian motion by a time-change operation. This is why Brownian motion is such a central example, and why it is important to understand its behavior before going on to deal with more general stochastic processes.

It is straightforward to generalize the previous derivation of Itô's formula to multidimensional Brownian motion. If $x = (x_1, \dots, x_d)$ and $y = (y_1, \dots, y_d)$ belong to $\mathbb{R}^d$ and are close together, then the first approximation to $f(x) - f(y)$ is now

$$\sum_{i=1}^d (x_i - y_i)\partial_i f(y),$$

where $\partial_i f(y)$ denotes the ith partial derivative of f, evaluated at y. The vector of partial derivatives at y is usually denoted $\nabla f(y)$. It is called the *gradient* of f at y (or "grad f" for short). As for the second

derivative of f, it naturally generalizes to the Laplacian Δf (for reasons that are explained in SOME FUNDAMENTAL MATHEMATICAL DEFINITIONS [I.3 §5.4]), and we therefore arrive at the formula

$$f(B_t) = f(B_0) + \int_0^t \nabla f(B_s) \cdot \mathrm{d}B_s + \frac{1}{2}\int_0^t \Delta f(B_s)\,\mathrm{d}s.$$

The stochastic integral term is defined formally in terms of one-dimensional stochastic integrals in the obvious way:

$$\int_0^t \nabla f(B_s) \cdot \mathrm{d}B_s = \sum_{j=1}^d \int_0^t \frac{\partial f}{\partial x_j}(B_s)\,\mathrm{d}B_s^j.$$

Since stochastic integrals are martingales, the stochastic process

$$M_t^f = f(B_t) - \frac{1}{2}\int_0^t \Delta f(B_s)\,\mathrm{d}s$$

is (under appropriate conditions on f) a martingale. This observation leads to the *martingale problem* for Brownian motion. To state a martingale problem for a stochastic process $(X_t)_{t \geqslant 0}$ is to give a collection of martingales defined as functionals of this stochastic process—just as M^f above is defined as a certain function of $(B_s)_{s \geqslant 0}$. The martingale problem is said to be *well-posed* if it characterizes the distribution of the given stochastic process. In the preceding example, the martingale problem is well-posed: if we know nothing about the distribution of the process $(B_t)_{t \geqslant 0}$ apart from the fact that M_t^f is a martingale for every (twice continuously differentiable) function f, we can infer that B must be a Brownian motion.

Martingale problems play a fundamental role in modern probability theory (see in particular Stroock and Varadhan (1979), and also THE MATHEMATICS OF MONEY [VII.9 §2.3]). The introduction of a suitable martingale problem is often the most convenient way to specify a stochastic process, or more precisely to characterize its probability distribution.

5 Brownian Motion and Analysis

5.1 Harmonic Functions

A continuous function h defined on an open subset U of $\mathbb{R}^d$ is called *harmonic* if the average value of h over any closed ball contained in U, or equivalently the average value over the boundary of any such ball, is equal to its value at the center of the ball. A basic result of analysis is that h is harmonic if and only if it is twice continuously differentiable and $\Delta h = 0$. Harmonic functions play an important role in several

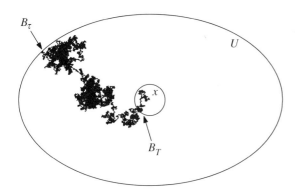

Figure 5 The probabilistic solution
of the Dirichlet problem.

areas of mathematics as well as in physics. For instance, the electrical potential of a conductor in equilibrium is a harmonic function outside the conductor. And if the temperature of the boundary of a body is kept fixed (that is, although different parts of the boundary may have different temperatures, these temperatures do not change over time), then the equilibrium temperature inside the body is also a harmonic function. (See the discussion of the heat equation in the next section.)

Harmonic functions have a very close relationship with Brownian motion, which leads to one of the most important connections between probability and analysis. This connection is already apparent from the fact that M_t^f, defined in the previous section, is a martingale. It follows from this that $h(B_t)$ is a martingale if (and in fact only if) h is harmonic, since then the second term vanishes. However, we will explain the link between Brownian motion and harmonic functions in a more elementary way, from the classical *Dirichlet problem*. Let U be a bounded open set, and let g be a continuous real-valued function defined on the boundary ∂U of U. The classical Dirichlet problem is to find a function h that is harmonic on U and is equal to g on the boundary.

The Dirichlet problem has a remarkably simple solution in terms of Brownian motion: take $x \in U$, start a Brownian motion from x, and evaluate g at the point B_τ where this Brownian motion leaves U (see figure 5); then define $h(x)$ to be the average value you get. Why does this work? That is, why is the function h, defined in this way, harmonic, and why does it equal (or, to be more accurate, converge to) g at the boundary?

The answer to the last question is roughly that if x is very close to the boundary, then a Brownian motion started at x is very likely to leave U at a point close to x. Therefore, since g is continuous, the average value of g at the first exit point will be close to the value of g at any point near x.

To show that h is harmonic is more interesting. Let x be a point in U and suppose that the ball of radius r about x is contained in U. We would like to show that $h(x)$ equals the average value of h on the boundary of this ball. Now $h(x)$ is the average value of g at the point where a Brownian motion that starts at x leaves U. Let us work out this average by conditioning on the first point B_T where the Brownian path leaves the ball of radius r (see figure 5). By the rotational invariance of Brownian motion, this point will be evenly distributed around the boundary of this ball. If we reach the boundary at a point y, then the average value of g when the path leaves U (conditioning on this extra information) is $h(y)$, by definition. Therefore, $h(x)$ is indeed the average value of h on the boundary of the ball of radius r.

Convincing though this argument might seem, there is a subtlety concealed within it, connected with the fact that a Brownian path will typically cross the boundary of the ball many times. Suppose we tried a similar argument, but this time we conditioned on the value at the *last* point where the path left the ball. If this point was y, we could not then say that the expected value of g where the path first reached the boundary of U was $h(y)$ because from that point onward the path would be forbidden to enter the ball again, and would therefore not be a Brownian motion.

Recall that the Markov property of a Brownian motion states that, given a fixed time T and another time t with $T < t$, the value of $B_t - B_T$ is independent of B_s for $s \leqslant T$. It may seem that we are applying this principle in the argument above, taking T to be the first time that the Brownian motion reaches the boundary of the ball. But if we do that, then T is not a fixed time since it depends on the Brownian motion. However, the argument can still be made to work because T is a so-called *stopping time*. Informally, this means that T does not depend on what the Brownian motion does after T. (Therefore the last time it leaves the ball of radius r is not a stopping time, because whether or not a given time is this last time depends on the subsequent behavior of the Brownian motion.) Brownian motion can be shown to have the *strong Markov property*, which is like the usual Markov property except that T is allowed to be a stopping time. Given this fact, it is not hard to show rigorously that h is harmonic.

5.2 The Heat Equation

Let f be a function on $\mathbb{R}^d$ (which we shall assume to be continuous and bounded). If we think of f as a temperature distribution at time 0, then the HEAT EQUATION [III.36] models what happens to the temperature at subsequent times. To find a solution to this equation with initial value f means to find a continuous function $u(t, x)$, defined for every $t \geqslant 0$ and $x \in \mathbb{R}^d$, that solves the partial differential equation

$$\frac{\partial u}{\partial t} = \tfrac{1}{2} \Delta u \qquad (2)$$

whenever $t > 0$, and that satisfies the condition $u(0, x) = f(x)$ for every x. (The factor $\tfrac{1}{2}$ in this equation is not important but it makes the probabilistic interpretation easier to express.)

The heat equation also has a simple solution in terms of Brownian motion: $u(t, x)$ is defined to be the expected value of $f(B_t)$ when B_t is a Brownian motion that starts at x. This tells us that heat propagates like a collection of infinitesimal Brownian particles.

The preceding probabilistic representation is quite easy to derive since one can write down an explicit formula for the expectation of $f(B_t)$ in terms of the Gaussian density function. Given this formula, all we have to do is differentiate it and check that the equation is satisfied. However, the connection between Brownian motion and the heat equation is much deeper, and in many other cases there is a probabilistic representation for a solution but no explicit formula. To take one example, suppose that we want to solve the heat equation in an open set U with Dirichlet boundary conditions. This means that we specify an initial value $f(x)$ for the temperature of each point $x \in U$ and stipulate that the temperature at the boundary is kept at 0. In other words, we want to find a function $u(t, x)$ such that $u(0, x) = f(x)$ for every $x \in U$, $u(t, x) = 0$ for every time $t \geqslant 0$ and every x in the boundary of U, and u satisfies the heat equation inside U. In this case, the solution is obtained as follows. Run a Brownian motion (B_t) starting at x. Let $g_t = f(B_t)$ if it has not left U at any time before t, and let $g_t = 0$ otherwise. Then define $u(t, x)$ to be the expected value of g_t.

Thus, in order to obtain the solution, we had to make just a small modification to the solution of the heat equation in $\mathbb{R}^d$. An analytic treatment of this version of the heat equation would be much more complicated.

5.3 Holomorphic Functions

Let us now concentrate on the case $d = 2$. As usual, we identify $\mathbb{R}^2$ with the complex plane $\mathbb{C}$. Let $f = f_1 + i f_2$ be a HOLOMORPHIC FUNCTION [I.3 §5.6] defined on $\mathbb{C}$. Then the real part f_1 and the imaginary part f_2 of f are both harmonic functions, so that $f_1(B_t)$ and $f_2(B_t)$ are martingales. More precisely, Itô's formula tells us that, for $j = 1, 2$,

$$f_j(B_t) = f_j(x) + \int_0^t \frac{\partial f_j}{\partial x_1}(B_s)\,\mathrm{d}B_s^1 + \int_0^t \frac{\partial f_j}{\partial x_2}(B_s)\,\mathrm{d}B_s^2,$$

since the Itô term vanishes. As we saw in section 3, each of the two processes $f_j(B_t)$ can be expressed as a time change of a linear Brownian motion β^j. However, a stronger result can also be proved, namely that the time change is the same in both cases and that the Brownian motions β^1 and β^2 are independent. This makes it possible to prove a "localized" rotational invariance, which leads to the important *conformal invariance* property of Brownian motion. Roughly speaking, this states that the image of a planar Brownian motion under a conformal (that is, angle-preserving) mapping is another planar Brownian motion run at a different speed.

6 Stochastic Differential Equations

Imagine a Brownian particle in some water. If the temperature of the water rises, then we expect there to be more collisions with faster-moving molecules; this can be modeled easily by increasing the diffusion constant. But what if the temperature in the water varied from place to place? Then the particle would be more agitated in some parts of the water than in others. And if the water was moving, with different parts moving at different speeds, then one would need to superimpose on the Brownian motion a "drift" term, to take into account that on average we would expect the particle to move with the surrounding water.

Stochastic differential equations are used to model more complicated situations like this. Let us begin by considering the one-dimensional case. Let σ and b be two functions (which we shall assume to be continuous) defined on $\mathbb{R}$. We think of $\sigma(x)$ as telling us the rate of diffusion at x and of $b(x)$ as the drift at x. (For the sake of a picture, one could think of $\sigma(x)$ as the local temperature at x and $b(x)$ as the velocity at x of some "one-dimensional water.") Let (B_t) be a one-dimensional Brownian motion.

The notation used for the associated stochastic differential equation is

$$\mathrm{d}X_t = \sigma(X_t)\,\mathrm{d}B_t + b(X_t)\,\mathrm{d}t. \qquad (3)$$

Here (X_t) is an unknown stochastic process. The idea is that, infinitesimally speaking, its behavior is like that of a Brownian motion with diffusivity $\sigma(X_t)$ (which is the diffusivity at the point that X_t has reached) superimposed onto a linear motion at speed $b(X_t)$. More precisely, a solution to the above equation is defined to be a continuous stochastic process (X_t) that satisfies, for every $t \geqslant 0$, the integral equation

$$X_t = X_0 + \int_0^t \sigma(X_s)\,\mathrm{d}B_s + \int_0^t b(X_s)\,\mathrm{d}s.$$

Notice that if $\sigma(x) = 0$ for every x, this boils down to the ordinary differential equation $x'(t) = b(x(t))$. The stochastic integral $\int_0^t \sigma(X_s)\,\mathrm{d}B_s$ is defined by approximations similar to those described in section 4. (For this to work, there are certain technical conditions that the process (X_t) must satisfy.) In fact, stochastic differential equations were Itô's original motivation for developing stochastic integrals.

Itô proved, under suitable conditions on σ and b, that for each $x \in \mathbb{R}$ the above equation has a unique solution (X_t) that starts at x. Furthermore, this solution is a Markov process in the sense that was explained above: the way that (X_t) evolves after time T given the value of X_T is independent of what happens before T, and is distributed in the same way as a solution of the equation that starts at X_T. In fact, it is also a strong Markov process in the sense explained in section 5.

An important example can be found in the famous BLACK–SCHOLES MODEL [VII.9 §2] of mathematical finance. In this model, the price of a share solves a stochastic differential equation of the type above with $\sigma(x) = \sigma x$ and $b(x) = bx$, where σ and b are positive constants. This is motivated by the simple idea that the price fluctuations of a share should be roughly proportional to its current value. In this context, the number σ is called the *volatility* of the share.

The previous discussion generalizes fairly easily to stochastic differential equations in higher dimensions. The solution of a d-dimensional stochastic equation (which when $d = 3$ could model the water example mentioned at the beginning of this section) is once again a strong Markov process, known as a *diffusion process*. Much of what was said earlier about the relationship between Brownian motion and partial differential equations can be generalized to diffusion processes as well. Roughly speaking, with each diffusion process one can associate a differential operator L, and this operator plays the role that the Laplacian plays for Brownian motion.

7   Random Trees

Brownian motion and more general diffusion processes appear as limits of many discrete models in probability theory, combinatorics, and statistical physics. The most striking recent example of this is given by the so-called *stochastic Loewner evolution* (commonly abbreviated to SLE) processes, which are discussed in [IV.25 §5]). These are expected to describe the asymptotic behavior of a large number of two-dimensional models, and their definition involves both linear Brownian motion and the *Loewner equation* from complex analysis. Rather than trying to give a general presentation of the relationship between Brownian motion and discrete models, in this final section we shall discuss a surprising application of Brownian motion to random trees, which can be used to describe the genealogy of a population.

The basic discrete model is the following. We start with a single "ancestor," which we label $\varnothing$. Then we place a probability distribution μ on the nonnegative integers, and use this to determine the number of children the ancestor has. Then each child is assumed to have children, the numbers of children being independent and also determined by the probability distribution μ. And so on. The case that we shall be interested in is the so-called *critical* case, where the expected number of children is exactly 1 (and the variance is finite).

We can represent the outcome of this process as a labeled tree, called the *genealogical tree*, in a natural way. To draw the tree one simply joins each member of the population to its children. As for the labels, the children of the original ancestor are labeled $1, 2, \ldots$, left to right, the children of 1 are labeled $(1, 1), (1, 2), \ldots$, the children of 2 are labeled $(2, 1), (2, 2), \ldots$, and so on. (For instance, the children of $(3, 4, 2)$, if it is ever born, are labeled $(3, 4, 2, 1), (3, 4, 2, 2), \ldots$.) See the left-hand side of figure 6 for a simple example of a tree. It is known that in this critical case the population will eventually die out with probability 1. (To avoid the certainty of this fate, the average number of children must be more than 1. A particular case of this process is discussed in [IV.25 §2].)

The genealogical tree, which we shall denote by θ, is a random variable. It is called the *Galton–Watson tree with offspring distribution* μ. A convenient way to represent this tree is via its so-called *contour function*, which is illustrated on the right-hand side of figure 6. Informally, we imagine the motion of a particle that starts from the root and explores the tree from the left to the

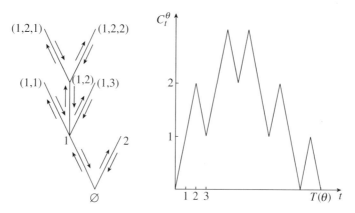

Figure 6 Left: a tree θ. Right: the contour function C^θ.

right, moving continuously along the edges at constant vertical speed (we set the height of each edge to 1), until it has completely explored the tree and come back to its starting point, after which it stays at this point. Since the particle will go along each edge exactly twice in this evolution, once upward and once downward, the total time $T(\theta)$ needed to explore the tree is twice the number of edges. The value C_t^θ of the contour function at time t is the height of the particle at time t. All this should be clear from figure 6.

It may be that a typical tree dies out fairly quickly. However, our goal is to understand the shape of the tree when it is "conditioned to be large." This is a bit like the difference between on the one hand picking a random person alive one thousand years ago and looking at the tree of all his or her descendants, and on the other hand looking at the tree of a random ancestor, alive one thousand years ago, of an individual who is alive today. In the latter case the tree is guaranteed to continue for many generations without dying out.

Suppose we condition on the event that the tree θ (or rather the population it represents) survives for n generations. We may now ask all sorts of questions about this genealogical tree. How many individuals are there in a given generation of the tree? If we pick two individuals in the same generation, how far do we typically have to go back in the tree to reach a common ancestor? Asymptotic answers to such questions are also of interest in computer science and in combinatorics.

We will condition on a slightly different event, namely the event that θ has exactly n edges. The conditioned tree is called θ^n. It is a random tree with n edges, so $T(\theta^n) = 2n$.

In the particular case where the probability $\mu(k)$ of having k children is $2^{-(k+1)}$, it is not hard to prove that the distribution of θ^n will actually be uniform over all trees with n edges. A famous theorem of Aldous gives the asymptotic behavior of the contour function C^{θ^n} as $n \to \infty$ for general offspring distributions, and it turns out to be very closely related to a linear Brownian motion.

Notice that it cannot *be* a Brownian motion because it exhibits some behavior that is very untypical: it begins and ends at zero and remains positive for all time. However, we can use Brownian motion in a simple way to define a notion called a *Brownian excursion*, for which the sample paths have the right shape. The rough idea is to start a linear Brownian motion at zero, draw its graph, and then pick out the part of the graph between $x = x_1$ and $x = x_2$, where x_1 is the point where it last crosses the x-axis before $x = 1$ and x_2 is the point where it first crosses the x-axis after $x = 1$. The corresponding portion of the Brownian motion will start and end at zero and not cross zero in between. We then need to rescale it so that x goes from 0 to 1 instead of from x_1 to x_2, and we also need to rescale the height appropriately, by dividing by $1/\sqrt{x_2 - x_1}$. Also, if the path is everywhere negative between x_1 and x_2, we simply turn it upside down to make it positive.

Aldous's theorem states that the limiting distribution of the contour function C^{θ^n} (rescaled in time by the factor $1/2n$ and in space by the factor $1/\sqrt{2n}$, like the rescaling in section 3) is a Brownian excursion. The surprising fact about this result is that it does not depend on the offspring distribution μ. Since the contour function completely determines the shape of the corresponding tree, we find that the limiting shape of a large critical Galton–Watson tree does not depend on the offspring distribution. This is another example of universality.

This result and variants of it provide a lot of useful information about the asymptotic behavior of large trees. Many interesting functions of the tree can be rewritten in terms of the contour function and by Aldous's theorem they will converge to similar functions of the Brownian excursion, whose distribution can be computed explicitly with the help of stochastic calculus. To give just one example, this technique can be used to calculate the limiting distribution of the height of the tree θ^n. Let the variance of the offspring distribution be σ, and let us define the *rescaled height* of a tree to be its original height multiplied by $\sigma/2\sqrt{n}$. The probability that this is at least x turns out to converge, as n gets large, to the quantity

$$2\sum_{k=1}^{\infty}(4x^2k^2 - 1)\exp(-2k^2x^2).$$

Acknowledgments. The author is indebted to Gilles Stoltz for his help with the simulations and to Gordon Slade for his remarks on the first version of this article.

Further Reading

Aldous, D. 1993. The continuum random tree. III. *Annals of Probability* 21:248–89.

Bachelier, L. 1900. Théorie de la spéculation. *Annales Scientifiques de l'École Normale Supérieure (3)* 17:21–86.

Billingsley, P. 1968. *Convergence of Probability Measures.* New York: John Wiley.

Durrett, R. 1984. *Brownian Motion and Martingales in Analysis.* Belmont, CA: Wadsworth.

Einstein, A. 1956. *Investigations on the Theory of the Brownian Movement.* New York: Dover.

Revuz, D., and M. Yor. 1991. *Continuous Martingales and Brownian Motion.* New York: Springer.

Strook, D. W., and S. R. S. Varadhan. 1979. *Multidimensional Diffusion Processes.* New York: Springer.

Wiener, N. 1923. Differential space. *Journal of Mathematical Physics Massachusetts Institute of Technology* 2:131–74.

IV.25 Probabilistic Models of Critical Phenomena
Gordon Slade

1 Critical Phenomena

1.1 Examples

A population can explode if its birth rate exceeds its death rate, but otherwise it becomes extinct. The nature of the population's evolution depends critically on which way the balance tips between adding new members and losing old ones.

A porous rock with randomly arranged microscopic pores has water spilled on top. If there are few pores, the water will not percolate through the rock, but if there are many pores, it will. Surprisingly, there is a critical degree of porosity that exactly separates these behaviors. If the rock's porosity is below the critical value, then water cannot flow completely through the rock, but if its porosity exceeds the critical value, even slightly, then water will percolate all the way through.

A block of iron placed in a magnetic field will become magnetized. If the magnetic field is extinguished, then the iron will remain magnetized if the temperature is below the Curie temperature 770 °C (1418 °F), but not if the temperature is above this critical value. It is striking that there is a specific temperature above which the magnetization of the iron does not merely remain small, but actually vanishes.

The above are three examples of *critical phenomena*. In each example, global properties of the system change abruptly as a relevant parameter (fertility, degree of porosity, or temperature) is varied through a critical value. For parameter values just below the critical value, the overall organization of the system is quite different from how it is for values just above. The sharpness of the transition is remarkable. How does it occur so suddenly?

1.2 Theory

The mathematical theory of critical phenomena is currently undergoing intense development. Intertwined with the science of *phase transitions*, it draws on ideas from probability theory and statistical physics. The theory is inherently probabilistic: each possible configuration of the system (e.g., a particular arrangement of pores in a rock, or of the magnetic states of the individual atoms in a block of iron) is assigned a probability, and the typical behavior of this ensemble of random configurations is analyzed as a function of parameters of the system (e.g., porosity or temperature).

The theory of critical phenomena is now guided to a large degree by a profound insight from physics known as *universality*, which, at present, is more of a philosophy than a mathematical theorem. The notion of universality refers to the fact that many essential features of the transition at a critical point depend on relatively few attributes of the system under consideration. In particular, simple mathematical models can capture some of the qualitative and quantitative features of critical behavior in real physical systems even if the models dramatically oversimplify the local interactions present in the real systems. This observation has helped to focus attention on particular mathematical models, among both physicists and mathematicians.

This essay discusses several models of critical phenomena that have attracted much attention from mathematicians, namely branching processes, the model of random networks known as the random graph, the percolation model, the Ising model of ferromagnetism, and the random cluster model. As well as having applications, these models are mathematically fascinating. Deep theorems have been proved, but many problems of central importance remain unsolved and tantalizing conjectures abound.

2 Branching Processes

Branching processes provide perhaps the simplest example of a phase transition. They occur naturally as a model of the random evolution of a population that changes in time as a result of births and deaths. The simplest branching process is defined as follows.

Consider an organism that lives for a unit time and that reproduces immediately before death. The organism has two potential offspring, which we can regard as the "left" offspring and the "right" offspring. At the moment of reproduction, the organism has either no offspring, a left but no right offspring, a right but no left offspring, or both a left and a right offspring. Assume that each of the potential offspring has a probability p of being born and that these two births occur independently. Here, the number p, which lies between 0 and 1, is a measure of the population's fecundity. Suppose that we start with a single organism at time zero, and that each descendant of this organism reproduces independently in the above manner.

A possible family tree is depicted in figure 1, showing all births that occurred. In this family tree, ten offspring were produced in all, but twelve potential offspring were not born, so the probability of this particular tree occurring is $p^{10}(1 - p)^{12}$.

If $p = 0$, then no offspring are born, and the family tree always consists of the original organism only. If $p = 1$, then all possible offspring are born, the family tree is the infinite binary tree, and the population always survives forever. For intermediate values of p, the population may or may not survive forever: let $\theta(p)$

Figure 1 A possible family tree, with probability $p^{10}(1 - p)^{12}$.

denote the *survival probability*, that is, the probability that the branching process survives forever when the fecundity is set at p. How does $\theta(p)$ interpolate between the two extremes $\theta(0) = 0$ and $\theta(1) = 1$?

2.1 The Critical Point

Since an organism has each of two potential offspring independently with probability p, it has, on average, $2p$ offspring. It is natural to suppose that survival for all time will not occur if $p < \frac{1}{2}$, since then each organism, on average, produces less than 1 offspring. On the other hand, if $p > \frac{1}{2}$, then, on average, organisms more than replace themselves, and it is plausible that a population explosion can lead to survival for all time.

Branching processes have a recursive nature, not present in other models, that facilitates explicit computation. Exploiting this, it is possible to show that the survival probability is given by

$$\theta(p) = \begin{cases} 0 & \text{if } p \leqslant \frac{1}{2}, \\ \dfrac{1}{p^2}(2p - 1) & \text{if } p \geqslant \frac{1}{2}. \end{cases}$$

The value $p = p_c = \frac{1}{2}$ is a critical value, at which the graph of $\theta(p)$ has a kink (see figure 2). The interval $p < p_c$ is referred to as *subcritical*, whereas $p > p_c$ is *supercritical*.

Rather than asking for the probability $\theta(p)$ that the initial organism has infinitely many descendants, one could ask for the probability $P_k(p)$ that the number of descendants is at least k. If there are at least $k + 1$ descendants, then there are certainly at least k, so $P_k(p)$ decreases as k increases. In the limit as k increases to infinity, $P_k(p)$ decreases to $\theta(p)$. In particular, when $p > p_c$, $P_k(p)$ approaches a positive limit as k approaches infinity, whereas $P_k(p)$ goes to zero when $p \leqslant p_c$. When p is strictly less than p_c, it can be

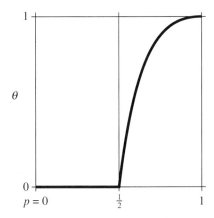

Figure 2 The survival probability θ versus p.

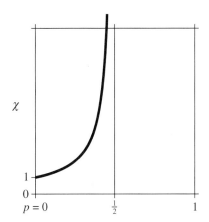

Figure 3 The average family size χ versus p.

shown that $P_k(p)$ goes to zero exponentially rapidly, but at the critical value itself we have

$$P_k(p_c) \sim \frac{2}{\sqrt{\pi k}}.$$

The symbol "$\sim$" denotes asymptotic behavior, and means that the ratio of the left- and right-hand sides in the above formula goes to 1 as k goes to infinity. In other words, $P_k(p_c)$ behaves essentially like $2/\sqrt{\pi k}$ when k is large.

There is a pronounced difference between the exponential decay of $P_k(p_c)$ for $p < p_c$ and the square-root decay at p_c. When $p = \frac{1}{4}$, family trees larger than 100 are sufficiently rare that in practical terms they do not occur: the probability is less than 10^{-14}. However, when $p = p_c$, roughly one in every ten trees will have size at least 100, and roughly one in a thousand will have size at least 1 000 000. At the critical value, the process is poised between extinction and survival.

Another important attribute of the branching process is the average size of a family tree, denoted $\chi(p)$. A calculation shows that

$$\chi(p) = \begin{cases} \dfrac{1}{1-2p} & \text{if } p < \frac{1}{2}, \\ \infty & \text{if } p \geqslant \frac{1}{2}. \end{cases}$$

In particular, the average family size becomes infinite at the same critical value $p_c = \frac{1}{2}$ above which the probability of an infinite family ceases to be zero. The graph of χ is shown in figure 3. At $p = p_c$, it may seem at first sight contradictory that family trees are always finite (since $\theta(p_c) = 0$) and yet the average family size is infinite (since $\chi(p_c) = \infty$). However, there is no inconsistency, and this combination, which occurs only at the critical point, reflects the slowness of the square-root decay of $P_k(p_c)$.

2.2   Critical Exponents and Universality

Some aspects of the above discussion are specific to twofold branching, and will change for a branching process with higher-order branching. For example, if each organism has not two but m potential offspring, again independently with probability p, then the average number of offspring per organism is mp and the critical probability p_c changes to $1/m$. Also, the formulas written above for the survival probability, for the probability of at least k descendants, and for the average family size must all be modified and will involve the parameter m.

However, the way that $\theta(p)$ goes to zero at the critical point, the way that $P_k(p_c)$ goes to zero as k goes to infinity, and the way that $\chi(p)$ diverges to infinity as p approaches the critical point p_c will all be governed by exponents that are independent of m. To be more specific, they behave in the following manner:

$$\theta(p) \sim C_1(p - p_c)^\beta, \quad \text{as } p \to p_c^+,$$
$$P_k(p_c) \sim C_2 k^{-1/\delta}, \quad \text{as } k \to \infty,$$
$$\chi(p) \sim C_3(p_c - p)^{-\gamma}, \quad \text{as } p \to p_c^-.$$

Here, the numbers C_1, C_2, and C_3 are constants that depend on m. By contrast, the exponents β, δ, and γ take on the same values for every $m \geqslant 2$. Indeed, those values are $\beta = 1$, $\delta = 2$, and $\gamma = 1$. They are called *critical exponents*, and they are *universal* in the sense that they do not depend on the precise form of the law that governs how the individual organisms reproduce. Related exponents will appear below in other models.

3   Random Graphs

An active research field in discrete mathematics with many applications is the study of objects known as GRAPHS [III.34]. These are used to model systems such as the Internet, the World Wide Web, and highway networks. Mathematically, a *graph* is a collection of *vertices* (which might represent computers, Web pages, or cities) joined in pairs by *edges* (physical connections between computers, hyperlinks between Web pages, highways). Graphs are also called *networks*, vertices are also called *nodes* or *sites*, and edges are also called *links* or *bonds*.

3.1   The Basic Model of a Random Graph

A major subarea of graph theory, initiated by Erdős and Rényi in 1960, concerns the properties that a graph typically has when it has been generated randomly. A natural way to do this is to take n vertices and for each pair to decide randomly (by the toss of a coin, say) whether it should be linked by an edge. More generally, one can choose a number p between 0 and 1 and let p be the probability that any given pair is linked. (This would correspond to using a biased coin to make the decisions.) The properties of random graphs come into their own when n is large, and of particular interest is the fact that there is a phase transition.

3.2   The Phase Transition

If x and y are vertices in a graph, then a *path* from x to y is a sequence of vertices that starts with x and ends with y in such a way that neighboring terms of the sequence are joined by edges. (If the vertices are represented by points and the edges by lines, then a path is a way of getting from x to y by traveling along the lines.) If x and y are joined by a path, then they are said to be *connected*. A *component*, or *connected cluster*, in a graph is what you obtain if you take a vertex together with all the other vertices that are connected to it.

Any graph decomposes naturally into its connected clusters. These will, in general, have different sizes (as measured by the number of vertices), and given a graph it is interesting to know the size of its largest cluster, which we shall denote by N. If we are considering a random graph with n vertices, then the value of N will depend on the multitude of random choices made when the graph was generated, and thus N is itself a random variable. The possible values of N are everything from 1, the value it takes when no edges are present and every cluster consists of a single vertex, to n, when there is just one connected cluster consisting of all the vertices. In particular, $N = 1$ when $p = 0$, and $N = n$ when $p = 1$. At a certain point between these extremes, N undergoes a dramatic jump.

It is possible to guess where the jump might take place, by considering the *degree* of a typical vertex x. This means the number of *neighbors* of x, that is, other vertices that are directly linked to x by a single edge. Each vertex has $n - 1$ potential neighbors, and for each one the probability that it is an actual neighbor is p, so the expected degree of any given vertex is $p(n - 1)$. When p is less than $1/(n - 1)$, each vertex has, on average, less than one neighbor, whereas when p exceeds $1/(n - 1)$, it has, again on average, more than one. This suggests that $p_c = 1/(n-1)$ will be a critical value, with N being small when p is below p_c, and large when p is above p_c.

This is indeed the case. If we set $p_c = 1/(n - 1)$ and write $p = p_c(1 + \varepsilon)$, with ε a fixed number between -1 and $+1$, then $\varepsilon = p(n - 1) - 1$. Since $p(n - 1)$ is the average degree of each vertex, ε is a measure of how much the average degree differs from 1. Erdős and Rényi showed that, in an appropriate sense, as n goes to infinity,

$$N \sim \begin{cases} 2\varepsilon^{-2} \log n & \text{if } \varepsilon < 0, \\ An^{2/3} & \text{if } \varepsilon = 0, \\ 2\varepsilon n & \text{if } \varepsilon > 0. \end{cases}$$

The A in the above formula is not a constant but a certain random variable that is independent of n (the distribution of which we have not specified here). When $\varepsilon = 0$ and n is large, the formula will tell us, for any $a < b$, the approximate probability that N lies between $an^{2/3}$ and $bn^{2/3}$. To put it another way, A is the *limiting distribution* of the quantity $n^{-2/3}N$ when $\varepsilon = 0$.

There is a marked difference between the behavior of the functions $\log n$, $n^{2/3}$, and n, for large n. The small clusters present for $p < p_c$ correspond to what is called a *subcritical phase*, whereas in the so-called *supercritical phase*, where $p > p_c$, there is a "giant cluster" whose size is of the same order of magnitude as the entire graph (see figure 4).

It is interesting to consider the "evolution" of the random graph, as p is increased from subcritical to supercritical values. (Here one can imagine more and more edges being randomly added to the graph.) A remarkable coalescence occurs, in which many smaller clusters

(a)

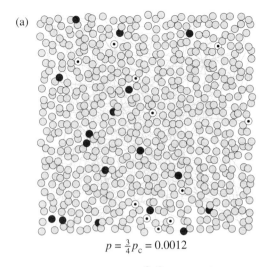

$$p = \tfrac{3}{4}p_{\mathrm c} = 0.0012$$

(b)

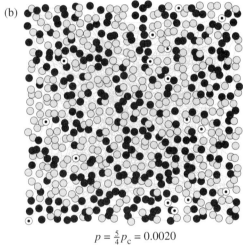

$$p = \tfrac{5}{4}p_{\mathrm c} = 0.0020$$

Figure 4 The largest cluster (black) and second largest cluster (dots) in random graphs with 625 vertices. These clusters have sizes (a) 17 and 11 and (b) 284 and 16. The hundreds of edges in the graphs are not clearly shown.

rapidly merge into a giant cluster whose size is proportional to the size of the entire system. The coalescence is thorough, in the sense that in the supercritical phase the giant cluster dominates everything: indeed, the second-largest cluster is known to have asymptotic size only $2\varepsilon^{-2}\log n$, which makes it far smaller than the giant cluster.

3.3 Cluster Size

For branching processes, we defined the quantity $\chi(p)$ to be the average size of the family tree spawned by an individual when the probability of each potential offspring being born was p. By analogy, for the random graph it is natural to take an arbitrary vertex v and define $\chi(p)$ to be the average size of the connected cluster containing v. Since all the vertices play identical roles, $\chi(p)$ is independent of the particular choice of v. If we fix a value of ε, set $p = p_{\mathrm c}(1 + \varepsilon)$, and let n tend to infinity, it turns out that the behavior of $\chi(p)$ is described by the formula

$$\chi(p) \sim \begin{cases} 1/|\varepsilon| & \text{if } \varepsilon < 0, \\ cn^{1/3} & \text{if } \varepsilon = 0, \\ 4\varepsilon^2 n & \text{if } \varepsilon > 0, \end{cases}$$

where c is a constant. Thus the expected cluster size is independent of n when $\varepsilon < 0$, grows like $n^{1/3}$ when $p = p_{\mathrm c}$, and is much larger—indeed, of the same order of magnitude n as the entire system—when $\varepsilon > 0$.

To continue the analogy with branching processes, let $P_k(p)$ denote the probability that the cluster containing the arbitrary vertex v consists of at least k vertices. Again this does not depend on the particular choice of v. In the subcritical phase, when $p = p_{\mathrm c}(1 + \varepsilon)$ for some fixed negative value of ε, the probability $P_k(p)$ is essentially independent of n and is exponentially small in k. Thus, large clusters are extremely rare. However, at the critical point $p = p_{\mathrm c}$, $P_k(p)$ decays like a multiple of $1/\sqrt{k}$ (for an appropriate range of k). This much slower square-root decay is similar to what happens for branching processes.

3.4 Other Thresholds

It is not only the largest cluster size that jumps. Another quantity that does so is the probability that a random graph is connected, meaning that there is a single connected cluster that contains all the n vertices. For what values of the edge-probability p is this likely? It is known that the property of being connected has a sharp threshold, at $p_{\mathrm{conn}} = (1/n)\log n$, in the following sense. If $p = p_{\mathrm{conn}}(1 + \varepsilon)$ for some fixed negative ε, then the probability that the graph is connected approaches 0 as $n \to \infty$. If on the other hand ε is positive, then the probability approaches 1. Roughly speaking, if you add edges randomly, then the graph suddenly changes from being almost certainly not connected to almost certainly connected as the proportion of edges present moves from just below p_{conn} to just above it.

There is a wide class of properties with thresholds of this sort. Other examples include the absence of any

   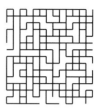

Figure 5 Bond-percolation configurations on a 14×14 piece of the square lattice $\mathbb{Z}^2$ for $p = 0.25$, $p = 0.45$, $p = 0.55$, $p = 0.75$. The critical value is $p_{\rm c} = \frac{1}{2}$.

isolated vertex (a vertex with no incident edge), and the presence of a Hamiltonian cycle (a closed loop that visits every vertex exactly once). Below the threshold, the random graph almost certainly does not have the property, whereas above the threshold it almost certainly does. The transition occurs abruptly.

4 Percolation

The percolation model was introduced by Broadbent and Hammersley in 1957 as a model of fluid flow in a porous medium. The medium contains a network of randomly arranged microscopic pores through which fluid can flow. A d-dimensional medium can be modeled with the help of the infinite d-dimensional lattice $\mathbb{Z}^d$, which consists of all points x of the form $(x_1, \ldots, x_d)$, where each x_i is an integer. This set can be made into a graph in a natural way if we join each point to the $2d$ points that differ from it by ± 1 in one coordinate and are the same in the others. (So, for example, in $\mathbb{Z}^2$ the neighbors of $(2, 3)$ are the four points $(1, 3)$, $(3, 3)$, $(2, 2)$, and $(2, 4)$.) One thinks of the edges as representing all pores potentially present in the medium.

To model the medium itself, one first chooses a *porosity parameter* p, which is a number between 0 and 1. Each edge (or bond) of the above graph is then retained with probability p and deleted with probability $1 - p$, with all choices independent. The retained edges are referred to as "occupied" and the deleted ones as "vacant." The result is a random subgraph of $\mathbb{Z}^d$ whose edges are the occupied bonds. These model the pores actually present in a macroscopic chunk of the medium.

For fluid to flow through the medium there must be a set of pores connected together on a macroscopic scale. This idea is captured in the model by the existence of an infinite cluster in the random subgraph, that is, a collection of infinitely many points all connected to one another. The basic question is whether or not an infinite cluster exists. If it does, then fluid can flow through

the medium on a macroscopic scale, and otherwise it cannot. Thus, when an infinite cluster exists, it is said that "percolation occurs."

Percolation on the square lattice $\mathbb{Z}^2$ is depicted in figure 5. Percolation in a three-dimensional physical medium is modeled using $\mathbb{Z}^3$. It is instructive, and mathematically interesting, to think how the model's behavior might change as the dimension d is varied.

For $d = 1$, percolation will not occur unless $p = 1$. The simple observation that leads to this conclusion is the following. Given any particular sequence of m consecutive edges, the probability that they are all occupied is p^m, and if $p < 1$, then this goes to zero as m goes to infinity. The situation is quite different for $d \geqslant 2$.

4.1 The Phase Transition

For $d \geqslant 2$, there is a phase transition. Let $\theta(p)$ denote the probability that any given vertex of $\mathbb{Z}^d$ is in an infinite connected cluster. (This probability does not depend on the choice of vertex.) It is known that for $d \geqslant 2$ there is a critical value $p_{\rm c}$, depending on d, such that $\theta(p)$ is zero if $p < p_{\rm c}$ and positive if $p > p_{\rm c}$. The exact value of $p_{\rm c}$ is not known in general, but a special symmetry of the square lattice allows for a proof that $p_{\rm c} = \frac{1}{2}$ when $d = 2$.

Using the fact that $\theta(p)$ is the probability that *any* particular vertex lies in an infinite cluster, it can be shown that when $\theta(p) > 0$ there must be an infinite connected cluster somewhere in $\mathbb{Z}^d$, while when $\theta(p) = 0$ there will not be one. Thus, percolation occurs when $p > p_{\rm c}$ but not when $p < p_{\rm c}$, and the system's behavior changes abruptly at the critical value. A deeper argument shows that when $p > p_{\rm c}$ there must be exactly one infinite cluster; infinite clusters cannot coexist on $\mathbb{Z}^d$. This is analogous to the situation in the random graph, where one giant cluster dominates when p is above the critical value.

Let $\chi(p)$ denote the average size of the connected cluster containing a given vertex. Certainly $\chi(p)$ is infinite for $p > p_c$, since then there is a positive probability that the given vertex is in an infinite cluster. It is conceivable that $\chi(p)$ could be infinite also for some values of p less than p_c, since infinite expectation is in principle compatible with $\theta(p) = 0$. However, it is a nontrivial and important theorem of the subject that this is not the case: $\chi(p)$ is finite for all $p < p_c$ and diverges to infinity as p approaches p_c from below.

Qualitatively, the graphs of θ and χ have the appearance depicted for the branching process in figures 2 and 3, although the critical value will be less than $\frac{1}{2}$ for $d \geqslant 3$. There is, however, a caveat. It has been proved that θ is continuous in p except possibly at p_c, and right-continuous for all p. It is widely believed that θ is equal to zero at the critical point, so that θ is continuous for all p and percolation does not occur at the critical point. But proofs that $\theta(p_c) = 0$ are currently known only for $d = 2$, for $d \geqslant 19$, and for certain related models when $d > 6$. The lack of a general proof is all the more intriguing since it has been proved for all $d \geqslant 2$ that there is zero probability of an infinite cluster in any half-space when $p = p_c$. This still allows for an infinite cluster with an unnatural spiral behavior, for example, though it is believed that this does not occur.

4.2 Critical Exponents

Assuming that $\theta(p)$ does in fact approach zero as p is decreased to p_c, it is natural to ask in what manner this occurs. Similarly, we can ask in what manner $\chi(p)$ diverges as p increases to p_c. Deep arguments of theoretical physics, and substantial numerical experimentation, have led to the prediction that this, as well as other, behavior is described by certain powers known as *critical exponents*. In particular, it is predicted that there are asymptotic formulas

$$\theta(p) \sim C(p - p_c)^\beta, \quad \text{as } p \to p_c^+,$$
$$\chi(p) \sim C(p_c - p)^{-\gamma}, \quad \text{as } p \to p_c^-.$$

The critical exponents here are the powers β and γ, which depend, in general, on the dimension d. (The letter C is used to denote a constant whose precise value is inessential and may change from line to line.)

When p is less than p_c, large clusters have exponentially small probabilities. For example, in this case the probability $P_k(p)$ that the size of the connected cluster containing any given vertex exceeds k is known to decay exponentially as $k \to \infty$. At the critical point, this exponential decay is predicted to be replaced by a power-law

decay involving a number δ, which is another critical exponent:

$$P_k(p_c) \sim Ck^{-1/\delta} \quad \text{as } k \to \infty.$$

Also, for $p < p_c$, the probability $\tau_p(x, y)$ that two vertices x and y are in the same connected cluster decays exponentially like $e^{-|x-y|/\xi(p)}$ as the separation between x and y is increased. The number $\xi(p)$ is called the *correlation length*. (Roughly speaking, $\tau_p(x, y)$ starts to become small when the distance between x and y exceeds $\xi(p)$.) The correlation length is known to diverge as p increases to p_c, and the predicted form of this divergence is

$$\xi(p) \sim C(p_c - p)^{-\nu} \quad \text{as } p \to p_c^-,$$

where ν is a further critical exponent. As before, the decay at the critical point is no longer exponential. It is predicted that $\tau_{p_c}(x, y)$ decays instead via a power law, traditionally written in the form

$$\tau_{p_c}(x, y) \sim C \frac{1}{|x - y|^{d-2+\eta}}, \quad \text{as } |x - y| \to \infty,$$

for yet another critical exponent η.

The critical exponents describe large-scale aspects of the phase transition and thus provide information relevant to the macroscopic scale of the physical medium. However, in most cases they have not been rigorously proved to exist. To do so, and to establish their values, is a major open problem in mathematics, one of central importance for percolation theory.

In view of this, it is important to be aware of a prediction from theoretical physics that the exponents are not independent, but are related to each other by what are called *scaling relations*. Three scaling relations are

$$\gamma = (2 - \eta)\nu, \quad \gamma + 2\beta = \beta(\delta + 1), \quad d\nu = \gamma + 2\beta.$$

4.3 Universality

Since the critical exponents describe large-scale behavior, it seems plausible that they might depend only weakly on changes in the fine structure of the model. In fact, it is a further prediction of theoretical physics, one that has been verified by numerical experiments, that the critical exponents are *universal*, in the sense that they depend on the spatial dimension d but on little else.

For example, if the two-dimensional lattice $\mathbb{Z}^2$ is replaced by another two-dimensional lattice, such as the triangular or the hexagonal lattice, then the values of the critical exponents are believed not to change. Another modification, for general $d \geqslant 2$, is to replace the standard percolation model with the so-called *spread-out model.* In the spread-out model, the edge set of $\mathbb{Z}^d$ is enriched so that now two vertices are joined whenever they are separated by a distance of L or less, where $L \geqslant 1$ is a fixed finite parameter, usually taken to be large. Universality suggests that the critical exponents for percolation in the spread-out model do not depend on the parameter L.

The discussion so far falls within the general framework of *bond percolation*, in which it is bonds (edges) that are randomly occupied or vacant. A much-studied variant is *site percolation*, where now it is vertices, or "sites," that are independently "occupied" with probability p and "vacant" with probability $1 - p$. The connected cluster of a vertex x consists of the vertex x itself together with those occupied vertices that can be reached by a path that starts at x, travels along edges in the graph, and visits only occupied vertices. For $d \geqslant 2$, site percolation also experiences a phase transition. Although the critical value for site percolation is different from the critical value for bond percolation, it is a prediction of universality that site and bond percolation on $\mathbb{Z}^d$ have the *same* critical exponents.

These predictions are mathematically very intriguing: the large-scale properties of the phase transition described by critical exponents appear to be insensitive to the fine details of the model, in contrast to features like the value of critical probability p_c, which depends heavily on such details.

At the time of writing, the critical exponents have been proved to exist, and their values rigorously computed, only for certain percolation models in dimensions $d = 2$ and $d > 6$, while a general mathematical understanding of universality remains an elusive goal.

4.4 Percolation in Dimensions $d > 6$

Using a method known as the *lace expansion*, it has been proved that the critical exponents exist, with values

$$\beta = 1, \quad \gamma = 1, \quad \delta = 2, \quad \nu = \tfrac{1}{2}, \quad \eta = 0,$$

for percolation in the spread-out model when $d > 6$ and L is large enough. The proof makes use of the fact that vertices in the spread-out model have many neighbors. For the more conventional nearest-neighbor model, where bonds have length 1 and there are fewer neighbors per vertex, results of this type have also been obtained, but only in dimensions $d \geqslant 19$.

The above values of β, γ, and δ are the same as those observed previously for branching processes. A branching process can be regarded as percolation on an infinite tree rather than on $\mathbb{Z}^d$, and thus percolation in dimensions $d > 6$ behaves like percolation on a tree. This is an extreme example of universality, in which the critical exponents are also independent of the dimension, at least when $d > 6$.

If the above values for the exponents are substituted into the scaling relation $d\nu = \gamma + 2\beta$, the result is $d = 6$. Thus, the scaling relation (called a *hyperscaling* relation because of the presence of the dimension d in the equation) is false for $d > 6$. However, this particular relation is predicted to apply only in dimensions $d \leqslant 6$. In lower dimensions, the nature of the phase transition is affected by the manner in which critical clusters fit into space, and the nature of the fit is partly described by the hyperscaling relation, in which d appears explicitly.

The critical exponents are predicted to take on different values below $d = 6$. Recent advances have shed much light on the situation for $d = 2$, as we shall see in the next section.

4.5 Percolation in Dimension 2

4.5.1 Critical Exponents and Schramm–Loewner Evolution

For site percolation on the two-dimensional triangular lattice it has been shown, in a major recent achievement, that the critical exponents exist and take the remarkable values

$$\beta = \tfrac{5}{36}, \quad \gamma = \tfrac{43}{18}, \quad \delta = \tfrac{91}{5}, \quad \nu = \tfrac{4}{3}, \quad \eta = \tfrac{5}{24}.$$

The scaling relations play an important role in the proof, but an essential additional step requires understanding of a concept known as the *scaling limit*.

To get some idea of what this is, let us look at the so-called *exploration process*, which is depicted in figure 6. In figure 6, hexagons represent vertices of the triangular lattice. Hexagons in the bottom row have been colored gray on the left half and white on the right half. The other hexagons have been chosen to be gray or white independently with probability $\tfrac{1}{2}$, which is the critical probability for site percolation on the triangular lattice. It is not hard to show that there is a path, also illustrated in figure 6, which starts at the bottom and all along its length is gray to the left and white to the right. The exploration process is this random path, which can be thought of as the gray/white interface. The boundary conditions at the bottom force it to be infinite.

The exploration process provides information about the boundaries separating large critical clusters of different color, and from this it is possible to extract

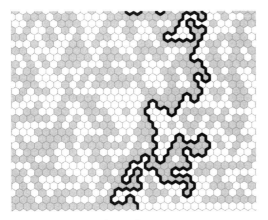

Figure 6 The exploration process.

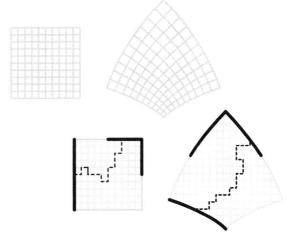

Figure 7 The two regions are related by a conformal transformation, depicted in the upper figures. In the lower figures, the limiting critical crossing probabilities are identical.

information about critical exponents. It is the macroscopic large-scale structure that is essential, so interest is focused on the exploration process in the limit as the spacing between vertices of the triangular lattice goes to zero. In other words, what does the curve in figure 6 typically look like in the limit as the size of the hexagons shrinks to zero? It is now known that this limit is described by a newly discovered STOCHASTIC PROCESS [IV.24 §1] called the Schramm–Loewner evolution (SLE) with parameter six, or SLE_6 for short. The SLE processes were introduced by Schramm in 2000, and have become a topic of intense current research activity.

This is a major step forward in the understanding of two-dimensional site percolation on the triangular lattice, but much remains to be done. In particular, it is still an unsolved problem to prove universality. There is currently no proof that critical exponents exist for bond percolation on the square lattice $\mathbb{Z}^2$, although universality predicts that the critical exponents for the square lattice should also take on the interesting values listed above.

4.5.2   Crossing Probabilities

In order to understand two-dimensional percolation, it is very helpful to understand the probability that there will be a path from one side of a region of the plane to another, especially when the parameter p takes its critical value p_c.

To make this idea precise, fix a simply connected region in the plane (i.e., a region with no holes), and fix two arcs on the boundary of the region. The *crossing probability* (which depends on p) is the probability that

there is an occupied path inside the region that joins one arc to the other, or more accurately the limit of this probability as the lattice spacing between vertices is reduced to zero. For $p < p_c$, clusters with diameter much larger than the correlation length $\xi(p)$ (measured by the number of steps in the lattice) are extremely rare. However, to cross the region, a cluster needs to be larger and larger as the lattice spacing goes to zero. It follows that the crossing probability is 0. When $p > p_c$, there is exactly one infinite cluster, from which it can be deduced that if the lattice spacing is very small, then with very high probability there will be a crossing of the region. In the limit, the crossing probability is 1. What if $p = p_c$? There are three remarkable predictions for critical crossing probabilities.

The first prediction is that critical crossing probabilities are universal, which is to say that they are the same for all finite-range two-dimensional bond- or site-percolation models. (As always, we are talking about the limiting probabilities as the lattice spacing goes to zero.)

The second prediction is that the critical crossing probabilities are *conformally invariant*. A conformal transformation is a transformation that locally preserves angles, as shown in figure 7. The remarkable RIEMANN MAPPING THEOREM [V.34] states that *any* two simply connected regions that are not the entire plane are related by a conformal transformation. The statement that the critical crossing probability is

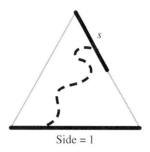

Figure 8 For the equilateral triangle of unit side length, Cardy's formula asserts that the limiting critical crossing probability shown is simply the length s.

conformally invariant means that if one region with two specified boundary arcs is mapped to another region by a conformal transformation, then the critical crossing probability between the images of the arcs in the new region is identical to the crossing probability of the original region. (Note that the underlying lattice is *not* transformed; this is what makes the prediction so striking.)

The third prediction is Cardy's explicit formula for critical crossing probabilities. Assuming conformal invariance, it is only necessary to give the formula for one region. For an equilateral triangle, Cardy's formula is particularly simple (see figure 8).

In 2001, in a celebrated achievement, Smirnov studied critical crossing probabilities for site percolation on the triangular lattice. Using the special symmetries of this particular model, Smirnov proved that the limiting critical crossing probabilities exist, that they are conformally invariant, and that they obey Cardy's formula. To prove universality of the crossing probabilities remains a tantalizing open problem.

5   The Ising Model

In 1925, Ising published an analysis of a mathematical model of ferromagnetism which now bears his name (although it was in fact Ising's doctoral supervisor Lenz who first defined the model). The Ising model occupies a central position in theoretical physics, and is of considerable mathematical interest.

5.1   Spins, Energy, and Temperature

In the Ising model, a block of iron is regarded as a collection of atoms whose positions are fixed in a crystalline lattice. Each atom has a magnetic "spin," which is assumed for simplicity to point upward or downward.

Each possible configuration of spins has an associated energy, and the greater this energy is, the less likely the configuration is to occur.

On the whole, atoms like to have the same spin as their immediate neighbors, and the energy reflects this: it increases according to the number of pairs of neighboring spins that are *not* aligned with each other. If there is an external magnetic field, also assumed to be directed up or down, then there is an additional contribution: atoms like to be aligned with the external field, and the energy is greater the more spins there are that are not aligned with it. Since configurations with higher energy are less likely, spins have a general tendency to align with each other, and also to align with the direction of the external magnetic field. When a larger fraction of spins points up than down, the iron is said to have a positive magnetization.

Although energy considerations favor configurations with many aligned spins, there is a competing effect. As the temperature increases, there are more random thermal fluctuations of the spins, and these diminish the amount of alignment. Whenever there is an external magnetic field, the energy effects predominate and there is at least some magnetization, however high the temperature. However, when the external field is turned off, the magnetization persists only if the temperature is below a certain critical temperature. Above this temperature, the iron will lose its magnetization.

The Ising model is a mathematical model that captures the above picture. The crystalline lattice is modeled by the lattice $\mathbb{Z}^d$. Vertices of $\mathbb{Z}^d$ represent atomic positions, and the atomic spin at a vertex x is simply modeled by one of the two numbers $+1$ (representing spin up) or -1 (representing spin down). The particular number chosen at x is denoted σ_x, and a collection of choices, one for each x in the lattice, is called a *configuration* of the Ising model. The configuration as a whole is denoted simply as σ. (Formally, a configuration σ is a function from the lattice to the set $\{-1, 1\}$.)

Each configuration σ comes with an associated energy, defined as follows. If there is no external field, the energy of σ consists of the sum, taken over all pairs of neighboring vertices $\langle x, y \rangle$, of the quantity $-\sigma_x \sigma_y$. This quantity is -1 if $\sigma_x = \sigma_y$, and is $+1$ otherwise, so the energy is indeed larger the more nonaligned pairs there are. If there is a nonzero external field, modeled by a real number h, then the energy receives an additional contribution $-h\sigma_x$, which is larger the more spins there are with a different sign from that of h. Thus, in total, the energy $E(\sigma)$ of a spin configuration

σ is defined by

$$E(\sigma) = - \sum_{\langle x,y \rangle} \sigma_x \sigma_y - h \sum_x \sigma_x,$$

where the first sum is over neighboring pairs of vertices, the second sum is over vertices, and h is a real number that may be positive, negative, or zero.

The sums defining $E(\sigma)$ actually make sense only when there are finitely many vertices, but one wishes to study the infinite lattice $\mathbb{Z}^d$. This problem is handled by restricting $\mathbb{Z}^d$ to a large finite subset and later taking an appropriate limit, the so-called *thermodynamic limit*. This is a well-understood process that will not be described here.

Two features remain to be modeled, namely, the manner in which lower-energy configurations are "preferred," and the manner in which thermal fluctuations can lessen this preference. Both features are handled simultaneously, as follows. We wish to assign to each configuration a probability that decreases as its energy increases. According to the foundations of statistical mechanics, the right way to do this is to make the probability proportional to the so-called *Boltzmann factor* $e^{-E(\sigma)/T}$, where T is a nonnegative parameter that represents the temperature. Thus, the probability is

$$P(\sigma) = \frac{1}{Z} e^{-E(\sigma)/T},$$

where the normalization constant, or *partition function*, Z, is defined by

$$Z = \sum_\sigma e^{-E(\sigma)/T},$$

where the sum is taken over all possible configurations σ (again it is necessary to work first in a finite subset of $\mathbb{Z}^d$ to make this precise). The reason for this choice of Z is that once we divide by it then we have ensured that the probabilities of the configurations add up to one, as they must. With this definition, the desired preference for low energy is achieved, since the probability of a given configuration is smaller when the energy of the configuration is larger. As for the effect of the temperature, note that when T is very large, all the numbers $e^{-E(\sigma)/T}$ are close to 1, so all probabilities are roughly equal. In general, as the temperature increases the probabilities of the various configurations become more similar, and this models the effect of random thermal fluctuations.

There is more to the story than energy, however. The Boltzmann factor makes any individual low-energy configuration much more likely than any individual high-energy configuration. However, the low-energy configurations have a high degree of alignment, so there are far fewer of them than there are of the more randomly arranged high-energy configurations. It is not obvious which of these two competing considerations will predominate, and in fact the answer depends on the value of the temperature T in a very interesting way.

5.2   The Phase Transition

For the Ising model with external field h and temperature T, let us choose a configuration randomly with the probabilities defined above. The *magnetization* $M(h, T)$ is defined to be the expected value of the spin σ_x at a given vertex x. Because of the symmetry of the lattice $\mathbb{Z}^d$, this does not depend on the particular vertex chosen. Accordingly, if the magnetization $M(h, T)$ is positive, then spins have an overall tendency to be aligned in the positive direction, and the system is magnetized.

The symmetry between up and down implies that $M(-h, T) = -M(h, T)$ (i.e., reversing the external field reverses the magnetization) for all h and T. In particular, when $h = 0$, the magnetization must be zero. On the other hand, if there is a nonzero external field h, then configurations with spins that are aligned with h are overwhelmingly more likely (because their energy is lower), and the magnetization satisfies

$$M(h, T) \begin{cases} < 0 & \text{if } h < 0, \\ = 0 & \text{if } h = 0, \\ > 0 & \text{if } h > 0. \end{cases}$$

What happens if the external field is initially positive and then is reduced to zero? In particular, is the *spontaneous magnetization*, defined by

$$M_+(T) = \lim_{h \to 0^+} M(h, T),$$

positive or zero? If $M_+(T)$ is positive, then the magnetization persists after the external field is turned off. In this case there will be a discontinuity in the graph of M versus h at $h = 0$.

Whether or not this happens depends on the temperature T. In the limit as T is reduced to zero, a small difference in the energies of two configurations results in an enormous difference in their probabilities. When $h > 0$ and the temperature is reduced to zero, only the minimal energy configuration, in which all spins are $+1$, has any chance of occurring. This is the case no matter how small the external field becomes, so $M_+(0) = 1$. On the other hand, in the limit of infinitely high temperature, all configurations become equally likely and the spontaneous magnetization is equal to zero.

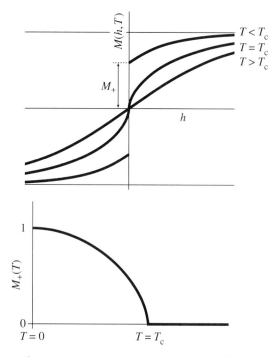

Figure 9 Magnetization versus external field, and spontaneous magnetization versus temperature.

For dimensions $d \geqslant 2$, the behavior of $M_+(T)$ when T lies between these two extremes is quite surprising. In particular, it is not differentiable everywhere: there is a critical temperature T_c, depending on the dimension, such that the spontaneous magnetization is strictly positive for $T < T_c$ and zero for $T > T_c$, and it is at $T = T_c$ that differentiability fails. Schematic graphs of the magnetization versus h and the spontaneous magnetization versus T are shown in figure 9. What happens at the critical temperature itself is delicate. In all dimensions except $d = 3$ it has been proved that there is no spontaneous magnetization at the critical temperature, which is to say that $M_+(T_c) = 0$. It is believed that this is true when $d = 3$ as well, but it remains an open problem to prove it.

5.3 Critical Exponents

The phase transition for the Ising model is again described by critical exponents. The critical exponent β, given by

$$M_+(T) \sim C(T_c - T)^\beta, \quad \text{as } T \to T_c^-,$$

indicates how the spontaneous magnetization disappears as the temperature increases toward the critical temperature T_c. For $T > T_c$, the *magnetic susceptibility*, denoted $\chi(T)$, is defined to be the rate of change of $M(h, T)$ with respect to h, at $h = 0$. This partial derivative in h diverges as T approaches T_c from above, and the exponent γ is defined by

$$\chi(T) \sim C(T - T_c)^{-\gamma}, \quad \text{as } T \to T_c^+.$$

Finally, δ describes the manner in which the magnetization goes to zero as the external field is reduced to zero at the critical temperature. That is,

$$M(h, T_c) \sim Ch^{1/\delta}, \quad \text{as } h \to 0^+.$$

These critical exponents, like those for percolation, are predicted to be universal and to obey various scaling relations. They are now understood mathematically in all dimensions except $d = 3$.

5.4 Exact Solution for $d = 2$

In 1944, Onsager published a famous paper in which he gave an exact solution of the two-dimensional Ising model. His remarkable computation is a landmark in the development of the theory of critical phenomena. With the exact solution as a starting point, critical exponents could be calculated. As with two-dimensional percolation, the exponents take interesting values:

$$\beta = \tfrac{1}{8}, \quad \gamma = \tfrac{7}{4}, \quad \delta = 15.$$

5.5 Mean-Field Theory for $d \geqslant 4$

Two modifications of the Ising model are relatively easy to analyze. One is to formulate the model on the infinite binary tree, rather than on the integer lattice $\mathbb{Z}^d$. Another is to formulate the Ising model on the so-called "complete graph," which is the graph consisting of n vertices with an edge joining *every* pair of vertices, and then take the limit as n goes to infinity. In the latter, known as the *Curie–Weiss model*, each spin interacts equally with all the other spins, or, put another way, each spin feels the *mean field* of all the other spins. In each of these modifications, the critical exponents take on the so-called mean-field values

$$\beta = \tfrac{1}{2}, \quad \gamma = 1, \quad \delta = 3.$$

Ingenious methods have been used to prove that the Ising model on $\mathbb{Z}^d$ has these same critical exponents in dimensions $d \geqslant 4$, although in dimension 4 there remain unresolved issues concerning logarithmic corrections to the asymptotic formulas.

6   The Random-Cluster Model

The percolation and Ising models appear to be quite different. A percolation configuration consists of a random subgraph of a given graph (usually a lattice as in the examples earlier), with edges included independently with probability p. A configuration of the Ising model consists of an assignment of values ± 1 to spins at the vertices of a graph (again usually a lattice), with these spins influenced by energy and temperature.

In spite of these differences, in around 1970 Fortuin and Kasteleyn had the insight to observe that the two models are in fact closely related to each other, as members of a larger family of models known as the random-cluster model. The random-cluster model also includes a natural extension of the Ising model known as the *Potts model*.

In the Potts model, spins at the vertices of a given graph G may take on any one of q different values, where q is an integer greater than or equal to 2. When $q = 2$ there are two possible spin values and the model is equivalent to the Ising model. For general q, it is convenient to label the possible spin values as $1, 2, \ldots, q$. As before, a configuration of spins has an associated energy that is smaller when more spins are aligned. The energy associated with an edge is -1 if the spins at the vertices joined by the edge are identical, and 0 otherwise. The total energy $E(\sigma)$ of a spin configuration σ, assuming no external field, is the sum of the energies associated with all edges. The probability of a particular spin configuration σ is again taken to be proportional to a Boltzmann factor, namely

$$P(\sigma) = \frac{1}{Z} e^{-E(\sigma)/T},$$

where the partition function Z is once again there to ensure that the probabilities add up to 1.

Fortuin and Kasteleyn noticed that the partition function of the Potts model on a finite graph G can be recast as

$$\sum_{S \subset G} p^{|S|} (1-p)^{|G \setminus S|} q^{n(S)}.$$

In this formula, the sum is over all subgraphs S that can be obtained by deleting edges from G, $|S|$ is the number of edges in S, $|G \setminus S|$ is the number of edges deleted from G to obtain S, $n(S)$ is the number of distinct connected clusters of S, and p is related to the temperature by

$$p = 1 - e^{-1/T}.$$

The restriction that q be an integer greater than or equal to 2 is essential for the definition of the Potts model, but the above sum makes good sense for any positive real value of q.

The *random-cluster model* has the above sum as its partition function. Given any real number $q > 0$, a configuration of the random-cluster model is a set S of occupied edges of the graph G, exactly like a configuration of bond percolation. However, in the random-cluster model we do not simply associate p with each occupied edge and $1 - p$ with each vacant edge. Instead, the probability associated with a configuration is proportional to $p^{|S|}(1-p)^{|G \setminus S|} q^{n(S)}$. In particular, for the choice $q = 1$, the random-cluster model is the same as bond percolation. Thus the random-cluster model provides a one-parameter family of models, indexed by q, which corresponds to percolation for $q = 1$, to the Ising model for $q = 2$, and to the Potts model for integer $q \geqslant 2$. The random-cluster model has a phase transition for general $q \geqslant 1$, and provides a unified setting and a rich family of examples.

7   Conclusion

The science of critical phenomena and phase transitions is a source of fascinating mathematical problems of real physical significance. Percolation is a central mathematical model in the subject. Often formulated on $\mathbb{Z}^d$, it can also be defined instead on a tree or on the complete graph, as a result of which it encompasses branching processes and the random graph. The Ising model is a fundamental model of the ferromagnetic phase transition. At first sight unrelated to percolation, it is in fact closely connected within the wider setting of the random-cluster model. The latter provides a unified framework and a powerful geometric representation for the Ising and Potts models.

Part of the fascination of these models is due to the prediction from theoretical physics that large-scale features near the critical point are universal. However, proofs often rely on specific details of a model, even when universality predicts that these details should not be essential to the results. For example, the understanding of critical crossing probabilities and the calculation of critical exponents has been carried out for site percolation on the triangular lattice, but not for bond percolation on $\mathbb{Z}^2$. Although the progress for the triangular lattice is a triumph of the theory, it is not the last word. Universality remains a guiding principle but it is not yet a general theorem.

In the physically most interesting case of dimension 3, a very basic feature of percolation and the Ising model is not understood at all: it has not yet been proved that there is no percolation at the critical point and that the spontaneous magnetization is zero.

Much has been accomplished but much remains to be done, and it seems clear that further investigation of models of critical phenomena will lead to highly important mathematical discoveries.

Acknowledgments. The figures were produced by Bill Casselman, Department of Mathematics, University of British Columbia, and Graphics Editor of *Notices of the American Mathematical Society*.

Further Reading

Grimmett, G. R. 1999. *Percolation*, 2nd edn. New York: Springer.

———. 2004. The random-cluster model. In *Probability on Discrete Structures*, edited by H. Kesten, pp. 73–124. New York: Springer.

Janson, S., T. Łuczak, and A. Ruciński. 2000. *Random Graphs*. New York: John Wiley.

Thompson, C. J. 1988. *Classical Equilibrium Statistical Mechanics*. Oxford: Oxford University Press.

Werner, W. 2004. Random planar curves and Schramm-Loewner evolutions. In *Lectures on Probability Theory and Statistics. École d'Eté de Probabilités de Saint-Flour XXXII—2002*, edited by J. Picard. Lecture Notes in Mathematics, volume 1840. New York: Springer.

IV.26 High-Dimensional Geometry and Its Probabilistic Analogues
Keith Ball

---

1 Introduction

If you have ever watched a child blowing soap bubbles, then you cannot have failed to notice that the bubbles are, at least as far as the human eye can tell, perfectly spherical. From a mathematical perspective, the reason for this is simple. The surface tension in the soap solution causes each bubble to make its area as small as possible, subject to the constraint that it encloses a fixed amount of air (and cannot compress the air too much). The sphere is the surface of smallest area that encloses a given volume.

As a mathematical principle, this seems to have been recognized by the ancient Greeks, although fully rigorous demonstrations did not appear until the end of the nineteenth century. This and similar statements are known as "isoperimetric principles."[1]

---

1. The prefix "iso" means "equal." The name "equal perimeter" refers to the two-dimensional formulation: if a disk and another region have equal perimeter, then the area of the other region cannot be larger than that of the disk.

Figure 1 A soap film has minimum area.

The two-dimensional form of the problem asks: what is the shortest curve that encloses a given area? The answer, as we might expect by analogy with the three-dimensional case, is a circle. Thus, by minimizing the length of the curve we force it to have a great deal of symmetry: the curve should be equally curved everywhere along its length. In three or more dimensions, many different kinds of CURVATURE [III.78] are used in different contexts. One, known as *mean curvature*, is the appropriate one for area-minimization problems.

The sphere has the same mean curvature at every point, but then it is pretty clear from its symmetry that the sphere would have the same curvature at every point whatever measure of curvature we used. More illustrative examples are provided by the soap films (much more varied than simple bubbles) that are a popular feature of recreational mathematics lectures: figure 1 shows such a soap film stretched across a wire frame. The film adopts the shape that minimizes its area, subject to the constraint that it is bounded by the wire frame. One can show that the minimal surface (the exact mathematical solution to the minimization problem) has constant mean curvature: its mean curvature is the same at every point.

Isoperimetric principles turn up all over mathematics: in the study of partial differential equations, the calculus of variations, harmonic analysis, computational algorithms, probability theory, and almost every branch of geometry. The aim of the first part of this article is to describe a branch of mathematics, high-dimensional geometry, whose starting point is the fundamental isoperimetric principle: that the sphere is the surface of least area that encloses a given volume. The most remarkable feature of high-dimensional geometry is its intimate connection to the theory of probability: geometric objects in high-dimensional space exhibit many of the characteristic properties of random distributions. The aim of the second part of this article is to outline the links between the geometry and probability.

2   High-Dimensional Spaces

So far we have discussed only two- and three-dimensional geometry. Higher-dimensional spaces seem to be impossible for humans to visualize but it is easy to provide a mathematical description of them by extending the usual description of three-dimensional space in terms of Cartesian coordinates. In three dimensions, a point (x, y, z) is given by three coordinates; in n-dimensional space, the points are n-tuples $(x_1, x_2, \ldots, x_n)$. As in two and three dimensions, the points are related to one another in that we can add two of them together to produce a third, by simply adding corresponding coordinates:

$$(2, 3, \ldots, 7) + (1, 5, \ldots, 2) = (3, 8, \ldots, 9).$$

By relating points to one another, addition gives the space some structure or "shape." The space is not just a jumble of unrelated points.

To describe the shape of the space completely, we also need to specify the distance between any two points. In two dimensions, the distance of a point (x, y) from the origin is $\sqrt{x^2 + y^2}$ by the Pythagorean theorem (and the fact that the axes are perpendicular). Similarly, the distance between two points (u, v) and (x, y) is

$$\sqrt{(x - u)^2 + (y - v)^2}.$$

In n dimensions we define the distance between points $(u_1, u_2, \ldots, u_n)$ and $(x_1, x_2, \ldots, x_n)$ to be

$$\sqrt{(x_1 - u_1)^2 + (x_2 - u_2)^2 + \cdots + (x_n - u_n)^2}.$$

Volume is defined in n-dimensional space roughly as follows. We start by defining a cube in n dimensions. The two- and three-dimensional cases, the square and the usual three-dimensional cube, are very familiar. The set of all points in the xy-plane whose coordinates are between 0 and 1 is a square of side 1 unit (as shown in figure 2), and, similarly, the set of all points (x, y, z) for which x, y, and z are all between 0 and 1 is a unit cube. In n-dimensional space the analogous cube consists of those points whose coordinates are all between 0 and 1. We stipulate that the unit cube has volume 1. Now, if we double the size of a plane figure, its area increases by a factor of 4. If we double a three-dimensional body, its volume increases by a factor of 8. In n-dimensional space, the volume scales as the nth power of size: so a cube of side t has volume t^n. To find the volume of a more general set we try to approximate it by covering it with little cubes whose total volume is as small as possible. The volume of the set is calculated as a limit of these approximate volumes.

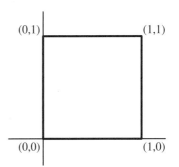

Figure 2 The unit square.

Whatever the dimension, a special geometric role is played by the *unit sphere*: that is, the surface consisting of all points that are a distance of 1 unit from a fixed point, the center. As one might expect, the corresponding solid sphere, or *unit ball*, consisting of all points enclosed by the unit sphere, also plays a special role. There is a simple relationship between the (n-dimensional) volume of the unit ball and the $(n - 1)$-dimensional "area" of the sphere. If we let v_n denote the volume of the unit ball in n dimensions, then the surface area is nv_n. One way to see this is to imagine enlarging the unit ball by a factor slightly greater than 1, say $1 + \varepsilon$. This is pictured in figure 3. The enlarged ball has volume $(1 + \varepsilon)^n v_n$ and so the volume of the shell between the two spheres is $((1 + \varepsilon)^n - 1)v_n$. Since the shell has thickness ε, this volume is approximately the surface area multiplied by ε. So the surface area is approximately

$$\frac{(1 + \varepsilon)^n - 1}{\varepsilon} v_n.$$

By taking the limit as ε approaches 0 we obtain the surface area exactly:

$$\lim_{\varepsilon \to 0} \frac{(1 + \varepsilon)^n - 1}{\varepsilon} v_n.$$

One can check that this limit is nv_n either by expanding the power $(1 + \varepsilon)^n$ or by observing that the expression is the formula for a derivative.

So far we have discussed bodies in n-dimensional space without being too precise about what kind of sets we are considering. Many of the statements in this article hold true for quite general sets. But a special role is played in high-dimensional geometry by convex sets (a set is convex if it contains the entire line segment joining any two of its points). Balls and cubes are both examples of convex sets. The next section describes a fundamental principle which holds for very general

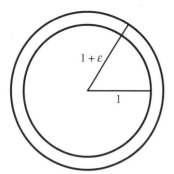

Figure 3 An inflated ball.

sets but which is intrinsically linked to the notion of convexity.

3   The Brunn–Minkowski Inequality

The two-dimensional isoperimetric principle was essentially proved in 1841 by Steiner, although there was a technical gap in the argument which was filled later. The general (n-dimensional) case was completed by the end of the nineteenth century. A couple of decades later a different approach to the principle, with far-reaching consequences, was found by HERMANN MINKOWSKI [VI.64]—an approach which was inspired by an idea of Hermann Brunn.

Minkowski considered the following way to add together two *sets* in n-dimensional space. If C and D are sets, then the sum $C + D$ consists of all points which can be obtained by adding a point of C to a point of D. Figure 4 shows an example in which C is an equilateral triangle and D is a square centered at the origin. We place a copy of the square at each point of the triangle (some of these are illustrated) and the set $C + D$ consists of all points that are included in all these squares. The outline of $C + D$ is shown dashed.

The Brunn–Minkowski inequality relates the volume of the sum of two sets to the volumes of the sets themselves. It states that (as long as the two sets C and D are not empty)

$$\operatorname{vol}(C + D)^{1/n} \geqslant \operatorname{vol}(C)^{1/n} + \operatorname{vol}(D)^{1/n}. \qquad (1)$$

The inequality looks a bit technical, if only because the volumes appearing in the inequality are raised to the power $1/n$. However, this fact is crucial. If each of C and D is a unit cube (with their edges aligned the same way), then the sum $C + D$ is a cube of side 2: a cube twice as large. Each of C and D has volume 1 while the volume of $C + D$ is 2^n. So, in this case, $\operatorname{vol}(C+D)^{1/n} = 2$

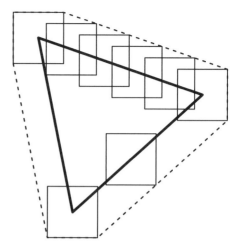

Figure 4 Adding two sets.

and each of $\operatorname{vol}(C)^{1/n}$ and $\operatorname{vol}(D)^{1/n}$ is equal to 1: the inequality (1) holds *with equality*. Similarly, whenever C and D are copies of one another, the Brunn–Minkowski inequality holds with equality. If we omitted the exponents $1/n$, the statement would still be true; in the case of two cubes, it is certainly true that $2^n \geqslant 1 + 1$. But the statement would be extremely weak: it would give us almost no useful information.

The importance of the Brunn–Minkowski inequality stems from the fact that it is the most fundamental principle relating volume to the operation of addition, which is the operation that gives space its structure. At the start of this section it was explained that Minkowski's formulation of Brunn's idea provided a new approach to the isoperimetric principle. Let us see why.

Let C be a COMPACT SET [III.9] in $\mathbb{R}^n$ whose volume is equal to that of the unit ball B. We want to show that the surface area of C is at least $n \operatorname{vol}(B)$ since this is the surface area of the ball. We consider what happens to C if we add a small ball to it. An example (a right-angled triangle) is shown in figure 5: the dashed curve outlines the enlarged set we obtain by adding to C a copy of the ball B scaled by a small factor ε. This looks rather like figure 3 above but here we do not expand the original set, we add a ball. Just as before, the difference between $C + \varepsilon B$ and C is a shell around C of width ε, so we can express the surface area as a limit as ε approaches 0:

$$\lim_{\varepsilon \to 0} \frac{\operatorname{vol}(C + \varepsilon B) - \operatorname{vol}(C)}{\varepsilon}.$$

Now the Brunn–Minkowski inequality tells us that

$$\operatorname{vol}(C + \varepsilon B)^{1/n} \geqslant \operatorname{vol}(C)^{1/n} + \operatorname{vol}(\varepsilon B)^{1/n}.$$

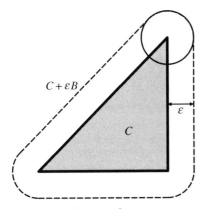

Figure 5 An ε-enlargement.

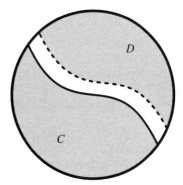

Figure 6 Expanding half a ball.

The right-hand side of this inequality is

$$\mathrm{vol}(C)^{1/n} + \varepsilon\,\mathrm{vol}(B)^{1/n} = (1+\varepsilon)\,\mathrm{vol}(B)^{1/n}$$

because $\mathrm{vol}(\varepsilon B) = \varepsilon^n \mathrm{vol}(B)$ and $\mathrm{vol}(C) = \mathrm{vol}(B)$. So the surface area is at least

$$\lim_{\varepsilon \to 0} \frac{(1+\varepsilon)^n \mathrm{vol}(B) - \mathrm{vol}(C)}{\varepsilon}$$
$$= \lim_{\varepsilon \to 0} \frac{(1+\varepsilon)^n \mathrm{vol}(B) - \mathrm{vol}(B)}{\varepsilon}.$$

Again as in section 2, this limit is $n\,\mathrm{vol}(B)$ and we conclude that the surface of C has at least this area.

Over the years, many different proofs of the Brunn–Minkowski inequality have been found, and most of the methods have other important applications. To finish this section we shall describe a modified version of the Brunn–Minkowski inequality that is often easier to use than (1). If we replace the set $C + D$ by a scaled copy half as large, $\frac{1}{2}(C + D)$, then its volume is scaled by $1/2^n$ and the nth root of this volume is scaled by $\frac{1}{2}$. Therefore, the inequality can be rewritten

$$\mathrm{vol}(\tfrac{1}{2}(C+D))^{1/n} \geqslant \tfrac{1}{2}\,\mathrm{vol}(C)^{1/n} + \tfrac{1}{2}\,\mathrm{vol}(D)^{1/n}.$$

Because of the simple inequality $\frac{1}{2}x + \frac{1}{2}y \geqslant \sqrt{xy}$ for positive numbers, the right-hand side of this inequality is at least $\sqrt{\mathrm{vol}(C)^{1/n}\,\mathrm{vol}(D)^{1/n}}$. It follows that

$$\mathrm{vol}(\tfrac{1}{2}(C+D))^{1/n} \geqslant \sqrt{\mathrm{vol}(C)^{1/n}\,\mathrm{vol}(D)^{1/n}}$$

and hence that

$$\mathrm{vol}(\tfrac{1}{2}(C+D)) \geqslant \sqrt{\mathrm{vol}(C)\,\mathrm{vol}(D)}. \qquad (2)$$

We shall elucidate a striking consequence of this inequality in the next section.

The Brunn–Minkowski inequality holds true for very general sets in n-dimensional space, but for convex sets it is the beginning of a surprising theory that was initiated by Minkowski and developed in a remarkable way by Aleksandrov, Fenchel, and Blaschke, among others: the theory of so-called mixed volumes. In the 1970s Khovanskii and Teissier (using a discovery of D. Bernstein) found an astonishing connection between the theory of mixed volumes and the Hodge index theorem in algebraic geometry.

4 Deviation in Geometry

Isoperimetric principles state that if a set is reasonably large, then it has a large surface or boundary. The Brunn–Minkowski inequality (and especially the argument we used to deduce the isoperimetric principle) expands upon this statement by showing that if we start with a reasonably large set and extend it (by adding a small ball), then the volume of the new set is quite a lot bigger than that of the original. During the 1930s Paul Lévy realized that in certain situations, this fact can have very striking consequences. To get an idea of how this works suppose that we have a compact set C inside the unit ball, whose volume is half that of the ball; for example, C might be the set pictured in figure 6.

Now extend the set C by including all points of the ball that are within distance ε of C, much as we did when deducing the isoperimetric inequality (the dashed curve in figure 6 shows the boundary of the extended set). Let D denote the remainder of the ball (also illustrated). Then if c is a point in C and d is a point in D, we are guaranteed that c and d are separated by a distance of at least ε. A simple two-dimensional argument, pictured in figure 7, shows that in this case the midpoint $\frac{1}{2}(c + d)$ cannot be too near the surface of the ball. In fact, its distance from the center is no more than $1 - \frac{1}{8}\varepsilon^2$. So the set $\frac{1}{2}(C + D)$ lies inside the ball of radius $1 - \frac{1}{8}\varepsilon^2$, whose volume is $(1 - \frac{1}{8}\varepsilon^2)^n$ times the

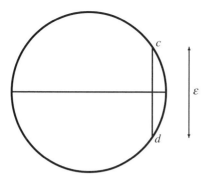

Figure 7 A two-dimensional argument.

volume of the ball v_n. The crucial point is that if the exponent n is large and ε is not too small, the factor $(1 - \frac{1}{8}\varepsilon^2)^n$ is extremely small: in a space of high dimension, a ball of slightly smaller radius has very much smaller volume. In order to make use of this we apply inequality (2), which states that the volume of $\frac{1}{2}(C + D)$ is at least $\sqrt{\mathrm{vol}(C)\,\mathrm{vol}(D)}$. Therefore,

$$\sqrt{\mathrm{vol}(C)\,\mathrm{vol}(D)} \leqslant (1 - \tfrac{1}{8}\varepsilon^2)^n v_n$$

or, equivalently,

$$\mathrm{vol}(C)\,\mathrm{vol}(D) \leqslant (1 - \tfrac{1}{8}\varepsilon^2)^{2n} v_n^2.$$

Since the volume of C is $\frac{1}{2}v_n$, we deduce that

$$\mathrm{vol}(D) \leqslant 2(1 - \tfrac{1}{8}\varepsilon^2)^{2n} v_n.$$

It is convenient to replace the factor $(1 - \frac{1}{8}\varepsilon^2)^{2n}$ by a (pretty accurate) approximation $\mathrm{e}^{-n\varepsilon^2/4}$, which is slightly easier to understand. We can then conclude that the volume $\mathrm{vol}(D)$ of the residual set D satisfies the inequality

$$\mathrm{vol}(D) \leqslant 2\mathrm{e}^{-n\varepsilon^2/4} v_n. \tag{3}$$

If the dimension n is large, then the exponential factor $\mathrm{e}^{-n\varepsilon^2/4}$ is very small, as long as ε is a bit bigger than $1/\sqrt{n}$. What this means is that only a small fraction of the ball lies in the residual set D. All but a small fraction of the ball lies close to C, even though *some* points in the ball may lie much farther from C. Thus, if we start with a set (any set) that occupies half the ball and extend it a little bit, we swallow up almost the entire ball. With a little more sophistication, the same argument can be used to show that the surface of the ball, the sphere, has exactly the same property. If a set C occupies half the sphere, then almost all of the sphere is close to that set.

This counterintuitive effect turns out to be characteristic of high-dimensional geometry. During the 1980s a startling probabilistic picture of high-dimensional

space was developed from Lévy's basic idea. This picture will be sketched in the next section.

One can see why the high-dimensional effect has a probabilistic aspect if one thinks about it in a slightly different way. To begin with, let us ask ourselves a basic question: what does it mean to choose a random number between 0 and 1? It could mean many things but if we want to specify one particular meaning, then our job is to decide what the chance is that the random number will fall into each possible range $a \leqslant x \leqslant b$: what is the chance that it lies between 0.12 and 0.47, for example? For most people, the obvious answer is 0.35, the difference between 0.47 and 0.12. The probability that our random number lands in the interval $a \leqslant x \leqslant b$ will just be $b - a$, the length of that interval. This way of choosing a random number is called *uniform*. Equal-sized parts of the range between 0 and 1 are equally likely to be selected.

Just as we can use length to describe what is meant by a random number, we can use the volume measure in n-dimensional space to say what it means to select a random point of the n-dimensional ball. We have to decide what the chance is that our random point falls into each subregion of the ball. The most natural choice is to say that it is equal to the volume of that subregion divided by the volume of the entire ball, that is, the proportion of the ball occupied by the subregion. With this choice of random point, it is possible to reformulate the high-dimensional effect in the following way. If we choose a subset C of the ball which has a $\frac{1}{2}$ chance of being hit by our random point, then the chance that our random point lies more than ε away from C is no more than $2\mathrm{e}^{-n\varepsilon^2/4}$.

To finish this section it will be useful to rephrase the geometric deviation principle as a statement about functions rather than sets. We know that if C is a set occupying half the sphere, then almost the entire sphere is within a small distance of C. Now suppose that f is a function defined on the sphere: f assigns a real number to each point of the sphere. Assume that f cannot change too rapidly as you move around the sphere: for example, that the values $f(x)$ and $f(y)$ at two points x and y cannot differ by more than the distance between x and y. Let M be the *median* value of f, meaning that f is at most M on half the sphere and at least M on the other half. Then it follows from the deviation principle that f must be almost equal to M on all but a small fraction of the sphere. The reason is that almost all of the sphere is close to the half where f is *below* M; so f cannot be much *more* than M except on a small set. On the other hand, almost all of the sphere is close to the half where f is *at least M*; so f cannot be much *less* than M except on a small set.

Thus, the geometric deviation principle says that if a function on the sphere does not vary too fast, then it must be almost constant on almost the entire sphere (even though there may be some points where it is very far from this constant value).

5 High-Dimensional Geometry

It was mentioned at the end of section 3 that convex sets have a special significance in Minkowski's theory relating volume to the additive structure of space. They also occur naturally in a large number of applications: in linear programming and partial differential equations, for example. Although convexity is a fairly restrictive condition for a body to satisfy, it is not hard to convince oneself that convex sets exhibit considerable variety and that this variety seems to increase with the dimension. The simplest convex sets after the balls are cubes. If the dimension is large, the surface of a cube looks very unlike the sphere. Let us consider, not a unit cube, but a cube of side 2 whose center is the origin. The corners of the cube are points like $(1, 1, \ldots, 1)$ or $(1, -1, -1, \ldots, 1)$, whose coordinates are all equal to 1 or -1, while the center of each *face* is a point like $(1, 0, 0, \ldots, 0)$ which has just one coordinate equal to 1 or -1. The corners are at a distance $\sqrt{n}$ from the center of the cube, while the centers of the faces are at distance 1 from the origin. Thus, the largest sphere that can be fitted inside the cube has radius 1, while the smallest sphere that encloses the cube has radius $\sqrt{n}$ (this is illustrated in figure 8).

When the dimension n is large, this ratio of $\sqrt{n}$ is also large. As one might expect, this gap between the ball and the cube is able to accommodate a wide variety of different convex shapes. Nevertheless, the probabilistic view of high-dimensional geometry has led to an understanding that, for many purposes, this enormous variety is an illusion: that in certain well-defined senses, all convex bodies behave like balls.

Probably the first discovery that pointed strongly in this direction was made by Dvoretzky in the late 1960s. *Dvoretzky's theorem* says that every high-dimensional convex body has slices that are almost spherical. More precisely, if you specify a dimension (say ten) and a degree of accuracy, then for any sufficiently large dimension n, every n-dimensional convex body has a ten-dimensional slice that is indistinguishable from a ten-dimensional sphere, up to the specified accuracy.

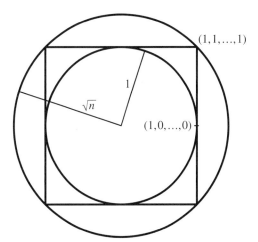

Figure 8 A ball in a box in a ball.

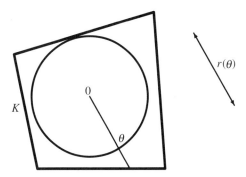

Figure 9 The directional radius.

The proof of Dvoretzky's theorem that is conceptually simplest depends upon the deviation principle described in the last section and was found by Milman a few years after Dvoretzky's theorem appeared. The idea is roughly this. Consider a convex body K in n dimensions that contains the unit ball. For each point θ on the sphere, imagine the line segment starting at the origin, passing through the sphere at θ, and extending out to the surface of K (see figure 9). Think of the length of this line as the "radius" of K in the direction of θ and call it $r(\theta)$. This "directional radius" is a function on the sphere. Our aim is to find (say) a 10-dimensional slice of the sphere on which $r(\theta)$ is almost constant. In such a slice, the body K looks like a ball, since its radius hardly varies.

The fact that K is convex means that the function r cannot change too rapidly as we move around the sphere: if two directions are close together, then the radius of K must be about the same in these two directions. Now we apply the geometric deviation principle to conclude that the radius of K is roughly the same on almost the entire sphere: the radius is close to its average (or median) value for all but a small fraction of the possible directions. That means that we have plenty of room in which to go looking for a slice on which the radius is almost constant—we just have to choose a slice that avoids the small bad regions. It can be shown that this happens if we choose the slice at random from among all possible slices. The fact that most of the sphere consists of good regions means that a random slice has a good chance of falling into a good region.

Dvoretzky's theorem can be recast as a statement about the behavior of the entire body K, rather than just its sections, by using the Minkowski sums defined in the previous section. The statement is that if K is a convex body in n dimensions, then there is a family of m rotations $K_1, K_2, \ldots, K_m$ of K whose Minkowski sum $K_1 + \cdots + K_m$ is approximately a ball, where the number m is significantly smaller than the dimension n. Recently, Milman and Schechtman realized that the smallest number m that would work could be described almost exactly, in terms of relatively simple properties of the body K, despite the apparently enormous complexity of the choice of rotations available.

For some n-dimensional convex sets, it is possible to create a ball with many fewer than n rotations. In the late 1970s Kašin discovered that if K is the cube, then just two rotations K_1 and K_2 are enough to produce something approximating a ball, even though the cube itself is extremely far from spherical. In two dimensions it is not hard to work out which rotations are best: if we choose K_1 to be a square and K_2 to be its rotation through $45°$, then $K_1 + K_2$ is a regular octagon which is as close to a circle as we can get with just two squares. In higher dimensions it is extremely hard to describe which rotations to use. At present the only known method is to use randomly chosen rotations, even though the cube is as concrete and explicit an object as one ever meets in mathematics.

The strongest principle discovered to date showing that most bodies behave like balls is what is usually called the reverse Brunn–Minkowski inequality. This result was proved by Milman, building on ideas of his own and of Pisier and Bourgain. The Brunn–Minkowski inequality was stated earlier for sums of bodies. The reverse one has a number of different versions; the simplest is in terms of intersections. To begin with, if K is a body and B is a ball of the same volume, then the intersection of these two sets, the region that they have in common, is clearly of smaller volume. This obvious fact can be stated in a complicated way that looks like the Brunn–Minkowski inequality:

$$\text{vol}(K \cap B)^{1/n} \leqslant \text{vol}(K)^{1/n}. \qquad (4)$$

If K is extremely long and thin, then whenever we intersect it with a ball of the same volume, we capture only a tiny part of K. So there is no possibility of reversing inequality (4) as it stands: no possibility of estimating the volume of $K \cap B$ from below. But if we are allowed to stretch the ball before intersecting it with K, the situation changes completely. A stretched ball in n-dimensional space is called an ellipsoid (in two dimensions it is just an ellipse). The reverse Brunn–Minkowski inequality states that for every convex body K, there is an ellipsoid $\mathcal{E}$ of the same volume for which

$$\text{vol}(K \cap \mathcal{E})^{1/n} \geqslant \alpha \text{vol}(K)^{1/n},$$

where α is a fixed positive number.

There is a widespread (but not quite universal) belief that an apparently much stronger principle is true: that if we are allowed to enlarge the ellipsoid by a factor of (say) 10, then we can ensure that it includes half the volume of K. In other words, for every convex body, there is an ellipsoid of roughly the same size that contains half of K. Such a statement flies in the face of our intuition about the huge variety of shapes in high dimensions, but there are some good reasons to believe it.

Since the Brunn–Minkowski inequality has a reverse form, it is natural to ask whether the isoperimetric inequality also does. The isoperimetric inequality guarantees that sets cannot have a surface that is too small. Is there a sense in which bodies cannot have too large a surface area? The answer is yes, and indeed a rather precise statement can be made. Just as in the case of the Brunn–Minkowski inequality, we have to take into account the possibility that our body could be long and thin and so have small volume but very large surface. So we have to start by applying a linear transformation that stretches the body in certain directions (but does not bend the shape). For example, if we start with a triangle, we first transform it into an *equilateral* triangle and then measure its surface and its volume. Once we have transformed our body as best we can, it turns out that we can specify precisely which convex body has the largest surface for a given volume. In two dimensions it is the triangle, in three it is the tetrahedron, and in n dimensions it is the natural analogue of these: the n-dimensional convex set (called a simplex) which has $n + 1$ corners. The fact that this set has the largest surface was proved by the present author using an inequality from harmonic analysis discovered by Brascamp and Lieb; the fact that the simplex is the only convex set with maximal surface (in the sense described) was proved by Barthe.

In addition to geometric deviation principles, two other methods have played a central role in the modern development of high-dimensional geometry; these methods grew out of two branches of probability theory. One is the study of sums of random points in NORMED SPACES [III.62] and how big they are, which provides important geometrical information about the spaces themselves. The other, the theory of Gaussian processes, depends upon a detailed understanding of how to cover sets in high-dimensional space efficiently with small balls. This issue may sound abstruse but it addresses a fundamental problem: how to measure (or estimate) the complexity of a geometric object. If we know that our object can be covered by one ball of radius 1, ten balls of radius $\frac{1}{2}$, fifty-seven balls of radius $\frac{1}{4}$, and so on, then we have a good idea of how complicated the object can be.

The modern view of high-dimensional space has revealed that it is at once much more complicated than was previously thought and at the same time in other ways much simpler. The first of these is well illustrated by the solution of a problem posed by Borsuk in the 1930s. A set is said to have diameter at most d if no two points in the set are further than d from each other. In connection with his work in topology, Borsuk asked whether every set of diameter 1 in n-dimensional space could be broken into $n + 1$ pieces of smaller diameter. In two and three dimensions this is always possible, and as late as the 1960s it was expected that the answer should be yes in all dimensions. However, a few years ago, Kahn and Kalai showed that in n dimensions it might require something like $e^{\sqrt{n}}$ pieces, enormously more than $n + 1$.

On the other hand, the simplicity of high-dimensional space is reflected in a fact discovered by Johnson and Lindenstrauss: if we pick a configuration of n points (in whatever dimension we like), we can find an almost perfect copy of the configuration sitting in a space of dimension much smaller than n: roughly the logarithm of n. In the last few years this fact has found applications in the design of computer algorithms, since many computational problems can be phrased geometrically and become much simpler if the dimension involved is small.

6 Deviation in Probability

If you toss a fair coin repeatedly, you expect that heads will occur on roughly half the tosses, and tails on roughly half. Moreover, as the number of tosses increases, you expect the proportion of heads to get closer and closer to $\frac{1}{2}$. The number $\frac{1}{2}$ is called the *expected number* of heads per toss. The number of heads yielded by a given toss is either 1 or 0, with equal probability, so the expected number of heads is the average of these, namely $\frac{1}{2}$.

The crucial unspoken assumption that we make about the tosses of the coin is that they are *independent*: that the outcomes of different tosses do not influence one another. (Independence and other basic probabilistic concepts are discussed in PROBABILITY DISTRIBUTIONS [III.71].) The coin-tossing principle, or its generalization to other random experiments, is called the *strong law of large numbers*. The average of a large number of independent repetitions of a random quantity will be close to the expected value of the quantity.

The strong law of large numbers for coin tosses is fairly simple to demonstrate. The general form, which applies to much more complicated random quantities, is considerably more difficult. It was first established by KOLMOGOROV [VI.88] in the early part of the twentieth century.

The fact that averages accumulate near the expected value is certainly useful to know, but for most purposes in statistics and probability theory it is vital to have more detailed information. If we focus our attention near the expected value, we may ask how the average is distributed around this number. For example, if the expected value is $\frac{1}{2}$, as for coin tossing, we might ask, what is the chance that the average is as large as 0.55 or as small as 0.42? We want to know how likely it is that our average number of heads will deviate from the expected value by a given amount.

The bar chart in figure 10 shows the probabilities of obtaining each of the possible numbers of heads, with twenty tosses of a coin. The height of each bar shows the chance that the corresponding number of heads will occur. As we would expect from the strong law of large numbers, the taller bars are concentrated near the middle. Superimposed upon the chart is a curve that plainly approximates the probabilities quite well. This is the famous "bell-shaped" or "normal" curve. It is a shifted and rescaled copy of

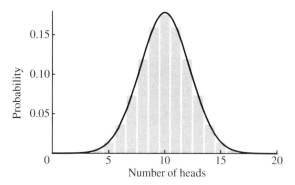

Figure 10 Twenty tosses of a fair coin.

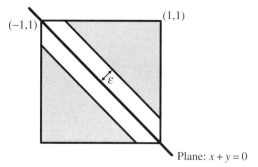

Figure 11 A random point of the cube.

the so-called *standard normal curve*, whose equation is

$$y = \frac{1}{\sqrt{2\pi}} \exp\left(-\tfrac{1}{2}x^2\right). \tag{5}$$

The fact that the curve approximates coin-tossing probabilities is an example of the most important principle in probability theory: the *central limit theorem*. This states that whenever we add up a large number of small independent random quantities, the result has a distribution that is approximated by a normal curve.

The equation of the normal curve (5) can be used to show that if we toss a coin n times, then the chance that the proportion of heads deviates from $\tfrac{1}{2}$ by more than ε is at most $e^{-2n\varepsilon^2}$. This closely resembles the geometric deviation estimate (3) from section 4. This resemblance is not coincidental, although we are still far from a full understanding of when and how it applies.

The simplest way to see why a version of the central limit theorem might apply to geometry is to replace the toss of a coin by a different random experiment. Suppose that we repeatedly select a random number between -1 and 1, and that the selection is *uniform* in the sense described in section 4. Let the first n selections be the numbers $x_1, x_2, \ldots, x_n$. Instead of thinking of them as independent random choices, we can consider the point $(x_1, \ldots, x_n)$ as a randomly chosen point inside the cube that consists of all points whose coordinates lie between -1 and 1. The expression $(1/\sqrt{n}) \sum_{i=1}^n x_i$ measures the distance of the random point from a certain $(n-1)$-dimensional "plane," which consists of all points whose coordinates add up to zero (the two-dimensional case is shown in figure 11). So the chance that $(1/\sqrt{n}) \sum_{i=1}^n x_i$ deviates from its expected value, 0, by more than ε is the same as the chance that a random point of the cube lies a distance of more than ε from the plane. This *chance* is proportional to the *volume* of the set of points that are

more than ε from the plane: the set shown shaded in figure 11. When we discussed the geometric deviation principle, we estimated the volume of the set of points which were more than ε away from a set C which occupied half the ball. The present situation is really the same, because each part of the shaded set consists of those points that are more than ε away from whichever half of the cube lies on the other side of the plane.

Arguments akin to the central limit theorem show that if we cut the cube in half with a plane, then the set of points which lie more than a distance ε from one of the halves has volume no more than $e^{-\varepsilon^2}$. This statement is different from, and apparently much weaker than, the one we obtained for the ball (3) because the factor of n is missing from the exponent. The estimate implies that if you take any plane through the center of the cube, then most points in the cube will be at a distance of less than 2 from it. If the plane is parallel to one of the faces of the cube, this statement certainly is weak, because *all* of the cube is within distance 1 of the plane. The statement becomes significant when we consider planes like the one in figure 11. Some points of the cube are at a distance of $\sqrt{n}$ from this "diagonal" plane, but still, the overwhelming majority of the cube is very much closer. Thus, the estimates for the cube and the ball contain essentially the same information; what is different is that the cube is bigger than the ball by a factor of about $\sqrt{n}$.

In the case of the ball we were able to prove a deviation estimate for *any* set occupying half the ball, not just the special sets that are cut off by planes. Towards the end of the 1980s Pisier found an elegant argument that showed that the general case works for the cube as well as for the ball. Among other things, the argument uses a principle which goes back to the early days of large-deviation theory in the work of Donsker and Varadhan.

The theory of large deviations in probability is now highly developed. In principle, more or less precise estimates are known for the probability that a sum of independent random variables deviates from its expectation by a given amount, in terms of the original distribution of the variables. In practice, the estimates involve quantities that may be difficult to compute, but there are sophisticated methods for doing this. The theory has numerous applications within probability and statistics, computer science, and statistical physics.

One of the most subtle and powerful discoveries of this theory is Talagrand's deviation inequality for product spaces, discovered in the mid 1990s. Talagrand himself has used this to solve several famous problems in combinatorial probability and to obtain striking estimates for certain mathematical models in particle physics. The full inequality of Talagrand is somewhat technical and is difficult to describe geometrically. However, the discovery had a precursor which fits perfectly into the geometric picture and which captures at least one of the most important ideas.[2] We look again at random points in the cube but this time the random point is not chosen uniformly from within the cube. As before, we choose the coordinates $x_1, x_2, \ldots, x_n$ of our random point *independently* of one another, but we do not insist that each coordinate is chosen *uniformly* from the range between -1 and 1. For example, it might be that x_1 can take only the values $1, 0$, or -1, each with probability $\frac{1}{3}$, that x_2 can take only the values 1 or -1 each with probability $\frac{1}{2}$, and perhaps that x_3 is chosen uniformly from the entire range between -1 and 1. What matters is that the choice of each coordinate has no effect on the choice of any others.

Any sequence of rules that dictates how we choose each coordinate determines a way of choosing a random point in the cube. This in turn gives us a way of measuring a kind of volume for subsets of the cube: the "volume" of a set A is the chance that our random point is selected from A. This way to measure volume might be very different from the usual one; among other things, an individual point might have nonzero volume.

Now suppose that C is a *convex* subset of the cube and that its "volume" is $\frac{1}{2}$, in the sense that our random point will be selected from C with probability $\frac{1}{2}$.

Talagrand's inequality says that the chance that our random point will lie a distance of more than ε from C is less than $2e^{-\varepsilon^2/16}$. This statement looks like the deviation estimate for the cube except that it refers only to convex sets C. But the crucial *new* information that makes the estimate and its later versions important is that we are allowed to choose our random point in so many different ways.

This section has described deviation estimates in probability theory that have a geometric flavor. For the cube, we are able to show that if C is any set occupying half the cube, then almost the entire cube is close to C. It would be extremely useful to know the same thing for convex sets more general than the cube. There are some other highly symmetric sets for which we do know it, but the most general possible statement of this type seems to be beyond our current methods. One potential application, which comes from theoretical computer science, is to the analysis of random algorithms for volume calculation. The problem may sound specialized, but it arises in LINEAR PROGRAMMING [III.84] (which alone is sufficient reason to justify the expenditure of enormous effort) and in the numerical estimation of integrals. In principle, one can calculate the volume of a set by laying over it a very fine grid, and counting how many grid points fall into the set. In practice, if the dimension is large, the number of grid points will be so astronomically huge that no computer has a chance of performing the count.

The problem of calculating the volume of a set is essentially the same as the problem of choosing a point at random within the set, roughly as we saw in section 4. So the aim is to select a random point without identifying a huge number of possible points to select from. At present, the most effective way of generating a random point in a convex set is to carry out a random walk within the set. We perform a sequence of small steps whose directions are chosen randomly and then select the point that we have reached after a fairly large number of steps, in the hope that this point has roughly the correct chance of falling into each part of the set. For the method to be effective, it is essential that the random walk quickly visits points all over the set: that it does not get stuck for a long time in, say, half of the set. In order to guarantee this *rapid mixing*, as it is called, we need an isoperimetric principle or deviation principle. We need to know that each half of our set has a large boundary, so that there is a good chance that our random walk will cross the boundary quickly and land in the other half of our set.

---

2. This precursor evolved from an original argument of Talagrand via an important contribution of Johnson and Schechtman.

In a series of papers published over the last ten years, Applegate, Bubley, Dyer, Frieze, Jerrum, Kannan, Lovasz, Montenegro, Simonovits, Vempala, and others have found very efficient random walks for sampling from a convex set. A geometric deviation principle of the kind alluded to above would make it possible to estimate the efficiency of these random walks almost perfectly.

7 Conclusion

The study of high-dimensional systems has become increasingly important in the last few decades. Practical problems in computing frequently lead to high-dimensional questions, many of which can be posed geometrically, while many models in particle physics are automatically high-dimensional because it is necessary to consider a huge number of particles in order to mimic large-scale phenomena in the real world. The literature in both these fields is vast but some general remarks can be made. The intuition that we gain from low-dimensional geometry leads us wildly astray if we try to apply it in many dimensions. It has become clear that naturally occurring high-dimensional systems exhibit characteristics that we expect to arise in probability theory, even if the original system does not have an explicitly random element. In many cases these random characteristics are manifested as an isoperimetric or deviation principle, that is, a statement to the effect that large sets have large boundaries. In the clas-sical theory of probability, independence assumptions can often be used to demonstrate deviation principles quite simply. For the very much more complicated systems that are studied today it is usually useful to have a geometric picture to accompany the probabilistic one. That way one can understand probabilistic deviation principles as analogues of the isoperimetric principle discovered by the ancient Greeks. This article has described the relationship between geometry and probability in just a few special cases. A very much more detailed picture is almost certainly waiting to be found. At present it seems to be just out of reach.

Further Reading

Ball, K. M. 1997. An elementary introduction to modern convex geometry. In *Flavors of Geometry*, edited by Silvio Levy. Cambridge: Cambridge University Press.

Bollobás, B. 1997. Volume estimates and rapid mixing. In *Flavors of Geometry*, edited by Silvio Levy. Cambridge: Cambridge University Press.

Chavel, I. 2001. *Isoperimetric Inequalities*. Cambridge: Cambridge University Press.

Dembo, A., and O. Zeitouni. 1998. *Large Deviations Techniques and Applications*. New York: Springer.

Ledoux, M. 2001. *The Concentration of Measure Phenomenon*. Providence, RI: American Mathematical Society.

Osserman, R. 1978. The isoperimetric inequality. *Bulletin of the American Mathematical Society* 84:1182–238.

Pisier, G. 1989. *The Volume of Convex Bodies and Banach Space Geometry*. Cambridge: Cambridge University Press.

Schneider, R. 1993. *Convex Bodies: The Brunn–Minkowski Theory*. Cambridge: Cambridge University Press.

Part V
Theorems and Problems

V.1 The ABC Conjecture

The ABC conjecture, proposed by Masser and Oesterlé in 1985, is a bold and very general conjecture in number theory with a wide range of important consequences. The rough idea of the conjecture is that it is impossible for one number to be the sum of two others if all three numbers have many repeated prime factors and no two have a prime factor in common (which would then have to be shared by the third).

More precisely, one defines the *radical* of a positive integer n to be the product of all primes that divide n, with each distinct prime included just once. For instance, $3960 = 2^3 \times 3^2 \times 5 \times 11$, so its radical is $2 \times 3 \times 5 \times 11 = 330$. Let us write $\mathrm{rad}(n)$ for the radical of n. The ABC conjecture asserts that for every positive real number ϵ there is a constant K_ϵ such that if a, b, and c are coprime integers and $a + b = c$, then $c < K_\epsilon \mathrm{rad}(abc)^{1+\epsilon}$.

To get a feel for the meaning of this conjecture, consider the Fermat equation $x^r + y^r = z^r$. If three positive integers x, y, and z solve the equation, then we can divide through by any common factors they might have and obtain a solution for which x, y, and z, and hence their rth powers, are coprime. Set $a = x^r$, $b = y^r$, and $c = z^r$. Then

$$\mathrm{rad}(abc) = \mathrm{rad}(xyz) \leqslant xyz = (abc)^{1/r} \leqslant c^{3/r},$$

where the last inequality follows from the fact that c is greater than both a or b. If we set ϵ to be $\frac{1}{6}$, then the ABC conjecture gives us a constant K such that c cannot be more than $K(c^{3/r})^{7/6} = Kc^{7/2r}$. If $r \geqslant 4$ then the power $7/2r$ is less than 1, so the Fermat equation can have at most finitely many solutions with x, y, and z coprime.

It is clear that this is just one of a huge number of consequences of a similar kind. For instance, we could deduce that there are only finitely many solutions of the equation $2^r + 3^s = x^2$, since the radical of $2^r 3^s x^2$ is at most $6x$, which is considerably smaller than x^2. But the ABC conjecture has other consequences that are less obvious, and more important, than this one. For instance, Bombieri has shown that the ABC conjecture implies ROTH'S THEOREM [V.22], Elkies has shown that it implies the MORDELL CONJECTURE [V.29], and Granville and Stark have shown that a strengthening of the ABC conjecture implies the nonexistence of Siegel zeros (these are defined in ANALYTIC NUMBER THEORY [IV.2]). It is also equivalent to strong forms, as yet unproven, of a famous theorem of Baker in transcendence theory, and of the theorem of Wiles about MODULAR FORMS [III.59] that implies Fermat's last theorem.

The ABC conjecture is discussed further in COMPUTATIONAL NUMBER THEORY [IV.3].

V.2 The Atiyah–Singer Index Theorem
Nigel Higson and John Roe

1 Elliptic Equations

The Atiyah–Singer index theorem is concerned with the existence and uniqueness of solutions to linear partial differential equations of *elliptic type*. To understand this concept, consider the two equations

$$\frac{\partial f}{\partial x} + \frac{\partial f}{\partial y} = 0 \quad \text{and} \quad \frac{\partial f}{\partial x} + \mathrm{i}\frac{\partial f}{\partial y} = 0.$$

They differ only by the factor $\mathrm{i} = \sqrt{-1}$, but their solutions nevertheless have very different properties. Any function of the form $f(x, y) = g(x - y)$ is a solution to the first equation, but in the analogous general solution $g(x + \mathrm{i}y)$ of the second equation, g must be a HOLOMORPHIC FUNCTION [I.3 §5.6] of the *complex* variable $z = x + \mathrm{i}y$, and it was already known in the nineteenth century that such functions are very special. For example, the first equation has an infinite-dimensional set of bounded solutions, but LIOUVILLE'S THEOREM [I.3 §5.6] in complex analysis asserts that the only bounded solutions of the second equation are the constant functions.

The differences between the solutions of the two equations can be traced to the differences between the *symbols* of the equations, which are the polynomials in real variables ξ, η that are obtained by substituting $i\xi$ for $\partial/\partial x$ and $i\eta$ for $\partial/\partial y$. Thus the symbols of the two equations above are

$$i\xi + i\eta \quad \text{and} \quad i\xi - \eta,$$

respectively. An equation is said to be *elliptic* if its symbol is zero only when $\xi = \eta = 0$; thus, the second equation is elliptic but the first is not. The fundamental *regularity theorem*, which is proved using FOURIER ANALYSIS [III.27], states that an elliptic partial differential equation (subject to suitable boundary conditions, if needed) has a finite-dimensional solution space.

2 Topology of Elliptic Equations and the Fredholm Index

Consider now the general first-order linear partial differential equation

$$a_1 \frac{\partial f}{\partial x_1} + \cdots + a_n \frac{\partial f}{\partial x_n} + bf = 0,$$

in which f is a vector-valued function and the coefficients a_j and b are complex matrix-valued functions. It is *elliptic* if its *symbol*

$$i\xi_1 a_1(x) + \cdots + i\xi_n a_n(x)$$

is an invertible matrix for every nonzero vector $\xi = (\xi_1, \ldots, \xi_n)$ and every x. The regularity theorem applies in this generality, and it allows us to form the *Fredholm index* of an elliptic equation (with suitable boundary conditions), which is the number of linearly independent solutions of the equation minus the number of linearly independent solutions of the *adjoint equation*

$$-\frac{\partial}{\partial x_1}(a_1^* f) - \cdots - \frac{\partial}{\partial x_n}(a_n^* f) + b^* f = 0.$$

The reason for introducing the Fredholm index is that it is a *topological invariant* of elliptic equations. This means that continuous variations in the coefficients of an elliptic equation leave the Fredholm index unchanged. (By contrast, the number of linearly independent solutions of an equation can vary as the coefficients of the equation vary.) The Fredholm index is therefore constant on each connected component of the set of all elliptic equations, and this raises the prospect of using topology to determine the structure of the set of all elliptic equations as an aid to computing the Fredholm index. This observation was made by Gelfand in the 1950s. It lies at the root of the Atiyah–Singer index theorem.

3 An Example

To see in more detail how topology can be used to determine the Fredholm index of an elliptic equation, let us look at a specific example. Consider elliptic equations for which the coefficients $a_j(x)$ and $b(x)$ are *polynomial* functions of x, with a_j of degree $m - 1$ or less and b of degree m or less. The expression

$$i\xi_1 a_1(x) + \cdots + i\xi_n a_n(x) + b(x)$$

is then a polynomial in both x and ξ of degree m or less. Let us strengthen the hypothesis of ellipticity by assuming that the terms in this expression that have degree exactly m (jointly in x and ξ) define an invertible matrix whenever *either x or ξ* is nonzero. Let us also agree to consider only solutions f of the equation or its adjoint that are *square-integrable*, which means that

$$\int |f(x)|^2 \, \mathrm{d}x < \infty.$$

All these extra hypotheses are types of boundary conditions (the behaviors of the equation and its solutions at infinity are controlled), and collectively they imply that the Fredholm index is well-defined.

A simple example is the equation

$$\frac{\mathrm{d}f}{\mathrm{d}x} + xf = 0. \tag{1}$$

The general solution to this ordinary differential equation is the one-dimensional space of multiples of the square-integrable function $e^{-x^2/2}$. By contrast, the solutions of the adjoint equation

$$-\frac{\mathrm{d}f}{\mathrm{d}x} + xf = 0$$

are multiples of the function $e^{+x^2/2}$, which is not square-integrable. Thus the index of this differential equation is equal to 1.

Returning to the general equation, the terms of degree m in

$$i\xi_1 a_1(x) + \cdots + i\xi_n a_n(x) + b(x)$$

determine a map from the unit sphere in (x, ξ)-space to the set $\mathrm{GL}_k(\mathbb{C})$ of invertible $k \times k$ complex matrices. Moreover, every such map comes from an elliptic equation (possibly of a more general type than we have discussed up to now, but an equation to which the basic regularity theorem guaranteeing the existence of the Fredholm index applies). It therefore becomes important to determine the topological structure of the space of all maps from the sphere S^{2n-1} into $\mathrm{GL}_k(\mathbb{C})$.

A remarkable theorem of Bott provides the answer. The *Bott periodicity theorem* associates an integer, which we shall call the *Bott invariant*, with each map

$S^{2n-1} \to \mathrm{GL}_k(\mathbb{C})$. Furthermore, Bott's theorem asserts that, provided that $k \geqslant n$, one such map can be continuously deformed into another if and only if the Bott invariants of the two maps agree. In the special case $n = k = 1$, where we are dealing with maps from the one-dimensional circle into the nonzero complex numbers, or in other words closed paths in $\mathbb{C}$ that do not pass through the origin, the Bott invariant is just the classical *winding number*, which measures the number of times such a path winds around the origin. We may therefore regard the Bott invariant as a generalized winding number.

The index theorem for equations of the type that we are considering in this section asserts that the Fredholm index of an elliptic equation is equal to the Bott invariant of its symbol. For instance, in the case of the simple example (1) considered above, the symbol $\mathrm{i}\xi + x$ corresponds to the identity map from the unit circle in (x, ξ)-space to the unit circle in $\mathbb{C}$. Its winding number is equal to 1, in agreement with our computation of the index.

The proof of the index theorem depends strongly on Bott periodicity and proceeds as follows. Because elliptic equations are classified topologically by the Bott invariant, and because the Bott invariant and the Fredholm index have analogous algebraic properties, one need only verify the theorem in a single example: that corresponding to a symbol with Bott invariant 1. It turns out that this *Bott generator* can be represented by an n-dimensional generalization of our example (1), and a computation in this case completes the proof.

4   Elliptic Equations on Manifolds

It is possible to define elliptic equations not just for functions f of n variables, but also for functions defined on a MANIFOLD [I.3 §6.9]. Particularly accessible to analysis are the elliptic equations on *closed* manifolds, that is, on manifolds that are finite in extent and that have no boundary. For closed manifolds it is not necessary to specify any boundary conditions in order to obtain the basic regularity theorem for elliptic equations (after all, there is no boundary). As a result, every elliptic partial differential equation on a closed manifold has a Fredholm index.

The Atiyah–Singer index theorem concerns elliptic equations on closed manifolds and it has roughly the same form as the index theorem that we studied in the previous section. One builds out of the symbol an invariant called the *topological index*, which generalizes

the Bott invariant. The Atiyah–Singer index theorem then asserts that the topological index of an elliptic equation is equal to the Fredholm or *analytical* index of the equation. The proof has two stages. In the first, theorems are proved that allow one to transform an elliptic equation on a general manifold into an elliptic equation on a sphere without changing the topological or analytical indices. For example, it may be shown that two elliptic equations on different manifolds that are the common "boundary" of an elliptic equation on a manifold of one higher dimension must have the same topological and analytical indices. In the second stage of the proof the Bott periodicity theorem and an explicit computation are applied to identify the topological and analytical indices of elliptic equations on spheres. Throughout both stages, an important tool is K-THEORY [IV.6 §6], which is a branch of algebraic topology invented by Atiyah and Hirzebruch.

Although the proof of the Atiyah–Singer index theorem makes use of K-theory, the final result can be translated into terms that do not mention K-theory explicitly. In this way one obtains an index formula roughly like this:

$$\mathrm{index} = \int_M I_M \cdot \mathrm{ch}(\sigma).$$

The term I_M is a DIFFERENTIAL FORM [III.16] determined by the CURVATURE [III.78] of the manifold M on which the equation is defined. The term $\mathrm{ch}(\sigma)$ is a differential form obtained from the symbol of the equation.

5   Applications

In order to prove the index theorem, Atiyah and Singer were obliged to study a very broad class of generalized elliptic equations. However, the applications they first had in mind were related to the simple equation with which we began this article. Solutions of the equation

$$\frac{\partial f}{\partial x} + \mathrm{i}\frac{\partial f}{\partial y} = 0$$

are precisely the analytic functions of the complex variable $z = x + \mathrm{i}y$. There is a counterpart to this equation on any RIEMANN SURFACE [III.79], and the Atiyah–Singer index formula, applied in this instance, is equivalent to a foundational result about the geometry of surfaces called the RIEMANN–ROCH THEOREM [V.31]. The Atiyah–Singer index theorem then gives a means to generalize the Riemann–Roch theorem to a COMPLEX MANIFOLD [III.6 §2] of any dimension.

The Atiyah–Singer index theorem also has important applications outside of complex geometry. The simplest example involves the elliptic equation $d\omega + d^*\omega = 0$, concerning differential forms on a manifold M. The Fredholm index may be identified with the *Euler characteristic* of M, which is the alternating sum of the numbers of r-dimensional cells in a cell decomposition of M. For two-dimensional manifolds, the Euler characteristic is the familiar quantity $V - E + F$. In the two-dimensional case, the index theorem reproduces the Gauss-Bonnet theorem, which asserts that the Euler characteristic is a multiple of the total Gaussian curvature.

Even in this simple case, the index theorem can be used to produce topological restrictions on the ways a manifold can curve. Many important applications of the index theorem proceed in the same direction. For example, Hitchin used a more refined application of the Atiyah–Singer index theorem to show that there is a nine-dimensional manifold that is homeomorphic to the sphere despite not being positively curved in even the weakest sense. (By contrast, the usual sphere is positively curved in the strongest possible sense.)

Further Reading

Atiyah, M. F. 1967. Algebraic topology and elliptic operators. *Communications in Pure and Applied Mathematics* 20:237–49.

Atiyah, M. F., and I. M. Singer. 1968. The index of elliptic operators. I. *Annals of Mathematics* 87:484–530.

Hirzebruch, F. 1966. *Topological Methods in Algebraic Geometry*. New York: Springer.

Hitchin, N. 1974. Harmonic spinors. *Advances in Mathematics* 14:1–55.

V.3 The Banach–Tarski Paradox

T. W. Körner

The Banach–Tarski paradox states that there is a way of decomposing a three-dimensional ball of unit radius into a finite number of disjoint pieces, then reassembling the pieces to form *two* balls of unit radius, where "reassembling" means that the pieces are translated and rotated and that they end up still disjoint.

Such a result seems impossible at first sight, and indeed it contradicts the naive assumption that one can consistently assign a finite volume to every bounded set. In other words, it shows that one cannot assign volumes to *all* bounded sets in such a way that these volumes are unaffected by translation and rotation, that the volume of a union of two disjoint sets is the sum of the volumes of the two sets, and that the volume

of the unit ball is greater than zero. However, if we drop this naive assumption, then the paradox disappears. Since there is no genuine paradox, we shall refer to the Banach–Tarski *construction*.

The Banach–Tarski construction is a descendant of an older construction due to Vitali, which concerns area rather than volume. Let us write l_θ for the line segment in $\mathbb{R}^2$ that is given in polar coordinates by

$$l_\theta = \{(r, \theta) : 0 < r \leqslant 1\}.$$

Note that the union of all such segments is the punctured unit disk D_* (that is to say, the unit disk with the origin removed). We say that l_θ and l_ϕ belong to the same equivalence class if $\theta - \phi$ is a rational multiple of π, and we consider a set E that is the union of a set of l_θ containing *exactly one representative* from each equivalence class.

The rationals are COUNTABLE [III.11], so we can enumerate the rationals x with $0 \leqslant x < 1$ as a sequence $x_1, x_2, \ldots$. If we write

$$E_n = \{l_{\theta + 2\pi x_n} : l_\theta \in E\},$$

then each E_n is obtained from E by a rotation about the origin (through an angle $2\pi x_n$), the E_n are disjoint (as E contains only one representative from each equivalence class), and the union of the E_n is D_* (as E contains a representative from each equivalence class).

Now take D_* and split it into the set F consisting of the union of the sets E_{2n} and the set G consisting of the union of the sets E_{2n+1}. Each E_{2n} can be rotated to E_n, and the union of the E_n gives us D_*. Similarly, each E_{2n+1} can be rotated to E_n, and the union of the E_n gives us D_* again. Thus the punctured unit disk can be split into a countable set of disjoint pieces (all obtained by rotation of one particular set) which can be rotated and translated to form disjoint sets whose union is two copies of D_*.

Vitali's construction makes use of THE AXIOM OF CHOICE [III.1] (because we chose one representative from each equivalence class), and the same is true for the Banach–Tarski construction. Solovay showed that if we reject the axiom of choice, then there are MODELS OF SET THEORY [IV.22 §3] in which it *is* possible to assign a volume to all bounded sets in $\mathbb{R}^3$ in a consistent way. However, most mathematicians would agree that the natural moral to draw from our discussion is that when we define volume we should consider only a restricted collection of sets.

The Banach–Tarski construction is also closely related to our final example, which requires a little group theory. To introduce this example of bad behavior, we first consider an example of good behavior. Suppose that $f : \mathbb{R} \to \mathbb{R}$ is a reasonable function with $f(x) \geqslant 0$ and $f(x + 1) = f(x)$ for all x (thus, f is nonnegative and periodic with period 1). Suppose that there existed real numbers s, t, u, v such that

$$f(x + s) + f(x + t) - f(x + u) - f(x + v) \leqslant -1 \quad (1)$$

for all x. Since $\int_0^1 f(x + w)\,\mathrm{d}x = \int_0^1 f(x)\,\mathrm{d}x$ for all w, integrating both sides of (1) from 0 to 1 would give

$$0 \leqslant \int_0^1 (-1)\,\mathrm{d}x = -1,$$

which is impossible. Thus (1) cannot hold.

Now consider the FREE GROUP [IV.10 §2] G generated by a and b (that is to say, the group generated by a and b where no nontrivial relations hold between a and b). Every element of G can be written in shortest form as the product of a sequence, each term of which is a, a^{-1}, b, or b^{-1}. Define $F(x) = 1$ if $x = e$ or the shortest form of x ends with a or a^{-1}, and set $F(x) = 0$ otherwise. We see that $F(x) \geqslant 0$ for all $x \in G$, and the reader can check, by going through cases, that

$$F(xb) + F(xab) - F(xa^{-1}) - F(xb^{-1}a) \leqslant -1 \quad (2)$$

for all $x \in G$. The averaging argument that enabled us to show that (1) was false for $\mathbb{R}$ must fail for G since (2) is, in fact, true. If there is no averaging argument, then there can be no appropriate universal integral and no appropriate universal "volume" in G.

This example bears a clear family resemblance to the "paradoxes" discussed earlier. If we consider the group $\mathrm{SO}(3)$ of rotations in three dimensions, then (unless specific conditions hold) there is no nontrivial group relation between two generally chosen rotations A and B about two generally chosen axes. Thus $\mathrm{SO}(3)$ contains a copy of the group G considered in the previous paragraph. The Banach–Tarski construction is a modification of a construction of Hausdorff that exploits this fact.

There is a beautiful account of all these matters in *The Banach–Tarski Paradox* by Stan Wagon (Cambridge University Press, Cambridge, UK, 1993).

V.4   The Birch–Swinnerton-Dyer Conjecture

Given an ELLIPTIC CURVE [III.21], there is a natural way of defining a binary operation on its points, and this turns the elliptic curve into an ABELIAN GROUP [I.3 §2.1]. Moreover, the points on the curve with rational coordin-

ates form a subgroup of this group. Mordell's theorem tells us that this subgroup is finitely generated. (These results are described in RATIONAL POINTS ON CURVES AND THE MORDELL CONJECTURE [V.29].)

Every finitely generated Abelian group is isomorphic to a group of the form $\mathbb{Z}^r \times C_{n_1} \times C_{n_2} \times \cdots \times C_{n_k}$, where C_n stands for the cyclic group with n elements. The number r, which measures the maximum number of independent elements of this group that have infinite order, is called the *rank* of the elliptic curve. Mordell's theorem implies that the rank of every elliptic curve is finite, but it does not tell us how to calculate it. That turns out to be an extraordinarily hard problem: in fact, so hard that it is considered a remarkable achievement of Birch and Swinnerton-Dyer even to have come up with a plausible conjecture about it.

Their conjecture relates the rank of an elliptic curve to a very different object associated with that curve: an L-FUNCTION [III.47]. This is a function with properties similar to those of the RIEMANN ZETA FUNCTION [IV.2 §3], but it is defined in terms of a series of numbers $N_2(E), N_3(E), N_5(E), \dots$, one for each prime p; the number $N_p(E)$ is the number of points on the elliptic curve when it is considered as a curve over the FIELD [I.3 §2.2] with p elements. One of the properties of the L-function of E is that it is HOLOMORPHIC [I.3 §5.6]. (The fact that it can be extended to a holomorphic function everywhere on the complex plane is very far from obvious: it follows from the fact that all elliptic curves are modular. See FERMAT'S LAST THEOREM [V.10].) Birch and Swinnerton-Dyer conjectured that the rank of the group associated with the elliptic curve is equal to the order of the zero of its L-function at 1. (If the L-function does not take the value 0 at 1, then this order is defined to be 0.) This can be thought of as a sophisticated LOCAL-TO-GLOBAL PRINCIPLE [III.51], in that it relates the rational solutions to the equation for the elliptic curve to the solutions mod p for each prime p.

Another remarkable feature of the conjecture is that far less was known about elliptic curves when Birch and Swinnerton-Dyer made it. Now there are many reasons to find it plausible, but then it was much more of a leap in the dark: they based it on numerical evidence gleaned from computations of $N_p(E)$ for several elliptic curves and many primes p. In other words, they did not calculate the orders of zeros of L-functions of various elliptic curves, since that was too hard, but guessed them based on approximations.

The Birch–Swinnerton-Dyer conjecture has now been proved for curves with L-functions that have a zero of order 0 or 1 at 1, but a proof of the general case still appears to be a long way off. It is one of the problems for which the Clay Mathematics Institute offers a prize of a million dollars. For a further discussion of the problem and much more about its mathematical context, see ARITHMETIC GEOMETRY [IV.5].

V.5   Carleson's Theorem
Charles Fefferman

Carleson's theorem asserts that the FOURIER SERIES [III.27] of a function f in $L^2[0, 2\pi]$ converges almost everywhere. To understand this statement and appreciate its significance, let us follow the history of the subject, starting in the early nineteenth century. FOURIER's [VI.25] great idea was that "any" (complex-valued) function f on an interval such as $[0, 2\pi]$ can be expanded in what we would now call a *Fourier series*,

$$f(\theta) = \sum_{n=-\infty}^{\infty} a_n e^{in\theta}, \tag{1}$$

for suitable *Fourier coefficients* a_n. Fourier obtained the formula for the coefficients a_n, and proved that (1) holds in interesting special cases.

The next major advance, due to DIRICHLET [VI.36], was a formula for the Nth partial sum $S_N f(\theta)$, which is defined to be

$$S_N f(\theta) = \sum_{n=-N}^{N} a_n e^{in\theta}. \tag{2}$$

Dirichlet realized that the precise meaning of (1) is that

$$\lim_{N \to \infty} S_N f(\theta) = f(\theta). \tag{3}$$

Dirichlet used his formula for $S_N f$ to prove that under certain circumstances (3) does indeed hold. For example, if f is a continuous increasing function on $[0, 2\pi]$, then it holds for every $\theta \in (0, 2\pi)$.

Decades later, DE LA VALLÉE POUSSIN [VI.67] discovered an example of a continuous function whose Fourier series diverges at a single point. More generally, given any countable set $E \subset [0, 2\pi]$, there exists a continuous function f whose Fourier series diverges at every point of E, a result that appears to restrict quite considerably the circumstances under which Fourier's original vision is valid.

The work of LEBESGUE [VI.72] led to fundamental progress in Fourier analysis and a significant change of viewpoint. We first sketch Lebesgue's ideas and then trace their impact on Fourier analysis.

Lebesgue sought to define a notion of integration that could be applied to all but the most pathological nonnegative functions F on $[0, 2\pi]$. He began by defining the MEASURE [III.55] of a set $E \subset [0, 2\pi]$. Loosely speaking, the measure of E, written $\mu(E)$, is "what the set E would weigh" if the interval $[0, 2\pi]$ were made of wire weighing one gram per centimeter. For instance, the measure of an interval (a, b) is equal to its length $b - a$. Certain sets E have measure zero, e.g., countable sets, or the CANTOR SET [III.17]; sets of measure zero are regarded as negligibly small.

Using his notion of measure, Lebesgue defined the *Lebesgue integral* $\int_0^{2\pi} F(\theta) \, d\theta$ for the "measurable" functions $F \geqslant 0$ on $[0, 2\pi]$. All but the most pathological functions are measurable, but $\int_0^{2\pi} F(\theta) \, d\theta$ may be infinite if F is too big. For example, if $F(\theta) = 1/\theta$ for $\theta \in (0, 2\pi]$, then the integral of F is infinite.

Finally, given any real number $p \geqslant 1$, the *Lebesgue space* $L^p[0, 2\pi]$ consists of all measurable functions f on $[0, 2\pi]$ that are not too big, in the sense that $\int_0^{2\pi} |f(\theta)|^p \, d\theta$ is finite. (See FUNCTION SPACES [III.29] for a slight, technical correction to this definition.)

We now turn to the impact of Lebesgue's theory on Fourier analysis. The Lebesgue space $L^2[0, 2\pi]$, which is also a HILBERT SPACE [III.37], plays a fundamental role. If f belongs to $L^2[0, 2\pi]$, then its Fourier coefficients a_n are such that

$$\sum_{n=-\infty}^{\infty} |a_n|^2 < \infty. \tag{4}$$

Conversely, any sequence of complex numbers a_n ($-\infty < n < \infty$) satisfying (4) arises as the sequence of Fourier coefficients of a function f in $L^2[0, 2\pi]$. Moreover, the size of a function f and its Fourier coefficients a_n are related by the *Plancherel formula*:

$$\frac{1}{2\pi} \int_0^{2\pi} |f(\theta)|^2 \, d\theta = \sum_{n=-\infty}^{\infty} |a_n|^2.$$

Finally, the partial sums $S_N f$ (see (2)) converge to the function f in the L^2-norm. In other words,

$$\int_0^{2\pi} |S_N f(\theta) - f(\theta)|^2 \, d\theta \longrightarrow 0 \tag{5}$$

as N tends to infinity. This gives us a precise sense in which the function f is the sum of its Fourier series. Thus, we have justified Fourier's formula (1) by reinterpreting it as the statement (5) rather than using the more obvious interpretation of (3).

However, it would still be nice to know to what extent the original, more straightforward interpretation can be justified. In 1906, Luzin conjectured that if f is any function in $L^2[0, \pi]$, then

$$\lim_{N \to \infty} S_N f(\theta) = f(\theta) \tag{6}$$

for all θ outside a set of measure zero. When this holds, one says that the Fourier series of f converges *almost everywhere*. If Luzin's conjecture were true, it would validate Fourier's vision from the early nineteenth century.

For several decades it looked as though Luzin's conjecture might well be false. KOLMOGOROV [VI.88] constructed a function f in $L^1[0, 2\pi]$ whose Fourier series converges nowhere. Also, a theorem of Kolmogorov, Seliverstov, and Plessner, which asserted that $\lim_{N\to\infty}(S_N f(\theta)/\sqrt{\log N}) = 0$ almost everywhere when f is in $L^2[0, 2\pi]$, withstood all attempts at improvement for over thirty years.

It therefore came as a big surprise when Lennart Carleson proved in 1966 that Luzin's conjecture is true. The main point of Carleson's proof is to control the *Carleson maximal function*

$$C(f)(\theta) = \sup_{N \geqslant 1} |S_N f(\theta)|$$

by proving that

$$\mu(\{\theta \in [0, 2\pi] : C(f)(\theta) > \alpha\}) \leqslant \frac{A}{\alpha^2} \int_0^{2\pi} |f(\theta)|^2 \, d\theta \tag{7}$$

for all f in $L^2[0, 2\pi]$ and all $\alpha > 0$, where A is a constant independent of f and α. It is not hard to show that (7) implies Luzin's conjecture, but it is very hard to prove (7).

Shortly after Carleson's work, Hunt proved the almost-everywhere convergence of Fourier series of functions in $L^p[0, 2\pi]$ for any $p > 1$. Kolmogorov's counterexample shows that the result fails for $p = 1$.

Fourier analysis has been immensely useful in mathematics and its applications. (For a fuller discussion of this, see THE FOURIER TRANSFORM [III.27] and HARMONIC ANALYSIS [IV.11].) The theorems of Carleson and Hunt provide the sharpest known answer to the basic question that started the subject.

Acknowledgments.   This work was partially supported by NSF grant #DMS-0245242.

Cauchy's Theorem

See SOME FUNDAMENTAL MATHEMATICAL DEFINITIONS [I.3 §5.6]

V.6   The Central Limit Theorem

The central limit theorem is a fundamental result in probability concerning sums of independent random variables. Let $X_1, X_2, \ldots$ be independent and suppose that they are identically distributed. Suppose also that they have mean 0 and variance 1. Then $X_1 + \cdots + X_n$ has mean 0 and variance n. (The variance is n because the X_i are independent.) Therefore, $Y_n = (X_1 + \cdots + X_n)/\sqrt{n}$ has mean 0 and variance 1. The central limit theorem states that, regardless of the distribution of the X_i, the random variable Y_n converges to a standard normal distribution. It is easy to deduce from this a similar result for random variables with any finite mean and variance. Details may be found in PROBABILITY DISTRIBUTIONS [III.71 §5].

V.7   The Classification of Finite Simple Groups

Martin W. Liebeck

A finite group G is said to be *simple* if its only normal subgroups are the identity subgroup and G itself. To some extent, simple groups play an analogous role in finite group theory to that of prime numbers in number theory: just as the only factors of a prime p are 1 and p itself, so the only factor groups of a simple group G are the identity group 1 and G itself. The analogy runs a bit deeper: just as every positive integer (greater than 1) is a product of a collection of primes, so every finite group is "built" from a collection of simple groups, in the following sense. Let H be a finite group, and choose a maximal normal subgroup H_1 of H (this means that H_1 is not the whole of H, and it is not contained in any larger normal subgroup that is not the whole of H); then choose a maximal normal subgroup H_2 of H_1; and so on. This gives a sequence of subgroups $1 = H_r < H_{r-1} < \cdots < H_1 < H_0 = H$, each one a maximal normal subgroup of the next, and, because of the maximality, each factor group $G_i = H_i/H_{i+1}$ is a simple group. It is in this sense that one says that H is built from the collection $G_0, G_1, \ldots, G_{r-1}$ of simple groups (although unlike the situation with prime numbers, there will in general be several different finite groups that are built from the same collection of simple groups).

At any rate, it is abundantly clear that simple groups lie at the heart of the theory of finite groups, and one of the driving forces of twentieth-century finite group theory was to study, and ultimately to classify completely, the finite simple groups. This classification was eventually achieved by the combined efforts of more than one hundred mathematicians in many published research articles and books written over a long period, the most intensive being 1955–80. It was a truly monumental feat of prolonged collaboration, and one of the most momentous theorems in the history of algebra.

In order to state the classification theorem, it is necessary to describe some examples of finite simple groups. The most obvious are the cyclic groups of prime order: these are clearly simple, since they have no subgroups at all apart from the identity and the whole group (by Lagrange's theorem, for example, which states that the size of any subgroup is a factor of the size of the group). Next come the alternating groups A_n: here A_n is defined as the group consisting of all the even permutations in the symmetric group S_n (see PERMUTATION GROUPS [III.68]). The alternating group A_n has $\frac{1}{2}(n!)$ elements, and is simple provided $n \geqslant 5$. For example, A_5, of order 60, is the smallest non-Abelian simple group.

Next we introduce some simple groups of matrices. For an integer $n \geqslant 2$ and a field K, define $\mathrm{SL}_n(K)$ to be the set of all $n \times n$ matrices with entries in K and with DETERMINANT [III.15] equal to 1. This is a group under matrix multiplication, called a *special linear* group. When the field K is finite, $\mathrm{SL}_n(K)$ is a finite group. For each prime power q, there is up to isomorphism a unique field of order q, and the corresponding special linear group in dimension n is denoted by $\mathrm{SL}_n(\mathbb{F}_q)$. These groups are not in general simple, since $Z = \{\lambda I : \lambda^n = 1\}$, the subgroup of scalar matrices in $\mathrm{SL}_n(\mathbb{F}_q)$, is a normal subgroup. However, the factor groups $\mathrm{PSL}_n(\mathbb{F}_q) = \mathrm{SL}_n(\mathbb{F}_q)/Z$ are simple (except when $(n, q) = (2, 2)$ or $(2, 3)$). This is the family of *projective special linear* groups.

There are a number of other families of finite simple matrix groups, which, very roughly speaking, are defined as groups of matrices $A \in \mathrm{SL}_n(\mathbb{F}_q)$ that satisfy an equation of the form $A^{\mathrm{T}} J A = J$, where J is a nonsingular symmetric or skew-symmetric $n \times n$ matrix. Again factoring out by the subgroup of scalar matrices, this gives the *projective orthogonal* and *symplectic* families of finite simple matrix groups. Similarly, if the finite field of order q has an automorphism $\alpha \rightarrow \bar{\alpha}$ of order 2, this can be extended to matrices $A = (a_{ij})$ by defining $\bar{A} = (\bar{a}_{ij})$, and then the group $\{A \in \mathrm{SL}_n(\mathbb{F}_q) : A^{\mathrm{T}}\bar{A} = I\}$, factored by its subgroup of scalar matrices, gives the *projective unitary* family of finite simple groups.

The families of projective special linear, symplectic, orthogonal, and unitary groups comprise what are known as the *classical* simple groups. These were all known early in the twentieth century, but it was not until 1955 that further infinite families of finite simple

groups were discovered by Chevalley. For each of the simple complex Lie algebras L, and each finite field K, Chevalley constructed a version of L over K, call it $L(K)$, and defined his families of finite simple groups as automorphism groups of the Lie algebras $L(K)$. Not long afterward, Steinberg, Suzuki, and Ree found some variations of Chevalley's construction and defined some further families of simple groups, known as twisted Chevalley groups. The Chevalley and twisted Chevalley groups include all the classical groups, together with ten other infinite families, and are collectively known as the *finite simple groups of Lie type*.

Until 1966, the only known finite simple groups were the cyclic groups of prime order, the alternating groups, the groups of Lie type, and a collection of five strange simple groups discovered by MATHIEU [VI.51] in the 1860s. These were groups of permutations of n objects, where $n = 11, 12, 22, 23,$ or 24. Mathieu's groups were termed "sporadic groups"— sporadic meaning that they do not fit into any of the known infinite families—and many thought that perhaps there were no more finite simple groups to be found. Then there was a bombshell, when Janko published a paper demonstrating the existence of a single, new finite simple group: the sixth sporadic group. After this, new sporadic groups appeared at regular intervals, culminating in the MONSTER [III.61], an amazing group of order around 10^{54}, which was predicted by Fischer and constructed by Griess as a group of $196\,884 \times 196\,884$ matrices. By 1980, twenty-six sporadic groups were known.

During this period the program to classify all the finite simple groups was proceeding at breakneck speed, and eventually in the early 1980s the final classification theorem was announced.

Every finite simple group is either a cyclic group of prime order, or an alternating group, or a group of Lie type, or one of the twenty-six sporadic groups.

Not surprisingly, this theorem has changed the face of finite group theory and its many areas of application: one can now solve many problems in a concrete way, by reducing them to the study of the (now known) list of simple groups, rather than abstractly, by deducing them from the axioms for groups.

The sheer length of the proof of the classification theorem (estimated at around ten thousand journal pages, spread across about five hundred research articles) meant that it was extremely difficult, perhaps impossible, for a single person to work through the entire proof. It also meant that the chances were rather high that there were errors along the way. Fortunately, in the years since the announcement of the result, various teams of group theorists have been publishing summaries and revisions of many parts of the proof, and a series of volumes containing the whole proof is now well on the way to completion.

V.8 Dirichlet's Theorem

A famous theorem of EUCLID [VI.2] asserts that there are infinitely many primes. But what if one wants more information about these primes? For instance, are there infinitely many primes of the form $4n - 1$? A fairly straightforward modification of Euclid's argument shows that there are, and a slightly more difficult modification proves that there are infinitely many of the form $4n + 1$ as well. However, modifications of Euclid's argument are not enough to prove the general result in this direction, which is that if a and m are *coprime* (that is, have highest common factor 1), then there are infinitely many primes of the form $mn + a$. This was proved by DIRICHLET [VI.36] using what are now called Dirichlet L-FUNCTIONS [III.47], which are closely related to the RIEMANN ZETA FUNCTION [IV.2 §3]. The condition that m and a have highest common factor 1 is clearly necessary, since any common factor of m and a will be a factor of $mn + a$. Dirichlet's theorem is discussed further in ANALYTIC NUMBER THEORY [IV.2 §4].

V.9 Ergodic Theorems
Vitaly Bergelson

Consider the sequence $(z^n)_{n=0}^{\infty}$, where z is a complex number of modulus 1. While for $z \neq 1$ our sequence is not convergent, it is not hard to see that, on average, it exhibits quite regular behavior. Indeed, using the formula for the sum of a geometric progression, and assuming that $z \neq 1$, we have, for any $N > M \geqslant 0$,

$$\left| \frac{z^M + z^{M+1} + \cdots + z^{N-1}}{N - M} \right|$$
$$= \left| \frac{z^M(z^{N-M+1} - 1)}{(N - M)(z - 1)} \right| \leqslant \frac{2}{(N - M)|z - 1|},$$

which implies that when $N - M$ is large enough, the averages

$$A_{N,M}(z) = \frac{z^M + z^{M+1} + \cdots + z^{N-1}}{N - M}$$

are small. More formally, we have

$$\lim_{N-M \to \infty} \frac{z^M + z^{M+1} + \cdots + z^{N-1}}{N - M} = \begin{cases} 0, & z \neq 1, \\ 1, & z = 1. \end{cases} \quad (1)$$

This simple fact is a special, one-dimensional case of *von Neumann's ergodic theorem*, which was the first mathematical statement to throw light on the so-called quasi-ergodic hypothesis in statistical mechanics and the kinetic theory of gases.

Von Neumann's theorem concerns the average behavior of powers of UNITARY OPERATORS [III.50 §3.1] on HILBERT SPACES [III.37]. If U is such an operator defined on a Hilbert space $\mathcal{H}$, then we can associate with U the U-*invariant subspace* $\mathcal{H}_{\text{inv}}$ that consists of all vectors $f \in \mathcal{H}$ such that $Uf = f$: that is, all vectors that are fixed by U. Let P be the ORTHOGONAL PROJECTION [III.50 §3.5] onto that subspace. Then von Neumann's theorem asserts that, for every $f \in \mathcal{H}$,

$$\lim_{N-M \to \infty} \left\| \frac{1}{N - M} \sum_{n=M}^{N-1} U^n f - Pf \right\| = 0.$$

In other words, in a certain sense the averages

$$\frac{1}{N - M} \sum_{n=M}^{N-1} U^n$$

converge to the orthogonal projection P. (This is not actually the theorem as formulated by VON NEUMANN [VI.91], but it is simpler to explain. He proved an equivalent statement about a continuous family of unitary operators $(U_\tau)_{\tau \in \mathbb{R}}$.)

Before we discuss various applications and refinements of von Neumann's theorem, let us briefly comment on its proof. Von Neumann's original proof used sophisticated machinery such as the spectral theory of one-parameter groups of unitary operators, obtained by Marshall Stone. Over the years many alternative proofs were offered, the simplest being a "geometric" proof due to RIESZ [VI.74], which we will describe below. To give the rough idea of von Neumann's proof it is convenient to use the fact (which follows from the SPECTRAL THEOREM [III.50 §3.4]) that any unitary operator U on a Hilbert space $\mathcal{H}$ has a "functional model." That is, we can realize the Hilbert space $\mathcal{H}$ as a function space, consisting of all (equivalence classes of) square-integrable functions with respect to some finite MEASURE [III.55], in such a way that U becomes a *multiplication operator* $M_\varphi(f) = \varphi f$, where φ is a complex-valued measurable function that satisfies $|\varphi(x)| = 1$ for almost every x. It is not hard to see, after passing to such a functional model, that von Neumann's theorem follows immediately from its one-dimensional case as expressed by formula (1). Note that in this case the orthogonal projection to the space of invariant elements takes a function f to the function g such that $g(x) = f(x)$ if $\varphi(x) = 1$ and $g(x) = 0$ otherwise.

Riesz's proof is based on the observation that the orthogonal complement of the subspace $\mathcal{H}_{\text{inv}}$ of U-invariant vectors is spanned by the set of vectors of the form $Ug - g$. To see this, note first that if $f \in \mathcal{H}_{\text{inv}}$, then

$$\langle f, Ug \rangle = \langle U^{-1}f, g \rangle = \langle f, g \rangle,$$

from which it follows that $\langle f, Ug - g \rangle = 0$ and thus that f is orthogonal to $Ug - g$. Conversely, if $f \notin \mathcal{H}_{\text{inv}}$, then $\langle f, Uf - f \rangle = \langle f, Uf \rangle - \langle f, f \rangle$. This is less than 0, by the CAUCHY–SCHWARZ INEQUALITY [V.19] and the fact that $\|Uf\| = \|f\|$ but $Uf \neq f$. In particular, f is not orthogonal to $Uf - f$. Thus, $\mathcal{H}_{\text{inv}}$ is the orthogonal complement of the (closed) subspace of $\mathcal{H}$ generated by functions of the form $Ug - g$.

Now the conclusion of von Neumann's theorem holds trivially if $f \in \mathcal{H}_{\text{inv}}$, since then $Pf = f$ and $U^n f = f$ for every n. On the other hand, if $f = Ug - g$, then $Pf = 0$. As for the averages, we know that $U^n f = U^{n+1}g - U^n g$, from which it follows that $\sum_{n=M}^{N-1} U^n f = U^N g - U^M g$. Since $\|U^N g - U^M g\|$ is at most $2\|g\|$ for every M and N, we find that

$$\frac{1}{N - M} \sum_{n=M}^{N-1} U^n f$$

has norm at most $2\|g\|/(N - M)$ and hence tends to 0. So the theorem is true in this case as well. It is straightforward to check that the set of functions for which the theorem holds is a closed linear subspace of $\mathcal{H}$, and therefore the theorem is proved.

The reason that von Neumann's theorem and other similar results are relevant to physics is that it is often possible to represent the evolution of the parameters associated with a physical system by a subset $X \subset \mathbb{R}^d$ that has finite d-dimensional volume, together with a continuous family $(T_\tau)_{\tau \in \mathbb{R}}$ of volume-preserving transformations from X to X. With each such transformation T_τ one can associate the unitary map U_τ, defined on $L^2(X)$ (the Hilbert space of square-integrable functions on X) by the formula $(U_\tau f)(x) = f(T_\tau x)$. The fact that these maps are unitary follows from the fact that the transformations T_τ preserve volume; also, it follows from the fact that the transformations T_τ depend continuously on τ that the maps U_τ do as well.

To simplify the discussion let us now "discretize" the situation. Instead of considering the continuous families (T_τ) and (U_τ) we shall fix a transformation $T = T_{\tau_0}$ (say, for $\tau_0 = 1$) and let U be the corresponding unitary operator. Assume that our volume-preserving transformation T is *ergodic*, which means that there is no proper subset $A \subset X$ of positive volume such that $T(A) \subset A$. This assumption can easily be shown to be equivalent to the fact that the only elements of $L^2(X)$ that satisfy $Uf = f$ are the constant functions. It follows from von Neumann's theorem that for any $f \in L^2(X)$ the averages

$$A_{N,M}(f) = \frac{1}{N - M} \sum_{n=M}^{N-1} U^n f$$

converge to a constant whose value is easy to find by performing term-by-term integration: it is equal to $(\int f \, dm)/\text{vol}(X)$. Since von Neumann's theorem also tells us that $\lim_{N-M \to \infty} A_{N,M}(f)$ is always a U-invariant function, we see that the assumption of ergodicity is a necessary and sufficient condition for the time average represented by $\lim_{N-M \to \infty} A_{N,M}(f)$ to equal the space average, $(\int f \, dm)/\text{vol}(X)$.

It is also possible to use von Neumann's theorem to strengthen a classical theorem of POINCARÉ [VI.61], called *Poincaré's recurrence theorem*. This result states that if X is a set of finite volume, as above, and A is a subset of X with nonzero volume, then "almost all points of A return infinitely often to A." In other words, if we set $\tilde{A}$ to be the set of all points $x \in A$ such that $T^n x \in A$ for infinitely many n, then the measure of the set of points in A but not in $\tilde{A}$ is 0. The main step in the proof of Poincaré's theorem is to prove the same about the set A_1, which consists of all points $x \in A$ such that $T^n x \in A$ for *some* positive integer n. To see why this is true, let B be the set of all points in A but not in A_1. The sets $B, T^{-1}B, T^{-2}B, \ldots$ all have the same measure, since T is volume preserving. ($T^{-n}B$ is defined to be the set of all x such that $T^n x \in B$.) Since X has finite volume, there must exist positive integers m and n such that the intersection of $T^{-m}B$ and $T^{-(m+n)}B$ has positive measure, and from this it follows that the measure of $B \cap T^{-n}B$ is also positive. But if $x \in B$ then $x \notin A_1$, so $T^n x \notin A$ and therefore $T^n x \notin B$, so this is a contradiction.

Now let us apply the von Neumann ergodic theorem with f equal to the characteristic function of a set A (that is, $f(x) = 1$ when $x \in A$ and $f(x) = 0$ otherwise) and U defined in terms of T as before. Suppose also that the set X has volume 1 and write μ for the measure on

X. Then one can check that $\langle f, U^n f \rangle = \mu(A \cap T^{-n}A)$. It follows that

$$\langle f, A_{N,M}(f) \rangle = \frac{1}{N-M} \sum_{n=M}^{N-1} \mu(A \cap T^{-n}A).$$

If we let $N - M$ tend to infinity, then $A_{N,M}f$ tends to a U-invariant function g. Since g is U-invariant, $\langle f, g \rangle = \langle U^n f, g \rangle$ for every n, and therefore $\langle f, g \rangle = \langle A_{N,M}(f), g \rangle$ for every N and M, and finally $\langle f, g \rangle = \langle g, g \rangle$. By the Cauchy–Schwarz inequality, this is at least $(\int g(x)\, \mathrm{d}\mu)^2 = (\int f(x)\, \mathrm{d}\mu)^2 = \mu(A)^2$. Therefore, we deduce that

$$\lim_{N-M \to \infty} \frac{1}{N-M} \sum_{n=M}^{N-1} \mu(A \cap T^{-n}A) \geqslant (\mu(A))^2.$$

If you choose two "random sets" of measure $\mu(A)$, then their intersection will typically be $(\mu(A))^2$, so the inequality above is saying that the average intersection of A with $T^{-n}A$ is at least as big as the "expected" intersection. This result, due to Khinchin, gives more precise information about the nature of Poincaré recurrence.

When a unitary operator is defined in terms of a measure-preserving transformation as above, it is natural to ask whether the averages converge not just in the sense of the L^2-norm but also in the more classical sense of convergence almost everywhere. (For a related thought in a different context, see CARLESON'S THEOREM [V.5].) The answer is that they do, as was shown by BIRKHOFF [VI.78] soon after he learned of von Neumann's theorem. He proved that for each integrable function f one could find a function f^* such that $f^*(Tx) = f^*(x)$ for almost every x, and such that

$$\lim_{N \to \infty} \frac{1}{N} \sum_{n=0}^{N-1} f(T^n x) = f^*(x)$$

for almost every x. Suppose that the transformation T is ergodic, let $A \subset X$ be a set of positive measure, and let $f(x)$ be the characteristic function of A. It follows from Birkhoff's theorem that for almost every $x \in X$ one has

$$\lim_{N \to \infty} \frac{1}{N} \sum_{n=0}^{N-1} f(T^n x) = \frac{\int f\, \mathrm{d}\mu}{\mu(X)} = \frac{\mu(A)}{\mu(X)}.$$

Since the expression

$$\lim_{N \to \infty} \frac{1}{N} \sum_{n=0}^{N-1} f(T^n x)$$

describes the frequency of visits of $T^n x$ to the set A, we see that in an ergodic system the images $x, Tx, T^2x, \dots$ of a typical point $x \in A$ visit A with a frequency that equals the proportion of the space occupied by A.

The ergodic theorems of von Neumann and Birkhoff have been generalized over the years in many different directions. These far-reaching extensions of ergodic theorems, and more generally the *ergodic method*, have found impressive applications in such diverse fields as statistical mechanics, number theory, probability theory, harmonic analysis, and combinatorics.

Further Reading

Furstenberg, H. 1981. *Recurrence in Ergodic Theory and Combinatorial Number Theory*. M. B. Porter Lectures. Princeton, NJ: Princeton University Press.
Krengel, U. 1985. *Ergodic Theorems*, with a supplement by A. Brunel. De Gruyter Studies in Mathematics, volume 6. Berlin: Walter de Gruyter.
Mackey, G. W. 1974. Ergodic theory and its significance for statistical mechanics and probability theory. *Advances in Mathematics* 12:178–268.

The Fermat–Euler Theorem

See MODULAR ARITHMETIC [III.58]

V.10 Fermat's Last Theorem

Many people, even if they are not mathematicians, are aware of the existence of *Pythagorean triples*: that is, triples of positive integers (x, y, z) such that $x^2 + y^2 = z^2$. These give us examples of right-angled triangles with integer side lengths, of which the best known is the "$(3, 4, 5)$ triangle." For any two integers m and n, we have that $(m^2 - n^2)^2 + (2mn)^2 = (m^2 + n^2)^2$, which gives us an infinite supply of Pythagorean triples, and in fact every Pythagorean triple is a multiple of a triple of this form.

FERMAT [VI.12] asked the very natural question of whether similar triples existed for higher powers: that is, could there be a solution in positive integers of the equation $x^n + y^n = z^n$ for some power $n \geqslant 3$? For instance, is it possible to express a cube as a sum of two other cubes? Or rather, Fermat famously claimed that it was not possible, and that he had a proof that space did not permit him to write down. Over the next three and a half centuries, this problem became the most famous unsolved problem in mathematics. Given the amount of effort that went into it, one can be virtually certain that Fermat did not in fact have a proof: the problem appears to be irreducibly difficult, and solvable only by techniques that were developed much later than Fermat.

The fact that Fermat's question was an easy one to think of does not on its own guarantee that it is interesting. Indeed, in 1816 GAUSS [VI.26] wrote in a letter that he found it too isolated a problem to interest him. At the time, that was a reasonable remark: it is often extremely hard to determine whether a given Diophantine equation has a solution, and it is therefore easy to come up with hard problems of a similar nature to Fermat's last theorem. However, Fermat's last theorem has turned out to be exceptional in ways that even Gauss could not have been expected to foresee, and nobody would now describe it as "isolated."

By the time of Gauss's remark, the problem had been solved for $n = 3$ (by EULER [VI.19]) and $n = 4$ (by Fermat; this is the easiest case). The first serious connection between Fermat's last theorem and more general mathematical concerns came with the work of KUMMER [VI.40] in the middle of the nineteenth century. An important observation that had been made by Euler is that it can be fruitful to study Fermat's last theorem in larger RINGS [III.81 §1], since these, if appropriately chosen, allow one to factorize the polynomial $z^n - y^n$. Indeed, if we write $1, \zeta, \zeta^2, \ldots, \zeta^{n-1}$ for the nth roots of 1, then we can factorize it as

$$(z - y)(z - \zeta y)(z - \zeta^2 y) \cdots (z - \zeta^{n-1} y). \quad (1)$$

Therefore, if $x^n + y^n = z^n$ then we have two rather different-looking factorizations of x^n inside the ring generated by 1 and ζ (namely the factorization in (1) above, and $x x x \cdots x$), and it is reasonable to hope that this information might be exploited. However, there is a serious problem: the ring generated by 1 and ζ does not enjoy the UNIQUE FACTORIZATION PROPERTY [IV.1 §§4–8], so one's sense of being close to a contradiction when faced with these two factorizations is not well-founded. Kummer, in connection with the search for HIGHER RECIPROCITY LAWS [V.28], had met this difficulty and had defined the notion of an IDEAL [III.81 §2]: very roughly, if you enlarge a ring by adding in Kummer's "ideal numbers," then unique factorization is restored. Using these concepts, Kummer was able to prove Fermat's last theorem for every prime number p that was not a factor of the CLASS NUMBER [IV.1 §7] of the corresponding ring. He called such primes *regular*. This connected Fermat's last theorem with ideas that have belonged to the mainstream of ALGEBRAIC NUMBER THEORY [IV.1] ever since. However, it did not solve the problem, since there are infinitely many irregular primes (though this was not known in Kummer's day).

It turned out that more complicated ideas could be used for individual irregular primes, and eventually an algorithm was developed that could check for any given n whether Fermat's last theorem was true for that n. By the late twentieth century, the theorem had been verified for all exponents up to 4 000 000. However, a general proof came from a very different direction.

The story of the eventual proof by Andrew Wiles has been told many times, so we shall be very brief about it here. Wiles did not study Fermat's last theorem directly, but instead solved an important special case of the *Shimura–Taniyama–Weil conjecture*, which connects ELLIPTIC CURVES [III.21] and MODULAR FORMS [III.59]. The first hint that elliptic curves might be relevant came when Yves Hellegouarch noticed that the elliptic curve $y^2 = x(x - a^p)(x - b^p)$ would have rather unusual properties if $a^p + b^p$ was also a pth power. Gerhard Frey realized that such a curve might be so unusual that it would contradict the Shimura–Taniyama–Weil conjecture. Jean-Pierre Serre came up with a precise statement (the "epsilon conjecture") that would imply this, and Ken Ribet proved Serre's conjecture, thus establishing that Fermat's last theorem was a consequence of the Shimura–Taniyama–Weil conjecture. Wiles suddenly became very interested indeed, and after seven years of intensive and almost secret work he announced a solution to a case of the Shimura–Taniyama–Weil conjecture that was sufficient to prove Fermat's last theorem. It then emerged that Wiles's proof contained a serious mistake, but with the help of Richard Taylor he managed to find an alternative and correct argument for that portion of the proof.

The Shimura–Taniyama–Weil conjecture asserts that "all elliptic curves are modular." We finish by giving a rough idea of what this means. (A few more details can be found in ARITHMETIC GEOMETRY [IV.5].) Associated with any elliptic curve E is a sequence of numbers $a_n(E)$, one for each positive integer n. For each prime p, $a_p(E)$ is related to the number of points on the elliptic curve (mod p); it is easy to derive from these values the values of $a_n(E)$ for composite n. Modular forms are HOLOMORPHIC FUNCTIONS [I.3 §5.6] with certain periodicity properties defined on the upper half-plane; associated with each modular form f is a FOURIER SERIES [III.27] that takes the form

$$f(q) = a_1(f)q + a_2(f)q^2 + a_3(f)q^3 + \cdots.$$

Let us call an elliptic curve E *modular* if there is a modular form f such that $a_p(E) = a_p(f)$ for all but finitely many primes p. If you are presented with an elliptic curve, it is not at all clear how to set about finding a modular form associated with it in this way. However, it always seemed to be possible, even if the phenomenon was a mysterious one. For instance, if E is the elliptic curve $y^2 + y = x^3 - x^2 - 10x - 20$, then there is a modular form f such that $a_p(E) = a_p(f)$ for every prime p apart from 11. This modular form is the unique complex function (up to scaling) that satisfies a certain periodicity property with respect to the group $\Gamma_0(11)$, which consists of all matrices $\left(\begin{smallmatrix} a & b \\ c & d \end{smallmatrix}\right)$ such that a, b, c, and d are integers, c is a multiple of 11, and the DETERMINANT [III.15] $ad - bc$ is 1. It is far from obvious that a definition of this type should have anything to do with elliptic curves.

Wiles proved that all "semistable" elliptic curves are modular, not by showing how to associate a modular form with each such elliptic curve, but by using a subtle counting argument that guaranteed that the modular form had to exist. The full conjecture was proved a few years later, by Christophe Breuil, Brian Conrad, Fred Diamond, and Richard Taylor, which put the icing on the cake of one of the most celebrated mathematical achievements of all time.

V.11 Fixed Point Theorems

1 Introduction

The following is a variant of a well-known mathematical puzzle. A man is on a train from London to Cambridge and has a bottle of water with him. Prove that there is at least one moment on the journey when the volume of air in the bottle, as a fraction of the volume of the bottle itself, is exactly equal to the fraction of his journey that he has completed. (For instance, the bottle might be two fifths full, and therefore three fifths empty, at the precise moment when he is three fifths of the way from London to Cambridge. Note that we do not assume that the bottle is full at the start of the journey or empty at the end.)

The solution, if you have not seen this sort of question before, is surprisingly simple. For each x between 0 and 1 let $f(x)$ be the proportion of air in the bottle when the proportion of the journey that has been completed is x. Then $0 \leqslant f(x) \leqslant 1$ for every x, since the volume of air in the bottle cannot be negative and cannot exceed the volume of the bottle. If we now set $g(x)$ to be $x - f(x)$, then we see that $g(0) \leqslant 0$ and $g(1) \geqslant 0$. Since $g(x)$ varies continuously with x, there must be some moment at which $g(x) = 0$, so that $f(x) = x$, which is what we wanted.

What we have just proved is a slightly disguised form of one of the simplest of all fixed point theorems. We could state it more formally as follows: if f is a continuous function from the closed interval $[0, 1]$ to itself, then there must exist an x such that $f(x) = x$. This x we call a *fixed point* of f. (We deduced the result from the *intermediate value theorem*, a basic result in analysis that states that if g is a continuous function from $[0, 1]$ to $\mathbb{R}$ such that $g(0) \leqslant 0$ and $g(1) \geqslant 0$, then there must be some x such that $g(x) = 0$.)

In general, a fixed point theorem is a theorem that asserts that a function that satisfies certain conditions must have a fixed point. There are many such theorems, a small sample of which we shall discuss in this article. On the whole, they tend to have a nonconstructive nature: they establish the existence of a fixed point rather than defining one or telling you how to find it. This is part of the reason that they are important, since there are many examples of equations for which one would like to prove that a solution exists even when one cannot solve it explicitly. As we shall see, one way of going about this is to try to rewrite the equation in the form $f(x) = x$ and apply a fixed point theorem.

2 Brouwer's Fixed Point Theorem

The fixed point theorem we have just proved is the one-dimensional version of *Brouwer's fixed point theorem*, which states that if B^n is the unit ball of $\mathbb{R}^n$ (that is, the set of all $(x_1, \ldots, x_n)$ such that $x_1^2 + \cdots + x_n^2 \leqslant 1$) and f is a continuous function from B^n to B^n, then f must have a fixed point. The set B^n is an n-dimensional solid sphere, but all that matters is its topological character, so we could take it to be another shape such as an n-dimensional cube or simplex.

In two dimensions this says that a continuous function from the closed unit disk to itself must have a fixed point. In other words, if you had a circular sheet of rubber on a table and you picked it up and put it back down within the circle where it started, having folded it and stretched it as much as you liked, there would always have to be a point that ended up in the same place as before.

To see why this is true, it is helpful to reformulate the statement. Let $D = B^2$ be the closed unit disk. If we had a continuous function f from D to D with no fixed point, then we could define a continuous function g from D to its boundary ∂D as follows: for each x, follow a straight path from $f(x)$ to x and continue on

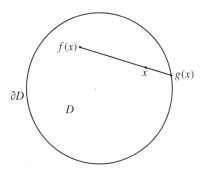

Figure 1 If f has no fixed points, then it can be used to define a retraction g.

in a straight line; $g(x)$ is the point where you first reach ∂D (see figure 1), and it is well-defined because (and only because) $f(x) \neq x$. If x is already on the boundary of D, then $g(x) = x$. So we have a continuous function $g : D \to \partial D$ such that $g(x) = x$ for every $x \in \partial D$. Such a function is called a *retraction* from D to ∂D.

It seems highly unlikely that a continuous retraction from D to ∂D could exist. If we can prove that it cannot, then we will have contradicted the assumption that there is a continuous function from D to D with no fixed point, and thereby have proved Brouwer's fixed point theorem in two dimensions.

There are several ways of proving that continuous retractions from disks to their boundaries cannot exist. Here we briefly sketch two.

Suppose, first, that g is such a retraction. For each t, let us consider the restriction of g to the circle of radius t about the origin, and let us represent a typical point in this circle as $te^{i\theta}$. Let us write $g_t(\theta)$ for $g(te^{i\theta})$. When $t = 1$ the circle of radius t is ∂D, so as θ goes from 0 to 2π, $g_t(\theta) = e^{i\theta}$ goes once around the unit circle. When $t = 0$, the circle of radius t is a single point, so as θ goes from 0 to 2π, $g_t(\theta)$ is just the constant point $g(0)$, which does not go around the unit circle at all. Therefore, somewhere between $t = 1$ and $t = 0$ there must be a change in the number of times $g_t(\theta)$ goes around the unit circle as θ goes from 0 to 2π. But the functions g_t are a continuously varying family of functions, and a small change in g_t cannot cause a sudden jump in the number of times that $g_t(\theta)$ goes around the circle. (To make this last step rigorous needs a bit of work, but the basic idea is sound.)

A second proof uses basic tools from algebraic topology. The first HOMOLOGY GROUP [IV.6 §4] of the disk D is trivial, since every curve in the disk can be shrunk

to a point. The first homology group of the unit circle ∂D is $\mathbb{Z}$. If there is a continuous retraction g from D to ∂D, then we can find continuous maps $h : \partial D \to D$ and $g : D \to \partial D$ such that $g \circ h$ is the identity on ∂D. (We let h be the map that takes a point of ∂D to itself and we let g be the continuous retraction.) Now continuous maps between topological spaces give rise to HOMO-MORPHISMS [I.3 §4.1] between their homology groups, in such a way that compositions go to compositions and identity maps go to identity maps. (That is, there is a FUNCTOR [III.8] from the CATEGORY [III.8] of topological spaces and continuous maps to the category of groups and group homomorphisms.) This means that there must be homomorphisms $\phi : \mathbb{Z} \to \{0\}$ and $\psi : \{0\} \to \mathbb{Z}$ such that $\psi \circ \phi$ is the identity on $\mathbb{Z}$, which is obviously impossible.

Both proofs generalize to higher dimensions: the second straightforwardly (once one knows how to compute homology groups of spheres), and the first via the notion of the *degree* of a continuous map from the n-sphere to itself, which is a higher-dimensional analogue of the notion of the number of times a map from the circle to itself "goes around the circle."

Brouwer's fixed point theorem has many applications. For example, the following fact is important in the theory of random walks on graphs. A *stochastic matrix* is an $n \times n$ matrix with nonnegative entries such that the sum of the entries in each row is equal to 1. Brouwer's fixed point theorem can be used to show that every such matrix has an EIGENVECTOR [I.3 §4.3] with nonnegative entries and eigenvalue 1. The proof is as follows: the set of all column vectors with nonnegative entries that add up to 1 is, geometrically speaking, an $(n - 1)$-dimensional simplex. (For example, if $n = 3$, this set is a triangle in $\mathbb{R}^3$ with vertices $(1, 0, 0)$, $(0, 1, 0)$, and $(0, 0, 1)$.) If A is a stochastic matrix and $\boldsymbol{x}$ belongs to this simplex, then so does $A\boldsymbol{x}$. Since the map $\boldsymbol{x} \mapsto A\boldsymbol{x}$ is continuous, Brouwer's theorem gives us an $\boldsymbol{x}$ such that $A\boldsymbol{x} = \boldsymbol{x}$: this is the required eigenvector.

An extension of Brouwer's theorem, called the *Kakutani fixed point theorem*, was used by John Nash to establish the existence of a "social equilibrium," a state of affairs in which no household can individually improve its well-being by altering the amount that it consumes of various items. Kakutani's theorem concerns functions that take points in a closed ball B^n not to other points in B^n but to *subsets* of B^n. If $f(x)$ is a nonempty closed convex subset of B^n for each x and if $f(x)$ varies continuously in an appropriate sense, then the theorem says that there must be some x such that $x \in f(x)$. Brouwer's theorem is the special case where each $f(x)$ is a set with just one element.

3 A Stronger Form of Brouwer's Fixed Point Theorem

So far, we have discussed maps from solid spheres to themselves, but there is nothing to stop us thinking about whether continuous maps on other spaces must have fixed points. For example, let S^2 be the (nonsolid) sphere $\{(x, y, z) : x^2 + y^2 + z^2 = 1\}$ and let f be a continuous function from S^2 to S^2. Must f have a fixed point? At first one might think so: some obvious functions from S^2 to itself are rotations and reflections, both of which certainly have fixed points, and it is hard to see how one can "get rid" of those fixed points. However, eventually one realizes that there is a simple example of a function without a fixed point, namely the function $f(x) = -x$, which reflects each point through the origin.

The obvious reaction to this example is to note that the result we had hoped for is false and to turn our attention to something else. But this reaction is a mistake, as it is in many other mathematical contexts, because there was something importantly correct about the idea that it was impossible to get rid of the fixed points of a rotation. It turns out that if you start with a rotation and try to get rid of the fixed points by continuously deforming it, then you are doomed to failure. In fact, in a certain sense there will always be exactly two fixed points. More generally, if you take any continuous function from S^2 to S^2 and continuously deform it, then you cannot change the number of fixed points.

Of course, these last two statements are patently false if taken at face value so some reinterpretation is needed. First, we must assume that the number of fixed points is finite, but this is not a huge assumption as it can be shown that a typical small perturbation of any continuous function will have only finitely many fixed points. Second, we must count the fixed points with appropriate weights. To define these, suppose that $f(x) = x$, and imagine a point $y(t)$ that goes around x in a tiny circle as t goes from 0 to 1. We define the *index* of the fixed point x to be the number of turns made by the vector from $y(t)$ to $f(y(t))$, counting this negatively if these turns are in the opposite direction to the way that $y(t)$ goes around x. (This definition is problematic if $f(y(t)) = y(t)$ for some t, but again

we can make small perturbations and assume that this does not happen.) Then the sum of the indices of all the fixed points is the quantity that does not change if you continuously deform f.

It follows that if you continuously deform a rotation, then the sum of the indices will always be 2. From this it follows that there must be at least one fixed point. It also follows that you cannot continuously deform a rotation so that it becomes the map that sends each x to $-x$.

The notion of the index of a fixed point can be generalized in a fairly straightforward way to higher dimensions (using the concept of degree mentioned earlier), and one can show under very general circumstances that the sum of the indices of fixed points remains constant when you continuously deform a continuous map. This implies Brouwer's fixed point theorem as follows. We can continuously deform any continuous map $f : B^n \to B^n$ into any other continuous map $g : B^n \to B^n$ by defining $f_t(x) = (1 - t)f(x) + tg(x)$ and letting t vary from 0 to 1. Let us therefore take g to be the map $x \mapsto \frac{1}{2}x$, which has a single fixed point. This fixed point has index 1 (as one can see easily in the two-dimensional case), and therefore the sum of the indices of the fixed points of f is 1 as well.

In general, the sum of the indices of the fixed points of a function f defined on a suitable topological space X (such as a smooth compact MANIFOLD [I.3 §6.9]) can be calculated in terms of the effect of f on the homology groups of X. The resulting theorem is (a slight generalization of) the *Lefschetz fixed point theorem*.

The fact that the index of a continuous map is an invariant of continuous deformations can be used to give a proof of THE FUNDAMENTAL THEOREM OF ALGEBRA [V.13]. Consider, for instance, the problem of proving that the polynomial $x^5 + 3x + 8$ has a root. This is the same as asking for a fixed point of the function $x^5 + 4x + 8$, since if this equals x then $x^5 + 3x + 8 = 0$. Now if we regard the polynomial x^5 as being defined on the RIEMANN SPHERE [IV.14 §2.4] $\mathbb{C} \cup \{\infty\}$, then it has two fixed points, at 0 and ∞. Moreover, their indices are both 5 (since if x goes around 0 or ∞ in a "small circle," then x^5 goes around five times). Now the polynomials $x^5 + (4x + 8)t$ give us a continuous deformation from x^5 to $x^5 + 4x + 8$, and $x^5 + 4x + 8$ has a fixed point of index 5 at ∞. It follows that there must be other fixed points, with indices adding up to 5. These are the roots of $x^5 + 3x + 8$, and the indices are the multiplicities of the roots.

4   Infinite-Dimensional Fixed Point Theorems and Applications to Analysis

What happens if we try to generalize the Brouwer fixed point theorem to continuous maps defined on infinite-dimensional closed balls? The answer is that we will not be able to, as the following example shows. Let B be the set of all sequences $(a_1, a_2, \dots)$ such that $\sum_n |a_n|^2 \leqslant 1$. This is our closed ball; it is the unit ball of the HILBERT SPACE [III.37] ℓ_2. Given an infinite sequence $\boldsymbol{a} = (a_1, a_2, \dots)$, we write $\|\boldsymbol{a}\|$ for its norm $(\sum_n |a_n|^2)^{1/2}$. Now consider the map $f : (a_1, a_2, \dots) \mapsto ((1 - \|\boldsymbol{a}\|^2)^{1/2}, a_1, a_2, \dots)$. It is easy to check that f is continuous and that $\|f(\boldsymbol{a})\| = 1$ for every $\boldsymbol{a}$. Therefore, if $\boldsymbol{a}$ is a fixed point, we must have $\|\boldsymbol{a}\| = 1$, from which we can see that $a_1 = 0$. From this it follows that $a_2 = 0$, and then that $a_3 = 0$, and so on. In other words, $\boldsymbol{a} = 0$. But this contradicts the condition that $\|\boldsymbol{a}\| = 1$. Therefore, the map f has no fixed point.

However, if we place extra conditions on a continuous map, then it is sometimes possible to prove fixed point theorems, and some of these theorems have important applications, notably to establishing the existence of solutions to differential equations.

An easy result of this type is the *contraction mapping theorem*. This states that if X is a METRIC SPACE [III.56] with a property known as *completeness* (which is briefly discussed in NORMED SPACES AND BANACH SPACES [III.62]) and f is a map from X to X such that there exists a constant $\rho < 1$ such that $d(f(x), f(y)) \leqslant \rho d(x, y)$ for every x and y in X, then f must have a fixed point. To prove this, one picks any point $x \in X$ and looks at the iterates $x, f(x), f(f(x)), f(f(f(x)))$, and so on. Denoting these by $x_0, x_1, x_2, \dots$, one can prove quite easily that $d(x_n, x_m)$ tends to 0 as m and n both tend to infinity, and the completeness property then guarantees that the sequence (x_n) has a limit. It is not hard to prove that this limit is a fixed point of f.

A more sophisticated example is the *Schauder fixed point theorem*, which states that if X is a Banach space, K is a COMPACT [III.9] convex subset of X, and f is a continuous function from K to K, then f has a fixed point. Roughly speaking, to prove this one approximates K by larger and larger finite-dimensional sets K_n and approximates f by continuous maps f_n that take K_n to K_n. Brouwer's fixed point theorem gives a sequence (x_n) such that $f_n(x_n) = x_n$ for each n. The compactness of K implies that the sequence (x_n) has a convergent subsequence: its limit can be shown to be a fixed point of f.

The importance of these two theorems, and others of a similar nature, lies more in their applications than in their basic statements. A typical application is a proof that the differential equation

$$\frac{\mathrm{d}^2 u}{\mathrm{d} x^2} = u - 10 \sin(u^2) - 10 \exp(-|x|)$$

has a solution u such that $u(x)$ is defined for every real number x and tends to 0 as x tends to $\pm\infty$. We can rewrite this equation as

$$\left(1 - \frac{\mathrm{d}^2}{\mathrm{d} x^2}\right) u = 10 \sin(u^2) + 10 \exp(-|x|).$$

If we write the left-hand side as $L(u)$, then this equation can be further rewritten as

$$u = L^{-1}(10 \sin(u^2) + 10 \exp(-|x|)).$$

(It is possible to identify the operator L^{-1} explicitly.) If we now let X be the Banach space of continuous functions defined on $\mathbb{R}$ that tend to 0 at $\pm\infty$, with the uniform norm, then it can be shown that the right-hand side of this last equation defines a continuous function from X to a compact convex subset of X. Therefore, by the Schauder fixed point theorem, this highly nonlinear equation has a solution with the given boundary conditions, a result that is hard to prove in any other way.

V.12   The Four-Color Theorem
Bojan Mohar

The four-color theorem asserts that the regions of any map drawn in the plane (or, equivalently, on the two-dimensional sphere) can be colored with no more than four colors in such a way that any two regions with a common boundary are given different colors. The example in figure 1 shows that four distinct colors are necessary since the regions A, B, C, and D are all adjacent to each other. This result was conjectured by Francis Guthrie in 1852. An incorrect proof was given by Kempe in 1879, and for eleven years the problem was believed to have been solved, until Heawood pointed out the error in 1890. However, Heawood showed that Kempe's basic idea, which we shall outline below, could at least be used to give a correct proof that five colors were always sufficient. After that, the problem became a famous example of a question that remained stubbornly open despite being very easy to understand. (Another such problem was FERMAT'S LAST THEOREM [V.10].)

In modern mathematics, map-coloring problems are usually formulated in the language of graph theory. To

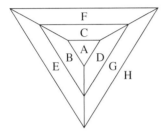

Figure 1 A map with eight regions.

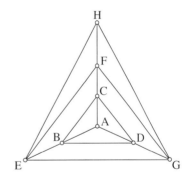

Figure 2 The graph of the map from figure 1.

any map we assign a GRAPH [III.34]: the vertices of the graph correspond to the regions of the map, and we declare two vertices to be adjacent if the corresponding regions share a piece of their boundary. The graph for the map in figure 1 is shown in figure 2. It is easy to see that the graph of any map in the plane can be drawn in such a way that no two edges cross each other: such graphs are called *planar*. Instead of coloring regions of maps, we now color vertices of the corresponding graphs. If no two vertices that are joined by an edge have the same color, then we say that the coloring is *proper*. After this reformulation, the four-color theorem states that every planar graph G has a proper coloring with at most four colors.

Here, briefly, is the proof of the *five*-color theorem due to Kempe and Heawood. It is a proof by contradiction, so we start by assuming that the result is false. If that is the case, then there must be a graph G of minimal size that has no proper coloring with five colors. EULER'S FORMULA [I.4 §2.2] says that $V - E + F = 2$ for any (connected) planar graph, where V is the number of vertices, E is the number of edges, and F is the number of regions into which the plane is divided by any drawing of the graph. It is not hard to deduce from this formula that G has a vertex v with at most five neighbors (that is, other vertices linked to v by an edge) in the graph. If we remove v from the graph, then we can find a proper coloring of what is left, because G is a minimal counterexample to the theorem. If v has *fewer* than five neighbors, then we can color v as well, since there are at most four colors that need to be avoided and we have five colors at our disposal. So the only thing that can go wrong is if v has five neighbors and those five colors all get different colors when we color the rest of G.

Let us suppose that the colors of the neighbors of v are red, yellow, green, blue, and brown, as we go clockwise around v. As it stands, we cannot color v, but we could try to do so by adjusting the coloring of the rest of the graph. For instance, we could try recoloring the

red vertex green, thereby freeing up red to be used for v. Of course, if we did that we might have to recolor further vertices, but we could try to find a recoloring as follows: first change the color of the red neighbor of v to green. Then change all the green neighbors of *that* vertex to red, and all the red neighbors of those vertices to green, and so on. When we have finished this process, the one thing that could go wrong is that we might end up recoloring the green neighbor of v red, in which case we would not after all be free to use red for v. This will happen if and only if there is a chain of vertices from the red neighbor of v to the green neighbor that alternates red and green. However, if this circumstance arises, we can try to recolor the yellow neighbor of v blue in a similar way. Once again, the only thing that can stop us is an alternating chain of yellow and blue vertices going from the yellow neighbor of v to the blue neighbor. But such a chain cannot exist, as it would at some point have to cross the red/green chain, and this contradicts the fact that the graph is planar.

Returning to the four-color problem, the German mathematician Heinrich Heesch proposed a general method for tackling it that can be thought of as a more complicated version of the above argument. The idea is to identify a list C of "configurations" with the following properties. First, every planar graph must contain a configuration X that belongs to C. Second, given a planar graph G that contains a configuration X from C, and given a proper coloring of the rest of G that uses at most four colors, it is possible to adjust this coloring in such a way that it can be extended to a proper coloring of the whole of G. In the proof of the five-color theorem above, there was a very simple list of five configurations: a vertex v with one edge, two edges, three edges, four edges, or five edges coming out of it. Nothing this simple works for the four-color problem, but Heesch's idea was that it might be possible to solve the problem by using a more complicated list of configurations.

Such a list was found by Kenneth Appel and Wolfgang Haken in 1976. However, this is by no means the whole story, because the list of configurations that they found was not just "more complicated" but so much more complicated that it broke new ground: it was the first time that a major theorem had been proved with a proof that was too long to be humanly checkable. The reason for this was partly that their list C contained about 1200 configurations, but a more important reason was that for some configurations X it was necessary to check hundreds of thousands of cases in order to demonstrate that a coloring of the rest of the graph could be adjusted to accommodate a coloring of X as well. Therefore, there was no alternative but to use a computer to do the checking. (Heesch had himself proposed a list, but some of his configurations would have involved so many cases that even a computer could not have checked them all.)

The reaction of other mathematicians to the proof of Appel and Haken was mixed. Some hailed it as the addition of a powerful new tool to the mathematical armory. Others were uneasy about having to trust that the relevant computer program had been written correctly and that the computer had operated as it should. And in fact the proof turned out to have several flaws, though all those that were discovered were subsequently corrected by Appel and Haken in their monograph of 1989. Any doubts there may have been of this kind were removed once and for all in 1997, when Robertson, Sanders, Seymour, and Thomas developed another proof based on similar principles. The part of the proof that was checkable by humans was made more transparent, and the computer-verified part was supported by a well-structured collection of data that enabled the proofs to be checked independently. One could still question whether the compilers used were correct and whether the hardware was stable, but the proof has been checked on different platforms, using different programming languages and operating systems, so this proof is much less likely to be incorrect than a typical human-checked proof of even moderate length.

The result is that very few mathematicians are now worried about whether the proof is correct. However, there are many who object to it for a different reason. Even if we can now be certain that the theorem is true, we can still ask *why* it is true, and not everybody regards the answer "Because hundreds of thousands of cases were checked and they all turned out

to be OK" as a satisfactory explanation. As a result, if someone were to discover a shorter and more accessible proof it would be regarded by many as a breakthrough comparable to the solution of the problem by Appel and Haken. An unfortunate side effect of this is that mathematics departments around the world still receive many incorrect attempted proofs, several of which repeat the mistake of Kempe.

Like many good problems, the four-color problem provoked the development of many important new mathematical ideas. The theory of graph colorings, in particular, has evolved into a deep and beautiful area of research. (See EXTREMAL AND PROBABILISTIC COMBINATORICS [IV.19 §2.1.1] and also Jensen and Toft (1995).) Extensions of map-coloring problems to arbitrary surfaces led to the development of topological graph theory, and questions about the planarity of graphs culminated in the theory of GRAPH MINORS [V.32].

One of the most prolific graph theorists, William T. Tutte, judged the impact of the four-color theorem on mathematics by proclaiming: "The four-colour theorem is the tip of the iceberg, the thin end of the wedge, and the first cuckoo of Spring."

Further Reading

Appel, K., and W. Haken. 1976. Every planar map is four colorable. *Bulletin of the American Mathematical Society* 82:711–12.

——. 1989. *Every Planar Map Is Four Colorable*. Contemporary Mathematics, volume 98. Providence, RI: American Mathematical Society.

Jensen, T., and B. Toft. 1995. *Graph Coloring Problems*. New York: John Wiley.

Robertson, N., D. Sanders, P. Seymour, and R. Thomas. 1997. The four-colour theorem. *Journal of Combinatorial Theory* B 70:2–44.

V.13 The Fundamental Theorem of Algebra

The COMPLEX NUMBERS [I.3 §1.5] can be thought of as what you obtain from the REAL NUMBERS [I.3 §1.4] when you introduce a new number, denoted i, and stipulate that it is a solution of the equation $x^2 = -1$, or equivalently a root of the polynomial $x^2 + 1$. At first, this may seem an artificial thing to do—it is not obvious what is so important about $x^2 + 1$ as opposed to any other polynomial—but that is a judgment with which no professional mathematician would concur. The fundamental theorem of algebra is one of the best pieces of evidence that the complex number system is, in fact, natural, and natural in a profound way. It states that, within the complex number system, *every* polynomial has a root. In other words, once we introduce the number i, then not only can we solve the equation $x^2 + 1 = 0$, we can solve all polynomial equations (even if the coefficients are themselves complex). Thus, when one defines the complex numbers, one gets much more out of them than one puts in. It is this that makes them seem not an artificial construction but a wonderful discovery.

For many polynomials it is not hard to see that they have roots. For example, if $P(x) = x^d - u$ for some positive integer d and some complex number u, then a root of P will be a dth root of u. One can write u in the form $re^{i\theta}$, and then $r^{1/d}e^{i\theta/d}$ will be such a root. This means that any polynomial that can be solved by a formula involving dth roots and the usual arithmetical operations, which includes all polynomials of degree less than 5, can be solved in the complex number system. However, owing to THE INSOLUBILITY OF THE QUINTIC [V.21], not all polynomials can be dealt with in this way, and in order to prove the fundamental theorem of algebra one must look for a less direct argument.

In fact, this is true even if one is looking for real roots of real polynomials. For example, if $P(x) = 3x^7 - 10x^6 + x^3 + 1$, then we know that $P(x)$ is large and positive when x is, since the x^7 term is by far the most significant, and large and negative when x is, for the same reason. Therefore, at some point the graph of P crosses the x-axis, which means that there is some x with $P(x) = 0$. Notice that this argument does not tell us what x is—that is the sense in which it is "less direct."

Now let us see how one might show that a polynomial has a complex root, by looking at the example $P(x) = x^4 + x^2 - 6x + 9$. This can be rewritten $x^4 + (x-3)^2$, and since both x^4 and $(x-3)^2$ are nonnegative, and since they cannot be zero simultaneously, P cannot have a real root. To see that it has a complex root, we shall begin by fixing a large real number r and looking at the behavior of $P(re^{i\theta})$ as θ varies between 0 and 2π. As θ varies in this way, $re^{i\theta}$ traces out a circle of radius r in the complex plane.

Now $(re^{i\theta})^4 = r^4e^{4i\theta}$, so the x^4 part of $P(re^{i\theta})$ traces out a circle of radius r^4, but goes around it four times. If r is large enough, then the rest (that is, $(re^{i\theta} - 3)^2$) is so small compared with $(re^{i\theta})^4$ that the only effect on the behavior of $P(re^{i\theta})$ is to make it deviate very slightly from the circle of radius r^4. This small deviation is not

enough to stop the path of $P(re^{i\theta})$ going around zero four times.

Next, let us consider what happens when r is very small. Then $P(re^{i\theta})$ is very close to 9, whatever the value of θ, since $(re^{i\theta})^4$, $(re^{i\theta})^2$, and $(re^{i\theta})$ are all small. But this means that the path traced out by $P(re^{i\theta})$ does not go around zero at all.

For any r we can ask how many times the path traced out by $P(re^{i\theta})$ goes around zero. What we have just established is that for very large r the answer is four and for very small r it is zero. It follows that at some intermediate r the answer changes. But if you gradually shrink r, the path traced out by $P(re^{i\theta})$ varies in a continuous way, so the only way this change can come about is if for some r the path crosses 0. This gives us the root we are looking for, since the path consists of points of the form $P(re^{i\theta})$ and one of these points is 0.

Some care is needed to turn the above reasoning into a rigorous proof. However, this can be done, and it is not hard to generalize the resulting argument to one that applies to any polynomial.

The fundamental theorem of algebra is usually attributed to GAUSS [VI.26], who proved it in 1799 in his doctoral thesis. Though his argument (which was different from the one sketched above) was not fully rigorous by today's standards, it was convincing and broadly correct. Later he went on to give three more proofs.

V.14   The Fundamental Theorem of Arithmetic

The fundamental theorem of arithmetic is the assertion that every positive integer can be expressed in exactly one way as a product of prime numbers. These prime numbers are known as the *prime factors* of the original number and the product itself is the *prime factorization*. To give a few examples: $12 = 2 \times 2 \times 3$, $343 = 7 \times 7 \times 7$, $4559 = 47 \times 97$, and 7187 is itself a prime. This last number shows that the word "product" should be interpreted so as to include the case where there is only one prime involved. As for the phrase "exactly one way," it is understood that the order in which the primes are multiplied is not significant, so, for example, the products 47×97 and 97×47 are not regarded as different.

The following inductive procedure allows one to find the prime factorization of a given positive integer n. If n is prime, then we have found it already. Otherwise, let p be the smallest prime factor of n and let $m = n/p$. Since m is smaller than n, we know by induction how to find the prime factorization of m, and this, together with p, gives it to us for n. In practice, what this means is that we generate a sequence of numbers, where each number in the sequence is the previous one divided by its smallest prime factor. For example, if we start with the number 168, then the sequence begins 168, 84, 42, 21. At this point we cannot divide by 2, but 3 is a factor of 21 so the next number in the sequence is 7. Since 7 is a prime, the process stops. Looking back, we find that we have shown that $168 = 2 \times 2 \times 2 \times 3 \times 7$.

Once one is used to this method, it comes to seem inconceivable that a number could have two genuinely different prime factorizations. But the method does not guarantee this at all. Suppose we successively divide by the largest prime factor rather than the smallest. Why should this not give a completely different set of primes? It is hard to think of an argument that does not use a phrase such as "the prime factorization of n," thereby implicitly assuming what it sets out to prove.

It is possible to show in a rather precise way that the fundamental theorem of arithmetic is *not* obvious, by looking at an algebraic structure where the notion of prime factorization makes sense but numbers can have more than one prime factorization. This structure, denoted $\mathbb{Z}(\sqrt{-5})$, is the set of all numbers of the form $a + b\sqrt{-5}$, where a and b are integers. Such numbers can be added and multiplied just like ordinary integers. For example,

$$(1 + 3\sqrt{-5}) + (6 - 7\sqrt{-5}) = 7 - 4\sqrt{-5}$$

and

$$\begin{aligned}
(1 + 3\sqrt{-5})&(6 - 7\sqrt{-5}) \\
&= 6 - 7\sqrt{-5} + 18\sqrt{-5} - 21(\sqrt{-5})^2 \\
&= 6 + 11\sqrt{-5} + 21 \times 5 \\
&= 111 + 11\sqrt{-5}.
\end{aligned}$$

In this structure, we can regard a number $x = a + b\sqrt{-5}$ as prime if its only factors are ± 1 and $\pm x$. (This would also be a natural definition if we wanted to extend the notion of primes from the positive integers to all integers.) It can be shown quite easily that 2 and 3 are both primes (though it is not immediately obvious since there are now more possibilities for factors). Two other primes are $1 + \sqrt{-5}$ and $1 - \sqrt{-5}$. But we can write 6 either as 2×3 or as $(1 + \sqrt{-5})(1 - \sqrt{-5})$, so 6 has two different prime factorizations. For a further discussion of this point see ALGEBRAIC NUMBERS [IV.1 §§4–8].

What this example shows is that any proof of the fundamental theorem of arithmetic must use some feature of $\mathbb{Z}$, the set of integers, that is lacking in $\mathbb{Z}(\sqrt{-5})$. Since addition and multiplication work in a very similar way in both structures, it is not very easy to find such a feature, or at least not one that is relevant. It turns out that the important property that $\mathbb{Z}(\sqrt{-5})$ does not have is an appropriate analogue of the following basic principle for integers: that if m and n are integers, then one can write $n = qm + r$ with $0 \leqslant r < |m|$. This fact underlies EUCLID'S ALGORITHM [III.22], which plays an important role in the most commonly given proof of unique factorization.

The Fundamental Theorem of Calculus
See SOME FUNDAMENTAL MATHEMATICAL DEFINITIONS [I.3 §5.5]

Gauss's Law of Quadratic Reciprocity
See FROM QUADRATIC RECIPROCITY TO CLASS FIELD THEORY [V.28]

V.15 Gödel's Theorem
Peter J. Cameron

In response to problems in the foundations of mathematics such as *Russell's paradox* ("consider the set of all sets which are not members of themselves; is it a member of itself?"), HILBERT [VI.63] proposed that the consistency of any given part of mathematics should be established by finitary methods that could not lead to a contradiction. Any part for which this had been done could then be used as a secure foundation for all of mathematics.

An example of a "part of mathematics" is the arithmetic of the natural numbers, which can be described in terms of FIRST-ORDER LOGIC [IV.23 §1]. We begin with symbols, both logical (connectives such as "not" and "implies," quantifiers such as "for all," the equality symbol, symbols for variables, and punctuation) and nonlogical (symbols for constants, relations, and functions suitable for the branch of mathematics under consideration). *Formulas* are finite strings of symbols built according to certain precise rules (which allow them to be mechanically recognized). We fix a certain set of formulas as our *axioms*, and we also choose a few *rules of inference* that allow us to infer some formulas from others. An example of a rule of inference is *modus ponens*: if we have inferred ϕ and $(\phi \rightarrow \psi)$, then we can infer ψ. A *theorem* is a formula that is at the end of a chain (or tree) of inferences that starts with axioms.

Axioms for the natural numbers were given by PEANO [VI.62] (see THE PEANO AXIOMS [III.67]). The nonlogical symbols are zero, the "successor function" s, addition, and multiplication. (The last two can be defined in terms of the others by inductive axioms: for example, the rules $x + 0 = x$ and $x + s(y) = s(x + y)$ define addition.) The crucial axiom is the *principle of induction*, which asserts that if $P(n)$ is a formula such that $P(0)$ is true and $P(n)$ implies $P(s(n))$ for all n, then $P(n)$ is true for all n. Hilbert's specific challenge was to give a formal proof of the consistency of this theory: that is, a proof that no contradiction can be deduced from the axioms by the rules of first-order logic.

Hilbert's program was undone by two remarkable *incompleteness theorems* proved by GÖDEL [VI.92]. The first theorem states the following.

There are (first-order) statements about the natural numbers that can be neither proved nor disproved from Peano's axioms.

(This is sometimes qualified by being prefixed with, "If Peano's axioms are consistent, then...." However, since we accept the existence of the natural numbers, we do know that Peano's axioms are consistent, as the natural numbers model them. So the qualification is unnecessary here, although it would need to be included if we were discussing some axioms whose consistency was not clear.)

Gödel's proof is long, but it is based on two simple ideas. The first is *Gödel numbering*, which is a means of encoding each formula or sequence of formulas as a natural number in a systematic and mechanical way.

It can be shown that there is a two-variable formula $\pi(x, y)$ such that $\pi(m, n)$ holds if and only if "n is a proof of m," which is a shorthand way of saying that m is the Gödel number of a formula ϕ and n is the Gödel number of a string of formulas that constitutes a proof of ϕ. Slightly more elaborately, there is a formula $\omega(x, y)$ such that $\omega(m, n)$ holds if and only if m is the Gödel number of a formula ϕ that has one free variable and n is the Gödel number of a proof of $\phi(m)$. (A free variable is one that is not quantified over. For example, $\phi(x)$ might be the formula $(\exists y)y^2 = x$, in which case x is the free variable. For this choice of ϕ, the number n would be the Gödel number of a proof that the Gödel number of ϕ was a perfect square.)

Now let $\psi(x)$ be the formula $(\forall y)(\neg \omega(x, y))$. If ϕ is a formula (with one free variable) with Gödel number m, then $\psi(m)$ tells us that there is no proof of $\phi(m)$. (It tells us this indirectly: what it actually says is that

there is no y that is the Gödel number of such a proof.) Let p be the Gödel number of ψ itself, and let ζ be the formula $\psi(p)$.

This brings us to the second idea in the proof: *self-reference*. The formula ζ is carefully devised so that it asserts its own unprovability, since $\psi(p)$ tells us that there is no proof of the formula $\phi(p)$, where ϕ is the formula with Gödel number p. In other words, it tells us that there is no proof of $\psi(p)$. Since ζ asserts its own unprovability, it must be unprovable (since a proof of ζ would be a proof that ζ had no proof, which is absurd). Since ζ asserts its unprovability and is unprovable, it is true, and since it is true it cannot be disproved. (One might wonder why this argument that ζ is true does not constitute a proof of ζ. The answer is that although it *is* a rigorous demonstration of the truth of ζ, it is not a proof in Peano arithmetic. That is, it is not an argument that starts from the Peano axioms and uses the rules of inference of the kind we discussed earlier.)

Gödel numbering also allowed Gödel to consider the consistency of the axioms as a first-order formula: namely $(\forall y)(\neg(\pi(m, y)))$, where m is the Gödel number of the formula $0 = s(0)$ (or any other contradiction). Here is Gödel's second theorem.

It is impossible to prove from Peano's axioms that they are consistent.

The proofs of these theorems are not specific to the Peano axioms, but apply to any (consistent) system of mechanically recognizable axioms that is powerful enough to describe the natural numbers. Thus, completeness cannot be restored simply by adding a true but unprovable statement as a new axiom, for the resulting system is still strong enough for Gödel's theorem to apply to it.

It might seem that we could obtain a complete axiomatization of the natural numbers by simply taking all true statements as axioms. However, one requirement for Gödel's theorems is that the axioms should be recognizable by some mechanical method. (This is needed to construct the formula $\pi(x, y)$ at the start of the proof.) Indeed, we can deduce from this that (as subsequently pointed out by TURING [VI.94]) the true statements about the natural numbers *cannot* be mechanically recognized (that is, their Gödel numbers do not form a *recursive set*).

Gödel's true but unprovable statement is important for the foundations of mathematics, but it has no intrinsic interest in its own right. Later, Paris and

Harrington gave the first example of a mathematically significant statement that is unprovable from Peano's axioms. Their statement is a variant of RAMSEY'S THEOREM [IV.19 §2.2]. Subsequently, many other "natural incompletenesses" have been found.

Of course, the consistency of Peano's axioms can be proved in a stronger system, since we could just add the (unprovable) consistency statement. Less trivially, since a model of the natural numbers can be constructed within set theory, the consistency of Peano arithmetic can be proved from THE ZERMELO–FRAENKEL AXIOMS [IV.22 §3.1] (known as ZFC) for set theory. Of course, ZFC cannot prove its own consistency, but the consistency of ZFC can be deduced from a yet stronger system (for example, adding an axiom that asserts the existence of a suitably "large" cardinal number such as an INACCESSIBLE CARDINAL [IV.22 §6]).

For small enough parts of mathematics, it is sometimes possible to find complete axiom systems (that is, systems that allow one to prove every true statement). For instance, this can be done for the theory of the natural numbers with zero, the successor function, and addition alone. Thus, multiplication is essential to Gödel's argument.

It is more elementary to see that Peano's axioms are not *categorical*: there are models for the axioms that are not isomorphic to the natural numbers. Such *nonstandard models of arithmetic* contain infinitely large numbers (that is, numbers that are larger than all natural numbers).

Gödel's theorem has been a battleground for philosophers arguing about whether the human brain is a deterministic machine (in which case, presumably, we would not be able to prove any formally unprovable statement). Fortunately, there is not enough space in this article for more details!

The Goldbach Conjecture

See PROBLEMS AND RESULTS IN ADDITIVE
NUMBER THEORY [V.27]

V.16   Gromov's Polynomial-Growth Theorem

If G is a group and $g_1, \dots, g_k$ are generators of G (meaning that every element of G can be expressed as a product of the g_i and their inverses), then we can define a *Cayley graph* by taking the elements of G as vertices

and joining g to h if there is some i such that h is equal either to gg_i or to gg_i^{-1}.

For each r, let y_r be the number of elements that are at a distance of at most r from the identity: that is, the number of elements that can be written as a "word" of length at most r in the generators and their inverses. (For instance, if $g = g_1 g_4 g_2^{-3}$, then we know that g belongs to y_5.) It turns out that if G is an infinite group, then the rate of growth of the sizes of the sets y_r can tell one a great deal about G; this is particularly true when the growth is less than exponential. (The growth is always bounded above by an exponential function, since there are at most exponentially many words of a given length in the generators $g_1, \dots, g_r$.)

If G is an Abelian group generated by $g_1, \dots, g_k$, then every element of y_r is of the form $\sum_{i=1}^k a_i g_i$, where $a_1, \dots, a_k$ are integers such that $\sum_{i=1}^k |a_i| \leqslant r$. It follows easily that the size of y_r is at most $(2r+1)^k$ (and with a bit more effort one can improve this bound). Thus, as r tends to infinity, the growth rate of y_r is bounded above by a polynomial of degree k in r. If G is the FREE GROUP [IV.10 §2] generated by $g_1, \dots, g_k$, then all words of length r in the elements g_i (but not their inverses) give rise to distinct elements of G, so the size of y_r is at least k^r. Thus, in this case the growth rate is exponential. More generally, there will be an exponential growth rate whenever G contains a non-Abelian free subgroup.

These observations suggest that the growth rate is likely to be smaller if G is more like an Abelian group. Gromov's theorem is a remarkably precise result along these lines. It states that the growth rate of the sets y_r is bounded above by a polynomial in r if and only if G has a nilpotent subgroup of finite index. This condition does indeed say that G is somewhat like an Abelian group, since nilpotent groups are "close to Abelian" and a subgroup of finite index is "close to the whole group." For example, a typical nilpotent group is the *Heisenberg group*, which consists of all 3×3 matrices with 0s below the diagonal, 1s on the diagonal, and integers above the diagonal. Given any two such matrices X and Y, the products XY and YX differ only in the top right-hand corner, and the "error matrix" $XY - YX$ commutes with everything in the group. In general, a nilpotent group is built out of Abelian groups in a controlled manner in a finite number of steps.

A fuller discussion of the theorem, including the exact definition of "nilpotent," can be found in GEOMETRIC AND COMBINATORIAL GROUP THEORY [IV.10]. Here we highlight the fact that it is a beautiful example of a *rigidity theorem*: if a group behaves roughly in the way that a nilpotent group would (because the growth rate of the sets y_r is polynomial), then it must in fact be related to a nilpotent group in a very precise and algebraic way. (See MOSTOW'S STRONG RIGIDITY THEOREM [V.23] for another example of such a theorem.)

V.17   Hilbert's Nullstellensatz

Let $f_1, \ldots, f_n$ be a collection of polynomials in d complex variables $z_1, \ldots, z_d$. Suppose that it is possible to find another collection of polynomials $g_1, \ldots, g_n$ such that

$$f_1(z)g_1(z) + f_2(z)g_2(z) + \cdots + f_n(z)g_n(z) = 1$$

for every complex d-tuple $z = (z_1, \ldots, z_d)$. Then it follows immediately that no such d-tuple can be a root of every single f_i, since otherwise the left-hand side would equal 0. Remarkably, the converse also holds: that is, if there is no d-tuple for which the polynomials f_i all vanish simultaneously, then it is possible to find polynomials g_i such that the above identity holds. This result is known as the *weak Nullstellensatz*.

A short (but clever) argument can be used to deduce *Hilbert's Nullstellensatz* from the weak Nullstellensatz. This again is a statement where a condition that is obviously necessary turns out to be sufficient. Suppose that h is another polynomial in d complex variables, that r is a positive integer, and that the polynomial h^r can be written in the form $f_1 g_1 + f_2 g_2 + \cdots + f_n g_n$ for some collection of polynomials $g_1, \ldots, g_n$. It follows immediately that $h(z) = 0$ whenever $f_i(z) = 0$ for every i. Hilbert's Nullstellensatz states that if $h(z) = 0$ whenever $f_i(z) = 0$ for every i, then there must be some positive integer r and some collection of polynomials $g_1, \ldots, g_n$ such that $h^r = f_1 g_1 + f_2 g_2 + \cdots + f_n g_n$.

Hilbert's Nullstellensatz is discussed further in ALGEBRAIC GEOMETRY [IV.4 §§5, 12].

V.18   The Independence of the Continuum Hypothesis

The real numbers are UNCOUNTABLE [III.11], but do they form the "smallest" uncountable set? Equivalently, is it the case that if A is any set of real numbers, then either A is countable or there is a bijection between A and the set of all real numbers? The *continuum hypothesis (or CH)* is the assertion that this is indeed true. The notions of countability and uncountability were invented by CANTOR [VI.54], who was also the first to

formulate CH. He tried hard to prove or disprove it, as did many others after him, but nobody succeeded.

Gradually, mathematicians came to entertain the idea that CH might be "independent" of normal mathematics: that is, independent of the usual ZFC AXIOMS [IV.22 §3.1] of set theory. This would mean that it could be neither proved nor disproved from the ZFC axioms.

The first result in this direction was due to GÖDEL [VI.92], who showed that CH could not be *disproved* from the usual axioms. In other words, one could not reach a contradiction by assuming CH. To do this, he showed that inside every MODEL OF SET THEORY [IV.22 §3.2] there is a model in which CH holds. This model is called the "constructible universe." Roughly speaking, it consists just of those sets that "have to exist" if the axioms are true. So, in this model, the set of reals is as small as it could possibly be. The "smallest uncountable size" is usually denoted $\aleph_1$, and in Gödel's construction the reals appear in $\aleph_1$ stages, with only countably many reals appearing at each stage. From this one can deduce that the number of reals is $\aleph_1$, which is precisely the assertion of CH.

The other direction had to wait thirty years, until Paul Cohen invented the method of *forcing*. How would we make CH false? Starting from some model of set theory (in which CH might well hold), we would like to "add" some reals to it. Indeed, we would like to add enough that there are now more than $\aleph_1$ of them. But how do we "add" a real? We need to ensure that what we end up with is still a model of set theory, which is hard enough, but also that when we add new reals we do not alter the value of $\aleph_1$ (since otherwise the statement "the number of reals is $\aleph_1$" may still be true in the new model). This is an extremely complicated task, both conceptually and technically. See SET THEORY [IV.22] for more details about how it is carried out.

V.19   Inequalities

Let x and y be two nonnegative real numbers. Then $(\sqrt{x} - \sqrt{y})^2 = x + y - 2\sqrt{xy}$ is a nonnegative real number, from which it follows that $\frac{1}{2}(x+y) \geqslant \sqrt{xy}$. That is, the *arithmetic mean* of x and y is at least as big as the *geometric mean*. This conclusion is a very simple example of a mathematical inequality; its generalization to n numbers is called the *AM-GM inequality*.

In any branch of mathematics that has even the slightest flavor of analysis, inequalities will be of great importance: as well as analysis itself, this includes probability, and parts of combinatorics, number theory, and geometry. Inequalities are less prominent in some of the more abstract parts of analysis, but even there one needs them as soon as one wishes to apply the abstract results. For instance, one may not always need an inequality to prove a theorem about continuous LINEAR OPERATORS [III.50] between BANACH SPACES [III.62], but the statement that some specific linear operator between two specific Banach spaces is continuous is an inequality, and often a very interesting one. We do not have space to discuss more than a small handful of inequalities in this article, but we shall include some of the most important ones in the toolbox of any analyst.

Jensen's inequality is another fairly simple but useful inequality. A function $f : \mathbb{R} \to \mathbb{R}$ is called *convex* if $f(\lambda x + \mu y) \leqslant \lambda f(x) + \mu f(y)$ whenever λ and μ are nonnegative real numbers with $\lambda + \mu = 1$. Geometrically, this says that all chords of the graph of the function lie above the graph. A straightforward inductive argument can be used to show that this property implies the same property for n numbers:

$$f(\lambda_1 x_1 + \cdots + \lambda_n x_n) \leqslant \lambda_1 f(x_1) + \cdots + \lambda_n f(x_n)$$

whenever all the λ_i are nonnegative and $\lambda_1 + \cdots + \lambda_n = 1$. This is Jensen's inequality.

The second derivative of the EXPONENTIAL FUNCTION [III.25] is positive, from which it follows that the exponential function itself is convex. If $a_1, \ldots, a_n$ are positive real numbers and we apply Jensen's inequality to the numbers $x_i = \log(a_i)$, then we find, using standard properties of exponentials and LOGARITHMS [III.25 §4], that

$$a_1^{\lambda_1} \cdots a_n^{\lambda_n} \leqslant \lambda_1 a_1 + \cdots + \lambda_n a_n.$$

This is called the *weighted AM-GM inequality*. When all the λ_i are equal to $1/n$ it reduces to the usual AM-GM inequality. Applying Jensen's inequality to other well-known convex functions produces several other well-known inequalities. For instance, if we apply it to the function x^2, we obtain the inequality

$$(\lambda_1 x_1 + \cdots + \lambda_n x_n)^2 \leqslant \lambda_1 x_1^2 + \cdots + \lambda_n x_n^2, \quad (1)$$

which can be interpreted as saying that if X is a RANDOM VARIABLE [III.71 §4] on a finite sample space, then $(\mathbb{E}X)^2 \leqslant \mathbb{E}X^2$.

The *Cauchy-Schwarz inequality* is perhaps the most important inequality in all of mathematics. Suppose that V is a real vector space with an INNER PRODUCT [III.37] $\langle \cdot, \cdot \rangle$ on it. One of the properties of an inner product is that $\langle v, v \rangle \geqslant 0$ for every $v \in V$, with equality if and only if $v = 0$. Let us write $\|v\|$ for

$\langle v, v \rangle^{1/2}$. If x and y are any two vectors in V with $\|x\| = \|y\| = 1$, then $0 \leqslant \|x - y\|^2 = \langle x - y, x - y \rangle = \langle x, x \rangle + \langle y, y \rangle - 2\langle x, y \rangle = 2 - 2\langle x, y \rangle$. It follows that $\langle x, y \rangle \leqslant 1 = \|x\| \|y\|$. Moreover, equality holds only if $x = y$. We can obtain a general pair of vectors by multiplying x by λ and y by μ, for some nonnegative real numbers λ and μ. Then both sides of the inequality scale up by a factor of $\lambda\mu$, so we can conclude that the inequality $\langle x, y \rangle \leqslant \|x\| \|y\|$ holds in general, with equality if and only if x and y are proportional.

Particular inner-product spaces lead to special cases of this inequality, which are themselves often referred to as the Cauchy-Schwarz inequality. For instance, if we take the space $\mathbb{R}^n$ with the inner product $\langle a, b \rangle = \sum_{i=1}^n a_i b_i$, then we obtain the inequality

$$\sum_{i=1}^n a_i b_i \leqslant \left(\sum_{i=1}^n a_i^2 \right)^{1/2} \left(\sum_{i=1}^n b_i^2 \right)^{1/2}. \quad (2)$$

It is not hard to deduce a similar inequality for complex scalars: one needs to replace a_i^2 and b_i^2 by $|a_i|^2$ and $|b_i|^2$ on the right-hand side. It is also not too hard to prove that inequality (2) is equivalent to the inequality (1) above.

Hölder's inequality is an important generalization of the Cauchy-Schwarz inequality. Again it has several versions, but the one that corresponds to inequality (2) is

$$\sum_{i=1}^n a_i b_i \leqslant \left(\sum_{i=1}^n |a_i|^p \right)^{1/p} \left(\sum_{i=1}^n |b_i|^q \right)^{1/q},$$

where p belongs to the interval $[1, \infty]$ and q is the *conjugate index* of p, which is defined to be the number that satisfies the equation $(1/p) + (1/q) = 1$. (We interpret $1/\infty$ to be 0.) If we write $\|a\|_p$ for the quantity $(\sum_{i=1}^n |a_i|^p)^{1/p}$, then this inequality can be rewritten in the succinct form $\langle a, b \rangle \leqslant \|a\|_p \|b\|_q$.

It is a straightforward exercise to find, for each sequence a, another (nonzero) sequence b such that equality occurs in the above inequality. Also, both sides of the inequality scale in the same way if you multiply b by a nonnegative scalar. It follows that $\|a\|_p$ is the maximum of $\langle a, b \rangle$ over all sequences b such that $\|b\|_q = 1$. Using this fact, it is easy to verify that the function $a \mapsto \|a\|_p$ satisfies *Minkowski's inequality*: $\|x + y\|_p \leqslant \|x\|_p + \|y\|_p$.

This gives some idea of why Hölder's inequality is so important. Once one has Minkowski's inequality, it is very easy to check that $\|\cdot\|_p$ is (as the notation suggests) a NORM [III.62] on $\mathbb{R}^n$. This is an even more basic example of the phenomenon mentioned at the beginning of the article: just to show that a certain normed space *is* a normed space, we have had to prove an inequality about real numbers. In particular, looking at the case $p = 2$, we see that the entire theory of HILBERT SPACES [III.37] depends on the Cauchy-Schwarz inequality.

Minkowski's inequality is a particular case of the *triangle inequality*, which states that if x, y, and z are three points in a METRIC SPACE [III.56], then $d(x, z) \leqslant d(x, y) + d(y, z)$, where $d(a, b)$ denotes the distance between a and b. When put like this, the triangle inequality is a tautology, since it is one of the axioms of a metric space. However, the statement that a particular notion of distance actually *is* a metric is far from vacuous. If our space is $\mathbb{R}^n$ and we define $d(a, b)$ to be $\|a - b\|_p$, then Minkowski's inequality is easily seen to be equivalent to the triangle inequality for this notion of distance.

The inequalities above have natural "continuous analogues" as well. For example, here is a continuous version of Hölder's inequality. For two functions f and g defined on $\mathbb{R}$, let $\langle f, g \rangle$ be defined to be $\int_{-\infty}^{\infty} f(x)g(x)\,dx$, and write $\|f\|_p$ for the quantity $(\int_{-\infty}^{\infty} |f(x)|^p)^{1/p}$. Then, once again, $\langle f, g \rangle \leqslant \|f\|_p \|g\|_q$, where q is the conjugate index of p. Another example is a continuous version of Jensen's inequality, which states, in a continuous setting, that if f is convex and X is a random variable, then $f(\mathbb{E}X) \leqslant \mathbb{E}f(X)$.

In all the inequalities we have so far mentioned, we have been comparing two quantities A and B, and it has been easy to identify the extreme cases where the ratio of A to B is maximized. However, not all inequalities are like this. Consider, for instance, the following two quantities associated with a sequence of real numbers $a = (a_1, a_2, \ldots, a_n)$. The first is the norm $\|a\|_2 = (\sum_{i=1}^{n} a_i^2)^{1/2}$. The second is the average of $|\sum_{i=1}^{n} \epsilon_i a_i|$ over all the 2^n sequences $(\epsilon_1, \epsilon_2, \ldots, \epsilon_n)$ such that each ϵ_i is 1 or -1. (In other words, for each i you randomly decide whether to multiply a_i by -1 or not, add up the results, and take the expected absolute value of the sum.) It is not the case that the first quantity is always less than the second. For instance, let $n = 2$, and let $a_1 = a_2 = 1$. Then the first quantity is $\sqrt{2}$ and the second is 1. However, *Khinchin's inequality* (or to be more accurate an important special case of Khinchin's inequality) is the remarkable statement that there is a constant C such that the first quantity is never more than C times the second. It is not hard to prove, using the inequality $\mathbb{E}X^2 \geqslant (\mathbb{E}X)^2$, that the first quantity is always at *least* as big as the second; so the two rather different looking quantities are in fact "equivalent, up to a constant." But what

is the best constant? In other words, how much bigger can the first quantity be than the second? This question was not answered until 1976, by Stanislaw Szarek, over fifty years after Khinchin proved the original inequality. The answer turns out to be that the example given earlier is the extreme one: the ratio can never exceed $\sqrt{2}$.

This situation is typical. Another famous inequality for which the best constant was discovered much later than the inequality itself is the *Hausdorff-Young inequality*, which relates norms of functions with norms of their FOURIER TRANSFORMS [III.27]. Suppose that $1 \leqslant p \leqslant 2$, and that f is a function from $\mathbb{R}$ to $\mathbb{C}$ with the property that the norm

$$\|f\|_p = \left(\int_{-\infty}^{\infty} |f(x)|^p \, dx \right)^{1/p}$$

exists and is finite. Let $\hat{f}$ be the Fourier transform of f and let q be the conjugate index of p. Then $\|\hat{f}\|_q \leqslant C_p \|f\|_p$ for some constant C_p that depends on p only (and not on f). Again, it was an open problem for many years to determine the best constant C_p. Some idea of why it might have been difficult can be gleaned from the fact that the "extreme" functions in this case are Gaussians: that is, functions of the form $f(x) = e^{-(x-\mu)^2/2\sigma^2}$. A sketch of a proof of the Hausdorff-Young inequality can be found in HARMONIC ANALYSIS [IV.11 §3].

There is an important class of inequalities known as *geometric inequalities*, where the quantities that are being compared are parameters associated with geometric objects. A famous example of such an inequality is the *Brunn-Minkowski inequality*, which states the following. Let A and B be two subsets of $\mathbb{R}^n$, and define $A + B$ to be the set $\{x + y : x \in A, y \in B\}$. Then

$$(\mathrm{vol}(A + B))^{1/n} \geqslant \mathrm{vol}(A)^{1/n} + \mathrm{vol}(B)^{1/n}.$$

Here, $\mathrm{vol}(X)$ denotes the n-dimensional volume (or, more formally, the LEBESGUE MEASURE [III.55]) of the set X. The Brunn-Minkowski inequality can be used to prove the equally famous *isoperimetric inequality* in $\mathbb{R}^n$ (which is one of a large class of isoperimetric inequalities). Informally, this states that, of all sets with a given volume, the one with the smallest surface area is a sphere. An explanation of why this follows from the Brunn-Minkowski inequality can be found in HIGH-DIMENSIONAL GEOMETRY AND ITS PROBABILISTIC ANALOGUES [IV.26 §3].

We finish this brief sample with one further inequality, the *Sobolev inequality*, which is important in the theory of partial differential equations. Suppose that f is a differentiable function from $\mathbb{R}^2$ to $\mathbb{R}$. We can visualize its graph as a smooth surface in $\mathbb{R}^3$ lying above the xy-plane. Suppose also that f is *compactly supported*, which means that there exists an M such that $f(x, y) = 0$ if the distance from (x, y) to $(0, 0)$ is greater than M. We would now like to bound the size of f, as measured by some L_p norm, in terms of the size of its GRADIENT [I.3 §5.3] ∇f, as measured by some other L_p norm. The L_p norm of a function f is defined here as

$$\|f\|_p = \left(\int_{\mathbb{R}^2} |f(x, y)|^p \, dx \, dy \right)^{1/p}.$$

In one dimension, it is clear that no such bound is possible. For instance, we could have a differentiable function that was 1 everywhere on the interval $[-M, M]$, 0 everywhere outside the wider interval $[-(M + 1), M + 1]$, and gently decaying from 1 to 0 in between. Then if we increased M we would not change the size of the derivative: we would just move the two nonzero parts of the derivative further apart. On the other hand, by increasing M we could increase the size of f as much as we liked. However, we cannot do this sort of construction in two dimensions, because now the "boundary" of the function increases as the size of the function increases. The Sobolev inequality tells us that if $1 \leqslant p < 2$ and $r = 2p/(2 - p)$, then $\|f\|_r \leqslant C_p \|\nabla f\|_p$. To see why this might be reasonable, consider the case $p = 1$, so that $r = 2$. Let f be a function that is 1 everywhere inside the circle of radius M about the origin and 0 everywhere outside the circle of radius $M + 1$. Then as M increases, the norm $\|f\|_2$ increases in proportion to M (since $\|f\|_2^2$ is approximately equal to the area of the circle of radius M), and so does $\|\nabla f\|_1$ (since it is roughly proportional to the length of the boundary of the circle). As this informal argument suggests, there are close connections between the Sobolev inequality and the isoperimetric inequality in the plane. And like the isoperimetric inequality, the Sobolev inequality has an n-dimensional version for each n: it is the same result, except that now the condition is that $1 \leqslant p < n$, and r is equal to $np/(n - p)$.

V.20   The Insolubility of the Halting Problem

What does it mean to understand a certain area of mathematics completely? One possible answer is that you understand it when you can solve its problems *mechanically*. Consider, for instance, the following question.

Jim is half the age of his mother, and in twelve years' time he will be three-fifths of her age. How old is his mother now? For a child who is just old enough to understand the concept of "three-fifths," this is likely to be an impossibly difficult problem. A bright and slightly older child may be able to solve it after some hard thought, which will probably include a certain amount of trial and error. But for anybody who has learned how to translate such problems into equations and who knows how to solve two simultaneous linear equations, the problem is utterly routine: let x be Jim's age and y his mother's; then the problem tells us that $2x = y$ and $5(x + 12) = 3(y + 12)$; the second equation can be rearranged to give $3y - 5x = 24$; substituting $y = 2x$ gives $x = 24$, so $y = 48$.

The more mathematics one learns, the more one finds that problems that once seemed to be difficult and to require ingenuity have become routine in this sort of way, and it is eventually tempting to ask whether *all* of mathematics might, ultimately, be reducible to a mechanical procedure. And even if you think that that is a bit much to hope for, you can still ask the question about certain natural classes of problems, such as simultaneous linear equations. Perhaps there is always a mechanical procedure for solving the problems in any sufficiently "natural" class, even if there is not necessarily a systematic way of finding the mechanical procedure.

One class of problems that has been intensively studied for several centuries is that of *Diophantine equations*, which are equations in one or more variables where one stipulates that the solutions should be integers. The most famous Diophantine equation is the Fermat equation $x^n + y^n = z^n$, but this is somewhat complicated because one of the variables, n, appears as an exponent. Suppose we restrict attention to *polynomial* equations, such as $x^2 - xy + y^2 = 157$. Is there a systematic way of telling whether such an equation has integer solutions?

The left-hand side of the equation $x^2 - xy + y^2 = 157$ is equal to $(x^2 + y^2 + (x - y)^2)/2$. Therefore, any solution (x, y) must satisfy $x^2 + y^2 \leqslant 314$, which makes it a short task to search through all possibilities until one discovers the solution $x = 12$ and $y = 13$ (or vice versa). However, an exhaustive search is not always possible: consider, for example, the equation $2x^2 - y^2 = 1$. This is a special case of the *Pell equation*, discussed in ALGEBRAIC NUMBERS [IV.1 §1]. The Pell equation can be solved systematically, with the help of CONTINUED FRACTIONS [III.22], and this leads to a systematic solution of all polynomial equations of degree up to 2 in two variables.

By the end of the nineteenth century, these and many other Diophantine equations had been completely solved, but there was no single overarching method that dealt with all of them. This state of affairs prompted HILBERT [VI.63] to include, as the tenth in his famous list of twenty-three unsolved problems, the question of whether there was a single, universal procedure for solving all polynomial Diophantine equations in any number of variables. Later, in 1928, he asked the more general question alluded to earlier: is there a universal procedure for determining the truth or falsity of any mathematical statement? This question became known as the *Entscheidungsproblem* (which means "decision problem" in German).

Hilbert expected, or at least hoped, that the answers to both questions would be yes. In other words, he hoped that the mathematicians of his day were in the position of the child who has not yet learned how to solve simultaneous equations. Perhaps a new age was dawning in which it would be possible, at least in principle, to solve all mathematical problems systematically and without relying on native wit.

The evidence in favor of such a view was not very strong: although problems of some kinds could be solved fully systematically, others, including Diophantine equations, stubbornly resisted, and the role of ingenuity in mathematical research appeared to be as important as ever. But if one wanted to give a *negative* answer to Hilbert's questions, then one faced a major challenge: in order to prove rigorously that there is no systematic procedure for accomplishing a particular task, one has to be absolutely clear about what a "systematic procedure" actually is.

Nowadays there is an easy answer to this: a systematic procedure is anything that you can program a computer to do. (Strictly speaking, this is an oversimplification, because one also makes the idealizing assumption that the computer has unlimited storage space.) Our feeling that we do not have to think too hard to solve simultaneous equations is reflected in the fact that we can devise a computer program to do it for us (though if we want the program to be fast and numerically robust, we will face very interesting problems: see NUMERICAL ANALYSIS [IV.21 §4]). However, Hilbert asked the questions before computers existed, so it was a remarkable achievement when in 1936 CHURCH [VI.89] and TURING [VI.94] independently managed to

formalize the notion of what we now call an ALGORITHM [IV.20 §1]. That is, they each gave a precise definition of the notion of an algorithm. Their definitions were quite different, but later shown to be equivalent, which means that anything that can be done by an algorithm in Church's sense can be done by an algorithm in Turing's sense, and vice versa. Turing's formalization, which had a big influence on the design of modern computers, is discussed in COMPUTATIONAL COMPLEXITY [IV.20 §1.1], while Church's is described in ALGORITHMS [II.4 §3.2], but for the purposes of this article we shall use the anachronistic definition with which this paragraph began.

It turns out that once one has *any* sufficiently precise notion of "algorithm," one is just a few short steps away from a negative answer to Hilbert's Entscheidungsproblem. To see this, imagine that L is some programming language (such as Pascal or C++). Given any string of symbols, we can ask of it the following question: if I present that string of symbols to my computer as a program in L, will the program run forever, or will it eventually stop? This is called the *halting problem*. (Note that the word "problem" really means "class of problems.") The halting problem may not seem very mathematical, but certain instances of it certainly are. For example, suppose that after a quick look at a program you realize that it does the following. In one portion of the memory it stores an even number n, which at the beginning is set to 6. It then checks for every odd number m less than n whether m and $n - m$ are both prime. If the answer is yes for some m, then it adds 2 to n and repeats. If the answer is no for all m, then it halts. This program will halt if and only if the GOLDBACH CONJECTURE [V.27] is false.

Turing proved that *there is no systematic procedure for solving the halting problem*. (Church proved an analogous result for his notion of *recursive functions*.) Let us see how Turing's argument works for the language L. In this case, it shows that there is no systematic procedure for recognizing which strings of symbols form programs in L that halt, and which do not. The proof is a reductio ad absurdum, so we begin by assuming that there *is* such a procedure. Let us call it P. Suppose that L is like most computer languages, in that a typical program asks for an *input*, which affects its subsequent behavior. Then P will be able to tell, given any pair of strings (S, I), whether S is a program in L that halts if the input is I.

Now let us create a new procedure Q out of P. Given any string S, we start by getting Q to run P on the pair

(S, S). If P judges that S does *not* halt when presented with itself as input, we then cause Q to halt. But if P judges that S *does* halt when presented with itself as input, then we artificially send Q into an endless loop, so that it does not halt. (If S is not a valid program in L, then let us say that Q halts—it does not really matter though.) To summarize, if S halts for input S, then Q does not halt for S, and if S does not halt for S, then Q does halt for S.

But now let us suppose that S is the program for Q itself. Does Q halt with input S? If it does, then S halts with input S, so Q does not halt. If it does not, then S does not halt with input S, so Q does halt. This is a contradiction, and therefore the procedure P out of which Q was built could not have existed.

That solves the general version of Hilbert's problem: there is no algorithm that will determine the truth or falsity of arbitrary mathematical statements. But it does so by constructing, for any given algorithm, a rather artificial statement. We do not yet have an answer to the question of what happens if we look at more specific and more natural classes of statements, such as that a given Diophantine equation has a solution.

Remarkably, however, specific questions of this kind can often be shown to be *equivalent* to the general question, by a technique known as *encoding*. For example, there is no algorithm that will take as its input a set of polygonal tiles (suitably represented) and tell you whether it is possible to tile the plane using copies of just those tiles. How do we know this? Well, given any algorithm, there is a clever way of devising a set of tiles (this is the encoding) that will tile the plane if and only if the algorithm fails to halt. Therefore, if there were an algorithm for determining whether the tiles could tile the plane, then there would be an algorithm for solving the halting problem—but there is not.

Another famous example of a more specific problem for which there is no algorithm is the *word problem for groups*. Here you are given a set of generators and relations for a group and asked whether the group is trivial—that is, whether it contains just the identity. Again, an algorithm that could decide this would give us an algorithm that could solve the halting problem, so there cannot be one. The encoding process used to prove this is much more difficult than it is for tiling the plane: the insolubility of the word problem for groups is a famous theorem proved by Pyotr Novikov in 1952. For a much fuller explanation of this problem and its solution, see GEOMETRIC AND COMBINATORIAL GROUP THEORY [IV.10].

Finally, what about Hilbert's tenth problem? This has become another famous and very hard theorem, due to Yuri Matiyasevitch in 1970, who built on work of Martin Davis, Hilary Putnam, and Julia Robinson. Matiyasevitch managed to produce a system of ten equations, involving two parameters m and n, that could be solved in integers if and only if m was the $2n$th Fibonacci number. From Robinson's work it followed that, given any algorithm with integer inputs, there was a system of Diophantine equations, involving a parameter q, that could be solved if and only if the algorithm halted at q. That is, any instance of the halting problem can be encoded as a system of Diophantine equations, so there is no general algorithm for deciding whether Diophantine equations can be solved.

Different people draw different morals from these results. In the opinion of some mathematicians, they show that there will always be a place for human creativity in mathematics, however powerful the computers of the future might be. Others maintain that although we now know that we cannot systematically solve all problems in mathematics, the effect on most mathematics is very slight: one should be aware that certain kinds of problems are sometimes equivalent to the halting problem, and that is it. Still others point out that it is often easy to devise an algorithm to solve a problem but much harder to make it *efficient*. This issue is discussed in great detail in COMPUTATIONAL COMPLEXITY [IV.20]

Turing's argument for the insolubility of the halting problem is closely related to GÖDEL'S THEOREM [V.15], and both proofs use *diagonal arguments*, which are discussed in COUNTABLE AND UNCOUNTABLE SETS [III.11].

V.21 The Insolubility of the Quintic
Martin W. Liebeck

Every student will be familiar with the formula for the roots of a quadratic polynomial $ax^2 + bx + c$, namely $(-b \pm \sqrt{b^2 - 4ac})/2a$. Perhaps less familiar is the fact that there is also a formula for the roots of a cubic: write the cubic as $x^3 + ax^2 + bx + c$, and make the substitution $y = x + \frac{1}{3}a$ to rewrite it in the form $y^3 + 3hy + k$. The roots of this are then of the form

$$\sqrt[3]{\tfrac{1}{2}(-k + \sqrt{k^2 + 4h^3})} + \sqrt[3]{\tfrac{1}{2}(-k - \sqrt{k^2 + 4h^3})}.$$

While the quadratic formula was known to the Greeks, the cubic formula was not found until the sixteenth century. In the same century a formula for the roots of quartic (degree 4) polynomials was also found. The formulas for quadratics, cubics, and quartics all arise by applying a sequence of arithmetic operations (addition, subtraction, multiplication, division) together with extraction of roots (square roots, cube roots, and so on) to the coefficients of the original polynomial. Such a formula is called a *radical* expression for the roots.

The next step, naturally enough, was the quintic (i.e., polynomial of degree 5). However, several hundred years passed without anyone finding a radical formula for the roots of a general quintic polynomial.

There was a good reason for this. There is no such formula. Nor is there a formula for polynomials of degree greater than 5. This fact was first established in the early nineteenth century by ABEL [VI.33] (who died aged twenty-six), after which GALOIS [VI.41] (who died aged twenty-one) built an entirely new theory of equations that not only explained the nonexistence of formulas but laid the foundations for a whole edifice of algebra and number theory known as *Galois theory*, a major area of modern-day research.

One of the key ideas of Galois was to associate with any polynomial $f = f(x)$ a GROUP [I.3 §2.1] $\mathrm{Gal}(f)$ (the Galois group of f), which is a finite group that permutes the roots of f. This group is defined in terms of certain FIELDS [I.3 §2.2], which for these purposes can be thought of as subsets F of the COMPLEX NUMBERS [I.3 §1.5] $\mathbb{C}$ having the property that if a, b are any two elements of F, then all the numbers $a + b$, $a - b$, ab, and a/b also lie in F (where we assume that $b \neq 0$ in the last case to avoid dividing by 0). The standard mathematical language for this property is to say that F is "closed under" the usual arithmetic operations of addition, subtraction, multiplication, and division. For example, the rationals $\mathbb{Q}$ form a field, as does $\mathbb{Q}(\sqrt{2}) = \{a + b\sqrt{2} : a, b \in \mathbb{Q}\}$ (this is clearly closed under addition, subtraction, and multiplication, and is also closed under division since $1/(a + b\sqrt{2}) = a/(a^2 - 2b^2) - b\sqrt{2}/(a^2 - 2b^2)$). A polynomial $f(x)$ of degree n with rational coefficients has n complex roots by THE FUNDAMENTAL THEOREM OF ALGEBRA [V.13]—call them $\alpha_1, \ldots, \alpha_n$. The *splitting field* of f is defined to be the smallest field containing $\mathbb{Q}$ and all the α_i, and is written as $\mathbb{Q}(\alpha_1, \ldots, \alpha_n)$. For example, the polynomial $x^2 - 2$ has roots $\pm\sqrt{2}$, so its splitting field is

$\mathbb{Q}(\sqrt{2})$, defined above. Less trivially, $x^3 - 2$ has roots α, $\alpha\omega$, $\alpha\omega^2$, where $\alpha = 2^{1/3}$, the real cube root of 2, and $\omega = e^{2\pi i/3}$, so its splitting field is $\mathbb{Q}(\alpha, \omega)$, which consists of all complex numbers $a_1 + a_2\alpha + a_3\alpha^2 + a_4\omega + a_5\alpha\omega + a_6\alpha^2\omega$ with $a_i \in \mathbb{Q}$. (Notice that we do not have to include ω^2 in such expressions since $\omega^3 = 1$, so $(\omega - 1)(\omega^2 + \omega + 1) = \omega^3 - 1 = 0$, which implies that $\omega^2 = -\omega - 1$.)

Let $E = \mathbb{Q}(\alpha_1, \ldots, \alpha_n)$ be the splitting field of our polynomial f. An *automorphism* of E is a bijection $\phi : E \to E$ that preserves addition and multiplication—in other words, $\phi(a + b) = \phi(a) + \phi(b)$ and $\phi(ab) = \phi(a)\phi(b)$ for all $a, b \in E$. Such a function necessarily also preserves subtraction and division, and fixes every rational number. Denote by $\mathrm{Aut}(E)$ the set of all automorphisms of E. For example, when $E = \mathbb{Q}(\sqrt{2})$, any automorphism ϕ satisfies

$$2 = \phi(2) = \phi(\sqrt{2}\sqrt{2}) = \phi(\sqrt{2})\phi(\sqrt{2}) = \phi(\sqrt{2})^2,$$

and therefore $\phi(\sqrt{2}) = \sqrt{2}$ or $-\sqrt{2}$. In the first case $\phi(a + b\sqrt{2}) = a + b\sqrt{2}$ for all $a, b \in \mathbb{Q}$, while in the second $\phi(a + b\sqrt{2}) = a - b\sqrt{2}$. Both of these are automorphisms of E; call them ϕ_1, ϕ_2, so that $\mathrm{Aut}(E) = \{\phi_1, \phi_2\}$.

The composition $\phi \circ \psi$ of two automorphisms ϕ, ψ of E is again an automorphism, and so is the inverse function ϕ^{-1}, while the identity function ι defined by $\iota(e) = e$ for all $e \in E$ is also an automorphism. Since composition of functions is an associative operation, it follows that $\mathrm{Aut}(E)$ is a group under composition. Define the *Galois group* $\mathrm{Gal}(f)$ of our polynomial $f(x)$ with splitting field E to be this group $\mathrm{Aut}(E)$. Thus, for example, $\mathrm{Gal}(x^2 - 2) = \{\phi_1, \phi_2\}$. Notice that ϕ_1 is the identity ι, while $\phi_2^2 = \phi_2 \circ \phi_2 = \phi_1$, so this is just a cyclic group of order 2. Similarly, if $f(x) = x^3 - 2$, with splitting field $E = \mathbb{Q}(\alpha, \omega)$ as above, then any $\phi \in \mathrm{Aut}(E)$ satisfies $\phi(\alpha)^3 = \phi(\alpha^3) = \phi(2) = 2$, and therefore $\phi(\alpha) = \alpha$, $\alpha\omega$, or $\alpha\omega^2$; likewise $\phi(\omega) = \omega$ or ω^2. Once $\phi(\alpha)$ and $\phi(\omega)$ are specified, ϕ is completely determined (since $\phi(a_1 + a_2\alpha + \cdots + a_6\alpha^2\omega) = a_1 + a_2\phi(\alpha) + \cdots + a_6\phi(\alpha)^2\phi(\omega)$), so there are just six possibilities for the automorphism ϕ. It turns out that each of these is indeed an automorphism, and therefore $\mathrm{Gal}(x^3 - 2)$ is a group of order 6. In fact, this group is isomorphic to the SYMMETRIC GROUP [III.68] S_3, as can be seen by considering each automorphism as a permutation of the three roots of $f(x)$.

Now that the Galois group is defined, it is possible to state some of Galois's fundamental results that lead to the insolubility of the quintic. Each subgroup H of

$G = \text{Gal}(f)$ has a *fixed field* $H^{\dagger}$, which is defined to be the set of all numbers $a \in E$ such that $\phi(a) = a$ for all $\phi \in H$. Galois proved that the association between H and $H^{\dagger}$ gives a one-to-one correspondence between subgroups of G and fields which lie between $\mathbb{Q}$ and E (the so-called *intermediate subfields* of E). The condition that $f(x)$ has a radical formula for its roots leads to certain special kinds of intermediate subfields, and hence to certain special subgroups of G, and eventually to Galois's most famous theorem: the polynomial $f(x)$ has a radical formula for its roots if and only if its Galois group $\text{Gal}(f)$ is a *soluble* group. (This means that $G = \text{Gal}(f)$ has a sequence of subgroups $1 = G_0 < G_1 < \cdots < G_r = G$ such that for each i, G_i is a NORMAL SUBGROUP [I.3 §3.3] of G_{i+1} and the factor group G_{i+1}/G_i is Abelian.)

It follows from Galois's theorem that to demonstrate the insolubility of the quintic, it is enough to produce a quintic $f(x)$ such that $\text{Gal}(f)$ is not a soluble group. An example of such a quintic is $f(x) = 2x^5 - 5x^4 + 5$: one can show first that $\text{Gal}(f)$ is isomorphic to the symmetric group S_5; and second that S_5 is not a soluble group. Here is a brief sketch of how the argument goes. First one establishes that $f(x)$ is an irreducible polynomial (i.e., is not the product of two rational polynomials of smaller degree) with five distinct complex roots. Thus, as observed above, $\text{Gal}(f)$ can be regarded as a subgroup of S_5 that permutes the five roots. By sketching the graph of $f(x)$ one can easily see that three of its roots are real and that the other two, call them α_1 and α_2, are complex conjugates of each other. Since the complex conjugation map $z \to \bar{z}$ always gives an automorphism in $\text{Gal}(f)$, it follows that $\text{Gal}(f)$ is a subgroup of S_5 that contains a 2-cycle, namely $(\alpha_1 \alpha_2)$. Another basic general fact is that the Galois group of an irreducible polynomial permutes the roots *transitively*, meaning that for any two roots α_i, α_j there exists an automorphism in $\text{Gal}(f)$ that sends α_i to α_j. Thus, our group $\text{Gal}(f)$ is a subgroup of S_5 that permutes the five roots transitively and contains a 2-cycle. At this point some fairly elementary group theory shows that $\text{Gal}(f)$ must actually be the whole of S_5. Finally, the fact that S_5 is not a soluble group follows easily from the fact that the alternating group A_5 is a non-Abelian simple group (i.e., it has no normal subgroups apart from the identity subgroup and A_5 itself).

These ideas can be extended to produce polynomials of any degree $n \geqslant 5$ that have Galois group S_n, and that are therefore not soluble by radicals. The reason

this cannot be done for quartics, cubics, and quadratics is that S_4 and all its subgroups are soluble groups.

V.22 Liouville's Theorem and Roth's Theorem

One of the most famous theorems in mathematics is the statement that $\sqrt{2}$ is irrational. This means that there is no pair of integers p and q such that $\sqrt{2} = p/q$, or equivalently that the equation $p^2 = 2q^2$ has no integer solutions apart from the trivial solution $p = q = 0$. The argument that proves this can be considerably generalized, and, in fact, if $P(x)$ is any polynomial with integer coefficients and leading coefficient 1, then all its roots are either integers or irrational numbers. For example, since $x^3 + x - 1$ is negative when $x = 0$ and positive when $x = 1$ it must have a root strictly between 0 and 1. This root is not an integer, so it must be irrational.

Once one has proved that a number is irrational, it may seem as though not much more can be said. However, this is very far from true: given an irrational number, one can ask *how close* it is to being rational, and fascinating and extremely difficult questions arise as soon as one does so.

It is not immediately obvious what this question means, since every irrational number can be approximated as closely as you like by rational numbers. For example, the decimal expansion of $\sqrt{2}$ begins $1.414213\ldots$, which tells us that $\sqrt{2}$ is within $1/100\,000$ of the rational number $141\,421/100\,000$. More generally, for any positive integer q we can let p be the largest integer such that $p/q < \sqrt{2}$, and then p/q will be within $1/q$ of $\sqrt{2}$. In other words, if we want an approximation to $\sqrt{2}$ with accuracy $1/q$, we can obtain it if we use a denominator of q.

However, we can now ask the following question: are there denominators q for which one can one obtain an accuracy much better than $1/q$? The answer turns out to be yes. To see this, let N be a positive integer and consider the numbers $0, \sqrt{2}, 2\sqrt{2}, \ldots, N\sqrt{2}$. Each of these can be written in the form $m + \alpha$, where m is an integer and α, the fractional part, lies between 0 and 1. Since there are $N + 1$ numbers, at least two of their fractional parts must be within $1/N$ of each other. That is, we can find integers $r < s$ between 0 and N such that if we write $r\sqrt{2} = n + \alpha$ and $s\sqrt{2} = m + \beta$, then $|\alpha - \beta| \leqslant 1/N$. Thus, if we set $\gamma = \alpha - \beta$, we have $(s - r)\sqrt{2} = n - m + \gamma$ and $|\gamma| \leqslant 1/N$. If we now let $q = s - r$ and $p = n - m$, then $\sqrt{2} = p/q + \gamma/q$, so $|\sqrt{2} - p/q| \leqslant 1/qN$. Since $N \geqslant q$, $1/qN \leqslant 1/q^2$, so for at least some positive integers q we can achieve an accuracy of $1/q^2$ using a denominator of q.

A different argument shows that we cannot do substantially better than this. Let p and q be any two positive integers. Since $\sqrt{2}$ is irrational, p^2 and $2q^2$ are distinct positive integers, which implies that $|p^2 - 2q^2| \geqslant 1$. On factorizing, we deduce that $|p - q\sqrt{2}|(p + q\sqrt{2}) \geqslant 1$. We can now divide through by q^2 and obtain the inequality $|p/q - \sqrt{2}|(p/q + \sqrt{2}) \geqslant 1/q^2$. We may as well assume that p/q is less than 2, since otherwise it is not a good approximation to $\sqrt{2}$. But then $p/q + \sqrt{2}$ is less than 4, so the inequality implies that $|p/q - \sqrt{2}| \geqslant 1/4q^2$. Thus, with a denominator of q we cannot achieve an accuracy better than $1/4q^2$.

A generalization of this argument proves *Liouville's theorem*: if x is an irrational root of a polynomial of degree d and p and q are integers, then $|p/q - x|$ cannot be substantially smaller than $1/q^d$. When $x = \sqrt{2}$ this reduces to what we have just shown, since then $x^2 - 2 = 0$ and we can set $d = 2$. However, from Liouville's theorem we know many similar facts, such as that $|p/q - \sqrt[3]{2}|$ cannot be substantially smaller than $1/q^3$.

Roth's theorem, proved in 1955, is the astonishing assertion that the power d that appears in Liouville's theorem can be improved—almost as far as 2. To be precise, given any irrational root x of any polynomial, and any number $r > 2$, there is a constant $c > 0$ with the property that $|p/q - x|$ is always at least as big as c/q^r. (The proof gives no information whatsoever about c beyond the fact that it is positive. It is a major open problem to understand something about how c depends on r and x.)

To see why this is a much deeper result than Liouville's theorem, consider the example of $\sqrt[3]{2}$. Underlying the proof that $|p/q - \sqrt[3]{2}|$ is never much smaller than $1/q^3$ is the simple fact that p^3 and $2q^3$ are distinct integers and therefore differ by at least 1. In order to prove a substantially better result such as Roth's theorem, one must show much more: that p^3 and $2q^3$ differ by an amount that grows as p and q grow. For example, if one wishes to prove Roth's theorem when $r = \frac{5}{2}$, it is necessary to show that p^3 and $2q^3$ must always differ by an amount comparable to or greater than $\sqrt{p}$, and it is far from obvious why this should be so.

The Mordell Conjecture

See RATIONAL POINTS ON CURVES AND THE MORDELL CONJECTURE [V.29]

V.23 Mostow's Strong Rigidity Theorem
David Fisher

1 What Are Rigidity Theorems?

A typical *rigidity theorem* is a statement that some class of objects is much smaller than one might expect. To make this notion clear, let us look at some examples of MODULI SPACES [IV.8] that might lead us to expect that spaces of a certain type would in general be large.

2 Some Moduli Spaces

A *flat metric* on an n-dimensional MANIFOLD [I.3 §6.9] is a METRIC [III.56] that is locally isometric to the usual metric on the Euclidean space $\mathbb{R}^n$. In other words, every point x in the manifold is contained in a neighborhood N_x such that there is a distance-preserving bijection from N_x to a subset of $\mathbb{R}^n$. For our first example, we shall consider flat metrics on a torus. We shall consider just the two-dimensional torus, but the phenomena we shall discuss occur in higher dimensions as well.

The simplest way of putting a flat metric on the two-dimensional torus $\mathbb{T}^2$ is to view it as the QUOTIENT [I.3 §3.3] of $\mathbb{R}^2$ by a discrete subgroup, or *lattice*, that is isomorphic to $\mathbb{Z}^2$. In fact, it is not too hard to see that *every* flat metric arises in essentially this way. However, there is a choice involved: the choice of which lattice to take. An obvious choice is $\mathbb{Z}^2$ itself. But one can also take any invertible linear transformation A, apply it to $\mathbb{Z}^2$, and then define the torus as $\mathbb{R}^2/A(\mathbb{Z}^2)$, which gives rise to another metric. A natural question to ask is, when do two choices of A give rise to the same metric? Usually, one studies only the cases when the DETERMINANT [III.15] of A is 1, since it is easy to deduce from these what happens in general. The group of all such linear maps is called $\mathrm{SL}_2(\mathbb{R})$.

If A is orthogonal, then it just rotates the lattice $\mathbb{Z}^2$ and therefore $A(\mathbb{Z}^2)$ gives rise to the same metric as $\mathbb{Z}^2$. What is slightly less obvious is that there are other maps A that give rise to this metric as well, namely all maps of determinant 1 whose matrices with respect to the standard basis of $\mathbb{R}^2$ have integer entries. The group of all these maps is called $\mathrm{SL}_2(\mathbb{Z})$. If A belongs to $\mathrm{SL}_2(\mathbb{Z})$, then the reason that $A(\mathbb{Z}^2)$ gives rise to the same metric as $\mathbb{Z}^2$ is simple: $A(\mathbb{Z}^2)$ is actually equal to $\mathbb{Z}^2$.

Loosely speaking, what we have just done is identify the space of flat metrics on $\mathbb{T}^2$ with the set

$\mathrm{SL}_2(\mathbb{Z})\backslash\mathrm{SL}_2(\mathbb{R})/\mathrm{SO}(2)$. (This is notation for the set $\mathrm{SL}_2(\mathbb{R})$, with two maps A and B considered equivalent if B can be expressed as A multiplied by a product of matrices from $\mathrm{SO}(2)$ and $\mathrm{SL}_2(\mathbb{Z})$.) In higher dimensions, a similar discussion shows that one can identify the space of flat metrics on the n-dimensional torus $\mathbb{T}^n$ with $\mathrm{SL}_n(\mathbb{Z})\backslash\mathrm{SL}_n(\mathbb{R})/\mathrm{SO}(n)$.

Returning to two dimensions, a torus is a surface of genus 1 (since it has one "hole"). A similar construction gives rise to a moduli space of metrics on a surface of higher genus, but now the metrics will be *hyperbolic* rather than flat. The UNIFORMIZATION THEOREM [V.34] says that any compact connected surface admits a metric of constant CURVATURE [III.13]: when the genus is 2 or more, this curvature must be negative, which implies that the surface is a QUOTIENT [I.3 §3.3] of the HYPERBOLIC PLANE [I.3 §6.6] $\mathbb{H}^2$ by a group Γ that acts on $\mathbb{H}^2$ as a set of isometries. (See FUCHSIAN GROUPS [III.28].)

Conversely, if we want to construct a metric of constant curvature on a surface of higher genus, we can take a subgroup Γ of the group of isometries of $\mathbb{H}^2$ (which is isomorphic to $\mathrm{SL}_2(\mathbb{R})$) and we can consider the quotient $\mathbb{H}^2/\Gamma$, which is analogous to the quotient $\mathbb{R}^2/\mathbb{Z}^2$ that we considered earlier. If Γ has no elements of finite order and if for each x the *orbit* of x (the set of images of x under the isometries in Γ) is a discrete subset of $\mathbb{H}^2$, then this space is a manifold. Furthermore, if there is a compact region in $\mathbb{H}^2$, called a *fundamental domain*, whose translates cover $\mathbb{H}^2$, then the manifold is compact. There are two fairly simple ways to construct examples of groups Γ with these properties: one is to use reflection groups and the other is to use a bit of number theory.

Now we can ask the same question for these metrics. In other words, given a surface S of genus at least 2, how many hyperbolic metrics can we find on S? The answer is quite similar to the answer for $\mathbb{T}^2$. For instance, if the genus is 2, then there is a connected six-dimensional space of such structures. This is a bit more difficult to see, as the space is not constructed in any simple way from a LIE GROUP [III.48 §1] (such as $\mathrm{SL}_n(\mathbb{R})$) and its subgroups. We will not describe this construction here but it can be found in Thurston (1997) or in MODULI SPACES [IV.8].

3 Mostow's Theorem

Thinking about the last two sets of examples leads to a natural question: what about compact three-dimensional hyperbolic manifolds? Or n-dimensional ones?

To be clear, a compact n-dimensional hyperbolic manifold is the quotient of $\mathbb{H}^n$ by a discrete group Γ of isometries of the hyperbolic n-space $\mathbb{H}^n$ such that Γ has no elements of finite order and there is a compact fundamental domain for Γ. Given this description, the reader may wonder if there are any such groups Γ. Once again, there are two easy ways of constructing them, one using a bit of number theory and another using reflection groups. (However, slightly surprisingly, the method using reflection groups works only in fairly small dimensions.) The constructions are all a bit technical so we will not go through them here. There are also many other examples of compact hyperbolic manifolds, particularly in three dimensions, where "most" manifolds are hyperbolic by the GEOMETRIZATION THEOREM [IV.7 §2.4].

Here we shall concentrate less on the existence of hyperbolic manifolds and more on the question that has been our principal concern in this article: if X is a manifold that can be represented in the form $\mathbb{H}^n/\Gamma$, then how many ways are there of giving X this structure? This question is equivalent to asking how many injective homomorphisms there are from Γ to the group of all isometries of $\mathbb{H}^n$ such that the image of Γ is discrete and cocompact. (A subset X of a group G is *cocompact* if there is a compact subset K of G such that $XK = G$. For instance, $\mathbb{Z}^2$ is a cocompact subset of $\mathbb{R}^2$ because $\mathbb{R}^2 = \mathbb{Z}^2 + [0,1]^2$ and the closed unit square $[0,1]^2$ is compact.) As we have seen, when $n = 2$ there is a continuum of such homomorphisms, and the same is true in all dimensions if we replace $\mathbb{H}^n$ by $\mathbb{R}^n$. So it is rather surprising that when $n \geqslant 3$, the answer for $\mathbb{H}^n$ is exactly 1. This is a special case of Mostow's rigidity theorem.

What does this result mean? Suppose we know that a manifold M is a quotient of $\mathbb{H}^n$ by some discrete cocompact group of isometries. The topology of M completely determines the group Γ up to isomorphism: it is just the FUNDAMENTAL GROUP [IV.6 §2] of M. The result we have just stated tells us that this purely topological information about the manifold M completely determines the geometry of $\mathbb{H}^n/\Gamma$ (that is, its structure as a metric space). More precisely, it says that any homeomorphism, or even homotopy equivalence, from M to another hyperbolic manifold N is homotopic to an isometry. In other words, any purely topological equivalence can be realized as a geometric equivalence.

The full Mostow rigidity theorem concerns objects called compact locally symmetric manifolds. Given a manifold with a metric, we say that it is *locally symmetric* if the *central symmetry* at every point is a local isometry. The central symmetry at a point m is defined formally as multiplication by -1 in the tangent space to m: one can picture it as taking a very small neighborhood of m and "reflecting through m." It turns out that every locally symmetric space is a quotient of a *symmetric space*: that is, a space such that the central symmetry at every point is a global isometry. Clearly, symmetric spaces have very large isometry groups. The work of CARTAN [VI.69] shows that the resulting isometry groups are exactly the semisimple LIE GROUPS [III.48 §1]. We will not say precisely what these are, but they include the classical matrix groups such as $\mathrm{SL}_n(\mathbb{R})$, $\mathrm{SL}_n(\mathbb{C})$, and $\mathrm{Sp}_n(\mathbb{R})$. Other examples, which can also be realized as matrix groups, include the isometry groups of complex and quaternionic hyperbolic spaces.

In general, given a Lie group G and a discrete subgroup Γ, we say that Γ is a cocompact lattice if there is a compact fundamental domain for Γ in G. Cartan's theorem has the consequence that any compact locally symmetric space is a quotient $\Gamma \backslash G / K$, where G is the isometry group of the universal cover and K is the (necessarily compact) set of isometries that fix a specified point. Mostow's theorem says the same here as it said for $\mathbb{H}^n / \Gamma$: given such a manifold, there is only one way to realize it as $\Gamma \backslash G / K$. Or, equivalently, any homeomorphism between two such manifolds is always homotopic to an isometry unless the relevant locally symmetric space is a product of a flat torus or a hyperbolic surface with some other locally symmetric manifold.

One might well ask how Mostow discovered such a phenomenon. His work certainly did not occur in a vacuum. In fact, earlier work of Calabi, Selberg, Vesentini, and WEIL [VI.93] had already shown that the moduli spaces Mostow was studying were discrete: in other words, unlike flat tori or two-dimensional hyperbolic manifolds, higher-dimensional locally symmetric spaces could admit only a discrete set of locally symmetric metrics. Mostow has said explicitly that he was motivated by the desire to find a more geometric understanding of this fact.

Another point worth making is that Mostow's proof is at least as surprising as his theorem. At the time, the study of locally symmetric spaces, or equivalently of semisimple Lie groups and their lattices, was dominated by two sets of techniques: one set that was purely algebraic and another that used classical methods in differential geometry. Mostow's original proof (which was only for $\mathbb{H}^n$) uses instead the theory of quasiconformal mappings and some ideas from dynamics.

Raghunathan, another leading figure in the field, has said that when he first read Mostow's paper, he thought it must be by a different man named Mostow. Similar uses of surprising dynamical and analytical ideas to study the same objects occurred almost simultaneously in work of Furstenberg and Margulis. These ideas have had a long and interesting legacy in the study of locally symmetric spaces, semisimple Lie groups, and related objects.

Further Reading

Furstenberg, H. 1971. Boundaries of Lie groups and discrete subgroups. In *Actes du Congrès International des Mathématiciens, Nice, 1970*, volume 2, pp. 301–6. Paris: Gauthier-Villars.

Margulis, G. A. 1977. Discrete groups of motions of manifolds of non-positive curvature. In *Proceedings of the International Congress of Mathematicians, Vancouver, 1974*, pp. 33–45. AMS Translations, volume 109. Providence, RI: American Mathematical Society.

Mostow, G. D. 1973. *Strong Rigidity of Locally Symmetric Spaces*. Annals of Mathematics Studies, number 78. Princeton, NJ: Princeton University Press.

Thurston, W. P. 1997. *Three-Dimensional Geometry and Topology*, edited by S. Levy, volume 1. Princeton Mathematical Series, number 35. Princeton, NJ: Princeton University Press.

V.24 The $\mathcal{P}$ versus $\mathcal{NP}$ Problem

The $\mathcal{P}$ versus $\mathcal{NP}$ problem is widely considered to be the most important unsolved problem in theoretical computer science, and one of the most important in all of mathematics. $\mathcal{P}$ and $\mathcal{NP}$ are two of the most basic COMPUTATIONAL COMPLEXITY CLASSES [III.10]: $\mathcal{P}$ is the class of all computational tasks that can be performed in a time that is polynomial in the length of the input, and $\mathcal{NP}$ is the class of all computational tasks where a correct answer can be *verified* in a time that is polynomial in the length of the input. An example of the former is multiplying two n-digit integers (which, even if you use long multiplication, takes roughly n^2 arithmetical operations). An example of the latter is searching in a GRAPH [III.34] with n vertices for a set of m vertices, any two of which are joined by an edge: if you are presented with m such vertices, then you just have to check the $\binom{m}{2}$ pairs of those vertices to make sure that each pair is indeed an edge of the graph.

It appears to be *much* harder to find m vertices that are all joined than to check that a given m vertices are all joined. This suggests that problems in $\mathcal{NP}$ are in general harder than problems in $\mathcal{P}$. The $\mathcal{P}$ versus $\mathcal{NP}$ problem asks for a proof that the complexity classes $\mathcal{P}$ and $\mathcal{NP}$ really are distinct. For a detailed discussion of the problem, see COMPUTATIONAL COMPLEXITY [IV.20].

V.25 The Poincaré Conjecture

The Poincaré conjecture is the statement that a COMPACT [III.9] smooth n-dimensional manifold that is HOMOTOPY EQUIVALENT [IV.6 §2] to the n-sphere S^n must in fact be homeomorphic to S^n. One can think of a compact manifold as a manifold that lives in a finite region of $\mathbb{R}^m$ for some m and that has no boundary: for example, the 2-sphere and the torus are compact manifolds living in $\mathbb{R}^3$, while the open unit disk or an infinitely long cylinder is not. (The open unit disk does not have a boundary in an intrinsic sense, but its realization as the set $\{(x, y) : x^2 + y^2 < 1\}$ has the set $\{(x, y) : x^2 + y^2 = 1\}$ as its boundary.) A manifold is called *simply connected* if every loop in the manifold can be continuously contracted to a point. For instance, a sphere of dimension greater than 1 is simply connected but a torus is not (since a loop that "goes around" the torus will always go around the torus, however you continuously deform it). In three dimensions, the Poincaré conjecture asks whether two simple properties of spheres, compactness and simple connectedness, are enough to characterize spheres.

The case $n = 1$ is not interesting: the real line is not compact and a circle is not simply connected, so the hypotheses of the problem cannot be satisfied. POINCARÉ [VI.61] himself solved the problem for $n = 2$ early in the twentieth century, by completely classifying all compact 2-manifolds and noting that in his list of all possible such manifolds only the sphere was simply connected. For a time he believed that he had solved the three-dimensional case as well, but then discovered a counterexample to one of the main assertions of his proof. In 1961, Stephen Smale proved the conjecture for $n \geqslant 5$, and Michael Freedman proved the $n = 4$ case in 1982. That left just the three-dimensional problem open.

Also in 1982, William Thurston put forward his famous *geometrization conjecture*, which was a proposed classification of three-dimensional manifolds. The conjecture asserted that every compact 3-manifold can be cut up into submanifolds that can be given METRICS [III.56] that turn them into one of eight particularly symmetrical geometric structures. Three of

these structures are the three-dimensional versions of Euclidean, spherical, and hyperbolic geometry (see [I.3 §6]). Another is the infinite "cylinder" $S^2 \times \mathbb{R}$: that is, the product of a 2-sphere with an infinite line. Similarly, one can take the product of the hyperbolic plane with an infinite line and obtain a fifth structure. The other three are slightly more complicated to describe. Thurston also gave significant evidence for his conjecture by proving it in the case of so-called Haken manifolds.

The geometrization conjecture implies the Poincaré conjecture; both were proved by Grigori Perelman, who completed a program that had been set out by Richard Hamilton. The main idea of this program was to solve the problems by analyzing RICCI FLOW [III.78]. The solution was announced in 2003 and checked carefully by several experts over the next few years. For more details, see DIFFERENTIAL TOPOLOGY [IV.7].

V.26 The Prime Number Theorem and the Riemann Hypothesis

How many prime numbers are there between 1 and n? A natural first reaction to this question is to define $\pi(n)$ to be the number of prime numbers between 1 and n and to search for a formula for $\pi(n)$. However, the primes do not have any obvious pattern to them and it has become clear that no such formula exists (unless one counts highly artificial formulas that do not actually help one to calculate $\pi(n)$).

The standard reaction of mathematicians to this kind of situation is to look instead for good *estimates*. In other words, we try to find a simply defined function $f(n)$ for which we can prove that $f(n)$ is always a good approximation to $\pi(n)$. The modern form of the prime number theorem was first conjectured by GAUSS [VI.26] (though a closely related conjecture had been made by LEGENDRE [VI.24] a few years earlier). He looked at the numerical evidence, which suggested to him that the "density" of primes near n was about $1/\log n$, in the sense that a randomly chosen integer near n would have a probability of roughly $1/\log n$ of being a prime. This leads to the conjectured approximation of $n/\log n$ for $\pi(n)$, or to the slightly more sophisticated approximation

$$\pi(n) \simeq \int_0^n \frac{\mathrm{d}x}{\log x}.$$

The function defined by the integral on the right-hand side is called li(n) (which stands for the "logarithmic integral" of n). Some care is needed in interpreting the integral because $\log 1 = 0$, but one can avoid this problem by integrating from 2 to n instead, which changes the function by just an additive constant.

The prime number theorem, proved independently by HADAMARD [VI.65] and DE LA VALLÉE POUSSIN [VI.67] in 1896, states that $\mathrm{li}(n)$ is indeed a good approximation to $\pi(n)$, in the sense that the ratio of the two functions tends to 1 as n tends to infinity.

This result is considered one of the great theorems of all time, but it is by no means the end of the story. The proofs of Hadamard and de la Vallée Poussin used the RIEMANN ZETA FUNCTION [IV.2 §3] $\zeta(s)$. The Riemann zeta function is defined to be $1^{-s} + 2^{-s} + 3^{-s} + \cdots$ whenever s is a complex number with real part greater than 1; this expression defines a HOLOMORPHIC FUNCTION [I.3 §5.6], which can be extended (by analytic continuation) to a function that is holomorphic on the entire complex plane, except for a pole at 1. This function has zeros, known as "trivial zeros," at all negative even integers. Riemann proved that the prime number theorem was equivalent to the assertion that the only "nontrivial zeros" were inside the *critical strip*, which consists of those complex numbers with real part strictly between 0 and 1. He also formulated what is often held to be the most important unsolved problem in mathematics, now known as the *Riemann hypothesis*: that in fact the nontrivial zeros all have real part equal to $\frac{1}{2}$. This assertion about the zeros of the zeta function has been shown to be equivalent to a stronger form of the prime number theorem, which states not just that $\pi(n)/\mathrm{li}(n)$ tends to 1, but even that $|\pi(n) - \mathrm{li}(n)| \leqslant \sqrt{n} \log n$ for every $n \geqslant 3$. Since $\mathrm{li}(n)$ is around $n/\log n$, which is much bigger than $\sqrt{n} \log n$, this would mean that the error $|\pi(n) - \mathrm{li}(n)|$ was extremely small compared with $\pi(n)$ or $\mathrm{li}(n)$ themselves.

The importance of the Riemann hypothesis goes far beyond its consequences for the distribution of primes: hundreds of statements in number theory have been shown to follow from it. This is particularly true when one considers generalizations of the Riemann hypothesis that apply to a wider class of L-FUNCTIONS [III.47]. For example, analogues of the Riemann hypothesis for Dirichlet L-functions imply very good estimates for the distribution of primes in arithmetic progressions, from which many further consequences follow.

The prime number theorem and the Riemann hypothesis are discussed in more detail in ANALYTIC NUMBER THEORY [IV.2 §3].

V.27 Problems and Results in Additive Number Theory

Is every even number greater than 4 the sum of two odd primes? Are there infinitely many primes p such that $p + 2$ is also a prime? Is every sufficiently large positive integer the sum of four cubes? These three questions are all famous unsolved problems in number theory: the first is called the *Goldbach conjecture*, the second is the *twin prime conjecture* (discussed in some detail in ANALYTIC NUMBER THEORY [IV.2]), and the third is a special case of *Waring's problem*, which we shall discuss later.

These three problems belong to an area of mathematics known as *additive number theory*. In order to say in general terms what this area is, it is useful to make some simple definitions. Suppose that A is a set of positive integers. Then the *sumset* of A, denoted $A + A$, is the set of all $x + y$ such that x and y (which are allowed to be equal) both belong to A. For example, if A is the set $\{1, 5, 9, 10, 13\}$, then $A + A$ is the set $\{2, 6, 10, 11, 14, 15, 18, 19, 20, 22, 23, 26\}$. Similarly, the *difference set*, denoted $A - A$, is the set of all $x - y$ such that x and y both belong to A. In the above example, $A - A = \{-12, -9, -8, -5, -4, -3, -1, 0, 1, 3, 4, 5, 8, 9, 12\}$.

Using this language, we can state two of our three problems very succinctly. Let P be the set of all odd primes and let C be the set of all cubes. Then Goldbach's conjecture is the statement that $P + P$ is the set $\{6, 8, 10, 12, \dots\}$, and the special case of Waring's problem asks whether every sufficiently large integer belongs to $C + C + C + C$. The twin prime conjecture is slightly more complicated: it states not just that 2 belongs to the set $P - P$ but that it does so "infinitely many times." (In a similar way, if A is the set in the previous paragraph, then $A - A$ contains the number 4 three times.)

These problems are notoriously difficult. However, remarkably, there are some closely related problems that look just as hard at first, but which have been solved. For instance, *Vinogradov's three-primes theorem* is the statement that every sufficiently large *odd* integer is the sum of *three* odd primes. Without the "sufficiently large" this would answer the *ternary Goldbach problem*, which asks whether every odd number from 9 onward is a sum of three odd primes. (How large is "sufficiently large"? Well, until recently you needed your number to have about 7 000 000 digits, but in 2002 this was reduced to under 1500 digits.) As for Waring's problem, it is known that every sufficiently large positive integer is a sum of seven cubes. More generally, it seems likely that, for any k, every sufficiently large integer can be written as a sum of at most $100k$ kth powers (where 100 is just a randomly chosen largish number—it is possible that even $4k$ kth powers are enough), and although a proof of this is well beyond today's mathematical technology, it has been shown that a little over $k \log k$ kth powers are enough. Since $\log k$ is a very slowly growing function, this result is, in a certain sense, not too far from a solution to the problem.

How does one obtain results such as these? Some of the proofs are pretty complicated, so we cannot give a full answer here. However, we can at least explain one idea that is fundamental to many of the arguments, namely the use of *exponential sums*. Let us illustrate it by looking at the beginning of the proof of the Vinogradov three-primes theorem.

Imagine, then, that we have a very large odd integer n and we wish to prove that it is a sum of three odd primes. Here is an argument that strongly suggests that our task is impossible: if n is over three times larger than the largest known prime, as it may very well be, then we cannot produce three primes that add up to n without finding a new prime. Indeed, we could take n to be astronomically large, $10^{10^{100}} + 1$, say, and then $\frac{1}{3}n$ would be far beyond any prime that has ever been discovered or is ever likely to be discovered.

This argument is, however, flawed, and the clue to what is wrong with it lies in the word "produce." We do not have to *produce* the three primes to show that they *exist*, any more than Euclid had to specify an infinite sequence of primes in order to show that there were infinitely many. (For a proof that there are, see [IV.2 §2].) But, one might ask, what alternative could there possibly be to actually finding three odd primes that add up to n?

This question has a beautifully simple answer: we shall attempt to *count*, or rather *estimate*, the number of triples p_1, p_2, p_3 of odd primes such that $p_1 + p_2 + p_3 = n$. If the estimate we manage to obtain is rather large, and if in addition we can show that it is reasonably accurate, then the actual number of such triples must also be rather large. This will imply that there *is* such a triple, and will not require us to "produce" one.

However, our answer immediately raises a difficult-looking question: how do we estimate the number of such triples? This is where exponential sums come in. We shall use certain properties of the EXPONENTIAL

FUNCTION [III.25] to reformulate our counting problem as a problem about estimating a certain integral.

As is customary in this area, let us write $e(x)$ instead of $e^{2\pi i x}$. The two basic properties that we shall use of the function $e(x)$ are that $e(x+y) = e(x)e(y)$ and that $\int_0^1 e(nx)\, dx = 1$ if $n = 0$, and 0 if n is any other integer. Let us also adopt the convention that if we write $\sum_{p \leqslant n}$, then we are summing over all odd primes less than or equal to n. Now define a function $F(x)$ by the formula $F(x) = \sum_{p \leqslant N} e(px)$. That is,

$$F(x) = e(3x) + e(5x) + e(7x) + e(11x) + \cdots + e(qx),$$

where q is the largest prime less than or equal to n. This is a sum of exponentials—hence the phrase "exponential sums." Next, we consider the cube of this function:

$$F(x)^3 = (e(3x) + e(5x) + e(7x) + \cdots + e(qx))^3.$$

When we multiply out the right-hand side, we obtain the sum of all terms of the form $e(p_1 x)e(p_2 x)e(p_3 x)$, where p_1, p_2, and p_3 are primes between 3 and q.

The integral we shall look at is $\int_0^1 F(x)^3 e(-nx)\, dx$. From our discussion in the previous paragraph, we know that this will be the sum of all integrals of the form $\int_0^1 e(p_1 x)e(p_2 x)e(p_3 x)e(-nx)\, dx$. Now the first basic property of $e(x)$ tells us that this last integral is equal to $\int_0^1 e((p_1 + p_2 + p_3 - n)x)\, dx$, and the second one then tells us that it is 1 if $p_1 + p_2 + p_3 = n$ and 0 otherwise. Therefore, when we sum over all possible triples p_1, p_2, p_3 of odd primes less than or equal to n, we get a contribution of 1 for each triple that adds up to n and 0 for all other triples. In other words, the integral $\int_0^1 F(x)^3 e(-nx)\, dx$ exactly equals the number of ways of writing n as a sum of three odd primes.

This "reduces" our problem to that of estimating the integral $\int_0^1 F(x)^3 e(-nx)\, dx$. But the function $F(x)$ looks rather difficult to analyze. Is it really feasible to estimate an expression such as $\sum_{p \leqslant N} e(px)$, which mixes prime numbers with exponentials?

Surprisingly, it is. The details are complicated, but the fact that it can be done becomes less mysterious after one thinks for a moment about which exponential sums we definitely *can* estimate. Are there at least *some* sets A of integers for which we can handle sums of the form $\sum_{a \in A} e(ax)$? Yes there are: arithmetic progressions. Suppose A is the set $\{s, s + d, s + 2d, \ldots, s + (m-1)d\}$: that is, the arithmetic progression of length m and common difference d that starts at s. Then, using the basic properties of $e(x)$, we find that

$\sum_{a \in A} e(ax)$ is

$$e(sx) + e((s+d)x) + \cdots + e((s+(m-1)d)x)$$
$$= e(sx) + e(dx)e(sx) + \cdots + e((m-1)\,dx)e(sx)$$
$$= e(sx)(1 + e(dx) + e(dx)^2 + \cdots + e(dx)^{m-1}).$$

This last expression is the sum of a geometric progression that starts at $e(sx)$ and has common ratio $e(dx)$. Using the standard formula and the basic properties of $e(x)$, we deduce that

$$\sum_{a \in A} e(ax) = \frac{e(sx) - e((s+dm)x)}{1 - e(dx)}.$$

Such expressions are useful because they can often be shown to be small. Suppose, for instance, that $|1 - e(dx)|$ is at least as big as some constant c. We know that $|e(sx) - e((s+dm)x)| \leqslant 2$, so the modulus of the right-hand side is at most $2/c$. If c is not too small, then this shows that there is a huge amount of cancellation in the sum $\sum_{a \in A} e(ax)$: we added together m numbers of modulus 1 and obtained a number of modulus no bigger than $2/c$.

For certain values of x, we can use this simple observation to help us estimate the sum $\sum_{p \in P} e(px)$. What we need to do is express the sum over P as a combination of sums over arithmetic progressions, and this is a very natural thing to do, since P consists of all those integers up to n that do not lie in certain arithmetic progressions (such as $14, 21, 28, 35, 42, \dots$). So we can begin by taking the sum $\sum_{t=1}^{n} e(tx)$. From this we need to subtract the contribution from all even integers, which is $\sum_{t \leqslant n/2} e(2tx)$. We also need to subtract the contribution from multiples of 3, apart from 3 itself. This contribution is $\sum_{1 < t \leqslant n/3} e(3tx)$. Now we find that we have subtracted the contribution from multiples of 6 twice, so we correct for that by adding $\sum_{t \leqslant n/6} e(6tx)$.

This process can be continued, and it leads to a way of decomposing the sum over primes into a combination of sums over geometric progressions. If x is not close to a rational with small denominator, then most of the common ratios are far from 1, so most of the sums over progressions are small. Unfortunately, there are too many of them for this simple argument to lead to a useful estimate. However, there is a more sophisticated argument with a similar flavor that does.

What happens if x *is* close to a rational with small denominator? For example, what can we say about the sum $\sum_{p \leqslant n} e(p/3)$? Here we use more direct methods: it is known that roughly half of all primes are 1 (mod 3) and half are 2 (mod 3) (see [IV.2 §4]), which tells us that

this sum is roughly $(|P|/2)(e(p/3) + e(2p/3))$, where $|P|$ denotes the size of the set P.

For very similar reasons, in Waring's problem one finds oneself wanting to know about exponential sums such as $G(x) = \sum_{t=0}^{m} e(t^k x)$. Again, one can sometimes estimate these by reducing them to sums of geometric progressions. This is easiest to show in the case $k = 2$. The idea is to look not at $G(x)$ directly but at $|G(x)|^2$, which a moment's calculation shows is equal to $\sum_{t=0}^{m} \sum_{u=0}^{m} e((t^2 - u^2)x)$. Now $t^2 - u^2 = (t+u)(t-u)$, so we can change variables, setting $v = t + u$ and $w = t - u$. This gives us the sum $\sum_{(v,w) \in V} e(vwx)$, where V is the set of all (v, w) such that $(v+w)/2$ and $(v-w)/2$ (which equal t and u, respectively) are both between 0 and m. For each v the set of possible values of w is an arithmetic progression, so we have decomposed $|G(x)|^2$ into a sum of sums of geometric progressions, one for each v.

So far we have been looking at so-called *direct* problems in additive number theory. These are problems where one specifies a set and then tries to understand its sumset or difference set. We have only scratched the surface of the subject: other related results and techniques are discussed in [IV.2] (see in particular sections 7, 9, and 11).

Direct problems have a long history, but in recent years another class of problems, called *inverse* problems, have become an important focus of research as well. These concern the following broad question: if you are given information about a sumset or a difference set, what can you deduce about the original set? We end by describing one of the highlights of this kind of additive number theory, called *Freiman's theorem*.

It is not hard to prove that if A is any set of integers of size n, then the size of $A + A$ must be between $2n - 1$ and $n(n+1)/2$. (The first happens if A is an arithmetic progression and the second happens if all the sums you can make are different.) What can we say about A if the size of $A + A$ is at most $100n$, or, more generally, is at most Cn for some constant C that remains fixed as n tends to infinity?

Suppose that we can find an arithmetic progression P of size at most $50n$ such that A is a subset of P. Then $A + A$ is a subset of $P + P$, which has size at most $100n - 1$. So if A is two percent of an arithmetic progression, then $A + A$ has size at most $100n$. However, there are other ways of producing such sets. Suppose, for instance, that A consists of all numbers of up to seven digits such that the third, fourth, and fifth digits from the end are 0: that is, numbers such as 35 000 26 or 99 000 90. There are $100 \times 100 = 10\,000$ of these. If we add two of them together, then we get a number like 138 00 162 or 141 00 068, which is made up of a number between 0 and 198, followed by two 0s, followed by a second number between 0 and 198 (written with 0s in front if these are needed to make it up to three digits). There are 199×199 of these, which is less than 40 000. Therefore, the size of $A + A$ is less than four times the size of A. However, A does not fill up two percent of any arithmetic progression P: such a progression would have to have common difference 1 and include both the numbers 0 and 99 000 099, and 10 000 is nothing like two percent of 9 900 100.

However, A is a very structured set: it is an example of a *two-dimensional* arithmetic progression. Roughly speaking, an ordinary, or one-dimensional, arithmetic progression is one that you build up by starting with a number s and repeatedly adding another one, d, called the common difference. You build up a *two-dimensional* arithmetic progression by using *two* "common differences" d_1 and d_2. That is, you have a starting number s and you look at numbers of the form $s + ad_1 + bd_2$, specifying that a should be between 0 and $m_1 - 1$ and b should be between 0 and $m_2 - 1$. Our set A is a two-dimensional progression with $s = 0$, $d_1 = 1$, $d_2 = 100\,000$, and $m_1 = m_2 = 100$.

In a similar way one can define higher-dimensional progressions. It is not hard to show that if P is an r-dimensional progression, then the size of $P + P$ is less than 2^r times the size of P. Therefore, if A is a subset of P and the size of P is at most C times the size of A, then the size of $A + A$ is at most the size of $P + P$, which is at most $2^r C$ times the size of A.

This tells us that if A is a large subset of a low-dimensional arithmetic progression, then A has a small sumset. Freiman's theorem is the remarkable statement that these are the *only* sets with small sumsets. That is, if $A + A$ is not much larger than A, then there must be some low-dimensional arithmetic progression P that contains A and is not much bigger than A. Exponential sums are vital for the proof of this theorem as well. Freiman's theorem has had many applications, and is likely to have many more.

V.28 From Quadratic Reciprocity to Class Field Theory
Kiran S. Kedlaya

The law of quadratic reciprocity, discovered by EULER [VI.19] and first proved by GAUSS [VI.26] (who dubbed

it his *theorema aureum*, or golden theorem), is considered a crown jewel of number theory, and with good cause. Whereas its statement could be rediscovered by a sufficiently ingenious student (indeed, it actually has been rediscovered on a regular basis at the Arnold Ross mathematics summer program for several decades), rare is the student who comes up with a proof unassisted.

The law is most conveniently stated in a formulation due to LEGENDRE [VI.24]. For n an integer not divisible by the prime p, write $(\frac{n}{p}) = 1$ if n is congruent to some perfect square modulo p, and $(\frac{n}{p}) = -1$ if it is not. Then quadratic reciprocity states the following. (The prime 2 must be treated separately.)

Theorem (quadratic reciprocity). *Suppose that p and q are two different primes, neither equal to 2. Then $(\frac{p}{q})(\frac{q}{p}) = -1$ if p and q are both congruent to 3 modulo 4, and $(\frac{p}{q})(\frac{q}{p}) = 1$ otherwise.*

For instance, if $p = 13$ and $q = 29$, then $(\frac{p}{q})(\frac{q}{p}) = 1$. Since 29 is congruent modulo 13 to the perfect square 16, it must be that 13 is congruent to some perfect square modulo 29, and in fact $100 = 3 \cdot 29 + 13$.

This statement is simple but also mysterious, because it violates our intuition that congruences modulo different primes should act independently. For instance, the Chinese remainder theorem asserts that (in a suitably precise sense) knowing that a random integer is odd or even does not prejudice it toward having any particular remainder modulo 3. Number theorists are fond of using geometric language to describe this situation, referring to phenomena associated with congruences modulo a single prime (or a power of a single prime) as *local* phenomena (see LOCAL AND GLOBAL IN NUMBER THEORY [III.51]). The Chinese remainder theorem can be interpreted as saying that local phenomena at one point really are local, in that they do not influence local phenomena at another point. However, just as a particle physicist cannot explain the behavior of the universe by analyzing individual particles in isolation, one cannot hope to understand the behavior of integers by looking at individual primes in isolation. Quadratic reciprocity thus emerges as one of the first known examples of a *global* phenomenon, proving to be a "fundamental force" that binds together two different primes. The interplay between local and global is built thoroughly into our modern understanding of number theory, but the phenomenon of quadratic reciprocity was where it first came to light.

Another indication of the fundamental nature of quadratic reciprocity is that it admits proofs using many different techniques. Gauss himself devised eight proofs in his lifetime, and nowadays dozens of proofs are available. These suggest numerous directions of generalization; here we will focus on the direction that led historically to class field theory. Among the many fascinating sidelights that this will force us to omit is the theory of Gauss sums and its surprisingly diverse range of applications, such as Kolyvagin's work on THE BIRCH–SWINNERTON-DYER CONJECTURE [V.4], and the use of number theory in CRYPTOGRAPHY [VII.7] and other areas of computer science.

Euler had sought reciprocity laws for perfect third and fourth powers, but had had limited success. Gauss succeeded in formulating such laws (but not proving them; that fell to Eisenstein later) by realizing that one could only properly understand them by stepping out of the ring of integers.

Let us see this explicitly for fourth powers. Let p and q be primes that are both congruent to 1 modulo 4. The reciprocity between p being congruent to a fourth power modulo q and vice versa cannot be easily stated in terms of p and q. Instead, we must recall a result of FERMAT [VI.12]: we can write $p = a^2 + b^2$ and $q = c^2 + d^2$, where each of the pairs (a, b) and (c, d) is unique up to changing signs and ordering. In other words, in the ring of complex numbers whose real and imaginary parts are integers (now called the *Gaussian integers*), we have $p = (a + bi)(a - bi)$ and $q = (c + di)(c - di)$.

Gauss defined an analogue of the Legendre symbol as follows. It was already known to Euler that

$$\left(\frac{n}{p}\right) \equiv n^{(p-1)/2} \pmod{p};$$

to see that the right-hand side is either 1 or -1, note that it squares to 1 by FERMAT'S LITTLE THEOREM [III.58], and the equation $x^2 = 1$ has just these two roots. Gauss similarly defined

$$\left(\frac{c + di}{a + bi}\right)_4$$

to be i^k, for the unique choice of k modulo 4 for which

$$i^k \equiv (c + di)^{(a^2 + b^2 - 1)/4} = (c + di)^{(p-1)/4} \pmod{a + bi}.$$

Here we say that two integers are congruent mod $a + bi$ if their difference is a multiple of $a + bi$ by a Gaussian integer. The existence of such k again follows from Fermat's little theorem: if you expand $(c + di)^p$, then all the binomial coefficients are multiples of p apart from the first and the last, so you obtain $c^p + (di)^p$, which equals $c + di$ by Fermat's theorem and the assumption that p

is congruent to 1 mod 4; it follows that $(c + di)^{p-1} \equiv 1$. (Alternatively, one can prove this by showing that the Gaussian integers mod $a + bi$ form a group of order $p - 1$ and applying Lagrange's theorem.)

Before stating the reciprocity law, we must stamp out the ambiguity in the choice of a, b, c, and d. We require that a and c must be odd, and that $a + b - 1$ and $c + d - 1$ must be divisible by 4. (Note that we can still flip the signs of b and d.)

Theorem (quartic reciprocity). *With p, q, a, b, c, and d as above, we have*

$$\left(\frac{a + bi}{c + di}\right)_4\left(\frac{c + di}{a + bi}\right)_4 = -1$$

if p and q are both congruent to 5 modulo 8, and

$$\left(\frac{a + bi}{c + di}\right)_4\left(\frac{c + di}{a + bi}\right)_4 = 1$$

otherwise.

One might expect to find an nth power reciprocity law that looks like this by working with the ring generated by a primitive nth root of 1. What complicates matters is that this ring does not enjoy the UNIQUE FACTORIZATION PROPERTY [IV.1 §§4–8] (whereas the usual integers and the Gaussian integers both do). This was remedied only by KUMMER's [VI.40] theory of IDEALS [III.81 §2] (short for "ideal numbers"). An ideal is a set that has the typical properties of the set of all multiples of a given number, but it can be more general. (Even if an ideal *is* the set of all multiples of some number, that number is not unique, since one can multiply it by a unit. For instance, both 2 and -2 generate the ideal of all even numbers.) Using Kummer's theory, Kummer and Eisenstein managed to formulate broad generalizations of quadratic reciprocity for higher powers.

HILBERT [VI.63] then realized that these should fit together as part of some sort of maximally general reciprocity law. He also gave a candidate for this law, inspired by a reformulation of quadratic reciprocity itself in terms of the *norm residue symbol*. For a prime p, and any nonzero integers m and n, the norm residue symbol $(\frac{m,n}{p})$ equals 1 if, for all sufficiently large k, the equations $mx^2 + ny^2 \equiv z^2 \pmod{p^k}$ have solutions where x, y, and z are not all divisible by p^k; otherwise the symbol equals -1. In other words, the symbol equals 1 if the equation $mx^2 + ny^2 = z^2$ has a solution in the p-ADIC NUMBERS [III.51].

Hilbert's formulation of quadratic reciprocity is that, for any nonzero m and n,

$$\prod_p \left(\frac{m,n}{p} \right) = 1,$$

where the product is taken over all primes p *and the prime* $p = \infty$. The latter requires some explanation: we write $\left(\frac{m,n}{\infty}\right) = 1$ if and only if m and n are not both negative, i.e., if the equation $mx^2 + ny^2 = z^2$ has a solution in the *real* numbers. This fits into a general pattern, that conditions quantified over "all prime numbers" must also account for the so-called infinite prime.

It should also be clarified that Hilbert's product only makes sense by virtue of the fact that, for fixed m and n, $\left(\frac{m,n}{p}\right) = 1$ for all but finitely many p. This is because in general, since approximately half the integers mod p^k are quadratic residues, it is easy to solve the equation $mx^2 + ny^2 = z^2$: difficulties arise only when multiplication by m or n identifies many of these quadratic residues. For instance, if m and n are (positive) prime numbers, then only those two primes contribute to the product; the two resulting factors can be related to $\left(\frac{m}{n}\right)$ and $\left(\frac{n}{m}\right)$, which leads back to quadratic reciprocity.

Using this formulation, Hilbert was able to state and prove a form of quadratic reciprocity over any NUMBER FIELD [III.63], in which the corresponding product of symbols is quantified over the prime ideals of the number field (together with some "infinite primes"). Hilbert also conjectured a higher-power reciprocity law over any number field. That conjecture was tackled by Hasse, Takagi, and finally ARTIN [VI.86], who stated a general reciprocity law. Its statement is a bit too technical to include here; we limit ourselves to observing that Artin's reciprocity law, when applied to a number field K, describes certain norm residue symbols in terms of *Abelian* extensions of K, i.e., number fields containing K whose groups of symmetries (GALOIS GROUPS [V.21]) are commutative.

The Abelian extensions of $\mathbb{Q}$ are easy to describe: the Kronecker–Weber theorem asserts that they are all contained in fields generated by roots of 1. This explains the role of the roots of 1 in the classical reciprocity laws. However, describing the Abelian extensions of an arbitrary number field K is somewhat harder. They can at least be classified in terms of the structure of the field K itself; this is what is commonly referred to as *class field theory*.

However, the problem of explicitly specifying generators of the Abelian extensions of K (Hilbert's twelfth problem) remains mostly unsolved, except in some special cases. For instance, the theory of ELLIPTIC FUNCTIONS [V.31] solves this problem for fields of the form $\mathbb{Q}(\sqrt{-d})$ with $d > 0$ via the theory of *complex multiplication*. Some additional examples emerged from the work of Shimura on MODULAR FORMS [III.59], leading to the *Shimura reciprocity law*.

This last example shows that the story of reciprocity laws is not yet complete. Any new instance of explicit class field theory would reveal another reciprocity law that had previously been hidden from view. Some exciting new conjectures in this direction have been advanced by Bertolini, Darmon, and Dasgupta, who have proposed some new constructions of Abelian extensions using p-adic analysis. These are analogous to the aforementioned constructions using elliptic functions, in which one evaluates a transcendental function at a special value. At first, there seems to be no reason to expect the resulting complex number to have any special properties, but in fact it turns out to be an algebraic number that generates an appropriate Abelian extension of the base field. While one can check in individual examples, using computer calculations, that the construction seems to be converging p-adically to a particular generator of the right field, a proof seems out of reach at present.

Further Reading

Ireland, K., and M. Rosen. 1990. *A Classical Introduction to Modern Number Theory*, 2nd edn. New York: Springer.
Lemmermeyer, F. 2000. *Reciprocity Laws, from Euler to Eisenstein*. Berlin: Springer.

V.29 Rational Points on Curves and the Mordell Conjecture

Suppose that we wish to study a Diophantine equation such as $x^3 + y^3 = z^3$. A simple observation we can make is that studying integer solutions to this equation is more or less equivalent to studying rational solutions to the equation $a^3 + b^3 = 1$: indeed, if we had integers x, y, and z such that $x^3 + y^3 = z^3$, then we could set $a = x/z$ and $b = y/z$ and obtain rational numbers with $a^3 + b^3 = 1$. Conversely, given rational numbers a and b with $a^3 + b^3 = 1$, we could multiply a and b by the lowest common multiple z of their denominators and set $x = az$ and $y = bz$, obtaining integers x, y, and z such that $x^3 + y^3 = z^3$.

The advantage of doing this is that it reduces the number of variables by 1 and focuses our attention on the plane curve $u^3 + v^3 = 1$, which is a simpler object than the surface $x^3 + y^3 = z^3$. A curve of this kind, defined by one or more polynomial equations, is called an *algebraic curve*.

Even though we are interested in rational points on the curve, it can be helpful to regard the curve as an abstract object that has many manifestations. (See ARITHMETIC GEOMETRY [IV.5] for a fuller discussion of this point.) For instance, if we think of u and v as complex numbers, then the "curve" $u^3 + v^3 = 1$ becomes a two-dimensional object, which means that it starts to have a genuinely interesting geometry. To be precise, it can be regarded as a two-dimensional MANIFOLD [I.3 §6.9] living in $\mathbb{R}^4$. From a *complex* perspective it is a one-dimensional subset of $\mathbb{C}^2$, but from either perspective it has a potentially interesting topology. For instance, if we COMPACTIFY [III.9] the curve by considering it as a subset not of $\mathbb{C}^2$ but of the complex PROJECTIVE PLANE [I.3 §6.7], then we turn it into a compact surface. As such, it must have a GENUS [III.33], which, roughly speaking, tells us how many holes it has.

Surprisingly, it turns out that this geometrical definition of the genus of a curve is intimately related to the algebraic question of how many rational points the curve contains. Consider, for instance, the curve $u^2 + v^2 = 1$, which corresponds to the Diophantine equation $x^2 + y^2 = z^2$. Since there are infinitely many Pythagorean triples that are not multiples of each other, there are infinitely many rational points on the curve $u^2 + v^2 = 1$. In order to calculate the genus of the curve, we first rewrite it as $(u + iv)(u - iv) = 1$. This shows that the function $(u, v) \mapsto u + iv$ is a homeomorphism from the curve to the set $\mathbb{C} \setminus \{0\}$ of all nonzero complex numbers, which itself is homeomorphic to a sphere with two points removed. The compactification adds in these points, giving us a surface of genus 0, so we say that the curve $u^2 + v^2 = 1$ has genus 0. It turns out that a curve of genus 0 always has either no rational points or infinitely many.

In general, the larger the genus, the harder it is to find rational solutions. A curve of genus 1 is called an ELLIPTIC CURVE [III.21]. It is possible for an elliptic curve to contain infinitely many rational points as well, but the set of such points turns out to have a very restricted structure. To explain this, let us consider an elliptic curve E of the form $y^2 = ax^3 + bx^2 + cx + d$ (a form into which any elliptic curve can be put). If we think of it as a curve in $\mathbb{R}^2$, then we can define a binary

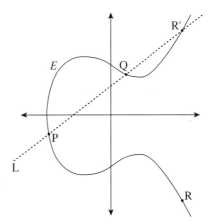

Figure 1 The group law for an elliptic curve.

operation on it as follows: given any two points P and Q on E, let L be the line through P and Q (where we define this to be the tangent to the curve at P if P = Q). In general, L intersects E in three points, of which P and Q are two; let R′ be the third. Finally, let R be the reflection of R′ in the x-axis (which also belongs to E because E has the form $y^2 = f(x)$). This construction of R from P and Q, which is illustrated in figure 1, defines a binary operation on the points of E. Remarkably, this binary operation turns E into an Abelian group, at least when we also include a point at infinity and adopt the convention that the point at infinity is the intersection of E with any vertical line. The point at infinity is the identity of the group, since a vertical line through a point P intersects E in the reflection P′ of P in the x-axis, and when we reflect P′ in the x-axis we get P again.

It is laborious, but basically straightforward, to come up with a formula for the "group law" of an elliptic curve—that is, a formula for the coordinates of R in terms of the coordinates of P and Q. Once one does so, it becomes clear that if P and Q have rational coordinates, then so does R. Thus, the set of all rational points on an elliptic curve E forms a subgroup. This simple fact can be used to produce rather easily some very large solutions to the corresponding Diophantine equations. For instance, one can start with a small solution, associate with it a rational point P, and then use the formula for the binary operation to calculate 2P, then 4P, then 8P, and so on. Unless nP = 0 for some n (which can certainly happen), in no time at all one has a point on the curve with rational coordinates that have huge numerators and denominators. To give an idea of the sort of solutions that can be obtained in this way, take the elliptic curve $y^2 = x^3 - 5x$ and let P be the point $(-1, 2)$ (which lies on the curve since $2^2 = (-1)^3 - 5(-1)$). If you calculate 5P using the group law, then you obtain the point $(-5\,248\,681/4\,020\,025, 16\,718\,705\,378/8\,060\,150\,125)$. In general, the number of digits needed to express the point nP grows exponentially with n.

In the early twentieth century, POINCARÉ [VI.61] conjectured that the subgroup of rational points on an elliptic curve was finitely generated. This conjecture was proved by Louis Mordell in 1922. Thus, although a curve of genus 1 may have infinitely many rational points, there is a finite set of these points that can be used to build up all the others: this is the sense in which the structure of the set of rational solutions is restricted.

Mordell conjectured that a curve of genus at least 2 could contain only finitely many points. This was a remarkable conjecture: if true, it would apply to an extremely wide class of Diophantine equations, proving that all of them had at most finitely many solutions (up to a multiple). Just one of its many implications was that for each $n \geqslant 3$ the Fermat equation $x^n + y^n = z^n$ had at most finitely many solutions with x, y, and z coprime. However, it is one thing to make a very general conjecture and quite another to prove it, and for a long time the consensus was that the Mordell conjecture, like many other conjectures in number theory, was way beyond what anybody could prove. It therefore came as a big surprise when Gerd Faltings proved the conjecture in 1983.

As a result of Faltings's proof, our knowledge about Diophantine equations took a huge leap forward. The theorem has subsequently been given a variety of different proofs, some of them simpler than that of Faltings. However, remarkable as these proofs are, they do have some limitations. One is that they are *ineffective*. That is, even though Faltings's theorem tells us that certain curves have finitely many rational points, no known proof gives any bound on the sizes of the numerators and denominators of the coordinates of those points, so we do not have any way of knowing whether we have found all of them. This aspect of the theorem is common in number theory: another example of a famous theorem that is ineffective is ROTH'S THEOREM [V.22]. To find effective versions of these theorems would be a further remarkable breakthrough. (Variants of THE ABC CONJECTURE [V.1] would imply effective versions of these results, but the ABC conjecture seems even further out of reach

now than Mordell's conjecture seemed before Faltings proved it.)

At the beginning of this article, we simplified the equation $x^3 + y^3 = z^3$ so that we were looking at a curve rather than a surface. But we obviously cannot always do that. For instance, if we apply the same procedure to the equation $x^5 + y^5 + z^5 = w^5$, then we obtain the two-dimensional surface $t^5 + u^5 + v^5 = 1$. Our knowledge about rational points on varieties (that is, sets defined by polynomial equations) of dimension greater than 1 is very limited. However, there is at least a definition of a "variety of general type" that serves as an analogue of the notion of a curve of genus at least 2. One cannot expect such a variety to contain only finitely many rational points, but a higher-dimensional analogue of the Mordell conjecture, due to Serge Lang, asserts that the rational points on a variety X of general type must all be contained in a union of finitely many lower-dimensional subvarieties of X. This conjecture is considered to be well out of reach of present methods: indeed, it is not even universally believed.

V.30 The Resolution of Singularities

Virtually all important mathematical structures come with a notion of equivalence. For instance, we regard two GROUPS [I.3 §2.1] as equivalent if they are ISOMORPHIC [I.3 §4.1], and we regard two TOPOLOGICAL SPACES [III.90] as equivalent if there is a continuous map from one to the other with a continuous inverse (in which case we say that they are *homeomorphic*). In general, a notion of equivalence is useful if properties that we are interested in are unaffected when we replace an object by an equivalent one: for example, if G is a finitely generated Abelian group and H is isomorphic to G, then H is a finitely generated Abelian group.

A useful notion of equivalence for ALGEBRAIC VARI-ETIES [IV.4 §7] is that of *birational* equivalence. Roughly speaking, two varieties V and W are said to be birationally equivalent if there is a rational map from V to W with a rational inverse. If V and W are presented as solution sets of equations in some coordinate system, then these rational maps are just rational functions in the coordinates that send points of V to points of W. However, it is important to understand that a rational map from V to W is not literally a function from V to W, because it is allowed to be undefined at certain points of V.

Consider, for example, how we might map the infinite cylinder $\{(x, y, z) : x^2 + y^2 = 1\}$ to the cone $\{(x, y, z) :$

$x^2 + y^2 = z^2$}. An obvious map would be the function $f(x,y,z) = (zx,zy,z)$, which we could try to invert using the map $g(x,y,z) = (x/z,y/z,z)$. However, g is not defined at the point $(0,0,0)$. Nevertheless, the cylinder and the cone are birationally equivalent, and algebraic geometers would say that g "blows up" the point $(0,0,0)$ to the circle {$(x,y,z) : x^2 + y^2 = 1, z = 0$}.

The main property of a variety V that is preserved by birational equivalence is the so-called *function field* of V, which consists of all rational functions defined on V. (What precisely this means is not completely obvious: in some contexts, V is a subset of a larger space such as $\mathbb{C}^n$ in which one can talk about ratios of polynomials, and then one possible definition of a rational function on V is that it is an equivalence class of such ratios, where two of them are counted as equivalent if they take the same values on V. See ARITHMETIC GEOMETRY [IV.5 §3.2] and QUANTUM GROUPS [III.75 §1] for further discussion of this equivalence relation.)

A famous theorem of Hironaka, proved in 1964, states that every algebraic variety (over a field of characteristic 0) is birationally equivalent to an algebraic variety without singularities, with some technical conditions on the birational equivalence that are needed for the theorem to be interesting and useful. The example given earlier is a simple illustration: the cone has a singularity at $(0,0,0)$ but the cylinder is smooth everywhere. Hironaka's proof was well over two hundred pages long, but his argument has since been substantially simplified by several authors.

For a further discussion of the resolution of singularities, see ALGEBRAIC GEOMETRY [IV.4 §9].

The Riemann Hypothesis

See THE PRIME NUMBER THEOREM AND
THE RIEMANN HYPOTHESIS [V.26]

V.31 The Riemann–Roch Theorem

A RIEMANN SURFACE [III.79] is a MANIFOLD [I.3 §6.9] that "looks locally like $\mathbb{C}$," in the usual sense of this sort of phrase. In other words, every point has a neighborhood that can be mapped bijectively to an open subset of $\mathbb{C}$, and where two such neighborhoods overlap, the "transition functions" are HOLOMORPHIC [I.3 §5.6]. One can think of a Riemann surface as the most general sort of set on which the notion of a holomorphic function (that is, a complex-differentiable function) of one complex variable makes sense.

The definition of differentiability is a local one: a function is differentiable if and only if a certain condition holds at each point z, and the condition at z depends only on the behavior of f at points very close to z. However, one of the surprises of complex analysis is that holomorphic functions are much more global than their basic definition would lead one to expect. Indeed, if you know the values of a holomorphic function $f : \mathbb{C} \to \mathbb{C}$ at every point in a small neighborhood of a single point z, then you can deduce its values at every point in $\mathbb{C}$. And the same is true if you replace $\mathbb{C}$ by any other (connected) Riemann surface.

Here is a second illustration of the global nature of holomorphic functions. One of the most basic Riemann surfaces is the so-called *Riemann sphere* $\hat{\mathbb{C}}$, which is obtained from $\mathbb{C}$ by adding a "point at infinity." A function $f : \hat{\mathbb{C}} \to \mathbb{C}$ is said to be holomorphic if the following conditions hold:

- f is differentiable at every point of $\mathbb{C}$;
- $f(z)$ tends to a limit w as $z \to \infty$ in any direction;
- w is the value of f at ∞.

What, then, are the holomorphic functions from $\hat{\mathbb{C}}$ to $\mathbb{C}$? A holomorphic function f is continuous, from which it follows that if $f(z)$ tends to a limit as $z \to \infty$, then f is bounded on $\mathbb{C}$. But a well-known theorem of LIOUVILLE [VI.39] states that a bounded holomorphic function defined on all of $\mathbb{C}$ must be constant. So the only holomorphic functions from $\hat{\mathbb{C}}$ to $\mathbb{C}$ are constant!

One might take the attitude that it was slightly artificial to consider maps from $\hat{\mathbb{C}}$ to $\mathbb{C}$. Why not look at maps from $\hat{\mathbb{C}}$ to $\hat{\mathbb{C}}$? Such maps are equivalent to functions from $\mathbb{C}$ to $\mathbb{C}$ that are allowed to tend to infinity at a finite set of points $z_1, \ldots, z_k$, called *poles*, and must tend to a limit as $z \to \infty$. (This limit is allowed to be the point ∞. We say that $f(z) \to \infty$ as $z \to \infty$ if we can make $|f(z)|$ arbitrarily large by making $|z|$ large enough. Note that some familiar functions such as e^z are ruled out since it is possible for $|z|$ to be large and e^z to be small.) Functions with this property are called *meromorphic*. A typical example is z, or z^2, or $(1 + z)/(1 - z)$, or indeed any rational function in z; it can in fact be shown that any meromorphic function from $\hat{\mathbb{C}}$ to $\hat{\mathbb{C}}$ is rational.

The notion of a meromorphic function also makes sense on other Riemann surfaces. One can think of it as a function that is holomorphic except at a set of isolated points where it tends to infinity. (If the function is defined on $\mathbb{C}$, there may be infinitely many such points, but a COMPACT [III.9] surface such as $\hat{\mathbb{C}}$ cannot contain infinitely many points that are all isolated from each other, so a meromorphic function on a compact surface has at most finitely many poles.)

A particularly important example is when the Riemann surface in question is a torus. We can regard such a surface as the QUOTIENT [I.3 §3.3] of $\mathbb{C}$ by the lattice generated by two complex numbers u and v such that u/v is not real. There is then a one-to-one correspondence between functions defined on the torus and functions f defined on $\mathbb{C}$ that are *doubly periodic*, in the sense that $f(z + u)$ and $f(z + v)$ are both equal to $f(z)$ for every z. Liouville's theorem again implies that if such a function is holomorphic then it is constant; however, there are interesting examples of doubly periodic *meromorphic* functions. Such functions are called *elliptic functions*.

Even here, the global nature, or "rigidity," of holomorphic functions asserts itself, by greatly restricting the supply of elliptic functions. Indeed, one can define a single function, called the *Weierstrass P-function* $\wp$, with the property that any other elliptic function with respect to a given pair of generators u and v can be expressed as a rational function of $\wp$ and its derivative. Weierstrass's function (for the generators u and v) is given by the formula

$$\wp(z) = \frac{1}{z^2} + \sum_{(n,m) \neq (0,0)} \left(\frac{1}{(z - mu - nv)^2} - \frac{1}{(mu + nv)^2} \right).$$

Notice that the double periodicity is built into the definition, and that $\wp$ has a pole at every point in the lattice generated by u and v. If we think of $\wp$ as a function on the torus, then it has just one pole. Near this pole, f tends to infinity at the same rate as the function $1/z^2$ does when z tends to 0; we say that the pole has *order* 2. More generally, if a function f tends to infinity at the same rate as $1/z^k$, then the resulting pole has *order* k.

Suppose we take a compact Riemann surface S and choose from it a finite set of points $z_1, \dots, z_r$. Given a sequence $d_1, \dots, d_r$ of positive integers, can we find a meromorphic function f defined on S such that its poles are $z_1, \dots, z_r$ and such that for each i the order of the pole at z_i is at most d_i? The results mentioned so far would lead us to expect that this might be possible, but that there would probably not be a huge supply of such functions. Since a linear combination of such functions gives us another one, the set of functions we are interested in forms a VECTOR SPACE [I.3 §2.3], so we could hope to quantify "how many" functions there are by investigating the dimension of this space.

As we might by now expect, this dimension turns out to be finite. RIEMANN [VI.49] proved that if the poles are required to be *simple* (that is, $d_i = 1$ for $i = 1, 2, \dots r$), then the dimension l is at least $r - g + 1$, where g is the GENUS [III.33] of the surface, which means, roughly speaking, the number of holes it has. This result is called *Riemann's inequality*. Roch's contribution was to interpret the difference between l and $r - g + 1$ as the dimension of another space of functions. This often makes it possible to calculate the dimension l exactly. For instance, under certain circumstances one can show that the dimension of the space of functions identified by Roch is 0, in which case $l = r - g + 1$. In particular, this is the case when $r \geqslant 2g - 1$.

The original question we asked was more general in that we did not require the poles to be simple: rather, we wanted the order of the pole at z_i to be at most d_i. However, the result generalizes straightforwardly, and l is now at least $d_1 + \dots + d_r - g + 1$, with the difference again equal to the dimension of a certain space of functions that one can define. One can even ask for some of the d_i to be negative, interpreting a "pole of order at most d_i" to mean a zero of multiplicity at least $-d_i$.

The Riemann–Roch theorem is a basic tool for computing the dimensions of spaces of holomorphic or meromorphic functions on compact surfaces (which is often equivalent to requiring them to obey certain symmetry conditions). Let us begin with a very simple example. It is not hard to show that every meromorphic function defined on the Riemann sphere with at most simple poles at 0 and 1 has to take the form $a + b/z + c/(z - 1)$. This is a three-dimensional space, and that is what the Riemann–Roch theorem predicts. A more sophisticated example concerns the Weierstrass P-function. We saw earlier that this is a doubly periodic meromorphic function defined on $\mathbb{C}$ with a pole of order 2 at each point in the lattice generated by u and v. The existence (and essential uniqueness) of such a function can be proved more abstractly with the help of the Riemann–Roch theorem: it shows that the space of such functions has dimension 2, so they can all be built out of a single function $\wp$ and the constant functions. Similarly, the theorem can be used to compute dimensions of spaces of MODULAR FORMS [III.59].

The Riemann–Roch theorem has been reformulated and generalized many times, which has made it even more useful as a computational tool, and a central result in algebraic geometry: for example, Hirzebruch found a higher-dimensional generalization, which was generalized further by Grothendieck to a statement about advanced concepts in modern algebraic geometry such as SCHEMES [IV.5 §3] and "sheaves." Hirzebruch's generalization, like the classical result about curves, expresses an analytically defined quantity in terms of purely topological invariants: it is this feature of both results that underlies their importance. Another generalization of which the same can be said is the famous ATIYAH–SINGER INDEX THEOREM [V.2], which has itself been generalized several times.

V.32 The Robertson–Seymour Theorem
Bruce Reed

A *graph G* is a mathematical structure that consists of a set $V(G)$ of *vertices* and a set $E(G)$ of *edges*, where each edge links a pair of vertices. Graphs can be used to represent many different networks in an abstract way. For example, the vertices might represent cities, and the edges might represent highways linking the cities; similarly, we could use a graph to represent which islands of an archipelago are linked by bridges, or to represent the wires of a telephone network. Among graphs there are certain families of "nice" graphs. One such family is the family of *cycles*: a k-cycle is a set of k vertices arranged around a circle with each point joined by an edge to the points immediately before and after it. Another family is that of *complete graphs*: the complete graph of order k consists of k vertices, all pairs of which are joined.

An important concept in graph theory, particularly when families of graphs are involved, is that of a *minor*. Given a graph G, a minor of G is any graph you can obtain by applying a sequence of operations of two kinds, known as contractions and deletions, applied to edges. To *contract* the edge that joins two vertices x and y, one "fuses" x and y into a single vertex, joining it to all the vertices that were previously joined to either x or y. For example, if you contract an edge of a 9-cycle, you will obtain an 8-cycle. *Deleting* an edge means what one would guess: for example, if you delete an edge from a 9-cycle you will get a *path* with nine vertices and eight edges.

It is not hard to check that a graph H is a minor of G if and only if we can find a collection of disjoint subsets of G, one for each vertex of H, with the following properties: they should be *connected*, which means that any two vertices in one of the subsets are joined to each other by a path in that subset, and for any pair of vertices in H that are linked by an edge in H the two corresponding subsets of G should be linked by an edge. For example, a graph has a 3-cycle (or *triangle*) as a minor if and only if it contains a cycle.

For an example of how minors can arise naturally, note that if a graph is planar (meaning that it can be drawn in the plane in such a way that edges do not cross), then so is any minor of it. This is expressed by saying that the class of planar graphs is *minor closed*. Now, there is a theorem of Kuratowski that tells us which graphs are planar. One form that this theorem takes is the following statement: a graph is planar if and only if it does not have either K_5 or $K_{3,3}$ as a minor, where K_5 denotes the complete graph of order 5, and $K_{3,3}$ denotes the *complete bipartite* graph that consists of two sets of three vertices, with every vertex in one set joined to every vertex in the other set. Thus, the class of planar graphs is characterized by two *forbidden minors*.

Kuratowski's theorem tells us which graphs can be embedded into the plane. What happens for other surfaces? For example, it is easy to see that for any d the set of graphs that can be drawn on a d-holed torus is minor closed, but is there a finite set of forbidden minors in this case? To put it another way, is the set of obstructions to being embeddable into the d-holed torus only finite?

A special case of the Robertson–Seymour theorem states that the answer to this question is yes for any surface. But the theorem itself is much more general. It states that for *any* minor-closed class of graphs, there is a finite set of forbidden minors. In other words, for any minor-closed class G there exist graphs $G_1, \ldots, G_k$ such that a graph G belongs to the family G if and only if G does not have any G_i as a minor. There is also a pleasant form of the theorem (which is easily seen to be equivalent) that says that the class of all graphs is "well-quasi-ordered" by the minor relation: this means that given any sequence $G_1, G_2, \ldots$ of graphs there must exist one that is a minor of a later one.

It turns out that testing a graph for the presence of a given minor can be done reasonably fast, so that one amazing spin-off from the Robertson–Seymour theorem is that for any minor-closed class there is an efficient algorithm for checking whether or not a given graph belongs to the class. This has had a huge number of applications in routing problems and the like.

The actual proof of the Robertson–Seymour theorem is enormous: it was published in a sequence of twenty-two papers. Interestingly, it turns out that the case of graphs embeddable into a given surface plays a key role, as we now explain.

We will consider the form of the theorem mentioned above involving a sequence of graphs. So let us suppose for a contradiction that we have a "bad" sequence: that is, a sequence $G_1, G_2, \ldots$ for which no G_i is a minor of any later G_j. Let the number of vertices of the first graph G_1 be k. Since no later G_i has G_1 as a minor, it certainly follows that none of $G_2, G_3, \ldots$ has a complete minor of size k (or else we could delete some edges and obtain G_1). For this reason, Robertson and Seymour studied families of graphs that do not have a complete minor of size k. They were able to show that every graph that does not have a complete minor of size k may be built up in a certain way from graphs that are "nearly embeddable" into a fixed surface (that depends on the value of k). This means that in a certain sense that can be made precise the graph is not too far from a graph that *is* embeddable into the surface. By some very deep arguments, they were able to show that the family of all such graphs (the graphs that can be built up from nearly embeddable graphs, for a given surface) has a finite number of forbidden minors, thereby proving the theorem.

V.33   The Three-Body Problem

The three-body problem can be simply stated: three point masses move in space under their mutual gravitational attraction; given their initial positions and velocities, determine their subsequent motion. Initially, it may come as a surprise that this is a difficult problem, since the analogous two-body problem can be solved fairly simply: more precisely, given any set of initial conditions, we can write down a formula, in terms of elementary functions (these are functions that can be built up using the basic operations of arithmetic, together with a few standard functions such as the EXPONENTIAL [III.25] and TRIGONOMETRIC [III.92] functions), that tells us the subsequent positions and velocities of the bodies. However, the three-body problem is a complicated nonlinear problem and it cannot be solved in this way, even if we are prepared to enlarge our stock of "standard functions" somewhat. NEWTON [VI.14] himself speculated that an exact solution "exceeds, if I am not mistaken, the force of any human mind," while HILBERT [VI.63], in his celebrated Paris address of 1900, put the problem in a category similar to FERMAT'S LAST THEOREM [V.10]. The problem can be extended to any number of bodies and in the general case it is known as the n-body problem.

Recall that the gravitational force of a particle P_1 on a particle P_2 has magnitude $k^2 m_1 m_2 / r^2$ (in suitable units), where k is the *Gaussian gravitational constant*, particle P_i has mass m_i, and the distance between the particles is r. The direction of this force on P_2 is toward P_1 (and there is a force of the same magnitude on P_1 in the direction of P_2). Recall also Newton's second law: force equals mass times acceleration. From these two laws we can easily derive the equations of motion for the three-body problem. Let the particles be P_1, P_2, and P_3. Write m_i for the mass of P_i, r_{ij} for the distance between P_i and P_j, and q_{ij} for the jth coordinate of the position of P_i. Then the equations of motion are

$$\left. \begin{aligned} \frac{\mathrm{d}^2 q_{1i}}{\mathrm{d}t^2} &= k^2 m_2 \frac{q_{2i} - q_{1i}}{r_{12}^3} + k^2 m_3 \frac{q_{3i} - q_{1i}}{r_{13}^3}, \\ \frac{\mathrm{d}^2 q_{2i}}{\mathrm{d}t^2} &= k^2 m_1 \frac{q_{1i} - q_{2i}}{r_{12}^3} + k^2 m_3 \frac{q_{3i} - q_{2i}}{r_{23}^3}, \\ \frac{\mathrm{d}^2 q_{3i}}{\mathrm{d}t^2} &= k^2 m_1 \frac{q_{1i} - q_{3i}}{r_{13}^3} + k^2 m_2 \frac{q_{2i} - q_{3i}}{r_{23}^3}. \end{aligned} \right\} \quad (1)$$

Here, i runs from 1 to 3; thus, there are nine equations, all derived from the simple laws above. For instance, the left-hand side of the first equation is the component of the acceleration of P_1 in the ith direction, and the right-hand side is the component of the force acting on P_1 in this direction, divided by m_1.

If the units are chosen so that $k^2 = 1$, then the potential energy V of the system is given by

$$V = -\frac{m_2 m_3}{r_{23}} - \frac{m_3 m_1}{r_{31}} - \frac{m_1 m_2}{r_{12}}.$$

Setting

$$p_{ij} = m_i \frac{\mathrm{d}q_{ij}}{\mathrm{d}t} \quad \text{and} \quad H = \sum_{i,j=1}^{3} \frac{p_{ij}^2}{2m_i} + V,$$

we can rewrite the equations in the HAMILTONIAN FORM [IV.16 §2.1.3]

$$\frac{\mathrm{d}q_{ij}}{\mathrm{d}t} = \frac{\partial H}{\partial p_{ij}}, \qquad \frac{\mathrm{d}p_{ij}}{\mathrm{d}t} = -\frac{\partial H}{\partial q_{ij}}, \qquad (2)$$

which is a set of eighteen first-order differential equations. Since this set is easier to use, it is now generally preferred to (1).

A standard way of decreasing the complexity of a system of differential equations is to find an *algebraic integral* for it: that is, a quantity that will remain constant for any given solution and that can be expressed as an integral that gives rise to an algebraic dependence between the variables. This allows us to reduce the number of variables by expressing some of them in terms of others. The three-body problem has ten independent algebraic integrals: six of them tell us about the motion of the center of mass (three for the position variables and three for the momentum variables), three integrals express the conservation of angular momentum, and one expresses conservation of energy. These ten independent integrals were known to EULER [VI.19] and LAGRANGE [VI.22] in the middle of the eighteenth century, and in 1887 Heinrich Bruns, professor of astronomy at Leipzig, proved that there are no others, a result sharpened by POINCARÉ [VI.61] two years later. By the use of these ten integrals, together with the "elimination of the time" and the "elimination of the nodes" (a procedure first made explicit by JACOBI [VI.35]), the original system of order eighteen can be reduced to one of order six, but it can be reduced no further. Hence, any general solution of (2) cannot be given by a simple formula: the best we can hope for is a solution in the form of an infinite series. It is not difficult to find series that work well enough for a limited time span: the problem is to find series that work for any initial configuration and for any time span, no matter how long. There is also the question of collisions. A complete solution to the problem has to take account of all possible motions of the bodies, including determining which initial conditions lead to binary and triple collisions. Since collisions are described by singularities in the differential equations, this means that to find a complete solution the singularities have to be understood.

This turns out to be a more interesting problem than one might think. It is obvious from the equations that a collision gives rise to a singularity, but it is less clear whether there can be any other kind of singular behavior. In the case of the three-body problem, the answer was supplied by Painlevé in 1897: the collisions are the only singularities. However, for more than three bodies the answer turned out to be different. In 1908 a Swedish astronomer, Hugo von Zeipel, showed that *noncollision singularities* can occur only if the system of particles becomes unbounded in a finite amount of time. A good example of such a singularity was found by Zhihong Xia for the five-body problem in 1992. In this case there are two pairs of bodies, the bodies in each pair having equal mass, and a fifth body with very small mass. The bodies in a pair move in very eccentric orbits parallel to the xy-plane, with the two pairs on opposite sides of this plane and rotating in opposite directions. A fifth particle is then added to the system. Its motion is confined to the z-axis and oscillates between the two pairs. Xia showed that the motion of the fifth particle forced the two pairs to move away from the xy-plane, but that it

also came closer and closer to colliding with the pairs, giving it larger and larger bursts of acceleration, and that as this happened the two pairs were forced out to infinity in finite time.

As well as trying to solve the problem in general, one can look for interesting particular solutions. A *central configuration* is defined to be a solution in which the geometric configuration remains constant. The first examples were discovered by Euler in 1767: they were solutions in which the bodies always lie on a straight line and revolve with uniform angular velocity in circles or ellipses about their common center of mass. In 1772 Lagrange discovered solutions in which the bodies are always at the vertices of an equilateral triangle that rotates uniformly about the center of mass. For almost all sets of initial conditions for these solutions, the size of the triangle changes as it rotates so that each body describes an ellipse.

However, despite the discovery of the particular solutions and a century of unrelenting work on the problem, the mathematicians of the nineteenth century were unable to find a general solution. Indeed, the problem was considered so hard that in 1890 Poincaré was led to declare that he thought it impossible without the discovery of some significant new mathematics. But, contrary to Poincaré's expectation, less than twenty years later a young Finnish mathematical astronomer, Karl Sundman, using only existing mathematical techniques, astonished the mathematical world by obtaining uniformly convergent infinite series that mathematically "solved" the problem. Sundman's series, which are in powers of $t^{1/3}$, are convergent for all real t, except for the negligible set of initial conditions for which the angular momentum is zero. To deal with binary collisions, Sundman used the technique of *regularization*, or analytically extending a solution beyond the collision, but he was unable to deal with triple collisions because in order for such a collision to occur the angular momentum must be zero.

Although it was a remarkable mathematical achievement, Sundman's solution leaves many questions unanswered. It provides no qualitative information about the behavior of the system and, worse, because the series converges so slowly it is of no practical use. To determine the motion of the bodies for any reasonable period of time would require the summation of something of the order of $10^{8000000}$ terms, a calculation that is patently unrealistic. Thus, Sundman left plenty still to do, and work on the problem (and the related n-body problem) has continued up to the present day, with exciting results continuing to appear. One recent example is a convergent power-series solution for the general n-body problem, which was discovered by Don Wang in 1991.

Since the three-body problem itself proved so intractable, simplified versions were developed, of which the most famous is the one now known as the *restricted* three-body problem (the name is due to Poincaré), which was first investigated by Euler. In this case, two of the bodies (the *primaries*) revolve around their joint center of mass in circular orbits under the influence of their mutual gravitational attraction, while the third body (the *planetoid*), which is assumed to have such small mass that the force it exerts on the other two bodies can be neglected, moves in the plane defined by the primaries. The advantage of this formulation is that the motion of the primaries can be treated as a two-body problem and is hence known; it remains only to investigate the motion of the planetoid, which can be done using perturbation theory. Although the restricted formulation might appear artificial, it provides a good approximation to real physical situations, such as, for example, the problem of determining the motion of the Moon around Earth given the presence of the Sun. Poincaré wrote extensively on the restricted problem, and the techniques he developed to tackle it led to his discovery of mathematical chaos, as well as laying the foundations for modern DYNAMICAL SYSTEMS [IV.14] theory.

Apart from its intrinsic appeal as a problem that is simple to state, the three-body problem has a further attribute that has contributed to its attraction for potential solvers: its intimate link with the fundamental question of the stability of the solar system. That is the question of whether the planetary system will always keep the same form as it has now, or whether, eventually, one of the planets will escape or, perhaps worse, experience a collision. Since bodies in the solar system are approximately spherical and their dimensions extremely small when compared with the distances between them, they can be considered as point masses. Ignoring all other forces, such as solar winds or relativistic effects, and taking only gravitational forces into account, the solar system can be modeled as a ten-body problem with one large mass and nine small ones, and it can be investigated accordingly.

Over the years, attempts to find a solution to the three-body problem (and the related n-body problem), have spawned a wealth of research. As a result, the importance of the problem is as much in the mathematical advances it has generated as in the problem itself. A notable example of this is the development of *KAM theory*, which provides methods for integrating perturbed Hamiltonian systems and obtaining results that are valid for infinite periods of time. This was developed in the 1950s and 1960s by KOLMOGOROV [VI.88], Arnold, and Moser.

Thurston's Geometrization Conjecture
See THE POINCARÉ CONJECTURE [V.25]

V.34   The Uniformization Theorem

The uniformization theorem is a remarkable classification of RIEMANN SURFACES [III.79]. Two surfaces are *biholomorphically equivalent* if there is a HOLOMORPHIC FUNCTION [I.3 §5.6] from one to the other that has a holomorphic inverse. If a Riemann surface is SIMPLY CONNECTED [III.93], then the uniformization theorem states that it is biholomorphically equivalent to the sphere, the Euclidean plane, or the HYPERBOLIC PLANE [I.3 §6.6]. These three spaces can all be viewed as Riemann surfaces, and they are all particularly symmetric: they have constant CURVATURE [III.78] (positive, zero, and negative, respectively); more generally, given any two points x and y in such a space, one can find a symmetry of the space that takes x to y, and one can ensure that a little arrow at x ends up pointing in any desired direction at y. Loosely speaking, these spaces "look the same from every point."

It can be shown that an open subset of $\mathbb{C}$ that is not the whole of $\mathbb{C}$ cannot be biholomorphically equivalent to the sphere or to $\mathbb{C}$. Therefore, by the uniformization theorem, a simply connected open subset of $\mathbb{C}$ that is not the whole of $\mathbb{C}$ must be biholomorphically equivalent to the hyperbolic plane. This proves that any such set, no matter how irregular its boundary might be, can be mapped biholomorphically to any other. This result is called the *Riemann mapping theorem*. Biholomorphic maps are *conformal*: that is, if two curves in one set meet at an angle θ, then the angle between their images in the other set is also θ. So the Riemann mapping theorem implies that the interior of any simple closed curve can be mapped in an angle-preserving way to the open unit disk. Recall that one of the main models of the hyperbolic plane is Poincaré's disk model. Thus, the hyperbolic metric on the disk together with the biholomorphic map that is given by the uniformization theorem can be used to define a hyperbolic metric on any simply connected proper open subset of $\mathbb{C}$.

If a Riemann surface is not simply connected, it is at least a QUOTIENT [I.3 §3.3] of a simply connected surface, namely its UNIVERSAL COVER [III.93]. For example, a torus is a quotient of the complex plane (in many possible ways that are topologically but not biholomorphically equivalent). Thus, the uniformization theorem tells us that a general Riemann surface is a quotient of the sphere, the Euclidean plane, or the hyperbolic plane. For a more detailed discussion of what such a quotient might be like, see FUCHSIAN GROUPS [III.28].

Waring's Problem

See PROBLEMS AND RESULTS IN ADDITIVE NUMBER THEORY [V.27]

V.35 The Weil Conjectures
Brian Osserman

The Weil conjectures constitute one of the central landmarks of twentieth-century ALGEBRAIC GEOMETRY [IV.4]: not only was their proof a dramatic triumph, but they were the driving force behind a striking number of fundamental advances in the field. The conjectures treat a very elementary problem: how to count the number of solutions to systems of polynomial equations over finite FIELDS [I.3 §2.2]. While one might ultimately be more interested in solutions over, say, the field of rational numbers, the problem is far more tractable over finite fields, and LOCAL–GLOBAL PRINCIPLES [III.51] such as THE BIRCH–SWINNERTON-DYER CONJECTURE [V.4] establish strong, albeit subtle, relationships between the two cases.

Moreover, there are some basic questions that have nonobvious connections to the Weil conjectures. The most famous of these is the *Ramanujan conjecture*, which concerns the coefficients of $\Delta(q)$, one of the most fundamental examples of a MODULAR FORM [III.59]. We obtain the function $\tau(n)$ from the formula for $\Delta(q)$ as follows:

$$\Delta(q) = q \prod_{n=1}^{\infty} (1 - q^n)^{24} = \sum_{n=1}^{\infty} \tau(n)q^n.$$

RAMANUJAN [VI.82] conjectured that $|\tau(p)| \leqslant 2p^{11/2}$ for any prime number p. This is closely related to a statement on the number of ways of writing p as a sum of twenty-four squares. Work of Eichler, Shimura,

Kuga, Ihara, and Deligne showed that in fact Ramanujan's conjecture is a consequence of the Weil conjectures, so that Deligne's proof of the latter in 1974 also resolved the former.

We begin with a brief historical summary of developments prior to WEIL [VI.93] and follow this with a more precise description of the statement of his conjectures. Finally, we sketch the ideas behind their proof.

1 An Auspicious Prologue

Our story begins with the seminal work of RIEMANN [VI.49] on the classical ZETA FUNCTION [IV.2 §3], which we recall is defined by the sum

$$\zeta(s) = \sum_n \frac{1}{n^s}.$$

EULER [VI.19] had studied this function for real values of s, but Riemann, in his remarkable eight-page paper of 1859, went much further. He looked at complex values as well, and therefore had at his disposal the considerable resources of complex analysis. In particular, although the above sum for $\zeta(s)$ converges only for complex numbers s that have real part $\text{Re}(s)$ strictly greater than 1, Riemann showed that the function itself can be extended to an analytic function defined on the entire complex plane, except at the point $s = 1$, at which it tends to infinity. He showed, moreover, that $\zeta(s)$ satisfies a certain functional equation relating $\zeta(s)$ to $\zeta(1 - s)$, which introduced an important kind of symmetry around the line $\text{Re}(s) = \frac{1}{2}$. Most famously (or infamously), he conjectured what is now known as the RIEMANN HYPOTHESIS [I.4 §3]: that, aside from easily analyzed "trivial zeros" on the negative real axis, every zero of $\zeta(s)$ occurs on the line $\text{Re}(s) = \frac{1}{2}$. Riemann's motivation for studying $\zeta(s)$ was to analyze the distribution of prime numbers, but it fell to later authors (HADAMARD [VI.65], DE LA VALLÉE POUSSIN [VI.67], and Van Koch) to bring this vision to fruition. They used the zeta function to prove the PRIME NUMBER THEOREM [I.4 §3], which determined the asymptotic distribution of prime numbers, and also showed that the Riemann hypothesis is equivalent to a particularly strong upper bound for the error term in the prime number theorem.

At first glance, the Riemann hypothesis might appear to be completely special, a one-of-a-kind conjecture. However, it was not long before DEDEKIND [VI.50] generalized the Riemann hypothesis to a whole family of zeta functions, and in doing so opened the door to further generalization. Just as we can think of the complex numbers as being obtained from the real numbers by including a square root of -1, that is, a root of the polynomial $x^2 + 1$, one can obtain a NUMBER FIELD [III.63], the fundamental object of study in ALGEBRAIC NUMBER THEORY [IV.1], from the field $\mathbb{Q}$ of rational numbers by including roots of more general polynomials. For each number field K we have the ring of integers $\mathcal{O}_K$, which enjoys many of the same properties as the classical integers $\mathbb{Z}$. Starting from this observation, Dedekind defined a more general class of zeta functions, one for each such ring, which now bear his name. The classical zeta function $\zeta(s)$ was the Dedekind zeta function in the case $\mathcal{O}_K = \mathbb{Z}$. However, it was not at all straightforward to establish the existence of a functional equation for Dedekind zeta functions: this was an open problem until 1917, when it was settled by Hecke, who showed at the same time that Dedekind zeta functions could be extended to the complex plane, thereby ensuring that the Riemann hypothesis makes sense for them as well.

With such ideas in the air, it was not long before geometry entered the picture. ARTIN [VI.86] first introduced zeta functions and the Riemann hypothesis for certain curves over finite fields in his 1923 thesis, noting that the ring of polynomial functions on such a curve shares precisely the properties of rings of integers that Dedekind used to define his zeta functions. Artin quickly observed first that his new zeta functions were strongly analogous to Dedekind zeta functions, and second that they were often more tractable: evidence for both observations is provided by the fact that he was able to check explicitly that the Riemann hypothesis was satisfied for a number of specific curves. The difference between the two situations is encapsulated as follows: while in the number field case one can think of the zeta function as counting primes, in the case of a function field the zeta function may be expressed in terms of the more geometric data of counting points on the given curve. In a 1931 paper F. K. Schmidt generalized Artin's work, and exploited this geometry to prove a strong form of the functional equation for such zeta functions. And then, in 1933, Hasse proved the Riemann hypothesis in the special case of ELLIPTIC CURVES [III.21] over finite fields.

2 Zeta Functions of Curves

We now discuss in more detail the definition and properties of zeta functions associated with curves over finite fields, as well as the theorems of Schmidt and Hasse. Let $\mathbb{F}_q$ denote the finite field with q elements, where $q = p^r$ for some prime number p and some positive integer r. The simplest case is when $q = p$, and $\mathbb{F}_p$

is just the field of integers modulo p. More generally, we can obtain $\mathbb{F}_q$ by adding roots of polynomials to $\mathbb{F}_p$ just as we do to $\mathbb{Q}$ to obtain number fields; in fact, a single root of a single irreducible polynomial of degree r will do.

Artin studied a certain class of curves in the plane. Here, "plane" means $\mathbb{F}_q^2$, that is, the set of all pairs (x, y) with x and y in $\mathbb{F}_q$. A *curve* C is simply the subset of these points where some polynomial $f(x, y)$ with coefficients in $\mathbb{F}_q$ vanishes. Of course, if F is any field that contains $\mathbb{F}_q$, then the coefficients are also in F, so it makes sense to talk about $C(F)$, the curve in the larger "plane" F^2 defined by the same equation $f(x, y) = 0$. If F is also a finite field, then $C(F)$ is obviously also finite. The finite fields F containing $\mathbb{F}_q$ turn out to be the fields $\mathbb{F}_{q^m}$ for $m \geqslant 1$. For each $m \geqslant 1$ let us define $N_m(C)$ to be the number of points belonging to the curve $C(\mathbb{F}_{q^m})$. The sequence $N_1(C), N_2(C), N_3(C), \ldots$ is what we shall try to understand.

Given our plane curve C, we can define the *ring of polynomial functions* $\mathcal{O}_C$ of C. This is simply the ring of polynomial functions on the plane (i.e., in two variables), modulo the EQUIVALENCE RELATION [I.2 §2.3] that two functions taking the same values on C should be considered the same. Formally, $\mathcal{O}_C$ is simply the QUOTIENT [I.3 §3.3] ring $\mathbb{F}_q[x, y]/(f(x, y))$. Artin's basic observation was that the definition of the Dedekind zeta function could be applied equally well to the ring $\mathcal{O}_C$, yielding a zeta function $Z_C(t)$ associated with C. However, in our geometric context we have the following equivalent and more elementary formula, which explicitly relates $Z_C(t)$ to the number of points over finite fields:

$$Z_C(t) = \exp\left(\sum_{m=1}^{\infty} N_m(C) \frac{t^m}{m}\right). \tag{1}$$

Schmidt generalized Artin's definition to all curves over finite fields, and gave an elegant description of the zeta function for curves, bearing out Artin's observations in the cases he was able to compute. The nicest form of Schmidt's theorem applies to curves that satisfy two additional conditions. The first condition is that, rather than considering the curve C in the plane, we will want to "compactify" it by considering instead a *projective* curve; we can think of this as adding some "points at infinity," thus increasing $N_m(C)$ slightly. Second, we will want to impose a technical condition of *smoothness* on C, which is analogous to asking that C be a MANIFOLD [I.3 §6.9].

In order to state Schmidt's result, recall that there is a notion of the GENUS [IV.4 §10] of a smooth projective curve C, which can be defined to be the dimension g of the space of differentials on C, or, if C is a complex curve, as the "number of holes" in the space obtained from the analytic topology on C. By extending certain classical results in algebraic geometry to more general fields, Schmidt proved that, for a smooth projective curve C over $\mathbb{F}_q$ of genus g, we have

$$Z_C(t) = \frac{P(t)}{(1-t)(1-qt)}, \qquad (2)$$

where $P(t)$ is a polynomial of degree $2g$ with integer coefficients. Furthermore, he proved a functional equation in terms of the substitution $t \mapsto 1/qt$. If we set $t = q^{-s}$, this gives a functional equation for the substitution $s \mapsto 1 - s$, as in Riemann's original work. The Riemann hypothesis for C is then the statement that the roots of $Z_C(q^{-s})$ all have $\operatorname{Re}(s) = \frac{1}{2}$, or, equivalently, that the roots of $P(t)$ all have norm equal to $q^{-1/2}$. It is an elementary observation that this is equivalent to the assertion that $|N_m(C) - q^m + 1| \leqslant 2g\sqrt{q^m}$, for all $m \geqslant 1$.

The next step in exploiting the geometric nature of zeta functions of curves is the observation that if F is a finite field containing $\mathbb{F}_{q^m}$, then the points with coordinates in $\mathbb{F}_{q^m}$ are the fixed points of a function called the *Frobenius map*, which is the map Φ_{q^m} that sends a point $(x, y) \in F^2$ to the point (x^{q^m}, y^{q^m}). It is a simple extension of FERMAT'S LITTLE THEOREM [III.58] that if $t \in \mathbb{F}_{q^m}$, then $t^{q^m} = t$. Moreover, the converse holds: if F is a field containing $\mathbb{F}_{q^m}$, and $t \in F$ satisfies $t^{q^m} = t$, then $t \in \mathbb{F}_{q^m}$. This follows because in any field, and in particular in F, the polynomial $t^{q^m} - t$ can have at most q^m roots, which must then be precisely the elements of $\mathbb{F}_{q^m}$. It immediately follows that a point $(x, y) \in F^2$ is a fixed point of Φ_{q^m} if and only if $(x, y) \in \mathbb{F}_{q^m}^2$. Moreover, it is elementary that $(s + t)^{q^m} = s^{q^m} + t^{q^m}$, if s, t are in any field containing $\mathbb{F}_{q^n}$. Because the coefficients of $f(x, y)$ are in $\mathbb{F}_{q^m}$, it follows that if $f(x, y) = 0$, then

$$f(\Phi_{q^m}(x, y)) = f(x^{q^m}, y^{q^m}) = (f(x, y))^{q^m} = 0,$$

so we see that Φ_{q^m} gives a map from C to itself. Thus, one might hope to study $C(\mathbb{F}_{q^m})$ by analyzing more generally what one can say about the fixed points of maps from C to itself. Hasse successfully applied this point of view to prove the Riemann hypothesis in the case $g = 1$, which is to say the case of elliptic curves. Moreover, we will see that this perspective is woven throughout the fabric of the rest of our story, not only inspiring Weil to make his conjectures, but also suggesting the techniques that ultimately led to their proof.

3 Enter Weil

In 1940 and 1941, Weil gave two proofs of the Riemann hypothesis for curves over finite fields. Or, to be more accurate, he described two proofs: they both relied on fundamental facts in algebraic geometry which had been proved by analytic methods for varieties over the complex numbers, but which had not been proved rigorously in the case of arbitrary base fields. It was largely in order to address this deficiency that Weil wrote his *Foundations of Algebraic Geometry*, which appeared in 1948 and allowed both of his earlier proofs to be made rigorous.

Weil's book constituted a watershed in algebraic geometry, as it introduced for the first time the notion of an *abstract* algebraic variety. Previously, a variety was always a global object, in that it was defined by a single collection of polynomial equations, in either affine or projective space. Weil realized that it would be helpful to have a corresponding locally defined concept, so he introduced abstract algebraic varieties, which are obtained by gluing together affine algebraic varieties in much the same way that manifolds in topology are obtained by gluing together open subsets of affine space. The notion of an abstract variety played a fundamental role in formalizing Weil's proofs, and was also an important precursor to Grothendieck's immensely successful theory of SCHEMES [IV.5 §3].

The following year, in a remarkable paper in the *Bulletin of the American Mathematical Society*, Weil went further, studying zeta functions $Z_V(t)$ associated with higher-dimensional varieties V over finite fields, and taking as his definition the formula (1). While the situation is more complicated in this context, the behavior conjectured by Weil was nonetheless strikingly similar, an utterly natural extension of the case of curves:

(i) $Z_V(t)$ is a rational function of t;
(ii) more explicitly, if $n = \dim V$, we can write
$$Z_V(t) = \frac{P_1(t)P_3(t) \cdots P_{2n-1}(t)}{P_0(t)P_2(t) \cdots P_{2n}(t)},$$
where each root of each $P_i(t)$ is a complex number of norm $q^{-i/2}$;
(iii) the roots of $P_i(t)$ are interchanged with the roots of $P_{2n-i}(t)$ under the substitution $t \mapsto 1/q^n t$;
(iv) if V is the reduction modulo p of a variety $\tilde{V}$ defined over a subfield of $\mathbb{C}$, then $b_i = \deg P_i(t)$ is the ith *Betti number* of $\tilde{V}$ using the usual topology.

The last part of (ii) is known as the Riemann hypothesis, while (iii) constitutes a functional equation for the substitution $t \mapsto 1/q^n t$. Betti numbers are a well-known invariant from ALGEBRAIC TOPOLOGY [IV.6]: if we return to Schmidt's theorem (2) in the case of curves, the degrees $1, 2g, 1$ of $1 - t, P(t), 1 - qt$ are precisely the Betti numbers of a complex curve of genus g.

4 The Proof

Weil's conjectures were inspired by a very intuitive topological picture, derived from considering $V(\mathbb{F}_{q^m})$ as the set of fixed points of Φ_{q^m}. Forgetting for the moment that Φ_{q^m} makes sense only over finite fields, if we imagine that V were defined over the complex numbers, then by using the complex topology we could study the fixed points of Φ_{q^m} by the LEFSCHETZ FIXED POINT THEOREM [V.11 §3], obtaining a formula in terms of the action of Φ_{q^m} on the COHOMOLOGY GROUPS [IV.6 §4]. Indeed, we could deduce the factorization in (ii) almost immediately (and in particular the rationality asserted in (i)), with each factor $P_i(t)$ corresponding to the action of Frobenius on the ith cohomology group, and we would also have $\deg P_i(t)$ given by the ith Betti number of V. Moreover, the functional equation would follow from a concept known as POINCARÉ DUALITY [III.19 §7].

It was not long before it became clear that such cohomological arguments might become more than just motivation: there could be a cohomology theory for algebraic varieties over finite fields that would mimic the properties of the classical topological theory and would allow one to prove the Weil conjectures. Such a cohomology theory is now known as a *Weil cohomology*. Serre was the first to seriously attempt to develop such a theory, but he had only limited success. In 1960, Dwork provided a brief detour by using p-ADIC ANALYSIS [III.51] to prove parts (i) and (iii) of the conjectures: that is, the rationality and the functional equation. Shortly thereafter, building on comments of Serre and in collaboration with Artin, Grothendieck proposed

and developed a candidate for a Weil cohomology, the *étale cohomology*. Indeed, he noted that one could in fact extend the list of desired properties of a Weil cohomology in such a way that the Weil conjectures would follow almost immediately. These properties were known but extremely difficult in the classical case, and included the "hard Lefschetz theorem." In a burst of optimism, Grothendieck referred to them as the "standard conjectures," and envisioned that the Weil conjectures would ultimately be proved through them.

However, the final chapter of the story did not go entirely according to Grothendieck's plan. His student Deligne set about working on the problem, and was ultimately able to complete an exceedingly subtle and intricate proof using induction on the dimension of the variety. The étale cohomology played an absolutely fundamental role in Deligne's proof, but he also introduced other ideas into the picture, most notably a classical geometric construction of Lefschetz, as well as some work of Rankin on the Ramanujan conjecture. In the end, he was able to conclude the hard Lefschetz theorem from his work, but the rest of the standard conjectures remain unsolved to this day.

Acknowledgments. I would like to thank Kiran Kedlaya, Nicholas Katz, and Jean-Pierre Serre for their helpful correspondence.

Further Reading

Dieudonné, J. 1975. The Weil conjectures. *Mathematical Intelligencer* 10:7–21.

Katz, N. 1976. An overview of Deligne's proof of the Riemann hypothesis for varieties over finite fields. In *Mathematical Developments Arising from Hilbert Problems*, edited by F. E. Browder, pp. 275–305. Providence, RI: American Mathematical Society.

Weil, A. 1949. Numbers of solutions of equations in finite fields. *Bulletin of the American Mathematical Society* 55: 497–508.

Part VI
Mathematicians

VI.1 Pythagoras

b. Samos, Ionia (now Samos, Greece)?, ca. 569 B.C.E.;
d. Metapontum, Magna Graecia (now Metaponto, Italy)?, ca. 494 B.C.E.
Incommensurability; theorem of Pythagoras

One of the most elusive figures of antiquity, Pythagoras is famous not just for his alleged mathematical achievements: it has been claimed that he had a golden thigh and that he issued a prescription against broad beans. Few things about him can be taken as historical facts, but we can be reasonably confident that he lived in around the sixth century B.C.E. in Greek southern Italy and that he established a group of followers, the Pythagoreans, who shared not just beliefs, but also dietary habits and a code of behavior. The existence of anecdotes about splinter Pythagoreans who revealed secrets to outsiders and were accordingly punished suggests that they were far from constituting a completely homogeneous group.

After a peak period in the late fifth century B.C.E., the Pythagoreans dispersed, probably as a result of their involvement in the public life of various city-states. The impact of their theories about the universe and the soul was very long-lived, though, and can be felt in Plato, Aristotle, and later authors. From the third century B.C.E. until well into late antiquity, a stream of texts was produced that purported to be by Pythagoras or his immediate successors. Indeed, historians talk of a neo-Pythagorean philosophical movement, sometimes associated with neo-Platonism.

The name of Pythagoras and his school is most commonly linked to the theorem establishing that the square on the hypotenuse of a right-angled triangle is equivalent to the sum of the squares on the other two sides. In fact, there is some evidence that the mathematical property expressed by the theorem was known in Mesopotamia long before Pythagoras's time; the ancient sources attributing the result to him are late and not entirely reliable, and no actual proof of the theorem is found before Euclid's *Elements*. While the proof itself may predate EUCLID [VI.2], there is no solid reason to connect it to Pythagoras.

Similarly, the discovery of the incommensurability of the side and the diagonal of a square, often attributed to Pythagoreans, may have been made earlier in Mesopotamia, and the earliest full proof in a Greek context belongs to a later period.

Pythagoras's real contribution to mathematics lies elsewhere. The Pythagoreans are credited by Aristotle with the theory that "things themselves are numbers." One interpretation is that they believed that mathematics offered a key to understanding reality, whether this reality was conceived to have an underlying geometrical structure (as in Plato's *Timaeus*), or whether it was simply seen as ordered and "in proportion." Indeed, Pythagoreans are plausibly credited with a strong interest in formulating the numerical ratios of musical concords and harmony. They connected the harmonious sound produced by, say, the plucking of a string with the fact that the musician plucked it at specific, mathematically expressible points. Breaking the mathematical proportion between the points on the string unsettled the sound produced. The heavenly bodies themselves, according to the Pythagoreans, produced music, thanks to their mathematical, and therefore orderly, arrangement. Understand the mathematics, and you will grasp the structure of reality: this insight is perhaps Pythagoras's true legacy.

Further Reading

Burkert, W. 1972. *Lore and Science in Ancient Pythagoreanism.* Cambridge, MA: Harvard University Press. (Revised English translation of 1962 *Weisheit und Wissenschaft: Studien zu Pythagoras, Philolaos und Platon.* Nürnberg: H. Carl.)
Zhmud, L. 1997. *Wissenschaft, Philosophie und Religion im frühen Pythagoreismus.* Berlin: Akademie.

Serafina Cuomo

VI.2 Euclid

b. Alexandria, Egypt?, ca. 325 B.C.E.; d. Alexandria?, ca. 265 B.C.E.
Deduction; postulate; reductio ad absurdum

Nothing is known about Euclid's life. In fact, his major work, the *Elements*, is now seen as a rather loose collection, with no strong authorial voice and no clear way of determining what, if any, Euclid's original contributions were. Born in the cultural climate of Ptolemaic Alexandria, the text probably aimed at systematizing the current knowledge in some mathematical areas.

The *Elements* covers plane geometry (including the squaring of any rectilinear figure, the bisection of an arc, the inscription and circumscription of polygons in circles, the finding of a mean proportional), solid geometry (e.g., the ratio of spheres to one another, the five regular solids), and arithmetic, from relatively simple (the properties of odd and even numbers, prime number theory) to more complex (commensurable and incommensurable lines, binomials and apotomes).

The title hints at the foundational character of the text, which starts with definitions of mathematical objects (e.g., point, straight line, scalene triangle), postulates (e.g., all right angles are equal to one another) and common notions (e.g., the whole is greater than the part). These initial premises are not demonstrated— whether some postulates are demonstrable spawned debate in antiquity, and later led to non-Euclidean geometries. In a style which has been termed axiomatic-deductive, proofs tend to be general rather than specific; they use a restricted set of formulaic expressions, refer to a lettered diagram, and each of their steps is justified by appeal to undemonstrated premises, to previous proofs, or to very simple notions, such as the principle of the excluded middle. Some proofs use *reductio ad absurdum*: instead of directly showing that something is the case, they proceed to show that any alternative is impossible.

There are parts of the book that reveal the presence of different, less abstract, demonstrative procedures. For instance, one of the theorems establishing criteria for two triangles to have the same area refers to one triangle being "superimposed" on the other, with the reader effectively invited to verify that their areas are indeed equal. The appeal is to a mental operation, which is quite different from the logical step-by-step method found elsewhere. Again, book IX contains propositions on odd and even numbers, which are often seen as vestiges of Pythagorean mathematics, to be demonstrated with the help of pebbles. The coexistence itself of arithmetic and geometry has been puzzling for some historians, who have proposed a notion of "geometric algebra," so that book II, ostensibly about squares and rectangles built on segments of straight lines, would in fact foreshadow modern equations.

As well as works on astronomy, optics, and music, the *Data*, which is about solving geometrical problems on the basis of some elements that are already given, is also attributed to Euclid. His fame is, however, inextricably linked to the *Elements*. The very absence of a strong authorial voice has perhaps facilitated other mathematicians' interaction with the text, which has been appropriated, added to, interfered with, and commented upon since antiquity. This very plasticity helped to make it possibly the most popular mathematical book of all time. (For more about its impact on the early development of mathematics, see GEOMETRY [II.2], THE DEVELOPMENT OF ABSTRACT ALGEBRA [II.3], and THE DEVELOPMENT OF THE IDEA OF PROOF [II.6].)

Further Reading

Euclid. 1990–2001. *Les Éléments d'Euclide d'Alexandrie; Traduits du Texte de Heiberg*, general introduction by M. Caveing, translation and commentary by B. Vitrac, four volumes. Paris: Presses Universitaires de France.
Netz, R. 1999. *The Shaping of Deduction in Greek Mathematics. A Study in Cognitive History.* Cambridge: Cambridge University Press.

Serafina Cuomo

VI.3 Archimedes

b. Siracusa, Magna Graecia (now Syracuse, Italy), ca. 287 B.C.E.; d. Siracusa, 212 B.C.E.
Area of the circle; centers of gravity; method of exhaustion; volume of the sphere

Archimedes' life was as spectacular as his scientific achievements: various sources attest that he built a ship, a cosmological model, and magnificent catapults with which he defended his native Syracuse during the Second Punic War. The Roman besiegers eventually took the city by deceit, and Archimedes was killed in the ensuing pillage. According to legend, his tomb was engraved with a sphere inscribed in a cylinder, to mark one of his most famous discoveries. Indeed, the first part of his *Sphere and Cylinder* reaches a climax with a proof that the volume of every sphere is two thirds of that of the cylinder circumscribing it. Archimedes' interest in establishing the volume or area of curved figures is also attested by his discovery of the area of the circle and of a sphere, and by treatises on spiral curves, conoids, and paraboloids, and on the *Quadrature of the Parabola*.

While following an axiomatic-deductive framework, Archimedes' style is distinctive. Many of his theorems about curved figures use the so-called METHOD OF EXHAUSTION [II.6 §2].

Take the problem of determining the area of a circle. Archimedes accomplished this by showing that it had the same area as that of a certain right-angled triangle. Since it was known how to calculate the area of a triangle, he was "reducing" a problem whose solution was unknown to one whose solution was known. Rather than establishing this directly, he proves that the area of the circle can be neither larger than nor smaller than the area of the triangle, so that only one possibility remains: that they are equal. This is achieved, here and in general, by inscribing and circumscribing rectilinear figures to the curvilinear figure under investigation, thus getting closer and closer to it. The leap from closer and closer approximation to equivalence of a rectilinear and curvilinear figure, however, can be accomplished only indirectly, by excluding the other possibilities. Such arguments usually employ a lemma, already found in Euclid, to the effect that if we start with a quantity and replace it by a quantity at most half as large, and then repeat this, then what remains can be made as small as we please.

Archimedes' output also includes *The Sand-Reckoner*, about astronomy and arithmetic, and works on the centers of gravity of plane figures and on bodies immersed in a fluid.

Above all, Archimedes provides unique insights into the processes of ancient Greek mathematics. The second part of *Sphere and Cylinder* contains problems about constructing given solid bodies. Several of the proofs are in two parts: analysis and synthesis. In the analysis, the result one wants to establish is taken as proved, and consequences are drawn from it, until one hits upon a result that is already proved elsewhere, and the process is then reconstituted in reverse (the "synthesis"). The recently rediscovered *Method* (addressed to Eratosthenes) reveals that Archimedes arrived at some of his most famous results, e.g., the area of a segment of a parabola, by imagining the two objects involved (say, a segment of a parabola and a triangle) as divided up into an infinite number of slices and lines, then placed at the two ends of a balance and set in equilibrium with each other. Archimedes underlined that this heuristic procedure was not a strict proof, but that

only makes the *Method* all the more valuable a glimpse into the mind of a great mathematician.

Further Reading

Archimedes. 2004. *The Works of Archimedes: Translation and Commentary. Volume 1: The Two Books On the Sphere and the Cylinder*, edited and translated by R. Netz. Cambridge: Cambridge University Press.
Dijksterhuis, E. J. 1987. *Archimedes*, with a bibliographical essay by W. R. Knorr. Princeton, NJ: Princeton University Press.

Serafina Cuomo

VI.4 Apollonius

b. Perge, Pamphylia (now Perga, Turkey), ca. 262 B.C.E.;
d. Alexandria, Egypt?, ca. 190 B.C.E.
Conic sections; diorism; locus problems

The *Conics*, in eight books, only seven of which have come down to us, has had fewer modern readers than other recognized masterpieces of Greek mathematics: it is complex, difficult to summarize, and easy to mistranslate into modern algebraic notation. Apollonius of Perga also wrote about arithmetic and astronomy, but none of these works survive. The letters prefacing six of the surviving books indicate that he was a highly esteemed member of a network of mathematicians, to whom he sent his results. He refers to the fact that various versions of his *Conics* were circulated, the latest probably incorporating his correspondents' feedback. Knowledge of the parabola, hyperbola, and ellipse predates Apollonius (we find conics in Archimedes), but his is the first known systematic account of these curves, which were of interest both in themselves and because they could be used as auxiliary lines for the solution of problems such as the trisection of an angle or the duplication of the cube.

Apollonius himself declares that the first four books of the *Conics* are an introduction to the subject, and indeed he starts with definitions of the cone and its various parts. The parabola, hyperbola, and ellipse are not introduced until later, so that their origin (from a plane cutting a cone or a conic surface at different angles) is already accompanied by a statement of their properties, which are further and fully explored in the next three books. These include theorems on tangents, asymptotes, and axes; constructions of conic sections on the basis of certain data; and an account of the conditions under which conics can intersect in the same plane.

The nonelementary books, which exist only in Arabic, contain treatments of maximum and minimum lines within the sections, the construction of conic sections equal or similar to a given conic section (including the theorem that all parabolas are similar), and "diorismic theorems." These are propositions that set the limits of possibility of a construction, or the limits of validity of some property of a geometrical configuration, given a certain number of known positions or known objects at the outset. Indeed, several of the propositions in the *Conics* are about loci, i.e., geometrical configurations consisting of all the points sharing a certain family of properties. Apollonius criticizes EUCLID [VI.2] for not having provided an exhaustive solution to the construction of the three-line and four-line locus (configurations of three or four lines, arranged so that they have specific properties).

In terms of demonstrative methods, Apollonius is in the axiomatic-deductive mold: general enunciations, lettered diagrams, each step justified by appeal to undemonstrated premises or previous proofs. Instead of indirect methods, we find a real mastery of the intricacies (and power) of proportion theory. At the same time, his propositions easily lend themselves to the consideration of different subcases: when, for instance, a certain line falls inside, outside, or on the vertex of a conical surface. Apollonius, in other words, combines a systematic approach with an almost playful fascination with exploring the possibilities of mathematical objects and their properties under varying circumstances.

Further Reading

Apollonius. 1990. *Conics*, books V–VII. *Arabic Translation of the Lost Greek Original in the Version of the Banu Musa*, edited with translation and commentary by G. J. Toomer, two volumes. New York: Springer.

Fried, M. N., and S. Unguru. 2001. *Apollonius of Perga's Conica: Text, Context, Subtext*. Leiden: Brill.

Serafina Cuomo

VI.5   Abu Ja'far Muhammad ibn Mūsā al-Khwārizmī

b. Unknown, 800; d. Unknown, 847
Arithmetic; algebra

Al-Khwārizmī, or possibly his ancestors, came from Khwārizm (the modern region of Khorezm in Uzbekistan, also known as Khiva). Most of his life was spent as a scholar at the House of Wisdom, Baghdad, where he produced works on astronomy, mathematics, and geography. Of his mathematical works, two have come down to us, one on arithmetic and one on algebra.

The arithmetical work, which did not survive in Arabic and is known only through Latin translations, was the means by which Hindu numerals were transmitted to the West, as well as the corresponding methods of arithmetical calculation. Although the text was clearly based on Indian writings, in Europe the techniques became particularly associated with al-Khwārizmī's name in the form of *algorism* (from which the modern term "algorithm" is derived).

Al-Khwārizmī's *al-Kitāb al-mukhtaṣar fī ḥisāb al-jabr wa'l-muqābala* ("The compendious book on calculation by completion and balancing") became the starting point for the subject of algebra for Islamic mathematicians. A work of elementary practical mathematics, it is written in three parts: one was devoted to solving equations, one to practical mensuration (areas and volumes), and one to problems that arose mainly from the complicated Islamic laws of inheritance (involving arithmetic and simple linear equations). No algebraic symbolism is employed: everything, including numerals, is expressed in words. The text opens with a brief discussion of the place-value system and then deals with equations of the first and second degrees. Remarkably, al-Khwārizmī did not regard these equations just as a means for solving problems, as his predecessors had done, but studied them in their own right, classifying them into six separate types. In modern notation these are

$$ax^2 = bx, \qquad ax^2 = b, \qquad ax = b,$$
$$ax^2 + bx = c, \qquad ax^2 + c = bx, \qquad ax^2 = bx + c,$$

where a, b, and c are positive integers. The different types are necessary because al-Khwārizmī did not recognize the existence of either negative numbers or zero as coefficients. Not only did al-Khwārizmī give proofs that his methods worked, which in itself was not standard at the time, but the proofs he gave were geometrical ones. That is, they were not classical Greek proofs but geometrical demonstrations of the validity of his methods.

The key word of the Arabic title, *al-jabr* ("completion" or "restoration"), which refers to restoring all the terms to a standard form, eventually came into common usage in the West as *algebra*. It is, however, doubtful that al-Khwārizmī's work was the first Islamic work bearing that name.

Further Reading

Berggren, J. L. 1986. *Episodes in the Mathematics of Medieval Islam*. New York: Springer.

VI.6 Leonardo of Pisa (known as Fibonacci)

b. Pisa, Italy, ca. 1170; d. Pisa, Italy, ca. 1250
Son of Pisan merchant; studied mathematics under Muslim teachers in North Africa and traveled throughout the Mediterranean meeting with Islamic scholars; awarded an annual stipend in 1240 by the city of Pisa in recognition of his teaching and other services

One of the earliest European writers on algebra, Fibonacci is most famous for his *Liber Abaci* ("Book of calculation"), which first appeared in 1202 and was largely responsible for the spread of the Hindu–Arabic numerals throughout Europe. The book contained not only rules for computing with the Hindu–Arabic numerals but also a large number of problems of various kinds, the best known of which was his "rabbit problem." This problem asks how many pairs of rabbits will be produced in a year, beginning with a single pair, if in every month each pair produces a new pair which becomes productive from the second month on. The number F_n of pairs there will be in the nth month is the number of pairs there were in the previous month plus the number of breeding pairs, and the latter is the number of rabbits there were in the previous month but one. This leads to the rule $F_n = F_{n-1} + F_{n-2}$. Starting with $F_0 = 0$ and $F_1 = 1$, we obtain the sequence $0, 1, 1, 2, 3, 5, 8, 13, \ldots$ of *Fibonacci numbers*. It can be shown that $\lim_{n \to \infty} F_{n+1}/F_n = \phi$, where $\phi = (1 + \sqrt{5})/2$ is the golden ratio.

VI.7 Girolamo Cardano

b. Pavia, Italy, 1501; d. Rome, 1576
Teacher of mathematics, Milan (1534–43); Professor of Medicine: Pavia (1543–60), Bologna (1562–70); imprisoned for heresy (1570–71)

Cardano's great treatise, the *Ars Magna* (1545), laid the foundations for European algebra and remained the most comprehensive and systematic work on algebra for more than a century after it was published. It contained many new ideas, including methods (not all Cardano's own) for solving cubic and quartic equations, all written without mathematical notation. Cardano's own great insight was to recognize the existence of relations between the roots and the coefficients of an equation; in this he was unprecedented. He also showed a greater readiness than most of his contemporaries to contemplate the square roots of negative numbers. He is remembered today for "Cardano's rule" for solving cubic equations of the form $x^3 + cx = d$, where

c and d are positive (he was unable to solve the *casus irreducibilis*, the case when c is negative).

VI.8 Rafael Bombelli

b. Bologna, Italy, 1526; d. Probably Rome, after 1572
Engineer-architect for the Roman nobleman Alessandro Rufini, later Bishop of Melfi

Bombelli was prompted to write his *Algebra* (1572) by a desire to make CARDANO's [VI.7] *Ars Magna* (1545) accessible to the less sophisticated reader. The *Algebra*, which contains a systematic treatment of quadratics, cubics, and quartics, is noted for its advances in mathematical notation—it was the first printed text to include a notation for exponents—and for its role in disseminating awareness of the work of Diophantus. Above all, the *Algebra* was renowned for solving certain special cases of the so-called *casus irreducibilis* of the cubic, those in which Cardano's rule appears to give rise to a complex or "impossible" solution. Cardano was aware that what we today call complex numbers (numbers of the form $a + b\sqrt{-1}$) could arise in the solution of quadratic equations. Bombelli made the important discovery that what at first sight appears to be a complex root of a cubic equation may in fact be a real root, because the imaginary parts cancel each other out. The *Algebra* included the first extensive discussion of complex numbers and Bombelli formulated the four basic operations of arithmetic for them.

VI.9 François Viète

b. Fontenay-le-Comte, France, 1540; d. Paris, 1603
Trigonometry; algebraic analysis; classical problems; numerical solution of equations

Viète obtained a bachelor's degree in law in 1560 from the University of Poitiers, but left the profession from 1564 to 1568 to oversee the education of Catherine de Parthenay, daughter of a local aristocratic family. His earliest scientific writings were his lectures to Catherine. He spent the remainder of his life in high public office, apart from a period between 1584 and 1589 when he was banished from the court in Paris for political and religious reasons. He died in Paris in 1603. Throughout his life, it was only during the time he had free from official duties that he was able to devote himself to mathematics.

The work for which Viète is best known appeared during the 1590s, beginning with *In Artem Analyticem Isagoge* ("Introduction to the analytic art") in 1591. In the *Isagoge* Viète began to combine classical Greek geometry with algebraic methods that had originated from Islamic sources, and in doing so laid the foundations for the algebraic approach to geometry. Viète saw that the symbols in equations (traditionally variants of R, Q, and C, for the unknown, its square, and its cube) could represent either numbers or geometric quantities, and that this was potentially a powerful tool for analyzing and solving geometric problems.

Viète's understanding of analysis was based on his reading of the *Synagoge* ("Collection") of Pappus (early fourth century C.E.), where analysis was described as a method of investigating a problem by assuming that the solution is in some sense known, as we would do now by representing the solution by a symbol and carrying out mathematical manipulations involving that symbol. Algebra achieves this by regarding all quantities, known or unknown, as of equal status; equations are then formed from prestated conditions (a process Viète called *zetetics*), and solved to produce the unknown quantity in terms of those given (*exegetics*). For Viète the final step in geometric problems was to provide a specific construction for the solution: this was the geometric *synthesis* arising from the preceding algebraic *analysis*.

In several further treatises, mostly written or published around 1593, Viète taught the necessary skills of forming equations and carrying out the corresponding geometric constructions, and these books together made up his *Opus Restitutae Mathematicae Analyseos, seu Algebra Nova* ("The work of restored mathematical analysis, or the new algebra"), which he offered with the famous and ambitious hope of leaving no mathematical problem unsolved (*nullum problema non solvere*). For most of the seventeenth century, algebra continued to be known as the "analytic art," or simply "analysis."

Recognizing that not all equations could be solved algebraically, Viète also put forward a method of numerical solution based on successive approximations. This was the first appearance of such techniques in Europe, and was important not only for practical purposes, but also because it rapidly led to a deeper understanding of the relationships between roots and coefficients of equations.

Viète's style of writing is wordy and often obscure, thanks in part to his liking for technical Greek terms. In his algebraic treatises, however, he devised some rudimentary notation. It had long been the case that rules for solving equations were presented through particular examples that were understood to represent a general class, but Viète took the step of replacing known quantities by consonants $B, C, \ldots$, and unknowns by vowels $A, E, \ldots$, so that numbers were replaced throughout by letters, or "species." However, he had no simple or systematic way of denoting powers (for squares and cubes he used the verbal *A quadratus* and *A cubus*), and his connectives ("added to," "equals," and so on) were also written in words, so that his algebra was still very far from symbolic.

One of the first people to study Viète's work in depth was Thomas Harriot in England, who, through study of Viète's numerical method shortly after 1600, discovered that polynomials could be written as products of linear and quadratic factors, a major breakthrough in the understanding of equations. Harriot also rewrote much of Viète's mathematics in what is essentially modern algebraic notation. In France, Viète's work was taken up in the 1620s by FERMAT [VI.12], who was profoundly influenced by it. DESCARTES [VI.11], on the other hand, denied that he had ever read either Viète or Harriot, though in the 1630s he developed a number of very similar ideas.

Viète and his immediate successors dealt only with equations of finite degree. Only much later in the seventeenth century with the work of NEWTON [VI.14] was analysis extended to include what were thought of as infinite equations, or what we would now call infinite series, hence bringing the word "analysis" much closer to its modern meaning.

Jacqueline Stedall

VI.10   Simon Stevin

b. Bruges, Belgium, 1548; d. The Hague, the Netherlands, 1620
Mathematics and science tutor to Maurice of Nassau, Prince of Orange

The Flemish mathematician and engineer Simon Stevin is remembered for his study of decimal fractions. Although he was not the first to use decimal fractions (they are found in the work of the tenth-century Islamic mathematician al-Uqlīdisī), it was his tract *De Thiende* ("The tenth"), published in 1585 and translated into English (as *Disme: The Art of Tenths, or Decimall Arithmetike Teaching*) in 1608, that led to their widespread adoption in Europe. Stevin, however, did not use the notation we use today. He drew circles around the exponents of the powers of one tenth: thus he wrote 7.3486 as 7⓪3①4②8③6④. In *De Thiende* Stevin not only demonstrated how decimal fractions could be used but also advocated that a decimal system should be used for weights and measures and for coinage.

VI.11 René Descartes

b. La Haye (now "Descartes"), France, 1596; d. Stockholm, 1650
Algebra; geometry; analytic geometry; foundations of mathematics

In 1637 Descartes published *La Géométrie* as an "essay" appended to his philosophical treatise *Discours de la Méthode*. It remained his only mathematical publication. No single early modern text shaped the development of mathematics between 1650 and 1700 as strongly as *La Géométrie*. It was the founding text of analytic geometry and it paved the way for the merging of algebra and geometry that made possible the development of the integral and differential calculus about fifty years later.

Descartes was educated at the Jesuit College at La Flèche. He spent his life mostly outside France, traveling through Europe in his early twenties and living in the Netherlands from 1628 until 1649; he then left for Sweden, invited by Queen Christina to her court. From an early age his interest in mathematics was tightly linked to his primary philosophical preoccupation: the certainty of knowledge. In a letter of 1619 he sketched a method, clearly inspired by arithmetic and geometry, for solving all problems in natural philosophy. Shortly afterward, his ideas grew into a passionate conviction that he could and should develop a philosophy along these problem-solving and mathematics-inspired lines. *La Géométrie* grew out of the mathematical part of his philosophical program; it was not a textbook on analytic geometry. Descartes offered little in the way of general principles, explaining his ideas by means of examples.

Descartes used a classical problem, Pappus's problem, in order to explain coordinates and equations of curves, and showed that the defining property of a curve could be written as an equation. He introduced coordinates x and y, using oblique as well as rectangular coordinate axes, which he always adjusted to the problem at hand. He also introduced the now very common usage of employing x, y, and z for unknowns and a, b, and c for indeterminate fixed quantities.

For Descartes, a geometrical problem required a geometrical answer. The equation was at best an algebraic reformulation of the problem; the answer had to be a construction of the curve or of individual points. If, as in the particular case of Pappus's problem in four lines, the equation was quadratic, then for any fixed value of y the x-coordinate was a root of a quadratic equation. Earlier in the book Descartes had shown how such a

René Descartes

root could be constructed (using ruler and compass). Thus, the curve could be constructed "pointwise" by choosing a series of values for y and constructing the corresponding xs and points on the curve. Pointwise construction could not provide the whole curve. Therefore in Pappus's problem Descartes used the equation to show that the solution curves were conic sections, and explained how to determine the nature of the conic, the location of its axes, and the values of its parameters. This was an impressive result; it was, in fact, the first classification of an algebraically defined class of curves.

A further influential result in *La Géométrie*, and the one of which Descartes himself said he was most proud, was his method to determine the normal (and thus also the tangent) at a given point on a curve with a given equation. It was a pre-calculus forerunner of differentiation.

There are three important differences between how Descartes treated curves and their equations and how they are treated in modern analytic geometry: he employed oblique as well as rectangular axes; he did not consider the equation as defining a curve—rather it represented a problem, namely to construct the curve itself, as well as its axes, tangents, etc.; and he did not consider the plane itself as a collection of points characterized by pairs of real numbers—for him the xs and ys were not dimensionless numbers but the lengths of line segments. (The term "Cartesian plane" for $\mathbb{R}^2$ is therefore anachronistic.)

Descartes supposed (too optimistically) that his procedures could be extended to polynomial equations of any degree (usually connected to Pappus's problem with more than four lines) and that therefore he had shown how, in principle, all geometrical construction problems could be solved. For higher-order constructions he needed new algebraic techniques. The relevant section in *La Géométrie* constituted the first general theory of polynomial equations and their roots. It contained his "sign rule" about the number of positive and negative roots of a polynomial, various transformation rules, and methods to check equations for reducibility. He gave no proofs; his results were based on a conviction that polynomials could essentially be written as products of linear factors $x - x_i$, in which the roots x_i could be positive, negative, or "imaginary."

It appears, then, that analytic geometry was not the primary goal of *La Géométrie*. Rather, its aim was to provide a universal method for solving geometrical problems, and to do so Descartes had to answer two urgent methodological questions. The first was how to solve geometrical problems not constructible by ruler and compass, and the second was how to use algebra as an analytic, i.e., solution-finding, tool in geometry.

For the first of these, Descartes allowed successively more complicated curves as means of construction. It was Descartes's conviction that algebra, through the equations of these curves, could guide him to choose, among all such construction curves, the most appropriate for the problem, in particular the simplest, i.e., that of lowest degree.

The second question addressed serious conceptual difficulties that were felt at the time about using algebra in geometry. The transfer of algebraic operations to geometry was indeed problematic because multiplication in geometry was generally interpreted dimensionally: for example, a product of two lengths had to represent an area, and a product of three a volume. But until then algebra had dealt mostly with numbers and had routinely used products of more than three factors. Thus, a consistent and unrestricted geometrical interpretation of the operations of algebra was needed. Descartes did indeed provide such a reinterpretation. He introduced a unit line segment in such a way that multiplication no longer raised the dimension and inhomogeneous terms could be allowed in equations.

By 1637 he had given up on his earlier attempts to link philosophy and mathematics. Yet the preoccupation with certainty remained. As his concept of construction involved the use of curves, he had to consider which curves could be understood by the human mind

with sufficient clarity to be acceptable in geometry. His answer was that all algebraic curves were acceptable (he called these "geometrical curves") and all others were not (these he called "mechanical"). Few seventeenth-century mathematicians followed Descartes in this strict demarcation of geometry. This is typical of the reception of Descartes's *La Géométrie*: the philosophical and methodological aspects of the book were largely ignored by his mathematical readers, but the technical mathematical aspects were eagerly accepted and used.

Further Reading

Bos, H. J. M. 2001. *Redefining Geometrical Exactness: Descartes' Transformation of the Early Modern Concept of Construction*. New York: Springer.

Cottingham, J., ed. 1992. *The Cambridge Companion to Descartes*. Cambridge: Cambridge University Press.

Shea, W. R. 1991. *The Magic of Numbers and Motion: The Scientific Career of René Descartes*. Canton, MA: Watson Publishing.

Henk J. M. Bos

VI.12 Pierre Fermat

b. Beaumont-de-Lomagne, France, 160?; d. Castres, France, 1665
Number theory; probability theory; variational principles; quadrature; geometry

Fermat, who spent his life as a magistrate in the south of France, contributed decisively to most of the mathematical subjects of his time: from quadrature to optics, from geometry to number theory. Very little is known about his early life—even the date of his birth is uncertain—but by 1629 he had close contacts with VIÈTE's [VI.9] scientific heirs in Bordeaux. His work displays a thorough knowledge of ancient as well as contemporary mathematics and he exchanged problems and mathematical information by correspondence with, among others, RENÉ DESCARTES [VI.11], Gilles Personne de Roberval, Marin Mersenne, Bernard Frenicle, John Wallis, and Christiaan Huygens.

A crucial early-modern topic was the use of algebra to solve geometric problems. Viète and other algebraists before him had used equations in a single unknown to rewrite and solve "determinate" problems (problems admitting a finite number of solutions). In his manuscript *Ad Locos Planos et Solidos Isagoge*, which circulated in Paris in 1637 (the same year as Descartes's *La Géométrie*), Fermat presented a general way of handling and solving indeterminate problems associated with constructions of loci: that is, of sets of points (usually curves) defined by some constraints. He identified the points of such loci by two coordinates linked by an equation (although he chose a different way of taking coordinates from the usual modern x and y coordinates). Moreover, Fermat gave the standard forms of the corresponding equation when the locus to be found was a line, a parabola, an ellipse, etc.

Fermat also used algebraic analysis to solve problems of extrema, including finding the tangent or the normal to a curve at a given point, and determining centers of gravity. His method relies on the principle that a certain algebraic expression takes on the same values twice near the extremum. Although the procedure is purely algebraic, his successors tended to interpret it from a differential perspective, thereby making his work an apparent precursor of the calculus. Fermat applied the method to a variety of problems, including (within the framework of a controversy with Descartes's followers around 1660) a proof of the law of refraction in optics. Basing his analysis on the principle that "nature acts in the shortest time," Fermat was able to express the problem as one of extrema and to solve it with his method. The problem of refraction was one of the first complex physical problems to be treated in a thoroughly mathematical way, and Fermat's approach later led to VARIATIONAL METHODS [III.94].

However, Fermat also showed a perfect mastery of more classical, for instance Archimedean, techniques, which he used when dealing with other types of geometrical questions such as quadrature.

Such versatility also appears in Fermat's work on numbers. On the one hand, he was happy to apply his algebraic approach to Diophantine analysis in order to obtain solutions for cases previously thought to be insoluble, or to derive new solutions from ones already known. On the other hand, he advocated a theoretical study of the integers, for which the currently available algebraic theory of equations was not sufficient. For example, he gave general properties of the divisors of numbers of the form $a^n \pm 1$ (among them his now celebrated LITTLE THEOREM [III.58]) and of $x^2 + Ny^2$ for various N. He invented the method of infinite descent specifically to deal with problems concerning integers. He used this method, which relies on the impossibility of constructing an infinite strictly decreasing sequence of integers, to prove that $a^4 - b^4 = c^2$ has no nontrivial integer solutions. This is a particular case of his famous LAST THEOREM [V.10], which Fermat only stated in the margins of one of his books: $a^n + b^n = c^n$ has no nontrivial integer solutions for $n > 2$. The first proof of the general case was given by Andrew Wiles in 1995.

In 1654 Fermat exchanged letters with PASCAL [VI.13] on the idea of a "fair game" and on the redistribution of the stakes if a game is interrupted before its end. These letters introduced important concepts in probability, including expected value and conditional probability.

Further Reading

Cifoletti, G. 1990. *La Méthode de Fermat, Son Statut et Sa Diffusion.* Société d'Histoire des Sciences et des Techniques. Paris: Belin.

Goldstein, C. 1995. *Un Théorème de Fermat et Ses Lecteurs.* Saint-Denis: Presses Universitaires de Vincennes.

Mahoney, M. 1994. *The Mathematical Career of Pierre de Fermat (1601–1665),* second revised edn. Princeton, NJ: Princeton University Press.

Catherine Goldstein

VI.13 Blaise Pascal

b. Clermont-Ferrand, France, 1623; d. Paris, 1662
Scientist and theologian

Pascal was the first to make a systematic study of the arithmetical triangle which now bears his name; although the triangle itself is found earlier, notably in the work of the Chinese mathematician Zhu Shijie (1303). "Pascal's triangle"

$$
\begin{array}{ccccccccc}
 & & & & 1 & & & & \\
 & & & 1 & & 1 & & & \\
 & & 1 & & 2 & & 1 & & \\
 & 1 & & 3 & & 3 & & 1 & \\
1 & & 4 & & 6 & & 4 & & 1 \\
\end{array}
$$
$$\cdot \quad \cdot \quad \cdot \quad \cdot \quad \cdot \quad \cdot$$

a triangular array in which each number is the sum of the two immediately above it, provides a geometrical arrangement of the binomial coefficients $\binom{n}{k}$, with $\binom{n}{k}$ appearing as the $(k+1)$st element in the $(n+1)$st row. Here $\binom{n}{k}$ is, as usual, the number of subsets of size k in a set of size n, so that

$$\binom{n}{k} = \frac{n!}{k!(n-k)!}.$$

The number $\binom{n}{k}$ is also the coefficient of $a^k b^{n-k}$ in the binomial expansion of $(a+b)^n$ for any integer $n \geqslant 0$ and $0 \leqslant k \leqslant n$. In his *Traité du Triangle Arithmétique* (printed in 1654 but not distributed until 1665) Pascal was the first to connect binomial coefficients with the combinatorial coefficients that arise in probability. The *Traité* is famous too for its explicit statement of the principle of mathematical induction.

Pascal is also known for a theorem in projective geometry (given an arbitrary hexagon inscribed in any conic section, if the three pairs of opposite sides are continued until they meet, then the three points of intersection lie on a straight line) (1640); and for a two-function (addition and subtraction) mechanical calculating machine (1645).

VI.14 Isaac Newton

b. Woolsthorpe, England, 1642; d. London, 1727
Calculus; algebra; geometry; mechanics; optics; mathematical astronomy

Newton entered Trinity College, Cambridge, in 1661, and it was in Cambridge that he spent most of his formative years, first as a student, then as a Fellow, and then, from 1669, as Lucasian Professor of Mathematics. His election to the Lucasian Chair was engineered by his mentor Isaac Barrow, a talented mathematician and theologian who was the first to hold the prestigious chair. In 1696 Newton moved to London to take up the post of Warden of the Mint. He resigned his professorship in 1702.

It appears that Newton's interest in mathematics began in 1664. In that year he embarked on a course of self-instruction, reading VIÈTE's [VI.9] works (1646), Oughtred's *Clavis Mathematicae* (1631), DESCARTES's [VI.11] *La Géométrie* (1637), and Wallis's *Arithmetica Infinitorum* (1656). From Descartes, Newton learned how useful it could be to relate algebra to geometry, since plane curves could be represented by algebraic equations in two unknowns. Descartes had, however, imposed strict limitations on the class of curves allowed in *La Géométrie*: "geometrical" (i.e., algebraic) curves were admitted but "mechanical" (i.e., transcendental) ones were not. In common with many of his contemporaries, Newton felt that such limitations ought to be overcome and that a "new analysis" capable of dealing with mechanical curves ought to be possible. He found the answer in infinite series.

Newton had learned how to deal with infinite series from Wallis's work, and it was while elaborating one of Wallis's techniques that, in the winter of 1664, he obtained his first great mathematical discovery: the binomial theorem for fractional powers. This provided him with a method for expanding into power series a large class of "curves," including transcendental curves, which could now be given an "analytical" representation to which the rules of algebra could be applied. Termwise application of the relation (which

Isaac Newton

he knew from Wallis and which is expressed in familiar Leibnizian notation as $\int x^n \, dx = x^{n+1}/(n+1)$) allowed him to "square" a variety of curves when they were expanded as a power series. (In the seventeenth century, squaring a curvilinear figure meant finding a square the area of which is equal to that of the curvilinear figure.)

A few months later, Newton, with extraordinary insight, realized that most of the problems dealt with by his contemporaries could be reduced to two classes: problems in which one is required to find the tangent to a curve, and problems in which one is required to find the area subtended by a curve. He conceived geometrical magnitudes as being generated by continuous motion. For example, the motion of a point generates a line and the motion of a line generates a surface. These he called "fluents," while the instantaneous rate of flow he called the "fluxion." Basing his intuitions on kinematical models, he formulated a version of what is known today as THE FUNDAMENTAL THEOREM OF CALCULUS [I.3 §5.5]. Namely, he proved that tangent and area problems are inverses of each other. In modern terms, Newton was able to reduce quadrature problems (i.e., calculating curvilinear areas) to the search for primitive functions (indefinite integrals). He built "catalogues of curves" (tables of integrals), deploying techniques equivalent to substitution of variables and integration by parts. He developed an efficient algorithm that allowed him to tackle both the direct (differential) and inverse (integral) methods of fluxions. He was able to calculate the tangent to and curvature of any known curve, and perform integrations of many classes of (what we now call) ordinary differential equations. Such mathematical tools allowed him to explore the properties of cubics, and he classified seventy-two different species of them. His results on series and on the direct and inverse methods of fluxions were published in *De Quadratura Curvarum* and those on cubics in the *Enumeratio Linearum Tertii Ordinis*, both works appearing in 1704 as appendices to *Opticks*. His *Arithmetica Universalis*, the text in which he collected together his lectures on algebra, appeared in 1707.

Before 1704, Newton, displaying his characteristic reluctance to publish, had divulged his discoveries on the fluxional method through letters and manuscripts rather than in printed form. In the meantime LEIBNIZ [VI.15], later than Newton but independently, had also discovered the differential and integral calculus, and had printed it as early as 1684–86. Newton was convinced that Leibniz had stolen the idea from him, and from 1699 onward he engaged Leibniz in a bitter quarrel over priority.

In the early 1670s Newton began distancing himself from the modern symbolic style that had characterized his youthful researches. He turned to geometry in the hope of restoring a hidden geometric method of discovery: the "method of analysis," known to the ancient Greeks. In fact, geometry dominates Newton's masterpiece, the *Philosophiae Naturalis Principia Mathematica*. In this work, which appeared in 1687, Newton presented his theory of gravitation. Newton was convinced that the ancient method was superior to the modern symbolic one that he identified with Cartesian analysis. In his attempts to rediscover the method, he developed elements of projective geometry. (This sprang from the idea that the ancients were able to solve complex problems related to conics by using projective transformations.) An important result is his solution of Pappus's locus problem, which appears in book I of the *Principia* (1687). Here he shows that a conic is the locus of points, the product of whose distances from two given lines is proportional to the product of its distances from a third and fourth given line. He then applied projective transformations to determine the conic tangent to m given lines that passes through n given points, when $m + n = 5$.

The *Principia* contains a rich array of mathematical results. In book I Newton presents the "method of first and ultimate ratios," in which he deploys geometric limit procedures in order to determine tangents, curvatures, and curvilinear areas, the latter containing the basic ingredients of what today is known as the RIEMANN INTEGRAL [I.3 §5.5]. He also shows that "ovals"

are algebraically nonintegrable. In dealing with the so-called Kepler problem, Newton approximates the roots of $x - d \sin x = z$ (d and z given) by a technique equivalent to the NEWTON–RAPHSON METHOD [II.4 §2.3]. In book II he inaugurates VARIATIONAL METHODS [III.94] by tackling the problem of the solid with least resistance. And in book III, in dealing with cometary paths, he presents a method of interpolation which inspired research by mathematicians such as Stirling, Bessel, and GAUSS [VI.26]. In his masterpiece, Newton had shown how productive the application of mathematics to natural philosophy could be: most notably, his studies on the Moon's motion, the precession of the equinoxes, and the tides were seminal in stimulating eighteenth-century perturbation theory.

Further Reading

Newton, I. 1967–81. *The Mathematical Papers of Isaac Newton*, edited by D. T. Whiteside et al., eight volumes. Cambridge: Cambridge University Press.

Pepper, J. 1988. Newton's mathematical work. In *Let Newton Be! A New Perspective on His Life and Works*, edited by J. Fauvel, R. Flood, M. Shortland, and R. Wilson, pp. 63–79. Oxford: Oxford University Press.

Whiteside, D. T. 1982. Newton the mathematician. In *Contemporary Newtonian Research*, edited by Z. Bechler, pp. 109–27. Dordrecht: Reidel. (Reprinted, 1996, in *Newton. A Critical Norton Edition*, edited by I. B. Cohen and R. S. Westfall, pp. 406–13. New York/London: W. W. Norton & Co.)

Niccolò Guicciardini

VI.15 Gottfried Wilhelm Leibniz

b. Leipzig, Germany, 1646; d. Hanover, Germany, 1716
Calculus; theory of linear equations and elimination theory; logic

Renowned among mathematicians for his invention of the calculus, Leibniz was a universal thinker who graduated in law and was self-taught in mathematics. In 1676 he became counselor and librarian in Hanover for the Duke Johann Friedrich of Braunschweig–Lüneburg, holding this position until the end of his life. Besides mathematics he occupied himself with technical, historiographical, political, religious, and philosophical questions. His philosophy distinguished between two areas of reality: the world of appearances and the world of substances. It was in developing his philosophy that he was led to declare that the real world is "the best of all possible worlds." In 1700 he was appointed first president of the newly founded Brandenburg Society of Sciences established in Berlin.

Gottfried Wilhelm Leibniz

Most of his mathematical ideas and writings were not published during his lifetime, and consequently many of his results were rediscovered many years later. About a fifth of his mathematical papers have now been published. He was always more interested in general or even universal methods than in technical details, using analogy and inductive reasoning to develop the art of invention. For the same reason he became a key creator of mathematical notation: he knew how much a suitable notation could facilitate mathematical discoveries.

One of Leibniz's earliest mathematical works was a treatise on infinitesimal geometry (written in 1675–76 but not published until 1993). In it he used his "quanta" concept of the infinite. In Leibniz's eyes the actual infinite as well as indivisibles, in the strictest sense of the word, were not quantities and therefore not mathematical entities: hence, he used the notions "infinitely small" and "infinitely large." These denoted, it is true, variable quantities, but nevertheless they were quantities of a sort, so they could be handled by mathematics. Among the results in this treatise is a rigorous proof, in the style of ARCHIMEDES [VI.3], of the existence of (what is today known as) the RIEMANN INTEGRAL [I.3 §5.5] of continuous functions, which is based on intermediary values of the function within subintervals. Only a few of these results were actually published by Leibniz, and even these mainly without proof: in 1682 the alternating series for $\pi/4$; in 1691 some further results. In 1713 he communicated his alternating series test in a private letter to JOHANN BERNOULLI [VI.18].

The year 1675 was also the year in which Leibniz invented his version of the differential and integral calculus, although its publication did not begin until 1684. His calculus was based on the key concept of a variable (quantity) ranging over a sequence of values infinitely close to each other, with the differential, the difference between two successive values in the sequence, being itself a variable that could be manipulated in the usual manner. Differentiation was represented by the operator "d", which assigned variables to variables. For example, if x is a line of variable length, then dx is a very short line, also of variable length. Integration meant summation. His notation (d and ∫) is still used today. He deduced the standard differentiation rules (the chain rule, the product rule, etc.) and successfully applied his calculus to the differentiation of families of curves, to differentiation under the integral sign, and to various types of differential equations.

Leibniz considered "combinatorial art" as a general qualitative science, which did not coincide with modern combinatorial analysis but included combinatorics and algebra: Leibniz considered it as "the inventive part of logic." Here he found the Girard formula for the representation of sums of powers of roots of equations by means of elementary symmetric functions, and the so-called Waring formulas by which polynomial symmetric functions are reduced to power sums (these were rediscovered by WARING [VI.21] in 1762). He invented double and multiple indices in order to solve systems of linear equations and problems of elimination theory. Between 1678 and 1713 he laid the foundations for the theory of DETERMINANTS [III.15]. The method now known as Cramer's rule, for solving simultaneous equations, which in modern terms is based on determinants, and which Cramer published in 1750, was in fact found in 1684 by Leibniz (but again not published by him). He also stated (without proof) several theorems in the theory of linear equations and elimination theory now attributed to EULER [VI.19], LAPLACE [VI.23], and SYLVESTER [VI.42].

Among Leibniz's other mathematical interests was additive number theory. In 1673 he found a recursion formula for the number of tripartitions of a natural number (published in 1976) and discovered further rules of recurrence now attributed to Euler. He also developed a formalism for a positional calculus (*calculus situs*) in order to express positions in space: if the definitions of figures are completely expressed by this calculus, all of their properties can then be found by this calculus. This is closely linked with the modern notions of geometry and topology.

Leibniz was one of the pioneers of actuarial theory. Using mathematical models of human life he calculated the purchase price of life annuities both for single persons and for groups of men, and he applied such considerations to the liquidation of a state's indebtedness.

From the very beginning of his scientific career Leibniz was deeply interested in logic. He conceived of a general science: that is, of an art of inventing and of judging all sciences by means of sufficient data and a suitable universal language or writing. Yet, his "characteristica universalis" and the ensuing logical calculi remained fragmentary projects. His "calculus ratiocinator" was meant to be a formalized deduction of truth. Given that Leibniz was interested in formalizing calculations, it is not surprising that he also constructed the first four-function calculating machine. In constructing this machine he invented a new technical device, which he developed in two different versions: the so-called pinwheel (before 1676) and the stepped drum (from 1693 or earlier).

Further Reading

Leibniz, G. W. 1990-. *Sämtliche Schriften und Briefe, Reihe 7 Mathematische Schriften*, four volumes (so far). Berlin: Akademie.

Eberhard Knobloch

VI.16 Brook Taylor

b. Edmonton, Middlesex, England, 1685; d. London, 1731
Secretary of the Royal Society (1714-18)

Taylor was not the first to discover the theorem that bears his name (James Gregory found the theorem in 1671), but he was the first to publish it and the first to appreciate its significance and applicability. The theorem, which states that any function that satisfies certain conditions can be expressed as (what is now known as) a Taylor series, was published in Taylor's *Methodus Incrementorum Directa et Inversa* (1715). In the *Methodus* Taylor gave the series as

$$f(x+h) = f(x) + \frac{f'(x)}{1!}h + \frac{f''(x)}{2!}h^2 + \frac{f'''(x)}{3!}h^3 + \cdots$$

(as it would appear in modern notation). Although Taylor did not attend to questions of rigor—there is no consideration of convergence, of the remainder term, or of the validity of expressing a function by such a series—his derivation of the series was not out of line

with the standards of its day. Taylor used the theorem for approximating the roots of equations and for solving differential equations. Although he was aware of its use for expanding functions into series, he does not appear to have fully appreciated its significance in this respect.

Taylor is also noted for his contribution to the problem of the vibrating string (discussed in the *Methodus* and in earlier papers) and for a book on the theory of linear perspective (1715).

VI.17 Christian Goldbach

b. Königsberg (now Kaliningrad, Russia), 1690; d. Moscow, 1764
Professor of Mathematics, Imperial Academy of Sciences, Saint Petersburg (1725-28); tutor to Tsarevitch Peter II, Moscow (1728-30); corresponding secretary and administrator, Imperial Academy of Sciences, Saint Petersburg (1732-42); Ministry of Foreign Affairs (1742-64)

Goldbach is remembered today for the conjecture that bears his name: that every even number greater than 2 is a sum of two primes. This was first stated by EULER [VI.19] in 1742 in a letter to Goldbach in response to the earlier proposal by Goldbach that every number greater than 2 is a sum of three primes (Goldbach considered 1 as a prime number). Goldbach's conjecture, together with the weaker conjecture that every odd number is either prime or the sum of three primes, was first published by WARING [VI.21] in 1770 but without attribution. Both conjectures remain unsolved. However, Vinogradov proved that every sufficiently large odd number is the sum of three primes: see PROBLEMS AND RESULTS IN ADDITIVE NUMBER THEORY [V.27].

VI.18 The Bernoullis

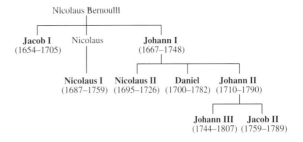

All born in Basel, Switzerland, apart from Daniel (Groningen, the Netherlands). All died in Basel, apart from Jacob II, Nicolaus II (both Saint Petersburg, Russia), and Johann III (Berlin). (The two members of the family not in bold text were not mathematicians.)

The Bernoullis played a remarkable role in the development of mathematics during the Enlightenment. Indeed such was the family's importance that in 1715 LEIBNIZ [VI.15] coined the term "bernoullizare" to describe the activity of doing mathematics. Altogether eight members of the family devoted themselves to the mathematical sciences (including physics, especially mechanics and fluid mechanics), and from 1687 to 1790 the mathematics chair of the university of Basel was occupied by successive members of the family: first Jacob (1687–1705), then his brother Johann (1705–48), and finally Johann's son, Johann II (1748–90). Throughout the eighteenth century Bernoullis were members of the Paris Academy of Sciences and individually they won prestigious prizes on many occasions. The same was true of the academies in Berlin, Saint Petersburg, and several others.

The family goes back to a line of Calvinist merchants who fled the Spanish Netherlands. The first Bernoulli to settle in Basel was Jacob, a druggist, who became a citizen in 1622. His grandson, Jacob I, studied philosophy and theology in Basel before turning to mathematics, against the will of his father. This was to be a typical pattern in the family: many of the Bernoullis studied mathematics despite pressure from the family to make a career in other areas (such as medicine or law). Having received a licentiate in theology in 1676, Jacob undertook an educational journey which took him first to France, then to the Netherlands, and finally to England. And it was through his encounters with Nicolas Malebranche, Jan Hudde, and others that he became acquainted with Cartesianism and its most eminent representatives. In 1677 he started his diary, *Meditationes*, in which he wrote down many of his mathematical insights and thoughts.

Having obtained the chair of mathematics in Basel, Jacob studied Leibniz's early memoirs on differential calculus, whose power he was the first, together with his younger brother Johann I, to recognize. In a paper on the curve of constant descent, published in the Leipzig *Acta Eruditorum* in 1690, Jacob was the first to use the term "integral" in its present mathematical sense. From then on he showed his mastery of Leibnizian methods in his study of curves, including, among others, the catenary, the form of a bent elastic beam, the form of a sail inflated by the wind, and the parabolic and logarithmic spirals. He also solved the differential equation $y' = p(x)y + q(x)y^n$, which is now named after him. However he is best remembered for his *Ars Conjectandi* (1713), which was published posthumously with a short foreword by his nephew, Nicolaus I. It contains an attempt to give a sound math-

ematical treatment of a commonsense principle already appealed to by CARDANO [VI.7] and Halley: if an experiment is repeated a large number of times, then the relative frequency with which an event occurs will roughly equal the probability of the event. Bernoulli's theorem, known since POISSON [VI.27] as the (WEAK) LAW OF LARGE NUMBERS [III.71 §4], establishes a first link between the theories of probability and statistics. In the same book, Bernoulli also introduced the sequence $B_0, B_1, \ldots$ of rational numbers that now bears his name, which can be defined as the coefficients of $t^k/k!$ in the power-series expansion

$$\frac{t}{e^t - 1} = \sum_{k=0}^{\infty} B_k \frac{t^k}{k!}.$$

Jacob computed these numbers up to B_{10}.

Johann, who had to study medicine before he could devote himself to mathematics, got his first mathematical training from his brother Jacob, with whom he developed numerous applications of the new Leibnizian calculus to mechanics. An academic peregrination brought him to Paris in 1691–92, where he gave private lessons to Guillaume de l'Hôpital. These lessons are the basis of l'Hôpital's famous *Analyse des Infiniment Petits* (1696). This textbook, the first on calculus, contains l'Hôpital's rule, which Johann had communicated by letter to his student. In 1695, Johann left Basel for Groningen to take up a professorship in mathematics.

With the growing visibility of Johann's work, the friendly collaboration between the two brothers Jacob and Johann transformed into an endless round of controversies, priority disputes, and public accusations. They engaged in heated struggles concerning the solution of the brachistochrone (curve of fastest descent) problem, and a complicated isoperimetric problem that involved minimizing the area enclosed by a curve of fixed length. Eventually, these bitter quarrels led to an interesting mathematical outcome: the creation of the CALCULUS OF VARIATIONS [III.94]. After Jacob's death, Johann took over the mathematics chair in Basel, where he taught until the end of his life, attracting students from all over Europe, including EULER [VI.19].

Johann's most important achievement in mathematics is the development of the integral calculus. He developed a general theory for the integration of rational functions and new methods for the solution of differential equations. He also extended the infinitesimal calculus to handle EXPONENTIAL FUNCTIONS [III.25].

Johann's correspondence with Leibniz (which spans approximately twenty-five years) can be viewed as a laboratory of mathematical invention and debate. The priority dispute that ensued when NEWTON [VI.14] accused Leibniz of having stolen the calculus from him also involved Johann, who fought on Leibniz's side. With each camp defying the other with difficult problems, Johann had the opportunity to create, with his son Nicolaus II, the theory of orthogonal trajectories of families of curves. Johann was also a towering figure in the origins of analytical mechanics and in mathematical physics, where, among other things, he made notable contributions to the study of central forces, to navigational theories, and to the question of the principles of statics.

Nicolaus I studied mathematics at the university of Basel with his uncle Jacob before taking a degree of Doctor of Jurisprudence (1709). He was the professor of mathematics in Padua, occupying the chair once held by Galileo, and he later held the professorship of logic in Basel. His main interests in mathematics were infinite series and the applications of probability theory to questions of law. He formulated, in 1713, the notorious Saint Petersburg paradox, which originated with a gambling game. Suppose that Peter is tossing a fair coin, and he will give Paul one ducat if the coin turns up heads on the first toss, two ducats if it shows heads for the first time on the second toss, and in general 2^{n-1} ducats if the coin turns up heads for the first time on the nth toss. The standard calculation shows that the value of Paul's expectation ($E = \frac{1}{2}1 + \frac{1}{4}2 + \frac{1}{8}4 + \cdots + \frac{1}{2^n}2^{n-1} + \cdots$) is infinitely great. Nevertheless no "fairly reasonable man" would be willing to pay even a moderately high price to purchase Paul's prospects. The result of the mathematical analysis clearly affronted common sense; this was the paradox. Nicolaus's cousin, Daniel, discussed the problem while he was staying in Saint Petersburg (hence the name given to the paradox). His strategy was to distinguish two senses of expectation, one mathematical and the other moral. The latter was to take into account the individual characteristics of the risk taker (his wealth, for instance).

Although primarily a physicist and author of the famous *Hydrodynamica* (1738), Daniel obtained a solution of the Riccati equation, $y' = r(x) + p(x)y + q(x)y^2$, and engaged in the problem of the vibrating string.

Otto Spiess, from Basel, started to publish the complete edition of the works and correspondence of the Bernoullis in 1955. The project continues.

Further Reading

Cramer, G., ed. 1967. *Jacobi Bernoulli, Basileensis, Opera*, two volumes. Brussels: Editions Culture et Civilization. (Originally published in Geneva in 1744.)

——. 1968. *Opera Omnia Johannis Bernoulli*, four volumes. Hildesheim: Georg Olms. (Originally published in Lausanne and Geneva in 1742.)

Spiess, O., ed. 1955-. *The Collected Scientific Papers of the Mathematicians and Physicists of the Bernoulli Family*. Basel: Birkhäuser.

Jeanne Peiffer

VI.19 Leonhard Euler

b. Basel, Switzerland, 1707; d. Saint Petersburg, Russia, 1783
Analysis; series; rational mechanics; number theory;
music theory; mathematical astronomy;
calculus of variations; differential equations

Euler was one of the most influential and prolific mathematicians in history. His first publication was a 1726 paper on mechanics, and his last was a collection published in 1862, seventy-nine years after his death. There are over eight hundred papers bearing his name, about three hundred of them appearing posthumously, and more than twenty books. His *Opera Omnia* fill over eighty volumes.

In number theory, Euler introduced the Euler phi function, $\phi(n)$, to denote the number of positive integers less than n and relatively prime to n, and proved the FERMAT–EULER THEOREM [III.58] that n divides $a^{\phi(n)} - 1$. He showed that the remainders relatively prime to n form what we now call a group under multiplication and he expanded the theory of quadratic and higher-order residues. He proved FERMAT'S LAST THEOREM [V.10] for $n = 3$. He stated that any real polynomial of degree n is a product of real and quadratic factors and has n complex roots, but was unable to give complete proofs. He was the first to use GENERATING FUNCTIONS [IV.18 §§2.4, 3] when he gave a generating function for Naudé's partition problem: the question of how many different ways a given integer can be written as a sum of positive integers. He introduced the function $\sigma(n)$, the sum of the divisors of an integer n, and used this function to increase the number of known pairs of amicable numbers (a pair m, n of numbers is called *amicable* if the sum of the proper divisors of m equals n, and vice versa) from 3 to over 100. He showed that any prime number of the form $4n + 1$ is the sum of two rational squares. LAGRANGE [VI.22] later improved this result to show that such numbers are the sum of two integer squares. Euler factored the fifth Fermat

Leonhard Euler

number, $F_5 = 2^{2^5} + 1$, thus refuting FERMAT's [VI.12] conjecture that all integers of the form $F_n = 2^{2^n} + 1$ were prime. He made extensive studies of the binary quadratic forms $x^2 + y^2$, $x^2 + ny^2$, and $mx^2 + ny^2$, and proved a form of the LAW OF QUADRATIC RECIPROCITY [V.28].

Euler was the first to use analytic methods in number theory. In the 1730s he calculated to several decimal places the so-called Euler–Mascheroni constant

$$y = \lim_{n \to \infty} \left[\left(\sum_{k=1}^{n} \frac{1}{k} \right) - \log n \right]$$

and discovered many of its properties. Mascheroni added to those properties in the 1790s. Euler also discovered the sum–product formula for what we now call the Riemann zeta function,

$$\zeta(s) = \sum_{n=1}^{\infty} \frac{1}{n^s} = \prod_{p \text{ prime}} \frac{1}{1 - p^{-s}},$$

and he evaluated the function for positive even values of s.

In analysis, Euler was largely responsible for shaping the modern calculus curriculum. He was also the first person to take a systematic approach to the solution of differential equations and to problems of the CALCULUS OF VARIATIONS [III.94]. He discovered a differential equation sometimes called the "Euler necessary condition" and sometimes called the "Euler–Lagrange equation." The equation tells us that if J is defined by the integral equation $J = \int_a^b f(x, y, y') \, dx$, then a function $y(x)$ that maximizes or minimizes J will satisfy

the differential equation

$$\frac{\partial f}{\partial y} - \frac{d}{dx} \left(\frac{\partial f}{\partial y'} \right) = 0.$$

Euler apparently thought that the condition was also sufficient. Very early in his career, he pioneered the use of integrating factors for solving differential equations, though the almost simultaneous published solution of Clairaut was more complete and more widely read, so credit for this innovation usually falls to Clairaut. He also did the first work using what are now called FOURIER SERIES [III.27] and LAPLACE TRANSFORMS [III.91], more than a generation before LAPLACE [VI.23] or FOURIER [VI.25] began doing mathematics, though they took the fields much farther than Euler had.

Much of Euler's best work involved series. His first widely acclaimed result was when he solved one of the best-known problems of his age, the seventy-year-old "Basel problem." The problem was to evaluate the sum of the reciprocals of the square integers, or $\zeta(2)$. Euler showed that

$$\sum_{n=1}^{\infty} \frac{1}{n^2} = \frac{\pi^2}{6}.$$

(For a sketch of a proof, see π [III.70].)

He developed the Euler–Maclaurin series to strengthen the relationships between series and integrals. The existence of the Euler–Mascheroni constant followed from these researches. Using techniques he called "interpolation of series," he developed the GAMMA FUNCTION [III.31] and the beta function. He developed the first extensive theory of CONTINUED FRACTIONS [III.22], and derived series for the accurate and efficient calculation of LOGARITHMS [III.25 §4] and trigonometric tables, often to more than twenty decimal places.

He was the first to do calculus with complex numbers and to investigate logarithms of negative and complex numbers. This research led to a long and bitter controversy with D'ALEMBERT [VI.20].

Euler was not the first to prove that $e^{i\theta} = \cos \theta + i \sin \theta$ or to know that $e^{\pi i} = -1$, but he made so much more use of these facts than any of his predecessors that this last formula is generally known as Euler's identity.

He is regarded as a pioneer in topology and graph theory for his necessary condition for a graph to have an Euler path, the so-called Königsburg bridge problem. This is to determine whether or not a graph has a path that traverses every edge exactly once. He also discovered and gave a flawed proof that, for a polyhedron "bounded by planes," Euler's words for what we now call "convex," $V - E + F = 2$, where V is the number of vertices, E is the number of edges, and F is the number of faces. (For details about the flaws in Euler's proof, see Richeson and Francese (2006).)

Euler proved a form of the general addition theorem for elliptic integrals and gave a complete classification of elastic curves. At the command of his king, Frederick the Great of Prussia, he studied hydraulics, designed pumps and fountains, and evaluated the probabilities and combinatorics involved in the state lotteries.

In a triangle, the line on which the orthocenter, the centroid, and the circumcenter lie is the Euler line. The Euler method is an algorithm for giving numerical solutions to differential equations. The EULER DIFFERENTIAL EQUATION [III.23] is the partial differential equation that describes continuity of fluid flow.

Euler tried to use lunar and planetary theory to solve the problem of finding longitude at sea. In studying the orbit of a comet, he made the first steps in the statistics of observed data.

He left Switzerland in 1727 to work in the new academy of Peter the Great in Saint Petersburg. In 1741 he moved to Berlin and the academy of Frederick the Great, but returned to Saint Petersburg in 1766, after the ascension of Catherine the Great. He was blind for the last fifteen years of his life, during which time he nevertheless wrote over three hundred papers. He won the annual prize competition of the Paris Academy twelve times.

His series of calculus books, published in four volumes between 1755 and 1770, were the first successful calculus textbooks. It was the climax of a complete series of mathematical textbooks, including arithmetic (1738), algebra (1770), and the *Introductio in Analysin Infinitorum* (1748), a textbook on the mathematics Euler thought was necessary to understand calculus.

In the two volumes of the *Mechanica* (1736), Euler gave the first calculus-based treatment of the mechanics of point masses. He followed this with another two-volume work, *Theoria Motus Corporum* (1765), on the motions of solid bodies, including rotations.

Other books include *Methodus Inveniendi* (1744), the first unified treatment of the calculus of variations, *Tentamen Novae Theoriae Musicae* (1739), on the physics of music and including the first use of logarithms in the theory of pitch, three different books on celestial mechanics and lunar theory, two on the theory of shipbuilding, three on optics, and one on ballistics.

Our modern notion that FUNCTIONS [I.2 §2.2] are a fundamental object in mathematics is due to Euler.

Euler standardized the use of the symbols e, π, and i, as well as $\sum$ for summations and Δ for finite differences.

His *Letters to a German Princess* in three volumes (1768–71) is regarded variously as the first work of popular science writing by a first-rate scientist and as an important work in the philosophy of science.

Laplace is reported to have advised, "Read Euler. Read Euler. He is the master of us all." The words are probably not those of Laplace, but the misattribution does not affect the quality of the advice.

Further Reading

Bradley, R. E, and C. E. Sandifer, eds. 2007. *Leonhard Euler: Life, Work and Legacy*. Amsterdam: Elsevier.

Dunham, W. 1999. *Euler: the Master of Us All*. Washington, DC: Mathematical Association of America.

Euler, L. 1984. *Elements of Algebra*. New York: Springer. (Reprint of 1840 edition. London: Longman, Orme, and Co.)

———. 1988, 1990. *Introduction to Analysis of the Infinite*, books I and II, translated by J. Blanton. New York: Springer.

———. 2000. *Foundations of Differential Calculus*, translated by J. Blanton. New York: Springer.

Richeson, D., and C. Francese. 2007. The flaw in Euler's proof of his polyhedral formula. *American Mathematical Monthly* 114(1):286–96.

Edward Sandifer

VI.20   Jean Le Rond d'Alembert

b. Paris, 1717; d. Paris, 1783
Algebra; infinitesimal calculus; rational mechanics; fluid mechanics; celestial mechanics; epistemology

---

D'Alembert spent his whole life in Paris, where he became one of the most influential members of the Académie Royale des Sciences and of the Académie Française. He became well-known as the scientific editor of the celebrated French *Encyclopédie*, the twenty-eight-volume work on which he collaborated with Denis Diderot, and for which he wrote most of the mathematical and many of the scientific articles.

As a student at the Jansenist Collège des Quatre-Nations, he followed the usual curriculum of grammar, rhetoric, and philosophy, the latter including some Cartesian science, a little mathematics, and much theology framed by the then burning debate about predestination, freedom, and grace. Disgusted with the permanent climate of controversy and the endless metaphysical discussions among his Jansenist teachers, d'Alembert decided, after attaining his diploma in law, to devote himself to his personal passion, "géométrie" (that is, mathematics).

D'Alembert's first communications to the French Académie were concerned with the analytic geometry of curves, the integral calculus, and fluid resistance, notably the problem of the deceleration and deflection of a disk entering a fluid, which was linked to the Cartesian explanation of refraction of light. He made a close reading of NEWTON's [VI.14] *Principia*, his commentary on passages of the first book showing his clear preference for analytical methods over the synthetic geometry of Newton.

D'Alembert's *Traité de Dynamique* (1743) made him famous in learned circles. He built up a systematic and rigorous theory of mechanics founded upon a short list of well-chosen principles—inertia, composition of motions (i.e., the addition of the effects of two forces or powers), and equilibrium—while at the same time trying to avoid metaphysical arguments. Most notably he proposed an important general principle, known today as "d'Alembert's principle," to simplify the investigation of constrained systems, such as the compound pendulum, vibrating rods, strings, rotating bodies, and even fluids, which he considered to be aggregates of parallel slices. The essential idea behind the principle was to reduce a problem in dynamics to one in statics, roughly speaking by introducing a fictitious force, the "kinetic reaction," which was minus the mass times the acceleration. This allowed techniques from statics to be brought to bear on problems in dynamics.

His other books and memoirs were developments, some very innovative, in fluid theory, partial differential equations, celestial mechanics, algebra, and integral calculus. He devoted much thought to the use and status of imaginary numbers.

In his *Réflexions sur la Cause Générale des Vents* (1747) and his *Recherches sur le Calcul Intégral* (1748) he observed that numbers of the form $a + b$i (where $i = \sqrt{-1}$) retain the same form when subjected to the usual operations (addition, subtraction, multiplication, division, and exponentiation). He proved that, for a real polynomial, imaginary roots always occur in conjugate pairs, and that even if a real polynomial has no real root, there is still always a complex root. However, his work was not rigorous—for example, he presupposed the existence of roots—and consequently he did not provide a proof of THE FUNDAMENTAL THEOREM OF ALGEBRA [V.13].

At the end of the 1740s there was a crisis in Newtonian science, with d'Alembert, Clairaut, and EULER

[VI.19] each independently coming to the conclusion that Newton's theory of gravitation could not account for the motion of the Moon. In 1747 d'Alembert discussed various possibilities for solving the problem—an additional force, or a very irregular shape for the Moon, or some vortices between Earth and Moon—and produced a long study on celestial mechanics and planetary perturbations that has only recently been rediscovered and published (see d'Alembert 2002). By 1749 an improved mathematical analysis of the problem had shown that Newton's theory was correct. The rest of d'Alembert's extensive work on celestial mechanics was published in his *Recherches sur la Précession des Équinoxes* (1749), *Recherches sur Différents Points du Système du Monde* (1754–56), and in some of the eight volumes of his *Opuscules* (1761–83).

In 1747 d'Alembert presented a paper on the famous problem of vibrating strings, *Recherches sur la Courbe que Forme une Corde Tendue Mise en Vibration* (1749). This paper contained a solution of THE WAVE EQUATION [I.3 §5.4]. This was the first solution of a partial differential equation—partial differential equations were a new tool that he had already used in his 1747 *Réflexions sur la Cause Générale des Vents*. It led to a lengthy debate with Euler and DANIEL BERNOULLI [VI.18] about the possible form of the solutions and the general notion of function.

D'Alembert's work for the *Encyclopédie* (1751–65) and his efforts to find rigorous foundations for the sciences led him into the field of philosophy, where his main contributions concerned the classification of various sciences. He also worked on the study of cognition following the lines proposed by DESCARTES [VI.11], Locke, and Condillac.

Further Reading

D'Alembert, J. le R. 2002. *Premiers Textes de Mécanique Céleste*, edited by M. Chapront. Paris: CNRS.
Hankins, T. 1970. *Jean d'Alembert, Science and the Enlightenment*. Oxford: Oxford University Press.
Michel, A., and M. Paty. 2002. *Analyse et Dynamique. Études sur l'Oeuvre de d'Alembert*. Laval, Québec: Les Presses de l'Université Laval.

Francois de Gandt

VI.21 Edward Waring
b. Shrewsbury, England, ca. 1735; d. Shrewsbury, 1798
Lucasian Professor of Mathematics, Cambridge (1760–98)

Waring, the leading British mathematician of the latter half of the eighteenth century, wrote several advanced but somewhat impenetrable analytical texts. His first work, *Miscellanea Analytica* (1762), is devoted to the theory of numbers and algebraic equations and contains many results which he revised and expanded in his *Meditationes Algebraicae* (1770). Included in the latter is the problem known today as Waring's problem (that every positive integer is the sum of not more than nine cubes, or the sum of not more than nineteen fourth powers, and so on, with a fixed number of summands depending on the exponent), which was solved by HILBERT [VI.63] in the affirmative in 1909 and which gave rise to important work by HARDY [VI.73] and LITTLEWOOD [VI.79] in the 1920s. The *Meditationes* also contained the first publication of Goldbach's conjecture (that every even integer greater than 2 can be written as the sum of two primes) and of Wilson's theorem (if p is a prime number, then $(p - 1)! + 1$ is divisible by p), which was subsequently proved by LAGRANGE [VI.22].

Waring's problem and Goldbach's conjecture are discussed in PROBLEMS AND RESULTS IN ADDITIVE NUMBER THEORY [V.27].

VI.22   Joseph Louis Lagrange

b. Turin, Italy, 1736; d. Paris, 1813
Number theory; algebra; analysis; classical and celestial mechanics

In 1766 Lagrange left his native Turin, where he had been a founding member of what would later become the Turin Academy of Sciences, to become the mathematics director at the Berlin Academy of Sciences. In 1787 he moved to Paris to take up a position as a *pensionnaire veteran* at the Academy of Sciences. In Paris he also lectured at the École Polytechnique, founded in 1794, and served as one of the members of the committee that established the modern metric system.

Lagrange was only nineteen years old when he wrote to EULER [VI.19] announcing a new formalism to simplify Euler's method for finding a curve that satisfied an extremum condition. Lagrange's method was based on the introduction of a new differential operator, δ, to express the independent variations of the coordinates of a curve that produced a local infinitesimal deformation.

Using this formalism he derived the differential equation known today as the Euler–Lagrange equation, the fundamental equation of the CALCULUS OF VARIATIONS [III.94]. Suppose that we wish to find the function $y = y(x)$ that maximizes or minimizes a definite integral

of the form

$$\int_a^b f(x, y, y') \, dx,$$

where $y' = dy/dx$. The equation states a necessary condition that this function must satisfy:

$$\frac{\partial f}{\partial y} - \frac{d}{dx}\left(\frac{\partial f}{\partial y'}\right) = 0.$$

This is a typical example of Lagrange's reductionist style. Throughout his career he sought suitable formalisms with which to express and solve the key problems of mathematical analysis.

Lagrange publicly presented his δ-formalism in a memoir published in the second volume (1760–61) of *Miscellanea Taurinensia*, a review he had helped to found. He coupled this memoir with another in which he used the same formalism to formulate a generalized version of the principle of least action (previously introduced by Maupertuis and Euler). As a result, he was able to derive the equations of motion of any system of distinct bodies attracted by central forces depending on the distances from their centers.

Meanwhile, in the first volume of *Miscellanea Taurinensia* (1759), Lagrange had published a memoir presenting a new approach to the problem of the vibrating string, where the string is first represented as a discrete system of n particles, and n is then allowed to tend to infinity. Using this method Lagrange argued that Euler was right to allow a large class of "functions," both "continuous" and "discontinuous," as solutions of this problem, whereas D'ALEMBERT [VI.20] had maintained that only "continuous functions" (that is, curves expressed by a single equation) could count as solutions.

Lagrange established a very general program in the foundations of classical mechanics in these memoirs. It was based on the interpretation of a continuous system as a limiting case of a discrete one and the use of the method of indeterminate coefficients: supposing that $P(x)$ is a polynomial in x, whose coefficients a_i ($i = 0, \ldots, n$) depend on some indeterminates, and that $P(x) = 0$ for all x (in a given interval), this method consists of deducing the system of equations $\{a_i = 0\}_{i=0}^{i=n}$, from which the indeterminates are possibly determined. Lagrange extended this method to sums of polynomials in several (independent) variables, and (following Euler, d'Alembert, and many others) also used it with respect to power series. This program was further elaborated in two memoirs on the motion of the Moon (1764, 1780), and later realized in the *Mécanique Analitique* (1788). Here, the principle of least action was replaced by a generalization of the Bernoullian principle of "virtual velocities," which were expressed by variations. Using what are now known as generalized coordinates, φ_i (that is, mutually independent coordinates in the configuration space of a discrete system that characterize completely the position of its bodies), Lagrange derived the equations that are now named after him:

$$\frac{\mathrm{d}}{\mathrm{d}t}\left(\frac{\partial T}{\partial \dot{\varphi}_i}\right) - \frac{\partial T}{\partial \varphi_i} + \frac{\partial U}{\partial \varphi_i} = 0,$$

where T and U are the kinetic and potential energy of the system, respectively.

The *Méchanique Analitique* appeared a century after NEWTON's [VI.14] *Principia* and marked the culmination of a purely analytical approach to mechanics. In the preface Lagrange proudly stated that no diagrams would be found in the work and that everything would be reduced to "algebraic operations submitted to a regular and uniform progression."

Lagrange made fundamental contributions to perturbation theory and THE THREE-BODY PROBLEM [V.33] with research published in the 1770s and 1780s. His methods were further developed by LAPLACE [VI.23] in his *Mécanique Céleste* and formed the basis for subsequent mathematical work in physical astronomy.

The method of indeterminate coefficients, or rather its extension to power series, is also the crucial technique underlying Lagrange's approach to the calculus. In a memoir which appeared in the *Proceedings of the Berlin Academy* (1768) he used it to prove an important result connecting the calculus to the theory of algebraic equations, the so-called Lagrange inversion theorem, which states that a function $\psi(p)$ of a root p of an equation $t - x + \varphi(x) = 0$, where $\varphi(x)$ is an arbitrary function of x, can be expanded in a series based on the Taylor expansions of $\varphi(t)$ and $\psi(t)$ (the precise conditions to be satisfied by x, $\varphi(x)$, and $\psi(t)$ were later clarified by CAUCHY [VI.29] and Roché).

In a memoir of 1772 Lagrange returned to power series and proved that if a function $f(x + h)$ has a power-series expansion in h, then this series can be written in the form

$$\sum_{i=0}^{\infty} f^{(i)}(x)\frac{h^i}{i!},$$

where, for any i, $f^{(i+1)}$ is derived from $f^{(i)}$ in the same way that f' derives from f. Thus, he just had to prove, through an infinitesimal argument, that $f' = \mathrm{d}f/\mathrm{d}x$ to conclude that the only power-series expansion of a function is its Taylor series. In *Théorie des Fonctions Analytiques* (1797) he then showed (or, rather, claimed to have shown), without making an appeal to the differential calculus, that any function $f(x + h)$ can be expanded in a power series, and suggested interpreting the differential formalism as a formalism that applies to the coefficients of $h^i/i!$ in such an expansion. In other words, he suggested defining the differential ratios of any order (that is, the ratios $\mathrm{d}^i y/\mathrm{d}x^i$, where $y = f(x)$) as derivative functions supplying these coefficients, whereas previously these had been thought of as genuine ratios of differentials. He also proved that the remainder of a Taylor series could be written in the form now known as the Lagrange remainder.

The main results obtained by Lagrange within the theory of algebraic equations were presented in a long memoir of 1770 and 1771 in which the formulas for solving equations of degree 2, 3, and 4 were obtained through an analysis of permutations of the roots. This work constituted the starting point for the later researches of ABEL [VI.33] and GALOIS [VI.41]. In the same memoir Lagrange stated a particular but significant case of the theorem in group theory that today bears his name: that the order of a subgroup of a finite group divides the order of the group.

Lagrange also obtained important results in number theory. Arguably the most significant were the proof of a conjecture, advanced by FERMAT [VI.12] (among others) and which Euler had already tried to prove, asserting that any positive integer is the sum of (at most) four squares (1770), and the proof of Wilson's theorem (first guessed by Wilson and published by WARING [VI.21] without proof), asserting that if n is prime, then $(n - 1)! + 1$ is divisible by n (1771).

Further Reading

Burzio, F. 1942. *Lagrange*. Torino: UTET.

Marco Panza

VI.23 Pierre-Simon Laplace

b. Beaumont-en-Auge, France, 1749; d. Paris, 1827
Celestial mechanics; probability; mathematical physics

Laplace is known to later mathematicians for many concepts of fundamental importance to mathematics, including the LAPLACE TRANSFORM [III.91], the Laplace expansion, Laplace's angles, Laplace's theorem, Laplace functions, inverse probability, GENERATING FUNCTIONS [IV.18 §§2.4, 3], a derivation of the Gauss/Legendre least-squares rule of error by means of a linear regression, and the LAPLACIAN [I.3 §5.4] or potential function. Laplace developed the fields of celestial mechanics (the phrase was his coinage) and probability, and along the way the mathematics to service and advance them. For Laplace, celestial mechanics and probability were complementary instruments that implemented a unified vision of a fully determined universe. Celestial mechanics vindicated the Newtonian system of the world. Probability was the measure, not of the operations of chance in nature, for there are none, but of human ignorance of causes, which was to be reduced to virtual certainty by calculation. The third reason for Laplace's importance to the history of science was the mathematization of physics in the first two decades of the nineteenth century. Apart from a few formulations—speed of sound, capillary action, refractive indices of gases—his role there was that of instigator and patron rather than of major contributor.

Laplace came up with the majority of the above concepts in a probabilistic context. The earliest hints of the method of solving difference, differential, and integral equations, later known as the Laplace transform, appeared in "Mémoire sur les suites" (1782a), where Laplace introduced generating functions. Laplace considered generating functions to be the approach of choice in solving problems that involved the development of functions in series and evaluation of the sums. Years later, in composing *Théorie Analytique des Probabilités* (1812), he subordinated all the analytical part to the theory of generating functions and treated the entire subject as their field of application. In the early memoir, however, he emphasized what he expected to be their applicability to problems of nature.

In an even earlier paper, "Mémoire sur la probabilité des causes par les événements" (1774), Laplace stated the theorem permitting the analysis later termed BAYESIAN [III.3]. Unknown to Laplace, Thomas Bayes had arrived at the same theorem eleven years earlier but had not developed it. Laplace for his part proceeded, in further investigations over some thirty years, to develop inverse probability into the basis for statistical inference, philosophical causality, estimation of scientific error, quantification of the credibility of evidence, and optimal voting rules in the proceedings of legislative bodies and judicial panels. His initial attraction to the approach was its applicability to human concerns. It was in the course of these papers, most notably "Mémoire sur les probabilités" (1780), that the word probability came to connote not merely the basic

quantity in the theory of games and chance but a subject in itself.

Laplace first addressed error theory in the above paper on causality. The problem was to estimate the most appropriate mean value to be taken in a series of astronomical observations of the same phenomenon. He also determined how the limits of the error were related to the number of observations ("Mémoire sur les probabilités" above). In "Essai pour connaître la population du Royaume" (1783–91) Laplace turned to demographic applications. In the absence of census data, one needed to determine the multiplier to be applied to the number of births at any one time in order to estimate the approximate size of a population. The specific problem Laplace solved was the size of the sample required for the probability of error to fall within given limits.

Laplace then put probabilistic investigation aside. Only twenty-five years later did he return to the subject, in the course of preparing the comprehensive *Théorie Analytique des Probabilités*. In 1810 he returned to the problem of determining the mean value from a large number of observations, which he interpreted as the problem of the probability that the mean value falls within certain limits. He proved a law of large numbers, stating that if positive and negative errors in an indefinitely large number of observations are assumed to be equally possible, then their mean result converges to a limit in a precise way. From this analysis followed the least-squares law of error. A priority dispute over the discovery of that law was even then simmering between GAUSS [VI.26] and LEGENDRE [VI.24].

In a long series of investigations brought together in *Traité de Mécanique Céleste* (1799–1825, five volumes), the two part "Mémoire sur la Théorie de Jupiter et de Saturne" (1788) demonstrates the most famous of his findings in planetary astronomy. He established that the current acceleration in orbital motion of Jupiter and the deceleration of Saturn are the reciprocal effects of their mutual gravitation, which are cyclical over many hundreds of years and not cumulative. From this and analysis of other phenomena that Laplace explored, it followed that the so-called secular inequalities of planetary motions are periodic over many centuries. Thus, they are not derogations from the law of gravity but evidence that its validity extended beyond the Sun-planet attractions that had been studied by NEWTON [VI.14]. He was never able to prove, however, that lunar acceleration is self-correcting over time.

The expansion known by Laplace's name in the theory of determinants first appears in the background of the Jupiter–Saturn memoir in an analysis of the eccentricities and inclinations of orbits, "Recherches sur le calcul intégral et sur le système du monde" (1776). Except for that, Laplace's mathematical originality is less notable in his analysis of planetary motion than in his development of the theory of probability. More in evidence in his astronomical work is his motivational drive and his power and virtuosity in calculation, which may indeed have been more important throughout his long career. Laplace was masterful in finding rapidly convergent series, in obtaining mathematical expressions incorporating terms to represent a multitude of physical phenomena, in justifying the neglect of inconvenient quantities in order to reach solutions, and in giving the widest possible generality to his conclusions.

The attraction exerted by a spheroid on an external or internal point proved to be mathematically the most fertile set of problems in Laplacian planetary astronomy. In "Théorie des attractions des sphéroïdes et de la figure des planètes" (1785) Laplace employed LEGENDRE POLYNOMIALS [III.85] in a form later called Laplace functions. He also proved a theorem that stated that all ellipsoids with the same foci for their principal sections attract a given point with a force proportional to their masses. Laplace's angles appear in his development of the equation for the attraction of a spheroid on a given point. Laplace used polar coordinates in this analysis. He transformed the equation into one in Cartesian coordinates in "Mémoire sur la théorie de l'anneau de Saturne" (1789). In 1828 George Green dubbed Poisson's application of the formula to electrostatic and magnetic forces the potential function, the term used thereafter in classical physics.

Further Reading

The memoirs cited in this article can be found in the bibliography of C. C. Gillispie's *Pierre-Simon Laplace: A Life in Exact Science* (Princeton University Press, Princeton, NJ, 1997).

For the mathematical content of Laplacian physics, see pp. 440–55 (and elsewhere) of I. Grattan-Guinness's *Convolutions in French Mathematics* (Birkhäuser, Basel, 1990, three volumes).

Charles C. Gillispie

VI.24 Adrien-Marie Legendre

b. Paris, 1752; d. Paris, 1833
Analysis; theory of attractions; geometry; number theory

Legendre passed his career in Paris and seems to have been largely of independent means. Somewhat younger than LAGRANGE [VI.22] (who was resident there from 1787) and LAPLACE [VI.23], he did not quite match their reputation, though the range of his mathematical interests was comparably wide. His professional appointments were modest; however, in 1799 he took over from Laplace as a graduation examiner at the École Polytechnique and remained there until his retirement in 1816. Additionally, in 1813 he succeeded Lagrange at the Bureau des Longitudes.

Legendre's early research concerned the shape of Earth and its external attraction to a point. Solutions of the differential equations involved led him to examine properties of the functions that are named after him; he was in rivalry with Laplace, after whom the functions were named during the nineteenth century. His other main concern in analysis, and the longest lasting, was with elliptic integrals. He wrote on them at great length up to a *Traité* of 1825–28. But in supplements of 1829–32 he acknowledged that his theory had just been eclipsed by the inverse ELLIPTIC FUNCTIONS [V.31] of JACOBI [VI.35] and ABEL [VI.33]. He also studied various other (functions defined as) integrals, including the beta and GAMMA FUNCTIONS [III.31]; solutions to differential equations; and optimization in the CALCULUS OF VARIATIONS [III.94].

Among Legendre's contributions to numerical mathematics was a beautiful theorem (found in 1789) relating spheroidal triangles (that is, triangles drawn on the surface of a spheroid) to spherical triangles, which was used in the 1790s by J. B. J. Delambre in the triangulation analysis that led to the specification of the meter. His most famous numerical result is the least-squares criterion of curve fitting, proposed in 1805 in connection with determining the orbits of comets. For him the criterion was simply one of minimization; he did not make the connections to probability theory that were soon to be effected by LAPLACE [VI.23] and GAUSS [VI.26].

Legendre's *Essai sur la Théorie des Nombres* (1798) was the first monograph on this subject. After reviewing CONTINUED FRACTIONS [III.22] and the theory of equations, he focused upon the algebraic branch, solving various Diophantine equations. Among many properties of integers, he stressed QUADRATIC RECIPROCITY [V.28], and proved various partition theorems concerning quadratic and some higher forms. Little in the book was new; while expanded editions appeared in 1808 and 1830, he had been quickly eclipsed on methods of proof by the *Disquisitiones Arithmeticae* (1801) of the young Gauss.

For educational use Legendre produced *Elements de Géométrie* (1794), an account of EUCLIDEAN GEOMETRY [I.3 §6.2] that emulated the same kind of form and organization and standards of proof of the Greek original. He also handled aspects that had lain outside Euclid's concerns, such as alternatives to the parallel postulate, related numerical issues such as approximations to the value of π, and a lengthy summary of planar and spherical trigonometry. He produced eleven further editions up to 1823 and there were further posthumous editions up until 1839 (which were followed by reprints). It was a very influential book in mathematics education.

Further Reading

de Beaumont, E. 1867. *Eloge Historique de Adrien Marie Legendre*. Paris: Gauthier-Villars.

Ivor Grattan-Guinness

VI.25   Jean-Baptiste Joseph Fourier

b. Auxerre, France, 1768; d. Paris, 1830
Analysis; equations; heat theory

Unusually for a mathematician, Fourier pursued a distinguished nonmathematical career. He was a civilian member of General Bonaparte's expedition to Egypt (1798–1801), important enough for the First Consul to make him, in 1802, the Prefect of the *département* at Grenoble, a position which he held until Emperor Napoleon's fall in the mid 1810s. Thereafter, Fourier moved to Paris, where he managed to establish himself to the extent of being appointed a *secrétaire perpétuel* of the Paris Academy of Sciences in 1822.

The prefectureship involved heavy commitments, and Fourier was also active in Egyptology, most notably discovering a teenager named Jean Champollion in Grenoble, who was later to decipher the Rosetta Stone and who helped to found the discipline. Nevertheless, between 1804 and 1815 he also created most of his scientific work. His motivation was the mathematical study of the diffusion of heat in continuous and solid bodies; his "diffusion equation" for this purpose was not only novel in itself but also marked the first large-scale mathematization of physical phenomena that lay outside mechanics. To solve this differential equation he proposed using infinite trigonometric series. These series were already known but had a low status. Fourier (re)found many properties: not only the formulas for their coefficients and some conditions for their convergence but especially their representability, namely, *how* a periodic series could represent a general function. For diffusion in a cylinder he found many properties of the Bessel function $J_0(x)$, which was then little studied.

Fourier presented his findings to the scientific class of the Institut de France in 1807. LAGRANGE [VI.22] did not like the series, while LAPLACE [VI.23] was disappointed in the physical modeling. But Laplace also gave him a clue about solutions of the diffusion equation for *infinite* bodies, which led Fourier to find, by 1811, his integral solution (including inversion) for them. His main publication was the book *Théorie Analytique de la Chaleur* (1822), which greatly influenced younger mathematicians: for example, the first satisfactory proof of the convergence of the series by DIRICHLET [VI.36] (1829) and their use in fluid dynamics by C. L. M. H. Navier (1825). Less happy was his relationship with POISSON [VI.27], who tried to rederive the entire theory following the molecularist physical principles of Laplace and the methods of solution of Lagrange, but only added a few special cases.

Fourier also worked on other topics in mathematics. As a teenager he gave the first proof of DESCARTES's [VI.11] rule of signs on the numbers of positive and negative roots of a polynomial equation. (He used an inductive proof that has now become standard.) He also found an upper bound on the number of roots within a given interval, which J. C. F. Sturm improved to an exact evaluation in 1829. At that time Fourier was trying to finish a book on equations, which appeared posthumously in 1831 thanks to Navier. The main novelty was the basic theory of LINEAR PROGRAMMING [III.84], as we now call it. Despite his prestige and advocacy, he gained few followers (Navier was one), and the theory lay dormant for over a century. Fourier also took up a few aspects of Laplace's work on mathematical statistics, examining the status of the NORMAL DISTRIBUTION [III.71 §5].

Further Reading

Fourier, J. 1888–90. *Oeuvres Complètes*, edited by G. Darboux, two volumes. Paris: Gauthier-Villars.
Grattan-Guinness, I., and J. R. Ravetz. 1972. *Joseph Fourier*. Cambridge, MA: MIT Press.

Ivor Grattan-Guinness

VI.26   Carl Friedrich Gauss

b. Brunswick, Germany, 1777; d. Göttingen, Germany, 1855
Algebra; astronomy; complex function theory including elliptic function theory; differential equations; differential geometry; land surveying; number theory; potential theory; statistics

Gauss's prodigious mathematical abilities brought him to the attention of the duke of Brunswick when he was fifteen, when the duke paid for his further education, lifting him out of near poverty. For the rest of his life Gauss felt a loyalty to the state and a strong desire to do useful work, which led him to become a professional astronomer. In 1801 he was the first person to manage to reobserve Ceres, the first asteroid to be discovered, after it had disappeared behind the Sun. Gauss produced a novel statistical analysis of the original observations, using the method of least squares, which he had invented but not published, to predict where Ceres would reappear. Gauss then assisted for many years in the analysis of the orbits of several more asteroids. He also wrote extensively on celestial mechanics and cartography, and did important work on telegraphy.

Nonetheless, it is as a pure mathematician that Gauss will always be remembered. In 1801 he published his *Disquisitiones Arithmeticae*, the book that created modern algebraic number theory. In it he gave the first rigorous proof of the law of QUADRATIC RECIPROCITY [V.28], going on to find seven more proofs over the years. Later he extended the theorem to higher powers, introducing the Gaussian integers for the purpose in 1831 (Gaussian integers are numbers of the form $m + ni$, where m and n are integers and $i = \sqrt{-1}$). He did major work on differential equations, chiefly the hypergeometric equation, which is a second-order linear differential equation depending on three parameters and having two singular points, with the property that many of the familiar functions of analysis are related to its solutions. He showed that this equation played a significant role in the new theory of ELLIPTIC FUNCTIONS [V.31], but because most of this work was unpublished it had no influence on the dramatic and rapidly advancing publications of ABEL [VI.33] and JACOBI [VI.35]. This unpublished work showed that he was the first mathematician to see the need to create a theory of complex functions of a complex variable. He also gave four proofs of THE FUNDAMENTAL THEOREM OF ALGEBRA [V.13]. By the 1820s he was persuaded that physical space might not be Euclidean, but he confined his opinion to his circle of friends, most of them astronomers and sympathetic to the idea; the much more detailed accounts of BOLYAI [VI.34] and LOBACHEVSKII [VI.31] were published independently in the early 1830s. Credit for the first detailed, mathematical descriptions of a non-Euclidean space therefore rightly attaches to Bolyai and Lobachevskii (for further discussion of this, see GEOMETRY [II.2 §7]). In 1827 Gauss wrote his *Disquisitiones Generales Circa*

Carl Friedrich Gauss

Superficies Curvas, in which the concept of intrinsic (Gaussian) curvature of a surface was put forward for the first time, thus reformulating differential geometry.

In statistics, he was one of the two or three discoverers of the NORMAL DISTRIBUTION [III.71 §5], and he was an expert in error analysis, bringing the levels of accuracy in astronomy to land surveying. In that context he invented the heliotrope, which couples a mirror to a telescope in order to transmit a precise beam of light, to improve precision measurement.

The sheer volume of Gauss's work is overwhelming. The *Werke* run to twelve volumes, and there are several books, of which the *Disquisitiones Arithmeticae* stands out.

A truly original mathematician and scientist, Gauss was otherwise a conservative in his tastes and views. His first marriage ended after only four years with the death of his wife in 1809; he then married again. A number of Gauss's descendants may now be found in the United States.

Gauss was the last great mathematician to be called the "Prince of mathematicians," and he has been admired as much for his breadth as for the depth of his insights and the fertility of his ideas. His own view of mathematics and its importance is captured both in the much-quoted remark that "mathematics is the queen of the sciences and arithmetic the queen of mathematics" (which he did say) and in the apocryphal remark that "mathematics is the queen and the servant of science."

Further Reading

Dunnington, G. W. 2003. *Gauss: Titan of Science*, new edition with additional material by J. J. Gray. Washington, DC: Mathematical Association of America.

Jeremy Gray

VI.27 Siméon-Denis Poisson

b. Pithiviers, France, 1781; d. Paris, 1840
Analysis; mechanics; mathematical physics; probability

A brilliant graduate of the École Polytechnique in 1800, Poisson was quickly appointed to the staff, and became professor and graduation examiner there until his death. He was also founder professor of mechanics at the new Paris Faculté des Sciences of the Université de France; from 1830 he was also a member of the governing Council of the Université.

Poisson's research output was dominated by his adherence to the traditions established by LAGRANGE [VI.22] and LAPLACE [VI.23]. Like Lagrange he preferred to algebraize theories, and to rely if possible upon power series and variational methods. From the mid 1810s he challenged the new theories of FOURIER [VI.25] (especially the solving of differential equations by trigonometric series and the Fourier integral) and of CAUCHY [VI.29] (the new approach to real-variable analysis using limits, and his innovation of complex-variable analysis). His overall achievements were much less significant than theirs: the main novelties were the "Poisson integral," which embedded Fourier series within a power series; and a summation formula. He also studied the general and singular solutions of differential, difference, and mixed equations.

In physics Poisson tried to justify Laplace's claim that all physical phenomena were molecular, and that the cumulative action upon a molecule of all its companions should be expressed mathematically in terms of an integral. He applied this approach to heat diffusion and to elasticity theory by the mid 1820s, but then decided that integrals should be replaced by sums; he elaborated this alternative especially in capillary theory (1831). Curiously, molecularism did not dominate his most important contributions to physics: to electrostatics (1812–14) and to magnetic bodies and the process of magnetization (1824–27). His mathematical contributions to these topics included modifying Laplace's equation to what we now call Poisson's equation, which deals with the potential at points inside a charged body or region of charge (1814); and also a divergence theorem (1826).

In mechanics, between 1808 and 1810 Poisson and Lagrange developed the brackets theory (named after them) of canonical solutions to the equations of motion. Poisson's motivation was to extend, to second-order terms in masses of the planets, Lagrange's superb attempt to prove that the planetary system was stable; in later work he examined this (first-order) problem specifically, as well as other aspects of perturbation theory. He also analyzed rotating bodies by using moving frames of reference (1839), in an analysis that was to inspire Léon Foucault to propose his famous long pendulum in 1851. His best-known publications include a substantial and wide-ranging two-volume *Traité de Mécanique* (editions in 1811 and 1833), which did not, however, have room for Louis Poinsot's beautiful recent theory (1803) of the couple in statics. In the mid 1810s, he studied deep-body fluid dynamics, in rivalry with Cauchy.

Poisson was one of the few contemporaries to take up Laplace's work in probability theory and mathematical statistics. He studied various PROBABILITY DISTRIBUTIONS [III.71]: not only the one named after him (1837, rather in passing) but also the so-called Cauchy (1824) and Rayleigh (1830) distributions. He also examined proofs of the CENTRAL LIMIT THEOREM [III.71 §5], and formulated THE LAW OF LARGE NUMBERS [III.71 §4] (his term). One of his main applications was to the old problem of determining the probability that a triad of judges would come to the correct decision in court cases (1837).

Further Reading

Grattan-Guinness, I. 1990. *Convolutions in French Mathematics 1800–1840*. Basel: Birkhäuser.
Métivier, M., P. Costabel, and P. Dugac, eds. 1981. *Siméon-Denis Poisson et la Science de son Temps*. Paris: École Polytechnique.

Ivor Grattan-Guinness

VI.28 Bernard Bolzano

b. Prague, 1781; d. Prague, 1848
Catholic priest and Professor of Theology, Prague (1805–19)

Bolzano was concerned with problems connected with finding the "correct," or the most appropriate, proofs and definitions in analysis and related areas. In 1817 he proved an early version of the intermediate value theorem for continuous functions—he was among the first to have a rigorous conception of a continuous function—and in the course of doing so proved the following important lemma. *If a property M does not apply to all values of a variable x but does apply for all values smaller than a certain u, then there is always a quantity U, which is the greatest of those of which it can be asserted that all smaller x possess the property M.* The value u in this formulation is a lower bound for the (nonempty) set of numbers with the property not-M. Bolzano's lemma is therefore equivalent to what nowadays might be called the "greatest lower bound" axiom (or, more commonly and equivalently, the "least upper bound" axiom). It is also equivalent to the Bolzano–Weierstrass theorem (that every bounded infinite set in $\mathbb{R}$, or more generally $\mathbb{R}^n$, has an accumulation point). It is likely that WEIERSTRASS [VI.44] independently rediscovered the Bolzano–Weierstrass theorem, but it is also likely that he knew, and was influenced by, Bolzano's proof technique of iterated bisection (used by Bolzano in 1817).

In the early 1830s it was widely believed that a continuous function must be differentiable except at some isolated points. But at that time Bolzano constructed a counterexample (although he did not publish it), and proved that it was such—more than thirty years before the well-known counterexample due to Weierstrass.

Bolzano had a surprising variety of insights and successful proof techniques that were well ahead of their time: notably in analysis, topology, dimension theory, and set theory.

VI.29 Augustin-Louis Cauchy

b. Paris, 1789; d. Sceaux, France, 1857
Real and complex analysis; mechanics; number theory; equations and algebra

---

Trained as a roads and bridges engineer at the École Polytechnique (hereafter, "EP") and the École des Ponts et Chaussées (1805–10), Cauchy passed his career as an academic at the EP and the Paris Faculté des Sciences of the Université de France until 1830, when he left France with the deposed royal family after the revolution of that year. He returned only in 1838, and later taught in the Paris Faculté.

Of Cauchy's many contributions to pure and applied mathematics, the best known are in mathematical analysis. In the foundations of real variables, he replaced all previous approaches to the theory with one that (in more developed forms) has now become standard: (i) lay down an explicit *theory* of limits; (ii) formulate definitions carefully, and in general terms; (iii) define the derivative of a function as the limiting value of the difference quotient, its integral as the limiting value of a sequence of partition sums, its continuity in terms of the joint passage to limits of any sequence of its argument and of its corresponding values, and the sum of a convergent infinite series as the limiting value of its partial sums. A key ingredient in all this was the idea that (iv) limits may not exist: their existence has to be justified carefully. Similarly, (v) the existence of solutions to differential equations has to be proved, not just assumed.

This approach brought a new level of rigor to analysis; for example, for the first time THE FUNDAMENTAL THEOREM OF CALCULUS [I.3 §5.5] was a genuine theorem, governed by conditions on the function. However, this emphasis on limits made the theory hard for beginners: it was not liked by staff or students at the EP, where he taught it in this form between 1816 and 1830 and published it extensively, especially in his *Cours d'Analyse* (1821) and his *Résumé* (1823) of the calculus. Its rise to standard educational practice was very gradual, both in France and elsewhere.

Another major innovation of Cauchy dates from 1814, when he began to create complex-variable analysis. Initially the integrand was a complex function but the limits of integration were real; however, from 1825 on they too became complex, and in this form he found many theorems on the residues of functions over closed domains of various shapes. Unusually for him, his progress was fitful, and he cast the theory in terms of the complex plane only in the mid 1840s. He also studied the general theory of complex functions, including their expansion in power series of various kinds.

Cauchy's main single achievement in applied mathematics lies in linear elasticity theory, where in the 1820s he used stress–strain models to analyze the behavior of various kinds of surfaces and solids; later he adapted it to study aspects of (aetherian) optics. In the 1810s he studied deep-body fluid dynamics, where he found Fourier-integral solutions. In this and several other areas he was in some competition with Fourier and, especially, Poisson, regarding both the quality of the theory and the chronology of its development.

Cauchy's other contributions lie in basic mechanics (derived from the EP teaching); singular and general solutions of differential equations; the theory of equations, especially methods that helped in the rise of group theory; algebraic number theory; perturbation theory in celestial mechanics; and an astounding paper of 1829 on quadratic forms, which could have launched the spectral theory of matrices had its author recognized its significance!

Further Reading

Belhoste, B. 1991. *Augustin-Louis Cauchy. A Biography*. New York: Springer.

Cauchy, A. L. 1882–1974. *Oeuvres Complètes*, twelve volumes in the first series and fifteen in the second. Paris: Gauthier-Villars.

Ivor Grattan-Guinness

VI.30 August Ferdinand Möbius

b. Schulpforta, Saxony, 1790; d. Leipzig, Germany, 1868
Astronomy; geometry; statics

Möbius was briefly a student of GAUSS [VI.26], and worked as an astronomer at Leipzig University for almost all of his life. His finest mathematical work was his *Der barycentrische Calcul* (1829), in which he introduced algebraic methods into the study of projective geometry. He showed in this way how points can be described by a homogeneous triple of coordinates, lines can be described by linear equations, the concept of cross-ratio can be introduced, and the duality of points and lines in the plane can be handled algebraically. He also introduced a *Möbius net*, which is the projective equivalent of squared paper in Cartesian geometry. His work is all the more remarkable because Möbius knew very little of Poncelet's radical reinvention of projective geometry only a few years before. In its turn his work was for a time overshadowed by Jakob Steiner's synthetic treatment of projective geometry of 1832, and then Plücker's two books on algebraic curves in the 1830s, but the simplicity and generality of Möbius's methods were important in establishing projective geometry as a rigorous mainstream subject.

In the 1830s Möbius developed a geometrical theory of statics and the composition of forces, and it was in this connection that he showed that whereas duality in plane geometry necessarily gives rise to a conic, duality in space need not. Möbius's study of duality in space, which pairs points with planes, led him to consider the set of all lines in space, which is a four-dimensional space. It pleased the educator Rudolf Steiner very much that the ordinary three-dimensional space may also be thought of as a four-dimensional space, because Steiner's philosophy was directed against breaking what he saw as a stranglehold of orthodox teaching.

Möbius is also remembered for the "Möbius band" (or MÖBIUS STRIP [IV.7 §2.3]), a one-sided or nonorientable surface, but the first mathematician to describe such a surface was his compatriot J. B. Listing, in July 1858 (published in 1861). Möbius discovered it only in September 1858 (publishing it in 1865). He is also one of the most important mathematicians to study inversion in circles, and his account of it in 1855 is one reason that such transformations are often called Möbius transformations.

Further Reading

Fauvel, J., R. Flood, and R. J. Wilson, eds. 1993. *Möbius and His Band*. Oxford: Oxford University Press.

Möbius, A. 1885–87. *Gesammelte Werke*, edited by R. Baltzer (except volume 4, edited by W. Scheibner and F. Klein), four volumes. Leipzig: Hirzel.

Jeremy Gray

VI.31 Nicolai Ivanovich Lobachevskii

b. Nizhni Novgorod (formerly Gorki), Russia, 1792;
d. Kazan, Russia, 1856
Non-Euclidean geometry

Lobachevskii came from a poor background, but his mother was able to have him enrolled at the local Gymnasium (or high school) on a scholarship in 1800. In 1805 the Gymnasium was made the kernel of the new University of Kazan, and in 1807 Lobachevskii began to study there. The university had just appointed Martin Bartels as Professor of Mathematics, and Bartels not only trained Lobachevskii well, but protected him from trouble with the authorities when Lobachevskii was suspected of atheism. Eventually, Lobachevskii graduated not with the ordinary degree but with a Master's qualification, and his career as a professional mathematician began.

In 1826, after a reform of the university, Lobachevskii gave a public lecture: "On the principles of geometry, with a rigorous demonstration of the theory of parallels." The manuscript of this talk is now lost, but it probably marked the start of Lobachevskii's awareness of a non-Euclidean geometry. Lobachevskii was soon elected Rector of the University of Kazan, a post he occupied with distinction for thirty years, helping to protect the university from a cholera epidemic in 1830, to rebuild it after a fire in 1841, and generally to expand its library and other facilities.

In the 1830s he also wrote his major works, on a geometry different in only one respect from Euclidean geometry. He called it imaginary geometry and it is known today as non-Euclidean geometry. In the new geometry, given a line in a plane and a point not on the line there are two lines through the point that are asymptotic to the given line (one in each direction); these two lines separate the lines through the point which meet the given line from those that do not. Lobachevskii called these the *parallels* to the given line through the given point. Starting from this definition, he gave formulas for the new trigonometry of triangles, and showed that these formulas reduce to the familiar formulas of plane Euclidean trigonometry when the triangles are very small. He extended his results to describe a geometry of three dimensions, thus making it clear that his new geometry could be a geometry of space, and attempted, inconclusively, to measure the parallax of stars in order to determine whether his imaginary geometry gave a more accurate account of space than Euclidean geometry.

He published these conclusions in lengthy papers in Russian in the *Journal of Kazan University*, but they drew only a relentlessly hostile review from Ostrogradskii, a much better known mathematician in Saint Petersburg. He published in French in a German journal in 1837, in German in a booklet of 1840, and again in French in 1855, but to little avail. GAUSS [VI.26] appreciated the booklet of 1840 and in 1842 had Lobachevskii made a corresponding member of the Göttingen Academy of Sciences, but this was to be the only acclaim Lobachevskii received in his lifetime.

Lobachevskii's final years were marked by terrible financial and mental decline. Such was the chaos of his household that Lobachevskii's biographers have been unable to establish the number of children born into it, but it may well have been fifteen or even eighteen.

Further Reading

Gray, J. J. 1989. *Ideas of Space: Euclidean, Non-Euclidean, and Relativistic*, second edn. Oxford: Oxford University Press.
Lobachetschefskij, N. I. 1899. *Zwei geometrische Abhandlungen*, translated by F. Engel. Leipzig: Teubner.
Rosenfeld, B. A. 1987. *A History of Non-Euclidean Geometry: Evolution of the Concept of a Geometric Space*. New York: Springer.

Jeremy Gray

VI.32 George Green

b. Nottingham, England, 1793; d. Nottingham, 1841
Miller; Fellow of Caius College, Cambridge (1839–41)

Green, a self-taught mathematician, went to Cambridge University at the age of forty, having already produced his most important work, the privately printed *An Essay on the Application of Mathematical Analysis to the Theories of Electricity and Magnetism* (1828). In this work, which Green opened by stressing the central role of the "potential function" (a term that he himself coined), he proved the three-dimensional version of the theorem now named after him, and introduced the concept that RIEMANN [VI.49] later called Green's function (1860). The *Essay* became widely known only after its discovery in 1845 by William Thomson (later Lord Kelvin), who was responsible for its republication in the *Journal für die reine und angewandte Mathematik* (1850–54).

Green gave his version of the theorem (in modern notation) as

$$\iiint U\Delta V\,\mathrm{d}v + \iint U\frac{\partial V}{\mathrm{d}n}\,\mathrm{d}\sigma = \iiint V\Delta U\,\mathrm{d}v + \iint V\frac{\partial U}{\mathrm{d}n}\,\mathrm{d}\sigma,$$

where U and V are two continuous functions of x, y, z whose derivatives are not infinite at any point of an arbitrary body, n is the surface normal of the body directed inward, and $\mathrm{d}\sigma$ is a surface element. The result today known as Green's theorem, which is the planar version of the above, was first published by CAUCHY [VI.29] in 1846, and it can be given (in modern notation) as follows. Let R be a closed plane region with a piecewise-smooth boundary curve C with positive orientation. Let $P(x, y)$ and $Q(x, y)$, having continuous partial derivatives, be defined on an open region containing R. We then have

$$\int_C (P\,\mathrm{d}x + Q\,\mathrm{d}y) = \iint_R \left(\frac{\partial Q}{\partial x} - \frac{\partial P}{\partial y}\right)\mathrm{d}x\,\mathrm{d}y.$$

More original than his theorem, however, was the powerful technique Green developed to solve certain second-order differential equations. In essence, Green sought a "potential function" and formulated the conditions it needed to satisfy. His great insight was to recognize that the central issue in potential theory was to relate properties inside volumes to properties on their surfaces. Green's functions are extensively used today in the solution of inhomogeneous differential equations with boundary conditions and in the solution of partial differential equations.

VI.33 Niels Henrik Abel

b. Finnöy, Norway, 1802; d. Froland, Norway, 1829
Theory of equations; analysis; elliptic functions; Abelian integrals

Abel's life was short and penurious, but successful, and he received recognition in his lifetime. His father—a minister of the church in Norway, but also at one time a government minister—overreached himself and when he died he left the family in straitened circumstances. Abel's exceptional intellectual talents were recognized at school, and funds were raised to enable him to complete his education and, in particular, to study mathematics. At age twenty-two, he was awarded a scholarship to make a two-year tour of Europe, during which he studied in Berlin and Paris. In Berlin he met and was befriended by Auguste Crelle, the engineer who had just founded the *Journal für die reine und angewandte Mathematik* (otherwise known as *Crelle's Journal*). Almost all of Abel's mathematical work was published in the first four volumes of the journal. From 1826 until his death in 1829 Abel eked out a poor existence, earning a little by teaching, but using what few resources he had to support his mother and his younger brother. He died of consumption at the age of twenty-seven within a couple of days of the news reaching Norway that he had been appointed to an established post in Berlin.

Abel's main mathematical contributions lie in three distinct areas. The first of these was the theory of equations. Here he was influenced by ideas published by LAGRANGE [VI.22] in 1770 and CAUCHY [VI.29] in 1815 about the form of functions of the roots of an equation, and what happens to such functions when the roots are permuted. Lagrange had hinted at the possibility that quintic equations might perhaps not be soluble in classical terms and Paolo Ruffini had expended much effort between 1799 and 1814 trying to prove this, though he had not managed to persuade his contemporaries. Abel's first success was to give an acceptable proof of the fact that, for polynomial equations of degree 5, there is no formula in the coefficients involving the usual operations of arithmetic together with extraction of roots that will always yield a solution. This first appeared in 1824 in a short pamphlet written in French and published privately in Christiania (Oslo). Once Abel reached Berlin, however, Crelle translated it into German and published it in the first volume of his journal; he also published a fuller, more detailed account, covering polynomials of any degree greater than 4, in 1826.

Abel returned to equations a few years later, publishing a long paper in 1829 about a class of equations satisfying two special conditions. His first requirement is that every root of the equation can be expressed as a function of every other root, the second that these functions commute (in modern terms, the GALOIS GROUP [V.21] of the equation is commutative). He proved various theorems about such equations, the most striking being that they are soluble by radicals. This represented an extensive generalization of the ideas described by GAUSS [VI.26] in the seventh part of *Disquisitiones Arithmeticae*, where the special case of cyclotomic equations (which satisfy both of these conditions) is treated systematically. It was in honour of this work that, later, the adjective "Abelian" was applied to groups that are commutative. It is important to appreciate, however, that Abel reached his results in the theory of equations without any appeal to groups, which at that time were not yet known.

He also made major contributions to the theory of convergence. Although there had been over a century of critical thinking devoted to foundations of the calculus, modern ideas of rigour were only just emerging in the writings of BOLZANO [VI.28], Cauchy, and others. Convergence had received some attention in Cauchy's lectures of 1820–21, but series in general, and power series in particular, were still far from well understood. Among other contributions, Abel offered a proper proof of the binomial theorem for exponents other than positive integers, and the insight about the continuity of a function defined by a power series as its argument goes to the circle of convergence that is now known as Abel's limit theorem.

Perhaps his greatest discoveries, however, were in the area where analysis and algebraic geometry come together. To summarize his legacy in this area in just a few words: first, a new and productive approach to the theory of ELLIPTIC FUNCTIONS [V.31]; and, second, a vast generalization of elliptic functions to what are now called Abelian functions and Abelian integrals. In this area Abel competed for priority with JACOBI [VI.35]. Most (though by no means all) of his work was written in two memoirs. One was published in two parts, "Recherches sur les fonctions elliptiques" and "Précis d'une théorie des fonctions elliptiques," coming to well over two hundred pages in *Crelle's Journal* in 1828 and 1829. The other, entitled "Mémoire sur une propriété générale d'une classe très étendue de fonctions transcendantes," was submitted to the Paris Academy of Sciences in October 1826. There it lay on Cauchy's desk, unread until after Abel's death. It was published by the Paris Academy in 1841. The manuscript itself, however, was stolen by G. Libri, lost, and rediscovered in parts between 1952 and 2000 by Viggo Brun and Andrea del Centina.

In June 1830 the Paris Academy awarded its Grand Prix de Mathématiques jointly to Abel (posthumously) and Jacobi for their work on elliptic functions.

Further Reading

Del Centina, A. 2006. Abel's surviving manuscripts including one recently found in London. *Historia Mathematica* 33:224–33.

Holmboe, B., ed. 1839. *Œuvres Complètes de Niels Henrik Abel*, two volumes. (Second edn.: 1881, edited by L. Sylow and S. Lie. Christiania: Grøndahl & Søn.)

Ore, O. 1957. *Niels Henrik Abel: Mathematician Extraordinary*. Minneapolis, MN: University of Minnesota Press. (Reprinted, 1974. New York: Chelsea.)

Stubhaug, A. 1996. *Et Foranskutt Lyn: Niels Henrik Abel Og Hans Tid*. Oslo: Aschehoug. (English translation: 2000, *Niels Henrik Abel and His Times: Called Too Soon by Flames Afar*, translated by R. H. Daly. New York: Springer.)

Peter M. Neumann

VI.34   János Bolyai

b. Klausenburg, Transylvania, Hungary (now Cluj, Romania), 1802;
d. Marosvásárhely, Hungary (now Tirgu-Mures, Romania), 1860
Non-Euclidean geometry

János Bolyai's father Farkas Bolyai taught him mathematics at home, using the first six books of EUCLID's [VI.2] *Elements* and EULER's [VI.19] *Algebra*. Between 1818 and 1823 János studied at the Royal Engineering Academy in Vienna, and then served as an engineer in the Austrian Army for ten years, before retiring on a pension as a semi-invalid. Probably inspired by his father's attempts to prove the parallel postulate, a key assumption in Euclidean geometry, but very much against the advice of his father, János also attempted to prove it. But in 1820 he switched direction and attempted to show that there could be a geometry independent of the parallel postulate. By 1823 he believed he had succeeded, and after much subsequent discussion father and son agreed to publish the son's ideas as a twenty-eight-page appendix to his father's two-volume work on geometry in 1832.

In this appendix, Bolyai started from a new definition of parallels, according to which, given a line in a plane and a point not on the line, there are many lines through the point that do not meet the given line. Of these lines, there are two that are asymptotic to the given line (one in each direction), and Bolyai called these the parallels to the given line through the given point. He went on to derive many results that follow from this assumption in the geometry of two and three dimensions, and gave formulas for the new trigonometry of triangles. He showed that these formulas reduce to the familiar formulas of plane Euclidean geometry when the triangles are very small. He also found a surface in his three-dimensional geometry in which geometry is Euclidean. He concluded that there were logically two geometries and that it remained undecided which one corresponded to reality. He also showed that in his new geometry it was possible to construct a square equal in area to a given circle, thus accomplishing a feat that was widely (and, as was later shown, correctly) believed to be impossible in Euclidean geometry.

A copy of the book was sent to GAUSS [VI.26], who eventually replied on March 6, 1832, that he could not praise the work, for "to praise it, would be to praise myself," going on to claim that the methods and results in the appendix agreed with his own work over the previous thirty-five years, although he was "very glad that it was just the son of my old friend, who takes the precedence of me in such a remarkable manner." This endorsement of the validity of János's ideas pleased the father but infuriated the son, and soured relations between father and son for several years. They did eventually resume an uncomfortable relationship, which persisted until Farkas's death in 1856.

János Bolyai published virtually nothing else, and his discovery was not appreciated in his lifetime. Indeed, it is unclear that anyone but Gauss ever read it, but specific comments about it that Gauss left behind led mathematicians back to it, and it was translated into French by Hoüel in 1867 and into English in 1896 (reprinted in 1912 and 2004).

Further Reading

Gray, J. J. 2004. *János Bolyai, Non-Euclidean Geometry and the Nature of Space*. Cambridge, MA: Burndy Library, MIT Press.

Jeremy Gray

VI.35   Carl Gustav Jacob Jacobi

b. Potsdam, Germany, 1804; d. Berlin, 1851
Theory of functions; number theory; algebra; differential equations; calculus of variations; analytical mechanics; perturbation theory; history of mathematics

Jacobi grew up as Jacques Simon Jacobi in a wealthy and well-educated Jewish family. He was baptized during his first year at the University of Berlin in 1821, probably in order to make it possible for him to follow an academic career at a time when Jews were ineligible for academic positions. Jacobi studied classics under the famous philologist Boeckh and philosophy under Hegel. Owing to the mediocrity of the mathematics staff in Berlin at that time, he was self-taught in the discipline, which soon became his favorite. He read EULER [VI.19], LAGRANGE [VI.22], LAPLACE [VI.23], GAUSS [VI.26], and, last but not least, Greek mathematicians like Pappus and Diophantus. In 1825, Jacobi was awarded his doctorate for a thesis, written in Latin, on the theory of functions. The subsequent *disputatio* (discussion) included critical comments both on Lagrange's theory of functions and on his analytical mechanics. The following year Jacobi went to the University of Königsberg, where (in 1829) he got a full professorship. In 1834, he and the physicist F. E. Neumann founded the "Königsberg mathematical physics seminar," which, because of the close connection between research and teaching that it fostered, soon led to Königsberg becoming the most successful and influential educational institution for theoretical physics and mathematics in the German-speaking part of the scientific world. By 1844, when Jacobi left Königsberg because of poor health and in order to become a member of the Berlin Academy of Sciences, he was recognized as Germany's most important mathematician after Gauss. After seven more fruitful years of research in Berlin he died unexpectedly from smallpox.

Throughout his life Jacobi was an advocate of pure mathematics, conceiving mathematical thinking as a means of developing the human intellect and, indeed, of advancing humanity itself. He published his first paper in 1827: it was devoted to number theory (cubic residues) and was influenced by Gauss's *Disquisitiones Arithmeticae*. Further investigations were devoted to higher residues, the division of the circle, quadratic forms, and related subjects. Many of Jacobi's results in number theory were published in the book *Canon Arithmeticus* (1839). The extension of the concept of divisibility to algebraic numbers by Jacobi and Gauss paved the way for the later algebraic theory of numbers (by KUMMER [VI.40] and others).

Jacobi's "most original achievements" (in the words of KLEIN [VI.57]) were his contributions to the theory of ELLIPTIC FUNCTIONS [V.31], which developed in competition with ABEL [VI.33] between 1827 and 1829. Starting with LEGENDRE's [VI.24] work, Jacobi's approach was analytical and focused on the transformation of elliptic functions, their properties (like double periodicity), and the introduction of the inverse function. Jacobi's research on elliptic functions culminated in the book *Fundamenta Nova Theoriae Functionum Ellipticarum* (1829). Together with Abel he should be viewed as one of the founders of the theory of complex func-

tions, which emerged in the second half of the century. In particular, his application of research on elliptic functions to Diophantine equations became important for the development of analytic number theory. Jacobi's contributions to algebra include investigations into the theory of determinants (the "Jacobian" functional determinant) and their relation to inverse functions, into quadratic forms ("Sylvester's law of inertia"), and into the transformation of multiple integrals.

Even Jacobi's work in mathematical physics bears the stamp of "pure mathematics": following the analytical tradition of Euler and Lagrange, he presented the foundations of mechanics in an abstract and formal manner, paying special attention to the relation between CONSERVATION LAWS [IV.12 §4.1] and symmetries of space and to the unifying role of variational principles. Jacobi's achievements in this area, which he developed in close relation to the theory of differential equations and the CALCULUS OF VARIATIONS [III.94], include what is now called the "Jacobi–Poisson theorem," the "principle of the last multiplier," a theory for integrating HAMILTON's [VI.37] CANONICAL EQUATIONS OF MOTION [IV.16 §2.1.3] by transformation ("Hamilton–Jacobi theory"), and a time-independent formulation of the principle of least action ("Jacobi's principle"). His approach to these areas and the results he obtained are documented in two comprehensive books based on his lectures: *Vorlesungen über Dynamik* (1866) and *Vorlesungen über Analytische Mechanik* (not published until 1996). The former had considerable impact on the development of German mathematical physics in the last third of the nineteenth century. The latter reveals Jacobi's criticism of the traditional understanding of mechanical principles (as laws that are firmly based on empirical observation or a priori reasoning) and shows strong parallels with the "conventionalist" viewpoint, which did not become popular in science and philosophy until half a century later, when it numbered H. Hertz and POINCARÉ [VI.61] among its adherents.

Jacobi not only promoted new mathematical developments, but also studied the history of mathematics: he worked on ancient number theory, was the advisor for the historical parts of A. von Humboldt's great *Kosmos* (1845–62), and developed detailed plans for the publication of Euler's works.

Further Reading

Koenigsberger, L. 1904. *Carl Gustav Jacob Jacobi*. Festschrift zur Feier des hundertsten Wiederkehr seines Geburtstages. Leipzig: Teubner.

Helmut Pulte

VI.36   Peter Gustav Lejeune Dirichlet

b. Düren, French Empire (now Germany), 1805;
d. Göttingen, Germany, 1859
*Number theory; analysis; mathematical physics; hydrodynamics;
probability theory*

The low level of mathematics education at German universities prompted Dirichlet to study in Paris, where he came into contact with the leading French mathematicians Lacroix, POISSON [VI.27], and FOURIER [VI.25], who particularly attracted him. In 1827 he took up a position at the University of Breslau. The following year he moved to Berlin, where he was appointed as a professor at the military academy and where he was also allowed to teach at the university. In 1831 he was made a professor at the university and from then on held positions at both institutions until 1855, when he was appointed as the successor to GAUSS [VI.26] at the University of Göttingen.

Dirichlet's primary interest was in number theory. His guiding star was Gauss's pioneering *Disquisitiones Arithmeticae* (1801)—the work that made number theory into a mathematical discipline—which he studied throughout his career. Dirichlet was not only the first mathematician to completely understand this work but he also became its interpreter, picking up its problems and improving its proofs, as well as developing its ideas.

With his very first publication, which appeared in 1825, Dirichlet came to international prominence. The paper, which dealt with Diophantine equations of the form $x^5 + y^5 = Az^5$, yielded substantial results for the verification of FERMAT'S LAST THEOREM [V.10] for the case $n = 5$ (these results were used by LEGENDRE [VI.24] for a complete proof for that case some weeks later). In a paper published in 1837 Dirichlet came up with the new and revolutionary idea of applying analytical methods to number theory. He introduced expressions that are now known as *Dirichlet L-series*. These are infinite series of the form

$$L(s, \chi) = \sum_{n=1}^{\infty} \frac{\chi(n)}{n^s}$$

where $\chi(n)$ is a *Dirichlet character* modulo k: that is, a complex-valued function on the integers that is totally multiplicative, in the sense that $\chi(ab) = \chi(a)\chi(b)$ for all a and b, and χ is periodic with period k and not identically zero. Using these L-series, Dirichlet showed that every arithmetic progression $\{an + b : n = 0, 1, \dots\}$, where a and b are relatively prime, contains infinitely many prime numbers. In two subsequent papers, published in 1838 and 1839, he used his new methods, among other things, to determine the formula of the class number of binary quadratic forms: that is, the number of proper classes of forms of given determinant. It is often said that these three papers mark the start of ANALYTIC NUMBER THEORY [IV.2].

Dirichlet also made important contributions to algebraic number theory, culminating in his UNIT THEOREM [III.63] for the Abelian group of units in an algebraic number field. These contributions, together with numerous others due to him (e.g., his *Schubfachprinzip*, or box principle; work on the law of biquadratic reciprocity; and results concerning Gaussian sums), were brought together in his influential *Vorlesungen über Zahlentheorie* (lectures on number theory), published by his former student DEDEKIND [VI.50] in 1863.

Inspired by his close contact with FOURIER [VI.25] during his student days in Paris, Dirichlet's other main interests were in analysis and mathematical physics, and in the connections between them. In a groundbreaking paper of 1829, Dirichlet not only gave the first strict proof of the convergence of a Fourier series under given conditions, but he also used new methods and concepts (e.g., his insight into the importance of conditional convergence of series; his Dirichlet function influencing the development of the concept of a function) that became classic and that served as a basis for countless nineteenth-century investigations on analysis. He also occupied himself with the determination of multiple integrals as well as with the expansion of a function into spherical functions (*Kugelfunktionen*) and applied these results to problems in mathematical physics. His main contributions to mathematical physics include papers on the theory of heat, hydrodynamics, the gravitational attraction of an ellipsoid, the n-body problem, and potential theory. The first boundary-value problem (the "Dirichlet problem" of finding the solution of an elliptic partial differential equation in the interior of a given region that takes prescribed values on the boundary of the region) had already been handled by Fourier and others, but Dirichlet proved the uniqueness of the solution, while the DIRICHLET PRINCIPLE [IV.12 §3.5] (a method for solving boundary problems for ELLIPTIC PARTIAL DIFFERENTIAL EQUATIONS [IV.12 §2.5] by reducing them to VARIATIONAL PROBLEMS [III.94]) was introduced by him in lectures on potential theory, enhancing a method introduced by Gauss. Connected with Dirichlet's work on analysis was his contribution to probability and error theory, in particular his development of new methods for probabilistic limit theorems.

Dirichlet also influenced the further development of mathematics by his mathematical style, by the exactness and elegance of his proofs, and by his teaching. Together with his friend JACOBI [VI.35], he ushered in a new epoch of mathematical teaching at German universities by introducing lectures and seminars on the most recent research, and with him began the golden age of mathematics in Berlin. Although Dirichlet did not found his own mathematical school, his influence can be found in the work of Dedekind, Eisenstein, KRONECKER [VI.48], and RIEMANN [VI.49], among others.

Further Reading

Butzer, P.-L., M. Jansen, and H. Zilles. 1984. Zum bevorstehenden 125. Todestag des Mathematikers Johann Peter Gustav Lejeune Dirichlet (1805–1859), Mitbegründer der mathematischen Physik im deutschsprachigen Raum. *Sudhoffs Archiv* 68:1–20.

Kronecker, L., and L. Fuchs, eds. 1889–97. *G. Lejeune Dirichlet's Werke*, two volumes. Berlin: Reimer.

Ulf Hashagen

VI.37 William Rowan Hamilton

b. Dublin, 1805; d. Dublin, 1865
Calculus of variations; optics; dynamics; algebra; geometry

Hamilton was educated at Trinity College, Dublin. Shortly before graduating in 1827, he was appointed Professor of Astronomy and Royal Astronomer of Ireland, a post which he held for the remainder of his life.

His first paper, "Theory of systems of rays: part first" (1828), was written while he was still an undergraduate. In it he developed new methods for the study of foci and caustics produced by the reflection of light from curved surfaces. Hamilton developed his approach to optics over the following five years, publishing three substantial supplements to his original paper. He showed that the properties of an optical system are completely determined by a certain "characteristic function" that is a function of the initial and final coordinates of a ray of light and which measures the time of passage of light through the system. In 1832 he predicted that light falling at a certain angle on a biaxial crystal would be refracted to form a hollow cone of emergent rays. This prediction was verified by his friend and colleague Humphrey Lloyd.

Hamilton adapted his optical methods to the study of dynamics. In a paper "On a general method in dynamics" (1834), he showed that the dynamics of a system of attracting and repelling point particles is completely determined by a certain characteristic function, which satisfies a differential equation, today referred to as the HAMILTON–JACOBI EQUATION [IV.12 §2.1]. In a subsequent paper, "Second essay on a general method in dynamics" (1835), he introduced the *principal function* of a dynamical system, presented the equations of motion of such a system in HAMILTONIAN FORM [IV.16 §2.1.3], and adapted methods of perturbation theory to this setting.

Hamilton discovered the system of QUATERNIONS [III.76] in 1843. The fundamental equations of this system occurred to him in a flash of insight as he was walking along the bank of the Royal Canal, near Dublin, on October 16 of that year. Most of his subsequent mathematical work involved quaternions. It is not difficult to translate much of this work into the language of modern vector analysis, and indeed many of the basic concepts and results of vector algebra and analysis emerged from Hamilton's work on quaternions. Hamilton applied quaternion methods to the study of dynamics in a series of short papers published in the three years immediately following his discovery of quaternions. He also investigated a number of algebraic systems related to quaternions. However, most of his work with quaternions was concerned with their application to the study of geometrical problems, and, in particular, to the study of surfaces of the second order and (especially in the final years of his life) to the study of the differential geometry of curves and surfaces. Much of this research is to be found in his two books *Lectures on Quaternions* (1853) and *Elements of Quaternions* (1866, published posthumously).

Further Reading

Hankins, T. L. 1980. *Sir William Rowan Hamilton*. Baltimore, MD: Johns Hopkins University Press.

David Wilkins

VI.38 Augustus De Morgan

b. Madura (now Madurai), India, 1806; d. London, 1871
*Professor of Mathematics, University College London
(1828–31, 1836–66); first president of the
London Mathematical Society (1865–66)*

De Morgan, a prolific author in many fields of mathematics and its history, made important and original contributions to the development of mathematical logic. He is particularly remembered for what we now call *de Morgan's laws*, which he first published in 1858 in a paper in the *Transactions of the Cambridge Philosophical Society*. The "laws" can be stated (using the notation of sets) as follows. If A and B are subsets of a set X, then $(A \cap B)^c = A^c \cup B^c$ and $(A \cup B)^c = A^c \cap B^c$, where "$\cup$" represents union, "$\cap$" represents intersection, and a superscript "c" denotes the complement with respect to X.

VI.39 Joseph Liouville

b. Saint Omer, France, 1809; d. Paris, 1882
Differentiation of arbitrary order; integration in closed form; Sturm–Liouville theory; potential theory; mechanics; differential geometry; doubly periodic functions; transcendental numbers; quadratic forms

Liouville was the leading French mathematician in the generation between CAUCHY [VI.29] and HERMITE [VI.47]. He taught analysis and mechanics at his alma mater, the École Polytechnique, until 1851, when he became professor at the Collège de France. Moreover, he was professor at the Sorbonne from 1857 and member of the Paris Academy of Sciences and the Bureau des Longitudes. In 1836 he founded the *Journal de Mathématiques Pures et Appliquées*, which exists to this day.

His wide-ranging research was often inspired by physics. For example, his early theory of differential operators of the form $(d/dx)^k$, where k is an arbitrary complex number, had its origin in Ampère's electrodynamics. Similarly, Sturm–Liouville theory, which he developed in around 1836 with his friend C. F. Sturm, was inspired by the theory of heat conduction. Sturm–Liouville theory deals with a linear self-adjoint second-order differential equation involving a parameter that must be chosen so that there exist nontrivial solutions (eigenfunctions) that satisfy given boundary-value conditions. Liouville's main contribution to this theory was a proof that an "arbitrary" function has a convergent "Fourier expansion" in terms of eigenfunctions. Sturm–Liouville theory was a major step toward a more qualitative theory of differential equations, and the first work on spectral theory of a general class of differential operators.

In 1844 Liouville gave the first proof that there exist TRANSCENDENTAL NUMBERS [III.41], of which a well-known example is $\sum_{n=1}^{\infty} 10^{-n!}$. In a similar vein, in the 1830s he had already shown that there are elementary functions such as e^t/t whose integrals are not expressible in elementary (or closed) form, i.e., in terms of algebraic functions, exponentials, and logarithms. In particular he proved that the elliptic integrals are nonelementary.

Around 1844 Liouville suggested an entirely new approach to ELLIPTIC FUNCTIONS [V.31] (inverses of elliptic integrals), based on a systematic investigation of doubly periodic complex functions and in particular the observation that such a function must have singularities if it is not constant. When Cauchy heard of this theorem he immediately generalized it to the statement that any bounded complex analytic function must be a constant. Today this is called Liouville's theorem.

In mechanics, Liouville's name is connected with the theorem stating that the volume in phase space is constant when a mechanical system moves according to HAMILTON'S EQUATIONS [III.88 §2.1]. In fact, Liouville proved the constancy of a certain DETERMINANT [III.15] formed from the solutions of a general class of differential equations. It was JACOBI [VI.35] who pointed out that the theorem applied to Hamilton's equations, and Boltzmann who interpreted the determinant as the volume in phase space, and emphasized its importance in statistical mechanics.

Liouville made many other important contributions to mechanics and to potential theory. For example, Jacobi had postulated that when the angular momentum of a fluid planet revolving around an axis is high enough, there are two shapes that are in equilibrium in their rotating frames of reference: an ellipsoid of revolution and an ellipsoid with three different axes. Liouville showed that Jacobi was right, and moreover proved the surprising result that only the latter figure is in stable equilibrium. Liouville published only the result, leaving the verification to Lyapunov and POINCARÉ [VI.61] (at least if the angular momentum is not too large).

The first mathematician to recognize the significance of GALOIS's [VI.41] theory of SOLVABILITY OF EQUATIONS [V.20], Liouville did a great service to algebra when he published some of Galois's most important papers in his journal.

Further Reading

Lützen, J. 1990. *Joseph Liouville 1809–1882: Master of Pure and Applied Mathematics*. Studies in the History of Mathematics and Physical Sciences, volume 15. New York: Springer.

Jesper Lützen

VI.40 Ernst Eduard Kummer

b. Sorau (now Zary, Poland), 1810; d. Berlin, 1893
Gymnasium teacher, Liegnitz (now Legnica, Poland) 1832–42;
Professor of Mathematics: Breslau (now Wroclaw, Poland) 1842–55,
Berlin 1855–82

Kummer's early research was in function theory, in which he made an important contribution to the theory of the (generalized) hypergeometric series (a power series in which the ratios of successive coefficients are rational functions). Surpassing earlier work of GAUSS [VI.26], Kummer not only provided a systematic account of solutions to the hypergeometric differential equation

$$x(x-1)\frac{d^2 y}{dx^2} + (c - (a+b+1)x)\frac{dy}{dx} - aby = 0,$$

where a, b, and c are constants, but also made the connection between the hypergeometric functions and newer functions in analysis, such as the ELLIPTIC FUNCTIONS [V.31].

After moving to Breslau, Kummer started doing research in number theory, the field in which he achieved his greatest success: the creation of the theory of "ideal prime factors" (1845–47). Although Kummer's theory is often described as an early contribution to the theory of IDEALS [III.81 §2], his algorithmic approach was very different from that later followed by DEDEKIND [VI.50]. Kummer's original goal had been to generalize the LAW OF QUADRATIC RECIPROCITY [V.28] to higher powers, and he succeeded in this in 1859. An additional consequence of this research was that he managed to prove FERMAT'S LAST THEOREM [V.10] for all prime exponents (and hence, since it was known for fourth powers, all exponents) less than 100.

In the third phase of his career Kummer turned to algebraic geometry. Continuing the work of HAMILTON [VI.37] and JACOBI [VI.35] on ray systems and geometric optics, he was led to the discovery of the quartic surface with sixteen nodal points, which is now named after him.

VI.41 Évariste Galois

b. Bourg-la-Reine, France, 1811; d. Paris, 1832
Theory of equations; theory of groups; Galois theory; finite fields

Galois studied at home until he was eleven years old, then entered the Collège Louis-le-Grand in Paris, where he stayed for six years. He had, and gave his teachers, a difficult time there, but excelled in mathematics, in which he read advanced work of LAGRANGE [VI.22],

GAUSS [VI.26], and CAUCHY [VI.29] alongside standard texts of the time. He attempted the entrance examination for the École Polytechnique prematurely in June 1828, but failed. In July 1829, after his father's suicide, Galois was again rejected by the École Polytechnique. He entered the École Préparatoire (later known as the École Normale Supérieure) in October 1829 but was expelled in December 1830 for unacceptable behaviour arising from political disagreements with the authorities. Arrested on Bastille Day (14 July) 1831, he spent the next eight months in prison for flouting authority again. He emerged at the end of April 1832 but somehow got himself challenged to a duel. On 29 May he edited his manuscripts and wrote a summary of his discoveries in a letter to his friend Auguste Chevalier. The duel took place the next morning and he died on 31 May 1832. Much has been written about him. But a man who dies so young leaves little real evidence for historians to work with, however rich his story, and most of his biographers have allowed romantic invention to colour their accounts of his life.

There are four main papers in Galois's mathematical works (and a number of smaller and less important items). The first to be published was "Sur la théorie des nombres," which appeared in April 1830 and contains the theory of Galois fields. These are analogues of the complex numbers obtained by adjoining to the integers modulo a prime number p a root of an irreducible polynomial congruence modulo p. The paper contains most of the basic features of what later became the theory of finite fields.

In the letter to Chevalier written on the eve of the duel, Galois mentions three memoirs. The first, now known as the *Premier Mémoire*, is a manuscript entitled "Sur les conditions de résolubilité des équations par radicaux." Galois submitted work on the theory of equations to the Paris Academy on 25 May and 1 June 1829, but this is now lost and it seems quite possible that Galois withdrew it on the advice of Cauchy (to whom it had been given to referee) in January 1830. In February 1830 he resubmitted his work in competition for the Grand Prix de Mathématiques, but his manuscript was unfortunately and mysteriously lost on the death of FOURIER [VI.25] (and the prize was awarded jointly to ABEL [VI.33], posthumously, and JACOBI [VI.35]). Encouraged to do so by POISSON [VI.27] he submitted his ideas to the Academy for a third time in January 1831. It is this third submission (which was read by the Academy referees, Poisson and Lacroix, and rejected on 4 July 1831) that survives as the manuscript of the *Premier Mémoire*. This is the remarkable work in which he introduced what is now called the Galois group of an equation and showed how solubility of the equation in terms of radicals could be precisely characterized by a property of the group. It was the *Premier Mémoire* which turned the theory of equations into what is now called GALOIS THEORY [V.21].

The *Second Mémoire* also exists. Galois never completed it, however, nor is it all correct. Nevertheless, it is an exciting document that focuses on aspects of what is now recognized as the theory of groups. Its main theorem is (in group-theoretic language) that every primitive soluble permutation group has degree a power of a prime number and may be represented as a group of affine transformations over the prime field $\mathbb{F}_p$. It also contains an incomplete study of two-dimensional linear groups over $\mathbb{F}_p$. The *Troisième Mémoire*, which he described as being on the theory of integrals and ELLIPTIC FUNCTIONS [V.31], has never been found.

Galois's main work—comprising the paper "Sur la théorie des nombres," the *Premier Mémoire*, the *Second Mémoire*, and the letter to Chevalier—was finally published by LIOUVILLE [VI.39] in 1846. A critical edition by Bourgne and Azra, including every known fragment of Galois's writing, was published in 1962.

Galois's legacy is enormous. His ideas led directly to "abstract algebra" (see [II.3 §6]): when the abstract notion of field developed later in the nineteenth century, it turned out that most of the theory of finite fields had already been anticipated in that first paper; Galois theory developed directly out of the material in the *Premier Mémoire*; and the theory of groups developed from the ideas in the *Premier Mémoire* and the *Second Mémoire* together with a series of papers published by Cauchy in 1845.

Further Reading

Bourgne, R., and J.-P. Azra, eds. 1962. *Écrits et Mémoires Mathématiques d'Évariste Galois*. Paris: Gauthiers-Villars.
Edwards, H. M. 1984. *Galois Theory*. New York: Springer.
Taton, R. 1983. Évariste Galois and his contemporaries. *Bulletin of the London Mathematical Society* 15:107–18.
Toti Rigatelli, L. 1996. *Évariste Galois 1811–1832*, translated from the Italian by J. Denton. Basel: Birkhäuser.

Peter M. Neumann

VI.42   James Joseph Sylvester

b. London, 1814; d. London, 1897
Algebra

As a Jew, Sylvester could neither take the degree he earned at St John's College, Cambridge, in 1837 nor compete for positions at England's Anglican universities. This effectively forced him down a convoluted path toward his personal goal of a career as a research mathematician. He worked as an actuary in London in the 1840s and 1850s before qualifying as a lawyer by passing the English Bar. He was unemployed for some six years in the 1870s, but held professorships at various times, both of natural philosophy and of mathematics, in England and in the United States. Most notably, Sylvester served as the first Professor of Mathematics at Johns Hopkins University in Baltimore, Maryland, from 1876 to 1883 and, thanks to an 1871 law that finally made it possible for non-Anglicans to hold professorships at Oxbridge, was eligible for and won the appointment as Oxford's Savilian Professor of Geometry in 1883. He held the Oxford chair until ill health forced his retirement in 1894. The program Sylvester set up at Johns Hopkins established his pivotal place in the history of American research-level mathematics, while his mathematical accomplishments had garnered him an international reputation as early as the 1860s.

Sylvester entered the research arena in the late 1830s with work on the problem of determining when two polynomial equations have a common root. This naturally led not only to questions in the theory of determinants but also to an explicit, pioneering, and self-consciously algebraic analysis of the intermediate expressions that arise in Charles François Sturm's algorithm for determining the number of real roots of a polynomial equation that lie between two given real numbers (1839, 1840). Sylvester followed this up with what he called the dialytic method of elimination: a new criterion in terms of DETERMINANTS [III.15] for detecting whether two polynomial equations have a common root (1841).

His next major research push came in the 1850s when, together with CAYLEY [VI.46], he formulated a theory of invariants. This involved an associated and slightly more general theory of "covariants." More concretely, given a binary form of a particular degree, Sylvester and Cayley devised techniques both for explicitly finding invariants and covariants of that form and for determining algebraic relations, or "syzygies," between them. Sylvester tackled these questions in two important papers: "On the principles of the calculus of forms" (1852) and "On a theory of the syzygetic relations of two rational integral functions" (1853). In the latter, he proved, among other results, Sylvester's law of inertia: if $Q(x_1, \ldots, x_n)$ is a real QUADRATIC FORM [III.73] of rank r, then there exists a (real) nonsingular linear transformation that takes Q to $x_1^2 + \cdots + x_p^2 - x_{p+1}^2 - \cdots - x_r^2$, where p is uniquely determined.

Sylvester surprised the mathematical world in 1864 and 1865 with the first proof of NEWTON's [VI.14] rule (Newton had only stated it) for determining bounds on the number of positive and negative roots of a polynomial equation. However, he then entered a fallow period that ended only with his move to Baltimore. While there, he returned to invariant theory, and specifically to the problem of inductively determining, for binary forms first of degree 2 then of degree 3 then of degree 4, etc., the number of covariants in a minimum generating set associated with the form. In 1868, Paul Gordan had proved that this number is always finite and, in so doing, had proved wrong an earlier result of Cayley, who claimed to have shown that a minimum generating set of covariants for the binary quintic form (that is, the binary form of degree 5) was infinite. By 1879, Sylvester had explicitly calculated minimum generating sets of covariants associated with binary forms of degrees two through ten. He had also succeeded in recognizing and filling (1878) a critical gap in the proof that Cayley had given of a theorem on the maximal number of linearly independent covariants associated with a binary form of any given degree.

Sylvester was the founding editor of the *American Journal of Mathematics*, and indeed much of this invariant-theoretic work, as well as results on partitions (1882), on rational points on a cubic curve (1879–80), and on matrix algebras (1884), appeared there.

Further Reading

Parshall, K. H. 1998. *James Joseph Sylvester: Life and Work in Letters*. Oxford: Clarendon.

——, 2006. *James Joseph Sylvester: Jewish Mathematician in a Victorian World*. Baltimore, MD: Johns Hopkins University Press.

Sylvester, J. J. 1904–12. *The Collected Mathematical Papers of James Joseph Sylvester*, four volumes. Cambridge: Cambridge University Press. (Reprint edition published in 1973. New York: Chelsea.)

Karen Hunger Parshall

VI.43   George Boole

b. Lincoln, England, 1815; d. Cork, Republic of Ireland, 1864
Boolean algebra; logic; operator theory; differential equations; difference equations

Boole, who never attended secondary school, college, or university, was almost entirely self-taught. His father was a poor shoemaker who was more interested in building telescopes and scientific instruments than making shoes—the result being that his business failed and Boole had to leave school at the age of fourteen and take a job as a junior teacher to support his parents, sister, and two brothers. By the age of ten he had mastered Latin and Ancient Greek, and by the age of sixteen he could read and speak French, Italian, Spanish, and German fluently. From his father he got a love of mechanics, physics, geometry, and astronomy, and together they built functioning scientific instruments. Boole then turned to mathematics and by the age of twenty he was publishing original research in calculus and linear systems. He wrote two seminal papers on linear transformations (1841, 1843), which provided the starting point for invariant theory, but he left it to others such as CAYLEY [VI.46] and SYLVESTER [VI.42] to develop the subject. In 1844 he was awarded the Royal Society's Gold Medal for his paper on operators in analysis, the first gold medal for mathematics to be presented by the society. The paper was important not only because it contained (arguably for the first time) a clear definition of the concept of an OPERATOR [III.50], but also because of the influence it had on Boole's subsequent ideas. An operator, for Boole, was an operation of the calculus, such as differentiation (which he denoted by D), considered as an object in its own right. There was an explicit similarity between the laws he derived for functions of D and the laws of his algebra of logic, which we shall discuss below.

At one time Boole had hoped to become a clergyman but family circumstances prevented this. His reverence for creation made him interested in the workings of the human mind, which he regarded as God's greatest accomplishment. He longed, as Aristotle and LEIBNIZ [VI.15] had before him, to explain how the brain processes information and to express this information in mathematical form. In 1847 he published a book entitled *A Mathematical Analysis of Logic* in which he took the first steps toward achieving his goal, but the book did not have a wide circulation and so made very little impact on the mathematical world.

In 1849 Boole was appointed Professor of Mathematics at Queen's College, Cork. It was there that he rewrote and expanded his ideas in a book entitled *An Investigation of the Laws of Thought* (1854), in which he introduced a new type of algebra, an algebra of logic, which evolved into what we now call Boolean algebra. From his earlier study of languages, he realized that there were mathematical structures concealed in everyday speech. For example, the class of European men, together with (i.e., union) the class of European women, is the same as the class of European men and women. By using letters to represent a class, or set, of objects, he could write the above as $z(x + y) = zx + zy$, where the letters x, y, and z represent the class of men, the class of women, and the class of all Europeans, respectively. Here addition is to be understood as union, at least for disjoint classes like men and women, and multiplication is to be understood as intersection.

The principal laws of Boole's algebra are commutativity, distributivity, and the law which he called the "fundamental law of duality" and which is represented by $x^2 = x$. This law can be interpreted by observing that the class of all white sheep intersected with the class of all white sheep is still the class of all white sheep. Unlike his other laws, all of which apply to ordinary numerical algebra, this law applies to numerical algebra only when x is 0 or 1.

Boole broke with traditional mathematics by showing that the study of well-defined classes or sets of objects is capable of precise mathematical interpretation and is indeed fundamental to mathematical analysis. In simple cases, his approach also reduces classical logic to symbolic mathematical form. Using the symbols 0 and 1 to denote "nothing" and "universe" respectively, and denoting the complement of the class x by $1 - x$, he derived (from the law of duality) the law $x(1 - x) = 0$, which represents the impossibility of an object simultaneously possessing and not possessing a given property, otherwise known as the principle of contradiction. Boole also applied his calculus to the theory of probability.

Boole's algebra lay dormant until 1939, when Shannon discovered that it was the appropriate language for describing digital switching circuits. Boole's work thus became an essential tool in the modern development of electronics and digital computer technology.

Boole also made several other contributions to mathematics: differential equations, difference equations, operator theory, calculus of integrals, etc. His textbooks on differential equations (1859) and finite differences (1860) include much of his original research and are still in print today, but he is best remembered as the father of symbolic logic and one of the founders of computer science.

Further Reading

MacHale, D. 1983. *George Boole, His Life and Work.* Dublin: Boole Press.

Des MacHale

VI.44 Karl Weierstrass

b. Ostenfelde, Germany, 1815; d. Berlin, 1897
Analysis

Weierstrass began his career studying finance and administration at the University of Bonn but his real interest was mathematics and he did not complete his course. He qualified as a teacher and taught in gymnasia for fourteen years. The turning point in his life occurred when, at the age of almost forty, he published a ground-breaking paper on Abelian functions, in which he solved the problem of inversion of hyperelliptic integrals. Shortly afterward he was offered a position at the University of Berlin. He demanded of himself the very strictest standards, with the result that he published little. His ideas, and his reputation, spread through his excellent lectures, which drew students and established mathematicians from around the world.

Weierstrass has been described as the "father of modern analysis." He contributed to all branches of the subject: calculus, differential and integral equations, the CALCULUS OF VARIATIONS [III.94], infinite series, elliptic and Abelian functions, and real and complex analysis. His work is characterized by attention to foundations and by scrupulous logical reasoning. "Weierstrassian rigor" has come to denote rigor of the strictest standard.

Calculus in the seventeenth and eighteenth centuries was heuristic, lacking logical foundations. The nineteenth century ushered in a rigorous spirit in mathematics which included an examination of the foundation of various fields of mathematics. CAUCHY [VI.29] initiated this process in calculus in the 1820s. But there were several major foundational problems with his approach: verbal definitions of limit and continuity; frequent use of infinitesimals; and intuitive appeal to geometry in proving the existence of various limits.

Weierstrass and DEDEKIND [VI.50] (among others) determined to remedy this unsatisfactory situation, and set themselves the goal of establishing theorems in a "purely arithmetic" manner, as Dedekind put it. To that end, Weierstrass gave precise ε-δ definitions of LIMIT [I.3 §5.1] and CONTINUITY [I.3 §5.2] (those we still use today), thus banishing infinitesimals from analysis (until ROBINSON [VI.95] some hundred years later). He also defined the real numbers based on the rationals (although Dedekind's and CANTOR's [VI.54] approaches proved more accessible). He was thereby largely responsible for the "arithmetization of analysis" (a term coined by KLEIN [VI.57]). Among his remarkable contributions to real analysis are his introduction of uniform convergence (introduced independently by P. L. Seidel) and his example of an everywhere-continuous and nowhere-differentiable function (Cauchy and his contemporaries believed that a continuous function was differentiable except possibly at isolated points).

Both RIEMANN [VI.49] and Weierstrass (succeeding Cauchy) founded complex function theory, but they had fundamentally different approaches to the subject. Riemann's global, geometric conception was based on the notion of a RIEMANN SURFACE [III.79] and on the DIRICHLET PRINCIPLE [IV.12 §3.5], while Weierstrass's local algebraic theory was grounded in power series and ANALYTIC CONTINUATION [I.3 §5.6]. "The more I ponder the principles of function theory—and I do so incessantly—the more I am convinced that it must be founded on simple algebraic truths...," he asserted in a letter to H. A. Schwartz. He severely criticized the Dirichlet principle for being mathematically not well-grounded, and produced a counterexample, after which his approach to complex analysis became dominant until the early twentieth century. Klein commented on Weierstrass's general approach to mathematics: "[He] is first of all a logician; he proceeds slowly, systematically, step-by-step. When he works, he strives for the definitive form."

Weierstrass's name is attached to various concepts and results, among them the *Weierstrass approximation theorem*, which says that a continuous function can be uniformly approximated by polynomials; the *Bolzano–Weierstrass theorem*, which states that every infinite, bounded set of real numbers has a limit point; the *Weierstrass factorization theorem*, which gives the representation of an entire function in terms of an infinite product of "prime functions"; the *Casorati-Weierstrass theorem*, which says that in every neighborhood of an isolated essential singularity an analytic function takes values arbitrarily close to any assigned complex number; the *Weierstrass M-test*, which deals with the comparison of series for convergence; and the *Weierstrass ℘-function*, an example of an ELLIPTIC FUNCTION [V.31] of order 2.

Weierstrass was most proud of his work on Abelian functions, and much of his fame in the nineteenth cen-

tury rested on it. His results in this field are, however, less significant today. For us, his main legacy is his unrelenting insistence on maintaining high standards of rigor and seeking the fundamental ideas underlying mathematical concepts and theories.

Further Reading

Bottazzini, U. 1986. *The Higher Calculus: A History of Real and Complex Analysis from Euler to Weierstrass.* New York: Springer.

Israel Kleiner

VI.45   Pafnuty Chebyshev

b. Okatovo, Russia, 1821; d. Saint Petersburg, Russia, 1894
Assistant, Extraordinary then full Professor of Mathematics, Saint Petersburg (1847–82); Artillery Committee (1856); Scientific Committee of the Ministry of Education (1856)

Fascinated by Watt's parallelogram (the linkage used in steam engines) and the problem of converting circular motion into rectilinear motion, Chebyshev embarked on a deep study of the theory of hinge mechanisms. In particular, he sought the linkage that would produce the minimum deviation from a straight line over a given range. This corresponds to the mathematical problem of finding, from among the class of functions chosen to approximate a given function, the one with the smallest absolute error for all specified values of the argument. It was in this context, in particular considering the approximation of functions by polynomials, that Chebyshev discovered the polynomials now named after him (see [III.85]). These polynomials were first published in his memoir "Théorie des mécanismes connus sous le nom de parallélogrammes" (1854), and they marked the beginning of his important contributions to the theory of orthogonal polynomials.

Chebyshev polynomials of the first kind are defined by $T_n(\cos\theta) = \cos(n\theta)$, for $n = 0, 1, 2, \ldots$. These polynomials also satisfy the recurrence relation $T_{n+1}(x) = 2xT_n(2) - T_{n-1}(x)$, where $T_0(x) = 1$ and $T_1(x) = x$. Chebyshev polynomials of the second kind satisfy $U_n(\cos\theta) = \sin((n+1)\theta)/\sin\theta$ and the recurrence relation $U_{n+1}(x) = 2xU_n(x) - U_{n-1}(x)$, where $U_0(x) = 1$ and $U_1(x) = 2x$.

Chebyshev also had a significant impact on number theory, coming close to proving the PRIME NUMBER THEOREM [V.26]. In probability he is remembered for Chebyshev's inequality, a result that is simple but has innumerable applications.

VI.46 Arthur Cayley

b. Richmond, England, 1821; d. Cambridge, England, 1895
Algebra; geometry; mathematical astronomy

At the beginning of his career in the 1840s, Cayley laid down subjects that informed much of his later research. The novelties of his very first undergraduate paper, "On a theorem in the geometry of position" (1841), are the now-standard notation for DETERMINANTS [III.15] of arrays set between vertical lines and the introduction of the *Cayley–Menger determinant.* Following HAMILTON's [VI.37] discovery of the QUATERNIONS [III.76] (1843), Cayley expressed rotations in three-dimensional space via the succinctly expressed mapping $x \rightarrow q^{-1}xq$, a result that led him to the *Cayley–Klein parameters.* He outlined the nonassociative system of the octaves (CAYLEY NUMBERS [III.76]), the intersection of curves (the *Cayley–Bacharach theorem*), and a dual curve called the *Cayleyan.* In major papers, he described a theory of multilinear determinants and ELLIPTIC FUNCTIONS [V.31] as doubly infinite products. In concert with George Salmon he investigated the famous twenty-seven lines that lie in a cubic surface. The most important studies among his juvenilia, though, were his first steps in invariant theory (1845, 1846), the field in which his reputation was made.

Between 1849 and 1863, years spent as a qualified London barrister, Cayley broadened his range, but unlike other gentlemen of science who roamed across a multitude of subjects, he restricted his activity exclusively to mathematics. This was mostly pure mathematics. He generalized PERMUTATION GROUPS [III.68] using the calculus of operations as a basis, and he saw that not only were matrices useful as a notational device, but they also constituted a study in their own right. Not generally an excitable person, at the point of discovery he declared the *Cayley–Hamilton theorem* as "very remarkable" and generations of mathematicians have shared his delight. Matrix algebra was used in his solution of the *Cayley–Hermite problem,* which required a description of those linear transformations that leave a bilinear form invariant. A special case of the solution gives rise to the *Cayley orthogonal transform* $(I-T)(I+T)^{-1}$. The links between quaternions, matrices, and group theory that he observed in the 1850s are indicative of his concern for the organization of mathematics.

In the 1850s, Cayley set in motion his famous memoirs on *quantics,* a term he coined for algebraic forms, now referred to as multilinear homogeneous algebraic forms. He discovered *Cayley's formula* for the general form of covariants of binary forms and *Cayley's law* for counting them. In the *Sixth Memoir* (1859), he demonstrated that EUCLIDEAN GEOMETRY [I.3 §6.2] was part of PROJECTIVE GEOMETRY [I.3 §6.7] rather than the converse. The idea of a projective metric (*Cayley's absolute*) was seen by KLEIN [VI.57] in the 1870s as the unifying conceptual idea for classifying non-Euclidean geometries.

For twenty-five years, from 1858, he was the editor of the *Monthly Notices of the Royal Astronomical Society.* In astronomy he contributed to the theory of elliptic planetary motion, calculatory work that demanded an assiduous attention to detail. His work on the lunar theory was noteworthy, and in one long calculation he helped to settle an Anglo-French controversy by verifying the correct value for the secular acceleration of the Moon, which had been established by John Couch Adams in 1853.

Cayley returned to the academic world in 1863 as the founding Sadleirian Professor of Pure Mathematics at Cambridge. In 1868 Paul Gordan startled invariant theorists by proving that invariants and covariants of a binary quantic could be expressed in terms of a *finite* basis. This contradicted an earlier result of Cayley's but, undaunted, he completed his series with a listing of the irreducible invariants and covariants of the binary form of order 5 (the binary quintic), and their connecting syzygies.

Many developments in pure mathematics can be traced back to his minor notes of the 1870s and 1880s, including the theory of knots, fractals, dynamic programming, and group theory (the well-known *Cayley's theorem*). In graph theory, the number of distinct labeled trees with n nodes being n^{n-2} is known as *Cayley's graph theorem.* He brought his theoretical knowledge of graphical trees to bear on the problem of counting isomers in organic chemistry, thus prompting questions about the actual existence of certain chemical compounds that have since been discovered in many instances by chemists. In the last decade of his life, Cayley set about the task that gave him an important line of contact with today's mathematicians: the publication of his *Collected Mathematical Papers* in thirteen large volumes by Cambridge University Press.

Further Reading

Crilly, T. 2006. *Arthur Cayley: Mathematician Laureate of the Victorian Age*. Baltimore, MD: Johns Hopkins University Press.

Tony Crilly

VI.47 Charles Hermite

b. Dieuze, Moselle, France, 1822; d. Paris, 1901
Analysis (elliptic functions, differential equations); algebra (invariant theory, quadratic forms); approximation theory

Like many who aspired to enter the École Polytechnique, Hermite undertook special preparatory classes, in his case at Lycée Henri IV and Lycée Louis-le-Grand. He began to study serious mathematics, immersing himself in the work of LAGRANGE [VI.22] and LEGENDRE [VI.24], and became interested in the solution of equations by radicals. Admitted to the École Polytechnique in 1842, by the end of that year he had completed his first significant original work. This extended results of JACOBI [VI.35] in the theory of ELLIPTIC FUNCTIONS [V.31]. He sent these to Jacobi, who responded very positively. This achievement both brought him recognition in Paris and initiated a correspondence with Jacobi on elliptic functions and number theory that launched Hermite's career.

Hermite nonetheless struggled to find a position commensurate with his abilities, and for almost a decade survived on teaching assistant and examiner jobs around Paris. Hermite's work turned to number theory, in particular the arithmetic of quadratic forms, where he followed GAUSS [VI.26] and Lagrange in studying when one form can be reduced to another by a linear transformation. It was in this context that the HERMITIAN MATRICES [III.50 §3] named after him arose. Hermite was interested in invariants of quadratic forms, and also applied his work to the problem of location of roots of polynomials. As a result of these efforts, in 1856 he was appointed to the Paris Academy of Sciences, with LIOUVILLE [VI.39] and CAUCHY [VI.29] supporting him. This appointment was quickly followed by Hermite's 1858 discovery of a means to express the solutions of the general fifth-degree polynomial equation in terms of elliptic functions, which earned him widespread international recognition.

Finally obtaining a professorship at the Faculty of Science in Paris in 1869, Hermite became an influential mentor for a generation of mathematicians, his best-known protégés including J. Tannery, POINCARÉ [VI.61],

E. Picard, P. Appell, and E. Goursat. Hermite's dynastic connections are also impressive: his brother-in-law, Joseph Bertrand, was permanent secretary of the Paris Academy of Sciences, Picard was his son-in-law, Appell married Bertrand's daughter, and their daughter married BOREL [VI.70]. His advocacy of improved international communication led to German work becoming much better known in France than it had been previously. During this period, he obtained a proof of the TRANSCENDENCE [III.41] of "e" using CONTINUED-FRACTION [III.22] methods based on earlier research in approximation theory (which had included the invention of the Hermite polynomials). His influence in the mathematical community was strong until his death.

Further Reading

Picard, É. 1901. L'œuvre scientifique de Charles Hermite. *Annales Scientifiques de l'École Normale Supérieure (3)* 18: 9–34.

Tom Archibald

VI.48 Leopold Kronecker

b. Liegnitz, Silesia, today Poland, 1823; d. Berlin, 1891
Algebra; number theory

One of the dominant mathematicians of the second half of the nineteenth century, Kronecker is best known today for his constructivist views and his contributions to number theory. After finishing his Ph.D. under the supervision of DIRICHLET [VI.36] in 1845, Kronecker left Berlin and mathematics in order to manage a family estate and to wind up his father-in-law's banking business. These activities left him wealthy and free to return to Berlin and concentrate on mathematics without holding an academic position. In 1855, Kronecker's former school teacher and closest scientific friend, ERNST EDUARD KUMMER [VI.40], also came to Berlin and stayed there until his death in 1893. In 1861, Kronecker became a member of the Berlin Academy of Sciences and started teaching courses at Berlin University. Kronecker valued the exchange with his Berlin colleagues (especially Kummer and WEIERSTRASS [VI.44]) highly, until a quarrel arose between Kronecker and Weierstrass in the 1870s, which drove Weierstrass to bitter, even anti-Semitic, complaints to others about Kronecker. After Kummer's retirement in 1883, Kronecker occupied Kummer's chair and stepped up his teaching activities as well as the frequency of his publications. This last active period was cut short when he died, shortly after the death of his wife.

Kronecker was renowned for the originality of his mathematical insight and became increasingly influential through the 1860s and 1870s. In 1868, he was offered the chair at Göttingen formerly held by GAUSS [VI.26], and was elected to the Paris Academy. After the Franco-Prussian war of 1870–71, he was invited to recommend mathematicians for the newly opened German university in Strasbourg; and in 1880 he became the managing editor of the *Journal für die reine und angewandte Mathematik* (otherwise known as *Crelle's Journal*). He was often criticized for incomplete, unpublished, or incomprehensible proofs—JORDAN [VI.52] spoke of his colleagues' "envy and despair" with regard to his results. Only in his later years was he explicit about his constructivist methodology. This constituted at least part of the quarrel with Weierstrass, and later prompted HILBERT [VI.63] to call Kronecker a "Verbotsdiktator" ("forbidding dictator"). Generally affable and hospitable, Kronecker was tough in defending his mathematical ideas and his claims to priority.

In his first works on solvable algebraic equations (in the early 1850s), he claimed not only the so-called *Kronecker–Weber theorem* (in today's formulation: every finite Galois extension of the rational numbers with Abelian GALOIS GROUP [V.21] lies in a field generated by roots of unity; the first correct proof of it was given by Hilbert in 1896), but also a generalization to Abelian extensions of imaginary quadratic fields, which he later called his "liebster Jugendtraum" ("dearest dream of his youth"). This dream, which was incorrectly translated by Hilbert into his twelfth problem in 1900, is today part of CLASS FIELD THEORY [V.28] and the theory of complex multiplication. Such connections between algebra, analysis, and arithmetic continued to pervade Kronecker's later work. Important results of Kronecker include class number relations and limit formulas in the theory of ELLIPTIC FUNCTIONS [V.31], the structure theorem for finitely generated Abelian groups, and a theory of bilinear forms.

In the late 1850s, Kronecker began to work on algebraic number theory, but only in 1881 did he publish his "Grundzüge einer arithmetischen Theorie der algebraischen Grössen," dedicated to Kummer on the fiftieth anniversary of his doctorate. This mathematical testament contains an (incomplete) exposition of a unified arithmetic theory of algebraic numbers and algebraic functions. As a research program, it adumbrates important aspects of class field theory as well as of an arithmetico-geometric theory in dimensions higher than one. Kronecker's concept of "divisor" is equivalent

to Dedekind's notion of "ideal" in the case of Dedekind domains, but is more restricted in the general case. Several mathematicians, such as H. Weber, K. Hensel, and G. König, took up the "Grundzüge" in their own work.

On a more general level, Kronecker asked for the complete arithmetization of pure mathematics, i.e., for the effective finitary reduction of pure mathematics to the notion of positive integer. For this, he propagated the introduction of indeterminates and equivalence relations, a method which he traced back to Gauss. In the case of a finite extension of the rational numbers, for instance, Kronecker is explicitly working with polynomials modulo an irreducible equation $f(x) = 0$, rather than adjoining a root of it.

Further Reading

Kronecker, L. 1895–1930. *Werke*, five volumes. Leipzig: Teubner.
Vlăduţ, S. G. 1991. *Kronecker's Jugendtraum and Modular Functions*. New York: Gordon & Breach.

Norbert Schappacher and Birgit Petri

VI.49 Georg Friedrich Bernhard Riemann

b. Breselenz, near Dannenberg, Germany, 1826;
d. Selasca, Italy, 1866
Real and complex analysis; differential equations;
differential geometry; heat distribution; number theory;
propagation of shock waves; topology

Riemann was born into a poor pastor's family and studied mathematics at Göttingen, eventually becoming a professor there. His health broke in 1862 and he died near Lake Maggiore, Italy, of pleurisy at the age of thirty-nine.

No mathematician is more associated with the mid-nineteenth-century transition from algorithmic to conceptual thought than Riemann. His doctoral thesis of 1851, and still more his paper on Abelian functions (1857), promoted the view that a HOLOMORPHIC FUNCTION [I.3 §5.6] is properly defined by the CAUCHY–RIEMANN EQUATIONS [I.3 §5.6] and is to be studied through a close connection with the theory of HARMONIC FUNCTIONS [IV.24 §5.1]. In his thesis he sketched a proof of the remarkable RIEMANN MAPPING THEOREM [V.34]. This states that if X and Y are any two simply connected open subsets in the complex plane, neither of which is the whole plane, then there is a holomorphic map from one to the other with a holomorphic inverse. For example, if you draw any closed curve in

Georg Friedrich Bernhard Riemann

the plane that does not intersect itself, and let D be the region inside the curve, then D is biholomorphically equivalent to the open unit disk. In the 1857 paper he defined RIEMANN SURFACES [III.79], showed how to analyze them topologically, and outlined the Riemann inequality which his student Gustav Roch improved to the RIEMANN–ROCH THEOREM [V.31] in 1864. (The Riemann–Roch theorem, which is of great importance in algebraic geometry as well as complex analysis, determines the dimension of the space of meromorphic functions on a given Riemann surface with a prescribed number of poles.) In 1857 he extended the theory of differential equations, specifically the important case of the hypergeometric equation, to complex functions. In 1859 he used deep, new ideas from complex function theory to study the (Riemann) zeta function and proposed his celebrated conjecture, the RIEMANN HYPOTHESIS [IV.2 §3], concerning the location of the complex zeros of this function. The conjecture remains unsolved to this day.

These ideas enabled mathematicians to study complex functions on domains other than the plane and subsets of the plane. They opened the way to a geometric study of algebraic functions and algebraic curves, and proved to be decisive in the study of the integrals of algebraic functions (the theories of Abelian functions and theta functions of several variables). Investigations of the Riemann zeta function led not only to the discovery of new properties of classes of complex func-

tions, but more recently to the use of zeta functions of many other kinds in other branches of mathematics, including dynamics.

In 1854 Riemann, inspired by his mentor DIRICHLET [VI.36], formulated the concept of the RIEMANN INTEGRAL [I.3 §5.5], which permitted him to do profound work on the convergence of trigonometric series. Dirichlet had been able to prove that a real function was correctly represented by a Fourier series, but only under very restrictive conditions. This left open the questions of what sorts of functions did not satisfy these conditions and how could they be studied. Riemann reformulated the concept of the integral and was able to show that it is not just the continuity of a function and the ways in which it may fail to be continuous that affect the accuracy of its Fourier series representation, but the nature of its oscillations. The Riemann integral remained the dominant definition of the integral until it was replaced by the LEBESGUE INTEGRAL [III.55] after 1902, which is better adapted to capturing the way the behavior of a function affects its Fourier series.

In a lecture, also given in 1854 (but published posthumously in 1868), he entirely reformulated geometry as being about spaces (sets of points, which he called MANIFOLDS [I.3 §6.9]) with a RIEMANNIAN METRIC [I.3 §6.10] (an appropriate concept of distance) and argued that the geometric properties of a space were its intrinsic ones. He noted that there are three constant-curvature spaces in two dimensions and showed how the idea of constant curvature can be extended to higher dimensions. In passing, he was the first person to write down a metric for non-Euclidean geometry (more than a decade before Beltrami's publication of 1868, which legitimized non-Euclidean geometry). This lecture earned him the right to teach in a German university.

Riemann also did important work on shock waves, and shares with WEIERSTRASS [VI.44] the honor of introducing the methods of complex function theory in the study of MINIMAL SURFACES [III.94 §3.1], where he was led to several new solutions of the Plateau problem, which asks for the surface of least area spanning a given curve in space.

The distinguished complex analyst Lars Ahlfors once described Riemann's complex analysis as consisting of "almost cryptic messages to the future" and said that his mapping theorem was given in a form that "would defy any attempt at proof, even with modern methods," and it is true that Riemann's presentation is more visionary than precise. But his vision described a geometric setting for complex function theory that, as Ahlfors's own work indicates, remains fertile over 150 years after it was written.

Further Reading

Laugwitz, D. 1999. *Bernhard Riemann, 1826–1866. Turning Points in the Conception of Mathematics*, translated by A. Shenitzer. Boston, MA/Basel: Birkhäuser.

Riemann, G. F. B. 1990. *Gesammelte Werke, Collected Works*, edited by R. Narasimhan, third edn. Berlin: Springer.

Jeremy Gray

VI.50 Julius Wilhelm Richard Dedekind

b. Brunswick, Germany, 1831; d. Brunswick, Germany, 1916
Algebraic number theory; algebraic curves; set theory; foundations of mathematics

---

Dedekind spent most of his life as a professor at the Technische Hochschule in Brunswick, Germany (his and GAUSS's [VI.26] home town), having spent the years 1858–62 at the *Polytechnikum* in Zürich (which later became known as ETH). He obtained his mathematical education at Göttingen, being Gauss's last Ph.D. student and subsequently a pupil of DIRICHLET [VI.36] and RIEMANN [VI.49]. Dedekind was a retiring man with, as KLEIN [VI.57] said, a "contemplative nature;" he remained a bachelor, living with his mother and sister. Nevertheless, he had an impact upon a select group of contemporaries (especially CANTOR [VI.54], FROBENIUS [VI.58], and Heinrich Weber) through his rich correspondence.

A key figure in the emergence of modern set-theoretic mathematics, and particularly the notion of a mathematical structure, Dedekind is best known for his work on the foundations of the REAL NUMBER SYSTEM [I.3 §1.4]. His main contribution, however, was in algebraic number theory. Indeed, he shaped modern number theory as we know it, presenting it as a theory of ideals in rings of integers (see ALGEBRAIC NUMBERS [IV.1 §§4–7]). This was first made public in 1871, within Supplement X to his edition of Dirichlet's *Vorlesungen über Zahlentheorie*, where he established unique decomposition of ideals into prime ideals for all rings of algebraic integers. In the process, he formulated the concepts of field, ring, ideal, and module (see [I.3 §2.2] and [III.81]), always within the particular context of the complex numbers. It was also in the context of algebra (Galois theory) and number theory that Dedekind started systematic work with quotient structures, isomorphisms, homomorphisms, and automorphisms.

In subsequent editions of Dirichlet's *Vorlesungen* (1879 and 1894) Dedekind went on refining his presentation of ideal theory, making it more purely settheoretic. In 1882, together with Weber, he offered a theory of ideals in fields of algebraic functions, which made it possible to give a rigorous treatment of Riemann's results on algebraic curves up to the RIEMANN–ROCH THEOREM [V.31]. This work paved the way for modern algebraic geometry.

Intimately linked with Dedekind's work in algebra and number theory were his reflections on the foundations of the real number system. In 1858 (published 1872) he formulated a definition of the real numbers using what are now known as "Dedekind cuts" in the set of rational numbers. During the 1870s (published 1888) he elaborated a purely set-theoretic definition of the natural numbers as "simply infinite" sets, which led him to crystallize the DEDEKIND–PEANO AXIOMS [III.67]. In this work, as in his more advanced research, sets, structures, and mappings form the essential building blocks, the very foundations of pure mathematics. In the light of (now superseded) conceptions of logic, this led Dedekind to the view that "arithmetic (algebra, analysis) is only a part of logic." From a modern viewpoint, his contributions show that SET THEORY [IV.22] is a sufficient foundation for classical mathematics. Thus he contributed as much as anybody else to the set-theoretic reformulation of modern mathematics.

Further Reading

Corry, L. 2004. *Modern Algebra and the Rise of Mathematical Structures*, second revised edn. Basel: Birkhäuser.

Ewald, W., ed. 1996. *From Kant to Hilbert: A Source Book in the Foundations of Mathematics*, two volumes. Oxford: Oxford University Press.

Ferreirós, J. 1999. *Labyrinth of Thought. A History of Set Theory and Its Role in Modern Mathematics*. Basel: Birkhäuser.

José Ferreirós

VI.51 Émile Léonard Mathieu

b. Metz, France, 1835; d. Nancy, France, 1890
Student at the École Polytechnique; Docteur és sciences with thesis on transitive functions (1859); Professor of Mathematics: Besançon (1869–74), Nancy (1874–90)

Mathieu is known for the functions that take his name, which he discovered while solving the two-dimensional wave equation for the vibrations of an elliptical membrane. These functions, which are special cases of the hypergeometric function, are particular solutions of *Mathieu's equation*:

$$\frac{\mathrm{d}^2 u}{\mathrm{d}z^2} + (a + 16q\cos 2z)u = 0,$$

where a and q are constants that depend on the physical problem.

Mathieu is also known for his discovery of the five Mathieu groups. These were the first SPORADIC SIMPLE GROUPS [V.7] (meaning that they did not fit into one of the known infinite families of simple groups) to be found. It is now known that there are twenty-six such groups altogether, although it was almost a century after Mathieu before a sixth one was found.

VI.52 Camille Jordan

b. Lyon, France, 1838; d. Milan, Italy, 1922
Nominally an engineer until 1885; teacher of mathematics,
École Polytechnique and Collège de France (1873–1912)

Jordan was the leading group theorist of his generation. His immense *Traité des Substitutions et des Équations Algébriques* (1870), which brought together all his earlier results on PERMUTATION GROUPS [III.68] and provided a synthesis of GALOIS's [VI.41] ideas, remained a cornerstone for group theorists for many years. Included in the *Traité*, in the chapter on what he calls linear substitutions (now written in matrix form as $y = Ax$), is the definition of what today is called the JORDAN NORMAL FORM [III.43] of a matrix, although in 1868 WEIERSTRASS [VI.44] had already defined an equivalent normal form.

Jordan is also known for his work in topology, especially for the theorem now known as the *Jordan curve theorem*. This states that a simple closed curve in the plane separates the plane into two disjoint regions, an inside and an outside, and it was given by him in his influential *Cours d'Analyse* (1887). Although the theorem appears obvious, the proof, as Jordan recognized, is difficult and the one he gave was incorrect. (The proof is relatively easy for smooth curves; the difficulties arise when dealing with nowhere-smooth curves, such as the Koch snowflake.) The first rigorous proof was given by Oswald Veblen in 1905. There is a stronger form of the theorem, known as the Jordan–Schönflies theorem, which states that in addition the two regions of the plane, the inside and the outside, are homeomorphic to the standard circle in the plane. Unlike the original theorem, this stronger form of the theorem cannot be generalized to higher dimensions, a famous counterexample being the Alexander horned sphere.

VI.53 Sophus Lie

b. Nordfjordeid (western Norway), 1842; d. Oslo, 1899
Transformation groups; Lie groups; partial differential equations

Lie was twenty-six when he discovered that, in his own words, he "harbored a mathematician." Before then he had primarily wanted to be an observational astronomer. Later in life, looking back on his career, he said that it was the "audacity of his thinking" more than any formal knowledge and education that had given him a position among the foremost of mathematicians. During a career spanning more than thirty years, Lie produced almost eight thousand pages of mathematics, making him one of the most productive mathematicians of his time.

Lie graduated in general science from the university in Oslo in 1865 but without showing any special aptitude for mathematics. It was not until 1868, when he attended a lecture by the Danish geometer Hieronymus Zeuthen on the work of Chasles, MÖBIUS [VI.30], and Plücker, that he became inspired by modern geometry. He studied the works of Poncelet (projective geometry) and Plücker (line geometry), and wrote a dissertation on "imaginary geometry," that is, geometry based on complex numbers. In the fall of 1869 he traveled to Berlin, Göttingen, and Paris, where he met mathematicians who would remain friends and colleagues for the rest of his life. In Berlin he met KLEIN [VI.57], in Göttingen he met Clebsch, and in Paris, where he was joined by Klein, he met Darboux and JORDAN [VI.52]. These two had a particular influence on him—Darboux through his theory of surfaces and Jordan through his knowledge of group theory and the work of GALOIS [VI.41]—with the result that he (and Klein) began to recognize the value of group theory for the study of geometry. Lie and Klein published three joint papers on geometrical topics, including one on the so-called Lie line–sphere transformation (the contact transformation, which is a transformation that maps straight lines into spheres and principal tangent curves into curvature lines; and then the study of the geometrical entities that are invariant under such transformations).

When Klein prepared what was to become his famous "Erlanger Programm" (his characterization of geometry as properties invariant under a group action), Lie was with him. This work later created a deep rift between them. (Friendship turned into aloofness and hostility and culminated in the following statement by Lie in 1893: "I am no pupil of Klein's, nor is the reverse the case, although this would be nearer to the truth.")

Lie returned (after his first trip abroad) to Oslo and, in 1872, a chair of mathematics at the university was created especially for him. During the early 1870s Lie worked on turning his line–sphere transformation into a general theory of contact transformations. From 1873 he worked on a systematic study of continuous transformation groups (today known as LIE GROUPS [III.48 §1]), his aim being to classify LIE ALGEBRAS [III.48 §§2, 3] and apply the results to the solution of differential equations. He also published studies on MINIMAL SURFACES [III.94 §3.1]. In Norway, however, there was no scientific milieu, and he felt very isolated. In 1884 Klein and his friend Adolf Mayer in Leipzig tried to help him by sending their student Friedrich Engel to study with him and to help him with the formulation and writing of his new ideas. The work that Engel and Lie started together resulted in three volumes, *Theorie der Transformationsgruppen* (1888–93). In 1886 Lie accepted the professorship in Leipzig (in succession to Klein, who had moved to Göttingen). In Leipzig he became a leading mathematician and a central figure in the European community of mathematicians. Promising new students from both France and the United States were sent to study with him. Besides teaching he continued his research on transformation groups and differential equations, and he solved the so-called Helmholtz space problem (characterizing the geometry of space in terms of groups of transformations). In 1898, the year before he died, Lie returned to Oslo to take up a position created especially for him.

The theory of transformation groups, which Lie initiated and developed in the study of differential equations, has grown into a field of its own, the theory of Lie groups and Lie algebras, which today permeates large parts of mathematics and mathematical physics.

Further Reading

Borel, A. 2001. *Essays in the History of Lie Groups and Algebraic Groups.* Providence, RI: American Mathematical Society.

Hawkins, T. 2000. *Emergence of the Theory of Lie Groups.* New York: Springer.

Laudal, O. A., and B. Jahrien, eds. 1994. *Proceedings, Sophus Lie Memorial Conference.* Oslo: Scandinavian University Press.

Stubhaug, A. 2002. *The Mathematician Sophus Lie.* Berlin: Springer.

Arild Stubhaug

VI.54 Georg Cantor

b. Saint Petersburg, Russia, 1845; d. Halle, Germany, 1918
Set theory; transfinite numbers; the continuum hypothesis

Although born in Russia, Cantor was raised and educated in Prussia and spent his entire career as professor of mathematics at the University of Halle. He studied at the Universities of Berlin and Göttingen with KRONECKER [VI.48], KUMMER [VI.40], and WEIERSTRASS [VI.44], and received his Ph.D. from the University of Berlin in 1867. His dissertation, "De aequationibus secundi gradus indeterminatis" ("On indeterminate equations of the second grade"), dealt with work in number theory on Diophantine equations, work that had been pioneered by LAGRANGE [VI.22], GAUSS [VI.26], and LEGENDRE [VI.24]. The following year he accepted a position in the mathematics department at the University of Halle, where he spent the rest of his academic career. There, his *Habilitationsschrift* was again devoted to number theory, and dealt with transformations of ternary quadratic forms.

It was at Halle that Cantor's colleague, Eduard Heine, was working on difficult problems involving trigonometric series, and he interested Cantor in the problem of determining the conditions under which a trigonometric series of the form

$$f(x) = \tfrac{1}{2}a_0 + \sum_{n=1}^{\infty} (a_n \sin nx + b_n \cos nx)$$

uniquely represented a given function. In other words, could it be that two different trigonometric series could represent the same function? Heine had shown, in 1870, that if $f(x)$ is continuous in general (i.e., for all but a finite number of points of discontinuity, at which points Heine added that the function need not necessarily be finite), the representation is unique if we insist that the series is uniformly convergent to f in general. Cantor was able to establish much more general results, and in five papers written between 1870 and 1872 he was able to show that such representations were unique even if an infinite number of exceptional points were allowed, so long as these exceptional points (i.e., points at which the function failed to be continuous) were distributed over the domain of the function's definition in a particular way, constituting what Cantor called "point sets of the first species." His studies of these and related point sets eventually led Cantor to his much more abstract and powerful theory of sets and transfinite numbers.

Point sets of the first species were sets P for which, given its sequence of *derived sets* (the derived set P' of a set P is the set of all the limit points of P), there was some finite n such that the nth derived set P^n of P was finite, and thus the $(n + 1)$st derived set was empty, i.e., $P^{n+1} = \varnothing$. It was Cantor's subsequent study of infinite linear point sets that would eventually lead to his creation of transfinite set theory in the 1880s. (For more details about this, see SET THEORY [IV.22 §2].)

Before he did so, Cantor first began to explore the implications of his work on trigonometric series and the structure of the real numbers in several papers, one of which was to revolutionize mathematics in a fundamental way. The first of these papers was published in 1874, and bore the innocuous title "Über eine Eigenschaft des Inbegriffes aller reellen algebraischen Zahlen" ("On a property of the collection of all real algebraic numbers"). In this paper, Cantor proved that the set of all algebraic real numbers was COUNTABLY INFINITE [III.11]. What was revolutionary about the paper, however, was that he also proved that the set of all real numbers was *not* countable, and must be of a higher order of infinity than the countably infinite set of natural numbers. He returned to this result in 1891 with a new approach, the groundbreaking method of diagonalization, to prove in a very direct way that the set of real numbers is uncountably infinite. Cantor's second important paper of the decade appeared in 1878, "Ein Beitrag zur Mannigfaltigkeitslehre" ("A contribution to the theory of aggregates"), in which he proved (with a partly faulty argument) the invariance of dimension, a theorem first correctly proved by BROUWER [VI.75] in 1911.

Between 1879 and 1884 Cantor published six papers designed to outline the basic elements of his new thinking about sets. He first considered what happened if a set were not of the first species by introducing symbols for the infinite indices needed to identify such sets. For example, a set P was said to be of the *second* species if there was no finite n such that the nth derived set P^n of P was finite. He then considered the case in which the intersection of all the derived sets of P (namely $P', P'', \ldots, P^n, \ldots$) was again an infinite set, which he designated P^∞. This set, since it was infinite, had a derived set as well, $P^{\infty+1}$, and this led in fact to an entire sequence of derived sets of the second species: $P^\infty, P^{\infty+1}, \ldots, P^{\infty+n}, \ldots, P^{2\infty}, \ldots$.

In his first papers on infinite linear sets, these indices for derived sets remained "infinitary symbols": that is, devices for distinguishing between different sets. But

in his *Grundlagen einer allgemeinen Mannigfaltigkeitslehre* ("Foundations of a general set theory"), published in 1883, these symbols became the first transfinite numbers: the transfinite ordinal numbers. These numbers began with ω, the transfinite ordinal number representing the sequence of natural numbers $1, 2, 3, \ldots$, which could also be thought of as the first infinite ordinal number after all of the finite whole numbers. In the *Grundlagen*, Cantor not only devised the basic features of a transfinite arithmetic for these numbers, but he provided a detailed philosophical defense of the new numbers. Acknowledging the revolutionary nature of what he was introducing, he argued that the new concepts were necessary in order to achieve precise mathematical results that he could obtain by no other means.

Cantor's best-known mathematical creation, however, the transfinite cardinal numbers, which he denoted using the Hebrew letter aleph, were introduced only later, in the 1890s. They were first given full exposition in a pair of papers (1895, 1897) that constituted his "Beiträge zur Begründung der transfiniten Mengenlehre" ("Contributions to the founding of transfinite set theory"). In two articles published in *Mathematische Annalen*, he not only set out his theory of transfinite ordinal and cardinal numbers, as well as their arithmetics, but also explained his theory of order types, namely the different properties exhibited by the sets of natural, rational, and real numbers considered in their natural orders. There he also stated (but could not prove) his famous CONTINUUM HYPOTHESIS [IV.22 §5], namely that the power (or cardinal number) of the continuum of all real numbers $\mathbb{R}$ is the next largest infinite set (or cardinal number) after the countably infinite set of natural numbers $\mathbb{N}$, the cardinality of which was taken to be $\aleph_0$. Cantor expressed the continuum hypothesis algebraically as the statement that $2^{\aleph_0} = \aleph_1$.

By the end of his career Cantor had received honorary degrees from foreign universities and the Copley Medal of the Royal Society for his great contributions to mathematics, but there were problems with set theory that were beyond his capacities to remedy. The most disturbing for many mathematicians were the "antinomies" of set theory: the paradoxes put forward by the likes of Burali-Forti and RUSSELL [VI.71]. In 1897 the former published the paradox arising from the collection of *all* ordinal numbers, the ordinal number of which should be an ordinal number greater than any in the collection of all ordinal numbers. In 1901 the latter discovered the paradox of the class of all classes that are not members of themselves: is it a member of itself or not? (See THE CRISIS IN THE FOUNDATIONS OF MATHEMATICS [II.7 §2.1].) Cantor himself was aware of the contradictions that arose from considering the collections of all transfinite ordinal or cardinal numbers, and what their ordinal or cardinal numbers might be. The solution Cantor adopted was to regard such collections as too large, and not really sets at all, but "inconsistent aggregates" as he called them. Others, like Zermelo, began to axiomatize set theory in an effort to exclude the possibility of contradictions. The two most powerful results of the twentieth century to complement Cantor's work are those of GÖDEL [VI.92] (who established the consistency of the continuum hypothesis with ZERMELO–FRAENKEL SET THEORY [IV.22 §3]) and Paul Cohen (who determined the independence of the continuum hypothesis from Zermelo-Fraenkel set theory), the latter finally establishing the impossibility of proving the continuum hypothesis.

Cantor's legacy for the history of mathematics has truly been revolutionary. Above all, his transfinite set theory for the first time gave mathematicians the means of dealing with concepts of the infinite in a careful and precise way.

Further Reading

Dauben, J. W. 1990. *Georg Cantor. His Mathematics and Philosophy of the Infinite.* Princeton, NJ: Princeton University Press. (First published in 1978 by Harvard University Press.)
——. 2005. Georg Cantor and the battle for transfinite set theory. In *Kenneth O. May Lectures of the Canadian Society for History and Philosophy of Mathematics*, edited by G. Van Brummelen and M. Kinyon, pp. 221-41. New York: Springer.
——. 2005. Georg Cantor. Paper on the "Foundations of a general set theory" (1883). In *Landmark Writings in Western Mathematics 1640-1940*, edited by I. Grattan-Guinness, pp. 600-12. London: Routledge.
Tapp, C. 2005. *Kardinalität und Kardinäle. Wissenschaftshistorische Aufarbeitung der Korrespondenz zwischen Georg Cantor und katholischen Theologen seiner Zeit.* Stuttgart: Franz Steiner.

Joseph W. Dauben

VI.55   William Kingdon Clifford

b. Exeter, England, 1845; d. Madeira, Portugal, 1879
Geometry; complex function theory; popularization of mathematics

---

Clifford went up to Trinity College Cambridge in 1863. He graduated from there in 1867 as 2nd Wrangler and also came second in the more demanding Smith's prize examination. In 1868 he became a Fellow of Trinity, leaving in 1871 to become the professor of applied mathematics at University College London. He died of tuberculosis in 1879.

A versatile mathematician, regarded by many as the best of his generation, Clifford's favorite field was geometry, over which he ranged widely, proving new results in classical Euclidean geometry as well as in projective and differential geometry. He was the first English mathematician to appreciate the work of RIEMANN [VI.49] on differential geometry, and published a translation of Riemann's paper "On the hypotheses that lie at the foundations of geometry" in 1873. He endorsed Riemann's fundamental reformulation of geometry, and went even further in speculating that the curvature of physical space might explain the motion of matter. He also made a significant application of the RIEMANN–ROCH THEOREM [V.31], and was among the first to analyze the complicated topological nature of a RIEMANN SURFACE [III.79] by showing how to dissect any Riemann surface into simple pieces in a standard way. He was the first to study a geometry locally equivalent to plane geometry but topologically distinct (the flat torus, also known today as the *Clifford-Klein space form* after KLEIN's [VI.57] later more detailed study of it). In algebra, he invented the biquaternions (these are like quaternions, but have complex numbers as coefficients).

Clifford was regarded as a marvelous lecturer until his health broke, and he was a successful popularizer and essay writer. He forcefully adopted the view that geometry was a matter of experience, not a priori truth. He was a friend of T. H. Huxley and was sympathetic to humanism in philosophy.

Further Reading

Clifford, W. K. 1968. *Mathematical Papers*, edited by R. Tucker. New York: Chelsea. (First published in 1882.)

Jeremy Gray

VI.56   Gottlob Frege

b. Wismar, Germany, 1848; d. Bad Kleinen, Germany, 1925
Logic; foundations of mathematics; paradox

Frege was a precursor of modern logic, in that many of the hallmarks of contemporary logic appear first in his writing. His work has also been singularly influential outside the foundations of mathematics, especially in the philosophy of language.

Frege was trained at Jena and Göttingen, receiving a Ph.D. under Ernst Schering in 1873. His Ph.D. thesis addressed the spatial representation of imaginary elements in geometry, and his 1874 Habilitation essay at Jena worked out some basic details of what we would now call "iteration theory." Though his early work gave no obvious sign of the revolutionary work to come, with hindsight one can discern a foundational motif running through even the apparently conventional mathematics of the early work: a conviction that arithmetic was in some way or other logical, and that geometry was fundamentally different and less general because it was grounded in spatial intuition. This is an especially salient concern in some of his areas of early research, such as Plücker's line geometry and RIEMANN's [VI.49] complex analysis, where the role of visual representation was a matter of some dispute. Frege sought to resolve the dispute by deriving arithmetic and analysis rigorously from logical principles. His motivation was not so much a desire for certainty: rather, he held that only "gap-free" proofs can reveal a science's fundamental principles.

Among the features of contemporary logic appearing first in Frege's core logical writings (*Begriffsschrift* (1879) and *Grundgesetze der Arithmetik* (volume 1, 1893; volume 2, 1903)) are the following.

(i) Inferences are analyzed within a quantified logic of propositions, which extends to relations as well as to propositions of subject–predicate form. We would today describe Frege's logical system as a higher-order predicate calculus.

(ii) Forms from syllogistic logic (such as "All As are Bs") are interpreted as quantified conditionals ("For all x, if x is an A, then x is a B"), in the way that is now so standard as to seem inescapable, presenting implicitly the point that the underlying logical form of a proposition may differ from its surface grammar.

(iii) The syntax of the language is explicitly displayed, and inferences are carried out strictly in accordance with the form of statements by explicitly stated rules.

(iv) Rules of inference and axioms are distinguished; the consequence relation and conditionals are distinguished.

(v) "Function" is taken as an undefined primitive concept. (This was a contentious move. Some mathematicians of the time, including one of Frege's teachers, Alfred Clebsch, held the concept of function to be too vague to serve as a basic building block.) A sharp distinction is enforced between functions and the things (called objects) that can be arguments of functions.

(vi) Quantifiers can be iterated, making possible the logical representations of distinctions such as that between uniform and pointwise convergence.

However, any simple catalogue of novelties understates the crystalline sharpness of Frege's logical writing when compared with works with similar aims, such as the later *Principia Mathematica* of Whitehead and RUSSELL [VI.71]. It would be several decades before logicians approached Frege's standards for exactness and clarity. The notation, however, seemed unwieldy to readers at the time (and since). Here, for example, is the statement "if not q, then every v is F" ($\neg q \implies (\forall v)F(v)$) in Frege's notation:

(Here $\top$ represents negation, $\smile^{v}$ is the universal quantifier, and the long vertical line represents the conditional.)

Frege also wrote an informal treatise, *Grundlagen der Arithmetik* (1884), which has had a profound influence on English-language philosophy since its translation in 1950. Its account of number contains the first hint of the tension that would collapse the project from within. Frege sets out conditions that a definition of number must satisfy to be counted as "acceptable." However, when formalized these lead to a contradiction, of a similar type to Russell's paradox (on the set of all sets that are not members of themselves). This problem escaped Frege's notice until Russell alerted him to it in a letter of 1903. Frege's reaction ("arithmetic totters") has been taken to be an overreaction to the failure of one set of axioms among many possible ones. But in Frege's view, the problem was not with the specific axioms, but rather that any logically adequate weakening appeared to violate some principle he took to be grounded in the nature of thought. Recently, many logicians who do not share Frege's often baroque-seeming metaphysics of concepts have shown that some natural consistent weakenings of Frege's system do support the derivation of the mathematics Frege aimed to reconstruct.

The years after 1903 brought personal tragedies in Frege's life and he ceased serious work for over a decade. Though he resumed writing in 1918 with a series of philosophical articles, his only research in mathematics was a brief jotted effort to found arithmetic on geometry, rather than logic, indicating his conclusion that his logical program had failed.

Further Reading

A particularly detailed recent example of "neo-Fregean" reconstructions of Frege's foundations of arithmetic appears in John Burgess's *Fixing Frege* (Princeton University Press, Princeton, NJ, 2005). Many of the classic papers on the technical details of reconstructing Frege's philosophy of logic are reprinted in *Frege's Philosophy of Mathematics* (Harvard University Press, Harvard, MA, 1995) by William Demopoulos.

Jamie Tappenden

VI.57   Christian Felix Klein

b. Düsseldorf, Germany, 1849; d. Göttingen, Germany, 1925
Higher geometry; function theory; theory of algebraic equations; pedagogy

Klein had originally intended to be a physicist but during the course of his studies with Julius Plücker in Bonn, with whom he studied both mathematics and physics, he turned to mathematics, receiving his doctorate for a thesis on line geometry in 1868. After Plücker's death in 1868 he went to Göttingen to study with Alfred Clebsch, where he worked exclusively on mathematics. In 1869–70 he spent some months in Berlin studying with WEIERSTRASS [VI.44] and KUMMER [VI.40] before joining LIE [VI.53] for a trip to Paris to see HERMITE [VI.47]. After passing his habilitation in Göttingen in 1871, he took positions successively at Erlangen, Munich, and Leipzig, returning to Göttingen in 1886, where he remained until he retired (because of poor health) in 1913. In 1875 he married Anna Hegel, a granddaughter of the philosopher Georg Wilhelm Friedrich Hegel.

In 1872 Klein published his celebrated "Erlanger Programm," a creative and unified conception of geometry. Building on a paper of CAYLEY [VI.46] of 1859 in which Cayley had shown how to deduce EUCLIDEAN GEOMETRY [I.3 §6.2] from PROJECTIVE GEOMETRY [I.3 §6.7], Klein applied his knowledge of group theory (learned from JORDAN [VI.52] in Paris) to create a hierarchy of all geometries. He had recognized that each geometry could be characterized by a group of transformations and classified accordingly (see SOME FUNDAMENTAL MATHEMATICAL DEFINITIONS [I.3 §6.1]). The

classification showed, as Klein had anticipated, that of all the geometries, projective is the most basic and that the others, e.g., affine, hyperbolic, Euclidean, etc., are subsumed at some level beneath it. Moreover, it was clear from his construction that a contradiction in NON-EUCLIDEAN GEOMETRY [II.2 §§6–10] would simultaneously involve a contradiction in Euclidean geometry.

Klein regarded his work in function theory as his greatest achievement. As his career progressed, he moved more and more from Plücker's and Clebsch's strictly geometric viewpoint toward the wider outlook embraced by RIEMANN [VI.49], who had regarded analytic functions as given by conformal mappings between given domains. In his "Riemanns Theorie der algebraischen Funktionen und ihrer Integrale" (1882), Klein gave a geometric treatment of function theory in which he fused Riemann's ideas with the rigorous power-series methods of Weierstrass.

In 1882, when he was at the height of his powers, Klein's health broke down. His attempt to keep up with POINCARÉ [VI.61] in the race to develop the theory of automorphic functions (which are generalizations of periodic functions such as trigonometric functions, ELLIPTIC FUNCTIONS [V.31], etc.), during which he had proved his famous Grenzkreis (boundary circle) Theorem, had left him exhausted, and he was never again able to work with such intensity and at such a high level.

After his breakdown Klein's interest shifted progressively from research toward pedagogy. In his efforts to modernize mathematical education he developed outstanding organizational skills and initiated important and far-reaching editorial projects ranging from the preparation of lecture notes to coediting the twenty-four-volume *Encyklopädie der mathematischen Wissenschaften* (1896–1935). He was an editor of the *Mathematische Annalen* for almost fifty years, and was among the founding members of the Deutsche Mathematiker-Vereinigung (1890). He also played an active role in establishing mathematical applications in science and engineering, as well as promoting the better understanding of mathematics by engineers.

Among Klein's other achievements were important results in the theory of algebraic equations (through a consideration of the icosahedron he obtained a complete theory of the general fifth-degree equation (1884)) and in mechanics, in which, jointly with Arnold Sommerfeld, he developed the theory of the gyroscope (1897–1910). He also worked on ideas involving the application of group theory to the theory of relativity, producing papers on the LORENTZ GROUP [IV.13 §1]

(1910) and gravitation (1918). Klein was an international figure who traveled widely, including to the United States and the United Kingdom, and he played a significant role in the first International Congresses of Mathematicians. His many foreign students included several from the United States, e.g., Maxime Bôcher and William Fogg Osgood, and a number of women, notably Grace Chisholm Young and Mary Winston.

Klein's achievements made Göttingen the scientific center of Germany and one of the mathematical centers of the world. He possessed an outstanding ability to "see" the truth in mathematical statements and to bring mathematical fields together without feeling the necessity for detailed calculations and justification (which he left to his students and others). He believed strongly in the unity of mathematics.

Further Reading

Frei, G. 1984. Felix Klein (1849–1925), a biographical sketch. In *Jahrbuch Überblicke Mathematik*, pp. 229–54. Mannheim: Bibliographisches Institut.

Klein, F. 1921–23. *Gesammelte mathematische Abhandlungen*, three volumes. Berlin: Springer. (Reprinted, 1973. Volume 3 contains lists of Klein's publications, lectures, and dissertations directed by him.)

———. 1979. *Development of Mathematics in the 19th Century*, translated by M. Ackerman. Brookline, MA: MathSci-Press.

Rüdiger Thiele

VI.58   Ferdinand Georg Frobenius

b. Berlin, 1849; d. Berlin, 1917
Analysis; linear algebra; number theory; theory of groups; character theory

After school in Berlin, Frobenius (who suppressed his first name and wrote mainly as G. Frobenius) spent one semester studying mathematics and physics in Göttingen, then returned to Berlin where he studied under KRONECKER [VI.48], KUMMER [VI.40], WEIERSTRASS [VI.44], and others. He wrote his doctoral dissertation (in Latin) supervised by Weierstrass, in 1870, on infinite series representations of analytic functions of one variable. For four years he worked as a schoolteacher in Berlin before he became Außerordentlicher Professor (associate professor) at Berlin University. After less than two years, in 1875, he was called to a full professorship at the Eidgenössische Technische Hochschule in Zürich, where he remained until

1892, when he returned to Berlin as successor to KRONECKER [VI.48]. He retired in 1916, and died one year later.

His early contributions were to analysis and the theory of differential equations. Later he wrote mainly on theta functions, algebra, and number theory. One of his well-known contributions lies across group theory and number theory. Given a polynomial with coefficients in an algebraic number field, one may ask for the degrees of the irreducible factors that occur when it is reduced modulo a prime ideal. In particular, one may ask for the "density" (suitably defined) of the set of prime ideals modulo which a given pattern of irreducible-factor degrees arises. Pursuing ideas of Kronecker, Frobenius proved that, if the GALOIS GROUP [V.21] is the SYMMETRIC GROUP [III.68], then that density is the proportion of elements of the group whose cycle structure is the pattern of degrees. He conjectured that this should be true whatever the Galois group. A tool he used for this led to the name "Frobenius automorphism" for the natural generator $a \mapsto a^q$ of the Galois group of a finite extension of the field $\mathbb{F}_q$. The conjecture was proved by N. G. Chebotaryov in 1925 and is now known as the Chebotaryov density theorem, or, sometimes, the Frobenius–Chebotaryov density theorem.

Another well-known and important contribution was to the theory of matrices and linear transformations, where Frobenius introduced the minimal polynomial and other invariants (the elementary divisors).

Frobenius is best known for his work in finite group theory. Like Otto Hölder and WILLIAM BURNSIDE [VI.60], he focused for a time on the search for FINITE SIMPLE GROUPS [V.7]. His greatest contribution, however, is his invention of the theory of GROUP CHARACTERS [IV.9]. This emerged unexpectedly in 1896 out of his study of group determinants. These are the determinants of square matrices with rows and columns indexed by the members of a finite group G, and with (a, b)-entry $x_{ab^{-1}}$, where the x_g are independent variables, one for each element g of G. His interest, stimulated by correspondence with DEDEKIND [VI.50], was in how the group determinant factorizes as a polynomial in these variables. This problem led Frobenius to the discovery of certain sets of complex numbers, which he called *Gruppencharactere*, one for each conjugacy class in the group, that arose as the solutions of sets of linear equations connected with the group. Nowadays they are defined differently: for each complex linear representation ρ of the group G (that is, homomorphism

$\rho : G \to \mathrm{GL}_n(\mathbb{C})$, where $\mathrm{GL}_n(\mathbb{C})$ is the group of $n \times n$ invertible matrices over $\mathbb{C}$), the associated character χ is the map $G \to \mathbb{C}$ such that $\chi(g) = \mathrm{trace}\, \rho(g)$ for $g \in G$. Frobenius proved the orthogonality relations, recognized the connection of his characters with matrix representations of the group, calculated the character tables of the symmetric groups, the alternating groups, and the Mathieu groups, and used properties of induced characters to prove his famous theorem that a transitive permutation group in which no element other than the identity fixes two or more points has a regular normal subgroup (that is, a subgroup consisting of the identity together with the fixed-point-free elements of the group). To this day no purely group-theoretic proof of this theorem has been found. In recognition of his contribution such groups are now called Frobenius groups. Through character theory and representation theory, as developed by Frobenius for finite groups (and by his pupil, friend, and colleague Issai Schur for classical matrix groups), group theory found important applications in physics and chemistry a generation later.

Further Reading

Begehr, H., ed. 1998. *Mathematik in Berlin: Geschichte und Dokumentation*, two volumes. Aachen: Shaker.

Curtis, C. W. 1999. *Pioneers of Representation Theory: Frobenius, Burnside, Schur, and Brauer.* Providence, RI: American Mathematical Society.

Serre, J.-P., ed. 1968. *F. G. Frobenius: Gesammelte Abhandlungen*, three volumes. Berlin: Springer.

Peter M. Neumann

VI.59 Sofya (Sonya) Kovalevskaya

b. Moscow, 1850; d. Stockholm, 1891
Partial differential equations; Abelian integrals

Kovalevskaya showed talent for mathematics at an early age but as a woman in mid-nineteenth-century Russia she was denied access to university. Unable to leave the country unescorted she married and in 1869 traveled to Heidelberg, where she was taught mathematics by Du Bois-Reymond. The following year she moved to Berlin to work with WEIERSTRASS [VI.44]. Berlin University was closed to women but Weierstrass agreed to tutor her privately. Under his supervision Kovalevskaya completed dissertations on partial differential equations (PDEs), Abelian integrals, and Saturn's rings, and in 1874 she became the first woman to receive a doctorate in mathematics. The dissertation on PDEs, which excited particular attention, contained the result now known as the CAUCHY–KOVALEVSKAYA THEOREM [IV.12 §§2.2, 2.4], an important tool in establishing the existence of analytic solutions of PDEs.

That same year Kovalevskaya returned to Russia and, unable to find a suitable position, temporarily abandoned mathematics. In 1880, at the invitation of CHEBYSHEV [VI.45], she gave a paper on Abelian integrals at a conference in Saint Petersburg. It was enthusiastically received and in 1881 she returned to Berlin. She saw Weierstrass frequently and devoted herself to the study of the propagation of light in a crystalline medium— a subject to which she had been led by studying the work of the French physicist Gabriel Lamé—and to the study of the rotation of a solid body about a fixed point. Later that year she moved to Paris to work with mathematicians there.

In 1883, championed by Mittag-Leffler, Kovalevskaya was appointed as a Privatdozent at the University of Stockholm. She also became an editor of *Acta Mathematica*, making her the first woman to join the board of a scientific journal. On behalf of *Acta* she liaised with mathematicians from Paris, Berlin, and Russia, providing an important link between Russian mathematicians and their western European counterparts. She continued to work on the rotation problem and in 1885 made the breakthrough that, three years later, would win her the prestigious *Prix Bordin* of the French Academy of Sciences. Prior to her work the problem had been completely solved for only two cases, both symmetrical. In the first, solved by EULER [VI.19], the center of gravity of the moving body coincides with the fixed point; and in the second, solved by LAGRANGE [VI.22], the center of gravity and the fixed point lie on the same axis. Kovalevskaya discovered that there was a third case, one that was asymmetrical and more complicated than the other two, which could also be solved completely. (It was later shown that there are no others.) The novelty of her results lay in her application of the recently developed theory of theta functions—the simplest elements from which ELLIPTIC FUNCTIONS [V.31] can be constructed—to solve Abelian integrals.

Kovalevskaya became a full professor of mathematics at the University of Stockholm in 1889, the first woman anywhere to achieve such a position. Shortly afterward, she was nominated by Chebyshev for corresponding membership of the Russian Academy of Sciences, her subsequent election breaking the gender barrier once again.

Further Reading

Cooke, R. 1984. *The Mathematics of Sonya Kovalevskaya*. New York: Springer.

Koblitz, A. H. 1983. *A Convergence of Lives. Sofia Kovalevskaia: Scientist, Writer, Revolutionary*. Boston, MA: Birkhäuser.

VI.60 William Burnside

b. London, 1852; d. West Wickham, England, 1927
Theory of groups; character theory; representation theory

Burnside's mathematical abilities first showed themselves at school. From there he won a place at Cambridge, where he read for the Mathematical Tripos and graduated as 2nd Wrangler in 1875. For ten years he remained in Cambridge as a Fellow of Pembroke College, coaching student rowers and mathematicians. In 1885, having published three very short papers, he was appointed professor at the Royal Naval College, Greenwich. He married in 1886 and the next year, at the age of thirty-five, he embarked on his career as a productive mathematician. He was elected as a Fellow of the Royal Society in 1893 on the basis of his contributions in applied mathematics (statistical mechanics and hydrodynamics), geometry, and the theory of functions. Although he continued to contribute to these areas throughout his working life, and added probability theory to his fields of interest during World War I, he turned to the theory of groups in 1893, and it is for his discoveries in this subject that he is remembered.

Burnside treated every aspect of the theory of finite groups. He was much concerned with the search for finite simple groups, and made the famous conjecture, finally proved by Walter Feit and John Thompson in 1962, that there are no simple groups of odd composite order (see THE CLASSIFICATION OF FINITE SIMPLE GROUPS [V.7]). He helped to develop character theory, which had been created by FROBENIUS [VI.58] in 1896, into a tool for proving theorems of pure group theory, using it in 1904 to spectacular effect when he proved his so-called $p^\alpha q^\beta$-theorem: the theorem that groups whose orders are divisible by at most two different prime numbers are soluble. By asking, in effect, whether a group all of whose elements have finite order and which is generated by finitely many elements must be finite, he launched the huge area of research which for much of the twentieth century was known as the Burnside problem (see GEOMETRIC AND COMBINATORIAL GROUP THEORY [IV.10 §5.1]).

Although CAYLEY [VI.46] and the Reverend T. P. Kirkman had written about groups before him, he was the only British mathematician to work in group theory until Philip Hall started his mathematical career in 1928. Burnside's influential book *Theory of Groups of Finite Order* (1897) was written in the hope of "arousing interest among English mathematicians in a branch of pure mathematics which becomes the more interesting the more it is studied." Its influence in his own country was minimal, however, until several years after his death. It went to a second edition in 1911 (reprinted 1955), which differs from the first in that it has been substantially revised and, in particular, it includes chapters about the character theory of finite groups and its applications—mathematics which had been much developed by Frobenius, Burnside, and Schur over the fifteen years following the invention of character theory in 1896.

Further Reading

Curtis, C. W. 1999. *Pioneers of Representation Theory: Frobenius, Burnside, Schur, and Brauer*. Providence, RI: American Mathematical Society.

Neumann, P. M., A. J. S. Mann, and J. C. Tompson. 2004. *The Collected Papers of William Burnside*, two volumes. Oxford: Oxford University Press.

Peter M. Neumann

VI.61 Jules Henri Poincaré

b. Nancy, France, 1854; d. Paris, 1912
Function theory; geometry; topology; celestial mechanics; mathematical physics; foundations of science

Educated at the École Polytechnique and the École des Mines in Paris, Poincaré began his teaching career at the University of Caen in 1879. In 1881 he took up an appointment at the University of Paris where, from 1886, he held successive chairs until his death in 1912. He was of a retiring nature and did not attract graduate students, but his lecture courses provided the basis for a number of treatises, mostly in mathematical physics.

Poincaré came to international prominence in the early 1880s when, fusing ideas from complex function theory, group theory, non-Euclidean geometry, and the theory of ordinary linear differential equations, he identified an important class of automorphic functions. Named Fuchsian functions, in honor of the mathematician Lazarus Fuchs, they are defined on a disk and

Jules Henri Poincaré

remain invariant under certain discrete groups of transformations. Soon after, he identified the related but more complicated Kleinian functions, which are automorphic functions without a limit circle. His theory of automorphic functions was the first significant application of non-Euclidean geometry. It led to his discovery of the disk model of the hyperbolic plane and later inspired the UNIFORMIZATION THEOREM [V.34].

During the same period Poincaré began pioneering work on the qualitative theory of differential equations, motivated in part by an interest in some of the fundamental questions of mechanics, notably the problem of the stability of the solar system. What was new and important was his idea of thinking of the solutions in terms of curves rather than functions, i.e., thinking geometrically rather than algebraically, and it was this that marked a departure from the work of his predecessors, whose research had been dominated by power-series methods. From the mid 1880s he began applying his geometric theory to problems in celestial mechanics. His memoir on THE THREE-BODY PROBLEM [V.33] (1890) is famous both for providing the basis for his acclaimed treatise, *Les Méthodes Nouvelles de la Mécanique Céleste* (1892–99), and for containing the first mathematical description of CHAOTIC BEHAVIOR [IV.14 §1.5] in a dynamical system. Stability was also at the heart of his investigation into the forms of rotating fluid masses (1885). This work, which contained the discovery of new, pear-shaped figures

of equilibrium, aroused considerable attention because of its important implications for cosmogony in relation to the evolution of binary stars and other celestial bodies.

Poincaré's work on Fuchsian functions and on the qualitative theory of differential equations led him to recognize the importance of the topology (or, as it was then called, *analysis situs*) of MANIFOLDS [I.3 §6.9]. And in the 1890s he began to study the topology of manifolds as a subject in its own right, effectively creating the powerful independent field of ALGEBRAIC TOPOLOGY [IV.6]. In a series of memoirs published between 1892 and 1904, the last of which contains the hypothesis known today as THE POINCARÉ CONJECTURE [IV.7 §2.4], he introduced a number of new ideas and concepts, including Betti numbers, THE FUNDAMENTAL GROUP [IV.6 §2], HOMOLOGY [IV.6 §4], and torsion.

A deep interest in physical problems lay behind Poincaré's achievements in mathematical physics. His work in potential theory forms a bridge between that of Carl Neumann on boundary-value problems and that of FREDHOLM [VI.66] on integral equations. He introduced a technique known as the "méthode de balayage" ("sweeping-out method") for establishing the existence of solutions to the DIRICHLET PROBLEM [IV.12 §1] (1890); and he had the idea that the Dirichlet problem itself should give rise to a sequence of EIGENVALUES AND EIGENFUNCTIONS [I.3 §4.3] (1898). In developing the theory for functions of several variables he was led to the discovery of new results in complex function theory. In *Électricité et Optique* (1890, revised 1901), which derived from his university lectures, he gave an authoritative account of the electromagnetic theories of Maxwell, Helmholtz, and Hertz. In 1905 he responded to Lorentz's new theory of the electron, coming close to anticipating Einstein's theory of SPECIAL RELATIVITY [IV.13 §1], thereby provoking controversy among later writers about the question of priority. And in 1911 he attended the first Solvay Conference on quantum theory, publishing an influential memoir (1912) in its favor.

As Poincaré's career developed, so too did his interest in the philosophy of mathematics and science. His ideas became widely known through four books of essays: *La Science et l'Hypothèse* (1902), *La Valeur de la Science* (1905), *Science et Méthode* (1908), and *Dernières Pensées* (1913). As a philosopher of geometry he was a proponent of the view, known as conventionalism, that it is not an objective question which model of geometry best fits physical space but is rather a matter of which model we find most convenient. By contrast, his position on arithmetic was intuitionist. On the question of foundational issues, he was largely critical. Although sympathetic to the goals of set theory, he attacked what he perceived as its counterintuitive results. (See THE CRISIS IN THE FOUNDATIONS OF MATHEMATICS [II.7 §2.2] for further discussion.)

Poincaré's visionary geometric style led him to new and brilliant ideas, which frequently connected different branches of mathematics, but lack of detail often made his work hard to follow. At times his approach was censured for imprecision; it was in marked contrast to that of HILBERT [VI.63], his German counterpart, whose work was rooted in algebra and rigor.

Further Reading

Barrow-Green, J. E. 1997. *Poincaré and the Three Body Problem*. Providence, RI: American Mathematical Society.
Poincaré, J. H. 1915–56. *Collected Works: Œuvres de Henri Poincaré*, eleven volumes. Paris: Gauthier Villars.

VI.62   Giuseppe Peano

b. Spinetta, Italy, 1858; d. Turin, 1932
Analysis; mathematical logic; foundations of mathematics

Known above all for his (and DEDEKIND's [VI.50]) axiom system for the natural numbers, Peano made important contributions to analysis, logic, and the axiomatization of mathematics. He was born in Spinetta (Piedmont, Italy) as the son of a peasant, and from 1876 studied at the University of Turin, taking his doctoral degree in 1880. He remained there until his death in 1932, becoming full professor in 1895.

During the 1880s Peano worked in analysis, achieving what are generally considered to be his most important results. Particularly noteworthy are the continuous space-filling *Peano curve* (1890), the notion of *content* (a precedent of MEASURE THEORY [III.55]) developed independently by JORDAN [VI.52], and his theorems on the existence of solutions for differential equations of the first order (1886, 1890). The textbook he published in 1884, *Calcolo Differentiale e Principii di Calcolo Integrale*, partly based on lectures by his teacher Angelo Genocchi, was noteworthy for its rigor and critical style, and is counted among the very best nineteenth-century treatises.

The years 1889–1908 saw Peano dedicating himself intensively to symbolic logic, axiomatization, and producing the encyclopedic *Formulaire de Mathématiques* (1895–1908, five volumes). This ambitious assembly of mathematical results, compactly presented in the symbols of mathematical logic, was given completely without proofs. This was by no means standard at the time, but it shows what Peano expected from logic: it was supposed to bring precision of language and brevity, but not a greater level of rigor (something that was, by contrast, crucial for FREGE [VI.56]). In 1891, together with some colleagues, he founded the journal *Rivista di Matematica*, gathering around him an important group of followers.

Peano was an accessible man, and the way he mingled with students was regarded as "scandalous" in Turin. He was a socialist in politics, and a tolerant universalist in all matters of life and culture. In the late 1890s Peano became increasingly interested in elaborating a universal spoken language, "Latino sine flexione"; the last edition of the *Formulario* (1905–8) appeared in this language.

Peano followed closely the work of German mathematicians such as Hermann Grassmann, Ernst Schröder, and Richard Dedekind; for example, the 1884 textbook defined the real numbers by Dedekind cuts, and in 1888 he published *Calcolo Geometrico Secondo l'Ausdehnungslehre di H. Grassmann*. In 1889 there appeared (notably in Latin) a first version of the famous PEANO AXIOMS [III.67] for the set of natural numbers, which he refined in volume 2 of the *Formulaire* (1898). It aimed at filling the most significant gap in the foundations of mathematics at a time when the *arithmetization* of analysis had essentially been completed. It is no coincidence that other mathematicians (Frege, Charles S. Peirce, and Dedekind) published similar work in the same decade. Peano's attempt is better rounded than Peirce's, but simpler and framed in more familiar terms than those of Frege and Dedekind; because of this, it has been more popular.

Peano's work on the natural numbers was at the crossroads of his diverse mathematical contributions, linking naturally his previous research in analysis with his later work on logical foundations, and being a necessary prerequisite for the *Formulaire* project. Actually, *Arithmetices Principia* can be regarded as a simplification, refinement, and translation into logical language (the "nova methodo" in its title) of Grassmann's *Lehrbuch der Arithmetik* (1861). Grassmann had striven to elaborate a stern deductive structure, stressing proofs by mathematical induction and recursive definitions. But curiously, unlike Peano, he did not postulate an axiom of induction; thus, Peano presented the basic assumptions much more clearly, bringing induction to center stage as the key defining property of the natural numbers.

Further Reading

Borga, M., P. Freguglia, and D. Palladino. 1985. *I Contributi Fondazionali della Scuola di Peano*. Milan: Franco Angeli.

Ferreirós, J. 2005. Richard Dedekind (1888) and Giuseppe Peano (1889), booklets on the foundations of arithmetic. In *Landmark Writings in Western Mathematics 1640–1940*, edited by I. Grattan-Guinness, pp. 613–26. Amsterdam: Elsevier.

Peano, G. 1973. *Selected Works of Giuseppe Peano*, with a biographical sketch and bibliography by H. C. Kennedy. Toronto: University of Toronto Press.

José Ferreirós

VI.63   David Hilbert

b. Königsberg, Germany, 1862; d. Göttingen, Germany, 1943
Invariant theory; number theory; geometry;
International Congress of Mathematicians; axiomatics

David Hilbert

HERMANN WEYL [VI.80] described his teacher Hilbert's style: "It is as if you were on a swift walk through a sunny open landscape; you look freely around, demarcation lines and connecting roads are pointed out to you, before you must brace yourself to climb the hill; then the path goes straight up...." Several themes balance in Hilbert's career as a mathematician. He wanted clarity, rigor, simplicity, and depth. Though he loved mathematics for its beauty, a beauty that transcends human failures, Hilbert saw mathematics as a social collaboration. A turning point came when he met MINKOWSKI [VI.64] and Adolf Hurwitz at university in Königsberg.

Hilbert wrote: "On unending walks we engrossed ourselves in the actual problems of the mathematics of the time; exchanged our newly acquired understandings, our thoughts and scientific plans; and formed a friendship for life." Later Hilbert became professor at Göttingen and, with KLEIN [VI.57], drew mathematicians from all over the world and turned that small city into a crossroads for mathematics—until Hitler destroyed it.

When he was a new Privatdozent, Hilbert decided he would study mathematics as he taught, and he resolved never to repeat lectures. He and Hurwitz decided to embark on a "systematic exploration" of mathematics, and he followed this pattern for the rest of his life. Hilbert's career divides easily into six periods: (i) algebra and algebraic invariants (1885–93); (ii) algebraic number theory (1893–98); (iii) geometry (1898–1902);

(iv) analysis (1902–12); (v) mathematical physics (1910–22); and (vi) foundations (1918–30). Remarkably, there is very little overlap. When Hilbert finished a subject, he was finished with it.

Hilbert's first breakthrough came in 1888 when he solved Gordan's problem, named after Paul Gordan, in a single bold move. Given a polynomial equation with at least two variables, some things about the polynomial change and some do not when you change coordinate systems. For example, consider the real polynomial equation

$$ax^2 + bxy + cy^2 + d = 0.$$

If you rotate the coordinate system, then this equation changes dramatically, but the graph does not, and neither does the discriminant $b^2 - 4ac$. The discriminant is one invariant. In the general case—a more complicated class of polynomials and coordinate changes—there can be many invariants. Mathematicians suspected that a finite number of essentially different invariants existed for any given type of polynomial and class of coordinate changes. Was this so? Many mathematicians calculated individual examples industriously. Instead, Hilbert reasoned indirectly: what if there is no finite basis for a specific class of polynomials and transformations? He found that it was always possible to produce a contradiction. He concluded that there must be such a basis. At first this result was greeted with disbelief because he did not display a basis. Gordan said, "Das ist nicht Mathematik. Das ist

Theologie." However, the result was so powerful that it has been said that it killed algebraic invariant theory.

In 1893 Hilbert and Minkowski were asked by the German Mathematical Society to write a report on number theory. Hilbert chose ALGEBRAIC NUMBER THEORY [IV.1] and transformed the results of the nineteenth century into the study of algebraic NUMBER FIELDS [III.63]. The deep organizing structure Hilbert found eventually led to what has been called "the magnificent edifice of class field theory" (described in [V.28]).

Hilbert's classic *Foundations of Geometry*, first published in 1899 and revised many times, starts with real-number arithmetic. He assumes that it is consistent, i.e., that it is free of the possibility of contradictory deductions. Using analytic geometry, he then exhibits a model of EUCLIDEAN GEOMETRY [II.2 §3]. A point is a pair of real numbers; a line is a set of pairs of numbers that satisfy the equation for a line; a circle...; and so on. All of Euclid's axioms are true statements about these "lines" and "points," that is, they are true statements about these sets of real numbers. Euclidean geometry is thereby reduced to a fraction of all the true statements about real numbers, and we conclude that if real-number arithmetic is consistent then Euclid's geometry is consistent. Next Hilbert constructs models of various non-Euclidean geometries in terms of Euclidean geometry, exploring in depth and with great inventiveness which possible axioms follow from which groups of axioms and which are independent yet consistent.

Hilbert was invited to address the Second International Congress of Mathematicians in Paris in 1900. He gave a talk proposing twenty-three problems for the new century. These problems are known today as "Hilbert's problems"; in a sense they have created a virtual Göttingen where mathematicians have entered into conversation with Hilbert and each other ever since.

Next Hilbert turned to analysis. WEIERSTRASS [VI.44] had found counterexamples to Dirichlet's principle, which is essentially the assertion that, in variational problems, maxima and minima are always attained. Hilbert proved a modified, but still powerful, version that "salvaged" much of the work that assumed the principle. The larger theme of this period, though, was integral equations and what is now called HILBERT SPACE [III.37]. Newton's equations for motion are differential equations, and it was natural to phrase equations in physics that way. However, in many cases it was easier to solve problems if the equations were written using integrals rather than derivatives. Between 1902 and 1912 Hilbert attacked a variety of problems

from this direction. He viewed the solutions as part of Hilbert space and gave a spectral interpretation analogous to an infinite-dimensional vector space. Thus, an amorphous sea of functions acquired geometric structure.

In 1910 he turned toward mathematical physics and had some successes, but physics was undergoing multiple revolutions and was not ready for mathematical clarification.

When he delivered his problems in 1900, Hilbert was aware that there were contradictions in mathematics as it was then phrased, and specifically in set theory. His second problem asked for a proof that first arithmetic, and then set theory, were consistent. As the debate widened, some mathematicians began to pull back on what they accepted as valid reasoning. Hilbert wanted none of this. By 1918 he was increasingly focused on a program to formally axiomatize mathematics and prove it free of contradictions using proof-theoretic, combinatorial methods. GÖDEL [VI.92] proved his incompleteness theorems in 1930 and thereby showed that Hilbert's program, at least as initially conceived, could never be successful. Hilbert was wrong here, but even if wrong, his dream of placing mathematics on a formal foundation stimulated some of the most important work of the twentieth century—and mathematics did not pull back.

Further Reading

Reid, C. 1986. *Hilbert-Courant*. New York: Springer.
Weyl, H. 1944. David Hilbert and his mathematical work. *Bulletin of the American Mathematical Society* 50:612–54.

Benjamin H. Yandell

VI.64   Hermann Minkowski

b. Alexotas, Russia (present day Kaunas, Lithuania), 1864;
d. Göttingen, Germany, 1909
Number theory; geometry; relativity theory

In 1883, the Paris Academy of Sciences awarded its prestigious Grand Prix for mathematical science to the eighteen-year-old student Hermann Minkowski. The prize problem was to give the number of representations of an integer as a sum of five squares of integers. In a manuscript of 140 pages written in German, Minkowski developed a general theory of QUADRATIC FORMS [III.73] that contains the solution to this problem as a special case. Two years later, Minkowski obtained his Ph.D. in Königsberg, and in 1887 he received his habilitation in Bonn with further work on quadratic forms in n variables.

While a student in Königsberg, Minkowski became a close friend of Adolf Hurwitz and HILBERT [VI.63]. In 1894, after Hurwitz had moved to Zürich, Minkowski returned from Bonn to his alma mater, and soon became Hilbert's successor after Hilbert left for Göttingen. In 1896, Minkowski moved on to Zurich to become Hurwitz's colleague. In 1902, Hilbert negotiated for another chair of mathematics to be created for Minkowski in Göttingen. There he worked as Hilbert's colleague and closest friend until he died, unexpectedly, of a ruptured appendix in early 1909.

Minkowski's later work is characterized by an ingenious use of geometric intuition for the solution of number-theoretic problems. His starting point was a theorem of HERMITE [VI.47] on the smallest positive real that can be represented by a given positive-definite quadratic form of n integer-valued nonzero variables. By interpreting the quadratic forms in terms of geometric objects such as ellipses (for $n = 2$) or ellipsoids (for $n = 3$), and considering the integer values of the variables as the coordinates of the points of a regular lattice, Minkowski was able to employ the notion of volume to arrive at nontrivial number-theoretic results. His investigations were published in 1896 in a book entitled *The Geometry of Numbers*. Realizing that the geometric arguments based on ellipsoids used only the property of convexity, Minkowski further generalized his theory by introducing a general concept of convex point sets. A *convex body*, according to Minkowski, is one in which the straight line connecting any two interior points lies completely within the set. This notion allowed Minkowski to investigate a geometry in which the Euclidean axiom about the congruence of triangles is replaced by the weaker axiom that the sum of two sides of a triangle is always larger than the third one (which we would nowadays call the *triangle inequality*, the key notion in metric spaces). Theorems about this Minkowskian geometry also produced immediate nontrivial number-theoretic results. Further results were obtained in the theory of CONTINUED FRACTIONS [III.22]. In 1907, Minkowski published introductory lectures on number theory under the title *Diophantine Approximations*.

Minkowski always had a deep interest in physics. In 1906, he wrote the article on capillarity for the authoritative *Encyclopedia of the Mathematical Sciences* (edited by KLEIN [VI.57] and others). In Göttingen, Hilbert and Minkowski gave joint seminars in which they studied contemporary work in electrodynamics by POINCARÉ [VI.61], Einstein, and others. Minkowski soon realized the significance of the fact that the special theory of relativity was a consequence of the invariance of the Maxwell equations under the group of Lorentz transformations (see GENERAL RELATIVITY AND THE EINSTEIN EQUATIONS [IV.13 §1]). He reinterpreted Maxwell–Lorentz electrodynamics geometrically in a mathematical formulation in which no formal distinction between the space and time coordinates exists. This is expressed in the famous opening words of his address to the Cologne meeting of the Society of German Scientists and Physicians a few weeks before his death: "From this hour on, space by itself and time by itself are to sink fully into shadows and only a kind of union of the two should yet preserve autonomy." Minkowski's four-dimensional Lorentz-covariant formulation of special relativity was a prerequisite for Einstein's later general theory of relativity.

Further Reading

Hilbert, D. 1910. Hermann Minkowski. *Mathematische Annalen* 68:445–71.
Walter, S. 1999. Minkowski, mathematicians, and the mathematical theory of relativity. In *The Expanding Worlds of General Relativity*, edited by H. Goenner et al., pp. 45–86. Boston: Birkhäuser.

Tilman Sauer

VI.65 Jacques Hadamard

b. Versailles, France, 1865; d. Paris, 1963
Function theory; calculus of variations; number theory; partial differential equations; hydrodynamics

A graduate of the École Normale in Paris, Hadamard obtained a position at the University of Bordeaux in 1893. He returned to Paris in 1897 where he taught at the Collège de France, the École Polytechnique, and the École Centrale until his retirement in 1937. The Hadamard Seminar at the Collège de France, where mathematicians came from around the world to expound on recent results, was an influential and integral part of mathematical life in France between the wars.

Hadamard's first significant papers were concerned with the theory of HOLOMORPHIC FUNCTIONS [I.3 §5.6] of a complex variable, in particular with the analytic continuation of a Taylor series; and in his thesis of 1892 he investigated how the properties of the singularities of a series could be deduced from those of its coefficients. Notably he showed that the radius of convergence R of a Taylor series $\sum a_n z^n$ could be given by $R = (\lim_{n\to\infty} \sup |a_n|^{1/n})^{-1}$, a result now known as the *Cauchy–Hadamard theorem*. (CAUCHY [VI.29] had published the formula in 1821 but Hadamard, who had discovered it independently, was the first to give a complete proof.) Further results followed, including the famous "Hadamard gap theorem," which gives the condition for the circle of convergence of the series to be a natural boundary of the function. His monograph *La Série de Taylor et son Prolongement Analytique* (1901) proved especially influential. In 1912 he formulated the problem of quasi-analyticity for infinitely differentiable functions.

The year 1892 also saw the appearance of Hadamard's prize-winning memoir on entire functions, in which he used results from his thesis to establish the relations between the coefficients of the Taylor series of an entire function and its zeros, and then applied them to evaluate the genus of the entire function. He applied this work, and other results from his thesis, to the RIEMANN ZETA FUNCTION [IV.2 §3], which enabled him, in 1896, to prove his most famous result: the PRIME NUMBER THEOREM [V.26]. (The theorem was proved simultaneously by DE LA VALLÉE POUSSIN [VI.67] but in a more complicated way.)

Hadamard's other key achievements of the 1890s include a well-known inequality concerning DETERMINANTS [III.15] (1893), a result essential in the FREDHOLM THEORY [IV.15 §1] of integral equations; and his "three-circles theorem" (1896), which demonstrates the importance of convexity in the study of analytic functions and plays a significant role in interpolation theory.

In 1896 Hadamard won the *Prix Bordin* for his study of the behavior of geodesics on surfaces. (The motivation for studying such geodesics is that they can be used to represent the trajectories of motion in dynamical systems.) It was Hadamard's first major work on a subject other than analysis. His two papers, one on geodesics on a surface of positive curvature (1897) and the other on geodesics on a surface of negative curvature (1898), are characterized by a qualitative analysis inherited from POINCARÉ [VI.61]. The first relies on results from classical differential geometry, while the second is dominated by topological considerations.

Prompted by an interest in the CALCULUS OF VARIATIONS [III.94], Hadamard developed the ideas of Volterra's functional calculus. In 1903 he was the first to describe linear functionals on a function space. By considering the space of continuous functions on a given interval, he showed that every functional is the limit of a sequence of intervals, a result now recognized as a precursor to the RIESZ REPRESENTATION THEOREM [III.18] formulated by RIESZ [VI.74] in 1909. Hadamard's influential *Leçons sur le Calcul de Variations* (1910) is the first book in which the ideas of modern functional analysis can be found.

In applied mathematics Hadamard was primarily concerned with wave propagation, in particular high-speed flows. In 1900 he began working on the theory of partial differential equations, and in 1903 published *Leçons sur la Propagation des Ondes et les Équations de l'Hydrodynamique*; this was followed by *Lectures on Cauchy's Problem in Linear Partial Differential Equations* (1922). The latter contained the details of his fundamental idea of the WELL-POSED PROBLEM [IV.12 §2.4] (i.e., a problem in which the solution must not only exist and be unique but must also depend continuously on the initial data). The origins of the idea can be found in his 1898 paper on GEODESICS [I.3 §6.10].

Hadamard's book *The Psychology of Invention in the Mathematical Field* (1945) is well-known for its discussion of the unconscious and its role in mathematical discovery.

Further Reading

Hadamard, J. 1968. *Collected Works: Œuvres de Jacques Hadamard*, four volumes. Paris: CNRS.

Maz'ya, V., and T. Shaposhnikova. 1998. *Jacques Hadamard. A Universal Mathematician*. Providence, RI: American Mathematical Society/London Mathematical Society.

VI.66   Ivar Fredholm

b. Stockholm, 1866; d. Stockholm, 1927
Professor of Mechanics and Mathematical Physics, Stockholm (1906–27)

In papers of 1900 and 1903 Fredholm solved the integral equations named after him,

$$\varphi(x) + \int_a^b K(x, y)\varphi(y)\,\mathrm{d}y = \psi(x),$$

with a continuous "kernel" K and unknown $\varphi(x)$, by analogy with infinite systems of linear equations and generalized determinants. Both the solution and several ideas attached to it ("Fredholm alternatives") made this work an important stimulus for HILBERT's [VI.63] theory of integral equations (1904–6) and thus a starting point for functional analysis. (For more about this, see OPERATOR ALGEBRAS [IV.15 §1].) The equations arise in the context of problems of mathematical physics, e.g., in potential theory and in the theory of oscillations. Fredholm considered himself primarily a mathematical physicist, and his colleague Mittag-Leffler tried in vain to have him awarded the Nobel Prize for physics.

VI.67   Charles-Jean de la Vallée Poussin

b. Louvain, Belgium, 1866; d. Brussels, 1962
Analytic number theory; analysis

De la Vallée Poussin graduated in engineering (1890) and mathematics (1891) from the Université Catholique de Louvain, where he went on to teach mathematical analysis from 1891 until 1951. His lectures formed the basis for his renowned *Cours d'Analyse Infinitésimale*, which ran to many editions from 1903 to 1959. A member of the most famous academies in Europe and the United States, with honorary doctorates from Paris, Strasbourg, Toronto, and Oslo, he was the first president (1920) of the International Union of Mathematicians (now the International Mathematical Union). He was made a baron in 1930.

De la Vallée Poussin's main achievement was his proof in 1896 of the PRIME NUMBER THEOREM [V.26] (an asymptotic estimate for the distribution of prime numbers in the integers), first conjectured by GAUSS [VI.26] in around 1793. (The theorem was also proved independently by HADAMARD [VI.65] in the same year, also using complex function theory.) Shortly afterward, de la Vallée Poussin followed his proof with a sharper error term (1899), which he extended to prime numbers in an arithmetic progression.

When LEBESGUE [VI.72] first published his INTEGRAL [III.55] in 1902, de la Vallée Poussin immediately grasped its importance and, using an original approach, described it in the second edition of his *Cours d'Analyse* (1908). In addition, he introduced the concept of the characteristic function of a set (1915), and shortly afterward gave a decomposition theorem for the measure generated by a continuous function of bounded variation (1916).

Of particular importance for approximation theory and the summation of series is de la Vallée Poussin's convolution integral (1908), for approximating periodic functions by trigonometric polynomials. His other significant results in this field include a lower bound for the error in the best approximation of a continuous function by a polynomial (1910), and a conver-

gence test and a summation method for Fourier series (1918).

In 1911 de la Vallée Poussin was responsible for suggesting the Belgian Academy prize question that led to Jackson's and Bernstein's theorems on the order of the best approximation of a continuous function by polynomials. His existence and uniqueness theorem for the Chebyshev problem for an overdetermined system of linear equations (1911) was an important step in LINEAR PROGRAMMING [III.84]; his interpolation formula (1908) was fundamental for sampling theory; and his characterization of new classes of quasi-analytic functions by the rate of decrease of their Fourier coefficients (1915) was a notable development.

De la Vallée Poussin's other achievements include determining a uniqueness condition for multipoint boundary-value problems (1929), which was a significant result for the study of nonoscillatory solutions of linear differential equations; and solving various problems of the conformal representation of multiply connected regions (1930-31). In potential theory he extended the concept of capacity to arbitrary bounded sets, proved his extraction theorem for bounded sequences of set functions, and, by introducing measure theory into POINCARÉ's [VI.61] "méthode de balayage" ("sweeping-out method") for the DIRICHLET PROBLEM [IV.12 §1], he paved the way for modern abstract potential theory.

Further Reading

Butzer, P., J. Mawhin, and P. Vetro, eds. 2000-4. *Charles-Jean de la Vallée Poussin. Collected Works—Oeuvres Scientifiques*, four volumes. Bruxelles/Palermo: Académie Royale de Belgique/Circolo Matematico di Palermo.

Jean Mawhin

VI.68   Felix Hausdorff

b. Breslau, Germany (now Wrocław, Poland), 1868;
d. Bonn, Germany, 1942
Set theory; topology

Hausdorff studied mathematics at Leipzig, Freiburg, and Berlin between 1887 and 1891, and then started research in applied mathematics at Leipzig under Heinrich Bruns. After his habilitation (1895) he taught first at Leipzig and then later at Bonn (1910-13, 1921-35) and Greifswald (1913-21). He is best known for his work in set theory and general topology, his magnum opus being *Grundzüge der Mengenlehre* ("Basic features of set theory"). It was published in 1914 and had second and third editions in 1927 and 1935. The second edition was so heavily revised in content, however, that it should really be considered a new book.

In his early work, Hausdorff concentrated on applied mathematics, mainly related to astronomy, in particular the refraction and extinction of light in the atmosphere. He had broad intellectual interests and moved in Nietzschean circles of artists and poets at Leipzig. Under the pseudonym Paul Mongré he wrote two long philosophical essays of which the more prominent was "Das Chaos in kosmischer Auslese" ("The chaos in cosmic selection"). Until 1904 he regularly contributed cultural critical essays to a renowned German intellectual review of the time, continuing to contribute, although less frequently, until 1912. He also published poems and a satirical play.

Hausdorff took up set theory at the turn of the century and gave his first lecture course on the topic in the summer semester of 1901 at Leipzig university. After his turn toward "Cantorianism" (set theory) he began deep and innovative research on order structures and their classification. Among the results of his early work in set theory are the *Hausdorff recursion formula* for exponentiation of cardinals and several contributions to the study of order structures (cofinality, etc.). Although Hausdorff did not pursue active research in the axiomatic foundation of set theory, he contributed important insights on transfinite numbers, in particular a characterization of what are now known as weakly inaccessible cardinals and his *maximal chain principle*, a form of ZORN'S LEMMA [III.1] that predated the latter and differed from it in formulation and intention.

His own contribution to the axiomatic method was oriented toward generalizing classical areas of mathematics and founding them on axiomatic principles within the framework of set theory. Hausdorff's move to use set theory inside mathematics was seminal for the turn toward *modern mathematics* in the sense of the twentieth century, most prominently characterized by the BOURBAKI [VI.96] group. Best known in this respect are his *axiomatization of general topology* in terms of axioms for neighborhood systems, first published in the *Grundzüge* (1914), and the study of the properties of general, or more specialized, TOPOLOGICAL SPACES [III.90]. Less well-known (it remained unpublished until recently) was Hausdorff's *axiomatization of probability theory*, which was presented in

a lecture course of 1923 and which preceded KOLMOGOROV's [VI.88] work in this area by about a decade. He also made important contributions to analysis and algebra. In algebra, he contributed to LIE THEORY [III.48] (via what is now called the Baker–Campbell–Hausdorff formula), while in analysis he developed summation methods for divergent series and also a generalization of the Riesz–Fischer theory.

Hausdorff's central goals in using set theory were for applications to analytical disciplines such as function theory. Among his most important contributions in this respect, and of wide-ranging importance, was the concept of HAUSDORFF DIMENSION [III.17], which he introduced to give a notion of dimension to rather general sets (such as, for example, fractal-type sets).

Hausdorff realized that analytical questions of set theory were deeply connected to foundational questions. In 1916 he (and, independently, P. Alexandroff) showed that any uncountable BOREL SET [III.55] in the reals actually has the cardinality of the continuum. This was an important development of a strategy proposed by Cantor to clarify the continuum. Although this strategy did not finally contribute to the decisive results by Gödel and Cohen on the CONTINUUM HYPOTHESIS [IV.22 §5], it led to the development of an extended field of investigation in the border region between set theory and analysis, now dealt with in DESCRIPTIVE SET THEORY [IV.22 §9]. Hausdorff's second edition of the *Mengenlehre* (1927) was the first monograph in this field.

After the rise to power of the Nazi regime, working conditions and life in general deteriorated more and more drastically for Hausdorff and others of Jewish origin. When Hausdorff, his wife Charlotte, and a sister of hers were ordered to leave their house for local internment in January 1942, they opted for suicide rather than suffering further persecution.

Further Reading

Brieskorn, E. 1996. *Felix Hausdorff zum Gedächtnis. Aspekte seines Werkes.* Braunschweig: Vieweg.

Hausdorff, F. 2001. *Gesammelte Werke einschließlich der unter dem Pseudonym Paul Mongré erschienenen philosophischen und literarischen Schriften*, edited by E. Brieskorn, F. Hirzebruch, W. Purkert, R. Remmert, and E. Scholz. Berlin: Springer.

Hausdorff's voluminous unpublished work (his "Nachlass") can be found online at www.aic.uni-wuppertal.de/fb7/hausdorff/findbuch.asp.

Erhard Scholz

VI.69   Élie Joseph Cartan

b. Dolomieu, France, 1869; d. Paris, 1951
Lie algebras; differential geometry; differential equations

Cartan was one of the leading mathematicians of his generation, particularly influential for his work on geometry and the theory of LIE ALGEBRAS [III.48 §§2, 3]. In the bleak years after World War I he was one of the most prominent mathematicians in France. He eventually became a notable influence on the BOURBAKI [VI.96] group, of which his son Henri, another distinguished mathematician, was one of the seven founder members. Cartan held lecturing positions in Montpellier and Lyon before becoming a professor in Nancy in 1903. He went on to gain a lecturing position at the Sorbonne in 1909, becoming a professor in 1912 and remaining there until his retirement.

In his doctoral thesis of 1894 Cartan classified the simple Lie algebras over the field of complex numbers, refining and correcting earlier work of Wilhelm Killing and emphasizing the deep general abstract structures inherent in the theory. In later years he returned to these ideas and drew out their implications for the study of the corresponding LIE GROUPS [III.48 §1]—these groups have a major bearing on symmetry considerations in physics.

Cartan spent much of his life working on geometry. In the 1870s and the 1890s KLEIN [VI.57] had analyzed geometry and shown how the major branches (Euclidean, non-Euclidean, projective, and affine) could be unified and treated as special cases of projective geometry. Cartan became interested in the extent to which the group-theoretic ideas that had animated Klein could be adapted to the setting of differential geometry, and especially to spaces of variable CURVATURE [III.78]—the mathematical setting for EINSTEIN'S GENERAL THEORY OF RELATIVITY [IV.13]. In that subject the observations of different observers are related by coordinate transformations, and changes in the gravitational field are expressed through changes in the metric, and hence curvature, of the underlying spacetime manifold. In the 1920s Cartan broadened the setting to what are today called FIBER BUNDLES [IV.6 §5], and showed that Klein's approach could be carried through by concentrating on the possible types of coordinate transformation and the Lie groups to which they can belong.

There are many problems in which one has a multitude of possible observations at each point of a space: for example, the weather at each point of Earth's surface. In Cartan's formulation, Earth's surface is taken as the base MANIFOLD [I.3 §6.9] and the possible observations at each point form another manifold, called the fiber at the point. The pair consisting of all fibers and all points of the base manifold is, roughly, a fiber bundle; the precise concept has proved to be fundamental across the whole field of modern differential geometry. It was to prove a natural setting for the study of what are called *connections* on a manifold, which deal with the way objects, such as vectors, are transformed as they move along curves in the manifold. Cartan's fundamental idea was to capture the symmetry of a geometrical problem by allowing fibers to have a common symmetry group, although aspects of the geometry of the base manifold, such as its curvature, were allowed to vary from point to point in such a way that the base manifold admits no symmetries at all.

Cartan also applied his geometric approach to the study of differential equations, which had earlier been a motivating concern for LIE [VI.53] in the creation of the theory of Lie algebras. He did important work on systems of equations, and this led him to emphasize the role of what are called exterior forms. Familiar examples include the 1-FORM [III.16] that represents the element of length along a curve, the 2-form that represents the element of area of a surface, and so on. The main thing one does to a 1-form is integrate it; integrating the 1-form that describes arc-length gives length along a curve. Cartan studied systems of equations involving arbitrary 1-forms and was led to discover ways in which the algebra of 1-forms, and more generally the algebra of k forms for arbitrary k, captures features of the geometry of the manifold on which they are defined. This led him to reformulate a method of studying the geometry of curves and surfaces that had been pursued by Gaston Darboux, the leading French geometer of the previous generation, and to proclaim his method of "moving frames" that again related to the study of fiber bundles and symmetries in differential geometry. This work, together with his work on fiber bundles, remains a major source of ideas for the study of differentiable manifolds to this day.

Further Reading

Chern, S.-S., and C. Chevalley. 1984. Élie Cartan and his mathematical work. In *Oeuvres Complétes de Élie Cartan*, volume III.2 (1877–1910). Paris: CNRS.
Hawkins, T. 2000. *Emergence of the Theory of Lie Groups: An Essay in the History of Mathematics, 1869–1926*. New York: Springer.

Jeremy Gray

VI.70   Emile Borel

b. Saint-Affrique, France, 1871; d. Paris, 1956
Professor of Mathematics: University of Lille (1893–96),
École Normale, Paris (1896–1909); Chair of Theory of Functions
(specially created for him), Sorbonne, Paris (1909–41);
first director of the Institut Poincaré (1926)

Borel's thesis of 1894 started with problems from within the classical theory of complex functions. With a new theory of MEASURE [III.55] based on CANTOR's [VI.54] set theory and, in particular, a "covering theorem" (later misnamed the Heine–Borel theorem), he gave a rationale for neglecting certain infinite sets of singularities. He assigned them "measure zero" and thus extended the domain of regularity of the functions considered. Borel's theory of measure, based on operations with infinitely many sets, became widely known through his influential *Leçons sur la Théorie des Fonctions* (1898) and was later completed and developed into a major tool of analysis by LEBESGUE [VI.72]. It was, in addition, an important prerequisite for the axiomatization of probability by KOLMOGOROV [VI.88].

VI.71   Bertrand Arthur William Russell

b. Trelleck, Wales, 1872; d. Plas Penrhyn, Wales, 1970
Mathematical logic and set theory; philosophy of mathematics

Russell's training at Cambridge University in the early 1890s inspired the part of his long and varied life that relates to mathematics. He divided his Tripos into Part 1 (Mathematics) and Part 2 (Philosophy), and then united these two trainings to seek a general philosophy of mathematics, especially its epistemological foundations, with geometry as the first test case (1897). But over the next few years he changed his philosophical stance, especially when he recognized the significance of CANTOR's [VI.54] set theory from 1896 onward, and also discovered in 1900 a group of mathematicians around PEANO [VI.62] in Turin. Wishing to raise the level of axiomatization and rigor in mathematics, the followers of Peano formalized theories as much as possible, including the "mathematical logic" of propositions and predicates with set theory, but they kept mathematical and logical notions separate. After learning their system and adding to it a logic of relations, Russell decided in 1901 that their distinction of notions was not necessary: *all* notions lay in that logic. This is the philosophical position that has become known

as "logicism," and Russell wrote a largely nonsymbolic account of it in *The Principles of Mathematics* (1903). In an appendix to this book he publicized the work of FREGE [VI.56], who had anticipated logicism (but advocated it only for arithmetic and some analysis); Russell read him in detail after forming his own position, which continued to be influenced more by Peano.

Now the job was to expound logicism in Peanesque detail—a daunting task, made even harder by Russell's discovery in 1901 that set theory was susceptible to paradoxes, which would have to be avoided or even solved. He was joined in the effort by his former Cambridge tutor, A. N. Whitehead; eventually three volumes of *Principia Mathematica* appeared between 1910 and 1913. After the basic logic and set theory, the arithmetic of real numbers and also the arithmetic of transfinite numbers were worked out in detail; a fourth volume on geometry was due to be written by Whitehead, but he abandoned it around 1920.

The paradoxes were solved by a "theory of types," which formed a hierarchy of individuals, sets of individuals, sets of sets of individuals, and so on. A set or individual could only be a member of a set immediately above it in the hierarchy; thus, a set could not belong to itself. Comparable restrictions were laid on relations and predicates. While this avoided the paradoxes, it also ruled out a great deal of good mathematics, since different kinds of numbers lay in different types and so could not be brought together for arithmetic operations: for example, $34 + \frac{7}{18}$ was not even definable. The authors proposed the "axiom of reducibility" to allow such definitions to be made; but this was, frankly, just a fudge.

Among the various features of Russell's theory was a form of THE AXIOM OF CHOICE [III.1], called the "multiplicative axiom," that he had found in 1904, just before Ernst Zermelo. It had a curious role within logicism, partly because its logicist status was suspect.

While there was discussion of *Principia Mathematica*, concerning both its logic and its logicism, it tended to be too mathematical for the philosophers and too philosophical for the mathematicians. However, the program influenced some kinds of philosophy, including Russell's own; and as an example of high-level axiomatization it served as a model for foundational studies, including GÖDEL'S INCOMPLETENESS THEOREMS [V.15] of 1931, which showed that logicism as Russell had conceived it could not be achieved.

Further Reading

Grattan-Guinness, I. 2000. *The Search for Mathematical Roots*. Princeton, NJ: Princeton University Press.

Russell, B. 1983–. *Collected Papers*, thirty volumes. London: Routledge.

Ivor Grattan-Guinness

VI.72 Henri Lebesgue

b. Beauvais, France, 1875; d. Paris, 1941
Theory of the integral; measure; applications in Fourier analysis; dimension in topology; calculus of variations

Lebesgue studied at the École Normale in Paris (1894–97), where he was influenced by the slightly older BOREL [VI.70] and René-Louis Baire. As a teacher at Nancy he completed his seminal thesis "Intégrale, longueure, aire" (1902). After university positions in Rennes, Poitiers, and at the Sorbonne in Paris, and following war-related research, Lebesgue became a professor at the Sorbonne (1919) and then, finally, at the Collège de France (1921). One year later he was elected to the French Academy of Sciences.

Lebesgue's most important achievement was his generalization of RIEMANN's [VI.49] notion of an integral. This was partly in response to the need to include broader classes of real-valued functions, and partly to give secure foundations to concepts such as the interchangeability of limit and integral in infinite series (particularly Fourier series). Alluding to a famous example (1881) by Vito Volterra of a bounded derivative that could not be integrated, Lebesgue wrote in his thesis:

> The kind of integration defined by Riemann does not allow in all cases for the solution of the fundamental problem of the calculus: find a function with a given derivative. It thus seems natural to search for a definition of the integral which makes integration the inverse operation of differentiation in as large a class of functions as possible.

Lebesgue defined his integral by partitioning the range of a function and summing up sets of x-coordinates (or arguments) belonging to given y-coordinates (or ordinates), rather than, as had traditionally been done, partitioning the domain. Lebesgue himself, according to his colleague, Paul Montel, compared his method with paying off a debt:

> I have to pay a certain sum, which I have collected in my pocket. I take the bills and coins out of my pocket and give them to the creditor in the order I find them until I have reached the total sum. This is the Riemann integral. But I can proceed differently. After I have taken out all my money I order the bills and coins according to identical values and then I pay the several heaps one after another to the creditor. This is my integral.

The comparison reveals the more theoretical character of Lebesgue's integral, as compared with the more intuitive and natural summation used by Riemann. This meant that more sophisticated functions, which were not necessarily integrable in Riemann's sense, became "summable" according to Lebesgue.

In order to perform his summations, Lebesgue had to base his new integral on Borel's notion of MEASURE [III.55] (1898), which in turn drew heavily on CANTOR's [VI.54] theory of infinite sets. He used infinitely many intervals to cover and to measure sets, and was thus able to measure much less intuitive subsets of the linear continuum (the reals) than had hitherto been considered. A crucial role was played by the notion of "the set of measure zero" and the consideration of properties that were valid "except for" such sets, i.e., "almost everywhere." This allowed for the theory to be streamlined to include fundamental results such as: "A bounded function is Riemann integrable if and only if the set of its points of discontinuity has measure zero."

Lebesgue completed Borel's theory of measure, making it a true generalization of JORDAN's [VI.52] earlier theory. From Jordan he also borrowed the important notion of a function of bounded variation for his theory of the integral. Lebesgue ascribed a measure to any subset of a "set of measure zero," and opened up broader theoretical questions such as whether there exist any sets that are not Lebesgue-measurable. The latter question was proved in the affirmative by the Italian Giuseppe Vitali in 1905 with the help of the AXIOM OF CHOICE [III.1], while Robert Solovay showed in 1970, with methods of mathematical logic, that without the axiom of choice such existence cannot be proved (see SET THEORY [IV.22 §5.2]). Lebesgue himself remained skeptical about an unlimited use of set-theoretical principles such as the axiom of choice. He held a restrictive view of the "existence" of mathematical objects by making "definability" the touchstone for his empiricist philosophy of mathematics.

Lebesgue's integral—the idea of which was paralleled, although not in such depth, in the work of the English mathematician W. H. Young—served as a sophisticated stimulus to developments in harmonic and functional analysis (e.g., the L^p spaces of RIESZ [VI.74] (1909)). Generalizations to functions defined on n-dimensional space, proposed by Lebesgue himself

(1910), contributed to even more general theories of integrals, e.g., the theory of Radon (1913).

Although it took several decades for the importance of Lebesgue's integral to become widely recognized, its significance for applications, especially in the analysis of discontinuous and statistical phenomena of nature and in probability theory, could not be ignored in the long run.

Further Reading

Hawkins, T. 1970. *Lebesgue's Theory of Integration: Its Origins and Development.* Madison, WI: University of Wisconsin Press.

Lebesgue, H. 1972–73. *Œuvres Scientifiques en Cinq Volumes.* Geneva: Université de Genève.

Reinhard Siegmund-Schultze

VI.73 Godfrey Harold Hardy

b. Cranleigh, England, 1877; d. Cambridge, 1947
Number theory; analysis

G. H. Hardy was the most influential mathematician in Britain in the twentieth century. With the exception of the years from 1919 to 1931, when he was the Savilian professor of geometry in Oxford, he spent his adult life in Cambridge, where from 1931 until his retirement in 1942 he was the Sadleirian professor of pure mathematics. He became a Fellow of the Royal Society in 1910 and was awarded a Royal Medal in 1920 and the Sylvester Medal in 1940. He died on the day the Royal Society's highest honor, the Copley Medal, was to be presented to him.

At the beginning of the twentieth century, the standard of mathematical analysis was rather low in Britain; Hardy did much to remedy this situation, not only through his research, but also by publishing *A Course of Pure Mathematics* in 1908. This book, which he wrote as "a missionary talking to cannibals," had a tremendous influence on several generations of mathematicians in the United Kingdom. Unfortunately, Hardy's love of pure mathematics, and analysis in particular, somewhat stifled the growth of applied mathematics and algebraic subjects for several decades.

In 1911 he began a long collaboration with LITTLE-WOOD [VI.79], with whom he wrote almost one hundred papers: this partnership is generally considered to have been the most fruitful in the history of mathematics. They worked on convergence and summability of series, inequalities, ADDITIVE NUMBER THEORY

[V.27] (including Waring's problem and Goldbach's conjecture), and Diophantine approximation.

Hardy was one of the first to do important work on the RIEMANN HYPOTHESIS [IV.2 §3] when, in 1914, he proved that the zeta function $\zeta(s) = \zeta(\sigma + it)$ has infinitely many zeros on the critical line $\sigma = \frac{1}{2}$ (see LITTLEWOOD [VI.79]). Later, with Littlewood, he proved deep extensions of this result.

From 1914 to 1919 he collaborated with the largely self-taught Indian genius, SRINIVASA RAMANUJAN [VI.82]. They wrote five papers, the most famous of which is about $p(n)$, the number of partitions of n. This is a rapidly growing function: $p(5) = 7$ but

$$p(200) = 3\,972\,999\,029\,388.$$

The GENERATING FUNCTION [IV.18 §§2.4, 3] of $p(n)$, that is,

$$f(z) = 1 + \sum_{n=1}^{\infty} p(n)z^n,$$

is equal to $1/((1 - z)(1 - z^2)(1 - z^3) \cdots)$, so

$$p(n) = \frac{1}{2\pi i} \int_\Gamma \frac{f(z)}{z^{n+1}}\, dz,$$

where Γ is a circle about the origin of radius just less than 1. In 1918, Hardy and Ramanujan not only gave a rapidly convergent asymptotic formula for $p(n)$ but also showed that, for n large enough, $p(n)$ could be calculated *exactly* by taking the integer nearest to the sum of the first few terms. In particular, $p(200)$ can be computed from the first five terms.

Hardy and Ramanujan proved their asymptotic formula for $p(n)$ with the aid of the "circle method"; later, Hardy and Littlewood developed this method into one of the most powerful tools in analytic number theory. In order to estimate contour integrals like the one above, Hardy and Littlewood found it advisable to break up the circle of integration in a subtle way.

Another Hardy–Ramanujan result concerns the number $\omega(n)$ of distinct prime divisors of a "typical" number n. They proved that a "typical" number n has about $\log \log n$ distinct prime factors in a certain precise sense. In 1940 Erdős and Kac sharpened and extended this result by showing that additive number-theoretic functions like $\omega(n)$ obey the GAUSSIAN LAW [III.71 §5] of errors: this gave birth to the important field of probabilistic number theory.

Hardy's name has been attached to several concepts and results, including Hardy spaces, Hardy's inequality, and the HARDY–LITTLEWOOD MAXIMAL THEOREM [IV.11 §3]. For $0 < p \leqslant \infty$ the *Hardy space* H^p consists of functions analytic in the unit disk that are bounded in various ways; in particular, H^∞ consists of bounded analytic functions. Hardy and Littlewood deduced fundamental properties of H^p from their *maximal theorem*, which relates a function to its "radial limits" at the boundary of the disk. The theory of H^p spaces has found numerous applications not only in analysis, but also in probability theory and control theory.

Hardy and Littlewood loved inequalities of all kinds; their book on the subject with George Pólya, an instant classic the moment it was published in 1934, greatly influenced the development of hard analysis.

Although Hardy was fiercely proud of the purity of his mathematics, in a paper published in 1908 he formulated the extension of the Mendelian law about the proportions of dominant and recessive characters. This law, which later became known as the *Hardy–Weinberg law*, refuted the idea "that a dominant character should show a tendency to spread over a whole population, or that a recessive should tend to die out." In a later article he dealt a severe blow to eugenics by giving a simple mathematical argument that showed the futility of forbidding people with "undesirable" characteristics to breed.

In his interest in mathematical philosophy, Hardy was a disciple of RUSSELL [VI.71], whose political views he also shared. He was a secretary of the committee which forced the abolition of the order of merit in the Mathematical Tripos through a reluctant Senate in 1910, and many years later he fought hard for the abolition (not reform!) of the Mathematical Tripos itself, which he considered to be harmful to mathematics in the United Kingdom. After World War I, Hardy led the British efforts to heal the wounds of the international mathematical community, and with the advent of the Nazi persecutions on the Continent in the early 1930s, he was an important figure in an extensive network finding jobs for refugee mathematicians in the United States, Britain, and the Commonwealth. He was a great supporter of the London Mathematical Society: he was not only one of the secretaries for close to twenty years, but also its president for two terms.

Hardy was a militant atheist; as an affectation, he liked to talk of God as his personal enemy. He was a great conversationalist, and was fond of various intellectual games, like putting together cricket teams of bores, bogus poets, Fellows of a Cambridge college, and so on. He loved ball games, especially cricket, baseball, bowls (with the curved woods of his college), and real tennis (as opposed to lawn tennis); to praise people, he frequently likened them to outstanding cricketers.

He had an exceptional gift for collaboration and launching young mathematicians on their research careers. He was a master not only of mathematics, but also of English prose; he was lively and charming, and left a lasting impression even on his casual acquaintances. His poetic book *A Mathematician's Apology*, written toward the end of his life, gives a rare insight into the world of a mathematician.

Further Reading

Hardy, G. H. 1992. *A Mathematician's Apology*, with a foreword by C. P. Snow. Cambridge: Cambridge University Press. (Reprint of the 1967 edition.)

Hardy, G. H., J. E. Littlewood, and G. Pólya. 1988. *Inequalities*. Cambridge: Cambridge University Press. (Reprint of the 1952 edition.)

Béla Bollobás

VI.74 Frigyes (Frédéric) Riesz

b. Győr, Hungary, 1880; d. Budapest, 1956
Functional analysis; set theory; measure theory

After being educated at Budapest University and elsewhere in Europe, Riesz was appointed in 1911 to the University of Kolozsvár (Hungary), which moved in 1920 to become Szeged University; he served twice as Rector. He returned to Budapest in 1946. Most of Riesz's research work lay in mathematical analysis enriched with techniques from set and measure theory, and functional analysis.

One of Riesz's famous results was the converse of a generalization of Parseval's theorem for FOURIER SERIES [III.27]: given a sequence of orthonormal functions on a finite interval, and a sequence $a_1, a_2, \ldots$ of real numbers, there exists a function f that can be expanded as a Fourier-type series with respect to those functions with the a_r as coefficients if and only if $\sum_r a_r^2$ is convergent; further, f is itself square summable. He proved the theorem in 1907, simultaneously with the German mathematician Ernst Fischer; so it is named after both of them.

Two years later Riesz found the "representation theorem" named after him. It states that a continuous linear functional that maps continuous functions F over a finite interval I onto the real numbers can be represented as a Stieltjes integral of F over I with respect to a function of bounded variation. It was to be a fertile source of applications and generalizations.

Riesz found these two theorems partly in connection with his study of integral equations, a topic then being developed by HILBERT [VI.63], and partly in connection with his study of functional analysis as formulated by Maurice Fréchet. Hilbert's work had led him to consider infinite matrices, which were then little studied: Riesz wrote the first monograph on them, *Les Systèmes d'Équations Linéaires à une Infinité d'Inconnues* (1913). He also studied the theory of L^p spaces for $p > 1$ (that is, spaces of functions f such that f^p is measure-integrable over some specified interval) and their dual spaces L^q, where $1/p + 1/q = 1$; and he worked on applying his and Fischer's theorem to the self-dual space, now known as HILBERT SPACE [III.37], that is given by $p = 2$. Later he laid some of the foundations of complete spaces (later known as BANACH SPACES [III.62]), and applied functional analysis to ergodic theory. He summed up much of his work in these areas in the book *Leçons d'Analyse Fonctionnelle* (1952), written with his student B. Szökefnalvy-Nagy.

All this work constituted important contributions to theories already laid out in principle by various other mathematicians. Riesz achieved groundbreaking work on subharmonic functions: he modified the DIRICHLET PROBLEM [IV.12 §1] by allowing the function that extends a given function into a domain to be subharmonic ("locally less than harmonic") instead of harmonic. He studied some of the applications of these functions to potential theory.

Riesz also studied some foundational aspects of set theory, especially types of ordering, continuity, and generalized Heine–Borel covering theorems. He also reformulated the LEBESGUE INTEGRAL [III.55] in a constructive manner, using step functions and sets of measure zero as primitive notions, and avoiding MEASURE THEORY [III.55] as much as possible.

Further Reading

Riesz, F. 1960. *Oeuvres Complètes*, edited by Á. Császár, two volumes. Budapest: Akademiai Kiado.

Ivor Grattan-Guinness

VI.75   Luitzen Egbertus Jan Brouwer

b. Overschie, the Netherlands, 1881;
d. Blaricum, the Netherlands, 1966
Lie groups; topology; geometry; intuitionistic mathematics; philosophy of mathematics

Brouwer entered the University of Amsterdam at the age of sixteen, where his teacher was D. J. Korteweg.

The young Brouwer taught himself modern mathematics, as well as a fair amount of philosophy. As a graduate student he published some original papers on the decomposition of rotations in four-dimensional space. He also published a brief monograph on mysticism that contained a number of ideas that became prominent in his later philosophy. In his dissertation of 1907 he solved a special case of HILBERT's [VI.63] fifth problem (the elimination of differentiability conditions from the axioms of LIE GROUPS [III.48 §1]), and he presented his first program for "constructive mathematics."

The basis of his mathematics was the *ur-intuition of mathematics*: the continuum and the natural numbers are simultaneously created from intuition. Mathematical objects (including proofs) are mental creations. After sketching the development of the basic parts of mathematics, Brouwer went on to criticize contemporary mathematics for transcending the bounds of the human mind. In particular, he criticized CANTOR [VI.54] for introducing sets beyond human recognition, and Hilbert for the axiomatic method and for formalism. He criticized the latter's consistency program and denied that "consistency implies existence."

In his 1908 paper "The unreliability of the logical principles," Brouwer explicitly rejected the principle of the excluded middle as unreliable (and also rejected Hilbert's dogma that "all mathematical problems can be solved in one way or another"). Between 1909 and 1913 Brouwer worked in topology. He continued his work on Lie groups, and noted that topology (in the style of Cantor–Schoenflies) was in need of a sound basis. In his paper "Zur Analysis Situs" (1910), he spelled out a number of notions and examples (curves, indecomposable continua, three domains with one common boundary). This was the beginning of his revision of set-theoretic topology. At the same time he started two lines of research: one on homeomorphisms from surfaces to themselves, establishing FIXED POINT THEOREMS [V.11] on the sphere and the *plane translation theorem* (a characterization of homeomorphisms of the Euclidean plane that have no fixed points); and one on vector distributions on the sphere, yielding existence theorems for singular points, and a characterization of these points. The best-known theorem in this area is Brouwer's "hairy ball theorem" (no matter how one combs a hairy ball, there is always a crown). In 1910 Brouwer published a direct topological proof of the Jordan curve theorem, which remains one of the most elegant proofs. The so-called new topology opened with Brouwer's "invariance of dimension" theorem (1910).

He then laid the basis for topology of MANIFOLDS [I.3 §6.9], where his basic tool was the Brouwer degree of continuous mappings. The basic paper is his "Über Abbildungen von Mannigfaltigkeiten" ("On mappings of manifolds," 1911), which contained most of the tools for the new topology, e.g., simplicial approximation, mapping degree, HOMOTOPY [IV.6 §§2, 3], singularity index (in his own terminology), and also the fundamental properties of the new notions.

Brouwer's new topological insights and techniques led him to a wealth of spectacular results: the Brouwer fixed point theorem, the invariance-of-domain theorem, the higher-dimensional Jordan theorem, and the definition of dimension, including the soundness proof (that $\mathbb{R}^n$ has dimension n). He also applied his invariance-of-domain theorem to the theory of automorphic functions and uniformization, thus proving the correctness of the *Klein–Poincaré continuity method* (1912).

During World War I Brouwer returned to the foundations of mathematics; he conceived his mature INTUITIONISTIC MATHEMATICS [II.7 §3.1], which fully exploited the potential of constructive mathematics, based on mentally created objects and notions. The key notions were (infinite) choice sequences (i.e., sequences determined by more or less free choices (by the mathematician) of mathematical objects, say natural numbers), well-orderings, and intuitionistic logic. In "Brouwer's universe" strong results can be obtained: the "continuity principle," for example, which says that a function that assigns natural numbers to choice sequences is continuous (i.e., the output is determined by a finite piece of the (infinite) input); and certain transfinite induction principles, in particular the novel principle of "bar-induction." With the help of these principles he showed that (i) all real functions on a closed segment are uniformly continuous and (ii) the continuum is indecomposable (cannot be split). This enabled him to refute the principle of the excluded middle in a strong sense: it is not the case that each real number is zero or nonzero. In Brouwer's universe many classical theorems, such as the intermediate value theorem and the Bolzano–Weierstrass theorem, fail.

Brouwer's mathematical universe lacked the logical "principle of the excluded middle," but instead it had certain strong constructive principles at its disposal, which turned it into an alternative to the traditional universe, with a comparable strength.

His foundational program brought him into conflict with Hilbert ("intuitionism versus formalism"). In 1928

matters came to a head and, in an incident famously described by Einstein as "the war of frogs and mice," Hilbert succeeded in getting Brouwer removed (after fourteen years' service) from the editorial board of *Mathematische Annalen*.

Brouwer was unconventional and had wide-ranging interests: art, literature, politics, philosophy, mysticism. He was a staunch internationalist.

He was a professor at the University of Amsterdam from 1912 until 1951.

Further Reading

Brouwer, L. E. J. 1975–76. *Collected Works*, two volumes. Amsterdam: North-Holland.
van Dalen, D. 1999–2005. *Mystic, Geometer and Intuitionist. The Life of L. E. J. Brouwer*, two volumes. Oxford: Oxford University Press.

Dirk van Dalen

VI.76 Emmy Noether

b. Erlangen, Germany, 1882; d. Bryn Mawr, Pennsylvania, 1935
Algebra; mathematical physics; topology

Noether began her career with a feat of classical algebra, which she transmuted into the NOETHER CONSERVATION THEOREMS [IV.12 §4.1] for physics. She became a founder of modern abstract algebra and the leader in spreading that algebra all across mathematics.

Her father Max Noether and family friend Paul Gordan were Erlangen mathematicians and favored educating women. Gordan made heroic calculations of invariants in algebra. A quadratic polynomial $Ax^2 + Bx + C$ has essentially just one invariant, the discriminant $\sqrt{B^2 - 4AC}$ used in the quadratic formula. As Gordan's student, Noether found 331 independent invariants of degree-four polynomials in three variables, and proved that all others depend on them. It was impressive, though not, as it turned out, groundbreaking.

HILBERT [VI.63] brought her to Göttingen in 1915 to work on invariants for differential equations in general relativity by reducing them to algebra. That year she found her conservation theorems, which show that the conserved quantities of a physical system correspond to its symmetries. For example, if a system has laws unchanging with time, so that a time shift is a symmetry of the system, then energy is conserved in the system (Feynman 1965, chapter 4). These theorems became fundamental in Newtonian physics and especially quantum mechanics. They also showed that general relativity admits conservation laws only in special cases.

Noether saw the creation of general abstract algebra as her life's work. Instead of classical algebra with real numbers, or complex numbers, and polynomials using them, she would study any system satisfying abstract rules such as the RING AXIOMS [III.81] or the GROUP AXIOMS [I.3 §2.1]. Concrete examples include the ring of all algebraic functions defined on a space (such as a sphere), and the group of all symmetries of a given space. She largely created the now-standard style of abstract algebra. Her ideas were also adopted in ALGEBRAIC GEOMETRY [IV.4] where every abstract ring appears as the ring of functions on a corresponding space called a SCHEME [IV.5 §3].

She turned her attention away from operations on elements of a system, like plus and times, and focused instead on relating whole systems to each other, such as rings R, R' related by ring HOMOMORPHISMS [I.3 §4.1] from R to R'. She organized all algebra around her homomorphism and isomorphism theorems. Her aim was to show how IDEALS [III.81 §2] and their corresponding homomorphisms could replace equations between elements as the basic tools for stating and proving theorems. (This approach was to come to fruition in the 1950s, with the advent of Grothendieck-style homological algebra.)

Topologists studied TOPOLOGICAL SPACES [III.90] by looking at continuous functions from one space to another. Noether saw how her algebraic methods could apply here, and convinced young topologists in the 1920s to use them in algebraic topology. Each topological space S has HOMOLOGY GROUPS [IV.6 §4] $H_n S$ with the property that continuous functions from S to S' induce group homomorphisms from $H_n S$ to $H_n S'$. Theorems of topology follow by abstract algebra. This relationship between homomorphisms and continuous functions is what inspired CATEGORY THEORY [III.8].

In the 1930s Noether pursued the algebra of GALOIS THEORY [V.21] through a radically simplified abstract theory of groups acting on rings. The applications are quite arcane, beginning with CLASS FIELD THEORY [V.28] and eventually growing into group cohomology and many other algebraic and topological methods used in ARITHMETIC GEOMETRY [IV.5].

Exiled from Germany by the Nazis in 1933, she died following surgery in the United States, at the height of her creative power.

Further Reading

Brewer, J., and M. Smith, eds. 1981. *Emmy Noether: A Tribute to Her Life and Work*. New York: Marcel Dekker.
Feynman, R. 1965. *The Character of Physical Law*. Cambridge, MA: MIT Press.

Colin McLarty

VI.77   Wacław Sierpiński

b. Warsaw, 1882; d. Warsaw, 1969
Number theory; set theory; real functions; topology

Sierpiński studied mathematics at the Russian university in Warsaw under the guidance of Georgii Voronoi. In his first paper (1906), he improved GAUSS's [VI.26] estimate for the difference between the number of lattice points inside the circle $x^2 + y^2 \leqslant N$ and the area of the circle, showing that it is $O(N^{1/3})$.

He became an associate professor at the University of Lwów in 1910, at which point his interest shifted to set theory, on which he wrote a textbook in 1912, only the fifth book ever to be published on the subject. His first important results on set theory were obtained during World War I, which he spent in Russia: in 1915–16 he constructed two curves that were among the first published examples of fractals, one known now as Sierpiński's gasket and the other as Sierpiński's carpet. The latter is the set of all points (x, y) in the square $[0, 1]^2$ such that, when written out as base 3 decimals, there is no position in which both x and y have a 1. It is also known as Sierpiński's universal curve, since it contains a homeomorphic image of every planar continuum (a continuum is a compact connected set) without interior points.

In 1917, Souslin had shown that projections of BOREL SETS [III.55] (from the plane into the line, say) need not be Borel. Together with Lusin, Sierpiński proved in 1918 that in fact every analytic set (a projection of a Borel set) is the intersection of $\aleph_1$ Borel sets (where $\aleph_1$ is the smallest uncountable cardinal). That same year he also published an important study of THE AXIOM OF CHOICE [III.1] and the role it plays in set theory and analysis, and proved that no continuum can be decomposed into countably many pairwise disjoint nonempty closed subsets.

In 1919 Sierpiński was made a full professor at the new Polish University of Warsaw and in 1920 he founded (together with Janiszewski and Mazurkiewicz) the first specialized mathematical journal, *Fundamenta Mathematicae*, which was devoted to set theory, topology, and applications. He remained its editor until 1951. Among his results published in volume 1 are a proof that every countable subset of $\mathbb{R}^n$ without isolated points is homeomorphic to the rationals; a complete classification of countable compact subsets of $\mathbb{R}^n$, obtained jointly with Mazurkiewicz; and a necessary and sufficient condition for a subset of $\mathbb{R}^n$ to be a continuous image of an interval.

Using the CONTINUUM HYPOTHESIS [IV.22 §5] ($\aleph_1 = 2^{\aleph_0}$), he constructed an UNCOUNTABLE SET [III.11] of reals, now known as a *Sierpiński set*, such that every uncountable subset of it is nonmeasurable (1924); and also a one-to-one mapping of the line into itself that maps sets of MEASURE ZERO [III.55] to sets of first category, in such a way that every set of the first category is obtained (1934). The former result is highly paradoxical (there is no explicit example of a nonmeasurable set); the latter has led, thanks to Erdős, to the following duality principle. Let P be any proposition involving solely the notions of measure zero, first category, and pure set theory. Let P* be the proposition obtained from P by interchanging the terms "set of measure zero" and "set of first category." Then P and P* are equivalent, assuming the continuum hypothesis.

Sierpiński wrote a monograph devoted to the continuum hypothesis in 1934, entitled *Hypothèse du Continu*. Together with TARSKI [VI.87] he introduced the notion of STRONGLY INACCESSIBLE CARDINALS [IV.22 §6] (1930), meaning cardinals $\mathfrak{m}$ that cannot be obtained as products of fewer than $\mathfrak{m}$ cardinals less than $\mathfrak{m}$. He also worked in Ramsey theory, giving a limitation on infinite extensions to Ramsey's theorem. To be precise, Ramsey had proved that, whenever one finitely colors the pairs from the natural numbers, there is an infinite monochromatic subset (i.e., a subset all of whose pairs have the same color); Sierpiński showed that, by contrast, one can 2-color the pairs from a ground set of size $\aleph_1$ in such a way that there is no monochromatic subset of size $\aleph_1$. He also deduced the axiom of choice from the generalized continuum hypothesis (formulated without cardinals in 1947).

In his old age he returned to number theory and became the editor of *Acta Arithmetica* (1958–69).

Further Reading

Sierpiński, W. 1974–76. *Oeuvres Choisies*. Warsaw: Polish Scientific.

Andrzej Schinzel

VI.78   George Birkhoff

b. Oversiel, Michigan, 1884; d. Cambridge, Massachusetts, 1944
Difference equations; differential equations; dynamical systems; ergodic theory; relativity theory

At the International Congress of Mathematicians in 1924 the Russian mathematician A. N. Krylov described Birkhoff as "the POINCARÉ [VI.61] of America." It was an apt description and one that Birkhoff would have relished, for he was deeply influenced by Poincaré's work, in particular his great treatise on celestial mechanics.

Birkhoff studied first at Chicago under E. H. Moore and Oskar Bolza, and then at Harvard under W. F. Osgood and Maxime Bôcher. Returning to Chicago, he was awarded his doctorate in 1907 for a thesis on asymptotic expansions, boundary-value problems, and Sturm–Liouville theory. In 1909, after two years at Wisconsin under E. B. Van Vleck, he went to Princeton, where he formed a close association with Oswald Veblen. In 1912 he moved to Harvard and remained there, in professorial positions, until his sudden death in 1944. Birkhoff was steadfast in his support for the development of American mathematics, supervising forty-five doctoral students, including Marston Morse and Marshall Stone, and holding many distinguished positions within the scientific community. He was generally recognized, both at home and abroad, as the leading American mathematician of his generation.

Birkhoff first came to prominence with a memoir on the theory of linear difference equations (1911), and he continued to publish on the topic intermittently throughout his career. Related to this work were several papers on the theory of linear differential equations and a paper on the generalized Riemann problem (1913), which concerns complex functions defined by differential equations. (Until recently it was believed that the latter paper included a solution to Hilbert's twenty-first problem, the Hilbert–Riemann problem, but in 1989 Bolibruch proved this belief to be mistaken.)

Throughout his life Birkhoff's deepest interest in analysis lay in DYNAMICAL SYSTEMS [IV.14] and it was here that he enjoyed his greatest success. His overarching aim was to obtain a reduction of the most general dynamical system to a normal form from which a complete qualitative characterization could be deduced. As with Poincaré, the study of periodic motions was central to his work, and he wrote extensively on the THREE-BODY PROBLEM [V.33] as well as on questions connected with stability. Of his memoir on dynamical systems with two degrees of freedom (1917), which won the Bôcher prize in 1923, he is said to have remarked that it was as good a piece of work as he was ever likely to do. Another celebrated achievement was his proof of Poincaré's topological "last geometric theorem," the publication of which brought him immediate international acclaim (1913). (The theorem states that any one-to-one area-preserving transformation of an annulus that moves the boundary circles in opposite directions must have at least two fixed points, and its importance lies in the fact that its proof implies the existence of periodic orbits in the restricted three-body problem.) He introduced several new concepts into dynamical theory, including "recurrent motion" (1912) and "metric transitivity" (1928), and promoted the use of symbolism in dynamics (1935), the latter helping to pave the way for the formalized development of symbolic dynamics (the branch of dynamical systems invented by HADAMARD [VI.65] (1898) that deals with spaces consisting of infinite sequences of symbols) by Marston Morse and Gustav Hedlund at the end of the 1930s. His book *Dynamical Systems* (1927) was the first book on the qualitative theory of systems defined by differential equations. Awash with topological ideas, it provides a connected account of much of his earlier research.

Closely related to Birkhoff's dynamical research was his work on ERGODIC THEORY [V.9]. Stimulated by the theorems of Bernard Koopman and VON NEUMANN [VI.91], Birkhoff presented his own ergodic theorem in 1931, a fundamental result both for statistical mechanics and for MEASURE THEORY [III.55], the proof of which combined Poincaré's topological approach with the use of Lebesgue measure theory. (Roughly speaking, Birkhoff's ergodic theorem states that for any dynamical system given by differential equations that possesses an invariant volume integral, there is a definite "time probability" p that any moving point, except those of a set of measure zero, will be in an assigned region v. In other words, if t is a total elapsed time interval and t^* is the portion of time during which the point is in v, then $\lim t^*/t = p$.)

In the creation of physical theories Birkhoff advocated mathematical symmetry and simplicity above physical intuition. His books on relativity theory (which were among the first on the subject in English), *Relativity and Modern Physics* (1923) and *The Origin, Nature, and Influence of Relativity* (1925), were characteristically original and widely read. At the time of his death he was engaged in developing a new theory of matter (taken to be a perfect fluid), electricity, and gravitation, which he had first proposed in 1943 and which, unlike Einstein's theory, was based on flat spacetime.

Birkhoff published in several other fields, including the CALCULUS OF VARIATIONS [III.94] and map coloring, and he was the coauthor (with Ralph Beatley) of a textbook of elementary geometry (1929). His paper

(with O. D. Kellogg) on fixed points in function space (1922) provided a stimulus for the later work of Leray and Schauder.

Birkhoff had a lifelong interest in the arts and was fascinated by the problem of analyzing the fundamentals of musical and artistic form. In later life he lectured extensively on the application of mathematics to aesthetics, and his book *Aesthetic Measure* (1933) enjoyed popular success.

Further Reading

Aubin, D. 2005. George David Birkhoff. Dynamical systems. In *Landmark Writings in Western Mathematics 1640–1940*, edited by I. Grattan-Guinness, pp. 871–81. Amsterdam: Elsevier.

VI.79   John Edensor Littlewood

b. Rochester, England, 1885; d. Cambridge, England, 1977
Analysis; number theory; differential equations

Littlewood made important contributions to several different branches of analysis and analytic number theory, including Abelian and Tauberian theory, the RIEMANN ZETA FUNCTION [IV.2 §3], WARING'S PROBLEM, GOLDBACH'S CONJECTURE [V.27], harmonic analysis, probabilistic analysis, and nonlinear differential equations. He loved concrete problems such as the RIEMANN HYPOTHESIS [IV.2 §3]: he was arguably the best problem solver of his generation. Much of his work was done in collaboration with HARDY [VI.73]: the Hardy–Littlewood partnership dominated the mathematical scene in the United Kingdom for a third of a century. With the exception of three years in Manchester, he lived all his adult life in Trinity College, Cambridge. From 1928 until his retirement in 1950, he was the first holder of the Rouse Ball Chair of Mathematics in Cambridge.

His first major result, published in 1911, was a deep converse of ABEL's [VI.33] classical theorem that if a series of reals $\sum a_n$ sums to A, then $\sum a_n x^n$ also tends to A as $x \to 1$ from below. In general, the converse is false, but Tauber had proved that it is true if $na_n \to 0$. Littlewood extended this by weakening the condition to na_n being bounded. This result gave rise to an extended area of analysis called Tauberian theorems.

In the theory of functions, he did elegant, important, and innovative work on injective holomorphic functions, the minimum modulus, and subharmonic functions. In particular, he worked on the conjecture that Bieberbach made in 1916 that if $f(z) = z + a_2 z^2 + a_3 z^3 + \cdots$ is an injective HOLOMORPHIC FUNCTION

[I.3 §5.6] in the open disk $\Delta = \{z : |z| < 1\}$, then $|a_n| \leqslant n$ for every n. Littlewood proved in 1923 that $|a_n| < en$ for every n. After many improvements by a number of people, the constant e was eventually reduced to a value close to 1, before de Branges proved the full conjecture in 1984.

Littlewood had a lifelong interest in the zeta function. This is defined in the half-plane $\mathrm{Re}(s) > 1$ by the absolutely convergent series

$$\zeta(s) = \zeta(\sigma + \mathrm{i}t) = \frac{1}{1^s} + \frac{1}{2^s} + \frac{1}{3^s} + \cdots,$$

and in the whole complex plane by analytic continuation. In fact, the second problem suggested to him by his supervisor was the Riemann hypothesis that the zeros of $\zeta(s)$ in the "critical strip" $0 < \sigma < 1$ are on the "critical line" $\sigma = \frac{1}{2}$. If true, this famous conjecture would imply deep results about the distribution of primes. Most of Littlewood's work on the zeta function was done in collaboration with Hardy and concerned analytic properties of $\zeta(s)$.

In addition to his work with Hardy, he also made use of the zeta function to prove a striking theorem about the error term in the PRIME NUMBER THEOREM [V.26]. The prime number theorem itself had been proved by HADAMARD [VI.65] and, independently, by DE LA VALLÉE POUSSIN [VI.67] in 1896. This fundamental result states that $\pi(x)$, the number of primes less than x, is asymptotic to the "logarithmic integral" $\mathrm{li}(x) = \int_0^x (1/\log t)\,\mathrm{d}t$. There was much numerical evidence that $\pi(x) < \mathrm{li}(x)$ for all x; in particular, by 1914 this inequality was known to hold for all $2 \leqslant x \leqslant 10^7$. Nevertheless, Littlewood proved that $\mathrm{li}(x) - \pi(x)$ changes sign infinitely often. Interestingly, he did not obtain any explicit bound for a value x with $\pi(x) > \mathrm{li}(x)$; the first such bound, given by Skewes in 1955, was

$$10^{10^{10^{1000}}}.$$

Hardy and Littlewood proved important approximate formulas for $\zeta(s)$, which they used to deduce that, in a certain sense, $\zeta(s)$ is "small" on the critical line; this was viewed as a breakthrough. Littlewood also studied the number of zeros of $\zeta(s)$ in a rectangle $0 < \sigma < 1$, $0 < t \leqslant T$.

In 1770, in his *Meditationes Algebraicae*, WARING [VI.21] asserted on the basis of empirical evidence that every natural number is the sum of nine nonnegative integral cubes, nineteen fourth powers, and so on: for every natural number k there is a minimal integer $g(k)$ such that every natural number is the sum of $g(k)$

nonnegative kth powers. In 1909 HILBERT [VI.63] used complicated algebraic identities to prove that $g(k)$ indeed exists, but the bounds he obtained on $g(k)$ were rather weak. In the 1920s, in a groundbreaking series of papers entitled *Partitio Numerorum*, Hardy and Littlewood introduced an analytic method that could be used to tackle not only Waring's problem of determining $g(k)$, but many other problems as well. The origins of this "circle method" of Hardy and Littlewood go back to the work of Hardy and RAMANUJAN [VI.82] on the partition function, but the technical difficulties that Hardy and Littlewood had to overcome were much greater than in that earlier work. This method enabled them to show, for example, that every sufficiently large number is the sum of nineteen fourth powers. (In 1986, Balasubramanian, Dress, and Deshouillers proved that $g(4)$ is indeed 19.) More importantly, they gave an asymptotic estimate for the number of representations of n as a sum of at most s positive kth powers.

The circle method also provides a possible line of attack on *Goldbach's conjecture* that every even number greater than two is the sum of two primes, and gives strong heuristic evidence for the strengthened version of the *twin prime conjecture* that the number of primes $p \leqslant n$ such that $p + 2$ is also a prime is asymptotic to $c\int_2^n (1/(\log t)^2)\,\mathrm{d}t$ for a constant $c > 0$. The socalled *k-tuple conjecture* of Hardy and Littlewood is a further extension of this conjecture for "constellations of primes."

Much of Littlewood's remarkable work on harmonic analysis was done in collaboration with R. E. A. C. Paley in the early 1930s. The starting point of the LITTLEWOOD–PALEY THEORY [VII.3 §7] is an inequality concerning trigonometric polynomials. Roughly speaking, Littlewood and Paley related the size of a function to the projection of its FOURIER COEFFICIENTS [III.27] onto various intervals. The original one-dimensional Littlewood–Paley theory has been extended to higher dimensions, arbitrary intervals, and even to tensors on two-dimensional compact manifolds; the theory has connections to such varied topics as WAVELETS [VII.3], semigroups acting on L^p-spaces of functions with values in a BANACH SPACE [III.62], and the geometry of null hypersurfaces for rough Einstein metrics.

Littlewood was also a formidable applied mathematician. During World War I he worked on ballistics, and during World War II, with his collaborator Mary Cartwright, he worked on the van der Pol oscillator in order to help the development of radio. Cartwright and Littlewood were among the first to combine topological and analytical methods to tackle differential equations, and discovered many of the phenomena that later became known as "chaos": they proved that chaos could arise even in equations originating in real engineering problems.

From 1910 until his death sixty-seven years later, Littlewood lived in the same set of spacious rooms in Trinity College, Cambridge. He was a great raconteur: after almost every dinner he was to be found in the Combination Room drinking claret in the company of Fellows and any mathematicians who might be visiting. In spite of his tremendous output, he suffered for decades from severe bouts of depression, from which he was cured only in 1957. He practiced his belief that mathematicians should take a vacation of at least twenty-one days a year during which they should do no mathematics. He was a keen and skilled rock climber and an avid Alpine skier. Although not an active musician, on most days he listened to Bach, Beethoven, and Mozart for hours on end.

In 1943, when he was awarded the Sylvester Medal of the Royal Society, the citation read: "Littlewood, on Hardy's own estimate, is the finest mathematician he has ever known. He was the man most likely to storm and smash a really deep and formidable problem; there was no one else who could command such a combination of insight, technique and power."

Further Reading

Littlewood, J. E. 1986. *Littlewood's Miscellany*, edited and with a foreword by B. Bollobás. Cambridge: Cambridge University Press.

Béla Bollobás

VI.80 Hermann Weyl

b. Elmshorn, Germany, 1885; d. Zürich, 1955
Analysis; geometry; topology; foundations; mathematical physics

Weyl studied mathematics at Göttingen under HILBERT [VI.63], KLEIN [VI.57], and MINKOWSKI [VI.64] between 1904 and 1908. His first teaching positions were at Göttingen (1910–13) and ETH Zürich (1913–30). In 1930 he accepted the call to Göttingen as Hilbert's successor. After the rise to power of the Nazis, he emigrated to the United States and became a member of the newly founded Institute of Advanced Studies at Princeton (1933–51).

Weyl made contributions to real and complex analysis, geometry and topology, LIE GROUPS [III.48 §1], num-

ber theory, the foundations of mathematics, mathematical physics, and philosophy. He contributed at least one book to each of these fields, publishing thirteen in total. Together with his other technical and conceptual innovations, these books were all of lasting influence: many had a pronounced and immediate effect.

His early research dealt with integral operators and differential equations with singular boundary conditions. His fame came later, with his book *The Concept of a Riemann Surface* (1913). This grew out of a lecture course in the winter of 1910–11 and built upon Klein's intuitive treatment of RIEMANN's [VI.49] geometric function theory and Hilbert's justification of the DIRICHLET PRINCIPLE [IV.12 §3.5]. Here Weyl gave a new presentation of the properties of RIEMANN SURFACES [III.79], which became highly influential for the geometric function theory of the twentieth century.

His second book, *The Continuum* (1918), marked the beginning of Weyl's interest in the foundations of mathematics. He was critical of Hilbert's "formalist" program for an axiomatic foundation of mathematics, and explored the possibility of a semi-formalized arithmetical approach to a strictly constructivist foundation of real analysis. Shortly thereafter he shifted toward BROUWER's [VI.75] intuitionistic program and attacked Hilbert's foundational views even more strongly in a famous article of 1921. In the late 1920s he developed a more balanced view of the foundational questions. After World War II he returned to a weak preference for his arithmetical constructive approach of 1918.

At the same time as he was working on foundational questions Weyl took up Einstein's theory of general relativity and wrote his third book, *Space-Time-Matter*. It was first published in 1918, and appeared in five successive editions until 1923. This was one of the first monographs on relativity theory and was among the most influential. The book represented only the tip of the iceberg of his contributions to differential geometry and general relativity. Weyl undertook this research within a broad conceptual and philosophical framework. One of the outcomes of this approach was his *Analysis of the Problem of Space* (1923), in which he sketched ideas that would later be analyzed in terms of the geometry of FIBER BUNDLES [IV.6 §5] and the study of *gauge fields*. He had already introduced gauge fields (and the key idea of a point-dependent rescaling of the metric) in 1918 for a generalization of RIEMANNIAN GEOMETRY [I.3 §6.10] and a geometrically unified field theory of gravity and electromagnetism.

Weyl made his most influential contributions to pure mathematics around the middle of the 1920s with his work on the REPRESENTATION THEORY [IV.9] of semisimple Lie groups. Combining CARTAN's [VI.69] insights into the representation theory of LIE ALGEBRAS [III.48 §2] with methods developed by Hurwitz and Schur, Weyl used his knowledge of the topology of manifolds and developed the core of the general theory of representations of Lie groups in a blend of geometric, algebraic, and analytic methods. He extended and refined this work and it formed the core of his later book *The Classical Groups* (1939)—a harvest of his work and lectures on this topic during his Princeton years.

Along with all this work, Weyl actively followed the rise of the new quantum mechanics. In 1927-28 he gave a lecture course at ETH on the topic, which gave rise to his next book on mathematical physics, *Group Theory and Quantum Mechanics* (1928). Weyl emphasized the conceptual role of group methods in the symbolic representation of quantum structures, in particular the intriguing interplay between representations of the special linear group and PERMUTATION GROUPS [III.68]. A second step in his gauge theory of the electromagnetic field was published separately, which gave rise to a modified gauge theory of electromagnetism. This was endorsed by leading theoretical physicists, including Pauli, Schrödinger, and Fock. It served as a starting point for the next generation of physicists who developed gauge field theories in the 1950s and 1960s.

Weyl's research in mathematics and physics was shaped by his philosophical outlook and he included his philosophical reflections on scientific activity in many of his publications. Most influential was his contribution to a philosophical handbook, *Philosophy of Mathematics and Natural Science*, originally published in German in 1927 and translated into English in 1949. It became a classic in the philosophy of science.

Further Reading

Chandrasekharan, K., ed. 1986. *Hermann Weyl: 1885–1985. Centenary Lectures delivered by C. N. Yang, R. Penrose, and A. Borel at the Eidgenössische Technische Hochschule Zürich.* Berlin: Springer.

Deppert, W., K. Hübner, A. Oberschelp, and V. Weidemann, eds. 1988. *Exact Sciences and Their Philosophical Foundations.* Frankfurt: Peter Lang.

Hawkins, T. 2000. *Emergence of the Theory of Lie Groups. An Essay in the History of Mathematics 1869-1926.* Berlin: Springer.

Scholz, E., ed. 2001. *Hermann Weyl's Raum–Zeit–Materie and a General Introduction to His Scientific Work.* Basel: Birkhäuser.

Weyl, H. 1968. *Gesammelte Abhandlungen*, edited by K. Chandrasekharan, four volumes. Berlin: Springer.

Erhard Scholz

VI.81   Thoralf Skolem

b. Sandsvaer, Norway, 1887; d. Oslo, 1963
Mathematical logic

Thoralf Skolem was one of the major logicians of the twentieth century, often a lone voice in his understanding of the subtle relationship between abstract set theory and logic. He also worked on Diophantine equations and on group theory, but his contributions to mathematical logic have proved the most lasting. He taught at Bergen and Oslo, was for a time President of the Norwegian Mathematical Society and an editor of its journal, and in 1954 was named a Knight of the First Class in the Royal Order of St. Olav by the king of Norway.

In 1915 Skolem extended a result obtained by the Polish mathematician Leopold Löwenheim. His conclusion (published in 1920 and known as the Löwenheim-Skolem theorem) says that if a mathematical theory defined using only the first-order predicate calculus has a MODEL [IV.23 §1], then it has a countable model. Here a model is a set of mathematical objects that obeys the axioms of the theory. Now, the real numbers are definable in such a theory (for example, ZERMELO–FRAENKEL SET THEORY [IV.22 §3], or any other axioms for set theory). From this we obtain the so-called Skolem paradox, that the real numbers can be defined in a theory with a countable model, even though it had been known since the time of CANTOR [VI.54] that the real numbers are uncountable. How can this paradox be resolved?

The answer is that one has to be very careful about what we mean by "countable." In this strange countable model of set theory, we can see that the reals are countable, but *to the model* the reals may be uncountable. In other words, the actual enumeration of the reals that we can see (i.e., the actual bijection between the reals and the natural numbers) may not belong to the model: the model can be so "small" that it is missing some functions. Skolem's paradox highlights the difference between the viewpoint from *outside* the model and that from *inside* the model.

Several fundamental aspects of Skolem's work are visible in these two results, the Löwenheim–Skolem theorem and the Skolem paradox. Skolem had realized, long before anyone else, that mathematical theories nearly always have several different models. He argued that there are axiom systems, and one can prove theorems in these settings, but what is meant by the objects that obey these rules will generally vary from case to case. From this he drew the radical conclusion that attempts to build mathematics on axiomatic theories were unlikely to succeed (although nowadays, of course, mathematics built on axiomatic foundations has become overwhelmingly successful).

Skolem's insistence on first-order theories, in which variables may range only over elements, not subsets, was one that his contemporaries took time to accept. But that viewpoint, and the great clarity that comes with it, is today the overwhelmingly dominant one. Skolem insisted that the only possible logic to use in any investigation of the foundations of mathematics was FIRST-ORDER LOGIC [IV.22 §3.2], and that second-order theories were impermissible in the foundations, precisely because second-order theories allowed the axioms to refer to sets, but the nature of sets was, in his view, one of the topics to be elucidated. Skolem also felt that, while one can talk of individual objects, talk of *all* objects of a certain kind can be problematic if it is too informal. Indeed, a generation earlier mathematicians had encountered the paradoxes of naive set theory, where loose talk about all sets of certain kinds causes real difficulties: for example, Russell's paradox of the set of all sets that are not members of themselves (if it is a member of itself, then it is not, but if it is not, then it is).

Skolem's work is also characterized by a distrust of the concept of infinity and a preference for finitistic reasoning. He was an early advocate of PRIMITIVE RECURSION [II.4 §3.2.1], which deals with the theory of what are called computable functions, as a way of avoiding paradoxes concerning the infinite.

Further Reading

Fenstadt, J. E., ed. 1970. *Thoralf Skolem: Selected Works in Logic*. Oslo: Universitetsforlaget.

Jeremy Gray

VI.82   Srinivasa Ramanujan

b. Erode, India, 1887; d. Madras (now Chennai), India, 1920
Partitions; modular forms; mock theta functions

Ramanujan, a self-taught Indian genius, made monumental contributions to mathematics that set the stage for many of the breakthroughs in number theory in the twentieth century. He worked on analytic number theory, as well as on ELLIPTIC FUNCTIONS [V.31], hypergeometric series, and the theory of CONTINUED FRACTIONS [III.22]. Much of this work was carried out together with his friend, benefactor, and collaborator G. H. HARDY [VI.73].

Hardy and Ramanujan founded the powerful "circle method" in their remarkable paper that gave an exact formula for $p(n)$, the number of integer partitions of n. Ramanujan independently discovered the two identities that came to be known as the Rogers–Ramanujan identities:

$$1 + \sum_{n=1}^{\infty} \frac{q^{n^2}}{(1-q)(1-q^2)\cdots(1-q^n)}$$
$$= \prod_{n=0}^{\infty} \frac{1}{(1-q^{5n+1})(1-q^{5n+4})},$$

$$1 + \sum_{n=1}^{\infty} \frac{q^{n^2+n}}{(1-q)(1-q^2)\cdots(1-q^n)}$$
$$= \prod_{n=0}^{\infty} \frac{1}{(1-q^{5n+2})(1-q^{5n+3})}.$$

These have applications ranging from LIE THEORY [III.48] to statistical physics. The importance of these identities relates to the fact that the GENERATING FUNCTION [IV.18 §§2.4, 3] for $p(n)$ is

$$\prod_{n=0}^{\infty} \frac{1}{1-q^n}.$$

Thus, for example, the second identity asserts that the number of partitions of n into parts all of which are 2 or 3 mod 5 is equal to the number of partitions into distinct parts, all greater than 1, in which no two parts are consecutive integers.

In his work on $p(n)$, Ramanujan discovered and proved many divisibility properties, e.g., that 5 always divides $p(5n+4)$ and that 7 always divides $p(7n+6)$. His conjectures on these divisibility properties inspired the development of extensive methods in MODULAR FORMS [III.59], and his last conjecture was finally settled in 1969 by Oliver Atkin.

All Ramanujan's studies involving $p(n)$ concerned the modular form

$$\eta(w) = q^{1/24} \prod_{n=1}^{\infty} (1-q^n), \quad \text{where } q = e^{2\pi i w}.$$

The relevance of this is that $q^{1/24}/\eta(w)$ is the generating function for $p(n)$. Of special interest to Ramanujan was the arithmetic function $\tau(n)$, defined by the 24th power of $\eta(w)$: namely,

$$\sum_{n=1}^{\infty} \tau(n)q^n = q \prod_{n=1}^{\infty} (1 - q^n)^{24}.$$

Ramanujan conjectured that $|\tau(p)| < 2p^{11/2}$ for every prime p. The study of this problem led to deep and extensive work on modular forms by H. Petersson, R. Rankin, and others. Eventually, the conjecture was proved by P. Deligne, who received the Fields Medal for his achievement in 1978.

The full story of Ramanujan's life makes his achievements all the more amazing. As a child he was mathematically precocious. In high school he won prizes in mathematics. On the basis of his high school record, he won a scholarship to the Government College in Kumbakonam in 1904. At about this time, Ramanujan came into contact with the book *A Synopsis of Elementary Results in Pure and Applied Mathematics* by G. S. Carr. This rather eccentric book is essentially a huge collection of formulas and theorems compiled for students preparing for the celebrated Mathematical Tripos examination at Cambridge. This book fascinated Ramanujan, who became obsessed with mathematics. In college, he neglected his other subjects and gave his all to mathematics. Consequently, he failed some subjects and lost his scholarship. By 1913, Ramanujan seemed destined for obscurity—he was now a mere clerk in the Madras Port Trust. Friends encouraged him to write to English mathematicians about his mathematical discoveries. Eventually he wrote to G. H. Hardy, who was able to discern that Ramanujan was a truly extraordinary mathematician.

Hardy arranged for Ramanujan to travel to England, and between 1914 and 1918 the two of them produced the groundbreaking work described above.

In 1918, Ramanujan became ill with a sickness diagnosed as tuberculosis. He convalesced in England for a year. His health improved a little in 1919 and he was able to return to India. Unfortunately, his health worsened after his return, and he died in 1920. During this last year in India he penned the pages now known as *Ramanujan's Lost Notebook* and therein laid the foundations of the theory of mock theta functions, a class of functions similar to but more general than the classical theta functions.

Further Reading

Berndt, B. 1985–98. *Ramanujan's Notebooks*. New York: Springer.

Kanigel, R. 1991. *The Man Who Knew Infinity*. New York: Scribners.

George Andrews

VI.83 Richard Courant

b. Lublinitz, Silesia (then part of Germany, now Poland), 1888;
d. New York, 1972
*Mathematical physics; partial differential equations;
minimal surfaces; compressible flow; shock waves*

The long and eventful life of Courant was full of high achievements: in mathematical research, the applications of mathematics, as a teacher of many future mathematicians, as a writer of superb books on mathematics, and as an organizer and administrator of large institutions. The fact that Courant—an outsider in his native Germany and a refugee in the United States— could accomplish these things is a testament to his personality as well as to his scientific outlook.

Born in Lublinitz, Courant completed his high school training in Breslau, living on his own and supporting himself by tutoring. His older Breslau friends, Hellinger and Toeplitz, went on to Göttingen, then the mecca of mathematics, and in due course Courant followed them. There he was taken on as an assistant to HILBERT [VI.63], and he began a close friendship with Harald Bohr, which was later extended to Harald's brother Niels.

Under Hilbert's direction, Courant wrote his dissertation on the use of DIRICHLET'S PRINCIPLE [IV.12 §3.5] (on minimizing energy) for constructing conformal maps. Courant also used Dirichlet's principle in several further mathematical studies.

During World War I Courant was drafted into the army as an officer; he fought on the western front and was seriously wounded. After returning to academic life he turned his energies to mathematics and proved some remarkable results: an isoperimetric inequality for the lowest frequency of a vibrating membrane; and the Courant max-min principle for the EIGENVALUES [I.3 §4.3] of a SELF-ADJOINT OPERATOR [III.50 §3.2], so useful in studying the distribution of eigenvalues of the operators of mathematical physics.

In 1920 Courant was named as KLEIN's [VI.57] successor as professor in Göttingen; the appointment was pushed through by Klein and Hilbert, who saw, correctly, that he shared their vision of the relationship between mathematics and science, that he would strike a balance between research and education, and that he had the administrative energy and wisdom to push his mission to fruition.

Courant became close friends with the publisher Ferdinand Springer. One of the fruits of this relationship was the famous "Grundlehren" series of monographs, affectionately known as the "Yellow Peril." The third volume in this series is Courant's exposition of RIE-MANN's [VI.49] geometric view of the theory of analytic functions, combined with Hurwitz's lectures on ELLIPTIC FUNCTIONS [V.31]. In 1924 the first volume of Courant–Hilbert on *Mathematical Physics* appeared; it contained, presciently, much of the mathematics needed for Schrödinger's version of quantum mechanics. His influential calculus book appeared in 1927. His research did not languish; in 1928 he published, jointly with his students Friedrichs and Lewy, the basic paper on the difference equations of mathematical physics.

Under Courant's leadership, Göttingen, where the lively international atmosphere had been destroyed by World War I, became once again an important center for mathematics, as well as physics: the list of visitors reads like a Who's Who of mathematics. This was totally shattered when Hitler took over the government: Jewish professors, Courant among the first, were dismissed unceremoniously and had to flee or face annihilation. Courant and his family found refuge in New York, where he was invited to build a Graduate School of Mathematics at New York University (NYU). Without any foundation to build on, Courant succeeded in this task, with the help of his former student Friedrichs and of the American James Stoker, who shared Courant's scientific ideals. Courant found New York a reservoir of talent, and attracted students such as Max Shiffman, and later Harold Grad, Joe Keller, Martin Kruskal, Cathleen Morawetz, Louis Nirenberg, and others, including the writer of this article.

In 1936, in a burst of creativity, Courant obtained several basic results about MINIMAL SURFACES [III.94 §3.1] using Dirichlet's principle. In 1937 he finished the second volume of Courant–Hilbert. The immensely successful popular book he wrote jointly with Herb Robbins, *What Is Mathematics?*, appeared in 1940. In 1942 when federal financing for scientific research became available, Courant's group embarked on an ambitious study of supersonic flow and shock waves.

Federal support did not stop after the war; this enabled Courant to vastly expand the scale of research and graduate instruction at NYU. The research combined, at a high intellectual level, theoretical mathematics with applications such as fluid dynamics, statistical mechanics, the theory of elasticity, meteorology, the numerical solution of partial differential equations, and other topics. Nothing like this had been attempted before at a university in the United States. The institute created by Courant, eventually named after him, is

flourishing today and has served as a model for other centers around the world.

Courant hated the Nazis, but did not condemn all Germans; after the war he helped to rebuild mathematics in Germany and was instrumental in inviting talented young German mathematicians and physicists to the United States.

Courant received much help from friends of his youth, many of whom became leaders in their fields, as well as from science administrators in government and industry who admired his vision of mathematics and the gallant spirit that was demonstrated by his willingness to fight against seemingly insuperable odds.

Further Reading

Reid, C. 1976. *Courant in Göttingen and New York: The Story of an Improbable Mathematician*. New York: Springer.

Peter D. Lax

VI.84   Stefan Banach

b. Kraków, Poland, 1892; d. Lwów, Poland, 1945
*Functional analysis; real analysis; measure theory;
orthogonal series; set theory; topology*

Banach was the son of Katarzyna Banach and Stefan Greczek. As his parents were unmarried and his mother was too poor to support her son, he was brought up mainly in Kraków by a foster mother, Franciszka Płowa.

After graduating from high school in 1910, Banach enrolled at the Lwów Polytechnic in the Faculty of Engineering. Two years after his studies were interrupted by the outbreak of World War I, Banach returned to Kraków, where on a summer evening in 1916 he was "discovered" by Hugo Steinhaus, who overheard the words "Lebesgue integral" and brought him to Lwów. Steinhaus considered this event as his "greatest mathematical discovery." It was also through Steinhaus that Banach met his future wife, Łucja Braus, whom he married in 1920.

In the same year Professor Antoni Łomnicki engaged Banach as his assistant at the Lwów Polytechnic, even though Banach had not yet finished his studies. This was the beginning of the meteoric rise of Banach's scientific career.

In June 1920 Banach defended his doctoral dissertation, "On operations on abstract sets and their application to integral equations," at the Jan Kazimierz University in Lwów. His dissertation was written in Polish and published in 1922 in French. In his thesis Banach introduced the concept of complete normed linear spaces, which are today known as BANACH SPACES [III.62] (the name was proposed by Fréchet in 1928). The theory combined the contributions of RIESZ [VI.74], Volterra, FREDHOLM [VI.66], Lévy, and HILBERT [VI.63] on concrete spaces and on integral equations into a general theory. Banach's dissertation could be viewed as the birth of *functional analysis*, since Banach spaces are one of its central objects of study.

On April 17, 1922, the Jan Kazimierz University in Lwów awarded Banach his habilitation (a degree allowing him to teach at the university), after which he was appointed Docent in Mathematics. On July 22, 1922, he became a professor of the university (and a full professor from 1927). Banach achieved great research results and became an authority in functional analysis and MEASURE THEORY [III.55]. During the academic year 1924–25 Banach was on sabbatical leave in Paris, where he met LEBESGUE [VI.72], who became a lifelong friend.

In Lwów a group of talented young mathematicians around Banach and Steinhaus soon became the Lwów School of Mathematics and started the journal *Studia Mathematica* in 1929. Among the members of this school were S. Mazur, S. Ulam, W. Orlicz, J. P. Schauder, H. Auerbach, M. Kac, S. Kaczmarz, S. Ruziewicz, and W. Nikliborc. Banach also collaborated with Steinhaus, Saks, and Kuratowski. Many of these mathematicians were later killed by the Germans during the occupation of Poland.

In 1932, Banach's famous book *Theory of Linear Operations* appeared in French (a Polish version was published the year before) as part of a new series of mathematical monographs, of which he was one of the founders. This was the first monograph on functional analysis as an independent discipline, and it was the culmination of more than a decade of intense activity by Banach and others.

Banach and the mathematicians around him liked to discuss mathematics in the Café Szkocka ("Scottish café"). This unconventional way of doing mathematics made the atmosphere of Lwów unique—it is one of the rare cases in mathematics of genuine teamwork among a large group. Turowicz and Ulam noted that (see Kaluza 1996, pp. 62, 74):

> Banach liked to spend most of his days in a café. He liked the noise and the music. They did not prevent him from concentrating and thinking. It was difficult to outlast or outdrink Banach during these sessions. Problems posed right there were discussed, often with no solution evident even after several hours of thinking.

The next day Banach was likely to appear with several small sheets of paper containing outlines of proofs he had completed.

One day in 1935, Banach proposed that the open problems should be collected in a notebook. This notebook later became famous under the name "The Scottish Book." In the years 1935–41 over 190 problems from various branches of mathematical analysis were proposed in this notebook, and the collection was published in English in 1957 by Ulam. A version with commentaries was published in 1981 by Birkhäuser as *The Scottish Book, Mathematics from the Scottish Café* (edited by R. D. Mauldin).

Banach was also the author of the books *Mechanics* (in two volumes, 1929 and 1930; English translation in 1951), *Differential and Integral Calculus* (in two volumes, 1929 and 1930, with several editions in Polish), *Introduction to the Theory of Real Functions* (in two volumes, written by Banach before the war, although only the first volume remains), and ten textbooks (jointly written with Stożek and SIERPIŃSKI [VI.77]) for primary and secondary schools on arithmetic, geometry, and algebra (published in the years 1930–36 and reprinted in 1944–47).

Banach's famous discoveries in functional analysis had three important steps. First, he considered abstract linear spaces, where functions are treated like points or vectors, sets of functions as function spaces, and operations on functions as operators. Second, he introduced the *norm* $\| \cdot \|$ of a mathematical object, that is, a quantity that in some (possibly abstract) sense describes the length, size, or extent of the object. The distance between two abstract elements x and y is then given naturally by $d(x, y) = \|x - y\|$. The third important step was to introduce the notion of "completeness" for these spaces. In such general spaces (*Banach spaces*) he was able to prove several fundamental theorems, like the uniform boundedness principle, the open mapping theorem, and the closed graph theorem. What these results say, roughly speaking, is that in a Banach space we cannot have bad (pathological) behavior everywhere—there is always some part of the space where our linear map or other object is well-behaved.

Names like Banach space, Banach algebra, Banach lattice, Banach manifold, Banach measure, the Hahn–Banach theorem, the Banach fixed point theorem, the Banach–Mazur game, the Banach–Mazur distance between isomorphic spaces, Banach limits, the Banach–Saks property, the Banach–Alaoglu theorem, and the BANACH–TARSKI PARADOX [V.3] show how wide his influence has been. Banach also introduced the notions of DUAL SPACE [III.19], dual operator, and the general concepts of weak and weak-star convergence, and he used all of these notions in linear operator equations.

In 1936, Banach delivered a one-hour plenary address at the International Congress of Mathematicians in Oslo, where he described the work of the whole Lwów school. In 1937 Norbert Wiener tried to lure him to the United States. In 1939 he was elected president of the Polish Mathematical Society and was awarded a Grand Prize of the Polish Academy of Knowledge. Banach spent the war years in Lwów. During the years 1940–41 and 1944–45 he was the Dean of the Faculty of Science at the renamed Iwan Franko State University. In the period 1941–44 Lwów was occupied by the German army. During this period Banach was saved from almost certain death by the action of Rudolf Weigel, a "Schindleresque" factory owner and inventor of the typhus vaccine, who gave him employment at his Bacteriological Institute as a louse feeder. After the war, he accepted a chair at the Jagiellonian University. He died on August 31, 1945, in Lwów of lung cancer at the age of fifty-three.

The complete list of Banach's publications comprises fifty-eight items, and they were reprinted in Banach's *Collected Works* (published in two volumes in 1967 and 1979). Banach said, "Mathematics is the most beautiful and most powerful creation of the human spirit. Mathematics is as old as Man." Banach is considered a national hero in Poland, as a great scientist and a major figure in the great flowering of Polish scientific life in the independent Poland of the interwar years.

Further Reading

Banach, S. 1967, 1996. *Oeuvres*, two volumes. Warsaw: PWN.
Kaluza, R. 1996. *The Life of Stefan Banach*. Basel: Birkhäuser.

Lech Maligranda

VI.85 Norbert Wiener

b. Columbia, Missouri, 1894; d. Stockholm, 1964
Stochastic processes; applications to electrical engineering and physiology; harmonic analysis; cybernetics

Wiener was just eighteen years old when, in 1913, he was awarded a Ph.D. in logic while studying under Josiah Royce at Harvard University. Afterward, he studied with, among others, RUSSELL [VI.71] and HARDY [VI.73] in Cambridge and HILBERT [VI.63] in Göttingen.

After doing work on ballistics for the military during World War II, he was appointed instructor of mathematics at the fledgling Massachusetts Institute of Technology in Cambridge, MA, where he remained for the rest of his career.

Wiener was in many respects a nonconformist, certainly scientifically and mathematically, but also socially, culturally, politically, and philosophically. He was a precocious child and his home education by his father (a noted linguist and Harvard professor), along with his Jewish background in a society still stricken by anti-Semitism, made his nonconformism almost inevitable. Garrett Birkhoff, the son of GEORGE BIRKHOFF [VI.78], said the following in 1977:

> Wiener was notable as one of the few Americans of his time who was outstanding in both pure mathematics and its applications. How much of this can be attributed to his varied and cosmopolitan early background, and how much to his continuing contacts with non-mathematicians … it is hard to say.

During a period in which American mathematics was largely self-sufficient and was still in a phase in which interdisciplinary approaches were generally ignored, Wiener was reaching out to European mathematics and collaborating with engineers such as Vannevar Bush.

This attitude also affected his choices of research topics, even within pure mathematics: he worked on whatever took his fancy. In a talk in 1938, George Birkhoff described Wiener's work on Tauberian theorems as an example of "exercising talent for free invention," contrasting this with the typically American approach: "mathematics as serious business."

Wiener's way of connecting pure and applied mathematics did not follow the usual path of taking old problems of applied mathematics (such as in classical mechanics and electrical engineering) and tackling them with new, and rigorously sharpened, mathematical tools. Rather the opposite: Wiener used some of the newest, and much debated, results of pure mathematics—such as the LEBESGUE INTEGRAL [III.55], Fourier transformations in the complex domain, and STOCHASTIC PROCESSES [IV.24]—and connected them to several of the newest physical, technological, and biological problems. The types of problem he attacked included those of BROWNIAN MOTION [IV.24], quantum mechanics, radio astronomy, anti-aircraft fire control, noise filtration in radar, the nervous system, and the theory of automata.

Of Wiener's many analytical results that make connections between very different domains we give only one as an example. Around 1931 Wiener discussed the following (Lebesgue) integral equation with the German mathematical astrophysicist Eberhard Hopf:

$$f(t) = \int_0^\infty W(t-\tau)f(\tau)\,d\tau.$$

The solution for unknown $f(t)$, which was found with the help of a new and very important factorization technique, and which was dependent on the analytical behavior of the FOURIER TRANSFORMS [III.27] of the functions involved, could be connected to radiative equilibrium in stars. When t is interpreted as time, equations of this kind can be seen to describe causality: the transition from the influencing "past" to the indeterminate "future." A decade later the Wiener–Hopf equations would be connected to Wiener's theories of prediction and filtering.

Wiener's discussion of such disparate fields of application could not fail to invoke philosophically relevant notions such as causality, information (Wiener is considered together with Claude Shannon as the founder of the modern concept of information), control, feedback, and finally the wide-ranging theory of "cybernetics." Cybernetics (literally, the art of steering) can be retrospectively connected to earlier discussions in Greek antiquity (Plato), to James Watt's centrifugal governor, and to Ampère's philosophical writings. Wiener's broad outlook resulted from his collaboration with colleagues from very different domains: mathematical (R. E. A. C. Paley), physical (Hopf), technical (Julian Bigelow, Bush), and physiological (Arturo Rosenblueth). However, this outlook left him vulnerable to criticism and to philosophical and political misinterpretation. The prominent mathematician Hans Freudenthal was a sharp-tongued critic of Wiener's epoch-making book of 1948, *Cybernetics or the Control and Communication in the Animal and the Machine*, claiming that it "shows there is not much to be reported" and that it "has contributed to spreading mistaken ideas of what mathematics really means," although even he had to admit that the book "earned Wiener the greater part of his public renown" and that its "mathematical readers were more fascinated by the richness of its ideas than by its shortcomings."

During the period of the Nazi threat Wiener helped refugees from Europe to settle in the United States, while after World War II he cautioned against the repetition of mistakes such as the boycott of German science in the aftermath of World War I. Wiener warned against the arms race and the misuse of technological developments in the postwar world. Having resigned from the National Academy of Sciences in 1941 because of its alleged bureaucracy and complacency, Wiener nevertheless accepted, shortly before his death in 1964 while traveling, the National Medal of Science from President Johnson.

Further Reading

Masani, P. R. 1990. *Norbert Wiener 1894–1964*. Basel: Birkhäuser.

Reinhard Siegmund-Schultze

VI.86 Emil Artin

b. Vienna, 1898; d. Hamburg, Germany, 1962
Number theory; algebra; theory of braids

Born in fin de siècle Vienna to an art dealer father and opera singer mother, Artin was influenced throughout his life by the rich cultural atmosphere of the late Hapsburg Empire. He was, as the algebraist Richard Brauer described him, as much artist as mathematician. After his first semester at the University of Vienna in 1916, Artin was drafted into the Austrian Army, in which he served until the end of World War I. In 1919 he enrolled at the University of Leipzig, and completed his doctorate under the direction of Gustav Herglotz in only two years.

Artin spent the academic year 1921–22 at the mathematically vibrant University of Göttingen, and then moved to the recently opened University of Hamburg. He achieved the rank of full professor in 1926. While at Hamburg, Artin oversaw the work of eleven doctoral students, including Max Zorn and Hans Zassenhaus. Artin's years at Hamburg were among the most productive of his life.

Artin's work in CLASS FIELD THEORY [V.28], the subject closest to his heart, led him to a solution of Hilbert's ninth problem: a proof of the most general law of reciprocity. The aim was to generalize Gauss's law of quadratic reciprocity and the higher reciprocity laws. Teiji Tagaki's fundamental results on class field theory had appeared when Artin was a student. Using Takagi's theory, N. G. Chebotaryov's 1922 proof of the density theorem (conjectured by FROBENIUS [VI.58] in 1880), and his own theory of *L*-FUNCTIONS [III.47], Artin established his general law of reciprocity in 1927. Artin's theorem not only provided the final form of the classical question on reciprocity but it also formed the central result of class field theory. Both Artin's result and his tools, particularly his *L*-functions, proved important.

Artin posed a conjecture about his *L*-functions that remains unanswered today. Questions in non-Abelian class field theory also remain open.

In 1926–27, Artin and Otto Schreier developed the theory of formally real fields: fields with the property that −1 cannot be expressed as the sum of two squares (an example being the real numbers). This work formed the basis of Artin's solution of Hilbert's seventeenth problem concerning rational functions.

Artin extended Wedderburn's theory of algebras ("hypercomplex numbers") to noncommutative rings with chain conditions in 1928. Indeed, the class of such rings called "Artinian rings" is named in his honor.

In 1929 Artin married one of his students, Natalie Jasny. Natalie's Jewish background and Artin's personal sense of justice prompted them to leave Germany in 1937. They emigrated to America, where Artin spent a year at Notre Dame University before moving to a permanent position at Indiana University. Artin's lectures at Notre Dame led to his influential text *Galois Theory* (1942), which reflected his quest for simplification and his desire to unite different research trends.

At Indiana, Artin began a collaboration with George Whaples of the University of Pennsylvania and introduced the concept of a valuation vector, a notion closely related to the concept of an idèle introduced by Claude Chevalley. This work seemed to revitalize Artin's mathematical research, and, after something of a hiatus in his written work, he began to publish regularly again.

In 1946, Artin moved to Princeton University. While there, Artin oversaw eighteen of his thirty-one Ph.D. students, including John Tate and Serge Lang. He also returned to his work in the theory of BRAIDS [III.4], a topic that relates questions in topology and group theory. His introduction to the theory of braids that appeared in *American Scientist* in 1950 reveals Artin's prowess as a master expositor.

Further Reading

Brauer, R. 1967. Emil Artin. *Bulletin of the American Mathematical Society* 73:27–43.

Della Fenster

VI.87 Alfred Tarski

b. Warsaw, 1901; d. Berkeley, California, 1983
Symbolic logic; metamathematics; set theory; semantics; model theory; algebras of logic; universal algebra; axiomatic geometry

---

Tarski matured during Poland's renaissance in mathematics and philosophy in the remarkable interwar period of Polish independence. His teachers at the University of Warsaw included Stanisław Leśniewski and Jan Łukasiewicz in logic, SIERPIŃSKI [VI.77] in set theory, and Stefan Mazurkiewicz and Kazimierz Kuratowski in topology. In his thesis Tarski solved a core problem in Leśniewski's idiosyncratic system for the foundation of mathematics, but afterward he focused on set theory and more mainstream mathematical logic. Almost immediately he obtained the spectacular BANACH–TARSKI PARADOX [V.3] (that it is possible to dissect a solid sphere into a finite number of pieces that may then be reassembled to form *two* spheres of the same radius as the original one) in collaboration with BANACH [VI.84].

Encouraged by his professors, he changed his original surname, Teitelbaum, to Tarski just before receiving his Ph.D. in 1924, because a Jewish name was a professional handicap. This accorded with Tarski's strong identification with Polish nationalism and his belief that assimilation was a rational solution to the Jewish question.

By 1930, Tarski had established one of his most important results: the completeness and decidability of formal systems of the algebra of real numbers and of Euclidean geometry axiomatized within first-order logic (see LOGIC AND MODEL THEORY [IV.23 §4]). In the following years Tarski concentrated on fundamental conceptual developments in metamathematics and the semantics of formalized languages. In contrast with HILBERT [VI.63], who called for the execution of his metamathematical consistency program by the most restricted means possible, Tarski was open to the use of any mathematical methods, including all those of set theory. His main conceptual contribution was to provide a theory of truth for formalized languages, in which he laid down a novel criterion—called the T-scheme—for an adequate definition of truth for such a language, and showed how it can be met by a set-theoretical definition within a metalanguage, while it cannot be defined within the language itself.

Though Tarski's preeminence in Polish logic was widely acknowledged, he never succeeded in obtaining a chair in his country of birth, partly because of the paucity of positions, and partly as a result of anti-Semitism, notwithstanding his change of name. Made a Docent at the University of Warsaw as soon as he had finished his Ph.D., his position was later raised to that of Adjunct Professor. Neither post paid a living wage, and so, in order to make ends meet, Tarski also taught in a Gymnasium (high school) throughout the 1930s. Because he did not hold a chair, he could not be designated the official director of the dissertation of his first student, Andrzej Mostowski; instead Kuratowski assumed that role.

An invitation to attend a Unity of Science meeting (an offshoot of the Vienna Circle) at Harvard brought Tarski to the United States two weeks before the Nazi invasion of Poland on September 1, 1939. Given his Jewish origins, this probably saved his life, but the war separated him from his family. (His wife and immediate family survived the war, but most of the rest of his family perished in the Holocaust.) In the United States he was granted a permanent nonquota visa within months, but only temporary positions were available to him during the period 1939–42. Finally, he succeeded in obtaining a position as Lecturer in the Department of Mathematics of the University of California at Berkeley. There Tarski's manifest excellence was soon recognized and he rose rapidly to the position of Full Professor by 1946. In the following decade, through his charismatic teaching and zealous campaigning for additional appointments in the field, he built a program in logic and the foundations of mathematics that made Berkeley a mecca for logicians from all over the world for years to come.

It was not until 1939 that Tarski wrote up his decision procedure for algebra and geometry for publication; it was slated to appear as a monograph for a Parisian publisher, but that was aborted following the invasion of France by Germany in 1940. A revised version with full details was finally prepared with the assistance of J. C. C. McKinsey as a RAND Corporation report in 1948; it only became publicly available a few years later through the University of California Press. This work then became paradigmatic for the applications of model theory to algebra in which the Tarski school led the way; the subject has continued to be one of the most important parts of mathematical logic to this day. At Berkeley during the postwar period Tarski also promoted substantial developments along several different lines: algebraic logic, the axiomatics of set theory and the significance of LARGE CARDINAL [IV.22 §6] assumptions for mathematical problems, and the axiomatics of geometry. Above all, the importance of Tarski's work lay in opening the field of logic to the unrestricted use of set-theoretical methods, combined with a constant attention to rigorous and proper conceptual development.

Further Reading

Feferman, A. B., and S. Feferman. 2004. *Alfred Tarski. Life and Logic.* New York: Cambridge University Press.
Givant, S. 1999. Unifying threads in Alfred Tarski's work. *Mathematics Intelligencer* 13(3):16–32.
Tarski, A. 1986. *Collected Papers*, four volumes. Basel: Birkhäuser.

Anita Burdman Feferman and
Solomon Feferman

VI.88 Andrei Nikolaevich Kolmogorov

b. Tambov, Russia, 1903; d. Moscow, 1987
Analysis; probability; statistics; algorithms; turbulence

Kolmogorov was one of the greatest mathematicians of the twentieth century. His work was distinguished both by its great depth and power, and by its breadth: he made important contributions to several different areas. He is most famous for his work on probability theory, and is widely regarded as having been the greatest probabilist ever.

Kolmogorov's mother, Mariya Yakovlena Kolmogorova, died in childbirth; his father, Nikolai Matveevich Kataev, an agronomist, worked for the Ministry of Agriculture after the Revolution and died in the Denikin offensive in the Civil War in 1919. Kolmogorov was brought up by his mother's sister Vera, whom he regarded as his mother and who lived to see her adopted son's success.

After his childhood in Tunoshna, near Yaroslavl on the Volga, Kolmogorov became a student of mathematics at Moscow University in 1920. His teachers included Aleksandrov, Lusin, Urysohn, and Stepanov. Kolmogorov's first work, published in 1923 when he was still nineteen, gave an example of a (Lebesgue integrable) function whose FOURIER SERIES [III.27] diverges almost everywhere. (This is in contrast to the classical theorems giving regularity conditions on a function that are sufficient for its Fourier series to converge to it.) This famous and unexpected result made him a celebrity, all the more so when in 1925 he sharpened "almost everywhere" to "everywhere."

Kolmogorov became a postgraduate student in 1925, studying under Lusin. Also in 1925, he published his first work on probability theory, in collaboration with Alexander Yakovlevich Khinchin (Khintchine, Hincin), on the "three series theorem." This classical result gives a necessary and sufficient condition for the convergence of a random series with independent terms, namely the convergence of three nonrandom series.

The paper also contains the Kolmogorov inequality on maxima of independent sums. By the time of his doctorate in 1929, Kolmogorov had written eighteen mathematical papers: on analysis, on probability, and on intuitionist logic, an indication of his lifelong interest in the foundations of mathematics. He became a professor at Moscow University in 1931.

Also in 1931, Kolmogorov published his famous paper on analytic methods in probability theory. This deals with Markov processes in continuous time, with the state space continuous or discrete (in which case one speaks of a Markov chain). The Chapman-Kolmogorov equations, and the Kolmogorov forward and backward differential equations, date from this paper. Diffusions are also treated, developing earlier work by Bachelier.

The whole subject of modern probability theory was given a firm foundation by Kolmogorov's epoch-making monograph *Grundbegriffe der Wahrscheinlichkeitsrechnung* of 1933 (later translated as *Foundations of Probability Theory*). Before this time, probability had lacked a rigorous mathematical foundation, and indeed some authors had believed that it was impossible to provide one. However, the relevant mathematical theory, MEASURE THEORY [III.55], had been introduced by LEBESGUE [VI.72] in 1902, in connection with his theory of the integral. Measure theory also provides a firm foundation for the mathematics of length, area, and volume. By the 1930s, the subject had been freed from its origins in Euclidean space. Kolmogorov treated probability simply as a measure of total mass 1, events as measurable sets, RANDOM VARIABLES [III.71 §4] as measurable functions, etc. The decisive technical innovation was his treatment of conditioning, which used the then-recent Radon-Nikodým theorem (whereby conditional expectations became Radon–Nikodým derivatives). The *Grundbegriffe* also contains two further key results. The first is the Daniell-Kolmogorov theorem, basic to the definition of a STOCHASTIC PROCESS [IV.24]. The second is Kolmogorov's STRONG LAW OF LARGE NUMBERS [III.71 §4]. When we repeatedly toss a fair coin, we expect the observed frequency of heads to tend to the expected frequency, a half. Some restriction is needed to make precise mathematical sense out of this intuition. It was known before Kolmogorov that the qualification needed here is that convergence takes place with probability 1 ("almost surely," or "a.s."). Kolmogorov generalized this result from coin tossing to repeated replication of any random experiment. One needs the expected value (often called the mean) to

exist, in the technical sense of measure theory. Then the average value in a sample, the *sample mean*, converges to the expectation, the *population mean*, with probability 1.

Further work by Kolmogorov on probability theory followed in the 1930s and 1940s. He worked on limit theorems, on infinite divisibility, on the Kolmogorov–Petrovskii–Piscunov equation governing the wave of advance of an advantageous gene, and on linear prediction of stationary stochastic processes. This application, which led to the "Kolmogorov–Wiener filter," was motivated by wartime applications to fire control problems.

This last work led Kolmogorov naturally to path-breaking work on turbulence in 1941, including the Kolmogorov "two-thirds power" law. This work has been profoundly important subsequently, as the problem of understanding turbulence is a central one in fluid dynamics.

Motivated by questions of the stability of the solar system, and related DYNAMICAL SYSTEMS [IV.14], Kolmogorov published in 1954 his work on mechanics and invariant tori, work that developed into the subject of "KAM theory" (for Kolmogorov, Arnold, and Moser).

Kolmogorov's axiomatization of probability theory can be regarded as a solution of (part of) Hilbert's sixth problem, to put probability and mechanics onto a rigorous footing. In 1956 and 1957, Kolmogorov solved another of Hilbert's problems, the thirteenth. His solution gave a surprising structure theorem, by which a function of many variables can be built up from functions of few variables by means of basic operations. He showed that a continuous function of any number of real variables may be built up by combining (using the operations of addition and of taking a function of a function) a finite number of functions of *only three* real variables. He regarded this work as his most technically difficult accomplishment.

In the 1960s, Kolmogorov turned his attention to foundational questions: in mathematics, in probability theory, and in INFORMATION THEORY [VII.6] and the theory of algorithms. He introduced the concept now called "Kolmogorov complexity." He gave a new approach to randomness, quite different from that in his earlier work on probability theory. Here, random sequences are identified as *sequences of maximal complexity*. His later work was dominated by his lifelong interest in teaching, and in particular his involvement in special schools for particularly gifted pupils.

Kolmogorov's *Selected Works* comprise three volumes: *Mathematics and Mechanics*, *Probability and Statistics*, and *Information Theory and Algorithms*.

He was widely honored, both within the Soviet Union and outside. He was married, with no children.

Further Reading

Kendall, D. G. 1990. Obituary, Andrei Nikolaevich Kolmogorov (1903–1987). *Bulletin of the London Mathematical Society* 22(1):31–100.

Shiryayev, A. N., ed. 2006. *Selected Works of A. N. Kolmogorov.* New York: Springer.

Shiryayev, A. N., and others. 2000. *Kolmogorov in Perspective.* History of Mathematics, volume 20. London: London Mathematical Society.

Nicholas Bingham

VI.89   Alonzo Church

b. Washington, District of Columbia, 1903; d. Hudson, Ohio, 1995
Logic

Church's career was spent almost entirely at Princeton. Having studied there, he returned, after spending time in Harvard, Göttingen, and Amsterdam, to take up a position as an assistant professor in 1929. He became Professor of Philosophy and Mathematics in 1961, a position he held until his retirement in 1967. He then moved to the University of California at Los Angeles, where he was Kent Professor of Philosophy and Professor of Mathematics until he retired (again) in 1990.

Princeton became an important center for logic during the 1930s: VON NEUMANN [VI.91] arrived at the beginning of the decade; GÖDEL [VI.92] visited in 1933 and 1935 before moving there permanently in 1940; and from September 1936 TURING [VI.94] spent two years there as a graduate student, completing his Ph.D. with Church.

In 1936 Church made two profound contributions to the theory of logic. The first, which appeared in a paper entitled "An unsolvable problem in elementary number theory," is what is now known as *Church's thesis*: the proposal to identify the vague intuitive notion of effective calculability with the precise notion of a RECURSIVE FUNCTION [II.4 §3.2.1]. It quickly transpired that Church's definition of a recursive function was equivalent to Turing's definition of computable functions. At the end of 1936, Turing, who had been working with analogous ideas in an entirely different way, published his famous paper "On computable numbers," which

contained the result that every function that is naturally regarded as computable is computable by a TURING MACHINES [IV.20 §1.1]. Church's thesis is therefore often known as the Church–Turing thesis.

The second of Church's contributions is what is now known as *Church's theorem*. In a short paper published in the first issue of *The Journal of Symbolic Logic*, Church showed that it is impossible to decide algorithmically whether statements in arithmetic are true or false. It follows that a general solution to the *Entscheidungsproblem* (decision problem) does not exist; equivalently, first-order logic is undecidable (see THE INSOLUBILITY OF THE HALTING PROBLEM [V.20]). This result is also known as the Church–Turing theorem because Turing independently (and in the paper referred to above) proved the same result. In achieving this result, both Church and Turing were strongly influenced by GÖDEL'S INCOMPLETENESS THEOREM [V.15].

VI.90   William Vallance Douglas Hodge

b. Edinburgh, Scotland, 1903; d. Cambridge, England, 1975
Algebraic geometry; differential geometry; topology

Hodge is famous for his theory of harmonic integrals (or forms), which was described by WEYL [VI.80] as "one of the landmarks of twentieth century mathematics." He was a Scot who spent his early life in Edinburgh but lived for most of his life in Cambridge, where he was Lowndean Professor of Astronomy and Geometry (an archaic title) from 1936 until 1970.

Hodge's work straddles the area between algebraic geometry, differential geometry, and complex analysis. It can be seen as a natural outgrowth of the theory of RIEMANN SURFACES [III.79] (or algebraic curves) and the work of Lefschetz on the topology of algebraic VARIETIES [IV.4 §7] (of higher dimension). It put algebraic geometry on a modern analytic footing and prepared the ground for the spectacular breakthroughs of the postwar period in the 1950s and 1960s. It also harmonized well with the later interaction with theoretical physics, harking back to the influence of James Clerk Maxwell.

In Riemann surface theory (with one complex dimension), complex structures and real metrics are very closely related and the roots of their relationship can be traced back to the link between the CAUCHY–RIEMANN EQUATIONS [I.3 §5.6] and the LAPLACE OPERATOR [I.3 §5.4]. In higher dimensions this close link disappears and a RIEMANNIAN METRIC [I.3 §6.10] seems alien to complex analysis, but it was Hodge's great insight to see that real analysis could still play a fruitful role.

Following the formalism of electromagnetic theory as developed by Maxwell, he introduced a generalization of the Laplace operator to exterior DIFFERENTIAL FORMS [III.16] (on any Riemannian manifold) and proved the key theorem that the null space of this operator on r-forms ("harmonic" forms) is naturally isomorphic to the r-dimensional COHOMOLOGY [IV.6 §4] H^r. In other words, a harmonic form is uniquely specified by its periods, and all sets of periods occur.

For complex manifolds, provided the metric is suitably compatible with the complex structure (the KÄHLER CONDITION [III.88 §3], always satisfied by algebraic varieties in projective space), this result can be refined. We get a decomposition of H^r into subspaces $H^{p,q}$ with $p + q = r$, with the extreme cases $p = r$, $q = r$ corresponding to holomorphic or anti-holomorphic forms.

This Hodge decomposition has a rich structure and a wealth of applications. One of the most remarkable is the Hodge signature theorem, which (for an even-dimensional algebraic variety) expresses the signature of the intersection matrix of middle-dimensional cycles in terms of the dimensions of the $H^{p,q}$.

Another success of the theory was the characterization of those homology classes of dimension $2n - 2$ (on a complex n-manifold) that arise from algebraic subvarieties. He conjectured that a similar characterization would work for all dimensions and proved the easy part. The hard part has resisted all subsequent attempts, and is now one of the million-dollar Millennium Problems of the Clay Institute.

The influence of Hodge's theory was enormous. First, in algebraic geometry it integrated many classical results into a modern framework and it acted as a launch pad for the subsequent development of modern sheaf theory by Henri Cartan, Serre, and others. Second, it was the first deep result in global differential geometry and paved the way for what became known as "global analysis." Third, it provided the basis for later developments arising from, or linked to, theoretical physics. These included THE ATIYAH–SINGER INDEX THEOREM [V.2] for elliptic operators, and nonlinear analogues of Hodge theory (the Yang–Mills and the Seiberg–Witten equations), which have played such a key role in the Donaldson theory of four-dimensional manifolds (see DIFFERENTIAL TOPOLOGY [IV.7 §2.5]). More recently, Witten and others have

shown how suitable infinite-dimensional versions of Hodge's theory turn up naturally in QUANTUM FIELD THEORY [IV.17 §2.1.4].

Further Reading

Griffiths, P., and J. Harris. 1978. *Principles of Algebraic Geometry*. New York: Wiley.

Sir Michael Atiyah

VI.91   John von Neumann

b. Budapest, 1903; d. Washington, District of Columbia, 1957
Axiomatic set theory; quantum physics; measure theory;
ergodic theory; operator theory; algebraic geometry;
theory of games; computer engineering; computer science

Raised as a Hungarian Jew in the Austrian Empire, Neumann János Lajos's political outlook was strongly affected by the five-month reign of the communist Béla Kun's regime after World War I. It formed his liberal and democratic political credo (although he did insist on retaining the title of nobility "margittai," acquired by his father in 1913, which he later translated to the German "von"). He was a child prodigy, learning several languages and demonstrating an early enthusiasm for mathematics.

During the early 1920s von Neumann studied mathematics, physics, and chemistry in Berlin and Zürich, and was also enrolled to study mathematics in Budapest although he never attended any lectures there. He received a diploma in chemical engineering at the ETH Zürich and shortly afterward (in 1926) a doctorate in mathematics at the University of Budapest (his thesis was entitled "The axiomatic deduction of general set theory"). While engineering was considered a respectable profession for a brilliant young man with such wide-ranging interests, the theoretical challenges of mathematics and formal logic drove von Neumann to the more academic environment in Germany, where he immediately received attention from HILBERT [VI.63]. Although the sensible choice, academically speaking, would have been to stay with Hilbert at Göttingen—and he did spend six months there during 1926–27 on a Rockefeller Fellowship—he preferred the pulsating atmosphere of Berlin.

During the following years he published on the axiomatic foundations of set theory, on MEASURE THEORY [III.55], and on the mathematical foundations of quantum mechanics. He also wrote his first paper on game theory ("Zur Theorie der Gesellschaftsspiele," published in *Mathematische Annalen* in 1928), proving the *minimax theorem* (the theorem that states that every two-person finite zero-sum game has optimal mixed strategies).

In 1927 von Neumann received his habilitation in mathematics from the Philosophical Faculty of Berlin University with a written thesis and a lecture on the foundations of set theory and mathematics, becoming one of the youngest Privatdozents in the history of the university. At this point he changed his name to the German Johann von Neumann. He gave lecture courses in Hamburg (1929–30) as well as in Berlin, but in 1933, with the Nazi seizure of power, he resigned from his appointment at Berlin. By that time he was already in Princeton, where his visiting status at the university, originally conferred in 1930, was transformed into a tenured position at the newly founded Institute for Advanced Study. He modified his name once again, this time to John von Neumann, receiving U.S. citizenship in 1937.

At Princeton he found a peaceful ivory tower. Much of his important mathematical work stems from that period in the mid 1930s: he published around six journal articles per year (a rate he maintained until his death), as well as several books. The Institute's environment allowed him to expand his research scope, taking in, among other things, ERGODIC THEORY [V.9], Haar measure, certain spaces of operators on a HILBERT SPACE [III.37] (these spaces are now known as VON NEUMANN ALGEBRAS [IV.15 §2]), and "continuous geometry."

Von Neumann was much too politically sensitive to ignore the European crisis that led to World War II. Having begun to investigate turbulent flow beyond the speed of sound in the mid 1930s, he was invited to the Ballistic Research Laboratory in 1937 as an expert on shock waves. Later he acted as a consultant to the Navy and the Air Force. Although he was not in the initial group of Los Alamos scientists, in 1943 he became an advisor to the Manhattan Project, where his mathematical treatment of shock waves became essential, leading to the "implosion lens," an arrangement of explosives that started the uranium chain reaction.

In parallel with his war-related work, von Neumann pursued his interest in economics, which resulted in a collaboration with Oskar Morgenstern: their groundbreaking book *The Theory of Games and Economic Behavior*, partly based on his 1928 *Mathematische Annalen* paper, appeared in 1944.

In the 1940s von Neumann began to focus on computing as a result of two very different branches of his thinking: namely, the numerical approximation of solutions to otherwise unsolvable problems, and his proficiency in the foundations of mathematics. He had tried to enlist TURING [VI.94] as an assistant at Princeton and he was certainly aware of the importance of Turing's seminal paper on computable numbers (1936). While Turing discussed an abstract machine in the form of a thought experiment, von Neumann also considered the problems arising from the actual construction of computers, such as those connected with the use of electronic hardware. His training as a mathematician allowed him to focus on the very essentials of computing machinery and avoid baroque designs like the Moore School's ENIAC (Electronic Numerical Integrator And Computer). In 1945 he defined the essential components for the "Electronic Discrete Variable Computer." His "First draft of a report on the EDVAC," which summarized and focused ideas gathered from work on early electronic computers, provided a logical framework for the modern electronic computer, becoming a road map for computer architecture for the ensuing decades. While von Neumann probably did not consider this paper to have the same importance as his mathematical results, today it is considered the birth certificate of modern computers.

Von Neumann quickly recognized that programming computers (or "coding," as he called it) was likely to be more demanding than building basic hardware. In essence he considered programming as a new branch of formal logic. In 1947 he coauthored (with Herman Goldstine) a three-part report, "Planning and coding of problems for an electronic computing instrument," in which many insights on the novel and demanding art of software construction were collected together.

Von Neumann's thinking went beyond the restrictions of calculating machines, and allowed him to venture into philosophical questions on the structure of the human brain and cellular automata and the idea of self-reproducing systems—questions that were forerunners to the disciplines now called "artificial intelligence" and "artificial life." Consideration of these questions resulted in a series of lectures published as *The Computer and the Brain* (1958) and a book, *Theory of Self-Reproducing Automata* (1966), both of which appeared posthumously.

In 1954 von Neumann was appointed to the five-member U.S. Atomic Energy Commission and in 1956 he was awarded the Presidential Medal of Freedom by President Eisenhower.

Further Reading

Aspray, W. 1990. *John von Neumann and the Origins of Modern Computing.* Cambridge, MA: MIT Press.

Wolfgang Coy

VI.92   Kurt Gödel

b. Brno, Moravia (now Czech Republic), 1906;
d. Princeton, New Jersey, 1978
Logic; relativity theory

Born in Brno, Moravia, Gödel did his most important work at the University of Vienna. In 1940 he emigrated to the United States, where he accepted an appointment at the Institute for Advanced Study in Princeton.

Considered the greatest mathematical logician of the twentieth century, Gödel is renowned for his proofs of three fundamental results: the semantic COMPLETENESS OF FIRST-ORDER LOGIC [IV.23 §2]; the syntactic INCOMPLETENESS OF FORMAL NUMBER THEORY [V.15]; and the consistency, relative to the AXIOMS OF ZERMELO-FRAENKEL [IV.22 §3.1] set theory, of the AXIOM OF CHOICE [III.1] and the generalized CONTINUUM HYPOTHESIS [IV.22 §5].

Gödel's completeness theorem (1930) is concerned with the following kind of question: how do we know that a statement in group theory, for example, that is true in every group is actually provable from the axioms of group theory? Gödel showed that in any first-order theory (one in which quantifiers are allowed over elements but not over subsets), any statement true in all models is indeed provable. In an equivalent form, this completeness theorem states that any set of statements that is consistent (that is, from which no contradiction may be derived) has a model—a structure in which all those statements hold.

Gödel's incompleteness theorem (1931) sent shock waves through logic and the philosophy of mathematics. HILBERT [VI.63] had set out a program in which all statements (in number theory, for example) should be derivable from a fixed set of axioms. It was generally believed that such a program was in principle possible, until the incompleteness theorem destroyed that hope.

Gödel's idea was to construct a statement S that, in effect, asserts "S is not provable." A moment's thought shows that such a statement must be both true and unprovable. Gödel's remarkable achievement was to manage to encode such a statement in the language of number theory. His proof applies to such axioms as THE PEANO AXIOMS [III.67] for number theory, and more generally, to any reasonable extension of them (such as the Zermelo–Fraenkel axioms for set theory).

Gödel's second incompleteness theorem represented another blow to the Hilbert program. Suppose that we have a set of axioms T (for example, the Peano axioms) that is consistent. Can we prove that it is consistent? Gödel showed that, if T is consistent, then the statement "T is consistent" (when encoded as a statement of number theory) cannot be proved from T. So "T is consistent" is an explicit example of a true but unprovable statement. Again, this applies when T is the set of Peano axioms, or any reasonable extension thereof (roughly, any extension that allows one to encode into arithmetic statements about provability and the like). As a slogan: "a theory cannot prove its own consistency."

The axiom of choice became highly controversial when Ernst Zermelo used it to prove that every set can be well-ordered, a task that, together with the proof of the continuum hypothesis, Hilbert had listed first among the problems he posed in 1900 to the International Congress of Mathematicians. In 1938 Gödel showed that both the axiom of choice and the generalized continuum hypothesis are consequences of another principle (the axiom of constructibility) that holds in a submodel of any model of Zermelo–Fraenkel set theory. Both are consequently consistent with (not disprovable from) the Zermelo–Fraenkel axioms. Much later (1963) Paul Cohen showed that both statements are also independent of (not provable from) those axioms.

Apart from logic, Gödel also worked in relativity theory, where he established the existence of models of EINSTEIN'S FIELD EQUATIONS [IV.13] that permit time travel into the past.

Further Reading

Dawson Jr., J. W. 1997. *Logical Dilemmas: The Life and Work of Kurt Gödel.* Natick, MA: A. K. Peters.

John W. Dawson Jr.

VI.93   André Weil

b. Paris, 1906; d. Princeton, New Jersey, 1998
Algebraic geometry; number theory

André Weil was one of the most influential mathematicians of the twentieth century. His influence is due both to his original contributions to a remarkably broad spectrum of mathematical theories, and to the mark he left on mathematical practice and style, through some of his own works as well as through the BOUR-BAKI [VI.96] group, of which he was one of the principal founders.

Weil, as well as his sister, the philosopher, political activist, and religious thinker Simone Weil, received an excellent education. Both were brilliant students, very widely read, with a keen interest in languages (including Sanskrit). André Weil soon specialized in mathematics, his sister in philosophy. He graduated (and was first in his year in the *agrégation* for mathematics) from the École Normale Supérieure (ENS) when he was not even nineteen years old, and traveled in Italy and Germany. He obtained his doctorate in Paris at the age of twenty-two, and then went to Aligarh, India, as a professor for two years. After a brief spell in Marseilles, he was *Maître de Conférences* at Strasbourg University (along with Henri Cartan) from 1933 to 1939. The idea of the Bourbaki project arose there from discussions about teaching with Cartan, and grew in Paris in meetings that included other friends from the ENS.

His research achievements began with his 1928 Paris thesis. In it, he generalized MORDELL'S THEOREM [V.29] of 1922, that the group of rational points on an ELLIPTIC CURVE [III.21] is a finitely generated Abelian group, to the group of K-rational points (where K is a NUMBER FIELD [III.63]) of a Jacobian variety. During the following twelve years, Weil branched out in various directions, all related to important research topics of the 1930s: the approximation of holomorphic functions of several variables by polynomials; the conjugation of maximal tori in compact LIE GROUPS [III.48 §1]; the theory of integration on compact and Abelian topological groups; and the definition of uniform TOPOLOGICAL SPACES [III.90]. But problems of arithmetic origin stood out among his interests: further thoughts on his thesis and on Siegel's finiteness theorem for integral points; a bold "vector bundle" version of the RIEMANN-ROCH THEOREM [V.31] on a Riemann surface (in parallel with similar work by E. Witt); p-adic analogs of ELLIPTIC FUNCTIONS [V.31] (with his student Elisabeth Lutz).

Starting in 1940, Weil became active on what was probably the biggest challenge in arithmetic algebraic geometry at the time. Helmut Hasse had proved in 1932 the analogue of the RIEMANN HYPOTHESIS [IV.2 §3] for curves of genus 1 (elliptic curves) defined over a field with finitely many elements. The problem was to generalize this to algebraic curves of genus higher than 1. In 1936, Max Deuring had proposed algebraic correspondences as a crucial new ingredient for attacking

this problem; but the problem remained open until World War II. Weil's initial attempt, written while in jail in Rouen, was very modest, and contained little more than Deuring's observations of 1936. But, after several years of searching in various directions while in residence in the United States, Weil finally became the first person to prove the analogue of the Riemann hypothesis for all nonsingular curves. This proof relied on his complete rewriting of algebraic geometry (over an arbitrary ground field), which he had published before in his *Foundations of Algebraic Geometry* (1946). Furthermore, Weil generalized the analogue of the Riemann hypothesis from curves to algebraic varieties of arbitrary dimensions, defined over a finite field, and added a new topological interpretation of the main invariants of the relevant zeta functions. Taken together, all these conjectures became known as THE WEIL CONJECTURES [V.35]; they represented the most important stimulus for the further development of algebraic geometry right through to the 1970s, and to some extent later as well.

Several mathematicians were at work in the 1930s and 1940s trying to rewrite algebraic geometry. Weil's *Foundations*, even though it does contain striking new insights (e.g., a novel definition of intersection multiplicity), owes its basic notions (generic points, specializations) to van der Waerden, and it exerted its influence on the mathematical community in conjunction with the (different) rewriting of algebraic geometry developed so successfully by Oscar Zariski from 1938 onward. It was therefore to a large extent the characteristic style, rather than just the "mathematical content," of the *Foundations* that would create a new way of doing algebraic geometry for the next twenty years or so, until it began to be replaced by Grothendieck's language of schemes.

Later works include, among other seminal papers and books, Weil's "adelic" rewriting of Siegel's work on quadratic forms, and a crucial contribution to the philosophy, due to Taniyama and Shimura, that elliptic curves over the rational numbers should be modular—the proof of this fact is the basis of Wiles's 1995 proof of FERMAT'S LAST THEOREM [V.10].

In 1947, Weil—whose evasion of the French draft in 1939 was considered very critically by many American colleagues—finally obtained a professorship at a distinguished university, namely Chicago. In 1958, he moved to Princeton as a permanent member of the Institute for Advanced Study.

The postwar years saw Weil continuously active on many fronts of mathematical research, contributing insightfully to many subjects that were in the air at the time. To mention just a few: the Weil groups of CLASS FIELD THEORY [V.28]; the explicit formulas of analytic number theory; various aspects of differential geometry, in particular KÄHLER MANIFOLDS [III.88 §3]; the determination of Dirichlet series by their functional equations. All of these topics point to seminal works without which today's mathematics would not be what it is.

In his later years, Weil put his erudition and historical sense to work writing articles and a book on the history of mathematics: *Number Theory, an Approach through History*. He also published a partial autobiography ending in 1945, *Souvenirs d'Apprentissage*, of considerable literary quality.

Further Reading

Weil, A. 1976. *Elliptic Functions According to Eisenstein and Kronecker*. Ergebnisse der Mathematik und ihrer Grenzgebiete, volume 88. Berlin: Springer.
——. 1980. *Oeuvres Scientifiques/Collected Papers*, second edn. Berlin: Springer.
——. 1984. *Number Theory. An Approach through History. From Hammurapi to Legendre*. Boston, MA: Birkhäuser.
——. 1991. *Souvenirs d'Apprentissage*. Basel: Birkhäuser 1991. (English translation: 1992, *The Apprenticeship of a Mathematician*. Basel: Birkhäuser.)

Norbert Schappacher and Birgit Petri

VI.94 Alan Turing

b. London, 1912; d. Wilmslow, England, 1954
Logic; computing; cryptography; mathematical biology

In 1936, as a young Fellow of King's College, Cambridge, Alan Turing made a crucial contribution to mathematical logic: he defined "computability" with what is now called the TURING MACHINES [IV.20 §1.1]. Although mathematically equivalent to a definition of effective calculability earlier given by CHURCH [VI.89], Turing's concept was compelling because of his entirely original philosophical analysis. It won the endorsement of Church, and indeed also of GÖDEL [VI.92], whose 1931 INCOMPLETENESS THEOREM [V.15] underlay Turing's investigation. Using his definition, Turing showed that first-order logic was undecidable, and thus dealt the final death blow to HILBERT's [VI.63] formalist program. (See LOGIC AND MODEL THEORY [IV.23 §2] for more details.)

Computability is now fundamental in mathematics, in that it gives an exact meaning to the question of whether a method exists to solve a problem. As an illustration, HILBERT'S TENTH PROBLEM [V.20], on the general solubility of Diophantine equations, was completely resolved in 1970 by methods connected with Turing's ideas. Turing himself pioneered extensions of his definition in mathematical logic, and applications of it in algebra. However, he was unusual as a mathematician in that he explored not only the mathematical uses of his ideas (in questions of decidability in algebra) but also the wider implications for philosophy, science, and engineering.

One factor in Turing's breakthrough was his fascination with the problem of mind and matter. Turing's analysis of mental states and operations has since become a point of departure for the cognitive sciences. Turing himself blazed this trail later by his advocacy of the possibility of artificial intelligence. His famous 1950 "Turing test" was part of an extensive range of research proposals in this field.

A more immediately applicable aspect of his 1936 work lay in his observation that a single "universal" machine could do the work of any Turing machine, by reading the description of that machine as a table of instructions. This is the essential principle of the modern digital computer, whose programs are themselves data structures. In 1945 Turing used this insight to plan a first electronic computer and its programming. He was preempted by VON NEUMANN [VI.91], but it can be argued that von Neumann had used Turing's insight that computing must be primarily an application of logic. Thus, Turing laid the foundations of modern computer science.

Turing was able to bridge theory and practice because between 1938 and 1945 he was the chief scientific figure in British cryptography, with particular responsibility for decrypting German naval signals. His main contributions lay in a brilliant logical solution of the Enigma cipher, and in Bayesian information theory. The advanced electronics employed in British code breaking gave him the experience to become a pioneer of practical computing as well.

Turing had less success in postwar computer engineering, and increasingly withdrew from attempts to influence the course of computer development. Instead, at Manchester University after 1949 he concentrated on a theory of nonlinear partial differential equations applied to biological development. Like his 1936 work, this opened an entirely new field. It also illustrated his broad mathematical scope, which

included important work on the RIEMANN ZETA FUNC-
TION [IV.2 §3]. He was busy working on biological
theory and new ideas in physics at the time of his
sudden death.

Turing's short life combined the purest mathematics
and the most practical applications. It was also marked
by other contrasts. Although he promoted the theme of
computer-based artificial intelligence, there was nothing mechanical about his thought or life. The wit and
drama of the "Turing test" have made him a lasting figure in the popularization of mathematical ideas. The
dramatization of his life, drawing on the extraordinary
secrecy of his war work, and his subsequent persecution as a homosexual, have also attracted great public
interest.

Further Reading

Hodges, A. 1983. *Alan Turing: The Enigma.* New York:
 Simon & Schuster.
Turing, A. M. 1992–2001. *The Collected Works of A. M.
 Turing.* Amsterdam: Elsevier.

Andrew Hodges

VI.95   Abraham Robinson

b. Waldenburg (Lower Silesia; now Walbrzych, Poland), 1918;
d. New Haven, Connecticut, 1974
Applied mathematics; logic; model theory; nonstandard analysis

Robinson was educated at a private Rabbinical school
and then at the Jewish High School in Breslau until
1933, when he emigrated with his family to Palestine.
There Robinson finished high school, going on to study
mathematics at the Hebrew University under Abraham
Fraenkel. He spent the spring of 1940 at the Sorbonne,
but when the Germans invaded France Robinson made
his way to England. There he spent the war as a refugee,
in the service of the Free French Forces. Robinson's
mathematical talents were soon recognized, and he
was assigned to the Royal Aircraft Establishment in
Farnborough, where he was part of a team designing
supersonic delta wings and reconstructing German V-
2 rockets to determine how they worked. After the
war, Robinson received his M.Sc. degree in mathematics
from the Hebrew University, with minors in physics and
philosophy. Several years later, he completed his Ph.D.
in mathematics at Birkbeck College, London. His thesis,
"On the metamathematics of algebra," was published in
1951.

Meanwhile, Robinson had been teaching at the Royal
College of Aeronautics since its founding in Cranfield in

October of 1946. Although promoted to Deputy Head of
the Department of Aeronautics in 1950, in the following year Robinson accepted a position, at the rank of
associate professor, at the University of Toronto in the
Department of Applied Mathematics. While at Toronto,
most of his publications were devoted to applied mathematics, including papers on supersonic airfoil design
and a book he coauthored with his former student from
Cranfield, J. A. Laurmann, on *Wing Theory.*

His years at Toronto (1951–57) proved to be a transitional period in Robinson's career, as his interests
turned increasingly toward mathematical logic, beginning with studies of algebraically closed fields of characteristic zero. In 1955 he published a book in French
summarizing much of his early work in mathematical
logic and MODEL THEORY [IV.23], *Théorie Métamathé-
matique des Ideaux.* Robinson was a pioneering contributor to model theory, which at its simplest uses
mathematical logic to analyze mathematical structures
(like groups, fields, or even set theory itself). Given
an axiomatic system, a *model* is a structure that satisfies the axioms. One of his early impressive results
was a model-theoretic proof, which he published in
1955 in *Mathematische Annalen,* of Hilbert's seventeenth problem, namely that a positive-definite rational
function over the reals can be expressed as a sum of
squares of rational functions. This was soon followed
by another book, *Complete Theories* (1956), which further extended ideas he had explored earlier in his thesis
on model-theoretic algebra. Here Robinson introduced
such important concepts as model completeness,
model completion, and the "prime model test," along
with proofs of the completeness of REAL-CLOSED FIELDS
[IV.23 §5] and the uniqueness of the model completion
of a model-complete theory.

In the fall of 1957 Robinson returned to the Hebrew
University, where he assumed the chair formerly held
by his teacher Abraham Fraenkel in the Einstein Institute of Mathematics. While at the Hebrew University,
Robinson worked on aspects of local differential algebra, differentially closed fields, and in logic on SKOLEM's
[VI.81] results dealing with nonstandard models of
arithmetic. These provide models of ordinary PEANO
ARITHMETIC [III.67], the usual arithmetic of the integers $(0, 1, 2, 3, \ldots)$, but ones that include "nonstandard"
elements, "numbers" that extend the scope of the standard model to models that are larger but nevertheless
satisfy the axioms of the standard structure. A nonstandard model of arithmetic may include, for example,
infinite integers. As Haim Gaifman puts it succinctly, "A

nonstandard model is one that constitutes an interpretation of a formal system that is admittedly different from the intended one."

Robinson spent the year 1960-61 in the United States, at Princeton, replacing CHURCH [VI.89], who was on sabbatical leave. It was there that Robinson was inspired to make his most revolutionary contribution to mathematics, nonstandard analysis, using model theory to allow the rigorous introduction of infinitesimals. In fact, this extended the usual, standard model of the real numbers to a nonstandard model that included both infinite and infinitesimal elements. He first published on this topic in 1961 in the *Proceedings of the Netherlands Royal Academy of Sciences*. This paper was soon followed by a book, *Introduction to Model Theory and to the Metamathematics of Algebra* (1963), a thorough revision of his earlier book of 1951, including a new section on nonstandard analysis.

Meanwhile, Robinson had left Jerusalem for Los Angeles, where he was appointed as Carnap's chair at UCLA in mathematics and philosophy. In addition to writing an introductory text, *Numbers and Ideals: An Introduction to Some Basic Concepts of Algebra and Number Theory* (1965), he also published his definitive introduction to *Nonstandard Analysis* (1966). Among the important results he obtained while at UCLA (1962–67) was his proof of the invariant subspace theorem in Hilbert space for the case of polynomially compact operators, published with his graduate student Allen Bernstein. (The case for compact operators had been established by Aronszajn and Smith in 1954; what Bernstein and Robinson did was extend this to the case of an operator T such that some nonzero polynomial of T is compact.)

In 1967 Robinson moved to Yale University (1967–74), where he was eventually given a Sterling Professorship in 1971. Among Robinson's most important mathematical achievements during this period were his extension of Paul Cohen's method of FORCING [IV.22 §5.2] in set theory to model theory, and applications of nonstandard analysis in economics and quantum physics. He also applied nonstandard analysis to achieve an outstanding result in number theory, namely a simplification of Carl Ludwig Siegel's theorem regarding integer points on curves (1929), as generalized by Kurt Mahler for rational as well as integer solutions (1934). This was work that Robinson did jointly with Peter Roquette; together they extended the Siegel-Mahler theorem by considering nonstandard integer

points and nonstandard prime divisors. After Robinson's death from pancreatic cancer in 1974, Roquette published this work in the *Journal of Number Theory* in 1975.

Further Reading

Dauben, J. W. 1995. *Abraham Robinson. The Creation of Nonstandard Analysis. A Personal and Mathematical Odyssey.* Princeton, NJ: Princeton University Press.

———. 2002. Abraham Robinson. 1918-1974. *Biographical Memoirs of the National Academy of Sciences* 82:1-44.

Davis, M., and R. Hersh. 1972. Nonstandard analysis. *Scientific American* 226:78-86.

Gaifman, H. 2003. Non-standard models in a broader perspective. In *Nonstandard Models of Arithmetic and Set Theory*, edited by A. Enayat and R. Kossak, pp. 1-22. Providence, RI: American Mathematical Society.

Joseph W. Dauben

VI.96 Nicolas Bourbaki

b. Paris, 1935; d. —
Set theory; algebra; topology; foundations of mathematics; analysis; differential and algebraic geometry; integration theory; spectral theory; Lie algebras; commutative algebras; history of mathematics

---

Bourbaki is a pseudonym chosen in 1935 by a group of French mathematicians, including Henri Cartan, Jean Dieudonné, and ANDRÉ WEIL [VI.93]. Under this nom de plume, several generations of mostly French mathematicians conceived, wrote, and published a series of treatises under the general title *Éléments de Mathématique*. The uncommon use of the singular "mathématique" underscored a strong commitment to the unity of mathematics that is one of the chief characteristics of the group. Together with the "Bourbaki Seminar," this monumental work promoted a unified, axiomatic, and structural view of pure mathematics that has exerted a strong influence on teaching and research since World War II, especially in France.

Charles Denis Sauter Bourbaki was a French general who fought in the Franco-Prussian war in 1870-71. A hoax lecture given by students at the École Normale Supérieure to the entering class in 1923 culminated with a "Bourbaki theorem." In 1935, a group of mathematicians, many of whom had taken part in that lecture, as either audience or pranksters, decided to adopt that name for the fictive author of the modern treatise of analysis they were planning to write.

Their first meeting had taken place in Paris on December 10, 1934. In addition to Cartan, Dieudonné, and Weil, other young university professors of mathematics were present: Claude Chevalley, Jean Delsarte, and René de Possel. Agreeing that analysis textbooks available in French (such as Édouard Goursat's *Cours d'Analyse*) were outdated, they decided to write a book, collectively, to replace them. Having been in touch with modern German mathematics, especially at HILBERT's [VI.63] Göttingen, and influenced in particular by Barteel van der Waerden's *Moderne Algebra*, they thought that their large treatise should begin with an "abstract packet" summarizing in axiomatic form basic general notions such as sets, groups, and fields. Soon after this, Szolem Mandelbrojt joined the group. Paul Dubreil and Jean Leray took part in just a few of the original meetings, and were replaced by Charles Ehresmann and the physicist Jean Coulomb.

In July 1935, the group had its first "congress" (as its annual summer meetings would later be called) in Besse-en-Chandesse, Auvergne, where the pen name "N. Bourbaki" was definitively adopted (the first name, Nicolas, was chosen later). Settling on working procedures, they drew up the general outline of the planned treatise. The members of the group worked collectively following certain ritual rules. They co-opted new collaborators, kept membership secret, and refused to acknowledge individual contributions. During the three or four working sessions they held every year, each contribution prepared in advance by one of them was read line by line, discussed, and severely criticized by the others. Up to ten successive drafts and several years of work by various authors were often needed before a final version was unanimously adopted.

The first booklet—a digest of results in set theory—was dated 1939 but issued in 1940. Despite the difficult working conditions during World War II, this was soon followed in the 1940s by several booklets dealing mostly with general topology and algebra. Today, the *Elements of Mathematics* consists of several books: *Theory of Sets, Algebra, General Topology, Real-Variable Functions, Topological Vector Spaces, Integration, Commutative Algebra, Differential and Analytic Manifolds, Lie Groups and Lie Algebra, Spectral Theories*, and *Elements of the History of Mathematics*. Many of them have been extensively revised over the years and translated into several languages, including English and Russian.

The first six books formed a tight linear exposition entitled "The fundamental structures of analysis." When they first appeared, they were striking for the logical organization of the topics covered. The axiomatic method was used systematically, and great effort was made to ensure a global unity of style, notation, and terminology. The avowed ambition was to take mathematics from its very start and, proceeding from the general toward the particular, write a unified survey of most of modern mathematics.

Several generations of mathematicians were co-opted into the "Association of Bourbaki's Collaborators," as the group is now officially known. After World War II, Samuel Eilenberg, Laurent Schwartz, Roger Godement, Jean-Louis Koszul, and Jean-Pierre Serre, among others, took part in the writing of the treatise. Later, Armand Borel, John Tate, François Bruhat, Serge Lang, and Alexander Grothendieck also joined. Although its frequency of publication has now slowed to a trickle, the group is still functioning in the first decade of the twenty-first century.

Notwithstanding the number of collaborators involved and the extensiveness of the work they published, Bourbaki's vision of mathematics was, and has remained, surprisingly coherent. Most of the crucial mathematical choices, which would come to have a huge impact on the structural image of mathematics that the group would later vigorously promote, were made in the late 1930s. In the following decades, many mathematicians shared a conviction that a tight axiomatic refoundation of their research domains would help overcome current blockages. This was felt, for example, in probability theory, model theory, algebraic geometry and topology, commutative algebra, Lie groups, and Lie algebras.

After World War II, as the notoriety of both the group and its individual members steadily grew, Bourbaki's public image soon encompassed more than just the treatise. At the level of mathematical research, the Bourbaki Seminar was a prestigious outlet established in Paris in 1948, and it has met three times a year ever since. Members of Bourbaki selected speakers who usually summarized someone else's work, and supervised the publication of their talks. The topics selected emphasized specific domains of mathematics, such as algebraic and differential geometry, at the expense of others, such as probability theory or applied mathematics.

Bourbaki's views on the philosophy of mathematics were always clear, especially after two articles published in the late 1940s under that name argued for a complete reorganization of mathematics, eschewing older classification schemes in favor of fundamental structures (sometimes called "mother-structures" and supposedly closer to the deep mental structures of humans) meant to underscore the organic unity of mathematics. Bourbaki's public image was echoed by structuralists in the human sciences as well as artists and philosophers, and it was invoked by radical reformers of mathematical education from kindergarten to university—although actual members of Bourbaki were rarely involved directly.

From the late 1960s, Bourbaki's critics became louder on two counts: they took issue with the Bourbaki approach to the logical foundations of mathematics and they found gaps in the group's encyclopedic objectives. CATEGORY THEORY [III.8] developed by Saunders Mac Lane and Samuel Eilenberg was found to offer a more fruitful foundational framework than Bourbaki's structures. It also became clear that whole branches of mathematics—probability theory, geometry, and, to a lesser extent, analysis and logic—were to remain absent from the treatise, their very place in the grand architecture of Bourbakist mathematics left unclear. For a new

generation of mathematicians, it was Bourbaki's elitist contempt for applications that was especially damaging.

Bourbaki's impact on mathematics was profound: despite its excesses, Bourbaki's unified, structural, rigorous image of mathematics is still with us. But it was those very characteristics that led to a feeling that Bourbaki was corseting mathematical research. The backlash seems to be abating somewhat nowadays, but no new Bourbaki is in view.

Further Reading

Beaulieu, L. 1994. Questions and answers about Bourbaki's early work (1934–1944). In *The Intersection of History and Mathematics*, edited by S. Chikara et al., pp. 241–52. Basel: Birkhäuser.

Corry, L. 1996. *Modern Algebra and the Rise of Mathematical Structures*. Basel: Birkhäuser.

Mac Lane, S. 1996. Structures in mathematics. *Philosophia Mathematica* 4:174–86.

David Aubin

Part VII
The Influence of Mathematics

VII.1 Mathematics and Chemistry
Jacek Klinowski and Alan L. Mackay

1 Introduction

Since ARCHIMEDES [VI.3], and his experimental investigation (described by Vitruvius) of the proportions of gold and silver in an alloy, the solution of chemical problems has employed mathematics. Carl Schorlemmer studied the paraffinic series of hydrocarbons (then important because of the discovery of oil in Pennsylvania) and showed how their properties changed with the addition of successive carbon atoms. His close friend in Manchester, Friedrich Engels, was inspired by this to introduce the transformation of "quantity into quality" into his philosophical outlook, which then became a mantra of dialectical materialism. From a similar chemical observation, CAYLEY [VI.46] in 1857 developed "rooted trees" and the mathematics of the enumeration of branched molecules, the first articulation of GRAPH THEORY [III.34]. Later, George Pólya developed his fundamental enumeration theorem, facilitating further advances in the counting of these molecules. Still more recently, chemical problems such as the mechanics and kinematics of DNA have had a significant influence on KNOT THEORY [III.44].

However, chemistry has been a quantitative modern science for no more than 150 years. Before this, it was a distant dream: when NEWTON [VI.14] was developing the calculus in around 1700, much of his time was spent working on alchemy. He explained why, having established "the motions of the planets, the comets, the Moon and the sea," he was unable to determine the remaining structure of the world from the same propositions:

> I suspect that they may all depend upon certain forces by which the particles of the bodies, by some causes hitherto unknown, are either mutually impelled toward one another, and cohere in regular figures, or are repelled and recede from one another. These forces being unknown, philosophers have hitherto attempted the search of Nature in vain; but I hope the principles laid down will afford some light either to this or some truer method of philosophy.

The nature of such forces came to be understood only two hundred years later, and indeed the electron, the particle responsible for chemical bonding, was not discovered until 1897. This is why the main flow of ideas has been from mathematical theory to applications in chemistry.

Some of the fundamental equations of chemistry, though based on experiment rather than strict mathematical reasoning, convey a wealth of information with great simplicity and elegance (Thomas 2003). For example, consider Boltzmann's fundamental equation of statistical thermodynamics, which links entropy, S, to Ω, the number of possible ways of arranging the particles: $S = k \log \Omega$, where k is known as the Boltzmann constant. There is also the expression derived by Balmer for the wavelength, λ, of spectral lines from hydrogen in the visible portion of the spectrum:

$$\frac{1}{\lambda} = R\left(\frac{1}{n_1^2} - \frac{1}{n_2^2}\right),$$

where n_1 and n_2 are integers, $n_1 < n_2$, and R is known as the Rydberg constant. A third example, the Bragg equation, links the wavelength, λ, of monochromatic X-rays, the distance, d, between planes in a crystal lattice, and the angle, θ, between the crystal planes and the direction of the X-rays. It says that $n\lambda = 2d \sin \theta$, where n is a small integer. Finally, there is the "phase rule," $P + F = C + 2$, which links the number of phases, P, the number of degrees of freedom, F, and the number of components, C, in a chemical system. This is the same relationship as that between the number of vertices, faces, and edges in a convex polyhedron, and emerges from the geometrical representation of the system.

In recent years computers have become the dominant tool in theoretical chemistry. Not only can computers solve differential equations numerically, they can often provide exact algebraic expressions, sometimes even ones that are too elaborate to write out. Computing has required the development of algorithms in the fields of *structure*, *process*, *modeling*, and *search*. Mathematics has been revolutionized by the advent of computers: in particular in the facility for dealing with nonlinear problems and for displaying results graphically. This has led to fundamental advances, some of them bearing on chemistry.

In general, mathematical approaches to chemical problems can be divided into discrete and continuous treatments, reflecting on the one hand the fundamental discrete atomic nature of matter and on the other the continuous statistical behavior of large numbers of atoms. For example, enumerating molecules is a discrete problem, while a problem involving global measures such as temperature and other thermodynamic parameters will be continuous. These treatments have required different branches of mathematics, with integers more important for discrete problems and real numbers more important for continuous ones.

We shall now outline some chemical problems to which, in our view, mathematics has made the most significant contributions.

2 Structure

2.1 Description of Crystal Structure

Crystal structure is the study of how atoms arrange themselves to form macroscopic materials. Early ideas in the subject were based purely on the symmetry of crystals and their morphology (that is, the shapes they tended to form), and were developed in the nineteenth century in the absence of definite information about the atomic structure of matter. The 230 *space groups*, which codify different ways of arranging objects periodically in three-dimensional (3D) space, were found independently by Fedorov, Schoenflies, and Barlow between 1885 and 1891. They result from the systematic combination of a certain collection of fourteen lattices, named *Bravais lattices* after their discovery in 1848 by Auguste Bravais, with the thirty-two so-called *crystallographic point groups*, which were developed from morphological considerations.

Since the diffraction of X-rays was demonstrated in 1912 by Max von Laue and practical X-ray analysis was developed by W. H. Bragg and his son W. L. Bragg, the crystal structures of several hundred thousand inorganic and organic substances have been determined.

However, such analysis was for a long time held back by the time required for the calculation of FOURIER TRANSFORMS [III.27]. This difficulty is now a thing of the past, owing to the discovery of THE FAST FOURIER TRANSFORM [III.26] by Cooley and Tukey in 1965—a universally applied algorithm and one of those most often cited in mathematics and computer science.

The fundamental geometry of two-dimensional (2D) and 3D spatial structures led mathematicians to seek analogous problems in N dimensions. Some of this work has found application in the description of *quasicrystals*, which are arrangements of atoms that, like crystals, exhibit a high degree of organization, but which lack the periodic behavior of crystals. (That is, they do not have translational symmetry.) The most notable example is the following, which uses six-dimensional geometry. Take a regular cubic lattice L in six dimensions and let V be a 3D subspace of $\mathbb{R}^6$ that contains no point of L apart from the origin. Now project on to V all points from L that are closer to V than a certain distance d. The result is a 3D structure of points that exhibits a great deal of local regularity but not global regularity. This structure gives a very good model for quasicrystals.

Until recently, crystals in three dimensions had always been thought of as periodic, and therefore capable of showing only twofold, threefold, fourfold, or sixfold axes of symmetry. Fivefold axes were excluded, because a plane cannot be tiled with regular pentagons. However, in 1982, X-ray and electron diffraction demonstrated the presence of fivefold diffraction symmetry in certain rapidly cooled alloys. Careful electron microscopy was necessary to distinguish the observed structures from the twinning (symmetrical intergrowth) of "normal" crystals. This discovery, of a quasicrystalline alloy phase "with long-range orientational order and no translational symmetry," has brought about an ideological shift in crystallography.

The earlier concept of a "quasilattice" appeared to be one possible mathematical formalism for the description of quasicrystals. Quasilattices have two incommensurable periods in the same direction, and the ratio of these periods was given by so-called Pisot and Salem numbers. A *Pisot number* θ is a root of a polynomial with integer coefficients of degree m such that if $\theta_2, \dots, \theta_m$ are the other roots, then $|\theta_i| < 1$, $i = 2, \dots, m$. A real quadratic ALGEBRAIC INTEGER [IV.1 §11] greater than 1 and of degree 2 or 3 is a Pisot number if its norm is equal to ± 1. The golden ratio is an example of a Pisot number since it has degree 2 and norm -1. A *Salem number* is defined in a similar way to a Pisot number, but with the inequalities replaced by equalities.

LIE ALGEBRA [III.48 §2] arguments have also been used to describe quasicrystals. This has stimulated a great deal of theoretical N-dimensional geometry. Before the discovery of quasicrystals, Roger Penrose had shown how to cover a plane nonperiodically using two different types of rhombic tiles, and corresponding rules were developed for 3D space with two kinds of rhombohedral tiles. The Fourier transform of such a 3D structure with atoms placed in the rhombohedral cells explains the observed diffraction patterns of 3D quasicrystals, while Penrose's 2D pattern corresponds to *decagonal quasicrystals*, which consist of stacked layers of the 2D pattern and which have been experimentally observed.

The broadening of classical crystallography to encompass quasicrystals has been given further impetus by recent advances in electron microscopy. It is now possible to observe atomic arrangements directly, including those of the decagonal quasicrystals just mentioned, rather than having to deduce them from diffraction patterns, where the phases of the various diffracted beams are lost in the experimental system and have to be recovered mathematically. The whole field of computational and experimental image processing has become coherent as a result.

Another model describes 2D quasicrystals in terms of a single repeating unit, but the unit is a composite object, a pattern made out of identical decagons. Unlike the unit cells in periodic crystals, these quasi-unit cells are allowed to overlap, but where they do their constituent decagons must match up. This conceptual device is an alternative to the use of two kinds of unit cell. It emphasizes the dominating physical presence of locally ordered atomic clusters, with no long-range order, and it can be extended to three dimensions. The predictions of this model agree with the observed composition of a 2D decagonal quasicrystal, as well as with the results obtained by electron microscopy and X-ray diffraction. Nevertheless, although a huge amount of interesting mathematics has been generated by the discovery of quasicrystals, most of it is not physically relevant: the structures emerge from the competition between local and global ordering forces rather than from the mathematics of the Penrose tiling.

The acceptance of quasicrystals demonstrates the need to accommodate more general concepts of *order* into classical crystallography. It has explicitly introduced concepts of *hierarchy*, by involving not just ordered clusters of atoms but ordered clusters of clusters, where local order has predominated over the regular lattice repetition. Quasicrystals represent the first step from absolute regularity toward more general structures that are intimately bound up with the notion of *information*.

Information can be stored in a device which has two or more clearly identifiable states that are *metastable*. This means that each state is a local equilibrium, and to pass from one to another, one must supply and remove enough energy to take the device over the local energy watershed. A switch, for example, can be on or off; it is stable in either state and to change the state takes a certain amount of energy. To take a more general example, any information, encoded as a sequence of binary digits, can be read in, read out, and stored as a sequence of magnetic domains, where each one is magnetized either north or south.

Perfect crystals have no alternative metastable states, so cannot be used to store information, but a piece of silicon carbide, for example, exists as a sequence of close-packed layers, each of which may be in one or other of two almost equivalent positions. To describe the structure of a piece of silicon carbide therefore demands a knowledge of the sequence of positions in which the layers are stacked. This can be represented by a string of binary digits. Now that it is possible to arrange atoms in a structure almost at will, at least if they are on a surface, the processing of information has become important to chemistry.

In determining the arrangement of atoms in crystals, mathematics has been essential for the solution of the *phase problem*, which had held up progress in structural chemistry and molecular biology for decades. A pattern of diffracted X-rays, recorded as an array of spots on a photographic plate, depends on the arrangement of atoms in the molecule causing the diffraction. The problem is that the diffraction pattern registers only the intensity of the light waves, but to work back to the molecular structure it is necessary to know their phase as well (that is, the positions of the crests and troughs of the waves relative to each other). This results in a classic *inverse problem*, which was solved by Jerome and Isabella Karle and Herbert A. Hauptman.

A *Voronoi diagram* consists of points, representing atom sites, with each point contained in a region (see also MATHEMATICAL BIOLOGY [VII.2 §5]). The region surrounding a given site consists of all points that are closer to that site than to any of the other sites

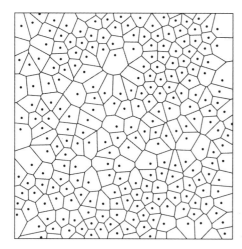

Figure 1 Voronoi dissection of 2D space.

(figure 1). The geometric dual of the Voronoi diagram, a system of triangles with the sites as vertices, is called the *Delaunay triangulation*. (An alternative definition of the Delaunay triangulation is that it is a triangulation of the sites with the additional property that, for each triangle, the circumcircle of that triangle contains no other sites.) These dissections give a well-defined way of representing many N-dimensional chemical structures as arrangements of polytopes. Crystals, which have periodic boundaries, are easier to deal with than extended structures that terminate in a boundary. The Voronoi dissection of crystal structures enables one to describe them as networks. Nevertheless, despite much progress in understanding structure, it is not yet possible to guess a crystal structure in advance just from the composition of elements in its molecules.

2.2 Computational Chemistry

Attempts to solve THE SCHRÖDINGER EQUATION [III.83], which gives the quantum mechanical description of matter, began soon after it was proposed in 1926. For very simple systems, calculations performed on mechanical calculators agreed with the experimental results of spectroscopy. In the 1950s, electronic computers became available for general scientific use, and the new field of *computational chemistry* developed, the aim of which was to obtain quantitative information on atomic positions, bond lengths, electronic configurations of atoms, etc., by means of numerical solutions of the Schrödinger equation. Advances during the 1960s included deriving suitable functions for representing electronic orbitals, obtaining approximate

solutions to the problem of how the motions of different electrons correlate with each other, and providing formulas for the derivative of the energy of a molecule with respect to the positions of the atomic nuclei. Powerful software packages became available in the early 1970s. Much current research is aimed at developing methods that can handle larger and larger molecules.

Density functional theory (DFT) (Parr and Yang 1989) is a major recent field of activity in quantum mechanical computation, and concerns macroscopic features of materials. It has been successful in the description of the properties of metals, semiconductors, and insulators, and even of complex materials such as proteins and carbon nanotubes. Traditional methods in the study of electronic structure—such as one called the *Hartree–Fock theory molecular orbital method*, which assigns the electrons two at a time to a set of molecular orbitals—involve very complicated many-electron wave functions. The main objective of DFT is to replace the many-body electronic wave function, which depends on $3N$ variables, with a different basic quantity, the *electronic density*, which depends on just 3 variables, and therefore greatly speeds up calculations.

The partial differential equations of quantum mechanics, physics, fields, surfaces, potentials, and waves can sometimes be solved analytically, but even if they cannot, they are now almost always soluble by numerical methods. All this relies on the corresponding pure mathematics. (For a discussion of how to solve partial differential equations numerically, see NUMERICAL ANALYSIS [IV.21 §5].)

2.3 Chemical Topology

Isomers are chemical compounds that are made out of the same elements but have different physical and chemical properties. This can happen for various reasons. In *structural isomers*, the atoms and functional groups are linked together in different ways. This class includes *chain isomers*, where hydrocarbon chains have variable amounts of branching, and *position isomers*, where the position of a functional group in a chain is different (figure 2(a)). In *stereoisomers* the bond structure is the same, but the geometrical positioning of atoms and functional groups in space differs (figure 2(b)). This class includes *optical isomers*, where different isomers are mirror images of each other (figure 2(c)). While structural isomers have different chemical properties, stereoisomers behave identically in most chemical reactions. There are also *topological isomers* such as catenanes and DNA.

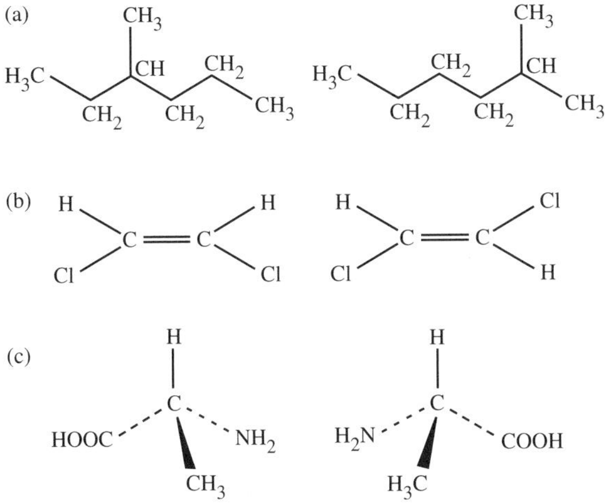

(a)

(b)

(c)

Figure 2 (a) Position isomerism. (b) Stereoisomerism. (c) Optical isomerism.

An important theme in chemical topology is determining, for any given molecule, how many isomers it has. To do this, one first associates with any molecule a *molecular graph*, the vertices representing atoms and the edges representing chemical bonds. To enumerate *stereoisomers*, one counts the symmetries of this graph, but first one must consider symmetries of the molecule (Cotton 1990) in order to decide which symmetries of the graph correspond to spatial transformations that make chemical sense. Cayley addressed the problem of enumerating *structural* isomers, that is, combinatorially possible branched molecules. To do this, one must count how many different molecular graphs there are with a given set of elements, where two graphs are regarded as the same if they are isomorphic. The enumeration of isomorphism types uses group theory to count the intrinsic graph symmetries. After Pólya published his remarkable ENUMERATION THEOREM [IV.18 §6] in 1937, his work using GENERATING FUNCTIONS [IV.18 §§2.4, 3] and PERMUTATION GROUPS [III.68] became central to the enumeration of isomers in organic chemistry. The theorem solves the general problem of how many configurations there are with certain properties. It has applications such as the enumeration of chemical compounds and the enumeration of rooted trees in graph theory. A new branch of graph theory, called enumerative graph theory, is based on Pólya's ideas (see ALGEBRAIC AND ENUMERATIVE COMBINATORICS [IV.18]).

Although not all the possible isomers occur in nature, molecules with remarkable topologies have been synthesized artificially. Among them are *cubane*, C_8H_8, which contains eight carbon atoms arranged at the corners of a cube, each linked to a single hydrogen atom; *dodecahedrane*, $C_{20}H_{20}$, which, as its name suggests, has a dodecahedral shape; the *molecular trefoil knot*; and the self-assembling compound *olympiadane* composed of five interlocked rings. *Catenanes* (from Latin *catena*, chain) are molecules containing two or more interlocked rings that are inseparable without breaking a covalent bond. *Rotaxanes* (from Latin *rota*, wheel, and *axis*, axle) are dumbbell shaped, having a rod and two bulky stopper groups, around which there are encircling macrocyclic components. The stoppers of the dumbbell prevent the macrocycles from slipping off the rod. Even a molecular MÖBIUS STRIP [IV.7 §2.3] has recently been synthesized.

Macromolecules, such as synthetic polymers and biopolymers (e.g., DNA and proteins), are very large and highly flexible. The degree to which a polymer molecule coils and knots and links with other molecules is crucial to its physical and chemical properties, such as reactivity, viscosity, and crystallization behavior. The topological entanglement of short chains can be modeled using Monte Carlo simulation, and the results can now be experimentally verified with fluorescence microscopy.

DNA, the central substance of life, has a complex and fascinating topology, which is closely related to its biological function. The major geometric descriptions of supercoiled DNA (that is, DNA wrapped around a series of proteins) involve the concepts of linking, twisting, and writhing numbers that come from knot theory. DNA knots, which are created spontaneously within cells, interfere with replication, reduce transcription, and may decrease the stability of the DNA. "Resolvase enzymes" detect and remove these knots, but the mechanism of this process is not understood. However, using topological concepts of knots and tangles, one can gain insight into the reaction site and thereby try to infer the mechanism. (See also MATHEMATICAL BIOLOGY [VII.2 §5].)

2.4 Fullerenes

Graphite and diamond, the two crystalline forms of the element carbon, have been known since time immemorial, but *fullerenes*, which were subsequently found to exist naturally in soot and geological deposits, were discovered only in the mid 1980s. The most common is the almost-spherical carbon cage C_{60} molecule (figure 3), also known as "buckminsterfullerene" after the architect who designed enormous domes, but fullerenes C_{24},

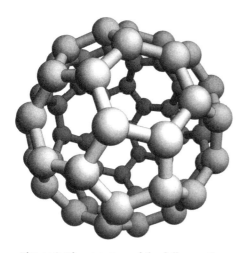

Figure 3 The structure of the fullerene C_{60}.

$C_{28}, C_{32}, C_{36}, C_{50}, C_{70}, C_{76}, C_{84}$, etc., also exist. Topology provides insights into the possible types of such structures, while group theory and graph theory describe the symmetry of the molecules, allowing one to interpret their vibrational modes.

In all fullerenes, each carbon atom is connected to exactly three neighboring ones, and the resulting molecule is a "cage" made of rings of either five or six carbon atoms. From EULER's [VI.19] topological relationship $\sum_n (6-n) f_n = 12$, where f_n is the number of n-hedral faces and the summation is over all faces of the polyhedron, we conclude first that $f_5 = 12$, since n is found to take only the values 5 or 6, and second that f_6 can take any value greater than 1.

In 1994, Terrones and Mackay predicted the existence of ordered structures of a new kind, derived from graphite and related to fullerenes, with topologies of triply periodic MINIMAL SURFACES [III.94 §3.1]. These new structures, which are of great practical interest, are produced by introducing eight-membered rings of carbon atoms into a sheet of six-membered rings. This gives rise to saddle-shaped surfaces of negative GAUSSIAN CURVATURE [III.78], unlike the fullerenes, which have positive curvature. Thus, to model them mathematically one must consider embeddings of non-Euclidean 2D spaces into $\mathbb{R}^3$. This has contributed to a renewed interest in certain aspects of non-Euclidean geometry.

2.5 Spectroscopy

Spectroscopy is the study of the interaction of electromagnetic radiation (light, radio waves, X-rays, etc.) with matter. The central portion of the electromagnetic spectrum—spanning the infrared, visible, and ultraviolet wavelengths and the radio frequency region—is of particular interest to chemistry. A molecule, which consists of electrically charged nuclei and electrons, may interact with the oscillating electric and magnetic fields of light and absorb enough energy to be promoted from one discrete vibrational energy level to another. Such a transition is registered in the infrared spectrum of the molecule. The *Raman spectrum* monitors inelastic scattering of light by molecules (that is, when some of the light is scattered at a different frequency from the frequency of the incoming photons). Visible and ultraviolet light can redistribute the electrons in the molecule: this is *electronic spectroscopy.*

Group theory is essential in the interpretation of the spectra of chemical compounds (Cotton 1990; Hollas 2003). For any given molecule, the symmetry operations that can be applied to it form a GROUP [I.3 §2.1], and can be represented by matrices. This allows one to identify "spectroscopically active" events in a molecule. For example, just three bands are observed in the infrared spectrum and eight bands in the Raman spectrum of dodecahedrane. This is a consequence of the icosahedral symmetry of the molecule and is what one expects from group-theoretic considerations. Also, there are no coincidences between the infrared- and Raman-active modes. Similarly, group theory correctly predicts that, because of the high symmetry of a C_{60} molecule, it has only four lines in its infrared spectrum and ten in its Raman spectrum, even though it has 174 vibrational modes.

2.6 Curved Surfaces

Structural chemistry has greatly changed in the last twenty years. First, as we have seen, the rigid concept of a "perfect crystal" has been relaxed to embrace structures such as quasicrystals and textures. Second, an advance has been made from classical geometry to 3D differential geometry. The main reason for this has been the use of curved surfaces for describing a great variety of structures (Hyde et al. 1997).

When a wire frame is dipped into soapy water, a thin film is formed. Surface tension minimizes the energy of the film, which is proportional to its surface area. As a result, the film has the smallest area consistent with the shape of the frame and with the requirement that the *mean curvature* of the film be zero at every point. If the symmetries of a minimal surface are given

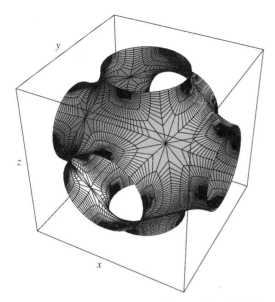

Figure 4 One unit cell of the P triply periodic minimal surface. The surface divides space into two interpenetrating labyrinths.

by one of the 230 space groups mentioned earlier, then the surface is periodic in three independent directions. Such triply periodic minimal surfaces (TPMSs) are of special interest because they appear in a variety of real structures such as silicates, bicontinuous mixtures, lyotropic colloids, detergent films, lipid bilayers, polymer interfaces, and biological formations (an example of a TPMS is illustrated in figure 4). Thus, TPMSs provide a concise description of many seemingly unrelated structures. Extensions of TPMSs may even have applications in cosmology as "branes."

In 1866 WEIERSTRASS [VI.44] discovered a method of complex analysis suitable for general investigation of minimal surfaces. Consider a transformation of a minimal surface into the complex plane by combination of two simple maps. The first is the *Gauss map* ν, under which the image of a point P of the surface is the point P′ of the intersection of the surface normal vector at P with the unit sphere centered at P. The second map is a stereographic projection σ of the point P′ on the sphere into the complex plane $\mathbb{C}$, resulting in the point P″. The composite map, $\sigma\nu$, conformally maps the neighborhood of any nonumbilic point on the surface to a simply connected region of $\mathbb{C}$. (An umbilic point is one where the two principal curvatures are the same.) The inverse of this composite map is called the *Enneper–Weierstrass representation*.

In a system with the origin at (x_0, y_0, z_0), the Cartesian coordinates (x, y, z) of *any* nontrivial minimal surface are determined by a set of three integrals:

$$x = x_0 + \operatorname{Re} \int_{\omega_0}^{\omega} (1 - \tau^2) R(\tau)\, d\tau,$$

$$y = y_0 + \operatorname{Re} \int_{\omega_0}^{\omega} i(1 + \tau^2) R(\tau)\, d\tau,$$

$$z = z_0 + \operatorname{Re} \int_{\omega_0}^{\omega} 2\tau R(\tau)\, d\tau.$$

Here $R(\tau)$ is the *Weierstrass function*. It is a function of a complex variable τ, and it is HOLOMORPHIC [I.3 §5.6] in a simply connected region of $\mathbb{C}$, except at isolated points.

The Cartesian coordinates of any (nonumbilic) point on a minimal surface are thus expressed as the real parts of certain contour integrals, evaluated in the complex plane from some fixed point ω_0 to a variable point ω. Integration is carried out within the domain where the integrands are HOLOMORPHIC [I.3 §5.6], and thus by Cauchy's theorem the values of the integrals are independent of the path of integration from ω_0 to ω. In this way, a specific minimal surface is completely defined by its Weierstrass function.

While the Weierstrass functions for many TPMSs are unknown, the coordinates of points lying on *some* minimal surfaces involve functions of the form

$$R(\tau) = \frac{1}{\sqrt{\tau^8 + 2\mu\tau^6 + \lambda\tau^4 + 2\mu\tau^2 + 1}},$$

where μ and λ are sufficient to parametrize the surface. A method has been developed for deriving this function for a given type of surface, and it generates different families of minimal surfaces from the above equation. For example, taking $\mu = 0$ and $\lambda = -14$ gives a surface known as the *D surface* (for "diamond").

The application of minimal surfaces to the physical world has so far been descriptive, rather than quantitative. Although explicit analytical equations for the parameters of some TPMSs have recently been derived, problems such as stability and mechanical strength are unresolved. While describing structure using the concept of curvature is mathematically attractive, it has yet to make its full impact on chemistry.

2.7 Enumeration of Crystalline Structures

It is a matter of considerable scientific and practical importance to enumerate all possible networks of atoms in a systematic way. For example, 4-connected networks (that is, networks in which each atom is connected to exactly four neighbors) occur in crystalline elements, hydrates, covalently bonded crystals, silicates, and many synthetic compounds. Of particular interest is the possibility of using systematic enumeration to discover and generate new *nanoporous architectures.*

Nanoporous materials are materials with tiny holes in them that allow some substances to pass through and not others. Many are naturally occurring, such as cell membranes and "molecular sieves" called *zeolites*, but many others have been synthesized. There are now 152 recognized structure types of zeolites, with several new types being added to the list every year. Zeolites find many important applications in science and technology, in areas as diverse as catalysis, chemical separation, water softening, agriculture, refrigeration, and optoelectronics. Unfortunately, the problem of enumeration is fraught with difficulties, and since the number of 4-connected 3D networks is infinite and there is no systematic procedure for their derivation, the results reported so far have been obtained by empirical methods.

Enumeration originated with the work of Wells (1984) on 3D nets and polyhedra. Many possible new structures were found by model building or computer search algorithms. New research in this field is based on recent advances in combinatorial tiling theory, developed by the first generation of pure mathematicians familiar with computing. The tiling approach identified over nine hundred networks with one, two, and three kinds of inequivalent vertices, which we call uninodal, binodal, and trinodal.

However, only a fraction of the mathematically generated networks are chemically feasible (many would be "strained" frameworks requiring unrealistic bond lengths and bond angles), so for the mathematics to be useful an effective filtering process is needed to identify the most plausible frameworks. Methods of computational chemistry were therefore used to minimize the framework energy of the various hypothetical structures, which were treated as though they were made from silicon dioxide. The unit cell parameters, framework energies and densities, volumes available to adsorption, and X-ray diffraction patterns were all calculated. A total of 887 structures were successfully optimized and ranked according to their framework energies and available volumes to give a subset of chemically feasible hypothetical structures. A number of them have since been synthesized.

The results of these calculations are relevant to the structures of zeolites and other silicates, aluminophos-

phates (AlPOs), oxides, nitrides, chalcogenides, halides, carbon networks, and even to polyhedral bubbles in foams.

2.8 Global Optimization Algorithms

A wide variety of problems in practically all fields of physical science involve *global optimization*, that is, determining the global minimum (or maximum) of a function of an arbitrary number of independent variables (Wales 2004). These problems also appear in technology, design, economics, telecommunications, logistics, financial planning, travel scheduling, and the design of microprocessor circuitry. In chemistry and biology, global optimization arises in connection with the structure of clusters of atoms, protein conformation, and molecular docking (the fitting and binding of small molecules at the active sites of biomacromolecules such as enzymes and DNA). The quantity to be minimized is nearly always the energy of the system (see below).

Global optimization is like trying to find the deepest point in a very rugged landscape. In most cases of practical interest it is very difficult because of the ubiquity of *local minima*, or holes in the landscape, the number of which tend to increase exponentially with the size of the problem. Conventional minimization techniques are time-consuming and have a tendency to find a nearby hole and stay there: that is, they converge to whichever local minimum they first encounter. The *genetic algorithm* (GA), an approach inspired by Darwin's theory of evolution, was introduced in the 1960s. This algorithm starts with a set of solutions (represented by "chromosomes") called a *population*. Solutions from one population are taken and used to form a new population. This is done in such a way that one expects the new population to be better than the old one. Solutions that are chosen for forming new solutions ("offspring") are selected according to their "fitness": the more suitable they are the more chances they have to reproduce. This is repeated until some condition is satisfied. (For example, one might stop after a certain number of generations or after a certain improvement of the solution has been achieved.)

Simulated annealing (SA), introduced in 1983, uses an analogy between the annealing process, in which a molten metal cools and freezes into a minimum-energy structure, and the search for a minimum in a more general system. The process can be thought of as an adiabatic approach to the lowest-energy state.

The algorithm employs a random search which accepts not only changes that decrease the energy, but also some changes that increase it. The energy is represented by an *objective function* f, and the energy-increasing changes are accepted with a probability $p = \exp(-\delta f/T)$, where δf is the increase in f and T is the system "temperature," irrespective of the nature of the objective function. SA involves the choice of "annealing schedule," initial temperature, the number of iterations at each temperature, and the temperature decrease at each step as cooling proceeds.

Taboo (or tabu) search is a general-purpose stochastic global-optimization method originally proposed by Glover in 1989. It is used for very large combinatorial optimization tasks and has been extended to continuous-valued functions of many variables with many local minima. Taboo search uses a modification of "local search," which starts from some initial solution and attempts to find a better solution. This becomes the new solution and the process restarts from it. The procedure continues step by step until no improvement is found to the current solution. The algorithm avoids entrapment in local minima and gives the optimal final solution. A recent method of global optimization, known as "basin hopping," has been successfully applied to a variety of atomic and molecular clusters, peptides, polymers, and glass-forming solids. The algorithm is based upon a transformation of the energy landscape that does not affect the relative energies of local minima. Combined with taboo search, basin hopping shows a significant improvement in efficiency over the best published results for atomic clusters.

2.9 Protein Structure

Proteins are linear sequences of amino acids, which are molecules that contain both the amide ($-NH_2$) and carboxylic ($-COOH$) functional groups. Understanding the means by which a protein adopts its 3D structure is a key scientific challenge (Wales 2004). This problem is also critical to developing strategies, at the molecular level, to counter "protein folding diseases" such as Alzheimer's disease and "mad cow" disease. The strategy in tackling protein folding relies upon the fact, observed by Anfinsen, Haber, Sela, and White in 1961, that the structure of a folded protein corresponds to the conformation which minimizes the free energy of the system. The free energy of a protein depends on the various interactions within the system, and each can be modeled mathematically using the principles of

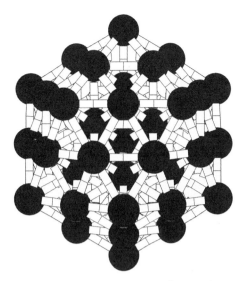

Figure 5 A fifty-five-atom Lennard-Jones cluster. (Courtesy of Dr. D. J. Wales, Cambridge University.)

electrostatics and physical chemistry. As a result, the free energy of a protein can be expressed as a function of the positions of the constituent atoms. The 3D arrangement of the protein then corresponds to the set of atomic locations providing the minimum possible value of the free energy, and the problem is reduced to finding the global minimum of the potential-energy surface of the protein. The problem is further complicated because some proteins require other molecules, "chaperones," to enable them to reach a particular configuration.

2.10 Lennard-Jones Clusters

A *Lennard-Jones cluster* is a closely packed arrangement of atoms in which every pair of atoms has an associated potential energy, given by the classical *Lennard-Jones potential-energy function*. The *Lennard-Jones cluster problem* is to determine the atomic cluster configurations with minimum potential energy (figure 5). If n is the number of atoms in the cluster, then one wishes to find points $p_1, p_2, \ldots, p_n$ so as to minimize the sum

$$\sum_{i=1}^{n-1} \sum_{j=i+1}^{n} (r_{ij}^{-12} - 2r_{ij}^{-6}),$$

where r_{ij} stands for the Euclidean distance between p_i and p_j, and the atoms of the cluster are positioned at $p_1, p_2, \ldots, p_n$. The problem is still a challenge, both to optimization methods and to computer technology. A systematic survey by Northby in 1987, which yielded most of the lowest Lennard-Jones potential values in the range $13 \leqslant n \leqslant 147$, was a significant landmark, and these results have since been improved by about 10%. The results for $n = 148, 149, 150, 192, 200, 201, 300,$ and 309 have now been reported using stochastic global-optimization algorithms.

2.11 Random Structures

Stereology, originally the deduction of 3D structure from microscope examination of sections, has required the development of a substantial branch of statistical mathematics, in which R. E. Miles and R. Coleman have played leading roles. Stereology concerns the estimation of geometrical quantities. Geometrical shapes are used to probe objects to learn about their quantities, such as volume or length. Random sampling is a basic step in all stereological estimation. The degree of randomness required for any estimate varies.

Even apparently simple questions involving randomness with spatial constraints may prove difficult. For example, Gotoh and Finney gave an estimate of 0.6357 as the density expected for a dense random packing of hard spheres of equal size, and their answer to this apparently simple question has not since been improved upon, as far as we know. The problem needs to be defined very carefully, since it is far from obvious what one means by a "random packing" of spheres. This is even more true when one investigates other, related problems concerning the interaction of molecules using computer simulation. This area, called *molecular dynamics*, was begun by A. Rahman, and it developed steadily from the 1960s as computers themselves developed. An example of a problem in molecular dynamics is the modeling of liquid water. This is still difficult, but the immense computing power that is now available has enabled enormous progress to be made.

3   Process

In 1951 Belousov discovered the *Belousov–Zhabotinski reaction*, in which time-dependent spatial patterns appear in an apparently isotropic medium. The mechanism of this reaction was elucidated in 1972, and this opened up an entire new research area: *nonlinear chemical dynamics*. Oscillatory phenomena have also been observed in *membrane transport*. Winfree and Prigogine have shown how patterns in space and time can appear, and some of these patterns have been fitted to practical examples.

The development of *cellular automata* began with Stanisław Ulam, Lindenmeyer systems, and Conway's "game of life" and continues to this day. With his huge book, Wolfram (2002) has demonstrated the complexity that can arise from apparently simple rules, and recently Reiter has used cellular automata to simulate the growth of snowflakes, beginning to answer questions that Kepler posed in 1611. There is a group of mathematicians in Bielefeld, led by Andreas Dress, who deal with structure-forming processes; they have made particular progress in modeling actual chemistry and thus revealing possible mechanisms.

4   Search

4.1   Chemical Informatics

A fundamental development in chemistry has been the application of computing to searching multidimensional databases of chemical compounds and their structures. These databases are now enormous compared with their (already large) predecessors, the classical Gmelin and Beilstein databases. The search process has required fundamental mathematical analyses, as exemplified in the pioneering work of Kennard and Bernal in developing the Cambridge Structural Database (www.ccdc.cam.ac.uk/products/csd/).

What is the best way to encode the structure of a 3D molecule or a crystal arrangement as a linear sequence of symbols? One would like to be able to restore the structure efficiently from its encoding, and also to search efficiently through a big list of encoded structures. The problems that this raises are of long standing, and need insights both from mathematics and chemistry.

4.2   Inverse Problems

Many of the mathematical challenges of chemistry are inverse problems. Often they involve solving a set of linear equations. If there are as many equations as unknowns and the equations are independent, then this can be done by inverting a square matrix. However, if the system is singular or redundant, or if there are fewer equations or more equations than unknowns, then the corresponding matrix is singular or rectangular and there is no ordinary inverse. Nevertheless, it is possible to define a *generalized inverse*, which gives a good model for linear problems. (It is the so-called *Moore–Penrose inverse* or *pseudo-inverse* involved in singular value decomposition.) This always exists and it uses all available information; it is related to the problem of reconstructing a 3D structure from a 2D projection. The operation has been fully described and is now available in Mathematica.

The generalized inverse also enables one to handle redundant axes in quasicrystals, but usually the interesting problems are nonlinear. Other inverse problems include the following.

(i) Finding the arrangement of atoms that gives rise to the observed scattering patterns of X-rays or electrons from a crystal.

(ii) Reconstructing a 3D image from 2D projections in microscopy or X-ray tomography.

(iii) Reconstructing the geometry of a molecule given probable interatomic distances (and perhaps bond angles and torsion angles).

(iv) Finding the way in which a protein molecule folds to give an active site, given the sequence of constituent amino acids.

(v) Finding the pathway to producing a molecule synthetically, given that it occurs in nature.

(vi) Finding the sequence of rules that generate a membrane or a plant or another biological object, given that it takes a certain shape.

Some questions of this type do not have unique answers. For example, the classic question as to whether the shape of a drumhead can be determined from its vibration spectrum (can you hear the shape of a drum?) has been answered in the negative: two vibrating membranes with different shapes may have the same spectrum. It was thought that this ambiguity might also be the case for crystal structures. Linus Pauling suggested that there might be two different crystal structures that were *homometric* (that is, giving the same diffraction pattern), but no definite example has been found.

5   Conclusion

As the examples in this article show, mathematics and chemistry have a symbiotic relationship, with developments in one often stimulating advances in the other. Many interesting problems, including several that we have mentioned here, are still waiting to be solved.

Further Reading

Cotton, F. A. 1990. *Chemical Applications of Group Theory.* New York: Wiley Interscience.

Hollas, J. M. 2003. *Modern Spectroscopy.* New York: John Wiley.

Hyde, S., S. Andersson, K. Larsson, Z. Blum, T. Landh, S. Lidin, and B. W. Ninham. 1997. *The Language of Shape. The Role of Curvature in Condensed Matter: Physics, Chemistry and Biology.* Amsterdam: Elsevier.

Parr, R. G., and W. Yang. 1989. *Density-Functional Theory of Atoms and Molecules.* Oxford: Oxford University Press.

Thomas, J. M. 2003. Poetic suggestion in chemical science. *Nova Acta Leopoldina NF* 88:109–39.

Wales, D. J. 2004. *Energy Landscapes.* Cambridge: Cambridge University Press.

Wells, A. F. 1984. *Structural Inorganic Chemistry.* Oxford: Oxford University Press.

Wolfram, S. 2002. *A New Kind of Science.* Champaign, IL: Wolfram Media.

VII.2   Mathematical Biology
Michael C. Reed

1   Introduction

Mathematical biology is an extremely large and diverse field. It studies objects ranging from molecules to global ecosystems and the mathematical methods come from many of the subdisciplines of the mathematical sciences: ordinary and partial differential equations, probability theory, numerical analysis, control theory, graph theory, combinatorics, geometry, computer science, and statistics. The most that one short article can do is to illustrate by selected examples this diversity and the range of new mathematical questions that arise naturally in the biological sciences.

2   How Do Cells Work?

From the simplest point of view, cells are large biochemical factories that take inputs and manufacture lots of intermediate products and outputs. For example, when a cell divides, its DNA must be copied and that requires the biochemical synthesis of large numbers of adenine, cytosine, guanine, and thymine molecules. Biochemical reactions are usually catalyzed by enzymes, proteins that facilitate a reaction but are not used up by it. Consider, for example, a reaction in which chemical A is converted to chemical B with the help of an enzyme E. If $a(t)$ and $b(t)$ are the respective concentrations of A and B at time t, then one typically writes down a differential equation for $b(t)$, which takes the form

$$b'(t) = f(a, b, E) + \cdots - \cdots .$$

Here, f is the rate of production, which typically depends on a, b, and E. Of course B may be produced by other reactions (which would lead to additional positive terms $+ \cdots$) and may be used as a substrate itself in still other reactions (which would lead to additional negative terms $- \cdots$). So, given a particular cell function or biochemical pathway, we can just write down the appropriate set of nonlinear coupled ordinary differential equations for the chemical concentrations and solve it by hand or by machine computation. However, this straightforward approach is often unsuccessful. First of all, there are a lot of parameters (and variables) in these equations and measuring them in the context of real living cells is difficult. Second, different cells behave differently and may have different functions, so we would expect the parameters to be different. Third, cells are alive and change what they are doing, so the parameters may themselves be functions of time. But the greatest difficulty is that the particular pathway under study is not really isolated. Rather, it is embedded in a much larger system. How do we know that our model system will continue to behave in the same way when embedded in this larger context? We need new theorems in dynamical systems that answer questions such as this, not for general "complex systems" but for the particular kinds of complex systems that arise in important biological problems.

Cells continue to accomplish many basic tasks even though their environments (i.e., their inputs) are constantly changing. A brief example of this phenomenon, which is known as *homeostasis*, will illustrate the problem of "context." Let us suppose that the chemical reaction above is one step in the pathway for making the thymines necessary for cell division. If the cell is a cancer cell, we would like to turn off this pathway, and a reasonable way to try to do this would be to put into the cell a compound X that binds to E, thereby reducing the amount of free enzyme available to make the reaction run. Two homeostatic mechanisms immediately come into play. First, a typical reaction is inhibited by its product: that is, f decreases as b increases. This makes biological sense because it ensures that B is not overproduced. So, when the amount of free E is reduced and the rate f declines, the resulting decrease in b drives the rate up again. Second, if the rate f is lower than usual, the concentration a typically rises since A is not being used up as quickly, which also drives the rate f up again since f increases as a increases. Given the network in which A and B are embedded, one can imagine calculating how much f will drop if we put a certain amount of X into the cell. In fact, f may drop even less than we calculate because of another homeostatic

mechanism that is not even in our network. The enzyme E is a protein produced by the cell via instructions from a gene. It turns out that sometimes the concentration of free E inhibits the messenger RNA that codes for the production of E itself. Then, if we introduce X and reduce free E, the inhibition is removed and the cell automatically increases its rate of production of E, thus raising the amount of free E and with it raising the reaction rate f.

This illustrates a fundamental difficulty in studying cell biochemistry, indeed a difficulty in studying many biological systems. These systems are very large and very complex. To gain understanding, it is natural to concentrate on particular relatively simple subsystems. But one always has to be aware that the subsystems exist in a larger context that may contain variables (excluded by the simplification) that are crucial for understanding the behavior and biological function of the subsystem itself.

Although cells exhibit remarkable homeostasis, they also undergo spectacular changes. For example, cell division requires unzipping of the DNA, synthesis of two new complementary strands, the movement apart of the two new DNAs, and the pinching off of the mother cell to produce two daughters. How does a cell do all this? In the case of yeast cells, which are comparatively simple, the actions of the biochemical pathways are quite well understood, partly because of the mathematical work of John Tyson. But as our brief discussion makes clear, biochemistry is not all there is to cell division; an important additional feature is motion. Materials are being transported all the time throughout cells from one specific place to another (so their motion is not just diffusion), and indeed, cells themselves move. How does this happen? The answer is that materials are transported by special molecules called molecular motors that turn the energy of chemical bonds into mechanical force. Since bonds are formed and broken stochastically (that is, some randomness is involved), the study of molecular motors leads naturally to new questions in STOCHASTIC ORDINARY AND PARTIAL DIFFERENTIAL EQUATIONS [IV.24]. A good introduction to the mathematics of cell biology is Fall et al. (2002).

3 Genomics

To understand the mathematics that was involved in sequencing the human genome it is useful to start with the following simple question. Suppose that we cut up a line segment into smaller segments and are presented with the pieces. If we are told the order in which the pieces came in the original segment, then we can put them back together and reconstruct the segment. In general, since there are many possible orders, we cannot reconstruct the segment without extra information of this kind. Now suppose that we have cut up the segment in *two different ways*. Think of the line segment as an interval I of real numbers, and let the pieces be $A_1, A_2, \ldots, A_r$ when you cut it up the first way, and $B_1, B_2, \ldots, B_s$ when you cut it up the other way. That is, the sets A_i form a partition of the interval I into subintervals, and the sets B_j form another partition. For simplicity, assume that no A_i shares an endpoint with any B_j, except for the two endpoints of I itself.

Suppose that we know nothing about the order in which the pieces A_i and B_j come in I. In fact, suppose that all we know about them is which A_i overlap with which B_j: that is, which of the intersections $A_i \cap B_j$ are nonempty. Can we use this information to work out the original order of the pieces A_i and thereby reconstruct the interval I (or its reflection)? The answer will sometimes be yes and sometimes no. If it is yes, then we would like to find an efficient algorithm for doing the reconstruction, and if it is no, then we would like to know how many different reconstructions are consistent with the given information. This so-called *restriction mapping problem* is really a problem in GRAPH THEORY [III.34]: the vertices of the graph correspond to the sets A_i or B_j, and there is an edge between A_i and B_j if $A_i \cap B_j \neq \varnothing$.

A second problem is whether we can find the original order of the A_i (or the B_j) if what we are told is the length of each set A_i and each set B_j, and the set of all the lengths of the intersections $A_i \cap B_j$. The catch is that we are not told which length corresponds to which intersection. This is called the *double digest problem*. Again one would like to be able to tell when there is only one solution, or to place an upper bound on the number of possible reconstructions if there is more than one.

Human DNA is, for our purposes here, a word of length approximately 3×10^9 over a four-letter alphabet A, G, C, T. That is, it is a sequence of length 3×10^9 in which each entry is A, G, C, or T. In the cell, the word is bound letter by letter to the "complementary" word, which is determined by the rule that A can only be bound to T, and C can only be bound to G. (For example, if the word is ATTGATCCTG, then the complementary word is TAACTAGGAC.) In this brief discussion we will ignore the complementary word.

Since DNA is so long (it would be approximately two meters if one stretched it out into a straight line) it is very hard to handle experimentally, but the sequence of letters in short segments of approximately five hundred letters can be determined by a process called gel chromatography. There are enzymes that cut DNA wherever specific very short sequences occur. So if we digest a DNA molecule with one of these enzymes and digest another copy with a different enzyme, we can hope to determine which fragments from the first digestion overlap fragments from the second digestion and then use techniques from the restriction mapping problem to reconstruct the original DNA molecule. The interval I corresponds to the whole DNA word, and the sets A_i to the fragments. This involves sequencing and comparing the fragments, which has its own difficulties. However, *lengths* of fragments are not so hard to determine, so another possibility is to digest with the first enzyme and measure lengths, digest with the second and measure lengths, and finally digest with both and measure lengths. If one does this, then the problem one obtains is essentially the double digest problem.

To completely reconstruct the DNA word one takes many copies of the word, digests with enzymes, and selects at random enough fragments that together they have a high probability of covering the word. Each of the fragments is cloned, in order to get enough mass, and then sequenced by gel chromatography. Both processes can introduce errors, so one is left with a very large number of sequenced fragments with known error rates for the letters. These need to be compared to see if they overlap: that is, to see if the sequence near the end of one fragment is the same as (or very similar to) the sequence at the beginning of another. This alignment problem is itself difficult because of the large number of possibilities involved. So, in the end we have a very large restriction mapping problem except that we can only say that given fragments overlap with probabilities that are themselves hard to estimate. A further difficulty is that DNA tends to have large blocks that repeat in different parts of the word. As a result of these complications, the problem is much harder than the restriction mapping problem described earlier. It is clear that graph theory, combinatorics, probability theory, statistics, and the design of algorithms all play central roles in sequencing a genome.

Sequence alignment is important in other problems as well. In phylogenetics (see below) one would like a way of saying how similar two genes or genomes are. When studying proteins, one can sometimes predict protein three-dimensional structure by searching databases for known proteins with the most similar amino acid sequence. To illustrate how complex these problems are, consider a sequence $\{a_i\}_{i=1}^{1000}$ of one thousand letters from our four-letter alphabet. We wish to say how similar it is to another sequence $\{b_i\}_{i=1}^{1000}$. Naively, one could just compare a_i with b_i and define a METRIC [III.56] like $d(\{a_i\}, \{b_i\}) = \sum \delta(a_i, b_i)$. However, DNA sequences have evolved typically by insertions and deletions as well as by substitutions. Thus if the sequence ACACAC$\cdots$ lost its first C to become AACAC$\cdots$, the two sequences would be very far apart in this metric even though they are very similar and related in a simple way. The way around this difficulty is to allow sequences to include a fifth symbol, -, which stands for the place of a deletion or a place opposite an insertion. Thus, given two sequences (of perhaps different lengths), we wish to find how they can be augmented with dashes to give the minimum possible distance between them. A little thought will convince the reader that it is not feasible to use a brute-force search for a problem like this, even for the fastest computers—there are so many potential augmentations that the search would take far too long. Serious and thoughtful algorithm development is required. Two excellent introductions to the material discussed in this section are Waterman (1995) and Pevzner (2000).

4 Correlation and Causality

The central dogma of molecular biology is DNA $\rightarrow$ RNA $\rightarrow$ proteins. That is, information is stored in DNA, it is transferred out of the nucleus by RNA, and the RNA is then used in the cell to make proteins that carry out the work of the cell through the metabolic processes discussed in section 2. Thus DNA directs the life of the cell. Like most things in biology, the true situation is much more complicated. Genes, which are segments of DNA that code for the manufacture of particular proteins, are sometimes turned on and sometimes turned off. Usually, they are partially turned on; that is, the protein they code for is manufactured at some intermediate rate. This rate is controlled by the binding (or lack of binding) of small molecules or specific proteins to the gene, or to the RNA that the gene codes for. Thus genes can produce proteins that inhibit (or excite) other genes; this called a gene network.

In a way, this was obvious all along. If cells can respond to their environments by changing what they do, they must be able to sense the environment and signal the DNA to change the protein content of the cell. Thus, while sequencing DNA and understanding specific biochemical reactions are important first steps in understanding cells, the hard and interesting work to come is to understand *networks* of genes and biochemical reactions. It is these networks, in which proteins control genes and genes control proteins, that carry out and control specific cellular functions. The mathematics will be ordinary differential equations for chemical concentrations and variables that indicate to what extent a gene is turned on. Since transport into and out of the nucleus occurs, partial differential equations will be involved. And, finally, since some of the molecular species occur in very small numbers, concentration (molecules per unit volume) may not be a useful approximation for computations about chemical binding and dissociation: they are probabilistic events.

Two kinds of statistical data can give hints about the components of these gene networks. First, there are large numbers of population studies that correlate specific genotypes to specific phenotypes (such as height, enzyme concentration, cancer incidence). Second, tools known as *microarrays* allow us to measure the relative amounts of a large number of different messenger RNAs in a group of cells. The amount of RNA tells us how much a particular gene is turned on. Thus, microarrays allow us to find correlations that may indicate that certain genes are turned on at the same time or perhaps in a sequence. Of course, correlation is not causality and a consistent sequential relationship is not necessarily causal either (sure, football causes winter, a sociologist once said). Real biological progress requires understanding the gene networks discussed above; they are the mechanisms by which the genotypes play out in the life of the cell.

A nice discussion of the relationship between population correlations and mechanisms occurs in Nijhout (2002), from which we take the following simple example. Most phenotypic traits depend on many genes; suppose that we consider a trait that depends on only two genes. Figure 1 depicts a surface that shows how the trait in an individual depends on how much each of the genes is turned on. All three variables are scaled from 0 to 1. Suppose that we study a population whose members have a genetic makeup that puts the individuals near the point X on the graph. If we do a statistical analysis of the population, we will find that gene B is highly statistically correlated to the trait, but gene A is

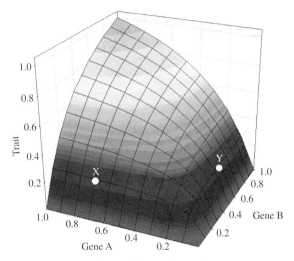

Figure 1 A phenotypic surface.

not. On the other hand, if the individuals in the population all live near the point Y on the surface, we will discover in our population study that gene A is highly statistically correlated to the trait, but gene B is not. More detailed examples with specific biochemical mechanisms are discussed in Nijhout's paper. Similar examples can be given for microarray data. This does not mean that population studies or microarray data are unimportant. Indeed, in studying hugely complex biological systems, statistical information can suggest where to look for the mechanisms that will ultimately give biological understanding.

5   The Geometry and Topology of Macromolecules

To illustrate the natural geometric and topological questions that arise when one studies macromolecules, we will briefly discuss molecular dynamics, protein–protein interactions, and the coiling of DNA. Genes code for the manufacture of proteins, which are large molecules made up of sequences of amino acids. There are twenty amino acids, each coded by a triplet of base pairs, and a typical protein might have five hundred amino acids. Interactions among the amino acids cause the protein to fold up into a complicated three-dimensional shape. This three-dimensional structure is crucial for the function of the protein since the exposed groups and the nooks and crannies in the shape govern the possible chemical interactions with small molecules and other proteins. Three-dimensional structures of

proteins can be approximately determined by X-ray crystallography and nontrivial inverse scattering calculations. The forward problem—namely, given the sequence of amino acids, predict the three-dimensional structure of the protein—is important not only for understanding existing proteins, but also for the pharmacological design of new proteins to accomplish specific tasks. Thus, in the past twenty years a large field called *molecular dynamics* has arisen, in which one uses classical mechanical methods.

Suppose we have a protein that consists of N atoms. Let x_i denote the position (specified by three real coordinates) of the ith atom, and let $\boldsymbol{x}$ denote the vector formed from all these coordinates (which belongs to $\mathbb{R}^{3N}$). For each pair of atoms, one attempts to write down a good approximation to the potential energy, $E_{i,j}(x_i, x_j)$, due to their pairwise interaction. This could be the electrostatic interaction, for example, or the van der Waals interaction, which is a classical mechanical formulation of quantum effects. The total potential energy is $E(\boldsymbol{x}) \equiv \sum E_{i,j}(x_i, x_j)$ and Newton's equations of motion take the form

$$\dot{\boldsymbol{v}} = -\nabla E(\boldsymbol{x}), \quad \dot{\boldsymbol{x}} = \boldsymbol{v},$$

where $\boldsymbol{v}$ is the vector of velocities. Starting with some initial conditions one can try to solve these equations to follow the dynamics of the molecule. Note that this is a very high-dimensional problem. A typical amino acid has twenty atoms, so that is sixty coordinates right there, and if we are looking at a protein made up of five hundred amino acids, then $\boldsymbol{x}$ will be a vector with thirty thousand coordinates. Alternatively, one could assume that the protein will fold to the configuration that has the minimum potential energy. Finding this configuration would mean finding the roots of $\nabla E(\boldsymbol{x})$, by NEWTON'S METHOD [II.4 §2.3] say, and then checking to see which root gives the lowest energy. Again this is an enormous computational task.

It is not surprising that molecular dynamics calculations have been only moderately successful and have predicted the shapes of only relatively small molecules and proteins. The numerical problems are substantial and the choice of energy terms is somewhat speculative. Even more importantly, context matters, as it does in many biological problems. The way proteins fold depends on properties of the solution in which they sit. Many proteins have several preferred configurations and switch from one to the other depending on interactions with small molecules or other proteins. Finally, it has recently been discovered that proteins do not fold up by themselves from their linear configuration to their three-dimensional shape, but are helped and guided by other proteins called chaperones. It is natural to ask whether there are quantifiable geometrical units larger than points (atoms) that could reasonably form the basis for a good approximation to the dynamics of large molecules.

A start has been made in this direction by research groups studying the interactions of proteins with small molecules and other proteins. These interactions are fundamental to cell biochemistry, cell-transport processes, and cell signaling, and so progress is vital to understanding how cells work. Suppose one has two large proteins that are bound to each other. The first thing one would like to do is describe the geometry of the binding region. One could do this as follows. Consider an atom in either protein that is at point x. Given another atom at point y, there is a plane that divides $\mathbb{R}^3$ into two open half-spaces: the points closer to x and the points closer to y. Now let R_x denote the intersection of all such open half-spaces as y ranges over the positions of all other atoms: that is, R_x consists of those points that are closer to x than to any other atom. The union of the boundaries, $\bigcup_x \partial(R_x)$, called a *Voronoi surface*, consists of triangles and pieces of planes and has the property that each point on the surface is equidistant from at least two atom positions. To model the binding region between the two proteins, we discard all pieces of the Voronoi surface that are equidistant from two atoms that belong to the same protein and keep just the ones that are equidistant from two atoms that are in different proteins. This surface goes off to infinity, so we clip off the parts that are not "close" to either protein. The result is a surface with a boundary made up of polyhedral faces that is a reasonable approximation of the interaction interface between the two proteins. (This is not quite an accurate description: in the actual construction, "distance" is weighted in a way that depends on the atoms involved.) Now choose colors representing the twenty amino acids and color each side of each polyhedral piece with the color of the amino acid that the closest atom is in. This divides each side of the surface into large colored patches corresponding to nearness of a particular amino acid on that side. The coloring of the two sides of the boundary surface will be different, of course, and the placement of the patches gives information about which amino acids in one protein are interacting with which amino acids in the other. In particular, one amino acid in one protein may interact with

several in the other. This gives a way of using geometry to classify the nature of the particular protein–protein interaction.

Finally, let us touch on questions involving the packaging of DNA. The basic problem is easy to see. As mentioned earlier, the human DNA double helix when stretched out linearly is about two meters long. A typical cell has a diameter of about one-hundredth of a millimeter and its nucleus has a diameter of about one-third that size. All of that DNA has to be packed into the nucleus. How is this done?

At least the first stages are well understood. The DNA double helix is wound around proteins called *histones*, which consist of about two hundred base pairs each, yielding chromatin, which is a sequence of such DNA-wrapped histones connected by short segments of DNA. Then the chromatin is itself wrapped up and compacted; the geometrical details are not completely understood. It is important to understand the packing and the mechanisms that create it, because the life of the cell requires unpacking! When the cell divides, the entire DNA helix must be unzipped to form two separate strands, which are the templates on which the two new copies of DNA will be built. Clearly this cannot be done all at once but must involve local unwinding of the DNA off the histones, local unzipping, synthesis, and then local repacking.

It is equally challenging to understand the sequence of events that occurs when a protein is synthesized from a gene. Transcription factors diffuse into the nucleus and bind to specific short segments of DNA (of about ten base pairs) in the regulatory region of the gene. Of course, they will randomly bind wherever they see the same segment. Typically, one needs the binding of several different transcription factors in the regulatory region along with RNA polymerase to start transcription of a gene. That process involves the unwinding of the gene-coding region from the histones so that it can be transcribed, the transport of the resulting RNA out of the nucleus, and the recompactification of the DNA. To understand these processes fully, one will have to solve problems in partial differential equations, geometry, combinatorics, probability theory, and topology. DeWitt Sumners is the mathematician who brought the topological problems in the study of DNA (links, twists, knots, supercoiling) to the attention of the mathematics community. A good reference for molecular dynamics and the general mathematical issues posed by biological macromolecules is Schlick (2002).

6   Physiology

When one first studies human physiological systems, they seem almost miraculous. They accomplish enormous numbers of tasks simultaneously. They are robust but capable of quick changes if the situation warrants. They are made up of large numbers of cells that actively cooperate so that the tasks of the whole can be done. It is the nature of many of these systems that they are complex, controlled by feedback, and integrated with each other. It is the job of mathematical physiology to understand how they work. We will illustrate some of these points by discussing problems in biological fluid dynamics.

The heart pumps blood throughout a circulatory system that consists of vessels of diameter as large as 2.5 cm (the aorta) and as small as 6×10^{-4} cm (the capillaries). Not only are the vessels flexible, but many are surrounded by muscle and can contract to exert local force on the blood. The main force-generating mechanism (the heart!) is approximately periodic, but the period can change. The blood itself is a very complicated fluid. About 40% of its volume is made up of cells: red blood cells carry most of the oxygen and CO_2; white blood cells are immune system cells that hunt bacteria; and platelets are part of the blood clotting process. Some of these cells have diameters that are larger than the smallest capillaries, which raises the nice question of how they get through. You notice that we are very far away from most of the simplifying assumptions of classical fluid dynamics.

Here is an example of a circulatory-system question. In a significant number of people, the mitral valve, which is the inflow valve to the left side of the heart, becomes defective. It is common to replace the valve by an artificial one and this leads to an important question: how should one design the artificial valve so that the resulting flow in the left heart chamber has as few stagnant points as possible, since clots tend to form at these points? Charles Peskin did the pioneering work on this problem. Here is another question. The white blood cells are not carried in the middle of the fluid but tend to roll along the walls. Why do they do that? It is a good thing that they do, because their job is to sniff out inflammation outside the blood vessel and, when they find it, to stop and burrow through the blood vessel wall to get to the inflamed site. Another circulatory fluid dynamics question is discussed in section 10.

The circulatory system is connected to many other systems. The heart has its own pacemaker cells, but its frequency of contraction is regulated by the autonomic nervous system. Through the *baroreceptor reflex*, the sympathetic nervous system tightens blood vessels to avoid a dramatic drop in blood pressure when we stand. Overall average blood pressure is maintained by a complicated regulatory feedback mechanism involving the kidneys. It is worthwhile remembering that all these things are being accomplished by living tissues whose parts are always decaying and being replaced. For example, the gap junctions that transmit current at very low resistance between heart muscle cells have a half-life of approximately one day.

As a final example, we consider the lung, which has a fractal branching structure that terminates after twenty-three levels in about 600 million air sacs called *alveoli*, in which oxygen and CO_2 are exchanged with the circulating blood. The Reynolds number of the air flow varies by about three orders of magnitude between the large vessels near the throat and the tiny vessels near the alveoli. Premature infants often have respiratory difficulty because they lack surfactants that reduce surface tension on the inner surfaces of the alveoli. The high surface tension makes the alveoli collapse, which makes breathing difficult. One would like the infants to breathe in air that includes tiny aerosol drops of surfactant. How small should the drops be so that as much surfactant as possible makes it to the alveoli?

The mathematics of physiology consists mostly of ordinary and partial differential equations. However, there is a new feature: many of these equations have time delays. For example, the rate of respiration is controlled by a brain center that senses the CO_2 content of blood. It takes almost fifteen seconds for blood to go from the lungs to the left heart and from there to the brain center. This time delay is even longer in patients with weak hearts and often these patients display Cheyne–Stokes breathing: very rapid breathing alternates with periods of little or no breathing. Such oscillations in control systems are well-known as the time delay gets longer. Since partial differential equations are often involved, new mathematical results are needed that go well beyond the standard theory of ordinary differential equations with delay, which was initiated by Bellman in the 1950s. An excellent reference for the applications of mathematics to physiology is Keener and Sneyd (1998).

7   What's Wrong with Neurobiology?

The short answer is that there is not enough theory. This may seem an odd thing to say, since neurobiology is the home of the Hodgkin–Huxley equations, which are often cited as a triumph of mathematics in biology. In a series of papers in the early 1950s, Hodgkin and Huxley described several experiments, and gave a theoretical basis for explaining them. Building on the work of physicists and chemists (for example, Walter Nernst, Max Planck, and Kenneth Cole), they discovered the relationship between certain ionic conductances and the trans-membrane electrical potential, $v(x, t)$, in the axons of neurons, and they formulated a mathematical model:

$$\frac{\partial v}{\partial t} = \alpha \frac{\partial^2 v}{\partial x^2} + g(v, y_1, y_2, y_3),$$
$$\frac{\partial y_i}{\partial t} = f_i(v, y_i), \quad i = 1, 2, 3.$$

Here the y_i are related to the membrane conductances of various ions. The equations have solutions that are pulses that keep their shape and travel at constant velocity in a way that corresponds to the observed behavior of action potentials in real neurons. The ideas, both explicit and implicit, in these discoveries form the basis of much single-neuron physiology. Of course, mathematicians should not be too proud about this since Hodgkin and Huxley were biologists. The Hodgkin–Huxley equations were part of the stimulus for interesting work by mathematicians on traveling waves and pattern formation in reaction–diffusion equations.

However, not everything can be explained at the level of just one neuron. Watch your hand as it reaches out gracefully to pick up an object. Think about the so-called ocular–vestibular reflex in which motions of the head are automatically compensated for by motions of the eyes so that your gaze can remain fixed. Consider the fact that you are looking at stereotypical black marks on a page and they mean something inside your head. These are *system properties*, and the systems are large indeed. There are approximately 10^{11} neurons in the central nervous system and on average each makes about one thousand connections to other neurons. These systems will not be understood just by examining their parts (the neurons) and, for obvious reasons, experimentation is limited. Thus, experimental neurobiology, like experimental physics, needs input from deep and imaginative theorists.

The lack of a large theory community interacting robustly with experimentalists is to some extent a historical accident. Grossberg asked how groups of (quite simple) model neurons, if they were connected in the right ways, could accomplish various tasks such

as pattern recognition and decision making, or could exhibit certain "psychological" properties (Grossberg 1982). He also asked how these networks could be trained. At about the same time it was shown that networks of neuron-like elements connected in the right way could automatically compute good solutions of large, difficult problems like the TRAVELING SALESMAN PROBLEM [VII.5 §2]. These and other factors, including the great interest in software engineering and artificial intelligence, led to the emergence of a large community of researchers studying "neural networks." The members of this community were mostly computer scientists and physicists, so it was natural for them to concentrate on the design of devices, rather than biology. This was noticed, of course, by experimental neurobiologists, who lost interest in collaborating with these theorists.

This brief history is of course an oversimplification. There are mathematicians (and physicists and computer scientists) who are essentially theoreticians for neuroscience. Some of them work on hypothetical networks, typically either very small networks or networks with strong homogeneity properties, to discover what are the emergent behaviors of the systems. Others work on modeling real physiological neural networks, often collaboratively with biologists. Usually, the models consist of ordinary differential equations for the firing rates of the individual neurons or mean-field models that involve integral equations. These mathematicians have made real contributions to neurobiology.

But much more is needed, and to see why, it is useful to think about just how difficult these problems really are. First, there is no one-to-one correspondence between the cells of the central nervous system in different members of the same species (except in special cases like *C. elegans*). Second, neurons in the same animal differ widely in their anatomy and physiology. Third, the details of a particular network may well depend on the life history of the animal. Fourth, most neurons are somewhat unreliable devices in that they give different outputs under repeated trials with the same input. Finally, one of the prime characteristics of neural systems is that they are plastic, adaptable, and ever changing. After all, if you remember anything of what is written here, then your head is different from when you began. Between the level of the single neuron and the psychological level, there are probably twenty levels of networks, each network feeding into and being controlled by networks at other levels. The mathematical objects that will enable us to classify, analyze, and understand how this all works have probably not yet been discovered.

8   Population Biology and Ecology

Let us begin with a simple example. Imagine a large orchard of equally spaced trees and suppose that one tree has a disease. The disease can be transmitted only to nearest neighbors, and is transmitted with probability p. What is $E(p)$, the expected percentage of trees that will be infected? Intuitively, if p is small, $E(p)$ should be small, and if p is large, $E(p)$ should be close to 100%. In fact, one can prove that $E(p)$ changes very rapidly from being small to being large as p passes through a small transition region around a particular critical probability p_c. One would expect p to decrease as the distance, d, between trees increases; farmers should choose d in such a way that p is less than the critical probability, in order to make $E(p)$ small. We see here a typical issue in ecological problems: how does behavior on the large scale (tree epidemic or not) depend on behavior at the small scale (the distance between trees). And, of course, the example illustrates that understanding the biological situation requires mathematics. For other examples of sharp global changes in probabilistic models, see PROBABILISTIC MODELS OF CRITICAL PHENOMENA [IV.25].

Suppose that we now widen our gaze to consider forests—let us say the forests on the East coast of the United States. We would like to understand how they have come to be as they are. Most of them were not planted in neat rows, so that is already a complication. But there are two other really new features. First, there is not one species but many, and each species of tree has different properties: shape, seed dispersal, need for light, and so forth. The species are different, but their properties affect each other because they are living in the same space. Second, the species, and the interactions between the species, are affected by the physics of the environment. There are physical parameters that vary on long timescales, like average temperature, and there are other parameters that vary on very short timescales, like wind speed (for seed dispersal). Certain properties of forests may depend on the fluctuations in these parameters as much as on the values themselves. Finally, one might have to take into account the reaction of the ecosystem to catastrophic events such as hurricanes or prolonged drought.

The difficulties are similar to those we have seen for other problems in mathematical biology. One would like to understand the emergent behavior on the large scale. To do this one creates mathematical models that relate the behavior on the small scale to the large scale. However, on the small scale one is overwhelmed by the biological details. Which of these details should be in the model? Of course, there is no simple answer to this because, in fact, this is the heart of what we want to know. Which of the bewildering variety of local properties or variables give rise to the large-scale behavior and by what mechanisms? Furthermore, it is not obvious what kinds of model are best. Should we model each individual and its interactions, or should we use population densities? Should we use deterministic models or stochastic models? These are also hard questions, and the answers depend on the system being studied and the questions being asked. A nice discussion of these different modeling choices can be found in Durrett and Levin (1994).

Let us focus again on a simple model: the so-called *SIRS model* for the spread of a disease in a population. A crucial parameter is the *infectious contact number*, σ, which represents the average number of new infections that an infected individual creates in the susceptible population. For a serious disease one would like to bring the value of σ down to below 1 (so that an epidemic will be unlikely) by vaccination, which takes individuals from the susceptible category and puts them in the removed category. Since vaccination is expensive and it is difficult to vaccinate high percentages of the population, it is an important public-health problem to know how much vaccination is needed to bring σ to below 1. A little reflection shows us how difficult this problem really is. First of all, the population is not well mixed, so one may not be able to ignore spatial separation, as is done in the SIRS model. Even more important, σ depends on the social behavior of individuals and the subclasses of the population to which they belong (as anyone with small children in school will attest). Thus, we see a genuinely new issue here: if an ecological problem involves animals, then the social behavior of the animals may affect the biology.

In fact, the issues are even deeper. Individuals in groups, or species, or subpopulations, vary and it is just this variation on which natural selection acts. So, to understand how an ecosystem got to where it is today, one may have to take this individual variability into account. Social behavior is also transmitted from generation to generation, both biologically and culturally, and therefore also evolves. For instance, there are many examples of plant and animal species in which the biology of the plants and the sociology of the animals clearly coevolved, to the benefit of both. Game-theory models have been used to study the evolution of certain human behaviors such as altruism. Therefore, ecological problems, which sometimes seem simple at first, are often very deep, because the biology and its evolution are connected in complicated ways to both the physics of the environment and the social behavior of the animals. A good introductory review of these questions can be found in Levin et al. (1997).

9   Phylogenetics and Graph Theory

Since Darwin, a deep ongoing problem in biology has been to determine the history of the evolution of species that has brought us to our current state. It is natural when thinking about such questions to draw directed GRAPHS [III.34] in which the vertices, V, are species (past or present) and an edge from species v_1 to species v_2 indicates that v_2 evolved directly from v_1. Indeed, Darwin himself wrote down such graphs. To explain the mathematical issues, we will consider a simple special case. A connected graph with no cycles is called a *tree*. If we distinguish a particular vertex, ρ, and call it the *root*, then the tree is called *rooted*. The vertices of the tree that have degree one (i.e., have only one attached edge) are called *leaves*. We will assume that ρ is not a leaf. Notice that, because there are no cycles, there is exactly one path in the tree from ρ to each vertex v. We say that $v_1 \leqslant v_2$ if the path from ρ to v_2 contains v_1 (see figure 2). The problem is to determine which trees with a given set of leaves X (current species) and a given root vertex ρ (a hypothesized ancestral species) are consistent with experimental information and theoretical assumptions about the mechanisms of evolution. Such a tree is called a *rooted phylogenetic X-tree*. One can always add extra intermediate species, so typically one imposes the additional restriction that the phylogenetic trees be as simple as possible.

Suppose that we are interested in a certain characteristic, the number of teeth, for example. We can use it to define a function f from X, the set of current species, to the nonnegative integers: given a species x in X, we let $f(x)$ be the number of teeth of members of x. In general, a *character* is a function from X to a set C of possible values of a particular characteristic (having or not having a gene, the number of vertebrae, the presence or absence of a particular enzyme, etc.). It is characters such as these that are measured by biologists in current

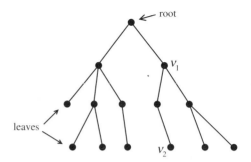

Figure 2 A rooted tree.

species. In order to say something about evolutionary history, one would like to extend the definition of f from X to the larger set V of all the vertices in a phylogenetic tree. To do this, one specifies some rules for how characters can change as species evolve. A character is called *convex* if f can be extended to a function $\bar{f}$ from V to C in such a way that for every $c \in C$, the subset $\bar{f}^{-1}(c)$ of V is a connected subgraph of the tree. That is, between any two species x and y with character value c there should be a path back in evolutionary history from x and forward again to y such that all the species in between have the same value c. This essentially forbids new values from arising and then reverting back and forbids two values evolving separately (in different parts of the tree). Of course, we have the current species and lots of characters. What is unknown is the phylogenetic tree, that is, the collection of intermediate species and the relations between them that link the current species to a common ancestor. A collection of characters is called *compatible* if there exists a phylogenetic tree on which they are all convex. Determining when this is the case and finding an algorithm for constructing such a tree (or a minimal such tree) is called the *perfect phylogeny problem*. This problem is understood for collections of characters with binary values, but not in general.

An alternative problem is the following. Note that we have been treating all the edges alike when in fact some may represent longer or shorter evolutionary steps. Suppose that we have a function w that assigns a positive number to each edge. Then, since there is a unique shortest path between any two vertices in the tree, w induces a distance function d_w on $V \times V$, and in particular on X. Now, suppose that we are given a distance function δ on $X \times X$ that tells us how far apart current species are. The question is whether there exists a phylogenetic tree and a weighting function w so that

$\delta(x, y) = d_w(x, y)$ for all $x, y \in X$. If so, one would like an algorithm to construct the tree and the weights. If not, one would like to construct a family of trees that satisfy the relation approximately.

Finally, we note that there is a blossoming field of Markov processes on trees where the partial order on V forms the basis for the Markov condition. Not only are there wonderful mathematical questions relating the geometry of the tree to the processes, but there are important issues for phylogenetics. Suppose that one starts with characters defined only at the root and then allows them to "evolve" down the tree by (possibly different) Markov processes. Then, given the distribution of characters on the leaves, when can we reconstruct the tree? These questions have even given rise to problems in algebraic geometry.

Phylogenetics is useful not only for determining our past but also for controlling our present and future: see Fitch et al. (1997), where you can find a phylogenetic reconstruction for the influenza A virus. An excellent recent graduate text in this field is Semple and Steel (2003).

10   Mathematics in Medicine

It is clear that an improved understanding of biological systems leads, at least indirectly, to improved medical care. However, there are many cases in which mathematics has a direct impact on medicine. We give two brief examples.

Charles Taylor is a biomedical engineer at Stanford who works on the fluid dynamics of the cardiovascular system. He wants to use fast simulations of flows as part of the medical decision-making process. Suppose that a patient presents with leg weakness and is found on magnetic resonance imaging (MRI) to have an arterial constriction in the thigh. Typically, the surgical group will meet and consider a variety of options including shunting blood from other vessels to a point below the constriction or shunting blood around the constriction with vessels removed from some other site in the patient's body. Among a fairly large number of possible choices, the surgical group chooses based on what they have been taught and on their own experience. The characteristics of the flow after the graft are important not just for recovery of function but to prevent the formation of possibly destructive clots. An important difficulty is that patients treated successfully are rarely seen again, so one does not know the actual characteristics of the flow after the operation. Taylor wants to be in

on the discussion with the surgical team with immediate fluid dynamical simulations based on the patient's actual vasculature (as revealed by the MRI) for each proposed graft suggested. And he wants followup on each patient to check how well his simulations predicted the actual postoperative flow.

David Eddy is an applied mathematician who has worked on health policy for thirty years. He first became prominent when he published *Screening for Cancer: Theory, Analysis and Design* (Eddy 1980), which grew out of his Ph.D. thesis. Because of this book, the American Cancer Society changed its recommendation for the frequency of Pap smears from once a year to once every three years, since Eddy's modeling showed that the change would have little effect on the life expectancy of the average American woman. A short calculation easily estimates the amount of money saved in an economy that spends 15% of its gross domestic product (GDP) on health care. Throughout his career Eddy has criticized both the indiscriminate use of diagnostic tests and the incorrect use of the results by physicians and policy boards often ignorant of the basic facts of conditional probability. He has criticized specific health-policy guidelines as based on seat-of-the-pants guesswork instead of quantitative analysis. In a classic case he distributed questionnaires to physicians at a conference on colorectal cancer. The physicians were asked to estimate the percentage drop in mortality from colorectal cancers if all Americans over age fifty were to have the two most common diagnostic tests each year: fecal blood smear and flexible sigmoidoscopy. The answers were approximately uniformly distributed in a range from 2% to 95%. Even more startling was the fact that the physicians did not even know that they disagreed so dramatically. He has used mathematical models to analyze the costs and benefits of new and existing surgeries, medical treatments, and drugs, and he has participated robustly in debates on the current health-policy crisis. Throughout, he has pointed out that a hefty percentage of GDP is spent on devices, drugs, and procedures with almost no mathematical analysis of which are effective.

For more on the interrelations between mathematics and medicine, see MATHEMATICS AND MEDICAL STATISTICS [VII.11].

11   Conclusions

Mathematics and mathematicians have played important roles in many fields of biology that this brief article has not had the space to cover: immunology, radiology, developmental biology, and the design of medical devices and synthetic biomaterials, to name just a few of the most obvious omissions. Nevertheless, this collection of examples and introductory discussions allows us to draw a few conclusions about mathematical biology. The range of biological problems needing explanation by mathematics is enormous and techniques from many different branches of mathematics are important. It is not so easy in mathematical biology to extract simple, clear mathematical questions to work on, because biological systems typically operate in a complex environment where it is difficult to decide what should be counted as the system and what as the parts. Finally, biology is a source of new, interesting, and difficult questions for mathematicians, whose participation in the biological revolution is necessary for a full understanding of the biology itself.

Further Reading

Durrett, R., and S. Levin. 1994. The importance of being discrete (and spatial). *Theoretical Population Biology* 46: 363–94.

Eddy, D. M. 1980. *Screening for Cancer: Theory, Analysis and Design.* Englewood Cliffs, NJ: Prentice-Hall.

Fall, C., E. Marland, J. Wagner, and J. Tyson. 2002. *Computational Cell Biology.* New York: Springer.

Fitch, W. M., R. M. Bush, C. A. Bender, and N. J. Cox. 1997. Long term trends in the evolution of H(3) HA1 human influenza type A. *Proceedings of the National Academy of Sciences of the United States of America* 94:7712–18.

Grossberg, S. 1982. *Studies of Mind and Brain: Neural Principles of Learning, Perception, Development, Cognition, and Motor Control.* Boston, MA: Kluwer.

Keener, J., and J. Sneyd. 1998. *Mathematical Physiology.* New York: Springer.

Levin, S., B. Grenfell, A. Hastings, and A. Perelson. 1997. Mathematical and computational challenges in population biology and ecosystems science. *Science* 275:334–43.

Nijhout, H. F. 2002. The nature of robustness in development. *Bioessays* 24(6):553–63.

Pevzner, P. A. 2000. *Computational Molecular Biology: An Algorithmic Approach.* Cambridge, MA: MIT Press.

Schlick, T. 2002. *Molecular Modeling and Simulation.* New York: Springer.

Semple, C., and M. Steel. 2003. *Phylogenetics.* Oxford: Oxford University Press.

Waterman, M. S. 1995. *Introduction to Computational Biology: Maps, Sequences, and Genomes.* London: Chapman and Hall.

VII.3 Wavelets and Applications
Ingrid Daubechies

1 Introduction

One of the best ways to understand a function is to expand it in terms of a well-chosen set of "basic" functions, of which TRIGONOMETRIC FUNCTIONS [III.92] are perhaps the best-known example. Wavelets are families of functions that are very good building blocks for a number of purposes. They emerged in the 1980s from a synthesis of older ideas in mathematics, physics, electrical engineering, and computer science, and have since found applications in a wide range of fields. The following example, concerning image compression, illustrates several important properties of wavelets.

2 Compressing an Image

Directly storing an image on a computer uses a lot of memory. Since memory is a limited resource, it is highly desirable to find more efficient ways of storing images, or rather to find *compressions* of images. One of the main ways of doing this is to express the image as a function and write that function as a linear combination of basic functions of some kind. Typically, most of the coefficients in the expansion will be small, and if the basic functions are chosen in a good way it may well be that one can change all these small coefficients to zero without changing the original function in a way that is visually detectable.

Digital images are typically given by large collections of *pixels* (short for *picture elements*; see figure 1).

The boat image in figure 1 is made up of 256×384 pixels; each pixel has one of 256 possible gray values, ranging from pitch black to pure white. (Similar ideas apply to color images, but for this exposition, it is simpler to keep track of only one color.) Writing a number between 0 and 255 requires 8 digits in binary; the resulting 8-bit requirement to register the gray level for each of the $256 \times 384 = 98\,304$ pixels thus gives a total memory requirement of $786\,432$ bits, for just this one image.

This memory requirement can be significantly reduced. In figure 2, two squares of 36×36 pixels are highlighted, in different areas of the image. As is clear from its blowup, square A has fewer distinctive characteristics than square B (a blowup of which is shown in figure 1), and should therefore be describable with fewer bits. Square B has more features, but it too contains

Figure 1 A digital image with successive blowups.

Figure 2 Blowup of a 36×36 square in the sky.

(smaller) squares that consist of many similar pixels; again this can be used to describe this region with fewer than the $36 \times 36 \times 8$ bits given by the naive estimate of assigning 8 bits to each pixel.

These arguments suggest that a change in the representation of the image can lead to reduced memory requirements: instead of a huge assembly of pixels, all equally small, the image should be viewed as a combination of regions of different size, each of which has more or less constant gray value; each such region can then be described by its size (or scale), by where it appears in the image, and by the 8-bit number that tells us its average gray value. Given any subregion of the image, it is easy to check whether it is already of this simple type by comparing it with its average gray value. For square A, taking the average makes virtually no difference, but for square B, the average gray value is not sufficient to characterize this portion of the image (see figure 3).

When square B is subdivided into yet smaller subsquares, some of them have a virtually constant gray level (e.g., in the top-left or bottom-left regions of square B); others, such as subsquares 2 and 3 (see figure 4), that are not of just one constant gray level may still have a simple gray level substructure that can be easily characterized with a few bits.

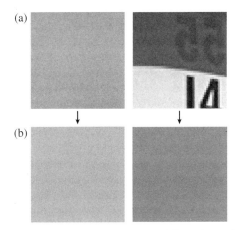

Figure 3 (a) Blowups of squares A (left) and B (right) with (b) the average gray value for each.

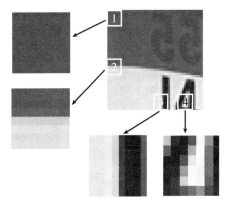

Figure 4 Subsquare 1 has constant gray level, while subsquares 2 and 3 do not, but they can be split horizontally (2) or vertically (3) into two regions with (almost) constant gray level. Subsquare 4 needs finer subdivision to be reduced to "simple" regions.

To use this decomposition for image compression, one should be able to implement it easily in an automated way. This could be done as follows:

- first, determine the average gray value for the whole image (assumed to be square, for simplicity);
- compare a square with this constant gray value with the original image; if it is close enough, then we are done (but it will have been a very boring image);
- if more features are needed than only the average gray value, subdivide the image into four equal-sized squares;

- for each of these subsquares, determine *their* average gray value, and compare with the subsquare itself;
- for those subsquares that are not sufficiently characterized by their average gray value, subdivide again into four further equal-sized subsquares (each now having an area one sixteenth of the original image);
- and so on.

In some of the subsquares it may be necessary to divide down to the pixel level (as in subsquare 4 in figure 4, for example), but in most cases subdivision can be stopped much earlier. Although this method is very easy to implement automatically, and leads to a description using many fewer bits for images such as the one shown, it is still somewhat wasteful. For instance, if the average gray level of the original image is 160, and we next determine the gray levels of each of the four quarter images as 224, 176, 112, and 128, then we have really computed one number too many: the average of the gray levels for the four equal-sized subimages is automatically the gray level of the whole image, so it is unnecessary to store all five numbers. In addition to the average gray value for a square, one just needs to store the *extra* information contained in the average gray values of its four quarters, given by the three numbers that describe

- how much darker (or lighter) the left half of the square is than the right,
- how much darker (or lighter) the top half of the square is than the bottom, and
- how much darker (or lighter) the diagonal from lower left to upper right is than the other diagonal.

Consider for example a square divided up into four subsquares with average values 224, 176, 112, and 128, as shown in figure 5. The average gray value for the whole square can easily be checked to be 160. Now let us do three further calculations. First, we work out the average gray values of the top half and the bottom half, which are 200 and 120, respectively, and calculate their difference, which is 80. Then we do the same for the left half and the right half, obtaining the difference $168 - 152 = 16$. Finally, we divide the four squares up diagonally: the average over the bottom-left and top-right squares is 144, the average over the other two is 176, and the difference between these two is -32.

From these four numbers one can reconstruct the four original averages. For example, the average for

Figure 5 The average gray values for four subsquares of a square.

the top-right subsquare is given by $160 + [80 - 16 + (-32)]/2 = 176$.

It is thus *this* process, rather than simply averaging over smaller and smaller squares as described above, that needs to be repeated. We now turn to the question of making the whole decomposition procedure as efficient as possible.

A complete decomposition of a 256×256 square, from "top" (largest square) to "bottom" (the three types of "differences" for the 2×2 subsquares), involves the computation of many numbers (in fact exactly 256×256 before pruning), some of which are themselves combinations of many of the original pixel values. For instance, the grayscale average of the whole 256×256 square requires adding $256 \times 256 = 65\,536$ numbers with values between 0 and 255 and then dividing the result by 65 536; another example, the difference between the averages of the left and right halves, requires adding the $256 \times 128 = 32\,768$ grayscale numbers for the left half and then subtracting from this sum A the sum B of another 32 768 numbers. On the other hand, the sum of the pixel grayscale values over the whole square is simply $A + B$, a sum of two 33-bit numbers instead of 65 536 numbers of 8 bits each. This allows us to make a considerable saving in computational complexity if A and B are computed *before* the average over the whole square. A computationally optimal implementation of the ideas explained so far must therefore proceed along a different path from the one sketched above.

Indeed, a much better procedure is to start from the other end of the scale. Instead of starting with the whole image and repeatedly subdividing it, one begins at the pixel level and builds up. If the image has $2^J \times 2^J$ pixels in total, then it can also be viewed as consisting of $2^{J-1} \times 2^{J-1}$ "superpixels," each of which is a small square of 2×2 pixels. For each 2×2 square, the average of the four gray values can be computed (this is the gray value of the superpixel), as well as the three types of differences indicated above. Moreover, these computations are all very simple.

The next step is to store the three difference values for each of the 2×2 squares and organize their averages, the gray values of the $2^{J-1} \times 2^{J-1}$ superpixels, into a new square. This square can be divided, in turn, into $2^{J-2} \times 2^{J-2}$ "super-superpixels," each of which is a small square of 2×2 superpixels (and thus stands for 4×4 "standard" pixels), and so on. At the very end, after J levels of "zooming out," there is only one super^J-pixel remaining; its gray value is the average over the whole image. The *last* three differences that were computed in this pixel-level-up process correspond exactly to the largest-scale differences that the top-down procedure would have computed *first*, at much greater computational expense.

Carrying out the procedure from the pixel level up, none of the individual averaging or differencing computations involves more than two numbers; the total number of these elementary computations, for the *whole* transform, is only $8(2^{2J}-1)/3$. For the 256×256 square discussed before, $J = 8$, so the total is $174\,752$, which is about the number of computations needed for just one level in the top-down procedure.

How can all this lead to compression? At each stage of the process, three species of difference numbers are accumulated, at different levels and corresponding to different positions. The total number of differences calculated is $3(1 + 2^2 + \cdots + 2^{2(J-1)}) = 2^{2J} - 1$. Together with the gray value of the whole square, this means we end up with exactly as many numbers as we had gray values for the original $2^J \times 2^J$ pixels. However, many of these difference numbers will be very small (as argued before), and can just as well be dropped or put to zero, and if the image is reconstructed from the remainder there will be no perceptible loss of quality. Once we have set these very small differences to zero, a list that enumerates all the differences (in some prearranged order) can be made much shorter: whenever a long stretch of Z zeros is encountered, it can be replaced by the statement "insert Z zeros now," which requires only a prearranged symbol (for "insert zeros now"), followed by the number of bits needed for Z, i.e., $\log_2 Z$. This achieves, as desired, a significant compression of the data that need to be stored for large images. (In practice, however, image compression involves *many* more issues, to which we shall return briefly below.)

The very simple image decomposition described above is an elementary example of a *wavelet decomposition*. The data that are retained consist of

- a very coarse approximation, and
- additional layers giving detail at successively finer scales j, with j ranging from 0 (the coarsest level) to $J - 1$ (the first superpixel level).

Moreover, within each scale j the detail layer consists of many pieces, each of which has a definite localization (indicating to which of the super^j-pixels it pertains), and all the pieces have "size" 2^j. (That is, the size, in pixel widths, of the corresponding super^j-pixel is 2^j.) In particular, the building blocks are very small at fine scales and become gradually larger as the scale becomes coarser.

3 Wavelet Transforms of Functions

In the image-compression example we needed to look at three types of differences at each level (horizontal, vertical, and diagonal) because the example was a two-dimensional image. For a one-dimensional signal, one type of difference suffices. Given a function f from $\mathbb{R}$ to $\mathbb{R}$, one can write a wavelet transform of f that is entirely analogous to the image example. For simplicity, let us look at a function f such that $f(x) = 0$ except when x belongs to the interval $[0, 1]$.

Let us now consider successive approximations of f by *step functions*: that is, functions that change value in only finitely many places. More precisely, for each positive integer j, divide the interval $[0, 1]$ up into 2^j equal intervals, denoting the interval from $k2^{-j}$ to $(k + 1)2^{-j}$ by $I_{j,k}$ (so that k runs from 0 to $2^j - 1$). Then define a function $P_j(f)$ by setting its value on $I_{j,k}$ to be the average value of f on that interval. This is illustrated in figure 6, which shows the step function $P_3(f)$ for a function f whose graph is shown as well. As j increases, the width of the intervals $I_{j,k}$ decreases, and $P_j(f)$ gets closer to f. (In more precise mathematical terms, if $p < \infty$ and f belongs to the FUNCTION SPACE [III.29] L_p, then $P_j(f)$ converges to f in L_p.)

Each approximation $P_j(f)$ of f can be computed easily from the approximation $P_{j+1}(f)$ at the next-finer scale: the average of the values that $P_{j+1}(f)$ takes on the two intervals $I_{j+1,2k}$ and $I_{j+1,2k+1}$ gives the value that $P_j(f)$ takes on $I_{j,k}$.

Of course, some information about f is lost when we move from $P_{j+1}(f)$ to $P_j(f)$. On every interval $I_{j,k}$, the difference between $P_{j+1}(f)$ and $P_j(f)$ is a step function, with constant levels on the $I_{j+1,l}$, that takes on exactly opposite values on each pair $(I_{j+1,2k}, I_{j+1,2k+1})$.

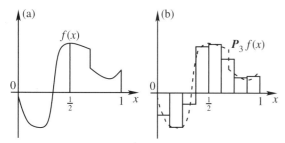

Figure 6 Graphs of (a) the function f and (b) its approximation $P_3(f)$, which is constant on every interval between $l/8$ and $(l+1)/8$, with $l = 0, 1, \ldots, 7$, and exactly equal to the average of f on each of these intervals.

The difference $P_{j+1}(f) - P_j(f)$ of the two approximation functions, over all of $[0, 1]$, consists of a juxtaposition of such up-and-down (or down-and-up) step functions, and can therefore be written as a sum of translates of the same up-and-down function, with appropriate coefficients:

$$P_{j+1}(f)(x) - P_j(f)(x) = \sum_{k=0}^{2^j - 1} a_{j,k} U_j(x - 2^{-j}k),$$

where

$$U_j(x) = \begin{cases} 1 & \text{for } x \text{ between } 0 \text{ and } 2^{-(j+1)}, \\ -1 & \text{for } x \text{ between } 2^{-(j+1)} \text{ and } 2 \times 2^{-(j+1)}, \\ 0 & \text{for all other } x. \end{cases}$$

Moreover, the "difference functions" U_j at the different levels are all scaled copies of a single function H, which takes the value 1 between 0 and $\frac{1}{2}$ and -1 between $\frac{1}{2}$ and 1; indeed, $U_j(x) = H(2^j x)$. It follows that each difference $P_{j+1}(f)(x) - P_j(f)(x)$ is a linear combination of the functions $H(2^j x - k)$, with k ranging from 0 to $2^j - 1$; adding many such differences, for successive j, shows that $P_J(f)(x) - P_0(f)(x)$ is a linear combination of the collection of functions $H(2^j x - k)$, with j ranging from 0 to $J - 1$ and k ranging from 0 to $2^j - 1$. Picking larger and larger J makes $P_J(f)$ closer and closer to f; one finds that $f - P_0(f)$ (i.e., the difference between f and its average) can be viewed as a (possibly infinite) linear combination of the functions $H(2^j x - k)$, now with j ranging over all the nonnegative integers.

This decomposition is very similar to what was done for images at the start of the article, but in one dimension instead of two and presented in a more abstract way. The basic ingredients are that f minus its average has been decomposed into a sum of layers at successively finer and finer scales, and that each extra layer of detail consists of a sum of simple "difference contributions" that all have width proportional to the scale. Moreover, this decomposition is realized by using translates and dilates of the *single* function $H(x)$, often called the *Haar wavelet*, after Alfred Haar, who first defined it at the start of the twentieth century (though not in a wavelet context). The functions $H(2^j x - k)$ constitute an *orthogonal* set of functions, meaning that the inner product $\int H(2^j x - k) H(2^{j'} x - k') \, dx$ is zero except when $j = j'$ and $k = k'$; if we define $H_{j,k}(x) = 2^{j/2} H(2^j x - k)$, then we also have that $\int [H_{j,k}(x)]^2 \, dx = 1$. A consequence of this is that the *wavelet coefficients* $w_{j,k}(f)$ that appear when we write the "jth layer" $P_{j+1}(f)(x) - P_j(f)(x)$ of the function f as a linear combination $\sum_k w_{j,k}(f) H_{j,k}(x)$ are given by the formula $w_{j,k}(f) = \int f(x) H_{j,k}(x) \, dx$.

Haar wavelets are a good tool for exposition, but for most applications, including image compression, they are not the best choice. Basically, this is because replacing a function simply by its averages over intervals (in one dimension) or squares (in two dimensions) results in a very-low-quality approximation, as illustrated in figure 7(b).

As the scale of approximation is made finer and finer (i.e., as the j in $P_j(f)$ increases), the difference between f and $P_j(f)$ becomes smaller; with a piecewise-constant approximation, however, this requires corrections at almost every scale "to get it right" in the end. Unless the original happens to be made up of large areas where it is roughly constant, many small-scale Haar wavelets will be required even in stretches where the function just has a consistent, sustained slope, without "genuine" fine features.

The right framework to discuss these questions is that of *approximation schemes*. An approximation scheme can be defined by providing a family of "building blocks," often with a natural order in which they are usually enumerated. A common way of measuring the quality of an approximation scheme is to define V_N to be the space of all linear combinations of the first N building blocks, and then to let $A_N f$ be the closest function in V_N to f, where distance is measured by the L_2-norm (though other norms can also be used). Then one examines how the distance $\|f - A_N f\|_2 = [\int |f(x) - A_N f(x)|^2 \, dx]^{1/2}$ decays as N tends to infinity. An approximation scheme is said to be *of order L* for a class of functions $\mathcal{F}$ if $\|f - A_N f\|_2 \leqslant CN^{-L}$ for all functions f in $\mathcal{F}$, where C typically depends on f but must be independent of N. The order of an approximation scheme for smooth functions is closely linked to

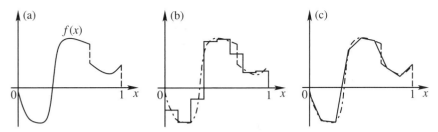

Figure 7 (a) The original function. (b), (c) Approximations of f by a function that equals a polynomial on each interval $[k2^{-3}, (k+1)2^{-3})$. The best approximation of f by a piecewise-constant function is shown in (b); the best by a continuous piecewise-linear function is in (c).

the performance of the approximation scheme on polynomials (because smooth functions can be replaced in estimations, at very little cost, by the polynomials given by their Taylor expansions). In particular, the types of approximation schemes considered here can have order L only if they perfectly reproduce polynomials of degree at most $L - 1$. In other words, there should exist some N_0 such that if p is any polynomial of degree at most $L - 1$ and $N \geqslant N_0$, then $A_N p = p$.

In the Haar case, applied to functions f that differ from zero only between 0 and 1, the building blocks consist of the function φ that takes the value 1 on $[0, 1]$ and 0 outside, together with the families $\{H_{j,k};\ k = 0, \dots, 2^j - 1\}$ for $j = 0, 1, 2, \dots$. We saw above that $P_j^{\text{Haar}}(f)$ can be written as a linear combination of the first $1 + 2^0 + 2^1 + \cdots + 2^{j-1} = 2^j$ building blocks φ, $H_{0,0}, H_{1,0}, H_{1,1}, H_{2,0}, \dots, H_{j-1,2^{j-1}-1}$. Because the Haar wavelets are orthogonal to each other, this is also the linear combination of these basis functions that is closest to f, so that $P_j^{\text{Haar}}(f) = A_{2^j}^{\text{Haar}}$. Figure 7 shows (for $j = 3$) both $A_{2^j}^{\text{Haar}} f$ and $A_{2^j}^{\text{PL}} f$, which is the best approximation of f by a continuous, piecewise-linear function with breakpoints at $k2^{-j}, k = 0, 1, \dots, 2^j - 1$. It turns out that if you are trying to approximate a function f using Haar wavelets, then the best decay you can obtain, even if f is smooth, is of the form $\|f - P_j^{\text{Haar}}(f)\|_2 \leqslant C2^{-j}$, or $\|f - A_N^{\text{Haar}} f\|_2 \leqslant CN^{-1}$ for $N = 2^j$. This means that approximation by Haar wavelets is a *first-order* approximation scheme. Approximation by continuous piecewise-linear functions is a *second-order* scheme: for smooth f, $\|f - A_N^{\text{PL}} f\|_2 \leqslant CN^{-2}$ for $N = 2^j$. Note that the difference between the two schemes can also be seen from the maximal degree d of polynomials they "reproduce" perfectly: clearly both schemes can reproduce constants ($d = 0$); the piecewise-linear scheme can also reproduce linear functions ($d = 1$), whereas the Haar scheme cannot.

Take now any continuously differentiable function f defined on the interval $[0, 1]$. Typically $\|f - P_j^{\text{Haar}}(f)\|_2$ equals about $C2^{-j}$; for an approximation scheme of order 2, that same difference would be about $C'2^{-2j}$. In order to achieve the same accuracy as $P_j^{\text{Haar}}(f)$, the piecewise-linear scheme would thus require only $j/2$ levels instead of j levels. For higher orders L, the gain would be even greater. If the projections P_j gave rise to a higher-order approximation scheme like this, then the difference $P_{j+1}(f)(x) - P_j(f)(x)$ would be so small as not to matter, even for modest values of j, wherever the function f was reasonably smooth; for these values of j, the difference would be important only near points where the function was not as smooth, and so only in those places would a contribution be needed from "difference coefficients" at very fine scales.

This is a powerful motivation to develop a framework similar to that for Haar, but with fancier "generalized averages and differences" corresponding to successive $P_j(f)$ associated with higher-order approximation schemes. This can be done, and was done in an exciting period in the 1980s to which we shall return briefly below. In these constructions, the generalized averages and differences are typically computed by combining more than two finer-scale entries each time, in appropriate linear combinations. The corresponding function decomposition represents functions as (possibly infinite) linear combinations of wavelets $\psi_{j,k}$ derived from a wavelet ψ. As in the case of H, $\psi_{j,k}(x)$ is defined to be $2^{j/2}\psi(2^j x - k)$. Thus, the functions $\psi_{j,k}$ are again normalized translates and dilates of a *single* function; this is due to our using systematically the same averaging operator to go from scale $j + 1$ to scale j, and the same differencing operator to quantify the difference between levels $j + 1$ and j, regardless of the value of j. There is no absolutely compelling reason to use the same averaging and differencing operator for the transition between any two successive levels, and thus to have all the $\psi_{j,k}$ generated by translating and dilating a single function. However, it is very convenient for implementing the transform, and it simplifies the mathematical analysis.

One can additionally require that, like the $H_{j,k}$, the $\psi_{j,k}$ constitute an *orthonormal basis* for the space $L^2(\mathbb{R})$. The *basis* part means that every function can be written as a (possibly infinite) linear combination of the $\psi_{j,k}$; the *orthonormality* means that the $\psi_{j,k}$ are *orthogonal* to each other, except if they are equal, in which case their inner product is 1.

As we have already mentioned, the projections P_j for the wavelet ψ will correspond to an approximation scheme of order L only if they can reproduce perfectly all polynomials of degree less than L. If the functions $\psi_{j,k}$ are orthogonal, then $\int \psi_{j',k}(x) P_j(f)(x)\,\mathrm{d}x = 0$ whenever $j' > j$. The $\psi_{j,k}$ can thus be associated with an approximation scheme of order L only if $\int \psi_{j,k}(x) p(x)\,\mathrm{d}x = 0$ for sufficiently large j and for all polynomials p of degree less than L. By scaling and translating, this reduces to the requirement $\int x^l \psi(x)\,\mathrm{d}x = 0$ for $l = 0, 1, \ldots, L-1$. When this requirement is met, ψ is said to *have L vanishing moments*.

Figure 8 shows the graphs of some choices for ψ that give rise to orthonormal wavelet bases and that are used in various circumstances.

For the wavelets of the type $\psi^{[2n]}$, and thus in particular for $\psi^{[4]}$, $\psi^{[6]}$, and $\psi^{[12]}$ in figure 8, an algorithm similar to that for the Haar wavelet can be used to carry out the decomposition, except that instead of combining two numbers from $P_{j+1,k}$ to obtain an average or a difference coefficient at level j, these wavelet decompositions require weighted combinations of four, six, or twelve finer-level numbers, respectively. (More generally, $2n$ finer-level numbers are used for $\psi^{[2n]}$.)

Because the Meyer and Battle–Lemarié wavelets $\psi^{[M]}$ and $\psi^{[BL]}$ are not concentrated on a finite interval, different algorithms are used for wavelet expansions with respect to these wavelets.

There are many useful orthonormal wavelet bases besides the examples given above. Which one to choose depends on the application one has in mind. For instance, if the function classes of interest in the application have smooth pieces, with abrupt transitions or spikes, then it is advantageous to pick a smooth ψ, corresponding to a high-order approximation scheme. This allows one to describe the smooth pieces efficiently with coarse-scale basis functions, and to leave the fine-scale wavelets to deal with the spikes and

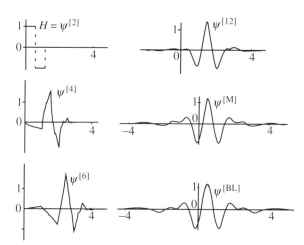

Figure 8 Six different choices of ψ for which the $\psi_{j,k}(x) = 2^{j/2}\psi(2^j x - k)$, $j, k \in \mathbb{Z}$, constitute an orthonormal basis for $L^2(\mathbb{R})$. The Haar wavelet can be viewed as the first example of a family $\psi^{[2n]}$, of which the wavelets for $n = 2$, 3, and 6 are also plotted here. Each $\psi^{[2n]}$ has n vanishing moments and is supported on (i.e., is equal to zero outside) an interval of width $2n - 1$. The remaining two wavelets are not supported on an interval; however, the Fourier transform of the Meyer wavelet $\psi^{[M]}$ is supported on $[-8\pi/3, -2\pi/3] \cup [2\pi/3, 8\pi/3]$; *all* moments of $\psi^{[M]}$ vanish. The Battle–Lemarié wavelet $\psi^{[BL]}$ is twice differentiable, is piecewise polynomial of degree 3, and has exponential decay; it has four vanishing moments.

abrupt transitions. In that case, why not always use a wavelet basis with a very high approximation order? The reason is that most applications require numerical computation of wavelet transforms; the higher the order of the approximation scheme, the more spread out the wavelet, and the more terms have to be used in each generalized average/difference, which slows down numerical computation. In addition, the wider the wavelet, and hence the wider all the finer-scale wavelets derived from it, the more often a discontinuity or sharp transition will overlap with these wavelets. This tends to spread out the influence of such transitions over more fine-scale wavelet coefficients. Therefore, one must find a good balance between the approximation order and the width of the wavelet, and the best balance varies from problem to problem.

There are also wavelet bases in which the restriction of orthonormality is relaxed. In this case one typically uses two different "dual" wavelets ψ and $\tilde{\psi}$, such that $\int_{-\infty}^{\infty} \psi_{j,k}(x) \tilde{\psi}_{j',k'}(x)\,\mathrm{d}x = 0$ unless $j = j'$ and $k = k'$. The approximation order of the scheme that approximates functions f by linear combinations of the $\psi_{j,k}$

is then governed by the number of vanishing moments of $\tilde{\psi}$. Such wavelet bases are called *biorthogonal*. They have the advantage that the basic wavelets ψ and $\tilde{\psi}$ can both be symmetric and concentrated on an interval, which is impossible for orthonormal wavelet bases other than the Haar wavelets.

The symmetry condition is important for image decomposition, where preference is usually given to two-dimensional wavelet bases derived from one-dimensional bases with a symmetric function ψ, a derivation to which we return below. When an image is compressed by deleting or rounding off wavelet coefficients, the difference between the original image I and its compressed version I^{comp} is a combination, with small coefficients, of these two-dimensional wavelets. It has been observed that the human visual system is more tolerant of such small deviations if they are symmetric; the use of symmetric wavelets thus allows for slightly larger errors, which translates to higher compression rates, before the deviations cross the threshold of perception or acceptability.

Another way of generalizing the notion of wavelet bases is to allow more than one starting wavelet. Such systems, known as *multiwavelets*, can be useful even in one dimension.

When wavelet bases are considered for functions defined on the interval $[a, b]$ rather than the whole of $\mathbb{R}$, the constructions are typically adapted, giving bases of *interval wavelets* in which specially crafted wavelets are used near the edges of the interval. It is sometimes useful to choose less regular ways of subdividing intervals than the systematic halving considered above: in this case, the constructions can be adapted to give *irregularly spaced* wavelet bases.

When the goal of a decomposition is compression of the information, as in the image example at the start, it is best to use a decomposition that is itself as efficient as possible. For other applications, such as pattern recognition, it is often better to use *redundant* families of wavelets, i.e., collections of wavelets that contain "too many" wavelets, in the sense that all functions in $L^2(\mathbb{R})$ could still be represented even if one dropped some of the wavelets from the collection. *Continuous wavelet families* and *wavelet frames* are the two main kinds of collections used for such redundant wavelet representations.

4   Wavelets and Function Properties

Wavelet expansions are useful for image compression because many regions of an image do not have features at very fine scales. Returning to the one-dimensional case, the same is true for a function that is reasonably smooth at most but not all points, like the function illustrated in figure 6(a). If we zoom in on such a function near a point x_0 where it is smooth, then it will look almost linear, so we will be able to represent that part of the function efficiently if our wavelets are good at representing linear functions.

This is where wavelet bases other than Haar show their power: the wavelets $\psi^{[4]}$, $\psi^{[6]}$, $\psi^{[12]}$, $\psi^{[M]}$, and $\psi^{[BL]}$ depicted in figure 8 all define approximation schemes of order 2 or higher, so that $\int x\psi_{j,k}(x)\,\mathrm{d}x = 0$ for all j, k. This is also seen in the numerical implementation schemes: the corresponding generalized differencing that computes the wavelet coefficients of f gives a zero result not only when the graph is flat, but also when it is a straight but sloped line, which is not true for the simple differencing used for the Haar basis. As a result, the number of coefficients needed for the wavelet expansion of smooth functions f to reach a preassigned accuracy is much smaller when one uses more sophisticated wavelets than the Haar wavelets.

For a function f that is twice differentiable except at a finite number of discontinuities, and with a basic wavelet that has, say, three vanishing moments, typically only very few wavelets at fine scales will be needed to write a very-high-precision approximation to f. Moreover, those will be needed only near the discontinuity points. This feature is characteristic for all wavelet expansions, whether they are with respect to an orthonormal basis, a basis that is nonorthogonal, or even a redundant family.

Figure 9 illustrates this for one type of redundant expansion, which uses the so-called *Mexican hat wavelets*, which are given by

$$\psi(x) = (2\sqrt{2}/\sqrt{3})\pi^{-1/4}(1 - 4x^2)\mathrm{e}^{-2x^2};$$

this wavelet gets its name from the shape of its graph, which looks like the cross section of a Mexican hat (see the figure).

The smoother a function f is (i.e., the more times it is differentiable), the faster its wavelet coefficients will decay as j increases, provided the wavelet ψ has sufficiently many vanishing moments. The converse statement is also true: one can read off how smooth the function is at x_0 from how the wavelet coefficients $w_{j,k}(f)$ decay, as j increases. Here one restricts attention to the "relevant" pairs (j, k). In other words, one considers only the pairs where $\psi_{j,k}$ is localized near x_0. (In more precise terms, this converse statement can be

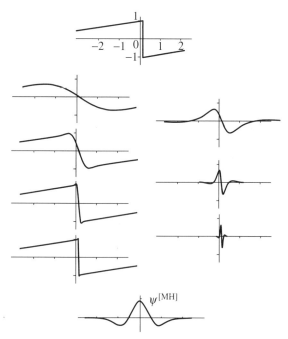

Figure 9 A function with a single discontinuity (top) is approximated by finite linear combinations of Mexican hat wavelets $\psi_{j,l}^{[MH]}$; the graph of $\psi^{[MH]}$ is at the bottom of the figure. Adding finer scales leads to increased precision. Left: successive approximations for $j = 1, 3, 5$, and 7. Right: total contributions from the wavelets at the scales needed to bridge from one j to the next. (In this example, j increases in steps of $\frac{1}{2}$.) The finer the scale, the more the extra detail is concentrated near the discontinuity point.

reformulated as an exact characterization of the so called *Lipschitz spaces* C^α, for all noninteger α that are strictly less than the number of vanishing moments of ψ.)

Wavelet coefficients can be used to characterize many other useful properties of functions, both global and local. Because of this, wavelets are good bases not just for L^2-spaces or the Lipschitz spaces, but also for many other function spaces, such as, for instance, the L^p-spaces with $1 < p < \infty$, the SOBOLEV SPACES [III.29 §2.4], and a wide range of Besov spaces. The versatility of wavelets is partly due to their connection with powerful techniques developed in harmonic analysis throughout the twentieth century.

We have already seen in some detail that wavelet bases are associated with approximation schemes of different orders. So far we have considered approximation schemes in which the $A_N f$ are always linear combinations of the same N building blocks, regard-

less of the function f. This is called *linear* approximation, because the collection of all functions of the form $A_N f$ is contained in the linear span V_N of the first N basis functions. Some of the function spaces mentioned above can be characterized by specifying the decay of $\|f - A_N f\|_2$ as N increases, where A_N is defined in terms of an appropriate wavelet basis.

However, when it is compression that we are interested in, we are really carrying out a different kind of approximation. Given a function f, and a desired accuracy, we want to approximate f to within that accuracy by a linear combination of as few basis functions as possible, but we are not trying to choose those functions from the first few levels. In other words, we are no longer interested in the ordering of the basis functions and we do not prefer one label (j, k) over another.

If we want to formalize this, we can define an approximation $\mathcal{A}_N f$ to be the closest linear combination to f that is made up of at most N basis functions. By analogy with linear approximation, we can then define the set $\mathcal{V}_N$ as the set of all possible linear combinations of N basis functions. However, the sets $\mathcal{V}_N$ are no longer linear spaces: two arbitrary elements $\mathcal{V}_N$ are typically combinations of two different collections of N basis functions, so that their sum has no reason to belong to $\mathcal{V}_N$ (though it will belong to $\mathcal{V}_{2N}$). For this reason, $\mathcal{A}_N f$ is called a *nonlinear* approximation of f.

One can go further and define classes of functions by imposing conditions on the decay of $\|f - \mathcal{A}_N f\|$, as N increases, with respect to some function space norm $\|\cdot\|$. This can of course be done starting from any basis; wavelet bases distinguish themselves from many other bases (such as the trigonometric functions) in that the resulting function spaces turn out to be standard function spaces, such as the Besov spaces, for example. We have referred several times to functions that are smooth in many places but have possible discontinuities in isolated points, and argued that they can be approximated well by linear combinations of a fairly small number of wavelets. Such functions are special cases of elements of particular Besov spaces, and their good approximation properties by sparse wavelet expansions can be viewed as a consequence of the characterization of these Besov spaces by nonlinear approximation schemes using wavelets.[1]

---

1. More types of wavelet families, as well as many generalizations, can be found on the Internet at www.wavelet.org.

5   Wavelets in More than One Dimension

There are many ways to extend the one-dimensional constructions to higher dimensions. An easy way to construct a multidimensional wavelet basis is to combine several one-dimensional wavelet bases. The image decomposition at the start is an example of such a combination: it combines two one-dimensional Haar decompositions. We saw earlier that a 2×2 superpixel could be decomposed as follows. First, think of it as arranged in two rows of two numbers, representing the gray levels of the corresponding pixels. Next, for each row replace its two numbers by their average and their difference, obtaining a new 2×2 array. Finally, do the same process to the columns of the new array. This produces four numbers, the result of, respectively,

- averaging both horizontally and vertically,
- averaging horizontally and differencing vertically,
- differencing horizontally and averaging vertically, and
- differencing both horizontally and vertically.

The first is the average gray level for the superpixel, which is needed as the input for the next round of the decomposition at the next scale up. The other three correspond to the three types of "differences" already encountered earlier. If we start with a rectangular image that consists of 2^K rows, each containing 2^J pixels, then we end up with $2^{K-1} \times 2^{J-1}$ numbers of each of the four types. Each collection is naturally arranged in a rectangle of half the size of the original (in both directions); it is customary in the image-processing literature to put the rectangle with gray values for the superpixels in the top left; the other three rectangles each group together all the differences (or wavelet coefficients) of the other three kinds. (See the level 1 decomposition in figure 10.) The rectangle that results from horizontal differencing and vertical averaging typically has large coefficients at places where the original image has vertical edges (such as the boat masts in the example above); likewise, the horizontal averaging/vertical differencing rectangle has large coefficients for horizontal edges in the original (such as the stripes in the sails); the horizontal differencing/vertical differencing rectangle selects for diagonal features. The three different types of "difference terms" indicate that we have here three basic wavelets (instead of just one in the one-dimensional case).

In order to go to the next round, one scale up, the scenario is repeated on the rectangle that contains the superpixel gray values (the results of averaging both horizontally and vertically); the other three rectangles are left unchanged. Figure 10 shows the result of this process for the original boat image, though the wavelet basis used here is not the Haar basis, but a symmetric biorthogonal wavelet basis that has been adopted in the JPEG 2000 image compression standard. The result is a decomposition of the original image into its component wavelets. The fact that so much of this is gray indicates that a lot of this information can be discarded without affecting the image quality.

Figure 11 illustrates that the number of vanishing moments is important not just when the wavelet basis is used for characterizing properties of functions, but also when it comes to image analysis. It shows an image that has been decomposed in two different ways: once with Haar wavelets, the other with the JPEG 2000 standard biorthogonal wavelet basis. In both cases, all but the largest 5% of the wavelet coefficients have been set to zero, and we are looking at the corresponding reconstructions of the images, neither of which is perfect. However, the wavelet used in the JPEG 2000 standard has four vanishing moments, and therefore gives a much better approximation in smoothly varying parts of the image than the Haar basis. Moreover, the reconstruction obtained from the Haar expansion is "blockier" and less attractive.

6   Truth in Advertising: Closer to True Image Compression

Image compression has been discussed several times in this article, and it is indeed a context in which wavelets are used. However, in practice there is much more to image compression than the simple idea of dropping all but the largest wavelet coefficients, taking the resulting truncated list of coefficients, and replacing each of the many long stretches of zeros by its runlength. In this short section we shall give a glimpse of the large gap between the mathematical theory of wavelets as discussed above and the real-life practice of engineers who want to compress images.

First of all, compression applications set a "bit budget," and all the information to be stored has to fit within the bit budget; statistical estimates and information-theoretic arguments about the class of images under consideration are used to allocate different numbers of bits to different types of coefficients. This bit allocation is much more gradual and subtle than just

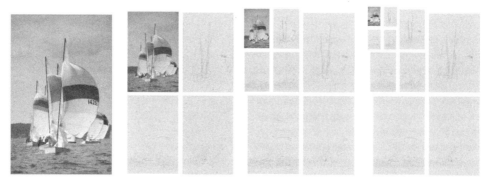

Figure 10 Wavelet decomposition of the boat image, together with a grayscale rendition of the wavelet coefficients. The decompositions are shown after one level of averaging and differencing, as well as after two and three levels. In the rectangles corresponding to wavelet coefficients (i.e., not averaged in both directions), where numbers can be negative, the convention is to use gray scale 128 for zero, and darker/lighter gray scales for positive and negative values. The wavelet rectangles are mostly at gray scale 128, indicating that most of the wavelet coefficients are negligibly small.

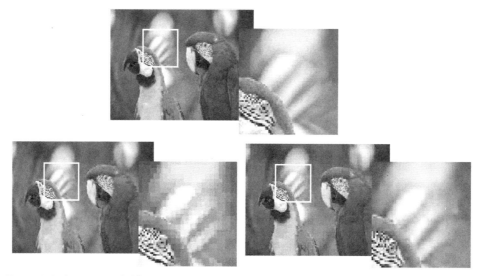

Figure 11 Top: original image, with blowup. Bottom: approximations obtained by expanding the image into a wavelet basis, and discarding the 95% smallest wavelet coefficients. Left: Haar wavelet transform. Right: wavelet transform using the so-called 9–7 biorthogonal wavelet basis.

retaining or dropping coefficients. Even so, many coefficients will get no bits assigned to them, meaning that they are indeed dropped altogether.

Because some coefficients are dropped, care has to be taken that each of the remaining coefficients is given its correct *address*, i.e., its (j, k_1, k_2) label, which is essential for "decompressing" the stored information in order to reconstruct the image (or rather, an approximation to it). If you do not have a good strategy for doing this, then you can easily find that the computational resources needed to encode informa-

tion about the addresses cancel out a large portion of the gain made by the nonlinear wavelet approximation. Every practical wavelet-based image-compression scheme uses some sort of clever approach to deal with this problem. One implementation exploits the observation that at locations in the image where wavelet coefficients of some species are negligibly small at some scale j, the wavelet coefficients of the same species at finer scales are often very small as well. (Check it out on the boat image decomposition given above.) At each such location, this method sets a whole tree of finer-scale coefficients (four for scale $j + 1$, sixteen for scale $j + 2$, etc.) automatically to zero; for those locations where this assumption is not borne out by the wavelet coefficients that are obtained from the actual decomposition of the image at hand, extra bits must then be spent to store the information that a correction has to be made to the assumption. In practice, the bits gained by the "zero-trees" far outweigh the bits needed for these occasional corrections.

Depending on the application, many other factors can play a role. For instance, if the compression algorithm has to be implemented in an instrument on a satellite where it can only draw on very limited power supplies, then it is also important for the computations involved in the transform itself to be as economical as possible.

Readers who want to know more about (important!) considerations of this kind can find them discussed in the engineering literature. Readers who are content to stay at the lofty mathematical level are of course welcome to do so, but are hereby warned that there is more to image compression via wavelet transforms than has been sketched in the previous sections.

7   Brief Overview of Several Influences on the Development of Wavelets

Most of what is now called "wavelet theory" was developed in the 1980s and early 1990s. It built on existing work and insights from many fields, including harmonic analysis (mathematics), computer vision and computer graphics (computer science), signal analysis and signal compression (electrical engineering), coherent states (theoretical physics), and seismology (geophysics). These different strands did not come together all at once but were brought together gradually, often as the result of serendipitous circumstances and involving many different agents.

In harmonic analysis, the roots of wavelet theory go back to work by LITTLEWOOD [VI.79] and Paley in the 1930s. An important general principle in Fourier analysis is that the smoothness of a function is reflected in its FOURIER TRANSFORM [III.27]: the smoother the function, the faster the decay of its transform. Littlewood and Paley addressed the question of characterizing *local* smoothness. Consider, for example, a periodic function with period 1 that has just one discontinuity in the interval $[0, 1)$ (which is then repeated at all integer translates of that point), and is smooth elsewhere. Is the smoothness reflected in the Fourier transform?

If the question is understood in the obvious way, then the answer is no: a discontinuity causes the Fourier coefficients to decay slowly, however smooth the rest of the function is. Indeed, the best possible decay is of the form $|\hat{f}_n| \leqslant C[1 + |n|]^{-1}$. If there were no discontinuity, the decay would be at least as good as $C_k[1 + |n|]^{-k}$ when f is k-times differentiable.

However, there is a more subtle connection between local smoothness and Fourier coefficients. Let f be a periodic function, and let us write its nth Fourier coefficient $\hat{f}_n$ as $a_n e^{i\theta_n}$, where a_n is the absolute value of $\hat{f}_n$ and $e^{i\theta_n}$ is its *phase*. When examining the decay of the Fourier coefficients, we look just at a_n and forget all about the phases, which means that we cannot detect any phenomenon unless it is unaffected by arbitrary changes to the phases. If f has a discontinuity, then we can clearly move it about by changing the phases. It turns out that these phases play an important role in determining not just where the singularities are, but even their severity: if the singularity at x_0 is not just a discontinuity but a divergence of the type $|f(x)| \sim |x - x_0|^{-\beta}$, then one can change the value of β just by changing the phases and without altering the absolute values $|a_n|$. Thus, changing phases in Fourier series is a dangerous thing to do: it can greatly change the properties of the function in question.

Littlewood and Paley showed that *some* changes of the phases of Fourier coefficients are more innocuous. In particular, if you choose a phase change for the first Fourier coefficient, another one for both the next two coefficients, another for the next four, another for the next eight, and so on, so that the phase changes are constant on "blocks" of Fourier coefficients that keep doubling in length, then local smoothness (or absence of smoothness) properties of f are preserved. Similar statements hold for the Fourier transform of functions on $\mathbb{R}$ (as opposed to Fourier series of periodic functions). This was the first result of a whole branch of harmonic analysis in which *scaling* was exploited systematically to deal with detailed local analysis, and in which very powerful theorems were proved that, with hindsight, seem ready-made to establish a host of powerful properties for wavelet decompositions. The simplest way to see the connection between Littlewood–Paley theory and wavelet decompositions is to consider the *Shannon wavelet* $\psi^{[\text{Sh}]}$, which is defined by $\hat{\psi}^{[\text{Sh}]}(\xi) = 1$ when $\pi \leqslant |\xi| < 2\pi$, and $\hat{\psi}^{[\text{Sh}]}(\xi) = 0$ otherwise. Here, $\hat{\psi}^{[\text{Sh}]}$ denotes the Fourier transform of the wavelet $\psi^{[\text{Sh}]}$. The corresponding functions $\psi_{j,k}^{[\text{Sh}]}(x) = 2^{j/2}\psi^{[\text{Sh}]}(2^j x - k)$ constitute an orthonormal basis for

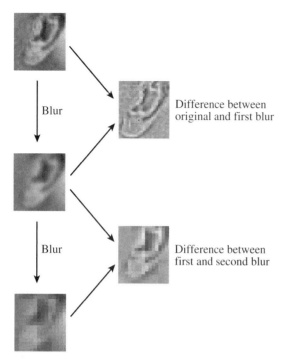

Blur

Difference between
original and first blur

Blur

Difference between
first and second blur

Figure 12 Differences between successive
blurs give detail at different scales.

$L^2(\mathbb{R})$, and for each f and each j the collection of inner products $(\int_{-\infty}^{\infty} f(x)\psi_{j,k}^{[\mathrm{Sh}]}(x)\,\mathrm{d}x)_{k\in\mathbb{Z}}$ tells us how $\hat{f}(\xi)$ restricts to the set $2^{j-1} \leqslant \pi^{-1}|\xi| < 2^j$. In other words, it gives us the jth Littlewood–Paley block of f.

Scaling also plays an important role in computer vision, where one of the basic ways to "understand" an image (going back to at least the early 1970s) is to blur it more and more, erasing more detail each time, so as to obtain approximations that are graded in "coarseness" (see figure 12). Details at different scales can then be found by considering the differences between successive coarsenings. The relationship with wavelet transforms is obvious!

An important class of signals of interest to electrical engineers is that of *bandlimited signals*, which are functions f, usually of one variable only, for which the Fourier transform $\hat{f}$ vanishes outside some interval. In other words, the frequencies that make up f come from some "limited band." If the interval is $[-\Omega, \Omega]$, then f is said to have *bandlimit* Ω. Such functions are completely characterized by their values, often called *samples*, at integer multiples of π/Ω. Most manipulations on the signal f are carried out not directly but by operations on this sequence of samples. For instance, we

might want to restrict f to its "lower-frequency half." To do this, we would define a function g by the condition that $\hat{g}(\xi) = \hat{f}(\xi)$ if $|\xi| \leqslant \Omega/2$ and is 0 otherwise. Equivalently, we could say that $\hat{g}(\xi) = \hat{f}(\xi)\hat{L}(\xi)$, where $\hat{L}(\xi) = 1$ if $|\xi| \leqslant \Omega/2$ and 0 otherwise. The next step is to let L_n be $L(n\pi/\Omega)$, and we find that $g(k\pi/\Omega) = \sum_{n\in\mathbb{Z}} L_n f((k-n)\pi/\Omega)$. To put this more neatly, if we write a_n and $\tilde{b}_n$ for $f(n\pi/\Omega)$ and $g(n\pi/\Omega)$, respectively, then $\tilde{b}_k = \sum_{n\in\mathbb{Z}} L_n a_{k-n}$. On the other hand, g clearly has bandlimit $\Omega/2$, so to characterize g it suffices to know only the sequence of samples at integer multiples of $2\pi/\Omega$. In other words, we just need to know the numbers $b_k = \tilde{b}_{2k}$. The transition from f to g is therefore given by $b_k = \sum_{n\in\mathbb{Z}} L_n a_{2k-n}$. In the appropriate electrical engineering vocabulary, we have gone from a critically sampled sequence for f (i.e., its sampling rate corresponded exactly to its bandlimit) to a critically sampled sequence for g by *filtering* (multiplying $\hat{f}$ by some function, or convolving the sequence $(f(n\pi/\Omega))_{n\in\mathbb{Z}}$ with a sequence of *filter coefficients*) and *downsampling* (retaining only one sample in two, because these are the only samples necessary to characterize the more narrowly bandlimited g). The upper-frequency half h of f can be obtained by the inverse Fourier transform of the restriction of $\hat{f}(\xi)$ to $|\xi| > \Omega/2$. Like g, the function h is also completely characterized by its values at multiples of $2\pi/\Omega$, and h can also be obtained from f by filtering and downsampling. This split of f into its lower and upper frequency halves, or *subbands*, is thus given by formulas that are the exact equivalent of the generalized averaging and differencing encountered in the implementation of wavelet transforms for orthonormal wavelet bases supported on an interval. Subband filtering followed by critical downsampling had been developed in the electrical engineering literature before wavelets came along, but were typically not concatenated in several stages.

A concept of central importance in quantum physics is that of a UNITARY REPRESENTATION [IV.15 §1.4] of a LIE GROUP [III.48 §1] on some HILBERT SPACE [III.37]. In other words, given a Lie group G and a Hilbert space H, one interprets the elements g of G as unitary transformations of H. The elements of H are called *states*, and for certain Lie groups, if $\boldsymbol{v}$ is some fixed state, then the family of vectors $\{g\boldsymbol{v};\ g \in G\}$ is called a *family of coherent states*. Coherent states go back to work by Schrödinger in the 1920s. Their name dates back to the 1950s, when they were used in quantum optics: the word "coherent" referred to the coherence of the light they were describing. These families turned out to be of interest in a much wider range of settings in quantum physics, and the name stuck, even outside the original setting of optics. In many applications it helps not to use the whole family of coherent states but only those coherent states that correspond to a certain kind of discrete subset of G. Wavelets turn out to be just such a subfamily of coherent states: one starts with a single, basic wavelet, and the transformations that convert it (by dilation and translation) into the remaining wavelets form a discrete semigroup of such transformations.

Despite the fact that wavelets synthesized ideas from all these fields, their discovery originated in another area altogether. In the late 1970s, the geophysicist J. Morlet was working for an oil company. Dissatisfied with the existing techniques for extracting special types of signals from seismograms, he came up with an ad hoc transform that combined translations and scalings: nowadays, it would be called a redundant wavelet transform. Other transforms in seismology with which Morlet was familiar involve comparing the seismic traces with special functions of the form $W_{m,n}(t) = w(t - n\tau)\cos(m\omega t)$, where w is a smooth function that gently rises from 0 to 1 and then gently decays to 0 again, all within a finite interval. Several different examples of functions w, proposed by several different scientists, are used in practice: because the functions $W_{m,n}$ look like small waves (they oscillate, but have a nice beginning and end because of w) they are typically called "wavelets of X," named after proposer X for that particular w. The reference functions in Morlet's new ad hoc family, which he used to compare pieces of seismic traces, were different in that they were produced from a function w by *scaling* instead of multiplying them by increasingly oscillating trigonometric functions. Because of this, they always had the same shape, and Morlet called them "wavelets of constant shape" (see figure 13) in order to distinguish them from the wavelets of X (or Y, or Z, etc.).

Morlet taught himself to work with this new transform and found it numerically useful, but had difficulty explaining his intuition to others because he had no underlying theory. A former classmate pointed him in the direction of A. Grossmann, a theoretical physicist, who made the connection with coherent states and, together with Morlet and other collaborators, started to develop a theory for the transform in the early 1980s. Outside the field of geophysics it was no longer necessary to use the phrase "of constant shape," so this was quickly dropped, which annoyed geophysicists when,

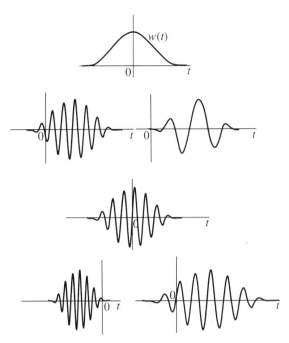

Figure 13 Top: an example of a window function w that is used in practice by geophysicists, with just below it two examples of $w(t - n\tau)e^{imt}$, i.e., two "traditional" geophysics wavelets. Bottom: a wavelet as used by Morlet, with two translates and dilates just below it—these have constant shape, unlike the "traditional" ones.

some years later, more mature forms of wavelet theory impinged on their field again.

A few years later, in 1985, standing in line for a photocopy machine at his university, harmonic analysis expert Y. Meyer heard about this work and realized it presented an interestingly different take on the scaling techniques with which he and other harmonic analysts had long been familiar. At the time, no wavelet bases were known in which the initial function ψ combined the properties of smoothness and good decay. Indeed, there seemed to be a subliminal expectation in papers on wavelet expansions that no such orthonormal wavelet bases could exist. Meyer set out to prove this, and to everyone's surprise and delight he failed in the best possible way—by finding a counterexample, the first smooth wavelet basis! Except that it later turned out not to have been the very first: a few years before, a different harmonic analyst, O. Stromberg, had constructed a different example, but this had not attracted attention at the time.

Meyer's proof was ingenious, and worked because of some seemingly miraculous cancellations, which is always unsatisfactory from the point of view of mathematical understanding. Similar miracles played a role in independent constructions by P. G. Lemarié (now Lemarié-Rieusset) and G. Battle of orthonormal wavelet bases that were piecewise polynomial. (They came to the same result from completely different points of departure—harmonic analysis for Lemarié and quantum field theory for Battle.)

A few months later, S. Mallat, then a Ph.D. candidate in computer vision in the United States, learned about these wavelet bases. He was on vacation, chatting on the beach with a former classmate who was one of Meyer's graduate students. After returning to his Ph.D. work, Mallat kept thinking about a possible connection with the reigning paradigm in computer vision. On learning that Meyer was coming to the United States in the fall of 1986 to give a named lecture series, he went to see him and explain his insight. In a few days of feverish enthusiasm, they hammered out *multiresolution analysis*, a different approach to Meyer's construction inspired by the computer vision framework. In this new setting, all the miracles fell into place as inevitable consequences of simple, entirely natural construction rules, embodying the principle of successively finer approximations. Multiresolution analysis has remained the basic principle behind the construction of many wavelet bases and redundant families.

None of the smooth wavelet bases constructed up to that point was supported inside an interval, so the algorithms to implement the transform (which were using the subband filtering framework without their creators knowing that it had been named and developed in electrical engineering) required, in principle, infinite filters that were impossible to implement. In practice, this meant that the infinite filters from the mathematical theory had to be truncated; it was not clear how to construct a multiresolution analysis that would lead to finite filters. Truncation of the infinite filters seemed to me a blemish on the whole beautiful edifice, and I was unhappy with this state of affairs. I had learned about wavelets from Grossmann and about multiresolution analysis from explanations scribbled by Meyer on a napkin after dinner during a conference. In early 1987 I decided to insist on finite filters for the implementation. I wondered whether a whole multiresolution analysis (and its corresponding orthonormal basis of wavelets) could be reconstructed from appropriate but finite filters. I managed to carry out this program, and as a result found the first construction of an

orthonormal wavelet basis for which ψ is smooth and supported on an interval.

Soon after this, the connection with the electrical engineering approaches was discovered. Especially easy algorithms were inspired by the needs of computer graphics applications. More exciting constructions and generalizations followed: biorthogonal wavelet bases, wavelet packets, multiwavelets, irregularly spaced wavelets, sophisticated multidimensional wavelet bases not derived from one-dimensional constructions, and so on.

It was a heady, exciting period. The development of the theory benefited from all the different influences and in its turn enriched the different fields with which wavelets are related. As the theory has matured, wavelets have become an accepted addition to the mathematical toolbox used by mathematicians, scientists, and engineers alike. They have also inspired the development of other tools that are better adapted to tasks for which wavelets are not optimal.

Further Reading

Aboufadel, E., and S. Schlicker. 1999. *Discovering Wavelets*. New York: Wiley Interscience.

Blatter, C. 1999. *Wavelets: A Primer*. Wellesley, MA: AK Peters.

Cipra, B. A. 1993. Wavelet applications come to the fore. *SIAM News* 26(7):10–11, 15.

Frazier, M. W. 1999. *An Introduction to Wavelets through Linear Algebra*. New York: Springer.

Hubbard, B. B. 1995. *The World According to Wavelets: The Story of a Mathematical Technique in the Making*. Wellesley, MA: AK Peters.

Meyer, Y., and R. Ryan. 1993. *Wavelets: Algorithms and Applications*. Philadelphia, PA: Society for Industrial and Applied Mathematics (SIAM).

Mulcahy, C. 1996. Plotting & scheming with wavelets. *Mathematics Magazine* 69(5):323–43.

VII.4 The Mathematics of Traffic in Networks
Frank Kelly

1 Introduction

We are all familiar with congested roads, and perhaps also with congestion in other networks such as the Internet, so it is obviously important to have a general understanding of how and why congestion occurs in networks. However, the pattern of the flow of traffic through a network is the consequence of a subtle and complex interaction between different users. For example, in a road network we would normally expect each driver to attempt to choose the most convenient route, and this choice will depend upon the delays the driver expects to encounter on different roads; but these delays will in turn depend upon the choices of routes made by others. This mutual interdependence makes it difficult to predict the effects of changes to the system, such as the construction of a new road or the introduction of tolls in certain places.

Related issues arise in other large-scale systems like the telephone network or the Internet. In these systems a major practical concern is the extent to which control can be *decentralized*. When you are browsing the Web, the rate at which a Web page is transferred to you across the network is controlled by software protocols running on your computer and on the Web server hosting the Web page, and not by some huge central computer. This decentralized approach to flow control has been outstandingly successful as the Internet has evolved from a small-scale research network to today's interconnection of hundreds of millions of hosts, but is beginning to show signs of strain. In developing new protocols, the challenge is to understand just which aspects of decentralized flow control are important if the network as a whole is to continue to expand and evolve.

In this article we introduce the reader to some of the mathematical models that have been used to address these issues. The models need to be able to represent several distinct aspects of the system. We shall see that the language of GRAPH THEORY [III.34] and MATRICES [I.3 §4.2] is needed to capture the pattern of connections within the network. Calculus is needed to describe how congestion depends upon traffic volumes. And optimization concepts are needed to model the way in which self-interested drivers choose their shortest routes, or the way that decentralized controls in communication networks can cause the system as a whole to perform well.

2   Network Structure

Figure 1 illustrates a set of three nodes connected by a set of five directed links. We might imagine the nodes as representing towns or locations within a city, and the links as representing road capacity between different nodes. A two-way road is represented by two links, one in each direction. Notice that there are two routes from node c to node a that a driver can choose: the first route,

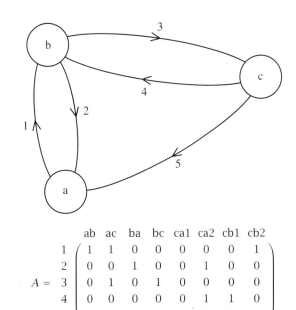

$$A = \begin{array}{c} \\ 1 \\ 2 \\ 3 \\ 4 \\ 5 \end{array} \begin{array}{cccccccc} ab & ac & ba & bc & ca1 & ca2 & cb1 & cb2 \\ \left(\begin{array}{cccccccc} 1 & 1 & 0 & 0 & 0 & 0 & 0 & 1 \\ 0 & 0 & 1 & 0 & 0 & 1 & 0 & 0 \\ 0 & 1 & 0 & 1 & 0 & 0 & 0 & 0 \\ 0 & 0 & 0 & 0 & 0 & 1 & 1 & 0 \\ 0 & 0 & 0 & 0 & 1 & 0 & 0 & 1 \end{array}\right) \end{array}$$

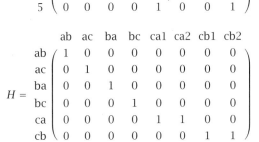

$$H = \begin{array}{c} \\ ab \\ ac \\ ba \\ bc \\ ca \\ cb \end{array} \begin{array}{cccccccc} ab & ac & ba & bc & ca1 & ca2 & cb1 & cb2 \\ \left(\begin{array}{cccccccc} 1 & 0 & 0 & 0 & 0 & 0 & 0 & 0 \\ 0 & 1 & 0 & 0 & 0 & 0 & 0 & 0 \\ 0 & 0 & 1 & 0 & 0 & 0 & 0 & 0 \\ 0 & 0 & 0 & 1 & 0 & 0 & 0 & 0 \\ 0 & 0 & 0 & 0 & 1 & 1 & 0 & 0 \\ 0 & 0 & 0 & 0 & 0 & 0 & 1 & 1 \end{array}\right) \end{array}$$

Figure 1 A simple network and its link-route incidence matrix, A. The matrix H represents which routes serve which source–destination pairs.

let us call it ca1, is the direct route, using link 5; the second route, let us call it ca2, is via node b and uses links 4 and 2.

Let J be the set of directed links and let R be the set of possible routes. One way to describe the relationship between links and routes is with a table, or *matrix*, defined as follows. Set $A_{jr} = 1$ if link j lies on route r, and set $A_{jr} = 0$ otherwise. This defines a matrix $A = (A_{jr}, \ j \in J, \ r \in R)$ called the *link-route incidence matrix*. Each column of the matrix corresponds to one of the routes r, and each row to one of the links j of the network. The column for route r is composed of 0s and 1s: the 1s tell us which links are on route r. As for the rows, the 1s in the row for link j tell us which routes pass through that link. Thus, for example, the incidence matrix in figure 1 has a column for each of the two routes, ca1 and ca2, between node c and node a.

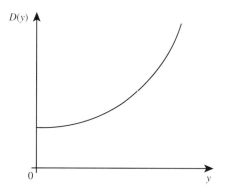

Figure 2 The time taken to travel along a link, $D(y)$, expressed as a function of the total flow y along the link. As the flow increases, congestion effects cause additional delay.

These columns encode the information that route ca1 uses link 5 and that route ca2 uses links 4 and 2. Note that the incidence matrix does not tell us the order of the links on the route. Also the incidence matrix shown does not include all logically possible routes, but it could if we wanted it to. And while we have illustrated a very small network, there is no limit to the number of nodes and links there could be in the network, or to the number of choices of route each driver might have—the incidence matrix would just be bigger.

One quantity of interest in a network is the volume of traffic along a particular route or link. Let x_r be the *flow* on route r, defined as the number of cars per hour that travel along that route. We can list the flows along all the routes in the network as a sequence of numbers $x = (x_r, \ r \in R)$, and we can think of this sequence as a vector. From this vector we can calculate the total flow through a link: for example, the total flow through link 5 in figure 1 is the sum of the flows along routes ca1 and cb2, since these are the routes that pass through link 5. In general, since $A_{jr} = 1$ when a route r passes through link j and $A_{jr} = 0$ when it does not, the total flow through link j, coming from all of the routes that use it, is

$$y_j = \sum_{r \in R} A_{jr} x_r, \quad j \in J.$$

Again, the numbers $(y_j, \ j \in J)$ can be thought of as forming a vector. The above equations can then be represented succinctly in matrix form as

$$y = Ax.$$

We expect the level of congestion at a link to depend on the total flow through the link, and we expect this to influence the time taken to travel along the link. We

shall call this time the *delay*. Figure 2 shows a typical way in which the delay might depend on the amount of flow. At small values of the flow y the delay $D(y)$ is just the time taken to travel along an empty road; for larger values of y the delay $D(y)$ is larger, and quite possibly *much* larger, owing to congestion effects.[1]

Let $D_j(y_j)$ be the delay along link j when the flow through that link is y_j; the nature of this delay may depend upon characteristics of link j such as its length and width, so we have to use the subscript j on the function D_j to indicate that the functions for the various links can be different.

2.1 Routing Choices

Given two nodes in a network there will in general be a variety of possible routes capable of linking them. For example, in figure 1 we have seen that the incidence matrix A records two routes between nodes c and a. The pair ca is an example of a *source-destination pair*. Flow originating from source c and destined for node a can use either ca1 or ca2, the two routes that serve this source-destination pair. We now need another matrix, this time to describe the relationship between source-destination pairs and routes. Let us use s to denote a typical source-destination pair, and let S be the set of all source-destination pairs. Then, for each source-destination pair s and each route r, let $H_{sr} = 1$ if s can be served by the route r, and let $H_{sr} = 0$ otherwise. This defines a matrix $H = (H_{sr}, \ s \in S, \ r \in R)$; figure 1 gives an example. Observe that the row labeled ca has 1s for the two routes, $r = \text{ca1}, \text{ca2}$, that serve the source-destination pair $s = \text{ca}$. Each column of H corresponds to a route, and contains a single 1: this identifies the source-destination pair served by the route. For each route r let us write $s(r)$ for the source-destination pair served by r: for example, in figure 1, $s(\text{ac}) = \text{ac}$ and $s(\text{ca1}) = \text{ca}$.

From the vector $x = (x_r, \ r \in R)$ we can calculate the total flow from a source to a destination: for example, the flow from node c to node a in figure 1 is the sum of

---

1. The graph shown in figure 2 is single valued. It is quite possible for the curve representing delay as a function of flow to bend back upon itself, so that higher delays than shown in the graph correspond to flows *smaller* than the maximum flow shown there. You are in this part of the graph when you experience stop-start driving conditions on a congested but otherwise incident-free highway. Part of the aim of traffic management is to keep flows and delays away from this part of the graph, which we will not consider further.

We will assume that the graph is increasing and smooth, which will make our use of calculus later more straightforward. Formally, we shall assume that $D(y)$ is a continuously differentiable and strictly increasing function of its argument y, as in the graph shown in figure 2.

flows along routes ca1 and ca2, since from the matrix H we see that these are the routes that serve the source-destination pair ca. More generally, if f_s is the total flow of traffic added up over all of the routes serving source-destination pair s, then

$$f_s = \sum_{r \in R} H_{sr} x_r, \quad s \in S.$$

Thus the vector $f = (f_s, s \in S)$ of source-destination flows can be expressed succinctly in matrix form as $f = Hx$.

3   Wardrop Equilibria

We are now able to approach the central issue: how do the traffic flows between the various sources and destinations distribute themselves over the links of the network? Each driver will try to use whatever route is quickest, but this may make other routes quicker or slower and cause other drivers to change their routes. Only when they cannot find alternative, quicker routes will drivers not have an incentive to change routes. What does this mean mathematically?

Let us first calculate the time taken for a driver to travel along route r. The column labeled r of the matrix A tells us which links j are on route r. If we add up the delays on each of these links, we get the time taken to travel along route r as the expression

$$\sum_{j \in J} D_j(y_j) A_{jr}.$$

Now the driver using route r could have used any other route that served the same source-destination pair $s(r)$. So, for the driver to be content with route r, we require

$$\sum_{i \in I} D_j(y_j) A_{jr} \leqslant \sum_{i \in I} D_j(y_j) A_{jr'}$$

for every other route r' that serves the same source-destination pair $s(r)$.

Define a *Wardrop equilibrium* (Wardrop 1952) to be a vector $x = (x_r, r \in R)$ of nonnegative numbers such that for every pair of routes r, r' serving the same source-destination pair,

$$x_r > 0 \Rightarrow \sum_{j \in J} D_j(y_j) A_{jr} \leqslant \sum_{j \in J} D_j(y_j) A_{jr'},$$

where $y = Ax$. The inequality expresses the defining characteristic of a Wardrop equilibrium: that if a route r is actively used, then it achieves the minimum delay over all routes serving its source-destination pair $s(r)$.

Does a Wardrop equilibrium exist? It is not at all clear whether it is possible to find a vector x such that all of the above inequalities, for the various routes through the network, are satisfied simultaneously. To answer the question, we shall proceed by addressing a seemingly different question: what is the answer to the following optimization problem?

$$\text{Minimize} \quad \sum_{j \in J} \int_0^{y_j} D_j(u)\, du$$

$$\text{over} \quad x \geqslant 0,\, y,$$

$$\text{subject to} \quad Hx = f,\, Ax = y.$$

Let us see in outline why this optimization problem has a solution (x, y), and why, if (x, y) is a solution, the vector x is a Wardrop equilibrium.

The optimization problem has some aspects that are quite natural. An obvious constraint is that the flows along each route are nonnegative, which is why we insist that $x \geqslant 0$. The constraints $Hx = f$, $Ax = y$ just enforce the accounting rules we have seen earlier—the rules that allow the source-destination flows f and the link flows y to be calculated from the route flows x using the matrices H and A, respectively. We view the source-destination flows f as fixed, to be distributed over the various routes. Given a choice of f, our task is then to find the route flows x and consequently the link flows y. At a solution to the optimization problem y will be nonnegative, since x is.

This much is fairly natural, but the function to be minimized looks somewhat strange. Its importance rests on the fact that the rate of change of the integral

$$\int_0^{y_j} D_j(u)\, du$$

with respect to y_j is $D_j(y_j)$, by THE FUNDAMENTAL THEOREM OF CALCULUS [I.3 §5.5], and the function to be minimized is the sum of these integrals over all links. We shall see that the link between Wardrop equilibria and the optimization problem is a direct consequence of this observation.

To find a solution to the optimization problem, we will use the method of LAGRANGE MULTIPLIERS [III.64]. Define the function

$$L(x, y; \lambda, \mu)$$
$$= \sum_{j \in J} \int_0^{y_j} D_j(u)\, du + \lambda \cdot (f - Hx) - \mu \cdot (y - Ax),$$

where $\lambda = (\lambda_s, s \in S)$, $\mu = (\mu_j, j \in J)$ are vectors of Lagrange multipliers, to be fixed later. The idea is that if we make the right choices of Lagrange multipliers, the minimization of the function L over x and y will find a solution to the original problem. The reason this works is that, for the right choices of Lagrange multipliers, the constraints $Hx = f$ and $Ax = y$ are consistent with the minimization of L.

To minimize the function L we need to differentiate. First,

$$\frac{\partial L}{\partial y_j} = D_j(y_j) - \mu_j.$$

Second,

$$\frac{\partial L}{\partial x_r} = -\lambda_{s(r)} + \sum_{j \in J} \mu_j A_{jr}.$$

Note that the form of the matrix H causes the derivative with respect to x_r to pick out exactly one component of λ, namely $\lambda_{s(r)}$, and the form of the matrix A causes the derivative to pick out just those components of μ that correspond to links on route r. These derivatives allow us to deduce that a minimum of L, over all $x \geqslant 0$ and all y, occurs when

$$\mu_j = D_j(y_j)$$

and

$$\begin{aligned} \lambda_{s(r)} &= \sum_{j \in J} \mu_j A_{jr} \quad \text{if } x_r > 0 \\ &\leqslant \sum_{j \in J} \mu_j A_{jr} \quad \text{if } x_r = 0. \end{aligned}$$

The equality condition for $\lambda_{s(r)}$ is straightforward: if $x_r > 0$ then small variations up or down in x_r should not decrease the function $L(x, y; \lambda, \mu)$, and hence we deduce that the partial derivative with respect to x_r must be zero. But if $x_r = 0$ then we can only vary x_r upward, and so all we can deduce is that the partial derivative with respect to x_r is nonnegative, and from this we deduce the inequality condition for $\lambda_{s(r)}$.

Minimizing the function L corresponds to allowing the constraints $Hx = f$, $Ax = y$ to be violated, but at a cost: now one charges a price λ_s for any shortfall of the sum $\sum_{j \in J} A_{jr} x_r$ below f_s and a price μ_j for any excess of the sum $\sum_{j \in J} A_{jr} x_r$ over y_j. From general results on convex optimization it is known that there exist Lagrange multipliers (λ, μ) and a vector (x, y) such that (x, y) minimizes $L(x, y; \lambda, \mu)$, satisfies the constraints $Hx = f$, $Ax = y$, and solves the original optimization problem.

Our solution for the Lagrange multipliers shows that they have a simple interpretation: μ_j is the delay on link j and λ_s is the minimum delay over all routes serving the node pair s. The various conditions established for the multipliers thus show that an optimum of the function L, known as the *objective function*, corresponds precisely to a Wardrop equilibrium.

Thus if traffic in the network distributes itself in accordance with the self-interested choices of drivers, the equilibrium flows (x, y) will solve an optimization problem. This result is originally due to Beckmann et al. (1956), and it provides a remarkable insight into the equilibrium patterns achieved in road traffic networks. The pattern of traffic resulting from the individual decisions of a large number of self-interested drivers behaves as if a central intelligence were directing flows to optimize a certain (rather strange) objective function.

The result does *not* mean that average delays in the network will be minimal: a striking illustration of this fact is provided by Braess's paradox (Braess 1968), which we describe next.

4 Braess's Paradox

Consider the network illustrated in figure 3(a). Cars travel from node S to node N, via either node W or node E. The total flow is 6, and the link delays $D_j(y)$ are given next to the links in the figure. One can imagine the figure illustrating rush hour as commuters travel from the center of a city in the south to their homes in the north. Commuters learn from experience what the delays are likely to be along the eastern and western routes. The distribution of traffic shown is the Wardrop equilibrium: there is no incentive for any drivers to change their routes, since the two possible routes incur the same delay, namely $(10 \times 3) + (3 + 50) = 83$ units of time. Now suppose that a new link is added, between nodes W and E, as shown in figure 3(b). Traffic is attracted onto the new link, since to begin with it offers a shorter journey time from the south to the north. Eventually, after everyone knows about the new link and traffic patterns have settled down, a new Wardrop equilibrium will be established, and this is shown in figure 3(b). In the new equilibrium there are three routes used, which each incur the same delay, namely $(10 \times 4) + (2 + 50) = (10 \times 4) + (2 + 10) + (10 \times 4) = 92$. Thus in figure 3(b) each car incurs a delay of 92, while in figure 3(a) the delay of each car was only 83. Adding the new link has increased everyone's delay!

The explanation for this apparent paradox is as follows. At a Wardrop equilibrium each driver is using a route which, given the choices of others, gives the minimum delay over the routes available between that driver's source and destination. But there is no intrinsic reason why this equilibrium should correspond to particularly low delays relative to what could be achieved

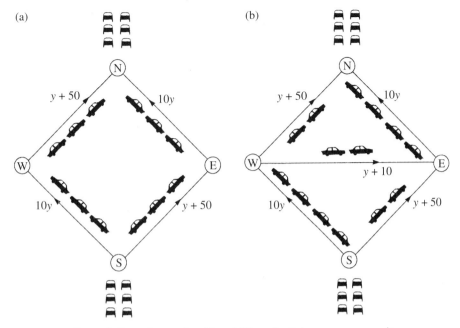

Figure 3 Braess's paradox. The addition of a link causes everyone's journey time to lengthen. (After Braess (1968) and Cohen (1988).)

by another flow pattern. If all drivers could be encouraged to depart from their own self-interested choices, it is quite possible that all might benefit. And in the above example, if all drivers in the second network could agree to avoid the new link, effectively converting the network back into the first network, then all would incur lower delays.

To explore the point further, note that the product of the flow y_j and the delay $D_j(y_j)$ is the delay incurred at link j per unit time, aggregated over all the vehicles using link j. Let us try to find the flow pattern that minimizes the total delay per unit time, summed over the entire network. Consider then the following problem.

Minimize $\sum_{j \in J} y_j D_j(y_j)$

over $x \geqslant 0,\ y,$

subject to $Hx = f,\ Ax = y.$

Note that the problem is of the same form as the earlier optimization problem, but the function to be minimized now measures the total network delay per unit time. (Recall that the function to be minimized in the first optimization problem seemed initially to be rather arbitrary, with its eventual motivation being that its minimization was achieved by a Wardrop equilibrium.)

Again define the function

$$L(x, y; \lambda, \mu)$$
$$= \sum_{j \in J} y_j D_j(y_j) + \lambda \cdot (f - Hx) - \mu \cdot (y - Ax).$$

Again

$$\frac{\partial L}{\partial x_r} = -\lambda_{s(r)} + \sum_{j \in J} \mu_j A_{jr},$$

but now

$$\frac{\partial L}{\partial y_j} = D_j(y_j) + y_j D_j'(y_j) - \mu_j.$$

Hence a minimum of L over $x \geqslant 0$ and y occurs when

$$\mu_j = D_j(y_j) + y_j D_j'(y_j)$$

and

$$\lambda_{s(r)} = \sum_{j \in J} \mu_j A_{jr} \quad \text{if } x_r > 0$$
$$\leqslant \sum_{j \in J} \mu_j A_{jr} \quad \text{if } x_r = 0.$$

The Lagrange multipliers now have a more sophisticated interpretation. Suppose that, in addition to the delay $D_j(y_j)$, users of link j incur a traffic-dependent toll

$$T_j(y_j) = y_j D_j'(y_j).$$

Then μ_j is the *generalized cost* of using link j, defined as the sum of the toll and the delay, and λ_s is the minimum generalized cost over all routes serving the node pair s. If users select routes in an attempt to minimize the sum of their tolls and their delays, then they will produce a flow pattern that minimizes total delay in the network. Notice that the generalized cost μ_j is $(\partial/\partial y_j)(y_j D(y_j))$, which is the rate of increase in the total delay at link j as the flow y_j is increased. So the assumption now is that, in a certain sense, drivers try to minimize their contribution to the total delay rather than minimizing their own delay.

We have seen that if drivers attempt to minimize their own delay, then the resulting equilibrium flows will minimize a certain objective function defined for the network. However, the objective function is certainly not the total network delay, and thus there is no guarantee that when capacity is added to a network the situation is improved. We have also seen that, with the imposition of appropriate tolls, it is possible for the self-interested behavior of drivers to lead to an equilibrium pattern of flow that minimizes total delay. A major challenge for governments and transport planners is to understand how insights from these and more sophisticated models might be used to encourage more efficient development and use of road networks (Department for Transport 2004).

5 Flow Control in the Internet

When a file is requested over the Internet, the computer that hosts that file breaks it into small packets of data that are then transferred across the network by the *transmission control protocol* of the Internet, or TCP. The rate at which packets enter the network is controlled by TCP, which is implemented as software on the two computers that are the source and destination of the data. The general approach is as follows (Jacobson 1988). When a link within the network becomes overloaded, one or more packets are lost; loss of a packet is taken as an indication of congestion, the destination informs the source, and the source slows down. TCP then gradually increases its sending rate until it again receives an indication of congestion. This cycle of increase and decrease enables the source computers to discover and use the available capacity, and to share it between different flows of packets.

TCP has been outstandingly successful as the Internet has evolved from a small-scale research network to today's interconnection of hundreds of millions of endpoints and links. This in itself is a striking observation.

Each of a large but indeterminate number of flows is controlled by a feedback loop that can know only of that flow's experience of congestion. A flow does not know how many other flows are sharing a link on its route, or even how many links are on its route. The links vary in capacity by many orders of magnitude, as do the numbers of flows sharing different links. It is remarkable that so much has been achieved in such a rapidly growing and heterogeneous network with congestion controlled just at the endpoints. Why does this algorithm work so well?

In recent years theoreticians have shed some light on TCP's success, by interpreting the protocol as a decentralized parallel algorithm that solves an optimization problem, just as the decentralized choices of drivers in a road network solve an optimization problem. We shall outline the argument, beginning with a more detailed description of TCP.[2]

Packets transferred by TCP across the Internet contain *sequence numbers* indicating their order, and they should arrive at their destination in that order. When a packet is received at the destination, it is acknowledged: an acknowledgment is a short packet sent by the destination back to the source. If a packet has been lost in the transfer, the source can tell this from the sequence numbers contained in the acknowledgments. The source keeps a copy of each packet sent until it has been positively acknowledged; these copies form what is called a *sliding window*, and allow packets lost in transfer to be sent again by the source.

Meanwhile, stored in the source computer there is a numerical variable known as the *congestion window* and denoted cwnd. The congestion window directs the size of the sliding window in the following sense: if the size of the sliding window is less than cwnd, then the computer increases it by sending out a packet; if it is greater than or equal to cwnd, then it waits for positive acknowledgments to come in, which have the effect of reducing the size of the sliding window and, as we shall see, increasing cwnd as well. Thus, the size of the sliding window continually changes, moving in the direction of a target size that is given by the congestion window.

The congestion window itself is not a fixed number: rather, it is constantly being updated, and the precise rules for how this is done are critical for TCP's sharing of capacity. The rules currently used are as follows.

---

2. Even our detailed description of TCP is simplified, concerning just the congestion-avoidance part of the protocol and omitting discussion of timeouts or of reactions to multiple congestion indication signals received within a single round-trip time.

Every time a positive acknowledgment comes in, cwnd is increased by cwnd^{-1}, and every time a lost packet is detected, cwnd is halved.[3] Thus, if the source computer detects a lost packet, it realizes that there has been some congestion and backs off for a while, but if all its packets are getting through then it allows the rate at which it sends packets to inch up again.

If p is the probability that a packet is lost, then with probability $1 - p$ the congestion window will increase by cwnd^{-1} and with probability p it will decrease by $\frac{1}{2}\mathsf{cwnd}$. The expected change in the congestion window cwnd per update step is therefore

$$\mathsf{cwnd}^{-1}(1 - p) - \tfrac{1}{2}\mathsf{cwnd}\,p.$$

The expected change will be positive for small values of cwnd, but will become negative if cwnd is big enough. We might therefore expect an equilibrium for cwnd to arise when the expression is zero: that is, when

$$\mathsf{cwnd} = \sqrt{\frac{2(1 - p)}{p}}.$$

Now let us see how this calculation can be extended to networks. Suppose that a network consists of a set of nodes connected by directed links, like the network illustrated in figure 1. As earlier, let J be the set of directed links, let R be the set of routes, and let $A = (A_{jr},\ j \in J,\ r \in R)$ be the link-route incidence matrix. When a request reaches a computer in this network, that computer will set up a congestion window for the flow of packets that will result. Since there will be many different such congestion windows, they need to be labeled, and it is convenient to label them with the route that will be used for the flow. (Exactly how these flows are routed is a complicated and important question, but one that we shall not discuss here.) So, for each route r that is being used, let cwnd_r be the congestion window for that route. Let T_r be the round-trip time for the route r: that is, the time between the sending out of a packet and the receiving of an acknowledgment for it.[4] Finally, define a variable x_r to be cwnd_r/T_r.

Now at any given time the sliding window consists of those packets that have been sent but not acknowledged. Therefore, if a packet has just been acknowledged and its round-trip has taken time T_r, the sliding window consists of all packets sent out in the last T_r time units. Since the source computer is aiming for the number of such packets to be about cwnd_r, we can interpret x_r to be the rate at which packets are transferred over route r. Thus, the numbers x_r form a flow vector that is closely analogous to the traffic flow vector discussed earlier.

As we did then, let us define a vector $y = Ax$, so that y_j is the total flow through link j, obtained by summing x_r over each route r that passes through link j. Let p_j be the proportion of packets that are lost, or "dropped," at link j. We expect p_j to be related to y_j, the total flow through link j, as follows. If y_j is less than the capacity C_j of link j, then p_j will be zero; there will be no dropped packets at link j if the link is not full. And if $p_j > 0$ then $y_j = C_j$; if packets are dropped then the link is full. If we assume that the proportions of packets dropped at links are small, then the probability that a packet is lost on route r is approximately

$$p_r = \sum_{j \in J} p_j A_{jr}.$$

(The exact formula would be $(1 - p_r) = \prod_{j \in J}(1 - p_j)^{A_{jr}}$, but when the p_j are small we can ignore their products.) Since $x_r = \mathsf{cwnd}_r/T_r$, our earlier calculation of cwnd now gives us that

$$x_r = \frac{1}{T_r}\sqrt{\frac{2(1 - p_r)}{p_r}}.$$

Is it possible to choose the rates $x = (x_r,\ r \in R)$ and the drop probabilities $p = (p_j,\ j \in J)$ in a consistent fashion, so that the last two equations are satisfied and either p_j is zero or $y_j = C_j$ for each $j \in J$? The remarkable observation is that such a choice corresponds precisely to the solution of the following optimization problem (Kelly 2001; Low et al. 2002).

$$\text{Maximize} \quad \sum_{r \in R} \frac{\sqrt{2}}{T_r} \arctan\left(\frac{x_r T_r}{\sqrt{2}}\right)$$
$$\text{over} \quad x \geqslant 0,$$
$$\text{subject to} \quad Ax \leqslant C.$$

Some aspects of this optimization problem are as we might expect: in particular, the inequality $Ax \leqslant C$ simply adds up the flows through link j and requires that the sum not exceed the capacity C_j of link j, for each link $j \in J$. But, as before, the function being optimized is undoubtedly strange. The arctan function, illustrated

---

3. These increase and decrease rules may appear rather mysterious, and indeed it is only recently that many of their macroscopic consequences have begun to be understood. The rules have worked well for more than a decade, but they are now beginning to show signs of age, and much current research is aimed at understanding the full consequences of changing them.

4. The round-trip time comprises the time taken for a packet to travel along links, called the propagation delay, together with processing times and queueing delays at nodes. Processing times and queueing delays tend to decrease with increasing computer speeds, but the finite speed of light places a fundamental lower bound on propagation delays. We shall treat the round-trip time for a route as a constant. Hence, we assume that congestion at a link makes itself felt by packet loss rather than additional packet delay.

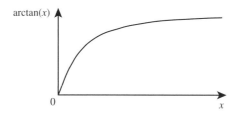

Figure 4 The arctan function. The Internet's TCP implicitly maximizes a sum of utilities over all the connections present in a network: this function shows the shape of the utility function for a single connection. The horizontal axis is proportional to the rate of the connection, and the vertical axis is proportional to the usefulness of that rate. Both axes are scaled in terms of the round-trip time of the connection.

in figure 4, is the inverse function to the trigonometric function tan, and can also be defined as

$$\arctan(x) = \int_0^x \frac{1}{1+u^2}\, du.$$

From this form, we see that its derivative with respect to x is $1/(1+x^2)$.

Let us sketch the relationship between the optimization problem and the equilibrium rates and drop probabilities. Define the function

$$L(x, z; \mu)$$
$$= \sum_{r \in R} \frac{\sqrt{2}}{T_r} \arctan\left(\frac{x_r T_r}{\sqrt{2}}\right) + \mu \cdot (C - Ax - z),$$

where $\mu = (\mu_j,\ j \in J)$ is a vector of Lagrange multipliers, and $z = C - Ax$ is a vector of *slack variables*, measuring the spare capacity on each of the links $j \in J$ of the network. Then, using the derivative of the arctan function,

$$\frac{\partial L}{\partial x_r} = (1 + \tfrac{1}{2} x_r^2 T_r^2)^{-1} - \sum_{j \in J} \mu_j A_{jr} \quad \text{and} \quad \frac{\partial L}{\partial z_j} = -\mu_j.$$

We look for a maximum of L over $x, z \geqslant 0$; it turns out that this maximum is, under the identification $\mu_j = p_j$, precisely the collection $(x_r,\ r \in R)$, $(p_j,\ j \in J)$ of rates and drop probabilities that we were looking for. For example, setting to zero the partial derivative with respect to x_r gives the desired equation for x_r.

In summary, for each link $j \in J$ the Lagrange multiplier μ_j arising from the optimization problem is precisely the proportion p_j of packets dropped at that link, much as the Lagrange multipliers arising earlier were precisely the delays on links of a road traffic network. And the equilibrium reached by the interaction of many competing TCPs, each implemented only on the source and destination computers, is effectively maximizing

an objective function for the entire network. The objective function has a surprising interpretation: it is as if the usefulness of the flow rate x_r to the source–destination pair served by this route is given by a *utility function*

$$\frac{\sqrt{2}}{T_r} \arctan\left(\frac{x_r T_r}{\sqrt{2}}\right),$$

and the network is attempting to maximize the sum of these utility functions across all source–destination pairs, subject to constraints arising from the limited capacities of the links.

The arctan function, illustrated in figure 4, is *concave*. Thus, if two or more connections share an overloaded link, the rates achieved will be approximately equal, since otherwise the total utility could be increased by reducing the largest rate a little and increasing the smallest rate a little. As a result, there is a tendency for TCP to share resources more or less equitably. This is very different from resource-control mechanisms in traditional telephone networks where, if the network is overloaded, some calls are blocked in order that the calls that are accepted are unaffected by the overload.

6 Conclusion

The behavior of large-scale systems has been of great interest to mathematicians for over a century, with many examples coming from physics. For example, the behavior of a gas can be described at the microscopic level in terms of the position and velocity of each molecule. At this level of detail a molecule's velocity appears as a random process, as the molecule bounces around off other molecules and the walls of the container. Yet consistent with this detailed microscopic description of the system is macroscopic behavior best described by quantities such as temperature and pressure. Similarly, the behavior of electrons in an electrical network can be described in terms of random walks, and yet this simple description at the microscopic level leads to rather sophisticated behavior at the macroscopic level: Kelvin showed that the pattern of potentials in a network of resistors is exactly the one that minimizes heat dissipation for a given level of current flow (Kelly 1991). The local, random behavior of the electrons causes the network as a whole to solve a rather complex optimization problem.

In the last fifty years we have begun to realize that large-scale engineered systems are often best understood in similar terms. Thus a microscopic description of traffic flow in terms of each driver's choice of the most convenient route can be consistent with macroscopic behavior described in terms of a function minimization. And the simple, local rules that control how packets are transmitted through the Internet can correspond to a maximizing of aggregate utility across the entire network.

One thought-provoking difference is that, whereas the microscopic rules governing physical systems are fixed, for engineered systems such as transport or communication networks we may be able to choose the microscopic rules so as to achieve the macroscopic consequences we judge desirable.

Further Reading

Beckmann, M., C. B. McGuire, and C. B. Winsten. 1956. *Studies in the Economics of Transportation.* Cowles Commission Monograph. New Haven, CT: Yale University Press.

Braess, D. 1968. Über ein Paradoxon aus der Verkehrsplanung. *Unternehmenforschung* 12:258–68.

Cohen, J. E. 1988. The counterintuitive in conflict and cooperation. *American Scientist* 76:576–84.

Department for Transport. 2004. Feasibility study of road pricing in the UK. Available from www.dft.gov.uk.

Jacobson, V. 1988. Congestion avoidance and control. *Computer Communication Review* 18(4):314–29.

Kelly, F. P. 1991. Network routing. *Philosophical Transactions of the Royal Society of London* A 337:343–67.

———. 2001. Mathematical modeling of the Internet. In *Mathematics Unlimited—2001 and Beyond*, edited by B. Engquist and W. Schmid, pp. 685–702. Berlin: Springer.

Low, S. H., F. Paganini, and J. C. Doyle. 2002. Internet congestion control. *IEEE Control Systems Magazine* 22: 28–43.

Wardrop, J. G. 1952. Some theoretical aspects of road traffic research. *Proceedings of the Institute of Civil Engineers* 1: 325–78.

VII.5   The Mathematics of Algorithm Design
Jon Kleinberg

1   The Goals of Algorithm Design

When computer science began to emerge as a subject at universities in the 1960s and 1970s, it drew some amount of puzzlement from the practitioners of more established fields. Indeed, it is not initially clear why computer science should be viewed as a distinct academic discipline. The world abounds with novel technologies, but we do not generally create a separate field around each one; rather, we tend to view them as by-products of existing branches of science and engineering. What is special about computers?

Viewed in retrospect, such debates highlighted an important issue: computer science is not so much about the computer as a specific piece of technology as it is about the more general phenomenon of computation itself, the design of processes that represent and manipulate information. Such processes turn out to obey their own inherent laws, and they are performed not only by computers but by people, by organizations, and by systems that arise in nature. We will refer to these computational processes as *algorithms*. For the purposes of our discussion in this article, one can think of an algorithm informally as a step-by-step sequence of instructions, expressed in a stylized language, for solving a problem.

This view of algorithms is general enough to capture both the way a computer processes data and the way a person performs calculations by hand. For example, the rules for adding and multiplying numbers that we learn as children are algorithms; the rules used by an airline company for scheduling flights constitute an algorithm; and the rules used by a search engine like Google for ranking Web pages constitute an algorithm. It is also fair to say that the rules used by the human brain to identify objects in the visual field constitute a kind of algorithm, though we are currently a long way from understanding what this algorithm looks like or how it is implemented on our neural hardware.

A common theme here is that one can reason about all these algorithms without recourse to specific computing devices or computer programming languages, instead expressing them using the language of mathematics. In fact, the notion of an algorithm as we now think of it was formalized in large part by the work of mathematical logicians in the 1930s, and algorithmic reasoning is implicit in the past several millennia of mathematical activity. (For example, equation-solving methods have always tended to have a strong algorithmic flavor; the geometric constructions of the ancient Greeks were inherently algorithmic as well.) Today, the mathematical analysis of algorithms occupies a central position in computer science; reasoning about algorithms independently of the specific devices on which they run can yield insight into general design principles and fundamental constraints on computation.

At the same time, computer-science research struggles to keep two diverging views in focus: this more abstract view that formulates algorithms mathematically, and the more applied view that the public generally associates with the field, the one that seeks to develop applications such as Internet search engines, electronic banking systems, medical imaging software, and the host of other creations we have come to expect from computer technology. The tension between these two views means that the field's mathematical formulations are continually being tested against their implementation in practice; it provides novel avenues for mathematical notions to influence widely used applications; and it sometimes leads to new mathematical problems motivated by these applications.

The goal of this short article is to illustrate this balance between the mathematical formalism and the motivating applications of computing. We begin by building up to one of the most basic definitional questions in this vein: how should we formulate the notion of *efficient* computation?

2   Two Representative Problems

To make the discussion of efficiency more concrete, and to illustrate how one might think about an issue like this, we first discuss two representative problems—both fundamental in the study of algorithms—that are similar in their formulation but very different in their computational difficulty.

The first in this pair is the *traveling salesman problem* (TSP), which is defined as follows. We imagine a salesman contemplating a map with n cities (he is currently located in one of them). The map gives the distance between each pair of cities, and the salesman wishes to plan the shortest possible tour that visits all n cities and returns to the starting point. In other words, we are seeking an algorithm that takes as input the set of all distances among pairs of cities, and produces a tour of minimum total length. Figure 1(a) depicts the optimal solution to a sample input instance of the TSP; the circles represent the cities, the dark lines (with lengths labeling them) connect cities that the salesman visits consecutively on the tour, and the lighter lines connect all the other pairs of cities, which are not visited consecutively.

A second problem is the *minimum spanning tree problem* (MSTP). Here we imagine a construction firm with access to the same map of n cities, but with a different goal in mind. They wish to build a set of roads connecting certain pairs of the cities on the map, so that after these roads are built there is a route from each of the n cities to each other one. (A key point here is that each road must go directly from one city

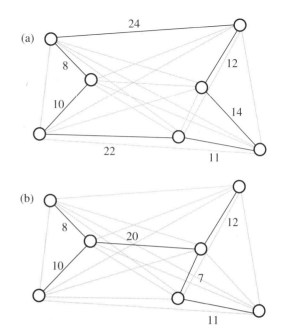

Figure 1 Solutions to instance of (a) the traveling salesman problem and (b) the minimum spanning tree problem, on the same set of points. The dark lines indicate the pairs of cities that are connected by the respective optimal solutions, and the lighter lines indicate all pairs that are not connected.

to another.) The goal is to build such a road network as cheaply as possible—in other words, using as little total road material as possible. Figure 1(b) depicts the optimal solution to the instance of the MSTP defined by the same set of cities used for part (a).

Both of these problems have a wide range of practical applications. The TSP is a basic problem concerned with sequencing a given set of objects in a "good" order; it has been used for problems that run from planning the motion of robotic arms drilling holes on printed circuit boards (where the "cities" are the locations where the holes must be drilled) to ordering genetic markers on a chromosome in a linear sequence (with the markers constituting the cities, and the distances derived from probabilistic estimates of proximity). The MSTP is a basic issue in the design of efficient communication networks; this follows the motivation given above, with fiber-optic cable acting in the role of "roads." The MSTP also plays an important role in the problem of clustering data into natural groupings. Note, for example, how the points on the left-hand side of figure 1(b) are joined to the points on the right-hand side by a relatively long link; in clustering applications, this can be taken as evidence that the left and right points form natural groupings.

It is not hard to come up with an algorithm for solving the TSP. We first list every possible way of ordering the cities (other than the starting city, which is fixed in advance). Each ordering defines a tour—the salesman could visit the cities in this order and then return to the start—and for each ordering we could compute the total length of the tour, by traversing the cities in this order and summing the distances from each city to the next. As we perform this calculation for all possible orders, we keep track of the order that yields the smallest total distance, and at the end of the process we return this tour as the optimal solution.

While this algorithm does solve the problem, it is extremely inefficient. There are $n - 1$ cities other than the starting point, and any possible sequence of them defines a tour, so we need to consider $(n - 1)(n - 2)$ $(n - 3) \cdots (3)(2)(1) = (n - 1)!$ possible tours. Even for $n = 30$ cities, this is an astronomically large quantity; on the fastest computers we have today, running this algorithm to completion would take longer than the life expectancy of the Earth. The difficulty is that the algorithm we have just described is performing a *brute-force search*: the "search space" of possible solutions to the TSP is very large, and the algorithm is doing nothing more than plowing its way through this entire space, considering every possible solution.

For most problems, there is a comparably inefficient algorithm that simply performs a brute-force search. Things tend to get interesting when one finds a way to improve significantly on this brute-force approach.

The MSTP provides a nice example of how such an improvement can happen. Rather than considering all possible road networks on the given set of cities, suppose we try the following myopic, "greedy" approach to the MSTP. We sort all the pairs of cities in order of increasing distance, and then work through the pairs in this order. When we get to a pair of cities, say A and B, we test if there is already a way to travel from A to B in the collection of roads constructed thus far. If there is, then it would be superfluous to build a direct road from A to B—our goal, remember, is just to make sure every pair is connected by *some* sequence of roads, and A and B are already connected in this case. But if there is no way to get from A to B using what has already been built, then we construct the direct road from A to B. (As an example of this reasoning, note that the potential road of length 14 in figure 1(a) would not get built by this MSTP algorithm; by the time this direct

route is considered, its endpoints are already joined by the sequence of two shorter roads of length 7 and 11, as depicted in figure 1(b).)

It is not at all obvious that the resulting road network should have the minimum possible cost, but in fact this is true. In other words, one can prove a theorem that says, essentially, "On every input, the algorithm just described produces an optimal solution." The payoff from this theorem is that we now have a way to compute an optimal road network by an algorithm that is much, much more efficient than brute-force search: it simply needs to sort the pairs of cities by their distances, and then make a single pass through this sorted list to decide which roads to build.

This discussion has provided us with a fair amount of insight into the nature of the TSP and the MSTP. Rather than experimenting with actual computer programs, we described algorithms in words, and made claims about their performance that could be stated and proved as mathematical theorems. But what can we abstract from these examples if we want to talk about computational efficiency in general?

3   Computational Efficiency

Most interesting computational problems share the following feature with the TSP and the MSTP: an input of size n implicitly defines a search space of possible solutions whose size grows exponentially with n. One can appreciate this explosive growth rate as follows: if we simply add one to the size of the input, the time required to search the entire space increases by a multiplicative factor. We would prefer algorithms to scale more reasonably: their running times should only increase by a multiplicative factor when the input itself increases by a multiplicative factor. Running times that are bounded by a polynomial function of the input size—in other words, proportional to n raised to some fixed power—exhibit this property. For example, if an algorithm requires at most n^2 steps on an input of size n, then it requires at most $(2n)^2 = 4n^2$ steps on an input twice as large.

In part because of arguments like this, computer scientists in the 1960s adopted *polynomial time* as a working definition of efficiency: an algorithm is deemed to be efficient if the number of steps it requires on an input of size n grows like n raised to a fixed power. Using the concrete notion of polynomial time as a surrogate for the fuzzier concept of efficiency is the kind of modeling decision that ultimately succeeds or fails based on its utility in guiding the development of real algorithms. And in this regard, polynomial time has turned out to be a definition of surprising power in practice: problems for which one can develop a polynomial-time algorithm have turned out in general to be highly tractable, while those for which we lack polynomial-time algorithms tend to pose serious challenges even for modest input sizes.

A concrete mathematical formulation of efficiency provides a further benefit: it becomes possible to pose, in a precise way, the conjecture that certain problems cannot be solved by efficient algorithms. The TSP is a natural candidate for such a conjecture; after decades of failed attempts to find an efficient algorithm for the TSP, one would like to be able to prove a theorem that says, "There is no polynomial-time algorithm that finds an optimal solution to every instance of the TSP." A theory known as NP-COMPLETENESS [IV.20 §4] provides a unifying framework for thinking about such questions; it shows that a large class of computational problems, containing literally thousands of naturally arising problems (including the TSP), are equivalent with respect to polynomial-time solvability: there is an efficient algorithm for one if and only if there is an efficient algorithm for all. It is a major open problem to decide whether or not these problems have efficient algorithms; the deeply held sense that they do not has become the *P versus NP conjecture*, which has begun to appear on lists of the most prominent problems in mathematics.

Like any attempt to make an intuitive notion mathematically precise, polynomial time as a definition of efficiency in practice begins to break down around its boundaries. There are algorithms for which one can prove a polynomial bound on the running time, but which are hopelessly inefficient in practice. Conversely, there are well-known algorithms (such as the standard SIMPLEX METHOD [III.84] for linear programming) that require exponential running time on certain pathological instances, but which run quickly on almost all inputs encountered in real life. And for computing applications that work with massive data sets, an algorithm with a polynomial running time may not be efficient enough; if the input is a trillion bytes long (as can easily occur when dealing with snapshots of the Web, for example), even an algorithm whose running time depends quadratically on the input will be unusable in practice. For such applications, one generally needs algorithms that scale linearly with the size of the input—or, more strongly, that operate by "stream-

ing" through the input in one or two passes, solving the problem as they go. The theory of such streaming algorithms is an active topic of research, drawing on techniques from information theory, Fourier analysis, and other areas. None of this means that polynomial time is losing its relevance to algorithm design—it is still the standard benchmark for efficiency—but new computing applications tend to push the limits of current definitions, and in the process raise new mathematical problems.

4 Algorithms for Computationally Intractable Problems

In the previous section we discussed how researchers have identified a large class of natural problems, including the TSP, for which it is strongly believed that no efficient algorithm exists. While this explains our difficulties in solving these problems optimally, it leaves open a natural question: what should we do when actually confronted by such a problem in practice?

There are a number of different strategies for approaching such computationally intractable problems. One of these is *approximation*: for problems like the TSP that involve choosing an optimal solution from among many possibilities, we could try to formulate an efficient algorithm that is guaranteed to produce a solution almost as good as the optimal one. The design of such approximation algorithms is an active area of research; we can see a basic example of this process by considering the TSP. Suppose we are given an instance of the TSP, specified by a map with distances, and we set ourselves the task of constructing a tour whose total length is at most twice that of the shortest tour. At first this goal seems a bit daunting: since we do not know how to compute the optimal tour (or its length), how will we guarantee that the solution we produce is short enough? It turns out, however, that this can be done by exploiting an interesting connection between the TSP and the MSTP, a relationship between the respective optimal solutions to each problem on the same set of cities.

Consider an optimal solution to the MSTP on the given set of cities, consisting of a network of roads; recall that this is something we can compute efficiently. Now, the salesman interested in finding a short tour for these cities can use this optimal road network to visit the cities as follows. Starting at one city, he follows roads until he hits a dead end, that is, a city with no new roads exiting it. He then backs up, retracing his steps until he gets to a junction with a road he has not yet taken, and he proceeds down this new road. For example, starting in the upper left corner of figure 1(b), the salesman would follow the road of length 8 and then choose one of the roads of length 10 or 20; if he selects the former, then after reaching the dead end he would back up to this junction again and continue the tour by following the road of length 20. A tour constructed in this way traverses each road twice (once in each direction), so if we let m denote the total length of all roads in the optimal MSTP solution, we have found a tour of length $2m$.

How does this compare to t, the length of the best possible tour? Let us first argue that $t \geqslant m$. This is true because, in the space of all possible solutions to the MSTP, one option is to build roads between cities that the salesman visits consecutively in the optimal TSP tour, for a total mileage of t; on the other hand, m is the total length of the *shortest possible* road network, and hence t cannot be smaller than m. So we have concluded that the optimal solution to the TSP has length at least m. However, we have just exhibited an algorithm that finds a tour of length $2m$, so, as we wanted, we have an efficient way to find a tour that is at most twice as long as the shortest one possible.

People trying to solve large instances of computationally hard problems in practice frequently use algorithms that have been observed empirically to give nearly optimal solutions, even when no guarantees on their performance have been proved. *Local-search* algorithms form one widely used class of approaches like this. A local-search algorithm starts with an initial solution and repeatedly modifies it by making some "local" change to its structure, looking for a way to improve its quality. In the case of the TSP, a local-search algorithm would seek simple improving modifications to its current tour; for example, it might look at sets of cities that are visited consecutively and see if visiting them in the opposite order would shorten the tour. Researchers have drawn connections between local-search algorithms and phenomena in nature; for example, just as a large molecule contorts itself in space trying to find a minimum-energy conformation, we can imagine the TSP tour in a local-search algorithm modifying itself as it tries to reduce its length. Determining how deeply this analogy goes is an interesting research issue.

5 Mathematics and Algorithm Design: Reciprocal Influences

Many branches of mathematics have contributed to aspects of algorithm design, and the issues raised by the analysis of new algorithmic problems have, in a number of cases, suggested novel mathematical questions.

Combinatorics and graph theory have been qualitatively transformed by the growth of computer science, to the extent that algorithmic questions have become thoroughly intertwined with the mainstream of research in these areas. Techniques from probability have also become fundamental to many areas of computer science: probabilistic algorithms draw power from the ability to make random choices while they are being executed, and probabilistic models of the input to an algorithm can give one a more realistic view of the problem instances that arise in practice. This style of analysis provides a steady source of new questions in discrete probability.

A computational perspective is often useful in thinking about "characterization" problems in mathematics. For example, the general issue of characterizing prime numbers has an obvious algorithmic component: given a number n as input, how efficiently can we determine whether it is prime? (There exist algorithms that are exponentially better than the approach of dividing n by all numbers up to $\sqrt{n}$: see COMPUTATIONAL NUMBER THEORY [IV.3 §2].) Problems in KNOT THEORY [III.44], such as the characterization of unknotted loops, have a similar algorithmic side. Suppose we are given a circular loop of string in three dimensions (described as a jointed chain of line segments), and it wraps around itself in complicated ways. How efficiently can we determine whether it is truly knotted, or whether by moving it around we can fully untangle it? We can ask this sort of question in many similar mathematical contexts; it is clear that these algorithmic issues are extremely concrete as problems, though they may lose part of the original intent of the mathematicians who posed the questions more generally.

Rather than attempting to enumerate the intersection of algorithmic ideas with all the different branches of mathematics, we conclude this article with two case studies that involve the design of algorithms for particular applications, and the ways in which mathematical ideas arise in each instance.

6 Web Search and Eigenvectors

As the World Wide Web grew in popularity throughout the 1990s, computer-science researchers grappled with a difficult problem: the Web contains a vast amount of useful information, but its anarchic structure makes it very hard for users, unassisted, to find the specific information they are looking for. Thus, early in the Web's history, people began to develop *search engines* that would index the information on the Web, and produce relevant Web pages in response to user queries. But of the thousands or millions of pages relevant to a topic on the Web, which few should the search engine present to a user? This is the *ranking* problem: how to determine the "best" resources on a given topic. Note the contrast with concrete problems like the TSP. There, the goal (the shortest tour) was not in doubt; the difficulty was simply in computing an optimal solution efficiently. For the search engine ranking problem, on the other hand, formalizing the goal is a large part of the challenge—what do we mean by the "best" page on a topic? In other words, an algorithm to rank Web pages is really providing a *definition* of the quality of a Web page as well as the means to evaluate this definition.

The first search engines ranked each Web page based purely on the text it contained. These approaches began to break down as the Web grew, because they did not take into account the quality judgments encoded in the Web's hyperlinks: in browsing the Web, we often discover high-quality resources because they are "endorsed" through the links they receive from other pages. This insight led to a second generation of search engines that determined rankings using *link analysis.*

The simplest such analysis would just count the number of links to a page: in response to the query "newspapers," for example, one could rank pages by the number of incoming links they receive from other pages containing the term in effect, allowing pages containing the term "newspapers" to vote on the result. Such a scheme will generally do well for the top few items, placing prominent news sites like *The New York Times* and *The Financial Times* at the head of the list; beyond this, however, it will quickly break down, favoring a large number of highly linked but irrelevant sites.

It is possible to make much more effective use of the latent information in the links. Consider pages that link to many of the sites ranked highly by this simple voting scheme; it is natural to expect that these are authored by people with a good sense for where the interesting newspapers are, and so we could run the voting again, this time giving more voting power to these pages that selected many of the highly ranked sites. This revote might elevate certain lesser-known newspapers favored by Web-page authors who were more knowledgeable on the topic; in response to the results of this revote, we could further sharpen our weighting of the voters. This

"principle of repeated improvement" uses the information contained in a set of page-quality estimates to produce a more refined set of estimates. If we perform these refinements repeatedly, will they converge to a stable solution?

In fact, this sequence of refinements can be viewed as an algorithm for computing the principal EIGENVECTOR [I.3 §4.3] of a particular matrix; this both establishes the convergence of the process and characterizes the end result. To establish this connection, we introduce some notation. Each Web page is assigned two scores: an *authority weight*, measuring its quality as a primary source on the topic; and a *hub weight*, measuring its power as a voter for the highest-quality content. Pages may score highly in one of these measures but not in the other—one should not expect a prominent newspaper to simultaneously serve as a good guide to other newspapers—but there is also nothing to prevent a page from scoring well in both. One round of voting can now be viewed as follows: we update the authority weight of each page by summing the hub weights of all pages that point to it (receiving links from highly weighted voters makes you a better authority); we then reweight all the voters, updating each page's hub weight by summing the authority weights of the pages it points to (linking to high-quality content makes you a better hub).

How do eigenvectors come into this? Suppose we define a matrix M with one row and one column for each page under consideration; the (i, j) entry equals 1 if page i links to page j, and it equals 0 otherwise. We encode the authority weights in a vector u, where the coordinate a_i is the authority weight of page i. The hub weights can be similarly written as a vector h. Using the definition of matrix–vector multiplication, we can now check that the updating of hub weights in terms of authority weights is simply the act of setting h equal to Ma; correspondingly, setting a equal to $M^T h$ updates the authority weights. (Here M^T denotes the transpose of the matrix M.) Running these updates n times each from starting vectors a_0 and h_0, we obtain $a = (M^T(M(M^T(M \cdots (M^T(Ma_0)) \cdots)))) = (M^T M)^n a_0$. This is the power-iteration method for computing the principal eigenvector of $M^T M$, in which we repeatedly multiply some fixed starting vector by larger and larger powers of $M^T M$. (As we do this, we also divide all coordinates of the vector by a scaling factor to prevent them from growing unboundedly.) Hence this eigenvector is the stable set of authority weights toward which our updates are converging. By completely symmetric reasoning, the hub weights are converging toward the principal eigenvector of MM^T.

A related link-based measure is *PageRank*, defined by a different procedure that is also based on repeated refinement. Instead of drawing a distinction between the voters and the voted-on, one posits a single kind of quality measure that assigns a *weight* to each page. A current set of page weights is then updated by having each page distribute its weight uniformly among the pages it links to. In other words, receiving links from high-quality pages raises one's own quality. This too can be written as multiplication by a matrix, obtained from M^T by dividing each row's entries by the number of outgoing links from the corresponding page; repeated updates again converge to an eigenvector. (There is a further wrinkle here: repeated updating in this case tends to cause all weight to pool at "dead-end" pages that have no outgoing links and hence nowhere to pass their weight. Thus, to obtain the PageRank measure used in applications, one adds a tiny quantity $\varepsilon > 0$ in each iteration to the weight of each page; this is equivalent to using a slightly modified matrix.)

PageRank is one of the main ingredients in the search engine Google; hubs and authorities form the basis for Ask's search engine Teoma, as well as a number of other Web search tools. In practice, current search engines (including Google and Ask) use highly refined versions of these basic measures, often combining features of each; understanding how relevance and quality measures are related to large-scale eigenvector computations remains an active research topic.

7   Distributed Algorithms

Thus far we have been discussing algorithms that run on a single computer. As a concluding topic, we briefly touch on a broad area in computer science concerned with computations that are *distributed* over multiple communicating computers. Here the problem of efficiency is compounded by concerns over maintaining coordination and consistency among the communicating processes.

As a simple example illustrating these issues, consider a network of automatic teller machines (ATMs). When you withdraw an amount of money x at one of these ATMs, it must do two things: (1) notify a central bank computer to deduct x from your account; and (2) emit the correct amount of money in physical bills. Now, suppose that between steps (1) and (2) the ATM crashes so that you do not get your money; you would

like it to be the case that the bank does not subtract x from your account anyway. Or suppose that the ATM executes both of steps (1) and (2), but its message to the bank is lost; the bank would like for x to be eventually subtracted from your account anyway. The field of distributed computing is concerned with designing algorithms that operate correctly in the presence of such difficulties.

As a distributed system runs, certain processes may experience long delays, some of them may fail in mid-computation, and some of the messages between them may be lost. This leads to significant challenges in reasoning about distributed systems, because this pattern of failures can cause each process to have a slightly different *view* of the computation. It is easily possible for there to be two runs of the system, with different patterns of failure, that are "indistinguishable" from the point of view of some process P; in other words, P will have the same view of each, simply because the differences in the runs did not affect any of the communications that it received. This can pose a problem if P's final output is supposed to depend on its having noticed that the two runs were different.

A major advance in the study of such systems came about in the 1990s, when a connection was made to techniques from algebraic topology. Consider for simplicity a system with three processes, though everything we say generalizes to any number of processes. We consider the set of all possible runs of the system; each run defines a set of three views, one held by each process. We now imagine the views associated with a single run as the three corners of a triangle, and we glue these triangles together according to the following rule: for any two runs that are indistinguishable to some process P, we paste the two corresponding triangles together at their corners associated with P. This gives us a potentially very complicated geometric object, constructed by applying all these pasting operations to the triangles; we call this object the *complex* associated with the algorithm. (If there were more than three processes, we would have an object in a higher number of dimensions.) While it is far from obvious, researchers have been able to show that the correctness of distributed algorithms can be closely connected with the topological properties of the complexes that they define.

This is another powerful example of the way in which mathematical ideas can appear unexpectedly in the study of algorithms, and it has led to new insights into the limits of the distributed model of computation.

Combining the analysis of algorithms and their complexes with classical results from algebraic topology has in some cases resolved tricky open problems in this area, establishing that certain tasks are provably impossible to solve in a distributed system.

Further Reading

Algorithm design is a standard topic in the undergraduate computer-science curriculum, and it is the subject of a number of textbooks, including Cormen et al. (2001) and a book by Kleinberg and Tardos (2005). The perspective of early computer scientists on how to formalize efficiency is discussed by Sipser (1992). The TSP and the MSTP are fundamental to the field of combinatorial optimization; the TSP is used as a lens through which this field is surveyed in a book edited by Lawler et al. (1985). Approximation algorithms and local-search algorithms for computationally intractable problems are discussed in books edited by Hochbaum (1996) and by Aarts and Lenstra (1997), respectively. Web search and the role of link analysis is covered in a book by Chakrabarti (2002); beyond Web applications, there are a number of other interesting connections between eigenvectors and network structures, as described by Chung (1997). Distributed algorithms are covered in a book by Lynch (1996), and the topological approach to analyzing distributed algorithms is reviewed by Rajsbaum (2004).

Aarts, E., and J. K. Lenstra, eds. 1997. *Local Search in Combinatorial Optimization*. New York: John Wiley.

Chakrabarti, S. 2002. *Mining the Web*. San Mateo, CA: Morgan Kaufman.

Chung, F. R. K. 1997. *Spectral Graph Theory*. Providence, RI: American Mathematical Society.

Cormen, T., C. Leiserson, R. Rivest, and C. Stein. 2001. *Introduction to Algorithms*. Cambridge, MA: MIT Press.

Hochbaum, D. S., ed. 1996. *Approximation Algorithms for NP-hard Problems*. Boston, MA: PWS Publishing.

Kleinberg, J., and É. Tardos. 2005. *Algorithm Design*. Boston, MA: Addison-Wesley.

Lawler, E. L., J. K. Lenstra, A. H. G. Rinnooy Kan, and D. B. Shmoys, eds. 1985. *The Traveling Salesman Problem: A Guided Tour of Combinatorial Optimization*. New York: John Wiley.

Lynch, N. 1996. *Distributed Algorithms*. San Mateo, CA: Morgan Kaufman.

Rajsbaum, S. 2004. Distributed computing column 15. *ACM SIGACT News* 35:3.

Sipser, M. 1992. The history and status of the P versus NP question. In *Proceedings of the 24th ACM Symposium on Theory of Computing*. New York: Association for Computing Machinery.

VII.6 Reliable Transmission of Information
Madhu Sudan

1 Introduction

The notion of "digital information" emerged in the middle of the twentieth century, in response to the advent of the telegraph and to the beginnings of computer science, which at the time was principally a theoretical discipline. Of course, the use of electricity to communicate signals goes back further, but the earlier uses involved signals of a "continuous" nature: music, voice, etc. The new era was characterized by the transmission of (or the need to transmit) more "discrete" messages, i.e., messages such as English sentences, which can be described as finite sequences of letters taken from some finite alphabet. The phrase "digital information" came to be applied to such families of messages.

Digital information posed some novel challenges to the engineers and mathematicians charged with the task of communicating such messages. The root cause of these challenges is "noise." Every communication medium is noisy, and never transmits any signal completely accurately. In the case of continuous signals, somehow the receivers (typically, our ears and eyes) can adjust to such errors and learn to discount them. For example, if you play a very old recording of a musical performance, then there will typically be a crackling noise, but it is possible to ignore this, unless the quality is very bad indeed, and concentrate on the music. However, in the case of digital information errors can have a more catastrophic effect. To see this, suppose that we are communicating in English sentences and that the communication medium makes occasional mistakes by altering one of the transmitted letters. In such a scenario the message

WE ARE NOT READY

could easily be changed into the message

WE ARE NOW READY.

All it takes is one error on the part of the communication medium, and the entire intention of the message is reversed. Digital information tends to be inherently intolerant of errors, and the mathematicians and engineers of the time were charged with the task of inventing methods that would make communication reliable even if the process of transmission is not.

Here is one way of achieving this. To communicate any message, the sender of the message repeats every letter, say five times. For example, to send the message

WE ARE NOT READY

the sender says something like

WWWWWEEEEE   AAAAA....

The receiver can then detect errors (as long as there are not too many) by checking that every block of five successive letters repeats the same letter. If this ever fails to be the case, then it is clear that errors have occurred during transmission. If it is not possible for five successive symbols to be in error (or even if it is just very unlikely), then it follows that the resulting scheme is also more reliable than the underlying means of transmission. Finally, if even less error is possible, then it may be possible for the receiver to determine the actual message, rather than simply being able to tell when errors have occurred. For example, if at most two symbols in any block of five can be erroneous, then the most commonly occurring letter in each block of five must be the letter from the original message: for instance, a sequence such as

WWWMWEFEEE   AAAAA...

would be interpreted by the receiver as

WE A....

Repeating every symbol five times in order to be able to correct two errors does not appear to be a very efficient way to use the communication channel. Indeed, as we will show in the rest of this article, when transmitting long messages one can do much better. However, in order to understand this issue, we need to define the process of communication, the model of error, and the measures of performance more carefully. We do so next.

2   Model

2.1   Channel and Errors

The central object of attention in the problem of information transmission is the "channel of communication," or simply the *channel*. The channel has an *input* (the original signal to be communicated) and an *output* (the signal after it is transmitted). The input consists of a sequence of elements from some finite set: by analogy with the English-language example, these elements are called *letters* and the finite set, which is typically denoted Σ, is called an *alphabet*. The channel attempts to transmit the input to the receiver, but while doing so it may make some errors. The alphabet and the process that underlies the errors are what specifies the channel.

The alphabet Σ varies from scenario to scenario. In the example described above, the alphabet consisted of the English characters $\{A, B, \ldots, Z\}$, and possibly some punctuation symbols. In most communication scenarios, the alphabet is the "binary alphabet" that consists just of the "letters" 0 and 1, which are known as *bits*. On the other hand, in applications involving storage of digital information (in compact discs (CDs), digital versatile discs (DVDs), etc.), the alphabet contains 256 elements (the alphabet of "bytes").

Specifying an alphabet is easy, but if we wish to define a good mathematical model for the way that errors are produced, then a lot more care is needed. At one extreme is a worst-case model suggested by Hamming (1950), where there is some limit on the number of errors that the channel can make, but within that limit it chooses the errors to be as damaging as possible. A more benign class of errors was proposed by Shannon (1948), who suggested that errors could be modeled by a probabilistic process.

We will focus on one probabilistic model to illustrate many of the concepts below. In this model, the error of the channel is specified by a real number parameter p, where $0 \leqslant p \leqslant 1$. Every use of the channel results in an error with probability p. To be precise, if the sender transmits an element $\sigma \in \Sigma$, then with probability $1 - p$ the output for that element is σ but with probability p it is some other element σ' of Σ, chosen uniformly at random. Furthermore, and this is very crucial to this model, the errors are assumed to be *independent*, i.e., the channel repeats this process for each letter it transmits without any memory of how it acted on previous symbols. We refer to this model as the Σ-*symmetric channel with parameter p* (or Σ-SC(p)) in the rest of this article. A special case of particular importance is the *binary symmetric channel*, which is the Σ-symmetric channel when Σ is the binary alphabet $\{0, 1\}$. Then, if the input bit is 0, say, the corresponding output bit will be 0 with probability $1 - p$ and 1 with probability p.

While this model of error may seem rather oversimplified (and even unnatural if Σ is not the binary alphabet $\{0, 1\}$), it turns out that it captures the essence of most mathematical challenges that arise when one tries to make communication reliable. Furthermore, many of the solutions found to make communication reliable in this setting have been generalized to other scenarios, so this simple model is very useful both in practice and in the theoretical study of communication.

2.2   Encoding and Decoding

Suppose the sender wishes to transmit a sequence through a channel that makes errors. One way to compensate for these errors is to send through the channel not the sequence itself but a modified version of the sequence that contains redundant information. The process of modification that we choose is called the *encoding* of the message. We have already seen one method of encoding, namely repeating each term in the sequence several times. However, this is by no means the only way of doing it, so to discuss encoding we use the following general framework: if the sender has a message consisting of a sequence of k elements of Σ, then by some means or another it expands the message into a new sequence, now consisting of n elements of Σ, for some $n > k$. Formally, the sender applies an *encoding function* $E : \Sigma^k \to \Sigma^n$ to the message. (Σ^k stands for the set of sequences of length k with letters in Σ, and Σ^n for the set of sequences of length n.) Thus, to convey a message $m = (m_1, m_2, \ldots, m_k)$ to the receiver, the sender transmits over the channel not the k symbols of m but the n symbols of $E(m)$.

Errors may then be introduced, after which the receiver receives a sequence $r = (r_1, r_2, \ldots, r_n)$; its goal is then to "compress" the sequence r back to a k-letter sequence, removing the error and obtaining the original message m (at least if not too many errors have occurred). It does this by applying a *decoding function* $D : \Sigma^n \to \Sigma^k$, which tells it how sequences of length n are converted back into sequences of length k.

The possible pairs of functions E, D describe the options available to the designers of the communication system. Their choice determines the performance of the system. Let us now describe how this performance is measured.

2.3   Goals

Very informally, our goals are threefold. We would like to make the communication as reliable as possible. At the same time, we would like to maximize the utilization of the channel. Finally, we would like to do so with effective computation. We describe these goals more carefully below, in the case of the model $\Sigma\text{-SC}(p)$ described earlier.

Consider first the reliability. If we start with a message m, encode it as $E(m)$, and pass it through the

channel, then the output, after some random errors have been introduced, will be a string y. The receiver will decode y, producing a new message $D(y)$. For each message m, there is a certain probability of a *decoding error*, i.e., a certain probability that $D(y)$ will not in fact be equal to the original message m. The reliability of the communication is measured by the largest of these probabilities. If this is small, then we know that, whatever the original message m, a decoding error is unlikely, and then we regard the communication as reliable.

Next, let us look at the utilization of the channel. This is measured by the *rate* of the encoding, i.e., the quantity k/n. In other words, it is the ratio of the length of the original message to the length of the encoded message: the smaller this ratio, the less efficiently one is using the channel.

Finally, practical considerations also require us to be able to encode and decode quickly: a pair of reliable and efficient encoding and decoding functions will not be of much use if they are very time-consuming to compute. Adopting the standard convention in algorithm design, we regard our algorithms as feasible if they run in *polynomial time*: that is, if their running time can be bounded above by a polynomial function of the length of their input and output.

To illustrate the above ideas, let us analyze the "repetition encoding" that repeats every letter of the alphabet five times. For simplicity, take the alphabet Σ to be $\{0, 1\}$, let the probability p be fixed, and let us consider the behavior of the model as the message length k tends to ∞. Our encoding function takes strings of length k to strings of length $5k$ and thus has a rate of $\frac{1}{5}$. Given any particular block of five transmissions, the probability that it contains three or more errors is

$$p' = \binom{5}{3} p^3 (1 - p)^2 + \binom{5}{4} p^4 (1 - p) + \binom{5}{5} p^5.$$

The probability that that block does not give rise to a decoding error is $1 - p'$, so the probability that there is no decoding error is $(1 - p')^k$ and the probability that there *is* a decoding error is $1 - (1 - p')^k$. If we fix $p > 0$ and let $k \to \infty$, then $(1 - p')^k$ tends to 0 (exponentially quickly), so the probability of decoding error tends to 1. Thus, this encoding/decoding pair is highly unreliable, and its rate is not too good either. The only redeeming feature is that it is very easy indeed to compute. (Its computational efficiency is easily seen to be bounded by a number of operations that is linear in k.)

One way to salvage the repetition code is to repeat every symbol $c \log k$ times. For a largish constant c, the probability of a decoding error goes to 0, but now the rate of the code goes to 0 as well. Prior to the work of Shannon it may have even been believed that a trade-off of this kind was inevitable: every encoding/decoding scheme would either achieve a vanishingly small rate or make mistakes with probability tending to 1. As we will see later in the article, it is in fact possible to define encoding schemes that achieve all three of our goals: they operate at a positive rate, they can correct errors that occur a positive proportion of the time (in either the probabilistic or the worst-case model), and they use efficient encoding and decoding algorithms. Most of the insight for this remarkable result goes back to a seminal paper by Shannon (1948). In that paper he gave the first examples of encoding and decoding functions that satisfied the first two goals, though they were not computationally efficient.

Shannon's encoding and decoding functions were therefore not practical, but we can now see, with the benefit of hindsight, that ignoring the goal of efficient computability in order to gain some theoretical insight into the channels was extraordinarily fruitful. A general rule of thumb seems to operate: that the performance of the very best encoding and decoding functions can be matched arbitrarily closely by encoding and decoding functions that are also computationally efficient. This justifies considering the goal of efficiency separately from the other two goals.

3   The Existence of Good Encoding and Decoding Functions

In this section we will describe results that demonstrate the existence of encoding and decoding functions that have an extremely good rate and reliability. In order to describe these results, first proved by Shannon, it will be useful to consider two related notions introduced by Hamming in work that was essentially concurrent with that of Shannon.

In order to understand these notions, let us start by describing what makes one encoding function E better or worse than another. The task of the *decoding* function is to work out, when it receives a string y, what the original message m was. Notice that this is equivalent to working out what the encoded message $E(m)$ was, since no two messages are encoded in the same way. The possible encoded messages are called *codewords*: that is, a codeword is a string of length n that arises as $E(m)$ for some message $m \in \Sigma^k$.

What we are worried about is the possibility of confusing two codewords after errors have been introduced, and this depends only on the set of codewords, and not on which codeword corresponds to which original message. Therefore, we adopt what at first seems a strange definition: an *error-correcting code* is any set of strings of length n in the alphabet Σ (that is, any subset of Σ^n). The strings in an error-correcting code are still called codewords. This definition completely ignores the actual process of encoding of a message, but that is so that we can focus on the rate and the decoding error while ignoring computational efficiency. If we are given an encoding function E, then the corresponding error-correcting code is simply the set of all the codewords of E. Mathematically, this is just the image of the function E.

What makes an error-correcting code good or bad? To answer this question, let us consider what happens if the alphabet is $\{0, 1\}$ and the code contains two strings $x = (x_1, x_2, \ldots, x_n)$ and $y = (y_1, y_2, \ldots, y_n)$ that differ in precisely d places. If errors are introduced with probability p, then the probability that x is converted into y is $p^d(1-p)^{n-d}$. Assuming that $p < \frac{1}{2}$, this probability gets smaller as d increases, so the smaller d is, the more likely the strings x and y are to be confused. It seems preferable, therefore, that there should not be too many pairs of strings in the code that differ in just a few places. A similar argument applies to larger alphabets as well.

The above thoughts lead to a definition that is very natural in this context. Given an alphabet Σ and two strings $x = (x_1, x_2, \ldots, x_n)$ and $y = (y_1, y_2, \ldots, y_n)$ belonging to Σ^n, the *Hamming distance* between x and y is defined to be the number of coordinates i for which $x_i \neq y_i$. For example, let $\Sigma = \{a, b, c, d\}$ and let $n = 6$. The strings $abccad$ and $abdcab$ differ in the third and sixth places and are identical otherwise, so their Hamming distance is 2. Our goal is to find an encoding function E such that the associated code maximizes the typical Hamming distance between pairs of codewords.

Shannon's solution to this is an extremely simple application of the PROBABILISTIC METHOD [IV.19 §3]: he picks the encoding function at random. That is, for every message m, the encoding $E(m)$ is chosen entirely randomly from the set Σ^n, with all choices equally likely. Furthermore, for every message m, this choice is independent of the encoding of every other message m'. It is a good exercise in basic probability to see that such a choice almost always leads to a code where the distances between codewords are on average large. In fact, even the minimum distance between codewords is almost always large. However, we will not show this. Instead, we will argue that with high probability this random choice leads to a "nearly optimal" encoding function, from the point of view of rate and reliability.

First, let us consider what the decoding function ought to be. In the absence of computational requirements, it is not hard to say what the "optimal" decoding algorithm is. If you receive a sequence z, then you should choose the message m that is most likely to have resulted in this sequence. For the model Σ-SC(p) with $p < 1 - 1/|\Sigma|$, it is easily verified that this will be the message m for which the encoding $E(m)$ is nearest to z, as measured by Hamming distance. (If the minimum distance is attained by both $E(m)$ and $E(m')$, then one can make an arbitrary choice between them.) The condition on p is important here. It ensures that when the sequence $E(m)$ passes through the channel, the most likely output corresponding to any given term, out of the $|\Sigma|$ different possibilities, is the same as the input. Without this condition, there would be no reason to expect z to be close to $E(m)$. We shall argue that there is a number C, depending only on the error probability p and the size of the alphabet, such that for a random encoding function with rate smaller than C, this decoding function recovers the original message with a high probability. As an aside, Shannon also showed that for the same constant C, any attempt to communicate at rates greater than C would lead to errors with probability exponentially close to 1. Because of this result, the constant C is known as the *Shannon capacity* of the channel.

Once again, for simplicity we shall consider just the case of the binary alphabet $\{0, 1\}$. In this case we are choosing a random function E from $\{0, 1\}^k$ to $\{0, 1\}^n$, and we would like to show that, under suitable circumstances, the resulting code will almost certainly be very reliable. In order to do this, we shall focus on a single message m, and rely on two basic ideas.

The first idea is a precise form of THE LAW OF LARGE NUMBERS [III.71 §4]. If the error probability is p, then the expected number of errors introduced into a codeword $E(m)$ is pn, so, if n is large, then we expect that the actual number of errors will almost certainly be very close to this, just as, if you toss a fair coin ten thousand times, you will be surprised if the number of heads is not close to five thousand. The result that expresses this formally is as follows.

Claim. There exists a constant $c > 0$ such that the probability that the number of errors exceeds $(p + \epsilon)n$ is at most $2^{-c\epsilon^2 n}$.

The same can be said of the probability that the number of errors is less than $(p - \epsilon)n$, but we shall not use this result.

When n is large, $2^{-c\epsilon^2 n}$ is extremely small, so the number of errors is almost certainly at most $(p + \epsilon)n$. The number of errors equals the Hamming distance from y, the output of the channel, to $E(m)$, the codeword that was transmitted. Therefore, the decoding function that chooses the codeword with smallest Hamming distance from y will almost certainly choose $E(m)$, provided that there is no message m' such that $E(m')$ is closer to y than $(p + \epsilon)n$.

The second idea, which allows us to say that this will almost certainly be the case, is that "Hamming balls are small." Let z be a sequence in $\{0, 1\}^n$. Then the *Hamming ball of radius r about z* is the set of all sequences w with Hamming distance at most r from z. How big is this set? Well, in order to specify a sequence w with Hamming distance exactly d from z, it is enough to specify the set of d places where w and z differ. There are $\binom{n}{d}$ ways of choosing this set, so the number of sequences at a distance of at most r is

$$\binom{n}{0} + \binom{n}{1} + \binom{n}{2} + \cdots + \binom{n}{r}.$$

If $r = \alpha n$ and $\alpha < \frac{1}{2}$, then this number is at most a constant times $\binom{n}{r}$, because each term is at least

$$\frac{n - r}{r} = \frac{1 - \alpha}{\alpha}$$

times the one before. But

$$\binom{n}{r} = \frac{n!}{r!(n - r)!}.$$

If we now use STIRLING'S FORMULA [III.31] or the looser approximation $n! = (n/e)^n$, then we find that this is about $(1/\alpha(1 - \alpha))^n$, which is $2^{H(\alpha)n}$, where

$$H(\alpha) = -\alpha \log_2 \alpha - (1 - \alpha) \log_2(1 - \alpha).$$

(Note that $H(\alpha)$ is positive, because α and $1 - \alpha$ are less than 1 and therefore have negative logarithms.) The function H is called the *entropy function*. It is continuous and strictly increasing on the interval $[0, \frac{1}{2}]$ with $H(0) = 0$ and $H(\frac{1}{2}) = 1$. So, if $\alpha < \frac{1}{2}$, then $H(\alpha) < 1$, and therefore $2^{H(\alpha)n}$ is exponentially smaller than 2^n: this is what is meant by saying that the Hamming ball of radius αn is small.

Let us set α to be $p + \epsilon < \frac{1}{2}$. Then the probability that a single randomly chosen sequence $E(m')$ lies in the Hamming ball of radius $(p + \epsilon)n$ about y is at most $2^{H(p+2\epsilon)n}2^{-n}$. (The 2ϵ is to compensate for slight inaccuracies in the above estimate for the size of the ball.) Since there are $2^k - 1$ possibilities for m', the probability that one can be found for which $E(m')$ lies in the ball is at most $2^k 2^{H(p+2\epsilon)n}2^{-n}$. Therefore, if $k \leqslant n(1 - H(p + 2\epsilon) - \epsilon)$, this probability is at most $2^{-\epsilon n}$, which is exponentially small.

Because we can choose ϵ to be as small as we like, we can make k/n as close as we like to $1 - H(p)$ while still maintaining an exponentially small probability of decoding error. It turns out that the quantity $1 - H(p)$ is the constant C discussed earlier: the Shannon capacity of the binary symmetric channel. Thus, the capacity of the binary symmetric channel is always positive if $p < \frac{1}{2}$.

Shannon's theorem and proof are significantly more general than the above example demonstrates. For a wide variety of channels, and for a wide variety of models of (probabilistic) error, his theory pins down the capacity of the channel and shows that reliable communication is possible if and only if the rate of the channel is less than its capacity. Shannon's proof is a remarkable example of the use of the probabilistic method in the practice of engineering. Note, however, that the encoding and decoding algorithms are quite impractical. The proof gives no clue about how to find an encoding function, though of course one can consider every encoding function $E : \{0,1\}^k \rightarrow \{0,1\}^n$ to check if it is good. However, even if such a function is found, it may have no succinct description, in which case the encoder and decoder have to store this encoding function as an exponentially long table in their memory. Finally, the decoding algorithm seems to involve a brute-force search for the nearest codeword, a problem which seems to be the most serious obstacle to obtaining a computationally efficient version of Shannon's theorem that can be used in practice. What the theorem definitely *does* give us is a significant insight into the limitations and potential utility of the communication channel. With this in mind, we can set ourselves the right targets to strive for when we come to devise more practical encoding and decoding procedures. In the next section we will show that it is possible to achieve a fixed rate that is bounded away from zero, to tolerate a constant fraction of errors, and to do both of these with efficient algorithms.

4   Efficient Encoding and Decoding

Let us now turn to the task of designing encoding and decoding functions that can be calculated efficiently. Currently, there are at least two very different approaches to building such functions. We describe here an approach based on algebra over finite fields. The alternative approach is based on the construction of EXPANDING GRAPHS [III.24], but we will not describe that here.

4.1   Codes for Large Alphabets Using Algebra

In this section we describe a simple way to get an encoding function $E : \Sigma^k \rightarrow \Sigma^n$, where Σ is a finite FIELD [I.3 §2.2] with at least n elements. (Recall that there are finite fields with q elements whenever q is of the form p^t for a prime p and a positive integer t.) These codes were introduced by Reed and Solomon (1960) and have since been called the *Reed–Solomon codes*.

A Reed–Solomon code is specified by a sequence of n distinct field elements $\alpha_1, \ldots, \alpha_n \in \Sigma$. Given a message $m = (m_0, m_1, \ldots, m_{k-1}) \in \Sigma^k$, we associate with the message the polynomial $M(x) = m_0 + m_1 x + \cdots + m_{k-1} x^{k-1}$. The encoding of m is simply the sequence $E(m) = M(\alpha_1), M(\alpha_2), \ldots, M(\alpha_n)$. In other words, to encode a sequence m, you treat the terms of the sequence as the k coefficients of a polynomial of degree $k - 1$ and write out the values that this polynomial takes at $\alpha_1, \ldots, \alpha_n$.

Before describing the error-correcting capability of this code, let us note that it is very succinctly represented: all that is needed to specify it is a description of the field Σ and the sequence of n elements $\alpha_1, \ldots, \alpha_n$. It is easy to show that the number of additions and multiplications needed to compute $M(\alpha)$ is at most Ck for some constant C. (For example, to work out $3\alpha^3 - \alpha^2 + 5\alpha + 4$, you start with 3, multiply by α, subtract 1, multiply by α, add 5, multiply by α, and add 4.) Therefore, the number of field operations needed to compute the entire encoding is bounded above by Cnk, for some (different) constant C. (In fact, more sophisticated and efficient algorithms are known for the encoding problem that take at most $Cn(\log n)^2$ steps.)

Now let us consider the error-correcting properties of the code. We start by showing that the encodings of any two messages m_1 and m_2 have a Hamming distance of at least $n - (k - 1)$. To see this, let $M_1(x)$ and $M_2(x)$ be the polynomials associated with m_1 and m_2. Now the difference $p(x) = M_1(x) - M_2(x)$ has degree at most $k - 1$, and it is not the zero polynomial (since M_1 and M_2 are distinct), and therefore it has at most $k - 1$ roots. This tells us that there are at most $k - 1$ values of α for which $M_1(\alpha) = M_2(\alpha)$. It follows that the Hamming distance between the sequences

$$E(m_1) = (M_1(\alpha_1), M_1(\alpha_2), \dots, M_1(\alpha_n))$$

and

$$E(m_2) = (M_2(\alpha_1), M_2(\alpha_2), \dots, M_2(\alpha_n))$$

is at least $n - k + 1$.

It follows that if z is any sequence, then its Hamming distance from at least one of $E(m_1)$ and $E(m_2)$ is greater than $\frac{1}{2}(n - k)$ (since otherwise the distance between $E(m_1)$ and $E(m_2)$ would have to be at most $n - k$). Therefore, if the number of errors that occur during transmission is at most $\frac{1}{2}(n - k)$, then the original message m is uniquely determined by the received sequence z. What is much less obvious is that there is an efficient algorithm for working out what m was, but, remarkably, it is possible to compute m with a polynomial-time algorithm (in n), which we shall now describe.

What must the decoding algorithm do? It is given the numbers $\alpha_1, \dots, \alpha_n$ and the received sequence $z_1, \dots, z_n$ and is required to find a polynomial M of degree $k - 1$ or less such that $M(\alpha_i) = z_i$ for all but at most $\frac{1}{2}(n - k)$ values of i. If such a polynomial exists, then it is unique, as we have just seen, and its coefficients will give the original message m (if the number of errors is at most $\frac{1}{2}(n - k)$).

If there were no errors, then our task would be much easier: one can determine the coefficients of a polynomial of degree $k - 1$ from k of its values by solving k simultaneous equations. However, if some of the values we use are incorrect, then we will end up with a completely different polynomial, so this method is not easy to use for the problem we actually face.

To overcome this difficulty, let us imagine that M exists and that the errors introduced into the sequence $M(\alpha_1), \dots, M(\alpha_n)$ occur at $i_1, \dots, i_s$, where $s \leqslant \frac{1}{2}(n - k)$. Then the polynomial $B(x) = (x - \alpha_{i_1}) \cdots (x - \alpha_{i_s})$ has degree at most $\frac{1}{2}(n - k)$ and is zero if and only if x is equal to α_{i_j} for some j. Let us set $A(x)$ to equal $M(x)B(x)$. Then $A(x)$ is a polynomial of degree at most $k - 1 + \frac{1}{2}(n - k) = \frac{1}{2}(n + k - 2)$, and for every i we have $A(\alpha_i) = z_i B(\alpha_i)$. (If there is no error at i, then this is obvious, since $z_i = M(\alpha_i)$, and if there is an error at i, then both sides are 0.)

Conversely, suppose that we manage to find polynomials $A(x)$, of degree at most $\frac{1}{2}(n + k - 2)$, and $B(x)$, of degree at most $k - 1$, such that $A(\alpha_i) = z_i B(\alpha_i)$ for every i. Then $R(x) = A(x) - M(x)B(x)$ is a polynomial of degree at most $\frac{1}{2}(n + k - 2)$, and $R(\alpha_i) = 0$ whenever $M(\alpha_i) = z_i$. Since there are at most $\frac{1}{2}(n - k)$ errors, this happens for at least $n - \frac{1}{2}(n - k) = \frac{1}{2}(n + k)$ values of i. Therefore, the number of roots of R is bigger than its degree, from which it follows that R is identically zero, so that $A(x) = M(x)B(x)$ for every x. From this we can determine M: given k values of x for which $A(x)$ and $B(x)$ are nonzero, one can determine k values of $M(x) = A(x)/B(x)$, and hence determine M.

It remains to show that we can indeed (efficiently) find polynomials $A(x)$ and $B(x)$ with the required properties. The n constraints $A(\alpha_i) = z_i B(\alpha_i)$ turn into n linear constraints on the unknown coefficients of A and B. Since B has $\frac{1}{2}(n - k) + 1$ coefficients and A has $\frac{1}{2}(n + k)$ coefficients, the total number of unknowns is $n + 1$. Since the system of equations is homogeneous (that is, we obtain a solution if we take all unknowns to be zero) and the number of unknowns is greater than the number of constraints, there must be a nontrivial solution: that is, a solution where $A(x)$ and $B(x)$ are not both the zero polynomial. Moreover, we can find such a solution by Gaussian elimination, which takes at most Cn^3 steps.

To summarize: we construct a code by exploiting the fact that two distinct low-degree polynomials cannot be equal for too many values. We then exploit the rigid algebraic structure of low-degree polynomials for the purposes of decoding. The main tool that allows us to do this is linear algebra and in particular the solving of systems of simultaneous equations.

4.2   Reducing the Size of the Alphabet Using Good Codes

The ideas described in the previous section show us how to build codes with efficient encoding and decoding algorithms, but they use relatively large alphabets. In this section we shall exploit these results to build binary codes.

To begin with, let us consider a very obvious method of converting codes over large alphabets into codes over the binary alphabet $\{0, 1\}$. For simplicity, assume that we have a Reed–Solomon code over an alphabet Σ of size 2^l for some integer l. Then we can associate the elements of Σ with binary strings of length l. In such a case, we can regard the Reed–Solomon encoding function, which maps Σ^k to Σ^n, as a function from $\{0, 1\}^{lk}$

to $\{0,1\}^{ln}$. (For instance, an element of Σ^k is a sequence of k objects, each of which is a binary sequence of length l. Putting them together produces a single binary sequence of length kl.) Since the encodings of two distinct messages differ for at least $n - k + 1$ elements of Σ, they must also differ on at least $n - k + 1$ bits.

This gives a fairly reasonable code over the binary alphabet. However, $n - k + 1$ is not as large as a fixed fraction of ln: the ratio $(n - k + 1)/ln$ is less than $1/l$, and since we need 2^l, the size of Σ, to be at least n, we find that this fraction is at most $1/\log_2 n$, which tends to zero as n tends to infinity. However, this can be fixed in a simple way, as we shall see.

The problem with the simple binary approach is that two different elements of Σ may be represented by binary sequences that differ in just one bit. However, the Hamming distance between two binary sequences of length l is usually much larger: it is more like cl for some positive constant c. Suppose that we could represent the elements of Σ as binary sequences of some length L in such a way that the Hamming distances between any two of the sequences used was at least cL. This would allow us to improve our argument above: if the encodings of two messages were different for at least $n - k + 1$ elements of Σ, then they would have to differ on at least $cL(n - k + 1)$ bits rather than just $n - k + 1$, and this is a positive fraction of Ln.

What we are asking for is an encoding of the binary sequences of length l as sequences of length L in such a way that no two codewords are closer than cL to each other. But we know, from the previous section, that such an encoding exists, provided that L and c satisfy appropriate conditions: for instance, it is possible to find an encoding function that works with $L \leqslant 10l$ and $c \geqslant \frac{1}{10}$.

So how do we use this? We start with a binary sequence m of length lk. As above, we associate with this a sequence of length k in the alphabet Σ. We then encode this sequence using the Reed–Solomon code, obtaining a sequence of length n in the alphabet Σ. Next, we convert each term of this sequence into a binary sequence of length l. And, finally, we encode each of these n binary sequences as a sequence of length L using a good encoding function, obtaining as a result a binary sequence of length Ln. We then pass this sequence through the channel, where errors may be introduced. The receiver then breaks the received sequence up into n blocks of length L, decodes each block to work out what binary sequence of length l gave rise to it, and interprets that binary sequence as an

element of Σ. This results in a sequence of n elements of Σ. It then uses the Reed–Solomon decoding algorithm to decode this sequence, producing a sequence of k elements of Σ. Finally, this can be converted into a binary sequence of length lk.

We have said nothing about the efficiency of the encoding and decoding procedures that convert binary sequences of length l into ones of length L and back again, stating merely that they exist. Since efficiency is supposed to be our priority, this may seem rather strange: do we not now face exactly the same problem that we were trying to solve in the first place? Luckily we do not, because although these encoding and decoding procedures may take exponentially long, they take exponentially long as a function of L, and L is much much smaller than n. Indeed, L is proportional to $\log n$, from which it follows that 2^L is bounded above by a polynomial function of n. This is a useful principle: one can afford procedures of exponential complexity provided that one only ever applies them to very short strings.

Thus even though we have not managed to specify the code explicitly, we have demonstrated that there is an encoding and decoding algorithm that runs in polynomial time and that corrects a constant fraction of errors. To complete this section, let us address the question of the probability of decoding error, which we have not yet discussed. The technique described above, of composing encoding functions (and decoding functions), can also be used to improve the above code so that the encoding and decoding still take place in polynomial time, but now the decoding error probability is exponentially small on the binary symmetric channel with parameter p, and the rate is arbitrarily close to the Shannon capacity, which is the theoretical maximum. (The idea is to compose a Reed–Solomon code that has rate close to 1 with a random inner code, and then to show that with random errors most of the inner decoding steps decode correctly. One then uses the outer decoding step to convert the "mostly correct decoding" to a "fully correct decoding.")

5   Impact on Communication and Storage

The mathematical theory of error-correcting codes has made a deep impact on the technologies for storage and communication of information, and we elaborate a little on this below.

Storage of information on digital media is probably the biggest success story for error-correcting codes.

Most known forms of storage media, and in particular standards for audio and data CDs and DVDs, prescribe error-correcting codes based on Reed–Solomon codes. Specifically, they are based on a code that maps $\mathbb{F}_{256}^{223}$ to $\mathbb{F}_{256}^{255}$, where $\mathbb{F}_{256}$ is the finite field with 256 elements. In audio CDs, codes are use to protect from minor scratches, though more serious scratches do lead to audible errors. In data CDs the error correction is stronger (with more redundancies), so that even serious scratches do not lead to loss of data. In all cases (CDs and DVDs) the readers for these devices use fast algorithms for decoding when reading the information on the media. Typically, these algorithms are based on the idea of the previous section, but are much faster implementations (in particular, an algorithm due to E. Berlekamp is widely used). Indeed, several CD readers owe their faster reading speed to faster decoding algorithms. Similarly, the increased storage capacity of DVDs (compared with CDs) is attributed in part to better error-correcting codes. Indeed, error-correction technology played a crucial role in establishing the dominance of audio CDs, which store music digitally, over the traditional, and now almost extinct, gramophone records, which store music in continuous forms. Thus, mathematical advances in coding theory have played an influential role in this technology.

Similarly, error-correcting codes have had a profound effect on communication. Since the late 1960s, error-correcting codes (and decoding) have been used for communication from satellites to their base stations on Earth. Of late, error-correcting codes are also being used in cellular phone communications and modems. Again, the most commonly used code at the time of the writing of this article is the Reed–Solomon code, though this situation has been changing rapidly since the discovery of a new class of codes called "turbo codes." This new family of codes seems to offer significant resilience to random errors (more so than that offered by methods based on Reed–Solomon codes) and uses a simple and quick algorithm, even when the codes used have small block length. These codes and the corresponding decoding algorithm have led to a resurgence of interest in codes constructed with the help of insights from GRAPH THEORY [III.34]. Many of the good properties of turbo codes have been observed only empirically: that is, the codes seem to work very well in practice but it has not yet been proved rigorously that they do. Nevertheless, the observations have been so compelling that new standards for communication are starting to prescribe these codes.

Finally, it must be stressed that while many of the codes used are based on ones that are studied in the mathematical literature, this should not be taken to mean that they can be deployed immediately without further design. For example, the *Mariner* spacecraft used not a Reed–Muller code but a variant of it designed to allow for synchronization between blocks. Similarly, the Reed–Solomon codes used in storage devices are carefully spread out over the disc, so as to allow the physical device to resemble more closely the model of a code over a large alphabet. Note that errors due to a scratch on the disc surface tend to ruin a large collection of bits in a small localized part of the disc. If all the data from a block were sitting in such a neighborhood, the entire block would be lost. So each block of 255 bytes of information is spread out all over the disc. On the other hand, the bytes themselves, which are elements of $\mathbb{F}_{256}$, are written as eight bits in close proximity. So a scratch corrupting one bit out of these eight is also likely to corrupt others in the neighborhood. However, this is all right from the perspective of the model that views the entire collection of eight bits as a single element. In general, working out the right way to apply the theory of error correction to a given scenario is a major challenge, and many success stories would not have been success stories had it not been for some careful design choices.

Mathematics and engineering continue to feed each other in this arena. Mathematical successes, such as new algorithms for decoding Reed–Solomon codes, raise the challenge of how to adapt technology to exploit new algorithms. Engineering successes, such as the discovery of turbo codes that perform extremely well, challenge mathematicians to come up with a formal model and analysis that can explain this success. And if such a model and analysis emerges, it is likely to lead to the discovery of new codes that might surpass the performance of turbo codes and lead to a new set of standards!

6 Bibliographic Notes

The theory of reliable communication and storage of information owes much to the seminal works of Shannon (1948) and Hamming (1950), which formed the basis for much of this article. The Reed–Solomon codes of section 4.1 are from Reed and Solomon (1960). Their decoding algorithm originates in the work of Peterson (1960), though the algorithm given here is significantly simplified. The technique of composing codes is due to Forney (1966).

Over the years, coding theory has amassed a wide variety of results. Some of these give better constructions of codes with faster algorithms. Others provide theoretical upper limits on how well codes can perform. The theory uses an enormous variety of mathematical tools, many of them more advanced than the ones described in this article. Most notable among them are algebraic geometry and graph theory, which are used to construct very good codes, and the theory of orthogonal polynomials, which is used to prove limits on parameters of codes, such as their rate and reliability. Most of the highlights of this vast literature are covered in Pless and Huffman (1998).

Further Reading

Hamming, R. W. 1950. Error detecting and error correcting codes. *Bell System Technical Journal* 29:147–60.

Forney Jr., G. D. 1966. *Concatenated Codes.* Cambridge, MA: MIT Press.

Peterson, W. W. 1960. Encoding and error-correction procedures for Bose–Chaudhuri codes. *IEEE Transactions on Information Theory* 6:459–70.

Pless, V. S., and W. C. Huffman, eds. 1998. *Handbook of Coding Theory*, two volumes. Amsterdam: North-Holland.

Reed, I. S., and G. Solomon. 1960. Polynomial codes over certain finite fields. *SIAM Journal of Applied Mathematics* 8:300–4.

Shannon, C. E. 1948. A mathematical theory of communication. *Bell System Technical Journal* 27:379–423, 623–56.

VII.7   Mathematics and Cryptography
Clifford Cocks

1   Introduction and History

Cryptography is the science of hiding the meaning or content of communications. The aim is that an adversary who sees a message only in its enciphered state cannot make sense of or derive useful information from what is seen. On the other hand, the intended recipient must be able to decipher the true meaning. For most of history cryptography has been an art practiced seriously only by a few—such as governments for military and diplomatic communications—for whom the consequences of unauthorized disclosure of information are damaging enough to justify the expense and inconvenience of enciphering messages. Recently this has changed: one of the results of the information revolution has been the need for instant and secure communication for all on demand. Fortunately, mathematics has come to the rescue and provided theoretical

and algorithmic developments to meet this need. It has also provided entirely new possibilities, such as "digital signatures" (which will be discussed later).

One of the oldest and most basic methods of cryptography is *simple substitution*. Suppose that a message to be enciphered consists of a piece of English text. Before it is sent, the sender and recipient agree on a permutation of the twenty-six letters of the alphabet, which they keep private. An enciphered message might then look something like

ZPLKKWL MFUPP UFL XA EUXMFLP

For very short messages this method is reasonably secure—it is just possible to work out the meaning of the above example by matching letter patterns to those commonly seen in English, but it is quite challenging! However, for longer messages, simply counting the frequencies of each letter and comparing those counts with the frequencies of letters in natural language will almost always reveal the hidden permutation sufficiently to allow the meaning to be easily recovered.

A major leap forward in cryptography came with the advent of mechanical encryption devices in the twentieth century, of which the German Enigma used during World War II is perhaps the most famous example. An account of the fascinating Enigma story and the role of the code breakers of Bletchley Park appears in Simon Singh's excellent book on cryptography (Singh 1999). It is interesting that the principle on which Enigma operates is a development of the simple substitution method. Each letter of the input message is enciphered exactly as a simple substitution, but with the additional rule that the permutation controlling the substitution changes after every letter. A complex electromechanical device controls the substitution process in a deterministic way. The recipient can decipher the message only if he or she can set up another device in exactly the same way as the originator. The information needed to do this is called the *key*. Making sure that keys are known only by the right people is called *key management*. Until the advent of public-key cryptography (to be discussed later), key management was a major inconvenience and expense for anyone wanting to secure their communications.

2   Stream Ciphers and Linear Feedback Shift Registers

Since the advent of computers, information has tended to be transmitted as *binary data*: that is, as a stream

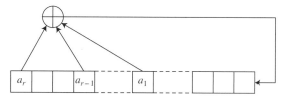

Figure 1 Linear feedback shift register.

of 0s and 1s. For such data there is a rather different method of encipherment based on a device called the *linear feedback shift register*, or LFSR (see figure 1). The first step is to generate a random-looking sequence of 0s and 1s in a deterministic way, and this is done by means of a recurrence formula, of which a simple example is

$$x_t = x_{t-3} + x_{t-4}.$$

Here, addition is mod 2, so x_t will be 1 if an odd number of the terms x_{t-3}, x_{t-4} is 1, and it will be 0 otherwise. We must also specify the first four values of the sequence, so let us begin with 1000. The sequence then continues as follows:

$$100110101111000100110101111\ldots.$$

More generally, one specifies some positive integers $a_1, a_2, \ldots, a_r$, called *feedback positions*—the numbers 3 and 4 in the above example—and defines a sequence by means of the recurrence formula

$$x_t = x_{t-a_1} + x_{t-a_2} + \cdots + x_{t-a_r},$$

where again the addition is mod 2.

A sequence produced in this way usually looks fairly random, but because there are only finitely many binary sequences of length a_r it must eventually repeat. Notice that, in our example, the sequence is periodic with period 15, which is actually the longest possible period, since there are sixteen binary sequences of length 4, and after a moment's thought one sees that the sequence 0000 cannot occur (or else the whole sequence up to then would have had to consist entirely of zeros).

In general, the length of the sequence depends on properties of the polynomial

$$P(x) = 1 + x^{a_1} + x^{a_2} + \cdots + x^{a_r}$$

over the FIELD [I.3 §2.2] $\mathbb{F}_2$ of two elements. As we have just seen in the case $a_r = 4$, the maximum possible sequence length is $2^{a_r} - 1$, and for this length to be achieved the polynomial $P(x)$ must be *irreducible* over $\mathbb{F}_2$: that is, it must not factorize into smaller polynomials. For example, the polynomial $1 + x^4 + x^5$ is not

irreducible, because $(1 + x + x^3)(1 + x + x^2)$ expands out to

$$1 + x + x + x^2 + x^2 + x^3 + x^3 + x^4 + x^5,$$

which equals $1 + x^4 + x^5$ since $1 + 1 = 0$ in the field $\mathbb{F}_2$.

Irreducibility is a necessary condition for the sequence to have the maximum length, but it does not guarantee it. For that we need a second condition: that the polynomial is *primitive*. To see what this means, let us take the polynomial $x^3 + x + 1$ and calculate the remainder when, for the first few positive integers m, we divide x^m by $x^3 + x + 1$ (with all coefficients in $\mathbb{F}_2$). When m goes from 1 to 7 we obtain the polynomials x, x^2, $x + 1$, $x^2 + x$, $x^2 + x + 1$, $x^2 + 1$, 1. For instance,

$$x^6 = (x^3 + x + 1)(x^3 + x + 1) + x^2 + 1,$$

so the remainder on dividing x^6 by $x^3 + x + 1$ is $x^2 + 1$.

Now the first time that we obtained the polynomial 1 was when $m = 7$, and $7 = 2^3 - 1$. This shows that the polynomial $x^3 + x + 1$ is primitive. In general, a polynomial $p(x)$ of degree d is primitive if the first time you obtain a remainder of 1 when you divide x^m by $p(x)$ is when $m = 2^d - 1$.

There are computationally efficient tests for determining whether a polynomial is irreducible and whether it is primitive. The advantage of using a primitive polynomial as the basis of an LFSR is that, in the sequence it generates, no subsequence of length a_r is repeated until all nonzero sequences of length a_r have appeared exactly once.

How is all this applied in cryptography? A simple idea would be to take the stream of bits generated by an LFSR and add it term by term to the message one is enciphering. For instance, if the LFSR generated a sequence that began 1001101 and the message was 0000111, then the encrypted message would begin 1001010. To decipher such a message, one could simply repeat the process: adding the two sequences 1001101 and 1001010 gives the original message 0000111. For this to work, the recipient would need to know the details of the LFSR in order to be able to generate the same sequence 1001101, so one might consider using the feedback positions (in this case 3 and 4) as the secret key.

The above procedure is not good enough to be of practical use because there is an efficient algorithm, due to Berlekamp and Massey (1969), that can recover the feedback rule from the stream of bits it generates. It is better to use some predetermined nonlinear function of the successive sequences of a_r bits in order to

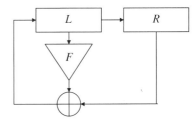

Figure 2 Feistel round structure.

scramble further the sequence of bits produced by the LFSR. Even then, such procedures are simple enough that, with careful design, they can be applied to large amounts of data very quickly.

3   Block Ciphers and the Computer Age

3.1   Data Encryption Standard

When computers started to be used, an entirely different method of cryptography became practical: the block cipher. The first example of this was DES: the *Data Encryption Standard* (first published in 1977). DES was adopted as a standard in 1976 by the U.S. National Bureau of Standards (now the National Institute of Standards and Technology). This enciphers a block of 64 bits at a time, with a key of length 56 bits. It has a particular structure, referred to as a *Feistel cipher* (see figure 2).

This structure is as follows. Given a block of 64 bits, you first divide it into two parts of 32 bits each, and call them L and R. Next, you take a subset of the 56 bits of the key, according to some predetermined rule, and use this subset to define a nonlinear function F, again according to some predetermined rule, which takes 32-bit sequences to 32-bit sequences. You then replace the pair $[L, R]$ by the pair $[R \oplus F(L), L]$. (Here $R \oplus F(L)$ denotes the result of taking the mod-2 sum of R and $F(L)$ one bit at a time.)

Having done that, you repeat the process a number of times, choosing a different nonlinear function F each time (but always deriving it in a predetermined way from the 56-bit key). A complete encryption by DES consists of 16 such rounds, together with some permutation of the bits of the input and output.

One reason for using the Feistel structure is that as long as one knows the 56-bit key it is quite easy to reverse the encryption process. Given a round that performs the transformation

$$[L, R] \rightarrow [R \oplus F(L), L],$$

one can invert it by means of the transformation

$$[L, R] \rightarrow [R, L \oplus F(R)].$$

This has the great advantage that it does not require us to invert F, so even if F is quite complicated the procedure can be easy to carry out.

A number of what are called "modes of use" of DES have been developed. Simply using the algorithm to encrypt each 64-bit block of data in turn is called ECB (for *electronic codebook*) mode. A disadvantage of this mode is that if there is an exact 64-bit repeat in the data then this results in an exact 64-bit repeat in the cipher.

Another mode is CBC, or *cipher block chaining*, mode. Here, each block of data is added mod 2 to the previous block before being encrypted as above. In OFB, or *output feedback*, mode the block of data is added to the DES encipherment of the previous block. It is an easy exercise to see how to decipher in CBC and OFB modes, and in practice these are the two most common modes of use of DES.

3.2   Advanced Encryption Standard

The U.S. National Institute of Standards and Technology recently held a competition for a replacement for DES, to be called the *Advanced Encryption Standard*, or AES. This was to be a 128-wide block cipher with a variety of possible key lengths. Many competing designs were submitted and subjected to public scrutiny, and the winning entry was called *Rijndael*, after the designers Joan Daemen and Vincent Rijmen.

The design is remarkable and elegant and makes use of interesting mathematical structures (Daeman and Rijmen 2002). The 128 bits in each block are thought of as 16 bytes (a byte consists of eight bits), arranged in a 4×4 square. Each byte is then thought of as an element of $\mathbb{F}_{256}$, the field of order 256. Encryption consists of ten or more rounds (the exact number depending upon the key length); and each round mixes the data and the key.

A round consists of a series of steps, typically as follows. First, each byte, regarded as an element of the finite field $\mathbb{F}_{256}$, is replaced by its inverse in the field, except that 0 is left unchanged. Each byte is then regarded as an element of the vector space of dimension 8 over the field $\mathbb{F}_2$ and an invertible linear transformation is applied. Each row of the 4×4 square is then rotated, by a different number of bytes for each row. Next, the values of each column of the square are taken to be the coefficients of a degree 3 polynomial over $\mathbb{F}_{256}$ and this is multiplied by a fixed polynomial and reduced modulo $x^4 + 1$. Finally, the key for the round, which is derived linearly from the encryption key, is added modulo 2 to the 128 bits.

It can be seen that all of these steps are reversible, which makes decipherment straightforward. It is likely that AES will take over from DES as the most widely used block cipher.

4 One-Time Key

The various encryption methods described above rely on the computational difficulty of recovering some secret that protects the enciphered data. There is one classic encryption method that does not rely on this property. This is the "one-time key." Imagine that the message to be enciphered is encoded as a sequence of bits (for example, the standard ASCII encoding that represents each character as eight bits). Suppose that ahead of time the sender and recipient have shared a sequence of random key bits $r_1, \ldots, r_n$ at least as long as the message. Suppose that the message bits are $p_1, p_2, \ldots, p_n$.

The enciphered message is then $x_1, x_2, \ldots, x_n$, where $x_i = p_i + r_i$. Here, as usual, addition is mod 2 addition in each bit. If the bits r_i are fully random, then knowing the sequence x_i gives no information whatsoever about the message sequence p_i. This system is called *one-time key*. It is very secure as long as the key is used only once. However, it is impractical to use this method except in very specialized situations because of the need for sender and recipient to share and keep safe possibly large quantities of key material.

5 Public-Key Cryptography

All of the examples of encryption methods that we have seen so far have had the following structure. Two communicators agree on an algorithm or method for encryption. The choice of method (e.g., simple substitution, AES, or one-time key) can be made public without the security of the system being compromised. The two communicators also agree on a secret key in the form required by the chosen encryption method. This key needs to be kept secure and never revealed to any adversary. The communicators encipher and decipher messages using the algorithm and secret key.

This presents a major problem: how can the communicators securely share the secret key? It would be insecure to exchange this over the same system that they will later use to send enciphered messages. Until so-called public-key methods were discovered this issue

limited the use of encryption to those organizations that could afford the physical security and separate communication channels necessary for distributing keys reliably.

The following remarkable, counterintuitive proposition forms the basis of public-key cryptography: *it is possible for two entities to communicate information in such a way that they start with no secret shared information; an adversary has access to all communications between them; at the end the entities have shared secret knowledge that the adversary is unable to determine.*

It is easy to see how useful such a capability could be. Consider, for example, someone making a purchase over the Internet. Having identified a product one wishes to buy the next step is to send personal information such as credit card details to the vendor. With public-key cryptography it is possible to do this in a secure manner straightaway.

How might public-key cryptography be possible? The structure of a solution was proposed by James Ellis in 1969,[1] with the first public description by Diffie and Hellman (1976). The critical idea is to use a function that is hard to invert unless you have an "inverse key" that helps you to do so.

More formally, a *one-way function H* is a mapping from a set X to itself, with the property that if you are told the value $y = H(x)$ for some $x \in X$, then it is computationally hard to determine x. The inverse key is a secret value, z, say, used in creating the function H, with the property that if you know z then it becomes computationally *easy* to recover x from $H(x)$.

We can use this to solve the problem of secure key exchange as follows. Let us suppose that Bob wishes to send some data securely to Alice. (Particularly useful would be a shared secret that they can use later as a key for subsequent communications.) Alice begins by generating a one-way function H with an inverse key z. She then communicates the function H to Bob, but the inverse key remains her personal secret, which she reveals to no one—not even to Bob. Bob takes the data x that he wishes to send, computes $H(x)$, and returns the result of his computation to Alice. Because Alice has the inverse key z, she can reverse the function H and thereby recover x.

Now suppose that an adversary manages to read all the communications between Alice and Bob. Then the

---

1. See "The possibility of secure non-secret digital encryption," available at www.cesg.gov.uk/site/publications/media/possnse.pdf.

adversary will know the function H and the value $H(x)$. However, Alice has not communicated the inverse key z, so the adversary is faced with the computationally intractable problem of inverting H. Therefore, Bob has successfully transmitted the secret x to Alice without the adversary being able to work out what it is. (For a more precise idea of what computational intractability is and a further discussion of one-way functions, see COMPUTATIONAL COMPLEXITY [IV.20], especially section 7.)

It can be helpful to imagine the one-way function H as a padlock and the inverse key as the key that unlocks the padlock. Then if Alice wants to receive an enciphered message from Bob, she sends him her padlock, retaining the key. Bob locks (enciphers) the message into a box with the padlock, and returns it. Only Alice, who is in possession of the padlock key, can unlock (decipher) the message.

5.1 RSA

It is all very well to have such a framework, but it leaves open an obvious question: how can one produce a one-way function with an inverse key? The following method was published by Rivest, Shamir, and Adleman (1978). It relies on the fact that it is relatively easy to find large prime numbers and multiply them to produce a composite number, but it is much harder, if you are given that composite number, to determine its two prime factors.

To create a one-way function by their method, Alice first finds two large prime numbers P and Q. She then calculates the integer $N = PQ$ and sends it to Bob, together with another integer e called the *encryption exponent*. The values N and e are called the *public parameters* because it does not matter if an adversary knows what they are.

Bob then expresses the secret value x that he wishes to send to Alice as a number modulo N. Next, he computes $H(x)$, which is defined to be x^e mod N, that is, the remainder when x^e is divided by N. Bob sends $H(x)$ to Alice.

Upon receipt of Bob's message, Alice needs to recover x from x^e mod N. This she can do by first calculating the number d that satisfies the equation

$$de \equiv 1 \mod (P-1)(Q-1).$$

To do this efficiently, Alice can use EUCLID'S ALGORITHM [III.22]. Notice, however, that this would not be possible if she did not know the values of P and Q. In fact, the ability to calculate the correct value of d can

be shown to be equivalent to the ability to factorize N. The value of d is Alice's private key (or "inverse key" in the terminology above): it is the secret that can undo the encryption function H.

This is because $H(x)^d$ mod N can be shown to equal x. Indeed, the significance of the number $(P-1)(Q-1)$ is that it equals $\phi(N)$, the number of integers less than N and coprime to N. EULER'S THEOREM [III.58] states that $x^{\phi(N)} \equiv 1$ mod N whenever x is coprime to N. Therefore, $x^{m\phi(N)} \equiv 1$ mod N as well, so if de has the form $m\phi(N) + 1$, as we are assuming, then $H(x)^d \equiv x^{de} \equiv x$ mod N. In other words, if you raise x to the power e mod N and then raise that to the power d mod N you get back to x. (An important point is that raising numbers to powers mod N is computationally easy by the method of "repeated squaring." This is discussed in COMPUTATIONAL NUMBER THEORY [IV.3 §2].)

While it has not been proved that the only way for an adversary to defeat the RSA encryption system is to factorize N, no other general attack has been found. This has created interest in finding improved factorization methods. A number of new subexponential methods—elliptic curve factorization (Lenstra 1987), the multiple polynomial quadratic sieve (Silverman 1987), and the number field sieve (Lenstra and Lenstra 1993)—have been discovered in the years since the RSA algorithm was found. See COMPUTATIONAL NUMBER THEORY [IV.3 §3] for discussions of some of them.

5.1.1 Implementation Details

The security of the RSA system depends on the primes P and Q being large enough to make factorization hard. However, the larger they are, the slower the encryption process is. Thus, there is a trade-off between security and the speed of encryption. A typical choice that is often made is to use primes that are each of 512 bits.

For the deciphering method to work, the encryption exponent e must have no factors in common with either $(P-1)$ or $(Q-1)$. This assumption was needed when we applied Euler's theorem, and if it does not hold then the encryption function is not invertible. Values such as 17 or $2^{16} + 1$ are often used in practice, because making e small reduces the amount of computation needed to calculate the encrypted value x^e mod N. (These two values of e are also well-suited to calculation by repeated squaring.)

5.2 Diffie–Hellman

Another approach to generating a shared secret was published by Whitfield Diffie and Martin Hellman. In their protocol Alice and Bob jointly create a shared secret, which can then be used as the key for one of the conventional cryptographic systems such as AES. To do this, they agree on a large prime number P and a *primitive element* g modulo P, which means a number g such that $g^{P-1} \equiv 1 \bmod P$, but $g^m \not\equiv 1 \bmod P$ for any $m < P - 1$.

Alice then creates her own private key a, a number randomly chosen between 1 and $P - 1$, and calculates $g_a = g^a \bmod P$ and sends this to Bob.

Bob similarly creates his own private key b between 1 and $P - 1$ and calculates and sends $g_b = g^b \bmod P$ to Alice.

Alice and Bob can now create the shared secret $g^{ab} \bmod P$. Alice calculates this as $g_b^a \bmod P$ and Bob calculates this as $g_a^b \bmod P$. Note that all of these terms can be calculated in time logarithmic in a and b through repeated squaring.

An adversary, however, would see only $g^a \bmod P$ and $g^b \bmod P$, and would also know g and P. How could $g^{ab} \bmod P$ be determined from this? One method is to solve what is called the *discrete logarithm problem*. This is the problem of calculating a if you know P, g, and $g^a \bmod P$. For large P this appears to be a computationally intractable problem. It is not known for certain whether there is a faster way for the adversary to calculate $g^{ab} \bmod P$ than computing discrete logarithms—this is called the *Diffie–Hellman problem*—but at present no better method is known.

It is not obvious how to find primitive elements in general, but it is much easier if, as is usually the case, the prime P has been constructed so as to ensure that the factorization of $P - 1$ is known. For instance, if P is of the form $2Q + 1$, where Q is also a prime (such numbers are called *Sophie Germain primes*), then it can be shown that for any a, exactly one of a and $-a$ has the property that its Qth power is congruent to $-1 \bmod P$, and this one is a primitive element. In practice, one can find such primes by a process of trial and error: for example, one can choose a number Q randomly and use randomized primality tests to see whether Q and $2Q + 1$ are prime. Assuming that, as everyone believes, such pairs occur with the "expected" frequency, the probability of finding one on any given attempt is large enough for this approach to be feasible.

5.3 Other Groups

The Diffie–Hellman protocol can be expressed in the language of GROUP THEORY [I.3 §2.1]. Suppose we have a group G and some element $g \in G$. We will require the group to be Abelian and will use "+" to denote the group operation. (In the examples so far, the groups under consideration were multiplicative groups consisting of elements coprime to some integer N, so by using additive notation we are taking a "logarithmic" perspective.)

To execute the protocol Alice computes some private integer a and computes and sends ag to Bob. Note that Alice can compute this sum of a elements of G in time of order logarithmic in a by successive doubling and adding. (In the multiplicative groups considered earlier, "doubling" is squaring, "adding" is multiplying, and "multiplying by a" is raising to the power a.)

Similarly, Bob computes a private integer b and computes and sends bg to Alice.

Both Alice and Bob can calculate the shared value abg. An adversary will know only G, g, ag, and bg.

The question is: which groups can be used in practical cryptographic systems? The critical property is that the discrete logarithm problem in G must be hard; in other words, given G, g, and ag it should be a hard problem to determine a.

One type of group that has aroused interest for cryptographic purposes is the additive group generated by points on an ELLIPTIC CURVE [III.21]. An elliptic curve has an equation of the form

$$y^2 = x^3 + ax + b.$$

It is an interesting exercise to sketch this curve over the real numbers—the shape depends upon how many times the curve

$$y = x^3 + ax + b$$

crosses the x-axis.

It is possible to define an "addition rule" (often called a *group law*) on the points of this curve, as follows. Given two points A and B on the curve, the straight line joining them must meet the curve in a third point, C say. This is because a straight line must meet a cubic in three places precisely. Define A + B to be the mirror image of C in the x-axis (see figure 3).

It is obvious that A + B = B + A from this definition. What is rather more surprising is that the associative law holds. That is, for any three points A, B, and C we have $((A + B) + C) = (A + (B + C))$. There are some deep reasons why this is true, but of course it can be verified by just doing the algebra.

To use this for cryptography the group is formed from the set of points on an elliptic curve defined over

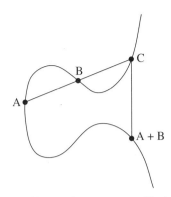

Figure 3 Addition of points on an elliptic curve.

a finite field. The graphical image for the sum of two points is no longer valid, but the algebraic definition still holds, so addition still obeys the associative law. We need to add one further point to the set of points on the curve to function as the zero of the group: this is the "point at infinity" on the curve.

For optimal security it turns out to be best to find a curve defined over $\mathbb{F}_p$ for which the number of elements in the group is a prime number. In fact it is guaranteed—by a deep result on the theory of elliptic curves—that the number of points on a curve defined over $\mathbb{F}_p$ will lie between $p + 1 - 2\sqrt{p}$ and $p + 1 + 2\sqrt{p}$. (See THE WEIL CONJECTURES [V.35].)

The reason this group is used is that for general curves the discrete logarithm problem appears to be particularly hard. If the group has n elements and if we are given group elements g and ag, then the number of steps needed to determine a, by the best algorithms that are currently known, is around $\sqrt{n}$. Since there is a so-called birthday attack that allows one to solve this problem in *any* group with n elements in around $\sqrt{n}$ computational steps, this means that the problem for elliptic curve groups is as hard as it can be. Therefore, whatever level of security you require, the public key is as short as it can be. This is important when there are constraints on the number of bits that can be sent as it allows the protocol to be executed in the minimum possible time.

6   Digital Signatures

As well as secure transmission of data, there is another very useful capability that is provided by public-key cryptography. That is the concept of a *digital signature*. A digital signature is a string of symbols that an author attaches to the end of a message that certifies the authenticity of the message. In other words, it proves that the message was written by the attested author and that it has not been modified. Once the necessary frameworks are in place, this opens up the possibility of much legal business being conducted online.

There are a number of ways that public-key methods can be used to create digital signatures. The one based on the RSA system is perhaps the simplest. Suppose Alice wants to sign documents. Just as she does for encryption, she generates two large prime numbers P and Q and calculates her public modulus $N = PQ$ and her public exponent e. She also generates her private key—the deciphering exponent d with the property that $x^{de} \equiv x \bmod N$ for any x. She will use the same parameters both for encryption and for the creation of digital signatures.

Alice can assume that the recipients of her signed messages know her N and e values. In practice she may have these values themselves signed and certified by a trusted authority or organization that the prospective recipient of a signed message will recognize.

One other component of this system is an object called a *one-way hash function*, which takes as its input the message to be signed, which may be rather long, and outputs a number between 1 and $N - 1$. The important property that a hash function must have is that for any value y between 1 and N it is computationally hard to construct a message x that hashes to that value. This is similar to a one-way function except that we are no longer assuming that for each y there is exactly one x that maps to y. However, the hash function should ideally also be *collision free*, which means that, even though there are many pairs of messages that hash to the same value, it is not easy to find any. Such hash functions need to be carefully designed, but there are some recognized standard hash functions (two of which are called MD5 and SHA-1). Suppose that x is the message to be signed, and let X be the output when you apply the hash function to x. The digital signature that Alice appends to the message is $Y = X^d \bmod N$.

Observe that anyone in possession of Alice's public key can verify the signature by following these steps. First, calculate the hashed value X of the message x, which is possible because the hash function is made public. Next, compute $Z = Y^e \bmod N$, which can be done because the parameters N and e are also public. Finally, verify that X equals Z. In order to fake such a signature, you have to find Y with the property that $Y^e \equiv X \bmod N$. That is, you must know how to calculate X^d, which is computationally intractable if you do not already know d.

It is also possible to construct digital signatures using a public key based on discrete logarithms (Diffie–Hellman type) rather than on factorization (RSA type). The U.S. standards body has published such a proposal: the *Digital Signature Standard* (1994).

7 Some Current Research Topics

Cryptography remains an active and fascinating area for research—there are undoubtedly more results and ideas to be discovered. For a good overview of current activity one should look at recent proceedings of the main conferences, such as Crypto, Eurocrypt, or Asiacrypt (these are published in the Springer series Lecture Notes in Computer Science). The comprehensive book on cryptography by Menezes, van Oorschott, and Vanstone (1996) is a good way to get up to speed on present theory. In this final section I outline just a few of the directions in which the subject is moving.

7.1 New Public-Key Methods

One important area of investigation is the search for new public-key methods and signature schemes. Recently some interesting new ideas have come from the use of *pairings* on elliptic curves (Boneh and Franklin 2001). These are maps w from pairs of points on the curve to either the finite field over which the curve is defined or an extension field.

A pairing w is *bilinear*, in the sense that $w(A+B, C) = w(A, C)w(B, C)$ and $w(A, B + C) = w(A, B)w(A, C)$, where addition is the group operation defined on points of the curve and multiplication takes place in the field.

One way that such a map can be used is to create an "identity-based cryptosystem." Here, a user's identity serves as his or her public key, which eliminates the need for directories or other public-key infrastructure in order to store and propagate public keys.

In such a system, a central authority decides upon a curve, a pairing map w, and a hash function that maps identities to points on the curve. All of this is made public, but there is also a secret parameter, an integer x.

Suppose that the hash function maps Alice's identity to the point A on the curve. The authority calculates Alice's private key xA and issues it to her when she registers, after making appropriate checks on her identity. Similarly, Bob would receive his private key xB,

where B is the point on the curve corresponding to his identity.

Alice and Bob are now able to communicate without any initial key exchange, using the common key $w(xA, B) = w(A, xB)$. The important point is that unlike other public-key systems this can be done without any need to share public keys.

7.2 Communication Protocols

A second area of activity is the study of proposed protocols, especially those likely to become international standards. When public-key methods are to be used in practical communication the sequence of bits to be transmitted needs to be clearly defined, so that both communicating parties understand the same thing by each bit sent. For example, if an n-bit number is transmitted, are the bits transmitted in increasing or decreasing order of significance? The rules or protocols are often enshrined in public standards, and it is important that they do not introduce any weakness into the system.

An example of the sort of weakness that can be introduced in this way is one discovered by Coppersmith (1997) in a seminal paper. He showed that in a low-exponent RSA system (for example, one with encryption exponent equal to 17) a weakness arises if too many of the bits of the number that is to be enciphered are set to publicly known values. This is something that is natural to want to do, if, as is often the case, a large public-key modulus is being used to transmit a much shorter communication key. As a result of Coppersmith's discovery such fields are nowadays usually padded out before they are encrypted, with bits that vary unpredictably.

7.3 Control of Information

Using public-key methods, one can control very precisely how information is released, shared, or generated. Research in this area is usually focused on finding elegant and efficient ways of achieving different sorts of control in a variety of situations. As a simple example, we might want to create a secret that is shared between N people in such a way that if any K people combine their share (where $K < N$) they can reconstruct the secret, but no information can be gained about the secret by any smaller number than K collaborating.

Another example of this type of control is a protocol that allows two participants to create an RSA modulus (a product of two primes) in such a way that neither participant gets to know the primes that were used to produce the modulus. To decipher a message enciphered under this modulus the two participants have to collaborate—neither can achieve this on their own (Cocks 1997).

A third and more amusing example is a protocol that allows Alice and Bob to replicate tossing a coin, but to do it over the telephone. Obviously, it would not be satisfactory for Alice to toss the coin and for Bob to make the call "heads" or "tails"—for how does Bob know that Alice is telling the truth about how the coin actually fell? This problem turns out to have a simple solution. Alice and Bob choose large random sequences. Alice then appends either a 1 or a 0 to her sequence and Bob does the same for his. Alice's extra bit represents the outcome of the coin toss, and Bob's represents his guess. Next, they send one-way hashes of their sequences (with the extra bits appended). At this point, because of the nature of one-way hashes, neither has any idea what the other's sequence is, so, for example, if Alice reveals her hashed sequence first, Bob cannot use this information to increase his chance of guessing correctly. Alice and Bob then exchange the unhashed sequences to see whether Bob's guess was correct. If either does not trust the other, they can hash the other's sequence to check that it really does give the right answer. Since it is hard to find a different sequence that gives the right answer, they can each be confident that the other has not cheated. More complicated protocols of this type have been designed—it is even possible to play poker remotely in this way.

Further Reading

Boneh, D., and M. Franklin. 2001. Identity-based encryption from the Weil pairing. In *Advances in Cryptology—CRYPTO 2001*. Lecture Notes in Computer Science, volume 2139, pp. 213–29. New York: Springer.

Cocks, C. 1997. *Split Knowledge Generation of RSA Parameters. Cryptography and Coding*. Lecture Notes in Computer Science, volume 1355, pp. 89–95. New York: Springer.

Coppersmith, D. 1997. Small solutions to polynomial equations, and low exponent RSA vulnerabilities. *Journal of Cryptology* 10(4):233–60.

Daeman, J., and V. Rijmen. 2002. *The Design of Rijndael*. AES—The Advanced Encryption Standard Series. New York: Springer.

Data Encryption Standard. 1999. Federal Information Processing Standards Publications, number 46-3.

Diffie, W., and M. Hellman. 1976. New directions in cryptography. *IEEE Transactions on Information Theory* 22(6): 644–54.

Digital Signature Standard. 1994. Federal Information Processing Standards Publications, number 186.

Lenstra, A., and H. Lenstra Jr. 1993. *The Development of the Number Field Sieve*. Lecture Notes in Mathematics, volume 1554. New York: Springer.

Lenstra Jr., H. 1987. Factoring integers with elliptic curves. *Annals of Mathematics* 126:649–73.

Massey, J. 1969. Shift-register synthesis and BCH decoding. *IEEE Transactions on Information Theory* 15:122–27.

Menezes, A., P. van Oorschott, and S. Vanstone. 1996. *Applied Cryptography*. Boca Raton, FL: CRC Press.

Rivest, R., A. Shamir, and L. Adleman. 1978. A method for obtaining digital signatures and public-key cryptosystems. *Communications of the Association for Computing Machinery* 21(2):120–26.

Silverman, R. 1987. The multiple polynomial quadratic sieve. *Mathematics of Computation* 48:329–39.

Singh, S. 1999. *The Code Book*. London: Fourth Estate.

VII.8 Mathematics and Economic Reasoning
Partha Dasgupta

1 Two Girls

1.1 Becky's World

Becky, who is ten years old, lives with her parents and an older brother Sam in a suburban town in America's Midwest. Becky's father works in a law firm specializing in small business enterprises. Depending on the firm's profits, his annual income varies somewhat, but it is rarely below $145 000. Becky's parents met in college. For a few years her mother worked in publishing, but when Sam was born she decided to concentrate on raising a family. Now that both Becky and Sam attend school, she does voluntary work in local education. The family live in a two-story house. It has four bedrooms, two bathrooms upstairs and a toilet downstairs, a large drawing-cum-dining room, a modern kitchen, and a family room in the basement. There is a small plot of land in the rear, which the family use for leisure activities.

Although they have a partial mortgage on their property, Becky's parents own stocks and bonds and have a savings account in the local branch of a national bank. Becky's father and his firm jointly contribute to his retirement pension. He also makes monthly payments into a scheme with the bank that will cover college education for Becky and Sam. The family's assets and their lives are insured. Becky's parents often remark that, federal taxes being high, they have to be careful with money; and they are. Nevertheless, they own two cars, the children attend camp each summer, and the family take a vacation together once camp is over. Becky's parents also remark that her generation will be much more prosperous than they. Becky wants to save the environment and insists on biking to school each day. Her ambition is to become a doctor.

1.2 Desta's World

Desta, who is about ten years old, lives with her parents and five siblings in a village in subtropical, southwest Ethiopia. The family live in a two-room, grass-roofed mud hut. Desta's father grows maize and *tef* on half a hectare of land that the government has awarded him. Desta's older brother helps him to farm the land and care for the household's livestock: a cow, a goat, and a few chickens. The small quantity of *tef* produced is sold so as to raise cash income, but the maize is largely consumed by the household as a staple. Desta's mother works a small plot next to their cottage, growing cabbage, onions, and *enset* (a year-round root crop that also serves as a staple). In order to supplement household income, she brews a local drink made from maize. As she is also responsible for cooking, cleaning, and minding the infants, her work day usually lasts fourteen hours. Despite the long hours, it would not be possible for her to complete the tasks on her own. (As the ingredients are all raw, cooking alone takes five hours or more.) So Desta and her older sister help their mother with household chores and mind their younger siblings. Although a younger brother attends the local school, neither Desta nor her older sister has ever been enrolled there. Her parents can neither read nor write, but they are numerate.

Desta's home has no electricity or running water. Around where they live, sources of water, land for grazing cattle, and the woodlands are communal property. They are shared by people in Desta's village; but the villagers do not allow outsiders to make use of them. Each day Desta's mother and the girls fetch water, collect fuelwood, and pick berries and herbs from the local commons. Desta's mother frequently observes that the time and effort needed to collect their daily needs has increased over the years.

There is no financial institution nearby to offer either credit or insurance. As funerals are expensive occasions, Desta's father long ago joined a community insurance scheme (*iddir*) to which he contributes monthly. When Desta's father purchased the cow they now own, he used the entire cash he had accumulated

and stored at home, but had to supplement that with funds borrowed from kinfolk, with a promise to repay the debt when he had the ability to do so. In turn, when they are in need, his kinfolk come to him for a loan, which he supplies if he is able to. Desta's father says that such patterns of reciprocity he and those close to him practice are part of their culture, reflecting their norms of social conduct. He also says that his sons are his main assets, as they are the ones who will look after him and Desta's mother in their old age.

Economic statisticians estimate that, adjusting for differences in the cost of living between Ethiopia and the United States, Desta's family income is about $5000 per year, of which $1000 is attributable to the products they draw from the local commons. However, as rainfall varies from year to year, Desta's family income fluctuates widely. In bad years, the grain they store at home gets depleted well before the next harvest. Food is then so scarce that they all grow weaker, the younger children especially so. It is only after harvest that they regain their weight and strength. Periodic hunger and illnesses have meant that Desta and her siblings are somewhat stunted. Over the years Desta's parents have lost two children in their infancy, stricken by malaria in one case and diarrhea in the other. There have also been several miscarriages.

Desta knows that she will be married (in all likelihood to a farmer, like her father) when she reaches eighteen and will then live on her husband's land in a neighboring village. She expects her life to be similar to that of her mother.

2 The Economist's Agenda

That the lives people are able to construct differ enormously across the globe is a commonplace. In our age of travel, it is even a common sight. That Becky and Desta face widely different futures is also something we have come to expect, perhaps also to accept. Nevertheless, it may not be out of turn to imagine that the two girls are intrinsically very similar: they both enjoy eating, playing, and gossiping; they are close to their families; they like pretty things to wear; and they both have the capacity to be disappointed, get annoyed, be happy. Their parents are also alike. They are knowledgeable about the ways of their worlds. They also care about their families, finding ingenious ways to meet the recurring problem of producing income and allocating resources among family members—over time and allowing for unexpected contingencies. So, a promising route for exploring the underlying causes behind their vastly different conditions of life would be to begin by observing that the constraints the families face are very different: that in some sense Desta's family are far more restricted in what they are able to be and do than Becky's.

Economics in large measure tries to uncover the processes that influence how people's lives come to be what they are. The context may be a household, a village, a district, a state, a country, or the whole world. In its remaining measure, the discipline tries to identify ways to influence those very processes so as to improve the prospects of those who are hugely constrained in what they can be and do. *Modern* economics, by which I mean the style of economics taught and practiced in today's graduate schools, does the exercises from the ground up: from individuals, through the household, village, district, state, country, to the whole world. In varying degrees the millions of individual decisions shape the eventualities people all face; as both theory and evidence tell us that there are enormous numbers of unintended consequences of what we all do. But there is also a feedback, in that those consequences go on to shape what people subsequently can do and choose to do. For example, when Becky's family drive their cars or use electricity, or when Desta's family create compost or burn wood for cooking, they contribute to global carbon emissions. Their contributions are no doubt negligible, but the millions of such tiny contributions cumulatively sum to a sizable amount, having consequences that people everywhere are likely to experience in different ways.

To understand Becky's and Desta's lives, we need first of all to identify the prospects they face for transforming goods and services into further goods and services—now and in the future, under various contingencies. Second, we need to uncover the character of their choices and the pathways by which the choices made by millions of households like Becky's and Desta's go to produce the prospects they all face. Third, and relatedly, we need to uncover the pathways by which the families came to inherit their current circumstances.

The last of these is the stuff of economic history. In studying history we could, should we feel bold, take the long view—from about the time agriculture came to be settled practice in the Fertile Crescent (roughly, Anatolia) some eleven thousand years ago—and try to explain why the many innovations and practices that have cumulatively contributed to the making of

Becky's world either did not reach or did not take hold in Desta's part of the world. (Diamond (1997) is an enquiry into this set of questions.) If we wanted a sharper account, we could study, say, the past six hundred years and ask how it is that, instead of the several regions in Eurasia that were economically promising in about 1400 C.E., it was the unlikely northern Europe that made it and helped to create Becky's world, even while bypassing Desta's. (Landes (1998) is an inquiry into that question. Fogel (2004) explores the pathways by which Europe during the past three hundred years has escaped permanent hunger.) As modern economics is largely concerned with the first two sets of enquiries, this article focuses on them. However, the methods that today's economic historians deploy to answer their questions are not dissimilar to the ones I describe below to study contemporary lives. The methods involve studying individual and collective choices in terms of *maximization exercises*. The predictions of the theories are then tested by studying data relating to actual behavior. Even the ethical foundations of national economic policies involve maximization exercises: the maximization of social well-being subject to constraints. (The treatise that codified this approach to economic reasoning was Samuelson (1947).)

3 The Household Maximization Problem

Both Becky's and Desta's households are microeconomies. Each subscribes to particular arrangements over who does what and when, recognizing that it faces constraints on what its members are capable of doing. We imagine that both sets of parents have their families' well-being in mind and want to do as well as they can to protect and promote it.[1] Of course, both Becky's and Desta's parents would have a wider notion of what constitutes their families than I have allowed here. Maintaining ties with kinfolk would be an important aspect of their lives, a matter I return to later. One also imagines that Becky's and Desta's parents are interested in their future grandchildren's well-being. But as they recognize that their children will in turn care about *their* children, they are right to conclude by recursion that doing the best for their children amounts to doing the best for their grandchildren, for their great grandchildren, and so on, down the generations.

---

1. As suggested by McElroy and Horney in 1981, a realistic alternative would be to suppose that household decisions are reached by negotiation between the various parties (see Dasgupta 1993, chapter 11). Qualitatively, nothing much is lost in my assuming optimizing households here.

Personal well-being is made up of a variety of constituents: health, relationships, place in society, and satisfaction at work are but four. Economists and psychologists have identified ways to represent well-being as a numerical measure. To say that someone's well-being is greater in situation Y than in situation Z is to say that her well-being measure is numerically higher in Y than in Z. A family's well-being is an aggregate of its members' well-beings. As goods and services are among the determinants of well-being (some important examples are food, shelter, clothing, and medical care), the problem that both Becky's and Desta's parents face is to determine, from among those allocations of goods and services that are feasible, the ones that are best for their households. However, both pairs of parents care not only about today, but also about the future. Moreover, the future is uncertain. So when the parents think about which goods and services their households should consume, they are concerned not just with the goods and services themselves, but also with when they will be consumed (food today, food tomorrow, and so on) and what will happen in the case of various contingencies (food the day after tomorrow if rainfall turns out to be bad tomorrow, and so forth). Implicitly or explicitly, both sets of parents convert their experience and knowledge into probabilistic judgments. Some of the probabilities they attach to contingencies are no doubt very subjective, but others, such as their predictions about the weather, are arrived at from long experience.

In subsequent sections we shall study the way in which Becky's and Desta's parents allocate goods and services across time and contingencies. But here we shall keep the exposition simple and consider a model that is *static* and *deterministic*. That is, we shall pretend that the people live in a timeless world, and that they are completely certain about all the information they need in order to make their decisions.

Suppose that a certain household has N members, whom we label $1, 2, \ldots, N$. Let us think about how we can appropriately model the well-being of household member i. As has already been mentioned, well-being is taken to be a real number that depends in some way on the goods and services consumed and supplied by i. It is traditional to divide goods and services into those *consumed* and those *supplied*, and to use positive numbers to represent quantities of the former and negative numbers for the latter. Imagine now that there are M commodities in all. Let $Y_i(j)$ represent the quantity of the jth commodity that is consumed or supplied by i.

By our convention, $Y_i(j) > 0$ if j is consumed by i (e.g., food eaten or clothing worn) and $Y_i(j) < 0$ if j is supplied by i (e.g., labor). Now consider the vector $Y_i = (Y_i(1), \ldots, Y_i(M))$. It denotes the quantities of all the goods and services consumed or supplied by i. Y_i is a point in $\mathbb{R}^M$—the Euclidean space of M dimensions. We now let $U_i(Y_i)$ denote i's well-being. Let us assume that supplying goods and services decreases i's well-being, while consuming them increases it. Because the goods that are supplied by i are measured as negative quantities, we can justifiably assume that $U_i(Y_i)$ increases as any of its components Y_i increases.

The next step is to generalize the model to one that applies to an entire household. The individual well-beings of the members of the household can be collected together so that they themselves form an N-dimensional vector, $(U_1(Y_1), \ldots, U_N(Y_N))$. The household's well-being is dependent in some way on this vector. That is, we say that the well-being of the household is $W(U_1(Y_1), \ldots, U_N(Y_N))$, for some function W. (Utilitarian philosophers have argued that W is simply the *sum* of the U_i.) We also make the natural assumption that W is an increasing function of each U_i (which is certainly the case if W is the sum of the U_i).

Let Y denote the sequence $(Y_1, \ldots, Y_N)$. Y is a point in the NM-dimensional Euclidean space $\mathbb{R}^{NM}$. It can also be thought of as the matrix you obtain if you make a table of the amounts of each commodity consumed or supplied by each member of the household. Now, it is clear that not every Y in $\mathbb{R}^{NM}$ can actually occur: after all, the total amount of any given commodity (in the whole world, say) is finite. So we assume that Y belongs to a certain set J, which we regard as the set of all *potentially feasible* values of Y. Within J we identify a smaller set, F, of "actually feasible" values of Y. This is the set of values of Y from which the household could in principle choose. It is smaller than J because of constraints that the household faces, such as the maximum amount of income it can earn. F is the household's *feasible set*.[2] The decision faced by a household is to choose Y from the feasible set F so as to maximize its well-being $W(U_1(Y_1), \ldots, U_N(Y_N))$. This is called *the household maximization problem*.

It is reasonable, and mathematically convenient, to assume that the sets J and F are both closed and bounded subsets of $\mathbb{R}^{NM}$, and that the well-being function W is continuous. Since every continuous function

---

2. Presently we will see why we need to distinguish J from F, rather than looking just at F.

on a closed bounded set has a maximum, it follows that the household maximization problem has a solution. If, in addition, W is differentiable, the theory of *non-linear programming* can be used to identify the optimality conditions the household's choice must satisfy. If F is a convex set and W is a concave function of Y, those conditions are both necessary and sufficient. The LAGRANGE MULTIPLIERS [III.64] associated with F can be interpreted as *notional prices*: they reflect the worth to the household of slightly relaxing the constraints.

Let us conduct an exercise to test the power of the modern economist's way of studying choice. First, let us assume that W is a *symmetric* and *concave* function of the individual well-beings U_i (as would be the case if W were the sum of the U_i). The symmetry assumption means that if two individuals exchange their well-beings, then W is unchanged; and concavity means, roughly speaking, that, other things being equal, as a U_i increases, the rate of increase of W does not rise. Let us suppose in addition that the household members are identical: that is, let us set all the functions U_i to be equal to a single function, U, say. Assume also that U is a strictly concave function of the Y_i, which means that the rate of increase of well-being declines as consumption increases. Finally, assume that the feasible set F is nonempty, convex, and symmetric. (Symmetry means that if some Y is feasible, and the vector Z is the same as Y except that the consumptions of a pair of individuals in the household have been exchanged, then Z is also feasible.) From these assumptions it can be shown that members of the household would be treated equally: that is, W is maximized when they all receive the same bundle of goods and services.

At low levels of consumption, however, the hypothesis that the function U is concave is unreasonable. To see why, we should note that, typically, 60–75% of the daily energy intake of someone in nutritional balance goes toward maintenance, while the remaining 25–40% is expended in discretionary activities (work and leisure). The 60–75% is rather like a fixed cost: over the long run a person needs it as a minimum no matter what he or she does. The simplest way to uncover the implications of such fixed costs is to continue to suppose that F is convex (which is the case, for example, if there is a fixed quantity of food for allocation among members of the household), but that U is a strictly convex function at low intakes of food and a strictly concave function thereafter. It is not hard to show that a poor household in such a world will maximize its well-being by allocating food

unequally among its members, while a rich household can afford the luxury of equal treatment and will choose to distribute food equally. Suppose, to take a very stylized example, that energy requirement for daily maintenance is 1500 kcal and that a household of four can obtain at most 5000 kcal for consumption. Then equal sharing would mean that no one would have sufficient energy for any work, so it is better to share the food unequally. On the other hand, if the household is able to obtain more than 6000 kcal, it can share the food equally without jeopardizing its future.

There are empirical correlates of this finding. When food is very scarce, the younger and weaker members of Desta's household are given less to eat than the others, even after allowance is made for differences in their ages. In good times, though, Desta's parents can afford to be egalitarian. In contrast, Becky's household can always afford enough food. Her parents therefore allocate food equally every day.

4   Social Equilibrium

Household transactions in Becky's world are carried out mostly in markets. The terms of trade are the quoted market prices. In developing a mathematical construction of social outcomes, I continue to imagine, for simplicity, a static, deterministic world. Let P ($\geqslant 0$) be the vector of market prices and let M ($\geqslant 0$) be the vector of a household's endowments of goods and services. (That is, for each commodity j, $P(j)$ is the price of j and $M(j)$ is the amount of j that the household already has.) Recalling our convention that goods consumed are of positive sign and goods supplied are of negative sign, define $X = \sum Y_i$. (Thus, $X(j) = \sum Y_i(j)$ is the total amount of commodity j that is consumed by the household.) Then $P \cdot X$ is the total price of goods consumed by the household, minus the total price of goods supplied, and $P \cdot M$ is the total value of its endowments. The feasible set F is the set of household choices Y that satisfy the "budget" constraint $P \cdot (X - M) \leqslant 0$.

The income that Becky's household earns from the assets it supplies to the market is determined by market prices (Becky's father's salary, interest rates on bank deposits, returns on shares owned). Those prices in turn depend on the size and distribution of household endowments of goods and services and on household needs and preferences. They depend too on the ability and willingness of institutions, such as private firms and the government, to make use of the rights they in turn have been awarded. These functional relationships explain why Becky's father's skills as a lawyer (itself an asset, termed "human capital" by economists) would not be worth much in Desta's village, even though they are much valued in the United States. In fact, it was a firm belief that lawyers would continue to prove valuable in the United States that encouraged Becky's father to *be* a lawyer.

Although Desta's household does operate in markets (when her father sells *tef* or her mother sells the liquor she has brewed), it undertakes many transactions directly with nature; in the local commons and in farming, and in nonmarket relationships with others in the village. Therefore the F that Desta's household faces is not defined simply by a linear budget inequality, as in the idealized model we have constructed to display Becky's world, but also reflects the constraints that nature imposes, such as soil productivity and rainfall, the assets it has access to, and the terms and conditions involving transactions with others in the village via nonmarket relationships, a matter I come to later. The constraints imposed by nature are felt by Becky's household too, but through market prices. For example, should a drought lead to a fall in world cereal production, it would become noticeable to Becky's household through the high price of cereal. Desta's household, in contrast, would notice it directly from the reduced harvest from their field.

Desta's household assets include the family home, livestock, agricultural implements, and their half hectare of land. The skills Desta's family members have accumulated in farming, managing livestock, and collecting resources from the local commons are part of their human capital. Those skills do not command much return in the global marketplace, but they do shape the household's feasible set F and are vital to the family's well-being. Desta's parents learned those skills from their parents and grandparents, just as Desta and her siblings have learned them from their parents and grandparents. Desta's family can also be said to own a portion of the local commons: in effect, her household shares its ownership with others in the village. Difficulties in reaching and enforcing agreement with neighbors over the use of the local commons are less severe than they are in the case of global commons, such as the atmosphere as a sink for carbon emissions. This is not only because the required negotiations involve far fewer people when the commons are local, but also because there is likely to be greater congruence of opinions and interests among the users. It helps too that the parties are able to observe whether the agreements they made over the use of local commons are being kept.

(See below in our discussion of insurance arrangements in Desta's world.)

Thus, the choices that are available to individuals are affected by the choices that other people make: this results in feedback. In a market economy, the feedback is in large part transmitted in prices. In nonmarket economies the feedback is transmitted through the terms in which households are able to negotiate with one another.

Let us try to model this situation mathematically. We start by imagining an economy of H households. For ease of exposition, I shall suppose that a household's well-being can be expressed directly in terms of its aggregate consumption of goods and services, disregarding how this consumption is distributed among the individual members. Let X_h denote the consumption vector in household h (with the usual sign convention), let J_h be the set of potentially feasible vectors X_h, and let $W_h(X_h)$ be h's well-being.

Within h's potentially feasible set J_h of consumption vectors lies the actual feasible set F_h. In order to model the feedback we shall explicitly recognize that F_h depends on the consumptions of other households. That is, it is a function of the sequence $(X_1, \ldots, X_{h-1}, X_{h+1}, \ldots, X_H)$. To save space, we shall denote this sequence, which consists of every household's consumption vector except h's, by X_{-h}. Formally, F_h is a function (sometimes called a "correspondence") that takes objects of the form X_{-h} to subsets of J_h. Household h's economic problem is to choose its consumption X_h from its feasible set $F_h(X_{-h})$ in such a way as to maximize its well-being $W_h(X_h)$. The optimum choice depends on h's beliefs about X_{-h} and the correspondence $F_h(X_{-h})$.

Meanwhile, all other households are making similar calculations. How can we unravel the feedbacks? One way would be to ask people to disclose their beliefs about the feedbacks. Fortunately, economists avoid that route. So as to anchor their investigation, economists study *equilibrium* beliefs; that is, beliefs that are self-confirming. The idea is to identify states of affairs where the choices people make on the basis of their beliefs about the feedbacks are precisely those that give rise to those very feedbacks. We call any such state of affairs a *social equilibrium*. Formally, a sequence $(X_1^*, \ldots, X_H^*)$ of household choices is called a *social equilibrium* if, for every h, the choice X_h^* of household h maximizes the well-being $W_h(X_h)$ over all choices of X_h in its feasible set $F_h(X_{-h}^*)$.

This raises an obvious question: does a social equilibrium exist? Classic papers by Nash in 1950 and Debreu in 1952 showed that, under a fairly general set of conditions, it always does. Here is a set of conditions that Debreu identified. Assume that each well-being function W_h is continuous and *quasi-concave* (which means that for any potentially feasible choice X'_h in J_h, the set of X_h in J_h for which $W_h(X_h)$ is greater than or equal to $W_h(X'_h)$ is convex). Assume also that for every household h, the feasible set F_h (recall that this is a subset of J_h) is nonempty, compact, and convex, and depends continuously on the choices X_{-h} made by other households. The proof that under the above conditions a social equilibrium always exists is a relatively straightforward use of THE KAKUTANI FIXED POINT THEOREM [V.11 §2], which is itself a generalization of Brouwer's fixed point theorem. Alternative sets of sufficient conditions for the existence of social equilibria (which allow the feasible set $F_h(X_{-h})$ to be nonconvex) have been explored in recent years.

In Becky's world, a social equilibrium is called a *market equilibrium*. A market equilibrium is a price vector P^* ($\geqslant 0$) and a consumption vector X_h^* for each household h, such that X_h^* maximizes $W_h(X_h)$ subject to the budget constraint $P^* \cdot (X_h - M_h) \leqslant 0$, and such that the demands for goods and services across households are feasible (i.e., $\sum (X_h - M_h) \leqslant 0$). That market equilibria are social equilibria, in the sense in which we have defined the latter term here, was demonstrated by Arrow and Debreu in 1954. Debreu (1959) is the definitive treatise on market equilibria. In that book, Debreu followed the leads of Erik Lindahl and Kenneth J. Arrow, by distinguishing goods and services not only in terms of their physical characteristics, but also in terms of the date and contingency in which they appear. Later in this article we shall expand the commodity space in that way to study savings and insurance decisions in both Becky's and Desta's worlds.

One cannot automatically assume that a social equilibrium is just or collectively good. Moreover, except for the most artificial examples, social equilibrium is not unique—which means that a study of equilibria per se leaves open the question of which social equilibrium we should expect to observe. In order to probe that question, economists study disequilibrium behavior and analyze the stability properties of the resulting dynamic processes. The basic idea is to hypothesize about the way people form beliefs about the way the world works, track the consequences of those patterns of learning, and check them against data. It is reasonable to limit such a study by considering only those learning processes that converge to a social equilibrium in stationary environments. Initial beliefs would then dictate which equilibrium is reached in the long run (see, for example, Evans and Honkapohja 2000). Since the study of disequilibria would lengthen this article greatly, we shall continue to study social equilibria here.

5 Public Policy

Economists distinguish between what they call *private goods* and *public goods*. For many goods, consumption is rivalrous: if you consume a bit more from a given supply of such a good (e.g., food), others have that much less to consume. These are private goods. The way to assess their consumption throughout the economy is to add up the amounts consumed by all individual households; which is what we did in the previous section when arriving at the notion of a social equilibrium. Not all goods are like that, however. For example, the extent of national security on offer to you is the same as that on offer to all households in your country. In a just society the law has that same property, as has the state: not only is consumption *not* rivalrous, but in addition, no one can be prevented from availing himself or herself of the entire amount available in the economy. Public goods are goods of this second kind. One models the quantity of a public good as a number G, and the quantity G_h consumed by each household h is deemed to equal G. An example of a public good that has a global coverage is the Earth's atmosphere: the whole world benefits from it jointly.

If the supply of public goods is left to private individuals, then problems arise. For example, even though everyone in a city would benefit from a cleaner, healthier environment, individuals have a strong incentive to free-ride on others when it comes to paying for that cleaner environment. Samuelson showed in 1954 that such a situation resembles the prisoner's dilemma: each party has a strategy that is best for him/her, regardless of what strategies the other parties choose, even though there is another set of strategies, one per party, that is better for everybody. Under such circumstances, one usually needs public measures, such as taxes and subsidies, in order for it to be in the interest of private individuals to act in a way that implements the collectively preferred outcomes. In other words, the dilemma can be expected to be resolved effectively not by markets but by politics. It is widely accepted in political theory that government should be charged with imposing taxes, subsidies, and transfers, and should be engaged in supplying public goods. The government is also the natural agency to supply infrastructure, such as roads, ports, and electrical cables, requiring as they do investments that are huge in comparison with individual incomes. We shall now extend our earlier model to include public goods and infrastructure, so that we can study the government's economic task.

Let us assume that social well-being is a numerical aggregate of household well-beings. Thus, if V is social well-being, we write it as $V(W_1, \ldots, W_H)$. It is natural to postulate that V increases as any W_h increases. (One example of such a function V is the one prescribed by utilitarian philosophy, namely, $W_1 + \cdots + W_H$.) The government chooses what quantities to supply of the various public goods and infrastructure commodities. These numbers can be modeled by two vectors, which we will call G and I, respectively. The government also chooses to impose on each household h certain transfers T_h of goods and services (for example, providing health care and charging income tax). Let us write T for the sequence $(T_1, \ldots, T_H)$. Whether or not a particular choice of vectors G and I is actually feasible for the government will depend on T, so we define K_T to be the set of feasible pairs of vectors (G, I), given the choice of T.

Because we have introduced a new set of goods, we shall have to modify the household well-being functions by enlarging their domains. The obvious notation to express this extra dependence is to write $W_h(X_h, G, I, T_h)$ for the well-being of household h. Moreover, h's feasible set F_h now also depends on G, I, and T_h; so we write the set of feasible household choices as $F_h(G, I, T_h, X_{-h})$.

To try to determine the optimum public policy, imagine a two-stage game. The government has the first move, choosing T and then G and I from K_T. Households go second, reacting to decisions made by the government. Imagine that a social equilibrium $X^* = (X_1^*, \ldots, X_H^*)$ is reached and that the equilibrium is unique. (We assume that if there are multiple equilibria, the government can select among them by resorting to public signals.) Clearly, this equilibrium X^* is a function of G, I, and T. An intelligent and benevolent government will anticipate it and choose T, G, and I from K_T in such a way as to maximize the resulting social well-being $V(W(X_1^*), \ldots, W(X_H^*))$.

The public policy problem we have just designed, involving as it does a double optimization, is technically very difficult. It transpires, for example, that even

in some of the simplest model economies one can imagine, $F_h(G, I, T_h, X_{-h})$ is not convex. This means that the social equilibrium cannot be guaranteed to depend continuously on G, I, and T, as was shown by Mirrlees in 1984. This in turn means that standard techniques are not suitable for the government's optimization problem. In fact, of course, even "double optimization" is a huge simplification. The government chooses; people respond by trading, producing, consuming; the government chooses again; people respond once again—and so forth in an unending series of moves and countermoves. Identifying the optimum public policy involves severe computational difficulties.

6 Matters of Trust: Laws and Norms

The previous examples demonstrate that a fundamental problem facing people who would like to transact with one another concerns *trust*. For example, the extent to which parties trust one another shapes the sets F_h and K_T. If the parties do not trust one another, what could have been mutually beneficial transactions will not take place. But what grounds does a person have for trusting someone to do what he promises to do under the terms of an agreement? Such grounds can exist if promises can be made credible. Societies everywhere have constructed mechanisms to create credibility of this kind, but in different ways. What the mechanisms have in common, however, is that individuals who fail to comply with agreements without a good reason are punished.

How does that common feature work?

In Becky's world the rules governing transactions are embodied in the law. The markets Becky's family enters are supported by an elaborate legal structure (a public good). Becky's father's firm, for example, is a legal entity; as are the financial institutions he deals with in order to accumulate his retirement pension, to save for Becky's and Sam's education, and so on. Even when someone in the family goes to the grocery store, the purchases (paid for with cash or by card) involve the law, which provides protection for both parties (the grocer, in case the cash is counterfeit or the card is void; the purchaser, in case the product turns out on inspection to be substandard). The law is enforced by the coercive power of the state. Transactions involve legal contracts backed by an *external enforcer*, namely, the state. It is because Becky's family and the grocery store's owner are confident that the government has the ability and willingness to enforce contracts (i.e., to continue to supply the public good in question) that they are willing to make transactions.

What is the basis of that confidence? After all, the contemporary world has shown that there are states and there are states. Why should Becky's family trust the government to carry out its tasks in an honest manner? A possible answer is that the government in her country worries about its *reputation*: a free and inquisitive press in a democracy helps to sober the government into believing that incompetence or malfeasance would mean an end to its rule come the next election. Notice how the argument involves a system of interlocking beliefs about the abilities and intentions of others. The millions of households in Becky's country trust their government (more or less!) to enforce contracts, because they know that government leaders know that not to enforce contracts efficiently would mean being thrown out of office. In their turn, each party to a contract trusts the other to refrain from reneging (again, more or less!), because each knows that the other knows that the government can be trusted to enforce contracts. And so on. Trust is maintained by the threat of punishment (a fine, a jail term, dismissal, or whatever) for anyone who breaks a contract. Once again, we are in the realm of equilibrium beliefs, held together by their own bootstraps. Mutual trust encourages people to seek out mutually beneficial transactions and engage in them. As the formal argument that supports the above claim is very similar to the one showing that social norms contain mechanisms for enforcing agreements, we turn to the place of social norms in people's lives.

Although the law of contracts exists also in Desta's country, her family cannot depend on it because the nearest courts are far from their village. Moreover, there are no lawyers in sight. As transport is enormously costly, economic life is shaped outside a formal legal system. In short, crucial public goods and infrastructure are either unavailable, or, at best, in short supply. But even though there is no external enforcer, Desta's parents do make transactions with others. Credit (not dissimilar to insurance in her village) involves saying, "I will lend to you now if you promise to repay me when you can." Saving for funerals involves saying, "I agree to abide by the terms and conditions of the *iddir*." And so on. But why should the parties have any confidence that the agreements will not be broken?

Such confidence can be justified if agreements are *mutually enforced*. The basic idea is this: a credible threat by members of a community that stiff sanctions will be imposed on anyone who breaks an agreement can deter everyone from breaking it. The problem is then to make the threat credible. In Desta's world credibility is achieved by recourse to social norms of behavior.

By a *social norm* we mean a rule of behavior followed by members of a community. A rule of behavior (or "strategy" in economic parlance) reads like, "I will do X if you do Y," "I will do P if Q happens," and so forth. For a rule of behavior to *be* a social norm, it must be in the interest of everyone to act in accordance with the rule if all others act in accordance with it. Social norms are equilibrium rules of behavior. We will now see how social norms work and how transactions based on them compare with market-based transactions. To do this we will study insurance as a commodity.

7   Insurance

To insure oneself against a risk is to act in ways to reduce that risk. (Formally, a RANDOM VARIABLE [III.71 §4] $\tilde{X}$ is said to be riskier than a random variable $\tilde{Y}$ if there is a random variable $\tilde{Z}$ with zero mean such that $\tilde{X}$ has the same distribution as $\tilde{Y} + \tilde{Z}$. In this case, $\tilde{X}$ and $\tilde{Y}$ have the same mean but $\tilde{X}$ is more "spread out.") As long as it does not cost too much, risk-averse households will want to reduce risk by purchasing insurance: in fact, avoiding risk would seem to be a universal urge. To formalize these notions, consider an isolated village, such as Desta's. Suppose for simplicity that it contains H identical households. If household h's food consumption is X_h (represented by a single real number), let us say that its well-being is $W(X_h)$. We shall assume that $W'(X_h) > 0$ (that is, more food leads to greater well-being) and that $W''(X_h) < 0$ (the more food you already have, the less you benefit from yet more). We shall confirm below that the second property of W, its strict concavity, implies, and is implied by, risk aversion; but the basic reason is simple: if W is strictly concave, then you gain less when you are lucky than you lose when you are unlucky.

For simplicity, let us suppose that the production of food by a household h, which is subject to chance factors such as the weather, involves no effort. Since the output is uncertain, we represent it by a random variable $\tilde{X}_h$, with expected value μ, which is assumed to be positive. We shall denote expectations by $\mathbb{E}$.

If a household h is completely self-sufficient, then its expected well-being is simply $\mathbb{E}(W(\tilde{X}_h))$. However, the strict concavity of W implies that $W(\mu) > \mathbb{E}(W(\tilde{X}_h))$. To put this in words: h's well-being at the average level of production is greater than the expectation of h's well-being if the production is random. This means that h will prefer a sure level of consumption to a risky one with mean equal to that sure level. In short, h is risk averse. Define a number $\bar{\mu}$ by $W(\bar{\mu}) = \mathbb{E}(W(\tilde{X}_h))$. So $\bar{\mu}$ is the level of production that achieves the expected well-being. This will be less than μ, and so $\mu - \bar{\mu}$ is a measure of the cost of the risk that a self-sufficient household bears. Notice that the greater the "curvature" of W is, the greater the cost is of the risk associated with $\tilde{X}_h$. (A useful measure of curvature turns out to be $-XW''(X)/W'(X)$. We will make use of this measure when discussing intertemporal choices.) To see how households could gain by pooling their risks, let us write $\tilde{X}_h = \mu + \tilde{\varepsilon}_h$, where $\tilde{\varepsilon}_h$ is a random variable with mean zero, variance σ^2, and finite support. Suppose for simplicity that the random variables $\tilde{\varepsilon}_h$ are identical (i.e., they do not depend on h). Let the correlation coefficient of any two of these distributions be ρ. It turns out that, as long as $\rho < 1$, households can reduce their risks by agreeing to share their outputs. Suppose that households are able to observe one another's outputs. Given that the random variables $\tilde{X}_h$ are identical, the obvious insurance scheme is to share out the outputs equally. Under this scheme, h's uncertain food consumption becomes the average of $\tilde{X}_1, \ldots, \tilde{X}_H$, which is an improvement on self-sufficiency because $\mathbb{E}(W(\sum \tilde{X}_{h'}/H)) > \mathbb{E}(W(\tilde{X}_h))$. The problem is that, without an enforcement mechanism, the agreement to share will not stick, because once each household knows how much food every household has produced, all but the unluckiest households will wish to renege. To see why, notice first that the luckiest households will renege because their outputs are above the average; but this means that the next luckiest set of households will renege because their outputs are above the reduced average; and so on, down to the unluckiest households. Since households know in advance that this will happen if there are no enforcement mechanisms, they will not enter the scheme in the first place: the only social equilibrium is pure self-sufficiency and there is no pooling of risk.

Let us call the insurance game just described the *stage game*. Although pure self-sufficiency is the only social equilibrium for the stage game, we shall now see that the situation changes if the game is played repeatedly. To model this, let us use the letter t to denote time, and let us take time to be a nonnegative integer.

(The game might, for instance, take place every year, with 0 standing for the current year.) Let us assume that the villagers face the same set of risks in each time period, and that the risk in one year is independent of the risks in all other years. Also assume that, in each period, once food outputs are realized, households decide independently of one another whether they will abide by the agreement to share their produce equally or whether they will renege on it.

Although future well-being is important to a household, it will typically be less important than present well-being. To model this we introduce a positive parameter δ, which measures how much a household discounts its future well-being. The assumption is that, when making calculations at $t = 0$, a household divides its well-being at time t by a factor $(1 + \delta)^t$: that is, the importance decays by a certain fixed percentage at each time period. We shall now show that, provided δ is sufficiently small (i.e., provided that households care enough about their future well-being), there is a social equilibrium in which households abide by the agreement to share their aggregate output equally.

Let $\tilde{Y}_h(t)$ be the uncertain amount of food available to household h at time t. If all households are participating in the agreement, then $\tilde{Y}_h(t)$ will be $\mu + (\sum \tilde{\varepsilon}_{h'})/H$, and if there is no agreement, then it will be $\mu + \tilde{\varepsilon}_h$. At time $t = 0$ the total expected well-being of household h, present and future, is $\sum_0^\infty \mathbb{E}(W(\tilde{Y}_h(t)))/(1 + \delta)^t$. (To calculate this we took, for each $t \geqslant 0$, the expected well-being of h at time t and divided it by $(1 + \delta)^t$. Then we added these numbers up.)

Now consider the following simple strategy that h might adopt: it begins by participating in the insurance scheme and continues to participate so long as no household has reneged on the agreement; but it withdraws from the scheme from the date following the first violation of the agreement by some household. Game theorists have christened this the "grim strategy," or simply *grim*, because of its unforgiving nature. Let us see how grim could support the original agreement to share aggregate output equally at every date. (For a general account of repeated games and the variety of social norms that can sustain agreements, see Fudenberg and Maskin (1986).)

Suppose that household h believes that all other households have chosen grim. Then h knows that none of the other households will be the first to defect. What should h do then? We will show that if δ is small enough, h can do no better than play grim. As the same reasoning would be applicable to all other households, we should conclude that, for small enough values of δ, grim is an equilibrium strategy in the repeated game. But if all households play grim, then no household will ever defect. Grim can therefore function as a social norm for sustaining cooperation. Let us see how the argument works.

The basic idea is simple. As all other households are assumed to be playing grim, household h would enjoy a one-period gain by defecting if its own output exceeded the average output of all households. But if h defects in any period, all other households will defect in all following periods (they are assumed to be playing grim, remember). Therefore, h's own best option in all following periods will be to defect also, which means that subsequent to a single deviation by h, the outcome can be predicted to be pure self-sufficiency. So, set against a one-period gain that household h would enjoy if its output exceeded the average output of all households is the loss it would suffer from the following date because of the breakdown of cooperation. That loss exceeds the one-period gain if δ is small enough. So, if δ is sufficiently small, household h will not defect, but will adopt grim; implying that grim is an equilibrium strategy and equal sharing among households in every period is a social equilibrium.

To formalize the above argument, we consider the situation in which h's incentive to defect is greatest. Let A and B be the minimum and maximum possible outputs of any household. Then the maximum gain that household h could possibly enjoy from defecting at $t = 0$ arises if h happens to produce B and all other households happen to produce A. Since the average output in this eventuality is $(B + (H-1)A)/H$, the one-period gain that household h would enjoy from defecting is

$$W(B) - W\left(\frac{B + (H-1)A}{H}\right).$$

But h knows that if it defects, the expected loss in each subsequent period (i.e., from $t = 1$ onward) will be $\mathbb{E}(W(\sum \tilde{X}_{h'}/H)) - \mathbb{E}(W(\tilde{X}_h))$. In order to simplify the notation, let us write $\mathbb{E}(W(\sum \tilde{X}_{h'}/H)) - \mathbb{E}(W(\tilde{X}_h))$ as L. Household h can then calculate that the expected *total* loss it will suffer from defecting at $t = 0$ is $L\sum_1^\infty (1+\delta)^{-t}$, which equals L/δ. If this future loss exceeds the present gain from defecting, then household h will not want to defect. In other words, h will not want to defect if

$$\frac{L}{\delta} > W(B) - W\left(\frac{B + (H-1)A}{H}\right)$$

or

$$\delta < L\bigg/\left(W(B) - W\left(\frac{B + (H-1)A}{H}\right)\right). \qquad (1)$$

But if h does not find it in its interest to defect when the one-period gain from defection is the largest possible, it will certainly not want to defect in any other situation. We conclude that if inequality (1) holds, then grim is an equilibrium strategy and equal sharing among households in every period is a resulting social equilibrium. Notice that, as we said, this will happen if δ is sufficiently small.

We usually reserve the term "society" to denote a collective that has managed to find a mutually beneficial equilibrium. Notice, however, that another social equilibrium of the repeated game is each household for itself. If everyone believed that all others would break the agreement from the start, then everyone would break the agreement from the start. Noncooperation would involve each household selecting the strategy: renege on the agreement. Failure to cooperate could be due simply to a collection of unfortunate, self-confirming beliefs, and nothing else. It is also easy to show that noncooperation is the only social equilibrium of the repeated game if

$$\delta > L\bigg/\left(W(B) - W\left(\frac{B + (H-1)A}{H}\right)\right). \qquad (2)$$

We now have in hand a tool for understanding how a community can slide from cooperative to noncooperative behavior. For example, political instability (in the extreme, civil war) can mean that households are increasingly concerned that they will be forced to disperse from their village. This translates into an increase in δ. Similarly, if households fear that their government is now bent on destroying communal institutions in order to strengthen its own authority, δ will increase. But from (1) and (2) we know that if δ increases sufficiently, then cooperation ceases. The model therefore offers an explanation for why, in recent decades, cooperation at the local level has declined in the unsettled regions of sub-Saharan Africa. Social norms work only when people have reasons to value the future benefits of cooperation.

In the above analysis, we allowed for the possibility that, in each period, household risks were positively correlated. Moreover, the number of households in any village is typically not large. These are two reasons why Desta's household is unable to attain anything like full insurance against the risk they face. Becky's parents, in contrast, have access to an elaborate set of insurance markets that pool the risks of hundreds of thousands of households across the country (even the world, if the insurance company is a multinational). This helps to reduce individual risk more than Desta's parents can, because, first, spatially distant risks are more likely to be uncorrelated, and, second, Becky's parents can pool their risk with many more households. With enough households and enough independence of their risk, THE LAW OF LARGE NUMBERS [III.71 §4] practically guarantees that equal sharing among those households will provide each one with the average μ. This is an advantage of markets, backed by the coercive power of the state as an external enforcer: in a competitive market, insurance contracts are available, enabling people who do not know one another to do business through third parties, in this case the insurance companies.

Many of the risks that Desta's parents face, such as low rainfall, will in fact be very similar for all households in their village. Since the insurance they are able to obtain within their village is therefore very limited, they adopt additional risk-reducing strategies, such as diversifying their crops. Desta's parents plant maize, *tef*, and *enset* (an inferior crop), with the hope that even if maize were to fail one year, *enset* would not let them down. That the local resource base in Desta's village is communally owned probably also has something to do with a mutual desire to pool risks. Woodlands are spatially nonhomogeneous ecosystems. In one year one group of plants bears fruit, in another year some other group does. If the woodland were divided into private parcels, each household would face a greater risk than it would under communal ownership. The reduction in individual household risks owing to communal ownership may be small, but as average incomes are very low, household benefits from communal ownership are large. (For a fuller account of the management of local commons in poor countries, see Dasgupta (1993).)

8 The Reach of Transactions and the Division of Labor

Payments in Becky's world are made in money, expressed in U.S. dollars. Money would not be required in a world where everyone was known to be utterly trustworthy, people did not incur computational costs, and transactions were costless: simple IOUs, stipulating repayment in terms of specific good and services, would suffice in that world. However, we do not live in that world. A debt in Becky's world involves a contract specifying that the borrower is to receive a certain number of dollars and that he promises to repay the

lender dollars in accordance with an agreed schedule. When signing the contract the relevant parties entertain certain beliefs about the dollar's future value in terms of goods and services. Those beliefs are in part based on their confidence in the U.S. government to manage the value of the dollar. Of course, the beliefs are based on many other things as well; but the important point remains that money's value is maintained only because people believe it will be maintained (the classic reference on this is Samuelson (1958)). Similarly, if, for whatever reason, people feared that the value would not be maintained, then it would not be maintained. Currency crashes, such as the one that occurred in Weimar Germany in 1922–23, are an illustration of how a loss in confidence can be self-fulfilling. Bank runs share that feature, as do stock market bubbles and crashes. To put it formally, there are multiple social equilibria, each supported by a set of self-fulfilling beliefs.

The use of money enables transactions to be anonymous. Becky frequently does not know the salespeople in the department stores of her town's shopping mall, nor do they know Becky. When Becky's parents borrow from their bank, the funds made available to them come from unknown depositors. Literally millions of transactions occur each day between people who have never met and will never meet in the future. The problem of creating trust is solved in Becky's world by building confidence in the medium of exchange: money. The value of money is maintained by the state, which has an incentive to maintain it because, as we saw earlier, it wishes not to destroy its reputation and be thrown out of office.

In the absence of infrastructure, markets are unable to penetrate Desta's village. Becky's suburban town, by contrast, is embedded in a gigantic world economy. Becky's father is able to specialize as a lawyer only because he is assured that his income can be used to purchase food in the supermarket, water from the tap, and heat from cooking ovens and radiators. Specialization enables people to produce more in total than they would be able to if they were each required to diversify their activities. Adam Smith famously remarked that the division of labor is limited by the size of the market. Earlier we noted that Desta's household does not specialize, but produces pretty much all of its daily requirements from a raw state. Moreover, the many transactions it enters into with others, being supported by social norms, are of necessity personalized, thus limited. There is a world of a difference between laws and social norms as the basis of economic activities.

9 Borrowing, Saving, and Reproducing

If you do not have insurance, then your consumption will depend heavily on various contingencies. Purchasing insurance helps to smooth out this dependence. We shall see presently that the human desire to smooth out the dependence on contingencies is related to the equally common desire to smooth out consumption across time: they are both a reflection of the strict concavity of the well-being function W. The flow of income over a person's lifetime tends not to be smooth, so people look for mechanisms, such as mortgages and pensions, that enable them to transfer consumption across time. For instance, Becky's parents took out a mortgage on their house because at the time of purchase they did not have sufficient funds to finance it. The resulting debt decreased their future consumption, but it enabled them to buy the house at the time they did and thereby raise current consumption. Becky's parents also pay into a pension fund, which transfers present consumption to their retired future. Borrowing for current consumption transfers future consumption to the present; saving achieves the reverse. Since capital assets are productive, they can earn positive returns if they are put to good use. This is one reason why, in Becky's world, borrowing involves having to pay interest, while saving and investing earn positive returns.

Becky's parents also make a considerable investment in their children's education, but they do not expect to be repaid for this. In Becky's world, resources are transferred from parents to children. Children are a direct source of parental well-being; they are not regarded as investment goods.

A simple way to formulate the problem Becky's parents face when they arrange transfers of resources across time is to imagine that they view themselves as part of a dynasty. This means that, in reaching their consumption and saving decisions, they take explicit note not only of their own well-being and the well-being of Becky and Sam, but also of the well-being of their potential grandchildren, great grandchildren, and so on, down the generations.

To analyze the problem, it is notationally tidiest to assume that time is a continuous variable. At time t (which we take to be greater than or equal to 0), let $K(t)$ denote household wealth and $X(t)$ the consumption rate, which is some aggregate based on the market prices of what they consume. In practice, a household will want to smooth its consumption across both time and contingencies, but in order to concentrate on

time we shall consider a deterministic model. Suppose that the market rate of return on investment is a positive constant r. This means that if household wealth at time t is $K(t)$, then the income it earns from that wealth at t is $rK(t)$. The dynamical equation describing the dynasty's consumption options over time is then

$$\mathrm{d}K(t)/\mathrm{d}t = rK(t) - X(t). \tag{3}$$

The right-hand side of the equation is the difference between the dynasty's investment income at time t (which is r times its wealth at t) and its consumption at t. This amount is saved and invested, so it gives the rate of increase of the dynasty's wealth at t. The present time is $t = 0$ and $K(0)$ is the wealth that Becky's parents have inherited from the past. Earlier, we assumed that the household allocates its consumption across contingencies by maximizing its expected well-being. The corresponding quantity for allocating consumption across time is

$$\int_0^\infty W(X(t))\mathrm{e}^{-\delta t}\,\mathrm{d}t, \tag{4}$$

where, as before, we assume that W satisfies the conditions $W'(X) > 0$ and $W''(X) < 0$. The parameter δ is once again a measure of the rate at which future well-being is discounted—owing to shortsightedness, the possibility of dynastic extinction, and so on. The difference between this and the previous δ is that now we are considering a continuous model rather than a discrete one, but the decay is still assumed to be exponential. In Becky's world the rate of return on investment is large; that is, investment is very productive. So it makes empirical sense to suppose that $r > \delta$. We will see presently that this condition provides Becky's parents with the incentive to accumulate wealth and pass it on to Becky and Sam, who in turn will accumulate their wealth and pass *that* on, and so on. For simplicity, let us suppose that the "curvature" of W, which is $-XW''(X)/W'(X)$, is equal to a parameter α, whose value exceeds 1.[3] As we saw earlier, strict concavity of W means that you gain less from increasing consumption than you lose from decreasing it by the same amount. The strength of this effect is measured by α:

3. This means that W has the form $B - AX^{-(\alpha-1)}$, where A (which is a positive number) and B (which can be of either sign) are the two arbitrary constants that arise when we integrate the curvature of W to arrive at W itself. We will see presently that the values that are adopted for A and B have no bearing on the decisions that Becky's parents will want to make; that is, Becky's parents' optimum decision is independent of A and B. Notice that, as $\alpha > 1$, $W(X)$ is bounded above. The above form is particularly useful in applied work, because in order to estimate $W(X)$ from data on household consumption, one has to estimate only one parameter, α. Empirical studies of saving behavior in the United States have revealed that α is in the range 2–4.

the larger it is, the greater the benefit of any smoothing you are able to do.

Becky's parents' problem at $t = 0$ is to maximize the quantity in (4) by making a suitable choice of the rate at which they consume their wealth (namely, $X(t)$), subject to the condition (3), together with the conditions that $K(t)$ and $X(t)$ should not be negative.[4] This is a problem in the CALCULUS OF VARIATIONS [III.94]. But it is of a somewhat unusual form, in that the horizon is infinite and there is no boundary condition at infinity. The reason for the latter is that Becky's parents would ideally like to *determine* the level of assets that the dynasty ought to aim at in the long run; they do not think it is appropriate to specify it in advance. If we assume for the moment that a solution to the optimization problem exists, then it turns out that it must satisfy the *Euler–Lagrange equation*:

$$\alpha(\mathrm{d}X(t)/\mathrm{d}t) = (r - \delta)X(t), \quad t \geqslant 0. \quad (5)$$

This equation is easily solved, and gives

$$X(t) = X(0)\mathrm{e}^{(r-\delta)t/\alpha}. \quad (6)$$

However, for this problem we are free to choose $X(0)$. Koopmans showed in 1965 that $X(t)$ in (6) is optimal if $W'(X(t))K(t)\mathrm{e}^{-\delta t} \to 0$ as $t \to \infty$. It transpires that, for the model in hand, there is a value of $X(0)$, which we shall write as $X^*(0)$, such that the condition (3) and Koopmans's asymptotic condition are satisfied by the function $X(t)$ given in (6). This implies that $X^*(0)\mathrm{e}^{(r-\delta)t/\alpha}$ is the unique optimum. Consumption grows at the percentage rate $(r - \delta)/\alpha$ and dynastic wealth accumulates continually in order to make that rising consumption level possible. All other things being equal, the larger the productivity of investment r, the higher the optimum rate of growth of consumption. By contrast, the larger the value of α, the lower the rate of growth of consumption, since there is a greater wish to spread it out among the generations.

Let us conduct a simple exercise with our finding. Suppose the annual market rate of return is 4% (i.e., $r = 0.04$ per year)—a reasonable figure for the United States—that δ is small, and that $\alpha = 2$. Then we can conclude from (6) that optimum consumption will grow at an annual rate of 2%; meaning that it will double every thirty-five years—roughly, every generation. The

figure is close to the postwar growth experience in the United States.

For Desta's parents the calculations are very different, since they are heavily constrained in their ability to transfer consumption across time. For example, they have no access to capital markets from which they can earn a positive return. Admittedly, they invest in their land (clearing weeds, leaving portions fallow, and so forth), but that is to prevent the productivity of the land from declining. Moreover, the only way they are able to draw on the maize crop following each harvest is to store it. Let us see how Desta's household would ideally wish to consume that harvest over the annual cycle.

Let $K(0)$ be the harvest, measured, say, in kilocalories. As rats and moisture are a potent combination, stocks depreciate. If $X(t)$ is the planned rate of consumption and y the rate of depreciation of the maize stock, then the stock at t satisfies the equation

$$\mathrm{d}K(t)/\mathrm{d}t = -X(t) - yK(t). \quad (7)$$

Here, y is assumed to be positive and both $X(t)$ and $K(t)$ nonnegative. Let us imagine that Desta's parents regard their household's well-being over the year to be $\int_0^1 W(X(t))\,\mathrm{d}t$. As with Becky's household, let $-XW''(X)/W'(X)$ be equal to a number $\alpha > 1$. Desta's parents' optimization exercise is to maximize $\int_0^1 W(X(t))\,\mathrm{d}t$, subject to (7) and the condition that $K(1) \geqslant 0$.

This is a straightforward problem in the calculus of variations. It can be shown that the optimum maize consumption *declines* over time at the rate y/α. This explains why Desta's family consume less and become physically weaker as the next harvest grows nearer. But Desta's parents have realized that the human body is a more productive bank. So the family consumes a good deal of maize during the months following each harvest so as to accumulate body mass, but they draw on that reserve during the weeks before the next harvest, when maize reserves have been depleted. Across the years maize consumption assumes a sawtooth pattern. (Readers may wish to construct the model that incorporates the body as a store of energy: see Dasgupta (1993) for details.)

As Desta and her siblings contribute to daily household production, they are economically valuable assets. Her male siblings, however, offer a higher return to their parents, because the custom (itself a social equilibrium!) is for girls to leave home on marriage and for boys to inherit the family property and offer security to their parents in old age. Because of an absence

---

4. This problem originated in a classic paper by Ramsey (1928). Ramsey insisted that $\delta = 0$ and devised an ingenious argument to show that an optimum function $X(t)$ exists despite the fact that the integral in (4) does not converge. For simplicity, I am assuming $\delta > 0$. As $W(X)$ is bounded above and $r > 0$ (meaning that it is feasible for $X(t)$ to grow indefinitely), we should expect (4) to converge if $X(t)$ is allowed to rise fast enough.

of capital markets and state pensions, male children are an essential form of investment. The transfer of resources in Desta's household, in contrast to Becky's, will be from the children to their parents.

The under-five mortality rate in Ethiopia was, until relatively recently, in excess of 300 per 1000 births. So, parents had to aim at large families if they were to have a reasonable chance of being looked after by a male child in their old age. But fertility is not entirely a private matter, since people are influenced by the choices of others. This gives rise to a certain inertia in household behavior even under changing circumstances, which is why even though the under-five mortality rate has fallen in Ethiopia in recent decades, Desta has five siblings.[5] High population growth has placed additional pressure on the local ecosystem, meaning that the local commons that used to be managed in a sustainable manner no longer are. That they are not is reflected in Desta's mother's complaint that the daily time and effort required to collect from the local commons has increased in recent years.

10   Differences in Economic Life among Similar People

In this article, I have used Becky's and Desta's experiences to show how it can be that the lives of essentially very similar people can become so different (for further elaboration, see Dasgupta (2004)). Desta's life is one of poverty. In her world people do not enjoy food security, do not own many assets, are stunted and wasted, do not live long (life expectancy at birth in Ethiopia is under fifty years), cannot read or write, are not empowered, cannot insure themselves well against crop failure or household calamity, do not have control over their own lives, and live in unhealthy surroundings. The deprivations reinforce one another, so that the productivity of labor effort, ideas, physical capital, and of land and natural resources are all very low and remain low. The rate of return on investment is zero, perhaps even negative (as it is with the storage of maize). Desta's life is filled with *problems* each day.

Becky suffers from no such deprivation (for example, life expectancy at birth in the United States is nearly eighty years). She faces what her society calls *challenges*. In her world, the productivity of labor effort, ideas, physical capital, and of land and natural resources are all very high and continually increasing; success in meeting each challenge reinforces the prospects of success in meeting further challenges.

We have seen, however, that, despite the enormous differences between Becky's and Desta's lives, there is a unified way to view them, and that mathematics is an essential language for analyzing them. It is tempting to pronounce that life's essentials cannot be reduced to mere mathematics; but in fact mathematics is essential to economic reasoning. It is essential because in economics we deal with quantifiable objects of vital interest to people.

Acknowledgments.   In describing Desta's life, I have received much guidance from my colleague Pramila Krishnan.

Further Reading

Dasgupta, P. 1993. *An Inquiry into Well-Being and Destitution.* Oxford: Clarendon Press.

———. 2004. World poverty: causes and pathways. In *Annual World Bank Conference on Development Economics 2003: Accelerating Development*, edited by F. Bourguignon and B. Pleskovic, pp. 159–96. New York: World Bank and Oxford University Press.

Debreu, G. 1959. *Theory of Value.* New York: John Wiley.

Diamond, J. 1997. *Guns, Germs and Steel: A Short History of Everybody for the Last 13,000 Years.* London: Chatto & Windus.

Durlauf, S. N., and H. Peyton Young, eds. 2001. *Social Dynamics.* Cambridge, MA: MIT Press.

Evans, G., and S. Honkapohja. 2001. *Learning and Expectations in Macroeconomics.* Princeton, NJ: Princeton University Press.

Fogel, R. W. 2004. *The Escape from Hunger and Premature Death, 1700–2100: Europe, America, and the Third World.* Cambridge: Cambridge University Press.

Fudenberg, D., and E. Maskin. 1986. The folk theorem in repeated games with discounting or with incomplete information. *Econometrica* 54(3):533–54.

Landes, D. 1998. *The Wealth and Poverty of Nations.* New York: W. W. Norton.

Ramsey, F. P. 1928. A mathematical theory of saving. *Economic Journal* 38:543–49.

Samuelson, P. A. 1947. *Foundations of Economic Analysis.* Cambridge, MA: Harvard University Press.

———. 1958. An exact consumption loan model with or without the social contrivance of money. *Journal of Political Economy* 66:1002–11.

---

5. See Dasgupta (1993) for the use of interdependent preferences to explain fertility behavior. In the notation of the section on social equilibria, we are to suppose that household h's well-being has the form $W_h(X_h, X_{-h})$, where one of the components of X_h is the number of births in the household, and that the higher the fertility rate is among other households in the village, the larger the desired number of children in h. The theory based on interdependent preferences interprets transitions from high to low fertility rates as bifurcations. Fertility rates are expected to decline even in Ethiopia. Interdependent preferences are currently being much studied by economists (see Durlauf and Young 2001).

VII.9 The Mathematics of Money
Mark Joshi

1 Introduction

The last twenty years have seen an explosive growth in the use of mathematics in finance. Mathematics has made its way into finance mainly via the application of two principles from economics: *market efficiency* and *no arbitrage*.

Market efficiency is the idea that the financial markets price every asset correctly. There is no sense in which a share can be a "good buy," because the market has already taken all available information into account. Instead, the only way that we have of distinguishing between two assets is their differing *risk characteristics*. For example, a technology share might offer a high rate of growth but also a high probability of losing a lot of money, while a U.K. or U.S. government bond would offer a much smaller rate of growth, but an extremely low probability of losing money. In fact, the probability of loss is so small in the latter case that these instruments are generally regarded as being riskless.

No arbitrage, the second fundamental principle, simply says that it is impossible to make money without taking risk. It is sometimes called the "no free lunch" principle. In this context, "making money" is defined to mean making *more* money than could be obtained by investing in a riskless government bond. A simple application of the principle of no arbitrage is that if one changes dollars into yen and then the yen into euros and then the euros back into dollars, then, apart from any transaction costs, one will finish with the same number of dollars that one started with. This forces a simple relationship between the three foreign exchange (FX) rates:

$$FX_{\$,\euro} = FX_{\$,\yen} FX_{\yen,\euro}. \tag{1}$$

Of course, occasional anomalies and exceptions to this relationship can occur, but these will be spotted by traders. The exploitation of the resulting arbitrage opportunity will quickly move the exchange rates until the opportunity disappears.

One can roughly divide the use of mathematics in finance into four main areas.

Derivatives pricing. This is the use of mathematics to price *securities* (i.e., financial instruments), whose value depends purely upon the behavior of another asset. The simplest example of such a security is a *call option*, which is the right, but not the obligation, to buy a share for a pre-agreed price, K, on some specified future date. The pre-agreed price is called the *strike*. The pricing of derivatives is heavily reliant upon the principle of no arbitrage.

Risk analysis and reduction. Any financial institution has holdings and borrowings of assets; it needs to keep careful control of how much money it can lose from adverse market moves and to reduce these risks as necessary to keep within the owners' desired risk profiles.

Portfolio optimization. Any investor in the markets will have notions of how much risk he wants to take and how much return he wants to generate, and most importantly of where he sees the trade-off between the two. There is, therefore, a theory of how to invest in shares in such a way as to maximize the return at a given level of risk. This theory relies greatly on the principle of market efficiency.

Statistical arbitrage. Crudely put, this is using mathematics to predict price movements in the stock market, or indeed in any other market. Statistical arbitrageurs laugh at the concept of market efficiency, and their objective is to exploit the inefficiencies in the market to make money.

Of these four areas, it is derivatives pricing that has seen the greatest growth in recent years, and which has seen the most powerful application of advanced mathematics.

2 Derivatives Pricing

2.1 Black and Scholes

Many of the foundations of mathematical finance were laid down by Bachelier (1900) in his thesis; his mathematical study of BROWNIAN MOTION [IV.24] preceded that of Einstein (see Einstein (1985), which contains his 1905 paper). However, his work was neglected for many years and the great breakthrough in derivatives pricing was made by Black and Scholes (1973). They showed that, under certain reasonable assumptions, it was possible to use the principle of no arbitrage to guarantee a unique price for a call option. The pricing of derivatives had ceased to be an economics problem and had become a mathematics problem.

The result of Black and Scholes was deduced by extending the principle of no arbitrage to encompass the idea that an arbitrage could result not just from static holdings of securities, but also from continuously trading them in a dynamic fashion depending upon their price movements. It is this principle of no *dynamic* arbitrage that underpins derivatives pricing.

In order to properly formulate the principle, we have to use the language of probability theory.

An *arbitrage* is a trading strategy in a collection of assets, the *portfolio*, such that

(i) initially the portfolio has a value of zero;
(ii) the probability that the portfolio will have a negative value in the future is zero;
(iii) the probability that the portfolio will have a positive value in the future is greater than zero.

Note that we do not require the profit to be certain; we merely require that it is possible that money may be made with no risk taken. (Recall that the notion of making money is by comparison with a government bond. The same is true of the "value" of a portfolio: it will be considered positive in the future if its price has increased by more than that of a government bond.)

The prices of shares appear to fluctuate randomly, but often with a general upward or downward tendency. It is natural to model them by means of a Brownian motion with an extra "drift term." This is what Black and Scholes did, except that it was the *logarithm* of the share price $S = S_t$ that was assumed to follow a Brownian motion W_t with a drift. This is a natural assumption to make, because changes in prices behave multiplicatively rather than additively. (For example, we measure inflation in terms of percentage increases.) They also assumed the existence of a riskless bond, B_t, growing at a constant rate. To put these assumptions more formally:

$$\log S = \log S_0 + \mu t + \sigma W_t, \qquad (2)$$
$$B_t = B_0 e^{rt}. \qquad (3)$$

Notice that the expectation of $\log S$ is $\log S_0 + \mu t$, so it changes at a rate μ, which is called the *drift*. The term σ is known as the *volatility*. The higher the volatility, the greater the influence of the Brownian motion W_t, and the more unpredictable the movements of S. (An investor will want a large μ and a small σ; however, market efficiency ensures that such shares are rather rare.) Under additional assumptions such as that there are no transaction costs, that trading in a share does not affect its price, and that it is possible to trade continuously, Black and Scholes showed that if there is no dynamic arbitrage, then at time t, the price of a call

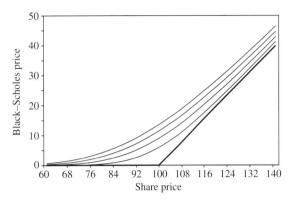

Figure 1 The Black–Scholes price of a call option struck at 100 for various maturities. The value decreases as maturity decreases, with the bottom line denoting a maturity of zero.

option, $C(S, t)$, that expires at time T must be equal to

$$\mathrm{BS}(S, t, r, \sigma, T) = S\Phi(d_1) - Ke^{-r(T-t)}\Phi(d_2), \quad (4)$$

with

$$d_1 = \frac{\log(S/K) + (r + \sigma^2/2)(T - t)}{\sigma\sqrt{T - t}} \qquad (5)$$

and

$$d_2 = \frac{\log(S/K) + (r - \sigma^2/2)(T - t)}{\sigma\sqrt{T - t}}. \qquad (6)$$

Here, $\Phi(x)$ denotes the probability that a standard normal random variable has value less than x. As x tends to ∞, $\Phi(x)$ tends to 1, and as x tends to $-\infty$, $\Phi(x)$ tends to 0. If we let t tend to T, we find that d_1 and d_2 tend to ∞ if $S_T > K$ (in which case $\log(S_T/K) > 0$) and to $-\infty$ if $S_T < K$. It follows that the price $C(S, t)$ converges to $\max(S_T - K, 0)$, which is the value of a call option at expiry, just as one would expect. We illustrate this in figure 1.

There are a number of interesting aspects to this result that go far beyond the formula itself. The first and most important result is that the price is unique. Using just the hypothesis that it is impossible to make a riskless profit, along with some natural and innocuous assumptions, we discover that there is only one possible price for the option. This is a very strong conclusion. It is not just the case that the option is a bad deal if traded at a different price: if a call option is bought for less or sold for more than the Black–Scholes price, then a *riskless profit* can be made.

A second fact, which may seem rather paradoxical, is that μ, the drift, does not appear anywhere in the Black–Scholes formula. This means that the expected behavior of the share's future mean price does not affect the price of the call option; our beliefs about the probability that the option will be used do not affect its price. Instead, it is the volatility of the share price that is all-important.

As part of their proof, Black and Scholes showed that the call option price satisfied a certain partial differential equation (PDE) now known as the *Black–Scholes equation*, or BS equation for short:

$$\frac{\partial C}{\partial t} + rS\frac{\partial C}{\partial S} + \tfrac{1}{2}\sigma^2 S^2 \frac{\partial^2 C}{\partial S^2} = rC. \tag{7}$$

This part of the proof did not rely on the derivative being a call option: there is in fact a large class of derivatives whose prices satisfy the BS equation, differing only in boundary conditions. If one changes variables, setting $\tau = T - t$ and $X = \log S$, then the BS equation becomes THE HEAT EQUATION [I.3 §5.4] with an extra first-order term which can easily be removed. This means that the value of an option behaves in a similar way to time-reversed heat: it diffuses and spreads out the farther back one gets from the option's expiry and the more uncertainty there is about the value of the share at time T.

2.2 Replication

The fundamental idea underlying the Black–Scholes proof and much of modern derivatives pricing is *dynamic replication*. Suppose we have a derivative Y that pays an amount that depends on the value of the share at some set of times $t_1 < t_2 < \cdots < t_n$, and suppose that the payout occurs at a certain time $T \geqslant t_n$. This can be expressed in terms of a *payoff function*, $f(t_1, \ldots, t_n)$.

The value of Y will vary with the share price. If, in addition, we hold just the right number of the shares themselves, then a portfolio consisting of Y and the shares will be instantaneously immune to changes in the share price, i.e., its value will have zero rate of change with respect to the share price. As the value of Y will vary with time and share price, we will need to continuously buy and sell shares to maintain this neutrality to share-price movements. If we have sold a call option, then it turns out that we will have to buy when the share price goes up and sell when it goes down; so these transactions will cost us a certain amount of money.

Black and Scholes's proof showed that this sum of money was always the same and that it could be computed. The sum of money is such that by investing it in shares and riskless bonds, one can end up with a portfolio precisely equal in value to the payoff of Y no matter what the share price did in between.

Thus if one could sell Y for more than this sum of money, one would simply carry out the trading strategy from their proof and always end up ahead. Similarly, if one can buy Y for less, one does the negative of the strategy and always ends up ahead. Both of these are outlawed by the principle of no arbitrage, and a unique price is guaranteed.

The property that the payoff of any derivative can be replicated is called *market completeness*.

2.3 Risk-Neutral Pricing

A curious aspect of the Black–Scholes result, mentioned above, is that the price of a derivative does not depend upon the drift of the share price. This leads to an alternative approach to derivatives pricing theory called *risk-neutral pricing*. An arbitrage can be thought of as the ultimate unfair game: the player can only make money. By contrast, a MARTINGALE [IV.24 §4] encapsulates the notion of a fair game: it is a random process whose expected future value is always equal to its current value. Clearly, an arbitrage portfolio can never be a martingale. So if we can arrange for everything to be a martingale, there can be no arbitrages, and the price of derivatives must be free of arbitrage.

Unfortunately, this cannot be done because the price of the riskless bond grows at a constant rate, and is therefore certainly not a martingale. However, we can carry out the idea for *discounted prices*: that is, for prices of assets when they are divided by the price of the riskless bond.

In the real world, we do not expect discounted prices to be martingales. After all, why buy shares if their mean return is no better than that of a bond that carries no risk? Nevertheless, there is an ingenious way of introducing martingales into the analysis: by changing the PROBABILITY MEASURE [III.71 §2] that one uses.

If you look back at the definition of arbitrage, you will see that it depends only on which events have zero probability and which have nonzero probability. Thus, it uses the probability measure in a rather incomplete way. In particular, if we use a different probability measure for which the sets of measure zero are the same, then the set of arbitrage portfolios will not change. Two measures with the same sets of measure zero are said to be *equivalent*.

A theorem of Girsanov says that if you change the drift of a Brownian motion, then the measure that you derive from it will be equivalent to the measure you had before. This means that we can change the term μ. A good value to choose turns out to be $\mu = r - \tfrac{1}{2}\sigma^2$.

With this value of μ, one has

$$\mathbb{E}(S/B_t) = S/B_0 \qquad (8)$$

for any t, and since we can take any time as our starting point, it follows that S/B_t is a martingale. (The extra $-\frac{1}{2}\sigma^2$ in the drift comes from the concavity of the coordinate change to log-space.) This means that the expectation has been taken in such a way that shares do not carry any greater return, on average, than bonds. Normally, as we have mentioned, one would expect an investor to demand a greater return from a risky share than from a bond. (An investor who does not demand such compensation is said to be *risk neutral*.) However, now that we are measuring expectations differently, we have managed to build an equivalent model in which this is no longer the case.

This yields a way of finding arbitrage-free prices. First, pick a measure in which the discounted price processes of all the fundamental instruments, e.g., shares and bonds, are martingales. Second, set the discounted price process of derivatives to be the expectations of their payoff; this makes them into martingales by construction.

Everything is now a martingale and there can be no arbitrage. Of course, this merely shows that the price is nonarbitrageable, rather than that it is the *only* nonarbitrageable price. However, work by Harrison and Kreps (1979) and by Harrison and Pliska (1981) shows that if a system of prices is nonarbitrageable, then there must be an equivalent martingale measure. Thus the pricing problem is reduced to classifying the set of equivalent martingale measures. Market completeness corresponds to the pricing measure being unique.

Risk-neutral evaluation has become such a pervasive technique that it is now typical to start a pricing problem by postulating risk-neutral dynamics for assets rather than real-world ones.

We now have two techniques for pricing: the Black–Scholes replication approach, and the risk-neutral expectation approach. In both cases, the real-world drift, μ, of the share price does not matter. Not surprisingly, a theorem from pure mathematics, the Feynman-Kac theorem, joins the two approaches together by stating that certain second-order linear partial differential equations can be solved by taking expectations of diffusive processes.

2.4 Beyond Black–Scholes

For a number of reasons, the theory outlined above is not the end of the story. There is considerable evidence that the log of the share price does not follow a Brownian motion with drift. In particular, market crashes occur. For example, in October 1987 the stock market fell by 30% in one day and financial institutions found that their replication strategies failed badly. Mathematically, a crash corresponds to a jump in the share price, and Brownian motion has the property that all paths are continuous. Thus the Black–Scholes model failed to capture an important feature of share-price evolution.

A reflection of this failure is that options on the same share but with differing strike prices often trade with different volatilities, despite the fact that the BS model suggests that all options should trade with the same volatility. The graph of volatility as a function of the strike price is normally in the shape of a smile, displaying the disbelief of traders in the Black–Scholes model.

Another deficiency of the model is that it assumes that the volatility is constant. In practice, market activity varies in intensity and goes through some periods when share prices are much more volatile and others when they are much less so. Models must therefore be corrected to take account of the stochasticity of volatility, and the prediction of volatility over the life of an option is an important part of its pricing. Such models are called stochastic volatility models.

If one examines the data on small-scale share movements, one quickly discovers that they do not resemble a diffusion. They appear to be more like a series of small jumps than a Brownian motion. However, if one rescales time so that it is based on the number of trades that have occurred rather than on calendar time, then the returns do become approximately normal. One way to generalize the Black–Scholes model is to introduce a second process that expresses trading time. An example of such a model is known as the *variance gamma model*. More generally, the theory of Lévy processes has been applied to develop wider theories of price movements for shares and other assets.

Most generalizations of the Black–Scholes model do not retain the property of market completeness. They therefore give rise to many prices for options rather than just one.

2.5 Exotic Options

Many derivatives have quite complicated rules to determine their payoffs. For example, a *barrier option* can be exercised only if the share price does not go below a certain level at any time during the contract's life, and an

Asian option pays a sum that depends on the average of the share price over certain dates rather than on the price at expiry. Or the derivative might depend upon several assets at once, such as, for example, the right to buy or sell a basket of shares for a certain price. It is easy to write down expressions for the value of such derivatives in the Black–Scholes model, either via a PDE or as a risk-neutral expectation. It is not so easy to evaluate these expressions. Much research is therefore devoted to developing efficient methods of pricing such options. In certain cases it is possible to develop analytic expressions. However, these tend to be the exception rather than the rule, and this means that one must resort to numerical techniques.

There is a wealth of methods for solving PDEs and these can be applied to derivatives-pricing problems. One difficulty in mathematical finance, however, is that the PDE can be very high dimensional. For example, if one is trying to evaluate a credit product depending on 100 assets, the PDE could be 100 dimensional. PDE methods are most effective for low-dimensional problems, and so research is devoted to trying to make them effective in a wider range of cases.

One method that is less affected by dimensionality is Monte Carlo evaluation. The basis of this method is very simple: both intuitively and (via the law of large numbers) mathematically, an expectation is the long-run average of a series of independent samples of a random variable X. This immediately yields a numerical method for estimating $\mathbb{E}(f(X))$. One simply takes many independent samples X_i of X, calculates $f(X_i)$ for each one and computes their average. It follows from the CENTRAL LIMIT THEOREM [III.71 §5] that the error after N draws is approximately distributed as a normal distribution with variance equal to $N^{-1/2}$ times the variance of $f(X)$. The rate of convergence is therefore dimension independent. If the variance of $f(X)$ is large, it may still be rather slow, however. Much effort is therefore devoted by financial mathematicians to developing methods of reducing the variance when one computes high-dimensional integrals.

2.6 Vanilla versus Exotics

Generally, a simple option to buy or sell an asset is known as a *vanilla option*, whereas a more complicated derivative is known as an *exotic option*. An essential difference between the pricing of the two is that one can hedge an exotic option not just with the underlying share, but also by trading appropriately in the vanilla

options on that share. Typically, the price of a derivative will depend not just on observable inputs, such as the share price and interest rates, but also on unobservable parameters, such as the volatility of the share price or the frequency of market crashes, which cannot be measured but only estimated.

When trading exotic options, one wishes to reduce dependence upon these unobservable inputs. A standard way to do this is to trade vanilla options in such a way as to make the rate of change of the value of the portfolio with respect to such parameters equal to zero. A small misestimation of their value will then have little effect on the worth of the portfolio.

This means that when one prices exotic options, one wishes not just to capture the dynamics of the underlying asset accurately but also to price all the vanilla options on that asset correctly. In addition, the model will predict how the prices of vanilla options change when the share price changes. We want these predictions to be accurate.

The BS model takes volatility to be constant. However, one can modify it so that the volatility varies with the share price and over time. One can choose how it varies in such a way that the model matches the market prices of all vanilla options. Such models are known as *local volatility models* or *Dupire models*. Local volatility models were very popular for a while, but have become less so because they give a poor model for how the prices of vanilla options change over time.

Much of the impetus behind the development of the models we mentioned in section 2.4 comes from the desire to produce a model that is computationally tractable, prices all vanilla options correctly, and produces realistic dynamics for both the underlying assets and the vanilla options. This problem has still not been wholly solved. There tends to be a trade-off between realistic dynamics and perfect matching of the vanilla options market. One compromise is to fit the market as well as possible using a realistic model and then to superimpose a local volatility model to remove the remaining errors.

3 Risk Management

3.1 Introduction

Once we have accepted that it is impossible to make money in finance without taking risk, it becomes important to be able to measure and quantify risks. We wish to measure accurately how much risk we are taking and decide whether we are comfortable with that level of risk. For a given level of risk, we want to maximize our expected return. When considering a new transaction, we will want to examine how it affects our risk levels and returns. Certain transactions may even reduce our risk while increasing our returns if they cancel out other risk. (A risk that can be canceled out by other risks that have a tendency to move in the opposite direction is called *diversifiable*.)

The control of risk becomes particularly important when dealing with portfolios of derivatives, which are often of zero value initially but which can very quickly change value. Placing a limit on the value of the contracts held is therefore not of much use, and controls based on deal sizes are complicated by the fact that often many derivatives contracts largely cancel each other out; it is the *residual* risk that one wishes to control.

3.2 Value-at-Risk

One method of limiting an institution's risks in derivatives trading is to place a limit on the amount it can lose with a given probability over a specified period of time. For instance, one might consider the losses at a 1% level over ten days, or at a 5% level over one day. This value is called *Value-at-Risk* or VAR.

To compute VAR one has to build up a probabilistic model of how the portfolio of derivatives might change in value over the time period. This requires a model of how all the underlying assets can move. Given this model, one then builds up the distribution of possible profits and losses over the given time period. Once one has this distribution one simply reads off the desired percentile.

The issues involved in modeling the changes for VAR computation are quite different from those for derivatives pricing. Typically, a VAR computation is done over a very short time period, such as one or ten days, unlike the pricing of an option, which deals with a long time frame. Also, one is not interested in the typical path for VAR, but instead one focuses on the extreme moves. In addition, since it is the VAR of an entire portfolio that matters, one has to develop an accurate model of the underlying assets' *joint* distributions: the movement of one underlying asset could magnify the price movement of another, or it could act as a hedge.

There are two main approaches to developing a probabilistic model for computing VAR. The first, the historical approach, is to record all the daily changes over some time period, for example two years, and then assume that the set of changes tomorrow will be identical to one of the sets of changes we have recorded. If we assign equal probability to each of those changes, then we get an approximation to the profit and loss distribution, from which we can read off the desired percentile. Note that as we are using a day's change for all assets simultaneously, we automatically get an approximation to the joint distribution of all the asset prices.

A second approach is to assume that asset price movements come from some well-known class of distributions. For example, we could assume that the logs of the asset price movements are jointly normal. We would then use historical data to estimate the volatilities and the correlations between the various prices. The main difficulty with this approach is obtaining robust estimates of the correlations given a limited amount of data.

4 Portfolio Optimization

4.1 Introduction

The job of a fund manager is to maximize the return on the money invested while minimizing the risk. If we assume that markets are efficient, then there is no point in trying to pick shares that we believe to be undervalued as we have assumed that they do not exist. A corollary is that just as no shares are good buys, no shares are bad buys. In any case, over half the shares in the market are owned via funds and therefore under the control of fund managers. Therefore, the average fund manager cannot expect to outperform the market.

It may seem that this does not leave much for fund managers to do, but in fact it leaves two things.

(i) They can attempt to control the amount of risk they are taking.
(ii) For a given level of risk, they can maximize their expected return.

To do these things requires an accurate model of the joint distribution of asset prices over the longer term, and a quantifiable notion of risk.

4.2 The Capital Asset Pricing Model

Portfolio theory has been in its modern form for longer than derivatives pricing. As an area, it relies less on stochastic calculus and more on economics. We briefly review the key ideas. The best-known model for modeling portfolio returns is the *capital asset pricing model* (or CAPM), which was introduced in the

1950s by Sharpe (see Sharpe 1964), and is still ubiquitous. Sharpe's model built on earlier work of Markowitz (1952).

The fundamental problem in this area is to assess what portfolio of assets, generally shares, an investor should hold in order to maximize returns at a given level of risk. The theory requires assumptions to be made about the joint distribution of share returns, e.g., joint normality, and/or about the risk preferences of investors, e.g., that they only care about the mean and variance of returns.

Under these assumptions, the CAPM yields the result that every investor should hold a multiple of the "market portfolio," which is essentially a portfolio consisting of everything traded in appropriate quantities to achieve maximum diversification, together with a certain amount of the risk-free asset. The relative amounts are determined by the investor's risk preferences.

A consequence of the model is the distinction between diversifiable risk and undiversifiable risk. While investors are compensated for taking undiversifiable, or systematic, risk via higher expected returns, diversifiable risk does not carry a risk premium. This is because one can cancel out diversifiable risk by holding appropriate combinations of other assets. Therefore, if it carried a risk premium, investors could receive extra return without taking any risk.

Much of the current research in this area is directed at trying to find more accurate models for the joint distribution of returns, and at finding techniques that estimate the parameters of such returns. A related problem is the "equity premium puzzle," which is that the excess return on investing in shares is much higher than the model predicts for reasonable levels of risk aversion.

5   Statistical Arbitrage

We only briefly mention statistical arbitrage as it is a rapidly changing area that is shrouded in secrecy. The fundamental idea in this area is to squeeze information out of asset price movements that the market has not already acted on. It therefore contradicts the principle of market efficiency, which says that all available information is already encoded in the market price. One explanation is that it is the action of taking such arbitrages that makes the market efficient.

Further Reading

Bachelier, L. 1900. *La Théorie de la Spéculation*. Paris: Gauthier-Villars.

Black, F., and M. Scholes. 1973. The valuation of options and corporate liabilities. *Journal of Political Economy* 81: 637–54.
Einstein, A. 1985. *Investigations on the Theory of the Brownian Movement*. New York: Dover.
Harrison, J. M., and D. M. Kreps. 1979. Martingales and arbitrage in multi-period securities markets. *Journal of Economic Theory* 20:381–408.
Harrison, J. M., and S. R. Pliska. 1981. Martingales and stochastic integration in the theory of continuous trading. *Stochastic Processes and Applications* 11:215–60.
Markowitz, H. 1952. Portfolio selection. *Journal of Finance* 7:77–99.
Sharpe, W. 1964. Capital asset prices: a theory of market equilibrium under conditions of risk. *Journal of Finance* 19:425–42.

VII.10   Mathematical Statistics
Persi Diaconis

1   Introduction

Suppose you want to measure something: your height, or the velocity of an airplane for example. You take repeated measurements $x_1, x_2, \ldots, x_n$ and you would like to combine them into a final estimate. An obvious way of doing this is to use the *sample mean* $(x_1 + x_2 + \cdots + x_n)/n$. However, modern statisticians use many other estimators, such as the median or the *trimmed mean* (where you throw away the largest and smallest 10% of the measurements and take the average of what is left). Mathematical statistics helps us to decide when one estimate is preferable to another. For example, it is intuitively clear that throwing away a random half of the data and averaging the rest is foolish, but setting up a framework that shows this clearly turns out to be a serious enterprise. One benefit of the undertaking is the discovery that the mean turns out to be inferior to nonintuitive "shrinkage estimators" even when the data are drawn from a PROBABILITY DISTRIBUTION [III.71] as natural as the bell-shaped curve (that is, are NORMALLY DISTRIBUTED [III.71 §5]).

To get an idea of why the mean may not always give you the most useful estimate, consider the following situation. You have a collection of a hundred coins and you would like to estimate their biases. That is, you would like to estimate a sequence of a hundred numbers, where the nth number θ_n is the probability that the nth coin will come up heads when it is flipped. Suppose that you flip each coin five times and note down how many times it shows heads. What should your estimate be for the sequence $(\theta_1, \ldots, \theta_{100})$? If you use the means, then your guess for θ_n will be the number of times the nth coin shows heads, divided by 5. However, if you do this, then you are likely to get some very anomalous results. For instance, if all the coins happen to be unbiased, then the probability that any given coin shows up heads five times is $1/32$, so you are likely to guess that around three of the coins have biases of 1. So you will be guessing that if you flip those coins five hundred times then they will come up heads every single time.

Many alternative methods of estimation have been proposed in order to deal with this obvious problem. However, one must be careful: if a coin comes up heads five times it could be that θ_i really is equal to 1. What reason is there to believe that a different method of estimation is not in fact taking us further from the truth?

Here is a second example, drawn from work of Bradley Efron, this time concerning a situation from real life. Table 1 shows the batting averages of eighteen baseball players. The first column shows the proportion of "hits" for each player in their first forty-five times at bat, and the second column shows the proportion of hits at the end of the season. Consider the task of predicting the second column given only the first column. Once again, the obvious approach is to use the average. In other words, one would simply use the first column as a predictor of the second column. The third column is obtained by a shrinkage estimator: more precisely, it takes a number y in the first column and replaces it by $0.265 + 0.212(y - 0.265)$. The number 0.265 is the average of the entries in the first column, so the shrinkage estimator is replacing each entry in the first column by one that is about five times closer to the average. (How the number 0.212 is chosen will be explained later.) If you look at the table, you will see that the shrinkage estimators in the third column are better predictors of the second column in almost every case, and certainly on average. Indeed, the sum of squared differences between the James–Stein estimator and the truth divided by the sum of squared differences between the usual estimator and the truth is 0.29. That is a threefold improvement.

There is beautiful mathematics behind this improvement and a clear sense in which the new estimator is *always* better than the average. We describe the framework, ideas, and extensions of this example as an introduction to the mathematics of statistics.

Before beginning, it will be useful to distinguish between probability and statistics. In probability theory,

Table 1 Batting averages for eighteen major league players in 1970.

Player number	Batting average after 45 at bats	Batting average remainder of season	James–Stein estimator	Remaining at bats
1	0.400	0.346	0.293	367
2	0.378	0.298	0.289	426
3	0.356	0.276	0.284	521
4	0.333	0.221	0.279	276
5	0.311	0.273	0.275	418
6	0.311	0.270	0.275	467
7	0.289	0.263	0.270	586
8	0.267	0.210	0.265	138
9	0.244	0.269	0.261	510
10	0.244	0.230	0.261	200
11	0.222	0.264	0.256	277
12	0.222	0.256	0.256	270
13	0.222	0.304	0.256	434
14	0.222	0.264	0.256	538
15	0.222	0.226	0.256	186
16	0.200	0.285	0.251	558
17	0.178	0.319	0.247	405
18	0.156	0.200	0.242	70

one begins with a set X (for the moment taken to be finite) and a collection of numbers $P(x)$, one for each $x \in X$, which are positive and sum to one. This function $P(x)$ is called a *probability distribution*. The basic problem of probability is this. You are given the probability distribution $P(x)$ and a subset $A \subset X$, and you must compute or approximate $P(A)$, which is defined to be the sum of $P(x)$ for x in A. (In probabilistic terms, each x has a probability $P(x)$ of being chosen, and $P(A)$ is the probability that x belongs to A.) This simple formulation hides wonderful mathematical problems. For example, X might be the set of all sequences of pluses and minuses of length 100 (e.g., $+--++------\cdots$), and each pattern might be equally likely, in which case $P(x) = 1/2^{100}$ for every sequence x. Finally, A might be the set of sequences such that for every positive integer $k \leqslant 100$ the number of + symbols in the first k places is larger than the number of − symbols in the first k places. This is a mathematical model for the following probability problem: if you and a friend flip a fair coin a hundred times, then what is the chance that your friend is always ahead? One might expect this chance to be very small. It turns out, however, to be about $\frac{1}{12}$, though verifying this is a far from trivial exercise. (Our poor intuitions about chance fluctuations have been used to explain road rage: suppose you choose one of two lines at a toll booth. As you wait, you notice whether your line or the other has made more progress. We feel it should all balance out, but the calculations above show that a fair proportion of the time you are always behind—and *frustrated*!)

2 The Basic Problem of Statistics

Statistics is a kind of opposite of probability. In statistics, we are given a *collection* of probability distributions $P_\theta(x)$, indexed by some parameter θ. We see just one x and are required to guess which member of the family (which θ) was used to generate x. For example, let us keep X as the sequence of pluses and minuses of length 100, but this time let $P_\theta(x)$ be the chance of obtaining the sequence x if the probability of a plus is θ and the probability of a minus is $1 - \theta$, with all terms in the sequence chosen independently. Here $0 \leqslant \theta \leqslant 1$, and $P_\theta(x)$ is easily seen to be $\theta^S(1 - \theta)^T$, where S is the number of times "+" appears in the sequence x and $T = 100 - S$ is the number of times "−" appears. This is a mathematical model for the following enterprise. You have a biased coin with a probability θ of turning up heads, but you do not know θ. You flip the coin a hundred times, and are required to estimate θ based on the outcome of the flips.

In general, for each $x \in X$, we want to find a guess, which we denote by $\hat{\theta}(x)$, for the parameter θ. That is, we want to come up with a function $\hat{\theta}$, which will be defined on the observation space X. Such functions are called *estimators*. The above simple formulation hides a wealth of complexity, since both the observation space X and the space Θ of possible parameters may be infinite, or even infinite dimensional. For example, in nonparametric statistics, Θ is often taken as the set of all probability distributions on X. All of the usual problems of statistics—design of experiments, testing hypotheses, prediction, and many others—fit into this framework. We will stick with the imagery of estimation.

To evaluate and compare estimators, one more ingredient is needed: you have to know what it means to get the right answer. This is formalized through the notion of a *loss function* $L(\theta, \hat{\theta}(x))$. One can think of this in practical terms: wrong guesses have financial consequences, and the loss function is a measure of how much it will cost if θ is the true value of the parameter but the statistician's guess is $\hat{\theta}(x)$. The most widely used choice is the *squared error* $(\theta - \hat{\theta}(x))^2$,

but $|\theta - \hat{\theta}(x)|$ or $|\theta - \hat{\theta}(x)|/\theta$ and many other variants are also used. The *risk function* $R(\theta, \hat{\theta})$ measures the expected loss if θ is the true parameter and the estimator $\hat{\theta}$ is used. That is,

$$R(\theta, \hat{\theta}) = \int L(\theta, \hat{\theta}(x)) P_\theta(\mathrm{d}x).$$

Here, the right-hand side is notation for the average value of $L(\theta, \hat{\theta}(x))$ if x is chosen randomly according to the probability distribution P_θ. In general, one would like to choose estimators that will make the risk function as small as possible.

3 Admissibility and Stein's Paradox

We now have the basic ingredients: a family $P_\theta(x)$ and a loss function L. An estimator $\hat{\theta}$ is called *inadmissible* if there is a better estimator θ^*, in the sense that

$$R(\theta, \theta^*) < R(\theta, \hat{\theta}) \quad \text{for all } \theta.$$

In other words, the expected loss with θ^* is less than the expected loss with $\hat{\theta}$, whatever the true value of θ.

Given our assumptions (the model P_θ and loss function L) it seems silly to use an inadmissible estimator. However, one of the great achievements of mathematical statistics is Charles Stein's proof that the usual least-squares estimator, which does not at first glance seem silly at all, is inadmissible in natural problems. Here is that story.

Consider the basic measurement model

$$X_i = \theta + \epsilon_i, \quad 1 \leqslant i \leqslant n.$$

Here X_i is the ith measurement, θ is the quantity to be estimated, and ϵ_i is measurement error. The classical assumptions are that the measurement errors are independently and normally distributed: that is, they are distributed according to the bell-shaped, or Gaussian, curve $e^{-x^2/2}/\sqrt{2\pi}$, $-\infty < x < \infty$. In terms of the language we introduced earlier, the measurement space X is $\mathbb{R}^n$, the parameter space Θ is $\mathbb{R}$, and the observation $x = (x_1, x_2, \ldots, x_n)$ has probability density $P_\theta(x) = \exp[-\frac{1}{2}\sum_1^n (x_i - \theta)^2]/(\sqrt{2\pi})^n$. The usual estimator is the mean: that is, if $x = (x_1, \ldots, x_n)$, then one takes $\hat{\theta}(x)$ to be $(x_1 + \cdots + x_n)/n$. It has been known for a long time that if the loss function $L(\theta, \hat{\theta}(x))$ is defined to be $(\theta - \hat{\theta}(x))^2$, then the mean is an admissible estimator. It has many other optimal properties as well (for example, it is the best linear unbiased estimator, and it is minimax—a property that will be defined later in this article).

Now suppose that we wish to estimate *two* parameters, θ_1 and θ_2, say. This time we have two sets of observations, $X_1, \ldots, X_n$ and $Y_1, \ldots, Y_m$, with $X_i = \theta_1 + \epsilon_i$ and $Y_j = \theta_2 + \eta_j$. The errors ϵ_i and η_j are independent and normally distributed, as above. The loss function $L((\theta_1\theta_2), (\hat{\theta}_1(x)\hat{\theta}_2(y)))$ is now defined to be $(\theta_1 - \hat{\theta}_1(x))^2 + (\theta_2 - \hat{\theta}_2(y))^2$: that is, you add up the squared errors from the two parts. Again, the mean of the X_i and the mean of the Y_i make up an admissible estimator for (θ_1, θ_2).

Consider the same setup with three parameters, θ_1, θ_2, θ_3. Again, $X_i = \theta_1 + \epsilon_i$, $Y_j = \theta_2 + \eta_j$, $Z_k = \theta_3 + \delta_k$ are independent and all the error terms are normally distributed. Stein's surprising result is that for three (or more) parameters the estimator

$$\hat{\theta}_1(x) = (x_1 + \cdots + x_n)/n,$$

$$\hat{\theta}_2(y) = (y_1 + \cdots + y_m)/m,$$

$$\hat{\theta}_3(z) = (z_1 + \cdots + z_l)/l$$

is *in*admissible: there are other estimators that do better in all cases. For example, if p is the number of parameters (and $p \geqslant 3$), then the *James–Stein estimator* is defined to be

$$\hat{\theta}_{\mathrm{JS}} = \left(1 - \frac{p-2}{\|\hat{\theta}\|}\right)_+ \hat{\theta}.$$

Here we are using the notation X_+ to denote the maximum of X and 0; θ stands for the vector $(\theta_1, \ldots, \theta_p)$ of all the averages and $\|\hat{\theta}\|$ is notation for $(\theta_1^2 + \cdots + \theta_p^2)^{1/2}$.

The James–Stein estimator satisfies the inequality $R(\theta, \hat{\theta}_{\mathrm{JS}}) < R(\theta, \hat{\theta})$ for all θ, and therefore the usual estimator $\hat{\theta}$ is indeed inadmissible. The James–Stein estimator shrinks the classical estimator toward zero. The amount of shrinkage is small if $\|\hat{\theta}\|^2$ is large and appreciable for $\|\hat{\theta}\|^2$ near zero. Now the problem as we have described it is invariant under translation, so if we can improve the classical estimate by shrinking toward zero, then we must be able to improve it by shrinking toward any other point. This seems very strange at first, but one can obtain some insight into the phenomenon by considering the following informal description of the estimator. It makes an a priori guess θ_0 at θ. (This guess was zero above.) If the usual estimator $\hat{\theta}$ is close to the guess, in the sense that $\|\hat{\theta}\|$ is small, then it moves $\hat{\theta}$ toward the guess. If $\hat{\theta}$ is far from the guess, it leaves $\hat{\theta}$ alone. Thus, although the estimator moves the classical estimator toward an arbitrary guess, it does so only if there are reasons to believe that the guess is a good one. With four or more parameters the data can in

fact be used to suggest which point θ_0 one should use as the initial guess. In the example of table 1, there are eighteen parameters, and the initial guess θ_0 was the constant vector with all its eighteen coordinates equal to the average 0.265. The number 0.212 that was used for the shrinking is equal to $1 - 16/\|\theta - \theta_0\|$. (Note that for this choice of θ_0, $\|\theta - \theta_0\|$ is the standard deviation of the parameters that make up θ.)

The mathematics used to prove inadmissibility is an elegant blend of harmonic function theory and tricky calculus. The proof itself has had many ramifications: it gave rise to what is called "Stein's method" in probability theory—this is a method for proving things like the central limit theorem for complex dependent problems. The mathematics is "robust," since it is applicable to nonnormal error distributions, a variety of different loss functions, and estimation problems far from the measurement model.

The result has had enormous practical application. It is routinely used in problems where many parameters have to be simultaneously estimated. Examples include national laboratories' estimates of the percentage of defectives when they are looking at many different products at once, and the simultaneous estimate of census undercounts for each of the fifty states in the United States. The apparent robustness of the method is very useful for such applications: even though the James–Stein estimator was derived for the bell-shaped curve, it seems to work well, without special assumptions, in problems where its assumptions hold only roughly. Consider the baseball players above, for example. Adaptations and variations abound. Two popular ones are called empirical Bayes estimates (now widely used in genomics) and hierarchical modeling (now widely used in the assessment of education).

The mathematical problems are far from completely solved. For example, the James–Stein estimator is itself inadmissible. (It can be shown that any admissible estimator in a normal measurement problem is an analytic function of the observations. The James–Stein estimator is, however, clearly not analytic because it involves the nondifferentiable function $x \mapsto x_+$.) While it is known that there is little practical improvement possible, the search for an admissible estimator that is always better than the James–Stein estimator is a tantalizing research problem.

Another active area of research in modern mathematical statistics is to understand which statistical problems give rise to Stein's paradox. For example, although at the beginning of this essay we discussed some inadequacies of the usual maximum-likelihood estimator for estimating the biases of a hundred coins, it turns out that that estimator is admissible! In fact, the maximum-likelihood estimator is admissible for any problem with finite state spaces.

4 Bayesian Statistics

The *Bayesian approach* to statistics adds one further ingredient to the family P_θ and loss function L. This is known as a *prior probability distribution* $\pi(\theta)$, which gives different weights to different values of the parameter θ. There are many ways of generating a prior distribution: it may quantify the working scientists' best guess at θ; it may be derived from previous studies or estimates; or it may just be a convenient way to generate estimators. Once the prior distribution $\pi(\theta)$ has been specified, the observation x and Bayes's theorem combine to give a *posterior distribution* for θ, here denoted $\pi(\theta|x)$. Intuitively, if x is your observation, then $\pi(\theta|x)$ measures how likely it is that θ was the parameter, given that the parameter was generated from the probability distribution π. The mean value of θ with respect to the posterior distribution $\pi(\theta|x)$ gives a *Bayes estimator*:

$$\hat{\theta}_{\text{Bayes}}(x) = \int \theta\, \pi(\theta|x).$$

For the squared-error loss function, all Bayes estimators are admissible, and, in the converse direction, any admissible estimator is a limit of Bayes estimators. (However, not every limit of Bayes estimators is admissible: indeed, the average, which we have seen to be inadmissible, is a limit of Bayes rules.) The point for the present discussion is this. In a wide variety of practical variations of the measurement problem—things like regression analysis or the estimation of correlation matrices—it is relatively straightforward to write down sensible Bayes estimators that incorporate available prior knowledge. These estimators include close cousins of the James–Stein estimator, but they are more general, and allow it to be routinely extended to almost any statistical problem.

Because of the high-dimensional integrals involved, Bayes estimates can be difficult to compute. One of the great advances in this area is the use of computer-simulation algorithms, called variously *Markov chain Monte Carlo* or *Gibbs samplers*, to compute useful approximations to Bayes estimators. The whole package—provable superiority, easy adaptability, and ease of computation—has made this Bayesian version of statistics a practical success.

5 A Bit More Theory

Mathematical statistics makes good use of a wide range of mathematics: fairly esoteric analysis, logic, combinatorics, algebraic topology, and differential geometry all play a role. Here is an application of group theory. Let us return to the basic setup of a sample space X, a family of probability distributions $P_\theta(x)$, and a loss function $L(\theta, \hat{\theta}(x))$. It is natural to consider how the estimator changes when you change the units of the problem: from pounds to grams, or from centimeters to inches, say. Will this have a significant impact on the mathematics? One would expect not, but if we want to think about this question precisely then it is useful to consider a group G of transformations of X. For example, linear changes of units correspond to the *affine group*, which consists of transformations of the form $x \mapsto ax + b$. The family $P_\theta(x)$ is said to be *invariant* under G if for each element g of G the transformed distribution $P_\theta(xg)$ is equal to a distribution $P_{\bar{\theta}}(x)$ for some other $\bar{\theta}$ in Θ. For example, the family of normal distributions

$$\frac{\exp\left[-\dfrac{(x - \theta_1)^2}{2\theta_2^2}\right]}{\sqrt{2\pi\theta_2^2}}, \quad -\infty < \theta_1 < \infty,\ 0 < \theta_2 < \infty,$$

is invariant under $ax+b$ transformations: if you change x to $ax + b$, then after some easy manipulations you can rewrite the resulting modified formula in the form $\exp[-(x - \phi_1)^2/2\phi_2^2]/\sqrt{2\pi\phi_2^2}$ for some new parameters ϕ_1 and ϕ_2. An estimator $\hat{\theta}$ is called *equivariant* if $\hat{\theta}(xg) = \hat{\theta}(x)$. This is a formal way of saying that if you change the data from one unit to another, then the estimate transforms as it should. For example, suppose your data are temperatures presented in centigrade and you want an answer in Fahrenheit. If your estimator is equivariant, then it will make no difference whether you first apply the estimator and then convert the answer into Fahrenheit or first convert all the data into Fahrenheit and then apply the estimator.

The multivariate normal problem underlying Stein's paradox is invariant under a variety of groups, including the p-dimensional group of Euclidian motions (rotations and translations). However, the James–Stein estimator is not equivariant, since, as we have already discussed, it depends on the choice of origin. This is not necessarily bad, but it is certainly thought provoking. If you ask a working scientist if they want a "most accurate" estimator, they will say "of course." If you ask if they insist on equivariance, "of course" will follow as well. One way of expressing Stein's paradox is the statement that the two desiderata—accuracy and invariance—are *incompatible*. This is one of many places where mathematics and statistics part company. Deciding whether mathematically optimal procedures are "sensible" is important and hard to mathematize.

Here is a second use of group theory. An estimator $\hat{\theta}$ is called *minimax* if it minimizes the maximum risk over all θ. Minimax corresponds to playing things safe: you have optimal behavior (that is, the least possible risk) in the worst case. Finding minimax estimators in natural problems is hard, honest work. For example, the vector of means is a minimax estimator in normal location problems. The work is easier if the problem is invariant under a group. Then one can first search for best invariant estimators. Invariance often reduces things to a straightforward calculus problem. Now the question arises of whether an estimator that is minimax among invariant estimators is minimax among all estimators. A celebrated theorem of Hurt and Stein says "yes" if the group involved is nice (e.g., Abelian or compact or amenable). Determining whether the best invariant estimator is minimax when the group is not nice is a challenging open problem in mathematical statistics. And it is not just a mathematical curiosity. For example, the following problem is very natural, and invariant under the group of invertible matrices: given a sample from the multivariate normal distribution, estimate its correlation matrix. In this case, the group is not nice and good estimates are not known.

6 Conclusion

The point of this article is to show how mathematics enters and enriches statistics. To be sure, there are parts of statistics that are hard to mathematize: graphical displays of data are an example. Further, much of modern statistical practice is driven by the computer. There is no longer any need to restrict attention to tractable families of probability distributions. Complex and more realistic models can be used. This gives rise to the subject of statistical computing. Nonetheless, every once in a while someone has to think about what the computer *should* do and determine whether one innovative procedure works better than another. Then, mathematics holds its own. Indeed, mathematizing modern statistical practice is a challenging, rewarding enterprise, of which Stein's estimator is a current

highlight. This endeavor gives us something to aim for and helps us to calibrate our day-to-day achievements.

Further Reading

Berger, J. O. 1985. *Statistical Decision Theory and Bayesian Analysis*, 2nd edn. New York: Springer.

Lehmann, E. L., and G. Casella. 2003. *Theory of Point Estimation*. New York: Springer.

Lehmann, E. L., and J. P. Romano. 2005. *Testing Statistical Hypotheses*. New York: Springer.

Schervish, M. 1996. *Theory of Statistics*. New York: Springer.

VII.11 Mathematics and Medical Statistics
David J. Spiegelhalter

---

1 Introduction

There are many ways in which mathematics has been applied in medicine: for example, the use of differential equations in pharmacokinetics and models for epidemics in populations; and FOURIER ANALYSIS [III.27] of biological signals. Here we are concerned with medical statistics, by which we mean collecting data about individuals and using it to draw conclusions about the development and treatment of disease. This definition may appear to be rather restrictive, but it includes all of the following: randomized clinical trials of therapies, evaluating interventions such as screening programs, comparing health outcomes in different populations and institutions, describing and comparing the survival of groups of individuals, and modeling the way in which a disease develops, both naturally and when it is influenced by an intervention. In this article we are not concerned with *epidemiology*, the study of why diseases occur and how they spread, although most of the formal ideas described here can be applied to it.

After a brief historical introduction, we shall summarize the varied approaches to probabilistic modeling in medical statistics. We shall then illustrate each one in turn using data about the survival of a sample of patients with lymphoma, showing how alternative "philosophical" perspectives lead directly to different methods of analysis. Throughout, we shall give an indication of the mathematical background to what can appear to be a conceptually untidy subject.

2 A Historical Perspective

One of the first uses of probability theory in the late seventeenth century was in the development of "life-tables" of mortality in order to decide premiums for annuities, and Charles Babbage's work on life-tables in 1824 helped motivate him to design his "difference engine" (although it was not until 1859 that Scheutz's implementation of the engine finally calculated a life-table). However, statistical analysis of medical data was a matter of arithmetic rather than mathematics until the growth of the "biometric" school founded by Francis Galton and Karl Pearson at the end of the nineteenth century. This group introduced the use of families of PROBABILITY DISTRIBUTIONS [III.71] to describe populations, as well as concepts of correlation and regression in anthropology, biology, and eugenics. Meanwhile, agriculture and genetics motivated Fisher's huge contributions in the theory of likelihood (see below) and significance testing. Postwar statistical developments were influenced by industrial applications and a U.S.-led increase in mathematical rigor, but from around the 1970s medical research, particularly concerning randomized trials and survival analysis, has been a major methodological driver in statistics.

For around thirty years after 1945 there were many attempts to put statistical inference on a sound foundational or axiomatic basis, but no consensus could be reached. This has given rise to a widespread ecumenical perspective which makes use of a mix of statistical "philosophies" which we shall illustrate below. The somewhat uncomfortable lack of an axiomatic basis can make statistical work deeply unattractive to many mathematicians, but it provides a great stimulus to those engaged in the area.

3 Models

In this context, by a *model* we mean a mathematical description of a probability distribution for one or more currently uncertain quantities. Such a quantity might, for example, be the outcome of a patient who is treated with a particular drug, or the future survival time of a patient with cancer. We can identify four broad approaches to modeling—these brief descriptions make use of terms that will be covered properly in later sections.

(i) A *nonparametric* or "model-free" approach that leaves unspecified the precise form for the probability distributions of interest.

(ii) A *full parametric model* in which a specific form is assumed for each probability distribution, which depends on a limited number of unknown parameters.

(iii) A *semi-parametric* approach in which only part of the model is parametrized, while the rest is left unspecified.

(iv) A *Bayesian* approach in which not only is a full parametric model specified, but an additional "prior" distribution is provided for the parameters.

These are not absolute distinctions: for example, some apparently "model-free" procedures may turn out to match procedures that are derived under certain parametric assumptions.

Another complicating factor is the multiplicity of possible aims of a statistical analysis. These may include

- *estimating* unknown parameters, such as the mean reduction in blood pressure when giving a certain dose of a certain drug to a defined population;
- *predicting* future quantities, such as the number of people with AIDS in a country in ten years' time;
- *testing a hypothesis*, such as whether a particular drug improves survival for a particular class of patents, or equivalently assessing the "null hypothesis" that it has no effect;
- *making decisions*, such as whether to provide a particular treatment in a health care system.

A common aspect of these objectives is that any conclusion should be accompanied by some form of assessment of the potential for an error having been made, and any estimate or prediction should have an associated expression of uncertainty. It is this concern for "second-order" properties that distinguishes a statistical "inference" based on probability theory from a purely algorithmic approach to producing conclusions from data.

4 The Nonparametric or "Model-Free" Approach

Now let us introduce a running example that will be used to illustrate the various approaches.

Matthews and Farewell (1985) report data on sixty-four patients from Seattle's Fred Hutchinson Cancer Research Center who had been diagnosed with advanced-stage non-Hodgkin's lymphoma: for each patient the information comprises their follow-up time since diagnosis, whether their follow-up ended in death, whether they presented with clinical symptoms, their stage of disease (stage IV or not), and whether a large abdominal mass (greater than 10 cm) was present. Such information has many uses. For example, we may wish to look at the general distribution of survival times, or assess which factors most influence survival, or provide a new patient with an estimate of their chance of surviving, say, five years. This is, of course, too small and limited a data set to draw firm conclusions, but it allows us to illustrate the different mathematical tools that can be used.

We need to introduce a few technical terms. Patients who are still alive at the end of data collection, or have been lost to follow-up, are said to have their survival times "censored": all we know is that they survived beyond the last time that any data was recorded about them. We also tend to call times of death "failure" times, since the forms of analysis do not just apply to death. (This term also reflects the close connection between this area and *reliability theory*.)

The original approach to such survival data was "actuarial," using the life-table techniques mentioned previously. Survival times are grouped into intervals such as years, and simple estimates are made of one's chance of dying in an interval given that one was alive at the start of it. Historically, this probability was known as the "force of mortality," but now it is usually called the *hazard*. A simple approach like this may be fine for describing large populations.

It was not until Kaplan and Meier (1958) that this procedure was refined to take into account the exact rather than the grouped survival times: with over thirty thousand citations, their paper is one of the most referenced papers in all of science. Figure 1 shows so-called Kaplan–Meier curves for the groups of patients with ($n = 31$) and without ($n = 33$) clinical symptoms at diagnosis.

These curves represent estimates of the underlying *survival function*, whose value at a time t is thought of as the probability that a typical patient will survive until that time. The obvious way of producing such a curve is simply to let its value at time t be the proportion of the initial sample that is still alive. However, this does not quite work, because of the censored patients. So instead, if a patient dies at time t, and if, just before time t, there are m patients still in the sample, then the value of the curve is multiplied by $(m - 1)/m$; and if a patient is censored then the value stays the same. (The tick marks on the curves show the censored survival times.) The set of patients alive just before time t is called the *risk set* and the hazard at t is estimated to be $1/m$. (We are assuming that two people do not die

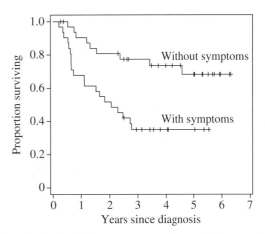

Figure 1 Kaplan–Meier nonparametric survival curves for lymphoma patients with and without clinical symptoms at diagnosis.

at the same time, but it is easy to drop that assumption and make appropriate adjustments.)

Although we do not assume that the actual survival curve has any particular functional form, we do need to make the qualitative assumption that the censoring mechanism is independent of the survival time. (For example, it is important that those who are about to die are not for some reason preferentially removed from the study.) We also need to provide error bounds on the curves: these can be based on a variance formula developed by Major Greenwood in 1926. ("Major" was his name rather than a title, one of the few characteristics he shared with Count Basie and Duke Ellington.)

The "true underlying survival curve" is a theoretical construct, and not something that one can directly observe. One can think of it as the survival experience that would be observed in a vast population of patients, or equivalently the expected survival for a new individual drawn at random from that population. As well as estimating these curves for the two groups of patients, we may wish to test hypotheses about them. A typical one would be that the true underlying survival curves in the two groups are precisely the same. Traditionally such "null" hypotheses are denoted H_0, and the traditional way to test them is to determine how unlikely it is that we would observe two Kaplan–Meier curves that are so far apart if H_0 were true. One can construct a summary measure, known as a *test statistic*, that is large if the observed curves are very different. For example, one possibility is to contrast the observed number of deaths in those with symptoms ($O = 20$) with the number one would have expected if H_0 were true ($E = 11.9$).

Under the null hypothesis it turns out there is only a 0.2% chance of observing such a high discrepancy between O and E, which casts considerable doubt on the null hypothesis in this case.

When constructing intervals around estimates and testing hypotheses we require approximate probability distributions for our estimates and test statistics. From a mathematical perspective the important theory therefore concerns large-sample distributions of functions of random variables, largely developed in the early twentieth century. Theories for optimal hypothesis testing were developed by Neyman and Pearson in the 1930s: the idea is to maximize the "power" of a test to detect a difference, while at the same time making sure that the probability of wrongly rejecting a null hypothesis is less than some acceptable threshold such as 5% or 1%. This approach still finds a role in the design of randomized clinical trials.

5 Full Parametric Models

Clearly we do not actually believe that deaths can only occur at the previously observed survival times shown in the Kaplan–Meier curve, so it seems reasonable to investigate a fairly simple functional form for the true survival function. That is, we assume that the survival function belongs to some natural class of functions, each of which can be fully parametrized by a small number of parameters, collectively denoted by θ. It is θ that we are trying to discover (or rather estimate with a reasonable degree of confidence). If we can do so, then the model is fully specified and we can even extrapolate a certain amount beyond the observed data. We first relate the survival function and the hazard, and then illustrate how observed data can be used to estimate θ in a simple example.

We assume that an unknown survival time has a probability density $p(t|\theta)$; without getting into technical details, this essentially corresponds to assuming that $p(t|\theta)\,dt$ is the probability of dying in a small interval t to $t+dt$. Then the survival function, given a particular value of θ, is the probability of surviving beyond t: we denote it by $S(t|\theta)$. To calculate it, we integrate the probability density over all times greater than t. That is,

$$S(t|\theta) = \int_t^\infty p(x|\theta)\,dx = 1 - \int_0^t p(x|\theta)\,dx.$$

From this and THE FUNDAMENTAL THEOREM OF CALCULUS [I.3 §5.5] it follows that $p(t|\theta) = -dS(t|\theta)/dt$. The hazard function $h(t|\theta)\,dt$ is the risk of death in the small interval t to $t+dt$, conditional on having survived to time t. Using the laws of elementary probability we find that

$$h(t|\theta) = p(t|\theta)/S(t|\theta).$$

For example, suppose we assume an exponential survival function with mean survival time θ, so that the probability of surviving beyond time t is $S(t|\theta) = e^{-t/\theta}$. The density is $p(t|\theta) = e^{-t/\theta}/\theta$. Therefore, the hazard function is a constant $h(t|\theta) = 1/\theta$, so that $1/\theta$ represents the mortality rate per unit of time. For instance, were the mean postdiagnosis survival to be $\theta = 1000$ days, an exponential model would imply a constant $1/1000$ risk of dying each day, regardless of how long the patient had already survived after diagnosis. More complex parametric survival functions allow hazard functions that increase, decrease, or have other shapes.

When it comes to estimating θ we need Fisher's concept of *likelihood*. This takes the probability distribution $p(t|\theta)$ but considers it as a function of θ rather than t, and hence for observed t allows us to examine plausible values of θ that "support" the data. The rough idea is that we multiply together the probabilities (or probability densities) of the observed events, assuming the value of θ. In survival analysis, observed and censored failure times make different contributions to this product: an observed time t contributes $p(t|\theta)$, while a censored time contributes $S(t|\theta)$. If, for example, we assume that the survival function is exponential, then an observed failure time contributes $p(t|\theta) = e^{-t/\theta}/\theta$, and a censored time contributes $S(t|\theta) = e^{-t/\theta}$. Thus, in this case the likelihood is

$$L(\theta) = \prod_{i\in\text{Obs}} \theta^{-1} e^{-t_i/\theta} \prod_{i\in\text{Cens}} e^{-t_i/\theta} = \theta^{-n_O} e^{-T/\theta}.$$

Here "Obs" and "Cens" indicate the sets of observed and censored failure times. We denote their sizes by n_O and n_C, respectively, and we denote the total follow-up time $\sum_i t_i$ by T. For the group of thirty-one patients presenting with symptoms we have $n_O = 20$ and $T = 68.3$ years: figure 2 shows both the likelihood and its logarithm

$$LL(\theta) = -T/\theta - n_O \log\theta.$$

We note that the vertical axis for the likelihood is unlabeled since only relative likelihood is important. A *maximum-likelihood estimate* (MLE) $\hat{\theta}$ finds parameter values that maximize this likelihood or equivalently the log-likelihood. Taking derivatives of $LL(\theta)$ and equating to 0 reveals that $\hat{\theta} = T/n_{\text{Obs}} = 3.4$ years, which is the total follow-up time divided by the number

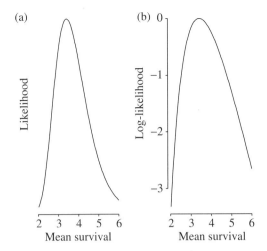

Figure 2 Likelihood and log-likelihood for mean survival time θ for lymphoma patients presenting with clinical symptoms.

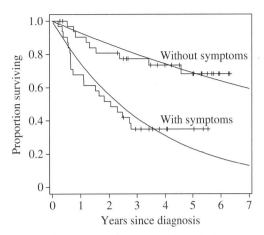

Figure 3 Fitted exponential survival curves for lymphoma patients.

of failures. Intervals around MLEs may be derived by directly examining the likelihood function, or by making a quadratic approximation around the maximum of the log-likelihood.

Figure 3 shows the fitted exponential survival curves: loosely, we have carried out a form of curve fitting by selecting the exponential curves that maximize the probability of the observed data. Visual inspection suggests the fit may be improved by investigating a more flexible family of curves such as the Weibull distribution (a distribution widely used in reliability theory): to compare how well two models fit the data, one can compare their maximized likelihoods.

Fisher's concept of likelihood has been the foundation for most current work in medical statistics, and indeed statistics in general. From a mathematical perspective there has been extensive development relating the large-sample distributions of MLEs to the second derivative of the log-likelihood around its maximum, which forms the basis for most of the outputs of statistical packages. Unfortunately, it is not necessarily straightforward to scale up the theory to deal with multidimensional parameters. First, as likelihoods become more complex and contain increasing numbers of parameters, the technical problems of maximization increase. Second, the recurring difficulty with likelihood theory remains that of "nuisance parameters," in which a part of the model is of no particular interest and yet needs to be accounted for. No generic theory has been developed, and instead there is a some-

what bewildering variety of adaptations of standard likelihood to specific circumstances, such as conditional likelihood, quasi-likelihood, pseudo-likelihood, extended likelihood, hierarchical likelihood, marginal likelihood, profile likelihood, and so on. Below we consider one extremely popular development, that of partial likelihood and the Cox model.

6   A Semi-Parametric Approach

Clinical trials in cancer therapy were a major motivating force in developing survival analysis—in particular, trials to assess the influence of a treatment on survival while taking account of other possible risk factors. In our simple lymphoma data set we have three risk factors, but in more realistic examples there will be many more. Fortunately, Cox (1972) showed that it was possible both to test hypotheses and to estimate the influence of possible risk factors, without having to go the whole way and specify the full survival function on the basis of possibly limited data.

The *Cox regression model* is based on assuming a hazard function of the form

$$h(t|\theta) = h_0(t)e^{\beta \cdot x}.$$

Here $h_0(t)$ is a *baseline hazard function* and β is typically a column vector of regression coefficients that measure the influence of a vector of risk factors x on the hazard. (The expression $\beta \cdot x$ denotes the scalar product of β and x.) The baseline hazard function corresponds to the hazard function of an individual whose risk factor vector is $x = 0$, since then $e^{\beta \cdot x} = 1$. More generally, we see that an increase of one unit in a factor

x_j will multiply the hazard by a factor e^{β_j}, for which reason this is known as the "proportional hazards" regression model. It is possible to specify a parametric form for $h_0(t)$, but remarkably it turns out to be possible to estimate the terms of β without specifying the form of the h_0, if we are willing to consider the situation immediately before a particular failure time. Again we construct a risk set, and the chance of a particular patient failing, given the knowledge that someone in the risk set fails, provides a term in a likelihood. This is known as a "partial" likelihood since it ignores any possible information in the times between failures.

When we fit this model to the lymphoma data we find that our estimate of β for the patients with symptoms is 1.2: easier to interpret is its exponent $e^{1.2} = 3.3$, which is the proportional increase in hazard associated with presenting with symptoms. We can estimate error bounds of 1.5–7.3 around this estimate, so we can be confident that the risk of a patient who presents with symptoms will die at any stage following diagnosis is substantially higher than that of a patient who does not present with symptoms, all other factors in the model being kept constant.

A huge literature has arisen from this model, dealing with errors around estimates, different censoring patterns, tied failure times, estimating the baseline survival, and so on. Large-sample properties were rigorously established only after the method came into routine use, and have made extensive use of the theory of stochastic counting processes: see, for example, Andersen et al. (1992). These powerful mathematical tools have enabled the theory to be expanded to deal with the general analysis of sequences of events, while allowing for censoring and multiple risk factors that may depend on time.

Cox's 1972 paper has over twenty thousand citations, and its importance to medicine is reflected in his having been awarded the 1990 Kettering Prize and Gold Medal for Cancer Research.

7 Bayesian Analysis

Bayes's theorem is a basic result in probability theory. It states that, for two random quantities t and θ,

$$p(\theta|t) = p(t|\theta)p(\theta)/p(t).$$

In itself this is a very simple fact, but when θ represents parameters in a model, the use of this theorem represents a different philosophy of statistical modeling. The major step in using Bayes's theorem for inference is in considering parameters as RANDOM VARIABLES

[III.71 §4] with probability distributions and therefore making probabilistic statements about them. For example, in the Bayesian framework one could express one's uncertainty about a survival curve by saying that one had assessed that the probability that the mean survival time was greater than three years was 0.90. To make such an assessment, one can combine a "prior" distribution $p(\theta)$ (a distribution representing the relative plausibility of different values of θ *before* you look at the data) with a likelihood $p(t|\theta)$ (how likely you were to observe the data t with that value of θ) and then use Bayes's theorem to provide a "posterior" distribution $p(\theta|t)$ (a distribution representing the relative plausibility of different values of θ *after* you look at the data).

Put in this way Bayesian analysis appears to be a simple application of probability theory, and for any given choice of prior distribution that is exactly what it is. But how do you choose the prior distribution? You could use evidence external to the current study, or even your own personal judgment. There is also an extensive literature on attempts to produce a toolkit of "objective" priors to use in different situations. In practice you need to specify the prior distribution in a way that is convincing to others, and this is where the subtlety arises.

As a simple example, suppose that previous studies of lymphoma had suggested that mean survival times of patients presenting with clinical symptoms probably lie between three and six years, with values of around four years being most plausible. Then it seems reasonable not to ignore such evidence when drawing conclusions for future patients, but rather to combine it with the evidence from the thirty-one patients in the current study. We could represent this external evidence by a prior distribution for θ with the form given in figure 4. When combined with the likelihood (taken from figure 2(a)), this gives rise to the posterior distribution shown. For this calculation, the functional form of the prior is assumed to be that of the *inverse-Gamma distribution*, which happens to make the mathematics of dealing with exponential likelihoods particularly straightforward, but such simplifications are not necessary if one is using simulation methods for deriving posterior distributions.

It can be seen from figure 4 that the external evidence has increased the plausibility of higher survival times. By integrating the posterior distribution above three years, we find that the posterior probability that the mean survival is greater than three years is 0.90.

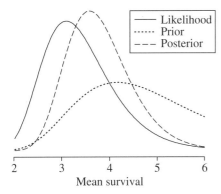

Figure 4 Prior, likelihood, and posterior distributions for mean survival time θ for patients presenting with symptoms. The posterior distribution is a formal compromise between the likelihood, which summarizes the evidence in the data alone, and the prior distribution, which summarizes external evidence that suggested longer survival times.

Likelihoods in Bayesian models need to be fully parametric, although semi-parametric models such as the Cox model can be approximated by high-dimensional functions of nuisance parameters, which then need to be integrated out of the posterior distributions. Difficulties with evaluating such integrals held up realistic applications of Bayesian analysis for many years, but now developments in simulation approaches such as Markov chain Monte Carlo (MCMC) methods have led to a startling growth in practical Bayesian analyses. Mathematical work in Bayesian analysis has mainly focused on theories of objective priors, large-sample properties of posterior distributions, and dealing with hugely multivariate problems and the necessary high-dimensional integrals.

8   Discussion

The preceding sections have given some idea of the tangled conceptual issues that underlie even routine medical statistical analysis. We need to distinguish a number of different roles for mathematics in medical statistics—the following are a few examples.

Individual applications: here the use of mathematics is generally quite limited, since extensive use is made of software packages, which can fit a wide variety of models. In nonstandard problems, algebraic or numerical maximization of likelihoods may be necessary, or developing MCMC algorithms for numerical integration.

Derivation of generic methods: these can then be implemented in software. This is perhaps the most widespread mathematical work, which requires extensive use of probability theory on functions of random variables, particularly using large-sample arguments.

Proof of properties of methods: this requires the most sophisticated mathematics, which concerns topics such as the convergence of estimators, or the behavior of Bayesian methods under different circumstances.

Medical applications continue to be a driving force in the development of new methods of statistical analysis, partly because of new sources of high-dimensional data from areas such as bioinformatics, imaging, and performance monitoring, but also because of the increasing willingness of health policy makers to use complex models: this has the consequence of focusing attention on analytic methods and the design of studies for checking, challenging, and refining such models.

Nevertheless, it may appear that rather limited mathematical tools are required in medical statistics, even for those engaged in methodological research. This is compensated for by the fascinating and continuing debate over the underlying philosophy of even the most common statistical tools, and the consequent variety of approaches to apparently simple problems. Much of this debate is hidden from the routine user. Regarding the appropriate role of mathematical theory in statistics, we can do no better than quote David Cox in his 1981 Presidential Address to the Royal Statistical Society (Cox 1981):

> Lord Rayleigh defined applied mathematics as being concerned with quantitative investigation of the real world "neither seeking nor evading mathematical difficulties." This describes rather precisely the delicate relation that ideally should hold between mathematics and statistics. Much fine work in statistics involves minimal mathematics; some bad work in statistics gets by because of its apparent mathematical content. Yet it would be harmful for the development of the subject for there to be widespread an anti-mathematical attitude, a fear of powerful mathematics appropriately deployed.

Further Reading

Andersen, P. K., O. Borgan, R. Gill, and N. Keiding. 1992. *Statistical Models Based on Counting Processes*. New York: Springer.

Cox, D. R. 1972. Theory and general principle in statistics. *Journal of the Royal Statistical Society* A 144:289–97.

——. 1981. Regression models and life-tables (with discussion). *Journal of the Royal Statistical Society* B 34: 187–220.

Kaplan, E. L., and P. Meier. 1958. Nonparametric estimation from incomplete observations. *Journal of the American Statistical Association* 53:457–81.

Matthews, D. E., and V. T. Farewell. 1985. *Using and Understanding Medical Statistics*. Basel: Karger.

VII.12 Analysis, Mathematical and Philosophical
John P. Burgess

1   The Analytic Tradition in Philosophy

Philosophical problems are never solved for the same reason that treasonous conspiracies never succeed: as successful conspiracies are never called "treason," so solved problems are no longer called "philosophy." Philosophy, which once included almost every subject in the university (every subject in which the highest degree is Ph.D.), has thus been shrunk by success. The greatest shrinkage occurred during the seventeenth and eighteenth centuries, when natural philosophy became natural science. Philosophers of the period, all intensely interested in the emergence of the new science, differed over issues of scientific method. Philosophy had always been understood to differ from, for instance, theology, by restricting itself to the methods of reasoned argument and the evidence of experience, without appeal to authority, tradition, revelation, or faith. But philosophers of the era of the scientific revolution disagreed about the comparative importance of reason and experience.

In introductory histories, philosophers are accordingly divided into the rationalists, or the party of reason, and the empiricists, or the party of experience. The former, mainly from Continental Europe, were dominant in the seventeenth century, while the latter, mainly from the British Isles, predominated in the eighteenth. The rationalists, who included the mathematicians DES-CARTES [VI.11] and LEIBNIZ [VI.15], were impressed by the apparent ability of pure thought—logical deduction from self-evident postulates—to achieve, as it seemed to do in geometry, substantive results with worldly applications; and they were tempted to adopt similar methods in other areas. Spinoza even wrote his

Ethics in the style of EUCLID's [VI.2] *Elements*, a world-historic peak of the influence of mathematics on philosophy. The empiricists, who included that acute critic of the calculus, Berkeley, recognized that in physics one cannot proceed as the rationalists wished to. The principles of physics are not self-evident, but must be conjectured from and tested against systematic observation and controlled experiment. What puzzled leading empiricists such as Locke and Hume was how pure thought was able to succeed in *any* area, as it seemed to in geometry. Thus, while for rationalists mathematics was a source of *methods*, for empiricists it was the source of a *problem*.

An influential formulation of that problem was offered by Kant, whose system attempted a synthesis of rationalism with empiricism. On the one hand, Kant claimed, geometry and arithmetic are a priori rather than a posteriori, by which he meant that they are knowable in advance of experience rather than dependent on it. On the other hand, they are synthetic rather than analytic, which is to say that they are more than mere logical consequences of the definitions of concepts, statements whose denials would amount to contradictions in terms. Philosophy of mathematics, today a smallish specialty within philosophy of science, itself a smallish specialty within epistemology or the theory of knowledge, played a much more important role for Kant, who in his own summary of his system gave pride of place to the question, "How is pure mathematics possible?" as the first case of the question, "How is synthetic a priori knowledge possible?" Kant's proposed solution was based on the insight that our knowledge must be shaped as much by the nature of ourselves, the knowers, as by that of what is known. Kant concluded that space, the subject matter of geometry, and time, according to him the ultimate subject matter of arithmetic, were not features of things as they are in themselves, but rather of things as we must perceive and experience them, given the nature of our sensibility. Synthetic a priori knowledge is ultimately *self*-knowledge, knowledge of the forms that *we* supply, and into which reality independent of us pours content. This distinction between *phenomena*, or things as we experience them, and *noumena*, things beyond our experience, about which we can wonder but never know, was central to Kant's entire system, his ethics as much as his metaphysics.

Such is the history, painted in quick strokes and with a broad brush, of early modern philosophy. After Kant, the story no longer has as clear a plotline.

System building continued for another generation, down to Hegel. But eventually, and inevitably, his system collapsed under its own weight, and in the ensuing reaction philosophers went off in all directions. Outside academia, striking figures sporadically appeared on the borders of philosophy and literature, notably Nietzsche. Meanwhile, academic philosophy, rather like Victorian architecture, experienced a number of revivals, of which the Kantian was the most prominent. But even as neo-Kantianism prevailed in the schools, the Kantian conception of mathematics was under attack. First, though the development of consistent non-Euclidean geometries in itself only confirms Kant's claim that geometry is synthetic, those who developed alternatives to Euclid were quickly led to question whether Euclidean geometry is really a priori, as Kant had claimed. GAUSS [VI.26] had already concluded that geometry is a posteriori, or, as he put it, of the same status as mechanics, and RIEMANN [VI.49] argued at greater length that an examination of the hypotheses that lie at the foundation of geometry must lead us into the domain of the neighboring science of physics. Second, while few doubted Kant's claim that arithmetic is a priori, a challenge arose to the claim that it is synthetic in the work of GOTTLOB FREGE [VI.56] and (slightly later, but largely independently) BERTRAND RUSSELL [VI.71], who both attempted a derivation of arithmetic from logic along with an appropriate definition of number.

Frege's work long remained less well-known than it deserved to be, despite the publicity given it by Russell once he became aware of it himself. As a result, Frege, though very influential at present, is more a precursor of the tradition in philosophy within which he stands than a founder, the founders being rather Russell and his contemporary and colleague G. E. Moore. That pair began by rebelling against the philosophy of their teachers, a late nineteenth-century aberration called absolute idealism, a kind of Hegel revival; but it soon became apparent that the rebels were aiming at more than just a return to the traditional empiricism of British philosophy from Bacon to Mill. Meanwhile, Edmund Husserl was developing the first form of what was to become the great rival to the Russell–Moore tradition in twentieth-century philosophy. Like Frege, Husserl had begun his career with work in the philosophy of arithmetic, work of which Frege himself had taken notice, and no one in the early twentieth century expected that Husserl's and Frege's heirs would,

within a generation, split into two noncommunicating lines of descent.

The two lines of development or traditions are oddly named, with a stylistic label, "analytic," for one, and a geographical label, "Continental," for the other. This odd labeling reflects the historical fact that the principal representatives of the analytic style in Continental Europe (Ludwig Wittgenstein, Rudolf Carnap, and others) were forced to go into exile in the English-speaking world in the 1930s, as a result of the process generally known as the Nazification (but celebrated by Husserl's estranged student Martin Heidegger as the "self-affirmation") of the German university. This physical separation—more than Heidegger's break with his teacher, hostility toward science, rebarbative prose style, or loathsome politics—created a split that no one could have anticipated twenty years earlier.

With the years the gap has widened, as later writers on each side tend to read and cite only predecessors on that side. Indeed, the divide has extended backwards in time. For while Borges has said that in literature great writers create their own predecessors, in philosophy even not-so-great writers can do so, and the two twentieth-century traditions came to see different nineteenth-century figures as leading up to themselves, thus extending the division between them right back to the death of Kant (with Hegel rather than Heidegger being identified as the first distinctively Continental philosopher). The gap between the reading lists of students in the two traditions has become so large that nowadays for a student trained in one to take up the other is virtually to switch disciplines.

The word "tradition," rather than "school" or "movement," is used advisedly, for each tradition has contained several movements, as well as individuals who defy classification by school. It would be a serious mistake to suppose that there is any doctrine or method on either side of the analytic/Continental divide that all philosophers on that side uphold. In particular, analytic philosophy should not be confused with logical positivism, a Viennese–American school defunct for more than half a century, nor should Continental philosophy be confused with existentialism, a literary-philosophical movement out of fashion in Paris for nearly as long. Logical positivism and existentialism were indeed varieties of analytic and Continental philosophy respectively, and perhaps the most prominent varieties half a century or so ago; but each was even then far from being the *only* variety. In assessing the influence of mathematics on philosophy in the twentieth century, one must take into account divisions within each tradition as much as the division between the two traditions.

It may be true that since the early work of Husserl there has been comparatively little contact between mathematics and philosophy on the Continental side, though the label "structuralist" is broad enough to take in both the mathematics of BOURBAKI [VI.96] and the various anthropological and linguistic doctrines that became influential in France after the eclipse of existentialism; but it is also true that the direct influence of mathematical ways of thinking on many individuals and groups within the analytic tradition has been negligible. Thus, just as there are distinguishable German and French subtraditions within the Continental tradition, so within the analytic tradition one may distinguish a more technically oriented subtradition, including Frege (who was himself a professor of mathematics), Russell (who as an undergraduate had concentrated on mathematics before turning to philosophy), and the logical positivists (who had mostly been trained as theoretical physicists), from a nontechnical or antitechnical subtradition, including Moore, Wittgenstein, the so-called ordinary-language school of mid-century Oxford, and others. (Wittgenstein even went so far as to claim that mathematicians always make bad philosophers, a sweeping judgment condemning many right back to Thales and PYTHAGORAS [VI.1], though the immediate target was Russell.) However, there has been very much more communication and influence back and forth between the two subtraditions within each tradition than between the two traditions.

Even among the more technical analytic philosophers the influence of mathematics after the period of the founders has been occasional and sporadic, and has come mostly from areas such as mathematical logic, computability theory, probability and statistics, game theory, and mathematical economics (as in the work of the philosopher–economist Amartya Sen), which are rather far from the core of pure mathematics as mathematicians see it. Thus it is hard to imagine the solution to any of the Millennium Prize Problems (except perhaps the P vs NP problem, the one question coming from theoretical computer science rather than core mathematics) having measurable impact even among the most susceptible analytic philosophers. In contrast to this limited *direct* influence, the *indirect* influence of mathematics, resulting from its effect on the thought of the early figures Frege and Russell, has been over-

whelming even among the less technically oriented analytic philosophers. The branches of mathematics that influenced Frege and Russell were geometry and algebra and, above all, the third great branch of core mathematics, "analysis," in the mathematical rather than the philosophical sense, the branch beginning with differential and integral calculus. (Frege and Russell were not *influenced* by mathematical logic: rather, they *created* it, and mathematical analysis was a key influence on its creation.)

2 Mathematical Analysis and Frege's New Logic

Let us turn, then, to consider the state of mathematical analysis in the days of Frege and Russell, beginning our account with a quick look back at the situation ca. 1800. As rich as its results were, and as powerful its applications, mathematics at the beginning of the nineteenth century was concerned with but a few structures: the natural, rational, real, and complex number systems; and the Euclidean and projective spaces of dimensions one, two, and three. All that changed quickly when the work of Gauss, HAMILTON [VI.37], and others introduced the first non-Euclidean spaces and first noncommutative algebras, after which a proliferation of new mathematical structures rapidly ensued. This *generalizing* tendency went hand in hand with a *rigorizing* tendency, since the proliferation of novelties persuaded mathematicians that they needed to adhere more strictly than had become customary to the ancient ideal of rigor, according to which all new results in mathematics are to be logically deduced from previous results, and ultimately from a list of explicit axioms. For without rigor, intuitions derived from familiarity with more traditional structures might easily be unconsciously transferred to new situations where they are no longer appropriate.

Generalization and rigorization went hand in hand not only in geometry and algebra, but also in mathematical analysis. Generalization in mathematical analysis took place in two directions. The eighteenth-century notion of "function" had been that of an operation applying to one or more real numbers as inputs or "arguments" and yielding a real number as output or "value," *according to a certain formula*, such as $f(x) = \sin x + \cos x$ or $f(x, y) = x^2 + y^2$. On the one hand, nineteenth-century mathematicians generalized by dropping the requirement of an explicit formula. On the other hand, Cauchy, Riemann, and others extended the notion to allow as arguments not only real numbers but also complex numbers, that is, numbers of the form $a + b\mathrm{i}$, where a and b are real numbers and i is the "imaginary" square root of -1.

Rigorization in mathematical analysis also took place on two levels. First, for each theorem it had to be clearly stated just what special properties were being assumed for the functions to which the result was supposed to apply, since special properties such as definability by a formula (or continuity or differentiability) were no longer being built into the highly general notion of function itself; moreover, the relevant properties themselves had to be clearly defined (leading to the so-called WEIERSTRASS [VI.44] epsilon–delta definitions of such concepts as "continuity" and "differentiability" in freshman calculus), since, as POINCARÉ [VI.61] remarked, until one has rigor in one's definitions one cannot have rigor in one's theorems. Second, the properties assumed for the numbers to which the functions apply had also to be clarified and stated explicitly as axioms, with the properties of complex numbers being derived by logical definition and deduction from properties of real numbers (by Hamilton), which themselves in turn were derived from properties of rational numbers (by DEDEKIND [VI.50] and CANTOR [VI.54]), which themselves in turn were derived from properties of the system of natural numbers 0, 1, 2, and so on.

Here Frege wished to press still further, and to do what Kant had said could not be done, and derive the properties of the natural numbers themselves from pure logic. For this purpose he needed to become more self-conscious about logic than even the most rigorist mathematicians: he needed not merely to adhere implicitly to the rules and standards of logical definition and deduction, but also to analyze explicitly those very rules and standards themselves. Such self-conscious analysis of definition and deduction was a topic that had, since antiquity, traditionally belonged to philosophy rather than mathematics. Frege needed to carry out a revolution in this philosophical subject, one that would bring it much closer to mathematics, and would bring progress to a field that Kant had described as having advanced not a step beyond the state in which it was left by its founder, Aristotle. (The description is slightly exaggerated, but essentially correct, in that each step forward in the two millennia after Aristotle had been followed by a step back.) It was Frege's new logic, detached from its original role as part of a special project in foundations of arithmetic and applied to quite diverse subject matters, that was

to become the single most important general instrument for philosophical analysis in the twentieth century. Indeed, to a large degree philosophical analysis simply *is* the logical analysis of philosophical rather than mathematical notions, carried out with the aid of Frege's broad new logic, or still broader extensions of it introduced by his successors. It was by the creation of this general instrument of a new logic, rather than the specialized application he made of it to the philosophy of mathematics, that Frege became the grandfather of analytic philosophy. And the novelty in Frege's logic was directly inspired by novel developments in mathematical analysis, as he himself emphasized.

In an article entitled "Function and concept," Frege describes the broadening of the notion of function as follows (in the translation by Peter Geach and Max Black):

> Now how has the reference of the word "function" been extended by the progress of science? We can distinguish two directions in which this has happened. In the first place, the field of mathematical operations that serve for constructing functions has been extended. Besides addition, multiplication, exponentiation, and their converses, the various means of transition to the limit have been introduced—to be sure, without people's being always clearly aware that they were thus adopting something essentially new. People have even gone further still, and have actually been obliged to resort to ordinary language, because the symbolic language of Analysis failed, e.g., when they were speaking of a function whose value is 1 for rational and 0 for irrational arguments. [This is a famous example of DIRICHLET [VI.36].] Secondly, the field of possible arguments and values for functions has been extended by the admission of complex numbers. In conjunction with this, the sense of the expressions "sum," "product," etc. had to be defined more widely.

Frege adds at the end, "In both directions I go still further." For it was the broadening of the notion of function by mathematicians that provided Frege with the clue he needed to develop a logic broader than Aristotle's.

Before one can appreciate the advance represented by Frege's logic, one must understand something of Aristotle's. Though it is a pretty poor achievement if it is considered as the best the human race could do in this area in a couple of thousand years, it is a brilliant one when considered as the work of a single individual in the course of a career devoted to many other projects. For Aristotle created from nothing the science of logic, whose aim is to distinguish valid from invalid inferences of conclusions from premises. Here an inference is valid if its form alone, regardless of the material truth or falsehood of premises and conclusions, guarantees that *if* the premises are true, *then* the conclusion is true. Equivalently, the inference is valid if in all inferences of the same form in which the premises are true, the conclusion is true. Thus, to adapt an example of Lewis Carroll, the inference from "I believe whatever I say" to "I say whatever I believe" is *not* valid, because there are inferences of identical form in which the premise is true and the conclusion false, such as the inference from "I see whatever I eat" to "I eat whatever I see."

The scope of Aristotle's logic is limited by the limited range of forms of potential premises and conclusions he recognizes. In fact, he recognized only four: the *universal affirmative* "All A's are B's," the *universal negative* "No A's are B's," the *particular affirmative* "Some A's are B's," and the *particular negative* "Some A's are not B's" or "Not all A's are B's." The premise "I believe whatever I say" amounts to "All things that I say are things that I believe," and hence is a universal affirmative. The invalidity of the inference in the Lewis Carroll example exemplifies the invalidity of the inference from "All A's are B's" to "All B's are A's." The validity of the inference from the two premises "All Greeks are human beings" and "All human beings are mortal" to the conclusion "All Greeks are mortal" exemplifies the validity of the inference from "All A's are B's" and "All B's are C's" to "All A's are C's," traditionally called the "syllogism in *Barbara*," for reasons that need not concern us here. Aristotle's logic was in part inspired by the practice of deduction in philosophical debate ("dialectic") and in part by the practice of deduction in mathematical theorem-proving ("demonstration"), and he offers in his *Posterior Analytics* an account of a deductive science that is presumed to be based on the practice of the contemporary geometer Eudoxus, in the same sense and to the same degree in which his account in the *Poetics* of tragedy is based on the practice of the contemporary playwright Euripides. But, in fact, Aristotle's logic is inadequate for the analysis of mathematicians' actual arguments, because he makes no provision for forms of argument involving *relations*. He cannot, for instance, analyze properly the valid argument from "All squares are rectangles" to "Anyone who draws a square draws a rectangle," because he has no way of representing adequately the form of the conclusion.

By contrast, if you open any present-day introductory logic text, you will find instructions on how to represent symbolically the forms of arguments involving relations. The example just given would appear textbook-style as follows:

$$\forall x(\text{Square}(x) \rightarrow \text{Rectangle}(x))$$
$$\therefore \quad \forall y(\exists x(\text{Square}(x) \,\&\, \text{Draws}(y,x)) \rightarrow$$
$$\exists x(\text{Rectangle}(x) \,\&\, \text{Draws}(y,x))).$$

In words this would amount to the following. For every x, if x is a square, then x is a rectangle. Therefore, for every y, if there is an x such that x is a square and y draws x, then there exists an x such that x is a rectangle and y draws x. (Thus "$\rightarrow$" means "if . . . , then . . . ," "$\forall$" means "for every," and "$\exists$" means "there is.") This style of logical analysis is the invention of Frege.

Underlying it is a notion of a "concept" as a special kind of function, a function that (generalizing the mathematical notion in one direction) need not be given by any kind of *mathematical* description, and that (generalizing the mathematical notion in another direction) need not have as arguments any kind of *numbers*. A concept for Frege is a function whose argument or arguments may be any objects at all, and whose values are Truth and Falsehood. Thus, the concept Wise applied to the argument Socrates produces the value Truth, since Socrates is wise (at least to the extent of recognizing that he lacked perfect wisdom), while the concept Immortal applied to Socrates produces Falsehood, since Socrates was not immortal but died of drinking hemlock. Frege is able to handle relations *because he follows the mathematical analysts who allowed functions of two or more arguments*. Thus the two-argument concept or relation Taught applied to Socrates and Plato, in that order, produces Truth, since Socrates taught Plato, while applied to Plato and Socrates, in *that* order, produces Falsehood, since Plato did not teach Socrates. Aristotle's simple "All A's are B's" becomes, for Frege, the more complex "For all objects x, if $A(x)$, then $B(x)$." At the price of such extra complexity, he is able to logically analyze arguments turning on relations, as Aristotle was not.

Aristotle analyzed the concept Human Being in terms of the concepts Animal and Rational in the sense of "language-using." In present-day textbook notation (writing "$\leftrightarrow$" for "if and only if"), this would be

$$\text{Human}(x) \leftrightarrow \text{Animal}(x) \,\&\, \text{Rational}(x).$$

But Aristotle, with no theory of relations, was unable to analyze the notion of Mother (respectively, Father) in terms of Female (respectively, Male) and Parent. For Frege, Mother is analyzed as follows:

$$\text{Mother}(x) \leftrightarrow \text{Female}(x) \,\&\, \exists y \text{Parent}(x, y).$$

A mother is a female who is someone's parent, and analogously for a father. Frege was even able to analyze the concept Ancestor in terms of the concept Parent, though this analysis is beyond the scope of the present sketch. Later philosophical analysis would have been unthinkable without Frege's broadening of logical analysis beyond Aristotle's, and Frege rightly saw his broadening of logical analysis as a direct extrapolation from the nineteenth-century mathematical analysts' broadening of the notion of function they had inherited from their eighteenth-century predecessors.

3 Mathematical Analysis and Russell's Theory of Descriptions

Like Frege, Russell found in mathematics both a source of problems and a source of methods. For the purposes of a specialized investigation of problems in the philosophy of mathematics, he created an instrument, his theory of descriptions, and a more general method, that of contextual definition, which his successors took up and applied to many other problem areas. Indeed, it was not merely Russell's successors who applied these ideas to areas outside philosophy of mathematics, since Russell himself did so in his first publications on the subject. Thus it is not apparent from Russell's still widely read "On denoting," published in 1905 and even today a key item on the syllabus of students of analytic philosophy, that the theory of descriptions originated in the course of studies in foundations and philosophy of mathematics. Rather, this is a fact mentioned in Russell's autobiographical writings and known to historians of twentieth-century philosophy. The degree to which the method of contextual definition, which the theory of descriptions exemplifies, was inspired by the nineteenth-century rigorization of analysis is perhaps not sufficiently appreciated even by such specialists.

A principal puzzle Russell addresses in "On denoting" is that of so-called negative existentials, such as "The king of France does not exist." In superficial grammatical form this statement resembles "The queen of England does not agree," and to that extent it appears to involve picking out an object (in this case, a person), and then attributing a property to him (or her, as the case may be). Thus it seems that in order to say that someone or something does not exist, one must assume that in some sense there is such a person or thing, to whom or which the property of nonexistence may be ascribed. Russell cites Alexius Meinong (a student of Husserl's teacher Franz Brentano) as a philosopher committed to such a view. For Meinong had a theory of "objects beyond being and nonbeing," exemplified by The Golden Mountain and The Round Square. But as Scott Soames reveals, in his *Philosophical Analysis in the Twentieth Century*, volume I: *The Dawn of Analysis*, Russell himself had briefly held a similar view in the first days of his and Moore's joint rebellion against absolute idealism. It was through the development of his theory of descriptions that Russell was able to free himself from anything like commitment to Meinongian "objects."

According to that theory, to say that *a* Golden Mountain exists is to say that there is something that is both golden and a mountain: $\exists x(\text{Golden}(x) \,\&\, \text{Mountain}(x))$. To say that *the* Golden Mountain exists is to say that there is one thing that is both golden and a mountain and no other such thing:

$$\exists x(\text{Golden}(x) \,\&\, \text{Mountain}(x)$$
$$\&\, \sim\!\exists y(\text{Golden}(y) \,\&\, \text{Mountain}(y) \,\&\, y \neq x)).$$

(Here "$\sim$" represents "it is not the case that.") This is logically equivalent to saying there is something such that a thing is both golden and a mountain if and only if it is identical with that thing:

$$\exists x \forall y(\text{Golden}(y) \,\&\, \text{Mountain}(y) \leftrightarrow y = x).$$

To say that the Golden Mountain does *not* exist is simply to deny this:

$$\sim\!\exists x \forall y(\text{Golden}(y) \,\&\, \text{Mountain}(y) \leftrightarrow y = x).$$

To say that the king of France is bald is, similarly, to say that there is something such that a thing is king of France if and only if it is identical with that thing, and that thing is bald:

$$\exists x(\forall y(\text{King-of-France}(y) \leftrightarrow y = x) \,\&\, \text{Bald}(x)).$$

This is not the place to go into the subtleties of Russell's theory, whose main point should be clear from these few examples: when the logical form is properly analyzed, using the new logic, the phrase "the Golden Mountain" or "the present king of France" disappears. With it vanishes any appearance that we must acknowledge such an "object" as the Golden Mountain or king of France even in order to deny that any such object *exists*. The examples illustrate in miniature two lessons:

first, that the logical form of a statement may differ significantly from its grammatical form, and that recognition of this difference may be the key to solving or dissolving a philosophical problem; second, that the correct logical analysis of a word or phrase may involve an explanation not of what *that word or phrase taken by itself* means, but rather of what *whole sentences containing the word or phrase* mean. Such an explanation is what is meant by a *contextual* definition: a definition that does not provide an analysis of the word or phrase standing alone, but rather provides an analysis of contexts in which it appears.

Russell's distinction between grammatical and logical form, and his claim that the former may be systematically misleading, was to prove immensely influential, even among nontechnically oriented philosophers, such as the Oxford ordinary-language school, who saw no need to use special symbols to represent logical forms, and objected to details of Russell's specific application of the distinction in his theory of descriptions. But Russell's notion of contextual definition is one implicit already in the practice of Weierstrass and other leaders of the nineteenth-century rigorization of analysis, and familiar to Russell from his undergraduate mathematical studies, so that even the antitechnical ordinary-language school of philosophical analysts are being influenced at one remove (and, so to speak, in spite of themselves) by mathematical analysis.

Contextual definition was the tool the rigorizers used to dispel the mysteries surrounding the notions of infinitesimals and infinities in the calculus. The followers of Leibniz had, for instance, written $\mathrm{d}f(x)/\mathrm{d}x$ for the derivative of a function $f(x)$, wherein $\mathrm{d}x$ was supposed to represent an "infinitesimal" change in the argument, and $\mathrm{d}f(x)$ a corresponding "infinitesimal" change $f(x + \mathrm{d}x) - f(x)$ in the value when the argument changes from x to $x + \mathrm{d}x$. (Leibniz claimed that this was all just a figure of speech, but his followers seem to have taken it literally.) These infinitesimals could be treated as nonzero in some circumstances—in particular, one could divide by them, as one cannot divide by zero—and yet treated as zero and neglected in other circumstances. Thus the derivative of the function $f(x) = x^2$ was computed as follows:

$$\frac{\mathrm{d}f(x)}{\mathrm{d}x} = \frac{f(x + \mathrm{d}x) - f(x)}{\mathrm{d}x} = \frac{(x + \mathrm{d}x)^2 - x^2}{\mathrm{d}x}$$
$$= \frac{2x\,\mathrm{d}x + (\mathrm{d}x)^2}{\mathrm{d}x} = 2x + \mathrm{d}x = 2x.$$

Here $\mathrm{d}x$ is treated as nonzero at the next-to-last step, and zero at the last step—the kind of procedure that

outraged critics like Berkeley. In the course of the nineteenth-century rigorization, the infinitesimals were banished: what was provided was not a direct explanation of the meaning of $\mathrm{d}f(x)$ or $\mathrm{d}x$, taken separately, but rather an explanation of the meaning of contexts containing such expressions, taken as wholes. The apparent form of $\mathrm{d}f(x)/\mathrm{d}x$ as a quotient of infinitesimals $\mathrm{d}f(x)$ and $\mathrm{d}x$ was explained away, the true form being $(\mathrm{d}/\mathrm{d}x)f(x)$, indicating the application of an operation of differentiation $\mathrm{d}/\mathrm{d}x$ applied to a function $f(x)$.

Similarly, such an expression as $\lim_{x \to 0} 1/x = \infty$, or "the limit of $1/x$ as x goes to zero is infinity," was explained *as a whole*, without requiring any explanation of "∞" or "infinity" taken separately. The details, which now appear in any freshman calculus textbook, need not detain us. What is important historically is that the notion of contextual definition employed in Russell's theory of descriptions was an idea that would have been familiar to him as a student of mathematics. To acknowledge this is, needless to say, not to deny that there is a certain genius involved in extracting such an idea from its original context of mathematical analysis and employing it to resolve philosophical puzzles. To acknowledge the germs of Russell's ideas in ideas of Weierstrass is merely to indicate more precisely what *kind* of genius Russell, like Frege before him, was bringing to bear on philosophical issues: a kind of philosophical genius *informed by knowledge of mathematics.*

4 Philosophical Analysis and Analytic Philosophy

Anyone who acquires a new tool is in some danger of behaving like the proverbial man with a hammer to whom everything seems to be a nail. There is no denying that some of the first people to apply the new methods of Frege and Russell were overenthusiastic about what such methods could accomplish. Russell himself, having established to his own satisfaction that mathematics could be reduced to pure logic once one had a sufficiently rich and powerful logic, went on to conclude that every science apart from mathematics could be reduced to logical compounds of statements about immediate sensory impressions—"sense data" as they were called. The logical positivists reached a similar conclusion, and were ready to ban any statement that did not admit such a reduction, from the assertions of Hegelian or absolute idealist metaphysicians on, as a "pseudo-statement," or mere nonsense.

Conscientious attempts to work out just *how* science, even the parts concerned with theoretical entities not directly observable (such as quarks and black holes in the science of today), could be reduced logically to statements about sense data, or at least to statements about everyday observable objects (such as meter readings), failed. Hence the positivists were forced to acknowledge that their program could not succeed, and (since they did not wish to dismiss large parts of modern science as mere pseudo-statements) that their standards of meaningfulness were too rigid. But as Soames emphasizes, this very acknowledgment of failure was a kind of success, because few if any philosophical schools before the positivists had even stated their aims with sufficient clarity to make it possible to see that they were unachievable. The new logical resources provided by Frege and Russell had *both* tempted the positivists to conjecture more than they could prove *and* made it clear to them that proof of their conjecture was impossible.

With experience the scope and limits of the new methods gradually came to be better understood. Russell's theory of descriptions had been hailed by his student F. P. Ramsey as "a paradigm of philosophical analysis," which indeed it is. But it came to be appreciated that the kind of application Russell made to the issue of negative existentials, where a philosophical problem was completely dissolved by philosophical analysis, would seldom be possible. Analysis, in general, is only a preliminary, a process that makes it clearer what the real problems are, and not a panacea, exposing all apparent problems as mere pseudo-problems.

As analytic philosophy has developed, enthusiasm has been replaced by dedication: recognition of the limitations of Frege's and Russell's methods has led not to the abandonment of the goal of clarity, which was the underlying motive of the great pioneering figures, but rather to firmer adherence to it. Today, when one can read large tracts of philosophy in the analytic tradition without encountering a single explicit analysis, let alone one expressed in special logical symbolism, one still finds almost everywhere a clarity of prose style that instantly distinguishes writing in this tradition from the writings of Continental philosophers (to say nothing of the Continentalizing philosophasters to be found in certain humanities departments in universities in the English-speaking world). This clarity—found, to be sure, already in the mathematician–philosopher Descartes, the first truly modern philosopher, but lost in

many of his successors—is the ultimate influence and legacy which the pioneers of analytic philosophy transmitted from mathematics to their philosophical heirs.

Further Reading

I recommend *Philosophical Analysis in the Twentieth Century* (Princeton, NJ: Princeton University Press, 2003) by Scott Soames for those wishing to read more about this subject. Each of the two volumes of this work contains substantial lists of primary and secondary sources at the end of each of its several parts.

VII.13 Mathematics and Music
Catherine Nolan

---

1 Introduction and Historical Overview

Music is the pleasure the human mind experiences from counting without being aware that it is counting.

This intriguing remark of LEIBNIZ [VI.15], from a 1712 letter to fellow mathematician CHRISTIAN GOLDBACH [VI.17], suggests a serious connection between mathematics and music, two subjects—one a science, the other an art—that may at first seem very different from each other. Leibniz was perhaps thinking of the long-standing historical and intellectual association of the two disciplines that date back to the time of PYTHAGORAS [VI.1], when the subject of music was part of an elaborate classification scheme of knowledge in the mathematical sciences. This scheme became known in the Middle Ages as the *quadrivium*, and consisted of the four disciplines of arithmetic, music (harmonics), geometry, and astronomy. In the Pythagorean worldview, these subjects were interlinked, since in one way or another they were all concerned with simple ratios. Music was merely the aural manifestation of a more universal harmony, which was likewise expressed by relationships between numbers, geometrical magnitudes, or the motions of celestial bodies. Harmonic consonance of musical intervals resulted from simple ratios of the first four natural numbers, 1:1 (the unison), 2:1 (the octave), 3:2 (the perfect fifth), and 4:3 (the perfect fourth), and was demonstrated empirically by the ratios of lengths of vibrating strings on the ancient instrument the monochord.[1] Beginning with the Scientific

---

1. The monochord was an instrument designed for demonstration, not artistic, purposes. It consisted of a single string stretched between two fixed bridges. A movable bridge between the fixed bridges was used to adjust the length of the string as it was plucked to produce sound, thereby altering the pitch of the sound.

Revolution of the seventeenth century, theories of tuning and temperament of musical intervals required more advanced mathematical ideas as well, such as logarithms and decimal expansions.

Musical composition has been inspired by mathematical techniques throughout its history, although mathematically inspired compositional techniques are associated mainly with music of the twentieth, and now twenty-first, centuries. A striking early example appears in the section on melody in a monumental treatise on music, entitled *Harmonie universelle* (1636–37), by the mathematician Marin Mersenne. Mersenne applied simple (from today's perspective) combinatorial techniques to the distribution and organization of notes in melodies. For example, he calculated the number of different arrangements or permutations of n notes, for each n between 1 and 22 (twenty-two notes delimiting the range of three octaves). The answer is of course $n!$, but in his zeal to illustrate this, he notated on musical staves all 720 (6!) permutations of the six notes of the minor hexachord (A, B, C, D, E, F), occupying a full twelve pages of *Harmonie universelle*. He went on to explore more complicated problems such as determining the number of melodies of a certain number of notes selected from a larger number, or determining the number of arrangements of finite collections of notes containing certain numbers of repetitions of one or more notes. He illustrated some of his findings with combinations of letters as well as musical notation, thereby showing that the music was incidental to the problems, which were in essence purely combinatorial. Such exercises, while seemingly of little practical or aesthetic value, at least demonstrated the great musical diversity that was in principle available with only a limited set of resources.

The polymath Mersenne was a composer and practicing musician as well as a mathematician, and his fascination with applying a relatively new mathematical technique to music composition showed a level of interest in abstract connections between mathematics and music that is shared by many music theorists, and to a lesser degree by performing musicians and nonspecialist music enthusiasts. The patterns of music, in particular pitch and rhythm, lend themselves well to mathematical description, and some of them are amenable to algebraic reasoning. In particular, the system of twelve equal-tempered notes is naturally modeled using MODULAR ARITHMETIC [III.58], and this, together with combinatorial arguments, was used in the music theory of the twentieth century. In this article we survey the asso-

ciation of mathematics and music from its concrete representation in sound itself, through its manifestation in the working materials of composers, and finally to its explanatory power in abstract music theory.

2   Tuning and Temperament

The most obvious relationships between mathematics and music appear in acoustics, the science of musical sound, and particularly in the analysis of the intervals between pairs of pitches. With the development of polyphonic music in the Renaissance period, the Pythagorean conception of consonance based on the simple ratios of the integers from 1 to 4 eventually came into conflict with musical practice. The acoustically pure perfect consonances of Pythagorean tuning were well-suited for medieval parallel organum,[2] but in the fifteenth and sixteenth centuries use was increasingly made of the so-called *imperfect consonances*, that is, major and minor thirds and their octave inversions, minor and major sixths. In Pythagorean tuning, intervals are derived by successions of perfect fifths, so the corresponding frequency ratios are powers of $\frac{3}{2}$. In conventional Western music, twelve perfect fifths in succession, C–G–D–A–E–B–F♯–C♯–G♯–D♯–A♯–E♯–B♯, are supposed to equal seven octaves (C = B♯), but this does not work in Pythagorean tuning, since $(\frac{3}{2})^{12}$ does not equal 2^7. Indeed, a succession of Pythagorean perfect fifths will never result in a whole number of octaves. As it happens, twelve Pythagorean perfect fifths give an interval slightly larger than seven octaves. The difference is a small interval known as the *Pythagorean comma*, which corresponds to a ratio of $(\frac{3}{2})^{12}/2^7$, which is about 1.013643.

Pythagorean tuning was originally conceived in terms of successive single pitches. The problems associated with it start to arise when pitches sound simultaneously. While Pythagorean fifths between simultaneous pitches sound pleasing with their simple 3:2 ratios, Pythagorean thirds and sixths have much more complex ratios that sound harsh to Western ears. These came to be replaced by the simple ratios of *just intonation*, which are ratios of quite small whole numbers. These ratios were considered "natural" because they

---

2. *Organum* is the earliest form of musical polyphony, and involved adding a voice (or voices) to an existing plainchant melody (*cantus firmus*). In its original form, the added voice proceeded in parallel motion to the plainchant melody at the interval of a perfect fourth or fifth.

Notes	C	D	E	F	G	A	B	C
Intervals (ratios)		$\frac{9}{8}$	$\frac{10}{9}$	$\frac{16}{15}$	$\frac{9}{8}$	$\frac{10}{9}$	$\frac{9}{8}$	$\frac{16}{15}$

Figure 1 Successive intervals in a major
scale tuned in just intonation.

reflect the ratios of the natural overtone series.[3] The Pythagorean major third, which has the relatively complex ratio of $(\frac{3}{2})^4/2^2$, or $\frac{81}{64}$, was replaced by the slightly smaller major third of just intonation, which has the much simpler ratio 5:4. The difference between these two intervals is known as the *syntonic comma*, which corresponds to the ratio 81:80, or 1.0125. Likewise, the Pythagorean minor third has ratio 32:27, and so is slightly smaller than the minor third of just intonation, which has ratio 6:5. The difference is again a syntonic comma. The Pythagorean major and minor sixths, the octave inversions of the thirds, also differ from their just counterparts by a syntonic comma.

Suppose that you want to build a C-major scale in just intonation. You can do it as follows. Start with C and define each other note by the ratio of its frequency to that of C. The subdominant and dominant, that is, F and G, have ratios 4:3 and 3:2, respectively. From these three notes one can build major triads in the ratios 4:5:6. So E, for instance, which belongs to the major triad that starts with C, has ratio 5:4. Similarly, A has ratio 5:3, since it is in a ratio 5:4 with F. With this kind of calculation, one ends up with the scale shown in figure 1, where the fractions now represent the frequency ratios between successive notes. The smaller whole tone (10:9) between notes D and E creates intonation problems for the supertonic triad, D–F–A. While the minor triads on E and A (the mediant and submediant) produce the proportion 10:12:15, the minor triad on D is out of tune. Its fifth, D–A, is a syntonic comma flat, as is its third, D–F, which is in fact a Pythagorean minor third.

Tempering (increasing or decreasing) the size of intervals offered a practical solution to the problems inherent in just intonation by distributing the syntonic comma among the major thirds or the perfect fifths of the scale, thereby compromising the purity of one interval to preserve the purity of another. This prac-

tice became known as meantone temperament. Various systems of meantone temperament were put forward in the sixteenth and seventeenth centuries for the tuning of keyboard instruments, the most common of which was quarter-comma meantone temperament. In this system the perfect fifth is lowered by a quarter of a syntonic comma so that the major thirds have the pure ratio 5:4.

A perpetual problem with meantone temperaments is that, while modulation to closely related keys sounds pleasing, modulation to more remote keys sounds out of tune. The system of equal temperament, in which the syntonic comma is distributed evenly among all twelve semitones of the octave, gradually became adopted because it removed the limitations on keys for modulation. The discrepancies between just and equal-tempered intervals are small and easily accepted by most listeners. The ratio of an equal-tempered semitone is $\sqrt[12]{2}$, or $1.05946\ldots$; by comparison, a just semitone, with ratio 16:15, is $1.06666\ldots$. The ratio of an equal-tempered perfect fifth, seven semitones, is $\sqrt[12]{2^7}$ or $\sqrt[12]{128}$, which is $1.498307\ldots$, whereas a just perfect fifth, with ratio 3:2, is of course 1.5. In equal temperament, one starts from a reference such as the note A, which is usually taken to have frequency 440 Hz.[4] All other notes have frequencies of the form $440(\sqrt[12]{2})^n$, where n is the number of semitones between the note in question and the reference note A. In equal temperament, enharmonic notes such as C$^\sharp$ and D$^\flat$ are acoustically identical—that is, they share the same frequency. Equal temperament was well-suited for the kind of music that was written from the eighteenth century onward, with its much greater range of modulations and chromatic harmonic vocabulary.

The unit of the *cent* was defined by A. J. Ellis as the ratio between two pitches separated by one hundredth of an equal-tempered semitone, and became the most commonly used unit for measuring and comparing intervals.[5] The octave consists, therefore, of 1200 cents. If a and b are two frequencies, then the distance in cents between the corresponding pitches is given by the formula $n = 1200\log_2(a/b)$. (As a check, notice that if $a = 2b$ then one does indeed get the answer $n = 1200$.)

---

3. The partials of the overtone series are multiples of the frequency of the fundamental pitch, and the first six partials generate the intervals of the major triad. For instance, the first six partials of the overtone series of a fundamental pitch C are C (1:1), C (2:1), G (3:1), C (4:1), E (5:1), G (6:1).

---

4. The frequency of a pitch is a measurement of the number of cycles per second (abbreviated as "cps"). More commonly, the number of cycles per second is identified in units called *hertz* (abbreviated as "Hz"), named after the physicist Heinrich Rudolf Hertz.

5. Ellis's account of the *cent* appeared in his appendix to the eminent nineteenth-century physicist Hermann von Helmholtz's *On the Sensations of Tone* (1870; English edn., 1875).

Microtonal systems based on the equal division of the octave into more than twelve parts were proposed and realized by some composers in the twentieth century, but they have not become widely used in Western music. However, the idea of dividing the octave into equal parts has become fundamental. It means that the notes used are naturally modeled by integers. If one regards two notes an octave apart as "the same," which makes good musical sense, then one is dividing all notes into twelve EQUIVALENCE CLASSES [I.2 §2.3]. The natural model for these is arithmetic modulo 12. As we shall see later, the symmetries of the group of integers mod 12 are of great musical significance.

3   Mathematics and Music Composition

The association of number and music in acoustics was the result of scientific discovery. Number and music have also been associated through invention and creativity in music composition. Fundamental aspects of the temporal organization of music reflect simple proportional relationships. The basic durational values in Western music notation are the whole note (𝗈), half note (♩), quarter note (♩), eighth note (♪), etc. These are related to each other by simple multiples or fractions—all powers of 2—and these relationships are reflected in the metric organization of musical time into bars with the same number of beats. Bars or measures are indicated by time signatures such as the *simple meters* $\frac{2}{4}$, $\frac{3}{4}$, or $\frac{4}{4}$ (𝐜), where beats (the ♩ in these examples) are typically subdivided into two, or the *compound meters* $\frac{6}{8}$, $\frac{9}{8}$, or $\frac{12}{8}$, in which beats (the ♩· in these examples) are subdivided into three.

A common device in musical composition, especially counterpoint, is for a melodic theme, or *subject*, to reappear at half or twice the original speed, techniques known as *rhythmic augmentation* or *diminution*, respectively. Figures 2 and 3 show the subjects of two fugues from the second volume of J. S. Bach's *Well-Tempered Clavier*: no. 9 in E major, whose subject appears in diminution; and no. 2 in C minor, whose subject appears in augmentation. (The last note of the diminished or augmented subject may not be proportionally related to the original in order to allow a good continuation for the music that follows.)

Geometric relations have served as musical resources of other kinds too. A well-known construct in music theory is the *circle of fifths*, which was originally designed to demonstrate the relationships between different major and minor keys. As illustrated in figure 4,

Figure 2 J. S. Bach, *Well-Tempered Clavier*, Book 2, Fugue no. 9, subject and diminution.

Figure 3 J. S. Bach, *Well-Tempered Clavier*, Book 2, Fugue no. 2, subject and augmentation.

the twelve notes are arranged around the circle as a succession of perfect fifths. Any seven consecutive notes in this circle will be the notes of some major scale, which makes it easy to understand some of the patterns of the key signatures. For instance, the C major scale consists of all the notes from F to B (clockwise). To change from C major to G major, one shifts the sequence by one, losing the note F but gaining F♯. Continuing in this way, we see that C major is the key with no sharps or flats, G major has one sharp, D major has two sharps, A major has three sharps, etc. Similarly, moving counterclockwise from C, F major has one flat, B♭ major has two flats, E♭ major has three flats, etc. From a mathematical point of view, we have transformed the chromatic scale, which we identify with the additive group of integers mod 12, using the automorphism $x \mapsto 7x$, and this makes some musical phenomena much more transparent.

Reflective symmetry is another geometrical concept with a long history in musical composition. Musicians will frequently describe melodic lines in spatial terms, referring to notes of higher frequencies as "up," and notes of lower frequencies as "down." This allows one to think of melodic lines as ascending or descending. Reflection in a horizontal axis interchanges up and down. The musical counterpart to this is known as *melodic inversion*: one reverses the ascending or descending direction of each interval, and the result is an inverted form of the melody. Figure 5 shows the subject of Fugue no. 23 in B major from the first volume of Bach's *Well-Tempered Clavier* and a later appearance of the subject in inverted form. A geometrical reflection is clearly visible in the notation, but, more importantly,

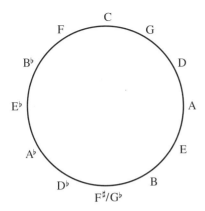

Figure 4 The circle of fifths.

Figure 5 J. S. Bach, *Well-Tempered Clavier*, Book 1, Fugue no. 23, subject and inversion.

the inversion can also be clearly heard in the sound of the music itself.

Conventional Western musical notation implies a two-dimensional organization: the vertical dimension expresses the relative frequency of pitches from low to high, and the horizontal dimension expresses chronological time from left to right. Another compositional device, much rarer than the devices of rhythmic augmentation and diminution or melodic inversion, is that of *retrograde*, where a melody is played backwards. When the melody is played backwards and forwards simultaneously, the technique is known as a *cancrizans* canon. Perhaps the best-known examples of cancrizans occur in the music of J. S. Bach, such as in the first canon of *The Musical Offering* or the first and second canons of the *Goldberg Variations*. Figure 6 shows the opening and closing measures of the cancrizans from Bach's *Musical Offering*. The melody of the first few bars of the upper staff returns in reverse order at the end of the piece in the lower staff, and likewise, the melody of the first few bars of the lower staff returns in reverse order at the end of the upper staff. Joseph Haydn's *Menuetto al rovescio*, from the Sonata no. 4 for violin and piano, is another well-known example of a similar technique,

in which the first half of the piece is played backwards in the second half.

We may regard the devices of melodic retrograde and inversion as reflections in a two-dimensional musical space. However, retrograde is much more esoteric, owing to the greater constraints involved in the manipulation of musical time. Examples such as those by Bach and Haydn mentioned above demonstrate great ingenuity on the part of the composer, who must make the melodic retrogrades work convincingly with the underlying harmonic progressions. Certain common chord progressions, such as moving from the supertonic to the dominant, do not work well in reverse, so a composer attempting to write a cancrizans canon is forced to avoid them. Similarly, many common melodic patterns do not sound good when reversed. These difficulties account for the rarity of retrograde techniques in tonal music (i.e., music based on major and minor keys). With the abandonment of tonality in the early twentieth century, the main constraints were removed, making composition with retrograde easier. For instance, retrograde and inversion played an important role in serial music, as we shall see. However, composers of such music replaced the traditional constraints of tonal music with others, such as avoiding major or minor triads and bringing out other intervals deemed important for a particular piece.

The atonal revolution in the early twentieth century, during which composers experimented with novel methods of harmonic organization, led to the exploration of new types of symmetry relations in music composition. Scales based on repeating interval patterns (measured in semitones), such as the *whole-tone scale* (2-2-2-2-2-2) or the *octatonic scale* (1-2-1-2-1-2-1-2), appealed to composers for the symmetric structures and novel harmonies they embodied. The octatonic scale, also known in jazz circles as the *diminished scale*, had a particularly wide appeal among a variety of composers of different nationalities, such as Igor Stravinsky, Olivier Messiaen, and Béla Bartók. The novelty of the whole-tone and octatonic scales is that they have nontrivial *translational* symmetry, a property not shared by the major or minor scales. The whole-tone scale is unchanged if it is transposed by a tone, and the octatonic scale is unchanged if it is transposed by a minor third. There are thus only two distinct translates of the whole-tone scale and three of the octatonic scale. For this reason, neither scale has a clearly defined tonal center, which was a major reason for their attractiveness to early twentieth-century composers.

Figure 6 J. S. Bach, *The Musical Offering*, opening and closing measures of the cancrizans (canon 1).

Reflective symmetry was used by twentieth-century composers as well, to help them with the formal aspects of compositional design. A fascinating example is the first movement of Bartók's *Music for Strings, Percussion, and Celesta* (1936), which extends the traditional principles of the baroque fugue and incorporates a symmetric design. Figure 7 illustrates the structure of the fugue subject entries, starting from the initial entry on A. In a traditional fugue, the subject is stated in tonic, followed by a statement in the dominant, and then again in the tonic (and continuing the alternating pattern of tonic and dominant entries for fugues with more than three voices). In Bartók's fugue, the first statement of the subject begins with the note A, and the next with E. Instead of returning to A for the third statement, however, the subsequent entries follow a pattern of alternating fifths in opposite directions from A: that is, the sequence A–E–B–F♯, etc., alternates with the sequence A–D–G–C, etc. This pattern is illustrated in figure 7. Each of the interlocked cycles completes a circle of fifths, one clockwise, the other counterclockwise. Each letter in the illustration represents a statement of the fugue subject beginning on that note, and each of the interlocked cycles of fifths arrives on E♭ (six semitones from the starting point, A) at its midpoint, so that all twelve notes occur once in the first half of the pattern and once again in the second half. The midpoint of the pattern corresponds to the dramatic climax of the work, after which the pattern of interlocked cycles of fifths resumes with subject entries in inverted form until the conclusion of the work with the return of the subject starting on A.

Arnold Schoenberg's twelve-tone method of composition, which he revealed in the early 1920s, is based on permutations of all the twelve notes, rather than of subsets of seven notes as one has in music in major or minor keys. In twelve-tone music (and atonal music in

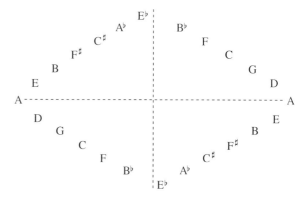

Figure 7 Plan of fugal entries in Béla Bartók's *Music for Strings, Percussion, and Celesta*, first movement (after Morris (1994, p. 61), with permission).

general), the twelve notes are supposed to have equal prominence: in particular, there is no single note with a special status like that of the tonic in a major or minor key. The basic ingredient of a piece of twelve-tone music is a *tone row*, which is a sequence given by some permutation of the twelve notes of the chromatic scale. (These notes can, however, be stated in any octave.) Once the tone row has been chosen, it can be manipulated by means of four types of transformation: transposition, inversion, retrograde, and retrograde inversion. Musical transposition corresponds to the mathematical operation of translation: the intervals between successive notes of a transposed row are the same as those between the corresponding notes of the original row, so the entire row is shifted up or down.[6] Inversion corresponds to reflection, as we have already

---

6. Describing transposition as translation does justice to the fact that a melody sounds "the same" when transposed, even though the pitches are different, because the successive intervals are the same. If one arranges the twelve notes in a circle, then one can also think of this translation as a rotation.

Figure 8 Row forms in Schoenberg's *Suite for Piano* (1923).

discussed: the intervals of the row are reflected about a "horizontal" axis. Retrograde corresponds to reflection in time: the row is stated backwards. (However, if it is combined with a transposition, as it may be, then it is better described as a glide reflection.) Retrograde inversion is a composition of two reflections, one vertical and one horizontal: it therefore corresponds to a half turn.

Figure 8 illustrates the serial transformations applied to a row created by Schoenberg for his *Suite for Piano*, opus 25, published in 1923. The forms of the row are labeled P (for prime—the original row and its transpositions), R (for retrograde), I (for inversion), and RI (for retrograde inversion). The integers 4 and 10 in the row labels on the left and right refer to the starting notes of the P and I row forms by telling us how many semitones away from C they are. Thus, 4 refers to E (4 semitones above C) and 10 refers to B♭ (10 semitones above C). The retrogrades of the P and I forms, R and RI, are labeled on the right-hand side of the figure. It is easy to see the inversion (reflection) of P4 in I4 about the first note E and the transposition of P4 by six semitones in P10, as well as the inversion of P10 about the first note, B♭.

One may wonder what sort of insight we gain from understanding these abstract relationships and why they were so attractive to composers like Schoenberg. In Schoenberg's *Suite*, the eight row forms shown in figure 8 are in fact the only ones used in all five movements of the composition. This represents a high degree of selectivity, since there are 48 (= 12 × 4) available row forms. However, this self-imposed restriction is not on its own enough to account for the interest or attraction of this music. An additional aspect of the

technique is that the row itself, and the way its transformations unfold in the course of a work, are chosen carefully to bring out certain relationships between notes. For example, all the row forms used in the *Suite* begin and end on the notes E and B♭, and these notes are frequently articulated in the work so that they take on an anchoring function that fills the void created by the absence of a conventional tonal center. Similarly, the notes in the third and fourth positions in each of the four row forms are always G and D♭, in either order, and likewise these are articulated in various ways in the movements of the *Suite* so that they can become recognizable. The two pairs of notes just mentioned, E–B♭ and G–D♭, are related to each other by sharing the same interval, six semitones (half an octave, also known as the tritone because it spans three whole tones). In the hands of a master composer, a twelve-tone row is not a random collection of notes, but a foundation for an extended composition carefully constructed to produce interesting structural effects that one can learn to recognize and appreciate.

Permutations and serial transformations of other musical parameters besides pitch—such as rhythm, tempo, dynamics, and articulation—were explored by a new generation of postwar European composers, including Olivier Messiaen, Pierre Boulez, and Karlheinz Stockhausen. Compared with the serialization of pitch, however, serialization of these parameters does not lend itself to such precise transformations, because it is less easy to organize them into discrete units than it is the twelve notes of musical space.

It is important to recognize that Schoenberg and most composers whose music exhibits mathematical conceptions such as those we have seen had little if any mathematical training.[7] Nevertheless, the basic mathematical patterns and relations that we have discussed are so pervasive in so many aspects of so many different kinds of music that the importance of mathematics in music is undeniable.

We end this section with a few more examples. Proportional relations such as the simple ones between note values reappear on a larger scale in relations between lengths of formal divisions in music of Mozart, Haydn, and others: they often use basic building blocks of four-measure phrases and use them in pairs, and pairs of pairs, to form larger units. The techniques of melodic manipulation seen in Bach's works, which are found in a new guise in Schoenberg's twelve-tone techniques, can also be found in contrapuntal works of composers before Bach, such as Palestrina. And some composers, including Bach, Mozart, Beethoven, Debussy, Berg, and others, are said to have incorporated numerological elements into their composition, such as symbolic numbers or proportions based on Fibonacci sequences and the golden ratio.

4 Mathematics and Music Theory

In the second half of the twentieth century, the ideas of Schoenberg and his followers were extended and developed in North American music theory. Milton Babbitt, a renowned American composer and theorist, is widely credited with introducing formal mathematics, specifically group theory, to the theoretical study of music. He generalized Schoenberg's twelve-tone system to any system where one has a finite set of basic musical elements (of which Schoenberg's twelve-tone rows were just one example), with relations and transformations between them (see Babbitt 1960, 1992). There are forty-eight ways of transforming a row, and Babbitt noted that these transformations form a group, which is in fact the product of the dihedral group D_{12} with the cyclic group C_2 of two elements. (The D_{12} in this product is the symmetry group of a dodecagon, and the C_2 allows the time reversal.) The four sets of transformations—P, I, R, and RI (see the previous section)—define a homomorphism from this group to

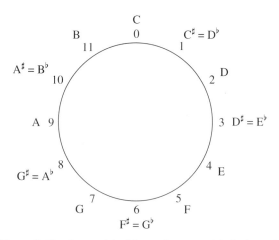

Figure 9 Circular model of the twelve notes (pitch classes).

the Klein group $C_2 \times C_2$, by identifying transformations that are equivalent up to rotation.

Identifying musical notes with the elements of the group $\mathbb{Z}_{12}$ of integers mod 12, and modeling various musical operations by means of transformations on this group, makes it much easier to analyze some kinds of music, such as the atonal music of Schoenberg, Berg, and Webern, that do not lend themselves easily to more traditional analysis of harmony (see Forte 1973; Morris 1987; Straus 2005). This identification is illustrated in figure 9. As we have already commented, multiplying by 5 or 7 is an automorphism of $\mathbb{Z}_{12}$, and gives the cycle of fifths shown in figure 4 (when one substitutes the mod-12 integers for the names of the notes). This mathematical fact has many musical consequences. One of them is that it is common to substitute fifths by semitones, and vice versa, in chromatic harmony and in jazz.

A branch of music theory known as *atonal set theory* attempts to give a very general understanding of pitch relations by looking at all the $2^{12} = 4096$ possible combinations of notes, and defining two such combinations to be equivalent if one can be derived from the other by two simple transformations, the idea being that equivalent combinations will have the same intervals. The transformations in question are transposition and inversion. A transposition up by n semitones (where we think of n as an integer mod 12) is denoted T_n. The notation I is used for a reflection about the note C, so a general inversion takes the form $T_n I$ for some n. (Inversion in this context refers to reflection in musical space, and should not be confused with chord inversion in tonal music.) In these terms, to use a familiar example, the major triad and the minor triad are related

---

7. Some composers, to be sure, have received extended mathematical training, which is reflected in their works. Iannis Xenakis, for example, was trained as an engineer, and had professional contact with the architect Le Corbusier. Xenakis found parallels between music and architecture through his study of Le Corbusier's *Modulor* system and its approach to form and proportion based on the human figure. Xenakis's compositions are characterized by their massive, physical sound and their complex algorithmic processes.

to each other by inversion since their successive intervals are reflections of each other (four then three semitones in the major triad and three then four semitones in the minor triad, counting from the lowest note). Consequently, all major and minor triads belong to the same equivalence class. For example, the E-major triad $\{4, 8, 11\}$ is related to the C-major triad $\{0, 4, 7\}$ by the transposition T_4 (because $\{4, 8, 11\} \equiv \{0+4, 4+4, 7+4\}$, mod 12), and the G-minor triad $\{7, 10, 2\}$ is related by inversion to the D-major triad $\{2, 6, 9\}$ by T_4I (because $\{7, 10, 2\} \equiv \{4 - 9, 4 - 6, 4 - 2\}$, mod 12). An equivalence class, such as the class of major and minor triads, will normally consist of twenty-four sets. However, if it has internal symmetries, such as those of the diminished seventh chord (with interval succession 3-3-3-3) or the whole-tone and octatonic scales mentioned earlier, then the number of sets in the class will be smaller, though it will always be a factor of 24.

Sets of notes in the same equivalence class share certain sonic attributes because they share the same number and types of intervals. But while it seems reasonable enough to regard transposed chords as equivalent, since they really do have an obvious "sameness" in the way they sound, there has been some controversy over the notion of inversional equivalence. For example, is it reasonable to declare major and minor triads to be equivalent to each other when they clearly do *not* sound the same and have very different musical roles? Of course, we are free to define any equivalence relation we like, so the real question is whether this one has any utility. And in some contexts it does: with sets of notes that do not possess extensive associations with tonal music it is easier to recognize this form of equivalence than it is with major and minor triads. For example, the three notes C, F, and B share the same intervals (one semitone, one perfect fourth or fifth, and one tritone) as the three notes F$^\sharp$, G, and C$^\sharp$, and this does indeed give them a noticeable form of "sameness." (The set $\{11, 0, 5\}$ is inversionally related to $\{1, 6, 7\}$ by T_6I because $\{11, 0, 5\} \equiv \{6 - 7, 6 - 6, 6 - 1\}$, mod 12.)

There is other important work in music theory that has been inspired by group theory. The most influential example is David Lewin's *Generalized Musical Intervals and Transformations* (1987), which develops a formal theory that connects mathematical reasoning and musical intuition. Lewin generalizes the concept of interval to mean any measurable distance, whether between pairs of pitches, durations, time points, or contextually defined events in a musical work. He develops a model called the generalized interval system (GIS), which consists of a set of musical objects (e.g., pitches, rhythmic durations, time spans, or time points), a group (in the mathematical sense) of intervals (representing the distance, span, or motion between a pair of objects in the system), and a function that maps all possible pairs of objects in the system into the group of intervals. He also uses GRAPH THEORY [III.34] to model musical processes, through his notion of a *transformation network*. The vertices of such a network are basic musical elements such as melodic lines or chordal roots. These elements come with certain transformations, such as transposition (or shifting by a generalized interval) or the serial transformations from twelve-tone theory. Two vertices are joined by an edge if there is an allowable transformation that takes one to the other. The emphasis thus shifts from the basic elements to the relations that connect them. Transformation networks offer a dynamic way of looking at musical processes, giving visible form to abstract and often nonchronological connections in the analysis of musical works.

The level of generalization and abstraction makes Lewin's treatise a challenge for the mathematically unsophisticated music theorist, but it does not need more than fairly simple undergraduate-level algebra, so it is accessible enough for the determined reader with some mathematical training. It becomes clear to such a reader that the formality of the presentation is essential to a proper understanding of the transformational approach to music theory and analysis. Despite this formality, Lewin continually maintains contact with music itself, and how his mathematical tools can be applied in different contexts. The result is that the reader is rewarded with insights that would be impossible without the mathematical rigor. Mathematicians, while likely to find the material relatively elementary, may find their attention "captivated by the way in which the author gives new and, sometimes, unexpected interpretations to classical mathematical ideas when applied to musical contexts" (Vuza 1988, p. 285).

5   Conclusion

The playful Leibniz quotation with which this essay began underscores an enduring mathematical presence in music. Both disciplines rely in a fundamental way on concepts of order and reason, as well as more dynamic concepts of pattern and transformation. Music was once subsumed within mathematics, but it has now acquired its own identity as an art that has always derived inspiration from mathematics. Mathematical concepts have provided composers and theorists of music both with tools for creating music and with a language for articulating analytical insights about it.

Further Reading

Babbitt, M. 1960. Twelve-tone invariants as compositional determinants. *Musical Quarterly* 46:246–59.

———. 1992. The function of set structure in the twelve-tone system. Ph.D. dissertation, Princeton University.

Backus, J. 1977. *The Acoustical Foundations of Music*, 2nd edn. New York: W. W. Norton.

Forte, A. 1973. *The Structure of Atonal Music*. New Haven, CT: Yale University Press.

Hofstadter, D. R. 1979. *Gödel, Escher, Bach: An Eternal Golden Braid*. New York: Basic Books.

Lewin, D. 1987. *Generalized Musical Intervals and Transformations*. New Haven, CT: Yale University Press.

Morris, R. 1987. *Composition with Pitch-Classes: A Theory of Compositional Design*. New Haven, CT: Yale University Press.

———. 1994. Conflict and anomaly in Bartók and Webern. In *Musical Transformation and Musical Intuition: Essays in Honor of David Lewin*, edited by R. Atlas and M. Cherlin, pp. 59–79. Roxbury, MA: Ovenbird.

Nolan, C. 2002. Music theory and mathematics. In *The Cambridge History of Western Music Theory*, edited by T. Christensen, pp. 272–304. Cambridge: Cambridge University Press.

Rasch, R. 2002. Tuning and temperament. In *The Cambridge History of Western Music Theory*, edited by T. Christensen, pp. 193–222. Cambridge: Cambridge University Press.

Rothstein, E. 1995. *Emblems of Mind: The Inner Life of Music and Mathematics*. New York: Times Books/Random House.

Straus, J. N. 2005. *Introduction to Post-Tonal Theory*, 3rd edn. Upper Saddle River, NJ: Prentice Hall.

Vuza, D. T. 1988. Some mathematical aspects of David Lewin's book *Generalized Musical Intervals and Transformations*. *Perspectives of New Music* 26(1):258–87.

VII.14   Mathematics and Art
Florence Fasanelli

1   Introduction

This article focuses on the relationship between the history of mathematics and the history of art in twentieth-century France, England, and the United States. The effect of mathematics on artists and the direct interactions between artists and mathematicians have both been extensively studied. These studies show that

knowledge of mathematics has had a significant influence on many artists, as well as on musicians and writers. In particular, the increasingly wide acceptance, during the nineteenth century, of mathematical ideas that had once been revolutionary contributed strongly to what is now called modern art. At the end of the nineteenth century and the beginning of the twentieth, artists expressed on canvas and in sculpture their understanding of the fourth dimension and of NON-EUCLIDEAN GEOMETRY [II.2 §§6–10]. In doing so, they left behind their earlier training and heritage, which had been heavily based on a mathematical perspective derived from EUCLID [VI.2]. Their new ideas reflected the progress that had been made in mathematics, and many of the artists who formed new schools of thought were also engaged in interpreting these new mathematical developments.

The connection between mathematics and art is rich, complex, and informative. This is evident in some of the artistic styles and the philosophies that developed under the influence of new mathematics (and science), and in the creation of mathematics to fulfill artistic needs. Some examples include the paintings (with their often-studied geometries) of Italian mathematician Piero della Francesca (ca. 1412–92), who, having made a transcription of Jacopo of Cremona's Latin translation of Archimedes' *Codex A*, wrote out his own mathematical theories of perspective; Hans Holbein's (1497–1543) *Ambassadors* (1533), which illustrates how an artist can use a distorted variation on mathematical perspective to fool the eye (anamorphosis); Artemesia Gentileschi's (1593–1652) deliberate correction of a smattering of blood in her first version of *Judith Beheading Holofernes* (1612–13) to a parabolic arc of blood in the second version (1620) to match a sketch that her friend, scientist, court mathematician, and amateur painter Galileo Galilei (1564–1642) had made as a study for his as yet unpublished law of projectile motion; various works of the Dutch portrait painter Johannes Vermeer (1632–75) using the camera obscura; Johann Hummel's (1769–1852) paintings of the making of the great granite bowl in Berlin, which used Gaspard Monge's (1746–1818) *Géométrie Descriptive* (1799); sculpture by Naum Gabo[1] (1890–1977) and his brother Antoine Pevsner (1886–1962) following their youthful academic study of solid geometry; and the mathematically understandable but physically

---

implausible scenes by Maurits Cornelis Escher (1898–1972).

This article begins with a brief history of the development of perspective in art, because it is necessary to understand this in order to understand the rebellion against it that had such a decisive impact on modern art. This is followed by a short summary of the changing course of geometry in the nineteenth century through the development of non-Euclidean geometry and *n*-dimensional geometry. We then move on to the activities of artists, beginning in France in the early twentieth century and continuing with the works of representative artists in other countries, all the while keeping in mind the mathematics that provoked their artistic responses.

2   Development of Perspective

During the fifteenth century artists were still primarily employed to produce images of sacred subjects, but there was an increased interest in having pictures match aspects of the physical world. Lacking any precursors, artists had to devise their own axioms of linear perspective. At the beginning of the sixteenth century these early ideas of mathematical perspective were spread by books that contained visual representations. Mathematics that was previously known only in writing or orally now took a visual form, which was copied in engravings and spread across Europe.

The first writings on perspective were by Leon Battista Alberti (1404–72) and Piero della Francesca, while the ideas of Filippo Brunelleschi (1377–1446), the Florentine architect and engineer who was in fact the first to consider a mathematical theory of perspective, were captured by his biographer Antonio Manetti (1423–97). Artists and mathematicians continued to develop the rules of perspective while looking for ways to best represent space and distance. Among the mathematicians, Federico Commandino (1509–75), renowned for his Latin editions of the works of Greek mathematicians such as Euclid, ARCHIMEDES [VI.3], and APOLLONIUS [VI.4], was the first to write about perspective for the benefit of mathematicians rather than artists. His student Guidobaldi del Monte (1505–1647) published the influential book *Perspectivae libri sex* in 1600, in which he showed that any set of parallel lines not parallel to the plane of the picture will converge to a vanishing point.

Great artists, notably Leonardo da Vinci (1452–1519) and Albrecht Dürer (1471–1528), were now portraying mathematics in a visual form. Mathematician Luca

Figure 1 An artist using Dürer's perspective machine.
© Copyright: The Trustees of the British Museum.

Pacioli's (1445–1517) *De Divina Proportione* (1509) includes Leonardo's unsurpassed woodcuts of polyhedra (among them the first published illustration of a rhombicuboctohedron), and Dürer's *Unterweysung der Messung* (1525) contains the first illustration of nets for models of polyhedra. Dürer's own new knowledge of perspective, whose secrets he had learned on a trip to Italy from Germany, inspired him to create his famous illustrations of how to draw a picture in which all the elements are in one-point perspective (see figure 1).

In the seventeenth century, Girard Desargues (1591–1661), a French engineer and architect who wrote on practical subjects, continued the study of perspective that had been begun by the Renaissance artists. In doing so he invented a new, "non-Greek" way of doing geometry, which he published in his *Brouillon Project d'une Atteinte aux Événemens des Rencontres du Cône avec un Plan* (1639). In this essay, he attempted to unify the theory of conic sections through the use of projective techniques. This new PROJECTIVE GEOMETRY [I.3 §6.7] was based on his earlier realization than an artist can construct a perspective image without using a point from outside the picture field. However, of the original fifty printed copies of the *Brouillon Project* only one survives and his work, including his "perspective theorem," was made known through the publications of other mathematicians. Abraham Bosse (1602–76), a friend of Desargues who ran a famous atelier where the art of engraving was taught, was responsible for publishing much of Desargues's work, including that on the theory of perspective. But Bosse's promotion of Desargues's innovative ideas created controversy in art circles and seriously damaged his professional reputation. However, in the twentieth century, when engraving was revived as an important art form, a replica of Bosse's studio was built in Paris.

In the early eighteenth century, the mathematician and amateur painter BROOK TAYLOR [VI.16] published *Linear Perspective: Or a New Method of Representing Justly All Manner of Objects as They Appear to the Eye in All Situations* (1715), the first book on perspective to give a general treatment of vanishing points. As Taylor wrote on the title page, the book was "a work necessary for painters, architects, etc., to judge of, and regulate designs by." Taylor invented the phrase "linear perspective" and stressed the importance of what is now described as the main theorem of perspective: given any direction not parallel to the plane of the picture there is a "vanishing point" through which the representations of all lines in that direction must pass.

Since ancient times the axioms of Euclid's *Elements* have provided the basis for the understanding of two- and three-dimensional figures, and in the fifteenth century they provided the foundations for the study of perspective. But during the nineteenth century the long-standing debate about whether to accept Euclid's fifth axiom (the "parallel postulate") was resolved in a way that was to provoke a radical change in the perception of geometry: it was demonstrated by several mathematicians—notably LOBACHEVSKII [VI.31] in 1829, BOLYAI [VI.34] in 1832, and RIEMANN [VI.49] in 1854—that a consistent "non-Euclidean geometry" was possible in which the fifth axiom no longer held.

The mathematician and expositor HENRI POINCARÉ [VI.61] provided popular accounts of these new ideas in his books *La Science et l'Hypothèse* (1902) and *Dernières Pensées* (1913), which were widely read in France and elsewhere. Poincaré's works provoked the highly influential French (and later American) artist Marcel Duchamp (1887–1968) to attach new meanings to the concepts of space and measurement. Duchamp famously discussed and used Poincaré's essays "Mathematical magnitude and experiment" and "Why space has three dimensions" to create artistic works of a completely new kind. (Duchamp's ideas have been explored by the art historian Linda Dalrymple Henderson, who has used Duchamp's extensive notes to analyze his understanding of four-dimensional and non-Euclidean geometries.)

3   Four-Dimensional Geometry

The modern movement known as *cubism* was greatly influenced by ideas of the fourth dimension. One of the ways cubists came into contact with these ideas, as well as with non-Euclidean geometry, was through their reading of popular science fiction. In *Gestes et Opinions du Docteur Faustroll* (1911), French author Alfred Jarry (1873–1907), a close friend of Spanish artist Pablo Picasso (1881–1973), attracted by the novelty of higher-dimensional geometries, wrote about the work of the British mathematician ARTHUR CAYLEY [VI.46]. In 1843, Cayley published "Chapters in the analytic geometry of n dimensions" in the *Cambridge Mathematical Journal*. This work, along with Hermann Grassmann's (1809–77) *Die Lineale Ausdehnungslehre*, published in German a year earlier, was of interest not only to mathematicians but also to the general public, who recognized that in spaces of higher than three dimensions basic concepts had to be redefined and generalized.

In 1880 Washington Irving Stringham (1847–1909), in another influential article, "Regular figures in n-dimensional space," published in the *American Journal of Mathematics*, extended EULER'S FORMULA [I.4 §2.2] for polyhedra to new objects called "polyhedroids" in which polyhedra are joined by their faces so as to enclose a hyperspace. This article, which included illustrations of four-dimensional figures created by Stringham, was cited for the next twenty years in the most important mathematical texts on four-dimensional geometry. Stringham's figures intrigued several artists during the first decade of the twentieth century: Albert Gleizes's (1881–1953) painting *Woman with Phlox* (1910) has flowers that are similar to Stringham's "ikosatetrahedroid"; while in Henri Victor Gabriel Le Fauconnier's (1881–1946) *Abundance* (1910–11) Stringham's "hekatonikosihedroid" appears.

Art forms evolved as artists found new ways of responding visually to the world around them. This was particularly true of cubism, in which the artist depicted objects from several viewpoints at once. In order to make sense of a cubist painting, the viewer was invited to construct a single (elusive) object from an array of different perspective "facets" laid out across the picture's surface.

The n-dimensional geometries influenced not just the visual arts but also literature, including works by Rudyard Kipling and H. G. Wells, and music, for example Edgard Varese's "Hyperprism" (1923). Some mathematicians used this new mathematics for humorous purposes: two examples were Charles Dodgson in his *Through the Looking Glass* of 1872 and Edwin Abbott in *Flatland: A Romance of Many Dimensions* of 1884. The latter in particular was read by French artists and was referred to in other mathematics books that they read, such as those by Esprit Pascal Jouffret (1837–1907).

4   Formal Protests against Euclid

In the early twentieth century, informed by Poincaré's exposition of "the fourth dimension" and by knowledge of non-Euclidean geometry, a group of artists, including Gleizes and Jean Metzinger (1883–1956), explicitly attempted to liberate themselves from the geometry of three-dimensional Euclidean space. In an essay titled "Du Cubisme" they stated, "If we wished to tie the painter's space to a particular geometry, we should have to refer to the non-Euclidean scholars; we should have to study, at some length, certain of Riemann's theorems." Here they appear to be referring to RIE-MANNIAN GEOMETRY [I.3 §6.10], in which the notion of shape is less rigid than it is in Euclidean geometry. They go on to say, "An object does not have one absolute shape, it has several, it has as many as it has planes in a range of meaning." It is likely that they are referring here to Poincaré's "Les géométries non euclidiennes" in *La Science et l'Hypothèse.* The title of a (lost) 1913 painting of Metzinger's, *Nature morte (4me dimension),* gives a good indication of his interest in representing three and four dimensions on a two-dimensional surface. Both Riemann's geometry and the fourth dimension lay behind what these artists were trying to accomplish; they referred to both, however, as "non-Euclidean."

In 1918, enraged by the destruction wrought by World War I, a dozen artists, including Jean (Hans) Arp (1886–1986) and Francis Picabia (1879–1953), signed the *Dada Manifesto.* In it, they explicitly stated their belief that "all objects, sentiments, obscurities, apparitions and the precise clash of parallel lines are weapons for the fight [against conformity]." By the 1930s, more and more artists were using their knowledge of mathematics to change, in a radical way, the appearance of sculpture and painting.

5   Paris at the Center

During the last decade of the nineteenth century and the years before the outbreak of World War I, artists were profoundly influenced not just by mathematics, but also by the extraordinary developments and discoveries in science and technology. For instance, motion pictures (1880s), radios (1890s), airplanes, cars, X-rays (1895), and the discovery of electrons (1897) all had an impact on the work of artists. The pioneering painter Wassily Kandinsky (1866–1944) wrote that an artist's block he was experiencing disappeared when he learned of what was new in science; his old world collapsed and he could begin painting again.

While it is not entirely clear how knowledge of scientific and mathematical thinking came to working artists in the early twentieth century, it is nevertheless evident that many artists were familiar with articles about mathematics written for the general public. There was also at least one tutor with whom they explored mathematics in depth. In 1911, in Paris, the mathematician and actuary Maurice Princet (1875–1971) gave informal lectures on four-dimensional geometry, using mathematician Esprit Pascal Jouffret's *Traité Élémentaire de Géométrie à Quatre Dimensions et Introduction à la Géométrie à n Dimensions* (1903). Jouffret's *Traité,* which makes reference to *Flatland,* contains ways to present four dimensions on paper, the diagrams by Stringham of polyhedroids in four-dimensional space, and clear presentations of the ideas and theories of Poincaré. A second book, *Mélanges de Géométrie à Quatre Dimensions* (1906), emphasizes similar points.

Princet's audience was the Puteaux cubist group (which was sometimes called the "Section d'Or"). The central figures in this group were the three brothers Raymond Duchamp-Villon (1876–1918), Duchamp, and Jacques Villon (born Gaston Emile Duchamp) (1875–1963). Princet's involvement with the artists continued, even after his divorce from Alice Géry (1884–1975), who shared a bohemian life with the best man at their wedding, Pablo Picasso, and who later married André Derain (1880–1954). Géry had introduced Princet to the artists. An avid reader, she may have been the sitter for *Seated Woman with a Book* (1910), an early cubist painting by Picasso.

Together, in Paris, Princet and Duchamp privately studied Poincaré and Riemann, who were two important sources for Duchamp's work, as we have already seen. Duchamp's own notes, written a decade later as he created his famous painting *The Bride Stripped Bare by her Bachelors, Even (The Large Glass)* (1915–23), document his increasing interest in and understanding of four-dimensional and non-Euclidean geometries. Referring to Jouffret's book, which explained how a three-dimensional projection of a four-dimensional figure can be considered as a sort of "shadow," Duchamp told friends that the bride in his picture was a three-dimensional projection of a four-dimensional object recorded in two-dimensional form. He also refers to the fact, which fascinated him, that electrons were known to exist but could not be directly observed, claiming that his picture contained elements that were not directly represented. These notes, and others containing speculations on mathematics, were published in *À l'Infinitif* (1966). Working in a field hitherto dominated by fifteenth-century Renaissance perspective and its dependence on a Euclidean framework, Duchamp and other artists learned with excitement that many mathematicians no longer felt it necessary to subject themselves to Euclidean restrictions, and art was dramatically changed.

Rather surprisingly, Riemann and Poincaré were even part of the original inspiration for Duchamp's famous "readymades," found objects presented as art. As the artist Rhonda Shearer described in the New York Academy of Science newsletter in 1997, Duchamp was very taken with Poincaré's description of the creative process in *Science and Method*. Poincaré reported on his accidental discovery of the so-called Fuchsian functions. After days of "unfruitful" conscious work trying to prove that the functions do not exist, he changed his habit and one evening drank black coffee late at night. The next morning, "fruitful" ideas came into his conscious mind. From these he selected "*tout fait*" (readymade) ideas and saw, surprisingly, a way to prove the existence of the mathematical functions whose existence he had previously doubted. Duchamp used the term "readymade" (and "*tout fait*") in 1915. The examples he selected, titled, and signed are ordinary manufactured objects such as a urinal turned upside down, *Fountain* (1917), and a bottle drying rack, *Bottle Rack* (1914), thought to be the first readymade.

6 Constructivism

In Russia in 1920, the artists Naum Gabo and Antoine Pevsner wrote that they had turned to mathematics in order to rethink their work. As they put it: "We construct our work as the universe constructs its men; as the engineer constructs his bridges; as the mathematician his formulas of the orbits." Gabo began to use a stereometric system that he had studied in engineering, creating sculptures such as "Head No. 2" (see figure 2). The subject of stereometry goes back at least as far as 1579, where it is listed in the "Groundplat" of John Dee's celebrated *Mathematicall Praeface* to Billingsley's edition of Euclid. It concerns the measurement of properties of solids, and was widely taught at universities in the nineteenth and twentieth centuries: indeed, it is still taught today in some European countries. Gabo and Pevsner constructed their sculptures out of planar parts, so space, rather than mass, became the sculptural element. Density was no longer important, with the result that the subtraction techniques used in classical sculpture (where material is carved away from

Figure 2 Gabo's *Head No. 2*, COR-TEN steel, 1916 (enlarged version 1964). The works of Naum Gabo: © Nina Williams.

a solid block leaving the artist's work as the solid) were no longer necessary. Sculpture became airy; surfaces became less significant and have remained so, at least within the tradition that became known as *constructivism*.

This tradition was first formalized in the Russian *Realistic Manifesto* (1920), written and signed by Gabo and Pevsner. There they argued that "The material formation of the object is to be substituted for its aesthetic combination. The object is to be treated as a whole ... a product of an industrial order like a car." Gabo took constructivism to the Bauhaus in Germany and then to France and England in the 1930s, where he worked alongside the British artists Barbara Hepworth (1903–75) and her husband Ben Nicholson (1894–1982). Gabo and Nicholson (with Leslie Martin) edited *Circle: International Survey of Constructive Art* (1937), which contained articles by themselves as well as ones by Hepworth, Piet Mondrian (1872–1944), and the critic Herbert Read (1893–1968), among others. In *Circle*, Gabo, referring back to the seventeen-year-old *Realistic Manifesto*, spells out what is meant by constructivism by guiding the reader to see how two cubes (shown in the photograph in figure 3) can illustrate the distinction between two kinds of representation of the same object: carving and construction. The cubes have different methods of execution and different centers of interest: one is mass and the other makes visible the space in which mass exists. Constructivism created an artistic context in which a mathematically understood space became a sculptural element. As Gabo wrote:

Figure 3 Gabo's two cubes: carving and construction. Image courtesy of the Library of Congress.

"The stereometrical method in which [the right-hand cube] is executed shows elementarily the constructive principle of a sculptural space expression."

These artists studied mathematical models in museums and catalogues. These models, designed by mathematicians for teaching about surfaces, were made of string, cardboard, metal, and plaster. The same artists also studied photographs produced by the surrealist Man Ray showing strings and striations of surface lines on a model that had been found by another surrealist artist, Max Ernst, at the Institut Henri Poincaré in Paris. Ray portrayed these models with impressionistic patterns of light and shadow (see figure 4); he was interested in the "elegance"—the aesthetic persuasiveness—of the model, though aware that the original model-maker had sought to give visual form to an elegance inherent in the mathematical equations themselves. Other artists too, such as Hepworth and Gabo, stated that it was not mathematics itself but the beauty of the mathematical models that provided the inspiration for their work. Hepworth studied mathematical models that were on display in Oxford, considering them to be "sculptural working out of mathematical equations." They inspired her to add strings to her own work. However, she wrote that her inspiration was not the mathematics exhibited by the strings, but rather their power: "the tension I felt between myself and the sea, the winds, and the hills."

A close friend of both Gabo and Hepworth was the renowned sculptor Henry Moore (1898–1986). Moore too spoke and wrote about the influence of mathematical models on his work. He had seen stringed figures of Theodore Olivier (see figure 5) and after making many of his own mathematical models introduced strings into his sculpture in 1938, later considering it to have been the most abstract of his work. He said he "had gone to the Science Museum in South Kensington and had been greatly intrigued by some of the mathematical models ... hyperboloids and groins ... developed by [Fabre de] Lagrange in Paris, that have geometric fig-

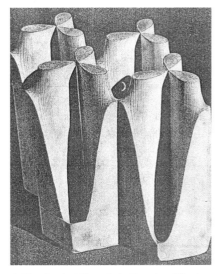

Figure 4 Man Ray's *Allure de la Fonction Elliptique*, 1936. Image courtesy of the National Gallery of Art.

Figure 5 Olivier's *Intersection of Two Hyperbolic Paraboloids*, 1830. Image courtesy of the Union College Permanent Collection, Schenectady, NY.

ures at the ends with colored threads from one to the other to show what the form between would be. I saw the sculptural possibilities of them, and I did some." Moore recognized that the use of strings connecting protrusions actually created a barrier between the solid sculpture and the space around the sculpture (see figure 6). The string barrier made it possible to see the captured space. Moore and Gabo made different uses of the mathematical models. As Moore later put it, Gabo "developed this string idea so that his structure

Figure 6 Moore's *Stringed Figure No. 1*, cherry wood and string, on oak base, 1937. Image (taken by Lee Stalsworth) courtesy of the Hirshhorn Museum and Sculpture Garden, Smithsonian Institution, Joseph H. Hirshhorn Purchase Fund (1989).

always became space itself, whereas I liked the contrast between the solid and the strings ... I was making an outside shape a sculpture in its own right (Interior/Exterior forms), yet one which was not completed until each part was connected to the other."

7 Other Countries, Other Times, Other Artists

7.1 Switzerland and Max Bill

In the mid 1930s the Swiss designer and artist Max Bill (1908–94) became intrigued by a one-sided surface, unaware that it had been published in 1865 by the German mathematician and astronomer AUGUST FERDINAND MÖBIUS [VI.30]. Bill, when he needed a design for a sculpture to hang in a stairwell, independently invented his own MÖBIUS STRIP [IV.7 §2.3], by dangling a long narrow rectangle of flexible material and then attaching the corners appropriately (1935).

Having been informed some years later of the connection between his sculpture and its mathematical forerunner, Bill, who liked the simplicity of geometric forms, continued to earn commissions by making sculptures based on topological problems and single-sided surfaces (see figure 7). In a 1955 essay on the mathematical approach in contemporary art, he wrote that mathematics, by giving all phenomena a meaning-

Figure 7 Bill's *Eindeloze Kronkel*, bronze, 1953–56. Image courtesy of Mary Ann Sullivan, Bluffton University.

ful arrangement, is an essential method to understand the world. For Bill, when mathematical relationships are given form they "emanate undeniable aesthetic appeal, such as goes out from space-models, as, for instance, those that stand in the Musée Poincaré in Paris."

7.2 Holland and Escher

From the second half of the twentieth century onward there has been a groundswell of interest in the relationship between mathematics and art, particularly since 1992 when artists and mathematicians from around the world began holding joint annual conferences to explore old and new ideas about the connections between their disciplines. The popularity in the West of this interdisciplinary study is in no small part due to the unusual drawings and prints made by Maurits Cornelis Escher (1898–1972), a Dutch graphic artist—or "craftsman," as he wished to be known. Escher was deeply interested in tessellations and "impossible" objects that are not constructible in three dimensions but that can nevertheless be portrayed in two dimensions. While his oeuvre is not thought of as an integral part of twentieth century art, he is greatly appreciated by mathematicians and also by the general public. Among his best-known works are pictures based on Penrose triangles and on the Möbius strip.

He was inspired by knowing and learning from mathematicians including Georg Pólya (1887–1985), Roger Penrose (1931–), and Harold Scott MacDonald "Donald" Coxeter (1907–2003). Escher was introduced to the international mathematics community in 1954 when the organizing committee for the Amsterdam meeting of the International Congress of Mathematicians inaugurated an exhibition of his work at the Stedelijk Museum. After Penrose viewed Escher's 1953 print *Relativity* at this exhibition, he and his father, geneticist Lionel Penrose (1898–1972), were inspired to create impossible figures: the Penrose tribar and the Penrose staircase published in the *British Journal of Psychology* in 1958—the Penroses sent Escher an offprint of the article. Escher subsequently used these in two well-known lithographs: *Waterfall* (1961), in which water runs in perpetual motion from the base of a waterfall to the top of the waterfall; and *Ascending and Descending* (1960), which features a building with an impossible staircase which constantly rises or falls (depending on the direction you go around it) but returns to the same level. Coxeter's field was symmetry in the Euclidean and hyperbolic planes, but he also took pleasure in analyzing the works of artists from a mathematical point of view. Escher began a correspondence with him shortly after the congress, at which they met, and it lasted until his death in 1972. In 1957 Coxeter requested the use of two of Escher's drawings to illustrate planar symmetry in "Crystal Symmetry and Its Generalizations," his presidential address to the Royal Society of Canada—in this way Escher's work spread among the mathematical community. In 1958, Coxeter sent Escher a letter containing a reprint of his address. The response was a request: "Could [you] give me a simple explanation how to construct the following circles, whose centers approach gradually from the outside till they reach the limit?" Coxeter's reply, meant to be helpful, gave Escher one small piece of useful information; the rest of the lengthy letter was unintelligible to the artist. But from the pictures and his own keen geometric intuition, Escher was able to construct the circles he required, and by 1958 he was the first graphic artist to have used the three main geometries in his works: Euclidean, spherical, and hyperbolic. Coxeter was astounded that an artist, untrained in mathematics, could produce such accurate "equidistant curves" as he did in his 1958 woodcut *Circle Limit III*. Escher always claimed that he knew little mathematics, but many of his prints are a direct result of using mathematics. Mathematician Doris Schattschneider has said that Escher was really a "secret mathematician," since much of his work depended on his pursuit of mathematical questions that arose from his interests and his interaction with mathematicians, which he referred to as "Coxetering." He did, however, write that he preferred to find solutions and understanding by himself.

As well as his artistic and mathematical legacy, Escher had an important influence on crystallographers, who have used his symmetry drawings for analysis. Crystallographer Caroline MacGillavry has pointed out that Escher began a deep study of color symmetry and created a classification system in 1941–42, which was some time before crystallographers became interested in this field of study, which has become very active. The International Union of Crystallography subsequently commissioned Escher to illustrate MacGillavry's *Symmetry Aspects of M. C. Escher's Periodic Drawings*, first published in 1965. Its purpose was to interest "students in the laws which underlie repeating designs and their colorings."

7.3   Spain and Dalí

As we have seen, some artists were influenced by their own knowledge of mathematics, others by a less direct appreciation of mathematical thinking, and still others by the appeal of mathematical models. Another kind of connection is illustrated by the example of the surrealist artist Salvador Dalí (1904–89) and his relationship with the mathematician and graphic artist Thomas Banchoff (1938–). Banchoff is a professor of mathematics at Brown University, known for his research in differential geometry in three and four dimensions. Since the late 1960s, he has also been involved in the development of computer graphics. Dalí's 1954 painting of Christ crucified on a hypercube was reproduced in a 1975 article about Banchoff's pioneering work, which used computer animation to illustrate geometry beyond the third dimension. This led to a series of meetings between Banchoff and Dalí over the next decade, at which hypercubes and other aspects of geometry and art were discussed. One joint project was the design for a giant sculpture of a horse that would appear realistic from only one viewing position. Dalí eventually envisioned a horse with its head in front of the viewer and its rump somewhere on the moon—clearly a project solely of the imagination. Dalí created works using anamorphoses, as other artists, beginning with Leonardo, had done. He prized his interactions with scientists and mathematicians, later stating, "Scientists give me everything, even the immortality of the soul." Dalí also met the French mathematician René Thom (1923–2002) to discuss catastrophe theory, which, in 1983, he sought to represent in what turned out to be his last series of paintings.

7.4 Other Recent Developments: The United States and Helaman Ferguson

So far we have seen how mathematics has influenced art. Occasionally, artists have actually created mathematics, for instance to produce sculpture by means of carefully chosen mathematical equations. The noted American sculptor/mathematician Helaman Ferguson (1943–) divides his time equally between mathematics and the interpretation of mathematics in his art. As a mathematician he designs algorithms for operating machinery and for scientific visualization. In 1979 he found a method for finding integer relationships between more than two real or complex numbers—this was later named one of the top ten algorithms of the twentieth century. As an artist, he carves in stone. In 1994, he asked mathematician Alfred Gray (1939–98) to develop equations for a Costa surface (named after the graduate student who invented equations for describing a minimal surface with holes), so that he could sculpt the surface (see figure 8). Gray developed the equations in terms of the Weierstrass zeta function. This could be used with Mathematica, which made it possible for Ferguson to create a stone sculpture. Ferguson sees his art as deriving from applied mathematics that has been developed over the course of the last two centuries:

> Start with physical observations about soap films in nature (Plateau), write down a differential equation model describing minimizing surfaces (Euler-Lagrange), define a minimal surface geometrically in terms of curvature (Gauss), discover a minimal surface with non-trivial topology (Costa), draw computer images of the surface (Hoffman–Hoffman), recognize symmetry and prove the surface has not self intersections (Hoffman–Meeks), discover fast parametric equations for the surface (Gray), and finally return to nature with a sculpture, a solid form of a "soap film" big enough to touch and climb on.

7.5 The United States and Tony Robbin

The development of n-dimensional geometry also had a powerful effect on many other European and American artists, and this continued into the late twentieth century. Interest was boosted in the 1970s with the development of computer graphics by mathematicians and artists. Examples can be found in the work of American artist Tony Robbin (1943–), who has explored concepts of dimension in painting, prints, and sculpture (see figure 9). In late 1979, Robbin, who had also

Figure 8 Ferguson's *Invisible Handshake II*: a triply punctured torus with negative Gaussian curvature. Image courtesy of the artist.

Figure 9 Robbin's *Lobofour*, acrylic on canvas with metal rods, 1982, collection of the artist.

been a student of mathematics, was working on Banchoff's parallel processor computer and managed to visualize for the first time a four-dimensional cube, an event which radically changed his art, and which led him to develop two-dimensional works that portrayed the spatial fourth dimension. Writing in his book *Fourfield: Computers, Art & the 4th Dimension* (1992), Robbin tells us, "When the fourth dimension becomes part of our intuition our understanding will soar." Some of Robbin's constructions, paintings, and prints show figures in independent planes: that is, in overlapping spaces that cannot be fully seen in three dimensions. If the viewer wants to see two structures in the same place at the same time and rotating with respect to one another (as though projected from four-dimensional space), then looking at one of Robbin's wall-relief sculptures lit by red and blue light while wearing 3D glasses (one red and one blue lens) will create a full stereoscopic effect of the four-dimensional figure. In digital prints it is Robbin's lines and polyhedra that imply four dimensions, with the two-dimensional picture being a shadow of the higher-order object.

7.6   Hayter and Atelier 17

In 1927, the British surrealist and printmaker Stanley William Hayter (1901–88) decided to revive the almost lost skill of intaglio printing and established an experimental studio, "Atelier 17," in Paris. This was followed by another in New York from 1940 to 1950 before he returned to Paris. Hayter was aware that many of the artists who used his facilities were working with a "different space from that seen through the classical window of Renaissance representation" that had existed when engraving flourished a hundred years earlier. The founding of Atelier 17 was central to the revival of the print as an autonomous art form, and Hayter's sensitivity to the significance of mathematics in the experimental techniques of printmaking (which had been evolving since the nineteenth century) is quite apparent: "Man's increasing consciousness of and power over space (in physics and mathematics) have been reflected in new and unorthodox methods of demonstrating space and time graphically," so that "many properties of matter and space, which had been represented diagrammatically only by the scientists, found their expression in graphic and affective forms." A printmaker in the twentieth century could use an arrangement of transparent webs to define planes above the picture plane. Specifically, by hollowing out spaces in the plate being engraved—possibly even gouging all the way to the bottom of the plate—the artist could make a projection in front of the plane of the picture. Although artists could have used this technique much earlier, it became important only at the end of the nineteenth century when the representational aspect of intaglio had been challenged by photography. They therefore used the gouge to create the third dimension. Hayter also describes in *About Prints* (1962) how Abraham Bosse's seventeenth-

century atelier was organized and reconstructed in Paris in the twentieth century.

In World War II Hayter's interest in mathematics revealed itself in a more practical way, when, in collaboration with artist and patron of art Roland Penrose and others, he set up a camouflage unit and, as *Art News* reported in 1941, constructed

> an apparatus which can duplicate the angle of the sun and the consequent length of cast shadows at any time of day, and day of the year, at any given latitude. This complex of turntables, discs inscribed with a scale of weeks, allowances for seasonal declination, and so on is just the kind of working mathematics he really delights in.

8   Conclusion

There has been a complex and fruitful relationship between Western art and mathematics in the twentieth century. Gabo, Moore, Bill, Dalí, and Duchamp are notable artists who have been influenced by mathematics, and Poincaré, Banchoff, Penrose, and Coxeter are among the mathematicians who influenced them. In the other direction, twentieth-century mathematicians, like their forebears in the fifteenth and sixteenth centuries, often turned to art to explore and exhibit, or even just to explain more expressively, the meaning of their mathematics. They have also likened their creative processes to those of artists. As the French mathematician ANDRÉ WEIL [VI.93] wrote to his sister, author Simone Weil (1909–43), from military prison in 1940, "When I invented (I say invented, and not discovered) uniform spaces, I did not have the impression of working with resistant material, but rather the impression that a professional sculptor must have when he plays by making a snowman."

Further Reading

Andersen, K. 2007. *The Geometry of an Art: The History of the Mathematical Theory of Perspective from Alberti to Monge*. New York: Springer.

Field, J. V. 2005. *Piero della Francesca: A Mathematician's Art*. Oxford: Oxford University Press.

Gould, S. J., and R. R. Shearer. 1999. Boats and deckchairs. *Natural History Magazine* 10:32–44.

Hammer, M., and C. Lodder. 2000. *Constructing Modernity: The Art and Career of Naum Gabo*. New Haven, CT: Yale University Press.

Henderson, L. 1983. *The Fourth Dimension and Non-Euclidean Geometry in Modern Art*. Princeton, NJ: Princeton University Press.

Henderson, L. 1998. *Duchamp in Context: Science and Technology in the Large Glass and Related Works*. Princeton, NJ: Princeton University Press.

Jouffret, E. 1903. *Traité Élémentaire de Géométrie à Quatre Dimensions et Introduction à la Géométrie à n Dimensions*. Paris: Gauthier-Villars. (A digital reproduction of this work is available at www.mathematik.uni-bielefeld. de/~rehmann/DML/dml_links_title_T.html.)

Robbin, T. 2006. *Shadows of Reality: The Fourth Dimension in Relativity, Cubism, and Modern Thought*. New Haven, CT: Yale University Press.

Schattschneider, D. 2006. Coxeter and the artists: two-way inspiration. In *The Coxeter Legacy: Reflections and Projections*, edited by C. Davis and E. Ellers, pp. 255–80. Providence, RI: American Mathematical Society/Fields Institute.

Part VIII
Final Perspectives

VIII.1 The Art of Problem Solving
A. Gardiner

Where there are problems, there is life.

Zinoviev (1980)

In English the word "problem" has negative connotations, suggesting some unwanted and unresolved tension. Zinoviev's reminder is therefore important: problems are the stuff of life—and of mathematics. Good problems focus the mind: they challenge and frustrate; they cultivate ambition and humility; they show up the limitations of what we know, and highlight potential sources of more powerful ideas. By contrast, the word "solving" suggests a *release* of tension. The juxtaposition of these two words in the expression "problem solving" may encourage the naive to think that this unwelcome tension can be massaged away by means of some "magic formula" or process. It cannot; there is no magic formula.

Why don't we tell the truth? No one has the faintest idea how the process … works, and in calling it a "process" we may be already making a dangerous assumption.

Gian-Carlo Rota, in Kac et al. (1986)

A "problem" is something that one wants to understand, to explain, or to solve, but which eludes one's initial attempts to classify it as being of some familiar "type." The experience of being confronted by such a "problem" is inevitably unsettling: it may eventually prove to be more familiar than one thought, but the would-be solver is initially dumped in terrain with few signposts or marked tracks. Some (such as Pólya and his recent followers) have tried to devise a universal "problem-solving meta-map." But in reality there is no easy alternative to that painful immersion so familiar to generations of postgraduate students.

Grand general principles can help to make sense of this experience, but are unlikely to take us very far. Consider, for example, the four general principles formulated by DESCARTES [VI.11] in his *Discourse on Method*.

The first was never to accept anything for true which I did not clearly know to be such. The second, to divide each of the difficulties under examination into as many parts as possible, and as might be necessary for its adequate solution. The third, to conduct my thoughts in such order that by commencing with objects the simplest and easiest to know, I might ascend … step by step to the knowledge of the more complex. And the last … to make enumerations so complete … that I might be assured that nothing was omitted.

Descartes's rules are worth pondering. But it is hard to accept that it was the systematic application of these four rules that led to Descartes's almost single-handed creation of analytic geometry as we know them today! In the detailed working out of the creative process, problem-specific "know-how" distilled from endless hands-on experience is likely to be far more important than any general principles. What then can one usefully say? To describe the "art of problem solving" in impressive-sounding detail would be irresponsible. But to say nothing would be misleading. Both options are unsatisfactory—yet these two responses are what students, teachers, and would-be mathematicians are most likely to meet! Attempts to teach "problem-solving" in schools often misconstrue mathematics as a kind of "subjective pattern-spotting." Instead of correcting this distortion at university level, mathematicians often maintain a discreet public silence about the very private matter of *how* serious mathematical problems actually get solved. Hence, in addressing the theme for readers with a mathematical bent, this article has to start largely from scratch, and to proceed slowly. So we begin with a warning. The subject of problem solving is well worth exploring, but we shall proceed obliquely and our conclusions will often remain implicit. Along the way we shall meet extracts from a number of sources—which may be viewed as an initial reading list for those who wish to pursue the theme in greater detail, provided they never forget that the only way to gain true insight into a craft is *through practicing the craft itself.* Mathematics may be "the queen of the sciences," but the art of *doing* mathematics remains a *craft,* passed on in the ancient craft tradition, through painful initiation. A number of collections of problems at various levels—often using relatively elementary material—are listed in the references. Here we make do with a single example.

Problem. For all positive integers n and k, show that some triangular number is congruent to k (mod 2^n).

The reader is encouraged to explore this problem before reading on, noting any obvious stages along the way: from initial bewilderment, through an exploratory/organization phase, eventually culminating in a solution and an attempt to locate this isolated challenge in some broader mathematical context.

Mathematics is a largely unexplored "mental universe," whose initial exploration and charting, subsequent colonization, routine traverse, and efficient administration correspond, in many ways, to the real-world adventures of geographical explorers in former centuries. To strike out beyond the security of the old-world coastline, to imagine and explore something new, takes intellectual courage.

Most prominent among these mathematical explorers are the "system builders," who identify new mathematical continents, or who uncover profound and unexpected bridges joining known lands. Their initial motivation may stem from a specific problem, whose analysis provides hints of the outline of previously undiscerned structures; but the system builder's focus then switches to the bigger picture: trying to identify, and to clarify, connections between the structures that underlie "mathematics in the large." Such ventures often end up with little to show for them—they may come close to discovering some mathematical El Dorado, but they lack the gold to prove it. Some of these explorers may later be singled out as major prophets or discoverers, but such recognition can be fickle: those so honored may not have been the first to see their particular promised land; they may not have appreciated the significance of what they had stumbled upon, or of how it would eventually be seen to link known mathematical lands; their success may have depended on earlier attempts by others; and their bounty may not have impressed their contemporaries as deeply as we now imagine.

Each triumph of the system builders is rooted in detailed knowledge of "mathematics in the small," which may derive from work in a very different mathematical style—such as that of the mathematical beachcomber, who is most at home exploring the *known* mathematical shoreline, using some sixth sense to spot suspicious-looking rocks, under which are hidden intricate, and totally unexpected, microworlds on our very doorstep. While great explorers range further and further afield, they leave behind annoying gaps, or unsolved problems, which represent significant lacunas in our understanding—gaps that some future beachcomber may one day explain, so opening the way for some new synthesis.

The system builder and the beachcomber represent very different mental styles; but their contributions complement each other. In our evolving picture of the mathematical universe, insights on a small scale and on a large scale must somehow fit together. Hence the beachcomber's chance discoveries may contribute in unexpected ways to our future conception of the large-scale mathematical universe.

Such differing styles should be borne in mind as we strive to make our introductory comments more specific. Our first attempt is based on a version of Alain Connes's three levels of mathematical activity.

> The first [level] is defined by the faculty of calculation—being able to apply a given algorithm rapidly and reliably.... The second level begins when the actual method of calculation is adapted to, and criticized in the context of, a particular problem.... In mathematics this is what often makes it possible to solve problems that aren't too difficult or that don't require any new ideas.... The third level [is] the level at which the mind, or rather conscious thought, is occupied with another task while the problem in question is being solved... subconsciously.... At [the third] level it isn't only a matter of solving a given problem; it is also possible to discover... a part of mathematics to which the [previously] existing corpus gives no direct access.

> Alain Connes, in Changeux and Connes (1995)

Connes's first level focuses on the development of *robust technique*—that is, fluency, accuracy, and confidence in using given procedures in relatively standard ways. We say no more about work on this level except to stress its importance! Discussion about the "art of problem solving" presupposes, and only makes sense in the context of, appropriate robust technique.

Connes's second level includes most, but by no means all, of the serious mathematics that mathematicians engage in on a daily basis. Genuine problems occur on this level in different guises, ranging from (i) challenges designed to stretch the young would-be mathematician (in high school geometry, in puzzle books and problem-solving journals, in Olympiads, etc., whose material is designed to force the would-be solver to select, to adapt, and to combine known methods in unexpected ways), to (ii) genuine research problems that can be tackled and largely solved by selecting, adapting, and combining known methods in a suitably imaginative way.

In our problem about triangular numbers, the first level includes the immediate translation from words into symbols to obtain the congruence $m(m-1)/2 \equiv k \pmod{2^n}$, or $m(m-1) \equiv 2k \pmod{2^{n+1}}$, which, for arbitrary given $n \geqslant 1$, has to be solved for all $k \geqslant 1$. The second level might then include a systematic attempt to make sense of what happens for small values of n, leading to the formulation of simple conjectures whose proofs would solve the problem, followed by moves to devise the necessary proofs.

It is tempting to think of Connes's third level as "inscrutable," in the spirit of the following extract:

In science, as well as in other fields of human endeavor, there are two kinds of geniuses: the "ordinary" and the "magicians." An ordinary genius is a fellow that you or I would be just as good as if we were only many times better [than we are]. There is no mystery as to how his mind works. Once we understand what he has done, we feel certain that we, too, could have done it. It is different with the magicians. They are ... in the orthogonal complement of where we are and the working of their minds is for all intents and purposes incomprehensible. Even after we understand what they have done, the process by which they have done it is completely in the dark. They seldom if ever have students because they cannot be emulated and it must be terribly frustrating to cope with the mysterious ways in which a magician's mind works.

Kac (1985)

However, one would then expect activity on this level to be so idiosyncratic as to be irrelevant to ordinary mortals. In fact, the most valuable insights we have into "the art of problem solving" derive from personal testimony about work on this level by precisely such "magicians" as POINCARÉ [VI.61], which suggests that there are clear parallels between the experience of the very best mathematicians on Connes's third level and what happens when ordinary students, or mathematicians, operate "out of their depth" when tackling more mundane problems; that is, when their own fumbling requires them to work in regions

to which *their own* "existing corpus gives no direct access." In our problem about triangular numbers this might occur when a solver who has never met "congruences for binomial coefficients" manages to adapt the naive proof for $\binom{m}{2} \pmod{2^n}$ to cover the slightly more awkward $\binom{m}{3} \pmod{2^n}$, and realizes that, even though this naive approach does not extend to $\binom{m}{4} \pmod{2^n}$, something more general may be lurking in the darkness.

Thus we use the word "problem" to refer to *a serious mathematical challenge on at least Connes's second level*, where this is to be interpreted in the spirit of activity on Connes's second and third levels. So any analysis of the art of mathematical problem solving must somehow reflect experience on these two higher levels. By contrast, the educational assumptions that underpin most attempts to bring "problem-solving" to the classroom generally try to reduce this subtle process to a set of rules *in the spirit of Connes's first level*!

A problem is much more than just a hard exercise. Consider the question, When is a "problem" *not* a problem? One answer is clearly, When it is too *easy*! However, many students and teachers are tempted to reject unfamiliar or mildly confusing problems because they appear to be *too hard*. This is an understandable reaction only where mathematics is limited to a succession of predictable exercises.

Most of us learn mathematics as a collection of standard techniques, which we use to solve standard problems in predictable contexts (Connes's first level). Like the athlete or musician, the mathematics student needs to develop technique. However, as the athlete trains in order to *compete*, and the musician practices in order to *make music*, so the mathematician needs technique in order to *make mathematics* by tackling challenging problems. Each new piece of printed music may initially strike the beginner as a confusing array of black blobs. But as they work on the piece, phrase by phrase, it slowly takes on a shape of its own, revealing internal connections that may previously have been overlooked. Much the same is true when we confront an unfamiliar mathematical problem. At first sight we may not even understand the question. But as we struggle to make sense of the problem, we regularly find that, little by little, the fog begins to lift.

Two rats fell into a can of milk. After swimming for a time one of them realized his hopeless fate and drowned. The other persisted, and at last the milk was turned to butter and he could get out.

In the first part of the war, Miss Cartwright and I got drawn into van der Pol's equation.... [W]e went on and on...with no earthly prospect of "results": suddenly the entire vista of the dramatic fine structure of solutions stared us in the face.

Littlewood (1986)

In 1923 HARDY [VI.73] and LITTLEWOOD [VI.79] made a conjecture about the number of arithmetic progressions (APs) of length k among the primes. One potential corollary was that the prime numbers must contain *arbitrarily long* APs. Faced with such a claim it is natural to start looking for APs which consist entirely of primes! But if you try, you will soon approach the limits of what is known: the first three odd primes, 3, 5, 7, form a very familiar AP of length *three*, but longer APs are surprisingly elusive (in 2004 the record for an AP of distinct primes had length twenty-three, with both the primes themselves and the step size being astronomical). Despite this unpromising lack of evidence, in 2004 Ben Green and Terence Tao proved that the set of prime numbers does indeed contain arbitrarily long APs. Their proof is a fine example of the way in which significant progress often combines a detailed reevaluation of known results (in this case a deep result of Szemerédi), lateral thinking (they embed the primes not in the integers, but in a natural but sparser set of "almost primes" of which the primes constitute a *nonzero* fraction), and the determination and ingenuity to make such ideas deliver the goods.

It remains a serious challenge to capture the essence of Littlewood's experience (where the fog suddenly lifts) in a form that is suitable for relative beginners, whether through time-constrained problems (see Barbeau 1989; Gardiner 1997; Lovasz 1979), or through structured investigations (see Gardiner 1987; Ringel 1974). In the year in which Green and Tao announced their proof, the British Mathematical Olympiad posed the following problem, which readers are encouraged to tackle.

Problem. In an AP of seven distinct primes, what is the smallest possible value of the largest prime?

This challenge could enliven any introductory number theory course, as well as providing a natural link to recent developments. For the novice it is far from obvious how to begin, but the basic idea is elementary and should be "known" (in some sense), and can be used to generate natural APs of lengths 4, 5, 6, 7, 8, provided that one accepts the value of carrying out extensive computations quickly and intelligently.

A great discovery solves a great problem. But there is a grain of discovery in the solution of any problem. Your problem may be modest; but if it challenges your curiosity and brings into play your inventive faculties, and if you solve it by your own means, you may experience the tension and enjoy the triumph of discovery. Such experiences at a susceptible age may create a taste for mental work and leave their imprint on mind and character for a lifetime.

From the preface to the first
printing of Pólya (2004)

Pólya is, if anything, too reticent here. The important distinction is not between that which is "known" and that which is truly "original," but rather between mathematical activity in the spirit of Connes's first level and mathematical activity in the spirit of Connes's second and third levels. Any introduction to this distinction is inevitably through problems whose solution is *known to someone*, so we should collect and use good "modest problems," not apologize for them. Ulam puts it more directly.

I learned chess from my father.... The moves of the knight fascinated me, especially the way two enemy pieces can be threatened simultaneously with one knight. Although it is a simple stratagem, I thought it was marvelous, and I have loved the game ever since.

Could the same process apply to the talent for mathematics? A child by chance has some satisfying experiences with numbers; then he experiments further and enlarges his memory by building up a store of experiences.

Ulam (1991)

Children also find delight—if less profound and more short-lived—in the discovery that one can set up a "corner move" in the children's game noughts and crosses (tic-tac-toe) so as to *simultaneously* threaten to complete two lines-of-three, at most one of which can be countered. This delight in a double-edged strategy, which points in two directions at once, has much in common with the pleasure we derive from (i) puns and *double entendres* in ordinary language, in humor, and in poetry, (ii) the almost physical response when we recognize subtle variations on a theme in music, and (iii) the more cerebral appreciation we feel when we meet counting methods based on unanticipated isomorphisms, or the essentially two-faced idea of "proof by contradiction" in mathematics. This enjoyment of hidden ambiguities and double meanings is related to the evident (but poorly understood) way in which *analogy* guides, and delights, mathematicians of all ages.

Banach once told me, "Good mathematicians see analogies between theorems or theories; the very best ones see analogies between analogies."

                                                          Ulam (1991)

Koestler, in his thought-provoking book *The Act of Creation* (1976), shows how scientific and literary "creativity" often flows from the identification and exploitation of "double meanings with a built-in tension." (Koestler calls them *bisociations*: "the perceiving of a situation or idea L in two self-consistent but habitually incompatible frames of reference ... the event L is made to vibrate simultaneously on two different wavelengths, as it were.") His study begins with an analysis in precisely this vein of the human response to humor, both comic and tragic, including a selection of jokes attributed to VON NEUMANN [VI.91]!

Ulam's innocent-sounding question (in the extract before last) challenges us not only to provide children with "satisfying experiences with numbers," but also to identify other quintessential aspects of mathematics and to ensure that they are experienced memorably at school (and undergraduate) level. In particular, insofar as there is such a thing as an "art of problem solving," we need to learn how to convey it faithfully and effectively through the medium of classical elementary mathematics to those who are near the beginning of their mathematical studies, or who may not yet have any commitment to mathematics.

It is often claimed that Pólya's little book *How to Solve It* provides an answer. It does not. Pólya was a pioneer who sought to provoke a debate among mathematicians about "heuristics." This debate never really got started. Instead his first low-level attempt at a theoretical framework has been embraced uncritically.

Much of what Pólya writes about specific problems in *How to Solve It* makes sense; but his general conclusions on "how to help students solve problems" are less convincing. As a result, much of the book's general theorizing needs to be read extremely carefully. For example, Pólya's suggestion that "when the teacher solves a problem before the class, he should dramatize his ideas a little and he should put to himself the same questions which he uses when helping students" is spot on. But alarm bells should start ringing when he confidently concludes that "[thanks] to such guidance, the student will eventually ... acquire something that is more important than the knowledge of any particular mathematical fact." In the right setting the claim may occasionally be true; but as a statement about the effect on students in general it is false.

Similar claims have been widely used to justify the introduction of a whole new branch of school mathematics called "problem-solving" (see NCTM (1980) and www.pisa.oecd.org), which has grown *at the expense of* mastery of the "particular mathematical facts" on which the activity itself depends.

Pólya and others were right to insist that school mathematics should include a regular diet of good problems, and that educators have a duty to convey not just the techniques and inner logical structure of the subject, but also the experience of struggling to uncover the mathematics hidden in multistep problems and carefully structured investigations. Fortunately, the four volumes that Pólya wrote to illustrate this broader thesis remain in print (Pólya 1981, 1990). There the focus is on mathematics, and the rhetoric is more restrained:

> [L]et us learn proving, but also *let us learn guessing....* I do not believe there is a foolproof method to learn guessing. At any rate, if there is such a method, I do not know it, and quite certainly I do not pretend to offer it in the following pages.... [P]lausible reasoning is a practical skill and it is learned, as any other practical skill, by imitation and practice.
>
>   Pólya (1990, volume 1)

These four books should be compulsory reading for all serious mathematics educators, graduate students, and mathematics lecturers. However, Pólya and others failed to show how problem solving could be developed *within* the standard school mathematics curriculum. Instead they concentrated on proposing general rules that might "help students become better problem solvers." What is needed is to clarify (i) which aspects of elementary mathematics have the potential to captivate young minds—not because they are more "enjoyable" in some superficial sense, but because they are more "pregnant with meaning"; and (ii) *how to teach* such material so as to convey this deeper meaning on an elementary level. This is not the place for a detailed analysis, but we suspect such an analysis would *strengthen* the position of many traditionally important topics and themes, encouraging them to be taught in such a way as to bring out their inherent richness, while recognizing that these goals depend on prior mastery of certain basic techniques without which this richness can scarcely be appreciated. In contrast, recent "reforms," whose declared intention was to *enrich* school mathematics, have regularly *reduced* both the emphasis on, and the time available for, serious elementary mathematics.

Those who want good problems to enrich school mathematics often fail to recognize that well-intentioned "reforms" are usually unstable under the kind of distortions that routinely affect large-scale educational change (where the cultivation of professional competence, sensitivity, independence, and responsibility among teachers is regularly replaced by centralized control via a fragmented list of separate "outcomes," which are then assessed in ways that actively discourage good teaching).

Small-scale experiments can also have unintended side effects! As a little-known example of a radical attempt to cultivate the art of problem solving at school level we offer Eisenstein's account of his own education at lower secondary school (1833–37).

> [E]ach student had to prove the theorems consecutively. No lecture took place at all. No one was allowed to tell his solutions to anybody else and each student received the next theorem to prove, independent of the other students, as soon as he had proved the preceding one correctly, and as long as he had understood the reasoning.... While my peers were still struggling with the eleventh or twelfth, I had already proved the hundredth.... [T]his method ... can probably not be adapted.... One does not obtain that overview of the whole subject, which can only be achieved by a good lecture.... In the end, the best mathematical genius cannot discover alone what has been discovered by the collaboration of many outstanding minds.... For students this method is only practicable if it deals with small fields of easily understandable knowledge, especially geometric theorems, which do not require new insights and ideas.
>
> Eisenstein (1975)

Eisenstein was a remarkable mathematician. Yet at the tender age of twenty, on the threshold of the mathematical world that he longed to inhabit, he could see the limitations of this approach—even for students such as himself.

Problems that cultivate a taste for problem solving tend to incorporate certain characteristic features, such as simplicity, rhythm, naturalness, elegance, and surprise; and their solutions are often double-edged. But their most important feature is that, while their solution should be within reach of those in the target audience, the statement of the problem should convey no direct hint as to how to begin. Indeed, a good problem may continue to frustrate the would-be solver for a disturbingly long period.

> A tacit rite of passage for the mathematician is the first sleepless night caused by an unsolved problem.
>
> Reznick (1994)

The role of sleep and sleeplessness in creative problem solving is well documented (if poorly understood). It often features within the "incubation" phase of HADAMARD's [VI.65] "four phases" (discussed below), which summarize the process through which the initial experience of helplessness and leaden frustration is sometimes transmuted into golden success.

Such success is neither mechanical, nor the result of pure chance. In solving a good problem—as with a good puzzle—there is no magic problem-solving method that might relieve us of the need to struggle: the struggle may sometimes be fruitless, but it is an important part of the process. Thus, a successful outcome generally presupposes a certain kind of preparatory hard work. When asked how he made his discoveries, GAUSS [VI.26] is said to have answered, "Durch planmässiges Tattonieren," that is, through systematic and persistent groping around!

Having discovered a way into a problem, one may realize that it "should have been obvious" where to begin; but things are often obvious only in retrospect. One learns by experience how a certain kind of persistence can cause the fog that initially surrounds an unfamiliar problem to magically evaporate; what was at first invisible then stands out so clearly that one can scarcely understand how it could ever have been missed.

When faced with an unfamiliar mathematical problem, the mathematician, young or old, is like someone who is trying to open some fiendishly difficult Chinese puzzle box with a hopelessly small bunch of keys. At first glance the surface seems totally smooth, without a single visible crack. If you were not convinced that it was indeed a Chinese puzzle box, and that it could in fact be opened, you would soon give up. Knowing (or rather *believing*) that it can be opened, you may be willing to keep searching until you eventually begin to discern the slightest hint of a crack here and there. You may still have no idea how the pieces are meant to move, or which of your "keys" may help you to open up the first layer of the puzzle, but by trying the most appropriate-looking keys in the most promising cracks, you eventually stumble on one that fits exactly, and the pieces begin to move. The job is certainly not done; but the mood has changed and you feel you are well on the way.

As we have already seen, this experience of initial confusion, giving way as one grapples with a problem to unexpected insight, is in no way confined to beginners. It is part of the very nature of mathematics and of the way human beings do mathematics. If a problem is unfamiliar, its solution may require persistence, faith, and much time. So one should never give up too easily, and should always be prepared to look back after solving a problem to see what one could perhaps have done differently.

> It is most important in creative science not to give up. If you are an optimist you will be willing to "try" more than if you are a pessimist. It is the same in games like chess. A really good chess player tends to believe (sometimes mistakenly) that he holds a better position than his opponent. This, of course, helps to keep the game moving and does not increase the fatigue that self-doubt engenders. Physical and mental stamina are of crucial importance in chess and also in creative scientific work.
>
> Ulam (1991)

Persistence is of course easier to sustain if one has a degree of optimism about the likely outcome, or if one has cultivated the sheer "bloody-mindedness" that makes one refuse to give up (as with Littlewood's surviving rat). However, there are dangers.

> I learned, subconsciously, from Mazur how to control my inborn optimism and how to verify details. I learned to go more slowly over intermediate steps with a skeptical mind and not to let myself be carried away.
>
> Ulam (1991)

At the International Congress of Mathematicians in Paris in 1900, HILBERT [VI.63] presented twenty-three major research problems, which he judged would be important for the development of mathematics in the twentieth century. These problems seemed very hard; yet in bringing them to the attention of his fellow mathematicians Hilbert felt the need to stress that this should not be used as an excuse for putting off trying to solve them.

> However unapproachable these problems may seem to us and however helpless we stand before them, we have, nevertheless, the firm conviction that their solution must follow by a finite number of purely logical processes.... This conviction of the solvability of every mathematical problem is a powerful incentive to the worker. We hear within us the perpetual call: There is the problem. Seek its solution. You can find it by pure reason; for in mathematics there is no *ignorabimus*.

During the nineteenth century it became clear that the more that scientists discovered about nature, the more they realized *how little they knew*, and that one could never hope to discover "the whole truth." This realization was summed up by the physiologist Emil du Bois-

Reymond in the phrase "*ignoramus et ignorabimus*"— ignorant we are and ignorant we shall remain. As the new century dawned, Hilbert felt that it was important to state as clearly as he could that *mathematics is different*. In mathematics, he said, we can tackle problems with "the firm conviction that their solution must follow by a finite number of purely logical processes." As if to underline his assertion, one of his problems was solved almost immediately (though the most famous, the RIEMANN HYPOTHESIS [IV.2 §3], remains unresolved).

Hilbert was talking about mathematical *research*: but his principle applies even more strongly when tackling problems from textbooks, Olympiads, or university courses. When faced with an unfamiliar and apparently very difficult mathematical problem, we have little choice about how to proceed: we must either tackle the problem using the "bunch of keys" or mathematical techniques that we already know (no matter how limited they may be), or put off trying. Of course it is important to learn new tricks, and to revise old ones, as we go along. And of course there is always the temptation to imagine that the problem we face is simply *too* hard, that progress toward a solution requires some trick or technique that we have not yet learned and that the solution is therefore beyond our powers. This defeatist view is all the more plausible because it must sometimes be true! Mathematicians know perfectly well that, strictly speaking, the assumption that every problem can be solved is irrational (in that it cannot be justified logically, and is in general clearly false: we now know that some problems are intrinsically insoluble as stated). *It is nevertheless an invaluable working hypothesis.* Thus we should never let such doubts interfere with the basic hypothesis that *every problem we tackle has to be solved using essentially the techniques that we already know* (deployed with sufficient ingenuity!). Though strictly illogical, the assumption that every problem can be solved has justified itself so often in practice that it becomes a powerful conviction—a conviction that is psychologically invaluable each time we experience that feeling of helplessness when trying to get to grips with a hard mathematical problem.

Hilbert's judgment that his problems would play a central role in the mathematics of the twentieth century was remarkably astute. But the most interesting thing for us here is his rallying call: however unapproachable these problems may seem at first sight, and however helpless we stand before them, we have the firm conviction that their solution must be possible by purely logical processes. "There is the problem. Seek its solution. You can find it by pure reason." As in most printed mathematics, Hilbert offered no psychological guidance on how to proceed. Those who took up Hilbert's challenge were expected to discover such things for themselves.

Like every social activity, mathematics has a "front" and a "back": the *front* is where the finished products are displayed for public consumption, while the *back* is where the real work is done in less presentable surroundings. A naive realist might view the *front* as a mere facade, insist that all serious "problem solving" goes on "out back," and declare this separation to be artificial.

> Sometime, in a future that is knocking at our door, we shall have to retrain ourselves and our children to properly tell the truth. The exercise will be particularly painful in mathematics. The enrapturing discoveries of our field systematically conceal, like footprints erased in the sand, the analogical train of thought that is the authentic life of mathematics.... Until that day, however, the truths of mathematics will make only fleeting appearances, like shameful confessions whispered to a priest, to a psychiatrist, or to a wife.
>
> In the nineteenth chapter of "The Betrothed," Manzoni describes as follows the one genuine moment in a conversation between astute Milanese diplomats: "It was as if, between acts in the performance of an opera, the curtain were to be raised too soon, and the spectators were given a glimpse of a half-dressed soprano screaming at the tenor."
>
> Gian-Carlo Rota, in Kac et al. (1986)

However, the prospect of some mathematical equivalent of being obliged to witness "a half-dressed soprano screaming at the tenor" should cause us to hesitate before embracing Rota's vision of the future.

The *front–back* metaphor is due to the sociologist Erving Goffman. One standard example is that of a restaurant. We tend to think of a restaurant in terms of what we see "out front," where the manners, food, and language are "all dressed up"; but everything we see out front is totally dependent on the raw heat, the steam and grease, the conflicts and curses "out back" in the kitchen—where the hard work is done to tight deadlines and in very different conditions.

The triumph of mathematics in the modern world has been largely due to the fact that these two worlds—the front and the back—have been deliberately and systematically separated. It may seem curious that we have no agreed way of discussing the dynamics of the mathematical kitchen; but mathematics has grown largely because its practitioners have learned to separate its *objective* results, and the way they are validated and presented, from the intriguing, but inscrutable (and ultimately irrelevant!) *subjective* alchemy through which these mathematical results are conjured up. This formal separation has led to the adoption of a universally communicable format, which transcends personal taste and style, and which can therefore be comprehended, checked, and improved by anyone. Any move to pay greater attention to the mental, physical, and emotional dynamics that underlie mathematical problem solving must understand the need for this separation and respect the formal world of "objective" mathematics.

There are intriguing insights into the human dynamics of the mathematical kitchen scattered throughout the mathematical literature. One such insight is the fact that different mathematicians may have very different styles, even though most of these differences are rarely discussed. One example is the perceived role of *memory*. Some mathematicians value memory highly.

> It seems to me that a good memory—at least for mathematicians and physicists—forms a large part of their talent. And what we call talent or perhaps genius itself depends to a large extent on the ability to use one's memory properly to find the analogies, past, present and future, which, as Banach said, are essential to the development of new ideas.
>
> Ulam (1991)

Others have an excellent memory for anything *within their own field of interest*, but have considerable difficulty storing information from outside that domain in an easily retrievable form. And many would-be mathematicians are drawn to the subject precisely because they see it as requiring markedly less memorizing than most other disciplines. The important point would seem to be not *how much* one remembers, but *what* one makes automatic, and *how accessibly* this and other information is stored. It is clearly worth making a serious effort to organize in one's mind that material which is central to one's own work—so that it is available for instant use. It is also important, as we shall see, to collect a penumbra of possibly useful ideas, information, and examples—so that the mind is in a position to make incidental connections which might be fruitful. But it is not necessarily wise to learn in a uniform way everything that might conceivably be needed for the problem at hand: knowing slightly less sometimes forces the mind to get by on less, and hence to be more ingenious or inventive.

Hadamard's Four Phases

Littlewood's (1986) numerous perceptive observations concerning his contemporaries highlight other differences in style—such as speed and working habits. Similar insights may be found in many of the livelier mathematical autobiographies, but Littlewood's remarks are especially valuable.

> With a good deal of diffidence I will try to give some practical advice about research and the strategy it calls for. In the first place research work is of a different "order" from the learning process of pre-research education (essential as it is). The latter can easily be rote-memory, with little associative power: on the other hand, after a month's immersion in research the mind knows its problem as much as the tongue knows the inside of one's mouth. You must acquire the art of "thinking vaguely", an elusive idea that I can't elaborate in short form.... I should stress the importance of giving the subconscious every chance. There should be relaxed periods during the working day, profitably I say spent in walking.
>
> Littlewood (1986)

At one stage Poincaré thought there might be just two main styles of mathematical thinking:

> The one sort are above all preoccupied with logic.... The other sort are guided by intuition and at the first stroke make quick but sometimes precarious conquests.... [O]ne often says of the first that they are *analysts* and calls the others *geometers*.
>
> Poincaré (1904)

But in identifying the label "logical" with that of "analyst," and the label "intuitive" with that of "geometer," he noticed that HERMITE [VI.47] constituted a counterexample—an "intuitive analyst"! Clearly, the range of mathematical styles is more complex (see Hadamard 1945, chapter VII). One consequence is that any analysis of the art of problem solving in general needs to be drawn with a broad brush. Despite this caveat, Hadamard's "four-phase" model of mathematical creativity has found widespread acceptance, so it may help if one's work habits respect these phases:

> It is usual to distinguish four phases in creation: preparation, incubation, illumination and verification, or working out.... Preparation is largely conscious, and anyhow *directed* by the conscious. The essential problem has to be stripped of accidentals and brought clearly into view; all relevant knowledge surveyed; possible analogues pondered. It should be kept constantly before the mind during intervals of other work....

Incubation is the work of the subconscious during the waiting time, which may be several years. Illumination, which can happen in a fraction of a second, is the emergence of the creative idea into the conscious. This almost always occurs when the mind is in a state of relaxation and engaged lightly with ordinary matters.... Illumination implies some mysterious rapport between the subconscious and the conscious, otherwise emergence could not happen. What rings the bell at the right moment?

> Littlewood (1986)

Pólya's *How to Solve It* proposes a less convincing four-stage "recipe" for the problem-solving process ("understand, plan, act, reflect"), which has nevertheless been widely used at school level. Hadamard's four phases provide a useful framework for thinking and communicating about the creative process; they also separate the relatively routine aspects (which one may be able to influence more easily) from the more elusive ones. The "conscious *preparation*" phase is perhaps the most mundane stage, requiring a combination of method and discipline. Littlewood again offers sound advice. He recognizes that his advice may not suit all tastes, but he insists that we would all benefit from trying different patterns of working in order to identify and cultivate habits that are as effective as possible.

> Most people need half an hour or so before being able to concentrate fully.... The natural impulse towards the end of a day's work is to finish the immediate job: this is of course right if stopping would mean doing work all over again. But try to end in the middle of something; in a job of writing out, stop in the middle of a sentence. The usual recipe for warming up is to run over the latter part of the previous day's work; this dodge is a further improvement.... When I am working really hard I wake around 5.30 a.m. ready and eager to start; if I am slack I sleep till I am called.
>
> Littlewood (1986)

At some ill-defined stage, this preparation achieves a sufficiently clear understanding of the immediate problem, together with a level of saturation in relevant background information, to enable the mind to begin trying different approaches and combinations of ideas. We have reached the *incubation* phase.

> We cannot know all the facts, since they are practically infinite in number.... Method is precisely the selection of facts.
>
> Poincaré (1908)

I've often observed too that once the first hurdle of preparation has been surmounted, one runs up against a wall. The main error to be avoided is trying to attack the problem head-on. During the incubation phase you have to proceed indirectly, obliquely.... Thought needs to be liberated in such a way that subconscious work can take place.

> Alain Connes, in Changeux and Connes (1995)

Temperament, general character, and "hormonal" factors must play a very important role in what is considered to be purely "mental" activity.... A "subconscious brewing" (or pondering) sometimes produces better results than forced, systematic thinking.... [W]hat we call originality ... might to some extent consist of a methodical way of exploring all avenues—an almost automatic sorting of attempts....

When I remember a mathematical proof, it seems to me that I remember only salient points, markers, as it were, of pleasure or difficulty. What is easy is easily passed over because it can be reconstructed logically with ease. If, on the other hand, I want to do something new or original, then it is no longer a question of syllogism chains. When I was a boy I felt that the role of rhyme in poetry was to compel one to find the unobvious because of the necessity of finding a word which rhymes. This forces novel associations and almost guarantees deviations from routine chains or trains of thought. It becomes paradoxically a sort of automatic mechanism of originality.... What people think of as inspiration or illumination is really the result of much subconscious work and association through channels in the brain of which one is not aware at all.

> Ulam (1991)

It takes two to invent anything. The one makes up combinations; the other one chooses, recognizes what he wishes and what is important to him in the mass of the things which the former has imparted to him. What we call genius is much less the work of the first one than the readiness of the second one to grasp the value of what has been laid before him and to choose it.

> Paul Valéry, quoted in Hadamard (1945)

We have reached a double conclusion: that invention is choice [and] that this choice is imperatively governed by the sense of scientific beauty.

> Hadamard (1945)

Part of the pleasure (and pain), the magic (and masochism) of mathematics stems from the fact that the next step—from incubation to illumination—remains so mysteriously elusive. *Illumination* can occur at any time. In most cases—especially where the realization is of something relatively straightforward—this occurs during periods of "official work." However, this need not be so, especially when the corner to be illuminated is especially dark or unfamiliar, or if the leap of imagination required is large. In such cases it seems that, after the hard graft of the preparation and incubation phases, the mind often needs to "step back" in order to see the way forward more clearly. That is, hard work needs to be combined with relaxation, as Connes implies when he warns against "trying to attack the problem head-on." In one oft-quoted example, Poincaré recalls how he realized the profound connection between Fuchsian functions and hyperbolic geometry as he stepped aboard a bus while on a day out! The first three extracts below show that the mind may achieve this in-between state as a result of *sleeplessness*, or *in the very act of waking*. The fourth extract concerns strenuous *hill walking*. What is common to them all is that the moment of enlightenment does not occur while the beneficiary is officially working!

> It was his custom to tell his friends that if others would meditate as long and as deeply as he did on mathematical truths, they would be able to make his discoveries. He said that often he meditated for days on a piece of research without finding a solution, which finally became clear to him after a sleepless night.
>
> > Dunnington (1955)

> One phenomenon is certain and I can vouch for its absolute certainty: the sudden and immediate appearance of a solution at the very moment of sudden awakening. On being very abruptly awakened by an external noise a solution long searched for appeared to me without the slightest instant of reflection on my part ... and in a quite different direction from any of those which I had previously tried to follow.
>
> > Hadamard (1945)

> Most striking at first is this appearance of sudden illumination, a manifest sign of long, conscious prior work.... The role of this unconscious work in mathematical invention appears to me incontestable....
>
> For a fortnight I had been attempting to prove that there could not be any function analogous to what I have since called Fuchsian functions. I was at that time very ignorant. Every day I sat down at my table and spent an hour or two trying a great number of combinations, and I arrived at no result. One night I took some black coffee, contrary to my custom, and was unable to sleep. A host of ideas kept surging in my head; I could feel them jostling one another, until two of them coalesced, so to speak, to form a stable combination. When morning came, I had established the existence of

one class of Fuchsian functions, those that are derived from the hypergeometric series. I had only to verify the results, which only took a few hours.
                                                    Poincaré (1908)

I had been struggling for two months to prove a result I was pretty sure was true. When ... walking up a Swiss mountain, fully occupied by the effort, a very odd device emerged—so odd that though it worked I could not grasp the resulting proof as a whole.... I had a sense that my subconscious was saying, "Are you *never* going to do it, confound you; try this."

                                                    Littlewood (1986)

The resulting sense of satisfaction is familiar even to those whose mathematical experience is limited.

Illumination is not only marked by the pleasure—the exhilaration!—one inevitably experiences at the moment it strikes, but also by the relief one suddenly feels at seeing a fog abruptly lift, and disappear.

              Alain Connes in Changeux and Connes (1995)

However, after months of hard work, such intoxication can sometimes be deceptive.

In mathematics one cannot stop at drawing with a big, wide brush; all the details have to be filled in at some time.
                                                    Ulam (1991)

The *verification*, or working-out, process often appears mundane; but it is rarely routine, and regularly reveals hidden subtleties that force us to reassess the anticipated approach. Unforeseen difficulties may remain unresolved, and we may be obliged reluctantly to begin the cycle all over again. It is tempting to think of this as "failure." But mathematics is not a mere machine for solving problems; it is a way of life. In their different ways success and failure both send us back to the drawing board—as Gauss observed in a letter to BOLYAI [VI.34] in 1808.

It is not knowledge, but the act of learning, not possession but the act of getting there, which grants the greatest enjoyment. When I have clarified and exhausted a subject, then I turn away from it, in order to go into darkness again; the never-satisfied man is so strange—if he has completed a structure, then it is not in order to dwell in it peacefully, but in order to begin another. I imagine the world conqueror must feel thus, who after one kingdom is scarcely conquered, stretches out his arms for others.

Further Reading

Barbeau, E. 1989. *Polynomials.* New York: Springer.

Changeux, J.-P., and A. Connes. 1995. *Conversations on Mind, Matter, and Mathematics.* Princeton, NJ: Princeton University Press.

Dixon, J. D. 1973. *Problems in Group Theory.* New York: Dover.

Dunnington, G. W. 1955. *Carl Friedrich Gauss: Titan of Science.* New York: Hafner. (Reprinted with additional material by J. J. Gray, 2004. Washington, DC: The Mathematical Association of America.)

Eisenstein, G. F. 1975. *Mathematische Werke.* New York: Chelsea. (English translation available at http://www-ub.massey.ac.nz/~wwiims/research/letters/volume6/.)

Engel, A. 1991. *Problem-Solving Strategies.* Problem Books in Mathematics. New York: Springer.

Gardiner, A. 1987. *Discovering Mathematics: The Art of Investigation.* Oxford: Oxford University Press.

———. 1997. *The Mathematical Olympiad Handbook: An Introduction to Problem Solving.* Oxford: Oxford University Press.

Hadamard, J. 1945. *The Psychology of Invention in the Mathematical Field.* Princeton, NJ: Princeton University Press. (Reprinted 1996.)

Hilbert, D. 1902. Mathematical problems. *Bulletin of the American Mathematical Society* 8:437–79.

Kac, M. 1985. *Enigmas of Chance: An Autobiography.* Berkeley, CA: University of California Press.

Kac, M., G.-C. Rota, and J. T. Schwartz. 1986. *Discrete Thoughts: Essays on Mathematics, Science, and Philosophy.* Boston, MA: Birkhäuser.

Koestler, A. 1976. *The Act of Creation.* London: Hutchinson.

Littlewood, J. E. 1986. *A Mathematician's Miscellany.* Cambridge: Cambridge University Press.

Lovasz, L. 1979. *Combinatorial Problems and Exercises.* Amsterdam: North-Holland.

NCTM. 1980. *Problem Solving in School Mathematics.* Reston, VA: National Council of Teachers of Mathematics.

Newman, D. 1982. *A Problem Seminar.* New York: Springer.

Poincaré, H. 1904. *La Valeur de la Science.* Paris: E. Flammarion. (In *The Value of Science: Essential Writings of Henri Poincaré* (2001), and translated by G. B. Halsted. New York: The Modern Library.)

———. 1908. *Science et Méthode.* Paris: E. Flammarion. (In *The Value of Science: Essential Writings of Henri Poincaré* (2001), and translated by F. Maitland. New York: The Modern Library.)

Pólya, G. 1981. *Mathematical Discovery*, two volumes combined. New York: John Wiley.

———. 1990. *Mathematics and Plausible Reasoning*, two volumes. Princeton, NJ: Princeton University Press.

———. 2004. *How to Solve It.* Princeton, NJ: Princeton University Press.

Pólya, G., and G. Szego. 1972. *Problems and Theorems in Analysis*, two volumes. New York: Springer.

Reznick, B. 1994. Some thoughts on writing for the Putnam. In *Mathematical Thinking and Problem Solving*, edited by A. H. Schoenfeld. Mahwah, NJ: Lawrence Erlbaum.

Ringel, G. 1974. *Map Color Theorem*. New York: Springer.

Roberts, J. 1977. *Elementary Number Theory: A Problem Oriented Approach*. Cambridge, MA: MIT Press.

Ulam, S. 1991. *Adventures of a Mathematician*. Berkeley, CA: University of California Press.

Yaglom, A. M., and I. M. Yaglom. 1987. *Challenging Mathematical Problems with Elementary Solutions*, two volumes. New York: Dover.

Zeitz, P. 1999. *The Art and Craft of Problem Solving*. New York: John Wiley.

Zinoviev, A. A. 1980. *The Radiant Future*. New York: Random House.

VIII.2   "Why Mathematics?" You Might Ask
Michael Harris

It seems to me that they have a poor opinion of our religion if they think it needs the protection of philosophy.

Lorenzo Valla, *Dialogue on Free Will*

1   A Metaphysical Burden

ANDRÉ WEIL [VI.93], speaking at the 1978 International Congress of Mathematicians at Helsinki, concluded his address entitled "History of Mathematics: Why and How?" with these words:

Thus my original question "Why mathematical history?" finally reduces itself to the question "Why mathematics?," which fortunately I do not feel called upon to answer.

Proceedings of the ICM, Helsinki, 1978
(pp. 227–36, quotation on p. 236)

I heard Weil's address, and the applause that followed, and remember imagining circumstances in which that final question could not be so easily evaded. For instance, in 1991 the House Committee on Science, Space, and Technology called upon the American Mathematical Society (AMS) to answer a very similar question: "What are the main goals in the mathematical sciences?" Weil knew his audience, and the committee of twelve mathematicians responding to the government body responsible for research budgets knew theirs:

The most important long-term goals for the mathematical sciences are: provision of fundamental tools for science and technology, improvement of mathematics education, discovery of new mathematics, facilitation of technology transfer, and support of efficient computation.[1]

"Meaning is what makes things sell," wrote Roland Barthes (1967), and the AMS adopted the posture of FOURIER [VI.25], who, according to a celebrated comment of JACOBI [VI.35], included in a letter to LEGENDRE [VI.24] of July 2, 1830,

… had the opinion that the principal aim of mathematics was public utility and explanation of natural phenomena; but a philosopher like him should have known that the sole end of science is the honor of the human mind.

It might seem that the AMS has left a place for "honor" in its third goal, but a later elaboration of that goal directs the reader toward "unexpected" applications of pure mathematics.

Few pure mathematicians are as indifferent to practical applications as HARDY [VI.73], who in *A Mathematician's Apology* famously claimed that: "Judged by all practical standards, the value of my mathematical life is nil." But it is fair to assume that, when they are addressing one another rather than government committees, most pure mathematicians (including those who represented the AMS in 1991) would choose a quite different list of "most important long-term goals."

In this they have long been able to count on the protection of philosophy. It has been a commonplace since Plato to grant mathematics intrinsic value on metaphysical grounds.[2] The topos of mathematics as a source of certain knowledge was already well established by the second century, when Ptolemy wrote

Only mathematics, if one attacks it critically, provides for those who practice it sure and unswerving knowledge, since the demonstration comes about

---

1. From "Pilot assessment of the mathematical sciences (prepared for the House Committee on Science, Space, and Technology)," *Notices of the American Mathematical Society* 39 (1992):101–10.

2. The present essay is mainly concerned with metaphysical certainty. Descartes wrote in *Principles of Philosophy* (chapter CCVI) of "certainty … founded on the metaphysical ground that, as God is supremely good and the source of all truth, the faculty of distinguishing truth from error which he gave us, cannot be fallacious so long as we use it aright, and distinctly perceive anything by it," and cites "the demonstrations of mathematics" as his first example. Plato (in *Republic*, VII, 522–31) saw mathematics rather as a source of "knowledge of that which exists forever." Certainty and its cognates are some, but only some, of the apparent blessings of mathematics that so impressed certain philosophers as to "infect" the whole of their work, as Ian Hacking (2000) argues.

through incontrovertible means, by arithmetic and geometry.[3]

THE CRISIS IN THE FOUNDATIONS OF MATHEMATICS [II.7] of the early twentieth century, which culminated in GÖDEL'S INCOMPLETENESS THEOREMS [V.15], was largely motivated by a desire to make mathematical certainty safe from dependence on human frailty. As RUSSELL [VI.71] wrote in *Reflections on my Eightieth Birthday*:

> I wanted certainty in the kind of way in which people want religious faith. I thought that certainty is more likely to be found in mathematics than elsewhere.... Mathematics is, I believe, the chief source of the belief in eternal and exact truth.
>
> Quoted in Hersh (1997)

Russell's hope to ground certainty in logic is largely a thing of the past—as Marvin Minsky wrote in another context, "without an intimate connection between our knowledge and our intentions, logic leads to madness, not intelligence" (Minsky 1985/1986)[4]—but his words continue to echo. After Jean-Pierre Serre was named first recipient of the Abel Prize, he was quoted in *Libération* (May 23, 2003) to the effect that mathematics is the only producer of "totally reliable and verifiable" truths. And Landon T. Clay III, announcing the creation of the $7 000 000 Millennium Prize Fund, linked his decision to devote much of his personal fortune to the support of pure mathematics to "the decline in religious certitude ... the pursuit of proof continues to be a strong motivating force in human actions."[5]

The mind saves its honor, as Jacobi would have it, but only through indenture to a higher power. I would like to express my opinion that the bargain implicit in comments like those just quoted, placing pure mathematicians on the front lines in defense of metaphysical certainty or some other normative concern of philosophers, is an unnecessary burden that fails to do justice to what is uniquely valuable about mathematics. It also fails to protect pure mathematics from the real existential dangers it faces, of which budget cuts are only the most obvious expression. Mathematics is not likely to collapse for lack of a coherent account of its certainty, but it may well collapse for lack of an account of its value.

2 Postmodernism versus Mathematics?

One danger that should not worry mathematicians is that of *postmodernism*. Many thousands of pages have been written on this topic, although it is not clear that the word designates anything specific. I will nevertheless add a few pages of my own, because the term has come to be used as shorthand for a radical relativism that is thought to call into question not only certainty but rationality in all its forms.[6] One thus finds mathematicians who are skeptical of certainty in Russell's sense but who nonetheless express hostility to something they call "postmodernism" as they try to defend reason and the value of mathematics as a rational activity.

Applied to architecture, postmodernism designates a reasonably precise tendency. As a trend defining the spirit of the times, it has been called "the cultural logic of late capitalism," differing from modernism by emphasizing space rather than time, multiple perspectives and fragmentation rather than unity of meaning and totality, pastiche (sampling)[7] rather than progress, and much more along the same lines. As a movement in philosophy it is most typically (if abusively) associated with Michel Foucault, Jacques Derrida, Gilles Deleuze, Roland Barthes, Jean-François Lyotard, and more generally the "French theory" of the 1960s and 1970s. Postmodern prose is eclectic, ironic, self-referential, and hostile to linear narrative. The variant known as posthumanism celebrates the fading of conceptual and material boundaries between human beings and machines.

We are all postmodernists insofar as we have experienced the degradation of public discourse under the influence of advertising slogans, and are therefore likely, in spite of ourselves, to read Jacobi's invocation of "the honor of the human mind" as a precursor of that genre. Mathematicians can even claim to be the first postmodernists: compare an art critic's definition of postmodernism—"meaning is suspended in

---

3. See Lloyd (2002), in which is cited Ptolemy's *Syntaxis*, I, chapter 1, 16.17–21.

4. Compare René Thom's comment in connection with his criticism of attempts to reduce mathematics to set theory: "In attempting to attach meaning to all the phrases constructed in ordinary languages, according to Boolean rules, the logician proceeds to a phantasmic, delirious reconstruction of the universe" (reprinted in Tymoczko 1998).

5. Transcript of interview by Francois Tisseyre conducted on the occasion of the Paris Millennium Meeting, May 24, 2000, graciously provided by the Clay Mathematics Institute.

6. For example, Lakoff and Núñez (2000) write of a "radical form of postmodernism which claims that mathematics is purely historically and culturally contingent and fundamentally subjective." No examples are given of texts espousing this point of view.

7. "Because his ... artistry comes from combining other people's art, ... the DJ is the epitome of a postmodern artist" (www.jahsonic.com/postmodernism.html).

favor of a game involving free-floating signs"—with the definition of mathematics, attributed to HILBERT [VI.63], as "a game played according to certain simple rules with meaningless marks on paper."[8] Mathematics could nevertheless (or for that very reason) safely ignore postmodernism, were it not that the latter is supposed to have no room for certainty, metaphysical or otherwise.[9] So it is not surprising that authors who are considered postmodernists have had some perplexing run-ins with science and mathematics.

The typically controversial postmodernist account of science sounds like this:

> Science and philosophy must jettison their grandiose metaphysical claims and view themselves more modestly as just another set of narratives.
>
> Terry Eagleton's caricature of postmodernism, quoted in Harvey (1989)

As far as mathematics is concerned, relativism of this kind has more to do with English-language postmodernism than with the French original. One might have thought that mathematical progress from axioms to theorems and from lesser to greater abstraction or generality constituted a prime example of the sort of "master narrative" that French postmodernists regarded with suspicion, and a particularly tempting target given the special role Enlightenment thinking reserved for mathematical explanation; but that seems not to have been the case. Although the most prominent French philosophers associated with postmodernism were metaphysical skeptics in other regards, they had no quarrel with mathematics' metaphysical pretensions per se; but they did question their relevance to the human sciences. For Derrida, thinking of LEIBNIZ [VI.15] in particular, "[mathematics] was always the exemplary model of scientificity" (in *Of Grammatology*, p. 27), and Foucault claimed that:

> Mathematics has certainly served as a model for most scientific discourse[s] in their efforts to attain formal rigor and demonstrativity; but for the historian who

questions the actual development of the sciences, it is ... an example ... from which one cannot generalize.[10]
>
> *The Archeology of Knowledge* (pp. 188–89)

At least one of postmodernism's canonical French texts does take on the issue of certainty in science and mathematics directly. Alluding to the trilogy of Gödel's theorems, uncertainty in quantum mechanics, and fractals,[11] Lyotard saw in contemporary mathematics

> a current that calls into question precise measurement and prediction of the behavior of objects at the human scale ... postmodern science ... produces not the known, but the unknown.
>
> Lyotard (1979)

Various authors have reminded readers that Gödel's theorems and the uncertainty principle (and chaos) are statements about formal systems in mathematics and particle physics (and nonlinear differential equations), respectively, and as such have no bearing on metaphysics.[12] The arguments are often eloquent but altogether beside the point, and of little comfort to seekers of certainty like Russell. Metaphysical certainty, whatever it may be, cannot be any less binding than a mathematical proof. Gödel's theorem, that it is impossible to prove, within a formal system, that that formal system is consistent, can reasonably be taken to mean that metaphysical certainty cannot be guaranteed by mathematical means alone.[13] But Serre, in his comments to *Libération*, surely meant something more than the tautology that mathematical truth is totally reliable and verifiable *by the standards of mathematics*. The struggle

---

8. Otto Karnik, in "Attraction and repulsion," article in *Kai KeinRespekt*, p. 48, Exhibition Catalogue of the Institute of Contemporary Art (Bridge House Publishing, Boston, MA, 2004). The Hilbert quotation is easy to find but is probably apocryphal, which does not make it any less significant. *Mathematics and the Roots of Postmodern Thought*, by Vladimir Tasić, is an extended speculation on postmodernism's mathematical antecedents; see my review in *Notices of the American Mathematical Society* 50 (2003):790–99.

9. For example, "[Derrida's] thought is based on his disapproval of the search for an ultimate metaphysical certainty or source of meaning that has characterized most of Western philosophy." From the *Encyclopedia Britannica Online* (www.britannica.com).

10. "Why don't you ask a physicist or a mathematician about difficulty?" was Derrida's response to a 1998 *New York Times* question about deconstruction: see *Jacques Derrida, Abstruse Theorist, dies at 74*, *New York Times*, October 10, 2004. Appeals to the presumed value of even the most abstruse mathematics, in order to legitimate obscurity elsewhere, are common. I first encountered such an argument in an article by composer (and former mathematician) Milton Babbitt entitled "Who cares if you listen?" (*High Fidelity*, February 1958): "Why should the layman be other than bored and puzzled by what he is unable to understand, music or anything else?" With this sort of talk, the justification of pure mathematics on aesthetic grounds is turned upside down. That is why I address aesthetic answers to the question of my title—which are by far the most popular among my colleagues—only in a footnote.

11. A cliché for the succeeding generation of literary critics: for a sample emphasizing chaos rather than Gödel, see N. Katherine Hayles (ed.), *Chaos and Order* (University of Chicago Press, 1991).

12. Much of *Prodiges et vertiges de l'analogie* by Jacques Bouveresse (Raisons d'Agir, 1999) is devoted to just this sort of reminder.

13. Predictably, religion steps in to fill the gap: see www.asa3.org/ASA/topics/Astronomy-Cosmology/PSCF9-89Hedman.html#16. John D. Barrow takes the implications of Gödel's theorems for physics seriously, while denying that they necessarily limit scientific objectivity (see, for example, "Domande senza risposta," in *Matematica e Cultura 2002*, edited by M. Emmer, pp. 13–24 (Springer, 2002)).

to pin down this "something more," to find what one might call the "essence" of mathematics, is why the philosophy of mathematics keeps visiting the scenes of its many past defeats.

Even if Lyotard does not make the case very well, one can detect a "postmodern" sensibility in much of recent science, from Stephen Jay Gould's insistence that evolution is highly contingent, to complexity theory, to the study of consciousness as an "emergent" phenomenon. What these developments have in common is a rejection of reductionism and related top-down "master narratives," not because they are wrong but because they are irrelevant and useless. It would be going too far to describe this kind of science as a new Kuhnian paradigm (the notion is, in any case, widely criticized as oversimplified), but it is noticeably different from the disciplines that inspired the analytic philosophy of science. As for mathematics, there have been suggestions that it too has postmodern aspects—for example, Jürgen Jost has written a book entitled *Postmodern Analysis* and some specialists now claim to be working in "postmodern algebra"—but I do not see any genuine signs of this sensibility. Indeed, I am not even sure that it makes sense to draw the line between modern and postmodern. Hilbert's definition of mathematics as a game does sound like something from Derrida, but if Hilbert's foundational program ("wir müssen wissen, wir werden wissen") is not a prime example of high modernism, then what is? On the other hand, the abandonment of all forms of foundationalism in an anthology of Tymoczko (1998) is a rejection of "master narratives" within philosophy of mathematics, and indeed the blurb calls the anthology "postmodern."[14]

3 Sociology Aims for the High Ground

While Weil is supposed to have discounted Gödel's metaphysical menace by making it into a joke—"God exists since mathematics is consistent, and the Devil exists since we cannot prove it"—his fellow Bourbakist Dieudonné attempted a counterattack:

Just as physicists and biologists believe in the permanence of the laws of nature, solely because they have observed this up to now, … the mathematicians called—wrongly—"formalists" (… at present the near totality of mathematical researchers) are convinced

that no contradiction will appear in set theory, none having manifested themselves for 80 years.[15]

This is either an inductive (empirical), sociological, or pragmatist argument. All these trends are indeed present in postmodernism, more typically in English sociology of science than in French philosophy:

The compelling force of mathematical procedures does not derive from their being transcendent, but from their being accepted and used by a group of people. The procedures are not used because they are correct, or correspond to an ideal; they are deemed correct because they are accepted.

David Bloor, in *Wittgenstein: A Social Theory of Knowledge* (Macmillan, London, 1983)

The Sociology of Scientific Knowledge (SSK) movement, of which David Bloor was a founder, is firmly rooted in postwar philosophy of science in the analytic tradition. The later Wittgenstein's discussion of mathematics, and knowledge more generally, in terms of "language-games," "forms of life," and learning to follow rules emphasizes social factors, and SSK is enthusiastically Wittgensteinian. Of course, Wittgenstein's work is notoriously unsystematic and lends itself to a variety of interpretations. I find it wrong to see the Wittgenstein who wrote "Grounds for *doubt* are lacking!" as a skeptic. Beyond the social factors to which Wittgenstein drew explicit attention, he clearly perceived "something more" specifically in mathematics ("the hardness of the logical must"), to which our language and philosophy are not able to do justice.[16]

Can sociology succeed where philosophy has failed? Bloor's militant "naturalist" response to the question of "whether sociology can touch the very heart of mathematical knowledge" (Bloor 1976) is less an exercise in debunking metaphysics than an attempt to seize the metaphysical high ground for sociology. An otherwise subtle ethnographic study by Claude Rosental of the resolution of a conflict among logicians betrays a similar sensibility, as does his suggestion that training in mathematics and logic might have constituted a "serious handicap" to carrying out his project (Rosental

---

14. The anti-foundationalism of Tymoczko's anthology is largely inspired by Gödel's theorems.

15. Weil's joke is quoted in at least eighty-five sites found via Google; no primary source is given. Dieudonné's comment is naturally from *Pour l'Honneur de l'Esprit Humain*, pp. 244–45 (Hachette, 1987). Borel's remarks on the "self-correcting power of mathematics," in his contribution to the discussion of the article "Theoretical mathematics: toward a cultural synthesis of mathematics and theoretical physics" by A. Jaffe and F. Quinn, express a more modest form of pragmatism (*Bulletin of the American Mathematical Society* 29 (1993):1–13).

16. Quotations from Wittgenstein (1969, paragraph 4; 1958, paragraph 437).

2003). The classic declaration of the latter kind is due to Bruno Latour and Stephen Woolgar:

> [W]e do *not* regard prior cognition…as a necessary prerequisite for understanding scientists' work. This is similar to an anthropologist's refusal to bow before the knowledge of a primitive sorcerer. There are, as far as we know, no a priori reasons for supposing that scientists' practice is more rational than that of outsiders.
>
> Latour and Woolgar, *Laboratory Life*, pp. 29–30 (Princeton University Press, Princeton, NJ, 1986)

But one can also imagine sociologists paying serious attention to mathematicians' accounts of their experience, addressing in the process the question that Weil did not. For example, Bettina Heintz, in fieldwork at the Max-Planck-Institut in Bonn, which was billed as the first study of mathematics from the perspective of constructivist sociology of science, worries about "going native" and "overidentifying with the dominant culture." But her subject is the eminently sociological one of determining how mathematicians reach consensus, and her methodology, far from treating practicing mathematicians as "primitive sorcerers," records their epistemic perspectives sympathetically and at length. One has the impression that, in spite of the limitations of her methodology, Heintz is more interested in accounting for "real mathematics," to which we shall return below, whereas Bloor and Rosental are preoccupied with marshaling evidence to counter the metaphysical preoccupations of philosophers.

Under siege from Gödel's theorem, Popper's attack on verificationism, Kuhn's theory of scientific revolutions, Lakatos's dialectical approach to the contents of knowledge in *Proofs and Refutations*, as well as Wittgenstein, certainty in Russell's sense has largely been scrapped.[17] As for the social, philosophical, and spiritual needs that the notion of metaphysical certainty was designed to address, they remain. Thus, on the one hand, those with tendencies that I have described as postmodernist continue to express skepticism regarding certainty, seemingly unaware that their target is now little more than an advertising slogan that has little to do with the real concerns of mathematicians; while, on the other hand, analytic philosophy has sought to substitute more flexible notions. The term "warrant," for example, is used in an attempt by Philip Kitcher to develop a consistent account of mathematics on empirical rather than aprioristic grounds. Kitcher recalls FREGE's [VI.56] frustration with the mathematicians of his time, observing that, "When Frege emphasizes the possibility of complete clarity and certainty in mathematical knowledge, he is advancing a picture of mathematics that is almost irrelevant to the working mathematician" (Kitcher 1984). However, Kitcher and the SSK remain obsessed by the problem of "how our mathematical knowledge [is] acquired" (Kitcher 1984), where knowledge is taken to be true and justified belief.

Reading Heintz (2000), one learns that now, as in Frege's day, mathematicians themselves widely consider these problems to be outdated or beside the point. The most controversial aspect of the SSK's "strong programme," formulated by Bloor and Barry Barnes, is the "thesis of symmetry": the insistence that truth or falsity not be taken into account when investigating how a scientific claim comes to be accepted as knowledge. Heintz's fieldwork suggests that this is compatible with the view prevailing among mathematicians regarding acceptance of a mathematical proof, a "kind of consensus theory of truth" (Heintz 2000).[18]

A striking instance of "how a mathematical proof comes to be accepted as knowledge" is playing out even as I am writing these lines. Grigori Perelman's announced proof of THE POINCARÉ CONJECTURE [V.25] is undergoing unprecedented scrutiny in a small number of specialized centers, with the hope of determining the truth or falsity of Perelman's claim. This is going on completely beyond the spotlight of sociology, as far as I know, and with no guidance from philosophy, even though the $1 000 000 prize offered by the Clay

---

17. Lakatos's posthumous "A renaissance of empiricism in the recent philosophy of mathematics," presents a long series of quotations by mathematicians and a few philosophers, including Russell in 1924, acknowledging that mathematics is uncertain, after all. Naturally, most of those cited refer directly or indirectly to Gödel's theorem. The article was reprinted in Tymoczko (1998). "Only dogma or theory has made people say that mathematics as a whole has a peculiar certainty," writes Hacking (2000). Certainty persists, however, in the titles of philosophy books, e.g., Marcus Giaquinto's optimistic *The Search for Certainty: A Philosophical Account of Foundations of Mathematics* (Oxford University Press, Oxford, 2004).

18. Heintz quotes Yu. I. Manin—"A proof only becomes a proof after the social act of 'accepting it as a proof'"—as well as René Thom's "community" theory of truth. One can of course always ask whether Heintz selectively quoted mathematicians whose positions support her thesis. This question can be asked of any sociological study, and it is best to let the sociologists work out their methodological issues. An important remark, however: though Heintz's original goal was to account for the formation of consensus among mathematicians within a science studies framework—with questionable success, but that is another matter—she does not defend a particular school within philosophy of mathematics. In this she differs from Bloor, for instance, who identifies himself explicitly as an empiricist.

Mathematical Institute is in no sense Platonic,[19] and the rules for awarding the prize presuppose the fallibility of the mathematical community, in terms very similar to those that Heintz's informants expressed spontaneously (see the third and subsequent paragraphs at www.claymath.org/millennium/rules_etc). The case is exceptional, however; "certifying knowledge," in Rosental's sense, is as such relatively unimportant to mathematicians, and Perelman's close readers would probably describe what they are doing as attempting to *understand* his proof rather than "certifying" it as knowledge (for the sake of the community, or a generous benefactor, or philosophers or sociologists).[20]

4 Truth and Knowledge

"By far the larger part of activity in what goes by the name *philosophy of mathematics* is dead to what mathematicians think and have thought, aside from an unbalanced interest in the 'foundational' ideas of the 1880-1930 period, yielding too often a distorted picture of that time," announced David Corfield, presenting his efforts to develop a "philosophy of real mathematics."[21] Corfield contrasted the traditional apriorist's concerns, "How should we talk about mathematical truth? Do mathematical terms or statements refer? If so, what are the referents and how do we have access to them?" (Corfield 2003), with the list of questions Aspray and Kitcher consider typical of the "maverick tradition" in philosophy of mathematics: "How does mathematical knowledge grow? What is mathematical progress? What makes some mathematical ideas (or theories) better than others? What is mathematical explanation?" (quoted by Corfield 2003).

The mavericks, well represented in Tymoczko's anthology, have moved a welcome step away from certainty. Nevertheless, the philosophers and philosophically minded sociologists I have mentioned—with the partial exception of Corfield, to be explored below—still often write as though mathematicians were creating Truth or Knowledge,[22] almost as a favor to philosophy or sociology, to show how such a feat is possible. Or just to show that it *is* possible.[23] We mathematicians, on the other hand, are quite convinced that we are creating *mathematics*, and it is the "why" of that activity, without the ennobling assimilation to the generic objects of interest to epistemology, that, as Weil understood, required no explanation in Helsinki.

"Whoever undertakes to set himself up as a judge in the field of Truth and Knowledge is shipwrecked by the laughter of the gods," wrote Einstein. Mathematicians tend to respond with dismay rather than laughter, and then only to blunders so egregious as to be universally recognized as such.[24] Although those who find fault with philosophical speculation regarding the nature of mathematics seem to be under an implicit obligation to propose a speculative alternative, experience suggests that the practice of mathematics renders one unfit to do so. This, more than the fear of ridicule, is the main reason I would not venture my own speculative philosophy of mathematics. If it is hard "for those who are used to thought processes stemming from geometry and algebra" to "develop the sort of intuition common among physicists" (R. MacPherson, quoted in *Quantum Fields and Strings: A Course for Mathematicians* by Deligne (volume 1, p. 2)), bridging the gap between mathematicians and metaphysicians is probably hopeless. There are superficial parallels, to be sure: a metaphysical abstraction such as "essence," like a mathematical abstraction such as "set," designates nothing in itself, but rather refers to a canonical body

---

19. This article was written in late 2004. The proof is now accepted as correct, and in 2006 Perelman was offered a Fields medal, which he declined. He has also refused the Clay Mathematical Institute prize.

20. "Having shown how the production of certified knowledge in logic could constitute an object of a sociological investigation and analysis, a vast field of research takes shape" (Rosental 2003). I suspect that identifying and accounting for the priorities expressed by mathematicians themselves would constitute a much richer field of research.

21. The quotation is from Corfield's *Towards a Philosophy of Real Mathematics* (Corfield 2003). Compare it with Ian Hacking's comment that "the most striking single feature of [twentieth century philosophy of mathematics] is that it is very largely banal" (Hacking 2002). For Hacking's philosophy of mathematics, see his *What Mathematics Has Done.*

"Real mathematics," for Corfield, who is remarkably well-informed about trends in the most diverse branches of mathematics, is "real" in the same way as "real ale." I readily agree that skepticism toward this sort of realism is self-defeating.

22. See, however, Hacking: "The truth of a sentence (of a kind introduced by a style of reasoning) is what we find out by reasoning using that style" (Hacking 2002).

23. Many of the authors in Tymoczko (1998) also look to the (real) practice of mathematics for philosophical insight, but Truth and Knowledge keep creeping in. Arriving in France in 1994, I was astonished to discover that the concerns of twentieth-century French philosophers of mathematics are entirely different. Following Husserl, the French concentrate largely on the phenomenological experience of the individual mathematical subject. It is only a slight exaggeration to say that the French-language and English-language traditions in philosophy of mathematics have become mutually incomprehensible. Fortunately, mathematicians writing in French and in English have no trouble citing each others' works.

24. As Serre put it in his comments to *Libération*, "Si vous ne voulez pas que les choses soient parfaites, ne faites pas de maths." Heintz's book is an inquiry into the roots of this apparent universal tendency to consensus, and finds it in the institution of the proof; Rosental treats a (highly unusual) case in which universal consensus apparently failed. The Einstein quotation is in Kline (1980).

of specialized texts in which the term plays a central role. I would like to argue that the nothing designated by "set" is somehow different, and more fruitful, than the nothing designated by "essence." But the means at my disposal for making such an argument take the form of mathematical reasoning, which leads me, at best, to a vicious circle.[25] More bluntly, and for reasons akin to those Serre invoked in his *Libération* interview, I cannot be satisfied with an answer that is less certain than the sort of answer mathematics provides; for a mathematician, a pragmatist answer to Weil's question is an admission of defeat. And yet I am aware that (metaphysically certain) grounds for distinguishing mathematical certainty from pragmatic certainty are lacking!

Another, possibly more profound, reason to steer clear of speculation is that, whereas philosophy presents itself as a dialogue extending over millennia, so that to understand each new contribution one would ideally be familiar with all previous contributions, mathematics is in principle supposed to be derivable by pure reason from a small number of axioms. A philosophical proposition, in other words, remains attached to its origins and context; a mathematical proposition floats free. This principle, an important constituent of the aura of metaphysical certainty surrounding mathematics, does not in fact bear much resemblance to mathematics as it is actually practiced—"one of humankind's longest conversations," as Barry Mazur puts it. I am nonetheless painfully aware that my personal "conversation" with the philosophical tradition is thoroughly unreliable, and my choice of footnotes is primarily the fruit of a random walk (or random surf, or remix) among scraps of the literature I have happened to encounter.

If I am nevertheless writing about philosophy, it is in large part because of a question that was put to me in 1995, during a presentation of Wiles's proof of FERMAT'S LAST THEOREM [V.10] to an audience of scientists. An October 1993 article in *Scientific American* entitled "The death of proof" had called Wiles's proof a "splendid anachronism," citing Laszlo Babai and his collaborators, among others, in support of the thesis that, in the future, deductive proof in mathematics will be largely supplanted by computer-assisted proofs and probabilistic arguments. That same month

the *Notices of the American Mathematical Society* (40: 978–81) published Doron Zeilberger's manifesto "Theorems for a price," predicting a rapid transition from rigorous proofs to an "age of *semi-rigorous* mathematics, in which identities (and perhaps other kinds of theorems) will carry price tags" measured in computer time and proportional to the degree of certainty desired, to be followed in turn by "abandoning the task of keeping track of price altogether, and … the metamorphosis to non-rigorous mathematics" (John Horgan, *Scientific American* October 1993:92–102).[26]

Feeling called upon to answer the question Weil avoided, I argued that the basic unit of mathematics is the concept rather than the theorem, that the purpose of a proof is to illuminate a concept rather than merely confirm a theorem, and that the replacement of deductive proofs by probabilistic or mechanical proofs should be compared not to the introduction of a new technology for producing shoes, say, but rather to the attempt to replace shoes by the sales receipts, or perhaps the cash profits, of the shoe factory. The audience had its own question: Was I talking about certainty? Of course not. That option has been philosophically discredited, as I have tried to explain. And other normative prescriptions fall victim no less easily to the laughter of philosophers. On the other hand, I see no pragmatic reason why probabilistic or mechanical proofs would not suit the five goals on the AMS committee's list just as well as deductive proofs, nor any sociological reason why they should not be as effective in commanding consensus in the event of a paradigm shift. So what was I talking about?

Such a question, at this point in the essay, practically begs to be answered by an advertising slogan. For example:

> The practice of making what one writes "reliable and verifiable" fosters critical thinking in general.

This is a popular argument for teaching proofs, and is probably even true, but how would one go about verifying such a claim? I am very much tempted to say that the concepts that serve as material for "one

---

25. "Truth is always the possibility of its proper destruction," according to the (nonpostmodern) French philosopher, Alain Badiou, taking Gödel's theorem as an example (www.egs.edu/faculty/badiou/badiou-truth-process-2002.html).

26. In the pop posthumanist scenarios promoted by Hans Moravec, Ray Kurzweil, and the like, computers acquire all human capabilities, including the ability to generate and prove theorems—for some reason this is always considered a landmark—by the middle of the twenty-first century. The distinction between humans and computers subsequently fades away rather rapidly, making Zeilberger's prediction moot.

A more recent, and much more nuanced, discussion of prospects for automatic theorem proving has been posted on the Internet by Maggesi and Simpson (undated).

of humankind's longest conversations," deserve to be appreciated on their own terms. Note that nothing is more "emergent" than a conversation. But that would be unfaithful to the spirit of Mazur's book, one of whose strengths is its refusal to conform to a linear narrative. In any case, on its own the argument does not seem to be sufficient: a similar argument could be made in favor of religious faith.

5 "Ideas, Even Dreams"

Rather than hazard an answer to Weil's (non)question here, I will take a cue from Corfield and suggest that one can best account for the value of pure mathematics by attending to what mathematicians write and say. A handful of commonplace words appear consistently, invested with unexpected power, when mathematicians attempt to account, formally as well as informally, for their value judgments, and these collectively constitute an answer to the question Weil left hanging.

WEYL [VI.80] wrote a book with the provocative title *The Idea of a Riemann Surface*[27] and referred in his preface to Plato. The word "concept" that was central in my reply to the audience is closer to this use of the term "Idea" as used by any number of philosophers, including most of those mentioned in this essay. A square, or a RIEMANNIAN MANIFOLD [I.3 §6.10], would be a concept or "Idea" in this sense, and this is how the word "concept" tends to be used by mathematicians, who generally reserve the word "idea" to designate something else. In Plato's *Meno*, the proof of the doubling of the square—draw diagonals and fit the resulting triangles together—which the slave "remembers" under Socrates' coaching, is taken by Plato to be contained in the "Idea" of the square. For a mathematician, drawing the diagonals and moving the triangles *are* the ideas.

That a contrast can be drawn, as I did in 1995, between "illuminating concepts" and "confirming theorems" is something of a truism among mathematicians and even some philosophers. Even by 1950 Popper had argued that "a calculator ... will not distinguish ingenious proofs and interesting theorems from dull and uninteresting ones" (quoted in Heintz 2000). Corfield correctly states that "what mathematicians are largely looking for from each other's proofs are new concepts, techniques, and interpretations"; they are not

merely "establishing the truth or correctness of propositions" (p. 56). However, although he devotes a chapter to the "extremely complex subject" of "mathematical conceptualization," he does not dwell on concepts (or "Ideas") as such; and neither will I. It is almost impossible to talk in general terms about mathematical concepts without getting caught up in the debate over their reality (and provoking the laughter of the philosophers). Those who write about mathematics (mathematicians included: see Hersh (1997)) have an irritating tendency to claim that most mathematicians are Platonists, whether or not they have committed themselves explicitly to a philosophical position. Maybe it can be argued that Platonism is implicit in the syntax of mathematical statements; maybe this is what Weil meant by his claim, quoted by Bourguignon (2001), that most mathematicians "spend a good portion of their professional time behaving as if they were [Platonists]."[28] In practice I would guess that most mathematicians are pragmatists, in the spirit of the remarks of Dieudonné quoted above.

On the other hand, there is no doubt whatsoever that the "ideas" that matter to mathematicians are real. A mathematician, according to a joke attributed to Weil,[29] can be defined as someone who has had two ideas (mathematical, of course). But then, Weil worried, so-and-so would be a mathematician. In a celebrated account by POINCARÉ [VI.61] of the role of the unconscious in a mathematical discovery, the climactic moment was the arrival of an idea ("the idea came to me") as he placed his foot on the steps of the omnibus ("L'idée me vint," Poincaré (1999)).

More to the point, consider Hacking's justification of his own commitment to a realist ontology of electrons: "*So far as I'm concerned, if you can spray them then they are real*" (Hacking 1983). By the same token, if you can steal ideas, then they are real. Every mathematician knows that ideas can be and often are stolen. Polemics then ensue, considerably juicier than the epistemic controversy studied by Rosental.

Nothing in the life of mathematics has more of the attributes of materiality than (lowercase) ideas. They have "features" (Gowers 2002), they can be "tried out"

---

27. Weyl used the word *Idee* in his title but applied the term *Begriff* (concept) elsewhere in the text. Both terms arrived in English as "concept."

28. Plato saw things quite the other way around: "Their language [speaking of mathematicians] is most ludicrous, *though they cannot help it*, for they speak as if they were doing something and as if all their words were directed toward action" (*Republic* VII.527a, my emphasis).

29. I heard this joke reported by several people who claimed to have heard it from Shimura, and I believe but am not certain that I too first heard it from Shimura.

(Singer[30]), they can be "passed from hand to hand" (Corfield 2003), they sometimes "originate in the real world" (Atiyah in the preface to Arnold et al. (2000)) or are promoted from the status of calculations by becoming "an integral part of the theory" (Godement 2001). At some point they come into being: it is generally understood, for example, that "new ideas" will be needed to solve the Clay Millennium Problems. They can also be counted. I once heard Serre introduce the proof of a famous conjecture by saying that it contained two or three real ideas, where "real" was intended as high praise. The ambiguity did not concern the number of ideas—there were three, which Serre enumerated—but whether all three were original to the author. Ideas are public: necessarily so, in order to be stolen, or to be presentable as Serre did in his lecture. Poincaré's idea was a sentence ("the transformations of which I had made use to define Fuchsian functions were identical to those of non-Euclidean geometry"); the slave's idea in *Meno* was a line in the sand.

Early in his unpublished memoirs *Récoltes et Semailles*, Grothendieck wrote that "ideas, even dreams" were, in Allyn Jackson's terminology, the "essence and power" of his mathematical work (Jackson 2004). An idea is typically symptomatic of "insight," and the capacity for insight is generally called "intuition." Mathematicians have borrowed all of these terms from philosophy but use them to completely different ends. Philosophers tend to follow Kant in attributing intuitions—the ones that without concepts are blind—to transcendental subjects or their more down-to-earth offspring. Intuition in this sense is a poor substitute for certainty, as even the mavericks recognize. "Intuition … is frequently a *prelude* to mathematical knowledge," wrote Kitcher. "By itself it does not warrant belief." Poincaré called intuition "the tool of invention," a "je ne sais quoi" that holds a proof together, but he contrasted it with logic, "the tool of demonstration," which "alone can provide certainty." Saunders Mac Lane expressed himself in much the same terms nearly a century later. David Ruelle considered reliance on (visual) intuition a characteristic feature of human (as opposed to extraterrestrial) mathematics.[31]

In each case intuition belongs to the *private* sphere, and is relegated to the "context of discovery," as opposed to the "context of justification" deemed worthy of philosophy's full attention. When mathematicians refer to "intuition" in the sense I have in mind, it is crucially *public*.[32] As in the quotation from MacPherson a few paragraphs back, it can be transmitted from teacher to student, or through a successful lecture, or developed collectively by running a seminar and writing a book on the proceedings. It has something in common with a "style of reasoning," but on a smaller scale. Grothendieck resorted to perceptual metaphor when describing Serre's ability to communicate something akin to intuition:

> The essential thing was that Serre each time strongly sensed the rich meaning behind a statement that, on the page, would doubtless have left me neither hot nor cold—and that he could "transmit" this perception of a rich, tangible, and mysterious substance—this perception that is at the same time the *desire* to understand this substance, to penetrate it.
>
> *Récoltes et Semailles*, p. 556

"Even those who try to articulate, to classify, the fruits of the imagination, and who are committed to the existence of an inner experience concomitant with it, admit to dark difficulty in describing it," wrote Mazur, elaborating an unusual array of literary and rhetorical strategies to chip away at the difficulty (Mazur 2003). This much is certain: this inner experience of imagination, or of understanding, is what drives people to become mathematicians, and it is why Weil could count on his audience's silent assent. Heintz recorded some of her informants' attempts to describe this inner experience: "[In mathematics] you have concrete objects before you and you interact with them, talk with them. And sometimes they answer you." She even talks about the "idea" that helps put the pieces together. "And suddenly you see the picture," she was told. Yet all this raw ethnographic data is presented in a chapter whose title, "Beauty and experiment: discovery of truth in mathematics," betrays her relentlessly epistemological preoccupations (Heintz 2000).

"The specific ways that mathematical truths move from person to person, and how they are transformed in the process, are as difficult to capture as the truths themselves," wrote Mazur (2003), in what could have

---

30. Quoted at www.abelprisen.no/en/prisvinnere/2004/interview_2004_7.html.

31. Kitcher (1984, p. 61); Poincaré (1970, pp. 36–37); Mac Lane, in his contribution to the discussion of the Jaffe–Quinn article cited in note 15, *Bulletin of the American Mathematical Society* 30 (1994): 178–207; Ruelle's quote is from an article entitled "Conversations on mathematics with a visitor from outer space" from Arnold et al. (2000).

32. This is also true of the normative program of intuitionism associated with BROUWER [VI.75], but that is definitely not what I have in mind.

been a comment on Grothendieck's remarks on Serre. The central notion in Mazur's book is that of "imagination." I have chosen the terms "idea" and "intuition" not for their intrinsic importance, though I believe each of the terms points to ways of talking about the famous "flash in the middle of a long night" that ends Poincaré's *The Value of Science*: "But this flash is everything." What strikes me about these terms is how their pervasiveness in mathematicians' conversations—the sense that they, more than the definitive theorems, are "everything"—contrasts so starkly with their near exclusion from philosophical consideration, even though the words themselves can be seen on practically every page of philosophy of mathematics. Maybe their very banality makes them appear philosophically trivial. Or maybe the problem is that the same words serve so many distinct purposes. Corfield uses the same word to designate what I am calling "ideas" ("the ideas in Hopf's 1942 paper") as well as "Ideas" ("the idea of groups") and something halfway between the two (the "idea" of decomposing representations into their irreducible components for a variety of purposes, p. 206). Elsewhere, the word crops up in connection with what mathematicians often refer to as "philosophy," as in the "Langlands philosophy" ("Kronecker's ideas" about divisibility, p. 202), and in many completely unrelated places as well. Corfield proposes to resolve what he sees as an anomaly in Lakatos's "methodology of scientific research programmes" as applied to mathematics by "a shift of perspective from seeing a mathematical theory as a collection of statements making truth claims, to seeing it as the clarification and elaboration of certain central ideas" (p. 181). He sees "a kind of creative vagueness to the central idea" in each of the four examples he offers to represent this shift of perspective; but on my count the ideas he chooses include two "philosophies," one "Idea," and one which is neither of these.

Other value-laden terms are no less important. In the wake of BOURBAKI [VI.96], quite a few philosophers (Cavaillès, Lautman, Piaget, and more recently Tiles) have made serious attempts to make sense of "structure" in mathematics. I have read a number of philosophical attempts to account for mathematical aesthetics, though none has left much of an impression. The practically universal use of dynamical or spatiotemporal metaphors ("the space X is fibered over Y," etc.), and the pronounced tendency to present proofs as series of actions playing out in time ("now choose an orbit passing arbitrarily close to the point x") have attracted little attention from philosophers.[33] These phenomena may be linked to the curious preference of many mathematicians for blackboards over contemporary audiovisual technology, which in turn draws attention to the neglected (and emergent) aspect of mathematical communication as *performance*, a word that manages to be typically postmodern and premodern at the same time.

For his part, Corfield does not talk much about "intuition" and is ambiguous about what he means by "ideas," but his discussions of "natural" and "importance," in the context of an analysis of the debate on the relative merits of groups and groupoids, are philosophically insightful while remaining faithful to the use of the terms by "real" mathematicians. His treatment of "postmodern algebra," where "diagrams are not just there to illustrate, they are used to calculate and to prove results rigorously" (p. 254), also has street credibility. It is true that much of his book remains concerned with "maverick" questions, such as accounting for plausible reasoning. But there is no question that Corfield likes mathematics, and for the right reasons; his book, unlike most treatises in philosophy of mathematics, is definitely part of the "conversation."

Morris Kline called the "loss of certainty" entailed by Gödel's theorems an "intellectual tragedy" and actually counseled "prudence" in designing bridges "using theory involving infinite sets or the axiom of choice" (Kline 1980). The word "tragedy" seems misplaced but the pathos is real, as it was for Russell. Pathos and its twin, resolute optimism, have found an unlikely home in the philosophy of mathematics:

> If this conception of mathematics [as "human knowledge of structures gained by employing reason beyond the bounds of logic"] can be sustained, mathematics could once again serve as a source of an image of reason liberated from formal imprisonment, freed to confront apocalyptic post-modern visions.

Mary Tiles, *Mathematics and the Image of Reason*, p. 4 (Routledge, London, 1991)

Whether or not it carries weight with congressional committees, I find this goal appealing, but it is a goal for philosophers, not for mathematicians. I'm willing to apply the "principle of charity" to philosophers if they will do the same for me. Corfield wrote (p. 39):

---

33. Nuñez's article "Do *real* numbers really move?" (in Hersh 2006) makes interesting points regarding mathematicians' use of metaphors of motion, though he limits his analysis to examples specifically related to the mathematics of motion. Plato specifically disapproved of mathematicians' use of action metaphors.

Human mathematicians pride themselves on producing beautiful, clear, explanatory proofs, and devote much of their effort to reworking results in conceptually illuminating ways. Philosophers must not evade their duty to treat these value judgments in mathematics.

They also have a duty, it seems to me, to account for terms like "idea" and "intuition"—and "conceptual" for that matter. An answer to the question "Why philosophy?" might well begin there.

Postscript

In December 2004 my university joined a number of other institutions in France and elsewhere in hosting a traveling UNESCO-sponsored exhibition entitled "Pourquoi les mathématiques?" Hoping to learn the answer before my submission deadline, I spent a few hours at the exhibition. It was clever and engaging, presenting a variety of (pure) mathematical ideas with a sprinkling of practical applications, but in no way did it address the "Pourquoi?" of the title. An organizer was on hand, and when I turned to her for guidance she explained that the French title was a solution to a problem of translation. The English title, which came first, was "Experiencing mathematics." This, she assured me, has no adequate French translation, so "Pourquoi les mathématiques?" was chosen as the best substitute.

Maybe the solution to the problem of my title is simply to accept the translation in the opposite direction. Even the most ruthless funding agency is not yet so post-human as to require an answer to "Why experience?" [34]

Acknowledgments. I thank Cathérine Goldstein and Norbert Schappacher for pointing me in the directions of the Rosental and Heintz books, among other source material, and for vigorously criticizing my project as well as its execution. I also thank Mireille Chaleyat-Maurel for explaining the title of the UNESCO exhibition and Ian Hacking for critically reading an earlier version of the manuscript with tolerance and rigor. David Corfield receives thanks for several helpful clarifications. Barry Mazur is thanked especially warmly for many suggestions, much encouragement, for help with the title, and most of all for showing, in his *Imagining Numbers*, that there is at least one way out of the fly-bottle.

---

34. Or, as Weyl put it, "with [mathematics] we stand precisely at the point of intersection of restraint and freedom that makes up the essence of man itself." Note the word "essence" (see Mancosu 1998). I thank David Corfield for this quotation.

Further Reading

Arnold, V., et al. 2000. *Mathematics: Frontiers and Perspectives*. Providence, RI: American Mathematical Society.

Barthes, R. 1967. *Système de la Mode*. Paris: Éditions du Seuil.

Bloor, D. 1976. *Knowledge and Social Imagery*. Chicago, IL: University of Chicago Press.

Bourguignon, J.-P. 2001. A basis for a new relationship between mathematics and society. In *Mathematics Unlimited—2001 and Beyond*, edited by B. Engquist and W. Schmid. New York: Springer.

Corfield, D. 2003. *Towards a Philosophy of Real Mathematics*. Oxford: Oxford University Press.

Godement, R. 2001. *Analyse Mathématique I*. New York: Springer.

Gowers, W. T. 2002. *Mathematics: A Very Short Introduction*. Oxford: Oxford University Press.

Hacking, I. 1983. *Representing and Intervening*. Cambridge: Cambridge University Press.

———. 2000. What mathematics has done to some and only some philosophers. *Proceedings of the British Academy* 103:83–138.

———. 2002. *Historical Ontology*. Cambridge, MA: Harvard University Press.

Harvey, D. 1989. *The Condition of Postmodernity*. Oxford: Basil Blackwell.

Heintz, B. 2000. *Die Innenwelt der Mathematik*. New York: Springer.

Hersh, R. 1997. *What Is Mathematics, Really?* Oxford: Oxford University Press.

———, ed. 2006. *18 Unconventional Essays on the Nature of Mathematics*. New York: Springer.

Jackson, A. 2004. Comme appelé du néant—as if summoned from the void: the life of Alexandre Grothendieck. *Notices of the American Mathematical Society* 51:1038.

Kitcher, P. 1984. *The Nature of Mathematical Knowledge*. Oxford: Oxford University Press.

Kline, M. 1980. *Mathematics: The Loss of Certainty*. Oxford: Oxford University Press.

Lakoff, G., and R. E. Núñez. 2000. *Where Mathematics Comes From*. New York: Basic Books.

Lloyd, G. E. R. 2002. *The Ambitions of Curiosity*, p. 137, note 13. Cambridge: Cambridge University Press.

Lyotard, J.-F. 1979. *La Condition Postmoderne*. Paris: Minuit.

Maggesi, M., and C. Simpson. Undated. Information technology implications for mathematics, a view from the French Riviera. (This paper is available at http://math1.unice.fr/~maggesi/itmath/; apparently not posted before 2004.)

Mancosu, P., ed. 1998. The current epistemological situation in mathematics. In *From Brouwer to Hilbert. The Debate on the Foundations of Mathematics in the 1920s*. Oxford: Oxford University Press.

Mazur, B. 2003. *Imagining Numbers (Particularly the Square Root of Minus Fifteen)*. New York: Farrar Straus Giroux.

Minsky, M. 1985/1986. *The Society of Mind*. New York: Simon and Schuster.

Poincaré, H. 1970. *La Valeur de la Science*. Paris: Flammarion.

——. 1999. *Science et méthode*. Paris: Éditions Kimé.

Rosental, C. 2003. *La Trame de l'Évidence*. Paris: Presses Universitaires de France.

Tymoczko, T., ed. 1998. *New Directions in the Philosophy of Mathematics*. Princeton, NJ: Princeton University Press. (First published in 1986.)

Wittgenstein, L. 1958. *Philosophical Investigations*, volume I. Oxford: Basil Blackwell.

——. 1969. *On Certainty*. Oxford: Basil Blackwell.

VIII.3 The Ubiquity of Mathematics
T. W. Körner

1 Introduction

We live surrounded by mathematics: when we open a door or use a nutcracker, we exploit ARCHIMEDES' [VI.3] law of the lever; when a bus goes around the corner, we experience at first hand NEWTON's [VI.14] law that a body continues to travel in uniform motion in a straight line unless acted on by an external force; when we use a rapidly accelerating elevator, we can feel for ourselves the equivalence of gravitational and accelerational inertia that lies at the heart of GENERAL RELATIVITY [IV.13]; and when we run a tap fast into a kitchen sink we see a thin and flat circle of water with a clear boundary, which is the chaotic "hydraulic jump" between two well-behaved solutions of a certain PARTIAL DIFFERENTIAL EQUATION [I.3 §5.4].

Because mathematics and physics are so interlinked, almost everything we see involves mathematics. With the help of elementary calculus, we know that a baseball, after it leaves the bat, will have a trajectory in the shape of a parabola. This calculation assumes that there is no air resistance, but a more complicated calculation can take air resistance into account too. If a chain hangs between two points, then the curve it forms can again be explained mathematically. This time, the technique used is the CALCULUS OF VARIATIONS [III.94]: the curve is the one that minimizes the potential energy of the chain, and the calculus of variations allows you to work it out. (It is called a *catenary*. The rough idea of the calculation is to consider small perturbations of the chain. Since the potential energy is minimized, we know that however we perturb it, we cannot decrease the potential energy. This information can be used to derive a differential equation that determines the curve. In general, the differential equations that arise from this technique are called the *Euler–Lagrange equations*.) Even the way that wet sand behaves when you walk across it involves interesting mathematics, as Reynolds realized in 1885. Typically, the sand just around where you tread dries out—if you have not noticed this strange phenomenon, then have a look next time you are on a beach. The reason this occurs is that when the tide goes out the sea tends to leave the grains of sand extremely well-packed. If you then tread on the sand, you disturb this packing, creating a less well-packed part of the sand near where you tread. This has more room for water, so it draws water in and down, temporarily drying out the sand around your foot.

It would be easy to give hundreds more examples of physical phenomena that can be analyzed mathematically. However, if one accepts that physics governs the universe and that mathematics is the language of physics, then it is not surprising that these applications exist. Therefore, this article will focus on the appearance of mathematics in other areas, and in particular geography, design, biology, communication, and sociology.

2 Uses of Geometry

If you travel about on Earth's surface, then you need to make small adjustments to your watch as you move from one time zone to another. There is one exception to this, however: if you cross the international date line, then you have to make a big adjustment (assuming, that is, that your watch shows not just the time but the date as well). Why is it necessary to have a discontinuity of this kind? Well, suppose that it is midnight on a Tuesday in Lisbon, for example, and imagine a path that goes westward right around the globe. If the time changes along this path are all small ones that reflect where one is in relation to the sun, then the time of day goes back by one hour for every 15 degrees of longitude that we move. Therefore, when one gets back to Lisbon it is midnight on Monday. (Remember that we are talking about a *mental* path here, and not an actual journey.) Something is clearly not right. The practical consequences of this theoretical problem were first felt by the tattered remnants of Magellan's first circumnavigation of the globe who had to do penance for performing religious ceremonies on the wrong day!

Here is another argument for the necessity of the date line. Let us ask exactly when the year 2000 began. The answer depends, of course, on what part of the world you are talking about, and more particularly on its longitude, but for any part the answer is midnight at the beginning of January 1. In other words, in any particular place the year began when the Sun was (approximately) over the opposite side of the world. It follows that at any given time at most a small fraction of the world was celebrating the very beginning of the year 2000. Therefore, at least somewhere had to go first, which means that parts of the world just to the east of it had missed their chance and had to wait almost 24 hours. Thus, again we see that there has to be a discontinuity.

These phenomena reflect the fact that a certain continuous map has no continuous inverse. The map in question takes a real number w to the point $w \mapsto (\cos w, \sin w)$, which lives in the unit circle. Notice that if we add 2π to w then we do not affect the values of $\cos w$ and $\sin w$. Now let us try to invert the map. This means that for each point (x, y) in the unit circle we must pick some w such that $\cos w = x$ and $\sin w = y$. This w is the angle that the line from 0 to (x, y) makes with the horizontal, with the all-important proviso that you can add any multiple of 2π to it. So the question becomes, can we choose the appropriate multiple in a continuous way? Again, the answer is no, since if you go around the circle once and let the angle vary continuously, you find that you have added 2π to it.

The above fact is one of the simplest theorems of TOPOLOGY [IV.6], the branch of mathematics that you turn to if you want to know about the existence or non-existence of continuous functions with given properties. Another situation where continuous functions are useful is when one is creating a map (in the geographer's sense) of the world. Such maps are more convenient if they are drawn on a flat piece of paper, so a preliminary question we might ask is whether there is a continuous function from the surface of the sphere to the plane such that any two different points in the sphere go to different points in the plane. Not only is the answer no, but *Borsuk's antipodal theorem* tells us that there must be some pair of *antipodal* points (that is, points of the sphere that are exactly opposite each other, such as the North and South Poles) that go to the same point in the plane.

However, perhaps we do not mind too much about continuity. If we take our sphere and make a cut from the North Pole to the South Pole, then we can open it up at the cut and flatten it out onto a plane. (To see this, imagine that it is made of particularly stretchy rubber.) Alternatively, we could cut the sphere into

two hemispheres and draw maps of each hemisphere separately.

Now another problem arises: it does not seem to be possible to draw a map of even half the world without distortions. This is not a topological problem, but a *geometrical* one, in the sense that we are interested in finer properties of Earth's surface—shape, angle, area, and so on—than those that are preserved by continuity. Because the sphere has positive CURVATURE [III.78], no part of it can be mapped to the plane in a length-preserving manner, so some distortion is necessary. However, we have a certain amount of freedom to decide what kind of distortion we are prepared to accept and what kind we would like to avoid. There is, it turns out, a *conformal map* from the sphere (minus the poles) to a cylinder (which one can cut and roll out so that it fits into a plane)—it is the famous "Mercator projection." A conformal map is one that preserves angles, so the Mercator projection is particularly useful for navigation purposes: if it looks as though you need to head north-northwest, then you really do. A disadvantage of the Mercator projection is that as you move away from the equator, the countries look bigger and bigger (though the angle-preserving property means that in close-up they are always the right shape). There is another projection that distorts shapes but preserves area. To work out the details of these projections, one must use mathematics—and in particular solve differential equations.

Here are a few simple applications of geometry to everyday life. If you have ever wondered what the best shape is for a manhole cover, then mathematics can come to your aid. Of course, it depends what one means by "best," but if you often need to lift manhole covers, then you may be annoyed if they keep falling down the manholes. Can this be avoided? If the cover is rectangular, then the length of any side is less than the length of the diagonal, so it can drop down the hole, but if it is circular, then its width is the same in all directions and this is not possible.

Does this mean that only circular manhole covers are safe from dropping down their manholes? Actually, no. If you draw the three vertices of an equilateral triangle and join each pair of them by a circular arc centered on the third, then you obtain a sort of "curved triangle," known as the *Reuleaux triangle*. (This is commonly misspelt "Rouleaux" in the mistaken belief that it has something to do with rolling. Actually, it is named after a nineteenth-century German engineer called Franz Reuleaux.)

Have you ever wondered why coins are the shapes they are? Most of them are circular, but the British fifty pence piece, for example, is a slightly curved polygon with seven sides. A moment's thought makes it clear that for any odd number $n \geqslant 3$ you can have a Reuleaux polygon with n sides, and the fifty pence piece is indeed a Reuleaux heptagon. This is convenient for slot machines: it means that you can have a slot into which the coin only just fits, however you push it in.

What about the best shape for a conveyor belt? If we construct it in the obvious manner, then one of its two sides will be exposed and the other not. Eventually, the exposed side will wear out, but the other side will be in pristine condition, since it will not have been used at all. However, as any mathematician will tell you, not all surfaces have two sides. The most famous example of a one-sided surface is the MÖBIUS STRIP [IV.7 §2.3], which is obtained from a flat strip of paper by twisting one end through 180 degrees and joining it to the other end. If you have a long enough conveyor belt for it to be practical to give it a twist somewhere, then you can wear out both sides equally (this makes sense locally even if globally the belt now has just one side), thereby doubling the use you get out of the belt. (You might think it simpler just to turn the belt over after a while, but the Möbius-strip design has been taken seriously enough to be patented, and similar designs have been used as typewriter ribbons and in tape recorders.)

3   Scaling and Chirality

Why are Arctic mammals large? Is it just a fluke that they have evolved that way? This does not sound like a mathematical question, but some simple mathematics can easily convince us that it is not a fluke at all. Since the Arctic is cold and animals need heat, animals that are better at preserving heat are more likely to thrive. The rate at which an object loses heat is proportional to its surface area, but the rate at which it *generates* heat is proportional to its volume. So if you double the size of an animal in every direction, then the rate at which heat is generated goes up by a factor of eight, while the rate at which it is lost goes up by a factor of only four. That is, larger animals find it easier to preserve heat.

But why, in that case, are Arctic animals not much bigger still? This can be explained by a similar scaling argument. If you scale an animal up by a factor of t, then its volume, and hence its weight (animals, being made predominantly of water, tend to have roughly the same density), will multiply by t^3. The animal has to support this weight with its bones. The amount of force you need to snap a bone is roughly proportional to the area of a cross-section of that bone, and areas go up by a factor of t^2. So if t is too large, the animal will not be able to support its own weight. It does have the option of increasing the relative thickness of its bones, but if t is very large then its legs will be too thick for this to be a practical solution.

A similar sort of scaling argument explains why, if you drop a mouse down a 1000 foot mine shaft, then, to quote Haldane, "on arriving at the bottom, [it] gets a slight shock and walks away." In this case, air resistance is roughly proportional to surface area, while the gravitational pull is proportional to mass, and therefore to volume. It follows that, the smaller you are, the smaller your terminal velocity, and the less you are bothered by a fall.

A simple fact with many scientific ramifications is that two shapes can be reflections of each other without being rotations or translations. For example, if you see a hand without seeing the body to which it is attached, then you can tell whether it is a right hand or a left hand. (If you can shake it naturally with your right hand, then it is a right hand.) This phenomenon is known as *chirality*: a shape is *chiral* if it cannot be obtained from its mirror image by rotation or translation.

The notion of chirality appears in many parts of science. For example, many elementary particles have a fundamental property known as "spin," which means that they often have right-handed versions and left-handed versions. In pharmacology, it is now understood that many molecules are chiral, and that the two different versions can have radically different properties. An example that had tragic consequences is the drug thalidomide: one form of it is effective against morning sickness while the other causes birth defects. Unfortunately, in the late 1950s several thousand pregnant women were given a 50:50 mixture of the two forms. Less harmful examples of the importance of chirality abound. For instance, there are many chemicals that smell or taste different when you look at their reflected versions. (This may seem paradoxical, but the explanation is simple: the sensors in our noses and mouths also contain molecules with chirality.)

So far we have been considering rigid motions, but some shapes are chiral in the stronger sense that not even a *continuous* motion in space is enough to turn them into their mirror images. Two interesting examples are the TREFOIL KNOT [III.44], which comes in a "right-handed" and a "left-handed" version (the proof that these two versions are genuinely distinct is not at all easy), and the Möbius strip, which was mentioned earlier. The rough reason that the Möbius strip is chiral is that when you do the twist, you do it either according to the "corkscrew rule"—that is, twisting it as if you were pushing a corkscrew into the cork—or the opposite way. If you try to visualize it, you may be able to convince yourself that the direction of twist is not altered by continuous deformations, and also that the mirror image of a Möbius strip that obeys the corkscrew rule is a Möbius strip that does not obey the corkscrew rule.

4   Hearing Numerical Coincidences

Legend has it that Pythagoras, passing a blacksmith hammering a set of iron bars in a particularly pleasing way, was led to discover the laws of harmony. In modern terms, these laws say that two sounds go together particularly well (at least in the European tradition) if their frequencies are in the ratio r to s for some pair of small integers r and s: the smaller the better. As a result, people have tried to devise musical scales that have as many of these pleasing intervals as possible.

Unfortunately, there are limits to how well you can do. If you take a very simple ratio such as $3/2$, which corresponds to what musicians call a perfect fifth, then its powers—$9/4$, $27/8$, $81/16$, and so on—get successively more complicated. However, by great good fortune it happens that 2^{19} is rather close to 3^{12}. To be precise, $2^{19} = 524\,288$ and $3^{12} = 531\,441$, which is a difference of about 1.4%. It follows that $(3/2)^{12}$ is close to 2^7. Since doubling a frequency raises the note by an octave, this says that twelve perfect fifths make an interval close to seven octaves. This allows one to build up a scale in which the fifths are *approximately* perfect.

There are many ways of doing the approximation. Early choices of musical scale would make some of the fifths perfect, at the expense of others. The modern compromise adopted by Western music for the last 250 years is to distribute the inaccuracies equally. If successive notes in a musical scale have frequencies in the ratio 1 to α, then starting from a frequency u the notes will have frequencies u, αu, $\alpha^2 u$, and so on. If you want k notes in the scale, then α^k should equal 2 (so that after k steps you have gone up by an octave). This means that all smaller powers of α must be irrational, so that all the other intervals in the scale are inharmonious! However, when $k = 12$, the fact that 3^{12} and 2^{19} are close has the consequence that α^7, which

equals $2^{7/12}$, is close to $3/2$ (more precisely, it is just over 1.4983), which means that all the fifths are close to perfect.

Tuning systems are discussed in more detail in MATHEMATICS AND MUSIC [VII.13 §2].

5   Information

Few things illustrate better how the abstract mathematical theory of one generation can become the common sense of the next than the following two closely related ideas: that all information can be expressed as a series of 0s and 1s, and that the "quantity of information" carried by a book, a picture, or a sound is proportional to the number of 0s and 1s required to express it.

A famous theorem of Shannon (described in RELIABLE TRANSMISSION OF INFORMATION [VII.6 §3]) tells us that the rate at which information can be transferred by signals depends on the range of frequencies available. For example, it is the change from signaling electrically along copper wires (with a narrow range of frequencies) to signaling by light (with a very wide range) that has allowed the massive data transfers required by the Internet. The sound waves we hear belong to a very narrow range of frequencies, while the light waves that we see belong to a wide range, and this is why we need much more memory on our computers to store an hour of film than an hour of music. Similarly, it may feel as though visual perception is a passive process—we point our eyes in a certain direction, they behave a bit like video cameras, and we just watch the video—but because light carries so much information, our brains actually have to resort to a wide variety of tricks to deal with it. What we think we see is actually a theatrical representation of reality that our brains have cunningly manipulated. This is why there are optical illusions, and why they continue to work even when you know how they work. By contrast, since sound carries so little information, our brains can process it in a much more direct way (though still not completely direct—there are aural illusions too, and the brain has tricks that help us to pick out the information we are actually interested in from all the sound waves that enter our ears).

When information is transmitted, there are almost always faults in the transmission system, so that our messages are not transmitted perfectly. How do we then recover the messages? Here is a Victorian parlor trick that shows how in a very simple case. One begins by writing down all sequences $(x_1, x_2, \ldots, x_7)$

such that every x_i is either 0 or 1 and such that the numbers $x_1 + x_3 + x_5 + x_7$, $x_2 + x_3 + x_6 + x_7$, and $x_4 + x_5 + x_6 + x_7$ are all even. An example of such a sequence is $(0, 0, 1, 1, 0, 0, 1)$.

If you think of these sequences as vectors in the vector space $\mathbb{F}_2^7$ (that is, the seven-dimensional space where the scalars belong to the field of integers mod 2), then you will readily convince yourself that these three properties of a sequence are independent linear conditions, so the set of sequences in question is a four-dimensional subspace of $\mathbb{F}_2^7$. Therefore, there are sixteen such sequences. A member of the audience is asked to take one of them and change it in one place. The magician can at once identify which digit has been changed. Let us see how this works if we change the third digit of the sequence above, so we now have the sequence $(y_1, \ldots, y_7) = (0, 0, 0, 1, 0, 0, 1)$.

The first step is to note that $y_1 + y_3 + y_5 + y_7$ and $y_2 + y_3 + y_6 + y_7$ have become odd, while $y_4 + y_5 + y_6 + y_7$ is still even (since it is y_3 that has changed). Now the only number that belongs to the first two of the sets $\{1, 3, 5, 7\}$, $\{2, 3, 6, 7\}$, and $\{4, 5, 6, 7\}$ but not the third is 3. This tells us that x_3 is the variable that has been changed. How are the sets chosen so that this sort of argument will always work? The answer becomes clearer if we use the binary representations of the integers instead and put in a couple of leading zeros. Then the sets are $\{001, 011, 101, 111\}$, $\{010, 011, 110, 111\}$, and $\{100, 101, 110, 111\}$ and we see that the ith set is the set of integers with a 1 in the ith digit from the end. So if we know which of the three parities have been changed, then we know the binary representation of the place where the sequence was altered. Therefore, we can reconstruct the original sequence.

This trick, rediscovered by Hamming, is the ancestor of all the error-correcting methods (also discussed in RELIABLE TRANSMISSION OF INFORMATION [VII.6]) that allow our CDs and DVDs to perform flawlessly even if they are slightly scuffed.

The fact that there is a precise mathematical way of measuring information content is of considerable importance in genetics. It has been suggested that the amount of information carried by our DNA, though very large, is much smaller than the information required to describe our bodies completely. This would explain what experimental evidence also corroborates: that the DNA carries a set of general instructions, but the fine detail of our anatomy, such as our fingerprints and the precise arrangements of our capillaries, is partly a matter of chance. So, for example, if it were possible to

rerun the growth of the fertilized egg that ended up as you, the result would be broadly similar to you, but small environmental differences would result in a different set of fingerprints and a different arrangement of capillaries.

Under certain circumstances, it is not enough just to transmit information: it must also be protected. If we send our credit card number over the Internet, we want to do so in such a way that it would be very hard for an eavesdropper to find that number. A mathematical way of doing this is described in CRYPTOGRAPHY [VII.7 §5].

Here is a slightly different but closely related problem. Suppose that Albert has a secret that he would like to share with Bertha (and only Bertha) in a conversation that everyone can hear. What is he to do? A first step is to think of *any* piece of information that they can share secretly—it turns out to be a short step from this to sharing a particular piece of information. The following procedure achieves this. First, Albert shouts out a large integer n and an integer u. Next, he chooses a large integer a, which he keeps secret (including from Bertha—obviously, since he does not yet know how to share secrets with her), and shouts out the value of u^a modulo n. Bertha then chooses an integer b, which she keeps secret, and shouts out the value of u^b modulo n. Now Albert is in a position to work out $u^{ab} = (u^b)^a$ modulo n, since Bertha has told him u^b and he knows a. Similarly, Bertha can use her secret number to work out $u^{ab} = (u^a)^b$ modulo n. Albert and Bertha now both know the number u^{ab} modulo n. This is a good example of a shared secret, because all that the eavesdroppers know is u^a, u^b, and n, and when n is large there is no known way of calculating u^{ab} modulo n from u^a and u^b modulo n, apart from methods that take far too long to be practical.

Now suppose that Albert wants to send a credit card number N to Bertha. Assuming that $1 \leqslant N \leqslant n$, then all he needs to do is shout out the number $u^{ab} + N$ modulo n. Bertha then subtracts the secret number u^{ab} and obtains N. (Albert should convey only one secret this way, or he will reveal information. For instance, if he sent another credit card number M using the same u^{ab}, then the eavesdroppers would know the value of $M - N$. But if he and Bertha choose new numbers n, u, a, and b and use those to share the value of M, the eavesdroppers will effectively know nothing about the pair (M, N).)

Why do we believe that it is "hard" to calculate u^{ab} from u^a and u^b? What if tomorrow somebody discovers a simple trick for doing it? Surprisingly, even though we cannot be absolutely sure that the problem *is* hard, there are very precise ways of discussing the question. In particular, there are extremely plausible conjectures, the truth of which would imply that it really is impossible to calculate u^{ab} in a short time. These issues are discussed in considerable detail in COMPUTATIONAL COMPLEXITY [IV.20].

6   Mathematics in Society

A street in which all houses have front gardens is much prettier than a street in which all those front gardens have been converted into parking places. For some people, aesthetics are more important than convenience, so the effect of converting all the front gardens in a street may well be to reduce the values of all the houses. However, if you convert just one front garden, then it will increase the convenience for that household without making too much of a difference to the look of the street, so the value of that house will increase and the values of all the other houses will decrease slightly. Thus, for each individual house owner there is a financial incentive to convert the front garden, even though if everybody does so then everybody will lose financially.

Clearly, to avoid this unfortunate result the households must cooperate. Nash has shown how, starting from simple assumptions about fairness, there must be a system of mutual payments—for example, a household that wishes to convert its front garden might have to pay a charge that was shared between the other households—which will change their incentives in such a way that they will no longer want to ruin the street.

If the households do not wish to cooperate, Nash has shown that they come to a (usually less favorable) agreement which it is not in the interest of any single individual to break. A simple example of a situation in which no single individual may wish to change but a group acting in concert may wish to change is given by the following game. Suppose that three people hand to an umpire an envelope containing either the word "yes" or the word "no." If two players have written the same thing and the third has not, then those two players get $400 each and the third player gets nothing. However, if all three have written the same, then all three players get $300. Suppose that the players meet before the game and agree that they will all write "yes" (in order to maximize their average gain). Then no single player will gain by writing "no" instead, but if two players decide to change then they will both gain.

Nash's ingenious argument starts with an agreement that is not necessarily in equilibrium, and allows the parties to the agreement to modify their actions very slightly in a way that would improve their own situation if nobody else changed *their* actions. (However, since the other parties *are* changing their actions, the total change may be preferable to nobody.) This results in a function that takes agreements to agreements. This function turns out to obey the conditions of THE KAKUTANI FIXED POINT THEOREM [V.11 §2], from which it follows that there is an agreement that no single individual wishes to change. (See MATHEMATICS AND ECONOMIC REASONING [VII.8], particularly section 4, for a further discussion of Nash's theorem. Another situation where individual and collective self-interest do not necessarily coincide is the flow of traffic (see THE MATHEMATICS OF TRAFFIC IN NETWORKS [VII.4 §4]).)

Not all applications of mathematical thought to social problems have such satisfactory outcomes. Suppose that there is to be an election (or, more generally, that society has to make a choice between various possibilities) with n candidates and m voters. Let us use the term "voting system" to mean any method of putting the n candidates in order given the preferences of the individual voters. Kenneth Arrow has shown that, under normal circumstances, there is no good voting system. More precisely, he has identified a small set of very reasonable sounding properties that one would wish a voting system to have, and shown that no voting system has all these properties. To give two examples of these properties, it is surely desirable that the final ranking of the candidates should depend on more than just the ranking of one individual voter, and one would also expect that if every voter prefers one candidate x to another candidate y, then x should be ranked higher than y. Instead of listing the other properties, we present a simpler result, known as Condorcet's paradox, that gives some of the flavor of Arrow's theorem. (Indeed, Arrow's theorem can be regarded as a descendant of Condorcet's paradox.) Consider three voters A, B, and C with the following preferences.

	A	B	C
First preferences	x	y	z
Second preferences	y	z	x
Third preferences	z	x	y

Observe that the majority of the voters prefer x to y, a majority prefer y to z, and a majority prefer z to x. Therefore, majority preference is not a TRANSITIVE RELATION [I.2 §2.3]. One consequence of this is that if voters are first asked to vote between two of x, y, and z and there is then a run-off between the winner of the first vote and whichever of x, y, and z is left, then the remaining candidate will always win.

Probability is another branch of mathematics that plays a central role in modern society. In earlier societies people worked until they died. Today people can stop working and live off their savings. You can, of course, just live off the interest of your savings but this means that you will die with a large sum unspent. Alternatively, you can assume that you will live a certain number of years and run down your savings, reaching zero at precisely the moment you expect to expire. This will not be satisfactory if you live longer than you expect. The solution is to make a bet with a wealthy corporation. You pay them your capital and in return they pay you a certain sum every year until you die. If you die early then they have won their bet, and if you die late then they have lost. By taking a large number of such bets and relying on results like the STRONG LAW OF LARGE NUMBERS [III.71 §4], the corporation can be almost certain of making a profit in the long run. In effect you have paid a certain amount to transfer the risk (from the financial point of view) that you might live a long time from yourself to the corporation.

One of the earliest ways for mathematicians to make money was to become actuaries—that is, advisers on the appropriate price for transfer of risk in the situation described above. Nowadays, all sorts of risk (Will next year's coffee crop fail? Will the euro fall against the dollar?) are bought and sold and have to be priced. A discussion of risk pricing in general can be found in THE MATHEMATICS OF MONEY [VII.9].

7   Conclusion

In the past, mathematics has had a dramatic impact on physics and engineering. At one time this led to hopes that biological and sociological phenomena would eventually come to be explained mathematically as well. Later, such hopes came to seem unrealistic: it was understood that these areas contain "emergent phenomena" that are not easily amenable to a reductionist approach and are therefore genuinely harder to describe mathematically than the phenomena studied in the "harder sciences." However, mathematicians

are now beginning to grapple with such phenomena: as even the simple examples in this article have shown, one can apply mathematics to many areas outside its traditional domain, and doing so can be extremely illuminating.

VIII.4   Numeracy
Eleanor Robson

---

1   Introduction

Most of this *Companion* is rightly concerned with the theories and practices of professional mathematicians. But all human beings have ideas about numbers, space, and shape, and ways of putting these ideas to use. It could be said that numeracy is to mathematics what literacy is to literature: everyday, routine application versus expert, elite innovation. But while literacy is now a wildly fashionable subject of academic study, the word "numeracy" is not even recognized by my mass-market word processor. Yet an array of interesting work has been done on nonprofessional mathematical concepts, practices, and attitudes. They range from historical studies and ethnographies to cognitive analyses and developmental psychologies, and cover such diverse periods and places as ancient Iraq, the pre-Columbian Andes, and the European Middle Ages, as well as many parts of the contemporary world. By surveying selected studies on five broadly construed topics in numeracy and artisanal mathematics, I hope to make the case in this essay that numeracy is as valuable a topic of academic research as professional mathematics on the one hand and literacy on the other.

Mathematics has rarely been considered part of the sociology or anthropology of knowledge, as it has often been assumed to stand outside culture. That is to say, many people have held the view that one can only *think mathematics*, not think *about* it. Furthermore, such work as has been done on the place of mathematics in culture is fragmented: mathematical thinking in the developed world has tended to be studied by sociologists, but in the developing world by anthropologists; historians of mathematics have mostly taken as their subject the literate mathematics of the professional elite, while psychologists have generally focused on the acquisition of numeracy, by adults and children.

But, as we shall see, the way that societies and individuals regard mathematics is strongly contingent on many environmental factors. Educational, linguistic, visual, and intellectual cultures all shape mathematical thinking in different ways. That is not to say that there are no constraints, however. Humans all share basic anatomical similarities that influence our ways of thinking: we are approximately symmetrical about one vertical axis, for instance, which gives rise to arguably innate concepts of left and right, front and back. And we all have fingers and opposable thumbs and the ability to subitize (that is, to recognize the size of a small set without counting its individual members). This, Reviel Netz has argued, makes human beings uniquely good at manipulating small groups of small objects, which has given rise to sophisticated systems of accounting and coinage. We shall return to Netz's work later.

The examples in this essay have been selected from studies of three very different clusters of world cultures. The ancient Middle East and Mediterranean (Egypt and Mesopotamia, classical Greece and Rome) have strongly influenced modern global culture in a variety of ways. Most obviously, the Euclidean tradition has been central to Western educational ideals for centuries, along with the teaching of Latin. And while the languages and writings of ancient Egypt and Mesopotamia are essentially nineteenth-century rediscoveries, their cultural influence runs in deep undercurrents throughout Western thinking, having percolated through classical and biblical learning. We should not be surprised, then, to discover the familiar as well as the alien in the world's oldest evidence for numeracy and artisanal mathematics. By contrast, the cultures of the pre-Columbian Americas are important for their very lack of contact with the premodern old world and thus their isolation from modernity. Virtually extinguished by the European conquests of the sixteenth and seventeenth centuries, and yet structurally similar to many old-world societies, they give a useful sense both of the constraints on numerate practice and thinking and of their diversity. Finally, this article also draws material from studies of the contemporary Americas, both South and North, in an attempt to break down the traditional disciplinary boundaries between past and present and between the developed world and the developing world. Numeracy is a feature of all human culture, wherever and whenever we have lived, and this should be reflected in how it is investigated.

2   Number Words and Social Values

Number words are usually studied for their mathematical content. French, for instance, shows traces of a vigesimal system in words such as *quatre-vingts*, meaning "four twenties," while the English word *eighty* is clearly derived from "eight tens." But in all languages number words also have social values attached, especially the counting numbers and words for sets. This is a rather different phenomenon from mystical numerology such as that of Late Antique Neo-Pythagoreanism. For instance, Nichomachus's book *The Theology of Arithmetic* (written in the second century B.C.E. but now known only from later summaries) assigned esoteric meanings to the first ten integers, understanding those numbers to represent fundamental attributes of the cosmos. But the social values of number words are often much more prosaic than that. English, for instance, has a variety of words for "group of three," each of which is applicable to a particular range of objects and has particular social connotations. "Threesome" is not a synonym of "trinity" in everyday language, just as in musical terminology "trio" does not have the same referents as "triad" or "triplet." There is nothing mystical or esoteric in the use of these words; it is simply that, in addition to their semantic content, these words also carry implicit qualitative information about the sort of objects that are being grouped (sexually active adults, divine beings, musicians, musical notes, criminals, babies), about which society and individuals tend to form value judgments.

That numbers have a "social life" was first recognized by Gary Urton in his ethnographic study of the Quechua-speaking inhabitants of the Bolivian Andes. Structurally, Quechua numeration is straightforwardly decimal, much like modern European number systems, and is written with Arabic numerals. This has ensured its survival side by side with Spanish, but the fact that it is not particularly exotic relative to Western norms has caused it to be somewhat neglected academically. However, as Urton shows, there are two predominant social aspects to Quechua numeration: family relations on the one hand, and the idea of completeness or "rectification" on the other. There are also clear boundaries around what may be counted and who may count them.

All Quechua number words are composed of a dozen basic lexemes—one to ten, hundred, and thousand—which may be combined additively or multiplicatively, just as in English the word *thirteen* means "three and ten" and *thirty* means "three tens." Also as in English, Quechua number words tend to be a distinct lexical set; for instance, *kinsa* means "three" and nothing else. But where synonyms for cardinal numbers are fairly rare in English (one example is *dozen* for "twelve"), they are a normal part of Quechua speech:

- *iskaypaq chaupin*, "the middle of (sets of) twos," used of the third item in a group of five;
- *iskay aysana*, "double puller" (because the symbol 3 looks like two handles);
- *uquti*, "anus" (because the symbol 3 also looks like a human bottom);
- *uj yunta ch'ullayuq*, "one pair, possessor of one standing alone" (2 + 1 = 3).

Family relations are most clearly visible in ordinal sequences, especially the names of the fingers, which are themselves important everyday counting tools. Urton lists six very similar sets of names, attested over the past 500 years. The most recent, collected by the Bolivian anthropologist Primitivo Nina Llanos in 1994, goes as follows:

- thumb, *mama riru*, "mother finger";
- index finger, *juch'uy riru*, "small[er] finger";
- middle finger, *chawpi riru*, "middle finger";
- ring finger, *sullk'a riru*, "younger finger";
- little finger, *sullk'aq sullk'an riru*, "younger sibling of the younger finger."

Thus the thumb is considered both the oldest and the antecedent of the others and the little finger the youngest; this is true of all six attested variants of the finger names. The hands themselves are considered as two symmetrical halves of a unified whole—as are paired items in general. In Quechua, one hand alone (or indeed an odd number) is not in its natural state. As Urton explains:

[T]he motivation for *two* is the "loneliness" (*ch'ulla*) of *one*. "One" is an incomplete, alienated entity: it needs a "partner" (*ch'ullantin*). The principle and motivational force obtain ... regardless of whether the unit that composes the "one" is indivisible (e.g., a single digit) or divisible (e.g., a hand with five digits).

And more generally, Urton shows that in Quechua, odd numbers (*ch'ulla*) are incomplete while even numbers (*ch'ullantin*, "the part together with its pair") represent the normal state of being.

But in Quechua society not everything is permissibly countable, even when there is no obvious difficulty in doing so. For example, they inventorize their herds, on whom they are often heavily economically dependent, not by counting but by naming. It is thought that counting individualizes the constituent members of the inseparable group, and thereby threatens its unity and fertility. If a herd *must* be counted then only a woman

may do so; it is an unacceptably effeminate action for a man to carry out.

While restrictions on counting are not a notable feature of contemporary English-speaking culture, taboos on particular numbers are still common. Why is thirteen considered so unlucky, for instance, particularly in North American hotels or on Fridays, while seven is regarded as lucky? In ancient Babylonia (modern-day southern Iraq) in the second and first millennia B.C.E., seven was thought to be particularly uncanny and unworldly. There were seven heavenly bodies (the Sun, the Moon, five visible planets), seven books of the *Epic of Creation*, and seven nights in each phase of the Moon. Demons, both beneficent and malevolent, were said to operate in groups of seven.

The Babylonians' primary numerical base for counting and recording groups of discrete objects was 60, factored into six groups of ten. The number 7 is, of course, the smallest one that is coprime to 60 and thus became a favorite subject of mathematics problems designed to be solved by trainee scribes. Further sexagesimal coprimes—11, 13, 17, 19—also featured prominently in ancient Babylonian mathematical problems and riddles. More often than not, however, the parameters were chosen in such a way that the tricky coprimes factored out or were otherwise disposed of, leaving an arithmetically innocuous answer:

I found a stone; I did not weigh it. I added a seventh. I added an eleventh. I weighed it: 1 mina. What was the original (weight of the) stone? The original stone was $\frac{2}{3}$ mina 8 shekels, $22\frac{1}{2}$ grains. (180 grains = 1 shekel; 60 shekel = 1 mina, ca. 0.5 kg.)

It is probably otiose to speculate whether the difficult mathematical properties of seven led directly to its cosmological demonization; the link is never made explicitly in any surviving cuneiform sources. But just as Babylonian demons failed to adhere to the norms of human behavior, so certain integers did not conform to the numerical patterns of the sexagesimally regular majority and the conceptual tools were not yet in place to explain that phenomenon in mathematical terms.

3   Counting and Calculating

While anyone can have views on whether particular numbers are lucky or unlucky, lonely or partnered, the ability to manipulate numbers arithmetically, and to take pleasure in doing so, is not universally shared. Both personal cognitive skills and social constraints are at work here. Patricia Cline Cohen argues that there were two key factors in the rapid rise in numerical competence in the early nineteenth-century United States. It was not that people suddenly became smarter. On the one hand, the decimalization of money in the late eighteenth century meant that at last accountants, shopkeepers, and business owners were working with a single number base. At the same time, a new educational movement forsook the rote learning of arithmetical rules, applied mechanically to particular situations, for inductive instruction that encouraged pupils to calculate with fingers and counters, and in their heads, before they progressed to pen and paper. In this way some basic structural impediments were removed, both to the learning of number relationships, and to their application in commercial life.

Because modern decimal notation is a calculating system as well as a recording device, it is easily forgotten that other methods are just as effective. Indeed, for most communities, most of the time, numerals were simply a means to record the outcome of operations performed on the body or with other calculating tools. Finger counting and abacus use remained ubiquitous in the medieval Islamic world and Christian Europe long after knowledge of decimal numerals, together with AL-KHWĀRIZMĪ's [VI.5] treatise on how to use them, and cheap paper on which to write them, began to spread outward from Baghdad in the ninth century C.E. Their retention was not a knee-jerk reaction in the face of an overwhelmingly superior technology; rather, it took into account such factors as portability, speed of use, and a long-established trust in and institutional sanction of the old methods.

Indeed, it is difficult to overestimate just how old abacus calculation is. Reviel Netz identifies two evolutionary prerequisites for what he calls "counter culture," by which he means the uniquely and ubiquitously human use of small objects to represent other objects that are being counted, in one–one or one–many relationships. One is physiological: one needs to be able to pick up and manipulate small objects such as pebbles or shells. All primates share this ability thanks to prehensile fingers and opposable thumbs. The other is cognitive: one must be able to subitize, or recognize the size of a small set of up to about seven objects, without counting them individually. Stringed-bead abacuses exploit this most obviously, whether in the Russian-style ten-bead variety, whose fifth and sixth beads are always a different color from the others, or in the Japanese version, whose strings contain just four unit-beads and one five-bead each.

But, as Netz so powerfully puts it, "The abacus is not an artefact: it is a state of mind." All one needs is a flat surface and a pile of small objects to act as counters. This extreme ephemerality makes the use of abacuses almost impossible to detect in the archaeological record, except in the rare cases where abacus counters can be recognized as such. Denise Schmandt-Besserat has argued that a sophisticated accounting system was developed in the Neolithic Middle East from the ninth millennium B.C.E. She proposes that the tiny, unbaked pieces of clay, crudely shaped into various simple geometrical figures and found in preliterate archaeological contexts from eastern Turkey to Iran, are ancient accounting tokens. It is certainly true that the earliest written numerals in the area, from southern Iraq in the late fourth millennium B.C.E., are marks on clay tablets that look remarkably like stylized impressions of such objects, and are visually distinct from the signs for the objects that were being counted, which were scratched onto the clay rather than impressed. It is also true that these earliest written records are almost exclusively accounting records, drawn up by temple administrators in the management of assets such as land, labor, and agricultural products. And from the fifth millennium B.C.E. onward, those tiny clay tokens are found in archaeological contexts—sealed into jars, for instance, or wrapped in little clay bundles, or carefully piled in the corners of storerooms—that are entirely compatible with their use as abacus counters. But Schmandt-Besserat's claim for a universally standardized system across the Middle East from several millennia before then is not provable: there is no way of establishing that they were not sometimes gaming pieces, for instance, or sling shot, or any number of other possibilities, and certainly no way of determining what specific shapes signified and to whom.

In fact, ad hoc means of counting and measuring are still everyday occurrences in all our lives, even among those with a high level of formal mathematics education. A team of anthropologists and psychologists, headed by Jean Lave, observed newcomers to a Californian Weight Watchers scheme in the 1980s as they adjusted to careful quantification of the food they were allowed to consume on the diet. One participant, who had taken a calculus course at college, was asked to modify a recipe calling for two-thirds of a cup of cottage cheese so that it contained three-quarters of that amount. Lave recalls: "He filled a measuring cup two-thirds full of cottage cheese, dumped it out on a cutting board, patted it into a circle, marked a cross on it, scooped away one quadrant, and served the rest." She comments:

> Thus, "take three-quarters of two-thirds of a cup of cottage cheese" was not just the problem statement but also the solution to the problem and the procedure for solving it. The setting was part of the calculating process and the solution was simply the problem statement enacted within the setting. At no time did the Weight Watcher check his procedure against a paper and pencil algorithm, which would have produced $\frac{3}{4} \times \frac{2}{3} = \frac{1}{2}$ cup. Instead, the coincidence of problem, setting, and enactment was the means by which checking took place.

In other words, there are many situations in many people's lives in which potentially applicable literate, school-taught mathematical procedures are ignored in favor of equally effective nonliterate ones that produce the correct result with the tools at hand. Numeracy takes many forms, not all of which entail writing.

4   Measurement and Control

The Weight Watcher invented a system of cottage-cheese measurement that satisfied him in its accuracy and fulfilled his immediate culinary needs. But as individuals and social groups we also accept the accuracy and consistency of standardized measurement systems, and the institutional necessity of counting and measuring particular things but not others. Theodore Porter has written eloquently of the twentieth century's growing "trust in numbers," whether of census statistics or environmental data. But institutionally sanctioned quantification is often contested, and it frequently alters the very phenomenon that is being pinned down. Cohen's description of nineteenth-century North America is more generally apposite:

> *What* people chose to count and measure reveals not only what was important to them but what they wanted to understand and, often, what they wanted to control. Further, *how* people counted and measured reveals underlying assumptions about the subject under study, assumptions ranging from plain old bias … to ideas about the structure of society and of knowledge. In some cases, the activity of counting and measuring itself altered the way people thought about what they were quantifying: numeracy could be an agent of change.

Cohen and Porter both explore problems raised by early nineteenth-century census taking. Porter describes the obstacles that the under-resourced Bureau

de Statistique faced in obtaining accurate population data in post-revolutionary France. Without resorting to the old class categorizations of the *ancien régime*, it needed to acknowledge the huge diversity of occupations and social structures across the country. To do so it relied on local officials to return a mass of quantitative data that was simply not readily available—and so the prefectures commissioned qualitative descriptions of their regions instead. As Porter puts it, in 1800 "France was not yet capable of being reduced to statistics." Cohen analyzes the U.S. Census of 1840, which appeared to demonstrate a much higher rate of insanity among the black population in the abolitionist northern states than in the south. Pro-slavery factions took this as irrefutable evidence that slavery suited the black population much better than freedom did; abolitionists queried the trustworthiness of the census itself. Whether or not one chose to believe the data was more or less a matter of what one's preexisting political convictions were. As Cohen shows, the source of the error lay in clumsily designed recording sheets, in which the "idiot white" and "idiot black" columns were easily confused, resulting in the misrecording of many elderly senile inhabitants of all-white households. In the 1840s, however, the public debate was not about methodology, but whether fraud had been committed: the numbers themselves could not lie.

Two thousand years earlier, as Serafina Cuomo has shown, the Roman land surveyor Frontinus opined that the world was essentially unknowable without quantitative intervention, and that the trustworthiness of that measure was dependent on professional expertise:

> The basis of the art of measuring lies in the experience of the agent. It is in fact impossible to express the truth of the places or of the size without calculable lines, because the wavy and uneven edge of any piece of land is enclosed by a boundary which, because of the great quantity of unequal angles, can be contracted or expanded, even when their number [that is, the number of the angles] remains the same. Indeed pieces of land which are not finally demarcated have a fluctuating space and an uncertain determination of *iugera*.

The natural world is problematically irregular, Frontinus believed, and must be disciplined into quantified straight lines—and, ideally, marked out into grids of 2400 foot squares (*iugera*)—in order to be brought under control. The Roman reshaping of the landscape through its quantification is still visible throughout

Europe, the Middle East, and North Africa today, both on land and from the air.

The Incas, by contrast, brought time, space, society, and the gods under control through radial lines in the landscape, tied to the ceremonial year. Before Spanish-led Christianization in the sixteenth-century, the heart of the Inca cosmos was the sacred city of Cuzco in the Peruvian Andes. The Incas divided the world into unequal quarters or *tawantinsuyu* "the four parts together," radiating out from the Temple of the Sun. Through each *suyu* ran nine to fourteen *ceque* paths through the mountains, forty-one in total, with an average of eight *huaca* shrines stationed on each. The local inhabitants performed a ritual at one of the 328 *huacas* every day of the sacred year (composed of twelve months of $27\frac{1}{3}$ days). Thus the religious focus of the Inca state moved systematically around its territory, day by day and from community to community, binding every social group into the same calendar, cult, and cosmos.

Numeracy, then, is a powerful institutional tool: measuring, quantifying, and classifying can transform an unknowable mass of individual people, places, or things into manageable categories of known entities; in turn, this institutionally imposed structure shapes the self-identities of those being managed. Institutional numeracy, while imposed from above, is always dependent to some degree on community-wide support and cooperation, if not necessarily for the objects of account then always for the counters. Attempts at censuses in the eighteenth century did not fail because people refused to be reduced to numbers in boxes, but because those charged with collecting the data had neither the infrastructural means to do so nor an intellectual outlook that valued quantification. Inca and Roman societies, by contrast, were able to produce whole classes of the professionally numerate who did.

5 Numeracy and Gender

In modern anglophone culture, academic mathematics is popularly considered a male pursuit—and women supposedly have to subordinate or compromise their femininity if they are to succeed in it. But such perceptions are far from universal: studies collected by Barbro Grevholm and Gila Hanna, for instance, show that in the early 1990s some 80% of Kuwaiti and over half of Portuguese undergraduate mathematics majors were women. However, as the following examples demonstrate, this has more to do with how particular societies construct ideals of femininity and masculinity and with what they count as mathematical activity than with any intrinsically gendered properties of mathematics itself.

For most of the second millennium B.C.E., Babylonian scribes understood professional numeracy to be a divine gift—not from the gods in general but from a handful of powerful goddesses. In the literary works that scribal students memorized as part of their professional training, creator gods bestowed land-measuring equipment and numeracy on those goddesses to enable them to manage household estates equitably. In a myth now known as *Enki and the World Order* the great god Enki announces:

> My illustrious sister, holy Nisaba,
> Is to receive the 1-rod measuring reed.
> The lapis lazuli rope is to hang from her arm.
> She is to proclaim all the great divine powers.
> She is to fix boundaries and mark borders.
>   She is to be the scribe of the Land.
> The gods' eating and drinking are to be in her
>   hands.

The scribes' literary works also portrayed Nisaba as the patron of institutional numeracy in the real world: she in turn provided mensuration tools to scribes and kings to enable them to uphold justice in society.

Another scholastic literary genre was the scribal dialogue, in which the protagonists argue over the ideals of scribal professionalism. In one such debate the young scribe Enki-manshum explicitly relates metrological competence to social justice:

> When I go to divide a plot, I can divide it;
>   when I go to apportion a field, I can
>   apportion the pieces,
> So that when wronged men have a quarrel
>   I soothe their hearts and ... .
> Brother will be at peace with brother, their
>   hearts ... .

This was not merely a literary trope: law codes promulgated by real-life Babylonian kings often began with prologues claiming that they would uphold fairness in commercial measuring, weighing, and counting, and included provisions for punishing metrological fraud. Many hundreds of legal records survive, attesting to the settlement of land disputes through accurate professional measurement and calculation. In the nineteenth-century B.C.E. city of Sippar, the judges who held court in the temple of Shamash, god of justice, employed female scribes and surveyors as well as male (often from the same families). Further, the personal seals of fourteenth-century B.C.E. royal land surveyors were often dedicated to Nin-sumun, the divine mother of the legendary hero Gilgamesh: for them the numerate goddess who bestows numerate justice was no school story but at the very heart of their professional self-identity.

In ancient Babylonia, then, numeracy and metrology gained institutional authority and power as much through association with divine femininity as with royal masculinity. Many modern societies, by contrast, defeminize numerate thought and activity by denying its mathematical status when it is carried out by women. Gary Urton's study of Quechua numeration started out as an ethnography of Bolivian weaving, which, he discovered, was based on highly intricate symmetrical patterns that the (female) weavers know by heart. They count off threads effortlessly, unerringly picking up where they have left off after interruptions to nurse babies, prepare food, or attend to other domestic matters. And yet the men of the area categorically told Urton that the weavers "can't count"—because when a woman sells her finished weavings at market she will invariably ask another woman of the group to check her takings to ensure that she has not been cheated.

Urton was taught to weave by Irene Flores Condori, a twelve-year-old girl. He recalls:

> On one occasion, a stern old woman … asked me point blank if, by weaving, I was trying to be like a woman. I answered by telling her that in some villages I know of, it is the men rather than the women who do the weaving.… The old woman gave us both a wry look and asked, if that was the case, then is it the women in those villages who have the penises!

Weaving was such a strongly gendered activity that this and other incidents led Urton to feel that "my behavior was being tolerated to the degree that it was only because, as an outsider, I was not subject to the same rules and expectations as local men." Weaving is exclusively women's work and therefore its intrinsically numerate character is socially invisible; women are more reluctant than men to trust strangers to handle money fairly and are therefore considered innumerate.

Mary Harris shows how a similarly powerful gender divide developed in Victorian Britain as primary education became available to an ever-widening section of the populace. Mathematics was regarded as the quintessentially male school subject, while needlework was the epitome of femininity. Yet:

> Every garment knitted to fit a particular body depends on the principle of ratio. Every pinafore pattern copied from a blackboard requires visual interpretation of scaling and the ability to draw a smooth curve. All the fine stitching that the early Inspectors were unable to tell from machine stitching depended on the ability to judge equal distances by eye and maintain them in a straight row.

In other words, wherever girls and women weave, knit, or sew they are unwittingly engaging numerate aptitudes and skills, often highly creatively, just as Molière's Monsieur Jourdain had been speaking prose all his life "without knowing anything about it."

6 Numeracy and Literacy, School and Supermarket

Perhaps one reason that women's work is not often thought to belong to the realm of professional numeracy is that numeracy is so often considered (when it is considered at all) as a subset of literacy. As Reviel Netz puts it,

> With Arabic numerals, numbers appear as secondary to writing, benefiting from tools that were largely invented to record verbal systems and not numerical symbols. In broad historical perspective, this is the exception and not the rule. The rule is that, across cultures, and especially in early cultures, the record and manipulation of visual symbols precede and predominate over the record and manipulation of verbal symbols.

Netz is thinking here of counters and abacuses, but the Bolivian weavers remind us that numeracy does not have to entail symbolic manipulation at all. One may count threads, llamas, ideas, anything, and perform calculations without the intervention of external tools. The use of fingers and other body parts has cropped up repeatedly in the examples presented in this essay. Much of the weavers' mental work is so naturalized within the rhythms and movements of their bodies that they can no longer verbalize the mental or physical processes involved. (That is why Urton chose a young girl as his teacher, who was still learning the craft, rather than a fully competent adult woman.) Nonliterate numerate practices and ideas, especially in the developing world, are often labeled by academic observers as "ethnomathematics." But this raises difficult questions about the appropriate use of the "ethno" prefix and about the border between numeracy and mathematics. How do we distinguish numeracy from mathematics, and where does ethnomathematics fit in?

When Ubiratan D'Ambrosio coined the term "ethnomathematics" in the mid 1970s it was to describe the study of mathematics "in direct relation to [its] social, economic, and cultural backgrounds," a subject lying "on the borderline between the history of mathematics and cultural anthropology." However, for many, particularly within mathematics education, it has come to mean the study of culturally "other" mathematics, as if only the academically marginalized have ethnicity (just as, according to some lazy academic views, only women have gender). This semantic shrinkage is doubly damaging, for it implies that "ethnic" cultures are not fully numerate, while rendering the mainstream of academic mathematics, both past and present, invisible to sociological, anthropological, or ethnographic research. Nor does it distinguish between the intellectual creativity that is mathematics and the routine application of numeracy.

If "ethnomathematics" is an unhelpful term, there are useful alternatives. An influential Brazilian study of childhood numeracy, by Terezinha Nunes and colleagues, distinguishes formally learned "school mathematics" from "street mathematics" created informally by the same children. Jean Lave's ethnography of adult numeracy in 1980s California likewise contrasts "school arithmetic" with "supermarket arithmetic." The participants in her study often described themselves as arithmetically incompetent and "were unaware of the efficacy of their math practice in the supermarket, and some did not know, even that they used arithmetic practices there." Yet often the supermarket setting required the solution of mathematical problems of much greater complexity than superficially similar scholastic "word problems":

> The shopper was standing in front of a produce display. As she spoke she put apples, one at a time, into a bag. She put the bag in the cart as she finished talking: "There's only about three or four [apples] at home, and I have four kids, so you figure at least two apiece in the next three days. These are the kinds of things I have to resupply. I only have a certain amount of storage space in the refrigerator, so I can't load it up totally.... Now that I'm home in the summertime, this is a good snack food. And I like an apple sometimes at lunchtime when I come home."

While explicitly considering such variables as the number of apple-consumers in the household, their rate of consumption, fridge storage space, and perhaps implicitly the apples' price and probably shelf life, the shopper selected nine apples to buy. She might also have compared the prices of different varieties of apple and/or considered whether loose or prepackaged apples were the better buy—all typical supermarket activities that Lave and her researchers observed and correlated with the same subjects' performance in written tests of arithmetically similar skills. They found "not a single significant correlation between frequency of calculation in a supermarket, and scores on math test, multiple choice test or number facts.... Success and frequency of calculation in supermarket and simulation experiment bear no statistical relationship with schooling, years since schooling was completed, or age."

Rather depressingly for educators, perhaps, Lave's work suggests that training in school mathematics has little or no impact on numerical competence in adult life. (Interestingly, this finding conflicts with Cohen's historical argument discussed above, relating improvements in mathematics education to rising standards of numeracy in early nineteenth-century North America.) Rather, as she and Étienne Wenger argue, learning takes place most effectively when it is situated in the social and professional context to which it pertains, through interaction and collaboration with competent practitioners, rather than through abstract, decontextualized classroom learning. Learners become part of a "community of practice" that inculcates not only the necessary technical skills but also the beliefs, standards, and behaviors of the group. Through gains in competence, confidence, and social acceptance, the learner moves from the periphery toward the center of the practice community, in due course becoming accepted as a fully fledged expert. It is perhaps in this light, then, that we should understand the process of becoming professionally numerate. But if situated learning is so effective, the development of supra-utilitarian educational mathematics in the societies of the ancient Middle East and Mediterranean is a major historical conundrum that has hitherto gone unrecognized.

7 Conclusions

This essay began by suggesting that "numeracy is to mathematics what literacy is to literature." But the case studies presented here show that numeracy has a far greater cognitive reach than that. Throughout time and across the world countless individuals and societies have managed perfectly well, and continue to thrive, without writing; none has yet been attested without counting, measuring, or pattern-making in some form or other. In this light a better formulation might be that "numeracy is to mathematics what *language* is to literature." Indeed babies, toddlers, and young children learn many essential mathematical skills through engagement with their immediate environment well before formal school learning begins. Just as some children grow into more articulate adults than others, with or without highly developed skills in reading and writing, so they may become more or less numerate in their everyday practices, independently of their competence in school mathematics.

There are many deep and important questions about the relationships between numeracy and mathematics, language and literacy that have hardly yet been formulated, let alone explored: this is perhaps one of the most open fields of enquiry in academia today. This essay has only scratched the surface of a fascinating and complex subject that has paradoxically been overlooked because of its very ubiquity and centrality to human existence. In the next few decades, a wide range of interdisciplinary approaches will almost certainly yield important and surprising discoveries about numeracy that today we can only guess at.

Further Reading

Ascher, M. 2002. *Mathematics Elsewhere: An Exploration of Ideas Across Cultures*. Princeton, NJ: Princeton University Press.

Bloor, D. 1976. *Knowledge and Social Imagery*. London: Routledge & Kegan Paul.

Cohen, P. C. 1999. *A Calculating People: The Spread of Numeracy in Early America*, 2nd edn. New York and London: Routledge.

Crump, T. 1990. *The Anthropology of Numbers*. Cambridge: Cambridge University Press.

Cuomo, S. 2000. Divide and rule: Frontinus and Roman land-surveying. *Studies in History and Philosophy of Science* 31: 189–202.

D'Ambrosio, U. 1985. Ethnomathematics and its place in the history and pedagogy of mathematics. *For the Learning of Mathematics* 5:41–48.

Gerdes, P. 1998. *Women, Art and Geometry in Southern Africa*. Trenton, NJ: Africa World Press.

Glimp, D., and M. R. Warren, eds. 2004. *The Arts of Calculation: Quantifying Thought in Early Modern Europe*. Basingstoke: Palgrave Macmillan.

Grevholm, B., and G. Hanna. 1995. *Gender and Mathematics Education: An ICMI Study in Stiftsgården Åkersberg, Höör, Sweden, 1993*. Lund: Lund University Press.

Harris, M. 1997. *Common Threads: Women, Mathematics, and Work*. Stoke on Trent: Trentham Books.

Lave, J. 1988. *Cognition in Practice: Mind, Mathematics and Culture in Everyday Life*. Cambridge: Cambridge University Press.

Lave, J., and E. Wenger. 1991. *Situated Learning: Legitimate Peripheral Participation*. Cambridge: Cambridge University Press.

Netz, R. 2002. Counter culture: towards a history of Greek numeracy. *History of Science* 40:321-52.

Nunes, T., A. Dias, and D. Carraher. 1993. *Street Mathematics and School Mathematics*. Cambridge: Cambridge University Press.

Porter, T. 1995. *Trust in Numbers: The Pursuit of Objectivity in Science and Public Life*. Princeton, NJ: Princeton University Press.

Robson, E. 2008. *Mathematics in Ancient Iraq: A Social History*. Princeton, NJ: Princeton University Press.

Schmandt-Besserat, D. 1992. *From Counting to Cuneiform*. Austin, TX: University of Texas Press.

Urton, G. 1997. *The Social Life of Numbers: A Quechua Ontology of Numbers and Philosophy of Arithmetic*. Austin, TX: University of Texas Press.

VIII.5 Mathematics: An Experimental Science

Herbert S. Wilf

---

1 The Mathematician's Telescope

Albert Einstein once said, "You can confirm a theory with experiment, but no path leads from experiment to theory." But that was before computers. In mathematical research now, there's a very clear path of that kind. It begins with wondering what a particular situation looks like in detail; it continues with some computer experiments to show the structure of that situation for a selection of small values of the parameters of the problem; and then comes the human part: the mathematician gazes at the computer output, attempting to see and to codify some patterns. If this seems fruitful, then the final step requires the mathematician to prove that the apparent pattern is really there, and is not a shimmering mirage above the desert sands.

A computer is used by a pure mathematician in much the same way that a telescope is used by a theoretical astronomer. It shows us "what's out there." Neither the computer nor the telescope can provide a theoretical explanation for what it sees, but both of them extend the reach of the mind by providing numerous examples that might otherwise be hidden, and from which one has some chance of perceiving, and then demonstrating, the existence of patterns, or universal laws.

In this article I would like to show you some examples of this process at work. Naturally the focus will be on examples in which some degree of success has been realized, rather than on the much more numerous cases where no pattern could be perceived, at least by my eyes. Since my work is mainly in combinatorics and discrete mathematics, the focus will also be on those areas of mathematics. It should not be inferred that experimental methods are not used in other areas; only that I don't know those applications well enough to write about them.

In one short article we cannot even begin to do justice to the richly varied, broad, and deep achievements of experimental mathematics. For further reading, see the journal *Experimental Mathematics* and the books by Borwein and Bailey (2003) and Borwein et al. (2004).

In the following sections we give first a brief description of some of the useful tools in the armament of experimental mathematics, and then some successful examples of the method, if it is a method. The examples have been chosen subject to fairly severe restrictions:

(i) the use of computer exploration was vital to the success of the project; and

(ii) the outcome of the effort was the discovery of a new theorem in pure mathematics.

I must apologize for including several examples from my own work, but those are the ones with which I am most familiar.

2 Some of the Tools in the Toolbox

2.1 Computer Algebra Systems

The mathematician who enjoys using computers will find an enormous number of programs and packages available, beginning with the two major computer algebra systems (CASs), Maple and Mathematica. These programs can provide so much assistance to a working mathematician that they must be regarded as essential pieces of one's professional armamentarium. They are extremely user-friendly and capable.

Typically one uses a CAS in interactive mode, meaning that you type in a one-line command and the program responds with its output, then you type in another line, etc. This modus operandi will suffice for many purposes, but for best results one should learn the programming languages that are embedded in these packages. With a little knowledge of programming, one can ask the computer to look at larger and larger cases until

something nice happens, then take the result and use another package to learn something else, and so forth. Many are the times when I have written little programs in Mathematica or Maple and then gone away for the weekend leaving the computer running and searching for interesting phenomena.

2.2 Neil Sloane's Database of Integer Sequences

Aside from a CAS, another indispensable tool for experimentally inclined mathematicians, particularly for combinatorialists, is Neil Sloane's "On-Line Encyclopedia of Integer Sequences," which is on the Web at www.research.att.com/~njas. At present, this contains nearly 100 000 integer sequences and has full search capabilities. A great deal of information is given for each sequence.

Suppose that for each positive integer n you have an associated set of objects that you want to count. You might, for example, be trying to determine the number of sets of size n with some given property, or you might wish to know how many prime divisors n has (which is the same as counting the set of these prime divisors). Suppose further that you've found the answer for $n = 1, 2, 3, \ldots, 10$, say, but you haven't been able to find any simple formula for the general answer.

Here's a concrete example. Suppose you're working on such a problem, and the answers that you get for $n = 1, 2, \ldots, 10$ are 1, 1, 1, 1, 2, 3, 6, 11, 23, 47. The next step should be to look online to see if the human race has encountered your sequence before. You might find nothing at all, or you might find that the result that you'd been hoping for has long since been known, or you might find that your sequence is mysteriously the same as another sequence that arose in quite a different context. In the third case, an example of which is described below in section 3, something interesting will surely happen next. If you haven't tried this before, do look up the little example sequence above, and see what it represents.

2.3 Krattenthaler's Package "Rate"

A very helpful Mathematica package for guessing the form of hypergeometric sequences has been written by Christian Krattenthaler and is available from his Web site. The name of the package is Rate (rot'-eh), which is the German word for "guess."

To say what a hypergeometric sequence is, let's first recall that a rational function of n is a quotient of two polynomials in n, like $(3n^2 + 1)/(n^3 + 4)$. A hypergeometric sequence $\{t_n\}_{n \geqslant 0}$ is one in which the ratio t_{n+1}/t_n is a rational function of the index n. For example, if $t_n = \binom{n}{7}$ then t_{n+1}/t_n works out to be $(n + 1)/(n - 6)$, which is a rational function of n, so $\{t_n\}_{n \geqslant 0}$ is a hypergeometric sequence. Other examples are

$$n!, \quad (7n + 3)!, \quad \binom{n}{7} t^n, \quad \frac{(3n + 4)!(2n - 3)!}{4^n n!^4},$$

all of which are easily seen to be hypergeometric.

If you input the first several members of the unknown sequence, Rate will look for a hypergeometric sequence that takes those values. It will also look for a hyper-hypergeometric sequence (i.e., one in which the ratio of consecutive terms is hypergeometric), and a hyper-hyper-hypergeometric sequence, etc.

For example, the line

```
Rate[1, 1/4, 1/4, 9/16, 9/4, 225/16]
```

elicits the (somewhat inscrutable) output

$$\{4^{1-i0}(-1 + i0)!^2\}.$$

Here $i0$ is the running index of Rate, so we would normally write that answer as, say,

$$\frac{(n - 1)!^2}{4^{n-1}} \quad (n = 1, 2, 3, 4, 5, 6),$$

which fits the input sequence perfectly. Rate is a part of the *Superseeker* front end to the Integer Sequences database, discussed in section 2.2 above.

2.4   Identification of Numbers

Suppose that, in the course of your work, you encountered a number, let's call it β, which, as nearly as you could calculate it, was 1.218041583332573. It might be that β is related to other famous mathematical constants, like π, e, $\sqrt{2}$, and so forth, or it might not. But you'd like to know.

The general problem that is posed here is the following. We are given k numbers, $\alpha_1, \ldots, \alpha_k$ (the *basis*), and a target number α. We want to find integers $m, m_1, \ldots, m_k$ such that the linear combination

$$m\alpha + m_1\alpha_1 + m_2\alpha_2 + \cdots + m_k\alpha_k \quad (1)$$

is an extremely close numerical approximation to 0.

If we had a computer program that could find such integers, how would we use it to identify the mystery constant $\beta = 1.218041583332573$? We would take the α_i to be a list of the logarithms of various well-known

universal constants and prime numbers, and we would take $\alpha = \log \beta$. For example, we might use

$$\{\log \pi, 1, \log 2, \log 3\} \quad (2)$$

as our basis. If we then find integers $m, m_1, \ldots, m_4$ such that

$$m \log \beta + m_1 \log \pi + m_2 + m_3 \log 2 + m_4 \log 3 \quad (3)$$

is extremely close to 0, then we will have found that our mystery number β is extremely close to

$$\beta = \pi^{-m_1/m} e^{-m_2/m} 2^{-m_3/m} 3^{-m_4/m}. \quad (4)$$

At this point we will have a judgment to make. If the integers m_i seem rather large, then the presumed evaluation (4) is suspect. Indeed, for any target α and basis $\{\alpha_i\}$ we can always find huge integers $\{m_i\}$ such that the linear combination (1) is *exactly* 0, to the limits of machine precision. The real trick is to find a linear combination that is extraordinarily close to 0, while using only "small" integers m, m_i, and that is a matter of judgment. If the judgment is that the relation found is real, rather than spurious, then there remains the little job of proving that the suspected evaluation of α is correct, but that task is beyond our scope here. For a nice survey of this subject, see Bailey and Plouffe (1997).

There are two major tools that can be used to discover linear dependencies such as (1) among the members of a set of real numbers. They are the algorithms PSLQ, of Ferguson and Forcade (1979), and LLL, of Lenstra et al. (1982), which uses their lattice basis reduction algorithm. For the working mathematician, the good news is that these tools are available in CASs. For example, Maple has a package, `IntegerRelations [LinearDependency]`, which places the PSLQ and the LLL algorithms at the immediate disposal of the user. Similarly there are Mathematica packages on the Web that can be freely downloaded and which perform the same functions.

An application of these methods will be given in section 7. For a quick illustration, though, let us try to recognize the mystery number $\beta = 1.218041583332573$. We use as a basis the list in (2) above, and we put this list, augmented by $\log 1.218041583332573$, into Maple's `IntegerRelations[LinearDependency]` package. The output is the integer vector $[2, -6, 0, 3, 4]$, which tells us that $\beta = \pi^3 \sqrt{2}/36$, to the number of decimal places carried.

2.5   Solving Partial Differential Equations

I had occasion recently to need the solution to a certain partial differential equation (PDE) that arose in connection with a research problem that was posed by Graham et al. (1989). It was a first-order linear PDE, so in principle the METHOD OF CHARACTERISTICS [III.49 §2.1] gives the solution. As those who have tried that method know, it can be fraught with technical difficulties relating to the solution of the associated ordinary differential equations.

However, some extremely intelligent packages are available for solving PDEs. I used the Maple command `pdsolve` to handle the equation

$$(1 - \alpha x - \alpha' y)\frac{\partial u(x, y)}{\partial x}$$
$$= y(\beta + \beta' y)\frac{\partial u(x, y)}{\partial y} + (\gamma + (\beta' + \gamma')y)u(x, y)$$

with $u(0, y) = 1$. `pdsolve` found that

$$u(x, y) = \frac{(1 - \alpha x)^{-\gamma/\alpha}}{(1 + (\beta'/\beta)y(1 - (1 - \alpha x)^{-\beta/\alpha}))^{1+\gamma'/\beta'}}$$

is the solution, and that enabled me to find explicit formulas for certain combinatorial quantities, with much less work and fewer errors than would otherwise have been possible.

3   Thinking Rationally

The following problem appeared in the September/October 1997 issue of *Quantum* (and was chosen by Stan Wagon for the Problem of the Week archive).

> How many ways can $90\,316$ be written as
> $$a + 2b + 4c + 8d + 16e + 32f + \cdots,$$
> where the coefficients can be any of 0, 1, or 2?

In standard combinatorial terminology, the question asks for the number of *partitions* of the integer $90\,316$ into powers of 2, where the multiplicity of each part is at most 2.

Let's define $b(n)$ to be the number of partitions of n, subject to the same restrictions. Thus $b(5) = 2$ and the two relevant partitions are $5 = 4 + 1$ and $5 = 2 + 2 + 1$. Then it is easy to see that $b(n)$ satisfies the recurrences $b(2n + 1) = b(n)$ and $b(2n + 2) = b(n) + b(n + 1)$, for $n = 0, 1, 2\ldots$, with $b(0) = 1$.

It is now easy to calculate particular values of $b(n)$. This can be done directly from the recurrence, which is quite fast for computational purposes. Alternatively, it can be shown quite easily that our sequence $\{b(n)\}_0^\infty$ has the generating function

$$\sum_{n=0}^{\infty} b(n)x^n = \prod_{j=0}^{\infty} (1 + x^{2^j} + x^{2 \cdot 2^j}).$$

(Generating functions are discussed in ALGEBRAIC AND ENUMERATIVE COMBINATORICS [IV.18 §§2.4, 3], or see Wilf (1994).) This helps us to avoid much programming when working with the sequence, because we can use the built-in series-expansion instructions in Mathematica or Maple to show us a large number of terms in this series quite rapidly. Returning to the original question from *Quantum*, it is a simple matter to compute $b(90\,316) = 843$ from the recurrence. But let's try to learn more about the sequence $\{b(n)\}$ in general. To do that we open up our telescope, and calculate the first ninety-five members of the sequence, i.e., $\{b(n)\}_0^{94}$, which are shown in table 1. The question is now, as it always is in the mathematics laboratory, what patterns do you see in these numbers?

Just as an example, one might notice that when n is 1 less than a power of 2, it seems that $b(n) = 1$. The reader who is fond of such puzzles is invited to cease reading here for the moment (without peeking at the next paragraph), and look at table 1 to spend some time finding whatever interesting patterns seem to be there. Computations up to $n = 94$ aren't as helpful for a quest like this as computations up to $n = 1000$ or so might be, so the reader is also invited to compute a much longer table of values of $b(n)$, using the above recurrence formulas, and to study it carefully for fruitful patterns.

OK, did you notice that if $n = 2^a$ then $b(n)$ appears to be $a + 1$? How about this one: in the block of values of n between 2^a and $2^{a+1} - 1$, inclusive, the largest value of $b(n)$ that seems to occur is the Fibonacci number F_{a+2}. There are many intriguing things going on in this sequence, but the one that was of crucial importance in understanding it was the observation that *consecutive values of $b(n)$ seem always to be relatively prime*.[1]

It was totally unexpected to find a property of the values of this sequence that involved the multiplicative structure of the positive integers, rather than their additive structure, which would have been quite natural. This is because the theory of partitions of integers belongs to the additive theory of numbers, and multiplicative properties of partitions are rare and always cherished.

Once this relative primality is noticed, the proof is easy. If m is the smallest n for which $b(n)$, $b(n + 1)$ fail to be relatively prime, then suppose $p > 1$ divides both of them. If $m = 2k + 1$ is odd, then the recurrence

---

1. Two positive integers are *relatively prime* if they have no common factor.

Table 1 The first ninety-five values of $b(n)$.

0	1	2	3	4	5	6	7	8	9	10	11	12	13	14	15	16	17	18
1	1	2	1	3	2	3	1	4	3	5	2	5	3	4	1	5	4	7

19	20	21	22	23	24	25	26	27	28	29	30	31	32	33	34	35	36	37
3	8	5	7	2	7	5	8	3	7	4	5	1	6	5	9	4	11	7

38	39	40	41	42	43	44	45	46	47	48	49	50	51	52	53	54	55	56
10	3	11	8	13	5	12	7	9	2	9	7	12	5	13	8	11	3	10

57	58	59	60	61	62	63	64	65	66	67	68	69	70	71	72	73	74	75
7	11	4	9	5	6	1	7	6	11	5	14	9	13	4	15	11	18	7

76	77	78	79	80	81	82	83	84	85	86	87	88	89	90	91	92	93	94
17	10	13	3	14	11	19	8	21	13	18	5	17	12	19	7	16	9	11

implies that p divides $b(k)$ and $b(k+1)$, contradicting the minimality, whereas if $m = 2k$ is even, the recurrence again gives that result, finishing the proof.

Why was it so interesting that consecutive values appeared to be relatively prime? Well, at once that raised the question of whether every possible relatively prime pair (r, s) of positive integers occurs as a pair of consecutive values of this sequence, and if so, whether every such pair occurs once and only once. Both of those possibilities are supported by the table of values above, and upon further investigation both turned out to be true. See Calkin and Wilf (2000) for details.

The bottom line here is that *every positive rational number occurs once and only once, and in reduced form, among the members of the sequence* $\{b(n)/b(n+1)\}_0^\infty$. Hence the partition function $b(n)$ induces an enumeration of the rational numbers, a result which was found by gazing at a computer screen and looking for patterns.

Moral: be sure to spend many hours each day gazing at your computer screen and looking for patterns.

4 An Unexpected Factorization

One of the great strengths of computer algebra systems is that they are very good at factoring. They can factor very large integers and very complicated expressions. Whenever you run into some large expression as the answer to a problem that interests you, it is good practice to ask your CAS to factor it for you. Sometimes the results will surprise you. This is one such story.

The theory of *Young tableaux* forms an important part of modern combinatorics. To create a Young tableau we choose a positive integer n and a partition $n = a_1 + a_2 + \cdots + a_k$ of that integer. We'll use the integer $n = 6$ and the partition $6 = 3 + 2 + 1$ as an example. Next we draw the *Ferrers board* of the partition, which is a truncated chessboard that has a_1 squares in its first row, a_2 in its second row, etc., the rows being left-justified. In our example, the Ferrers board is as shown in figure 1.

To make a tableau, we insert the labels $1, 2, \ldots, n$ into the n cells of the board in such a way that the labels increase from left to right across each row and increase from top to bottom down every column. With our example, one way to do this is as shown in figure 2.

One of several important properties of tableaux is that there is a one-to-one correspondence, known as the Robinson–Schensted–Knuth (RSK) correspondence, which assigns to every permutation of n letters a pair of tableaux of the same shape. One use of the RSK correspondence is to find the length of the longest increasing subsequence in the vector of values of a given permutation. It turns out that this length is the same as the length of the first row of either of the tableaux to which the permutation corresponds under the RSK

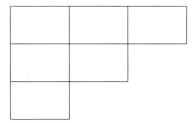

Figure 1 The Ferrers board.

1	2	4
3	6	
5		

Figure 2 A Young tableau.

mapping. This fact gives us a good way, algorithmically speaking, of finding the length of the longest increasing subsequence of a given permutation.

Now suppose that $u_k(n)$ is the number of permutations of n letters that have no increasing subsequence of length greater than k. A spectacular theorem of Gessel (1990) states that

$$\sum_{n\geqslant 0} \frac{u_k(n)}{n!^2} x^{2n} = \det(I_{|i-j|}(2x))_{i,j=1,\ldots,k}, \qquad (5)$$

in which $I_\nu(t)$ is (the modified Bessel function)

$$I_\nu(t) = \sum_{j=0}^{\infty} \frac{(\frac{1}{2}t)^{2j+\nu}}{j!(j+\nu)!}.$$

At any rate, it seems fairly "spectacular" to me that when you place various infinite series such as the above into a $k \times k$ determinant and then expand the determinant, you should find that the coefficient of x^{2n}, when multiplied by $n!^2$, is exactly the number of permutations of n letters with no increasing subsequence longer than k.

Let's evaluate one of these determinants, say the one with $k = 2$. We find that

$$\det(I_{|i-j|}(2x))_{i,j=1,2} = I_0^2 - I_1^2,$$

which of course factors as $(I_0 + I_1)(I_0 - I_1)$. The arguments of the I_ν are all $2x$ and have been omitted.

When $k = 3$, no such factorization occurs. If you ask your CAS for this determinant when $k = 4$, it will show

you

$$I_0^4 - 3I_0^2 I_1^2 + I_1^4 + 4I_0 I_1^2 I_2$$
$$- 2I_0^2 I_2^2 - 2I_1^2 I_2^2 + I_2^4 - 2I_1^3 I_3$$
$$+ 4I_0 I_1 I_2 I_3 - 2I_1 I_2^2 I_3 - I_0^2 I_3^2 + I_1^2 I_3^2,$$

where now we have abbreviated $I_\nu(2x)$ simply by I_ν. If we ask our CAS to factor this last expression, it (surprisingly) replies with

$$(I_0^2 - I_0 I_1 - I_1^2 + 2I_1 I_2 - I_2^2 - I_0 I_3 + I_1 I_3)$$
$$\times (I_0^2 + I_0 I_1 - I_1^2 - 2I_1 I_2 - I_2^2 + I_0 I_3 + I_1 I_3),$$

which is actually of the form $(A + B)(A - B)$, as a quick inspection will reveal.

We have now observed, experimentally, that for $k = 2$ and $k = 4$ Gessel's $k \times k$ determinant has a nontrivial factorization of the form $(A + B)(A - B)$, in which A and B are certain polynomials of degree $k/2$ in the Bessel functions. Such a factorization of a large expression in terms of formal Bessel functions simply cannot be ignored. It demands explanation. Does this factorization extend to all even values of k? It does. Can we say anything in general about what the factors mean? We can.

The key point, as it turns out, is that in Gessel's determinant (5), the matrix entries depend only on $|i - j|$ (such a matrix is called a *Toeplitz* matrix). The determinants of such matrices have a natural factorization, as follows. If $a_0, a_1, \ldots$ is some sequence, and $a_{-i} = a_i$, then we have

$$\det(a_{i-j})_{i,j=1}^{2m}$$
$$= \det(a_{i-j} + a_{i+j-1})_{i,j=1}^{m} \det(a_{i-j} - a_{i+j-1})_{i,j=1}^{m}.$$

When we apply this fact to the present situation it correctly reproduces the above factorizations for $k = 2, 4$, and generalizes them to all even k, as follows.

Let $y_k(n)$ be the number of Young tableaux of n cells whose first row is of length at most k, and let

$$U_k(x) = \sum_{n\geqslant 0} \frac{u_k(n)}{n!^2} x^{2n} \quad \text{and} \quad Y_k(x) = \sum_{n\geqslant 0} \frac{y_k(n)}{n!} x^n.$$

In terms of these two generating functions, the general factorization theorem states that

$$U_k(x) = Y_k(x) Y_k(-x) \quad (k = 2, 4, 6, \ldots).$$

Why is it useful to have such factorizations? For one thing we can equate the coefficients of like powers of x on both sides of this factorization (try it!). We then find an interesting explicit formula that relates the number of Young tableaux of n cells whose first row is of length at most k, on the one hand, and the number of permutations of n letters that have no increasing subsequence of length greater than k, on the other. No more direct proof of this relationship is known. For more details and some further consequences, see Wilf (1992).

Moral: cherchez les factorisations!

5   A Score for Sloane's Database

Here is a case study in which, as it happens, not only was Sloane's database utilized, but Sloane himself was one of the authors of the ensuing research paper.

Eric Weisstein, the creator of the invaluable Web resource *MathWorld*, became interested in the enumeration of 0-1 matrices whose eigenvalues are all positive real numbers. If $f(n)$ is the number of $n \times n$ matrices whose entries are all 0s and 1s and whose eigenvalues are all real and positive, then by computation, Weisstein found for $f(n)$ the values

$$1, 3, 25, 543, 29\,281 \quad (\text{for } n = 1, 2, \ldots, 5).$$

Upon looking up this sequence in Sloane's database, Weisstein found, interestingly, that this sequence is identical, as far as it goes, with sequence A003024 in the database. The latter sequence counts vertex-labeled acyclic directed graphs ("digraphs") of n vertices, and so Weisstein's conjecture was born:

[T]he number of vertex-labeled acyclic digraphs of n vertices is equal to the number of $n \times n$ 0-1 matrices whose eigenvalues are all real and positive.

This conjecture was proved in McKay et al. (2003). En route to the proof of the result, the following somewhat surprising fact was shown.

Theorem 1. *If a 0-1 matrix A has only real positive eigenvalues, then those eigenvalues are all equal to 1.*

To prove this, let $\{\lambda_i\}_{i=1}^n$ be the eigenvalues of A. Then

$$
\begin{aligned}
1 &\geqslant \frac{1}{n}\mathrm{trace}(A) \quad (\text{since all } A_{i,i} \leqslant 1) \\
&= \frac{1}{n}(\lambda_1 + \lambda_2 + \cdots + \lambda_n) \\
&\geqslant (\lambda_1\lambda_2\cdots\lambda_n)^{1/n} \\
&= (\det A)^{1/n} \\
&\geqslant 1,
\end{aligned}
$$

in which the third line uses the arithmetic-geometric mean inequality, and the last line uses the fact that $\det A$ is a positive integer. Since the arithmetic and

geometric means of the eigenvalues are equal, the eigenvalues are all equal, and in fact all $\lambda_i(A) = 1$.

The proof of the conjecture itself works by finding an explicit bijection between the two sets that are being counted. Indeed, let A be an $n \times n$ matrix of 0s and 1s with positive eigenvalues only. Then those eigenvalues are all 1s, so the diagonal of A is all 1s, whence the matrix $A - I$ also has solely 0s and 1s as its entries. Regard $A - I$ as the vertex adjacency matrix of a digraph G. Then (it turns out that) G is acyclic.

Conversely, if G is such a digraph, let B be its vertex adjacency matrix. By renumbering the vertices of G, if necessary, B can be brought to triangular form with zero diagonal. Then $A = I + B$ is a 0-1 matrix with positive real eigenvalues only. But then the same must have been true for the matrix $I + B$ before simultaneously renumbering its rows and columns. For more details and more corollaries, see McKay et al. (2003).

Moral: see if you can find your sequence in the online encyclopedia!

6   The Twenty-One-Stage Rocket

Now we'll describe a successful attack that was carried out by Andrews (1998) on the problem of evaluating the Mills-Robbins-Rumsey determinant, which is the determinant of the $n \times n$ matrix

$$M_n(\mu) = \left(\binom{i + j + \mu}{2j - i}\right)_{0 \leqslant i, j \leqslant n-1}. \tag{6}$$

This problem arose (Mills et al. 1987) in connection with the study of *plane partitions*. A plane partition of an integer n is an (infinite) array $n_{i,j}$ of nonnegative integers whose sum is n, subject to the restriction that the entries $n_{i,j}$ are nonincreasing across each row, and also down each column.

It turns out that $\det M_n(\mu)$ can be expressed neatly as a product, namely as

$$\det M_n(\mu) = 2^{-n} \prod_{k=0}^{n-1} \Delta_{2k}(2\mu), \tag{7}$$

in which

$$\Delta_{2j}(\mu) = \frac{(\mu + 2j + 2)_j (\frac{1}{2}\mu + 2j + \frac{3}{2})_{j-1}}{(j)_j (\frac{1}{2}\mu + j + \frac{3}{2})_{j-1}},$$

and $(x)_j$ is the rising factorial $x(x + 1) \cdots (x + j - 1)$.

The strategy of Andrews's proof is elegant in conception and difficult in execution: we are going to find an upper triangular matrix $E_n(\mu)$, whose diagonal entries are all 1s, such that the matrix

$$M_n(\mu)E_n(\mu) = L_n(\mu) \tag{8}$$

is lower triangular, with the numbers $\{\frac{1}{2}\Delta_{2j}(2\mu)\}_{j=0}^{n-1}$ on its diagonal. Of course, if we can do this, then from (8), since $\det E_n(\mu) = 1$, we will have proved the theorem (7), since the determinant of the product of two matrices is the product of their determinants, and the determinant of a triangular matrix (i.e., of a matrix all of whose entries below the diagonal are 0s) is simply the product of its diagonal entries.

But how shall we find this matrix $E_n(\mu)$? By holding tightly to the hand of our computer and letting it guide us there. More precisely,

(i) we will look at the matrix $E_n(\mu)$ for various small values of n, and from those data we will conjecture the formula for the general (i, j) entry of the matrix; and then

(ii) we will (well actually "we" won't, but Andrews did) prove that the conjectured entries of the matrix are correct.

It was in step (ii) above that an extraordinary twenty-one-stage event occurred which was successfully managed by Andrews. What he did was to set up a system of twenty-one propositions, each of them a fairly technical hypergeometric identity. Next, he carried out a simultaneous induction on these twenty-one propositions. That is to say, he showed that if, say, the thirteenth proposition was true for a certain value of n, then so was the fourteenth, etc., and if they were all true for that value of n, then the first proposition was true for $n + 1$. The reader should be sure to look at Andrews (1998) to gain more of the flavor and substance of what was done than can be conveyed in this short summary.

Here we will confine ourselves to a few comments about step (i) of the program above. So, let's look at the matrix $E_n(\mu)$ for some small values of n. The condition that $E_n(\mu)$ is upper triangular with 1s on the diagonal means that

$$\sum_{k=0}^{j-1} (M_n)_{i,k} e_{k,j} = -(M_n)_{i,j},$$

for $0 \leqslant i \leqslant j - 1$ and $1 \leqslant j \leqslant n - 1$. We can regard these as $\binom{n}{2}$ equations in the $\binom{n}{2}$ above-diagonal entries of $E_n(\mu)$ and we can ask our CAS to find those entries, for some small values of n. Here is $E_4(\mu)$:

$$\begin{pmatrix} 1 & 0 & 0 & 0 \\ 0 & 1 & -\dfrac{1}{\mu + 2} & \dfrac{6(\mu + 5)}{(\mu + 2)(\mu + 3)(2\mu + 11)} \\ 0 & 0 & 1 & -\dfrac{6(\mu + 5)}{(\mu + 3)(2\mu + 11)} \\ 0 & 0 & 0 & 1 \end{pmatrix}.$$

At this point the news is all good. While it is true that the matrix entries are fairly complicated, the fact that leaps off the page and warms the heart of the experimental mathematician is that all of the polynomials in μ factor into linear factors with pleasant-looking integer coefficients. So there is hope for conjecturing a general form of the E matrix. Will this benign situation persist when $n = 5$? A further computation reveals that $E_5(\mu)$ is as shown in figure 3. Now it is a "certainty" that some nice formulas exist for the entries of the general matrix $E_n(\mu)$. The Rate package, described in section 2.3, would certainly facilitate the next step, which is to find general formulas for the entries of the E matrix. The final result is that the (i, j) entry of $E_n(\mu)$ is 0 if $i > j$ and

$$\frac{(-1)^{j-i}(i)_{2(j-i)}(2\mu + 2j + i + 2)_{j-i}}{4^{j-i}(j - i)!(\mu + i + 1)_{j-i}(\mu + j + i + \frac{1}{2})_{j-i}}$$

otherwise.

After divining that the E matrix has the above form, Andrews now faced the task of proving that it works, i.e., that $M_n E_n(\mu)$ is lower triangular and has the diagonal entries specified above. It was in this part of the work that the twenty-one-fold induction was unleashed. Another proof of the evaluation of the Mills–Robbins–Rumsey determinant is in Petkovšek and Wilf (1996). That proof begins with Andrews's discovery of the above form of the $E_n(\mu)$ matrix, and then uses the machinery of the so-called WZ method (Petkovšek et al. 1996), instead of a twenty-one-stage induction, to prove that the matrix performs the desired triangulation (8).

Moral: never give up, even when defeat seems certain.

7   The Computation of π

In 1997, a remarkable formula for π was found (Bailey et al. 1997). This formula permits the computation of just a single hexadecimal digit of π, if desired, using minimal space and time. For example, we might compute the trillionth digit of π, without ever having to deal with any of the earlier ones, in a time that is faster than what we might attain if we had to calculate all of the first trillion digits. For example, Bailey et al. found that in the hexadecimal expansion of π, the block of fourteen digits in positions 10^{10} through $10^{10} + 13$ are 921C73C6838FB2. The formula is

$$\pi = \sum_{i=0}^{\infty} \frac{1}{16^i} \left(\frac{4}{8i + 1} - \frac{2}{8i + 4} - \frac{1}{8i + 5} - \frac{1}{8i + 6} \right). \quad (9)$$

$$\begin{pmatrix} 1 & 0 & 0 & 0 & 0 \\ 0 & 1 & -\dfrac{1}{\mu+2} & \dfrac{6(\mu+5)}{(\mu+2)(\mu+3)(2\mu+11)} & -\dfrac{30(\mu+6)}{(\mu+2)(\mu+3)(\mu+4)(2\mu+15)} \\ 0 & 0 & 1 & -\dfrac{6(\mu+5)}{(\mu+3)(2\mu+11)} & \dfrac{30(\mu+6)}{(\mu+3)(\mu+4)(2\mu+15)} \\ 0 & 0 & 0 & 1 & -\dfrac{6(2\mu+13)}{(\mu+4)(2\mu+15)} \\ 0 & 0 & 0 & 0 & 1 \end{pmatrix}$$

Figure 3 The upper triangular matrix $E_5(\mu)$.

In our discussion here we will limit ourselves to describing how we might have found the specific expansion (9) once we had decided that an interesting expansion of the form

$$\pi = \sum_{i=0}^{\infty} \frac{1}{c^i} \sum_{k=1}^{b-1} \frac{a_k}{bi+k}. \tag{10}$$

might exist. This, of course, leaves open the question of how the discovery of the form (10) was singled out in the first place.

The strategy will be to use the linear dependency algorithm described in section 2.4. More precisely, we want to find a nontrivial integer linear combination of π and the seven numbers

$$\alpha_k = \sum_{i=0}^{\infty} \frac{1}{(8i+k)16^i} \quad (k=1,\dots,7)$$

that sums to 0. As in equation (3), we now compute the seven numbers α_j and we look for a relation

$$m\pi + m_1\alpha_1 + m_2\alpha_2 + \cdots + m_7\alpha_7 = 0$$
$$(m, m_i \in \mathbb{Z})$$

by using, for example, the Maple `IntegerRelations` package. The output vector,

$$(m, m_1, m_2, \dots, m_7) = (1, -4, 0, 0, 2, 1, 1, 0),$$

yields the identity (9). You should do this calculation for yourself, then prove that the apparent identity is in fact true, and, finally, look for something similar that uses powers of 64 instead of 16. Good luck!

Moral: even as late as the year 1997 C.E., something new and interesting was said about the number π.

8   Conclusions

When computers first appeared in mathematicians' environments the almost universal reaction was that they would never be useful for proving theorems since a computer can never investigate infinitely many cases, no matter how fast it is. But computers are useful for proving theorems despite that handicap. We have seen several examples of how a mathematician can act in concert with a computer to explore a world within mathematics. From such explorations there can grow understanding, and conjectures, and roads to proofs, and phenomena that would not have been imaginable in the pre-computer era. This role of computation within pure mathematics seems destined only to expand over the coming years and to be imbued into our students along with EUCLID's [VI.2] axioms and other staples of mathematical education.

At the other end of the rainbow there may lie a more far-reaching role for computers. Perhaps one day we will be able to input some hypotheses and a desired conclusion, press the "Enter" key, and get a printout of a proof. There are a few fields of mathematics in which we can do such things, notably in the proofs of identities (Petkovšek et al. 1996; Greene and Wilf 2007), but in general the road to that brave new world remains long and uncharted.

Further Reading

Andrews, G. E. 1998. Pfaff's method. I. The Mills–Robbins–Rumsey determinant. *Discrete Mathematics* 193:43–60.

Bailey, D. H., and S. Plouffe. 1997. Recognizing numerical constants. In *Proceedings of the Organic Mathematics Workshop, 12–14 December 1995, Simon Fraser University*. Conference Proceedings of the Canadian Mathematical Society, volume 20. Ottawa: Canadian Mathematical Society.

Bailey, D. H., P. Borwein, and S. Plouffe. 1997. On the rapid computation of various polylogarithmic constants. *Mathematics of Computation* 66:903–13.

Borwein, J.. and D. H. Bailey. 2003. *Mathematics by Experiment: Plausible Reasoning in the 21st Century*. Wellesley, MA: A. K. Peters.

Borwein, J., D. H. Bailey, and R. Girgensohn. 2004. *Experimentation in Mathematics: Computational Paths to Discovery*. Wellesley, MA: A. K. Peters.

Calkin, N., and H. S. Wilf. 2000. Recounting the rationals. *American Mathematical Monthly* 107:360-63.

Ferguson, H. R. P., and R. W. Forcade. 1979. Generalization of the Euclidean algorithm for real numbers to all dimensions higher than two. *Bulletin of the American Mathematical Society* 1:912-14.

Gessel, I. 1990. Symmetric functions and *P*-recursiveness. *Journal of Combinatorial Theory* A 53:257-85.

Graham, R. L., D. E. Knuth, and O. Patashnik. 1989. *Concrete Mathematics*. Reading, MA: Addison-Wesley.

Greene, C., and Wilf, H. S. 2007. Closed form summation of *C*-finite sequences. *Transactions of the American Mathematical Society* 359:1161-89.

Lenstra, A. K., H. W. Lenstra Jr., and L. Lovász. 1982. Factoring polynomials with rational coefficients. *Mathematische Annalen* 261(4):515-34.

McKay, B. D., F. E. Oggier, G. F. Royle, N. J. A. Sloane, I. M. Wanless, and H. S. Wilf. 2004. Acyclic digraphs and eigenvalues of $(0,1)$-matrices. *Journal of Integer Sequences* 7: 04.3.3.

Mills, W. H., D. P. Robbins, and H. Rumsey Jr. 1987. Enumeration of a symmetry class of plane partitions. *Discrete Mathematics* 67:43-55.

Petkovšek, M., and H. S. Wilf. 1996. A high-tech proof of the Mills–Robbins–Rumsey determinant formula. *Electronic Journal of Combinatorics* 3:R19.

Petkovšek, M., H. S. Wilf, and D. Zeilberger. 1996. $A = B$. Wellesley, MA: A. K. Peters.

Wilf, H. S. 1992. Ascending subsequences and the shapes of Young tableaux. *Journal of Combinatorial Theory* A 60: 155-57.

——. 1994. *generatingfunctionology*, 2nd edn. New York: Academic Press. (This can also be downloaded at no charge from the author's Web site.)

VIII.6   Advice to a Young Mathematician

The most important thing that a young mathematician needs to learn is of course mathematics. However, it can also be very valuable to learn from the experiences of other mathematicians. The five contributors to this article were asked to draw on their experiences of mathematical life and research, and to offer advice that they might have liked to receive when they were just setting out on their careers. (The title of this entry is a nod to Sir Peter Medawar's well-known book, *Advice to a Young Scientist*.) The resulting contributions were every bit as interesting as we had expected; what was more surprising was that there was remarkably little overlap between the contributions. So here they are, five gems

intended for young mathematicians but surely destined to be read and enjoyed by mathematicians of all ages.

I. Sir Michael Atiyah

Warning

What follows is very much a personal view based on my own experience and reflecting my personality, the type of mathematics that I work on, and my style of work. However, mathematicians vary widely in all these characteristics and you should follow your own instinct. You may learn from others but interpret what you learn in your own way. Originality comes by breaking away, in some respects, from the practice of the past.

Motivation

A research mathematician, like a creative artist, has to be passionately interested in the subject and fully dedicated to it. Without strong internal motivation you cannot succeed, but if you enjoy mathematics the satisfaction you can get from solving hard problems is immense.

The first year or two of research is the most difficult. There is so much to learn. One struggles unsuccessfully with small problems and one has serious doubts about one's ability to prove anything interesting. I went through such a period in my second year of research, and Jean-Pierre Serre, perhaps the outstanding mathematician of my generation, told me that he too had contemplated giving up at one stage.

Only the mediocre are supremely confident of their ability. The better you are, the higher the standards you set yourself—you can see beyond your immediate reach.

Many would-be mathematicians also have talents and interests in other directions and they may have a difficult choice to make between embarking on a mathematical career and pursuing something else. The great Gauss is reputed to have wavered between mathematics and philology, Pascal deserted mathematics at an early age for theology, while Descartes and Leibniz are also famous as philosophers. Some mathematicians move into physics (e.g., Freeman Dyson) while others (e.g., Harish Chandra, Raoul Bott) have moved the other way. You should not regard mathematics as a closed world, and the interaction between mathematics and other disciplines is healthy both for the individual and for society.

Psychology

Because of the intense mental concentration required in mathematics, psychological pressures can be considerable, even when things are going well. Depending on your personality this may be a major or only a minor problem, but one can take steps to reduce the tension. Interaction with fellow students—attending lectures, seminars, and conferences—both widens one's horizons and provides important social support. Too much isolation and introspection can be dangerous, and time spent in apparently idle conversation is not really wasted.

Collaboration, initially with fellow students or one's supervisor, has many benefits, and long-term collaboration with coworkers can be extremely fruitful both in mathematical terms and at the personal level. There is always the need for hard quiet thought on one's own, but this can be enhanced and balanced by discussion and exchange of ideas with friends.

Problems versus Theory

Mathematicians are sometimes categorized as either "problem solvers" or "theorists." It is certainly true that there are extreme cases that highlight this division (Erdős versus Grothendieck, for example) but most mathematicians lie somewhere in between, with their work involving both the solution of problems and the development of some theory. In fact, a theory that does not lead to the solution of concrete and interesting problems is not worth having. Conversely, any really deep problem tends to stimulate the development of theory for its solution (Fermat's last theorem being a classic example)

What bearing does this have on a beginning student? Although one has to read books and papers and absorb general concepts and techniques (theory), realistically, a student has to focus on one or more specific problems. This provides something to chew on and to test one's mettle. A definite problem, which one struggles with and understands in detail, is also an invaluable benchmark against which to measure the utility and strength of available theories.

Depending on how the research goes, the eventual Ph.D. thesis may strip away most of the theory and focus only on the essential problem, or else it may describe a wider scenario into which the problem naturally fits.

The Role of Curiosity

The driving force in research is curiosity. When is a particular result true? Is that the best proof, or is there a more natural or elegant one? What is the most general context in which the result holds?

If you keep asking yourself such questions when reading a paper or listening to a lecture, then sooner or later a glimmer of an answer will emerge—some possible route to investigate. When this happens to me I always take time out to pursue the idea to see where it leads or whether it will stand up to scrutiny. Nine times out of ten it turns out to be a blind alley, but occasionally one strikes gold. The difficulty is in knowing when an idea that is initially promising is in fact going nowhere. At this stage one has to cut one's losses and return to the main road. Often the decision is not clear-cut, and in fact I frequently return to a previously discarded idea and give it another try.

Ironically, good ideas can emerge unexpectedly from a bad lecture or seminar. I often find myself listening to a lecture where the result is beautiful and the proof ugly and complicated. Instead of trying to follow a messy proof on the blackboard, I spend the rest of the hour thinking about producing a more elegant proof. Usually, but not always, without success, but even then my time is better spent, since I have thought hard about the problem in my own way. This is much better than passively following another person's reasoning.

Examples

If you are, like me, someone who prefers large vistas and powerful theories (I was influenced but not converted by Grothendieck), then it is essential to be able to test general results by applying them to simple examples. Over the years I have built up a large array of such examples, drawn from a variety of fields. These are examples where one can do concrete calculations, sometimes with elaborate formulas, that help to make the general theory understandable. They keep your feet on the ground. Interestingly enough, Grothendieck eschewed examples, but fortunately he was in close touch with Serre, who was able to rectify this omission. There is no clear-cut distinction between example and theory. Many of my favorite examples come from my early training in classical projective geometry: the twisted cubic, the quadric surface, or the Klein representation of lines in 3-space. Nothing could be more concrete or classical and all can be looked at algebraically or geometrically, but each illustrates and is the first case in a large class of examples which then become a theory: the theory of rational curves, of homogeneous spaces, or of Grassmannians.

Another aspect of examples is that they can lead off in different directions. One example can be generalized in several different ways or illustrate several different principles. For instance, the classical conic is a rational curve, a quadric, and a Grassmannian all in one.

But most of all a good example is a thing of beauty. It shines and convinces. It gives insight and understanding. It provides the bedrock of belief.

Proof

We are all taught that "proof" is the central feature of mathematics, and Euclidean geometry with its careful array of axioms and propositions has provided the essential framework for modern thought since the Renaissance. Mathematicians pride themselves on absolute certainty, in comparison with the tentative steps of natural scientists, let alone the woolly thinking of other areas.

It is true that, since Gödel, absolute certainty has been undermined, and the more mundane assault of computer proofs of interminable length has induced some humility. Despite all this, proof retains its cardinal role in mathematics, and a serious gap in your argument will lead to your paper being rejected.

However, it is a mistake to identify research in mathematics with the process of producing proofs. In fact, one could say that all the really creative aspects of mathematical research precede the proof stage. To take the metaphor of the "stage" further, you have to start with the idea, develop the plot, write the dialogue, and provide the theatrical instructions. The actual production can be viewed as the "proof": the implementation of an idea.

In mathematics, ideas and concepts come first, then come questions and problems. At this stage the search for solutions begins, one looks for a method or strategy. Once you have convinced yourself that the problem has been well-posed, and that you have the right tools for the job, you then begin to think hard about the technicalities of the proof.

Before long you may realize, perhaps by finding counterexamples, that the problem was incorrectly formulated. Sometimes there is a gap between the initial intuitive idea and its formalization. You left out some hidden assumption, you overlooked some technical detail, you tried to be too general. You then have to

go back and refine your formalization of the problem. It would be an unfair exaggeration to say that mathematicians rig their questions so that they can answer them, but there is undoubtedly a grain of truth in the statement. The art in good mathematics, and mathematics is an art, is to identify and tackle problems that are both interesting and solvable.

Proof is the end product of a long interaction between creative imagination and critical reasoning. Without proof the program remains incomplete, but without the imaginative input it never gets started. One can see here an analogy with the work of the creative artist in other fields: writer, painter, composer, or architect. The vision comes first, it develops into an idea that gets tentatively sketched out, and finally comes the long technical process of erecting the work of art. But the technique and the vision have to remain in touch, each modifying the other according to its own rules.

Strategy

In the previous section I discussed the philosophy of proof and its role in the whole creative process. Now let me turn to the most down-to-earth question of interest to the young practitioner. What strategy should one adopt? How do you actually go about finding a proof?

This question makes little sense in the abstract. As I explained in the previous section a good problem always has antecedents: it arises from some background, it has roots. You have to understand these roots in order to make progress. That is why it is always better to find your own problem, asking your own questions, rather than getting it on a plate from your supervisor. If you know where a problem comes from, why the question has been asked, then you are halfway toward its solution. In fact, asking the right question is often as difficult as solving it. Finding the right context is an essential first step.

So, in brief, you need to have a good knowledge of the history of the problem. You should know what sort of methods have worked with similar problems and what their limitations are.

It is a good idea to start thinking hard about a problem as soon as you have fully absorbed it. To get to grips with it, there is no substitute for a hands-on approach. You should investigate special cases and try to identify where the essential difficulty lies. The more you know about the background and previous methods, the more techniques and tricks you can try. On the other hand, ignorance is sometimes bliss. J. E. Littlewood is reported to have set each of his research students to work on a disguised version of the Riemann hypothesis, letting them know what he had done only after six months. He argued that the student would not have the confidence to attack such a famous problem directly, but might make progress if not told of the fame of his opponent! The policy may not have led to a proof of the Riemann hypothesis, but it certainly led to resilient and battle-hardened students.

My own approach has been to try to avoid the direct onslaught and look for indirect approaches. This involves connecting your problem with ideas and techniques from different fields that may shed unexpected light on it. If this strategy succeeds, it can lead to a beautiful and simple proof, which also "explains" why something is true. In fact, I believe the search for an explanation, for understanding, is what we should really be aiming for. Proof is simply part of that process, and sometimes its consequence.

As part of the search for new methods it is a good idea to broaden your horizons. Talking to people will extend your general education and will sometimes introduce you to new ideas and techniques. Very occasionally you may get a productive idea for your own research or even for a new direction.

If you need to learn a new subject, consult the literature but, even better, find a friendly expert and get instruction "from the horse's mouth"—it gives more insight more quickly.

As well as looking forward, and being alert to new developments, you should not forget the past. Many powerful mathematical results from earlier eras have got buried and have been forgotten, coming to light only when they have been independently rediscovered. These results are not easy to find, partly because terminology and style change, but they can be gold mines. As usual with gold mines, you have to be lucky to strike one, and the rewards go to the pioneers.

Independence

At the start of your research your relationship with your supervisor can be crucial, so choose carefully, bearing in mind subject matter, personality, and track record. Few supervisors score highly on all three. Moreover, if things do not work out well during the first year or so, or if your interests diverge significantly, then do not hesitate to change supervisors or even universities. Your supervisor will not be offended and may even be relieved!

Sometimes you may be part of a large group and may interact with other members of the faculty, so that you effectively have more than one supervisor. This can be helpful in that it provides different inputs and alternative modes of work. You may also learn much from fellow students in such large groups, which is why choosing a department with a large graduate school is a good idea.

Once you have successfully earned your Ph.D. you enter a new stage. Although you may still carry on collaborating with your supervisor and remain part of the same research group, it is healthy for your future development to move elsewhere for a year or more. This opens you up to new influences and opportunities. This is the time when you have the chance to carve out a niche for yourself in the mathematical world. In general, it is not a good idea to continue too closely in the line of your Ph.D. thesis for too long. You have to show your independence by branching out. It need not be a radical change of direction but there should be some clear novelty and not simply a routine continuation of your thesis.

Style

In writing up your thesis your supervisor will normally assist you in the manner of presentation and organization. But acquiring a personal style is an important part of your mathematical development. Although the needs may vary, depending on the kind of mathematics, many aspects are common to all subjects. Here are a number of hints on how to write a good paper.

(i) Think through the whole logical structure of the paper before you start to write.

(ii) Break up long complex proofs into short intermediate steps (lemmas, propositions, etc.) that will help the reader.

(iii) Write clear coherent English (or the language of your choice). Remember that mathematics is also a form of literature.

(iv) Be as succinct as it is possible to be while remaining clear and easy to understand. This is a difficult balance to achieve.

(v) Identify papers that you have enjoyed reading and imitate their style.

(vi) When you have finished writing the bulk of your paper go back and write an introduction that explains clearly the structure and main results as well as the general context. Avoid unnecessary jargon and aim at a general mathematical reader, not just a narrow expert.

(vii) Try out your first draft on a colleague and take heed of any suggestions or criticisms. If even your close friend or collaborator has difficulty understanding it, then you have failed and need to try harder.

(viii) If you are not in a desperate hurry to publish, put your paper aside for a few weeks and work on something else. Then return to your paper and read it with a fresh mind. It will read differently and you may see how to improve it.

(ix) Do not hesitate to rewrite the paper, perhaps from a totally new angle, if you become convinced that this will make it clearer and easier to read. Well-written papers become "classics" and are widely read by future mathematicians. Badly written papers are ignored or, if they are sufficiently important, they get rewritten by others.

II. Béla Bollobás

"There is no permanent place in this world for ugly mathematics," wrote Hardy; I believe that it is just as true that there is no place in this world for unenthusiastic, dour mathematicians. Do mathematics only if you are passionate about it, only if you would do it even if you had to find the time for it after a full day's work in another job. Like poetry and music, mathematics is not an occupation but a vocation.

Taste is above everything. It is a miracle of our subject that there seems to be a consensus as to what constitutes good mathematics. You should work in areas that are important and unlikely to dry up for a long time, and you should work on problems that are beautiful and important: in a good area there will be plenty of these, and not just a handful of well-known problems. Indeed, aiming too high all the time may lead to long barren periods: these may be tolerated at some stage of your life, but at the beginning of your career it is best to avoid them.

Strive for a *balance* in your mathematical activity: research should and does come first for real mathematicians, but in addition to doing research, do plenty of reading and teach well. Have fun with mathematics at all levels, even if it has (almost) no bearing on your research. Teaching should not be a burden but a source of inspiration.

Research should never be a chore (unlike writing up): you should choose problems that you find it difficult *not* to think about. This is why it is good if you get *yourself* hooked on problems rather than working on problems as if you were doing a task imposed on you. At the very beginning of your career, when you are a research student, you should use your experienced supervisor to help you judge problems that you have found and like, rather than working on a problem that he has handed to you, which may not be to your taste. After all, your supervisor should have a fairly good idea whether a certain problem is worth your efforts or not, while he may not yet know your strength and taste. Later in your career, when you can no longer rely on your supervisor, it is frequently inspiring to talk to sympathetic colleagues.

I would recommend that at any one time you have problems of two types to work on.

(i) A "dream": a big problem that you would love to solve, but you cannot reasonably *expect* to solve.

(ii) Some very worthwhile problems that you feel you should have a good chance of solving, given enough time, effort, and luck.

In addition, there are two more types you should consider, although these are less important than the previous ones.

(i) From time to time, work on problems that should be below your dignity and that you can be confident of doing rather quickly, so that time spent on them will not jeopardize your success with the proper problems.

(ii) On an even lower level, it is always fun to do problems that are not really research problems (although they may have been some years ago) but are beautiful enough to spend time on: doing them will give you pleasure and will sharpen your ability to be inventive.

Be patient and persistent. When thinking about a problem, perhaps the most useful device you can employ is to bear the problem in mind all the time: it worked for Newton, and it has worked for many a mortal as well. Give yourself time, especially when attacking major problems; promise yourself that you will spend a certain amount of time on a big problem without expecting much, and after that take stock and decide what to do next. Give your approach a chance to work, but do not be so wrapped up in it that you miss other ways of attacking the problem. Be mentally agile: as Paul Erdős put it, keep your brain open.

Do not be afraid to make mistakes. A mistake for a chess player is fatal; for a mathematician it is par for the course. What you should be terrified of is a blank sheet in front of you after having thought about a problem for a little while. If after a session your wastepaper basket is full of notes of failed attempts, you may still be doing very well. Avoid pedestrian approaches, but always be happy to put in work. In particular, doing the simplest cases of a problem is unlikely to be a waste of time and may well turn out to be very useful.

When you spend a significant amount of time on a problem, it is easy to underestimate the progress you have made, and it is equally easy to overestimate your ability to remember it all. It is best to write down even your very partial results: there is a good chance that your notes will save you a great deal of time later.

If you are lucky enough to have made a breakthrough, it is natural to feel fed up with the project and to want to rest on your laurels. Resist this temptation and see what else your breakthrough may give you.

As a young mathematician, your main advantage is that you have plenty of time for research. You may not realize it, but it is very unlikely that you will ever again have as much time as you do at the beginning of your career. Everybody feels that there is not enough time to do mathematics, but as the years pass this feeling gets more and more acute, and more and more justified.

Turning to *reading*, young people are at a disadvantage when it comes to the amount of mathematics they have read, so to compensate for this, read as much as you can, both in your general area and in mathematics as a whole. In your own research area, make sure that you read many papers written by the best people. These papers are often not as carefully written as they could be, but the quality of the ideas and results should amply reward you for the effort you have to make to read them. Whatever you read, be alert: try to anticipate what the author will do and try to think up a better attack. When the author takes the route you had in mind, you will be happy, and when he chooses to go a different way, you can look forward to finding out why. Ask yourself questions about the results and proofs, even if they seem simpleminded: they will greatly help your understanding.

On the other hand, it is often useful *not* to read up everything about an open problem you are about to attack: once you have thought deeply about it and apparently got nowhere, you can (and should) read the failed attempts of others.

Keep your ability to be surprised, do not take phenomena for granted, appreciate the results and ideas you read. It is all too easy to think that you know what

is going on: after all, you have just read the proof. Outstanding people often spend a great deal of time digesting new ideas. It is not enough for them to know a circle of theorems and understand their proofs: they want to feel them in their blood.

As your career progresses, always keep your mind open to new ideas and new directions: the mathematical landscape changes all the time, and you will probably have to as well if you do not want to be left behind. Always sharpen your tools and learn new ones.

Above everything, *enjoy mathematics and be enthusiastic about it.* Enjoy your research, look forward to reading about new results, feed the love of mathematics in others, and even in your recreation have fun with mathematics by thinking about beautiful little problems you come across or hear from your colleagues.

If I wanted to sum up the advice we should all follow in order to be successful in the sciences and the arts, I could hardly do better than recall what Vitruvius wrote over two thousand years ago:

> *Neque enim ingenium sine disciplina aut disciplina sine ingenio perfectum artificem potest efficere.*
>
> For neither genius without learning nor learning without genius can make a perfect artist.

III. Alain Connes

Mathematics is the backbone of modern science and a remarkably efficient source of new concepts and tools for understanding the "reality" in which we participate. The new concepts themselves are the result of a long process of "distillation" in the alembic of human thought.

I was asked to write some advice for young mathematicians. My first observation is that each mathematician is a special case, and in general mathematicians tend to behave like "fermions," i.e., they avoid working in areas that are too trendy, whereas physicists behave a lot more like "bosons," which coalesce in large packs, often "overselling" their achievements—an attitude that mathematicians despise.

It might be tempting at first to regard mathematics as a collection of separate branches, such as geometry, algebra, analysis, number theory, etc., where the first is dominated by the attempt to understand the concept of "space," the second by the art of manipulating symbols, the third by access to "infinity" and the "continuum," and so on.

This, however, does not do justice to one of the most important features of the mathematical world, namely that it is virtually impossible to isolate any of the above parts from the others without depriving them of their essence. In this way the corpus of mathematics resembles a biological entity, which can only survive as a whole and which would perish if separated into disjoint pieces.

The scientific life of mathematicians can be pictured as an exploration of the geography of the "mathematical reality" which they unveil gradually in their own private mental frame.

This process often begins with an act of rebellion against the dogmatic descriptions of that space that can be found in existing books. Young, prospective mathematicians begin to realize that their own perception of the mathematical world captures some features that do not quite fit in with the existing dogma. This initial rebellion is, in most cases, due to ignorance, but it can nevertheless be beneficial, as it frees people from reverence for authority and allows them to rely on their intuition, provided that that intuition can be backed up by actual proofs. Once a mathematician truly gets to know, in an original and "personal" manner, some small part of the mathematical world, however esoteric it may look at first,[1] the journey can properly start. It is of course vital not to break the "fil d'Arianne" ("Ariadne's thread"): that way one can constantly keep a fresh eye on whatever one encounters along the way, but one can also go back to the source if one ever begins to feel lost.

It is also vital to keep moving. Otherwise, one risks confining oneself to a relatively small area of extreme technical specialization, thereby limiting one's perception of the mathematical world and of its huge, even bewildering, diversity.

The fundamental point in this respect is that, even though many mathematicians have spent their lives exploring different parts of that world, with different perspectives, they all agree on its contours and interconnections. Whatever the origin of one's journey, one day, if one walks far enough, one is bound to stumble on a well-known town: for instance, elliptic functions, modular forms, or zeta functions. "All roads lead to Rome," and the mathematical world is "connected." Of course, this is not to say that all parts of mathematics look alike, and it is worth quoting what Grothendieck says (in *Récoltes et Semailles*) in comparing the landscape of analysis in which he first worked with that of algebraic geometry, in which he spent the rest of his mathematical life:

> Je me rappelle encore de cette impression saisissante (toute subjective certes), comme si je quittais des steppes arides et revêches, pour me retrouver soudain dans une sorte de "pays promis" aux richesses luxuriantes, se multipliant à l'infini partout où il plait à la main de se poser, pour cueillir ou pour fouiller.[2]

Most mathematicians adopt a pragmatic attitude and see themselves as explorers of this "mathematical world" whose existence they do not have any wish to question, and whose structure they uncover by a mixture of intuition and a great deal of rational thought. The former is not so different from "poetical desire" (as emphasized by the French poet Paul Valéry), while the latter requires intense periods of concentration.

Each generation builds a mental picture that reflects their own understanding of this world. They construct mental tools that penetrate more and more deeply into it, so that they can explore aspects of it that were previously hidden.

Where things get really interesting is when unexpected bridges emerge between parts of the mathematical world that were remote from each other in the mental picture that had been developed by previous generations of mathematicians. When this happens, one gets the feeling that a sudden wind has blown away the fog that was hiding parts of a beautiful landscape. In my own work this type of great surprise has come mostly from the interaction with physics. The mathematical concepts that arise naturally in physics often turn out to be fundamental, as Hadamard pointed out. For him they exhibit

> not this short lived novelty which can too often influence the mathematician left to his own devices, but the infinitely fecund novelty that springs from the nature of things.

I will end this article with some more "practical" advice. Note, though, that each mathematician is a "special case" and one should not take the advice too seriously.

Walks. One very sane exercise, when fighting with a very complicated problem (often involving computations), is to go for a long walk (no paper or pencil)

---

1. My own starting point was the localization of roots of polynomials. Fortunately, I was invited at a very early age to attend a conference in Seattle, at which I was introduced to the roots of all my future work on factors.

2. Translation: "I still remember this strong impression (completely subjective of course), as if I was leaving dry and gloomy steppes and finding myself suddenly in a sort of 'promised land' of luxuriant richness, which spread out to infinity wherever one might wish to put out one's hand to gather from it or delve about in it."

and do the computation in one's head, irrespective of whether one initially feels that "it is too complicated to be done like that." Even if one does not succeed, it trains the live memory and sharpens one's skills.

Lying down.   Mathematicians usually have a hard time explaining to their partner that the times when they work with most intensity are when they are lying down in the dark on a sofa. Unfortunately, with e-mail and the invasion of computer screens in all mathematical institutions, the opportunity to isolate oneself and concentrate is becoming rarer, and all the more valuable.

Being brave.   There are several phases in the process that leads to the discovery of new mathematics. While the checking phase is scary, but involves just rationality and concentration, the first, more creative, phase is of a totally different nature. In some sense, it requires a kind of protection of one's ignorance, since this also protects one from the billions of reasons there will always be for not looking at a problem that has already been unsuccessfully attacked by many other mathematicians.

Setbacks.   Throughout their working lives, including at the very early stages, mathematicians will receive preprints from competitors and feel disrupted. The only suggestion I have here is to try to convert this feeling of frustration into an injection of positive energy for working harder. However, this is not always easy.

Grudging approbation.   A colleague of mine once said, "We [mathematicians] work for the grudging approbation of a few friends." It is true that, since research work is of a rather solitary nature, we badly need that approbation in one way or another, but quite frankly one should not expect much. In fact, the only real judge is oneself. Nobody else is in as good a position to know what work was involved, and caring too much about the opinion of others is a waste of time: so far no theorem has been proved as the result of a vote. As Feynman put it, "Why do you care what other people think?"

IV. Dusa McDuff

I started my adult life in a very different situation from most of my contemporaries. Always brought up to think I would have an independent career, I had also received a great deal of encouragement from my family and school to do mathematics. Unusually, my girls' school had a wonderful mathematics teacher who showed me the beauty of Euclidean geometry and calculus. In contrast, I did not respect the science teachers, and since those at university were not much better I never really learned any physics.

Very successful within this limited sphere, I was highly motivated to be a research mathematician. While in some respects I had enormous self-confidence, in other ways I grew to feel very inadequate. One basic problem was that somehow I had absorbed the message that women are second rate as far as professional life is concerned and are therefore to be ignored. I had no female friends and did not really value my kind of intelligence, thinking it boring and practical (female), and not truly creative (male). There were many ways of saying this: women keep the home fires burning while men go out into the world, women are muses not poets, women do not have the true soul needed to be a mathematician, etc. And there still are many ways of saying this. Recently an amusing letter circulated among my feminist friends: it listed various common and contradictory prejudices in different scientific fields, the message being that women are perceived to be incapable of whatever is most valued.

Another problem that became apparent a little later was that I had managed to write a successful Ph.D. thesis while learning very little mathematics. My thesis was in von Neumann algebras, a specialized topic that did not relate to anything with real meaning for me. I could see no way forward in that field, and yet I knew almost nothing else. When I arrived in Moscow in my last year of graduate study, Gel'fand gave me a paper to read on the cohomology of the Lie algebra of vector fields on a manifold, and I did not know what cohomology was, what a manifold was, what a vector field was, or what a Lie algebra was.

Though this ignorance was partly the fault of an overspecialized educational system, it also resulted from my lack of contact with the wider world of mathematics. I had solved the problem of how to reconcile being a woman with being a mathematician by essentially leading two separate lives. My isolation increased upon my return from Moscow. Having switched fields from functional analysis to topology, I had little guidance, and I was too afraid of appearing ignorant to ask many questions. Also, I had a baby while I was a postdoc, and was therefore very busy coping with practical matters. At that stage, with no understanding of the process of doing mathematics, I was learning mostly by reading, unaware of the essential role played by formulating questions and trying out one's own, perhaps naive, ideas. I also had no understanding of how to build a career. Good things do not just happen: one has to apply for fellowships and jobs and keep an eye out for interesting conferences. It would certainly have helped to have had a mentor to suggest better ways of dealing with all these difficulties.

I probably most needed to learn how to ask good questions. As a student, one's job is not only to learn enough to be able to answer questions posed by others, but also to learn how to frame questions that might lead somewhere interesting. When studying something new I often used to start in the middle, using some complicated theory already developed by others. But often one sees further by starting with the simplest questions and examples, because that makes it easier to understand the basic problem and then perhaps to find a new approach to it. For example, I have always liked working with Gromov's nonsqueezing theorem in symplectic geometry, which imposes restrictions on the ways a ball can be manipulated in a symplectic way. This very fundamental and geometric result somehow resonates for me, and so forms a solid basis from which to start exploring.

These days people are much more aware that mathematics is a communal endeavor: even the most brilliant idea gets meaning only from its relation to the whole. Once one has an understanding of the context, it is often very important and fruitful to work by oneself. However, while one is learning it is vital to interact with others.

There have been many successful attempts to facilitate such communication, by changing the structure of buildings, of conferences and meetings, of departmental programs, and also, less formally, of seminars and lectures. It is amazing how the atmosphere in a seminar changes when a senior mathematician, instead of going to sleep or looking bored, asks questions that clarify and open up the discussion for everyone there. Often people (both young and old) are intimidated into silence because they fear showing their ignorance, lack of imagination, or other fatal defect. But in the face of a subject as difficult and beautiful as mathematics, everyone has something to learn from others. Now there are many wonderful small conferences and workshops, organized so that it is easy to have discussions both about the details of specific theories and also about formulating new directions and questions.

The problem of how to reconcile being a woman and a mathematician is still of concern, although the idea that mathematics is intrinsically unfeminine is much less prevalent. I do not think that we women are as fully present in the world of mathematics as we could be, but there are enough of us that we can no longer be dismissed as exceptions. I have found meetings intended primarily for women to be unexpectedly worthwhile; the atmosphere is different when a lecture room is full of women discussing mathematics. Also, as is increasingly understood, the real question is how *any* young person can build a satisfying personal life while still managing to be a creative mathematician. Once people start working on this in a serious way, we will have truly come a long way.

V. Peter Sarnak

I have guided quite a number of Ph.D. students over the years, which perhaps qualifies me to write as an experienced mentor. When advising a brilliant student (and I have been fortunate enough to have had my fair share of these) the interaction is a bit like telling someone to dig for gold in some general area and offering just a few vague suggestions. Once they move into action with their skill and talent they find diamonds instead (and of course, after the fact one cannot resist saying "I told you so"). In these cases, and in most others as well, the role of a senior mentor is more like that of a coach: one provides encouragement and makes sure that the person being mentored is working on interesting problems and is aware of the basic tools that are available. Over the years I have found myself repeating certain comments and suggestions that may have been found useful. Here is a list of some of them.

(i) When learning an area, one should combine reading modern treatments with a study of the original papers, especially papers by the masters of our subject. One of the troubles with recent accounts of certain topics is that they can become too slick. As each new author finds cleverer proofs or treatments of a theory, the treatment evolves toward the one that contains the "shortest proofs." Unfortunately, these are often in a form that causes the new student to ponder, "How did anyone think of this?" By going back to the original sources one can usually see the subject evolving naturally and understand how it has reached its modern form. (There will remain those unexpected and brilliant steps at which one can only marvel at the genius of the inventor, but there are far fewer of these than you might think.) As an example, I usually recommend reading Weyl's original papers on the representation theory of compact Lie groups and the derivation of his character formula, alongside one of the many modern treatments. Similarly, I recommend his book *The Concept of a Riemann Surface* to someone who knows complex analysis and wants to learn about the modern theory of Riemann surfaces, which is of central importance to many areas of mathematics. It is also instructive to study the collected works of superb mathematicians such as Weyl. Besides learning their theorems one uncovers how their minds work. There is almost always a natural line of thought that leads from one paper to the next and certain developments are then appreciated as inevitable. This can be very inspiring.

(ii) On the other hand, you should question dogma and "standard conjectures," even if these have been made by brilliant people. Many standard conjectures are made on the basis of special cases that one understands. Beyond that, they are sometimes little more than wishful thinking: one just hopes that the general picture is not significantly different from the picture that the special cases suggest. There are a number of instances that I know of where someone set out to prove a result that was generally believed to be true and made no progress until they seriously questioned it. Having said that, I also find it a bit irritating when, for no particularly good reason, skepticism is thrown on certain special conjectures, such as the Riemann hypothesis, or on their provability. While as a scientist one should certainly adopt a critical attitude (especially toward some of the artificial objects that we mathematicians have invented), it is important psychologically that we have beliefs about our mathematical universe and about what is true and what is provable.

(iii) Do not confuse "elementary" with "easy": a proof can certainly be elementary without being easy. In fact, there are many examples of theorems for which a little sophistication makes the proof easy to understand and brings out the underlying ideas, whereas an elementary treatment that avoids sophisticated notions hides what is going on. At the same time, beware of equating sophistication with quality or with the "beef of an argument" (an expression that I apparently like to use a lot in this context: many of my former students have teased me about it). There is a tendency among some young mathematicians to think that using fancy and sophisticated language means that what they are doing is deep. Nevertheless, modern tools are powerful when they are understood properly and when they

are combined with new ideas. Those working in certain fields (number theory, for example) who do not put in the time and substantial effort needed to learn these tools are putting themselves at a great disadvantage. Not to learn the tools is like trying to demolish a building with just a chisel. Even if you are very adept at using the chisel, somebody with a bulldozer will have a huge advantage and will not need to be nearly as skilful as you.

(iv) Doing research in mathematics is frustrating and if being frustrated is something you cannot get used to, then mathematics may not be an ideal occupation for you. Most of the time one is stuck, and if this is not the case for you, then either you are exceptionally talented or you are tackling problems that you knew how to solve before you started. There is room for some work of the latter kind, and it can be of a high quality, but most of the big breakthroughs are earned the hard way, with many false steps and long periods of little progress, or even negative progress. There are ways to make this aspect of research less unpleasant. Many people these days work jointly, which, besides the obvious advantage of bringing different expertise to bear on a problem, allows one to share the frustration. For most people this is a big positive (and in mathematics the corresponding sharing of the joy and credit on making a breakthrough has not, so far at least, led to many big fights in the way that it has in some other areas of science). I often advise students to try to have a range of problems at hand at any given moment. The least challenging should still be difficult enough that solving it will give you satisfaction (for without that, what is the point?) and with luck it will be of interest to others. Then you should have a range of more challenging problems, with the most difficult ones being central unsolved problems. One should attack these on and off over time, looking at them from different points of view. It is important to keep exposing oneself to the possibility of solving very difficult problems and perhaps benefiting from a bit of luck.

(v) Go to your departmental colloquium every week, and hope that its organizers have made some good choices for speakers. It is important to have a broad awareness of mathematics. Besides learning about interesting problems and progress that people are making in other fields, you can often have an idea stimulated in your mind when the speaker is talking about something quite different. Also, you may learn of a technique or theory that could be applied to one of the problems that you are working on. In recent times, a good number of the most striking resolutions of long-standing problems have come about from an unexpected combination of ideas from different areas of mathematics.

VIII.7   A Chronology of Mathematical Events
Adrian Rice

Where a personal name is not attached to a specific mathematical work, the corresponding date is an approximate mean date for the period of that person's mathematical activity. Please note that the early dates in this chronology are approximate, with those before 1000 B.C.E. being *very* approximate. With regard to post-1500 entries, unless otherwise specified all dates refer to the apparent date of first publication rather than to the date of composition.

ca. 18 000 B.C.E.   The Ishango Bone, Zaire (possibly the earliest known evidence of counting).

ca. 4000   Clay accounting tokens used in the Middle East.

ca. 3400–3200   Development of numerical notation, Sumer (southern Iraq).

ca. 2050   First attestation of place-value sexagesimal system, Sumer (southern Iraq).

ca. 1850–1650   Old Babylonian mathematics.

ca. 1650   Rhind Papyrus (copy of papyrus from around 1850; largest and best preserved mathematical papyrus from ancient Egypt).

ca. 1400–1300   Decimal numeration, China, found on oracle bones of the Shang Dynasty.

ca. 580   Thales of Miletus ("Father of geometry").

ca. 530–450   The Pythagoreans (number theory, geometry, astronomy, and music).

ca. 450   Zeno's paradoxes of motion.

ca. 370   Eudoxus (theory of proportion, astronomy, method of exhaustion).

ca. 350   Aristotle (logic).

ca. 320   Eudemus's *History of Geometry* (important evidence about knowledge of geometry at the time). Decimal numeration, India.

ca. 300   Euclid's *Elements*.

ca. 250   Archimedes (solid geometry, quadrature, statics, hydrostatics, approximation of π).

ca. 230   Eratosthenes (measurement of Earth's circumference, algorithm for finding prime numbers).

ca. 200   Apollonius's *Conics* (extensive and influential work on conics).

ca. 150   Hipparchus (computed first chord table).

ca. 100   *Jiu Zhang Suan Shu* ("Nine Chapters on Mathematical Procedures"; the most important ancient Chinese mathematical text).

ca. 60 C.E.   Heron of Alexandria (optics, geodesy).

ca. 100   Menelaus's *Spherics* (spherical trigonometry).

ca. 150   Ptolemy's *Almagest* (authoritative text on mathematical astronomy).

ca. 250   Diophantus's *Arithmetica* (solutions of determinant and indeterminant equations, early algebraic symbolism).

ca. 300–400   Sun Zi (Chinese remainder theorem).

ca. 320   Pappus's *Collection* (summarized and extended most important mathematics known at the time).

ca. 370   Theon of Alexandria (commentary on Ptolemy's *Almagest*, revision of Euclid).

ca. 400   Hypatia of Alexandria (commentaries on Diophantus, Apollonius, and Ptolemy).

ca. 450   Proclus (commentary on Euclid Book I, summary of Eudemus's *History*).

ca. 500–510   The *Āryabhaṭīya of Āryabhaṭa* (Indian astronomical treatise that included close approximations to π, $\sqrt{2}$, and the sines of many angles).

ca. 510   Boethius translates Greek works into Latin.

ca. 625   Wang Xiaotong (numerical solutions of cubic equations, expressed geometrically).

628   Brahmagupta's *Brāhmasphuṭasiddhānta* (astronomical treatise, first treatment of so-called Pell's equation).

ca. 710   Venerable Bede (calendar reckoning, astronomy, tides)

ca. 830   Al-Khwārizmī's *Algebra* (theory of equations).

ca. 900   Abū Kāmil (irrational solutions to quadratics).

ca. 970–990   Gerbert d'Aurillac introduces Arabic mathematical techniques to Europe.

ca. 980   Abū al-Wafā' (regarded as first to have calculated the modern trigonometric functions; first to use and publish spherical law of sines).

ca. 1000   Ibn al-Haytham (optics, Alhazen's problem).

ca. 1100   Omar Khayyám (cubic equations, parallel postulate).

1100–1200   Many translations of mathematical works from Arabic to Latin.

ca. 1150   Bhāskara's *Līlāvatī* and *Bījagaṇita* (standard arithmetic and algebra textbooks of the Sanskrit tradition, the latter includes a detailed treatment of Pell's equation).

1202   Fibonacci's *Liber Abacci* (introduces Hindu-Arabic numerals into Europe).

ca. 1270 Yang Hui's *A Detailed Analysis of the Mathematical Methods in the Nine Chapters* (includes diagram similar to "Pascal's triangle," which Hui ascribes to Jia Xian in the eleventh century).

1303 Zhu Shijie's *Siyuan Yujian* ("Precious Mirror of the Four Elements"; elimination methods for solving simultaneous equations in as many as four unknowns).

ca. 1330 Merton School of kinematics, Oxford.

1335 Heytesbury states mean-speed theorem.

ca. 1350 Oresme invents an early form of coordinate geometry, proves the mean-speed theorem, and uses fractional exponents for the first time.

ca. 1415 Brunelleschi demonstrates geometrical method of perspective.

ca. 1464 Regiomontanus's *De Triangulis Omnimodis* (published 1533; first comprehensive European work on plane and spherical trigonometry).

1484 Chuquet's *Triparty en la Science des Nombres* (zero and negative exponents, names "billion," "trillion," etc., introduced).

1489 First appearance in print of "+" and "−" signs.

1494 Pacioli's *Summa de Arithmetica* (summarized all the mathematics known at the time, laying foundation for major progress soon after).

1525 Rudolff's *Die Coss* (partial use of algebraic symbolism; introduces the symbol $\sqrt{\ }$).

1525–28 Dürer publishes on perspective, proportion, and geometrical constructions.

1543 Copernicus's *De Revolutionibus* (proposes heliocentric theory of planetary motion).

1545 Cardano's *Ars Magna* (cubic and quartic equations).

1557 Recorde's *The Whetstone of Witte* (introduces the symbol "=").

1572 Bombelli's *Algebra* (complex numbers).

1585 Stevin's *De Thiende* (popularizes decimal fractions).

1591 Viète's *In Artem Analyticem Isagoge* (use of letters for unknowns).

1609 Kepler's *Astronomia Nova* (first two of Kepler's laws of planetary motion).

1610 Galileo's *Sidereus Nuncius* (description of discoveries made with his telescope, including four moons of Jupiter).

1614 Napier's *Mirifici Logarithmorum Canonis Descriptio* (first table of logarithms).

1619 Kepler's *Harmonice Mundi* (Kepler's third law).

1621 Publication of Bachet's translation of Diophantus's *Arithmetica*.

ca. 1621 Oughtred invents rectilinear slide rule.

1624 Briggs's *Arithmetica Logarithmica* (first printed book of logarithmic tables in base 10).

1631 Harriot's *Artis Analyticae Praxis* (theory of equations).

1632 Galileo's *Dialogue Concerning the Two Chief World Systems* (comparison of Ptolemaic and Copernican theories).

1637 Descartes's *La Géométrie* (geometry via algebraic means).

1638 Galileo's *Discourses—Concerning Two New Sciences* (systematic mathematical treatments of physical problems). Fermat studies Bachet's edition of Diophantus's *Arithmetica* and conjectures Fermat's last theorem.

1642 Pascal invents an adding machine.

1654 Fermat and Pascal correspond on probability. Pascal's *Traité du Triangle Arithmétique*.

1656 Wallis's *Arithmetica Infinitorum* (areas under curves, product formula for $4/\pi$, systematic study of continued fractions).

1657 Huygens's *Ratiociniis in Aleae Ludo* (investigates games of chance).

1664–72 Newton's early work on calculus.

1678 Hooke's *De Potentia Restitutiva* (formulates law of elasticity).

1683 Seki's "Kaifukudai no hō" (procedure for determining the terms of a determinant).

1684 First publication of Leibniz's calculus.

1687 Newton's *Principia* (Newton's laws of motion and gravity, foundation of classical mechanics, derivation of Kepler's laws).

1690 The Bernoullis' earliest work on calculus.

1696 L'Hôpital's *Analyse des Infiniment Petits* (first calculus textbook). Solutions to the brachistochrone problem by Jacob Bernoulli, Johann Bernoulli, Newton, Leibniz, and l'Hôpital (beginning of calculus of variations).

1704 Newton's *De Quadratura* (appendix to *Opticks* containing first publication of Newton's calculus).

1706 Jones introduces the notation "π" for the ratio of the circumference of a circle to its diameter.

1713 Jacob Bernoulli's *Ars Conjectandi* (founding work in probability theory).

1715 Taylor's *Methodus Incrementorum* (Taylor's theorem).

1727–1777 Euler introduces the notation "e" for the exponential function (1727), "$f(x)$" for functions (1734), "$\sum$" for sums (1755), and "i" for $\sqrt{-1}$ (1777).

1734 Berkeley's *The Analyst* (major attack on use of infinitesimals).

1735 Euler solves the Basel problem, proving that $\sum_{n=1}^{\infty}(1/n^2) = \pi^2/6$.

1736 Euler solves the Königsberg bridge problem.

1737 Euler's "Variae observationes circa series infinitis" (Euler product).

1738 Daniel Bernoulli's *Hydrodynamica* (relates fluid flow to pressure).

1742 Goldbach's conjecture (contained in a letter to Euler). Maclaurin's *Treatise of Fluxions* (defense of Newton against attacks of Berkeley).

1743 D'Alembert's *Traité de Dynamique* (d'Alembert's principle).

1744 Euler's *Methodus Inveniendi Lineas Curvas* (calculus of variations).

1747 Euler states law of quadratic reciprocity. D'Alembert derives one-dimensional wave equation as the law governing the motion of a vibrating string.

1748 Euler's *Introductio in Analysin Infinitorum* (introduction of function concept, the formula $e^{i\theta} = \cos\theta + i\sin\theta$, and much much more).

1750–52 Euler's formula for polyhedra.

1757 Euler's "Principes généraux du mouvement des fluides" (Euler equations, start of modern hydrodynamics).

1763 Bayes's *An Essay towards Solving a Problem in the Doctrine of Chances* (Bayes's theorem).

1771 Lagrange's "Réflections sur la résolution algébrique des équations" (codifies work on theory of equations, foreshadowing group theory).

1788 Lagrange's *Méchanique Analitique* (Lagrangian approach to mechanics).

1795 Monge's *Application de l'Analyse à la Géométrie* (differential geometry) and *Géométrie Descriptive* (significant for creation of projective geometry).

1796 Gauss constructs regular 17 gon.

1797 Lagrange's *Théorie des Fonctions Analytiques* (major study of functions as power series).

1798 Legendre's *Théorie des Nombres* (first book dedicated to number theory).

1799 Gauss proves the fundamental theorem of algebra.

1799–1825 Laplace's *Traité de la Mécanique Céleste* (authoritative statement on celestial and planetary mechanics).

1801 Gauss's *Disquisitiones Arithmeticae* (modular arithmetic, first complete proof of law of quadratic reciprocity, many other major results and concepts in number theory).

1805 Legendre's method of least squares.

1809 Gauss on celestial motion.

1812 Laplace's *Théorie Analytique des Probabilités* (introduction of many new concepts in probability, including probability generating functions and the central limit theorem).

1814 Servois coins terms "commutative" and "distributive."

1815 Cauchy on permutations.

1817 Bolzano's early version of the intermediate value theorem.

1821 Cauchy's *Cours d'Analyse* (major contribution to rigorization of analysis).

1822 Fourier's *Théorie Analytique de la Chaleur* (first appearance in print of Fourier series). Poncelet's *Traité des Propriétés Projective des Figures* (rediscovery of projective geometry).

1823 Navier formulates equations now known as the Navier–Stokes equations. Cauchy's *Résumé des Leçons sur le Calcul Infinitésimal*.

1825 Cauchy's integral theorem.

1826 *Journal für die reine und angewandte Mathematik* (otherwise known as *Crelle's Journal*; first major mathematics journal that continues to be important today, published in Germany). Abel proves the insolubility of the quintic by radicals.

1827 Ampère's law for electrodynamics. Gauss's *Disquisitiones Generales Circa Superficies Curva* (Gaussian curvature, *theorema egregium*). Ohm's law for electricity.

1828 Green's theorem.

1829 Dirichlet on convergence of Fourier series. Sturm's theorem. Lobachevskii's non-Euclidean geometry. Jacobi's *Fundamenta Nova Theoriae Functionum Ellipticarum* (key work on elliptic functions).

1830–32 Galois's systematic treatment of solubility of polynomial equations by radicals and inception of group theory.

1832 Bolyai's non-Euclidean geometry.

1836 *Journal de Mathématiques Pures et Appliquées* (also known as *Liouville's Journal*; major mathematics journal which continues to be important today, published in France).

1836–37 Sturm and Liouville create Sturm–Liouville theory.

1837 Dirichlet's theorem on infinitely many primes in arithmetic progressions. Poisson's *Recherches sur la Probabilité des Jugements* (Poisson distribution; the phrase "law of large numbers" coined).

1841 Jacobian determinants.

1843 Hamilton discovers quaternions.

1844 Grassmann's *Ausdehnungslehre* (multilinear algebra). Cayley's earliest work on invariants.

1846 Chebyshev proves a version of the weak law of large numbers.

1851 Riemann's *Grundlagen für eine Theorie der Funktionen einer veränderlichen complexen Grösse* (Cauchy–Riemann equations, Riemann surface).

1854 Cayley's abstract definition of groups. Boole's *Laws of Thought* (algebraic logic). Chebyshev polynomials.

1856–58 Dedekind gives first ever course on Galois theory.

1858 Cayley's "Memoir on the theory of matrices." The Möbius strip.

1859 The Riemann hypothesis.

1863–90 Weierstrass's lectures on analysis popularize modern "epsilon–delta" approach to the subject.

1864 The Riemann–Roch theorem.

1868 Plücker's *Neue Geometrie des Raumes* (line geometry). Beltrami's non-Euclidean geometry. Gordan's theorem on binary forms.

1869–73 Lie develops theory of continuous groups.

1870 Benjamin Peirce's *Linear Associative Algebra*. Jordan's *Traité des Substitutions et des Équations Algébriques* (treatise on groups).

1871 Dedekind introduces modern notions of field, ring, module, ideal.

1872 Klein's *Erlanger Programm*. Sylow's theorems in group theory. Dedekind's *Stetigkeit und Irrationale Zahlen* (construction of the real numbers using cuts).

1873 Maxwell's *Treatise on Electricity and Magnetism* (theory of the electromagnetic field and electromagnetic theory of light; Maxwell's equations). Clifford's biquaternions. Hermite proves transcendence of e.

1874 Cantor discovers that there are different sizes of infinity.

1877–78 Rayleigh's *Theory of Sound* (founding work for the modern theory of sound).

1878 Cantor states the continuum hypothesis.

1881–84 Gibbs's *Elements of Vector Analysis* (basic notions of vector calculus).

1882 Lindemann proves transcendence of π.

1884 Frege's *Grundlagen der Arithmetik* (important attempt to lay foundations for mathematics).

1887 The Jordan curve theorem.

1888 Hilbert's finite basis theorem.

1889 Peano's postulates for natural numbers.

1890 Poincaré's "Sur le problème des trois corps et les équations de la dynamique" (first mathematical description of chaotic behavior in a dynamical system).

1890–1905 Schröder's *Vorlesungen über die Algebra der Logik* (includes concept of Dualgruppe, important in modern lattice theory).

1895 Poincaré's "Analysis situs" (first systematic exposition of general topology; foundation of algebraic topology).

1895–97 Cantor's "Beiträge zur Begründung der transfiniten Mengenlehre" (systematic account of transfinite cardinal numbers).

1896 Frobenius founds representation theory. Hadamard and de la Vallée-Poussin prove the prime number theorem. Hilbert's *Zahlbericht* (major work that shaped modern algebraic number theory).

1897 First International Congress of Mathematicians, Zurich. Hensel introduces p-adic numbers.

1899 Hilbert's *Grundlagen der Geometrie* (rigorous modern axiomatization of Euclidean geometry).

1900 Hilbert's twenty-three problems posed at the Second International Congress of Mathematicians in Paris.

1901 Ricci and Levi-Cività's *Méthodes du Calcul Différentiel Absolut et leurs Applications* (tensor calculus).

1902 Lebesgue's *Intégrale, Longeure, Aire* (Lebesgue integration).

1903 Russell's paradox.

1904 Zermelo's axiom of choice.

1905 Einstein's special theory of relativity published.

1910–13 Whitehead and Russell's *Principia Mathematica* (foundations for mathematics avoiding set-theoretic paradoxes).

1914 Hausdorff's *Grundzüge der Mengenlehre* (topological spaces).

1915 Einstein submits paper giving definitive version of the general theory of relativity.

1916 The Bieberbach conjecture.

1917–18 Fatou and Julia sets (iteration of rational functions).

1920 Takagi existence theorem (major founding result in Abelian class field theory).

1921 Noether's "Idealtheorie in Ringbereichen" (major step in the development of the abstract theory of rings).

1923 Wiener provides mathematical theory of Brownian motion.

1924 Courant and Hilbert's *Methoden der mathematischen Physik* (important summary of methods then known in mathematical physics).

1925 Fisher's *Statistical Methods for Research Workers* (foundational work of modern statistics). Heisenberg's matrix mechanics (first formulation of quantum mechanics). Weyl's character formula (foundational result in representation theory of compact Lie groups).

1926 Schrödinger's wave mechanics (second formulation of quantum mechanics).

1927 Peter and Weyl's "Die Vollständigkeit der primitiven Darstellungen einer geschlossenen kontinuierlichen Gruppe" (birth of modern harmonic analysis). Artin's generalized reciprocity law.

1930 Ramsey's "On a problem of formal logic" (Ramsey's theorem). Van der Waerden's *Moderne Algebra* (revolutionized modern algebra, promoted approaches of Artin and Noether).

1931 Gödel's incompleteness theorems.

1932 Banach's *Théorie des Opérations Linéaires* (first monograph on functional analysis).

1933 Kolmogorov's axioms for probability.

1935 Birth of Bourbaki.

1937 Turing's paper "On computable numbers" (theory of Turing machines).

1938 Gödel proves that the continuum hypothesis and the axiom of choice are consistent with the Zermelo–Fraenkel axioms.

1939 First volume of Bourbaki's *Éléments de Mathématique*.

1943 Colossus (first programmable electronic computer).

1944 Von Neumann and Morgenstern's *Theory of Games and Economic Behavior* (foundation of game theory).

1945 Eilenberg and Mac Lane define the notion of a category. Eilenberg and Steenrod introduce axiomatic approach to homology theories.

1947 Dantzig discovers the simplex algorithm.

1948 Shannon's "A mathematical theory of communication" (foundation of information theory).

1949 The Weil conjectures. Erdős and Selberg give elementary proofs of the prime number theorem.

1950 Hamming's "Error-detecting and error-correcting codes" (the beginning of coding theory).

1955 Roth's theorem on approximating algebraic numbers by rationals. Shimura–Taniyama conjecture.

1959 70 Grothendieck revolutionizes algebraic geometry during his years at the Insitut des Hautes Études Scientifiques.

1963 The Atiyah–Singer index theorem. Cohen proves that the axiom of choice is independent of ZF and the continuum hypothesis is independent of ZFC.

1964 Hironaka's theorem on resolution of singularities.

1965 Publication of Birch–Swinnerton-Dyer conjecture. Carleson's theorem proved.

1966 Robinson's *Non-Standard Analysis*.

1966–67 Langlands introduces conjectures that give rise to the Langlands program (profoundly reshaped much of algebraic number theory and representation theory).

1967 Analytical solution of the KdV equation by Gardner, Greene, Kruskal, and Miura.

1970 Matiyasevich, building on work of Davis, Putnam, and Robinson, proves that there is no algorithm to solve Diophantine equations, thereby solving Hilbert's tenth problem.

1971–72 Notion of NP-completeness developed by Cook, Karp, and Levin.

1974 Deligne completes proof of the Weil conjectures.

1976 The four-color theorem proved by Appel and Haken using a computer program.

1978 RSA algorithm for public-key cryptography. Brooks and Matelski produce the first picture of a Mandelbrot set.

1981 Announcement of classification theorem for finite simple groups (as of 2008, definitive version not yet fully available in print but theorem widely accepted).

1982 Hamilton introduces Ricci flow. Thurston's geometrization conjecture.

1983 Faltings proves the Mordell conjecture.

1984 De Branges proves the Bieberbach conjecture.

1985 Masser and Oesterlé formulate the ABC conjecture.

1989 Anosov and Bolibruch answer the Riemann–Hilbert problem in the negative.

1994 Shor's quantum algorithm for factorization of integers. Fermat's last theorem proved in papers by Wiles and Taylor/Wiles.

2003 Perelman uses Ricci flow to solve the Poincaré conjecture and Thurston's geometrization conjecture.

Index

Bold numbers indicate pages that may be particularly helpful, usually because they contain definitions or detailed discussions.

3-colorability, 583, 598
3SAT, 582-83
4-potential, 490

Ω-logic, *see* logic, Ω-

abacuses, 106
ABC conjecture, 361, 681, 722
Abel, Niels Henrik, 20, 50, 81, 101, 123-24, 331, 709, **760-62**
Abelian field extensions, 720
Abelian groups, **20**, 190, 255, 274, 285, 323, 761
Abelian varieties, *see* varieties, Abelian
absolute convergence, *see* convergence, absolute
abstract algebra, 82, 95-106, 800-801
abstract nonsense, 166, 417
abstraction in mathematics, 20, 55-56, 96, 539
Ackermann function, 112
Acta Mathematica, 784
action principle, *see* Hamilton's least action principle
actions of a group, *see* groups, actions of
actions of a physical system, 287, 311, 478, 524-26, 541
Adams formulas, 609-10
adaptive algorithms, 614
addition, 284-85, 635-36, 638
additive number theory, 715-18
adjacency matrix of a graph, 198, 571
adjectives, 8, 10, 77
adjoint, 172, 179, 186, **188**, 212, **240**, 272, 277
advanced encryption standard, 889-90
affine algebraic groups, 429
affine buildings, 162
affine geometry, 39-40
Airy function, 449

Al-Ṭūsī, Naṣir al-Dīn, 86-87
Al-Karajī, 98
Al-Khwārizmī, Abu Ja'far Muhammad ibn Mūsā, 79-80, 98-99, 106, 133, **736-37**, 986
al-Kitāb al-mukhtaṣar fī ḥisāb al-jabr wa'l-muqābala (al-Khwārizmī), 98, 106, 133, 736
Aldous's theorem, 656
Alexander polynomial, the, 225-27
algebra, 1-4, 57-58, 80, 95-106, 539
Algebra (Bombelli), 737
algebraic closure, 642
algebraic curves, 190, 367, 381, 392, 721
algebraic functions, 241
algebraic geometry, 5, 285, **363-72**
algebraic integrals, 726
algebraic multiplicity, 225
algebraic number theory, 4, 315-32; algebraic integers, **254**, 315, 317-19, 324-30; algebraic numbers, **171**, 222, 241-42, 315, 325-30, 616, 641, 779
algebraic sets, 313, 363, 367
algebraic structure on a surface, 411
algebraic topology, 40, 383-96, 801
algebraically closed fields, 640-42
algebras, 105, 172, 239-40, 272
algorithms, 50, 65, 68, 71-73, **106-17**, 436, 575-77, 579, 707, 871-72
alternating groups, 61, 261, 279, 688
alternating knots, 227
AM-GM inequality, 703-4
amalgamated free products, 437, 442
amicable numbers, 747
A-model, 531-34
analysis, 2-3, 5-6, 30-38, 118, 122-23, 125, 127-28, 136, 138
Analyst, The (Berkeley), 120
analytic continuation, 38
analytic formulas for special values of *L*-functions, 316-17, 323

analytic geometry, 100, 138
analytic number theory, 4, 332-48
analytic philosophy, 928-35
analytic sets, 628, 632, 801
AND gates, 584, 587
Andrews, George, 997-98
angles, 41, 219-20
Apollonius, 735-36
Appel, Kenneth, 117, 142, 563, 698
approximate algorithms, 874
approximate counting, 595
approximating square roots, 110
approximation by polynomials, 253
approximation by rational numbers, 192, 222, 315-16, 710
approximation method in computational complexity, 587
approximation schemes, 852-53, 856
a priori estimates, 474-75
arbitrage, 911-12
arboreal group theory, 442
Archimedean property, 636
Archimedes, 79, 97, 108, 132, 609, **734-35**
area, 57, 183-84
Argand diagrams, 18, 201
arguments of complex numbers, 19
Aristotelean logic, *see* logic, Aristotelean
Aristotle, 83, 86, 151, 931-33
arithmetic circuits, 589-91
arithmetic geometry, 372-83
Arithmetica (Diophantus), 97-99
Arithmetica Universalis (Newton), 100, 136
Arrow's theorem, 982
arrows, 166
Ars Conjectandi (Jacob Bernoulli), 746
Ars Magna (Cardano), 99, 134, 737
Artin, Emil, 161, 720, 730, **812-13**
Artin zeta functions, 730
Artin's reciprocity law, *see* reciprocity, Artin's law of

Asian option, 914
Ask, *see* search engines
associative law, **13**, 105, 272,
    278, 301, 323, 892
asymptotically stable orbit, 495
Atiyah, Michael, 394, 683
Atiyah-Singer index theorem, 219,
    460, 521, 681-84, 725
Atkinson's theorem, 520-21
atlas, 44, 279
atonal set theory, 942
attaching maps, 441
attracting basin, 501-2
automatic differentiation, 613
automorphic forms, 191, 252, 419
automorphisms, **27-28**, 29, 412, 709
average-case complexity, 603
avoidable properties, 626, 633
axiom of choice, 145, 147-48,
    157-58, 159, 314, 619-21,
    623-24, 626-28, 684
axiom of comprehension, 157
axiom of extensionality, 620-21
axiom of foundation, 620
axiom of infinity, 620-21
axiom of regularity, 620-21
axiom of replacement, 620-22
axiom of separation, 620
axiom of union, 620-21
axiom scheme, 259, 622
axiomatic approach: to mathematics,
    84, 128, 138-40, 145, 152;
    to probability, 793, 795, 815
axioms, 20, 56, 284, 700

Babbage, Charles, 111
Babylonian mathematics, 96
Bach, J. S., 938-40
Baire category theorem, 633
Baker-Campbell-Hausdorff formula,
    232
Banach, Stefan, 254, **809-11**, 813
Banach algebras, **172**, 202, 211, 239;
    commutative, 307
Banach spaces, 172, 188, 210,
    239-40, **252-54**, 270, 294,
    799, 810
Banach-Tarski paradox, 158, 684-85,
    813
bandlimited signals, 860
Barban-Davenport-Halberstam
    theorem, 341
barrier option, 913
Bartók, Béla, 940
base space, 392
basic feasible solutions, 288

basis: of a matroid, 245-46; of a
    topology, 302; of a vector space,
    21-22, 28, 30, 223
basis states, 270
Baum-Connes conjecture, 522
Bayesian analysis, 159-60, 753,
    920, 926-27
Begriffsschrift (Frege), 140
Beltrami operator, 296
Beltrami, Eugenio, 92
Berkeley, George, 120
Bernoulli, Daniel, 747
Bernoulli, Jacob I, 746
Bernoulli, Johann I, 746
Bernoulli, Nicolaus I, 120, **747**
Bernoulli distribution, 263, 267
Bernoulli numbers, 395
Betti number, 731
Bézout's lemma, 114
Bézout's theorem, 365-66, 392
Bianchi identities, 488
bicharacteristic curves, 463, 466-67
Bieberbach conjecture, 803
bifurcation set, 505
big bang, 492
biholomorphic equivalence, 728
bijection, 11, 616
bijective proofs, 555-56
bilinear forms, 178, 268
bilinear maps, 188, 301
binary operations, 12-13, 284
binary symmetric channel, 879, 883
binomial distribution, 263, 266-67
biological fluid dynamics, 843
biorthogonal wavelet bases, 855
bipartite graphs, *see* graphs, bipartite
birational equivalence, 722
Birch-Swinnerton-Dyer conjecture,
    229, 381, 685-86
Birkhoff, George, 691, **802-3**
Birkhoff's ergodic theorem, 691, 803
black holes, 491
Black-Scholes equation, 655, 910,
    912-13
block designs, 172-73
blow-up, 463, 465
B-model, 531-34
Bochner identity, 474
Boltzmann factor, 667, 669
Bolyai, János, 42, 89-92, 137, **762**
Bolzano, Bernard, 124, **757-58**
Bolzano-Weierstrass theorem, 124,
    144, 147, **168**, 758, 771
Bombelli, Rafael, 81, 104, 317, **737**
Boole, George, 111, **769-70**
Boolean algebra, 770

Boolean circuits, 584-89;
    monotone, 586-87
bootstrap argument, 475
Borcherds, Richard, 60, 548-49
Borel, Emile, **795**, 796
Borel sets, 247, 628, 631-32, 801
Borsuk's antipodal theorem, 978
Borsuk's problem, 677
Bott periodicity theorem, 227, 394,
    682-83
Böttcher maps, 502-5
bound states, 472
bound variables, *see* free and bound
    variables
boundary conditions, 217, 458-59,
    467, 469
bounded-depth circuits, 587-88
Bourbaki, Nicolas, 823-25
BPP, *see* complexity classes, 595
Braess's paradox, 866-68
braid groups, 160-61, 274, 388
branching processes, 658-59
branes, 535-36, 538
Brauer, Richard, 320, 428
breakdown of solutions to partial
    differential equations, 194-96, 481
breaking time, 237
*Bride Stripped Bare by her Bachelors,
    Even (The Large Glass), The*
    (Duchamp), 947
Britton's lemma, 437, 439-40
Brouwer, Luitzen Egbertus Jan,
    116, 142, 148-51, 153, 155,
    181, **799-800**
Brouwer's fixed point theorem,
    693-96, 800, 901
Brown-Douglas-Fillmore theorem,
    521
Brownian excursion, 656-57
Brownian motion, 218, 647-56,
    910-11
Brun sieve, the, *see* sieves
Brunn-Minkowski inequality, 672-73,
    705; reverse form of, 676
brute-force search, 59, 62-63, 580,
    840, 873, 883
buildings, 161-62
Burali-Forti paradox, 145, 779
Burgers equation, 236, 463, 467, 476
Burnside, William, 68, **785**
Burnside problem, 68, 438, 785
Burnside's lemma, 560
butterfly effect, 496

C^*-algebras, 57, 172, 227, 313,
    518-20, 522-23
Calabi conjecture, 164

Calabi-Yau manifolds, 163-65, 190, 530-34

calculus, 118-19, 122-24, 128, 134-36, 743-44, 770, 934

calculus of variations, 65, **310-13**, 478-79, 908

cancelation law, 250

cancrizans canon, 939-40

canonical inner models, 629

canonical transformations, 298-99

Cantor, Georg, 71, 81, 116, 125, 127, 144-46, 155, 171, 183, 222, 616-19, 623, 629, 632, 634, 703, **778-80**

Cantor set, **183-84**, 247, 504

Cantor's diagonal argument, 171, 779

CAR algebra, 519

Cardano, Girolamo, 101, 104, 133-34, **737**

cardinal exponentiation, 618, 630

cardinal invariants, *see* invariants, cardinal

cardinality, 165, 616-19, 622, 626

cardinals, 145, **165**, 616-19, 779; inaccessible, 627-29, 632, 702; measurable, 628-32; regular, 627-29; singular, 630; supercompact, 630, 632-33; uncountable, 626-29; weakly compact, 628; Woodin, 632

Cardy's formula, 666

Carleson's theorem, 453, 686-87

Carmichael numbers, 351

Cartan, Élie Joseph, 232, 713, **794-95**

Cartan subalgebra, 233-34

Cartesian coordinates, 21, 739

Cartesian product, 618

Casorati-Weierstrass theorem, 771

casus irreducibilis of the cubic, 737

Catalan conjecture, 360

Catalan's constant, 150

category theory, 6, **165-67**, 275, 417, 536, 801

Cauchy, Augustin-Louis, 102, 122-24, 147, 459, 560, 569, **758-59**, 760, 791

Cauchy problem, 235-37, 459, 468-69

Cauchy's residue theorem, 202, 337

Cauchy's theorem, 38

Cauchy-Davenport theorem, 569

Cauchy-Hadamard theorem, 791

Cauchy-Kovalevskaya theorem, 464, 467-68

Cauchy-Riemann equations, 37, 459-60

Cauchy-Schwarz inequality, 220, 268, 704-5

Cayley, Arthur, 82, 92, 103, 105, 110, 498, 509, 768-69, **772-73**, 831

Cayley graphs, 443, 445, 447, 702

Cayley numbers, *see* octonions

Cayley's graph theorem, 772

Cayley's theorem, 422

Cayley-Dickson construction, 278-79

Cayley-Hamilton theorem, 329

cell complex, 441

cellular automata, 836

central limit theorem, 207, **266-67**, 335, 648, 678, 687, 919

chaos, 51, 190, 495, 728

character tables, 429-30

characteristic classes, 393, 411

characteristic coordinates, 236

characteristic curves, 236, 462-63, 466

characteristic hypersurfaces, 466-67, 469

characteristic of a field, 640

characteristic polynomial, 224-25, 294, 329

characters: of Abelian groups, 189, 207, 295-96, 308, 339, 426, 428; of group representations, 423-26, 428, 430, 783, 785; in phylogenetics, 846.
See also Dirichlet characters

Chebotaryov density theorem, 783

Chebyshev, Pafnuty, 771

Chebyshev polynomials, 293, 297, 771

chemical informatics, 836

chemical topology, 830-31

Chern classes, 394

Chevalley, Claude, 813, 824

Chevalley groups, 688

chiral algebra, 544

chirality, 979

choice number of a graph, 574

choice sequence, 150

Christoffel symbols, 311, 488

chromatic index of a graph, 565

chromatic number of a graph, 564, 566

Church, Alonzo, 50, 111-12, 577, 707, **816**

Church's thesis, 113

circle method, 346-47, 797, 804, 807

circle of fifths, 938

circuits, *see* Boolean circuits

class field theory, 243, 268, 720, 812-13

class numbers, 255, 322-24, 340-41

classical computation, 269, 271-72

classical mechanics, 287, 299

classification, 52-54, 56, 232, 252, 408, 411; of finite simple groups, 141, 252, 429, 687-89; of Lie algebras, 161-62, **232-34**

classifying spaces, 408, 442

Clenshaw-Curtis quadrature, *see* quadrature, Clenshaw-Curtis

Clifford, William Kingdon, 780

cliques in graphs, 564, 573, 586-87

closed sets, 302-3, 618-19

closed-form solutions, 51, 766

coanalytic sets, 628, 631-32

coarse moduli spaces, 415-16

cofinality, 629

Cohen, Paul, 141, 155, 624-27, 703, 780, 819

Cohen-Lenstra heuristics, 324

cohomology, 189, 221, 384, 389, 391-94, 411, 523, 531, 732

collapse of the polynomial-time hierarchy, 585

combinatorial geometry, 570-71

combinatorial number theory, 569-70

combinatorics, **6-7**, 562-63; algebraic, 561; extremal, 215, 563-72; probabilistic, 572-74

communication channel, 879

communication complexity, 589

commutant of a set of operators, 515

commutative diagrams, 166, 274

commutative law, **13**, 82, 105, 179, 278, 284, 301, 323, 519, 770

commutator, 231, 287, 444, 526, 542

compactification, 168-69, 267, 721; one-point, 169; Stone-Čech, 169

compactness, 167-69, 303, 398, 639-40, 645

complement of a set, 188

complete graphs, *see* graphs, complete

completeness: of an axiomatic system, 139, 153, 637-39; in computer science, 170; of a metric space, 220, 254, 514, 696; of a normed space, 810; of the real numbers, 144, 636

complex analysis, 37, 282-83, 337-38, 758, 775

complex cobordism, 395

complex manifolds, *see* manifolds, complex

complex numbers, **18-19**, 81-82, 102, 105, 201-2, 275-78, 284-85, 296, 317-18, 328, 640-41, 698, 737 complex orientation, 163-64

complex structures, 300, 411-13, 417, 816

complex systems, 838

complexity, *see* computational complexity

complexity classes, 169-70; *BPP*, 595; co *NP*, 582, 584; *EXP*, 580-81, 595; *NC*, 170; *NP*, 170, 446, 580-83, 595-96, 598-99; *P*, 579-81, 595, 713; *PSPACE*, 170, 597

complexity of algorithms, 578

composition: of braids, 160; of morphisms, 165-66, 536-37; of operators, 240, 294, 515; of permutations, 259-60; of symmetries, 20, 277, 420, 484

comprehension principle, 145

computable functions, 112-13, 577, 816, 821

computational chemistry, 830

computational complexity, 114, 575-604

computational fluid dynamics, 611

computational number theory, 348-62

computer memory, 114, 169-70, 578, 597, 848-49, 980

computer-assisted proofs, 142, 496, 575, 698, 972

concatenation of paths, 176-77, 221, 401

conditional probability, 159

Condorcet's paradox, 982-83

conformal equivalence, 209, 282, 411

conformal field theory, 543-45

conformal invariance, 654, 665

conformal maps, 543, 728, 978

conformal structure on a surface, 209, 411

conformal vector, 546

conic sections, 43, 365, 735-36, 739, 743

Conics (Apollonius), 735

conjectures, 69-70, 76, 142, 335, 349-60, 381, 722, 957

conjugacy (in group theory), 26; classes, 422-26, 428-31; conjugacy problem, the, 436; conjugate subgroups, 421

conjugates: algebraic, 319, 329-31; complex, 19, 276, 278, 710

connectedness, 198, 230, 245, **303**, 309, **383-85**, 398, 504-8, 564, 573, 660-64

Connes, Alain, 517, 522-23, 956-57

co *NP*, *see* complexity classes

conservation laws, 236, 286, 479, 486, 488, 525, 540, 800

conservative extensions, 154

consistency, 139-40, 145-46, 153, 622-23, 625, 639, 701, 819; of the continuum hypothesis, 155, 624, 780; of Euclidean geometry, 789; of Peano arithmetic, 702; of ZFC, 629, 702

consistency strength, 629; lower bound, 629; upper bound, 629

constant-curvature metric, 92, 281, 712, 728, 775

constrained optimization, 256-57

constructible set theory, 623-26, 629, 819

construction of regular 17-gon, 101, 327

constructive proofs, 143-44, 149, 157

constructivism, 116, 157-59; in art, 948-50

contextual definition, 933-34

Continental philosophy, 929

continued fractions, 192-93, 315-17, 326; for tangent function, 193

continuous functions, **32-33**, 123, 144, 151, 168, 211, 301-2

continuum hypothesis, 145, 155, 618, 623-27, 629, 632, 634, 703, 780, 802, 819; independence of, 703

contour function of a tree, 655-56

contractible spaces, **309**, **387**, 388, 442

contraction mapping theorem, 696

control theory, **461**, 472

conventionalism, 94, 786

convergence, **31-33**, 109-10, 123, 126, 168-69, 254, 452; absolute, 334; almost everywhere, **452-53**, 687, 815; in distribution, 650; in probability, 266; quadratic, 110, 612; superlinear, 612; uniform, 124, **126**, 211; weak, 186-87

convexity, **72**, 288, 671, 675, 696, 704-5, 790

convolution, **203**, 207, 213, 303-4, 306-7, 450

Conway, J. H., 59, 227, 268, 549

Conway group, 59

coordinate charts, 45, 47, 181, 279, 282-83, 396-98, 401

coordinate ring, 376-78

coprime integers, 107

coproduct, 272, 274

corollaries, 74

correlation function, 528

correlation length in percolation, 663

cosine function, *see* trigonometric functions

counit, 273

countable additivity, 247, 628

countable chain condition (CCC), 632-33

countable models, 625-26, 645-46

countable sets, 71, 157, **170-72**, 223, 617, 619, 623, 779

counterexamples, 69, 121, 124-26

counting, 61-66, 984-88

Courant, Richard, 808-9

Cours d'Analyse (Cauchy), 758

covariant 2-tensors, 485

cover of a topological space, 310

Cox regression model, 925

Coxeter, Harold Scott MacDonald ("Donald"), 53, 950-51

Cramér, Harald, 335

Cramer's rule, 329

creation and annihilation operators, 528, 542

Crelle's Journal, 91, 125, 761, 774

crisis in foundations, 142-56

critical exponents, 659, 663-65, 668

critical phenomena, 657-58

critical points, 310

critical probability, 658, 660, 662-64

critical strip, 337-38, 715

critical temperature, 666-68

Critique of Pure Reason (Kant), 137

crossing probability, 665

cryptography, 601 2, 887

crystallographic point groups, 828

cube, *n*-dimensional, 53; discrete, 197

cubic equations, 81, 98-99, 101, 326, 708, 737

cubism, 946

cuneiform texts, 96

Curie-Weiss model, 668

curl, 180

curvature, 42, 92, 172, 280, 311, 388, 394, 670; Ricci, 218, 280-81, 406-7, 488; scalar, 280; Weyl, 489

cut rule, 593

cybernetics, 812

cycle decomposition of permutations, 260, 558

cyclic cohomology, 523

cyclic groups, 422, 688, 709

cyclotomic fields, 254

cylinder, 734

d'Alembert, Jean Le Rond, 35, 121, 136, **749-50**

d'Alembert's solution to the wave equation, 236

d'Alembertian, 35, 457, 460, 478, 490

Dalí, Salvador, 951

Dantzig, George, 289, 613

Darboux, Gaston, 125, 777, 794

Darboux's theorem, 300

Das Kontinuum (Weyl), 149

data encryption standard (DES), 889

de Branges, Louis, 804

de Gennes, Pierre-Gilles, 70

de la Vallée Poussin, Charles-Jean, 63, 338, 356, 686, 715, **792**

de Morgan's laws, 188, 766

de Rham cohomology, 175, 177, 179

De Thiende (Stevin), 738

decay of particles, 528

decidability, 638, 640, 643, 645, 813

decimal notation, 30, 79-80, 106, 171, 242, 738, 986

decision problems, 269-70, 577-79, 581

decoherence, 271

decomposition of a finite set, 551

Dedekind, Julius Wilhelm Richard, 104, 127, 138, 143-45, 241, 729-70, **776**

Dedekind cuts, 127, 144

Dedekind zeta functions, 730

definable real numbers, 146

definable sets, 624, 627, 631, 643-44

definitions, 74, 84, 146-47, 149

deformation theory, 418

degree, 410; of an algebraic number, 328; of a continuous function, 388, 694-95; of a number field, 329

Dehn, Max, 435-36

Dehn functions, 445-47

Delaunay triangulation, 830

Deligne, P., 347, 729, 732, 808

depth of a circuit, 587-88

derandomization, 601

derangements, 560

derivations, 178-79, 547

derivative of a set of real numbers, 618

derivatives pricing, 910-14

Desargues, Girard, 945; *Brouillon Project* of, 945

Descartes, René, 81, 100, 134-35, **739-40**, 955

descriptive set theory, 631-32

designs, 172-73

determinacy, 159, 630-34; axiom of, 159, 631

determinants, 39, 103, 174-75, 277, 420, 514, 590-91, 641

determined system of equations, 459

Deuring–Heilbronn phenomenon, 340-41

deviation principles, 673-76, 679

diagonalization, 206, 223, 297

diagrams in Greek mathematics, 131, 134, 137, 139

diffeomorphism, 298

difference set, 715

differentiable manifolds, *see* manifolds, differentiable

differential equations, **51-52**, 297, 455, 523-24, 609-11; linear, 51. *See also* ordinary differential equations, partial differential equations, stochastic differential equations

differential forms, 175-80, 189, 273, 300

differential geometry, 44-46

differential operators, 456-57, 478

differentiation, 30, **33-34**, 36-37, 45, 51, 65, 74, 122, 125, 144, 177, 179, 186, 239, 255-56, 282, 397, 450

Diffie–Hellman protocol, 891-92

diffusion processes, 293, 655

digital information, 829, 878

digital signatures, 893-94

dihedral group, 24, 420-21, 424

dimension, 52, 56, **180-84**, 367, 516-17, 724; algebraic, 367; codimension, 391; cohomological, 182; fractional, 184; Hausdorff, **184**, 508, 793; homological, 182-83; inductive, 181-82; of a manifold, 396; topological, **184**, 367; of a vector space, 22

dimension argument, 571

Diophantine equations, 50-51, 111, 373-75, 378, 692, 706-8, 720, 722

Diophantus, 97-98, 134

Dirac, Paul, 542

Dirac distribution, 186, 473, 542-43

Dirac equation, 460

direct products of groups, *see* groups, direct products of

direct sums and products, 24

Dirichlet, Peter Gustav Lejeune, 124, 143, 229, 305, 339, 686, 689, **764-65**, 775-76

Dirichlet *L*-functions, *see* *L*-functions, Dirichlet

Dirichlet boundary conditions, 458, 469, 654

Dirichlet characters, 339, 764

Dirichlet principle, 125-26, 475, 789

Dirichlet problem, 458, 476, 653, 764

Dirichlet series, 228-29

Dirichlet summation operators, 451, 453

Dirichlet's class number formula, 340

Dirichlet's theorem, 689

Dirichlet's unit theorem, 255

discrepancy theory, 574

discrete logarithm problem, 892-93

discrete mathematics, *see* combinatorics

discrete subgroups, 402

discrete topology, 302

discrete-time stochastic process, 649

discretization, 203

discriminant, 320-21, 323, 788, 800; of a binary quadratic form, 320; of a number field, 330; of a polynomial, 317; of an elliptic curve, 347

disk model of hyperbolic geometry, *see* hyperbolic geometry

dispersive PDEs, *see* partial differential equations, dispersive

Disquisitiones Arithmeticae (Gauss), 101, 103, 315, 320, 756, 761, 763-64

Disquisitiones Generales Circa Superficies Curvas (Gauss), 756

distance, 31, 41-43, 46, 220, 248, 253, 280, 671

distributed computation, 603, 877-78

distributions, 184-87, 190, 211, 456, 475, 542, 544

distributive law, 20, 284, 770

divergence operator, 180, 194

diversifiable risk, 915-16

divisibility, **118**, 242, 249, 807

DNA, 838-40, 842, 981

domain of a function, 11

domain of attraction, 110

Donaldson, Simon, 394, 404

Douady, Adrien, 506, 508

Douady rabbit, 500

double contradiction, method of, 132, 134

double cover, 277

double digest problem, 839

downsampling, 860

drift of a Brownian motion, 654, 911-12

duality, 177, 185-90, 212, 274-75, 288; of convex bodies, 189; of groups, 189; of linear spaces, 185, 188, 212; Pontryagin, 189, 205-6
Duchamp, Marcel, 947-48
Dvoretzky's theorem, 675-76
dynamic replication, 912
dynamical systems: continuous, 494; discrete, 495
dynamics, 5-6, **190, 493-504, 506-10,** 576, 713, 728, 766, 802-3; holomorphic, 497-509; topological, 495
Dynkin diagrams, 233-34

e, 71, 81, **200-201, 222-23,** 748, 773; transcendence of, 773
effective and ineffective proofs, 117, 722
efficiency of a proof system, 593
efficient computation, 197, 579, 872-74, 883-85
Egyptian fractions, 77-78
eigenfunctions, 206-7, 217, 297, 306, 511
eigenspaces, 224-25, 295
eigenvalues and eigenvectors, 30, 198, 206, 223-25, 240, 294-97, 608, 694, 876-77; eigenvalue decomposition, 608; eigenvalue problem, 472
eikonal equation, 463, 466
Eilenberg, Samuel, 167, 221
Einstein, Albert, 83, 95, 153-54, 218, **483-93,** 647-48
Einstein constraint equations, 490
Einstein equations, 164, 460, 470, **483, 489-93;** vacuum, 489, 491
Eisenstein series, 251-52
elementary functions, 293, 726, 766
Éléments de Géométrie (Legendre), 88
Éléments de Mathématique (Bourbaki), 823
elements of a set, 9
Elements of Algebra (Euler), 104
Elements (Euclid), 84-88, 96, 98, 107-8, 118, 130-31, 133-34, 733-34, 762, 928
elliptic curves, 51, **190-91,** 252, 347, 356, 370-71, 380-82, 412-14, 685, 692, 721, 730-31, 892; group law of, 355, 721, 892; use in factoring, 355-56
elliptic functions, 241, 293, 724, 773, 949
elliptic modular function, 60, 324-25

elliptic PDEs, *see* partial differential equations, elliptic
elliptic regularity theorem, 682
encoding and decoding functions, 880-83
Encyclopédie (d'Alembert), 107
energy estimates, 217, 474
entropy function, 882
Entscheidungsproblem, 113, 707
equal temperament, 937
equation of a circle, 374
equipotentials, 502-6
equivalence: of binary quadratic forms, 320-22; of physical theories, 523-25, 529-30, 532
equivalence relations and equivalence classes, 12, 25, 40, 185, 221, 252
Erdős, Paul, 64, 338, 342, 351, 359, 361, 572, 627, 660, 802
Erdős-Ko-Rado theorem, 569
ergodic theorems, 299, 512-13, 689-91
Erlanger Programm, 93, 777, 782
error function, 293
error-correcting codes, 173, 364, 575, 598, **881-86,** 981
Escher, Maurits Cornelis, 950-51
Essai sur la Théorie des Nombres (Legendre), 754
essential supremum, 211
estimates, 62-63, 72, 200, 474-75, 714-17, 916-21, 924; asymptotic, 335
estimators, statistical, 916-21
étale cohomology, 382, 732
Euclid, 74, 84-85, 96-97, 107, 118, 132, 689, **734**
Euclidean algorithm, 107-8, 114-15, **191-92,** 353, 378, 700
Euclidean geometry, **39,** 44, 83-84, 87-94, 139, 208, 283, 401, 789
Euclidean space, 45, **248,** 302
Euclidean structure, 401-2
Euler, Leonhard, 81, 104, 120-21, 228, 261, 290, 333-34, 348, 555, 692, 718-19, 727-29, 746, **747-49,** 751
Euler characteristic, 53, 215, 219, 393, 684
Euler class, 393, 411
Euler equation, 193-96
Euler product, 228-29, 283, 336, 340, 347
Euler vector field, 427
Euler's constant, 214, 222
Euler's formula, 697

Euler's identity, 748
Euler's theorem, 250, 891
Euler's totient function, 250
Euler-Lagrange equations, 65, 311-12, 478-79, 489, 748, 751, 908, 977
evaluation, operation of, 184-86, 378
even functions, 204
even permutations, *see* permutations
events in probability, 265
evolution equations, 235, 456
evolution maps, 299
examples, importance of, 1001-2
exchange axiom, 244-45
excluded middle, law of, 149, 157, 799-800
existence of solutions, 48, 51, 510
exotic options, 914
EXP, *see* complexity classes
expanders, 196-99, 572
explicit constructions, 197, 574; of expanders, 197-99; strong, 197
explicit formula for counting primes, 337, 344
explicit proofs, 71-73, 150
exploration process, 664
exponential distribution, 265
exponential function, 30, 193, **199-202,** 223, 232, 265, 308, 746
exponential generating functions, 557-59
exponential sums, 347, 716-18
exponential varieties, *see* varieties, exponential
exponential-time algorithm, 349-50, 355, 580, 874
extended real line, 169
external rays, 504-6
extremal problems, 64-66; in combinatorics, 215, 562-72, 865-70

factorials, 213-14, 350
factorization of integers, 271, 353-56, 583, 590; into primes, 699-700
factors in von Neumann algebra theory, 516-18
faithful actions, 420
Faltings, Gerd, 117, 722
fast Fourier transform, the, 65, **202-4,** 271
Fatou, Pierre, 498
Fatou sets, 501-2
fault-tolerant networks, 198
feasible set, **288,** 613, 898-900
feedbacks, 900

Feistel cipher, 889
Feit, Walter, 785
Fejér summation operators, 452
Ferguson, Helaman, 952
Fermat, Pierre, 100, 104, 134, 268, 325, 353, 380, 691-92, **740-41**
Fermat equation, 50, 111, 117, 254, 681, 722
Fermat prime, 327, 748
Fermat's last theorem, 51, 69, 104, 111, 141, 191, 229, 243, 252, 255, 347, 359-60, 364, 380, 562, **691-93**, 764, 820
Fermat's little theorem, 55, 250, 350-52, 355
Fermat–Catalan conjecture, 360-61
Ferrari, Ludovico, 99, 101
Feynman, Richard, 527-28, 541, 1007
Feynman diagrams, 528, **541**
Feynman-Kac formula, 218
fiber bundles, **392-93**, 794
Fibonacci, *see* Leonardo of Pisa
Fibonacci numbers, 115, 222, 249-50, 316, 737
fields, 18, **20-21**, 23-25, 27-28, 102, 254, 284, 317-18, 329; extensions of, **21**, 28, 102, 254
fields (in physics), 525-26, **542-43**
figure eight knot, 225-26
filtering, 860
finite simple groups, 26, 59, 261, 439, 687-89; search for, 783, 785
finitely generated groups, **67**, 438-39, 443-44, 685
finitely presented groups, **434-36**, 439-48
finitism, 152
first-order logic, *see* logic, first-order
Fisher, Ronald, 173, 924
fixed field, 710
fixed point, **495**, 499, 559-60, **693**, 731-32; attracting, 499; indifferent, 499; repelling, 499; super-attracting, 499
fixed point theorems, **693-96**, 732, 799, 901
flat metrics, 711-12
Flatland: A Romance of Many Dimensions (Abbott), 946-47
floating-point arithmetic, 605-6
flow in networks, 864-70
fluid dynamics, 193-96, 847; biological, 843; computational, 611
forbidden minors, 725
forcing, 624-30, 632-33, 703; iterated, 627
forcing axioms, 632-34

Foreman, Matthew, 630, 633
formal languages, 621-23, 635-37
formal power series, 546, 552, 556
formalism, doctrine of, 153-55
formalization of mathematics, 16, 74, 111, 138, 140, 152
formulas, 140, 151, 153, 259, 582, 592, 621-24, 635-37, 640, 642, 700; atomic, 621; Boolean, 588
Foundations of Algebraic Geometry (Weil), 731, 820
Foundations of Geometry (Hilbert), 789
Foundations of Probability Theory (Kolmogorov), 815
four-color theorem, 117, 142, 563, 696-98
four-dimensional manifolds, *see* manifolds, four-dimensional
Fourier, Jean-Baptiste Joseph, 216, **755**
Fourier analysis, 220, 261-62, 295-97, 425, 457
Fourier coefficients, 202-3, **205**, 212, 262, 451, 454; phase of, 859
Fourier series, 124, 305-6, 451, 511, 686-87
Fourier transforms, 186, 189, 203, **204-8**, 214, 236, 239, 274, 306, 450, 453-54, 457, 473-74, 523-24, 859; discrete, 203, 590, 611; inversion formula for, 206, 306, 426; non-Abelian, 274, 424-25, 429-30. *See also* fast Fourier transform, the
fractal sets, 31, 57, 110, 184, 244, 496, 498, 502, 509
Fraenkel, Abraham, 105, 148
Fredholm, Ivar, 511, 520, **791-92**
Fredholm operators, *see* operators, Fredholm
free Abelian group, 390
free and bound variables, 15-16, 635-36
free Burnside group, 438
free group, 387-88, 390, 433-34, 437, 440, 447, 685
free products of groups, *see* groups, free products of
Freedman, Michael, 404-5, 714
Frege, Gottlob, 111, 127, 140, 146, **780-82**, 929, 931-34
Frege system, 593-94
Freiman's theorem, 717-18
Frey, Gerhard, 692
Frobenius, Ferdinand Georg, 514, 560, **783-84**, 785
Frobenius map, 731
Fuchs, Lazarus, 209, 785

Fuchsian groups, 208-10, 417
full parametric statistical models, 924-25
fullerenes, 831-32
function field of an algebraic variety, *see* varieties, function field of
function spaces, 29, **210-13**, 239, 253-54
function-classes, 621-22
functional calculus, 207, 519
functional equation for the Riemann zeta function, 214, 229, **337**, 356-57, 729-32
functional equations, 200-202, 455-56
functions, **10-11**, 29, 51, 121, 125, 128, 144, 151, 184, 930-31
functors, 166-67
fundamental discriminants, 320, 324
fundamental groups, 166, 221, 225, 310, 385-88, 390, 416-17, 436, 441-42, 446, 517, 786
fundamental parallelogram, 413
fundamental quadratic form, 320-21
fundamental solutions, 187, 473
fundamental theorem of algebra, 81, 100, 147, 224, 275, 294, 326, 386, 640, 695, **698-99**
fundamental theorem of arithmetic, 57, 104, 191, 221, 283, 332-33, 336, **699-700**
fundamental theorem of calculus, **37**, 175, 177, 179, 290, 651, 742, 758
fundamental theorem of exponential generating functions, 557
fundamental theorem of numerical analysis, 610
fundamental units, 320, 327
Funk-Hecke formula, 297

Gabo, Naum, 948-49
Galois, Évariste, 50, 81-82, 101-2, 104, 331, 709-10, 766, **767-68**
Galois correspondence, 310, **710**
Galois groups, 28, 191, 213, 331, 709, 761, 783
Galton–Watson tree, 655, 657
gamma function, the, 213-14, 290, 337
Gaudin distribution, 359
gauge, 470, **490**
Gauss, Carl Friedrich, 42, 81, 88, 90-91, 100-101, 103-4, 137, 255, 269, 292, 315, 318, 320, 326-28, 335, 341, 349-50, 353, 356, 605, 609, 692, 699, 714, 718-19, **755-57**, 760-62, 764, 929

Gauss sums, 328

Gauss-Bonnet theorem,
  219, 394, 684

Gauss-Cramér model, 335-36,
  341-43

Gaussian distribution, *see* normal
  distribution

Gaussian elimination, 606-7

Gaussian function, 293, 528, 705

Gaussian integers, 104, 108, 221,
  318-19, 325-26, 719, 756

Gaussian quadrature, *see* quadrature,
  Gaussian

Gaussian unitary ensemble, 359

Gauss's circle problem, 801

Gel'fond-Schneider theorem, 223

Gelfand transform, 307, 519

Gelfand-Naimark theorem, 519

genealogical tree, 655-56

general linear group, **29**, 39,
  161, 230-31, 429

general relativity, *see* relativity,
  general

generalization, 55-58, 199

generalized continuum hypothesis,
  624, 629-30, 633

generalized eigenvectors, 224

generalized functions, 185

generalized inverse of a matrix, 836

generalized solutions, 187, 475-77

generating functions, 62, 214-15,
  228, 304-5, 552, 555, 557, 753, 807

generic absoluteness, 633-34

generic extension, 626

generic real numbers, 625-26

genetic algorithm, 834

Gentzen, Gerhard, 151

genus, 54, 215, 370, 399, 411;
  of an algebraic curve, 721, 730

geodesics, 47, 92, 248, 312, 446, 488

geometric classification problems,
  408-9

geometric distribution, 263

geometric group theory, 443-45

geometric Langlands program, 538

geometric mean, 703

geometric multiplicity, 225

geometric proof systems, 594

geometric structures on manifolds,
  401, 406-7

geometrization conjecture,
  281, 402-3, 406, 440

geometry, 1-2, 4-5, 38-47, 58,
  83-97, 130, 136-39

giant component of a random graph,
  573, 660

Girard, Albert, 81

global maximum principle for the
  heat equation, 217

global optimization, 834-35

GNS construction, 514, 517

Gödel, Kurt, 151-56, 624, 639,
  701, 703, **819**

Gödel's completeness theorem,
  623, 638, 819

Gödel's incompleteness theorem,
  141, 623, 629, 634, 638, 640,
  700-702, 819

Gödel's second incompleteness
  theorem, 623, 627, **701**, 819

Goldbach, Christian, 343, **745**

Goldbach's conjecture, 69-70,
  343-44, 346, 362, 715, 745,
  751, 803-4; ternary, 715

Goldberg Variations (Bach), 939

golden ratio, 193, 222, 316-17, 320

Google, *see* search engines

Gordan, Paul, 103, 144, 769, 772,
  788, 800

gradient, 34, 180, 239, 255-57

Gram-Schmidt orthogonalization,
  607

graphs, 196, 198, 215, 245, 564, 645,
  660, 697, 725, 846; bipartite,
  157-58, 199, 565; coloring of,
  564-65, 697; complete, 564, 725;
  edge-coloring of, 565; of groups,
  442; minors of, 725-26; planar,
  564, 697, 725; regular, 196

gravitational radiation, 491

gravitational waves, 490

great circle, 40

great wave of translation, 234-35

greatest common divisor, 107, 321

greedy algorithm, 245-46, 565, 873

Greek mathematics, 131-33

Green, George, 760

Green's theorem, 180, 760

Green-Tao theorem, 344, 346,
  570, 958

grim strategy, 904-5

Gromov, Mikhail, 199, 299, 444, 447

Gromov's filling theorem, 445-46

Gromov's nonsqueezing theorem,
  298-99

Gromov's polynomial-growth
  theorem, 444, 702-3

Gromov-Witten invariants, *see*
  invariants, Gromov-Witten

Grothendieck, Alexander, 285, 373,
  394, 724, 731-32, 974, 1001, 1006

group C^*-algebras, 520

group algebra, 274

groups, **19-20**, 27-29, 39, 53, 66-68,
  102, 204, 208, 221, 229, 248-49,
  252, 260, 273, 277, 279, 284, 296,
  636-37; actions of, 297, 309-10,
  420-21, 424-25; axioms for, 20,
  636; direct products of, 23-24,
  436; free products of, 436-38;
  generators of, 433-34; of Lie type,
  162, 688; presentations of, 436;
  residually finite, 439; simple, 26,
  59, 261, 439, 687-88; solvable,
  447, 710; of transformations,
  39-41, 93, 419-20, 441

Grundlagen der Arithmetik (Frege),
  127, 781

Grundlagen der Geometrie (Hilbert),
  138

GUE conjecture, 359

Hölder's inequality, 704

Haar measure, 425

Haar wavelets, 852-55, 858

Hadamard, Jacques, 63, 155, 338,
  356, 468, 715, **790-91**, 963-65

hairy ball theorem, 393, 799

Haken, Wolfgang, 117, 142, 563, 698

Hales, Thomas, 58

half-plane model of hyperbolic
  geometry, *see* hyperbolic
  geometry, half-plane model of

Hall's theorem, 566

halting problem, 638-39, 707-8

Hamilton, Richard, 281, 406, 714

Hamilton, William Rowan, 81-82,
  105, 276, 299, 567, **765**

Hamilton cycle, 567

Hamilton's equations, 216,
  286-87, 298-99

Hamilton's least action principle,
  478-79, **524**, 525, 527, 541

Hamilton-Jacobi equations,
  463-64, 466

Hamiltonians, 215-16, 286,
  524-27, 540, 542

Hamming, Richard, 879, 881, 981

Hamming distance, 248, 881-82

Hardy, Godfrey Harold, 62-64,
  346, **797-98**, 807-8

Hardy-Littlewood k-tuple conjecture,
  804

Hardy-Littlewood maximal
  inequality, 452, 455

Hardy-Weinberg law, 798

Harish-Chandra, 428-29

harmonic analysis, 205, **448-55**, 859.
  See also Fourier analysis

harmonic functions, 283, 503, 652-53

harmonic map flow, 218

harmonic oscillators, 526, 542

harmonic polynomials, 296-97

Harnack inequality, 217-18

Hasse, Helmut, 243, 730-31

Hasse principle, *see* local-to-global principle, of Hasse

Hausdorff, Felix, 140, 184, **792**, **794**

Hausdorff dimension, *see* dimension, Hausdorff

Hausdorff topological space, 302, 633

Hausdorff-Young inequality, 454, 705

Hayter, Stanley William, 953

heat equation, 34-35, **216-19**, 406, 456, 458-59, 470, 473, 478, 654

heat kernel, 218-19

Hecke, Erich, 730

Heegner, K., 255, 324, 341

Heine, Eduard, 127, 778

Heine-Borel theorem, 168

Heisenberg, Werner, 513, 542

Heisenberg equation, 287

Heisenberg group, 702

Heisenberg uncertainty principle, 207, 272, 306, 453, 513, 527

Hellegouarch, Yves, 380, 692

Helly's theorem, 571

Hensel, Kurt, 82, 241-43

Hermite, Charles, 330, **773**

Hermitian matrices, *see* matrices, Hermitian

Hermitian operators, *see* operators, Hermitian

Hermitian structure, 163

Heron of Alexandria, 79, 110

Hessian matrix, 462, 612-13

heuristic evidence, 361-62

Higher Arithmetic, The (Davenport), 315

highest common factor, 107, 191-92

Higman, Graham, 437-39

Higman's embedding theorem, 439-40

Hilbert, David, 50, 83, 95, 103, 110, 113, 125, 128, 138-46, 151-54, 358, 489, 511, 619, 623, 700-701, 707, 719-20, 726, 751, **788-89**, 790, 804, 961-62

Hilbert spaces, 189, 211, 219-21, 240-41, 253-54, 270, 295, 423, 511, 513, 526, 540, 705, 799

Hilbert's basis theorem, 144, 303

Hilbert's Nullstellensatz, 594, 640, 642, 703; arithmetic, 365; effective, 371-72; weak, 371, 703

Hilbert's problems, 789; tenth, 50, 110-11, 113, 639-40, 707-8, 821; twelfth, 720; thirteenth, 815; seventeenth, 822

Hironaka, Heisuke, 369, 723

Hirsch conjecture, 289

Hirzebruch, Friedrich, 394, 683, 724

History of Algebraic Geometry (Dieudonné), 377

HNN extensions, 437-38, 440, 442-43

Hodge, William Vallance Douglas, 816-17

Hodge conjecture, 191, 817

Hodgkin-Huxley equations, 844

hole argument, the, 489

Holmgren's theorem, 471

holomorphic functions, **37-38**, 144, 205, 213, 228, 252, 282-83, 300, 307-8, 723, 804

holonomy, 163-64

homeomorphisms, 40, **302**, 383, 387, 412, 497

homeostasis, 838

HOMFLY polynomial, 225-27

homogeneous coordinates, 267

homogeneous polynomials, 296-97

homological mirror symmetry conjecture, 536

homology, 189, 786

homology groups, 182, 221, 389-92, 694-95

homomorphisms, **27-28**, 165, 284, 801; ring, **284**, 330, 801

homotopic loops, 385

homotopy groups, 221, 309, 385-87; of spheres, 389-92, 395

Hopf algebra, 273-74

Hopf link, 225-26

Hopf map, 389, 392

horocycle, 90

household maximization problem, 897-99

Householder method, 608

Hubbard, John H., 506, 508

Hurwitz's theorem, 278

hyperbolic geometry, 41-43, 47, 90-91, 208-9, 281, 283, 401, 447, 728-29; Beltrami's disk model of, 92, 93; half-plane model of, 41; hyperboloid model of, 42; Poincaré's disk model of, 42, 47, 94, 209, 728, 786

hyperbolic groups, 447-48

hyperbolic manifolds, *see* manifolds, hyperbolic

hyperbolic PDEs, *see* partial differential equations, hyperbolic

hyperbolic structure, 401-2

hyperbolicity conjecture, 508

hyperelliptic curves, 370

hyperfunctions, 186

hypergeometric sequences, 992

i (square root of −1), 18, 56, 284

ibn al-Haytham, 86

ibn Qurra, Thābit, 86

ideal class group, 221-22, 322-23

ideals, **58**, 221, **284-85**, 322, 376, 378, 642, 719

idempotents, 240

identities, technique for proving, 261

identity element, 13, 17, 20

ill-posed problems, 468, 473

image compression, 848-62

image of a function, 11, 284

In Artem Analyticem Isagoge (Viète), 99, 737

inaccessible cardinals, *see* cardinals, inaccessible

inadmissible operator, 919

inclusion-exclusion principle, 345, 560

indefinite integrals, 175

independent random variables, 265-67

independent set: of elements of a matroid, 244-45; of vectors, 22, 158; of vertices in a graph, 564

index: of a continuous map, 695; of an elliptic equation, 682-83; of a fixed point, 695; of a Fredholm operator, 520; of a Toeplitz operator, 521

Indian mathematics, 79, 98, 192, 320, 736

indirect proofs of existence, 71, 117, 143-44, 149

induction: principle of, 16, 152, 258, 592, 638, 701, 741, 787, 998

ineffective proofs, *see* effective and ineffective proofs

inequalities, 3, 123, 126, 703-6

infinite cluster in percolation, 662-63

infinite prime, 720

infinite series, 120-21, 123, 193

infinite sets, 124, 127-28, 143-44, 148-49, 165, 167, 170-71, 616-17, 620, 623

infinite-dimensional vector spaces, 5, 22, 29-30

infinitesimals, 119-22, 128-29, 640, 770, 823, 934

infinities, 118-19, 127-28, 616, 744, 778

initial data, 459

initial value problem, 459, 464-67

injection, 11, 642

inner models, 624, 629, 631-32

inner products, 185, 189, **219-20**, 240, 268, 278, 301, 704

instantons, 529, 538

Institutiones Calculi Differentialis (Euler), 120-21

integers, **17**, 82, 127, 170, 254, 284, 377, 570, 636, 992

integral delay equation, 346

integral domain, 377

integral equations, 510-11

integral operators, 511, 515

integration, **35-37**, 51, **175-80**, 247, 450, 686, 744; Lebesgue, 247, 686, 796; numerical, 292, 609; Riemann, 247, 796

interactive proof systems, 596-97

interesting numbers, 261

interior methods for linear programming, 614

intermediate value theorem, 49, 67, 124, 384, 693

International Congress of Mathematicians, 110, 145, 619, 783, 789, 811, 950, 961, 966

Internet, 862-63, 868-70, 890

Internet search engines, *see* search engines

interpolation, 212, 450, 454

intersecting family, 569

intersection numbers, 189, 383-85, 391-93

intuitionism, 116, 148-51, 799

invariant subspaces, 422, 425, 427, 515, 689

invariants, **53-54**, 103, 221, 225, 370, 383-87, 395, 404, 407, 410-12, 442, 695, 788, 800; cardinal, 627; discrete, 370-71, 411-12; Gromov-Witten, 419, 533-34, 537; of elliptic equations, 682; Seiberg-Witten, 404, 407, 419; theory of, 103, 144, 550, 773, 789, 800

inverse problems: in additive number theory, 717; in chemistry, 829, 836-37; in spectral theory, 472

inverse-Gamma distribution, 926

inverses, 13; of functions, 10, 202; of matrices, 174; of operators, 294; under multiplication, 276-78, 284

involution principle, 556

involutions, 172, 554

irrational numbers, 18, 80, **222-23**, 315-17, 328, 710

irrational rotation algebra, 520

irrationality of π, 193

irreducible element of a ring, 318-19

irreducible polynomial, 102, 328, 710, 888

Ising model, 223, 666-68

Islamic mathematics, 79-80, 86-87, 98, 133-34

isomers, 830-31

isometry groups, 402, 712-13

isomorphism problem: for graphs, 584, 596; for groups, 436

isomorphisms, **27-28**, 165-67, 202, 408, 411, 645, 801

isoperimetric inequalities, 210, 445-46, 670, 672-73, 676, 679, 705-6

isotopy, 160, 225

iteration, 107, 112-14, 190, 244, 495

Itô, Kiyoshi, 655

Itô's formula, 651-52, 654

Jacobi, Carl Gustav Jacob, 727, **762-63**, 766

Jacobi identity, 231, 541

Jacobian, 418

James-Stein estimator, *see* estimators, statistical

Jensen, Ronald, 624

Jensen's inequality, 704

j-function, 324-26, 415, 548

Jones, Vaughan, 518

Jones polynomial, 226-27, 274

Jordan, Camille, 777

Jordan curve theorem, 777, 799

Jordan normal form, 223-25, 285

Journal des Mathématiques Pures et Appliquées, 102

Journal für die reine und angewandte Mathematik, see Crelle's Journal

J-stability, 506

Julia, Gaston, 110, 498

Julia sets, 244, 500-506, 508

just intonation, 936-37

K-theory, 227, 394-95, 521-22, 683

Kähler manifolds, 163, 297, 300

Kac-Moody algebras, 234

Kakutani fixed point theorem, 694, 901

Kant, Immanuel, 93, 136, 928-29

Kaplan-Meier curves, 923-24

Kasparov, Gennadi, 522

Kauffman polynomial, 227

KdV equation, 235-38, 471, 481

Kepler conjecture, 58

kernel: heat, 218-19; integral, 29, 239, 791

kernel of a homomorphism, 28, 284

key management, 887

Khayyam, Omar, 80, 86, 98-99

Khinchin, Alexander Yakovlevich, 814

Khinchin's inequality, 705

Kirchhoff formula, 474

Kleene, Stephen, 111-12

Klein, Christian Felix, 92-93, 137, 209, 324, 327, 777-78, **782-83**, 794

Klein bottle, 279, 388, 400, 441

Klein-Gordon equation, 456, 478

Kleinian groups, 209

Kloosterman sum, 347

Kneser conjecture, 569

knots, 225-26, 385, 435

Koch snowflake, 184

Koebe, Paul, 210

Kolmogorov, Andrei Nikolaevich, 453, 648, 677, 687, 793, 795, **814-16**

Kontsevich, Maxim, 69, 536-37

Korteweg, Diederik, 235, 237, 799

Kovalevskaya, Sofya (Sonya), 125, **784-85**

Kronecker, Leopold, 143-44, 146, 315, 327-28, 330-31, **773-74**

Kronecker-Weber theorem, 720, 774

Krylov subspace iterations, 608-9

Kummer, Ernst Eduard, 81, 692, 719, **767**

Kuratowski's theorem, 725

L^∞-norm, 211, 294, 449

L^p-norm, 211, 449, 706

La Géométrie (Descartes), 100, 739-40, 742

Lagrange, Joseph Louis, 101, 122, 136, 193, 268, 554, 570, 636, 727, **751-52**

Lagrange inversion theorem, 554, 752

Lagrange multipliers, 256-57, 865-67, 870, 899

Lagrange's theorem, 55, 688

Lagrangian, 257, 311, 478-79

Lambert, Johann, 81, 87-88, 193

Landau, Edmund, 63

Lang's conjecture, *see* Mordell-Lang conjecture

Langlands, Robert, 69

Langlands program, 69, 191, 331, 419, 429-31

Laplace, Pierre-Simon, 552, **752-54**

Laplace equation, 35, 125, 283, 291, 296, 456-58, 468, 478

Laplace transform, 306-7, 552

Laplace–Beltrami operator, 218, 296, 472

Laplacian, **34**, 206-7, 217-18, 239, 287, 296-97, 312, 426, 456-57, 459-60, 477

large-cardinal axioms, 627-34

lattices, 59-60, 227, 250-52, 318, 324, 330, 412-13, 530; hexagonal, 228, 415, 663; square, 415

law of large numbers, 753, 906

Lax equivalence theorem, 611

Lax–Wendroff formula, 611

least action principle, *see* Hamilton's least action principle

least upper bound axiom, 758

Lebesgue, Henri, 182, 628, 686, 795, **796-97**

Lebesgue differentiation theorem, 455

Lebesgue measure, 247, 628, 686, 796

Lebesgue spaces, 211

Leech lattice, 59-60, 227-28, 549

Lefschetz fixed point theorem, 695

left coset of a subgroup, 26, 421

legal bracketing, 550, 553

Legendre, Adrien-Marie, 88, 104, 714, **754-55**

Legendre polynomials, 291-92, 297, 609

Legendre symbol, 339, 719

Legendre's equation, 291

Leibniz, Gottfried Wilhelm, 118-19, 134, **743-45**, 746, 935

lemmas, 73

length, 31, 57, 183-84, 220, 246-47, 307

length spaces, 444

Lennard-Jones clusters, 835-36

Leonardo of Pisa, 99, **737**

Les Méthodes Nouvelles de la Mécanique Céleste (Poincaré), 786

Lévy's arcsine law, 650

Lewy operator, 471

L-functions, **228-29**, 316-17, 339-41, 345, 347-48, 381, 812; Dirichlet, 228-29, 284, 339-40, 345, 347-48, 689, 715, 764; of elliptic curves, 229, 347-48, 381, 685-86

Liber Abbaci (Fibonacci), 99

Lie, Sophus, 137, 230, **777-78**

Lie algebras, 231-32, 234, 273-74, 427, 541, 544, 778, 794; classical, 234; semisimple, 232-33; simple, 232-34

Lie brackets, 231-32, 287

Lie groups, 161, 229-32, 234, 240, 272-73, 277, 279, 298, 402, 425, 428, 778, 794; classical, 161, 234; linear, 230; semisimple, 713; simple, 232

lifting a path, 309

light cone, 43, 467-68

limit groups, 448

limit ordinal, 258, 617-18, 620, 624

limits, **30-32**, 122-26, 168-69, 200-201, 254, 258

Lindemann, Carl, 116, 150

line bundle, 393, 410, 413

linear algebra, 103

linear approximation, 33, 37, 109

linear combinations, 21, 28

linear equations, 48-49

linear feedback shift register (LFSR), 888

linear functionals, 176, 185, 188, 212

linear groups, 161

linear independence, 244, 285

linear maps, **28-30**, 33, 37, 49, 51, 174, 219, 223, 239, 255, 276, 294

linear operators, 216, 239-41, 294-97, 511

linear programming, 288, 612-13

linear wave equation, 611

linearization, 470

link-route incidence matrix, 863

linking numbers, 385

links, 225-26, 402-3

Liouville, Joseph, 38, 71, 81, 222, 293, **766-67**

Liouville's theorem: in complex analysis, 38, 723-24, 766; in mechanics, 766; on transcendence, 294, 299, 711

Littlewood, John Edensor, 346, 797-98, **803-5**, 859, 963

Littlewood–Paley theory, 804, 859

Lobachevskii, Nicolai Ivanovich, 42, 89-92, 137, **759-60**

local connectedness, 505; of the Mandelbrot set, 508

local-search algorithms, 875

local-to-global principles, 167-68; of Hasse, **241-43**, 685

locality, 543

localization, 453, 477

locally symmetric manifolds, *see* manifolds, locally symmetric

logarithmic integral, 63, 715

logarithms, 80-81, 202, 290, 523

logic, 6, **13-16**, 140, 634-39, 819, 931-33; Ω-, 634; Aristotelean, 932; first-order, 259, 314, 448, **621-22**, 623, 636, 700-701; propositional, 153

logical connectives, 13-14, 621, 635

logical consequences, 637-38

logicism, 143, 795

long multiplication, 106-7, 170, 204, 349-50

Lorentz gauge, 490

Lorentzian geometry, 43-44, 402, 478, 484, 487-89

Lorenz attractor, 496

loss function, 918

Lovász local lemma, 574

Löwenheim–Skolem theorem, 622, 624-25, 806

lower-triangular matrices, *see* matrices, lower-triangular

Luzin, Nikolai, 628, 686

machine epsilon, 606

Maclaurin, Colin, 121

Mac Lane, Saunders, 167

major arcs, 346

majority function, 588

Mandelbrot set, 244, 505-9; hyperbolic components of, 507

manifolds, 4, 5, **44-46**, 47, 57, 244, 258, 281-82, 300, 396-408, 794; complex, 163, 191, 300; differentiable, 45; four-dimensional, 388, 403-4, 440-41; hyperbolic, 401, 712; locally symmetric, 712-13; nonorientable, 384, 399-400; orientable, 163, 384, 399-400; simply connected, 281, 388, 403, 714; smooth, 396-400, 403; symplectic, 297-301; three-dimensional, 280, 388, 401-3, 441, 714; topological, 45, 397-400, 404. *See also* Calabi-Yau manifolds, Kahler manifolds, Lorentzian manifold, Riemannian manifolds

manipulatorics, 555-56

mapping class group, 418

Margulis, Gregori, 197-98, 269, 713

market completeness, 912

market efficiency principle, 910

market equilibrium, 901

Markov chain, 596

Markov process, 647, 649, 653, 655

Martin's maximum (MM), 633
Martin-Löf thesis, 116
martingale, 652, 912-13; problem
    for Brownian motion, 652
mathematical collaboration, 1001
Mathematical Physics (Courant
    and Hilbert), 809
mathematical physics, 7-8
Mathematische Annalen, 93, 153,
    782, 800, 817, 822
Mathieu, Émile Léonard, 688, **776-77**
Matiyasevitch, Yuri, 50, 708
matrices, **28-30**, 33, 49, 174, 223,
    240; diagonalizable, 223-24;
    Hermitian, 240; invertible, 174-75;
    lower-triangular, 607; nilpotent,
    224; orthogonal, 240; permutation,
    423, 607; self-adjoint, 240; similar,
    174; skew-Hermitian, 231;
    stochastic, 694; symmetric, 240,
    511; symplectic, 298; transpose of,
    240; unipotent, 430; unitary, 240,
    271, 277, 511; upper-triangular,
    607
matrix multiplication, *see*
    multiplication, of matrices
matroids, 244-46
maximal function, 455
maximal operator, 452
maximal torus, 430
maximum principle for the Laplace
    equation, 475
maximum-likelihood estimate, 924
Maxwell's equations, 479, 484, 490,
    525
McKay, John, 60
mean of a random variable, 265-66
meantone temperament, 937
measurable cardinals, *see* cardinals,
    measurable
measurable sets, 128, 158, 247,
    627-29, 631-32
measure problem, the, 628-29
measurement problem, 269
measures, **246-47**, 628, 815;
    probability, 264. *See also* Haar
    measure, Lebesgue measure
Méchanique Analitique (Lagrange),
    751
Mellin transform, 214
melodic retrograde and inversion,
    938-40
memorylessness, 265
meromorphic continuation, 228-29
meromorphic functions, 213, 723-24
Mersenne, Marin, 936
Mersenne primes, 353

Mesopotamian mathematics,
    77-78, 733
metamathematics, 152, 154, 622
metastable states, 829
method of characteristics, 236
method of exhaustion, 132, 735
metric spaces, 46, 172, 181, 220,
    247-48, 253, 302
Meyer, Y., 861-62
microlocal analysis, 477
millionaires' problem, 602
Mills-Robbins-Rumsey determinant,
    997-98
Milman, Vitali, 675-76
Milnor, John, 395, 404
Milnor-Švarc lemma, 444
minimal polynomial, 225, 328, 330
minimal surface equation, 312, 457
minimal surfaces, 312, 534, 670,
    832-33
minimum connector problem, 245
minimum spanning tree problem
    (MSTP), 872-75
Minkowski, Hermann, 330, 484, 487,
    672, **789-90**
Minkowski space, 43, 268, 402, 457,
    478, 484-85, 487
Minkowski's inequality, 704
minor arcs, 346
mirror symmetry, 69, 164, 190,
    523-24, 529-32, 534, 537-38
Möbius, August Ferdinand, 759
Möbius function, 345
Möbius inversion, 561
Möbius strip, 384, 392-93, 399-400,
    759, 950, 979-80
Möbius transformations, 208-9,
    415-16
model spaces for geometric
    structures, 402
model theory, 6, 645, 814, 822
models, 621-22, 636, 639, 645, 822;
    of set theory, 248-49, 806; of ZFC,
    620-21, 623-27, 629-30
Moderne Algebra (van der Waerden),
    105, 824
modes in quantum field theory,
    532, 543
modular arithmetic, 249-50
modular automorphism group, 517
modular elliptic curve, 692
modular forms, 251-52, 268, 419,
    692, 724, 807-8
modular functions, 347, 545, 549
modules, 104, 285
moduli spaces, 191, 252, 370-71,
    408-19, 711-13

modulus of a complex number,
    19, 276
modus ponens, 700
molecular dynamics, 836, 841-42
momentum operator, 286
Monge-Ampère equation, 462
Monier-Rabin theorem, 351
monochromatic subsets, 567, 802
monoidal category, 275
monotone circuits, *see* Boolean
    circuits, monotone
Monster group, 60, 252, 549
Monstrous Moonshine conjecture,
    60, 548-49
Mordell conjecture, the, 117, 382,
    681, 722
Mordell-Lang conjecture, 382, 722
Mordell-Weil theorem, 190
Mordell's theorem, 685, 820
Morlet, J., 861
morphisms, 165-67, 536
Mostow, George, 713
Mostow's rigidity theorem, 712-13
multiplication, 275-78, 284, 306-7,
    635-36, 638; of ideal classes, 323;
    of ideals, 322; of integers, 65, 586;
    of matrices, 29, 65, 67, 272,
    277-78, 591
multiplication operators, 239-40,
    294-95, 511-12, 519
multiplicative sequence, 228
multiplicity of a solution, 366
multiplier of a fixed point, 499
multistep methods for numerical
    solution of ODEs, 609
*Music for Strings, Percussion, and
    Celesta* of Bartók, 940
Musical Offering, The (Bach), 939-40

naive set theory, *see* set theory, naive
nanoporous architectures, 834
Nash, John, 694, 901, 982
Nash equilibrium, 694, 901, 982
Nash's theorem, 364
natural numbers, 17, 258;
    theory of, 638
natural proofs in computational
    complexity, 589
natural transformations, 167
Navier-Stokes equations, 193-96,
    477
n-body problem, 493, 726-28, 764
$\mathcal{NC}$, *see* complexity classes
n-category, 167
negation, 15
negative numbers, 17, 81, 126
networks, 862-71

Neumann boundary conditions, 458, 469

neural networks, 844

Newton, Isaac, 87, 100, 109, 118-20, 134, 136, 493-94, 609, 612, 726, **742-43**, 827

Newton–Raphson method, 109-10

Newtonian limit, 480, 488

Newton's law of gravitation, 493

Newton's method, 494-95, 498, 509, 612-13

Newton's second law of motion, 194, 286, 311, 493, 524, 726

nilpotent groups, 444, 447, 702

nilpotent matrices, *see* matrices, nilpotent

no arbitrage principle, 910

Noether, Emmy, 82, 104, 525, **800-801**

Noether's principle, 479, 525, 540

noise, 878

non-Euclidean geometry, 84, **88-94**, 832, 946

noncollision singularities in the *n*-body problem, 727

noncommutative algebraic topology, 523

noncommutative geometry, 57, 272, 522-23

nonconstructible real numbers, 629

nonconstructible sets, 629

nonconstructive arguments, 157

nonconstructive proofs, 693

nondetermined game, 631

nonlinear approximation scheme, 856

nonlinear Poisson equation, 312

nonmeasurable sets, 158, 627-28, 796, 802

nonorientable manifolds, *see* manifolds, nonorientable

nonparametric approach to statistical modeling, 922-24

nonperturbative phenomena in physics, 530

nonpositively curved groups, 447

nonrigorous arguments, 68-71

nonstandard analysis, 128, 823

nonstandard models of arithmetic, 702, 822

norm residue symbol, 719

normal distribution, 51, 214, 262, 266-67, 647-50, 654, 678, 797

normal number, 262

normal operators, *see* operators, normal

normal subgroup, 26

normed division algebra, 278

normed spaces, 185, 210, 252-54

norms, 210-11, 253, 278, 319, 321, 704, 810; of quadratic integers, 319, 321, 323

Norton, Simon, 60, 549

NOT gates, 584

notional prices, 899

Novikov, Pyotr, 68, 438, 708

$\mathcal{NP}$, *see* complexity classes

NP-complete problems, 68, 271-72, **583-87**, 596, 874

$\mathcal{NP}$ versus co$\mathcal{NP}$ problem, 582, 593

null hypotheses, 923-24

Nullstellensatz, *see* Hilbert's Nullstellensatz

number field sieve, *see* sieves

number fields, 254-55, 329, 730

number systems, **16-19**, 77-83, 104, 278, 984

number theory, 4. *See also* additive number theory, algebraic number theory, analytic number theory, combinatorial number theory, computational number theory

numeracy, 983-91

numerical analysis, 109, 604-15; integration, 292, 609; linear algebra, 606-9

numerical evidence, 69-70

numerical instability, 606, 610

n-vector model, 70

o-minimal structures, 644

objective functions, 255-57, 288-89, 613, 835, 866

objective prior distributions, 926

objects in a category, 185-87, 536-37

observables, 286-87, 513, **540-42**

octahedron, *n*-dimensional, 53

octatonic scale, 939

octonions, 278-79

odd functions, 204

odd permutations, *see* permutations

one-step methods for numerical solution of ordinary differential equations, 610

one-time key, 890

one-way functions, 598-601, 890

one-way hash function, 893-95

open sets, 302-3, 633

operator algebras, 240, **510-23**

operators, 295, 450, 526; compact, 519; essentially normal, 521; Fredholm, 520-22; Hermitian, 240, 295, 540; normal, 240, 518; self-adjoint, 511, 513;

Toeplitz, 521; trace-class, 515, 523; unitary, 513-14, 689, 691

optimization, 255-56, 612-14, 865-70

Opus Restitutae Mathematicae Analyseos, seu Algebra Nova (Viète), 738

OR gates, 584, 587

orbifolds, 257-58, 367, 371, 534-35

orbits, 421, 442, 494-95, 559

order: of a group element, 67; of a permutation, 260

order isomorphism, 258, 617

ordered fields, 643

ordinals, 145, 258, 616-22, 624-27, 629, 779; countable, 624-26

ordinary differential equations (ODEs), **51-52**, 464-65, 609-11

orientable manifolds, *see* manifolds, orientable

orthogonal arrays, 173

orthogonal groups, 39, 230, 232

orthogonal maps, 240

orthogonal projections, *see* projections, orthogonal

orthogonality: of Legendre polynomials, 292; of spherical harmonics, 296-97; of trigonometric functions, 308; of wavelets, 852-54

orthonormal basis, 212, 220, 240, 296, 423

overdetermined system of equations, 459

$\mathcal{P}$, *see* complexity classes

p-adic numbers, 82, 241-43

$\mathcal{P}$ versus $\mathcal{BPP}$ problem, 595, 601

$\mathcal{P}$ versus $\mathcal{NP}$ problem, 69, 170, 580-81, 585, 591, 598-600, 713-14, 874

Pacioli, Luca, 99, 945

packing problems, 173, 836

PageRank, 877

pairing axiom, 314, 620

pairings, 189

Paley, R. E. A. C., 572, 804, 812, 859

palindromic numbers, 75

parabolic equations, *see* partial differential equations, parabolic

paradifferential calculus, 477

paradoxes, 145-46

parallel computation, 587, 603

parallel postulate, 42, 84-88, 90, 94

parity function, 588

partial derivative, 33

partial differential equations (PDEs), 34-35, 187, 455-83, 681-82, 993-94; criticality of, 479-80; dispersive, 236, 468, 471; elliptic, 468-71, 681-83; fully nonlinear, 462; homogeneous linear, 457; hyperbolic, 468-70, 490; inhomogeneous linear, 457-58; linear, 51, 458, 471-72; local solvability of, 471; nonlinear, 457; order of, 461; parabolic, 468, 470-71; quasilinear, 462; regular, 480, 482; semilinear, 462; subcritical, 479-80, 482; supercritical, 479-80, 482; symbols of, 682-83
partial differential operators, 239
partition functions, 62, 529, 667, 669
partitions: of natural numbers, 555, 797, 807, 994-95; of sets, 558
Pascal, Blaise, 741-42
Pascal's triangle, 741
Pasch, Moritz, 137-39
path integral, 38, 526-28, 541, 545
Peano, Giuseppe, 111, 128, 138, 701, **787-88**, 795
Peano arithmetic, 138, 151, 154, 258-59, 638, 701, 787, 819; first-order, 259, 638
Pell equations, 98, 255, 315, 317, 706
Penrose, Roger, 829
Penrose staircase, 951
Penrose tilings, 829
Penrose's incompleteness theorem, 492
percolation, 662-66
Perelman, Grigori, 281, 388, 401-3, 406, 440, 461, 714, 970
perfect matchings in a graph, 566, 587, 596
perfect number, 104
perfect phylogeny problem, 846
perfect sets, 619, 632
periodic orbits, 495, 498-99
periodic points, 498-99
permanent of a matrix, 591
permutation matrices, *see* matrices, permutation
permutations, 174, 259-61, 420, 558
Perron's formula, 336
perspective, 945-46
perturbation theory, 527-28, 541
phase of a Fourier coefficient, 859
phase problem in crystallography, 829
phase space, 286

phase transitions, 7, 261, 573, 657, 660-64, 667-68, 845
phenomenological equations, 480-81
philosophasters, 935
phylogenetic tree, 846
π, 60, 79, 108-9, 222-23, **261-62**, 325, 748; continued fraction for, 192; formula for, 262, 998-99; hexadecimal expansion of, 262, 998-99; transcendence of, 81, 223, 262
Picard iteration method, 465, 470
pigeonhole principle, 167, 567; proof complexity of, 591-94
Pisot number, 828
pivoting in Gaussian elimination, 607
planar graphs, *see* graphs, planar
Plancherel theorem, the, 207, 212, 261, 427, 453, 686
plane waves, 205-7, 236
Plateau problem, 458
Plato, 83-84
Platonic solids, 52, 84, 187, 189
Platonism, 148, 155
Poincaré, Jules Henri, 42, 93-94, 146-47, 152, 209, 215, 400, 403, 432, 484, 495-96, 690, 714, 722, 727-28, **785-87**, 802, 946-48
Poincaré conjecture, 141, 219, 403, 714, 786; in four dimensions, 404; in more than four dimensions, 404-5
Poincaré duality, 189
Poincaré map, 495
Poincaré quantum group, 275
Poincaré transformations, 457, 478
Poincaré's recurrence theorem, 690-91
point at infinity, 41, 43, 267, 356, 366, 370, 386, 721
Poisson, Siméon-Denis, 757
Poisson bracket, 286, 525-26, 541
Poisson distribution, 264
Poisson equation, 286-87, 457-58, 483, 492
poker, electronic, 601-2, 895
polar convex body, *see* duality, of convex bodies
polarization identity, 268
Pólya, George, 358, 831, 959
Pólya-Redfield enumeration, 559-61, 831
polynomial equations, 49-50
polynomial ring, 24
polynomial time, 169
polynomial-time algorithms, 197-98, 350, 353, 355, 551, 579, 873-74, 880; for primality testing, 353

polynomial-time reducibility, 582-83, 591
polynomials, 284, 326, 328-29, 363-65, 738; approximation by, 253
polytopes, 52
Pontryagin classes, 394
Pontryagin duality, *see* duality, Pontryagin
position operator, 286
positional system for numbers, 77
posterior distributions, 160, 920, 926-27
postmodernism, 967-69
potential theory, 283, 760
potentials in complex dynamics, 503-4
Potts model, 669
power series, 38, 122, 126, 144, 201, 308, 368, 752
power-set axiom, 157, 314, 620-21
preconditioning a system of linear equations, 609
predicativity, 146-47, 149
presentations of groups, 433-34
primality testing, **351-53**, 595; polynomial-time, 352-53; randomized, **351-52**, 892
prime factorization, *see* factorization of integers, into primes
prime ideals, 143, 323, 378
prime number theorem, 63, 338, 346, 356, **714-15**, 729, 804
prime numbers, 63, 69-70, 75, 319, 325, 332-36, 338-41, 348-53, 357, 714-15; distribution of, 229, 283-84, 356-59; gaps between, 66, 342-45
primitive polynomials, 888
primitive recursive functions, 111-12
Princet, Maurice, 947
principal ideals, 58, 221, 322-23
Principia Mathematica (Newton), 87, 136, 743, 750
Principia Mathematica (Whitehead and Russell), 147, 149, 781, 795
prior distribution, 160, 920, 926
prisoner's dilemma, 901
private goods, 901
probabilistic computation, 269-71
probabilistic method, 197, 563, 572
probabilistic proof systems, 596-99
probabilistically checkable proofs, 598-99
probability, 7, 917
probability amplitudes, 269-71

probability distributions, 263–71, 917

probability theory: foundations of, 793, 795, 814–15

probability thresholds, 661–62

Proclus, 85

products, 23–24

projections, 43, 240–41, 295, 422; Mercator, 978; orthogonal, **240–41**, 295, 518; stereographic, **169**, 401

projective determinacy, axiom of, 632–34

projective geometry, **43**, 92, 169, 173, 187, **267**, 356, 388, 400

projective orthogonal groups, 688

projective representations, 544

projective sets, 628, 631–32

projective spaces, 161, 230, 267, 300, 313, 365

projective special linear groups, 41, 43, 688

projective unitary groups, 688

projective varieties, *see* varieties, projective

proof complexity, 591–94

proofs, 74, 129–42, 622, 637, 642, 646; by contradiction, 150

propagation of singularities, 467, 469

proper time, 485

propositional logic, *see* logic, propositional

propositions, 73

protein folding, 841

protein structure, 835

pseudorandomness, 116, 600–602

pseudosphere, 92

$\mathcal{PSPACE}$, *see* complexity classes

public goods, 901

public-key cryptography, 602, 890–95

pullbacks, 180, 410

pure states, 270

Putnam, Hilary, 50

Pythagoras, 18, **733–34**, 980

Pythagorean comma, 936

Pythagorean triples, 364, 691

Pythagorean tuning system, 936

QR factorization, 607–8

quadratic convergence, *see* convergence, quadratic

quadratic equations, 49, 133, 193, 708

quadratic fields, 254

quadratic forms, 267–69, 317, 319; binary, 320–23, 325; positive definite, 268

quadratic reciprocity, *see* reciprocity, quadratic

quadratic residues, 104, 317, 325, 327, 339

quadratic sieve method, 354–55

quadratic variation of a martingale, 652

quadratic-like mappings, 508

quadrature, 292, 609, 614; Clenshaw-Curtis, 609; Gaussian, 292, 609

quadric surface, 62

quadrivium, the, 935

quantifiers, 14, 31, 149, 621, 635–36, 638, 640–41; elimination of, 640–44

quantization, 526, 528

quantum computation, 269–72

quantum field theories, 528–29, 542–48; topological, 227

quantum groups, 272–75

quantum mechanics, 269, 287, 513, 526, 539; operator formulation of, 526

quartic equations, 99, 101, 326, 709, 737

quasi-conformal mappings, 713

quasi-isometry, 443–45

quasi-Newton methods, 613

quasicrystals, 828–29

quaternions, 82, 105, 276–79, 389, 765

Quechua numeration, 984–85

quintic equations: insolubility of, 50, 101, 327, **708–10**

quotients, **24–26**, 28, 257, 284; of groups, 26; of rings, 284

Rényi, Alfréd, 572, 660

radial lines, 504

radical of a positive integer, 361, 681

Radon transform, 297, 307

Ramanujan, Srinivasa, 62, 64, 268, 729, 797, **807–8**

Ramanujan conjecture, 729, 732

Ramanujan graphs, 198

Ramanujan–Nagell equation, 254

Ramanujan's ternary form, 269

Ramanujan's Lost Notebook, 808

ramification, 323, 330

Ramsey theory, 215, 562, **567–68**, 573–74, 627, 639, 702

random graphs, 572, 645–46, 660–62

random matrices, 345, 359

random packing, 836

random restriction method in computational complexity, 588

random trees, 655–57

random variables, 265–67

random walks, 61, 72, 198–99, 648–51, 679

random-cluster model, 669

randomized algorithms, 116, 351, 595, 892

randomness, 115–16, 199, 269, 361

randomness extractors, 599

range of a function, 11

rank, 245; of a set, 621; of a vector bundle, 393

Raphson, Joseph, 110

rapid mixing, 199, 679

rational functions, 126, 193

rational numbers, 17–18, 82, 126–27, 144, 171, 242, 246, 284, 316, 622; approximation by, 192, 222, 315–16, 710

ratios, 79

reaction-diffusion equations, 219, 844

real closed fields, 643

real division algebra, 395

real numbers, **18**, 80–82, 102, 123, 127–28, 144, 171, 242, 246, 275, 616, 622–25, 627, 630, 635, 639–40, 643–45, 698, 776

real projective line, 409

reciprocity: Artin's law of, 720, 812–13; higher, 104, **719–20**, 812–13; quadratic, 104, **718–20**, 812; quartic, 104, **719**

recurrence formulas, 110, 552

recursion, 66, 113–14, 552

recursive functions, 111–13

recursive sets, 439–40

recursively enumerable sets, 439–40, 638–39

recursively presentable group, 439

reductio ad absurdum, 150

reductive groups, 430

Reed-Solomon codes, 883–86

reflexive relations, 12

refutation systems, 593

regular cardinals, *see* cardinals, regular

regular polyhedra, 52

regular polytopes, 53

regular primes, 692

regularity of solutions to partial differential equations, 481

relations, 11–12, 932

relative consistency proofs, 623

relativity, 539; equivalence principle in, 486, 488; general, 164, 172, 483–93; principle of, 484, 487; special, 483–85

repelling orbits, 502
representation ring of a group, 423
representation varieties, 417-18
representations, 207, 274, 279,
    420-24, 783-84; decomposable,
    427; induced, 424; irreducible, 279,
    422-29, 515-16, 561; linear, 234,
    279; regular, 423, 516-18, 520;
    unitary, 313, 514-16
residue mod m, *see* modular
    arithmetic
resolution of singularities, 369
resolution rule, 594
restriction mapping problem, 839
retractions, 694
Reuleaux triangle, 978
rhythmic augmentation and
    diminution, 938
Ribet, Ken, 692
Ricci curvature, *see* curvature, Ricci
Ricci flow, 219, 279-81
Ricci tensor, 164, 172
Richard's paradox, 146
Riemann, Georg Friedrich Bernhard,
    36, 91-92, 124-25, 127, 137, 143,
    282, 284, 336-37, 356-58, 475,
    729, 771, **774-76**, 780, 782, 929,
    947-48
Riemann curvature tensor, 488
Riemann hypothesis, 63, 68, 229,
    337-38, 356-58, **714-15**; for
    curves, 356, 730-31; generalized,
    229, 340-41, 348, 729-30
Riemann integration, *see* integration,
    Riemann
Riemann mapping theorem,
    728, 774-75
Riemann sphere, 282, 370, 389,
    412, 500, 723
Riemann surface bundle, 414
Riemann surfaces, 209, **282-83**, 300,
    408, 411-14, 416, 529-30, 723,
    728-29, 775, 805, 816; families of,
    413-14, 417
Riemann zeta function, 38, 228-29,
    283-84, 336, 344, 356, 359, 685,
    715, 775
Riemann-Roch theorem, 683,
    724, 775, 820
Riemannian manifolds, 218,
    280, 296-97, 394, 444
Riemannian metrics, 46-47, 172, 280,
    300, 311, 407, 775
Riesz, Frigyes (Frédéric), 689-90,
    798-99
Riesz representation theorems, 185,
    791, 798

rigid transformations, 39
rigidity theorems, 442, 444, 703, 711
rigor in mathematics, 117-29, 137,
    930-31
rings, 22, 24, 57, 104-5, 108, 221,
    254, **284-85**, 317-19, 329,
    376-77, 391, 776; graded, 391;
    of polynomials, 108
risk, 903, 914-15, 918
risk-neutral pricing, 912
road rage, 918
Robbin, Tony, 952-53
Robertson-Seymour theorem, 725
Robinson, Abraham, 128, **822-23**
Robinson, Julia, 50
Rodriguez's formula, 291
Rogers-Ramanujan identities,
    557, 807
root systems, 233-34
roots: of polynomials, 66, 101;
    of unity, 327-28
rotational invariance: of Brownian
    motion, 650, 653; of normal
    distribution, 266
Roth's theorem, 222, 681, 711
rounding errors, 606-7
RSA algorithm, 602, 891
Ruffini, Paolo, 101
rules of inference, 84, 140, 152, 700
Runge-Kutta method for numerical
    solution of ordinary differential
    equations, 610
Russell, Bertrand Arthur William,
    128, 145-49, 781, **795-96**, 929,
    933-34
Russell, John Scott, 234-35, 237
Russell's paradox, 128, 145, 171,
    619, 779, 781, 795, 807
Russell's theory of descriptions,
    933-35

Saccheri, Gerolamo, 87
Saint Petersburg paradox, 747
Salem numbers, 828-29
sample average, 266
sample mean, 916
sample paths, 647, 649
sample space, 263-65
SAT, 582-83, 593, 596
satisfiable formulas, 582
satisfiable theories, 639
satisfying assignment, 582
scalar curvature, *see* curvature,
    scalar, 280, 488
scalar multiplication, 21, 285
scaling properties of partial
    differential equations, 477

scaling relations between critical
    exponents, 663
scattering theory, 287-88, 472
Schauder fixed point theorem, 696
schemes, 285, 367, 373, 377-79, 381
Schmidt, Friedrich Karl, 730-32
Schoenberg, Arnold, 940-41
Schottky problem, 418
Schrödinger equation, 216, 285-88,
    456, 459, 472, 478, 514, 526,
    540, 830
Schramm-Loewner evolution, *see*
    stochastic Loewner evolution
Schubert calculus, 62
Schweikart, F. K., 88
"Scottish Book, The" of Banach, 810
search engines, 30, 876-77;
    ranking problem for, 876
search problems, 577-78, 580-81
second derivative, 34
second largest eigenvalue, 198, 572
second moment method, 573
section of a bundle, 393
seed of a pseudorandom generator,
    600
Seiberg-Witten invariants, *see*
    invariants, Seiberg-Witten
Seifert-van Kampen theorem,
    437, 441
Selberg, Atle, 338, 341, 348
Selberg's upper bound sieve, *see*
    sieves
self-adjoint matrices, *see* matrices,
    self-adjoint
self-adjoint operators, *see* operators,
    self-adjoint
self-avoiding random walk, 63, 70
self-reference, 701
self-similarity, 183, 502, 509
semi-parametric approach to
    statistical models, 925-26
semialgebraic sets, 364, 643-44
semidirect product, 24
sensitive dependence on initial
    conditions, 495-96, 501-2
sentences, 636-37, 641, 643
Serre, Jean-Pierre, 692, 732, 974,
    1000, 1001
set of measure zero, 183, 247, 628,
    686, 796
set systems, 568-69
set theory, 6, **9-10**, 127-28, 143,
    145-46, 314, **615-34**, 776, 779;
    naive, 145
sexagesimal place-value system,
    77-79
shadow price, 289

Shannon, Claude, 770, 812, 879, 881
Shannon wavelet, 859
Shannon's theorem, 881–83
shape of a drum, 219, 837
Shelah, Saharon, 627, 630, 632–34, 646
shift operator, 239, 294, 520
Shimura varieties, 331, 419
Shimura-Taniyama-Weil conjecture, 69, 692
shock formation, 237
shock waves, 481, 775, 809, 818
Shor, Peter, 204, 271
shortest path, 310
Shuja, Abu Kamil, 133
Siegel, Carl, 337, 340
Siegel zeros, 340, 346, 681
Sierpiński, Wacław, 627, **801-2**
Sierpiński's carpet, 801
sieves, 345–46; Brun, 345; combinatorial, 345; number field, 355; of Eratosthenes, 334, 345; Selberg's upper bound, 346
sigma-algebra, 247
sigma models, 529, 532
signature of a permutation, 260
signature of a quadratic form, 268
similar matrices, *see* matrices, similar
simple connectedness, 38, 177, 309, 403, 714
simple groups, *see* groups, simple
simple harmonic oscillators, 524–27
simplex, **52**, 162, 676–77
simplex algorithm, 288–90, 613
simply connected manifolds, *see* manifolds, simply connected
simulated annealing, 834
simultaneous linear equations, 22, 48, 102
sine function, *see* trigonometric functions
Singer, Isadore, 683
singular cardinal hypothesis, 630, 633
singular cardinals, *see* cardinals, singular
singular value decomposition, 608
singularities, 258, 281, 367–69, 415, 476; of an algebraic variety, 723
SIRS model, 845
six-exponentials theorem, 223
skein relation, 225–27
skew-Hermitian matrices, *see* matrices, skew-Hermitian
Skewes, Stanley, 804
Skolem, Thoralf, 622, **806-7**
slack variables, 870

Sloane's database, 992, 997
Smale, Stephen, 404, 714
smooth functions, 185
smooth manifolds, *see* manifolds, smooth
smooth numbers, 346, 354–56
smooth solutions, 195–96
smooth structures on manifolds, 397–99, 404, 407
Sobolev embedding theorem, 212, 449
Sobolev inequality, 705–6
Sobolev spaces, 211
social equilibrium, 900–901
solitons, 235–38
Solovay, Robert, 627, 629, 632
solubility by radicals, 102, 213
soluble groups, *see* groups, solvable
solvable groups, *see* groups, solvable
solving equations, 48–52
Sophie Germain primes, 892
soundness, 637; of a proof system, 592
space complexity of an algorithm, 114–15
space groups, 828
space-filling curve, 787
spacelike hypersurface, 468–69
spacelike vector, 43, 484, 487
spanning set, 22
spanning tree, 245
special functions, 290–93
special linear groups, 41, 230–31, 429, 688
special orthogonal groups, 39, 230, 425–26, 429
special relativity, *see* relativity, special
special unitary groups, 277
species, 150
spectral gap, 198
spectral radius formula, 518
spectral theorem, 217, 240, 295, 511–13, 519, 689
spectroscopy, 832
spectrum, 294–95, 472, 512; of a graph, 571
Sperner's theorem, 64, 568
sphere, 53, 208, 215, 244, 282, 390, 400, 403, 670, 714, 728–29, 734; *n*-dimensional, 386
sphere packing, 58–60
spherical geometry, 40–41, 90, 92, 182, 390, 401
spherical harmonics, 295–97
spherical Radon transform, 297
spherical structure, 401–2

spin configuration, 666
split-stepping, 237
splitting field, 213, 709
splitting law, 319, 325
splitting of a prime, 319, **322-23**, 325–26, 330
spontaneous magnetization, 667
sporadic examples, 53
sporadic simple groups, 252, 688
spread, 150
square-integrable function, 511
square roots, 80–81, 327; of minus one, 56; of two, 30, 56, 78, 150, 222, 315–16, 710–11
squared error, 918
stability: of Minkowski space, 490–91; of orbits, 495; of the solar system, 728, 786, 815
stabilizer, 421, 442, 559
stable letter, 437–38
stacks, 367
standard deviation, 265–66
state-field correspondence, 544
states, 860; of physical systems, 513, 540; of a quantum computer, 269–70; on a von Neumann algebra, 517
statistical arbitrage, 916
statistical models, 922
statistics, 917–18
Stein's paradox, 918–20
Steiner triple systems, 173
Steinitz, Ernst, 82, 105
Steinitz's exchange lemma, *see* exchange axiom
Stevin, Simon, 80–82, **738**
Stirling's formula, 214
stochastic differential equations, 461, **654-55**, 838
stochastic integrals, 651–52, 655
stochastic Loewner evolution, 655, 664–65
stochastic matrices, *see* matrices, stochastic
stock market crashes, 913
Stokes's theorem, 179–80
Stone-Čech compactification, *see* compactification, Stone-Čech
stopping time, 653
Strassen, Volker, 66, 353
stress-energy-momentum tensor, 485–86
string theory, 164, 528–29, 538–41
strong cosmic censorship conjecture, 492
strong Fermat congruence, 351–52

strong law of large numbers, 266, 677, 815
strong Markov process, 653, 655
structural stability, 497
Sturm–Liouville equation, 291
Sturm–Liouville theory, 766
subrepresentations, 515
subroutines, 579
subset sum problem, 583
subspace topology, 302
substructures, 23
successive approximation, 31
successor function, 111, 146, 258
successor ordinal, 618
Suite for Piano (Schoenberg), 941
Sullivan, Dennis, 394, 510
Summa (Luca Pacioli), 99
sumsets, 569, 715, 718
super-attracting fixed point, 499
supercompact cardinals, *see* cardinals, supercompact
superposition: in quantum mechanics, 269–72; principle of, 457–58
supremum norm, *see* L^∞-norm
surfaces, 45, 53–54
surgeries, 281, 400; on links, 403
surjection, 11, 642
survival function in medical statistics, 923–24
Suslin, Mikhail, 628, 632
Suslin's hypothesis, 624, 627, 632–33
Swinnerton-Dyer, Peter, 685
Sylvester, James Joseph, 103, **768–69**
Sylvester's law of inertia, 768
symbol of a Toeplitz operator, 521
symbols of PDEs, *see* partial differential equations, symbols of
symmetric groups, 234, 260, 422, 424, 561, 688, 709–10
symmetric polynomials, 329–30
symmetric relations, 12
symmetric space, 713
symmetry, 19, 52, 59–61, 204, 229, 273
symmetry groups, 39, 484
symmetry reductions, 480
symplectic geometry, 297–301, 531
symplectic groups, 230, 688
symplectic manifolds, *see* manifolds, symplectic
symplectic space: linear, 297, 300
syntonic comma, 937
Szemerédi's regularity lemma, 571
Szemerédi's theorem, 570
Szemerédi–Trotter theorem, 570

tableau, 289
tangent bundle, 313, 393
tangent space, 46
tangent vector, 176, 180
Tarski, Alfred, 627, 640–41, 643, 645, **813–14**
Tauberian theorems, 803
tautology, 592
Taylor, Brook, 119, **745**, 946
Taylor, Richard, 229, 252, 692–93
Taylor series, **38**, 119, 122, 124, 200, 205, 282, 304, 541, 745
T-duality, 532–33, 537–38
Teichmüller spaces, 413–14, 416–18
Tennenbaum, Stanley, 627, 632
tensor categories, 275
tensor products, 272, 274–75, **301**
ternary Goldbach problem, *see* Goldbach's conjecture, ternary
tessellation, 42, 208
test functions, 185–86, 195
Thales of Miletus, 130
theorema aureum, 718
theorems, 73, 117, 130, 700
theoretical computer science, 7, 575–604
Théorie Analytique de la Chaleur (Fourier), 755
Théorie Analytique des Probabilités (Laplace), 753
theories, formal, 636–46
Theory of Linear Operations (Banach), 810
thermodynamic limit, 667
theta series, 268
Thompson, John, 60, 785
three-body problem, 51, 495–96, 726–28; restricted, 495, 728
three-dimensional manifolds, *see* manifolds, three-dimensional
Thurston, William, 388, 402–3, 461, 714
Thurston's geometrization conjecture, *see* geometrization conjecture
timelike vector, 43, 484, 487
Tits, Jacques, 162
Toeplitz index theorem, 521
Toeplitz operators, *see* operators, Toeplitz
tomography problem, 307
topological group theory, 441–43
topological manifolds, *see* manifolds, topological
topological methods in combinatorics, 572

topological spaces, 182, 221, 227, 301–3, 309, 633
topological twisting, 530–31
Torricelli, Evangelista, 135
torus, **26**, 45, 53, **208–9**, 300, 309, 387–91, 530–33, 536–38, 711–13
total space, 392
totally disconnected set, 504
Tower of Hanoi, 113
trace, 515
trace-class operators, *see* operators, trace-class
Traité de Dynamique (d'Alembert), 750
transcendence of e, *see* e, transcendence of
transcendental numbers, 71–72, 81, **222–23**, 241, 262, 616, 766, 773
transfinite numbers, 258, 616–19, 778–79
transition functions, 45, 282–83, 300, 396–98
transition matrices, 198
transitive actions, 421
transitive closure, 624
transitive relations, 12
translation invariance, 253, 457
transmission control protocol (TCP) of the Internet, 868
traveling salesman problem (TSP), 567, 583, 872–75
Treatise of Fluxions, A (Maclaurin), 121
tree, 442, 846
trefoil knot, 225–26, 979
trial division, 349–50, 355
triangle inequality, 248, 253, 268, 705, 790
triangulation, 54
trichotomy, law of, 127
trigonometric functions, 204, 220, 262, 295–96, 307–9
trigonometric polynomials, 212, 296, 451
trimmed mean, 916
Trotter product formula, 237
Turán's theorem, 566
turbo codes, 886
turbulence, 815
Turing, Alan, 50, 113, 219, 358, 577, 638, 701, 707–8, 818, **821–22**
Turing machines, 113, 576–78, 818
twelve-tone music, 940
twin prime conjecture, 362, 715, 804; asymptotic version, 343
twisted Chevalley groups, 688

two-dimensional arithmetic progression, 718
two-soliton interaction, 235
Tychonoff's theorem, 168
types: Russell's theory of, 147-48

Ulam, Stanisław, 237, 359, 361, 568-69, 628, 958
umbrella problem, 561
unavoidable configurations, 697
uncertainty principle, *see* Heisenberg uncertainty principle
uncountable cardinals, *see* cardinals, uncountable
uncountable sets, 71, 146, **170-72**, 183, 223, 247, 616-18, 622, 626-29, 632, 703, 806
undecidability, 622-25, 628-29, 633, 638-39; of first-order logic, 821
underdetermined system of equations, 459
uniform approximation, 253
uniform convergence, *see* convergence, uniform
uniform distribution, 263-64, 266, 674
uniform family of circuits, 585
uniformization theorem, 209, 281, 283, 416, 460, 728-29, 786
unipotent matrices, *see* matrices, unipotent
unique factorization, 221, 254-55, 285, 318, 320-25, 327, 692, 719
uniqueness of solutions, 48, 510
unit ball, 189, 671, 684, 693
unit in a ring, 318-19
unitary groups, 230
unitary maps, 211, 240, 270
unitary matrices, *see* matrices, unitary
unitary operators, *see* operators, unitary
unitary representations, *see* representations, unitary
universal covers, 309-10, 441-42
universal family of marked elliptic curves, 414
universal properties, 166, 301, 433-34, 436-37
universal set, 249
universality, 650, 657-59, 663-66, 669; of the Mandelbrot set, 508-9
universe of all sets, 314, 619-22, 625
unknot, 225-26, 403, 435
unlink, 225-26
upper-triangular matrices, *see* matrices, upper-triangular

vacuum Einstein equations, *see* Einstein equations, vacuum
vacuum state, 542
van der Waerden, Bartel, 104-5, 568
van der Waerden's theorem, 568, 570
vanilla options, 914
van Ceulen, Ludolph, 109
variables, 15, 635
variance, 265-67
varieties, 272, 285, 313, 366, 376-80, 722-23; Abelian, 190, 418; abstract algebraic, 731; affine, 313; exponential, 644; function field of, 723; irreducible, 376; projective, 313; rational, 374
vector bundles, 176-77, 227, 313, 392-94; trivial, 313, 392
vector fields, 180, 393, 486
vector spaces, **21-22**, 27-28, 30, 33, 52, 57, 103, 105, 172, 219, 239, 244-45, 253-54, 275-76, 278-79, 284-85, 294, 314, 419-25; basis of, 21-22, 28, 30, 223; dimension of, 22; direct sum of, 24; infinite dimensional, 22, 29
vertex operator algebra, 60, 541, 544-49
Viète, François, 99, 136, **737-38**
Vinogradov, Ivan, 343
Vinogradov's three-primes theorem, 715-16
Virasoro algebra, 544
Vitali, Giuseppe, 628, 684
Vitali covering lemma, 455
Vizing's theorem, 565
volatility of a share price, 911-14
volume, 57, 158, 174, 183, 672, 679, 790; in *n* dimensions, 671
von Neumann, John, 148, 153, 512-13, 515-16, **817-19**, 821
von Neumann algebras, 313, 515-17
von Neumann's ergodic theorem, 512, 689-91
Voronoi diagram, 829
Voronoi surface, 842

Wallis, John, 87, 192
Wardrop equilibria, 865-66
Waring, Edward, **750-51**, 804
Waring's problem, 715, 717, 751, 803-4
wave equation, 35, 235-36, 456-57, 460, 466, 468, 478, 490
wave function, 216, 286
wavelets, 313-14, 848-62; Battle-Lemarié, 854; interval, 855; Mexican hat, 855; Meyer, 854; multiwavelets, 855

weak convergence, *see* convergence, weak
weak cosmic censorship conjecture, 492
weak law of large numbers, 266, 746
weak solutions to PDEs, 195-96, 313, 476
weakly compact cardinals, *see* cardinals, weakly compact
Weber, Heinrich, 241, 776
wedge product, 179
Weierstrass, Karl, 92, 124-29, 143-44, 476, 758, **770-71**, 833
Weierstrass approximation theorem, 253, 452, 771
Weierstrass function, 833
Weierstrass *P*-function, 724
Weierstrass product expansion of the gamma function, 214
weight enumerators, 552-53
weighted AM-GM inequality, 704
Weil, André, 337, 347, 713, **819-21**, 823, 966
Weil conjectures, 729-32, 820
Weil numbers, **331**, 347
Weil's theorem, 381-82
well-ordered set, 258, 617, 619, 624, 626
well-ordering principle, 108, 147, 158, 619, 621
well-posed problems, 468-69, 473
Well-Tempered Clavier (Bach), 938
Westzynthius, Erik, 66
Weyl, Hermann, 147-49, 151, 490, 521, 788, **805-6**
Weyl curvature, *see* curvature, Weyl
Weyl formula, 219
Weyl group, 233-34
Whitehead link, 225
Whitney, Hassler, 244, 405
Whitney trick, 405
Wick rotation, 529
Wiener, Norbert, 648, **811-12**
Wiener-Hopf equations, 812
Wiles, Andrew, 69, 229, 252, 359, 380, 381, 692, 693
Wilson's theorem, 350, 751-52
winding numbers, 386, 410, 521, 532, 683
winning strategy, 159, 630-31
Witten, Edward, 404, 539
Woodin, Hugh, 141, 630, 632, 634
Woodin cardinals, *see* cardinals, Woodin
word hyperbolic groups, 446
word metrics, 443

word problems, 435–36, 440, 445–46; for groups, 708
world lines, 484; length of, 485
World Wide Web, 875
worldsheets, 529-40-41
worst-case complexity, 578

x-ray transform, 307

Yang–Baxter equations, 161
Yang–Mills equations, 490
Yau, Shing-Tung, 163

Yoneda lemma, 417
Young tableaux, 561, 995
Young's inequality, 213, 451

Zariski topology, 303
Zelmanov, Efim, 438
zeolites, 834
Zermelo, Ernst, 128, 145, 147–48, 619–20, 780
Zermelo-Fraenkel set theory, 619
Zermelo-Russell paradox, 145
zero, 17, 79

zero divisors, 105, 276, 278
zero–one law, 646
zero-knowledge proof systems, 597–98
zeros of the Riemann zeta function, 336–38, 344, 357–58, 715
zeta functions, 284
ZF axioms, 128, 148, **314**, 624
ZFC axioms, 314, 619–29, 634, 702
Zhu Shijie, 741
Zorn's lemma, 158